Handbook of Data Compression

Fifth Edition

David Salomon
Giovanni Motta

With Contributions by David Bryant

Handbook of Data Compression

Fifth Edition

Previous editions published under the title
"Data Compression: The Complete Reference"

Springer

Prof. David Salomon (emeritus)
Computer Science Dept.
California State University, Northridge
Northridge, CA 91330-8281
USA
dsalomon@csun.edu

Dr. Giovanni Motta
Personal Systems Group, Mobility Solutions
Hewlett-Packard Corp.
10955 Tantau Ave.
Cupertino, Califormia 95014-0770
gim@ieee.org

ISBN 978-1-84882-902-2 ISBN 978-1-84882-903-9
DOI 10.1007/10.1007/978-1-84882-903-9
Springer London Dordrecht Heidelberg New York

British Library Cataloguing in Publication Data
A catalogue record for this book is available from the British Library

Library of Congress Control Number: 2009936315

Cover design: eStudio Calamar S.L.

Printed on acid-free paper

Springer is part of Springer Science+Business Media (www.springer.com)

To users of data compression everywhere

I love being a writer. What I can't stand is the paperwork.
—Peter De Vries

Preface to the New Handbook

GENTLE READER. The thick, heavy volume you are holding in your hands was intended to be the fifth edition of *Data Compression: The Complete Reference.* Instead, its title indicates that this is a handbook of data compression. What makes a book a handbook? What is the difference between a textbook and a handbook? It turns out that "handbook" is one of the many terms that elude precise definition. The many definitions found in dictionaries and reference books vary widely and do more to confuse than to illuminate the reader. Here are a few examples:

- A concise reference book providing specific information about a subject or location (but this book is not concise).

- A type of reference work that is intended to provide ready reference (but every reference work should provide ready reference).

- A pocket reference is intended to be carried at all times (but this book requires big pockets as well as deep ones).

- A small reference book; a manual (definitely does not apply to this book).

- General information source which provides quick reference for a given subject area. Handbooks are generally subject-specific (true for this book).

Confusing; but we will use the last of these definitions. The aim of this book is to provide a quick reference for the subject of data compression. Judging by the size of the book, the "reference" is certainly there, but what about "quick?" We believe that the following features make this book a quick reference:

- The detailed index which constitutes 3% of the book.

- The glossary. Most of the terms, concepts, and techniques discussed throughout the book appear also, albeit briefly, in the glossary.

- The particular organization of the book. Data is compressed by removing redundancies in its original representation, and these redundancies depend on the type of data. Text, images, video, and audio all have different types of redundancies and are best compressed by different algorithms which in turn are based on different approaches. Thus, the book is organized by different data types, with individual chapters devoted to image, video, and audio compression techniques. Some approaches to compression, however, are general and work well on many different types of data, which is why the book also has chapters on variable-length codes, statistical methods, dictionary-based methods, and wavelet methods.

The main body of this volume contains 11 chapters and one appendix, all organized in the following categories, basic methods of compression, variable-length codes, statistical methods, dictionary-based methods, methods for image compression, wavelet methods, video compression, audio compression, and other methods that do not conveniently fit into any of the above categories. The appendix discusses concepts of information theory, the theory that provides the foundation of the entire field of data compression.

In addition to its use as a quick reference, this book can be used as a starting point to learn more about approaches to and techniques of data compression as well as specific algorithms and their implementations and applications. The broad coverage makes the book as complete as practically possible. The extensive bibliography will be very helpful to those looking for more information on a specific topic. The liberal use of illustrations and tables of data helps to clarify the text.

This book is aimed at readers who have general knowledge of computer applications, binary data, and files and want to understand how different types of data can be compressed. The book is not for dummies, nor is it a guide to implementors. Someone who wants to implement a compression algorithm A should have coding experience and should rely on the original publication by the creator of A.

In spite of the growing popularity of Internet searching, which often locates quantities of information of questionable quality, we feel that there is still a need for a concise, reliable reference source spanning the full range of the important field of data compression.

New to the Handbook

The following is a list of the new material in this book (material not included in past editions of *Data Compression: The Complete Reference*).

- The topic of compression benchmarks has been added to the Introduction.

- The paragraphs titled "How to Hide Data" in the Introduction show how data compression can be utilized to quickly and efficiently hide data in plain sight in our computers.

- Several paragraphs on compression curiosities have also been added to the Introduction.

- The new Section 1.1.2 shows why irreversible compression may be useful in certain situations.

- Chapters 2 through 4 discuss the all-important topic of variable-length codes. These chapters discuss basic, advanced, and robust variable-length codes. Many types of VL codes are known, they are used by many compression algorithms, have different properties, and are based on different principles. The most-important types of VL codes are prefix codes and codes that include their own length.

- Section 2.9 on phased-in codes was wrong and has been completely rewritten.

- An example of the start-step-stop code $(2, 2, \infty)$ has been added to Section 3.2.

- Section 3.5 is a description of two interesting variable-length codes dubbed recursive bottom-up coding (RBUC) and binary adaptive sequential coding (BASC). These codes represent compromises between the standard binary (β) code and the Elias gamma codes.

- Section 3.28 discusses the original method of interpolative coding whereby dynamic variable-length codes are assigned to a strictly monotonically increasing sequence of integers.

- Section 5.8 is devoted to the compression of PK (packed) fonts. These are older bitmaps fonts that were developed as part of the huge TeX project. The compression algorithm is not especially efficient, but it provides a rare example of run-length encoding (RLE) without the use of Huffman codes.

- Section 5.13 is about the Hutter prize for text compression.

- PAQ (Section 5.15) is an open-source, high-performance compression algorithm and free software that features sophisticated prediction combined with adaptive arithmetic encoding. This free algorithm is especially interesting because of the great interest it has generated and because of the many versions, subversions, and derivatives that have been spun off it.

- Section 6.3.2 discusses LZR, a variant of the basic LZ77 method, where the lengths of both the search and look-ahead buffers are unbounded.

- Section 6.4.1 is a description of LZB, an extension of LZSS. It is the result of evaluating and comparing several data structures and variable-length codes with an eye to improving the performance of LZSS.

- SLH, the topic of Section 6.4.2, is another variant of LZSS. It is a two-pass algorithm where the first pass employs a hash table to locate the best match and to count frequencies, and the second pass encodes the offsets and the raw symbols with Huffman codes prepared from the frequencies counted by the first pass.

- Most LZ algorithms were developed during the 1980s, but LZPP, the topic of Section 6.5, is an exception. LZPP is a modern, sophisticated algorithm that extends LZSS in several directions and has been inspired by research done and experience gained by many workers in the 1990s. LZPP identifies several sources of redundancy in the various quantities generated and manipulated by LZSS and exploits these sources to obtain better overall compression.

- Section 6.14.1 is devoted to LZT, an extension of UNIX compress/LZC. The major innovation of LZT is the way it handles a full dictionary.

- LZJ (Section 6.17) is an interesting LZ variant. It stores in its dictionary, which can be viewed either as a multiway tree or as a forest, *every* phrase found in the input. If a phrase is found n times in the input, only one copy is stored in the dictionary. Such behavior tends to fill the dictionary up very quickly, so LZJ limits the length of phrases to a preset parameter h.

- The interesting, original concept of antidictionary is the topic of Section 6.31. A dictionary-based encoder maintains a list of bits and pieces of the data and employs this list to compress the data. An antidictionary method, on the other hand, maintains a list of strings that do not appear in the data. This generates negative knowledge that allows the encoder to predict with certainty the values of many bits and thus to drop those bits from the output, thereby achieving compression.

- The important term "pixel" is discussed in Section 7.1, where the reader will discover that a pixel is not a small square, as is commonly assumed, but a mathematical point.

- Section 7.10.8 discusses the new HD photo (also known as JPEG XR) compression method for continuous-tone still images.

- ALPC (*adaptive linear prediction and classification*), is a lossless image compression algorithm described in Section 7.12. ALPC is based on a linear predictor whose coefficients are computed for each pixel individually in a way that can be mimiced by the decoder.

- Grayscale Two-Dimensional Lempel-Ziv Encoding (GS-2D-LZ, Section 7.18) is an innovative dictionary-based method for the lossless compression of grayscale images.

- Section 7.19 has been partially rewritten.

- Section 7.40 is devoted to spatial prediction, a combination of JPEG and fractal-based image compression.

- A short historical overview of video compression is provided in Section 9.4.

- The all-important H.264/AVC video compression standard has been extended to allow for a compressed stream that supports temporal, spatial, and quality scalable video coding, while retaining a base layer that is still backward compatible with the original H.264/AVC. This extension is the topic of Section 9.10.

- The complex and promising VC-1 video codec is the topic of the new, long Section 9.11.

- The new Section 11.6.4 treats the topic of syllable-based compression, an approach to compression where the basic data symbols are syllables, a syntactic form between characters and words.

- The commercial compression software known as stuffit has been around since 1987. The methods and algorithms it employs are proprietary, but some information exists in various patents. The new Section 11.16 is an attempt to describe what is publicly known about this software and how it works.

- There is now a short appendix that presents and explains the basic concepts and terms of information theory.

We would like to acknowledge the help, encouragement, and cooperation provided by Yuriy Reznik, Matt Mahoney, Mahmoud El-Sakka, Pawel Pylak, Darryl Lovato, Raymond Lau, Cosmin Truţa, Derong Bao, and Honggang Qi. They sent information, reviewed certain sections, made useful comments and suggestions, and corrected numerous errors.

A special mention goes to David Bryant who wrote Section 10.11.

Springer Verlag has created the Springer Handbook series on important scientific and technical subjects, and there can be no doubt that data compression should be included in this category. We are therefore indebted to our editor, Wayne Wheeler, for proposing this project and providing the encouragement and motivation to see it through.

The book's Web site is located at `www.DavidSalomon.name`. Our email addresses are `dsalomon@csun.edu` and `gim@ieee.org` and readers are encouraged to message us with questions, comments, and error corrections.

Those interested in data compression in general should consult the short section titled "Joining the Data Compression Community," at the end of the book, as well as the following resources:

- `http://compression.ca/`,
- `http://www-isl.stanford.edu/~gray/iii.html`,
- `http://www.hn.is.uec.ac.jp/~arimura/compression_links.html`, and
- `http://datacompression.info/`.

(URLs are notoriously short lived, so search the Internet.)

David Salomon Giovanni Motta

The preface is usually that part of a book which can most safely be omitted.

—William Joyce, *Twilight Over England* (1940)

Preface to the Fourth Edition

(This is the Preface to the 4th edition of *Data Compression: The Complete Reference*, the predecessor of this volume.) I was pleasantly surprised when in November 2005 a message arrived from Wayne Wheeler, the new computer science editor of Springer Verlag, notifying me that he intends to qualify this book as a Springer major reference work (MRW), thereby releasing past restrictions on page counts, freeing me from the constraint of having to compress my style, and making it possible to include important and interesting data compression methods that were either ignored or mentioned in passing in previous editions.

> These fascicles will represent my best attempt to write a comprehensive account, but computer science has grown to the point where I cannot hope to be an authority on all the material covered in these books. Therefore I'll need feedback from readers in order to prepare the official volumes later.
>
> I try to learn certain areas of computer science exhaustively; then I try to digest that knowledge into a form that is accessible to people who don't have time for such study.
>
> —Donald E. Knuth, `http://www-cs-faculty.stanford.edu/~knuth/` (2006)

Naturally, all the errors discovered by me and by readers in the third edition have been corrected. Many thanks to all those who bothered to send error corrections, questions, and comments. I also went over the entire book and made numerous additions, corrections, and improvements. In addition, the following new topics have been included in this edition:

- Tunstall codes (Section 2.6). The advantage of variable-size codes is well known to readers of this book, but these codes also have a downside; they are difficult to work with. The encoder has to accumulate and append several such codes in a short buffer, wait until n bytes of the buffer are full of code bits (where n must be at least 1), write the n bytes on the output, shift the buffer n bytes, and keep track of the location of the last bit placed in the buffer. The decoder has to go through the reverse process.

The idea of Tunstall codes is to construct a set of fixed-size codes, each encoding a variable-size string of input symbols. As an aside, the "pod" code (Table 10.29) is also a new addition.

- Recursive range reduction (3R) (Section 1.7) is a simple coding algorithm due to Yann Guidon that offers decent compression, is easy to program, and its performance is independent of the amount of data to be compressed.

- LZARI, by Haruhiko Okumura (Section 6.4.3), is an improvement of LZSS.

- RAR (Section 6.22). The popular RAR software is the creation of Eugene Roshal. RAR has two compression modes, general and special. The general mode employs an LZSS-based algorithm similar to ZIP Deflate. The size of the sliding dictionary in RAR can be varied from 64 Kb to 4 Mb (with a 4 Mb default value) and the minimum match length is 2. Literals, offsets, and match lengths are compressed further by a Huffman coder. An important feature of RAR is an error-control code that increases the reliability of RAR archives while being transmitted or stored.

- 7-z and LZMA (Section 6.26). LZMA is the main (as well as the default) algorithm used in the popular `7z` (or `7-Zip`) compression software [7z 06]. Both `7z` and LZMA are the creations of Igor Pavlov. The software runs on Windows and is free. Both LZMA and `7z` were designed to provide high compression, fast decompression, and low memory requirements for decompression.

- Stephan Wolf made a contribution to Section 7.34.4.

- H.264 (Section 9.9). H.264 is an advanced video codec developed by the ISO and the ITU as a replacement for the existing video compression standards H.261, H.262, and H.263. H.264 has the main components of its predecessors, but they have been extended and improved. The only new component in H.264 is a (wavelet based) filter, developed specifically to reduce artifacts caused by the fact that individual macroblocks are compressed separately.

- Section 10.4 is devoted to the `WAVE` audio format. `WAVE` (or simply Wave) is the native file format employed by the Windows opearting system for storing digital audio data.

- FLAC (Section 10.10). FLAC (free lossless audio compression) is the brainchild of Josh Coalson who developed it in 1999 based on ideas from Shorten. FLAC was especially designed for audio compression, and it also supports streaming and archival of audio data. Coalson started the FLAC project on the well-known sourceforge Web site [sourceforge.flac 06] by releasing his reference implementation. Since then many developers have contributed to improving the reference implementation and writing alternative implementations. The FLAC project, administered and coordinated by Josh Coalson, maintains the software and provides a reference codec and input plugins for several popular audio players.

- WavPack (Section 10.11, written by David Bryant). WavPack [WavPack 06] is a completely open, multiplatform audio compression algorithm and software that supports three compression modes, lossless, high-quality lossy, and a unique hybrid compression mode. It handles integer audio samples up to 32 bits wide and also 32-bit IEEE floating-point data [IEEE754 85]. The input stream is partitioned by WavPack into blocks that can be either mono or stereo and are generally 0.5 seconds long (but the length is actually flexible). Blocks may be combined in sequence by the encoder to handle multichannel audio streams. All audio sampling rates are supported by WavPack in all its modes.

- Monkey's audio (Section 10.12). Monkey's audio is a fast, efficient, free, lossless audio compression algorithm and implementation that offers error detection, tagging, and external support.

- MPEG-4 ALS (Section 10.13). MPEG-4 Audio Lossless Coding (ALS) is the latest addition to the family of MPEG-4 audio codecs. ALS can input floating-point audio samples and is based on a combination of linear prediction (both short-term and long-term), multichannel coding, and efficient encoding of audio residues by means of Rice codes and block codes (the latter are also known as block Gilbert-Moore codes, or BGMC [Gilbert and Moore 59] and [Reznik 04]). Because of this organization, ALS is not restricted to the encoding of audio signals and can efficiently and losslessly compress other types of fixed-size, correlated signals, such as medical (ECG and EEG) and seismic data.

- AAC (Section 10.15). AAC (advanced audio coding) is an extension of the three layers of MPEG-1 and MPEG-2, which is why it is often called `mp4`. It started as part of the MPEG-2 project and was later augmented and extended as part of MPEG-4. Apple Computer has adopted AAC in 2003 for use in its well-known iPod, which is why many believe (wrongly) that the acronym AAC stands for apple audio coder.

- Dolby AC-3 (Section 10.16). AC-3, also known as Dolby Digital, stands for Dolby's third-generation audio coder. AC-3 is a perceptual audio codec based on the same principles as the three MPEG-1/2 layers and AAC. The new section included in this edition concentrates on the special features of AC-3 and what distinguishes it from other perceptual codecs.

- Portable Document Format (PDF, Section 11.13). PDF is a popular standard for creating, editing, and printing documents that are independent of any computing platform. Such a document may include text and images (graphics and photos), and its components are compressed by well-known compression algorithms.

- Section 11.14 (written by Giovanni Motta) covers a little-known but important aspect of data compression, namely how to compress the differences between two files.

- Hyperspectral data compression (Section 11.15, partly written by Giovanni Motta) is a relatively new and growing field. Hyperspectral data is a set of data items (called pixels) arranged in rows and columns where each pixel is a vector. A home digital camera focuses visible light on a sensor to create an image. In contrast, a camera mounted on a spy satellite (or a satellite searching for minerals and other resources) collects and measures radiation of many wavelegths. The intensity of each wavelength is converted into a number, and the numbers collected from one point on the ground form a vector that becomes a pixel of the hyperspectral data.

Another pleasant change is the great help I received from Giovanni Motta, David Bryant, and Cosmin Truţa. Each proposed topics for this edition, went over some of the new material, and came up with constructive criticism. In addition, David wrote Section 10.11 and Giovanni wrote Section 11.14 and part of Section 11.15.

I would like to thank the following individuals for information about certain topics and for clearing up certain points. Igor Pavlov for help with `7z` and LZMA, Stephan Wolf for his contribution, Matt Ashland for help with Monkey's audio, Yann Guidon for his help with recursive range reduction (3R), Josh Coalson for help with FLAC, and Eugene Roshal for help with RAR.

> In the first volume of this biography I expressed my gratitude to those individuals and corporate bodies without whose aid or encouragement it would not have been undertaken at all; and to those others whose help in one way or another advanced its progress. With the completion of this volume my obligations are further extended. I should like to express or repeat my thanks to the following for the help that they have given and the premissions they have granted.
>
> Christabel Lady Aberconway; Lord Annan; Dr Igor Anrep; ...
>
> —Quentin Bell, *Virginia Woolf: A Biography* (1972)

Currently, the book's Web site is part of the author's Web site, which is located at `http://www.ecs.csun.edu/~dsalomon/`. Domain `DavidSalomon.name` has been reserved and will always point to any future location of the Web site. The author's email address is `dsalomon@csun.edu`, but email sent to ⟨*anyname*⟩`@DavidSalomon.name` will be forwarded to the author.

Those interested in data compression in general should consult the short section titled "Joining the Data Compression Community," at the end of the book, as well as the following resources:

- `http://compression.ca/`,
- `http://www-isl.stanford.edu/~gray/iii.html`,
- `http://www.hn.is.uec.ac.jp/~arimura/compression_links.html`, and
- `http://datacompression.info/`.

(URLs are notoriously short lived, so search the Internet).

> People err who think my art comes easily to me.
>
> —Wolfgang Amadeus Mozart

Lakeside, California David Salomon

Contents

Content comes first... yet excellent design can catch people's eyes and impress the contents on their memory.

—Hideki Nakajima

Introduction

Giambattista della Porta, a Renaissance scientist sometimes known as the professor of secrets, was the author in 1558 of *Magia Naturalis* (Natural Magic), a book in which he discusses many subjects, including demonology, magnetism, and the camera obscura [della Porta 58]. The book became tremendously popular in the 16th century and went into more than 50 editions, in several languages beside Latin. The book mentions an imaginary device that has since become known as the "sympathetic telegraph." This device was to have consisted of two circular boxes, similar to compasses, each with a magnetic needle. Each box was to be labeled with the 26 letters, instead of the usual directions, and the main point was that the two needles were supposed to be magnetized by the *same lodestone*. Porta assumed that this would somehow coordinate the needles such that when a letter was dialed in one box, the needle in the other box would swing to point to the same letter.

Needless to say, such a device does not work (this, after all, was about 300 years before Samuel Morse), but in 1711 a worried wife wrote to the *Spectator*, a London periodical, asking for advice on how to bear the long absences of her beloved husband. The adviser, Joseph Addison, offered some practical ideas, then mentioned Porta's device, adding that a pair of such boxes might enable her and her husband to communicate with each other even when they "were guarded by spies and watches, or separated by castles and adventures." Mr. Addison then added that, in addition to the 26 letters, the sympathetic telegraph dials should contain, when used by lovers, "several entire words which always have a place in passionate epistles." The message "I love you," for example, would, in such a case, require sending just three symbols instead of ten.

> A woman seldom asks advice before
> she has bought her wedding clothes.
> —Joseph Addison

This advice is an early example of *text compression* achieved by using short codes for common messages and longer codes for other messages. Even more importantly, this shows how the concept of data compression comes naturally to people who are interested in communications. We seem to be preprogrammed with the idea of sending as little data as possible in order to save time.

Data compression is the process of converting an input data stream (the source stream or the original raw data) into another data stream (the output, the bitstream, or the compressed stream) that has a smaller size. A stream can be a file, a buffer in memory, or individual bits sent on a communications channel.

The decades of the 1980s and 1990s saw an exponential decrease in the cost of digital storage. There seems to be no need to compress data when it can be stored inexpensively in its raw format, yet the same two decades have also experienced rapid progress in the development and applications of data compression techniques and algorithms. The following paragraphs try to explain this apparent paradox.

- Many like to accumulate data and hate to throw anything away. No matter how big a storage device one has, sooner or later it is going to overflow. Data compression is useful because it delays this inevitability.

- As storage devices get bigger and cheaper, it becomes possible to create, store, and transmit larger and larger data files. In the old days of computing, most files were text or executable programs and were therefore small. No one tried to create and process other types of data simply because there was no room in the computer. In the 1970s, with the advent of semiconductor memories and floppy disks, still images, which require bigger files, became popular. These were followed by audio and video files, which require even bigger files.

- We hate to wait for data transfers. When sitting at the computer, waiting for a Web page to come in or for a file to download, we naturally feel that anything longer than a few seconds is a long time to wait. Compressing data before it is transmitted is therefore a natural solution.

- CPU speeds and storage capacities have increased dramatically in the last two decades, but the speed of mechanical components (and therefore the speed of disk input/output) has increased by a much smaller factor. Thus, it makes sense to store data in compressed form, even if plenty of storage space is still available on a disk drive. Compare the following scenarios: (1) A large program resides on a disk. It is read into memory and is executed. (2) The same program is stored on the disk in compressed form. It is read into memory, decompressed, and executed. It may come as a surprise to learn that the latter case is faster in spite of the extra CPU work involved in decompressing the program. This is because of the huge disparity between the speeds of the CPU and the mechanical components of the disk drive.

- A similar situation exists with regard to digital communications. Speeds of communications channels, both wired and wireless, are increasing steadily but not dramatically. It therefore makes sense to compress data sent on telephone lines between fax machines, data sent between cellular telephones, and data (such as web pages and television signals) sent to and from satellites.

The field of data compression is often called *source coding*. We imagine that the input symbols (such as bits, ASCII codes, bytes, audio samples, or pixel values) are emitted by a certain information source and have to be coded before being sent to their destination. The source can be *memoryless*, or it can have memory. In the former case, each symbol is independent of its predecessors. In the latter case, each symbol depends on some of its predecessors and, perhaps, also on its successors, so they are correlated. A memoryless source is also termed "independent and identically distributed" or IIID.

Data compression has come of age in the last 20 years. Both the quantity and the quality of the body of literature in this field provide ample proof of this. However, the need for compressing data has been felt in the past, even before the advent of computers, as the following quotation suggests:

> I have made this letter longer than usual because I lack the time to make it shorter.
>
> —Blaise Pascal

There are many known methods for data compression. They are based on different ideas, are suitable for different types of data, and produce different results, but they are all based on the same principle, namely they compress data by removing *redundancy* from the original data in the source file. Any nonrandom data has some structure, and this structure can be exploited to achieve a smaller representation of the data, a representation where no structure is discernible. The terms *redundancy* and *structure* are used in the professional literature, as well as *smoothness*, *coherence*, and *correlation*; they all refer to the same thing. Thus, redundancy is a key concept in any discussion of data compression.

◇ **Exercise Intro.1:** (Fun) Find English words that contain all five vowels "aeiou" in their original order.

In typical English text, for example, the letter `E` appears very often, while `Z` is rare (Tables Intro.1 and Intro.2). This is called *alphabetic redundancy*, and it suggests assigning variable-length codes to the letters, with `E` getting the shortest code and `Z` getting the longest code. Another type of redundancy, *contextual redundancy*, is illustrated by the fact that the letter `Q` is almost always followed by the letter `U` (i.e., that in plain English certain digrams and trigrams are more common than others). Redundancy in images is illustrated by the fact that in a nonrandom image, adjacent pixels tend to have similar colors.

Section A.1 discusses the theory of information and presents a rigorous definition of redundancy. However, even without a precise definition for this term, it is intuitively clear that a variable-length code has less redundancy than a fixed-length code (or no redundancy at all). Fixed-length codes make it easier to work with text, so they are useful, but they are redundant.

The idea of compression by reducing redundancy suggests the *general law* of data compression, which is to "assign short codes to common events (symbols or phrases) and long codes to rare events." There are many ways to implement this law, and an analysis of any compression method shows that, deep inside, it works by obeying the general law.

Compressing data is done by changing its representation from inefficient (i.e., long) to efficient (short). Compression is therefore possible only because data is normally represented in the computer in a format that is longer than absolutely necessary. The reason that inefficient (long) data representations are used all the time is that they make it easier to process the data, and data processing is more common and more important than data compression. The ASCII code for characters is a good example of a data

Letter	Freq.	Prob.	Letter	Freq.	Prob.
A	51060	0.0721	E	86744	0.1224
B	17023	0.0240	T	64364	0.0908
C	27937	0.0394	I	55187	0.0779
D	26336	0.0372	S	51576	0.0728
E	86744	0.1224	A	51060	0.0721
F	19302	0.0272	O	48277	0.0681
G	12640	0.0178	N	45212	0.0638
H	31853	0.0449	R	45204	0.0638
I	55187	0.0779	H	31853	0.0449
J	923	0.0013	L	30201	0.0426
K	3812	0.0054	C	27937	0.0394
L	30201	0.0426	D	26336	0.0372
M	20002	0.0282	P	20572	0.0290
N	45212	0.0638	M	20002	0.0282
O	48277	0.0681	F	19302	0.0272
P	20572	0.0290	B	17023	0.0240
Q	1611	0.0023	U	16687	0.0235
R	45204	0.0638	G	12640	0.0178
S	51576	0.0728	W	9244	0.0130
T	64364	0.0908	Y	8953	0.0126
U	16687	0.0235	V	6640	0.0094
V	6640	0.0094	X	5465	0.0077
W	9244	0.0130	K	3812	0.0054
X	5465	0.0077	Z	1847	0.0026
Y	8953	0.0126	Q	1611	0.0023
Z	1847	0.0026	J	923	0.0013

Frequencies and probabilities of the 26 letters in a previous edition of this book. The histogram in the background illustrates the byte distribution in the text.

Most, but not all, experts agree that the most common letters in English, in order, are `ETAOINSHRDLU` (normally written as two separate words `ETAOIN SHRDLU`). However, [Fang 66] presents a different viewpoint. The most common digrams (2-letter combinations) are TH, HE, AN, IN, HA, OR, ND, RE, ER, ET, EA, and OU. The most frequently appearing letters *beginning* words are S, P, and C, and the most frequent final letters are E, Y, and S. The 11 most common letters in French are `ESARTUNILOC`.

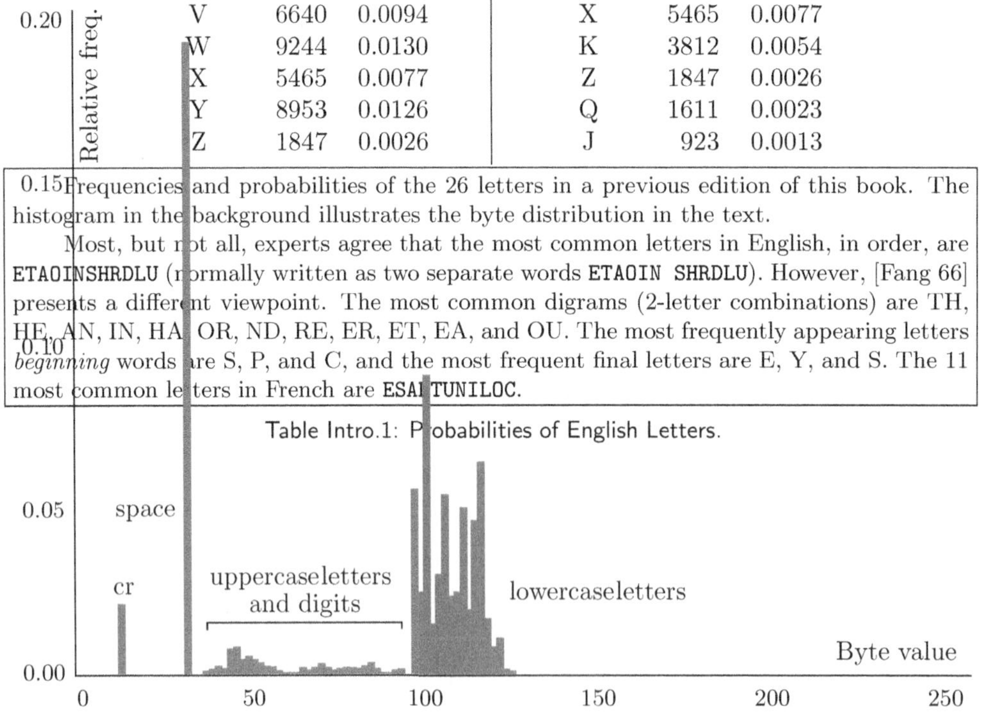

Table Intro.1: Probabilities of English Letters.

Char.	Freq.	Prob.	Char.	Freq.	Prob.	Char.	Freq.	Prob.
e	85537	0.099293	x	5238	0.006080	F	1192	0.001384
t	60636	0.070387	|	4328	0.005024	H	993	0.001153
i	53012	0.061537	-	4029	0.004677	B	974	0.001131
s	49705	0.057698	)	3936	0.004569	W	971	0.001127
a	49008	0.056889	(	3894	0.004520	+	923	0.001071
o	47874	0.055573	T	3728	0.004328	!	895	0.001039
n	44527	0.051688	k	3637	0.004222	#	856	0.000994
r	44387	0.051525	3	2907	0.003374	D	836	0.000970
h	30860	0.035823	4	2582	0.002997	R	817	0.000948
l	28710	0.033327	5	2501	0.002903	M	805	0.000934
c	26041	0.030229	6	2190	0.002542	;	761	0.000883
d	25500	0.029601	I	2175	0.002525	/	698	0.000810
m	19197	0.022284	^	2143	0.002488	N	685	0.000795
\	19140	0.022218	:	2132	0.002475	G	566	0.000657
p	19055	0.022119	A	2052	0.002382	j	508	0.000590
f	18110	0.021022	9	1953	0.002267	@	460	0.000534
u	16463	0.019111	[	1921	0.002230	Z	417	0.000484
b	16049	0.018630	C	1896	0.002201	J	415	0.000482
.	12864	0.014933	]	1881	0.002183	O	403	0.000468
1	12335	0.014319	’	1876	0.002178	V	261	0.000303
g	12074	0.014016	S	1871	0.002172	X	227	0.000264
0	10866	0.012613	_	1808	0.002099	U	224	0.000260
,	9919	0.011514	7	1780	0.002066	?	177	0.000205
&	8969	0.010411	8	1717	0.001993	K	175	0.000203
y	8796	0.010211	‘	1577	0.001831	%	160	0.000186
w	8273	0.009603	=	1566	0.001818	Y	157	0.000182
$	7659	0.008891	P	1517	0.001761	Q	141	0.000164
}	6676	0.007750	L	1491	0.001731	>	137	0.000159
{	6676	0.007750	q	1470	0.001706	*	120	0.000139
v	6379	0.007405	z	1430	0.001660	<	99	0.000115
2	5671	0.006583	E	1207	0.001401	”	8	0.000009

Frequencies and probabilities of the 93 most-common characters in a prepublication previous edition of this book, containing 861,462 characters. See Figure Intro.3 for the Mathematica code.

Table Intro.2: Frequencies and Probabilities of Characters.

representation that is longer than absolutely necessary. It uses 7-bit codes because fixed-size codes are easy to work with. A variable-size code, however, would be more efficient, since certain characters are used more than others and so could be assigned shorter codes.

In a world where data is always represented by its shortest possible format, there would therefore be no way to compress data. Instead of writing books on data compression, authors in such a world would write books on how to determine the shortest format for different types of data.

```
fpc = OpenRead["test.txt"];
g = 0; ar = Table[{i, 0}, {i, 256}];
While[0 == 0,
 g = Read[fpc, Byte];
 (* Skip space, newline & backslash *)
 If[g==10||g==32||g==92, Continue[]];
 If[g==EndOfFile, Break[]];
 ar[[g, 2]]++] (* increment counter *)
Close[fpc];
ar = Sort[ar, #1[[2]] > #2[[2]] &];
tot = Sum[
ar[[i,2]], {i,256}] (* total chars input *)
Table[{FromCharacterCode[ar[[i,1]]],ar[[i,2]],ar[[i,2]]/N[tot,4]},
 {i,93}] (* char code, freq., percentage *)
TableForm[%]
```

Figure Intro.3: Code for Table Intro.2.

A Word to the Wise . . .

The main aim of the field of data compression is, of course, to develop methods for better and faster compression. However, one of the main dilemmas of the *art* of data compression is when to stop looking for better compression. Experience shows that fine-tuning an algorithm to squeeze out the last remaining bits of redundancy from the data gives diminishing returns. Modifying an algorithm to improve compression by 1% may increase the run time by 10% and the complexity of the program by more than that. A good way out of this dilemma was taken by Fiala and Greene (Section 6.10). After developing their main algorithms A1 and A2, they modified them to produce less compression at a higher speed, resulting in algorithms B1 and B2. They then modified A1 and A2 again, but in the opposite direction, sacrificing speed to get slightly better compression.

The principle of compressing by removing redundancy also answers the following question: Why is it that an already compressed file cannot be compressed further? The answer, of course, is that such a file has little or no redundancy, so there is nothing to remove. An example of such a file is random text. In such text, each letter occurs with equal probability, so assigning them fixed-size codes does not add any redundancy. When such a file is compressed, there is no redundancy to remove. (Another answer is that if it were possible to compress an already compressed file, then successive compressions would reduce the size of the file until it becomes a single byte, or even a single bit. This, of course, is ridiculous since a single byte cannot contain the information present in an arbitrarily large file.)

In spite of the arguments above and the proof below, claims of recursive compression appear from time to time in the Internet. These are either checked and proved wrong or disappear silently. Reference [Barf 08], is a joke intended to amuse (and temporarily confuse) readers. A careful examination of this "claim" shows that any gain achieved by recursive compression of the Barf software is offset (perhaps more than offset) by the long name of the output file generated. The reader should also consult page 1132 for an interesting twist on the topic of compressing random data.

Definition of barf (verb): to vomit; purge; cast; sick; chuck; honk; throw up.

Since random data has been mentioned, let's say a few more words about it. Normally, it is rare to have a file with random data, but there is at least one good example—an already compressed file. Someone owning a compressed file normally knows that it is already compressed and would not attempt to compress it further, but there may be exceptions and one of them is data transmission by modems. Modern modems include hardware to automatically compress the data they send, and if that data is already compressed, there will not be further compression. There may even be expansion. This is why a modem should monitor the compression ratio "on the fly," and if it is low, it should stop compressing and should send the rest of the data uncompressed. The V.42bis protocol (Section 6.23) is a good example of this technique.

Before we prove the impossibility of recursive compression, here is an interesting twist on this concept. Several algorithms, such as JPEG, LZW, and MPEG, have long become de facto standards and are commonly used in web sites and in our computers. The field of data compression, however, is rapidly advancing and new, sophisticated methods are continually being developed. Thus, it is possible to take a compressed file, say JPEG, decompress it, and recompress it with a more efficient method. On the outside, this would look like recursive compression and may become a marketing tool for new, commercial compression software. The Stuffit software for the Macintosh platform (Section 11.16) does just that. It promises to compress already-compressed files and in many cases, it does!

The following simple argument illustrates the essence of the statement "Data compression is achieved by reducing or removing redundancy in the data." The argument shows that most data files cannot be compressed, no matter what compression method is used. This seems strange at first because we compress our data files all the time. The point is that most files cannot be compressed because they are random or close to random and therefore have no redundancy. The (relatively) few files that can be compressed are the ones that we *want* to compress; they are the files we use all the time. They have redundancy, are nonrandom, and are therefore useful and interesting.

Here is the argument. Given two different files A and B that are compressed to files C and D, respectively, it is clear that C and D must be different. If they were identical, there would be no way to decompress them and get back file A or file B.

Suppose that a file of size n bits is given and we want to compress it efficiently. Any compression method that can compress this file to, say, 10 bits would be welcome. Even compressing it to 11 bits or 12 bits would be great. We therefore (somewhat arbitrarily) assume that compressing such a file to half its size or better is considered good compression. There are 2^n n-bit files and they would have to be compressed into 2^n different files of sizes less than or equal to $n/2$. However, the total number of these files is

$$N = 1 + 2 + 4 + \cdots + 2^{n/2} = 2^{1+n/2} - 1 \approx 2^{1+n/2},$$

so only N of the 2^n original files have a chance of being compressed efficiently. The problem is that N is much smaller than 2^n. Here are two examples of the ratio between these two numbers.

For $n = 100$ (files with just 100 bits), the total number of files is 2^{100} and the number of files that can be compressed efficiently is 2^{51}. The ratio of these numbers is the ridiculously small fraction $2^{-49} \approx 1.78\times 10^{-15}$.

For $n = 1000$ (files with just 1000 bits, about 125 bytes), the total number of files is 2^{1000} and the number of files that can be compressed efficiently is 2^{501}. The ratio of these numbers is the incredibly small fraction $2^{-499} \approx 9.82\times 10^{-91}$.

Most files of interest are at least some thousands of bytes long. For such files, the percentage of files that can be efficiently compressed is so small that it cannot be computed with floating-point numbers even on a supercomputer (the result comes out as zero).

The 50% figure used here is arbitrary, but even increasing it to 90% isn't going to make a significant difference. Here is why. Assuming that a file of n bits is given and that $0.9n$ is an integer, the number of files of sizes up to $0.9n$ is

$$2^0 + 2^1 + \cdots + 2^{0.9n} = 2^{1+0.9n} - 1 \approx 2^{1+0.9n}.$$

For $n = 100$, there are 2^{100} files and $2^{1+90} = 2^{91}$ of them can be compressed well. The ratio of these numbers is $2^{91}/2^{100} = 2^{-9} \approx 0.00195$. For $n = 1000$, the corresponding fraction is $2^{901}/2^{1000} = 2^{-99} \approx 1.578\times 10^{-30}$. These are still extremely small fractions.

It is therefore clear that no compression method can hope to compress all files or even a significant percentage of them. In order to compress a data file, the compression algorithm has to examine the data, find redundancies in it, and try to remove them. The redundancies in data depend on the type of data (text, images, audio, etc.), which is why a new compression method has to be developed for each specific type of data and it performs best on this type. There is no such thing as a universal, efficient data compression algorithm.

Data compression has become so important that some researchers (see, for example, [Wolff 99]) have proposed the SP theory (for "simplicity" and "power"), which suggests that all computing is compression! Specifically, it says: Data compression may be interpreted as a process of removing unnecessary complexity (redundancy) in information, and thereby maximizing simplicity while preserving as much as possible of its nonredundant descriptive power. SP theory is based on the following conjectures:

- All kinds of computing and formal reasoning may usefully be understood as information compression by pattern matching, unification, and search.

- The process of finding redundancy and removing it may always be understood at a fundamental level as a process of searching for patterns that match each other, and merging or unifying repeated instances of any pattern to make one.

This book discusses many compression methods, some suitable for text and others for graphical data (still images or video) or for audio. Most methods are classified into four categories: run length encoding (RLE), statistical methods, dictionary-based (sometimes called LZ) methods, and transforms. Chapters 1 and 11 describe methods based on other principles.

Before delving into the details, we discuss important data compression terms.

- A *compressor* or *encoder* is a program that compresses the raw data in the input stream and creates an output stream with compressed (low-redundancy) data. A *decompressor* or *decoder* converts in the opposite direction. Note that the term *encoding* is very general and has several meanings, but since we discuss only data compression, we use the name *encoder* to mean data compressor. The term *codec* is often used to describe both the encoder and the decoder. Similarly, the term *companding* is short for "compressing/expanding."

- The term *stream* is used throughout this book instead of *file*. *Stream* is a more general term because the compressed data may be transmitted directly to the decoder, instead of being written to a file and saved. Also, the data to be compressed may be downloaded from a network instead of being input from a file.

- For the original input stream, we use the terms *unencoded*, *raw*, or *original* data. The contents of the final, compressed, stream are considered the *encoded* or *compressed* data. The term *bitstream* is also used in the literature to indicate the compressed stream.

The Gold Bug

Here, then, we have, in the very beginning, the groundwork for something more than a mere guess. The general use which may be made of the table is obvious—but, in this particular cipher, we shall only very partially require its aid. As our predominant character is 8, we will commence by assuming it as the "e" of the natural alphabet. To verify the supposition, let us observe if the 8 be seen often in couples—for "e" is doubled with great frequency in English—in such words, for example, as "meet," "fleet," "speed," "seen," "been," "agree," etc. In the present instance we see it doubled no less than five times, although the cryptograph is brief.

—Edgar Allan Poe

- A *nonadaptive* compression method is rigid and does not modify its operations, its parameters, or its tables in response to the particular data being compressed. Such a method is best used to compress data that is all of a single type. Examples are the Group 3 and Group 4 methods for facsimile compression (Section 5.7). They are specifically designed for facsimile compression and would do a poor job compressing any other data. In contrast, an *adaptive* method examines the raw data and modifies its operations and/or its parameters accordingly. An example is the adaptive Huffman method of Section 5.3. Some compression methods use a 2-pass algorithm, where the first pass reads the input stream to collect statistics on the data to be compressed, and the second pass does the actual compressing using parameters determined by the first pass. Such a method may be called *semiadaptive*. A data compression method can also be *locally adaptive*, meaning it adapts itself to local conditions in the input stream and varies this adaptation as it moves from area to area in the input. An example is the move-to-front method (Section 1.5).

- *Lossy/lossless compression:* Certain compression methods are lossy. They achieve better compression at the price of losing some information. When the compressed stream is decompressed, the result is not identical to the original data stream. Such a method makes sense especially in compressing images, video, or audio. If the loss of data is small, we may not be able to tell the difference. In contrast, text files, especially files containing computer programs, may become worthless if even one bit gets modified. Such files should be compressed only by a lossless compression method. [Two points should be mentioned regarding text files: (1) If a text file contains the source code of a program, consecutive blank spaces can often be replaced by a single space. (2) When the output of a word processor is saved in a text file, the file may contain information about the different fonts used in the text. Such information may be discarded if the user is interested in saving just the text.]

- *Cascaded compression:* The difference between lossless and lossy codecs can be illuminated by considering a cascade of compressions. Imagine a data file A that has been compressed by an encoder X, resulting in a compressed file B. It is possible, although pointless, to pass B through another encoder Y, to produce a third compressed file C. The point is that if methods X and Y are lossless, then decoding C by Y will produce an exact B, which when decoded by X will yield the original file A. However, if any of the compression algorithms is lossy, then decoding C by Y may produce a file B' different from B. Passing B' through X may produce something very different from A and may also result in an error, because X may not be able to read B'.

- *Perceptive compression:* A lossy encoder must take advantage of the special type of data being compressed. It should delete only data whose absence would not be detected by our senses. Such an encoder must therefore employ algorithms based on our understanding of psychoacoustic and psychovisual perception, so it is often referred to as a perceptive encoder. Such an encoder can be made to operate at a constant compression ratio, where for each x bits of raw data, it outputs y bits of compressed data. This is convenient in cases where the compressed stream has to be transmitted at a constant rate. The trade-off is a variable subjective quality. Parts of the original data that are difficult to compress may, after decompression, look (or sound) bad. Such parts may require more than y bits of output for x bits of input.

- *Symmetrical compression* is the case where the compressor and decompressor employ basically the same algorithm but work in "opposite" directions. Such a method makes sense for general work, where the same number of files are compressed as are decompressed. In an asymmetric compression method, either the compressor or the decompressor may have to work significantly harder. Such methods have their uses and are not necessarily bad. A compression method where the compressor executes a slow, complex algorithm and the decompressor is simple is a natural choice when files are compressed into an archive, where they will be decompressed and used very often. The opposite case is useful in environments where files are updated all the time and backups are made. There is a small chance that a backup file will be used, so the decompressor isn't used very often.

> Like the ski resort full of girls hunting for husbands and husbands hunting for girls, the situation is not as symmetrical as it might seem.
>
> —Alan Lindsay Mackay, lecture, Birckbeck College, 1964

◇ **Exercise Intro.2:** Give an example of a compressed file where good compression is important but the speed of both compressor and decompressor isn't important.

- Many modern compression methods are asymmetric. Often, the formal description (the standard) of such a method specifies the decoder and the format of the compressed stream, but does not discuss the operation of the encoder. Any encoder that generates a correct compressed stream is considered *compliant*, as is also any decoder that can read and decode such a stream. The advantage of such a description is that anyone is free to develop and implement new, sophisticated algorithms for the encoder. The implementor need not even publish the details of the encoder and may consider it proprietary. If a compliant encoder is demonstrably better than competing encoders, it may become a commercial success. In such a scheme, the encoder is considered *algorithmic*, while the decoder, which is normally much simpler, is termed *deterministic*. A good example of this approach is the MPEG-1 audio compression method (Section 10.14).

- A data compression method is called *universal* if the compressor and decompressor do not know the statistics of the input stream. A universal method is *optimal* if the compressor can produce compression ratios that asymptotically approach the entropy of the input stream for long inputs.

- The term *file differencing* refers to any method that locates and compresses the differences between two files. Imagine a file A with two copies that are kept by two users. When a copy is updated by one user, it should be sent to the other user, to keep the two copies identical. Instead of sending a copy of A, which may be big, a much smaller file containing just the differences, in compressed format, can be sent and used at the receiving end to update the copy of A. Section 11.14.2 discusses some of the details and shows why compression can be considered a special case of file differencing. Note that the term *differencing* is used in Section 1.3.1 to describe an entirely different compression method.

- Most compression methods operate in the *streaming mode*, where the codec inputs a byte or several bytes, processes them, and continues until an end-of-file is sensed. Some methods, such as Burrows-Wheeler transform (Section 11.1), work in the *block mode*, where the input stream is read block by block and each block is encoded separately. The block size should be a user-controlled parameter, since its size may significantly affect the performance of the method.

■ Most compression methods are *physical.* They look only at the bits in the input stream and ignore the meaning of the data items in the input (e.g., the data items may be words, pixels, or audio samples). Such a method translates one bitstream into another, shorter bitstream. The only way to make sense of the output stream (to decode it) is by knowing how it was encoded. Some compression methods are *logical.* They look at individual data items in the source stream and replace common items with short codes. A logical method is normally special purpose and can operate successfully on certain types of data only. The pattern substitution method described on page 35 is an example of a logical compression method.

■ *Compression performance*: Several measures are commonly used to express the performance of a compression method.

1. The *compression ratio* is defined as

$$\text{Compression ratio} = \frac{\text{size of the output stream}}{\text{size of the input stream}}.$$

A value of 0.6 means that the data occupies 60% of its original size after compression. Values greater than 1 imply an output stream bigger than the input stream (negative compression). The compression ratio can also be called bpb (bit per bit), since it equals the number of bits in the compressed stream needed, on average, to compress one bit in the input stream. In modern, efficient text compression methods, it makes sense to talk about bpc (bits per character)—the number of bits it takes, on average, to compress one character in the input stream.

Two more terms should be mentioned in connection with the compression ratio. The term *bitrate* (or "bit rate") is a general term for bpb and bpc. Thus, the main goal of data compression is to represent any given data at low bit rates. The term *bit budget* refers to the functions of the individual bits in the compressed stream. Imagine a compressed stream where 90% of the bits are variable-size codes of certain symbols, and the remaining 10% are used to encode certain tables. The bit budget for the tables is 10%.

2. The inverse of the compression ratio is called the *compression factor*:

$$\text{Compression factor} = \frac{\text{size of the input stream}}{\text{size of the output stream}}.$$

In this case, values greater than 1 indicate compression and values less than 1 imply expansion. This measure seems natural to many people, since the bigger the factor, the better the compression. This measure is distantly related to the sparseness ratio, a performance measure discussed in Section 8.6.2.

3. The expression $100 \times (1 - \text{compression ratio})$ is also a reasonable measure of compression performance. A value of 60 means that the output stream occupies 40% of its original size (or that the compression has resulted in savings of 60%).

4. In image compression, the quantity bpp (bits per pixel) is commonly used. It equals the number of bits needed, on average, to compress one pixel of the image. This quantity should always be compared with the bpp before compression.
5. The *compression gain* is defined as

$$100 \log_e \frac{\text{reference size}}{\text{compressed size}},$$

where the reference size is either the size of the input stream or the size of the compressed stream produced by some standard lossless compression method. For small numbers x, it is true that $\log_e(1+x) \approx x$, so a small change in a small compression gain is very similar to the same change in the compression ratio. Because of the use of the logarithm, two compression gains can be compared simply by subtracting them. The unit of the compression gain is called *percent log ratio* and is denoted by $\overset{\circ}{\circ}$.
6. The speed of compression can be measured in cycles per byte (CPB). This is the average number of machine cycles it takes to compress one byte. This measure is important when compression is done by special hardware.
7. Other quantities, such as mean square error (MSE) and peak signal to noise ratio (PSNR), are used to measure the distortion caused by lossy compression of images and movies. Section 7.4.2 provides information on those.
8. Relative compression is used to measure the compression gain in lossless audio compression methods, such as MLP (Section 10.7). This expresses the quality of compression by the number of bits each audio sample is reduced.

Name	Size	Description	Type
bib	111,261	A bibliography in UNIX *refer* format	Text
book1	768,771	Text of T. Hardy's *Far From the Madding Crowd*	Text
book2	610,856	Ian Witten's *Principles of Computer Speech*	Text
geo	102,400	Geological seismic data	Data
news	377,109	A Usenet news file	Text
obj1	21,504	VAX object program	Obj
obj2	246,814	Macintosh object code	Obj
paper1	53,161	A technical paper in *troff* format	Text
paper2	82,199	Same	Text
pic	513,216	Fax image (a bitmap)	Image
progc	39,611	A source program in C	Source
progl	71,646	A source program in LISP	Source
progp	49,379	A source program in Pascal	Source
trans	93,695	Document teaching how to use a terminal	Text

Table Intro.4: The Calgary Corpus.

■ The *Calgary Corpus* is a set of 18 files traditionally used to test data compression algorithms and implementations. They include text, image, and object files, for a total of more than 3.2 million bytes (Table Intro.4). The corpus can be downloaded from [Calgary 06].

■ The *Canterbury Corpus* (Table Intro.5) is another collection of files introduced in 1997 to provide an alternative to the Calgary corpus for evaluating lossless compression methods. The following concerns led to the new corpus:

1. The Calgary corpus has been used by many researchers to develop, test, and compare many compression methods, and there is a chance that new methods would unintentionally be fine-tuned to that corpus. They may do well on the Calgary corpus documents but poorly on other documents.
2. The Calgary corpus was collected in 1987 and is getting old. "Typical" documents change over a period of decades (e.g., html documents did not exist in 1987), and any body of documents used for evaluation purposes should be examined from time to time.
3. The Calgary corpus is more or less an arbitrary collection of documents, whereas a good corpus for algorithm evaluation should be selected carefully.

The Canterbury corpus started with about 800 candidate documents, all in the public domain. They were divided into 11 classes, representing different types of documents. A representative "average" document was selected from each class by compressing every file in the class using different methods and selecting the file whose compression was closest to the average (as determined by statistical regression). The corpus is summarized in Table Intro.5 and can be obtained from [Canterbury 06].

Description	File name	Size (bytes)
English text (*Alice in Wonderland*)	alice29.txt	152,089
Fax images	ptt5	513,216
C source code	fields.c	11,150
Spreadsheet files	kennedy.xls	1,029,744
SPARC executables	sum	38,666
Technical document	lcet10.txt	426,754
English poetry ("Paradise Lost")	plrabn12.txt	481,861
HTML document	cp.html	24,603
LISP source code	grammar.lsp	3,721
GNU manual pages	xargs.1	4,227
English play (*As You Like It*)	asyoulik.txt	125,179
Complete genome of the *E. coli* bacterium	E.Coli	4,638,690
The King James version of the Bible	bible.txt	4,047,392
The CIA World Fact Book	world192.txt	2,473,400

Table Intro.5: The Canterbury Corpus.

The last three files constitute the beginning of a random collection of larger files. More files are likely to be added to it.

■ The Calgary challenge [Calgary challenge 08], is a contest to compress the Calgary corpus. It was started in 1996 by Leonid Broukhis and initially attracted a number of contestants. In 2005, Alexander Ratushnyak achieved the current record of 596,314 bytes, using a variant of PAsQDa with a tiny dictionary of about 200 words.

Currently (late 2008), this challenge seems to have been preempted by the bigger prizes offered by the Hutter prize, and has featured no activity since 2005.

Here is part of the original challenge as it appeared in [Calgary challenge 08].

> I, Leonid A. Broukhis, will pay the amount of $(759,881.00 - X)/333$ US dollars (but not exceeding $1001, and no less than $10.01 "ten dollars and one cent") to the first person who sends me an archive of length X bytes, containing an executable and possibly other files, where the said executable file, run repeatedly with arguments being the names of other files contained in the original archive file one at a time (or without arguments if no other files are present) on a computer with no permanent storage or communication devices accessible to the running process(es) produces 14 new files, so that a 1-to-1 relationship of bytewise identity may be established between those new files and the files in the original Calgary corpus. (In other words, "solid" mode, as well as shared dictionaries/models and other tune-ups specific for the Calgary Corpus are allowed.)
>
> I will also pay the amount of $(777,777.00 - Y)/333$ US dollars (but not exceeding $1001, and no less than $0.01 "zero dollars and one cent") to the first person who sends me an archive of length Y bytes, containing an executable and exactly 14 files, where the said executable file, run with standard input taken directly (so that the `stdin` is seekable) from one of the 14 other files and the standard output directed to a new file, writes data to standard output so that the data being output matches one of the files in the original Calgary corpus and a 1-to-1 relationship may be established between the files being given as standard input and the files in the original Calgary corpus that the standard output matches. Moreover, after verifying the above requirements, an arbitrary file of size between 500 KB and 1 MB will be sent to the author of the decompressor to be compressed and sent back. The decompressor must handle that file correctly, and the compression ratio achieved on that file must be not worse that within 10% of the ratio achieved by gzip with default settings. (In other words, the compressor must be, roughly speaking, "general purpose.")

▪ The *probability model*. This concept is important in statistical data compression methods. In such a method, a model for the data has to be constructed before compression can begin. A typical model may be built by reading the entire input stream, counting the number of times each symbol appears (its frequency of occurrence), and computing the probability of occurrence of each symbol. The data stream is then input again, symbol by symbol, and is compressed using the information in the probability model. A typical model is shown in Table 5.42, page 266.

Reading the entire input stream twice is slow, which is why practical compression methods use estimates, or adapt themselves to the data as it is being input and compressed. It is easy to scan large quantities of, say, English text and calculate the frequencies and probabilities of every character. This information can later serve as an approximate model for English text and can be used by text compression methods to compress any English text. It is also possible to start by assigning equal probabilities to all the symbols in an alphabet, then reading symbols and compressing them, and, while doing that, also counting frequencies and changing the model as compression progresses. This is the principle behind the various *adaptive compression methods*.

- Source. A source of data items can be a file stored on a disk, a file that is input from outside the computer, text input from a keyboard, or a program that generates data symbols to be compressed or processed in some way. In a memoryless source, the probability of occurrence of a data symbol does not depend on its context. The term i.i.d. (independent and identically distributed) refers to a set of sources that have the same probability distribution and are mutually independent.

- Alphabet. This is the set of symbols that an application has to deal with. An alphabet may consist of the 128 ASCII codes, the 256 8-bit bytes, the two bits, or any other set of symbols.

- Random variable. This is a function that maps the results of random experiments to numbers. For example, selecting many people and measuring their heights is a random variable. The number of occurrences of each height can be used to compute the probability of that height, so we can talk about the probability distribution of the random variable (the set of probabilities of the heights). A special important case is a discrete random variable. The set of all values that such a variable can assume is finite or countably infinite.

- Compressed stream (or encoded stream). A compressor (or encoder) compresses data and generates a compressed stream. This is often a file that is written on a disk or is stored in memory. Sometimes, however, the compressed stream is a string of bits that are transmitted over a communications line.

[End of data compression terms.]

The concept of *data reliability and integrity* (page 247) is in some sense the opposite of data compression. Nevertheless, the two concepts are often related since any good data compression program should generate reliable code and so should be able to use error-detecting and error-correcting codes.

Compression benchmarks

Research in data compression, as in many other areas of computer science, concentrates on finding new algorithms and improving existing ones. In order to prove its value, however, an algorithm has to be implemented and tested. Thus, every researcher, programmer, and developer compares a new algorithm to older, well-established and known methods, and draws conclusions about its performance.

In addition to these tests, workers in the field of compression continually conduct extensive benchmarks, where many algorithms are run on the same set of data files and the results are compared and analyzed. This short section describes a few independent compression benchmarks.

Perhaps the most-important fact about these benchmarks is that they generally restrict themselves to compression ratios. Thus, a winner in such a benchmark may not be the best choice for general, everyday use, because it may be slow, may require large memory space, and may be expensive or protected by patents. Benchmarks for compression speed are rare, because it is difficult to accurately measure the run time of an executable program (i.e., a program whose source code is unavailable). Another drawback of benchmarking is that their data files are generally publicly known, so anyone interested in record breaking for its own sake may tune an existing algorithm to the particular data files used by a benchmark and in this way achieve the smallest (but nevertheless meaningless) compression ratio.

From the Dictionary

Benchmark: A standard by which something can be measured or judged; "his painting sets the benchmark of quality."

In computing, a benchmark is the act of running a computer program, a set of programs, or other operations, in order to assess the relative performance of an object, normally by running a number of standard tests and trials against it.

The term benchmark originates from the chiseled horizontal marks that surveyors made in stone structures, into which an angle-iron could be placed to form a "bench" for a leveling rod, thus ensuring that a leveling rod could be accurately repositioned in the same place in future.

The following independent benchmarks compare the performance (compression ratios but generally not speeds, which are difficult to measure) of many algorithms and their implementations. Surprisingly, the results often indicate that the winner comes from the family of context-mixing compression algorithms. Such an algorithm employs several models of the data to predict the next data symbol, and then combines the predictions in some way to end up with a probability for the next symbol. The symbol and its computed probability are then sent to an adaptive arithmetic coder, to be encoded. Included in this family of lossless algorithms are the many versions and derivatives of PAQ (Section 5.15) as well as many other, less well-known methods.

- The Maximum Compression Benchmark, managed by Werner Bergmans. This suite of tests (described in [Bergmans 08]) was started in 2003 and is still frequently updated. The goal is to discover the best compression ratios for several different types of data, such as text, images, and executable code. Every algorithm included in the tests is first tuned by setting any switches and parameters to the values that yield best performance. The owner of this benchmark prefers command line (console) compression programs over GUI ones. At the time of writing (late 2008), more than 150 programs have been tested on several large collections of test files. Most of the top-ranked algorithms are of the context mixing type. Special mention goes to PAsQDa 4.1b and WinRK 2.0.6/pwcm. The latest update to this benchmark reads as follows:

> 28-September-2008: Added PAQ8P, 7-Zip 4.60b, FreeARC 0.50a (June 23 2008), Tornado 0.4a, M1 0.1a, BZP 0.3, NanoZIP 0.04a, Blizzard 0.24b and WinRAR 3.80b5 (MFC to do for WinRAR and 7-Zip). PAQ8P manages to squeeze out an additional 12 KB from the BMP file, further increasing the gap to the number 2 in the SFC benchmark; newcomer NanoZIP takes 6th place in SFC!. In the MFC benchmark PAQ8 now takes a huge lead over WinRK 3.0.3, but WinRK 3.1.2 is on the todo list to be tested. To be continued....

- Johan de Bock started the UCLC (ultimate command-line compressors) benchmark project [UCLC 08]. A wide variety of tests are performed to compare the latest state-of-the-art command line compressors. The only feature being compared is the compression ratio; run-time and memory requirements are ignored. More than 100 programs have been tested over the years, with WinRK and various versions of PAQ declared the best performers (except for audio and grayscale images, where the records were achieved by specialized algorithms).

- The EmilCont benchmark [Emilcont 08] is managed by Berto Destasio. At the time of writing, the latest update of this site dates back to March 2007. EmilCont tests hundreds of algorithms on a confidential set of data files that include text, images, audio, and executable code. As usual, WinRK and PAQ variants are among the record holders, followed by SLIM.

> An important feature of these benchmarks is that the test data will not be released to avoid the unfortunate practice of compression writers tuning their programs to the benchmarks.
>
> —From [Emilcont 08]

- The Archive Comparison Test (ACT), maintained by Jeff Gilchrist [Gilchrist 08], is a set of benchmarks designed to demonstrate the state of the art in lossless data compression. It contains benchmarks on various types of data for compression speed, decompression speed, and compression ratio.

  The site lists the results of comparing 162 DOS/Windows programs, eight Macintosh programs, and 11 JPEG compressors. However, the tests are old. Many were performed in or before 2002 (except the JPEG tests, which took place in 2007).

- In [Ratushnyak 08], Alexander Ratushnyak reports the results of hundreds of speed tests done in 2001.

- Site [squeezemark 09] is devoted to benchmarks of lossless codecs. It is kept up to date by its owner, Stephan Busch, and has many pictures of compression pioneers.

How to Hide Data

Here is an unforeseen, unexpected application of data compression. For a long time I have been thinking about how to best hide a sensitive data file in a computer, while still having it ready for use at a short notice. Here is what we came up with. Given a data file A, consider the following steps:

1. Compress A. The result is a file B that is small and also seems random. This has two advantages (1) the remaining steps encrypt and hide small files and (2) the next step encrypts a random file, thereby making it difficult to break the encryption simply by checking every key.

2. Encrypt B with a secret key to obtain file C. A would-be codebreaker may attempt to decrypt C by writing a program that loops and tries every key, but here is the rub. Each time a key is tried, someone (or something) has to check the result. If the result looks meaningful, it may be the decrypted file B, but if the result seems random, the loop should continue. At the end of the loop; frustration.

3. Hide C inside a cover file D to obtain a large file E. Use one of the many steganographic methods for this (notice that many such methods depend on secret keys). One reference for steganography is [Salomon 03], but today there may be better texts.

4. Hide E in plain sight in your computer by changing its name and placing it in a large folder together with hundreds of other, unfamiliar files. A good idea may be to change the file name to `msLibPort.dll` (or something similar that includes MS and other familiar-looking terms) and place it in one of the many large folders created and used exclusively by Windows or any other operating system. If files in this folder are visible, do not make your file invisible. Anyone looking inside this folder will see hundreds of unfamiliar files and will have no reason to suspect `msLibPort.dll`. Even if this happens, an opponent would have a hard time guessing the three steps above (unless he has read these paragraphs) and the keys used. If file E is large (perhaps more than a few Gbytes), it should be segmented into several smaller files and each hidden in plain sight as described above. This step is important because there are utilities that identify large files and they may attract unwanted attention to your large E.

For those who require even greater privacy, here are a few more ideas. (1) A password can be made strong by including in it special characters such §, ¶, †, and ‡. These can be typed with the help of special modifier keys found on most keyboards. (2) Add a step between steps 1 and 2 where file B is recompressed by any compression method. This will not decrease the size of B but will defeat anyone trying to decompress B into meaningful data simply by trying many decompression algorithms. (3) Add a step between steps 1 and 2 where file B is partitioned into segments and random data inserted between the segments. (4) Instead of inserting random data segments, swap segments to create a permutation of the segments. The permutation may be determined by the password used in step 2.

> Until now, the US government's default position has been: If you can't keep data secret, at least hide it on one of 24,000 federal Websites, preferably in an incompatible or obsolete format.
>
> —*Wired*, July 2009.

Compression Curiosities

People are curious and they also like curiosities (not the same thing). It is easy to find curious facts, behavior, and objects in many areas of science and technology. In the early days of computing, for example, programmers were interested in programs that print themselves. Imagine a program in a given programming language. The source code of the program is printed on paper and can easily be read. When the program is executed, it prints its own source code. Curious! The few compression curiosities that appear here are from [curiosities 08].

- We often hear the following phrase "I want it all and I want it now!" When it comes to compressing data, we all want the best compressor. So, how much can a file possibly be compressed? Lossless methods routinely compress files to less than half their size and can go down to compression ratios of about 1/8 or smaller. Lossy algorithms do much better. Is it possible to compress a file by a factor of 10,000? Now that would be a curiosity.

The Iterated Function Systems method (IFS, Section 7.39) can compress certain files by factors of thousands. However, such files must be especially prepared (see four examples in Section 7.39.3). Given an arbitrary data file, there is no guarantee that IFS will compress it well.

The 115-byte RAR-compressed file (see Section 6.22 for RAR) found at [curio.test 08] swells, when decompressed, to 4,884,863 bytes, a compression factor of 42,477! The hexadecimal listing of this (obviously contrived) file is

```
526172211A07003BD07308000D00000000000000CB5C74C0802800300000007F894A0002F4EE5
A3DA39B29351D35080020000000746573742E747874A1182FD22D77B8617EF782D70000000000
00000000000000000000000000000093DE30369CFB76A800BF8867F6A9FFD4C43D7B00400700
```

■ Even if we do not obtain good compression, we certainly don't expect a compression program to *increase* the size of a file, but this is precisely what happens sometimes (even often). Expansion is the bane of compression and it is easy to choose the wrong data file or the wrong compressor for a given file and so end up with an expanded file (and high blood pressure into the bargain). If we try to compress a text file with a program designed specifically for compressing images or audio, often the result is expansion, even considerable expansion. Trying to compress random data very often results in expansion.

> The Ring was cut from Sauron's hand by Isildur at the slopes of Mount Doom, and he in turn lost it in the River Anduin just before he was killed in an Orc ambush. Since it indirectly caused Isildur's death by slipping from his finger, it was known in Gondorian lore as Isildur's Bane.
>
> —From Wikipedia

■ Another contrived file is the 99,058-byte `strange.rar`, located at [curio.strange 08]. When decompressed with RAR, the resulting file, `Compression Test.gba`, is 1,048,576-bytes long, indicating a healthy compression factor of 10.59. However, when `Compression Test.gba` is compressed in turn with zip, there is virtually no compression. Two high-performance compression algorithms yield very different results when processing the same data. Curious behavior!

■ Can a compression algorithm compress a file to itself? Obviously, the trivial algorithm that does nothing, achieves precisely that. Thus, we better ask, given a well-known, nontrivial compression algorithm, is it possible to find (or construct) a file that the algorithm will compress to itself? The surprising answer is yes. File `selfgz.gz` that can be found at `http://www.maximumcompression.com/selfgz.gz` yields, when compressed by gzip, itself!

> "Curiouser and curiouser!" Cried Alice (she was so much surprised, that for the moment she quite forgot how to speak good English).
>
> —Lewis Carroll, *Alice's Adventures in Wonderland* (1865)

The Ten Commandments of Compression

1. Redundancy is your enemy, eliminate it.

2. Entropy is your goal, strive to achieve it.

3. Read the literature before you try to publish/implement your new, clever compression algorithm. Others may have been there before you.

4. There is no universal compression method that can compress any file to just a few bytes. Thus, refrain from making incredible claims. They will come back to haunt you.

5. The G-d of compression prefers free and open source codecs.

6. If you decide to patent your algorithm, make sure anyone can understand your patent application. Others might want to improve on it. Talking about patents, recall the following warning about them (from D. Knuth) "there are better ways to earn a living than to prevent other people from making use of one's contributions to computer science."

7. Have you discovered a new, universal, simple, fast, and efficient compression algorithm? Please don't ask others (especially these authors) to evaluate it for you for free.

8. The well-known saying (also from Knuth) "beware of bugs in the above code; I have only proved it correct, not tried it," applies also (perhaps even mostly) to compression methods. Implement and test your method thoroughly before you brag about it.

9. Don't try to optimize your algorithm/code to squeeze the last few bits out of the output in order to win a prize. Instead, identify the point where you start getting diminishing returns and stop there.

10. This is your own, private commandment. Grab a pencil, write it here, and obey it.

Why this book? Most drivers know little or nothing about the operation of the engine or transmission in their cars. Few know how cellular telephones, microwave ovens, or combination locks work. Why not let scientists develop and implement compression methods and have us use them without worrying about the details? The answer, naturally, is curiosity. Many drivers try to tinker with their car out of curiosity. Many weekend sailors love to mess about with boats even on weekdays, and many children spend hours taking apart a machine, a device, or a toy in an attempt to understand its operation. If you are curious about data compression, this book is for you.

The typical reader of this book should have a basic knowledge of computer science; should know something about programming and data structures; feel comfortable with terms such as *bit*, *mega*, *ASCII*, *file*, *I/O*, and *binary search*; and should be curious. The necessary mathematical background is minimal and is limited to logarithms, matrices, polynomials, differentiation/integration, and the concept of probability. This book is not intended to be a guide to software implementors and has few programs.

The following URLs have useful links and pointers to the many data compression resources available on the Internet and elsewhere:

`http://www.hn.is.uec.ac.jp/~arimura/compression_links.html`,

`http://cise.edu.mie-u.ac.jp/~okumura/compression.html`,
`http://compression-links.info/`, `http://compression.ca/` (mostly comparisons),
`http://datacompression.info/`. This URL has a wealth of information on data compression, including tutorials, links, and lists of books. The site is owned by Mark Nelson.
`http://directory.google.com/Top/Computers/Algorithms/Compression/` is also a growing, up-to-date, site.

Reference [Okumura 98] discusses the history of data compression in Japan.

Data Compression Resources

A vast number of resources on data compression are available. Any Internet search under "data compression," "lossless data compression," "image compression," "audio compression," and similar topics returns at least tens of thousands of results. Traditional (printed) resources range from general texts and texts on specific aspects or particular methods, to survey articles in magazines, to technical reports and research papers in scientific journals. Following is a short list of (mostly general) books, sorted by date of publication.

Khalid Sayood, *Introduction to Data Compression*, Morgan Kaufmann, 3rd edition (2005).

Ida Mengyi Pu, *Fundamental Data Compression*, Butterworth-Heinemann (2005).

Darrel Hankerson, *Introduction to Information Theory and Data Compression*, Chapman & Hall (CRC), 2nd edition (2003).

Peter Symes, *Digital Video Compression*, McGraw-Hill/TAB Electronics (2003).

Charles Poynton, *Digital Video and HDTV Algorithms and Interfaces*, Morgan Kaufmann (2003).

Iain E. G. Richardson, *H.264 and MPEG-4 Video Compression: Video Coding for Next Generation Multimedia*, John Wiley and Sons (2003).

Khalid Sayood, *Lossless Compression Handbook*, Academic Press (2002).

Touradj Ebrahimi and Fernando Pereira, *The MPEG-4 Book*, Prentice Hall (2002).

Adam Drozdek, *Elements of Data Compression*, Course Technology (2001).

David Taubman and Michael Marcellin, (eds), *JPEG2000: Image Compression Fundamentals, Standards and Practice*, Springer Verlag (2001).

Kamisetty R. Rao, *The Transform and Data Compression Handbook*, CRC (2000).

Ian H. Witten, Alistair Moffat, and Timothy C. Bell, *Managing Gigabytes: Compressing and Indexing Documents and Images*, Morgan Kaufmann, 2nd edition (1999).

Peter Wayner, *Compression Algorithms for Real Programmers*, Morgan Kaufmann (1999).

John Miano, *Compressed Image File Formats: JPEG, PNG, GIF, XBM, BMP*, ACM Press and Addison-Wesley Professional (1999).

Mark Nelson and Jean-Loup Gailly, *The Data Compression Book*, M&T Books, 2nd edition (1995).

William B. Pennebaker and Joan L. Mitchell, *JPEG: Still Image Data Compression Standard*, Springer Verlag (1992).

Timothy C. Bell, John G. Cleary, and Ian H. Witten, *Text Compression*, Prentice Hall (1990).

James A. Storer, *Data Compression: Methods and Theory*, Computer Science Press.

John Woods, ed., *Subband Coding*, Kluwer Academic Press (1990).

Notation

- The symbol "␣" is used to indicate a blank space in places where spaces may lead to ambiguity.
- The acronyms MSB and LSB refer to most-significant-bit and least-significant-bit, respectively.
- The notation 1^i0^j indicates a bit string of i consecutive 1's followed by j zeros.

Some readers called into question the title of the predecessors of this book. What does it mean for a work of this kind to be complete, and how complete is this book? Here is our opinion on the matter. We like to believe that if the entire field of data compression were (heaven forbid) to disappear, a substantial part of it could be reconstructed from this work. Naturally, we don't compare ourselves to James Joyce, but his works provide us with a similar example. He liked to claim that if the Dublin of his time were to be destroyed, it could be reconstructed from his works.

Readers who would like to get an idea of the effort it took to write this book should consult the Colophon.

The authors welcome any comments, suggestions, and corrections. They should be sent to `dsalomon@csun.edu` or `gim@ieee.org`.

> The days just prior to marriage are like a snappy introduction to a tedious book.
>
> —Wilson Mizner

1
Basic Techniques

1.1 Intuitive Compression

Data is compressed by reducing its redundancy, but this also makes the data less reliable, more prone to errors. Increasing the integrity of data, on the other hand, is done by adding check bits and parity bits, a process that increases the size of the data, thereby increasing redundancy. Data compression and data reliability are therefore opposites, and it is interesting to note that the latter is a relatively recent field, whereas the former existed even before the advent of computers. The sympathetic telegraph, discussed in the Preface, the Braille code of 1820 (Section 1.1.1), the shutter telegraph of the British admiralty [Bell et al. 90], and the Morse code of 1838 (Table 2.1) use simple, intuitive forms of compression. It therefore seems that reducing redundancy comes naturally to anyone who works on codes, but increasing it is something that "goes against the grain" in humans. This section discusses simple, intuitive compression methods that have been used in the past. Today these methods are mostly of historical interest, since they are generally inefficient and cannot compete with the modern compression methods developed during the last several decades.

1.1.1 Braille

This well-known code, which enables the blind to read, was developed by Louis Braille in the 1820s and is still in common use today, after having been modified several times. Many books in Braille are available from the National Braille Press. The Braille code consists of groups (or cells) of 3×2 dots each, embossed on thick paper. Each of the six dots in a group may be flat or raised, implying that the information content of a group is equivalent to six bits, resulting in 64 possible groups. The letters (Table 1.1), digits, and common punctuation marks do not require all 64 codes, which is why the remaining

D. Salomon, G. Motta, *Handbook of Data Compression*, 5th ed.
DOI 10.1007/978-1-84882-903-9_1, © Springer-Verlag London Limited 2010

groups may be used to code common words—such as `and`, `for`, and `of`—and common strings of letters—such as `ound`, `ation` and `th` (Table 1.2).

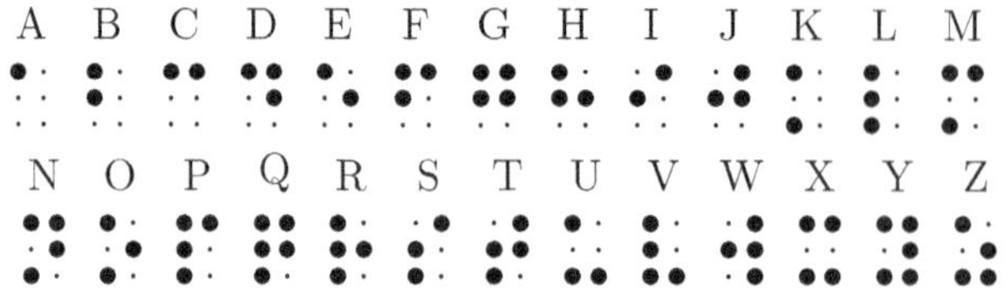

Table 1.1: The 26 Braille Letters.

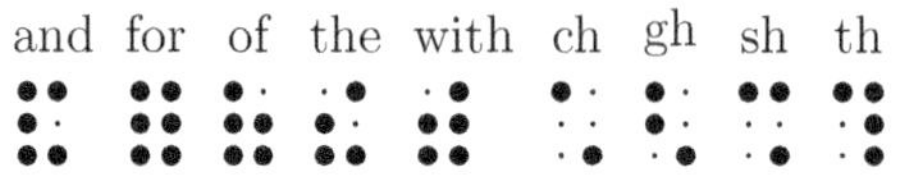

Table 1.2: Some Words and Strings in Braille.

Redundancy in Everyday Situations

Even though we don't unnecessarily increase redundancy in our data, we use redundant data all the time, mostly without noticing it. Here are some examples:

All natural languages are redundant. A Portuguese who does not speak Italian may read an Italian newspaper and still understand most of the news because he recognizes the basic form of many Italian verbs and nouns and because most of the text he does not understand is superfluous (i.e., redundant).

PIN is an acronym for "Personal Identification Number," but banks always ask you for your "PIN number." SALT was an acronym for "Strategic Arms Limitations Talks," but TV announcers in the 1970s kept talking about the "SALT Talks." These are just two examples illustrating how natural it is to be redundant in everyday situations. More examples can be found at URL
`http://www.corsinet.com/braincandy/twice.html`

◇ **Exercise 1.1:** Find a few more everyday redundant phrases.

The amount of compression achieved by Braille is small but significant, because books in Braille tend to be very large (a single group covers the area of about ten printed letters). Even this modest compression comes with a price. If a Braille book is mishandled or becomes old and some dots are flattened, serious reading errors may result since every possible group is used.

(*Windots2*, from [windots 06] and *Duxbury Braille Translator*, from [afb 06], are current programs for those wanting to experiment with Braille.)

1.1.2 Irreversible Text Compression

Sometimes it is acceptable to "compress" text by simply throwing away some information. This is called *irreversible text compression* or *compaction*. The decompressed text will not be identical to the original, so such methods are not general purpose; they can only be used in special cases.

A run of consecutive blank spaces may be replaced by a single space. This may be acceptable in literary texts and in most computer programs, but it should not be used when the data is in tabular form.

In extreme cases all text characters except letters and spaces may be thrown away, and the letters may be case flattened (converted to all lower- or all uppercase). This will leave just 27 symbols, so a symbol can be encoded in 5 instead of the usual 8 bits. The compression ratio is $5/8 = .625$, not bad, but the loss may normally be too great. (An interesting example of similar text is the last chapter of *Ulysses* by James Joyce. In addition to letters, digits, and spaces, this long chapter contains only a few punctuation marks.)

◇ **Exercise 1.2:** A character set including the 26 uppercase letters and the space can be coded with 5-bit codes, but that would leave five unused codes. Suggest a way to use them.

1.1.3 Irreversible Compression

The following method achieves unheard-of compression factors. We consider the input data a stream of bytes, read it byte by byte, and count the number of times each byte value occurs. There are 256 byte values, so we end up with 256 integers. After being encoded with variable-length codes, these integers become the output file. The resulting compression factor is astounding, especially if the file is large. For example, a 4 Gbyte file (about the size of a DVD) may be compressed by a factor of 650,000! Here is why. In one extreme case, all 256 counts would be roughly equal, so each would equal approximately $4G/256 = 2^2 \times 2^{30}/2^8 = 2^{24}$ and each would require about 24 bits. The total compressed size would be about $24 \times 256 = 6{,}144$ bits, yielding a compression factor of $4 \times 2^{30}/6{,}144 \approx 651{,}000$. In the other extreme case, all the bytes in the input would be identical. One count would become 4G (a 32-bit number), and the other 255 counts would be zeros (and would require one or two bits each to encode). The total number of bits would be $32 + 255 \times 2 = 543$, leading to another fantastic compression factor.

With such an excellent algorithm, why do researchers still attempt to develop other methods? Answer, because the compressed file cannot be decompressed; the method is irreversible. However, we live in such a complex world that even this seemingly useless scheme may find applications. Here is an interesting example.

Today, when a scientist discovers or develops something new, they rush to publish it, but in the past scientists often kept their discoveries secret for various reasons. The great mathematician Gauss did not consider much of his work worth publishing. Isaac Newton did not publish his work on the calculus until Gottfried Leibniz came up with his version. Charles Darwin kept his theory of evolution private for fear of violent reaction by the church. Niccolo Tartaglia discovered how to solve cubic equations but kept his method secret in order to win a competition (in 1535).

So what do you do if you don't want to publish a discovery, but still want to be able to claim priority in case someone else discovers the same thing? Simple; you write down a description of your achievement, compress it with the method above, keep the original in a safe place, and give the compressed version to a third party, perhaps a lawyer. When someone else claims the discovery, you show your description, compress it in the presence of witnesses, and show that the result is identical to what the third party has received from you in the past. Clever!

1.1.4 Ad Hoc Text Compression

Here are some simple, intuitive ideas for cases where the compression must be reversible (lossless).

■ If the text contains many spaces but they are not clustered, they may be removed and their positions indicated by a bit-string that contains a 0 for each text character that is not a space and a 1 for each space. Thus, the text

`Here␣are␣some␣ideas`

is encoded as the bit-string `0000100010000100000` followed by the text

`Herearesomeideas.`

If the number of blank spaces is small, the bit-string will be sparse, and the methods of Section 11.5 can be employed to compress it efficiently.

■ Since ASCII codes are essentially 7 bits long, the text may be compressed by writing 7 bits per character instead of 8 on the output stream. This may be called *packing*. The compression ratio is, of course, $7/8 = 0.875$.

■ The numbers $40^3 = 64{,}000$ and $2^{16} = 65{,}536$ are not very different and satisfy the relation $40^3 < 2^{16}$. This can serve as the basis of an intuitive compression method for a small set of symbols. If the data to be compressed is text with at most 40 different characters (such as the 26 letters, 10 digits, a space, and three punctuation marks), then this method produces a compression factor of $24/16 = 1.5$. Here is how it works.

Given a set of 40 characters and a string of characters from the set, we group the characters into triplets. Each character can take one of 40 values, so a trio of characters can have one of $40^3 = 64{,}000$ values. Such values can be expressed in 16 bits each, because 40^3 is less than 2^{16}. Without compression, each of the 40 characters requires one byte, so our intuitive method produces a compression factor of $3/2 = 1.5$. (This is one of those rare cases where the compression factor is constant and is known in advance.)

■ If the text includes just uppercase letters, digits, and some punctuation marks, the old 6-bit CDC display code (Table 1.3) may be used. This code was commonly used in second-generation computers (and even a few third-generation ones). These computers did not need more than 64 characters because they did not have any display monitors and they sent their output to printers that could print only a limited set of characters.

■ Another old code worth mentioning is the Baudot code (Table 1.4). This was a 5-bit code developed by J. M. E. Baudot in about 1880 for telegraph communication. It became popular and by 1950 was designated the International Telegraph Code No. 1. It was used in many first- and second-generation computers. The code uses 5 bits per character but encodes more than 32 characters. Each 5-bit code can be the code of two characters, a letter and a figure. The "letter shift" and "figure shift" codes are used to shift between letters and figures.

Using this technique, the Baudot code can represent $32 \times 2 - 2 = 62$ characters (each code can have two meanings except the LS and FS codes). The actual number of

Bits	Bit positions 210							
543	0	1	2	3	4	5	6	7
0		A	B	C	D	E	F	G
1	H	I	J	K	L	M	N	O
2	P	Q	R	S	T	U	V	W
3	X	Y	Z	0	1	2	3	4
4	5	6	7	8	9	+	-	*
5	/	(	)	$	=	sp	,	.
6	≡	[	]	:	≠	_	∨	∧
7	↑	↓	<	>	≤	≥	¬	;

Table 1.3: The CDC Display Code.

characters, however, is smaller than that, because five of the codes have one meaning each, and some codes are not assigned.

The Baudot code is not reliable because it does not employ a parity bit. A bad bit transforms a character into another character. In particular, a bad bit in a shift character causes a wrong interpretation of all the characters following, up to the next shift.

Letters	Code	Figures	Letters	Code	Figures
A	10000	1	Q	10111	/
B	00110	8	R	00111	-
C	10110	9	S	00101	SP
D	11110	0	T	10101	na
E	01000	2	U	10100	4
F	01110	na	V	11101	'
G	01010	7	W	01101	?
H	11010	+	X	01001	,
I	01100	na	Y	00100	3
J	10010	6	Z	11001	:
K	10011	(	LS	00001	LS
L	11011	=	FS	00010	FS
M	01011	)	CR	11000	CR
N	01111	na	LF	10001	LF
O	11100	5	ER	00011	ER
P	11111	%	na	00000	na

LS, Letter Shift; FS, Figure Shift.
CR, Carriage Return; LF, Line Feed.
ER, Error; na, Not Assigned; SP, Space.

Table 1.4: The Baudot Code.

- If the data includes just integers, each decimal digit may be represented in 4 bits, with two digits packed in a byte. Data consisting of dates may be represented as the number of days since January 1, 1900 (or some other convenient start date). Each date may be stored as a 16-bit or 24-bit number (2 or 3 bytes). If the data consists of date/time pairs, a possible compressed representation is the number of seconds since a convenient start date. If stored as a 32-bit number (4 bytes) such a representation can be sufficient for about 136 years.

- Dictionary data (or any list sorted lexicographically) can be compressed using the concept of *front compression*. This is based on the observation that adjacent words in such a list tend to share some of their initial characters. A word can therefore be compressed by dropping the n characters it shares with its predecessor in the list and replacing them with the numeric value of n.

The 9/19/89 Syndrome

How can a date, such as 11/12/71, be represented inside a computer? One way to do this is to store the number of days since January 1, 1900 in an integer variable. If the variable is 16 bits long (including 15 magnitude bits and one sign bit), it will overflow after $2^{15} = 32K = 32{,}768$ days, which is September 19, 1989. This is precisely what happened on that day in several computers (see the January, 1991 issue of the *Communications of the ACM*). Notice that doubling the size of such a variable to 32 bits would have delayed the problem until after $2^{31} = 2$ giga days have passed, which would occur sometime in the fall of year 5,885,416.

Table 1.5 shows a short example taken from a word list used to create anagrams. It is clear that it is easy to obtain significant compression with this simple method (see also [Robinson and Singer 81] and [Nix 81]).

a	a
aardvark	1ardvark
aback	1back
abaft	3ft
abandon	3ndon
abandoning	7ing
abasement	3sement
abandonment	3ndonment
abash	3sh
abated	3ted
abate	5
abbot	2bot
abbey	3ey
abbreviating	3reviating
abbreviate	9e
abbreviation	9ion

Table 1.5: Front Compression.

■ The MacWrite word processor [Young 85] used a special 4-bit code to encode the most-common 15 characters "␣etnroaisdlhcfp" plus an escape code. Any of these 15 characters is encoded by 4 bits. Any other character is encoded as the escape code followed by the 8 bits of the ASCII code of the character; a total of 12 bits. Each paragraph is coded separately, and if this results in expansion, the paragraph is stored as plain ASCII. One more bit is added to each paragraph to indicate whether or not it uses compression.

> The principle of parsimony values a theory's ability to compress a maximum of information into a minimum of formalism. Einstein's celebrated $E = mc^2$ derives part of its well-deserved fame from the astonishing wealth of meaning it packs into its tiny frame. Maxwell's equations, the rules of quantum mechanics, and even the basic equations of the general theory of relativity similarly satisfy the parsimony requirement of a fundamental theory: They are compact enough to fit on a T-shirt. By way of contrast, the human genome project, requiring the quantification of hundreds of thousands of genetic sequences, represents the very antithesis of parsimony.
>
> —Hans C. von Baeyer, *Maxwell's Demon*, 1998

1.2 Run-Length Encoding

The idea behind this approach to data compression is this: If a data item d occurs n consecutive times in the input stream, replace the n occurrences with the single pair nd. The n consecutive occurrences of a data item are called a *run length* of n, and this approach to data compression is called *run-length encoding* or RLE. We apply this idea first to text compression and then to image compression.

1.3 RLE Text Compression

Just replacing `2.␣all␣is␣too␣well` with `2.␣a2␣is␣t2␣we2` will not work. Clearly, the decompressor should have a way to tell that the first 2 is part of the text while the others are repetition factors for the letters `o` and `l`. Even the string `2.␣a2l␣is␣t2o␣we2l` does not solve this problem (and also does not produce any compression). One way to solve this problem is to precede each repetition with a special escape character. If we use the character `@` as the escape character, then the string `2.␣a@2l␣is␣t@2o␣we@2l` can be decompressed unambiguously. However, this string is longer than the original string, because it replaces two consecutive letters with three characters. We have to adopt the convention that only three or more repetitions of the same character will be replaced with a repetition factor. Figure 1.6a is a flowchart for such a simple run-length text compressor.

After reading the first character, the repeat-count is 1 and the character is saved. Subsequent characters are compared with the one already saved, and if they are identical to it, the repeat-count is incremented. When a different character is read, the operation depends on the value of the repeat count. If it is small, the saved character is written on the compressed file and the newly-read character is saved. Otherwise, an `@` is written, followed by the repeat-count and the saved character.

Decompression is also straightforward. It is shown in Figure 1.6b. When an `@` is read, the repetition count n and the actual character are immediately read, and the character is written n times on the output stream.

The main problems with this method are the following:

1. In plain English text there are not many repetitions. There are many "doubles" but a "triple" is rare. The most repetitive character is the space. Dashes or asterisks may sometimes also repeat. In mathematical texts, digits may repeat. The following "paragraph" is a contrived example.

> The abbott from Abruzzi accedes to the demands of all abbesses from Narragansett and Abbevilles from Abyssinia. He will accommodate them, abbreviate his sabbatical, and be an accomplished accessory.

2. The character "`@`" may be part of the text in the input stream, in which case a different escape character must be chosen. Sometimes the input stream may contain every possible character in the alphabet. An example is an object file, the result of compiling a program. Such a file contains machine instructions and can be considered a string of bytes that may have any values. The MNP5 method described below and in Section 5.4 provides a solution.
3. Since the repetition count is written on the output stream as a byte, it is limited to counts of up to 255. This limitation can be softened somewhat when we realize that the existence of a repetition count means that there is a repetition (at least three identical consecutive characters). We may adopt the convention that a repeat count of 0 means three repeat characters, which implies that a repeat count of 255 means a run of 258 identical characters.

> There are three kinds of lies: lies, damned lies, and statistics.
> (Attributed by Mark Twain to Benjamin Disraeli)

The MNP class 5 method was used for data compression in old modems. It has been developed by Microcom, Inc., a maker of modems (MNP stands for Microcom Network Protocol), and it uses a combination of run-length and adaptive frequency encoding. The latter technique is described in Section 5.4, but here is how MNP5 solves problem 2 above.

When three or more identical consecutive bytes are found in the input stream, the compressor writes three copies of the byte on the output stream, followed by a repetition count. When the decompressor reads three identical consecutive bytes, it knows that the next byte is a repetition count (which may be 0, indicating just three repetitions). A disadvantage of the method is that a run of three characters in the input stream results in four characters written to the output stream; expansion! A run of four characters results in no compression. Only runs longer than four characters get compressed. Another slight

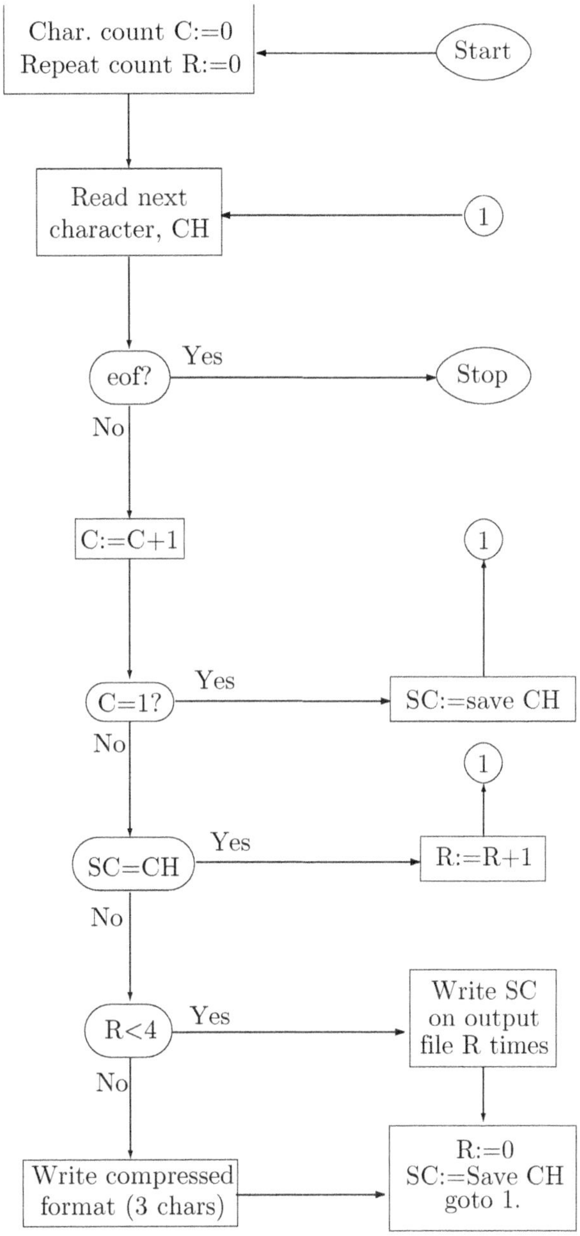

(a)

Figure 1.6: RLE. Part I: Compression.

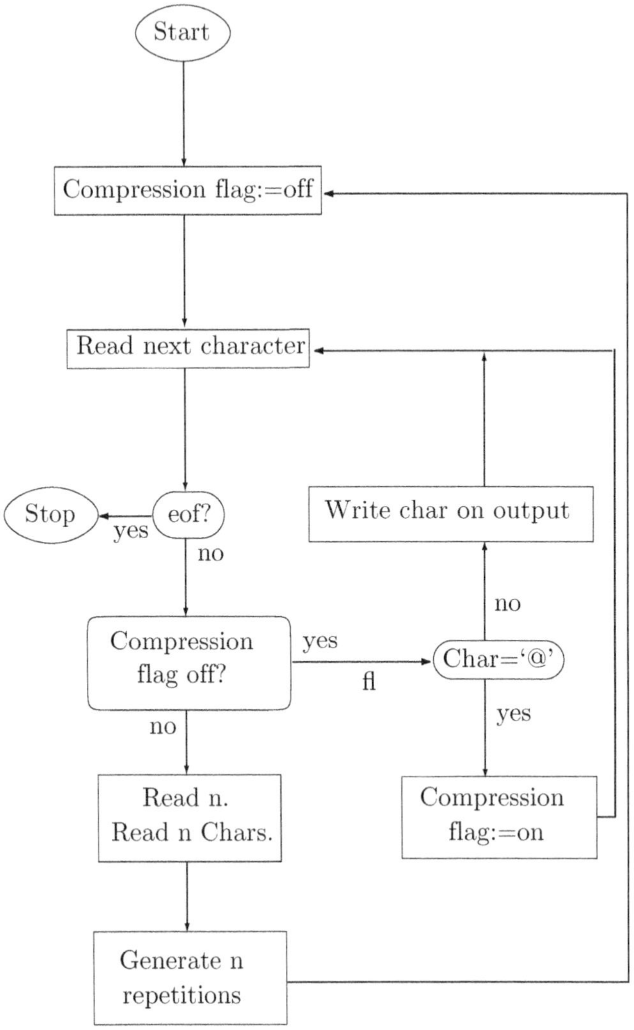

(b)

Figure 1.6 RLE. Part II: Decompression.

problem is that the maximum count is artificially limited in MNP5 to 250 instead of 255.

To get an idea of the compression ratios produced by RLE, we assume a string of N characters that needs to be compressed. We assume that the string contains M repetitions of average length L each. Each of the M repetitions is replaced by 3 characters (escape, count, and data), so the size of the compressed string is $N - M \times L + M \times 3 = N - M(L-3)$ and the compression factor is

$$\frac{N}{N - M(L-3)}.$$

(For MNP5 just substitute 4 for 3.) Examples: $N = 1000, M = 10, L = 3$ yield a compression factor of $1000/[1000 - 10(4-3)] = 1.01$. A better result is obtained in the case $N = 1000, M = 50, L = 10$, where the factor is $1000/[1000 - 50(10-3)] = 1.538$.

A variant of run-length encoding for text is *digram encoding*. This method is suitable for cases where the data to be compressed consists only of certain characters, e.g., just letters, digits, and punctuation marks. The idea is to identify commonly-occurring pairs of characters and to replace a pair (a digram) with one of the characters that cannot occur in the data (e.g., one of the ASCII control characters). Good results can be obtained if the data can be analyzed beforehand. We know that in plain English certain pairs of characters, such as `E␣`, `␣T`, `TH`, and `␣A`, occur often. Other types of data may have different common digrams. The sequitur method of Section 11.10 is an example of a method that compresses data by locating repeating digrams (as well as longer repeated phrases) and replacing them with special symbols.

A similar variant is *pattern substitution*. This is suitable for compressing computer programs, where certain words, such as `for`, `repeat`, and `print`, occur often. Each such word is replaced with a control character or, if there are many such words, with an escape character followed by a code character. Assuming that code `a` is assigned to the word `print`, the text `m:␣print,b,a;` will be compressed to `m:␣@a,b,a;`.

1.3.1 Relative Encoding

This is another variant, sometimes called *differencing* (see [Gottlieb et al. 75]). It is used in cases where the data to be compressed consists of a string of numbers that do not differ by much, or in cases where it consists of strings that are similar to each other. An example of the former is telemetry. The latter case is used in facsimile data compression described in Section 5.7 and also in LZW compression (Section 6.13.4).

In telemetry, a sensing device is used to collect data at certain intervals and transmit it to a central location for further processing. An example is temperature values collected every hour. Successive temperatures normally do not differ by much, so the sensor needs to send only the first temperature, followed by differences. Thus the sequence of temperatures $70, 71, 72.5, 73.1, \ldots$ can be compressed to $70, 1, 1.5, 0.6, \ldots$. This compresses the data, because the differences are small and can be expressed in fewer bits.

Notice that the differences can be negative and may sometimes be large. When a large difference is found, the compressor sends the actual value of the next measurement instead of the difference. Thus, the sequence $110, 115, 121, 119, 200, 202, \ldots$ can be compressed to $110, 5, 6, -2, 200, 2, \ldots$. Unfortunately, we now need to distinguish between a difference and an actual value. This can be done by the compressor creating an extra bit (a flag) for each number sent, accumulating those bits, and sending them to the decompressor from time to time, as part of the transmission. Assuming that each difference is sent as a byte, the compressor should follow (or precede) a group of eight bytes with a byte consisting of their eight flags.

Another practical way to send differences mixed with actual values is to send pairs of bytes. Each pair is either an actual 16-bit measurement (positive or negative) or two 8-bit signed differences. Thus, actual measurements can be between 0 and ±32K and differences can be between 0 and ±255. For each pair, the compressor creates a flag: 0 if the pair is an actual value, 1 if it is a pair of differences. After 16 pairs are sent, the compressor sends the 16 flags.

Example: The sequence of measurements $110, 115, 121, 119, 200, 202, \ldots$ is sent as (110), $(5, 6)$, $(-2, -1)$, (200), $(2, \ldots)$, where each pair of parentheses indicates a pair of bytes. The -1 has value 11111111_2, which is ignored by the decompressor (it indicates that there is only one difference in this pair). While sending this information, the compressor prepares the flags $01101\ldots$, which are sent after the first 16 pairs.

Relative encoding can be generalized to the lossy case, where it is called *differential encoding*. An example of a differential encoding method is differential pulse code modulation (or DPCM, Section 7.30).

1.4 RLE Image Compression

RLE is a natural candidate for compressing graphical data. A digital image consists of small dots called *pixels*. (For information on pixels and their history, see Section 7.1 as well as the lively references [Lyon 09] and [Smith 09].) Each pixel can be either one bit, indicating a black or a white dot, or several bits, indicating one of several colors or shades of gray. We assume that the pixels are stored in an array called a *bitmap* in memory, so the bitmap is the input stream for the image. Pixels are normally arranged in the bitmap in scan lines, so the first bitmap pixel is the dot at the top-left corner of the image, and the last pixel is the one at the bottom-right corner.

Compressing an image using RLE is based on the observation that if we select a pixel in the image at random, there is a good chance that its neighbors will have the same color (see also Sections 7.34 and 7.36). The compressor therefore scans the bitmap row by row, looking for runs of pixels of the same color. If the bitmap starts, e.g., with 17 white pixels, followed by 1 black pixel, followed by 55 white ones, etc., then only the numbers 17, 1, 55, ... need be written on the output stream.

The compressor assumes that the bitmap starts with white pixels. If this is not true, then the bitmap starts with zero white pixels, and the output stream should start with the run length 0. The resolution of the bitmap should also be saved at the start of the output stream.

The size of the compressed stream depends on the complexity of the image. The more detail, the worse the compression. However, Figure 1.7 shows how scan lines pass through a uniform region. A line enters through one point on the perimeter of the region and exits through another point, and these two points are not "used" by any other scan lines. It is now clear that the number of scan lines traversing a uniform region is roughly equal to half the length (measured in pixels) of its perimeter. Since the region is uniform, each scan line contributes two runs to the output stream for each region it crosses. The compression ratio of a uniform region therefore roughly equals the ratio

$$\frac{2\times\text{half the length of the perimeter}}{\text{total number of pixels in the region}}=\frac{\text{perimeter}}{\text{area}}.$$

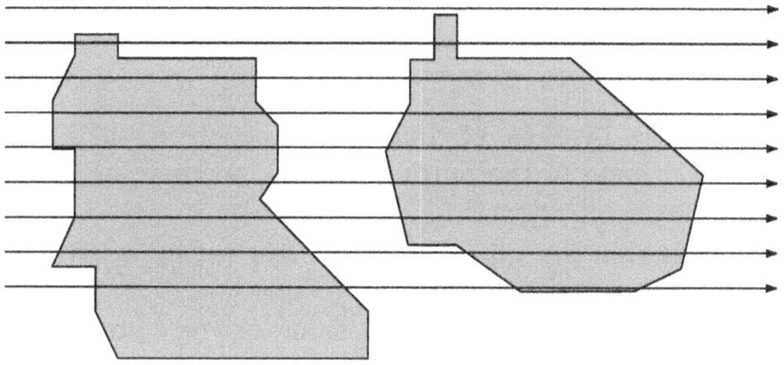

Figure 1.7: Uniform Areas and Scan Lines.

◇ **Exercise 1.3:** What would be the compressed file in the case of the following 6×8 bitmap?

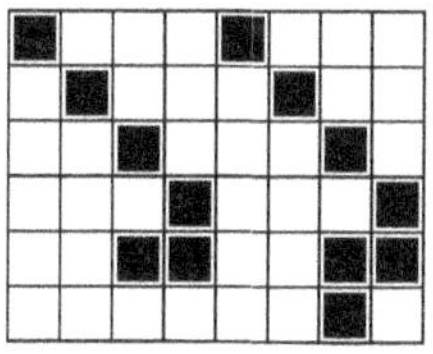

RLE can also be used to compress grayscale images. Each run of pixels of the same intensity (gray level) is encoded as a pair (run length, pixel value). The run length usually occupies one byte, allowing for runs of up to 255 pixels. The pixel value occupies several bits, depending on the number of gray levels (typically between 4 and 8 bits).

Example: An 8-bit-deep grayscale bitmap that starts with

$$12,12,12,12,12,12,12,12,12,35,76,112,67,87,87,87,5,5,5,5,5,5,1,\ldots$$

is compressed into $\boxed{9}$,12,35,76,112,67,$\boxed{3}$,87, $\boxed{6}$,5,1,..., where the boxed numbers indicate counts. The problem is to distinguish between a byte containing a grayscale value (such as 12) and one containing a count (such as $\boxed{9}$). Here are some solutions (although not the only possible ones):

1. If the image is limited to just 128 grayscales, we can devote one bit in each byte to indicate whether the byte contains a grayscale value or a count.
2. If the number of grayscales is 256, it can be reduced to 255 with one value reserved as a flag to precede every byte with a count. If the flag is, say, 255, then the sequence

above becomes

$$255, 9, 12, 35, 76, 112, 67, 255, 3, 87, 255, 6, 5, 1, \ldots .$$

3. Again, one bit is devoted to each byte to indicate whether the byte contains a grayscale value or a count. This time, however, these extra bits are accumulated in groups of 8, and each group is written on the output stream preceding (or following) the 8 bytes it "corresponds to."

Example: the sequence $\boxed{9}$,12,35,76,112,67,$\boxed{3}$,87,$\boxed{6}$,5,1,... becomes

$$\boxed{10000010},9,12,35,76,112,67,3,87,\boxed{100.....},6,5,1,\ldots .$$

The total size of the extra bytes is, of course, 1/8 the size of the output stream (they contain one bit for each byte of the output stream), so they increase the size of that stream by 12.5%.

4. A group of m pixels that are all different is preceded by a byte with the negative value $-m$. The sequence above is encoded by
$9, 12, -4, 35, 76, 112, 67, 3, 87, 6, 5, ?, 1, \ldots$ (the value of the byte with ? is positive or negative depending on what follows the pixel of 1). The worst case is a sequence of pixels (p_1, p_2, p_2) repeated n times throughout the bitmap. It is encoded as $(-1, p_1, 2, p_2)$, four numbers instead of the original three! If each pixel requires one byte, then the original three bytes are expanded into four bytes. If each pixel requires three bytes, then the original three pixels (which constitute nine bytes) are compressed into $1 + 3 + 1 + 3 = 8$ bytes.

Three more points should be mentioned:

1. Since the run length cannot be 0, it makes sense to write the [run length minus one] on the output stream. Thus, the pair $(3, 87)$ denotes a run of *four* pixels with intensity 87. This way, a run can be up to 256 pixels long.

2. In color images it is common to have each pixel stored as three bytes, representing the intensities of the red, green, and blue components of the pixel. In such a case, runs of each color should be encoded separately. Thus the pixels $(171, 85, 34)$, $(172, 85, 35)$, $(172, 85, 30)$, and $(173, 85, 33)$ should be separated into the three sequences $(171, 172, 172, 173, \ldots)$, $(85, 85, 85, 85, \ldots)$, and $(34, 35, 30, 33, \ldots)$. Each sequence should be run-length encoded separately. This means that any method for compressing grayscale images can be applied to color images as well.

3. It is preferable to encode each row of the bitmap individually. Thus if a row ends with four pixels of intensity 87 and the following row starts with 9 such pixels, it is better to write $\ldots, 4, 87, 9, 87, \ldots$ on the output stream rather than $\ldots, 13, 87, \ldots$. It is even better to write the sequence $\ldots, 4, 87, \text{eol}, 9, 87, \ldots$, where "eol" is a special end-of-line code. The reason is that sometimes the user may decide to accept or reject an image just by examining its general shape, without any details. If each line is encoded individually, the decoding algorithm can start by decoding and displaying lines $1, 6, 11, \ldots$, continue with lines $2, 7, 12, \ldots$, etc. The individual rows of the image are interlaced, and the image is constructed on the screen gradually, in several steps. This way, it is possible to get an idea of what is in the image at an early stage. Figure 1.8c shows an example of such a scan.

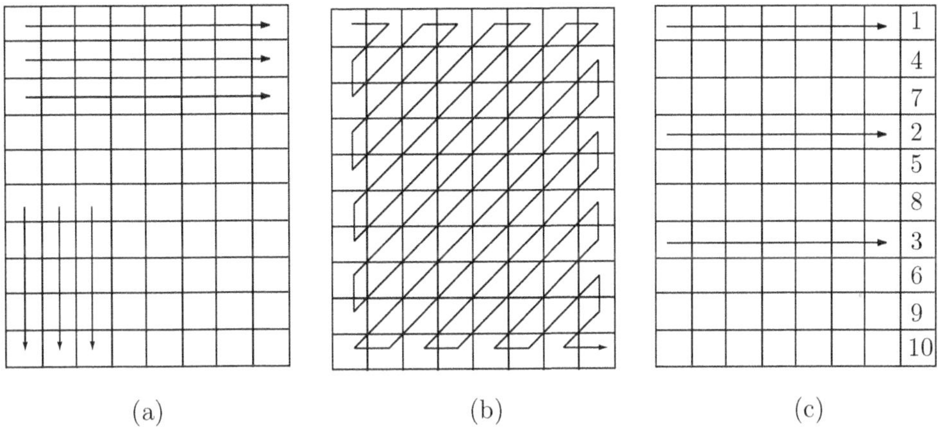

Figure 1.8: RLE Scanning.

Another advantage of individual encoding of rows is to make it possible to extract just part of an encoded image (such as rows k through l). Yet another application is to merge two compressed images without having to decompress them first.

If this idea (encoding each bitmap row individually) is adopted, then the compressed stream must contain information on where each bitmap row starts in the stream. This can be done by writing a header at the start of the stream that contains a group of 4 bytes (32 bits) for each bitmap row. The kth such group contains the offset (in bytes) from the start of the stream to the start of the information for row k. This increases the size of the compressed stream but may still offer a good trade-off between space (size of compressed stream) and time (time to decide whether to accept or reject the image).

◇ **Exercise 1.4:** There is another, obvious, reason why each bitmap row should be coded individually. What is it?

Figure 1.9a lists Matlab code to compute run lengths for a bi-level image. The code is very simple. It starts by flattening the matrix into a one-dimensional vector, so the run lengths continue from row to row.

Disadvantage of image RLE: When the image is modified, the run lengths normally have to be completely redone. The RLE output can sometimes be bigger than pixel-by-pixel storage (i.e., an uncompressed image, a raw dump of the bitmap) for complex pictures. Imagine a picture with many vertical lines. When it is scanned horizontally, it produces very short runs, resulting in very bad compression, or even in expansion. A good, practical RLE image compressor should be able to scan the bitmap by rows, columns, or in zigzag (Figure 1.8a,b) and it may automatically try all three ways on every bitmap to achieve the best compression.

◇ **Exercise 1.5:** Given the 8×8 bitmap of Figure 1.9b, use RLE to compress it, first row by row, then column by column. Describe the results in detail.

```
% Returns the run lengths of
% a matrix of 0s and 1s
function R=runlengths(M)
[c,r]=size(M);
for i=1:c;
 x(r*(i-1)+1:r*i)=M(i,:);
end
N=r*c;
y=x(2:N);
u=x(1:N-1);
z=y+u;
j=find(z==1);
i1=[j N];
i2=[0 j];
R=i1-i2;
```

the test

```
M=[0 0 0 1; 1 1 1 0; 1 1 1 0]
runlengths(M)
```

produces

```
3    4    1    3    1
```

(a)

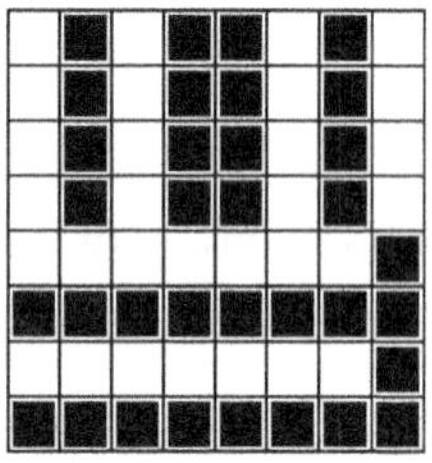

(b)

Figure 1.9: (a) Matlab Code To Compute Run Lengths. (b) A Bitmap.

1.4.1 Lossy Image Compression

It is possible to get even better compression if short runs are ignored. Such a method loses information when compressing an image, but sometimes this is acceptable to the user. (Images that permit no loss are medical X-rays and pictures taken by large telescopes, where the price of an image is astronomical.)

A lossy run-length encoding algorithm should start by asking the user for the longest run that should still be ignored. If the user specifies, for example, 3, then the program merges all runs of 1, 2, or 3 identical pixels with their neighbors. The compressed data "6,8,1,2,4,3,11,2" would be saved, in this case, as "6,8,7,16" where 7 is the sum $1+2+4$ (three runs merged) and 16 is the sum $3 + 11 + 2$. This makes sense for large high-resolution images where the loss of some detail may be invisible to the eye, but may significantly reduce the size of the output stream (see also Chapter 7).

1.4.2 Conditional Image RLE

Facsimile compression (Section 5.7) uses a modified Huffman code, but it can also be considered a modified RLE. This section discusses another modification of RLE, proposed in [Gharavi 87]. Assuming an image with n grayscales, the method starts by assigning an n-bit code to each pixel depending on its near neighbors. It then concatenates the n-bit codes into a long string and computes run lengths. The run lengths are assigned prefix codes (Huffman or other, Section 2.2) that are written on the compressed stream.

The method considers each scan line in the image a second-order Markov model. In such a model the value of the current data item depends on just two of its past neighbors, not necessarily the two immediate ones. Figure 1.10 shows the two neighbors A and B used by our method to predict the current pixel X (compare this with the lossless mode of JPEG, Section 7.10.5). A set of training images is used to count—for each possible pair of values of the neighbors A, B—how many times each value of X occurs. If A and B have similar values, it is natural to expect that X will have a similar value. If A and B have very different values, we expect X to have many different values, each with a low probability. The counts therefore produce the conditional probabilities $P(X|A,B)$ (the probability of the current pixel having value X if we already know that its neighbors have values A and B). Table 1.11 lists a small part of the results obtained by counting this way several training images with 4-bit pixels.

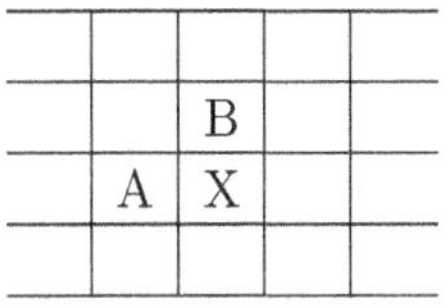

Figure 1.10: Neighbors Used To Predict X.

A	B		W1	W2	W3	W4	W5	W6	W7 ...
2	15	value:	4	3	10	0	6	8	1 ...
		count:	21	6	5	4	2	2	1 ...
3	0	value:	0	1	3	2	11	4	15 ...
		count:	443	114	75	64	56	19	12 ...
3	1	value:	1	2	3	4	0	5	6 ...
		count:	1139	817	522	75	55	20	8 ...
3	2	value:	2	3	1	4	5	6	0 ...
		count:	7902	4636	426	264	64	18	6 ...
3	3	value:	3	2	4	5	1	6	7 ...
		count:	33927	2869	2511	138	93	51	18 ...
3	4	value:	4	3	5	2	6	7	1 ...
		count:	2859	2442	240	231	53	31	13 ...

Table 1.11: Conditional Counting of 4-Bit Pixels.

Each pixel in the image to be compressed is assigned a new 4-bit code depending on its conditional probability as indicated by the table. Imagine a current pixel X with value 1 whose neighbors have values $A = 3$, $B = 1$. The table indicates that the conditional probability $P(1|3,1)$ is high, so X should be assigned a new 4-bit code with few runs (i.e., codes that contain consecutive 1's or consecutive 0's). On the other hand, the same $X = 1$ with neighbors $A = 3$ and $B = 3$ can be assigned a new 4-bit code with many runs, since the table indicates that the conditional probability $P(1|3,3)$ is low. The method therefore uses conditional probabilities to detect common pixels (hence the name *conditional RLE*), and assigns them codes with few runs.

Examining all 16 four-bit codes W_1 through W_{16}, we find that the two codes 0000 and 1111 have one run each, while 0101 and 1010 have four runs each. The codes should be arranged in order of increasing runs. For 4-bit codes we end up with the four groups

1. W_1 to W_2 : $0000, 1111,$
2. W_3 to W_8 : $0001, 0011, 0111, 1110, 1100, 1000,$
3. W_9 to W_{14} : $0100, 0010, 0110, 1011, 1101, 1001,$
4. W_{15} to W_{16} : $0101, 1010.$

The codes of group i have i runs each. The codes in each group are selected such that the codes in the second half of a group are the complements of those in the first half.

◇ **Exercise 1.6:** Apply this principle to construct the 32 five-bit codes.

The method can now be described in detail. The image is scanned in raster order. For each pixel X, its neighbors A and B are located, and the table is searched for this triplet. If the triplet is found in column i, then code W_i is selected. The first pixel has no neighbors, so assuming that its value is i, code W_i is selected for it. If X is located at the top row of the image, it has an A neighbor but not a B neighbor, so this case is handled differently. Imagine a pixel $X = 2$ at the top row with a neighbor $A = 3$. All the rows in the table with $A = 3$ are examined, in this case, and the one with the largest count for $X = 2$ is selected. In our example this is the row with count 7902, so code W_1 is selected. Pixels X with no A neighbors (on the left column) are treated similarly.

Rule of complementing: After the code for X has been selected, it is compared with the preceding code. If the least-significant bit of the preceding code is 1, the current code is complemented. This is supposed to reduce the number of runs. As an example, consider the typical sequence of 4-bit codes W_2, W_4, W_1, W_6, W_3, W_2, W_1

$$1111, 0011, 0000, 1110, 0001, 1111, 0000.$$

When these codes are concatenated, the resulting 28-bit string has eight runs. After applying the rule above, the codes become

$$1111, 1100, 0000, 1110, 0001, 0000, 0000,$$

a string with just six runs.

◇ **Exercise 1.7:** Do the same for the code sequence W_1, W_2, W_3, W_6, W_1, W_4, W_2.

A variation of this method uses the counts of Table 1.11 but not its codes and run lengths. Instead, it assigns a prefix code to the current pixel X depending on its neighbors A and B. Table 1.12 is an example. Each row has a different set of prefix codes constructed according to the counts of the row.

A	B		W1	W2	W3	W4	W5	W6	W7 ...
2	15	value:	4	3	10	0	6	8	1 ...
		code:	01	00	111	110	1011	1010	10010 ...
3	0	value:	0	1	3	2	11	4	15 ...
		code:	11	10	00	010	0110	011111	011101 ...
3	1	value:	1	2	3	4	0	5	6 ...
		code:	0	11	100	1011	10100	101010	10101111 ...
3	2	value:	2	3	1	4	5	6	0 ...
		code:	0	11	100	1011	10100	101010	10101111 ...
3	3	value:	3	2	4	5	1	6	7 ...
		code:	0	11	100	1011	101001	1010000	10101000 ...
3	4	value:	4	3	5	2	6	7	1 ...
		code:	11	10	00	0111	0110	0100	010110 ...

Table 1.12: Prefix Codes For 4-Bit Pixels.

1.4.3 The BinHex 4.0 Format

BinHex 4.0 is an old file format for reliable file transfers, designed by Yves Lempereur for use on the Macintosh platform. Before delving into the details of the format, the reader should understand why such a format is useful. ASCII is a 7-bit code. Each character is coded as a 7-bit number, which allows for 128 characters in the ASCII table. The ASCII standard recommends adding an eighth bit as parity to every character for increased reliability. However, the standard does not specify odd or even parity, and many computers simply ignore the extra bit or even set it to 0. As a result, when files are transferred in a computer network, some transfer programs may ignore the eighth bit and transfer just seven bits per character. This isn't so bad when a text file is being transferred but when the file is binary, no bits should be ignored. This is why it is safer to transfer text files, rather than binary files, over computer networks.

The idea of BinHex is to translate any file to a text file. The BinHex program reads an input file (text or binary) and produces an output file with the following format:

1. The comment:

```
(This␣file␣must␣be␣converted␣with␣BinHex␣4.0)
```

2. A header including the items listed in Table 1.13.
3. The input file is then read and RLE is used as the first step. Character 90_{16} is used as the RLE marker, and the following examples speak for themselves:

Source string	Packed string
00 11 22 33 44 55 66 77	00 11 22 33 44 55 66 77
11 22 22 22 22 22 22 33	11 22 90 06 33
11 22 90 33 44	11 22 90 00 33 44

(The character 00 indicates no run.) Runs of lengths 3 thorugh 255 characters are encoded in this way.

Field	Size
Length of FileName (1–63)	byte
FileName	(“Length” bytes)
Version	byte
Type	long
Creator	long
Flags (And $F800)	word
Length of Data Fork	long
Length of Resource Fork	long
CRC	word
Data Fork	(“Data Length” bytes)
CRC	word
Resource Fork	(“Rsrc Length” bytes)
CRC	word

Table 1.13: The BinHex Header.

◇ **Exercise 1.8:** How is the string “11 22 90 00 33 44” encoded?

4. Encoding into 7-bit ASCII characters. The input file is considered a stream of bits. As the file is being read, it is divided into blocks of 6 bits, and each block is used as a pointer to the BinHex table below. The character that's pointed to in this table is written on the output file. The table is

`!"#$%&'()*+,-012345689@ABCDEFGHIJKLMNPQRSTUVXYZ[‘abcdefhijklmpqr`

The output file is organized in “lines” of 64 characters each (except, perhaps, the last line). Each line is preceded and followed by a pair of colons “`:`”. The following is a quotation from the designer:

> “The characters in this table have been chosen for maximum noise protection.”

◇ **Exercise 1.9:** Manually convert the string “`123ABC`” to BinHex. Ignore the comment and the file header.

1.4.4 BMP Image Files

BMP is the native format for image files in the Microsoft Windows operating system. It has been modified several times since its inception, but has remained stable from version 3 of Windows. BMP is a palette-based graphics file format for images with 1, 2, 4, 8, 16, 24, or 32 bitplanes. It uses a simple form of RLE to compress images with 4 or 8 bitplanes. The format of a BMP file is simple. It starts with a file header that contains the two bytes `BM` and the file size. This is followed by an image header with the width, height, and number of bitplanes (there are two different formats for this header). Following the two headers is the color palette (that can be in one of three formats) which is followed by the image pixels, either in raw format or compressed by RLE. Detailed information on the BMP file format can be found in, for example, [Miano 99] and [Swan 93]. This section discusses the particular version of RLE used by BMP to compress pixels.

For images with eight bitplanes, the compressed pixels are organized in pairs of bytes. The first byte of a pair is a count C, and the second byte is a pixel value P which is repeated C times. Thus, the pair 04_{16} 02_{16} is expanded to the four pixels 02_{16} 02_{16} 02_{16} 02_{16}. A count of 0 acts as an escape, and its meaning depends on the byte that follows. A zero byte followed by another zero indicates end-of-line. The remainder of the current image row is filled with pixels of 00 as needed. A zero byte followed by 01_{16} indicates the end of the image. The remainder of the image is filled up with 00 pixels. A zero byte followed by 02_{16} indicates a skip to another position in the image. A 00_{16} 02_{16} pair must be followed by two bytes indicating how many columns and rows to skip to reach the next nonzero pixel. Any pixels skipped are filled with zeros. A zero byte followed by a byte C greater than 2 indicates C raw pixels. Such a pair must be followed by the C pixels. Assuming a 4×8 image with 8-bit pixels, the following sequence

$$04_{16}\,02_{16},\ 00_{16}\,04_{16}\,a35b1247_{16},\ 01_{16}\,f5_{16},\ 02_{16}\,e7_{16},\ 00_{16}\,02_{16}\,0001_{16},$$
$$01_{16}\,99_{16},\ 03_{16}\,c1_{16},\ 00_{16}\,00_{16},\ 00_{16}\,04_{16}\,08926bd7_{16},\ 00_{16}\,01_{16}$$

is the compressed representation of the 32 pixels

```
02 02 02 02 a3 5b 12 47
f5 e7 e7 00 00 00 00 00
00 00 99 c1 c1 c1 00 00
08 92 6b d7 00 00 00 00
```

Images with four bitplanes are compressed in a similar way but with two exceptions. The first exception is that a pair (C, P) represents a count byte and a byte of two pixel values that alternate. The pair 05_{16} $a2_{16}$, for example, is the compressed representation of the five 4-bit pixels a, 2, a, 2, and a, while the pair 07_{16} ff_{16} represents seven consecutive 4-bit pixels of f_{16}. The second exception has to do with pairs $(0, C)$ where C is greater than 2. Such a pair is followed by C 4-bit pixel values, packed two to a byte. The value of C is normally a multiple of 4, implying that a pair $(0, C)$ specifies pixels to fill up an integer number of words (where a word is two bytes). If C is not a multiple of 4, the remainder of the last word is padded with zeros. Thus, 00_{16} 08_{16} $a35b1247_{16}$ specifies eight pixels and fills up four bytes (or two words) with $a35b1247_{16}$, whereas 00_{16} 06_{16} $a35b12_{16}$ specifies six pixels and also fills up four bytes but with $a35b1200_{16}$.

1.5 Move-to-Front Coding

The basic idea of this method [Bentley 86] is to maintain the alphabet A of symbols as a list where frequently-occurring symbols are located near the front. A symbol `s` is encoded as the number of symbols that precede it in this list. Thus if A=(`t`, `h`, `e`, `s`,...) and the next symbol in the input stream to be encoded is the `e`, it will be encoded as 2, since it is preceded by two symbols. There are several possible variants to this method; the most basic of them adds one more step: After symbol `s` is encoded, it is moved to the front of list A. Thus, after encoding the `e`, the alphabet is modified to A=(`e`, `t`, `h`, `s`,...). This move-to-front step reflects the expectation that once `e` has been read from the input stream, it will be read many more times and will, at least for a while, be a common symbol. The move-to-front method is *locally adaptive*, since it adapts itself to the frequencies of symbols in local areas of the input stream.

The method produces good results if the input stream satisfies this expectation, i.e., if it contains concentrations of identical symbols (if the local frequency of symbols changes significantly from area to area in the input stream). We call this *the concentration property*. Here are two examples that illustrate the move-to-front idea. Both assume the alphabet A=(`a`, `b`, `c`, `d`, `m`, `n`, `o`, `p`).

1. The input stream `abcddcbamnopponm` is encoded as $C = (0, 1, 2, 3, 0, 1, 2, 3, 4, 5, 6, 7, 0, 1, 2, 3)$ (Table 1.14a). Without the move-to-front step it is encoded as $C' = (0, 1, 2, 3, 3, 2, 1, 0, 4, 5, 6, 7, 7, 6, 5, 4)$ (Table 1.14b). Both C and C' contain codes in the same range $[0, 7]$, but the elements of C are smaller on average, since the input starts with a concentration of `abcd` and continues with a concentration of `mnop`. (The average value of C is 2.5, while that of C' is 3.5.)

a	abcdmnop	0	a	abcdmnop	0	a	abcdmnop	0	a	abcdmnop	0
b	abcdmnop	1	b	abcdmnop	1	b	abcdmnop	1	b	abcdmnop	1
c	bacdmnop	2	c	abcdmnop	2	c	bacdmnop	2	c	abcdmnop	2
d	cbadmnop	3	d	abcdmnop	3	d	cbadmnop	3	d	abcdmnop	3
d	dcbamnop	0	d	abcdmnop	3	m	dcbamnop	4	m	abcdmnop	4
c	dcbamnop	1	c	abcdmnop	2	n	mdcbanop	5	n	abcdmnop	5
b	cdbamnop	2	b	abcdmnop	1	o	nmdcbaop	6	o	abcdmnop	6
a	bcdamnop	3	a	abcdmnop	0	p	onmdcbap	7	p	abcdmnop	7
m	abcdmnop	4	m	abcdmnop	4	a	ponmdcba	7	a	abcdmnop	0
n	mabcdnop	5	n	abcdmnop	5	b	aponmdcb	7	b	abcdmnop	1
o	nmabcdop	6	o	abcdmnop	6	c	baponmdc	7	c	abcdmnop	2
p	onmabcdp	7	p	abcdmnop	7	d	cbaponmd	7	d	abcdmnop	3
p	ponmabcd	0	p	abcdmnop	7	m	dcbaponm	7	m	abcdmnop	4
o	ponmabcd	1	o	abcdmnop	6	n	mdcbapon	7	n	abcdmnop	5
n	opnmabcd	2	n	abcdmnop	5	o	nmdcbapo	7	o	abcdmnop	6
m	nopmabcd	3	m	abcdmnop	4	p	onmdcbap	7	p	abcdmnop	7
	mnopabcd						ponmdcba				
	(a)			(b)			(c)			(d)	

Table 1.14: Encoding With and Without Move-to-Front.

2. The input stream `abcdmnopabcdmnop` is encoded as $C = (0, 1, 2, 3, 4, 5, 6, 7, 7, 7, 7, 7, 7, 7, 7, 7)$ (Table 1.14c). Without the move-to-front step it is encoded as $C' = (0, 1, 2, 3, 4, 5, 6, 7, 0, 1, 2, 3, 4, 5, 6, 7)$ (Table 1.14d). The average of C is now 5.25, greater than that of C', which is 3.5. The move-to-front rule creates a worse result in this case, because the input does not contain concentrations of identical symbols (it does not satisfy the concentration property).

Before getting into further details, it is important to understand the advantage of having small numbers in C. This feature makes it possible to efficiently encode C with either Huffman or arithmetic coding (Chapter 5). Here are four ways to do this:

1. Assign Huffman codes to the integers in the range $[0, n]$ such that the smaller integers get the shorter codes. Here is an example of such a code for the integers 0 through 7:

0—0, 1—10, 2—110, 3—1110, 4—11110, 5—111110, 6—1111110, 7—1111111.

2. Assign codes to the integers such that the code of integer $i \geq 1$ is its binary code preceded by $\lfloor \log_2 i \rfloor$ zeros. This is the well-known Elias gamma code (Section 3.4). Table 1.15 lists some examples.

i	Code	Size
1	1	1
2	010	3
3	011	3
4	00100	5
5	00101	5
6	00110	5
7	00111	5
8	0001000	7
9	0001001	7
⋮	⋮	⋮
15	0001111	7
16	000010000	9

Table 1.15: Examples of Variable-Length Codes.

◊ **Exercise 1.10:** What is the total size of the code of i in this case.

3. Use adaptive Huffman coding (Section 5.3).

4. For maximum compression, perform two passes over C, the first pass counts frequencies of codes and the second pass performs the actual encoding. The frequencies counted in pass 1 are used to compute probabilities and assign Huffman codes to be used later by pass 2.

It can be shown that the move-to-front method performs, in the worst case, slightly worse than Huffman coding. At best, it performs significantly better.

As has been mentioned earlier, it is easy to come up with variations of the basic idea of move-to-front. Here are a few.

1. Move-ahead-k. The element of A matched by the current symbol is moved ahead k positions instead of all the way to the front of A. The parameter k can be specified by the user, with a default value of either n or 1. This tends to reduce performance (i.e., to increase the average size of the elements of C) for inputs that satisfy the concentration property, but it works better for other inputs. Notice that assigning $k = n$ is identical to move-to-front. The case $k = 1$ is especially simple, since it only requires swapping an element of A with the one preceding it.

◊ **Exercise 1.11:** Use move-ahead-k to encode each of the strings `abcddcbamnopponm` and `abcdmnopabcdmnop` twice, with $k = 1$ and $k = 2$.

2. Wait-c-and-move. An element of A is moved to the front only after it has been matched c times to symbols from the input stream (not necessarily c consecutive times). Each element of A should have a counter associated with it, to count the number of matches. This method makes sense in implementations where moving and rearranging elements of A is slow.
3. Normally, a symbol read from the input is a byte. If the input stream consists of text, however, it may make sense to treat each *word*, not each character, as a symbol. Consider the simple case where the input consists of just lowercase letters, spaces, and one end-of-text marker at the end. We can define a word as a string of letters followed by a space or by the end-of-text marker. The number of words in this case can be huge, so the alphabet list A should start empty, and words should be added as they are being input and encoded. We use the text

`the␣boy␣on␣my␣right␣is␣the␣right␣boy`

as an example.

The first word input is `the`. It is not found in A, since A is empty, so it is added to A. The encoder emits 0 (the number of words preceding `the` in A) followed by `the`. The decoder also starts with an empty A. The 0 tells it to select the first word in A, but since A is empty, the decoder knows to expect the 0 to be followed by a word. It adds this word to A.

The next word is `boy`. It is added to A, so A=(`the`, `boy`) and the encoder emits `1boy`. The word `boy` is moved to the front of A, so A=(`boy`, `the`). The decoder reads the 1, which refers to the second word of A, but the decoder's A has only one word in it so far. The decoder thus knows that a new word must follow the 1. It reads this word and adds it to the front of A. Table 1.16 summarizes the encoding steps for this example.

List A may grow very large in this variant, but any practical implementation has to limit its size. This is why the last item of A (the least-recently-used item) has to be deleted when A exceeds its size limit. This is another difference between this variant and the basic move-to-front method.

◇ **Exercise 1.12:** Decode `the␣boy␣on␣my␣right␣is␣the␣right␣boy` and summarize the steps in a table.

Word	A (before adding)	A (after adding)	Code emitted
the	()	(the)	0the
boy	(the)	(the, boy)	1boy
on	(boy, the)	(boy, the, on)	2on
my	(on, boy, the)	(on, boy, the, my)	3my
right	(my, on, boy, the)	(my, on, boy, the, right)	4right
is	(right, my, on, boy, the)	(right, my, on, boy, the, is)	5is
the	(is, right, my, on, boy, the)	(is, right, my, on, boy, the)	5
right	(the, is, right, my, on, boy)	(the, is, right, my, on, boy)	2
boy	(right, the, is, my, on, boy)	(right, the, is, my, on, boy)	5
	(boy, right, the, is, my, on)		

Table 1.16: Encoding Multiple-Letter Words.

> "I've been looking," Lana Lee said, becoming a grave personnel manager, "for the right boy for this job for several days." She put her hands in the pockets of her leather overcoat and looked into the sunglasses.
>
> —John Kennedy Toole, *A Confederacy of Dunces*

1.6 Scalar Quantization

The dictionary definition of the term "quantization" is "to restrict a variable quantity to discrete values rather than to a continuous set of values." In the field of data compression, quantization is employed in two ways:

1. If the data to be compressed is in the form of large numbers, quantization is used to convert it to small numbers. Small numbers take less space than large ones, so quantization generates compression. On the other hand, small numbers generally contain less information than large numbers, so quantization results in lossy compression.
2. If the data to be compressed is analog (i.e., a voltage that changes with time) quantization is used to digitize it into small numbers. The smaller the numbers the better the compression, but also the greater the loss of information. This aspect of quantization is exploited by several speech compression methods.

In the discussion here we assume that the data to be compressed is in the form of numbers, and that it is input, number by number, from an input stream (or a source). Section 7.19 discusses a generalization of discrete quantization to cases where the data consists of sets (called vectors) of numbers rather than of individual numbers.

The first example is naive discrete quantization of an input stream of 8-bit numbers. We can simply delete the least-significant four bits of each data item. This is one of those rare cases where the compression factor (=2) is known in advance and does not depend on the data. The input data consists of 256 different symbols, while the output data consists of just 16 different symbols. This method is simple but not very practical because too much information is lost in order to get the unimpressive compression factor of 2.

In order to develop a better approach we assume again that the data consists of 8-bit numbers, and that they are unsigned. Thus, input symbols are in the range $[0, 255]$ (if the input data is signed, input symbols have values in the range $[-128, +127]$). We select a spacing parameter s and compute the sequence of uniform quantized values 0, s, $2s$,...,ks, such that $(k+1)s > 255$ and $ks \leq 255$. Each input symbol S is quantized by converting it to the nearest value in this sequence. Selecting $s = 3$, e.g., produces the uniform sequence 0,3,6,9,12,...,252,255. Selecting $s = 4$ produces 0,4,8,12,...,252,255 (since the next multiple of 4, after 252, is 256).

A similar approach is to select quantized values such that any number in the range $[0, 255]$ will be no more than d units distant from one of the data values that are being quantized. This is done by dividing the range into segments of size $2d + 1$ and centering them on the range $[0, 255]$. If, e.g., $d = 16$, then the range $[0, 255]$ is divided into seven segments of size 33 each, with 25 numbers remaining. We can thus start the first segment 12 numbers from the start of the range, which produces the 10-number sequence 12, 33, 45, 78, 111, 144, 177, 210, 243, and 255. Any number in the range $[0, 255]$ is at most

16 units distant from any of these numbers. If we want to limit the quantized sequence to just eight numbers (so each can be expressed in 3 bits) we can apply this method to compute the sequence 8, 41, 74, 107, 140, 173, 206, and 239.

The quantized sequences above make sense in cases where each symbol appears in the input data with equal probability (cases where the source is *memoryless*). If the input data is not uniformly distributed, the sequence of quantized values should be distributed in the same way as the data.

bbbbbbbb		bbbbbbbb	
1	1	.	
10	2	.	
11	3	.	
100	4	100|000	32
101	5	101|000	40
110	6	110|000	48
111	7	111|000	56
100|0	8	100|0000	64
101|0	10	101|0000	80
110|0	12	110|0000	96
111|0	14	111|0000	112
100|00	16	100|00000	128
101|00	20	101|00000	160
110|00	24	110|00000	192
111|00	28	111|00000	224

Table 1.17: A Logarithmic Quantization Table.

Imagine, e.g., an input stream of 8-bit unsigned data items most of which are zero or close to zero and few are large. A good sequence of quantized values for such data should have the same distribution, i.e., many small values and few large ones. One way of computing such a sequence is to select a value for the length parameter l and to construct a "window" of the form

$$1\underbrace{b\ldots bb}_{l},$$

where each b is a bit, and place it under each of the 8 bit positions of a data item. If the window sticks out on the right, some of the l bits are truncated. As the window is shifted to the left, bits of 0 are appended to it. Table 1.17 illustrates this construction with $l = 2$. It is easy to see how the resulting quantized values start with initial spacing of one unit, continue with spacing of two units and four units, until the last four values are spaced by 32 units. The numbers 0 and 255 should be manually added to such a quasi-logarithmic sequence to make it more general.

Scalar quantization is an example of a lossy compression method, where it is easy to control the trade-off between compression ratio and the amount of loss. However, because it is so simple, its use is limited to cases where much loss can be tolerated. Many image compression methods are lossy, but scalar quantization is not suitable for image compression because it creates annoying artifacts in the decompressed image. Imagine an image with an almost uniform area where all pixels have values 127 or 128. If 127 is quantized to 111 and 128 is quantized to 144, then the result, after decompression, may resemble a checkerboard where adjacent pixels alternate between 111 and 144. This is why practical algorithms use *vector quantization*, instead of scalar quantization, for lossy (and sometimes lossless) compression of images and sound. See also Section 7.3.

1.7 Recursive Range Reduction

In their original 1977 paper [Ziv and Lempel 77], Lempel and Ziv have proved that their dictionary method can compress data to the entropy, but they pointed out that vast quantities of data would be needed to approach ideal compression. Other algorithms, most notably PPM (Section 5.14), suffer from the same problem. Often, a user is willing to sacrifice compression performance in favor of easy implementation and the knowledge that the performance of the algorithm is independent of the quantity of the data. The recursive range reduction (3R) method described here generates decent compression, is easy to implement, and its performance is independent of the amount of data to be compressed. These features make it an attractive candidate for compression in embedded systems, low-cost microcontrollers, and other applications where space is limited or resources are constrained. The method is the brainchild of Yann Guidon who described it in [3R 06]. The method is distantly related to variable-length codes.

The method is described first for a sorted list of integers, where its application is simplest and no recursion is required. We term this version "range reduction" (RR). We then show how RR can be extended to a recursive version (RRR or 3R) that can be applied to any set of integers.

Given a list of integers, we first eliminate the effects of the sign bit. We either rotate each integer to move the sign bit to the least-significant position (a folding) or add an offset to all the integers, so they become nonnegative. The list is then sorted in nonincreasing order. The first element of the list is the largest one, and we assume that its most-significant bit is a 1 (i.e., it has no extra zeros on the left).

It is obvious that the integers that follow the first element may have some most-significant zero bits, and the heart of the RR method is to eliminate most of those bits while leaving enough information for the decoder to restore them. The first item in the compressed stream is a header with the length L of the largest integer. This length is stored in a fixed-size field whose size is sufficient for any data that may be encountered. In the examples here, this size is four bits, allowing for integers up to 16 bits long. Once the decoder inputs L, it knows the length of the first integer. The MSB of this integer is a 1, so this bit can be eliminated and the first integer can be emitted in $L-1$ bits. The next integer is written on the output as an L-bit number, thereby allowing the decoder to read it unambiguously. The decoder then checks the most-significant bits of the second integer. If the leftmost k bits are zeros, then the decoder knows that the next integer (i.e., the third one) was written without its k leftmost zeros and thus occupies the next $L-k$ bits in the compressed stream. This is how RR compresses a sorted list of nonnegative integers. The compression efficiency depends on the data, but not on the amount of data available for compression. The method does not improve by applying it to vast quantities of data.

Table 1.18 is an example. The data consists of ten 7-bit integers. The 4-bit length field is set to 6, indicating 7-bit integers (notice that L cannot be zero, so 6 indicates a length of seven bits). The first integer is emitted minus its leftmost bit and the next integer is output in its entirety. The third integer is also emitted as is, but its leftmost zero (listed in bold) indicates to the decoder that the following (fourth) integer will have only six bits.

Data	RR code	Rice code
	0110 = 6	
1011101	011101	000001 1101
1001011	1001011	00001 1011
0110001	**0**110001	0001 0001
0101100	101100	001 1100
0001110	**00**1110	1 1110
0001101	1101	1 1101
0001100	1100	1 1100
0001001	1001	1 1001
0000010	**00**10	1 0010
0000001	01	1 0001
70	53	64 bits

Table 1.18: Example of Range Reduction.

The total size of the ten integers is 70 bits, and this should be compared with the 53 bits created by RR and the 64 bits resulting from the Rice codes (with $n = 4$) of the same integers (see Section 10.9 for Rice codes). Compression is not impressive, but it is obvious that it is not affected by the amount of data.

The limited experience available with RR seems to indicate (but a rigorous proof is still needed) that this method performs best on data that decreases exponentially, where each integer is about half the size of its predecessor. In such a list, each number is one bit shorter than its predecessor. Worst results should be obtained with data that either decreases slowly, such as (900, 899, 898, 897, 896), or that decreases fast, such as (100000, 1000, 10, 1). These cases are illustrated in Table 1.19. Notice that the header is the only side information sent to the decoder and that RR never expands the data.

Now for unsorted data. Given an unsorted list of integers, we can sort it, compress it with RR, and then include side information about the sort, so that the decoder can unsort the data correctly after decompressing it. An efficient method is required to specify the relation between a list of numbers and its sorted version, and the method described here is referred to as recursive range reduction or 3R.

The 3R algorithm involves the creation of a binary tree where each path from the root to a leaf is a nonincreasing list of integers. Each path is encoded with 3R separately, which requires a recursive algorithm (hence the word "recursive" in the algorithm's name). Figure 1.20 shows an example. Given the five unsorted integers `A` through `E`, a binary tree is constructed where each pair of consecutive integers becomes a subtree at the lowest level. Each of the five paths from the root to a leaf is a sorted list, and

Data	RR code	Data	RR code
	0110 = 6		1101 = 13
1000010	000010	10011100010000	0011100010000
1000001	1000001	00001111101000	**0000**1111101000
1000000	1000000	00000001100100	**000**1100100
0111111	**0**111111	00000000001010	**000**1010
0111110	111110	00000000000001	0001
0111101	111101		
0111100	111101		
49	49	70	52

Table 1.19: Range Reduction Worst Performance.

it is compressed with 3R. It is obvious that writing the compressed results of all five lists on the output would normally cause expansion, so only certain nodes are actually written. The rule is to follow every path from the root and select the nodes along edges that go to the left (or only those that go to the right). Thus, in our example, we write the following values on the output: (1) a header with the length of the root, (2) the 3R-encoded path (root, A+B, A), (3) the value of node C [after the path (root, C+D+E, C) is encoded], and (4) the value of node D [after the path (root, C+D+E, D+E, D) is encoded].

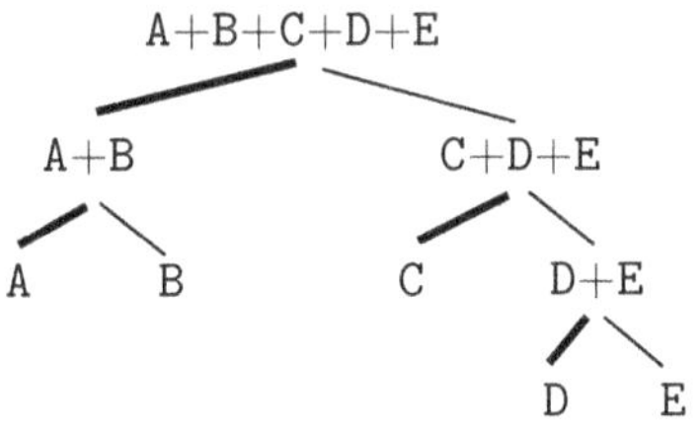

Figure 1.20: A Binary Tree Encoded With 3R.

The decoder reads the header and first path (root, A+B, A). It decodes the path to obtain the three values and subtracts (A + B) − A to obtain B. It then subtracts Root − (A + B) to obtain C+D+E. It inputs C and now it can decode the path (root, C+D+E, C). Next, the decoder subtracts C from C+D+E to obtain D+E, it inputs D, and decodes path (root, C+D+E, D+E, D). One more subtraction yields E.

This sounds like much work for very little compression, and it is! The 3R method is not the most efficient compression algorithm, but it may find its applications. The nice feature of the tree structure described here is that the number of nodes written on the output equals the size of the original data. In our example there are five data items and five 3R codes written on the output. It's easy to see why this is generally true. Given a list of n data items, we observe that they end up as the leaves of the binary tree. What is eventually written on the output is half the nodes at each level of the tree. A complete binary tree with n leaves has $2n$ nodes, so half of this number, n nodes, ends up on the output.

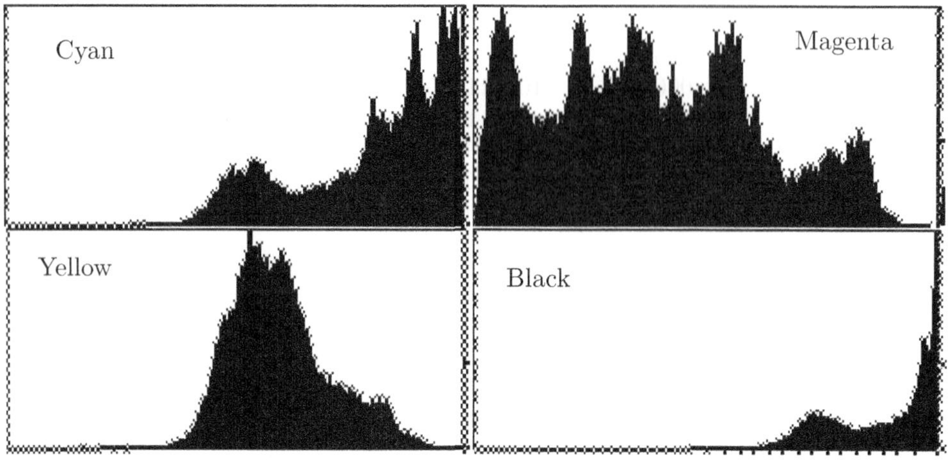

Figure 1.21: A Histogram.

It seems that 3R would be suitable for applications where the data does not fit any of the standard statistical distributions and yet is not random. One example of such data is the histogram of an image (Figure 1.21 shows the CMYK histograms of the Lena image, Figure 7.55). On the other hand, 3R is restricted to nonnegative integers and the encoder requires two passes, one for constructing the tree and one for traversing it and collecting nodes.

It has already been mentioned that RR never expands the data. However, when fed with random data, 3R generates output that grows by almost one bit per data item. Because of this feature, its developer refers to 3R as a nearly entropy encoder, not a real entropy encoder.

> Explanation. Because of the summations, random data are "averaged" so after a few tree levels, this is like almost-monotonous data. These data are then further summed together, still adding one bit of size per level but halving the number of sums each time, so in the end it is equivalent to one bit per input element.
>
> —Yann Guidon (private communication)

> Compression algorithms are often described as squeezing, squashing, crunching or imploding data, but these are not very good descriptions of what is actually happening.
>
> —James D. Murray and William Vanryper (1994)

2
Basic VL Codes

The dates of most of the important historical events are known, but not always very precisely. We know that Kublai Khan, grandson of Ghengis Khan, founded the Yuan dynasty in 1280 (it lasted until 1368), but we don't know precisely (i.e., the month, day and hour) when this act took place. A notable exception to this state of affairs is the modern age of telecommunications, a historical era whose birth is known precisely, up to the minute. On Friday, 24 May 1844, at precisely 9:45 in the morning, Samuel Morse inaugurated the age of modern telecommunications by sending the first telegraphic message in his new code. The message was sent over an experimental line funded by the American Congress from the Supreme Court chamber in Washington, DC to the B & O railroad depot in Baltimore, Maryland. Taken from the Bible (*Numbers* 23:23), the message was "What hath God wrought?" It had been suggested to Morse by Annie Ellsworth, the young daughter of a friend. It was prerecorded on a paper tape, was sent to a colleague in Baltimore, and was then decoded and sent back by him to Washington. An image of the paper tape can be viewed at [morse-tape 06].

Morse was born near Boston and was educated at Yale. We would expect the inventor of the telegraph (and of such a sophisticated code) to have been a child prodigy who tinkered with electricity and gadgets from an early age (the electric battery was invented when Morse was nine years old). Instead, Morse became a successful portrait painter with more than 300 paintings to his credit. It wasn't until 1837 that the 46 year old Morse suddenly quit his painting career and started thinking about communications and tinkering with electric equipment. It is not clear why he made such a drastic career change at such an age, but it is known that two large, wall-size paintings that he made for the Capitol building in

D. Salomon, G. Motta, *Handbook of Data Compression*, 5th ed.
DOI 10.1007/978-1-84882-903-9_2,

Washington, DC were ignored by museum visitors and rejected by congressmen. It may have been this disappointment that gave us the telegraph and the Morse code.

Given this background, it is easy to imagine how the 53-year-old Samuel Morse felt on that fateful day, Friday, 24 May 1844, as he sat hunched over his mysterious apparatus, surrounded by a curious crowd of onlookers, some of whom had only a vague idea of what he was trying to demonstrate. He must have been very anxious, because his telegraph project, his career, and his entire future depended on the success of this one test. The year before, the American Congress awarded him $30,000 to prepare this historical test and prove the value of the electric telegraph (and thus also confirm the ingenuity of yankees), and here he is now, dependent on the vagaries of his batteries, on the new, untested 41-mile-long telegraph line, and on a colleague in Baltimore.

Fortunately, all went well. The friend in Baltimore received the message, decoded it, and resent it within a few minutes, to the great relief of Morse and to the amazement of the many congressmen assembled around him.

The Morse code, with its quick dots and dashes (Table 2.1), was extensively used for many years, first for telegraphy, and beginning in the 1890s, for early radio communications. The development of more advanced communications technologies in the 20th century displaced the Morse code, which is now largely obsolete. Today, it is used for emergencies, for navigational radio beacons, land mobile transmitter identification, and by continuous wave amateur radio operators.

A	.-	N	-.	1	.----	Period	.-.-.-
B	-...	O	---	2	..---	Comma	--..--
C	-.-.	P	.--.	3	...--	Colon	---...
Ch	----	Q	--.-	4	....-	Question mark	..--..
D	-..	R	.-.	5	.....	Apostrophe	.----.
E	.	S	...	6	-....	Hyphen	-....-
F	..-.	T	-	7	--...	Dash	-..-.
G	--.	U	..-	8	---..	Parentheses	-.--.-
H	....	V	...-	9	----.	Quotation marks	.-..-.
I	..	W	.--	0	-----		
J	.---	X	-..-				
K	-.-	Y	-.--				
L	.-..	Z	--..				
M	--						

Table 2.1: The Morse Code for English.

Our interest in the Morse code is primarily with a little-known aspect of this code. In addition to its advantages for telecommunications, the Morse code is also an early example of text compression. The various dot-dash codes developed by Morse (and possibly also by his associate, Alfred Vail) have different lengths, and Morse intuitively assigned the short codes (a single dot and a single dash) to the letters E and T, the longer, four dots-dashes, he assigned to Q, X, Y, and Z. The even longer, five dots-dashes codes, were assigned to the 10 digits, and the longest codes (six dots and dashes) became those of the punctuation marks. Morse also specified that the signal for error is eight consecutive dots, in response to which the receiving operator should delete the last word received.

It is interesting to note that Morse was not the first to think of compression (in terms of time saving) by means of a code. The well-known Braille code for the blind was developed by Louis Braille in the 1820s and is still in common use today. It consists of groups (or cells) of 3×2 dots each, embossed on thick paper. Each of the six dots in a group may be flat or raised, implying that the information content of a group is equivalent to six bits, resulting in 64 possible groups. The letters, digits, and common punctuation marks do not require all 64 codes, which is why the remaining groups may be used to code common words—such as `and`, `for`, and `of`—and common strings of letters—such as `ound`, `ation`, and `th`.

The Morse code has another feature that makes it relevant to us. Because the individual codes have different lengths, there must be a way to identify the end of a code. Morse solved this problem by requiring accurate relative timing. If the duration of a dot is taken to be one unit, then that of a dash is three units, the space between the dots and dashes of one character is one unit, the space between characters is three units, and the interword space is six units (five for automatic transmission). This chapter and the two following it are concerned with the use of variable-length codes to compress digital data. With these codes, it is important not to have any extra spaces. In fact, there is no such thing as a space, because computers use only zeros and 1's. Thus, when a string of data symbols is compressed by assigning codewords to the symbols, the codewords (whose lengths vary) are concatenated into a long binary string without any spaces or separators. Such variable-length codes must therefore be designed to allow for unambiguous reading. Somehow, the decoder should be able to read bits and identify the end of each codeword. Such codes are referred to as uniquely decodable or uniquely decipherable (UD).

Variable-length codes have become important in many areas of computer science. This chapter and the two following it are related and constitute a survey of this important topic. They present the principles underlying this type of codes and describe the important classes of variable-length codes. Many examples illustrate the applications of these codes to data compression. These three related chapters are devoted to the codes, which is why they describe the codes rather than actual compression algorithms. The remainder of the book is devoted to many actual methods, algorithms, and techniques for compressing data.

The term *representation* is central to our discussion. A number can be represented in decimal, binary, or any other number base (or number system, see Section 3.19). Mathematically, a representation is a bijection (or a bijective function) of an infinite, countable set S_1 of strings onto another set S_2 of strings (in practice, S_2 consists of binary strings, but it may also be ternary or based on other number systems), such that any concatenation of any elements of S_2 is UD. The elements of S_1 are called data symbols and those of S_2 are codewords. Set S_1 is an alphabet and set S_2 is a code. An interesting example is the standard binary notation. We normally refer to it as the binary representation of the integers, but according to the definition above it is not a representation because it is not UD. It is easy to see, for example, that a string of binary codewords that starts with 11 can be either two consecutive 1's or the code of 3.

A function $f : X \Rightarrow Y$ is said to be bijective, if for every $y \in Y$, there is exactly one $x \in X$ such that $f(x) = y$.

Figure 4.19 and Table 4.22 list several variable-length UD codes assigned to the 26 letters of the English alphabet.

This group of three related chapters starts with several introductory sections (Sections 2.1 through 2.6) that discuss information theory concepts such as entropy and redundancy, and concepts that are used throughout the text, such as prefix codes, complete codes, and universal codes.

The remainder of the group deals mostly with block-to-variable codes, although its first part deals with the Tunstall codes and other variable-to-block codes. It concentrates on the codes themselves, not on the compression algorithms. Thus, the individual sections describe various variable-length codes and classify them according to their structure and organization. The main techniques employed to design variable-length codes are the following:

- The phased-in codes (Section 2.9) are a minor extension of fixed-length codes and may contribute a little to the compression of a set of consecutive integers by changing the representation of the integers from fixed n bits to either n or $n-1$ bits (recursive phased-in codes are also described).

- Self-delimiting codes. These are intuitive variable-length codes—mostly due to Gregory Chaitin, the originator of algorithmic information theory—where a code signals its end by means of extra flag bits. The self-delimiting codes of Section 2.12 are inefficient and are not used in practice.

- Prefix codes. Such codes can be read unambiguously (they are uniquely decodable, or UD codes) from a long string of codewords because they have a special property (the prefix property) which is stated as follows: Once a bit pattern is assigned as the code of a symbol, no other codes can start with that pattern. The most common example of prefix codes are the Huffman codes (Section 5.2). Other important examples are the unary, start-step-stop, and start/stop codes (Sections 3.2 and 3.3, respectively).

- Codes that include their own length. One way to construct a UD code for the integers is to start with the standard binary representation of an integer and prepend to it its length L_1. The length may also have variable length, so it has to be encoded in some way or have its length L_2 prepended. The length of an integer n equals approximately $\log n$ (where the logarithm base is the same as the number base of n), which is why such methods are often called logarithmic ramp representations of the integers. The most common examples of this type of codes are the Elias codes (Section 3.4), but other types are also presented. They include the Levenstein code (Section 3.6), Even–Rodeh code (Section 3.7), punctured Elias codes (Section 3.8), the ternary comma code (Section 3.10), Stout codes (Section 3.12), Boldi–Vigna (zeta) codes (Section 3.13), and Yamamoto's recursive code (Section 3.14).

- Suffix codes (codes that end with a special flag). Such codes limit the propagation of an error and are therefore robust. An error in a codeword affects at most that codeword and the one or two codewords following it. Most other variable-length codes sacrifice data integrity to achieve short codes, and are fragile because a single error can propagate indefinitely through a sequence of concatenated codewords. The taboo codes of Section 3.16 are UD because they reserve a special string (the taboo) to indicate the end of the code. Wang's flag code (Section 3.17) is also included in this category.

Note. The term "suffix code" is ambiguous. It may refer to codes that end with a special bit pattern, but it also refers to codes where no codeword is the suffix of another codeword (the opposite of prefix codes). The latter meaning is used in Section 4.5, in connection with bidirectional codes.

- Flag codes. A true flag code differs from the suffix codes in one interesting aspect. Such a code may include the flag inside the code, as well as at its right end. The only example of a flag code is Yamamoto's code, Section 3.18.

- Codes based on special number bases or special number sequences. We normally use decimal numbers, and computers use binary numbers, but any integer greater than 1 can serve as the basis of a number system and so can noninteger (real) numbers. It is also possible to construct a sequence of numbers (real or integer) that act as weights of a numbering system. The most important examples of this type of variable-length codes are the Fibonacci (Section 3.20), Goldbach (Section 3.22), and additive codes (Section 3.23).

- The Golomb codes of Section 3.24 are designed in a special way. An integer parameter m is selected and is used to encode an arbitrary integer n in two steps. In the first step, two integers q and r (for quotient and remainder) are computed from n such that n can be fully reconstructed from them. In the second step, q is encoded in unary and is followed by the binary representation of r, whose length is implied by the parameter m. The Rice code of Section 3.25 is a special case of the Golomb codes where m is an integer power of 2. The subexponential code (Section 3.26) is related to the Rice codes.

- Codes ending with "1" are the topic of Section 3.27. In such a code, all the codewords end with a 1, a feature that makes them the natural choice in special applications.

- Variable-length codes are designed for data compression, which is why implementors select the shortest possible codes. Sometimes, however, data reliability is a concern, and longer codes may help detect and isolate errors. Thus, Chapter 4 discusses robust codes. Section 4.3 presents synchronous prefix codes. These codes are useful in applications where it is important to limit the propagation of errors. Bidirectional (or reversible) codes (Sections 4.5 and 4.6) are also designed for increased reliability by allowing the decoder to read and identify codewords either from left to right or in reverse.

The discussion in this chapter starts with codes, prefix codes, and information theory concepts. This is followed by a description of basic codes such as variable-to-block codes, phased-in codes, and the celebrated Huffman code.

2.1 Codes, Fixed- and Variable-Length

A code is a symbol that stands for another symbol. At first, this idea seems pointless. Given a symbol S, what is the use of replacing it with another symbol Y? However, it is easy to find many important examples of the use of codes. Here are a few.

- Any language and any system of writing are codes. They provide us with symbols Y that we use in order to express our thoughts S.

- Acronyms and abbreviations can be considered codes. Thus, the string IBM is a symbol that stands for the much longer symbol "International Business Machines" and the well-known French university École Supérieure D'électricité is known to many simply as Supélec.

- Cryptography is the art and science of obfuscating messages. Before the age of computers, a message was typically a string of letters and was encrypted by replacing each letter with another letter or with a number. In the computer age, a message is a binary string (a bitstring) in a computer, and it is encrypted by replacing it with another bitstring, normally of the same length.

- Error control. Messages, even secret ones, are often transmitted over communications channels and may become damaged, corrupted, or garbled on their way from transmitter to receiver. We often experience low-quality, garbled telephone conversations. Even experienced pharmacists often find it difficult to read and understand a handwritten prescription. Computer data stored on magnetic disks may become corrupted because of exposure to magnetic fields or extreme temperatures. Music and movies recorded on optical discs (CDs and DVDs) may become unreadable because of scratches. In all these cases, it helps to augment the original data with error-control codes. Such codes—formally titled channel codes, but informally known as error-detecting or error-correcting codes—employ redundancy to detect and even correct, certain types of errors.

- ASCII and Unicode. These are character codes that make it possible to store characters of text as bitstrings in a computer. The ASCII code, which dates back to the 1960s [ascii-wiki 09], assigns 7-bit codes to 128 characters including 26 letters (upper- and lowercase), the 10 digits, certain punctuation marks, and several control characters. The Unicode project assigns 16-bit codes to many characters, and has a provision for even longer codes. The long codes make it possible to store and manipulate many thousands of characters, taken from many languages and alphabets (such as Greek, Cyrillic, Hebrew, Arabic, and Indic), and including punctuation marks, diacritics, mathematical symbols, technical symbols, arrows, and dingbats.

The last example illustrates the use of codes in the field of computers and computations. Mathematically, a code is a mapping. Given an alphabet of symbols, a code maps individual symbols or strings of symbols to codewords, where a codeword is a string of bits, a bitstring. The process of mapping a symbol to a codeword is termed encoding and the reverse process is known as decoding.

Codes can have a fixed or variable length, and can be static or adaptive (dynamic). A static code is constructed once and never changes. ASCII and Unicode are examples of such codes. A static code can also have variable length, where short codewords are assigned to the commonly-occurring symbols. A variable-length, static code is normally designed based on the probabilities of the individual symbols. Each type of data has different probabilities and may benefit from a different code. The Huffman method (Section 5.2) is an example of an excellent variable-length, static code that can be constructed once the probabilities of all the symbols in the alphabet are known. In general, static codes that are also variable-length can match well the lengths of individual codewords to the probabilities of the symbols. Notice that the code table must normally be included in the compressed file, because the decoder does not know the symbols' probabilities (the model of the data) and so has no way to construct the codewords independently.

A dynamic code varies over time, as more and more data is read and processed and more is known about the probabilities of the individual symbols. The dynamic (adaptive) Huffman algorithm [Section 5.3] is a method that employs such a code.

Fixed-length codes are known as block codes. They are easy to implement in software. It is easy to replace an original symbol with a fixed-length code, and it is equally easy to start with a string of such codes and break it up into individual codes that are then replaced by the original symbols.

There are cases where variable-length codes (VLCs) have obvious advantages. As their name implies, VLCs are codes that have different lengths. They are also known as variable-length codes. A set of such codes consists of short and long codewords. The following is a short list of important applications where such codes are commonly used.

- Data compression (or source coding). Given an alphabet of symbols where certain symbols occur often in messages, while other symbols are rare, it is possible to compress messages by assigning short codes to the common symbols and long codes to the rare symbols. This is an important application of variable-length codes.

- The Morse code for telegraphy, originated in the 1830s by Samuel Morse and Alfred Vail, exploits the same idea. It assigns short codes to commonly-occurring letters (the code of `E` is a dot and the code of `T` is a dash) and long codes to rare letters and punctuation marks (`--.-` to `Q`, `--..` to `Z`, and `--..--` to the comma).

- Processor design. Part of the architecture of any computer is an instruction set and a processor that fetches instructions from memory and executes them. It is easy to handle fixed-length instructions, but modern computers normally have instructions of different sizes. It is possible to reduce the overall size of programs by designing the instruction set such that commonly-used instructions are short. This also reduces the processor's power consumption and physical size and is especially important in embedded processors, such as processors designed for digital signal processing (DSP).

- Country calling codes. ITU-T recommendation E.164 is an international standard that assigns variable-length calling codes to many countries such that countries with many telephones are assigned short codes and countries with fewer telephones are assigned long codes. These codes also obey the prefix property (Section 2.2) which means that once a calling code C has been assigned, no other calling code will start with C.

- The International Standard Book Number (ISBN) is a unique number assigned to a book, to simplify inventory tracking by publishers and bookstores. The ISBN numbers are assigned according to an international standard known as ISO 2108 (1970). One component of an ISBN is a country code, that can be between one and five digits long. This code also obeys the prefix property. Once C has been assigned as a country code, no other country code will start with C.

- VCR Plus+ (also known as G-Code, VideoPlus+, and ShowView) is a prefix, variable-length code for programming video recorders. A unique number, a VCR Plus+, is computed for each television program by a proprietary algorithm from the date, time, and channel of the program. The number is published in television listings in newspapers and on the Internet. To record a program on a VCR, the number is located in a newspaper and is typed into the video recorder. This programs the recorder to record the correct channel at the right time. This system was developed by Gemstar-TV Guide International [Gemstar 06].

> I gave up on new poetry myself thirty years ago, when most of it began to read like coded messages passing between lonely aliens on a hostile world.
>
> —Russell Baker

2.2 Prefix Codes

Encoding a string of symbols a_i with VLCs is easy. No clever methods or algorithms are needed. The software reads the original symbols a_i one by one and replaces each a_i with its binary, variable-length code c_i. The codes are concatenated to form one (normally long) bitstring. The encoder either includes a table with all the pairs (a_i, c_i) or it executes a procedure to compute code c_i from the bits of symbol a_i.

Decoding is slightly more complex, because of the different lengths of the codes. When the decoder reads the individual bits of VLCs from a bitstring, it has to know either how long each code is or where each code ends. This is why a set of variable-length codes has to be carefully selected and why the decoder has to be taught about the codes. The decoder either has to have a table of all the valid codes, or it has to be told how to identify valid codes.

We start with a simple example. Given the set of four codes $a_1 = 0$, $a_2 = 01$, $a_3 = 011$, and $a_4 = 111$ we easily encode the message $a_2a_3a_3a_1a_2a_4$ as the bitstring 01|011|011|0|01|111. This string can be decoded unambiguously, but not easily. When the decoder inputs a 0, it knows that the next symbol is either a_1, a_2, or a_3, but the decoder has to input more bits to find out how many 1's follow the 0 before it can identify the next symbol. Similarly, given the bitstring 011...111, the decoder has to read the entire string and count the number of consecutive 1's before it finds out how many 1's (zero, one, or two 1's) follow the single 0 at the beginning. We say that such codes are not instantaneous.

In contrast, the following set of VLCs $a_1 = 0$, $a_2 = 10$, $a_3 = 110$, and $a_4 = 111$ is similar and is also instantaneous. Given a bitstring that consists of these codes, the decoder reads consecutive 1's until it has read three 1's (an a_4) or until it has read another 0. Depending on how many 1's precede the 0 (zero, one, or two 1's), the decoder knows whether the next symbol is a_1, a_2, or a_3. The 0 acts as a separator, which is why instantaneous codes are also known as comma codes. The rules that drive the decoder can be considered a finite automaton or a decision tree.

The next example is similar. We examine the set of VLCs $a_1 = 0$, $a_2 = 10$, $a_3 = 101$, and $a_4 = 111$. Only the code of a_3 is different, but a little experimenting shows that this set of VLCs is bad because it is not uniquely decodable (UD). Given the bitstring 0101111..., it can be decoded either as $a_1a_3a_4\ldots$ or $a_1a_2a_4\ldots$.

This observation is crucial because it points the way to the construction of large sets of VLCs. The set of codes above is bad because 10, the code of a_2, is also the prefix of the code of a_3. When the decoder reads 10..., it often cannot tell whether this is the code of a_2 or the start of the code of a_3.

Thus, a useful, practical set of VLCs has to be instantaneous and has to satisfy the following *prefix property*. Once a code c is assigned to a symbol, no other code should start with the bit pattern c. Prefix codes are also referred to as prefix-free codes, prefix condition codes, or instantaneous codes.

The following results can be proved: (1) A code is instantaneous if and only if it is a prefix code. (2) The set of UD codes is larger than the set of instantaneous codes (i.e., there are UD codes that are not instantaneous). (3) There is an instantaneous variable-length code with codeword lengths L_i if and only if there is a UD code with these codeword lengths.

The last of these results indicates that we cannot reduce the average word length of a variable-length code by using a UD code rather than an instantaneous code. Thus, there is no loss of compression performance if we restrict our selection of codes to instantaneous codes.

A UD code that consists of r codewords of lengths l_i must satisfy the Kraft inequality (Section 2.5), but this inequality does not require a prefix code. Thus, if a code satisfies the Kraft inequality it is UD, but if it is also a prefix code, then it is instantaneous. This feature of a UD code being also instantaneous, comes for free, because there is no need to add bits to the code and make it longer.

A prefix code (a set of codewords that satisfy the prefix property) is UD. Such a code is also *complete* if adding any codeword to it turns it into a non-UD code. A complete code is the largest UD code, but it also has a downside; it is less robust. If even a single bit is accidentally modified or deleted (or if a bit is somehow added) during storage or transmission, the decoder will lose synchronization and the rest of the transmission will be decoded incorrectly (see the discussion of robust codes in Chapter 4).

While discussing UD and non-UD codes, it is interesting to note that the Morse code is non-UD (because, for example, the code of `I` is ".." and the code of `H` is "...."), so Morse had to make it UD by requiring accurate relative timing.

2.3 VLCs, Entropy, and Redundancy

Understanding data compression and its codes must start with understanding information, because the former is based on the latter. Hence this short section that introduces a few important concepts from information theory (see the more detailed discussion in the Appendix).

Information theory is the creation, in the late 1940s, of Claude Shannon. Shannon tried to develop means for measuring the amount of information stored in a symbol without considering the meaning of the information. He discovered the connection between the logarithm function and information, and showed that the information content (in bits) of a symbol with probability p is $-\log_2 p$. If the base of the logarithm is e, then the information is measured in units called nats. If the base is 3, the information units are trits, and if the base is 10, the units are referred to as Hartleys.

Information theory is concerned with the transmission of information from a sender (termed a source), through a communications channel, to a receiver. The sender and receiver can be persons or machines and the receiver may, in turn, act as a sender and send the information it has received to another receiver. The information is sent in units called symbols (normally bits, but in verbal communications the symbols are spoken words) and the set of all possible data symbols is an alphabet.

The most important single factor affecting communications is noise in the communications channel. In verbal communications, this noise is literally noise. When trying to talk in a noisy environment, we may lose part of the discussion. In electronic communications, the channel noise is caused by imperfect hardware and by factors such as lightning, voltage fluctuations—old, high-resistance wires—sudden surge in temperature, and interference from machines that generate strong electromagnetic fields.

The presence of noise implies that special codes should be used to increase the reliability of transmitted information. This is referred to as channel coding or, in everyday language, error-control codes.

The second most important factor affecting communications is sheer volume. Any communications channel has a limited capacity. It can transmit only a limited number of symbols per time unit. An obvious way to increase the amount of data transmitted is to compress it before it is sent (in the source). Methods to compress data are therefore known as source coding or, in everyday language, data compression. The feature that makes it possible to compress data is the fact that individual symbols appear in our data files with different probabilities. One important principle (although not the only one) used to compress data is to assign variable-length codes to the individual data symbols such that short codes are assigned to the common symbols.

Two concepts from information theory, namely entropy and redundancy, are needed in order to fully understand the application of VLCs to data compression.

Roughly speaking, the term "entropy" as defined by Shannon is proportional to the minimum number of yes/no questions needed to reach the answer to some question. Another way of looking at entropy is as a quantity that describes how much information is included in a signal or event. Given a discrete random variable X that can have n values x_i with probabilities P_i, the entropy $H(X)$ of X is defined as

$$H(X) = -\sum_{i=1}^{n} P_i \log_2 P_i.$$

A detailed discussion of information theory is outside the scope of this book. Interested readers are referred to the many texts on this subject. Here, we will only show intuitively why the logarithm function plays such an important part in measuring information.

Imagine a source that emits symbols a_i with probabilities p_i. We assume that the source is memoryless, i.e., the probability of a symbol being emitted does not depend on what has been emitted so far. We want to define a function $I(a_i)$ that will measure the amount of information gained when we discover that the source has emitted symbol a_i. Function I will also measure our uncertainty as to whether the next symbol will be a_i. Alternatively, $I(a_i)$ corresponds to our surprise in finding that the next symbol is a_i. Clearly, our surprise at seeing a_i emitted is inversely proportional to the probability p_i (we are surprised when a low-probability symbol is emitted, but not when we notice a high-probability symbol). Thus, it makes sense to require that function I satisfies the following conditions:

1. $I(a_i)$ is a decreasing function of p_i, and returns 0 when the probability of a symbol is 1. This reflects our feeling that high-probability events convey less information.

2. $I(a_i a_j) = I(a_i) + I(a_j)$. This is a result of the source being memoryless and the probabilities being independent. Discovering that a_i was immediately followed by a_j, provided us with the same information as knowing that a_i and a_j were emitted independently.

Even those with a minimal mathematical background will immediately see that the logarithm function satisfies the two conditions above. This is the first example of the relation between the logarithm function and the quantitative measure of information. The next few paragraphs illustrate other connections between the two.

Consider the case of person A selecting at random an integer N between 1 and 64 and person B having to guess N. What is the minimum number of yes/no questions that are needed for B to guess N? Those familiar with the technique of binary search know how to guess efficiently. Using this technique, B should divide the interval 1–64 in two, and should start by asking "is N between 1 and 32?" If the answer is no, then N is in the interval 33 to 64. This interval is then divided by two and B's next question should be "is N between 33 and 48?" This process continues until the interval selected by B shrinks to a single number.

It does not take much to see that exactly six questions are necessary to determine N. This is because 6 is the number of times 64 can be divided in half. Mathematically, this is equivalent to writing $6 = \log_2 64$, which is why the logarithm is the mathematical function that quantifies information.

> What we call reality arises in the last analysis from the posing of yes/no questions. All things physical are information-theoretic in origin, and this is a participatory universe.
>
> —John Wheeler

Another approach to the same problem is to consider a nonnegative integer N and ask how many digits does it take to express it. The answer, of course, depends on N. The greater N, the more digits are needed. The first 100 nonnegative integers (0 through 99) can be expressed by two decimal digits. The first 1000 such integers can be expressed by three digits. Again it does not take long to see the connection. The number of digits required to represent N equals approximately $\log N$. The base of the logarithm is the same as the base of the digits. For decimal digits, use base 10; for binary digits (bits), use base 2. If we agree that the number of digits it takes to express N is proportional to the information content of N, then again the logarithm is the function that gives us a measure of the information. As an aside, the precise length, in bits, of the binary (β) representation of a positive integer n is

$$1 + \lfloor \log_2 n \rfloor \tag{2.1}$$

or, alternatively, $\lceil \log_2(n+1) \rceil$. When n is represented in any other number base b, its length is given by the same formula, but with the logarithm in base b instead of 2.

Here is another observation that illuminates the relation between the logarithm and information. A 10-bit string can have $2^{10} = 1{,}024$ values. We say that such a string may contain one of 1,024 messages, or that the length of the string is the logarithm of the number of possible messages the string can convey.

The following example sheds more light on the concept of entropy and will prepare us for the definition of redundancy. Given a set of two symbols a_1 and a_2, with probabilities P_1 and P_2, respectively, we compute the entropy of the set for various values of the probabilities. Since $P_1+P_2=1$, the entropy of the set is $-P_1 \log_2 P_1 - (1-P_1)\log_2(1-P_1)$ and the results are summarized in Table 2.2.

When $P_1 = P_2$, at least one bit is required to encode each symbol, reflecting the fact that the entropy is at its maximum, the redundancy is zero, and the data cannot be compressed. However, when the probabilities are very different, the minimum number of bits required per symbol drops significantly. We may not be able to conceive a compression method that expresses each symbol in just 0.08 bits, but we know that when $P_1 = 99\%$, such compression is theoretically possible.

P_1	P_2	Entropy
0.99	0.01	0.08
0.90	0.10	0.47
0.80	0.20	0.72
0.70	0.30	0.88
0.60	0.40	0.97
0.50	0.50	1.00

Table 2.2: Probabilities and Entropies of Two Symbols.

In general, the entropy of a set of n symbols depends on the individual probabilities P_i and is largest when all n probabilities are equal. Data representations often include redundancies and data can be compressed by reducing or eliminating these redundancies. When the entropy is at its maximum, the data has maximum information content and therefore cannot be further compressed. Thus, it makes sense to define redundancy as a quantity that goes down to zero as the entropy reaches its maximum.

> The fundamental problem of communication is that of reproducing at one point either exactly or approximately a message selected at another point.
>
> —Claude Shannon (1948)

To understand the definition of redundancy, we start with an alphabet of symbols a_i, where each symbol appears in the data with probability P_i. The data is compressed by replacing each symbol with an l_i-bit-long code. The average code length is the sum $\sum P_i l_i$ and the entropy (the smallest number of bits required to represent the symbols) is $\sum[-P_i \log_2 P_i]$. The redundancy R of the set of symbols is defined as the average code length minus the entropy. Thus,

$$R = \sum_i P_i l_i - \sum_i [-P_i \log_2 P_i]. \tag{2.2}$$

The redundancy is zero when the average code length equals the entropy, i.e., when the codes are the shortest and compression has reached its maximum.

Given a set of symbols (an alphabet), we can assign binary codes to the individual symbols. It is easy to assign long codes to symbols, but most practical applications require the shortest possible codes.

Consider the four symbols a_1, a_2, a_3, and a_4. If they appear in our data strings with equal probabilities (= 0.25), then the entropy of the data is $-4(0.25 \log_2 0.25) = 2$. Two is the smallest number of bits needed, on average, to represent each symbol in this case. We can simply assign our symbols the four 2-bit codes 00, 01, 10, and 11. Since the probabilities are equal, the redundancy is zero and the data cannot be compressed below two bits/symbol.

Next, consider the case where the four symbols occur with different probabilities as shown in Table 2.3, where a_1 appears in the data (on average) about half the time, a_2 and a_3 have equal probabilities, and a_4 is rare. In this case, the data has entropy $-(0.49 \log_2 0.49 + 0.25 \log_2 0.25 + 0.25 \log_2 0.25 + 0.01 \log_2 0.01) \approx -(-0.050 - 0.5 - 0.5 - 0.066) = 1.57$. The smallest number of bits needed, on average, to represent each symbol has dropped to 1.57.

Symbol	Prob.	Code1	Code2
a_1	.49	1	1
a_2	.25	01	01
a_3	.25	010	000
a_4	.01	001	001

Table 2.3: Variable-Length Codes.

If we again assign our symbols the four 2-bit codes 00, 01, 10, and 11, the redundancy would be $R = -1.57 + \log_2 4 = 0.43$. This suggests assigning variable-length codes to the symbols. Code1 of Table 2.3 is designed such that the most common symbol, a_1, is assigned the shortest code. When long data strings are transmitted using Code1, the average size (the number of bits per symbol) is $1 \times 0.49 + 2 \times 0.25 + 3 \times 0.25 + 3 \times 0.01 = 1.77$, which is very close to the minimum. The redundancy in this case is $R = 1.77 - 1.57 = 0.2$ bits per symbol. An interesting example is the 20-symbol string $a_1 a_3 a_2 a_1 a_3 a_3 a_4 a_2 a_1 a_1 a_2 a_2 a_1 a_1 a_3 a_1 a_1 a_2 a_3 a_1$, where the four symbols occur with approximately the right frequencies. Encoding this string with Code1 yields the 37 bits:

1|010|01|1|010|010|001|01|1|1|01|01|1|1|010|1|1|01|010|1

(without the vertical bars). Using 37 bits to encode 20 symbols yields an average size of 1.85 bits/symbol, not far from the calculated average size. (The reader should bear in mind that our examples are short. To obtain results close to the best that's theoretically possible, an input stream with at least thousands of symbols is needed.)

However, the conscientious reader may have noticed that Code1 is bad because it is not a prefix code. Code2, in contrast, is a prefix code and can be decoded uniquely. Notice how Code2 was constructed. Once the single bit 1 was assigned as the code of a_1, no other codes could start with 1 (they all had to start with 0). Once 01 was assigned as the code of a_2, no other codes could start with 01. This is why the codes of a_3 and a_4 had to start with 00. Naturally, they became 000 and 001.

Designing variable-length codes for data compression must therefore take into account the following two principles: (1) assign short codes to the more frequent symbols and (2) obey the prefix property. Following these principles produces short, unambiguous codes, but not necessarily the best (i.e., shortest) codes. In addition to these principles, an algorithm is needed to generate a set of shortest codes (ones with the minimum average size). The only input to such an algorithm is the frequencies of occurrence (or alternatively the probabilities) of the symbols of the alphabet. The well-known Huffman algorithm (Section 5.2) is such a method. Given a set of symbols whose probabilities of occurrence are known, this algorithm constructs a set of shortest prefix codes for the symbols. Notice that such a set is normally not unique and there may be several sets of codes with the shortest length.

> The beauty of code is much more akin to the elegance, efficiency and clean lines of a spiderweb. It is not the chaotic glory of a waterfall, or the pristine simplicity of a flower. It is an aesthetic of structure, design and order.
>
> —Charles Gordon

Notice that a UD code does not have to be a prefix code. It is possible, for example, to designate the string 111 as a separator (a comma) to separate individual codewords of different lengths, provided that no codeword contains 111. Other examples of a non-prefix, variable-length codes are the C^3 code (page 146) and the generalized Fibonacci C_2 code (page 149).

2.4 Universal Codes

Mathematically, a code is a mapping. It maps source symbols into codewords. Mathematically, a source of messages is a pair (M, P) where M is a (possibly infinite) set of messages and P is a function that assigns a nonzero probability to each message. A message is mapped into a long bitstring whose length depends on the quality of the code and on the probabilities of the individual symbols. The best that can be done is to compress a message to its entropy H. A code is universal if it compresses messages to codewords whose average length is bounded by $C1(H + C2)$ where $C1$ and $C2$ are constants greater than or equal to 1, i.e., an average length that is a constant multiple of the entropy plus another constant. A universal code with large constants isn't very useful. A code with $C1 = 1$ is called asymptotically optimal.

A Huffman code often performs better than a universal code, but it can be used only when the probabilities of the symbols are known. In contrast, a universal code can be used in cases where only the ranking of the symbols' probabilities is known. If we know that symbol a_5 has the highest probability and a_8 has the next largest one, we can assign the shortest codeword to a_5 and the next longer codeword to a_8. Thus, universal coding amounts to ranking of the source symbols. After ranking, the symbol with index 1 has the largest probability, the symbol with index 2 has the next highest one, and so on. We can therefore ignore the actual symbols and concentrate on their new indexes. We can assign one codeword to index 1, another codeword to index 2, and so on, which is why variable-length codes are often designed to encode integers (Section 3.1). Such a set of variable-length codes can encode any number of integers with codewords that have increasing lengths.

Notice also that a set of universal codes is fixed and so doesn't have to be constructed for each set of source symbols, a feature that simplifies encoding and decoding. However, if we know the probabilities of the individual symbols (the probability distribution of the alphabet of symbols), it becomes possible to tailor the code to the probability, or conversely, to select a known code whose codewords fit the known probability distribution. In all cases, the code selected (the set of codewords) must be uniquely decodable (UD). A non-UD code is ambiguous and therefore useless.

2.5 The Kraft–McMillan Inequality

The Kraft–McMillan inequality is concerned with the existence of a uniquely decodable (UD) code. It establishes the relation between such a code and the lengths L_i of its codewords.

One part of this inequality, due to [McMillan 56], states that given a UD variable-length code, with n codewords of lengths L_i, the lengths must satisfy the relation

$$\sum_{i=1}^{n} 2^{-L_i} \le 1. \tag{2.3}$$

The other part, due to [Kraft 49], states the opposite. Given a set of n positive integers $(L_1, L_2, \ldots, L_n)$ that satisfy Equation (2.3), there exists an instantaneous variable-length code such that the L_i are the lengths of its individual codewords.

Together, both parts say that there is an instantaneous variable-length code with codeword lengths L_i if and only if there is a UD code with these codeword lengths. The two parts do not say that a variable-length code is instantaneous or UD if and only if the codeword lengths satisfy Equation (2.3). In fact, it is easy to check the three individual code lengths of the code $(0, 01, 011)$ and verify that $2^{-1} + 2^{-2} + 2^{-3} = 7/8$. This code satisfies the Kraft–McMillan inequality and yet it is not instantaneous, because it is not a prefix code. Similarly, the code $(0, 01, 001)$ also satisfies Equation (2.3), but is not UD. A few more comments on this inequality are in order:

- If a set of lengths L_i satisfies Equation (2.3), then there exist instantaneous and UD variable-length codes with these lengths. For example $(0, 10, 110)$.

■ A UD code is not always instantaneous, but there exists an instantaneous code with the same codeword lengths. For example, code $(0, 01, 11)$ is UD but not instantaneous, while code $(0, 10, 11)$ is instantaneous and has the same lengths.

■ The sum of Equation (2.3) corresponds to the part of the complete code tree that has been used for codeword selection. This is why the sum has to be less than or equal to 1. This intuitive explanation of the Kraft–McMillan relation is explained in the next paragraph.

We can gain a deeper understanding of this useful and important inequality by constructing the following simple prefix code. Given five symbols a_i, suppose that we decide to assign 0 as the code of a_1. Now all the other codes have to start with 1. We therefore assign 10, 110, 1110, and 1111 as the codewords of the four remaining symbols. The lengths of the five codewords are 1, 2, 3, 4, and 4, and it is easy to see that the sum

$$2^{-1}+2^{-2}+2^{-3}+2^{-4}+2^{-4}=\frac{1}{2}+\frac{1}{4}+\frac{1}{8}+\frac{2}{16}=1$$

satisfies the Kraft–McMillan inequality. We now consider the possibility of constructing a similar code with lengths 1, 2, 3, 3, and 4. The Kraft–McMillan inequality tells us that this is impossible, because the sum

$$2^{-1}+2^{-2}+2^{-3}+2^{-3}+2^{-4}=\frac{1}{2}+\frac{1}{4}+\frac{2}{8}+\frac{1}{16}$$

is greater than 1, and this is easy to understand when we consider the code tree. Starting with a complete binary tree of height 4, such as the tree of Figure 4.17, it is obvious that once 0 was assigned as a codeword, we have "used" one half of the tree and all future codes would have to be selected from the other half of the tree. Once 10 was assigned, we were left with only 1/4 of the tree. Once 110 was assigned as a codeword, only 1/8 of the tree remained available for the selection of future codes. Once 1110 has been assigned, only 1/16 of the tree was left, and that was enough to select and assign code 1111. However, once we select and assign codes of lengths 1, 2, 3, and 3, we have exhausted the entire tree and there is nothing left to select the last (4-bit) code from.

The Kraft–McMillan inequality can be related to the entropy by observing that the lengths L_i can always be written as $L_i = -\log_2 P_i + E_i$, where E_i is simply the amount by which L_i is greater than the entropy (the extra length of code i).

This implies that

$$2^{-L_i}=2^{(\log_2 P_i - E_i)}=2^{\log_2 P_i}/2^{E_i}=P_i/2^{E_i}.$$

In the special case where all the extra lengths are the same ($E_i = E$), the Kraft–McMillan inequality says that

$$1\geq\sum_{i=1}^{n}P_i/2^E=\frac{\sum_{i=1}^{n}P_i}{2^E}=\frac{1}{2^E}\Longrightarrow 2^E\geq 1\Longrightarrow E\geq 0.$$

An unambiguous code has nonnegative extra length, meaning its length is greater than or equal to the length determined by its entropy.

Here is a simple example of the use of this inequality. Consider the simple case of n equal-length binary codewords. The size of each codeword is $L_i = \log_2 n$, and the Kraft–McMillan sum is

$$\sum_1^n 2^{-L_i} = \sum_1^n 2^{-\log_2 n} = \sum_1^n \frac{1}{n} = 1.$$

The inequality is satisfied, so such a code is UD.

A more interesting example is the case of n symbols where the first one is compressed and the second one is expanded. We set $L_1 = \log_2 n - a$, $L_2 = \log_2 n + e$, and $L_3 = L_4 = \cdots = L_n = \log_2 n$, where a and e are positive. We show that $e > a$, which means that compressing a symbol by a factor a requires expanding another symbol by a larger factor. We can benefit from this only if the probability of the compressed symbol is greater than that of the expanded symbol.

$$\begin{aligned}\sum_1^n 2^{-L_i} &= 2^{-L_1} + 2^{-L_2} + \sum_3^n 2^{-\log_2 n}\\ &= 2^{-\log_2 n+a} + 2^{-\log_2 n-e} + \sum_1^n 2^{-\log_2 n} - 2\times 2^{-\log_2 n}\\ &= \frac{2^a}{n} + \frac{2^{-e}}{n} + 1 - \frac{2}{n}.\end{aligned}$$

The Kraft–McMillan inequality requires that

$$\frac{2^a}{n} + \frac{2^{-e}}{n} + 1 - \frac{2}{n} \le 1, \quad \text{or} \quad \frac{2^a}{n} + \frac{2^{-e}}{n} - \frac{2}{n} \le 0,$$

or $2^{-e} \le 2 - 2^a$, implying $-e \le \log_2(2 - 2^a)$, or $e \ge -\log_2(2 - 2^a)$.

The inequality above implies $a \le 1$ (otherwise, $2 - 2^a$ is negative) but a is also positive (since we assumed compression of symbol 1). The possible range of values of a is therefore $(0, 1]$, and in this range e is greater than a, proving the statement above. (It is easy to see that $a = 1 \to e \ge -\log_2 0 = \infty$, and $a = 0.1 \to e \ge -\log_2(2 - 2^{0.1}) \approx 0.10745$.)

It can be shown that this is just a special case of a general result that says, given an alphabet of n symbols, if we compress some of them by a certain factor, then the others must be expanded by a greater factor.

> One of my most productive days was throwing away 1000 lines of code.
> —Kenneth Thompson

2.6 Tunstall Code

The main advantage of variable-length codes is their variable lengths. Some codes are short, others are long, and a clever assignment of codes to symbols can produce compression. On the downside, variable-length codes are difficult to work with. The encoder has to construct each code from individual bits and pieces, has to accumulate and append several such codes in a short buffer, wait until n bytes of the buffer are full of code bits (where n must be at least 1), write the n bytes onto the output, shift the buffer n bytes, and keep track of the location of the last bit placed in the buffer. The decoder has to go through the reverse process. It is definitely easier to deal with fixed-length codes, and the Tunstall codes described here are an example of how such codes can be designed. The idea is to construct a set of fixed-length codes, each encoding a variable-length string of input symbols. As a result, these codes are also known as variable-to-fixed (or variable-to-block) codes, in contrast to the variable-length codes which are also referred to as fixed-to-variable.

Imagine an alphabet that consists of two symbols A and B where A is more common. Given a typical string from this alphabet, we expect substrings of the form AA, AAA, AB, AAB, and B, but rarely strings of the form BB. We can therefore assign fixed-length codes to the following five substrings as follows. $AA = 000$, $AAA = 001$, $AB = 010$, $ABA = 011$, and $B = 100$. A rare occurrence of two consecutive Bs will be encoded by 100100, but most occurrences of B will be preceded by an A and will be coded by 010, 011, or 100.

This example is both bad and inefficient. It is bad, because $AAABAAB$ can be encoded either as the four codes AAA, B, AA, B or as the three codes AA, ABA, AB; encoding is not unique and may require several passes to determine the shortest code. This happens because our five substrings don't satisfy the prefix property. This example is inefficient because only five of the eight possible 3-bit codewords are used. An n-bit Tunstall code should use all 2^n codewords. Another point is that our codes were selected without considering the relative frequencies of the two symbols, and as a result we cannot be certain that this is the best code for our alphabet.

Thus, an algorithm is needed to construct the best n-bit Tunstall code for a given alphabet of N symbols, and such an algorithm is given in [Tunstall 67]. Given an alphabet of N symbols, we start with a code table that consists of the symbols. We then iterate as long as the size of the code table is less than or equal to 2^n (the number of n-bit codes). Each iteration performs the following steps:

- Select the symbol with largest probability in the table. Call it S.
- Remove S and include the N substrings Sx where x goes over all the N symbols. This step increases the table size by $N - 1$ symbols (some of them may be substrings). Thus, after iteration k, the table size will be $N + k(N - 1)$ elements.
- If $N + (k + 1)(N - 1) \le 2^n$, perform another iteration (iteration $k + 1$).

It is easy to see that the elements (symbols and substrings) of the table satisfy the prefix property and thus ensure unique encodability. If the first iteration adds element AB to the table, it must have removed element A. Thus, A, the prefix of AB, is not a code. If the next iteration creates element ABR, then it has removed element AB, so AB is not a prefix of ABR. This construction also minimizes the average number of bits per alphabet symbol because of the requirement that each iteration select the element (or an element) of maximum probability. This requirement is similar to the way a Huffman code is constructed, and we illustrate it by an example.

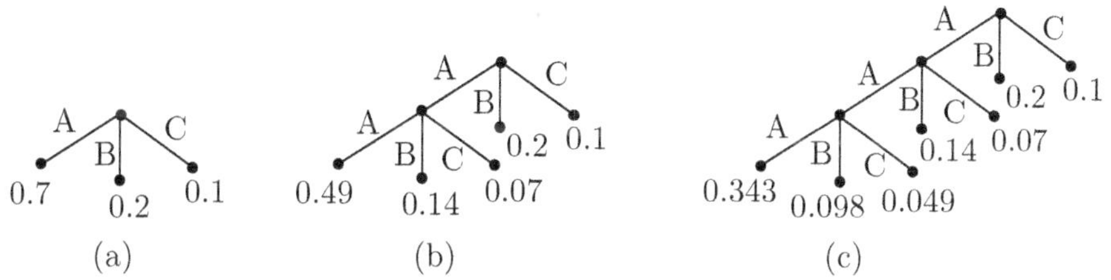

Figure 2.4: Tunstall Code Example.

Given an alphabet with the three symbols A, B, and C ($N = 3$), with probabilities 0.7, 0.2, and 0.1, respectively, we decide to try to construct a set of 3-bit Tunstall codes (thus, $n = 3$). We start our code table as a tree with a root and three children (Figure 2.4a). In the first iteration, we select A and turn it into the root of a subtree with children AA, AB, and AC with probabilities 0.49, 0.14, and 0.07, respectively (Figure 2.4b). The largest probability in the tree is that of node AA, so the second iteration converts it to the root of a subtree with nodes AAA, AAB, and AAC with probabilities 0.343, 0.098, and 0.049, respectively (Figure 2.4c). After each iteration we count the number of leaves of the tree and compare it to $2^3 = 8$. After the second iteration there are seven leaves in the tree, so the loop stops. Seven 3-bit codes are arbitrarily assigned to elements AAA, AAB, AAC, AB, AC, B, and C. The eighth available code should be assigned to a substring that has the highest probability and also satisfies the prefix property.

The average bit length of this code is easily computed as

$$\frac{3}{3(0.343 + 0.098 + 0.049) + 2(0.14 + 0.07) + 0.2 + 0.1} = 1.37 \text{ bits/symbol}.$$

In general, let p_i and l_i be the probability and length of tree node i. If there are m nodes in the tree, the average bit length of the Tunstall code is $n/\sum_{i=1}^{m} p_i l_i$. The entropy of our alphabet is $-(0.7 \times \log_2 0.7 + 0.2 \times \log_2 0.2 + 0.1 \times \log_2 0.1) = 1.156$, so the Tunstall codes do not provide the best compression.

The tree of Figure 2.4 is referred to as a parse tree, not a code tree. It is complete in the sense that every interior node has N children. Notice that the total number of nodes of this tree is $3 \times 2 + 1$ and in general $a(N - 1) + 1$. A parse tree defines a set of substrings over the alphabet (seven substrings in our example) such that any string of symbols from the alphabet can be broken up (subdivided) into these substrings (except that the last part may be only a prefix of such a substring) in one way only. The subdivision is unique because the set of substrings defined by the parse tree is proper, i.e., no substring is a prefix of another substring.

An important property of the Tunstall codes is their reliability. If one bit becomes corrupt, only one code will get bad. Normally, variable-length codes are not robust. One bad bit may corrupt the decoding of the remainder of a long sequence of such codes. It is possible to incorporate error-control codes in a string of variable-length codes, but this increases its size and reduces compression.

Section 5.2.1 illustrates how a combination of the Tunstall algorithm with Huffman coding can improve compression in a two-step, dual tree process.

A major downside of the Tunstall code is that both encoder and decoder have to store the complete code (the set of substrings of the parse tree).

> There are 10 types of people in this world: those who understand binary and those who don't.
>
> —Author unknown

2.7 Schalkwijk's Coding

One of the many contributions of Isaac Newton to mathematics is the well-known binomial theorem. It states

$$(a+b)^n = \sum_{i=0}^{n} \binom{n}{i} a^i b^{n-i},$$

where the term

$$\binom{n}{i} = \frac{n!}{i!(n-i)!}$$

is pronounced "n over i" and is referred to as a binomial coefficient.

Blaise Pascal (Section 3.24), a contemporary of Newton, discovered an elegant way to compute these coefficients without the lengthy calculations of factorials, multiplications, and divisions. He came up with the famous triangle that now bears his name (Figure 2.5) and showed that the general element of this triangle is a binomial coefficient.

```
1
1 1
1 2 1
1 3 3 1
1 4 6 4 1
1 5 10 10 5 1
1 6 15 20 15 6 1
1 7 21 35 35 21 7 1
1 8 28 56 70 56 28 8 1
1 9 36 84 126 126 84 36 9 1
1 10 45 120 210 252 210 120 45 10 1
1 11 55 165 330 462 462 330 165 55 11 1
1 12 66 220 495 792 924 792 495 220 66 12 1
1 13 78 286 715 1287 1716 1716 1287 715 286 78 13 1
1 14 91 364 1001 2002 3003 3432 3003 2002 1001 364 91 14 1
1 15 105 455 1365 3003 5005 6435 6435 5005 3003 1365 455 105 15 1
1 16 120 560 1820 4368 8008 11440 12870 11440 8008 4368 1820 560 120 16 1
```

Figure 2.5: Pascal Triangle.

> Quand on voit le style naturel, on est tout ètonnè et ravi, car on s'attendait de voir un auteur, et on trouve un homme. (When we see a natural style, we are quite surprised and delighted, for we expected to see an author and we find a man.)
>
> —Blaise Pascal, *Pensèes* (1670)

The Pascal triangle is an infinite triangular matrix that's constructed from the edges inwards. First fill the left and right edges with 1's, then compute each interior element as the sum of the two elements directly above it. The construction is simple and it is trivial to derive an explicit expression for the general element of the triangle and show that it is a binomial coefficient. Number the rows from 0 starting at the top, and number the columns from 0 starting on the left. A general element is denoted by $\binom{i}{j}$. Now observe that the top two rows (corresponding to $i = 0$ and $i = 1$) consist of 1's and that every other row can be obtained as the sum of its predecessor and a shifted version of its predecessor. For example,

$$\begin{array}{rccccc} & 1 & 3 & 3 & 1 & \\ + & & 1 & 3 & 3 & 1 \\ \hline & 1 & 4 & 6 & 4 & 1 \end{array}$$

This shows that the elements of the triangle satisfy

$$\binom{i}{0} = \binom{i}{i} = 1, \qquad i = 0, 1, \ldots,$$
$$\binom{i}{j} = \binom{i-1}{j-1} + \binom{i-1}{j}, \qquad i = 2, 3, \ldots, \quad j = 1, \ldots, (i-1).$$

From this, it is easy to obtain the explicit expression

$$\begin{aligned}\binom{i}{j} &= \binom{i-1}{j-1} + \binom{i-1}{j} \\ &= \frac{(i-1)!}{(j-1)!(i-j)!} + \frac{(i-1)!}{j!(i-1-j)!} \\ &= \frac{j(i-1)!}{j!(i-j)!} + \frac{(i-j)(i-1)!}{j!(i-j)!} \\ &= \frac{i!}{j!(i-j)!}.\end{aligned}$$

And this is Newton's binomial coefficient $\binom{i}{j}$.

The Pascal triangle has many interesting and unexpected properties, some of which are listed here.

- The sum of the elements of row i is 2^i.
- If the second element of a row is a prime number, all the elements of the row (except the 1's) are divisible by it. For example, the elements 7, 21, and 35 of row 7 are divisible by 7.
- Select any diagonal and any number of consecutive elements on it. Their sum will equal the number on the row below the end of the selection and off the selected diagonal. For example, $1 + 6 + 21 + 56 = 84$.

- Select row 7, convert its elements 1, 7, 21, 35, 35, 21, 7, and 1 to the single number 19487171 by concatenating the elements, except that a multidigit element is first carried over, such that 1, 7, 21, 35,... become $1(7+2)(1+3)(5+3)\ldots = 1948\ldots$. This number equals 11^7 and this magic-11 property holds for any row.

- The third column from the right consists of the triangular numbers 1, 3, 6, 10,....

- Select all the odd numbers on the triangle and fill them with black. The result is the Sierpinski triangle (a well-known fractal).

Other unusual properties can be found in the vast literature that exists on the Pascal triangle. The following is a quotation from Donald Knuth:

"There are so many relations in Pascal's triangle, that when someone finds a new identity, there aren't many people who get excited about it anymore, except the discoverer."

The Pascal triangle is the basis of the unusual coding scheme described in [Schalkwijk 72]. This method starts by considering all the finite bit strings of length n that have exactly w 1's. The set of these strings is denoted by $T(n, w)$. If a string t consists of bits t_1 through t_n, then we define weights w_1 through w_n as the partial sums

$$w_k = \sum_{i=k}^{n} t_i.$$

Thus, if $t = 010100$, then $w_1 = w_2 = 2$, $w_3 = w_4 = 1$, and $w_5 = w_6 = 0$. Notice that w_1 always equals w.

We now define, somewhat arbitrarily, a ranking $i(t)$ on the $\binom{n}{w}$ strings in set $T(n, w)$ by

$$i(t) = \sum_{k=1}^{n} t_k \binom{n-k}{w_k}.$$

(The binomial coefficient $\binom{i}{j}$ is defined only for $i \geq j$, so we set it to 0 when $i < n$.) The rank of $t = 010100$ becomes

$$0 + 1\binom{6-2}{2} + 0 + 1\binom{6-4}{1} + 0 + 0 = 6 + 2 = 8.$$

It can be shown that the rank of a string t in set $T(n, w)$ is between 0 and $\binom{n}{w} - 1$.

The following table lists the ranking for the $\binom{6}{2} = 15$ strings of set $T(6, 2)$.

0	000011	3	001001	6	010001	9	011000	12	100100
1	000101	4	001010	7	010010	10	100001	13	101000
2	000110	5	001100	8	010100	11	100010	14	110000

The first version of the Schalkwijk coding algorithm is not general. It is restricted to data symbols that are elements t of $T(n, w)$. We assume that both encoder and decoder know the values of n and w. The method employs the Pascal triangle to determine the rank $i(t)$ of each string t, and this rank becomes the code of t. The maximum value of the rank is $\binom{n}{w} - 1$, so it can be expressed in $\lceil \log_2 \binom{n}{w} \rceil$ bits. Thus, this method compresses each n-bit string (of which w bits are 1's) to $\lceil \log_2 \binom{n}{w} \rceil$ bits.

Consider a source of bits that emits a 0 with probability q and a 1 with probability $p = 1 - q$. The entropy of this source is $H(p) = -p \log_2 p - (1-p) \log_2 (1-p)$. In our strings, $p = w/n$, so the compression performance of this method is measured by the ratio

$$\frac{\lceil \log_2 \binom{n}{w} \rceil}{n}$$

and it can be shown that this ratio approaches $H(w/n)$ when n becomes very large. (The proof employs the famous Stirling formula $n! \approx \sqrt{2\pi n}\, n^n e^{-n}$.)

Figure 2.6a illustrates the operation of both encoder and decoder. Both know the values of n and w and they construct in the Pascal triangle a coordinate system tilted as shown in the figure and with its origin at element w of row n of the triangle.

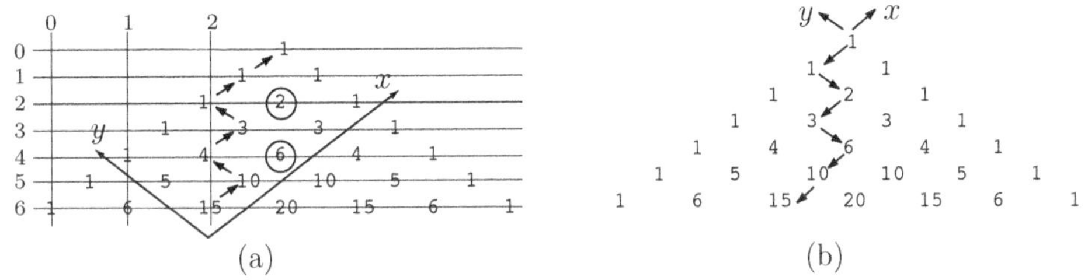

Figure 2.6: First Version.

As an example, suppose that $(n, w) = (6, 2)$. This puts the origin at the element 15 as shown in part (a) of the figure. The encoder starts at the origin, reads bits from the input string, and moves one step in the x direction for each 0 read and one step in the y direction for each 1 read. In addition, before moving in the y direction, the encoder saves the next triangle element in the x direction (the one it will not go to). Thus, given the string 010100, the encoder starts at the origin (15), moves to 10, 4, 3, 1, 1, and 1, while saving the values 6 (before it moves from 10 to 4) and 2 (before it moves from 3 to 1). The sum $6 + 2 = 8 = 1000_2$ is the 4-bit rank of the input string and it becomes the encoder's output.

The decoder also knows the values of n and w, so it constructs the same coordinate system in the triangle and starts at the origin. Given the 4-bit input $1000_2 = 8$, the decoder compares it to the next x value 10, and finds that $8 < 10$. It therefore moves in the x direction, to 10, and emits a 0. The input 8 is compared to the next x value 6, but it is not less than 6. The decoder responds by subtracting $8 - 6 = 2$, moving in the y direction, to 4, and emitting a 1. The current input, 2, is compared to the next x value 3, and is found to be smaller. The decoder therefore moves in the x direction, to 3, and emits a 0. When the input 2 is compared to the next x value 2, it is not smaller, so the decoder: (1) subtracts $2 - 2 = 0$, (2) moves in the y direction to 1, and (3) emits a 1. The decoder's input is now 0, so the decoder finds it smaller than the values on the x axis. It therefore keeps moving in the x direction, emitting two more zeros until it reaches the top of the triangle.

A similar variant is shown in Figure 2.6b. The encoder always starts at the apex of the triangle, moves in the $-x$ direction for each 0 and in the $-y$ direction for each 1, where it also records the value of the next element in the $-x$ direction. Thus, the two steps in the $-y$ direction in the figure record the values 1 and 3, whose sum 4 becomes the encoded value of string 010100. The decoder starts at 15 and proceeds in the opposite direction toward the apex. It is not hard to see that it ends up decoding the string 001010, which is why the decoder's output in this variant has to be reversed before it is used.

This version of Schalkwijk coding is restricted to certain bit strings, and is also block-to-block coding. Each block of n bits is replaced by a block of $\lceil \log_2 \binom{n}{w} \rceil$ bits. The next version is similar, but is variable-to-block coding. We again assume a source of bits that emits a 0 with probability q and a 1 with probability $p = 1 - q$. A string of n bits from this source may often have close to pn 1's and qn zeros, but may sometimes have different numbers of zeros and 1's. We select a convenient value for n, a value that is as large as possible and where both pn and qn are integers or very close to integers. If $p = 1/3$, for example, then $n = 12$ may make sense, because it results in $np = 4$ and $nq = 8$. We again employ the Pascal triangle and take a rectangular block of $(pn + 1)$ rows and $(qn + 1)$ columns such that the top of the triangle will be at the top-right corner of the rectangle (Figure 2.7).

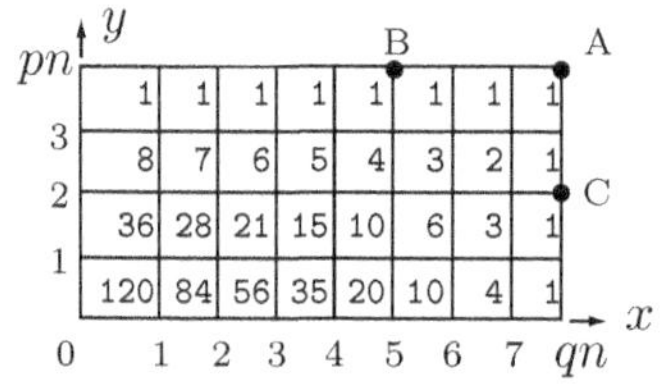

Figure 2.7: Second Version.

As before, we start at the bottom-left corner of the array and read bits from the source. For each 0 we move a step in the x direction and for each 1 we move in the y direction. If the next n bits have exactly pn 1's, we will end up at point "A," the top-right corner of the array, and encode n bits as before. If the first n bits happen to have more than pn 1's, then the top of the array will be reached (after we have read np 1's) early, say at point "B," before we have read n bits. We cannot read any more source bits, because any 1 would take us outside the array, so we append several dummy zeros to what we have read, to end up with n bits (of which np are 1's). This is encoded as before. Notice that the decoder can mimic this operation. It operates as before, but stops decoding when it reaches the top boundary. If the first n bits happen to have many zeros, the encoder will end up at the right boundary of the array, say, at point "C," after it has read qn zeros but before it has read n bits. In such a case, the encoder appends several 1's, to end up with exactly n bits (of which precisely pn are 1's), and encodes as before. The decoder can mimic this operation by simply stopping when it reaches the right boundary.

Any string that has too many or too few 1's degrades the compression, because it encodes fewer than n bits in the same $\lceil \log_2 \binom{n}{pn} \rceil$ bits. Thus, the method may not be very effective, but it is an example of a variable-to-block encoding.

The developer of this method points out that the method can be modified to employ the Pascal triangle for block-to-variable coding. The value of n is determined and it remains fixed. Blocks of n bits are encoded and each block is preceded by the number of 1's it contains. If the block contains w 1's, it is encoded by the appropriate part of the Pascal triangle. Thus, each block of n bits may be encoded by a different part of the triangle, thereby producing a different-length code. The decoder can still work in lockstep with the encoder, because it first reads the number w of 1's in a block. Knowing n and w tells it what part of the triangle to use and how many bits of encoding to read. It has been pointed out that this variant is similar to the method proposed by [Lynch 66] and [Davisson 66]. This variant has also been extended by [Lawrence 77], whose block-to-variable coding scheme is based on a Pascal triangle where the boundary points are defined in a special way, based on the choice of a parameter S.

2.8 Tjalkens–Willems V-to-B Coding

The little-known variable-to-block coding scheme presented in this section is due to [Tjalkens and Willems 92] and is an extension of earlier work described in [Lawrence 77]. Like the Schalkwijk's codes of Section 2.7 and the Lawrence algorithm, this scheme employs the useful properties of the Pascal triangle. The method is based on the choice of a positive integer parameter C. Once a value for C has been selected, the authors show how to construct a set L of M variable-length bitstrings that satisfy the following:

1. Set L is complete. Given any infinite bitstring (in practice, a string of M or more bits), L contains a prefix of the string.
2. Set L is proper. No segment in the set is a prefix of another segment.

Once L has been constructed, it is kept in lexicographically-sorted order, so each string in L has an index between 0 and $M-1$. The input to be encoded is a long bitstring. It is broken up by the encoder into segments of various lengths that are members of L. Each segment is encoded by replacing it with its index in L. Note that the index is a $(\log_2 M)$-bit number. Thus, if $M=256$, each segment is encoded in eight bits. The main task is to construct set L in such a way that the encoder will be able to read the input bit by bit, stop when it has read a bit pattern that is a string in L, and determine the code of the string (its index in L). The theory behind this method is complex, so only the individual steps and tests are summarized here.

Given a string s of a zeros and b 1's, we define the function

$$Q(s)=(a+b+1)\binom{a+b}{b}.$$

(The authors show that $1/Q$ is the probability of string s.) We denote by s_{-1} the string s without its last (rightmost) bit. String s is included in set L if it satisfies

$$Q(s_{-1})<C\le Q(s). \tag{2.4}$$

The authors selected $C=82$ (because this results in the convenient size $M=256$). Once C is known, it is easy to decide whether a given string s with a zeros and b 1's is a member of set L (i.e., whether s satisfies Equation (2.4)). If s is in L, then point (a, b) in the Pascal triangle (i.e., element b of row a, where row and column numbering starts at 0) is considered a boundary point. Figure 2.8a shows the boundary points (underlined) for $C = 82$.

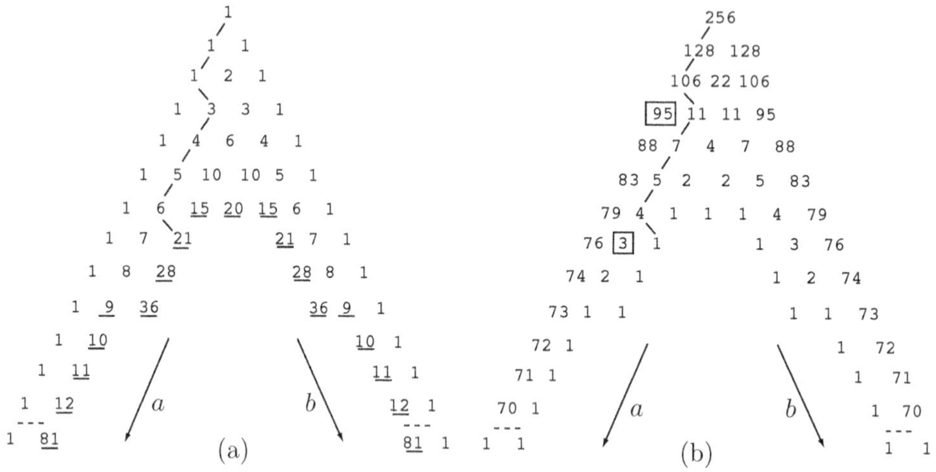

Figure 2.8: (a) Boundary Points. (b) Coding Array.

The inner parts of the triangle are not used in this method and can be removed. Also, The lowest boundary points are located on row 81 and lower parts of the triangle are not used. If string s is in set L, then we can start at the apex of the triangle, move in the a direction for each 0 and in the b direction for each 1 in s, and end up at a boundary point. The figure illustrates this walk for the string 0010001, where the boundary point reached is 21.

Setting up the partial Pascal triangle of Figure 2.8a is just the first step. The second step is to convert this triangle to a coding triangle $M(a, b)$ of the same size, where each walk for a string s can be used to determine the index of s in set L and thus its code. The authors show that element $M(a, b)$ of this triangle must equal the number of distinct ways to reach a boundary point after processing a zeros and b 1's (i.e., after moving a steps in the a direction and b steps in the b direction). Triangle $M(a, b)$ is constructed according to

$$M(a, b) = \begin{cases} 1, & \text{if } (a, b) \text{ is a boundary point,} \\ M(a+1, b) + M(a, b+1), & \text{otherwise.} \end{cases}$$

The coding array for $C = 82$ (constructed from the bottom up) is shown in Figure 2.8b. Notice that its apex, $M(0, 0)$, equals the total number of strings in L. Once this triangle is available, both encoding and decoding are simple and are listed in Figure 2.9a,b. The former inputs individual bits and moves in $M(a, b)$ in the a or b directions according to the inputs. The end of the current input string is signalled when a node with a 1 is reached in the coding triangle. For each move in the b direction, the next element in the a direction (the one that will not be reached) is added to the index. At the end, the index is the code of the current string. Figure 2.8b shows the moves for 0010001 and how the nodes 95 and 3 are selected and added to become code 98. The decoder starts with the code of a string in variable `index`. It compares `index` to the sum $(I + M(a+1,b))$ and moves in the a or b directions according to the result, generating one output bit as it moves. Decoding is complete when the decoder reaches a node with a 1.

```
index:=0; a:=0; b:=0;
while M(a,b) ≠ 1 do
  if next_input = 0
  then a:=a+1
  else index:=index+M(a+1,b);
       b:=b+1
  endif
endwhile
```

```
I:=0; a:=0; b:=0;
while M(a,b) ≠ 1 do
  if index < (I + M(a+1,b))
  then next_output:=0; a:=a+1;
  else next_output:=1;
       I := I+M(a+1,b); b:=b+1
  endif
endwhile
```

Figure 2.9: (a) Encoding and (b) Decoding.

> Extraordinary how mathematics help you. . . .
> —Samuel Beckett, *Molloy* (1951)

2.9 Phased-In Codes

Many of the prefix codes described here were developed for the compression of specific types of data. These codes normally contain codewords of various lengths and they are suitable for the compression of data where individual symbols have widely different probabilities. Data where symbols have equal probabilities cannot be compressed by VLCs and is normally assigned fixed-length codes. The codes of this section (also called phased-in binary codes, see Appendix A-2 in [Bell et al. 90]) constitute a compromise. A set of phased-in codes consists of codewords of two lengths and may contribute something (although not much) to the compression of data where symbols have roughly equal probabilities. Section 5.3.1 and page 376 discuss applications of these codes.

Given n data symbols, where $n = 2^m$ (implying that $m = \log_2 n$), we can assign them m-bit codewords. However, if $2^{m-1} < n < 2^m$, then $\log_2 n$ is not an integer. If we assign fixed-length codes to the symbols, each codeword would be $\lceil \log_2 n \rceil$ bits long, but not all the codewords would be used. The case $n = 1{,}000$ is a good example. In this case, each fixed-length codeword is $\lceil \log_2 1{,}000 \rceil = 10$ bits long, but only 1,000 out of the 1,024 possibe codewords are used.

In the approach described here, we try to assign two sets of codes to the n symbols, where the codewords of one set are $m-1$ bits long and may have several prefixes and the codewords of the other set are m bits long and have different prefixes. The average length of such a code is between $m-1$ and m bits and is shorter when there are more short codewords.

A little research and experimentation leads to the following method of constructing the two sets of codes. Given a set of n data symbols (or simply the integers 0 through $n-1$) where $2^m \le n < 2^{m+1}$ for some integer m, we denote $n = 2^m + p$ where $0 \le p < 2^m - 1$ and also define $P \stackrel{\text{def}}{=} 2^m - p$. We construct $2p$ long (i.e., m-bit) codewords and P short, $(m-1)$-bit codewords. The total number of codewords is always $2p + P = 2p + 2^m - p = (n - 2^m) + 2^m = n$ and there is an even number of long codewords. The first P integers 0 through $P-1$ receive the short codewords and the remaining $2p$ integers P through $n-1$ are assigned the long codewords. The short codewords are the integers 0 through $P-1$, each encoded in $m-1$ bits. The long codewords consist of p pairs, where each pair starts with the m-bit value P, $P+1, \ldots,$ $P+p-1$, followed by an extra bit, 0 or 1, to distinguish between the two codewords of a pair.

The following tables illustrate both the method and its compression performance. Table 2.10 lists the values of m, p, and P for $n = 7$, 8, 9, 15, 16, and 17. Table 2.11 lists the actual codewords.

n	m	$p = n - 2^m$	$P = 2^m - p$
7	2	3	1
8	3	0	8
9	3	1	7
15	3	7	1
16	4	0	16
17	4	1	15

Table 2.10: Parameters of Phased-In Codes.

i	$n = 7$	8	9	15	16	17
0	00	000	000	000	0000	0000
1	010	001	001	0010	0001	0001
2	011	010	010	0011	0010	0010
3	100	011	011	0100	0011	0011
4	101	100	100	0101	0110	0100
5	110	101	101	0110	0101	0101
6	111	110	110	0111	0110	0110
7		111	1110	1000	0111	0111
8			1111	1001	1000	1000
9				1010	1001	1001
10				1011	1010	1010
11				1100	1011	1011
12				1101	1100	1100
13				1110	1101	1101
14				1111	1110	1110
15					1111	11110
16						11111

Table 2.11: Six Phased-In Codes.

Table 2.11 is also the key to estimating the compression ratio produced by these codes. The table shows that when n approaches a power of 2 (such as $n = 7$ and $n = 15$), there are few short codewords and many long codewords, indicating low efficiency. When n is a power of 2, all the codewords are long and the method does not produce any compression (the compression ratio is 1). When n is slightly greater than a power of 2 (such as $n = 9$ and 17), there are many short codewords and therefore better compression.

The total size of the n phased-in codewords for a given n is

$$P \times m + 2p(m+1) = (2^{m+1} - n)m + 2(n - 2^m)(m+1)$$

where $m = \lfloor \log_2 n \rfloor$, whereas the ideal total size of n fixed-length codewords (keeping in mind the fact that $\log_2 n$ is normally a noninteger), is $n\lceil \log_2 n \rceil$. Thus, the ratio of these two quantities is the compression ratio of the phased-in codes. It is illustrated in Figure 2.12 for n values 2 through 256.

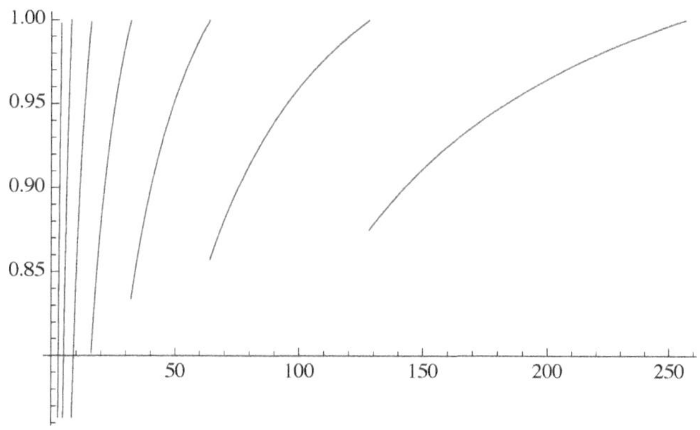


```
m:=Floor[Log[2,n]];
Plot[(n m+2 n-2^(m+1))/(n Ceiling[ Log[2,n]]), {n,2,256}]
```

Figure 2.12: Efficiency of Phased-In Codes.

It is obvious that the compression ratio goes up toward 1 (indicating worse compression) as n reaches a power of 2, then goes down sharply (indicating better compression) as we pass that value.

See `http://f-cpu.seul.org/whygee/phasing-in_codes/phasein.html` for other examples and a different way to measure the efficiency of these codes.

Another advantage of phased-in codes is the ease of encoding and decoding them. Once the encoder is given the value of n, it computes m, p, and P. Given an integer i to be encoded, if it is less than P, the encoder constructs an m-bit codeword with the value i. Otherwise, the encoder constructs an $(m+1)$-bit codeword where the first m bits (the prefix) have the value $P + (i - P) \div 2$ and the rightmost bit is $(i - P) \bmod 2$. Using $n = 9$ as an example, we compute $m = 3$, $p = 1$, and $P = 7$. If $i = 6$, then

$i < P$, so the 3-bit codeword 110 is prepared. If $i = 8$, then the prefix is the three bits $7 + (8-7) \div 2 = 7 = 111_2$ and the rightmost bit is $(8-7) \bmod 2 = 1$.

The decoder also starts with the given value of n and computes m, p, and P. It inputs the next m bits into i. If this value is less than P, then this is the entire codeword and it is returned by the decoder as the result. Otherwise, the decoder inputs the next bit b and appends it to $2(i - P) + P$. Using $n = 9$ as an example, if the decoder inputs the three bits 111 (equals P) into i, it inputs the next bit into b and appends it to $2(7-7) + 7 = 111_2$, resulting in either 1110 or 1111, depending on b.

The phased-in codes are closely related to the minimal binary code of Section 3.13.

See [seul.org 06] for a *Mathematica* notebook to construct phased-in codes.

Two variations: In the basic phased-in codes algorithm, the shorter codes are assigned to the first symbols. This makes sense if it is known (or suspected) that these symbols are more common. In some cases, however, we may know that the middle symbols are the most common, and such cases can benefit from rotating the original sequence of phased-in codes such that the shorter codes are located in the middle. This variation is termed centered phased-in codes. Table 2.11 shows that for $n = 9$ there are seven 3-bit codes and two 4-bit codes. The sequence of code lengths is therefore $(3,3,3,3,3,3,3,4,4)$ and it can be rotated one position to the right to produce $(4,3,3,3,3,3,3,3,4)$.

The FELICS image compression algorithm (Section 7.23) is an example of common symbols in the middle of a range. The values in the middle region of part (b) of Figure 7.133 are more common than their extreme neighbors and should therefore be assigned shorter codes.

Another useful version, the reverse centered phased-in codes, is obtained when the longer codes are placed in the middle. Rotating the sequence of code lengths $(3,3,3,3,3,3,3,4,4)$ three places to the left results in $(3,3,3,3,4,4,3,3,3)$.

It is also possible to construct suffix phased-in codes. We start with a set of fixed-length codes and convert it to two sets of codewords by removing the leftmost bit of some codewords. This bit is removed if it is a 0 and if its removal does not create any ambiguity. Table 2.13 (where the removed bits are in italics) illustrates an example for the first 24 nonnegative integers. The fixed-length representation of these integers requires five bits, but each of the eight integers 8 through 15 can be represented by only four bits because 5-bit codes can represent 32 symbols and we have only 24 symbols. A simple check verifies that, for example, coding the integer 8 as 1000 instead of 01000 does not introduce any ambiguity, because none of the other 23 codes ends with 1000. One-third of the codewords in this example are one bit shorter, but if we consider only the 17 integers from 0 to 16, about half will require four bits instead of five. The efficiency of this code depends on where n (the number of symbols) is located in the interval $[2^m, 2^{m+1} - 1)$.

00000	00001	00010	00011	00100	00101	00110	00111
*0*1000	*0*1001	*0*1010	*0*1011	*0*1100	*0*1101	*0*1110	*0*1111
10000	10001	10010	10011	10100	10101	10110	10111

Table 2.13: Suffix Phased-In Codes.

The suffix phased-in codes are suffix codes (if c has been selected as a codeword, no other codeword will end with c). Suffix codes can be considered the complements of prefix codes and are also mentioned in Section 4.5.

2.10 Redundancy Feedback (RF) Coding

The interesting and original method of redundancy feedback (RF) coding is the brainchild of Eduardo Enrique González Rodríguez who hasn't published it formally. As a result, information about it is hard to find. At the time of writing (early 2007), there is a discussion in file `new-entropy-coding-algorithm-312899.html` at web site `http://archives.devshed.com/forums/compression-130/` and some information (and source code) can also be obtained from the authors of this book.

The method employs phased-in codes, but is different from other entropy coders. It may perhaps be compared with static arithmetic coding. Most entropy coders assign variable-length codes to data symbols such that the length of the code for a symbol is inversely proportional to the symbol's frequency of occurrence. The RF method, in contrast, starts by assigning several fixed-length (block) codes to each symbol according to its probability. The method then associates a phased-in code (that the developer terms "redundant information") with each block codeword. Encoding is done in reverse, from the end of the input stream. Each symbol is replaced by one of its block codes B in such a way that the phased-in code associated with B is identical to some bits at the start (the leftmost part) of the compressed stream. Those bits are deleted from the compressed stream (which generates compression) and B is prepended to it. For example, if the current block code is 010111 and the compressed stream is 0111|0001010..., then the result of prepending the code and removing identical bits is 01|0001010....

We start with an illustrative example. Given the four symbols A, B, C, and D, with probabilities 37.5%, 25%, 12.5%, and 25%, respectively, we assign each symbol several block codes according to its probability. The total number of codes must be a power of 2, so A is assigned three codes, each of B and D gets two codes, and C becomes the "owner" of one code, for a total of eight codes. Naturally, the codes are the 3-bit numbers 0 through 7. Table 2.14a lists the eight codes and their redundant information (the associated phased-in codes). Thus, e.g., the three codes of A are associated with the phased-in codes 0, 10, and 11, because these are the codes for $n = 3$. (Section 2.9 shows that we have to look for the integer m that satisfies $2^m \leq n < 2^{m+1}$. Thus, for $n = 3$, m is 1. The first $2^m = 2$ symbols are assigned the 2-bit numbers $0 + 2$ and $1 + 2$ and the remaining $3 - 2$ symbol is assigned the 1-bit number $i - 2^m = 2 - 2 = 0$.) Similarly, the two phased-in codes associated with B are 0 and 1. Symbol D is associated with the same two codes, and the single block code 5 of C has no associated phased-in codes because there are no phased-in codes for a set of one symbol. Table 2.14b is constructed similarly and lists 16 4-bit block codes and their associated phased-in codes for the three symbols A, B, and C with probabilities 0.5, 0.2, and 0.2, respectively.

First, a few words on how to determine the number of codes per symbol from the number n of symbols and their frequencies f_i. Given an input string of F symbols (from an alphabet of n symbols) such that symbol i appears f_i times (so that $\sum f_i = F$), we first determine the number of codes. This is simply the power m of 2 that satisfies $2^{m-1} < n \le 2^m$. We now multiply each f_i by $2^m/F$. The new sum satisfies $\sum f_i \times 2^m/F = 2^m$. Next, we round each term of this sum to the nearest integer, and if any is rounded down to zero, we set it to 1. Finally, if the sum of these integers is slightly different from 2^m, we increment (or decrement) each of the largest ones by 1 until the sum equals 2^m.

Code	Symbol	Redundant Info
0	A	$0/3 \to 0$
1	A	$1/3 \to 10$
2	A	$2/3 \to 11$
3	B	$0/2 \to 0$
4	B	$1/2 \to 1$
5	C	$0/1 \to -$
6	D	$0/2 \to 0$
7	D	$1/2 \to 1$

(a)

Code	Symbol	Redundant Info
0 0000	A	$0/10 \to 000$
1 0001	A	$1/10 \to 001$
2 0010	A	$2/10 \to 010$
3 0011	A	$3/10 \to 011$
4 0100	A	$4/10 \to 100$
5 0101	A	$5/10 \to 101$
6 0110	A	$6/10 \to 1100$
7 0111	A	$7/10 \to 1101$
8 1000	A	$8/10 \to 1110$
9 1001	A	$8/10 \to 1111$
10 1010	B	$0/3 \to 0$
11 1011	B	$1/3 \to 10$
12 1100	B	$2/3 \to 11$
13 1101	C	$0/3 \to 0$
14 1110	C	$1/3 \to 10$
15 1111	C	$2/3 \to 11$

(b)

Table 2.14: Eight and 16 RF Codes.

As an example, consider a 6-symbol alphabet and an input string of $F = 47$ symbols, where the six symbols appear 17, 6, 3, 12, 1, and 8 times. We first determine that $2^2 < 6 \le 2^3$, so we need 8 codes. Multiplying $17\times 8/47 = 2.89 \to 3$, $6\times 8/47 = 1.02 \to 1$, $3\times 8/47 = 0.51 \to 1$, $12\times 8/47 = 2.04 \to 2$, $1\times 8/47 = 0.17 \to 0$, and $8\times 8/47 = 1.36 \to 1$. The last step is to increase the 0 to 1, and make sure the sum is 8 by decrementing the largest count, 3, to 2.

The codes of Table 2.14b are now used to illustrate RF encoding. Assume that the input is the string $AABCA$. It is encoded from end to start. The last symbol A is easy to encode as we can use any of its block codes. We therefore select 0000. The next symbol, C, has three block codes, and we select $13 = 1101$. The associated phased-in code is 0, so we start with 0000, delete the leftmost 0, and prepend 1101, to end up with 1101|000. The next symbol is B and we select block code $12 = 1100$ with associated phased-in code 11. Encoding is done by deleting the leftmost 11 and prepending 1100, to end up with 1100|01|000. To encode the next A, we select block code $6 = 0110$ with associated phased-in code 1100. Again, we delete 1100 and prepend 0110 to end up with 0110||01|000. Finally, the last (i.e., leftmost) symbol A is reached, for which we select block code $3 = 0011$ (with associated phased-in code 011) and encode by deleting 011 and prepending 0011. The compressed stream is 0011|0||01|000.

The RF encoding principle is simple. Out of all the block codes assigned to the current symbol, we select the one whose associated phased-in code is identical to the prefix of the compressed stream. This results in the deletion of the greatest number of bits and thus in maximum compression.

Decoding is the opposite of encoding. The decoder has access to the table of block codes and their associated codes (or it starts from the symbols' probabilities and constructs the table as the encoder does). The compressed stream is 0011001000 and the first code is the leftmost four bits $0011 = 3 \to A$. The first decoded symbol is A and the decoder deletes the 0011 and prepends 011 (the phased-in code associated with 3) to end up with 011001000. The rest of the decoding is straightforward.

Experiments with this method verify that its performance is generally almost as good as Huffman coding. The main advantages of RF coding are as follows:

1. It works well with a 2-symbol alphabet. We know that Huffman coding fails in this situation, even when the probabilities of the symbols are skewed, because it simply assigns the two 1-bit codes to the symbols. In contrast, RF coding assigns the common symbol many (perhaps seven or 15) codes, while the rare symbol is assigned only one or two codes, thereby producing compression even in such a case.

2. The version of RF presented earlier is static. The probabilities of the symbols have to be known in advance in order for this version to work properly. It is possible to extend this version to a simple dynamic RF coding, where a buffer holds the most-recent symbols and code assignment is constantly modified. This version is described below.

3. It is possible to replace the phased-in codes with a simple form of arithmetic coding. This slows down both encoding and decoding, but results in better compression.

Dynamic RF coding is slower than the static version above, but is more efficient. Assuming that the probabilities of the data symbols are unknown in advance, this version of the basic RF scheme is based on a long sliding buffer. The buffer should be long, perhaps 2^{15} symbols or longer, in order to reflect the true frequencies of the symbols. A common symbol will tend to appear many times in the buffer and will therefore be assigned many codes. For example, given the alphabet A, B, C, D, and E, with probabilities 60%, 10%, 10%, 15%, and 5%, respectively, the buffer may, at a certain point in the encoding, hold the following

raw	A	B	A	A	B	A	C	A	A	D	A	A	B	A	A	A	C	A	A	D	A	C	A	E	A	A	A	D	A	A	C	A	A	B	A	A	coded
←	36	35	34	33	32	31	30	29	28	27	26	25	24	23	22	21	20	19	18	17	16	15	14	13	12	11	10	9	8	7	6	5	4	3	2	1	←

On the left, there is raw (unencoded) text and on the right there is text that has already been encoded. We can imagine the text being stationary and the buffer sliding to the left. If the buffer is long enough, the text inside it will reflect the true probabilities of the symbols and each symbol will have a number of codes proportional to its probability. At any time, the symbol immediately to the right of the buffer is encoded by selecting one of its codes in the buffer, and then moving the buffer one symbol to the left. If the buffer happens to contain no occurrences of the symbol to be encoded, then the code of all zeros is selected (which is why the codes in the buffer start at 1) and is output, followed by the raw (normally ASCII) code of the symbol. Notice that sliding the buffer modifies the codes of the symbols, but the decoder can do this in lockstep with the encoder. Once a code has been selected for a symbol, the code is prepended to the compressed stream after its associated phased-in code is used to delete identical bits, as in the static version.

We illustrate this version with a 4-symbol alphabet and the string `ABBACBBBAADA`. We assume a buffer with seven positions (so the codes are between 1 and 7) and place the buffer initially such that the rightmost `A` is immediately to its right, thus `ABBA[CBBBAAD]A`.

The initial buffer position and codes (both the 3-bit RF codes and the associated phased-in codes) are shown here. Symbol `A` is immediately to the right of the buffer and it can be encoded as either 2 or 3. We arbitrarily select 2, ending up with a compressed stream of 010.

A	B	B	A	C	B	B	B	A	A	D	A
				7	6	5	4	3	2	1	
				111	110	101	100	011	010	001	
				1:-	3:2	3:1	3:0	2:1	2:0	1:-	
				-	11	10	0	1	0	-	

The buffer is slid, as shown below, thereby changing all the codes. This is why the dynamic version is slow. Symbol `D` is now outside the buffer and must be encoded as the pair (000, `D`) because there are no occurrences of `D` inside the buffer. The compressed stream becomes 000: 01000100|010.

A	B	B	A	C	B	B	B	A	A	D	A
			7	6	5	4	3	2	1		
			111	110	101	100	011	010	001		
			3:2	1:-	3:2	3:1	3:0	3:1	3:0		
			11	-	11	10	0	10	0		

Now comes another `A` that can be encoded as either 1 or 6. Selecting the 1 also selects its associated phased-in code 0, so the leftmost 0 is deleted from the compressed stream and 001 is prepended. The result is 001|00: 01000100|010.

A	B	B	A	C	B	B	B	A	A	D	A
		7	6	5	4	3	2	1			
		111	110	101	100	011	010	001			
		4:3	2:1	1:-	4:2	4:1	4:0	2:0			
		11	1	-	10	01	00	0			

The next symbol to be encoded is the third `A` from the right. The only available code is 5, which has no associated phased-in code. The output therefore becomes 101|001|00: 01000100|010.

A	B	B	A	C	B	B	B	A	A	D	A
	7	6	5	4	3	2	1				
	111	110	101	100	011	010	001				
	5:4	5:3	1:-	1:-	5:2	5:1	5:0				
	111	110	-	-	10	01	00				

Next in line is the `B`. Four codes are available, of which the best choice is 5, with associated phased-in code 10. The string 101|1|001|00: 01000100|010 now becomes the current output.

A	B	B	A	C	B	B	B A A D A
7	6	5	4	3	2	1	
111	110	101	100	011	010	001	
2:1	4:3	4:2	2:0	1:-	4:1	4:0	
1	11	10	0	-	01	00	

Encoding continues in this way even though the buffer is now only partially full. The next B is encoded with only three Bs in the buffer, and with each symbol encoded, fewer symbols remain in the buffer. Each time a symbol s is encoded that has no copies left in the buffer, it is encoded as a pair of code 000 followed by the ASCII code of s. As the buffer gradually empties, more and more pairs are prepended to the output, thereby degrading the compression ratio. The last symbol (which is encoded with an empty buffer) is always encoded as a pair.

Thus, the decoder starts with an empty buffer, and reads the first code (000) which is followed by the ASCII code of the first (leftmost) symbol. That symbol is shifted into the buffer, and decoding continues as the reverse of encoding.

2.11 Recursive Phased-In Codes

The recursive phased-in codes were introduced in [Acharya and JáJá 95] and [Acharya and JáJá 96] as an enhancement to the well-known LZW (Lempel Ziv Welch) compression algorithm [Section 6.13]. These codes are easily explained in terms of complete binary trees, although their practical construction may be simpler with the help of certain recursive formulas conceived by Steven Pigeon.

The discussion in Section 3.19 shows that any positive integer N can be written uniquely as the sum of certain powers of 2. Thus, for example, 45 is the sum $2^5 + 2^3 + 2^2 + 2^0$. In general, we can write $N = \sum_{i=1}^{s} 2^{a_i}$, where $a_1 > a_2 > \cdots > a_s \geq 0$ and $s \geq 1$. For $N = 45$, for example, these values are $s = 4$ and $a_1 = 5$, $a_2 = 3$, $a_3 = 2$, and $a_4 = 0$. Once a value for N has been selected and the values of all its powers a_i determined, a set of N variable-length recursive phased-in codes can be constructed from the tree shown in Figure 2.15. For each power a_i, this tree has a subtree that is a complete binary tree of height a_i. The individual subtrees are connected to the root by $2s - 2$ edges labeled 0 or 1 as shown in the figure.

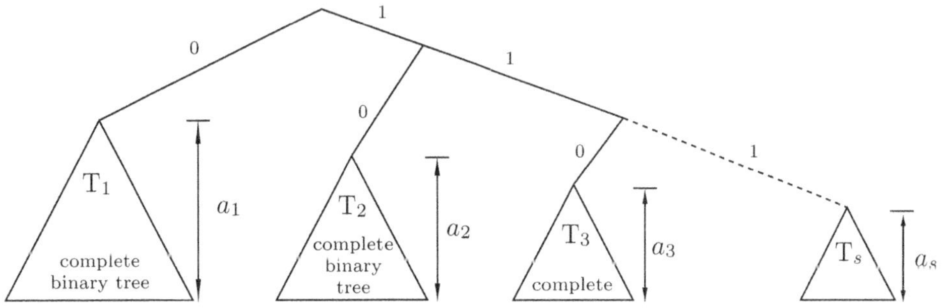

Figure 2.15: A Tree for Recursive Phased-In Codes.

The tree for $N = 45$ is shown in Figure 2.16. It is obvious that the complete binary tree for a_i has a_i leaves and the entire tree therefore has a total of N leaf nodes. The codes are assigned by sliding down the tree from the root to each leaf, appending a 0 to the code each time we slide to the left and a 1 each time we slide to the right. Some codes are also shown in the figure. Thus, the 45 recursive phased-in codes for $N = 45$ are divided into the four sets $0xxxxx$, $10xxx$, $110xx$, and 111, where the x's stand for bits. The first set consists of the 32 5-bit combinations prepended by a 0, the second set includes eight 5-bit codes that start with 10, the third set has four codes, and the last set consists of the single code 111. As we scan the leaves of each subtree from left to right, we find that the codewords in each set are in ascending order. Even the codewords in different sets appear sorted, when scanned from left to right, if we append enough zeros to each so they all become six bits long. The codes are prefix codes, because, for example, once a code of the form $0xxxxx$ has been assigned, no other codes will start with 0.

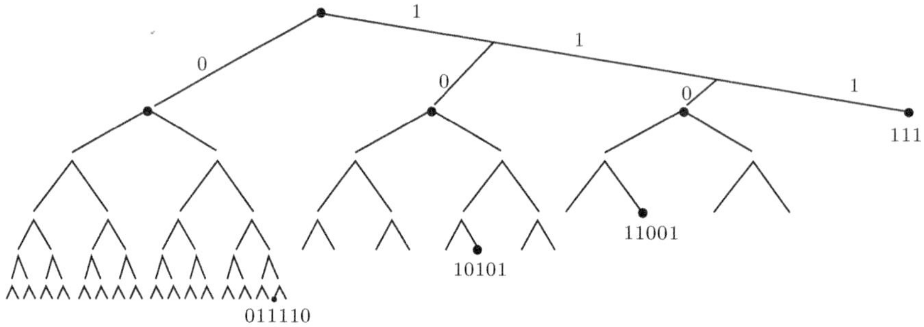

Figure 2.16: A Tree for $N = 45$.

In practice, these codes can be constructed in two steps, one trivial and the other one simple, as follows:

1. (This step is trivial.) Given a positive integer N, determine its powers a_i. Given, for example, $45 = \ldots 000101101$ we first locate its leftmost 1. The position of this 1 (its distance from the right end) is s. We then scan the bits from right to left while decrementing an index i from s to 1. Each 1 found designates a power a_i.

2. There is no need to actually construct any binary trees. We build the set of codes for a_1 by starting with the prefix 0 and appending to it all the a_1-bit numbers. The set of codes for a_2 is similarly built by starting with the prefix 10 and appending to it all the a_2-bit numbers.

In his PhD thesis [Pigeon 01b], Steven Pigeon proposes a recursive formula as an alternative to the steps above. Following Elias (Section 3.4), we denote by $\beta_k(n)$ the k-bit binary representation of the integer n. Given N, we find its largest power k, so $N = 2^k + b$ where $0 \le b < 2^k$ (k equals a_1 above). The N recursive phased-in codes $C_N(n)$ for $n = 0, 1, \ldots, N-1$ are computed by

$$C_N(n) = \begin{cases} 0 : C_b(n), & \text{if } 0 \le n < b, \\ \beta_k(n), & \text{if } b = 0, \\ 1 : \beta_k(n-b), & \text{otherwise.} \end{cases}$$

Their lengths $L_N(n)$ are given by

$$L_N(n) = \begin{cases} 1 + L_b(n), & \text{if } 0 \le n < b, \\ k, & \text{if } b = 0, \\ 1 + k, & \text{otherwise.} \end{cases}$$

Table 2.17 lists the resulting codes for $N = 11$, 13, and 18. It is obvious that these are slightly different from the original codes of Acharya and JáJá. The latter code for $N = 11$, for example, consists of the sets $0xxx$, $10x$, and 11, while Pigeon's formula generates the sets $1xxx$, $01x$, and 00.

n	11	13	18
0	00	00	00
1	010	0100	01
2	011	0101	10000
3	1000	0110	10001
4	1001	0111	10010
5	1010	1000	10011
6	1011	1001	10100
7	1100	1010	10101
8	1101	1011	10110
9	1110	1100	10111
10	1111	1101	11000
11		1110	11001
12		1111	11010
13			11011
14			11100
15			11101
16			11110
17			11111

Table 2.17: Codes for 11, 13, and 18.

The recursive phased-in codes bear a certain resemblance to the start-step-stop codes of Section 3.2, but a quick glance at Table 3.3 shows the difference between the two types of codes. A start-step-stop code consists of sets of codewords that start with 0, 10, 110, and so on and get longer and longer, while the recursive phased-in codes consist of sets that start with the same prefixes but get shorter.

"The grand aim of all science is to cover the greatest number of empirical facts by logical deduction from the smallest number of hypotheses or axioms," he [Einstein] maintained. The same principle is at work in Ockham's razor, in Feynman's panegyric upon the atomic doctrine, and in the technique of data compression in information technology—all three of which extol economy of expression, albeit for different reasons.

—Hans Christian von Baeyer, *Information, The New Language of Science* (2004)

2.12 Self-Delimiting Codes

Before we look at the main classes of VLCs, we list in this short section a few simple techniques (the first few of which are due to [Chaitin 66]) to construct *self-delimiting codes*, codes that have variable lengths and can be decoded unambiguously.

1. Double each bit of the original message, so the message becomes a set of pairs of identical bits, then append a pair of different bits. Thus, the message 01001 becomes the bitstring 00|11|00|00|11|01. This is simple but is obviously too long. It is also fragile, because one bad bit will confuse any decoder (computer or human). A variation of this technique precedes each bit of the number with an intercalary bit of 1, except the last bit, which is preceded with a 0. Thus, 01001 become $\bar{1}0\bar{1}1\bar{1}0\bar{1}0\bar{0}1$. We can also concentrate the intercalary bits together and have them followed by the number, as in 11110|01001 (which is the number itself preceded by its unary code).

2. Prepare a header with the length of the message and prepend it to the message. The size of the header depends on the size of the message, so the header should be made self-delimiting using method 1 above. Thus, the 6-bit message 010011 becomes the header 00|11|11|00|01 followed by 010011. It seems that the result is still very long (16 bits to encode six bits), but this is because our message is so short. Given a 1-million bit message, its length requires 20 bits. The self-delimiting header is therefore 42 bits long, increasing the length of the original message by 0.0042%.

3. If the message is extremely long (trillions of bits) its header may become too long. In such a case, we can make the header itself self-delimiting by writing it in raw format and preceding it with its own header, which is made self-delimiting with method 1.

4. It is now clear that there may be any number of headers. The first header is made self-delimiting with method 1, and all the other headers are concatenated to it in raw format. The last component is the (very long) original binary message.

5. A decimal variable-length integer can be represented in base 15 (quindecimal) as a string of nibbles (groups of four bits each), where each nibble is a base-15 digit (i.e., between 0 and 14) and the last nibble contains $16 = 1111_2$. This method is sometimes referred to as nibble code or byte coding. Table 3.25 lists some examples.

6. A variation on the nibble code is to start with the binary representation of the integer n (or $n-1$), prepend it with zeros until the total number of bits is divisible by 3, break it up into groups of three bits each, and prefix each group with a 0, except the leftmost (or alternatively, the rightmost) group, which is prepended by a 1. The length of this code for the integer n is $4\lceil(\log_2 n)/3\rceil$, so it is ideal for a distribution of the form

$$2^{-4\lceil(\log_2 n)/3\rceil} \approx \frac{1}{n^{4/3}}. \tag{2.5}$$

This is a power law distribution with a parameter of 3/4. A natural extension of this code is to k-bit groups. Such a code fits power law distributions of the form

$$\frac{1}{n^{1+\frac{1}{k}}}. \tag{2.6}$$

7. If the data to be compressed consists of a large number of small positive integers, then a word-aligned packing scheme may provide good (although not the best) compression combined with fast decoding. The idea is to pack several integers into fixed-length

fields of a computer word. Thus, if the word size is 32 bits, 28 bits may be partitioned into several k-bit fields while the remaining four bits become a selector that indicates the value of k.

The method described here is due to [Anh and Moffat 05] who employ it to compress inverted indexes. The integers they compress are positive and small because they are differences of consecutive pointers that are in sorted order. The authors describe three packing schemes, of which only the first, dubbed simple-9, is discussed here.

Simple-9 packs several small integers into 28 bits of a 32-bit word, leaving the remaining four bits as a selector. If the next 28 integers to be compressed all have values 1 or 2, then each can fit in one bit, making it possible to pack 28 integers in 28 bits. If the next 14 integers all have values of 1, 2, 3, or 4, then each fits in a 2-bit field and 14 integers can be packed in 28 bits. At the other extreme, if the next integer happens to be greater than $2^{14} = 16,384$, then the entire 28 bits must be devoted to it, and the 32-bit word contains just this integer. The choice of 28 is particularly fortuitous, because 28 is divisible by 1, 2, 3, 4, 5, 7, 9, 14, and itself. Thus, a 32-bit word packed in simple-9 may be partitioned in nine ways. Table 2.18 lists these nine partitions and shows that at most three bits are wasted (in row **e**).

Selector	Number of codes	Code length	Unused bits
a	28	1	0
b	14	2	0
c	9	3	1
d	7	4	0
e	5	5	3
f	4	7	0
g	3	9	1
h	2	14	0
i	1	28	0

Table 2.18: Summary of the Simple-9 Code.

Given the 14 integers 4, 6, 1, 1, 3, 5, 1, 7, 1, 13, 20, 1, 12, and 20, we encode the first nine integers as **c**|011|101|000|000|010|100|000|110|000|b and the following five integers as **e**|01100|10011|00000|01011|10011|bbb, for a total of 64 bits, where each b indicates an unused bit. The originators of this method point out that the use of a Golomb code would have compressed the 14 integers into 58 bits, but the small loss of compression efficiency of simple-9 is often more than compensated for by the speed of decoding. Once the leftmost four bit of a 32-bit word are examined and the selector value is determined, the remaining 28 bits can be unpacked with a few simple operations.

Allocating four bits for the selector is somewhat wasteful, because only nine of the 16 possible values are used, but the flexibility of the simple-9 code is the result of the many (nine) factors of 28. It is possible to give up one selector value, cut the selector size to three bits and increase the data segment to 29 bits, but 29 is a prime number, so a 29-bit segment cannot be partitioned into equal-length fields. The authors propose dividing a 32-bit word into a 2-bit selector and a 30-bit segment for packing data. The integer

30 has 10 factors, so a table of the simple-10 code, similar to Table 2.18, would have 10 rows. The selector field, however, can specify only four different values, which is why the resulting code (not described here) is more complex and is denoted by relative-10 instead of simple-10.

Intercalary: Inserted between other elements or parts; interpolated.

In Japan, the basic codes are the Civil Code, the Commercial Code, the Penal Code, and procedural codes such as the Code of Criminal Procedure and the Code of Civil Procedure.

—Roger E. Meiners, *The Legal Environment of Business*

3 Advanced VL Codes

We start this chapter with codes for the integers. This is followed by many types of variable-length codes that are based on diverse principles, have been developed by different approaches, and have various useful properties and features.

3.1 VLCs for Integers

Following Elias, it is customary to denote the standard binary representation of the integer n by $\beta(n)$. This representation can be considered a code (the beta code), but it does not satisfy the prefix property (because, for example, $2 = 10_2$ is the prefix of $4 = 100_2$). The beta code has another disadvantage. Given a set of integers between 0 and n, we can represent each in $1 + \lfloor \log_2 n \rfloor$ bits, a fixed-length representation. However, if the number of integers in the set is not known in advance (or if the largest integer is unknown), a fixed-length representation cannot be used and the natural solution is to assign variable-length codes to the integers. Any variable-length codes for integers should satisfy the following requirements:

1. Given an integer n, its codeword should be as short as possible and should be constructed from the magnitude, length, and bit pattern of n, without the need for any table lookups or other mappings.
2. Given a bitstream of variable-length codes, it should be easy to decode the next codeword and obtain an integer n even if n hasn't been seen before.

We will see that in many VLCs for integers, part of the binary representation of the integer is included in the codeword, and the rest of the codeword is side information indicating the length or precision of the encoded integer.

Several codes for the integers are described in the first few sections of this chapter. Some can code only nonnegative integers and others can code only positive integers. A

D. Salomon, G. Motta, *Handbook of Data Compression*, 5th ed.
DOI 10.1007/978-1-84882-903-9_3,

VLC for positive integers can be extended to encode nonnegative integers by incrementing the integer before it is encoded and decrementing the result produced by decoding. A VLC for arbitrary integers can be obtained by a bijection, a mapping of the form

$$\begin{array}{cccccccccccc} 0 & -1 & 1 & -2 & 2 & -3 & 3 & -4 & 4 & -5 & 5 & \cdots \\ 1 & 2 & 3 & 4 & 5 & 6 & 7 & 8 & 9 & 10 & 11 & \cdots \end{array}$$

A function is bijective if it is one-to-one and onto.

Perhaps the simplest variable-length code for integers is the well-known unary code. The unary code of the positive integer n is constructed as $n-1$ bits of 1 followed by a single 0, or alternatively as $n-1$ zeros followed by a single 1 (the three left columns of Table 3.1). The length of the unary code for the integer n is therefore n bits. The two rightmost columns of Table 3.1 show how the unary code can be extended to encode the nonnegative integers (which makes the codes one bit longer). The unary code is simple to construct and is useful in many applications, but it is not universal. Stone-age people indicated the integer n by marking n adjacent vertical bars on a stone, which is why the unary code is sometimes known as a stone-age binary and each of its n or $(n-1)$ 1's (or n or $(n-1)$ zeros) is termed a stone-age bit.

Stone Age Binary?

n	Code	Reverse	Alt. code	Alt. reverse
0	–	–	0	1
1	0	1	10	01
2	10	01	110	001
3	110	001	1110	0001
4	1110	0001	11110	00001
5	11110	00001	111110	000001

Table 3.1: Some Unary Codes.

It is easy to see that the unary code satisfies the prefix property, so it is instantaneous and can be used as a variable-length code. Since its length L satisfies $L = n$, we get $2^{-L} = 2^{-n}$, so it makes sense to use this code in cases where the input data consists of integers n with exponential probabilities $P(n) \approx 2^{-n}$. Given data that lends itself to the use of the unary code (i.e., a set of integers that satisfy $P(n) \approx 2^{-n}$), we can assign unary codes to the integers and these codes will be as good as the Huffman codes with the advantage that the unary codes are trivial to encode and decode. In general, the unary code is used as part of other, more sophisticated, variable-length codes.

Example: Table 3.2 lists the integers 1 through 6 with probabilities $P(n) = 2^{-n}$, except that $P(6) = 2^{-5} \approx 2^{-6}$. The table lists the unary codes and Huffman codes for the six integers, and it is obvious that these codes have the same lengths (except the code of 6, because this symbol does not have the correct probability).

Every positive number was one of Ramanujan's personal friends.
—J. E. Littlewood

n	Prob.	Unary	Huffman
1	2^{-1}	0	0
2	2^{-2}	10	10
3	2^{-3}	110	110
4	2^{-4}	1110	1110
5	2^{-5}	11110	11110
6	2^{-5}	111110	11111

Table 3.2: Six Unary and Huffman Codes.

3.2 Start-Step-Stop Codes

The unary code is ideal for compressing data that consists of integers n with probabilities $P(n) \approx 2^{-n}$. If the data to be compressed consists of integers with different probabilities, it may benefit from one of the general unary codes (also known as start-step-stop codes). Such a code, proposed by [Fiala and Greene 89], depends on a triplet (start, step, stop) of nonnegative integer parameters. A set of such codes is constructed subset by subset as follows:

1. Set $n = 0$.

2. Set $a = \text{start} + n \times \text{step}$.

3. Construct the subset of codes that start with n leading 1's, followed by a single intercalary bit (separator) of 0, followed by a combination of a bits. There are 2^a such codes.

4. Increment n by step. If $n < \text{stop}$, go to step 2. If $n > \text{stop}$, issue an error and stop. If $n = \text{stop}$, repeat steps 2 and 3 but without the single intercalary 0 bit of step 3, and stop.

This construction makes it obvious that the three parameters have to be selected such that start $+ n \times$ step will reach "stop" for some nonnegative n. The number of codes for a given triplet is normally finite and depends on the choice of parameters. It is given by

$$\frac{2^{\text{stop}+\text{step}} - 2^{\text{start}}}{2^{\text{step}} - 1}.$$

Notice that this expression increases exponentially with parameter "stop," so large sets of these codes can be generated even with small values of the three parameters. Notice that the case step $= 0$ results in a zero denominator and thus in an infinite set of codes.

Tables 3.3 and 3.4 show the 680 codes of (3,2,9) and the 2044 codes of (2,1,10). Table 7.109 lists the number of codes of each of the general unary codes $(2, 1, k)$ for $k = 2, 3, \ldots, 11$. This table was calculated by the *Mathematica* command `Table[2^(k+1)-4,{k,2,11}]`.

Examples:

1. The triplet $(n, 1, n)$ defines the standard (beta) n-bit binary codes, as can be verified by direct construction. The number of such codes is easily seen to be

$$\frac{2^{n+1} - 2^n}{2^1 - 1} = 2^n.$$

n	$a = 3 + n \cdot 2$	nth codeword	Number of codewords	Range of integers
0	3	$0xxx$	$2^3 = 8$	0–7
1	5	$10xxxxx$	$2^5 = 32$	8–39
2	7	$110xxxxxxx$	$2^7 = 128$	40–167
3	9	$111xxxxxxxxx$	$2^9 = 512$	168–679
		Total	680	

Table 3.3: The General Unary Code (3,2,9).

n	$a = 2 + n \cdot 1$	nth codeword	Number of codewords	Range of integers
0	2	$0xx$	4	0–3
1	3	$10xxx$	8	4–11
2	4	$110xxxx$	16	12–27
3	5	$1110xxxxx$	32	28–59
	...	...		...
8	10	$\underbrace{11...1}_{8}\underbrace{xx...x}_{10}$	1024	1020–2043
		Total	2044	

Table 3.4: The General Unary Code (2,1,10).

k :	2	3	4	5	6	7	8	9	10	11
$(2,1,k)$:	4	12	28	60	124	252	508	1020	2044	4092

Table 3.5: Number of General Unary Codes $(2,1,k)$ for $k = 2, 3, \ldots, 11$.

2. The triplet $(0, 0, \infty)$ defines the codes 0, 10, 110, 1110,... which are the unary codes but assigned to the integers 0, 1, 2,... instead of 1, 2, 3,... .
3. The triplet $(0, 1, \infty)$ generates a variance of the Elias gamma code.
4. The triplet (k, k, ∞) generates another variance of the Elias gamma code.
5. The triplet $(k, 0, \infty)$ generates the Rice code with parameter k.
6. The triplet $(s, 1, \infty)$ generates the exponential Golomb codes of page 198.
7. The triplet $(1, 1, 30)$ produces $(2^{30} - 2^1)/(2^1 - 1) \approx$ a billion codes.
8. Table 3.6 shows the general unary code for (10,2,14). There are only three code lengths since "start" and "stop" are so close, but there are many codes because "start" is large.

The start-step-stop codes are very general. Often, a person gets an idea for a variable-length code only to find out that it is a special case of a start-step-stop code. Here is a typical example. We can design a variable-length code where each code consists of triplets. The first two bits of a triplet go through the four possible values 00, 01, 10, and 11, and the rightmost bit acts as a stop bit. If it is 1, we stop reading the code, otherwise we read another triplet of bits. Table 3.7 lists the first 22 codes.

n	$a = 10 + n \cdot 2$	nth codeword	Number of codewords	Range of integers
0	10	$0\underbrace{x...x}_{10}$	$2^{10} = 1K$	0–1023
1	12	$10\underbrace{xx...x}_{12}$	$2^{12} = 4K$	1024–5119
2	14	$11\underbrace{xx...xx}_{14}$	$2^{14} = 16K$	5120–21503
		Total		21504

Table 3.6: The General Unary Code (10,2,14).

n	Codes	n	Codes
0	001	11	010 111
1	011	12	100 001
2	101	13	100 011
3	111	14	100 101
4	000 001	15	100 111
5	000 011	16	110 001
6	000 101	17	110 011
7	000 111	18	110 101
8	010 001	19	110 111
9	010 011	20	000 000 001
10	010 101	21	000 000 011

Table 3.7: Triplet-Based Codes.

n	Codes	n	Codes
0	1 00	11	01 01 11
1	1 01	12	01 10 00
2	1 10	13	01 10 01
3	1 11	14	01 10 10
4	01 00 00	15	01 10 11
5	01 00 01	16	01 11 00
6	01 00 10	17	01 11 01
7	01 00 11	18	01 11 10
8	01 01 00	19	01 11 11
9	01 01 01	20	001 00 00 00
10	01 01 10	21	001 00 00 01

Table 3.8: Same Codes Rearranged.

To see why this is a start-step-stop code, we rewrite it by placing the stop bits on the left end, followed by pairs of code bits, as listed in Table 3.8. Examining these 22 codes shows that the code for the integer n starts with a unary code of j zeros followed by a single 1, followed by $j+1$ pairs of code bits. There are four codes of length 3 (a 1-bit unary followed by a pair of code bits, corresponding to $j=0$), 16 6-bit codes (a 2-bit unary followed by two pairs of code bits, corresponding to $j=1$), 64 9-bit codes (corresponding to $j=2$), and so on. A little tinkering shows that this is the start-step-stop code $(2, 2, \infty)$.

3.3 Start/Stop Codes

The start-step-stop codes are flexible. By carefully adjusting the values of the three parameters it is possible to construct sets of codes of many different lengths. However, the lengths of these codes are restricted to the values $n + 1 + \text{start} + n \times \text{step}$ (except for the last subset where the codes have lengths stop + start + stop × step). The start/stop codes of this section were conceived by Steven Pigeon and are described in [Pigeon 01a,b], where it is shown that they are universal. A set of these codes is fully specified by an array of nonnegative integer parameters $(m_0, m_1, \ldots, m_t)$ and is constructed in subsets, similar to the start-step-stop codes, in the following steps:

1. Set $i = 0$ and $a = m_0$.

2. Construct the subset of codes that start with i leading 1's, followed by a single separator of 0, followed by a combination of a bits. There are 2^a such codes.

3. Increment i by 1 and set $a \leftarrow a + m_i$.

4. If $i < t$, go to step 2. Otherwise $(i = t)$, repeat step 2 but without the single 0 intercalary, and stop.

Thus, the parameter array $(0, 3, 1, 2)$ results in the set of codes listed in Table 3.9.

i	a	Codeword	# of codes	Length
0	2	$0xx$	4	3
1	5	$10xxxxx$	32	7
2	6	$110xxxxxx$	64	9
3	8	$111xxxxxxxx$	256	11

Table 3.9: The Start/Stop Code (2,3,1,2).

The maximum code length is $t + m_0 + \cdots + m_t$ and on average the start/stop code of an integer s is never longer than $\lceil \log_2 s \rceil$ (the length of the binary representation of s). If an optimal set of such codes is constructed by an encoder and is used to compress a data file, then the only side information needed in the compressed file is the value of t and the array of $t + 1$ parameters m_i (which are mostly small integers). This is considerably less than the size of a Huffman tree or the side information required by many other compression methods.

The start/stop codes can also encode an indefinite number of arbitrarily large integers. Simply set all the parameters to the same value and set t to infinity.

Steven Pigeon, the developer of these codes, shows that the parameters of the start/stop codes can be selected such that the resulting set of codes will have an average length shorter than what can be achieved with the start-step-stop codes for the same probability distribution. He also shows how the probability distribution can be employed to determine the best set of parameters for the code. In addition, the number of codes in the set can be selected as needed, in contrast to the start-step-stop codes that often result in more codes than needed.

> The Human Brain starts working the moment you are born and never stops until you stand up to speak in public!
>
> —George Jessel

3.4 Elias Codes

In his pioneering work [Elias 75], Peter Elias described three useful prefix codes. The main idea of these codes is to prefix the integer being encoded with an encoded representation of its order of magnitude. For example, for any positive integer n there is an integer M such that $2^M \le n < 2^{M+1}$. We can therefore write $n = 2^M + L$ where L is at most M bits long, and generate a code that consists of M and L. The problem is to determine the length of M and this is solved in different ways by the various Elias codes. Elias denoted the unary code of n by $\alpha(n)$ and the standard binary representation of n, from its most-significant 1, by $\beta(n)$. His first code was therefore designated γ (gamma).

Elias gamma code $\gamma(n)$ for positive integers n is simple to encode and decode and is also universal.

Encoding. Given a positive integer n, perform the following steps:

1. Denote by M the length of the binary representation $\beta(n)$ of n.
2. Prepend $M-1$ zeros to it (i.e., the $\alpha(n)$ code without its terminating 1).

Step 2 amounts to prepending the length of the code to the code, in order to ensure unique decodability.

The length M of the integer n is, from Equation (2.1), $1 + \lfloor \log_2 n \rfloor$, so the length of $\gamma(n)$ is

$$2M - 1 = 2\lfloor \log_2 n \rfloor + 1. \tag{3.1}$$

We later show that this code is ideal for applications where the probability of n is $1/(2n^2)$.

An alternative construction of the gamma code is as follows:

1. Find the largest integer N such that $2^N \le n < 2^{N+1}$ and write $n = 2^N + L$. Notice that L is at most an N-bit integer.
2. Encode N in unary either as N zeros followed by a 1 or N 1's followed by a 0.
3. Append L as an N-bit number to this representation of N.

Section 3.8 describes yet another way to construct the same code. Section 3.15 shows a connection between this code and certain binary search trees.

Peter Elias

$1 = 2^0 + 0 = 1$	$10 = 2^3 + 2 = 0001\mathit{010}$
$2 = 2^1 + 0 = 01\mathit{0}$	$11 = 2^3 + 3 = 0001\mathit{011}$
$3 = 2^1 + 1 = 01\mathit{1}$	$12 = 2^3 + 4 = 0001\mathit{100}$
$4 = 2^2 + 0 = 001\mathit{00}$	$13 = 2^3 + 5 = 0001\mathit{101}$
$5 = 2^2 + 1 = 001\mathit{01}$	$14 = 2^3 + 6 = 0001\mathit{110}$
$6 = 2^2 + 2 = 001\mathit{10}$	$15 = 2^3 + 7 = 0001\mathit{111}$
$7 = 2^2 + 3 = 001\mathit{11}$	$16 = 2^4 + 0 = 00001\mathit{0000}$
$8 = 2^3 + 0 = 0001\mathit{000}$	$17 = 2^4 + 1 = 00001\mathit{0001}$
$9 = 2^3 + 1 = 0001\mathit{001}$	$18 = 2^4 + 2 = 00001\mathit{0010}$

Table 3.10: 18 Elias Gamma Codes.

Table 3.10 lists the first 18 gamma codes, where the L part is in italics (see also Table 10.29 and the C_1 code of Table 3.20).

In his 1975 paper, Elias describes two versions of the gamma code. The first version (titled γ) is encoded as follows:

1. Generate the binary representation $\beta(n)$ of n.
2. Denote the length $|\beta(n)|$ of $\beta(n)$ by M.
3. Generate the unary $u(M)$ representation of M as $M-1$ zeros followed by a 1.
4. Follow each bit of $\beta(n)$ by a bit of $u(M)$.
5. Drop the leftmost bit (the leftmost bit of $\beta(n)$ is always 1).

Thus, for $n=13$ we prepare $\beta(13)=\bar{1}\bar{1}\bar{0}\bar{1}$, so $M=4$ and $u(4)=0001$, resulting in $\bar{1}0\bar{1}0\bar{0}0\bar{1}1$. The final code is $\gamma(13)=0\bar{1}0\bar{0}0\bar{1}1$. In general, the length of the gamma code for the integer i is $1+2\lfloor\log_2 i\rfloor$.

The second version, dubbed γ', moves the bits of $u(M)$ to the left. Thus $\gamma'(13)=0001|\bar{1}\bar{0}\bar{1}$. The gamma codes of Table 3.10 are Elias's γ' codes. Both gamma versions are universal.

Decoding is also simple and is done in two steps:

1. Read zeros from the code until a 1 is encountered. Denote the number of zeros by N.

2. Read the next N bits as an integer L. Compute $n=2^N+L$.

It is easy to see that this code can be used to encode positive integers even in cases where the largest integer is not known in advance. Also, this code grows slowly (see Figure 3.38), so it is a good candidate for compressing integer data where small integers are common and large ones are rare.

It is easy to understand why the gamma code (and in general, all variable-length codes) should be used only when it is known or estimated that the distribution of the integers to be encoded is close to the ideal distribution for the given code. The simple computation here assumes that the first n integers are given and are distributed uniformly (i.e., each appears with the same probability). We compute the average gamma code length and show that it is much larger than the length of the fixed-size binary codes of the same integers.

The length of the Elias gamma code for the integer i is $1+2\lfloor\log_2 i\rfloor$. Thus, the average of the gamma codes of the first n integers is

$$a=\frac{n+2\sum_{i=1}^{n}\lfloor\log_2 i\rfloor}{n}.$$

The length of the fixed-size codes of the same integers is $b=\lceil\log_2 n\rceil$. Figure 3.11 is a plot of the ratio a/b and it is easy to see that this value is greater than 1, indicating that for flat distributions of the integers the fixed-length (beta) code is better than the gamma code. The curve in the figure generally goes up, but features sudden drops (where it remains greater than 1) at points where $n=2^k+1$ for integer k.

Elias delta code. In his gamma code, Elias prepends the length of the code in unary (α). In his next code, δ (delta), he prepends the length in binary (β). Thus, the Elias delta code, also for the positive integers, is slightly more complex to construct.

Encoding a positive integer n, is done in the following steps:

1. Write n in binary. The leftmost (most-significant) bit will be a 1.

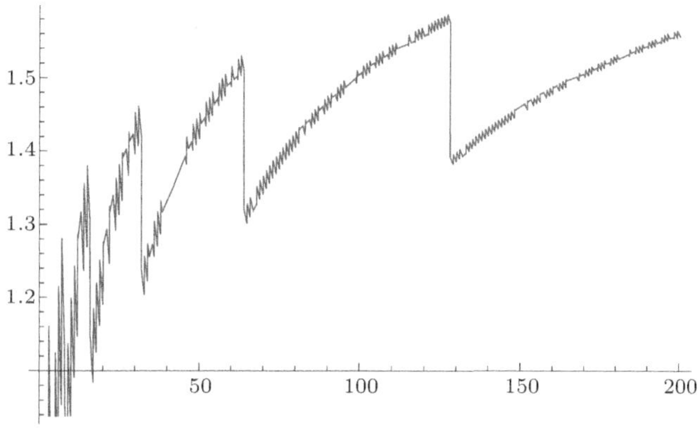


```
gamma[i_] := 1. + 2 Floor[Log[2, i]];
Plot[Sum[gamma[j], {j,1,n}]/(n Ceiling[Log[2,n]]), {n,1,200}]
```

Figure 3.11: Gamma Code Versus Binary Code.

2. Count the bits, remove the leftmost bit of n, and prepend the count, in binary, to what is left of n after its leftmost bit has been removed.

3. Subtract 1 from the count of step 2 and prepend that number of zeros to the code.

When these steps are applied to the integer 17, the results are: $17 = 10001_2$ (five bits). Remove the leftmost 1 and prepend $5 = 101_2$ yields 101|0001. Three bits were added, so we prepend two zeros to obtain the delta code 00|101|0001.

To compute the length of the delta code of n, we notice that step 1 generates (from Equation (2.1)) $M = 1 + \lfloor \log_2 n \rfloor$ bits. For simplicity, we omit the $\lfloor$ and $\rfloor$ and observe that

$$M = 1 + \log_2 n = \log_2 2 + \log_2 n = \log_2(2n).$$

The count of step 2 is M, whose length C is therefore $C = 1+\log_2 M = 1+\log_2(\log_2(2n))$ bits. Step 2 therefore prepends C bits and removes the leftmost bit of n. Step 3 prepends $C - 1 = \log_2 M = \log_2(\log_2(2n))$ zeros. The total length of the delta code is therefore the 3-part sum

$$\underbrace{\log_2(2n)}_{\text{step 1}} + \underbrace{[1 + \log_2 \log_2(2n)] - 1}_{\text{step 2}} + \underbrace{\log_2 \log_2(2n)}_{\text{step 3}} = 1 + \lfloor \log_2 n \rfloor + 2\lfloor \log_2 \log_2(2n) \rfloor. \tag{3.2}$$

Figure 3.38 illustrates the length graphically. We show below that this code is ideal for data where the integer n occurs with probability $1/[2n(\log_2(2n))^2]$.

An equivalent way to construct the delta code employs the gamma code:

1. Find the largest integer N such that $2^N \ge n < 2^{N+1}$ and write $n = 2^N + L$. Notice that L is at most an N-bit integer.

2. Encode $N + 1$ with the Elias gamma code.

3. Append the binary value of L, as an N-bit integer, to the result of step 2.

When these steps are applied to $n = 17$, the results are: $17 = 2^N + L = 2^4 + 1$. The gamma code of $N + 1 = 5$ is 00101, and appending $L = 0001$ to this yields 00101|0001.

Table 3.12 lists the first 18 delta codes, where the L part is in italics. See also the related code C_3 of Table 3.20, which has the same length.

$1 = 2^0 + 0 \to \lvert L\rvert = 0 \to 1$	$10 = 2^3 + 2 \to \lvert L\rvert = 3 \to 00100\mathit{010}$
$2 = 2^1 + 0 \to \lvert L\rvert = 1 \to 010\mathit{0}$	$11 = 2^3 + 3 \to \lvert L\rvert = 3 \to 00100\mathit{011}$
$3 = 2^1 + 1 \to \lvert L\rvert = 1 \to 010\mathit{1}$	$12 = 2^3 + 4 \to \lvert L\rvert = 3 \to 00100\mathit{100}$
$4 = 2^2 + 0 \to \lvert L\rvert = 2 \to 011\mathit{00}$	$13 = 2^3 + 5 \to \lvert L\rvert = 3 \to 00100\mathit{101}$
$5 = 2^2 + 1 \to \lvert L\rvert = 2 \to 011\mathit{01}$	$14 = 2^3 + 6 \to \lvert L\rvert = 3 \to 00100\mathit{110}$
$6 = 2^2 + 2 \to \lvert L\rvert = 2 \to 011\mathit{10}$	$15 = 2^3 + 7 \to \lvert L\rvert = 3 \to 00100\mathit{111}$
$7 = 2^2 + 3 \to \lvert L\rvert = 2 \to 011\mathit{11}$	$16 = 2^4 + 0 \to \lvert L\rvert = 4 \to 00101\mathit{0000}$
$8 = 2^3 + 0 \to \lvert L\rvert = 3 \to 00100\mathit{000}$	$17 = 2^4 + 1 \to \lvert L\rvert = 4 \to 00101\mathit{0001}$
$9 = 2^3 + 1 \to \lvert L\rvert = 3 \to 00100\mathit{001}$	$18 = 2^4 + 2 \to \lvert L\rvert = 4 \to 00101\mathit{0010}$

Table 3.12: 18 Elias Delta Codes.

Decoding is done in the following steps:

1. Read bits from the code until you can decode an Elias gamma code. Call the decoded result $M + 1$. This is done in the following substeps:

1.1 Count the leading zeros of the code and denote the count by C.

1.2 Examine the leftmost $2C + 1$ bits (C zeros, followed by a single 1, followed by C more bits). This is the decoded gamma code $M + 1$.

2. Read the next M bits. Call this number L.

3. The decoded integer is $2^M + L$.

In the case of $n = 17$, the delta code is 001010001. We skip two zeros, so $C = 2$. The value of the leftmost $2C + 1 = 5$ bits is $00101 = 5$, so $M + 1 = 5$. We read the next $M = 4$ bits 0001, and end up with the decoded value $2^M + L = 2^4 + 1 = 17$.

To better understand the application and performance of these codes, we need to identify the type of data it compresses best. Given a set of symbols a_i, where each symbol occurs in the data with probability P_i and the length of its code is l_i bits, the average code length is the sum $\sum P_i l_i$ and the entropy (the smallest number of bits required to represent the symbols) is $\sum[-P_i \log_2 P_i]$. The redundancy (Equation (2.2)) is the difference $\sum_i P_i l_i - \sum_i[-P_i \log_2 P_i]$ and we are looking for probabilites P_i that will minimize this difference.

In the case of the gamma code, $l_i = 1 + 2\lfloor \log_2 i \rfloor$. If we select symbol probabilities $P_i = 1/(2i^2)$ (a power law distribution of probabilities, where the first 10 values are 0.5, 0.125, 0.0556, 0.03125, 0.02, 0.01389, 0.0102, 0.0078, 0.00617, and 0.005, see also Table 3.21), both the average code length and the entropy become the identical sums

$$\sum_i \frac{1 + 2\log i}{2i^2},$$

indicating that the gamma code is asymptotically optimal for this type of data. A power law distribution of values is dominated by just a few symbols and especially by the first. Such a distribution is very skewed and is therefore handled very well by the gamma code which starts very short. In an exponential distribution, in contrast, the small values have similar probabilities, which is why data with this type of statistical distribution is compressed better by a Rice code (Section 3.25).

In the case of the delta code, $l_i = 1 + \log i + 2\log\log(2i)$. If we select symbol probabilities $P_i = 1/[2i(\log(2i))^2]$ (where the first five values are 0.5, 0.0625, 0.025, 0.0139, and 0.009), both the average code length and the entropy become the identical sums

$$\sum_i \frac{\log 2 + \log i + 2\log\log(2i)}{2i(\log(2i))^2},$$

indicating that the redundancy is zero and the delta code is therefore asymptotically optimal for this type of data.

Section 3.15 shows a connection between a variant of the delta code and certain binary search trees.

> The phrase "working mother" is redundant.
> —Jane Sellman

The Elias omega code. Unlike the previous Elias codes, the omega code uses itself recursively to encode the prefix M, which is why it is sometimes referred to as a recursive Elias code. The main idea is to prepend the length of n to n as a group of bits that starts with a 1, then prepend the length of the length, as another group, to the result, and continue prepending lengths until the last length is 2 or 3 (and therefore fits in two bits). In order to distinguish between a length group and the last, rightmost group (of n itself), the latter is followed by a delimiter of 0, while each length group starts with a 1.

Encoding a positive integer n is done recursively in the following steps:

1. Initialize the code-so-far to 0.
2. If the number to be encoded is 1, stop; otherwise, prepend the binary representation of n to the code-so-far. Assume that we have prepended L bits.
3. Repeat step 2, with the binary representation of $L-1$ instead of n.

The integer 17 is therefore encoded by (1) a single 0, (2) prepended by the 5-bit binary value 10001, (3) prepended by the 3-bit value of $5-1 = 100_2$, and (4) prepended by the 2-bit value of $3-1 = 10_2$. The result is 10|100|10001|0.

Table 3.13 lists the first 18 omega codes (see also Table 3.17). Note that $n = 1$ is handled as a special case.

Decoding is done in several nonrecursive steps where each step reads a group of bits from the code. A group that starts with a zero signals the end of decoding.

1. Initialize n to 1.
2. Read the next bit. If it is 0, stop. Otherwise read n more bits, assign the group of $n+1$ bits to n, and repeat this step.

1	0	10	11 1010 0
2	10 0	11	11 1011 0
3	11 0	12	11 1100 0
4	10 100 0	13	11 1101 0
5	10 101 0	14	11 1110 0
6	10 110 0	15	11 1111 0
7	10 111 0	16	10 100 10000 0
8	11 1000 0	17	10 100 10001 0
9	11 1001 0	18	10 100 10010 0

Table 3.13: 18 Elias Omega Codes.

Some readers may find it easier to understand these steps rephrased as follows.

1. Read the first group, which will either be a single 0, or a 1 followed by n more digits. If the group is a 0, the value of the integer is 1; if the group starts with a 1, then n becomes the value of the group interpreted as a binary number.

2. Read each successive group; it will either be a single 0, or a 1 followed by n more digits. If the group is a 0, the value of the integer is n; if it starts with a 1, then n becomes the value of the group interpreted as a binary number.

Example. Decode 10|100|10001|0. The decoder initializes $n = 1$ and reads the first bit. It is a 1, so it reads $n = 1$ more bit (0) and assigns $n = 10_2 = 2$. It reads the next bit. It is a 1, so it reads $n = 2$ more bits (00) and assigns the group 100 to n. It reads the next bit. It is a 1, so it reads four more bits (0001) and assigns the group 10001 to n. The next bit read is 0, indicating the end of decoding.

The omega code is constructed recursively, which is why its length $|\omega(n)|$ can also be computed recursively. We define the quantity $l^k(n)$ recursively by $l^1(n) = \lfloor \log_2 n \rfloor$ and $l^{i+1}(n) = l^1(l^i(n))$. Equation (2.1) tells us that $|\beta(n)| = l^1(n) + 1$ (where β is the standard binary representation), and this implies that the length of the omega code is given by the sum

$$|\omega(n)| = \sum_{i=1}^{k} \beta(l^{k-i}(n)) + 1 = 1 + \sum_{i=1}^{k} (l^i(n) + 1),$$

where the sum stops at the k that satisfies $l^k(n) = 1$. From this, Elias concludes that the length satisfies $|\omega(n)| \le 1 + \frac{5}{2}\lfloor \log_2 n \rfloor$.

A quick glance at any table of these codes shows that their lengths fluctuate. In general, the length increases slowly as n increases, but when a new length group is added, which happens when $n = 2^{2^k}$ for any positive integer k, the length of the code increases suddenly by several bits. For k values of 1, 2, 3, and 4, this happens when n reaches 4, 16, 256, and 65,536. Because the groups of lengths are of the form "length," "log(length)," "log(log(length))," and so on, the omega code is sometimes referred to as a logarithmic-ramp code.

Table 3.14 compares the length of the gamma, delta, and omega codes. It shows that the delta code is asymptotically best, but if the data consists mostly of small numbers

Values	Gamma	Delta	Omega
1	1	1	2
2	3	4	3
3	3	4	4
4	5	5	4
5–7	5	5	5
8–15	7	8	6–7
16–31	9	9	7–8
32–63	11	10	8–10
64–88	13	11	10
100	13	11	11
1000	19	16	16
10^4	27	20	20
10^5	33	25	25
10^5	39	28	30

Table 3.14: Lengths of Three Elias Codes.

(less than 8) and there are only a few large integers, then the gamma code performs better.

> Beware of bugs in the above code; I have only proved it correct, not tried it.
> —Donald Knuth

3.5 RBUC, Recursive Bottom-Up Coding

The RBUC (recursive bottom-up coding) method discussed here [Moffat and Anh 06] is based on the notion of a selector. We know from Equation (2.1) that the length b of the integer n is $1+\lfloor\log_2 n\rfloor = \lceil\log_2(1+n)\rceil$. Thus, given a set S of nonnegative integers, we can use the standard binary (β) code to encode each integer in $b=\lceil\log_2(1+x)\rceil$ bits, where x is the largest integer in the set. We can consider b the *selector* for S, and since the same b is used for the entire set, it can be termed a global selector. This type of encoding is simple but extremely ineffectual because the codes of many integers smaller than x will have redundant leading zeros.

In the other extreme we find the Elias Gamma code (Section 3.4). Given a positive integer n, its gamma code is constructed in the following steps:

1. Denote by b the length of the binary representation $\beta(n)$ of n.
2. Prepend $b-1$ zeros to $\beta(n)$.

Thus, each gamma code includes its own selector b. This code is more efficient than β but takes more steps to construct.

Once we look at the β and γ codes as extreme approaches to selectors, it is natural to search for a more flexible approach, and RBUC is an important step in this direction. It represents a compromise between the β and γ codes and it results in good compression when applied to the right type of data.

Given a set of m integers, the RBUC encoder partitions it into segments of size s (except the last segment, which may be shorter), finds the largest integer in each segment i, uses it to determine a selector b_i for the segment, and employs the standard β code to encode the elements of the segment in b_i bits each.

It is obvious that RBUC can yield good compression if each segment is uniform or close to uniform, and one example of such data is inverted indexes [Witten et al. 99]. An inverted index is a data structure that stores a mapping from an index item (normally a word) to its locations in a set of documents. Thus, the inverted index for index-item f may be the pair $[f, (2, 5, 6, 8, 11, 12, 13, 18)]$ where the integers 2, 5, 6, and so on are document numbers. It is known experimentally that such lists of document numbers exhibit local uniformity and are therefore natural candidates for RBUC encoding.

Even with ideal data, there is a price to pay for the increased performance of RBUC. Compressing each segment with a separate selector b_i must be followed by recursively compressing the set of selectors. There are $\lceil m/s \rceil$ selectors in this set and they are encoded by partitioning the set into subsets of size s each, determining a selector c_i for each, and encoding with β. The set of selectors c_i has to be encoded too, and this process continues recursively $\lceil \log_2 m \rceil$ times until a set with just one selector is obtained. This selector is encoded with an Elias code, and the encoder also has to encode the value of m, as overhead, for the decoder's use.

Here is an example. Given the $m = 13$ set of integers $(15, 6, 2, 3, 0, 0, 0, 0, 4, 5, 1, 7, 8)$ we select $s = 2$ and partition the set into seven subsets of two integers each (except the last subset). Examining the elements of each subset, it is easy to determine the set of selectors $b_i = (4, 3, 0, 0, 3, 3, 4)$ and encode the 13 integers. Notice that the four zeros are encoded into zero bits, thereby contributing to compression effectiveness. The process iterates four times (notice that $\log_2 m = 3.7$) and is summarized in Table 3.15. The last selector, 2, is encoded as the gamma code 010, and the encoder also has to encode $m = 13$, perhaps also as a gamma code. The codes are then collected from the top of the table and are concatenated to become the compressed stream. Notice that the original message and all the iterations must be computed and kept in memory until the moment of output, which is the main drawback of this method. We can say that RBUC is an offline algorithm.

010	$e_i = (2)$
10|10	$d_i = (2, 2)$
11|00|01|11	$c_i = (3, 0, 2, 2)$
100|011|011|011|100	$b_i = (4, 3, 0, 0, 3, 3, 4)$
1111|0110|010|011|100|101|001|111|1000	15, 6, 2, 3, 0, 0, 0, 0, 4, 5, 1, 7, 8

Table 3.15: An RBUC Example with $m = 13$ and $s = 2$.

Assuming that the largest integer in set i is a, the selector b_i for the set will be $\lceil \log_2(1+a) \rceil$. Thus, the set of selectors (b_i) consists of integers that are the logarithms of the original data symbols and are therefore much smaller. The set of selectors (c_i) will be much smaller still, and so on. Since all the data symbols and selectors are nonnegative integers, consecutive sets of selectors become logarithmically more uniform and reduce any nonhomogeneity in the original data.

The need to iterate $\lceil \log_2 m \rceil$ times and compress sets of selectors reduces the effectiveness and increases the execution time of RBUC. However, the sets of selectors become more and more uniform, which suggests the use of larger sets (i.e., larger values of s) in each iteration and thus fewer iterations. If we double s in each iteration, the number of iterations reduces to $\sqrt{\log_2 m}$, and if we square s each time, the number of iterations shrinks to the much smaller $\log_2 \log_2 m$.

Another minor improvement can be incorporated when we observe that the set of data items 1, 7 are encoded in three bits to become 001 and 111 (Table 3.15), but can also be encoded to 001 and 11. Once we see the 3-bit codeword 001, we know that the other codeword in this set must start with a 1, so this 1 can be omitted. In general, if $s-1$ codewords of a set start with a zero, then the remaining codeword must start with a 1 which can be omitted.

3.5.1 BASC, Binary Adaptive Sequential Coding

RBUC is an offline algorithm, which is why its developers also came up with a similar encoding method that is both online and adaptive. ("Online" means that the codewords can be emitted as soon as they are computed and there is no need to save large sets of numbers in memory.) They termed it binary adaptive sequential coding (BASC).

The principle of BASC is to determine a selector b for each data symbol based on the selector of the preceding symbol. Once b has been determined, the current symbol is encoded, the codeword is emitted, and only b need be saved, for use with the next data symbol.

The principle of BASC is to determine a selector b' for each data symbol based on the selector b of the preceding symbol. Once b' has been determined, the current symbol is encoded, the codeword is emitted, and b is set to b'. Only b need be saved, for use with the next data symbol.

Specifically, the BASC algorithm operates as follows: The initial value of b is either input by the user or is determined from the first data symbol. Given the current data symbol x_i, its length $b' = \lceil \log_2(1+x_i) \rceil$ is computed. If $b' \le b$, then a zero is emitted, followed by the β code of x_i as a b-bit number. Otherwise, the encoder emits $(b'-b)$ 1's, followed by a single zero, followed by the least-significant $b'-1$ bits of x_i (the most-significant bit of x_i must be a 1, and so can be ignored). This process is illustrated by the 13-symbol message $(15, 6, 2, 3, 0, 0, 0, 0, 4, 5, 1, 7, 8)$.

Assume that the initial b is 5. The first symbol, 15, yields $b' = 4$, which is less than b. The encoder generates and outputs a zero followed by 15 as a b-bit number 0|01111 and b is set to 4. The second symbol, 6, yields $b' = 3$, so 0|0110 is emitted and b is set to 3. The third symbol, 2, yields $b' = 2$, so 0|010 is output and b is set to 2. The next symbol, 3, results in $b' = 2$ and 0|11 is output, followed by $b \leftarrow 2$. The first of the four zeros yields $b' = 0$. The codeword 0|00 is emitted and b is set to 0. Each successive zero produces $b' = 0$ and is output as 0. The next symbol, 4, yields $b' = 3$ which is greater than b. The encoder emits $b' - b = 3 - 0 = 3$ 1's, followed by a zero, followed by the two least-significant bits 00 of the 4, thus 111|0|00. The current selector b is set to 3 and the rest of the example is left as an easy exercise.

The decoder receives the values of m (13) and the initial b (5) as overhead, so it knows how many codewords to read and how to read the first one. If a codeword starts with a zero, the decoder reads b bits, converts them to an integer n, computes the number of bits required for n, and sets b to that number. If a codeword starts with a 1, the decoder reads consecutive 1s until the first zero, skips the zero, and stores the number of 1s in c. This number is $b' - b$. The decoder has b from the previous step, so it can compute $b' = c + b$ and read $b' - 1$ bits from the current codeword. The decoder prepends a 1 to those bits, outputs the result as the next symbol, and sets b to b'.

The codewords produced by BASC are similar to the Elias gamma codes for the integers, the main difference is the adaptive step where b is set to b'. The developers of this method propose two variations to this simple step. The first variation is to set b to the average of several previous b's. This can be thought of as smoothing any large differences in consecutive b's because of uneveness in the input data. The second, damping, variation tries to achieve the same goal by limiting the changes in b to one unit at a time.

The developers of RBUC and BASC list results of tests comparing the performance of these algorithms to the Elias gamma codes, Huffman codes, and the interpolative coding algorithm of Section 3.28. The results indicate that BASC and RBUC offer new tradeoffs between compression effectiveness and decoding and are therefore serious competitors to the older, well-known, and established compression algorithms.

3.6 Levenstein Code

This little-known code for the nonnegative integers was conceived in 1968 by Vladimir Levenshtein [Levenstein 06]. Both encoding and decoding are multistep processes.

Encoding. The Levenstein code of zero is a single 0. To code a positive number n, perform the following:

1. Set the count variable C to 1. Initialize the code-so-far to the empty string.
2. Take the binary value of n without its leading 1 and prepend it to the code-so-far.
3. Let M be the number of bits prepended in step 2.
4. If $M \neq 0$, increment C by 1, go to and execute step 2 with M instead of n.
5. If $M = 0$, prepend C 1's followed by a 0 to the code-so-far and stop.

n	Levenstein code	n	Levenstein code
0	0	9	1110 1 001
1	10	10	1110 1 010
2	110 0	11	1110 1 011
3	110 1	12	1110 1 100
4	1110 0 00	13	1110 1 101
5	1110 0 01	14	1110 1 110
6	1110 0 10	15	1110 1 111
7	1110 0 11	16	11110 0 00 0000
8	1110 1 000	17	11110 0 00 0001

Table 3.16: 18 Levenstein Codes.

Table 3.16 lists some of these codes. Spaces have been inserted to indicate the individual parts of each code. As an exercise, the reader may verify that the Levenstein codes for 18 and 19 are 11110|0|00|0010 and 11110|0|00|0011, respectively.

Decoding is done as follows:

1. Set count C to the number of consecutive 1's preceding the first 0.
2. If $C = 0$, the decoded value is zero, stop.
3. Set $N = 1$, and repeat step 4 $(C - 1)$ times.
4. Read N bits, prepend a 1, and assign the resulting bitstring to N (thereby erasing the previous value of N). The string assigned to N in the last iteration is the decoded value.

The Levenstein code of the positive integer n is always one bit longer than the Elias omega code of n.

3.7 Even–Rodeh Code

The principle behind the Elias omega code occurred independently to S. Even and M. Rodeh [Even and Rodeh 78]. Their code is similar to the omega code, the main difference being that lengths are prepended until a 3-bit length is reached and becomes the leftmost group of the code. For example, the Even–Rodeh code of 2761 is 100|1100|101011001001|0 (4, 12, 2761, and 0).

The authors prove, by induction on n, that every nonnegative integer can be encoded in this way. They also show that the length $l(n)$ of their code satisfies $l(n) \leq 4 + 2L(n)$ where $L(n)$ is the length of the binary representation (beta code) of n, Equation (2.1).

The developers show how this code can be used as a comma to separate variable-length symbols in a string. Given a string of symbols a_i, precede each a_i with the Even–Rodeh code R of its length l_i. The codes and symbols can then be concatenated into a single string

$$R(l_1)a_1R(l_2)a_2\ldots R(l_m)a_m000$$

that can be uniquely separated into the individual symbols because the codes act as separators (commas). Three zeros act as the string delimiter.

The developers also prove that the extra length of the codes shrinks asymptotically to zero as the string becomes longer. For strings of length 2^{10} bits, the overhead is less than 2%. Table 3.17 (after [Fenwick 96a]) lists several omega and Even–Rodeh codes.

Table 3.18 (after [Fenwick 96a]) illustrates the different points where the lengths of these codes increase. The length of the omega code increases when n changes from the form $2^m - 1$ (where the code is shortest relative to n) to 2^m (where it is longest). The Even–Rodeh code behaves similarly, but may be slightly shorter or longer for various intervals of n.

n	Omega	Even–Rodeh
0	—	000
1	0	001
2	10 0	010
3	11 0	011
4	10 100 0	100 0
7	10 111 0	111 0
8	11 1000 0	100 1000 0
15	11 1111 0	100 1111 0
16	10 100 10000 0	101 10000 0
32	10 101 100000 0	110 100000 0
100	10 110 1100100 0	111 1100100 0
1000	10 110 1100100 0	110 1100100 0

Table 3.17: Several Omega and Even–Rodeh Codes.

Values	ω	ER
1	1	3
2–3	3	3
4–7	6	4
8–15	7	8
16–31	11	9
32–63	12	10
64–127	13	11
128–255	14	17
256–512	21	18

Table 3.18: Different Lengths.

3.8 Punctured Elias Codes

The punctured Elias codes for the integers were designed by Peter Fenwick in an attempt to improve the performance of the Burrows–Wheeler transform [Burrows and Wheeler 94]. The codes are described in the excellent, 15-page technical report [Fenwick 96a]. The term "punctured" comes from the field of error-control codes. Often, a codeword for error-detection or error-correction consists of the original data bits plus a number of check bits. If some check bits are removed, to shorten the codeword, the resulting code is referred to as punctured.

We start with the Elias gamma code. Section 3.4 describes two ways to construct this code, and here we consider a third approach to the same construction. Write the binary value of n, reverse its bits so its rightmost bit is now a 1, and prepend flags for the bits of n. For each bit of n, create a flag of 0, except for the last (rightmost) bit (which is a 1), whose flag is 1. Prepend the flags to the reversed n and remove the rightmost bit. Thus, $13 = 1101_2$ is reversed to yield 1011. The four flags 0001 are prepended and the rightmost bit of 1011 is removed to yield the final gamma code 0001|101.

The punctured code eliminates the flags for zeros and is constructed as follows. Write the binary value of n, reverse its bits, and prepend flags to indicate the number of 1's in n. For each bit of 1 in n we prepare a flag of 1, and terminate the flags with a single 0. Thus, $13 = 1101_2$ is reversed to 1011. It has three 1's, so the flags 1110 are prepended to yield 1110|1011. We call this punctured code $P1$ and notice that it starts with a 1 (there is at least one flag, except for the $P1$ code of $n = 0$) and also ends with a 1 (because the original n, whose MSB is a 1, has been reversed). We can therefore construct another punctured code $P2$ such that $P2(n)$ equals $P1(n+1)$ with its most-significant 1 removed.

Table 3.19 lists examples of $P1$ and $P2$. The feature that strikes us most is that the codes are generally getting longer as n increases, but are also getting shorter from time to time, for n values that are 1 less than a power of 2. For small values of n, these codes are often a shade longer than the gamma code, but for large values they average about $1.5\log_2 n$ bits, shorter than the $2\lfloor\log_2 n\rfloor + 1$ bits of the gamma code.

One design consideration for these codes was their expected behavior when applied to data with a skewed distribution and more smaller values than larger values. Smaller binary integers tend to have fewer 1's, so it was hoped that the punctures would reduce the average code length below $1.5\log_2 n$ bits. Later work by Fenwick showed that this hope did not materialize.

n	$P1$	$P2$	n	$P1$	$P2$
0	0	01	11	11101101	100011
1	101	001	12	1100011	1101011
2	1001	1011	13	11101011	1100111
3	11011	0001	14	11100111	11101111
4	10001	10101	15	111101111	000001
5	110101	10011	16	1000001	1010001
6	110011	110111	...		
7	1110111	00001	31	11111011111	0000001
8	100001	101001	32	10000001	10100001
9	1101001	100101	33	110100001	10010001
10	1100101	1101101			

Table 3.19: Two Punctured Elias Codes.

> And as they carried him away
> Our punctured hero was heard to say,
> When in this war you venture out,
> Best never do it dressed as a Kraut!
>
> —Stephen Ambrose, *Band of Brothers*

3.9 Other Prefix Codes

Four prefix codes, C_1 through C_4, are presented in this section. We denote by $B(n)$ the binary representation of the integer n (the beta code). Thus, $|B(n)|$ is the length, in bits, of this representation. We also use $\overline{B}(n)$ to denote $B(n)$ without its most significant bit (which is always 1).

Code C_1 consists of two parts. To code the positive integer n, we first generate the unary code of $|B(n)|$ (the size of the binary representation of n), then append $\overline{B}(n)$ to it. An example is $n = 16 = 10000_2$. The size of B(16) is 5, so we start with the unary code 11110 (or 00001) and append $\overline{B}(16) = 0000$. Thus, the complete code is 11110|0000 (or 00001|0000). Another example is $n = 5 = 101_2$ whose code is 110|01. The length of $C_1(n)$ is $2\lfloor\log_2 n\rfloor + 1$ bits. Notice that this code is identical to the general unary code $(0, 1, \infty)$ and is closely related to the Elias gamma code.

Code C_2 is a rearrangement of C_1 where each of the $1 + \lfloor\log_2 n\rfloor$ bits of the first part (the unary code) of C_1 is followed by one of the bits of the second part. Thus, code $C_2(16) = 101010100$ and $C_2(5) = 10110$.

Code C_3 starts with $|B(n)|$ coded in C_2, followed by $\overline{B}(n)$. Thus, 16 is coded as $C_2(5) = 10110$ followed by $\overline{B}(16) = 0000$, and 5 is coded as code $C_2(3) = 110$ followed by $\overline{B}(5) = 01$. The length of $C_3(n)$ is $1 + \lfloor \log_2 n \rfloor + 2\lfloor \log_2(1 + \lfloor \log_2 n \rfloor)\rfloor$ (same as the length of the Elias delta code, Equation (3.2)).

Code C_4 consists of several parts. We start with $B(n)$. To its left we prepend the binary representation of $|B(n)| - 1$ (one less than the length of n). This continues recursively, until a 2-bit number is written. A 0 is then appended to the right of the entire code, to make it uniquely decodable. To encode 16, we start with 10000, prepend $|B(16)| - 1 = 4 = 100_2$ to the left, then prepend $|B(4)| - 1 = 2 = 10_2$ to the left of that, and finally append a 0 on the right. The result is 10|100|10000|0. To encode 5, we start with 101, prepend $|B(5)| - 1 = 2 = 10_2$ to the left, and append a 0 on the right. The result is 10|101|0. Comparing with Table 3.13 shows that C_4 is the omega code.

(The 0 on the right make the code uniquely decodable because each part of C_4 is the standard binary code of some integer, so it starts with a 1. A start bit of 0 is therefore a signal to the decoder that this is the last part of the code.)

Table 3.20 shows examples of the four codes above, as well as $B(n)$ and $\overline{B}(n)$. The lengths of the four codes shown in the table increase as $\log_2 n$, in contrast with the length of the unary code, which increases as n. These codes are therefore good choices in cases where the data consists of integers n with probabilities that satisfy certain conditions. Specifically, the length L of the unary code of n is $L = n = \log_2 2^n$, so it is ideal for the case where $P(n) = 2^{-L} = 2^{-n}$. The length of code $C_1(n)$ is $L = 1 + 2\lfloor \log_2 n \rfloor = \log_2 2 + \log_2 n^2 = \log_2(2n^2)$, so it is ideal for the case where

$$P(n) = 2^{-L} = \frac{1}{2n^2}.$$

The length of code $C_3(n)$ is

$$L = 1 + \lfloor \log_2 n \rfloor + 2\lfloor \log_2(1 + \lfloor \log_2 n \rfloor)\rfloor = \log_2 2 + 2\lfloor \log \log_2 2n \rfloor + \lfloor \log_2 n \rfloor,$$

so it is ideal for the case where

$$P(n) = 2^{-L} = \frac{1}{2n(\log_2 n)^2}.$$

Table 3.21 shows the ideal probabilities that the first eight positive integers should have for the unary, C_1, and C_3 codes to be used.

More prefix codes for the positive integers, appropriate for special applications, may be designed by the following general approach. Select positive integers v_i and combine them in a list V (which may be finite or infinite according to needs). The code of the positive integer n is prepared in the following steps:

1. Find k such that

$$\sum_{i=1}^{k-1} v_i < n \le \sum_{i=1}^{k} v_i.$$

2. Compute the difference

$$d = n - \left[\sum_{i=1}^{k-1} v_i\right] - 1.$$

n	$B(n)$	$\overline{B}(n)$	C_1	C_2	C_3	C_4
1	1		0|	0	0|	0
2	10	0	10|0	100	100|0	10|0
3	11	1	10|1	110	100|1	11|0
4	100	00	110|00	10100	110|00	10|100|0
5	101	01	110|01	10110	110|01	10|101|0
6	110	10	110|10	11100	110|10	10|110|0
7	111	11	110|11	11110	110|11	10|111|0
8	1000	000	1110|000	1010100	10100|000	11|1000|0
9	1001	001	1110|001	1010110	10100|001	11|1001|0
10	1010	010	1110|010	1011100	10100|010	11|1010|0
11	1011	011	1110|011	1011110	10100|011	11|1011|0
12	1100	100	1110|100	1110100	10100|100	11|1100|0
13	1101	101	1110|101	1110110	10100|101	11|1101|0
14	1110	110	1110|110	1111100	10100|110	11|1110|0
15	1111	111	1110|111	1111110	10100|111	11|1111|0
16	10000	0000	11110|0000	101010100	10110|0000	10|100|10000|0
31	11111	1111	11110|1111	111111110	10110|1111	10|100|11111|0
32	100000	00000	111110|00000	10101010100	11100|00000	10|101|100000|0
63	111111	11111	111110|11111	11111111110	11100|11111	10|101|111111|0
64	1000000	000000	1111110|000000	1010101010100	11110|000000	10|110|1000000|0
127	1111111	111111	1111110|111111	1111111111110	11110|111111	10|110|1111111|0
128	10000000	0000000	11111110|0000000	101010101010100	1010100|0000000	10|111|10000000|0
255	11111111	1111111	11111110|1111111	111111111111110	1010100|1111111	10|111|11111111|0

Table 3.20: Some Prefix Codes.

n	Unary	C_1	C_3
1	0.5	0.5000000	
2	0.25	0.1250000	0.2500000
3	0.125	0.0555556	0.0663454
4	0.0625	0.0312500	0.0312500
5	0.03125	0.0200000	0.0185482
6	0.015625	0.0138889	0.0124713
7	0.0078125	0.0102041	0.0090631
8	0.00390625	0.0078125	0.0069444

Table 3.21: Ideal Probabilities of Eight Integers for Three Codes.

The largest value of n is $\sum_1^k v_i$, so the largest value of d is $\sum_i^k v_i - \left[\sum_1^{k-1} v_i\right] - 1 = v_k - 1$, a number that can be written in $\lceil \log_2 v_k \rceil$ bits. The number d is encoded, using the standard binary code, either in this number of bits, or if $d < 2^{\lceil \log_2 v_k \rceil} - v_k$, it is encoded in $\lfloor \log_2 v_k \rfloor$ bits.
3. Encode n in two parts. Start with k encoded in some prefix code, and concatenate the binary code of d. Since k is coded in a prefix code, any decoder would know how many bits to read for k. After reading and decoding k, the decoder can compute the value $2^{\lceil \log_2 v_k \rceil} - v_k$ which tells it how many bits to read for d.

A simple example is the infinite sequence $V = (1, 2, 4, 8, \ldots, 2^{i-1}, \ldots)$ with k coded in unary. The integer $n = 10$ satisfies

$$\sum_{i=1}^{3} v_i < 10 \leq \sum_{i=1}^{4} v_i,$$

so $k = 4$ (with unary code 1110) and $d = 10 - \left[\sum_{i=1}^{3} v_i\right] - 1 = 2$. Our values v_i are powers of 2, so $\log_2 v_i$ is an integer and $2^{\log_2 v_k}$ equals v_i. Thus, the length of d in our example is $\log_2 v_i = \log_2 8 = 3$ and the code of 10 is 1110|010.

3.10 Ternary Comma Code

Binary (base 2) numbers are based on the two bits 0 and 1. Similarly, ternary (base 3) numbers are based on the three digits (trits) 0, 1, and 2. Each trit can be encoded in two bits, but two bits can have four values. Thus, it makes sense to work with a ternary number system where each trit is represented by two bits and in addition to the three trits there is a fourth symbol, a comma (c). Once we include the c, it becomes easy to construct the ternary comma code for the integers. The comma code of n is simply the ternary representation of $n - 1$ followed by a c. Thus, the comma code of 8 is 21c (because $7 = 2 \cdot 3 + 1$) and the comma code of 18 is 122c (because $17 = 1 \cdot 9 + 2 \cdot 3 + 2$).

Table 3.22 (after [Fenwick 96a]) lists several ternary comma codes (the columns labeled L are the length of the code, in bits). These codes start long (longer than most of the other codes described here) but grow slowly. Thus, they are suitable for applications where large integers are common. These codes are also easy to encode and decode and their principal downside is the comma symbol (signalling the end of a code) that requires two bits. This inefficiency is not serious, but becomes more so for comma codes based on larger number bases. In a base-15 comma code, for example, each of the 15 digits requires four bits and the comma is also a 4-bit pattern. Each code ends with a 4-bit comma, instead of with the theoretical minimum of one bit, and this feature renders such codes inefficient. (However, the overall redundancy per symbol decreases for large number bases. In a base-7 system, one of eight symbols is sacrificed for the comma, while in a base 15 it is one of 16 symbols.)

The ideas above are simple and easy to implement, but they make sense only for long bitstrings. In data compression, as well as in many other applications, we normally need short codes, ranging in size from two to perhaps 12 bits. Because they are used for compression, such codes have to be self-delimiting without adding any extra bits. The

Value	Code	L	Value	Code	L
0	c	2	11	101c	8
1	0c	4	12	102c	8
2	1c	4	13	110c	8
3	2c	4	14	111c	8
4	10c	6	15	112c	8
5	11c	6	16	120c	8
6	12c	6	17	121c	8
7	20c	6	18	122c	8
8	21c	6	19	200c	8
9	22c	6	20	201c	8
...			...		
64	2100c	10	1,000	1101000c	16
128	11201c	12	3,000	11010002c	18
256	100110c	14	10,000	111201100c	20
512	200221c	14	65,536	10022220020c	24

Table 3.22: Ternary Comma Codes and Their Lengths.

solution is to design sets of prefix codes, codes that have the prefix property. Among these codes, the ones that yield the best performance are the universal codes.

In 1840, Thomas Fowler, a self-taught English mathematician and inventor, created a unique ternary calculating machine. Until recently, all detail of this machine was lost. A research project begun in 1997 uncovered sufficient information to enable the recreation of a physical concept model of Fowler's machine. The next step is to create a historically accurate replica.

—[Glusker et al 05]

3.11 Location Based Encoding (LBE)

Location based encoding (LBE) is a simple method for assigning variable-length codes to a set of symbols, not necessarily integers. The originators of this method are P. S. Chitaranjan, Arun Shankar, and K. Niyant [LBE 07]. The LBE encoder is essentially a two-pass algorithm. Given a data file to be compressed, the encoder starts by reading the file and counting symbol frequencies. Following the first pass, it (1) sorts the symbols in descending order of their probabilities, (2) stores them in a special order in a matrix, and (3) assigns them variable-length codes. In between the passes, the encoder writes the symbols, in their order in the matrix, on the compressed stream, for the use of the decoder.

The second pass simply reads the input file again symbol by symbol and replaces each symbol with its code. The decoder starts by reading the sorted sequence of symbols from the compressed stream and constructing the codes in lockstep with the encoder. The decoder then reads the codes off its input and replaces each code with the original symbol. Thus, decoding is fast.

The main idea is to assign short codes to the common symbols by placing the symbols in a matrix in a special diagonal order, vaguely reminiscent of the zigzag order used by JPEG, such that high-probability symbols are concentrated at the top-left corner of the matrix. This is illustrated in Figure 3.23a, where the numbers 1, 2, 3, ... indicate the most-common symbol, the second most-common one, and so on. Each symbol is assigned a variable-length code that indicates its position (row and column) in the matrix. Thus, the code of 1 (the most-common symbol) is 11 (row 1 column 1), the code of 6 is 0011 (row 3 column 1), and the code of 7 is 10001 (row 1 column 4). Each code has two 1's, which makes it trivial for the decoder to read and identify the codes. The length of a code depends on the diagonal (shown in gray in the figure), and the figure shows codes with lengths (also indicated in gray) from two to eight bits.

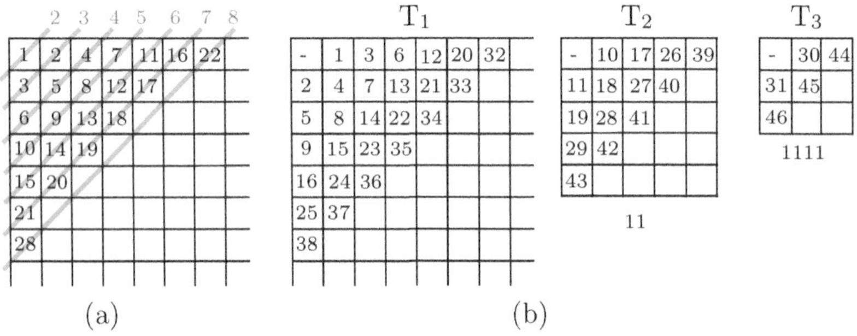

Figure 3.23: LBE Code Tables.

We know that Huffman coding fails for very small alphabets. For a 2-symbol alphabet, the Huffman algorithm assigns 1-bit codes to the two symbols regardless of their probabilities, so no compression is achieved. Given the same alphabet, LBE assigns the two codes 11 and 101 to the symbols, so the average length of its codes is between 2 and 3 (the shorter code is assigned to the most-common symbol, which is why the average code length is in the interval $(2, 2.5]$, still bad). Thus, LBE performs even worse in this case. For alphabets with small numbers of symbols, such as three or four, LBE performs better, but still not as good as Huffman.

The ASCII code assigns 7-bit codes to 128 symbols, and this should be compared with the average code length achievable by LBE. LBE assigns codes of seven or fewer bits to the 21 most-common symbols. Thus, if the alphabet consists of 21 or fewer symbols, replacing them with LBE codes generates compression. For larger alphabets, longer codes are needed, which may cause expansion (i.e., compression where the average code length is greater than seven bits).

It is possible to improve LBE's performance somewhat by constructing several tables, as illustrated in Figure 3.23b. Each code is preceded by its table number such that the codes of table n are preceded by $(2n - 2)$ 1's. The codes of table 1 are not preceded by any 1's, those in table 2 are preceded by 11, those of table 3 are preceded by 1111, and so on. Because of the presence of pairs of 1's, position $(1, 1)$ of a table cannot be used and the first symbol of a table is placed at position $(1, 2)$. The codes are placed in table 1 until a code can become shorter if placed in table 2. This happens with the 10th most-common symbol. Placing it in table 1 would have assigned it the code 100001, whereas placing it in table 2 (which is empty so far) assigns it the shorter code 11|101. The next example is the 12th symbol. Placing it in table 2 would have assigned it code 11|1001, but placing it in table 1 gives it the same-length code 100001. With this scheme it is possible to assign short codes (seven bits or fewer) to the 31 most-common symbols.

Notice that there is no need to actually construct any tables. Both encoder and decoder can determine the codes one by one simply by keeping track of the next available position in each table. Once a code has been computed, it can be stored with other codes in an array or another convenient data structure.

The decoder reads pairs of consecutive 1's until it reaches a 0 or a single 1. This provides the table number, and the rest of the code is read until two 1's (not necessarily consecutive) are found.

The method is simple but not very efficient, especially because the compressed stream has to start with a list of the symbols in ascending order of probabilities. The main advantage of LBE is simple implementation, simple encoding, and fast decoding.

> Die Mathematiker sind eine Art Franzosen: redet man zu ihnen, so uebersetzen sie es in ihre Sprache, und dann ist es also bald ganz etwas anderes.
> (The mathematicians are a sort of Frenchmen: when you talk to them, they immediately translate it into their own language, and right away it is something entirely different.)
>
> —Johann Wolfgang von Goethe

3.12 Stout Codes

In his 1980 short paper [Stout 80], Quentin Stout introduced two families R_l and S_l of recursive, variable-length codes for the integers, similar to and more general than Elias omega and Even–Rodeh codes. The Stout codes are universal and asymptotically optimal. Before reading ahead, the reader is reminded that the length L of the integer n is given by $L = 1 + \lfloor \log_2 n \rfloor$ (Equation (2.1)).

The two families of codes depend on the choice of an integer parameter l. Once a value (greater than or equal to 2) for l has been selected, a codeword in the first family consists of a prefix $R_l(n)$ and a suffix $0n$.

The suffix is the binary value of n in L bits, preceded by a single 0 separator. The prefix $R_l(n)$ consists of length groups. It starts with the length L of n, to which is prepended the length L_1 of L, then the length L_2 of L_1, and so on until a length L_i is reached that is short enough to fit in an l-bit group. Notice that the length groups (except perhaps the leftmost one) start with 1. Thus, the single 0 separator indicates the end of the length groups.

As an example, we select $l = 2$ and encode a 985-bit integer n whose precise value is irrelevant. The suffix is the 986-bit string $0n$ and the prefix starts with the group $L = 985 = 1111011001_2$. The length L_1 of this group is $10_{10} = 1010_2$, the length L_2 of the second group is $4 = 100_2$, and the final length L_3 is $3 = 11_2$. The complete codeword is therefore $11|100|1010|1111011001|0|n$. Notice how the length groups start with 1, which implies that each group is followed by a 1, except the last length group which is followed by a 0.

Decoding is simple. The decoder starts by reading the first l bits. These are the length of the next group. More and more length groups are read, until a group is found that is followed by a 0. This indicates that the next group is n itself.

With this background, the recursive definition of the R_l prefixes is easy to read and understand. We use the notation $L = 1+\lfloor\log_2 n\rfloor$ and denote by $B(n,l)$ the l-bit binary representation (beta code) of the integer n. Thus, $B(12,5) = 01100$. For $l \geq 2$, the prefixes are defined by

$$R_l(n) = \begin{cases} B(n,l), & \text{for } 0 \leq n \leq 2^l - 1, \\ R_l(L)B(n,L), & \text{for } n \geq 2^l. \end{cases}$$

Those familiar with the Even–Rodeh code (Section 3.7) may already have realized that this code is identical to R_3. Furthermore, the Elias omega code (Section 3.4) is in-between R_2 and R_3 with two differences: (1) the omega code encodes the quantities $L_i - 1$ and (2) the 0 separator is placed to the right of n.

$R_2(985) =$ 11 100 1010 1111011001	$R_2(31{,}925) =$ 11 100 1111 111110010110101
$R_3(985) =$ 100 1010 1111011001	$R_3(31{,}925) =$ 100 1111 111110010110101
$R_4(985) =$ 1010 1111011001	$R_4(31{,}925) =$ 1111 111110010110101
$R_5(985) =$ 01010 1111011001	$R_5(31{,}925) =$ 01111 111110010110101
$R_6(985) =$ 001010 1111011001	$R_6(31{,}925) =$ 001111 111110010110101

The short table lists the $R_l(n)$ prefixes for $2 \leq l \leq 6$ and 985-bit and 31,925-bit integers. It is obvious that the shortest prefixes are obtained for parameter values $l = L = 1+\lfloor\log_2 n\rfloor$. Larger values of l result in slightly longer prefixes, while smaller values require more length groups and therefore result in much longer prefixes. Elementary algebraic manipulations indicate that the range of best lengths L for a given parameter l is given by $[2^s, 2^e - 1]$ where $s = 2^{l-1}$ and $e = 2^l - 1 = 2s - 1$ (this is the range of best lengths L, not best integers n). The following table lists these intervals for a few l values and makes it clear that small parameters, such as 2 and 3, are sufficient for most practical applications of data compression. Parameter $l = 2$ is best for integers that are 2 to 7 bits long, while $l = 3$ is best for integers that are 8 to 127 bits long.

l	s	e	2^s	$2^e - 1$
2	2	3	2	7
3	4	7	8	127
4	8	15	128	32,767
5	9	31	32,768	2,147,483,647

The second family of Stout codes is similar, but with different prefixes that are denoted by $S_l(n)$. For small l values, this family offers some improvement over the R_l codes. Specifically, it removes the slight redundancy that exists in the R_l codes because a length group cannot be 0 (which is why a length group in the omega code encodes $L_i - 1$ and not L_i). The S_l prefixes are similar to the R_l prefixes with the difference that a length group for L_i encodes $L_i - 1 - l$. Thus, $S_2(985)$, the prefix of a 985-bit integer n, starts with the 10-bit length group 1111011001 and prepends to it the length group for $10-1-2=7=111_2$. To this is prepended the length group for $3-1-2=0$ as the two bits 00. The result is the 15-bit prefix 00|111|1111011001, shorter than the 19 bits of $R_2(985)$. Another example is $S_3(985)$, which starts with the same 1111011001 and prepends to it the length group for $10-1-3=6=110_2$. The recursion stops at this point because 110 is an l-bit group. The result is the 13-bit codeword 110|1111011001, again shorter than the 17 bits of $R_3(985)$. The $S_l(n)$ prefixes are defined recursively by

$$S_l(n) = \begin{cases} B(n,l), & \text{for } 0 \le n \le 2^l - 1, \\ R_l(L-1-l)B(n,L), & \text{for } n \ge 2^l. \end{cases}$$

Table 3.24 lists some $S_2(n)$ and $S_3(n)$ prefixes and illustrates their regularity. Notice that the leftmost column lists the values of L, i.e., the lengths of the integers being encoded, and not the integers themselves. A length group retains its value until the group that follows it becomes all 1's, at which point the group increments by 1 and the group that follows is reset to $10\ldots0$. All the length groups, except perhaps the leftmost one, start with 1. This regular behavior is the result of the choice $L_i - 1 - l$.

L	$S_2(n)$	$S_3(n)$
1	01	001
2	10	010
3	11	011
4	00 100	100
5	00 101	101
6	00 110	110
7	00 111	111
8	01 1000	000 1000
15	01 1111	000 1111
16	10 10000	001 10000
32	11 100000	010 100000
64	00 100 1000000	011 1000000
128	00 101 10000000	100 10000000
256	00 110 100000000	101 100000000
512	00 111 1000000000	110 1000000000
1024	01 1000 10000000000	111 10000000000
2048	01 1001 100000000000	000 1000 100000000000

Table 3.24: $S_2(n)$ and $S_3(n)$ Codes.

The prefix $S_2(64)$, for example, starts with the 7-bit group $1000000 = 64$ and prepends to it $S_2(7-1-2) = S_2(4) = 00|100$. We stress again that the table lists only the prefixes, not the complete codewords. Once this is understood, it is not hard to see that the second Stout code is a prefix code. Once a codeword has been assigned, it will not be the prefix of any other codeword. Thus, for example, the prefixes of all the codewords for 64-bit integers start with the prefix 00 100 of the 4-bit integers, but any codeword for a 4-bit integer has a 0 following the 00 100, whereas the codewords for the 64-bit integers have a 1 following 00 100.

3.13 Boldi–Vigna (ζ) Codes

The World Wide Web (WWW) was created in 1990 and currently, after only 20 years of existence, it completely pervades our lives. It is used by many people and is steadily growing and finding more applications. Formally, the WWW is a collection of interlinked, hypertext documents (web pages) that can be sent over the Internet. Mathematically, the pages and hyperlinks may be viewed as nodes and edges in a vast directed graph (a webgraph) which itself has become an important object of study. Currently, the graph of the entire WWW consists of hundreds of millions of nodes and over a billion links. Experts often claim that these numbers grow exponentially with time. The various web search engines crawl over the WWW, collecting pages, resolving their hypertext links, and creating large webgraphs. Thus, a search engine operates on a vast data base and it is natural to want to compress this data.

One approach to compressing a webgraph is outlined in [Randall et al. 01]. Each URL is replaced by an integer. These integers are stored in a list that includes, for each integer, a pointer to an adjacency list of links. Thus, for example, URL 104 may be a web page that links to URLs 101, 132, and 174. These three integers become the adjacency list of 104, and this list is compressed by computing the differences of adjacent links and replacing the differences with appropriate variable-length codes. In our example, the differences are $101 - 104 = -3$, $132 - 101 = 31$, and $174 - 132 = 42$.

Experiments with large webgraphs indicate that the differences (also called gaps or deltas) computed for a large webgraph tend to be distributed according to a power law whose parameter in normally in the interval $[1.1, 1.3]$, so the problem of compressing a webgraph is reduced to the problem of finding a variable-length code that corresponds to such a distribution.

A power law distribution has the form

$$Z_\alpha[n] = \frac{P}{n^\alpha}$$

where $Z_\alpha[n]$ is the distribution (the number of occurrences of the value n), α is a parameter of the distribution, and P is a constant of proportionality. Codes for power law distributions have already been mentioned. The length of the Elias gamma code (Equation (3.1)) is $2\lfloor\log_2 n\rfloor + 1$, which makes it a natural choice for a distribution of the form $1/(2n^2)$. This is a power law distribution with $\alpha = 2$ and $P = 1/2$. Similarly, the Elias delta code is suitable for a distribution of the form $1/[2n(\log_2(2n))^2]$. This is very close to a power law distribution with $\alpha = 1$. The nibble code and its variations (page 92) correspond to power law distributions of the form $1/n^{1+\frac{1}{k}}$, where the parameter is $1 + \frac{1}{k}$.

The remainder of this section describes the zeta (ζ) code, also known as Boldi–Vigna code, introduced by Paolo Boldi and Sebastiano Vigna as a family of variable-length codes that are best choices for the compression of webgraphs. The original references are [Boldi and Vigna 04a] and [Boldi and Vigna 04b]. The latest reference is [Boldi and Vigna 05].

We start with Zipf's law, an empirical power law [Zipf 07] introduced by the linguist George K. Zipf. It states that the frequency of any word in a natural language is roughly inversely proportional to its position in the frequency table. Intuitively, Zipf's law states that the most frequent word in any language is twice as frequent as the second most-frequent word, which in turn is twice as frequent as the third word, and so on.

For a language having N words, the Zipf power law distribution with a parameter α is given by

$$Z_\alpha[n] = \frac{1/n^\alpha}{\sum_{i=1}^{N} \frac{1}{i^\alpha}}$$

If the set is infinite, the denominator becomes the well-known Riemann zeta function

$$\zeta(\alpha) = \sum_{i=1}^{\infty} \frac{1}{i^\alpha}$$

which converges to a finite value for $\alpha > 1$. In this case, the distribution can be written in the form

$$Z_\alpha[n] = \frac{1}{\zeta(\alpha)n^\alpha}$$

The Boldi–Vigna zeta code starts with a positive integer k that becomes the shrinking factor of the code. The set of all positive integers is partitioned into the intervals $[2^0, 2^k - 1]$, $[2^k, 2^{2k} - 1]$, $[2^{2k}, 2^{3k} - 1]$, and in general $[2^{hk}, 2^{(h+1)k} - 1]$. The length of each interval is $2^{(h+1)k} - 2^{hk}$.

Next, a minimal binary code is defined, which is closely related to the phased-in codes of Section 2.9. Given an interval $[0, z-1]$ and an integer x in this interval, we first compute $s = \lceil \log_2 z \rceil$. If $x < 2^s - z$, it is coded as the xth element of the interval, in $s-1$ bits. Otherwise, it is coded as the $(x - z - 2^s)$th element of the interval in s bits.

With this background, here is how the zeta code is constructed. Given a positive integer n to be encoded, we employ k to determine the interval where n is located. Once this is known, the values of h and k are used in a simple way to construct the zeta code of n in two parts, the value of $h+1$ in unary (as h zeros followed by a 1), followed by the minimal binary code of $n - 2^{hk}$ in the interval $[0, 2^{(h+1)k} - 2^{hk} - 1]$.

Example. Given $k = 3$ and $n = 16$, we first determine that n is located in the interval $[2^3, 2^6 - 1]$, which corresponds to $h = 1$. Thus, $h + 1 = 2$ and the unary code of 2 is 01. The minimal binary code of $16 - 2^3 = 8$ is constructed in steps as follows. The length z of the interval $[2^3, 2^6 - 1]$ is 56. This implies that $s = \lceil \log_2 56 \rceil = 6$. The value 8 to be encoded satisfies $8 = 2^6 - 56$, so it is encoded as $x - z - 2^s = 8 - 56 - 2^6 = 16$ in six bits, resulting in 010000. Thus, the ζ_3 code of $n = 16$ is 01|010000.

Table 3.25 lists ζ codes for various shrinking factors k. For $k = 1$, the ζ_1 code is identical to the γ code. The nibble code of page 92 is also shown.

A nibble (more accurately, nybble) is the popular term for a 4-bit number (or half a byte). A nibble can therefore have 16 possible values, which makes it identical to a single hexadecimal digit

The length of the zeta code of n with shrinking factor k is $\lfloor 1 + (\log_2 n)/k \rfloor (k+1) + \tau(n)$ where

$$\tau(n) = \begin{cases} 0, & \text{if } (\log_2 n)/k - \lfloor (\log_2 n)/k \rfloor \in [0, 1/k), \\ 1, & \text{otherwise.} \end{cases}$$

n	$\gamma = \zeta_1$	ζ_2	ζ_3	ζ_4	δ	Nibble
1	1	10	100	1000	1	1000
2	010	110	1010	10010	0100	1001
3	011	111	1011	10011	0101	1010
4	00100	01000	1100	10100	01100	1011
5	00101	01001	1101	10101	01101	1100
6	00110	01010	1110	10110	01110	1101
7	00111	01011	1111	10111	01111	1110
8	0001000	011000	0100000	11000	00100000	1111
9	0001001	011001	0100001	11001	00100001	00011000
10	0001010	011010	0100010	11010	00100010	00011001
11	0001011	011011	0100011	11011	00100011	00011010
12	0001100	011100	0100100	11100	00100100	00011011
13	0001101	011101	0100101	11101	00100101	00011100
14	0001110	011110	0100110	11110	00100110	00011101
15	0001111	011111	0100111	11111	00100111	00011110
16	000010000	00100000	01010000	010000111	001011001	00011111

Table 3.25: ζ-Codes for $1 \le k \le 4$ Compared to γ, δ and Nibble Codes.

Thus, this code is ideal for integers n that are distributed as

$$\frac{1+\tau(n)}{n^{1+\frac{1}{k}}}.$$

This is very close to a power law distribution with a parameter $1+\frac{1}{k}$. The developers show that the zeta code is complete, and they provide a detailed analysis of the expected lengths of the zeta code for various values of k. The final results are summarized in Table 3.26 where certain codes are recommended for various ranges of the parameter α of the distribution.

α :	<1.06	$[1.06, 1.08]$	$[1.08, 1.11]$	$[1.11, 1.16]$	$[1.16, 1.27]$	$[1.27, 1.57]$	$[1.57, 2]$
Code:	δ	ζ_6	ζ_5	ζ_4	ζ_3	ζ_2	$\gamma = \zeta_1$

Table 3.26: Recommended Ranges for Codes.

3.14 Yamamoto's Recursive Code

In 2000, Hirosuke Yamamoto came up with a simple, ingenious way [Yamamoto 00] to improve the Elias omega code. As a short refresher on the omega code, we start with the relevant paragraph from Section 3.4.

> The omega code uses itself recursively to encode the prefix M, which is why it is sometimes referred to as a recursive Elias code. The main idea is to prepend the length of n to n as a group of bits that starts with a 1, then prepend the length of the length, as another group, to the result, and continue prepending lengths until the last length is 2 or 3 (and it fits in two bits). In order to distinguish between a length group and the last, rightmost group (that of n itself), the latter is followed by a delimiter of 0, while each length group starts with a 1.

The decoder of the omega code reads the first two bits (the leftmost length group), interprets its value as the length of the next group, and continues reading groups until a group is found that is followed by the 0 delimiter. That group is the binary representation of n.

This scheme makes it easy to decode the omega code, but is somewhat wasteful, because the MSB of each length group is 1, and so the value of this bit is known in advance and it acts only as a separator.

Yamamoto's idea was to select an f-bit delimiter a (where f is a small positive integer, typically 2 or 3) and use it as the delimiter instead of a single 0. In order to obtain a UD code, none of the length groups should start with a. Thus, the length groups should be encoded with special variable-length codes that do not start with a. Once a has been selected and its length f is known, the first step is to prepare all the binary strings that do not start with a and assign each string as the code $\tilde{B}_{a,f}()$ of one of the positive integers. Now, if a length group has the value n, then $\tilde{B}_{a,f}(n)$ is placed in the codeword instead of the binary value of n.

We start with a simpler code that is denoted by $B_{a,f}(n)$. As an example, we select $f=2$ and the 2-bit delimiter $a=00$. The binary strings that do not start with 00 are the following: the two 1-bit strings 0 and 1; the three 2-bit strings 01, 10, and 11; the six 3-bit strings 010, 011, 100, 101, 110, and 111; the 12 4-bit strings 0100, 1001, through 1111 (i.e., the 16 4-bit strings minus the four that start with 00), and so on. These strings are assigned to the positive integers as listed in the second column of Table 3.27. The third column of this table lists the first 26 B codes for $a=100$.

One more step is needed. Even though none of the $B_{a,f}(n)$ codes starts with a, it may happen that a short $B_{a,f}(n)$ code (a code whose length is less than f) followed by another $B_{a,f}(n)$ code will accidentally contain the pattern a. Thus, for example, the table shows that $B_{100,3}(5)=10$ followed by $B_{100,3}(7)=000$ becomes the string 10|000 and may be interpreted by the decoder as 100|00.... Thus, the $B_{100,3}(n)$ codes have to be made UD by eliminating some of the short codes, those that are shorter than f bits and have the values 1 or 10. Similarly, the single short $B_{00,2}(1)=0$ code should be discarded. The resulting codes are designated $\tilde{B}_{a,f}(n)$ and are listed in the last two columns of the table.

If the integers to be encoded are known not to exceed a few hundred or a few thousand, then both encoder and decoder can have built-in tables of the $\tilde{B}_{a,f}(n)$ codes

n	$B_{a,f}(n)$		$\tilde{B}_{a,f}(n)$	
	$a=00$	$a=100$	$a=00$	$a=100$
1	0	0	1	0
2	1	1	01	00
3	01	00	10	01
4	10	01	11	11
5	11	10	010	000
6	010	11	011	001
7	011	000	100	010
8	100	001	101	011
9	101	010	110	101
10	110	011	111	110
11	111	101	0100	111
12	0100	110	0101	0000
13	0101	111	0110	0001
14	0110	0000	0111	0010
15	0111	0001	1000	0011
16	1000	0010	1001	0100
17	1001	0011	1010	0101
18	1010	0100	1011	0110
19	1011	0101	1100	0111
20	1100	0110	1101	1010
21	1101	0111	1110	1011
22	1110	1010	1111	1100
23	1111	1011	01000	1101
24	01000	1100	01001	1110
25	01001	1101	01010	1111
26	01010	1110	01011	00000

Table 3.27: Some $B_{a,f}(n)$ and $\tilde{B}_{a,f}(n)$ Yamamoto Codes.

for all the relevant integers. In general, these codes have to be computed "on the fly" and the developer provides the following expression for them (the notation $K^{[j]}$ means the integer K expressed in j bits, where some of the leftmost bits may be zeros).

$$\tilde{B}_{a,f}(n) = \begin{cases} [n - M(j,f) + L(j,f)]^{[j]}, \\ \qquad \text{if } M(j,f) - L(j,f) \le n < M(j,f) - L(j,f) + N(j,f,a), \\ [n - M(j,f-1) + L(j+1,f)]^{[j]}, \\ \qquad \text{if } M(j,f) - L(j,f) + N(j,f,a) \le n < M(j+1,f) - L(j+1,f), \end{cases}$$

where $M(j,f) = 2^j - 2^{(j-f)_+}$, $(t)_+ = \max(t,0)$, $L(j,f) = (f-1) - (f-j)_+$, and $N(j,f,a) = \lfloor 2^{j-f}a \rfloor$. Given a positive integer n, the first step in encoding it is to determine the value of j by examining the inequalities. Once j is known, functions M and L can be computed.

Armed with the $\tilde{B}_{a,f}(n)$ codes, the recursive Yamamoto code $C_{a,f}(n)$ is easy to describe. We select an f-bit delimiter a. Given a positive integer n, we start with the group $\tilde{B}_{a,f}(n)$. Assuming that this group is n_1 bits long, we prepend to it the length group $\tilde{B}_{a,f}(n_1 - 1)$. If this group is n_2 bits long, we prepend to it the length group $\tilde{B}_{a,f}(n_2 - 1)$. This is repeated recursively until the last length group is $\tilde{B}_{a,f}(1)$. (This code is always a single bit because it depends only on a, and the decoder knows this bit because it knows a. Thus, in principle, the last length group can be omitted, thereby shortening the codeword by one bit.) The codeword is completed by appending the delimiter a to the right end. Table 3.28 lists some codewords for $a = 00$ and for $a = 100$.

n	$a = 00$	$a = 100$
1	1 00	0 100
2	1 01 00	0 00 100
3	1 10 00	0 01 100
4	1 11 00	0 11 100
5	1 01 010 00	0 00 000 100
6	1 01 011 00	0 00 001 100
7	1 01 100 00	0 00 010 100
8	1 01 101 00	0 00 011 100
9	1 01 110 00	0 00 101 100
10	1 01 111 00	0 00 110 100
11	1 10 0100 00	0 00 111 100
12	1 10 0101 00	0 01 0000 100
13	1 10 0110 00	0 01 0001 100
14	1 10 0111 00	0 01 0010 100
15	1 10 1000 00	0 01 0011 100
16	1 10 1001 00	0 01 0100 100
17	1 10 1010 00	0 01 0101 100
18	1 10 1011 00	0 01 0110 100
19	1 10 1100 00	0 01 0111 100
20	1 10 1101 00	0 01 1010 100
21	1 10 1110 00	0 01 1011 100
22	1 10 1111 00	0 01 1100 100
23	1 11 01000 00	0 01 1101 100
24	1 11 01001 00	0 01 1110 100
25	1 11 01010 00	0 01 1111 100
26	1 11 01011 00	0 11 00000 100

Table 3.28: Yamamoto Recursive Codewords.

The developer, Hirosuke Yamamoto, also proves that the length of a codeword $C_{a,f}(n)$ is always less than or equal to $\log_2^* n + F(f)w_f(n) + c_f + \delta(n)$. More importantly, for infinitely many integers, the length is also less than or equal to $\log_2^* n + (1 - F(f))w_f(n) + c_f + 2\delta(n)$, where $F(f) = -\log_2(1 - 2^{-f})$, $c_f = 5(f-2)_+ + f + 5F(f)$,

$$\delta(n) \le \frac{\log_2 e}{n}\left[1 + \frac{(w(n)-1)(\log_2 e)^{w(n)-1}}{\log_2 n}\right] \le \frac{4.7}{n},$$

and the log-star function $\log_2^* n$ is the finite sum

$$\log_2 n + \log_2 \log_2 n + \cdots + \underbrace{\log_2 \log_2 \ldots \log_2 n}_{w(n)}$$

where $w(n)$ is the largest integer for which the compound logarithm is still nonnegative.

The important point (at least from a theoretical point of view) is that this recursive code is shorter than $\log_2^* n$ for infinitely many integers. This code can also be extended to other number bases and is not limited to base-2 numbers.

3.15 VLCs and Search Trees

There is an interesting association between certain unbounded searches and prefix codes. A search is a classic problem in computer science. Given a data structure, such as an array, a list, a tree, or a graph, where each node contains an item, the problem is to search the structure and find a given item in the smallest possible number of steps. The following are practical examples of computer searches.

1. A two-player game. Player A chooses a positive integer n and player B has to guess n by asking only the following type of question: Is the integer i less than n? Clearly, the number of questions needed to find n depends on n and on the strategy (algorithm) used by B.

2. Given a function $y = f(x)$, find all the values of x for which the function equals a given value Y (typically zero).

3. An online business stocks many items. Customers should be able to search for an item by its name, price range, or manufacturer. The items are the nodes of a large structure (normally a tree), and there must be a fast search algorithm that can find any item among many thousands of items in a few seconds.

4. An Internet search engine faces a similar problem. The first part of such an engine is a crawler that locates a vast number of Web sites, inputs them into a data base, and indexes every term found. The second part is a search algorithm that should be able to locate any term (out of many millions of terms) in just a few seconds. (There is also the problem of ranking the results.) Internet users search for terms (such as "variable-length codes") and naturally prefer search engines that provide results in just a few seconds.

Searching a data base, even a very large one, is considered a bounded search, because the search space is finite. Bounded searches have been a popular field of research for many years, and many efficient search methods are known and have been analyzed in detail. In contrast, searching for a zero of a function is an example of unbounded search, because the domain of the function is normally infinite. In their work [Bentley and Yao 76], Jon Bentley and Andrew Yao propose several algorithms for searching a linearly ordered unbounded table and show that each algorithm is associated with a binary string that can be considered a codeword in a prefix code. This creates an interesting and unexpected relation between searching and variable-length codes, two hitherto unrelated areas of scientific endeavor. The authors restrict themselves to algorithms that search for an integer n in an ordered table $F(i)$ using only elementary tests of the form $F(i) < n$.

The simplest search method, for both bounded and unbounded searches, is linear search. Given an ordered array of integers $F(i)$ and an integer n, a linear search performs the tests $F(1) < n$, $F(2) < n$,... until a match is found. The answers to the tests are No, No,..., Yes, which can be expressed as the bitstring 11...10. Thus, a linear search (referred to by its developers as algorithm B_0) corresponds to the unary code.

The next search algorithm (designated B_1) is an unbounded variation of the well-known binary search. The first step of this version is a linear search that determines an interval. This step starts with an interval $[F(a), F(b)]$ and performs the test $F(b) < n$. If the answer is No, then n is in this interval, and the algorithm executes its second step, where n is located in this interval with a bounded binary search. If the answer is Yes, then the algorithm computes another interval that starts at $F(b+1)$ and is twice as wide as the previous interval. Specifically, the algorithm computes $F(2^i - 1)$ for $i = 1$, 2,... and each interval is $[F(2^{i-1}), F(2^i - 1)]$.

Figure 3.29 (compare with Figures 2.15 and 2.16) shows how these intervals can be represented as a binary search tree where the root and each of its right descendants correspond to an index $2^i - 1$. The left son of each descendant is the root of a subtree that contains the remaining indexes less than or equal to $2^i - 1$. Thus, the left son of the root is index 1. The left son of node 3 is the root of a subtree containing the remaining indexes that are less than or equal to 3, namely 2 and 3. The left son of 7 is the root of a subtree with 4, 5, 6, and 7, and so on.

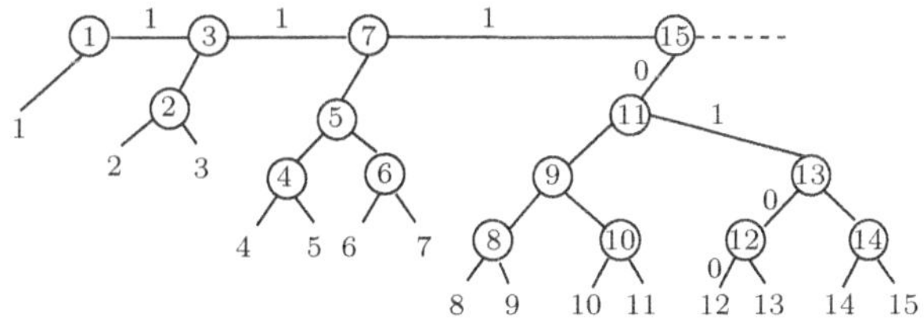

Figure 3.29: Gamma Code and Binary Search.

The figure also illustrates how $n = 12$ is searched for and located by the B_1 method. The algorithm compares 12 to 1, 3, 7, and 15 (a linear search), with results No, No, No, and Yes, or 1110. The next step performs a binary search to locate 12 in the interval [8, 15] with results 100. These results are listed as labels on the appropriate edges. The final bitstring associated with 12 by this method is therefore 1110|100. This result is the Elias gamma code of 12, as produced by the alternative construction method of page 101 (this construction produces codes different from those of Table 3.10, but the code lengths are the same).

If m is the largest integer such that $2^{m-1} \le n < 2^m$, then the first step (a linear search to determine m) requires $m = \lfloor \log_2 n \rfloor + 1$ tests. The second step requires $\log_2 2^{m-1} = m - 1 = 2\lfloor \log_2 n \rfloor$ tests. The total number of tests is therefore $2\lfloor \log_2 n \rfloor + 1$, which is the length of the gamma code (Equation (3.1)).

The third search method proposed by Bentley and Yao (their algorithm B_2) is a double binary search. The second step is the same, but the first step is modified to determine m by constructing intervals that become longer and longer. The intervals computed by method B_1 start at $2^i - 1$ for $i = 1, 2, 3, \ldots$, or $1, 3, 7, 15, 31, \ldots$. Method B_2 constructs intervals that start with numbers of the form $2^{2^i-1} - 1$ for $i = 1, 2, 3, \ldots$. These numbers are 1, 7, 127, 32,767, $2^{31} - 1$, and so on. They become the roots of subtrees as shown in Figure 3.30. There are fewer search trees, which speeds up step 1, but they grow very fast. The developers show that the first step requires $1 + 2\lfloor \log_2 n \rfloor$ tests (compared to the $m = \lfloor \log_2 n \rfloor + 1$ tests required by B_1) and the second step requires $\lfloor \log_2 n \rfloor$ tests (same as B_1), for a total of

$$\lfloor \log_2 n \rfloor + 2\lfloor \log_2(\lfloor \log_2 n \rfloor + 1) \rfloor + 1 = 1 + \lfloor \log_2 n \rfloor + 2\lfloor \log_2 \log_2(2n) \rfloor.$$

The number of tests is identical to the length of the Elias delta code, Equation (3.2). Thus, the variable-length code associated with this search tree is a variant of the delta code. The codewords are different from those listed in Table 3.12, but the code is equivalent because of the identical lengths.

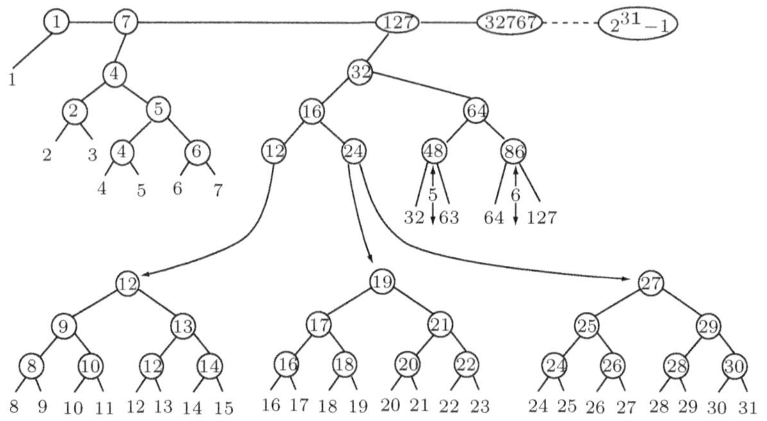

Figure 3.30: Delta Code and Binary Search.

Bentley and Yao go on to propose unbounded search algorithms B_k (k-nested binary search) and U (ultimate). The latter is shown by [Ahlswede et al. 97] to be associated with a modified Elias omega code. The authors of this paper also construct the binary search trees for the Stout codes (Section 3.12).

3.16 Taboo Codes

The taboo approach to variable-length codes, as well as the use of the term "taboo," are the brainchilds of Steven Pigeon. The two types of taboo codes are described in [Pigeon 01a,b] where it is shown that they are universal (more accurately, they can be made as close to universal as desired by the choice of a parameter). The principle of the taboo codes is to select a positive integer parameter n and reserve a pattern of n bits to indicate the end of a code. This pattern should not appear in the code itself, which is the reason for the term taboo. Thus, the taboo codes can be considered suffix codes.

The first type of taboo code is block-based and its length is a multiple of n. The block-based taboo code of an integer is a string of n-bit blocks, where n is a user-selected parameter and the last block is a taboo bit pattern that cannot appear in any of the other blocks. An n-bit block can have 2^n values, so if one value is reserved for the taboo pattern, each of the remaining code blocks can have one of the remaining $2^n - 1$ bit patterns. In the second type of taboo codes, the total length of the code is not restricted to a multiple of n. This type is called unconstrained and is shown to be related to the n-step Fibonacci numbers.

We use the notation $\langle n \rangle : t$ to denote a string of n bits concatenated with the taboo string t. Table 3.31 lists the lengths, number of codes, and code ranges as the codes get longer when more blocks are added. Each row in this table summarizes the properties of a *range* of codes. The number of codes in the kth range is $(2^n - 1)^k$, and the total number of codes $g_n(k)$ in the first k ranges is obtained as the sum of a geometric progression

$$g_n(k) = \sum_{i=1}^{k} (2^n - 1)^i = \frac{[(2^n - 1)^k - 1](2^n - 1)}{2^n - 2}.$$

Codes	Length	# of values	Range
$\langle n:t \rangle$	$2n$	$2^n - 1$	0 to $(2^n - 1) - 1$
$\langle 2n:t \rangle$	$3n$	$(2^n - 1)^2$	$(2^n - 1)$ to $(2^n - 1) + (2^n - 1)^2 - 1$
$\vdots$			$\vdots$
$\langle kn:t \rangle$	$(k+1)n$	$(2^n - 1)^k$	$\sum_{i=1}^{k-1} (2^n - 1)^i$ to $-1 + \sum_{i=1}^{k} (2^n - 1)^i$
$\vdots$			$\vdots$

Table 3.31: Structure of Block-Based Taboo Codes.

The case $n = 1$ is special and uninteresting. A 1-bit block can either be a 0 or a 1. If we reserve 0 as the taboo pattern, then all the other blocks of codes cannot contain any zeros and must be all 1's. The result is the infinite set of codewords **10**, 11**0**, 111**0**,..., which is the unary code and therefore not universal (more accurately, the unary code is ∞-universal). The first interesting case is $n = 2$. Table 3.32 lists some codes for this

m	Code	m	Code	m	Code	m	Code
0	01 **00**	4	01 10 **00**	8	10 11 **00**	12	01 01 01 **00**
1	10 **00**	5	01 11 **00**	9	11 01 **00**	13	01 01 10 **00**
2	11 **00**	6	10 01 **00**	10	11 10 **00**	14	01 01 11 **00**
3	01 01 **00**	7	10 10 **00**	11	11 11 **00**	...	

Table 3.32: Some Block-Based Taboo Codes for $n = 2$.

case. Notice that the pattern of all zeros is reserved for the taboo, but any other n-bit pattern would do as well.

In order to use this code in practice, we need simple, fast algorithms to encode and decode it. Given an integer m, we first determine the number of n-bit blocks in the code of m. Assuming that m appears in the kth range of Table 3.31, we can write the basic relation $m \le g_n(k) - 1$. This can be written explicitly as

$$m \le \frac{[(2^n-1)^k - 1](2^n-1)}{2^n-2} - 1$$

or

$$k \ge \frac{\log_2\left[m + 2 - \frac{m+1}{2^n-1}\right]}{\log_2(2^n-1)}.$$

The requirement that m be less than $g_n(k)$ yields the value of k (which depends on both n and m) as

$$k_n(m) = \left\lceil \frac{\log_2\left[m + 2 - \frac{m+1}{2^n-1}\right]}{\log_2(2^n-1)} \right\rceil.$$

Thus, the code of m consists of $k_n(m)$ blocks of n bits each plus the n-bit taboo block, for a total of $[k_n(m)+1]n$ bits. This code can be considered the value of the integer $c \stackrel{\text{def}}{=} m - g_n(k_n(m) - 1)$ represented in base $(2^n - 1)$, where each "digit" is n bits long.

Encoding: Armed with the values of n, m, and c, we can write the individual values of the n-bit blocks b_i that constitute the code of m as

$$b_i = \left[\lceil c(2^n-1)^i \rceil \bmod (2^n-1)\right] + 1, \quad i = 0, 1, \ldots, k_n(m) - 1. \tag{3.3}$$

The taboo pattern is the $(k_n(m)+1)$th block, block $b_{k_n(m)}$. Equation (3.3) is the basic relation for encoding integer m.

Decoding: The decoder knows the value of n and the taboo pattern. It reads n-bit blocks b_i until it finds the taboo block. The number of blocks read is k (rather $k_n(m)$) and once k is known, decoding is done by

$$m = g_n(k-1) + \sum_{i=0}^{k-1} (b_i - 1)(2^n-1)^i. \tag{3.4}$$

Equations (3.3) and (3.4) lend themselves to fast, efficient implementation. In particular, decoding is easy. The decoder can read n-bit blocks b_i and add terms to the partial sum of Equation (3.4) until it reaches the taboo block, at which point it only has to compute and add $g_n\bigl(k_n(m)-1\bigr)$ to the sum in order to obtain m.

The block-based taboo code is longer than the binary representation (beta code) of the integers by $n+k\bigl[n-\log_2(2^n-1)\bigr]$ bits. For all values of n except the smallest ones, $2^n-1\approx 2^n$, so this difference of lengths (the "waste" of the code) equals approximately n. Even for $n=4$, the waste is only $n+0.093k$. Thus, for large integers m, the waste is very small compared to m.

The developer of these codes, Steven Pigeon, proves that the block-based taboo code can be made as close to universal as desired.

> Let us overthrow the totems, break the taboos. Or better, let us consider them cancelled. Coldly, let us be intelligent.
>
> —Pierre Trudeau

Now, for the unconstrained taboo codes. The idea is to have a bitstring of length l, the actual code, followed by a taboo pattern of n bits. The length l of the code is independent of n and can even be zero. Thus, the total length of such a code is $l+n$ for $l\geq 0$. The taboo pattern should not appear inside the l code bits and there should be a separator bit at the end of the l-bit code part, to indicate to the decoder the start of the taboo. In the block-based taboo codes, any bit pattern could serve as the taboo, but for the unconstrained codes we have to choose a bit pattern that will prevent ambiguous decoding. The two ideal patterns for the taboo are therefore all zeros or all 1's, because a string of all zeros can be separated from the preceding l bits by a single 1, and similarly for a string of all 1's. Other strings require longer separators. Thus, the use of all zeros or 1's as the taboo string leads to the maximum number of l-bit valid code strings and we will therefore assume that the taboo pattern consists of n consecutive zeros.

As a result, an unconstrained taboo code starts with l bits (where $l\geq 0$) that do not include n consecutive zeros and that end with a separator bit of 1 (except when $l=0$, where the code part consists of no bits and there is no need for a separator). The case $n=1$ is special. In this case, the taboo is a single zero, the l bits preceding it cannot include any zeros, so the unconstrained code reduces to the unary code of l 1's followed by a zero.

In order to figure out how to encode and decode these codes, we first have to determine the number of valid bit patterns of l bits, i.e., patterns that do not include any n consecutive zeros. We distinguish three cases.

1. The case $l=0$ is trivial. The number of bit patterns of length zero is one, the pattern of no bits.

2. The case $0<l<n$ is simple. The string of l bits is too short to have any n consecutive zeros. The last bit must be a 1, so the remaining $l-1$ bits can have 2^{l-1} values.

3. When $l\geq n$, we determine the number of valid l-bit strings (i.e., l-bit strings that do not contain any substrings of n zeros) recursively. We denote the number of valid strings by $\langle\!\langle {l \atop n} \rangle\!\rangle$ and consider the following cases. When a valid l-bit string starts with a 1 followed by $l-1$ bits, the number of valid strings is $\langle\!\langle {l-1 \atop n} \rangle\!\rangle$. When such a string start with a 01, the number of valid strings is $\langle\!\langle {l-2 \atop n} \rangle\!\rangle$. We continue in this way until

we reach strings that start with $\underbrace{00\ldots0}_{n-1}1$, where the number of valid strings is $\langle\!\langle {l-n \atop n} \rangle\!\rangle$.

Thus, the number of valid l-bit strings in this case is the sum shown in the third case of Equation (3.5).

$$\left\langle\!\!\left\langle {l \atop n} \right\rangle\!\!\right\rangle = \begin{cases} 1 & \text{if } l = 0, \\ 2^{l-1} & \text{if } l < n, \\ \sum_{i=1}^{n} \langle\!\langle {l-i \atop n} \rangle\!\rangle & \text{otherwise.} \end{cases} \tag{3.5}$$

Table 3.33 lists some values of $\langle\!\langle {l \atop n} \rangle\!\rangle$ and it is obvious that the second column (the values of $\langle\!\langle {l \atop 2} \rangle\!\rangle$) consists of the well-known Fibonacci numbers. Thus, $\langle\!\langle {l \atop 2} \rangle\!\rangle = F_{l+1}$. A deeper look at the table shows that $\langle\!\langle {l \atop 3} \rangle\!\rangle = F^{(3)}_{l+1}$, where $F^{(3)}_{l} = F^{(3)}_{l-1} + F^{(3)}_{l-2} + F^{(3)}_{l-3}$. This column consists of the n-step Fibonacci numbers of order 3 (tribonaccis), a sequence that starts with $F^{(3)}_1 = 1$, $F^{(3)}_2 = 1$, and $F^{(3)}_3 = 2$. Similarly, the 4th column consists of tetrabonacci numbers, and the remaining columns feature the pentanacci, hexanacci, heptanacci, and other n-step "polynacci" numbers.

	$\langle\!\langle {l \atop 1} \rangle\!\rangle$	$\langle\!\langle {l \atop 2} \rangle\!\rangle$	$\langle\!\langle {l \atop 3} \rangle\!\rangle$	$\langle\!\langle {l \atop 4} \rangle\!\rangle$	$\langle\!\langle {l \atop 5} \rangle\!\rangle$	$\langle\!\langle {l \atop 6} \rangle\!\rangle$	$\langle\!\langle {l \atop 7} \rangle\!\rangle$	$\langle\!\langle {l \atop 8} \rangle\!\rangle$
$\langle\!\langle {0 \atop n} \rangle\!\rangle$	1	1	1	1	1	1	1	1
$\langle\!\langle {1 \atop n} \rangle\!\rangle$	1	1	1	1	1	1	1	1
$\langle\!\langle {2 \atop n} \rangle\!\rangle$	1	2	2	2	2	2	2	2
$\langle\!\langle {3 \atop n} \rangle\!\rangle$	1	3	4	4	4	4	4	4
$\langle\!\langle {4 \atop n} \rangle\!\rangle$	1	5	7	8	8	8	8	8
$\langle\!\langle {5 \atop n} \rangle\!\rangle$	1	8	13	15	16	16	16	16
$\langle\!\langle {6 \atop n} \rangle\!\rangle$	1	13	24	29	31	32	32	32
$\langle\!\langle {7 \atop n} \rangle\!\rangle$	1	21	44	56	61	63	64	64
$\langle\!\langle {8 \atop n} \rangle\!\rangle$	1	34	81	108	120	125	127	128
$\langle\!\langle {9 \atop n} \rangle\!\rangle$	1	55	149	208	236	248	253	255
$\langle\!\langle {10 \atop n} \rangle\!\rangle$	1	89	274	401	464	492	504	509

Table 3.33: The First Few Values of $\langle\!\langle {l \atop n} \rangle\!\rangle$.

The n-step Fibonacci numbers of order k are defined recursively by

$$F^{(k)}_l = \sum_{i=1}^{k} F(k)_{l-i}, \tag{3.6}$$

with initial conditions $F(k)_1 = 1$ and $F(k)_i = 2^{i-2}$ for $i = 2, 3, \ldots, k$. (As an aside, the limit $\lim_{k\to\infty} F^{(k)}$ is a sequence that starts with an infinite number of 1's, followed by the powers of 2. The interested reader should also look for anti-Fibonacci numbers, generalized Fibonacci numbers, and other sequences and numbers related to Fibonacci.) We therefore conclude that $\langle\!\langle {l \atop n} \rangle\!\rangle = F^{(n)}_{l+1}$, a relation that is exploited by the developer of these codes to prove that the unconstrained taboo codes are universal (more accurately, can be made as close to universal as desired by increasing n).

Table 3.34 lists the organization, lengths, number of codes, and code ranges of some unconstrained taboo codes, and Table 3.35 lists some of these codes for $n = 3$ (the taboo string is shown in boldface).

Codes	Length	# of values	Range
t	n	$\langle\langle {0 \atop n} \rangle\rangle$	0 to $\langle\langle {0 \atop n} \rangle\rangle - 1$
$\langle 1\rangle : t$	$1+n$	$\langle\langle {1 \atop n} \rangle\rangle$	$\langle\langle {0 \atop n} \rangle\rangle$ to $\langle\langle {0 \atop n} \rangle\rangle + \langle\langle {1 \atop n} \rangle\rangle - 1$
$\vdots$			$\vdots$
$\langle l\rangle : t$	$l+n$	$\langle\langle {l \atop n} \rangle\rangle$	$\sum_{i=0}^{l-1} \langle\langle {i \atop n} \rangle\rangle$ to $-1 + \sum_{i=0}^{l} \langle\langle {i \atop n} \rangle\rangle$
$\vdots$			$\vdots$

Table 3.34: Organization and Features of Unconstrained Taboo Codes.

m	Code	m	Code	m	Code	m	Code
0	**000**	7	111**000**	14	1111**000**	21	10011**000**
1	1**000**	8	0011**000**	15	00101**000**	22	10101**000**
2	01**000**	9	0101**000**	16	00111**000**	23	10111**000**
3	11**000**	10	0111**000**	17	01001**000**	24	11001**000**
4	001**000**	11	1001**000**	18	01011**000**	25	11011**000**
5	011**000**	12	1011**000**	19	01101**000**	26	11101**000**
6	101**000**	13	1101**000**	20	01111**000**	27	...

Table 3.35: Unconstrained Taboo Codes for $n = 3$.

> And I promise you, right here and now, no subject will ever be taboo...except, of course, the subject that was just under discussion.
>
> —Quentin Tarantino, *Kill Bill*, Vol. 1

The main reference [Pigeon 01b] describes the steps for encoding and decoding these codes.

3.17 Wang's Flag Code

Similar to the taboo code, the flag code of Muzhong Wang [Wang 88] is based on a flag of f zeros appended to the codewords. The name *suffix code* is perhaps more appropriate, because the flag must be the suffix of a codeword and cannot appear anywhere inside it. However, this type of suffix code is not the opposite of a prefix code.

The principle is to scan the positive integer n that is being encoded and append a single 1 to each sequence of $f-1$ zeros found in it. This effectively removes any occurrences of substrings of f consecutive zeros. Before this is done, n is reversed, so that its LSB becomes a 1. This guarantees that the flag will be preceded by a 1 and will therefore be easily recognized by the decoder.

We use the notation 0^f for a string of f zeros, select a value $f \geq 2$, and look at two examples. Given $n = 66 = 1000010_2$ and $f = 3$, we reverse n to obtain 0100001 and scan it from left to right, appending a 1 to each string of two consecutive zeros. The result is 010010011 to which is appended the flag. Thus, 010010011|000. Given $n = 288 = 100100000_2$, we reverse it to 000001001, scan it and append 001001010011, and append the flag 001001010011|000.

It is obvious that such codewords are easy to decode. The decoder knows the value of f and looks for a string of $f-1$ zeros. If such a string is followed by a 0, then it is the flag. The flag is removed and the result is reversed. If such a string is followed by a 1, the 1 is removed and the scan continues.

This code, as originally proposed by Wang, is slow to implement in software because reversing an m-bit string requires $\lfloor m/2 \rfloor$ steps where a pair of bits is swapped in each step. A possible improvement is to move the MSB (which is always a 1) to the right end, where it acts as an intercalary bit to separate the code bits from the flag.

Reference [Wang 88] shows that this code is universal, and for large values of n its asymptotic efficiency, defined by Wang as the limit

$$\Gamma = \lim_{n\to\infty} \sup \frac{\log_2(n+1)}{L(n)}, \tag{3.7}$$

(where $L(n)$ is the length of the Wang code of n) approaches 1.

The codeword length $L(n)$ increases with n, but not monotonically because it depends on the number of consecutive zeros in n and on the relation of this number to f. Recall that the binary representation of a power of 2 is a single 1 followed by several consecutive zeros. Thus, integers of the form 2^k or slightly bigger tend to have many consecutive zeros. As a result, the Wang codewords of 2^k and of its immediate successors tend to be longer than the codewords of its immediate predecessors because of the many intercalary 1's. On the other hand, the binary representation of the integer $2^k - 1$ consists of k consecutive 1's, so such numbers and their immediate predecessors tend to have few consecutive zeros, which is why their Wang codewords tend to be shorter. We therefore conclude that the lengths of Wang codewords feature jumps at n values that are powers of 2.

It is easy to figure out the length of the Wang code for two types of numbers. The integer $n = 2^k - 1 = 1^k$ has no zeros in its binary representation, so its Wang code consists of the original k bits plus the f-bit flag. Its successor $n + 1 = 2^k = 10^k$, on the other hand, has k consecutive zeros, so $\lfloor k/f \rfloor$ intercalary bits have to be inserted. The length of the codeword is therefore the original $k+1$ bits, plus the extra $\lfloor k/f \rfloor$ bits, plus the flag, for a total of $k + 1 + \lfloor k/f \rfloor + f$. Thus, when n passes through a value 2^k, the codeword length increases by $\lfloor k/f \rfloor + 1$ bits.

Reference [Yamamoto and Ochi 91] shows that in the special case where each bit b_j of n (except the MSB b_0, which is effectively unused by this code) is selected independently with probability $P(b_j = 0) = P(b_j = 1) = 1/2$ for $j = 1, 2, \ldots, M$, the average codeword length $\overline{L(n)}$ depends on M and f in a simple way (compare with Equation (3.9))

$$\overline{L(n)} = \begin{cases} M + 1 + f & \text{if } M \leq f - 2, \\ M + 1 + \sum_{j=1}^{M-f+2} l_j + f, & \text{if } M \geq f - 1, \end{cases}$$

where the quantity l_j is defined recursively by

$$l_j = \begin{cases} 0, & j \leq 0, \\ \frac{1}{2^{f-1}}, & j = 1, \\ \frac{1}{2^{f-1}}\left[\frac{1}{2} + l_{j-(f-1)}\right], & 2 \leq j \leq M - f + 2. \end{cases} \tag{3.8}$$

3.18 Yamamoto Flag Code

Wang's code requires reversing the bits of the integer n before it is encoded and reversing it again as the last step in its decoding. This code can be somewhat improved if we simply move the MSB of n (which is always a 1) to its right end, to separate the codeword from the flag. The fact that the MSB of n is included in the codeword of n introduces a slight redundancy because this bit is known to be a 1. This bit is needed in Wang's code, because it helps to identify the flag at the end of a codeword. The flag code of this section, due to Hirosuke Yamamoto and Hiroshi Ochi [Yamamoto and Ochi 91], is more complex to understand and implement, but is faster to encode and decode and does not include the MSB of n. It is therefore slightly shorter on average. In addition, this code is a true flag code, because it is UD even though the bit pattern of the flag may appear inside a codeword. We first explain the encoding process in the simple case $f = 3$ (a 3-bit flag).

We start with a positive integer $n = b_0b_1b_2\ldots b_M$ and the particular 3-bit flag $\mathbf{p} = p_1p_2p_3 = 100$. We ignore b_0 because it is a 1. Starting at b_1, we compare overlapping pairs of bits b_ib_{i+1} to the fixed pair $p_1p_2 = 10$. If $b_ib_{i+1} \neq 10$, we append b_i to the codeword-so-far and continue with the pair $b_{i+1}b_{i+2}$. If, however, $b_ib_{i+1} = 10$, we append the triplet 101 (which is $b_ib_{i+1}\overline{p}_f$) to the codeword-so-far and continue with the pair $b_{i+2}b_{i+3}$. Notice that $b_ib_{i+1}\overline{p}_f = 101$ is different from the flag, a fact exploited by the decoder. At the end, when all the pairs b_ib_{i+1} have been checked, the LSB b_M of n may be left over. This bit is appended to the codeword-so-far, and is followed by the flag.

We encode $n = 325 = 101000101_2$ as an example. Its nine bits are numbered $b_0b_1\ldots b_8$ with $b_0 = 1$. The codeword-so-far starts as an empty string. The first pair is $b_1b_2 = 01 \neq 10$, so $b_1 = 0$ is appended to the codeword-so-far. The next pair is $b_2b_3 = 10$, so the triplet 101 is appended to the codeword-so-far, which becomes 0|101. The next pair is $b_4b_5 = 00 \neq 10$, so $b_4 = 0$ is appended to the codeword-so-far. The next pair is $b_5b_6 = 01 \neq 10$, so $b_5 = 0$ is appended to the codeword-so-far, which becomes 0|101|0|0. The next pair is $b_6b_7 = 10$, so the triplet 101 is appended to the codeword-so-far, which becomes 0|101|0|0|101. Once b_6 and b_7 have been included in the codeword-so-far, only $b_8 = 1$ remains and it is simply appended, followed by the flag bits. The resulting codeword is 0|10**1**|0|0|10**1**|1|100 where the two bits in boldface are the complement of p_3. Notice that the flag pattern 100 also appears inside the codeword.

It is now easy to see how such a codeword can be decoded. The decoder initializes the decoded bitstring to 1 (the MSB of n). Given a string of bits $a_1a_2\ldots$ whose length is unknown, the decoder scans it, examining overlapping triplets of bits and looking for the pattern $a_ia_{i+1}\overline{p}_f = 101$. When such a triplet is found, its last bit is discarded and the first two bits 10 are appended to the decoded string. The process stops when the flag is located.

In our example, the decoder appends $a_1 = 0$ to the decoded string. It then locates $a_2a_3a_4 = 101$, discards a_4, and appends a_3a_4 to the decoded string. Notice that the codeword contains the flag pattern 100 in bits $a_4a_5a_6$ but this pattern disappears once a_4 has been discarded. The only potential problem in decoding is that an f-bit pattern of the form $\ldots a_{j-1}a_j|p_1p_2\ldots$ (i.e., some bits from the end of the codeword, followed by some bits from the start of the flag) will equal the flag. This problem is solved by selecting the special flag 100. For the special case $f = 3$, it is easy to verify that bit patterns of the form $xx|1$ or $x|10$ cannot equal 100. In the general case, a flag of the form 10^{f-1} (a 1 followed by $f - 1$ zeros) is selected, and again it is easy to see that no f-bit string of the form $xx\ldots x|100\ldots 0$ can equal the flag.

We are now ready to present the general case, where f can have any value greater than 1. Given a positive integer $n = b_0b_1b_2\ldots b_M$ and the f-bit flag $\mathbf{p} = p_1p_2\ldots p_f = 10^{f-1}$, the encoder initializes the codeword-so-far C to the empty string and sets a counter t to 1. It then performs the following loop:

1. If $t > M - f + 2$ then [if $t \le M$, then $C \leftarrow C + b_tb_{t+1}\ldots b_M$ endif],
   $C \leftarrow C + p_1p_2\ldots p_f$, Stop.
   endif

2. If $b_tb_{t+1}\cdots b_{t+(f-2)} \ne p_1p_2\ldots p_{f-1}$
   then [$C \leftarrow C + b_t$, $t \leftarrow t + 1$],
   else [$C \leftarrow C + b_tb_{t+1}\ldots b_{t+(f-2)}\overline{p_f}$, $t \leftarrow t + (f-1)$]
   endif
   Go to step 1.

Step 1 should read: If no more tuples remain [if some bits remain, append them to C, endif], append the flag to C. Stop.

Step 2 should read: If a tuple (of $f - 1$ bits from n) does not equal the most-significant $f-1$ bits of the flag, append the next bit b_t to C and increment t by 1. Else, append the entire tuple to C, followed by the complement of p_f, and increment t by $f - 1$. Endif. Go to step 1.

The developers of this code prove that the choice of the bit pattern 10^{f-1}, or equivalently 01^{f-1}, for the flag guarantees that no string of the form $xx\ldots x|p_1p_2\ldots p_j$ can equal the flag. They also propose flags of the form 1^20^{f-2} for cases where $f \ge 4$.

The decoder initializes the decoded string D to 1 (the MSB of n) and a counter s to 1. It then iterates the following step until it finds the flag and stops.

If $a_sa_{s+1}\ldots a_{s+(f-2)} \ne p_1p_2\ldots p_{f-1}$
 then $D \leftarrow D + a_s$, $s \leftarrow s + 1$,
 else
  [if $a_{s+(f-1)} = p_f$
    then Stop,
    else $D \leftarrow D + a_sa_{s+1}\ldots a_{s+(f-2)}$, $s \leftarrow s + f$
  endif]
endif

The developers also show that this code is universal and is almost asymptotically optimal in the sense of Equation (3.7). The codeword length $L(n)$ increases with n, but not monotonically, and is bounded by

$$\lfloor \log_2 n \rfloor + f \le L(n) \le \lfloor \log_2 n \rfloor + \frac{\lfloor \log_2 n \rfloor}{f-1} + f \le \frac{f}{f-1}\log_2 n + f.$$

In the special case where each bit b_j of n (except the MSB b_0, which is not used by this code) is selected independently with probability $P(b_j = 0) = P(b_j = 1) = 1/2$ for $j = 1, 2, \ldots, M$, the average codeword length $\overline{L(n)}$ depends on M and f in a simple way (compare with Equation (3.8))

$$\overline{L(n)} = \begin{cases} M + f & \text{if } M \le f - 2, \\ M + \frac{M-f+2}{2^{f-1}} + f, & \text{if } M \ge f - 1. \end{cases} \tag{3.9}$$

Table 3.36 lists some of the Yamamoto codes for $f = 3$ and $f = 4$. These are compared with the similar $S(r+1, 01^r)$ codes of Capocelli (Section 3.21.1 and Table 3.41).

Recall that the encoder inserts the intercalary bit $\overline{p_f}$ whenever it finds the pattern $p_1p_2 \ldots p_{f-1}$ in the integer n that is being encoded. Thus, if f is small (a short flag), large values of n may be encoded into long codewords because of the many intercalary bits. On the other hand, large f (a long flag) results in long codewords for small values of n because the flag has to be appended to each codeword. This is why a scheme where the flag starts small and becomes longer with increasing n seems ideal. Such a scheme, dubbed dynamically-variable-flag-length (DVFL), has been proposed by Yamamoto and Ochi as an extension of their original code.

The idea is to start with an initial flag length f_0 and increment it by 1 at certain points. A function $T(f)$ also has to be chosen that satisfies $T(f_0) \ge 1$ and $T(f+1) - T(f) \ge f - 1$ for $f \ge f_0$. Given a large integer $n = b_0b_1 \ldots b_M$, the encoder (and also the decoder, working in lockstep) will increment f when it reaches bits whose indexes equal $T(f_0)$, $T(f_0+1)$, $T(f_0+2)$, and so on. Thus, bits $b_1b_2 \ldots b_{T(f_0)}$ of n will be encoded with a flag length f_0, bits $b_{T(f_0)+1}b_{T(f_0)+2} \ldots b_{T(f_0+1)}$ will be encoded with a flag length f_0+1, and so on. In the original version, the encoder maintains a counter t and examines the $f-1$ bits starting at bit b_t. These are compared to the first $f-1$ bits 10^{f-2} of the flag. In the extended version, f is determined by the counter t and the function T by solving the inequality $T(f-1) < t \le T(f)$. In the last step, the flag 10^{f_M-1} is appended to the codeword, where f_M is determined by the inequality $T(f_M - 1) < M + 1 \le T(f_M)$. The encoding steps are as follows:

1. Initialize $f \leftarrow f_0$, $t \leftarrow 1$, and the codeword C to the empty string.

2. If $t > M - f + 2$
   then [if $t \le M$ then $C \leftarrow C + b_tb_{t+1} \ldots b_M$ endif],
   [if $M + 1 > T(f)$ then $f \leftarrow f + 1$ endif],
   $C \leftarrow C + 10^{f-1}$, Stop.
   endif.

3. If $b_tb_{t+1} \ldots\ b_{t+(f-2)} \ne 10^{f-2}$

n	$S(3,011)$	$Y_3(n)$	$S(4,0111)$	$Y_4(n)$
1	011	011	0111	0111
2	0 011	0 011	0 0111	0 0111
3	1 011	1 011	1 0111	1 0111
4	00 011	00 011	00 0111	00 0111
5	01 011	010 011	01 0111	01 0111
6	10 011	10 011	10 0111	10 0111
7	11 011	11 011	11 0111	11 0111
8	000 011	000 011	000 0111	000 0111
9	001 011	0010 011	001 0111	001 0111
10	010 011	0100 011	010 0111	010 0111
11	100 011	0101 011	011 0111	0110 0111
12	101 011	100 011	100 0111	100 0111
13	110 011	1010 011	101 0111	101 0111
14	111 011	110 011	110 0111	110 0111
15	0000 011	111 011	111 0111	111 0111
16	0001 011	0000 011	0000 0111	0000 0111
17	0010 011	00010 011	0001 0111	0001 0111
18	0100 011	00100 011	0010 0111	0010 0111
19	0101 011	00101 011	0011 0111	00110 0111
20	1000 011	01000 011	0100 0111	0100 0111
21	1001 011	010010 011	0101 0111	0101 0111
22	1010 011	01010 011	0110 0111	01100 0111
23	1100 011	01011 011	1000 0111	01101 0111
24	1101 011	1000 011	1001 0111	1000 0111
25	1110 011	10010 011	1010 0111	1001 0111
26	1111 011	10100 011	1011 0111	1010 0111
27	00000 011	10101 011	1100 0111	10110 0111
28	00001 011	1100 011	1101 0111	1100 0111
29	00010 011	11010 011	1110 0111	1101 0111
30	00100 011	1110 011	1111 0111	1110 0111
31	00101 011	1111 011	00000 0111	1111 0111
32	01000 011	00000 011	00001 0111	00000 0111

Table 3.36: Some Capocelli and Yamamoto Codes for $f = 3$ and $f = 4$.

```
    then C ← C + b_t, t ← t + 1,
         [if t > T(f) then f ← f + 1 endif]
    else C ← C + b_t b_{t+1} ... b_{t+(f-2)} 1,
         t ← t + (f - 1),
         [if t > T(f) then f ← f + 1 endif]
endif, Go to step 2.
```

The decoding steps are the following:

1. Initialize $f \leftarrow f_0$, $t \leftarrow 1$, $s \leftarrow 1$, $D \leftarrow 1$.

```
2. If $a_s a_{s+1} \ldots a_{s+(f-2)} \neq 10^{f-2}$
     then $D \leftarrow D + a_s$, $s \leftarrow s+1$, $t \leftarrow t+1$,
          [if $t > T(f)$ then $f \leftarrow f+1$ endif],
     else [if $a_{s+(f-1)} = 0$ then Stop
                              else
 $D \leftarrow D + a_s a_{s+1} \ldots a_{s+(f-2)}$, $s \leftarrow s+f$,
 $t \leftarrow t+(f-1)$, [if $t > T(f)$ then $f \leftarrow f+1$ endif]
            endif]
   endif, Go to step 2.
```

The choice of function $T(f)$ is the next issue. The authors show that a function of the form $T(f) = Kf(f-1)$, where K is a nonnegative constant, results in an asymptotically optimal DVFL. There are other functions that guarantee the same property of DVFL, but the point is that varying f, while generating short codewords, also eliminates one of the chief advantages of any flag code, namely its robustness. If the length f of the flag is increased during the construction of a codeword, then any future error may cause the decoder to lose synchronization and may propagate indefinitely through the string of codewords. It seems that, in practice, the increased reliability achieved by synchronization with a fixed f overshadows the savings produced by a dynamic encoding scheme that varies f.

3.19 Number Bases

This short section is a digression to prepare the reader for the Fibonacci and Goldbach codes that follow. Decimal numbers use base 10. The number 2037_{10}, for example, has a value of $2 \times 10^3 + 0 \times 10^2 + 3 \times 10^1 + 7 \times 10^0$. We can say that 2037 is the sum of the digits 2, 0, 3, and 7, each weighted by a power of 10. Fractions are represented in the same way, using negative powers of 10. Thus, $0.82 = 8 \times 10^{-1} + 2 \times 10^{-2}$ and $300.7 = 3 \times 10^2 + 7 \times 10^{-1}$.

Binary numbers use base 2. Such a number is represented as a sum of its digits, each weighted by a power of 2. Thus, $101.11_2 = 1\times2^2 + 0\times2^1 + 1\times2^0 + 1\times2^{-1} + 1\times2^{-2}$.

Since there is nothing special about 10 or 2 (actually there is, because 2 is the smallest integer that can be a base for a number system and 10 is the number of our fingers), it should be easy to convince ourselves that any positive integer $n > 1$ can serve as the basis for representing numbers. Such a representation requires n "digits" (if $n > 10$, we use the ten digits and the letters A, B, C, ...) and represents the number $d_3d_2d_1d_0.d_{-1}$ as the sum of the digits d_i, each multiplied by a power of n, thus $d_3n^3 + d_2n^2 + d_1n^1 + d_0n^0 + d_{-1}n^{-1}$. The base of a number system does not have to consist of powers of an integer but can be any *superadditive* sequence that starts with 1.

Definition: A superadditive sequence a_0, a_1, $a_2, \ldots$ is one where any element a_i is greater than the sum of all its predecessors. An example is 1, 2, 4, 8, 16, 32, 64, ... where each element equals 1 plus the sum of all its predecessors. This sequence consists of the familiar powers of 2, so we know that any integer can be expressed by it using just the digits 0 and 1 (the two bits). Another example is 1, 3, 6, 12, 24, 50, ..., where each element equals 2 plus the sum of all its predecessors. It is easy to see that any integer can be expressed by it using just the digits 0, 1, and 2 (the three trits).

Given a positive integer k, the sequence $1,\ 1+k,\ 2+2k,\ 4+4k,\ldots,2^i(1+k)$ is superadditive, because each element equals the sum of all its predecessors plus k. Any nonnegative integer can be represented uniquely in such a system as a number $x\ldots xxy$, where x are bits and y is a single digit in the range $[0,k]$.

In contrast, a general superadditive sequence, such as 1, 8, 50, 3102 can be used to represent integers, but not uniquely. The number 50, e.g., equals $8\times 6+1+1$, so it can be represented as $0062 = 0\times 3102+0\times 50+6\times 8+2\times 1$, but also as $0100 = 0\times 3102+1\times 50+0\times 8+0\times 1$.

It can be shown that $1+r+r^2+\cdots+r^k$ is less than r^{k+1} for any real number $r>1$. This implies that the powers of any real number $r>1$ can serve as the base of a number system using the digits $0,1,2,\ldots,d$ for some $d<r$.

The number $\phi=\frac{1}{2}(1+\sqrt{5})\approx 1.618$ is the well-known golden ratio. It can serve as the base of a number system, with 0 and 1 as the digits. Thus, for example, $100.1_\phi = \phi^2+\phi^{-1}\approx 3.23_{10}$.

Some real bases have special properties. For example, any positive integer R can be expressed as $R=b_1F_1+b_2F_2+b_3F_3+b_4F_5+b_5F_8+b_6F_{13}\cdots$, where b_i are either 0 or 1, and F_i are the Fibonacci numbers 1, 2, 3, 5, 8, 13, ... This representation has the interesting property, known as Zeckendorf's theorem [Zeckendorf 72], that the string $b_1b_2\ldots$ does not contain any adjacent 1's. This useful property, which is the basis for the Goldbach codes (Section 3.22) is easy to prove. If an integer I in this representation has the form $\ldots 01100\ldots$, then because of the definition of the Fibonacci numbers, I can be written $\ldots 00010\ldots$.

Examples are the integer 5, whose Fibonacci representation is 0001 and $33=1+3+8+21$, which is expressed in the Fibonacci base as the 7-bit number 1010101. Section 3.16 discusses the n-step Fibonacci numbers, defined by Equation (3.6).

The Australian Aboriginals use a number of languages, some of which employ binary or binary-like counting systems. For example, in the Kala Lagaw Ya language, the numbers 1 through 6 are urapon, ukasar, ukasar-urapon, ukasar-ukasar, ukasar-ukasar-urapon, and ukasar-ukasar-ukasar.

The familiar terms "dozen" (12) and "gross" (twelve dozen) originated in old duodecimal (base 12) systems of measurement.

Computers use binary numbers mostly because it is easy to design electronic circuits with just two states, as opposed to ten states.

Leonardo Pisano Fibonacci (1170–1250)

Leonard of Pisa (or Fibonacci), was an Italian mathematician, often considered the greatest mathematician of the Middle Ages. He played an important role in reviving ancient mathematics and also made significant original contributions. His book, *Liber Abaci*, introduced the modern decimal system and the use of Arabic numerals and the zero into Europe.

Leonardo was born in Pisa around 1170 or 1180 (he died in 1250). His father Guglielmo was nicknamed Bonaccio (the good natured or simple), which is why Leonardo is known today by the nickname Fibonacci (derived from filius Bonacci, son of Bonaccio). He is also known as Leonardo Pisano, Leonardo Bigollo, Leonardi Bigolli Pisani, Leonardo Bonacci, and Leonardo Fibonacci.

The father was a merchant (and perhaps also the consul for Pisa) in Bugia, a port east of Algiers (now Bejaïa, Algeria), and Leonardo visited him there while still a boy. It seems that this was where he learned about the Arabic numeral system.

Realizing that representing numbers with Arabic numerals rather than with Roman numerals is easier and greatly simplifies the arithmetic operations, Fibonacci traveled throughout the Mediterranean region to study under the leading Arab mathematicians of the time. Around 1200 he returned to Pisa, where in 1202, at age 32, he published what he had learned in *Liber Abaci*, or Book of Calculation.

Leonardo became a guest of the Emperor Frederick II, who enjoyed mathematics and science. In 1240, the Republic of Pisa honored Leonardo, under the name Leonardo Bigollo, by granting him a salary.

Today, the Fibonacci sequence is one of the best known mathematical objects. This sequence and its connection to the golden ratio $\phi = \frac{1}{2}(1+\sqrt{5}) \approx 1.618$ have been studied extensively and the mathematics journal *Fibonacci Quarterly* is dedicated to the Fibonacci and similar sequences. The publication [Grimm 73] is a short biography of Fibonacci.

3.20 Fibonacci Code

The Fibonacci code, as its name suggests, is closely related to the Fibonacci representation of the integers. The Fibonacci code of the positive integer n is the Fibonacci representation of n with an additional 1 appended to the right end. Thus, the Fibonacci code of 5 is 0001|1 and that of 33 is 1010101|1. It is obvious that such a code ends with a pair 11, and that this is the only such pair in the codeword (because the Fibonacci representation does not have adjacent 1's). This property makes it possible to decode such a code unambiguously, but also causes these codes to be long, because not having adjacent 1's restricts the number of possible binary patterns.

Table 3.37 lists the first 12 Fibonacci codes.

1	11	7	01011
2	011	8	000011
3	0011	9	100011
4	1011	10	010011
5	00011	11	001011
6	10011	12	101011

Table 3.37: Twelve Fibonacci Codes.

Decoding. Skip bits of the code until a pair 11 is reached. Replace this 11 by 1. Multiply the skipped bits by the values ..., 13, 8, 5, 3, 2, 1 (the Fibonacci numbers), and add the products. Obviously, it is not necessary to do any multiplication. Simply use the 1 bits to select the proper Fibonacci numbers and add.

The Fibonacci codes are long, but have the advantage of being more robust than most other variable-length codes. A corrupt bit in a Fibonacci code may change a pair of consecutive bits from 01 or 10 to 11 or from 11 to 10 or 01. In the former case, a code may be read as two codes, while in the latter case two codes will be decoded as a single code. In either case, the slippage will be confined to just one or two codewords and will not propagate further.

The length of the Fibonacci code for n is less than or equal to $1 + \lfloor \log_\phi \sqrt{5}n \rfloor$ where ϕ is the golden ratio (compare with Equation (2.1)).

Figure 3.38 represents the lengths of two Elias codes and the Fibonacci code graphically and compares them to the length of the standard binary (beta) code.

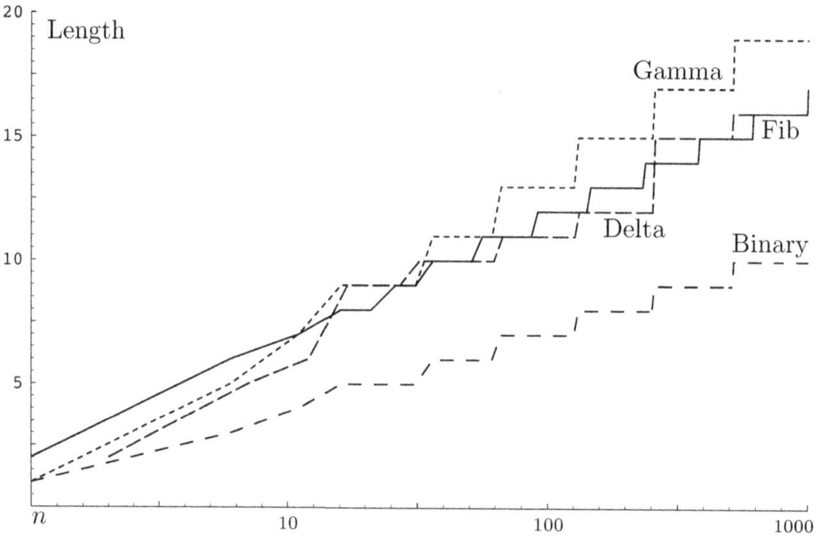


```
(* Plot the lengths of four codes
   1. staircase plots of binary representation *)
bin[i_] := 1 + Floor[Log[2, i]];
Table[{Log[10, n], bin[n]}, {n, 1, 1000, 5}];
g1 = ListPlot[%, AxesOrigin -> {0, 0}, PlotJoined -> True,
     PlotStyle -> { AbsoluteDashing[{5, 5}]}]
(* 2. staircase plot of Fibonacci code length *)
fib[i_] := 1 + Floor[ Log[1.618, Sqrt[5] i]];
Table[{Log[10, n], fib[n]}, {n, 1, 1000, 5}];
g2 = ListPlot[%, AxesOrigin -> {0, 0}, PlotJoined -> True]
(* 3. staircase plot of gamma code length*)
gam[i_] := 1 + 2Floor[Log[2, i]];
Table[{Log[10, n], gam[n]}, {n, 1, 1000, 5}];
g3 = ListPlot[%, AxesOrigin -> {0, 0}, PlotJoined -> True,
     PlotStyle -> { AbsoluteDashing[{2, 2}]}]
(* 4. staircase plot of delta code length*)
del[i_] := 1 + Floor[Log[2, i]] + 2Floor[Log[2, Log[2, i]]];
Table[{Log[10, n], del[n]}, {n, 2, 1000, 5}];
g4 = ListPlot[%, AxesOrigin -> {0, 0}, PlotJoined -> True,
     PlotStyle -> { AbsoluteDashing[{6, 2}]}]
Show[g1, g2, g3, g4, PlotRange -> {{0, 3}, {0, 20}}]
```

Figure 3.38: Lengths of Binary, Fibonacci and Two Elias Codes.

In [Fraenkel and Klein 96] (and also [Fraenkel and Klein 85a]), the authors denote this code by C^1 and show that it is universal, with $c1 = 2$ and $c2 = 3$, but is not asymptotically optimal because $c1 > 1$. They also prove that for any length $r \geq 1$, there are F_r codewords of length $r+1$ in C^1. As a result, the total number of codewords of length up to and including r is $\sum_{i=1}^{r-1} F_i = F_{r+1} - 1$. (See also Figure 3.45.)

A P-code is a set of codewords that end with the same pattern P (the pattern is the suffix of the codewords) and where no codeword includes the pattern anywhere else. Given a k-bit binary pattern P, the set of all binary strings of length $\geq k$ in which P occurs only as a suffix is called the set generated by P and is denoted by $\mathcal{L}(P)$. In [Berstel and Perrin 85] this set is called a semaphore code. All codewords in the C^1 code end with the pattern $P = 11$, so this code is $\mathcal{L}(11)$.

The next Fibonacci code proposed by Fraenkel and Klein is denoted by C^2 and is constructed from C^1 as follows:

1. Each codeword in C^1 ends with two consecutive 1's; delete one of them.
2. Delete all the codewords that start with 0.

Thus, the first few C^2 codewords, constructed with the help of Table 3.37, are 1, 101, 1001, 10001, 10101, 100001, and 101001. An equivalent procedure to construct this code is the following:

1. Delete the rightmost 1 of every codeword in C^1.
2. Prepend 10 to every codeword.
3. Include 1 as the first codeword.

A simple check verifies the equivalence of the two constructions. Code C^2 is not a straightforward Fibonacci code as is C^1, but it can be termed a Fibonacci code, because the interior bits of each codeword correspond to Fibonacci numbers. The code consists of one codeword of length 1, no codewords of length 2, and F_{r-2} codewords of length r for any $r \geq 3$.

The C^2 code is not a prefix code, but is UD. Individual codewords are identified by the decoder because each starts and ends with a 1. Thus, two consecutive 1's indicate the boundary between codewords. The first codeword introduces a slight complication, but can be handled by the decoder. A string of the form $\ldots 011110 \ldots$ is interpreted by the decoder as $\ldots 01|1|1|10 \ldots$, i.e., two consecutive occurrences of the codeword 1.

The C^2 code is also more robust than C^1. A single error cannot propagate far because the decoder is looking for the pattern 11. The worst case is a string of the form $\ldots xyz \ldots = \ldots 1|10 \ldots 01|1|10 \ldots 01|1 \ldots$ where the middle 1 gets corrupted to a 0. This results in $\ldots 1|10 \ldots 01\mathbf{0}10 \ldots 01|1 \ldots$ which is interpreted by the decoder as one long codeword. The three original codewords xyz are lost, but the error does not propagate any further. Other single errors (corrupted or lost bits) result in the loss of only two codewords.

The third Fibonacci code described in [Fraenkel and Klein 96] is denoted by C^3 and is constructed from C^1 as follows:

1. Delete the rightmost 1 of every codeword of C^1.
2. For every $r \geq 1$, select the set of C^1 codewords of length r, duplicate the set, and distinguish between the two copies by prepending a 10 to all the codewords in one copy and a 11 to all the codewords in the other copy.

This results in the codewords 101, 111, 1001, 1101, 10001, 10101, 11001, 11101, 100001, 101001, 100101, 110001, ... It is easy to see that every codeword of C^3 starts with a 1, has at most three consecutive 1's (and then they appear in the prefix), and every codeword except the second ends with 01. The authors show that for any $r \geq 3$ there are $2F_{r-2}$ codewords. It is easy to see that C^3 is not a prefix code because, for example, codeword 111 is also the prefix of 11101. However, the code is UD. The decoder first checks for the pattern 111 and interprets it depending on the bit that follows. If that bit is 0, then this is a codeword that starts with 111; otherwise, this is the codeword 111 followed by another codeword. If the current pattern is not 111, the decoder checks for 011. Every codeword except 111 ends with 01, so the pattern 01|1 indicates the end of a codeword and the start of the next one. This pattern does not appear inside any codeword.

Given an r-bit codeword $y_1 y_2 \ldots y_r$ (where $y_r = 1$), the developers show that its index (i.e., the integer whose codeword it is) is given by

$$2F_{r-1} - 2 + y_2 F_{r-2} + \sum_{i=3}^{r} y_i F_{i-1} - F_{r-1} + 1$$
$$= \sum_{i=3}^{r+2} y_i F_{i-1} + (y_2 - 1)F_{r-2} - 1,$$

where $y_{r+1} = 1$ is the leftmost bit of the next codeword.

The developers compare the three Fibonacci codes with the Huffman code using English text of 100,000 words collected from many different sources. Letter distributions were computed and were used to assign Huffman, C^1, C^2, and C^3 codes to the 26 letters. The average lengths of the codes were 4.185, 4.895, 5.298, and 4.891 bits/character, respectively. The Huffman code has the shortest length, but is vulnerable to storage and transmission errors. The conclusion is that the three Fibonacci codes are good candidates for compressing data in applications where robustness is more important than optimal compression.

It is interesting to note that the Fibonacci sequence can be generalized by adding a parameter. An m-step Fibonacci number equals the sum of its m immediate predecessors (Equation (3.6)). Thus, the common Fibonacci numbers are 2-step. The m-step Fibonacci numbers can be used to construct order-m Fibonacci codes (Section 3.21), but none of these codes is asymptotically optimal.

Section 11.5.4 discusses an application of this type of variable-length code to the compression of sparse strings.

The anti-Fibonacci numbers are defined recursively as $f(1) = 1$, $f(2) = 0$, and $f(k+2) = f(k) - f(k+1)$. The sequence starts with 1, 0, 1, −1, 2, −3, 5, −8, ... This sequence is also obtained when the Fibonacci sequence is extended backward from 0. Thus,

$$\ldots - 8, 5, -3, 2, -1, 1, 0, 1, 1, 2, 3, 5, 8, \ldots.$$

In much the same way that the ratios of successive Fibonacci numbers converge to ϕ, the ratios of successive anti-Fibonacci numbers converge to $1/\phi$.

3.21 Generalized Fibonacci Codes

The Fibonacci code of Section 3.20 is elegant and robust. The generalized Fibonacci codes presented here, due to Alberto Apostolico and Aviezri Fraenkel [Apostolico and Fraenkel 87], are also elegant, robust, and UD. They are based on m-step Fibonacci numbers (sometimes also called generalized Fibonacci numbers). The authors show that these codes are easy to encode and decode and can be employed to code integers as well as arbitrary, unbound bit strings. (The MSB of an integer is 1, but the MSB of an arbitrary string can be any bit.)

The sequence $F^{(m)}$ of m-step Fibonacci numbers $F_n^{(m)}$ is defined as follows:

$$F_n^{(m)} = F_{n-1}^{(m)} + F_{n-2}^{(m)} + \cdots + F_{n-m}^{(m)}, \text{ for } n \geq 1,$$

and

$$F_{-m+1}^{(m)} = F_{-m+2}^{(m)} = \cdots = F_{-2}^{(m)} = 0, \quad F_{-1}^{(m)} = F_0^{(m)} = 1.$$

Thus, for example, $F_1^{(m)} = 2$ for any m, while for $m = 3$ we have $F_{-2}^{(3)} = 0$, $F_{-1}^{(3)} = F_0^{(3)} = 1$, which implies $F_1^{(3)} = F_0^{(3)} + F_{-1}^{(3)} + F_{-2}^{(3)} = 2$, $F_2^{(3)} = F_1^{(3)} + F_0^{(3)} + F_{-1}^{(3)} = 4$, and $F_3^{(3)} = F_2^{(3)} + F_1^{(3)} + F_0^{(3)} = 7$.

The generalized Fibonacci numbers can serve as the basis of a numbering system. Any positive integer N can be represented as the sum of several distinct m-step Fibonacci numbers and this sum does not contain m consecutive 1's. Thus, if we represent N in this number basis, the representation will not have a run of m consecutive 1's. An obvious conclusion is that such a run can serve as a comma, to separate consecutive codewords.

The two generalized Fibonacci codes proposed by Apostolico and Fraenkel are pattern codes (P-codes) and are also universal and complete (a UD code is complete if the addition of any codeword turns it into a non-UD code). A P-code is a set of codewords that end with the same pattern P (the pattern is the suffix of the codewords) and where no codeword includes the pattern anywhere else. For example, if $P = 111$, then a P-code is a set of codewords that all end with 111 and none has 111 anywhere else. A P-code is a prefix code because once a codeword $c = x \ldots x111$ has been selected, no other codeword will start with c because no other codeword has 111 other than as a suffix.

Given a P-code with the general pattern $P = a_1 a_2 \ldots a_p$ and an arbitrary codeword $x_1 x_2 \ldots x_n a_1 a_2 \ldots a_p$, we consider the string $a_2 a_3 \ldots a_p | x_1 x_2 \ldots x_n a_1 a_2 \ldots a_{p-1}$ (the end of P followed by the start of a codeword). If P happens to appear anywhere in this string, then the code is not synchronous and a transmission error may propagate over many codewords. On the other hand, if P does not appear anywhere inside such a string, the code is synchronous and is referred to as an SP-code (it is also comma-free). Such codes are useful in applications where data integrity is important. When an error occurs during storage or transmission of an encoded message, the decoder loses synchronization, but regains it when it sees the next occurrence of P (and it sees it when it reads the next codeword). As usual, there is a price to pay for this useful property. In general, SP-codes are not complete.

The authors show that the Elias delta code is asymptotically longer than the generalized Fibonacci codes (of any order) for small integers, but becomes shorter at a certain point that depends on the order m of the generalized code. For $m = 2$, this transition point is at $F_{27}^{(2)} - 1 = 514,228$. For $m = 3$ it is at $(F_{63}^{(3)} + F_{631}^{(3)} - 1)/2 = 34,696,689,675,849,696 \approx 3.47 \times 10^{16}$, and for $m = 4$ it is at $(F_{231}^{(4)} + F_{229}^{(4)} + F_{228}^{(4)} - 1)/3 \approx 4.194 \times 10^{65}$.

The first generalized Fibonacci code, $C_1^{(m)}$, employs the pattern 1^m. The code is constructed as follows:

1. The $C_1^{(m)}$ code of $N = 1$ is 1^m and the $C_1^{(m)}$ code of $N = 2$ is 01^m.

2. All other $C_1^{(m)}$ codes have the suffix 01^m and a prefix of $n-1$ bits, where n starts at 2. For each n, there are $F_n^{(m)}$ prefixes which are the $(n-1)$-bit $F^{(m)}$ representations of the integers 0, 1, 2, ...

Table 3.39 lists examples of $C_1^{(3)}$. Once the codes of $N = 1$ and $N = 2$ are in place, we set $n = 2$ and construct the 1-bit $F^{(3)}$ representations of the integers 0, 1, 2, ... There are only two such representations, 0 and 1, and they become the prefixes of the next two codewords (for $N = 3$ and $N = 4$). We increment n to 3, and construct the 2-bit $F^{(3)}$ representations of the integers 0, 1, 2, ... There are four of them, namely 0, 1, 10, and 11. Each is extended to two bits to form 00, 01, 10, and 11, and they become the prefixes of $N = 5$ through $N = 8$.

String M	$C_1^{(3)}$ 7421|0111	$F^{(3)}$ ${}^{1}_{3}7421$	N
		0	0
0	111	1	1
00	0111	10	2
000	0|0111	11	3
1	1|0111	100	4
0000	00|0111	101	5
01	01|0111	110	6
2	10|0111	1000	7
3	11|0111	1001	8
00000	000|0111	1010	9
001	001|0111	1011	10
02	010|0111	1100	11
03	011|0111	1101	12
4	100|0111	10000	13
5	101|0111	10001	14
6	110|0111	10010	15
000000	0000|0111	10011	16

Table 3.39: The Generalized Fibonacci Code $C_1^{(3)}$.

$C_2^{(3)}$ ${}^{1}_{3}7421$|011	$C_2^{(2)}$ ${}^{1}_{3}85321$|01	N
11	1	1
1|011	1|01	2
10|011	10|01	3
11|011	100|01	4
100|011	101|01	5
101|011	1000|01	6
110|011	1001|01	7
1000|011	1010|01	8
1001|011	10000|01	9
1010|011	10001|01	10
1011|011	10010|01	11
1100|011	10100|01	12
1101|011	10101|01	13
10000|011	100000|01	14
10001|011	100001|01	15
10010|011	100010|01	16

Table 3.40: Generalized Codes $C_2^{(3)}$ $C_2^{(2)}$.

Notice that $N_1 < N_2$ implies that the C_1 code of N_1 is lexicographically smaller than the C_1 code of N_2 (where a blank space is considered lexicographically smaller than 0). The C_1 code is a P-code (with 1^m as the pattern) and is therefore a prefix code.

The code is partitioned into groups of $m+n$ codewords each. In each group, the prefix of the codewords is $n-1$ bits long and the suffix is 01^m. Each group consists of $F_{n-1}^{(m)}$ codewords of which $F_{n-2}^{(m)}$ codewords start with 0, $F_{n-3}^{(m)}$ start with 10, $F_{n-4}^{(m)}$ start with 110, and so on, up to $F_{n-m-1}^{(m)}$ codewords that start with $1^{m-1}0$.

A useful, interesting property of this code is that it can be used for arbitrary bitstrings, not just for integers. The difference is that an integer has a MSB of 1, whereas a binary string can start with any bit. The idea is to divide the set of bitstrings into those strings that start with 1 (integers) and those that start with 0. The former strings are assigned C_1 codes that start with 1 (in Table 3.39, those are the codes of 4, 7, 8, 13, 14, 15, ...). The latter are assigned C_1 codes that start with 0 in the following way. Given a bitstring of the form 0^iN (where N is a nonnegative integer), assign to it the code 0^iN01^m. Thus, the string 001 is assigned the code 00|1|0111 (which happens to be the code of 10 in the table).

The second generalized Fibonacci code, $C_2^{(m)}$, employs the pattern 1^{m-1}. The code is constructed as follows:

1. The $C_2^{(m)}$ code of $N=1$ is 1^{m-1}.

2. All other $C_2^{(m)}$ codes have the suffix 01^{m-1} and prefixes that are the $F^{(m)}$ representations of the positive integers, in increasing order.

Table 3.40 shows examples of $C_2^{(3)}$ and $C_2^{(2)}$. Once the code of $N=1$ has been constructed, we prepare the $F^{(m)}$ representation of $n=1$ and prepend it to 01^{m-1} to obtain the C_2 code of $N=2$. We then increment n by 1, and construct the other codes in the same way. Notice that C_2 is not a prefix code because, for example, $C_2^{(3)}(2)=1011$ is a prefix of $C_2^{(3)}(11)=1011011$. However, C_2 is a UD code because the string $01^{m-1}|1$ separates any two codewords (each ends with 01^{m-1} and starts with 1). The C_2 codes become longer with N and are organized in length groups for $n=0,1,2,\ldots$. Each group has $F_{n-1}^{(m)}-F_{n-2}^{(m)}$ codewords of length $m+n-1$ that can be partitioned as follows: $F_{n-3}^{(m)}$ codewords that start with 10, $F_{n-4}^{(m)}$ codewords that start with 110, and so on, up to $F_{n-m-1}^{(m)}$ codewords that start with $1^{m-1}0$.

The authors provide simple procedures for encoding and decoding the two codes. Once a value for m has been selected, the procedures require tables of many m-step Fibonacci numbers (if even larger numbers are needed, they have to be computed on the fly).

A note on robustness. It seems that a P-code is robust. Following an error, the decoder will locate the pattern P very quickly and will resynchronize itself. However, the term "robust" is imprecise and at least code $C_1^{(3)}$ has a weak point, namely the codeword 111. In his short communication [Capocelli 89], Renato Capocelli points out a case where the decoder of this code can be thrown completely off track because of this codeword. The example is the message 41^n3, which is encoded in $C_1^{(3)}$ as $10111|(111)^n|00111$. If the second MSB becomes 1 because of an error, the decoder will not sense any error and will decode this string as $(111)^{n+1}|1100111$, which is the message $1^{n+1}(15)$.

3.21.1 A Related Code

The simple code of this section, proposed by Renato Capocelli [Capocelli 89], is prefix, complete, universal, and also synchronizable (see also Section 4.3 for more synchronous codes). It is not a generalized Fibonacci code, but it is related to the C_1 and C_2 codes of Apostolico and Fraenkel. The code depends on a parameter r and is denoted by $S(r+1, 01^r)$. Once r has been selected, two-part codewords are constructed with a suffix 01^r and a prefix that is a binary string that does not contain the suffix. Thus, for example, if $r=2$, the suffix is 011 and the prefixes are all the binary strings, starting with the empty string, that do not contain 011. Table 3.41 lists a few examples of $S(3, 011)$ (see also Table 3.36) and it is easy to see how the prefixes are the empty string, 0, 1, 00, 01, and so on, but they include no strings with 011. The codeword of $N=9$ is 010|011, but the codeword of $N=10$ has the prefix 100 and not 011, so it is 100|011. In general a codeword $x\ldots x0101y\ldots y|011$ will be followed by $x\ldots x1000y\ldots y|011$ instead of by $x\ldots x0110y\ldots y|011$. Such codewords have either the form $0\beta|011$ (where β does not contain two consecutive 1's) or the form $1\gamma|011$ (where γ does not contain 011). For example, only 12 of the 16 4-bit prefixes can be used by this code, because the four prefixes 0011, 0110, 0111, and 1011 contain the pattern 011. In general, the number of codewords of length $N+3$ in $S(3, 011)$ is $F_{N+3}-1$. For $N=4$ (codewords of a 4-bit prefix and a 3-bit suffix), the number of codewords is $F_{4+3}-1=F_7-1=12$.

N	$S(3, 011)$	BS
0	011	0
1	0011	1
2	1011	00
3	00011	01
4	01011	10
5	10011	11
6	11011	000
7	000011	001
8	001011	010
9	010011	011
10	100011	100
11	101011	101
12	110011	110
13	111011	111

Table 3.41: Code $S(3, 011)$ for Integers N and Strings BS.

The table also illustrates how $S(3, 011)$ can be employed to encode arbitrary bit-strings, not just integers. Simply write all the binary strings in lexicographic order and assign them the $S(3, 011)$ codewords in increasing order.

The special suffix 01^r results in synchronized codewords. If the decoder loses synchronization, it regains it as soon as it recognizes the next suffix. What is perhaps more interesting is that the C_1 and C_2 codes of Apostolico and Fraenkel are special cases of this code. Thus, $C_1^{(3)}$ is obtained from $S(4, 0111)$ by replacing all the sequences that start with 11 with the sequence 11 and moving the rightmost 1 to the left end of the codeword. Also, $C_2^{(2)}$ is obtained from $S(3, 011)$ by replacing all the sequences that start with 1 with the codeword 1 and moving the rightmost 1 to the left end of the codeword.

The developer provides algorithms for encoding and decoding this code.

Fibonacci couldn't sleep— Counted rabbits instead of sheep. —Katherine O'Brien

3.22 Goldbach Codes

The Fibonacci codes of Section 3.20 have the rare property, termed Zeckendorf's theorem, that they do not have consecutive 1's. If a binary number of the form $1xx\ldots x$ does not have any consecutive 1's, then reversing its bits and appending a 1 results in a number of the form $xx\ldots x1|1$ that ends with two consecutive 1's, thereby making it easy for a decoder to read individual, variable-length Fibonacci codewords from a long stream of such codes. The variable-length codes described in this section are based on a similar property that stems from the Goldbach conjecture.

Goldbach's conjecture: Every even integer n greater than 2 is the sum of two primes.

A prime number is an integer that is not divisible by any other integer (other than trivially by itself and by 1). The integer 2 is prime, but all other primes are odd. Adding two odd numbers results in an even number, so mathematicians have always known that the sum of two primes is even. History had to wait until 1742, when it occurred to Christian Goldbach, an obscure German mathematician, to ask the "opposite" question. If the sum of any two primes is even, is it true that every even integer is the sum of two primes? Goldbach was unable to prove this simple-looking problem, but neither was he able to find a counter-example. He wrote to the famous mathematician Leonhard Euler and received the answer "There is little doubt that this result is true." However, even Euler was unable to furnish a proof, and at the time of this writing (late 2006), after more than 260 years of research, the Goldbach conjecture has almost, but not quite, been proved. It should be noted that many even numbers can be written as the sum of two primes in several ways. Thus $42 = 23 + 19 = 29 + 13 = 31 + 11 = 37 + 5$, 1,000,000 can be partitioned in 5,402 ways, and 100,000,000 has 291,400 Goldbach partitions. Reference [pass 06] is an online calculator that can compute the Goldbach partitions of even integers.

Christian Goldbach was a Prussian mathematician. Born in 1690, the son of a pastor, in Königsberg (East Prussia), Goldbach studied law and mathematics. He traveled widely throughout Europe and met with many well-known mathematicians, such as Gottfried Leibniz, Leonhard Euler, and Nicholas (I) Bernoulli. He went to work at the newly opened St Petersburg Academy of Sciences and became tutor to the later Tsar Peter II. The following quotation, from [Mahoney 90] reflects the feelings of his superiors in Russia "...a superb command of Latin style and equal fluency in German and French. Goldbach's polished manners and cosmopolitan circle of friends and acquaintances assured his success in an elite society struggling to emulate its western neighbors."

Goldbach is remembered today for Goldbach's conjecture. He also studied and proved some theorems on perfect powers. He died in 1764.

In 2001, Peter Fenwick had the idea of using the Goldbach conjecture (assuming that it is true) to design an entirely new class of codes, based on prime numbers [Fenwick 02]. The prime numbers can serve as the basis of a number system, so if we write an even integer in this system, its representation will have exactly two 1's. Thus, the even number 20 equals 7 + 13 and can therefore be written 10100, where the five bits are assigned the prime weights (from left to right) 13, 11, 7, 5, and 3. Now reverse this bit pattern so that its least-significant bit becomes 1, to yield 00101. Such a number is easy to read and extract from a long bitstring. Simply stop reading at the second 1. Recall that the unary code (a sequence of zeros terminated by a single 1) is read by a similar rule: stop at the first 1. Thus, the Goldbach codes can be considered an extension of the simple unary code.

Goldbach's original conjecture (sometimes called the "ternary" Goldbach conjecture), written in a June 7, 1742 letter to Euler [dartmouth 06], states "at least it seems that every integer that is greater than 2 is the sum of three primes." This made sense because Goldbach considered 1 a prime, a convention that is no longer followed. Today, his statement would be rephrased to "every integer greater than 5 is the sum of three primes." Euler responded with "There is little doubt that this result is true," and coined the modern form of the conjecture (a form currently referred to as the "strong" or "binary" Goldbach conjecture) which states "all positive even integers greater than 2 can be expressed as the sum of two primes." Being honest, Euler also added in his response that he regarded this as a fully certain theorem ("ein ganz gewisses Theorema"), even though he was unable to prove it.

Reference [utm 06] has information on the history of this and other interesting mathematical conjectures.

---

n	$2(n+3)$	Primes	Codeword
1	8	$3+5$	11
2	10	$3+7$	101
3	12	$5+7$	011
4	14	$3+11$	1001
5	16	$5+11$	0101
6	18	$7+11$	0011
7	18	$7+13$	00101
8	22	$5+17$	010001
9	24	$11+13$	00011
10	26	$7+19$	0010001
11	28	$11+17$	000101
12	30	$13+17$	000011
13	32	$13+19$	0000101
14	34	$11+23$	00010001

Table 3.42: The Goldbach G0 Code.

The first Goldbach code is designated G0. It encodes the positive integer n by examining the even number $2(n+3)$ and writing it as the sum of two primes in reverse (with its most-significant bit at the right end). The G0 codes listed in Table 3.42 are based on the primes (from right to left) 23, 19, 17, 13, 11, 7, 5, and 3. It is obvious that the codes increase in length, but not monotonically, and there is no known expression for the length of $\mathrm{G0}(n)$ as a function of n

The G0 codes are efficient for small values of n because (1) they are easy to construct based on a table of primes, (2) they are easy to decode, and (3) they are about as long as the Fibonacci codes or the standard binary representation (the β code) of the integers. However, for large values of n, the G0 codes become too long because, as Table 3.43 illustrates, the primes are denser than the powers of 2 or the Fibonacci numbers. Also, a large even number can normally be written as the sum of two primes in many different ways. For large values of n, a large table of primes is therefore needed and it may take the encoder a while to determine the pair of primes that yields the shortest code for a given large integer n. (The shortest code is normally produced by two primes of similar sizes. Writing $n = a + b$ where a is small and b is large, results in a long G0 code. The best Goldbach partition for 11,230, for example, is $2003 + 9227$ and the worst one is $17 + 11213$.)

n:	1	2	3	4	5	6	7	8	9	10	11	12
P_n:	1	3	5	7	11	13	17	19	23	29	31	37
2^{n-1}:	1	2	4	8	16	32	64	128	256	512	1024	2048
F_n:	1	1	2	3	5	8	13	21	34	55	89	144

Table 3.43: Growth of Primes, Powers of 2, and Fibonacci Numbers.

Thus, the G0 code of a large integer n is long, and since it has only two 1's, it must have many zeros and may have one or two runs of zeros. This property is the basis for the Goldbach G1 code. The principle of G1 is to determine two primes P_i and P_j (where $i \le j$) whose sum yields a given integer n, and encode the pair $(i, j-i+1)$ with two gamma codes. Thus, for example, $n = 100$ can be written as the following sums $3+97$, $11+89$, $17+83$, $29+71$, $41+59$, and $47+53$. We select $47+53 = P_{15} + P_{16}$, yielding the pair $(15, 16-15+1 = 2)$ and the two gamma codes 0001111:010 that are concatenated to form the G1 code of 100. For comparison, selecting the Goldbach partition $100 = 3 + 97$ yields the indexes 2 and 25, the pair $(2, 25-2+1)$, and the two gamma codes 010:000011000, two bits longer. Notice that i may equal j, which is why the pair $(i, j-i+1)$ and not $(i, j-i)$ is encoded. The latter may result in a second element of 0.

Table 3.44 lists several G1 codes of even integers. It is again obvious that the lengths of the G1 codes increase but not monotonically. The lengths of the corresponding gamma codes are also listed for comparison, and it is clear that the G1 code is the winner in most cases. Figure 3.45 illustrates the lengths of the G1, Elias gamma code, and the C^1 code of Fraenkel and Klein (Section 3.20).

The last two rows of Table 3.44 list two alternatives for the G1 code of 40, and they really are two bits longer than the best code of 40. However, selecting the Goldbach partition with the most similar primes does not always yield the shortest code, as in the

n	Sum	Indexes	Pair	Codeword	Len.	$\lvert\gamma(n)\rvert$
2	1+1	1,1	1,1	1:1	2	3
4	1+3	1,2	1,2	1:010	4	5
6	3+3	2,2	2,1	010:1	4	5
8	3+5	2,3	2,2	010:010	6	7
10	3+7	2,4	2,3	010:011	6	7
12	5+7	3,4	3,2	011:010	6	7
14	7+7	4,4	4,1	00100:1	6	7
16	5+11	3,5	3,3	011:011	6	9
18	7+11	4,5	4,2	00100:011	8	9
20	7+13	4,6	4,3	00100:011	8	9
30	13+17	6,7	6,2	00110:010	8	9
40	17+23	7,9	7,3	00111:011	8	11
50	13+37	6,12	6,7	00110:00111	10	11
60	29+31	10,11	10,2	0001010:010	10	11
70	29+41	10,13	10,4	0001010:00100	12	13
80	37+43	12,14	12,3	0001100:011	10	13
90	43+47	14,15	14,2	0001110:010	10	13
100	47+53	15,16	15,2	0001111:010	10	13
40	3+37	2,12	2,11	010:0001011	10	11
40	11+29	5,10	5,6	00101:00110	10	11

Table 3.44: The Goldbach G1 Code.

case of $50 = 13+37 = 19+31$. Selecting the first pair (where the two primes differ most) yields the indexes 6, 12 and the pair (6, 7) for a total gamma codes of 10 bits. Selecting the second pair (the most similar primes), on the other hand, results in indexes 8, 11 and a pair (8, 4) for a total of 12 bits of gamma codes.

In order for it to be useful, the G1 code has to be extended to arbitrary positive integers, a task that is done in two ways as follows:

1. Map the positive integer n to the even number $N = 2(n+3)$ and encode N in G1. This is a natural extension of G1, but it results in indexes that are about 60% larger and thus generate long codes. For example, the extended code for 20 would be the original G1 code of $46 = 17 + 29$. The indexes are 7, 10, the pair is (7, 4), and the codes are 00111:00100, two bits longer than the original G1 code of 20.

2. A better way to extend G1 is to consider various cases and handle each differently so as to obtain the best codes in every case. The developer, Peter Fenwick, refers to the result as the G2 code. The cases are as follows:

2.1. The integers 1 and 2 are encoded as 110 and 111, respectively (no other codes will start with 11).

2.2. The even integers are encoded as in G1, but with a small difference. Once it is determined that $n = P_i + P_j$, we encode the pair $(i+1, j-i+1)$ instead of $(i, j-i+1)$. Thus, if $i = 1$, it is encoded as 010, the gamma code of 2. This guarantees that the G2 code of an even integer will not start with a 1 and will always have the form 0. . . :0. . . .

2.3. If n is the prime P_i, it is encoded as the gamma code of $(i+1)$ followed by a

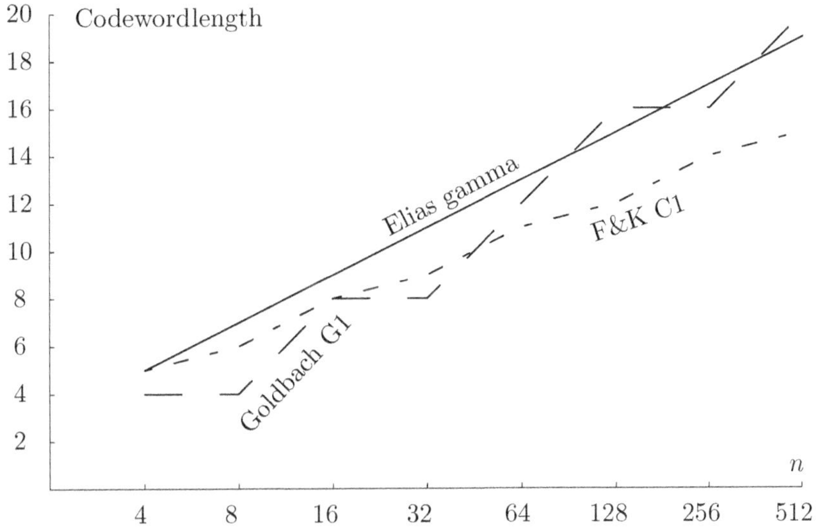


```
Clear[g1, g2, g3];
n=Table[2^i, {i,2,9}];
 (* ELias gamma *)
Table[{i, IntegerPart[1+2Log[2, n[[i]]]]}, {i,1,8}];
g1=ListPlot[%, PlotJoined->True];
 (* Goldbach G2 *)
m[i_]:=4IntegerPart[Log[2, (1.15n[[i]]/2)/Log[E, n[[i]]/2]]];
Table[{i, m[i]}, {i,1,8}];
g2=ListPlot[%, PlotJoined->True, PlotStyle->{Dashing[{0.05, 0.05}]}];
 (* Fraenkel Klein C^1 *)
Table[{i, IntegerPart[Log[GoldenRatio, n[[i]] Sqrt[5]]+1]}, {i,1,8}];
g3=ListPlot[%, PlotJoined->True,
 PlotStyle->{Dashing[{0.01,0.03,0.02,0.03}]}];
Show[g1, g2, g3,  PlotRange->{{0, 8},{0, 20}},
  Ticks->{{{1,"4"},{2,"8"},{3,"16"},{4,"32"},
          {5,"64"},{6,"128"},{7,"256"},{8,"512"}},
          {0, 2, 4, 6, 8, 10, 12, 14, 16, 18, 20}},
 TextStyle->{FontFamily->"cmr10", FontSize->10}]
```

Figure 3.45: Codeword Length Versus n for Three Codes.

single 1 to yield 0. . . :1.

2.4. If n is odd but is not a prime, its G2 code starts with a single 1 followed by the G2 code of the even number $n-1$. The resulting gamma codes have the form 1:0. . . :0. . . .

Table 3.46 lists some examples of the G2 code and compares their lengths to the lengths of the gamma and omega codes. In most cases, G2 is the shortest of the three, but in some cases, most notably when n is a power of 2, G2 is longer than the other codes.

Given an even integer n, there is no known formula to locate any of its Goldbach partitions. Thus, we cannot determine the precise length of any Goldbach code, but we can provide an estimate, and the G1 code length is the easiest to estimate. We assume that the even number $2n$ is the sum of two primes, each of order n. According to the prime number theorem, the number $\pi(n)$ of primes less than n is approximately proportional to $n/\ln n$. Extensive experiments carried out by Peter Fenwick, the developer of

n	Codeword	Len.	$\|\gamma(n)\|$	$\|\omega(n)\|$
1	110	3	1	1
2	111	3	3	3
3	0101	4	3	3
4	010010	6	5	6
5	0111	4	5	6
6	011010	6	5	6
7	001001	6	5	6
8	011011	6	7	7
9	1011011	7	7	7
10	01100100	8	7	7
11	001011	6	7	7
12	00101010	8	7	7
13	001101	6	7	7
14	00110010	8	7	7
15	100110010	9	7	7
16	0010100100	10	9	11
17	001111	6	9	11
18	00111010	8	9	11
19	00010001	8	9	11
20	00111011	8	9	11
30	0001010010	10	9	11
40	0001100011	10	11	12
50	0001111011	10	11	12
60	000010001010	12	11	12
70	000010011011	12	13	13
80	000010110010	12	13	13
90	000011000010	12	13	13
100	000011001011	12	13	13

Table 3.46: The Goldbach G2 Code.

these codes, indicate that for values up to 1000, a good constant of proportionality is 1.15. Thus, we estimate index i at $i \approx 1.15n/\ln n$, resulting in a gamma code of

$$2\left\lfloor \log_2 \frac{1.15n}{\ln n} \right\rfloor + 1 \stackrel{\text{def}}{=} 2L + 1 \quad \text{bits.} \tag{3.10}$$

We estimate the second element of the pair $(i, j - i + 1)$ at $i/2$ (Table 3.44 shows that it is normally smaller), which implies that the second gamma code is two bits shorter than the first. The total length of the G1 code is therefore $(2L + 1) + (2L - 1) = 4L$, where L is given by Equation (3.10). Direct computations show that the G1 code is generally shorter than the gamma code for values up to 100 and longer for larger values.

The length of the G2 code is more difficult to estimate, but a simple experiment that computed these codes for n values from 2 to 512 indicates that their lengths (which often vary by 2–3 bits from code to code) can be approximated by the smooth function

$2 + \frac{13}{8}\log_2 n$. For $n = 512$, this expression has the value 16.625, which is 1.66 times the length of the binary representation of 512 (10 bits). This should be contrasted with the gamma code, where the corresponding factor is 2.

> I can envision an abstract of a paper, circa 2100, that reads: "We can show, in a certain precise sense, that the Goldbach conjecture is true with probability larger than 0.99999, and that its complete truth could be determined with a budget of \$10B.")
>
> —Doron Zeilberger (1993)

3.23 Additive Codes

The fact that the Fibonacci numbers and the Goldbach conjecture lead to simple, efficient codes suggests that other number sequences may be employed in the same way to create similar, and perhaps even better, codes. Specifically, we may extend the Goldbach codes if we find a sequence S of "basis" integers such that any integer n can be expressed as the sum of two elements of S (it is possible to search for similar sequences that can express any n as the sum of three, four, or more elements, but we restrict our search to sums of two sequence elements). Given such a sequence, we can conceive a code similar to G0, but for all the nonnegative integers, not just the even integers. Given several such sequences, we can call the resulting codes "additive codes."

One technique to generate a sequence of basis integers is based on the sieve principle (compare with the sieve of Eratosthenes). We start with a short sequence (a basis set) whose elements a_i are sufficient to represent each of the first few positive integers as a sum $a_i + a_j$. This initial sequence is then enlarged in steps. In a general step, we have just added a new element a_k and ended up with the sequence $S = (0, a_1, a_2, \ldots, a_k)$ such that each integer $1 \le n \le a_k$ can be represented as a sum of two elements of S. We now generate all the sums $a_i + a_k$ for i values from 1 to k and check to verify that they include $a_k + 1$, $a_k + 2$, and so on. When we find the first missing integer, we append it to S as element a_{k+1}. We now know that each integer $1 \le n \le a_{k+1}$ can be represented as a sum of two elements from S, and we can continue to extend sequence S of basis integers.

This sieve technique is illustrated with the basis set $(0, 1, 2)$. Any of 0, 1, and 2 can be represented as the sum of two elements from this set. We generate all the possible sums $a_i + a_k = a_i + 2$ and obtain 3 and 4. Thus, 5 is the next integer that cannot be generated, and it becomes the next element of the set. Adding $a_i + 5$ yields 5, 6, and 7, so the next element is 8. Adding $a_i + 8$ in the sequence $(0, 1, 2, 5, 8)$ yields 8, 9, 10, 13, and 16, so the next element should be 11. Adding $a_i + 11$ in the sequence $(0, 1, 2, 5, 8, 11)$ yields 11, 12, 13, 16, 19, and 22, so the next element should be 14. Continuing in this way, we obtain the sequence 0, 1, 2, 5, 8, 11, 14, 16, 20, 23, 26, 29, 33, 46, 50, 63, 67, 80, 84, 97, 101, 114, 118, 131, 135, 148, 152, 165, 169, 182, 186, 199, 203, 216, 220, 233, and 237 whose 37 elements can represent every integer from 0 to 250 as the sum of two elements. The equivalent Goldbach code for integers up to 250 requires 53 primes, so it has the potential of being longer.

The initial basis set may be as small as $(0, 1)$, but may also contain other integers (seeds). In fact, the seeds determine the content and performance of the additive sequence. Starting with just the smallest basic set $(0, 1)$, adding $a_i + 1$ yields 1 and 2, so the next element should be 3. Adding $a_i + 3$ in the sequence $(0, 1, 3)$ yields 3, 4, and 6, so the next element should be 5. Adding $a_i + 5$ in the sequence $(0, 1, 3, 5)$ yields 5, 6, 8, and 10, so the next element should be 7. We end up with a sequence of only odd numbers, which is why it may be a good idea to start with the basic sequence plus some even seeds. Table 3.47 lists some additive codes based on the additive sequence $(0, 1, 3, 5, 7, 9, 11, 12, 25, 27, 29, 31, 33, 35, \ldots)$ and compares their lengths to the lengths of the corresponding gamma codes.

n	Sum	Indexes	Pair	Codeword	Len.	$\lvert\gamma(n)\rvert$
10	$3+7$	3,5	3,3	011:011	6	7
11	$0+11$	1,7	1,7	1:00111	6	7
12	$1+11$	2,7	2,6	010:00110	8	7
13	$1+12$	2,8	2,7	010:00111	8	7
14	$7+7$	5,5	5,1	00101:1	6	7
15	$3+12$	3,8	3,6	011:00110	8	7
16	$7+9$	5,6	5,2	00101:010	8	9
17	$5+12$	4,8	4,5	00100:00101	10	9
18	$7+11$	5,7	5,3	00101:011	8	9
20	$9+11$	6,7	6,2	00110:010	8	9
30	$5+25$	4,9	4,6	00100:00110	10	9
40	$11+29$	7,11	7,5	00111:00101	10	11
50	$25+25$	9,9	9,1	0001001:1	8	11
60	$29+31$	11,12	11,2	0001011:010	10	11
70	$35+35$	14,14	14,1	0001110:1	8	13
80	$0+80$	1,20	1,20	1:000010100	10	13
90	$29+61$	11,17	11,7	0001011:00111	14	13
100	$3+97$	3,23	3,21	011:000010101	14	13

Table 3.47: An Additive Code.

The developer of these codes presents a summary where the gamma, delta, Fibonacci, G1, and additive codes are compared for integers n from 1 to 256, and concludes that for values from 4 to 30, the additive code is the best of the five, and for the entire range, it is better than the gamma and delta codes.

Given integer data in a certain interval, it seems that the best additive codes to compress the data are those generated by the shortest additive sequence. Determining this sequence can be done by a brute force approach that tries many sets of seeds and computes the additive sequence generated by each set.

Ulam Sequence. The Ulam sequence (u, v) is defined by $a_1 = u$, $a_2 = v$, and a_i for $i > 2$ is the smallest integer that can be expressed *uniquely* as the sum of two distinct earlier elements. The numbers generated this way are sometimes called u-numbers or Ulam numbers. A basic reference is [Ulam 06].

The first few elements of the $(1, 2)$-Ulam sequence are 1, 2, 3, 4, 6, 8, 11, 13, 16,.... The element following the initial $(1, 2)$ is 3, because $3 = 1 + 2$. The next element is $4 = 1 + 3$. (There is no need to worry about $4 = 2 + 2$, because this is a sum of two identical elements instead of two distinct elements.) The integer 5 is not an element of the sequence because it is not uniquely representable, but $6 = 2 + 4$ is.

These simple rules make it possible to generate Ulam sequences for any pair (u, v) of integers. Table 3.48 lists several examples (the "Sloane" labels in the second column refer to integer sequences from [Sloane 06]).

(u, v)	Sloane	Sequence
$(1, 2)$	A002858	1, 2, 3, 4, 6, 8, 11, 13, 16, 18, ...
$(1, 3)$	A002859	1, 3, 4, 5, 6, 8, 10, 12, 17, 21, ...
$(1, 4)$	A003666	1, 4, 5, 6, 7, 8, 10, 16, 18, 19, ...
$(1, 5)$	A003667	1, 5, 6, 7, 8, 9, 10, 12, 20, 22, ...
$(2, 3)$	A001857	2, 3, 5, 7, 8, 9, 13, 14, 18, 19, ...
$(2, 4)$	A048951	2, 4, 6, 8, 12, 16, 22, 26, 32, 36, ...
$(2, 5)$	A007300	2, 5, 7, 9, 11, 12, 13, 15, 19, 23, ...

Table 3.48: Some Ulam Sequences.

It is clear that Ulam sequences are additive and can be used to generate additive codes. The following *Mathematica* code to generate such a sequence is from [Sloane 06], sequence A002858.

```
Ulam4Compiled = Compile[{{nmax, _Integer}, {init, _Integer, 1}, {s, _Integer}},
 Module[{ulamhash = Table[0, {nmax}], ulam = init},
 ulamhash[[ulam]] = 1;
 Do[ If[Quotient[Plus @@ ulamhash[[i - ulam]], 2] == s, AppendTo[ulam, i];
 ulamhash[[i]] = 1], {i, Last[init] + 1, nmax}]; ulam]];
Ulam4Compiled[355, {1, 2}, 1]
```

Stanislaw Marcin Ulam (1909–1984)

One of the most prolific mathematicians of the 20th century, Stanislaw Ulam also had interests in astronomy and physics. He worked on the hydrogen bomb at the Los Alamos National Laboratory, proposed the Orion project for nuclear propulsion of space vehicles, and originated the Monte-Carlo method (and also coined this term). In addition to his important original work, mostly in point set topology, Ulam was also interested in mathematical recreations, games, and oddities. The following quotation, by his friend and colleague Gian-Carlo Rota, summarizes Ulam's unusual personality and talents.

"Ulam's mind is a repository of thousands of stories, tales, jokes, epigrams, remarks, puzzles, tongue-twisters, footnotes, conclusions, slogans, formulas, diagrams, quotations, limericks, summaries, quips, epitaphs, and headlines. In the course of a normal conversation he simply pulls out of his mind the fifty-odd relevant items, and presents them in linear succession. A second-order memory prevents him from repeating himself too often before the same public."

There is another Ulam sequence that is constructed by a simple rule. Start with any positive integer n and construct a sequence as follows:

1. If $n = 1$, stop.
2. If n is even, the next number is $n/2$; go to step 1.
3. If n is odd, the next number is $3n + 1$; go to step 1.

Here are some examples for the first few integers: (1), (10, 5, 16, 8, 4, 2, 1), (2, 1), (16, 8, 4, 2, 1), (3, 10, 5, ...) (22, 11, 34, 17, 52, 26, 13, 40, 20, 10, 5, ...).

These sequences are sometimes known as the $3x + 1$ problem, because no one has proved that they always reach the value 1. However, direct checking up to 100×2^{50} suggests that this may be so.

3.24 Golomb Code

The seventeenth-century French mathematician Blaise Pascal is known today mostly for his contributions to the field of probability, but during his short life he made important contributions to many areas. It is generally agreed today that he invented (an early version of) the game of roulette (although some believe that this game originated in China and was brought to Europe by Dominican monks who were trading with the Chinese). The modern version of roulette appeared in 1842.

The roulette wheel has 37 shallow depressions (known as slots) numbered 0 through 36 (the American version has 38 slots numbered 00, 0, and 1 through 36). The dealer (croupier) spins the wheel while sending a small ball rolling in the opposite direction inside the wheel. Players can place bets during the spin until the dealer says "no more bets." When the wheel stops, the slot where the ball landed determines the outcome of the game. Players who bet on the winning number are paid according to the type of bet they placed, while players who bet on the other numbers lose their entire bets to the house. [Bass 92] is an entertaining account of an attempt to compute the result of a roulette spin in real time.

The simplest type of bet is on a single number. A player winning this bet is paid 35 times the amount bet. Thus, a player who plays the game repeatedly and bets \$1 each time expects to lose 36 games and win one game out of every set of 37 games on average. The player therefore loses on average \$37 for every \$35 won.

The probability of winning a game is $p = 1/37 \approx 0.027027$ and that of losing a game is the much higher $q = 1 - p = 36/37 \approx 0.972973$. The probability $P(n)$ of winning once and losing $n - 1$ times in a sequence of n games is the product $q^{n-1}p$. This probability is normalized because

$$\sum_{n=1}^{\infty} P(n) = \sum_{n=1}^{\infty} q^{n-1}p = p\sum_{n=0}^{\infty} q^n = \frac{p}{1-q} = \frac{p}{p} = 1.$$

As n grows, $P(n)$ shrinks slowly because of the much higher value of q. The values of $P(n)$ for $n = 1, 2, \ldots, 10$ are 0.027027, 0.026297, 0.025586, 0.024895, 0.024222, 0.023567, 0.022930, 0.022310, 0.021707, and 0.021120.

The probability function $P(n)$ is said to obey a *geometric distribution.* The reason for the name "geometric" is the resemblance of this distribution to the geometric sequence. A sequence where the ratio between consecutive elements is a constant q is called geometric. Such a sequence has elements a, aq, aq^2, aq^3, ... The (infinite) sum of these elements is a geometric series $\sum_{i=0}^{\infty} aq^i$. The interesting case is where q satisfies $-1 < q < 1$, in which the series converges to $a/(1-q)$. Figure 3.49 shows the geometric distribution for $p = 0.2$, 0.5, and 0.8.

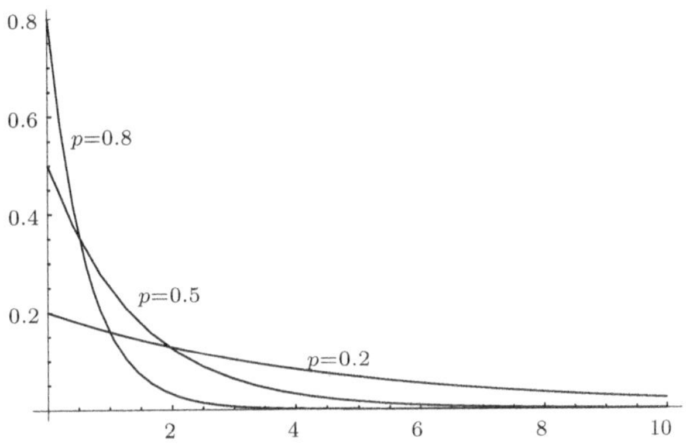

Figure 3.49: Geometric Distributions for $p = 0.2$, 0.5, and 0.8.

Certain data compression methods are based on run-length encoding (RLE). Imagine a binary string where a 0 appears with probability p and a 1 appears with probability $1-p$. If p is large, there will be runs of zeros, suggesting the use of RLE to compress the string. The probability of a run of n zeros is p^n, and the probability of a run of n zeros followed by a 1 is $p^n(1-p)$, indicating that run lengths are distributed geometrically. A naive approach to compressing such a string is to compute the probability of each run length and apply the Huffman method to obtain the best prefix codes for the run lengths. In practice, however, there may be a large number of run lengths and this number may not be known in advance. A better approach is to construct an infinite family of optimal prefix codes, such that no matter how long a run is, there will be a code in the family to encode it. The codes in the family must depend on the probability p, so we are looking for an infinite family of *parametrized* prefix codes. The Golomb codes presented here [Golomb 66] are such codes and they are the best ones for the compression of data items that are distributed geometrically.

Let's first examine a few numbers to see why such codes must depend on p. For $p = 0.99$, the probabilities of runs of two zeros and of 10 zeros are $0.99^2 = 0.9801$ and $0.99^{10} = 0.9$, respectively (both large). In contrast, for $p = 0.6$, the same run lengths have the much smaller probabilities of 0.36 and 0.006. The ratio 0.9801/0.36 is 2.7225, but the ratio 0.9/0.006 is the much greater 150. Thus, a large p implies higher probabilities for long runs, whereas a small p implies that long runs will be rare.

Two relevant statistical concepts are the mean and median of a sequence of run lengths. They are illustrated by the binary string

$$00000100110001010000001110100010000010001001000110100001001 \qquad (3.11)$$

that has the 18 run lengths 5, 2, 0, 3, 1, 6, 0, 0, 1, 3, 5, 3, 2, 3, 0, 1, 4, and 2. Its mean is the average $(5+2+0+3+1+6+0+0+1+3+5+3+2+3+0+1+4+2)/18 \approx 2.28$. Its median m is the value such that about half the run lengths are shorter than m and about half are equal to or greater than m. To find m, we sort the 18 run lengths to obtain 0, 0, 0, 0, 1, 1, 1, 2, 2, 2, 3, 3, 3, 3, 4, 5, 5, and 6 and find that the median (the central number) is 2.

We are now ready for a description of the Golomb code. The main feature of this code is its coding efficiency when the data consists of two asymmetric events, one common and the other one rare, that are interleaved.

Encoding. The Golomb code for nonnegative integers n depends on the choice of a parameter m (we'll see later that for RLE, m should depend on the probability p and on the median of the run lengths). Thus, it is a parametrized prefix code, which makes it especially useful in cases where good values for the parameter can be computed or estimated. The first step in constructing the Golomb code of the nonnegative integer n is to compute the three quantities q (quotient), r (remainder), and c by

$$q = \left\lfloor \frac{n}{m} \right\rfloor, \quad r = n - qm, \text{ and } c = \lceil \log_2 m \rceil,$$

following which the code is constructed in two parts; the first is the value of q, coded in unary, and the second is the binary value of r coded in a special way. The first $2^c - m$ values of r are coded, as unsigned integers, in $c-1$ bits each, and the rest are coded in c bits each (ending with the biggest c-bit number, which consists of c 1's). The case where m is a power of 2 ($m = 2^c$) is special because it requires no $(c-1)$-bit codes. We know that $n = r + qm$; so once a Golomb code is decoded, the values of q and r can be used to easily reconstruct n. The case $m = 1$ is also special. In this case, $q = n$ and $r = c = 0$, implying that the Golomb code of n is its unary code.

Examples. Choosing $m = 3$ produces $c = 2$ and the three remainders 0, 1, and 2. We compute $2^2 - 3 = 1$, so the first remainder is coded in $c - 1 = 1$ bit to become 0, and the remaining two are coded in two bits each ending with 11_2, to become 10 and 11. Selecting $m = 5$ results in $c = 3$ and produces the five remainders 0 through 4. The first three ($2^3 - 5 = 3$) are coded in $c-1 = 2$ bits each, and the remaining two are each coded in three bits ending with 111_2. Thus, 00, 01, 10, 110, and 111. The following simple rule shows how to encode the c-bit numbers such that the last of them will consist of c 1's. Denote the largest of the $(c-1)$-bit numbers by b, then construct the integer $b+1$ in $c-1$ bits, and append a zero on the right. The result is the first of the c-bit numbers and the remaining ones are obtained by incrementing.

Table 3.50 shows some examples of m, c, and $2^c - m$, as well as some Golomb codes for $m = 2$ through 13.

For a somewhat longer example, we select $m = 14$. This results in $c = 4$ and produces the 14 remainders 0 through 13. The first two ($2^4 - 14 = 2$) are coded in $c - 1 = 3$ bits each, and the remaining 12 are coded in four bits each, ending with

m	2	3	4	5	6	7	8	9	10	11	12	13	14	15	16
c	1	2	2	3	3	3	3	4	4	4	4	4	4	4	4
2^c-m	0	1	0	3	2	1	0	7	6	5	4	3	2	1	0

<table>
<tr><th>m/n</th><th>0</th><th>1</th><th>2</th><th>3</th><th>4</th><th>5</th><th>6</th><th>7</th><th>8</th><th>9</th><th>10</th><th>11</th><th>12</th></tr>
<tr><td>2</td><td>0|0</td><td>0|1</td><td>10|0</td><td>10|1</td><td>110|0</td><td>110|1</td><td>1110|0</td><td>1110|1</td><td>11110|0</td><td>11110|1</td><td>111110|0</td><td>111110|1</td><td>1111110|0</td></tr>
<tr><td>3</td><td>0|0</td><td>0|10</td><td>0|11</td><td>10|0</td><td>10|10</td><td>10|11</td><td>110|0</td><td>110|10</td><td>110|11</td><td>1110|0</td><td>1110|10</td><td>1110|11</td><td>11110|0</td></tr>
<tr><td>4</td><td>0|00</td><td>0|01</td><td>0|10</td><td>0|11</td><td>10|00</td><td>10|01</td><td>10|10</td><td>10|11</td><td>110|00</td><td>110|01</td><td>110|10</td><td>110|11</td><td>11110|00</td></tr>
<tr><td>5</td><td>0|00</td><td>0|01</td><td>0|10</td><td>0|110</td><td>0|111</td><td>10|00</td><td>10|01</td><td>10|10</td><td>10|110</td><td>10|111</td><td>110|00</td><td>110|01</td><td>110|10</td></tr>
<tr><td>6</td><td>0|00</td><td>0|01</td><td>0|100</td><td>0|101</td><td>0|110</td><td>0|111</td><td>10|00</td><td>10|01</td><td>10|100</td><td>10|101</td><td>10|110</td><td>10|111</td><td>110|00</td></tr>
<tr><td>7</td><td>0|00</td><td>0|010</td><td>0|011</td><td>0|100</td><td>0|101</td><td>0|110</td><td>0|111</td><td>10|00</td><td>10|010</td><td>10|011</td><td>10|100</td><td>10|101</td><td>10|110</td></tr>
<tr><td>8</td><td>0|000</td><td>0|001</td><td>0|010</td><td>0|011</td><td>0|100</td><td>0|101</td><td>0|110</td><td>0|111</td><td>10|000</td><td>10|001</td><td>10|010</td><td>10|011</td><td>10|100</td></tr>
<tr><td>9</td><td>0|000</td><td>0|001</td><td>0|010</td><td>0|011</td><td>0|100</td><td>0|101</td><td>0|110</td><td>0|1110</td><td>0|1111</td><td>10|000</td><td>10|001</td><td>10|010</td><td>10|011</td></tr>
<tr><td>10</td><td>0|000</td><td>0|001</td><td>0|010</td><td>0|011</td><td>0|100</td><td>0|101</td><td>0|1100</td><td>0|1101</td><td>0|1110</td><td>0|1111</td><td>10|000</td><td>10|001</td><td>10|010</td></tr>
<tr><td>11</td><td>0|000</td><td>0|001</td><td>0|010</td><td>0|011</td><td>0|100</td><td>0|1010</td><td>0|1011</td><td>0|1100</td><td>0|1101</td><td>0|1110</td><td>0|1111</td><td>10|000</td><td>10|001</td></tr>
<tr><td>12</td><td>0|000</td><td>0|001</td><td>0|010</td><td>0|011</td><td>0|1000</td><td>0|1001</td><td>0|1010</td><td>0|1011</td><td>0|1100</td><td>0|1101</td><td>0|1110</td><td>0|1111</td><td>10|000</td></tr>
<tr><td>13</td><td>0|000</td><td>0|001</td><td>0|010</td><td>0|0110</td><td>0|0111</td><td>0|1000</td><td>0|1001</td><td>0|1010</td><td>0|1011</td><td>0|1100</td><td>0|1101</td><td>0|1110</td><td>0|1111</td></tr>
</table>

Table 3.50: Some Golomb Codes for $m = 2$ Through 13.

1111_2 (and as a result starting with 0100_2). Thus, we have 000, 001, followed by the 12 values 0100, 0101, 0110, 0111, ..., 1111. Table 3.51 lists several detailed examples and Table 3.52 lists 48 codes for $m = 14$ and for $m = 16$. The former starts with two 4-bit codes, followed by sets of 14 codes each that are getting longer by one bit. The latter is simpler because 16 is a power of 2. The Golomb codes for $m = 16$ consist of sets of 16 codes each that get longer by one bit. The Golomb codes for the case where m is a power of 2 have been conceived by Robert F. Rice and are called Rice codes. They are employed by several algorithms for lossless audio compression.

n	0	1	2	3	...	13	14	15	16	17	...	27	28	29	30
$q=\lfloor\frac{n}{14}\rfloor$	0	0	0	0	...	0	1	1	1	1	...	1	2	2	2
unary(q)	0	0	0	0	...	0	10	10	10	10	...	10	110	110	110
r	000	001	0100	0101	...	1111	000	001	0100	0101	...	1111	000	001	0100

Table 3.51: Some Golomb Codes for $m = 14$.

Tables 3.51 and 3.52 illustrate the effect of m on the code length. For small values of m, the Golomb codes start short and rapidly increase in length. They are appropriate for RLE in cases where the probability p of a 0 bit is small, implying very few long runs. For large values of m, the initial codes (for $n = 1, 2, \ldots$) are long, but their lengths increase slowly. Such codes make sense for RLE when p is large, implying that many long runs are expected.

Decoding. The Golomb codes are designed in this special way to facilitate their decoding. We first demonstrate the decoding for the simple case $m = 16$ (m is a power of 2). To decode, start at the left end of the code and count the number A of 1's preceding the first 0. The length of the code is $A + c + 1$ bits (for $m = 16$, this is $A + 5$ bits). If we denote the rightmost five bits of the code by R, then the value of the code is $16A + R$. This simple decoding reflects the way the code was constructed. To encode n with $m = 16$, start by dividing it by 16 to get $n = 16A + R$, then write A 1's followed by a single 0, followed by the 4-bit representation of R.

$m = 14$

n	Code	n	Code
0	0000	24	101100
1	0001	25	101101
		26	101110
2	00100	27	101111
3	00101	28	110000
4	00110	29	110001
5	00111		
6	01000	30	1100100
7	01001	31	1100101
8	01010	32	1100110
9	01011	33	1100111
10	01100	34	1101000
11	01101	35	1101001
12	01110	36	1101010
13	01111	37	1101011
14	10000	38	1101100
15	10001	39	1101101
		40	1101110
16	100100	41	1101111
17	100101	42	1110000
18	100110	43	1110001
19	100111		
20	101000	44	11100100
21	101001	45	11100101
22	101010	46	11100110
23	101011	47	11100111

$m = 16$

n	Code	n	Code
0	00000	24	101000
1	00001	25	101001
2	00010	26	101010
3	00011	27	101011
4	00100	28	101100
5	00101	29	101101
6	00110	30	101110
7	00111	31	101111
8	01000		
9	01001	32	1100000
10	01010	33	1100001
11	01011	34	1100010
12	01100	35	1100011
13	01101	36	1100100
14	01110	37	1100101
15	01111	38	1100110
		39	1100111
16	100000	40	1101000
17	100001	41	1101001
18	100010	42	1101010
19	100011	43	1101011
20	100100	44	1101100
21	100101	45	1101101
22	100110	46	1101110
23	100111	47	1101111

Table 3.52: The First 48 Golomb Codes for $m = 14$ and $m = 16$.

For m values that are not powers of 2, decoding is slightly more involved. Assuming again that a code begins with A 1's, start by removing them and the zero immediately following them. Denote the $c-1$ bits that follow by R. If $R < 2^c - m$, then the total length of the code is $A+1+(c-1)$ (the A 1's, the zero following them, and the $c-1$ bits that follow) and its value is $m\times A+R$. If $R \geq 2^c - m$, then the total length of the code is $A+1+c$ and its value is $m\times A+R'-(2^c-m)$, where R' is the c-bit integer consisting of R and the bit that follows R.

An example is the code $0001xxx$, for $m = 14$. There are no leading 1's, so A is 0. After removing the leading zero, the $c-1 = 3$ bits that follow are $R = 001$. Since $R < 2^c - m = 2$, we conclude that the length of the code is $0+1+(4-1) = 4$ and its value is 001. Similarly, the code $00100xxx$ for the same $m = 14$ has $A = 0$ and $R = 010_2 = 2$. In this case, $R \geq 2^c - m = 2$, so the length of the code is $0+1+c = 5$, the value of R' is $0100_2 = 4$, and the value of the code is $14\times 0+4-2 = 2$.

The JPEG-LS method for lossless image compression (recommendation ISO/IEC CD 14495) employs the Golomb code.

The family of Golomb codes has a close relative, the exponential Golomb codes which are described on page 198.

It is now clear that the best value for m depends on p, and it can be shown that this value is the integer closest to $-1/\log_2 p$ or, equivalently, the value that satisfies

$$p^m \approx 1/2. \tag{3.12}$$

It can also be shown that in the case of a sequence of run lengths, this integer is the median of the run lengths. Thus, for $p = 0.5$, m should be $-1/\log_2 0.5 = 1$. For $p = 0.7$, m should be 2, because $-1/\log_2 0.7 \approx 1.94$, and for $p = 36/37$, m should be 25, because $-1/\log_2(36/37) \approx 25.29$.

It should also be mentioned that Gallager and van Voorhis [Gallager and van Voorhis 75] have refined and extended Equation (3.12) into the more precise relation

$$p^m + p^{m+1} \le 1 < p^m + p^{m-1}. \tag{3.13}$$

They proved that the Golomb code is the best prefix code when m is selected by their inequality. We first show that for a given p, inequality (3.13) has only one solution m. We manipulate this inequality in four steps as follows:

$$\begin{aligned} p^m(1+p) &\le 1 < p^{m-1}(1+p), \\ p^m &\le \frac{1}{(1+p)} < p^{m-1}, \\ m &\ge \frac{1}{\log p} \log \frac{1}{1+p} > m-1, \\ m &\ge -\frac{\log(1+p)}{\log p} > m-1, \end{aligned}$$

from which it is clear that the unique value of m is

$$m = \left\lceil -\frac{\log_2(1+p)}{\log_2 p} \right\rceil. \tag{3.14}$$

Three examples are presented here to illustrate the performance of the Golomb code in compressing run lengths. The first example is the binary string (3.11), which has 41 zeros and 18 ones. The probability of a zero is therefore $41/(41+18) \approx 0.7$, yielding $m = \lceil -\log 1.7/\log 0.7 \rceil = \lceil 1.487 \rceil = 2$. The sequence of run lengths 5, 2, 0, 3, 1, 6, 0, 0, 1, 3, 5, 3, 2, 3, 0, 1, 4, and 2 can therefore be encoded with the Golomb codes for $m = 2$ into the string of 18 codes

1101|100|00|101|01|11100|00|00|01|101|1101|101|100|101|00|01|1100|100.

The result is a 52-bit string that compresses the original 59 bits. There is almost no compression because p isn't large. Notice that string (3.11) has three short runs of 1's, which can be interpreted as four empty runs of zeros. It also has three runs (of zeros) of length 1. The next example is the 94-bit string

00000000001000000000100000001000000000001000000001000000000000100000000100000001000000000010000000,

which is sparser and therefore compresses better. It consists of 85 zeros and 9 ones, so $p = 85/(85+9) = 0.9$. The best value of m is therefore $m = \lceil -\log(1.9)/\log(0.9) \rceil = \lceil 6.09 \rceil = 7$. The 10 runs of zeros have lengths 10, 9, 7, 11, 8, 12, 8, 7, 10, and 7. When encoded by the Golomb codes for $m = 7$, the run lengths become the 47-bit string

10100|10011|1000|10101|10010|10110|10010|1000|10100|1000,

resulting in a compression factor of $94/47 = 2$.

The third, extreme, example is a really sparse binary string that consists of, say, 10^6 bits, of which only 100 are ones. The probability of zero is $p = 10^6/(10^6 + 10^2) = 0.9999$, implying $m = 6932$. There are 101 runs, each about 10^4 zeros long. The Golomb code of 10^4 for $m = 6932$ is 14 bits long, so the 101 runs can be compressed to 1414 bits, yielding the impressive compression factor of 707!

In summary, given a binary string, we can employ the method of run-length encoding to compress it with Golomb codes in the following steps: (1) count the number of zeros and ones, (2) compute the probability p of a zero, (3) use Equation (3.14) to compute m, (4) construct the family of Golomb codes for m, and (5) for each run-length of n zeros, write the Golomb code of n on the compressed stream.

In order for the run lengths to be meaningful, p should be large. Small values of p, such as 0.1, result in a string with more 1's than zeros and thus in many short runs of zeros and long runs of 1's. In such a case, it is possible to use RLE to compress the runs of 1's. In general, we can talk about a binary string whose elements are r and s (for run and stop). For r, we should select the more common element, but it has to be very common (the distribution of r and s should be skewed) for RLE to produce good compression. Values of p around 0.5 result in runs of both zeros and 1's, so regardless of which bit is selected for the r element, there will be many runs of length zero. For example, the string 00011100110000111101000111 has the following run lengths of zeros 3, 0, 0, 2, 0, 4, 0, 0, 0, 1, 3, 0, 0 and similar run lengths of 1's 0, 0, 3, 0, 2, 0, 0, 0, 4, 1, 0, 0, 3. In such a case, RLE is not a good choice for compression and other methods should be considered.

Another approach to adaptive RLE is to use the binary string input so far to estimate p and from it to estimate m, and then use the new value of m to encode the next run length (not the current one because the decoder cannot mimic this). Imagine that three runs of 10, 15, and 21 zeros have been input so far, and the first two have already been compressed. The current run of 21 zeros is first compressed with the current value of m, then a new p is computed as $(10 + 15 + 21)/[(10 + 15 + 21) + 3]$ and is used to update m either from $-1/\log_2 p$ or from Equation (3.14). (The 3 is the number of 1's input so far, one for each run.) The new m is used to compress the next run. The algorithm accumulates the lengths of the runs in variable L and the number of runs in N. Figure 3.53 is a simple pseudocode listing of this method. (A practical implementation should halve the values of L and N from time to time, to prevent them from overflowing.)

> In addition to the codes, Solomon W. Golomb has his "own" Golomb constant: 0.624329988543550870992936383100837244179642620180529286.

3.25 Rice Codes

The Golomb codes constitute a family that depends on the choice of a parameter m. The case where m is a power of 2 ($m = 2^k$) is special and results in a Rice code (sometimes also referred to as Golomb–Rice code), so named after its originator, Robert F. Rice ([Rice 79], [Rice 91], and [Fenwick 96a]). The Rice codes are also related to the subexponential code of Section 3.26. A Rice code depends on the choice of a base

```
L = 0; % initialize
N = 0;
m = 1; % or ask user for m
% main loop
for each run of r zeros do
  construct Golomb code for r using current m.
  write it on compressed stream.
  L = L + r:            % update L, N, and m
  N = N + 1;
  p = L/(L + N);
  m = ⌊-1/log₂ p + 0.5⌋;
endfor;
```

Figure 3.53: Simple Adaptive Golomb RLE Encoding.

k and is computed in the following steps: (1) Separate the sign bit from the rest of the number. This is optional and the bit becomes the most-significant bit of the Rice code. (2) Separate the k LSBs. They become the LSBs of the Rice code. (3) Code the remaining $j = \lfloor n/2^k \rfloor$ bits as either j zeros followed by a 1 or j 1's followed by a 0 (similar to the unary code). This becomes the middle part of the Rice code. Thus, this code is computed with a few logical operations, which is faster than computing a Huffman code, a process that requires sliding down the Huffman tree while collecting the individual bits of the code. This feature is especially important for the decoder, which has to be simple and fast. Table 10.28 shows examples of this code for $k = 2$, which corresponds to $m = 4$ (the column labeled "No. of ones" lists the number of 1's in the middle part of the code). Notice that the codes of this table are identical (except for the extra, optional, sign bit) to the codes on the 3rd row (the row for $m = 4$) of Table 3.50.

i	Binary	Sign	LSB	No. of ones	Code	i	Code
0	0	0	00	0	0|0|00		
1	1	0	01	0	0|0|01	−1	1|0|01
2	10	0	10	0	0|0|10	−2	1|0|10
3	11	0	11	0	0|0|11	−3	1|0|11
4	100	0	00	1	0|10|00	−4	1|10|00
5	101	0	01	1	0|10|01	−5	1|10|01
6	110	0	10	1	0|10|10	−6	1|10|10
7	111	0	11	1	0|10|11	−7	1|10|11
8	1000	0	00	2	0|110|00	−8	1|110|00
11	1011	0	11	2	0|110|11	−11	1|110|11
12	1100	0	00	3	0|1110|00	−12	1|1110|00
15	1111	0	11	3	0|1110|11	−15	1|1110|11

Table 3.54: Various Positive and Negative Rice Codes.

The length of the (unsigned) Rice code of the integer n with parameter k is $1+k+\lfloor n/2^k \rfloor$ bits, indicating that these codes are suitable for data where the integer n appears with a probability $P(n)$ that satisfies $\log_2 P(n) = -(1+k+n/2^k)$ or $P(n) \propto 2^{-n}$, an exponential distribution, such as the Laplace distribution. (See [Kiely 04] for a detailed analysis of the optimal values of the Rice parameter.) The Rice code is instantaneous, once the decoder reads the sign bit and skips to the first 0 from the left, it knows how to generate the left and middle parts of the code. The next k bits should be read and appended to that.

Rice codes are ideal for data items with a Laplace distribution, but other prefix codes exist that are easier to construct and to decode and that may, in certain circumstances, outperform the Rice codes. Table 10.29 lists three such codes. The "pod" code, due to Robin Whittle [firstpr 06], codes the number 0 with the single bit 1, and codes the binary number $1\underbrace{b\ldots b}_{k}$ as $\underbrace{0\ldots 0}_{k+1}1\underbrace{b\ldots b}_{k}$. In two cases, the pod code is one bit longer than the Rice code, in four cases it has the same length, and in all other cases it is shorter than the Rice codes. The Elias gamma code [Fenwick 96a] is identical to the pod code minus its leftmost zero. It is therefore shorter, but does not provide a code for zero (see also Table 3.10). The biased Elias gamma code corrects this fault in an obvious way, but at the cost of making some codes one bit longer.

Number Dec	Number Binary	Pod	Elias gamma	Biased Elias gamma
0	00000	1		1
1	00001	01	1	010
2	00010	0010	010	011
3	00011	0011	011	00100
4	00100	000100	00100	00101
5	00101	000101	00101	00110
6	00110	000110	00110	00111
7	00111	000111	00111	0001000
8	01000	00001000	0001000	0001001
9	01001	00001001	0001001	0001010
10	01010	00001010	0001010	0001011
11	01011	00001011	0001011	0001100
12	01100	00001100	0001100	0001101
13	01101	00001101	0001101	0001110
14	01110	00001110	0001110	0001111
15	01111	00001111	0001111	000010000
16	10000	0000010000	000010000	000010001
17	10001	0000010001	000010001	000010010
18	10010	0000010010	000010010	000010011

Table 3.55: Pod, Elias Gamma, and Biased Elias Gamma Codes.

There remains the question of what base value n to select for the Rice codes. The base determines how many low-order bits of a data symbol are included directly in the

Rice code, and this is linearly related to the variance of the data symbol. Tony Robinson, the developer of the Shorten method for audio compression [Robinson 94], provides the formula $n = \log_2[\log(2)E(|x|)]$, where $E(|x|)$ is the expected value of the data symbols. This value is the sum $\sum |x|p(x)$ taken over all possible symbols x.

Figure 3.56 lists the lengths of various Rice codes and compares them to the length of the standard binary (beta) code.

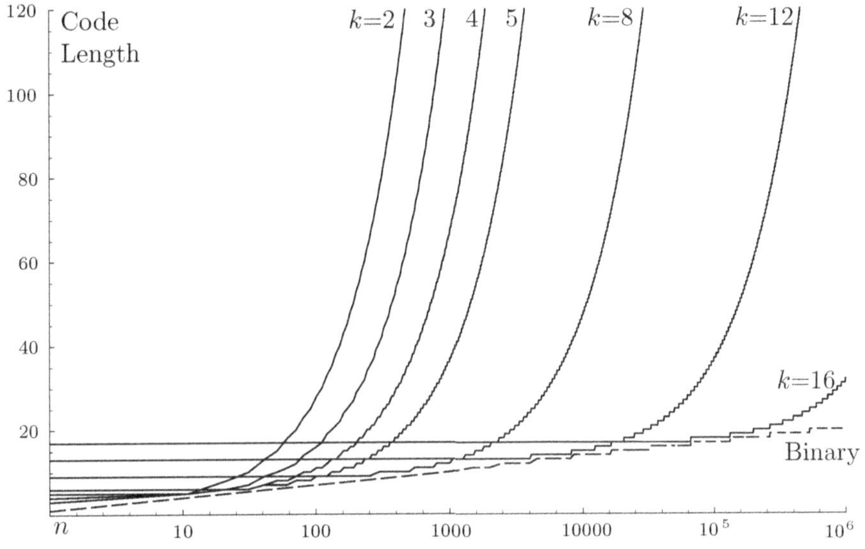


```
(* Lengths of binary code and 7 Rice codes *)
bin[i_] := 1 + Floor[Log[2, i]];
Table[{Log[10, n], bin[n]}, {n, 1, 1000000, 500}];
gb = ListPlot[%, AxesOrigin -> {0, 0}, PlotJoined -> True,
     PlotStyle -> { AbsoluteDashing[{6, 2}]}]
rice[k_, n_] := 1 + k + Floor[n/2^k];
k = 2; Table[{Log[10, n], rice[k, n]}, {n, 1, 10000, 10}];
g2 = ListPlot[%, AxesOrigin -> {0, 0}, PlotJoined -> True]
k = 3; Table[{Log[10, n], rice[k, n]}, {n, 1, 10000, 10}];
g3 = ListPlot[%, AxesOrigin -> {0, 0}, PlotJoined -> True]
k = 4; Table[{Log[10, n], rice[k, n]}, {n, 1, 10000, 10}];
g4 = ListPlot[%, AxesOrigin -> {0, 0}, PlotJoined -> True]
k = 5; Table[{Log[10, n], rice[k, n]}, {n, 1, 10000, 10}];
g5 = ListPlot[%, AxesOrigin -> {0, 0}, PlotJoined -> True]
k = 8; Table[{Log[10, n], rice[k, n]}, {n, 1, 100000, 50}];
g8 = ListPlot[%, AxesOrigin -> {0, 0}, PlotJoined -> True]
k = 12; Table[{Log[10, n], rice[k, n]}, {n, 1, 500000, 100}];
g12 = ListPlot[%, AxesOrigin -> {0, 0}, PlotJoined -> True]
k = 16; Table[{Log[10, n], rice[k, n]}, {n, 1, 1000000, 100}];
g16 = ListPlot[%, AxesOrigin -> {0, 0}, PlotJoined -> True]
Show[gb, g2, g3, g4, g5, g8, g12, g16, PlotRange -> {{0, 6}, {0, 120}}]
```

Figure 3.56: Lengths of Various Rice Codes.

3.26 Subexponential Code

The subexponential code of this section is related to the Rice codes. Like the Golomb codes and the Rice codes, the subexponential code depends on a parameter $k \ge 0$. The main feature of the subexponential code is its length. For integers $n < 2^{k+1}$, the code length increases linearly with n, but for larger values of n, it increases logarithmically. The subexponential code of the nonnegative integer n is computed in two steps. In the first step, values b and u are calculated by

$$b = \begin{cases} k, & \text{if } n < 2^k, \\ \lfloor \log_2 n \rfloor, & \text{if } n \ge 2^k, \end{cases} \quad \text{and} \quad u = \begin{cases} 0, & \text{if } n < 2^k, \\ b - k + 1, & \text{if } n \ge 2^k. \end{cases}$$

In the second step, the unary code of u (in $u+1$ bits) is followed by the b least-significant bits of n to become the subexponential code of n. Thus, the total length of the code is

$$u + 1 + b = \begin{cases} k + 1, & \text{if } n < 2^k, \\ 2\lfloor \log_2 n \rfloor - k + 2, & \text{if } n \ge 2^k. \end{cases}$$

Table 7.139 lists examples of the subexponential code for various values of n and k. It can be shown that for a given n, the code lengths for consecutive values of k differ by at most 1.

n	$k=0$	$k=1$	$k=2$	$k=3$	$k=4$	$k=5$
0	0|	0|0	0|00	0|000	0|0000	0|00000
1	10|	0|1	0|01	0|001	0|0001	0|00001
2	110|0	10|0	0|10	0|010	0|0010	0|00010
3	110|1	10|1	0|11	0|011	0|0011	0|00011
4	1110|00	110|00	10|00	0|100	0|0100	0|00100
5	1110|01	110|01	10|01	0|101	0|0101	0|00101
6	1110|10	110|10	10|10	0|110	0|0110	0|00110
7	1110|11	110|11	10|11	0|111	0|0111	0|00111
8	11110|000	1110|000	110|000	10|000	0|1000	0|01000
9	11110|001	1110|001	110|001	10|001	0|1001	0|01001
10	11110|010	1110|010	110|010	10|010	0|1010	0|01010
11	11110|011	1110|011	110|011	10|011	0|1011	0|01011
12	11110|100	1110|100	110|100	10|100	0|1100	0|01100
13	11110|101	1110|101	110|101	10|101	0|1101	0|01101
14	11110|110	1110|110	110|110	10|110	0|1110	0|01110
15	11110|111	1110|111	110|111	10|111	0|1111	0|01111
16	111110|0000	11110|0000	1110|0000	110|0000	10|0000	0|10000

Table 3.57: Some Subexponential Codes.

Subexponential codes are used in the progressive FELICS method for image compression [Howard and Vitter 94a].

3.27 Codes Ending with "1"

In general, the particular bits that constitute a code are irrelevant. Given a set of codewords that have the desired properties, we don't check the individual bits of a code and object if there is a dearth of zeros or too few 1's. Similarly, we don't complain if most or all of the codes start with a 1 or end with a 0. The important requirements of variable-length codes are (1) to have a set of codes that feature the shortest average length for a given statistical distribution of the source symbols and (2) to have a UD code. However, there may be applications where it is advantageous to have codes that start or end in a special way, and this section presents prefix codes that end with a 1. The main contributors to this line of research are R. Capocelli, A. De Santis, T. Berger, and R. Yeung (see [Capocelli and De Santis 94] and [Berger and Yeung 90]). They discuss special applications and coin the term "feasible codes" for this type of prefix codes. They also show several ways to construct such codes and prove the following bounds on their average lengths. The average length E of an optimum feasible code for a discrete source with entropy H satisfies $H + p_N \leq E \leq H + 1.5$, where p_N is the smallest probability of a symbol from the source.

This section discusses a simple way to construct a set of feasible codes from a set of Huffman codes. The main idea is to start with a set of Huffman codes and append a 1 to each code that ends with a 0. Such codes are called "derived codes." They are easy to construct, but are not always the best feasible codes (they are not optimal). The construction of the codes is as follows: Given a set of symbols and their probabilities, construct their Huffman codes. The subset of codes ending with a 0 is denoted by C_0 and the subset of codes ending with a 1 is denoted by C_1. We also denote by $p(C_0)$ and $p(C_1)$ the sum of probabilities of the symbols in the two subsets, respectively. Notice that $p(C_0)+p(C_1)=1$, which implies that either $p(C_0)$ or $p(C_1)$ is less than or equal to $1/2$ and the other one is greater than or equal to $1/2$. If $p(C_0) \leq 1/2$, the derived code is constructed by appending a 1 to each codeword in C_0; the codewords in C_1 are not modified. If, on the other hand, $p(C_0) > 1/2$, then the zeros and 1's are interchanged in the entire set of Huffman codes, resulting in a new $p(C_0)$ that is less than or equal to $1/2$, and the derived code is constructed as before.

As an example, consider the set of six symbols with probabilities 0.26, 0.24, 0.14, 0.13, 0.12, and 0.11. The entropy of this set is easily computed at approximately 2.497 and one set of Huffman codes for these symbols (from high to low probabilities) is 11, 01, 101, 100, 001, and 000. The average length of this set is

$$0.26\times 2 + 0.24\times 2 + 0.14\times 3 + 0.13\times 3 + 0.12\times 3 + 0.11\times 3 = 2.5 \text{ bits}.$$

It is easy to verify that $p(C_0) = 0.13 + 0.11 = 0.24 < 0.5$, so the derived code becomes 11, 01, 101, **1001**, 001, and **0001**, with an average size of 2.74 bits. On the other hand, if we consider the set of Huffman codes 00, 10, 010, 011, 110, and 111 for the same symbols, we compute $p(C_0) = 0.26 + 0.24 + 0.14 + 0.12 = 0.76 > 0.5$, so we have to interchange the bits and append a 1 to the codes of 0.13 and 0.11. This again results in a derived code with an average length E of 2.74, which satisfies $E < H + 1.5$.

It is easy to see how the extra 1's added to the codes increase its average size. Suppose that subset C_0 includes codes 1 through k. Originally, the average length of these codes was $E_0 = p_1 l_1 + p_2 l_2 + \cdots + p_k l_k$ where l_i is the length of code i. After a 1 is appended to each of the k codes, the new average length is

$$E = p_1(l_1+1) + p_2(l_2+1) + \cdots + p_k(l_k+1) = E_0 + (p_1 + p_2 + \cdots + p_k) \le E_0 + 1/2.$$

The average length has increased by less than half a bit.

3.28 Interpolative Coding

The interpolative coding method of this section ([Moffat and Stuiver 00] and [Moffat and Turpin 02]) is an unusual and innovative way of assigning dynamic codes to data symbols. It is different from the other variable-length codes described here because the codes it assigns to individual symbols are not static and depend on the entire message rather than on the symbols and their probabilities. We can say (quoting Aristotle) that the entire message encoded in this way becomes more than the sum of its individual parts, a behavior described by the term *holistic*.

> Holism is a Greek word meaning all, entire, total. It is the idea that all the attributes of a given system (physical, biological, chemical, social, or other) cannot be determined or explained by the sum of its component parts alone. Instead, the system as a whole determines in a significant way how the parts behave.

The entire message being encoded must be available to the encoder. Encoding is done by scanning the message in a special order, not linearly from start to end, and assigning codewords to symbols as they are being encountered. As a result, the codeword assigned to a symbol depends on the order of the symbols in the message.

As a result of this unusual code assignment, certain codewords may be zero bits long; they may be empty! Even more, when a message is encoded to a string S and then a new symbol is appended to the message, the augmented message is often reencoded to a string of the same length as S!

A binary tree is an important data structure that is used by many algorithms. Once such a tree has been constructed, it often has to be traversed. To traverse a tree means to go over it in a systematic way and visit each node once. Traversing a binary tree is often done in preorder, inorder, or postorder (there are other, less important, traversal methods). All three methods are recursive and are described in every text on data structures. The principle of inorder traversal is to (1) traverse the left subtree recursively in inorder, (2) visit the root, and (3) traverse the right subtree recursively in inorder. As an example, the inorder traversal of the tree of Figure 3.58a is A, B, C, D, E, F, G, H, and I.

Such an unusual method is best described by an example. Given a four-symbol alphabet, we arrange the symbols in order of decreasing (more accurately, nonincreasing) probabilities and replace them with the digits 1, 2, 3, and 4. Given the 12-symbol message $M = (1,1,1,2,2,2,1,3,4,2,1,1)$, the encoder starts by computing the sequence of 12 cumulative sums $L = (1,2,3,5,7,9,10,13,17,19,20,21)$ and it is these elements that are encoded. (Once the decoder decodes the cumulative sums, it can easily recreate the original message M.) Notice that L is a strictly monotonically increasing sequence because the symbols of the alphabet have been replaced by positive integers. Thus, element $L(i+1)$ is greater than $L(i)$ for any i and this fact is used by both encoder and decoder.

We denote the length of the message (12) by m and the last cumulative sum (21) by B. These quantities have to be transmitted to the decoder as overhead. The encoder considers L the inorder traversal of a binary tree (the tree itself is shown in Figure 3.58b) and it encodes its elements in the order 9, 3, 1, 2, 5, 7, 17, 10, 13, 20, 19, and 21.

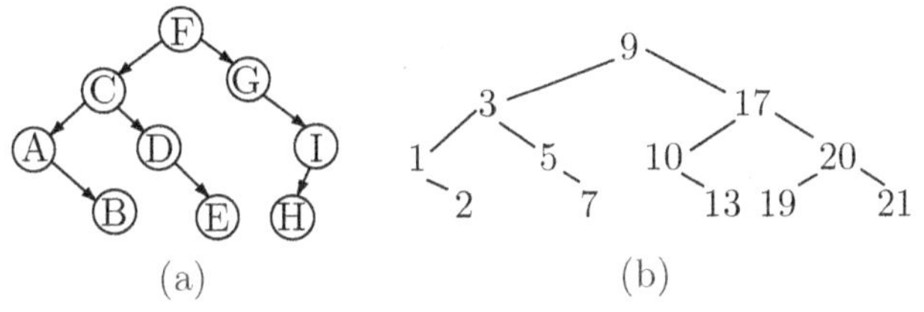

Figure 3.58: Binary Trees.

Encoding is done with reference to the position of each element in L. Thus, when the encoder processes element $L(6) = 9$, it assigns it a code based on the fact that this element is preceded by five elements and followed by six elements. Let's look at the encoding of $L(6)$ from the point of view of the decoder. The decoder has the values $m = 12$ and $B = 21$. It knows that the first code is that of the root of the tree which, with 12 elements, is element six (we could also choose element 7 as the root, but the choice has to be consistent). Thus, the first codeword to be decoded is that of $L(6)$. Later, the decoder will have to decode five codes for the elements preceding $L(6)$ and six codes for the elements following $L(6)$. The decoder also knows that the last element is 21 and that the elements are cumulative sums (i.e., they form a strictly monotonically increasing sequence). Since the root is element 6 and is preceded by five cumulative sums, its smallest value must be 6 (the five cumulative sums preceding it must be at least 1, 2, 3, 4, and 5). Since the root is followed by six cumulative sums, the last of which is 21, its largest value must be 15 (the six cumulative sums that follow it must be at least 16, 17, 18, 19, 20, and 21). Thus, the decoder knows that $L(6)$ must be an integer in the interval $[6, 15]$.

The encoder also knows this, and it also knows that the value of $L(6)$ is 9, i.e., it is the third integer in the sequence of 10 integers $(6, 7, \ldots, 15)$. The simplest choice for the encoder is to use fixed-length codes and assign $L(6)$ the code 0010, the third of the 16 4-bit codes. However, we have to encode one of only 10 integers, so one bit may be saved if we use the phased in codes of Section 2.9. The encoder therefore assigns $L(6)$ the codeword 010, the third code in the set of 10 phased-in codes.

Next to be encoded is element $L(3) = 3$, the root of the left subtree of $L(6)$. Cumulative sums in this subtree must be less than the root $L(6) = 9$, so they can be only between 1 and 8. The decoder knows that $L(3)$ is preceded by two elements, so its smallest value must be 3. Element $L(3)$ is also followed by two elements in its subtree, so two codes must be reserved. Thus, the value of $L(3)$ must be between 3 and 6. The encoder also knows this, and it also knows that $L(3)$ is 3. Thus, the encoder assigns $L(3)$ the phased-in code 00, the first code in the set of four phased-in codes.

Next comes $L(1) = 1$. The decoder knows that this is the root of the left subtree of $L(3)$. Cumulative sums in this subtree must be less than $L(3) = 3$, so they can be 1 and 2. The decoder also knows that $L(1)$ is not preceded by any other elements, so its smallest value can be 1. Element $L(1)$ is followed by the single element $L(2)$, so the second code in the interval $[1, 2]$ must be reserved for $L(2)$. Thus, the decoder knows that $L(1)$ must be 1. The encoder knows this too, so it assigns $L(1)$ a null (i.e., zero bits) codeword. Similarly, the decoder can figure out that $L(2)$ must be 2, so this element is also assigned a null codeword.

It is easy to see that elements $L(4)$ must be 5 and $L(5)$ must be 7, so these elements are also assigned null codewords. It is amazing how much information can be included in an empty codeword!

Next comes element $L(9) = 17$. The decoder knows that this is the right subtree of $L(6) = 9$, so elements (cumulative sums) in this subtree have values that start at 10 and go up to 21. Element $L(9)$ is preceded by two elements and is followed by three elements. Thus, its value must be between 12 and 18. The encoder knows this and also knows that the value of $L(9)$ is 17, the sixth integer in the seven-integer interval $[12, 18]$. Thus, $L(9)$ is assigned codeword 110, the sixth in the set of seven phased-in codes.

At this point the decoder can determine that the value of $L(7)$ must be between 10 and 15. Since it is 10, the encoder assigns it codeword 0, the first in the set of six phased-in codes. The value of $L(8)$ must be between 11 and 16. Since it is 13, the encoder assigns it codeword 011, the third in the set of six phased-in codes.

Next is element $L(11) = 20$. This is the root of the right subtree of $L(9) = 17$. The value of this element must be between $L(9) + 1 = 18$ and 20 (since $L(12)$ is 21). The encoder finds that $L(11)$ is 20, so its code becomes 11, the third code in the set of three phased-in codes. Next is $L(10)$. It must be between $L(9) + 1 = 18$ and $L(11) - 1 = 19$. It is 19, so it is assigned codeword 1, the second code in the set of two codes. Finally, the decoder knows that the value of the last cumulative sum $L(m)$ must be $B = 21$, so there is no need to assign this element any bits. These codewords are summarized in Table 3.59.

index	value	interval	code	index	value	interval	code
6	9	6–15	010	9	17	12–18	110
3	3	3–6	00	7	10	10–15	0
1	1	1–1	null	8	13	11–16	00
2	2	2–2	null	11	20	18–20	11
4	5	5–5	null	10	19	18–19	1
5	7	7–7	null	12	21	21–21	null

Table 3.59: Indexes, Values, Intervals, and Codes for $B = 21$.

The following pseudocode summarizes this recursive algorithm.

```
BinaryInterpolativeCode(L,f,lo,hi)
/* L[1...f] is a sorted array of symbols with values
   in the interval [lo, hi] */
1. if f = 0 return
2. if f = 1 call BinaryCode(L[1],lo,hi), return
3.  Otherwise compute h ← (f + 1) ÷ 2, f1 ← h − 1,
   f2 ← f − h, L1 ← L[1...(h − 1)], L2 ← L[(h + 1)...f].
4. call BinaryCode(L[h],lo + f1,hi − f2)
5. call BinaryInterpolativeCode(L1,f1,lo,L[h] − 1)
6. call BinaryInterpolativeCode(L2,f2,L[h] + 1,hi)
7. return
```

It is the cumulative sums that have been encoded, but in order to estimate the efficiency of the method, we have to consider the original message. The message $M = (1,1,1,2,2,2,1,3,4,2,1,1)$ was encoded into the 14-bit string 010|00|110|0|00|11|1. Is this a good result? The entropy of the message is easy to compute. The symbol 1 appears six out of 12 times, so its probability is 6/12, and similarly for the other three symbols. The entropy of our message is therefore

$$-\left[\frac{6}{12}\log_2\frac{6}{12}+\frac{4}{12}\log_2\frac{4}{12}+\frac{1}{12}\log_2\frac{1}{12}+\frac{1}{12}\log_2\frac{1}{12}\right]=1.62581.$$

Thus, to encode 12 symbols, the minimum number of bits needed is $12 \times 1.62581 = 19.5098$. Using interpolative coding, the message was compressed into fewer bits, but we shouldn't forget the overhead, in the form of $m = 12$ and $B = 21$. After these two numbers are encoded with suitable variable-length codes, the result will be at least 20 bits. In practice, messages may be hundreds or thousands of symbols long, so the overhead is negligible.

The developers of this method recommend the use of centered phased-in codes (Section 2.9) instead of the original phased-in codes. The original codes assign the shortest codewords to the first few symbols (for example, the sequence of nine phased-in codes starts with seven short codes and ends with two long codes). The centered codes are a rotated version where the shortest codes are in the middle (in the case of nine codes, the first and last ones are long, while the remaining seven codes are short).

This algorithm is especially efficient when the message is a bitstring with clusters of 1's, a practical example of which is inverted indexes [Witten et al. 99]. An inverted index is a data structure that stores a mapping from an index item (normally a word) to its locations in a set of documents. Thus, the inverted index for index item f may be the list $(f,2,5,6,8,11,12,13,18)$ where the integers 2, 5, 6, and so on are document numbers. Assuming a set of 20 documents, the document numbers can be written as the compact bitstring 01001101001110000100. Experience indicates that such bitstrings tend to contain clusters of 1's and it is easy to show that interpolative coding performs extremely well in such cases and produces many null codes.

As an extreme example of clustering, consider the 12-bit message 111111111111. The list of cumulative sums is $L = (1,2,3,4,5,6,7,8,9,10,11,12)$ and it is easy to verify that each element of L is encoded in zero bits, because its value $L(i)$ equals its index i; only the overhead $m = 12$ and $B = 12$ is needed!

The elements encoded by this method do not have to be cumulative sums as long as they form a strictly monotonically increasing sequence. As a voluntary exercise, the interested reader might want to encode the message $(1, 4, 5, 6, 7, 17, 25, 27, 28, 29)$ and compare the results with Table 3.60 (after [Moffat 00]).

index	value	interval	code	index	value	interval	code
5	7	5–24	00010	8	27	10–27	01111
2	4	2–4	10	6	17	8–25	01001
1	1	1–3	00	7	25	18–26	0111
3	5	5–5	null	9	28	28–28	null
4	6	6–6	null	10	29	29–29	null

Table 3.60: Indexes, Values, Intervals, and Codes for $B = 29$.

Any sufficiently advanced bug is indistinguishable from a feature.

—Arthur C. Clarke, paraphrased by Rich Kulawiec

4 Robust VL Codes

The many codes included in this chapter have a common feature; thet are robust. Any errors that creep into a string of such codewords can either be detected (or even corrected automatically) or have only limited effects and do not propagate indefinitely. The chapter starts with a discussion of the principles of error-control codes.

4.1 Codes For Error Control

When data is stored or transmitted, it is often encoded. Modern data communications is concerned with the following types of codes:

- Reliability. The detection and removal of errors caused by noise in the communications channel (this is also referred to as channel coding or error-control coding).
- Efficiency. The efficient encoding of the information in a small number of bits (source coding or data compression).
- Security. Protecting data against eavesdropping, intrusion, or tampering (the fields of cryptography and data hiding).

This section and the next cover the chief aspects of error-control as applied to fixed-length codes. Section 4.3 returns to variable-length codes and the remainder of this chapter is concerned with the problem of constructing reliable (or robust) codes.

Every time information is transmitted, over any channel, it may get corrupted by noise. In fact, even when information is stored in a storage device, it may become bad, because no piece of hardware is absolutely reliable. This important fact also applies to non-computer information. Text written or printed on paper fades over time and the paper itself degrades and may crumble. Audio and video data recorded on magnetic media fades over time. Speech sent on the air becomes corrupted by noise, wind, and

D. Salomon, G. Motta, *Handbook of Data Compression*, 5th ed.
DOI 10.1007/978-1-84882-903-9_4, © Springer-Verlag London Limited 2010

fluctuations of temperature and humidity. Speech, in fact, is a good starting point for understanding the principles of error-control. Imagine a noisy cocktail party where everybody talks simultaneously, on top of blaring music. We know that even in such a situation it is possible to carry on a conversation, but more attention than normal is needed.

The properties that make natural languages so robust and immune to errors are redundancy and context.

- Our language is redundant because only a very small fraction of all possible words are valid. A huge number of words can be constructed with the 26 letters of the English alphabet. Just the number of 7-letter words, for example, is $26^7 \approx 8.031$ billion. Yet only about 50,000 words are commonly used in daily conversations, and even the Oxford English Dictionary lists "only" about 500,000 words. When we hear a garbled word, our brain searches through many similar words for the "closest" valid word. Computers are very good at such searches, which is why redundancy is the basis of error-control codes.

- Our brain works by associations. This is why we humans excel at using the context of a message to locate and correct errors in the message. In receiving a sentence with a garbled word or a word that doesn't belong, such as "pass the thustard please," we first search our memory to find words that are associated with "thustard," then we use our accumulated life experience to select, among perhaps many possible candidates, the word that best fits in the present context. If we are driving on the highway, we pass the bastard in front of us; if we are at dinner, we pass the mustard (or custard). Another example is the (corrupted) written sentence `a*l n*tu*al l**gua*es a*e red***ant`, which we can easily complete. Computers don't have much life experience and are notoriously bad at such tasks, which is why context is not used in digital codes. In extreme cases, where much of the sentence is bad, even we may not be able to correct it, and we have to ask for a retransmission "say it again, Sam."

The idea of using redundancy to add reliability to information is due to Claude Shannon, the founder of information theory. It is not an obvious idea, since we are conditioned against it. Most of the time, we try to *eliminate* redundancy in digital data, in order to save space and compress the data. This habit is also reflected in our everyday, nondigital behavior. We shorten Michael to Mike and Samuel to Sam, and it is a rare Bob who insists on being a Robert.

There are several approaches to robust codes, but this section deals only with the concept of *check bits*, because this leads to the important concept of *Hamming distance*. Section 4.7 shows how this concept can be extended to variable-length codes.

Imagine a text message encoded in m-bit words (perhaps ASCII or Unicode) and then stored or transmitted. We can make the message robust by adding several check bits to the m data bits of each word of the message. We assume that k check bits are appended to the original m information bits, to produce a codeword of $n = m + k$ bits. Such a code is referred to as an (n, m) code. The codewords are then transmitted and decoded at the receiving end. Only certain combinations of the information bits and check bits are valid, in analogy with a natural language. The decoder knows what the valid codewords are. If a nonvalid codeword is received, the decoder considers it an error. By adding more check bits, the decoder can also correct certain errors, not just detect them. The principle of error correction, not just detection, is that, on receiving a bad codeword, the receiver selects the valid codeword that is the "closest" to it.

Example: Assume a set of 128 symbols (i.e., $m = 7$). If we select $k = 4$, we end up with 128 valid codewords, each 11 bits long. This is an $(11, 7)$ code. The valid codewords are selected from a total of $2^{11} = 2048$ possible codewords, so there remain $2048 - 128 = 1920$ nonvalid codewords. The big difference between the number of valid (128) and nonvalid (1920) codewords implies that if a codeword gets corrupted, chances are that it will change to a nonvalid one.

It may, of course, happen that a valid codeword gets modified, during transmission, to another valid codeword. Thus, our codes are not absolutely reliable, but can be made more and more robust by adding more check bits and by selecting the valid codewords carefully. The noisy channel theorem, one of the basic theorems of information theory, states that codes can be made as reliable as desired, even in the presence of much noise, by adding check bits and thus separating our codewords further and further, as long as the rate of transmission does not exceed a certain quantity referred to as the channel's capacity.

It is important to understand the meaning of the word "error" in data processing and communications. When an n-bit codeword is transmitted over a channel, the decoder may receive the same n bits, it may receive n bits, some of which are bad, but it may also receive fewer than or more than n bits. Thus, bits may be added, deleted, or changed (substituted) by noise in the communications channel. In this section we consider only substitution errors. A bad bit simply changes its value, either from 0 to 1, or from 1 to 0. This makes it relatively easy to correct the bit. If the error-control code tells the receiver which bits are bad, the receiver corrects those bits by inverting them.

Parity bits represent the next step in error-control. A parity bit can be added to a group of m information bits to complete the total number of 1's to an odd number. Thus, the (odd) parity of the group 10110 is 0, since the original group plus the parity bit has an odd number (3) of 1's. It is also possible to use even parity, and the only difference between odd and even parity is that, in the case of even parity, a group of all zeros is valid, whereas, with odd parity, any group of bits with a parity bit added, cannot be all zeros.

Parity bits can be used to design simple, but not very efficient, error-correcting codes. To correct 1-bit errors, the message can be organized as a rectangle of dimensions $(r - 1) \times (s - 1)$. A parity bit is added to each row of $s - 1$ bits, and to each column of $r - 1$ bits. The total size of the message (Table 4.1) is now $s \times r$.

0	1	0	0	1
1	0	1	0	0
0	1	1	1	1
0	0	0	0	0
1	1	0	1	1
0	1	0	0	1

Table 4.1.

0	1	0	0	**1**
1	0	1	**0**	
0	1	**0**		
0	**0**			
1				

Table 4.2.

If only one bit becomes bad, a check of all $s-1+r-1$ parity bits will detect the error, since only one of the $s-1$ parities and only one of the $r-1$ parities will be bad.

The overhead of a code is defined as the number of parity bits divided by the number of information bits. The overhead of the rectangular code is, therefore,

$$\frac{(s-1+r-1)}{(s-1)(r-1)} \approx \frac{s+r}{s\times r-(s+r)}.$$

A similar, slightly more efficient code is a triangular configuration, where the information bits are arranged in a triangle, with the parity bits placed on the diagonal (Table 4.2). Each parity bit is the parity of all the bits in its row and column. If the top row contains r information bits, the entire triangle has $r(r+1)/2$ information bits and r parity bits. The overhead is therefore

$$\frac{r}{r(r+1)/2} = \frac{2}{r+1}.$$

It is also possible to arrange the information bits in a number of two-dimensional planes, to obtain a three-dimensional cube, three of whose six outer surfaces consist of parity bits. It is not obvious how to generalize these methods to more than 1-bit error correction.

Hamming Distance and Error Detecting. In the 1950s, Richard Hamming conceived the concept of distance as a general way to use check bits for error detection and correction.

Symbol	code_1	code_2	code_3	code_4	code_5
A	0000	0000	001	001001	01011
B	1111	1111	010	010010	10010
C	0110	0110	100	100100	01100
D	0111	1001	111	111111	10101
k:	2	2	1	4	3

Table 4.3: Codes with $m=2$.

To illustrate this idea, we start with a simple example of four symbols A, B, C, and D. Only two information bits are required, but the codes of Table 4.3 add some check bits, for a total of 3–6 bits per symbol. The first of these codes, code_1, is simple. Its four codewords were selected from the 16 possible 4-bit numbers, and are not the best possible ones. When the receiver receives one of them, say, 0111, it assumes that there is no error and the symbol received is D. When a nonvalid codeword is received, the receiver signals an error. Since code_1 is not the best possible, not every error is detected. Even if we limit ourselves to single-bit errors, this code is not very good. There are 16 possible single-bit errors in its four 4-bit codewords, and of those, the following four cannot be detected: a 0110 changed during transmission to 0111, a 0111 modified to 0110, a 1111 corrupted to 0111, and a 0111 changed to 1111. Thus, the error detection rate is 12 out of 16, or 75%. In contrast, code_2 does a much better job. It can detect every single-bit error, because when only a single bit is changed in any of its codewords, the result is not any of the other codewords. We say that the four codewords of code$_2$ are sufficiently distant from one another. The concept of distance of codewords is easy to describe.

1. Two codewords are a Hamming distance d apart if they differ in exactly d of their n bits.

2. A code has a Hamming distance of d if every pair of codewords in the code is at least a Hamming distance d apart.

(For mathematically-inclined readers.) These definitions have a simple geometric interpretation. Imagine a hypercube in n-dimensional space. Each of its 2^n corners can be numbered by an n-bit number (Figure 4.4) such that each of the n bits corresponds to one of the n dimensions of the cube. In such a cube, points that are directly connected (near neighbors) have a Hamming distance of 1, points with a common neighbor have a Hamming distance of 2, and so on. If a code with a Hamming distance of 2 is desired, only points that are not directly connected should be selected as valid codewords.

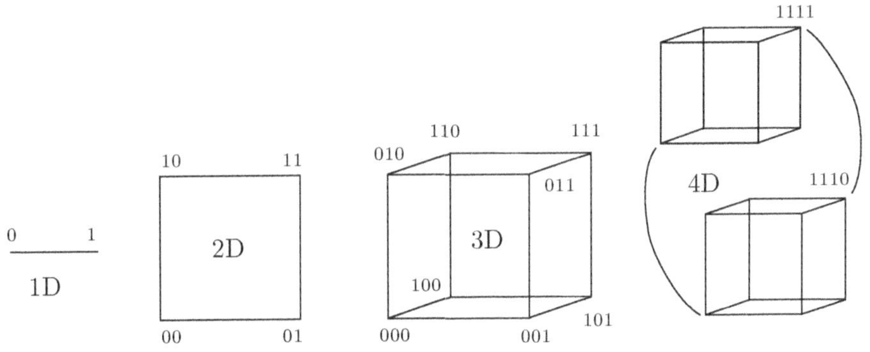

Figure 4.4: Cubes of Various Dimensions and Corner Numbering.

The "distance" approach to reliable communications is natural and has been employed for decades by many organizations—most notably armies, but also airlines and emergency services—in the form of phonetic alphabets. We know from experience that certain letters—such as `F` and `S`, or `B`, `D`, and `V`—sound similar over the telephone and often cause misunderstanding in verbal communications (the pair `8` and `H` also comes to mind). A phonetic alphabet replaces each letter by a word that starts with that letter, selecting words that are "distant" in some sense, in order to remove any ambiguity.

Thus, replacing `F` by **`foxtrot`** and `S` by **`sierra`** guarantees reliable conversations, because these words sound very different and will not be mistaken even under conditions of noise and garbled communications.

The NATO phonetic alphabet is Alfa Bravo Charlie Delta Echo Foxtrot Golf Hotel India Juliett Kilo Lima Mike November Oscar Papa Quebec Romeo Sierra Tango Uniform Victor Whiskey X-ray Yankee Zulu. For other phonetic alphabets, see [uklinux 07]. Reference [codethatword 07] may also be of interest to some.

The reason code_2 can detect all single-bit errors is that it has a Hamming distance of 2. The distance between valid codewords is 2, so a 1-bit error always changes a valid codeword into a nonvalid one. When two bits go bad, a valid codeword is moved to another codeword at distance 2. If we want that other codeword to be nonvalid, the code must have at least distance 3.

In general, a code with a Hamming distance of $d+1$ can detect all d-bit errors. In comparison, code_3 has a Hamming distance of 2 and can therefore detect all 1-bit errors even though it is short ($n=3$). Similarly, code_4 has a Hamming distance of 4, which is more than enough to detect all 2-bit errors. It is now obvious that we can increase the reliability of our data, but this feature does not come free. As always, there is a tradeoff, or a price to pay, in the form of overhead. Our codes are much longer than m bits per symbol because of the added check bits. A measure of the price is $n/m = \frac{m+k}{m} = 1 + k/m$, where the quantity k/m is the overhead of the code. In the case of code_1 the overhead is 2, and in the case of code_3 it is 3/2.

Example: A code with a single check bit, that is a parity bit (even or odd). Any single-bit error can easily be detected since it creates a nonvalid codeword. Such a code therefore has a Hamming distance of 2. Notice that code_3 uses a single, odd, parity bit.

Example: A 2-bit error-detecting code for the same four symbols. It must have a Hamming distance of at least 3, and one way of generating it is to duplicate code_3 (which results in code_4 with a distance of 4).

Unfortunately, the Hamming distance cannot be easily extended to variable-length codes, because it is computed between codes of the same length. Nevertheless, there are ways to extend it to variable-length codes and the most common of these is discussed in Section 4.2.

Error-Correcting Codes. The principle of error-correcting codes is to separate the codewords even farther by increasing the code's redundancy (i.e., adding more check bits). When an invalid codeword is received, the receiver corrects the error by selecting the valid codeword that is closest to the one received. An example is code_5, which has a Hamming distance of 3. When one bit is modified in any of its four codewords, that codeword is one bit distant from the original, but is still two bits distant from any of the other codewords. Thus, if there is only one error, the receiver can always correct it.

In general, when d bits go bad in a codeword C_1, it turns into an invalid codeword C_2 at a distance d from C_1. If the distance between C_2 and the other valid codewords is at least $d+1$, then C_2 is closer to C_1 than it is to any other valid codeword. This is why a code with a Hamming distance of $d+(d+1) = 2d+1$ can correct all d-bit errors.

How are the codewords selected? The problem is to select a good set of 2^m codewords out of the 2^n possible ones. The simplest approach is to use brute force. It is easy to write a computer program that will examine all the possible sets of 2^m codewords, and will stop at the first one that has the right distance. The problems with this approach are (1) the time and storage required at the receiving end to verify and correct the codes received, and (2) the amount of time it takes to examine all the possibilities.

Problem 1. The receiver must have a list of all the 2^n possible codewords. For each codeword, it must have a flag indicating whether it is valid, and if not, which valid codeword is the closest to it. Every codeword received has to be searched for and found in this list in order to verify it.

Problem 2. In the case of four symbols, only four codewords need be selected. For code_1 and code_2, these four codewords had to be selected from among 16 possible numbers, which can be done in $\binom{16}{4} = 7280$ ways. It is possible to write a simple program that will systematically select sets of four codewords until it finds a set with the required distance. In the case of code_4, however, the four codewords had to be selected from a set of 64 numbers, which can be done in $\binom{64}{4} = 635{,}376$ ways. It is still possible to write a program that will systematically explore all the possible codeword selections. In practical cases, however, where sets of hundreds of symbols are involved, the number of possibilities of selecting codewords is too large even for the fastest computers to handle in reasonable time.

This is why sophisticated methods are needed to construct sets of error-control codes. Such methods are out of the scope of this book but are discussed in many books and publications on error control codes. Section 4.7 discusses approaches to developing robust variable-length codes.

4.2 The Free Distance

The discussion above shows that the Hamming distance is an important metric (or measure) of the reliability (robustness) of an error-control code. This section shows how the concept of distance can be extended to variable-length codes. Given a code with s codewords c_i, we first construct the length vector of the code. We assume that there are s_1 codewords of length L_1, s_2 codewords of length L_2, and so on, up to s_m codewords of length L_m. We also assume that the lengths L_i are sorted such that L_1 is the shortest length and L_m is the longest. The length vector is $(L_1, L_2, \ldots, L_m)$.

The first quantity defined is b_i, the minimum block distance for length L_i. This is simply the minimum Hamming distance of the s_i codewords of length L_i (where i goes from 1 to m). The overall minimum block length $b_{\min}$ is defined as the smallest b_i.

We next look at all the possible pairs of codewords (c_i, c_j). The two codewords of a pair may have different lengths, so we first compute the distances of their prefixes. Assume that the lengths of c_i and c_j are 12 and 4 bits, respectively, we examine the four leftmost bits of the two, and compute their Hamming distance. The minimum of all these distances, over all possible pairs of codewords, is called the minimum diverging distance of the code and is denoted by $d_{\min}$.

Next, we do the same for the postfixes of the codewords. Given a pair of codewords of lengths 12 and 4 bits, we examine their last (rightmost) four bits and compute their Hamming distance. The smallest of these distances is the minimum converging distance of the code and is denoted by $c_{\min}$.

The last step is more complex. We select a positive integer N and construct all the sequences of codewords whose total length is N. If there are many codewords, there may be many sequences of codewords whose total length is N. We denote those sequences by f_1, f_2, and so on. The set of all the N-bit sequences f_i is denoted by F_N. We compute the Hamming distances of all the pairs (f_i, f_j) in F_N for different i and j and select the minimum distance. We repeat this for all the possible values of N (from 1 to infinity) and select the minimum distance. This last quantity is termed the free distance of the code [Bauer and Hagenauer 01] and is denoted by d_{free}. The free distance of a variable-length code is the single most important parameter determining the robustness of the code. This metric is considered the equivalent of the Hamming distance, and [Buttigieg and Farrell 95] show that it is bounded by

$$d_{\text{free}} \geq \min(b_{\min}, d_{\min} + c_{\min}).$$

4.3 Synchronous Prefix Codes

Errors are a fact of life. They are all around us, are found everywhere, and are responsible for many glitches and accidents and for much misery and misunderstanding. Unfortunately, computer data is not an exception to this rule and is not immune to errors. Digital data written on a storage device such as a disk, CD, or DVD is subject to corruption. Similarly, data stored in the computer's internal memory can become bad because of a sudden surge in the electrical voltage, a stray cosmic ray hitting the memory circuits, or an extreme variation of temperature. When binary data is sent over a communications channel, errors may creep up and damage the bits. This is why error-detecting and error-correcting codes (also known as error-control codes or channel codes) are so important and are used in many applications. Data written on CDs and DVDs is made very reliable by including sophisticated codes that detect most errors and can even correct many errors automatically. Data sent over a network between computers is also often protected in this way. However, error-control codes have a serious downside, they work by adding extra bits to the data (parity bits or check bits) and thus increase both the redundancy and the size of the data. In this sense, error-control (data reliability and integrity) is the opposite of data compression. The main goal of compression is to eliminate redundancy from the data, but this inevitably decreases data reliability and opens the door to errors and data corruption.

> We are built to make mistakes, coded for error.
> —Lewis Thomas

The problem of errors in communications (in scientific terms, the problem of noisy communications channels or of sources of noise) is so fundamental, that the first figure in Shannon's celebrated 1948 papers [Shannon 48] shows a source of noise (Figure 4.5).

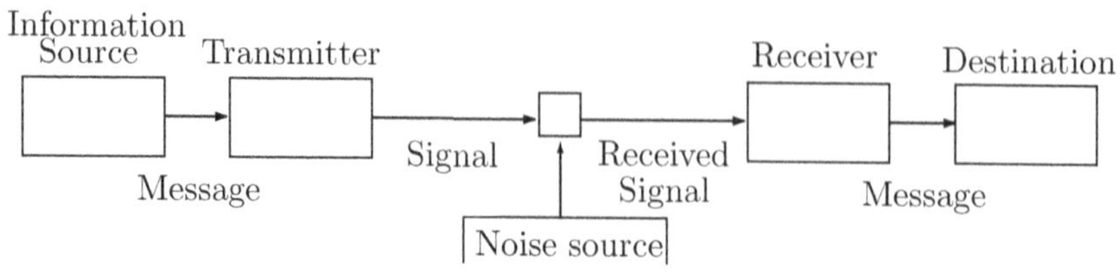

Figure 4.5: The Essence of Communications.

As a result, reliability or lack thereof is the chief downside of variable-length codes. A single error that creeps into a compressed stream can propagate and trigger many consecutive errors when the stream is decompressed. We say that the decoder loses synchronization with the data or that it slips while decoding. The first solution that springs to mind, when we consider this problem, is to add an error-control code to the compressed data. This solves the problem and results in reliable receiving and decoding, but it requires the introduction of many extra bits (often about 50% of the size of the data). Thus, this solution makes sense only in special applications, where data reliability and integrity is more important than size. A more practical solution is to develop *synchronous codes*. Such a code has one or more synchronizing codewords, so it does not artificially increase the size of the compressed data. On the other hand, such a code is not as robust as an error-control code and it allows for a certain amount of slippage for each error. The idea is that an error will propagate and will affect that part of the compressed data from the point it is encountered (from the bad codeword) until the first synchronizing codeword is input and decoded. The corrupted codeword and the few codewords following it (the slippage region) will be decoded into the wrong data symbols, but the synchronizing codeword will synchronize the decoder and stop the propagation of the error. The decoder will recognize either the synchronizing codeword or the codeword that follows it and will produce correct output from that point. Thus, a synchronous code is not as reliable as an error-control code. It allows for a certain amount of error propagation and should be used only in applications where that amount of bad decoded data is tolerable. There is also the important consideration of the average length of a synchronous code. Such a code may be a little longer than other variable-length codes, but it shouldn't be much longer.

The first example shows how a single error can easily propagate through a compressed file that's being decoded and cause a long slippage. We consider the simple feasible code 0001, 001, 0101, 011, 1001, 101, and 11. If we send the string of codewords 00**0**1|101|011|101|011|101|011|... and the third bit (in boldface) gets corrupted to a 1, then the decoder will decode this as the string 001|11|0101|11|0101|11|..., resulting in a long slippage.

However, if we send 01**0**1|011|11|101|0001|... and the third bit goes bad, the decoder will produce 011|101|11|11|010001.... The last string (010001) serves as an error indicator for the decoder. The decoder realizes that there is a problem and the best way for it to resynchronize itself is to skip a bit and try to decode first 10001... (which fails) and then 0001 (a success). The damage is limited to a slippage of a few symbols, because the codeword 0001 is synchronizing. Similarly, 1001 is also a synchronizing codeword, so a string of codewords will synchronize itself after an error when 1001 is read. Thus 01**0**1|011|11|101|1001|... is decoded as 011|101|11|11|011|001..., thereby stopping the slippage.

> Slippage: The act or an instance of slipping, especially movement away from an original or secure place.

The next example is the 4-ary code of Table 4.9 (after [Capocelli and De Santis 92]), where the codewords flagged by an "*" are synchronizing. Consider the string of codewords **0**000|100|002|3|... where the first bit is damaged and is input by the decoder as 1. The decoder will generate 100|01|0000|23|..., thereby limiting the slippage to four symbols. The similar string **0**000|100|002|23|... will be decoded as 100|01|0000|22|3..., and the error in the code fragment **0**000|100|002|101|... will result in 100|01|0000|21|01.... In all cases, the damage is limited to the substring from the position of the error to the location of the synchronizing codeword.

The material that follows is based on [Ferguson and Rabinowitz 84]. In order to understand how a synchronizing codeword works, we consider a case where an error has thrown the decoder off the right track. If the next few bits in the compressed file are 101110010..., then the best thing for the decoder to do is to examine this bit pattern and try to locate a valid codeword in it. The first bit is 1, so the decoder concentrates on all the codewords that start with 1. Of these, the decoder examines all the codewords that start with 10. Of these, it concentrates on all the codewords that starts with 101, and so on. If 101110010... does not correspond to any codeword, the decoder discards the first bit and examines the string 01110010... in the same way.

Based on this process, it is easy to identify the three main cases that can occur at each decoding step, and through them to figure out the properties that a synchronizing codeword should have. The cases are as follows:

1. The next few bits constitute a valid codeword. The decoder is synchronized and can proceed normally.

2. The next few bits are part of a long codeword a whose suffix is identical to a synchronizing codeword b. Suppose that $a = 1|101110010$ and $b = 010$. Suppose also that because of an error the decoder has become unsynchronized and is positioned after the first bit of a (at the vertical bar). When the decoder examines bit patterns as discussed earlier and skips bits, it will eventually arrive at the pattern 010 and identify it as the valid codeword b. We know that this pattern is the tail (suffix) of a and is not b, but the point is that b has synchronized the decoder and the error no longer propagates. Thus, we know that for b to be a synchronizing codeword, it has to satisfy the following: If b is a substring of a longer codeword a, then b should be the suffix of a and should not reappear elsewhere in a.

3. The decoder is positioned in front of the string 100|11001... where the 100 is the tail (suffix) of a codeword x that the decoder cannot identify because of a slippage, and 11001 is a synchronizing codeword c. Assume that the bit pattern 100|110 (the suffix of x followed by the prefix of c) happens to be a valid codeword. If the suffix 01 of c is a valid codeword a, then c would do its job and would terminate the slippage (although it wouldn't be identified by the decoder).

Based on these cases, we list the two properties that a codeword should have in order to be synchronizing. (1) If b is a synchronizing codeword and it happens to be the suffix of a longer codeword a, then b should not occur anywhere else in a. (2) If a prefix of b is identical to a suffix of another codeword x, then the remainder of b should be either a valid codeword or identical to several valid codewords. These two properties can be considered the definition of a synchronizing codeword.

Now assume that C is a synchronous code (i.e., a set of codewords, one or more of which are synchronizing) and that c is a synchronizing codeword in C. Codeword c is the variable-length code assigned to a symbol s of the alphabet, and we denote the probability of s by p. Given a long string of data symbols, s will appear in it $1/p$ percent of the time. Equivalently, every $(1/p)$th symbol will on average be s. When such a string is encoded, codeword c will appear with the same frequency and will therefore limit the length of any slippage to a value proportional to $1/p$. If a code C includes k synchronizing codewords, then on average every $(1/\sum_i^k p_i)$th codeword will be a synchronizing codeword. Thus, it is useful to have synchronizing codewords assigned to common (high probability) symbols. The principle of compression with variable-length codes demands that common symbols be assigned short codewords, leading us to conclude that the best synchronous codes are those that have many synchronizing short codewords. The shortest codewords are single bits, but since 0 and 1 are always suffixes, they cannot satisfy the definition above and cannot serve as synchronizing codewords (however, in a nonbinary code, such as a ternary code, one-symbol codewords can be synchronizing as illustrated in Table 4.9).

Table 4.6 lists four Huffman codes. Code C_1 is nonsynchronous and code C_2 is synchronous. Figure 4.7 shows the corresponding Huffman code trees. Starting with the data string `ADABCDABCDBECAABDECA`, we repeat the string indefinitely and encode it in C_1. The result starts with $\mathbf{0}1|000|01|10|11|000|01|10|11|000|10|001|\ldots$ and we assume that the second bit (in boldface) gets corrupted. The decoder, as usual, inputs the string $0000001101100001101100010001\ldots$, proceeds normally, and decodes it as the bit string $000|000|11|01|10|000|11|01|10|001|000|1\ldots$, resulting in unbounded slippage. This is a specially-contrived example, but it illustrates the risk posed by a nonsynchronous code.

Symbol	A	B	C	D	E
Prob.	0.3	0.2	0.2	0.2	0.1
C_1	01	10	11	000	001
C_2	00	10	11	010	011

Symbol	A	B	C	D	E	F
Prob.	0.3	0.3	0.1	0.1	0.1	0.1
C_3	10	11	000	001	010	011
C_4	01	11	000	001	100	101

Table 4.6: Synchronous and Nonsynchronous Huffman Codes.

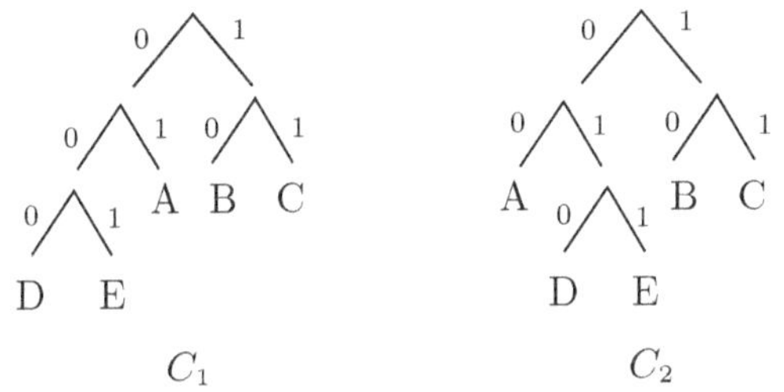

Figure 4.7: Strongly Equivalent Huffman Trees.

In contrast, code C_2 is synchronized, with the two synchronizing codewords 010 and 011. Codeword 010 is synchronizing because it satisfies the two parts of the definition above. It satisfies part 1 by default (because it is the longest codeword) and it satisfies part 2 because its suffix 10 is a valid codeword. The argument for 011 is similar. Thus, since both codes are Huffman codes and have the same average code length, code C_2 is preferable.

In [Ferguson and Rabinowitz 84], the authors concentrate on the synchronization of Huffman codes. They describe a twisting procedure for turning a nonsynchronous Huffman code tree into an equivalent tree that is synchronous. If the procedure can

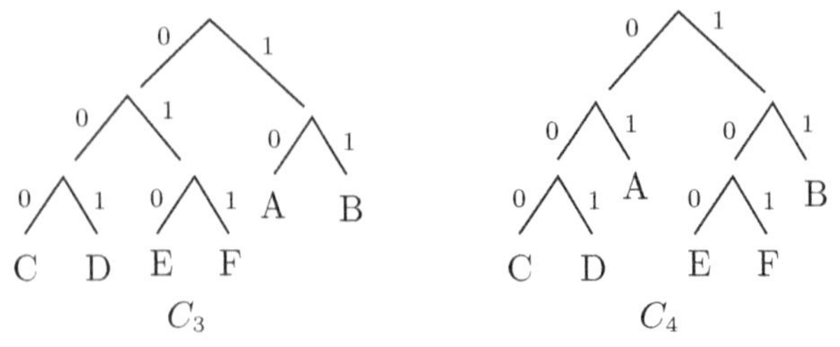

Figure 4.8: Weakly Equivalent Huffman Trees.

be carried out, then the two codes are said to be strongly equivalent. However, this procedure cannot always be performed. Code C_3 of Table 4.6 is nonsynchronous, while code C_4 is synchronous (with 101 as its synchronizing codeword). The corresponding Huffman code trees are shown in Figure 4.8 but C_3 cannot be twisted into C_4, which is why these codes are termed weakly equivalent. The same reference also proves the following criteria for nonsynchronous Huffman codes:

1. Given a Huffman code where the codewords have lengths l_i, denote by $(l_1, \ldots, l_j)$ the set of all the different lengths. If the greatest common divisor of this set does not equal 1, then the code is nonsynchronous and is not even weakly equivalent to a synchronous code.

2. For any $n \geq 7$, there is a Huffman code with n codewords that is nonsynchronous and is not weakly equivalent to a synchronous code.

3. A Huffman code in which the only codeword lengths are l, $l+1$, and $l+3$ for some $l > 2$, is nonsynchronous. Also, for any $n \geq 12$, there is a source where all the Huffman codes are of this type.

> The greatest common divisor (gcd) of a set of nonzero integers l_i is the largest integer that divides each of the l_i evenly (without remainders). If the gcd of a set is 1, then the members of the set are relatively prime. An example of relatively prime integers is $(9, 28, 29)$.

These criteria suggest that many Huffman codes are nonsynchronous and cannot even be modified to become synchronous. However, there are also many Huffman codes that are either synchronous or can be twisted to become synchronous. We start with the concept of quantized source. Given a source S of data symbols with known probabilities, we construct a Huffman code for the source. We denote the length of the longest codeword by L and denote by α_i the number of codewords with length i. The vector $S = (\alpha_1, \alpha_2, \ldots, \alpha_L)$ is called the quantized representation of the source S. A source is said to be gapless if the first nonzero α_i is followed only by nonzero α_j (i.e., if all the zero α's are concentrated at the start of the quantized vector). The following are criteria for synchronous Huffman codes:

1. If α_1 is positive (i.e., there are some codewords with length 1), then any Huffman code for S is synchronous.

2. Given a gapless source S where the minimum codeword length is 2, then S has a synchronous Huffman code, unless its quantized representation is one of the following $(0, 4)$, $(0, 1, 6)$, $(0, 1, 1, 10)$, and $(0, 1, 1, 1, 18)$.

The remainder of this section is based on [Capocelli and De Santis 92], a paper that offers an advanced discussion of synchronous prefix codes, with theorems and proofs. We present only a short summary of the many results discovered by these authors. Given a feasible code (Section 3.27) where the length of the longest codeword is L, all the codewords of the form $\underbrace{00\ldots0}_{L-1}1$, $\underbrace{00\ldots0}_{L-2}1$, and $1\underbrace{00\ldots0}_{L-2}1$, are synchronizing codewords. If there are no such codewords, it is possible to modify the code as follows. Select a codeword of length L with the most number of consecutive zeros on the left $\underbrace{00\ldots0}_{j}1x\ldots x$ and change the 1 to a 0 to obtain $\underbrace{00\ldots0}_{j+1}x\ldots x$. Continue in this way, until all the x bits except the last one have been changed to 1's and the codeword has the format $\underbrace{00\ldots0}_{L-1}1$. This codeword is now synchronizing.

As an example, given the feasible code **0001**, **001**, 0101, 011, **1001**, 101, and 11, $L = 4$ and the three codewords in boldface are synchronizing. This property can be extended to nonbinary codes, with the difference that instead of a 1, any digit other than 0 can be substituted. Thus, given the feasible ternary code 00001, 0001, 0002, 001, 002, 01, 02, 1001, 101, 102, 201, 21, and 22, L equals 5 and the first three codewords are synchronizing. Going back to the definition of a synchronizing codeword, it is easy to show by direct checks that the first 11 codewords of this code satisfy this definition and are therefore synchronizing.

Table 4.9 (after [Capocelli and De Santis 92]) lists a 4-ary code for the 21-character Latin alphabet. The rules above imply that all the codewords that end with 3 are synchronizing, while our previous definition shows that 101 and 102 are also synchronizing.

Letter	Freq.	Code	Letter	Freq.	Code	Letter	Freq.	Code
h	5	0000	d	17	*101	r	67	20
x	6	0001	l	21	*102	s	68	21
v	7	0002	p	30	*103	a	72	22
f	9	100	c	33	11	t	72	*03
b	12	001	m	34	12	u	74	*13
q	13	002	o	44	01	e	92	*23
g	14	*003	n	60	02	i	101	*3

Table 4.9: Optimal 4-ary Synchronous Code for the Latin Alphabet.

See also the code of Section 3.21.1.

4.4 Resynchronizing Huffman Codes

Because of the importance and popularity of Huffman codes, much research has gone into every aspect of these codes. The work described here, due to [Rudner 71], shows how to identify a resynchronizing Huffman code (RHC), a set of codewords that allows the decoder to always synchronize itself following an error. Such a set contains at least one synchronizing codeword (SC) with the following property: any bit string followed by an SC is a sequence of valid codewords.

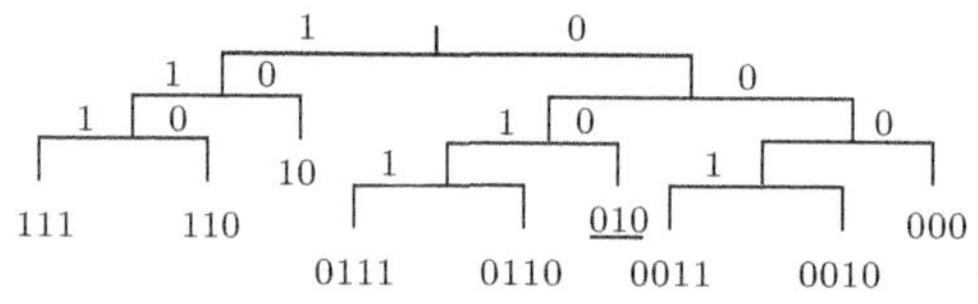

Figure 4.10: A Resynchronizing Huffman Code Tree.

Figure 4.10 is an example of such a code. It is a Huffman code tree for nine symbols having probabilities of occurrence of 1/4, 1/8, 1/8, 1/8, 1/8, 1/16, 1/16, 1/16, and 1/16. Codeword 010 (underlined) is an SC. A direct check verifies that any bit string followed by 010 is a sequence of valid codewords from this tree. For example, 001010001|010 is the string of four codewords 0010|10|0010|10. If an encoded file has an error at point A and an SC appears later, at point B, then the bit string from A to B (including the SC) is a string of valid codewords (although not the correct codewords) and the decoder becomes synchronized at B (although it may miss decoding the SC itself). Reference [Rudner 71] proposes an algorithm to construct an RHC, but the lengths of the codewords must satisfy certain conditions.

Given a set of prefix codewords, we denote the length of the shortest code by m. An integer $q < m$ is first determined by a complex test (listed below) that goes over all the nonterminal nodes in the code tree and counts the number of codewords of the same length that are descendants of the node. Once q has been computed, the SC is the codeword $0^q 1^{m-q} 0^q$. It is easy to see that $m = 2$ for the code tree of Figure 4.10. The test results in $q = 1$, so the SC is 010.

In order for an SC to exist, the code must either have $m = 1$ (in which case the SC has the form 0^{2q}) or must satisfy the following conditions:

1. The value of m is 2, 3, or 4.
2. The code must contain codewords with all the lengths from m to $2m - 1$.
3. The value of q that results from the test must be less than m.

Figure 4.11 is a bigger example with 20 codewords. The shortest codeword is three bits long, so $m = 3$. The value of q turns out to be 2, so the SC is 00100. A random bit string followed by the SC, such as 01000101100011100101010100010101|00100, becomes the string of nine codewords 0100|01011|00011|100|1010|1010|00101|0100|100.

The test to determine the value of q depends on many quantities that are defined in [Rudner 71] and are too specialized to be described here. Instead, we copy this test verbatim from the reference.

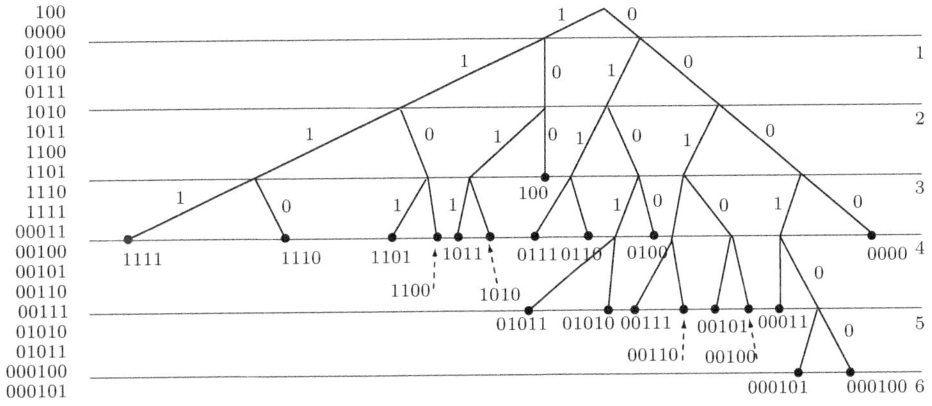

Figure 4.11: A Resynchronizing Huffman Code Tree With 20 Codewords.

"Let index I have $2 \le m \le 4$ and $p < m$. Let

$$Q_j(h) = (h + m - j) + \sum_{x=j-m+2}^{j} c_{x-h-1,x}$$

where $c_{a,b} = 0$ for $a \le 0$. Let q be the smallest integer $p \le q < m$, if any exists, for which $n_j \ge Q_j(q)$ for all j, $m \le j \le M$. If no such integer exists, let $q = \max(m, p)$. Then $r_0 \ge q + m$."

The operation of an SC is especially simple when it follows a nonterminal node in the code tree. Checking Figure 4.11 verifies that when any interior node is followed by the SC, the result is either one or two codewords. In the former case, the SC is simply a suffix of the single resulting codeword, while in the latter case, some of the leftmost zeros of the SC complete the string of the interior node, and the remainder of the SC (a string of the form $0^k 1^{m-q} 0^q$, where k is between 0 and $q - 1$) is a valid codeword that is referred to as a reset word.

In the code tree of Figure 4.11, the reset words are 0100 and 100. When the interior node 0101 is followed by the SC, the result is 01010|0100. When 0 is followed by the SC, the result is 000100, and when 01 is followed by 00100, we obtain 0100|100. Other codewords may also have this property (i.e., they may reset certain interior nodes). This happens when a codeword is a suffix of another codeword, such as 0100, which is a suffix of 000100 in the tree of Figure 4.11.

Consider the set that consists of the SC (whose length is $m + q$ bits) and all the reset words (which are shorter). This set can reset all the nonterminal nodes, which implies a certain degree of nonuniformity in the code tree. The tree of Figure 4.11 has levels from 1 to 6, and a codeword at level D is D bits long. Any node (interior or not) at level D must therefore have at least one codeword of length L that satisfies either $D + 1 \le L \le D + q - 1$ (this corresponds to the case where appending the SC to the node results in two codewords) or $L = D + q + m$ (this corresponds to the case where appending the SC to the node results in one codeword). Thus, the lengths of codewords in the tree are restricted and may not truely reflect the data symbols' probabilities.

Another aspect of this restriction is that the number of codewords of the same length that emanate from a nonterminal node is at most 2^q. Yet another nonuniformity is that short codewords tend to concentrate on the 0-branch of the tree. This downside of the RHC has to be weighed against the advantage of having a resynchronizing code.

Another important factor affecting the performance of an RHC is the expected time to resynchronize. From the way the Huffman algorithm works, we know that a codeword of length L appears in the encoded stream with probability $1/2^L$. The length of the SC is $m+q$, so if we consider only the resynchronization provided by the SC, and ignore the reset words, then the expected time τ to resynchronize (the inverse of the probability of occurrence of the SC) is 2^{m+q}. This is only an upper bound on τ, because the reset words also contribute to synchronization and reduce the expected time τ. If we assume that the SC and all the reset words reset all the nonterminal (interior) nodes, then it can be shown that

$$\frac{1}{2^{m-1}-2^{m+q}}$$

is a lower bound on τ. Better estimates of τ can be computed for any given code tree by going over all possible errors, determining the recovery time τ_i for each possible error i, and computing the average τ.

An RHC allows the decoder to resynchronize itself as quickly as possible following an error, but there still are two points to consider: (1) the number of data symbols decoded while the decoder is not synchronized may differ from the original number of symbols and (2) the decoder does not know that a slippage occurred. In certain applications, such as run-length encoding of bi-level images, where runs of identical pixels are decoded and placed consecutively in the decoded image, these points may cause a noticeable error when the decoded image is later examined.

A possible solution, due to [Lam and Kulkarni 96], is to take an RHC and modify it by including an extended synchronizing codeword (ESC), which is a bit pattern that cannot appear in any concatenation of codewords. The resulting code is no longer an RHC, but is still an RVLC (reversible VLC, page 195). The idea is to place the ESC at regular intervals in the encoded bitstream, say, after every 10 symbols or at the end of each line of text or each row of pixels. When the decoder decodes an ESC, it knows that 10 symbols should have been decoded since the previous ESC. If this is not true, the decoder can issue an error message, warning the user of a potential synchronization problem.

In order to understand the particular construction of an ESC, we consider all the cases in which a codeword c can be decoded incorrectly when there are errors preceding it. Luckily, there are only two such cases, illustrated in Figure 4.12.

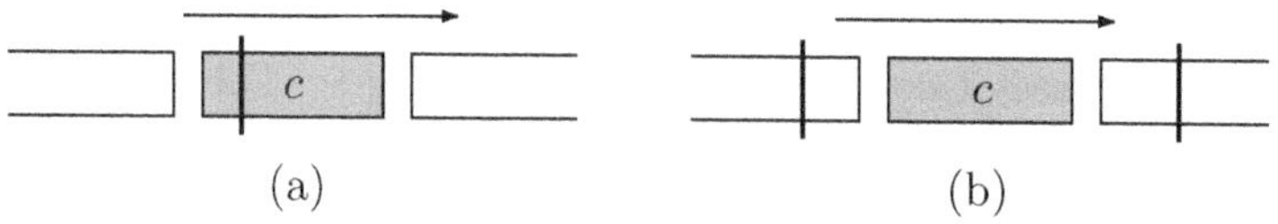

Figure 4.12: Two Ways to Incorrectly Decode a Codeword.

In part (a) of the figure, a prefix of c is decoded as the suffix of some other codeword and a suffix of c is decoded as the prefix of another codeword (in between, parts of c may be decoded as several complete codewords). In part (b), codeword c is concatenated with some bits preceding it and some bits following it, to form a valid codeword. This simple analysis points the way toward an ESC, because it is now clear that this bit pattern should satisfy the following conditions:

1. If the ESC can be written as the concatenation of two bit strings $\alpha\beta$ where α is a suffix of some codeword, then β should not be the prefix of any other codeword. Furthermore, if β itself can be written as the concatenation of two bit strings $\gamma\delta$, then γ can be empty or a concatenation of some codewords, and δ should be nonempty, should not be the prefix of any codeword, and no codeword should be a prefix of δ. This takes care of case (a).

2. The ESC should not be a substring of any codeword. This takes care of case (b).

If the ESC satisfies these conditions, then the decoder can recognize it even in the presence of errors preceding it. Once the decoder recognizes the ESC, it reads the bits that follow until it can continue the decoding. As can be expected, the life of a researcher in the field of coding is never that simple. It may happen that an error corrupts part of the encoded bit stream into an ESC. Also, an error can corrupt the ESC pattern itself. In these cases, the decoder loses synchronization, but regains it when the next ESC is found later. It is also possible to append a fixed-length count to each ESC, such that the first ESC is followed by a count of 1, the second ESC is followed by a 2, and so on. If the latest ESC had a count $i-1$ and the next ESC has a count of $i+1$, then the decoder knows that the ESC with a count of i has been missed because of such an error. If the previous ESC had a count of i and the current ESC has a completely different count, then the current ESC is likely spurious.

Given a Huffman code tree, how can an ESC be constructed? We first observe that it is always possible to rearrange a Huffman code tree such that the longest codeword (the codeword of the least-probable symbol) consists of all 1's. If this codeword is 1^k, then we construct the ESC as the bit pattern $1^k1^{k-1}0 = 1^{2k-1}0$. This pattern is longer than the longest codeword, so it cannot be a substring of any codeword, thereby satisfying condition 2 above. For condition 1, consider the following. All codeword suffixes that consist of consecutive 1's can have anywhere from no 1's to $(k-1)$ 1's. Therefore, if any prefix of the ESC is the suffix of a codeword, the remainder of the ESC will be a string of the form 1^j0 where j is between k and $2k-1$, and such a string is not the prefix of any codeword, thereby satisfying condition 1.

4.5 Bidirectional Codes

This book is enlivened by the use of quotations and epigraphs. Today, it is easy to search the Internet and locate quotations on virtually any topic. Many web sites maintain large collections of quotations, there are many books whose full text can easily be searched, and the fast, sophisticated Internet search engines can locate many occurrences of any word or phrase.

Before the age of computers, the task of finding words, phrases, and full quotations required long visits to a library and methodical reading of many texts. For this reason, scholars sometimes devoted substantial parts of their working careers to the preparation of a concordance.

A concordance is an alphabetical list of the principal words found in a book or a body of work, with their locations (page and line numbers) and immediate contexts. Common words such as "of" and "it" are excluded. The task of manually compiling and publishing a concordance was often gargantuan, which is why concordances were generated only for highly-valued works, such as the Bible, the writings of St. Thomas Aquinas, the Icelandic sagas, and the works of Shakespeare and Wordsworth.

The emergence of the digital computer, in the late 1940s and early 1950s, has given a boost to the art of concordance compiling. Suddenly, it became possible to store an entire body of literature in the computer and point to all the important words in it. It was also in the 1950s that Hans Peter Luhn, a computer scientist at IBM, conceived the technique of KWIC indexing. The idea was to search a concordance for a given word and display or print a condensed list of all the occurrences of the word, each with its location (page and line numbers) and immediate context. KWIC indexing remained popular for many years until it became superseded by the full text search that so many of us take for granted.

KWIC is an acronym for Key Word In Context, but like so many acronyms it has other meanings such as Kitchen Waste Into Compost, Kawartha World Issues Center, and Kids' Well-being Indicators Clearinghouse.

The following KWIC example lists the first 15 results of searching for the word **may** in a large collection (98,000 items) of excerpts from academic textbooks and introductory books located at the free concordance [amu 06].

```
No: Line:                    Concordance
 1:   73:  in the National Grid - which may accompany the beginning or end of
 2:   98: of a country, however much we may analyze it into separate rules,
 3:   98: and however much the analysis may be necessary to our understanding
 4:  104: y to interpret an Act a court may sometimes alter considerably the
 5:  110: ectly acquainted with English may know the words of the statute, but
 6:  114: ime. Therefore in a sense one may speak of the Common Law as unwritten
 7:  114: e for all similar cases which may arise in the future. This binding
 8:  120:  court will not disregard. It may happen that a question has never
 9:  120: ven a higher court, though it may think a decision of a lower court
10:  122: ween the parties. The dispute may be largely a question of fact.
11:  122: nts must be delivered, and it may often be a matter of doubt how far
12:  138:  of facts. The principles, it may be, give no explicit answer to the
13:  138:  the conclusions of a science may be involved in its premisses, and
14:  144: at conception is that a thing may change and yet remain the same thing.
15:  152: ions is unmanageable, Statute may undertake the work of codification,
```

Thus, a computerized concordance that supports KWIC searches consists of a set of texts and a dictionary. The dictionary lists all the important words in the texts, each with pointers to all its occurrences. Because of their size, the texts are normally stored in compressed format where each character is assigned a variable-length code (perhaps a Huffman code). The software inputs a search term, such as **may**, from the user, locates it in the dictionary, follows the first pointer to the text, and reads the immediate context of the first occurrence of **may**. The point is that this context both precedes and follows the word, but Huffman codes (and variable-length codes in general) are designed to be read and decoded from left to right (from early to late). In general, when reading a group of variable-length codes, it is impossible to reverse direction and read the preceding code, because there is no way to tell its length. We say that variable-length codes are unidirectional, but there are applications where bidirectional (or reversible) codes are needed.

Computerized concordances and KWIC indexing is one such application. Another application, no less important, is data integrity. We already know, from the discussion of synchronous codes in Section 4.3, that errors may occur and they tend to propagate through a sequence of variable-length codes. One way to limit error propagation (slippage) is to organize a compressed file in records, where each record consists of several variable-length codes. The record is read and decoded from beginning to end, but if an error is discovered, the decoder tries to read the record from end to start. If there is only one error in the record, the two readings and decodings can sometimes be combined to isolate the error and perhaps even to correct it

There are other, much less important applications of bidirectional codes. In the early days of the digital computer, magnetic tapes were the main input/output devices. A tape is sequential storage that lends itself to linear reading and writing. If we want to read an early record from a tape, we generally have to rewind the tape and then skip forward to the desired record. The ability to read a tape backward can speed up tape input/output and make the tape look more like a random-access I/O device. Another example of the use of bidirectional codes is a little-known data structure called deque (short for double-ended queue, and sometimes spelled "dequeue"). This is a linear data structure where elements can be added to or deleted from either end. (This is in contrast to a queue, where elements can only be added to the head and deleted from the tail.) If the elements of the deque are compressed by means of variable-length codes, then bidirectional codes allow for easy access of the structure from either end.

> A good programmer is someone who always looks both ways before crossing a one-way street.
>
> —Doug Linder

It should be noted that fixed-length codes are bidirectional, but they generally do not provide any compression (the Tunstall code of Section 2.6 is an exception). We say that fixed-length codes are trivially bidirectional. Thus, we need to develop variable-length codes that are as short as possible on average but are also bidirectional. The latter requirement is important, because the bidirectional codes are going to be used for compression (where they are often called reversible VLCs or RVLCs). Anyone with even little experience in the field of variable-length codes will immediately realize that it is easy to design bidirectional codes. Simply dedicate a bit pattern p to be both the prefix and suffix of all the codes, and make sure that p does not appear inside a code. Thus, if $p = 101$, then 101|10011001001|101 is an example of such a code. It is easy to see that such codes can be read in either direction, but it is also obvious that they are too long, because the code bits between the suffix and prefix are restricted to patterns that do not contain p. A little thinking shows that p doesn't even have to appear twice in each code. A code where each codeword ends with p is bidirectional. A string of such codes has the form $papbpc\ldots pypz$ and it is obvious that it can be read in either direction. The taboo codes of Section 3.16 are an example of this type of variable-length code.

The average length of a code is therefore important even in the case of bidirectional codes. Recall that the most common variable-length codes can be decoded easily and uniquely because they are prefix codes and therefore instantaneous. This suggests the idea of constructing a suffix code, a code where no codeword is the suffix of another codeword. A suffix code can be read in reverse and decoded uniquely in much the same way that a prefix code is uniquely decoded. Thus, a code that is both a prefix code and a suffix code is bidirectional. Such a code is termed *affix* (although some authors use the terms "biprefix" and "never-self-synchronizing").

The next few paragraphs are based on [Fraenkel and Klein 90], a work that analyses Huffman codes, looking for ways to construct affix Huffman codes or modify existing Huffman codes to make them affix and thus bidirectional. The authors first show how to start with a given affix Huffman code C and double its size. The idea is to take every codeword c_i in C and create two new codewords from it by appending a bit to it. The two new codewords are therefore c_i0 and c_i1. The resulting set of codewords is affix as can be seen from the following argument.

1. No other codeword in C starts with c_i (c_i is not the prefix of any other codeword). Therefore, no other codeword in the new code starts with c_i. As a result, no other codeword in the new code starts with c_i0 or c_i1.

2. Similarly, c_i is not the suffix of any other codeword in C, therefore neither c_i0 nor c_i1 are suffixes of a codeword in the new code.

Given the affix Huffman code 01, 000, 100, 110, 111, 0010, 0011, 1010, and 1011, we apply this method to double it and construct the code 010, 0000, 1000, 1100, 1110, 00100, 00110, 10100, 011, 0001, 1001, 1101, 1111, 00101, 00111, and 10101. A simple check verifies that the new code is affix. The conclusion is that there are infinitely many affix Huffman codes.

On the other hand, there are cases where affix Huffman codes do not exist. Consider, for example, codewords of length 1. In the trivial case where there are only two codewords, each is a single bit. This case is trivial and is also a fixed-length code. Thus, a variable-length code can have at most one codeword of length 1 (a single bit). If a code has such a codeword, then it is the suffix of other codewords, because in a complete prefix code, codewords must end with both 0 and 1. Thus, a code one of whose codewords is of length 1 cannot be affix. Such Huffman codes exist for sets of symbols with skewed probabilities. In fact, it is known that the existence of a codeword c_i of length 1 in a Huffman code implies that the probability of the symbol that corresponds to c_i must be greater than 1/3.

The authors then describe a complex algorithm (not listed here) to construct affix Huffman codes for cases where such codes exist.

There are many other ways to construct RVLCs from Huffman codes. Table 4.13 (after [Laković and Villasenor 03], see also Table 4.22) lists a set of Huffman codes for the 26 letters of the English alphabet together with three RVLC codes for the same symbols. These codes were constructed by algorithms proposed by [Takishima et al. 95], [Tsai and Wu 01a], and [Laković and Villasenor 03].

The remainder of this section describes the extension of Rice codes and exponential Golomb (EG) codes to bidirectional codes (RVLCs). The resulting bidirectional codes have the same average length as the original, unidirectional Rice and EG codes. They have been adopted by the International Telecommunications Union (ITU) for use

	p	Huffman	Takishima	Tsai	Laković
E	0.14878	001	001	000	000
T	0.09351	110	110	111	001
A	0.08833	0000	0000	0101	0100
O	0.07245	0100	0100	1010	0101
R	0.06872	0110	1000	0010	0110
N	0.06498	1000	1010	1101	1010
H	0.05831	1010	0101	0100	1011
I	0.05644	1110	11100	1011	1100
S	0.05537	0101	01100	0110	1101
D	0.04376	00010	00010	11001	01110
L	0.04124	10110	10010	10011	01111
U	0.02762	10010	01111	01110	10010
P	0.02575	11110	10111	10001	10011
F	0.02455	01111	11111	001100	11110
M	0.02361	10111	111101	011110	11111
C	0.02081	11111	101101	100001	100010
W	0.01868	000111	000111	1001001	100011
G	0.01521	011100	011101	0011100	1000010
Y	0.01521	100110	100111	1100011	1000011
B	0.01267	011101	1001101	0111110	1110111
V	0.01160	100111	01110011	1000001	10000010
K	0.00867	0001100	00011011	00111100	10000011
X	0.00146	00011011	000110011	11000011	11100111
J	0.00080	000110101	0001101011	100101001	100000010
Q	0.00080	0001101001	00011010011	0011101001	1000000010
Z	0.00053	0001101000	000110100011	1001011100	1000000111
Avg. length		4.15572	4.36068	4.30678	4.25145

Table 4.13: Huffman and Three RVLC Codes for the English Alphabet.

in the video coding parts of MPEG-4, and especially in the H.263v2 (also known as H.263+ or H.263 1998) and H263v3 (also known as H.263++ or H.263 2000) video compression standards [T-REC-h 06]. The material presented here is based on [Wen and Villasenor 98].

The Rice codes (Section 3.25) are a special case of the more general Golomb code, where the parameter m is a power of 2 ($m = 2^k$). Once the base k has been chosen, the Rice code of the unsigned integer n is constructed in two steps: (1) Separate the k least-significant bits (LSBs) of n. They become the LSBs of the Rice code. (2) Code the remaining $j = \lfloor n/2^k \rfloor$ bits in unary as either j zeros followed by a 1 or j 1's followed by a 0. This becomes the most-significant part of the Rice code. This code is therefore easily constructed with a few logical operations.

Decoding is also simple and requires only the value of k. The decoder scans the most-significant 1's until it reaches the first 0. This gives it the value of the most-significant part of n. The least-significant part of n is the k bits following the first 0. This simple decoding points the way to designing a bidirectional Rice code. The second part is always k bits long, so it can be read in either direction. To also make the first (unary) part bidirectional, we change it from 111...10 to 100...01, unless it is a single bit, in which case it becomes a single 0. Table 4.14 lists several original and bidirectional Rice codes. It should be compared with Table 10.28.

n	Binary	LSB	No. of 1's	Rice Code	Rev. Rice
0	0	00	0	0|00	0|00
1	1	01	0	0|01	0|01
2	10	10	0	0|10	0|10
3	11	11	0	0|11	0|11
4	100	00	1	10|00	11|00
5	101	01	1	10|01	11|01
6	110	10	1	10|10	11|10
7	111	11	1	10|11	11|11
8	1000	00	2	110|00	101|00
11	1011	11	2	110|11	101|11
12	1100	00	3	1110|00	1001|00
15	1111	11	3	1110|11	1001|11

Table 4.14: Original and Bidirectional Rice Codes.

An interesting property of the Rice codes is that there are 2^k codes of each length and the lengths start at $k+1$ (the prefix is at least one bit, and there is a k-bit suffix). Thus, for $k=3$ there are eight codes of length 4, eight codes of length 5, and so on. For certain probability distributions, we may want the number of codewords of length L to grow exponentially with L, and this feature is offered by a parametrized family of codes known as the exponential Golomb codes. These codes were first proposed in [Teuhola 78] and are also identical to the triplet $(s, 1, \infty)$ of start-step-stop codes. They perform well for probability distributions that are exponential but are taller than average and have a wide tail.

The exponential Golomb codes depend on the choice of a nonnegative integer parameter s (that becomes the length of the suffix of the codewords). The nonnegative integer n is encoded in the following steps:

1. Compute $w = 1 + \lfloor n/2^s \rfloor$.

2. Compute $f_{2^s}(n) = \left\lfloor \log_2 \left[1 + \frac{n}{2^s}\right]\right\rfloor$. This is the number of bits following the leftmost 1 in the binary representation of w.

3. Construct the codeword EG(n) as the unary representation of $f_{2^s}(n)$, followed by the $f_{2^s}(n)$ least-significant bits in the binary representation of w, followed by the s least-significant bits in the binary representation of n.

Thus, the length of this codeword is the sum

$$l(n) = 1 + 2f_{2^s}(n) + s = 1 + 2\left\lfloor \log_2 \left[1 + \frac{n}{2^s}\right]\right\rfloor + s = P + s,$$

where P is the prefix (whose length is always odd) and s is the suffix of a codeword. Because the logarithm is truncated, the length increases by 2 each time the logarithm increases by 1, i.e., each time $1 + n/2^s$ is a power of 2.

(As a side note, it should be mentioned that the exponential Golomb codes can be further generalized by substituting an arbitrary positive integer parameter m for the expression 2^s. Such codes can be called generalized exponential Golomb codes.)

As an example, we select $s = 1$ and determine the exponential Golomb code of $n = 11_{10} = 1011_2$. We compute $w = 1 + \lfloor 11/2^s \rfloor = 6 = 110_2$, $f_2(11) = \lfloor \log_2(11/2) \rfloor = 2$, and construct the three parts 110|10|1.

Table 4.15 lists (in column 2) several examples of the exponential Golomb codes for $s = 1$. Each code has an s-bit suffix and the prefixes get longer with n. The table also illustrates (in column 3) how these codes can be modified to become bidirectional. The idea is to have the prefix start and end with 1's, to fill the odd-numbered bit positions of the prefix (except the two extreme ones) with zeros, and to fill the even-numbered positions with bit patterns that represent increasing integers. Thus, for $s = 1$, the 16 $(7+1)$-bit codewords (i.e., the codewords for $n = 14$ through $n = 29$) have a 7-bit prefix of the form $1x0y0z1$ where the bits xyz take the eight values 000 through 111. Pairs of consecutive codewords have the same prefix and differ in their 1-bit suffixes.

n	Exp. Golomb	Rev. exp.
0	0|0	0|0
1	0|1	0|1
2	100|0	1**0**1|0
3	100|1	1**0**1|1
4	101|0	1**1**1|0
5	101|1	1**1**1|1
6	11000|0	1**0**0**0**1|0
7	11000|1	1**0**0**0**1|1
8	11001|0	1**0**0**1**1|0
9	11001|1	1**0**0**1**1|1
10	11010|0	1**1**0**0**1|0
11	11010|1	1**1**0**0**1|1

Table 4.15: Original and Bidirectional Exponential Golomb Codes.

Here is how the decoder can read such codewords in reverse and identify them. The decoder knows the value of s, so it first reads the s-bit suffix. If the next bit (the rightmost bit of the prefix) is 0, then the entire prefix is this single bit and the suffix determines the value of n (between 0 and $2^s - 1$). Otherwise, the decoder reads bits from right to left, isolating the bits with even indexes (shown in boldface in Table 4.15) and concatenating them from right to left, until an odd-indexed bit of 1 is found. The total number of bits read is the length P of the prefix. All the prefixes of length P differ only in their even-numbered bits, and a P-bit prefix (where P is always odd) has $(P-1)/2$ such bits. Thus, there are $2^{(P-1)/2}$ groups of prefixes, each with 2^s identical prefixes. The decoder identifies the particular n in a group by the s-bit suffix.

The bidirectional exponential Golomb codes therefore differ from the (original) exponential Golomb codes in their construction, but the two types have the same lengths.

The magical exclusive-OR. The methods presented earlier are based on sets of Huffman, Rice, and exponential Golomb codes that are modified and restricted in order to become bidirectional. In contrast, the next few paragraphs present a method, due to [Girod 99], where any set of prefix codes B_i can be transformed to a bitstring C that can be decoded in either direction. The method is simple and requires only logical operations and string reversals. It is based on a well-known "magical" property of the exclusive-OR (XOR) logical operation, and its only downside is the addition of a few extra zero bits. First, a few words about the XOR operation and what makes it special.

The logical OR operation is familiar to many. It receives two bits a and b as its inputs and it outputs one bit. The output is 1 if a or b or both are 1's. The XOR operation is similar but it excludes the case where both inputs are 1's. Thus, the result of (1 XOR 1) is 0. Both logical operations are summarized in the table.

a	0011
b	0101
a OR b	0111
a XOR b	0110

The table also shows that the XOR of any bit a with 0 is a.

What makes the XOR so useful is the following property: if $c = a$ XOR b, then $b = a$ XOR c and $a = b$ XOR c. This property is easily verified and it has made the XOR very popular in many applications (see page 357 of [Salomon 03] for an interesting example).

This useful property of the XOR is now exploited as follows. The first idea is to start with a string of n data symbols and encode them with a prefix code to produce a string B of n codewords $B_1B_2\ldots B_n$. Now reverse each codeword B_i to become B'_i and construct the string $B' = B'_1B'_2\ldots B'_n$. Next, compute the final result $C = B$ XOR B'. The hope is that the useful property of the XOR will enable us to decode B from C with the relation $B = C$ XOR B'. This simple scheme does not work because the relation requires knowledge of B'. We don't know string B', but we know that its components are closely related to those of string B. Thus, the following trick becomes the key to this elegant method. Denote by L the length of the longest codeword B_i, append L zeros to B and prepend L zeros to B' (more than L zeros can be used, but not fewer). Now perform the operation $C = (B\underbrace{00\ldots0}_{L})$ XOR $(\underbrace{00\ldots0}_{L}B')$. It is clear that the first L bits of C are identical to the first L bits of B. Because of the choice of L, those L bits constitute at least the first codeword B_1 (there may be more than one codeword and there may also be a remainder), so we can immediately reverse it to obtain B'_1. Now we can read more bits from C and XOR them with B'_1 to obtain more bits with parts of codewords from B. This unusual decoding procedure is best illustrated by an example.

We start with five symbols a_1 through a_5 and assign them the variable-length prefix codes 0, 10, 111, 1101, and 1100. The string of symbols $a_2a_1a_3a_5a_4$ is compressed by this code to the string $B = 10|0|111|1100|1101$. Reversing each codeword produces $B' = 01|0|111|0011|1011$. The longest codeword is four bits long, so we select $L = 4$.

We append four zeros to B and prepend four zeros to B'. Exclusive-ORing the strings yields

$$\begin{array}{r}B = 1001\,1111\,0011\,010000\\ B' = 0000\,0101\,1100\,111011\\ \hline C = 1001\,1010\,1111\,101011\end{array}$$

Decoding is done in the following steps:

1. XOR the first L bits of C and B'. This results in $1001 \oplus 0000 = 1001 \rightarrow a_2a_11$. This step decodes the first two symbols and leaves a "remainder" of 1 for the next step.

2. A total of three bits were decoded in the previous step, so the current step XORs the next three bits of C and B'. This results in $101 \oplus 010 = 111$. Prepending the remainder from the previous step yields 1111, which is decoded to a_3 and a remainder of 1.

3. A total of three bits were decoded in the previous step, so the current step XORs the next three bits of C and B', which results in $011 \oplus 111 = 100$. Prepending the remainder from the previous step yields 1100 which is decoded to a_5 with no remainder.

4. Four bits were decoded in the previous step, so the current step XORs the next four bits of C and B', which results in $1110 \oplus 0011 = 1101$, which is decoded to a_4.

5. Four bits were decoded in the previous step, so the current step XORs the next four bits of C and B', which results in $1011 \oplus 1011 = 0000$. This indicates the successful end of the decoding procedure.

If any data becomes corrupted, the last step will produce something other than L zeros, indicating unsuccessful decoding.

Decoding in the reverse direction is identical, except that C is fed to the decoder from end to start (from right to left) to produce substrings of B, which are then decoded and also reversed (to become codewords of B') and sent to the XOR. The first step XORs the reverse of the last four bits of C with the the reverse of the last four bits of B (the four zeros) to produce 1101, which is decoded as a_4, reversed, and sent to the XOR.

The only overhead is the extra L bits, but L is normally a small number. Figure 4.16 shows the encoder and decoder of this method.

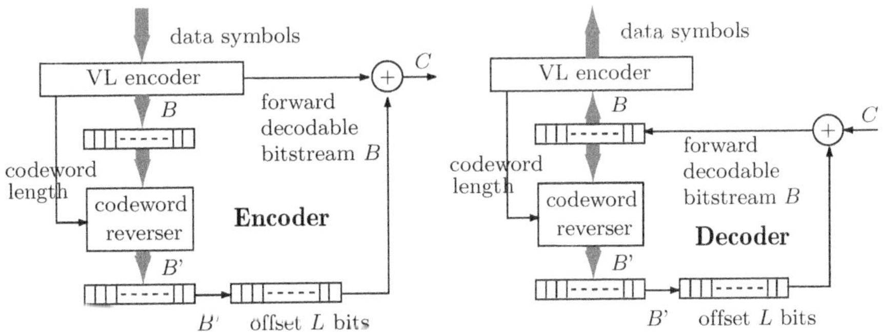

Figure 4.16: XOR-Based Encoding and Decoding.

4.6 Symmetric Codes

A symmetric code is one where every codeword is symmetric. Such a codeword looks the same when read in either direction, which is why symmetric codes are a special case of reversible codes. We can expect a symmetric code to feature higher average length compared with other codes, because the requirement of symmetry restricts the number of available bit patterns of any given length.

The material presented here describes one way of selecting a set of symmetric codewords. It is based on [Tsai and Wu 01b], which is an extension of [Takishima et al. 95]. The method starts from a variable-length prefix code, a set of prefix codewords of various lengths, and replaces the codewords with symmetric bit patterns that have the same or similar lengths and also satisfy the prefix property. Figure 4.17 is a good starting point. It shows a complete binary tree with four levels. The symmetric bit patterns at each level are underlined (as an aside, it can be proved by induction that there are $2^{\lfloor (i+1)/2 \rfloor}$ such patterns on level i) and it is clear that, even though the number of symmetric patterns is limited, every path from the root to a leaf passes through several such patterns. It is also clear from the figure that any bit pattern is the prefix of all the patterns below it on the same path. Thus, for example, selecting the pattern 00 implies that we cannot later select any of the symmetric patterns 000, 0000, or any other symmetric pattern below them on the same path. Selecting the 3-bit symmetric pattern 010, on the other hand, does not restrict the choice of 4-bit symmetric patterns. (It will restrict the choice of longer patterns, such as 01010 or 010010, but we are more interested in short patterns.) Thus, a clever algorithm is needed, to select those symmetric patterns at any level that will maximize the number of available symmetric patterns on the levels immediately below them.

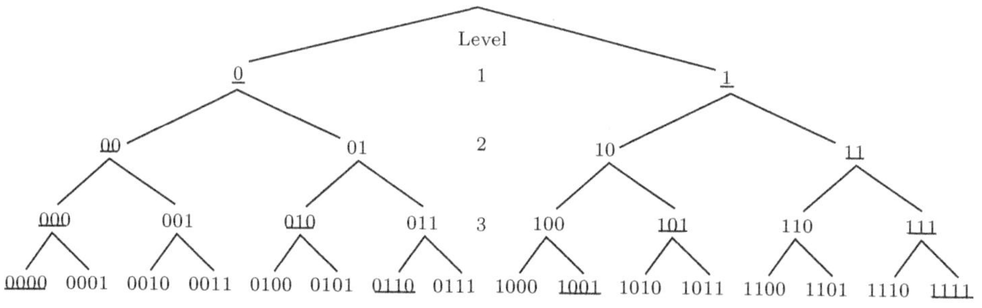

Figure 4.17: A 4-Level Full Binary Tree.

The analysis presented in [Tsai and Wu 01b] proposes the following procedure. Scan the complete binary tree level by level, locating the symmetric bit patterns on each level and ordering them in a special way as shown below. The ordered patterns are then used one by one to replace the original prefix codewords. However, if a symmetric pattern P violates the prefix property (i.e., if a previously-assigned pattern is a prefix of P), then P should be dropped.

To order the symmetric patterns of a level, go through the following steps:

1. Ignore the leftmost bit of the pattern.

2. Examine the remaining bits and determine the maximum number M of least-significant bits that are still symmetric (the maximum number of symmetric bit suffixes).

Table 4.18 lists the orders of the symmetric patterns on levels 3 through 6 of the complete binary tree. In this table, M stands for the maximum number of symmetric bit suffixes and CW is a symmetric pattern. For example, the 5-bit pattern 01110 has $M = 1$ because when we ignore its leftmost bit, the remaining four bits 1110 are asymmetric (only the rightmost bit is symmetric). On the other hand, pattern 10101 has $M = 3$ because the three rightmost bits of 0101 are symmetric.

The symmetric bit patterns on each level should be selected and assigned in the order shown in the table (order of increasing M). Thus, if we need 3-bit symmetric patterns, we should first select 010 and 101, and only then 000 and 111. Selecting them in this order maximizes the number of available symmetric patterns on levels 4 and 5.

Level 3		Level 4		Level 5		Level 6	
M	CW	M	CW	M	CW	M	CW
1	010	1	0110	1	01110	1	011110
1	101	1	1001	1	10001	1	100001
2	000	3	0000	2	00100	2	001100
2	111	3	1111	2	11011	2	110011
				3	01010	3	010010
				3	10101	3	101101
				4	00000	5	000000
				4	11111	5	111111

Table 4.18: Symmetric Codewords on Levels 3 Through 6, Ordered by Symmetric Bit Suffixes.

The following, incomplete, example illustrates the operation of this interesting algorithm. Figure 4.19 shows a Huffman code for the 26 letters of the English alphabet (the individual probabilities of the letters appear in Figure 5.5). The first Huffman codeword is 000. It happens to be symmetric, but Table 4.18 tells us to replace it with 010. It is immediately clear that this is a good choice (as would have been the choice of 101), because Figure 4.17 shows that the path that leads down from 010 to level 4 passes through patterns 0100 and 0101 which are not symmetric. Thus, the choice of 010 does not restrict the future choice of 4-bit symmetric codewords.

Figure 4.19 requires six 4-bit symmetric codewords, but there are only four such patterns. They are 0110, 1001, 0000, and 1111, selected in increasing number of symmetric bit suffixes. Pattern 010 is not a prefix of any of these patterns, so they are assigned as the new, symmetric codewords of T, A, O, and N. The remaining two symmetric codewords (of R and I) on this level will have to be longer (five bits each). Figure 4.19 requires 14 5-bit symmetric codewords (and two more, for R and I, are still needed), but there are only eight symmetric 5-bit patterns. Two of them cannot be used because of prefix violation. Pattern 010 is a prefix of 01010 and pattern 0000 is a prefix of 00000. The remaining six 5-bit codewords of Table 4.18 are selected and assigned to the letters R through L, while the letters C through V will obviously have to be assigned longer symmetric codewords.

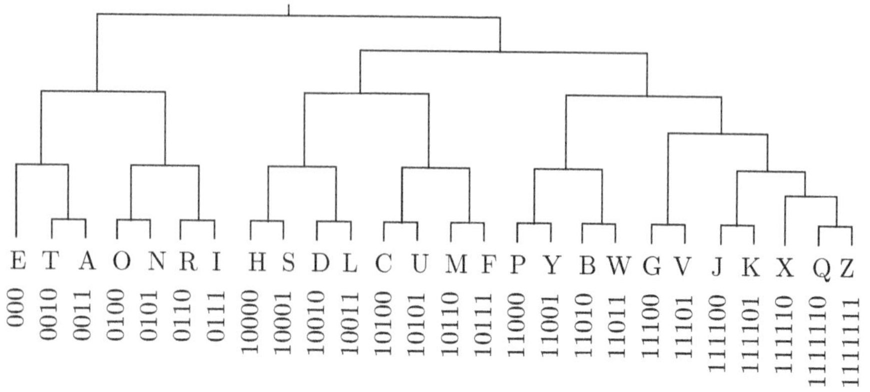

Figure 4.19: A Huffman Code for the 26-Letter Alphabet.

It seems that the symmetric codewords are longer on average than the original Huffman codewords, and this is true in general, because the number of available symmetric codewords of any given length is limited. However, as we go down the tree to longer and longer codewords, the situation improves, because every other level in the complete binary tree doubles the number of available symmetric codewords. Reference [Tsai and Wu 01b] lists Huffman codes and symmetric codes for the 26 letters with average lengths of 4.156 and 4.61 bits per letter, respectively.

The special ordering rules above seem arbitrary, but are not difficult to understand intuitively. As an example, we examine the eight symmetric 6-bit patterns. The two patterns of $M = 5$ become prefixes of patterns at level 7 simply by appending a 0 or a 1. These patterns should therefore be the last ones to be used because they restrict the number of available 7-bit symmetric patterns. The two patterns for $M = 3$ are better, because they are not the prefixes of any 7-bit or 8-bit symmetric patterns. It is obvious that appending one or two bits to 010010 or 101101 cannot result in a symmetric pattern. Thus, these two patterns are prefixes of 9-bit (or longer) symmetric patterns. When we examine the two patterns for $M = 2$, the situation is even better. Appending one, two, or even three bits to 100001 or 001100 cannot result in a symmetric pattern. These patterns are the prefixes of 10-bit (or longer) patterns. Similarly, the two 6-bit patterns for $M = 1$ are the best, because their large asymmetry implies that they can only be prefixes of 11-bit (or longer) patterns. Selecting these two patterns does not restrict the number of symmetric 7-, 8-, 9-, or 10-bit patterns.

4.7 VLEC Codes

The simplest approach to robust variable-length codes is to add a parity bit to each codeword. This approach combines compression (source coding) and reliability (channel coding), but keeps the two separate. The two aims are contradictory because the former removes redundancy while the latter adds redundancy. However, any approach to robust compressed data must necessarily reduce compression efficiency. This simple approach has an obvious advantage; it makes it easy to respond to statistical changes in both the source and the channel. If it turns out that the channel is noisy, several parity bits may be appended to each codeword, but the codewords themselves do not have to be modified. If it turns out that certain data symbols occur with higher probability than originally thought, their codewords can be replaced with shorter ones without having to modify the error-control scheme.

A different approach to the same problem is to construct a set of variable-length codewords that are sufficiently distant from one another (distant in the sense of Hamming distance). A set of such codewords can be called a variable-length error-correcting (VLEC) code. The downside of this approach is that any changes in source or channel statistics requires a new set of codewords, but the hope is that this approach may result in better overall compression. This approach may be justified if it meets the following two goals:

1. The average length of the codewords turns out to be shorter than the average length of the codewords with parity bits of the previous approach.

2. A decoding algorithm is found that can exploit the large distance between variable-length codewords to detect and even correct errors. Even better, such an algorithm should be able to recover synchronization in cases where bits get corrupted in the communications channel.

Several attempts to develop such codes are described here.

Alpha-Prompt Codes. The main idea of this technique is to associate each codeword c with a set $\alpha(c)$ of bit patterns of the same length that are at certain Hamming distances from c. If codeword c is transmitted and gets corrupted to c', the decoder checks the entire set $\alpha(c)$ and selects the pattern nearest c'. The problem is that the decoder doesn't know the length of the next codeword, which is why a practical method based on this technique should construct the sets $\alpha(c)$ in a special way.

We assume that there are s_1 codewords of length L_1, s_2 codewords of length L_2, and so on, up to length m. The set of all the codewords c and all the bit patterns in the individual sets $\alpha(c)$ is constructed as a prefix code (i.e., it satisfies the prefix property and is therefore instantaneous). The decoder starts by reading the next L_1 bits from the input. Let's denote this value by t. If t is a valid codeword c in s_1, then t is decoded to c. Otherwise, the decoder checks all the leftmost L_1 bits of all the patterns that are L_1 bits or longer and compares t to each of them. If t is identical to a pattern in set $\alpha(c_i)$, then t is decoded to c_i. Otherwise, the decoder again goes over the leftmost L_1 bits of all the patterns that are L_1 bits or longer and selects the one whose distance from t is minimal. If that pattern is a valid codeword, then t is decoded to it. Otherwise, the decoder reads a few more bits, for a total of L_2 bits, and repeats this process for the first L_2 bits of all the patterns that are L_2 bits long or longer.

This complex algorithm (proposed by [Buttigieg 95]) always decodes the input to some codeword if the set of all the codewords c and all the bit patterns in the individual sets $\alpha(c)$ is a prefix code.

VLEC Tree. We assume that our code has s_1 codewords of length L_1, s_2 codewords of length L_2, and so on, up to length m. The total number of codewords c_i is s. As a simple example, consider the code $a = 000$, $b = 0110$, and $c = 1011$ (one 3-bit and two 4-bit codewords).

The VLEC tree method attempts to correct errors by looking at an entire transmission and mapping it to a special VLEC tree. Each node in this tree is connected to s other nodes on lower levels with edges s that are labeled with the codewords c_i. Figure 4.20 illustrates an example with the code above. Each path of e edges from the root to level b is therefore associated with a different string of e codewords whose total length is b bits. Thus, node [1] in the figure corresponds to string $cba = 1011|0110|000$ (11 bits) and node [2] corresponds to the 12-bit string $bcc = 0110|1011|1011$. There are many possible strings of codewords, but it is also clear that the tree grows exponentially and quickly becomes very wide (the tree in the figure corresponds to strings of up to 12 bits).

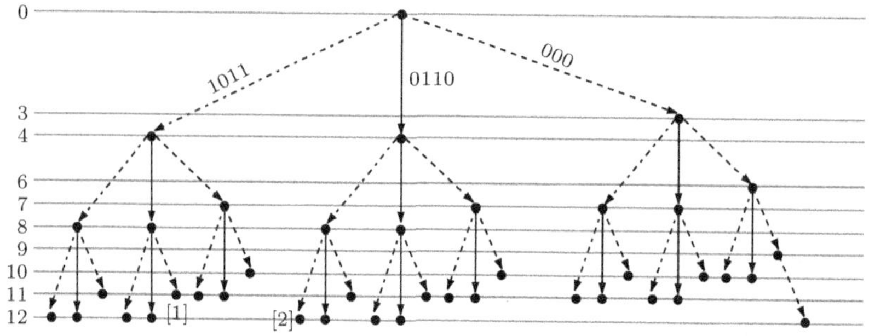

Figure 4.20: A VLEC Tree.

Assuming that the only transmission errors are corrupted bits (i.e., no bits are inserted to or deleted from an encoded string by noise in the communications channel), if an encoded string of b bits is sent, the decoder will receive b bits, some possibly bad. Any b-bit string corresponds to a node on level b, so error-correcting is reduced to the problem of selecting one of the paths that end on that level. The obvious solution is to measure the Hamming distances between the received b-bit string and all the paths that end on level b and select the path with the minimum distance. This simple solution, however, is impractical for the following reasons:

- Data symbols (and therefore their codewords) have different probabilities, which is why not all b-bit paths are equally likely. If symbol a is very probable, then paths with many occurrences of a are more likely and should be given more weight. Assume that symbol a in our example has probability 0.5, while symbols b and c occur with probability 0.25 each. A probable path such as $aabaca$ should be assigned the weight $4 \times 0.5 + 2 \times 0.25 = 2.5$, while a less probable path, such as $bbccca$ should be assigned the smaller weight $5 \times 0.25 + 0.5 = 1.75$. When the decoder receives a b-bit string S, it should (1) compute the Hamming distance between S and all the b-bit paths P, (2) divide each distance by the weight of the path, and (3) select the path with the smallest weighted distance. Thus, if two paths with weights 2.5 and 1.75 are at a Hamming distance of 2 from S, the decoder computes $2/2.5 = 0.8$ and $2/1.75 = 1.14$ and selects the former path.
- A VLEC tree for even a modest-size alphabet grows too wide very quickly and becomes unmanageable for even short messages. Thus, an algorithm is needed for the

decoder to decode an encoded message in short segments. In cases where many errors are rare, a possible approach may be to read the encoded string and encode it normally until the decoder loses synchronization (say, at point A). The decoder then skips bits until it regains synchronization (at point B), and then employs the VLEC tree to the input from A to B.

VLEC Trellis. The second point above, of the tree getting too big very quickly, can be overcome by replacing the tree with a trellis structure, as shown in Figure 4.21.

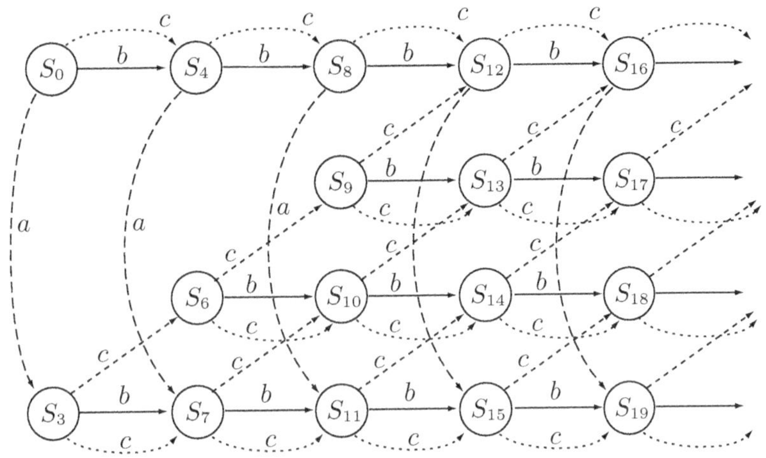

Figure 4.21: A VLEC Trellis.

The trellis is constructed as follows:

1. Start with state S_0.

2. From each state, construct m edges to other states, where each edge corresponds to one of the m codeword lengths L_i. Thus, state S_i will be connected to states S_{i+L_1}, S_{i+L_2}, up to S_{i+L_m}. If any of these states does not exist, it is created when needed.

> A trellis is a structure, usually made from interwoven pieces of wood, bamboo or metal that supports many types of climbing plant such as sweet peas, grapevines and ivy.
> —`wikipedia.com`

Notice that the states are arranged in Figure 4.21 in stages, where each stage is a vertical set of states. A comparison of Figures 4.20 and 4.21 shows their similarities, but also shows the important difference between them. Beyond a certain point (in our example, beyond the third stage), the trellis stages stop growing in size regardless of the length of the trellis. It is this difference that makes it possible to develop a reliable decoding algorithm (a modified Viterbi algorithm, described on page 69 of [Buttigieg 95], but not discussed here). An important feature of this algorithm is that it employs a metric (a weight assigned to each trellis edge), and it is possible to define this metric in a way that takes into account the different probabilities of the trellis edges.

4.7.1 Constructing VLEC Codes

	p	Huffman	Takishima	Tsai	Laković
E	0.14878	001	001	000	000
T	0.09351	110	110	111	111
A	0.08833	0000	0000	0101	0101
O	0.07245	0100	0100	1010	1010
R	0.06872	0110	1000	0010	0110
N	0.06498	1000	1010	1101	1001
H	0.05831	1010	0101	0100	0011
I	0.05644	1110	11100	1011	1100
S	0.05537	0101	01100	0110	00100
D	0.04376	00010	00010	11001	11011
L	0.04124	10110	10010	10011	01110
U	0.02762	10010	01111	01110	10001
P	0.02575	11110	10111	10001	010010
F	0.02455	01111	11111	001100	101101
M	0.02361	10111	111101	011110	100001
C	0.02081	11111	101101	100001	011110
W	0.01868	000111	000111	1001001	001011
G	0.01521	011100	011101	0011100	110100
Y	0.01521	100110	100111	1100011	0100010
B	0.01267	011101	1001101	0111110	1011101
V	0.01160	100111	01110011	1000001	0100010
K	0.00867	0001100	00011011	00111100	1101011
X	0.00146	00011011	000110011	11000011	10111101
J	0.00080	000110101	0001101011	100101001	010000010
Q	0.00080	0001101001	00011010011	0011101001	0100000010
Z	0.00053	0001101000	000110100011	1001011100	1011111101
Avg. length		4.15572	4.36068	4.30678	4.34534

Table 4.22: Huffman and Three RVLC Codes for the English Alphabet.

How can we construct an RVLC with a free distance greater than 1? One such algorithm, proposed by [Laković and Villasenor 02] is an extension of an older algorithm due to [Tsai and Wu 01a], which itself is based on the work of [Takishima et al. 95]. This algorithm starts from a set of Huffman codes. It then goes over these codewords level by level, from short to long codewords and replaces some of them with patterns taken from a complete binary tree, similar to what is described in Section 4.6, making sure that the prefix property is not violated while also ascertaining that the free distance of all the codewords that have so far been examined does not drop below the target free distance. If a certain pattern from the binary tree results in a free distance that is too small, that pattern is skipped.

Table 4.22 (after [Laković and Villasenor 02], see also the very similar Table 4.13) lists a set of Huffman codes for the 26 letters of the English alphabet together with three RVLC codes for the same symbols. The last of these codes has a free distance of 2. These codes were constructed by algorithms proposed by [Takishima et al. 95], [Tsai and Wu 01a], and [Laković and Villasenor 02].

Summary. Errors are a fact of life, which is why any practical method for coding, storing, or transmitting data should consider the use of robust codes. The various variable-length codes described in this chapter are robust, which makes them ideal choices for applications where both data compression and data reliability are needed.

Good programmers naturally write neat code when left to their own devices. But they also have an array of battle tactics to help write robust code on the front line.

—Pete Goodliffe, *Code Craft: The Practice of Writing Excellent Code*

5 Statistical Methods

Statistical data compression methods employ variable-length codes, with the shorter codes assigned to symbols or groups of symbols that appear more often in the data (have a higher probability of occurrence). Designers and implementors of variable-length codes have to deal with the two problems of (1) assigning codes that can be decoded unambiguously and (2) assigning codes with the minimum average size. The first problem is discussed in detail in Chapters 2 through 4, while the second problem is solved in different ways by the methods described here.

This chapter is devoted to statistical compression algorithms, such as Shannon-Fano, Huffman, arithmetic coding, and PPM. It is recommended, however, that the reader start with the short presentation of information theory in the Appendix. This presentation covers the principles and important terms used by information theory, especially the terms redundancy and entropy. An understanding of these terms leads to a deeper understanding of statistical data compression, and also makes it possible to calculate how redundancy is reduced, or even eliminated, by the various methods.

5.1 Shannon-Fano Coding

Shannon-Fano coding, named after Claude Shannon and Robert Fano, was the first algorithm to construct a set of the best variable-length codes.

We start with a set of n symbols with known probabilities (or frequencies) of occurrence. The symbols are first arranged in descending order of their probabilities. The set of symbols is then divided into two subsets that have the same (or almost the same) probabilities. All symbols in one subset get assigned codes that start with a 0, while the codes of the symbols in the other subset start with a 1. Each subset is then recursively divided into two subsubsets of roughly equal probabilities, and the second bit of all the codes is determined in a similar way. When a subset contains just two symbols,

D. Salomon, G. Motta, *Handbook of Data Compression*, 5th ed.
DOI 10.1007/978-1-84882-903-9_5,

their codes are distinguished by adding one more bit to each. The process continues until no more subsets remain. Table 5.1 illustrates the Shannon-Fano algorithm for a seven-symbol alphabet. Notice that the symbols themselves are not shown, only their probabilities.

Robert M. Fano was Ford Professor of Engineering, in the Department of Electrical Engineering and Computer Science at the Massachusetts Institute of Technology until his retirement. In 1963 he organized MIT's Project MAC (now the Computer Science and Artificial Intelligence Laboratory) and was its Director until September 1968. He also served as Associate Head of the Department of Electrical Engineering and Computer Science from 1971 to 1974.

Professor Fano chaired the Centennial Study Committee of the Department of Electrical Engineering and Computer Science whose report, "Lifelong Cooperative Education," was published in October, 1982.

Professor Fano was born in Torino, Italy, and did most of his undergraduate work at the School of Engineering of Torino before coming to the United States in 1939. He received the Bachelor of Science degree in 1941 and the Doctor of Science degree in 1947, both in Electrical Engineering from MIT. He has been a member of the MIT staff since 1941 and a member of its faculty since 1947.

During World War II, Professor Fano was on the staff of the MIT Radiation Laboratory, working on microwave components and filters. He was also group leader of the Radar Techniques Group of Lincoln Laboratory from 1950 to 1953. He has worked and published at various times in the fields of network theory, microwaves, electromagnetism, information theory, computers and engineering education. He is author of the book entitled *Transmission of Information*, and co-author of *Electromagnetic Fields, Energy and Forces* and *Electromagnetic Energy Transmission and Radiation*. He is also co-author of Volume 9 of the Radiation Laboratory Series.

The first step splits the set of seven symbols into two subsets, one with two symbols and a total probability of 0.45 and the other with the remaining five symbols and a total probability of 0.55. The two symbols in the first subset are assigned codes that start with 1, so their final codes are 11 and 10. The second subset is divided, in the second step, into two symbols (with total probability 0.3 and codes that start with 01) and three symbols (with total probability 0.25 and codes that start with 00). Step three divides the last three symbols into 1 (with probability 0.1 and code 001) and 2 (with total probability 0.15 and codes that start with 000).

The average size of this code is $0.25 \times 2 + 0.20 \times 2 + 0.15 \times 3 + 0.15 \times 3 + 0.10 \times 3 + 0.10 \times 4 + 0.05 \times 4 = 2.7$ bits/symbol. This is a good result because the entropy (the smallest number of bits needed, on average, to represent each symbol) is

$$-\big(0.25\log_2 0.25 + 0.20\log_2 0.20 + 0.15\log_2 0.15 + 0.15\log_2 0.15 \\ + 0.10\log_2 0.10 + 0.10\log_2 0.10 + 0.05\log_2 0.05\big) \approx 2.67.$$

Prob.		Steps				Final
1. 0.25	1	1				:11
2. 0.20	1	0				:10
3. 0.15	0	1	1			:011
4. 0.15	0	1	0			:010
5. 0.10	0	0	1			:001
6. 0.10	0	0	0	1		:0001
7. 0.05	0	0	0	0		:0000

Table 5.1: Shannon-Fano Example.

◇ **Exercise 5.1:** Repeat the calculation above but place the first split between the third and fourth symbols. Calculate the average size of the code and show that it is greater than 2.67 bits/symbol.

The code in the table in the answer to Exercise 5.1 has longer average size because the splits, in this case, were not as good as those of Table 5.1. This suggests that the Shannon-Fano method produces better code when the splits are better, i.e., when the two subsets in every split have very close total probabilities. Carrying this argument to its limit suggests that perfect splits yield the best code. Table 5.2 illustrates such a case. The two subsets in every split have identical total probabilities, yielding a code with the minimum average size (zero redundancy). Its average size is $0.25 \times 2 + 0.25 \times 2 + 0.125 \times 3 + 0.125 \times 3 + 0.125 \times 3 + 0.125 \times 3 = 2.5$ bits/symbols, which is identical to its entropy. This means that it is the theoretical minimum average size.

Prob.		Steps		Final
1. 0.25	1	1		:11
2. 0.25	1	0		:10
3. 0.125	0	1	1	:011
4. 0.125	0	1	0	:010
5. 0.125	0	0	1	:001
6. 0.125	0	0	0	:000

Table 5.2: Shannon-Fano Balanced Example.

The conclusion is that this method produces the best results when the symbols have probabilities of occurrence that are (negative) powers of 2.

◇ **Exercise 5.2:** Compute the entropy of the codes of Table 5.2.

The Shannon-Fano method is easy to implement but the code it produces is generally not as good as that produced by the Huffman method (Section 5.2).

5.2 Huffman Coding

David Huffman (1925–1999)

Being originally from Ohio, it is no wonder that Huffman went to Ohio State University for his BS (in electrical engineering). What is unusual was his age (18) when he earned it in 1944. After serving in the United States Navy, he went back to Ohio State for an MS degree (1949) and then to MIT, for a PhD (1953, electrical engineering).

That same year, Huffman joined the faculty at MIT. In 1967, he made his only career move when he went to the University of California, Santa Cruz as the founding faculty member of the Computer Science Department. During his long tenure at UCSC, Huffman played a major role in the development of the department (he served as chair from 1970 to 1973) and he is known for his motto "my products are my students." Even after his retirement, in 1994, he remained active in the department, teaching information theory and signal analysis courses.

Huffman made significant contributions in several areas, mostly information theory and coding, signal designs for radar and communications, and design procedures for asynchronous logical circuits. Of special interest is the well-known Huffman algorithm for constructing a set of optimal prefix codes for data with known frequencies of occurrence. At a certain point he became interested in the mathematical properties of "zero curvature" surfaces, and developed this interest into techniques for folding paper into unusual sculptured shapes (the so-called computational origami).

Huffman coding is a popular method for data compression. It serves as the basis for several popular programs run on various platforms. Some programs use just the Huffman method, while others use it as one step in a multistep compression process. The Huffman method [Huffman 52] is somewhat similar to the Shannon-Fano method. It generally produces better codes, and like the Shannon-Fano method, it produces the best code when the probabilities of the symbols are negative powers of 2. The main difference between the two methods is that Shannon-Fano constructs its codes top to bottom (from the leftmost to the rightmost bits), while Huffman constructs a code tree from the bottom up (builds the codes from right to left). Since its development, in 1952, by D. Huffman, this method has been the subject of intensive research into data compression.

Since its development in 1952 by D. Huffman, this method has been the subject of intensive research in data compression. The long discussion in [Gilbert and Moore 59] proves that the Huffman code is a minimum-length code in the sense that no other encoding has a shorter average length. An algebraic approach to constructing the Huffman code is introduced in [Karp 61]. In [Gallager 74], Robert Gallager shows that the redundancy of Huffman coding is at most $p_1 + 0.086$ where p_1 is the probability of the most-common symbol in the alphabet. The redundancy is the difference between the average Huffman codeword length and the entropy. Given a large alphabet, such as the set of letters, digits and punctuation marks used by a natural language, the largest symbol probability is typically around 15–20%, bringing the value of the quantity $p_1 + 0.086$ to around 0.1. This means that Huffman codes are at most 0.1 bit longer (per symbol) than an ideal entropy encoder, such as arithmetic coding.

The Huffman algorithm starts by building a list of all the alphabet symbols in descending order of their probabilities. It then constructs a tree, with a symbol at every leaf, from the bottom up. This is done in steps, where at each step the two symbols with smallest probabilities are selected, added to the top of the partial tree, deleted from the list, and replaced with an auxiliary symbol representing the two original symbols. When the list is reduced to just one auxiliary symbol (representing the entire alphabet), the tree is complete. The tree is then traversed to determine the codes of the symbols.

This process is best illustrated by an example. Given five symbols with probabilities as shown in Figure 5.3a, they are paired in the following order:

1. a_4 is combined with a_5 and both are replaced by the combined symbol a_{45}, whose probability is 0.2.
2. There are now four symbols left, a_1, with probability 0.4, and a_2, a_3, and a_{45}, with probabilities 0.2 each. We arbitrarily select a_3 and a_{45}, combine them, and replace them with the auxiliary symbol a_{345}, whose probability is 0.4.
3. Three symbols are now left, a_1, a_2, and a_{345}, with probabilities 0.4, 0.2, and 0.4, respectively. We arbitrarily select a_2 and a_{345}, combine them, and replace them with the auxiliary symbol a_{2345}, whose probability is 0.6.
4. Finally, we combine the two remaining symbols, a_1 and a_{2345}, and replace them with a_{12345} with probability 1.

The tree is now complete. It is shown in Figure 5.3a "lying on its side" with its root on the right and its five leaves on the left. To assign the codes, we arbitrarily assign a bit of 1 to the top edge, and a bit of 0 to the bottom edge, of every pair of edges. This results in the codes 0, 10, 111, 1101, and 1100. The assignments of bits to the edges is arbitrary.

The average size of this code is $0.4 \times 1 + 0.2 \times 2 + 0.2 \times 3 + 0.1 \times 4 + 0.1 \times 4 = 2.2$ bits/symbol, but even more importantly, the Huffman code is not unique. Some of the steps above were chosen arbitrarily, since there were more than two symbols with smallest probabilities. Figure 5.3b shows how the same five symbols can be combined differently to obtain a different Huffman code (11, 01, 00, 101, and 100). The average size of this code is $0.4 \times 2 + 0.2 \times 2 + 0.2 \times 2 + 0.1 \times 3 + 0.1 \times 3 = 2.2$ bits/symbol, the same as the previous code.

◇ **Exercise 5.3:** Given the eight symbols A, B, C, D, E, F, G, and H with probabilities 1/30, 1/30, 1/30, 2/30, 3/30, 5/30, 5/30, and 12/30, draw three different Huffman trees with heights 5 and 6 for these symbols and calculate the average code size for each tree.

◇ **Exercise 5.4:** Figure Ans.5d shows another Huffman tree, with height 4, for the eight symbols introduced in Exercise 5.3. Explain why this tree is wrong.

It turns out that the arbitrary decisions made in constructing the Huffman tree affect the individual codes but not the average size of the code. Still, we have to answer the obvious question, which of the different Huffman codes for a given set of symbols is best? The answer, while not obvious, is simple: The best code is the one with the

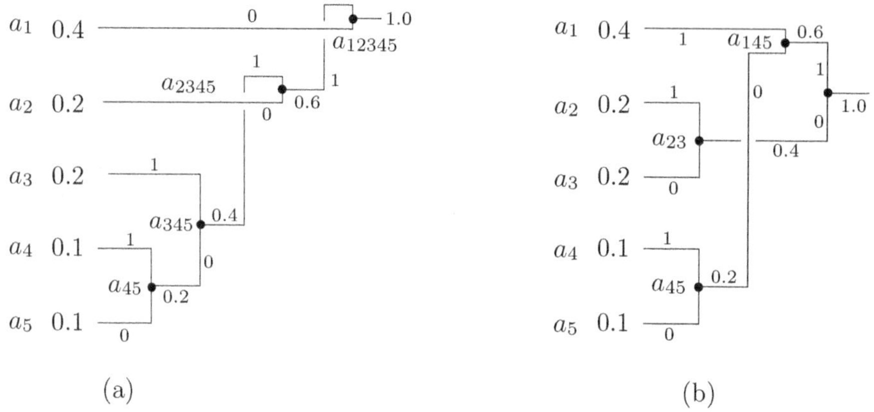

Figure 5.3: Huffman Codes.

smallest variance. The variance of a code measures how much the sizes of the individual codes deviate from the average size (see page 624 for the definition of variance). The variance of code 5.3a is

$$0.4(1-2.2)^2+0.2(2-2.2)^2+0.2(3-2.2)^2+0.1(4-2.2)^2+0.1(4-2.2)^2=1.36,$$

while the variance of code 5.3b is

$$0.4(2-2.2)^2+0.2(2-2.2)^2+0.2(2-2.2)^2+0.1(3-2.2)^2+0.1(3-2.2)^2=0.16.$$

Code 5.3b is therefore preferable (see below). A careful look at the two trees shows how to select the one we want. In the tree of Figure 5.3a, symbol a_{45} is combined with a_3, whereas in the tree of 5.3b it is combined with a_1. The rule is: When there are more than two smallest-probability nodes, select the ones that are lowest and highest in the tree and combine them. This will combine symbols of low probability with ones of high probability, thereby reducing the total variance of the code.

If the encoder simply writes the compressed stream on a file, the variance of the code makes no difference. A small-variance Huffman code is preferable only in cases where the encoder *transmits* the compressed stream, as it is being generated, over a communications line. In such a case, a code with large variance causes the encoder to generate bits at a rate that varies all the time. Since the bits have to be transmitted at a constant rate, the encoder has to use a buffer. Bits of the compressed stream are entered into the buffer as they are being generated and are moved out of it at a constant rate, to be transmitted. It is easy to see intuitively that a Huffman code with zero variance will enter bits into the buffer at a constant rate, so only a short buffer will be needed. The larger the code variance, the more variable is the rate at which bits enter the buffer, requiring the encoder to use a larger buffer.

The following claim is sometimes found in the literature:

> It can be shown that the size of the Huffman code of a symbol a_i with probability P_i is always less than or equal to $\lceil -\log_2 P_i \rceil$.

Even though it is correct in many cases, this claim is not true in general. It seems to be a wrong corollary drawn by some authors from the Kraft-MacMillan inequality, Equation (2.3). The authors are indebted to Guy Blelloch for pointing this out and also for the example of Table 5.4.

◇ **Exercise 5.5:** Find an example where the size of the Huffman code of a symbol a_i is greater than $\lceil -\log_2 P_i \rceil$.

P_i	Code	$-\log_2 P_i$	$\lceil -\log_2 P_i \rceil$
.01	000	6.644	7
*.30	001	1.737	2
.34	01	1.556	2
.35	1	1.515	2

Table 5.4: A Huffman Code Example.

◇ **Exercise 5.6:** It seems that the size of a code must also depend on the number n of symbols (the size of the alphabet). A small alphabet requires just a few codes, so they can all be short; a large alphabet requires many codes, so some must be long. This being so, how can we say that the size of the code of symbol a_i depends just on its probability P_i?

Figure 5.5 shows a Huffman code for the 26 letters.

As a self-exercise, the reader may calculate the average size, entropy, and variance of this code.

◇ **Exercise 5.7:** Discuss the Huffman codes for equal probabilities.

Exercise 5.7 shows that symbols with equal probabilities don't compress under the Huffman method. This is understandable, since strings of such symbols normally make random text, and random text does not compress. There may be special cases where strings of symbols with equal probabilities are not random and can be compressed. A good example is the string $a_1a_1 \ldots a_1a_2a_2 \ldots a_2a_3a_3 \ldots$ in which each symbol appears in a long run. This string can be compressed with RLE but not with Huffman codes.

Notice that the Huffman method cannot be applied to a two-symbol alphabet. In such an alphabet, one symbol can be assigned the code 0 and the other code 1. The Huffman method cannot assign to any symbol a code shorter than one bit, so it cannot improve on this simple code. If the original data (the source) consists of individual bits, such as in the case of a bi-level (monochromatic) image, it is possible to combine several bits (perhaps four or eight) into a new symbol and pretend that the alphabet consists of these (16 or 256) symbols. The problem with this approach is that the original binary data may have certain statistical correlations between the bits, and some of these correlations would be lost when the bits are combined into symbols. When a typical bi-level image (a painting or a diagram) is digitized by scan lines, a pixel is more likely to be followed by an identical pixel than by the opposite one. We therefore have a file that can start with either a 0 or a 1 (each has 0.5 probability of being the first bit). A zero is more likely to be followed by another 0 and a 1 by another 1. Figure 5.6 is a finite-state

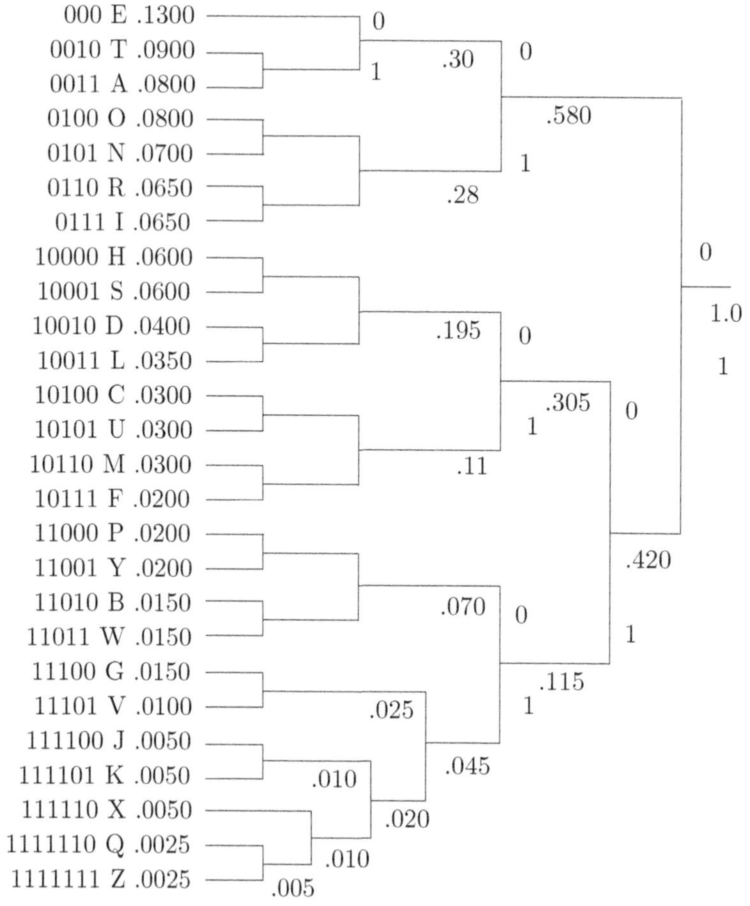

Figure 5.5: A Huffman Code for the 26-Letter Alphabet.

machine illustrating this situation. If these bits are combined into, say, groups of eight, the bits inside a group will still be correlated, but the groups themselves will not be correlated by the original pixel probabilities. If the input stream contains, e.g., the two adjacent groups 00011100 and 00001110, they will be encoded independently, ignoring the correlation between the last 0 of the first group and the first 0 of the next group. Selecting larger groups improves this situation but increases the number of groups, which implies more storage for the code table and longer time to calculate the table.

◇ **Exercise 5.8:** How does the number of groups increase when the group size increases from s bits to $s+n$ bits?

A more complex approach to image compression by Huffman coding is to create several complete sets of Huffman codes. If the group size is, e.g., eight bits, then several sets of 256 codes are generated. When a symbol S is to be encoded, one of the sets is selected, and S is encoded using its code in that set. The choice of set depends on the

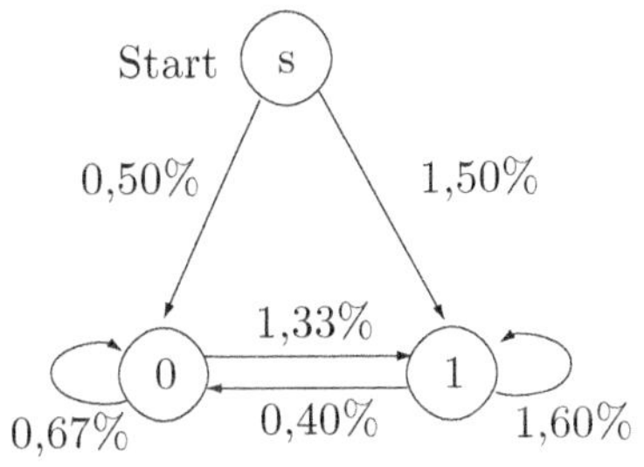

Figure 5.6: A Finite-State Machine.

symbol preceding S.

◇ **Exercise 5.9:** Imagine an image with 8-bit pixels where half the pixels have values 127 and the other half have values 128. Analyze the performance of RLE on the individual bitplanes of such an image, and compare it with what can be achieved with Huffman coding.

5.2.1 Dual Tree Coding

Dual tree coding, an idea due to G. H. Freeman ([Freeman 91] and [Freeman 93]), combines Tunstall and Huffman coding in an attempt to improve the latter's performance for a 2-symbol alphabet. The idea is to use the Tunstall algorithm to extend such an alphabet from 2 symbols to 2^k strings of symbols, and select k such that the probabilities of the strings will be close to negative powers of 2. Once this is achieved, the strings are assigned Huffman codes and the input stream is compressed by replacing the strings with these codes. This approach is illustrated here by a simple example.

Given a binary source that emits two symbols a and b with probabilities 0.15 and 0.85, respectively, we try to compress it in four different ways as follows:

1. We apply the Huffman algorithm directly to the two symbols. This simply assigns the two 1-bit codes 0 and 1 to a and b, so there is no compression.

2. We combine the two symbols to obtain the four 2-symbol strings aa, ab, ba, and bb, with probabilities 0.0225, 0.1275, 0.1275, and 0.7225, respectively. The four strings are assigned Huffman codes as shown in Figure 5.7a, and it is obvious that the average code length is $0.0225 \times 3 + 0.1275 \times 3 + 0.1275 \times 2 + 0.7225 \times 1 = 1.4275$ bits. On average, each 2-symbol string is compressed to 1.4275 bits, yielding a compression ratio of $1.4275/2 \approx 0.714$.

3. We apply Tunstall's algorithm to obtain the four strings bbb, bba, ba, and a with probabilities 0.614, 0.1084, 0.1275, and 0.15, respectively. The resulting parse tree is shown in Figure 5.7b. Tunstall's method compresses these strings by replacing each with a 2-bit code. Given a string of 257 bits with these probabilities, we expect the strings bbb, bba, ba, and a to occur 61, 11, 13, and 15 times, respectively, for a total of 100 strings. Thus, Tunstall's method compresses the 257 input bits to $2 \times 100 = 200$ bits, for a compression ratio of $200/257 \approx 0.778$.

4. We now change the probabilities of the four strings above to negative powers of 2, because these are the best values for the Huffman method. Strings bbb, bba,

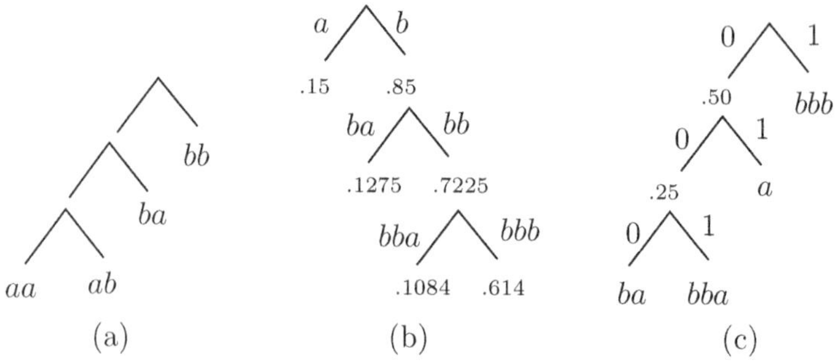

Figure 5.7: Dual Tree Coding.

ba, and a are thus assigned the probabilities 0.5, 0.125, 0.125, and 0.25, respectively. The resulting Huffman code tree is shown in Figure 5.7c and it is easy to see that the 61, 11, 13, and 15 occurrences of these strings will be compressed to a total of $61 \times 1 + 11 \times 3 + 13 \times 3 + 15 \times 2 = 163$ bits, resulting in a compression ratio of $163/257 \approx 0.634$, much better.

To summarize, applying the Huffman method to a 2-symbol alphabet produces no compression. Combining the individual symbols in strings as in 2 above or applying the Tunstall method as in 3, produce moderate compression. In contrast, combining the strings produced by Tunstall with the codes generated by the Huffman method, results in much better performance. The dual tree method starts by constructing the Tunstall parse tree and then using its leaf nodes to construct a Huffman code tree. The only (still unsolved) problem is determining the best value of k. In our example, we iterated the Tunstall algorithm until we had $2^2 = 4$ strings, but iterating more times may have resulted in strings whose probabilities are closer to negative powers of 2.

5.2.2 Huffman Decoding

Before starting the compression of a data stream, the compressor (encoder) has to determine the codes. It does that based on the probabilities (or frequencies of occurrence) of the symbols. The probabilities or frequencies have to be written, as side information, on the compressed stream, so that any Huffman decompressor (decoder) will be able to decompress the stream. This is easy, since the frequencies are integers and the probabilities can be written as scaled integers. It normally adds just a few hundred bytes to the compressed stream. It is also possible to write the variable-length codes themselves on the stream, but this may be awkward, because the codes have different sizes. It is also possible to write the Huffman tree on the stream, but this may require more space than just the frequencies.

In any case, the decoder must know what is at the start of the stream, read it, and construct the Huffman tree for the alphabet. Only then can it read and decode the rest of the stream. The algorithm for decoding is simple. Start at the root and read the first bit off the compressed stream. If it is zero, follow the bottom edge of the tree; if it is one, follow the top edge. Read the next bit and move another edge toward the leaves of the tree. When the decoder gets to a leaf, it finds the original, uncompressed code of the symbol (normally its ASCII code), and that code is emitted by the decoder. The process starts again at the root with the next bit.

This process is illustrated for the five-symbol alphabet of Figure 5.8. The four-symbol input string $a_4a_2a_5a_1$ is encoded into 1001100111. The decoder starts at the root, reads the first bit 1, and goes up. The second bit 0 sends it down, as does the third bit. This brings the decoder to leaf a_4, which it emits. It again returns to the root, reads 110, moves up, up, and down, to reach leaf a_2, and so on.

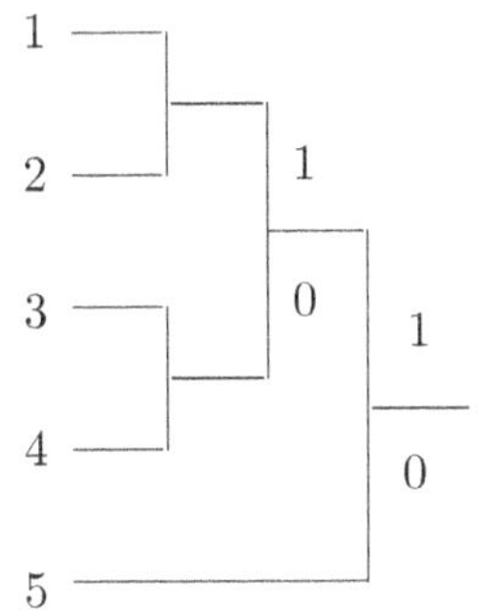

Figure 5.8: Huffman Codes for Equal Probabilities.

5.2.3 Fast Huffman Decoding

Decoding a Huffman-compressed file by sliding down the code tree for each symbol is conceptually simple, but slow. The compressed file has to be read bit by bit and the decoder has to advance a node in the code tree for each bit. The method of this section, originally conceived by [Choueka et al. 85] but later reinvented by others, uses preset partial-decoding tables. These tables depend on the particular Huffman code used, but not on the data to be decoded. The compressed file is read in chunks of k bits each (where k is normally 8 or 16 but can have other values) and the current chunk is used as a pointer to a table. The table entry that is selected in this way can decode several symbols and it also points the decoder to the table to be used for the next chunk.

As an example, consider the Huffman code of Figure 5.3a, where the five codewords are 0, 10, 111, 1101, and 1100. The string of symbols $a_1a_1a_2a_4a_3a_1a_5\ldots$ is compressed by this code to the string 0|0|10|1101|111|0|1100.... We select $k = 3$ and read this string in 3-bit chunks 001|011|011|110|110|0.... Examining the first chunk, it is easy to see that it should be decoded into a_1a_1 followed by the single bit 1 which is the prefix of another codeword. The first chunk is $001 = 1_{10}$, so we set entry 1 of the first table (table 0) to the pair $(a_1a_1, 1)$. When chunk 001 is used as a pointer to table 0, it points to entry 1, which immediately provides the decoder with the two decoded symbols a_1a_1 and also directs it to use table 1 for the next chunk. Table 1 is used when a partially-decoded chunk ends with the single-bit prefix 1. The next chunk is $011 = 3_{10}$, so entry 3 of table 1 corresponds to the encoded bits 1|011. Again, it is easy to see that these should be decoded to a_2 and there is the prefix 11 left over. Thus, entry 3 of table 1 should be $(a_2, 2)$. It provides the decoder with the single symbol a_2 and also directs it to use table 2 next (the table that corresponds to prefix 11). The next chunk is again $011 = 3_{10}$, so entry 3 of table 2 corresponds to the encoded bits 11|011. It is again obvious that these

```
i←0; output←null;
repeat
  j←input next chunk;
  (s,i)←Tablei[j];
  append s to output;
until end-of-input
```

Figure 5.9: Fast Huffman Decoding.

should be decoded to a_4 with a prefix of 1 left over. This process continues until the end of the encoded input. Figure 5.9 is the simple decoding algorithm in pseudocode.

Table 5.10 lists the four tables required to decode this code. It is easy to see that they correspond to the prefixes Λ (null), 1, 11, and 110. A quick glance at Figure 5.3a shows that these correspond to the root and the four interior nodes of the Huffman code tree. Thus, each partial-decoding table corresponds to one of the four prefixes of this code. The number m of partial-decoding tables therefore equals the number of interior nodes (plus the root) which is one less than the number N of symbols of the alphabet.

$T_0 = \Lambda$			$T_1 = 1$			$T_2 = 11$			$T_3 = 110$		
000	$a_1a_1a_1$	0	1|000	$a_2a_1a_1$	0	11|000	a_5a_1	0	110|000	$a_5a_1a_1$	0
001	a_1a_1	1	1|001	a_2a_1	1	11|001	a_5	1	110|001	a_5a_1	1
010	a_1a_2	0	1|010	a_2a_2	0	11|010	a_4a_1	0	110|010	a_5a_2	0
011	a_1	2	1|011	a_2	2	11|011	a_4	1	110|011	a_5	2
100	a_2a_1	0	1|100	a_5	0	11|100	$a_3a_1a_1$	0	110|100	$a_4a_1a_1$	0
101	a_2	1	1|101	a_4	0	11|101	a_3a_1	1	110|101	a_4a_1	1
110	–	3	1|110	a_3a_1	0	11|110	a_3a_2	0	110|110	a_4a_2	0
111	a_3	0	1|111	a_3	1	11|111	a_3	2	110|111	a_4	2

Table 5.10: Partial-Decoding Tables for a Huffman Code.

Notice that some chunks (such as entry 110 of table 0) simply send the decoder to another table and do not provide any decoded symbols. Also, there is a tradeoff between chunk size (and thus table size) and decoding speed. Large chunks speed up decoding, but require large tables. A large alphabet (such as the 128 ASCII characters or the 256 8-bit bytes) also requires a large set of tables. The problem with large tables is that the decoder has to set up the tables after it has read the Huffman codes from the compressed stream and before decoding can start, and this process may preempt any gains in decoding speed provided by the tables.

To set up the first table (table 0, which corresponds to the null prefix Λ), the decoder generates the 2^k bit patterns 0 through $2^k - 1$ (the first column of Table 5.10) and employs the decoding method of Section 5.2.2 to decode each pattern. This yields the second column of Table 5.10. Any remainders left are prefixes and are converted by the decoder to table numbers. They become the third column of the table. If no remainder is left, the third column is set to 0 (use table 0 for the next chunk). Each of the other partial-decoding tables is set in a similar way. Once the decoder decides that table 1 corresponds to prefix p, it generates the 2^k patterns $p|00\ldots0$ through $p|11\ldots1$ that become the first column of that table. It then decodes that column to generate the remaining two columns.

This method was conceived in 1985, when storage costs were considerably higher than today (early 2007). This prompted the developers of the method to find ways to cut down the number of partial-decoding tables, but these techniques are less important today and are not described here.

> Truth is stranger than fiction, but this is because fiction is obliged to stick to probability; truth is not.
>
> —Anonymous

5.2.4 Average Code Size

Figure 5.13a shows a set of five symbols with their probabilities and a typical Huffman tree. Symbol A appears 55% of the time and is assigned a 1-bit code, so it contributes $0.55\cdot 1$ bits to the average code size. Symbol E appears only 2% of the time and is assigned a 4-bit Huffman code, so it contributes $0.02\cdot 4 = 0.08$ bits to the code size. The average code size is therefore calculated to be

$$0.55\cdot 1 + 0.25\cdot 2 + 0.15\cdot 3 + 0.03\cdot 4 + 0.02\cdot 4 = 1.7 \text{ bits per symbol.}$$

Surprisingly, the same result is obtained by adding the values of the four internal nodes of the Huffman code-tree $0.05 + 0.2 + 0.45 + 1 = 1.7$. This provides a way to calculate the average code size of a set of Huffman codes without any multiplications. Simply add the values of all the internal nodes of the tree. Table 5.11 illustrates why this works.

$$\begin{aligned}
&\mathit{.05} = && .02+.03\\
&\mathit{.20} = \mathit{.05}+.15 && = .02+.03+.15\\
&\mathit{.45} = \mathit{.20}+.25 && = .02+.03+.15+.25\\
&\mathit{1.0} = \mathit{.45}+.55 && = .02+.03+.15+.25+.55
\end{aligned}$$

Table 5.11: Composition of Nodes.

$$\begin{aligned}
&\mathit{0.05} = && = 0.02+0.03+\cdots\\
&a_1 = \mathit{0.05}+\ldots && = 0.02+0.03+\cdots\\
&a_2 = a_1 \quad +\ldots && = 0.02+0.03+\cdots\\
&\vdots \quad = &&\\
&a_{d-2} = a_{d-3}+\ldots && = 0.02+0.03+\cdots\\
&\mathit{1.0} = a_{d-2}+\ldots && = 0.02+0.03+\cdots
\end{aligned}$$

Table 5.12: Composition of Nodes.

(Internal nodes are shown in italics in this table.) The left column consists of the values of all the internal nodes. The right columns show how each internal node is the sum of some of the leaf nodes. Summing the values in the left column yields 1.7, and summing the other columns shows that this 1.7 is the sum of the four values 0.02, the four values 0.03, the three values 0.15, the two values 0.25, and the single value 0.55.

This argument can be extended to the general case. It is easy to show that, in a Huffman-like tree (a tree where each node is the sum of its children), the weighted sum

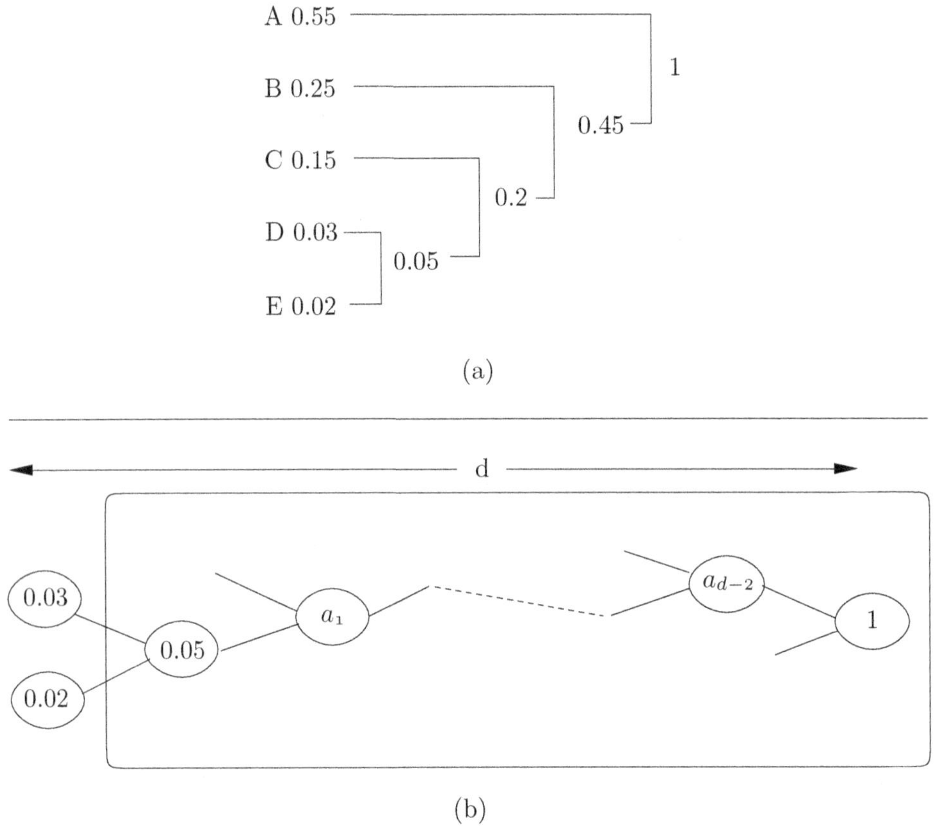

Figure 5.13: Huffman Code-Trees.

of the leaves, where the weights are the distances of the leaves from the root, equals the sum of the internal nodes. (This property has been communicated to us by John M. Motil.)

Figure 5.13b shows such a tree, where we assume that the two leaves 0.02 and 0.03 have d-bit Huffman codes. Inside the tree, these leaves become the children of internal node 0.05, which, in turn, is connected to the root by means of the $d-2$ internal nodes a_1 through a_{d-2}. Table 5.12 has d rows and shows that the two values 0.02 and 0.03 are included in the various internal nodes exactly d times. Adding the values of all the internal nodes produces a sum that includes the contributions $0.02 \cdot d + 0.03 \cdot d$ from the two leaves. Since these leaves are arbitrary, it is clear that this sum includes similar contributions from all the other leaves, so this sum is the average code size. Since this sum also equals the sum of the left column, which is the sum of the internal nodes, it is clear that the sum of the internal nodes equals the average code size.

Notice that this proof does not assume that the tree is binary. The property illustrated here exists for any tree where a node contains the sum of its children.

5.2.5 Number of Codes

Since the Huffman code is not unique, the natural question is: How many different codes are there? Figure 5.14a shows a Huffman code-tree for six symbols, from which we can answer this question in two different ways.

Answer 1. The tree of 5.14a has five interior nodes, and in general, a Huffman code-tree for n symbols has $n-1$ interior nodes. Each interior node has two edges coming out of it, labeled 0 and 1. Swapping the two labels produces a different Huffman code-tree, so the total number of different Huffman code-trees is 2^{n-1} (in our example, 2^5 or 32). The tree of Figure 5.14b, for example, shows the result of swapping the labels of the two edges of the root. Table 5.15a,b lists the codes generated by the two trees.

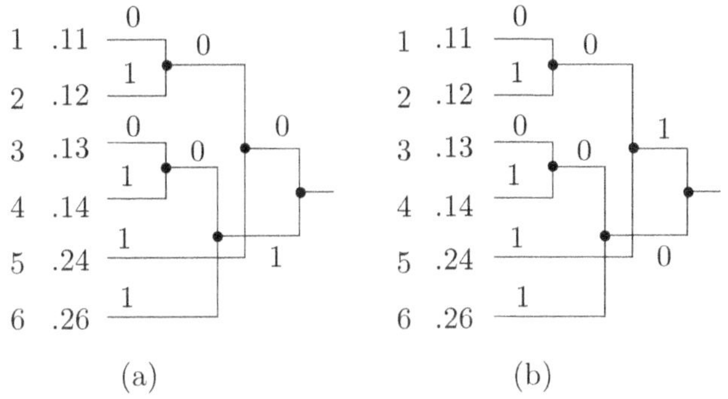

Figure 5.14: Two Huffman Code-Trees.

(a)	(b)	(c)
000	100	000
001	101	001
100	000	010
101	001	011
01	11	10
11	01	11

Table 5.15.

Answer 2. The six codes of Table 5.15a can be divided into the four classes $00x$, $10y$, 01, and 11, where x and y are 1-bit each. It is possible to create different Huffman codes by changing the first two bits of each class. Since there are four classes, this is the same as creating all the permutations of four objects, something that can be done in $4! = 24$ ways. In each of the 24 permutations it is also possible to change the values of x and y in four different ways (since they are bits) so the total number of different Huffman codes in our six-symbol example is $24 \times 4 = 96$.

The two answers are different because they count different things. Answer 1 counts the number of different Huffman code-trees, while answer 2 counts the number of different Huffman codes. It turns out that our example can generate 32 different code-trees but only 94 different codes instead of 96. This shows that there are Huffman codes that cannot be generated by the Huffman method! Table 5.15c shows such an example. A look at the trees of Figure 5.14 should convince the reader that the codes of symbols 5 and 6 must start with different bits, but in the code of Table 5.15c they both start with 1. This code is therefore impossible to generate by any relabeling of the nodes of the trees of Figure 5.14.

5.2.6 Ternary Huffman Codes

The Huffman code is not unique. Moreover, it does not have to be binary! The Huffman method can easily be applied to codes based on other number systems. Figure 5.16a shows a Huffman code tree for five symbols with probabilities 0.15, 0.15, 0.2, 0.25, and 0.25. The average code size is

$$2\times 0.25 + 3\times 0.15 + 3\times 0.15 + 2\times 0.20 + 2\times 0.25 = 2.3\,\text{bits/symbol}.$$

Figure 5.16b shows a ternary Huffman code tree for the same five symbols. The tree is constructed by selecting, at each step, three symbols with the smallest probabilities and merging them into one parent symbol, with the combined probability. The average code size of this tree is

$$2\times 0.15 + 2\times 0.15 + 2\times 0.20 + 1\times 0.25 + 1\times 0.25 = 1.5\,\text{trits/symbol}.$$

Notice that the ternary codes use the digits 0, 1, and 2.

◇ **Exercise 5.10:** Given seven symbols with probabilities .02, .03, .04, .04, .12, .26, and .49, we construct binary and ternary Huffman code-trees for them and calculate the average code size in each case.

5.2.7 Height Of A Huffman Tree

The height of the code-tree generated by the Huffman algorithm may sometimes be important because the height is also the length of the longest code in the tree. The Deflate method (Section 6.25), for example, limits the lengths of certain Huffman codes to just three bits.

It is easy to see that the shortest Huffman tree is created when the symbols have equal probabilities. If the symbols are denoted by `A`, `B`, `C`, and so on, then the algorithm combines pairs of symbols, such `A` and `B`, `C` and `D`, in the lowest level, and the rest of the tree consists of interior nodes as shown in Figure 5.17a. The tree is balanced or close to balanced and its height is $\lceil \log_2 n \rceil$. In the special case where the number of symbols n is a power of 2, the height is exactly $\log_2 n$. In order to generate the tallest tree, we need to assign probabilities to the symbols such that each step in the Huffman method will increase the height of the tree by 1. Recall that each step in the Huffman algorithm combines two symbols. Thus, the tallest tree is obtained when the first step combines two of the n symbols and each subsequent step combines the result of its predecessor with one of the remaining symbols (Figure 5.17b). The height of the complete tree is therefore $n-1$, and it is referred to as a lopsided or unbalanced tree.

It is easy to see what symbol probabilities result in such a tree. Denote the two smallest probabilities by a and b. They are combined in the first step to form a node whose probability is $a+b$. The second step will combine this node with an original symbol if one of the symbols has probability $a+b$ (or smaller) and all the remaining symbols have greater probabilities. Thus, after the second step, the root of the tree has probability $a+b+(a+b)$ and the third step will combine this root with one of the remaining symbols if its probability is $a+b+(a+b)$ and the probabilities of the remaining $n-4$ symbols are greater. It does not take much to realize that the symbols have to have probabilities $p_1 = a$, $p_2 = b$, $p_3 = a+b = p_1+p_2$, $p_4 = b+(a+b) = p_2+p_3$, $p_5 = (a+b)+(a+2b) = p_3+p_4$, $p_6 = (a+2b)+(2a+3b) = p_4+p_5$, and so on (Figure 5.17c). These probabilities form a Fibonacci sequence whose first two elements

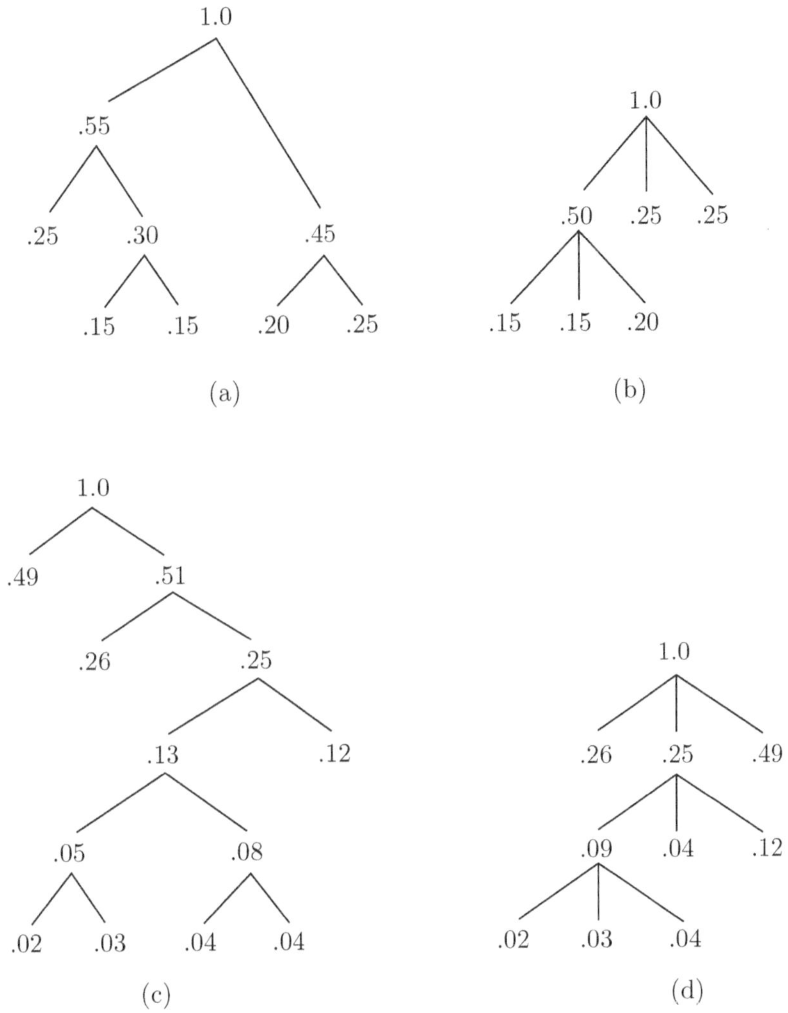

Figure 5.16: Binary and Ternary Huffman Code-Trees.

are a and b. As an example, we select $a = 5$ and $b = 2$ and generate the 5-number Fibonacci sequence 5, 2, 7, 9, and 16. These five numbers add up to 39, so dividing them by 39 produces the five probabilities 5/39, 2/39, 7/39, 9/39, and 15/39. The Huffman tree generated by them has a maximal height (which is 4).

In principle, symbols in a set can have any probabilities, but in practice, the probabilities of symbols in an input file are computed by counting the number of occurrences of each symbol. Imagine a text file where only the nine symbols A through I appear. In order for such a file to produce the tallest Huffman tree, where the codes will have lengths from 1 to 8 bits, the frequencies of occurrence of the nine symbols have to form a Fibonacci sequence of probabilities. This happens when the frequencies of the symbols are 1, 1, 2, 3, 5, 8, 13, 21, and 34 (or integer multiples of these). The sum of these

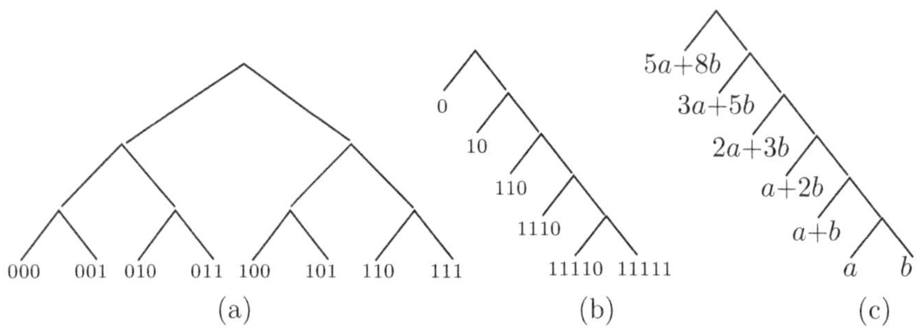

Figure 5.17: Shortest and Tallest Huffman Trees.

frequencies is 88, so our file has to be at least that long in order for a symbol to have 8-bit Huffman codes. Similarly, if we want to limit the sizes of the Huffman codes of a set of n symbols to 16 bits, we need to count frequencies of at least 4180 symbols. To limit the code sizes to 32 bits, the minimum data size is 9,227,464 symbols.

If a set of symbols happens to have the Fibonacci probabilities and therefore results in a maximal-height Huffman tree with codes that are too long, the tree can be reshaped (and the maximum code length shortened) by slightly modifying the symbol probabilities, so they are not much different from the original, but do not form a Fibonacci sequence.

5.2.8 Canonical Huffman Codes

The code of Table 5.15c has a simple interpretation. It assigns the first four symbols the 3-bit codes 0, 1, 2, 3, and the last two symbols the 2-bit codes 2 and 3. This is an example of a *canonical Huffman code*. The word "canonical" means that this particular code has been selected from among the several (or even many) possible Huffman codes because its properties make it easy and fast to use.

Table 5.18 shows a slightly bigger example of a canonical Huffman code. Imagine a set of 16 symbols (whose probabilities are irrelevant and are not shown) such that four symbols are assigned 3-bit codes, five symbols are assigned 5-bit codes, and the remaining seven symbols are assigned 6-bit codes. Table 5.18a shows a set of possible Huffman codes, while Table 5.18b shows a set of canonical Huffman codes. It is easy to see that the seven 6-bit canonical codes are simply the 6-bit integers 0 through 6. The five codes are the 5-bit integers 4 through 8, and the four codes are the 3-bit integers 3 through 6. We first show how these codes are generated and then how they are used.

The top row (length) of Table 5.19 lists the possible code lengths, from 1 to 6 bits. The second row (numl) lists the number of codes of each length, and the bottom row (first) lists the first code in each group. This is why the three groups of codes start with values 3, 4, and 0. To obtain the top two rows we need to compute the lengths of all the Huffman codes for the given alphabet (see below). The third row is computed by setting `first[6]:=0;` and iterating

```
for l:=5 downto 1 do first[l]:=⌈(first[l+1]+numl[l+1])/2⌉;
```

This guarantees that all the 3-bit prefixes of codes longer than three bits will be less than `first[3]` (which is 3), all the 5-bit prefixes of codes longer than five bits will be less than `first[5]` (which is 4), and so on.

	(a)	(b)		(a)	(b)
1:	000	011	9:	10100	01000
2:	001	100	10:	101010	000000
3:	010	101	11:	101011	000001
4:	011	110	12:	101100	000010
5:	10000	00100	13:	101101	000011
6:	10001	00101	14:	101110	000100
7:	10010	00110	15:	101111	000101
8:	10011	00111	16:	110000	000110

Table 5.18.

length:	1	2	3	4	5	6
numl:	0	0	4	0	5	7
first:	2	4	3	5	4	0

Table 5.19.

Now for the use of these unusual codes. Canonical Huffman codes are useful in cases where the alphabet is large and where fast decoding is mandatory. Because of the way the codes are constructed, it is easy for the decoder to identify the length of a code by reading and examining input bits one by one. Once the length is known, the symbol can be found in one step. The pseudocode listed here shows the rules for decoding:

```
l:=1; input v;
while v<first[l]
append next input bit to v; l:=l+1;
endwhile
```

As an example, suppose that the next code is 00110. As bits are input and appended to v, it goes through the values 0, 00=0, 001=1, 0011=3, 00110=6, while `l` is incremented from 1 to 5. All steps except the last satisfy `v<first[l]`, so the last step determines the value of `l` (the code length) as 5. The symbol itself is found by subtracting `v` − `first[5]` $= 6 - 4 = 2$, so it is the third symbol (numbering starts at 0) in group `l` $= 5$ (symbol 7 of the 16 symbols).

It has been mentioned that canonical Huffman codes are useful in cases where the alphabet is large and fast decoding is important. A practical example is a collection of documents archived and compressed by a *word-based* adaptive Huffman coder (Section 11.6.1). In an archive a slow encoder is acceptable, but the decoder should be fast. When the individual symbols are words, the alphabet may be huge, making it impractical, or even impossible, to construct the Huffman code-tree. However, even with a huge alphabet, the number of different code lengths is small, rarely exceeding 20 bits (just the number of 20-bit codes is about a million). If canonical Huffman codes are used, and the maximum code length is `L`, then the code length `l` of a symbol is found by the decoder in at most `L` steps, and the symbol itself is identified in one more step.

> He uses statistics as a drunken man uses lampposts—for support rather than illumination.
>
> —Andrew Lang, *Treasury of Humorous Quotations*

The last point to be discussed is the encoder. In order to construct the canonical Huffman code, the encoder needs to know the length of the Huffman code of every symbol. The main problem is the large size of the alphabet, which may make it impractical or even impossible to build the entire Huffman code-tree in memory. The algorithm described here (see [Hirschberg and Lelewer 90] and [Sieminski 88]) solves this problem. It calculates the code sizes for an alphabet of n symbols using just one array of size $2n$. One half of this array is used as a *heap*, so we start with a short description of this useful data structure.

A *binary tree* is a tree where every node has at most two children (i.e., it may have 0, 1, or 2 children). A *complete binary tree* is a binary tree where every node except the leaves has exactly two children. A *balanced binary tree* is a complete binary tree where some of the bottom-right nodes may be missing (see also page 276 for another application of those trees). A heap is a balanced binary tree where every leaf contains a data item and the items are ordered such that every path from a leaf to the root traverses nodes that are in sorted order, either nondecreasing (a max-heap) or nonincreasing (a min-heap). Figure 5.20 shows examples of min-heaps.

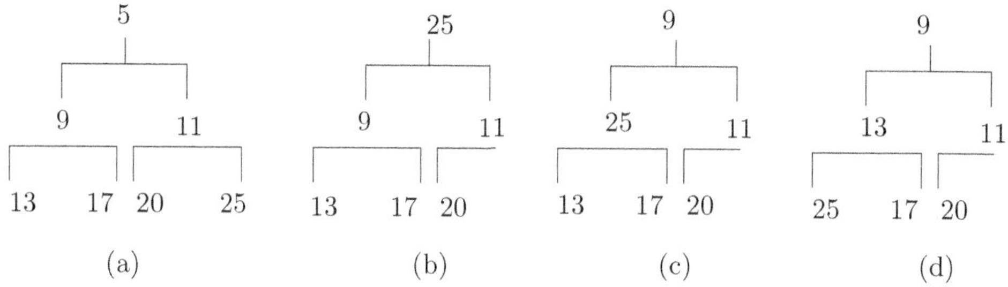

Figure 5.20: Min-Heaps.

A common operation on a heap is to remove the root and rearrange the remaining nodes to get back a heap. This is called *sifting* the heap. The four parts of Figure 5.20 show how a heap is sifted after the root (with data item 5) has been removed. Sifting starts by moving the bottom-right node to become the new root. This guarantees that the heap will remain a balanced binary tree. The root is then compared with its children and may have to be swapped with one of them in order to preserve the ordering of a heap. Several more swaps may be necessary to completely restore heap ordering. It is easy to see that the maximum number of swaps equals the height of the tree, which is $\lceil \log_2 n \rceil$.

The reason a heap must always remain balanced is that this makes it possible to store it in memory without using any pointers. The heap is said to be "housed" in an array. To house a heap in an array, the root is placed in the first array location (with index 1), the two children of the node at array location i are placed at locations $2i$ and $2i+1$, and the parent of the node at array location j is placed at location $\lfloor j/2 \rfloor$. Thus the heap of Figure 5.20a is housed in an array by placing the nodes 5, 9, 11, 13, 17, 20, and 25 in the first seven locations of the array.

The algorithm uses a single array `A` of size $2n$. The frequencies of occurrence of the n symbols are placed in the top half of `A` (locations $n+1$ through $2n$), and the bottom half of `A` (locations 1 through n) becomes a min-heap whose data items are pointers to the frequencies in the top half (Figure 5.21a). The algorithm then goes into a loop where in each iteration the heap is used to identify the two smallest frequencies and replace them with their sum. The sum is stored in the last heap position `A[h]`, and the heap shrinks by one position (Figure 5.21b). The loop repeats until the heap is reduced to just one pointer (Figure 5.21c).

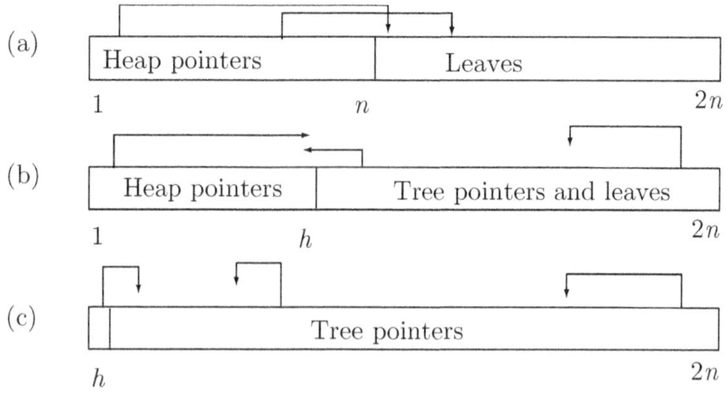

Figure 5.21: Huffman Heaps and Leaves in an Array.

We now illustrate this part of the algorithm using seven frequencies. The table below shows how the frequencies and the heap are initially housed in an array of size 14. Pointers are shown in italics, and the heap is delimited by square brackets.

1	2	3	4	5	6	7	8	9	10	11	12	13	14
[*14*	*12*	*13*	*10*	*11*	*9*	*8*]	25	20	13	17	9	11	5

The first iteration selects the smallest frequency (5), removes the root of the heap (pointer 14), and leaves `A[7]` empty.

1	2	3	4	5	6	7	8	9	10	11	12	13	14
[*12*	*10*	*13*	*8*	*11*	*9*]		25	20	13	17	9	11	5

The heap is sifted, and its new root (12) points to the second smallest frequency (9) in `A[12]`. The sum $5+9$ is stored in the empty location 7, and the three array locations `A[1]`, `A[12]`, and `A[14]` are set to point to that location.

1	2	3	4	5	6	7	8	9	10	11	12	13	14
[*7*	*10*	*13*	*8*	*11*	*9*]	5+9	25	20	13	17	*7*	11	*7*

The heap is now sifted.

1	2	3	4	5	6	7	8	9	10	11	12	13	14
[*13*	*10*	*7*	*8*	*11*	*9*]	14	25	20	13	17	*7*	11	*7*

The new root is 13, implying that the smallest frequency (11) is stored at `A[13]`. The root is removed, and the heap shrinks to just five positions, leaving location 6 empty.

1	2	3	4	5	6	7	8	9	10	11	12	13	14
[*10*	*11*	*7*	*8*	*9*]		14	25	20	13	17	*7*	11	*7*

The heap is now sifted. The new root is 10, showing that the second smallest frequency, 13, is stored at `A[10]`. The sum $11 + 13$ is stored at the empty location 6, and the three locations `A[1]`, `A[13]`, and `A[10]` are set to point to 6.

1	2	3	4	5	6	7	8	9	10	11	12	13	14
[*6*	*11*	*7*	*8*	*9*]	11+13	14	25	20	*6*	17	*7*	*6*	*7*

◇ **Exercise 5.11:** Continue this loop.

◇ **Exercise 5.12:** Complete this loop.

◇ **Exercise 5.13:** Find the lengths of all the other codes.

> Considine's Law. Whenever one word or letter can change the entire meaning of a sentence, the probability of an error being made will be in direct proportion to the embarrassment it will cause.
>
> —Bob Considine

5.2.9 Is Huffman Coding Dead?

The advantages of arithmetic coding are well known to users of compression algorithms. Arithmetic coding can compress data to its entropy, its adaptive version works well if fed the correct probabilities, and its performance does not depend on the size of the alphabet. On the other hand, arithmetic coding is slower than Huffman coding, its compression potential is not always utilized to its maximum, its adaptive version is very sensitive to the symbol probabilities and in extreme cases may even expand the data. Finally, arithmetic coding is not robust; a single error may propagate indefinitely and may result in wrong decoding of a large quantity of compressed data. (Some users may complain that they don't understand arithmetic coding and have no idea how to implement it, but this doesn't seem a serious concern, because implementations of this method are available for all major computing platforms.) A detailed comparison and analysis of both methods is presented in [Bookstein and Klein 93], with the conclusion that arithmetic coding has the upper hand only in rare situations.

In [Gallager 74], Robert Gallager shows that the redundancy of Huffman coding is at most $p_1 + 0.086$ where p_1 is the probability of the most-common symbol in the alphabet. The redundancy is the difference between the average Huffman codeword length and the entropy. Since arithmetic coding can compress data to its entropy, the quantity $p_1 + 0.086$ indicates by how much arithmetic coding outperforms Huffman coding. Given a two-symbol alphabet, the more probable symbol appears with probability 0.5 or more, but given a large alphabet, such as the set of letters, digits and punctuation marks used by a language, the largest symbol probability is typically around 15–20%, bringing the value of the quantity $p_1 + 0.086$ to around 0.1. This means that Huffman codes are at most 0.1 bit longer (per symbol) than arithmetic coding. For some (perhaps even many) applications, such a small difference may be insignificant, but those applications for which this difference is significant may be important.

Bookstein and Klein examine the two extreme cases of large and small alphabets. Given a text file in a certain language, it is often compressed in blocks. This limits the propagation of errors and also provides several entry points into the file. The authors examine the probabilities of characters of several large alphabets (each consisting of the letters and punctuations marks of a natural language), and list the average codeword length for Huffman and arithmetic coding (the latter is the size of the compressed file divided by the number of characters in the original file). The surprising conclusion is that the Huffman codewords are longer than the arithmetic codewords by less than one percent. Also, arithmetic coding performs better than Huffman coding only in large blocks of text. The minimum block size where arithmetic coding is preferable turns out to be between 269 and 457 characters. Thus, for shorter blocks, Huffman coding outperforms arithmetic coding.

The other extreme case is a binary alphabet where one symbol has probability e and the other has probability $1-e$. If $e=0.5$, no method will compress the data. If the probabilities are skewed, Huffman coding does a bad job. The Huffman codes of the two symbols are 0 and 1 independent of the symbols' probabilities. Each code is 1-bit long, and there is no compression. Arithmetic coding, on the other hand, compresses such data to its entropy, which is $-[e\log_2 e+(1-e)\log_2(1-e)]$. This expression tends to 0 for both small e (close to 0) and for large e (close to 1). However, there is a simple way to improve the performance of Huffman coding in this case. Simply group several bits into a word. If we group the bits in 4-bit words, we end up with an alphabet of 16 symbols, where the probabilities are less skewed and the Huffman codes do a better job, especially because of the Gallager bound.

Another difference between Huffman and arithmetic coding is the case of wrong probabilities. This is especially important when a compression algorithm employs a mathematical model to estimate the probabilities of occurrence of individual symbols. The authors show that, under reasonable assumptions, arithmetic coding is affected by wrong probabilities more than Huffman coding.

Speed is also an important consideration in many applications. Huffman encoding is fast. Given a symbol to encode, the symbol is used as a pointer to a code table, the Huffman code is read from the table, and is appended to the codes-so-far. Huffman decoding is slower because the decoder has to start at the root of the Huffman code tree and slide down, guided by the bits of the current codeword, until it reaches a leaf node, where it finds the symbol. Arithmetic coding, on the other hand, requires multiplications and divisions, and is therefore slower. (Notice, however, that certain versions of arithmetic coding, most notably the Q-coder, MQ-coder, and QM-coder, have been developed specifically to avoid slow operations and are not slow.)

Often, a data compression application requires a certain amount of robustness against transmission errors. Neither Huffman nor arithmetic coding is robust, but it is known from long experience that Huffman codes tend to synchronize themselves fairly quickly following an error, in contrast to arithmetic coding, where an error may propagate to the end of the compressed file. It is also possible to construct resynchronizing Huffman codes, as shown in Section 4.4.

The conclusion is that Huffman coding, being fast, simple, and effective, is preferable to arithmetic coding for most applications. Arithmetic coding is the method of choice only in cases where the alphabet has skewed probabilities that cannot be redefined.

5.3 Adaptive Huffman Coding

The Huffman method assumes that the frequencies of occurrence of all the symbols of the alphabet are known to the compressor. In practice, the frequencies are seldom, if ever, known in advance. One approach to this problem is for the compressor to read the original data twice. The first time, it just calculates the frequencies. The second time, it compresses the data. Between the two passes, the compressor constructs the Huffman tree. Such a method is called semiadaptive (page 10) and is normally too slow to be practical. The method that is used in practice is called adaptive (or dynamic) Huffman coding. This method is the basis of the UNIX `compact` program. (See also Section 11.6.1 for a word-based version of adaptive Huffman coding.) The method was originally developed by [Faller 73] and [Gallager 78] with substantial improvements by [Knuth 85].

The main idea is for the compressor and the decompressor to start with an empty Huffman tree and to modify it as symbols are being read and processed (in the case of the compressor, the word "processed" means compressed; in the case of the decompressor, it means decompressed). The compressor and decompressor should modify the tree in the same way, so at any point in the process they should use the same codes, although those codes may change from step to step. We say that the compressor and decompressor are synchronized or that they work in *lockstep* (although they don't necessarily work together; compression and decompression normally take place at different times). The term *mirroring* is perhaps a better choice. The decoder mirrors the operations of the encoder.

Initially, the compressor starts with an empty Huffman tree. No symbols have been assigned codes yet. The first symbol being input is simply written on the output stream in its uncompressed form. The symbol is then added to the tree and a code assigned to it. The next time this symbol is encountered, its current code is written on the stream, and its frequency incremented by one. Since this modifies the tree, it (the tree) is examined to see whether it is still a Huffman tree (best codes). If not, it is rearranged, which results in changing the codes (Section 5.3.2).

The decompressor mirrors the same steps. When it reads the uncompressed form of a symbol, it adds it to the tree and assigns it a code. When it reads a compressed (variable-length) code, it scans the current tree to determine what symbol the code belongs to, and it increments the symbol's frequency and rearranges the tree in the same way as the compressor.

The only subtle point is that the decompressor needs to know whether the item it has just input is an uncompressed symbol (normally, an 8-bit ASCII code, but see Section 5.3.1) or a variable-length code. To remove any ambiguity, each uncompressed symbol is preceded by a special, variable-length *escape code*. When the decompressor reads this code, it knows that the next 8 bits are the ASCII code of a symbol that appears in the compressed stream for the first time.

The trouble is that the escape code should not be any of the variable-length codes used for the symbols. These codes, however, are being modified every time the tree is rearranged, which is why the escape code should also be modified. A natural way to do this is to add an empty leaf to the tree, a leaf with a zero frequency of occurrence, that's always assigned to the 0-branch of the tree. Since the leaf is in the tree, it gets a variable-length code assigned. This code is the escape code preceding every uncompressed symbol. As the tree is being rearranged, the position of the empty leaf—and thus its code—change, but this escape code is always used to identify uncompressed symbols in the compressed stream. Figure 5.22 shows how the escape code moves and changes as the tree grows.

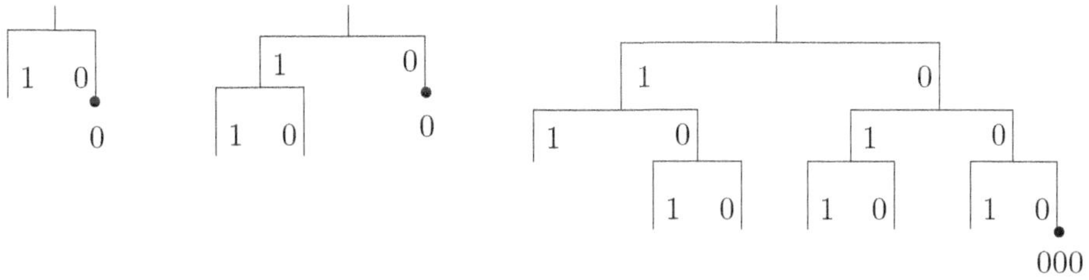

Figure 5.22: The Escape Code.

This method was used to compress/decompress data in the V.32 protocol for 14,400-baud modems.

> Escape is not his plan. I must face him. Alone.
>
> —David Prowse as Lord Darth Vader in *Star Wars* (1977)

5.3.1 Uncompressed Codes

If the symbols being compressed are ASCII characters, they may simply be assigned their ASCII codes as uncompressed codes. In the general case where there may be any symbols, the phased-in codes of Section 2.9 may be used.

Given n data symbols, where $n = 2^m$ (implying that $m = \log_2 n$), we can assign them m-bit codewords. However, if $2^{m-1} < n < 2^m$, then $\log_2 n$ is not an integer. If we assign fixed-length codes to the symbols, each codeword would be $\lceil \log_2 n \rceil$ bits long, but not all the codewords would be used. The case $n = 1{,}000$ is a good example. In this case, each fixed-length codeword is $\lceil \log_2 1{,}000 \rceil = 10$ bits long, but only 1,000 out of the 1,024 possibe codewords are used.

The idea of phased-in codes it to try to assign two sets of codes to the n symbols, where the codewords of one set are $m-1$ bits long and may have several prefixes and the codewords of the other set are m bits long and have different prefixes. The average length of such a code is between $m-1$ and m bits and is shorter when there are more short codewords. See Section 2.9 for more details and examples.

5.3.2 Modifying the Tree

The main idea is to check the tree each time a symbol is input. If the tree is no longer a Huffman tree, it should be updated. A glance at Figure 5.23a shows what it means for a binary tree to be a Huffman tree. The tree in the figure contains five symbols: A, B, C, D, and E. It is shown with the symbols and their frequencies (in parentheses) after 16 symbols have been input and processed. The property that makes it a Huffman tree is that if we scan it level by level, scanning each level from left to right, and going from the bottom (the leaves) to the top (the root), the frequencies will be in sorted, nondescending order. Thus, the bottom left node (A) has the lowest frequency, and the top right node (the root) has the highest frequency. This is called the *sibling property*.

◇ **Exercise 5.14:** Why is this the criterion for a tree to be a Huffman tree?

Here is a summary of the operations needed to update the tree. The loop starts at the current node (the one corresponding to the symbol just input). This node is a leaf that we denote by X, with frequency of occurrence F. Each iteration of the loop involves three steps as follows:

1. Compare X to its successors in the tree (from left to right and bottom to top). If the immediate successor has frequency $F+1$ or greater, the nodes are still in sorted order and there is no need to change anything. Otherwise, some successors of X have identical frequencies of F or smaller. In this case, X should be swapped with the last node in this group (except that X should not be swapped with its parent).
2. Increment the frequency of X from F to $F+1$. Increment the frequencies of all its parents.
3. If X is the root, the loop stops; otherwise, the loop repeats with the parent of node X.

Figure 5.23b shows the tree after the frequency of node A has been incremented from 1 to 2. It is easy to follow the three rules above to see how incrementing the frequency of A results in incrementing the frequencies of all its parents. No swaps are needed in this simple case because the frequency of A hasn't exceeded the frequency of its immediate successor B. Figure 5.23c shows what happens when A's frequency has been incremented again, from 2 to 3. The three nodes following A, namely, B, C, and D, have frequencies of 2, so A is swapped with the last of them, D. The frequencies of the new parents of A are then incremented, and each is compared with its successor, but no more swaps are needed.

Figure 5.23d shows the tree after the frequency of A has been incremented to 4. Once we decide that A is the current node, its frequency (which is still 3) is compared to that of its successor (4), and the decision is not to swap. A's frequency is incremented, followed by incrementing the frequencies of its parents.

In Figure 5.23e, A is again the current node. Its frequency (4) equals that of its successor, so they should be swapped. This is shown in Figure 5.23f, where A's frequency is 5. The next loop iteration examines the parent of A, with frequency 10. It should be swapped with its successor E (with frequency 9), which leads to the final tree of Figure 5.23g.

5.3.3 Counter Overflow

The frequency counts are accumulated in the Huffman tree in fixed-length fields, and such fields may overflow. A 16-bit unsigned field can accommodate counts of up to $2^{16}-1=65{,}535$. A simple solution to the counter overflow problem is to watch the count field of the root each time it is incremented, and when it reaches its maximum value, to *rescale* all the frequency counts by dividing them by 2 (integer division). In practice, this is done by dividing the count fields of the leaves, then updating the counts of the interior nodes. Each interior node gets the sum of the counts of its children. The problem is that the counts are integers, and integer division reduces precision. This may change a Huffman tree to one that does not satisfy the sibling property.

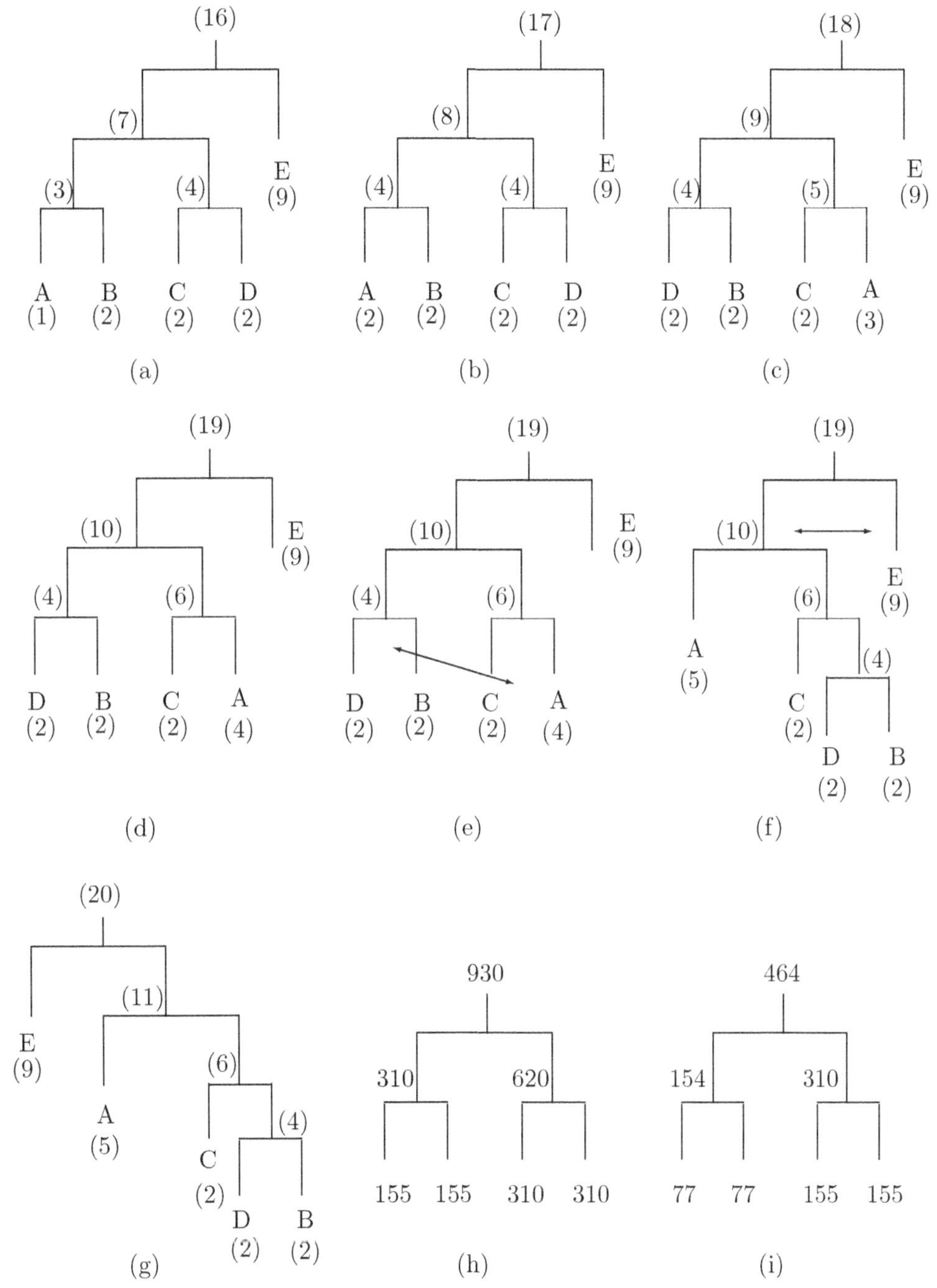

Figure 5.23: Updating the Huffman Tree.

A simple example is shown in Figure 5.23h. After the counts of the leaves are halved, the three interior nodes are updated as shown in Figure 5.23i. The latter tree, however, is no longer a Huffman tree, since the counts are no longer in sorted order. The solution is to rebuild the tree each time the counts are rescaled, which does not happen very often. A Huffman data compression program intended for general use should therefore have large count fields that would not overflow very often. A 4-byte count field overflows at $2^{32} - 1 \approx 4.3 \times 10^9$.

It should be noted that after rescaling the counts, the new symbols being read and compressed have more effect on the counts than the old symbols (those counted before the rescaling). This turns out to be fortuitous since it is known from experience that the probability of appearance of a symbol depends more on the symbols immediately preceding it than on symbols that appeared in the distant past.

5.3.4 Code Overflow

An even more serious problem is code overflow. This may happen when many symbols are added to the tree, and it becomes tall. The codes themselves are not stored in the tree, since they change all the time, and the compressor has to figure out the code of a symbol X each time X is input. Here are the details of this process:

1. The encoder has to locate symbol X in the tree. The tree has to be implemented as an array of structures, each a node, and the array is searched linearly.
2. If X is not found, the escape code is emitted, followed by the uncompressed code of X. X is then added to the tree.
3. If X is found, the compressor moves from node X back to the root, building the code bit by bit as it goes along. Each time it goes from a left child to a parent, a "1" is appended to the code. Going from a right child to a parent appends a "0" bit to the code (or vice versa, but this should be consistent because it is mirrored by the decoder). Those bits have to be accumulated someplace, since they have to be emitted in the *reverse order* in which they are created. When the tree gets taller, the codes get longer. If they are accumulated in a 16-bit integer, then codes longer than 16 bits would cause a malfunction.

One solution to the code overflow problem is to accumulate the bits of a code in a linked list, where new nodes can be created, limited in number only by the amount of available memory. This is general but slow. Another solution is to accumulate the codes in a large integer variable (perhaps 50 bits wide) and document a maximum code size of 50 bits as one of the limitations of the program.

Fortunately, this problem does not affect the decoding process. The decoder reads the compressed code bit by bit and uses each bit to move one step left or right down the tree until it reaches a leaf node. If the leaf is the escape code, the decoder reads the uncompressed code of the symbol off the compressed stream (and adds the symbol to the tree). Otherwise, the uncompressed code is found in the leaf node.

◇ **Exercise 5.15:** Given the 11-symbol string `sir␣sid␣is`, apply the adaptive Huffman method to it. For each symbol input, show the output, the tree after the symbol has been added to it, the tree after being rearranged (if necessary), and the list of nodes traversed left to right and bottom up.

5.3.5 A Variant

This variant of the adaptive Huffman method is simpler but less efficient. The idea is to calculate a set of n variable-length codes based on equal probabilities, to assign those codes to the n symbols at random, and to change the assignments "on the fly," as symbols are being read and compressed. The method is not efficient because the codes are not based on the actual probabilities of the symbols in the input stream. However, it is simpler to implement and also faster than the adaptive method described earlier, because it has to swap rows in a table, rather than update a tree, when updating the frequencies of the symbols.

Name	Count	Code	Name	Count	Code	Name	Count	Code	Name	Count	Code
a_1	0	0	a_2	1	0	a_2	1	0	a_4	2	0
a_2	0	10	a_1	0	10	a_4	1	10	a_2	1	10
a_3	0	110	a_3	0	110	a_3	0	110	a_3	0	110
a_4	0	111	a_4	0	111	a_1	0	111	a_1	0	111
	(a)			(b)			(c)			(d)	

Figure 5.24: Four Steps in a Huffman Variant.

The main data structure is an $n \times 3$ table where the three columns store the names of the n symbols, their frequencies of occurrence so far, and their codes. The table is always kept sorted by the second column. When the frequency counts in the second column change, rows are swapped, but only columns 1 and 2 are moved. The codes in column 3 never change. Figure 5.24 shows an example of four symbols and the behavior of the method when the string a_2, a_4, a_4 is compressed.

Figure 5.24a shows the initial state. After the first symbol a_2 is read, its count is incremented, and since it is now the largest count, rows 1 and 2 are swapped (Figure 5.24b). After the second symbol a_4 is read, its count is incremented and rows 2 and 4 are swapped (Figure 5.24c). Finally, after reading the last symbol a_4, its count is the largest, so rows 1 and 2 are swapped (Figure 5.24d).

The only point that can cause a problem with this method is overflow of the count fields. If such a field is k bits wide, its maximum value is $2^k - 1$, so it will overflow when incremented for the 2^kth time. This may happen if the size of the input stream is not known in advance, which is very common. Fortunately, we do not really need to know the counts, we just need them in sorted order, which makes it easy to solve this problem.

One solution is to count the input symbols and, after $2^k - 1$ symbols are input and compressed, to (integer) divide all the count fields by 2 (or shift them one position to the right, if this is easier).

Another, similar, solution is to check each count field every time it is incremented, and if it has reached its maximum value (if it consists of all ones), to integer divide all the count fields by 2, as mentioned earlier. This approach requires fewer divisions but more complex tests.

Naturally, whatever solution is adopted should be used by both the compressor and decompressor.

5.3.6 Vitter's Method

An improvement of the original algorithm, due to [Vitter 87], which also includes extensive analysis is based on the following key ideas:

1. A different scheme should be used to number the nodes in the dynamic Huffman tree. It is called *implicit numbering*, and it numbers the nodes from the bottom up and in each level from left to right.

2. The Huffman tree should be updated in such a way that the following will always be satisfied. For each weight w, all leaves of weight w precede (in the sense of implicit numbering) all the internal nodes of the same weight. This is an *invariant.*

These ideas result in the following benefits:

1. In the original algorithm, it is possible that a rearrangement of the tree would move a node down one level. In the improved version, this does not happen.

2. Each time the Huffman tree is updated in the original algorithm, some nodes may be moved up. In the improved version, at most one node has to be moved up.

3. The Huffman tree in the improved version minimizes the sum of distances from the root to the leaves and also has the minimum height.

A special data structure, called a *floating tree*, is proposed to make it easy to maintain the required invariant. It can be shown that this version performs much better than the original algorithm. Specifically, if a two-pass Huffman method compresses an input file of n symbols to S bits, then the original adaptive Huffman algorithm can compress it to at most $2S+n$ bits, whereas the improved version can compress it down to $S+n$ bits—a significant difference! Notice that these results do not depend on the size of the alphabet, only on the size n of the data being compressed and on its nature (which determines S).

> I think you're begging the question," said Haydock, "and I can see looming ahead one of those terrible exercises in probability where six men have white hats and six men have black hats and you have to work it out by mathematics how likely it is that the hats will get mixed up and in what proportion. If you start thinking about things like that, you would go round the bend. Let me assure you of that!
>
> —Agatha Christie, *The Mirror Crack'd*

5.4 MNP5

Microcom, Inc., a maker of modems, has developed a protocol (called MNP, for Microcom Networking Protocol) for use in its modems. Among other things, the MNP protocol specifies how to unpack bytes into individual bits before they are sent by the modem, how to transmit bits serially in the synchronous and asynchronous modes, and what modulation techniques to use. Each specification is called a *class*, and classes 5 and 7 specify methods for data compression. These methods (especially MNP5) have become very popular and were used by many modems in the 1980s and 1990s.

The MNP5 method is a two-stage process that starts with run-length encoding, followed by adaptive frequency encoding.

The first stage is described on page 32 and is repeated here. When three or more identical consecutive bytes are found in the source stream, the compressor emits three copies of the byte onto its output stream, followed by a repetition count. When the decompressor reads three identical consecutive bytes, it knows that the next byte is a repetition count (which may be zero, indicating just three repetitions). A downside of the method is that a run of three characters in the input stream results in four characters written to the output stream (expansion). A run of four characters results in no compression. Only runs longer than four characters get compressed. Another, slight, problem is that the maximum count is artificially limited to 250 instead of to 255.

The second stage operates on the bytes in the partially compressed stream generated by the first stage. Stage 2 is similar to the method of Section 5.3.5. It starts with a table of 256×2 entries, where each entry corresponds to one of the 256 possible 8-bit bytes 00000000 to 11111111. The first column, the frequency counts, is initialized to all zeros. Column 2 is initialized to variable-length codes, called *tokens*, that vary from a short "000|0" to a long "111|11111110". Column 2 with the tokens is shown in Table 5.25 (which shows column 1 with frequencies of zero). Each token starts with a 3-bit header, followed by some code bits.

The code bits (with three exceptions) are the two 1-bit codes 0 and 1, the four 2-bit codes 0 through 3, the eight 3-bit codes 0 through 7, the sixteen 4-bit codes, the thirty-two 5-bit codes, the sixty-four 6-bit codes, and the one hundred and twenty-eight 7-bit codes. This provides for a total of $2+4+8+16+32+64+128=254$ codes. The three exceptions are the first two codes "000|0" and "000|1", and the last code, which is "111|11111110" instead of the expected "111|11111111".

When stage 2 starts, all 256 entries of column 1 are assigned frequency counts of zero. When the next byte B is read from the input stream (actually, it is read from the output of the first stage), the corresponding token is written to the output stream, and the frequency of entry B is incremented by 1. Following this, tokens may be swapped to ensure that table entries with large frequencies always have the shortest tokens (see the next section for details). Notice that only the tokens are swapped, not the frequency counts. Thus, the first entry always corresponds to byte "00000000" and contains its frequency count. The token of this byte, however, may change from the original "000|0" to something longer if other bytes achieve higher frequency counts.

<table>
<tr><th>Byte</th><th>Freq.</th><th>Token</th><th>Byte</th><th>Freq.</th><th>Token</th><th>Byte</th><th>Freq.</th><th>Token</th><th>Byte</th><th>Freq.</th><th>Token</th></tr>
<tr><td>0</td><td>0</td><td>000|0</td><td>9</td><td>0</td><td>011|001</td><td>26</td><td>0</td><td>111|1010</td><td>247</td><td>0</td><td>111|1110111</td></tr>
<tr><td>1</td><td>0</td><td>000|1</td><td>10</td><td>0</td><td>011|010</td><td>27</td><td>0</td><td>111|1011</td><td>248</td><td>0</td><td>111|1111000</td></tr>
<tr><td>2</td><td>0</td><td>001|0</td><td>11</td><td>0</td><td>011|011</td><td>28</td><td>0</td><td>111|1100</td><td>249</td><td>0</td><td>111|1111001</td></tr>
<tr><td>3</td><td>0</td><td>001|1</td><td>12</td><td>0</td><td>011|100</td><td>29</td><td>0</td><td>111|1101</td><td>250</td><td>0</td><td>111|1111010</td></tr>
<tr><td>4</td><td>0</td><td>010|00</td><td>13</td><td>0</td><td>011|101</td><td>30</td><td>0</td><td>111|1110</td><td>251</td><td>0</td><td>111|1111011</td></tr>
<tr><td>5</td><td>0</td><td>010|01</td><td>14</td><td>0</td><td>011|110</td><td>31</td><td>0</td><td>111|1111</td><td>252</td><td>0</td><td>111|1111100</td></tr>
<tr><td>6</td><td>0</td><td>010|10</td><td>15</td><td>0</td><td>011|111</td><td>32</td><td>0</td><td>101|00000</td><td>253</td><td>0</td><td>111|1111101</td></tr>
<tr><td>7</td><td>0</td><td>010|11</td><td>16</td><td>0</td><td>111|0000</td><td>33</td><td>0</td><td>101|00001</td><td>254</td><td>0</td><td>111|1111110</td></tr>
<tr><td>8</td><td>0</td><td>011|000</td><td>17</td><td>0</td><td>111|0001</td><td>34</td><td>0</td><td>101|00010</td><td>255</td><td>0</td><td>111|11111110</td></tr>
</table>

18 to 25 and 35 to 246 continue in the same pattern.

Table 5.25: The MNP5 Tokens.

The frequency counts are stored in 8-bit fields. Each time a count is incremented, the algorithm checks to see whether it has reached its maximum value. If yes, all the counts are scaled down by dividing them by 2 (an integer division).

Another, subtle, point has to do with interaction between the two compression stages. Recall that each repetition of three or more characters is replaced, in stage 1, by three repetitions, followed by a byte with the repetition count. When these four bytes arrive at stage 2, they are replaced by tokens, but the fourth one does not cause an increment of a frequency count.

Example: Suppose that the character with ASCII code 52 repeats six times. Stage 1 will generate the four bytes `52, 52, 52, 3,` and stage 2 will replace each with a token, will increment the entry for `52` (entry 53 in the table) by 3, but will not increment the entry for `3` (which is entry 4 in the table). (The three tokens for the three bytes of `52` may all be different, since tokens may be swapped after each `52` is read and processed.)

The output of stage 2 consists of tokens of different sizes, from 4 to 11 bits. This output is packed in groups of 8 bits, which get written into the output stream. At the end, a special code consisting of 11 bits of 1 (the flush token) is written, followed by as many 1 bits as necessary, to complete the last group of 8 bits.

The efficiency of MNP5 is a result of both stages. The efficiency of stage 1 depends heavily on the original data. Stage 2 also depends on the original data, but to a smaller extent. Stage 2 tends to identify the most frequent characters in the data and assign them the short codes. A look at Table 5.25 shows that 32 of the 256 characters have tokens that are 7 bits or fewer in length, thereby resulting in compression. The other 224 characters have tokens that are 8 bits or longer. When one of these characters is replaced by a long token, the result is no compression, or even expansion.

The efficiency of MNP5 therefore depends on how many characters dominate the original data. If all characters occur with the same frequency, expansion will result. In the other extreme case, if only four characters appear in the data, each will be assigned a 4-bit token, and the compression factor will be 2.

◇ **Exercise 5.16:** Assuming that all 256 characters appear in the original data with the same probability (1/256 each), what will the expansion factor in stage 2 be?

5.4.1 Updating the Table

The process of updating the table of MNP5 codes by swapping rows can be done in two ways:

1. Sorting the entire table every time a frequency is incremented. This is simple in concept but too slow in practice, because the table is 256 entries long.
2. Using pointers in the table, and swapping pointers such that items with large frequencies will point to short codes. This approach is illustrated in Figure 5.26. The figure shows the code table organized in four columns labeled `F`, `P`, `Q`, and `C`. Columns `F` and `C` contain the frequencies and codes; columns `P` and `Q` contain pointers that always point to each other, so if `P[i]` contains index `j` (i.e., points to `Q[j]`), then `Q[j]` points to `P[i]`. The following paragraphs correspond to the nine different parts of the figure.

(a) The first data item `a` is read and `F[a]` is incremented from 0 to 1. The algorithm starts with pointer `P[a]` that contains, say, j. The algorithm examines pointer `Q[j-1]`, which initially points to entry `F[b]`, the one right above `F[a]`. Since F[a] $>$ F[b], entry `a` has to be assigned a short code, and this is done by swapping pointers `P[a]` and `P[b]` (and also the corresponding `Q` pointers).

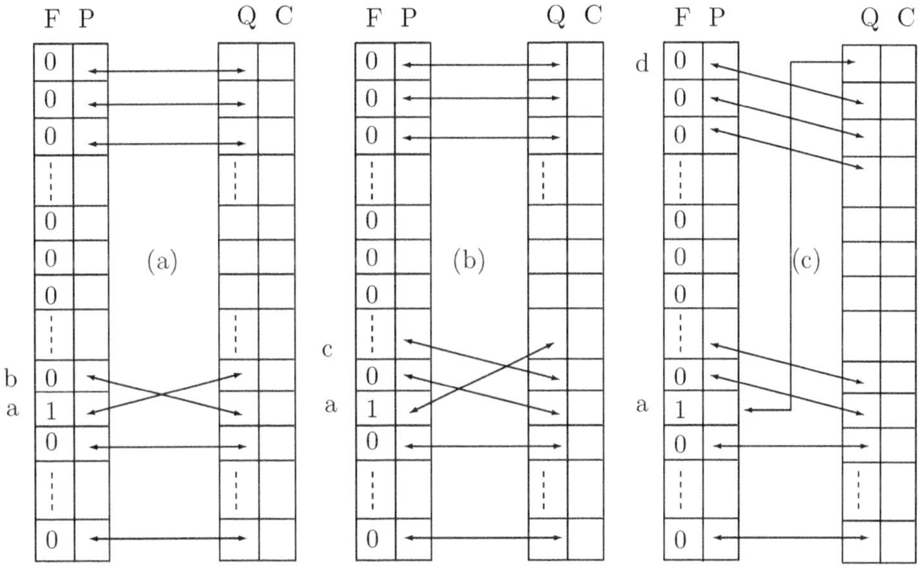

Figure 5.26: Swapping Pointers in the MNP5 Code Table (Part I).

(b) The same process is repeated. The algorithm again starts with pointer `P[a]`, which now points higher, to entry `b`. Assuming that `P[a]` contains the index `k`, the algorithm examines pointer `Q[k-1]`, which points to entry `c`. Since `F[a]` > `F[c]`, entry `a` should be assigned a code shorter than that of `c`. This again is done by swapping pointers, this time `P[a]` and `P[c]`.
(c) This process is repeated, and since `F[a]` is greater than all the frequencies above it, pointers are swapped until `P[a]` points to the top entry, `d`. At this point entry `a` has been assigned the shortest code.
(d) We now assume that the next data item has been input, and `F[m]` incremented to 1. Pointers `P[m]` and the one above it are swapped as in (a) above.
(e) After a few more swaps, `P[m]` is now pointing to entry `n` (the one just below `a`). The next step performs `j:=P[m]; b:=Q[j-1]`, and the algorithm compares `F[m]` to `F[b]`. Since `F[m]` > `F[b]`, pointers are swapped as shown in Figure 5.26f.
(f) After the swap, `P[m]` is pointing to entry `a` and `P[b]` is pointing to entry `n`.
(g) After a few more swaps, pointer `P[m]` points to the second entry `p`. This is how entry `m` is assigned the second-shortest code. Pointer `P[m]` is not swapped with `P[a]`, since they have the same frequencies.
(h) We now assume that the third data item has been input and `F[x]` incremented. Pointers `P[x]` and `P[y]` are swapped.
(i) After some more swaps, pointer `P[x]` points to the third table entry `z`. This is how entry `x` is assigned the third shortest code.

Assuming that `F[x]` is incremented next, the reader is invited to try to figure out how `P[x]` is swapped, first with `P[m]` and then with `P[a]`, so that entry `x` is assigned the shortest code.

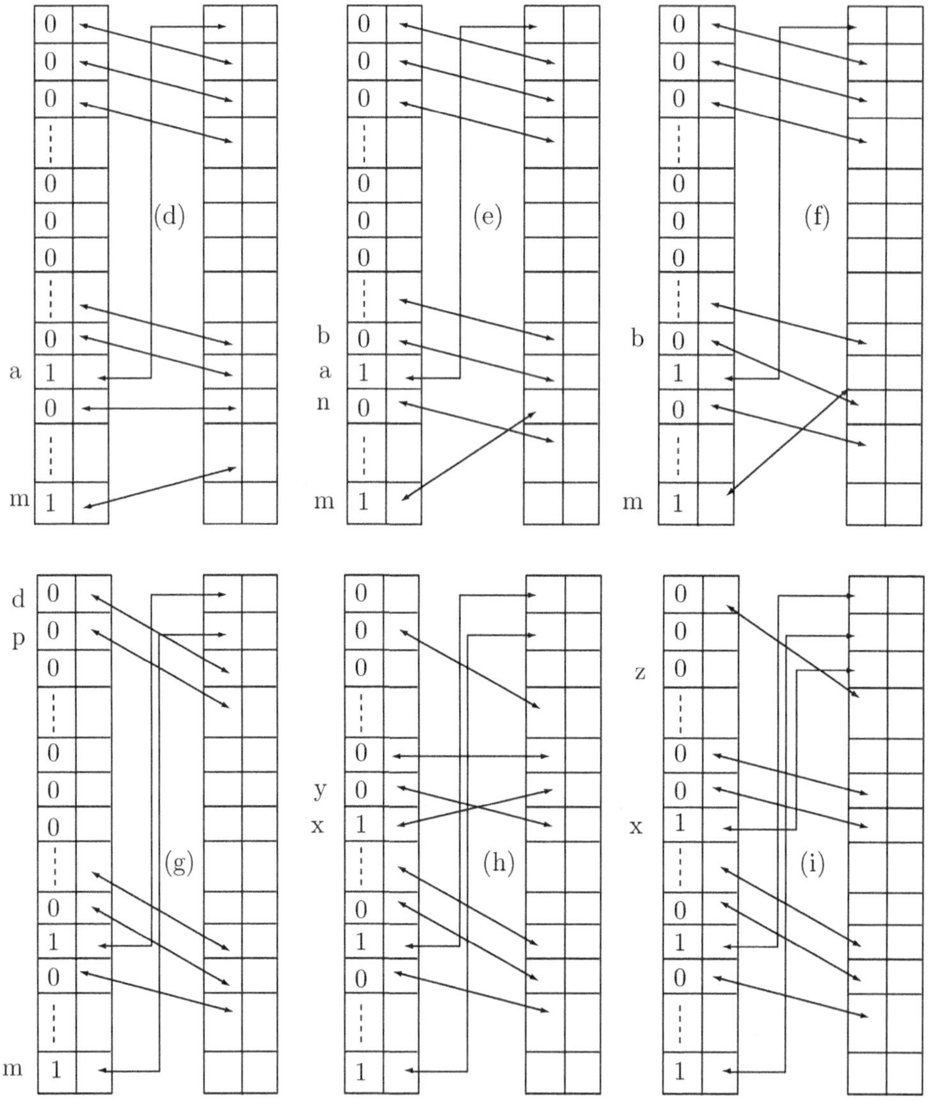

Figure 5.26 (Continued).

```
F[i]:=F[i]+1;
repeat forever
 j:=P[i];
 if j=1 then exit;
 j:=Q[j-1];
 if F[i]<=F[j] then exit
  else
   tmp:=P[i]; P[i]:=P[j]; P[j]:=tmp;
   tmp:=Q[P[i]]; Q[P[i]]:=Q[P[j]]; Q[P[j]]:=tmp
 endif;
end repeat
```

Figure 5.27: Swapping Pointers in MNP5.

The pseudo-code of Figure 5.27 summarizes the pointer swapping process.

Are no probabilities to be accepted, merely because they are not certainties?
—Jane Austen, *Sense and Sensibility*

5.5 MNP7

More complex and sophisticated than MNP5, MNP7 combines run-length encoding with a two-dimensional variant of adaptive Huffman coding. Stage 1 identifies runs and emits three copies of the run character, followed by a 4-bit count of the remaining characters in the run. A count of zero implies a run of length 3, and a count of 15 (the largest possible in a 4-bit nibble), a run of length 18. Stage 2 starts by assigning to each character a complete table with many variable-length codes. When a character C is read, one of the codes in its table is selected and output, depending on the *character preceding* C in the input stream. If this character is, say, P, then the frequency count of the pair (digram) PC is incremented by 1, and table rows may be swapped, using the same algorithm as for MNP5, to move the pair to a position in the table that has a shorter code.

MNP7 is therefore based on a first-order Markov model, where each item is processed depending on the item and one of its predecessors. In a k-order Markov model, an item is processed depending on itself and k of its predecessors (not necessarily the k immediate ones).

Here are the details. Each of the 256 8-bit bytes gets a table of codes assigned, of size 256×2, where each row corresponds to one of the 256 possible bytes. Column 1 of the table is initialized to the integers 0 through 255, and column 2 (the frequency counts) is initialized to all zeros. The result is 256 tables, each a double column of 256 rows (Table 5.28a). Variable-length codes are assigned to the rows, such that the first code is 1-bit long, and the others get longer towards the bottom of the table. The codes are stored in an additional code table that never changes.

When a character C is input (the current character to be compressed), its value is used as a pointer, to select one of the 256 tables. The first column of the table is

	Current character					
	0	1	2	...	254	255
	0 0	0 0	0 0	...	0 0	0 0
	1 0	1 0	1 0	...	1 0	1 0
	2 0	2 0	2 0	...	2 0	2 0
	3 0	3 0	3 0	...	3 0	3 0
Preced.	⋮	⋮	⋮		⋮	⋮
Char.	254 0	254 0	254 0	...	254 0	254 0
	255 0	255 0	255 0	...	255 0	255 0

(a)

...a	b	c	d	e...
t	l	h	o	d
h	e	o	a	r
c	u	r	e	s
⋮	⋮	⋮	⋮	⋮

(b)

Table 5.28: The MNP7 Code Tables.

searched, to find the row with the 8-bit value of the preceding character P. Once the row is found, the code from the same row in the code table is emitted and the count in the second column is incremented by 1. Rows in the table may be swapped if the new count of the digram PC is large enough.

After enough characters have been input and rows swapped, the tables start reflecting the true digram frequencies of the data. Table 5.28b shows a possible state assuming that the digrams `ta`, `ha`, `ca`, `lb`, `eb`, `ub`, `hc`, etc., are common. Since the top digram is encoded into 1 bit, MNP7 can be very efficient. If the original data consists of text in a natural language, where certain digrams are very common, MNP7 normally produces a high compression ratio.

His only other visitor was a woman: the woman who had attended his reading. At the time she had seemed to him to be the only person present who had paid the slightest attention to his words. With kittenish timidity she approached his table. Richard bade her welcome, and meant it, and went on meaning it as she extracted from her shoulder pouch a copy of a novel written not by Richard Tull but by Fyodor Dostoevsky. *The Idiot.* Standing beside him, leaning over him, her face awfully warm and near, she began to leaf through its pages, explaining. The book too was stained, not by gouts of blood but by the vying colors of two highlighting pens, one blue, one pink. And not just two pages but the whole six hundred. Every time the letters *h* and *e* appeared together, as in *the, then, there,* as in *forehead, Pashlishtchev, sheepskin,* they were shaded in blue. Every time the letters *s, h,* and *e* appeared together, as in *she, sheer, ashen, sheepskin,* etc., they were shaded in pink. And since every *she* contained a *he*, the predominance was unarguably and unsurprisingly masculine. Which was exactly her point. "You see?" she said with her hot breath, breath redolent of metallic medications, of batteries and printing-plates. "You see?"... The organizers knew all about this woman—this unfortunate recurrence, this indefatigable drag—and kept coming over to try and coax her away.

—Martin Amis, *The Information*

5.6 Reliability

The chief downside of variable-length codes is their vulnerability to errors. The prefix property is used to decode those codes, so an error in a single bit can cause the decompressor to lose synchronization and be unable to decode the rest of the compressed stream. In the worst case, the decompressor may even read, decode, and interpret the rest of the compressed data incorrectly, without realizing that a problem has occurred.

Example: Using the code of Figure 5.5 the string `CARE` is coded into 10100 0011 0110 000 (without the spaces). Assuming the error 10$\boxed{0}$00 0011 0110 000, the decompressor will not notice any problem but will decode the string as `HARE`.

◇ **Exercise 5.17:** What will happen in the case 11$\boxed{1}$11 0011 0110 000 ...(the string WARE ... with one bad bit)?

Users of Huffman codes have noticed long ago that these codes recover quickly from an error. However, if Huffman codes are used to code run lengths, then this property does not help, since all runs would be shifted after an error.

A simple way of adding reliability to variable-length codes is to break a long compressed stream, as it is being transmitted, into groups of 7 bits and add a parity bit to each group. This way, the decompressor will at least be able to detect a problem and output an error message or ask for a retransmission. It is, of course, possible to add more than one parity bit to a group of data bits, thereby making it more reliable. However, reliability is, in a sense, the opposite of compression. Compression is done by decreasing redundancy, while reliability is achieved by increasing it. The more reliable a piece of data is, the less compressed it is, so care should be taken when the two operations are used together.

Some Important Standards Groups and Organizations

ANSI (American National Standards Institute) is the private sector voluntary standardization system for the United States. Its members are professional societies, consumer groups, trade associations, and some government regulatory agencies (it is a federation). It collects and distributes standards developed by its members. Its mission is the enhancement of global competitiveness of U.S. business and the American quality of life by promoting and facilitating voluntary consensus standards and conformity assessment systems and promoting their integrity.

ANSI was founded in 1918 by five engineering societies and three government agencies, and it remains a private, nonprofit membership organization whose nearly 1,400 members are private and public sector organizations.

ANSI is located at 11 West 42nd Street, New York, NY 10036, USA. See also `http://web.ansi.org/`.

ISO (International Organization for Standardization, Organisation Internationale de Normalisation) is an agency of the United Nations whose members are standards organizations of some 100 member countries (one organization from each country). It develops a wide range of standards for industries in all fields.

Established in 1947, its mission is "to promote the development of standardization in the world with a view to facilitating the international exchange of goods and services, and to developing cooperation in the spheres of intellectual, scientific, technological and economic activity." This is the forum where it is agreed, for example, how thick your bank card should be, so every country in the world follows a compatible standard. The ISO is located at 1, rue de Varembé, CH-1211 Geneva 20, Switzerland, `http://www.iso.ch/`.

ITU (International Telecommunication Union) is another United Nations agency developing standards for telecommunications. Its members are mostly companies that make telecommunications equipment, groups interested in standards, and government agencies involved with telecommunications. The ITU is the successor of an organization founded by some 20 European nations in Paris in 1865.

IEC (International Electrotechnical Commission) is a nongovernmental international organization that prepares and publishes international standards for all electrical, electronic, and related technologies. The IEC was founded, in 1906, as a result of a resolution passed at the International Electrical Congress held in St. Louis (USA) in 1904. Its membership consists of more than 50 participating countries, including all the world's major trading nations and a growing number of industrializing countries.

The IEC's mission is to promote, through its members, international cooperation on all questions of electrotechnical standardization and related matters, such as the assessment of conformity to standards, in the fields of electricity, electronics and related technologies.

The IEC is located at 3, rue de Varembé, P.O. Box 131, CH-1211, Geneva 20, Switzerland, `http://www.iec.ch/`.

QIC (`http://www.qic.org/`) was an international trade association, incorporated in 1987, to encourage and promote the widespread use of quarter-inch tape cartridge technology. One hundred companies around the world were Members or Associates of QIC during its active years. The group became inactive in 1998 after more than 15 million QIC-compatible drives had been installed.

Most of the 15 million QIC tape drives in use worldwide were installed in business environments. QIC tape automation solutions enabled capacities well into the terabyte range, providing the hardware data compression and read-write features essential to network backup.

5.7 Facsimile Compression

Data compression is especially important when images are transmitted over a communications line because the user is often waiting at the receiver, eager to see something quickly. Documents transferred between fax machines are sent as bitmaps, so a standard data compression method was needed when those machines became popular. Several methods were developed and proposed by the ITU-T.

The ITU-T is one of four permanent parts of the International Telecommunications Union (ITU), based in Geneva, Switzerland (`http://www.itu.ch/`). It issues recommendations for standards applying to modems, packet switched interfaces, V.24 connectors, and similar devices. Although it has no power of enforcement, the standards it recommends are generally accepted and adopted by industry. Until March 1993, the ITU-T

CCITT: Can't Conceive Intelligent Thoughts Today

was known as the Consultative Committee for International Telephone and Telegraph (Comité Consultatif International Télégraphique et Téléphonique, or CCITT).

The first data compression standards developed by the ITU-T were T2 (also known as Group 1) and T3 (Group 2). These are now obsolete and have been replaced by T4 (Group 3) and T6 (Group 4). Group 3 is currently used by all fax machines designed to operate with the Public Switched Telephone Network (PSTN). These are the machines we have at home, and at the time of writing, they operate at maximum speeds of 9,600 baud. Group 4 is used by fax machines designed to operate on a digital network, such as ISDN. They have typical speeds of 64K baud. Both methods can produce compression factors of 10 or better, reducing the transmission time of a typical page to about a minute with the former, and a few seconds with the latter.

Some references for facsimile compression are [Anderson et al. 87], [Hunter and Robinson 80], [Marking 90], and [McConnell 92].

5.7.1 One-Dimensional Coding

A fax machine scans a document line by line, converting each line to small black and white dots called *pels* (from Picture ELement). The horizontal resolution is always 8.05 pels per millimeter (about 205 pels per inch). An 8.5-inch-wide scan line is therefore converted to 1728 pels. The T4 standard, though, recommends to scan only about 8.2 inches, thereby producing 1664 pels per scan line (these numbers, as well as those in the next paragraph, are all to within $\pm 1\%$ accuracy).

The vertical resolution is either 3.85 scan lines per millimeter (standard mode) or 7.7 lines/mm (fine mode). Many fax machines have also a very-fine mode, where they scan 15.4 lines/mm. Table 5.29 assumes a 10-inch-high page (254 mm), and shows the total number of pels per page, and typical transmission times for the three modes without compression. The times are long, illustrating how important data compression is in fax transmissions.

Scan lines	Pels per line	Pels per page	Time (sec.)	Time (min.)
978	1664	1.670M	170	2.82
1956	1664	3.255M	339	5.65
3912	1664	6.510M	678	11.3

Ten inches equal 254 mm. The number of pels is in the millions, and the transmission times, at 9600 baud without compression, are between 3 and 11 minutes, depending on the mode. However, if the page is shorter than 10 inches, or if most of it is white, the compression factor can be 10 or better, resulting in transmission times of between 17 and 68 seconds.

Table 5.29: Fax Transmission Times.

To derive the Group 3 code, the ITU-T counted all the run lengths of white and black pels in a set of eight "training" documents that they felt represent typical text and images sent by fax, and used the Huffman algorithm to assign a variable-length code to each run length. (The eight documents are described in Table 5.30. They are not shown because they are copyrighted by the ITU-T.) The most common run lengths were found to be 2, 3, and 4 black pixels, so they were assigned the shortest codes (Table 5.31). Next come run lengths of 2–7 white pixels, which were assigned slightly longer codes. Most run lengths were rare and were assigned long, 12-bit codes. Thus, Group 3 uses a combination of RLE and Huffman coding.

Image	Description
1	Typed business letter (English)
2	Circuit diagram (hand drawn)
3	Printed and typed invoice (French)
4	Densely typed report (French)
5	Printed technical article including figures and equations (French)
6	Graph with printed captions (French)
7	Dense document (Kanji)
8	Handwritten memo with very large white-on-black letters (English)

Table 5.30: The Eight CCITT Training Documents.

◇ **Exercise 5.18:** A run length of 1664 white pels was assigned the short code 011000. Why is this length so common?

Since run lengths can be long, the Huffman algorithm was modified. Codes were assigned to run lengths of 1 to 63 pels (they are the termination codes in Table 5.31a) and to run lengths that are multiples of 64 pels (the make-up codes in Table 5.31b). Group 3 is therefore a *modified Huffman code* (also called MH). The code of a run length is either a single termination code (if the run length is short) or one or more make-up codes, followed by one termination code (if it is long). Here are some examples:

1. A run length of 12 white pels is coded as 001000.
2. A run length of 76 white pels ($= 64 + 12$) is coded as 11011|001000.
3. A run length of 140 white pels ($= 128 + 12$) is coded as 10010|001000.
4. A run length of 64 black pels ($= 64 + 0$) is coded as 0000001111|0000110111.
5. A run length of 2561 black pels ($2560 + 1$) is coded as 000000011111|010.

◇ **Exercise 5.19:** There are no runs of length zero. Why then were codes assigned to runs of zero black and zero white pels?

◇ **Exercise 5.20:** An 8.5-inch-wide scan line results in 1728 pels, so how can there be a run of 2561 consecutive pels?

Each scan line is coded separately, and its code is terminated by the special 12-bit EOL code 000000000001. Each line also gets one white pel appended to it on the left when it is scanned. This is done to remove any ambiguity when the line is decoded on

the receiving end. After reading the EOL for the previous line, the receiver assumes that the new line starts with a run of white pels, and it ignores the first of them. Examples:
1. The 14-pel line  is coded as the run lengths 1w 3b 2w 2b 7w EOL, which become 000111|10|0111|11|1111|000000000001. The decoder ignores the single white pel at the start.
2. The line  is coded as the run lengths 3w 5b 5w 2b EOL, which becomes the binary string 1000|0011|1100|11|000000000001.

◇ **Exercise 5.21:** The group 3 code for a run length of five black pels (0011) is also the prefix of the codes for run lengths of 61, 62, and 63 white pels. Explain this.

The Group 3 code has no error correction, but many errors can be detected. Because of the nature of the Huffman code, even one bad bit in the transmission can cause the receiver to get out of synchronization, and to produce a string of wrong pels. This is why each scan line is encoded separately. If the receiver detects an error, it skips bits, looking for an EOL. This way, one error can cause at most one scan line to be received incorrectly. If the receiver does not see an EOL after a certain number of lines, it assumes a high error rate, and it aborts the process, notifying the transmitter. Since the codes are between 2 and 12 bits long, the receiver detects an error if it cannot decode a valid code after reading 12 bits.

Each page of the coded document is preceded by one EOL and is followed by six EOL codes. Because each line is coded separately, this method is a *one-dimensional coding* scheme. The compression ratio depends on the image. Images with large contiguous black or white areas (text or black and white images) can be highly compressed. Images with many short runs can sometimes produce negative compression. This is especially true in the case of images with shades of gray (such as scanned photographs). Such shades are produced by halftoning, which covers areas with many alternating black and white pels (runs of length one).

◇ **Exercise 5.22:** What is the compression ratio for runs of length one (i.e., strictly alternating pels)?

The T4 standard also allows for fill bits to be inserted between the data bits and the EOL. This is done in cases where a pause is necessary, or where the total number of bits transmitted for a scan line must be a multiple of 8. The fill bits are zeros.

Example: The binary string 000111|10|0111|11|1111|000000000001 becomes

000111|10|0111|11|1111|00|0000000001

after two zeros are added as fill bits, bringing the total length of the string to 32 bits ($= 8 \times 4$). The decoder sees the two zeros of the fill, followed by the 11 zeros of the EOL, followed by the single 1, so it knows that it has encountered a fill followed by an EOL.

See `http://www.doclib.org/rfc/rfc804.html` for a description of group 3.

At the time of writing, the T.4 and T.6 recommendations can also be found at URL `ftp://sunsite.doc.ic.ac.uk/` as files `7_3_01.ps.gz` and `7_3_02.ps.gz` (the precise subdirectory seems to change every few years and it is recommended to locate it with a search engine).

(a)

Run length	White codeword	Black codeword	Run length	White codeword	Black codeword
0	00110101	0000110111	32	00011011	000001101010
1	000111	010	33	00010010	000001101011
2	0111	11	34	00010011	000011010010
3	1000	10	35	00010100	000011010011
4	1011	011	36	00010101	000011010100
5	1100	0011	37	00010110	000011010101
6	1110	0010	38	00010111	000011010110
7	1111	00011	39	00101000	000011010111
8	10011	000101	40	00101001	000001101100
9	10100	000100	41	00101010	000001101101
10	00111	0000100	42	00101011	000011011010
11	01000	0000101	43	00101100	000011011011
12	001000	0000111	44	00101101	000001010100
13	000011	00000100	45	00000100	000001010101
14	110100	00000111	46	00000101	000001010110
15	110101	000011000	47	00001010	000001010111
16	101010	0000010111	48	00001011	000001100100
17	101011	0000011000	49	01010010	000001100101
18	0100111	0000001000	50	01010011	000001010010
19	0001100	00001100111	51	01010100	000001010011
20	0001000	00001101000	52	01010101	000000100100
21	0010111	00001101100	53	00100100	000000110111
22	0000011	00000110111	54	00100101	000000111000
23	0000100	00000101000	55	01011000	000000100111
24	0101000	00000010111	56	01011001	000000101000
25	0101011	00000011000	57	01011010	000001011000
26	0010011	000011001010	58	01011011	000001011001
27	0100100	000011001011	59	01001010	000000101011
28	0011000	000011001100	60	01001011	000000101100
29	00000010	000011001101	61	00110010	000001011010
30	00000011	000001101000	62	00110011	000001100110
31	00011010	000001101001	63	00110100	000001100111

(b)

Run length	White codeword	Black codeword	Run length	White codeword	Black codeword
64	11011	0000001111	1344	011011010	0000001010011
128	10010	000011001000	1408	011011011	0000001010100
192	010111	000011001001	1472	010011000	0000001010101
256	0110111	000001011011	1536	010011001	0000001011010
320	00110110	000000110011	1600	010011010	0000001011011
384	00110111	000000110100	1664	011000	0000001100100
448	01100100	000000110101	1728	010011011	0000001100101
512	01100101	0000001101100	1792	00000001000	same as
576	01101000	0000001101101	1856	00000001100	white
640	01100111	0000001001010	1920	00000001101	from this
704	011001100	0000001001011	1984	000000010010	point
768	011001101	0000001001100	2048	000000010011	
832	011010010	0000001001101	2112	000000010100	
896	011010011	0000001110010	2176	000000010101	
960	011010100	0000001110011	2240	000000010110	
1024	011010101	0000001110100	2304	000000010111	
1088	011010110	0000001110101	2368	000000011100	
1152	011010111	0000001110110	2432	000000011101	
1216	011011000	0000001110111	2496	000000011110	
1280	011011001	0000001010010	2560	000000011111	

Table 5.31: Group 3 and 4 Fax Codes: (a) Termination Codes, (b) Make-Up Codes.

5.7.2 Two-Dimensional Coding

Two-dimensional coding was developed because one-dimensional coding does not produce good results for images with gray areas. Two-dimensional coding is optional on fax machines that use Group 3 but is the only method used by machines intended to work in a digital network. When a fax machine using Group 3 supports two-dimensional coding as an option, each EOL is followed by one extra bit, to indicate the compression method used for the next scan line. That bit is 1 if the next line is encoded with one-dimensional coding, and 0 if it is encoded with two-dimensional coding.

The two-dimensional coding method is also called MMR, for *modified modified READ*, where READ stands for *relative element address designate*. The term "modified modified" is used because this is a modification of one-dimensional coding, which itself is a modification of the original Huffman method. The two-dimensional coding method works by comparing the current scan line (called the *coding line*) to its predecessor (which is called the *reference line*) and recording the differences between them, the assumption being that two consecutive lines in a document will normally differ by just a few pels. The method assumes that there is an all-white line above the page, which is used as the reference line for the first scan line of the page. After coding the first line, it becomes the reference line, and the second scan line is coded. As in one-dimensional coding, each line is assumed to start with a white pel, which is ignored by the receiver.

The two-dimensional coding method is less reliable than one-dimensional coding, since an error in decoding a line will cause errors in decoding all its successors and will propagate through the entire document. This is why the T.4 (Group 3) standard includes a requirement that after a line is encoded with the one-dimensional method, at most $K-1$ lines will be encoded with the two-dimensional coding method. For standard resolution $K=2$, and for fine resolution $K=4$. The T.6 standard (Group 4) does not have this requirement, and uses two-dimensional coding exclusively.

Scanning the coding line and comparing it to the reference line results in three cases, or modes. The mode is identified by comparing the next run length on the reference line [(b_1b_2) in Figure 5.33] with the current run length (a_0a_1) and the next one (a_1a_2) on the coding line. Each of these three runs can be black or white. The three modes are as follows (see also the flow chart of Figure 5.34):

1. **Pass mode**. This is the case where (b_1b_2) is to the left of (a_1a_2) and b_2 is to the left of a_1 (Figure 5.33a). This mode does not include the case where b_2 is above a_1. When this mode is identified, the length of run (b_1b_2) is coded using the codes of Table 5.32 and is transmitted. Pointer a_0 is moved below b_2, and the four values b_1, b_2, a_1, and a_2 are updated.
2. **Vertical mode**. (b_1b_2) overlaps (a_1a_2) by not more than three pels (Figure 5.33b1, b2). Assuming that consecutive lines do not differ by much, this is the most common case. When this mode is identified, one of seven codes is produced (Table 5.32) and is transmitted. Pointers are updated as in case 1 above. The performance of the two-dimensional coding method depends on this case being common.
3. **Horizontal mode**. (b_1b_2) overlaps (a_1a_2) by more than three pels (Figure 5.33c1, c2). When this mode is identified, the lengths of runs (a_0a_1) and (a_1a_2) are coded using the codes of Table 5.32 and are transmitted. Pointers are updated as in cases 1 and 2 above.

> . . . and you thought "impressive" statistics were 36–24–36.
>
> —Advertisement, *The American Statistician*, November 1979

Mode	Run length to be encoded	Abbreviation	Codeword
Pass	b_1b_2	P	0001+coded length of b_1b_2
Horizontal	a_0a_1, a_1a_2	H	001+coded length of a_0a_1 and a_1a_2
Vertical	$a_1b_1 = 0$	V(0)	1
	$a_1b_1 = -1$	VR(1)	011
	$a_1b_1 = -2$	VR(2)	000011
	$a_1b_1 = -3$	VR(3)	0000011
	$a_1b_1 = +1$	VL(1)	010
	$a_1b_1 = +2$	VL(2)	000010
	$a_1b_1 = +3$	VL(3)	0000010
Extension			0000001000

Table 5.32: 2D Codes for the Group 4 Method.

When scanning starts, pointer a_0 is set to an imaginary white pel on the left of the coding line, and a_1 is set to point to the first black pel on the coding line. (Recall that a_0 corresponds to an imaginary pel, which is why the first run length is $|a_0a_1| - 1$.) Pointer a_2 is set to the first white pel following that. Pointers b_1, b_2 are set to point to the start of the first and second runs on the reference line, respectively.

After identifying the current mode and transmitting codes according to Table 5.32, a_0 is updated as shown in the flow chart, and the other four pointers are updated relative to the new a_0. The process continues until the end of the coding line is reached. The encoder assumes an extra pel on the right of the line, with a color opposite that of the last pel.

The extension code in Table 5.32 is used to abort the encoding process prematurely, before reaching the end of the page. This is necessary if the rest of the page is transmitted in a different code or even in uncompressed form.

◇ **Exercise 5.23:** Manually figure out the code generated from the two lines below.

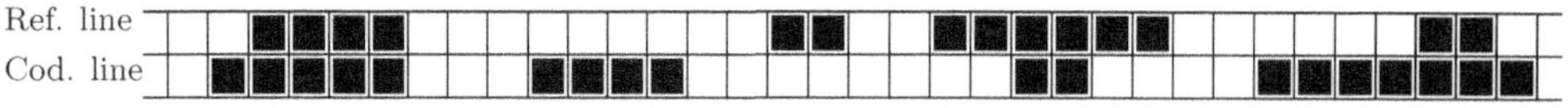

Table 5.36 summarizes the codes emitted by the group 4 encoder. Figure 5.35 is a tree with the same codes. Each horizontal branch corresponds to another zero and each vertical branch, to another 1.

> Teamwork is the ability to work as a group toward a common vision, even if that vision becomes extremely blurry.
>
> —Anonymous

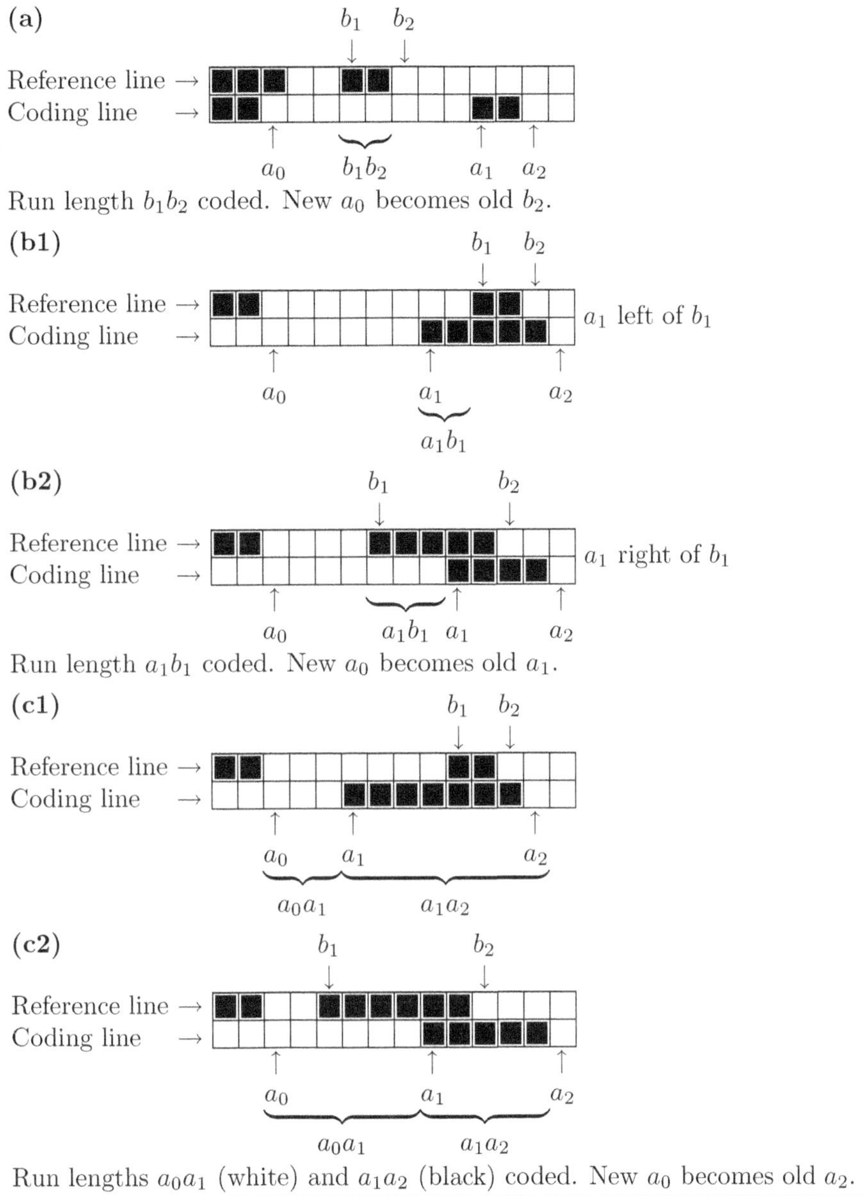

Notes:

1. a_0 is the first pel of a new codeword and can be black or white.
2. a_1 is the first pel to the right of a_0 with a different color.
3. a_2 is the first pel to the right of a_1 with a different color.
4. b_1 is the first pel on the reference line to the right of a_0 with a different color.
5. b_2 is the first pel on the reference line to the right of b_1 with a different color.

Figure 5.33: Five Run-Length Configurations: (a) Pass Mode, (b) Vertical Mode, (c) Horizontal Mode.

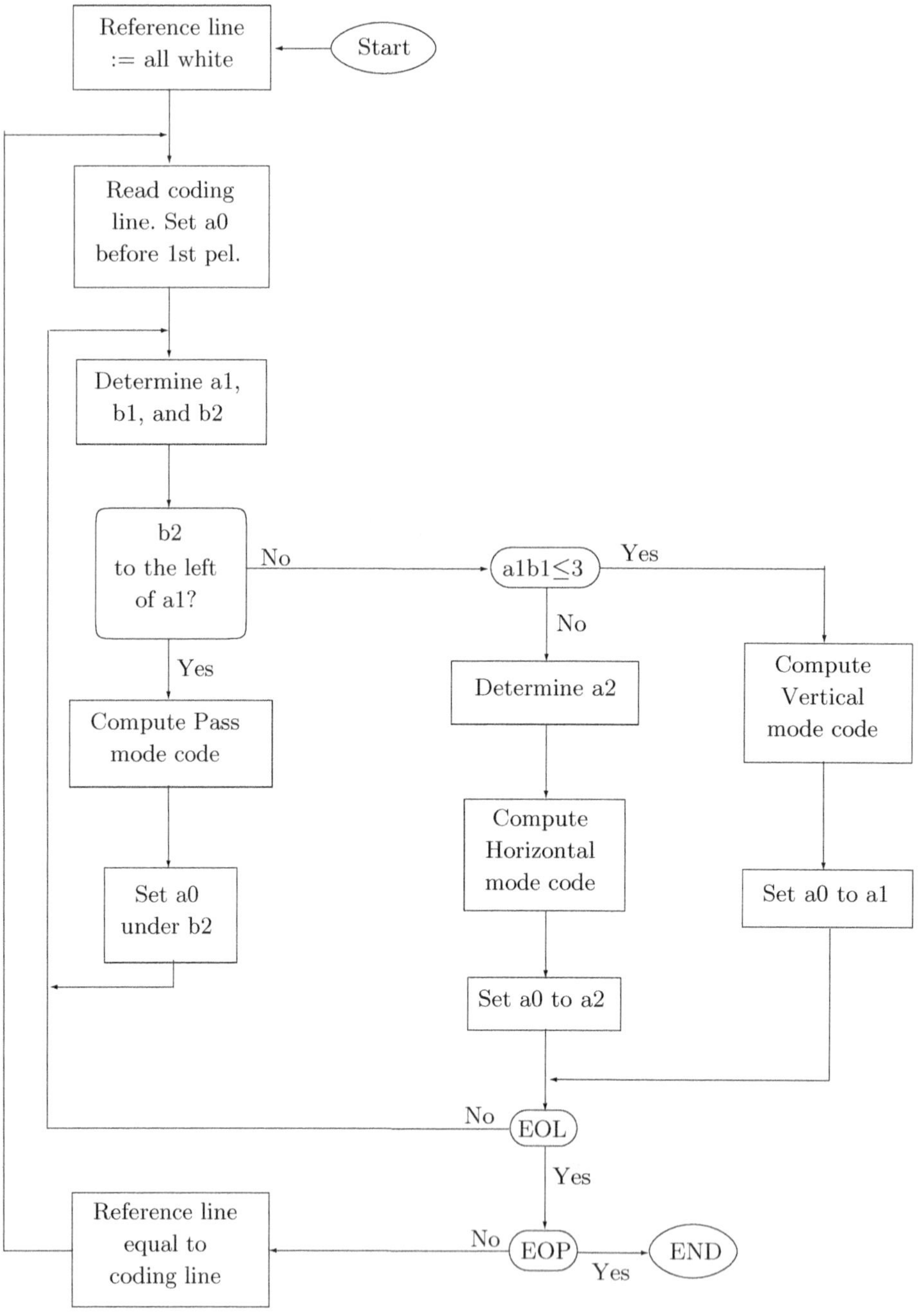

Figure 5.34: MMR Flow Chart.

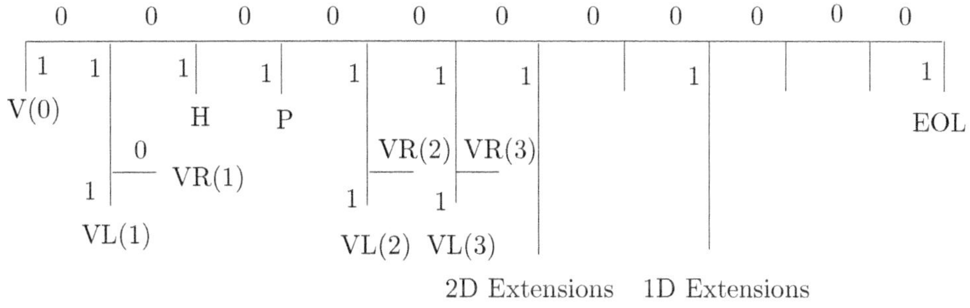

Figure 5.35: Tree of Group 3 Codes.

Mode	Elements to Be Coded		Notation	Codeword
Pass	b_1, b_2		P	0001
Horizontal	a_0a_1, a_1a_2		H	$001 + M(a_0a_1) + M(a_1a_2)$
Vertical	a_1 just under b_1	$a_1b_1 = 0$	V(0)	1
	a_1 to the right of b_1	$a_1b_1 = 1$	VR(1)	011
		$a_1b_1 = 2$	VR(2)	000011
		$a_1b_1 = 3$	VR(3)	0000011
	a_1 to the left of b_1	$a_1b_1 = 1$	VL(1)	010
		$a_1b_1 = 2$	VL(2)	000010
		$a_1b_1 = 3$	VL(3)	0000010
2D Extensions 1D Extensions				0000001xxx 000000001xxx
EOL				000000000001
1D Coding of Next Line 2D Coding of Next Line				EOL+‘1’ EOL+‘0’

Table 5.36: Group 4 Codes.

5.8 PK Font Compression

Obviously, these words are hard to read because the individual characters feature different styles and sizes. In a beautifully-typeset document, all the letters of a word and all the words of the document (with perhaps a few exceptions) should have the same size and style; they should belong to the same *font.*

Traditionally, the term "font" refers to a set of characters of type that are of the same size and style, such as Times Roman 12 point. A typeface is a set of fonts of different sizes but in the same style, like Times Roman. A typeface family is a set of typefaces in the same style, such as Times.

The size of a font is normally measured in points (more accurately, printer's points, where 72.27 points equal one inch). The style of a font describes its appearance. Traditional styles are roman, **boldface**, *italic*, *slanted*, `typewriter`, and sans serif.

Old operating systems did not support fonts. Even the DOS operating system, which was widely used on IBM-PC-compatible personal computers from 1980 to 1995, did not support fonts and was based on ASCII codes instead.

It was not until the mid 1980s that digital fonts became a part of many operating systems. In 1984, Adobe launched the PostScript language, which supported two types of digital fonts. In the same year, Apple released the first models of the Macintosh computer, whose operating system used fonts (Figure 5.37 illustrates some of the early fonts included in the Macintosh OS 6). In 1985, Apple announced the LaserWriter, one of the first laser printers intended for the mass market. These developments paved the way for future operating systems to support digital fonts, thus enabling users to create, print, and publish beautiful, professionally-looking documents.

Figure 5.37: Old Macintosh OS 6 Fonts.

A digital font is a set of symbols or characters that can be stored in the computer's memory or on an I/O device such as a disk. The symbols of the font are either displayed or printed. Each symbol consists of a glyph (the actual shape of the symbol) and bookkeeping information, such as the height, depth, and width of the symbol.

Modern fonts are referred to as outline fonts. A symbol in such a font is fully defined by a small set of control points that are connected by curves to form the outline of the symbol. The symbol π in Figure 5.38 is an example. It is easy to see how this symbol is fully specified by about 30 control points that are connected with a few smooth curves. An outline font is easy to store in a computer file simply by storing the coordinates of the control points. It is also easy to change the size of the font by scaling the coordinates of the points.

In contrast, early digital fonts were of the bitmap variety. The font designer would draw the glyph of each symbol on a grid of small squares (pixels) and then select the

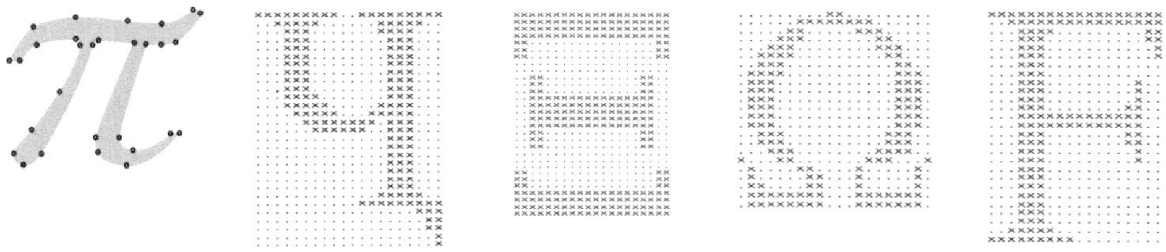

Figure 5.38: Outline and Bitmap Fonts.

best pixels for the symbol. Figure 5.38 shows four examples of bitmap symbols where the white pixels are represented by dots and the black pixels are shown as an $\times$. In the computer's memory, a bitmap symbol is stored as a map of bits, zeros for white pixels and 1's for black pixels. When such a font is written in raw format on a file, the file tends to be large and should be compressed. Also, a high-quality bitmap font has to be designed from scratch for each font size. Simply scaling a bitmap results in badly-looking characters.

The TEX project, started by Donald Knuth in 1975, is an early example of the use of bitmap fonts. The goal was to develop software for typesetting beautiful documents, especially documents with a lot of mathematics. Such software requires a set of fonts, even if the operating system does not support fonts. Thus, `METAFONT`, an application to generate fonts, was developed in parallel with the TEX application itself. `METAFONT` was then used to create a set of fonts that is referred to as computer modern (CM).

The CM set of fonts consists of 75 fonts that are described in volume E of [Knuth 86] as well as the "line," "circle," and symbol fonts associated with LaTeX.

The output of `METAFONT` is called a generic font (GF), to indicate that this file format does not follow the conventions of any font foundry. However, it is easy to convert GF font files to the special format required by most digital phototypesetting equipment.

In 1985, Tomas Rokicki, a student of Knuth's, developed the PK (for PacKed) font format. The idea was to develop a simple, fast compression method such that the fonts would be saved in small files and even the slow computers available at that time would be able to decompress a font, or any part of it, quickly. The main references for the format and organization of PK fonts are [Rokicki 95], [Rokicki 90], and [Haralambous 07].

The method selected for PK compression was based on run-length encoding (RLE) and on the special features of font bitmaps. The method is not very efficient, but it is described here because it offers an original approach to RLE, an approach that does not use Huffman codes and makes minimal use of variable-length codes (the packed numbers, described later, are the only variable-length code used by PK). A quick glance at the four bitmaps of Figure 5.38 shows two important features (1) they are narrow, resulting in mostly short runs of black and white pixels and (2) they tend to have consecutive identical rows of pixels.

This section discusses the compression of PK fonts, but a full understanding of this compression method requires a little knowledge of the organization of these fonts and of the way TEX employs font information.

When a font of type is designed, the designer determines, for each symbol in the font, its glyph and its dimensions. Symbols in a font are two-dimensional, but each has three dimensions, height, depth, and width. It is best to think of the symbol as if it is surrounded by a box (or a bitmap) as illustrated by the leftmost item of Figure 5.39. There is a baseline that separates the height and depth of the box, and there is a reference point at the left end of the baseline.

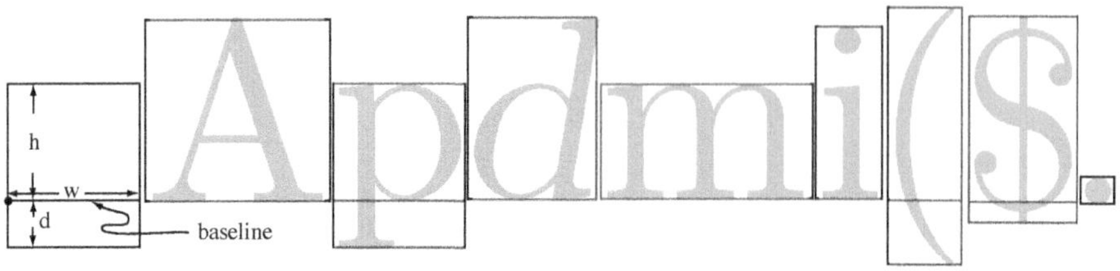

Figure 5.39: Three-Dimensional Character Boxes.

TEX uses the three dimensions of each font symbol but is not concerned with the actual shape of the symbol. As a result, symbols may stick out of their bitmap boxes, as illustrated by the italic d in the figure. TEX reads the input text and strings boxes horizontally to construct a line of text. The boxes butt together (in Figure 5.39 there are small spaces between boxes for better readability), so the font designer leaves spaces to the left and right of each symbol, as illustrated in the figure. (In those rare cases where a symbol, such as an m-dash, should touch their neighbors, the bitmap box is as wide as the symbol and no spaces are left.)

When the current line of text is ready, TEX starts on the next line. When that line is also ready, it is placed under the current line (and it then becomes the new current line) such that the baselines of the two lines are separated by a user-controlled parameter. If this causes the lines to overlap (because the top line has a symbol with a large depth and the bottom line has a symbol with a large height), TEX leaves some space (Figure 5.40) between the bottom of the upper line (the symbol with the largest depth) and the top of the lower line (the symbol with the largest height). This space can also be controlled by the user, but the font designer does not have to worry about it. Thus, there are spaces to the left and right of font symbols in their bitmap boxes but no spaces above and below them.

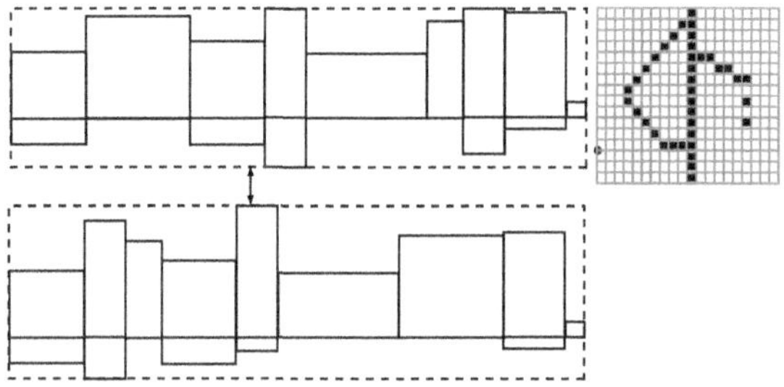

Figure 5.40: Two Lines of Text.

PK compression is based on run-length encoding. The method encodes the runs of black and white pixels in the bitmap of a symbol. In order to decrease the run lengths and thereby increase compression performance, the encoder first determines the bounding box of a bitmap, as illustrated by the 16×20 bitmap of Figure 5.40. The bounding box of a glyph bitmap is the smallest rectangle that encompasses all the black pixels of the glyph. It is easy to see that our bounding box is 15 pixels wide and 16 pixels tall. It starts on column 3 from the left and 13 of the 16 pixel rows are above the baseline. Once the bounding box is decompressed, the decoder needs the following dimensions to create the complete bitmap: The horizontal escapement (the width of the bitmap, 20), the width of the bounding box (15), the vertical escapement (the height of the bitmap, 16), the height of the bounding box (also 16), the horizontal offset (the distance between the reference point and leftmost column of the bounding box, 2), and the vertical offset (the distance from the reference point to the top of the bounding box, 13). Three of these dimensions are horizontal and three are vertical, but often only two vertical dimensions are needed, because the bitmap and bounding box tend to have the same height.

We are now ready to look at the details of the compression. A PK font file is organized in records, one for each symbol. A record starts with a character preamble that contains (in addition to other information) the five dimensions mentioned above and one bit that specifies the top-left pixel of the bounding box (1 for a black pixel and 0 for a white one). The character preamble is followed by the encoded run lengths of the symbol's pixels and by the preamble of the next character. Because of the use of nybbles to encode run lengths, the preamble may start in the middle of a byte. There are also a preamble and postamble for the entire file.

The first step of the encoder is to locate groups of repeating (i.e., identical consecutive) pixel rows in the bounding box. When a group of n such rows is located, $n-1$ of them are removed from the bounding box, and a repeat count of $[n-1]$ is suitably encoded somewhere between runs in the remaining row of the group. Figure 5.41 illustrates this process. Part (a) of the figure shows a 5×15 bitmap (note that this is not a bounding box) where the three middle rows repeat. In part (b), two of these rows have been deleted and part (c) shows the valid points between runs where the repeat count [2] may be inserted.

```
000001111000000  000001111000000  000001111000000  000001111000000
011100111101100  011100111101100  011100111101100  000000000000000
011100111101100  000111111000011   ↑  ↑ ↑   ↑↑ ↑   000000000000000
011100111101100                   000111111000011  000000000000000
000111111000011                                    000000000000000
                                                   110000000000000
      (a)              (b)              (c)              (d)
```

Figure 5.41: Repeating Rows in a Bitmap.

Thus, the run lengths of this example become the following sequence 5, 4, 7, [2], 3, 2, 4, 1, 2, 5, 6, 4, and 2. The repeat count was placed at the first valid point, but could have been placed at any of the other five points indicated in the figure. The repeat count for a group of identical rows is placed in the remaining row between runs of pixels, but this rule fails in bitmaps such as the one of Figure 5.41d, where the repeating rows are uniform and are preceded by a run of the same color pixels. In such a case, the encoder does not delete the repeating rows and ends up encoding a long run (66 white pixels in the figure), which reduces the overall compression somewhat.

It is now clear that we have to encode two types of data, run lengths of pixels and repeat counts. Since a typical bounding box is small, we expect most run lengths to be short and most repeat counts to be small. Thus, the principle of PK compression is to pack, whenever possible, two items in the two nybbles of a byte. Often, the first nybble is a flag and the second nybble is a data item. A nybble is four bits long and can have 16 values. We expect to have more run lengths than repeat counts, so we reserve nybble values 14 and 15 to indicate repeat counts. A nybble of 15 indicates a repeat count of 1 (the most common) and a nybble of 14 will be followed by a repeat count stored as a packed number (a term to be discussed shortly).

The remaining nybble values 0 through 13 indicate run lengths. Value 0 indicates a long run, occupying three or more nybbles. The length of the run is stored in as many nybbles as needed, encoded as a packed number. Values 1 through 13 indicate the short and medium run lengths. In order to use these 13 values to maximum effect, the encoder counts the run lengths of the bounding box of the current character, and uses their values, in a fast, one-pass algorithm, to compute a variable *dyn*. This variable is computed for each character of the font and is stored in the character's preamble. Variable *dyn* is used as follows: run lengths 1 through *dyn* are considered short and are stored in one nybble each. The medium run lengths are stored in two nybbles, the first of which, to be denoted by a, is between $dyn+1$ and 13 and the second one, b, can be any 4-bit value. The run length represented by a and b is defined as $16(a-dyn-1)+b+dyn+1$. The shortest run length is therefore $dyn+1$ (for $a=dyn+1$ and $b=0$) and the longest is $16(13-dyn)+dyn$ (for $a=13$ and $b=15$). This convention is best illustrated by examples.

We first select $dyn=4$. In this case, run lengths 1 through 4 are the short ones and are represented by one nybble each, thus 0001, 0010, 0011, and 0100. The medium run lengths are 5 (for $a=4+1$ and $b=0$) through $16(13-4)+4=148$ (for $a=13$ and $b=15$). Runs of more than 148 pixels are considered long and will start with a zero nybble.

We now select $dyn=12$. In this case, run lengths 1 through 12 are short and are represented by one nybble each. The medium run lengths are 13 (for $a=12+1$ and $b=0$) through $16(13-12)+12=28$ (for $a=13$ and $b=15$). Runs of more than 28 pixels are considered long.

It is clear that the value of *dyn* is critical. Small *dyn* values allow for a large range of medium run lengths, while large values of *dyn* should be used in cases where most run lengths are short.

Next, we discuss the format of a packed number. Given an integer i, the idea is to create its hexadecimal representation (let's say it occupies n nybbles), remove any leading zero nybbles and prepend $n-1$ zero nybbles. Thus, i values 1 through 15 occupy one nybble (nothing is prepended). Values $16\le i\le 255$ are two nybbles long, so one zero nybble should be prepended, for a total of three nybbles. Values $256\le i\le 4{,}095$ occupy three nybbles, so two nybbles should be prepended, for a total of five nybbles.

This variable-length format is used to encode repeat counts (indicated by a nybble flag of 14) and the long run lengths (indicated by a nybble flag of 0). However, the long run lengths are always greater than $16(13-dyn)+dyn$. The shortest of the long run lengths is therefore $s \stackrel{\text{def}}{=} 16(13-dyn)+dyn+1$, which is why it makes sense to subtract s from such a run length before it is encoded as a packed number. Recall that the long run lengths are indicated by a nybble flag of 0, and a packed integer in the interval $[16,255]$ is also preceded by a single zero nybble. Thus, it makes sense to add 16 to the long run lengths after s is subtracted.

Example. If $dyn = 4$, the longest medium run length is 148, so $s = 149$. Thus, a run length of 200 is long and is encoded by computing $200 - 149 + 16 = 67 = 43_{16}$ and constructing the 3-nybble packed number 043_{16}.

The algorithm for determining the optimal value of dyn (between 1 and 13) can now be described. This algorithm is executed after the sequence of run lengths has been determined. We start with a naive version of the algorithm. For each value of dyn between 1 and 13, it is easy to compute the ranges of the short, medium, and long runs. Each run in the sequence is examined, its type (short, medium, or long) is easily determined, and its length after being encoded is computed (short runs are one nybble long, medium runs are two nybbles, and long runs occupy three or more nybbles and their lengths take a bit more work to determine). The 13 total lengths for the values of dyn are saved and their minimum is then found.

A better version of this algorithm exploits the fact that as dyn is increased, the range of short run lengths increases while the range of medium run lengths decreases (for $dyn = 4$ this range is $[5,148]$ but for $dyn = 12$ it is only $[13,28]$). Thus, a run of pixels that was medium for a certain value of dyn may remain medium but may also become short or long when dyn is incremented by 1.

This version of the algorithm starts by setting dyn to zero. Thus, initially there are no short runs. The algorithm goes over all the run lengths. It determines the type and computes the encoded length of each run. The lengths are added into variable `total`. The algorithm then increments dyn by 1 and goes over all the runs again. If a run was short in the previous iteration, it will remain short. If it was medium and has now become short, it decreases the value of `total` by 1. If it was medium and has now become long, it increases the value of `total` by 1 (or in rare cases, by 2). After each iteration, the new value of `total` is saved in an array. After 13 iterations, the smallest `total` is located in the array and is used to determine the optimal value of dyn.

The PK compression method described so far is simple, but not very efficient. Given a bitmap with many short runs (such as a gray character, where black and white pixels alternate), this RLE-based method may produce large expansion. In such cases, the run length encoding described here is abandoned and the bitmap is simply written on the font file in raw format (one bit per pixel).

We end with a summarizing example. The middle bitmap of Figure 5.38 is the Greek letter Ξ (Xi). Its size is 20×29 pixels and it has several groups of repeating rows, although only five groups are not uniform and can employ repeat counts. Counting the runs of pixels produces the sequence 82, [2], 16, 2, 42, [2], 2, 12, 2, 4, [3], 16, 4, [2], 2, 12, 2, 62, [2], 2, 16, and 82. The optimal value of dyn turns out to be 8, but in order to make this example useful and have short, medium, and long runs, we assume that dyn is set to 12. Thus, the short runs are 1 to 12 pixels, the medium runs are 13 to 28 pixels, and the long runs are longer than 28 pixels.

The first run is 82; a long run. Subtracting $s = 29$ and adding 16 yields $69 = 45_{16}$. This is encoded as the three nybbles `045`. The repeat count of 2 is encoded as the nybble

$14 = \mathtt{E}_{16}$, followed by the packed number representation of 2, thus E2. The medium run of 16 is encoded in two nybbles ab, but since a is between $dyn + 1$ and 13, its value must be $13 = \mathtt{D}_{16}$, implying that $b = 4$ and the run of 16 is encoded as D4. The next run, 2, is short and is encoded as the single nybble 2. The run 42 is long. We first compute $42 - 29 + 16 = 29 = \mathtt{1D}_{16}$ and then encode 01D. The next two runs [2] and 2 are encoded as E22. So far we have 045E2D4201DE22. The next run, 12, is short and is encoded as the single nybble C. The remaining runs do not provide any new examples and should be encoded by the reader as an exercise.

5.9 Arithmetic Coding

The Huffman method is simple, efficient, and produces the best codes for the individual data symbols. However, Section 5.2 shows that the only case where it produces ideal variable-length codes (codes whose average size equals the entropy) is when the symbols have probabilities of occurrence that are negative powers of 2 (i.e., numbers such as 1/2, 1/4, or 1/8). This is because the Huffman method assigns a code with an integral number of bits to each symbol in the alphabet. Information theory shows that a symbol with probability 0.4 should ideally be assigned a 1.32-bit code, since $-\log_2 0.4 \approx 1.32$. The Huffman method, however, normally assigns such a symbol a code of 1 or 2 bits.

Arithmetic coding overcomes the problem of assigning integer codes to the individual symbols by assigning one (normally long) code to the entire input file. The method starts with a certain interval, it reads the input file symbol by symbol, and it uses the probability of each symbol to narrow the interval. Specifying a narrower interval requires more bits, so the number constructed by the algorithm grows continuously. To achieve compression, the algorithm is designed such that a high-probability symbol narrows the interval less than a low-probability symbol, with the result that high-probability symbols contribute fewer bits to the output.

An interval can be specified by its lower and upper limits or by one limit and width. We use the latter method to illustrate how an interval's specification becomes longer as the interval narrows. The interval $[0, 1]$ can be specified by the two 1-bit numbers 0 and 1. The interval $[0.1, 0.512]$ can be specified by the longer numbers 0.1 and 0.412. The very narrow interval $[0.12575, 0.1257586]$ is specified by the long numbers 0.12575 and 0.0000086.

The output of arithmetic coding is interpreted as a number in the range $[0, 1)$. [The notation $[a, b)$ means the range of real numbers from a to b, including a but not including b. The range is "closed" at a and "open" at b.] Thus the code 9746509 is be interpreted as 0.9746509, although the 0. part is not included in the output file.

Before we plunge into the details, here is a bit of history. The principle of arithmetic coding was first proposed by Peter Elias in the early 1960s and was first described in [Abramson 63]. Early practical implementations of this method were developed by [Rissanen 76], [Pasco 76], and [Rubin 79]. [Moffat et al. 98] and [Witten et al. 87] should especially be mentioned. They discuss both the principles and details of practical arithmetic coding and show examples.

The first step is to calculate, or at least to estimate, the frequencies of occurrence of each symbol. For best results, the exact frequencies are calculated by reading the entire input file in the first pass of a two-pass compression job. However, if the program can get good estimates of the frequencies from a different source, the first pass may be omitted.

The first example involves the three symbols a_1, a_2, and a_3, with probabilities $P_1 = 0.4$, $P_2 = 0.5$, and $P_3 = 0.1$, respectively. The interval $[0, 1)$ is divided among the three symbols by assigning each a subinterval proportional in size to its probability. The order of the subintervals is immaterial. In our example, the three symbols are assigned the subintervals $[0, 0.4)$, $[0.4, 0.9)$, and $[0.9, 1.0)$. To encode the string "$a_2a_2a_2a_3$", we start with the interval $[0, 1)$. The first symbol a_2 reduces this interval to the subinterval from its 40% point to its 90% point. The result is $[0.4, 0.9)$. The second a_2 reduces $[0.4, 0.9)$ in the same way (see note below) to $[0.6, 0.85)$, the third a_2 reduces this to $[0.7, 0.825)$, and the a_3 reduces this to the stretch from the 90% point of $[0.7, 0.825)$ to its 100% point, producing $[0.8125, 0.8250)$. The final code our method produces can be any number in this final range.

(Note: The subinterval $[0.6, 0.85)$ is obtained from the interval $[0.4, 0.9)$ by $0.4 + (0.9 - 0.4) \times 0.4 = 0.6$ and $0.4 + (0.9 - 0.4) \times 0.9 = 0.85$.)

With this example in mind, it should be easy to understand the following rules, which summarize the main steps of arithmetic coding:

1. Start by defining the "current interval" as $[0, 1)$.
2. Repeat the following two steps for each symbol s in the input stream:

2.1. Divide the current interval into subintervals whose sizes are proportional to the symbols' probabilities.

2.2. Select the subinterval for s and define it as the new current interval.

3. When the entire input stream has been processed in this way, the output should be any number that uniquely identifies the current interval (i.e., any number inside the current interval).

For each symbol processed, the current interval gets smaller, so it takes more bits to express it, but the point is that the final output is a single number and does not consist of codes for the individual symbols. The average code size can be obtained by dividing the size of the output (in bits) by the size of the input (in symbols). Notice also that the probabilities used in step 2.1 may change all the time, since they may be supplied by an adaptive probability model (Section 5.10).

> A theory has only the alternative of being right or wrong. A model has a third possibility: it may be right, but irrelevant.
>
> —Manfred Eigen, *The Physicist's Conception of Nature*

The next example is a little more involved. We show the compression steps for the short string `SWISS␣MISS`. Table 5.42 shows the information prepared in the first step (the *statistical model* of the data). The five symbols appearing in the input may be arranged in any order. For each symbol, its frequency is first counted, followed by its probability of occurrence (the frequency divided by the string size, 10). The range $[0, 1)$ is then divided among the symbols, in any order, with each symbol getting a chunk, or a subrange, equal in size to its probability. Thus `S` gets the subrange $[0.5, 1.0)$ (of size 0.5), whereas the subrange of I is of size 0.2 [0.2, 0.4). The cumulative frequencies column is used by the decoding algorithm on page 271.

Char	Freq	Prob.	Range	CumFreq
		Total CumFreq=		10
S	5	$5/10 = 0.5$	[0.5, 1.0)	5
W	1	$1/10 = 0.1$	[0.4, 0.5)	4
I	2	$2/10 = 0.2$	[0.2, 0.4)	2
M	1	$1/10 = 0.1$	[0.1, 0.2)	1
␣	1	$1/10 = 0.1$	[0.0, 0.1)	0

Table 5.42: Frequencies and Probabilities of Five Symbols.

The symbols and frequencies in Table 5.42 are written on the output stream before any of the bits of the compressed code. This table will be the first thing input by the decoder.

The encoding process starts by defining two variables, `Low` and `High`, and setting them to 0 and 1, respectively. They define an interval [`Low`, `High`). As symbols are being input and processed, the values of `Low` and `High` are moved closer together, to narrow the interval.

After processing the first symbol `S`, `Low` and `High` are updated to 0.5 and 1, respectively. The resulting code for the entire input stream will be a number in this range ($0.5 \leq \text{Code} < 1.0$). The rest of the input stream will determine precisely where, in the interval $[0.5, 1)$, the final code will lie. A good way to understand the process is to imagine that the new interval $[0.5, 1)$ is divided among the five symbols of our alphabet using the same proportions as for the original interval $[0, 1)$. The result is the five subintervals $[0.5, 0.55)$, $[0.55, 0.60)$, $[0.60, 0.70)$, $[0.70, 0.75)$, and $[0.75, 1.0)$. When the next symbol `W` is input, the third of those subintervals is selected and is again divided into five subsubintervals.

As more symbols are being input and processed, `Low` and `High` are being updated according to

```
NewHigh:=OldLow+Range*HighRange(X);
NewLow:=OldLow+Range*LowRange(X);
```

where `Range=OldHigh−OldLow` and `LowRange(X)`, `HighRange(X)` indicate the low and high limits of the range of symbol `X`, respectively. In the example above, the second input symbol is `W`, so we update $\text{Low} := 0.5 + (1.0 - 0.5) \times 0.4 = 0.70$, $\text{High} := 0.5 + (1.0-0.5) \times 0.5 = 0.75$. The new interval $[0.70, 0.75)$ covers the stretch $[40\%, 50\%)$ of the subrange of `S`. Table 5.43 shows all the steps involved in coding the string `SWISS␣MISS` (the first three steps are illustrated graphically in Figure 5.56a). The final code is the final value of `Low`, 0.71753375, of which only the eight digits `71753375` need be written on the output stream (but see later for a modification of this statement).

The decoder works in reverse. It starts by inputting the symbols and their ranges, and reconstructing Table 5.42. It then inputs the rest of the code. The first digit is `7`, so the decoder immediately knows that the entire code is a number of the form 0.7....

Char.		The calculation of low and high
S	L	$0.0 + (1.0 - 0.0) \times 0.5 = 0.5$
	H	$0.0 + (1.0 - 0.0) \times 1.0 = 1.0$
W	L	$0.5 + (1.0 - 0.5) \times 0.4 = 0.70$
	H	$0.5 + (1.0 - 0.5) \times 0.5 = 0.75$
I	L	$0.7 + (0.75 - 0.70) \times 0.2 = 0.71$
	H	$0.7 + (0.75 - 0.70) \times 0.4 = 0.72$
S	L	$0.71 + (0.72 - 0.71) \times 0.5 = 0.715$
	H	$0.71 + (0.72 - 0.71) \times 1.0 = 0.72$
S	L	$0.715 + (0.72 - 0.715) \times 0.5 = 0.7175$
	H	$0.715 + (0.72 - 0.715) \times 1.0 = 0.72$
␣	L	$0.7175 + (0.72 - 0.7175) \times 0.0 = 0.7175$
	H	$0.7175 + (0.72 - 0.7175) \times 0.1 = 0.71775$
M	L	$0.7175 + (0.71775 - 0.7175) \times 0.1 = 0.717525$
	H	$0.7175 + (0.71775 - 0.7175) \times 0.2 = 0.717550$
I	L	$0.717525 + (0.71755 - 0.717525) \times 0.2 = 0.717530$
	H	$0.717525 + (0.71755 - 0.717525) \times 0.4 = 0.717535$
S	L	$0.717530 + (0.717535 - 0.717530) \times 0.5 = 0.7175325$
	H	$0.717530 + (0.717535 - 0.717530) \times 1.0 = 0.717535$
S	L	$0.7175325 + (0.717535 - 0.7175325) \times 0.5 = 0.71753375$
	H	$0.7175325 + (0.717535 - 0.7175325) \times 1.0 = 0.717535$

Table 5.43: The Process of Arithmetic Encoding.

This number is inside the subrange $[0.5, 1)$ of `S`, so the first symbol is `S`. The decoder then eliminates the effect of symbol `S` from the code by subtracting the lower limit 0.5 of `S` and dividing by the width of the subrange of `S` (0.5). The result is 0.4350675, which tells the decoder that the next symbol is `W` (since the subrange of `W` is $[0.4, 0.5)$).

To eliminate the effect of symbol `X` from the code, the decoder performs the operation `Code:=(Code-LowRange(X))/Range`, where `Range` is the width of the subrange of `X`. Table 5.45 summarizes the steps for decoding our example string (notice that it has two rows per symbol).

The next example is of three symbols with probabilities as shown in Table 5.46a. Notice that the probabilities are very different. One is large (97.5%) and the others much smaller. This is a case of *skewed probabilities*.

Encoding the string $a_2a_2a_1a_3a_3$ produces the strange numbers (accurate to 16 digits) in Table 5.47, where the two rows for each symbol correspond to the `Low` and `High` values, respectively. Figure 5.44 lists the *Mathematica* code that computed the table.

At first glance, it seems that the resulting code is longer than the original string, but Section 5.9.3 shows how to figure out the true compression achieved by arithmetic coding.

Decoding this string is shown in Table 5.48 and involves a special problem. After eliminating the effect of a_1, on line 3, the result is 0. Earlier, we implicitly assumed that this means the end of the decoding process, but now we know that there are two more occurrences of a_3 that should be decoded. These are shown on lines 4, 5 of the table. This problem always occurs when the last symbol in the input stream is the one whose subrange starts at zero. In order to distinguish between such a symbol and the end of the input stream, we need to define an additional symbol, the end-of-input (or end-of-file, eof). This symbol should be added, with a small probability, to the frequency table (see Table 5.46b), and it should be encoded at the end of the input stream.

```
lowRange={0.998162,0.023162,0.};
highRange={1.,0.998162,0.023162};
low=0.; high=1.;
enc[i_]:=Module[{nlow,nhigh,range},
range=high-low;
nhigh=low+range highRange[[i]];
nlow=low+range lowRange[[i]];
low=nlow; high=nhigh;
Print["r=",N[range,25]," l=",N[low,17]," h=",N[high,17]]]
enc[2]
enc[2]
enc[1]
enc[3]
enc[3]
```

Figure 5.44: *Mathematica* Code for Table 5.47.

Tables 5.49 and 5.50 show how the string $a_3a_3a_3a_3$eof is encoded into the number 0.0000002878086184764172, and then decoded properly. Without the eof symbol, a string of all a_3s would have been encoded into a 0.

Notice how the low value is 0 until the eof is input and processed, and how the high value quickly approaches 0. Now is the time to mention that the final code does not have to be the final low value but can be any number between the final low and high values. In the example of $a_3a_3a_3a_3$eof, the final code can be the much shorter number 0.0000002878086 (or 0.0000002878087 or even 0.0000002878088).

◇ **Exercise 5.24:** Encode the string $a_2a_2a_2a_2$ and summarize the results in a table similar to Table 5.49. How do the results differ from those of the string $a_3a_3a_3a_3$?

If the size of the input stream is known, it is possible to do without an eof symbol. The encoder can start by writing this size (unencoded) on the output stream. The decoder reads the size, starts decoding, and stops when the decoded stream reaches this size. If the decoder reads the compressed stream byte by byte, the encoder may have to add some zeros at the end, to make sure the compressed stream can be read in groups of 8 bits.

Char.	Code−low	Range
S	$0.71753375 - 0.5 = 0.21753375$	$/0.5 = 0.4350675$
W	$0.4350675 - 0.4 = 0.0350675$	$/0.1 = 0.350675$
I	$0.350675 - 0.2 = 0.150675$	$/0.2 = 0.753375$
S	$0.753375 - 0.5 = 0.253375$	$/0.5 = 0.50675$
S	$0.50675 - 0.5 = 0.00675$	$/0.5 = 0.0135$
␣	$0.0135 - 0 = 0.0135$	$/0.1 = 0.135$
M	$0.135 - 0.1 = 0.035$	$/0.1 = 0.35$
I	$0.35 - 0.2 = 0.15$	$/0.2 = 0.75$
S	$0.75 - 0.5 = 0.25$	$/0.5 = 0.5$
S	$0.5 - 0.5 = 0$	$/0.5 = 0$

Table 5.45: The Process of Arithmetic Decoding.

Char	Prob.	Range
a_1	0.001838	[0.998162, 1.0)
a_2	0.975	[0.023162, 0.998162)
a_3	0.023162	[0.0, 0.023162)

(a)

Char	Prob.	Range
eof	0.000001	[0.999999, 1.0)
a_1	0.001837	[0.998162, 0.999999)
a_2	0.975	[0.023162, 0.998162)
a_3	0.023162	[0.0, 0.023162)

(b)

Table 5.46: (Skewed) Probabilities of Three Symbols.

a_2	$0.0 + (1.0 - 0.0) \times 0.023162 = 0.023162$
	$0.0 + (1.0 - 0.0) \times 0.998162 = 0.998162$
a_2	$0.023162 + .975 \times 0.023162 = 0.04574495$
	$0.023162 + .975 \times 0.998162 = 0.99636995$
a_1	$0.04574495 + 0.950625 \times 0.998162 = 0.99462270125$
	$0.04574495 + 0.950625 \times 1.0 = 0.99636995$
a_3	$0.99462270125 + 0.00174724875 \times 0.0 = 0.99462270125$
	$0.99462270125 + 0.00174724875 \times 0.023162 = 0.994663171025547$
a_3	$0.99462270125 + 0.00004046977554749998 \times 0.0 = 0.99462270125$
	$0.99462270125 + 0.00004046977554749998 \times 0.023162 = 0.994623638610941$

Table 5.47: Encoding the String $a_2a_2a_1a_3a_3$.

Char.	Code−low		Range	
a_2	0.99462270125 − 0.023162	= 0.97146170125	/0.975	= 0.99636995
a_2	0.99636995 − 0.023162	= 0.97320795	/0.975	= 0.998162
a_1	0.998162 − 0.998162	= 0.0	/0.00138	= 0.0
a_3	0.0 − 0.0	= 0.0	/0.023162	= 0.0
a_3	0.0 − 0.0	= 0.0	/0.023162	= 0.0

Table 5.48: Decoding the String $a_2a_2a_1a_3a_3$.

a_3	$0.0+(1.0-0.0)\times 0.0=0.0$
	$0.0+(1.0-0.0)\times 0.023162=0.023162$
a_3	$0.0+.023162\times 0.0=0.0$
	$0.0+.023162\times 0.023162=0.000536478244$
a_3	$0.0+0.000536478244\times 0.0=0.0$
	$0.0+0.000536478244\times 0.023162=0.000012425909087528$
a_3	$0.0+0.000012425909087528\times 0.0=0.0$
	$0.0+0.000012425909087528\times 0.023162=0.000000287808906285323$5
eof	$0.0+0.0000002878089062853235\times 0.999999=0.0000002878086184764172$
	$0.0+0.0000002878089062853235\times 1.0=0.0000002878089062853235$

Table 5.49: Encoding the String $a_3a_3a_3a_3$eof.

Char.	Code−low		Range
a_3	0.0000002878086184764172-0	=0.0000002878086184764172	/0.023162=0.00001242589666161891247
a_3	0.00001242589666161891247-0	=0.00001242589666161891247	/0.023162=0.000536477707521756
a_3	0.000536477707521756-0	=0.000536477707521756	/0.023162=0.023161976838
a_3	0.023161976838-0.0	=0.023161976838	/0.023162=0.999999
eof	0.999999-0.999999	=0.0	/0.000001=0.0

Table 5.50: Decoding the String $a_3a_3a_3a_3$eof.

5.9.1 Implementation Details

The encoding process described earlier is not practical, since it assumes that numbers of unlimited precision can be stored in `Low` and `High`. The decoding process described on page 267 ("The decoder then eliminates the effect of the `S` from the code by subtracting... and dividing ...") is simple in principle but also impractical. The code, which is a single number, is normally long and may also be very long. A 1 Mbyte file may be encoded into, say, a 500 Kbyte file that consists of a single number. Dividing a 500 Kbyte number is complex and slow.

Any practical implementation of arithmetic coding should use just integers (because floating-point arithmetic is slow and precision is lost), and they should not be very long (preferably just single precision). We describe such an implementation here, using two integer variables Low and High. In our example they are four decimal digits long, but in practice they might be 16 or 32 bits long. These variables hold the low and high limits of the current subinterval, but we don't let them grow too much. A glance at Table 5.43 shows that once the leftmost digits of Low and High become identical, they never change. We therefore shift such digits out of the two variables and write one digit on the output stream. This way, the two variables don't have to hold the entire code, just the most-recent part of it. As digits are shifted out of the two variables, a zero is shifted into the right end of Low and a 9 into the right end of High. A good way to understand this is to think of each of the two variables as the left end of an infinitely long number. Low contains $xxxx00\ldots$, and High= $yyyy99\ldots$.

One problem is that High should be initialized to 1, but the contents of Low and High should be interpreted as fractions less than 1. The solution is to initialize High to 9999..., since the infinite fraction 0.999... equals 1.

(This is easy to prove. If $0.999\ldots < 1$, then their average $a = (1+0.999\ldots)/2$ would be a number between 0.999... and 1, but there is no way to write a. It is impossible to give it more digits than to 0.999..., since the latter already has an infinite number of digits. It is impossible to make the digits any bigger, since they are already 9's. This is why the infinite fraction 0.999... must equal 1.)

◇ **Exercise 5.25:** Write the number 0.5 in binary.

Table 5.51 describes the encoding process of the string SWISS␣MISS. Column 1 shows the next input symbol. Column 2 shows the new values of Low and High. Column 3 shows these values as scaled integers, after High has been decremented by 1. Column 4 shows the next digit sent to the output stream. Column 5 shows the new values of Low and High after being shifted to the left. Notice how the last step sends the four digits 3750 to the output stream. The final output is 717533750.

Decoding is the opposite of encoding. We start with Low=0000, High=9999, and Code=7175 (the first four digits of the compressed stream). These are updated at each step of the decoding loop. Low and High approach each other (and both approach Code) until their most significant digits are the same. They are then shifted to the left, which separates them again, and Code is also shifted at that time. An index is calculated at each step and is used to search the cumulative frequencies column of Table 5.42 to figure out the current symbol.

Each iteration of the loop consists of the following steps:

1. Calculate index:=((Code-Low+1)x10-1)/(High-Low+1) and truncate it to the nearest integer. (The number 10 is the total cumulative frequency in our example.)
2. Use index to find the next symbol by comparing it to the cumulative frequencies column in Table 5.42. In the example below, the first value of index is 7.1759, truncated to 7. Seven is between the 5 and the 10 in the table, so it selects the S.
3. Update Low and High according to

```
Low:=Low+(High-Low+1)LowCumFreq[X]/10;
High:=Low+(High-Low+1)HighCumFreq[X]/10-1;
```

where LowCumFreq[X] and HighCumFreq[X] are the cumulative frequencies of symbol X and of the symbol above it in Table 5.42.

1	2	3	4	5
S	$L = 0+(1-0)\times 0.5 = 0.5$	5000		5000
	$H = 0+(1-0)\times 1.0 = 1.0$	9999		9999
W	$L = 0.5+(1-.5)\times 0.4 = 0.7$	7000	7	0000
	$H = 0.5+(1-.5)\times 0.5 = 0.75$	7499	7	4999
I	$L = 0+(0.5-0)\times 0.2 = 0.1$	1000	1	0000
	$H = 0+(0.5-0)\times 0.4 = 0.2$	1999	1	9999
S	$L = 0+(1-0)\times 0.5 = 0.5$	5000		5000
	$H = 0+(1-0)\times 1.0 = 1.0$	9999		9999
S	$L = 0.5+(1-0.5)\times 0.5 = 0.75$	7500		7500
	$H = 0.5+(1-0.5)\times 1.0 = 1.0$	9999		9999
␣	$L = 0.75+(1-0.75)\times 0.0 = 0.75$	7500	7	5000
	$H = 0.75+(1-0.75)\times 0.1 = 0.775$	7749	7	7499
M	$L = 0.5+(0.75-0.5)\times 0.1 = 0.525$	5250	5	2500
	$H = 0.5+(0.75-0.5)\times 0.2 = 0.55$	5499	5	4999
I	$L = 0.25+(0.5-0.25)\times 0.2 = 0.3$	3000	3	0000
	$H = 0.25+(0.5-0.25)\times 0.4 = 0.35$	3499	3	4999
S	$L = 0+(0.5-0)\times 0.5 = .25$	2500		2500
	$H = 0+(0.5-0)\times 1.0 = 0.5$	4999		4999
S	$L = 0.25+(0.5-0.25)\times 0.5 = 0.375$	3750		3750
	$H = 0.25+(0.5-0.25)\times 1.0 = 0.5$	4999		4999

Table 5.51: Encoding SWISS␣MISS by Shifting.

4. If the leftmost digits of `Low` and `High` are identical, shift `Low`, `High`, and `Code` one position to the left. `Low` gets a 0 entered on the right, `High` gets a 9, and `Code` gets the next input digit from the compressed stream.

Here are all the decoding steps for our example:

0. Initialize `Low`=0000, `High`=9999, and `Code`=7175.

1. `index`= $[(7175-0+1)\times 10-1]/(9999-0+1) = 7.1759 \to 7$. Symbol `S` is selected. `Low` $= 0+(9999-0+1)\times 5/10 = 5000$. `High` $= 0+(9999-0+1)\times 10/10-1 = 9999$.

2. `index`= $[(7175-5000+1)\times 10-1]/(9999-5000+1) = 4.3518 \to 4$. Symbol `W` is selected.
`Low` $= 5000+(9999-5000+1)\times 4/10 = 7000$. `High` $= 5000+(9999-5000+1)\times 5/10-1 = 7499$.
After the 7 is shifted out, we have `Low`=0000, `High`=4999, and `Code`=1753.

3. `index`= $[(1753-0+1)\times 10-1]/(4999-0+1) = 3.5078 \to 3$. Symbol `I` is selected. `Low` $= 0+(4999-0+1)\times 2/10 = 1000$. `High` $= 0+(4999-0+1)\times 4/10-1 = 1999$. After the 1 is shifted out, we have `Low`=0000, `High`=9999, and `Code`=7533.

4. `index`= $[(7533-0+1)\times 10-1]/(9999-0+1) = 7.5339 \to 7$. Symbol `S` is selected. `Low` $= 0+(9999-0+1)\times 5/10 = 5000$. `High` $= 0+(9999-0+1)\times 10/10-1 = 9999$.

5. index= $[(7533-5000+1)\times 10-1]/(9999-5000+1)=5.0678\rightarrow 5$. Symbol S is selected.
Low $=5000+(9999-5000+1)\times 5/10=7500$. High $=5000+(9999-5000+1)\times 10/10-1=9999$.

6. index= $[(7533-7500+1)\times 10-1]/(9999-7500+1)=0.1356\rightarrow 0$. Symbol $\sqcup$ is selected.
Low $=7500+(9999-7500+1)\times 0/10=7500$. High $=7500+(9999-7500+1)\times 1/10-1=7749$.
After the 7 is shifted out, we have Low=5000, High=7499, and Code=5337.

7. index= $[(5337-5000+1)\times 10-1]/(7499-5000+1)=1.3516\rightarrow 1$. Symbol M is selected.
Low $=5000+(7499-5000+1)\times 1/10=5250$. High $=5000+(7499-5000+1)\times 2/10-1=5499$.
After the 5 is shifted out we have Low=2500, High=4999, and Code=3375.

8. index= $[(3375-2500+1)\times 10-1]/(4999-2500+1)=3.5036\rightarrow 3$. Symbol I is selected.
Low $=2500+(4999-2500+1)\times 2/10=3000$. High $=2500+(4999-2500+1)\times 4/10-1=3499$.
After the 3 is shifted out we have Low=0000, High=4999, and Code=3750.

9. index= $[(3750-0+1)\times 10-1]/(4999-0+1)=7.5018\rightarrow 7$. Symbol S is selected.
Low $=0+(4999-0+1)\times 5/10=2500$. High $=0+(4999-0+1)\times 10/10-1=4999$.

10. index= $[(3750-2500+1)\times 10-1]/(4999-2500+1)=5.0036\rightarrow 5$. Symbol S is selected.
Low $=2500+(4999-2500+1)\times 5/10=3750$. High $=2500+(4999-2500+1)\times 10/10-1=4999$.

◇ **Exercise 5.26:** How does the decoder know to stop the loop at this point?

1	2	3	4	5
1	L=0+(1 − 0)×0.0 = 0.0	000000	0	000000
	H=0+(1 − 0)×0.023162= 0.023162	023162	0	231629
2	L=0+(0.231629 − 0)×0.0 = 0.0	000000	0	000000
	H=0+(0.231629 − 0)×0.023162= 0.00536478244	005364	0	053649
3	L=0+(0.053649 − 0)×0.0 = 0.0	000000	0	000000
	H=0+(0.053649 − 0)×0.023162= 0.00124261813	001242	0	012429
4	L=0+(0.012429 − 0)×0.0 = 0.0	000000	0	000000
	H=0+(0.012429 − 0)×0.023162= 0.00028788049	000287	0	002879
5	L=0+(0.002879 − 0)×0.0 = 0.0	000000	0	000000
	H=0+(0.002879 − 0)×0.023162= 0.00006668339	000066	0	000669

Table 5.52: Encoding $a_3a_3a_3a_3a_3$ by Shifting.

5.9.2 Underflow

Table 5.52 shows the steps in encoding the string $a_3a_3a_3a_3a_3$ by shifting. This table is similar to Table 5.51, and it illustrates the problem of underflow. `Low` and `High` approach each other, and since `Low` is always 0 in this example, `High` loses its significant digits as it approaches `Low`.

Underflow may happen not just in this case but in any case where `Low` and `High` need to converge very closely. Because of the finite size of the `Low` and `High` variables, they may reach values of, say, 499996 and 500003, and from there, instead of reaching values where their most significant digits are identical, they reach the values 499999 and 500000. Since the most significant digits are different, the algorithm will not output anything, there will not be any shifts, and the next iteration will only add digits beyond the first six ones. Those digits will be lost, and the first six digits will not change. The algorithm will iterate without generating any output until it reaches the eof.

The solution to this problem is to detect such a case early and *rescale* both variables. In the example above, rescaling should be done when the two variables reach values of 49xxxx and 50yyyy. Rescaling should squeeze out the second most significant digits, end up with 4xxxx0 and 5yyyy9, and increment a counter `cntr`. The algorithm may have to rescale several times before the most-significant digits become equal. At that point, the most-significant digit (which can be either 4 or 5) should be output, followed by `cntr` zeros (if the two variables converged to 4) or nines (if they converged to 5).

5.9.3 Final Remarks

All the examples so far have been in decimal, since the computations involved are easier to understand in this number base. It turns out that all the algorithms and rules described above apply to the binary case as well and can be used with only one change: Every occurrence of 9 (the largest decimal digit) should be replaced by 1 (the largest binary digit).

The examples above don't seem to show any compression at all. It seems that the three example strings `SWISS␣MISS`, $a_2a_2a_1a_3a_3$, and $a_3a_3a_3a_3$eof are encoded into very long numbers. In fact, it seems that the length of the final code depends on the probabilities involved. The long probabilities of Table 5.46a generate long numbers in the encoding process, whereas the shorter probabilities of Table 5.42 result in the more reasonable `Low` and `High` values of Table 5.43. This behavior demands an explanation.

> I am ashamed to tell you to how many figures I carried these computations, having no other business at that time.
>
> —Isaac Newton

To figure out the kind of compression achieved by arithmetic coding, we have to consider two facts: (1) In practice, all the operations are performed on binary numbers, so we have to translate the final results to binary before we can estimate the efficiency of the compression; (2) since the last symbol encoded is the eof, the final code does not have to be the final value of `Low`; it can be any value between `Low` and `High`. This makes it possible to select a shorter number as the final code that's being output.

Table 5.43 encodes the string `SWISS␣MISS` into the final Low and High values 0.71753375 and 0.717535. The approximate binary values of these numbers are

0.10110111101100000100101010111 and 0.10110111101100000101111111011, so we can select the number 10110111101100000100 as our final, compressed output. The ten-symbol string has thus been encoded into a 20-bit number. Does this represent good compression?

The answer is yes. Using the probabilities of Table 5.42, it is easy to calculate the probability of the string `SWISS␣MISS`. It is $P = 0.5^5 \times 0.1 \times 0.2^2 \times 0.1 \times 0.1 = 1.25 \times 10^{-6}$. The entropy of this string is therefore $-\log_2 P = 19.6096$. Twenty bits are therefore the minimum needed in practice to encode the string.

The symbols in Table 5.46a have probabilities 0.975, 0.001838, and 0.023162. These numbers require quite a few decimal digits, and as a result, the final Low and High values in Table 5.47 are the numbers 0.99462270125 and 0.994623638610941. Again it seems that there is no compression, but an analysis similar to the above shows compression that's very close to the entropy.

The probability of the string $a_2a_2a_1a_3a_3$ is $0.975^2 \times 0.001838 \times 0.023162^2 \approx 9.37361 \times 10^{-7}$, and $-\log_2 9.37361 \times 10^{-7} \approx 20.0249$.

The binary representations of the final values of Low and High in Table 5.47 are 0.111111101001111110010111111001 and 0.1111111010011111100100111101. We can select any number between these two, so we select 1111111010011111100, a 19-bit number. (This should have been a 21-bit number, but the numbers in Table 5.47 have limited precision and are not exact.)

◇ **Exercise 5.27:** Given the three symbols a_1, a_2, and eof, with probabilities $P_1 = 0.4$, $P_2 = 0.5$, and $P_{\text{eof}} = 0.1$, encode the string $a_2a_2a_2\text{eof}$ and show that the size of the final code equals the (practical) minimum.

The following argument shows why arithmetic coding can, in principle, be a very efficient compression method. We denote by s a sequence of symbols to be encoded, and by b the number of bits required to encode it. As s gets longer, its probability $P(s)$ gets smaller and b gets larger. Since the logarithm is the information function, it is easy to see that b should grow at the same rate that $\log_2 P(s)$ shrinks. Their product should therefore be constant, or close to a constant. Information theory shows that b and $P(s)$ satisfy the double inequality

$$2 \le 2^b P(s) < 4,$$

which implies

$$1 - \log_2 P(s) \le b < 2 - \log_2 P(s). \qquad (5.1)$$

As s gets longer, its probability $P(s)$ shrinks, the quantity $-\log_2 P(s)$ becomes a large positive number, and the double inequality of Equation (5.1) shows that in the limit, b approaches $-\log_2 P(s)$. This is why arithmetic coding can, in principle, compress a string of symbols to its theoretical limit.

For more information on this topic, see [Moffat et al. 98] and [Witten et al. 87].

5.10 Adaptive Arithmetic Coding

Two features of arithmetic coding make it easy to extend:

1. One of the main encoding steps (page 266) updates NewLow and NewHigh. Similarly, one of the main decoding steps (step 3 on page 271) updates Low and High according to

```
Low:=Low+(High-Low+1)LowCumFreq[X]/10;
High:=Low+(High-Low+1)HighCumFreq[X]/10-1;
```

This means that in order to encode symbol X, the encoder should be given the cumulative frequencies of the symbol and of the one above it (see Table 5.42 for an example of cumulative frequencies). This also implies that the frequency of X (or, equivalently, its probability) could be changed each time it is encoded, provided that the encoder and the decoder agree on how to do this.

2. The order of the symbols in Table 5.42 is unimportant. They can even be swapped in the table during the encoding process as long as the encoder and decoder do it in the same way.

With this in mind, it is easy to understand how adaptive arithmetic coding works. The encoding algorithm has two parts: the probability model and the arithmetic encoder. The model reads the next symbol from the input stream and invokes the encoder, sending it the symbol and the two required cumulative frequencies. The model then increments the count of the symbol and updates the cumulative frequencies. The point is that the symbol's probability is determined by the model from its *old* count, and the count is incremented only after the symbol has been encoded. This makes it possible for the decoder to mirror the encoder's operations. The encoder knows what the symbol is even before it is encoded, but the decoder has to decode the symbol in order to find out what it is. The decoder can therefore use only the old counts when decoding a symbol. Once the symbol has been decoded, the decoder increments its count and updates the cumulative frequencies in exactly the same way as the encoder.

The model should keep the symbols, their counts (frequencies of occurrence), and their cumulative frequencies in an array. This array should be kept in sorted order of the counts. Each time a symbol is read and its count is incremented, the model updates the cumulative frequencies, then checks to see whether it is necessary to swap the symbol with another one, to keep the counts in sorted order.

It turns out that there is a simple data structure that allows for both easy search and update. This structure is a balanced binary tree housed in an array. (A balanced binary tree is a complete binary tree where some of the bottom-right nodes may be missing.) The tree should have a node for every symbol in the alphabet, and since it is balanced, its height is $\lceil \log_2 n \rceil$, where n is the size of the alphabet. For $n = 256$ the height of the balanced binary tree is 8, so starting at the root and searching for a node takes at most eight steps. The tree is arranged such that the most probable symbols (the ones with high counts) are located near the root, which speeds up searches. Table 5.53a shows an example of a ten-symbol alphabet with counts. Table 5.53b shows the same symbols sorted by count.

The sorted array "houses" the balanced binary tree of Figure 5.55a. This is a simple, elegant way to build a tree. A balanced binary tree can be housed in an array without the use of any pointers. The rule is that the first array location (with index 1) houses

a_1	a_2	a_3	a_4	a_5	a_6	a_7	a_8	a_9	a_{10}
11	12	12	2	5	1	2	19	12	8

(a)

a_8	a_2	a_3	a_9	a_1	a_{10}	a_5	a_4	a_7	a_6
19	12	12	12	11	8	5	2	2	1

(b)

Table 5.53: A Ten-Symbol Alphabet With Counts.

the root, the two children of the node at array location i are housed at locations $2i$ and $2i+1$, and the parent of the node at array location j is housed at location $\lfloor j/2 \rfloor$. It is easy to see how sorting the array has placed the symbols with largest counts at and near the root.

In addition to a symbol and its count, another value is now added to each tree node, the total counts of its left subtree. This will be used to compute cumulative frequencies. The corresponding array is shown in Table 5.54a.

Assume that the next symbol read from the input stream is a_9. Its count is incremented from 12 to 13. The model keeps the array in sorted order by searching for the farthest array element left of a_9 that has a count smaller than that of a_9. This search can be a straight linear search if the array is short enough, or a binary search if the array is long. In our case, symbols a_9 and a_2 should be swapped (Table 5.54b). Figure 5.55b shows the tree after the swap. Notice how the left-subtree counts have been updated.

a_8	a_2	a_3	a_9	a_1	a_{10}	a_5	a_4	a_7	a_6
19	12	12	12	11	8	5	2	2	1
40	16	8	2	1	0	0	0	0	0

(a)

a_8	a_9	a_3	a_2	a_1	a_{10}	a_5	a_4	a_7	a_6
19	13	12	12	11	8	5	2	2	1
41	16	8	2	1	0	0	0	0	0

(b)

Tables 5.54: A Ten-Symbol Alphabet With Counts.

Finally, here is how the cumulative frequencies are computed from this tree. When the cumulative frequency for a symbol `X` is needed, the model follows the tree branches from the root to the node containing `X` while adding numbers into an integer `af`. Each time a right branch is taken from an interior node `N`, `af` is incremented by the two numbers (the count and the left-subtree count) found in that node. When a left branch is taken, `af` is not modified. When the node containing `X` is reached, the left-subtree count of `X` is added to `af`, and `af` then contains the quantity `LowCumFreq[X]`.

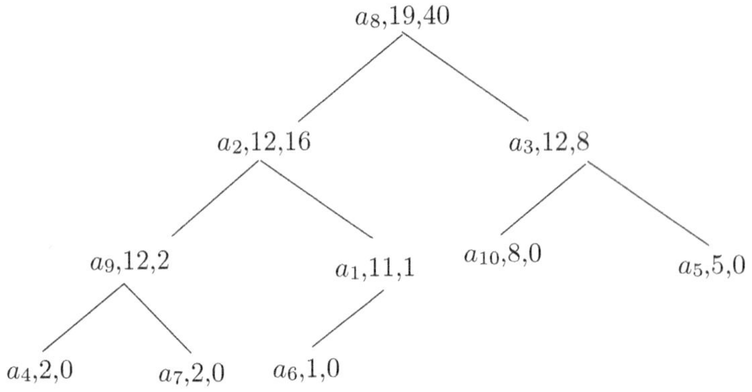

(a)

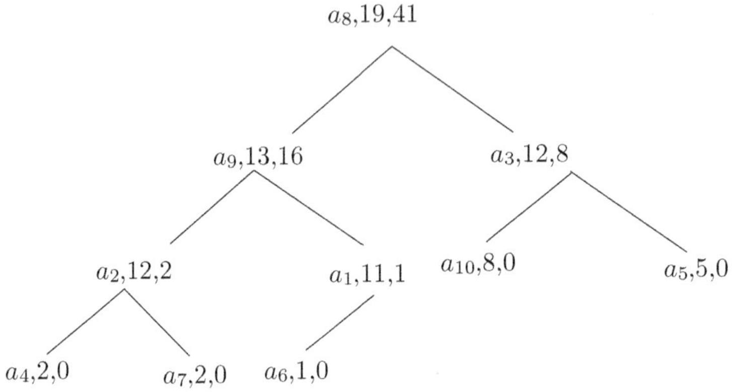

(b)

a_4	2	0—1
a_9	12	2—13
a_7	2	14—15
a_2	12	16—27
a_6	1	28—28
a_1	11	29—39
a_8	19	40—58
a_{10}	8	59—66
a_3	12	67—78
a_5	5	79—83

(c)

Figure 5.55: Adaptive Arithmetic Coding.

As an example, we trace the tree of Figure 5.55a from the root to symbol a_6, whose cumulative frequency is 28. A right branch is taken at node a_2, adding 12 and 16 to `af`. A left branch is taken at node a_1, adding nothing to `af`. When reaching a_6, its left-subtree count, 0, is added to `af`. The result in `af` is $12 + 16 = 28$, as can be verified from Figure 5.55c. The quantity `HighCumFreq[X]` is obtained by adding the count of a_6 (which is 1) to `LowCumFreq[X]`.

To trace the tree and find the path from the root to a_6, the algorithm performs the following steps:

1. Find a_6 in the array housing the tree by means of a binary search. In our example the node with a_6 is found at array location 10.
2. Integer-divide 10 by 2. The remainder is 0, which means that a_6 is the left child of its parent. The quotient is 5, which is the array location of the parent.
3. Location 5 of the array contains a_1. Integer-divide 5 by 2. The remainder is 1, which means that a_1 is the right child of its parent. The quotient is 2, which is the array location of a_1's parent.
4. Location 2 of the array contains a_2. Integer-divide 2 by 2. The remainder is 0, which means that a_2 is the left child of its parent. The quotient is 1, the array location of the root, so the process stops.

The PPM compression method, Section 5.14, is a good example of a statistical model that invokes an arithmetic encoder in the way described here.

> The driver held out a letter. Boldwood seized it and opened it, expecting another anonymous one—so greatly are people's ideas of probability a mere sense that precedent will repeat itself. "I don't think it is for you, sir," said the man, when he saw Boldwood's action. "Though there is no name I think it is for your shepherd."
>
> —Thomas Hardy, *Far From The Madding Crowd*

5.10.1 Range Encoding

The use of integers in arithmetic coding is a must in any practical implementation, but it results in slow encoding because of the need for frequent renormalizations. The main steps in any integer-based arithmetic coding implementation are (1) proportional range reduction and (2) range expansion (renormalization).

Range encoding (or range coding) is an improvement to arithmetic coding that reduces the number of renormalizations and thereby speeds up integer-based arithmetic coding by factors of up to 2. The main references are [Schindler 98] and [Campos 06], and the description here is based on the former.

The main idea is to treat the output not as a binary number, but as a number to another base (256 is commonly used as a base, implying that each digit is a byte). This requires fewer renormalizations and no bitwise operations. The following analysis may shed light on this method.

At any point during arithmetic coding, the output consists of four parts as follows:

1. The part already written on the output. This part will not change.

2. One digit (bit, byte, or a digit to another base) that may be modified by at most one carry when adding to the lower end of the interval. (There cannot be two carries

because when this digit was originally determined, the range was less than or equal to one unit. Two carries require a range greater than one unit.)

3. A (possibly empty) block of digits that passes on a carry (1 in binary, 9 in decimal, 255 for base-256, etc.) and are represented by a counter counting their number.

4. The low variable of the encoder.

The following states can occur while data is encoded:

- No renormalization is needed because the range is in the desired interval.
- The low end plus the range (this is the upper end of the interval) will not produce any carry. In this case the second and third parts can be output because they will never change.
- The digit produced will become part two, and part three will be empty. The low end has already produced a carry. In this case, the (modified) second and third parts can be output; there will not be another carry. Set the second and third part as before.
- The digit produced will pass on a possible future carry, so it is added to the block of digits of part three.

The difference between conventional integer-based arithmetic coding and range coding is that in the latter, part two, which may be modified by a carry, has to be stored explicitly. With binary output this part is always 0 since the 1's are always added to the carry-passing-block. Implementing that is straightforward.

More information and code can be found in [Campos 06]. Range coding is used in LZMA (Section 6.26).

5.11 The QM Coder

JPEG (Section 7.10) is an important image compression method. It uses arithmetic coding, but not in the way described in Section 5.9. The arithmetic coder of JPEG is called the QM-coder and is described in this section. It is designed for simplicity and speed, so it is limited to input symbols that are single bits and it uses an approximation instead of a multiplication. It also uses fixed-precision integer arithmetic, so it has to resort to *renormalization* of the probability interval from time to time, in order for the approximation to remain close to the true multiplication. For more information on this method, see [IBM 88], [Pennebaker and Mitchell 88a], and [Pennebaker and Mitchell 88b].

A slight confusion arises because the arithmetic coder of JPEG 2000 (Section 8.19) and JBIG2 (Section 7.15) is called the MQ-coder and is not the same as the QM-coder (we are indebted to Christopher M. Brislawn for pointing this out).

◇ **Exercise 5.28:** The QM-coder is limited to input symbols that are single bits. Suggest a way to convert an arbitrary set of symbols to a stream of bits.

The main idea behind the QM-coder is to classify each input symbol (which is a single bit) as either the more probable symbol (MPS) or the less probable symbol (LPS). Before the next bit is input, the QM-encoder uses a statistical model to determine whether a 0 or a 1 is more probable at that point. It then inputs the next bit and classifies it according to its actual value. If the model predicts, for example, that a 0 is more probable, and the next bit turns out to be a 1, the encoder classifies it as an LPS. It is important to understand that the only information encoded in the compressed stream is whether the next bit is MPS or LPS. When the stream is decoded, all that the decoder knows is whether the bit that has just been decoded is an MPS or an LPS. The decoder has to use the same statistical model to determine the current relation between MPS/LPS and 0/1. This relation changes, of course, from bit to bit, since the model is updated identically (in lockstep) by the encoder and decoder each time a bit is input by the former or decoded by the latter.

The statistical model also computes a probability Qe for the LPS, so the probability of the MPS is $(1 - Qe)$. Since Qe is the probability of the *less probable* symbol, it is in the range $[0, 0.5]$. The encoder divides the probability interval A into two subintervals according to Qe and places the LPS subinterval (whose size is $A \times Qe$) above the MPS subinterval [whose size is $A(1 - Qe)$], as shown in Figure 5.56b. Notice that the two subintervals in the figure are closed at the bottom and open at the top. This should be compared with the way a conventional arithmetic encoder divides the same interval (Figure 5.56a, where the numbers are taken from Table 5.43).

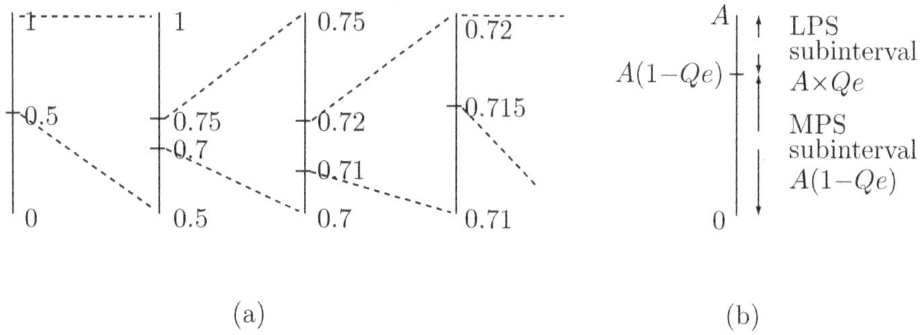

Figure 5.56: Division of the Probability Interval.

In conventional arithmetic coding, the interval is narrowed all the time, and the final output is any number inside the final subinterval. In the QM-coder, for simplicity, each step adds the bottom of the selected subinterval to the output-so-far. We denote the output string by C. If the current bit read from the input is the MPS, the bottom of the MPS subinterval (i.e., the number 0) is added to C. If the current bit is the LPS, the bottom of the LPS subinterval [i.e., the number $A(1 - Qe)$] is added to C. After C is updated in this way, the current probability interval A is shrunk to the size of the selected subinterval. The probability interval is always in the range $[0, A)$, and A gets smaller at each step. This is the main principle of the QM-encoder, and it is expressed by the rules

$$
\begin{aligned}
\text{After MPS:}\quad & C \text{ is unchanged}, \quad A \leftarrow A(1 - Qe),\\
\text{After LPS:}\quad & C \leftarrow C + A(1 - Qe), \quad A \leftarrow A \times Qe.
\end{aligned}
\tag{5.2}
$$

These rules set C to point to the bottom of the MPS or the LPS subinterval, depending on the classification of the current input bit. They also set A to the new size of the subinterval.

Table 5.57 lists the values of A and C when four symbols, each a single bit, are encoded. We assume that they alternate between an LPS and an MPS and that $Qe = 0.5$ for all four steps (normally, of course, the statistical model yields different values of Qe all the time). It is easy to see how the probability interval A shrinks from 1 to 0.0625, and how the output C grows from 0 to 0.625. Table 5.59 is similar, but uses $Qe = 0.1$ for all four steps. Again A shrinks, to 0.0081, and C grows, to 0.981. Figures 5.58 and 5.60 illustrate graphically the division of the probability interval A into an LPS and an MPS.

◇ **Exercise 5.29:** Repeat these calculations for the case where all four symbols are LPS and $Qe = 0.5$, then for the case where they are MPS and $Qe = 0.1$.

The principle of the QM-encoder is simple and easy to understand, but it involves two problems. The first is the fact that the interval A, which starts at 1, shrinks all the time and requires high precision to distinguish it from zero. The solution to this problem is to maintain A as an integer and double it every time it gets too small. This is called *renormalization*. It is fast, since it is done by a logical left shift; no multiplication is needed. Each time A is doubled, C is also doubled. The second problem is the multiplication $A \times Qe$ used in subdividing the probability interval A. A fast compression method should avoid multiplications and divisions and should try to replace them with additions, subtractions, and shifts. It turns out that the second problem is also solved by renormalization. The idea is to keep the value of A close to 1, so that Qe will not be very different from the product $A \times Qe$. The multiplication is *approximated* by Qe.

How can we use renormalization to keep A close to 1? The first idea that comes to mind is to double A when it gets just a little below 1, say to 0.9. The problem is that doubling 0.9 yields 1.8, closer to 2 than to 1. If we let A get below 0.5 before doubling it, the result will be less than 1. It does not take long to realize that 0.75 is a good minimum value for renormalization. If A reaches this value at a certain step, it is doubled, to 1.5. If it reaches a smaller value, such as 0.6 or 0.55, it ends up even closer to 1 when doubled.

If A reaches a value less than 0.5 at a certain step, it has to be renormalized by doubling it several times, each time also doubling C. An example is the second row of Table 5.59, where A shrinks from 1 to 0.1 in one step, because of a very small probability Qe. In this case, A has to be doubled three times, from 0.1 to 0.2, to 0.4, to 0.8, in order to bring it into the desired range $[0.75, 1.5)$. We conclude that A can go down to 0 (or very close to 0) and can be at most 1.5 (actually, less than 1.5, since our intervals are always open at the high end).

◇ **Exercise 5.30:** In what case does A always have to be renormalized?

Approximating the multiplication $A \times Qe$ by Qe changes the main rules of the QM-encoder to

$$\begin{aligned}
&\text{After MPS:}\quad C \text{ is unchanged},\quad A \leftarrow A(1-Qe) \approx A - Qe,\\
&\text{After LPS:}\quad C \leftarrow C + A(1-Qe) \approx C + A - Qe,\quad A \leftarrow A\times Qe \approx Qe.
\end{aligned}$$

Symbol	C	A
Initially	0	1
s1 (LPS)	$0+1(1-0.5)=0.5$	$1\times0.5=0.5$
s2 (MPS)	unchanged	$0.5\times(1-0.5)=0.25$
s3 (LPS)	$0.5+0.25(1-0.5)=0.625$	$0.25\times0.5=0.125$
s4 (MPS)	unchanged	$0.125\times(1-0.5)=0.0625$

Table 5.57: Encoding Four Symbols With $Qe=0.5$.

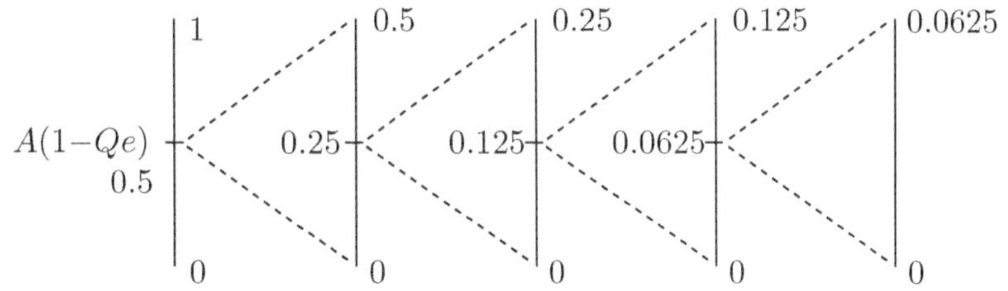

Figure 5.58: Division of the Probability Interval.

Symbol	C	A
Initially	0	1
s1 (LPS)	$0+1(1-0.1)=0.9$	$1\times0.1=0.1$
s2 (MPS)	unchanged	$0.1\times(1-0.1)=0.09$
s3 (LPS)	$0.9+0.09(1-0.1)=0.981$	$0.09\times0.1=0.009$
s4 (MPS)	unchanged	$0.009\times(1-0.1)=0.0081$

Table 5.59: Encoding Four Symbols With $Qe=0.1$.

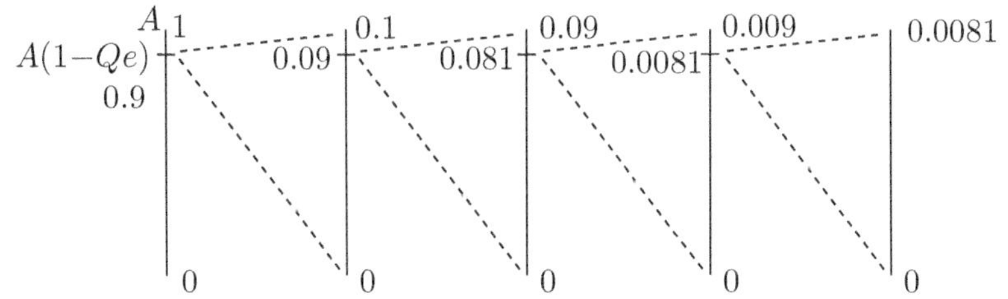

Figure 5.60: Division of the Probability Interval.

In order to include renormalization in these rules, we have to choose an integer representation for A where real values in the range $[0, 1.5)$ are represented as integers. Since many current and old computers have 16-bit words, it makes sense to choose a representation where 0 is represented by a word of 16 zero bits and 1.5 is represented by the smallest 17-bit number, which is

$$2^{16} = 65536_{10} = 10000_{16} = 1\underbrace{0\ldots0}_{16}{}_2.$$

This way we can represent 65536 real values in the range $[0, 1.5)$ as 16-bit integers, where the largest 16-bit integer, 65535, represents a real value slightly less than 1.5. Here are a few important examples of such values:

$$0.75 = 1.5/2 = 2^{15} = 32768_{10} = 8000_{16}, \quad 1 = 0.75(4/3) = 43690_{10} = AAAA_{16},$$
$$0.5 = 43690/2 = 21845_{10} = 5555_{16}, \quad 0.25 = 21845/2 = 10923_{10} = 2AAB_{16}.$$

(The optimal value of 1 in this representation is $AAAA_{16}$, but the way we associate the real values of A with the 16-bit integers is somewhat arbitrary. The important thing about this representation is to achieve accurate interval subdivision, and the subdivision is done by either $A \leftarrow A - Qe$ or $A \leftarrow Qe$. The accuracy of the subdivision depends, therefore, on the relative values of A and Qe, and it has been found experimentally that the average value of A is $B55A_{16}$, so this value, instead of $AAAA_{16}$, is associated in the JPEG QM-coder with $A = 1$. The difference between the two values $AAAA$ and $B55A$ is $AB0_{16} = 2736_{10}$. The JBIG QM-coder uses a slightly different value for 1.)

Renormalization can now be included in the main rules of the QM-encoder, which become

$$\begin{aligned} &\text{After MPS: } C \text{ is unchanged}, \quad A \leftarrow A - Qe, \\ &\qquad \text{if } A < 8000_{16} \text{ renormalize } A \text{ and } C. \\ &\text{After LPS: } C \leftarrow C + A - Qe, \quad A \leftarrow Qe, \\ &\qquad \text{renormalize } A \text{ and } C. \end{aligned} \tag{5.3}$$

Tables 5.61 and 5.62 list the results of applying these rules to the examples shown in Tables 5.57 and 5.59, respectively.

◇ **Exercise 5.31:** Repeat these calculations with renormalization for the case where all four symbols are LPS and $Qe = 0.5$. Following this, repeat the calculations for the case where they are all MPS and $Qe = 0.1$. (Compare this exercise with Exercise 5.29.)

The next point that has to be considered in the design of the QM-encoder is the problem of *interval inversion*. This is the case where the size of the subinterval allocated to the MPS becomes smaller than the LPS subinterval. This problem may occur when Qe is close to 0.5 and is a result of the approximation to the multiplication. It is illustrated in Table 5.63, where four MPS symbols are encoded with $Qe = 0.45$. In the third row of the table the interval A is doubled from 0.65 to 1.3. In the fourth row it is reduced to 0.85. This value is greater than 0.75, so no renormalization takes place; yet the subinterval allocated to the MPS becomes $A - Qe = 0.85 - 0.45 = 0.40$, which is smaller than the LPS subinterval, which is $Qe = 0.45$. Clearly, the problem occurs when $Qe > A/2$, a relation that can also be expressed as $Qe > A - Qe$.

Symbol	C	A	Renor. A	Renor. C
Initially	0	1		
s1 (LPS)	$0+1-0.5=0.5$	0.5	1	1
s2 (MPS)	unchanged	$1-0.5=0.5$	1	2
s3 (LPS)	$2+1-0.5=2.5$	0.5	1	5
s4 (MPS)	unchanged	$1-0.5=0.5$	1	10

Table 5.61: Renormalization Added to Table 5.57.

Symbol	C	A	Renor. A	Renor. C
Initially	0	1		
s1 (LPS)	$0+1-0.1=0.9$	0.1	0.8	$0.9\cdot 2^3=7.2$
s2 (MPS)	unchanged 7.2	$0.8-0.1=0.7$	1.4	$7.2\cdot 2=14.4$
s3 (LPS)	$14.4+1.4-0.1=15.7$	0.1	0.8	$15.7\cdot 2^3=125.6$
s4 (MPS)	unchanged	$0.8-0.1=0.7$	1.4	$125.6\cdot 2=251.2$

Table 5.62: Renormalization Added to Table 5.59.

Symbol	C	A	Renor. A	Renor. C
Initially	0	1		
s1 (MPS)	0	$1-0.45=0.55$	1.1	0
s2 (MPS)	0	$1.1-0.45=0.65$	1.3	0
s3 (MPS)	0	$1.3-0.45=0.85$		
s4 (MPS)	0	$0.85-0.45=0.40$	0.8	0

Table 5.63: Illustrating Interval Inversion.

The solution is to interchange the two subintervals whenever the LPS subinterval becomes greater than the MPS subinterval. This is called *conditional exchange*. The condition for interval inversion is $Qe > A - Qe$, but since $Qe \leq 0.5$, we get $A - Qe < Qe \leq 0.5$, and it becomes obvious that both Qe and $A - Qe$ (i.e., both the LPS and MPS subintervals) are less than 0.75, so renormalization must take place. This is why the test for conditional exchange is performed only *after* the encoder has decided that renormalization is needed. The new rules for the QM-encoder are shown in Figure 5.64.

The QM-Decoder: The QM-decoder is the reverse of the QM-encoder. For simplicity we ignore renormalization and conditional exchange, and we assume that the QM-encoder operates by the rules of Equation (5.2). Reversing the way C is updated in those rules yields the rules for the QM-decoder (the interval A is updated in the same way):

$$\begin{aligned} &\text{After MPS:} \quad C \text{ is unchanged}, \quad A \leftarrow A(1-Qe), \\ &\text{After LPS:} \quad C \leftarrow C - A(1-Qe), \quad A \leftarrow A\times Qe. \end{aligned} \tag{5.4}$$

These rules are demonstrated using the data of Table 5.57. The four decoding steps are as follows:

```
After MPS:
   C is unchanged
   A ← A − Qe;            % The MPS subinterval
   if A < 8000_16 then % if renormalization needed
    if A < Qe then       % if inversion needed
    C ← C + A;           % point to bottom of LPS
    A ← Qe               % Set A to LPS subinterval
    endif;
   renormalize A and C;
   endif;

After LPS:
   A ← A − Qe;            % The MPS subinterval
   if A ≥ Qe then % if interval sizes not inverted
    C ← C + A;           % point to bottom of LPS
    A ← Qe               % Set A to LPS subinterval
   endif;
   renormalize A and C;
```

Figure 5.64: QM-Encoder Rules With Interval Inversion.

Step 1: $C = 0.625$, $A = 1$, the dividing line is $A(1 - Qe) = 1(1 - 0.5) = 0.5$, so the LPS and MPS subintervals are $[0, 0.5)$ and $[0.5, 1)$. Since C points to the upper subinterval, an LPS is decoded. The new C is $0.625 - 1(1 - 0.5) = 0.125$ and the new A is $1 \times 0.5 = 0.5$.
Step 2: $C = 0.125$, $A = 0.5$, the dividing line is $A(1 - Qe) = 0.5(1 - 0.5) = 0.25$, so the LPS and MPS subintervals are $[0, 0.25)$ and $[0.25, 0.5)$, and an MPS is decoded. C is unchanged, and the new A is $0.5(1 - 0.5) = 0.25$.
Step 3: $C = 0.125$, $A = 0.25$, the dividing line is $A(1 - Qe) = 0.25(1 - 0.5) = 0.125$, so the LPS and MPS subintervals are $[0, 0.125)$ and $[0.125, 0.25)$, and an LPS is decoded. The new C is $0.125 - 0.25(1 - 0.5) = 0$, and the new A is $0.25 \times 0.5 = 0.125$.
Step 4: $C = 0$, $A = 0.125$, the dividing line is $A(1 - Qe) = 0.125(1 - 0.5) = 0.0625$, so the LPS and MPS subintervals are $[0, 0.0625)$ and $[0.0625, 0.125)$, and an MPS is decoded. C is unchanged, and the new A is $0.125(1 - 0.5) = 0.0625$.

◇ **Exercise 5.32:** Use the rules of Equation (5.4) to decode the four symbols encoded in Table 5.59.

Probability Estimation: The QM-encoder uses a novel, interesting, and little-understood method for estimating the probability Qe of the LPS. The first method that comes to mind in trying to estimate the probability of the next input bit is to initialize Qe to 0.5 and update it by counting the numbers of zeros and ones that have been input so far. If, for example, 1000 bits have been input so far, and 700 of them were zeros, then 0 is the current MPS, with probability 0.7, and the probability of the LPS is $Qe = 0.3$. Notice that Qe should be updated *before* the next input bit is read and encoded, since otherwise the decoder would not be able to mirror this operation (the decoder does not know what the next bit is). This method produces good results, but is slow, since Qe should be updated often (ideally, for each input bit), and the calculation involves a division (dividing 700/1000 in our example).

The method used by the QM-encoder is based on a table of preset Qe values. Qe is initialized to 0.5 and is modified when renormalization takes place, not for every input bit. Table 5.65 illustrates the process. The Qe index is initialized to zero, so the first value of Qe is $0AC1_{16}$ or very close to 0.5. After the first MPS renormalization, the Qe index is incremented by 1, as indicated by column "Incr MPS." A Qe index of 1 implies a Qe value of $0A81_{16}$ or 0.49237, slightly smaller than the original, reflecting the fact that the renormalization occurred because of an MPS. If, for example, the current Qe index is 26, and the next renormalization is LPS, the index is decremented by 3, as indicated by column "Decr LPS," reducing Qe to 0.00421. The method is not applied very often, and it involves only table lookups and incrementing or decrementing the Qe index: fast, simple operations.

Qe index	Hex Qe	Dec Qe	Decr LPS	Incr MPS	MPS exch	Qe index	Hex Qe	Dec Qe	Decr LPS	Incr MPS	MPS exch
0	0AC1	0.50409	0	1	1	15	0181	0.07050	2	1	0
1	0A81	0.49237	1	1	0	16	0121	0.05295	2	1	0
2	0A01	0.46893	1	1	0	17	00E1	0.04120	2	1	0
3	0901	0.42206	1	1	0	18	00A1	0.02948	2	1	0
4	0701	0.32831	1	1	0	19	0071	0.02069	2	1	0
5	0681	0.30487	1	1	0	20	0059	0.01630	2	1	0
6	0601	0.28143	1	1	0	21	0053	0.01520	2	1	0
7	0501	0.23456	2	1	0	22	0027	0.00714	2	1	0
8	0481	0.21112	2	1	0	23	0017	0.00421	2	1	0
9	0441	0.19940	2	1	0	24	0013	0.00348	3	1	0
10	0381	0.16425	2	1	0	25	000B	0.00201	2	1	0
11	0301	0.14081	2	1	0	26	0007	0.00128	3	1	0
12	02C1	0.12909	2	1	0	27	0005	0.00092	2	1	0
13	0281	0.11737	2	1	0	28	0003	0.00055	3	1	0
14	0241	0.10565	2	1	0	29	0001	0.00018	2	0	0

Table 5.65: Probability Estimation Table (Illustrative).

The column labeled "MPS exch" in Table 5.65 contains the information for the conditional exchange of the MPS and LPS definitions at $Qe = 0.5$. The zero value at the bottom of column "Incr MPS" should also be noted. If the Qe index is 29 and an MPS renormalization occurs, this zero causes the index to stay at 29 (corresponding to the smallest Qe value).

Table 5.65 is used here for illustrative purposes only. The JPEG QM-encoder uses Table 5.66, which has the same format but is harder to understand, since its Qe values are not listed in sorted order. This table was prepared using probability estimation concepts based on Bayesian statistics.

We now justify this probability estimation method with an approximate calculation that suggests that the Qe values obtained by this method will adapt to and closely approach the correct LPS probability of the binary input stream. The method updates

Qe each time a renormalization occurs, and we know, from Equation (5.3), that this happens every time an LPS is input, but not for all MPS values. We therefore imagine an ideal balanced input stream where for each LPS bit there is a sequence of consecutive MPS bits. We denote the true (but unknown) LPS probability by q, and we try to show that the Qe values produced by the method for this ideal case are close to q.

Equation (5.3) lists the main rules of the QM-encoder and shows how the probability interval A is decremented by Qe each time an MPS is input and encoded. Imagine a renormalization that brings A to a value A_1 (between 1 and 1.5), followed by a sequence of N consecutive MPS bits that reduce A in steps of Qe from A_1 to a value A_2 that requires another renormalization (i.e., A_2 is less than 0.75). It is clear that

$$N = \left\lfloor \frac{\Delta A}{Qe} \right\rfloor,$$

where $\Delta A = A_1 - A_2$. Since q is the true probability of an LPS, the probability of having N MPS bits in a row is $P = (1-q)^N$. This implies $\ln P = N \ln(1-q)$, which, for a small q, can be approximated by

$$\ln P \approx N(-q) = -\frac{\Delta A}{Qe}q, \text{ or } P \approx \exp\left(-\frac{\Delta A}{Qe}q\right). \tag{5.5}$$

Since we are dealing with an ideal balanced input stream, we are interested in the value $P = 0.5$, because it implies equal numbers of LPS and MPS renormalizations. From $P = 0.5$ we get $\ln P = -\ln 2$, which, when combined with Equation (5.5), yields

$$Qe = \frac{\Delta A}{\ln 2}q.$$

This is fortuitous because $\ln 2 \approx 0.693$ and ΔA is typically a little less than 0.75. We can say that for our ideal balanced input stream, $Qe \approx q$, providing one justification for our estimation method. Another justification is provided by the way P depends on Qe [shown in Equation (5.5)]. If Qe gets larger than q, P also gets large, and the table tends to move to smaller Qe values. In the opposite case, the table tends to select larger Qe values.

Qe index	Hex *Qe*	Next-Index LPS	Next-Index MPS	MPS exch
0	5A1D	1	1	1
1	2586	14	2	0
2	1114	16	3	0
3	080B	18	4	0
4	03D8	20	5	0
5	01DA	23	6	0
6	00E5	25	7	0
7	006F	28	8	0
8	0036	30	9	0
9	001A	33	10	0
10	000D	35	11	0
11	0006	9	12	0
12	0003	10	13	0
13	0001	12	13	0
14	5A7F	15	15	1
15	3F25	36	16	0
16	2CF2	38	17	0
17	207C	39	18	0
18	17B9	40	19	0
19	1182	42	20	0
20	0CEF	43	21	0
21	09A1	45	22	0
22	072F	46	23	0
23	055C	48	24	0
24	0406	49	25	0
25	0303	51	26	0
26	0240	52	27	0
27	01B1	54	28	0
28	0144	56	29	0
29	00F5	57	30	0
30	00B7	59	31	0
31	008A	60	32	0
32	0068	62	33	0
33	004E	63	34	0
34	003B	32	35	0
35	002C	33	9	0
36	5AE1	37	37	1
37	484C	64	38	0
38	3A0D	65	39	0
39	2EF1	67	40	0
40	261F	68	41	0
41	1F33	69	42	0
42	19A8	70	43	0
43	1518	72	44	0
44	1177	73	45	0
45	0E74	74	46	0
46	0BFB	75	47	0
47	09F8	77	48	0
48	0861	78	49	0
49	0706	79	50	0
50	05CD	48	51	0
51	04DE	50	52	0
52	040F	50	53	0
53	0363	51	54	0
54	02D4	52	55	0
55	025C	53	56	0
56	01F8	54	57	0
57	01A4	55	58	0
58	0160	56	59	0
59	0125	57	60	0
60	00F6	58	61	0
61	00CB	59	62	0
62	00AB	61	63	0
63	008F	61	32	0
64	5B12	65	65	1
65	4D04	80	66	0
66	412C	81	67	0
67	37D8	82	68	0
68	2FE8	83	69	0
69	293C	84	70	0
70	2379	86	71	0
71	1EDF	87	72	0
72	1AA9	87	73	0
73	174E	72	74	0
74	1424	72	75	0
75	119C	74	76	0
76	0F6B	74	77	0
77	0D51	75	78	0
78	0BB6	77	79	0
79	0A40	77	48	0
80	5832	80	81	1
81	4D1C	88	82	0
82	438E	89	83	0
83	3BDD	90	84	0
84	34EE	91	85	0
85	2EAE	92	86	0
86	299A	93	87	0
87	2516	86	71	0
88	5570	88	89	1
89	4CA9	95	90	0
90	44D9	96	91	0
91	3E22	97	92	0
92	3824	99	93	0
93	32B4	99	94	0
94	2E17	93	86	0
95	56A8	95	96	1
96	4F46	101	97	0
97	47E5	102	98	0
98	41CF	103	99	0
99	3C3D	104	100	0
100	375E	99	93	0
101	5231	105	102	0
102	4C0F	106	103	0
103	4639	107	104	0
104	415E	103	99	0
105	5627	105	106	1
106	50E7	108	107	0
107	4B85	109	103	0
108	5597	110	109	0
109	504F	111	107	0
110	5A10	110	111	1
111	5522	112	109	0
112	59EB	112	111	1

Table 5.66: The QM-Encoder Probability Estimation Table.

5.12 Text Compression

Before delving into the details of the next method, here is a general discussion of text compression. Most text compression methods are either statistical or dictionary based. The latter class breaks the text into fragments that are saved in a data structure called a dictionary. When a fragment of new text is found to be identical to one of the dictionary entries, a pointer to that entry is written on the compressed stream, to become the compression of the new fragment. The former class, on the other hand, consists of methods that develop statistical *models* of the text.

A common statistical method consists of a modeling stage followed by a coding stage. The model assigns probabilities to the input symbols, and the coding stage actually codes the symbols based on those probabilities. The model can be static or dynamic (adaptive). Most models are based on one of the following two approaches.

Frequency: The model assigns probabilities to the text symbols based on their frequencies of occurrence, such that commonly-occurring symbols are assigned short codes. A static model uses fixed probabilities, whereas a dynamic model modifies the probabilities "on the fly" while text is being input and compressed.

Context: The model considers the context of a symbol when assigning it a probability. Since the decoder does not have access to future text, both encoder and decoder must limit the context to past text, i.e., to symbols that have already been input and processed. In practice, the context of a symbol is the N symbols preceding it (where N is a parameter). We therefore say that a context-based text compression method uses the context of a symbol to *predict* it (i.e., to assign it a probability). Technically, such a method is said to use an "order-N" Markov model. The PPM method, Section 5.14, is an excellent example of a context-based compression method, although the concept of context can also be used to compress images.

Some modern context-based text compression methods perform a transformation on the input data and then apply a statistical model to assign probabilities to the transformed symbols. Good examples of such methods are the Burrows-Wheeler method, Section 11.1, also known as the Burrows-Wheeler transform, or *block sorting*; the technique of symbol ranking, Section 11.2; and the ACB method, Section 11.3, which uses an associative dictionary.

Reference [Bell et al. 90] is an excellent introduction to text compression. It also describes many important algorithms and approaches to this important problem.

5.13 The Hutter Prize

Among the different types of digital data, text is the smallest in the sense that text files are much smaller than audio, image, or video files. A typical 300–400-page book may contain about 250,000 words or about a million characters, so it fits in a 1 MB file. A raw (uncompressed) 1,024 × 1,024 color image, on the other hand, contains one mega pixels and therefore becomes a 3 MB file (three bytes for the color components of each pixel). Video files are much bigger. This is why many workers in the compression field concentrate their efforts on video and image compression, but there is a group of "hard core" researchers who are determined to squeeze out the last bit of redundancy from text and compress text all the way down to its entropy. In order to encourage this group of enthusiasts, scientists, and programmers, Marcus Hutter decided to offer the prize that now bears his name.

In August 2006, Hutter selected a specific 100-million-character text file with text from the English wikipedia, named it `enwik8`, and announced a prize with initial funding of 50,000 euros [Hutter 08]. Two other organizers of the prize are Matt Mahoney and Jim Bowery.

Specifically, the prize awards 500 euros for each 1% improvement in the compression of `enwik8`. The following is a quotation from [wikiHutter 08].

> The goal of the Hutter Prize is to encourage research in artificial intelligence (AI). The organizers believe that text compression and AI are equivalent problems. Hutter proved that the optimal behavior of a goal seeking agent in an unknown but computable environment is to guess at each step that the environment is controlled by the shortest program consistent with all interaction so far. Unfortunately, there is no general solution because Kolmogorov complexity is not computable. Hutter proved that in the restricted case (called AIXItl) where the environment is restricted to time t and space l, a solution can be computed in time $O(t \times 2^l)$, which is still intractable.
>
> The organizers further believe that compressing natural language text is a hard AI problem, equivalent to passing the Turing test. Thus, progress toward one goal represents progress toward the other. They argue that predicting which characters are most likely to occur next in a text sequence requires vast real-world knowledge. A text compressor must solve the same problem in order to assign the shortest codes to the most likely text sequences.

Anyone can participate in this competition. A competitor has to submit either an encoder and decoder or a compressed `enwik8` and a decoder (there are certain time and memory constraints on the decoder). The source code is not required and the compression algorithm does not have to be general; it may be optimized for `enwik8`. The total size of the compressed `enwik8` plus the decoder should not exceed 99% of the previous winning entry. For each step of 1% improvement, the lucky winner receives 500 euros.

Initially, the compression baseline was the impressive 18,324,887 bytes, representing a compression ratio of 0.183 (achieved by PAQ8F, Section 5.15).

So far, only Matt Mahoney and Alexander Ratushnyak, won prizes for improvements to the baseline, with Ratushnyak winning twice!

The awards are computed by the simple expression $Z(L - S)/L$, where Z is the total prize funding (starting at 50,000 euros and possibly increasing in the future), S is the new record (size of the encoder or size of the compressed file plus the decoder), and L is the previous record. The minimum award is 3% of Z.

5.14 PPM

The PPM method is a sophisticated, state of the art compression method originally developed by J. Cleary and I. Witten [Cleary and Witten 84], with extensions and an implementation by A. Moffat [Moffat 90]. The method is based on an encoder that maintains a statistical model of the text. The encoder inputs the next symbol S, assigns it a probability P, and sends S to an adaptive arithmetic encoder, to be encoded with probability P.

The simplest *statistical model* counts the number of times each symbol has occurred in the past and assigns the symbol a probability based on that. Assume that 1217 symbols have been input and encoded so far, and 34 of them were the letter `q`. If the next symbol is a `q`, it is assigned a probability of 34/1217 and its count is incremented by 1. The next time `q` is seen, it will be assigned a probability of $35/t$, where t is the total number of symbols input up to that point (not including the last `q`).

The next model up is a *context-based* statistical model. The idea is to assign a probability to symbol S depending not just on the frequency of the symbol but on the contexts in which it has occurred so far. The letter `h`, for example, occurs in "typical" English text (Table Intro.1) with a probability of about 5%. On average, we expect to see an `h` about 5% of the time. However, if the current symbol is `t`, there is a high probability (about 30%) that the next symbol will be `h`, since the digram `th` is common in English. We say that the model of typical English **predicts** an `h` in such a case. If the next symbol is in fact `h`, it is assigned a large probability. In cases where an `h` is the second letter of an unlikely digram, say `xh`, the `h` is assigned a smaller probability. Notice that the word "predicts" is used here to mean "estimate the probability of." A similar example is the letter `u`, which has a probability of about 2%. When a `q` is encountered, however, there is a probability of more than 99% that the next letter will be a `u`.

◇ **Exercise 5.33:** We know that in English, a `q` must be followed by a `u`. Why not just say that the probability of the digram `qu` is 100%?

A *static* context-based modeler always uses the same probabilities. It contains static tables with the probabilities of all the possible digrams (or trigrams) of the alphabet and uses the tables to assign a probability to the next symbol S depending on the symbol (or, in general, on the context) C preceding it. We can imagine S and C being used as indices for a row and a column of a static frequency table. The table itself can be constructed by accumulating digram or trigram frequencies from large quantities of text. Such a modeler is simple and produces good results on average, but has two problems. The first is that some input streams may be statistically very different from the data originally used to prepare the table. A static encoder may create considerable expansion in such a case. The second problem is zero probabilities.

What if after reading and analyzing huge amounts of English text, we still have never encountered the trigram `qqz`? The cell corresponding to `qqz` in the trigram frequency table will contain zero. The arithmetic encoder, Sections 5.9 and 5.10, requires all symbols to have nonzero probabilities. Even if a different encoder, such as Huffman, is used, all the probabilities involved must be nonzero. (Recall that the Huffman method works by combining two low-probability symbols into one high-probability symbol. If two zero-probability symbols are combined, the resulting symbol will have the same zero probability.) Another reason why a symbol must have nonzero probability is that its entropy (the smallest number of bits into which it can be encoded) depends on $\log_2 P$, which is undefined for $P = 0$ (but gets very large when $P \to 0$). This *zero-probability problem* faces any model, static or adaptive, that uses probabilities of occurrence of symbols to achieve compression. Two simple solutions are traditionally adopted for this problem, but neither has any theoretical justification.

1. After analyzing a large quantity of data and counting frequencies, go over the frequency table, looking for empty cells. Each empty cell is assigned a frequency count of 1, and the total count is also incremented by 1. This method pretends that every digram and trigram has been seen at least once.
2. Add 1 to the total count and divide this single 1 among all the empty cells. Each will get a count that's less than 1 and, as a result, a very small probability. This assigns a very small probability to anything that hasn't been seen in the training data used for the analysis.

An *adaptive* context-based modeler also maintains tables with the probabilities of all the possible digrams (or trigrams or even longer contexts) of the alphabet and uses the tables to assign a probability to the next symbol S depending on a few symbols immediately preceding it (its context C). The tables are updated all the time as more data is being input, which adapts the probabilities to the particular data being compressed.

Such a model is slower and more complex than the static one but produces better compression, since it uses the correct probabilities even when the input has data with probabilities much different from the average.

A text that skews letter probabilities is called a *lipogram*. (Would a computer program without any `goto` statements be considered a lipogram?) The word comes from the Greek stem $\lambda\epsilon\acute{\iota}\pi\omega$ (lipo or leipo) meaning to miss, to lack, combined with the Greek $\gamma\rho\acute{\alpha}\mu\mu\alpha$ (gramma), meaning "letter" or "of letters." Together they form $\lambda\iota\pi o\gamma\rho\acute{\alpha}\mu\mu\alpha\tau o\sigma$. There are not many examples of literary works that are lipograms:

1. Perhaps the best-known lipogram in English is *Gadsby*, a full-length novel [Wright 39], by Ernest V. Wright, that does not contain any occurrences of the letter E.
2. *Alphabetical Africa* by Walter Abish (W. W. Norton, 1974) is a readable lipogram where the reader is supposed to discover the unusual writing style while reading. This style has to do with the initial letters of words. The book consists of 52 chapters. In the first, all words begin with a; in the second, words start with either a or b, etc., until, in Chapter 26, all letters are allowed at the start of a word. In the remaining 26 chapters, the letters are taken away one by one. Various readers have commented on how little or how much they have missed the word "the" and how they felt on finally seeing it (in Chapter 20).
3. The novel *La Disparition* is a 1969 French lipogram by Georges Perec that does not contain the letter E (this letter actually appears several times, outside the main text, in words that the publisher had to include, and these are all printed in red). *La Disparition* has been translated to English, where it is titled *A Void*, by Gilbert Adair. Perec also wrote a univocalic (text employing just one vowel) titled *Les Revenentes* employing only the vowel E. The title of the English translation (by Ian Monk) is *The Exeter Text, Jewels, Secrets, Sex*. (Perec also wrote a short history of lipograms; see [Motte 98].)

A Quotation from the Preface to *Gadsby*

People as a rule will not stop to realize what a task such an attempt actually is. As I wrote along, in long-hand at first, a whole army of little E's gathered around my desk, all eagerly expecting to be called upon. But gradually as they saw me writing on and on, without even noticing them, they grew uneasy; and, with excited whisperings among themselves, began hopping up and riding on my pen, looking down constantly for a chance to drop off into some word; for all the world like sea birds perched, watching for a passing fish! But when they saw that I had covered 138 pages of typewriter size paper, they slid off unto the floor, walking sadly away, arm in arm; but shouting back: "You certainly must have a hodge-podge of a yarn there without Us! Why, man! We are in every story ever written, *hundreds and thousands of times!* This is the first time we ever were shut out!"

—Ernest V. Wright

4. Gottlob Burmann, a German poet, created our next example of a lipogram. He wrote 130 poems, consisting of about 20,000 words, without the use of the letter `R`. It is also believed that during the last 17 years of his life, he even omitted this letter from his daily conversation.
5. A Portuguese lipogram is found in five stories written by Alonso Alcala y Herrera, a Portuguese writer, in 1641, each suppressing one vowel.
6. Other examples, in Spanish, are found in the writings of Francisco Navarrete y Ribera (1659), Fernando Jacinto de Zurita y Haro (1654), and Manuel Lorenzo de Lizarazu y Berbinzana (also 1654).

An order-N adaptive context-based modeler reads the next symbol S from the input stream and considers the N symbols preceding S the current order-N context C of S. The model then estimates the probability P that S appears in the input data following the particular context C. Theoretically, the larger N, the better the probability estimate (the *prediction*). To get an intuitive feeling, imagine the case $N = 20{,}000$. It is hard to imagine a situation where a group of 20,000 symbols in the input stream is followed by a symbol S, but another group of the same 20,000 symbols, found later in the same input stream, is followed by a different symbol. Thus, $N = 20{,}000$ allows the model to predict the next symbol (to estimate its probability) with high accuracy. However, large values of N have three disadvantages:

1. If we encode a symbol based on the 20,000 symbols preceding it, how do we encode the first 20,000 symbols in the input stream? They may have to be written on the output stream as raw ASCII codes, thereby reducing the overall compression.
2. For large values of N, there may be too many possible contexts. If our symbols are the 7-bit ASCII codes, the alphabet size is $2^7 = 128$ symbols. There are therefore $128^2 = 16{,}384$ order-2 contexts, $128^3 = 2{,}097{,}152$ order-3 contexts, and so on. The number of contexts grows exponentially, since it is 128^N or, in general, A^N, where A is the alphabet size.

> I pounded the keys so hard that night that the letter `e` flew off the part of the machine that hits the paper. Not wanting to waste the night, I went next door to a neighbor who, I knew, had an elaborate workshop in his cellar. He attempted to solder my `e` back, but when I started to work again, it flew off like a bumblebee. For the rest of the night I inserted each `e` by hand, and in the morning I took the last dollars from our savings account to buy a new typewriter. Nothing could be allowed to delay the arrival of my greatest triumph.
>
> —Sloan Wilson, *What Shall We Wear to This Party*, (1976)

◇ **Exercise 5.34:** What is the number of order-2 and order-3 contexts for an alphabet of size $2^8 = 256$?

◇ **Exercise 5.35:** What would be a practical example of a 16-symbol alphabet?

3. A very long context retains information about the nature of old data. Experience shows that large data files tend to feature topic-specific content. Such a file may contain different distributions of symbols in different parts (a good example is a history book, where one chapter may commonly use words such as "Greek," "Athens," and "Troy," while the following chapter may use "Roman," "empire," and "legion"). Such data is termed *nonstationary*. Better compression can therefore be achieved if the model takes into account the "age" of data i.e, if it assigns less significance to information gathered from old data and more weight to fresh, recent data. Such an effect is achieved by a short context.

◇ **Exercise 5.36:** Show an example of a common binary file where different parts may have different bit distributions.

As a result, relatively short contexts, in the range of 2 to 10, are used in practice. Any practical algorithm requires a carefully designed data structure that provides fast search and easy update, while holding many thousands of symbols and strings (Section 5.14.5).

We now turn to the next point in the discussion. Assume a context-based encoder that uses order-3 contexts. Early in the compression process, the word `here` was seen several times, but the word `there` is now seen for the first time. Assume that the next symbol is the `r` of `there`. The encoder will not find any instances of the order-3 context `the` followed by `r` (the `r` has 0 probability in this context). The encoder may simply write `r` on the compressed stream as a literal, resulting in no compression, but we know that `r` was seen several times in the past following the order-2 context `he` (`r` has nonzero probability in this context). The PPM method takes advantage of this knowledge.

> "uvulapalatopharangoplasty" is the name of a surgical procedure to correct sleep apnea. It is rumored to be the longest (English?) word without any `e`'s.

5.14.1 PPM Principles

The central idea of PPM is to use this knowledge. The PPM encoder switches to a shorter context when a longer one has resulted in 0 probability. Thus, PPM starts with an order-N context. It searches its data structure for a previous occurrence of the current context C followed by the next symbol S. If it finds no such occurrence (i.e., if the probability of this particular C followed by this S is 0), it switches to order $N-1$ and tries the same thing. Let C' be the string consisting of the rightmost $N-1$ symbols of C. The PPM encoder searches its data structure for a previous occurrence of the current context C' followed by symbol S. PPM therefore tries to use smaller and smaller parts of the context C, which is the reason for its name. The name PPM stands for "prediction with partial string matching." Here is the process in some detail.

The encoder reads the next symbol S from the input stream, looks at the current order-N context C (the last N symbols read), and based on input data that has been seen in the past, determines the probability P that S will appear following the particular context C. The encoder then invokes an adaptive arithmetic coding algorithm to encode symbol S with probability P. In practice, the adaptive arithmetic encoder is a procedure that receives the quantities `HighCumFreq[X]` and `LowCumFreq[X]` (Section 5.10) as parameters from the PPM encoder.

As an example, suppose that the current order-3 context is the string `the`, which has already been seen 27 times in the past and was followed by the letters `r` (11 times), `s` (9 times), `n` (6 times), and `m` (just once). The encoder assigns these cases the probabilities 11/27, 9/27, 6/27, and 1/27, respectively. If the next symbol read is `r`, it is sent to the arithmetic encoder with a probability of 11/27, and the probabilities are updated to 12/28, 9/28, 6/28, and 1/28.

What if the next symbol read is `a`? The context `the` was never seen followed by an `a`, so the probability of this case is 0. This zero-probability problem is solved in PPM by switching to a shorter context. The PPM encoder asks; How many times was the order-2 context `he` seen in the past and by what symbols was it followed? The answer may be as follows: Seen 54 times, followed by `a` (26 times), by `r` (12 times), etc. The PPM encoder now sends the `a` to the arithmetic encoder with a probability of 26/54.

If the next symbol S was never seen before following the order-2 context `he`, the PPM encoder switches to order-1 context. Was S seen before following the string `e`? If yes, a nonzero probability is assigned to S depending on how many times it (and other symbols) was seen following `e`. Otherwise, the PPM encoder switches to order-0 context. It asks itself how many times symbol S was seen in the past, regardless of any contexts. If it was seen 87 times out of 574 symbols read, it is assigned a probability of 87/574. If the symbol S has never been seen before (a common situation at the start of any compression process), the PPM encoder switches to a mode called order -1 context, where S is assigned the fixed probability 1/(size of the alphabet).

> To predict is one thing. To predict correctly is another.
>
> —Unknown

Table 5.67 shows contexts and frequency counts for orders 4 through 0 after the 11-symbol string xyzzxyxyzzx has been input and encoded. To understand the operation of the PPM encoder, let's assume that the 12th symbol is z. The order-4 context is now yzzx, which earlier was seen followed by y but never by z. The encoder therefore switches to the order-3 context, which is zzx, but even this hasn't been seen earlier followed by z. The next lower context, zx, is of order 2, and it also fails. The encoder then switches to order 1, where it checks context x. Symbol x was found three times in the past but was always followed by y. Order 0 is checked next, where z has a frequency count of 4 (out of a total count of 11). Symbol z is therefore sent to the adaptive arithmetic encoder, to be encoded with probability 4/11 (the PPM encoder "predicts" that it will appear 4/11 of the time).

Order 4	Order 3	Order 2	Order 1	Order 0
xyzz→x 2	xyz→z 2	xy→z 2	x→y 3	x 4
yzzx→y 1	yzz→x 2	→x 1	y→z 2	y 3
zzxy→x 1	zzx→y 1	yz→z 2	→x 1	z 4
zxyx→y 1	zxy→x 1	zz→x 2	z→z 2	
xyxy→z 1	xyx→y 1	zx→y 1	→x 2	
yxyz→z 1	yxy→z 1	yx→y 1		

(a)

Order 4	Order 3	Order 2	Order 1	Order 0
xyzz→x 2	xyz→z 2	xy→z 2	x→y 3	x 4
yzzx→y 1	yzz→x 2	xy→x 1	→z 1	y 3
→z 1	zzx→y 1	yz→z 2	y→z 2	z 5
zzxy→x 1	→z 1	zz→x 2	→x 1	
zxyx→y 1	zxy→x 1	zx→y 1	z→z 2	
xyxy→z 1	xyx→y 1	→z 1	→x 2	
yxyz→z 1	yxy→z 1	yx→y 1		

(b)

Table 5.67: (a) Contexts and Counts for "xyzzxyxyzzx". (b) Updated After Another z Is Input.

Next, we consider the PPM decoder. There is a fundamental difference between the way the PPM encoder and decoder work. The encoder can always look at the next symbol and base its next step on what that symbol is. The job of the decoder is to find out what the next symbol is. The encoder decides to switch to a shorter context based on what the next symbol is. The decoder cannot mirror this, since it does not know what the next symbol is. The algorithm needs an additional feature that will make it possible for the decoder to stay in lockstep with the encoder. The feature used by PPM is to reserve one symbol of the alphabet as an *escape symbol*. When the encoder decides to switch to a shorter context, it first writes the escape symbol (arithmetically encoded) on the output stream. The decoder can decode the escape symbol, since it is encoded in the present context. After decoding an escape, the decoder also switches to a shorter context.

The worst that can happen with an order-N encoder is to encounter a symbol S for the first time (this happens mostly at the start of the compression process). The symbol hasn't been seen before in any context, not even in order-0 context (i.e., by itself). In such a case, the encoder ends up sending $N+1$ consecutive escapes to be arithmetically encoded and output, switching all the way down to order -1, followed by the symbol S encoded with the fixed probability 1/(size of the alphabet). Since the escape symbol may be output many times by the encoder, it is important to assign it a reasonable probability. Initially, the escape probability should be high, but it should drop as more symbols are input and decoded and more information is collected by the modeler about contexts in the particular data being compressed.

◇ **Exercise 5.37:** The escape is just a symbol of the alphabet, reserved to indicate a context switch. What if the data uses every symbol in the alphabet and none can be reserved? A common example is image compression, where a pixel is represented by a byte (256 grayscales or colors). Since pixels can have any values between 0 and 255, what value can be reserved for the escape symbol in this case?

Table 5.68 shows one way of assigning probabilities to the escape symbol (this is variant PPMC of PPM). The table shows the contexts (up to order 2) collected while reading and encoding the 14-symbol string `assanissimassa`. (In the movie "8 1/2," Italian children utter this string as a magic spell. They pronounce it `assa-neesee-massa`.) We assume that the alphabet consists of the 26 letters, the blank space, and the escape symbol, a total of 28 symbols. The probability of a symbol in order -1 is therefore 1/28. Notice that it takes 5 bits to encode 1 of 28 symbols without compression.

Each context seen in the past is placed in the table in a separate group together with the escape symbol. The order-2 context `as`, e.g., was seen twice in the past and was followed by `s` both times. It is assigned a frequency of 2 and is placed in a group together with the escape symbol, which is assigned frequency 1. The probabilities of `as` and the escape in this group are therefore 2/3 and 1/3, respectively. Context `ss` was seen three times, twice followed by `a` and once by `i`. These two occurrences are assigned frequencies 2 and 1 and are placed in a group together with the escape, which is now assigned frequency 2 (because it is in a group of 2 members). The probabilities of the three members of this group are therefore 2/5, 1/5, and 2/5, respectively.

The justification for this method of assigning escape probabilities is the following: Suppose that context `abc` was seen ten times in the past and was always followed by `x`. This suggests that the same context will be followed by the same `x` in the future, so the encoder will only rarely have to switch down to a lower context. The escape symbol can therefore be assigned the small probability 1/11. However, if every occurrence of context `abc` in the past was followed by a different symbol (suggesting that the data varies a lot), then there is a good chance that the next occurrence will also be followed by a different symbol, forcing the encoder to switch to a lower context (and thus to emit an escape) more often. The escape is therefore assigned the higher probability 10/20.

◇ **Exercise 5.38:** Explain the numbers 1/11 and 10/20.

Order 2			Order 1			Order 0		
Context	f	p	Context	f	p	Symbol	f	p
as→s	2	2/3	a→ s	2	2/5	a	4	4/19
esc	1	1/3	a→ n	1	1/5	s	6	6/19
			esc→	2	2/5	n	1	1/19
ss→a	2	2/5				i	2	2/19
ss→i	1	1/5	s→ s	3	3/9	m	1	1/19
esc	2	2/5	s→ a	2	2/9	esc	5	5/19
			s→ i	1	1/9			
sa→n	1	1/2	esc	3	3/9			
esc	1	1/2						
			n→ i	1	1/2			
an→i	1	1/2	esc	1	1/2			
esc	1	1/2						
			i→ s	1	1/4			
ni→s	1	1/2	i→ m	1	1/4			
esc	1	1/2	esc	2	2/4			
is→s	1	1/2	m→ a	1	1/2			
esc	1	1/2	esc	1	1/2			
si→m	1	1/2						
esc	1	1/2						
im→a	1	1/2						
esc	1	1/2						
ma→s	1	1/2						
esc	1	1/2						

Table 5.68: Contexts, Counts (f), and Probabilities (p) for "assanissimassa".

Order 0 consists of the five different symbols asnim seen in the input string, followed by an escape, which is assigned frequency 5. Thus, probabilities range from 4/19 (for a) to 5/19 (for the escape symbol).

> Wall Street indexes predicted nine out of the last five recessions.
> —Paul A. Samuelson, *Newsweek* (19 September 1966)

5.14.2 Examples

We are now ready to look at actual examples of new symbols being read and encoded. We assume that the 14-symbol string assanissimassa has been completely input and encoded, so the current order-2 context is "sa". Here are four typical cases:

1. The next symbol is n. The PPM encoder finds that sa followed by n has been seen before and has probability 1/2. The n is encoded by the arithmetic encoder with this

probability, which takes, since arithmetic encoding normally compresses at or close to the entropy, $-\log_2(1/2) = 1$ bit.
2. The next symbol is `s`. The PPM encoder finds that `sa` was not seen before followed by an `s`. The encoder therefore sends the escape symbol to the arithmetic encoder, together with the probability (1/2) predicted by the order-2 context of `sa`. It therefore takes 1 bit to encode this escape. Switching down to order 1, the current context becomes `a`, and the PPM encoder finds that an `a` followed by an `s` was seen before and currently has probability 2/5 assigned. The `s` is then sent to the arithmetic encoder to be encoded with probability 2/5, which produces another 1.32 bits. In total, $1 + 1.32 = 2.32$ bits are generated to encode the `s`.
3. The next symbol is `m`. The PPM encoder finds that `sa` was never seen before followed by an `m`. It therefore sends the escape symbol to the arithmetic encoder, as in Case 2, generating 1 bit so far. It then switches to order 1, finds that `a` has never been seen followed by an `m`, so it sends another escape symbol, this time using the escape probability for the order-1 `a`, which is 2/5. This is encoded in 1.32 bits. Switching to order 0, the PPM encoder finds `m`, which has probability 1/19 and sends it to be encoded in $-\log_2(1/19) = 4.25$ bits. The total number of bits produced is thus $1+1.32+4.25 = 6.57$.
4. The next symbol is `d`. The PPM encoder switches from order 2 to order 1 to order 0, sending two escapes as in Case 3. Since `d` hasn't been seen before, it is not found in order 0, and the PPM encoder switches to order -1 after sending a third escape with the escape probability of order 0, of 5/19 (this produces $-\log_2(5/19) = 1.93$ bits). The `d` itself is sent to the arithmetic encoder with its order -1 probability, which is 1/28, so it gets encoded in 4.8 bits. The total number of bits necessary to encode this first `d` is $1 + 1.32 + 1.93 + 4.8 = 9.05$, more than the five bits that would have been necessary without any compression.

◇ **Exercise 5.39:** Suppose that Case 4 has actually occurred (i.e., the 15th symbol to be input was a `d`). Show the new state of the order-0 contexts.

◇ **Exercise 5.40:** Suppose that Case 4 has actually occurred and the 16th symbol is also a `d`. How many bits would it take to encode this second `d`?

◇ **Exercise 5.41:** Show how the results of the above four cases are affected if we assume an alphabet size of 256 symbols.

5.14.3 Exclusion

When switching down from order 2 to order 1, the PPM encoder can use the information found in order 2 in order to exclude certain order-1 cases that are now known to be impossible. This increases the order-1 probabilities and thereby improves compression. The same thing can be done when switching down from any order. Here are two detailed examples.

In Case 2, the next symbol is `s`. The PPM encoder finds that `sa` was seen before followed by `n` but not by `s`. The encoder sends an escape and switches to order 1. The current context becomes `a`, and the encoder checks to see whether an `a` followed by an `s` was seen before. The answer is yes (with frequency 2), but the fact that `sa` was seen before followed by `n` implies that the current symbol cannot be `n` (if it were, it would be encoded in order 2).

The encoder can therefore *exclude* the case of an `a` followed by `n` in order-1 contexts [we can say that there is no need to reserve "room" (or "space") for the probability of this case, since it is impossible]. This reduces the total frequency of the order-1 group "a→" from 5 to 4, which increases the probability assigned to `s` from 2/5 to 2/4. Based on our knowledge from order 2, the `s` can now be encoded in $-\log_2(2/4) = 1$ bit instead of 1.32 (a total of two bits is produced, since the escape also requires 1 bit).

Another example is Case 4, modified for exclusions. When switching from order 2 to order 1, the probability of the escape is, as before, 1/2. When in order 1, the case of `a` followed by `n` is excluded, increasing the probability of the escape from 2/5 to 2/4. After switching to order 0, both `s` and `n` represent impossible cases and can be excluded. This leaves the order 0 with the four symbols `a`, `i`, `m`, and escape, with frequencies 4, 2, 1, and 5, respectively. The total frequency is 12, so the escape is assigned probability 5/12 (1.26 bits) instead of the original 5/19 (1.93 bits). This escape is sent to the arithmetic encoder, and the PPM encoder switches to order -1. Here it excludes all five symbols `asnim` that have already been seen in order 1 and are therefore impossible in order -1. The `d` can now be encoded with probability $1/(28-5) \approx 0.043$ (4.52 bits instead of 4.8) or $1/(256-5) \approx 0.004$ (7.97 bits instead of 8), depending on the alphabet size.

> Exact and careful model building should embody constraints that the final answer had in any case to satisfy.
>
> —Francis Crick, *What Mad Pursuit*, (1988)

5.14.4 Four PPM Variants

The particular method described earlier for assigning escape probabilities is called PPMC. Four more methods, titled PPMA, PPMB, PPMP, and PPMX, have also been developed in attempts to assign precise escape probabilities in PPM. All five methods have been selected based on the vast experience that the developers had with data compression. The last two are based on Poisson distribution [Witten and Bell 91], which is the reason for the "P" in PPMP (the "X" comes from "approximate," since PPMX is an approximate variant of PPMP).

Suppose that a group of contexts in Table 5.68 has total frequencies n (excluding the escape symbol). PPMA assigns the escape symbol a probability of $1/(n+1)$. This is equivalent to always assigning it a count of 1. The other members of the group are still assigned their original probabilities of x/n, and these probabilities add up to 1 (not including the escape probability).

PPMB is similar to PPMC with one difference. It assigns a probability to symbol S following context C only after S has been seen **twice** in context C. This is done by subtracting 1 from the frequency counts. If, for example, context `abc` was seen three times, twice followed by `x` and once by `y`, then `x` is assigned probability $(2-1)/3$, and `y` (which should be assigned probability $(1-1)/3 = 0$) is not assigned any probability (i.e., does not get included in Table 5.68 or its equivalent). Instead, the escape symbol "gets" the two counts subtracted from `x` and `y`, and it ends up being assigned probability 2/3. This method is based on the belief that "seeing twice is believing."

PPMP is based on a different principle. It considers the appearance of each symbol a separate Poisson process. Suppose that there are q different symbols in the input stream. At a certain point during compression, n symbols have been read, and symbol i has been input c_i times (so $\sum c_i = n$). Some of the c_is are zero (this is the zero-probability problem). PPMP is based on the assumption that symbol i appears according to a Poisson distribution with an expected value (average) λ_i. The statistical problem considered by PPMP is to estimate q by extrapolating from the n-symbol sample input so far to the entire input stream of N symbols (or, in general, to a larger sample). If we express N in terms of n in the form $N = (1+\theta)n$, then a lengthy analysis shows that the number of symbols that haven't appeared in our n-symbol sample is given by $t_1\theta - t_2\theta^2 + t_3\theta^3 - \cdots$, where t_1 is the number of symbols that appeared exactly once in our sample, t_2 is the number of symbols that appeared twice, and so on.

Hapax legomena: words or forms that occur only once in the writings of a given language; such words are extremely difficult, if not impossible, to translate.

In the special case where N is not the entire input stream but the slightly larger sample of size $n+1$, the expected number of new symbols is $t_1\frac{1}{n} - t_2\frac{1}{n^2} + t_3\frac{1}{n^3} - \cdots$. This expression becomes the probability that the next symbol is novel, so it is used in PPMP as the escape probability. Notice that when t_1 happens to be zero, this expression is normally negative and cannot be used as a probability. Also, the case $t_1 = n$ results in an escape probability of 1 and should be avoided. Both cases require corrections to the sum above.

PPMX uses the approximate value t_1/n (the first term of the sum) as the escape probability. This expression also breaks down when t_1 happens to be 0 or n, so in these cases PPMX is modified to PPMXC, which uses the same escape probability as PPMC.

Experiments with all five variants show that the differences between them are small. Version X is indistinguishable from P, and both are slightly better than A-B-C. Version C is slightly but consistently better than A and B.

It should again be noted that the way escape probabilities are assigned in the A-B-C variants is based on experience and intuition, not on any underlying theory. Experience with these variants indicates that the basic PPM algorithm is robust and is not affected much by the precise way of computing escape probabilities. Variants P and X are based on theory, but even they don't significantly improve the performance of PPM.

5.14.5 Implementation Details

The main problem in any practical implementation of PPM is to maintain a data structure where all contexts (orders 0 through N) of every symbol read from the input stream are stored and can be located fast. The structure described here is a special type of tree, called a *trie*. This is a tree in which the branching structure at any level is determined by just part of a data item, not by the entire item (page 357). In the case of PPM, an order-N context is a string that includes all the shorter contexts of orders $N-1$ through 0, so each context effectively adds just one symbol to the trie.

Figure 5.69 shows how such a trie is constructed for the string "zxzyzxxyzx" assuming $N = 2$. A quick glance shows that the tree grows in width but not in depth. Its depth remains $N+1 = 3$ regardless of how much input data has been read. Its width grows as more and more symbols are input, but not at a constant rate. Sometimes, no new nodes are added, such as in case 10, when the last x is read. At other times, up to three nodes are added, such as in cases 3 and 4, when the second z and the first y are added.

Level 1 of the trie (just below the root) contains one node for each symbol read so far. These are the order-1 contexts. Level 2 contains all the order-2 contexts, and so on. Every context can be found by starting at the root and sliding down to one of the leaves. In case 3, for example, the two contexts are xz (symbol z preceded by the order-1 context x) and zxz (symbol z preceded by the order-2 context zx). In case 10, there are seven contexts ranging from xxy and xyz on the left to zxz and zyz on the right.

The numbers in the nodes are context counts. The "z,4" on the right branch of case 10 implies that z has been seen four times. The "x,3" and "y,1" below it mean that these four occurrences were followed by x three times and by y once. The circled nodes show the different orders of the context of the last symbol added to the trie. In case 3, for example, the second z has just been read and added to the trie. It was added twice, below the x of the left branch and the x of the right branch (the latter is indicated by the arrow). Also, the count of the original z has been incremented to 2. This shows that the new z follows the two contexts x (of order 1) and zx (order 2).

It should now be easy for the reader to follow the ten steps of constructing the tree and to understand intuitively how nodes are added and counts updated. Notice that three nodes (or, in general, $N+1$ nodes, one at each level of the trie) are involved in each step (except the first few steps when the trie hasn't reached its final height yet). Some of the three are new nodes added to the trie; the others have their counts incremented.

The next point that should be discussed is how the algorithm decides which nodes to update and which to add. To simplify the algorithm, one more pointer is added to each node, pointing backward to the node representing the next shorter context. A pointer that points backward in a tree is called a *vine pointer*.

Figure 5.70 shows the first ten steps in the construction of the PPM trie for the 14-symbol string "assanissimassa". Each of the ten steps shows the new vine pointers (the dashed lines in the figure) constructed by the trie updating algorithm while that step was executed. Notice that old vine pointers are not deleted; they are just not shown in later diagrams. In general, a vine pointer points from a node X on level n to a node with the same symbol X on level $n-1$. All nodes on level 1 point to the root.

A node in the PPM trie therefore consists of the following fields:

1. The code (ASCII or other) of the symbol.
2. The count.
3. A down pointer, pointing to the leftmost child of the node. In Figure 5.70, Case 10, for example, the leftmost son of the root is "a,2". That of "a,2" is "n,1" and that of "s,4" is "a,1".
4. A right pointer, pointing to the next sibling of the node. The root has no right sibling. The next sibling of node "a,2" is "i,2" and that of "i,2" is "m,1".
5. A vine pointer. These are shown as dashed arrows in Figure 5.70.

◇ **Exercise 5.42:** Complete the construction of this trie and show it after all 14 characters have been input.

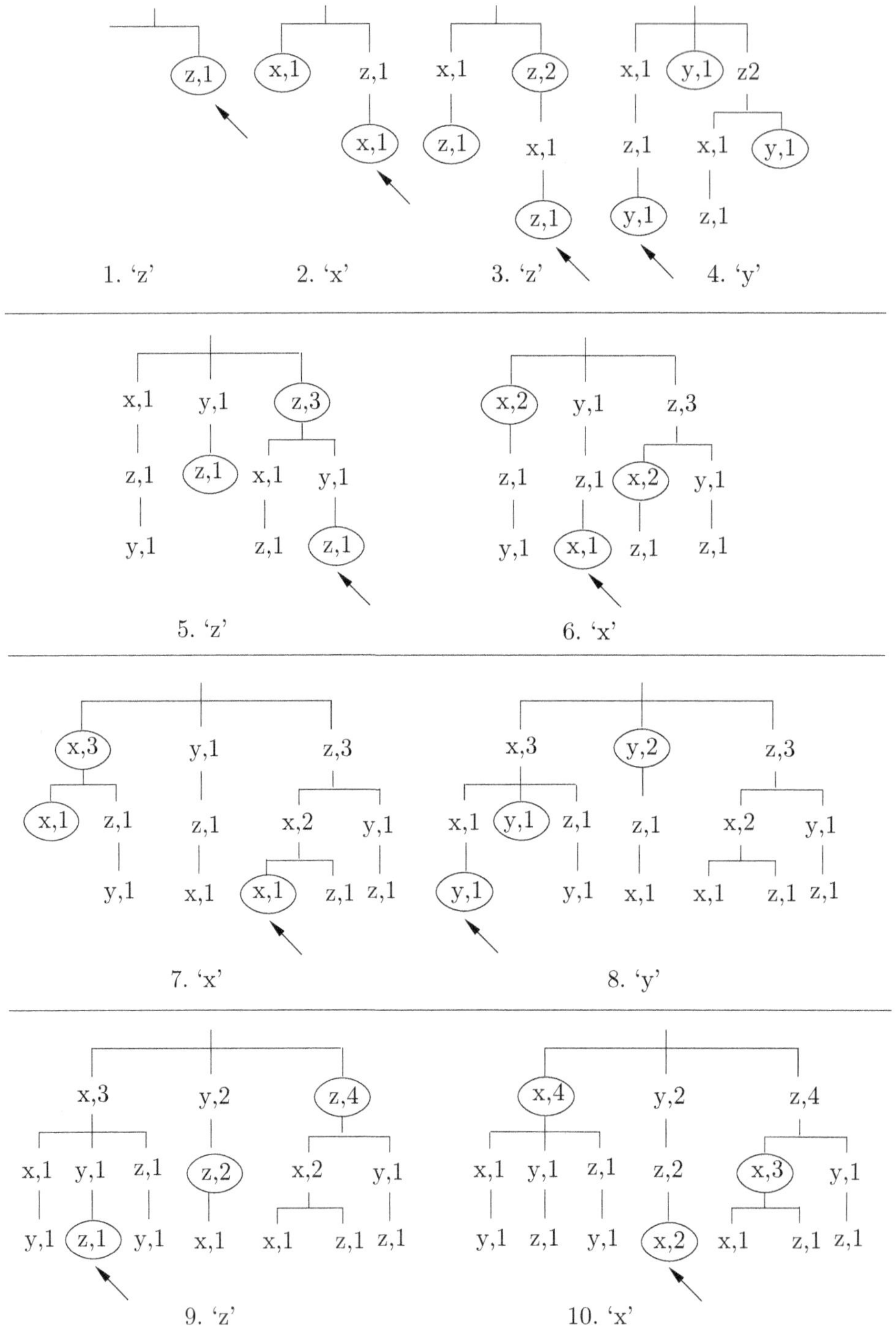

Figure 5.69: Ten Tries of "zxzyzxxyzx".

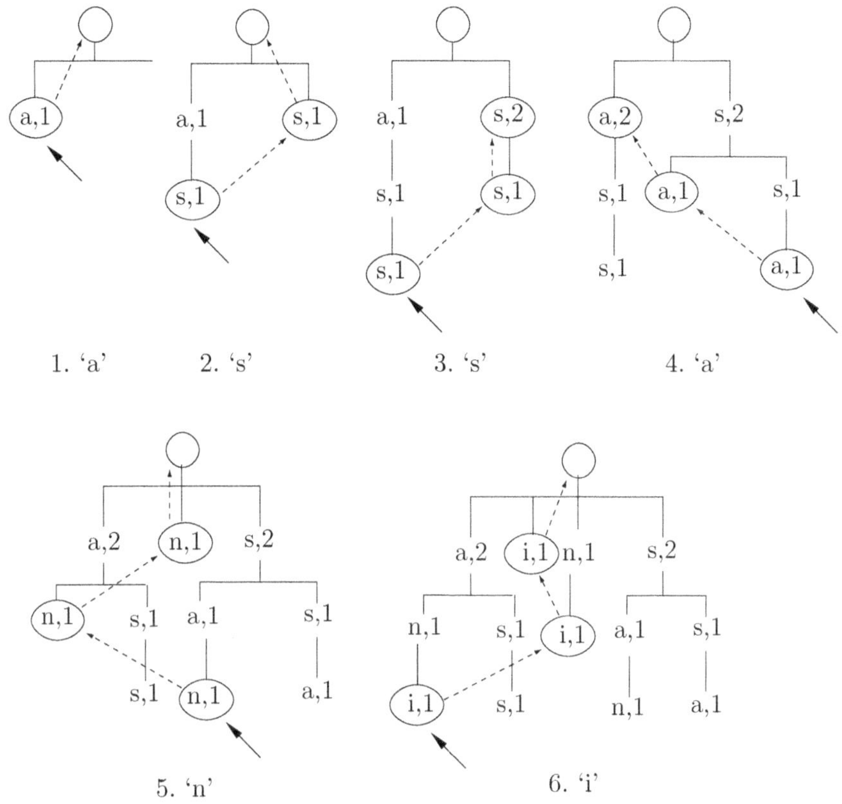

Figure 5.70: Part I. First Six Tries of "assanissimassa".

At any step during the trie construction, one pointer, called the *base*, is maintained that points to the last node added or updated in the previous step. This pointer is shown as a solid arrow in Figure 5.70. Suppose that symbol S has been input and the trie should be updated at this point. The algorithm for adding and/or updating nodes is as follows:

1. Follow the base pointer to node X. Follow the vine pointer from X to Y (notice that Y can be the root). Add S as a new child node of Y and set the base to point to it. However, if Y already has a child node with S, increment the count of that node by 1 (and also set the base to point to it). Call this node A.
2. Repeat the same step but without updating the base. Follow the vine pointer from Y to Z, add S as a new child node of Z, or update an existing child. Call this node B. If there is no vine pointer from A to B, install one. (If both A and B are old nodes, there will already be a vine pointer from A to B.)
3. Repeat until you have added (or incremented) a node at level 1.

During these steps, the PPM encoder also collects the counts that are needed to compute the probability of the new symbol S. Figure 5.70 shows the trie after the last

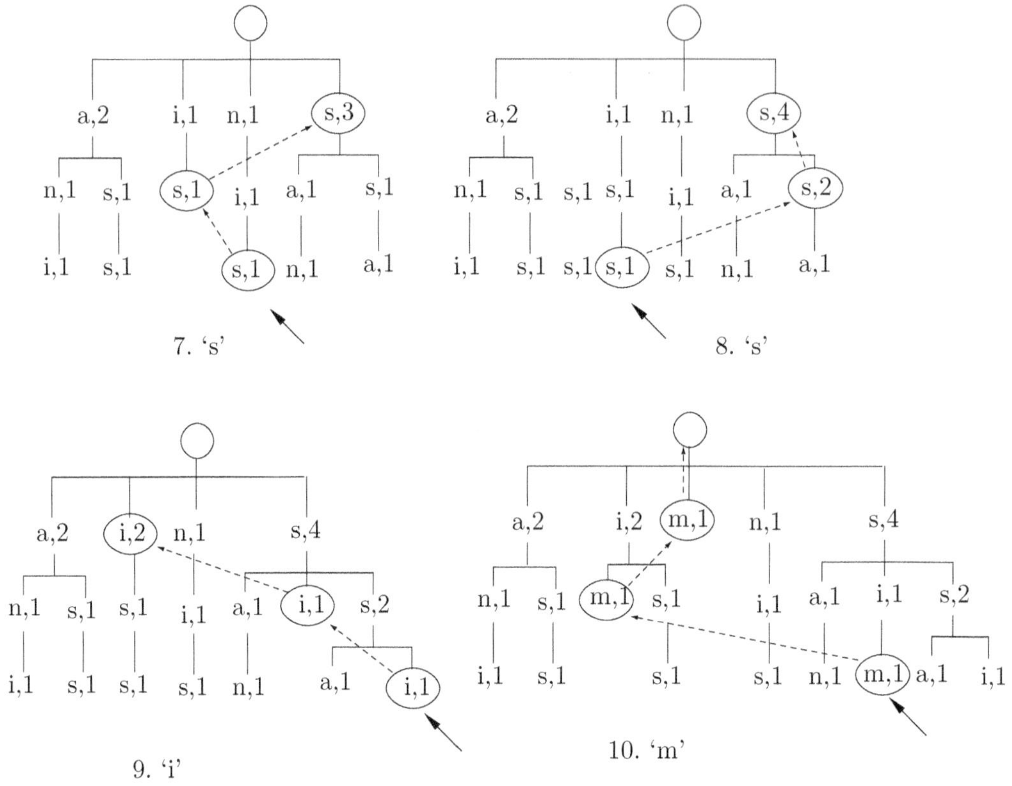

Figure 5.70: (Continued) Next Four Tries of "assanissimassa".

two symbols `s` and `a` were added. In Figure 5.70, Case 13, a vine pointer was followed from node "`s,2`", to node "`s,3`", which already had the two children "`a,1`" and "`i,1`". The first child was incremented to "`a,2`". In Figure 5.70, Case 14, the subtree with the three nodes "`s,3`", "`a,2`", and "`i,1`" tells the encoder that `a` was seen following context `ss` twice and `i` was seen following the same context once. Since the tree has two children, the escape symbol gets a count of 2, bringing the total count to 5. The probability of `a` is therefore 2/5 (compare with Table 5.68). Notice that steps 11 and 12 are not shown. The serious reader should draw the tries for these steps as a voluntary exercise (i.e., without an answer).

It is now easy to understand the reason why this particular trie is so useful. Each time a symbol is input, it takes the algorithm at most $N+1$ steps to update the trie and collect the necessary counts by going from the base pointer toward the root. Adding a symbol to the trie and encoding it takes $O(N)$ steps regardless of the size of the trie. Since N is small (typically 4 or 5), an implementation can be made fast enough for practical use even if the trie is very large. If the user specifies that the algorithm should use exclusions, it becomes more complex, since it has to maintain, at each step, a list of symbols to be excluded.

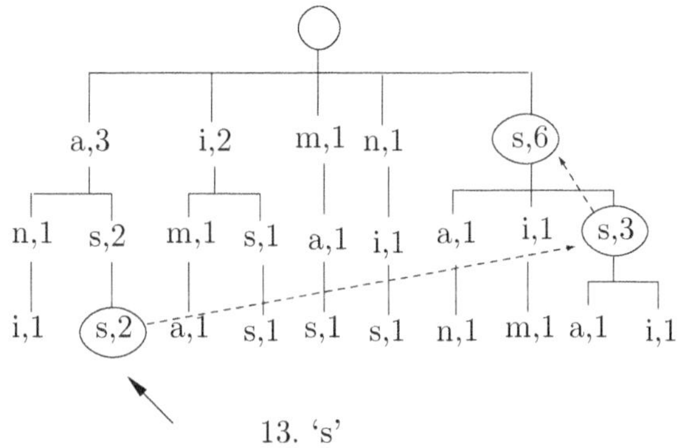

13. ‘s’

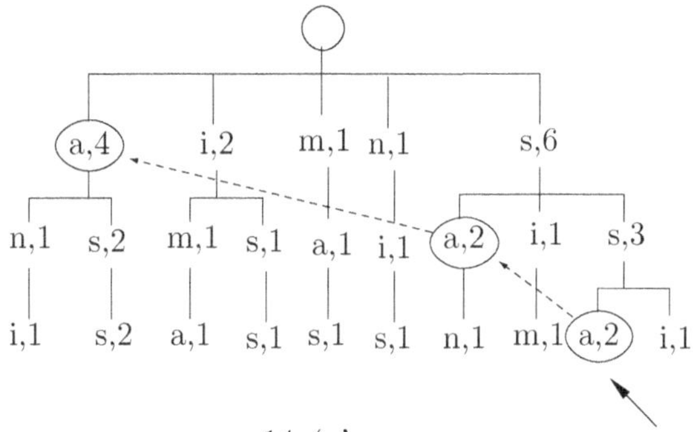

14. ‘a’

Figure 5.70: (Continued) Final Two Tries of "assanissimassa".

As has been noted, between 0 and 3 nodes are added to the trie for each input symbol encoded (in general, between 0 and $N+1$ nodes). The trie can therefore grow very large and fill up any available memory space. One elegant solution, adopted in [Moffat 90], is to discard the trie when it gets full and start constructing a new one. In order to bring the new trie "up to speed" fast, the last 2048 input symbols are always saved in a circular buffer in memory and are used to construct the new trie. This reduces the amount of inefficient code generated when tries are replaced.

5.14.6 PPM*

An important feature of the original PPM method is its use of a fixed-length, bounded initial context. The method selects a value N for the context length and always tries to predict (i.e., to assign probability to) the next symbol S by starting with an order-N context C. If S hasn't been seen so far in context C, PPM switches to a shorter context. Intuitively it seems that a long context (large value of N) may result in better prediction, but Section 5.14 explains the drawbacks of long contexts. In practice, PPM implementations tend to use N values of 5 or 6 (Figure 5.71).

The PPM* method, due to [Cleary et al. 95] and [Cleary and Teahan 97], tries to extend the value of N indefinitely. The developers tried to find ways to use unbounded

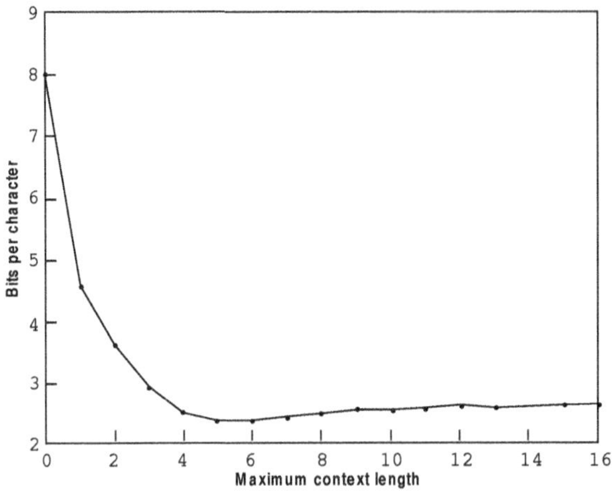

Figure 5.71: Compression Ratio as a Function of Maximum Context Length.

values for N in order to improve compression. The resulting method requires a new trie data structure and more computational resources than the original PPM, but in return it provides compression improvement of about 6% over PPMC.

(In mathematics, when a set S consists of symbols a_i, the notation S^* is used for the set of all the strings of symbols a_i.)

One problem with long contexts is the escape symbols. If the encoder inputs the next symbol S, starts with an order-100 context, and does not find any past string of 100 symbols that's followed by S, then it has to emit an escape and try an order-99 context. Such an algorithm may result in up to 100 consecutive escape symbols being emitted by the encoder, which can cause considerable expansion. It is therefore important to allow for contexts of various lengths, not only very long contexts, and decide on the length of a context depending on the current situation. The only restriction is that the decoder should be able to figure out the length of the context used by the encoder for each symbol encoded. This idea is the main principle behind the design of PPM*.

An early idea was to maintain a record, for each context, of its past performance. Such a record can be mirrored by the decoder, so both encoder and decoder can use, at any point, that context that behaved best in the past. This idea did not seem to work and the developers of PPM* were also faced with the task of having to explain why it did not work as expected.

The algorithm finally selected for PPM* depends on the concept of a deterministic context. A context is defined as deterministic when it gives only one prediction. For example, the context `this␣is␣my␣` is deterministic if every appearance of it so far in the input has been followed by the same symbol. Experiments indicate that if a context C is deterministic, the chance that when it is seen next time, it will be followed by a novel symbol is smaller than what is expected from a uniform prior distribution of the symbols. This feature suggests the use of deterministic contexts for prediction in the new version of PPM.

Based on experience with deterministic contexts, the developers have arrived at the following algorithm for PPM*. When the next symbol S is input, search all its contexts trying to find deterministic contexts of S. If any such contexts are found, use the shortest of them. If no deterministic contexts are found, use the longest nondeterministic context.

The result of this strategy for PPM* is that nondeterministic contexts are used most of the time, and they are almost always 5–6 symbols long, the same as those used by traditional PPM. However, from time to time deterministic contexts are used and they get longer as more input is read and processed. (In experiments performed by the developers, deterministic contexts started at length 10 and became as long as 20–25 symbols after about 30,000 symbols were input.) The use of deterministic contexts results in very accurate prediction, which is the main contributor to the slightly better performance of PPM* over PPMC.

A practical implementation of PPM* has to solve the problem of keeping track of long contexts. Each time a symbol S is input, all its past occurrences have to be checked, together with all the contexts, short and long, deterministic or not, for each occurrence. In principle, this can be done by simply keeping the entire data file in memory and checking back for each symbol. Imagine the symbol in position i in the input file. It is preceded by $i-1$ symbols, so $i-1$ steps are needed to search and find all its contexts. The total number of steps for n symbols is therefore $1+2+\cdots+(n-1)=n(n-1)/2$. For large n, this amounts to $O(n^2)$ complexity—too slow for practical implementations. This problem was solved by a special trie, termed a context-trie, where a leaf node points back to the input string whenever a context is unique. Each node corresponds to a symbol that follows some context and the frequency count of the symbol is stored in the node.

PPM* uses the same escape mechanism as the original PPM. The implementation reported in the PPM publications uses the PPMC algorithm to assign probabilities to the various escape symbols. Notice that the original PPM uses escapes less and less over time, as more data is input and more context predictions become available. In contrast, PPM* has to use escapes very often, regardless of the amount of data already input, particularly because of the use of deterministic contexts. This fact makes the problem of computing escape probabilities especially acute.

Compressing the entire (concatenated) Calgary corpus by PPM* resulted in an average of 2.34 bpc, compared to 2.48 bpc achieved by PPMC. This represents compression improvement of about 6% because 2.34 is 94.4% of 2.48.

5.14.7 PPMZ

The PPMZ variant, originated and implemented by Charles Bloom [Bloom 98], is an attempt to improve the original PPM algorithm. It starts from the premise that PPM is a powerful algorithm that can, in principle, compress data to its entropy, especially when presented with large amounts of input data, but performs less than optimal in practice because of poor handling of features such as deterministic contexts, unbounded-length contexts, and local order estimation. PPMZ attempts to handle these features in an optimal way, and it ends up achieving superior performance.

The PPMZ algorithm starts, similar to PPM*, by examining the maximum deterministic context of the current symbol. If no deterministic context is found, the PPMZ encoder executes a local-order-estimation (LOE) procedure, to compute an order in the interval $[0, 12]$ and use it to predict the current symbol as the original PPM algorithm does. In addition, PPMZ uses a secondary model to predict the probabilities of the various escape symbols.

The originator of the method noticed that the various PPM implementations compress data to about 2 bpc, where most characters are compressed to 1 bpc each, and the remaining characters represent either the start of the input stream or random data. The natural conclusion is that any small improvements in the probability estimation of the most common characters can lead to significant improvements in the overall performance of the method. We start by discussing the way PPMZ handles unbounded contexts.

Figure 5.72a shows a situation where the current character is `e` and its 12-order context is `ll␣assume␣th`. The context is hashed into a pointer P that points to a linked list. The nodes of the list point to all the 12-character strings in the input stream that happen to hash to the same pointer P. (Each node also has a count field indicating the number of times the string pointed to by the node has been a match.) The encoder follows the pointers, looking for a match whose minimum length varies from context to context. Assuming that the minimum match length in our case is 15, the encoder will find the 15-character match `e␣all␣assume␣th` (the preceding `w` makes this a 16-character match, and it may even be longer). The current character `e` is encoded with a probability determined by the number of times this match has been found in the past, and the match count (in the corresponding node in the list) is updated.

Figure 5.72b shows a situation where no deterministic match is found. The current character is again an `e` and its order-12 context is the same `ll␣assume␣th`, but the only string `ll␣assume␣th` in the data file is preceded by the three characters `y␣a`. The encoder does not find a 15-character match, and it proceeds as follows: (1) It outputs an escape to indicate that no deterministic match has been found. (2) It invokes the LOE procedure to compute an order. (3) It uses the order to predict the current symbol the way ordinary PPM does. (4) It takes steps to ensure that such a case will not happen again.

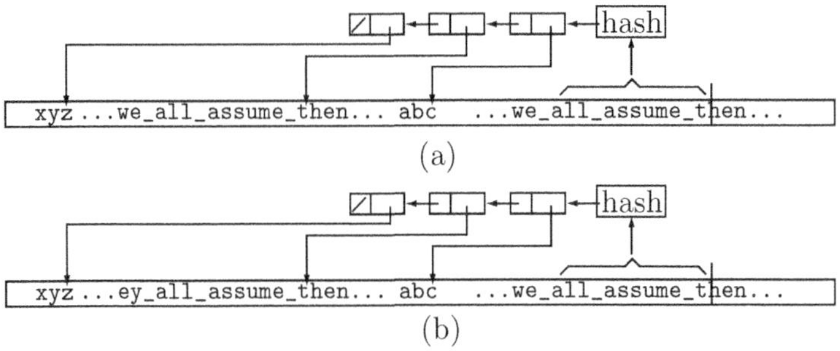

Figure 5.72: Unbounded-Length Deterministic Contexts in PPMZ.

In the first of these steps, the encoder appends a node to the list and sets it to point to the new 12-character context `ll␣assume␣th`. The second step increments the minimum match length of both contexts by 1 (i.e., to 16 characters). This ensures that these two contexts will be used in the future only when the encoder can match enough characters to distinguish between them.

This complex procedure is executed by the PPMZ encoder to guarantee that all the unbounded-length contexts are deterministic.

Local order estimation is another innovation of PPMZ. Traditional PPM uses the same value for N (the maximum order length, typically 5–6) for all input streams, but a more sophisticated version should attempt to estimate different values of N for different input files or even for different contexts within an input file. The LOE computation performed by PPMZ tries to decide which high-order contexts are unreliable. LOE finds a matching context, examines it in each order, and computes a *confidence rating* for each order.

At first, it seems that the best measure of confidence is the entropy of the context, because the entropy estimates the length of the output in bits. In practice, however, this measure turned out to underestimate the reliability of long contexts. The reason mentioned by the method's developer is that a certain symbol X may be common in an input stream; yet any specific context may include X only once.

The confidence measure finally selected for LOE is based on the probability P of the most probable character in the context. Various formulas involving P were tried, and all resulted in about the same performance. The conclusion was that the best confidence measure for LOE is simply P itself.

The last important feature of PPMZ is the way it estimates the escape probabilities. This is called secondary escape estimation or SEE. The main idea is to have an adaptive algorithm where not only the counts of the escapes but also the way the counts are computed are determined by the input stream. In each context, the PPMC method of counting the escapes is first applied. This method counts the number of novel characters (i.e., characters found that were nor predicted) in the context. This information is then used to construct an escape context which, in turn, is used to look up the escape probability in a table.

The escape context is a number constructed from the four fields listed here, each quantized to a few bits. Both linear and logarithmic quantization were tried. Linear quantization simply truncates the least-significant bits. Logarithmic quantization computes the logarithm of the number to be quantized. This quantizes large numbers more than small ones, with the result that small values remain distinguishable, while large values may become equal. The four components of the escape context are as follows:

1. The PPM order (which is between 0 and 8), quantized to two bits.
2. The escape count, quantized to two bits.
3. The number of successful matches (total count minus the escape count), quantized to three bits.
4. Ten bits from the two characters xy preceding the current symbol S. Seven bits are taken from x and three bits from y.

This number becomes the order-2 escape context. After deleting some bits from it, PPMZ also creates order-1 and order-0 contexts (15 bits and 7 bits long, respectively). The escape contexts are used to update a table of escape counts. Each entry in this table corresponds to matches coded from past PPM contexts that had the same escape contexts. The information in the table is then used in a complex way to construct the escape probability that is sent to the arithmetic coder to code the escape symbol itself.

The advantage of this complex method is that it combines the statistics gathered from the long (high order) contexts. These contexts provide high compression but are sparse, causing the original PPM to overestimate their escape probabilities.

Applied to the entire (concatenated) Calgary Corpus, PPMZ resulted in an average of 2.119 bpc. This is 10% better than the 2.34 bpc obtained by PPM* and 17% better than the 2.48 bpc achieved by PPMC.

5.14.8 Fast PPM

Fast PPM is a PPM variant developed and implemented by [Howard and Vitter 94b] as a compromise between speed and performance of PPM. Even though it is recognized as one of the best (if not *the* best) statistical compression method, PPM is not very popular because it is slow. Most general-purpose lossless compression software implementations select a dictionary-based method. Fast PPM attempts to isolate those features of PPM that contribute only marginally to compression performance and replace them by approximations. It was hoped that this would speed up execution to make this version competitive with common, commercial lossless compression products.

PPM has two main aspects: modeling and encoding. The modeling part searches the contexts of the current symbol to compute its probability. The fast version simplifies that part by eliminating the explicit use of escape symbols, computing approximate probabilities, and simplifying the exclusion mechanism. The encoding part of PPM uses adaptive arithmetic coding. The fast version speeds up this part by using quasi-arithmetic coding, a method developed by the same researchers [Howard and Vitter 92c] and not discussed here.

The modeling part of fast PPM is illustrated in Table 5.73. We assume that the input stream starts with the string `abcbabdbaeabbabe` and that the current symbol is the second `e` (the last character of the string). Part (a) of the table lists the contexts of this character starting with order 3. The order-3 context of this `e` is `bab`, which has been seen once in the past, but was followed by a `d`, so it cannot be used to predict the current `e`. The encoder therefore skips to order 2, where the context `ab` was seen three times, but never followed by an `e` (notice that the `d` following `ab` has to be excluded). Skipping to order 1, the encoder finds four different symbols following the order-1 context `b`. They are `c`, `a`, `d`, and `b`. Of these, `c`, `d`, and `b` have already been seen, following longer contexts, and are therefore excluded, and the `d` is designated NF (not found), because we are looking for an `e`. Skipping to order 0, the encoder finally finds `e`, following `a`, `b`, `c`, and `d`, which are all excluded. The point is that both the encoder and decoder of fast PPM can easily generate this table with the information available to them. All that the decoder needs in order to decode the `e` is the number of NFs (4 in our example) in the table.

Part (b) of the table illustrates the situation when the sixth `b` (there are seven `b`'s in all) is the current character. It shows that this character can be identified to the decoder by encoding three NFs and writing them on the compressed stream.

A different way of looking at this part of fast PPM is to imagine that the encoder generates a list of symbols, starting at the highest order and eliminating duplicates. The list for part (a) of the table consists of `dcbae` (four NFs followed by an F), while the list for part (b) is `cdab` (three NFs followed by an F).

Order	Context	Symbol	Count	Action
3	bab	d	1	NF, → 2
2	ab	c	1	NF
		d	1	exclude
		b	1	NF, → 1
1	b	c	1	exclude
		a	3	NF
		d	1	exclude
		b	1	exclude, → 0
0		a	5	exclude
		b	7	exclude
		c	1	exclude
		d	1	exclude
		e	1	found

(a)

Order	Context	Symbol	Count	Action
3	eab	–	–	→ 2
2	ab	c	1	NF
		d	1	NF, → 1
1	b	c	1	exclude
		a	2	NF
		d	1	exclude, → 0
0		a	4	exclude
		b	5	found

(b)

Figure 5.73: Two Examples of Fast PPM For abcbabdbaeabbabe.

Temporal reasoning involves both prediction and explanation. Prediction is projection forwards from causes to effects whilst explanation is projection backwards from effects to causes. That is, prediction is reasoning from events to the properties and events they cause, whilst explanation is reasoning from properties and events to events that may have caused them. Although it is clear that a complete framework for temporal reasoning should provide facilities for solving both prediction and explanation problems, prediction has received far more attention in the temporal reasoning literature than explanation.

—Murray Shanahan, *Proceedings IJCAI 1989*

Thus, fast PPM encodes each character by encoding a sequence of NFs, followed by one F (found). It therefore uses a binary arithmetic coder. For increased speed, quasi-arithmetic coding is used, instead of the more common QM coder of Section 5.11. For even faster operation, the quasi-arithmetic coder is used to encode the NFs only for symbols with highest probability, then use a Rice code (Section 10.9) to encode the symbol's (e or b in our example) position in the remainder of the list. Variants of fast PPM can eliminate the quasi-arithmetic coder altogether (for maximum speed) or use it all the way (for maximum compression).

The results of applying fast PPM to the Calgary corpus are reported by the developers and seem to justify its development effort. The compression performance is 2.341 bpc (for the version with just quasi-arithmetic coder) and 2.287 bpc (for the version with both quasi-arithmetic coding and Rice code). This is somewhat worse than the 2.074 bpc achieved by PPMC. However, the speed of fast PPM is about 25,000–30,000 characters per second, compared to about 16,000 characters per second for PPMC—a speedup factor of about 2!

5.15 PAQ

PAQ is an open-source high-performance compression algorithm and free software that features sophisticated prediction (modeling) combined with adaptive arithmetic encoding. PAQ encodes individual bits from the input data and its main novelty is its adaptive prediction method, which is based on mixing predictions (or models) obtained from several contexts. PAQ is therefore related to PPM (Section 5.14). The main references are [wikiPAQ 08] and [Mahoney 08] (the latter includes four pdf files with many details about PAQ and its varieties).

PAQ has been a surprise to the data compression community. Its algorithm naturally lends itself to all kinds of additions and improvements and this fact, combined with its being open source, inspired many researchers to extend and improve PAQ. In addition to describing the chief features of PAQ, this section discusses the main historical milestones in the rapid development of this algorithm, its versions, subversions, and derivatives. The conclusion at the end of this section is that PAQ may signal a change of paradigm in data compression. In the foreseeable future we may see more extensions of existing algorithms and fewer new, original methods and approaches to compression.

The original idea for PAQ is due to Matt Mahoney. He knew that PPM achieves high compression factors and he wondered whether the main principle of PPM, prediction by examining contexts, could be improved by including several predictors (i.e., models of the data) and combining their results in sophisticated ways. As always, there is a price to pay for improved performance, and in the case of PAQ the price is increased execution time and memory requirements.

Starting in 2002, Mahoney and others have come up with eight main versions of PAQ, each of which acquired several subversions (or variations), some especially tuned for certain types of data. As of late 2008, the latest version, PAQ8, has about 20 subversions.

The PAQ predictor predicts (i.e., computes a probability for) the next input bit by employing several context-based prediction methods (mathematical models of the data). An important feature is that the contexts do not have to be contiguous. Here are the main predictors used by the various versions of PAQ:

- An order-n context predictor, as used by PPM. The last n bits encoded are examined and the numbers of zeros and 1's are counted to estimate the probabilities that the next bit will be a 0 or a 1.

- Whole word order-n context. The context is the latest n whole words input. Nonalphabetic characters are ignored and upper- and lowercase letters are considered identical. This type of prediction makes sense for text files.

- Sparse contexts. The context consists of certain noncontiguous bytes preceding the current bit. This may be useful in certain types of binary files.

- A high-order context. The next bit is predicted by examining and counting the bits in the high-order parts of the latest bytes or 16-bit words. This has proved useful for the compression of multimedia files.

- Two-dimensional context. An image is a rectangle of correlated pixels, so we can expect consecutive rows of an image to be very similar. If a row is c pixels long, then the next pixel to be predicted should be similar to older pixels at distances of c, $2c$, and so on. This kind of context also makes sense for tables and spreadsheets.

- A compressed file generally looks random, but if we know the compression method, we can often see order in the file. Thus, common image compression methods, such as jpeg, tiff, and PNG, produce files with certain structures which can be exploited by specialized predictors.

The PAQ encoder examines the beginning of the input file, trying to determine its type (text, jpeg, binary, spreadsheet, etc.). Based on the result, it decides which predictors to use for the file and how to combine the probabilities that they produce for the next input bit. There are three main ways to combine predictions, as follows:

1. In version 1 through 3 of PAQ, each predictor produces a pair of bit counts (n_0, n_1). The pair for each predictor is multiplied by a weight that depends on the length of the context used by the predictor (the weight for an order-n context is $(n+1)^2$). The weighted pairs are then added and the results normalized to the interval $[0,1]$, to become meaningful as probabilities.

2. In versions 4 through 6, the weights assigned to the predictors are not fixed but are adjusted adaptively in the direction that would reduce future prediction mistakes. This adjustment tends to favor the more accurate predictors. Specifically, if the predicted bit is b, then weight w_i of the ith predictor is adjusted in the following steps

$$\begin{aligned} n_i &= n_{0i} + n_{1i}, \\ \text{error} &= b - p_1, \\ w_i &\leftarrow w_i + \big[(S \cdot n_{1i} - S_1 \cdot n_i)/(S_0 \cdot S_1)\big] \cdot \text{error}, \end{aligned}$$

where n_{0i} and n_{1i} are the counts of zeros and 1's produced by the predictor, p_1 is the probability that the next bit will be a 1, $n_i = n_{0i} + n_{1i}$, and S, S_0, and S_1 are defined by Equation (5.6).

3. Starting with version 7, each predictor produces a probability rather than a pair of counts. The individual probabilities are then combined with an artificial neural network in the following manner

$$\begin{aligned} x_i &= \text{stretch}\big(p_{i1}\big), \\ p_1 &= \text{squash}\Big(\sum_i w_i x_i\Big), \end{aligned}$$

where p_1 is the probability that the next input bit will be a 1, p_{i1} is the probability computed by the ith predictor, and the stretch and squash operations are defined as $\text{stretch}(x) = \ln\big(x/(1-x)\big)$ and $\text{squash}(x) = 1/(1+e^{-x})$. Notice that they are the inverse of each other.

After each prediction and bit encoding, the ith predictor is updated by adjusting the weight according to $w_i \leftarrow w_i + \mu x_i\big(b - p_1\big)$, where μ is a constant (typically 0.002 to 0.01) representing the adaptation rate, b is the predicted bit, and $\big(b - p_1\big)$ is the prediction error.

In PAQ4 and later versions, the weights are updated by

$$w_i \leftarrow \max[0, w_i + (b - p_i)(S\,n_{1i} - S_1 n_i)/S_0 S_1],$$

where $n_i = n_{0i} + n_{1i}$.

PAQ prediction (or modeling) pays special attention to the difference between stationary and nonstationary data. In the former type, the distribution of symbols (the statistics of the data), remains the same throughout the different parts of the input file. In the latter type, different regions of the input file discuss different topics and consequently feature different symbol distributions.

PAQ modeling recognizes that prediction algorithms for nonstationary data must perform poorly for stationary data and vice versa, which is why the PAQ encoder modifies predictions for stationary data to bring them closer to the behavior of nonstationary data.

Modeling stationary binary data is straightforward. The predictor initializes the two counts n_0 and n_1 to zero. If the next bit to be encoded is b, then count n_b is incremented by 1. At any point, the probabilities for a 0 and a 1 are $p_0 = n_0/(n_0 + n_1)$ and $p_1 = n_1/(n_0 + n_1)$. Modeling nonstationary data is more complex and can be done in various ways. The approach considered by PAQ is perhaps the simplest. If the next bit to be encoded is b, then increment count n_b by 1 and clear the other counter. Thus, a sequence of consecutive zeros will result in higher and higher values of n_0 and in $n_1 = 0$. Essentially, this is the same as predicting that the last input will repeat. Once the next bit of 1 appears in the input, n_0 is cleared, n_1 is set to 1, and we expect (or predict) a sequence of consecutive 1's.

In general, the input data may be stationary or nonstationary, so PAQ needs to combine the two approaches above; it needs a semi-stationary rule to update the two counters, a rule that will prove itself experimentally over a wide range of input files. The solution? Rather than keep all the counts (as in the stationary model above) or discard all the counts that disagree with the last input (as in the nonstationary model), the semi-stationary model discards about half the counts. Specifically, it keeps at least two counts and discards half of the remaining counts. This modeling has the effect of starting off as a stationary model and becoming nonstationary as more data is input and encoded and the counts grow. The rule is: if the next bit to be encoded is b, then increment n_b by 1. If the other counter, n_{1-b} is greater than 2, set it to $\lfloor 1 + n_{1-b}/2 \rfloor$.

The modified counts of the ith predictor are denoted by n_{0i} and n_{1i}. The counts from the predictors are combined as follows

$$S_0 = \epsilon + \sum_i w_i n_{0i}, \quad S_1 = \epsilon + \sum_i w_i n_{1i}, \tag{5.6}$$
$$S = S_0 + S_1, \quad p_0 = S_0/S, \quad p_1 = S_1/S,$$

where w_i is the weight assigned to the ith predictor, ϵ is a small positive parameter (whose value is determined experimentally) to make sure that S_0 and S_1 are never zeros, and the division by S normalizes the counts to probabilities.

The predicted probabilities p_0 and p_1 from the equation above are sent to the arithmetic encoder together with the bit to be encoded.

History of PAQ

The first version, PAQ1, appeared in early 2002. It was developed and implemented by Matt Mahoney and it employed the following contexts:

- Eight contexts of length zero to seven bytes. These served as general-purpose models. Recall that PAQ operates on individual bits, not on bytes, so each of these models also includes those bits of the current byte that precede the bit currently being predicted and encoded (there may be between zero and seven such bits). The weight of such a context of length n is $(n+1)^2$.

- Two word-oriented contexts of zero or one whole words preceding the currently predicted word. (A word consists of just the 26 letters and is case insensitive.) This constitutes the unigram and bigram models.

- Two fixed-length record models. These are specialized models for two-dimensional data such as tables and still images. This predictor starts with the byte preceding the current byte and looks for an identical byte in older data. When such a byte is located at a distance of, say, c, the model checks the bytes at distances $2c$ and $3c$. If all four bytes are identical, the model assumes that c is the row length of the table, image, or spreadsheet.

- One match context. This predictor starts with the eight bytes preceding the current bit. It locates the nearest group of eight (or more) consecutive bytes, examines the bit b that follows this context, and predicts that the current bit will be b.

All models except the last one (match) employ the semi-stationary update rule described earlier.

PAQ2 was written by Serge Osnach and released in mid 2003. This version adds an SSE (secondary symbol estimation) stage between the predictor and encoder of PAQ1. SSE inputs a short, 10-bit context and the current prediction and outputs a new prediction from a table. The table entry is then updated (in proportion to the prediction error) to reflect the actual bit value.

PAQ3 has two subversions, released in late 2003 by Matt Mahoney and Serge Osnach. It includes an improved SSE and three additional sparse models. These employ two-byte contexts that skip over intermediate bytes.

PAQ4, again by Matt Mahoney, came out also in late 2003. It used adaptive weighting of 18 contexts, where the weights are updated by

$$w_i \leftarrow \max[0, w_i + (b - p_i)(S\,n_{1i} - S_1 n_i)/S_0 S_1],$$

where $n_i = n_{0i} + n_{1i}$.

PAQ5 (December 2003) and PAQ6 (the same month) represent minor improvements, including a new analog model. With these versions, PAQ became competitive with the best PPM implementations and began to attract the attention of researchers and users. The immediate result was a large number (about 30) of small, incremental improvements and subversions. The main contributors were Berto Destasio, Johan de Bock, Fabio Buffoni, Jason Schmidt, and David A. Scott. These versions added a number of features as follows:

- A second mixer whose weights were selected by a 4-bit context (consisting of the two most-significant bits of the last two bytes).

- Six new models designed for audio (8 and 16-bit mono and stereo audio samples), 24-bit color images, and 8-bit data.

- Nonstationary, run-length models (in addition to the older, semi-stationary model).
- A special model for executable files with machine code for Intel x86 processors.
- Many more contexts, including general-purpose, match, record, sparse, analog, and word contexts.

In may, 2004, Alexander Ratushnyak threw his hat into the ring and entered the world of PAQ and incremental compression improvements. He named his design PAQAR and started issuing versions in swift succession (seven versions and subversions were issued very quickly). PAQAR extends PAQ by adding many new models, multiple mixers with weights selected by context, an SSE stage to each mixer output, and a preprocessor to improve the compression of Intel executable files (relative `CALL` and `JMP` operands are replaced by absolute addresses). The result was significant improvements in compression, at the price of slow execution and larger memory space.

Another spinoff of PAQ is PAsQDa, four versions of which were issued by Przemysław Skibiński in early 2005. This algorithm is based on PAQ6 and PAQAR, with the addition of an English dictionary preprocessor (termed a word reducing transform, or WRT, by its developer). WRT replaces English words with a code (up to three bytes long) from a dictionary. It also performs other transforms to model punctuations, capitalizations, and end-of-lines. (It should be noted that several specialized subversions of PAQ8 also employ text preprocessing to achieve minor improvements in the compression of certain text files.) PAsQDa achieved the top ranking on the Calgary corpus but not on most other benchmarks.

Here are a few words about one of the terms above. For the purposes of context prediction, the end-of-line (EOL) character is artificial in the sense that it intervenes between symbols that are correlated and spoils their correlation (or similarities). Thus, it makes sense to replace the EOL (which is ASCII character CR, LF, or both) with a space. This idea is due to Malcolm Taylor.

Another significant byproduct of PAQ6 is RK (or WinRK), by Malcolm Taylor. This variant of PAQ is based on a mixing algorithm named PWCM (PAQ weighted context mixing) and has achieved first place, surpassing PAQAR and PAsQDa, in many benchmarks.

In late 2007, Matt Mahoney released PAQ7, a major rewrite of PAQ6, based on experience gained with PAQAR and PAsQDa. The major improvement is not the compression itself but execution speed, which is up to three times faster than PAQAR. PAQ7 lacks the WRT dictionary and the specialized facilities to deal with Intel x86 executable code, but it does include predictors for jpg, bmp, and tiff image files. As a result, it does not compress executables and text files as well as PAsQDa, but it performs better on data that has text with embedded images, such as MS-excel, MS-Word and PDF. PAQ7 also employs a neural network to combine predictions.

PAQ8 was born in early 2006. Currently (early 2009), there are about 20 subversions, many issued in series, that achieve incremental improvements in performance by adding, deleting, and updating many of the features and facilities mentioned above. These subversions were developed and implemented by many of the past players in the PAQ arena, as well as a few new ones. The interested reader is referred to [wikiPAQ 08] for the details of these subversions. Only one series of PAQ subversions, namely the eight algorithms PAQ8HP1 through PAQ8HP8, will be mentioned here. They were released by Alexander Ratushnyak from August, 2006 through January, 2007 specifically to claim improvement steps for the Hutter Prize (Section 5.13).

> I wrote an experimental PAQ9A in Dec. 2007 based on a chain of 2-input mixers rather than a single mixer as in PAQ8, and an LZP preprocessor for long string matches. The goal was better speed at the cost of some compression, and to allow files larger than 2 GB. It does not compress as well as PAQ8, however, because it has fewer models, just order-n, word, and sparse. Also, the LZP interferes with compression to some extent.
>
> I wrote LPAQ1 in July 2007 based on PAQ8 but with fewer models for better speed. (Similar goal to PAQ9A, but a file compressor rather than an archiver). Alexander Ratushnyak wrote several later versions specialized for text which are now #3 on the large text benchmark.
>
> —Matt Mahoney to D. Salomon, Dec. 2008

A Paradigm Change

Many of the algorithms, techniques, and approaches described in this book have originated in the 1980s and 1990s, which is why these two decades are sometimes referred to as the golden age of data compression. However, the unexpected popularity of PAQ with its many versions, subversions, derivatives, and spinoffs (several are briefly mentioned below) seems to signal a change of paradigm. Instead of new, original algorithms, may we expect more and more small, incremental improvements of existing methods and approaches to compression? No one has a perfect crystal ball, and it is possible that there are clever people out there who have novel ideas for new, complex, and magical compression algorithms and are only waiting for another order of magnitude increase in processor speed and memory size to implement and test their ideas. We'll have to wait and see.

PAQ Derivatives

Another reason for the popularity of PAQ is its being free. The algorithm is not encumbered by a patent and existing implementations are free. This has encouraged several researchers and coders to extend PAQ in several directions. The following is a short list of the most notable of these derivatives.

- WinUDA 0.291, is based on PAQ6 but is faster [UDA 08].
- UDA 0.301, is based on the PAQ8I algorithm [UDA 08].
- KGB is essentially PAQ6v2 with a GUI (beta version supports PAQ7 compression algorithms). See [KGB 08].
- Emilcont, an algorithm based on PAQ6 [Emilcont 08].
- Peazip. This is a GUI frontend (for Windows and Linux) for LPAQ and various PAQ8* algorithms [PeaZip 08].
- PWCM (PAQ weighted context mixing) is an independently developed closed source implementation of the PAQ algorithm. It is used (as a plugin) in WinRK to win first place in several compression benchmarks.

5.16 Context-Tree Weighting

Whatever the input stream is, text, pixels, sound, or anything else, it can be considered a binary string. Ideal compression (i.e., compression at or very near the entropy of the string) would be achieved if we could use the bits that have been input so far in order to predict with certainty (i.e., with probability 1) the value of the next bit. In practice, the best we can hope for is to use history to estimate the probability that the next bit will be 1. The context-tree weighting (CTW) method [Willems et al. 95] starts with a given bit-string $b_1^t = b_1 b_2 \ldots b_t$ and the d bits that precede it $c_d = b_{-d} \ldots b_{-2} b_{-1}$ (the context of b_1^t). The two strings c_d and b_1^t constitute the input stream. The method uses a simple algorithm to construct a tree of depth d based on the context, where each node corresponds to a substring of c_d. The first bit b_1 is then input and examined. If it is 1, the tree is updated to include the substrings of $c_d b_1$ and is then used to calculate (or estimate) the probability that b_1 will be 1 given context c_d. If b_1 is zero, the tree is updated differently and the algorithm again calculates (or estimates) the probability that b_1 will be zero given the same context. Bit b_1 and its probability are then sent to an arithmetic encoder, and the process continues with b_2. The context bits themselves are written on the compressed stream in raw format.

The depth d of the context tree is fixed during the entire compression process, and it should depend on the expected correlation among the input bits. If the bits are expected to be highly correlated, a small d may be enough to get good probability predictions and thus good compression.

In thinking of the input as a binary string, it is customary to use the term "source." We think of the bits of the inputs as coming from a certain information source. The source can be *memoryless* or it can have memory. In the former case, each bit is independent of its predecessors. In the latter case, each bit depends on some of its predecessors (and, perhaps, also on its successors, but these cannot be used because they are not available to the decoder), so they are correlated.

We start by looking at a memoryless source where each bit has probability $P_a(1)$ of being a 1 and probability $P_a(0)$ of being a 0. We set $\theta = P_a(1)$, so $P_a(0) = 1 - \theta$ (the subscript a stands for "actual" probability). The probability of a particular string b_1^t being generated by the source is denoted by $P_a(b_1^t)$, and it equals the product

$$P_a(b_1^t) = \prod_{i=1}^{t} P_a(b_i).$$

If string b_1^t contains a zeros and b ones, then $P_a(b_1^t) = (1-\theta)^a \theta^b$.

Example: Let $t = 5$, $a = 2$, and $b = 3$. The probability of generating a 5-bit binary string with two zeros and three ones is $P_a(b_1^5) = (1-\theta)^2\theta^3$. Table 5.74 lists the values of $P_a(b_1^5)$ for seven values of θ from 0 to 1. It is easy to see that the maximum is obtained when $\theta = 3/5$. To understand these values intuitively we examine all 32 5-bit numbers. Ten of them consist of two zeros and three ones, so the probability of generating such a string is $10/32 = 0.3125$ and the probability of generating a particular string out of these 10 is 0.03125. This number is obtained for $\theta = 1/2$.

In real-life situations we don't know the value of θ, so we have to estimate it based on what has been input in the past. Assuming that the immediate past string b_1^t consists

θ:	0	1/5	2/5	1/2	3/5	4/5	5/5
$P_a(2,3)$:	0	0.00512	0.02304	0.03125	0.03456	0.02048	0

Table 5.74: Seven Values of $P_a(2,3)$.

of a zeros and b ones, it makes sense to estimate the probability of the next bit being 1 by

$$P_e(b_{t+1}=1|b_1^t)=\frac{b}{a+b},$$

where the subscript e stands for "estimate" (the expression above is read "the estimated probability that the next bit b_{t+1} will be a 1 given that we have seen string b_1^t is..."). The estimate above is intuitive and cannot handle the case $a=b=0$. A better estimate, due to Krichevsky and Trofimov [Krichevsky 81], is called KT and is given by

$$P_e(b_{t+1}=1|b_1^t)=\frac{b+1/2}{a+b+1}.$$

The KT estimate, like the intuitive estimate, predicts a probability of 1/2 for any case where $a=b$. Unlike the intuitive estimate, however, it also works for the case $a=b=0$.

The KT Boundary

All over the globe is a dark clay-like layer that was deposited around 65 million years ago. This layer is enriched with the rare element iridium. Older fossils found under this KT layer include many dinosaur species. Above this layer (younger fossils) there are no dinosaur species found. This suggests that something that happened around the same time as the KT boundary was formed killed the dinosaurs. Iridium is found in meteorites, so it is possible that a large iridium-enriched meteorite hit the earth, kicking up much iridium dust into the stratosphere. This dust then spread around the earth via air currents and was deposited on the ground very slowly, later forming the KT boundary.

This event has been called the "KT Impact" because it marks the end of the Cretaceous Period and the beginning of the Tertiary. The letter "K" is used because "C" represents the Carboniferous Period, which ended 215 million years earlier.

◇ **Exercise 5.43:** Use the KT estimate to calculate the probability that the next bit will be a zero given string b_1^t as the context.

Example: We use the KT estimate to calculate the probability of the 5-bit string 01110. The probability of the first bit being zero is (since there is no context)

$$P_e(0|\text{null})=P_e(0|_{a=b=0})=\left(1-\frac{0+1/2}{0+0+1}\right)=1/2.$$

The probability of the entire string is the product

$$\begin{aligned}
P_e(01110) &= P_e(2,3)\\
&= P_e(0|\text{null})P_e(1|0)P_e(1|01)P_e(1|011)P_e(0|0111)\\
&= P_e(0|_{a=b=0})P_e(1|_{a=1,b=0})P_e(1|_{a=b=1})P_e(1|_{a=1,b=2})P_e(0|_{a=1,b=3})\\
&= \left(1-\frac{0+1/2}{0+0+1}\right)\cdot\frac{0+1/2}{1+0+1}\cdot\frac{1+1/2}{1+1+1}\cdot\frac{2+1/2}{1+2+1}\cdot\left(1-\frac{3+1/2}{1+3+1}\right)\\
&= \frac{1}{2}\cdot\frac{1}{4}\cdot\frac{3}{6}\cdot\frac{5}{8}\cdot\frac{3}{10}=\frac{3}{256}\approx 0.01172.
\end{aligned}$$

In general, the KT estimated probability of a string with a zeros and b ones is

$$P_e(a,b)=\frac{1/2{\cdot}3/2\cdots(a-1/2){\cdot}1/2{\cdot}3/2\cdots(b-1/2)}{1{\cdot}2{\cdot}3\cdots(a+b)}. \qquad (5.7)$$

Table 5.75 lists some values of $P_e(a,b)$ calculated by Equation (5.7). Notice that $P_e(a,b)=P_e(b,a)$, so the table is symmetric.

	0	1	2	3	4	5
0	-	1/2	3/8	5/16	35/128	63/256
1	1/2	1/8	1/16	5/128	7/256	21/1024
2	3/8	1/16	3/128	3/256	7/1024	9/2048
3	5/8	5/128	3/256	5/1024	5/2048	45/32768
4	35/128	7/256	7/1024	5/2048	35/32768	35/65536
5	63/256	21/1024	9/2048	45/32768	35/65536	63/262144

Table 5.75: KT Estimates for Some $P_e(a,b)$.

Up until now we have assumed a memoryless source. In such a source the probability θ that the next bit will be a 1 is fixed. Any binary string, including random ones, is generated by such a source with equal probability. Binary strings that have to be compressed in real situations are generally not random and are generated by a nonmemoryless source. In such a source θ is not fixed. It varies from bit to bit, and it depends on the past context of the bit. Since a context is a binary string, all the possible past contexts of a bit can be represented by a binary tree. Since a context can be very long, the tree can include just some of the last bits of the context, to be called the *suffix*. As an example consider the 42-bit string

$$S=000101100111010110001101001011110010101100.$$

Let's assume that we are interested in suffixes of length 3. The first 3 bits of S don't have long enough suffixes, so they are written raw on the compressed stream. Next we examine the 3-bit suffix of each of the last 39 bits of S and count how many times each suffix is followed by a 1 and how many times by a 0. Suffix 001, for example, is followed twice by a 1 and three times by a 0. Figure 5.76a shows the entire suffix tree of depth 3 for this case (in general, this is not a complete binary tree). The suffixes are read from the leaves to the root, and each leaf is labeled with the probability of that suffix being followed by a 1-bit. Whenever the three most recently read bits are 001, the encoder starts at the root of the tree and follows the edges for 1, 0, and 0. It finds 2/5 at the leaf, so it should predict a probability of 2/5 that the next bit will be a 1, and a probability of $1 - 2/5$ that it will be a 0. The encoder then inputs the next bit, examines it, and sends it, with the proper probability, to be arithmetically encoded.

Figure 5.76b shows another simple tree of depth 2 that corresponds to the set of suffixes 00, 10, and 1. Each suffix (i.e., each leaf of the tree) is labeled with a probability θ. Thus, for example, the probability that a bit of 1 will follow the suffix $\ldots 10$ is 0.3. The tree is the *model* of the source, and the probabilities are the *parameters*. In practice, neither the model nor the parameters are known, so the CTW algorithm has to estimate them.

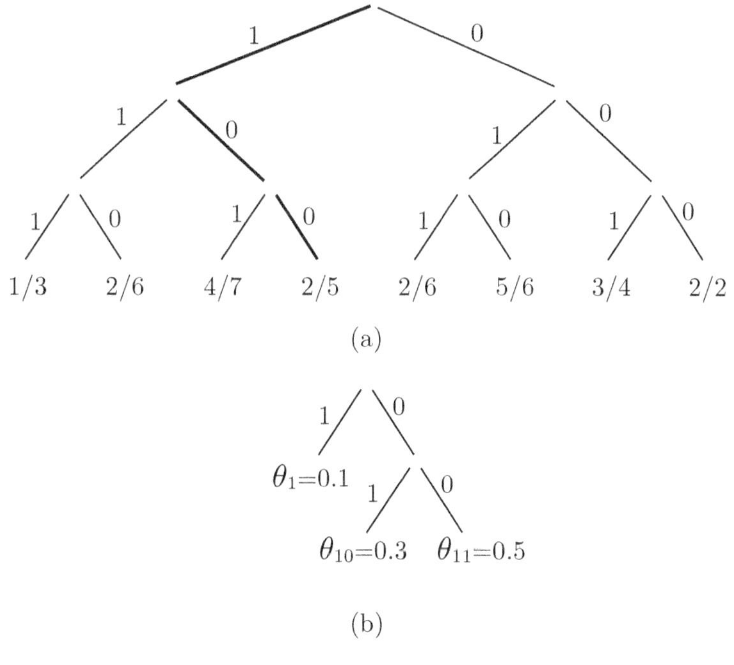

Figure 5.76: Two Suffix Trees.

Next we get one step closer to real-life situations. We assume that the model is known and the parameters are unknown, and we use the KT estimator to estimate the parameters. As an example we use the model of Figure 5.76b but without the probabilities. We use the string $10|0100110 = 10|b_1b_2b_3b_4b_5b_6b_7$, where the first two bits are the suffix, to illustrate how the probabilities are estimated with the KT estimator. Bits b_1 and b_4 have suffix 10, so the probability for leaf 10 of the tree is estimated as the KT probability of substring $b_1b_4 = 00$, which is $P_e(2,0) = 3/8$ (two zeros and no ones) from Table 5.75. Bits b_2 and b_5 have suffix 00, so the probability for leaf 00 of the tree is estimated as the KT probability of substring $b_2b_5 = 11$, which is $P_e(0,2) = 3/8$ (no zeros and two ones) from Table 5.75. Bits $b_3 = 0$, $b_6 = 1$, and $b_7 = 0$ have suffix 1, so the probability for leaf 1 of the tree is estimated as $P_e(2,1) = 1/16$ (two zeros and a single one) from Table 5.75. The probability of the entire string 0100110 given the suffix 10 is thus the product

$$\frac{3}{8} \cdot \frac{3}{8} \cdot \frac{1}{16} = \frac{9}{1024} \approx .0088.$$

◇ **Exercise 5.44:** Use this example to estimate the probabilities of the five strings 0, 00, 000, 0000, and 00000, assuming that each is preceded by the suffix 00.

In the last step we assume that the model, as well as the parameters, are unknown. We construct a binary tree of depth d. The root corresponds to the null context, and each node s corresponds to the substring of bits that were input following context s. Each node thus splits up the string. Figure 5.77a shows an example of a context tree for the string $10|0100110 = 10|b_1b_2b_3b_4b_5b_6b_7$.

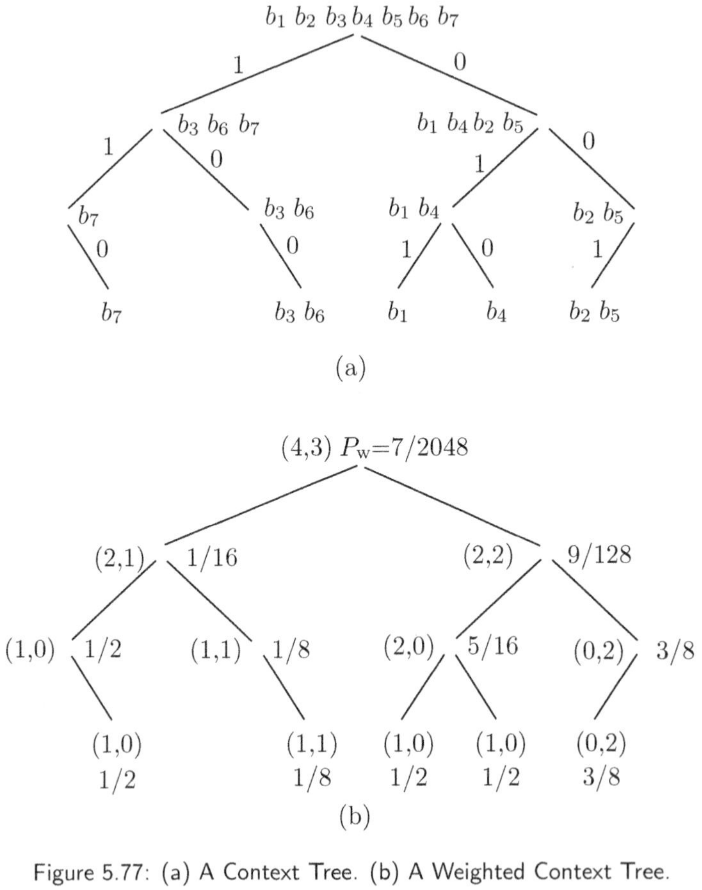

Figure 5.77: (a) A Context Tree. (b) A Weighted Context Tree.

Figure 5.77b shows how each node s contains the pair (a_s, b_s), the number of zeros and ones in the string associated with s. The root, for example, is associated with the entire string, so it contains the pair $(4,3)$. We still have to calculate or estimate a weighted probability P_w^s for each node s, the probability that should be sent to the arithmetic encoder to encode the string associated with s. This calculation is, in fact, the central part of the CTW algorithm. We start at the leaves because the only thing available in a leaf is the pair (a_s, b_s); there is no suffix. The best assumption that can therefore be made is that the substring consisting of a_s zeros and b_s ones that's associated with leaf s is memoryless, and the best weighted probability that can be defined for the node is the KT estimated probability $P_e(a_s, b_s)$. We therefore define

$$P_w^s \stackrel{\text{def}}{=} P_e(a_s, b_s) \quad \text{if depth}(s) = d. \tag{5.8}$$

Using the weighted probabilities for the leaves, we work our way recursively up the tree and calculate weighted probabilities for the internal nodes. For an internal node s we know a little more than for a leaf, since such a node has one or two children. The children, which are denoted by s_0 and s_1, have already been assigned weighted probabilities. We consider two cases. If the substring associated with suffix s is memoryless, then $P_e(a_s, b_s)$ is a good weighted probability for it. Otherwise the CTW method claims that the product $P_w^{s_0} P_w^{s_1}$ of the weighted probabilities of the child nodes is a good coding probability (a missing child node is considered, in such a case, to have weighted probability 1).

Since we don't know which of the above cases is true for a given internal node s, the best that we can do is to assign s a weighted probability that's the average of the two cases above, i.e.,

$$P_w^s \stackrel{\text{def}}{=} \frac{P_e(a_s, b_s) + P_w^{s_0} P_w^{s_1}}{2} \quad \text{if depth}(s) < d. \tag{5.9}$$

The last step that needs to be described is the way the context tree is updated when the next bit is input. Suppose that we have already input and encoded the string $b_1 b_2 \ldots b_{t-1}$. Thus, we have already constructed a context tree of depth d for this string, we have used Equations (5.8) and (5.9) to calculate weighted probabilities for the entire tree, and the root of the tree already contains a certain weighted probability. We now input the next bit b_t and examine it. Depending on what it is, we need to update the context tree for the string $b_1 b_2 \ldots b_{t-1} b_t$. The weighted probability at the root of the new tree will then be sent to the arithmetic encoder, together with bit b_t, and will be used to encode b_t.

If $b_t = 0$, then updating the tree is done by (1) incrementing the a_s counts for all nodes s, (2) updating the estimated probabilities $P_e(a_s, b_s)$ for all the nodes, and (3) updating the weighted probabilities $P_w(a_s, b_s)$ for all the nodes. If $b_t = 1$, then all the b_s should be incremented, followed by updating the estimated and weighted probabilities as above. Figure 5.78a shows how the context tree of Figure 5.77b is updated when $b_t = 0$.

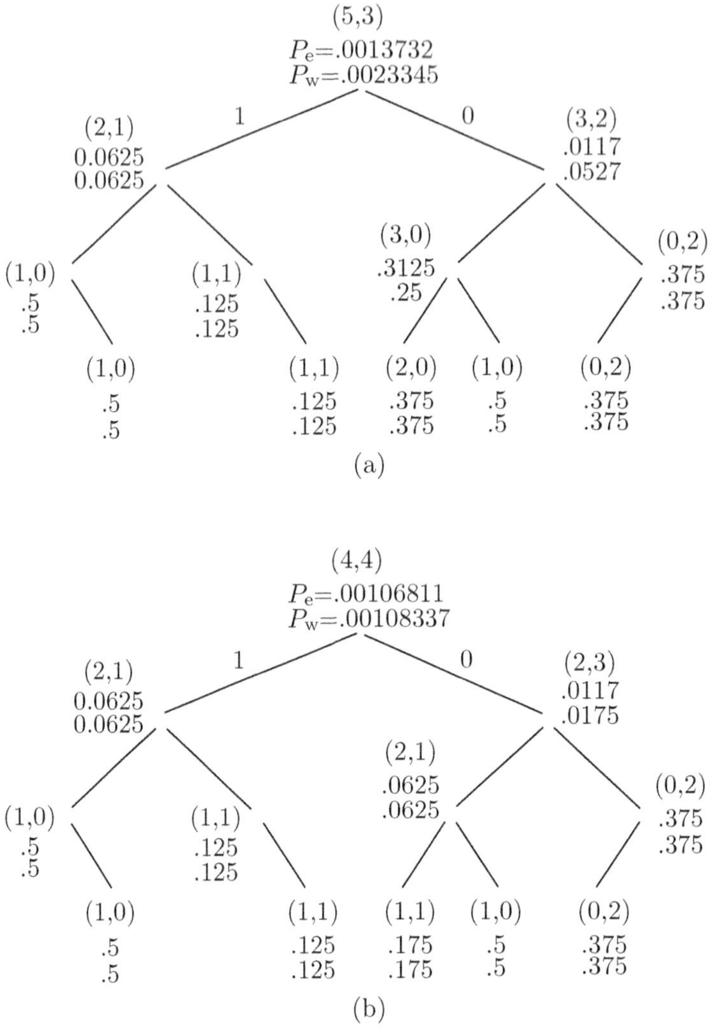

Figure 5.78: Context Trees for $b_t = 0, 1$.

Figure 5.78b shows how the context tree of Figure 5.77b is updated when $b_t = 1$.

◇ **Exercise 5.45:** Construct the context trees with depth 3 for the strings 000|0, 000|00, 000|1, and 000|11.

The depth d of the context tree is selected by the user (or is built into both encoder and decoder) and does not change during the compression job. The tree has to be updated for each input bit processed, but this requires updating at most $d + 1$ nodes. The number of operations needed to process n input bits is thus linear in n.

> I hoped that the contents of his pockets might help me to form a conclusion.
>
> —Arthur Conan Doyle, *Memoires of Sherlock Holmes*

5.16.1 CTW for Text Compression

The CTW method has so far been developed for compressing binary strings. In practice, we are normally interested in compressing text, image, and sound streams, and this section discusses one approach to applying CTW to text compression.

Each ASCII character consists of seven bits, and all 128 7-bit combinations are used. However, some combinations (such as `E` and `T`) are more common than others (such as `Z`, `<`, and certain control characters). Also, certain character pairs and triplets (such as `TH` and `THE`) appear more often than others. We therefore claim that if b_t is a bit in a certain ASCII character X, then the $t-1$ bits $b_1 b_2 \ldots b_{t-1}$ preceding it can act as context (even if some of them are not even parts of X but belong to characters preceding X). Experience shows that good results are obtained (1) with contexts of size 12, (2) when seven context trees are used, each to construct a model for one of the seven bits, and (3) if the original KT estimate is modified to the *zero-redundancy* estimate, defined by

$$P_e^z(a,b) \stackrel{\text{def}}{=} \frac{1}{2}P_e(a,b) + \frac{1}{4}\vartheta(a=0) + \frac{1}{4}\vartheta(b=0),$$

where $\vartheta(\text{true}) \stackrel{\text{def}}{=} 1$ and $\vartheta(\text{false}) \stackrel{\text{def}}{=} 0$.

Another experimental feature is a change in the definition of the weighted probabilities. The original definition, Equation (5.9), is used for the two trees on the ASCII borders (i.e., the ones for bits 1 and 7 of each ASCII code). The weighted probabilities for the five context trees for bits 2–6 are defined by $P_s^w = P_w^{s_0} P_w^{s_1}$.

This produces typical compression of 1.8 to 2.3 bits/character on the (concatenated) documents of the Calgary Corpus.

Paul Volf [Volf 97] has proposed other approaches to CTW text compression.

Statistics are like bikinis. What they reveal
is suggestive, but what they conceal is vital.

—Aaron Levenstein

6 Dictionary Methods

Statistical compression methods use a statistical model of the data, which is why the quality of compression they achieve depends on how good that model is. Dictionary-based compression methods do not use a statistical model, nor do they use variable-length codes. Instead they select strings of symbols and encode each string as a *token* using a dictionary. The dictionary holds strings of symbols, and it may be static or dynamic (adaptive). The former is permanent, sometimes permitting the addition of strings but no deletions, whereas the latter holds strings previously found in the input stream, allowing for additions and deletions of strings as new input is being read.

Given a string of n symbols, a dictionary-based compressor can, in principle, compress it down to nH bits where H is the entropy of the string. Thus, dictionary-based compressors are entropy encoders, but only if the input file is very large. For most files in practical applications, dictionary-based compressors produce results that are good enough to make this type of encoder very popular. Such encoders are also general purpose, performing on images and audio data as well as they perform on text.

The simplest example of a static dictionary is a dictionary of the English language used to compress English text. Imagine a dictionary containing perhaps half a million words (without their definitions). A word (a string of symbols terminated by a space or a punctuation mark) is read from the input stream and the dictionary is searched. If a match is found, an index to the dictionary is written into the output stream. Otherwise, the uncompressed word itself is written. (This is an example of *logical compression*.)

As a result, the output stream contains indexes and raw words, and it is important to distinguish between them. One way to achieve this is to reserve an extra bit in every item written. In principle, a 19-bit index is sufficient to specify an item in a $2^{19} = 524{,}288$-word dictionary. Thus, when a match is found, we can write a 20-bit token that consists of a flag bit (perhaps a zero) followed by a 19-bit index. When no match is found, a flag of 1 is written, followed by the size of the unmatched word, followed by the word itself.

D. Salomon, G. Motta, *Handbook of Data Compression*, 5th ed.
DOI 10.1007/978-1-84882-903-9_6, © Springer-Verlag London Limited 2010

Example: Assuming that the word `bet` is found in dictionary entry 1025, it is encoded as the 20-bit number 0|0000000010000000001. Assuming that the word `xet` is not found, it is encoded as 1|0000011|01111000|01100101|01110100. This is a 4-byte number where the 7-bit field 0000011 indicates that three more bytes follow.

Assuming that the size is written as a 7-bit number, and that an average word size is five characters, an uncompressed word occupies, on average, six bytes (= 48 bits) in the output stream. Compressing 48 bits into 20 is excellent, provided that it happens often enough. Thus, we have to answer the question how many matches are needed in order to have overall compression? We denote the probability of a match (the case where the word is found in the dictionary) by P. After reading and compressing N words, the size of the output stream will be $N[20P+48(1-P)] = N[48-28P]$ bits. The size of the input stream is (assuming five characters per word) $40N$ bits. Compression is achieved when $N[48-28P] < 40N$, which implies $P > 0.29$. We need a matching rate of 29% or better to achieve compression.

◇ **Exercise 6.1:** What compression factor do we get with $P = 0.9$?

As long as the input stream consists of English text, most words will be found in a 500,000-word dictionary. Other types of data, however, may not do that well. A file containing the source code of a computer program may contain "words" such as `cout`, `xor`, and `malloc` that may not be found in an English dictionary. A binary file normally contains gibberish when viewed in ASCII, so very few matches may be found, resulting in considerable expansion instead of compression.

This shows that a static dictionary is not a good choice for a general-purpose compressor. It may, however, be a good choice for a special-purpose one. Consider a chain of hardware stores, for example. Their files may contain words such as `nut`, `bolt`, and `paint` many times, but words such as `peanut`, `lightning`, and `painting` will be rare. Special-purpose compression software for such a company may benefit from a small, specialized dictionary containing, perhaps, just a few hundred words. The computers in each branch would have a copy of the dictionary, making it easy to compress files and send them between stores and offices in the chain.

In general, an adaptive dictionary-based method is preferable. Such a method can start with an empty dictionary or with a small, default dictionary, add words to it as they are found in the input stream, and delete old words because a big dictionary slows down the search. Such a method consists of a loop where each iteration starts by reading the input stream and breaking it up (parsing it) into words or phrases. It then should search the dictionary for each word and, if a match is found, write a token on the output stream. Otherwise, the uncompressed word should be written and also added to the dictionary. The last step in each iteration checks whether an old word should be deleted from the dictionary. This may sound complicated, but it has two advantages:

1. It involves string search and match operations, rather than numerical computations. Many programmers prefer that.
2. The decoder is simple (this is an asymmetric compression method). In statistical compression methods, the decoder is normally the exact opposite of the encoder (symmetric compression). In an adaptive dictionary-based method, however, the decoder has to read its input stream, determine whether the current item is a token or uncompressed data, use tokens to obtain data from the dictionary, and output the final, uncompressed

data. It does not have to parse the input stream in a complex way, and it does not have to search the dictionary to find matches. Many programmers like that, too.

J. Ziv

A. Lempel

The LZW Trio. Having one's name attached to a scientific discovery, technique, or phenomenon is considered a special honor in science. Having one's name associated with an entire field of science is even more so. This is what happened to Jacob Ziv and Abraham Lempel. In the late 1970s these two researchers developed the first methods, LZ77 and LZ78, for dictionary-based compression. Their ideas have been a source of inspiration to many researchers who generalized, improved, and combined them with RLE and statistical methods to form many popular lossless compression methods for text, images, and audio. Close to 30 such methods are described in this chapter, some in great detail. Of special interest is the popular LZW algorithm, partly devised by Terry Welch (Section 6.13), which has extended LZ78 and made it practical and popular.

T. Welch

> I love the dictionary, Kenny, it's the only book with the words in the right place.
>
> —Paul Reynolds as Colin Mathews in *Press Gang* (1989)

6.1 String Compression

In general, compression methods based on strings of symbols can be more efficient than methods that compress individual symbols. To understand this, the reader should first review Exercise A.4. This exercise shows that in principle, better compression is possible if the symbols of the alphabet have very different probabilities of occurrence. We use a simple example to show that the probabilities of strings of symbols vary more than the probabilities of the individual symbols constituting the strings.

We start with a 2-symbol alphabet a_1 and a_2, with probabilities $P_1 = 0.8$ and $P_2 = 0.2$, respectively. The average probability is 0.5, and we can get an idea of the variance (how much the individual probabilities deviate from the average) by calculating the sum of absolute differences $|0.8 - 0.5| + |0.2 - 0.5| = 0.6$. Any variable-length code would assign 1-bit codes to the two symbols, so the average length of the code is one bit per symbol.

We now generate all the strings of two symbols. There are four of them, shown in Table 6.1a, together with their probabilities and a set of Huffman codes. The average probability is 0.25, so a sum of absolute differences similar to the one above yields

$$|0.64 - 0.25| + |0.16 - 0.25| + |0.16 - 0.25| + |0.04 - 0.25| = 0.78.$$

String	Probability	Code
a_1a_1	$0.8 \times 0.8 = 0.64$	0
a_1a_2	$0.8 \times 0.2 = 0.16$	11
a_2a_1	$0.2 \times 0.8 = 0.16$	100
a_2a_2	$0.2 \times 0.2 = 0.04$	101

(a)

Str. size	Variance of prob.	Avg. size of code
1	0.6	1
2	0.78	0.78
3	0.792	0.728

(c)

String	Probability	Code
$a_1a_1a_1$	$0.8 \times 0.8 \times 0.8 = 0.512$	0
$a_1a_1a_2$	$0.8 \times 0.8 \times 0.2 = 0.128$	100
$a_1a_2a_1$	$0.8 \times 0.2 \times 0.8 = 0.128$	101
$a_1a_2a_2$	$0.8 \times 0.2 \times 0.2 = 0.032$	11100
$a_2a_1a_1$	$0.2 \times 0.8 \times 0.8 = 0.128$	110
$a_2a_1a_2$	$0.2 \times 0.8 \times 0.2 = 0.032$	11101
$a_2a_2a_1$	$0.2 \times 0.2 \times 0.8 = 0.032$	11110
$a_2a_2a_2$	$0.2 \times 0.2 \times 0.2 = 0.008$	11111

(b)

Table 6.1: Probabilities and Huffman Codes for a Two-Symbol Alphabet.

The average size of the Huffman code is $1 \times 0.64 + 2 \times 0.16 + 3 \times 0.16 + 3 \times 0.04 = 1.56$ bits per string, which is 0.78 bits per symbol.

In the next step we similarly create all eight strings of three symbols. They are shown in Table 6.1b, together with their probabilities and a set of Huffman codes. The average probability is 0.125, so a sum of absolute differences similar to the ones above yields

$$|0.512 - 0.125| + 3|0.128 - 0.125| + 3|0.032 - 0.125| + |0.008 - 0.125| = 0.792.$$

The average size of the Huffman code in this case is $1 \times 0.512 + 3 \times 3 \times 0.128 + 3 \times 5 \times 0.032 + 5 \times 0.008 = 2.184$ bits per string, which equals 0.728 bits per symbol.

As we keep generating longer and longer strings, the probabilities of the strings differ more and more from their average, and the average code size gets better (Table 6.1c). This is why a compression method that compresses strings, rather than individual symbols, can, in principle, yield better results. This is also the reason why the various dictionary-based methods are in general better and more popular than the Huffman method and its variants (see also Section 7.19). The above conclusion is a fundamental result of rate-distortion theory, that part of information theory that deals with data compression.

6.2 Simple Dictionary Compression

The topic of this section is a simple, two-pass method, related to me by Ismail Mohamed (see Preface to the 3rd edition of the complete reference). The first pass reads the source file and prepares a list of all the different bytes found. The second pass uses this list to actually compress the data bytes. Here are the steps in detail.

1. The source file is read and a list is prepared of the distinct bytes encountered. For each byte, the number of times it occurs in the source file (its frequency) is also included in the list.

2. The list is sorted in descending order of the frequencies. Thus, it starts with byte values that are common in the file, and it ends with bytes that are rare. Since the list consists of distinct bytes, it can have at most 256 elements.

3. The sorted list becomes the dictionary. It is written on the compressed file, preceded by its length (a 1-byte integer).

4. The source file is read again byte by byte. Each byte is located in the dictionary (by a direct search) and its index is noted. The index is a number in the interval $[0, 255]$, so it requires between 1 and 8 bits (but notice that most indexes will normally be small numbers because common byte values are stored early in the dictionary). The index is written on the compressed file, preceded by a 3-bit code denoting the index's length. Thus, code 000 denotes a 1-bit index, code 001 denotes a 2-bit index, and so on up to code 111, which denotes an 8-bit index.

The compressor maintains a short, 2-byte buffer where it collects the bits to be written on the compressed file. When the first byte of the buffer is filled, it is written on the file and the second byte is moved to the first byte.

Decompression is straightforward. The decompressor starts by reading the length of the dictionary, then the dictionary itself. It then decodes each byte by reading its 3-bit code, followed by its index value. The index is used to locate the next data byte in the dictionary.

Compression is achieved because the dictionary is sorted by the frequency of the bytes. Each byte is replaced by a quantity of between 4 and 11 bits (a 3-bit code followed by 1 to 8 bits). A 4-bit quantity corresponds to a compression ratio of 0.5, while an 11-bit quantity corresponds to a compression ratio of 1.375, an expansion. The worst case is a file where all 256 byte values occur and have a uniform distribution. The compression ratio in such a case is the average

$$\begin{aligned}&(2\times 4+2\times 5+4\times 6+8\times 7+16\times 8+32\times 9+64\times 10+128\times 11)/(256\times 8)\\&=2562/2048=1.2509765625,\end{aligned}$$

indicating expansion! (Actually, slightly worse than 1.25, because the compressed file also includes the dictionary, whose length in this case is 257 bytes.) Experience indicates typical compression ratios of about 0.5.

The probabilities used here were obtained by counting the numbers of codes of various sizes. Thus, there are two 4-bit codes 000|0 and 000|1, two 5-bit codes 001|10 and 001|11, four 6-bit codes 010|100, 010|101, 010|110 and 010|111, eight 7-bit codes 011|1000, 011|1001, 011|1010, 011|1011, 011|1100, 011|1101, 011|1110, and 011|1111, and so on, up to 128 11-bit codes.

The downside of the method is slow compression; a result of the two passes the compressor performs combined with the slow search (slow, because the dictionary is not sorted by byte values, so binary search cannot be used). Decompression, in contrast, is not slow.

◇ **Exercise 6.2:** Design a reasonable organization for the list maintained by this method.

6.3 LZ77 (Sliding Window)

An important source of redundancy in digital data is repeating phrases. This type of redundancy exists in all types of data—audio, still images, and video—but is easiest to visualize for text. Figure 6.2 lists two paragraphs from this book and makes it obvious that certain phrases, such as words, parts of words, and other bits and pieces of text repeat several times or even many times. Phrases tend to repeat in the same paragraph, but repetitions are often found in paragraphs that are distant from one another. Thus, phrases that occur again and again in the data constitute the basis of all the dictionary-based compression algorithms.

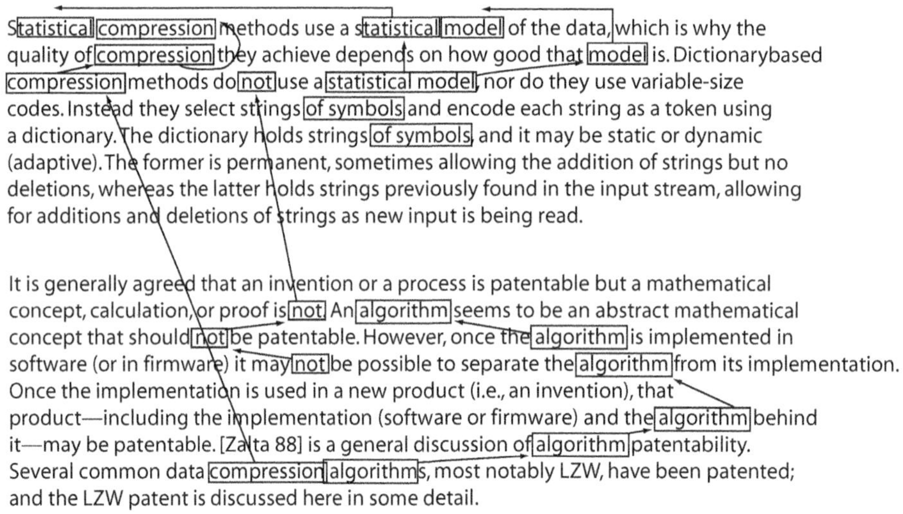

Figure 6.2: Repeating Phrases in Text.

The principle of LZ77 (sometimes also referred to as LZ1) [Ziv and Lempel 77] is to use part of the previously-seen input stream as the dictionary. The encoder maintains a window to the input stream and shifts the input in that window from right to left as strings of symbols are being encoded. Thus, the method is based on a *sliding window*. The window below is divided into two parts. The part on the left is the *search buffer*. This is the current dictionary, and it includes symbols that have recently been input and encoded. The part on the right is the *look-ahead buffer*, containing text yet to be encoded. In practical implementations the search buffer is some thousands of bytes long, while the look-ahead buffer is only tens of bytes long. The vertical bar between the `t` and the `e` below represents the current dividing line between the two buffers. We assume that the text `sir␣sid␣eastman␣easily␣t` has already been compressed, while the text `eases␣sea␣sick␣seals` still needs to be compressed.

← coded text... `sir␣sid␣eastman␣easily␣t|eases␣sea␣sick␣seals`... ← text to be read

The encoder scans the search buffer backwards (from right to left) looking for a match for the first symbol `e` in the look-ahead buffer. It finds one at the `e` of the word `easily`. This `e` is at a distance (offset) of 8 from the end of the search buffer. The encoder then matches as many symbols following the two `e`'s as possible. Three symbols `eas` match in this case, so the length of the match is 3. The encoder then continues the backward scan, trying to find longer matches. In our case, there is one more match, at the word `eastman`, with offset 16, and it has the same length. The encoder selects the longest match or, if they are all the same length, the last one found, and prepares the token (16, 3, `e`).

Selecting the last match, rather than the first one, simplifies the encoder, because it only has to keep track of the latest match found. It is interesting to note that selecting the first match, while making the program somewhat more complex, also has an advantage. It selects the smallest offset. It would seem that this is not an advantage, because a token should have room for the largest possible offset. However, it is possible to follow LZ77 with Huffman, or some other statistical coding of the tokens, where small offsets are assigned shorter codes. This method, proposed by Bernd Herd, is called LZH. Having many small offsets implies better compression in LZH.

◇ **Exercise 6.3:** How does the decoder know whether the encoder selects the first match or the last match?

In general, an LZ77 token has three parts: offset, length, and next symbol in the look-ahead buffer (which, in our case, is the *second* `e` of the word `teases`). The offset can also be thought of as the distance between the previous and the current occurrences of the string being compressed. This token is written on the output stream, and the window is shifted to the right (or, alternatively, the input stream is moved to the left) four positions: three positions for the matched string and one position for the next symbol.

`...sir␣|sid␣eastman␣easily␣tease|s␣sea␣sick␣seals...|...`

If the backward search yields no match, an LZ77 token with zero offset and length and with the unmatched symbol is written. This is also the reason a token has a third component. Tokens with zero offset and length are common at the beginning of any compression job, when the search buffer is empty or almost empty. The first five steps in encoding our example are the following:

	`sir␣sid␣eastman␣`	⇒	(0,0,"s")
`s`	`ir␣sid␣eastman␣e`	⇒	(0,0,"i")
`si`	`r␣sid␣eastman␣ea`	⇒	(0,0,"r")
`sir`	`␣sid␣eastman␣eas`	⇒	(0,0,"␣")
`sir␣`	`sid␣eastman␣easi`	⇒	(4,2,"d")

◇ **Exercise 6.4:** What are the next two steps?

Clearly, a token of the form $(0, 0, \ldots)$, which encodes a single symbol, does not provide good compression. It is easy to estimate its length. The size of the offset is $\lceil \log_2 S \rceil$, where S is the length of the search buffer. In practice, the search buffer may be a few thousand bytes long, so the offset size is typically 10–12 bits. The size of the "length" field is similarly $\lceil \log_2(L-1) \rceil$, where L is the length of the look-ahead buffer (see below for the -1). In practice, the look-ahead buffer is only a few tens of bytes long, so the size of the "length" field is just a few bits. The size of the "symbol" field is typically 8 bits, but in general, it is $\lceil \log_2 A \rceil$, where A is the alphabet size. The total size of the 1-symbol token $(0, 0, \ldots)$ may typically be $11 + 5 + 8 = 24$ bits, much longer than the raw 8-bit size of the (single) symbol it encodes.

Here is an example showing why the "length" field may be longer than the size of the look-ahead buffer:

`...Mr.␣|alf␣eastman␣easily␣grows␣alf|alfa␣in␣his␣|garden...` .

The first symbol `a` in the look-ahead buffer matches the five `a`'s in the search buffer. It seems that the two extreme `a`'s match with a length of 3 and the encoder should select the last (leftmost) of them and create the token (28,3,"a"). In fact, it creates the token (3,4,"␣"). The four-symbol string `alfa` in the look-ahead buffer is matched with the last three symbols `alf` in the search buffer **and** the first symbol `a` in the look-ahead buffer. The reason for this is that the decoder can handle such a token naturally, without any modifications. It starts at position 3 of its search buffer and copies the next four symbols, one by one, extending its buffer to the right. The first three symbols are copies of the old buffer contents, and the fourth one is a copy of the first of those three. The next example is even more convincing (and only somewhat contrived):

`...|alf␣eastman␣easily␣yells␣A|AAAAAAAAAA|AAAAAH...` .

The encoder creates the token (1,9,`A`), matching the first nine copies of `A` in the look-ahead buffer and including the tenth `A`. This is why, in principle, the length of a match can be up to the size of the look-ahead buffer minus 1.

The decoder is much simpler than the encoder (LZ77 is therefore an asymmetric compression method). It has to maintain a buffer, equal in size to the encoder's window. The decoder inputs a token, finds the match in its buffer, writes the match and the third token field on the output stream, and shifts the matched string and the third field into the buffer. This implies that LZ77, or any of its variants, is useful in cases where a file is compressed once (or just a few times) and is decompressed often. A rarely-used archive of compressed files is a good example.

At first it seems that this method does not make any assumptions about the input data. Specifically, it does not pay attention to any symbol frequencies. A little thinking, however, shows that because of the nature of the sliding window, the LZ77 method always compares the look-ahead buffer to the recently-input text in the search buffer and never to text that was input long ago (and has therefore been flushed out of the search buffer). Thus, the method implicitly assumes that patterns in the input data occur close together. Data that satisfies this assumption will compress well.

The basic LZ77 method was improved in several ways by researchers and programmers during the 1980s and 1990s. One way to improve it is to use variable-size "offset" and "length" fields in the tokens. Another way is to increase the sizes of both buffers. Increasing the size of the search buffer makes it possible to find better matches, but the trade-off is an increased search time. A large search buffer therefore requires a more sophisticated data structure that allows for fast search (Section 6.13.2). A third improvement has to do with sliding the window. The simplest approach is to move all the text in the window to the left after each match. A faster method is to replace the linear window with a *circular queue*, where sliding the window is done by resetting two pointers (Section 6.3.1). Yet another improvement is adding an extra bit (a flag) to each token, thereby eliminating the third field (Section 6.4). Of special notice is the hash table employed by the Deflate algorithm (Section 6.25.3) to search for matches.

6.3.1 A Circular Queue

The circular queue is a basic data structure. Physically, it is a linear array, but it is used as a circular array. Figure 6.3 illustrates a simple example. It shows a 16-byte array with characters appended at the "end" and deleted from the "start." Both the start and end positions move, and two pointers, `s` and `e`, point to them all the time. In (a) the queue consists of the eight characters `sid␣east`, with the rest of the buffer empty. In (b) all 16 bytes are occupied, and `e` points to the end of the buffer. In (c), the first letter `s` has been deleted and the `l` of `easily` inserted. Notice how pointer `e` is now located *to the left* of `s`. In (d), the two letters `id` have been deleted just by moving the `s` pointer; the characters themselves are still present in the array but have been effectively deleted. In (e), the two characters `y␣` have been appended and the `e` pointer moved. In (f), the pointers show that the buffer ends at `teas` and starts at `tman`. Inserting new characters into the circular queue and moving the pointers is thus equivalent to shifting the contents of the queue. No actual shifting or moving is necessary, though.

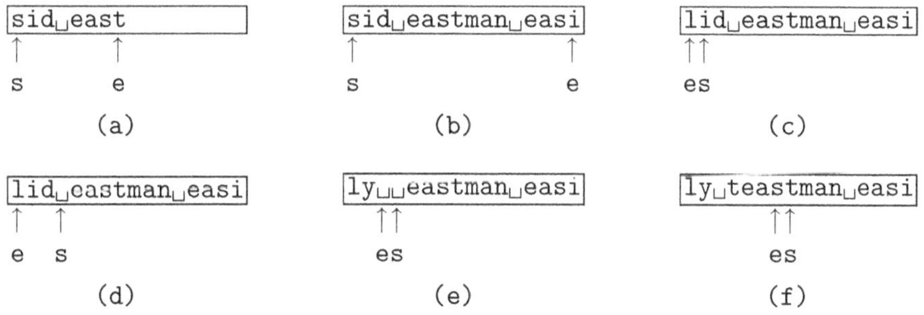

Figure 6.3: A Circular Queue.

More information on circular queues can be found in most texts on data structures.

From the dictionary
Circular. (1) Shaped like or nearly like a circle. (2) Defining one word in terms of another that is itself defined in terms of the first word. (3) Addressed or distributed to a large number of persons.

6.3.2 LZR

The data compression literature is full of references to LZ77 and LZ78. There is, however, also an LZ76 algorithm [Lempel and Ziv 76]. It is little known because it deals with the complexity of text, rather than with compressing the text. The point is that LZ76 measures complexity by looking for previously-found strings of text. It can therefore be extended to become a text compression algorithm. The LZR method described here, proposed by [Rodeh et al. 81], is such an extension.

LZR is a variant of the basic LZ77 method, where the lengths of both the search and look-ahead buffers are unbounded. In principle, such a method will always find the best possible match to the string in the look-ahead buffer, but it is immediately obvious that any practical implementation of LZR will have to deal with the finite memory space and run-time available.

The space problem can be handled by allocating more and more space to the buffers until no more space is left, and then either (1) keeping the buffers at their current sizes or (2) clearing the buffers and starting with empty buffers. The time constraint may be more severe. The basic algorithm of LZ76 uses linear search and requires $O(n^2)$ time to process an input string of n symbols. The developers of LZR propose to reduce these space and time complexities to $O(n)$ by employing the special suffix trees proposed by [McCreight 76], but readers of this chapter will have many opportunities to see that other LZ methods, such as LZ78 and its varieties, offer similar compression performance, while being simpler to implement.

The data structure developed by McCreight is based on a complicated multiway tree. Those familiar with tree data structures know that the basic operations on a tree are (1) adding a new node, (2) modifying an existing node, and (3) deleting a node. Of these, deletion is normally the most problematic since it tends to change the entire structure of a tree.

Thus, instead of deleting nodes when the symbols represented by them overflow off the left end of the search buffer, LZR simply marks a node as deleted. It also constructs several trees and deletes a tree when all its nodes have beed marked as deleted (i.e., when all the symbols represented by the tree have slid off the search buffer). As a result, LZR implementation is complex.

Another downside of LZR is the sizes of the offset and length fields of an output triplet. With buffers of unlimited sizes, these fields may become big. LZR handles this problem by encoding these fields in a variable-length code. Specifically, the Even–Rodeh codes of Section 3.7 are used to encode these fields. The length of the Even–Rodeh code of the integer n is close to $2\log_2 n$, so even for n values in the millions, the codes are about 20 bits long.

6.4 LZSS

LZSS is an efficient variant of LZ77 developed by Storer and Szymanski in 1982 [Storer and Szymanski 82]. It improves LZ77 in three directions: (1) It holds the look-ahead buffer in a circular queue, (2) it creates tokens with two fields instead of three, and (3) it holds the search buffer (the dictionary) in a binary search tree, (this data structure was incorporated in LZSS by [Bell 86]).

The second improvement of LZSS over LZ77 is in the tokens created by the encoder. An LZSS token contains just an offset and a length. If no match was found, the encoder emits the uncompressed code of the next symbol instead of the wasteful three-field token $(0, 0, \ldots)$. To distinguish between tokens and uncompressed codes, each is preceded by a single bit (a flag).

In practice, the search buffer may be a few thousand bytes long, so the offset field would typically be 11–13 bits. The size of the look-ahead buffer should be selected such that the total size of a token would be 16 bits (2 bytes). For example, if the search buffer size is 2 Kbyte ($= 2^{11}$), then the look-ahead buffer should be 32 bytes long ($= 2^5$). The offset field would be 11 bits long and the length field, 5 bits (the size of the look-ahead buffer). With this choice of buffer sizes the encoder will emit either 2-byte tokens or 1-byte uncompressed ASCII codes. But what about the flag bits? A good practical idea is to collect eight output items (tokens and ASCII codes) in a small buffer, then output one byte consisting of the eight flags, followed by the eight items (which are 1 or 2 bytes long each).

The third improvement has to do with binary search trees. A binary search tree is a binary tree where the left subtree of every node A contains nodes smaller than A, and the right subtree contains nodes greater than A. Since the nodes of our binary search trees contain strings, we first need to know how to compare two strings and decide which one is "bigger." This is easily understood by imagining that the strings appear in a dictionary or a lexicon, where they are sorted alphabetically. Clearly, the string `rote` precedes the string `said` since `r` precedes `s` (even though `o` follows `a`), so we consider `rote` smaller than `said`. This concept is called *lexicographic order* (ordering strings lexicographically).

What about the string `␣abc`? Most modern computers use ASCII codes to represent characters (although more and more use Unicode, discussed in Section 11.12, and some older IBM, Amdahl, Fujitsu, and Siemens mainframe computers use the old, 8-bit EBCDIC code developed by IBM), and in ASCII the code of a blank space precedes those of the letters, so a string that starts with a space will be smaller than any string that starts with a letter. In general, the *collating sequence* of the computer determines the sequence of characters arranged from small to big. Figure 6.4 shows two examples of binary search trees.

Notice the difference between the (almost) balanced tree in Figure 6.4a and the skewed one in Figure 6.4b. They contain the same 14 nodes, but they look and behave very differently. In the balanced tree any node can be found in at most four steps. In the skewed tree up to 14 steps may be needed. In either case, the maximum number of steps needed to locate a node equals the height of the tree. For a skewed tree (which is really the same as a linked list), the height is the number of elements n; for a balanced tree, the height is $\lceil \log_2 n \rceil$, a much smaller number. More information on the properties

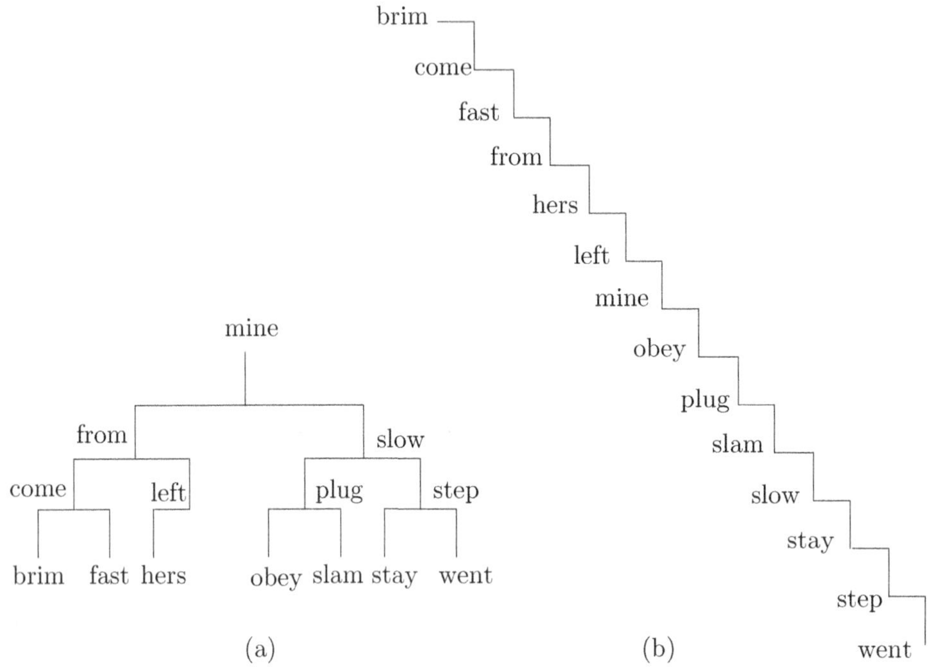

Figure 6.4: Two Binary Search Trees.

of binary search trees may be found in any text on data structures.

Here is an example showing how a binary search tree can be used to speed up the search of the dictionary. We assume an input stream with the short sentence `sid␣eastman␣clumsily␣teases␣sea␣sick␣seals`. To keep the example simple, we assume a window of a 16-byte search buffer followed by a 5-byte look-ahead buffer. After the first $16 + 5$ characters have been input, the sliding window is

`sid␣eastman␣clum|sily␣|teases␣sea␣sick␣seals`

with the string `teases␣sea␣sick␣seals` still waiting to be input.

The encoder scans the search buffer, creating the 12 five-character strings of Table 6.5 (12 since $16 - 5 + 1 = 12$), which are inserted into the binary search tree, each with its offset.

The first symbol in the look-ahead buffer is `s`, so the encoder searches the tree for strings that start with an `s`. Two are found, at offsets 16 and 10, and the first of them, `sid␣e` (at offset 16) provides a longer match.

(We now have to sidetrack and discuss the case where a string in the tree completely matches that in the look-ahead buffer. In that case the encoder should go back to the search buffer, to attempt to match longer strings. In principle, the maximum length of a match can be $L - 1$.)

In our example, the match is of length 2, and the 2-field token $(16, 2)$ is emitted. The encoder now has to slide the window two positions to the right, and update the tree. The new window is

sid␣e	16
id␣ea	15
d␣eas	14
␣east	13
eastm	12
astma	11
stman	10
tman␣	09
man␣c	08
an␣cl	07
n␣clu	06
␣clum	05

Table 6.5: Five-Character Strings.

`si|d␣eastman␣clumsi|ly␣te|ases␣sea␣sick␣seals`

The tree should be updated by deleting strings `sid␣e` and `id␣ea`, and inserting the new strings `clums` and `lumsi`. If a longer, k-letter, string is matched, the window has to be shifted k positions, and the tree should be updated by deleting k strings and adding k new strings, but which ones?

A little thinking shows that the k strings to be deleted are the first ones in the search buffer before the shift, and the k strings to be added are the last ones in it after the shift. A simple procedure for updating the tree is to prepare a string consisting of the first five letters in the search buffer, find it in the tree, and delete it. Then slide the buffer one position to the right (or shift the data to the left), prepare a string consisting of the last five letters in the search buffer, and append it to the tree. This should be repeated k times.

Since each update deletes and adds the same number of strings, the tree size never changes, except at the start and end of the compression. It always contains T nodes, where T is the length of the search buffer minus the length of the look-ahead buffer plus 1 ($T = S - L + 1$). The shape of the tree, however, may change significantly. As nodes are being added and deleted, the tree may change its shape between a completely skewed tree (the worst case for searching) and a balanced one, the ideal shape for searching.

When the encoder starts, it outputs the first L symbols of the input in raw format and shifts them from the look-ahead buffer to the search buffer. It then fills up the remainder of the search buffer with blanks. In our example, the initial tree will consist of the five strings `␣␣␣␣s`, `␣␣␣si`, `␣␣sid`, `␣sid␣`, and `sid␣e`. As compression progresses, the tree grows to $S - L + 1 = 12$ nodes, and stays at that size until it shrinks toward the end.

6.4.1 LZB

LZB is an extension of LZSS. This method, proposed in [Bell 87], is the result of evaluating and comparing several data structures and variable-length codes with an eye toward improving the performance of LZSS.

In LZSS, the length of the offset field is fixed and is large enough to accommodate any pointers to the search buffer. If the size of the search buffer is S, then the offset field is $s \stackrel{\text{def}}{=} \lceil \log_2 S \rceil$ bits long. However, when the encoder starts, the search buffer is empty, so the offset fields of the first tokens output do not use all the bits allocated to them. In principle, the size of the offset field should start at one bit. This is enough as long as there are only two symbols in the search buffer. The offset should then be increased to two bits until the search buffer contains four symbols, and so on. As more and more symbols are shifted into the search buffer and tokens are output, the size of the offset field should be increased. When the search buffer is between 25% and 50% full, the size of the offset should still be $s-1$ and should be increased to s only when the buffer is 50% full or more.

Even this sophisticated scheme can be improved upon to yield slightly better compression. Section 2.9 introduces phased-in codes and shows how to construct a set of pointers of two sizes to address an array of n elements even when n is not a power of 2. LZB employs phased-in codes to encode the size of the offset field at any point in the encoding. As an example, when there are nine symbols in the search buffer, LZB uses the seven 3-bit codes and two 4-bit codes 000, 001, 010, 011, 100, 101, 110, 1110, and 1111.

The second field of an LZSS token is the match length l. This is normally a fairly small integer, but can sometimes have large values. Recall that LZSS computes a value p and outputs a token only if it requires fewer bits than emitting p raw characters. Thus, the smallest value of l is $p+1$. LZB employs the Elias gamma code of Section 3.4 to encode the length field. A match of length i is encoded as the gamma code of the integer $i-p+1$. The length of the gamma code of the integer n is $1+2\lfloor \log_2 n \rfloor$ and it is ideal for cases where n appears in the input with probability $1/(2n^2)$. As a simple example, assuming that $p=3$, a match of five symbols is encoded as the gamma code of $5-3+1=3$ which is 011.

6.4.2 SLH

SLH is another variant of LZSS, described in [Brent 87]. It employs a circular buffer and it outputs raw symbols and pairs as does LZSS. The main difference is that SLH is a two-pass algorithm where the first pass employs a hash table to locate the best match and to count frequencies, and the second pass encodes the offsets and the raw symbols with Huffman codes prepared from the frequencies counted by the first pass.

We start with a few details about the hash table H. The first pass starts with an empty H. As strings are searched, matches are either found in H or are not found and are then added to H. What is actually stored in a location of H when a string is added, is a pointer to the string in the search buffer. In order to find a match for a string s of symbols, s is passed through a hashing function that returns a pointer to H. Thus, a match to s can be located (or verified as not found) in H in one step. (This discussion of hash tables ignores collisions, and detailed descriptions of this data structure can be found in many texts.)

We denote by p the break-even point, where a token representing a match of p or more symbols is shorter than the raw symbols themselves. At a general point in the encoding process, the search buffer has symbols a_1 through a_j and the look-ahead buffer has symbols a_{j+1} through a_n. We select the p-symbol string $s = a_{j+1}a_{j+2}\ldots a_{j+p-1}$, hash it to a pointer q and check $H[q]$. If a match is not found, then s is added to H (i.e., an offset P is stored in H and is updated as the buffers are shifted and s moves to the left). If a match is found, the offset P found in H and the match length are prepared as the token pair (P, p) representing the best match so far. This token is saved but is not immediately output. In either case, match found or not found, the match length is incremented by 1 and the process is repeated in an attempt to find a longer match.

In practice, the search buffer is implemented as a circular queue. Each time it is shifted, symbols are moved out of its "left" end and matches containing these symbols have to be deleted from H, which further complicates this algorithm.

The second pass of SLH encodes the offsets and the raw characters with Huffman codes. The frequencies counted in the first pass are first normalized to the interval $[0, 511]$ and a table with 512 Huffman codes is prepared (this table has to be written on the output, but its size is only a few hundred bytes). The implementation discussed in [Brent 87] assumes that the data symbols are bytes and that S, the size of the search buffer, is less than 2^{18}. Thus, an offset (a pointer to the search buffer) is at most 18 bits long. An 18-bit offset is broken up into two 9-bit parts and each part is Huffman encoded separately. Each raw character is also Huffman encoded.

The developer of this method does not indicate how an 18-bit offset should be split in two and precisely what frequencies are counted by the first pass. If that pass counts the frequencies of the raw input characters, then these characters will be efficiently encoded by the Huffman codes, but the same Huffman codes will not correspond to the frequencies of the 9-bit parts of the offset. It is possible to count the frequencies of the 9-bit parts and prepare two separate Huffman code tables for them, but other LZ methods offer the same or better compression performance, while being simpler to implement and evaluate.

The developer of this method does not explain the name SLH. The excellent book *Text Compression* [Bell et al. 90] refers to this method as LZH, but our book already uses LZH as the name of an LZ77 variant proposed by Bernd Herd, which leads to ambiguity. Even data compression experts cannot always avoid confusion.

> An expert is a man who tells you a simple thing in a confused way in such a fashion as to make you think the confusion is your own fault.
>
> —William Castle

6.4.3 LZARI

The following is quoted from [Okumura 98].

During the summer of 1988, I [Haruhiko Okumura] wrote another compression program, LZARI. This program is based on the following observation: Each output of LZSS is either a single character or a ⟨position,length⟩ pair. A single character can be coded as an integer between 0 and 255. As for the ⟨length⟩ field, if the range of ⟨length⟩ is 2 to 257, say, it can be coded as an integer between 256 and 511. Thus, I can say that there are 512 kinds of "characters," and the "characters" 256 through 511 are accompanied by a ⟨position⟩ field. These 512 "characters" can be Huffman-coded, or better still, algebraically coded. The ⟨position⟩ field can be coded in the same manner. In LZARI, I used an adaptive algebraic compression to encode the "characters," and static algebraic compression to encode the ⟨position⟩ field. (There were several versions of LZARI; some of them were slightly different from the above description.) The compression of LZARI was very tight, though rather slow.

6.5 LZPP

The original LZ methods were published in 1977 and 1978. They established the field of dictionary-based compression, which is why the 1980s were the golden age of this approach to compression, Most LZ algorithms were developed during this decade, but LZPP, the topic of this section, is an exception. This method [Pylak 03] is a modern, sophisticated algorithm that extends LZSS in several directions and has been inspired by research done and experience gained by many workers in the 1990s. LZPP identifies several sources of redundancy in the various quantities generated and manipulated by LZSS and exploits these sources to obtain better overall compression.

Virtually all LZ methods output offsets (also referred to as pointers or indexes) to a dictionary, and LZSS is no exception. However, many experiments indicate that the distribution of index values is not uniform and that most indexes are small, as illustrated by Figure 6.6. Thus, the set of indexes has large entropy (for the concatenated Calgary Corpus, [Pylak 03] gives this entropy as 8.954), which implies that the indexes can be compressed efficiently with an entropy encoder. LZPP employs a modified range encoder (Section 5.10.1) for this purpose. The original range coder has been modified by adding a circular buffer, an escape symbol, and an exclusion mechanism. In addition, each of the bytes of an index is compressed separately, which simplifies the range coder and allows for 3-byte indexes (i.e., a 16 MB dictionary).

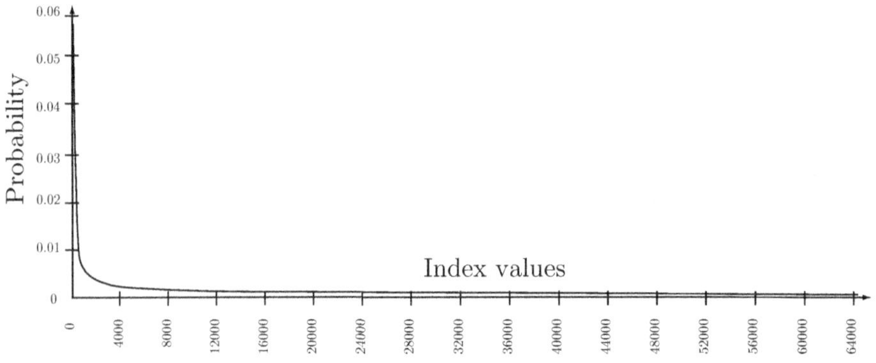

Figure 6.6: Distribution of Typical LZ Index Values (After [Pylak 03]).

LZSS outputs either raw characters or pairs (match length, index). It turns out that the distribution of match lengths is also far from uniform, which makes it possible to compress them too with an entropy encoder. Figure 6.7 shows the distribution of match lengths for the small, 21 KB file `obj1`, part of the Calgary Corpus. It is clear that short matches, of three and four characters, are considerably more common than longer matches, thereby resulting in large entropy of match lengths. The same range coder is used by LZPP to encode the lengths, allowing for match lengths of up to 64 KB.

The next source of redundancy in LZSS is the distribution of flag values. Recall that LZSS associates a flag with each raw character (a 0 flag) and each pair (a flag of 1) it outputs. Flag values of 2 and 3 are also used and are described below. The

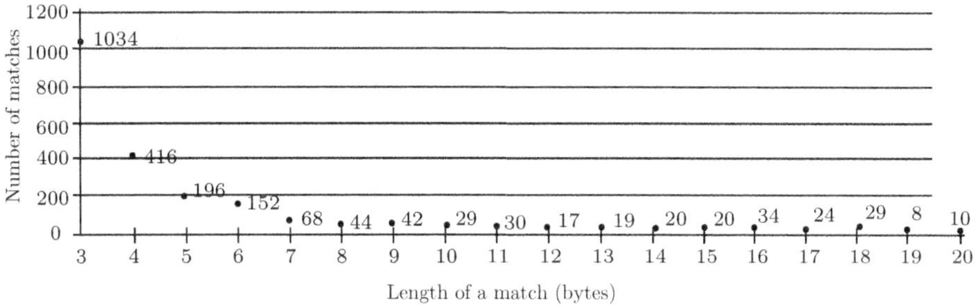

Figure 6.7: Distribution of Match Lengths (After [Pylak 03]).

redundancy in flag values exists because the probabilities of the two values are not 0.5 but vary considerably during encoding.

It is intuitively clear that when the encoder starts, few matches exist, many raw characters are output, and most flags are therefore 0. As encoding progresses, more and more matches are found, so more flags of 1 are output. Thus, the probability of a 0 flag starts high, and slowly drops to a low value, hopefully much less than 0.5.

This probability is illustrated in Figure 6.8, which shows how the probability of a 0 flag drops from almost 1 at the start, to around 0.2 after about 36,000 characters of the input have been read and processed. The input for the figure was file `paper1` of the Calgary Corpus.

Figure 6.8: Probability of a 0 Flag (After [Pylak 03]).

LZPP uses the same range encoder to compress the flags, which improves its compression pereformance by up to 2%. The description in this section shows that LZPP employs four different flags (as opposed to the two flags employed by LZSS), which makes the compression of flags even more meaningful.

Another source of redundancy in LZSS is the raw characters. LZSS outputs raw characters, while LZPP compresses these characters with the same range encoder.

It should be mentioned that the range encoder employed by LZPP is a variant of adaptive arithmetic coding. Its main input is an item to be encoded, but it must also receive a probability for the item. LZPP uses order-0 and order-1 contexts to estimate the probabilities of the indexes, flags, and characters it sends to the range encoder. The concept of contexts and their various orders is covered in Section 5.14, but here is what order-0 context means. Suppose that the current raw character is `q`. LZPP counts the number of times each character has been encoded and emitted as a raw character token. If `q` has been encoded and output 35 times as a raw character token, then the order-0 probability estimated for this occurrence of `q` is $35/t$, where t is the total number of raw character tokens output so far.

Figure 6.7 shows that match lengths of 3 are much more common than any other lengths, which is why LZPP treats 3-byte matches differently. It constructs and maintains a special data structure (a FIFO queue) D that holds a certain number of 3-byte matches, with the number of occurrences of each. When a new 3-byte match is found in the dictionary, the encoder searches D. If the match is already there, the encoder outputs the single item index3, where index3 is the index to the match in D. A flag of 2 is associated with this item. An index to D is encoded by an entropy coder before it is output, and the important point is that the occurrence count of the match is used by LZPP to compute a probability which is sent to the entropy coder together with index3.

The queue D is also exploited for a further small performance improvement. If LZPP outputs two consecutive raw characters, say, `t` and `h`, it knows that the next character x (the leftmost one in the look-ahead buffer) cannot be the last character of any of the 3-byte matches `th`y in D. Thus, if the trigrams `the`, `tha`, `thi`, and `tho` are in D, the next character cannot be `e`, `a`, or `i`. These characters are *excluded* when the x is encoded. When the encoder computes the probability of the x, it excludes the frequency counts for `e`, `a`, and `i`.

The following illustrates another type of exclusion. If the search and look-ahead buffers hold the following characters:

← coded text. . . | `... while␣these␣let␣` | `themselves␣out␣...` | . . . ← text to be read

and the strings `the` in the two buffers are matched, then LZPP knows that the character following `the` in the look-ahead buffer cannot be an `s`. Thus, the symbol `s` is excluded when LZPP computes a probability for encoding the length and index for string `the`.

It has already been mentioned that LZPP handles 3-byte matches in a special way. Experiments with LZPP indicate that other short matches, those of four and five bytes, should also be treated differently. Specifically, it has been noticed that when a 4-byte or a 5-byte match is distant (i.e., has a large index), the resulting pair (length, index) does not yield good compression. As a result, LZPP limits the indexes for short matches. If the smallest index of a 4-byte match exceeds 255, the match is ignored. Similarly, 5-byte matches are ignored if the smallest index exceeds 65,535.

The last source of redundancy in LZSS that is addressed by LZPP is symbol context. The concept of context is central to the PPM algorithm and is treated in detail in Section 5.14. The reader is advised to read the first few paragraphs of that section before continuing.

The idea is that characters read from the input file are not random. Certain characters (such as `e`, `t`, and `a` in English) occur more often than others (such as `j`, `q`, and `z`).

Also certain pairs and triplets (digrams and trigrams) of characters are common while others are rare. LZPP employs order-1 contexts to encode raw characters. It maintains a two-dimensional array C of 256×256 entries where it records the occurrences of digrams.

Thus, if the leftmost character in the look-ahead buffer is `h` and there are no 3-byte or longer matches for it, LZPP is supposed to output the `h` as a raw character (suitably encoded). Before it does that, it locates the rightmost character in the search buffer, say, `t`, and examines the frequency count in row `t` column `h` of array C. If the count c found there (the number of previous occurrences of the digram `th`) is nonzero, the `h` is encoded with probability c/a, where a is the number of characters read so far from the input file. A flag of 3 is associated with the resulting bits. If c is zero (the digram `th` hasn't been seen so far), the `h` is encoded and is output with a 0 flag. This is how LZPP handles the famous zero-probability problem in order-1 contexts.

The escape mechanism. An escape character is added to the alphabet, and is used to encode those raw characters that have zero frequencies. LZPP maintains two counts for each character of the alphabet. The main count is maintained in array C. It is the number of occurrences of the character so far and it may be zero. The secondary count is 1 plus the number of times the character has been coded with the escape mechanism. The secondary count is always nonzero and is used when a character is encoded with the escape. Notice that this count is maintained only for the most-recent 4,096 characters encoded with an escape.

When a character with zero main count is to be emitted, the escape flag is encoded first, followed by the character itself, encoded with a special probability that is computed by considering only the characters with zero occurrences and adding their secondary counts.

A 64 KB hash table is used by the encoder to locate matches in the dictionary. The hash function is the well-known CRC-32 (Section 6.32). The leftmost four characters in the look-ahead buffer are considered a 32-bit integer and the value returned by CRC-32 for this integer is truncated to 16 bits and becomes an index to the 64 MB dictionary. Further search in the dictionary is needed to find matches longer than four bytes.

Tests performed on the Calgary Corpus comparing several compression methods indicate that the compression ratios produced by LZPP are about 30% better than those of LZSS and about 1.5–2% worse than those of WinRAR (Section 6.22), a commercial product which is also based on LZSS. The developer of LZPP also notes that compression algorithms (including the commercial RK software) that are based on PPMZ (Section 5.14.7) outperform LZPP as well as any other dictionary-based algorithm.

6.5.1 Deficiencies

Before we discuss LZ78, let's summarize the deficiencies of LZ77 and its variants. It has already been mentioned that LZ77 uses the built-in implicit assumption that patterns in the input data occur close together. Data streams that don't satisfy this assumption compress poorly. A common example is text where a certain word, say `economy`, occurs often but is uniformly distributed throughout the text. When this word is shifted into the look-ahead buffer, its previous occurrence may have already been shifted out of the search buffer. A better algorithm would save commonly-occurring strings in the dictionary and not simply slide it all the time.

Another disadvantage of LZ77 is the limited size L of the look-ahead buffer. The size of matched strings is limited to $L-1$, but L must be kept small because the process of matching strings involves comparing individual symbols. If L were doubled in size, compression would improve, since longer matches would be possible, but the encoder would be much slower when searching for long matches. The size S of the search buffer is also limited. A large search buffer results in better compression but slows down the encoder, because searching takes longer (even with a binary search tree). Increasing the sizes of the two buffers also means creating longer tokens, thereby reducing compression efficiency. With two-byte tokens, compressing a two-character string into one token results in two bytes plus a flag. Writing the two characters as two raw ASCII codes results in two bytes plus two flags, a very small difference in size. The encoder should, in such a case, use the latter choice and write the two characters in uncompressed form, saving time and wasting just one bit. We say that the encoder has a two-byte *breakeven* point. With longer tokens, the breakeven point increases to three bytes.

6.6 Repetition Times

Frans Willems, one of the developers of context-tree weighting (Section 5.16), is also the developer of this original (although not very efficient) dictionary-based method. The input may consist of any symbols, but the method is described here and also in [Willems 89] for binary input. The input symbols are grouped into words of length L each that are placed in a sliding buffer. The buffer is divided into a look-ahead buffer with words still to be compressed, and a search buffer containing the B most-recently processed words. The encoder tries to match the leftmost word in the look-ahead buffer to the contents of the search buffer. Only one word in the look-ahead buffer is matched in each step. If a match is found, the distance (offset) of the word from the start of the match is denoted by m and is encoded by a 2-part prefix code that's written on the compressed stream. Notice that there is no need to encode the number of symbols matched, because exactly one word is matched. If no match is found, a special code is written, followed by the L symbols of the unmatched word in raw format.

The method is illustrated by a simple example. We assume that the input symbols are bits. We select $L = 3$ for the length of words, and a search buffer of length $B = 2^L - 1 = 7$ containing the seven most-recently processed bits. The look-ahead buffer contains just the binary data, and the commas shown here are used only to indicate word boundaries.

← coded input... `0100100|100,000,011,111,011,101,001`... ← input to be read

It is obvious that the leftmost word "100" in the look-ahead buffer matches the rightmost three bits in the search buffer. The repetition time (the offset) for this word is therefore $m = 3$. (The biggest repetition time is the length B of the search buffer, 7 in our example.) The buffer is now shifted one word (three bits) to the left to become

← ...010`0100100|000,011,111,011,101,001,...`... ← input to be read

The repetition time for the current word "000" is $m = 1$ because each bit in this word is matched with the bit immediately to its left. Notice that it is possible to match the leftmost 0 of the next word "011" with the bit to its left, but this method matches exactly one word in each step. The buffer is again shifted L positions to become

$\leftarrow \ldots 010010$|0100000|011,111,011,101,001,......|$\ldots \leftarrow$ input to be read

There is no match for the next word "011" in the search buffer, so m is set to a special value that we denote by 8* (meaning; greater than or equal 8). It is easy to verify that the repetition times of the remaining three words are 6, 4, and 8*.

Each repetition time is encoded by first determining two integers p and q. If $m = 8*$, then p is set to L; otherwise p is selected as the integer that satisfies $2^p \le m < 2^{p+1}$. Notice that p is located in the interval $[0, L-1]$. The integer q is determined by $q = m - 2^p$, which places it in the interval $[0, 2^p - 1]$. Table 6.9 lists the values of m, p, q, and the prefix codes used for $L = 3$.

m	p	q	Prefix	Suffix	Length
1	0	0	00	none	2
2	1	0	01	0	3
3	1	1	01	1	3
4	2	0	10	00	4
5	2	1	10	01	4
6	2	2	10	10	4
7	2	3	10	11	4
8*	3	—	11	word	5

Table 6.9: Repetition Time Encoding Table for $L = 3$.

Once p and q are known, a prefix code for m is constructed and is written on the compressed stream. It consists of two parts, a prefix and a suffix, that are the binary values of p and q, respectively. Since p is in the interval $[0, L-1]$, the prefix requires $\log(L+1)$ bits. The length of the suffix is p bits. The case $p = L$ is different. Here, the suffix is the raw value (L bits) of the word being compressed.

The compressed stream for the seven words of our example consists of the seven codes

$$01|1, 00, 11|011, 00, 10|10, 10|00, 11|001, \ldots,$$

where the vertical bars separate the prefix and suffix of a code. Notice that the third and seventh words (011 and 001) are included in the codes in raw format.

It is easy to see why this method generates prefix codes. Once a code has been assigned (such as 01|0, the code of $m = 2$), that code cannot be the prefix of any other code because (1) some of the other codes are for different values of p and thus do not start with 01, and (2) codes for the same p do start with 01 but must have different values of q, so they have different suffixes.

The compression performance of this method is inferior to that of LZ77, but it is interesting for the following reasons.

1. It is universal and optimal. It does not use the statistics of the input stream, and its performance asymptotically approaches the entropy of the input as the input stream gets longer.

2. It is shown in [Cachin 98] that this method can be modified to include data hiding (steganography).

6.7 QIC-122

QIC was an international trade association, incorporated in 1987, whose mission was to encourage and promote the widespread use of quarter-inch tape cartridge technology (hence the acronym QIC; see also `http://www.qic.org/html`). The association ceased to exist in 1998, when the technology it promoted became obsolete.

The QIC-122 compression standard is an LZ77 variant that has been developed by QIC for text compression on 1/4-inch data cartridge tape drives. Data is read and shifted into a 2048-byte ($= 2^{11}$) input buffer from right to left, such that the first character is the leftmost one. When the buffer is full, or when all the data has been read into it, the algorithm searches from left to right for repeated strings. The output consists of raw characters and of tokens that represent strings already seen in the buffer. As an example, suppose that the following data have been read and shifted into the buffer:

ABAAAAAACABABABA...............

The first character `A` is obviously not a repetition of any previous string, so it is encoded as a raw (ASCII) character (see below). The next character `B` is also encoded as raw. The third character `A` is identical to the first character but is also encoded as raw since repeated strings should be at least two characters long. Only with the fourth character `A` we do have a repeated string. The string of five `A`'s from position 4 to position 8 is identical to the one from position 3 to position 7. It is therefore encoded as a string of length 5 at offset 1. The offset in this method is the distance between the start of the repeated string and the start of the original one.

The next character `C` at position 9 is encoded as raw. The string `ABA` at positions 10–12 is a repeat of the string at positions 1–3, so it is encoded as a string of length 3 at offset $10 - 1 = 9$. Finally, the string `BABA` at positions 13–16 is encoded with length 4 at offset 2, since it is a repetition of the string at positions 10–13.

◇ **Exercise 6.5:** Suppose that the next four characters of data are `CAAC`

ABAAAAAACABABABACAAC...........

How will they be encoded?

A raw character is encoded as 0 followed by the 8 ASCII bits of the character. A string is encoded as a token that starts with 1 followed by the encoded offset, followed by the encoded length. Small offsets are encoded as 1, followed by 7 offset bits; large offsets are encoded as 0 followed by 11 offset bits (recall that the buffer size is 2^{11}). The length is encoded according to Table 6.10. The 9-bit string 110000000 is written, as an end marker, at the end of the output stream.

Bytes	Length	Bytes	Length
2	00	17	11 11 1001
3	01	18	11 11 1010
4	10	19	11 11 1011
5	11 00	20	11 11 1100
6	11 01	21	11 11 1101
7	11 10	22	11 11 1110
8	11 11 0000	23	11 11 1111 0000
9	11 11 0001	24	11 11 1111 0001
10	11 11 0010	25	11 11 1111 0010
11	11 11 0011	⋮	
12	11 11 0100	37	11 11 1111 1110
13	11 11 0101	38	11 11 1111 1111 0000
14	11 11 0110	39	11 11 1111 1111 0001
15	11 11 0111		etc.
16	11 11 1000		

Table 6.10: Values of the `<length>` Field.

◇ **Exercise 6.6:** How can the decoder identify the end marker?

When the search algorithm arrives at the right end of the buffer, it shifts the buffer to the left and inputs the next character into the rightmost position of the buffer. The decoder is the reverse of the encoder (symmetric compression).

> When I saw each luminous creature in profile, from the point of view of its body, its egglike shape was like a gigantic asymmetrical yoyo that was standing edgewise, or like an almost round pot that was resting on its side with its lid on. The part that looked like a lid was the front plate; it was perhaps one-fifth the thickness of the total cocoon.
>
> —Carlos Castaneda, *The Fire From Within* (1984)

Figure 6.11 is a precise description of the compression process, expressed in BNF, which is a metalanguage used to describe processes and formal languages unambiguously. BNF uses the following *metasymbols*:

`::=`	The symbol on the left is defined by the expression on the right.
`<expr>`	An expression still to be defined.
|	A logical OR.
`[]`	Optional. The expression in the brackets may occur zero or more times.
`()`	A comment.
`0,1`	The bits 0 and 1.

(Special applications of BNF may require more symbols.)

Table 6.12 shows the results of encoding `ABAAAAAACABABABA` (a 16-symbol string). The reader can easily verify that the output stream consists of the 10 bytes

`20 90 88 38 1C 21 E2 5C 15 80.`

```
(QIC-122 BNF Description)
<Compressed-Stream>::=[<Compressed-String>] <End-Marker>
<Compressed-String>::= 0<Raw-Byte> | 1<Compressed-Bytes>
<Raw-Byte>          ::=<b><b><b><b><b><b><b><b> (8-bit byte)
<Compressed-Bytes> ::=<offset><length>
<offset>            ::= 1<b><b><b><b><b><b><b> (a 7-bit offset)
                         |
         0<b><b><b><b><b><b><b><b><b><b><b> (an 11-bit offset)
<length>            ::= (as per length table)
<End-Marker>        ::=110000000 (Compressed bytes with offset=0)
<b>                 ::=0|1
```

Figure 6.11: BNF Definition of QIC-122.

Raw byte "A"	0 01000001
Raw byte "B"	0 01000010
Raw byte "A"	0 01000001
String "AAAAA" offset=1	1 1 0000001 1100
Raw byte "C"	0 01000011
String "ABA" offset=9	1 1 0001001 01
String "BABA" offset=2	1 1 0000010 10
End-Marker	1 1 0000000

Table 6.12: Encoding the Example.

6.8 LZX

In 1995, Jonathan Forbes and Tomi Poutanen developed an LZ variant (possibly influenced by Deflate) that they dubbed LZX [LZX 09]. The main feature of the first version of LZX was the way it encoded the match offsets, which can be large, by segmenting the size of the search buffer. They implemented LZX on the Amiga personal computer and included a feature where data was grouped into large blocks instead of being compressed as a single unit.

At about the same time, Microsoft devised a new installation media format that it termed, in analogy to a file cabinet, *cabinet files*. A cabinet file has an extension name of `.cab` and may consist of several data files concatenated into one unit and compressed. Initially, Microsoft used two compression methods to compress cabinet files, MSZIP (which is just another name for Deflate) and Quantum, a large-window dictionary-based encoder that employs arithmetic coding. Quantum was developed by David Stafford.

Microsoft later used cabinet files in its Cabinet Software Development Kit (SDK). This is a software package that provides software developers with the tools required to employ cabinet files in any applications that they implement.

In 1997, Jonathan Forbes went to work for Microsoft and modified LZX to compress cabinet files. Microsoft published an official specification for cabinet files, including MSZIP and LZX, but excluding Quantum. The LZX description contained errors to such an extent that it wasn't possible to create a working implementation from it.

LZX documentation is available in executable file `Cabsdk.exe` located at `http://download.microsoft.com/download/platformsdk/cab/2.0/w98nt42kmexp/en-us/`. After unpacking this executable file, the documentation is found in file `LZXFMT.DOC`.

LZX is also used to compress `chm` files. Microsoft Compiled HTML (`chm`) is a proprietary format initially developed by Microsoft (in 1997, as part of its Windows 98) for online help files. It was supposed to be the successor to the Microsoft WinHelp format. It seems that many Windows users liked this compact file format and started using it for all types of files, as an alternative to PDF. The popular `lit` file format is an extension of `chm`.

A `chm` file is identified by a `.chm` file name extension. It consists of web pages written in a subset of HTML and a hyperlinked table of contents. A `chm` file can have a detailed index and table-of-contents, which is one reason for the popularity of this format. Currently, there are `chm` readers available for both Windows and Macintosh.

LZX is a variant of LZ77 that writes on the compressed stream either unmatched characters or pairs (offset, length). What is actually written on the compressed stream is variable-length codes for the unmatched characters, offsets, and lengths. The size of the search buffer is a power of 2, between 2^{15} and 2^{21}. LZX uses static canonical Huffman trees (Section 5.2.8) to provide variable-length, prefix codes for the three types of data. There are 256 possible character values, but there may be many different offsets and lengths. Thus, the Huffman trees have to be large, but any particular cabinet file being compressed by LZX may need just a small part of each tree. Those parts are written on the compressed stream. Because a single cabinet file may consist of several data files, each is compressed separately and is written on the compressed stream as a block, including those parts of the trees that it requires. The other important features of LZX are listed here.

Repeated offsets. The developers of LZX noticed that certain offsets tend to repeat; i.e., if a certain string is compressed to a pair $(74, \text{length})$, then there is a good chance that offset 74 will be used again soon. Thus, the three special codes 0, 1, and 2 were allocated to encode three of the most-recent offsets. The actual offset associated with each of those codes varies all the time. We denote by $R0$, $R1$, and $R2$ the most-recent, second most-recent, and third most-recent offsets, respectively (these offsets must themselves be nonrepeating; i.e., none should be 0, 1, or 2). We consider $R0$, $R1$, and $R2$ a short list and update it similar to an LRU (least-recently used) queue. The three quantities are initialized to 1 and are updated as follows. Assume that the current offset is X, then

$$
\begin{gathered}
\text{if } X \neq R0 \text{ and } X \neq R1 \text{ and } X \neq R2, \text{ then } R2 \leftarrow R1,\ R1 \leftarrow R0,\ R0 \leftarrow X, \\
\text{if } X = R0, \text{ then nothing}, \\
\text{if } X = R1, \text{ then swap } R0 \text{ and } R1, \\
\text{if } X = R2, \text{ then swap } R0 \text{ and } R2.
\end{gathered}
$$

Because codes 0, 1, and 2 are allocated to the three special recent offsets, an offset of 3 is allocated code 5, and, in general, an offset x is assigned code $x+2$. The largest offset is the size of the search buffer minus 3, and its assigned code is the size of the search buffer minus 1.

Encoder preprocessing. LZX was designed to compress Microsoft cabinet files, which are part of the Windows operating system. Computers using this system are generally based on microprocessors made by Intel, and throughout the 1980s and 1990s, before the introduction of the pentium, these microprocessors were members of the well-known 80x86 family. The encoder preprocessing mode of LZX is selected by the user when the input stream is an executable file for an 80x86 computer. This mode converts 80x86 CALL instructions to use absolute instead of relative addresses.

Output block format. LZX outputs the compressed data in blocks, where each block contains raw characters, offsets, match lengths, and the canonical Huffman trees used to encode the three types of data. A canonical Huffman tree can be reconstructed from the path length of each of its nodes. Thus, only the path lengths have to be written on the output for each Huffman tree. LZX limits the depth of a Huffman tree to 16, so each tree node is represented on the output by a number in the range 0 to 16. A 0 indicates a missing node (a Huffman code that's not used by this block). If the tree has to be bigger, the current block is written on the output and compression resumes with fresh trees. The tree nodes are written in compressed form. If several consecutive tree nodes are identical, then run-length encoding is used to encode them. The three numbers 17, 18, and 19 are used for this purpose. Otherwise the difference (modulo 17) between the path lengths of the current node and the previous node is written. This difference is in the interval $[0, 16]$. Thus, what's written on the output are the 20 5-bit integers 0 through 19, and these integers are themselves encoded by a Huffman tree called a pre-tree. The pre-tree is generated dynamically according to the frequencies of occurrence of the 20 values. The pre-tree itself has to be written on the output, and it is written as 20 4-bit integers (a total of 80 bits) where each integer indicates the path length of one of the 20 tree nodes. A path of length zero indicates that one of the 20 values is not used in the current block.

The offsets and match lengths are themselves compressed in a complex process that involves several steps and is summarized in Figure 6.13. The individual steps involve many operations and use several tables that are built into both encoder and decoder. However, because LZX is not an important compression method, these steps are not discussed here.

6.9 LZ78

The LZ78 method (which is sometimes referred to as LZ2) [Ziv and Lempel 78] does not use any search buffer, look-ahead buffer, or sliding window. Instead, there is a dictionary of previously encountered strings. This dictionary starts empty (or almost empty), and its size is limited only by the amount of available memory. The encoder outputs two-field tokens. The first field is a pointer to the dictionary; the second is the code of a symbol. Tokens do not contain the length of a string, since this is implied in the dictionary. Each token corresponds to a string of input symbols, and that string is added to the dictionary after the token is written on the compressed stream. Nothing is ever deleted from the dictionary, which is both an advantage over LZ77 (since future strings can be compressed even by strings seen in the distant past) and a liability (because the dictionary tends to grow fast and to fill up the entire available memory).

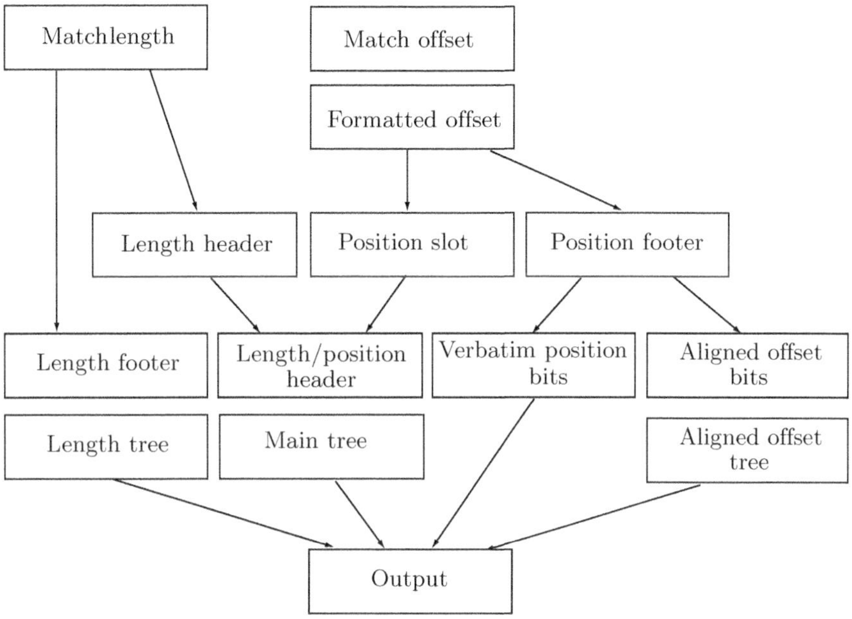

Figure 6.13: LZX Processing of Offsets and Lengths.

The dictionary starts with the null string at position zero. As symbols are input and encoded, strings are added to the dictionary at positions 1, 2, and so on. When the next symbol x is read from the input stream, the dictionary is searched for an entry with the one-symbol string x. If none are found, x is added to the next available position in the dictionary, and the token (0, x) is output. This token indicates the string "null x" (a concatenation of the null string and x). If an entry with x is found (at, say, position 37), the next symbol y is read, and the dictionary is searched for an entry containing the two-symbol string xy. If none are found, then string xy is added to the next available position in the dictionary, and the token (37, y) is output. This token indicates the string xy, since 37 is the dictionary position of string x. The process continues until the end of the input stream is reached.

In general, the current symbol is read and becomes a one-symbol string. The encoder then tries to find it in the dictionary. If the symbol is found in the dictionary, the next symbol is read and concatenated with the first to form a two-symbol string that the encoder then tries to locate in the dictionary. As long as those strings are found in the dictionary, more symbols are read and concatenated to the string. At a certain point the string is not found in the dictionary, so the encoder adds it to the dictionary and outputs a token with the last dictionary match as its first field, and the last symbol of the string (the one that caused the search to fail) as its second field. Table 6.14 shows the first 14 steps in encoding the string

`sir␣sid␣eastman␣easily␣teases␣sea␣sick␣seals`

◇ **Exercise 6.7:** Complete Table 6.14.

In each step, the string added to the dictionary is the one being encoded, minus

Dictionary		Token	Dictionary		Token
0	null				
1	"s"	(0, "s")	8	"a"	(0, "a")
2	"i"	(0, "i")	9	"st"	(1, "t")
3	"r"	(0, "r")	10	"m"	(0, "m")
4	"␣"	(0, "␣")	11	"an"	(8, "n")
5	"si"	(1, "i")	12	"␣ea"	(7, "a")
6	"d"	(0, "d")	13	"sil"	(5, "l")
7	"␣e"	(4, "e")	14	"y"	(0, "y")

Table 6.14: First 14 Encoding Steps in LZ78.

its last symbol. In a typical compression run, the dictionary starts with short strings, but as more text is being input and processed, longer and longer strings are added to it. The size of the dictionary can either be fixed or may be determined by the size of the available memory each time the LZ78 compression program is executed. A large dictionary may contain more strings and thus allow for longer matches, but the trade-off is longer pointers (and thus bigger tokens) and slower dictionary search.

A good data structure for the dictionary is a tree, but not a binary tree. The tree starts with the null string as the root. All the strings that start with the null string (strings for which the token pointer is zero) are added to the tree as children of the root. In the above example those are `s`, `i`, `r`, `␣`, `d`, `a`, `m`, `y`, `e`, `c`, and `k`. Each of them becomes the root of a subtree as shown in Figure 6.15. For example, all the strings that start with `s` (the four strings `si`, `sil`, `st`, and `s(eof)`) constitute the subtree of node `s`.

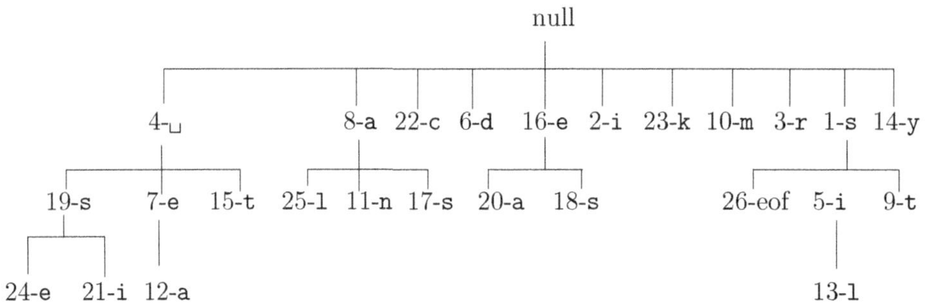

Figure 6.15: An LZ78 Dictionary Tree.

Assuming an alphabet with 8-bit symbols, there are 256 different symbols, so in principle, each node in the tree could have up to 256 children. The process of adding a child to a tree node should thus be dynamic. When the node is first created, it has no children and it should not reserve any memory space for them. As a child is added to the node, memory space should be claimed for it. Since no nodes are ever deleted, there is no need to reclaim memory space, which simplifies the memory management task somewhat.

Such a tree makes it easy to search for a string and to add strings. To search for `sil`, for example, the program looks for the child `s` of the root, then for the child `i` of `s`, and so on, going down the tree. Here are some examples:

1. When the `s` of `sid` is input in step 5, the encoder finds node "1-`s`" in the tree as a child of "null". It then inputs the next symbol `i`, but node `s` does not have a child `i` (in fact, it has no children at all at this point), so the encoder adds node "5-`i`" as a child of "1-`s`", which effectively adds the string `si` to the tree.

2. When the blank space between `eastman` and `easily` is input in step 12, a similar situation happens. The encoder finds node "4-␣", inputs `e`, finds "7-`e`", inputs `a`, but "7-`e`" does not have "`a`" as a child, so the encoder adds node "12-`a`", which effectively adds the string "␣`ea`" to the tree.

A tree of the type described here is called a *trie*. In general, a trie is a tree in which the branching structure at any level is determined by just part of a data item, not the entire item (Section 5.14.5). In the case of LZ78, each string added to the tree effectively adds just one symbol, and does that by adding a branch.

Since the total size of the tree is limited, it may fill up during compression. This, in fact, happens all the time except when the input stream is unusually small. The original LZ78 method does not specify what to do in such a case, so we list a few possible solutions.

1. The simplest solution is to freeze the dictionary at that point. No new nodes should be added, the tree becomes a static dictionary, but it can still be used to encode strings.

2. Delete the entire tree once it gets full and start with a new, empty tree. This solution effectively breaks the input into blocks, each with its own dictionary. If the content of the input varies from block to block, this solution will produce good compression, since it will eliminate a dictionary with strings that are unlikely to be used in the future. We can say that this solution implicitly assumes that future symbols will benefit more from new data than from old (the same implicit assumption used by LZ77).

3. The UNIX `compress` utility (Section 6.14) uses a more complex solution.

4. When the dictionary is full, delete some of the least-recently-used entries, to make room for new ones. Unfortunately there is no good algorithm to decide which entries to delete, and how many (but see the *reuse* procedure in Section 6.23).

The LZ78 decoder works by building and maintaining the dictionary in the same way as the encoder. It is therefore more complex than the LZ77 decoder. Thus, LZ77 decoding is simpler than its encoding, but LZ78 encoding and decoding involve the same complexity.

6.10 LZFG

Edward Fiala and Daniel Greene have developed several related compression methods [Fiala and Greene 89] that are hybrids of LZ77 and LZ78. All their methods are based on the following scheme. The encoder generates a compressed file with tokens and literals (raw ASCII codes) intermixed. There are two types of tokens: a *literal* and a *copy*. A literal token indicates that a string of literals follows, a copy token points to a string previously seen in the data. The string `the␣boy␣on␣my␣right␣is␣the␣right␣boy` produces, when encoded,

(literal 23)`the␣boy␣on␣my␣right␣is␣`(copy 4 23)(copy 6 13)(copy 3 29),

where the three copy tokens refer to the strings `the␣`, `right␣`, and "`boy`", respectively. The LZFG methods are best understood by considering how the decoder operates. The decoder starts with a large empty buffer in which it generates and shifts the decompressed stream. When the decoder inputs a (literal 23) token, it inputs the next 23 bytes as raw ASCII codes into the buffer, shifting the buffer such that the last byte input will be the rightmost one. When the decoder inputs (copy 4 23) it copies the string of length 4 that starts 23 positions from the right end of the buffer. The string is then appended to the buffer, while shifting it. Two LZFG variants, denoted by A1 and A2, are described here.

The A1 scheme employs 8-bit literal tokens and 16-bit copy tokens. A literal token has the format `0000nnnn`, where `nnnn` indicates the number of ASCII bytes following the token. Since the 4-bit `nnnn` field can have values between 0 and 15, they are interpreted as meaning 1 to 16. The longest possible string of literals is therefore 16 bytes. The format of a copy token is `sssspp...p`, where the 4-bit nonzero `ssss` field indicates the length of the string to be copied, and the 12-bit `pp...p` field is a displacement showing where the string starts in the buffer. Since the `ssss` field cannot be zero, it can have values only between 1 and 15, and they are interpreted as string lengths between 2 and 16. Displacement values are in the range $[0, 4095]$ and are interpreted as $[1, 4096]$.

The encoder starts with an empty search buffer, 4,096 bytes long, and fills up the look-ahead buffer with input data. At each subsequent step it tries to create a copy token. If nothing matches in that step, the encoder creates a literal token. Suppose that at a certain point the buffer contains

←text already encoded..|`..      ..xyz`|`abcd..      ..`|..← text yet to be input

The encoder tries to match the string `abc...` in the look-ahead buffer to various strings in the search buffer. If a match is found (of at least two symbols), a copy token is written on the compressed stream and the data in the buffers is shifted to the left by the size of the match. If a match is not found, the encoder starts a literal with the `a` and left-shifts the data one position. It then tries to match `bcd...` to the search buffer. If it finds a match, a literal token is output, followed by a byte with the `a`, followed by a match token. Otherwise, the `b` is appended to the literal and the encoder tries to match from `cd...` Literals can be up to 16 bytes long, so the string `the␣boy␣on␣my...` above is encoded as

(literal 16)`the␣boy␣on␣my␣ri`(literal 7)`ght␣is␣`(copy 4 23)(copy 6 13)(copy 3 29).

The A1 method borrows the idea of the sliding buffer from LZ77 but also behaves like LZ78, because it creates two-field tokens. This is why it can be considered a hybrid of the two original LZ methods. When A1 starts, it creates mostly literals, but when it gets up to speed (fills up its search buffer), it features strong adaptation, so more and more copy tokens appear in the compressed stream.

The A2 method uses a larger search buffer (up to 21K bytes long). This improves compression, because longer copies can be found, but raises the problem of token size. A large search buffer implies large displacements in copy tokens; long copies imply large "length" fields in those tokens. At the same time we expect both the displacement and the "length" fields of a typical copy token to be small, since most matches are found close to the beginning of the search buffer. The solution is to use a variable-length code for those fields, and A2 uses the general unary codes of Section 3.1. The "length" field of a copy token is encoded with a (2,1,10) code (Table 3.4), making it possible to match strings up to 2,044 symbols long. Notice that the (2,1,10) code is between 3 and 18 bits long.

The first four codes of the $(2, 1, 10)$ code are 000, 001, 010, and 011. The last three of these codes indicate match lengths of two, three, and four, respectively (recall that the minimum match length is 2). The first one (code 000) is reserved to indicate a literal. The length of the literal then follows and is encoded with code $(0, 1, 5)$. A literal can therefore be up to 63 bytes long, and the literal-length field in the token is encoded by between 1 and 10 bits. In case of a match, the "length" field is not 000 and is followed by the displacement field, which is encoded with the (10,2,14) code (Table 6.16). This code has 21K values, and the maximum code size is 16 bits (but see points 2 and 3 below).

n	$a = 10 + n \cdot 2$	nth codeword	Number of codewords	Range of integers
0	10	$0\underbrace{x...x}_{10}$	$2^{10} = 1K$	0–1023
1	12	$10\underbrace{xx...x}_{12}$	$2^{12} = 4K$	1024–5119
2	14	$11\underbrace{xx...xx}_{14}$	$2^{14} = 16K$	5120–21503
		Total	21504	

Table 6.16: The General Unary Code $(10, 2, 14)$.

Three more refinements are employed by the A2 method, to achieve slightly better (1% or 2%) compression.

1. A literal of maximum length (63 bytes) can immediately be followed by another literal or by a copy token of any length, but a literal of fewer than 63 bytes must be followed by a copy token matching *at least three symbols* (or by the end-of-file). This fact is used to shift down the (2,1,10) codes used to indicate the match length. Normally, codes 000, 001, 010, and 011 indicate no match, and matches of length 2, 3, and 4, respectively. However, a copy token following a literal token of fewer than 63 bytes uses codes 000, 001, 010, and 011 to indicate matches of length 3, 4, 5, and 6, respectively. This way the maximum match length can be 2,046 symbols instead of 2,044.

2. The displacement field is encoded with the $(10, 2, 14)$ code, which has 21K values and whose individual codes range in size from 11 to 16 bits. For smaller files, such large

displacements may not be necessary, and other general unary codes may be used, with shorter individual codes. Method A2 thus uses codes of the form $(10 - d, 2, 14 - d)$ for $d = 10, 9, 8, \ldots, 0$. For $d = 1$, code $(9, 2, 13)$ has $2^9 + 2^{11} + 2^{13} = 10{,}752$ values, and individual codes range in size from 9 to 15 bits. For $d = 10$ code $(0, 2, 4)$ contains $2^0 + 2^2 + 2^4 = 21$ values, and codes are between 1 and 6 bits long. Method A2 starts with $d = 10$ [meaning it initially uses code $(0, 2, 4)$] and a search buffer of size 21 bytes. When the buffer fills up (indicating an input stream longer than 21 bytes), the A2 algorithm switches to $d = 9$ [code $(1, 2, 5)$] and increases the search buffer size to 42 bytes. This process continues until the entire input stream has been encoded or until $d = 0$ is reached [at which point code (10,2,14) is used to the end]. A lot of work for a small gain in compression! (See the discussion of diminishing returns (*a word to the wise*) in the Preface.)

3. Each of the codes $(10 - d, 2, 14 - d)$ requires a search buffer of a certain size, from 21 up to 21K = 21,504 bytes, according to the number of codes it contains. If the user wants, for some reason, to assign the search buffer a different size, then some of the longer codes may never be used, which makes it possible to cut down a little the size of the individual codes. For example, if the user decides to use a search buffer of size 16K = 16,384 bytes, then code $(10, 2, 14)$ has to be used [because the next code $(9, 2, 13)$ contains just 10,752 values]. Code $(10, 2, 14)$ contains 21K = 21,504 individual codes, so the 5,120 longest codes will never be used. The last group of codes ("11" followed by 14 bits) in $(10, 2, 14)$ contains $2^{14} = 16{,}384$ different individual codes, of which only 11,264 will be used. Of the 11,264 codes the first 8,192 can be represented as "11" followed by $\lfloor \log_2 11{,}264 \rfloor = 13$ bits, and only the remaining 3,072 codes require $\lceil \log_2 11{,}264 \rceil = 14$ bits to follow the first "11". We thus end up with 8,192 15-bit codes and 3,072 16-bit codes, instead of 11,264 16-bit codes, a very small improvement.

These three improvements illustrate the great lengths that researchers are willing to go to in order to improve their algorithms ever so slightly.

> Experience shows that fine-tuning an algorithm to squeeze out the last remaining bits of redundancy from the data gives diminishing returns. Modifying an algorithm to improve compression by 1% may increase the run time by 10% (from the Introduction).

The LZFG "corpus" of algorithms contains four more methods. B1 and B2 are similar to A1 and A2 but faster because of the way they compute displacements. However, some compression ratio is sacrificed. C1 and C2 go in the opposite direction. They achieve slightly better compression than A1 and A2 at the price of slower operation. (LZFG has been patented, an issue that's discussed in Section 6.34.)

6.11 LZRW1

Developed by Ross Williams [Williams 91a] and [Williams 91b] as a simple, fast LZ77 variant, LZRW1 is also related to method A1 of LZFG (Section 6.10). The main idea is to find a match in one step, using a hash table. This is fast but not very efficient, since the match found is not always the longest. We start with a description of the algorithm, follow with the format of the compressed stream, and conclude with an example.

The method uses the entire available memory as a buffer and encodes the input stream in blocks. A block is read into the buffer and is completely encoded, then the next block is read and encoded, and so on. The length of the search buffer is 4K and that of the look-ahead buffer is 16 bytes. These two buffers slide along the input block in memory from left to right. Only one pointer, p_src, needs be maintained, pointing to the start of the look-ahead buffer. The pointer p_src is initialized to 1 and is incremented after each phrase is encoded, thereby moving both buffers to the right by the length of the phrase. Figure 6.17 shows how the search buffer starts empty, grows to 4K, and then starts sliding to the right, following the look-ahead buffer.

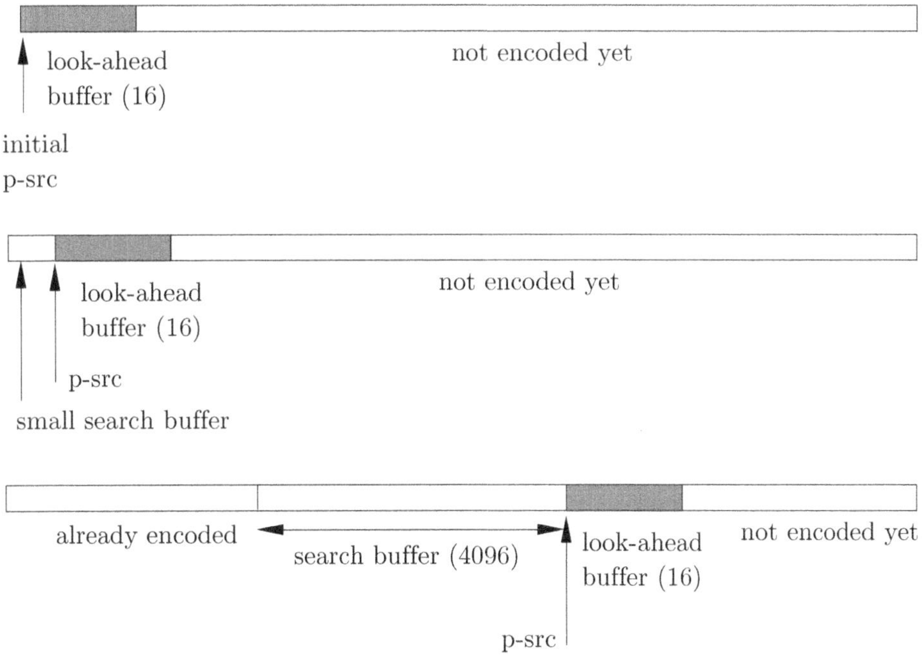

Figure 6.17: Sliding the LZRW1 Search- and Look-Ahead Buffers.

The leftmost three characters of the look-ahead buffer are hashed into a 12-bit number I, which is used to index an array of $2^{12} = 4{,}096$ pointers. A pointer P is retrieved and is immediately replaced in the array by p_src. If P points outside the search buffer, there is no match; the first character in the look-ahead buffer is output as a literal, and p_src is advanced by 1. The same thing is done if P points inside the search buffer but to a string that does not match the one in the look-ahead buffer. If P points to a match of at least three characters, the encoder finds the longest match (at most 16 characters), outputs a match item, and advances `p_src` by the length of the match. This process is depicted in Figure 6.19. An interesting point to note is that the array of pointers does not have to be initialized when the encoder starts, since the encoder checks every pointer. Initially, all pointers are random, but as they are replaced, more and more of them point to real matches.

The output of the LZRW1 encoder (Figure 6.20) consists of groups, each starting with a 16-bit *control word*, followed by 16 items. Each item is either an 8-bit literal or a 16-bit copy item (a match) consisting of a 4-bit length field b (where the length is $b+1$) and a 12-bit offset (the a and c fields). The length field indicates lengths between 3 and 16. The 16 bits of the control word flag each of the 16 items that follow (a 0 flag indicates a literal and a flag of 1 indicates a match item). Obviously, groups have different lengths. The last group may contain fewer than 16 items.

The decoder is even simpler than the encoder, since it does not need the array of pointers. It maintains a large buffer using a `p_src` pointer in the same way as the encoder. The decoder reads a control word from the compressed stream and uses its 16 bits to read 16 items. A literal item is decoded by appending it to the buffer and incrementing `p_src` by 1. A copy item is decoded by subtracting the offset from `p_src`, fetching a string from the search buffer, of length indicated by the length field, and appending it to the buffer. Then `p_src` is incremented by the length.

Table 6.18 illustrates the first seven steps of encoding "`that␣thatch␣thaws`". The values produced by the hash function are arbitrary. Initially, all pointers are random (indicated by "any") but they are replaced by useful ones very quickly.

◇ **Exercise 6.8:** Summarize the last steps in a table similar to Table 6.18 and write the final compressed stream in binary.

p_src	3 chars	Hash index	P	Output	Binary output
1	`tha`	4	any→1	`t`	01110100
2	`hat`	6	any→2	`h`	01101000
3	`at␣`	2	any→3	`a`	01100001
4	`t␣t`	1	any→4	`t`	01110100
5	`␣th`	5	any→5	`␣`	00100000
6	`tha`	4	4→1	6,5	0000|0011|00000101
10	`ch␣`	3	any→10	`c`	01100011

Table 6.18: First Seven Steps of Encoding `that thatch thaws`.

Tests done by the original developer indicate that LZRW1 performs about 10% worse than LZC (the UNIX `compress` utility) but is four times faster. Also, it performs about 4% worse than LZFG (the A1 method) but runs ten times faster. It is therefore suited for cases where speed is more important than compression performance. A 68000

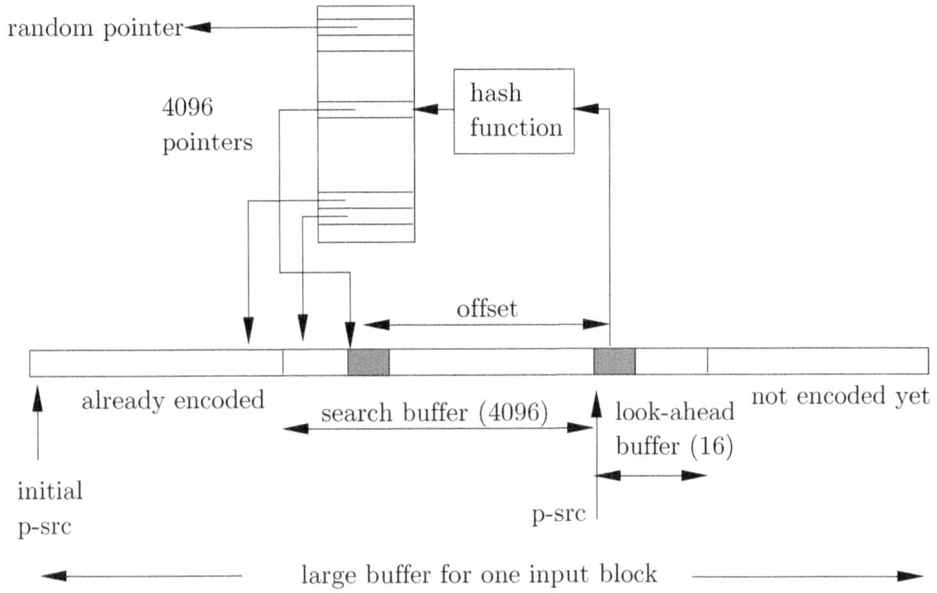

Figure 6.19: The LZRW1 Encoder.

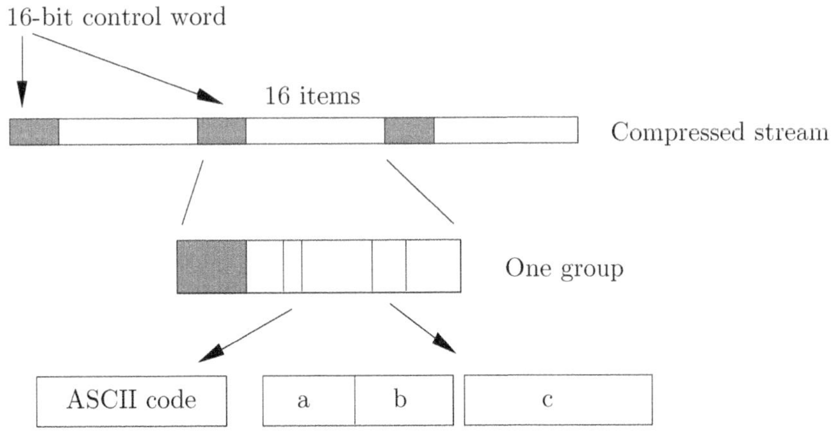

Figure 6.20: Format of the Output.

assembly language implementation has required, on average, the execution of only 13 machine instructions to compress, and four instructions to decompress, one byte.

◇ **Exercise 6.9:** Show a practical situation where compression speed is more important than compression ratio.

6.12 LZRW4

LZRW4 is a variant of LZ77, based on ideas of Ross Williams about possible ways to combine a dictionary method with prediction (Section 6.35). LZRW4 also borrows some ideas from LZRW1. It uses a 1 Mbyte buffer where both the search and look-ahead buffers slide from left to right. At any point in the encoding process, the order-2 context of the current symbol (the two most-recent symbols in the search buffer) is used to predict the current symbol. The two symbols constituting the context are hashed to a 12-bit number I, which is used as an index to a $2^{12} = 4{,}096$-entry array A of partitions. Each partition contains 32 pointers to the input data in the 1 Mbyte buffer (each pointer is therefore 20 bits long).

The 32 pointers in partition $A[I]$ are checked to find the longest match between the look-ahead buffer and the input data seen so far. The longest match is selected and is coded in 8 bits. The first 3 bits code the match length according to Table 6.21; the remaining 5 bits identify the pointer in the partition. Such an 8-bit number is called a *copy item*. If no match is found, a literal is encoded in 8 bits. For each item, an extra bit is prepared, a 0 for a literal and a 1 for a copy item. The extra bits are accumulated in groups of 16, and each group is output, as in LZRW1, preceding the 16 items it refers to.

3 bits:	000	001	010	011	100	101	110	111
length:	2	3	4	5	6	7	8	16

Table 6.21: Encoding the Length in LZRW4.

The partitions are updated all the time by moving "good" pointers toward the start of their partition. When a match is found, the encoder swaps the selected pointer with the pointer halfway toward the partition (Figure 6.22a,b). If no match is found, the entire 32-pointer partition is shifted to the left and the new pointer is entered on the right, pointing to the current symbol (Figure 6.22c).

> The Red Queen shook her head, "You may call it 'nonsense' if you like," she said, "but I've heard nonsense, compared with which that would be as sensible as a dictionary!"
>
> —Lewis Carroll, *Through the Looking Glass* (1872)

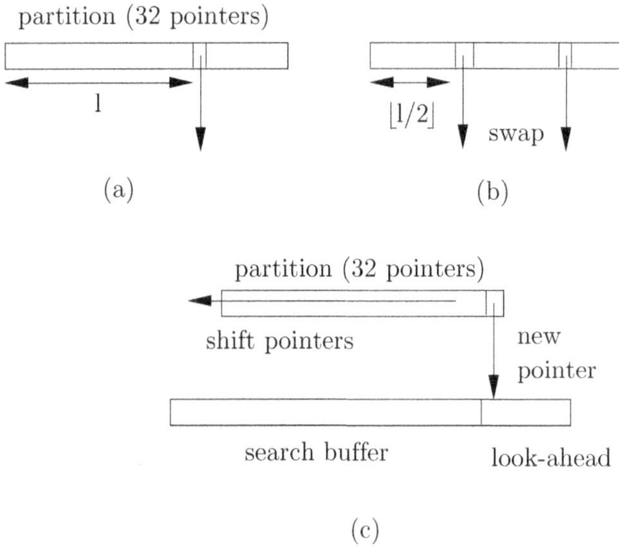

Figure 6.22: Updating an LZRW4 Partition.

6.13 LZW

This is a popular variant of LZ78, developed by Terry Welch in 1984 ([Welch 84] and [Phillips 92]). Its main feature is eliminating the second field of a token. An LZW token consists of just a pointer to the dictionary. To best understand LZW, we will temporarily forget that the dictionary is a tree, and will think of it as an array of variable-size strings. The LZW method starts by initializing the dictionary to all the symbols in the alphabet. In the common case of 8-bit symbols, the first 256 entries of the dictionary (entries 0 through 255) are occupied before any data is input. Because the dictionary is initialized, the next input character will always be found in the dictionary. This is why an LZW token can consist of just a pointer and does not have to contain a character code as in LZ77 and LZ78.

(LZW has been patented and for many years its use required a license. This issue is treated in Section 6.34.)

The principle of LZW is that the encoder inputs symbols one by one and accumulates them in a string `I`. After each symbol is input and is concatenated to `I`, the dictionary is searched for string `I`. As long as `I` is found in the dictionary, the process continues. At a certain point, adding the next symbol `x` causes the search to fail; string `I` is in the dictionary but string `Ix` (symbol `x` concatenated to `I`) is not. At this point the encoder (1) outputs the dictionary pointer that points to string `I`, (2) saves string `Ix` (which is now called a *phrase*) in the next available dictionary entry, and (3) initializes string `I` to symbol `x`. To illustrate this process, we again use the text string `sir␣sid␣eastman␣easily␣teases␣sea␣sick␣seals`. The steps are as follows:

0. Initialize entries 0–255 of the dictionary to all 256 8-bit bytes.
1. The first symbol `s` is input and is found in the dictionary (in entry 115, since this is

the ASCII code of `s`). The next symbol `i` is input, but `si` is not found in the dictionary. The encoder performs the following: (1) outputs 115, (2) saves string `si` in the next available dictionary entry (entry 256), and (3) initializes I to the symbol `i`.
2. The `r` of `sir` is input, but string `ir` is not in the dictionary. The encoder (1) outputs 105 (the ASCII code of `i`), (2) saves string `ir` in the next available dictionary entry (entry 257), and (3) initializes `I` to the symbol `r`.

Table 6.23 summarizes all the steps of this process. Table 6.24 shows some of the original 256 entries in the LZW dictionary plus the entries added during encoding of the string above. The complete output stream is (only the numbers are output, not the strings in parentheses) as follows:

115 (`s`), 105 (`i`), 114 (`r`), 32 (␣), 256 (`si`), 100 (`d`), 32 (␣), 101 (`e`), 97 (`a`), 115 (`s`), 116 (`t`), 109 (`m`), 97 (`a`), 110 (`n`), 262 (`␣e`), 264 (`as`), 105 (`i`), 108 (`l`), 121 (`y`), 32 (␣), 116 (`t`), 263 (`ea`), 115 (`s`), 101 (`e`), 115 (`s`), 259 (`␣s`), 263 (`ea`), 259 (`␣s`), 105 (`i`), 99 (`c`), 107 (`k`), 280 (`␣se`), 97 (`a`), 108 (`l`), 115 (`s`), eof.

Figure 6.25 is a pseudo-code listing of the algorithm. We denote by λ the empty string, and by `<<a,b>>` the concatenation of strings `a` and `b`.

The line "`append <<di,ch>> to the dictionary`" is of special interest. It is clear that in practice, the dictionary may fill up. This line should therefore include a test for a full dictionary, and certain actions for the case where it is full.

Since the first 256 entries of the dictionary are occupied right from the start, pointers to the dictionary have to be longer than 8 bits. A simple implementation would typically use 16-bit pointers, which allow for a 64K-entry dictionary (where $64\text{K} = 2^{16} = 65{,}536$). Such a dictionary will, of course, fill up very quickly in all but the smallest compression jobs. The same problem exists with LZ78, and any solutions used with LZ78 can also be used with LZW. Another interesting fact about LZW is that strings in the dictionary become only one character longer at a time. It therefore takes a long time to end up with long strings in the dictionary, and thus a chance to achieve really good compression. We can say that LZW adapts slowly to its input data.

◇ **Exercise 6.10:** Use LZW to encode the string `alf␣eats␣alfalfa`. Show the encoder output and the new entries added by it to the dictionary.

◇ **Exercise 6.11:** Analyze the LZW compression of the string "`aaaa...`".

`A dirty icon` (anagram of "dictionary")

6.13.1 LZW Decoding

To understand how the LZW decoder works, we recall the three steps the encoder performs each time it writes something on the output stream. They are (1) it outputs the dictionary pointer that points to string `I`, (2) it saves string `Ix` in the next available entry of the dictionary, and (3) it initializes string `I` to symbol `x`.

The decoder starts with the first entries of its dictionary initialized to all the symbols of the alphabet (normally 256 symbols). It then reads its input stream (which consists of pointers to the dictionary) and uses each pointer to retrieve uncompressed symbols from its dictionary and write them on its output stream. It also builds its dictionary in

I	in dict?	new entry	output	I	in dict?	new entry	output
s	Y			y	Y		
si	N	256-si	115 (s)	y␣	N	274-y␣	121 (y)
i	Y			␣	Y		
ir	N	257-ir	105 (i)	␣t	N	275-␣t	32 (␣)
r	Y			t	Y		
r␣	N	258-r␣	114 (r)	te	N	276-te	116 (t)
␣	Y			e	Y		
␣s	N	259-␣s	32 (␣)	ea	Y		
s	Y			eas	N	277-eas	263 (ea)
si	Y			s	Y		
sid	N	260-sid	256 (si)	se	N	278-se	115 (s)
d	Y			e	Y		
d␣	N	261-d␣	100 (d)	es	N	279-es	101 (e)
␣	Y			s	Y		
␣e	N	262-␣e	32 (␣)	s␣	N	280-s␣	115 (s)
e	Y			␣	Y		
ea	N	263-ea	101 (e)	␣s	Y		
a	Y			␣se	N	281-␣se	259 (␣s)
as	N	264-as	97 (a)	e	Y		
s	Y			ea	Y		
st	N	265-st	115 (s)	ea␣	N	282-ea␣	263 (ea)
t	Y			␣	Y		
tm	N	266-tm	116 (t)	␣s	Y		
m	Y			␣si	N	283-␣si	259 (␣s)
ma	N	267-ma	109 (m)	i	Y		
a	Y			ic	N	284-ic	105 (i)
an	N	268-an	97 (a)	c	Y		
n	Y			ck	N	285-ck	99 (c)
n␣	N	269-n␣	110 (n)	k	Y		
␣	Y			k␣	N	286-k␣	107 (k)
␣e	Y			␣	Y		
␣ea	N	270-␣ea	262 (␣e)	␣s	Y		
a	Y			␣se	Y		
as	Y			␣sea	N	287-␣sea	281 (␣se)
asi	N	271-asi	264 (as)	a	Y		
i	Y			al	N	288-al	97 (a)
il	N	272-il	105 (i)	l	Y		
l	Y			ls	N	289-ls	108 (l)
ly	N	273-ly	108 (l)	s	Y		
				s,eof	N		115 (s)

Table 6.23: Encoding sir sid eastman easily teases sea sick seals.

0	NULL	110	n	262	␣e	276	te
1	SOH	...		263	ea	277	eas
...		115	s	264	as	278	se
32	SP	116	t	265	st	279	es
...		...		266	tm	280	s
97	a	121	y	267	ma	281	␣se
98	b	...		268	an	282	ea␣
99	c	255	255	269	n␣	283	␣si
100	d	256	si	270	␣ea	284	ic
101	e	257	ir	271	asi	285	ck
...		258	r␣	272	il	286	k␣
107	k	259	␣s	273	ly	287	␣sea
108	l	260	sid	274	y␣	288	al
109	m	261	d␣	275	␣t	289	ls

Table 6.24: An LZW Dictionary.

```
for i:=0 to 255 do
  append i as a 1-symbol string to the dictionary;
append λ to the dictionary;
di:=dictionary index of λ;
repeat
  read(ch);
  if <<di,ch>> is in the dictionary then
    di:=dictionary index of <<di,ch>>;
  else
    output(di);
    append <<di,ch>> to the dictionary;
    di:=dictionary index of ch;
  endif;
until end-of-input;
```

Figure 6.25: The LZW Algorithm.

the same way as the encoder (this fact is usually expressed by saying that the encoder and decoder are *synchronized*, or that they work in *lockstep*).

In the first decoding step, the decoder inputs the first pointer and uses it to retrieve a dictionary item `I`. This is a string of symbols, and it is written on the decoder's output. String `Ix` needs to be saved in the dictionary, but symbol `x` is still unknown; it will be the first symbol in the next string retrieved from the dictionary.

In each decoding step after the first, the decoder inputs the next pointer, retrieves the next string `J` from the dictionary, writes it on the output, isolates its first symbol `x`, and saves string `Ix` in the next available dictionary entry (after checking to make sure string `Ix` is not already in the dictionary). The decoder then moves `J` to `I` and is ready for the next step.

In our `sir␣sid...` example, the first pointer that's input by the decoder is 115. This corresponds to the string `s`, which is retrieved from the dictionary, gets stored in `I`, and becomes the first item written on the decoder's output. The next pointer is 105, so string `i` is retrieved into `J` and is also written on the output. `J`'s first symbol is concatenated with `I`, to form string `si`, which does not exist in the dictionary, and is therefore added to it as entry 256. Variable `J` is moved to `I`, so `I` is now the string `i`. The next pointer is 114, so string `r` is retrieved from the dictionary into `J` and is also written on the output. `J`'s first symbol is concatenated with `I`, to form string `ir`, which does not exist in the dictionary, and is added to it as entry 257. Variable `J` is moved to `I`, so `I` is now the string `r`. The next step reads pointer 32, writes ␣ on the output, and saves string `r` .

◇ **Exercise 6.12:** Decode the string `alf␣eats␣alfalfa` by using the encoding results from Exercise 6.10.

◇ **Exercise 6.13:** Assume a two-symbol alphabet with the symbols `a` and `b`. Show the first few steps for encoding and decoding the string "`ababab...`".

6.13.2 LZW Dictionary Structure

Up until now, we have assumed that the LZW dictionary is an array of variable-size strings. To understand why a trie is a better data structure for the dictionary we need to recall how the encoder works. It inputs symbols and concatenates them into a variable `I` as long as the string in `I` is found in the dictionary. At a certain point the encoder inputs the first symbol `x`, which causes the search to fail (string `Ix` is not in the dictionary). It then adds `Ix` to the dictionary. This means that each string added to the dictionary effectively adds just one new symbol, `x`. (Phrased another way; for each dictionary string of more than one symbol, there exists a "parent" string in the dictionary that's one symbol shorter.)

A tree similar to the one used by LZ78 is therefore a good data structure, because adding string `Ix` to such a tree is done by adding one node with x. The main problem is that each node in the LZW tree may have many children (this is a multiway tree, not a binary tree). Imagine the node for the letter `a` in entry 97. Initially it has no children, but if the string `ab` is added to the tree, node 97 gets one child. Later, when, say, the string `ae` is added, node 97 gets a second child, and so on. The data structure for the tree should therefore be designed such that a node could have any number of children, but without having to reserve any memory for them in advance.

One way of designing such a data structure is to house the tree in an array of nodes, each a structure with two fields: a symbol and a pointer to the parent node. A node has no pointers to any child nodes. Moving down the tree, from a node to one of its children, is done by a *hashing process* in which the pointer to the node and the symbol of the child are hashed to create a new pointer.

Suppose that string `abc` has already been input, symbol by symbol, and has been stored in the tree in the three nodes at locations 97, 266, and 284. Following that, the encoder has just input the next symbol `d`. The encoder now searches for string `abcd`, or, more specifically, for a node containing the symbol `d` whose parent is at location 284. The encoder hashes the 284 (the pointer to string `abc`) and the 100 (ASCII code of `d`) to create a pointer to some node, say, 299. The encoder then examines node 299. There are three possibilities:

1. The node is unused. This means that `abcd` is not yet in the dictionary and should be added to it. The encoder adds it to the tree by storing the parent pointer 284 and ASCII code 100 in the node. The result is the following:

Node				
Address :	97	266	284	299
Contents :	(-:`a`)	(97:`b`)	(266:`c`)	(284:`d`)
Represents:	`a`	`ab`	`abc`	`abcd`

2. The node contains a parent pointer of 284 and the ASCII code of `d`. This means that string `abcd` is already in the tree. The encoder inputs the next symbol, say `e`, and searches the dictionary tree for string `abcde`.
3. The node contains something else. This means that another hashing of a pointer and an ASCII code has resulted in 299, and node 299 already contains information from another string. This is called a *collision*, and it can be dealt with in several ways. The simplest way to deal with a collision is to increment pointer 299 and examine nodes 300, 301,... until an unused node is found, or until a node with (284:`d`) is found.

In practice, we build nodes that are structures with three fields, a pointer to the parent node, the pointer (or index) created by the hashing process, and the code (normally ASCII) of the symbol contained in the node. The second field is necessary because of collisions. A node can therefore be illustrated by

parent
index
symbol

We illustrate this data structure using string `ababab...` of Exercise 6.13. The dictionary is an array `dict` where each entry is a structure with the three fields `parent`, `index`, and `symbol`. We refer to a field by, for example, `dict[pointer].parent`, where `pointer` is an index to the array. The dictionary is initialized to the two entries `a` and `b`. (To keep the example simple we use no ASCII codes. We assume that `a` has code 1 and `b` has code 2.) The first few steps of the encoder are as follows:

Step 0: Mark all dictionary locations from 3 on as unused.

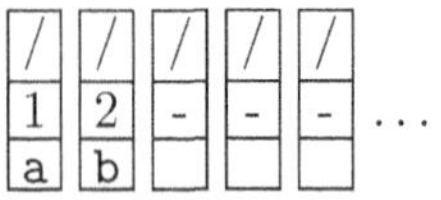

Step 1: The first symbol `a` is input into variable `I`. What is actually input is the code of `a`, which in our example is 1, so `I` = 1. Since this is the first symbol, the encoder assumes that it is in the dictionary and so does not perform any search.
Step 2: The second symbol `b` is input into `J`, so `J` = 2. The encoder has to search for string `ab` in the dictionary. It executes `pointer:=hash(I,J)`. Let's assume that the result is 5. Field `dict[pointer].index` contains "unused", since location 5 is still empty, so string `ab` is not in the dictionary. It is added by executing

```
dict[pointer].parent:=I;
dict[pointer].index:=pointer;
dict[pointer].symbol:=J;
```

with `pointer`=5. `J` is moved into `I`, so `I` = 2.

/	/	/	/	1	
1	2	-	-	5	...
a	b			b	

Step 3: The third symbol `a` is input into `J`, so `J` = 1. The encoder has to search for string `ba` in the dictionary. It executes `pointer:=hash(I,J)`. Let's assume that the result is 8. Field `dict[pointer].index` contains "unused", so string `ba` is not in the dictionary. It is added as before by executing

```
dict[pointer].parent:=I;
dict[pointer].index:=pointer;
dict[pointer].symbol:=J;
```

with `pointer`=8. `J` is moved into `I`, so `I` = 1.

/	/	/	/	1	/	/	2	/	
1	2	-	-	5	-	-	8	-	...
a	b			b			a		

Step 4: The fourth symbol `b` is input into `J`, so `J`=2. The encoder has to search for string `ab` in the dictionary. It executes `pointer:=hash(I,J)`. We know from step 2 that the result is 5. Field `dict[pointer].index` contains 5, so string `ab` is in the dictionary. The value of `pointer` is moved into `I`, so `I` = 5.
Step 5: The fifth symbol `a` is input into `J`, so `J` = 1. The encoder has to search for string `aba` in the dictionary. It executes as usual `pointer:=hash(I,J)`. Let's assume that the result is 8 (a collision). Field `dict[pointer].index` contains 8, which looks good, but field `dict[pointer].parent` contains 2 instead of the expected 5, so the hash function knows that this is a collision and string `aba` is not in dictionary entry 8. It increments `pointer` as many times as necessary until it finds a dictionary entry with `index`=8 and `parent`=5 or until it finds an unused entry. In the former case, string `aba` is in the dictionary, and `pointer` is moved to `I`. In the latter case `aba` is not in the dictionary, and the encoder saves it in the entry pointed at by `pointer`, and moves `J` to `I`.

/	/	/	/	1	/	/	2	5	/	
1	2	-	-	5	-	-	8	8	-	...
a	b			b			a	a		

Example: The 15 hashing steps for encoding the string `alf␣eats␣alfalfa` are shown below. The encoding process itself is illustrated in detail in the answer to Exercise 6.10. The results of the hashing are arbitrary; they are not the results produced by a real hash function. The 12 trie nodes constructed for this string are shown in Figure 6.26.

1. Hash(`l`,97) → 278. Array location 278 is set to (97, 278, `l`).
2. Hash(`f`,108) → 266. Array location 266 is set to (108, 266, `f`).
3. Hash(␣,102) → 269. Array location 269 is set to (102,269,␣).
4. Hash(`e`,32) → 267. Array location 267 is set to (32, 267, `e`).
5. Hash(`a`,101) → 265. Array location 265 is set to (101, 265, `a`).
6. Hash(`t`,97) → 272. Array location 272 is set to (97, 272, `t`).
7. Hash(`s`,116) → 265. A collision! Skip to the next available location, 268, and set it to (116, 265, `s`). This is why the index needs to be stored.
8. Hash(␣,115) → 270. Array location 270 is set to (115, 270, ␣).
9. Hash(`a`,32) → 268. A collision! Skip to the next available location, 271, and set it to (32, 268, `a`).
10. Hash(`l`,97) → 278. Array location 278 already contains index 278 and symbol `l` from step 1, so there is no need to store anything else or to add a new trie entry.
11. Hash(`f`,278) → 276. Array location 276 is set to (278, 276, `f`).
12. Hash(`a`,102) → 274. Array location 274 is set to (102, 274, `a`).
13. Hash(`l`,97) → 278. Array location 278 already contains index 278 and symbol `l` from step 1, so there is no need to do anything.
14. Hash(`f`,278) → 276. Array location 276 already contains index 276 and symbol `f` from step 11, so there is no need to do anything.
15. Hash(`a`,276) → 274. A collision! Skip to the next available location, 275, and set it to (276, 274, `a`).

Readers who have carefully followed the discussion up to this point will be happy to learn that the LZW decoder's use of the dictionary tree-array is simple and no hashing is needed. The decoder starts, like the encoder, by initializing the first 256 array locations. It then reads pointers from its input stream and uses each to locate a symbol in the dictionary.

In the first decoding step, the decoder inputs the first pointer and uses it to retrieve a dictionary item `I`. This is a symbol that is now written by the decoder on its output stream. String `Ix` needs to be saved in the dictionary, but symbol `x` is still unknown; it will be the first symbol in the next string retrieved from the dictionary.

In each decoding step after the first, the decoder inputs the next pointer and uses it to retrieve the next string `J` from the dictionary and write it on the output stream. If the pointer is, say 8, the decoder examines field `dict[8].index`. If this field equals 8, then this is the right node. Otherwise, the decoder examines consecutive array locations until it finds the right one.

Once the right tree node is found, the `parent` field is used to go back up the tree and retrieve the individual symbols of the string *in reverse order*. The symbols are then placed in `J` in the right order (see below), the decoder isolates the first symbol `x` of `J`, and saves string `Ix` in the next available array location. (String `I` was found in the previous step, so only one node, with symbol `x`, needs be added.) The decoder then moves `J` to `I` and is ready for the next step.

265	266	267	268	269	270	271	272	273	274	275	276	277	278
/	/	/	/	/	/	/	/	/	/	/	/	/	97
-	-	-	-	-	-	-	-	-	-	-	-	-	278
													l
/	108	/	/	/	/	/	/	/	/	/	/	/	97
-	266	-	-	-	-	-	-	-	-	-	-	-	278
	f												l
/	108	/	/	102	/	/	/	/	/	/	/	/	97
-	266	-	-	269	-	-	-	-	-	-	-	-	278
	f			␣									l
/	108	32	/	102	/	/	/	/	/	/	/	/	97
-	266	267	-	269	-	-	-	-	-	-	-	-	278
	f	e		␣									l
101	108	32	/	102	/	/	/	/	/	/	/	/	97
265	266	267	-	269	-	-	-	-	-	-	-	-	278
a	f	e		␣									l
101	108	32	/	102	/	/	97	/	/	/	/	/	97
265	266	267	-	269	-	-	272	-	-	-	-	-	278
a	f	e		␣			t						l
101	108	32	116	102	/	/	97	/	/	/	/	/	97
265	266	267	265	269	-	-	272	-	-	-	-	-	278
a	f	e	s	␣			t						l
101	108	32	116	102	115	/	97	/	/	/	/	/	97
265	266	267	265	269	270	-	272	-	-	-	-	-	278
a	f	e	s	␣	␣		t						l
101	108	32	116	102	115	32	97	/	/	/	/	/	97
265	266	267	265	269	270	268	272	-	-	-	-	-	278
a	f	e	s	␣	␣	a	t						l
101	108	32	116	102	115	32	97	/	/	/	278	/	97
265	266	267	265	269	270	268	272	-	-	-	276	-	278
a	f	e	s	␣	␣	a	t				f		l
101	108	32	116	102	115	32	97	/	102	/	278	/	97
265	266	267	265	269	270	268	272	-	274	-	276	-	278
a	f	e	s	␣	␣	a	t		a		f		l
101	108	32	116	102	115	32	97	/	102	276	278	/	97
265	266	267	265	269	270	268	272	-	274	274	276	-	278
a	f	e	s	␣	␣	a	t		a	a	f		l

Figure 6.26: Growing An LZW Trie for "alf eats alfalfa".

Retrieving a complete string from the LZW tree therefore involves following the pointers in the `parent` fields. This is equivalent to moving *up* the tree, which is why the hash function is no longer needed.

Example: The previous example describes the 15 hashing steps in the encoding of string `alf␣eats␣alfalfa`. The last step sets array location 275 to (276,274,a) and writes 275 (a pointer to location 275) on the compressed stream. When this stream is read by the decoder, pointer 275 is the last item input and processed by the decoder. The decoder finds symbol `a` in the `symbol` field of location 275 (indicating that the string stored at 275 ends with an `a`) and a pointer to location 276 in the `parent` field. The decoder then examines location 276 where it finds symbol `f` and parent pointer 278. In location 278 the decoder finds symbol `l` and a pointer to 97. Finally, in location 97 the decoder finds symbol `a` and a null pointer. The (reversed) string is therefore `afla`. There is no need for the decoder to do any hashing or to use the `index` fields.

The last point to discuss is string reversal. Two commonly-used approaches are outlined here:

1. Use a stack. A stack is a common data structure in modern computers. It is an array in memory that is accessed at one end only. At any time, the item that was last pushed into the stack will be the first one to be popped out (last-in-first-out, or LIFO). Symbols retrieved from the dictionary are pushed into the stack. When the last one has been retrieved and pushed, the stack is popped, symbol by symbol, into variable `J`. When the stack is empty, the entire string has been reversed. This is a common way to reverse a string.
2. Retrieve symbols from the dictionary and concatenate them into `J` *from right to left*. When done, the string will be stored in `J` in the right order. Variable `J` must be long enough to accommodate the longest possible string, but then it has to be long enough even when a stack is used.

◇ **Exercise 6.14:** What is the longest string that can be retrieved from the LZW dictionary during decoding?

(A reminder. The troublesome issue of software patents and licenses is treated in Section 6.34.)

6.13.3 LZW in Practice

The publication of the LZW algorithm, in 1984, has strongly affected the data compression community and has influenced many people to come up with implementations and variants of this method. Some of the most important LZW variants and spin-offs are described here.

6.13.4 Differencing

The idea of differencing, or relative encoding, has already been mentioned in Section 1.3.1. This idea turns out to be useful in LZW image compression, since most adjacent pixels don't differ by much. It is possible to implement an LZW encoder that computes the value of a pixel relative to its predecessor and then encodes this difference. The decoder should, of course, be compatible and should compute the absolute value of a pixel after decoding its relative value.

6.13.5 LZW Variants

A word-based LZW variant is described in Section 11.6.2.

LZW is an adaptive data compression method, but it is slow to adapt to its input, since strings in the dictionary become only one character longer at a time. Exercise 6.11 shows that a string of one million a's (which, of course, is highly redundant) produces dictionary phrases, the longest of which contains only 1,414 a's.

The LZMW method, Section 6.15, is a variant of LZW that overcomes this problem. Its main principle is this: Instead of adding I plus one character of the next phrase to the dictionary, add I plus the entire next phrase to the dictionary.

The LZAP method, Section 6.16, is yet another variant based on this idea: Instead of just concatenating the last two phrases and placing the result in the dictionary, place all prefixes of the concatenation in the dictionary. More specifically, if S and T are the last two matches, add St to the dictionary for every nonempty prefix t of T, including T itself.

Table 6.27 summarizes the principles of LZW, LZMW, and LZAP and shows how they naturally suggest another variant, LZY.

Increment	Add a string to the dictionary	
string by	per phrase	per input char.
One character:	LZW	LZY
Several chars:	LZMW	LZAP

Table 6.27: LZW and Three Variants.

LZW adds one dictionary string per phrase and increments strings by one symbol at a time. LZMW adds one dictionary string per phrase and increments strings by several symbols at a time. LZAP adds one dictionary string per input symbol and increments strings by several symbols at a time. LZY, Section 6.18, fits the fourth cell of Table 6.27. It is a method that adds one dictionary string per input symbol and increments strings by one symbol at a time.

6.14 UNIX Compression (LZC)

In the vast UNIX world, compress used to be the most common compression utility (although GNU gzip has become more popular because it is free from patent claims, is faster, and provides superior compression). This utility (also called LZC) uses LZW with a growing dictionary. It starts with a small dictionary of just $2^9 = 512$ entries (with the first 256 of them already filled up). While this dictionary is being used, 9-bit pointers are written onto the output stream. When the original dictionary fills up, its size is doubled, to 1024 entries, and 10-bit pointers are used from this point. This process continues until the pointer size reaches a maximum set by the user (it can be set to between 9 and 16 bits, with 16 as the default value). When the largest allowed dictionary fills up, the program continues without changing the dictionary (which then becomes static), but with monitoring the compression ratio. If the ratio falls below a predefined threshold, the dictionary is deleted, and a new 512-entry dictionary is started. This way, the dictionary never gets "too out of date."

Decoding is done by the `uncompress` command, which implements the LZC decoder. Its main task is to maintain the dictionary in the same way as the encoder.

Two improvements to LZC, proposed by [Horspool 91], are listed below:

1. Encode the dictionary pointers with the phased-in binary codes of Section 5.3.1. Thus if the dictionary size is $2^9 = 512$ entries, pointers can be encoded in either 8 or 9 bits.
2. Determine all the impossible strings at any point. Suppose that the current string in the look-ahead buffer is `abcd...` and the dictionary contains strings `abc` and `abca` but not `abcd`. The encoder will output, in this case, the pointer to `abc` and will start encoding a new string starting with `d`. The point is that after decoding `abc`, the decoder knows that the next string cannot start with an `a` (if it did, an `abca` would have been encoded, instead of `abc`). In general, if `S` is the current string, then the next string cannot start with any symbol `x` that satisfies "`Sx` is in the dictionary."

This knowledge can be used by both the encoder and decoder to reduce redundancy even further. When a pointer to a string should be output, it should be coded, and the method of assigning the code should eliminate all the strings that are known to be impossible at that point. This may result in a somewhat shorter code but is probably too complex to justify its use in practice.

6.14.1 LZT

LZT is an extension, proposed by [Tischer 87], of UNIX compress/LZC. The major innovation of LZT is the way it handles a full dictionary. Recall that LZC treats a full dictionary in a simple way. It continues without updating the dictionary until the compression ratio drops below a preset threshold, and then it deletes the dictionary and starts afresh with a new dictionary. In contrast, LZT maintains, in addition to the dictionary (which is structured as a hash table, as described in Section 6.13.2), a linked list of phrases sorted by the number of times a phrase is used. When the dictionary fills up, the LZT algorithm identifies the least-recently-used (LRU) phrase in the list and deletes it from both the dictionary and the list.

Maintaining the list requires several operations per phrase, which is why LZT is slower than LZW or LZC, but the fact that the dictionary is not completely deleted results in better compression performance (each time LZC deletes its dictionary, its compression performance drops significantly for a while).

There is some similarity between the simple LZ77 algorithm and the LZT encoder. In LZ77, the search buffer is shifted, which deletes the leftmost phrase regardless of how useful it is (i.e., how often it has appeared in the input). In contrast, the LRU approach of LZT deletes the least useful phrase, the one that has been used the least since its insertion into the dictionary. In both cases, we can visualize a window of phrases. In LZ77 the window is sorted by the order in which phrases have been input, while in LZT the window is sorted by the usefulness of the phrases. Thus, LZT adapts better to varying statistics in the input data.

Another innovation of LZT is its use of phased-in codes (Section 2.9). It starts with a dictionary of 512 entries, of which locations 0 through 255 are reserved for the 256 8-bit bytes (those are never deleted) and code 256 is reserved as an escape code. These 257 codes are encoded by 255 8-bit codes and two 9-bit codes. Once the first input phrase has been identified and added to the dictionary, it (the dictionary) has 258 codes and they are encoded by 254 8-bit codes and four 9-bit codes. The use of phased-in codes is especially useful at the beginning, when the dictionary contains only a few useful phrases. As encoding progresses and the dictionary fills up, these codes contribute less and less to the overall performance of LZT.

The LZT dictionary is maintained as a hash table where each entry contains a phrase. Recall that in LZW (and also in LZC), if a phrase sI is in the dictionary (where s is a string and I is a single character), then s is also in the dictionary. This is why each dictionary entry in LZT has three fields (1) an output code (the code that has been assigned to a string s), (2) the (ASCII) code of an extension character I, and (3) a pointer to the list. When an entry is deleted, it is marked as free and the hash table is reorganized according to algorithm R in volume 3, Section 6.4 of [Knuth 73].

Another difference between LZC and LZT is the hash function. In LZC, the hash function takes an output code, shifts it one position to the left, and exclusive-ors it with the bits of the extension character. The hash function of LZT, on the other hand, reverses the bits of the output code before shifting and exclusive-oring it, which leads to fewer collisions and an almost perfect hash performance.

6.15 LZMW

This LZW variant, developed by V. Miller and M. Wegman [Miller and Wegman 85], is based on two principles:

1. When the dictionary becomes full, the least-recently-used dictionary phrase is deleted. There are several ways to select this phrase, and the developers suggest that any reasonable way of doing so would work. One possibility is to identify all the dictionary phrases `S` for which there are no phrases `Sa` (nothing has been appended to `S`, implying that `S` hasn't been used since it was placed in the dictionary) and delete the oldest of them. An auxiliary data structure has to be constructed and maintained in this case, pointing to dictionary phrases according to their age (the first pointer always points to the oldest phrase). The first 256 dictionary phrases should never be deleted.
2. Each phrase added to the dictionary is a concatenation of two strings, the previous match (`S'` below) and the current one (`S`). This is in contrast to LZW, where each phrase added is the concatenation of the current match and the first symbol of the next match. The pseudo-code algorithm illustrates this:

```
Initialize Dict to all the symbols of alphabet A;
i:=1; S':=null;
while i <= input size
 k:=longest match of Input[i] to Dict;
 Output(k);
 S:=Phrase k of Dict;
 i:=i+length(S);
 If phrase S'S is not in Dict, append it to Dict;
 S':=S;
endwhile;
```

By adding the concatenation `S'S` to the LZMW dictionary, dictionary phrases can grow by more than one symbol at a time. This means that LZMW dictionary phrases are more "natural" units of the input (for example, if the input is text in a natural language, dictionary phrases will tend to be complete words or even several words in that language). This, in turn, implies that the LZMW dictionary generally adapts to the input faster than the LZW dictionary.

Table 6.28 illustrates the LZMW method by applying it to the string

`sir␣sid␣eastman␣easily␣teases␣sea␣sick␣seals`.

LZMW adapts to its input faster than LZW but has the following three disadvantages:

1. The dictionary data structure cannot be the simple LZW trie, because not every prefix of a dictionary phrase is included in the dictionary. This means that the one-symbol-at-a-time search method used in LZW will not work. Instead, when a phrase `S` is added to the LZMW dictionary, every prefix of `S` must be added to the data structure, and every node in the data structure must have a tag indicating whether the node is in the dictionary or not.
2. Finding the longest string may require backtracking. If the dictionary contains `aaaa` and `aaaaaaaa`, we have to reach the eighth symbol of phrase `aaaaaaab` to realize that we have to choose the shorter phrase. This implies that dictionary searches in LZMW are slower than in LZW. This problem does not apply to the LZMW decoder.
3. A phrase may be added to the dictionary twice. This again complicates the choice of data structure for the dictionary.

◇ **Exercise 6.15:** Use the LZMW method to compress the string `swiss miss`.

◇ **Exercise 6.16:** Compress the string `yabbadabbadabbadoo` using LZMW.

6.16 LZAP

LZAP is an extension of LZMW. The "AP" stands for "All Prefixes" [Storer 88]. LZAP adapts to its input fast, like LZMW, but eliminates the need for backtracking, a feature that makes it faster than LZMW. The principle is this: Instead of adding the concatenation `S'S` of the last two phrases to the dictionary, add all the strings `S't` where `t` is a prefix of `S` (including `S` itself). Thus if $\mathtt{S}' = \mathtt{a}$ and $\mathtt{S} = \mathtt{bcd}$, add phrases `ab`,`abc`, and `abcd` to the LZAP dictionary. Table 6.29 shows the matches and the phrases added to the dictionary for `yabbadabbadabbadoo`.

In step 7 the encoder concatenates `d` to the two prefixes of `ab` and adds the two phrases `da` and `dab` to the dictionary. In step 9 it concatenates `ba` to the three prefixes of `dab` and adds the resulting three phrases `bad`, `bada`, and `badab` to the dictionary.

LZAP adds more phrases to its dictionary than does LZMW, so it takes more bits to represent the position of a phrase. At the same time, LZAP provides a bigger selection of dictionary phrases as matches for the input string, so it ends up compressing slightly better than LZMW while being faster (because of the simpler dictionary data structure, which eliminates the need for backtracking). This kind of trade-off is common in computer algorithms.

Step	Input	Out-put	S	Add to dict.	S'
	sir sid eastman easily teases sea sick seals				
1	s	115	s	—	—
2	i	105	i	256-si	s
3	r	114	r	257-ir	i
4	-	32	␣	258-r␣	r
5	si	256	si	259-␣si	␣
6	d	100	d	260-sid	si
7	-	32	␣	261-d␣	d
8	e	101	e	262-␣e	␣
9	a	97	a	263-ea	e
10	s	115	s	264-as	a
11	t	117	t	265-st	s
12	m	109	m	266-tm	t
13	a	97	a	267-ma	m
14	n	110	n	268-an	a
15	-e	262	␣e	269-n␣e	n
16	as	264	as	270-␣eas	␣e
17	i	105	i	271-asi	as
18	l	108	l	272-il	i
19	y	121	y	273-ly	l
20	-	32	␣	274-y␣	y
21	t	117	t	275-␣t	␣
22	ea	263	ea	276-tea	t
23	s	115	s	277-eas	ea
24	e	101	e	278-se	s
25	s	115	s	279-es	e
26	-	32	␣	280-s␣	s
27	se	278	se	281-␣se	␣
28	a	97	a	282-sea	se
29	-si	259	␣si	283-a␣si	a
30	c	99	c	284-␣sic	␣si
31	k	107	k	285-ck	c
32	-se	281	␣se	286-k␣se	k
33	a	97	a	287-␣sea	␣se
34	l	108	l	288-al	a
35	s	115	s	289-ls	l

Table 6.28: LZMW Example

Step	Input	Match	Add to dictionary
	yabbadabbadabbadoo		
1	y	y	—
2	a	a	256-ya
3	b	b	257-ab
4	b	b	258-bb
5	a	a	259-ba
6	d	d	260-ad
7	ab	ab	261-da, 262-dab
8	ba	ba	263-abb, 264-abba
9	dab	dab	265-bad, 266-bada, 267-badab
10	bad	bad	268-dabb, 269-dabba, 270-dabbad
11	o	o	271-bado
12	o	o	272-oo

Table 6.29: LZAP Example.

6.17 LZJ

LZJ is an interesting LZ variant, originated by [Jakobsson 85]. It stores in its dictionary, which can be viewed either as a multiway tree or as a forest, *every* phrase found in the input. If a phrase is found n times in the input, only one copy is stored in the dictionary. Such behavior tends to fill the dictionary up very quickly, so LZJ limits the length of phrases to a preset parameter h. The developer of the method recommends h values of around 6. Thus, if $h = 4$, the alphabet includes the letters and the blank space, and the input is `Once␣upon␣a␣time...`, then the dictionary will start with all the symbols of the alphabet and the following strings will gradually be inserted in it: `on`, `onc`, `nc`, `ce`, `nce`, `Once`, `nce␣`, `ce␣u`, `e␣up`, `␣upo`, and so on.

(The value of h is critical. Small values of h cause the LZJ encoder to store only short phrases in the dictionary. This degrades the compression because every code in the compressed file will correspond to only a short string. Large values of h also degrade compression because only a few long strings occur often in the input.)

An indirect result of this principle and of the way LZJ operates is that LZJ encoding is simpler than its decoding. Thus, LZJ is an asymmetric compression method and is unusual in this respect because in other asymmetric methods decoding is simpler and faster than encoding.

Example. Given the alphabet `a`, `b`, and `c`, the value $h = 4$, and the input data `babbaaacba`, the left side of Figure 6.30 lists the 24 phrases whose length is h or shorter, together with their dictionary addresses. The right side shows the complete 3-way dictionary tree (as a forest of three trees) with all 24 phrases. Each node in the dictionary has four fields, the character itself, a pointer (shown as a solid arrow) to its leftmost child, a pointer (horizontal, in dashed) to its right sibling, if any, and a frequency count that is discussed below. If a node is a rightmost child, the sibling pointer is replaced by a pointer to the parent (dashed, pointing up).

The dictionary's size H is another parameter of LZJ and is preset in each implementation (in the mid 1980s, the developer recommended $H \approx 2^{13}$, but current imple-

a	0	bb	9	aac	17
b	1	bba	10	aacb	18
c	2	bbba	11	ac	19
ba	3	baa	12	acb	20
bab	4	baaa	13	acba	21
babb	5	aa	14	cb	22
ab	6	aaa	15	cba	23
abb	7	aaac	16		
abba	8				

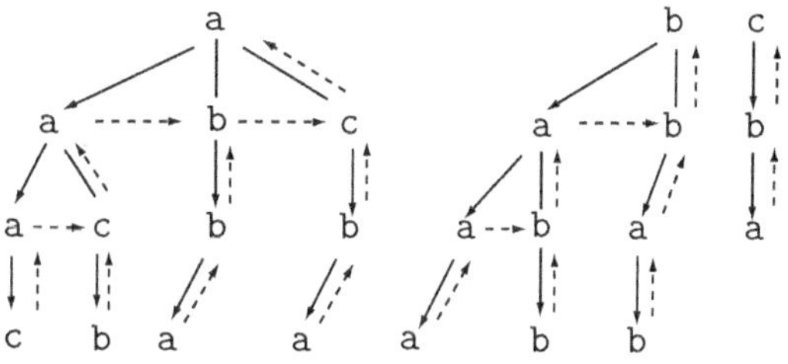

Figure 6.30: An LZJ Dictionary.

mentations can afford much larger values). Each input phrase found by the encoder in the dictionary is compressed by writing the dictionary address of the last character of the phrase on the output. This address is a number between 0 and $H-1$, written in $\lceil \log_2 H \rceil$ bits.

In order to understand the operation of the encoder, we assume $h=4$ and the input string `Once␣upon␣a␣time...`.

The encoder initializes the dictionary to the characters of the alphabet. At a general point in the encoding it reads the next input string, say `pon␣a␣time...`, selects the dictionary tree for `p`, and tries to match as many characters as possible. If `pon` is found in the tree, but `pon␣` isn't, the encoder outputs the dictionary address of the `n` of `pon` and updates the dictionary by inserting in it the strings `e␣up`, `␣upo`, and `upon`. They are placed in the trees whose roots are `e`, `␣`, and `u`, respectively. As each string is inserted, the encoder increments the counts of all the nodes that are on the path from the root to the leaf node of the string.

When the dictionary fills up, the encoder prunes the trees by deleting any nodes with a count of 1 (where instead of 1, an implementation can opt for a user-defined parameter). The roots of the trees correspond to the characters of the alphabet and should never be removed. The decoder can mimic this operation in lockstep with the encoder.

Pruning the trees results in free nodes, but consecutive prunings result in fewer and fewer free nodes, as illustrated by Figure 6.31. The figure shows that with $H=1{,}024$, many dictionary prunings are needed for the first 4,000 input characters. The developer of the method also shows that doubling the dictionary size to 2,048 nodes results in only three dictionary prunings for the same input data. Eventually, pruning becomes ineffective and LZJ continues (nonadaptively) without it. This behavior suggests that a large dictionary may not require any prunings at all and may therefore speed up LZJ (both encoding and decoding) considerably.

LZJ decoding is more time consuming than its encoding, because the decoder has to follow various pointers in a dictionary tree. This is best illustrated by an example (refer to Figure 6.30). Suppose that the decoder reads 17 from the compressed stream. Location 17 contains the character `c`, the last character in the phrase being decoded. The decoder has to follow the back pointers (the dashed pointers in the figure) until it arrives at the root of the tree. It then follows the forward pointers (both dashed horizontal and solid), collecting characters as it goes along, until it arrives back at location 17. This is

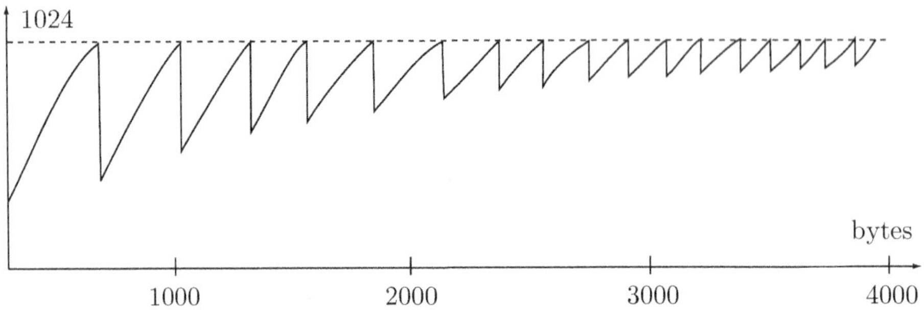

Figure 6.31: Effect of Pruning the LZJ Dictionary.

how the decoder constructs phrase aac. Once the phrase is decoded, the decoder uses it to update the dictionary by inserting phrases in lockstep with the encoder.

The last interesting feature of LZJ is the format of the compressed stream. If the dictionary size H is fixed, then each item in the compressed stream is a fixed-size, $\lceil \log_2 H \rceil$ bits pointer to the dictionary. In principle, however, the dictionary can be allowed to grow as needed and the algorithm may use one of the many variable-length codes for the integers to encode the dictionary pointers.

It turns out that this approach, which initially seems promising, is in fact worse than having fixed-length pointers. The reason is that each variable-length code is ideal for a certain distribution of the integers in the data, but in the case of LZJ nothing is known about this distribution. Initially, the dictionary is small, so all the pointers to it are small integers. When the dictionary grows, especially after it has been pruned, pointers to it may be small or large with the same probability. Given a set of n integers with a uniform distribution, the length of a fixed-length code for them is $\lceil \log_2 n \rceil$, but the average length of any variable-length code is generally longer (see Figure 3.11).

A good compromise is the use of phased-in codes (Section 2.9). If $H = 1{,}000$, the length of a fixed-size code is 10 bits, but only 1,000 out of the 1,024 possible 10-bit combinations are used. With phased-in codes, most codewords would be 10 bits long, but some codewords would be only nine bits long. The excellent book *Text Compression* [Bell et al. 90], refers to this compromise (on page 230) as LZJ$'$.

> "Heavens, Andrew!" said his wife; "what is a rustler?"
> It was not in any dictionary, and current translations of it were inconsistent. A man at Hoosic Falls said that he had passed through Cheyenne, and heard the term applied in a complimentary way to people who were alive and pushing. Another man had always supposed it meant some kind of horse. But the most alarming version of all was that a rustler was a cattle thief.
>
> —Owen Wister, *The Virginian—A Horseman of the Plains*

6.18 LZY

The LZY method is due to Dan Bernstein. The Y stands for Yabba, which came from the input string originally used to test the algorithm. The LZY dictionary is initialized to all the single symbols of the alphabet. For every symbol C in the input stream, the decoder looks for the longest string P that precedes C and is already included in the dictionary. If the string PC is not in the dictionary, it is added to it as a new phrase.

As an example, the input yabbadabbadabbadoo causes the phrases ya, ab, bb, ba, ad, da, abb, bba, ada, dab, abba, bbad, bado, ado, and oo to be added to the dictionary.

While encoding the input, the encoder keeps track of the list of matches-so-far L. Initially, L is empty. If C is the current input symbol, the encoder (before adding anything to the dictionary) checks, for every string M in L, whether string MC is in the dictionary. If it is, then MC becomes a new match-so-far and is added to L. Otherwise, the encoder outputs the number of L (its position in the dictionary) and adds C, as a new match-so-far, to L.

Here is a pseudo-code algorithm for constructing the LZY dictionary. The authors' personal experience suggests that implementing such an algorithm in a real programming language results in a deeper understanding of its operation.

```
Start with a dictionary containing all the symbols of the
 alphabet, each mapped to a unique integer.
M:=empty string.
Repeat
 Append the next symbol C of the input stream to M.
 If M is not in the dictionary, add it to the dictionary,
  delete the first character of M, and repeat this step.
Until end-of-input.
```

The output of LZY is not synchronized with the dictionary additions. Also, the encoder must be careful not to have the longest output match overlap itself. Because of this, the dictionary should consist of two parts, S and T, where only the former is used for the output. The algorithm is the following:

```
Start with S mapping each single character to a unique integer;
set T empty; M empty; and O empty.
Repeat
 Input the next symbol C. If OC is in S, set O:=OC;
 otherwise output S(O), set O:=C, add T to S,
  and remove everything from T.
 While MC is not in S or T, add MC to T (mapping to the next
   available integer), and chop off the first character of M.
 After M is short enough so that MC is in the dict., set M:=MC.
Until end-of-input.
Output S(O) and quit.
```

The decoder reads the compressed stream. It uses each code to find a phrase in the dictionary, it outputs the phrase as a string, then uses each symbol of the string to add a new phrase to the dictionary in the same way the encoder does. Here are the decoding steps:

```
Start with a dictionary containing all the symbols of the
 alphabet, each mapped to a unique integer.
M:=empty string.
Repeat
 Read D(O) from the input and take the inverse under D to find O.
 As long as O is not the empty string, find the first character C
  of O, and update (D,M) as above.
  Also output C and chop it off from the front of O.
Until end-of-input.
```

Notice that encoding requires two fast operations on strings in the dictionary: (1) testing whether string SC is in the dictionary if S's position is known and (2) finding S's position given CS's position. Decoding requires the same operations plus fast searching to find the first character of a string when its position in the dictionary is given.

Table 6.32 illustrates LZY for the input string abcabcabcabcabcabcabcx'. It shows the phrases added to the dictionary at each step, as well as the list of current matches.

The encoder starts with no matches. When it inputs a symbol, it appends it to each match-so-far; any results that are already in the dictionary become the new matches-so-far (the symbol itself becomes another match). Any results that are not in the dictionary are deleted from the list and added to the dictionary.

Before reading the fifth c, for example, the matches-so-far are bcab, cab, ab, and b. The encoder appends c to each match. bcabc doesn't match, so the encoder adds it to the dictionary. The rest are still in the dictionary, so the new list of matches-so-far is cabc, abc, bc, and c.

When the x is input, the current list of matches-so-far is abcabc, bcabc, cabc, abc, bc, and c. None of abcabcx, bcabcx, cabcx, abcx, bcx, or cx are in the dictionary, so they are all added to it, and the list of matches-so-far is reduced to just a single x.

> Airman stops coed
>
> Anagram of "data compression"

6.19 LZP

LZP is an LZ77 variant developed by Charles Bloom [Bloom 96] (the P stands for "prediction"). It is based on the principle of context prediction, which says, "if a certain string abcde has appeared in the input stream in the past and was followed by fg..., then when abcde appears again in the input stream, there is a good chance that it will be followed by the same fg...." Section 6.35 should be consulted for the relation between dictionary-based and prediction algorithms.

Figure 6.33 (part I) shows an LZ77 sliding buffer with fgh... as the current symbols (this string is denoted by S) in the look-ahead buffer, immediately preceded by abcde in the search buffer. The string abcde is called the *context* of fgh... and is denoted by C. In general, the context of a string S is the N-symbol string C immediately to the left of S. A context can be of any length N, and variants of LZP, discussed in Sections 6.19.3 and 6.19.4, use different values of N. The algorithm passes the context through a hash

Step	Input	Add to dict.	Current matches
	abcabcabcabcabcabcabcx		
1	a	—	a
2	b	256-ab	b
3	c	257-bc	c
4	a	258-ca	a
5	b	—	ab, b
6	c	259-abc	bc, c
7	a	260-bca	ca, a
8	b	261-cab	ab, b
9	c	—	abc, bc, c
10	a	262-abca	bca, ca, a
11	b	263-bcab	cab, ab, b
12	c	264-cabc	abc, bc, c
13	a	—	abca, bca, ca, a
14	b	265-abcab	bcab, cab, ab, b
15	c	266-bcabc	cabc, abc, bc, c
16	a	267-cabca	abca, bca, ca, a
17	b	—	abcab, bcab, cab, ab, b
18	c	268-abcabc	bcabc, cabc, abc, bc, c
19	a	269-bcabca	cabca, abca, bca, ca, a
20	b	270-cabcab	abcab, bcab, cab, ab, b
21	c	—	abcabc, bcabc, cabc, abc, bc, c
22	x	271-abcabcx	x
23		272-bcabcx	
24		273-cabcx	
25		274-abcx	
26		275-bcx	
27		276-cx	

Table 6.32: LZY Example.

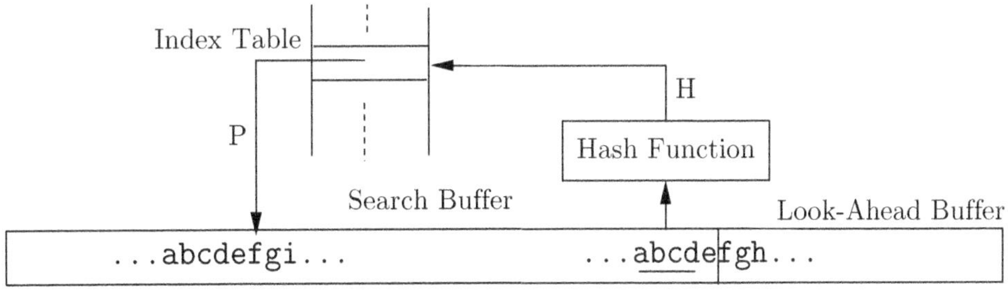

Figure 6.33: The Principle of LZP: Part I.

function and uses the result H as an index to a table of pointers called the *index table*. The index table contains pointers to various symbols in the search buffer. Index H is used to select a pointer P. In a typical case, P points to a previously seen string whose context is also abcde (see below for atypical cases). The algorithm then performs the following steps:

Step 1: It saves P and replaces it in the index table with a fresh pointer Q pointing to fgh... in the look-ahead buffer (Figure 6.33 Part II). An integer variable L is set to zero. It is used later to indicate the match length.

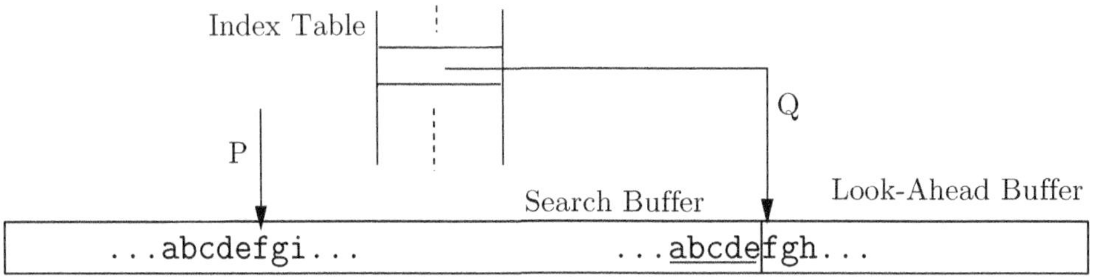

Figure 6.33: The Principle of LZP: Part II.

Step 2: If P is not a null pointer, the algorithm follows it and compares the string pointed at by P (string fgi... in the search buffer) to the string "fgh..." in the look-ahead buffer. Only two symbols match in our example, so the match length, L, is set to 2.
Step 3: If $L = 0$ (no symbols have been matched), the buffer is slid to the right (or, equivalently, the input is shifted to the left) **one position** and the first symbol of string S (the f) is written on the compressed stream as a raw ASCII code (a literal).
Step 4: If $L > 0$ (L symbols have been matched), the buffer is slid to the right L positions and the value of L is written on the compressed stream (after being suitably encoded).

In our example the single encoded value $L = 2$ is written on the compressed stream instead of the two symbols fg, and it is this step that produces compression. Clearly, the larger the value of L, the better the compression. Large values of L result when an N-symbol context C in the input stream is followed by the same long string S as a previous occurrence of C. This may happen when the input stream features high redundancy. In a random input stream each occurrence of the same context C is likely to be followed by another S, leading to $L = 0$ and therefore to no compression. An "average" input stream results in more literals than L values being written on the output stream (see also Exercise 6.18).

The decoder inputs the compressed stream item by item and creates the decompressed output in a buffer B. The steps are:

Step 1: Input the next item I from the compressed stream.
Step 2: If I is a raw ASCII code (a literal), it is appended to buffer B, and the data in B is shifted to the left one position.
Step 3: If I is an encoded match length, it is decoded, to obtain L. The present context C (the rightmost N symbols in B) is hashed to an index H, which is used to select a pointer P from the index table. The decoder copies the string of L symbols starting at B[P] and appends it to the right end of B. It also shifts the data in B to the left L positions and replaces P in the index table with a fresh pointer, to keep in lockstep with the encoder.

Two points remain to be discussed before we are ready to look at a detailed example.

1. When the encoder starts, it places the first N symbols of the input stream in the search buffer, to become the first context. It then writes these symbols, as literals, on the compressed stream. This is the only special step needed to start the compression. The decoder starts by reading the first N items off the compressed stream (they should be literals), and placing them at the rightmost end of buffer B, to serve as the first context.

2. It has been mentioned before that in the typical case, P points to a previously-seen string whose context is identical to the present context C. In an atypical case, P may be pointing to a string whose context is different. The algorithm, however, does not check the context and always behaves in the same way. It simply tries to match as many symbols as possible. At worst, zero symbols will match, leading to one literal written on the compressed stream.

6.19.1 Example

The input stream `xyabcabcabxy` is used to illustrate the operation of the LZP encoder. To simplify the example, we use $N = 2$.

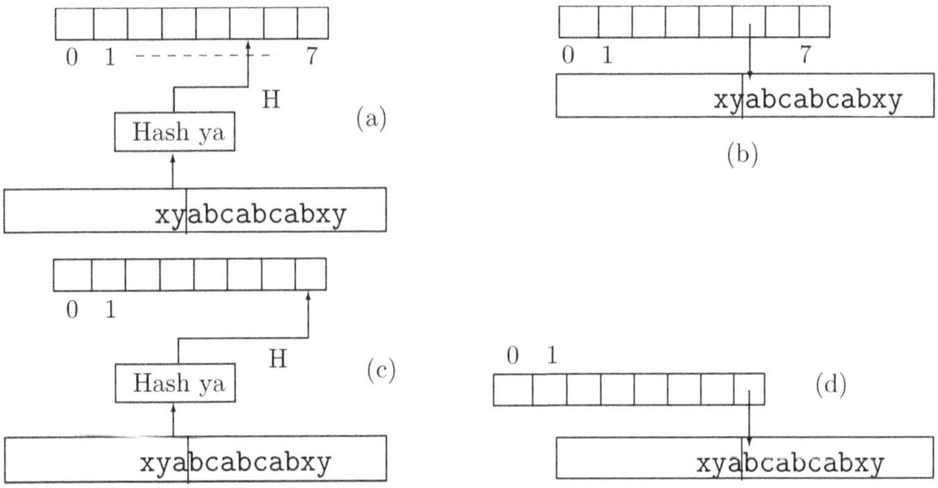

Figure 6.34: LZP Compression of `xyabcabcabxy`: Part I.

1. To start the operation, the encoder shifts the first two symbols `xy` to the search buffer and outputs them as literals. It also initializes all locations of the index table to the null pointer.

2. The current symbol is `a` (the first `a`), and the context is `xy`. It is hashed to, say, 5, but location 5 of the index table contains a null pointer, so P is null (Figure 6.34a). Location 5 is set to point to the first `a` (Figure 6.34b), which is then output as a literal. The data in the encoder's buffer is shifted to the left.

3. The current symbol is the first `b`, and the context is `ya`. It is hashed to, say, 7, but location 7 of the index table contains a null pointer, so P is null (Figure 6.34c). Location

7 is set to point to the first b (Figure 6.34d), which is then output as a literal. The data in the encoder's buffer is shifted to the left.
4. The current symbol is the first c, and the context is ab. It is hashed to, say, 2, but location 2 of the index table contains a null pointer, so P is null (Figure 6.34e). Location 2 is set to point to the first c (Figure 6.34f), which is then output as a literal. The data in the encoder's buffer is shifted to the left.
5. The same happens two more times, writing the literals a and b on the compressed stream. The current symbol is now (the second) c, and the context is ab. This context is hashed, as in step 4, to 2, so P points to "cabc...". Location 2 is set to point to the current symbol (Figure 6.34g), and the encoder tries to match strings cabcabxy and cabxy. The resulting match length is $L = 3$. The number 3 is written, encoded on the output, and the data is shifted three positions to the left.
6. The current symbol is the second x, and the context is ab. It is hashed to 2, but location 2 of the index table points to the second c (Figure 6.34h). Location 2 is set to point to the current symbol, and the encoder tries to match strings cabxy and xy. The resulting match length is, of course, $L = 0$, so the encoder writes x on the compressed stream as a literal and shifts the data one position.
7. The current symbol is the second y, and the context is bx. It is hashed to, say, 7. This is a hash collision, since context ya was hashed to 7 in step 3, but the algorithm does not check for collisions. It continues as usual. Location 7 of the index table points to the first b (or rather to the string bcabcabxy). It is set to point to the current symbol, and the encoder tries to match strings bcabcabxy and y, resulting in $L = 0$. The encoder writes y on the compressed stream as a literal and shifts the data one position.
8. The current symbol is the end-of-data, so the algorithm terminates.

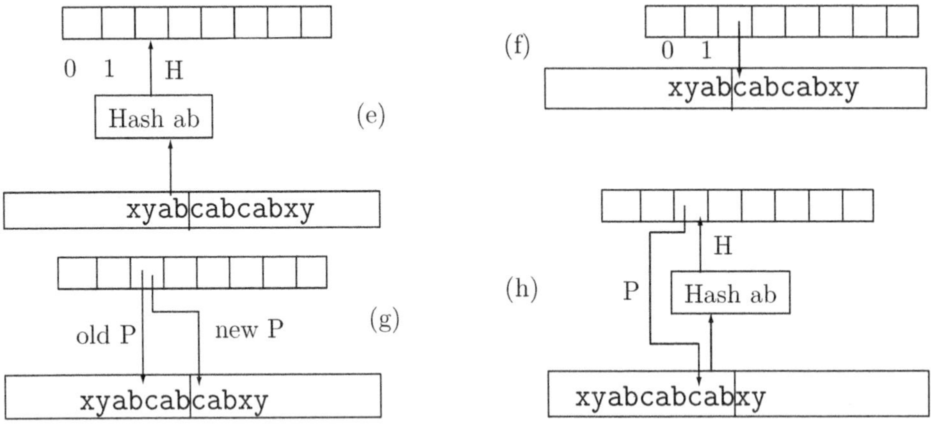

Figure 6.34 (Continued). LZP Compression of xyabcabcabxy: Part II.

◇ **Exercise 6.17:** Write the LZP encoding steps for the input string xyaaaa....

> Structures are the weapons of the mathematician.
>
> —Nicholas Bourbaki

6.19.2 Practical Considerations

Shifting the data in the buffer would require updating all the pointers in the index table. An efficient implementation should therefore adopt a better solution. Two approaches are described below, but other ones may also work.

1. Reserve a buffer as large as possible, and compress the input stream in blocks. Each block is input into the buffer and is never shifted. Instead, a pointer is moved from left to right, to point at any time to the current symbol in the buffer. When the entire buffer has been encoded, the buffer is filled up again with fresh input. This approach simplifies the program but has the disadvantage that the data at the end of a block cannot be used to predict anything for the next block. Each block is encoded independently of the other ones, leading to poorer compression.

2. Reserve a large buffer, and use it as a circular queue (Section 6.3.1). The data itself does not have to be shifted, but after encoding the current symbol the data is *effectively* shifted by updating the start and end pointers, and a new symbol is input and stored in the buffer. The algorithm is somewhat more complicated, but this approach has the advantage that the entire input is encoded as one stream. Every symbol benefits from the D symbols preceding it (where D is the total length of the buffer).

Imagine a pointer P in the index table pointing to some symbol `X` in the buffer. When the movement of the two pointers in the circular queue leaves `X` outside the queue, some new symbol `Y` will be input into the position occupied by `X`, and `P` will now be pointing to `Y`. When `P` is next selected by the hashing function and is used to match strings, the match will likely result in $L = 0$. However, the algorithm always replaces the pointer that it uses, so such a case should not degrade the algorithm's performance significantly.

6.19.3 LZP1 and LZP2

There are currently four versions of LZP, called LZP1 through LZP4. This section discusses the details of the first two. The context used by LZP1 is of order 3, i.e., it is the three bytes preceding the current one. The hash function produces a 12-bit index `H` and is best described by the following C code:

```
H=((C>>11)^C)&0xFFF.
```

Since `H` is 12 bits, the index table should be $2^{12} = 4{,}096$ entries long. Each entry is two bytes (16 bits), but only 14 of the 16 bits are used. A pointer `P` selected in the index table thus points to a buffer of size $2^{14} = 16\text{K}$.

The LZP1 encoder creates a compressed stream with literals and L values mixed together. Each item must therefore be preceded by a flag indicating its nature. Since only two flags are needed, the simplest choice would be 1-bit flags. However, we have already mentioned that an "average" input stream results in more literals than L values, so it makes sense to assign a short flag (less than one bit) to indicate a literal, and a long flag (a wee bit longer than one bit) to indicate a length. The scheme used by LZP1 uses 1 to indicate two literals, 01 to indicate a literal followed by a match length, and 00 to indicate a match length.

◇ **Exercise 6.18:** Let T indicate the probability of a literal in the compressed stream. For what value of T does the above scheme produce flags that are 1-bit long on average?

A literal is written on the compressed stream as an 8-bit ASCII code. Match lengths are encoded according to Table 6.35. Initially, the codes are 2 bits. When these are all used up, 3 bits are added, for a total of 5 bits, where the first 2 bits are 1's. When these are also all used, 5 bits are added, to produce 10-bit codes where the first 5 bits are 1's. From then on another group of 8 bits is added to the code whenever all the old codes have been used up. Notice how the pattern of all 1's is never used as a code and is reserved to indicate longer and longer codes. Notice also that a unary code or a general unary code (Section 3.1) might have been a better choice.

Length	Code	Length	Code
1	00	11	11|111|00000
2	01	12	11|111|00001
3	10	⋮	
4	11|000	41	11|111|11110
5	11|001	42	11|111|11111|00000000
6	11|010	⋮	
⋮		296	11|111|11111|11111110
10	11|110	297	11|111|11111|11111111|00000000

Table 6.35: Codes Used by LZP1 and LZP2 for Match Lengths.

The compressed stream consists of a mixture of literals (bytes with ASCII codes) and control bytes containing flags and encoded lengths. This is illustrated by the output of the example of Section 6.19.1. The input of this example is the string `xyabcabcabxy`, and the output items are `x`, `y`, `a`, `b`, `c`, `a`, `b`, 3, `x`, and `y`. The actual output stream consists of the single control byte 111 01|10 1 followed by nine bytes with the ASCII codes of `x`, `y`, `a`, `b`, `c`, `a`, `b`, `x`, and `y`.

◇ **Exercise 6.19:** Explain the content of the control byte 111 01|10 1.

Another example of a compressed stream is the three literals `x`, `y`, and `a` followed by the four match lengths 12, 12, 12, and 10. We first prepare the flags

1 (`x`, `y`) 01 (`a`, 12) 00 (12) 00 (12) 00 (12) 00 (10),

then substitute the codes of 12 and 10,

$1xy01a11|111|0000100|11|111|0000100|11|111|0000100|11|111|0000100|11|110,$

and finally group together the bits that make up the control bytes. The result is

10111111 `x`, `y`, `a`, 00001001 11110000 10011111 00001001 11110000 10011110.

Notice that the first control byte is followed by the three literals.

The last point to be mentioned is the case ...0 1yyyyyyy zzzzzzzz. The first control byte ends with the 0 of a pair 01, and the second byte starts with the 1 of the same pair. This indicates a literal followed by a match length. The match length is the yyy bits (at least some of them) in the second control byte. If the code of the match length is long, then the zzz bits or some of them may be part of the code. The literal is either the zzz byte or the byte following it.

LZP2 is identical to LZP1 except that literals are coded using nonadaptive Huffman codes. Ideally, two passes should be used; the first one counting the frequency of occurrence of the symbols in the input stream and the second pass doing the actual compression. In between the passes, the Huffman code table can be constructed.

6.19.4 LZP3 and LZP4

LZP3 is similar to both LZP1 and LZP2. It uses order-4 contexts and more sophisticated Huffman codes to encode both the match lengths and the literals. The LZP3 hash function is

```
H=((C>>15)^C)&0xFFFF,
```

so `H` is a 16-bit index, to be used with an index table of size $2^{16} = 64$ K. In addition to the pointers `P`, the index table contains also the contexts `C`. Thus if a context `C` is hashed to an index `H`, the encoder expects to find the same context `C` in location `H` of the index table. This is called *context confirmation*. If the encoder finds something else, or if it finds a null pointer, it sets `P` to null and hashes an order-3 context. If the order-3 context confirmation also fails, the algorithm hashes the order-2 context, and if that also fails, the algorithm sets `P` to null and writes a literal on the compressed stream. This method attempts to find the highest-order context seen so far.

LZP4 uses order-5 contexts and a multistep lookup process. In step 1, the rightmost four bytes `I` of the context are hashed to create a 16-bit index `H` according to the following:

```
H=((I>>15)^I)&0xFFFF.
```

Then `H` is used as an index to the index table that has 64K entries, each corresponding to one value of `H`. Each entry points to the start of a list linking nodes that have the same hash value. Suppose that the contexts `abcde`, `xbcde`, and `mnopq` hash to the same index `H` = 13 (i.e., the hash function computes the same index 13 when applied to `bcde` and `nopq`) and we are looking for context `xbcde`. Location 13 of the index table would point to a list with nodes for these contexts (and possibly others that have also been hashed to 13). The list is traversed until a node is found with `bcde`. This node points to a second list linking `a`, `x`, and perhaps other symbols that precede `bcde`. The second list is traversed until a node with `x` is found. That node contains a pointer to the most recent occurrence of context `xbcde` in the search buffer. If a node with `x` is not found, a literal is written to the compressed stream.

This complex lookup procedure is used by LZP4 because a 5-byte context does not fit comfortably in a single word in most current computers.

6.20 Repetition Finder

All the dictionary-based methods described so far have one thing in common: they use a large memory buffer as a dictionary that holds fragments of text found so far. The dictionary is used to locate strings of symbols that repeat. The method described here is different. Instead of a dictionary it uses a fixed-size array of integers to find previous occurrences of strings of text. The array size equals the square of the alphabet size, so it is not very large. The method is due to Hidetoshi Yokoo [Yokoo 91], who elected not to call it LZHY but left it nameless. The reason a name of the form LZxx was not used is that the method does not employ a traditional Ziv-Lempel dictionary. The reason it was left nameless is that it does not compress very well and should therefore be considered the first step in a new field of research rather than a mature, practical method.

The method alternates between two modes, normal and repeat. It starts in the normal mode, where it inputs symbols and encodes them using adaptive Huffman. When it identifies a repeated string it switches to the "repeat" mode where it outputs an escape symbol, followed by the length of the repeated string.

Assume that the input stream consists of symbols $x_1x_2\ldots$ from an alphabet A. Both encoder and decoder maintain an array REP of dimensions $|A|\times|A|$ that is initialized to all zeros. For each input symbol x_i, the encoder (and decoder) compute a value y_i according to $y_i = i - \text{REP}[x_{i-1}, x_i]$, and then update $\text{REP}[x_{i-1}, x_i] := i$. The 13-symbol string

x_i: X A B C D E Y A B C D E Z
i : 1 3 5 7 9 11 13

results in the following y values:

$i =$	1	2	3	4	5	6	7	8	9	10	11	12	13
$y_i =$	1	2	3	4	5	6	7	8	6	6	6	6	13
$x_{i-1}x_i$:	XA	AB	BC	CD	DE	EY	YA	AB	BC	CD	DE	EZ	

Table 6.36a shows the state of array REP after the eighth symbol has been input and encoded. Table 6.36b shows the state of REP after all 13 symbols have been input and encoded.

	A	B	C	D	E	...	X	Y	Z
A		3							
B			4						
C				5					
D					6				
E								7	
⋮									
X	2								
Y	8								
Z									

	A	B	C	D	E	...	X	Y	Z
A		9							
B			10						
C				11					
D					12				
E									13
⋮									
X	2								
Y	8								
Z									

Table 6.36: (a) REP at $i = 8$. (b) REP at $i = 13$.

Perhaps a better way to explain the way y is calculated is by the expression

$$y_i = \begin{cases} 1, & \text{for } i = 1,\\ i, & \text{for } i > 1 \text{ and first occurrence of } x_{i-1}x_i,\\ \min(k), & \text{for } i > 1 \text{ and } x_{i-1}x_i \text{ identical to } x_{i-k-1}x_{i-k}. \end{cases}$$

This shows that y is either i or is the distance k between the current string $x_{i-1}x_i$ and its most recent copy $x_{i-k-1}x_{i-k}$. However, recognizing a repetition of a string is done by means of array REP and without using any dictionary (which is the main point of this method).

When a string of length l repeats itself in the input, l consecutive identical values of y are generated, and this is the signal for the encoder to switch to the "repeat" mode. As long as consecutive different values of y are generated, the encoder stays in the "normal" mode, where it encodes x_i in adaptive Huffman and outputs it. When the encoder senses $y_{i+1} = y_i$, it outputs x_i in the normal mode, and then enters the "repeat" mode. In the example above this happens for $i = 9$, so the string XABCDEYAB is output in the normal mode.

Once in the "repeat" mode, the encoder inputs more symbols and calculates y values until it finds the first value that differs from y_i. In our example this happens at $i = 13$, when the Z is input. The encoder compresses the string "CDE" (corresponding to $i = 10, 11, 12$) in the "repeat" mode by emitting an (encoded) escape symbol, followed by the (encoded) length of the repeated string (3 in our example). The encoder then switches back to the normal mode, where it saves the y value for Z as y_i and inputs the next symbol.

The escape symbol must be an extra symbol, one that's not included in the alphabet A. Notice that only two y values, y_{i-1} and y_i, need be saved at any time. Notice also that the method is not very efficient, since it senses the repeating string "too late" and encodes the first two repeating symbols in the normal mode. In our example only three of the five repeating symbols are encoded in the "repeat" mode.

The decoder inputs and decodes the first nine symbols, decoding them into the string XABCDEYAB while updating array REP and calculating y values. When the escape symbol is input, i has the value 9 and y_i has the value 6. The decoder inputs and decodes the length, 3, and now it has to figure out the repeated string of length 3 using just the data in array REP, not any of the previously decoded input. Since $i = 9$ and y_i is the distance between this string and its copy, the decoder knows that the copy started at position $i - y_i = 9 - 6 = 3$ of the input. It scans REP, looking for a 3. It finds it at position REP[A,B], so it starts looking for a 4 in row B of REP. It finds it in REP[B,C], so the first symbol of the required string is C. Looking for a 5 in row C, the decoder finds it in REP[C,D], so the second symbol is D. Looking now for a 6 in row D, the decoder finds it in REP[D,E].

This is how a repeated string can be decoded without maintaining a dictionary.

Both encoder and decoder store values of i in REP, so an entry of REP should be at least two bytes long. This way i can have values of up to $64\text{K} - 1 \approx 65{,}500$, so the input has to be encoded in blocks of size 64K. For an alphabet of 256 symbols, the size of REP should therefore be $256 \times 256 \times 2 = 128$ Kbytes, not very large. For larger alphabets REP may have to be uncomfortably large.

In the normal mode, symbols (including the escape) are encoded using adaptive Huffman (Section 5.3). In the repeat mode, lengths are encoded in a recursive prefix code denoted $Q_k(i)$, where k is a positive integer (see Section 2.2 for prefix codes). Assuming that i is an integer whose binary representation is 1α, the prefix code $Q_k(i)$ of i is defined by

$$Q_0(i) = 1^{|\alpha|}0\alpha, \qquad Q_k(i) = \begin{cases} 0, & i = 1, \\ 1Q_{k-1}(i-1), & i > 1, \end{cases}$$

where $|\alpha|$ is the length of α and $1^{|\alpha|}$ is a string of $|\alpha|$ ones. Table 6.37 shows some of the proposed codes; however, any of the prefix codes of Section 2.2 can be used instead of the $Q_k(i)$ codes proposed here.

i	α	$Q_0(i)$	$Q_1(i)$	$Q_2(i)$
1	null	0	0	0
2	0	100	10	10
3	1	101	1100	110
4	00	11000	1101	11100
5	01	11001	111000	11101
6	10	11010	111001	1111000
7	11	11011	111010	1111001
8	000	1110000	111011	1111010
9	001	1110001	11110000	1111011

Table 6.37: Proposed Prefix Code.

The developer of this method, Hidetoshi Yokoo, indicates that compression performance is not very sensitive to the precise value of k, and he proposes $k = 2$ for best overall performance.

As mentioned earlier, the method is not very efficient, which is why it should be considered the start of a new field of research where repeated strings are identified without the need for a large dictionary.

6.21 GIF Images

GIF—the graphics interchange format—was developed by Compuserve Information Services in 1987 as an efficient, compressed graphics file format, which allows for images to be sent between different computers. The original version of GIF is known as GIF 87a. The current standard is GIF 89a and, at the time of writing, can be freely obtained as the file `http://delcano.mit.edu/info/gif.txt`. GIF is not a data compression method; it is a graphics file format that uses a variant of LZW to compress the graphics data (see [Blackstock 87]). This section reviews only the data compression aspects of GIF.

In compressing data, GIF is very similar to `compress` and uses a dynamic, growing dictionary. It starts with the number of bits per pixel b as a parameter. For a monochromatic image, $b = 2$; for an image with 256 colors or shades of gray, $b = 8$. The dictionary starts with 2^{b+1} entries and is doubled in size each time it fills up, until it reaches a size of $2^{12} = 4{,}096$ entries, where it remains static. At such a point, the encoder monitors the compression ratio and may decide to discard the dictionary at any point and start with a new, empty one. When making this decision, the encoder emits the value 2^b as the *clear code*, which is the sign for the decoder to discard its dictionary.

The pointers, which get longer by 1 bit from one dictionary to the next, are accumulated and are output in blocks of 8-bit bytes. Each block is preceded by a header that contains the block size (255 bytes maximum) and is terminated by a byte of eight zeros. The last block contains, just before the terminator, the eof value, which is $2^b + 1$. An interesting feature is that the pointers are stored with their least significant bit on the left. Consider, for example, the following 3-bit pointers 3, 7, 4, 1, 6, 2, and 5. Their binary values are 011, 111, 100, 001, 110, 010, and 101, so they are packed in 3 bytes |10101001|11000011|11110...|.

The GIF format is commonly used today by web browsers, but it is not an efficient image compressor. GIF scans the image row by row, so it discovers pixel correlations within a row, but not between rows. We can say that GIF compression is inefficient because GIF is one-dimensional while an image is two-dimensional. An illustrative example is the two simple images of Figure 7.4a,b (Section 7.3). Saving both in GIF89 has resulted in file sizes of 1053 and 1527 bytes, respectively.

Most graphics file formats use some kind of compression. For more information on those files, see [Murray and vanRyper 94].

6.22 RAR and WinRAR

The popular RAR software is the creation of Eugene Roshal, who started it as his university doctoral dissertation. RAR is an acronym that stands for Roshal ARchive (or Roshal ARchiver). The current developer is Eugene's brother Alexander Roshal. The following is a list of its most important features:

- RAR is currently available from [rarlab 06], as shareware, for Windows (WinRAR), Pocket PC, Macintosh OS X, Linux, DOS, and FreeBSD. WinRAR has a graphical user interface, whereas the other versions support only a command line interface.

- WinRAR provides complete support for RAR and ZIP archives and can decompress (but not compress) CAB, ARJ, LZH, TAR, GZ, ACE, UUE, BZ2, JAR, ISO, 7Z, and Z archives.

- In addition to compressing data, WinRAR can encrypt data with the advanced encryption standard (AES-128).

- WinRAR can compress files up to 8,589 billion Gb in size (approximately 9×10^{18} bytes).

- Files compressed in WinRAR can be self-extracting (SFX) and can come from different volumes.

- It is possible to combine many files, large and small, into a so-called "solid" archive, where they are compressed together. This may save 10–50% compared to compressing each file individually.

- The RAR software can optionally generate recovery records and recovery volumes that make it possible to reconstruct damaged archives. Redundant data, in the form of a Reed-Solomon error-correcting code, can optionally be added to the compressed file for increased reliability.

The user can specify the amount of redundancy (as a percentage of the original data size) to be built into the recovery records of a RAR archive. A large amount of redundancy makes the RAR file more resistant to data corruption, thereby allowing recovery from more serious damage. However, any redundant data reduces compression efficiency, which is why compression and reliability are opposite concepts.

- Authenticity information may be included for extra security. RAR software saves information on the last update and name of the archive.

- An intuitive wizard especially designed for novice users. The wizard makes it easy to use all of RAR's features through a simple question and answer procedure.

- Excellent drag-and-drop features. The user can drag files from WinRAR to other programs, to the desktop, or to any folder. An archive dropped on WinRAR is immediately opened to display its files. A file dropped on an archive that's open in WinRAR is immediately added to the archive. A group of files or folders dropped on WinRAR automatically becomes a new archive.

- RAR offers NTFS and Unicode support (see [ntfs 06] for NTFS).

- WinRAR is available in over 35 languages.

These features, combined with excellent compression, good compression speed, an attractive graphical user interface, and backup facilities, make RAR one of the best overall compression tools currently available.

The RAR algorithm is generally considered proprietary, and the following quote (from [donationcoder 06]) sheds some light on what this implies

> The fact that the RAR encoding algorithm is proprietary is an issue worth considering. It means that, unlike ZIP and 7z and almost all other compression algorithms, only the official WinRAR programs can create RAR archives (although other programs can decompress them). Some would argue that this is unhealthy for the software industry and that standardizing on an open format would be better for everyone in the long run. But for most users, these issues are academic; WinRAR offers great support for both ZIP and RAR formats.

Proprietary

A term that indicates that a party, or proprietor, exercises private ownership, control, or use over an item of property, usually to the exclusion of other parties.

—From `http://www.wikipedia.org/`

The following illuminating description was obtained directly from Eugene Roshal, the designer of RAR. (The source code of the RAR decoder is available at [unrarsrc 06], but only on condition that it is not used to reverse-engineer the encoder.)

RAR has two compression modes, general and special. The general mode employs an LZSS-based algorithm similar to ZIP Deflate (Section 6.25). The size of the sliding dictionary in RAR can be varied from 64 KB to 4 MB (with a 4 MB default value), and the minimum match length is 2. Literals, offsets, and match lengths are compressed further by a Huffman coder (recall that Deflate works similarly).

Starting at version 3.0, RAR also uses a special compression mode to improve the compression of text-like data. This mode is based on the PPMD algorithm (also known as PPMII) by Dmitry Shkarin.

RAR contains a set of filters to preprocess audio (in `wav` or `au` formats), true color data, tables, and executables (32-bit x86 and Itanium, see note below). A data-analyzing module selects the appropriate filter and the best algorithm (general or PPMD), depending on the data to be compressed.

(Note: The 80x86 family of processors was originally developed by Intel with a word length of 16 bits. Because of its tremendous success, its architecture was extended to 32-bit words and is currently referred to as IA-32 [Intel Architecture, 32-bit]. See [IA-32 06] for more information. The Itanium is an IA-64 microprocessor developed jointly by Hewlett-Packard and Intel.)

In addition to its use as Roshal ARchive, the acronym RAR has many other meanings, a few of which are listed here.
(1) Remote Access Router (a network device used to connect remote sites via private lines or public carriers). (2) Royal Anglian Regiment, a unit of the British Army. (3) Royal Australian Regiment, a unit of the Australian Army. (4) Resource Adapter, a specific module of the Java EE platform. (5) The airport code of Rarotonga, Cook Islands. (6) Revise As Required (editors' favorite). (7) Road Accident Rescue.
See more at [acronyms 06].

Rarissimo, by [softexperience 06], is a file utility intended to automatically compress and decompress files in WinRAR. Rarissimo by itself is useless. It can be used only if WinRAR is already installed in the computer. The Rarissimo user specifies a number of folders for Rarissimo to watch, and Rarissimo compresses or decompresses any files that have been modified in these folders. It can also move the modified files to target folders. The user also specifies how often (in multiples of 10 sec) Rarissimo should check the folders.

For each folder to be watched, the user has to specify RAR or UnRAR and a target folder. If RAR is specified, then Rarissimo employs WinRAR to compress each file that has been modified since the last check, and then moves that file to the target folder. If the target folder resides on another computer, this move amounts to an FTP transfer. If UnRAR has been specified, then Rarissimo employs WinRAR to decompress all the RAR-compressed files found in the folder and then moves them to the target folder.

An important feature of Rarissimo is that it preserves NTFS alternate streams (see [ntfs 06]). This means that Rarissimo can handle Macintosh files that happen to reside on the PC; it can compress and decompress them while preserving their data and resource forks.

6.23 The V.42bis Protocol

The V.42bis protocol is a set of rules, or a standard, published by the ITU-T (page 248) for use in fast modems. It is based on the existing V.32bis protocol and is supposed to be used for fast transmission rates, up to 57.6K baud. Thomborson [Thomborson 92] in a detailed reference for this standard. The ITU-T standards are recommendations, but they are normally followed by all major modem manufacturers. The standard contains specifications for data compression and error correction, but only the former is discussed here.

V.42bis specifies two modes: a *transparent* mode, where no compression is used, and a *compressed* mode using an LZW variant. The former is used for data streams that don't compress well and may even cause expansion. A good example is an already compressed file. Such a file looks like random data; it does not have any repetitive patterns, and trying to compress it with LZW will fill up the dictionary with short, two-symbol, phrases.

The compressed mode uses a growing dictionary, whose initial size is negotiated between the modems when they initially connect. V.42bis recommends a dictionary size of 2,048 entries. The minimum size is 512 entries. The first three entries, corresponding to pointers 0, 1, and 2, do not contain any phrases and serve as special codes. Code 0 (enter transparent mode—ETM) is sent when the encoder notices a low compression ratio, and it decides to start sending uncompressed data. (Unfortunately, V.42bis does not say how the encoder should test for low compression.) Code 1 is FLUSH, to flush data. Code 2 (STEPUP) is sent when the dictionary is almost full and the encoder decides to double its size. A dictionary is considered almost full when its size exceeds that of a special threshold (which is also negotiated by the modems).

When the dictionary is already at its maximum size and it becomes full, V.42bis recommends a *reuse* procedure. The least-recently-used phrase is located and deleted, to make room for a new phrase. This is done by searching the dictionary from entry 256 for the first phrase that is not a prefix to any other phrase. Suppose that the phrase `abcd` is found, and there are no phrases of the form `abcdx` for any `x`. This means that `abcd` has not been used since it was created, and that it is the oldest such phrase. It therefore makes sense to delete it, since it reflects an old pattern in the input stream. This way, the dictionary always reflects recent patterns in the input.

> ...there is an ever-increasing body of opinion which holds that The Ultra-Complete Maximegalon Dictionary is not worth the fleet of lorries it takes to cart its microstored edition around in. Strangely enough, the dictionary omits the word "floopily," which simply means "in the manner of something which is floopy."
>
> —Douglas Adams, *Life, the Universe, and Everything* (1982)

6.24 Various LZ Applications

ARC is a compression/archival/cataloging program developed by Robert A. Freed in the mid-1980s. It offers good compression and the ability to combine several files into one file, called an *archive*. Currently (early 2003) ARC is outdated and has been superseded by the newer PK applications.

PKArc is an improved version of ARC. It was developed by Philip Katz, who has founded the PKWare company [PKWare 03], which still markets the PKzip, PKunzip, and PKlite software. The PK programs are faster and more general than ARC and also provide for more user control. Past editions of this book have more information on these applications.

LHArc, from Haruyasu Yoshizaki, and LHA, by Haruhiko Okumura and Haruyasu Yoshizaki, use adaptive Huffman coding with features drawn from LZSS. The LZ window size may be up to 16K bytes. ARJ is another older compression tool by Robert K. Jung that uses the same algorithm, but increases the LZ window size to 26624 bytes. A similar program, called ICE (for the old MS-DOS operating system), seems to be a faked version of either LHarc or LHA. [Okumura 98] has some information about LHArc and LHA as well as a history of data compression in Japan.

Two newer applications, popular with Microsoft Windows users, are RAR/WinRAR [rarlab 06] (Section 6.22) and ACE/WinAce [WinAce 03]. They use LZ with a large search buffer (up to 4 Mb) combined with Huffman codes. They are available for several platforms and offer many useful features.

6.25 Deflate: Zip and Gzip

Deflate is a popular compression method that was originally used in the well-known Zip and Gzip software and has since been adopted by many applications, the most important of which are (1) the HTTP protocol ([RFC1945 96] and [RFC2616 99]), (2) the PPP compression control protocol ([RFC1962 96] and [RFC1979 96]), (3) the PNG (Portable Network Graphics, Section 6.27) and MNG (Multiple-Image Network Graphics) graphics file formats ([PNG 03] and [MNG 03]), and (4) Adobe's PDF (Portable Document File, Section 11.13) [PDF 01].

Phillip W. Katz was born in 1962. He received a bachelor's degree in computer science from the University of Wisconsin at Madison. Always interested in writing software, he started working in 1984 as a programmer for Allen-Bradley Co. developing programmable logic controllers for the industrial automation industry. He later worked for Graysoft, another software company, in Milwaukee, Wisconsin. At about that time he became interested in data compression and founded PKWare in 1987 to develop, implement, and market software products such as PKarc and PKzip. For a while, the company was very successful selling the programs as shareware.

Always a loner, Katz suffered from personal and legal problems, started drinking heavily, and died on April 14, 2000 from complications related to chronic alcoholism. He was 37 years old.

After his death, PKWare was sold, in March 2001, to a group of investors. They changed its management and the focus of its business. PKWare currently targets the corporate market, and emphasizes compression combined with encryption. Their product line runs on a wide variety of platforms.

Deflate was designed by Philip Katz as a part of the Zip file format and implemented in his PKZIP software [PKWare 03]. Both the ZIP format and the Deflate method are in the public domain, which allowed implementations such as Info-ZIP's Zip and Unzip (essentially, PKZIP and PKUNZIP clones) to appear on a number of platforms. Deflate is described in [RFC1951 96].

The most notable implementation of Deflate is zlib, a portable and free compression library ([zlib 03] and [RFC1950 96]) by Jean-Loup Gailly and Mark Adler who designed and implemented it to be free of patents and licensing requirements. This library (the source code is available at [Deflate 03]) implements the ZLIB and GZIP file formats ([RFC1950 96] and [RFC1952 96]), which are at the core of most Deflate applications, including the popular Gzip software.

Deflate is based on a variation of LZ77 combined with Huffman codes. We start with a simple overview based on [Feldspar 03] and follow with a full description based on [RFC1951 96].

The original LZ77 method (Section 6.3) tries to match the text in the look-ahead buffer to strings already in the search buffer. In the example

```
               search buffer           look-ahead
...old␣|..the␣a..then...there...|the␣new...|...more
```

the look-ahead buffer starts with the string `the␣`, which can be matched to one of three strings in the search buffer. The longest match has a length of 4. LZ77 writes tokens on the compressed stream, where each token is a triplet (offset, length, next symbol). The third component is needed in cases where no string has been matched (imagine having `che` instead of `the` in the look-ahead buffer) but it is part of every token, which reduces the performance of LZ77. The LZ77 algorithm variation used in Deflate eliminates the third component and writes a pair (offset, length) on the compressed stream. When no match is found, the unmatched character is written on the compressed stream instead of a token. Thus, the compressed stream consists of three types of entities: literals (unmatched characters), offsets (termed "distances" in the Deflate literature), and lengths. Deflate actually writes Huffman codes on the compressed stream for these entities, and it uses two code tables—one for literals and lengths and the other for distances. This makes sense because the literals are normally bytes and are therefore in the interval $[0, 255]$, and the lengths are limited by Deflate to 258. The distances, however, can be large numbers because Deflate allows for a search buffer of up to 32Kbytes.

When a pair (length, distance) is determined, the encoder searches the table of literal/length codes to find the code for the length. This code (we later use the term "edoc" for it) is then replaced by a Huffman code that's written on the compressed stream. The encoder then searches the table of distance codes for the code of the distance and writes that code (a special prefix code with a fixed, 5-bit prefix) on the compressed stream. The decoder knows when to expect a distance code, because it always follows a length code.

The LZ77 variant used by Deflate defers the selection of a match in the following way. Suppose that the two buffers contain

```
                search buffer          look-ahead
...old␣|..she␣needs..then...there...|the␣new...|...more input
```

The longest match is 3. Before selecting this match, the encoder saves the `t` from the look-ahead buffer and starts a secondary match where it tries to match `he␣new...` with the search buffer. If it finds a longer match, it outputs `t` as a literal, followed by the longer match. There is also a 3-valued parameter that controls this secondary match attempt. In the "normal" mode of this parameter, if the primary match was long enough (longer than a preset parameter), the secondary match is reduced (it is up to the implementor to decide how to reduce it). In the "high-compression" mode, the encoder always performs a full secondary match, thereby improving compression but spending more time on selecting a match. In the "fast" mode, the secondary match is omitted.

Deflate compresses an input data file in blocks, where each block is compressed separately. Blocks can have different lengths and the length of a block is determined by the encoder based on the sizes of the various prefix codes used (their lengths are limited to 15 bits) and by the memory available to the encoder (except that blocks in mode 1 are limited to 65,535 bytes of uncompressed data). The Deflate decoder must be able to decode blocks of any size. Deflate offers three modes of compression, and each block can be in any mode. The modes are as follows:

1. No compression. This mode makes sense for files or parts of files that are incompressible (i.e., random) or already compressed, or for cases where the compression software is asked to segment a file without compression. A typical case is a user who wants to move an 8 Gb file to another computer but has only a DVD "burner." The user may want to segment the file into two 4 Gb segments without compression. Commercial compression software based on Deflate can use this mode of operation to segment the file. This mode uses no code tables. A block written on the compressed stream in this mode starts with a special header indicating mode 1, followed by the length `LEN` of the data, followed by `LEN` bytes of literal data. Notice that the maximum value of `LEN` is 65,535.

2. Compression with fixed code tables. Two code tables are built into the Deflate encoder and decoder and are always used. This speeds up both compression and decompression and has the added advantage that the code tables don't have to be written on the compressed stream. The compression performance, however, may suffer if the data being compressed is statistically different from the data used to set up the code tables. Literals and match lengths are located in the first table and are replaced by a code (called "edoc") that is, in turn, replaced by a prefix code that's output to the compressed stream. Distances are located in the second table and are replaced by special prefix codes that are output to the compressed stream. A block written on the compressed stream in this mode starts with a special header indicating mode 2, followed by the compressed data in the form of prefix codes for the literals and lengths, and special prefix codes for the distances. The block ends with a single prefix code for end-of-block.

3. Compression with individual code tables generated by the encoder for the particular data that's being compressed. A sophisticated Deflate encoder may gather statistics about the data as it compresses blocks, and may be able to construct improved code tables as it proceeds from block to block. There are two code tables, for literals/lengths and for distances. They again have to be written on the compressed stream, and they are written in compressed format. A block written by the encoder on the compressed stream in this mode starts with a special header, followed by (1) a compressed Huffman code table and (2) the two code tables, each compressed by the Huffman codes that preceded them. This is followed by the compressed data in the form of prefix codes for the literals, lengths, and distances, and ends with a single code for end-of-block.

6.25.1 The Details

Each block starts with a 3-bit header where the first bit is 1 for the last block in the file and 0 for all other blocks. The remaining two bits are 00, 01, or 10, indicating modes 1, 2, or 3, respectively. Notice that a block of compressed data does not always end on a byte boundary. The information in the block is sufficient for the decoder to read all the bits of the compressed block and recognize the end of the block. The 3-bit header of the next block immediately follows the current block and may therefore be located at any position in a byte on the compressed stream.

The format of a block in mode 1 is as follows:

1. The 3-bit header 000 or 100.

2. The rest of the current byte is skipped, and the next four bytes contain `LEN` and the one's complement of `LEN` (as unsigned 16-bit numbers), where `LEN` is the number of data bytes in the block. This is why the block size in this mode is limited to 65,535 bytes.

3. `LEN` data bytes.

The format of a block in mode 2 is different:

1. The 3-bit header 001 or 101.

2. This is immediately followed by the fixed prefix codes for literals/lengths and the special prefix codes of the distances.

3. Code 256 (rather, its prefix code) designating the end of the block.

Mode 2 uses two code tables: one for literals and lengths and the other for distances. The codes of the first table are not what is actually written on the compressed stream, so in order to remove ambiguity, the term "edoc" is used here to refer to them. Each edoc is converted to a prefix code that's output to the compressed stream. The first table allocates edocs 0 through 255 to the literals, edoc 256 to end-of-block, and edocs 257–285 to lengths. The latter 29 edocs are not enough to represent the 256 match lengths of 3 through 258, so extra bits are appended to some of those edocs. Table 6.38 lists the 29 edocs, the extra bits, and the lengths represented by them. What is actually written on the compressed stream is prefix codes of the edocs (Table 6.39). Notice that edocs 286 and 287 are never created, so their prefix codes are never used. We show later that Table 6.39 can be represented by the sequence of code lengths

$$\underbrace{8,8,\ldots,8}_{144},\underbrace{9,9,\ldots,9}_{112},\underbrace{7,7,\ldots,7}_{24},\underbrace{8,8,\ldots,8}_{8}, \tag{6.1}$$

Code	Extra bits	Lengths	Code	Extra bits	Lengths	Code	Extra bits	Lengths
257	0	3	267	1	15,16	277	4	67–82
258	0	4	268	1	17,18	278	4	83–98
259	0	5	269	2	19–22	279	4	99–114
260	0	6	270	2	23–26	280	4	115–130
261	0	7	271	2	27–30	281	5	131–162
262	0	8	272	2	31–34	282	5	163–194
263	0	9	273	3	35–42	283	5	195–226
264	0	10	274	3	43–50	284	5	227–257
265	1	11,12	275	3	51–58	285	0	258
266	1	13,14	276	3	59–66			

Table 6.38: Literal/Length Edocs for Mode 2.

edoc	Bits	Prefix codes
0–143	8	00110000–10111111
144–255	9	110010000–111111111
256–279	7	0000000–0010111
280–287	8	11000000–11000111

Table 6.39: Huffman Codes for Edocs in Mode 2.

but any Deflate encoder and decoder include the entire table instead of just the sequence of code lengths. There are edocs for match lengths of up to 258, so the look-ahead buffer of a Deflate encoder can have a maximum size of 258, but can also be smaller.

Examples. If a string of 10 symbols has been matched by the LZ77 algorithm, Deflate prepares a pair (length, distance) where the match length 10 becomes edoc 264, which is written as the 7-bit prefix code 0001000. A length of 12 becomes edoc 265 followed by the single bit 1. This is written as the 7-bit prefix code 0001010 followed by 1. A length of 20 is converted to edoc 269 followed by the two bits 01. This is written as the nine bits 0001101|01. A length of 256 becomes edoc 284 followed by the five bits 11110. This is written as 11000101|11110. A match length of 258 is indicated by edoc 285 whose 8-bit prefix code is 11000110. The end-of-block edoc of 256 is written as seven zero bits.

The 30 distance codes are listed in Table 6.40. They are special prefix codes with fixed-size 5-bit prefixes that are followed by extra bits in order to represent distances in the interval $[1, 32768]$. The maximum size of the search buffer is therefore 32,768, but it can be smaller. The table shows that a distance of 6 is represented by 00100|1, a distance of 21 becomes the code 01000|101, and a distance of 8195 corresponds to code 11010|000000000010.

6.25.2 Format of Mode-3 Blocks

In mode 3, the encoder generates two prefix code tables, one for the literals/lengths and

Code	Extra bits	Distance	Code	Extra bits	Distance	Code	Extra bits	Distance
0	0	1	10	4	33–48	20	9	1025–1536
1	0	2	11	4	49–64	21	9	1537–2048
2	0	3	12	5	65–96	22	10	2049–3072
3	0	4	13	5	97–128	23	10	3073–4096
4	1	5,6	14	6	129–192	24	11	4097–6144
5	1	7,8	15	6	193–256	25	11	6145–8192
6	2	9–12	16	7	257–384	26	12	8193–12288
7	2	13–16	17	7	385–512	27	12	12289–16384
8	3	17–24	18	8	513–768	28	13	16385–24576
9	3	25–32	19	8	769–1024	29	13	24577–32768

Table 6.40: Thirty Prefix Distance Codes in Mode 2.

the other for the distances. It uses the tables to encode the data that constitutes the block. The encoder can generate the tables in any way. The idea is that a sophisticated Deflate encoder may collect statistics as it inputs the data and compresses blocks. The statistics are used to construct better code tables for later blocks. A naive encoder may use code tables similar to the ones of mode 2 or may even not generate mode 3 blocks at all. The code tables have to be written on the compressed stream, and they are written in a highly-compressed format. As a result, an important part of Deflate is the way it compresses the code tables and outputs them. The main steps are (1) Each table starts as a Huffman tree. (2) The tree is rearranged to bring it to a standard format where it can be represented by a sequence of code lengths. (3) The sequence is compressed by run-length encoding to a shorter sequence. (4) The Huffman algorithm is applied to the elements of the shorter sequence to assign them Huffman codes. This creates a Huffman tree that is again rearranged to bring it to the standard format. (5) This standard tree is represented by a sequence of code lengths which are written, after being permuted and possibly truncated, on the output. These steps are described in detail because of the originality of this unusual method.

Recall that the Huffman code tree generated by the basic algorithm of Section 5.2 is not unique. The Deflate encoder applies this algorithm to generate a Huffman code tree, then rearranges the tree and reassigns the codes to bring the tree to a standard form where it can be expressed compactly by a sequence of code lengths. (The result is reminiscent of the canonical Huffman codes of Section 5.2.8.) The new tree satisfies the following two properties:

1. The shorter codes appear on the left, and the longer codes appear on the right of the Huffman code tree.

2. When several symbols have codes of the same length, the (lexicographically) smaller symbols are placed on the left.

The first example employs a set of six symbols A–F with probabilities 0.11, 0.14, 0.12, 0.13, 0.24, and 0.26, respectively. Applying the Huffman algorithm results in a tree similar to the one shown in Figure 6.41a. The Huffman codes of the six symbols are 000, 101, 001, 100, 01, and 11. The tree is then rearranged and the codes reassigned to comply with the two requirements above, resulting in the tree of Figure 6.41b. The new codes of the symbols are 100, 101, 110, 111, 00, and 01. The latter tree has the advantage that it can be fully expressed by the sequence 3, 3, 3, 3, 2, 2 of the lengths of the codes of the six symbols. The task of the encoder in mode 3 is therefore to generate this sequence, compress it, and write it on the compressed stream.

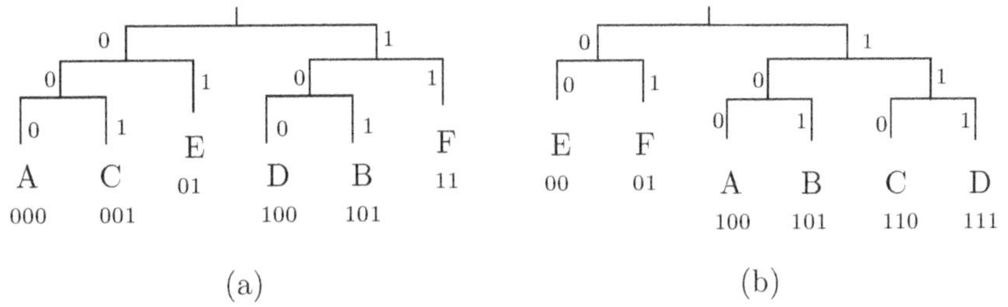

Figure 6.41: Two Huffman Trees.

The code lengths are limited to at most four bits each. Thus, they are integers in the interval [0, 15], which implies that a code can be at most 15 bits long (this is one factor that affects the Deflate encoder's choice of block lengths in mode 3).

The sequence of code lengths representing a Huffman tree tends to have runs of identical values and can have several runs of the same value. For example, if we assign the probabilities 0.26, 0.11, 0.14, 0.12, 0.24, and 0.13 to the set of six symbols A–F, the Huffman algorithm produces 2-bit codes for A and E and 3-bit codes for the remaining four symbols. The sequence of these code lengths is 2, 3, 3, 3, 2, 3.

The decoder reads a compressed sequence, decompresses it, and uses it to reproduce the standard Huffman code tree for the symbols. We first show how such a sequence is used by the decoder to generate a code table, then how it is compressed by the encoder.

Given the sequence 3, 3, 3, 3, 2, 2, the Deflate decoder proceeds in three steps as follows:

1. Count the number of codes for each code length in the sequence. In our example, there are no codes of length 1, two codes of length 2, and four codes of length 3.

2. Assign a base value to each code length. There are no codes of length 1, so they are assigned a base value of 0 and don't require any bits. The two codes of length 2 therefore start with the same base value 0. The codes of length 3 are assigned a base value of 4 (twice the number of codes of length 2). The C code shown here (after [RFC1951 96]) was written by Peter Deutsch. It assumes that step 1 leaves the number of codes for each code length `n` in `bl_count[n]`.

```
code = 0;
bl_count[0] = 0;
for (bits = 1; bits <= MAX_BITS; bits++) {
  code = (code + bl_count[bits-1]) << 1;
  next code[bits] = code;
}
```

3. Use the base value of each length to assign consecutive numerical values to all the codes of that length. The two codes of length 2 start at 0 and are therefore 00 and 01. They are assigned to the fifth and sixth symbols E and F. The four codes of length 3 start at 4 and are therefore 100, 101, 110, and 111. They are assigned to the first four symbols A–D. The C code shown here (by Peter Deutsch) assumes that the code lengths are in `tree[I].Len` and it generates the codes in `tree[I].Codes`.

```
for (n = 0; n <= max code; n++) {
  len = tree[n].Len;
  if (len != 0) {
    tree[n].Code = next_code[len];
    next_code[len]++;
  }
}
```

In the next example, the sequence 3, 3, 3, 3, 3, 2, 4, 4 is given and is used to generate a table of eight prefix codes. Step 1 finds that there are no codes of length 1, one code of length 2, five codes of length 3, and two codes of length 4. The length-1 codes are assigned a base value of 0. There are zero such codes, so the next group is assigned the base value of twice 0. This group contains one code, so the next group (length-3 codes) is assigned base value 2 (twice the sum $0+1$). This group contains five codes, so the last group is assigned base value of 14 (twice the sum $2+5$). Step 3 simply generates the five 3-bit codes 010, 011, 100, 101, and 110 and assigns them to the first five symbols. It then generates the single 2-bit code 00 and assigns it to the sixth symbol. Finally, the two 4-bit codes 1110 and 1111 are generated and assigned to the last two (seventh and eighth) symbols.

Given the sequence of code lengths of Equation (6.1), we apply this method to generate its standard Huffman code tree (listed in Table 6.39).

Step 1 finds that there are no codes of lengths 1 through 6, that there are 24 codes of length 7, 152 codes of length 8, and 112 codes of length 9. The length-7 codes are assigned a base value of 0. There are 24 such codes, so the next group is assigned the base value of $2(0+24)=48$. This group contains 152 codes, so the last group (length-9 codes) is assigned base value $2(48+152)=400$. Step 3 simply generates the 24 7-bit codes 0 through 23, the 152 8-bit codes 48 through 199, and the 112 9-bit codes 400 through 511. The binary values of these codes are listed in Table 6.39.

> "Must you deflate romantic rhetoric? Besides, the Astabigans have plenty of visitors from other worlds who will be viewing her."
>
> —Roger Zelazny, *Doorways in the Sand*

It is now clear that a Huffman code table can be represented by a short sequence (termed SQ) of code lengths (herein called CLs). This sequence is special in that it tends to have runs of identical elements, so it can be highly compressed by run-length encoding. The Deflate encoder compresses this sequence in a three-step process where the first step employs run-length encoding; the second step computes Huffman codes for the run lengths and generates another sequence of code lengths (to be called CCLs) for those Huffman codes. The third step writes a permuted, possibly truncated sequence of the CCLs on the compressed stream.

Step 1. When a CL repeats more than three times, the encoder considers it a run. It appends the CL to a new sequence (termed SSQ), followed by the special flag 16 and by a 2-bit repetition factor that indicates 3–6 repetitions. A flag of 16 is therefore preceded by a CL and followed by a factor that indicates how many times to copy the CL. Thus, for example, if the sequence to be compressed contains six consecutive 7's, it is compressed to 7, 16, 10_2 (the repetition factor 10_2 indicates five consecutive occurrences of the same code length). If the sequence contains 10 consecutive code lengths of 6, it will be compressed to 6, 16, 11_2, 16, 00_2 (the repetition factors 11_2 and 00_2 indicate six and three consecutive occurrences, respectively, of the same code length).

Experience indicates that CLs of zero are very common and tend to have long runs. (Recall that the codes in question are codes of literals/lengths and distances. Any given data file to be compressed may be missing many literals, lengths, and distances.) This is why runs of zeros are assigned the two special flags 17 and 18. A flag of 17 is followed by a 3-bit repetition factor that indicates 3–10 repetitions of CL 0. Flag 18 is followed by a 7-bit repetition factor that indicates 11–138 repetitions of CL 0. Thus, six consecutive zeros in a sequence of CLs are compressed to 17, 11_2, and 12 consecutive zeros in an SQ are compressed to 18, 01_2.

The sequence of CLs is compressed in this way to a shorter sequence (to be termed SSQ) of integers in the interval $[0, 18]$. An example may be the sequence of 28 CLs

4, 4, 4, 4, 4, 3, 3, 3, 6, 6, 6, 6, 6, 6, 6, 6, 6, 6, 0, 0, 0, 0, 0, 0, 2, 2, 2, 2

that's compressed to the 16-number SSQ

4, 16, 01_2, 3, 3, 3, 6, 16, 11_2, 16, 00_2, 17, 11_2, 2, 16, 00_2,

or, in decimal, 4, 16, 1, 3, 3, 3, 6, 16, 3, 16, 0, 17, 3, 2, 16, 0.

Step 2. Prepare Huffman codes for the SSQ in order to compress it further. Our example SSQ contains the following numbers (with their frequencies in parentheses): 0(2), 1(1), 2(1), 3(5), 4(1), 6(1), 16(4), 17(1). Its initial and standard Huffman trees are shown in Figure 6.42a,b. The standard tree can be represented by the SSQ of eight lengths 4, 5, 5, 1, 5, 5, 2, and 4. These are the lengths of the Huffman codes assigned to the eight numbers 0, 1, 2, 3, 4, 6, 16, and 17, respectively.

Step 3. This SSQ of eight lengths is now extended to 19 numbers by inserting zeros in the positions that correspond to unused CCLs.

Position:	0	1	2	3	4	5	6	7	8	9	10	11	12	13	14	15	16	17	18
CCL:	4	5	5	1	5	0	5	0	0	0	0	0	0	0	0	0	2	4	0

Next, the 19 CCLs are permuted according to

Position:	16	17	18	0	8	7	9	6	10	5	11	4	12	3	13	2	14	1	15
CCL:	2	4	0	4	0	0	0	5	0	0	0	5	0	1	0	5	0	5	0

The reason for the permutation is to end up with a sequence of 19 CCLs that's likely to have trailing zeros. The SSQ of 19 CCLs minus its trailing zeros is written on the compressed stream, preceded by its actual length, which can be between 4 and 19. Each CCL is written as a 3-bit number. In our example, there is just one trailing zero, so the 18-number sequence 2, 4, 0, 4, 0, 0, 0, 5, 0, 0, 0, 5, 0, 1, 0, 5, 0, 5 is written on the compressed stream as the final, compressed code of one prefix-code table. In mode 3,

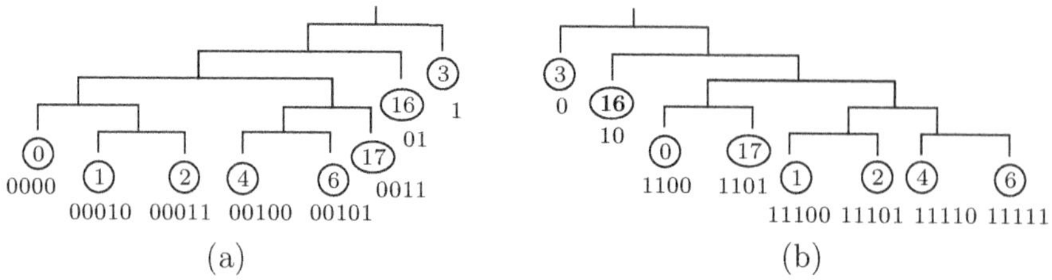

Figure 6.42: Two Huffman Trees for Code Lengths.

each block of compressed data requires two prefix-code tables, so two such sequences are written on the compressed stream.

A reader finally reaching this point (sweating profusely with such deep concentration on so many details) may respond with the single word "insane." This scheme of Phil Katz for compressing the two prefix-code tables per block is devilishly complex and hard to follow, but it works!

The format of a block in mode 3 is as follows:

1. The 3-bit header 010 or 110.

2. A 5-bit parameter `HLIT` indicating the number of codes in the literal/length code table. This table has codes 0–256 for the literals, code 256 for end-of-block, and the 30 codes 257–286 for the lengths. Some of the 30 length codes may be missing, so this parameter indicates how many of the length codes actually exist in the table.

3. A 5-bit parameter `HDIST` indicating the size of the code table for distances. There are 30 codes in this table, but some may be missing.

4. A 4-bit parameter `HCLEN` indicating the number of CCLs (there may be between 4 and 19 CCLs).

5. A sequence of `HCLEN` + 4 CCLs, each a 3-bit number.

6. A sequence SQ of `HLIT` + 257 CLs for the literal/length code table. This SQ is compressed as explained earlier.

7. A sequence SQ of `HDIST` + 1 CLs for the distance code table. This SQ is compressed as explained earlier.

8. The compressed data, encoded with the two prefix-code tables.

9. The end-of-block code (the prefix code of edoc 256).

Each CCL is written on the output as a 3-bit number, but the CCLs are Huffman codes of up to 19 symbols. When the Huffman algorithm is applied to a set of 19 symbols, the resulting codes may be up to 18 bits long. It is the responsibility of the encoder to ensure that each CCL is a 3-bit number and none exceeds 7. The formal definition [RFC1951 96] of Deflate does not specify how this restriction on the CCLs is to be achieved.

6.25.3 The Hash Table

This short section discusses the problem of locating a match in the search buffer. The buffer is 32 Kb long, so a linear search is too slow. Searching linearly for a match to any string requires an examination of the entire search buffer. If Deflate is to be able to compress large data files in reasonable time, it should use a sophisticated search method.

The method proposed by the Deflate standard is based on a hash table. This method is strongly recommended by the standard, but is not required. An encoder using a different search method is still compliant and can call itself a Deflate encoder. Those unfamiliar with hash tables should consult any text on data structures.

Instead of separate look-ahead and search buffers, the encoder should have one, 32 Kb buffer. The buffer is filled up with input data and initially all of it is a look-ahead buffer. In the original LZ77 method, once symbols have been examined, they are moved into the search buffer. The Deflate encoder, in contrast, does not move the data in its buffer and instead moves a pointer (or a separator) from left to right, to indicate the point where the look-ahead buffer ends and the search buffer starts. Short, 3-symbol strings from the look-ahead buffer are hashed and added to the hash table. After hashing a string, the encoder examines the hash table for matches. Assuming that a symbol occupies n bits, a string of three symbols can have values in the interval $[0, 2^{3n}-1]$. If $2^{3n}-1$ isn't too large, the hash function can return values in this interval, which tends to minimize the number of collisions. Otherwise, the hash function can return values in a smaller interval, such as 32 Kb (the size of the Deflate buffer).

We demonstrate the principles of Deflate hashing with the 17-symbol string

```
abbaabbaabaabaaaa
12345678901234567
```

Initially, the entire 17-location buffer is the look-ahead buffer and the hash table is empty

0	1	2	3	4	5	6	7	8
0	0	0	0	0	0	0	0	...

We assume that the first triplet abb hashes to 7. The encoder outputs the raw symbol a, moves this symbol to the search buffer (by moving the separator between the two buffers to the right), and sets cell 7 of the hash table to 1.

```
a|bbaabbaabaabaaaa
1 2345678901234567
```

0	1	2	3	4	5	6	7	8
0	0	0	0	0	0	0	1	...

The next three steps hash the strings bba, baa, and aab to, say, 1, 5, and 0. The encoder outputs the three raw symbols b, b, and a, moves the separator, and updates the hash table as follows:

```
abba|abbaabaabaaaa
1234 5678901234567
```

0	1	2	3	4	5	6	7	8
4	2	0	0	0	3	0	1	...

Next, the triplet abb is hashed, and we already know that it hashes to 7. The encoder finds 1 in cell 7 of the hash table, so it looks for a string that starts with abb at position 1 of its buffer. It finds a match of size 6, so it outputs the pair $(5-1, 6)$. The offset (4) is the difference between the start of the current string (5) and the start of the matching string (1). There are now two strings that start with abb, so cell 7 should point to both. It therefore becomes the start of a linked list (or chain) whose data items are 5 and 1. Notice that the 5 precedes the 1 in this chain, so that later searches of the chain will find the 5 first and will therefore tend to find matches with the smallest offset, because those have short Huffman codes.

```
abbaa|bbaabaabaaaa
12345 678901234567
```

0	1	2	3	4	5	6	7	8
4	2	0	0	0	3	0	↓	...

Cell 7 → | 5 | → | 1 | 0 |

Six symbols have been matched at position 5, so the next position to consider is $6+5=11$. While moving to position 11, the encoder hashes the five 3-symbol strings it finds along the way (those that start at positions 6 through 10). They are bba, baa, aab, aba, and baa. They hash to 1, 5, 0, 3, and 5 (we arbitrarily assume that aba hashes to 3). Cell 3 of the hash table is set to 9, and cells 0, 1, and 5 become the starts of linked chains.

abbaabbaab|aabaaaa
1234567890 1234567

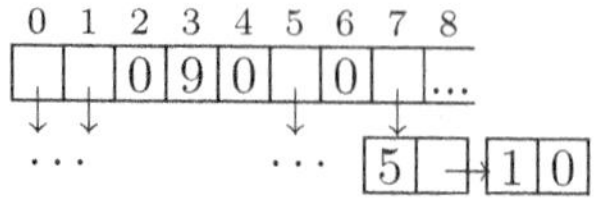

Continuing from position 11, string aab hashes to 0. Following the chain from cell 0, we find matches at positions 4 and 8. The latter match is longer and matches the 5-symbol string aabaa. The encoder outputs the pair $(11-8, 5)$ and moves to position $11+5=16$. While doing so, it also hashes the 3-symbol strings that start at positions 12, 13, 14, and 15. Each hash value is added to the hash table. (End of example.)

It is clear that the chains can become very long. An example is an image file with large uniform areas where many 3-symbol strings will be identical, will hash to the same value, and will be added to the same cell in the hash table. Since a chain must be searched linearly, a long chain defeats the purpose of a hash table. This is why Deflate has a parameter that limits the size of a chain. If a chain exceeds this size, its oldest elements should be truncated. The Deflate standard does not specify how this should be done and leaves it to the discretion of the implementor. Limiting the size of a chain reduces the compression quality but can reduce the compression time significantly. In situations where compression time is unimportant, the user can specify long chains.

Also, selecting the longest match may not always be the best strategy; the offset should also be taken into account. A 3-symbol match with a small offset may eventually use fewer bits (once the offset is replaced with a variable-length code) than a 4-symbol match with a large offset.

6.25.4 Conclusions

Deflate is a general-purpose lossless compression algorithm that has proved valuable over the years as part of several popular compression programs. The method requires memory for the look-ahead and search buffers and for the two prefix-code tables. However, the memory size needed by the encoder and decoder is independent of the size of the data or the blocks. The implementation is not trivial, but is simpler than that of some modern methods such as JPEG 2000 or MPEG. Compression algorithms that are geared for specific types of data, such as audio or video, may perform better than Deflate on such data, but Deflate normally produces compression factors of 2.5 to 3 on text, slightly less for executable files, and somewhat more for images. Most important, even in the worst case, Deflate expands the data by only 5 bytes per 32 Kb block. Finally, free implementations that avoid patents are available. Notice that the original method, as designed by Phil Katz, has been patented (United States patent 5,051,745, September 24, 1991) and assigned to PKWARE.

6.26 LZMA and 7-Zip

LZMA is the main (as well as the default) algorithm used in the popular `7z` (or `7-Zip`) compression software [7z 06]. Both `7z` and LZMA are the creations of Igor Pavlov. The software runs on Windows and is free. Both LZMA and `7z` were designed to provide high compression, fast decompression, and low memory requirements for decompression.

The main feature of `7z` is its open architecture. The software can currently compress data in one of six algorithms, and can in principle support any new compression methods. The current algorithms are the following:

1. LZMA. This is a sophisticated extension of LZ77
2. PPMD. A variant of Dmitry Shkarin's PPMdH
3. BCJ. A converter for 32-bit x86 executables (see note)
4. BCJ2. A similar converter
5. BZip2. An implementation of the Burrows-Wheeler method (Section 11.1)
6. Deflate. An LZ77-based algorithm (Section 6.25)

(Note: The 80x86 family of processors was originally developed by Intel with a word length of 16 bits. Because of its tremendous success, its architecture had been extended to 32-bit words and is currently referred to as IA-32 [Intel Architecture, 32-bit]. See [IA-32 06] for more information.)

Other important features of `7z` are the following:

- In addition to compressing and decompressing data in LZMA, the `7z` software can compress and decompress data in ZIP, GZIP, and BZIP2, it can pack and unpack data in the TAR format, and it can decompress data originally compressed in RAR, CAB, ARJ, LZH, CHM, Z, CPIO, RPM, and DEB.
- When compressing data in the ZIP or GZIP formats, `7z` provides compression ratios that are 2–10% better than those achieved by PKZip and WinZip.
- A data file compressed in `7z` includes a decompressor; it is self-extracting.
- The `7z` software is integrated with Windows Shell [Horstmann 06].
- It constitutes a powerful file manager.
- It offers a powerful command line version.
- It has a plugin for FAR Manager.
- It includes localizations for 60 languages.
- It can encrypt the compressed data with the advanced encryption standard (AES-256) algorithm, based on a 256-bit encryption key [FIPS197 01]. (The original AES algorithm, also known as Rijndael, was based on 128-bit keys.) The user inputs a text string as a passphrase, and `7z` employs a key-derivation function to obtain the 256-bit key from the passphrase. This function is based on the SHA-256 hash algorithm [SHA256 02] and it computes the key by generating a very long string based on the passphrase and using it as the input to the hash function. The 256 bits output by the hash function become the encryption key.

To generate the string, 7z starts with the passphrase, encoded in the UTF-16 encoding scheme (two bytes per character, see [UTF16 06]). It then generates and concatenates 256K $= 2^{18}$ pairs (passphrase, integer) with 64-bit integers ranging from 0 to $2^{18} - 1$. If the passphrase is p symbols long, then each pair is $2p + 8$ bytes long (the bytes are arranged in little endian), and the total length of the final string is $2^{18}(2p + 8)$ bytes; very long!

> The term Little Endian means that the low-order byte (the little end) of a number or a string is stored in memory at the lowest address (it comes first). For example, given the 4-byte number $b_3b_2b_1b_0$, if the least-significant byte b_0 is stored at address A, then the most-significant byte b_3 will be stored at address $A + 3$.

LZMA (which stands for Lempel-Ziv-Markov chain-Algorithm) is the default compression algorithm of 7z. It is an LZ77 variant designed for high compression ratios and fast decompression. A free software development kit (SDK) with the LZMA source code, in C, C++, C#, and Java, is available at [7z 06] to anyone who wants to study, understand, or modify this algorithm. The main features of LZMA are the following:

- Compression speeds of about 1 Mb/s on a 2 GHz processor. Decompression speeds of about 10–20 Mb/s are typically obtained on a 2 GHz CPU

- The size of the decompressor can be as little as 2 Kb (if optimized for size of code), and it requires only 8–32Kb (plus the dictionary size) for its data.

- The dictionary size is variable and can be up to 4 Gb, but the current implementation limits it to 1 Gb.

- It supports multithreading and the Pentium 4's hyperthreading. (Hyperthreading is a technology that allows resource sharing and partitioning while providing a multiprocessor environment in a unique CPU.) The current LZMA encoder can use one or two threads, and the LZMA decoder can use only one thread.

- The LZMA decoder uses only integer operations and can easily be implemented on any 32-bit processor (implementing it on a 16-bit CPU is more involved).

The compression principle of LZMA is similar to that of Deflate (Section 6.25), but uses range encoding (Section 5.10.1) instead of Huffman coding. This complicates the encoder, but results in better compression (recall that range encoding is an integer-based version of arithmetic coding and can compress very close to the entropy of the data), while minimizing the number of renormalizations needed. Range encoding is implemented in binary such that shifts are used to divide integers, thereby avoiding the slow "divide" operation.

Recall that LZ77 searches the search buffer for the longest string that matches the look-ahead buffer, then writes on the compressed stream a triplet (distance, length, next symbol) where "distance" is the distance from the string in the look-ahead buffer to its match in the search buffer. Thus, three types of data are written on the output, literals (the next symbol, often an ASCII code), lengths (relatively small numbers), and distances (which can be large numbers if the search buffer is large).

LZMA also outputs these three types. If nothing matches the look-ahead buffer, a literal (a value in the interval [0, 255]) is output. If a match is found, then a pair (length, distance) is output (after being encoded by the range coder). Because of the large size of the search buffer, a short, 4-entry, distance-history array is used that always contains the four most-recent distances that have been determined and output. If the distance of the current match equals one of the four distances in this array, then a pair (length, index) is output (after being encoded by the range coder), where "index" is a 2-bit index to the distance-history array.

Locating matches in the search buffer is done by hashing two bytes, the current byte in the look-ahead buffer and the byte immediately to its right (but see the next paragraph for more details). The output of the hash function is an index to an array (the hash-array). The size of the array is selected as the power of 2 that's closest to half the dictionary size, so the output of the hash function depends on the size of the dictionary. For example, if the dictionary size is 256 Mb (or 2^{28}), then the size of the array is 2^{27} and the hash function has to compute and output 27-bit numbers. The large size of the array minimizes hash collisions.

Experienced readers may have noticed the problem with the previous paragraph. If only two bytes are hashed, then the input to the hash function is only 16 bits, so there can be only $2^{16} = 65{,}536$ different inputs. The hash function can therefore compute only 65,536 distinct outputs (indexes to the hash-array) regardless of the size of the array. The full story is therefore more complex. The LZMA encoder can hash 2, 3, or 4 bytes and the number of items in the hash-array is selected by the encoder depending on the size of the dictionary. For example, a 1-Gb ($= 2^{30}$ bytes) dictionary results in a hash-array of size 512M $= 2^{29}$ items (where each item in the hash-array is a 32-bit integer). In order to take advantage of such a large hash-array, the encoder hashes four bytes. Four bytes constitute 32 bits, which provide the hash function with 2^{32} distinct inputs. The hash function converts each input to a 29-bit index. (Thus, many inputs are converted to the same index.)

Table 6.44 lists several user options and shows how the user can control the encoder by setting the Match Finder parameter to certain values. The value `bt4`, for example, specifies the binary tree mode with 2-3-4 bytes hashing. For simplicity, the remainder of this description talks about hashing two bytes (see Table 6.44 for various hashing options).

The output of the hash function is used as an index to a hash-array and the user can choose one of two search methods, hash-chain (the fast method) or binary tree (the efficient method).

In the fast method, the output of the hash function is an index to a hash-array of lists. Each array location is the start of a list of distances. Thus, if the two bytes hashed are `XY` and the result of the hash is index 123, then location 123 of the hash-array is a list of distances to pairs `XY` in the search buffer. These lists can be very long, so LZMA checks only the first 24 distances. These correspond to the 24 most-recent occurrences of `XY` (the number 24 is a user-controlled parameter). The best of the 24 matches is selected, encoded, and output. The distance of this match is then added to the start of the list, to become the first match found when array location 123 is checked again. This method is reminiscent of match searching in LZRW4 (Section 6.12).

In the binary tree method, the output of the hash function is an index to a hash-array of binary search trees (Section 6.4). Initially, the hash-array is empty and there are no trees. As data is read and encoded, trees are generated and grow, and each data byte becomes a node in one of the trees. A binary tree is created for every pair of consecutive bytes in the original data. Thus, if the data contains `good␣day`, then trees are generated for the pairs `go`, `oo`, `od`, `d␣`, `␣d`, `da`, and `ay`. The total number of nodes in those trees is eight (the size of the data). Notice that the two occurrences of `o` end up as nodes in different trees. The first resides in the tree for `oo`, and the second ends up in the tree of `od`.

The following example illustrates how this method employs the binary trees to find matches and how the trees are unpdated. The example assumes that the match-finder parameter (Table 6.44) is set to `bt2`, indicating 2-byte hashing.

We assume that the data to be encoded is already fully stored in a long buffer (the dictionary). Five strings that start with the pair `ab` are located at various points as shown

```
...abm...abcd2...abcx...abcd1...aby...
   11    24      30     57      78
```

We denote by `p(n)` the index (location) of string `n` in the dictionary. Thus, `p(abm...)` is 11 and `p(abcd2)` is 24. Each time a binary tree T is searched and a match is selected, T is rearranged and is updated according to the following two rules:

1. The tree must always remain a binary search tree.
2. If `p(n1)` $<$ `p(n2)`, then `n2` cannot be in any subtree of `n1`. This implies that (a) the latest string (the one with the greatest index) is always the root of the tree and (b) indexes decrease as we slide down the tree. The result is a tree where recent strings are located near the root.

We also assume that the pair `ab` of bytes is hashed to 62. Figure 6.43 illustrates how the binary tree for the pair `ab` is created, kept up to date, and searched. The following numbered items refer to the six parts of this figure.

1. When the LZMA encoder gets to location 11 and finds `a`, it hashes this byte and the `b` that follows, to obtain 62. It examines location 62 of the hash-array and finds it empty. It then generates a new binary tree (for the pair `ab`) with one node containing the pointer 11, and sets location 62 of the hash-array to point to this tree. There is no match, the byte `a` at 11 is output as a literal, and the encoder proceeds to hash the next pair `bm`.

2. The next pair `ab` is found by the LZMA encoder when it gets to location 24. Hashing produces the same 62, and location 62 of the hash-array is found to point to a binary tree whose root (which is so far its only node) contains 11. The encoder places 24 as the new root with 11 as its right subtree, because `abcd2...` is smaller than `abm...` (strings are compared lexicographically). The encoder matches `abcd2...` with `abm...` and the match length is 2. The encoder continues with `cd2...`, but before it does that, it hashes the pair `bc` and either appends it to the binary tree for `bc` (if such a tree exists) or generates a new tree.

3. The next pair `ab` is found at location 30. Hashing produces the same 62. The encoder places 30 (`abcx...`) as the new root with 24 (`abcd2...`) as its left subtree and 11 (`abm..`) as its right subtree. The better match is `abcd2...`, and the match length is 3. The next two pairs `bc` and `cx` are appended to their respective trees (if such trees exist), and the encoder continues with the pair `x..`

4. The next occurrence of `ab` is found at location 57 and is hashed to 62. It becomes the root of the tree, with 30 (`abcx...`) as its right subtree. The best match is with `abcd2...` where the match length is 4. The encoder continues with the string `1...` found at location 61.

5. In the last step of this example, the encoder finds a pair `ab` at location 78. This string (`aby...`) becomes the new root, with 57 as its left subtree. The match length is 2.

6. Now assume that location 78 contains the string `abk..` instead of `aby...` Since `abm..` is greater than `abk..`, it must be moved to the right subtree of `abk..`, resulting in a completely different tree.

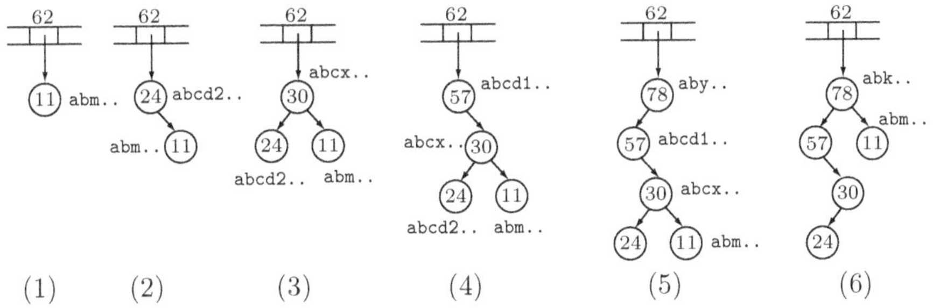

Figure 6.43: Binary Search Trees in LZMA.

Past versions of LZMA used a data structure called a Patricia trie (see page 357 for the definition of a trie), but this structure proved inefficient and has been eliminated.

We end this description with some of the options that can be specified by the user when LZMA is invoked.

- `-a{N}`. The compression mode. 0, 1, and 2 specify the fast, normal, and max modes, with 2 as the default. (The latest version, currently in its beta stage, does not support the max mode.)

- `-d{N}`. The Dictionary size. N is an integer between 0 and 30, and the dictionary size is set to 2^N. The default is $N = 23$ (an 8-Mb dictionary), and the current maximum is $N = 30$ (a 1-Gb dictionary).

- `-mf{MF_ID}`. The Match Finder. It specifies the method for finding matches and limits the number of bytes to be hashed. The memory requirements depend on this choice and on the dictionary size d. Table 6.44 lists the choices. The default value of `MF_ID` in the normal, max, and ultra modes is `bt4` and in the fast and fastest modes it is `hc4`.

We would like to thank Igor Pavlov for contributing important information and details to the material of this section.

MF_ID	Memory	Description
bt2	$d \times 9.5 + 1$ Mb	Binary Tree with 2 bytes hashing
bt3	$d \times 11.5 + 4$ Mb	Binary Tree with 3 bytes hashing
bt4	$d \times 11.5 + 4$ Mb	Binary Tree with 4 bytes hashing
hc4	$d \times 7.5 + 4$ Mb	Hash Chain with 4 bytes hashing

Table 6.44: LZMA Match Finder Options.

Notes for Table 6.44:

1. bt4 uses three hash tables with 2^{10} items for hashing two bytes, with 2^{16} items for hashing three bytes, and with a variable size to hash four bytes. Only the latter table points to binary trees. The other tables point to positions in the input buffer.

2. bt3 uses two hash tables, one with 2^{10} items for hashing two bytes and the other with a variable size to hash three bytes

3. bt2 uses only one hash table with 2^{16} items.

4. bt2 and bt3 also can find almost all the matches, but bt4 is faster.

6.27 PNG

The portable network graphics (PNG) file format has been developed in the mid-1990s by a group (the PNG development group [PNG 03]) headed by Thomas Boutell. The project was started in response to the legal issues surrounding the GIF file format (Section 6.21). The aim of this project was to develop a sophisticated graphics file format that will be flexible, will support many different types of images, will be easy to transmit over the Internet, and will be unencumbered by patents. The design was finalized in October 1996, and its main features are as follows:

1. It supports images with 1, 2, 4, 8, and 16 bitplanes.
2. Sophisticated color matching.
3. A transparency feature with very fine control provided by an alpha channel.
4. Lossless compression by means of Deflate combined with pixel prediction.
5. Extensibility: New types of meta-information can be added to an image file without creating incompatibility with existing applications.

Currently, PNG is supported by many image viewers and web browsers on various platforms. This subsection is a general description of the PNG format, followed by the details of the compression method it uses.

A PNG file consists of chunks that can be of various types and sizes. Some chunks contain information that's essential for displaying the image, and decoders must be able to recognize and process them. Such chunks are referred to as "critical chunks." Other chunks are ancillary. They may enhance the display of the image or may contain metadata such as the image title, author's name, creation and modification dates and times, etc. (but notice that decoders may choose not to process such chunks). New, useful types of chunks can also be registered with the PNG development group.

A chunk consists of the following parts: (1) size of the data field, (2) chunk name, (3) data field, and (4) a 32-bit cyclical redundancy code (CRC, Section 6.32). Each chunk has a 4-letter name of which (1) the first letter is uppercase for a critical chunk and lowercase for an ancillary chunk, (2) the second letter is uppercase for standard chunks (those defined by or registered with the PNG group) and lowercase for a private chunk (an extension of PNG), (3) the third letter is always uppercase, and (4) the fourth letter is uppercase if the chunk is "unsafe to copy" and lowercase if it is "safe to copy."

Any PNG-aware application will process all the chunks it recognizes. It can safely ignore any ancillary chunk it doesn't recognize, but if it finds a critical chunk it cannot recognize, it has to stop and issue an error message. If a chunk cannot be recognized but its name indicates that it is safe to copy, the application may safely read and rewrite it even if it has altered the image. However, if the application cannot recognize an "unsafe to copy" chunk, it must discard it. Such a chunk should not be rewritten on the new PNG file. Examples of "safe to copy" are chunks with text comments or those indicating the physical size of a pixel. Examples of "unsafe to copy" are chunks with gamma/color correction data or palette histograms.

The four critical chunks defined by the PNG standard are IHDR (the image header), PLTE (the color palette), IDAT (the image data, as a compressed sequence of filtered samples), and IEND (the image trailer). The standard also defines several ancillary chunks that are deemed to be of general interest. Anyone with a new chunk that may also be of general interest may register it with the PNG development group and have it assigned a public name (a second letter in uppercase).

The PNG file format uses a 32-bit CRC (Section 6.32) as defined by the ISO standard 3309 [ISO 84] or ITU-T V.42 [ITU-T 94]. The CRC polynomial is

$$x^{32}+x^{26}+x^{23}+x^{22}+x^{16}+x^{12}+x^{11}+x^{10}+x^{8}+x^{7}+x^{5}+x^{4}+x^{2}+x+1.$$

The particular calculation proposed in the PNG standard employs a precalculated table that speeds up the computation significantly.

A PNG file starts with an 8-byte signature that helps software to identify it as PNG. This is immediately followed by an IHDR chunk with the image dimensions, number of bitplanes, color type, and data filtering and interlacing. The remaining chunks must include a PLTE chunk if the color type is palette, and one or more adjacent IDAT chunks with the compressed pixels. The file must end with an IEND chunk. The PNG standard defines the order of the public chunks, whereas private chunks may have their own ordering constraints.

An image in PNG format may have one of the following five color types: RGB with 8 or 16 bitplanes, palette with 1, 2, 4, or 8 bitplanes, grayscale with 1, 2, 4, 8, or 16 bitplanes, RGB with alpha channel (with 8 or 16 bitplanes), and grayscale with alpha channel (also with 8 or 16 bitplanes). An alpha channel implements the concept of transparent color. One color can be designated transparent, and pixels of that color are not drawn on the screen (or printed). Instead of seeing those pixels, a viewer sees the background behind the image. The alpha channel is a number in the interval $[0, 2^{bp}-1]$, where bp is the number of bitplanes. Assuming that the background color is B, a pixel in the transparent color P is painted in color $(1-\alpha)B+\alpha P$. This is a mixture of $\alpha\%$ background color and $(1-\alpha)\%$ pixel color.

Perhaps the most intriguing feature of the PNG format is the way it handles interlacing. Interlacing makes it possible to display a rough version of the image on the screen, then improve it gradually until it reaches its final, high-resolution form. The special interlacing used in PNG is called Adam 7 after its developer, Adam M. Costello.

PNG divides the image into blocks of 8×8 pixels each, and displays each block in seven steps. In step 1, the entire block is filled up with copies of the top-left pixel (the one marked "1" in Figure 6.45a). In each subsequent step, the block's resolution is doubled by modifying half its pixels according to the next number in Figure 6.45a. This process is easiest to understand with an example, such as the one shown in Figure 6.45b.

1	6	4	6	2	6	4	6
7	7	7	7	7	7	7	7
5	6	5	6	5	6	5	6
7	7	7	7	7	7	7	7
3	6	4	6	3	6	4	6
7	7	7	7	7	7	7	7
5	6	5	6	5	6	5	6
7	7	7	7	7	7	7	7

(a) (b)

Figure 6.45: Interlacing in PNG.

PNG compression is lossless and is performed in two steps. The first step, termed *delta filtering* (or just *filtering*), converts pixel values to numbers by a process similar to the prediction used in the lossless mode of JPEG (Section 7.10.5). The filtering step calculates a "predicted" value for each pixel and replaces the pixel with the difference between the pixel and its predicted value. The second step employs Deflate to encode the differences. Deflate is the topic of Section 6.25, so only filtering needs be described here.

Filtering does not compress the data. It only transforms the pixel data to a format where it is more compressible. Filtering is done separately on each image row, so an intelligent encoder can switch filters from one image row to the next (this is called *adaptive filtering*). PNG specifies five filtering methods (the first one is simply no filtering) and recommends a simple heuristic for choosing a filtering method for each image row. Each row starts with a byte indicating the filtering method used in it. Filtering is done on bytes, not on complete pixel values. This is easy to understand in cases where a pixel consists of three bytes, specifying the three color components of the pixel. Denoting the three bytes by a, b, and c, we can expect a_i and a_{i+1} to be correlated (and also b_i and b_{i+1}, and c_i and c_{i+1}), but there is no correlation between a_i and b_i. Also, in a grayscale image with 16-bit pixels, it makes sense to compare the most-significant bytes of two adjacent pixels and then the least-significant bytes. Experiments suggest that filtering is ineffective for images with fewer than eight bitplanes, and also for palette images, so PNG recommends no filtering in such cases.

The heuristic recommended by PNG for adaptive filtering is to apply all five filtering types to the row of pixels and select the type that produces the smallest sum of absolute values of outputs. (For the purposes of this test, the filtered bytes should be considered signed differences.)

The five filtering types are described next. The first type (type 0) is no filtering. Filtering type 1 (sub) sets byte $B_{i,j}$ in row i and column j to the difference $B_{i,j} - B_{i-t,j}$, where t is the interval between a byte and its correlated predecessor (the number of bytes in a pixel). Values of t for the various image types and bitplanes are listed in Table 6.46. If $i - t$ is negative, then nothing is subtracted, which is equivalent to having a zero pixel on the left of the row. The subtraction is done modulo 256, and the bytes subtracted are considered unsigned.

Image type	Bit planes	Interval t
Grayscale	1, 2, 4, 8	1
Grayscale	16	2
Grayscale with alpha	8	2
Grayscale with alpha	16	4
Palette	1, 2, 4, 8	1
RGB	8	3
RGB	16	6
RGB with alpha	8	4
RGB with alpha	16	8

Figure 6.46: Interval Between Bytes.

Filtering type 2 (up) sets byte $B_{i,j}$ to the difference $B_{i,j} - B_{i,j-1}$. The subtraction is done as in type 1, but if j is the top image row, no subtraction is done.

Filtering type 3 (average) sets byte $B_{i,j}$ to the difference $B_{i,j} - \left[B_{i-t,j} + B_{i,j-1}\right] \div 2$. The average of the left neighbor and the top neighbor is computed and subtracted from the byte. Any missing neighbor to the left or above is considered zero. Notice that the sum $B_{i-t,j} + B_{i,j-1}$ may be up to 9 bits long. To guarantee that the filtering is lossless and can be reversed by the decoder, the sum must be computed exactly. The division by 2 is equivalent to a right shift and brings the 9-bit sum down to 8 bits. Following that, the 8-bit average is subtracted from the current byte $B_{i,j}$ modulo 256 and unsigned.

Example. Assume that the current byte $B_{i,j} = 112$, its left neighbor $B_{i-t,j} = 182$, and its top neighbor $B_{i,j-1} = 195$. The average is $(182 + 195) \div 2 = 188$. Subtracting $(112 - 188)$ mod 256 yields -76 mod 256 or $256 - 76 = 180$. Thus, the encoder sets $B_{i,j}$ to 180. The decoder inputs the value 180 for the current byte, computes the average in the same way as the encoder to end up with 188, and adds $(180 + 188)$ mod 256 to obtain 112.

Filtering type 4 (Paeth) sets byte $B_{i,j}$ to $B_{i,j}-\text{PaethPredict}\big[B_{i-t,j}, B_{i,j-1}, B_{i-t,j-1}\big]$. PaethPredict is a function that uses simple rules to select one of its three parameters, then returns that parameter. Those parameters are the left, top, and top-left neighbors. The selected neighbor is then subtracted from the current byte, modulo 256 unsigned. The PaethPredictor function is defined by the following pseudocode:

```
function PaethPredictor (a, b, c)
begin
; a=left, b=above, c=upper left
p:=a+b-c ;initial estimate
pa := abs(p-a) ; compute distances
pb := abs(p-b) ; to a, b, c
pc := abs(p-c)
; return nearest of a,b,c,
; breaking ties in order a,b,c.
if pa<=pb AND pa<=pc then return a
else if pb<=pc then return b
else return c
end
```

PaethPredictor must perform its computations exactly, without overflow. The order in which PaethPredictor breaks ties is important and should not be altered. This order (that's different from the one given in [Paeth 91]) is left neighbor, neighbor above, upper-left neighbor.

PNG is a single-image format, but the PNG development group has also designed an animated companion format named MNG (multiple-image network format), which is a proper superset of PNG.

We are indebted to Cosmin Truţa for reviewing and correcting this subsection.

Does the world really need yet another graphics format? We believe so. GIF is no longer freely usable,. . . it would not be all that much easier to implement than a whole new file format. (PNG is designed to be simple to implement, with the exception of the compression engine, which would be needed in any case.) We feel that this is an excellent opportunity to design a new format that fixes some of the known limitations of GIF.

From the PNG standard, RFC 2083, 1999

6.28 XML Compression: XMill

XMill is a special-purpose, efficient software application for the compression of XML (Extensible Markup Language) documents. Its description in this short section is based on [Liefke and Suciu 99], but more details and a free implementation are available from [XMill 03]. First, a few words about XML.

XML is a markup language for documents containing structured information. A markup language is a mechanism to identify structures in a document. A short story or a novel may have very little structure. It may be divided into chapters and may also include footnotes, an introduction, and an epilogue. A cooking recipe has more structure. It starts with the name of the dish and its class (such as salads, soups, etc.). This is followed by the two main structural items: the ingredients and the preparation. The recipe may then describe the proper way to serve the dish and may end with notes, reviews, and comments. A business card is similarly divided into a number of short items.

The XML standard [XML 03] defines a way to add markup to documents, and has proven very popular and successful. An XML file contains data represented as text and also includes tags that identify the types of (or that assign meaning to) various data items. HTML is also a markup language, familiar to many, but it is restrictive in that it defines only certain tags and their meanings. The `<H3>` tag, for example, is defined in HTML and has a certain meaning (a certain header), but anyone wanting to use a tag of, say, `<blah>`, has first to convince the WWW consortium to define it and assign it a meaning, then wait for the various web browsers to support it. XML, in contrast, does not specify a set of tags but provides a facility to define tags and structural relationships between them. The meaning of the tags (their semantics) is later defined by applications that process XML documents or by style sheets.

Here is a simple example of a business card in XML. Most tags specify the start and end of a certain data item; they are wrappers. A tag such as `<red_backgrnd/>` that has a trailing "/" is considered empty. It specifies something to the application that reads and processes the XML file, but that something does not require any extra data.

```
<card xmlns="http://businesscard.org">
   <name>Melvin Schwartzkopf</name>
   <title>Chief person, Monster Inc.</title>
   <email>mschwa@monster.com</email>
   <phone>(212)555-1414</phone>
   <logo url="widget.gif"/>
   <red_backgrnd/>
</card>
```

In summary, an XML file consists of markup and content. There are six types of markup that can occur in an XML document: elements, entity references, comments, processing instructions, marked sections, and document-type declarations. The contents can be any digital data.

The main aim of the developers of XMill was to design and implement a special-purpose XML encoder that will compress an XML file better than a typical compressor will compress just the data of that file. Given a data file A, suppose that a typical compressor, such as gzip, compresses it to X. Now add XML tags to A, converting it to B and increasing its size in the process. When B is compressed by XMill, the resulting file, Y, should be smaller than X. As an example, the developers had tested XMill on a 98-Mb data file taken from SwissProt, a data base for protein structure. The file was initially compressed by gzip down to 16 Mb. The original file was then converted to XML which increased its size to 165 Mb and was compressed by gzip (to 19 Mb) and by XMill (to 8.6 Mb, albeit after fine-tuning XMill for the specific data in that file). However, since XMill is a special-purpose encoder, it is not expected to perform well on arbitrary data files. The design of XMill is based on the following principles:

1. By itself, XMill is not a compressor. Rather, it is an extensible tool for specifying and applying existing compressors to XML data items. XMill analyzes the XML file, then invokes different compressors to compress different parts of the file. The main compressor used by XMill is gzip, but XMill includes other compressors and can also be linked by the user to any existing compressor.

2. Separate the structure from the raw data. The XML tags and attributes are separated by XMill from the data in the input file and are compressed separately. The data is the contents of XML elements and the values of attributes.

3. Group together items that are related. XMill uses the concept of a *container*. In the above example of business cards, all the URLs are grouped in one container, all the names are grouped in a second container, and so on. Also, all the XML tags and attributes are grouped in the structure container. The user can control the contents of the container by providing *container expressions* on the XMill command line.

4. Use semantic compressors. A container may include data items of a certain type, such as telephone numbers or airport codes. A sophisticated user may have an encoder that compresses such items efficiently. The user may therefore direct XMill to use certain encoders for certain containers, thereby achieving excellent compression for the entire XML input file. XMill comes with several built-in encoders, but any encoder available to the user may be linked to XMill and will be used by it to compress any specified container.

An important part of XMill is a concise language, termed the *container expressions*, that's used to group data items in containers and to specify the proper encoders for the various containers.

XMill was designed to prepare XML files for storage or transmission. Sometimes, an XML file is used in connection with a query processor, where the file has to be searched often. XMill is not a good choice for such an application. Another limitation of XMill is that it performs well only on large files. Any XML file smaller than about 20 Kb will be poorly compressed by XMill because XMill adds overhead in the form of bookkeeping to the compressed file. Small XML files, however, are common in situations where messages in XML format are exchanged between users or between applications.

> I do not know it—it is without name—it is a word unsaid, It is not in any dictionary, utterance, symbol.
>
> —Walt Whitman, *Leaves of Grass*, (1900)

6.29 EXE Compressors

The LZEXE program is freeware originally written in the late 1980s by Fabrice Bellard as a special-purpose utility to compress EXE files (PC executable files). The idea is that an EXE file compressed by LZEXE can be decompressed **and** executed with one command. The decompressor does not write the decompressed file on the disk but loads it in memory, relocates addresses, and executes it! The decompressor uses memory that's eventually used by the program being decompressed, so it does not require any extra RAM. In addition, the decompressor is very small compared with decompressors in self-extracting archives.

The algorithm is based on LZ. It uses a circular queue and a dictionary tree for finding string matches. The position and size of the match are encoded by an auxiliary algorithm based on the Huffman method. Uncompressed bytes are kept unchanged, since trying to compress them any further would have entailed a much more complex and larger decompressor. The decompressor is located at the end of the compressed EXE file and is 330 bytes long (in version 0.91). The main steps of the decoder are as follows:

1. Check the CRC (Section 6.32) to ensure data reliability.
2. Locate itself in high RAM; then move the compressed code in order to leave sufficient room for the EXE file.
3. Decompress the code, check that it is correct, and adjust the segments if bigger than 64K.
4. Decompress the relocation table and update the relocatable addresses of the EXE file.
5. Run the program, updating the CS, IP, SS, and SP registers.

The idea of EXE compressors, introduced by LZEXE, was attractive to both users and software developers, so a few more have been developed:

1. PKlite, from PKWare, is a similar EXE compressor that can also compress .COM files.
2. DIET, by Teddy Matsumoto, is a more general EXE compressor that can compress data files. DIET can act as a monitor, permanently residing in RAM, watching for applications trying to read files from the disk. When an application tries to read a DIET-compressed data file, DIET senses it and does the reading and decompressing in a process that's transparent to the application.

UPX is an EXE compressor started in 1996 by Markus Oberhumer and László Molnár. The current version (as of June 2006) is 2.01. Here is a short quotation from [UPX 03].

UPX is a free, portable, extendable, high-performance executable packer for several different executable formats. It achieves an excellent compression ratio and offers very fast decompression. Your executables suffer no memory overhead or other drawbacks.

6.30 Off-Line Dictionary-Based Compression

The dictionary-based compression methods described in this chapter are distinct, but have one thing in common; they generate the dictionary as they go along, reading data and compressing it. The dictionary is not included in the compressed file and is generated by the decoder in lockstep with the encoder. Thus, such methods can be termed "online." In contrast, the methods described in this section are also based on a dictionary, but can be considered "offline" because they include the dictionary in the compressed file.

BPE. The first method is byte pair encoding (BPE). This is a simple compression method, due to [Gage 94], that often features only mediocre performance. It is described here because (1) it is an example of a multipass method (two-pass compression algorithms are common, but multipass methods are generally considered too slow) and (2) it eliminates only certain types of redundancy and should therefore be applied only to data files that feature this redundancy. (The next method, by [Larsson and Moffat 00], does not suffer from these drawbacks and is much more efficient.) BPE is both an example of an offline dictionary-based compression algorithm and a simple example (perhaps the simplest) of a grammar-based compression method. In addition, the BPE decoder is very small, which makes it ideal for applications where memory size is restricted.

The BPE method is easy to describe. We assume that the data symbols are bytes and we use the term *bigram* for a pair of consecutive bytes. Each pass locates the most-common bigram and replaces it with an unused byte value. Thus, the method performs best on files that have many unused byte values, and one aim of this discussion is to point out the types of data that feature this kind of redundancy. First, however, a small example. Given the character set `A`, `B`, `C`, `D`, `X`, and `Y` and the data file `ABABCABCD` (where `X` and `Y` are unused bytes), the first pass identifies the pair `AB` as the most-common bigram and replaces each of its three occurrences with the single byte `X`. The result is `XXCXCD`. The second pass identifies the pair `XC` as the most-common bigram and replaces each of its two occurrences with the single byte `Y`. The result is `XYYD`, where no bigram occurs more than once. Bigrams that occur just once can also be replaced, if more unused byte values are available. However, each replacement rule must be appended to the dictionary and thus ends up being included in the compressed file. As a result, the BPE encoder stops when no bigram occurs more than once.

What types of data tend to have many unused byte values? The first type that comes to mind is text. Currently (in late 2008), most text files use the well-known ASCII codes to encode text. An ASCII code occupies a byte, but only seven bits constitue the actual code. The eighth bit can be used for a parity check, but is often simply set to zero. Thus, we can expect 128 byte values out of the 256 possible ones to be unused in a typical ASCII text file. A quick glance at an ASCII code table shows that codes 0 through 32 (as well as code 127) are control codes. They indicate commands such as backspace, carriage return, escape, delete, and blank space. It therefore makes sense to expect only a few of those to appear in any given text file.

The validity of these arguments can be checked by a simple test. The following Mathematica code prints the unused byte values in a given text file.

```
scn = ReadList["Hobbit.txt", Byte];
btc = Table[0, {256}];
```

```
Do[btc[[scn[[i]]]] = btc[[scn[[i]]]] + 1, {i, 1, Length[scn]}];
btc
dis = Table[0, {256}]; j = 0;
Do[If[btc[[i]] == 0, {j = j + 1, dis[[j]] = i - 1}], {i, 1, 256}];
Take[dis, j]
```

It was executed on three chapters from the book *Data Compression: The Complete Reference* (4th edition) and on Tolkien's *The Hobbit*. The following results were obtained:

Dcomp3: 0–7, 9–11, 13–30, 126–255.
Dcomp4.1: 0–11, 13–30, 126–255.
Dcomp5: 0–7, 9–11, 13–30, 126–255.
Hobbit: 0–7, 9–11, 13–30, 34–36, 42, 46, 60, 63, 87, 89–92, 95, 122–123, 125–255.

The results of this experiment are clear. More than half the byte values are unused. The 128 values 128 through 255 are unused and the only ASCII control characters used are BS, FF, US, and Space.

Today, more and more text files are encoded in Unicode, but even such files should have many unused byte values. (Notice that most Unicodes are 16 bits, but there are also 8-bit and 32-bit Unicodes.) A typical Unicode text file in an alphabetic language consists of letters, digits, punctuation marks, and accented letters, so the total number of codes is around 100–150, leaving many unused byte values. As an example, the 128 Unicodes for Greek and Coptic [Greek-Unicode 07] are `0370` through `03FF`, and these do not use byte values `00`, `01`, and `04` through `6F`. Naturally, text files in an ideograph-based language, such as Hangul or Cuneiforms, easily use all 256 byte values.

Grayscale images constitute another example of data where many byte values may be unused. A typical image in grayscale may have millions of pixels, each in one of 256 shades of gray, but many shades may be unused. An experiment with the Mathematica code above indicates 41 unused byte values in the well-known `lena` image (raw, 128×128, 1-byte pixels), and 35 unused shades of gray (out of 256) in the familiar, raw format, `baboon` image of the same resolution.

On the other hand, the same images in color (in raw format, with three bytes per pixels) use all the 256 byte values and it is easy to see why. A typical color image may consist of 6–7 million pixels (this is typical for today's digital cameras) and may use only a few thousand colors. However, each color occupies three bytes, so even if a certain color, say (r, g, b)=(108, 56, 213), is unused, there is a good chance that some pixels have color components 108, 56, or 213.

Thus, Gage's method makes sense for compressing text and grayscale images. If it is applied to other types of data files, a large file should be segmented into small sections with unused byte values in each.

At any given time, the method looks only at one pair of bytes, but this algorithm also indirectly takes advantage of longer repeating patterns. Thus, an input file of the form `abcdeabcdfabcdg` compresses quite efficiently. The most-common byte pairs are `ab`, `bc`, and `cd`. If the algorithm selects the first pair and replaces it with the unused byte `x`, the file becomes `xcdexcdfxcdg`. If `xc` is next selected and is replaced by `y`, the result is `ydeydfydg`. Now the pair `yd` is replaced by `z`, to produce `zezfzg`. The compression factor is $15/6 = 2.5$, comparable to (or even exceeding) more efficient methods.

The remainder of this section describes a possible implementation of this method (reference [Gage 94] includes C source code). To compress a file, the entire file must be input into a buffer in memory (if the file is too large, it should be compressed in segments). As an example, we consider an alphabet of the eight symbols a through h (coded as 0 through 7) and the input file ababcabcd (with symbols e through h unused).

The program constructs the following dictionary (also referred to as a pair-table or a phrase-table), where each zero in the bottom row indicates an unused character.

0	1	2	3	4	5	6	7
a	b	c	d	e	f	g	h
1	1	1	1	0	0	0	0

The first step is to locate the most-common bigram. The simplest approach is to set up a table of size 256 × 256, initialize it to all zeros, use the two bytes of the next input bigram as row and column indexes to the table, and increment the particular entry pointed to by this bigram. The implementation described in [Gage 94] is based on a hash table that hashes each pair of bytes into a 12-bit number. That number is used as an index to an array of size $2^{12} = 4{,}096$ and the array location pointed to by the index is incremented. This works if the number of bigrams in the data file is less than 4,096 (notice that it can be up to $256 \times 256 = 65{,}536$).

Once the most-common bigram is located (ab in our example), it is replaced by an unused character (h). The file in the buffer becomes hhchcd and the pair-table is updated to

0	1	2	3	4	5	6	7
a	b	c	d	e	f	g	a
1	1	1	1	0	0	0	b

The last entry indicates that byte-pair ab has been replaced by h (code 7).

The next (and last) pass identifies pair hc as the most common bigram and replaces it with unused symbol g. The file in the buffer becomes hggd and the pair-table is updated to

0	1	2	3	4	5	6	7
a	b	c	d	e	f	h	a
1	1	1	1	0	0	c	b

The 7th entry indicates that bigram hc has been replaced by g (code 6).

The pair-table consists of two types of entries. One type (type 1) has a binary flag in the bottom row indicating used or unused symbols (1 or 0 flags). This type is easy to identify because the element in the top row is identical to its index (thus, the first element a had code 0 and is in column 0, while the last element, also a, has code 0 but is in column 7). The other type (type 2) indicates pair substitutions, and entries of this type should be written on the compressed file. Notice that the two types of entries may be mixed and do not have to be contiguous. In our example, the pair-table consists of six type-1 entries followed by two type-2 entries and is written on the compressed file as the six bytes −6, h, c, 1, a, b, where the first three bytes indicate six irrelevant type-1 entries (corresponding to codes 0 through 5) followed by one type-2 entry hc (corresponding to code 6). The next three bytes indicate one type-2 entry ab (which corresponds to code 7). The encoding rule for this table is therefore the following: Each contiguous segment of n type-1 entries followed by a type-2 entry is encoded as $-n$ followed by the two bytes of the type-2 entry. Each segment of m consecutive type-2 entries is encoded as the byte m followed by the $2m$ bytes of the entries.

The last feature of the encoder has to do with replacing pairs of bytes in the buffer. When a pair of bytes `xy` is replaced by a single byte `p`, it (the pair) becomes `p-`, where the `-` indicates an empty byte. There may be several ways to handle empty bytes. The straightforward way is to move bytes and compact the buffer each time a pair is replaced by a single byte. This is extremely slow because it may involve many thousands of byte movements for each replacement. A better approach is to have an auxiliary array of flags that indicate which byte positions in the buffer are empty. If the size of the buffer is n bytes, the size of the auxiliary array should be n bits, one-eighth (or 12.5%) the buffer size. If the buffer contains the 16 bytes `the␣la-t␣fea-ure`, then the auxiliary array should have the two bytes 00000010|00001000, where the two 1's indicate empty bytes. When the compressed file is written from the buffer to the output, only those bytes that correspond to zero bits in the auxiliary array should be written. Another way to handle empty bytes is to organize the buffer as a linked list and establish another list (initially empty) of empty bytes. When a byte becomes empty, pointers are updated to take this byte out of the buffer and append it to the list of empty bytes.

It is now clear that encoding may not be very efficient, but on the other hand the method never expands the data (except for the overhead from the pair-table). If no byte values are unused, the data is simply written on the compressed file as is, with the addition of the pair-table. This should be compared to other, more efficient methods where the "wrong" type of data may cause significant expansion.

Encoding is slow, requiring multiple passes and a large buffer. Decoding, on the other hand, is fast. The decoder first reads and expands the pair-table, where only the type-2 entries are relevant. Bytes are then read from the compressed file. If a byte is literal (i.e., if it does not appear in the type-2 entries, such as byte `d` in our example), it is written to the final output as is. Otherwise, the byte is one of the type-2 entries and it represents a pair. The pair is constructed and is pushed into a stack. If the stack is not empty, the next byte is popped from the stack and handled as described above (which may cause another byte pair to be pushed into the stack). If the stack is empty, the next byte is read from the compressed file. Being so simple, the decoder is also very small, which is an advantage (as has been mentioned earlier).

Re-Pair, an efficient offline dictionary algorithm

The performance of BPE depends on the number of unused byte values in the original data. The next offline compression method (recursive pairing, or Re-Pair, by [Larsson and Moffat 00]) is much more sophisticated. It features high compression factors, a fast, small decoder, and it does not depend on the existence of unused byte values. It also employs sophisticated data structures that make it possible to select the most-common bigram in each pass without having to scan the entire input each time.

No unused byte values are required. In each pass, the most-common bigram is identified and is replaced by a *new* symbol. We first illustrate this idea symbolically, using the well-known line from the television film *Yabba-Dabba Do!* (based on the 1960–66 animated sitcom *The Flintstones*). The original data is shown in lowercase and the new symbols are in uppercase.

Pair	String
	yabba␣dabba,␣yabba␣dabba␣dabba␣do
A → ba	yabA␣dabA,␣yabA␣dabA␣dabA␣do
B → ab	yBA␣dBA,␣yBA␣dBA␣dBA␣do
C → BA	yC␣dC,␣yC␣dC␣dC␣do
D → C␣	yDdC,␣yDdDdDdo
E → Dd	yEC,␣yEEEo
F → yE	FC,␣FEEo

The compressed file consists of the final string FC,␣FEEo and the six-entry dictionary (or phrase-table). Notice that the final string contains one bigram, but replacing it with a new symbol would require an extra dictionary entry and would therefore contribute nothing to the compression (and may even cause expansion).

The main question is how to add new symbols. Assuming that the original data consists of bytes and that all 256 byte values are used, how can we add new symbols? We can start by placing each 8-bit byte in a pair of bytes (16 bits). Two-byte symbols can have values from 0 to $2^{16} - 1 = 65{,}535$, so there is room for the original 256 byte values and 65,280 new symbols. The bigram replacement algorithm is then executed and a phrase-table is constructed. Once this is done, the algorithm knows how many new symbols are needed, and the symbols (the 256 original ones and the new ones) are replaced with variable-length codes.

One way to assign reasonable variable-length codes to the symbols is to count the number of occurrences of each symbol in the final string and in the phrase-table and use this information to construct a set of Huffman codes. The table of Huffman codes becomes overhead that must be included in the compressed file for the decoder's use.

An alternative is to sort the list of symbols according to their frequencies of occurrence and assign them one of the many variable-length codes for the integers, such as the Golomb, Rice, or Elias codes. The most-common symbols are assigned the shortest codes and the overhead in this case is the sorting information. This information—which is a permutation of the symbols, each represented by its variable-length code—must be included in the compressed file for the decoder's use.

In either case, the compressed file consists of three parts, the final string (where each symbol is represented by its variable-length code), the phrase-table (which must also be compressed), and the overhead.

The Nakamura Murashima Method

The Nakamura Murashima method ([Nakamura and Murashima 91] and [Nakamura and Murashima 96]) also employs new symbols, but encodes the phrase-table differently. The authors propose two variants, dubbed SCT and SED. The former is straightforward. Given the 4-symbol alphabet a, b, c, and d and the source data string acabbadcaddbaaddcaaddb, SCT encodes it as follows:

Pair	String
	acabbadcaddbaaddcaaddb
A → ad	acabbAcAdbaAdcaAdb
B → Ad	acabbAcBbaBcaBb
C → ca	aCbbAcBbaBCBb
D → Bb	aCbbAcDaBCD

SCT encodes the phrase-table in a simple way and prepends it to the output. It generates a string that includes two symbols for each table entry. The names of the new symbols are not included in this string and are assumed to be A, B, C, and so on. Thus, our phrase-table is encoded as the 8-symbol string adAdcaBb and the complete encoder output is the string adAdcaBb|aCbbAcDaBCD, where the vertical bar is a special separator symbol. The decoder reconstructs the phrase-table easily. It reads pairs of symbols, assigns the first pair to the new symbol A, the second pair to new symbol B, and so on until it reaches the separator.

The SED method is more complex. It constructs a different phrase-table and its output is a string of (original and new) symbols where flags distinguish between symbols and phrase-table entries. Each symbol in the output string is preceded by such a flag (a bit). A single symbol (either from the original alphabet or a new symbol) is preceded by a zero, while a phrase-table entry is preceded by a 1. Thus, the already-familiar input string acabbadcaddbaaddcaaddb becomes 0a10c0a0b0b10a0d0c110β0d0b0a0γ0α0δ (where Greek letters stand for the new symbols). Here is why. The first input symbol a becomes 0a, but the following pair ca appears several times, so it becomes 10c0a, where the "1" indicates a new symbol, α. The next pair bb appears only once, so it becomes 0b0b, but the following pair ad repeats several times, so it becomes 10a0d, where the "1" indicates the next new symbol β. The single c that follows becomes 0c. Notice that the encoder does not process the pair ca, because the c is followed by the pair ad, which has already been replaced by β. Thus, the next substring addb becomes βdb, and then γb (where the new symbol γ is add), and finally the new symbol $\delta = \gamma\mathtt{b} = \beta\mathtt{db} = \mathtt{addb}$. This is encoded as the string 1(10β0d)0b (without the paretheses). The rest of the encoding is easy to follow.

There remains the question of how to write a hybrid string (with bits and symbols mixed up) such as 0a10c0a0b0b10a0d0c110β0d0b0a0γ0α0δ on the output. The authors of this algorithm mention three different methods that employ adaptive arithmetic coding and complete binary trees, but no details are given.

I would like to acknowledge the help provided by Hirofumi Nakamura in the preparation of this section.

6.31 DCA, Compression with Antidictionaries

A dictionary-based compression method is based on a dictionary; a collection of bits and pieces of data found in the input. If the encoder locates the next input string a in the dictionary, it compresses a by emitting a pointer to its location in the dictionary. An antidictionary method is based on an inverse kind of knowledge. Such a method maintains an antidictionary with strings that *do not* appear in the input. Using the antidictionary, the encoder can often predict the next data symbol a with certainty, so that a doesn't have to be output. This is how such a method results in compression. The decoder can mimic the steps of the encoder and can therefore determine many data symbols and place them in the output. This section is based on [Crochemore et al. 00], who dubbed their approach DCA (data compression, antidictionaries). Reference [Davidson and Ilie 04] takes the basic idea further and describes algorithms for fast encoding and decoding.

> Antidictionary: This book would explain the disadvantages and hazards of using the word you looked up, and try to persuade you not to have anything to do with it. For example, under "solution," it could say, "A term widely used by apish technology-floggers to avoid the need to think of actual descriptive words for products and services." Imagine what it could say under "country music."
>
> `http://www.halfbakery.com/idea/Anti-Dictionary#1096876321`

An antidictionary method employs a binary alphabet where the basic data symbols are 0 and 1. We denote the input data (a bitstring) by w and assume that there exists an antidictionary AD with bitstrings that are not found in w (forbidden bitstrings). Notice that AD does not have to contain *every* forbidden bitstring. The encoder scans the input w bit by bit from left to right. We denote the part that has been scanned so far (the prefix of the input) by v. Assume that at a certain point, v is the bitstring 10110 and the next (still unknown) bit to be scanned is b. The encoder examines all the suffixes s of v which are 0, 10, 110, 0110, and 10110. If any of the strings $s0$ (i.e., a suffix s followed by a zero) is found in the antidictionary, then it is a forbidden string and the encoder knows that b must be 1. Similarly, if any of the strings $s1$ is in the AD, then b must be 0. In either case, b can be omitted from the output because the decoder can determine its value by searching the antidictionary in lockstep with the encoder. If neither $s0$ nor $s1$ is located in AD, then b is read by the encoder, is appended to v, and is also output because the decoder will not be able to determine it.

Given the input string $w = 10110101$, it is easy to construct the set $F(w)$ of substrings in w (we denote the empty string by ε)

$$F(w) = (\varepsilon, 0, 1, 01, 10, 11, 011, 101, 110, 0101, 0110, 1010, 1011, 1101, \ldots, 10110101).$$

An antidictionary AD for w is any set of bitstrings that are not in $F(w)$. Such bitstrings are forbidden in w. For example, $(00, 000, 001, 010, 100, 111, 0000, 0001)$ is such a set. Assuming that such an antidictionary exists, both encoder and decoder can easily be described and understood. Figure 6.47 is a pseudocode listing of the encoder.

```
ADC_Encoder(w, AD, γ)
1. v ← ε; γ ← ε;
2. for a ← first to last bit of w
3.    if for every suffix s of v, neither s0 nor s1 is in AD
4.        then γ ← γ.a; /* no compression */
5.    endif;
6.    v ← v.a;
7. endfor
8. return γ;
```

Figure 6.47: A Simple DCA Encoder.

As an example, we assume the input string $w = 10110101$ and the antidictionary $AD = (00, 000, 111, 11011)$. Table 6.48 lists the eight encoding steps, the values of γ and v at the end of each step, and the suffixes that have to be checked against AD at each step. Notice that the suffixes checked in each step are those of v of the previous step, because v is updated (in step 6) after the suffix check. Each suffix found in the AD (i.e., each forbidden string) is underlined, and it is obvious that each step with an underlined suffix leaves γ unchanged and thus increases compression (the difference in length between γ and v) by one bit. At the end of the loop, variable v equals the original input w, which suggests that there is really no need to have v as an independent variable. v is always a prefix of w and it is enough to maintain a pointer to w that will point to the current bit (the rightmost bit of v) at each step.

	a	γ	v	Suffix pairs $s0$, $s1$ to examine
Initial		ε	ε	
1.	1	1	1	$\varepsilon 0$, $\varepsilon 1$
2.	0	10	10	10, 11
3.	1	10	101	$\underline{00}$, 01, 100, 101
4.	1	101	1011	10, 11, 010, 011, 1010, 1011
5.	0	101	10110	10, 11, 110, $\underline{111}$, 0110, 0111, 10110, 10111
6.	1	101	101101	$\underline{00}$, 01, 100, 101, 1100, 1101, 01100, 01101, 101100, ...
7.	0	101	1011010	10, 11, 010, 011, 1010, 1011, 11010, $\underline{11011}$, 011010, ...
8.	1	101	10110101	$\underline{00}$, 01, 100, 101, 0100, 0101, 10100, 10101, 110100, ...

Table 6.48: DCA Encoding of $w = 10110101$.

The compressed stream that is output by the encoder must consist of (1) the AD, (2) the length n of w, and (3) bitstring γ. Notice that n is needed by the decoder because different input strings w may produce the same γ when processed by the encoder of Figure 6.47 (this is illustrated by the decoding listing of Table 6.50). The antidictionary must also be included in the encoder's output, because it is static and has to be used by the decoder. (Notice that the entire input string w must be input in order to construct the antidictionary, which implies that DCA is a two-pass method.) Naturally, including

AD in the compressed stream reduces the effectiveness of this method, and suggests that an algorithm to construct an effective AD should be a crucial part of any practical DCA implementation. Alternatively, a dynamic approach is also possible, where the AD starts empty and is populated by forbidden strings as the encoder (and in lockstep, also the decoder) scans more and more input bits. At the very least, the AD should be compressed before it is written on the output. Notice that an AD is specific to the input data, which is why a general AD, based on training documents, cannot be used.

The DCA decoder is simple. It receives an AD, the length n of the original data w, and the compressed string γ. The decoder loops n times. In each iteration, it has a prefix v of w from previous iterations. It examines all the suffixes s of v. If a bitstring $s0$ is in the AD, then the next reconstructed bit must be a 1. Similarly, if $s1$ is found in the antidictionary, then the next bit must be 0. The next bit is then appended to v. (The notation $|v|$ denotes the length of string v, while $\bar{b}$ denotes the complement of bit b.) If neither $s0$ nor $s1$ is in the antidictionary, the next bit of γ is appended to v. In either case, v grows by one bit in the iteration. Figure 6.49 is a pseudocode listing of this simple decoder.

```
ADC_Decoder(AD, n, γ, v)
v ← ε;
while |v| < n
   if for a suffix s of v and a bit b, string s.b is in AD
     then v ← v.b̄;
     else v ← v.(next bit of γ);
   endif;
endwhile;
return v;
```

Figure 6.49: A Simple DCA Decoder.

Table 6.50 lists the steps performed by this decoder when it is given the same antidictionary and is asked to perform eight iterations with $\gamma = 101$ (when a bit of γ is used, it is underlined). Notice that the last (least-significant) bit of γ is used in iteration 4, which is why the decoder has to be given the length of the final, reconstructed bitstring.

Here are some observations regarding antidictionary design. If a substring s of the input w is forbidden, but a proper substring u of s is also forbidden, then u but not s should be included in the antidictionary. In general, only *minimal* forbidden strings (strings where no substring is also forbidden) should appear in an antidictionary. In our examples above, the antidictionary contains 000 but this string is never used because it is not minimal. Also, most matches between suffixes $s0$, $s1$ and antidictionary strings occur for short suffixes s. Thus, it is more important to include short forbidden strings in an AD, which keeps it short. In practice, the length of an input string may be many thousands of bits, whereas the length of the antidictionary may be a few hundred bits. Still, it is important to construct an antidictionary by an efficient algorithm and also to look for ways to compress the AD.

	v	$\lvert v \rvert$	γ	Suffix pairs $s0$, $s1$ to examine
1.	$\varepsilon \to 1$	0	$\underline{1}01$	none
2.	$1 \to 10$	1	$1\underline{0}1$	10, 11
3.	$10 \to 101$	2		$\underline{00}$, 01, 100, 101
4.	$101 \to 1011$	3	$10\underline{1}$	10, 11, 010, 011, 1010, 1011
5.	$1011 \to 10110$	4		10, 11, 110, $\underline{111}$, 0110, 0111, 10110, 10111
6.	$10110 \to 101101$	5		$\underline{00}$, 01, 100, 101, 1100, 1101, 01100, 01101, ...
7.	$101101 \to 1011010$	6		10, 11, 010, 011, 1010, 1011, 11010, $\underline{11011}$, ...
8.	$1011010 \to 10110101$	7		$\underline{00}$, 01, 100, 101, 0100, 0101, 10100, 10101, ...

Table 6.50: DCA Decoding of $\gamma = 101$ and $n = 8$.

The main reference of DCA is [Crochemore et al. 00] and it discusses ways to prune the antidictionary and to compress it. The developers observe the following. Given an antidictionary AD with only minimal forbidden strings, if we remove a string v from AD, the remaining strings constitute an antidictionary for v. Thus, an antidictionary can be compressed by removing each of its strings in turn and then using the remaining antidictionary to compress the string; a process that can be termed self compression.

The developers of this method also describe an algorithm to construct an AD. Unfortunately, this algorithm is complex, it is based on finite-state automata, and its construction is beyond the scope of this book. Instead, we show an example (after [Davidson and Ilie 04]) illustrating how the automaton that corresponds to an antidictionary for a given bitstring w can be used to create two state diagrams (also referred to as transducers) that simplify the encoding and decoding of w.

Given the input string $w = 11010001$, the AD-creation algorithm (not described here) produces the automaton of Figure 6.51a. Each state in such an automaton has two outgoing transitions, for 0 and 1. If one of the transitions of state A leads to a non-final state (a square), then A is referred to as a predict state and is shown in gray in the figure. Once this automaton has been obtained, it is used to create the antidictionary $AD = (0000, 111, 011, 0101, 1100)$ in the following way. We start at state 0 and work our way to one of the non-final states (shown as squares), collecting bits off the edges on our way and concatenating them to form a bitstring. When we get to a non-final state, the bitstring is added to AD as the next forbidden string. Once we have visited all the non-final states, the AD is complete.

With the AD in hand, the same automaton is used to construct a state diagram (a transducer) for encoding $w = 11010001$ (Figure 6.51b). This transducer has the following property: For each outgoing transition from a non-predict state (a white circle), the output equals the input. Thus, in our example, we enter state 0 with an input of 1 (because w starts with 1), so we have to leave it with the same output, which takes us to state 2. The input bit at this point is the second 1 of w, so we leave state 2 with an identical output and find ourselves in state 5. The next input bit is 0, so the output is ε and the transducer sends us to state 9. The entire encoding sequence can be summarized by

$$0 \xrightarrow{1/1} 2 \xrightarrow{1/1} 5 \xrightarrow{0/\varepsilon} 9 \xrightarrow{1/\varepsilon} 4 \xrightarrow{0/\varepsilon} 7 \xrightarrow{0/\varepsilon} 3 \xrightarrow{0/0} 6 \xrightarrow{1/\varepsilon} 4$$

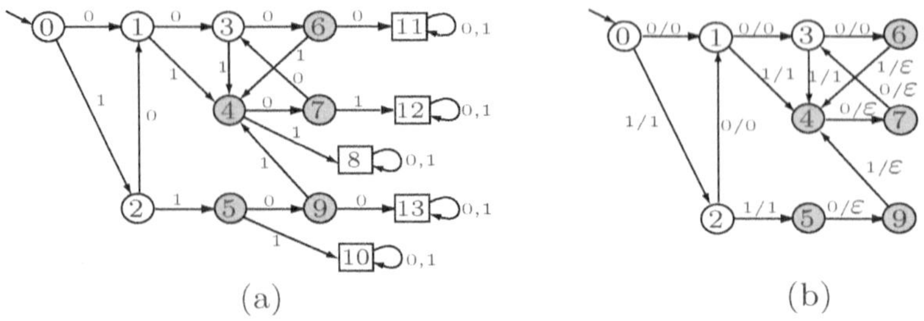

Figure 6.51: Automaton and Transducer for Fast Encoding and Decoding.

and the encoding produces (in states 0, 2, and 3) the output string 110.

Notice that state 5 has only one output transition (corresponding to an input of 0). The reason is that we can arrive at this state only after scanning the substring 11 from the input, and the AD tells us that this substring cannot be followed by a 0.

The same transducer is used for decoding. The only change needed is to swap the input and output bits on each transition edge. Thus, for example, the transition from state 5 to state 9 will now be labeled $\varepsilon/0$. Decoding string 111 is done in eight steps (notice that the decoder requires the length of w) as follows

$$0 \xrightarrow{1/1} 2 \xrightarrow{1/1} 5 \xrightarrow{\varepsilon/0} 9 \xrightarrow{\varepsilon/1} 4 \xrightarrow{\varepsilon/0} 7 \xrightarrow{\varepsilon/0} 3 \xrightarrow{0/0} 6 \xrightarrow{\varepsilon/1} 4.$$

It seems that the idea of using antidictionaries for compression has caught the attention of many researchers. Here are just two interesting contributions to this area. Reference [Crochemore and Navarro 02] introduces the notion of almost antifactors, which are strings that rarely appear in the text. More precisely, almost antifactors are strings that improve compression if they are considered forbidden and are added to the antidictionary. In [Ota and Morita 07], the authors present an algorithm (linear in space and time) to construct an antidictionary from a suffix tree.

6.32 CRC

The idea of a parity bit is simple, old, and familiar to most computer practitioners. A parity bit is the simplest type of error detecting code. It adds reliability to a group of bits by making it possible for hardware to detect certain errors that may occur when the group is stored in memory, is written on a disk, or is transmitted over communication lines between computers. A single parity bit does not make the group completely reliable. There are certain errors that cannot be detected with a parity bit, but experience shows that even a single parity bit can make data transmission reliable in most practical cases.

The parity bit is computed from a group of $n-1$ bits, then added to the group, making it n bits long. A common example is a 7-bit ASCII code that becomes 8 bits long after a parity bit is added. The parity bit p is computed by counting the number of 1's in the original group, and setting p to complete that number to either odd or even. The former is called odd parity, and the latter is called even parity.

Examples: Given the group of 7 bits 1010111, the number of 1's is five, which is odd. Assuming odd parity, the value of p should be 0, leaving the total number of 1's odd. Similarly, the group 1010101 has four 1's, so its odd parity bit should also be a 1, bringing the total number of 1's to five.

Imagine a block of data where the most significant bit (MSB) of each byte is an odd parity bit, and the bytes are written vertically (Table 6.52a).

1 01101001	1 01101001	1 01101001	1 01101001
0 00001011	0 00001011	0 00001011	0 00001011
0 11110010	0 11**0**10010	0 11**0**10**1**10	0 11**0**10**1**10
0 01101110	0 01101110	0 01101110	0 01101110
1 11101101	1 11101101	1 11101101	1 11101101
1 01001110	1 01001110	1 01001110	1 01001110
0 11101001	0 11101001	0 11101001	0 11101001
1 11010111	1 11010111	1 11010111	1 11010111
			0 00011100
(a)	(b)	(c)	(d)

Table 6.52: Horizontal and Vertical Parities.

When this block is read from a disk or is received by a computer, it may contain transmission errors, errors that have been caused by imperfect hardware or by electrical interference during transmission. We can think of the parity bits as *horizontal reliability*. When the block is read, the hardware can check every byte, verifying the parity. This is done by simply counting the number of 1's in the byte. If this number is odd, the hardware assumes that the byte is good. This assumption is not always correct, since two bits may get corrupted during transmission (Table 6.52c). A single parity bit is therefore useful (Table 6.52b) but does not provide full error detection capability.

A simple way to increase the reliability of a block of data is to compute vertical parities. The block is considered to be eight vertical columns, and an odd parity bit is computed for each column (Table 6.52d). If two bits in a byte go bad, the horizontal parity will not catch it, but two of the vertical ones will. Even the vertical bits do not provide complete error detection capability, but they are a simple way to significantly improve data reliability.

A CRC is a glorified vertical parity. CRC stands for Cyclical Redundancy Check (or Cyclical Redundancy Code) and is a rule that shows how to compute the vertical check bits (they are now called check bits, not just simple parity bits) from all the bits of the data. Here is how CRC-32 (one of the many standards developed by the CCITT) is computed. The block of data is written as one long binary number. In our example this will be the 64-bit number

101101001|000001011|011110010|001101110|111101101|101001110|011101001|111010111.

The individual bits are considered the coefficients of a *polynomial* (see below for definition). In our example, this will be the degree-63 polynomial

$$\begin{aligned}P(x) &= 1\times x^{63}+0\times x^{62}+1\times x^{61}+1\times x^{60}+\cdots+1\times x^2+1\times x^1+1\times x^0\\ &= x^{63}+x^{61}+x^{60}+\cdots+x^2+x+1.\end{aligned}$$

This polynomial is then divided by the standard CRC-32 *generating polynomial*

$$\mathrm{CRC}_{32}(x)=x^{32}+x^{26}+x^{23}+x^{22}+x^{16}+x^{12}+x^{11}+x^{10}+x^8+x^7+x^5+x^4+x^2+x^1+1.$$

When an integer M is divided by an integer N, the result is a quotient Q (which we will ignore) and a remainder R, which is in the interval $[0, N-1]$. Similarly, when a polynomial $P(x)$ is divided by a degree-32 polynomial, the result is two polynomials, a quotient and a remainder. The remainder is a degree-31 polynomial, which means that it has 32 coefficients, each a single bit. Those 32 bits are the CRC-32 code, which is appended to the block of data as four bytes. As an example, the CRC-32 of a recent version of the file with the text of this chapter is $\texttt{586DE4FE}_{16}$.

Selecting a generating polynomial is more an art than science. Page 196 of [Tanenbaum 02] is one of a few places where the interested reader can find a clear discussion of this topic.

The CRC is sometimes called the "fingerprint" of the file. Of course, since it is a 32-bit number, there are only 2^{32} different CRCs. This number equals approximately 4.3 billion, so, in theory, there may be different files with the same CRC, but in practice this is rare. The CRC is useful as an error detecting-code because it has the following properties:

1. Every bit in the data block is used to compute the CRC. This means that changing even one bit may produce a different CRC.
2. Even small changes in the data normally produce very different CRCs. Experience with CRC-32 shows that it is very rare that introducing errors in the data does not modify the CRC.
3. Any histogram of CRC-32 values for different data blocks is flat (or very close to flat). For a given data block, the probability of any of the 2^{32} possible CRCs being produced is practically the same.

Other common generating polynomials are $\mathrm{CRC}_{12}(x) = x^{12}+x^3+x+1$ and $\mathrm{CRC}_{16}(x) = x^{16}+x^{15}+x^2+1$. They generate the common CRC-12 and CRC-16 codes, which are 12 and 16 bits long, respectively.

Definition: A polynomial of degree n in x is the function

$$P_n(x)=\sum_{i=0}^{n} a_i x^i = a_0+a_1x+a_2x^2+\cdots+a_nx^n,$$

where a_i are the $n+1$ coefficients (in our case, real numbers).

Two simple, readable references on CRC are [Ramabadran and Gaitonde 88] and [Williams 93].

6.33 Summary

The dictionary-based methods presented here are different but are based on the same principle. They read the input stream symbol by symbol and add phrases to the dictionary. The phrases are symbols or strings of symbols from the input. The main difference between the methods is in deciding what phrases to add to the dictionary. When a string in the input stream matches a dictionary phrase, the encoder outputs the position of the match in the dictionary. If that position requires fewer bits than the matched string, compression results.

In general, dictionary-based methods, when carefully implemented, give better compression than statistical methods. This is why many popular compression programs are dictionary based or employ a dictionary as one of several compression steps.

6.34 Data Compression Patents

It is generally agreed that an invention or a process is patentable but a mathematical concept, calculation, or proof is not. An algorithm seems to be an abstract mathematical concept that should not be patentable. However, once the algorithm is implemented in software (or in firmware) it may not be possible to separate the algorithm from its implementation. Once the implementation is used in a new product (i.e., an invention), that product—including the implementation (software or firmware) and the algorithm behind it—may be patentable. [Zalta 88] is a general discussion of algorithm patentability. Several common data compression algorithms, most notably LZW, have been patented; and the LZW patent is discussed here in some detail.

The Sperry Corporation was granted a patent (4,558,302) on LZW in December 1985 (even though the inventor, Terry Welch, left Sperry prior to that date). When Unisys acquired Sperry in 1986 it became the owner of this patent and required users to obtain (and pay for) a license to use it.

When CompuServe designed the GIF format in 1987 it decided to use LZW as the compression method for GIF files. It seems that CompuServe was not aware at that point that LZW was patented, nor was Unisys aware that the GIF format uses LZW. After 1987 many software developers became attracted to GIF and published programs to create and display images in this format. GIF became widely accepted and was commonly used on the World-Wide Web, where it was one of the prime image formats for Web pages and browsers.

It was not until GIF had become a world-wide de facto standard that Unisys contacted CompuServe for a license. Naturally, CompuServe and other LZW users tried to challenge the patent. They applied to the United States Patent Office for a reexamination of the LZW patent, with the (perhaps surprising) result that on January 4, 1994, the patent was reconfirmed and CompuServe had to obtain a license (for an undisclosed sum) from Unisys later that year. Other important licensees of LZW (see [Rodriguez 95]) were Aldus (in 1991, for the TIFF graphics file format), Adobe (in 1990, for PostScript level II), and America Online and Prodigy (in 1995).

The Unisys LZW patent had significant implications for the World-Wide Web, where use of GIF format images was currently widespread. Similarly, the Unix `compress` utility uses LZW and therefore required a license. In the United States, the patent expired on 20 June 2003 (20 years from the date of first filing). In Europe (patent EP0129439) expired on 18 June 2004. In Japan, patents 2,123,602 and 2,610,084 expired on 20 June 2004, and in Canada, patent CA1223965 expired on 7 July 2004.

Unisys graciously exempted old software products (those written or modified before January 1, 1995) from a patent license. Also exempt was any noncommercial and nonprofit software, old and new. Commercial software (even shareware) or firmware created after December 31, 1994, had to be licensed if it supported the GIF format or implemented LZW. A similar policy was enforced with regard to TIFF, where the cutoff date is July 1, 1995. Notice that computer users could legally keep and transfer GIF and any other files compressed with LZW; only the compression/decompression software required a license.

For more information on the Unisys LZW patent and license see [unisys 03].

An alternative to GIF is the Portable Network Graphics, PNG (pronounced "ping," Section 6.27) graphics file format [Crocker 95], which was developed expressly to replace GIF, and avoid patent claims. PNG is simple, portable, with source code freely available, and is unencumbered by patent licenses. It has potential and promise in replacing GIF. However, any GIF-to-PNG conversion software required a Unisys license.

The GNU `gzip` compression software (Section 6.14) should also be mentioned here as a popular substitute for `compress`, since it is free from patent claims, is faster, and provides superior compression.

The LZW U.S. patent number is 4,558,302, issued on Dec. 10, 1985. Here is the abstract filed as part of it (the entire filing constitutes 50 pages).

> A data compressor compresses an input stream of data character signals by storing in a string table strings of data character signals encountered in the input stream. The compressor searches the input stream to determine the longest match to a stored string. Each stored string comprises a prefix string and an extension character where the extension character is the last character in the string and the prefix string comprises all but the extension character. Each string has a code signal associated therewith and a string is stored in the string table by, at least implicitly, storing the code signal for the string, the code signal for the string prefix and the extension character. When the longest match between the input data character stream and the stored strings is determined, the code signal for the longest match is transmitted as the compressed code signal for the encountered string of characters and an extension string is stored in the string table. The prefix of the extended string is the longest match and the extension character of the extended string is the next input data character signal following the longest match. Searching through the string table and entering extended strings therein is effected by a limited search hashing procedure. Decompression is effected by a decompressor that receives the compressed code signals and generates a string table similar to that constructed by the compressor to effect lookup of received code signals so as to recover the data character signals comprising a stored string. The decompressor string table is updated by storing a string having a prefix in accordance with a prior received code signal and an extension character in accordance with the first character of the currently recovered string.

Here are a few other patented compression methods, some of them mentioned elsewhere in this book:
1. "Textual Substitution Data Compression with Finite Length Search Windows," U.S. Patent 4,906,991, issued March 6, 1990 (the LZFG method).
2. "Search Tree Data Structure Encoding for Textual Substitution Data Compression Systems," U.S. Patent 5,058,144, issued Oct. 15, 1991.

The two patents above were issued to Edward Fiala and Daniel Greene.
3. "Apparatus and Method for Compressing Data Signals and Restoring the Compressed Data Signals." This is the LZ78 patent, assigned to Sperry Corporation by the inventors Willard L. Eastman, Abraham Lempel, Jacob Ziv, and Martin Cohn. U.S. Patent 4,464,650, issued August, 1984.

The following, from Ross Williams `http://www.ross.net/compression/` illustrates how thorny this issue of patents is.

> Then, just when I thought all hope was gone, along came some software patents that drove a stake through the heart of the LZRW algorithms by rendering them unusable. At last I was cured! I gave up compression and embarked on a new life, leaving behind the world of data compression forever.

Appropriately, Dr Williams maintains a list [patents 06] of data compression-related patents.

> Your patent application will be denied. Your permits will be delayed. Something will force you to see reason-and to sell your drug at a lower cost.
> Wu had heard the argument before. And he knew Hammond was right, some new bioengineered pharmaceuticals had indeed suffered inexplicable delays and patent problems.
> You don't even know exactly what you have done, but already you have reported it, patented it, and sold it.
>
> —Michael Crichton, *Jurassic Park* (1991)

6.35 A Unification

Dictionary-based methods and methods based on prediction approach the problem of data compression from two different directions. Any method based on prediction predicts (i.e., assigns probability to) the current symbol based on its order-N context (the N symbols preceding it). Such a method normally stores many contexts of different sizes in a data structure and has to deal with frequency counts, probabilities, and probability ranges. It then uses arithmetic coding to encode the entire input stream as one large number. A dictionary-based method, on the other hand, works differently. It identifies the next phrase in the input stream, stores it in its dictionary, assigns it a code, and continues with the next phrase. Both approaches can be used to compress data because each obeys the general law of data compression, namely, to assign short codes to common events (symbols or phrases) and long codes to rare events.

On the surface, the two approaches are completely different. A predictor deals with probabilities, so it can be highly efficient. At the same time, it can be expected to be slow, since it deals with individual symbols. A dictionary-based method deals with strings of symbols (phrases), so it gobbles up the input stream faster, but it ignores correlations between phrases, typically resulting in poorer compression.

The two approaches are similar because a dictionary-based method *does use* contexts and probabilities (although implicitly) just by storing phrases in its dictionary and searching it. The following discussion uses the LZW trie to illustrate this concept, but the argument is valid for any dictionary-based method, no matter what the details of its algorithm and its dictionary data structure.

Imagine the phrase `abcdef...` stored in an LZW trie (Figure 6.53a). We can think of the substring `abcd` as the order-4 context of `e`. When the encoder finds another occurrence of `abcde...` in the input stream, it will locate our phrase in the dictionary, parse it symbol by symbol starting at the root, get to node `e`, and continue from there, trying to match more symbols. Eventually, the encoder will get to a leaf, where it will add another symbol and allocate another code. We can think of this process as adding a new leaf to the subtree whose root is the `e` of `abcde...`. Every time the string `abcde` becomes the prefix of a parse, both its subtree and its code space (the number of codes associated with it) get bigger by 1. It therefore makes sense to assign node `e` a probability depending on the size of its code space, and the above discussion shows that the size of the code space of node `e` (or, equivalently, string `abcde`) can be measured by counting the number of nodes of the subtree whose root is `e`. This is how probabilities can be assigned to nodes in any dictionary tree.

The ideas of Glen Langdon in the early 1980s (see [Langdon 83] but notice that his equation (8) is wrong; it should read $P(y|s) = c(s)/c(s \cdot y)$; [Langdon 84] is perhaps more useful) led to a simple way of associating probabilities not just to nodes but also to edges in a dictionary tree. Assigning probabilities to edges is more useful, since the edge from node `e` to node `f`, for example, signifies an `f` whose context is `abcde`. The probability of this edge is thus the probability that an `f` will follow `abcde` in the input stream. The fact that these probabilities can be calculated in a dictionary tree shows that every dictionary-based data compression algorithm can be "simulated" by a prediction algorithm (but notice that the converse is not true). Algorithms based on prediction are, in this sense, more general, but the important fact is that these two seemingly different classes of compression methods can be unified by the observations listed here.

The process whereby a dictionary encoder slides down from the root of its dictionary tree, parsing a string of symbols, can now be given a different interpretation. We can visualize it as a sequence of making predictions for individual symbols, computing codes for them, and combining the codes into one longer code, which is eventually written on the compressed stream. It is as if the code generated by a dictionary encoder for a phrase is actually made up of small chunks, each a code for one symbol.

The rule for calculating the probability of the edge $e \to f$ is to count the number of nodes in the subtree whose root is `f` (including node `f` itself) and divide by the number of nodes in the subtree of `e`. Figure 6.53b shows a typical dictionary tree with the strings `aab`, `baba`, `babc`, and `cac`. The probabilities associated with every edge are also shown and should be easy for the reader to verify. Note that the probabilities of sibling subtrees don't add up to 1. The probabilities of the three subtrees of the root, for example, add up to 11/12. The remaining 1/12 is assigned to the root itself and

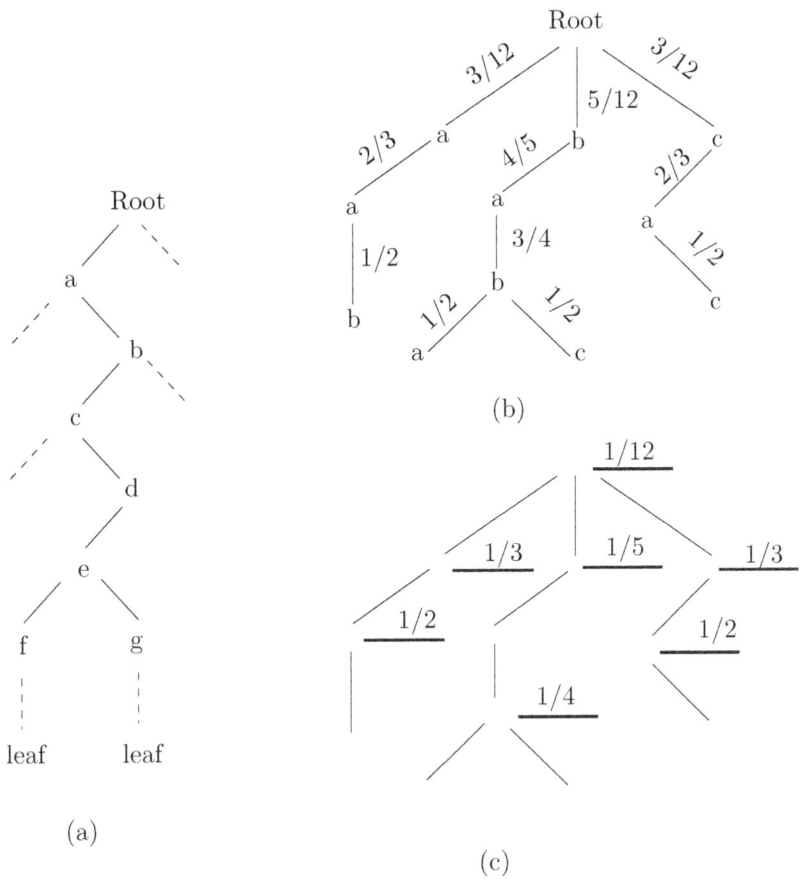

Figure 6.53: Defining Probabilities in a Dictionary Tree.

represents the probability that a fourth string will eventually start at the root. These "missing probabilities" are shown as horizontal lines in Figure 6.53c.

The two approaches, dictionary and prediction, can be combined in a single compression method. The LZP method of Section 6.19 is one example; the LZRW4 method (Section 6.12) is another. These methods work by considering the context of a symbol before searching the dictionary.

> The only place where housework comes
> before needlework is in the dictionary.
> —Mary Kurtz

7
Image Compression

The first part of this chapter discusses the basic features and types of digital images and the main approaches to image compression. This is followed by a description of about 30 different compression methods. We would like to start with the following observations:

1. Why were these particular methods included in the book, while others were left out? The simple answer is: Because of the documentation available to the authors. Image compression methods that are well documented were included. Methods that are proprietary, or whose documentation was not clear to the authors, were left out.
2. The treatment of the various methods is uneven. This, again, reflects the documentation available to the authors. Some methods have been documented by their developers in great detail, and this is reflected in this chapter. Where no detailed documentation was available for a compression algorithm, only its basic principles are outlined here.
3. There is no attempt to compare the various methods described here. This is because most image compression methods have been designed for a specific type of image, and also because of the practical difficulties of obtaining all the software and adapting it to run on the same platform.
4. The compression methods described in this chapter are not arranged in any particular order. After much thought and many trials, the authors gave up any hope of sorting the compression methods in any reasonable way. Readers looking for any particular method may consult the table of contents and the detailed index to easily locate it.

A digital image is a rectangular array of dots, or picture elements, arranged in m rows and n columns. The expression $m \times n$ is called the *resolution* of the image, and the dots are called *pixels* (except in the cases of fax images and video compression, where they are referred to as *pels*). The term "resolution" is sometimes also used to indicate the number of pixels per unit length of the image. Thus, dpi stands for dots per inch.

D. Salomon, G. Motta, *Handbook of Data Compression*, 5th ed.
DOI 10.1007/978-1-84882-903-9_7, © Springer-Verlag London Limited 2010

7.1 Pixels

Images are all around us. We see them in color and in high resolution. Many objects (especially artificial objects) seem perfectly smooth, with no jagged edges and no graininess. Computer graphics, on the other hand, deals with images that consist of small dots, pixels. The term pixel stands for "picture element" (see [Lyon 09] for a lively survey of the history of this important term). When we first hear of this feature of computer graphics, we tend to dismiss the entire field as trivial. It seems intuitively obvious that an image that consists of dots would always look pixelated, grainy, rough, and inferior to what we see with our eyes. Yet state-of-the-art computer-generated images are often difficult or impossible to distinguish from their real counterparts, even though they are discrete, made of pixels, and not continuous. (See also Page 526 for a discussion of human vision and the resolution of the eye.)

A similar dichotomy between discrete and continuous exists in art. Many painters try to mimic nature and employ wide, continuous brush strokes to paint smooth and continuous pictures, while others choose to be pointillists. They create a painting by placing many small dots on their canvas. The most important pointillist was the 19th century French impressionist Georges Seurat.

Seurat was a leader in the late 19th century neo-impressionism movement, a school of painting that uses tiny brushstrokes of contrasting colors to achieve a delicate play of light and create subtle changes in form. Seurat used this technique, which became known as pointillism or divisionism, to create large paintings made entirely of small dots of pure color. The dots are too small to be distinguished when looking at the work in its entirety, but they make his paintings shimmer with brilliance. His most well-known works are *Une Baignade* (1883–84) and *Un dimanche après-midi à l'Ile de la Grande Jatte* (1884–86). The art critic Arsène Alexandre had this to say about the latter painting: "Everything was so new in this immense painting—the conception was bold and the technique one that nobody had ever seen or heard before. This was the famous pointillism."

Most engineers, programmers, and users think of pixels as small squares, and this is generally true for pixels on computer monitors. Pixels in other digital output devices (displays or printers) may be rectangular or circular. However, in principle, a pixel should be considered a mathematical, dimensionless, point (see [Smith 09]). It seems impossible to reconstruct a continuous image from an array of discrete pixels, but this is precisely what the surprising Nyquist-Shannon sampling theorem [Wikipedia 03] tells us (in fact, what it guarantees). Section 10.2 (and also page 741) discuss the application of this theorem to digitized audio (a one-dimensional signal), while here we apply it to two-dimensional images.

Audio is a good starting point to understand the sampling theorem. Sound fed into a microphone is converted to an electrical voltage that varies with time; it becomes a wave. A wave has a frequency, and a wave that varies all the time consists of many frequencies. We denote the maximum frequency contained in a wave by B (cycles per second, or Hertz). The samping theorem says that it is possible to reconstruct the original wave if it is sampled at a rate greater than $2B$ samples per second.

An image is a rectilinear array of point samples (pixels). The sampling theorem guarantees that we'll be able to reconstruct the image (i.e., to compute the color of every mathematical point in the image) if we sample the image at a rate greater than $2B$ pixels per unit length, where B is the maximum pixel frequency in the image. Section 7.7 explains the meaning of the term "pixel frequencies in an image," but in practice, pixels, their values, and their frequencies depend on the accuracy of the capturing device. An ideal device should measure the color of an image at certain points, but image sensors (CCDs and CMOS) used in real devices (cameras and scanners) are often far from ideal. Because of physical limitations, manufacturing defects, and the need to capture enough light, an image sensor often measures the average color (or intensity) of a small area of the image, instead of the color at a point.

Assuming that we have enough pixels for a given digital image, we compute the color of a given point in the image by interpolation. Section 9.11.7 discusses bilinear interpolation and reference [Salomon 06] discusses this and other interpolation methods and illustrates them with examples. Here, we only touch on the principles of bilinear and bicubic interpolations. The discussion assumes a grayscale image, where each pixel is a number indicating a shade of gray (an intensity), but interpolation can easily be extended to color images, where a pixel is a triplet of primary colors.

Bilinear interpolation. Given enough pixels of an image and given a point Q in the image, we use bilinear interpolation to compute the intensity (grayscale) of the image at Q in the following steps:

1. We select the four pixels surrounding Q, denote their values by a, b, c, and d (Figure 7.1), and convert them to three-dimensional points by considering the value of a pixel the z coordinate of the point and adding appropriate x and y coordinates. We end up with the points $P_{00} = (0, 0, a)$, $P_{01} = (0, 1, b)$, $P_{11} = (1, 1, c)$, and $P_{10} = (1, 0, d)$.

2. We compute the parametric equations of the straight segment $L_1(u)$ running from P_{00} to P_{10} and the segment $L_2(u)$ running from P_{01} to P_{11}. These equations are the simple linear interpolations $L_1(u) = P_{00}(1-u)+P_{10}u$ and $L_2(u) = P_{01}(1-u)+P_{11}u$.

3. We compute the parametric equation $P_{u,w}$ of the bilinear surface generated by interpolating segments $L_1(u)$ and $L_2(u)$. The equation is

$$P(u, w) = L_1(u)(1-w)+L_2(u)w = P_{00}(1-u)(1-w)+P_{10}u(1-w)+P_{01}(1-u)w+P_{11}uw.$$

Figure 7.1 (see also Figure 9.63) is an example of such a surface. The entire surface is spanned when parameters u and w are varied independently in the interval $[0, 1]$.

4. We determine the values of u and w for the given point Q and compute its intensity as the z coordinate of surface $P(u, w)$ at these values.

Bicubic interpolation. The principle of bicubic interpolation is similar to that of bilinear interpolation and is discussed in Section 7.25.7. Given a point Q on the image, we select a group of 4×4 pixels, convert them to three-dimensional points, and compute the bicuic surface $P(u, w)$ that passes through these points. Once this surface is known,

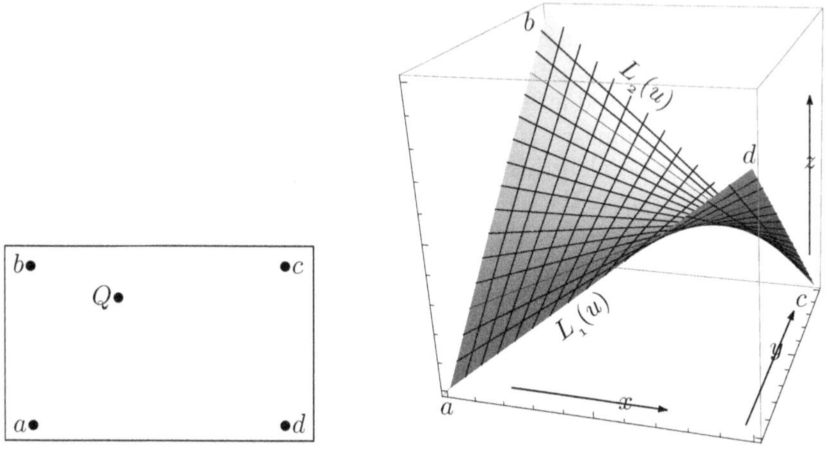

Figure 7.1: A Bilinear Surface.

the values of parameters u and w for Q are determined and the intensity of image point Q becomes the height of the surface at the point determined by these values.

7.2 Image Types

For the purpose of image compression it is useful to distinguish the following types of images: (For information on pixels and their history, see the lively references [Lyon 09] and [Smith 09].)

1. A *bi-level* (or monochromatic) image. This is an image where the pixels can have one of two values, normally referred to as black and white. Each pixel in such an image is represented by one bit, making this the simplest type of image.
2. A *grayscale* image. A pixel in such an image can have one of the n values 0 through $n-1$, indicating one of 2^n shades of gray (or shades of some other color). The value of n is normally compatible with a byte size; i.e., it is 4, 8, 12, 16, 24, or some other convenient multiple of 4 or of 8. The set of the most-significant bits of all the pixels is the most-significant bitplane. Thus, a grayscale image has n bitplanes.
3. A *continuous-tone* image. This type of image can have many similar colors (or grayscales). When adjacent pixels differ by just one unit, it is hard or even impossible for the eye to distinguish their colors. As a result, such an image may contain areas with colors that seem to vary continuously as the eye moves along the area. A pixel in such an image is represented by either a single large number (in the case of many grayscales) or three components (in the case of a color image). A continuous-tone image is normally a natural image (natural as opposed to artificial) and is obtained by taking a photograph with a digital camera, or by scanning a photograph or a painting. Figures 7.55 through 7.58 are typical examples of continuous-tone images. A general survey of lossless compression of this type of images is [Carpentieri et al. 00].

4. A *discrete-tone* image (also called a graphical image or a synthetic image). This is normally an artificial image. It may have a few colors or many colors, but it does not have the noise and blurring of a natural image. Examples are an artificial object or machine, a page of text, a chart, a cartoon, or the contents of a computer screen. (Not every artificial image is discrete-tone. A computer-generated image that's meant to look natural is a continuous-tone image in spite of its being artificially generated.) Artificial objects, text, and line drawings have sharp, well-defined edges, and are therefore highly contrasted from the rest of the image (the background). Adjacent pixels in a discrete-tone image often are either identical or vary significantly in value. Such an image does not compress well with lossy methods, because the loss of just a few pixels may render a letter illegible, or change a familiar pattern to an unrecognizable one. Compression methods for continuous-tone images often do not handle sharp edges very well, so special methods are needed for efficient compression of these images. Notice that a discrete-tone image may be highly redundant, since the same character or pattern may appear many times in the image. Figure 7.59 is a typical example of a discrete-tone image.
5. A *cartoon-like* image. This is a color image that consists of uniform areas. Each area has a uniform color but adjacent areas may have very different colors. This feature may be exploited to obtain excellent compression.

Whether an image is treated as discrete or continuous is usually dictated by the depth of the data. However, it is possible to force an image to be continuous even if it would fit in the discrete category. (From `www.genaware.com`)

It is intuitively clear that each type of image may feature redundancy, but they are redundant in different ways. This is why any given compression method may not perform well for all images, and why different methods are needed to compress the different image types. There are compression methods for bi-level images, for continuous-tone images, and for discrete-tone images. There are also methods that try to break an image up into continuous-tone and discrete-tone parts, and compress each separately.

7.3 Introduction

Modern computers employ graphics extensively. Window-based operating systems display the disk's file directory graphically. The progress of many system operations, such as downloading a file, may also be displayed graphically. Many applications provide a graphical user interface (GUI), which makes it easier to use the program and to interpret displayed results. Computer graphics is used in many areas in everyday life to convert many types of complex information to images. Thus, images are important, but they tend to be big! Modern hardware can display many colors, which is why it is common to have a pixel represented internally as a 24-bit number, where the percentages of red, green, and blue occupy 8 bits each. Such a 24-bit pixel can specify one of $2^{24} \approx 16.78$ million colors. As a result, an image at a resolution of 512×512 that consists of such pixels occupies 786,432 bytes. At a resolution of 1024×1024 it becomes four times as big, requiring 3,145,728 bytes. Videos are also commonly used in computers, making for even bigger images. This is why image compression is so important. An important feature of image compression is that it can be lossy. An image, after all, exists for people to look at, so, when it is compressed, it is acceptable to lose image features to which the eye is not sensitive. This is one of the main ideas behind the many lossy image compression methods described in this chapter.

In general, information can be compressed if it is redundant. It has been mentioned several times that data compression amounts to reducing or removing redundancy in the data. With lossy compression, however, we have a new concept, namely compressing by removing *irrelevancy*. An image can be lossy-compressed by removing irrelevant information even if the original image does not have any redundancy.

◇ **Exercise 7.1:** It would seem that an image with no redundancy is always random (and therefore uninteresting). It that so?

The idea of losing image information becomes more palatable when we consider how digital images are created. Here are three examples: (1) A real-life image may be scanned from a photograph or a painting and digitized (converted to pixels). (2) An image may be recorded by a digital camera that creates pixels and stores them directly in memory. (3) An image may be painted on the screen by means of a paint program. In all these cases, some information is lost when the image is digitized. The fact that the viewer is willing to accept this loss suggests that further loss of information might be tolerable if done properly.

(Digitizing an image involves two steps: *sampling* and *quantization*. Sampling an image is the process of dividing the two-dimensional original image into small regions: pixels. Quantization is the process of assigning an integer value to each pixel. Notice that digitizing sound involves the same two steps, with the difference that sound is one-dimensional.)

We present a simple process that can be employed to determine qualitatively the amount of data loss in a compressed image. Given an image A, (1) compress it to B, (2) decompress B to C, and (3) subtract $D = C - A$. If A was compressed without any loss and decompressed properly, then C should be identical to A and image D should be uniformly white. The more data was lost in the compression, the farther will D be from uniformly white.

How should an image be compressed? The compression techniques discussed in previous chapters are RLE, scalar quantization, statistical methods, and dictionary-based methods. By itself, none is very satisfactory for color or grayscale images (although they may be used in combination with other methods). Here is why:

Section 1.4.1 shows how RLE can be used for (lossless or lossy) compression of an image. This is simple, and it is used by certain parts of JPEG, especially by its lossless mode. In general, however, the other principles used by JPEG produce much better compression than does RLE alone. Facsimile compression (Section 5.7) uses RLE combined with Huffman coding and obtains good results, but only for bi-level images.

Scalar quantization has been mentioned in Section 1.6. It can be used to compress images, but its performance is mediocre. Imagine an image with 8-bit pixels. It can be compressed with scalar quantization by cutting off the four least-significant bits of each pixel. This yields a compression ratio of 0.5, not very impressive, and at the same time reduces the number of colors (or grayscales) from 256 to just 16. Such a reduction not only degrades the overall quality of the reconstructed image, but may also create bands of different colors, a noticeable and annoying effect that's illustrated here.

Imagine a row of 12 pixels with similar colors, ranging from 202 to 215. In binary notation these values are

11010111 11010110 11010101 11010011 11010010 11010001 11001111 11001110 11001101 11001100 11001011 11001010.

Quantization will result in the 12 4-bit values

1101 1101 1101 1101 1101 1101 1100 1100 1100 1100 1100 1100,

which will reconstruct the 12 pixels

11010000 11010000 11010000 11010000 11010000 11010000 11000000 11000000 11000000 11000000 11000000 11000000.

The first six pixels of the row now have the value $11010000_2 = 208$, while the next six pixels are $11000000_2 = 192$. If neighboring rows have similar pixels, the first six columns will form a band, distinctly different from the band formed by the next six columns. This banding, or contouring, effect is very noticeable to the eye, since our eyes are sensitive to edges and breaks in an image.

One way to eliminate this effect is called *improved grayscale* (IGS) *quantization.* It works by adding to each pixel a random number generated from the four rightmost bits of previous pixels. Section 7.4.1 shows that the least-significant bits of a pixel are fairly random, so IGS works by adding to each pixel randomness that depends on the neighborhood of the pixel.

The method maintains an 8-bit variable, denoted by `rsm`, that's initially set to zero. For each 8-bit pixel P to be quantized (except the first one), the IGS method does the following:

1. Set `rsm` to the sum of the eight bits of P and the four rightmost bits of `rsm`. However, if P has the form $1111xxxx$, set `rsm` to P.
2. Write the four leftmost bits of `rsm` on the compressed stream. This is the compressed value of P. IGS is thus not exactly a quantization method, but a variation of scalar quantization.

The first pixel is quantized in the usual way, by dropping its four rightmost bits. Table 7.2 illustrates the operation of IGS.

Pixel	Value	`rsm`	Compressed value
1	1010 0110	0000 0000	1010
2	1101 0010	1101 0010	1101
3	1011 0101	1011 0111	1011
4	1001 1100	1010 0011	1010
5	1111 0100	1111 0100	1111
6	1011 0011	1011 0111	1011

Table 7.2: Illustrating the IGS Method.

Vector quantization can be used more successfully to compress images. It is discussed in Section 7.19.

Statistical methods work best when the symbols being compressed have different probabilities. An input stream where all symbols have the same probability will not compress, even though it may not be random. It turns out that in a continuous-tone color or grayscale image, the different colors or shades of gray may often have roughly the same probabilities. This is why statistical methods are not a good choice for compressing such images, and why new approaches are needed. Images with color discontinuities, where adjacent pixels have widely different colors, compress better with statistical methods, but it is not easy to predict, just by looking at an image, whether it has enough color discontinuities.

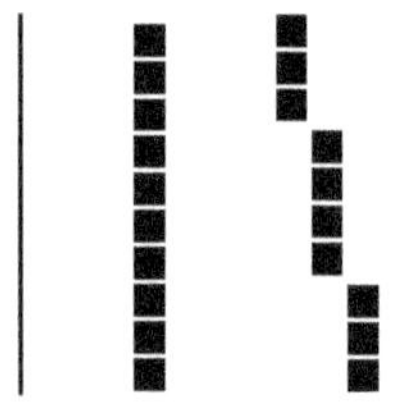

An ideal vertical rule is shown in (a). In (b), the rule is assumed to be perfectly digitized into ten pixels, stacked vertically. However, if the image is placed in the scanner slightly crooked, the scanning may be imperfect, and the resulting pixels might look as in (c).

(a) (b) (c)

Figure 7.3: Perfect and Imperfect Digitizing.

Dictionary-based compression methods also tend to be unsuccessful in dealing with continuous-tone images. Such an image typically contains adjacent pixels with similar colors, but does not contain repeating patterns. Even an image that contains repeated patterns such as vertical lines may lose them when digitized. A vertical line in the original image may become slightly crooked when the image is digitized (Figure 7.3), so the pixels in a scan row may end up having slightly different colors from those in neighboring rows, resulting in a dictionary with short strings. (This problem may also affect curved edges.)

Another problem with dictionary compression of images is that such methods scan the image row by row, and therefore may miss vertical correlations between pixels. An example is the two simple images of Figure 7.4a,b. Saving both in GIF89, a dictionary-based graphics file format (Section 6.21), has resulted in file sizes of 1053 and 1527 bytes, respectively, on the author's computer.

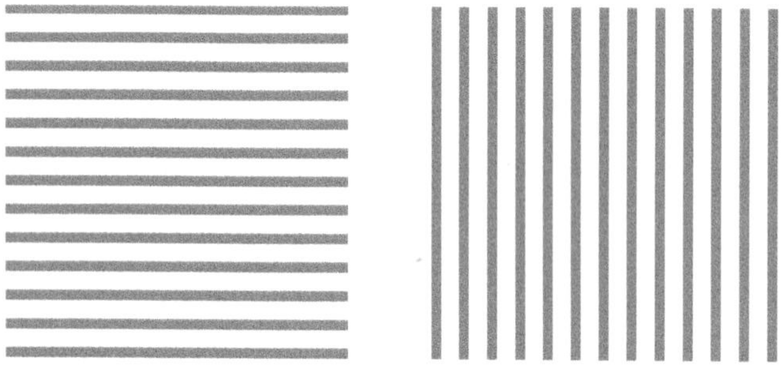

(a) (b)

Figure 7.4: Dictionary Compression of Parallel Lines.

Traditional methods are therefore unsatisfactory for image compression, so this chapter discusses novel approaches. They are all different, but they remove redundancy from an image by using the following principle (see also Section 1.4):

The Principle of Image Compression. If we select a pixel in an image at random, there is a good chance that its neighbors will have the same color or very similar colors.

Image compression is therefore based on the fact that neighboring pixels are *highly correlated.* This correlation is also called *spatial redundancy.*

Here is a simple example that illustrates what can be done with correlated pixels. The following sequence of values gives the intensities of 24 adjacent pixels in a row of a continuous-tone image:

12, 17, 14, 19, 21, 26, 23, 29, 41, 38, 31, 44, 46, 57, 53, 50, 60, 58, 55, 54, 52, 51, 56, 60.

Only two of the 24 pixels are identical. Their average value is 40.3. Subtracting pairs of adjacent pixels results in the sequence

12, 5, −3, 5, 2, 4, −3, 6, 11, −3, −7, 13, 4, 11, −4, −3, 10, −2, −3, 1, −2, −1, 5, 4.

The two sequences are illustrated in Figure 7.5.

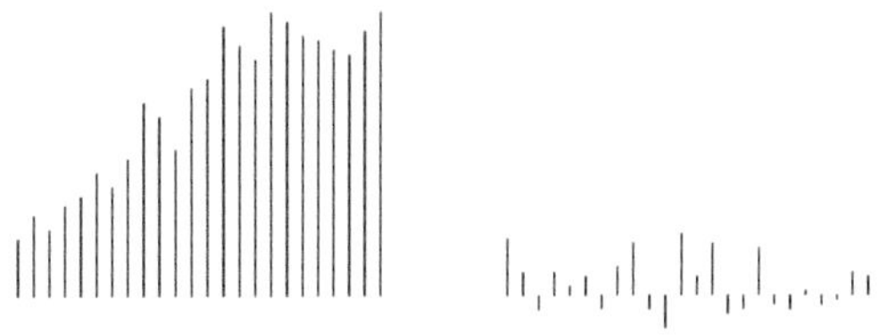

Figure 7.5: Values and Differences of 24 Adjacent Pixels.

The sequence of difference values has three properties that illustrate its compression potential: (1) The difference values are smaller than the original pixel values. Their average is 2.58. (2) They repeat. There are just 15 distinct difference values, so in principle they can be coded by four bits each. (3) They are *decorrelated*: adjacent difference values tend to be different. This can be seen by subtracting them, which results in the sequence of 24 second differences

12, −7, −8, 8, −3, 2, −7, 9, 5, −14, −4, 20, −11, 7, −15, 1, 13, −12, −1, 4, −3, 1, 6, 1.

They are larger than the differences themselves.

Figure 7.6 provides another illustration of the meaning of the words "correlated quantities." A 32×32 matrix A is constructed of random numbers, and its elements are displayed in part (a) as shaded squares. The random nature of the elements is clear. The matrix is then inverted and stored in B, which is shown in part (b). This time, there seems to be more structure to the 32×32 squares. A direct calculation using Equation (7.1) shows that the cross-correlation between the top two rows of A is 0.0412, whereas the cross-correlation between the top two rows of B is -0.9831. The elements of B are correlated since each depends on *all* the elements of A

$$R = \frac{n\sum x_i y_i - \sum x_i \sum y_i}{\sqrt{[n\sum x_i^2 - (\sum x_i)^2][n\sum y_i^2 - (\sum y_i)^2]}}. \tag{7.1}$$

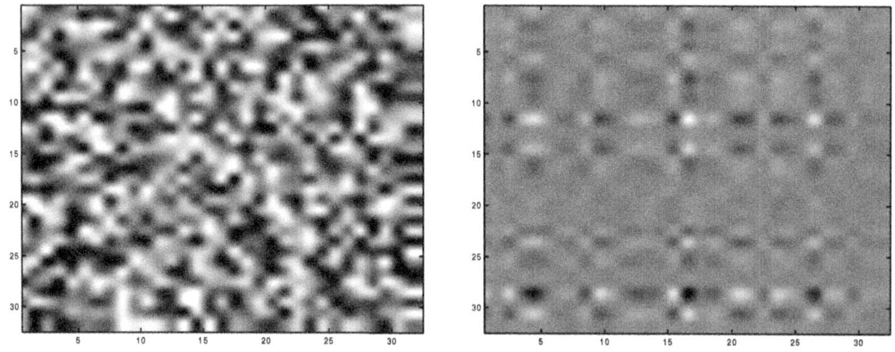

(a) (b)

```
n=32; a=rand(n); imagesc(a); colormap(gray)
b=inv(a); imagesc(b)
```

Figure 7.6: Maps of (a) a Random Matrix and (b) its Inverse.

◇ **Exercise 7.2:** Use mathematical software to illustrate the covariance matrices of (1) a matrix with correlated values and (2) a matrix with decorrelated values.

Once the concept of correlated quantities is grasped, we start looking for a correlation test. Given a matrix M, a statistical test is needed to determine whether its elements are correlated or not. The test is based on the statistical concept of covariance. If the elements of M are decorrelated (i.e., independent), then the covariance of any two different rows and any two different columns of M will be zero (the covariance of a row or of a column with itself is always 1). As a result, the covariance matrix of M (whether covariance of rows or of columns) will be diagonal. If the covariance matrix of M is not diagonal, then the elements of M are correlated. The statistical concepts of variance, covariance, and correlation are discussed in any text on statistics.

The principle of image compression has another aspect. We know from experience that the *brightness* of neighboring pixels is also correlated. Two adjacent pixels may have different colors. One may be mostly red, and the other may be mostly green. Yet if the red component of the first is bright, the green component of its neighbor will, in most cases, also be bright. This property can be exploited by converting pixel representations from RGB to three other components, one of which is the brightness, and the other two represent color. One such format (or *color space*) is YCbCr, where Y (the "luminance" component) represents the brightness of a pixel, and Cb and Cr define its color. This format is discussed in Section 7.10.1, but its advantage is easy to understand. The eye is sensitive to small changes in brightness but not to small changes in color. Thus, losing information in the Cb and Cr components compresses the image while introducing distortions to which the eye is not sensitive. Losing information in the Y component, on the other hand, is very noticeable to the eye.

7.4 Approaches to Image Compression

An image compression method is normally designed for a specific type of image, and this section lists various approaches to compressing images of different types. Only the general principles are discussed here; specific methods are described in the remainder of this chapter.

Approach 1: This is appropriate for bi-level images. A pixel in such an image is represented by one bit. Applying the principle of image compression to a bi-level image therefore means that the immediate neighbors of a pixel P tend to be *identical* to P. Thus, it makes sense to use run-length encoding (RLE) to compress such an image. A compression method for such an image may scan it in raster order (row by row) and compute the lengths of runs of black and white pixels. The lengths are encoded by variable-length codes and are written on the compressed stream. An example of such a method is facsimile compression, Section 5.7.

It should be stressed that this is just an approach to bi-level image compression. The details of specific methods vary. For instance, a method may scan the image column by column or in zigzag (Figure 1.8b), it may convert the image to a quadtree (Section 7.34), or it may scan it region by region using a space-filling curve (Section 7.36).

Approach 2: Also for bi-level images. The principle of image compression tells us that the neighbors of a pixel tend to be similar to the pixel. We can extend this principle and conclude that if the current pixel has color c (where c is either black or white), then pixels of the same color seen in the past (and also those that will be found in the future) tend to have the same immediate neighbors.

This approach looks at n of the near neighbors of the current pixel and considers them an n-bit number. This number is the *context* of the pixel. In principle there can be 2^n contexts, but because of image redundancy we expect them to be distributed in a nonuniform way. Some contexts should be common while others will be rare.

The encoder counts how many times each context has already been found for a pixel of color c, and assigns probabilities to the contexts accordingly. If the current pixel has color c and its context has probability p, the encoder can use adaptive arithmetic coding to encode the pixel with that probability. This approach is used by JBIG (Section 7.14).

Next, we turn to grayscale images. A pixel in such an image is represented by n bits and can have one of 2^n values. Applying the principle of image compression to a grayscale image implies that the immediate neighbors of a pixel P tend to be similar to P, but are not necessarily identical. Thus, RLE should not be used to compress such an image. Instead, two approaches are discussed.

Approach 3: Separate the grayscale image into n bi-level images and compress each with RLE and prefix codes. The principle of image compression seems to imply intuitively that two adjacent pixels that are similar in the grayscale image will be identical in most of the n bi-level images. This, however, is not true, as the following example makes clear. Imagine a grayscale image with $n = 4$ (i.e., 4-bit pixels, or 16 shades of gray). The image can be separated into four bi-level images. If two adjacent pixels in the original grayscale image have values 0000 and 0001, then they are similar. They are also identical in three of the four bi-level images. However, two adjacent pixels with values 0111 and 1000 are also similar in the grayscale image (their values are 7 and 8, respectively) but differ in all four bi-level images.

This problem occurs because the binary codes of adjacent integers may differ by several bits. The binary codes of 0 and 1 differ by one bit, those of 1 and 2 differ by two bits, and those of 7 and 8 differ by four bits. The solution is to design special binary codes such that the codes of any consecutive integers i and $i + 1$ will differ by one bit only. An example of such a code is the *reflected Gray codes* of Section 7.4.1.

Approach 4: Use the *context* of a pixel to *predict* its value. The context of a pixel is the values of some of its neighbors. We can examine some neighbors of a pixel P, compute an average A of their values, and predict that P will have the value A. The principle of image compression tells us that our prediction will be correct in most cases, almost correct in many cases, and completely wrong in a few cases. We can say that the predicted value of pixel P represents the redundant information in P. We now calculate the difference

$$\Delta \stackrel{\text{def}}{=} P - A,$$

and assign variable-length codes to the different values of Δ such that small values (which we expect to be common) are assigned short codes and large values (which are expected to be rare) are assigned long codes. If P can have the values 0 through $m - 1$, then values of Δ are in the range $[-(m-1), +(m-1)]$, and the number of codes needed is $2(m - 1) + 1$ or $2m - 1$.

Experiments with a large number of images suggest that the values of Δ tend to be distributed according to the Laplace distribution (Figure 7.143b). A compression method can, therefore, use this distribution to assign a probability to each value of Δ, and use arithmetic coding to encode the Δ values very efficiently. This is the principle of the MLP method (Section 7.25).

The context of a pixel may consist of just one or two of its immediate neighbors. However, better results may be obtained when several neighbor pixels are included in the context. The average A in such a case should be weighted, with near neighbors assigned higher weights (see, for example, Table 7.141). Another important consideration is the decoder. In order for it to decode the image, it should be able to compute the context of every pixel. This means that the context should employ only pixels that have already been encoded. If the image is scanned in raster order, the context should include only pixels located above the current pixel or on the same row and to its left.

Approach 5: Transform the values of the pixels and encode the transformed values. The concept of a transform, as well as the most important transforms used in image compression, are discussed in Section 7.6. Chapter 8 is devoted to the wavelet transform. Recall that compression is achieved by reducing or removing redundancy. The redundancy of an image is caused by the correlation between pixels, so transforming the pixels to a representation where they are decorrelated eliminates the redundancy. It is also possible to think of a transform in terms of the entropy of the image. In a highly correlated image, the pixels tend to have equiprobable values, which results in maximum entropy. If the transformed pixels are decorrelated, certain pixel values become common, thereby having large probabilities, while others are rare. This results in small entropy. Quantizing the transformed values can produce efficient lossy image compression. We want the transformed values to be independent because coding independent values makes it simpler to construct a statistical model.

We now turn to color images. A pixel in such an image consists of three color components, such as red, green, and blue. Most color images are either continuous-tone or discrete-tone.

Approach 6: The principle of this approach is to separate a continuous-tone color image into three grayscale images and compress each of the three separately, using approaches 3, 4, or 5.

For a continuous-tone image, the principle of image compression implies that adjacent pixels have similar, although perhaps not identical, colors. However, similar colors do not mean similar pixel values. Consider, for example, 12-bit pixel values where each color component is expressed in four bits. Thus, the 12 bits 1000|0100|0000 represent a pixel whose color is a mixture of eight units of red (about 50%, since the maximum is 15 units), four units of green (about 25%), and no blue. Now imagine two adjacent pixels with values 0011|0101|0011 and 0010|0101|0011. They have similar colors, since only their red components differ, and only by one unit. However, when considered as 12-bit numbers, the two numbers 001101010011 and 001001010011 are very different, since they differ in one of their most significant bits.

An important feature of this approach is to use a luminance chrominance color representation instead of the more common RGB. The concepts of luminance and chrominance are discussed in Section 7.10.1 and in [Salomon 99]. The advantage of the luminance chrominance color representation is that the eye is sensitive to small changes in luminance but not in chrominance. This allows the loss of considerable data in the chrominance components, while making it possible to decode the image without a significant visible loss of quality.

Approach 7: A different approach is needed for discrete-tone images. Recall that such an image contains uniform regions, and a region may appear several times in the image. A good example is a screen dump. Such an image consists of text and icons. Each character of text and each icon is a region, and any region may appear several times in the image. A possible way to compress such an image is to scan it, identify regions, and find repeating regions. If a region B is identical to an already found region A, then B can be compressed by writing a pointer to A on the compressed stream. The block decomposition method (FABD, Section 7.32) is an example of how this approach can be implemented.

Approach 8: Partition the image into parts (overlapping or not) and compress it by processing the parts one by one. Suppose that the next unprocessed image part is part number 15. Try to match it with parts 1–14 that have already been processed. If part 15 can be expressed, for example, as a combination of parts 5 (scaled) and 11 (rotated), then only the few numbers that specify the combination need be saved, and part 15 can be discarded. If part 15 cannot be expressed as a combination of already-processed parts, it is declared processed and is saved in raw format.

This approach is the basis of the various *fractal* methods for image compression. It applies the principle of image compression to image parts instead of to individual pixels. Applied this way, the principle tells us that "interesting" images (i.e., those that are being compressed in practice) have a certain amount of *self similarity*. Parts of the image are identical or similar to the entire image or to other parts.

Image compression methods are not limited to these basic approaches. This book discusses methods that use the concepts of context trees, Markov models (Section 11.8), and wavelets, among others. In addition, the concept of *progressive image compression* (Section 7.13) should be mentioned, since it adds another dimension to the field of image compression.

7.4.1 Gray Codes

An image compression method that has been developed specifically for a certain type of image can sometimes be used for other types. Any method for compressing bi-level images, for example, can be used to compress grayscale images by separating the bitplanes and compressing each individually, as if it were a bi-level image. Imagine, for example, an image with 16 grayscale values. Each pixel is specified by four bits, so the image can be separated into four bi-level images. The trouble with this approach is that it violates the general principle of image compression. Imagine two adjacent 4-bit pixels with values $7 = 0111_2$ and $8 = 1000_2$. These pixels have close values, but when separated into four bitplanes, the resulting 1-bit pixels are different in every bitplane! This is because the binary representations of the consecutive integers 7 and 8 differ in all four bit positions. In order to apply any bi-level compression method to grayscale images, a binary representation of the integers is needed where consecutive integers have codes differing by one bit only. Such a representation exists and is called *reflected Gray code* (RGC). This code is easy to generate with the following recursive construction:

Start with the two 1-bit codes $(0, 1)$. Construct two sets of 2-bit codes by duplicating $(0, 1)$ and appending, either on the left or on the right, first a zero, then a one, to the original set. The result is $(00, 01)$ and $(10, 11)$. We now reverse (reflect) the second set, and concatenate the two. The result is the 2-bit RGC $(00, 01, 11, 10)$; a binary code of the integers 0 through 3 where consecutive codes differ by exactly one bit. Applying the rule again produces the two sets $(000, 001, 011, 010)$ and $(110, 111, 101, 100)$, which are concatenated to form the 3-bit RGC. Note that the first and last codes of any RGC also differ by one bit. Here are the first three steps for computing the 4-bit RGC:

$$\text{Add a zero } (0000, 0001, 0011, 0010, 0110, 0111, 0101, 0100),$$
$$\text{Add a one } (1000, 1001, 1011, 1010, 1110, 1111, 1101, 1100),$$
$$\text{reflect } (1100, 1101, 1111, 1110, 1010, 1011, 1001, 1000).$$

Table 7.7 shows how individual bits change when moving through the binary codes of the first 32 integers. The 5-bit binary codes of these integers are listed in the odd-numbered columns of the table, with the bits of integer i that differ from those of $i - 1$ shown in boldface. It is easy to see that the least-significant bit (bit b_0) changes all the time, bit b_1 changes for every other number, and, in general, bit b_k changes every k integers. The even-numbered columns list one of the several possible reflected Gray codes for these integers. A recursive Matlab function to compute RGC is also listed.

43210	Gray	43210	Gray	43210	Gray	43210	Gray
00000	00000	01**000**	10010	1**0000**	00011	11**000**	10001
00001	00100	01001	10110	10001	00111	11001	10101
000**10**	01100	010**10**	11110	100**10**	01111	110**10**	11101
00011	01000	01011	11010	10011	01011	11011	11001
00**100**	11000	01**100**	01010	10**100**	11011	11**100**	01001
00101	11100	01101	01110	10101	11111	11101	01101
001**10**	10100	011**10**	00110	101**10**	10111	111**10**	00101
00111	10000	01111	00010	10111	10011	11111	00001

```
function b=rgc(a,i)
[r,c]=size(a);
b=[zeros(r,1),a; ones(r,1),flipud(a)];
if i>1, b=rgc(b,i-1); end;
```

Table 7.7: First 32 Binary and Reflected Gray Codes.

◇ **Exercise 7.3:** It is also possible to generate the reflected Gray code of an integer n with the following nonrecursive rule: Exclusive-OR n with a copy of itself that's logically shifted one position to the right. In the C programming language this is denoted by `n^(n>>1)`. Use this expression to construct a table similar to Table 7.7.

```
clear;
filename='parrots128'; dim=128;
fid=fopen(filename,'r');
img=fread(fid,[dim,dim])';
mask=1; % between 1 and 8


nimg=bitget(img,mask);
imagesc(nimg), colormap(gray)
```

Binary code

```
clear;
filename='parrots128'; dim=128;
fid=fopen(filename,'r');
img=fread(fid,[dim,dim])';
mask=1 % between 1 and 8
a=bitshift(img,-1);
b=bitxor(img,a);
nimg=bitget(b,mask);
imagesc(nimg), colormap(gray)
```

Gray code

Figure 7.8: Matlab Code to Separate Image Bitplanes.

The conclusion is that the most-significant bitplanes of an image obey the principle of image compression more than the least-significant ones. When adjacent pixels have values that differ by one unit (such as p and $p+1$), chances are that the least-significant bits are different and the most-significant ones are identical. Any image compression method that compresses bitplanes individually should therefore treat the least-significant bitplanes differently from the most-significant ones, or should use RGC instead of the binary code to represent pixels. Figures 7.10, 7.11, and 7.12 (prepared by the Matlab code of Figure 7.8) show the eight bitplanes of the well-known parrots image in both the binary code (the left column) and in RGC (the right column). The bitplanes are numbered 8 (the leftmost or most-significant bits) through 1 (the rightmost or least-significant bits). It is obvious that the least-significant bitplane doesn't show any correlations between the pixels; it is random or very close to random in both binary and RGC. Bitplanes 2 through 5, however, exhibit better pixel correlation in the Gray code. Bitplanes 6 through 8 look different in Gray code and binary, but seem to be highly correlated in either representation.

43210	Gray	43210	Gray	43210	Gray	43210	Gray
00000	00000	0**1000**	01100	**10000**	11000	**11000**	10100
00001	00001	01001	01101	10001	11001	11001	10101
000**10**	00011	010**10**	01111	100**10**	11011	110**10**	10111
00011	00010	01011	01110	10011	11010	11011	10110
00**100**	00110	01**100**	01010	10**100**	11110	11**100**	10010
00101	00111	01101	01011	10101	11111	11101	10011
001**10**	00101	011**10**	01001	101**10**	11101	111**10**	10001
00111	00100	01111	01000	10111	11100	11111	10000

```
a=linspace(0,31,32); b=bitshift(a,-1);
b=bitxor(a,b); dec2bin(b)
```

Table 7.9: First 32 Binary and Gray Codes.

Figure 7.13 is a graphic representation of two versions of the first 32 reflected Gray codes. Part (b) shows the codes of Table 7.7, and part (c) shows the codes of Table 7.9. Even though both are Gray codes, they differ in the way the bits in each bitplane alternate between 0 and 1. In part (b), the bits of the most-significant bitplane alternate four times between 0 and 1. Those of the second most-significant bitplane alternate eight times between 0 and 1, and the bits of the remaining three bitplanes alternate 16, two, and one times between 0 and 1. When the bitplanes are separated, the middle bitplane features the smallest correlation between the pixels, since the Gray codes of adjacent integers tend to have different bits in this bitplane. The Gray codes shown in Figure 7.13c, on the other hand, alternate more and more between 0 and 1 as we move from the most significant bitplanes to the least-significant ones. The least significant bitplanes of this version feature less and less correlation between the pixels and therefore tend to be random. For comparison, Figure 7.13a shows the binary code. It is obvious that bits in this code alternate more often between 0 and 1.

◇ **Exercise 7.4:** Even a cursory look at the Gray codes of Figure 7.13c shows that they exhibit some regularity. Examine these codes carefully and identify two features that may be used to compute the codes.

◇ **Exercise 7.5:** Figure 7.13 is a graphic representation of the binary codes and reflected Gray codes. Find a similar graphic representation of the same codes that illustrates the fact that the first and last codes also differ by one bit.

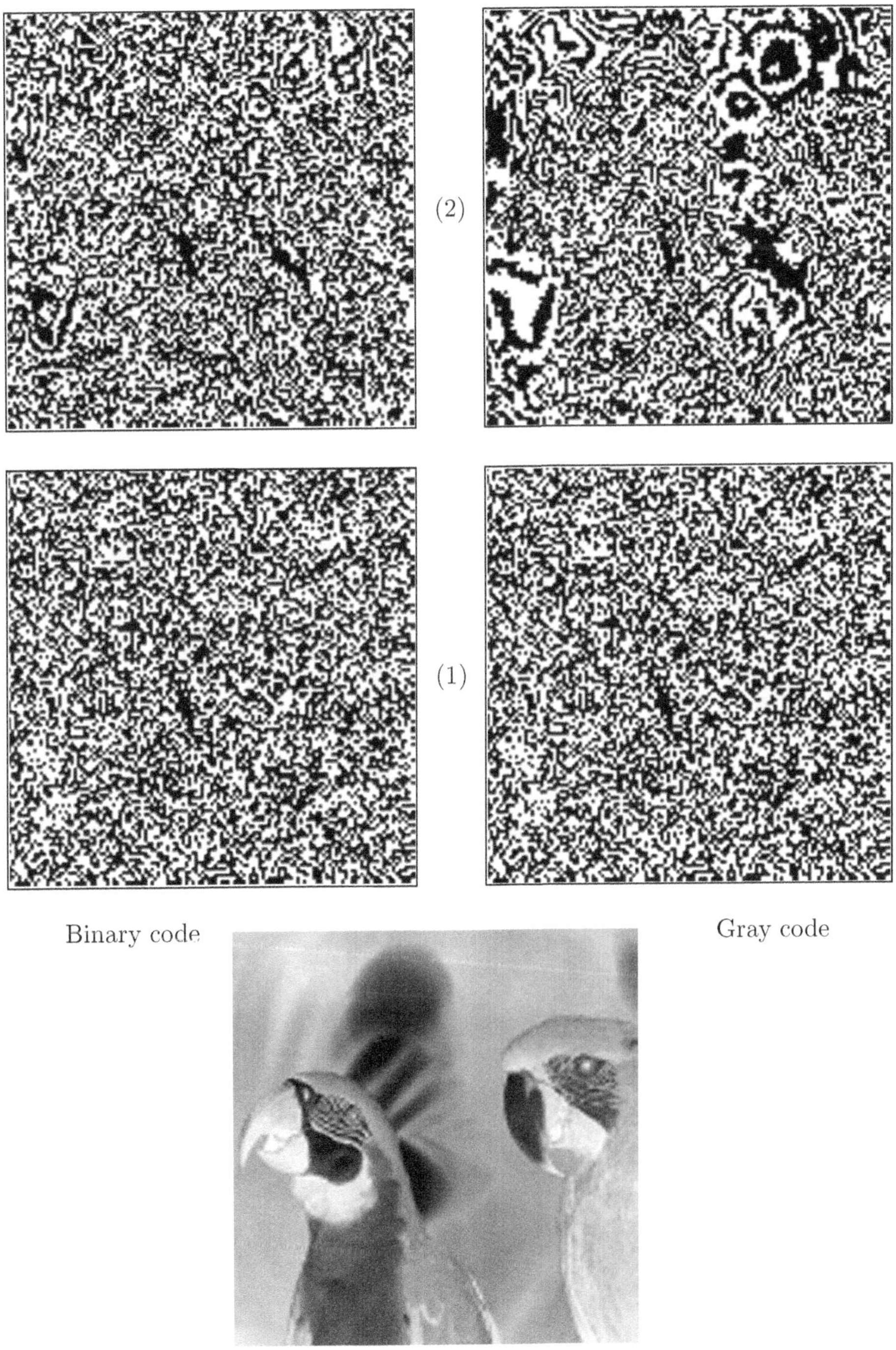

Figure 7.10: Bitplanes 1,2 of the Parrots Image.

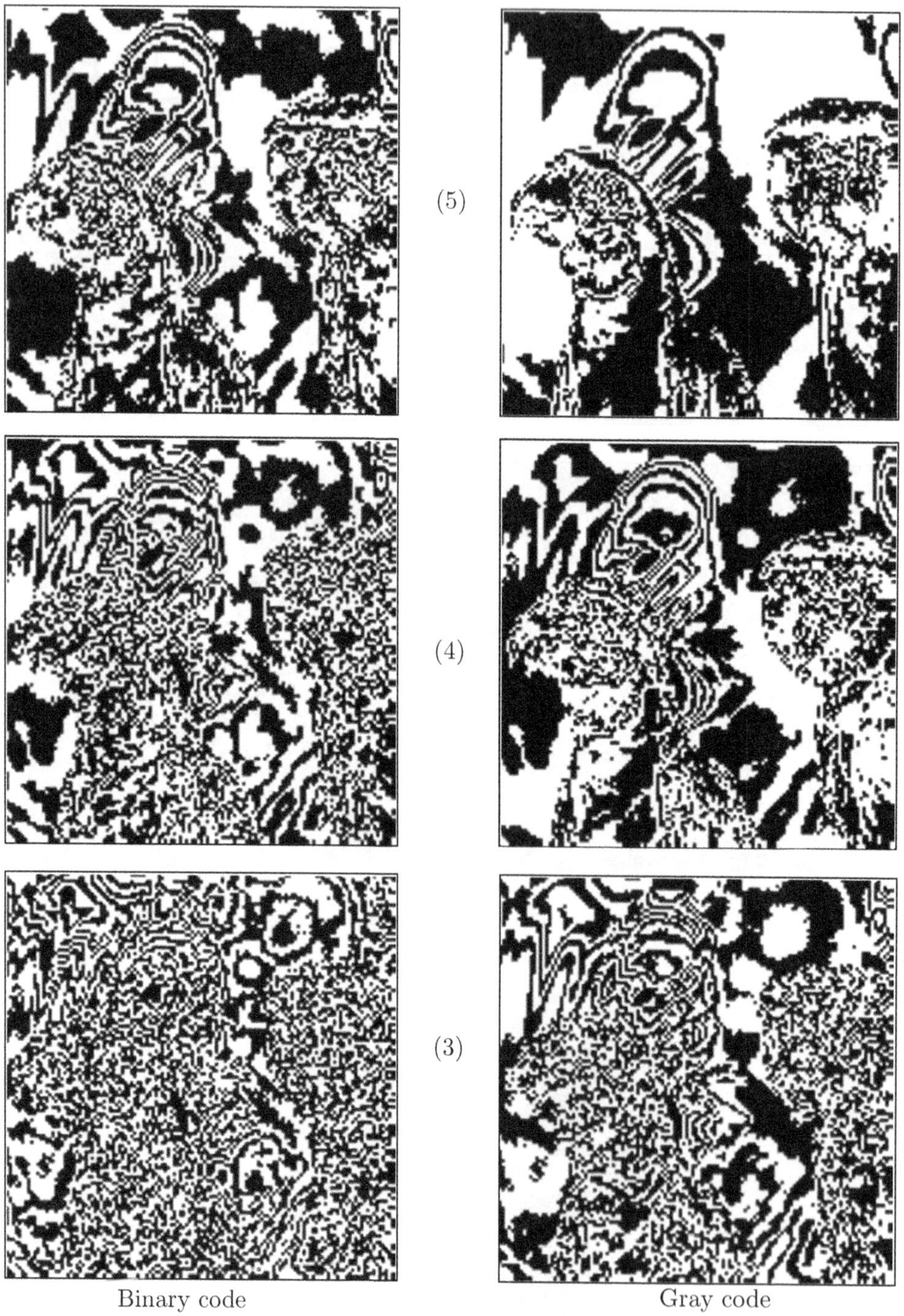

Figure 7.11: Bitplanes 3, 4, and 5 of the Parrots Image.

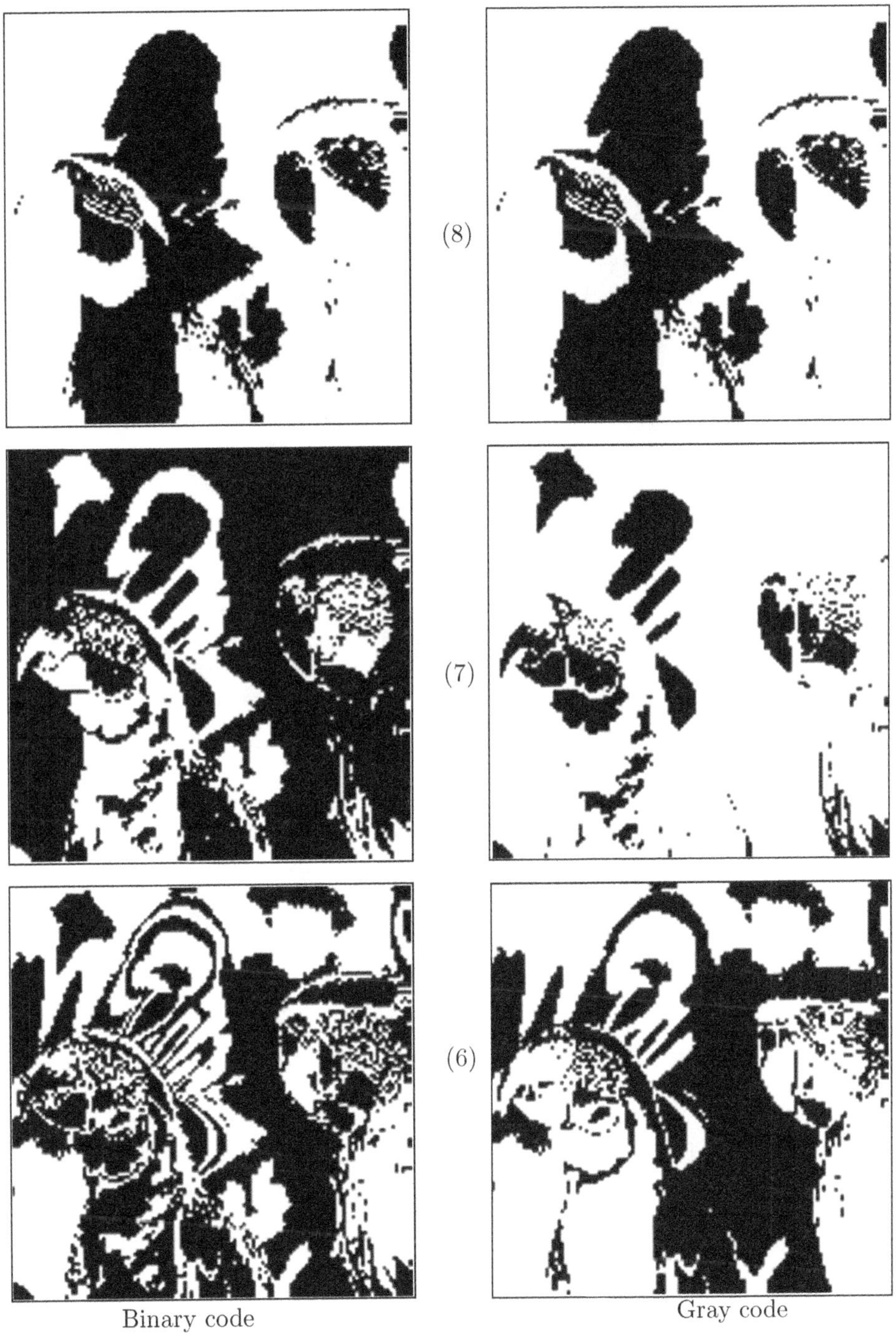

Figure 7.12: Bitplanes 6, 7, and 8 of the Parrots Image.

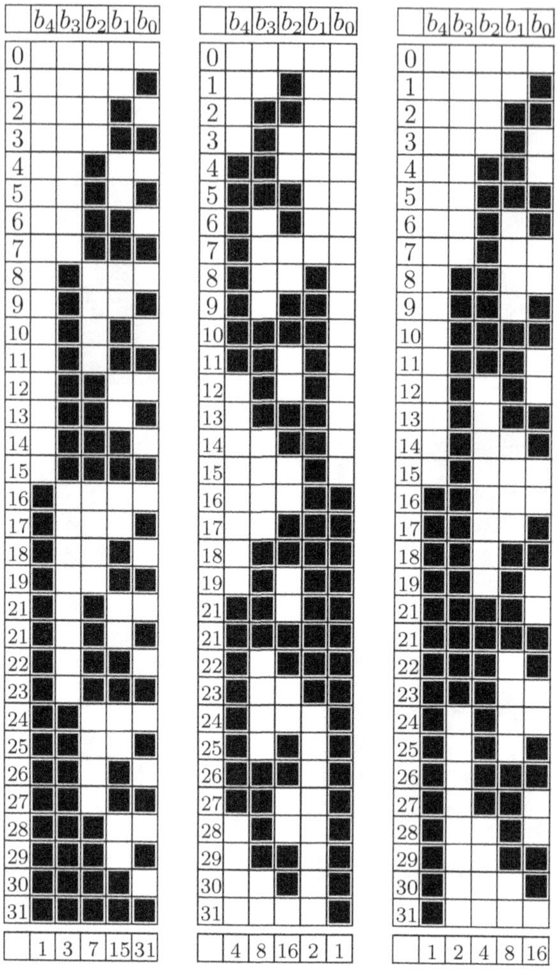

(a) (b) (c)

Table 7.13: First 32 Binary and Reflected Gray Codes.

> The binary Gray code is fun,
> For in it strange things can be done.
> Fifteen, as you know,
> Is one, oh, oh, oh,
> And ten is one, one, one, one.
>
> —Anonymous

Color images provide another example of using the same compression method across image types. Any compression method for grayscale images can be used to compress color images. In a color image, each pixel is represented by three color components (such as RGB). Imagine a color image where each color component is represented by one byte. A pixel is represented by three bytes, or 24 bits, but these bits should not be considered a single number. The two pixels 118|206|12 and 117|206|12 differ by just one unit in the first component, so they have very similar colors. Considered as 24-bit numbers, however, these pixels are very different, since they differ in one of their most-significant bits. Any compression method that treats these pixels as 24-bit numbers would consider these pixels very different, and its performance would suffer as a result. A compression method for grayscale images can be applied to compressing color images, but the color image should first be separated into three color components, and each component compressed individually as a grayscale image.

For an example of the use of RGC for image compression see Section 7.31.

History of Gray Codes

Gray codes are named after Frank Gray, who patented their use for shaft encoders in 1953 [Gray 53]. However, the work was performed much earlier, the patent being applied for in 1947. Gray was a researcher at Bell Telephone Laboratories. During the 1930s and 1940s he was awarded numerous patents for work related to television. According to [Heath 72] the code was first, in fact, used by J. M. E. Baudot for telegraphy in the 1870s (Section 1.1.4), though it is only since the advent of computers that the code has become widely known.

The Baudot code uses five bits per symbol. It can represent $32 \times 2 - 2 = 62$ characters (each code can have two meanings, the meaning being indicated by the `LS` and `FS` codes). It became popular and, by 1950, was designated the International Telegraph Code No. 1. It was used by many first- and second-generation computers.

The August 1972 issue of *Scientific American* contains two articles of interest, one on the origin of binary codes [Heath 72], and another [Gardner 72] on some entertaining aspects of the Gray codes.

7.4.2 Error Metrics

Developers and implementers of lossy image compression methods need a standard metric to measure the quality of reconstructed images compared with the original ones. The better a reconstructed image resembles the original one, the bigger should be the value produced by this metric. Such a metric should also produce a dimensionless number, and that number should not be very sensitive to small variations in the reconstructed image. A common measure used for this purpose is the *peak signal to noise ratio* (PSNR). It is familiar to workers in the field, it is also simple to calculate, but it has only a limited, approximate relationship with the perceived errors noticed by the human visual system. This is why higher PSNR values imply closer resemblance between the reconstructed and the original images, but they do not provide a guarantee that viewers will like the reconstructed image.

Denoting the pixels of the original image by P_i and the pixels of the reconstructed image by Q_i (where $1 \le i \le n$), we first define the *mean square error* (MSE) between the two images as

$$\text{MSE} = \frac{1}{n}\sum_{i=1}^{n}(P_i - Q_i)^2. \tag{7.2}$$

It is the average of the square of the errors (pixel differences) of the two images. The *root mean square error* (RMSE) is defined as the square root of the MSE, and the PSNR is defined as

$$\text{PSNR} = 20\log_{10}\frac{\max_i |P_i|}{\text{RMSE}}, \tag{7.3}$$

(but see Equation (9.4) for a different version). The absolute value is normally not needed, since pixel values are rarely negative. For a bi-level image, the numerator is 1. For a grayscale image with eight bits per pixel, the numerator is 255. For color images, only the luminance component is used.

Greater resemblance between the images implies smaller RMSE and, as a result, larger PSNR. The PSNR is dimensionless, since the units of both numerator and denominator are pixel values. However, because of the use of the logarithm, we say that the PSNR is expressed in *decibels* (dB, Section 10.1). The use of the logarithm also implies less sensitivity to changes in the RMSE. For example, dividing the RMSE by 10 multiplies the PSNR by 2. Notice that the PSNR has no absolute meaning. It is meaningless to say that a PSNR of, say, 25 is good. PSNR values are used only to compare the performance of different lossy compression methods or the effects of different parametric values on the performance of an algorithm. The MPEG committee, for example, uses an informal threshold of PSNR = 0.5 dB to decide whether to incorporate a coding optimization, since they believe that an improvement of that magnitude would be visible to the eye.

Typical PSNR values range between 20 and 40. Assuming pixel values in the range $[0, 255]$, an RMSE of 25.5 results in a PSNR of 20, and an RMSE of 2.55 results in a PSNR of 40. An RMSE of zero (i.e., identical images) results in an infinite (or, more precisely, undefined) PSNR. An RMSE of 255 results in a PSNR of zero, and RMSE values greater than 255 yield negative PSNRs.

◇ **Exercise 7.6:** If the maximum pixel value is 255, can the RMSE values be greater than 255?

Some authors define the PSNR as

$$\text{PSNR} = 10\log_{10}\frac{\max_i |P_i|^2}{\text{MSE}}.$$

In order for the two formulations to produce the same result, the logarithm is multiplied in this case by 10 instead of by 20, since $\log_{10} A^2 = 2\log_{10} A$. Either definition is useful, because only relative PSNR values are used in practice. However, the use of two different factors is confusing.

A related measure is *signal to noise ratio* (SNR). This is defined as

$$\text{SNR} = 20\log_{10}\frac{\sqrt{\frac{1}{n}\sum_{i=1}^{n} P_i^2}}{\text{RMSE}}.$$

The numerator is the root mean square of the original image.

Figure 7.14 is a Matlab function to compute the PSNR of two images. A typical call is `PSNR(A,B)`, where `A` and `B` are image files. They must have the same resolution and have pixel values in the range $[0, 1]$.

```
function PSNR(A,B)
if A==B
 error('Images are identical; PSNR is undefined')
end
max2_A=max(max(A)); max2_B=max(max(B));
min2_A=min(min(A)); min2_B=min(min(B));
if max2_A>1 | max2_B>1 | min2_A<0 | min2_B<0
   error('pixels must be in [0,1]')
end
differ=A-B;
decib=20*log10(1/(sqrt(mean(mean(differ.^2)))));
disp(sprintf('PSNR = +%5.2f dB',decib))
```

Figure 7.14: A Matlab Function to Compute PSNR.

Another relative of the PSNR is the *signal to quantization noise ratio* (SQNR) This is a measure of the effect of quantization on signal quality. It is defined as

$$\mathrm{SQNR} = 10\log_{10}\frac{\text{signal power}}{\text{quantization error}},$$

where the quantization error is the difference between the quantized signal and the original signal.

Another approach to the comparison of an original and a reconstructed image is to generate the difference image and judge it visually. Intuitively, the difference image is $D_i = P_i - Q_i$, but such an image is hard to judge visually because its pixel values D_i tend to be small numbers. If a pixel value of zero represents white, such a difference image would be almost invisible. In the opposite case, where pixel values of zero represent black, such a difference would be too dark to judge. Better results are obtained by calculating

$$D_i = a(P_i - Q_i) + b,$$

where a is a magnification parameter (typically a small number such as 2) and b is half the maximum value of a pixel (typically 128). Parameter a serves to magnify small differences, while b shifts the difference image from extreme white (or extreme black) to a more comfortable gray.

7.5 Intuitive Methods

It is easy to come up with simple, intuitive methods for compressing images. They are inefficient and are described here for the sake of completeness.

7.5.1 Subsampling

Subsampling is perhaps the simplest way to compress an image. One approach to subsampling is simply to delete some of the pixels. The encoder may, for example, ignore every other row and every other column of the image, and write the remaining pixels (which constitute 25% of the image) on the compressed stream. The decoder inputs the compressed data and uses each pixel to generate four identical pixels of the reconstructed image. This, of course, involves the loss of much image detail and is rarely acceptable. Notice that the compression ratio is known in advance.

A slight improvement is obtained when the encoder calculates the average of each block of four pixels and writes this average on the compressed stream. No pixel is totally deleted, but the method is still primitive, because a good lossy image compression method should lose only data to which the eye is not sensitive.

Better results (but worse compression) are obtained when the color representation of the image is changed from the original (normally RGB) to luminance and chrominance. The encoder subsamples the two chrominance components of a pixel but not its luminance component. Assuming that each component uses the same number of bits, the two chrominance components use 2/3 of the image size. Subsampling them reduces this to 25% of 2/3, or 1/6. The size of the compressed image is therefore 1/3 (for the uncompressed luminance component), plus 1/6 (for the two chrominance components) or 1/2 of the original size.

7.5.2 Quantization

Scalar quantization has been mentioned in Section 7.3. This is an intuitive, lossy method where the information lost is not necessarily the least important. Vector quantization can obtain better results, and an intuitive version of it is described here.

The image is partitioned into equal-size blocks (called *vectors*) of pixels, and the encoder has a list (called a *codebook*) of blocks of the same size. Each image block B is compared to all the blocks of the codebook and is matched with the "closest" one. If B is matched with codebook block C, then the encoder writes a pointer to C on the compressed stream. If the pointer is smaller than the block size, compression is achieved. Figure 7.15 shows an example.

The details of selecting and maintaining the codebook and of matching blocks are discussed in Section 7.19. Notice that vector quantization is a method where the compression ratio is known in advance.

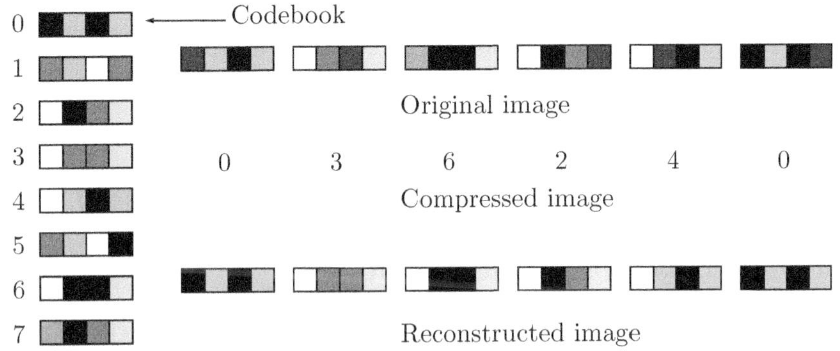

Figure 7.15: Intuitive Vector Quantization.

7.6 Image Transforms

The mathematical concept of a transform is a powerful tool that is employed in many areas and can also serve as an approach to image compression. Section 8.1 discusses this concept in general, as well as the Fourier transform. An image can be compressed by transforming its pixels (which are correlated) to a representation where they are *decorrelated*. Compression is achieved if the new values are smaller, on average, than the original ones. Lossy compression can be achieved by quantizing the transformed values. The decoder inputs the transformed values from the compressed stream and reconstructs the (precise or approximate) original data by applying the inverse transform. The transforms discussed in this section are *orthogonal*. Section 8.6.1 discusses *subband transforms*. One of several excellent references on transforms and their applications to data compression is [Rao and Yip 00].

The term *decorrelated* means that the transformed values are independent of one another. As a result, they can be encoded independently, which makes it simpler to construct a statistical model. An image can be compressed if its representation has redundancy. The redundancy in images stems from pixel correlation. If we transform the image to a representation where the pixels are decorrelated, we have eliminated the redundancy and the image has been fully compressed.

We start with a simple example, where we scan an image in raster order and group pairs of adjacent pixels. Because the pixels are correlated, the two pixels (x, y) of a pair normally have similar values. We now consider each pair of pixels a point in two-dimensional space, and we plot the points. We know that all the points of the form (x, x) are located on the 45° line $y = x$, so we expect our points to be concentrated around this line. Figure 7.17a shows the results of plotting the pixels of a typical image—where a pixel has values in the interval $[0, 255]$—in such a way. Most points form a cloud around this line, and only a few points are located away from it. We now transform the image by rotating all the points 45° clockwise about the origin, such that the 45° line now coincides with the x-axis (Figure 7.17b). This is done by the simple transformation [see Equation (7.63)]

$$(x^*, y^*) = (x, y)\begin{pmatrix} \cos 45^\circ & -\sin 45^\circ \\ \sin 45^\circ & \cos 45^\circ \end{pmatrix} = (x, y)\frac{1}{\sqrt{2}}\begin{pmatrix} 1 & -1 \\ 1 & 1 \end{pmatrix} = (x, y)\mathbf{R}, \tag{7.4}$$

where the rotation matrix $\mathbf{R}$ is orthonormal (i.e., the dot product of a row with itself is 1, the dot product of different rows is 0, and the same is true for columns). The inverse transformation is

$$(x, y) = (x^*, y^*)\mathbf{R}^{-1} = (x^*, y^*)\mathbf{R}^T = (x^*, y^*)\frac{1}{\sqrt{2}}\begin{pmatrix} 1 & 1 \\ -1 & 1 \end{pmatrix}. \tag{7.5}$$

(The inverse of an orthonormal matrix is its transpose.)

It is obvious that most points end up with y coordinates that are zero or close to zero, while the x coordinates don't change much. Figure 7.18a,b shows the distributions of the x and y coordinates (i.e., the odd-numbered and even-numbered pixels) of the $128 \times 128 \times 8$ grayscale Lena image before the rotation. It is clear that the two distributions don't differ by much. Figure 7.18c,d shows that the distribution of the x coordinates stays about the same (with greater variance) but the y coordinates are concentrated around zero. The Matlab code that generated these results is also shown. (Figure 7.18d shows that the y coordinates are concentrated around 100, but this is because a few were as small as -101, so they had to be scaled by 101 to fit in a Matlab array, which always starts at index 1.)

```
p={{5,5},{6, 7},{12.1,13.2},{23,25},{32,29}};
rot={{0.7071,-0.7071},{0.7071,0.7071}};
Sum[p[[i,1]]p[[i,2]], {i,5}]
q=p.rot
Sum[q[[i,1]]q[[i,2]], {i,5}]
```

Figure 7.16: Code for Rotating Five Points.

Once the coordinates of points are known before and after the rotation, it is easy to measure the reduction in correlation. A simple measure is the sum $\sum_i x_i y_i$, also called the *cross-correlation* of points (x_i, y_i).

◇ **Exercise 7.7:** Given the five points $(5,5)$, $(6,7)$, $(12.1, 13.2)$, $(23, 25)$, and $(32, 29)$, rotate them 45° clockwise and calculate their cross-correlations before and after the rotation.

We can now compress the image by simply writing the transformed pixels on the compressed stream. If lossy compression is acceptable, then all the pixels can be quantized (Sections 1.6 and 7.19), resulting in even smaller numbers. We can also write all the odd-numbered pixels (those that make up the x coordinates of the pairs) on the compressed stream, followed by all the even-numbered pixels. These two sequences are called the *coefficient vectors* of the transform. The latter sequence consists of small numbers and may, after quantization, have runs of zeros, resulting in even better compression.

It can be shown that the total variance of the pixels does not change by the rotation, because a rotation matrix is orthonormal. However, since the variance of the new y coordinates is small, most of the variance is now concentrated in the x coordinates. The variance is sometimes called the *energy* of the distribution of pixels, so we can say that

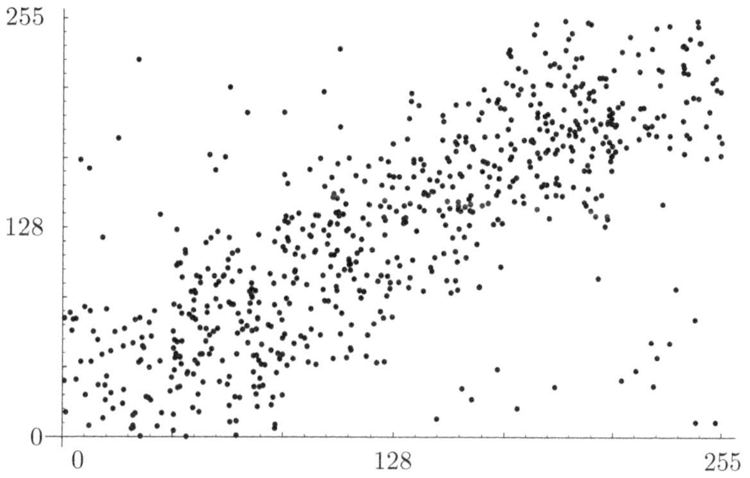

(a)

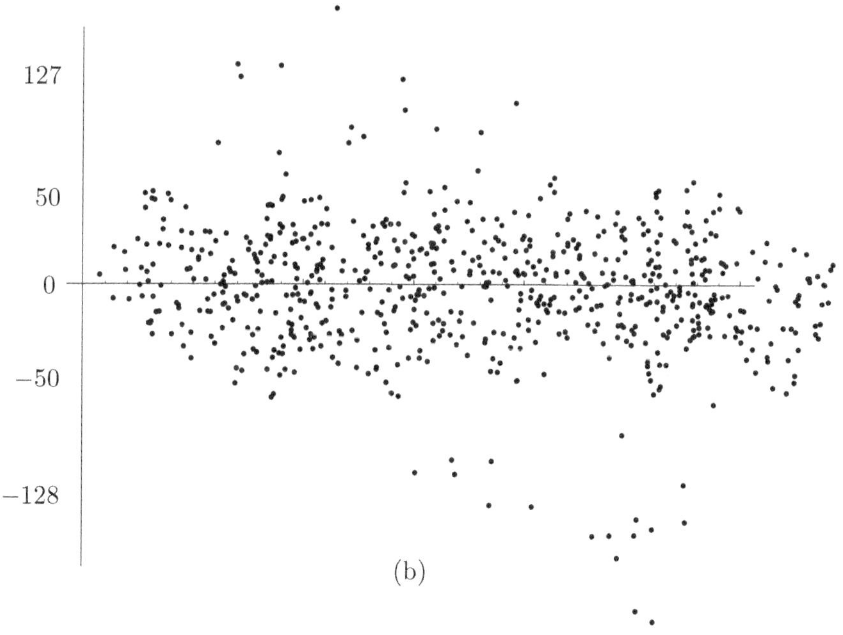

(b)

Figure 7.17: Rotating a Cloud of Points.

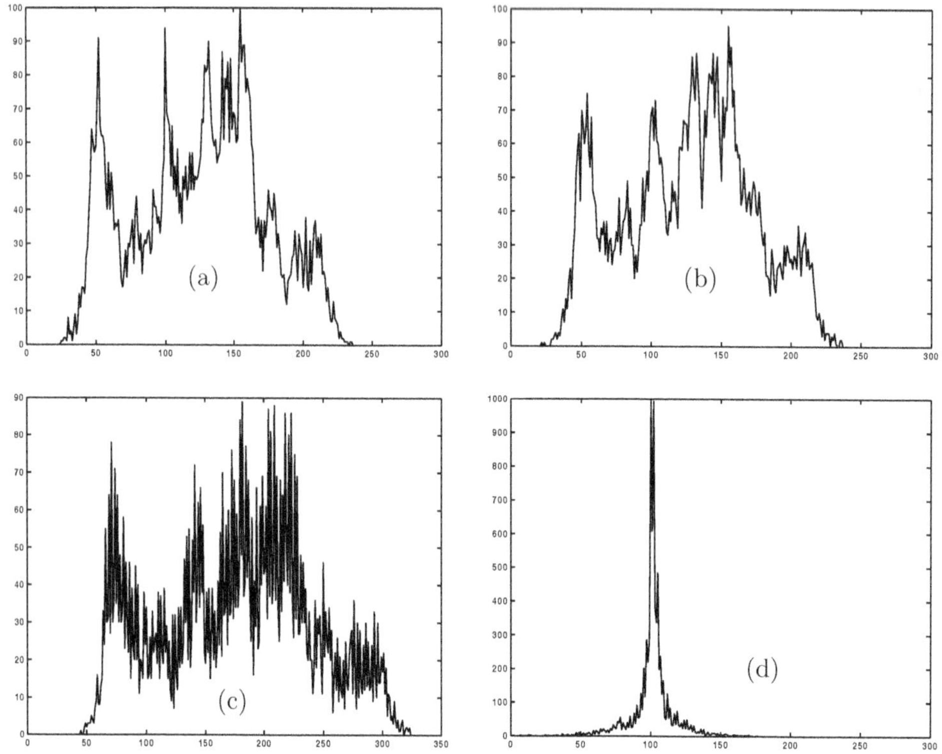


```
filename='lena128'; dim=128;
xdist=zeros(256,1); ydist=zeros(256,1);
fid=fopen(filename,'r');
img=fread(fid,[dim,dim])';
for col=1:2:dim-1
 for row=1:dim
  x=img(row,col)+1; y=img(row,col+1)+1;
  xdist(x)=xdist(x)+1; ydist(y)=ydist(y)+1;
 end
end
figure(1), plot(xdist), colormap(gray) %dist of x&y values
figure(2), plot(ydist), colormap(gray) %before rotation
xdist=zeros(325,1); % clear arrays
ydist=zeros(256,1);
for col=1:2:dim-1
 for row=1:dim
  x=round((img(row,col)+img(row,col+1))*0.7071);
  y=round((-img(row,col)+img(row,col+1))*0.7071)+101;
  xdist(x)=xdist(x)+1; ydist(y)=ydist(y)+1;
 end
end
figure(3), plot(xdist), colormap(gray) %dist of x&y values
figure(4), plot(ydist), colormap(gray) %after rotation
```

Figure 7.18: Distribution of Image Pixels Before and After Rotation.

the rotation has concentrated (or compacted) the energy in the x coordinate and has created compression this way.

Concentrating the energy in one coordinate has another advantage. It makes it possible to quantize that coordinate more finely than the other coordinates. This type of quantization results in better (lossy) compression.

The following simple example illustrates the power of this basic transform. We start with the point $(4,5)$, whose two coordinates are similar. Using Equation (7.4) the point is transformed to $(4,5)\mathbf{R} = (9,1)/\sqrt{2} \approx (6.36396, 0.7071)$. The energies of the point and its transform are $4^2+5^2 = 41 = (9^2+1^2)/2$. If we delete the smaller coordinate (4) of the point, we end up with an error of $4^2/41 = 0.39$. If, on the other hand, we delete the smaller of the two transform coefficients (0.7071), the resulting error is just $0.7071^2/41 = 0.012$. Another way to obtain the same error is to consider the reconstructed point. Passing $\frac{1}{\sqrt{2}}(9,1)$ through the inverse transform [Equation (7.5)] results in the original point $(4,5)$. Doing the same with $\frac{1}{\sqrt{2}}(9,0)$ results in the approximate reconstructed point $(4.5, 4.5)$. The energy difference between the original and reconstructed points is the same small quantity

$$\frac{\left[(4^2+5^2)-(4.5^2+4.5^2)\right]}{4^2+5^2} = \frac{41-40.5}{41} = 0.0012.$$

This simple transform can easily be extended to any number of dimensions. Instead of selecting pairs of adjacent pixels we can select triplets. Each triplet becomes a point in three-dimensional space, and these points form a cloud concentrated around the line that forms equal (although not 45°) angles with the three coordinate axes. When this line is rotated such that it coincides with the x axis, the y and z coordinates of the transformed points become small numbers. The transformation is done by multiplying each point by a 3×3 rotation matrix, and such a matrix is, of course, orthonormal. The transformed points are then separated into three coefficient vectors, of which the last two consist of small numbers. For maximum compression each coefficient vector should be quantized separately.

This can be extended to more than three dimensions, with the only difference being that we cannot visualize spaces of dimensions higher than three. However, the mathematics can easily be extended. Some compression methods, such as JPEG, divide an image into blocks of 8×8 pixels each, and rotate first each row then each column, by means of Equation (7.16), as shown in Section 7.8. This double rotation produces a set of 64 transformed values, of which the first—termed the "DC coefficient"—is large, and the other 63 (called the "AC coefficients") are normally small. Thus, this transform concentrates the energy in the first of 64 dimensions. The set of DC coefficients and each of the sets of 63 AC coefficients should, in principle, be quantized separately (JPEG does this a little differently, though; see Section 7.10.3).

7.7 Orthogonal Transforms

Image transforms are designed to have two properties: (1) reduce image redundancy by reducing the sizes of most pixels and (2) identify the less important parts of the image by isolating the various frequencies of the image. Thus, this section starts with a short discussion of frequencies. We intuitively associate a frequency with a wave. Water waves, sound waves, and electromagnetic waves have frequencies, but pixels in an image can also feature frequencies. Figure 7.19 shows a small, 5×8 bi-level image that illustrates this concept. The top row is uniform, so we can assign it zero frequency. The rows below it have increasing pixel frequencies as measured by the number of color changes along a row. The four waves on the right roughly correspond to the frequencies of the four top rows of the image.

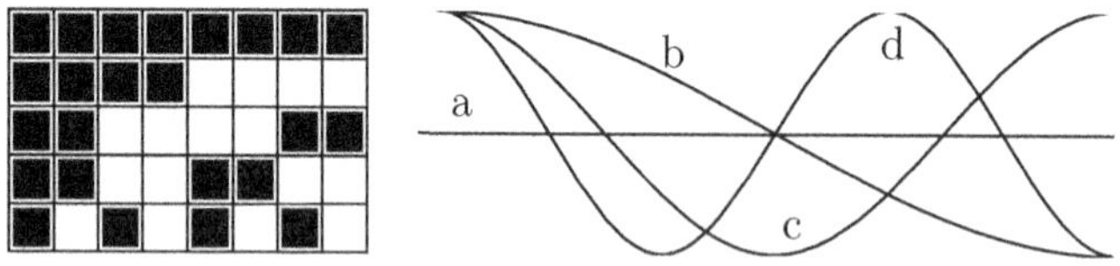

Figure 7.19: Image Frequencies.

Image frequencies are important because of the following basic fact: Low frequencies correspond to the important image features, whereas high frequencies correspond to the details of the image, which are less important. Thus, when a transform isolates the various image frequencies, pixels that correspond to high frequencies can be quantized heavily, while pixels that correspond to low frequencies should be quantized lightly or not at all. This is how a transform can compress an image very effectively by losing information, but only information associated with unimportant image details.

Practical image transforms should be fast and preferably also simple to implement. This suggests the use of *linear transforms.* In such a transform, each transformed value (or transform coefficient) c_i is a weighted sum of the data items (the pixels) d_j that are being transformed, where each item is multiplied by a weight w_{ij}. Thus, $c_i=\sum_j d_j w_{ij}$, for $i,j=1,2,\ldots,n$. For $n=4$, this is expressed in matrix notation as follows:

$$\begin{pmatrix} c_1\\ c_2\\ c_3\\ c_4 \end{pmatrix} = \begin{pmatrix} w_{11} & w_{12} & w_{13} & w_{14}\\ w_{21} & w_{22} & w_{23} & w_{24}\\ w_{31} & w_{32} & w_{33} & w_{34}\\ w_{41} & w_{42} & w_{43} & w_{44} \end{pmatrix} \begin{pmatrix} d_1\\ d_2\\ d_3\\ d_4 \end{pmatrix}.$$

For the general case, we can write $\mathbf{C}=\mathbf{W}\cdot\mathbf{D}$. Each row of $\mathbf{W}$ is called a "basis vector."

The only quantities that have to be computed are the weights w_{ij}. The guiding principles are as follows:

1. Reducing redundancy. The first transform coefficient c_1 can be large, but the remaining values $c_2, c_3, \ldots$ should be small.

2. Isolating frequencies. The first transform coefficient c_1 should correspond to zero pixel frequency, and the remaining coefficients should correspond to higher and higher frequencies.

The key to determining the weights w_{ij} is the fact that our data items d_j are not arbitrary numbers but pixel values, which are nonnegative and correlated.

The basic relation $c_i = \sum_j d_j w_{ij}$ suggests that the first coefficient c_1 will be large if all the weights of the form w_{1j} are positive. To make the other coefficients c_i small, it is enough to make half the weights w_{ij} positive and the other half negative. A simple choice is to assign half the weights the value $+1$ and the other half the value -1. In the extreme case where all the pixels d_j are identical, this will result in $c_i = 0$. When the d_j's are similar, c_i will be small (positive or negative).

This choice of w_{ij} satisfies the first requirement: to reduce pixel redundancy by means of a transform. In order to satisfy the second requirement, the weights w_{ij} of row i should feature frequencies that get higher with i. Weights w_{1j} should have zero frequency; they should all be $+1$'s. Weights w_{2j} should have one sign change; i.e., they should be $+1, +1, \ldots + 1, -1, -1, \ldots, -1$. This continues until the last row of weights w_{nj} should have the highest frequency $+1, -1, +1, -1, \ldots, +1, -1$. The mathematical discipline of vector spaces coins the term "basis vectors" for our rows of weights.

In addition to isolating the various frequencies of pixels d_j, this choice results in basis vectors that are orthogonal. The basis vectors are the rows of matrix $\mathbf{W}$, which is why this matrix and, by implication, the entire transform are also termed orthogonal.

These considerations are satisfied by the orthogonal matrix

$$\mathbf{W} = \begin{pmatrix} 1 & 1 & 1 & 1 \\ 1 & 1 & -1 & -1 \\ 1 & -1 & -1 & 1 \\ 1 & -1 & 1 & -1 \end{pmatrix}. \tag{7.6}$$

The first basis vector (the top row of $\mathbf{W}$) consists of all 1's, so its frequency is zero. Each of the subsequent vectors has two $+1$'s and two -1's, so they produce small transformed values, and their frequencies (measured as the number of sign changes along the basis vector) get higher. This matrix is similar to the Walsh-Hadamard transform [Equation (7.7)].

To illustrate how this matrix identifies the frequencies in a data vector, we multiply it by four vectors as follows:

$$\mathbf{W}\cdot\begin{bmatrix}1\\0\\0\\1\end{bmatrix} = \begin{bmatrix}2\\0\\2\\0\end{bmatrix},\quad \mathbf{W}\cdot\begin{bmatrix}0\\0.33\\-0.33\\-1\end{bmatrix} = \begin{bmatrix}0\\2.66\\0\\1.33\end{bmatrix},\quad \mathbf{W}\cdot\begin{bmatrix}1\\0\\0\\0\end{bmatrix} = \begin{bmatrix}1\\1\\1\\1\end{bmatrix},\quad \mathbf{W}\cdot\begin{bmatrix}1\\-0.8\\1\\-0.8\end{bmatrix} = \begin{bmatrix}0.4\\0\\0\\3.6\end{bmatrix}.$$

The results make sense when we discover how the four test vectors were determined

$$\begin{aligned}(1,0,0,1) &= 0.5(1,1,1,1) + 0.5(1,-1,-1,1),\\ (1,0.33,-0.33,-1) &= 0.66(1,1,-1,-1) + 0.33(1,-1,1,-1),\\ (1,0,0,0) &= 0.25(1,1,1,1) + 0.25(1,1,-1,-1) + 0.25(1,-1,-1,1) + 0.25(1,-1,1,\ \ 1),\\ (1,-0.8,1,-0.8) &= 0.1(1,1,1,1) + 0.9(1,-1,1,-1).\end{aligned}$$

The product of $\mathbf{W}$ and the first vector shows how that vector consists of equal amounts of the first and the third frequencies. Similarly, the transform $(0.4, 0, 0, 3.6)$ shows that vector $(1, -0.8, 1, -0.8)$ is a mixture of a small amount of the first frequency and nine times the fourth frequency.

It is also possible to modify this transform to conserve the energy of the data vector. All that's needed is to multiply the transformation matrix $\mathbf{W}$ by the scale factor 1/2. Thus, the product $(\mathbf{W}/2)\times(a, b, c, d)$ has the same energy $a^2 + b^2 + c^2 + d^2$ as the data vector (a, b, c, d). An example is the product of $\mathbf{W}/2$ and the correlated vector $(5, 6, 7, 8)$. It results in the transform coefficients $(13, -2, 0, -1)$, where the first coefficient is large and the remaining ones are smaller than the original data items. The energy of both $(5, 6, 7, 8)$ and $(13, -2, 0, -1)$ is 174, but whereas in the former vector the first component accounts for only 14% of the energy, in the transformed vector the first component accounts for 97% of the energy. This is how our simple orthogonal transform compacts the energy of the data vector.

Another advantage of $\mathbf{W}$ is that it also performs the inverse transform. The product $(\mathbf{W}/2)\cdot(13, -2, 0, -1)^T$ reconstructs the original data $(5, 6, 7, 8)$.

We are now in a position to appreciate the compression potential of this transform. We use matrix $\mathbf{W}/2$ to transform the (not very correlated) data vector $d = (4, 6, 5, 2)$. The result is $t = (8.5, 1.5, -2.5, 0.5)$. It's easy to transform t back to d, but t itself does not provide any compression. In order to achieve compression, we quantize the components of t, and the point is that even after heavy quantization, it is still possible to get back a vector very similar to the original d.

We first quantize t to the integers $(9, 1, -3, 0)$ and perform the inverse transform to get back $(3.5, 6.5, 5.5, 2.5)$. In a similar experiment, we completely delete the two smallest elements and inverse-transform the coarsely quantized vector $(8.5, 0, -2.5, 0)$. This produces the reconstructed data $(3, 5.5, 5.5, 3)$, still very close to the original values of d. The conclusion is that even this simple, intuitive transform is a powerful tool for "squeezing out" the redundancy in data. More sophisticated transforms produce results that can be quantized coarsely and still be used to reconstruct the original data to a high degree.

7.7.1 Two-Dimensional Transforms

Given two-dimensional data such as the 4×4 matrix

$$\mathbf{D} = \begin{pmatrix} 5 & 6 & 7 & 4 \\ 6 & 5 & 7 & 5 \\ 7 & 7 & 6 & 6 \\ 8 & 8 & 8 & 8 \end{pmatrix},$$

where each of the four columns is highly correlated, we can apply our simple one-dimensional transform to the columns of $\mathbf{D}$. The result is

$$\mathbf{C}' = \mathbf{W}\cdot\mathbf{D} = \begin{pmatrix} 1 & 1 & 1 & 1 \\ 1 & 1 & -1 & -1 \\ 1 & -1 & -1 & 1 \\ 1 & -1 & 1 & -1 \end{pmatrix}\cdot\mathbf{D} = \begin{pmatrix} 26 & 26 & 28 & 23 \\ -4 & -4 & 0 & -5 \\ 0 & 2 & 2 & 1 \\ -2 & 0 & -2 & -3 \end{pmatrix}.$$

Each column of $\mathbf{C}'$ is the transform of a column of $\mathbf{D}$. Notice how the top element of each column of $\mathbf{C}'$ is dominant, because the data in the corresponding column of

D is correlated. Notice also that the rows of $\mathbf{C}'$ are still correlated. $\mathbf{C}'$ is the first stage in a two-stage process that produces the two-dimensional transform of matrix **D**. The second stage should transform each *row* of $\mathbf{C}'$, and this is done by multiplying $\mathbf{C}'$ by the transpose $\mathbf{W}^T$. Our particular **W**, however, is symmetric, so we end up with $\mathbf{C} = \mathbf{C}'\cdot\mathbf{W}^T = \mathbf{W}\cdot\mathbf{D}\cdot\mathbf{W}^T = \mathbf{W}\cdot\mathbf{D}\cdot\mathbf{W}$ or

$$\mathbf{C} = \begin{pmatrix} 26 & 26 & 28 & 23 \\ -4 & -4 & 0 & -5 \\ 0 & 2 & 2 & 1 \\ -2 & 0 & -2 & -3 \end{pmatrix} \begin{pmatrix} 1 & 1 & 1 & 1 \\ 1 & 1 & -1 & -1 \\ 1 & -1 & -1 & 1 \\ 1 & -1 & 1 & -1 \end{pmatrix} = \begin{pmatrix} 103 & 1 & -5 & 5 \\ -13 & -3 & -5 & 5 \\ 5 & -1 & -3 & -1 \\ -7 & 3 & -3 & -1 \end{pmatrix}.$$

The elements of **C** are decorrelated. The top-left element is dominant. It contains most of the total energy of the original **D**. The elements in the top row and the leftmost column are somewhat large, while the remaining elements are smaller than the original data items. The double-stage, two-dimensional transformation has reduced the correlation in both the horizontal and vertical dimensions. As in the one-dimensional case, excellent compression can be achieved by quantizing the elements of **C**, especially those that correspond to higher frequencies (i.e., located toward the bottom-right corner of **C**).

This is the essence of orthogonal transforms. The remainder of this section discusses the following important transforms:

1. The Walsh-Hadamard transform (WHT, Section 7.7.2) is fast and easy to compute (it requires only additions and subtractions), but its performance, in terms of energy compaction, is lower than that of the DCT.
2. The Haar transform [Stollnitz et al. 96] is a simple, fast transform. It is the simplest wavelet transform and is discussed in Section 7.7.3 and in Chapter 8.
3. The Karhunen-Loève transform (KLT, Section 7.7.4) is the best one theoretically, in the sense of energy compaction (or, equivalently, pixel decorrelation). However, its coefficients are not fixed; they depend on the data to be compressed. Calculating these coefficients (the basis of the transform) is slow, as is the calculation of the transformed values themselves. Since the coefficients are data dependent, they have to be included in the compressed stream. For these reasons and because the DCT performs almost as well, the KLT is not generally used in practice.
4. The discrete cosine transform (DCT) is discussed in detail in Section 7.8. This important transform is almost as efficient as the KLT in terms of energy compaction, but it uses a fixed basis, independent of the data. There are also fast methods for calculating the DCT. This method is used by JPEG and MPEG audio.

7.7.2 Walsh-Hadamard Transform

As mentioned earlier, this transform has low compression efficiency, which is why it is not used much in practice. It is, however, fast, because it can be computed with just additions, subtractions, and an occasional right shift (to replace a division by a power of 2).

Given an $N\times N$ block of pixels P_{xy} (where N must be a power of 2, $N=2^n$), its two-dimensional WHT and inverse WHT are defined by Equations (7.7) and (7.8):

$$\begin{aligned} H(u,v) &= \sum_{x=0}^{N-1}\sum_{y=0}^{N-1} p_{xy}g(x,y,u,v) \\ &= \frac{1}{N}\sum_{x=0}^{N-1}\sum_{y=0}^{N-1} p_{xy}(-1)^{\sum_{i=0}^{n-1}[b_i(x)p_i(u)+b_i(y)p_i(v)]}, \end{aligned} \tag{7.7}$$

$$\begin{aligned} P_{xy} &= \sum_{u=0}^{N-1}\sum_{v=0}^{N-1} H(u,v)h(x,y,u,v) \\ &= \frac{1}{N}\sum_{u=0}^{N-1}\sum_{v=0}^{N-1} H(u,v)(-1)^{\sum_{i=0}^{n-1}[b_i(x)p_i(u)+b_i(y)p_i(v)]}, \end{aligned} \tag{7.8}$$

where $H(u,v)$ are the results of the transform (i.e., the WHT coefficients), the quantity $b_i(u)$ is bit i of the binary representation of the integer u, and $p_i(u)$ is defined in terms of the $b_j(u)$ by Equation (7.9):

$$\begin{aligned} p_0(u) &= b_{n-1}(u), \\ p_1(u) &= b_{n-1}(u)+b_{n-2}(u), \\ p_2(u) &= b_{n-2}(u)+b_{n-3}(u), \\ &\vdots \\ p_{n-1}(u) &= b_1(u)+b_0(u). \end{aligned} \tag{7.9}$$

(Recall that n is defined above by $N=2^n$.) As an example, consider $u=6=110_2$. Bits zero, one, and two of 6 are 0, 1, and 1, respectively, so $b_0(6)=0$, $b_1(6)=1$, and $b_2(6)=1$.

The quantities $g(x,y,u,v)$ and $h(x,y,u,v)$ are called the *kernels* (or *basis images*) of the WHT. These matrices are identical. Their elements are just $+1$ and -1, and they are multiplied by the factor $\frac{1}{N}$. As a result, the WHT transform consists in multiplying each image pixel by $+1$ or -1, summing, and dividing the sum by N. Since $N=2^n$ is a power of 2, dividing by it can be done by shifting n positions to the right.

The WHT kernels are shown, in graphical form, for $N=4$, in Figure 7.20, where white denotes $+1$ and black denotes -1 (the factor $\frac{1}{N}$ is ignored). The rows and columns of blocks in this figure correspond to values of u and v from 0 to 3, respectively. The rows and columns inside each block correspond to values of x and y from 0 to 3, respectively. The number of sign changes across a row or a column of a matrix is called the *sequency* of the row or column. The rows and columns in the figure are ordered in increased sequency. Some authors show similar but unordered figures, because this transform was defined by Walsh and by Hadamard in slightly different ways (see [Gonzalez and Woods 92] for more information).

Compressing an image with the WHT is done similarly to the DCT, except that Equations (7.7) and (7.8) are used, instead of Equations (7.16) and (7.17).

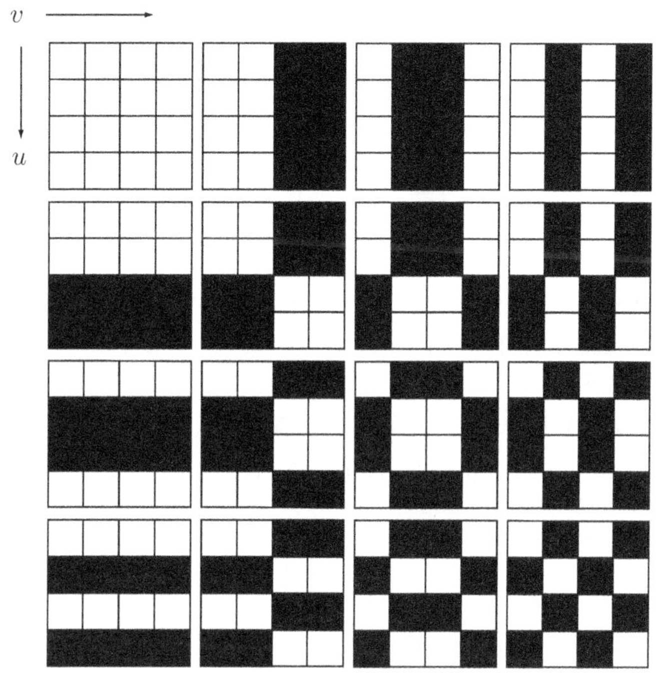

Figure 7.20: The Ordered WHT Kernel for $N = 4$.

◇ **Exercise 7.8:** Use appropriate mathematical software to compute and display the basis images of the WHT for $N = 8$.

7.7.3 Haar Transform

The Haar transform [Stollnitz et al. 92] is based on the Haar functions $h_k(x)$, which are defined for $x \in [0, 1]$ and for $k = 0, 1, \ldots, N-1$, where $N = 2^n$. Its application is also discussed in Chapter 8.

Before we discuss the actual transform, we have to mention that any integer k can be expressed as the sum $k = 2^p + q - 1$, where $0 \le p \le n-1$, $q = 0$ or 1 for $p = 0$, and $1 \le q \le 2^p$ for $p \ne 0$. For $N = 4 = 2^2$, for example, we get $0 = 2^0 + 0 - 1$, $1 = 2^0 + 1 - 1$, $2 = 2^1 + 1 - 1$, and $3 = 2^1 + 2 - 1$.

The Haar basis functions are now defined as

$$h_0(x) \stackrel{\text{def}}{=} h_{00}(x) = \frac{1}{\sqrt{N}}, \quad \text{for } 0 \le x \le 1, \tag{7.10}$$

and

$$h_k(x) \stackrel{\text{def}}{=} h_{pq}(x) = \frac{1}{\sqrt{N}} \begin{cases} 2^{p/2}, & \frac{q-1}{2^p} \le x < \frac{q-1/2}{2^p}, \\ -2^{p/2}, & \frac{q-1/2}{2^p} \le x < \frac{q}{2^p}, \\ 0, & \text{otherwise for } x \in [0, 1]. \end{cases} \tag{7.11}$$

The Haar transform matrix $\mathbf{A}_N$ of order $N\times N$ can now be constructed. A general element i,j of this matrix is the basis function $h_i(j)$, where $i=0,1,\ldots,N-1$ and $j=0/N,1/N,\ldots,(N-1)/N$. For example,

$$\mathbf{A}_2=\begin{pmatrix} h_0(0/2) & h_0(1/2)\\ h_1(0/2) & h_1(1/2)\end{pmatrix}=\frac{1}{\sqrt{2}}\begin{pmatrix}1 & 1\\ 1 & -1\end{pmatrix} \tag{7.12}$$

(recall that $i=1$ implies $p=0$ and $q=1$). Figure 7.21 shows code to calculate this matrix for any N, and also the Haar basis images for $N=8$.

◇ **Exercise 7.9:** Compute the Haar coefficient matrices $\mathbf{A}_4$ and $\mathbf{A}_8$.

Given an image block $\mathbf{P}$ of order $N\times N$ where $N=2^n$, its Haar transform is the matrix product $\mathbf{A}_N\mathbf{P}\mathbf{A}_N$ (Section 8.6).

7.7.4 Karhunen-Loève Transform

The Karhunen-Loève transform (also called the Hotelling transform) has the best efficiency in the sense of energy compaction, but for the reasons mentioned earlier, it has more theoretical than practical value. Given an image, we break it up into k blocks of n pixels each, where n is typically 64 but can have other values, and k depends on the image size. We consider the blocks vectors and denote them by $\mathbf{b}^{(i)}$, for $i=1,2,\ldots,k$. The average vector is $\overline{\mathbf{b}}=(\sum_i \mathbf{b}^{(i)})/k$. A new set of vectors $\mathbf{v}^{(i)}=\mathbf{b}^{(i)}-\overline{\mathbf{b}}$ is defined, causing the average $(\sum \mathbf{v}^{(i)})/k$ to be zero. We denote the $n\times n$ KLT transform matrix that we are seeking by $\mathbf{A}$. The result of transforming a vector $\mathbf{v}^{(i)}$ is the weight vector $\mathbf{w}^{(i)}=\mathbf{A}\mathbf{v}^{(i)}$. The average of the $\mathbf{w}^{(i)}$ is also zero. We now construct a matrix $\mathbf{V}$ whose columns are the $\mathbf{v}^{(i)}$ vectors and another matrix $\mathbf{W}$ whose columns are the weight vectors $\mathbf{w}^{(i)}$:

$$\mathbf{V}=\left(\mathbf{v}^{(1)},\mathbf{v}^{(2)},\ldots,\mathbf{v}^{(k)}\right),\quad \mathbf{W}=\left(\mathbf{w}^{(1)},\mathbf{w}^{(2)},\ldots,\mathbf{w}^{(k)}\right).$$

Matrices $\mathbf{V}$ and $\mathbf{W}$ have n rows and k columns each. From the definition of $\mathbf{w}^{(i)}$, we get $\mathbf{W}=\mathbf{A}\cdot\mathbf{V}$.

The n coefficient vectors $\mathbf{c}^{(j)}$ of the Karhunen-Loève transform are given by

$$\mathbf{c}^{(j)}=\left(w_j^{(1)},w_j^{(2)},\ldots,w_j^{(k)}\right),\quad j=1,2,\ldots,n.$$

Thus, vector $\mathbf{c}^{(j)}$ consists of the jth elements of all the weight vectors $\mathbf{w}^{(i)}$, for $i=1,2,\ldots,k$ ($\mathbf{c}^{(j)}$ is the jth coordinate of the $\mathbf{w}^{(i)}$ vectors).

We now examine the elements of the matrix product $\mathbf{W}\cdot\mathbf{W}^T$ (this is an $n\times n$ matrix). A general element in row a and column b of this matrix is the sum of products:

$$\left(\mathbf{W}\cdot\mathbf{W}^T\right)_{ab}=\sum_{i=1}^{k} w_a^{(i)}w_b^{(i)}=\sum_{i=1}^{k} c_i^{(a)}c_i^{(b)}=\mathbf{c}^{(a)}\bullet\mathbf{c}^{(b)},\quad \text{for } a,b\in[1,n]. \tag{7.13}$$

The fact that the average of each $\mathbf{w}^{(i)}$ is zero implies that a general diagonal element $(\mathbf{W}\cdot\mathbf{W}^T)_{jj}$ of the product matrix is the variance (up to a factor k) of the jth element

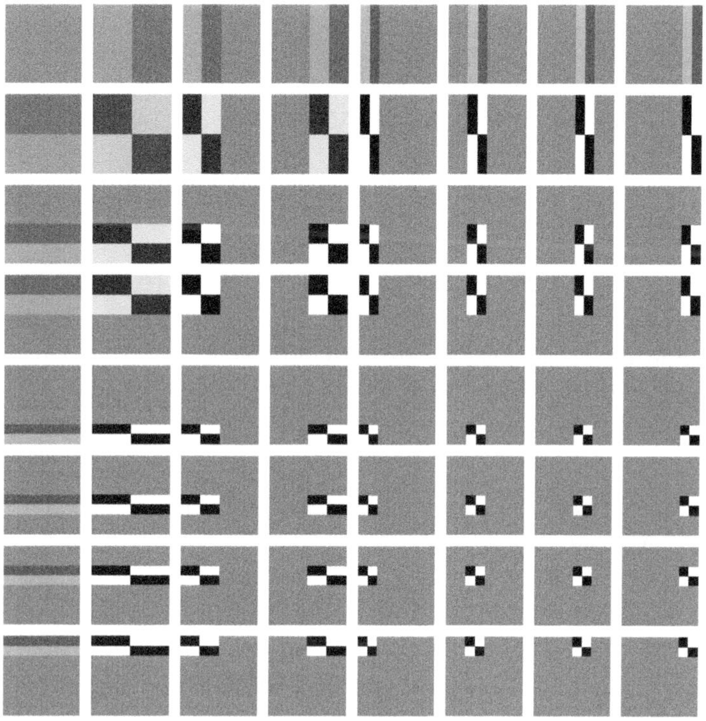

```
Needs["GraphicsImage`"] (* Draws 2D Haar Coefficients *)
n=8;
h[k_,x_]:=Module[{p,q}, If[k==0, 1/Sqrt[n],          (* h_0(x) *)
 p=0; While[2^p<=k ,p++]; p--; q=k-2^p+1; (* if k>0, calc. p, q *)
 If[(q-1)/(2^p)<=x && x<(q-.5)/(2^p),2^(p/2),
  If[(q-.5)/(2^p)<=x && x<q/(2^p),-2^(p/2),0]]]];
HaarMatrix=Table[h[k,x], {k,0,7}, {x,0,7/n,1/n}] //N;
HaarTensor=Array[Outer[Times, HaarMatrix[[#1]],HaarMatrix[[#2]]]&,
 {n,n}];
Show[GraphicsArray[Map[GraphicsImage[#, {-2,2}]&, HaarTensor,{2}]]]
```

Figure 7.21: The Basis Images of the Haar Transform for $n = 8$.

(or jth coordinate) of the $\mathbf{w}^{(i)}$ vectors. This, of course, is the variance of coefficient vector $\mathbf{c}^{(j)}$.

◇ **Exercise 7.10:** Show why this is true.

The off-diagonal elements of $(\mathbf{W}\cdot\mathbf{W}^T)$ are the covariances of the $\mathbf{w}^{(i)}$ vectors such that element $\left(\mathbf{W}\cdot\mathbf{W}^T\right)_{ab}$ is the covariance of the ath and bth coordinates of the $\mathbf{w}^{(i)}$'s. Equation (7.13) shows that this is also the dot product $\mathbf{c}^{(a)}\cdot\mathbf{c}^{(b)}$. One of the main aims of image transform is to decorrelate the coordinates of the vectors, and probability theory tells us that two coordinates are decorrelated if their covariance is zero (the other aim is energy compaction, but the two goals go hand in hand). Thus, our aim is to find a transformation matrix $\mathbf{A}$ such that the product $\mathbf{W}\cdot\mathbf{W}^T$ will be diagonal.

From the definition of matrix $\mathbf{W}$ we get

$$\mathbf{W}\cdot\mathbf{W}^T = (\mathbf{AV})\cdot(\mathbf{AV})^T = \mathbf{A}(\mathbf{V}\cdot\mathbf{V}^T)\mathbf{A}^T.$$

Matrix $\mathbf{V}\cdot\mathbf{V}^T$ is symmetric, and its elements are the covariances of the coordinates of vectors $\mathbf{v}^{(i)}$, i.e.,

$$\left(\mathbf{V}\cdot\mathbf{V}^T\right)_{ab} = \sum_{i=1}^{k} v_a^{(i)} v_b^{(i)}, \quad \text{for } a, b \in [1, n].$$

Since $\mathbf{V}\cdot\mathbf{V}^T$ is symmetric, its eigenvectors are orthogonal. We therefore normalize these vectors (i.e., make them orthonormal) and choose them as the rows of matrix $\mathbf{A}$. This produces the result

$$\mathbf{W}\cdot\mathbf{W}^T = \mathbf{A}(\mathbf{V}\cdot\mathbf{V}^T)\mathbf{A}^T = \begin{pmatrix} \lambda_1 & 0 & 0 & \cdots & 0 \\ 0 & \lambda_2 & 0 & \cdots & 0 \\ 0 & 0 & \lambda_3 & \cdots & 0 \\ \vdots & \vdots & & \vdots & \vdots \\ 0 & 0 & \cdots & 0 & \lambda_n \end{pmatrix}.$$

This choice of $\mathbf{A}$ results in a diagonal matrix $\mathbf{W}\cdot\mathbf{W}^T$ whose diagonal elements are the eigenvalues of $\mathbf{V}\cdot\mathbf{V}^T$. Matrix $\mathbf{A}$ is the Karhunen-Loève transformation matrix; its rows are the basis vectors of the KLT, and the energies (variances) of the transformed vectors are the eigenvalues $\lambda_1, \lambda_2, \ldots, \lambda_n$ of $\mathbf{V}\cdot\mathbf{V}^T$.

The basis vectors of the KLT are calculated from the original image pixels and are, therefore, data dependent. In a practical compression method, these vectors have to be included in the compressed stream, for the decoder's use, and this, combined with the fact that no fast method has been discovered for the calculation of the KLT, makes this transform less than ideal for practical applications.

7.8 The Discrete Cosine Transform

This important transform (DCT for short) has originated by [Ahmed et al. 74] and has been used and studied extensively since. Because of its importance for data compression, the DCT is treated here in detail. Section 7.8.1 introduces the mathematical expressions for the DCT in one dimension and two dimensions without any theoretical background or justifications. The use of the transform and its advantages for data compression are then demonstrated by several examples. Sections 7.8.2 and 7.8.3 cover the theory of the DCT and discuss its two interpretations as a rotation and as a basis of a vector space. Section 7.8.4 introduces the four DCT types, and Section 11.15.2 discusses the three-dimensional DCT. Section 7.8.5 describes ways to speed up the computation of the DCT, and Section 7.8.7 is a short discussion of the symmetry of the DCT and how it can be exploited for a hardware implementation. Several sections of important background material follow. Section 7.8.8 explains the QR decomposition of matrices. Section 7.8.9 introduces the concept of vector spaces and their bases. Section 7.8.10 shows how the rotation performed by the DCT relates to general rotations in three dimensions. Finally, the discrete sine transform is introduced in Section 7.8.11 together with the reasons that make it unsuitable for data compression.

For more information on this important transform, see [Ahmed et al. 74], [Rao and Yip 90], and [Britanak et al. 06].

It is possible to derive integer orthogonal transforms, where both the elements of the transform matrix and the resulting transform coefficients are integers. One such approach, based on the DCT and used for video compression, is discussed in detail in Section 9.11.9.

7.8.1 Introduction

The DCT in one dimension is given by

$$G_f = \sqrt{\frac{2}{n}} C_f \sum_{t=0}^{n-1} p_t \cos\left[\frac{(2t+1)f\pi}{2n}\right], \tag{7.14}$$

where

$$C_f = \begin{cases} \frac{1}{\sqrt{2}}, & f = 0, \\ 1, & f > 0, \end{cases} \quad \text{for } f = 0, 1, \ldots, n-1.$$

The input is a set of n data values p_t (pixels, audio samples, or other data), and the output is a set of n DCT *transform coefficients* (or weights) G_f. The first coefficient G_0 is called the DC coefficient, and the rest are referred to as the AC coefficients (these terms have been inherited from electrical engineering, where they stand for "direct current" and "alternating current"). Notice that the coefficients are real numbers even if the input data consists of integers. Similarly, the coefficients may be positive or negative even if the input data consists of nonnegative numbers only. This computation is straightforward but slow (Section 7.8.5 discusses faster versions). The decoder inputs the DCT coefficients in sets of n and uses the *inverse* DCT (IDCT) to reconstruct the original data values (also in groups of n). The IDCT in one dimension is given by

$$p_t = \sqrt{\frac{2}{n}} \sum_{j=0}^{n-1} C_j G_j \cos\left[\frac{(2t+1)j\pi}{2n}\right], \quad \text{for } t = 0, 1, \ldots, n-1. \tag{7.15}$$

The important feature of the DCT, the feature that makes it so useful in data compression, is that it takes correlated input data and concentrates its energy in just the first few transform coefficients. If the input data consists of correlated quantities, then most of the n transform coefficients produced by the DCT are zeros or small numbers, and only a few are large (normally the first ones). We will see that the early coefficients contain the important (low-frequency) image information and the later coefficients contain the less-important (high-frequency) image information. Compressing data with the DCT is therefore done by quantizing the coefficients. The small ones are quantized coarsely (possibly all the way to zero), and the large ones can be quantized finely to the nearest integer. After quantization, the coefficients (or variable-length codes assigned to the coefficients) are written on the compressed stream. Decompression is done by performing the inverse DCT on the quantized coefficients. This results in data items that are not identical to the original ones but are not much different.

In practical applications, the data to be compressed is partitioned into sets of n items each and each set is DCT-transformed and quantized individually. The value of n is critical. Small values of n such as 3, 4, or 6 result in many small sets of data items. Such a small set is transformed to a small set of coefficients where the energy of the original data is concentrated in a few coefficients, but there are only a few coefficients in such a set! Thus, there are not enough small coefficients to quantize. Large values of n result in a few large sets of data. The problem in such a case is that the individual data items of a large set are normally not correlated and therefore result in a set of transform coefficients where all the coefficients are large. Experience indicates that $n = 8$ is a good value, and most data compression methods that employ the DCT use this value of n.

The following experiment illustrates the power of the DCT in one dimension. We start with the set of eight correlated data items $\mathbf{p} = (12, 10, 8, 10, 12, 10, 8, 11)$, apply the DCT in one dimension to them, and find that it results in the eight coefficients

$$28.6375,\ 0.571202,\ 0.46194,\ 1.757,\ 3.18198,\ -1.72956,\ 0.191342,\ -0.308709.$$

These can be fed to the IDCT and transformed by it to precisely reconstruct the original data (except for small errors caused by limited machine precision). Our goal, however, is to compress the data by quantizing the coefficients. We first quantize them to $28.6, 0.6, 0.5, 1.8, 3.2, -1.8, 0.2, -0.3$, and apply the IDCT to get back

$$12.0254,\ 10.0233,\ 7.96054,\ 9.93097,\ 12.0164,\ 9.99321,\ 7.94354,\ 10.9989.$$

We then quantize the coefficients even more, to $28, 1, 1, 2, 3, -2, 0, 0$, and apply the IDCT to get back

$$12.1883,\ 10.2315,\ 7.74931,\ 9.20863,\ 11.7876,\ 9.54549,\ 7.82865,\ 10.6557.$$

Finally, we quantize the coefficients to $28, 0, 0, 2, 3, -2, 0, 0$, and still get back from the IDCT the sequence

$$11.236,\ 9.62443,\ 7.66286,\ 9.57302,\ 12.3471,\ 10.0146,\ 8.05304,\ 10.6842,$$

where the largest difference between an original value (12) and a reconstructed one (11.236) is 0.764 (or 6.4% of 12). The code that does all that is listed in Figure 7.22.

It seems magical that the eight original data items can be reconstructed to such high precision from just four transform coefficients. The explanation, however, relies on the following arguments instead of on magic: (1) The IDCT is given all eight transform coefficients, so it knows the positions, not just the values, of the nonzero coefficients. (2) The first few coefficients (the large ones) contain the important information of the original data items. The small coefficients, the ones that are quantized heavily, contain less important information (in the case of images, they contain the image details). (3) The original data is redundant because of pixel correlation.

```
n=8;
p={12.,10.,8.,10.,12.,10.,8.,11.};
c=Table[If[t==1, 0.7071, 1], {t,1,n}];
dct[i_]:=Sqrt[2/n]c[[i+1]]Sum[p[[t+1]]Cos[(2t+1)i Pi/16],{t,0,n-1}];
q=Table[dct[i],{i,0,n-1}] (* use precise DCT coefficients *)
q={28,0,0,2,3,-2,0,0}; (* or use quantized DCT coefficients *)
idct[t_]:=Sqrt[2/n]Sum[c[[j+1]]q[[j+1]]Cos[(2t+1)j Pi/16],{j,0,n-1}];
ip=Table[idct[t],{t,0,n-1}]
```

Figure 7.22: Experiments with the One-Dimensional DCT.

The following experiment illustrates the performance of the DCT when applied to decorrelated data items. Given the eight decorrelated data items −12, 24, −181, 209, 57.8, 3, −184, and −250, their DCT produces

$$-117.803,\ 166.823,\ -240.83,\ 126.887,\ 121.198,\ 9.02198,\ -109.496,\ -185.206.$$

When these coefficients are quantized to $(-120., 170., -240., 125., 120., 9., -110., -185)$ and fed into the IDCT, the result is

$$-12.1249,\ 25.4974,\ -179.852,\ 208.237,\ 55.5898,\ 0.364874,\ -185.42,\ -251.701,$$

where the maximum difference (between 3 and 0.364874) is 2.63513 or 88% of 3. Obviously, even with such fine quantization the reconstruction is not as good as with correlated data.

◇ **Exercise 7.11:** Compute the one-dimensional DCT [Equation (7.14)] of the eight correlated values 11, 22, 33, 44, 55, 66, 77, and 88. Show how to quantize them, and compute their IDCT from Equation (7.15).

An important relative of the DCT is the Fourier transform (Section 8.1), which also has a discrete version termed the DFT. The DFT has important applications, but it does not perform well in data compression because it assumes that the data to be transformed is periodic.

The following example illustrates the difference in performance between the DCT and the DFT. We start with the simple, highly correlated sequence of eight numbers $(8, 16, 24, 32, 40, 48, 56, 64)$. It is displayed graphically in Figure 7.23a. Applying the DCT to it yields $(100, -52, 0, -5, 0, -2, 0, 0.4)$. When this is quantized to $(100, -52, 0, -5, 0, 0, 0, 0)$ and transformed back, it produces $(8, 15, 24, 32, 40, 48, 57, 63)$, a sequence almost identical to the original input. Applying the DFT to the same input, on the other hand, yields $(36, 10, 10, 6, 6, 4, 4, 4)$. When this is quantized to $(36, 10, 10, 6, 0, 0, 0, 0)$ and is transformed back, it produces $(24, 12, 20, 32, 40, 51, 59, 48)$. This output is shown in Figure 7.23b, and it illustrates the tendency of the Fourier transform to produce a periodic result.

The DCT in one dimension can be used to compress one-dimensional data, such as audio samples. This chapter, however, discusses image compression which is based on the two-dimensional correlation of pixels (a pixel tends to resemble all its near neighbors,

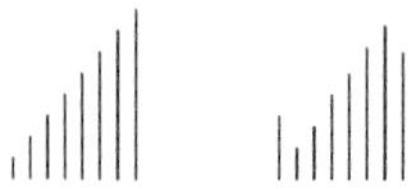

Figure 7.23: (a) One-Dimensional Input. (b) Its Inverse DFT.

not just those in its row). This is why practical image compression methods use the DCT in two dimensions. This version of the DCT is applied to small parts (data blocks) of the image. It is computed by applying the DCT in one dimension to each row of a data block, then to each column of the result. Because of the special way the DCT in two dimensions is computed, we say that it is separable in the two dimensions. Because it is applied to blocks of an image, we term it a "blocked transform." It is defined by

$$G_{ij}=\sqrt{\frac{2}{m}}\sqrt{\frac{2}{n}}C_iC_j\sum_{x=0}^{n-1}\sum_{y=0}^{m-1}p_{xy}\cos\left[\frac{(2y+1)j\pi}{2m}\right]\cos\left[\frac{(2x+1)i\pi}{2n}\right], \quad (7.16)$$

for $0\le i\le n-1$ and $0\le j\le m-1$ and for C_i and C_j defined by Equation (7.14). The first coefficient G_{00} is again termed the "DC coefficient," and the remaining coefficients are called the "AC coefficients."

The image is broken up into blocks of $n\times m$ pixels p_{xy} (with $n=m=8$ typically), and Equation (7.16) is used to produce a block of $n\times m$ DCT coefficients G_{ij} for each block of pixels. The coefficients are then quantized, which results in lossy but highly efficient compression. The decoder reconstructs a block of quantized data values by computing the IDCT whose definition is

$$p_{xy}=\sqrt{\frac{2}{m}}\sqrt{\frac{2}{n}}\sum_{i=0}^{n-1}\sum_{j=0}^{m-1}C_iC_jG_{ij}\cos\left[\frac{(2x+1)i\pi}{2n}\right]\cos\left[\frac{(2y+1)j\pi}{2m}\right], \quad (7.17)$$

where

$$C_f=\begin{cases}\frac{1}{\sqrt{2}}, & f=0\\ 1\ , & f>0,\end{cases}$$

for $0\le x\le n-1$ and $0\le y\le m-1$. We now show one way to compress an entire image with the DCT in several steps as follows:

1. The image is divided into k blocks of 8×8 pixels each. The pixels are denoted by p_{xy}. If the number of image rows (columns) is not divisible by 8, the bottom row (rightmost column) is duplicated as many times as needed.

2. The DCT in two dimensions [Equation (7.16)] is applied to each block B_i. The result is a block (we'll call it a vector) $W^{(i)}$ of 64 transform coefficients $w_j^{(i)}$ (where $j=0,1,\ldots,63$). The k vectors $W^{(i)}$ become the rows of matrix $\mathbf{W}$

$$\mathbf{W}=\begin{bmatrix} w_0^{(1)} & w_1^{(1)} & \ldots & w_{63}^{(1)}\\ w_0^{(2)} & w_1^{(2)} & \ldots & w_{63}^{(2)}\\ \vdots & \vdots & & \\ w_0^{(k)} & w_1^{(k)} & \ldots & w_{63}^{(k)}\end{bmatrix}.$$

3. The 64 columns of $\mathbf{W}$ are denoted by $C^{(0)}$, $C^{(1)}$, ..., $C^{(63)}$. The k elements of $C^{(j)}$ are $\left(w_j^{(1)}, w_j^{(2)}, \ldots, w_j^{(k)}\right)$. The first coefficient vector $C^{(0)}$ consists of the k DC coefficients.

4. Each vector $C^{(j)}$ is quantized separately to produce a vector $Q^{(j)}$ of quantized coefficients (JPEG does this differently; see Section 7.10.3). The elements of $Q^{(j)}$ are then written on the compressed stream. In practice, variable-length codes are assigned to the elements, and the codes, rather than the elements themselves, are written on the compressed stream. Sometimes, as in the case of JPEG, variable-length codes are assigned to runs of zero coefficients, to achieve better compression.

In practice, the DCT is used for lossy compression. For lossless compression (where the DCT coefficients are not quantized) the DCT is inefficient but can still be used, at least theoretically, because (1) most of the coefficients are small numbers and (2) there often are runs of zero coefficients. However, the small coefficients are real numbers, not integers, so it is not clear how to write them in full precision on the compressed stream and still have compression. Other image compression methods are better suited for lossless image compression.

The decoder reads the 64 quantized coefficient vectors $Q^{(j)}$ of k elements each, saves them as the columns of a matrix, and considers the k rows of the matrix weight vectors $W^{(i)}$ of 64 elements each (notice that these $W^{(i)}$'s are not identical to the original $W^{(i)}$'s because of the quantization). It then applies the IDCT [Equation (7.17)] to each weight vector, to reconstruct (approximately) the 64 pixels of block B_i. (Again, JPEG does this differently.)

We illustrate the performance of the DCT in two dimensions by applying it to two blocks of 8×8 values. The first block (Table 7.24a) has highly correlated integer values in the range $[8, 12]$, and the second block has random values in the same range. The first block results in a large DC coefficient, followed by small AC coefficients (including 20 zeros, Table 7.24b, where negative numbers are underlined). When the coefficients are quantized (Table 7.24c), the result, shown in Table 7.24d, is very similar to the original values. In contrast, the coefficients for the second block (Table 7.25b) include just one zero. When quantized (Table 7.25c) and transformed back, many of the 64 results are very different from the original values (Table 7.25d).

◇ **Exercise 7.12:** Show why the 64 values of Table 7.24 are correlated.

The next example illustrates the difference in the performance of the DCT when applied to a continuous-tone image and to a discrete-tone image. We start with the highly correlated pattern of Table 7.26. This is an idealized example of a continuous-tone image, since adjacent pixels differ by a constant amount except the pixel (underlined) at row 7, column 7. The 64 DCT coefficients of this pattern are listed in Table 7.27. It is clear that there are only a few dominant coefficients. Table 7.28 lists the coefficients after they have been coarsely quantized, so that only four nonzero coefficients remain! The results of performing the IDCT on these quantized coefficients are shown in Table 7.29. It is obvious that the four nonzero coefficients have reconstructed the original pattern to a high degree. The only visible difference is in row 7, column 7, which has changed from 12 to 17.55 (marked in both figures). The Matlab code for this computation is listed in Figure 7.34.

12	10	8	10	12	10	8	11
11	12	10	8	10	12	10	8
8	11	12	10	8	10	12	10
10	8	11	12	10	8	10	12
12	10	8	11	12	10	8	10
10	12	10	8	11	12	10	8
8	10	12	10	8	11	12	10
10	8	10	12	10	8	11	12

(a) Original data

81	0	0	0	0	0	0	0
0	1.57	0.61	1.90	0.38	$\underline{1.81}$	0.20	$\underline{0.32}$
0	$\underline{0.61}$	0.71	0.35	0	0.07	0	0.02
0	1.90	$\underline{0.35}$	4.76	0.77	$\underline{3.39}$	0.25	$\underline{0.54}$
0	$\underline{0.38}$	0	$\underline{0.77}$	8.00	0.51	0	0.07
0	$\underline{1.81}$	$\underline{0.07}$	$\underline{3.39}$	$\underline{0.51}$	1.57	0.56	0.25
0	$\underline{0.20}$	0	$\underline{0.25}$	0	$\underline{0.56}$	$\underline{0.71}$	0.29
0	$\underline{0.32}$	$\underline{0.02}$	$\underline{0.54}$	$\underline{0.07}$	0.25	$\underline{0.29}$	$\underline{0.90}$

(b) DCT coefficients

81	0	0	0	0	0	0	0
0	2	1	2	0	$\underline{2}$	0	0
0	$\underline{1}$	1	0	0	0	0	0
0	$\underline{2}$	0	5	1	$\underline{3}$	0	$\underline{1}$
0	0	0	$\underline{1}$	8	1	0	0
0	$\underline{2}$	0	$\underline{3}$	$\underline{1}$	2	1	0
0	0	0	0	0	$\underline{1}$	$\underline{1}$	0
0	0	0	$\underline{1}$	0	0	0	$\underline{1}$

(c) Quantized

12.29	10.26	7.92	9.93	11.51	9.94	8.18	10.97
10.90	12.06	10.07	7.68	10.30	11.64	10.17	8.18
7.83	11.39	12.19	9.62	8.28	10.10	11.64	9.94
10.15	7.74	11.16	11.96	9.90	8.28	10.30	11.51
12.21	10.08	8.15	11.38	11.96	9.62	7.68	9.93
10.09	12.10	9.30	8.15	11.16	12.19	10.07	7.92
7.87	9.50	12.10	10.08	7.74	11.39	12.06	10.26
9.66	7.87	10.09	12.21	10.15	7.83	10.90	12.29

(d) Reconstructed data (good)

Table 7.24: Two-Dimensional DCT of a Block of Correlated Values.

8	10	9	11	11	9	9	12
11	8	12	8	11	10	11	10
9	11	9	10	12	9	9	8
9	12	10	8	8	9	8	9
12	8	9	9	12	10	8	11
8	11	10	12	9	12	12	10
10	10	12	10	12	10	10	12
12	9	11	11	9	8	8	12

(a) Original data

79.12	0.98	0.64	$\underline{1.51}$	$\underline{0.62}$	$\underline{0.86}$	1.22	0.32
0.15	$\underline{1.64}$	$\underline{0.09}$	1.23	0.10	3.29	1.08	$\underline{2.97}$
$\underline{1.26}$	$\underline{0.29}$	$\underline{3.27}$	1.69	$\underline{0.51}$	1.13	1.52	1.33
$\underline{1.27}$	$\underline{0.25}$	$\underline{0.67}$	$\underline{0.15}$	1.63	$\underline{1.94}$	0.47	$\underline{1.30}$
$\underline{2.12}$	$\underline{0.67}$	$\underline{0.07}$	$\underline{0.79}$	0.13	$\underline{1.40}$	0.16	$\underline{0.15}$
$\underline{2.68}$	1.08	$\underline{1.99}$	$\underline{1.93}$	$\underline{1.77}$	$\underline{0.35}$	0	$\underline{0.80}$
1.20	2.10	$\underline{0.98}$	0.87	$\underline{1.55}$	$\underline{0.59}$	$\underline{0.98}$	2.76
$\underline{2.24}$	0.55	0.29	0.75	$\underline{2.40}$	$\underline{0.05}$	0.06	1.14

(b) DCT coefficients

79	1	1	$\underline{2}$	$\underline{1}$	$\underline{1}$	1	0
0	$\underline{2}$	0	1	0	3	1	$\underline{3}$
$\underline{1}$	0	$\underline{3}$	2	0	1	2	1
$\underline{1}$	0	$\underline{1}$	0	2	$\underline{2}$	0	$\underline{1}0$
$\underline{2}0$	$\underline{1}$	0	$\underline{1}$	0	$\underline{1}0$	0	0
$\underline{3}$	$\underline{1}$	$\underline{2}$	$\underline{2}$	$\underline{2}$	0	0	$\underline{1}$
1	2	$\underline{1}$	1	$\underline{2}$	$\underline{1}$	$\underline{1}$	3
$\underline{2}$	1	0	1	$\underline{2}$	0	0	1

(c) Quantized

7.59	9.23	8.33	11.88	7.12	12.47	6.98	8.56
12.09	7.97	9.3	11.52	9.28	11.62	10.98	12.39
11.02	10.06	13.81	6.5	10.82	8.28	13.02	7.54
8.46	10.22	11.16	9.57	8.45	7.77	10.28	11.89
9.71	11.93	8.04	9.59	8.04	9.7	8.59	12.14
10.27	13.58	9.21	11.83	9.99	10.66	7.84	11.27
8.34	10.32	10.53	9.9	8.31	9.34	7.47	8.93
10.61	9.04	13.66	6.04	13.47	7.65	10.97	8.89

(d) Reconstructed data (bad)

Table 7.25: Two-Dimensional DCT of a Block of Random Values.

Tables 7.30 through 7.33 show the same process applied to a Y-shaped pattern, typical of a discrete-tone image. The quantization, shown in Table 7.32, is light. The coefficients have only been truncated to the nearest integer. It is easy to see that the reconstruction, shown in Table 7.33, isn't as good as before. Quantities that should have been 10 are between 8.96 and 10.11. Quantities that should have been zero are as big as 0.86. The conclusion is that the DCT performs well on continuous-tone images but is less efficient when applied to a discrete-tone image.

7.8.2 The DCT as a Basis

The discussion so far has concentrated on how to use the DCT for compressing one-dimensional and two-dimensional data. The aim of this section and the next one is to show why the DCT works the way it does and how Equations (7.14) and (7.16) have been derived. This topic is approached from two different directions. The first interpretation of the DCT is as a special basis of an n-dimensional vector space. We

00	10	20	30	30	20	10	00
10	20	30	40	40	30	20	10
20	30	40	50	50	40	30	20
30	40	50	60	60	50	40	30
30	40	50	60	60	50	40	30
20	30	40	50	50	40	30	20
10	20	30	40	40	30	12	10
00	10	20	30	30	20	10	00

Table 7.26: A Continuous-Tone Pattern.

239	1.19	−89.76	−0.28	1.00	−1.39	−5.03	−0.79
1.18	−1.39	0.64	0.32	−1.18	1.63	−1.54	0.92
−89.76	0.64	−0.29	−0.15	0.54	−0.75	0.71	−0.43
−0.28	0.32	−0.15	−0.08	0.28	−0.38	0.36	−0.22
1.00	−1.18	0.54	0.28	−1.00	1.39	−1.31	0.79
−1.39	1.63	−0.75	−0.38	1.39	−1.92	1.81	−1.09
−5.03	−1.54	0.71	0.36	−1.31	1.81	−1.71	1.03
−0.79	0.92	−0.43	−0.22	0.79	−1.09	1.03	−0.62

Table 7.27: Its DCT Coefficients.

239	1	-90	0	0	0	0	0
0	0	0	0	0	0	0	0
-90	0	0	0	0	0	0	0
0	0	0	0	0	0	0	0
0	0	0	0	0	0	0	0
0	0	0	0	0	0	0	0
0	0	0	0	0	0	0	0
0	0	0	0	0	0	0	0

Table 7.28: Quantized Heavily to Just Four Nonzero Coefficients.

0.65	9.23	21.36	29.91	29.84	21.17	8.94	0.30
9.26	17.85	29.97	38.52	38.45	29.78	17.55	8.91
21.44	30.02	42.15	50.70	50.63	41.95	29.73	21.09
30.05	38.63	50.76	59.31	59.24	50.56	38.34	29.70
30.05	38.63	50.76	59.31	59.24	50.56	38.34	29.70
21.44	30.02	42.15	50.70	50.63	41.95	29.73	21.09
9.26	17.85	29.97	38.52	38.45	29.78	17.55	8.91
0.65	9.23	21.36	29.91	29.84	21.17	8.94	0.30

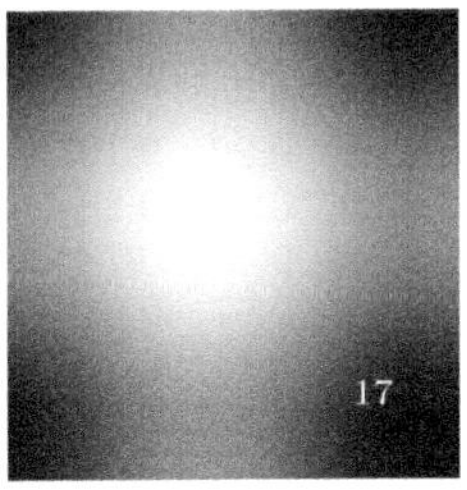

Table 7.29: Results of IDCT.

00	10	00	00	00	00	00	10
00	00	10	00	00	00	10	00
00	00	00	10	00	10	00	00
00	00	00	00	10	00	00	00
00	00	00	00	10	00	00	00
00	00	00	00	10	00	00	00
00	00	00	00	10	00	00	00
00	00	00	00	10	00	00	00

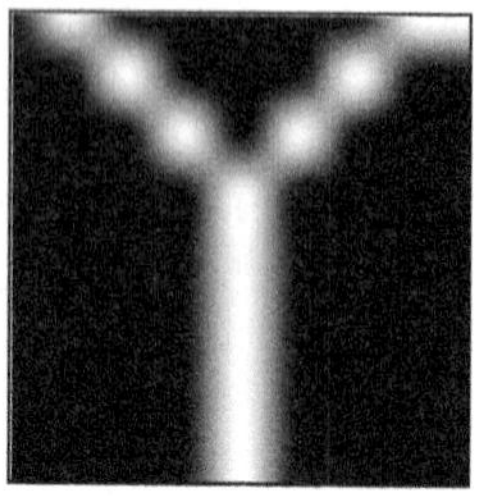

Table 7.30: A Discrete-Tone Image (Y).

13.75	−3.11	−8.17	2.46	3.75	−6.86	−3.38	6.59
4.19	−0.29	6.86	−6.85	−7.13	4.48	1.69	−7.28
1.63	0.19	6.40	−4.81	−2.99	−1.11	−0.88	−0.94
−0.61	0.54	5.12	−2.31	1.30	−6.04	−2.78	3.05
−1.25	0.52	2.99	−0.20	3.75	−7.39	−2.59	1.16
−0.41	0.18	0.65	1.03	3.87	−5.19	−0.71	−4.76
0.68	−0.15	−0.88	1.28	2.59	−1.92	1.10	−9.05
0.83	−0.21	−0.99	0.82	1.13	−0.08	1.31	−7.21

Table 7.31: Its DCT Coefficients.

13.75	−3	−8	2	3	−6	−3	6
4	−0	6	−6	−7	4	1	−7
1	0	6	−4	−2	−1	−0	−0
−0	0	5	−2	1	−6	−2	3
−1	0	2	−0	3	−7	−2	1
−0	0	0	1	3	−5	−0	−4
0	−0	−0	1	2	−1	1	−9
0	−0	−0	0	1	−0	1	−7

Table 7.32: Quantized Lightly by Truncating to Integer.

-0.13	8.96	0.55	-0.27	0.27	0.86	0.15	9.22
0.32	0.22	9.10	0.40	0.84	-0.11	9.36	-0.14
0.00	0.62	-0.20	9.71	-1.30	8.57	0.28	-0.33
-0.58	0.44	0.78	0.71	10.11	1.14	0.44	-0.49
-0.39	0.67	0.07	0.38	8.82	0.09	0.28	0.41
0.34	0.11	0.26	0.18	8.93	0.41	0.47	0.37
0.09	-0.32	0.78	-0.20	9.78	0.05	-0.09	0.49
0.16	-0.83	0.09	0.12	9.15	-0.11	-0.08	0.01

Table 7.33: The IDCT. Bad Results.

```
% 8x8 correlated values
n=8;
p=[00,10,20,30,30,20,10,00; 10,20,30,40,40,30,20,10; 20,30,40,50,50,40,30,20; ...
 30,40,50,60,60,50,40,30; 30,40,50,60,60,50,40,30; 20,30,40,50,50,40,30,20; ...
 10,20,30,40,40,30,12,10; 00,10,20,30,30,20,10,00];
figure(1), imagesc(p), colormap(gray), axis square, axis off
dct=zeros(n,n);
for j=0:7
  for i=0:7
    for x=0:7
      for y=0:7
dct(i+1,j+1)=dct(i+1,j+1)+p(x+1,y+1)*cos((2*y+1)*j*pi/16)*cos((2*x+1)*i*pi/16);
      end;
    end;
  end;
end;
dct=dct/4; dct(1,:)=dct(1,:)*0.7071; dct(:,1)=dct(:,1)*0.7071;
dct
quant=[239,1,-90,0,0,0,0,0; 0,0,0,0,0,0,0,0; -90,0,0,0,0,0,0,0; 0,0,0,0,0,0,0,0; ...
 0,0,0,0,0,0,0,0; 0,0,0,0,0,0,0,0; 0,0,0,0,0,0,0,0; 0,0,0,0,0,0,0,0];
idct=zeros(n,n);
for x=0:7
  for y=0:7
    for i=0:7
if i==0 ci=0.7071; else ci=1; end;
      for j=0:7
  if j==0 cj=0.7071; else cj=1; end;
idct(x+1,y+1)=idct(x+1,y+1)+ ...
              ci*cj*quant(i+1,j+1)*cos((2*y+1)*j*pi/16)*cos((2*x+1)*i*pi/16);
      end;
    end;
  end;
end;
idct=idct/4;
idct
figure(2), imagesc(idct), colormap(gray), axis square, axis off
```

Figure 7.34: Code for Highly Correlated Pattern.

show that transforming a given data vector **p** by the DCT is equivalent to representing it by this special basis that isolates the various frequencies contained in the vector. Thus, the DCT coefficients resulting from the DCT transform of vector **p** indicate the various frequencies in the vector. The lower frequencies contain the important information in **p**, whereas the higher frequencies correspond to the details of the data in **p** and are therefore less important. This is why they can be quantized coarsely.

The second interpretation of the DCT is as a rotation, as shown intuitively for $n = 2$ (two-dimensional points) in Figure 7.17. This interpretation considers the DCT a rotation matrix that rotates an n-dimensional point with identical coordinates $(x, x, \ldots, x)$ from its original location to the x axis, where its coordinates become $(\alpha, \epsilon_2, \ldots, \epsilon_n)$ where the various ϵ_i are small numbers or zeros. Both interpretations are illustrated for the case $n = 3$, because this is the largest number of dimensions where it is possible to visualize geometric transformations.

For the special case $n = 3$, Equation (7.14) reduces to

$$G_f = \sqrt{\frac{2}{3}} C_f \sum_{t=0}^{2} p_t \cos\left[\frac{(2t+1)f\pi}{6}\right], \quad \text{for } f = 0, 1, 2.$$

Temporarily ignoring the normalization factors $\sqrt{2/3}$ and C_f, this can be written in matrix notation as

$$\begin{bmatrix} G_0 \\ G_1 \\ G_2 \end{bmatrix} = \begin{bmatrix} \cos 0 & \cos 0 & \cos 0 \\ \cos\frac{\pi}{6} & \cos\frac{3\pi}{6} & \cos\frac{5\pi}{6} \\ \cos 2\frac{\pi}{6} & \cos 2\frac{3\pi}{6} & \cos 2\frac{5\pi}{6} \end{bmatrix} \begin{bmatrix} p_0 \\ p_1 \\ p_2 \end{bmatrix} = \mathbf{D}\cdot\mathbf{p}.$$

Thus, the DCT of the three data values $\mathbf{p} = (p_0, p_1, p_2)$ is obtained as the product of the DCT matrix $\mathbf{D}$ and the vector $\mathbf{p}$. We can therefore think of the DCT as the product of a DCT matrix and a data vector, where the matrix is constructed as follows: Select the three angles $\pi/6$, $3\pi/6$, and $5\pi/6$ and compute the three basis vectors $\cos(f\theta)$ for $f = 0$, 1, and 2, and for the three angles. The results are listed in Table 7.35 for the benefit of the reader.

θ	0.5236	1.5708	2.618
$\cos 0\theta$	1.	1.	1.
$\cos 1\theta$	0.866	0	−0.866
$\cos 2\theta$	0.5	−1	0.5

Table 7.35: The DCT Matrix for $n = 3$.

Because of the particular choice of the three angles, these vectors are orthogonal but not orthonormal. Their magnitudes are $\sqrt{3}$, $\sqrt{1.5}$, and $\sqrt{1.5}$, respectively. Normalizing them results in the three vectors $\mathbf{v}_1 = (0.5774, 0.5774, 0.5774)$, $\mathbf{v}_2 = (0.7071, 0, -0.7071)$, and $\mathbf{v}_3 = (0.4082, -0.8165, 0.4082)$. When stacked vertically, they produce the following 3×3 matrix

$$\mathbf{M} = \begin{bmatrix} 0.5774 & 0.5774 & 0.5774 \\ 0.7071 & 0 & -0.7071 \\ 0.4082 & -0.8165 & 0.4082 \end{bmatrix}. \tag{7.18}$$

[Equation (7.14) tells us how to normalize these vectors: Multiply each by $\sqrt{2/3}$, and then multiply the first by $1/\sqrt{2}$.] Notice that as a result of the normalization the columns of $\mathbf{M}$ have also become orthonormal, so $\mathbf{M}$ is an orthonormal matrix (such matrices have special properties).

The steps of computing the DCT matrix for an arbitrary n are as follows:

1. Select the n angles $\theta_j = (j+0.5)\pi/n$ for $j = 0, \ldots, n-1$. If we divide the interval $[0, \pi]$ into n equal-size segments, these angles are the centerpoints of the segments.

2. Compute the n vectors $\mathbf{v}_k$ for $k = 0, 1, 2, \ldots, n-1$, each with the n components $\cos(k\theta_j)$.

3. Normalize each of the n vectors and arrange them as the n rows of a matrix.

The angles selected for the DCT are $\theta_j = (j+0.5)\pi/n$, so the components of each vector $\mathbf{v}_k$ are $\cos[k(j+0.5)\pi/n]$ or $\cos[k(2j+1)\pi/(2n)]$. Section 7.8.4 covers three other ways to select such angles. This choice of angles has the following two useful properties: (1) The resulting vectors are orthogonal, and (2) for increasing values of k, the n vectors $\mathbf{v}_k$ contain increasing frequencies (Figure 7.36). For $n=3$, the top row of $\mathbf{M}$ [Equation (7.18)] corresponds to zero frequency, the middle row (whose elements become monotonically smaller) represents low frequency, and the bottom row (with three elements that first go down, then up) represents high frequency. Given a three-dimensional vector $\mathbf{v}=(v_1,v_2,v_3)$, the product $\mathbf{M}\cdot\mathbf{v}$ is a triplet whose components indicate the magnitudes of the various frequencies included in $\mathbf{v}$; they are *frequency coefficients*. [Strictly speaking, the product is $\mathbf{M}\cdot\mathbf{v}^T$, but we ignore the transpose in cases where the meaning is clear.] The following three extreme examples illustrate the meaning of this statement.

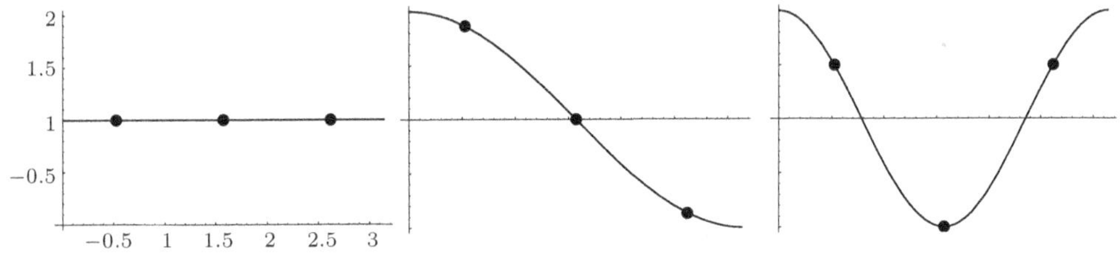

Figure 7.36: Increasing Frequencies.

The first example is $\mathbf{v}=(v,v,v)$. The three components of $\mathbf{v}$ are identical, so they correspond to zero frequency. The product $\mathbf{M}\cdot\mathbf{v}$ produces the frequency coefficients $(1.7322v,0,0)$, indicating no high frequencies. The second example is $\mathbf{v}=(v,0,-v)$. The three components of $\mathbf{v}$ vary slowly from v to $-v$, so this vector contains a low frequency. The product $\mathbf{M}\cdot\mathbf{v}$ produces the coefficients $(0,1.4142v,0)$, confirming this result. The third example is $\mathbf{v}=(v,-v,v)$. The three components of $\mathbf{v}$ vary from v to $-v$ to v, so this vector contains a high frequency. The product $\mathbf{M}\cdot\mathbf{v}$ produces $(0,0,1.6329v)$, again indicating the correct frequency.

These examples are not very realistic because the vectors being tested are short, simple, and contain a single frequency each. Most vectors are more complex and contain several frequencies, which makes this method useful. A simple example of a vector with two frequencies is $\mathbf{v}=(1.,0.33,-0.34)$. The product $\mathbf{M}\cdot\mathbf{v}$ results in $(0.572,0.948,0)$ which indicates a large medium frequency, small zero frequency, and no high frequency. This makes sense once we realize that the vector being tested is the sum $0.33(1,1,1)+0.67(1,0,-1)$. A similar example is the sum $0.9(-1,1,-1)+0.1(1,1,1)=(-0.8,1,-0.8)$, which when multiplied by $\mathbf{M}$ produces $(-0.346,0,-1.469)$. On the other hand, a vector with random components, such as $(1,0,0.33)$, typically contains roughly equal amounts of all three frequencies and produces three large frequency coefficients. The product $\mathbf{M}\cdot(1,0,0.33)$ produces $(0.77,0.47,0.54)$ because $(1,0,0.33)$ is the sum

$$0.33(1,1,1)+0.33(1,0,-1)+0.33(1,-1,1).$$

Notice that if $\mathbf{M}\cdot\mathbf{v} = \mathbf{c}$, then $\mathbf{M}^T\cdot\mathbf{c} = \mathbf{M}^{-1}\cdot\mathbf{c} = \mathbf{v}$. The original vector $\mathbf{v}$ can therefore be reconstructed from its frequency coefficients (up to small differences due to the limited precision of machine arithmetic). The inverse $\mathbf{M}^{-1}$ of $\mathbf{M}$ is also its transpose $\mathbf{M}^T$ because $\mathbf{M}$ is orthonormal.

A three-dimensional vector can have only three frequencies zero, medium, and high. Similarly, an n-dimensional vector can have n different frequencies, which this method can identify. We concentrate on the case $n = 8$ and start with the DCT in one dimension. Figure 7.37 shows eight cosine waves of the form $\cos(f\theta_j)$, for $0 \le \theta_j \le \pi$, with frequencies $f = 0, 1, \ldots, 7$. Each wave is sampled at the eight points

$$\theta_j = \frac{\pi}{16},\ \frac{3\pi}{16},\ \frac{5\pi}{16},\ \frac{7\pi}{16},\ \frac{9\pi}{16},\ \frac{11\pi}{16},\ \frac{13\pi}{16},\ \frac{15\pi}{16} \tag{7.19}$$

to form one basis vector $\mathbf{v}_f$, and the resulting eight vectors $\mathbf{v}_f$, $f = 0, 1, \ldots, 7$ (a total of 64 numbers) are shown in Table 7.38 (Equation (9.6) is the normalized form). They serve as the basis matrix of the DCT. Notice the similarity between this table and matrix $\mathbf{W}$ of Equation (7.6).

Because of the particular choice of the eight sample points, the $\mathbf{v}_i$'s are orthogonal. This is easy to check directly with appropriate mathematical software, but Section 7.8.4 describes a more elegant way of proving this property. After normalization, the $\mathbf{v}_i$'s can be considered either as an 8×8 transformation matrix (specifically, a rotation matrix, since it is orthonormal) or as a set of eight orthogonal vectors that constitute the basis of a vector space. Any vector $\mathbf{p}$ in this space can be expressed as a linear combination of the $\mathbf{v}_i$'s. As an example, we select the eight (correlated) numbers $\mathbf{p} = (0.6, 0.5, 0.4, 0.5, 0.6, 0.5, 0.4, 0.55)$ as our test data and express $\mathbf{p}$ as a linear combination $\mathbf{p} = \sum w_i\mathbf{v}_i$ of the eight basis vectors $\mathbf{v}_i$. Solving this system of eight equations yields the eight weights

$$\begin{array}{llll} w_0 = 0.506, & w_1 = 0.0143, & w_2 = 0.0115, & w_3 = 0.0439, \\ w_4 = 0.0795, & w_5 = -0.0432, & w_6 = 0.00478, & w_7 = -0.0077. \end{array}$$

Weight w_0 is not much different from the elements of $\mathbf{p}$, but the other seven weights are much smaller. This is how the DCT (or any other orthogonal transform) can lead to compression. The eight weights can be quantized and written on the compressed stream, where they occupy less space than the eight components of $\mathbf{p}$.

Figure 7.39 illustrates this linear combination graphically. Each of the eight $\mathbf{v}_i$'s is shown as a row of eight small, gray rectangles (a basis image) where a value of $+1$ is painted white and -1 is black. The eight elements of vector $\mathbf{p}$ are also displayed as a row of eight grayscale pixels.

To summarize, we interpret the DCT in one dimension as a set of basis images that have higher and higher frequencies. Given a data vector, the DCT separates the frequencies in the data and represents the vector as a linear combination (or a weighted sum) of the basis images. The weights are the DCT coefficients. This interpretation can be extended to the DCT in two dimensions. We apply Equation (7.16) to the case $n = 8$ to create 64 small basis images of 8×8 pixels each. The 64 images are then used as a basis of a 64-dimensional vector space. Any image B of 8×8 pixels can be expressed as

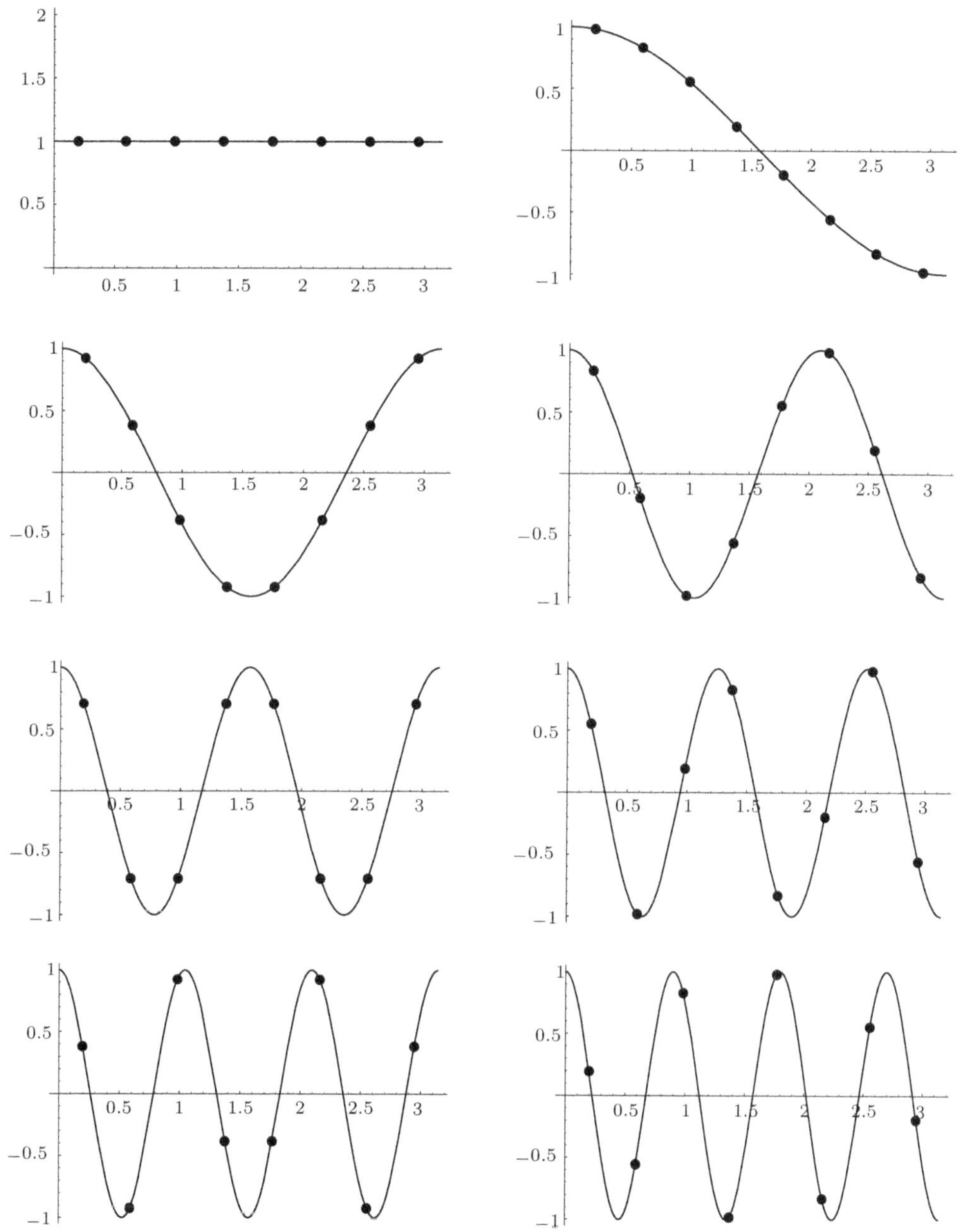

Figure 7.37: Angle and Cosine Values for an 8-Point DCT.

θ	0.196	0.589	0.982	1.374	1.767	2.160	2.553	2.945
$\cos 0\theta$	1.	1.	1.	1.	1.	1.	1.	1.
$\cos 1\theta$	0.981	0.831	0.556	0.195	−0.195	−0.556	−0.831	−0.981
$\cos 2\theta$	0.924	0.383	−0.383	−0.924	−0.924	−0.383	0.383	0.924
$\cos 3\theta$	0.831	−0.195	−0.981	−0.556	0.556	0.981	0.195	−0.831
$\cos 4\theta$	0.707	−0.707	−0.707	0.707	0.707	−0.707	−0.707	0.707
$\cos 5\theta$	0.556	−0.981	0.195	0.831	−0.831	−0.195	0.981	−0.556
$\cos 6\theta$	0.383	−0.924	0.924	−0.383	−0.383	0.924	−0.924	0.383
$\cos 7\theta$	0.195	−0.556	0.831	−0.981	0.981	−0.831	0.556	−0.195

Table 7.38: The Unnormalized DCT Matrix in One Dimension for $n = 8$.

```
Table[N[t],{t,Pi/16,15Pi/16,Pi/8}]
dctp[pw_]:=Table[N[Cos[pw t]],{t,Pi/16,15Pi/16,Pi/8}]
dctp[0]
dctp[1]
...
dctp[7]
```

Code for Table 7.38.

```
dct[pw_]:=Plot[Cos[pw t], {t,0,Pi}, DisplayFunction->Identity,
 AspectRatio->Automatic];
dcdot[pw_]:=ListPlot[Table[{t,Cos[pw t]},{t,Pi/16,15Pi/16,Pi/8}],
 DisplayFunction->Identity]
Show[dct[0],dcdot[0], Prolog->AbsolutePointSize[4],
 DisplayFunction->$DisplayFunction]
...
Show[dct[7],dcdot[7], Prolog->AbsolutePointSize[4],
 DisplayFunction->$DisplayFunction]
```

Code for Figure 7.37.

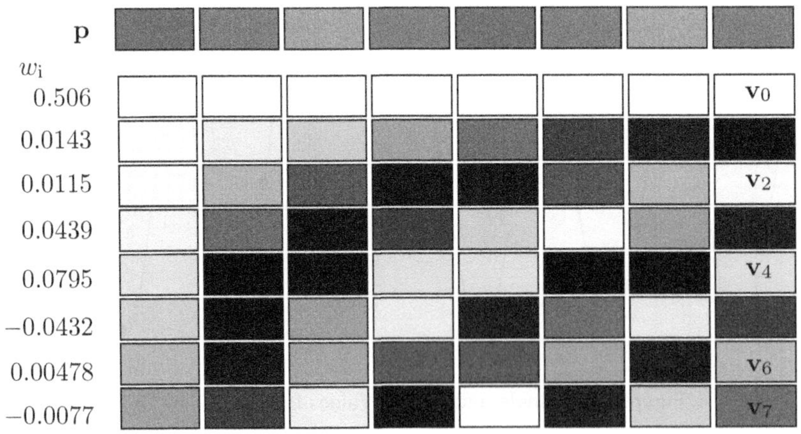

Figure 7.39: A Graphic Representation of the One-Dimensional DCT.

a linear combination of the basis images, and the 64 weights of this linear combination are the DCT coefficients of B.

Figure 7.40 shows the graphic representation of the 64 basis images of the two-dimensional DCT for $n = 8$. A general element (i, j) in this figure is the 8×8 image obtained by calculating the product $\cos(i \cdot s)\cos(j \cdot t)$, where s and t are varied independently over the values listed in Equation (7.19) and i and j vary from 0 to 7. This figure can easily be generated by the *Mathematica* code shown with it. The alternative code shown is a modification of code in [Watson 94], and it requires the `GraphicsImage.m` package, which is not widely available.

Using appropriate software, it is easy to perform DCT calculations and display the results graphically. Figure 7.41a shows a random 8×8 data unit consisting of zeros and ones. The same unit is shown in Figure 7.41b graphically, with 1 as white and 0 as black. Figure 7.41c shows the weights by which each of the 64 DCT basis images has to be multiplied in order to reproduce the original data unit. In this figure, zero is shown in neutral gray, positive numbers are bright (notice how bright the DC weight is), and negative numbers are shown as dark. Figure 7.41d shows the weights numerically. The *Mathematica* code that does all that is also listed. Figure 7.42 is similar, but for a very regular data unit.

◇ **Exercise 7.13:** . Imagine an 8×8 block of values where all the odd-numbered rows consist of 1's and all the even-numbered rows contain zeros. What can we say about the DCT weights of this block?

7.8.3 The DCT as a Rotation

The second interpretation of matrix $\mathbf{M}$ [Equation (7.18)] is as a rotation. We already know that $\mathbf{M}\cdot(v, v, v)$ results in $(1.7322v, 0, 0)$ and this can be interpreted as a rotation of point (v, v, v) to the point $(1.7322v, 0, 0)$. The former point is located on the line that makes equal angles with the three coordinate axes, and the latter point is on the x axis. When considered in terms of adjacent pixels, this rotation has a simple meaning. Imagine three adjacent pixels in an image. They are normally similar, so we start by examining the case where they are identical. When three identical pixels are considered the coordinates of a point in three dimensions, that point is located on the line $x = y = z$. Rotating this line to the x axis brings our point to that axis where its x coordinate hasn't changed much and its y and z coordinates are zero. This is how such a rotation leads to compression. Generally, three adjacent pixels p_1, p_2, and p_3 are similar but not identical, which locates the point (p_1, p_2, p_3) somewhat off the line $x = y = z$. After the rotation, the point will end up near the x axis, where its y and z coordinates will be small numbers.

This interpretation of $\mathbf{M}$ as a rotation makes sense because $\mathbf{M}$ is orthonormal and any orthonormal matrix is a rotation matrix. However, the determinant of a rotation matrix is 1, whereas the determinant of our matrix is -1. An orthonormal matrix whose determinant is -1 performs an *improper rotation* (a rotation combined with a reflection). To get a better insight into the transformation performed by $\mathbf{M}$, we apply the QR matrix decomposition technique (Section 7.8.8) to decompose $\mathbf{M}$ into the matrix

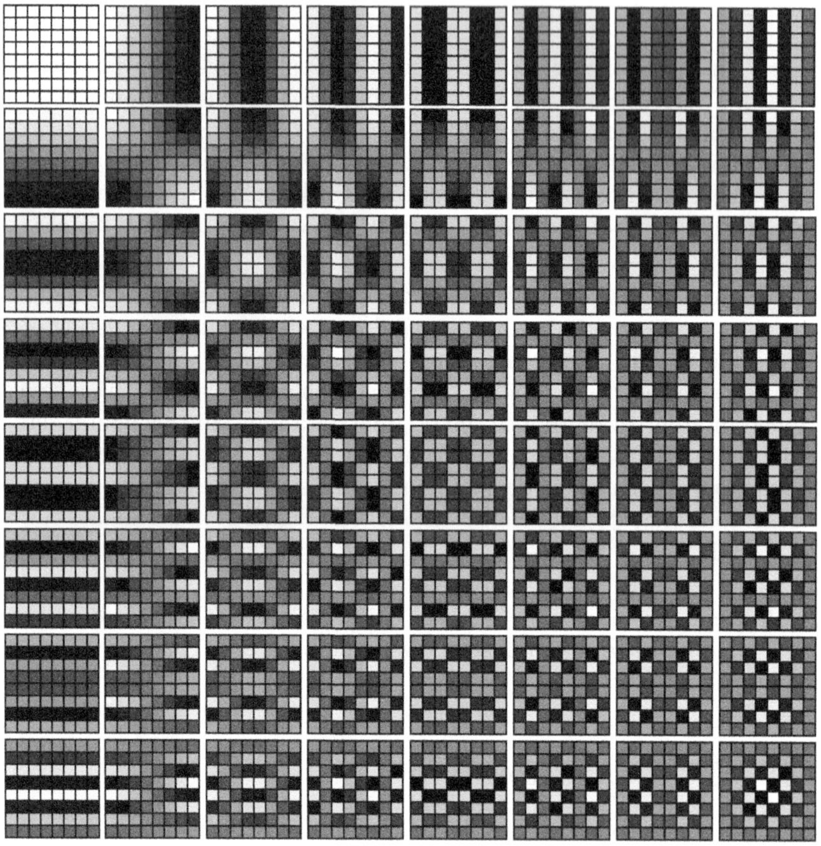

Figure 7.40: The 64 Basis Images of the Two-Dimensional DCT.

```
dctp[fs_,ft_]:=Table[SetAccuracy[N[(1.-Cos[fs s]Cos[ft t])/2],3],
 {s,Pi/16,15Pi/16,Pi/8},{t,Pi/16,15Pi/16,Pi/8}]//TableForm
dctp[0,0]
dctp[0,1]
...
dctp[7,7]
```

Code for Figure 7.40.

```
Needs["GraphicsImage`"] (* Draws 2D DCT Coefficients *)
DCTMatrix=Table[If[k==0,Sqrt[1/8],Sqrt[1/4]Cos[Pi(2j+1)k/16]],
 {k,0,7}, {j,0,7}] //N;
DCTTensor=Array[Outer[Times, DCTMatrix[[#1]],DCTMatrix[[#2]]]&,
 {8,8}];
Show[GraphicsArray[Map[GraphicsImage[#, {-.25,.25}]&, DCTTensor,{2}]]]
```

Alternative Code for Figure 7.40.

10011101
11001011
01100100
00010010
01001011
11100110
11001011
01010010

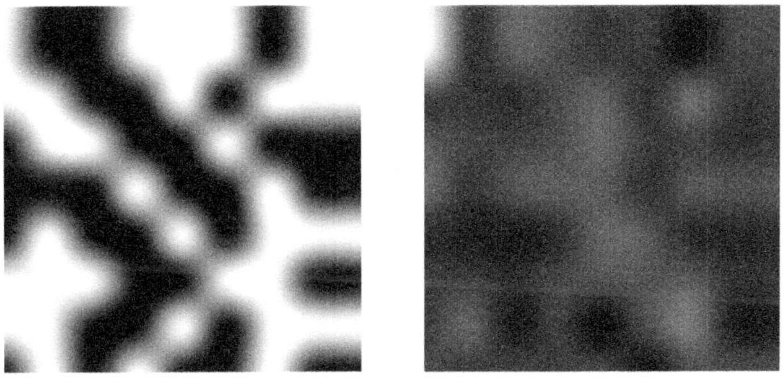

(a) (b) (c)

4.000	−0.133	0.637	0.272	−0.250	−0.181	−1.076	0.026
0.081	−0.178	−0.300	0.230	0.694	−0.309	0.875	−0.127
0.462	0.125	0.095	0.291	0.868	−0.070	0.021	−0.280
0.837	−0.194	0.455	0.583	0.588	−0.281	0.448	0.383
−0.500	−0.635	−0.749	−0.346	0.750	0.557	−0.502	−0.540
−0.167	0	−0.366	0.146	0.393	0.448	0.577	−0.268
−0.191	0.648	−0.729	−0.008	−1.171	0.306	1.155	−0.744
0.122	−0.200	0.038	−0.118	0.138	−1.154	0.134	0.148

(d)

Figure 7.41: An Example of the DCT in Two Dimensions.

```
DCTMatrix=Table[If[k==0,Sqrt[1/8],Sqrt[1/4]Cos[Pi(2j+1)k/16]],
 {k,0,7}, {j,0,7}] //N;
DCTTensor=Array[Outer[Times, DCTMatrix[[#1]],DCTMatrix[[#2]]]&,
 {8,8}];
img={{1,0,0,1,1,1,0,1},{1,1,0,0,1,0,1,1},
{0,1,1,0,0,1,0,0},{0,0,0,1,0,0,1,0},
{0,1,0,0,1,0,1,1},{1,1,1,0,0,1,1,0},
{1,1,0,0,1,0,1,1},{0,1,0,1,0,0,1,0}};
ShowImage[Reverse[img]]
dctcoeff=Array[(Plus @@ Flatten[DCTTensor[[#1,#2]] img])&,{8,8}];
dctcoeff=SetAccuracy[dctcoeff,4];
dctcoeff=Chop[dctcoeff,.001];
MatrixForm[dctcoeff]
ShowImage[Reverse[dctcoeff]]
```

Code for Figure 7.41.

01010101
01010101
01010101
01010101
01010101
01010101
01010101
01010101

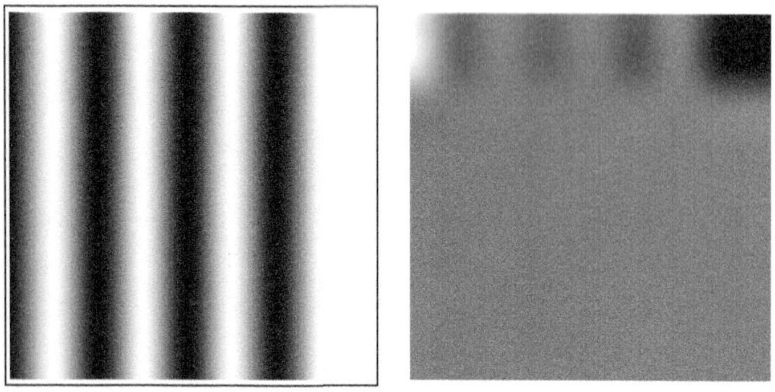

(a) (b) (c)

4.000	−0.721	0	−0.850	0	−1.273	0	−3.625
0	0	0	0	0	0	0	0
0	0	0	0	0	0	0	0
0	0	0	0	0	0	0	0
0	0	0	0	0	0	0	0
0	0	0	0	0	0	0	0
0	0	0	0	0	0	0	0
0	0	0	0	0	0	0	0

(d)

Figure 7.42: An Example of the DCT in Two Dimensions.

> Some painters transform the sun into a yellow spot; others transform a yellow spot into the sun.
>
> —Pablo Picasso

```
DCTMatrix=Table[If[k==0,Sqrt[1/8],Sqrt[1/4]Cos[Pi(2j+1)k/16]],
 {k,0,7}, {j,0,7}] //N;
DCTTensor=Array[Outer[Times, DCTMatrix[[#1]],DCTMatrix[[#2]]]&,
 {8,8}];
img={{0,1,0,1,0,1,0,1},{0,1,0,1,0,1,0,1},
 {0,1,0,1,0,1,0,1},{0,1,0,1,0,1,0,1},{0,1,0,1,0,1,0,1},
 {0,1,0,1,0,1,0,1},{0,1,0,1,0,1,0,1},{0,1,0,1,0,1,0,1}};
ShowImage[Reverse[img]]
dctcoeff=Array[(Plus @@ Flatten[DCTTensor[[#1,#2]] img])&,{8,8}];
dctcoeff=SetAccuracy[dctcoeff,4];
dctcoeff=Chop[dctcoeff,.001];
MatrixForm[dctcoeff]
ShowImage[Reverse[dctcoeff]]
```

Code for Figure 7.42.

product $T1\times T2\times T3\times T4$, where

$$T1=\begin{bmatrix}0.8165 & 0 & -0.5774\\ 0 & 1 & 0\\ 0.5774 & 0 & 0.8165\end{bmatrix},\quad T2=\begin{bmatrix}0.7071 & -0.7071 & 0\\ 0.7071 & 0.7071 & 0\\ 0 & 0 & 1\end{bmatrix},$$

$$T3=\begin{bmatrix}1 & 0 & 0\\ 0 & 0 & 1\\ 0 & -1 & 0\end{bmatrix},\quad T4=\begin{bmatrix}1 & 0 & 0\\ 0 & 1 & 0\\ 0 & 0 & -1\end{bmatrix}.$$

Each of matrices $T1$, $T2$, and $T3$ performs a rotation about one of the coordinate axes (these are called *Givens rotations*). Matrix $T4$ is a reflection about the z axis. The transformation $\mathbf{M}\cdot(1,1,1)$ can now be written as $T1\times T2\times T3\times T4\times(1,1,1)$, where $T4$ reflects point $(1,1,1)$ to $(1,1,-1)$, $T3$ rotates $(1,1,-1)$ 90° about the x axis to $(1,-1,-1)$, which is rotated by $T2$ 45° about the z axis to $(1.4142,0,-1)$, which is rotated by $T1$ 35.26° about the y axis to $(1.7321,0,0)$.

[This particular sequence of transformations is a result of the order in which the individual steps of the QR decomposition have been performed. Performing the same steps in a different order results in different sequences of rotations. One example is (1) a reflection about the z axis that transforms $(1,1,1)$ to $(1,1,-1)$, (2) a rotation of $(1,1,-1)$ 135° about the x axis to $(1,-1.4142,0)$, and (3) a further rotation of 54.74° about the z axis to $(1.7321,0,0)$.]

For an arbitrary n, this interpretation is similar. We start with a vector of n adjacent pixels. They are considered the coordinates of a point in n-dimensional space. If the pixels are similar, the point is located near the line that makes equal angles with all the coordinate axes. Applying the DCT in one dimension [Equation (7.14)] rotates the point and brings it close to the x axis, where its first coordinate hasn't changed much and its remaining $n-1$ coordinates are small numbers. This is how the DCT in one dimension can be considered a single rotation in n-dimensional space. The rotation can be broken up into a reflection followed by $n-1$ Givens rotations, but a user of the DCT need not be concerned with these details.

The DCT in two dimensions is interpreted similarly as a double rotation. This interpretation starts with a block of $n\times n$ pixels (Figure 7.43a, where the pixels are labeled `L`). It first considers each row of this block as a point $(p_{x,0},p_{x,1},\ldots,p_{x,n-1})$ in n-dimensional space, and it rotates the point by means of the innermost sum

$$G1_{x,j}=\sqrt{\frac{2}{n}}C_j\sum_{y=0}^{n-1}p_{xy}\cos\left(\frac{(2y+1)j\pi}{2n}\right)$$

of Equation (7.16). This results in a block $G1_{x,j}$ of $n\times n$ coefficients where the first element of each row is dominant (labeled `L` in Figure 7.43b) and the remaining elements are small (labeled `S` in that figure). The outermost sum of Equation (7.16) is

$$G_{ij}=\sqrt{\frac{2}{n}}C_i\sum_{x=0}^{n-1}G1_{x,j}\cos\left(\frac{(2x+1)i\pi}{2n}\right).$$

Here, the *columns* of $G1_{x,j}$ are considered points in n-dimensional space and are rotated. The result is one large coefficient at the top-left corner of the block (L in Figure 7.43c) and $n^2 - 1$ small coefficients elsewhere (S and s in that figure). This interpretation considers the two-dimensional DCT as two separate rotations in n dimensions; the first one rotates each of the n rows, and the second one rotates each of the n columns. It is interesting to note that $2n$ rotations in n dimensions are faster than one rotation in n^2 dimensions, since the latter requires an $n^2 \times n^2$ rotation matrix.

```
L L L L L L L L      L S S S S S S S      L S S S S S S S
L L L L L L L L      L S S S S S S S      S s s s s s s s
L L L L L L L L      L S S S S S S S      S s s s s s s s
L L L L L L L L      L S S S S S S S      S s s s s s s s
L L L L L L L L      L S S S S S S S      S s s s s s s s
L L L L L L L L      L S S S S S S S      S s s s s s s s
L L L L L L L L      L S S S S S S S      S s s s s s s s
L L L L L L L L      L S S S S S S S      S s s s s s s s

      (a)                  (b)                  (c)
```

Figure 7.43: The Two-Dimensional DCT as a Double Rotation.

7.8.4 The Four DCT Types

There are four ways to select n equally-spaced angles that generate orthogonal vectors of cosines. They correspond (after the vectors are normalized by scale factors) to four discrete cosine transforms designated DCT-1 through DCT-4. The most useful is DCT-2, which is normally referred to as *the* DCT. Equation (7.20) lists the definitions of the four types. The actual angles of the DCT-1 and DCT-2 are listed (for $n = 8$) in Table 7.44. Note that DCT-3 is the transpose of DCT-2 and DCT-4 is a shifted version of DCT-1. Notice that the DCT-1 has $n+1$ vectors of $n+1$ cosines each. In each of the four types, the n (or $n+1$) DCT vectors are orthogonal and become normalized after they are multiplied by the proper scale factor. Figure 7.45 lists *Mathematica* code to generate the normalized vectors of the four types and test for normalization.

$$\begin{aligned}
\mathrm{DCT}^1_{k,j} &= \sqrt{\frac{2}{n}}C_kC_j\cos\left[\frac{kj\pi}{n}\right], \quad k,j=0,1,\ldots,n,\\
\mathrm{DCT}^2_{k,j} &= \sqrt{\frac{2}{n}}C_k\cos\left[\frac{k(j+\frac{1}{2})\pi}{n}\right], \quad k,j=0,1,\ldots,n-1,\\
\mathrm{DCT}^3_{k,j} &= \sqrt{\frac{2}{n}}C_j\cos\left[\frac{(k+\frac{1}{2})j\pi}{n}\right], \quad k,j=0,1,\ldots,n-1,\\
\mathrm{DCT}^4_{k,j} &= \sqrt{\frac{2}{n}}\cos\left[\frac{(k+\frac{1}{2})(j+\frac{1}{2})\pi}{n}\right], \quad k,j=0,1,\ldots,n-1,
\end{aligned} \tag{7.20}$$

where the scale factor C_x is defined by

$$C_x = \begin{cases} 1/\sqrt{2}, & \text{if } x=0 \text{ or } x=n,\\ 1, & \text{otherwise.}\end{cases}$$

k	scale	DCT-1 Angles (9×9)								
0	$\frac{1}{2\sqrt{2}}$	0*	0	0	0	0	0	0	0	0*
1	$\frac{1}{2}$	0	$\frac{\pi}{8}$	$\frac{2\pi}{8}$	$\frac{3\pi}{8}$	$\frac{4\pi}{8}$	$\frac{5\pi}{8}$	$\frac{6\pi}{8}$	$\frac{7\pi}{8}$	$\frac{8\pi}{8}$
2	$\frac{1}{2}$	0	$\frac{2\pi}{8}$	$\frac{4\pi}{8}$	$\frac{6\pi}{8}$	$\frac{8\pi}{8}$	$\frac{10\pi}{8}$	$\frac{12\pi}{8}$	$\frac{14\pi}{8}$	$\frac{16\pi}{8}$
3	$\frac{1}{2}$	0	$\frac{3\pi}{8}$	$\frac{6\pi}{8}$	$\frac{9\pi}{8}$	$\frac{12\pi}{8}$	$\frac{15\pi}{8}$	$\frac{18\pi}{8}$	$\frac{21\pi}{8}$	$\frac{24\pi}{8}$
4	$\frac{1}{2}$	0	$\frac{4\pi}{8}$	$\frac{8\pi}{8}$	$\frac{12\pi}{8}$	$\frac{16\pi}{8}$	$\frac{20\pi}{8}$	$\frac{24\pi}{8}$	$\frac{28\pi}{8}$	$\frac{32\pi}{8}$
5	$\frac{1}{2}$	0	$\frac{5\pi}{8}$	$\frac{10\pi}{8}$	$\frac{15\pi}{8}$	$\frac{20\pi}{8}$	$\frac{25\pi}{8}$	$\frac{30\pi}{8}$	$\frac{35\pi}{8}$	$\frac{40\pi}{8}$
6	$\frac{1}{2}$	0	$\frac{6\pi}{8}$	$\frac{12\pi}{8}$	$\frac{18\pi}{8}$	$\frac{24\pi}{8}$	$\frac{30\pi}{8}$	$\frac{36\pi}{8}$	$\frac{42\pi}{8}$	$\frac{48\pi}{8}$
7	$\frac{1}{2}$	0	$\frac{7\pi}{8}$	$\frac{14\pi}{8}$	$\frac{21\pi}{8}$	$\frac{28\pi}{8}$	$\frac{35\pi}{8}$	$\frac{42\pi}{8}$	$\frac{49\pi}{8}$	$\frac{56\pi}{8}$
8	$\frac{1}{2\sqrt{2}}$	0*	$\frac{8\pi}{8}$	$\frac{16\pi}{8}$	$\frac{24\pi}{8}$	$\frac{32\pi}{8}$	$\frac{40\pi}{8}$	$\frac{48\pi}{8}$	$\frac{56\pi}{8}$	$\frac{64\pi}{8}$*

* the scale factor for these four angles is 4.

k	scale	DCT-2 Angles							
0	$\frac{1}{2\sqrt{2}}$	0	0	0	0	0	0	0	0
1	$\frac{1}{2}$	$\frac{\pi}{16}$	$\frac{3\pi}{16}$	$\frac{5\pi}{16}$	$\frac{7\pi}{16}$	$\frac{9\pi}{16}$	$\frac{11\pi}{16}$	$\frac{13\pi}{16}$	$\frac{15\pi}{16}$
2	$\frac{1}{2}$	$\frac{2\pi}{16}$	$\frac{6\pi}{16}$	$\frac{10\pi}{16}$	$\frac{14\pi}{16}$	$\frac{18\pi}{16}$	$\frac{22\pi}{16}$	$\frac{26\pi}{16}$	$\frac{30\pi}{16}$
3	$\frac{1}{2}$	$\frac{3\pi}{16}$	$\frac{9\pi}{16}$	$\frac{15\pi}{16}$	$\frac{21\pi}{16}$	$\frac{27\pi}{16}$	$\frac{33\pi}{16}$	$\frac{39\pi}{16}$	$\frac{45\pi}{16}$
4	$\frac{1}{2}$	$\frac{4\pi}{16}$	$\frac{12\pi}{16}$	$\frac{20\pi}{16}$	$\frac{28\pi}{16}$	$\frac{36\pi}{16}$	$\frac{44\pi}{16}$	$\frac{52\pi}{16}$	$\frac{60\pi}{16}$
5	$\frac{1}{2}$	$\frac{5\pi}{16}$	$\frac{15\pi}{16}$	$\frac{25\pi}{16}$	$\frac{35\pi}{16}$	$\frac{45\pi}{16}$	$\frac{55\pi}{16}$	$\frac{65\pi}{16}$	$\frac{75\pi}{16}$
6	$\frac{1}{2}$	$\frac{6\pi}{16}$	$\frac{18\pi}{16}$	$\frac{30\pi}{16}$	$\frac{42\pi}{16}$	$\frac{54\pi}{16}$	$\frac{66\pi}{16}$	$\frac{78\pi}{16}$	$\frac{90\pi}{16}$
7	$\frac{1}{2}$	$\frac{7\pi}{16}$	$\frac{21\pi}{16}$	$\frac{35\pi}{16}$	$\frac{49\pi}{16}$	$\frac{63\pi}{16}$	$\frac{77\pi}{16}$	$\frac{91\pi}{16}$	$\frac{105\pi}{16}$

Table 7.44: Angle Values for the DCT-1 and DCT-2.

Orthogonality can be proved either directly, by multiplying pairs of different vectors, or indirectly. The latter approach is discussed in detail in [Strang 99] and it proves that the DCT vectors are orthogonal by showing that they are the eigenvectors of certain symmetric matrices. In the case of the DCT-2, for example, the symmetric matrix is

$$\mathbf{A}=\begin{bmatrix} 1 & -1 & & & \\ -1 & 2 & -1 & & \\ & & \vdots & & \\ & & -1 & 2 & -1 \\ & & & -1 & 1 \end{bmatrix}. \qquad \text{For } n=3,\text{ matrix} \qquad \mathbf{A}_3=\begin{bmatrix} 1 & -1 & 0 \\ -1 & 2 & -1 \\ 0 & -1 & 1 \end{bmatrix}$$

has eigenvectors $(0.5774, 0.5774, 0.5774)$, $(0.7071, 0, -0.7071)$, $(0.4082, -0.8165, 0.4082)$ with eigenvalues 0, 1, and 3, respectively. Recall that these eigenvectors are the rows of matrix $\mathbf{M}$ of Equation (7.18).

```
(* DCT-1. Notice (n+1)x(n+1) *)
Clear[n, nor, kj, DCT1, T1];
n=8; nor=Sqrt[2/n];
kj[i_]:=If[i==0 || i==n, 1/Sqrt[2], 1];
DCT1[k_]:=Table[nor kj[j] kj[k] Cos[j k Pi/n], {j,0,n}]
T1=Table[DCT1[k], {k,0,n}]; (* Compute nxn cosines *)
MatrixForm[T1] (* display as a matrix *)
(* multiply rows to show orthonormality *)
MatrixForm[Table[Chop[N[T1[[i]].T1[[j]]]], {i,1,n}, {j,1,n}]]

(* DCT-2 *)
Clear[n, nor, kj, DCT2, T2];
n=8; nor=Sqrt[2/n];
kj[i_]:=If[i==0 || i==n, 1/Sqrt[2], 1];
DCT2[k_]:=Table[nor kj[k] Cos[(j+1/2)k Pi/n], {j,0,n-1}]
T2=Table[DCT2[k], {k,0,n-1}]; (* Compute nxn cosines *)
MatrixForm[T2] (* display as a matrix *)
(* multiply rows to show orthonormality *)
MatrixForm[Table[Chop[N[T2[[i]].T2[[j]]]], {i,1,n}, {j,1,n}]]

(* DCT-3. This is the transpose of DCT-2 *)
Clear[n, nor, kj, DCT3, T3];
n=8; nor=Sqrt[2/n];
kj[i_]:=If[i==0 || i==n, 1/Sqrt[2], 1];
DCT3[k_]:=Table[nor kj[j] Cos[(k+1/2)j Pi/n], {j,0,n-1}]
T3=Table[DCT3[k], {k,0,n-1}]; (* Compute nxn cosines *)
MatrixForm[T3] (* display as a matrix *)
(* multiply rows to show orthonormality *)
MatrixForm[Table[Chop[N[T3[[i]].T3[[j]]]], {i,1,n}, {j,1,n}]]

(* DCT-4. This is DCT-1 shifted *)
Clear[n, nor, DCT4, T4];
n=8; nor=Sqrt[2/n];
DCT4[k_]:=Table[nor Cos[(k+1/2)(j+1/2) Pi/n], {j,0,n-1}]
T4=Table[DCT4[k], {k,0,n-1}]; (* Compute nxn cosines *)
MatrixForm[T4] (* display as a matrix *)
(* multiply rows to show orthonormality *)
MatrixForm[Table[Chop[N[T4[[i]].T4[[j]]]], {i,1,n}, {j,1,n}]]
```

Figure 7.45: Code for Four DCT Types.

> From the dictionary
>
> Exegete (EK-suh-jeet), noun: A person who explains or interprets difficult parts of written works.

7.8.5 Practical DCT

Equation (7.16) can be coded directly in any higher-level language. Since this equation is the basis of several compression methods such as JPEG and MPEG, its fast calculation is essential. It can be speeded up considerably by making several improvements, and this section offers some ideas.

1. Regardless of the image size, only 32 cosine functions are involved. They can be precomputed once and used as needed to calculate all the 8×8 data units. Calculating the expression

$$p_{xy} \cos\left[\frac{(2x+1)i\pi}{16}\right] \cos\left[\frac{(2y+1)j\pi}{16}\right]$$

now amounts to performing two multiplications. Thus, the double sum of Equation (7.16) requires $64 \times 2 = 128$ multiplications and 63 additions.

◇ **Exercise 7.14:** (Proposed by V. Saravanan.) Why are only 32 different cosine functions needed for the DCT?

2. A little algebraic tinkering shows that the double sum of Equation (7.16) can be written as the matrix product $\mathbf{CPC}^T$, where $\mathbf{P}$ is the 8×8 matrix of the pixels, $\mathbf{C}$ is the matrix defined by

$$C_{ij} = \begin{cases} \frac{1}{\sqrt{8}}, & i = 0 \\ \frac{1}{2}\cos\left[\frac{(2j+1)i\pi}{16}\right], & i > 0, \end{cases} \tag{7.21}$$

and $\mathbf{C}^T$ is the transpose of $\mathbf{C}$. (The product of two matrices $\mathbf{A}_{mp}$ and $\mathbf{B}_{pn}$ is a matrix $\mathbf{C}_{mn}$ defined by

$$C_{ij} = \sum_{k=1}^{p} a_{ik} b_{kj}.$$

For other properties of matrices, see any text on linear algebra.)

Calculating one matrix element of the product $\mathbf{CP}$ therefore requires eight multiplications and seven (but for simplicity let's say eight) additions. Multiplying the two 8×8 matrices $\mathbf{C}$ and $\mathbf{P}$ requires $64\times 8 = 8^3$ multiplications and the same number of additions. Multiplying the product $\mathbf{CP}$ by $\mathbf{C}^T$ requires the same number of operations, so the DCT of one 8×8 data unit requires 2×8^3 multiplications (and the same number of additions). Assuming that the entire image consists of $n\times n$ pixels and that $n = 8q$, there are $q\times q$ data units, so the DCT of all the data units requires $2q^2 8^3$ multiplications (and the same number of additions). In comparison, performing one DCT for the entire image would require $2n^3 = 2q^3 8^3 = (2q^2 8^3)q$ operations. By dividing the image into data units, we reduce the number of multiplications (and also of additions) by a factor of q. Unfortunately, q cannot be too large, because that would mean very small data units.

Recall that a color image consists of three components (often RGB, but sometimes YCbCr or YPbPr). In JPEG, the DCT is applied to each component separately, bringing the total number of arithmetic operations to $3\times 2q^2 8^3 = 3{,}072q^2$. For a 512×512-pixel image, this implies $3072\times 64^2 = 12{,}582{,}912$ multiplications (and the same number of additions).

3. Another way to speed up the DCT is to perform all the arithmetic operations on fixed-point (scaled integer) rather than on floating-point numbers. On many computers, operations on fixed-point numbers require (somewhat) sophisticated programming techniques, but they are considerably faster than floating-point operations (except on supercomputers, which are optimized for floating-point arithmetic).

The DCT algorithm with smallest currently-known number of arithmetic operations is described in [Feig and Linzer 90]. Today, there are also various VLSI chips that perform this calculation efficiently.

7.8.6 The LLM Method

This section describes the Loeffler-Ligtenberg-Moschytz (LLM) method for the DCT in one dimension [Loeffler et al. 89]. Developed in 1989 by Christoph Loeffler, Adriaan Ligtenberg, and George S. Moschytz, this algorithm computes the DCT in one dimension with a total of 29 additions and 11 multiplications. Recall that the DCT in one dimension involves multiplying a row vector by a matrix. For $n = 8$, multiplying the row by one column of the matrix requires eight multiplications and seven additions, so the total number of operations required for the entire operation is 64 multiplications and 56 additions. Reducing the number of multiplications from 64 to 11 represents a savings of 83% and reducing the number of additions from 56 to 29 represents a savings of 49%—very significant!

Only the final result is listed here and the interested reader is referred to the original publication for the details. We start with the double sum of Equation (7.16) and claim that a little algebraic tinkering reduces it to the form $\mathbf{CPC}^T$, where $\mathbf{P}$ is the 8×8 matrix of the pixels, $\mathbf{C}$ is the matrix defined by Equation (7.21) and $\mathbf{C}^T$ is the transpose of $\mathbf{C}$. In the one-dimensional case, only one matrix multiplication, namely $\mathbf{PC}$, is needed. The originators of this method show that matrix $\mathbf{C}$ can be written (up to a factor of $\sqrt{8}$) as the product of seven simple matrices, as shown in Figure 7.47.

Even though the number of matrices has been increased, the problem has been simplified because our seven matrices are sparse and contain mostly 1's and −1's. Multiplying by 1 or by −1 does not require a multiplication, and multiplying something by 0 saves an addition. Table 7.46 summarizes the total number of arithmetic operations required to multiply a row vector by the seven matrices.

Matrix	Additions	Multiplications
$\mathbf{C}_1$	0	0
$\mathbf{C}_2$	8	12
$\mathbf{C}_3$	4	0
$\mathbf{C}_4$	2	0
$\mathbf{C}_5$	0	2
$\mathbf{C}_6$	4	0
$\mathbf{C}_7$	8	0
Total	26	14

Table 7.46: Number of Arithmetic Operations.

$$
\mathbf{C} = \begin{bmatrix}
1 & 0 & 0 & 0 & 0 & 0 & 0 & 0\\
0 & 0 & 0 & 0 & 1 & 0 & 0 & 0\\
0 & 0 & 1 & 0 & 0 & 0 & 0 & 0\\
0 & 0 & 0 & 0 & 0 & 0 & 1 & 0\\
0 & 1 & 0 & 0 & 0 & 0 & 0 & 0\\
0 & 0 & 0 & 0 & 0 & 1 & 0 & 0\\
0 & 0 & 0 & 1 & 0 & 0 & 0 & 0\\
0 & 0 & 0 & 0 & 0 & 0 & 0 & 1
\end{bmatrix}
$$

$$
\times \begin{bmatrix}
1 & 1 & 0 & 0 & 0 & 0 & 0 & 0\\
1 & -1 & 0 & 0 & 0 & 0 & 0 & 0\\
0 & 0 & \sqrt{2}\cos\frac{6\pi}{16} & \sqrt{2}\cos\frac{2\pi}{16} & 0 & 0 & 0 & 0\\
0 & 0 & -\sqrt{2}\cos\frac{2\pi}{16} & \sqrt{2}\cos\frac{6\pi}{16} & 0 & 0 & 0 & 0\\
0 & 0 & 0 & 0 & \sqrt{2}\cos\frac{7\pi}{16} & 0 & 0 & \sqrt{2}\cos\frac{\pi}{16}\\
0 & 0 & 0 & 0 & 0 & \sqrt{2}\cos\frac{3\pi}{16} & \sqrt{2}\cos\frac{5\pi}{16} & 0\\
0 & 0 & 0 & 0 & 0 & -\sqrt{2}\cos\frac{5\pi}{16} & \sqrt{2}\cos\frac{3\pi}{16} & 0\\
0 & 0 & 0 & 0 & -\sqrt{2}\cos\frac{\pi}{16} & 0 & 0 & \sqrt{2}\cos\frac{7\pi}{16}
\end{bmatrix}
$$

$$
\times \begin{bmatrix}
1 & 0 & 0 & 0 & 0 & 0 & 0 & 0\\
0 & 1 & 0 & 0 & 0 & 0 & 0 & 0\\
0 & 0 & 1 & 0 & 0 & 0 & 0 & 0\\
0 & 0 & 0 & 1 & 0 & 0 & 0 & 0\\
0 & 0 & 0 & 0 & 1 & -1 & 0 & 0\\
0 & 0 & 0 & 0 & 1 & 1 & 0 & 0\\
0 & 0 & 0 & 0 & 0 & 0 & -1 & 1\\
0 & 0 & 0 & 0 & 0 & 0 & 1 & 1
\end{bmatrix}
\times \begin{bmatrix}
1 & 0 & 0 & 0 & 0 & 0 & 0 & 0\\
0 & 1 & 0 & 0 & 0 & 0 & 0 & 0\\
0 & 0 & 1 & 0 & 0 & 0 & 0 & 0\\
0 & 0 & 0 & 1 & 0 & 0 & 0 & 0\\
0 & 0 & 0 & 0 & 1 & 0 & 0 & 0\\
0 & 0 & 0 & 0 & 0 & 1 & -1 & 0\\
0 & 0 & 0 & 0 & 0 & 1 & 1 & 0\\
0 & 0 & 0 & 0 & 0 & 0 & 0 & 1
\end{bmatrix}
$$

$$
\times \begin{bmatrix}
1 & 0 & 0 & 0 & 0 & 0 & 0 & 0\\
0 & 1 & 0 & 0 & 0 & 0 & 0 & 0\\
0 & 0 & 1 & 0 & 0 & 0 & 0 & 0\\
0 & 0 & 0 & 1 & 0 & 0 & 0 & 0\\
0 & 0 & 0 & 0 & 1 & 0 & 0 & 0\\
0 & 0 & 0 & 0 & 0 & 1/\sqrt{2} & 0 & 0\\
0 & 0 & 0 & 0 & 0 & 0 & 1/\sqrt{2} & 0\\
0 & 0 & 0 & 0 & 0 & 0 & 0 & 1
\end{bmatrix}
\times \begin{bmatrix}
1 & 0 & 0 & 1 & 0 & 0 & 0 & 0\\
0 & 1 & 1 & 0 & 0 & 0 & 0 & 0\\
0 & 1 & -1 & 0 & 0 & 0 & 0 & 0\\
1 & 0 & 0 & -1 & 0 & 0 & 0 & 0\\
0 & 0 & 0 & 0 & 1 & 0 & 0 & 0\\
0 & 0 & 0 & 0 & 0 & 1 & 0 & 0\\
0 & 0 & 0 & 0 & 0 & 0 & 1 & 0\\
0 & 0 & 0 & 0 & 0 & 0 & 0 & 1
\end{bmatrix}
$$

$$
\times \begin{bmatrix}
1 & 0 & 0 & 0 & 0 & 0 & 0 & 1\\
0 & 1 & 0 & 0 & 0 & 0 & 1 & 0\\
0 & 0 & 1 & 0 & 0 & 1 & 0 & 0\\
0 & 0 & 0 & 1 & 1 & 0 & 0 & 0\\
0 & 0 & 0 & 1 & -1 & 0 & 0 & 0\\
0 & 0 & 1 & 0 & 0 & -1 & 0 & 0\\
0 & 1 & 0 & 0 & 0 & 0 & -1 & 0\\
1 & 0 & 0 & 0 & 0 & 0 & 0 & -1
\end{bmatrix}
$$

$$
= \mathbf{C}_1\mathbf{C}_2\mathbf{C}_3\mathbf{C}_4\mathbf{C}_5\mathbf{C}_6\mathbf{C}_7.
$$

Figure 7.47: Product of Seven Matrices.

These surprisingly small numbers can be reduced further by the following observation. We notice that matrix $\mathbf{C}_2$ has three groups of four cosines each. One of the groups consists of (we ignore the $\sqrt{2}$) two $\cos\frac{6}{16}\pi$ and two $\cos\frac{2}{16}\pi$ (one with a negative sign). We use the trigonometric identity $\cos(\frac{\pi}{2}-\alpha)=\sin\alpha$ to replace the two $\pm\cos\frac{2}{16}\pi$ with $\pm\sin\frac{6}{16}\pi$. Multiplying any matrix by $\mathbf{C}_2$ now results in products of the form $A\cos(\frac{6}{16}\pi)-B\sin(\frac{6}{16}\pi)$ and $B\cos(\frac{6}{16}\pi)+A\sin(\frac{6}{16}\pi)$. It seems that computing these two elements requires four multiplications and two additions (assuming that a subtraction takes the same time to execute as an addition). The following computation, however, yields the same result with three additions and three multiplications:

$$T=(A+B)\cos\alpha,\quad T-B(\cos\alpha-\sin\alpha),\quad -T+A(\cos\alpha+\sin\alpha).$$

Thus, the three groups now require nine additions and nine multiplications instead of the original 6 additions and 12 multiplications (two more additions are needed for the other nonzero elements of $\mathbf{C}_2$), which brings the totals of Table 7.46 down to 29 additions and 11 multiplications.

> There is no national science just as there is no national multiplication table; what is national is no longer science.
>
> —Anton Chekhov

7.8.7 Hardware Implementation of the DCT

Table 7.44 lists the 64 angle values of the DCT-2 for $n=8$. When the cosines of those angles are computed, we find that because of the symmetry of the cosine function, there are only six distinct nontrivial cosine values. They are summarized in Table 7.48, where $a=1/\sqrt{2}$, $b_i=\cos(i\pi/16)$, and $c_i=\cos(i\pi/8)$. The six nontrivial values are b_1, b_3, b_5, b_7, c_1, and c_3.

1	1	1	1	1	1	1	1
b_1	b_3	b_5	b_7	$-b_7$	$-b_5$	$-b_3$	$-b_1$
c_1	c_3	$-c_3$	$-c_1$	$-c_1$	$-c_3$	c_3	c_1
b_3	$-b_7$	$-b_1$	$-b_5$	b_5	b_1	b_7	$-b_3$
a	$-a$	$-a$	a	a	$-a$	$-a$	a
b_5	$-b_1$	b_7	b_3	$-b_3$	$-b_7$	b_1	$-b_5$
c_3	$-c_1$	c_1	$-c_3$	$-c_3$	c_1	$-c_1$	c_3
b_7	$-b_5$	b_3	$-b_1$	b_1	$-b_3$	b_5	$-b_7$

Table 7.48: Six Distinct Cosine Values for the DCT-2.

This feature can be exploited in a fast software implementation of the DCT or to make a simple hardware device to compute the DCT coefficients G_i for eight pixel values p_i. Figure 7.49 shows how such a device may be organized in two parts, each computing four of the eight coefficients G_i. Part I is based on a 4×4 symmetric matrix whose elements are the four distinct b_i's. The eight pixels are divided into four groups of two pixels each. The two pixels of each group are subtracted, and the four differences become a row vector that's multiplied by the four columns of the matrix to produce the four DCT coefficients G_1, G_3, G_5, and G_7. Part II is based on a similar 4×4 matrix whose nontrivial elements are the two c_i's. The computations are similar except that the two pixels of each group are added instead of subtracted.

$$\bigl[(p_0-p_7),(p_1-p_6),(p_2-p_5),(p_3-p_4)\bigr]\begin{bmatrix} b_1 & b_3 & b_5 & b_7\\ b_3 & -b_7 & -b_1 & -b_5\\ b_5 & -b_1 & b_7 & b_3\\ b_7 & -b_5 & b_3 & -b_1\end{bmatrix}\to[G_1,G_3,G_5,G_7], \qquad \text{(I)}$$

$$\bigl[(p_0+p_7),(p_1+p_6),(p_2+p_5),(p_3+p_4)\bigr]\begin{bmatrix} 1 & c_1 & a & c_3\\ 1 & c_3 & -a & -c_1\\ 1 & -c_3 & -a & c_1\\ 1 & -c_1 & a & -c_3\end{bmatrix}\to[G_0,G_2,G_4,G_6]. \qquad \text{(II)}$$

Figure 7.49: A Hardware Implementation of the DCT-2.

Figure 7.50 (after [Chen et al. 77]) illustrates how such a device can be constructed out of simple adders, complementors, and multipliers. Notation such as $c(3\pi/16)$ in the figure refers to the cosine function.

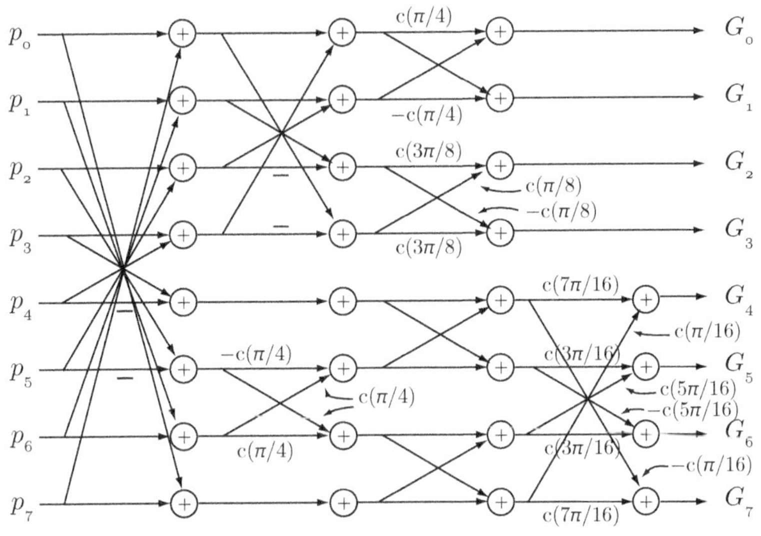

Figure 7.50: A Hardware Implementation of the DCT-2.

7.8.8 QR Matrix Decomposition

This section provides background material on the technique of QR matrix decomposition. It is intended for those already familiar with matrices who want to master this method.

Any matrix $\mathbf{A}$ can be factored into the matrix product $\mathbf{Q}\times\mathbf{R}$, where $\mathbf{Q}$ is an orthogonal matrix and $\mathbf{R}$ is upper triangular. If $\mathbf{A}$ is also orthogonal, then $\mathbf{R}$ will also be orthogonal. However, an upper triangular matrix that's also orthogonal must be diagonal. The orthogonality of $\mathbf{R}$ implies $\mathbf{R}^{-1}=\mathbf{R}^T$ and its being diagonal implies

$\mathbf{R}^{-1}\times\mathbf{R}=\mathbf{I}$. The conclusion is that if $\mathbf{A}$ is orthogonal, then $\mathbf{R}$ must satisfy $\mathbf{R}^T\times\mathbf{R}=\mathbf{I}$, which means that its diagonal elements must be $+1$ or -1. If $\mathbf{A}=\mathbf{Q}\times\mathbf{R}$ and $\mathbf{R}$ has this form, then $\mathbf{A}$ and $\mathbf{Q}$ are identical, except that columns i of $\mathbf{A}$ and $\mathbf{Q}$ will have opposite signs for all values of i where $\mathbf{R}_{i,i}=-1$.

The QR decomposition of matrix $\mathbf{A}$ into $\mathbf{Q}$ and $\mathbf{R}$ is done by a loop where each iteration converts one element of $\mathbf{A}$ to zero. When all the below-diagonal elements of $\mathbf{A}$ have been zeroed, it becomes the upper triangular matrix $\mathbf{R}$. Each element $\mathbf{A}_{i,j}$ is zeroed by multiplying $\mathbf{A}$ by a *Givens rotation* matrix $T_{i,j}$. This is an antisymmetric matrix where the two diagonal elements $T_{i,i}$ and $T_{j,j}$ are set to the cosine of a certain angle θ, and the two off-diagonal elements $T_{j,i}$ and $T_{i,j}$ are set to the sine and negative sine, respectively, of the same θ. The sine and cosine of θ are defined as

$$\cos\theta=\frac{\mathbf{A}_{j,j}}{D},\quad \sin\theta=\frac{\mathbf{A}_{i,j}}{D},\quad \text{where } D=\sqrt{\mathbf{A}^2_{j,j}+\mathbf{A}^2_{i,j}}.$$

Following are some examples of Givens rotation matrices:

$$\begin{bmatrix} c & s\\ -s & c\end{bmatrix},\quad \begin{bmatrix}1&0&0\\0&c&s\\0&-s&c\end{bmatrix},\quad \begin{bmatrix}1&0&0&0\\0&c&0&s\\0&0&1&0\\0&-s&0&c\end{bmatrix},\quad \begin{bmatrix}1&0&0&0&0\\0&c&0&s&0\\0&0&1&0&0\\0&-s&0&c&0\\0&0&0&0&1\end{bmatrix}. \tag{7.22}$$

Those familiar with rotation matrices will recognize that a Givens matrix [Givens 58] rotates a point through an angle whose sine and cosine are the s and c of Equation (7.22). In two dimensions, the rotation is done about the origin. In three dimensions, it is done about one of the coordinate axes [the x axis in Equation (7.22)]. In four dimensions, the rotation is about two of the four coordinate axes (the first and third in Equation (7.22)) and cannot be visualized. In general, an $n\times n$ Givens matrix rotates a point about $n-2$ coordinate axes of an n-dimensional space.

> J. Wallace Givens, Jr. (1910–1993) pioneered the use of plane rotations in the early days of automatic matrix computations. Givens graduated from Lynchburg College in 1928, and he completed his Ph.D. at Princeton University in 1936. After spending three years at the Institute for Advanced Study in Princeton as an assistant of Oswald Veblen, Givens accepted an appointment at Cornell University, but later moved to Northwestern University. In addition to his academic career, Givens was the director of the Applied Mathematics Division at Argonne National Lab and, like his counterpart Alston Householder at Oak Ridge National Laboratory, Givens served as an early president of SIAM. He published his work on the rotations in 1958.
>
> —Carl D. Meyer

Figure 7.51 is a Matlab function for the QR decomposition of a matrix $\mathbf{A}$. Notice how $\mathbf{Q}$ is obtained as the product of the individual Givens matrices and how the double loop zeros all the below-diagonal elements column by column from the bottom up.

```
function [Q,R]=QRdecompose(A);
% Computes the QR decomposition of matrix A
% R is an upper triangular matrix and Q
% an orthogonal matrix such that A=Q*R.
[m,n]=size(A); % determine the dimens of A
Q=eye(m); % Q starts as the mxm identity matrix
R=A;
for p=1:n
 for q=(1+p):m
  w=sqrt(R(p,p)^2+R(q,p)^2);
  s=-R(q,p)/w; c=R(p,p)/w;
  U=eye(m); % Construct a U matrix for Givens rotation
  U(p,p)=c; U(q,p)=-s; U(p,q)=s; U(q,q)=c;
  R=U'*R; % one Givens rotation
  Q=Q*U;
 end
end
```

Figure 7.51: A Matlab Function for the QR Decomposition of a Matrix.

"Computer!' shouted Zaphod, "rotate angle of vision through oneeighty degrees and don't talk about it!'

—Douglas Adams, *The Hitchhikers Guide to the Galaxy*

7.8.9 Vector Spaces

The discrete cosine transform can also be interpreted as a change of basis in a vector space from the standard basis to the DCT basis, so this section is a short discussion of vector spaces, their relation to data compression and to the DCT, their bases, and the important operation of change of basis.

An n-dimensional vector space is the set of all vectors of the form $(v_1, v_2, \ldots, v_n)$. We limit the discussion to the case where the v_i's are real numbers. The attribute of vector spaces that makes them important to us is the existence of *bases*. Any vector (a, b, c) in three dimensions can be written as the linear combination

$$(a, b, c) = a(1, 0, 0) + b(0, 1, 0) + c(0, 0, 1) = a\mathbf{i} + b\mathbf{j} + c\mathbf{k},$$

so we say that the set of three vectors $\mathbf{i}$, $\mathbf{j}$, and $\mathbf{k}$ forms a basis of the three-dimensional vector space. Notice that the three basis vectors are orthogonal; the dot product of any two of them is zero. They are also orthonormal; the dot product of each with itself is 1. It is convenient to have an orthonormal basis, but this is not a requirement. The basis does not even have to be orthogonal.

The set of three vectors $\mathbf{i}$, $\mathbf{j}$, and $\mathbf{k}$ can be extended to any number of dimensions. A basis for an n-dimensional vector space may consist of the n vectors $\mathbf{v}_i$ for $i = 1, 2, \ldots, n$, where element j of vector $\mathbf{v}_i$ is the Kronecker delta function δ_{ij}. This simple basis is the *standard basis* of the n-dimensional vector space. In addition to this basis, the n-dimensional vector space can have other bases. We illustrate two other bases for $n = 8$.

> God made the integers, all else is the work of man.
> —Leopold Kronecker

The DCT (unnormalized) basis consists of the eight vectors

$$(1,1,1,1,1,1,1,1), \quad (1,1,1,1,-1,-1,-1,-1),$$
$$(1,1,-1,-1,-1,-1,1,1), \quad (1,-1,-1,-1,1,1,1,-1),$$
$$(1,-1,-1,1,1,-1,-1,1), \quad (1,-1,1,1,-1,-1,1,-1),$$
$$(1,-1,1,-1,-1,1,-1,1), \quad (1,-1,1-1,1,-1,1,-1).$$

Notice how their elements correspond to higher and higher frequencies. The (unnormalized) Haar wavelet basis (Section 8.6) consists of the eight vectors

$$(1,1,1,1,1,1,1,1), \quad (1,1,1,1,-1,-1,-1,-1),$$
$$(1,1,-1,-1,0,0,0,0), \quad (0,0,0,0,1,1,-1,-1),$$
$$(1,-1,0,0,0,0,0,0), \quad (0,0,1,-1,0,0,0,0),$$
$$(0,0,0,0,1,-1,0,0), \quad (0,0,0,0,0,0,1,-1).$$

To understand why these bases are useful for data compression, recall that our data vectors are images or parts of images. The pixels of an image are normally correlated, but the standard basis takes no advantage of this. The vector of all 1's, on the other hand, is included in the above bases because this single vector is sufficient to express any uniform image. Thus, a group of identical pixels $(v, v, \ldots, v)$ can be represented as the single coefficient v times the vector of all 1's. [The discrete sine transform of Section 7.8.11 is unsuitable for data compression mainly because it does not include this uniform vector.] Basis vector $(1,1,1,1,-1,-1,-1,-1)$ can represent the energy of a group of pixels that's half dark and half bright. Thus, the group $(v,v,\ldots,v,-v,-v,\ldots,-v)$ of pixels is represented by the single coefficient v times this basis vector. Successive basis vectors represent higher-frequency images, up to vector $(1,-1,1,-1,1,-1,1,-1)$. This basis vector resembles a checkerboard and therefore isolates the high-frequency details of an image. Those details are normally the least important and can be heavily quantized or even zeroed to achieve better compression.

The vector members of a basis don't have to be orthogonal. In order for a set S of vectors to be a basis, it has to have the following two properties: (1) The vectors have to be linearly independent and (2) it should be possible to express any member of the vector space as a linear combination of the vectors of S. For example, the three vectors $(1,1,1)$, $(0,1,0)$, and $(0,0,1)$ are not orthogonal but form a basis for the three-dimensional vector space. (1) They are linearly independent because none of them can be expressed as a linear combination of the other two. (2) Any vector (a,b,c) can be expressed as the linear combination $a(1,1,1)+(b-a)(0,1,0)+(c-a)(0,0,1)$.

Once we realize that a vector space may have many bases, we start looking for good bases. A good basis for data compression is one where the inverse of the basis matrix is easy to compute and where the energy of a data vector becomes concentrated in a few coefficients. The bases discussed so far are simple, being based on zeros and ones.

The orthogonal bases have the added advantage that the inverse of the basis matrix is simply its transpose. Being fast is not enough, because the fastest thing we could do is to stay with the original standard basis. The reason for changing a basis is to get compression. The DCT base has the added advantage that it concentrates the energy of a vector of correlated values in a few coefficients. Without this property, there would be no reason to change the coefficients of a vector from the standard basis to the DCT basis. After changing to the DCT basis, many coefficients can be quantized, sometimes even zeroed, with a loss of only the least-important image information. If we quantize the original pixel values in the standard basis, we also achieve compression, but we lose image information that may be important.

Once a basis has been selected, it is easy to express any given vector in terms of the basis vectors. Assuming that the basis vectors are $\mathbf{b}_i$ and given an arbitrary vector $\mathbf{P} = (p_1, p_2, \ldots, p_n)$, we write $\mathbf{P}$ as a linear combination $\mathbf{P} = c_1\mathbf{b}_1 + c_2\mathbf{b}_2 + \cdots + c_n\mathbf{b}_n$ of the $\mathbf{b}_i$'s with unknown coefficients c_i. Using matrix notation, this is written $\mathbf{P} = \mathbf{c} \cdot \mathbf{B}$, where $\mathbf{c}$ is a row vector of the coefficients and $\mathbf{B}$ is the matrix whose rows are the basis vectors. The unknown coefficients can be computed by $\mathbf{c} = \mathbf{P} \cdot \mathbf{B}^{-1}$ and this is the reason why a good basis is one where the inverse of the basis matrix is easy to compute.

A simple example is the coefficients of a vector under the standard basis. We have seen that vector (a, b, c) can be written as the linear combination $a(1,0,0) + b(0,1,0) + c(0,0,1)$. Thus, when the standard basis is used, the coefficients of a vector $\mathbf{P}$ are simply its original elements. If we now want to compress the vector by changing to the DCT basis, we need to compute the coefficients under the new basis. This is an example of the important operation of *change of basis*.

Given two bases $\mathbf{b}_i$ and $\mathbf{v}_i$ and assuming that a given vector $\mathbf{P}$ can be expressed as $\sum c_i\mathbf{b}_i$ and also as $\sum w_i\mathbf{v}_i$, the problem of change of basis is to express one set of coefficients in terms of the other. Since the vectors $\mathbf{v}_i$ constitute a basis, any vector can be expressed as a linear combination of them. Specifically, any $\mathbf{b}_j$ can be written $\mathbf{b}_j = \sum_i t_{ij}\mathbf{v}_i$ for some numbers t_{ij}. We now construct a matrix $\mathbf{T}$ from the t_{ij} and observe that it satisfies $\mathbf{b}_i\mathbf{T} = \mathbf{v}_i$ for $i = 1, 2, \ldots, n$. Thus, $\mathbf{T}$ is a *linear transformation* that transforms basis $\mathbf{b}_i$ to $\mathbf{v}_i$. The numbers t_{ij} are the elements of $\mathbf{T}$ in basis $\mathbf{v}_i$.

For our vector $\mathbf{P}$, we can now write $(\sum c_i\mathbf{b}_i)\mathbf{T} = \sum c_i\mathbf{v}_i$, which implies

$$\sum_{j=1}^{n} w_j\mathbf{v}_j = \sum_j w_j\mathbf{b}_j\mathbf{T} = \sum_j w_j \sum_i \mathbf{v}_i t_{ij} = \sum_i \Big(\sum_j w_j t_{ij}\Big)\mathbf{v}_i.$$

This shows that $c_i = \sum_j t_{ij}w_j$; in other words, a basis is changed by means of a linear transformation $\mathbf{T}$ and the same transformation also relates the elements of a vector in the old and new bases.

Once we switch to a new basis in a vector space, every vector has new coordinates and every transformation has a different matrix.

A linear transformation $\mathbf{T}$ operates on an n-dimensional vector and produces an m-dimensional vector, Thus, $\mathbf{T}(\mathbf{v})$ is a vector $\mathbf{u}$. If $m = 1$, the transformation produces a scalar. If $m = n-1$, the transformation is a projection. Linear transformations satisfy the two important properties $\mathbf{T}(\mathbf{u}+\mathbf{v}) = \mathbf{T}(\mathbf{u}) + \mathbf{T}(\mathbf{v})$ and $\mathbf{T}(c\mathbf{v}) = c\mathbf{T}(\mathbf{v})$. In general, the linear transformation of a linear combination $\mathbf{T}(c_1\mathbf{v}_1 + c_2\mathbf{v}_2 + \cdots + c_n\mathbf{v}_n)$ equals the linear combination of the individual transformations $c_1\mathbf{T}(\mathbf{v}_1)+c_2\mathbf{T}(\mathbf{v}_2)+\cdots+c_n\mathbf{T}(\mathbf{v}_n)$. This implies that the zero vector is transformed to itself under any linear transformation.

Examples of linear transformations are projection, reflection, rotation, and differentiating a polynomial. The derivative of $c_1 + c_2x + c_3x^2$ is $c_2 + 2c_3x$. This is a transformation from the basis (c_1, c_2, c_3) in three-dimensional space to basis (c_2, c_3) in two-dimensional space. The transformation matrix satisfies $(c_1, c_2, c_3)\mathbf{T} = (c_2, 2c_3)$, so it is given by

$$\mathbf{T} = \begin{bmatrix} 0 & 0 \\ 1 & 0 \\ 0 & 2 \end{bmatrix}.$$

Examples of nonlinear transformations are translation, the length of a vector, and adding a constant vector v_0. The latter is nonlinear because if $\mathbf{T}(\mathbf{v}) = \mathbf{v} + \mathbf{v}_0$ and we double the size of $\mathbf{v}$, then $\mathbf{T}(2\mathbf{v}) = 2\mathbf{v} + \mathbf{v}_0$ is different from $2\mathbf{T}(\mathbf{v}) = 2(\mathbf{v} + \mathbf{v}_0)$. Transforming a vector $\mathbf{v}$ to its length $||v||$ is also nonlinear because $\mathbf{T}(-\mathbf{v}) \neq -\mathbf{T}(\mathbf{v})$. Translation is nonlinear because it transforms the zero vector to a nonzero vector.

In general, a linear transformation is performed by multiplying the transformed vector $\mathbf{v}$ by the transformation matrix $\mathbf{T}$. Thus, $\mathbf{u} = \mathbf{v}\cdot\mathbf{T}$ or $\mathbf{T}(\mathbf{v}) = \mathbf{v}\cdot\mathbf{T}$. Notice that we denote by $\mathbf{T}$ both the transformation and its matrix.

In order to describe a transformation uniquely, it is enough to describe what it does to the vectors of a basis. To see why this is true, we observe the following. If for a given vector $\mathbf{v}_1$ we know what $\mathbf{T}(\mathbf{v}_1)$ is, then we know what $\mathbf{T}(a\mathbf{v}_1)$ is for any a. Similarly, if for a given $\mathbf{v}_2$ we know what $\mathbf{T}(\mathbf{v}_2)$ is, then we know what $\mathbf{T}(b\mathbf{v}_1)$ is for any b and also what $\mathbf{T}(a\mathbf{v}_1 + b\mathbf{v}_2)$ is. Thus, we know how $\mathbf{T}$ transforms any vector in the plane containing $\mathbf{v}_1$ and $\mathbf{v}_2$. This argument shows that if we know what $\mathbf{T}(\mathbf{v}_i)$ is for all the vectors $\mathbf{v}_i$ of a basis, then we know how $\mathbf{T}$ transforms any vector in the vector space.

Given a basis $\mathbf{b}_i$ for a vector space, we consider the special transformation that affects the magnitude of each vector but not its direction. Thus, $\mathbf{T}(\mathbf{b}_i) = \lambda_i\mathbf{b}_i$ for some number λ_i. The basis $\mathbf{b}_i$ is the *eigenvector* basis of transformation $\mathbf{T}$. Since we know $\mathbf{T}$ for the entire basis, we also know it for any other vector. Any vector $\mathbf{v}$ in the vector space can be expressed as a linear combination $\mathbf{v} = \sum_i c_i\mathbf{b}_i$. If we apply our transformation to both sides and use the linearity property, we end up with

$$\mathbf{T}(\mathbf{v}) = \mathbf{v}\cdot\mathbf{T} = \sum_i c_i\mathbf{b}_i\cdot\mathbf{T}. \tag{7.23}$$

In the special case where $\mathbf{v}$ is the basis vector $\mathbf{b}_1$, Equation (7.23) implies $\mathbf{T}(\mathbf{b}_1) = \sum_i c_i\mathbf{b}_i\cdot\mathbf{T}$. On the other hand, $\mathbf{T}(\mathbf{b}_1) = \lambda_1\mathbf{b}_1$. We therefore conclude that $c_1 = \lambda_1$ and, in general, that the transformation matrix $\mathbf{T}$ is diagonal with λ_i in position i of its diagonal.

In the eigenvector basis, the transformation matrix is diagonal, so this is the perfect basis. We would love to have it in compression, but it is data dependent. It is called the Karhunen-Loève transform (KLT) and is described in Section 7.7.4.

7.8.10 Rotations in Three Dimensions

For those exegetes who want the complete story, the following paragraphs show how a proper rotation matrix (with a determinant of +1) that rotates a point (v, v, v) to the x axis can be derived from the general rotation matrix in three dimensions.

A general rotation in three dimensions is fully specified by (1) an axis $\mathbf{u}$ of rotation, (2) the angle θ of rotation, and (3) the direction (clockwise or counterclockwise as viewed from the origin) of the rotation about $\mathbf{u}$. Given a unit vector $\mathbf{u}=(u_x,u_y,u_z)$, matrix $\mathbf{M}$ of Equation (7.24) performs a rotation of $\theta°$ about $\mathbf{u}$. The rotation appears clockwise to an observer looking from the origin in the direction of $\mathbf{u}$. If $\mathbf{P}=(x,y,z)$ is an arbitrary point, its position after the rotation is given by the product $\mathbf{P}\cdot\mathbf{M}$.

$$\mathbf{M}= \tag{7.24}$$
$$\begin{pmatrix} u_x^2+\cos\theta(1-u_x^2) & u_xu_y(1-\cos\theta)-u_z\sin\theta & u_xu_z(1-\cos\theta)+u_y\sin\theta \\ u_xu_y(1-\cos\theta)+u_z\sin\theta & u_y^2+\cos\theta(1-u_y^2) & u_yu_z(1-\cos\theta)-u_x\sin\theta \\ u_xu_z(1-\cos\theta)-u_y\sin\theta & u_yu_z(1-\cos\theta)+u_x\sin\theta & u_z^2+\cos\theta(1-u_z^2) \end{pmatrix}.$$

The general rotation of Equation (7.24) can now be applied to our problem, which is to rotate the vector $\mathbf{D}=(1,1,1)$ to the x axis. The rotation should be done about the vector $\mathbf{u}$ that's perpendicular to both $\mathbf{D}$ and $(1,0,0)$. This vector is computed by the cross-product $\mathbf{u}=\mathbf{D}\times(1,0,0)=(0,1,-1)$. Normalizing it yields $\mathbf{u}=(0,\alpha,-\alpha)$, where $\alpha=1/\sqrt{2}$.

The next step is to compute the angle θ between $\mathbf{D}$ and the x axis. This is done by normalizing $\mathbf{D}$ and computing the dot product of it and the x axis (recall that the dot product of two unit vectors is the cosine of the angle between them). The normalized $\mathbf{D}$ is (β,β,β), where $\beta=1/\sqrt{3}$, and the dot product results in $\cos\theta=\beta$, which also produces $\sin\theta=-\sqrt{1-\beta^2}=-\sqrt{2/3}=-\beta/\alpha$. The reason for the negative sign is that a rotation from $(1,1,1)$ to $(1,0,0)$ about $\mathbf{u}$ appears counterclockwise to an observer looking from the origin in the direction of positive $\mathbf{u}$. The rotation matrix of Equation (7.24) was derived for the opposite direction of rotation. Also, $\cos\theta=\beta$ implies that $\theta=54.76°$. This angle, not 45°, is the angle made by vector $\mathbf{D}$ with each of the three coordinate axes. [As an aside, when the number of dimensions increases, the angle between vector $(1,1,\ldots,1)$ and any of the coordinate axes approaches 90°.]

Substituting $\mathbf{u}$, $\sin\theta$, and $\cos\theta$ in Equation (7.24) and using the relations $\alpha^2+\beta(1-\alpha^2)=(\beta+1)/2$ and $-\alpha^2(1-\beta)=(\beta-1)/2$ yields the simple rotation matrix

$$\begin{aligned}\mathbf{M}&=\begin{bmatrix}\beta & -\beta & -\beta\\ \beta & \alpha^2+\beta(1-\alpha^2) & -\alpha^2(1-\beta)\\ \beta & -\alpha^2(1-\beta) & \alpha^2+\beta(1-\alpha^2)\end{bmatrix}=\begin{bmatrix}\beta & -\beta & -\beta\\ \beta & (\beta+1)/2 & (\beta-1)/2\\ \beta & (\beta-1)/2 & (\beta+1)/2\end{bmatrix}\\ &\approx\begin{bmatrix}0.5774 & -0.5774 & -0.5774\\ 0.5774 & 0.7886 & -0.2115\\ 0.5774 & -0.2115 & 0.7886\end{bmatrix}.\end{aligned}$$

It is now easy to see that a point on the line $x=y=z$, with coordinates (v,v,v) is rotated by $\mathbf{M}$ to $(v,v,v)\mathbf{M}=(1.7322v,0,0)$. Notice that the determinant of $\mathbf{M}$ equals $+1$, so $\mathbf{M}$ is a rotation matrix, in contrast to the matrix of Equation (7.18), which generates improper rotations.

7.8.11 Discrete Sine Transform

Readers who made it to this point may raise the question of why the cosine function, and not the sine, is used in the transform? Is it possible to use the sine function in a similar way to the DCT to create a discrete sine transform? Is there a DST, and if not, why? This short section discusses the differences between the sine and cosine functions and shows why these differences lead to a very ineffective discrete sine transform.

A function $f(x)$ that satisfies $f(x) = -f(-x)$ is called *odd*. Similarly, a function for which $f(x) = f(-x)$ is called *even*. For an odd function, it is always true that $f(0) = -f(-0) = -f(0)$, so $f(0)$ must be 0. Most functions are neither odd nor even, but the trigonometric functions sine and cosine are important examples of odd and even functions, respectively. Figure 7.52 shows that even though the only difference between them is phase (i.e., the cosine is a shifted version of the sine), this difference is enough to reverse their parity. When the (odd) sine curve is shifted, it becomes the (even) cosine curve, which has the same shape.

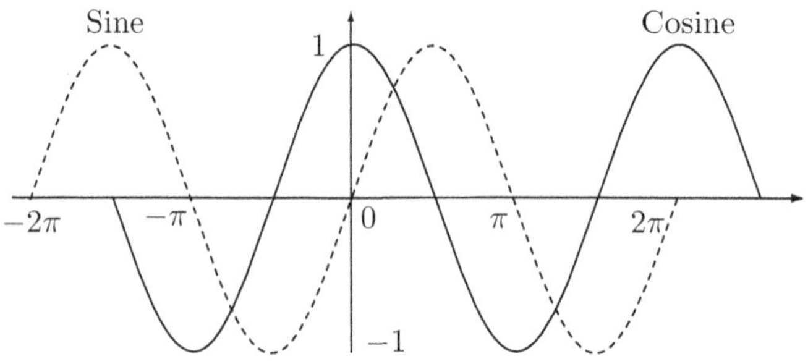

Figure 7.52: The Sine and Cosine as Odd and Even Functions, Respectively.

To understand the difference between the DCT and the DST, we examine the one-dimensional case. The DCT in one dimension, Equation (7.14), employs the function $\cos[(2t+1)f\pi/16]$ for $f = 0, 1, \ldots, 7$. For the first term, where $f = 0$, this function becomes $\cos(0)$, which is 1. This term is the familiar and important DC coefficient, which is proportional to the average of the eight data values being transformed. The DST is similarly based on the function $\sin[(2t+1)f\pi/16]$, resulting in a zero first term [since $\sin(0) = 0$]. The first term contributes nothing to the transform, so the DST does not have a DC coefficient.

The disadvantage of this can be seen when we consider the example of eight identical data values being transformed by the DCT and by the DST. Identical values are, of course, perfectly correlated. When plotted, they become a horizontal line. Applying the DCT to these values produces just a DC coefficient: All the AC coefficients are zero. The DCT compacts all the energy of the data into the single DC coefficient whose value is identical to the values of the data items. The IDCT can reconstruct the eight values perfectly (except for minor changes resulting from limited machine precision). Applying the DST to the same eight values, on the other hand, results in seven AC coefficients whose sum is a wave function that passes through the eight data points but oscillates between the points. This behavior, illustrated by Figure 7.53, has three disadvantages, namely (1) the energy of the original data values is not compacted, (2) the seven coefficients are not decorrelated (since the data values are perfectly correlated), and (3) quantizing the seven coefficients may greatly reduce the quality of the reconstruction done by the inverse DST.

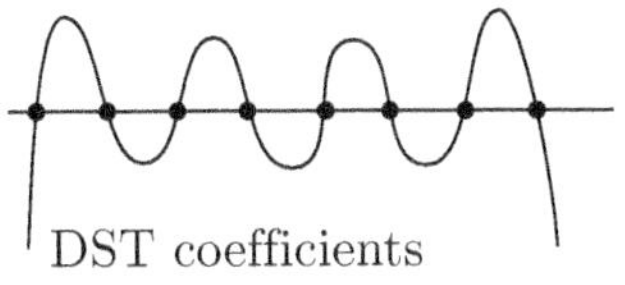

Figure 7.53: The DCT and DST of Eight Identical Data Values.

Example: Applying the DST to the eight identical values 100 results in the eight coefficients $(0, 256.3, 0, 90, 0, 60.1, 0, 51)$. Using these coefficients, the IDST can reconstruct the original values, but it is easy to see that the AC coefficients do not behave like those of the DCT. They are not getting smaller, and there are no runs of zeros among them. Applying the DST to the eight highly correlated values 11, 22, 33, 44, 55, 66, 77, and 88 results in the even worse set of coefficients

$$(0, 126.9, -57.5, 44.5, -31.1, 29.8, -23.8, 25.2).$$

There is no energy compaction at all.

These arguments and examples, together with the fact (discussed in [Ahmed et al. 74] and [Rao and Yip 90]) that the DCT produces highly decorrelated coefficients, argue strongly for the use of the DCT as opposed to the DST in data compression.

◇ **Exercise 7.15:** Use mathematical software to compute and display the 64 basis images of the DST in two dimensions for $n = 8$.

> We are the wisp of straw, the plaything of the winds. We think that we are making for a goal deliberately chosen; destiny drives us towards another. Mathematics, the exaggerated preoccupation of my youth, did me hardly any service; and animals, which I avoided as much as ever I could, are the consolation of my old age. Nevertheless, I bear no grudge against the sine and the cosine, which I continue to hold in high esteem. They cost me many a pallid hour at one time, but they always afforded me some first rate entertainment: they still do so, when my head lies tossing sleeplessly on its pillow.
>
> —J. Henri Fabre, *The Life of the Fly*

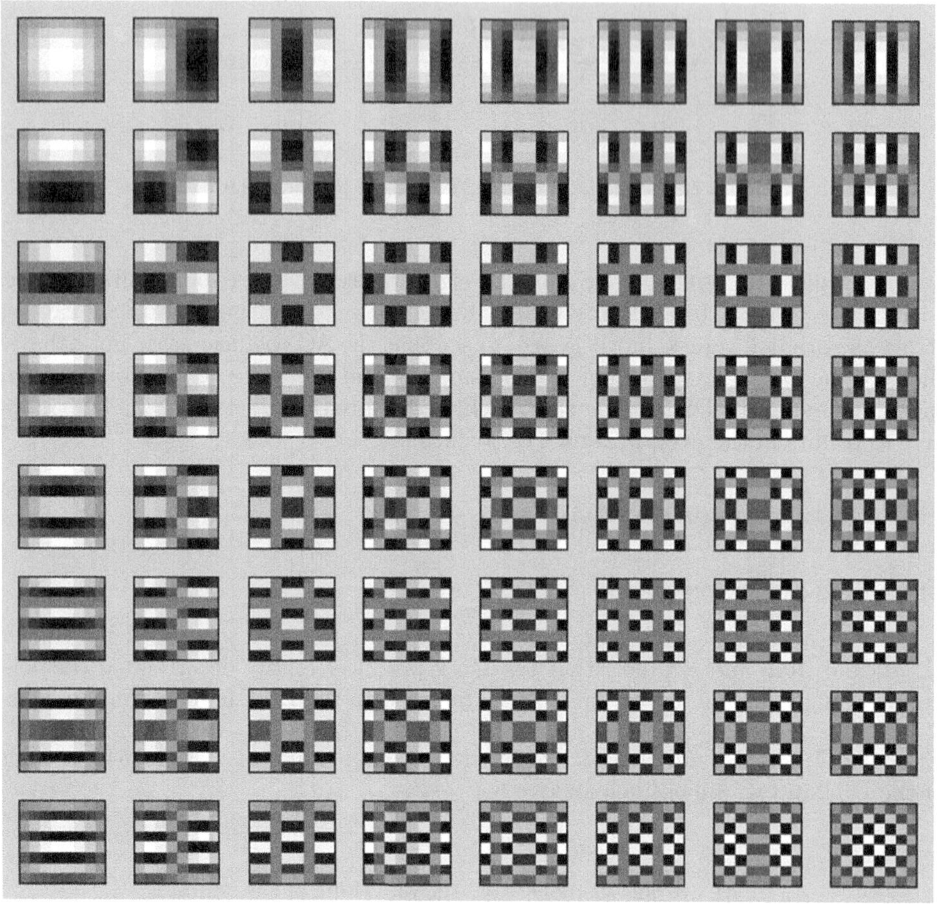

```
N=8;
m=[1:N]'*ones(1,N); n=m';
% can also use cos instead of sin
%A=sqrt(2/N)*cos(pi*(2*(n-1)+1).*(m-1)/(2*N));
A=sqrt(2/N)*sin(pi*(2*(n-1)+1).*(m-1)/(2*N));
A(1,:)=sqrt(1/N);
C=A';
for row=1:N
  for col=1:N
    B=C(:,row)*C(:,col).'; %tensor product
    subplot(N,N,(row-1)*N+col)
    imagesc(B)
    drawnow
  end
end
```

Figure 7.54: The 64 Basis Images of the DST in Two Dimensions.

7.9 Test Images

New data compression methods that are developed and implemented have to be tested. Testing different methods on the same data makes it possible to compare their performance both in compression efficiency and in speed. This is why there are standard collections of test data, such as the Calgary Corpus and the Canterbury Corpus (mentioned in the Preface), and the ITU-T set of eight training documents for fax compression (Section 5.7.1).

The need for standard test data has also been felt in the field of image compression, and there currently exist collections of still images commonly used by researchers and implementors in this field. Three of the four images shown here, namely *Lena*, *mandril*, and *peppers*, are arguably the most well known of them. They are continuous-tone images, although *Lena* has some features of a discrete-tone image.

Each image is accompanied by a detail, showing individual pixels. It is easy to see why the *peppers* image is continuous-tone. Adjacent pixels that differ much in color are fairly rare in this image. Most neighboring pixels are very similar. In contrast, the *mandril* image, even though natural, is a bad example of a continuous-tone image. The detail (showing part of the right eye and the area around it) shows that many pixels differ considerably from their immediate neighbors because of the animal's facial hair in this area. This image compresses badly under any compression method. However, the nose area, with mostly blue and red, is continuous-tone. The *Lena* image is mostly pure continuous-tone, especially the wall and the bare skin areas. The hat is good continuous-tone, whereas the hair and the plume on the hat are bad continuous-tone. The straight lines on the wall and the curved parts of the mirror are features of a discrete-tone image.

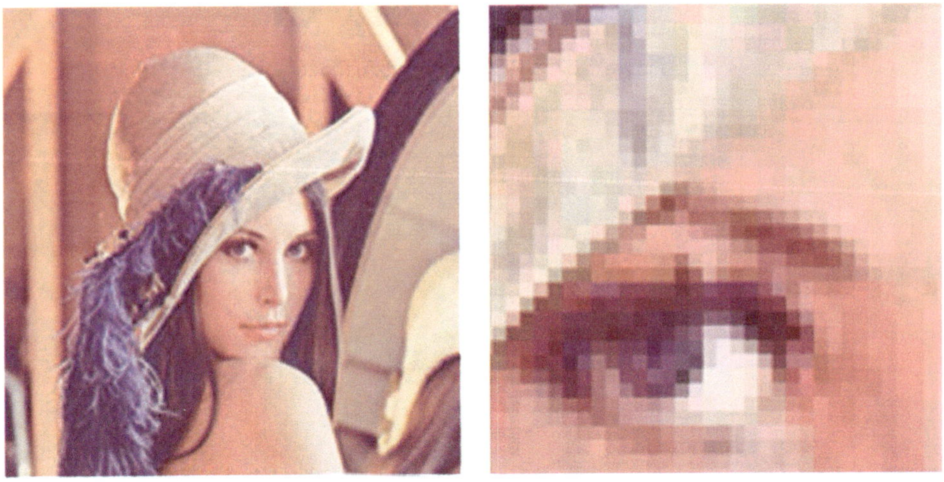

Figure 7.55: Lena and Detail.

The *Lena* image is widely used by the image processing community, in addition to being popular in image compression. Because of the interest in it, its origin and

Figure 7.56: Mandril and Detail.

Figure 7.57: JPEG Blocking Artifacts.

Figure 7.58: Peppers and Detail.

Figure 7.59: A Discrete-Tone Image.

Figure 7.60: A Discrete-Tone Image (Detail).

history have been researched and are well documented. This image is part of the *Playboy* centerfold for November, 1972. It features the Swedish playmate Lena Soderberg (née Sjooblom), and it was discovered, clipped, and scanned in the early 1970s by an unknown researcher at the University of Southern California for use as a test image for his image compression research. It has since become the most important, well-known, and commonly used image in the history of imaging and electronic communications. As a result, Lena is currently considered by many the First Lady of the Internet. *Playboy*, which normally prosecutes unauthorized users of its images, has found out about the unusual use of one of its copyrighted images, but decided to give its blessing to this particular "application."

Lena herself currently lives in Sweden. She was told of her "fame" in 1988, was surprised and amused by it, and was invited to attend the 50th Anniversary IS&T (the society for Imaging Science and Technology) conference in Boston in May 1997. At the conference she autographed her picture, posed for new pictures (available on the www), and gave a presentation (about herself, not compression).

The three images are widely available for downloading on the Internet.

Figure 7.59 shows a typical discrete-tone image, with a detail shown in Figure 7.60. Notice the straight lines and the text, where certain characters appear several times (a source of redundancy). This particular image has few colors, but in general, a discrete-tone image may have many colors.

> Lena, Illinois, is a community of approximately 2,900 people. Lena is considered to be a clean and safe community located centrally to larger cities that offer other interests when needed. Lena is 2-1/2 miles from Lake Le-Aqua-Na State Park. The park offers hiking trails, fishing, swimming beach, boats, cross country skiing, horse back riding trails, as well as picnic and camping areas. It is a beautiful well-kept park that has free admission to the public. A great place for sledding and ice skating in the winter! (From `http://www.villageoflena.com/`)

7.10 JPEG

JPEG is a sophisticated lossy/lossless compression method for color or grayscale still images (not videos). It does not handle bi-level (black and white) images very well. It also works best on continuous-tone images, where adjacent pixels have similar colors. An important feature of JPEG is its use of many parameters, allowing the user to adjust the amount of the data lost (and thus also the compression ratio) over a very wide range. Often, the eye cannot see any image degradation even at compression factors of 10 or 20. There are two operating modes, lossy (also called baseline) and lossless (which typically produces compression ratios of around 0.5). Most implementations support just the lossy mode. This mode includes progressive and hierarchical coding. A few of the many references to JPEG are [Pennebaker and Mitchell 92], [Wallace 91], and [Zhang 90].

JPEG is a compression method, not a complete standard for image representation. This is why it does not specify image features such as pixel aspect ratio, color space, or interleaving of bitmap rows.

JPEG has been designed as a compression method for continuous-tone images. The main goals of JPEG compression are the following:

1. High compression ratios, especially in cases where image quality is judged as very good to excellent.
2. The use of many parameters, allowing knowledgeable users to experiment and achieve the desired compression/quality trade-off.
3. Obtaining good results with any kind of continuous-tone image, regardless of image dimensions, color spaces, pixel aspect ratios, or other image features.
4. A sophisticated, but not too complex compression method, allowing software and hardware implementations on many platforms.
5. Several modes of operation: (a) A sequential mode where each image component (color) is compressed in a single left-to-right, top-to-bottom scan; (b) A progressive mode where the image is compressed in multiple blocks (known as "scans") to be viewed from coarse to fine detail; (c) A lossless mode that is important in cases where the user decides that no pixels should be lost (the trade-off is low compression ratio compared to the lossy modes); and (d) A hierarchical mode where the image is compressed at multiple resolutions allowing lower-resolution blocks to be viewed without first having to decompress the following higher-resolution blocks.

The name JPEG is an acronym that stands for Joint Photographic Experts Group. This was a joint effort by the CCITT and the ISO (the International Standards Organization) that started in June 1987 and produced the first JPEG draft proposal in 1991. The JPEG standard has proved successful and has become widely used for image compression, especially in Web pages.

The main JPEG compression steps are outlined here, and each step is then described in detail later.

1. Color images are transformed from RGB into a luminance/chrominance color space (Section 7.10.1; this step is skipped for grayscale images). The eye is sensitive to small changes in luminance but not in chrominance, so the chrominance part can later lose much data, and thus be highly compressed, without visually impairing the overall image quality much. This step is optional but important because the remainder of the algorithm works on each color component separately. Without transforming the color space, none of the three color components will tolerate much loss, leading to worse compression.
2. Color images are downsampled by creating low-resolution pixels from the original ones (this step is used only when hierarchical compression is selected; it is always skipped for grayscale images). The downsampling is not done for the luminance component. Downsampling is done either at a ratio of 2:1 both horizontally and vertically (the so-called `2h2v` or 4:1:1 sampling) or at ratios of 2:1 horizontally and 1:1 vertically (`2h1v` or 4:2:2 sampling). Since this is done on two of the three color components, `2h2v` reduces the image to $1/3 + (2/3) \times (1/4) = 1/2$ its original size, while `2h1v` reduces it to $1/3 + (2/3) \times (1/2) = 2/3$ its original size. Since the luminance component is not touched, there is no noticeable loss of image quality. Grayscale images don't go through this step.
3. The pixels of each color component are organized in groups of 8×8 pixels called *data units*, and each data unit is compressed separately. If the number of image rows or columns is not a multiple of 8, the bottom row and the rightmost column are duplicated

as many times as necessary. In the noninterleaved mode, the encoder handles all the data units of the first image component, then the data units of the second component, and finally those of the third component. In the interleaved mode the encoder processes the three top-left data units of the three image components, then the three data units to their right, and so on. The fact that each data unit is compressed separately is one of the downsides of JPEG. If the user asks for maximum compression, the decompressed image may exhibit blocking artifacts due to differences between blocks. Figure 7.57 is an extreme example of this effect.

4. The discrete cosine transform (DCT, Section 7.8) is then applied to each data unit to create an 8×8 map of frequency components (Section 7.10.2). They represent the average pixel value and successive higher-frequency changes within the group. This prepares the image data for the crucial step of losing information. Since DCT involves the transcendental function cosine, it must involve some loss of information due to the limited precision of computer arithmetic. This means that even without the main lossy step (step 5 below), there will be some loss of image quality, but it is normally small.

5. Each of the 64 frequency components in a data unit is divided by a separate number called its *quantization coefficient* (QC), and then rounded to an integer (Section 7.10.3). This is where information is irretrievably lost. Large QCs cause more loss, so the high-frequency components typically have larger QCs. Each of the 64 QCs is a JPEG parameter and can, in principle, be specified by the user. In practice, most JPEG implementations use the QC tables recommended by the JPEG standard for the luminance and chrominance image components (Table 7.63).

6. The 64 quantized frequency coefficients (which are now integers) of each data unit are encoded using a combination of RLE and Huffman coding (Section 7.10.4). An arithmetic coding variant known as the QM coder (Section 5.11) can optionally be used instead of Huffman coding.

7. The last step adds headers and all the required JPEG parameters, and outputs the result. The compressed file may be in one of three formats (1) the *interchange* format, in which the file contains the compressed image and all the tables needed by the decoder (mostly quantization tables and tables of Huffman codes), (2) the *abbreviated* format for compressed image data, where the file contains the compressed image and may contain no tables (or just a few tables), and (3) the *abbreviated* format for table-specification data, where the file contains just tables, and no compressed image. The second format makes sense in cases where the same encoder/decoder pair is used, and they have the same tables built in. The third format is used in cases where many images have been compressed by the same encoder, using the same tables. When those images need to be decompressed, they are sent to a decoder preceded by one file with table-specification data.

The JPEG decoder performs the reverse steps. (Thus, JPEG is a symmetric compression method.)

The progressive mode is a JPEG option. In this mode, higher-frequency DCT coefficients are written on the compressed stream in blocks called "scans." Each scan that is read and processed by the decoder results in a sharper image. The idea is to use the first few scans to quickly create a low-quality, blurred preview of the image, and then either input the remaining scans or stop the process and reject the image. The trade-off is that the encoder has to save all the coefficients of all the data units in a

In addition to scotopic and photopic visions, there is also mesopic vision. This is a combination of photopic and scotopic visions in low light but not full darkness.

Nothing in our world is perfect, and this includes people and other living beings. Color blindness in humans is not rare. It strikes about 8% of all males and about 0.5% of all females. It is caused by having just one or two types of cones (the other types may be completely missing or just weak). A color blind person with two types of cones can distinguish a limited range of colors, a range that requires only two numbers to specify a color. A person with only one type of cones senses even fewer colors and perceives a one-dimensional color space. With this in mind, can there be persons, animals, or aliens with four or more types of cones in their retina? The answer is yes, and such beings would perceive color spaces of more than three dimensions and would sense more colors that we do! This surprising conclusion stems from the fact that colors are not attributes of objects and exist only in our minds.

The famous "blind spot" also deserves mentioning. This is a point where the retina does not have any photoreceptors. Any image in this region is not sensed by the eye and is invisible. The blind spot is the point where the optic nerves come together and exit the eye on their way to the brain. To find your blind spot, look at this image,

close your left eye, and hold the page about 20 inches (or 50 cm) from you. Look at the square with your right eye. Slowly bring the page closer while looking at the square. At a certain distance, the cross will disappear. This occurs when the cross falls on the blind spot of your retina. Reverse the process. Close your right eye and look at the cross with your left eye. Bring the image slowly closer to you and the square will disappear.

Each of the photoreceptors sends a light sensation to the brain that's essentially a pixel, and the brain combines these pixels to a continuous image. Thus, the human eye is similar to a digital camera. Once this is understood, we naturally want to compare the resolution of the eye to that of a modern digital camera. Current (2009) digital cameras feature from 1–2 Mpixels (for a cell phone camera) to about ten Mpixels (for a good-quality camera).

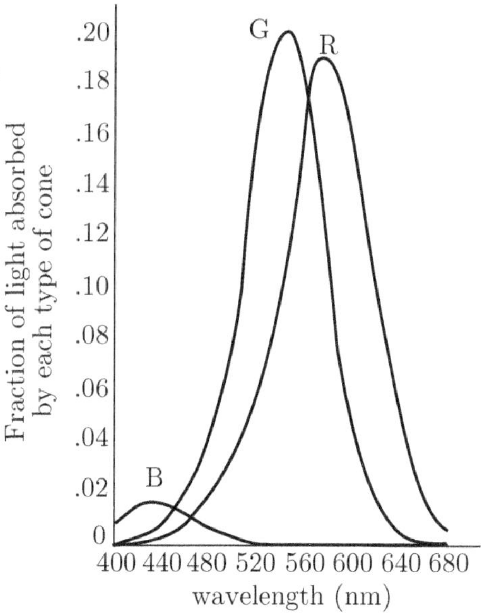

Figure 7.62: Sensitivity of the Cones.

Thus, the eye features a much higher resolution, but its effective resolution is even higher if we consider that the eye can move and refocus itself about three to four times a second. This means that in a single second, the eye can sense and send to the brain about half a billion pixels. Assuming that our camera takes a snapshot once a second, the ratio of the resolutions is about 100.

Certain colors—such as red, orange, and yellow—are psychologically associated with heat. They are considered *warm* and cause a picture to appear larger and closer than it actually is. Other colors—such as blue, violet, and green—are associated with cool things (air, sky, water, ice) and are therefore called *cool* colors. They cause a picture to look smaller and farther away.

7.10.2 DCT

The general concept of a transform is discussed in Section 7.6. The discrete cosine transform is discussed in much detail in Section 7.8. Other examples of important transforms are the Fourier transform (Section 8.1) and the wavelet transform (Chapter 8). Both have applications in many areas and also have discrete versions (DFT and DWT).

The JPEG committee elected to use the DCT because of its good performance, because it does not assume anything about the structure of the data (the DFT, for example, assumes that the data to be transformed is periodic), and because there are ways to speed it up (Section 7.8.5).

The JPEG standard calls for applying the DCT not to the entire image but to data units (blocks) of 8×8 pixels. The reasons for this are (1) Applying DCT to large blocks involves many arithmetic operations and is therefore slow. Applying DCT to small data units is faster. (2) Experience shows that, in a continuous-tone image, correlations between pixels are short range. A pixel in such an image has a value (color component or shade of gray) that's close to those of its near neighbors, but has nothing to do with the values of far neighbors. The JPEG DCT is therefore executed by Equation (7.16), duplicated here for $n=8$

$$
\begin{aligned}
G_{ij} &= \frac{1}{4}C_iC_j\sum_{x=0}^{7}\sum_{y=0}^{7}p_{xy}\cos\left(\frac{(2x+1)i\pi}{16}\right)\cos\left(\frac{(2y+1)j\pi}{16}\right),\\
&\text{where } C_f=\begin{cases}\frac{1}{\sqrt{2}}, & f=0,\\ 1, & f>0,\end{cases}\quad \text{and } 0\le i,j\le 7.
\end{aligned}
\tag{7.16}
$$

The DCT is JPEG's key to lossy compression. The unimportant image information is reduced or removed by quantizing the 64 DCT coefficients, especially the ones located toward the lower-right. If the pixels of the image are correlated, quantization does not degrade the image quality much. For best results, each of the 64 coefficients is quantized by dividing it by a different quantization coefficient (QC). All 64 QCs are parameters that can be controlled, in principle, by the user (Section 7.10.3).

The JPEG decoder works by computing the inverse DCT (IDCT), Equation (7.17), duplicated here for $n=8$

$$
\begin{aligned}
p_{xy} &= \frac{1}{4}\sum_{i=0}^{7}\sum_{j=0}^{7}C_iC_jG_{ij}\cos\left(\frac{(2x+1)i\pi}{16}\right)\cos\left(\frac{(2y+1)j\pi}{16}\right),\\
&\text{where } C_f=\begin{cases}\frac{1}{\sqrt{2}}, & f=0;\\ 1, & f>0.\end{cases}
\end{aligned}
\tag{7.17}
$$

It takes the 64 quantized DCT coefficients and calculates 64 pixels p_{xy}. If the QCs are the right ones, the new 64 pixels will be very similar to the original ones. Mathematically, the DCT is a one-to-one mapping of 64-point vectors from the image domain to the frequency domain. The IDCT is the reverse mapping. If the DCT and IDCT could be calculated with infinite precision and if the DCT coefficients were not quantized, the original 64 pixels would be exactly reconstructed.

7.10.3 Quantization

After each 8×8 data unit of DCT coefficients G_{ij} is computed, it is quantized. This is the step where information is lost (except for some unavoidable loss because of finite precision calculations in other steps). Each number in the DCT coefficients matrix is divided by the corresponding number from the particular "quantization table" used, and the result is rounded to the nearest integer. As has already been mentioned, three such tables are needed, for the three color components. The JPEG standard allows for up to four tables, and the user can select any of the four for quantizing each color component. The 64 numbers that constitute each quantization table are all JPEG parameters. In principle, they can all be specified and fine-tuned by the user for maximum compression. In practice, few users have the patience or expertise to experiment with so many parameters, so JPEG software normally uses the following two approaches:

1. Default quantization tables. Two such tables, for the luminance (grayscale) and the chrominance components, are the result of many experiments performed by the JPEG committee. They are included in the JPEG standard and are reproduced here as Table 7.63. It is easy to see how the QCs in the table generally grow as we move from the upper left corner to the bottom right corner. This is how JPEG reduces the DCT coefficients with high spatial frequencies.
2. A simple quantization table Q is computed, based on one parameter R specified by the user. A simple expression such as $Q_{ij} = 1 + (i + j) \times R$ guarantees that QCs start small at the upper-left corner and get bigger toward the lower-right corner. Table 7.64 shows an example of such a table with $R = 2$.

16	11	10	16	24	40	51	61
12	12	14	19	26	58	60	55
14	13	16	24	40	57	69	56
14	17	22	29	51	87	80	62
18	22	37	56	68	109	103	77
24	35	55	64	81	104	113	92
49	64	78	87	103	121	120	101
72	92	95	98	112	100	103	99

Luminance

17	18	24	47	99	99	99	99
18	21	26	66	99	99	99	99
24	26	56	99	99	99	99	99
47	66	99	99	99	99	99	99
99	99	99	99	99	99	99	99
99	99	99	99	99	99	99	99
99	99	99	99	99	99	99	99
99	99	99	99	99	99	99	99

Chrominance

Table 7.63: Recommended Quantization Tables.

If the quantization is done correctly, very few nonzero numbers will be left in the DCT coefficients matrix, and they will typically be concentrated in the upper-left region. These numbers are the output of JPEG, but they are further compressed before being

1	3	5	7	9	11	13	15
3	5	7	9	11	13	15	17
5	7	9	11	13	15	17	19
7	9	11	13	15	17	19	21
9	11	13	15	17	19	21	23
11	13	15	17	19	21	23	25
13	15	17	19	21	23	25	27
15	17	19	21	23	25	27	29

Table 7.64: The Quantization Table $1 + (i + j) \times 2$.

written on the output stream. In the JPEG literature this compression is called "entropy coding," and Section 7.10.4 shows in detail how it is done. Three techniques are used by entropy coding to compress the 8×8 matrix of integers:

1. The 64 numbers are collected by scanning the matrix in zigzags (Figure 1.8b). This produces a string of 64 numbers that starts with some nonzeros and typically ends with many consecutive zeros. Only the nonzero numbers are output (after further compressing them) and are followed by a special end-of block (EOB) code. This way there is no need to output the trailing zeros (we can say that the EOB is the run-length encoding of all the trailing zeros). The interested reader should also consult Section 11.5 for other methods to compress binary strings with many consecutive zeros.

◇ **Exercise 7.16:** Propose a practical way to write a loop that traverses an 8×8 matrix in zigzag.

2. The nonzero numbers are compressed using Huffman coding (Section 7.10.4).
3. The first of those numbers (the DC coefficient, page 471) is treated differently from the others (the AC coefficients).

> She had just succeeded in curving it down into a graceful zigzag, and was going to dive in among the leaves, which she found to be nothing but the tops of the trees under which she had been wandering, when a sharp hiss made her draw back in a hurry.
>
> —Lewis Carroll, *Alice in Wonderland* (1865)

7.10.4 Coding

We first discuss point 3 above. Each 8×8 matrix of quantized DCT coefficients contains one DC coefficient [at position $(0, 0)$, the top left corner] and 63 AC coefficients. The DC coefficient is a measure of the average value of the 64 original pixels, constituting the data unit. Experience shows that in a continuous-tone image, adjacent data units of pixels are normally correlated in the sense that the average values of the pixels in adjacent data units are close. We already know that the DC coefficient of a data unit is a multiple of the average of the 64 pixels constituting the unit. This implies that the DC coefficients of adjacent data units don't differ much. JPEG outputs the first one (encoded), followed by *differences* (also encoded) of the DC coefficients of consecutive data units. The concept of differencing is discussed in Section 1.3.1.

Example: If the first three 8×8 data units of an image have quantized DC coefficients of 1118, 1114, and 1119, then the JPEG output for the first data unit is 1118 (Huffman encoded, see below) followed by the 63 (encoded) AC coefficients of that data unit. The output for the second data unit will be $1114 - 1118 = -4$ (also Huffman encoded), followed by the 63 (encoded) AC coefficients of that data unit, and the output for the third data unit will be $1119 - 1114 = 5$ (also Huffman encoded), again followed by the 63 (encoded) AC coefficients of that data unit. This way of handling the DC coefficients is worth the extra trouble, because the differences are small.

Coding the DC differences is done with Table 7.65, so first here are a few words about this table. Each row has a row number (on the left), the unary code for the row (on the right), and several columns in between. Each row contains greater numbers (and also more numbers) than its predecessor but not the numbers contained in previous rows. Row i contains the range of integers $[-(2^i-1), +(2^i-1)]$ but is missing the middle range $[-(2^{i-1}-1), +(2^{i-1}-1)]$. Thus, the rows get very long, which means that a simple two-dimensional array is not a good data structure for this table. In fact, there is no need to store these integers in a data structure, since the program can figure out where in the table any given integer x is supposed to reside by analyzing the bits of x.

The first DC coefficient to be encoded in our example is 1118. It resides in row 11 column 930 of the table (column numbering starts at zero), so it is encoded as 111111111110|01110100010 (the unary code for row 11, followed by the 11-bit binary value of 930). The second DC difference is −4. It resides in row 3 column 3 of Table 7.65, so it is encoded as 1110|011 (the unary code for row 3, followed by the 3-bit binary value of 3).

◇ **Exercise 7.17:** How is the third DC difference, 5, encoded?

Point 2 above has to do with the precise way the 63 AC coefficients of a data unit are compressed. It uses a combination of RLE and either Huffman or arithmetic coding. The idea is that the sequence of AC coefficients normally contains just a few nonzero numbers, with runs of zeros between them, and with a long run of trailing zeros. For each nonzero number x, the encoder (1) finds the number Z of consecutive zeros preceding x; (2) finds x in Table 7.65 and prepares its row and column numbers (R and C); (3) the pair (R, Z) [that's (R, Z), not (R, C)] is used as row and column numbers for Table 7.68; and (4) the Huffman code found in that position in the table is concatenated to C (where C is written as an R-bit number) and the result is (finally) the code emitted by the JPEG encoder for the AC coefficient x and all the consecutive zeros preceding it.

The Huffman codes in Table 7.68 are not the ones recommended by the JPEG standard. The standard recommends the use of Tables 7.66 and 7.67 and says that up to four Huffman code tables can be used by a JPEG codec, except that the baseline mode can use only two such tables. The actual codes in Table 7.68 are thus arbitrary. The reader should notice the EOB code at position $(0,0)$ and the ZRL code at position $(0,15)$. The former indicates end-of-block, and the latter is the code emitted for 15 consecutive zeros when the number of consecutive zeros exceeds 15. These codes are the ones recommended for the luminance AC coefficients of Table 7.66. The EOB and ZRL codes recommended for the chrominance AC coefficients of Table 7.67 are 00 and 1111111010, respectively.

0:	0										0
1:	-1	1									10
2:	-3	-2	2	3							110
3:	-7	-6	-5	-4	4	5	6	7			1110
4:	-15	-14	...	-9	-8	8	9	10	...	15	11110
5:	-31	-30	-29	...	-17	-16	16	17	...	31	111110
6:	-63	-62	-61	...	-33	-32	32	33	...	63	1111110
7:	-127	-126	-125	...	-65	-64	64	65	...	127	11111110
⋮				⋮							
14:	-16383	-16382	-16381	...	-8193	-8192	8192	8193	...	16383	111111111111110
15:	-32767	-32766	-32765	...	-16385	-16384	16384	16385	...	32767	1111111111111110
16:	32768										1111111111111111

Table 7.65: Coding the Differences of DC Coefficients.

As an example consider the sequence

$$1118, 2, 0, -2, \underbrace{0, \ldots, 0}_{13}, -1, 0, \ldots.$$

The first AC coefficient 2 has no zeros preceding it, so Z = 0. It is found in Table 7.65 in row 2, column 2, so R = 2 and C = 2. The Huffman code in position (R, Z) = (2, 0) of Table 7.68 is 01, so the final code emitted for 2 is 01|10. The next nonzero coefficient, −2, has one zero preceding it, so Z = 1. It is found in Table 7.65 in row 2, column 1, so R = 2 and C = 1. The Huffman code in position (R, Z) = (2, 1) of Table 7.68 is 11011, so the final code emitted for 2 is 11011|01.

◇ **Exercise 7.18:** What code is emitted for the last nonzero AC coefficient, −1?

Finally, the sequence of trailing zeros is encoded as 1010 (EOB), so the output for the above sequence of AC coefficients is 01101101110111010101010. We saw earlier that the DC coefficient is encoded as 111111111110|1110100010, so the final output for the entire 64-pixel data unit is the 46-bit number

1111111111100111010001001101101110111010101010.

These 46 bits encode one color component of the 64 pixels of a data unit. Let's assume that the other two color components are also encoded into 46-bit numbers. If each pixel originally consists of 24 bits, then this corresponds to a compression factor of $64 \times 24/(46 \times 3) \approx 11.13$; very impressive!

(Notice that the DC coefficient of 1118 has contributed 23 of the 46 bits. Subsequent data units code differences of their DC coefficient, which may take fewer than 10 bits instead of 23. They may feature much higher compression factors as a result.)

The same tables (Tables 7.65 and 7.68) used by the encoder should, of course, be used by the decoder. The tables may be predefined and used by a JPEG codec as defaults, or they may be specifically calculated for a given image in a special pass preceding the actual compression. The JPEG standard does not specify any code tables, so any JPEG codec must use its own.

Readers who feel that this coding scheme is complex should take a look at the much more complex CAVLC coding method that is employed by H.264 (Section 9.9) to encode a similar sequence of 8×8 DCT transform coefficients.

Z	R				
	1 6	2 7	3 8	4 9	5 A
0	00 1111000	01 11111000	100 1111110110	1011 1111111110000010	11010 1111111110000011
1	1100 1111111110000100	11011 1111111110000101	11110001 1111111110000110	111110110 1111111110000111	11111110110 1111111110001000
2	11100 111111110001010	11111001 111111110001011	1111110111 111111110001100	111111110100 111111110001101	111111110001001 111111110001110
3	111010 1111111110010001	111110111 1111111110010010	111111110101 1111111110010011	1111111110001111 1111111110010100	1111111110010000 1111111110010101
4	111011 1111111110011001	1111111000 1111111110011010	1111111110010110 1111111110011011	1111111110010111 1111111110011100	1111111110011000 1111111110011101
5	1111010 1111111110100001	11111110111 1111111110100010	1111111110011110 1111111110100011	1111111110011111 1111111110100100	1111111110100000 1111111110100101
6	1111011 1111111110101001	111111110110 1111111110101010	1111111110100110 1111111110101011	1111111110100111 1111111110101100	1111111110101000 1111111110101101
7	11111010 1111111110110001	111111110111 1111111110110010	1111111110101110 1111111110110011	1111111110101111 1111111110110100	1111111110110000 1111111110110101
8	111111000 1111111110111001	111111111000000 1111111110111010	1111111110110110 1111111110111011	1111111110110111 1111111110111100	1111111110111000 1111111110111101
9	111111001 1111111111000010	1111111110111110 1111111111000011	1111111110111111 1111111111000100	1111111111000000 1111111111000101	1111111111000001 1111111111000110
A	111111010 1111111111001011	1111111111000111 1111111111001100	1111111111001000 1111111111001101	1111111111001001 1111111111001110	1111111111001010 1111111111001111
B	1111111001 1111111111010100	1111111111010000 1111111111010101	1111111111010001 1111111111010110	1111111111010010 1111111111010111	1111111111010011 1111111111011000
C	1111111010 1111111111011101	1111111111011001 1111111111011110	1111111111011010 1111111111011111	1111111111011011 1111111111100000	1111111111011100 1111111111100001
D	11111111000 1111111111100110	1111111111100010 1111111111100111	1111111111100011 1111111111101000	1111111111100100 1111111111101001	1111111111100101 1111111111101010
E	1111111111101011 1111111111110000	1111111111101100 1111111111110001	1111111111101101 1111111111110010	1111111111101110 1111111111110011	1111111111101111 1111111111110100
F	11111111001 1111111111111001	1111111111110101 1111111111111010	1111111111110110 1111111111111011	1111111111110111 1111111111111101	1111111111111000 1111111111111110

Table 7.66: Recommended Huffman Codes for Luminance AC Coefficients.

Z	R: 1 6	2 7	3 8	4 9	5 A
0	01 111000	100 1111000	1010 111110100	11000 1111110110	11001 111111110100
1	1011 111111110101	111001 1111111110001000	11110110 1111111110001001	111110101 1111111110001010	11111110110 1111111110001011
2	11010 1111111110001100	11110111 1111111110001101	1111110111 1111111110001110	111111110110 1111111110001111	111111111000010 1111111110010000
3	11011 1111111110010010	11111000 1111111110010011	1111111000 1111111110010100	111111110111 1111111110010101	1111111110010001 1111111110010110
4	111010 1111111110011010	111110110 1111111110011011	1111111110010111 1111111110011100	1111111110011000 1111111110011101	1111111110011001 1111111110011110
5	111011 1111111110100010	1111111001 1111111110100011	1111111110011111 1111111110100100	1111111110100000 1111111110100101	1111111110100001 1111111110100110
6	1111001 1111111110101010	11111110111 1111111110101011	1111111110100111 1111111110101100	1111111110101000 1111111110101101	1111111110101001 1111111110101110
7	1111010 1111111110110010	11111111000 1111111110110011	1111111110101111 1111111110110100	1111111110110000 1111111110110101	1111111110110001 1111111110110110
8	11111001 1111111110111011	1111111110110111 1111111110111100	1111111110111000 1111111110111101	1111111110111001 1111111110111110	1111111110111010 1111111110111111
9	111110111 1111111111000100	1111111111000000 1111111111000101	1111111111000001 1111111111000110	1111111111000010 1111111111000111	1111111111000011 1111111111001000
A	111111000 1111111111001101	1111111111001001 1111111111001110	1111111111001010 1111111111001111	1111111111001011 1111111111010000	1111111111001100 1111111111010001
B	111111001 1111111111010110	1111111111010010 1111111111010111	1111111111010011 1111111111011000	1111111111010100 1111111111011001	1111111111010101 1111111111011010
C	111111010 1111111111011111	1111111111011011 1111111111100000	1111111111011100 1111111111100001	1111111111011101 1111111111100010	1111111111011110 1111111111100011
D	11111111001 1111111111101000	1111111111100100 1111111111101001	1111111111100101 1111111111101010	1111111111100110 1111111111101011	1111111111100111 1111111111101100
E	11111111100000 1111111111110001	1111111111101101 1111111111110010	1111111111101110 1111111111110011	1111111111101111 1111111111110100	1111111111110000 1111111111110101
F	111111111000011 1111111111111010	111111111010110 1111111111111011	1111111111110111 1111111111111100	1111111111111000 1111111111111101	1111111111111001 1111111111111110

Table 7.67: Recommended Huffman Codes for Chrominance AC Coefficients.

R Z:	0	1	...	15
0:	1010			11111111001(ZRL)
1:	00	1100	...	1111111111110101
2:	01	11011	...	1111111111110110
3:	100	1111001	...	1111111111110111
4:	1011	111110110	...	1111111111111000
5:	11010	11111110110	...	1111111111111001
⋮	⋮			

Table 7.68: Coding AC Coefficients.

Some JPEG variants use a particular version of arithmetic coding, called the QM coder (Section 5.11), that is specified in the JPEG standard. This version of arithmetic coding is adaptive, so it does not need Tables 7.65 and 7.68. It adapts its behavior to the image statistics as it goes along. Using arithmetic coding may produce 5–10% better compression than Huffman for a typical continuous-tone image. However, it is more complex to implement than Huffman coding, so in practice it is rare to find a JPEG codec that uses it.

7.10.5 Lossless Mode

The lossless mode of JPEG uses differencing (Section 1.3.1) to reduce the values of pixels before they are compressed. This particular form of differencing is called *predicting*. The values of some near neighbors of a pixel are subtracted from the pixel to get a small number, which is then compressed further using Huffman or arithmetic coding. Figure 7.69a shows a pixel X and three neighbor pixels A, B, and C. Figure 7.69b shows eight possible ways (predictions) to combine the values of the three neighbors. In the lossless mode, the user can select one of these predictions, and the encoder then uses it to combine the three neighbor pixels and subtract the combination from the value of X. The result is normally a small number, which is then entropy-coded in a way very similar to that described for the DC coefficient in Section 7.10.4.

Predictor 0 is used only in the hierarchical mode of JPEG. Predictors 1, 2, and 3 are called one-dimensional. Predictors 4, 5, 6, and 7 are two-dimensional.

It should be noted that the lossless mode of JPEG has never been very successful. It produces typical compression factors of 2, and is therefore inferior to other lossless image compression methods. Because of this, many JPEG implementations do not even implement this mode. Even the lossy (baseline) mode of JPEG does not perform well when asked to limit the amount of loss to a minimum. As a result, some JPEG implementations do not allow parameter settings that result in minimum loss. The strength of JPEG is in its ability to generate highly compressed images that when decompressed are indistinguishable from the original. Recognizing this, the ISO has decided to come up with another standard for lossless compression of continuous-tone images. This standard is now commonly known as JPEG-LS and is described in Section 7.11.

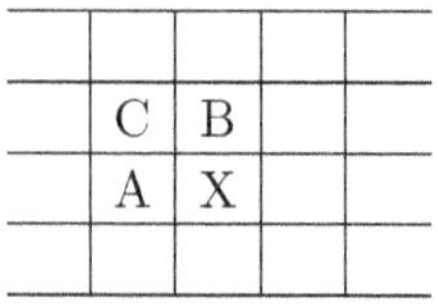

(a)

Selection value	Prediction
0	no prediction
1	A
2	B
3	C
4	A + B − C
5	A + ((B − C)/2)
6	B + ((A − C)/2)
7	(A + B)/2

(b)

Figure 7.69: Pixel Prediction in the Lossless Mode.

7.10.6 The Compressed File

A JPEG encoder outputs a compressed file that includes parameters, markers, and the compressed data units. The parameters are either four bits (these always come in pairs), one byte, or two bytes long. The markers serve to identify the various parts of the file. Each is two bytes long, where the first byte is `X'FF'` and the second one is not 0 or `X'FF'`. A marker may be preceded by a number of bytes with `X'FF'`. Table 7.71 lists all the JPEG markers (the first four groups are start-of-frame markers). The compressed data units are combined into MCUs (minimal coded unit), where an MCU is either a single data unit (in the noninterleaved mode) or three data units from the three image components (in the interleaved mode).

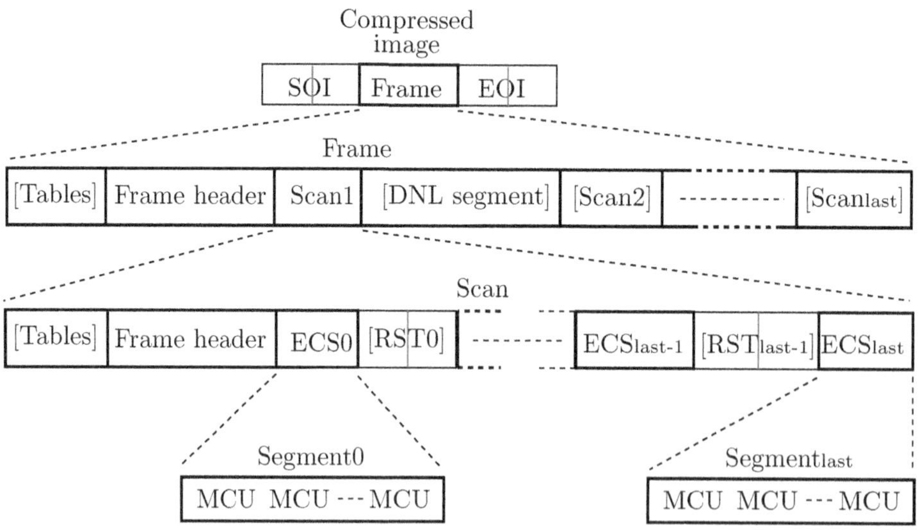

Figure 7.70: JPEG File Format.

Value	Name	Description
Nondifferential, Huffman coding		
FFC0	SOF_0	Baseline DCT
FFC1	SOF_1	Extended sequential DCT
FFC2	SOF_2	Progressive DCT
FFC3	SOF_3	Lossless (sequential)
Differential, Huffman coding		
FFC5	SOF_5	Differential sequential DCT
FFC6	SOF_6	Differential progressive DCT
FFC7	SOF_7	Differential lossless (sequential)
Nondifferential, arithmetic coding		
FFC8	JPG	Reserved for extensions
FFC9	SOF_9	Extended sequential DCT
FFCA	SOF_{10}	Progressive DCT
FFCB	SOF_{11}	Lossless (sequential)
Differential, arithmetic coding		
FFCD	SOF_{13}	Differential sequential DCT
FFCE	SOF_{14}	Differential progressive DCT
FFCF	SOF_{15}	Differential lossless (sequential)
Huffman table specification		
FFC4	DHT	Define Huffman table
Arithmetic coding conditioning specification		
FFCC	DAC	Define arith coding conditioning(s)
Restart interval termination		
FFD0–FFD7	RST_m	Restart with modulo 8 count m
Other markers		
FFD8	SOI	Start of image
FFD9	EOI	End of image
FFDA	SOS	Start of scan
FFDB	DQT	Define quantization table(s)
FFDC	DNL	Define number of lines
FFDD	DRI	Define restart interval
FFDE	DHP	Define hierarchical progression
FFDF	EXP	Expand reference component(s)
FFE0–FFEF	APP_n	Reserved for application segments
FFF0–FFFD	JPG_n	Reserved for JPEG extensions
FFFE	COM	Comment
Reserved markers		
FF01	TEM	For temporary private use
FF02–FFBF	RES	Reserved

Table 7.71: JPEG Markers.

Figure 7.70 shows the main parts of the JPEG compressed file (parts in square brackets are optional). The file starts with the SOI marker and ends with the EOI marker. In between these markers, the compressed image is organized in frames. In the hierarchical mode there are several frames, and in all other modes there is only one frame. In each frame the image information is contained in one or more scans, but the frame also contains a header and optional tables (which, in turn, may include markers). The first scan may be followed by an optional DNL segment (define number of lines), which starts with the DNL marker and contains the number of lines in the image that's represented by the frame. A scan starts with optional tables, followed by the scan header, followed by several entropy-coded segments (ECS), which are separated by (optional) restart markers (RST). Each ECS contains one or more MCUs, where an MCU is, as explained earlier, either a single data unit or three such units.

See Section 7.40 for an image compression method that combines JPEG with a fractal-based method.

7.10.7 JFIF

It has been mentioned earlier that JPEG is a compression method, not a graphics file format, which is why it does not specify image features such as pixel aspect ratio, color space, or interleaving of bitmap rows. This is where JFIF comes in.

JFIF (JPEG File Interchange Format) is a graphics file format that makes it possible to exchange JPEG-compressed images between computers. The main features of JFIF are the use of the YCbCr triple-component color space for color images (only one component for grayscale images) and the use of a *marker* to specify features missing from JPEG, such as image resolution, aspect ratio, and features that are application-specific.

The JFIF marker (called the APP0 marker) starts with the zero-terminated string `JFIF`. Following this, there is pixel information and other specifications (see below). Following this, there may be additional segments specifying JFIF extensions. A JFIF extension contains more platform-specific information about the image.

Each extension starts with the zero-terminated string `JFXX`, followed by a 1-byte code identifying the extension. An extension may contain application-specific information, in which case it starts with a different string, not `JFIF` or `JFXX` but something that identifies the specific application or its maker.

The format of the first segment of an APP0 marker is as follows:

1. APP0 marker (4 bytes): `FFD8FFE0`.
2. Length (2 bytes): Total length of marker, including the 2 bytes of the "length" field but excluding the APP0 marker itself (field 1).
3. Identifier (5 bytes): $\texttt{4A46494600}_{16}$. This is the `JFIF` string that identifies the APP0 marker.
4. Version (2 bytes): Example: 0102_{16} specifies version 1.02.
5. Units (1 byte): Units for the X and Y densities. 0 means no units; the Xdensity and Ydensity fields specify the pixel aspect ratio. 1 means that Xdensity and Ydensity are dots per inch, 2, that they are dots per cm.
6. Xdensity (2 bytes), Ydensity (2 bytes): Horizontal and vertical pixel densities (both should be nonzero).
7. Xthumbnail (1 byte), Ythumbnail (1 byte): Thumbnail horizontal and vertical pixel counts.

8. (RGB)n ($3n$ bytes): Packed (24-bit) RGB values for the thumbnail pixels. n = Xthumbnail × Ythumbnail.

The syntax of the JFIF extension APP0 marker segment is as follows:

1. APP0 marker.
2. Length (2 bytes): Total length of marker, including the 2 bytes of the "length" field but excluding the APP0 marker itself (field 1).
3. Identifier (5 bytes): $\texttt{4A46585800}_{16}$ This is the `JFXX` string identifying an extension.
4. Extension code (1 byte): 10_{16} = Thumbnail coded using JPEG. 11_{16} = Thumbnail coded using 1 byte/pixel (monochromatic). 13_{16} = Thumbnail coded using 3 bytes/pixel (eight colors).
5. Extension data (variable): This field depends on the particular extension.

> JFIF is the technical name for the image format better (but inaccurately) known as JPEG. This term is used only when the difference between the Image Format and the Image Compression is crucial. Strictly speaking, however, JPEG does not define an Image Format, and therefore in most cases it would be more precise to speak of JFIF rather than JPEG. Another Image Format for JPEG is SPIFF defined by the JPEG standard itself, but JFIF is much more widespread than SPIFF.
>
> —Erik Wilde, *WWW Online Glossary*

7.10.8 HD photo and JPEG XR

HD Photo is a complete specification (i.e., an algorithm and a file format) for the compression of continuous-tone photographic images. This method (formerly known as Windows Media Photo and also as Photon) was developed and patented by Microsoft as part of its Windows Media family. It supports both lossy and lossless compression, and is the standard image format for Microsoft's XPS documents. Microsoft official implementations of HD photo are part of the Windows Vista operating system and are also available for Windows XP.

The Joint Photographic Experts Group (the developer of JPEG) has decided to adopt HD photo as a JPEG standard, to be titled JPEG XR (extended range). The formal description of JPEG XR is ISO/IEC 29199-2.

An important feature of HD Photo is the use of integer operations only (even division is not required) in both encoder and decoder. The method supports bi-level, RGB, CMYK, and even n-channel color representation, with up to 16-bit unsigned integer representation, or up to 32-bit fixed-point or floating-point representation. The original color scheme is transformed to an internal color space and quantization is applied to achieve lossy compression. The algorithm has features for decoding just part of an image. Certain operations such as cropping, downsampling, horizontal or vertical flips, and 90° rotations can also be performed without decoding the entire image.

An HD Photo file has a format similar to TIFF (with all the advantages and limitations of this format, such as a maximum 4 GB file size). The file contains Image File Directory (IFD) tags, the image data (self-contained), optional alpha channel data (compressed as a separate image record), HD Photo metadata, optional XMP metadata, and optional Exif metadata, in IFD tags.

The compression algorithm of HD Photo is based on JPEG. The input image is converted to a luminance chrominance color space, the chroma components are optionally subsampled to increase compression, each of the three color components is partitioned into fixed-size blocks which are transformed into the frequency domain, and the frequency coefficients are quantized and entropy coded. The main differences between the original JPEG and HD photo are the following:

- JPEG supports 8-bit pixels (with some implementations also handling 12-bit pixels), while HD Photo supports pixels of up to 32 bits. HD Photo also includes features for the lossless compression of floating-point pixels.

- The color transformation of JPEG (from RGB to YCbCr) is reversible in principle, but in practice involves small losses of data because of roundoff errors. In contrast, HD Photo employs a lossless color transformation, that for RGB is given by

$$C_o = B - R, \quad C_g = R - G + \left\lceil \frac{C_o}{2} \right\rceil, \quad Y = G + \left\lfloor \frac{C_g}{2} \right\rfloor.$$

- JPEG uses blocks of 8×8 pixels for its DCT, while HD Photo also uses blocks of 4×4, 2×4, and 2×2 pixels. These block sizes can be selected by the encoder for image regions with lower-than-normal pixel correlation.

- The DCT is also reversible in principle and slightly lossy in practice because of the limited precision of computer arithmetic. The transform used by HD Photo is called the photo core transform (PCT). It is similar to the Walsh-Hadamard transform (Section 7.7.2), but is lossless because it employs only integers.

- Before applying the PCT, HD Photo offers an optional, lossless prefiltering step that has a visual blurring effect. This step filters 4×4 blocks that are offset by two pixels in each direction from the PCT blocks. Experiments indicate that this type of filtering reduces the annoying block-boundary JPEG artifacts that occur mostly at low bitrates. This step is normally skipped at high bitrates, where such artifacts are negligible or nonexistent.

- Section 7.10.4 explains how JPEG encodes the DCT coefficients (specifically, how the DC coefficients are encoded by left-prediction). In HD Photo, blocks are grouped into macroblocks of 16×16 pixels, and the DC components from each macroblock are themselves frequency transformed. Thus, HD photo has to encode three types of coefficients: the macroblock DC coefficients (called DC), macroblock-level AC coefficients (termed lowpass), and lower-level AC coefficients (referred to as AC).

- The encoding of these three types is far more complex in HD Photo than JPEG encoding. It consists of (1) a complex DC prediction algorithm, (2) an adaptive coefficient reordering (instead of simple zigzag ordering), followed by (3) a form of adaptive Huffman coding for the coefficients themselves.

- There is a single quantization coefficient for the DC transform coefficients in each color plane. The lowpass and the AC quantization coefficients, however, can vary from macroblock to macroblock.

- HD Photo produces lossless compression simply by setting all its quantization coefficients to 1. Because of the use of integer transform and special color space transformation, there are no other losses in HD photo. This is in contrast with JPEG, whose lossless mode is inefficient and often is not even implemented.

7.11 JPEG-LS

As has been mentioned in Section 7.10.5, the lossless mode of JPEG is inefficient and often is not even implemented. As a result, the ISO, in cooperation with the IEC, has decided to develop a new standard for the lossless (or near-lossless) compression of continuous-tone images. The result is recommendation ISO/IEC CD 14495, popularly known as JPEG-LS. The principles of this method are described here, but it should be noted that it is not simply an extension or a modification of JPEG. This is a new method, designed to be simple and fast. It does not use the DCT, does not employ arithmetic coding, and uses quantization in a restricted way, and only in its near-lossless option. JPEG-LS is based on ideas developed in [Weinberger et al. 96 and 00] and for their LOCO-I compression method. JPEG-LS examines several of the previously-seen neighbors of the current pixel, uses them as the *context* of the pixel, uses the context to predict the pixel and to select a probability distribution out of several such distributions, and uses that distribution to encode the prediction error with a special Golomb code. There is also a run mode, where the length of a run of identical pixels is encoded.

The context used to predict the current pixel x is shown in Figure 7.72. The encoder examines the context pixels and decides whether to encode the current pixel x in the *run mode* or in the *regular mode*. If the context suggests that the pixels y, z,... following the current pixel are likely to be identical, the encoder selects the run mode. Otherwise, it selects the regular mode. In the near-lossless mode the decision is slightly different. If the context suggests that the pixels following the current pixel are likely to be almost identical (within the tolerance parameter NEAR), the encoder selects the run mode. Otherwise, it selects the regular mode. The rest of the encoding process depends on the mode selected.

	c	b	d		
	a	x	y	z	

Figure 7.72: Context for Predicting x.

In the regular mode, the encoder uses the values of context pixels a, b, and c to predict pixel x, and subtracts the prediction from x to obtain the *prediction error*, denoted by *Errval*. This error is then *corrected* by a term that depends on the context (this correction is done to compensate for systematic biases in the prediction), and encoded with a Golomb code. The Golomb coding depends on all four pixels of the context and also on prediction errors that were previously encoded for the same context (this information is stored in arrays A and N, mentioned in Section 7.11.1). If near-lossless compression is used, the error is quantized before it is encoded.

In the run mode, the encoder starts at the current pixel x and finds the longest run of pixels that are identical to context pixel a. The encoder does not extend this run beyond the end of the current image row. Since all the pixels in the run are identical to a (and a is already known to the decoder) only the length of the run needs be encoded, and this is done with a 32-entry array denoted by J (Section 7.11.1). If near-lossless compression is used, the encoder selects a run of pixels that are close to a within the tolerance parameter NEAR.

The decoder is not substantially different from the encoder, so JPEG-LS is a nearly symmetric compression method. The compressed stream contains data segments (with the Golomb codes and the encoded run lengths), marker segments (with information needed by the decoder), and markers (some of the reserved markers of JPEG are used). A marker is a byte of all ones followed by a special code, signaling the start of a new segment. If a marker is followed by a byte whose most significant bit is 0, that byte is the start of a marker segment. Otherwise, that byte starts a data segment.

7.11.1 The Encoder

JPEG-LS is normally used as a lossless compression method. In this case, the reconstructed value of a pixel is identical to its original value. In the near lossless mode, the original and the reconstructed values may differ. In every case we denote the reconstructed value of a pixel p by Rp.

When the top row of an image is encoded, context pixels b, c, and d are not available and should therefore be considered zero. If the current pixel is located at the start or the end of an image row, either a and c, or d are not available. In such a case, the encoder uses for a or d the reconstructed value Rb of b (or zero if this is the top row), and for c the reconstructed value that was assigned to a when the first pixel of the previous line was encoded. This means that the encoder has to do part of the decoder's job and has to figure out the reconstructed values of certain pixels.

The first step in determining the context is to calculate the three *gradient* values

$$D1 = Rd - Rb, \quad D2 = Rb - Rc, \quad D3 = Rc - Ra.$$

If the three values are zero (or, for near-lossless, if their absolute values are less than or equal to the tolerance parameter NEAR), the encoder selects the run mode, where it looks for the longest run of pixels identical to Ra. Step 2 compares the three gradients Di to certain parameters and calculates three region numbers Qi according to certain rules (not shown here). Each region number Qi can take one of the nine integer values in the interval $[-4, +4]$, so there are $9 \times 9 \times 9 = 729$ different region numbers. The third step maps the region numbers Q_i to an integer Q in the interval $[0, 364]$. The tuple $(0, 0, 0)$ is mapped to 0, and the 728 remaining tuples are mapped to $[1, 728/2 = 364]$, such that (a, b, c) and $(-a, -b, -c)$ are mapped to the same value. The details of this calculation are not specified by the JPEG-LS standard, and the encoder can do it in any way it chooses. The integer Q becomes the context for the current pixel x. It is used to index arrays A and N in Figure 7.76.

After determining the context Q, the encoder predicts pixel x in two steps. The first step calculates the prediction Px based on *edge rules*, as shown in Figure 7.73.

The second step corrects the prediction, as shown in Figure 7.74, based on the quantity SIGN (determined from the signs of the three regions Qi), the correction values $C[Q]$ (derived from the bias and not discussed here), and parameter MAXVAL.

```
if(Rc>=max(Ra,Rb)) Px=min(Ra,Rb);
 else
  if(Rc<=min(Ra,Rb)) Px=max(Ra,Rb)
    else Px=Ra+Rb-Rc;
  endif;
endif;
```

Figure 7.73: Edge Detecting.

```
if(SIGN=+1) Px=Px+C[Q]
         else Px=Px-C[Q]
endif;
if(Px>MAXVAL) Px=MAXVAL
 else if(Px<0) Px=0 endif;
endif;
```

Figure 7.74: Prediction Correcting.

To understand the edge rules, let's consider the case where $b \leq a$. In this case the edge rules select b as the prediction of x in many cases where a vertical edge exists in the image just left of the current pixel x. Similarly, a is selected as the prediction in many cases where a horizontal edge exists in the image just above x. If no edge is detected, the edge rules compute a prediction of $a+b-c$, and this has a simple geometric interpretation. If we interpret each pixel as a point in three-dimensional space, with the pixel's intensity as its height, then the value $a+b-c$ places the prediction Px on the same plane as pixels a, b, and c.

Once the prediction Px is known, the encoder computes the prediction error $Errval$ as the difference $x - Px$ but reverses its sign if the quantity SIGN is negative.

In the near-lossless mode the error is quantized, and the encoder uses it to compute the reconstructed value Rx of pixel x the way the decoder will do it in the future. The encoder needs this reconstructed value to encode future pixels. The basic quantization step is

$$Errval \leftarrow \frac{Errval + \text{NEAR}}{2\times\text{NEAR}+1}.$$

It uses parameter NEAR, but it involves more details that are not shown here. The basic reconstruction step is

$$Rx \leftarrow Px + \text{SIGN}\times Errval\times(2\times\text{NEAR}+1).$$

The prediction error (after possibly being quantized) now goes through a range reduction (whose details are omitted here) and is finally ready for the important step of encoding.

The Golomb code was introduced in Section 3.24, where its main parameter was denoted by b. JPEG-LS denotes this parameter by m. Once m has been selected, the Golomb code of the nonnegative integer n consists of two parts, the unary code of the integer part of n/m and the binary representation of $n \bmod m$. These codes are ideal for integers n that are distributed geometrically (i.e., when the probability of n is $(1-r)r^n$, where $0 < r < 1$). For any such geometric distribution there exists a value m such that the Golomb code based on it yields the shortest possible average code length. The special case where m is a power of 2 ($m = 2^k$) leads to simple encoding/decoding operations. The code for n consists, in such a case, of the k least-significant bits of n, preceded by the unary code of the remaining most-significant bits of n. This particular Golomb code is denoted by $G(k)$.

As an example we compute the $G(2)$ code of $n = 19 = 10011_2$. Since $k = 2$, m is 4. We start with the two least-significant bits, 11, of n. They equal the integer 3, which is also $n \bmod m$ ($3 = 19 \bmod 4$). The remaining most-significant bits, 100, are also the integer 4, which is the integer part of the quotient n/m ($19/4 = 4.75$). The unary code of 4 is 00001, so the $G(2)$ code of $n = 19$ is $00001|11$.

In practice, we always have a finite set of nonnegative integers, where the largest integer in the set is denoted by I. The maximum length of $G(0)$ is $I + 1$, and since I can be large, it is desirable to limit the size of the Golomb code. This is done by the special Golomb code $LG(k, glimit)$ which depends on the two parameters k and $glimit$. We first form a number q from the most significant bits of n. If $q < glimit - \lceil \log I \rceil - 1$, the $LG(k, glimit)$ code is simply $G(k)$. Otherwise, the unary code of $glimit - \lceil \log I \rceil - 1$ is prepared (i.e., $glimit - \lceil \log I \rceil - 1$ zeros followed by a single 1). This acts as an escape code and is followed by the binary representation of $n - 1$ in $\lceil I \rceil$ bits.

Our prediction errors are not necessarily positive. They are differences, so they can also be zero or negative, but the various Golomb codes were designed for nonnegative integers. This is why the prediction errors must be mapped to nonnegative values before they can be coded. This is done by

$$MErrval = \begin{cases} 2Errval, & Errval \geq 0, \\ 2|Errval| - 1, & Errval < 0. \end{cases} \tag{7.25}$$

This mapping interleaves negative and positive values in the sequence

$$0, -1, +1, -2, +2, -3, \ldots.$$

Table 7.75 lists some prediction errors, their mapped values, and their $LG(2, 32)$ codes assuming an alphabet of size 256 (i.e., $I = 255$ and $\lceil \log I \rceil = 8$).

The next point to be discussed is how to determine the value of the Golomb code parameter k. This is done adaptively. Parameter k depends on the context, and the value of k for a context is updated each time a pixel with that context is found. The calculation of k can be expressed by the single C-language statement

```
for (k=0; (N[Q]<<k)<A[Q]); k++);
```

where A and N are arrays indexed from 0 to 364. This statement uses the context Q as an index to the two arrays. It initializes k to 0 and goes into a loop. In each iteration it shifts array element $N[Q]$ by k positions to the left and compares it to element $A[Q]$. If the shifted value of $N[Q]$ is greater than or equal to $A[Q]$, the current value of k is chosen. Otherwise, k is incremented by 1 and the test repeated.

After k has been determined, the prediction error $Errval$ is mapped, by means of Equation (7.25), to $MErrval$, which is encoded using code $LG(k, LIMIT)$. The quantity LIMIT is a parameter. Arrays A and N (together with an auxiliary array B) are then updated as shown in Figure 7.76 (RESET is a user-controlled parameter).

Encoding in the run mode is done differently. The encoder selects this mode when it finds consecutive pixels x whose values Ix are identical and equal to the reconstructed

Prediction error	Mapped value	Code
0	0	1 00
−1	1	1 01
1	2	1 10
−2	3	1 11
2	4	01 00
−3	5	01 01
3	6	01 10
−4	7	01 11
4	8	001 00
−5	9	001 01
5	10	001 10
−6	11	001 11
6	12	0001 00
−7	13	0001 01
7	14	0001 10
−8	15	0001 11
8	16	00001 00
−9	17	00001 01
9	18	00001 10
−10	19	00001 11
10	20	000001 00
−11	21	000001 01
11	22	000001 10
−12	23	000001 11
12	24	0000001 00
...		
50	100	000000000000 000000000001 01100011

Table 7.75: Prediction Errors, Their Mappings, and $LG(2,32)$ Codes.

```
B[Q]=B[Q]+Errval*(2*NEAR+1);
A[Q]=A[Q]+abs(Errval);
if(N[Q]=RESET) then
 A[Q]=A[Q]>>1; B[Q]=B[Q]>>1; N[Q]=N[Q]>>1
endif;
N[Q]=N[Q]+1;
```

Figure 7.76: Updating Arrays A, B, and N.

value Ra of context pixel a. For near-lossless compression, pixels in the run must have values Ix that satisfy

$$|Ix - Ra| \leq \text{NEAR}.$$

A run is not allowed to continue beyond the end of the current image row. The length of the run is encoded (there is no need to encode the value of the run's pixels, since it equals Ra), and if the run ends before the end of the current row, its encoded length is followed by the encoding of the pixel immediately following it (the pixel *interrupting* the run). The two main tasks of the encoder in this mode are (1) run scanning and run-length encoding and (2) run interruption coding. Run scanning is shown in Figure 7.77. Run-length encoding is shown in Figures 7.78 (for run segments of length rm) and 7.79 (for segments of length less than rm). Here are some of the details.

```
RUNval=Ra;
RUNcnt=0;
while(abs(Ix-RUNval)<=NEAR)
 RUNcnt=RUNcnt+1;
 Rx=RUNval;
 if(EOLine=1) break
  else GetNextSample()
 endif;
endwhile;
```

Figure 7.77: Run Scanning.

```
while(RUNcnt>=(1<<J[RUNindex]))
 AppendToBitStream(1,1);
 RUNcnt=RUNcnt-(1<<J[RUNindex]);
 if(RUNindex<31)
  RUNindex=RUNindex+1;
endwhile;
```

Figure 7.78: Run Encoding: I.

```
if(EOLine=0) then
 AppendToBitStream(0,1);
 AppendToBitStream
  (RUNcnt,J[RUNindex]);
 if(RUNindex>0)
  RUNindex=RUNindex-1;
 endif;
else if(RUNcnt>0)
 AppendToBitStream(1,1);
```

Figure 7.79: Run Encoding: II.

The encoder uses a 32-entry table J containing values that are denoted by rk. J is initialized to the 32 values

$$0, 0, 0, 0, 1, 1, 1, 1, 2, 2, 2, 2, 3, 3, 3, 3, 4, 4, 5, 5, 6, 6, 7, 7, 8, 9, 10, 11, 12, 13, 14, 15.$$

For each value rk, we use the notation $rm = 2^{rk}$. The 32 quantities rm are called *code-order*. The first four rms have values $2^0 = 1$. The next four have values $2^1 = 2$. The next four, $2^2 = 4$, up to the last rm, whose value is $2^{15} = 32768$. The encoder executes the procedure of Figure 7.77 to determine the run length, which it stores in variable `RUNlen`. This variable is then encoded by breaking it up into chunks whose sizes are the values of consecutive rms. For example, if `RUNlen` is 6, it can be expressed in terms of the *rms* as $1 + 1 + 1 + 1 + 2$, so it is equivalent to the first five *rms*. It is encoded by writing five bits of 1 on the compressed stream. Each of those bits is written by the statement `AppendToBitStream(1,1)` of Figure 7.78. Each time a 1 is written, the value of the corresponding *rm* is subtracted from `RUNlen`. If `RUNlen` is originally 6, it goes down to 5, 4, 3, 2, and 0.

It may happen, of course, that the length `RUNlen` of a run is not equal to an integer number of *rms*. An example is a `RUNlen` of 7. This is encoded by writing five bits of 1, followed by a *prefix* bit, followed by the remainder of `RUNlen` (in our example, a 1), written on the compressed stream as an *rk*-bit number (the current *rk* in our example is 1). This last operation is performed by the procedure call `AppendToBitStream(RUNcnt,J[RUNindex])` of Figure 7.79. The prefix bit is 0 if the run is interrupted by a different pixel. It is 1 if the run is terminated by the end of an image row.

The second main task of the encoder, encoding the interruption pixel, is similar to encoding the current pixel and is not discussed here.

7.12 Adaptive Linear Prediction and Classification

ALPC (*adaptive linear prediction and classification*), is a lossless image compression algorithm described in [Motta et al. 00]. ALPC uses a linear predictor that is trained on a selected part of the causal neighborhood of the pixel being encoded. The use of linear prediction in an image compressor is somewhat unusual, since most lossless compressors rely on non-linear predictors to model the discontinuities often present in a natural image. In ALPC, the non-linearity of the input image is modeled by a classification step that is performed before the prediction. The classification selects a causal subset of samples from which the weights of the predictor are determined.

Causality: A cause and effect relationship. The causality of two events describes to what extent one event is caused by the other. Causality implies a measure of predictability between the two events.

In a linear predictor, a pixel value is represented by a weighted sum of its context pixels. ALPC uses a context formed by the six causal pixels depicted in Figure 7.80. The value of the pixel $I(x, y)$ is predicted as

$$\begin{aligned}\hat{I}(x,y) = \text{round}\big(&w_0 \times I(x,y-2) + w_1 \times I(x-1,y-1) + w_2 \times I(x,y-1)\\ &+ w_3 \times I(x+1,y-1) + w_4 \times I(x-2,y) + w_5 \times I(x-1,y)\big).\end{aligned}$$

The shape of the context is fixed, while the weights w_0 through w_5 are adaptively determined for each pixel. Since the prediction is close to the value of the pixel, the prediction error $E(x, y) = I(x, y) - \hat{I}(x, y)$ exhibits a peaked distribution centered around zero. As a result, encoding the prediction error is more effective than encoding the pixel itself. The image is encoded in raster scan order, from top to bottom and from left to right, so the decoder can produce an exact reconstruction of the context pixels and

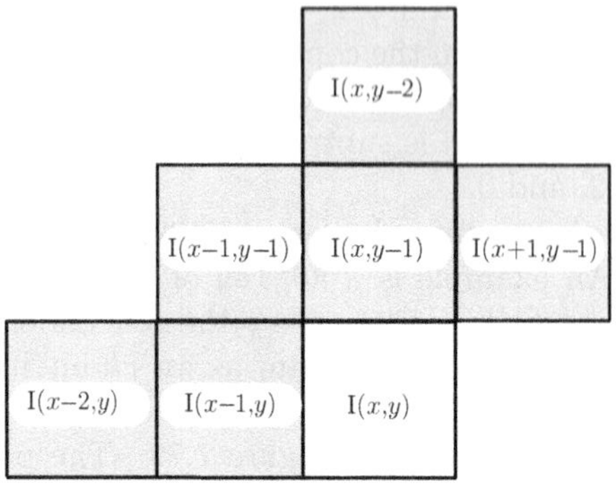

Figure 7.80: Context Used in ALPC Linear Predictors.

can mimic the encoder by computing $\hat{I}(x,y)$ and adding it to $E(x,y)$ to reconstruct the original value of $I(x,y)$.

The weights are computed on a pixel by pixel basis, in an attempt to model local characteristics of the image. The determination of the weights is based on the causal portion of a square window of radius R_p that is centered on the current pixel (Figure 7.81). This choice allows the use of *backward prediction*, a technique that does not require explicit transmission of the weights, because the decoder has enough information to determine them. The radius of the prediction window is a critical feature of ALPC. A small window will determine overspecialized predictors, while a large window will not be able to track the local variations in the image content. While it is possible to dynamically adapt R_p to the image content, in ALPC the radius of the window is fixed and is set equal to 10 pixels.

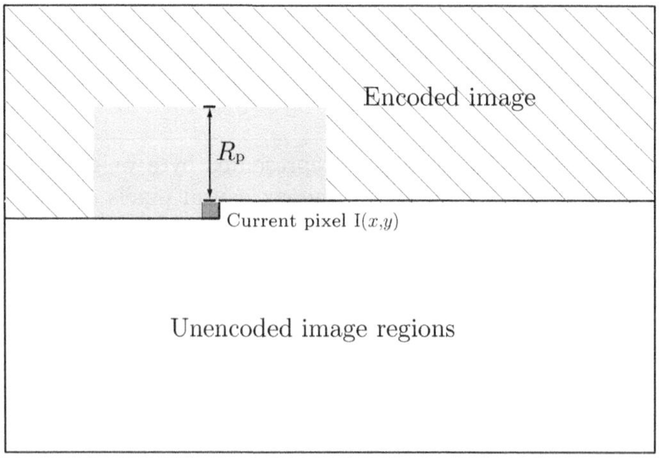

Figure 7.81: Prediction Window of Radius R_p Centered Around $I(x,y)$.

Not all samples in the window are used to compute the predictors. Each pixel in the window is first classified in one of n classes, according to its context. The number of classes n is predetermined and a classification algorithm such as LBG [Linde, Buzo, and Gray 80] can be used to partition the samples in the window. The context of the current pixel is then classified as well and the contexts and the pixels belonging to its class are used to determine the predictor's weights. While this classification method is not optimal, the selection of pixels having a context similar to the one being encoded is sufficient to improve the performance of the basic adaptive predictor.

In [Motta et al. 00] the weights are determined by using a *gradient descend* optimization that minimizes the energy of the error on the samples in the class. If the number of samples is too small, a fixed predictor is used instead.

The classified linear predictor is quite effective and the prediction error is distributed according to a skewed Laplacian-like distribution. In the experimental results reported in [Motta et al. 00], 95% of the error consistently falls into a very narrow range; we call these error values "typical." Two arithmetic encoders are used to encode error magnitudes, depending on whether the prediction error is typical or not. The sign of the error is encoded separately, with a third arithmetic encoder. The modeling of the arithmetic encoders is based on error statistics collected in a causal window of radius R_e, similar to the one used in the prediction. The use of multiple arithmetic encoders results in a small but consistent gain.

7.13 Progressive Image Compression

Most modern image compression methods are either progressive or optionally so. Progressive compression is an attractive choice when compressed images are transmitted over a communications line and are decompressed and viewed in real time. When such an image is received and is decompressed, the decoder can very quickly display the entire image in a low-quality format, and improve the display quality as more and more of the image is being received and decompressed. A user watching the image developing on the screen can normally recognize most of the image features after only 5–10% of it has been decompressed.

This should be compared to raster-scan image compression. When an image is raster scanned and compressed, a user normally cannot tell much about the image when only 5–10% of it has been decompressed and displayed. Images are supposed to be viewed by humans, which is why progressive compression makes sense even in cases where it is slower or less efficient than nonprogressive.

Perhaps a good way to think of progressive image compression is to imagine that the encoder compresses the most important image information first, then compresses less important information and appends it to the compressed stream, and so on. This explains why all progressive image compression methods have a natural lossy option; simply stop compressing at a certain point. The user can control the amount of loss by means of a parameter that tells the encoder how soon to stop the progressive encoding process. The sooner encoding is stopped, the better the compression ratio and the higher the data loss.

Another advantage of progressive compression becomes apparent when the compressed file has to be decompressed several times and displayed with different resolutions. The decoder can, in each case, stop the decompression when the image has reached the resolution of the particular output device used.

Progressive image compression has already been mentioned, in connection with JPEG (page 522). JPEG uses the DCT to break the image up into its spatial frequency components, and it compresses the low-frequency components first. The decoder can therefore display these parts quickly, and it is these low-frequency parts that contain the principal image information. The high-frequency parts contain image details. Thus, JPEG encodes spatial frequency data progressively.

It is useful to think of progressive decoding as the process of improving image features over time, and this can be achieved in three ways:

1. Encode spatial frequency data progressively. An observer watching such an image being decoded sees the image changing from blurred to sharp. Methods that work this way typically feature medium speed encoding and slow decoding. This type of progressive compression is sometimes called *SNR progressive* or *quality progressive*.
2. Start with a gray image and add colors or shades of gray to it. An observer watching such an image being decoded will see all the image details from the start, and will see them improve as more color is continuously added to them. Vector quantization methods (Section 7.19) use this kind of progressive compression. Such a method normally features slow encoding and fast decoding.
3. Encode the image in layers, where early layers consist of a few large low-resolution pixels, followed by later layers with smaller higher-resolution pixels. A person watching such an image being decoded will see more detail added to the image over time. Such a method thus adds detail (or resolution) to the image as it is being decompressed. This way of progressively encoding an image is called *pyramid coding* or *hierarchical coding*. Most progressive methods use this principle, so this section discusses general ideas for implementing pyramid coding. Figure 7.83 illustrates the three progressive methods mentioned here. It should be contrasted with Figure 7.82, which illustrates sequential decoding.

Assuming that the image size is $2^n \times 2^n = 4^n$ pixels, the simplest method that comes to mind, when trying to implement progressive compression, is to calculate each pixel of layer $i-1$ as the average of a group of 2×2 pixels of layer i. Thus layer n is the entire image, layer $n-1$ contains $2^{n-1}\times 2^{n-1} = 4^{n-1}$ large pixels of size 2×2, and so on, down to layer 1, with $4^{n-n} = 1$ large pixel, representing the entire image. If the image isn't too large, all the layers can be saved in memory. The pixels are then written on the compressed stream in reverse order, starting with layer 1. The single pixel of layer 1 is the "parent" of the four pixels of layer 2, each of which is the parent of four pixels in layer 3, and so on. The total number of pixels in the pyramid is 33% more than the original number!

$$4^0 + 4^1 + \cdots + 4^{n-1} + 4^n = (4^{n+1} - 1)/3 \approx 4^n(4/3) \approx 1.33\times 4^n = 1.33(2^n \times 2^n).$$

A simple way to bring the total number of pixels in the pyramid down to 4^n is to include only three of the four pixels of a group in layer i, and to compute the value of

Figure 7.82: Sequential Decoding.

the 4th pixel using the parent of the group (from the preceding layer, $i-1$) and its three siblings.

Example: Figure 7.84c shows a 4×4 image that becomes the third layer in its progressive compression. Layer two is shown in Figure 7.84b, where, for example, pixel 81.25 is the average of the four pixels 90, 72, 140, and 23 of layer three. The single pixel of layer one is shown in Figure 7.84a.

The compressed file should contain just the numbers

$$54.125,\ 32.5, 41.5, 61.25,\ 72, 23, 140,\ 33, 18, 21,\ 18, 32, 44,\ 70, 59, 16$$

(properly encoded, of course), from which all the missing pixel values can easily be determined. The missing pixel 81.25, e.g., can be calculated from $(x+32.5+41.5+61.25)/4=54.125$.

A small complication with this method is that averages of integers may be nonintegers. If we want our pixel values to remain integers we either have to lose precision or to keep using longer and longer integers. Assuming that pixels are represented by eight bits, adding four 8-bit integers produces a 10-bit integer. Dividing it by four, to create the average, reduces the sum back to an 8-bit integer, but some precision may be lost. If we don't want to lose precision, we should represent our second-layer pixels as 10-bit numbers and our first-layer (single) pixel as a 12-bit number. Figure 7.84d,e,f shows the results of rounding off our pixel values and thus losing some image information. The content of the compressed file in this case should be

$$54,\ 33, 42, 61,\ 72, 23, 140,\ 33, 18, 21,\ 18, 32, 44,\ 70, 59, 16.$$

The first missing pixel, 81, of layer three can be determined from the equation $(x+33+42+61)/4=54$, which yields the (slightly wrong) value 80.

Figure 7.83: Progressive Decoding.

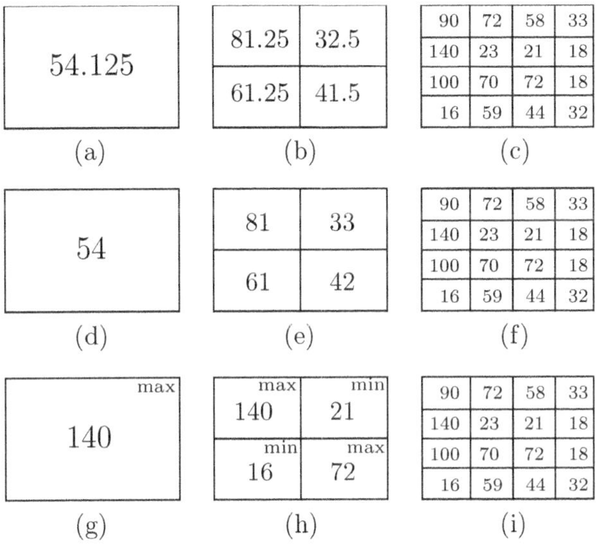

Figure 7.84: Progressive Image Compression.

◇ **Exercise 7.19:** Show that the sum of four n-bit numbers is an $(n+2)$-bit number.

A better method is to let the parent of a group help in calculating the values of its four children. This can be done by calculating the differences between the parent and its children, and writing the differences (suitably coded) in layer i of the compressed stream. The decoder decodes the differences, then uses the parent from layer $i-1$ to compute the values of the four pixels. Either Huffman or arithmetic coding can be used to encode the differences. If all the layers are calculated and saved in memory, then the distribution of difference values can be found and used to achieve the best statistical compression.

If there is no room in memory for all the layers, a simple adaptive model can be implemented. It starts by assigning a count of 1 to every difference value (to avoid the zero-probability problem). When a particular difference is computed, it is assigned a probability and is encoded according to its count, and its count is then updated. It is a good idea to update the counts by incrementing them by a value greater than 1, since this way the original counts of 1 become insignificant very quickly.

Some improvement can be achieved if the parent is used to help calculate the values of three child pixels, and then these three plus the parent are used to calculate the value of the fourth pixel of the group. If the four pixels of a group are a, b, c, and d, then their average is $v = (a+b+c+d)/4$. The average becomes part of layer $i-1$, and layer i need contain only the three differences $k = a-b$, $l = b-c$, and $m = c-d$. Once the decoder has read and decoded the three differences, it can use their values, together with the value of v from the previous layer, to compute the values of the four pixels of the group. Calculating v by a division by 4 still causes the loss of two bits, but this 2-bit quantity can be isolated before the division, and retained by encoding it separately, following the three differences.

The improvements mentioned above are based on the well-known fact that small integers are easy to compress (page 46).

The parent pixel of a group does not have to be its average. One alternative is to select the maximum (or the minimum) pixel of a group as the parent. This has the advantage that the parent is identical to one of the pixels in the group. The encoder has to encode just three pixels in each group, and the decoder decodes three pixels (or differences) and uses the parent as the fourth pixel, to complete the group. When encoding consecutive groups in a layer, the encoder should alternate between selecting the maximum and the minimum as parents, since always selecting the same creates progressive layers that are either too dark or too bright. Figure 7.84g,h,i shows the three layers in this case.

The compressed file should contain the numbers

$$140,\ (0), 21, 72, 16,\ (3), 90, 72, 23,\ (3), 58, 33, 18,\ (0), 18, 32, 44,\ (3), 100, 70, 59,$$

where the numbers in parentheses are two bits each. They tell where (in what quadrant) the parent from the previous layer should go. Notice that quadrant numbering is $\binom{0\,1}{3\,2}$.

Selecting the median of a group is a little slower than selecting the maximum or the minimum, but it improves the appearance of the layers during progressive decompression. In general, the median of a sequence $(a_1, a_2, \ldots, a_n)$ is an element a_i such that half the elements (or very close to half) are smaller than a_i and the other half are bigger. If the four pixels of a group satisfy $a < b < c < d$, then either b or c can be considered the median pixel of the group. The main advantage of selecting the median as the group's parent is that it tends to smooth large differences in pixel values that may occur because of one extreme pixel. In the group 1, 2, 3, 100, for example, selecting 2 or 3 as the parent is much more representative than selecting the average. Finding the median of four pixels requires a few comparisons, but calculating the average requires a division by 4 (or, alternatively, a right shift).

Once the median has been selected and encoded as part of layer $i-1$, the remaining three pixels can be encoded in layer i by encoding their (three) differences, preceded by a 2-bit code telling which of the four is the parent. Another small advantage of using the median is that once the decoder reads this 2-bit code, it knows how many of the three pixels are smaller and how many are bigger than the median. If the code says, for example, that one pixel is smaller, and the other two are bigger than the median, and the decoder reads a pixel that's smaller than the median, it knows that the next two pixels decoded will be bigger than the median. This knowledge changes the distribution of the differences, and it can be taken advantage of by using three count tables to estimate probabilities when the differences are encoded. One table is used when a pixel is encoded that the decoder will know is bigger than the median. Another table is used to encode pixels that the decoder will know are smaller than the median, and the third table is used for pixels where the decoder will not know in advance their relations to the median. This improves compression by a few percent and is another example of how adding more features to a compression method brings diminishing returns.

Some of the important progressive image compression methods used in practice are described in the rest of this chapter.

7.13.1 Growth Geometry Coding

The idea of growth geometry coding deserves its own section, since it combines image compression and progressive image transmission in an original and unusual way. This idea is due to Amalie J. Frank [Frank et al. 80]. The method is designed for progressive lossless compression of bi-level images. The idea is to start with some *seed* pixels and apply geometric rules to grow each seed pixel into a pattern of pixels. The encoder has the harder part of the job. It has to select the seed pixels, the growth rule for each seed, and the number of times the rule should be applied (the number of generations). Only this data is written on the compressed stream. Often, a group of seeds shares the same rule and the same number of generations. In such a case, the rule and number of generations are written once, following the coordinates of the seed pixels of the group. The decoder's job is simple. It reads the first group of seeds, applies the rule once, reads the next group and adds it to the pattern so far, applies the rule again, and so on. Compression is achieved if the number of seeds is small compared to the total image size, i.e., if each seed pixel is used to generate, on average, many image pixels.

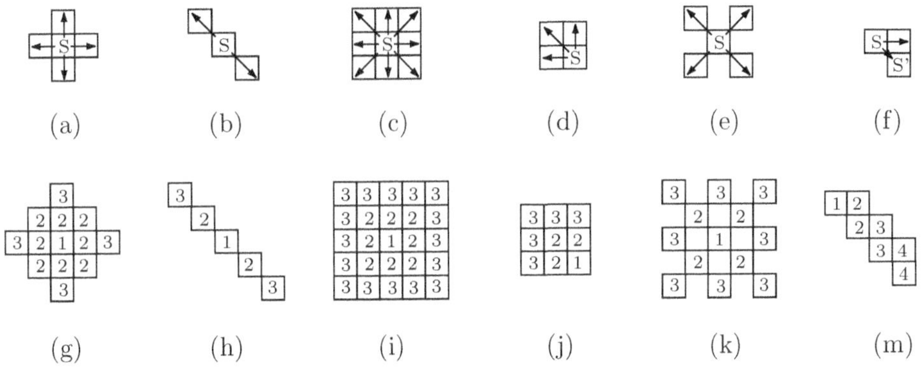

Figure 7.85: Six Growth Rules and Patterns.

Figure 7.85a–f shows six simple growth rules. Each adds some immediate neighbor pixels to the seed S, and the same rule is applied to these neighbors in the next generation (they become *secondary* seed pixels). Since a pixel can have up to eight immediate neighbors, there can be 256 such rules. Figure 7.85g–m shows the results of applying these rules twice (i.e., for two generations) to a single seed pixel. The pixels are numbered one plus the generation number. The growth rules can be as complex as necessary. For example, Figure 7.85m assumes that only pixel S' becomes a secondary seed. The resulting pattern of pixels may be solid or may have "holes" in it, as in Figure 7.85k.

Figure 7.86 shows an example of how a simple growth rule (itself shown in Figure 7.86a) is used to grow the disconnected pattern of Figure 7.86b. The seed pixels are shown in Figure 7.86c. They are numbered 1–4 (one more than the number of generations). The decoder first inputs the six pixels marked 4 (Figure 7.86d). The growth rule is applied to them, producing the pattern of Figure 7.86e. The single pixel marked 3 is then input, and the rule applied again, to pixels 3 and 4, to form the pattern of Figure 7.86f. The four pixels marked 2 are input, and the rule applied again, to form

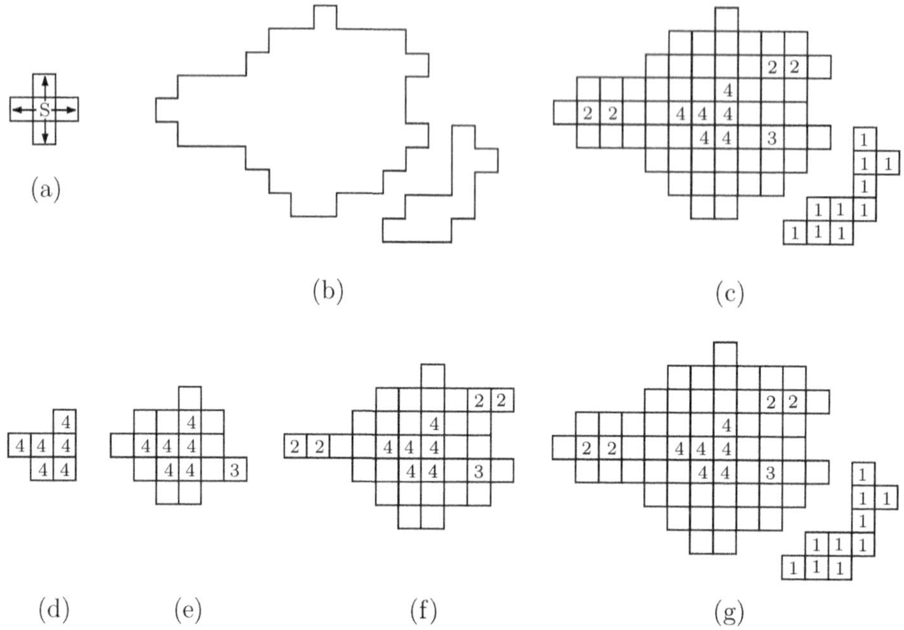

Figure 7.86: Growing a Pattern.

the pattern of Figure 7.86g. Finally, the ten pixels marked 1 are read, to complete the image. The number of generations for these pixels is zero; they are not used to grow any other pixels, and so they do not contribute to the image compression. Notice that the pixels marked 4 go through three generations.

In this example, the encoder wrote the seed pixels on the compressed stream in order of decreasing number of generations. In principle, it is possible to complicate matters as much as desired. It is possible to use different growth rules for different groups of pixels (in such a case, the growth rules must tell unambiguously who the parent of a pixel P is, so that the decoder will know what growth rule to use for P), to stop growth if a pixel comes to within a specified distance of other pixels, to change the growth rule when a new pixel bumps into another pixel, to reverse direction in such a case and start erasing pixels, or to use any other algorithm. In practice, however, complex rules require more bits to encode them, which is why simple rules may lead to better compression.

◇ **Exercise 7.20:** Simple growth rules have another advantage. What is it?

Figure 7.87 illustrates a simple approach to the design of a recursive encoder that identifies the seed pixels in layers. We assume that the only growth rule used is the one shown in Figure 7.87a. All the black (foreground) pixels are initially marked 1, and the white (background) pixels (not shown in the figure) are marked 0. The encoder scans the image and marks by 2 each 1-pixel that can grow. In order for a 1-pixel to grow it must be surrounded by 1-pixels on all four sides. The next layer consists of the resulting 2-pixels, shown in Figure 7.87c. The encoder next looks for 1-pixels that do not have at least one flanking 2-pixel. These are seed pixels. There are 10 of them, marked with a

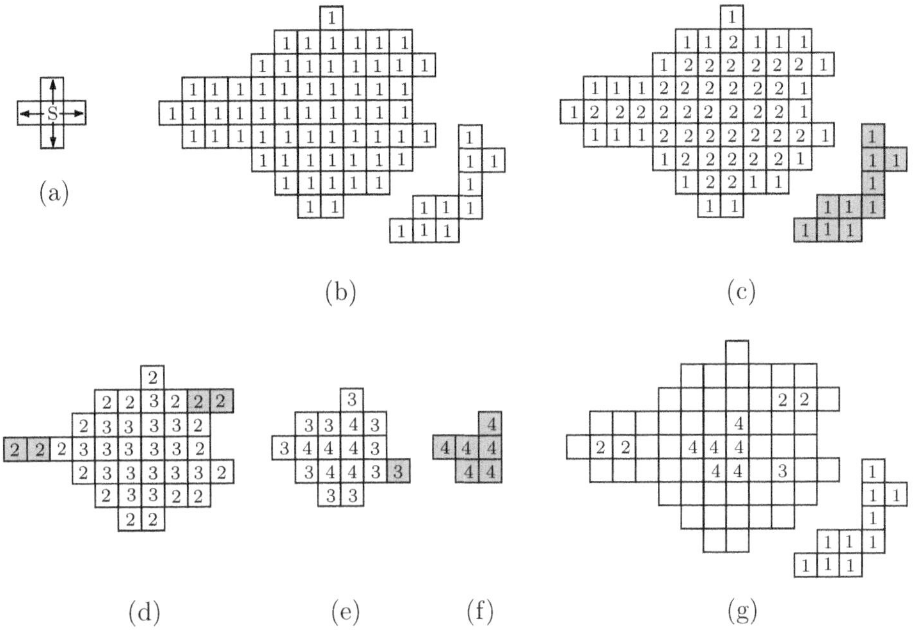

Figure 7.87: Recursive Encoding of a Pattern.

gray background in Figure 7.87c.

The same process is applied to the layer of 2-pixels. The encoder scans the image and marks by 3 each 2-pixel that can grow. The next (third) layer consists of the resulting 3-pixels, shown in Figure 7.87d. The encoder next looks for 2-pixels that do not have at least one flanking 3-pixel. These are seed pixels. There are four of them, marked with a gray background in Figure 7.87d.

Figure 7.87e shows the fourth layer. It consists of six 4-pixels and one seed 3-pixel (marked in gray). Figure 7.87f shows the six seed 4-pixels, and the final seed pixels are shown in Figure 7.87g.

This basic algorithm can be extended in a number of ways, most notably by using several growth rules for each layer, instead of just one.

7.14 JBIG

No compression method can efficiently compress every type of data. This is why new special-purpose methods are being developed all the time. JBIG [JBIG 03] is an example of a special-purpose method. It has been developed specifically for progressive compression of bi-level images. Such images, also called monochromatic or black and white, are common in applications where drawings (technical or artistic), with or without text, need to be saved in a database and retrieved. It is customary to use the terms "foreground" and "background" instead of black and white, respectively.

The term "progressive compression" means that the image is saved in several "layers" in the compressed stream, at higher and higher resolutions. When such an image is decompressed and viewed, the viewer first sees an imprecise, rough, image (the first layer) followed by improved versions of it (later layers). This way, if the image is the wrong one, it can be rejected at an early stage, without having to retrieve and decompress all of it. Section 7.13 shows how each high-resolution layer uses information from the preceding lower-resolution layer, so there is no duplication of data. This feature is supported by JBIG, where it is an option called *deterministic prediction*.

Even though JBIG was designed for bi-level images, where each pixel is one bit, it is possible to apply it to grayscale images by separating the bitplanes and compressing each individually, as if it were a bi-level image. RGC (*reflected Gray code*) should be used, instead of the standard binary code, as discussed in Section 7.4.1.

The name JBIG stands for Joint Bi-Level Image Processing Group. This is a group of experts from several international organizations, formed in 1988 to recommend such a standard. The official name of the JBIG method is ITU-T recommendation T.82. ITU is the International Telecommunications Union (part of the United Nations). The ITU-T is the telecommunication standardization sector of the ITU. JBIG uses multiple arithmetic coding to compress the image, and this part of JBIG, which is discussed below, is separate from the progressive compression part discussed in Section 7.14.1.

An important feature of the definition of JBIG is that the operation of the encoder is not defined in detail. The JBIG standard discusses the details of the decoder and the format of the compressed file. It is implied that any encoder that generates a JBIG file is a valid JBIG encoder. The JBIG2 method of Section 7.15 adopts the same approach, because this allows implementers to come up with sophisticated encoders that analyze the original image in ways that could not have been envisioned at the time the standard was developed.

One feature of arithmetic coding is that it is easy to separate the statistical model (the table with frequencies and probabilities) from the encoding and decoding operations. It is easy to encode, for example, the first half of a data stream using one model and the second half using another model. This is called *multiple arithmetic coding*, and it is especially useful in encoding images, since it takes advantage of any local structures and interrelationships that might exist in the image. JBIG uses multiple arithmetic coding with many models, each a two-entry table that gives the probabilities of a white and a black pixel. There are between 1,024 and 4,096 such models, depending on the resolution of the image that's being compressed.

A bi-level image is made up of foreground (black) and background (white) dots called *pixels*. The simplest way to compress such an image with arithmetic coding is to count the frequency of black and white pixels, and compute their probabilities. In practice, however, the probabilities vary from region to region in the image, and this fact can be used to produce better compression. Certain regions, such as the margins of a page, may be completely white, while other regions, such as the center of a complex diagram, or a large, thick rule, may be predominantly or completely black. Thus, pixel distributions in adjacent regions in an image can vary wildly.

Consider an image where 25% of the pixels are black. The entropy of this image is $-0.25\log_2 0.25 - 0.75\log_2 0.75 \approx 0.8113$. The best that we can hope for is to represent each pixel with 0.81 bits instead of the original 1 bit: an 81% compression ratio (or 0.81 bpp). Now assume that we discover that 80% of the image is predominantly white, with just 10% black pixels, and the remaining 20% have 85% black pixels. The entropies of these parts are $-0.1\log_2 0.1 - 0.9\log_2 0.9 \approx 0.47$ and $-0.85\log_2 0.85 - 0.15\log_2 0.15 \approx 0.61$, so if we encode each part separately, we can have 0.47 bpp 80% of the time and 0.61 bpp the remaining 20%. On average this results in 0.498 bpp, or a compression ratio of about 50%; much better than 81%!

We assume that a white pixel is represented by a 0 and a black one by a 1. In practice, we don't know in advance how many black and white pixels exist in each part of the image, so the JBIG encoder stops at every pixel and examines a *template* made of the 10 neighboring pixels marked (in Figures 7.88 and 7.89) "X" and "A" above it and to its left (those that have already been input; the ones below and to the right are still unknown). It interprets the values of these 10 pixels as a 10-bit integer which is then used as a pointer to a statistical model, which, in turn, is used to encode the current pixel (marked by a "?"). There are $2^{10} = 1{,}024$ 10-bit integers, so there should be 1,024 models. Each model is a small table consisting of the probabilities of black and white pixels (just one probability needs be saved in the table, since the probabilities add up to 1).

Figure 7.88a,b shows the two templates used for the lowest-resolution layer. The encoder decides whether to use the three-line or the two-line template and sets parameter `LRLTWO` in the compressed file to 0 or 1, respectively, to indicate this choice to the decoder. (The two-line template results in somewhat faster execution, while the three-line template produces slightly better compression.) Figure 7.88c shows the 10-bit template 0001001101, which becomes the pointer 77. The pointer shown in Figure 7.88d for pixel Y is 0000100101 or 37. Whenever any of the pixels used by the templates lies outside the image, the JBIG edge convention mentioned earlier should be used.

Figure 7.89 shows the four templates used for all the other layers. These templates reflect the fact that when any layer, except the first one, is decoded the low-resolution pixels from the preceding layer are known to the decoder. Each of the four templates is used to encode (and later decode) one of the four high-resolution pixels of a group. The context (pointer) generated in these cases consists of 12 bits, 10 taken from the template's pixels and two generated to indicate which of the four pixels in the group is being processed. The number of statistical models should therefore be $2^{12} = 4096$. The two bits indicating the position of a high-resolution pixel in its group are 00 for the top-left pixel (Figure 7.89a), 01 for the top-right (Figure 7.89b), 10 for the bottom-left (Figure 7.89c), and 11 for the bottom-right pixel (Figure 7.89d).

The use of these templates implies that the JBIG probability model is a 10th-order or a 12th-order Markov model.

The template pixels labeled A are called *adaptive pixels* (AT). The encoder is allowed to use as AT a pixel other than the one specified in the template, and it uses two parameters T_x and T_y (one byte each) in each layer to indicate to the decoder the actual location of the AT in that layer. (The AT is not allowed to overlap one of the X pixels in the template.) A sophisticated JBIG encoder may notice, for example, that the image uses halftone patterns where a black pixel normally has another black pixel three rows above it. In such a case the encoder may decide to use the pixel three rows above ? as the AT. Figure 7.90 shows the image area where AT may reside. Parameter M_x can vary in the range $[0, 127]$, while M_y can take the values 0 through 255.

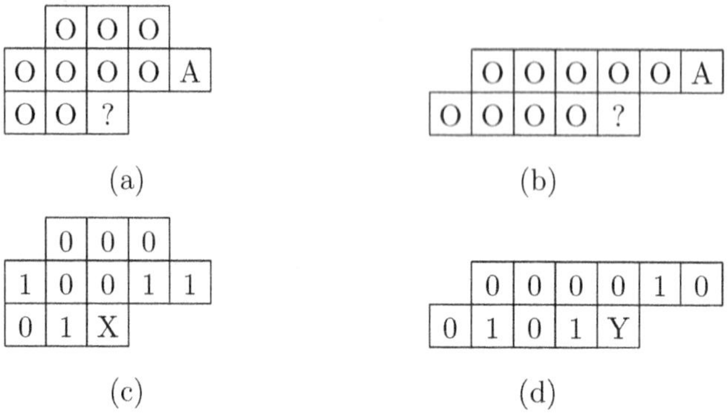

Figure 7.88: Templates for Lowest-Resolution Layer.

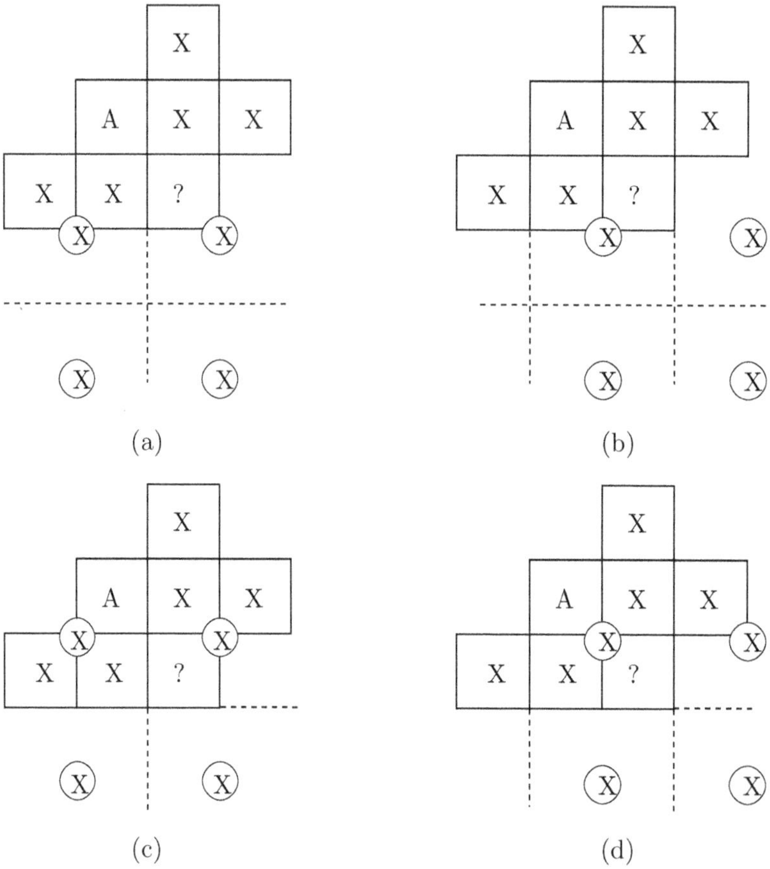

Figure 7.89: Templates for Other Layers.

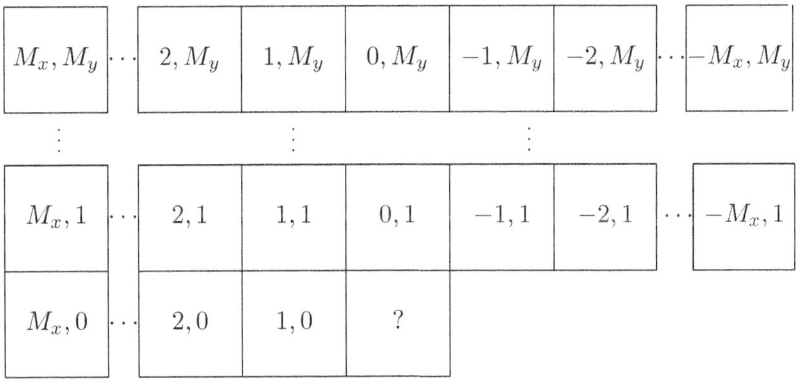

Figure 7.90: Coordinates and Allowed Positions of AT Pixels.

7.14.1 Progressive Compression

One advantage of the JBIG method is its ability to generate low-resolution versions (layers) of the image in the compressed stream. The decoder decompresses these layers progressively, from the lowest to the highest resolution.

We first look at the order of the layers. The encoder is given the entire image (the highest-resolution layer), so it is natural for it to construct the layers from high to low resolution and write them on the compressed file in this order. The decoder, on the other hand, has to start by decompressing and displaying the lowest-resolution layer, so it is easiest for it to read this layer first. As a result, either the encoder or the decoder should use buffering to reverse the order of the layers. If fast decoding is important, the encoder should use buffers to accumulate all the layers and then write them on the compressed file from low to high resolutions. The decoder then reads the layers in the right order. In cases where fast encoding is important (such as an archive that's being updated often but is rarely decompressed and used), the encoder should write the layers in the order in which they are generated (from high to low) and the decoder should use buffers. The JBIG standard allows either method, and the encoder has to set a bit denoted by `HITOLO` in the compressed file to either zero (layers are in low to high order) or one (the opposite case). It is implied that the encoder decides what the resolution of the lowest layer should be (it doesn't have to be a single pixel). This decision may be based on user input or on information built in the encoder about the specific needs of a particular environment.

Progressive compression in JBIG also involves the concept of *stripes*. A stripe is a narrow horizontal band consisting of L scan lines of the image, where L is a user-controlled JBIG parameter. As an example, if L is chosen such that the height of a stripe is 8 mm (about 0.3 inches), then there will be about 36 stripes in an 11.5-inch-high image. The encoder writes the stripes on the compressed file in one of two ways, setting parameter `SET` to indicate this to the decoder. Either all the stripes of a layer are written on the compressed file consecutively, followed by the stripes of the next layer, and so on (`SET`=0), or the top stripes of all the layers are first written, followed by the second stripes of all the layers, and so on (`SET`=1). If the encoder sets `SET`=0, then the decoder will decompress the image progressively, layer by layer, and in each layer stripe by stripe. Figure 7.91 illustrates the case of an image divided into four stripes and encoded in three layers (from low-resolution 150 dpi to high-resolution 600 dpi). Table 7.92 lists the order in which the stripes are output for the four possible values of parameters `HITOLO` and `SET`.

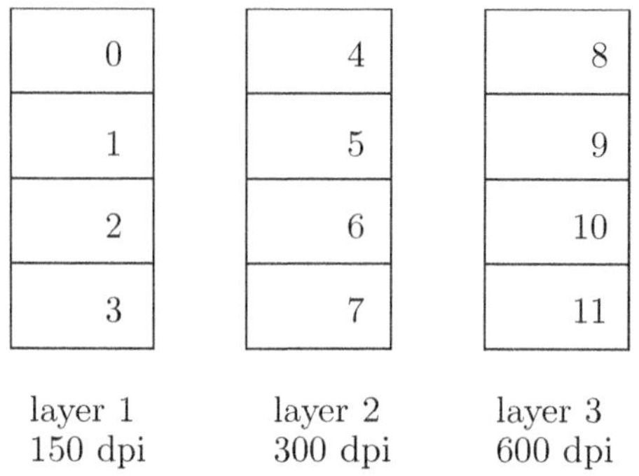

Figure 7.91: Four Stripes and Three Layers in a JBIG Image.

`HITOLO`	`SEQ`	Order
0	0	0,1,2,3,4,5,6,7,8,9,10,11
0	1	0,4,8,1,5,9,2,6,10,3,7,11
1	0	8,9,10,11,4,5,6,7,0,1,2,3
1	1	8,4,0,9,5,1,10,6,2,11,7,3

Table 7.92: The Four Possible Orders of Layers.

The basic idea of progressive compression is to group four high-resolution pixels into one low-resolution pixel, a process called *downsampling*. The only problem is to determine the value (black or white) of that pixel. If all four original pixels have the same value, or even if three are identical, the solution is obvious (follow the majority). When two pixels are black and the other two are white, we can try one of the following solutions:

1. Create a low-resolution pixel that's always black (or always white). This is a bad solution, since it may eliminate important details of the image, making it impractical or even impossible for an observer to evaluate the image by viewing the low-resolution layer.
2. Assign a random value to the new low-resolution pixel. This solution is also bad, since it may add too much noise to the image.
3. Give the low-resolution pixel the color of the top-left of the four high-resolution pixels. This prefers the top row and the left column, and has the drawback that thin lines can sometimes be completely missing from the low-resolution layer. Also, if the high-resolution layer uses halftoning to simulate grayscales, the halftone patterns may be corrupted.
4. Assign a value to the low-resolution pixel that depends on the four high-resolution pixels **and** on some of their nearest neighbors. This solution is used by JBIG. If most of

the near neighbors are white, a white low-resolution pixel is created; otherwise, a black low-resolution pixel is created. Figure 7.93a shows the 12 high-resolution neighboring pixels and the three low-resolution ones that are used in making this decision. A, B, and C are three low-resolution pixels whose values have already been determined. Pixels d, e, f, g, and, j are on top or to the left of the current group of four pixels. Pixel "?" is the low-resolution pixel whose value needs to be determined.

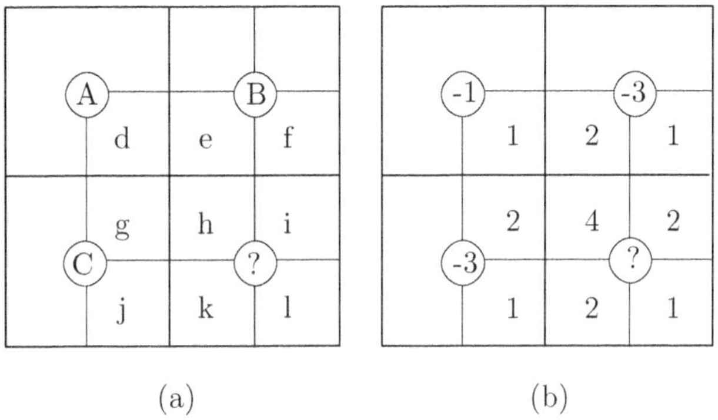

Figure 7.93: HiRes and LoRes Pixels.

Figure 7.93b shows the weights assigned to the various pixels involved in the determination of pixel "?". The weighted sum of the pixels can also be written as

$$\begin{aligned} &4h + 2(e + g + i + k) + (d + f + j + l) - 3(B + C) - A \\ &= 4h + 2(i + k) + l + (d - A) + 2(g - C) + (j - C) + 2(e - B) + (f - B). \end{aligned} \tag{7.26}$$

The second line of Equation (7.26) shows how the value of pixel "?" depends on differences such as $d - A$ (a difference of a high-resolution pixel and the adjacent low-resolution pixel). Assuming equal probabilities for black and white pixels, the first line of Equation (7.26) can have values between zero (when all 12 pixels are white) and 9 (when all are black), so the rule is to assign pixel "?" the value 1 (black) if expression (7.26) is greater than 4.5 (if it is 5 or more), and the value 0 (white) otherwise (if it is 4 or less). Thus, this expression acts as a filter that preserves the density of the high-resolution pixels in the low-resolution image.

◇ **Exercise 7.21:** We normally represent a black (or foreground) pixel with a binary 1, and a white (or background) pixel with a binary 0. How will the resolution reduction method above change if we use 0 to represent black and 1 to represent white?

Since the image is divided into groups of 4×4 pixels, it should have an even number of rows and columns. The JBIG *edge convention* says that an image can be extended if and when necessary by adding columns of 0 pixels at the left and right, rows of 0 pixels at the top, and by replicating the bottom row as many times as necessary.

The JBIG method of resolution reduction has been carefully designed and extensively tested. It is known to produce excellent results for text, drawings, halftoned grayscales, as well as for other types of images.

JBIG also includes exceptions to the above rule in order to preserve edges (132 exceptions), preserve vertical and horizontal lines (420 exceptions), periodic patterns in the image (10 exceptions, for better performance in regions of transition to and from periodic patterns), and dither patterns in the image (12 exceptions that help preserve very low density or very high density dithering, i.e., isolated background or foreground pixels). The last two groups are used to preserve certain shading patterns and also patterns generated by halftoning. Each exception is a 12-bit number. Table 7.94 lists some of the exception patterns.

The pattern of Figure 7.95b was developed in order to preserve thick horizontal, or near-horizontal, lines. A two-pixel-wide high-resolution horizontal line, e.g., will result in a low-resolution line whose width alternates between one and two low-resolution pixels. When the upper row of the high-resolution line is scanned, it will result in a row of low-resolution pixels because of Figure 7.95a. When the next high-resolution row is scanned, the two rows of the thick line will generate alternating low-resolution pixels because pattern 7.95b requires a zero in the bottom-left low-resolution pixel in order to complement the bottom-right low-resolution pixel.

Pattern 7.95b leaves two pixels unspecified, so it counts for four patterns. Its reflection is also used, to preserve thick vertical lines, so it counts for eight of the 420 line-preservation exceptions.

Figure 7.95c complements a low-resolution pixel in cases where there is one black high-resolution pixel among 11 white ones. This is common when a grayscale image is converted to black and white by halftoning. This pattern is one of the 12 dither preservation exceptions.

The JBIG method for reducing resolution seems complicated—especially since it includes so many exceptions—but is actually simple to implement and also executes very fast. The decision whether to paint the current low-resolution pixel black or white depends on 12 of its neighboring pixels, nine high-resolution pixels, and three low-resolution pixels. Their values are combined into a 12-bit number, which is used as a pointer to a 4,096-entry table. Each entry in the table is one bit, which becomes the value of the current low-resolution pixel. This way, all the exceptions are already included in the table and don't require any special treatment by the program.

It has been mentioned in Section 7.13 that compression is improved if the encoder writes on the compressed stream only three of the four pixels of a group in layer i. This is enough because the decoder can compute the value of the 4th pixel from its three siblings and the parent of the group (the parent is from the preceding layer, $i-1$). Equation (7.26) shows that in JBIG, a low-resolution pixel is not calculated as a simple average of its four high-resolution "children," so it may not be possible for the JBIG decoder to calculate the value of, say, high-resolution pixel k from the values of its three siblings h, i, and l and their parent. The many exceptions used by JBIG also complicate this calculation. As a result, JBIG uses special tables to tell both encoder and decoder which high-resolution pixels can be inferred from their siblings and parents. Such pixels are not compressed by the encoder, but all the other high-resolution pixels of the layer are. This method is referred to as *deterministic prediction* (DP), and is a JBIG option. If the encoder decides to use it, it sets parameter `DPON` to 1.

Figure 7.96 shows the pixel numbering used for deterministic prediction. Before it compresses high-resolution pixel 8, the encoder combines the values of pixels 0–7 to

The 10 periodic pattern preservation exceptions are (in hex)

5c7 36d d55 b55 caa aaa c92 692 a38 638.

The 12 dither pattern preservation exceptions are

fef fd7 f7d f7a 145 142 ebd eba 085 082 028 010.

The 132 edge-preservation exception patterns are

a0f 60f 40f 20f e07 c07 a07 807 607 407 207 007 a27 627 427 227 e1f 61f e17 c17 a17 617 e0f c0f 847 647 447 247 e37 637 e2f c2f a2f 62f e27 c27 24b e49 c49 a49 849 649 449 249 049 e47 c47 a47 e4d c4d a4d 84d 64d 44d 24d e4b c4b a4b 64b 44b e69 c69 a69 669 469 269 e5b 65b e59 c59 a59 659 ac9 6c9 4c9 2c9 ab6 8b6 e87 c87 a87 687 487 287 507 307 cf8 8f8 ed9 6d9 ecb ccb acb 6cb ec9 cc9 949 749 549 349 b36 336 334 f07 d07 b07 907 707 3b6 bb4 3b4 bb2 3a6 b96 396 d78 578 f49 d49 b49 ff8 df8 5f8 df0 5f0 5e8 dd8 5d8 5d0 db8 fb6 bb6

where the 12 bits of each pattern are numbered as in the diagram.

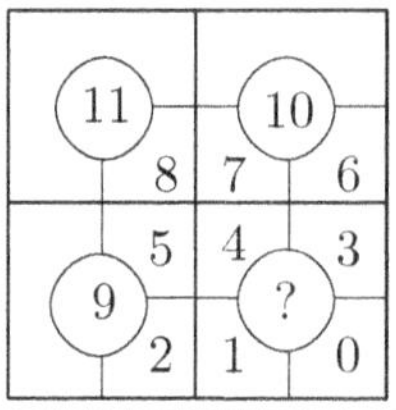

Table 7.94: Some JBIG Exception Patterns.

Figure 7.95 shows three typical exception patterns. A pattern of six zeros and three ones, as in Figure 7.95a, is an example exception, introduced to preserve horizontal lines. It means that the low-resolution pixel marked "C" should be complemented (assigned the opposite of its normal value). The normal value of this pixel depends, of course, on the three low-resolution pixels above it and to the left. Since these three pixels can have eight different values, this pattern covers eight exceptions. Reflecting 7.95a about its main diagonal produces a pattern (actually, eight patterns) that's natural to use as an exception to preserve vertical lines in the original image. Thus, Figure 7.95a corresponds to 16 of the 420 line-preservation exceptions.

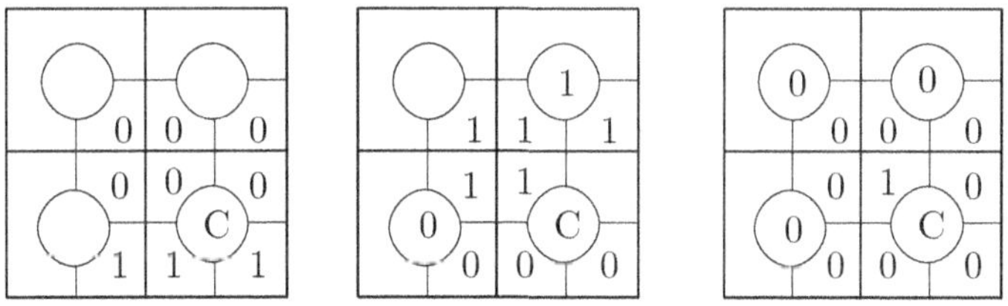

(a) (b) (c)

Figure 7.95: Some JBIG Exception Patterns.

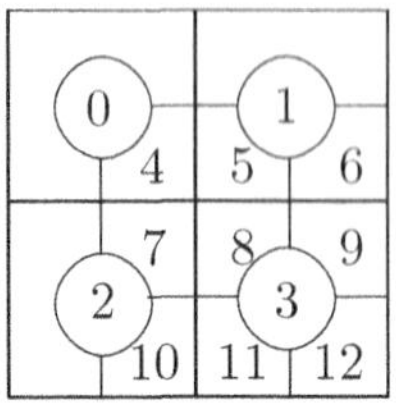

Figure 7.96: Pixel Numbering for Deterministic Prediction.

form an 8-bit pointer to the first DP table. If the table entry is 2, then pixel 8 should be compressed, since the decoder cannot infer it from its neighbors. If the table entry is 0 or 1, the encoder ignores pixel 8. The decoder prepares the same pointer (since pixels 0–7 are known when pixel 8 should be decoded). If the table entry is 0 or 1, the decoder assigns this value to pixel 8; otherwise (if the table entry is 2), the next pixel value is decoded from the compressed file and is assigned to pixel 8. Notice that of the 256 entries of Table 7.97 only 20 are not 2. Pixel 9 is handled similarly, except that the pointer used consists of pixels 0–8, and it points to an entry in the second DP table. This table, shown in 7.98, consists of 512 entries, of which 108 are not 2. The JBIG standard also specifies the third DP table, for pixel 10 (1024 entries, of which 526 are not 2), and the fourth table, for pixel 11 (2048 entries, of which 1044 are not 2). The total size of the four tables is 3840 entries, of which 1698 (or about 44%) actually predict pixel values.

Pointer	Value
0–63	02222222 22222222 22222222 22222222 02222222 22222222 22222222 22222222
64–127	02222222 22222222 22222222 22222222 00222222 22222222 22222222 22222222
128–191	02222222 22222222 00222222 22222222 02020222 22222222 02022222 22222222
192–255	00222222 22222222 22222222 22222221 02020022 22222222 22222222 22222222

Table 7.97: Predicting Pixel 8.

Pointer	Value
0–63	22222222 22222222 22222222 22000000 02222222 22222222 00222222 22111111
64–127	22222222 22222222 22222222 21111111 02222222 22111111 22222222 22112221
128–191	02222222 22222222 02222222 22222222 00222222 22222200 20222222 22222222
192–255	02222222 22111111 22222222 22222102 11222222 22222212 22220022 22222212
256–319	20222222 22222222 00220222 22222222 20000222 22222222 00000022 22222221
320–383	20222222 22222222 11222222 22222221 22222222 22222221 22221122 22222221
384–447	20020022 22222222 22000022 22222222 20202002 22222222 20220002 22222222
448–511	22000022 22222222 00220022 22222221 21212202 22222222 22220002 22222222

Table 7.98: Predicting Pixel 9.

7.15 JBIG2

They did it again! The Joint Bi-Level Image Processing Group has produced another standard for the compression of bi-level images. The new standard is called JBIG2, implying that the old standard should be called JBIG1. JBIG2 was developed in the late 1990s and was approved by the ISO and the ITU-T in 1999. Current information about JBIG2 can be found at [JBIG2 03] and [JBIG2 06]. The JBIG2 standard offers:

1. Large increases in compression performance (typically 3–5 times better than Group 4/MMR, and 2–4 times better than JBIG1).
2. Special compression methods for text, halftones, and other bi-level image parts.
3. Lossy and lossless compression.
4. Two modes of progressive compression. Mode 1 is quality-progressive compression, where the decoded image progresses from low to high quality. Mode 2 is content-progressive coding, where important image parts (such as text) are decoded first, followed by less-important parts (such as halftone patterns).
5. Multipage document compression.
6. Flexible format, designed for easy embedding in other image file formats.
7. Fast decompression: In some coding modes, images can be decompressed at over 250 million pixels/second in software.

The JBIG2 standard describes the principles of the compression and the format of the compressed file. It does not get into the operation of the encoder. Any encoder that produces a JBIG2 compressed file is a valid JBIG2 encoder. It is hoped that this will encourage software developers to implement sophisticated JBIG2 encoders. The JBIG2 decoder reads the compressed file, which contains dictionaries and image information, and decompresses it into the *page buffer*. The image is then displayed or printed from the page buffer. Several auxiliary buffers may also be used by the decoder.

A document to be compressed by JBIG2 may consist of more than one page. The main feature of JBIG2 is that it distinguishes between text, halftone images, and everything else on the page. The JBIG2 encoder should somehow scan the page before doing any encoding, and identify regions of three types:

1. Text regions. These contain text, normally arranged in rows. The text does not have to be in any specific language, and may consist of unknown symbols, dingbats, musical notation, or hieroglyphs. The encoder treats each symbol as a rectangular bitmap, which it places in a dictionary. The dictionary is compressed by either arithmetic coding (specifically, a version of arithmetic coding called the MQ-coder, [Pennebaker and Mitchell 88a,b]) or MMR (Section 5.7.2) and written, as a segment, on the compressed file. Similarly, the text region itself is encoded and written on the compressed file as another segment. To encode a text region the encoder prepares the relative coordinates of each symbol (the coordinates relative to the preceding symbol), and a pointer to the symbol's bitmap in the dictionary. (It is also possible to have pointers to several bitmaps and specify the symbol as an aggregation of those bitmaps, such as logical AND, OR, or XOR.) This data also is encoded before being written on the compressed file. Compression is achieved if the same symbol occurs several times. Notice that the symbol may occur in different text regions and even on different pages. The encoder should make sure that the dictionary containing this symbol's bitmap will be retained by

the decoder for as long as necessary. Lossy compression is achieved if the same bitmap is used for symbols that are slightly different.

2. Halftone regions. A bi-level image may contain a grayscale image done in halftone [Salomon 99]. The encoder scans such a region cell by cell (halftone cells, also called patterns, are typically 3×3 or 4×4 pixels). A halftone dictionary is created by collecting all the different cells and converting each to an integer (typically nine bits or 16 bits). The dictionary is compressed and is written on the output file. Each cell in the halftone region is then replaced by a pointer to the dictionary. The same dictionary can be used to decode several halftone dictionaries, perhaps located on different pages of the document. The encoder should label each dictionary so the decoder would know how long to retain it.

3. Generic regions. This is any region not identified by the encoder as text or halftone. Such a region may contain a large character, line art, mathematics, or even noise (specks and dirt). A generic region is compressed with either arithmetic coding or MMR. When the former is used, the probability of each pixel is determined by its context (i.e., by several pixels located above it and to its left that have already been encoded).

A page may contain any number of regions, and they may overlap. The regions are determined by the encoder, whose operation is not specified by the JBIG2 standard. As a result, a simple JBIG2 encoder may encode any page as one large generic region, while a sophisticated encoder may spend time analyzing the contents of a page and identifying several regions, thereby leading to better compression. Figure 7.99 is an example of a page with four text regions, one halftone region, and two generic region (a sheared face and fingerprints).

The fact that the JBIG2 standard does not specify the encoder is also used to generate lossy compression. A sophisticated encoder may decide that dropping certain pixels would not deteriorate the appearance of the image significantly. The encoder therefore drops those bits and creates a small compressed file. The decoder operates as usual and does not know whether the image that's being decoded is lossy or not. Another approach to lossy compression is for the encoder to identify symbols that are very similar and replace them by pointers to the same bitmap in the symbol dictionary (this approach is risky, since symbols that differ by just a few pixels may be completely different to a human reader, think of "e" and "c" or "i" and "ı"). An encoder may also be able to identify specks and dirt in the document and optionally ignore them.

JBIG2 introduces the concept of region *refinement*. The compressed file may include instructions directing the decoder to decode a region A from the compressed file into an auxiliary buffer. That buffer may later be used to refine the decoding of another region B. When B is found in the compressed file and is decoded into the page buffer, each pixel written in the page buffer is determined by pixels decoded from the compressed file *and* by pixels in the auxiliary buffer. One example of region refinement is a document where certain text regions contain the words "Top Secret" in large, gray type, centered as a background. A simple JBIG2 encoder may consider each of the pixels constituting "Top Secret" part of some symbol. A sophisticated encoder may identify the common background, treat it as a generic region, and compress it as a region refinement, to be applied to certain text regions by the decoder. Another example of region refinement is a document where certain halftone regions have dark background. A sophisticated encoder may identify the dark background in those regions, treat it as a generic region,

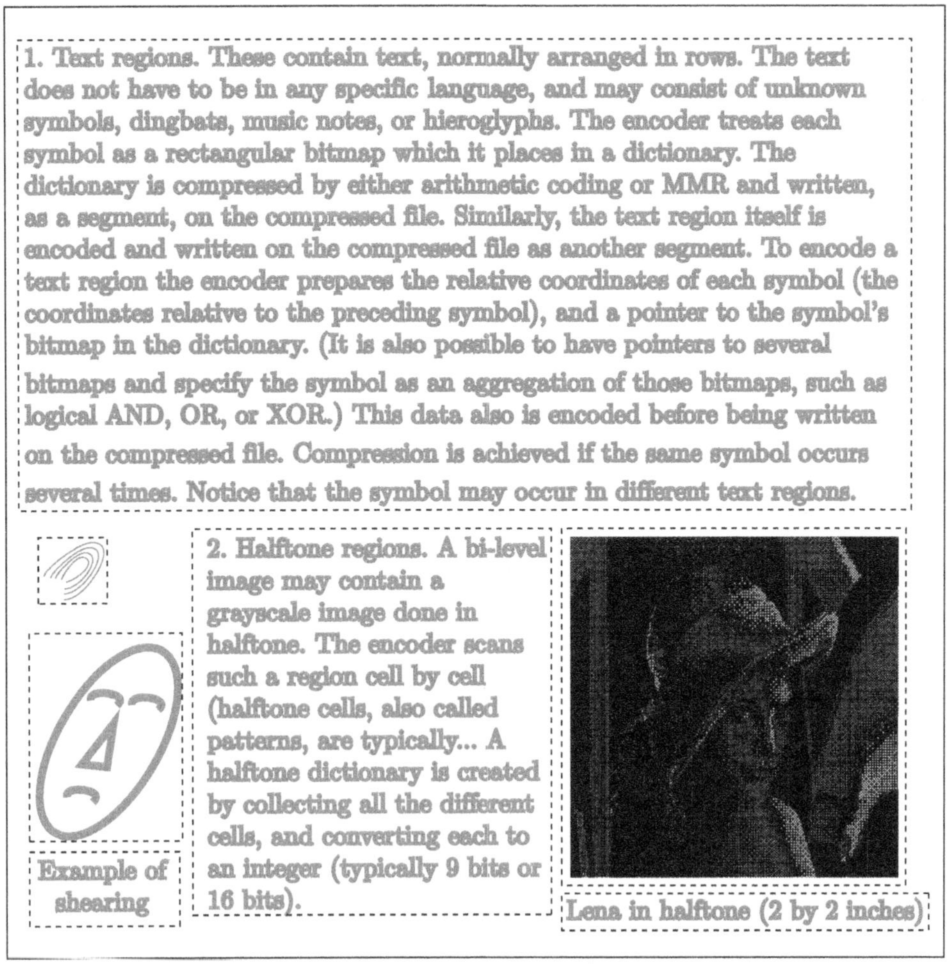

Figure 7.99: Typical JBIG2 Regions.

and place instructions in the compressed file telling the decoder to use this region to refine certain halftone regions. Thus, a simple JBIG2 encoder may never use region refinement, but a sophisticated one may use this concept to add special effects to some regions.

The decoder starts by initializing the page buffer to a certain value, 0 or 1, according to a code read from the compressed file. It then inputs the rest of the file segment by segment and executes each segment by a different procedure. There are seven main procedures.

1. The procedure to decode segment headers. Each segment starts with a header that includes, among other data and parameters, the segment's type, the destination of the decoded output from the segment, and which other segments have to be used in decoding this segment.

2. A procedure to decode a generic region. This is invoked when the decoder finds a segment describing such a region. The segment is compressed with either arithmetic coding or MMR, and the procedure decompresses it (pixel by pixel in the former case and runs of pixels in the latter). In the case of arithmetic coding, previously decoded pixels are used to form a prediction context. Once a pixel is decoded, the procedure does not simply store it in the page buffer, but combines it with the pixel already in the page buffer according to a logical operation (AND, OR, XOR, or XNOR) specified in the segment.
3. A procedure to decode a generic refinement region. This is similar to the above except that it modifies an auxiliary buffer instead of the page buffer.
4. A procedure to decode a symbol dictionary. This is invoked when the decoder finds a segment containing such a dictionary. The dictionary is decompressed and is stored as a list of symbols. Each symbol is a bitmap that is either explicitly specified in the dictionary or is specified as a refinement (i.e., a modification) of a known symbol (a preceding symbol from this dictionary or a symbol from another existing dictionary) or is specified as an aggregate (a logical combination) of several known symbols).
5. A procedure to decode a symbol region. This is invoked when the decoder finds a segment describing such a region. The segment is decompressed and it yields triplets. The triplet for a symbol contains the coordinates of the symbol relative to the preceding symbol and a pointer (index) to the symbol in the symbol dictionary. Since the decoder may keep several symbol dictionaries at any time, the segment should indicate which dictionary is to be used. The symbol's bitmap is brought from the dictionary, and the pixels are combined with the pixels in the page buffer according to the logical operation specified by the segment.
6. A procedure to decode a halftone dictionary. This is invoked when the decoder finds a segment containing such a dictionary. The dictionary is decompressed and is stored as a list of halftone patterns (fixed-size bitmaps).
7. A procedure to decode a halftone region. This is invoked when the decoder finds a segment describing such a region. The segment is decompressed into a set of pointers (indexes) to the patterns in the halftone dictionary.

Some of these procedures are described here in more detail.

7.15.1 Generic Region Decoding

This procedure reads several parameters from the compressed file (Table 7.100, where "I" stands for integer and "B" stands for bitmap), the first of which, **MMR**, specifies the compression method used in the segment about to be decoded. Either arithmetic coding or MMR can be used. The former is similar to arithmetic coding in JBIG1 and is described below. The latter is used as in fax compression (Section 5.7).

If arithmetic coding is used, pixels are decoded one by one and are placed row by row in the generic region being decoded (part of the page buffer), whose width and height are given by parameters **GBW** and **GBH**, respectively. They are not simply stored there but are logically combined with the existing background pixels. Recall that the decompression process in arithmetic coding requires knowledge of the probabilities of the items being decompressed (decoded). These probabilities are obtained in our case by generating a template for each pixel being decoded, using the template to generate an integer (the context of the pixel), and using that integer as a pointer to a table of

Name	Type	Size	Signed?	Description
MMR	I	1	N	MMR or arithcoding used
GBW	I	32	N	Width of region
GBH	I	32	N	Height of region
GBTEMPLATE	I	2	N	Template number
TPON	I	1	N	Typical prediction used?
USESKIP	I	1	N	Skip some pixels?
SKIP	B			Bitmap for skipping
GBATX$_1$	I	8	Y	Relative X coordinate of A_1
GBATY$_1$	I	8	Y	Relative Y coordinate of A_1
GBATX$_2$	I	8	Y	Relative X coordinate of A_2
GBATY$_2$	I	8	Y	Relative Y coordinate of A_2
GBATX$_3$	I	8	Y	Relative X coordinate of A_3
GBATY$_3$	I	8	Y	Relative Y coordinate of A_3
GBATX$_4$	I	8	Y	Relative X coordinate of A_4
GBATY$_4$	I	8	Y	Relative Y coordinate of A_4

Table 7.100: Parameters for Decoding a Generic Region.

probabilities. Parameter **GBTEMPLATE** specifies which of four types of templates should be used. Figure 7.101a–d shows the four templates that correspond to **GBTEMPLATE** values of 0–3, respectively (notice that the two 10-bit templates are identical to templates used by JBIG1, Figure 7.88b,d). The pixel labeled "O" is the one being decoded, and the "X" and the "A$_i$" are known pixels. If "O" is near an edge of the region, its missing neighbors are assumed to be zero.

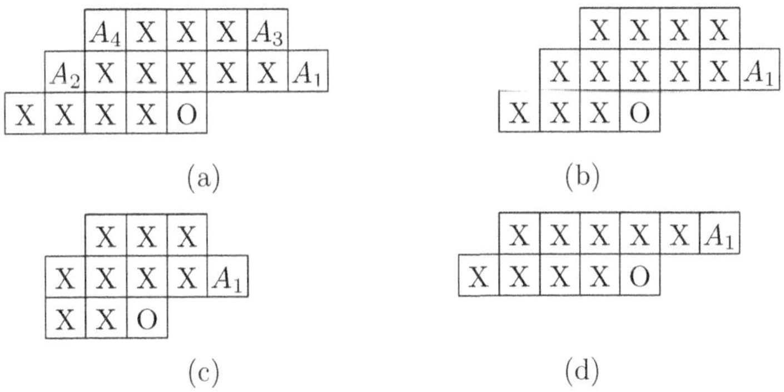

Figure 7.101: Four Templates for Generic Region Decoding

The values of the "X" and "A_i" pixels (one bit per pixel) are combined to form a context (an integer) of between 10 and 16 bits. It is expected that the bits will be collected top to bottom and left to right, but the standard does not specify that. The two templates of Figure 7.102a,b, e.g., should produce the contexts 1100011100010111 and

1000011011001, respectively. Once a context has been computed, it is used as a pointer to a probability table, and the probability found is sent to the arithmetic decoder to decode pixel "O."

(a) (b)

Figure 7.102: Two Templates.

The encoder specifies which of the four template types should be used. A simple rule of thumb is to use large templates for large regions.

An interesting feature of the templates is the use of the A_i pixels. They are called *adaptive* or AT pixels and can be located in positions other than the ones shown. Figure 7.101 shows their normal positions, but a sophisticated encoder may discover, for example, that the pixel three rows above the current pixel "O" is always identical to "O." The encoder may in such a case tell the decoder (by means of parameters $\mathbf{GBATX}_i$ and $\mathbf{GBATY}_i$) to look for A_1 at address $(0, -3)$ relative to "O." Figure 7.103 shows the permissible positions of the AT pixels relative to the current pixel, and their coordinates. Notice that an AT pixel may also be located at one of the "X" positions of the template.

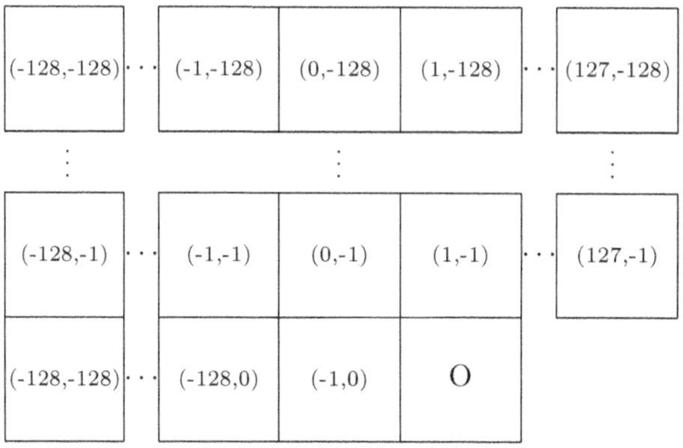

Figure 7.103: Coordinates and Allowed Positions of AT Pixels.

◇ **Exercise 7.22:** What are the relative coordinates of the AT pixels in the four types of templates shown in Figure 7.101?

The **TPON** parameter controls the so-called *typical prediction* feature. A typical row is defined as a row that is identical to its predecessor. If the encoder notices that certain rows of the generic region being encoded are typical, it sets **TPON** to 1 and then writes a code in the compressed file before each row, indicating whether or not it is typical. If the decoder finds **TPON** set to 1, it has to decode and check a special code preceding each row of pixels. When the code of a typical row is found, the decoder simply generates it as a copy of its predecessor.

The two parameters **USESKIP** and **SKIP** control the *skip* feature. If the encoder discovers that the generic region being encoded is sparse (i.e., most of its pixels are zero), it sets **USESKIP** to 1 and sets **SKIP** to a bitmap of size **GBW**×**GBH**, where each 1 bit indicates a zero-bit in the generic region. The encoder then compresses only the 1 bits of the generic region.

7.15.2 Symbol Region Decoding

This procedure is invoked when the decoder starts reading a new segment from the compressed file and identifies it as a symbol-region segment. The procedure starts by reading many parameters from the segment. It then inputs the coded information for each symbol and decodes it (using either arithmetic coding or MMR). This information contains the coordinates of the symbol relative to its row and the symbol preceding it, a pointer to the symbol in the symbol dictionary, and possibly also refinement information. The coordinates of a symbol are denoted by S and T. Normally, S is the x-coordinate and T is the y-coordinate of a symbol. However, if parameter **TRANSPOSED** has the value 1, the meaning of S and T is reversed. In general, T is considered the height of a row of text and S is the coordinate of a text symbol in its row.

◇ **Exercise 7.23:** What is the advantage of transposing S and T?

Symbols are encoded and written on the compressed file by strips. A strip is normally a row of symbols but can also be a column. The encoder makes that decision and sets parameter **TRANSPOSED** to 0 or 1 accordingly. The decoder therefore starts by decoding the number of strips, then the strips themselves. For each strip the compressed file contains the strip's T coordinate relative to the preceding strip, followed by coded information for the symbols constituting the strip. For each symbol this information consists of the symbol's S coordinate (the gap between it and the preceding symbol), its T coordinate (relative to the T coordinate of the strip), its ID (a pointer to the symbol dictionary), and refinement information (optionally). In the special case where all the symbols are aligned vertically on the strip, their T coordinates will be zero.

Once the absolute coordinates (x, y) of the symbol in the symbol region have been computed from the relative coordinates, the symbol's bitmap is retrieved from the symbol dictionary and is combined with the page buffer. However, parameter **REFCORNER** indicates which of the four corners of the bitmap should be placed at position (x, y). Figure 7.104 shows examples of symbol bitmaps aligned in different ways.

If the encoder decides to use MMR to encode a region, it selects one of 15 Huffman code tables defined by the JBIG2 standard and sets a parameter to indicate to the decoder which table is used. The tables themselves are built into both encoder and decoder. Table 7.105 shows two of the 15 tables. The OOB value (out of bound) is used to terminate a list in cases where the length of the list is not known in advance.

7.15.3 Halftone Region Decoding

This procedure is invoked when the decoder starts reading a new segment from the compressed file and identifies it as a halftone-region segment. The procedure starts by

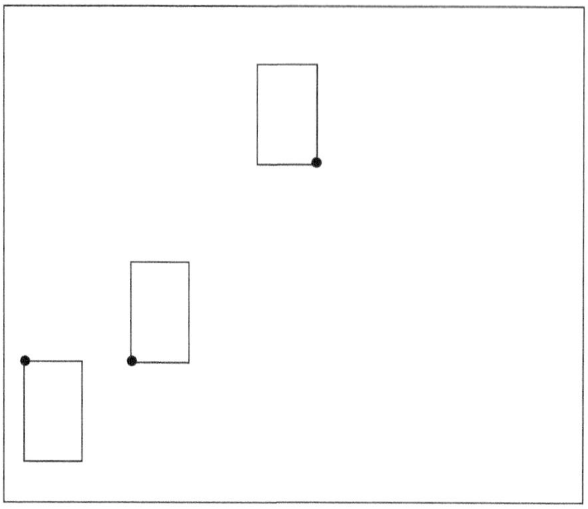

Figure 7.104: Three Symbol Bitmaps Aligned at Different Corners.

Value	Code
0–15	0+Value encoded as 4 bits
16–271	10+(Value − 16) encoded as 8 bits
272–65807	110+(Value − 272) encoded as 16 bits
65808–∞	111+(Value − 65808) encoded as 32 bits

Value	Code
0	0
1	10
2	110
3–10	1110+(Value − 3) encoded as 3 bits
11–74	11110+(Value − 11) encoded as 6 bits
75–∞	111110+(Value − 75) encoded as 32 bits
OOB	11111

Table 7.105: Two Huffman Code Tables for JBIG2 Decoding.

reading several parameters from the segment and setting all the pixels in the halftone region to the value of background parameter **HDEFPIXEL**. It then inputs the coded information for each halftone pattern and decodes it using either arithmetic coding or MMR. This information consists of a pointer to a halftone pattern in the halftone dictionary. The pattern is retrieved and is logically combined with the background pixels that are already in the halftone region. Parameter **HCOMBOP** specifies the logical combination. It can have one of the values REPLACE, OR, AND, XOR, and XNOR.

Notice that the region information read by the decoder from the compressed file does not include the positions of the halftone patterns. The decoder adds the patterns to the region at consecutive grid points of the *halftone grid*, which is itself defined by four parameters. Parameters **HGX** and **HGY** specify the origin of the grid relative to the origin of the halftone region. Parameters **HRX** and **HRY** specify the orientation of the grid by means of an angle θ. The last two parameters can be interpreted as the horizontal and vertical components of a vector $\mathbf{v}$. In this interpretation θ is the angle between $\mathbf{v}$ and the x axis. The parameters can also be interpreted as the cosine and sine of θ, respectively. (Strictly speaking, they are multiples of the sine and cosine, since $\sin^2\theta + \cos^2\theta$ equals unity, but the sum $\mathbf{HRX^2} + \mathbf{HRY^2}$ can have any value.) Notice that these parameters can also be negative and are not limited to integer values. They are written on the compressed file as integers whose values are 256 times their real values. Thus, for example, if **HRX** should be -0.15, it is written as the integer -38 because $-0.15 \times 256 = -38.4$. Figure 7.106a shows typical relations between a page, a halftone region, and a halftone grid. Also shown are the coordinates of three points on the grid.

The decoder performs a double loop in which it varies m_g from zero to $\mathbf{HGH} - \mathbf{1}$ and for each value of m_g it varies n_g from zero to $\mathbf{HGW} - \mathbf{1}$ (parameters **HGH** and **HGW** are the number of horizontal and vertical grid points, respectively). At each iteration the pair (n_g, m_g) constitutes the coordinates of a grid point. This point is mapped to a point (x, y) in the halftone region by the relation

$$\begin{aligned} x &= \mathbf{HGX} + m_g \times \mathbf{HRY} + n_g \times \mathbf{HRX}, \\ y &= \mathbf{HGY} + m_g \times \mathbf{HRX} - n_g \times \mathbf{HRY}. \end{aligned} \tag{7.27}$$

To understand this relation the reader should compare it to a rotation of a point (n_g, m_g) through an angle θ about the origin. Such a rotation is expressed by

$$(x, y) = (n_g, m_g)\begin{pmatrix} \cos\theta & -\sin\theta \\ \sin\theta & \cos\theta \end{pmatrix} = (n_g\cos\theta + m_g\sin\theta, m_g\cos\theta - n_g\sin\theta). \tag{7.28}$$

A comparison of Equations (7.27) and (7.28) shows that they are identical if we associate **HRX** with $\cos\theta$ and **HRY** with $\sin\theta$ (Equation (7.27) also adds the origin of the grid relative to the region, so it results in a point (x, y) whose coordinates are relative to the origin of the region).

It is expected that the grid would normally be identical to the region, i.e., its origin would be $(0, 0)$ and its rotation angle would be zero. This is achieved by setting **HGX**, **HGY**, and **HRY** to zero and setting **HRX** to the width of a halftone pattern. However, all four parameters may be used, and a sophisticated JBIG2 encoder may improve the overall compression quality by trying different combinations.

Once point (x, y) has been calculated, the halftone pattern is combined with the halftone region such that its top-left corner is at location (x, y) of the region. Since parts of the grid may lie outside the region, any parts of the pattern that lie outside the region are ignored (see gray areas in Figure 7.106b). Notice that the patterns themselves are oriented with the region, not with the grid. Each pattern is a rectangle with horizontal and vertical sides. The orientation of the grid affects only the position of the top-left corner of each pattern. As a result, the patterns added to the halftone region generally overlap by an amount that's greatly affected by the orientation and position of the grid. This is also illustrated by the three examples of Figure 7.106b.

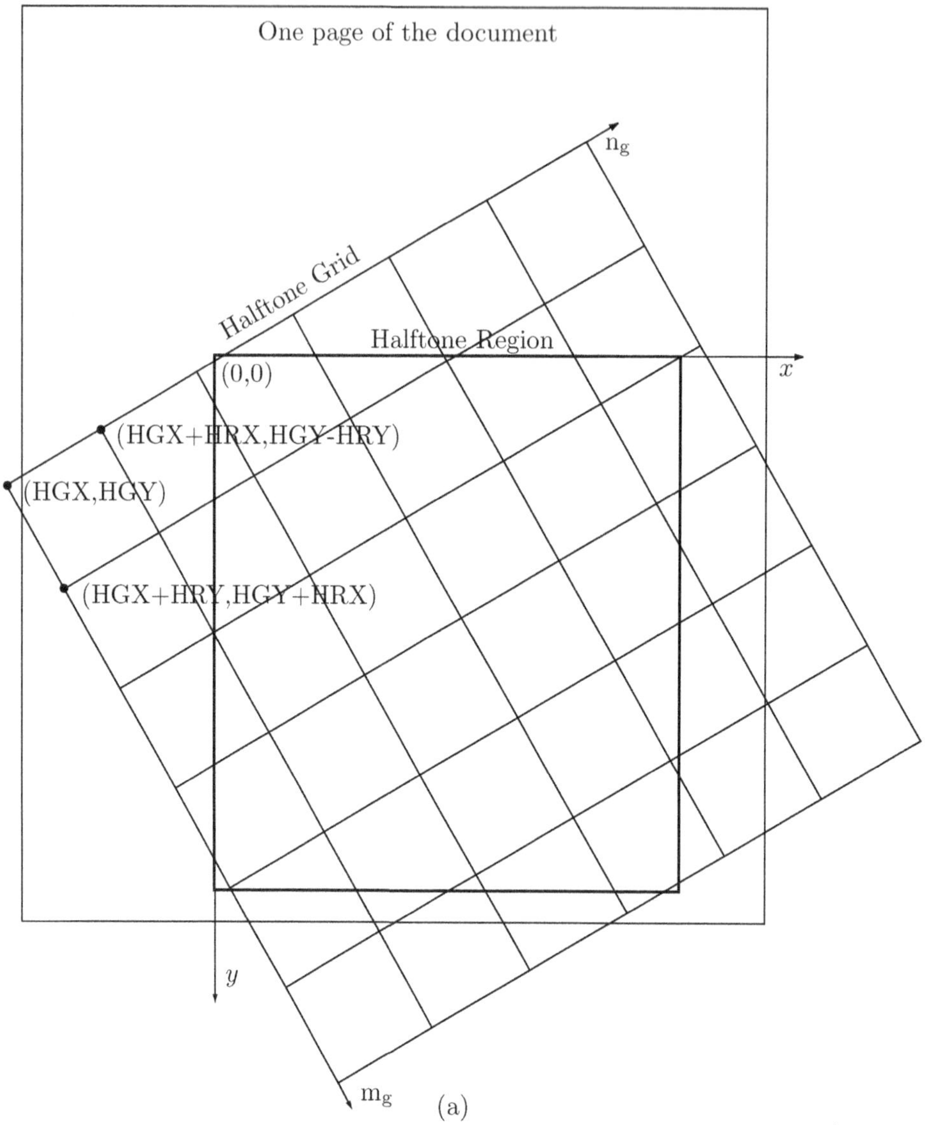

(a)

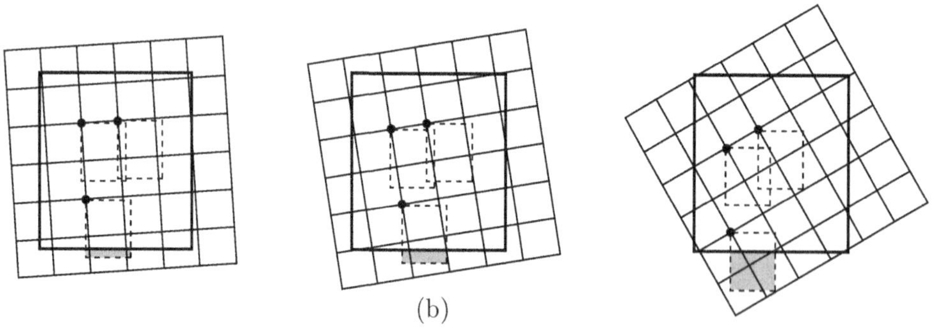

(b)

Figure 7.106: Halftone Grids and Regions.

7.15.4 The Overall Decoding Process

The decoder starts by reading general information for page 1 of the document from the compressed file. This information tells the decoder to what background value (0 or 1) to set the page buffer initially, and which of the four combination operators OR, AND, XOR, and XNOR to use when combining pixels with the page buffer. The decoder then reads segments from the compressed file until it encounters the page information for page 2 or the end of the file. A segment may specify its own combination operator, which overrides the one for the entire page. A segment can be a dictionary segment or an image segment. The latter has one of four types:

1. An immediate direct image segment. The decoder uses this type to decode a region directly into the page buffer.
2. An intermediate direct image segment. The decoder uses this type to decode a region into an auxiliary buffer.
3. An immediate refinement image segment. The decoder uses this type to decode an image and combine it with an existing region in order to refine that region. The region being refined is located in the page buffer, and the refinement process may use an auxiliary buffer (which is then deleted).
4. An intermediate refinement image segment. The decoder uses this type to decode an image and combine it with an existing region in order to refine that region. The region being refined is located in an auxiliary buffer.

We therefore conclude that an immediate segment is decoded into the page buffer and an intermediate segment involves an auxiliary buffer.

7.16 Simple Images: EIDAC

Image compression methods based on transforms perform best on continuous-tone images. There is, however, an important class of images where transform-based methods, such as JPEG (Section 7.10) and the various wavelet methods (Chapter 8), produce mediocre compression. This is the class of *simple images*. A simple image is one that uses a small fraction of all the possible grayscales or colors available to it. A common example is a bi-level image where each pixel is represented by eight bits. Such an image uses just two colors out of a palette of 256 possible colors. Another example is a grayscale image scanned from a bi-level image. Most pixels will be black or white, but some pixels may have other shades of gray. A cartoon is also an example of a simple image (especially a cheap cartoon, where just a few colors are used). A typical cartoon consists of uniform areas, so it may use a small number of colors out of a potentially large palette.

EIDAC is an acronym that stands for *embedded image-domain adaptive compression* [Yoo et al. 98]. This method is especially designed for the compression of simple images and combines high compression factors with lossless performance (although lossy compression is an option). The method is also progressive. It compresses each bitplane separately, going from the most-significant bitplane (MSBP) to the least-significant one (LSBP). The decoder reads the data for the MSBP first, and immediately generates a rough "black-and-white" version of the image. This version is improved each time a bitplane is read and decoded. Lossy compression is achieved if some of the LSBPs are not processed by the encoder.

The encoder scans each bitplane in raster order and uses several near neighbors of the current pixel X as the context of X, to determine a probability for X. Pixel X and its probability are then sent to an adaptive arithmetic encoder that does the actual encoding. Thus, EIDAC resembles JBIG and CALIC, but there is an important difference. EIDAC uses a *two-part context* for each pixel. The first part is an *intra* context, consisting of several neighbors of the pixel in the same bitplane. The second part is an *inter* context, whose pixels are selected from the already encoded bitplanes (recall that encoding is done from the MSBP to the LSBP).

To see why an inter context makes sense, consider a grayscale image scanned from a bi-level image. Most pixel values will be 255 ($1111\,1111_2$) or 0 ($0000\,0000_2$), but some pixels may have values close to these, such as 254 ($1111\,1110_2$) or 1 ($0000\,0001_2$). Figure 7.107a shows (in binary) the eight bitplanes of the six pixels 255 255 254 0 1 0. It is clear that there are spatial correlations between the bitplanes. A bit position that has a zero in one bitplane tends to have zeros in the other bitplanes. Another example is a pixel P with value "0101xxxx" in a grayscale image. In principle, the remaining four bits can have any of 16 different values. In a simple image, however, the remaining four bits will have just a few values. There is therefore a good chance that pixels adjacent to P will have values similar to P and will therefore be good predictors of P in all the bitplanes.

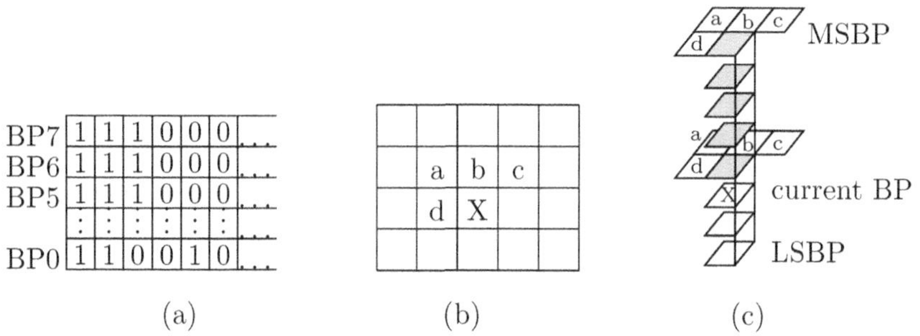

Figure 7.107: Pixel Contexts Used by EIDAC.

The intra context actually used by EIDAC is shown in Figure 7.107b. It consists of four of the already-seen nearest neighbors of the current pixel X. The inter context is shown in Figure 7.107c. It consists of (1) the same four neighbors in the bitplane immediately above the current bitplane, (2) the same four neighbors in the MSBP, and (3) the pixels in the same position as X in all the bitplanes above it (five shaded pixels in the figure). Thus, a pixel in the MSBP has no inter context, while the inter context of a pixel in the LSBP consists of $4+4+7=15$ pixels. Other contexts can be selected, which means that a general implementation of EIDAC should include three items of side information in the compressed stream. The first item is the dimensions of the image, the second item is the way the contexts are selected, and the third item is a flag indicating whether *histogram compaction* is used (if it is used, the flag should be followed by the new codes). This useful feature is described here.

A simple image is one that has a small number of colors or grayscales out of a large available palette of colors. Imagine a simple image with eight bits per pixel. If the image has just 27 colors, each can be assigned a 5-bit code instead of the original 8-bit value. This feature is referred to as histogram compaction. When histogram compaction is used, the new codes for the colors have to be included in the compressed stream, where they constitute overhead. Generally, histogram compaction improves compression, but in rare cases the overhead may be bigger than the savings due to histogram compaction.

7.17 Block Matching

The LZ77 sliding window method for compressing text (Section 6.3) can be applied to images as well. This section describes such an application for the lossless compression of images. It follows the ideas outlined in [Storer and Helfgott 97]. Figure 7.108a shows how a search buffer and a look-ahead buffer can be made to slide in raster order along an image. The principle of image compression (Section 7.3) says that the neighbors of a pixel P tend to have the same value as P or very similar values. Thus, if P is the leftmost pixel in the look-ahead buffer, there is a good chance that some of its neighbors on the left (i.e., in the search buffer) will have the same value as P. There is also a chance that a string of neighbor pixels in the search buffer will match P and the string of pixels that follow it in the look-ahead buffer. Experience also suggests that in any nonrandom image there is a good chance of having identical strings of pixels in several different locations in the image.

Since near-neighbors of a pixel are located also above and below it and not just on the left and right, it makes sense to use "wide" search and look-ahead buffers and to compare blocks of, say, 4×4 pixels instead of individual pixels (Figure 7.108b). This is the reason for the name *block matching*. When this method is used, the number of rows and columns of the image should be divisible by 4. If the image doesn't satisfy this, up to three artificial rows and/or columns should be added. Two reasonable *edge conventions* are (1) add columns of zero pixels on the left and rows of zero pixels on the top of the image, (2) duplicate the rightmost column and the bottom row as many times as needed.

Figure 7.108c,d,e shows three other ways of sliding the window in the image. The -45°-diagonal traversal visits pixels in the order (1,1), (1,2), (2,1), (1,3), (2,2), (3,1),.... In the rectilinear traversal pixels are visited in the order (1,1), (1,2), (2,2), (2,1), (1,3), (2,3), (3,3), (3,2), (3,1),.... The circular traversal visits pixels in order of increasing distance from the center.

The encoder proceeds as in the LZ77 algorithm. It finds all blocks in the search buffer that match the current 4×4 block in the look-ahead buffer and selects the longest match. A token is emitted (i.e., written on the compressed stream), and both buffers are advanced. In the original LZ77 the token consists of an offset (distance), match length, and next symbol in the look-ahead buffer (and the buffers are advanced by one plus the length of the match). The third item guarantees that progress will be made even in cases where no match was found. When LZ77 is extended to images, the third item of a token is the next pixel block in the look-ahead buffer. The method should therefore

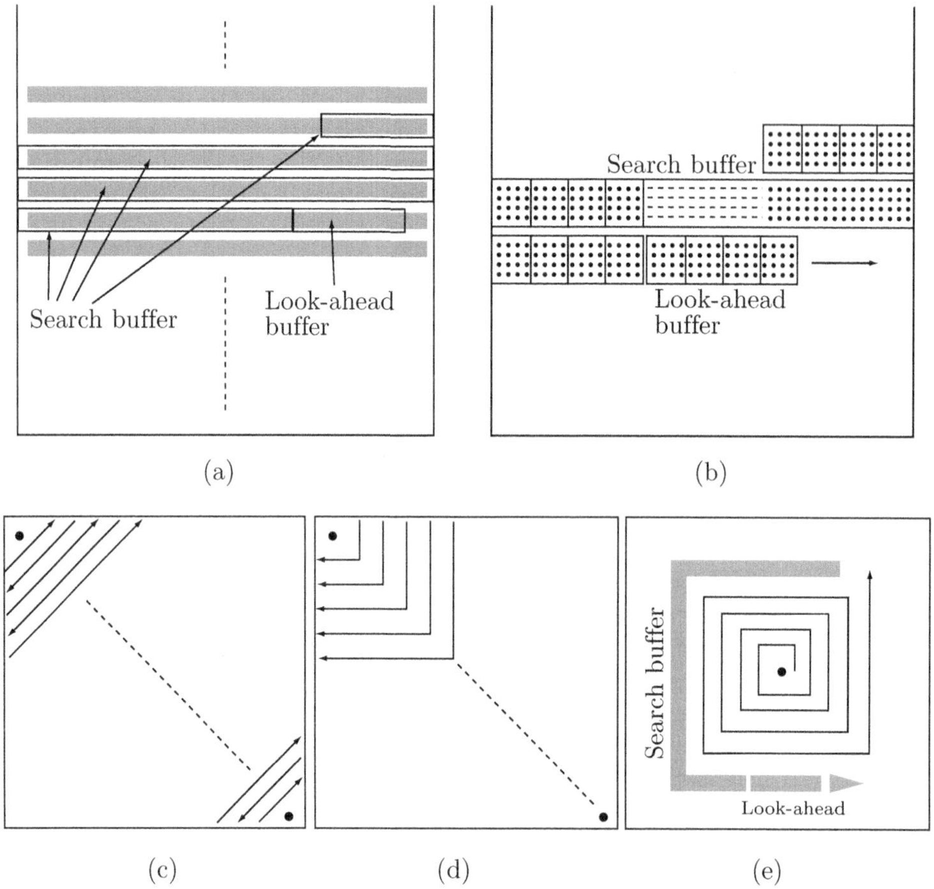

Figure 7.108: Block Matching (LZ77 for Images).

be modified to output tokens without this third item, and three ways of doing this are outlined here.

◇ **Exercise 7.24:** What's wrong with including the next block of pixels in the token?

1. The output stream consists of 2-field tokens (offset, length). When no match is found, a token of the form (0,0) is output, followed by the raw unmatched block of pixels (this is common at the start of the compression process, when the search buffer is empty or almost so). The decoder can easily mimic this.

2. The output stream consists of variable-size tags, some of which are followed by either a 2-field token or by a raw block of 16 pixels. This method is especially appropriate for bi-level images, where a pixel is represented by a single bit. We know from experience that in a "typical" bi-level image there are more uniform white areas than uniform black ones (the doubting reader should check the eighth of the ITU-T fax training documents, listed in Table 5.30, for an example of an atypical image). The tag can therefore be assigned one of the following four values: 0 if the current block in the look-ahead buffer

is all white, 10 if the current block is all black, 110 if a match was found (in which case the tag is followed by an offset and a length, encoded in either fixed size or variable size), and 111 if no match was found (in this case the tag is followed by the 16-bit raw pixel block).

3. This is similar to 2 above, but on discovering a current all-white or all-black pixel block, the encoder looks for a run of such blocks. If it finds n consecutive all-white blocks, it generates a tag of 0 followed by the encoded value of n. Similarly, a tag of 10 precedes an encoded count of a run of all-black pixel blocks. Since large values of n are rare, it makes sense to use a variable-length code such as the general unary codes of Section 3.1. A simple encoder can always use the same unary code, for example, the (2,1,10) code, which has 2044 values. A sophisticated encoder may use the image resolution to estimate the maximum size of a run of identical blocks, select a proper value for k based on this, and use code $(2,1,k)$. The value of k being used would have to be included in the header of the compressed file, for the decoder's use. Table 7.109 lists the number of codes of each of the general unary codes $(2,1,k)$ for $k=2,3,\ldots,11$. This table was calculated by the single *Mathematica* statement `Table[2^(k+1)-4,{k,2,11}]`.

k :	2	3	4	5	6	7	8	9	10	11
$(2,1,k)$:	4	12	28	60	124	252	508	1020	2044	4092

Table 7.109: Number of General Unary Codes $(2,1,k)$ for $k=2,3,\ldots,11$.

The offset can be a single number, the distance of the matched string of pixel blocks from the edge of the search buffer. This has the advantage that the maximum value of the distance depends on the size of the search buffer. However, since the search buffer may meander through the image in a complex way, the decoder may have to work hard to determine the image coordinates from the distance. An alternative is to use as offset the image coordinates (row and column) of the matched string instead of its distance. This makes it easy for the decoder to find the matched string of pixel blocks, but it decreases compression performance, since the row and column may be large numbers and since the offset does not depend on the actual distance (offsets for a short distance and for a long distance would require the same number of bits). An example may shed light on this point. Imagine an image with a resolution of $1\text{K}\times 1\text{K}$. The numbers of rows and columns are 10 bits each, so coding an offset as the pair (row, column) requires 20 bits. A sophisticated encoder, however, may decide, based on the image size, to select a search buffer of $256\text{K} = 2^{18}$ locations (25% of the image size). Coding the offsets as distances in this search buffer would require 18 bits, a small gain.

7.17.1 Implementation Details

The method described here for finding a match is just one of many possible methods. It is simple and fast, but is not guaranteed to find the best match. It is similar to the method used by LZP (Section 6.19) to find matches. Notice that finding matches is a task performed by the encoder. The decoder's task is to use the token to locate a string of pixel blocks, and this is straightforward.

The encoder starts with the current pixel block B in the look-ahead buffer. It hashes the values of the pixels of B to create a pointer p to an index table T. In $T[p]$ the encoder normally finds a pointer q pointing to the search buffer S. Block B is compared with $S[q]$. If they are identical, the encoder finds the longest match. If B and $S[q]$ are different, or if $T[p]$ is invalid (does not contain a pointer to the search buffer), there is no match. In any of these cases, a pointer to B is stored in $T[p]$, replacing its former content. Once the encoder finds a match or decides that there is no match, it encodes its findings using one of the methods outlined above.

The index table T is initialized to all invalid entries, so initially there are no matches. As more and more pointers to pixel blocks B are stored in T, the index table becomes more useful, since more of its entries contain meaningful pointers.

Another implementation detail has to do with the header of the compressed file. It should include side information that depends on the image or on the particular method used by the encoder. Such information is obviously needed by the decoder. The main items included in the header are the image resolution, number of bitplanes, the block size (4 in the examples above), and the method used for sliding the buffers. Other items that may be included are the sizes of the search and look-ahead buffers (they determine the sizes of the offset and the length fields of a token), the edge convention used, the format of the compressed file (one of the three methods outlined above), and whether the offset is a distance or a pair of image coordinates,

◇ **Exercise 7.25:** Indicate one more item that a sophisticated encoder may have to include in the header.

7.18 Grayscale LZ Image Compression

The dictionary-based compression methods described in Chapter 6 are one-dimensional. They were developed mostly for the compression of text, which is one-dimensional (although they do not perform badly on images). Images, however, are two-dimensional, and the algorithm described here is an extension of the traditional LZ methods to two dimensions. This method has been created specifically for the lossless compression of grayscale images and it is an LZ method, i.e., it is based on a dictionary of already-encoded data. This is why the name of this method is `GS-2D-LZ`. The main references are [Brittain and El-Sakka 05 and 07].

The algorithm scans the input image in raster order, encoding blocks of pixels as it goes along. In each step it tries to match a block E of yet-unencoded pixels to a block M of already-encoded pixels. The main novelty of the algorithm is that the matches do not have to be exact. Corresponding pixels in the two matched blocks are generally different, but the differences are within a threshold parameter. The differences (referred to as residuals) are small numbers and are efficiently encoded with an entropy encoder (the actual implementation employs PAQ6, Section 5.15, for this purpose).

Figure 7.110 illustrates the principle. The dark squares indicate encoded pixels. The $\times$ marks the current encoder's position. The white squares indicate unencoded pixels and the light-gray squares constitute the search region. This region is a rectangle of *search-height*$\times$ *search-width* pixels whose bottom-right corner is located at the encoder's position. Both *search-height* and *search-width* are small parameters whose best values need to be determined experimentally. Since they are small, it is reasonable to expect that the pixels inside the search region will be correlated with pixel $\times$ and its neighbors to the right and below.

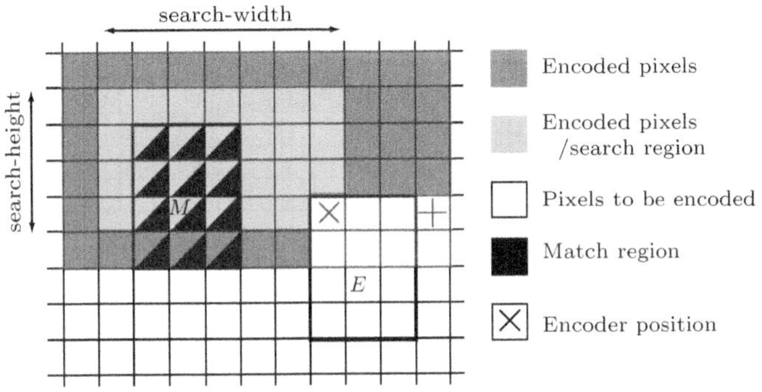

Figure 7.110: Matching a Region.

Figure 7.110 assumes that the rectangular 4×3 block M with black triangles has been (approximately) matched with the same-size block E whose top-left corner is at $\times$. The latter block is compressed by encoding (1) the location of the match relative to the block being encoded (five steps to the left and two steps up), (2) the dimensions of the match (4×3), and (3) the differences (residuals) between each pixel in the match block M and the corresponding pixel in the encoded block E. Once this is done, the encoder marks the 12 pixels of E as encoded and skips to the pixel marked $+$. The case where no match can be found in the search region is covered below.

Now for the details of (approximate) matching. The first encoder step is to scan the $S \stackrel{\text{def}}{=} (\textit{search-height} \times \textit{search-width}) - 1$ pixels in the search region and compare each to $\times$. If the difference between $\times$ and a pixel P is less than or equal to the *threshold* parameter, the encoder considers P an anchor and tries to match a maximal-size block M whose top-left corner is at P to the same-size block E whose top-left corner is $\times$. Once all S pixels in the search block have been checked in this way, the largest match M is selected and is used to encode block E as shown above.

For each anchor pixel P in the search region, the encoder compares P to $\times$. If their difference is within parameter *threshold*, the encoder compares the pixel to the right of P with the pixel to the right of $\times$. This is repeated until a pixel pair is found whose difference is greater than *threshold*. The encoder has now matched a sequence (a short row) of pixels. It moves down one row and checks the pixel pairs below those that have been matched. The next row may feature the same number of matches or fewer, but the encoder does not try to match a longer sequence. The process repeats row by row until a row is found where no pairs match. At this point, the encoder selects the largest rectangle in the matched region and considers it the matching block M found for anchor pixel P. After all S anchors have been examined in this way, the largest match block M is selected by the encoder and is employed to encode the corresponding block E. This

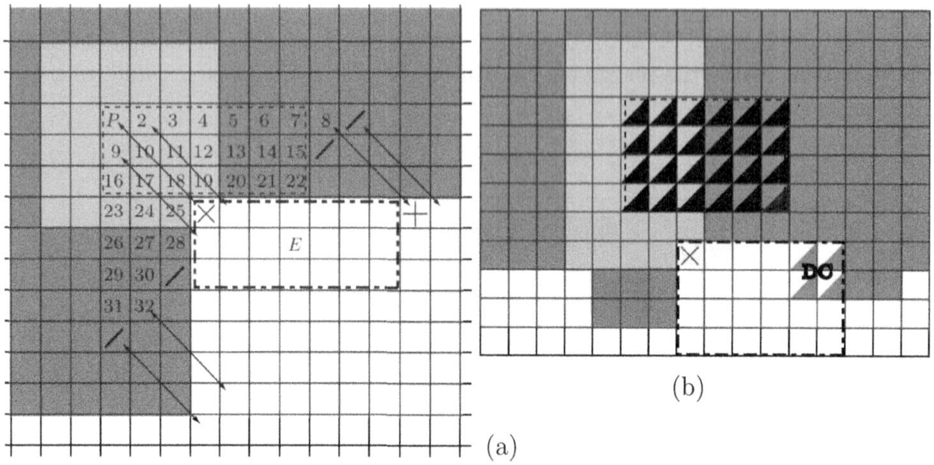

Figure 7.111: Matching a Region For An Anchor Pixel P.

process is illustrated in Figure 7.111a where the pixel pairs are numbered according to the search order.

The figure assumes that anchor pixel P has matched pixel $\times$. The encoder managed to match a sequence of seven pixels to the right of P (eight pixels including P), but only seven on the next two rows. Then came two rows of 3-pixel sequences and two more rows of short, 2-pixel sequences. The bold slashes indicate no match. No pixel matched on the next row, thereby signaling the end of the pixel-match loop. Two rectangles can be identified in the matched region, a horizontal 3×7 rectangle (shown in dashed) and a vertical 7×2 rectangle. The former is larger, so it is selected (if it satisfies the two rules below) as the match block M for anchor pixel P.

Once a match block M has been identified for each of the $6\times 6-1=35$ anchor pixels in the search region, the largest M that satisfies the two rules below is selected. Assuming that this is the 3×7 rectangle shown in Figure 7.111a in dashed, it is used to encode the 3×7 rectangle E whose top-left corner is at $\times$ (shown in bold dashed-dot). Once the differences between the 21 pixel pairs have been encoded, the encoder skips to the pixel indicated by $+$.

Matching involves two more rules. Rule 1. A block M can be a match to a block E only if the number of *new* pixels in E exceeds the value of parameter *minimum-match-size*. Figure 7.111b shows an example where block E contains four already-encoded pixels (marked DC, or don't care). Those pixels are not counted as new and also do not have to match their opposite numbers in M. Rule 2. The mean square error (MSE, Section 7.4.2) of blocks M and E must be less than the value of parameter *max-MSE.*

It is important to consider the case where no match is found. This may happen if the image region above and to the left of the encoder's position $\times$ is very different from the region below $\times$ and to its right. If the encoder cannot find a match block M for any anchor pixel, it selects a block E of dimensions (*no-match-block-height* $\times$ *no-match-block-width*) and with its top-left corner at $\times$. This block is then encoded with a simple prediction method, similar to the one used by CALIC (Section 7.28).

Both the context and the prediction rules are shown in Figure 7.112. This ensures that progress is made even in noisy image regions where pixels are not correlated.

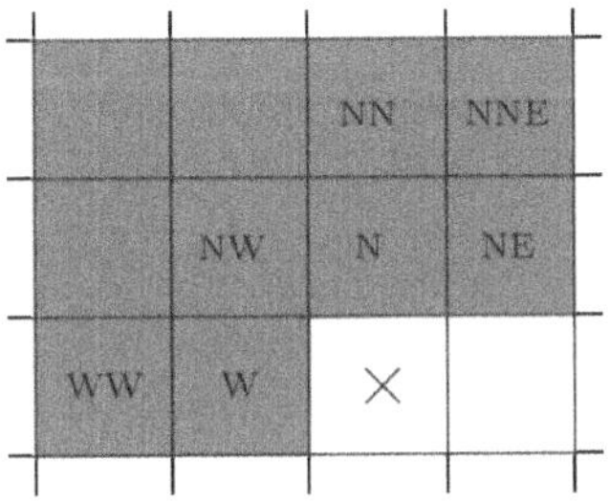

Dh = |W − WW| + |N − NW| + |NE − N|
Dv = |W − NW| + |N − NN| + |NE − NNE|
DiffDhDv = Dh − Dv
If (DiffDhDv > 80) Return N
If (DiffDhDv < −80) Return W
Pred = (N + W)/2 + (NE − NW)/4
If (DiffDhDv > 32) Return(Pred + N)/2
If (DiffDhDv < −32) Return(Pred + W)/2
If (DiffDhDv > 8) Return(3 × Pred + N)/4
If (DiffDhDv < −8) Return(3 × Pred + W)/4
Return(Pred)

Figure 7.112: Context And Rules For Predicting Pixels.

The main encoder loop of matching pixel pairs and selecting the best matching blocks M can lead to another complication, which is illustrated in Figure 7.113. We know that a data compressor can use only quantities that will be known to its decompressor at the time of decoding. The figure, however, shows a match that has extended to unencoded pixels. These pixels will be unknown to the decoder when it tries to decode pixel ×, but the surprising fact is that the decoder can handle this case recursively. Figure 7.113 shows a 3×4 matching block M (in dashed) four of whose pixels are also part of the matched block E (in bold dashed-dot). When the decoder first tries to decode E, it inputs the encoded differences between the 12 pixel pairs of M and E. The decoder first decodes the differences and then adds each difference to a pixel in M to obtain the corresponding pixel in E. In this way, it uses pixels 1–4 to decode pixels 7–10, pixels 5–8 to decode pixels 13–16, and pixels 11–14 to decode 17–20. The decoder now has the values of pixels 7, 8, 13, and 14, and it uses them to decode pixels 15, 16, 19, and 20. This neat process is the two-dimensional extension of LZW decoding (see exercise 6.10). Notice that this type of recursive decoding is possible only if the "problematic" pixels are located to the right and/or below the encoder's position.

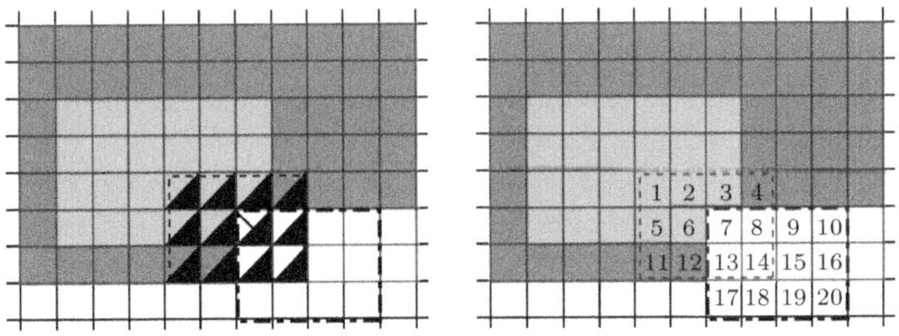

Figure 7.113: Decoding With Yet-Unencoded Pixels.

Once the encoder has decided on a match M to a block E (or discovered that there is no match at the current position), it prepares the data needed by the decoder and stores it in five tables as follows:

1. The *match-flag* table contains a boolean value for each encoding step, `true`, if a matching block was found; `false`, in cases of no match.

2. The *match-location* table, where the position of the matching block M relative to the encoded block E, is recorded. Thus, for the match shown in Figure 7.111b, the match location has coordinates $(2,5)$, while for the match shown in Figure 7.110, the match location has coordinates $(5,2)$.

3. The *match-dimensions* table contains the dimensions of the two matched blocks.

4. The differences between the pixel pairs of blocks M and E are recorded in table *residuals*.

5. An additional table, *prediction errors*, is used for those cases where no match was found (i.e., the entry in *match-flag* is `false`). Here are stored the differences between predicted and actual pixel values for each pixel in the block.

Data is prepared by the encoder and is appended to each table in each encoding step. Once the entire image has been scanned and encoded, each table is in turn encoded with an entropy encoder. After trying various statistical encoders, the developers of GS-2D-LZ decided to employ PAQ6 (Section 5.15) for this task.

Parameter settings

The algorithm depends on seven parameters whose values had to be determined experimentally for various types of data. It is intuitively clear that wrong values, whether too high or too low, can adversely affect the performance of the algorithm, and it is enlightening to see how.

1. The dimensions of a match region M are *search-width* and *search-height*. Large regions imply better approximate match, because the large number of matching pixel pairs normally results in lower MSE. However, it takes more bits to encode the dimensions of large regions.

2. The *threshold* parameter restricts the maximum allowable difference between the two pixels of a pair. Large values of this parameter may result in bigger match blocks, at the price of larger residuals and poorer compression.

3. The maximum MSE between blocks is specified by the *max-MSE* parameter. Small values of the parameter result in closer matches, but also in smaller matched blocks and more cases of no match.

4. The minimum number of new pixels in a matched block E is specified by parameter *min-match-size*. Large values imply larger matched blocks, but also fewer matches and more cases of no match.

5. Finally, parameters *no-match-block-width* and *no-match-block-height* determine the dimensions of blocks in cases of no match. Large values cause faster progress in such cases, but a smaller chance of employing the LZ principle for finding matches.

Experiments to determine the optimal values of the parameters were run with 24 images as test data. The images were divided into four classes—natural scenes (images of people, houses, birds, and a hill), geographic (taken from high altitudes, normally a satellite), graphic (hand drawn or computer generated), and medical (X-rays, ultrasound, and magnetic resonance)—with six images in each class. The results are summarized in Table 7.114 and show a clear preference of small blocks (typically 4×4 or 5×5) and large thresholds (about 25 out of 256 possible shades of grey).

	Natural scene	Geographic	Graphic	Medical
search-width	4	5	5	4
search-height	4	5	5	4
threshold	27	25	25	27
min-match-size	17	17	21	16
max-MSE	2.5	3.0	3.8	1.8
no-match-block-width	5	5	5	5
no-match-block-height	5	5	5	5

Table 7.114: Optimal Settings For Seven Parameters.

Finally, with the parameters set to their optimal values, the algorithm was tested on 112 grayscale images. It was found to perform as well or even outperform popular LZ methods (Table 7.115) and even state-of-the-art algorithms such as JPEG-LS, the lossless option of JPEG2000, and CALIC (Table 7.116).

Image class	GS-2D-LZ	PNG	GIF	UNIX Compress
Natural scene	4.50	4.73	6.83	6.21
Geographic	5.17	5.40	7.37	6.46
Graphic	1.61	1.77	2.73	2.48
Medical	3.25	3.42	4.96	4.58

Table 7.115: GS-2D-LZ Compared to Popular LZ Codecs.

Image name	GS-2D-LZ	BZIP2	JPEG2000	JPEG-LS	CALIC-h	CALIC-a
Natural scene	4.50	4.88	4.66	4.71	4.50	4.27
Geographic	5.17	5.24	5.31	5.24	5.23	5.06
Graphic	1.61	1.77	2.61	1.90	2.11	1.73
Medical	3.25	3.48	3.22	3.22	3.39	3.23

Table 7.116: GS-2D-LZ Compared to State-Of-The-Art Codecs.

I would like to thank Prof. El-Sakka for his help. He has read this section and provided useful comments and corrections.

7.19 Vector Quantization

Vector quantization is a generalization of the scalar quantization method (Section 1.6). It is used for both image and audio compression. In practice, vector quantization is commonly used to compress data that has been digitized from an analog source, such as audio samples and scanned images. Such data is called *digitally sampled analog data* (DSAD). Vector quantization is based on the fact that compression methods that compress strings (or blocks) rather than individual symbols (Section 6.1) can, in principle, produce better results. Such methods are called *block coding* and, as the block length increases, compression methods based on block coding can approach the entropy when compressing stationary information sources.

We start with a simple, intuitive vector quantization method for image compression. Given an image, we divide it into small blocks of pixels, typically 2×2 or 4×4. Each block is considered a vector. The encoder maintains a list (called a *codebook*) of vectors and compresses each block by writing on the compressed stream a pointer to the block in the codebook. The decoder has the easy task of reading pointers, following each pointer to a block in the codebook, and appending the block to the image-so-far. Thus, vector quantization is an asymmetric compression method.

In the case of 2×2 blocks, each block (vector) consists of four pixels. If each pixel is one bit, then a block is four bits long and there are only $2^4 = 16$ different blocks. It is easy to store such a small, permanent codebook in both encoder and decoder. However, a pointer to a block in such a codebook is, of course, four bits long, so there is no compression gain by replacing blocks with pointers. If each pixel is k bits, then each block is $4k$ bits long and there are 2^{4k} different blocks. The codebook grows very fast with k (for $k = 8$, for example, it is $256^4 = 2^{32} = 4$ Tera entries) but the point is that we again replace a block of $4k$ bits with a $4k$-bit pointer, resulting in no compression gain. This is true for blocks of any size.

Once it becomes clear that this simple method does not work, the next thing that comes to mind is that any given image may not contain every possible block. Given 8-bit pixels, the number of 2×2 blocks is $2^{2 \cdot 2 \cdot 8} = 2^{32} \approx 4.3$ billion, but any particular image may contain only a few million pixels and a few thousand different blocks. Thus, our next version of vector quantization starts with an empty codebook and scans the image block by block. The codebook is searched for each block. If the block is already in the codebook, the encoder outputs a pointer to the block in the (growing) codebook. If the block is not in the codebook, it is added to the codebook and a pointer is output.

The problem with this simple method is that each block added to the codebook has to be written on the compressed stream. This greatly reduces the effectiveness of the method and may lead to low compression and even to expansion. There is also the small added complication that the codebook grows during compression, so the pointers get longer, but this is not difficult for the decoder to handle.

These problems are the reason why image vector quantization is lossy. If we accept lossy compression, then the size of the codebook can be greatly reduced. Here is an intuitive lossy method for image compression by vector quantization. Analyze a large number of different "training" images and find the B most-common blocks. Construct a codebook with these B blocks and embed it into both encoder and decoder. Each entry of the codebook is a block. To compress an image, scan it block by block, and for each block find the codebook entry that best matches it, and output a pointer to that entry. The size of the pointer is, of course, $\lceil \log_2 B \rceil$, so the compression ratio (which is known in advance) is

$$\frac{\lceil \log_2 B \rceil}{\text{block size}}.$$

One problem with this approach is how to match image blocks to codebook entries. Here are a few common measures. Let $B = (b_1, b_2, \ldots, b_n)$ and $C = (c_1, c_2, \ldots, c_n)$ be a block of image pixels and a codebook entry, respectively (each is a vector of n bits). We denote the "distance" between them by $d(B, C)$ and measure it in three different ways as follows:

$$\begin{aligned} d_1(B,C) &= \sum_{i=0}^{n} |b_i - c_i|, \\ d_2(B,C) &= \sum_{i=0}^{n} (b_i - c_i)^2, \\ d_3(B,C) &= \mathrm{MAX}_{i=0}^{n} |b_i - c_i|. \end{aligned} \tag{7.29}$$

The third measure $d_3(B,C)$ is easy to interpret. It finds the component where B and C differ most, and it returns this difference. The first two measures are easy to visualize in the case $n = 3$. Measure $d_1(B,C)$ becomes the distance between the two three-dimensional vectors B and C when we move along the coordinate axes. Measure $d_2(B,C)$ becomes the Euclidean (straight line) distance between the two vectors. The quantities $d_i(B,C)$ can also be considered measures of distortion.

A distortion measure (often also called a metric) is a function m that measures "distance" between, or "closeness" of, two mathematical quantities a and b. Such a function should satisfy $m(a,b) \geq 0$, $m(a,a) = 0$, and $m(a,b) + m(b,c) \geq m(a,c)$.

Another problem with this approach is the quality of the codebook. In the case of 2×2 blocks with 8-bit pixels the total number of blocks is $2^{32} \approx 4.3$ billion. If we decide to limit the size of our codebook to, say, a million entries, it will contain only 0.023% of the total number of blocks (and still be 32 million bits, or about 4 Mb long). Using this codebook to compress an "atypical" image may result in a large distortion regardless of the distortion measure used. When the compressed image is decompressed, it may look so different from the original as to render our method useless. A natural way to solve this problem is to modify the original codebook entries in order to *adapt* them to the particular image being compressed. The final codebook will have to be included in the compressed stream, but since it has been adapted to the image, it may be small enough to yield a good compression ratio, yet close enough to the image blocks to produce an acceptable decompressed image.

Such an algorithm has been developed by Linde, Buzo, and Gray [Linde, Buzo, and Gray 80]. It is known as the LBG algorithm and it is the basis of many vector quantization methods for the compression of images and sound (see, for example, Section 7.37). Its main steps are the following:

Step 0: Select a threshold value ϵ and set $k = 0$ and $D^{(-1)} = \infty$. Start with an initial codebook with entries $C_i^{(k)}$ (where k is currently zero, but will be incremented in each iteration). Denote the image blocks by B_i (these blocks are also called *training vectors*, since the algorithm uses them to find the best codebook entries).

Step 1: Pick up a codebook entry $C_i^{(k)}$. Find all the image blocks B_m that are closer to C_i than to any other C_j. Phrased more precisely; find the set of all B_m that satisfy

$$d(B_m, C_i) < d(B_m, C_j) \quad \text{for all } j \neq i.$$

This set (or *partition*) is denoted by $P_i^{(k)}$. Repeat for all values of i. It may happen that some partitions will be empty, and we deal with this problem below.

Step 2: Select an i and calculate the distortion $D_i^{(k)}$ between codebook entry $C_i^{(k)}$ and the set of training vectors (partition) $P_i^{(k)}$ found for it in step 1. Repeat for all i, then calculate the average $D^{(k)}$ of all the $D_i^{(k)}$. A distortion $D_i^{(k)}$ for a certain i is calculated by computing the distances $d(C_i^{(k)}, B_m)$ for all the blocks B_m in partition $P_i^{(k)}$, then computing the average distance. Alternatively, $D_i^{(k)}$ can be set to the minimum of the distances $d(C_i^{(k)}, B_m)$.

Step 3: If $(D^{(k-1)} - D^{(k)})/D^{(k)} \leq \epsilon$, halt. The output of the algorithm is the last set of codebook entries $C_i^{(k)}$. This set can now be used to (lossy) compress the image with vector quantization. In the first iteration k is zero, so $D^{(k-1)} = D^{(-1)} = \infty > \epsilon$. This guarantees that the algorithm will not stop at the first iteration.

Step 4: Increment k by 1 and calculate new codebook entries $C_i^{(k)}$; each equals the average of the image blocks (training vectors) in partition $P_i^{(k-1)}$ that was computed in step 1. (This is how the codebook entries are adapted to the particular image.) Go to step 1.

A full understanding of such an algorithm calls for a detailed example, parts of which should be worked out by the reader in the form of exercises. In order to easily visualize the example, we assume that the image to be compressed consists of 8-bit pixels, and we divide it into small, two-pixel blocks. Normally, a block should be square, but the advantage of our two-pixel blocks is that they can be plotted on paper as two-dimensional points, thereby rendering the data (as well as the entire example) more visual. Examples of blocks are $(35, 168)$ and $(250, 37)$; we interpret the two pixels of a block as the (x, y) coordinates of a point.

Our example assumes an image consisting of 24 pixels, organized in the 12 blocks $B_1 = (32, 32)$, $B_2 = (60, 32)$, $B_3 = (32, 50)$, $B_4 = (60, 50)$, $B_5 = (60, 150)$, $B_6 = (70, 140)$, $B_7 = (200, 210)$, $B_8 = (200, 32)$, $B_9 = (200, 40)$, $B_{10} = (200, 50)$, $B_{11} = (215, 50)$, and $B_{12} = (215, 35)$ (Figure 7.117). It is clear that the 12 points are concentrated in four regions. We select an initial codebook with the four entries $C_1^{(0)} = (70, 40)$, $C_2^{(0)} = (60, 120)$, $C_3^{(0)} = (210, 200)$, and $C_4^{(0)} = (225, 50)$ (shown as x in the diagram). These entries were selected more or less at random but we show later how the LBG algorithm selects them methodically, one by one. Because of the graphical nature of the data, it is easy to determine the four initial partitions. They are $P_1^{(0)} = (B_1, B_2, B_3, B_4)$, $P_2^{(0)} = (B_5, B_6)$, $P_3^{(0)} = (B_7)$, and $P_4^{(0)} = (B_8, B_9, B_{10}, B_{11}, B_{12})$. Table 7.118 shows how the average distortion $D^{(0)}$ is calculated for the first iteration (we use the Euclidean

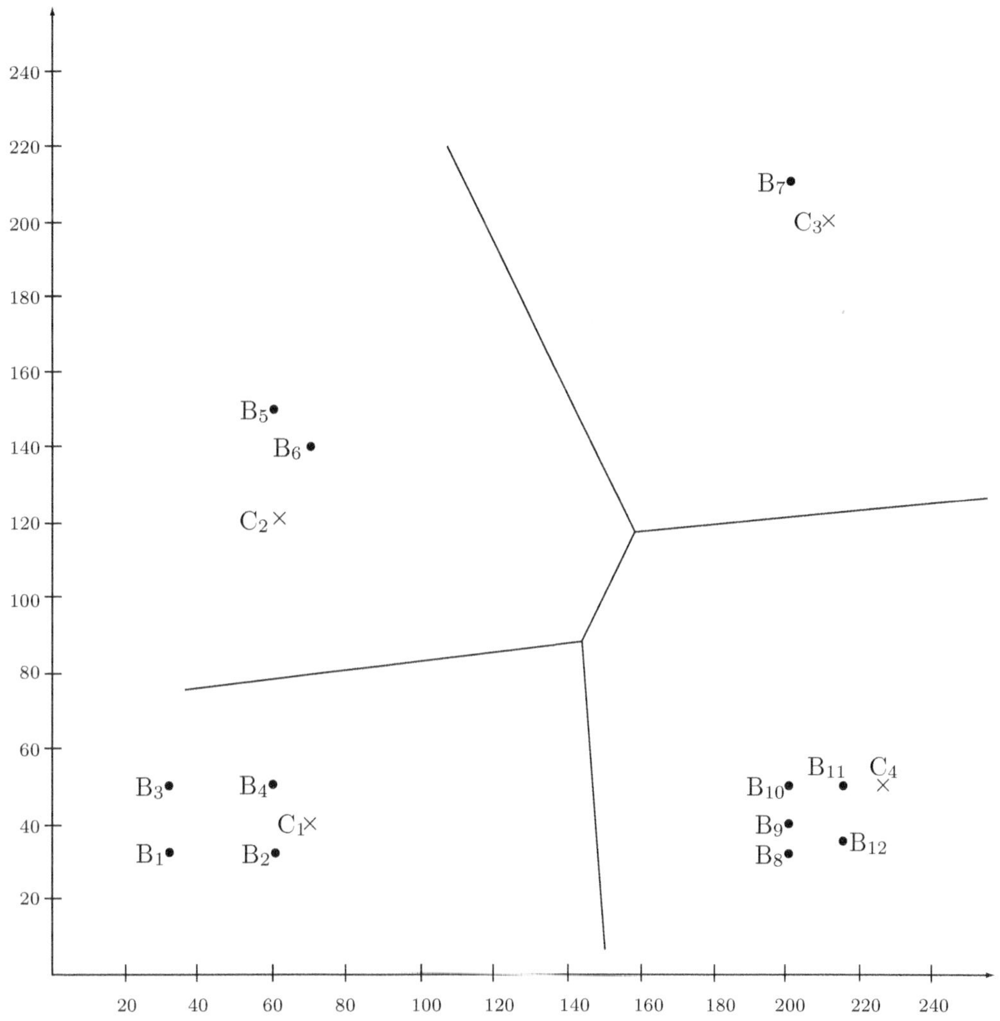

Figure 7.117: Twelve Points and Four Codebook Entries $C_i^{(0)}$.

I:	$(70-32)^2+(40-32)^2=1508,$	$(70-60)^2+(40-32)^2=164,$
	$(70-32)^2+(40-50)^2=1544,$	$(70-60)^2+(40-50)^2=200,$
II:	$(60-60)^2+(120-150)^2=\ 900,$	$(60-70)^2+(120-140)^2=500,$
III:	$(210-200)^2+(200-210)^2=\ 200,$	
IV:	$(225-200)^2+(50-32)^2=\ 449,$	$(225-200)^2+(50-40)^2=725,$
	$(225-200)^2+(50-50)^2=\ 625,$	$(225-215)^2+(50-50)^2=100,$
	$(225-215)^2+(50-35)^2=\ 325.$	

Table 7.118: Twelve Distortions for $k=0$.

distance function). The result is

$$D^{(0)} = (1508 + 164 + 1544 + 200 + 900 + 500 + 200 + 449 + 725 + 625 + 100 + 325)/12 = 603.33.$$

Step 3 indicates no convergence, since $D^{(-1)} = \infty$, so we increment k to 1 and calculate four new codebook entries $C_i^{(1)}$ (rounded to the nearest integer for simplicity)

$$\begin{aligned} C_1^{(1)} &= (B_1 + B_2 + B_3 + B_4)/4 = (46, 41), \\ C_2^{(1)} &= (B_5 + B_6)/2 = (65, 145), \\ C_3^{(1)} &= B_7 = (200, 210), \\ C_4^{(1)} &= (B_8 + B_9 + B_{10} + B_{11} + B_{12})/5 = (206, 41). \end{aligned} \tag{7.30}$$

They are shown in Figure 7.119.

◇ **Exercise 7.26:** Perform the next iteration.

◇ **Exercise 7.27:** Use the four codebook entries of Equation (7.30) to perform the next iteration of the LBG algorithm.

A slightly different interpretation of the LBG algorithm can provide intuitive understanding of why it works. Imagine a vector quantizer that consists of an encoder and a decoder. For a given input vector, the encoder finds the region with the most similar samples, the decoder outputs for a given region, the most representative vector for that region. The LBG algorithm optimizes the encoder (in step 1) and the decoder (in step 4) until no further improvement is possible.

Since the algorithm strictly reduces the average distortion from one iteration to the next, it does not guarantee that the codebook entries will converge to the optimum set.

In general, they will converge to a *local minimum*, a miminum that can be improved only by temporarily making the distortion of the solution worse (which we know is impossible for LBG). The quality of the solution and the speed of convergence depend heavily on the initial choice of codebook entries (i.e., on the values of $C_i^{(0)}$). We therefore discuss this aspect of the LBG algorithm next. The original LBG algorithm proposes a computationally intensive *splitting technique* where the initial codebook entries $C_i^{(0)}$ are selected in several steps as follows:

Step 0: Set $k = 1$ and select a codebook with k entries (i.e., one entry) $C_1^{(0)}$ that's an average of all the image blocks B_m.
Step 1: In a general iteration there will be k codebook entries $C_i^{(0)}$, $i = 1, 2, \ldots, k$. Split each entry (which is a vector) into the two similar entries $C_i^{(0)} \pm e$ where e is a fixed *perturbation vector*. Set $k \leftarrow 2k$. (Alternatively, we can use the two vectors $C_i^{(0)}$ and $C_i^{(0)} + e$, a choice that leads to smaller overall distortion.)
Step 2: If there are enough codebook entries, stop the splitting process. The current set of k codebook entries can now serve as the initial set $C_i^{(0)}$ for the LBG algorithm above.

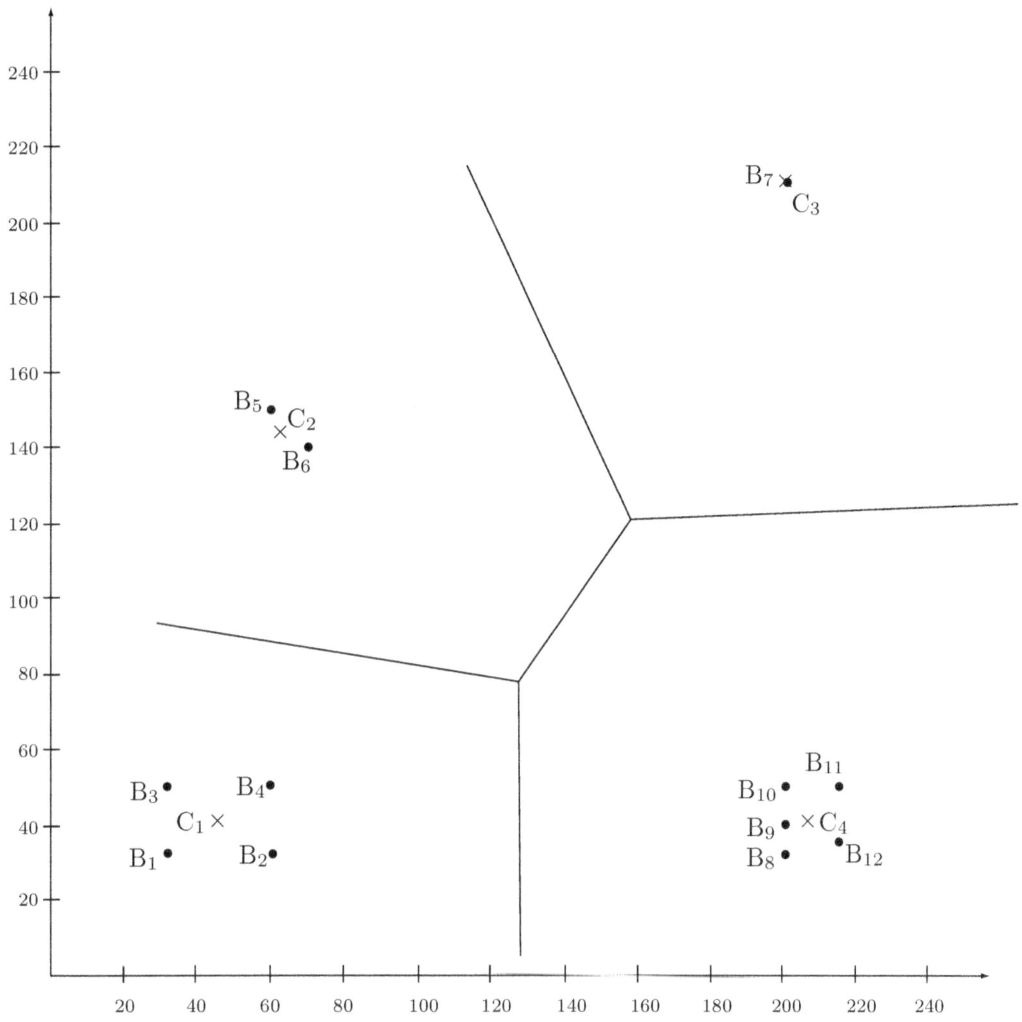

Figure 7.119: Twelve Points and Four Codebook Entries $C_i^{(1)}$.

I:	$(46-32)^2+(41-32)^2=277,$	$(46-60)^2+(41-32)^2=277,$
	$(46-32)^2+(41-50)^2=277,$	$(46-60)^2+(41-50)^2=277,$
II:	$(65-60)^2+(145-150)^2=\ 50,$	$(65-70)^2+(145-140)^2=\ 50,$
III:	$(210-200)^2+(200-210)^2=200,$	
IV:	$(206-200)^2+(41-32)^2=117,$	$(206-200)^2+(41-40)^2=\ 37,$
	$(206-200)^2+(41-50)^2=117,$	$(206-215)^2+(41-50)^2=162,$
	$(206-215)^2+(41-35)^2=117.$	

Table 7.120: Twelve Distortions for $k=1$.

If more entries are needed, execute the LBG algorithm on the current set of k entries, to converge them to a better set; then go to step 1.

This process starts with one codebook entry $C_1^{(0)}$ and uses it to create two entries, then four, eight, and so on. The number of entries is always a power of 2. If a different number is needed, then the last time step 1 is executed, it can split just some of the entries. For example, if 11 codebook entries are needed, then the splitting process is repeated until eight entries have been computed. Step 1 is then invoked to split just three of the eight entries, ending up with 11 entries.

One more feature of the LBG algorithm needs to be clarified. Step 1 of the algorithm says; "Find all the image blocks B_m that are closer to C_i than to any other C_j. They become partition $P_i^{(k)}$." It may happen that no image blocks are close to a certain codebook entry C_i, which creates an empty partition $P_i^{(k)}$. This is a problem, because the average of the image blocks included in partition $P_i^{(k)}$ is used to compute a better codebook entry in the next iteration. An empty partition is also bad because it "wastes" a codebook entry. The distortion of the current solution can be trivially reduced by replacing codebook entry C_i with any image block. An even better solution is to delete codebook entry C_i and replace it with a new entry chosen at random from one of the image blocks included in the partition $P_j^{(k)}$ having biggest distortion.

If an image block is an n-dimensional vector, then the process of constructing the partitions in step 1 of the LBG algorithm divides the n-dimensional space into Voronoi regions [Salomon 99] with a codebook entry C_i at the center of each region. Figures 7.118 and 7.119 show the Voronoi regions in the first two iterations of our example. In each iteration the codebook entries are moved, thereby changing the Voronoi regions.

Speeding up the codebook search: In order to quantize $V = (v_1, v_2, \ldots, v_n)$ with a vector quantizer having a k-entry codebook, the method described above requires the comparison of the vector V with each entry in the codebook. Each comparison involves the computation of the distortion between the vector and a codeword. This approach to vector quantization is called *Exhaustive Search Vector Quantizer* or ESVQ. An exhaustive search of high-dimensional vectors in a large codebook can be computationally prohibitive, so several algorithms have been designed to overcome this issue.

A divide and conquer search algorithm, very similar to the well-known binary search, can speed up the codebook search. For simplicity, we describe this method for the two-dimensional vector quantizer presented in Figure 7.119, however, it is possible to generalize it to quantizers of any dimension.

Line A in Figure 7.121 separates the two-dimensional space containing the codewords into two semi-planes A^+ and A^-. A similar subdivision is performed by lines B through E. For reasons that will become clear later, all lines in the figure mark one or more boundaries of the quantizer regions. If, instead of comparing vector V to each codeword, we check whether V lies to the left or to the right of a line, we can (almost) halve the size of the search space with a single comparison. By repeating the comparison with the other lines and intersecting the resulting semi-planes, we can pinpoint the region to which V belongs, and thus also the closest codeword in the codebook.

The sequence of decisions, illustrated in the form of a *decision tree*, is depicted in Figure 7.122. For example, to quantize V, we start comparing it to line A. V lies on the semi-plane A^+, so we proceed by comparing V to line B. Since V lies on the

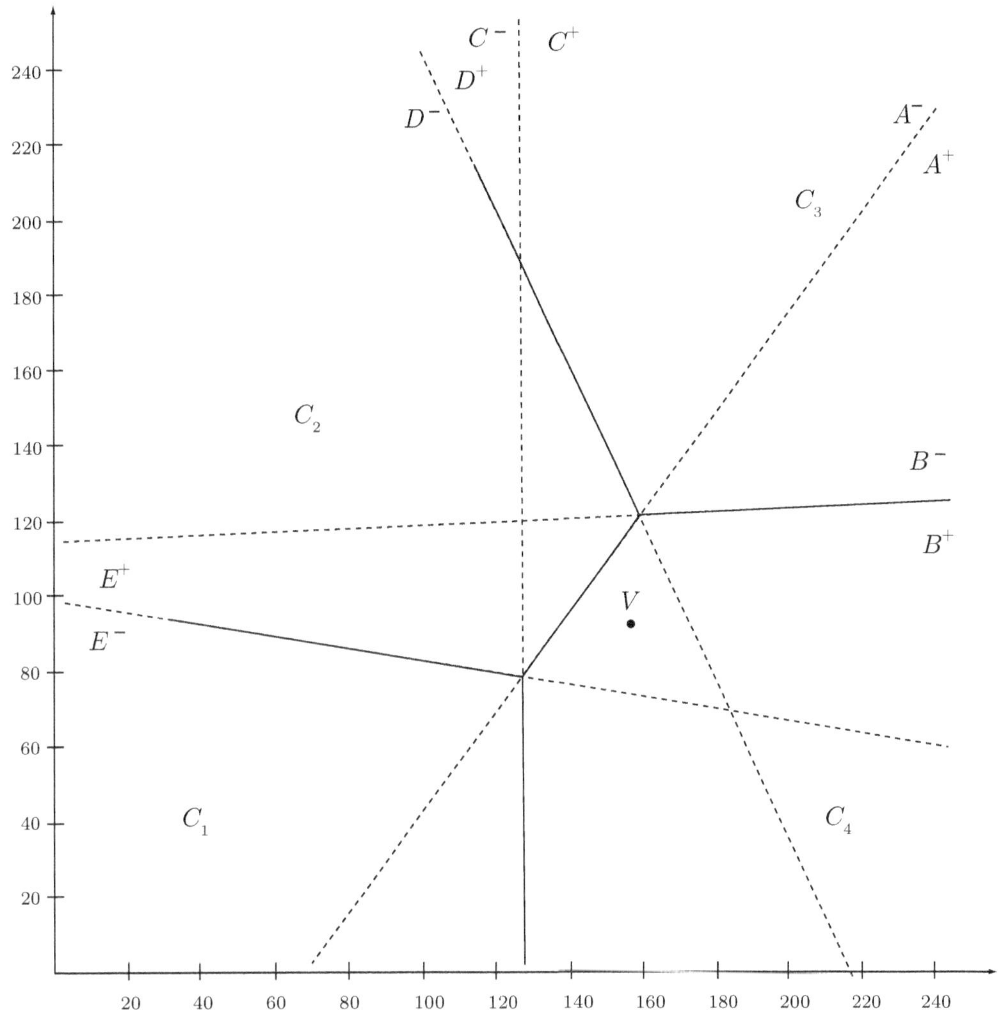

Figure 7.121: Quantizer Regions Partitioned into Semi-Planes.

semi-plane B^+, we continue by comparing V to C and conclude that V lies on the semi-plane C^+. The intersection of the semi-planes A^+, B^+, and C^+ is the quantizer region corresponding to codeword C_4.

From the depth of the decision tree it is evident, even with such a small example, that the number of comparisons is reduced from four to at most three. If the vector being quantized lies in the region associated with C_3, then only two comparisons are necessary.

It is important to notice that this method speeds up the codeword search without degrading the quality of the result. In the following paragraphs we introduce methods that are capable of even faster searches while marginally compromising the overall distortion of the quantizer. All the methods described impose some sort of structure on

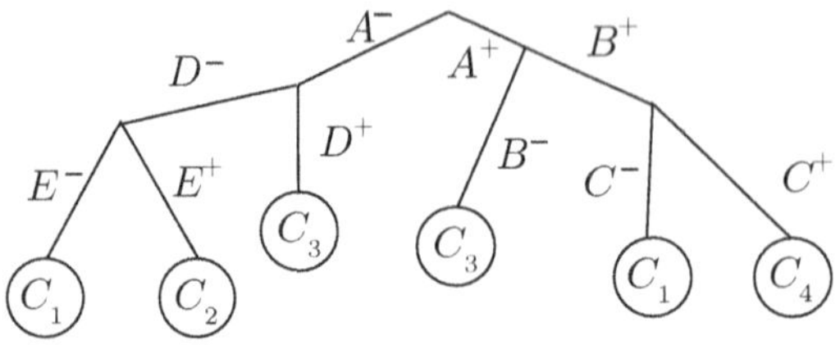

Figure 7.122: Decision Tree for the Two-Dimensional Vector Quantizer.

the codebook and exploit this structure to reduce its size and/or improve the speed of the search.

Tree-Structured VQ: An alternative to the previous method consists of structuring the codebook as a search tree. This strategy also allows a fast codebook search and, if desired, may provide additional functionality. The drawback of this approach is that, due to the structural constraints, the average distortion may be higher than that of a vector quantizer that uses an unstructured codebook.

In a *Tree-Structured Vector Quantizer*, or *TSVQ*, the quantization is performed by traversing a tree and comparing the input vector V to the vectors associated with the nodes along a path. The final result of the quantization is the vector associated with a leaf node. Figure 7.123 shows an example of such quantizer. The input vector V is compared first to the vectors y_1 and y_2. The quatization continues on the branch corresponding to the vector having minimum distortion (the vector "closer" to V). Assume that y_2 is such a vector. In the next step V is compared to y_5 and y_6 and since y_5 is closer, a comparison with C_5 and C_6 follows. The quantization ends at the leaf associated with C_6 and the index corresponding to this vector is output and sent to the decoder.

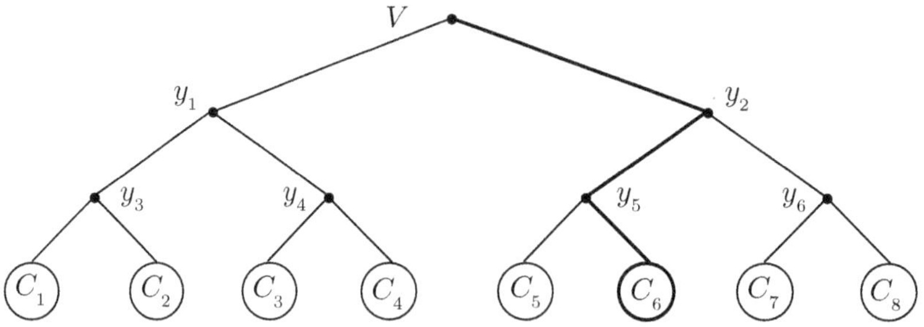

Figure 7.123: Tree-Structured Vector Quantizer.

Notice that in this example, the inner vectors are only used to speed up the search and that the decoder does not need to store them. On the other hand, vectors y_2 and y_5 encountered along the path can be considered a "coarser" quantization of V (with y_2, in general, less precise than y_5). By using this observation and choosing an indexing scheme that reflects the branching of the tree (such as the indexing used in a Huffman code), a decoder that stores the inner vectors can stop the quantization anywhere along the path and output a lower quality quantization of V. This method is particularly useful when implementing a *progressive* compression scheme, where decoding a part of the bitstream results in a lower-quality representation of the compressed signal.

Residual VQ: Progressive encoding can also be obtained with what is referred to as *Residual Vector Quantizer*. A residual vector quantizer consists of multiple cascaded stages, each quantizing the error (or residual) of the previous stage. Figure 7.124 illustrates such a quantizer.

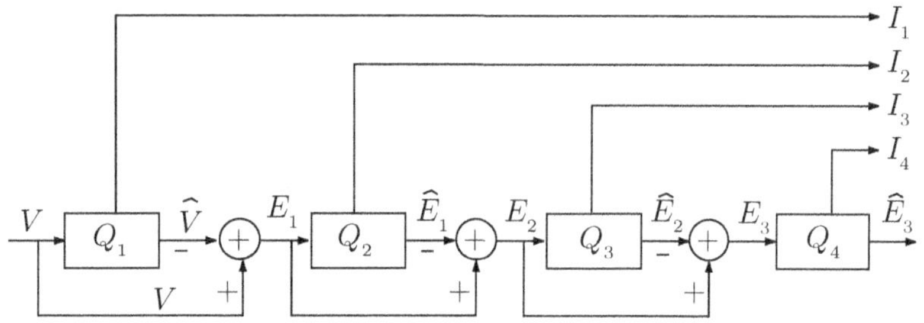

Figure 7.124: Four-Stages Residual Vector Quantizer.

A residual vector quantizer simplifies the search by using several small codebooks instead of a single large one. Instead of quantizing a vector with, say, 16 bits with a codebook containing $2^{16} = 65{,}536$ codevectors, a four-stage residual quantizer could use, for example, four 4-bit indexes and four codebooks of 16 codevectors each (the codebooks of the quantizers Q_1 through Q_4 in the figure). Even with an exhaustive search, comparing an input vector to 64 codevectors still results in a practical algorithm.

Each stage transmits a quantization index (I_1 through I_4 in the figure) that is independently dequantized at the decoder. To reconstruct the quantized vector, the decoder adds up the four codewords $\hat{V}$, $\hat{E}_1$, $\hat{E}_2$, and $\hat{E}_3$.

Product VQ: If a vector V has too many components to be quantized as a whole, a simple strategy is to partition it into two or more smaller vectors and quantize the smaller vectors separately. This is clearly worse than quantizing the entire vector, since otherwise vector quantization wouldn't be better than scalar quantization. However, if the sub-vectors are not strongly correlated, partitioning can result in an efficient and practical algorithm. The name *Product Vector Quantizer* derives from the fact that the codebook for the entire vector is the Cartesian product of the codebooks used to quantize the smaller vectors.

An algorithm that partitions long vectors into sub-vectors of possibly different dimensions is described in Section 11.15.3. In the following we discuss an approach (often used in image coding) called *Mean-Removed Vector Quantizer*.

Small pixel blocks in a digital image may repeat in the context of objects or areas having different luminosity. A vector quantizer would need different codewords to represent blocks that have the same pattern but different luminosity. For example, the vector representing a 4×4 image block containing a vertical edge will appear numerically different depending on whether the edge belongs to a well-lit object or to an object in a shaded area.

A mean-removed vector quantizer addresses this problem by decomposing an n-dimensional vector into a vector of dimension $(n-1)$ and a scalar (Figure 7.125). As the name suggests, the scalar represents the mean value of the pixels in the vector and the $(n-1)$-dimensional vector is obtained by subtracting the mean from each component and dropping a dimension. One of the dimensions can be discarded since the mean-removed vector has $n-1$ degrees of freedom (the sum of its components is always zero). The mean can now be quantized with a scalar quantizer and an $(n-1)$-dimensional codebook can be used on the mean-removed vectors.

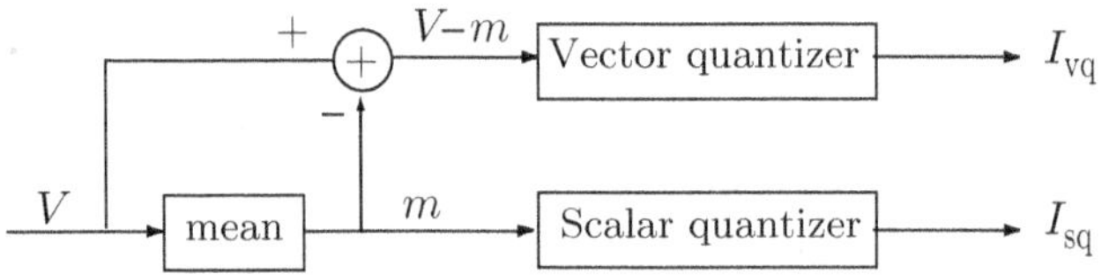

Figure 7.125: Mean-Removed Vector Quantizer.

The following illustrates the advantage of this approach. To quantize 4×4 image blocks with 16 bits, a traditional vector quantizer would require a codebook containing 65,536 vectors. A mean-removed quantizer would be much more practical since it could be designed with the mean quantized (very accurately) with seven bits, and the mean-removed vector with a codebook of only 512 entries.

The approaches to the design of structured quantizers presented here are not mutually exclusive. The tree and the residual structures, for example, are often combined so that paths along the tree encode quantization residuals. Similarly, the vector quantizer used in a mean-removed VQ can have a residual structure, and so on. An in-depth discussion of these methods and the presentation of many more can be found in the excellent text by Gersho and Gray [Gersho and Gray 92].

7.20 Adaptive Vector Quantization

The basic vector quantization method of Section 7.19 uses either a fixed codebook or the LBG (or similar) algorithm to construct the codebook as it goes along. In all these cases the codebook consists of fixed-size entries, identical in size to the image block. The adaptive method described here, due to Constantinescu and Storer [Constantinescu and Storer 94a,b], uses variable-size image blocks and codebook entries. The codebook is called *dictionary*, because this method bears a slight resemblance to the various LZ methods. The method combines the single-pass, adaptive dictionary used by the various dictionary-based algorithms with the distance measures used by the different vector quantization methods to obtain good approximations of data blocks.

At each step of the encoding process, the encoder selects an image block (a rectangle of any size), matches it to one of the dictionary entries, outputs a pointer to that entry, and updates the dictionary by adding one or more entries to it. The new entries are based on the current image block. The case of a full dictionary has to be taken care of, and is discussed below.

Some authors refer to such a method as *textual substitution*, since it substitutes pointers for the original data.

Extensive experimentation by the developers yielded compression ratios that equal or even surpass those of JPEG. At the same time, the method is fast (since it performs just one pass), simple (it starts with an empty dictionary, so no training is required), and reliable (the amount of loss is controlled easily and precisely by a single tolerance parameter). The decoder's operation is similar to that of a vector quantization decoder, so it is fast and simple.

The encoder has to select the image blocks carefully, making sure that they cover the entire image, that they cover it in a way that the decoder can mimic, and that there is not too much overlap between the different blocks. In order to select the blocks the encoder selects one or more *growing points* at each iteration, and uses one of them in the next iteration as the corner of a new image block. The block starts small and is increased in size (is "grown" from the corner) by the encoder as long as it can be matched with a dictionary entry. A growing point is denoted by G and is a triplet (x, y, q), where x and y are the coordinates of the point and q indicates the corner of the image block where the point is located. We assume that q values of 0, 1, 2, and 3 indicate the top-left, top-right, bottom-left, and bottom-right corners, respectively (thus, q is a 2-bit number). An image block B anchored at a growing point G is denoted by $B = (G, w, h)$, where w and h are the width and height of the block, respectively.

We list the main steps of the encoder and decoder, then discuss each in detail. The encoder's main steps are as follows:

Step 1: Initialize the dictionary D to all the possible values of the image pixels. Initialize the pool of growing points (GPP) to one or more growing points using one of the algorithms discussed below.
Step 2: Repeat Steps 3–7 until the GPP is empty.
Step 3: Use a growing algorithm to select a growing point G from GPP.
Step 4: Grow an image block B with G as its corner. Use a matching algorithm to match B, as it is being grown, to a dictionary entry with user-controlled fidelity.
Step 5: Once B has reached the maximum size where it still can be matched with a dictionary entry d, output a pointer to d. The size of the pointer depends on the size (number of entries) of D.
Step 6: Delete G from the GPP and use an algorithm to decide which new growing points (if any) to add to the GPP.
Step 7: If D is full, use an algorithm to delete one or more entries. Use an algorithm to update the dictionary based on B.

The operation of the encoder depends, therefore, on several algorithms. Each algorithm should be developed, implemented, and its performance tested with many test images. The decoder is much simpler and faster. Its main steps are as follows:

Step 1: Initialize the dictionary D and the GPP as in Step 1 of the encoder.
Step 2: Repeat Steps 3–5 until GPP is empty.
Step 3: Use the encoder's growing algorithm to select a growing point G from the GPP.

Step 4: Input a pointer from the compressed stream, use it to retrieve a dictionary entry d, and place d at the location and position specified by G.
Step 5: Update D and the GPP as in Steps 6–7 of the encoder.

The remainder of this section discusses the various algorithms needed, and shows an example.

Algorithm: Select a growing point G from the GPP. The simplest methods are LIFO (which probably does not yield good results) and FIFO (which is used in the example below), but the coverage methods described here may be better, since they determine how the image will be covered with blocks.

Wave Coverage: This algorithm selects from among all the growing points in the GPP the point $G = (x_s, y_s, 0)$ that satisfies

$$x_s + y_s \leq x + y, \quad \text{for any } G(x, y, 0) \text{ in } GPP$$

(notice that the growing point selected has $q = 0$, so it is located at the upper-left corner of an image block). When this algorithm is used, it makes sense to initialize the GPP with just one point, the top-left corner of the image. If this is done, then the wave coverage algorithm will select image blocks that cover the image in a wave that moves from the upper-left to the lower-right corners of the image (Figure 7.126a).

Circular Coverage: Select the growing point G whose distance from the center of the image is minimal among all the growing points G in the GPP. This covers the image with blocks located in a growing circular region around the center (Figure 7.126b).

Diagonal Coverage: This algorithm selects from among all the growing points in the GPP the point $G = (x_s, y_s, q)$ that satisfies

$$|x_s - y_s| \leq |x - y|, \quad \text{for any } G(x, y, p) \text{ in } GPP$$

(notice that the growing point selected may have any q). The result will be blocks that start around the main diagonal and move away from it in two waves that run parallel to it.

Algorithm: Matching an image block B (with a growing point G as its corner) to a dictionary entry. It seems that the best approach is to start with the smallest block B (just a single pixel) and try to match bigger and bigger blocks to the dictionary entries, until the next increase of B does not find any dictionary entry to match B to the tolerance specified by the user. The following parameters control the matching:

1. The distance measure. Any of the measures proposed in Equation (7.29) can be used.
2. The tolerance. A user-defined real parameter t that's the maximum allowed distance between an image block B and a dictionary entry.
3. The type of coverage. Since image blocks have different sizes and grow from the growing point in different directions, they can overlap (Figure 7.126a). Imagine that an image block B has been matched, but it partly covers some other blocks. We can compute the distance using just those parts of B that don't overlap any other blocks. This means that the first block that covers a certain image area will determine the quality of the decompressed image in that area. This can be called *first coverage*. The opposite is *last coverage*, where the distance between an image block B and a dictionary entry is computed using all of B (except those parts that lie outside the image), regardless of any

overlaps. It is also possible to have *average coverage*, where the distance is computed for all of B, but in a region of overlap, any pixel value used to calculate the distance will be the average of pixel values from all the image blocks in the region.
4. The elementary subblock size l. This is a "normalization" parameter that compensates for the different block sizes. We know that image blocks can have different sizes. It may happen that a large block will match a certain dictionary entry to within the tolerance, while some regions within the block will match poorly (and other regions will yield a very good match). An image block should therefore be divided into subblocks of size $l \times l$ (where the value of l depends on the image size, but is typically 4 or 5), and each dictionary entry should similarly be divided. The matching algorithm should calculate the distance between each image subblock and the corresponding dictionary entry subblock. The maximum of these distances should be selected as the distance between the image block and the dictionary entry.

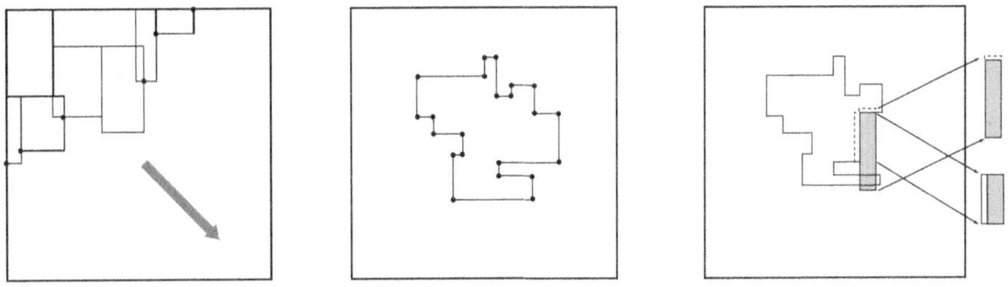

Figure 7.126: (a) Wave Coverage. (b) Growing Points. (c) Dictionary Update.

Algorithm: GPP update. A good policy is to choose as new growing points those points located at or near the borders of the partially encoded image (Figure 7.126b). This causes new image blocks to be adjacent to old ones. This policy makes sense because an old block has contributed to new dictionary entries, and those entries are perfect candidates to match a new, adjacent block, because adjacent blocks generally don't differ much. Initially, when the partially encoded image is small, consisting mainly of single-pixel blocks, this policy adds two growing points to the GPP for each small block added to the encoded image. Those are the points below and to the right of the image block (see example below). Another good location for new growing points is any new concave corners that have been generated by adding the recent image block to the partially encoded image (Figure 7.126a,b). We show below that when a new image block is grown from such a corner point, it contributes two or three new entries to the dictionary.

Algorithm: Dictionary update. This is based on two principles: (1) Each matched block added to the image-so-far should be used to add one or more dictionary entries; (2) the new entries added should contain pixels in the image-encoded-so-far (so the decoder can update its dictionary in the same way). Denoting the currently matched block by B, a simple policy is to add to the dictionary the two blocks obtained by adding one column of pixels to the left of B and one row of pixels above it. Sometimes, as Figure 7.126c shows, the row or column can be added to just part of B. Later experiments with the method added a third block to the dictionary, the block obtained by adding the column to the left of B *and* the row above it. Since image blocks B start small and square, this third dictionary entry is also square. Adding this third block improves compression performance significantly.

Algorithm: Dictionary deletion. The dictionary keeps growing but it can grow only so much. When it reaches its maximum size, we can select from among the following ideas: (1) Delete the least recently used (LRU) entry (except that the original entries, all the possible pixel values, should not be deleted); (2) freeze the dictionary (i.e., don't add any new entries); and (3) delete the entire dictionary (except the original entries) and start afresh.

◇ **Exercise 7.28:** Suggest another approach to dictionary deletion. (Hint: See Section 6.14.)

Experiments performed by the method's developers suggest that the basic algorithm is robust and does not depend much on the particular choice of the algorithms above, except for two choices that seem to improve compression performance significantly:

1. Wave coverage seems to offer better performance than circular or diagonal coverage.
2. Visual inspection of many decompressed images shows that the loss of data is more visible to the eye in image regions that are smooth (i.e., uniform or close to uniform). The method can therefore be improved by using a smaller tolerance parameter for such regions. A simple measure A of smoothness is the ratio V/M, where V is the variance of the pixels in the region (see page 624 for the definition of variance) and M is their mean. The larger A, the more active (i.e., the less smooth) is the region. The tolerance t is determined by the user, and the developers of the method suggest the use of smaller tolerance values for smooth regions as follows:

$$\text{if}\begin{cases} A \le 0.05, & \text{use } 0.4t,\\ 0.05 < A \le 0.1, & \text{use } 0.6t,\\ A > 0.1, & \text{use } t. \end{cases}$$

Example: We select an image that consists of 4-bit pixels. The nine pixels at the top-left corner of this image are shown in Figure 7.127a. They have image coordinates ranging from $(0,0)$ to $(2,2)$. The first 16 entries of the dictionary D (locations 0 through 15) are initialized to the 16 possible values of the pixels. The GPP is initialized to the top-left corner of the image, position $(0,0)$. We assume that the GPP is used as a (FIFO) queue. Here are the details of the first few steps.

Step 1: Point $(0,0)$ is popped out of the GPP. The pixel value at this position is 8. Since the dictionary contains just individual pixels, not blocks, this point can be matched only with the dictionary entry containing 8. This entry happens to be located in dictionary location 8, so the encoder outputs the pointer 8. This match does not have any concave corners, so we push the point on the right of the matched block, $(0,1)$, and the point below it, $(1,0)$, into the GPP.

Step 2: Point $(1,0)$ is popped out of the GPP. The pixel value at this position is 4. The dictionary still contains just individual pixels, not any bigger blocks, so this point can be matched only with the dictionary entry containing 4. This entry is located in dictionary location 4, so the encoder outputs the pointer 4. The match does not have any concave corners, so we push the point on the right of the matched block, (1, 1), and the point below it, (2, 0), into the GPP. The GPP now contains (from most recent to least recent) points (1, 1), (2, 0), and (0, 1). The dictionary is updated by appending to it (at location 16) the block $\boxed{8 \atop 4}$ as shown in Figure 7.127b.

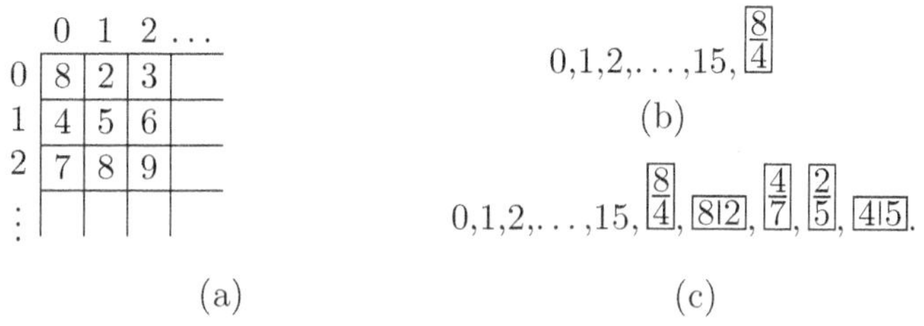

Figure 7.127: An Image and a Dictionary.

Step 3: Point (0, 1) is popped out of the GPP. The pixel value at this position is 2. The dictionary contains the 16 individual pixel values and one bigger block. The best match for this point is with the dictionary entry containing 2, which is at dictionary location 2, so the encoder outputs the pointer 2. The match does not have any concave corners, so we push the point on the right of the matched block, (0, 2), and the point below it, (1, 1), into the GPP. The GPP is now (from most recent to least recent) points (0, 2), (1, 1), and (2, 0). The dictionary is updated by appending to it (at location 17) the block $\boxed{8|2}$ (Figure 7.127c).

◇ **Exercise 7.29:** Do the next two steps.

7.21 Block Truncation Coding

Quantization is an important technique for data compression in general and for image compression in particular. Any quantization method should be based on a principle that determines what data items to quantize and by how much. The principle used by the *block truncation coding* (BTC) method and its variants is to quantize pixels in an image while preserving the first two or three *statistical moments.* The main reference is [Dasarathy 95], which includes an introduction and copies of the main papers in this area. In the basic BTC method, the image is divided into blocks (normally 4×4 or 8×8 pixels each). Assuming that a block contains n pixels with intensities p_1 through p_n, the first two moments are the mean and variance, defined as

$$\bar{p} = \frac{1}{n}\sum_{i=1}^{n} p_i, \tag{7.31}$$

and

$$\overline{p^2} = \frac{1}{n}\sum_{i=1}^{n} p_i^2, \tag{7.32}$$

respectively. The standard deviation of the block is

$$\sigma = \sqrt{\overline{p^2} - \bar{p}}. \tag{7.33}$$

The principle of the quantization is to select three values, a threshold p_{thr}, a high value p^+, and a low value p^-. Each pixel is replaced by either p^+ or p^-, such that the first two moments of the new pixels (i.e., their mean and variance) will be identical to the original moments of the pixel block. The rule of quantization is that a pixel p_i is quantized to p^+ if it is greater than the threshold, and is quantized to p^- if it is less than the threshold (if p_i equals the threshold, it can be quantized to either value). Thus,

$$p_i \leftarrow \begin{cases} p^+, & \text{if } p_i \geq p_{\text{thr}}, \\ p^-, & \text{if } p_i < p_{\text{thr}}. \end{cases}$$

Intuitively, it is clear that the mean $\bar{p}$ is a good choice for the threshold. The high and low values can be determined by writing equations that preserve the first two moments, and solving them. We denote by n^+ the number of pixels in the current block that are greater than or equal to the threshold. Similarly, n^- stands for the number of pixels that are smaller than the threshold. The sum $n^+ + n^-$ equals, of course, the number of pixels n in the block. Once the mean $\bar{p}$ has been computed, both n^+ and n^- are easy to calculate. Preserving the first two moments is expressed by the two equations

$$n\bar{p} = n^- p^- - n^+ p^+, \quad n\overline{p^2} = n^-(p^-)^2 - n^+(p^+)^2. \tag{7.34}$$

These are easy to solve even though they are nonlinear, and the solutions are

$$p^- = \bar{p} - \sigma\sqrt{\frac{n^+}{n^-}}, \quad p^+ = \bar{p} + \sigma\sqrt{\frac{n^-}{n^+}}. \tag{7.35}$$

These solutions are generally real numbers, but they have to be rounded to the nearest integer, which implies that the mean and variance of the quantized block may be somewhat different from those of the original block. Notice that the solutions are located on the two sides of the mean $\bar{p}$ at distances that are proportional to the standard deviation σ of the pixel block.

Example: We select the 4×4 block of 8-bit pixels

$$\begin{pmatrix} 121 & 114 & 56 & 47 \\ 37 & 200 & 247 & 255 \\ 16 & 0 & 12 & 169 \\ 43 & 5 & 7 & 251 \end{pmatrix}.$$

The mean is $\bar{p} = 98.75$, so we count $n^+ = 7$ pixels greater than the mean and $n^- = 16 - 7 = 9$ pixels less than the mean. The standard deviation is $\sigma = 92.95$, so the high and low values are

$$p^+ = 98.75 + 92.95\sqrt{\frac{9}{7}} = 204.14, \quad p^- = 98.75 - 92.95\sqrt{\frac{7}{9}} = 16.78.$$

They are rounded to 204 and 17, respectively. The resulting block is

$$\begin{pmatrix} 204 & 204 & 17 & 17 \\ 17 & 204 & 204 & 204 \\ 17 & 17 & 17 & 204 \\ 17 & 17 & 17 & 204 \end{pmatrix},$$

and it is clear that the original 4×4 block can be compressed to the 16 bits

$$\begin{pmatrix} 1 & 1 & 0 & 0 \\ 0 & 1 & 1 & 1 \\ 0 & 0 & 0 & 1 \\ 0 & 0 & 0 & 1 \end{pmatrix},$$

plus the two 8-bit values 204 and 17; a total of $16 + 2\times8$ bits, compared to the 16×8 bits of the original block. The compression factor is

$$\frac{16\times8}{16 + 2\times8} = 4.$$

◇ **Exercise 7.30:** Do the same for the 4×4 block of 8-bit pixels

$$\begin{pmatrix} 136 & 27 & 144 & 216 \\ 172 & 83 & 43 & 219 \\ 200 & 254 & 1 & 128 \\ 64 & 32 & 96 & 25 \end{pmatrix}.$$

It is clear that the compression factor of this method is known in advance, because it depends on the block size n and on the number of bits b per pixel. In general, the compression factor is

$$\frac{bn}{n + 2b}.$$

Figure 7.128 shows the compression factors for b values of 2, 4, 8, 12, and 16 bits per pixel, and for block sizes n between 2 and 16.

The basic BTC method is simple and fast. Its main drawback is the way it loses image information, which is based on pixel intensities in each block, and not on any properties of the human visual system. Because of this, images compressed under BTC tend to have a blocky character when decompressed and reconstructed. This led many researchers to develop enhanced and extended versions of the basic BTC, some of which are briefly mentioned here.

1. Encode the results before writing them on the compressed stream. In basic BTC, each original block of n pixels becomes a block of n bits plus two numbers. It is possible to

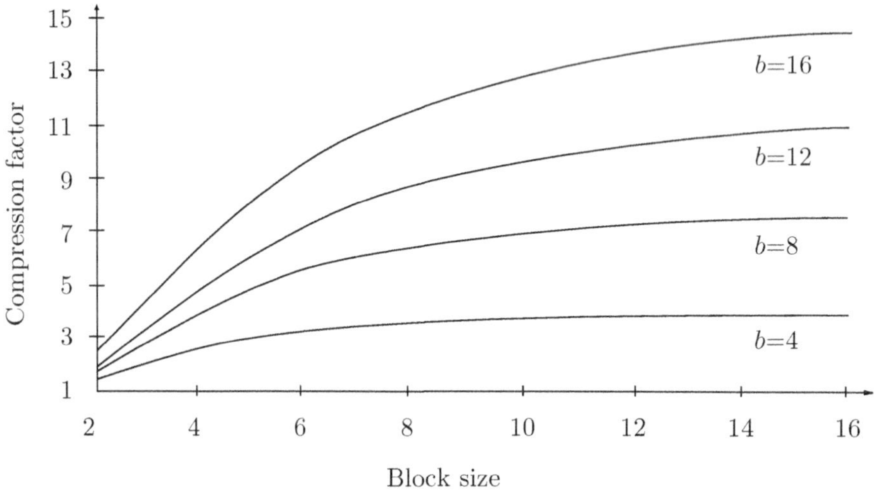

Figure 7.128: BTC Compression Factors for Two Preserved Moments.

accumulate B blocks of n bits each, and encode the Bn bits using run-length encoding (RLE). This should be followed by $2B$ values, and those can be encoded with prefix codes.

2. Sort the n pixels p_i of a block by increasing intensity values, and denote the sorted pixels by s_i. The first n^- pixels s_i have values that are less than the threshold, and the remaining n^+ pixels s_i have larger values. The high and low values, p^+ and p^-, are now the values that minimize the square-error expression

$$\sum_{i=1}^{n^- -1} \left(s_i - p^-\right)^2 + \sum_{i=n^-}^{n^+} \left(s_i - p^+\right)^2. \tag{7.36}$$

These values are

$$p^- = \frac{1}{n^-} \sum_{i=1}^{n^- -1} s_i \text{ and } p^+ = \frac{1}{n^+} \sum_{i=n^-}^{n^+} s_i.$$

The threshold in this case is determined by searching all $n-1$ possible threshold values and choosing the one that produces the lowest value for the square-error expression of Equation (7.36).

3. Sort the pixels as in 2 above, and change Equation (7.36) to a similar one using absolute values instead of squares:

$$\sum_{i=1}^{n^- -1} |s_i - p^-| + \sum_{i=n^-}^{n^+} |s_i - p^+|. \tag{7.37}$$

The high and low values p^+ and p^- should be, in this case, the medians of the low and high intensity sets of pixels, respectively. Thus, p^- should be the median of the set

$\{s_i|i = 1, 2, \ldots, n^- - 1\}$, and p^+ should be the median of the set $\{s_i|i = n^-, \ldots, n^+\}$. The threshold value is determined as in 2 above.

4. This is an extension of the basic BTC method, where *three* moments are preserved, instead of two. The three equations that result make it possible to calculate the value of the threshold, in addition to those of the high (p^+) and low (p^-) parameters. The third moment is

$$\overline{p^3} = \frac{1}{n}\sum_{i=1}^{n} p_i^3, \tag{7.38}$$

and its preservation yields the equation

$$n\overline{p^3} = n^- \left(p^-\right)^3 - n^+ \left(p^+\right)^3. \tag{7.39}$$

Solving the three equations (7.34) and (7.39) with p^-, p^+, and n^+ as the unknowns yields

$$n^+ = \frac{n}{2}\left(1 + \frac{\alpha}{\sqrt{\alpha^2+4}}\right),$$

where

$$\alpha = \frac{3\bar{p}(\overline{p^2}) - \overline{p^3} - 2(\bar{p})^3}{\sigma^3}.$$

And the threshold p_{thr} is selected as pixel p_{n^+}.

5. BTC is a lossy compression method where the data being lost is based on the mean and variance of the pixels in each block. This loss has nothing to do with the human visual system [Salomon 99], or with any general features of the image being compressed, so it may cause artifacts and unrecognizable regions in the decompressed image. One especially annoying feature of the basic BTC is that straight edges in the original image become jagged in the reconstructed image.

The *edge following* algorithm [Ronson and Dewitte 82] is an extension of the basic BTC, where this tendency is minimized by identifying pixels that are on an edge, and using an additional quantization level for blocks containing such pixels. Each pixel in such a block is quantized into one of *three* different values instead of the usual two. This reduces the compression factor but improves the visual appearance of the reconstructed image.

Another extension of the basic BTC is a three-stage process, proposed in [Dewitte and Ronson 83] where the first stage is basic BTC, resulting in a binary matrix and two numbers. The second stage classifies the binary matrix into one of three categories as follows:

a. The block has no edges (indicating a uniform or near-uniform region).
b. There is a single edge across the block.
c. There is a pair of edges across the block.

The classification is done by counting the number of transitions from 0 to 1 and from 1 to 0 along the four edges of the binary matrix. Blocks with zero transitions or with more than four transitions are considered category a (no edges). Blocks with two transitions are considered category b (a single edge), and blocks with four transitions are classified as category c (two edges).

Stage three matches the classified block of pixels to a particular model depending on its classification. Experiments with this method suggest that about 80% of the blocks belong in category a, 12% are category b, and 8% belong in category c. The fact that category a blocks are so common implies that the basic BTC can be used in most cases, and the particular models used for categories b and c improve the appearance of the reconstructed image without significantly degrading the compression time and compression factor.

6. The original BTC is based on the principle of preserving the first two statistical moments of a pixel block. The variant described here changes this principle to preserving the first two *absolute moments*. The first absolute moment is given by

$$\overline{p_a} = \frac{1}{n}\sum_{i=1}^{n}|p_i - \bar{p}|,$$

and it can be shown that this implies

$$p^+ = \bar{p} + \frac{n\overline{p_a}}{2n^+} \quad \text{and} \quad p^- = \bar{p} - \frac{n\overline{p_a}}{2n^-},$$

which shows that the high and low values are simpler than those used by the basic BTC. They are obtained from the mean by adding and subtracting quantities that are proportional to the absolute first moment $\overline{p_a}$ and inversely proportional to the numbers n^+ and n^- of pixels on both sides of this mean. Moreover, these high and low values can also be written as

$$p^+ = \frac{1}{n^+}\sum_{p_i \ge \bar{p}} p_i \quad \text{and} \quad p^- = \frac{1}{n^-}\sum_{p_i < \bar{p}} p_i,$$

simplifying their computation even more.

In addition to these improvements and extensions there are BTC variants that deviate significantly from the basic idea. The *three-level BTC* is a good example of the latter. It quantizes each pixel p_i into one of three values, instead of the original two. This, of course, requires two bits per pixel, thereby reducing the compression factor. The method starts by scanning all the pixels in the block and finding the range Δ of intensities

$$\Delta = \max(p_i) - \min(p_i), \text{ for } i = 1, 2, \ldots, n.$$

Two mean values, high and low, are defined by

$$\overline{p_h} = \max(p_i) - \Delta/3 \quad \text{and} \quad \overline{p_l} = \min(p_i) + \Delta/3.$$

The three quantization levels, high, low, and mid, are now given by

$$p^+ = \tfrac{1}{2}\left[\overline{p_h} + \max(p_i)\right], \quad p^- = \tfrac{1}{2}\left[\overline{p_l} + \min(p_i)\right], \quad p^{\rm m} = \tfrac{1}{2}\left[\overline{p_l} + \overline{p_h}\right].$$

A pixel p_i is compared to p^+, p^-, and $p^{\rm m}$ and is quantized to the nearest of these values. It takes two bits to express the result of quantizing p_i, so the original block of n b-bit pixels becomes a block of $2n$ bits. In addition, the values of $\max(p_i)$ and $\min(p_i)$ have to be written on the compressed stream as two b-bit numbers. The resulting compression factor is

$$\frac{bn}{2n+2b},$$

lower than in basic BTC. Figure 7.129 shows the compression factors for $b = 4, 8, 12$, and 16 and for n values in the range $[2, 16]$.

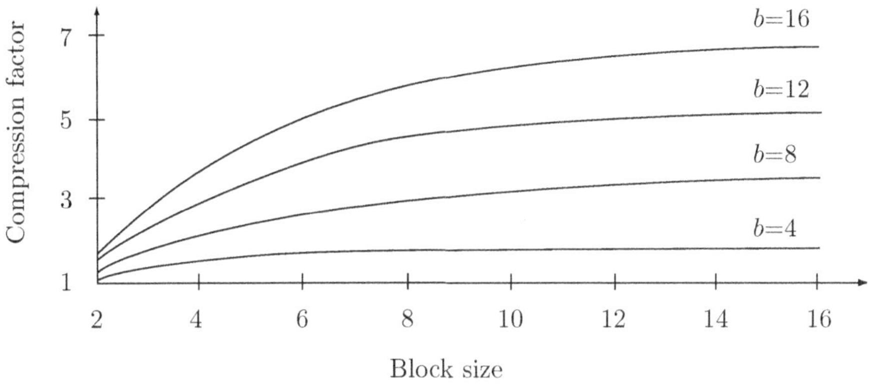

Figure 7.129: BTC Compression Factors for Three Quantization Levels.

In spite of worse compression, the three-level BTC has an overall better performance than the basic BTC because it requires simpler calculations (in particular, the second moment is not needed) and because its reconstructed images are of higher quality and don't have the blocky appearance typical of images compressed and decompressed under basic BTC.

7.22 Context-Based Methods

Most image-compression methods are based on the observation that for any randomly-selected pixel in the image, its near neighbors tend to have the same value as the pixel (or similar values). A context-based image compression method generalizes this observation. It is based on the idea that the context of a pixel can be used to predict (estimate the probability of) the pixel.

Such a method compresses an image by scanning it pixel by pixel, examining the context of every pixel, and assigning it a probability depending on how many times the same context was seen in the past. The pixel and its assigned probability are then sent to an arithmetic encoder that does the actual encoding. The methods described here are due to Langdon [Langdon and Rissanen 81] and Moffat [Moffat 91], and apply to monochromatic (bi-level) images. The context of a pixel consists of some of its neighbor pixels that have already been processed. The diagram below shows a possible seven-pixel context (the pixels marked P) made up of five pixels above and two on the left of the current pixel X. The pixels marked "?" haven't been input yet, so they cannot be included in the context.

·	·	P	P	P	P	P	·	·
·	·	P	P	X	?	?	?	?

The main idea is to use the values of the seven context pixels as a 7-bit index to a frequency table, and find the number of times a 0 pixel and a 1 pixel were seen in the past with the same context. Here is an example:

·	·	1	0	0	1	1	·	·
·	·	0	1	X	?	?	?	?

Since $1001101_2 = 77$, location 77 of the frequency table is examined. We assume that it contains a count of 15 for a 0 pixel and a count of 11 for a 1 pixel. The current pixel is therefore assigned probability $15/(15+11) \approx 0.58$ if it is 0 and $11/26 \approx 0.42$ if it is 1.

	77	
...	15	...
...	11	...

One of the counts at location 77 is then incremented, depending on what the current pixel is. Figure 7.130 shows ten possible ways to select contexts. They range from a 1-bit, to a 22-bit context. In the latter case, there may be $2^{22} \approx 4.2$ million contexts, most of which will rarely, if ever, occur in the image. Instead of maintaining a huge, mostly empty array, the frequency table can be implemented as a binary search tree or a hash table. For short, 7-bit, contexts, the frequency table can be an array, resulting in fast search.

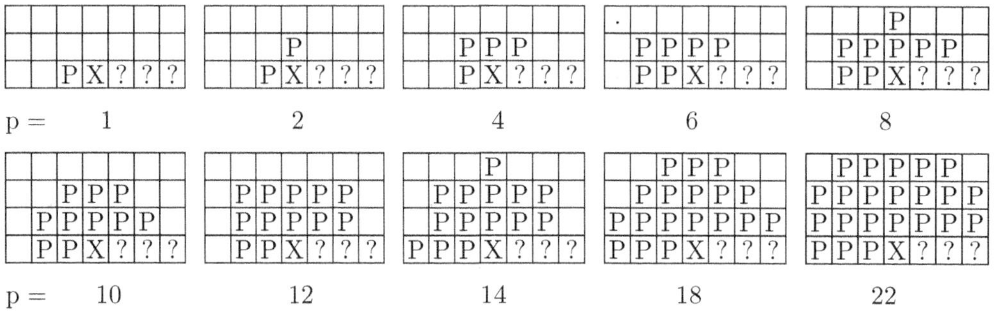

Figure 7.130: Context Pixel Patterns.

> "Bless me, bless me, Sept," returned the old lady, "you don't see the context! Give it back to me, my dear."
>
> —Charles Dickens, *The Mystery of Edwin Drood* (1870)

Experience shows that longer contexts result in better compression, up to about 12–14 bits. Contexts longer than that result in worse compression, indicating that a pixel in a typical image does not "relate" to distant pixels (correlations between pixels are typically limited to a distance of 1–2).

As usual for a context-based compression method, care should be taken to avoid zero probabilities. The frequency table should thus be initialized to nonzero values, and experience shows that the precise way of doing this does not affect the compression performance significantly. It seems best to initialize every table entry to 1.

When the process starts, the first pixels to be scanned don't have any neighbors above them or to the left. A simple way to handle this is to assume that any nonexistent neighbors are zeros. It is as if the image was extended by adding as many rows of zero pixels as necessary on top and as many zero columns on the left.

The 2-bit context of Figure 7.130 is now used, as an example, to encode the first row of the 4×4 image

$$\begin{array}{c|c} & 0\,0\,0\,0 \\ \hline 0 & 1\,0\,1\,1 \\ 0 & 0\,1\,0\,1 \\ 0 & 1\,0\,0\,1 \\ 0 & 0\,0\,1\,0 \end{array}$$

The results are summarized in Table 7.131.

Number	Pixel	Context	Counts	Probability	New counts
1	1	00=0	1, 1	$1/(1+1) = 1/2$	1, 2
2	0	01=1	1, 1	1/2	2, 1
3	1	00=0	1, 2	2/3	1, 3
4	1	01=1	2, 1	1/3	2, 2

Table 7.131: Counts and Probabilities for First Four Pixels.

◇ **Exercise 7.31:** Continue Table 7.131 for the next row of four pixels 0101.

The contexts of Figure 7.130 are not symmetric about the current pixel, since they must use pixels that have already been processed ("past" pixels). If the algorithm scans the image by rows, those will be the pixels above the current pixel or to its left. In practical work it is impossible to include "future" pixels in the context (because the decoder wouldn't have their values when decoding the current pixel), but for experiments, where there is no need to actually decode the compressed image, it is possible to store the entire image in memory so that the encoder can examine any pixel at any time. Experiments performed with symmetric contexts have shown that compression performance can improve by as much as 30%. (The MLP method, Section 7.25, provides an interesting twist to the question of a symmetric context.)

◇ **Exercise 7.32:** (Easy.) Why is it possible to use "future" pixels in an experiment but not in practice? It would seem that the image, or part of it, could be stored in memory and the encoder could use any pixel as part of a context?

One disadvantage of a large context is that it takes the algorithm longer to "learn" it. A 20-bit context, for example, allows for about a million different contexts. It takes many millions of pixels to collect enough counts for all those contexts, which is one reason large contexts do not result in better compression. One way to improve our method is to implement a *two-level* algorithm that uses a long context only if that context has already been seen Q times or more (where Q is a parameter, typically set to a small value such as 2 or 3). If a context has been seen fewer than Q times, it is deemed unreliable, and only a small subset of it is used to predict the current pixel. Figure 7.132 shows four such contexts, where the pixels of the subset are labeled S. The notation p, q means a two-level context of p bits with a subset of q bits.

$p, q =$ 12,6

	P	P	P	P	P	
	S	S	S	S	P	
	S	S	X	?	?	?

14,8

			P			
	P	P	S	P	P	
	S	S	S	S	S	
P	S	S	X	?	?	?

18,10

		P	P	P		
	P	S	S	S	P	
P	S	S	S	S	S	P
P	S	S	X	?	?	?

22,10

	P	P	P	P	P	
P	P	S	S	S	P	P
P	S	S	S	S	S	P
P	S	S	X	?	?	?

Figure 7.132: Two-Level Context Pixel Patterns.

Experience shows that the 18, 10 and 22, 10 contexts result in better, although not revolutionary, compression.

7.23 FELICS

FELICS is an acronym for Fast, Efficient, Lossless Image Compression System. It is a special-purpose compression method [Howard and Vitter 93] designed for grayscale images and it competes with the lossless mode of JPEG (Section 7.10.5). It is fast and it generally produces good compression. However, it cannot compress an image to below one bit per pixel, so it is not a good choice for bi-level or for highly redundant images.

The principle of FELICS is to code each pixel with a variable-length code based on the values of two of its previously seen neighbor pixels. Figure 7.133a shows the two known neighbors A and B of some pixels P. For a general pixel, these are the neighbors above it and to its left. For a pixel in the top row, these are its two left neighbors (except for the first two pixels of the image). For a pixel in the leftmost column, these are the first two pixels of the line above it. Notice that the first two pixels of the image don't have any previously seen neighbors, but since there are only two of them, they can be output without any encoding, causing just a slight degradation in the overall compression.

◇ **Exercise 7.33:** What model is used by FELICS to predict the current pixel?

Consider the two neighbors A and B of a pixel P. We use A, B, and P to denote both the three pixels and their intensities (grayscale values). We denote by L and H the neighbors with the smaller and the larger intensities, respectively. Pixel P should be assigned a variable-length code depending on where the intensity P is located relative to L and H. There are three cases:

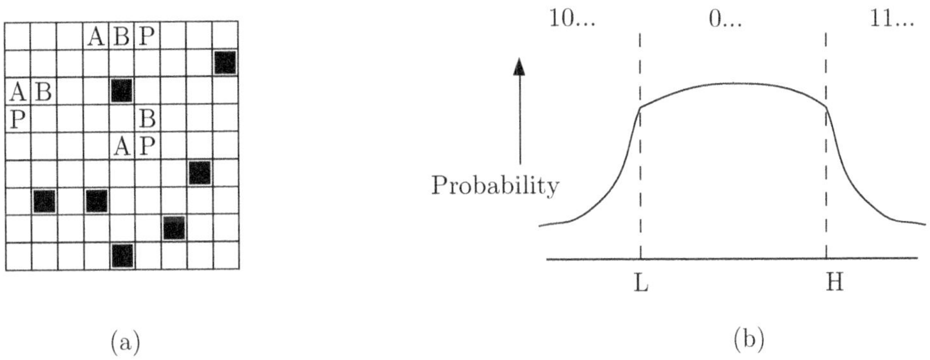

Figure 7.133: (a) The Two Neighbors. (b) The Three Regions.

1. The intensity of pixel P is between L and H (it is located in the central region of Figure 7.133b). This case is known experimentally to occur in about half the pixels, and P is assigned, in this case, a code that starts with 0. (A special case occurs when L = H. In such a case, the range [L, H] consists of one value only, and the chance that P will have that value is small.) The probability that P will be in this central region is almost, but not completely, flat, so P should be assigned a binary code that has about the same size in the entire region but is slightly shorter at the center of the region.
2. The intensity of P is lower than L (P is in the left region). The code assigned to P in this case starts with 10.
3. P's intensity is greater than H. P is assigned a code that starts with 11.

When pixel P is in one of the outer regions, the probability that its intensity will differ from L or H by much is small, so P can be assigned a long code in these cases.

The code assigned to P should therefore depend heavily on whether P is in the central region or in one of the outer regions. Here is how the code is assigned when P is in the central region. We need H − L + 1 variable-length codes that will not differ much in size and will, of course, satisfy the prefix property. We set $k = \lfloor \log_2(H - L + 1) \rfloor$ and compute integers a and b by

$$a = 2^{k+1} - (H - L + 1), \quad b = 2(H - L + 1 - 2^k).$$

Example: If H − L = 9, then $k = 3$, $a = 2^{3+1} - (9 + 1) = 6$, and $b = 2(9 + 1 - 2^3) = 4$. We now select the a codes $2^k - 1$, $2^k - 2, \ldots$ expressed as k-bit numbers, and the b codes $0, 1, 2, \ldots$ expressed as $(k + 1)$-bit numbers. In the example above, the a codes are $8 - 1 = 111$, $8 - 2 = 110$, through $8 - 6 = 010$, and the b codes, 0000, 0001, 0010, and 0011. The a short codes are assigned to values of P in the middle of the central region, and the b long codes are assigned to values of P closer to L or H. Notice that b is even, so the b codes can always be partitioned into two equal sets. Table 7.134 shows how ten such codes can be assigned in the case L = 15, H = 24.

When P is in one of the outer regions, say the upper one, the value P − H should be assigned a variable-length code whose size can grow quickly as P − H gets bigger. One way to do this is to select a small nonnegative integer m (typically 0, 1, 2, or 3) and to assign the integer n a two-part code. The second part is the lower m bits of n, and

Pixel P	Region code	Pixel code
L=15	0	0000
16	0	0010
17	0	010
18	0	011
19	0	100
20	0	101
21	0	110
22	0	111
23	0	0001
H=24	0	0011

Table 7.134: The Codes for the Central Region.

the first part is the unary code of [n without its lower m bits] (see Section 3.1 for this code). Example: If $m = 2$, then $n = 1101_2$ is assigned the code 110|01, since 110 is the unary code of 11. This code is a special case of the Golomb code (Section 3.24), where the parameter b is a power of 2 (2^m). Table 7.135 shows some examples of this code for $m = 0, 1, 2, 3$ and $n = 1, 2, \ldots, 9$. The value of m used in any particular compression job can be selected, as a parameter, by the user.

Pixel		Region	$m =$			
P	P–H	code	0	1	2	3
H+1=25	1	11	0	00	000	0000
26	2	11	10	01	001	0001
27	3	11	110	100	010	0010
28	4	11	1110	101	011	0011
29	5	11	11110	1100	1000	0100
30	6	11	111110	1101	1001	0101
31	7	11	1111110	11100	1010	0110
32	8	11	11111110	11101	1011	0111
33	9	11	111111110	111100	11000	10000
...	...	...	...		...	
...	...	...	...		...	

Table 7.135: The Codes for an Outer Region.

◇ **Exercise 7.34:** Given the 4×4 bitmap of Figure 7.136, calculate the FELICS codes for the three pixels with values 8, 7, and 0.

2	5	7	12
3	0	11	10
2	1	8	15
4	13	11	9

Figure 7.136: A 4×4 Bitmap.

7.24 Progressive FELICS

The original FELICS method can easily be extended to progressive compression of images because of its main principle. FELICS scans the image in raster order (row by row) and encodes a pixel based on the values of two of its (previously seen and encoded) neighbors. Progressive FELICS [Howard and Vitter 94a] works similarly, but it scans the pixels in levels. Each level uses the k pixels encoded in all previous levels to encode k more pixels, so the number of encoded pixels doubles after each level. Assuming that the image consists of $n\times n$ pixels, and the first level starts with just four pixels, consecutive levels result in

$$4, 8, \ldots, \frac{n^2}{8}, \frac{n^2}{4}, \frac{n^2}{2}, n^2$$

pixels. Thus, the number of levels is the number of terms, $2\log_2 n - 1$, in this sequence. (This property is easy to prove. The first term can be written

$$4 = \frac{n^2}{2^{-2}n^2} = \frac{n^2}{2^{2\log n-2}}.$$

Terms in the sequence therefore contain powers of 2 that go from 0 to $2\log_2 n - 2$, showing that there are $2\log_2 n - 1$ terms.)

Figure 7.137 shows the pixels encoded in most of the levels of a 16×16-pixel image. Figure 7.138 shows how the pixels of each level are selected. In Figure 7.138a there are $8\times 8 = 64$ pixels, one quarter of the final number, arranged in a square grid. Each group of four pixels is used to encode a new pixel, so Figure 7.138b has 128 pixels, half the final number. The image of Figure 7.138b is then rotated 45° and scaled by factors of $\sqrt{2}\approx 1.414$ in both directions, to produce Figure 7.138c, which is a square grid that looks exactly like Figure 7.138a. The next step (not shown in the figure) is to use every group of 4×4 pixels in Figure 7.138c to encode a pixel, thereby encoding the remaining 128 pixels. In practice, there is no need to actually rotate and scale the image; the program simply alternates between xy and diagonal coordinates.

Each group of four pixels is used to encode the pixel at its center. Notice that in early levels the four pixels of a group are far from each other and are therefore not correlated, thereby resulting in poor compression. However, the last two levels encode 3/4 of the total number of pixels, and these levels contain compact groups. Two of the four pixels of a group are selected to encode the center pixel, and are designated L and H. Experience shows that the best choice for L and H is the two median pixels (page 554), the ones with the middle values (i.e., not the maximum or the minimum pixels of the group). Ties can be resolved in any way, but it should be consistent. If the two medians in a group are the same, then the median and the minimum (or the median and the maximum) pixels can be selected. The two selected pixels, L and H, are

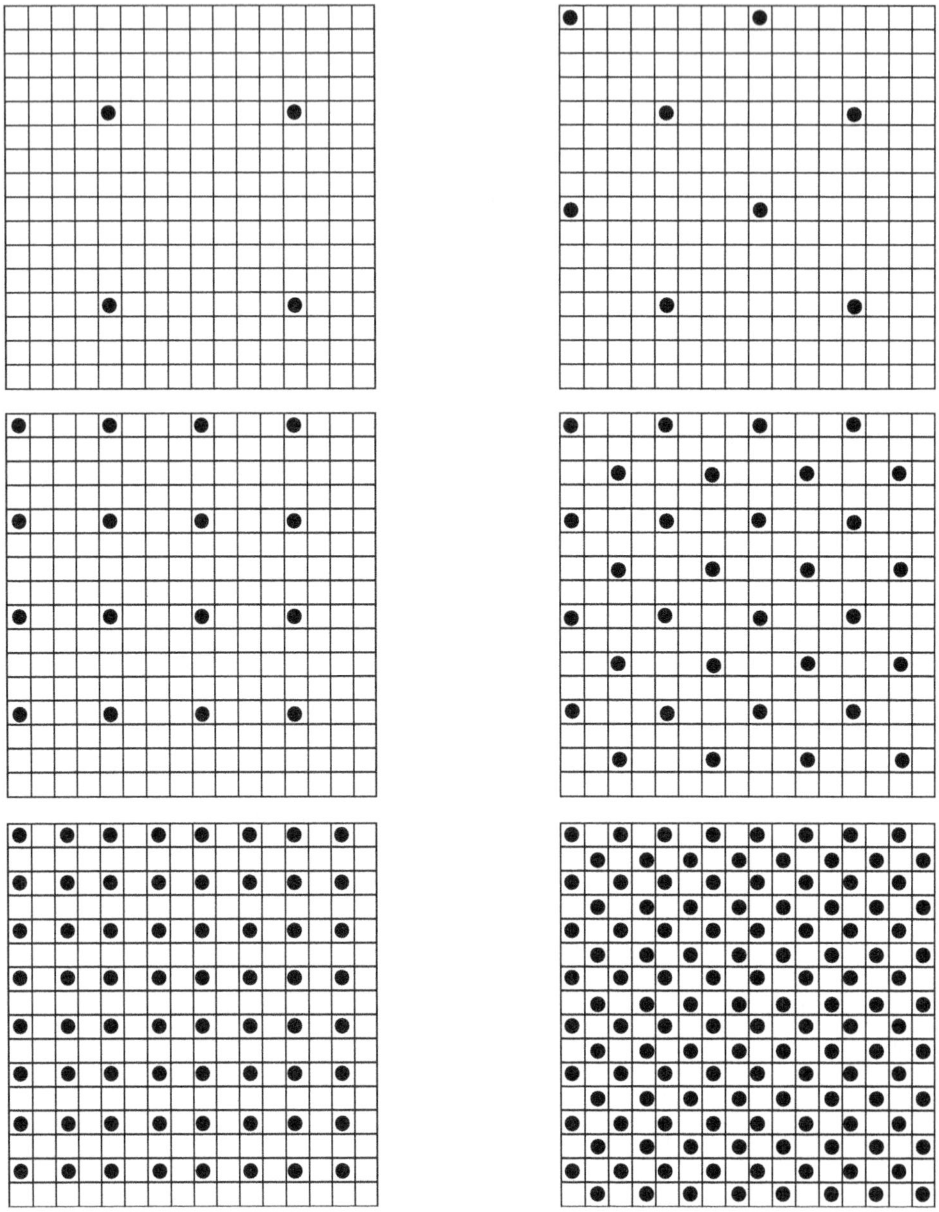

Figure 7.137: Some Levels of a 16×16 Image.

used to encode the center pixel in the same way FELICS uses two neighbors to encode a pixel. The only difference is that a new prefix code (Section 7.24.1) is used, instead of the Golomb code.

◇ **Exercise 7.35:** Why is it important to resolve ties in a consistent way?

7.24.1 Subexponential Code

In early levels, the four pixels used to encode a pixel are far from each other. As more levels are progressively encoded the groups get more compact, so their pixels get closer. The encoder uses the absolute difference between the L and H pixels in a group (the *context* of the group) to encode the pixel at the center of the group, but a given absolute difference means more for late levels than for early ones, because the groups of late levels are smaller, so their pixels are more correlated. The encoder should therefore scale the difference by a weight parameter s that gets heavier from level to level. The specific value of s is not critical, and experiments recommend the value 12.

The prefix code used by progressive FELICS is called *subexponential*. They are related to the Rice codes of Section 10.9. Like the Golomb code (Section 3.24), this new code depends on a parameter $k \geq 0$. The main feature of the subexponential code is its length. For integers $n < 2^{k+1}$ the code length increases linearly with n, but for larger n, it increases logarithmically. The subexponential code of the nonnegative integer n is computed in two steps. In the first step, values b and u are calculated by

$$b = \begin{cases} k & \text{if } n < 2^k, \\ \lfloor \log_2 n \rfloor & \text{if } n \geq 2^k, \end{cases} \quad \text{and} \quad u = \begin{cases} 0 & \text{if } n < 2^k, \\ b - k + 1 & \text{if } n \geq 2^k. \end{cases}$$

In the second step, the unary code of u (in $u+1$ bits), followed by the b least-significant bits of n, becomes the subexponential code of n. Thus, the total size of the code is

$$u + 1 + b = \begin{cases} k + 1 & \text{if } n < 2^k, \\ 2\lfloor \log_2 n \rfloor - k + 2 & \text{if } n \geq 2^k. \end{cases}$$

Table 7.139 shows examples of the subexponential code for various values of n and k. It can be shown that for a given n, the code lengths for consecutive values of k differ by at most 1.

If the value of the pixel to be encoded lies between those of L and H, the pixel is encoded as in FELICS. If it lies outside the range $[L, H]$, the pixel is encoded by using the subexponential code where the value of k is selected by the following rule: Suppose that the current pixel P to be encoded has context C. The encoder maintains a cumulative total, for some reasonable values of k, of the code length the encoder would have if it had used that value of k to encode all pixels encountered so far in context C. The encoder then uses the k value with the smallest cumulative code length to encode pixel P.

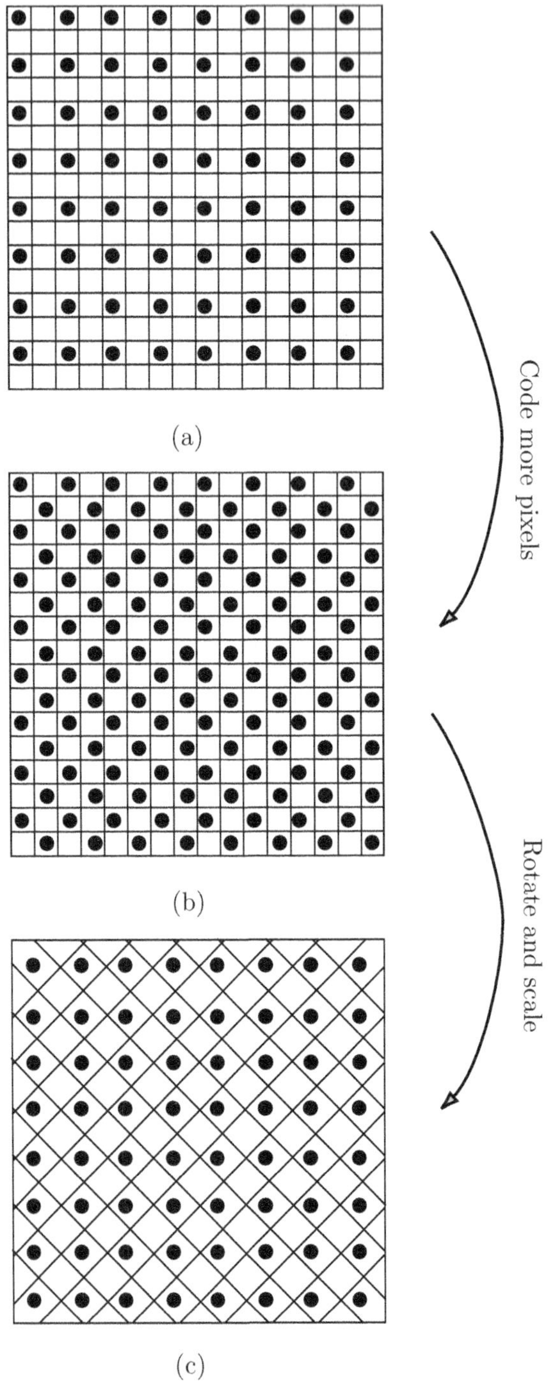

Figure 7.138: Rotation and Scaling.

n	$k=0$	$k=1$	$k=2$	$k=3$	$k=4$	$k=5$
0	0|	0|0	0|00	0|000	0|0000	0|00000
1	10|	0|1	0|01	0|001	0|0001	0|00001
2	110|0	10|0	0|10	0|010	0|0010	0|00010
3	110|1	10|1	0|11	0|011	0|0011	0|00011
4	1110|00	110|00	10|00	0|100	0|0100	0|00100
5	1110|01	110|01	10|01	0|101	0|0101	0|00101
6	1110|10	110|10	10|10	0|110	0|0110	0|00110
7	1110|11	110|11	10|11	0|111	0|0111	0|00111
8	11110|000	1110|000	110|000	10|000	0|1000	0|01000
9	11110|001	1110|001	110|001	10|001	0|1001	0|01001
10	11110|010	1110|010	110|010	10|010	0|1010	0|01010
11	11110|011	1110|011	110|011	10|011	0|1011	0|01011
12	11110|100	1110|100	110|100	10|100	0|1100	0|01100
13	11110|101	1110|101	110|101	10|101	0|1101	0|01101
14	11110|110	1110|110	110|110	10|110	0|1110	0|01110
15	11110|111	1110|111	110|111	10|111	0|1111	0|01111
16	111110|0000	11110|0000	1110|0000	110|0000	10|0000	0|10000

Table 7.139: Some Subexponential Codes.

7.25 MLP

Note. The MLP image compression method described in this section is different from and unrelated to the MLP (meridian lossless packing) audio compression method of Section 10.7. The identical acronyms are an unfortunate coincidence.

Text compression methods can use context to predict (i.e., to estimate the probability of) the next character of text. Context can also be used to predict the intensity of the next pixel in image compression, but this is more complex for two reasons: (1) An image is two-dimensional, allowing for many possible contexts, and (2) a digital image is often the result of digitizing an analog image. The intensity of any individual pixel is therefore determined by the details of scanning and may differ from the "ideal" intensity.

The multilevel progressive method (MLP) described here [Howard and Vitter 92a], is a computationally intensive, lossless method for compressing grayscale images. It uses context to predict the intensities of pixels, then uses arithmetic coding to encode the difference between the prediction and the actual value of a pixel (the error). The Laplace distribution is used to estimate the probability of the error. The method combines four separate steps: (1) pixel sequencing, (2) prediction (image modeling), (3) error modeling (by means of the Laplace distribution), and (4) arithmetically encoding the errors.

MLP is also progressive, encoding the image in levels, where the pixels of each level are selected as in progressive FELICS. When the image is decoded, each level adds details to the entire image, not just to certain parts, so a user can view the image as it is being decoded and decide in real time whether to accept or reject it. This feature is useful when an image has to be selected from a large archive of compressed images. The user can browse through images very fast, without having to wait for any image to be completely decoded and displayed. Another advantage of progressive compression is that it provides a natural lossy option. The encoder may be told to stop encoding before it reaches the last level (thereby encoding only half the total number of pixels) or before it reaches the next to last level (encoding only a quarter of the total number of pixels). Such an option results in excellent compression but a loss of image data. The decoder may be told to use interpolation to determine the intensities of any missing pixels.

Like any compression method for grayscale images, MLP can be used to compress color images. The original color image should be separated into three color components, and each component compressed individually as a grayscale image. Following is a detailed description of the individual MLP encoding steps.

> What we know is not much. What we do not know is immense.
> —(Allegedly Laplace's last words.)

7.25.1 Pixel Sequencing

Pixels are selected in levels, as in progressive FELICS, where each level encodes the same number of pixels as all the preceding levels combined, thereby doubling the number of encoded pixels. This means that the last level encodes half the number of pixels, the level preceding it encodes a quarter of the total number, and so on. The first level should start with at least four pixels, but may also start with 8, 16, or any desired power of 2.

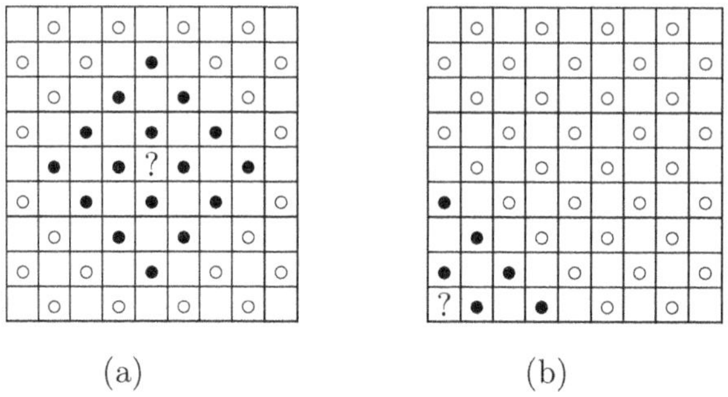

Figure 7.140: (a) Sixteen Neighbors. (b) Six Neighbors.

> Man follows only phantoms.
> —(Allegedly Laplace's last words.)

7.25.2 Prediction

A pixel is predicted by calculating a weighted average of 16 of its known neighbors. Keep in mind that pixels are not raster scanned but are encoded (and therefore also decoded) in levels. When decoding the pixels of level L, the MLP decoder has already decoded all the pixels of all the preceding levels, and it can use their values (gray scales) to predict values of pixels of L. Figure 7.140a shows the situation when the MLP encoder processes the last level. Half the pixels have already been encoded in previous levels, so they will be known to the decoder when the last level is decoded. The encoder can therefore use a diamond-shaped group of 4×4 pixels (shown in black) from previous levels to predict the pixel at the center of the group. This group becomes the *context* of the pixel. Compression methods that scan the image in raster order can use only pixels above and to the left of pixel P to predict P. Because of the progressive nature of MLP, it can use a symmetric context, which produces more accurate predictions. On the other hand, the pixels of the context are not near neighbors and may even be (in early levels) far from the predicted pixel.

Table 7.141 shows the 16 weights used for a group. They are calculated by bicubic polynomial interpolation (Section 7.25.4) and are normalized such that they add up to 1. (Notice that in Table 7.141a the weights are not normalized; they add up to 256. When these integer weights are used, the weighted sum should be divided by 256.) To predict a pixel near an edge, where some of the 16 neighbors are missing (as in Figure 7.140b), only those neighbors that exist are used, and their weights are renormalized, to bring their sum to 1.

			1			
		−9		−9		
	−9		81		−9	
1		81		81		1
	−9		81		−9	
		−9		−9		
			1			

			0.0039			
		−0.0351		−0.0351		
	−0.0351		0.3164		−0.0351	
0.0039		0.3164		0.3164		0.0039
	−0.0351		0.3164		−0.0351	
		−0.0351		−0.0351		
			0.0039			

Table 7.141: 16 Weights. (a) Integers. (b) Normalized.

◇ **Exercise 7.36:** Why do the weights have to add up to 1?

◇ **Exercise 7.37:** Show how to renormalize the six weights needed to predict the pixel at the bottom left corner of Figure 7.140b.

The encoder predicts all the pixels of a level by using the diamond-shaped group of 4×4 (or fewer) "older" pixels around each pixel of the level. This is the *image model* used by MLP.

> It's hard to predict, especially the future.
> —Niels Bohr

7.25.3 Error Modeling

Assume that the weighted sum of the 16 near-neighbors of pixel P equals R. Thus, R is the value predicted for P. The prediction error, E, is simply the difference R − P. Assuming an image with 16 gray levels (four bits per pixel) the largest value of E is 15 (when R = 15 and P = 0) and the smallest is −15. Depending on the image, we can expect most of the errors to be small integers, either zero or close to it. Few errors will be ±15 or close to that. Experiments with a large number of images (see, for example, [Netravali and Limb 80]) have produced the error distribution shown in Figure 7.143a. This is a symmetric, narrow curve, with a sharp peak, indicating that most errors are small and are therefore concentrated at the top. Such a curve has the shape of the well-known Laplace distribution (Section 10.2.1) with mean 0. This statistical distribution is similar to the normal (Gaussian) distribution but is narrower. It fits the error curve better than the normal distribution because most error values are small.

	x					
V	0	2	4	6	8	10
3:	0.408248	0.0797489	0.015578	0.00304316	0.00059446	0.000116125
4:	0.353553	0.0859547	0.020897	0.00508042	0.00123513	0.000300282
5:	0.316228	0.0892598	0.025194	0.00711162	0.00200736	0.000566605
1,000:	0.022360	0.0204475	0.018698	0.0170982	0.0156353	0.0142976

Table 7.142: Some Values of the Laplace Distribution with $V = 3$, 4, 5, and 1,000.

Table 7.142 shows some values for the Laplace distributions with $m = 0$ and $V = 3, 4, 5$, and 1,000. Figure 7.143b shows the graphs of the first three of those. It is clear that as V grows, the graph becomes lower and wider, with a less-pronounced peak.

The factor $1/\sqrt{2V}$ is included in the definition of the Laplace distribution in order to scale the area under the curve of the distribution to 1. Because of this, it is easy to use the curve of the distribution to calculate the probability of any error value. Figure 7.143c shows a gray strip, 1 unit wide, under the curve of the distribution, centered at an error value of k. The area of this strip equals the probability of any error E having the value k. Mathematically, the area is the integral

$$P_V(k) = \int_{k-.5}^{k+.5} \frac{1}{\sqrt{2V}} \exp\left(-\sqrt{\frac{2}{V}}|x|\right) dx, \tag{7.40}$$

and this is the key to encoding the errors. With 4-bit pixels, error values are in the range $[-15, +15]$. When an error k is obtained, the MLP encoder encodes it arithmetically with a probability computed by Equation (7.40). In practice, both encoder and decoder should have a table with all the possible probabilities precomputed.

The only remaining point to discuss is what value of the variance V should be used in Equation (7.40). Both encoder and decoder need to know this value. It is clear, from Figure 7.143b, that using a large variance (which corresponds to a low, flat distribution) results in too low a probability estimate for small error values k. The arithmetic encoder would produce an unnecessarily long code in such a case. On the

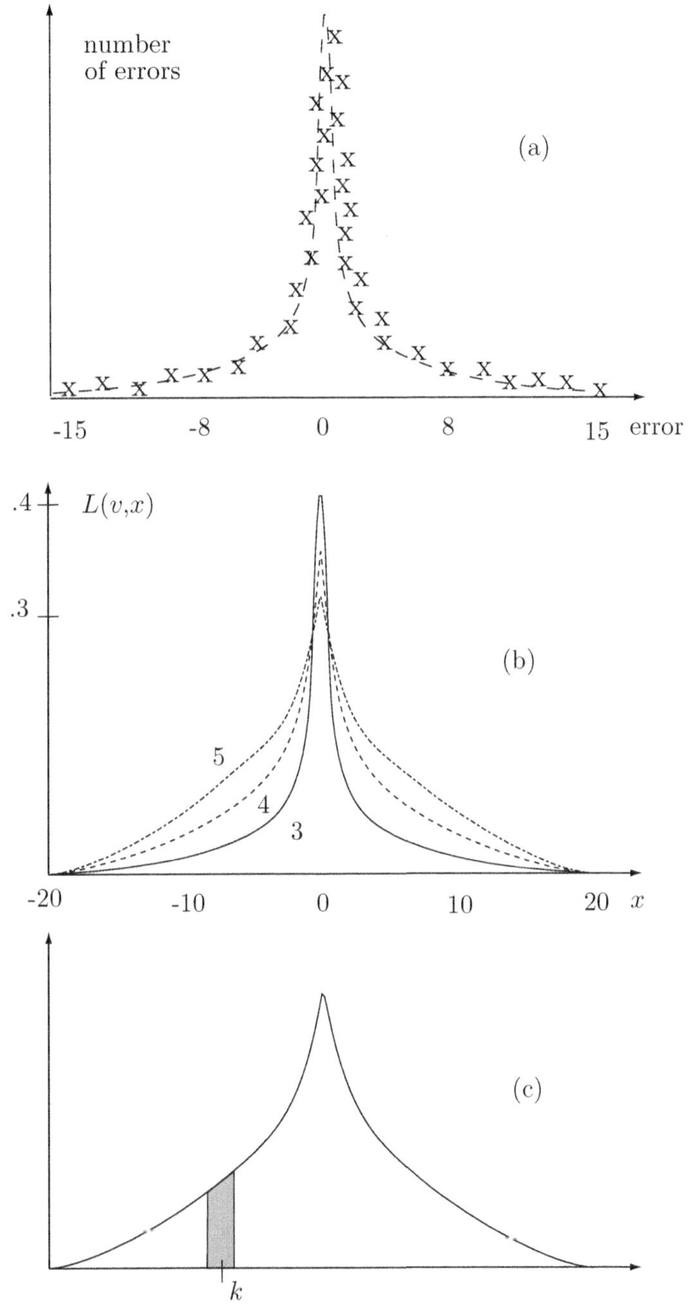

Figure 7.143: (a) Error Distribution. (b) Laplace Distributions. (c) Probability of k.

other hand, using a small variance (which corresponds to a high, narrow distribution) would allocate too low probabilities to large error values k. The choice of variance is therefore important. An ideal method to estimate the variance should assign the best variance value to each error and should involve no overhead (i.e., no extra data should be written on the compressed stream to help the decoder estimate the variances). Here are some approaches to variance selection:

1. The Laplace distribution was adopted as the MLP error distribution after many experiments with real images. The distribution obtained by all those images has a certain value of V, and this value should be used by MLP. This is a simple approach, which can be used in fast versions of MLP. However, it is not adaptive (since it always uses the same variance), and therefore does not result in best compression performance for all images.
2. (A two-pass compression job.) Each compression should start with a pass where all the error values are computed and their variance calculated (see below). This variance should be used, after the first pass, to compute a table of probabilities from Equation (7.40). The table should be used in the second pass where the error values are encoded and written on the compressed stream. Compression performance would be excellent, but any two-pass job is slow. Notice that the entire table can be written at the start of the compressed stream, thereby greatly simplifying the task of the decoder.

◇ **Exercise 7.38:** Show an example where this approach is practical (i.e., when slow encoding is unimportant but a fast decoder and excellent compression are important).

3. Every time an error is obtained and is about to be arithmetically coded, use some method to estimate the variance associated with that error. Quantize the estimate and use a number of precomputed probability tables, one for each quantized variance value, to compute the probability of the error. This is computationally intensive but may be a good compromise between approaches 1 and 2 above.

We now need to discuss how the variance of an error can be estimated, and we start with an explanation of the concept of variance. Variance is a statistical concept defined for a sequence of values $a_1, a_2, \ldots, a_n$. It measures how elements a_i vary by calculating the differences between them and the average A of the sequence, which is defined, as usual, as $A = (1/n)\sum a_i$. A small variance is the reason why the curves of Figure 7.143b that correspond to smaller variances are narrower; their values are concentrated near the average, which in this case is zero. The sequence $(5,5,5)$, for example, has average 5 and variance 0, because every element of the sequence equals the average. The sequence $(0,5,10)$ also has average 5 but should have a nonzero variance, because two of its elements differ from the average. In general, the variance of the sequence a_i is defined as the nonnegative quantity

$$V = \sigma^2 = E(a_i - A)^2 = \frac{1}{n}\sum_{1}^{n}(a_i - A)^2,$$

so the variance of $(0,5,10)$ is $[(0-5)^2+(5-5)^2+(10-5)^2]/3 = 50/3$. Statisticians also use a quantity called *standard deviation* (denoted by σ) that is defined as the square root of the variance.

We now discuss several ways to estimate the variance of a prediction error E.

3.1. Equation (7.40) gives the probability of an error E with value k, but this probability depends on V. We can consider $P_V(k)$ a function of the two variables V and k, and find the optimal value of V by solving the equation $\partial P_V(k)/\partial V = 0$. The solution is $V = 2k^2$, but this method is not practical, because the decoder does not know k (it is trying to decode k so it can find the value of pixel P), and thus cannot mirror the operations of the encoder. It is possible to write the values of all the variances on the compressed stream, but this would significantly reduce the compression ratio. This method can be used to encode a particular image in order to find the best compression ratio of the image and compare it to what is achieved in practice.

3.2. While the pixels of a level are being encoded, consider their errors E to be a sequence of numbers, and find its variance V. Use V to encode the pixels of the next level. The number of levels is never very large, so all the variance values can be written (arithmetically encoded) on the compressed stream, resulting in fast decoding. The variance used to encode the first level should be a user-controlled parameter whose value is not critical, because that level contains just a few pixels. Since MLP quantizes a variance to one of 37 values (see below), which is why each variance written on the compressed stream is encoded in just $\log_2 37 \approx 5.21$ bits, a negligible overhead. The obvious disadvantage of this method is that it disregards local concentrations of identical or very similar pixels in the same level.

3.3. (Similar to 3.2.) While the pixels of a level are being encoded, collect the prediction errors of each block of $b \times b$ pixels and use them to compute a variance that will be used to encode the pixels inside this block in the next level. The variance values for a level can also be written on the compressed stream following all the encoded errors for that level, so the decoder could use them without having to compute them. Parameter b should be adjusted by experiments, and the authors recommend the value $b = 16$. This method entails significant overhead and may therefore degrade compression performance.

3.4. (This is a later addition to MLP; see [Howard and Vitter 92b].) A *variability index* is computed, by both the encoder and decoder, for each pixel. This index should depend on the amount by which the pixel differs from its near neighbors. The variability indexes of all the pixels in a level are then used to adaptively estimate the variances for the pixels, based on the assumption that pixels with similar variability index should use Laplace distributions with similar variances. The method proceeds in the following steps:

1. Variability indexes are calculated for all the pixels of the current level, based on values of pixels in the preceding levels. This is done by the encoder and is later mirrored by the decoder. After several tries, the developers of MLP have settled on a simple way to calculate the variability index. It is calculated as the variance of the four nearest neighbors of the current pixel (the neighbors are from preceding levels, so the decoder can mirror this operation).

2. The variance estimate V is set to some initial value whose choice is not critical, as V is going to be updated often later. The decoder chooses this value in the same way.

3. The pixels of the current level are sorted in variability index order. The decoder can mirror this even though it still does not have the values of these pixels (the decoder has already calculated the values of the variability index in step 1, because they depend on pixels of previous levels).

4. The encoder loops over the sorted pixels in decreasing order (from large variability indexes to small ones). For each pixel:
4.1. The encoder calculates the error E of the pixel and sends E and V to the arithmetic encoder. The decoder mirrors this step. It knows V, so it can decode E.
4.2. The encoder updates V by

$$V \leftarrow f \times V + (1 - f)E^2,$$

where f is a smoothing parameter (experience suggests a large value, such as 0.99, for f). This is how V is adapted from pixel to pixel, using the errors E. Because of the large value of f, V is decreased in small steps. This means that latter pixels (those with small variability indexes) will get small variances assigned. The idea is that compressing pixels with large variability indexes is less sensitive to accurate values of V.

As the loop progresses, V gets assigned more accurate values, and these are used to compress pixels with small variability indexes, which are more sensitive to variance values. Notice that the decoder can mirror this step, since it has already decoded E in step 4.1. Notice also that the arithmetic encoder writes the encoded error values on the compressed stream in decreasing variability index order, not row by row. The decoder can mirror this too, since it has already sorted the pixels in this order in step 3.

This method gives excellent results but is even more computationally intensive than the original MLP (end of method 3.4).

Using one of the four methods above, variance V is estimated. Before using V to encode error E, V is quantized to one of 37 values as shown in Table 7.145. For example, if the estimated variance value is 0.31, it is quantized to 7. The quantized value is then used to select one of 37 precomputed probability tables (in our example Table 7, precomputed for variance value 0.290, is selected) prepared using Equation (7.40), and the value of error E is used to index that table. (The probability tables are not shown here.) The value retrieved from the table is the probability that's sent to the arithmetic encoder, together with the error value, to arithmetically encode error E.

MLP is therefore one of the many compression methods that implement a model to estimate probabilities and use arithmetic coding to do the actual compression.

Table 7.144 is a pseudo-code summary of MLP encoding.

7.25.4 Interpolating Polynomials

This section shows how to predict the value of a pixel from those of 16 of its near neighbors by means of a two-dimensional interpolating polynomial. The results are used in Table 7.141.

We start with an intuitive discussion of the term *interpolation*. Given two numbers a and b, their average $(a + b)/2$ is always located midway between them, so we can use the average to interpolate them. However, given four numbers a, b, c, and d, their average $(a+b+c+d)/4$ is not a good interpolation, because it is not located "midway" between the four. A simple example is the four numbers 1, 1, 1, and 100. Their average is close to 25, so it is nowhere "in the middle" of the four numbers. Interpolating four numbers is therefore done by (1) converting the numbers to two-dimensional points, (2) calculating a smooth curve that passes through the points, and (3) finding the midpoint of the curve.

```
for each level L do
 for every pixel P in level L do
  Compute a prediction R for P using a group from level L-1;
  Compute E=R-P;
  Estimate the variance V to be used in encoding E;
  Quantize V and use it as an index to select a Laplace table LV;
  Use E as an index to table LV and retrieve LV[E];
  Use LV[E] as the probability to arithmetically encode E;
 endfor;
Determine the pixels of the next level (rotate & scale);
endfor;
```

Table 7.144: MLP Encoding.

Variance range	Var. used	Variance range	Var. used	Variance range	Var. used
0.005–0.023	0.016	2.882–4.053	3.422	165.814–232.441	195.569
0.023–0.043	0.033	4.053–5.693	4.809	232.441–326.578	273.929
0.043–0.070	0.056	5.693–7.973	6.747	326.578–459.143	384.722
0.070–0.108	0.088	7.973–11.170	9.443	459.143–645.989	540.225
0.108–0.162	0.133	11.170–15.627	13.219	645.989–910.442	759.147
0.162–0.239	0.198	15.627–21.874	18.488	910.442–1285.348	1068.752
0.239–0.348	0.290	21.874–30.635	25.875	1285.348–1816.634	1506.524
0.348–0.502	0.419	30.635–42.911	36.235	1816.634–2574.021	2125.419
0.502–0.718	0.602	42.911–60.123	50.715	2574.021–3663.589	3007.133
0.718–1.023	0.859	60.123–84.237	71.021	3663.589–5224.801	4267.734
1.023–1.450	1.221	84.237–118.157	99.506	5224.801–7247.452	6070.918
1.450–2.046	1.726	118.157–165.814	139.489	7247.452–10195.990	8550.934
2.046–2.882	2.433				

Table 7.145: Thirty-Seven Quantized Variance Values.

Any numbers a, b, c, and d can be converted to the points $(1, a)$, $(2, b)$, $(3, c)$, and $(4, d)$. It is intuitively clear that the midpoint (x, y) of a smooth curve that passes through those points is a good candidate for the title "the interpolation of the four points." The y coordinate becomes the interpolation of the four numbers, and the x coordinate is ignored.

This method is called one-dimensional interpolation. It can be extended to more than four numbers, and also to pixels, where it becomes two-dimensional interpolation. As mentioned before, we want to use a group of 16 neighboring pixels to predict the value of a pixel at the center of the group. The main idea is to consider the 16 neighbor pixels a set of 4×4 equally-spaced points on a surface (where the value of a pixel is interpreted as the height of the surface) and to derive a polynomial function $\mathbf{P}(u, w)$ that passes through all 16 points. Graphically, $\mathbf{P}(u, w)$ can be thought of as a surface.

The value of the pixel at the center of the 4×4 group can then be predicted by calculating the height of the center point $\mathbf{P}(0.5, 0.5)$ of the surface. Mathematically, this surface is the two-dimensional polynomial interpolation of the 16 points.

7.25.5 One-Dimensional Interpolation

A surface can be viewed as an extension of a curve, so we start by deriving a one-dimensional polynomial (a curve) that interpolates four points, then extend it to a two-dimensional polynomial (a surface) that interpolates a grid of 4×4 points.

Given four points $\mathbf{P}_1$, $\mathbf{P}_2$, $\mathbf{P}_3$, and $\mathbf{P}_4$ we look for a polynomial that will pass through them. In general, a polynomial of degree n in x is defined (Section 6.32) as the function

$$P_n(x) = \sum_{i=0}^{n} a_i x^i = a_0 + a_1 x + a_2 x^2 + \cdots + a_n x^n, \tag{7.41}$$

where a_i are the $n+1$ coefficients of the polynomial and the parameter x is a real number. The one-dimensional interpolating polynomial that is of interest to us is special, and differs from the definition above in two respects

1. This polynomial goes from point $\mathbf{P}_1$ to point $\mathbf{P}_4$. Its length is finite, and it is therefore better to describe it as the function

$$P_n(t) = \sum_{i=0}^{n} a_i t^i = a_0 + a_1 t + a_2 t^2 + \cdots + a_n t^n; \text{ where } 0 \le t \le 1.$$

This is the *parametric representation* of a polynomial. We want this polynomial to go from $\mathbf{P}_1$ to $\mathbf{P}_4$ when the parameter t is varied from 0 to 1.

2. The only given data are the four points and we have to use them to calculate all $n+1$ coefficients of the polynomial. This suggests the value $n = 3$ (a polynomial of degree 3, a cubic polynomial; one that has four coefficients). The idea is to set up and solve four equations, with the four coefficients as the unknowns, and with the four points as known quantities. Thus, we use the notation (T indicates transpose)

$$\mathbf{P}(t) = \mathbf{a}t^3 + \mathbf{b}t^2 + \mathbf{c}t + \mathbf{d} = (t^3, t^2, t, 1)(\mathbf{a}, \mathbf{b}, \mathbf{c}, \mathbf{d})^T = \mathbf{T}(t) \cdot \mathbf{A}. \tag{7.42}$$

The four coefficients $\mathbf{a}, \mathbf{b}, \mathbf{c}, \mathbf{d}$ are shown in boldface because they are not numbers. Keep in mind that the polynomial has to pass through the given points, so the value of $\mathbf{P}(t)$ for any t must be the three coordinates of a point. Each coefficient should therefore be a triplet. $\mathbf{T}(t)$ is the row vector $(t^3, t^2, t, 1)$, and $\mathbf{A}$ is the column vector $(\mathbf{a}, \mathbf{b}, \mathbf{c}, \mathbf{d})^T$. Calculating the curve therefore involves finding the values of the four unknowns $\mathbf{a}$, $\mathbf{b}$, $\mathbf{c}$, and $\mathbf{d}$. $\mathbf{P}(t)$ is called a *parametric cubic* (or PC) polynomial.

It turns out that degree 3 is the smallest one that is still useful for an interpolating polynomial. A polynomial of degree 1 has the form $\mathbf{P}_1(t) = \mathbf{c}t + \mathbf{d}$ and is therefore a straight line, so it can be used only in special cases. A polynomial of degree 2 (quadratic) has the form $\mathbf{P}_2(t) = \mathbf{b}t^2 + \mathbf{c}t + \mathbf{d}$ and is a conic section, so it can take only a few different shapes. A polynomial of degree 3 (cubic) is therefore the simplest one that can take on complex shapes, and can also be a space curve.

◇ **Exercise 7.39:** Prove that a quadratic polynomial must be a plane curve.

Our ultimate problem is to interpolate pixels. Pixels are always spaced uniformly, so we assume that the two interior points $\mathbf{P}_2$ and $\mathbf{P}_3$ are equally spaced between $\mathbf{P}_1$ and $\mathbf{P}_4$. The first point $\mathbf{P}_1$ is the start point $\mathbf{P}(0)$ of the polynomial, the last point, $\mathbf{P}_4$ is the endpoint $\mathbf{P}(1)$, and the two interior points $\mathbf{P}_2$ and $\mathbf{P}_3$ are the two equally-spaced interior points $\mathbf{P}(1/3)$ and $\mathbf{P}(2/3)$ of the polynomial.

We therefore write $\mathbf{P}(0) = \mathbf{P}_1$, $\mathbf{P}(1/3) = \mathbf{P}_2$, $\mathbf{P}(2/3) = \mathbf{P}_3$, $\mathbf{P}(1) = \mathbf{P}_4$, or

$$\begin{aligned}
\mathbf{a}(0)^3 + \mathbf{b}(0)^2 + \mathbf{c}(0) + \mathbf{d} &= \mathbf{P}_1,\\
\mathbf{a}(1/3)^3 + \mathbf{b}(1/3)^2 + \mathbf{c}(1/3) + \mathbf{d} &= \mathbf{P}_2,\\
\mathbf{a}(2/3)^3 + \mathbf{b}(2/3)^2 + \mathbf{c}(2/3) + \mathbf{d} &= \mathbf{P}_3,\\
\mathbf{a}(1)^3 + \mathbf{b}(1)^2 + \mathbf{c}(1) + \mathbf{d} &= \mathbf{P}_4.
\end{aligned}$$

These equations are easy to solve, and the solutions are:

$$\begin{aligned}
\mathbf{a} &= -9/2\mathbf{P}_1 + 27/2\mathbf{P}_2 - 27/2\mathbf{P}_3 + 9/2\mathbf{P}_4,\\
\mathbf{b} &= 9\mathbf{P}_1 - 45/2\mathbf{P}_2 + 18\mathbf{P}_3 - 9/2\mathbf{P}_4,\\
\mathbf{c} &= -11/2\mathbf{P}_1 + 9\mathbf{P}_2 - 9/2\mathbf{P}_3 + \mathbf{P}_4,\\
\mathbf{d} &= \mathbf{P}_1.
\end{aligned}$$

Substituting into Equation (7.42) gives

$$\begin{aligned}
\mathbf{P}(t) =& (-9/2\mathbf{P}_1 + 27/2\mathbf{P}_2 - 27/2\mathbf{P}_3 + 9/2\mathbf{P}_4)t^3\\
&+ (9\mathbf{P}_1 - 45/2\mathbf{P}_2 + 18\mathbf{P}_3 - 9/2\mathbf{P}_4)t^2\\
&+ (-11/2\mathbf{P}_1 + 9\mathbf{P}_2 - 9/2\mathbf{P}_3 + \mathbf{P}_4)t + \mathbf{P}_1,
\end{aligned}$$

which, after rearranging, becomes

$$\begin{aligned}
\mathbf{P}(t) =& (-4.5t^3 + 9t^2 - 5.5t + 1)\mathbf{P}_1 + (13.5t^3 - 22.5t^2 + 9t)\mathbf{P}_2\\
&+ (-13.5t^3 + 18t^2 - 4.5t)\mathbf{P}_3 + (4.5t^3 - 4.5t^2 + t)\mathbf{P}_4\\
=& G_1(t)\mathbf{P}_1 + G_2(t)\mathbf{P}_2 + G_3(t)\mathbf{P}_3 + G_4(t)\mathbf{P}_4\\
=& \mathbf{G}(t) \cdot \mathbf{P},
\end{aligned} \tag{7.43}$$

where

$$\begin{aligned}
G_1(t) &= (-4.5t^3 + 9t^2 - 5.5t + 1), & G_2(t) &= (13.5t^3 - 22.5t^2 + 9t),\\
G_3(t) &= (-13.5t^3 + 18t^2 - 4.5t), & G_4(t) &= (4.5t^3 - 4.5t^2 + t);
\end{aligned} \tag{7.44}$$

$\mathbf{P}$ is the column $(\mathbf{P}_1, \mathbf{P}_2, \mathbf{P}_3, \mathbf{P}_4)^T$, and $\mathbf{G}(t)$ is the row vector

$$(G_1(t), G_2(t), G_3(t), G_4(t)).$$

The functions $G_i(t)$ are called *blending functions*, because they create any point on the curve as a blend of the four given points. Note that they add up to 1 for any value of

t. This property must be satisfied by any set of blending functions, and such functions are called *barycentric*. We can also write

$$G_1(t) = (t^3, t^2, t, 1)(-4.5, 9, -5.5, 1)^T$$

and, similarly, for $G_2(t)$, $G_3(t)$, and $G_4(t)$. In matrix notation this becomes

$$\mathbf{G}(t) = (t^3, t^2, t, 1)\begin{pmatrix} -4.5 & 13.5 & -13.5 & 4.5 \\ 9.0 & -22.5 & 18 & -4.5 \\ -5.5 & 9.0 & -4.5 & 1.0 \\ 1.0 & 0 & 0 & 0 \end{pmatrix} = \mathbf{T}(t)\cdot\mathbf{N}. \qquad (7.45)$$

The curve can now be written $\mathbf{P}(t) = \mathbf{G}(t)\cdot\mathbf{P} = \mathbf{T}(t)\cdot\mathbf{N}\cdot\mathbf{P}$. $\mathbf{N}$ is called the basis matrix, and $\mathbf{P}$ is the geometry vector. From Equation (7.42) we know that $\mathbf{P}(t) = \mathbf{T}(t)\cdot\mathbf{A}$, so we can write $\mathbf{A} = \mathbf{N}\cdot\mathbf{P}$. Equation (8.17) illustrates an application of this interpolating polynomial for image compression.

> The word *barycentric* is derived from *barycenter*, meaning "center of gravity," because such weights are used to calculate the center of gravity of an object. Barycentric weights have many uses in geometry in general, and in curve and surface design in particular.

Given the four points, the interpolating polynomial can be calculated in two steps:
1. Set-up the equation $\mathbf{A} = \mathbf{N}\cdot\mathbf{P}$ and solve it for $\mathbf{A} = (\mathbf{a}, \mathbf{b}, \mathbf{c}, \mathbf{d})^T$.
2. The polynomial is $\mathbf{P}(t) = \mathbf{T}(t)\cdot\mathbf{A}$.

7.25.6 Example

(This example is in two dimensions, each of the four points $\mathbf{P}_i$ and each of the four coefficients $\mathbf{a}$, $\mathbf{b}$, $\mathbf{c}$, and $\mathbf{d}$ is a pair. For three-dimensional curves, the method is the same, except that triplets should be used, instead of pairs.) Given the four two-dimensional points $\mathbf{P}_1 = (0,0)$, $\mathbf{P}_2 = (1,0)$, $\mathbf{P}_3 = (1,1)$, and $\mathbf{P}_4 = (0,1)$, we set up the equation

$$\begin{pmatrix} \mathbf{a} \\ \mathbf{b} \\ \mathbf{c} \\ \mathbf{d} \end{pmatrix} = \mathbf{A} = \mathbf{N}\cdot\mathbf{P} = \begin{pmatrix} -4.5 & 13.5 & -13.5 & 4.5 \\ 9.0 & -22.5 & 18 & -4.5 \\ -5.5 & 9.0 & -4.5 & 1.0 \\ 1.0 & 0 & 0 & 0 \end{pmatrix}\begin{pmatrix} (0,0) \\ (1,0) \\ (1,1) \\ (0,1) \end{pmatrix},$$

which is easy to solve

$$\begin{aligned}
\mathbf{a} &= -4.5(0,0) + 13.5(1,0) - 13.5(1,1) + 4.5(0,1) = (0,-9), \\
\mathbf{b} &= 19(0,0) - 22.5(1,0) + 18(1,1) - 4.5(0,1) = (-4.5, 13.5), \\
\mathbf{c} &= -5.5(0,0) + 9(1,0) - 4.5(1,1) + 1(0,1) = (4.5, -3.5), \\
\mathbf{d} &= 1(0,0) - 0(1,0) + 0(1,1) - 0(0,1) = (0,0).
\end{aligned}$$

Thus $\mathbf{P}(t) = \mathbf{T} \cdot \mathbf{A} = (0, -9)t^3 + (-4.5, 13.5)t^2 + (4.5, -3.5)t.$

It is now easy to calculate and verify that $\mathbf{P}(0) = (0, 0) = \mathbf{P}_1$, and

$$\mathbf{P}(1/3) = (0, -9)1/27 + (-4.5, 13.5)1/9 + (4.5, -3.5)1/3 = (1, 0) = \mathbf{P}_2,$$

$$\mathbf{P}(1) = (0, -9)1^3 + (-4.5, 13.5)1^2 + (4.5, -3.5)1 = (0, 1) = \mathbf{P}_4.$$

◇ **Exercise 7.40:** Calculate $\mathbf{P}(2/3)$ and verify that it is equal to $\mathbf{P}_3$.

◇ **Exercise 7.41:** Imagine the circular arc of radius one in the first quadrant (a quarter circle). Write the coordinates of the four points that are equally spaced on this arc. Use the coordinates to calculate a PC interpolating polynomial approximating this arc. Calculate point $\mathbf{P}(1/2)$. How far does it deviate from the midpoint of the true quarter circle?

The main advantage of this method is its simplicity. Given the four points, it is easy to calculate the PC polynomial that passes through them.

◇ **Exercise 7.42:** This method makes sense if the four points are (at least approximately) equally spaced along the curve. If they are not equally spaced, the following may be done: Instead of using 1/3 and 2/3 as the intermediate values, the user may specify values α, β such that $\mathbf{P}_2 = \mathbf{P}(\alpha)$ and $\mathbf{P}_3 = \mathbf{P}(\beta)$. Generalize Equation (7.45) such that it depends on α and β.

7.25.7 Two-Dimensional Interpolation

The PC polynomial, Equation (7.42), can easily be extended to two dimensions by means of a technique called *Cartesian product*. The polynomial is generalized from a cubic curve to a *bicubic* surface.

A one-dimensional PC polynomial has the form $\mathbf{P}(t) = \sum_{i=0}^{3} \mathbf{a}_i t^i$. Two such curves, $\mathbf{P}(u)$ and $\mathbf{P}(w)$, can be combined by means of this technique to form the surface:

$$\begin{aligned}
\mathbf{P}(u, w) &= \sum_{i=0}^{3}\sum_{j=0}^{3} \mathbf{a}_{ij} u^i w^j \\
&= \mathbf{a}_{33}u^3w^3 + \mathbf{a}_{32}u^3w^2 + \mathbf{a}_{31}u^3w + \mathbf{a}_{30}u^3 + \mathbf{a}_{23}u^2w^3 + \mathbf{a}_{22}u^2w^2 + \mathbf{a}_{21}u^2w + \mathbf{a}_{20}u^2 \\
&\quad + \mathbf{a}_{13}uw^3 + \mathbf{a}_{12}uw^2 + \mathbf{a}_{11}uw + \mathbf{a}_{10}u + \mathbf{a}_{03}w^3 + \mathbf{a}_{02}w^2 + \mathbf{a}_{01}w + \mathbf{a}_{00} \\
&= (u^3, u^2, u, 1) \begin{pmatrix} \mathbf{a}_{33} & \mathbf{a}_{32} & \mathbf{a}_{31} & \mathbf{a}_{30} \\ \mathbf{a}_{23} & \mathbf{a}_{22} & \mathbf{a}_{21} & \mathbf{a}_{20} \\ \mathbf{a}_{13} & \mathbf{a}_{12} & \mathbf{a}_{11} & \mathbf{a}_{10} \\ \mathbf{a}_{03} & \mathbf{a}_{02} & \mathbf{a}_{01} & \mathbf{a}_{00} \end{pmatrix} \begin{pmatrix} w^3 \\ w^2 \\ w \\ 1 \end{pmatrix}, \quad \text{where } 0 \le u, w \le 1.
\end{aligned} \tag{7.46}$$

This is a double cubic polynomial (hence the name *bicubic*) with 16 terms, where each of the 16 coefficients $\mathbf{a}_{ij}$ is a triplet. Note that the surface depends on all 16 coefficients. Any change in any of them produces a different surface. Equation (7.46) is the *algebraic representation* of a bicubic surface. In order to use it in practice, the

16 unknown coefficients have to be expressed in terms of the 16 known, equally-spaced points. We denote these points

$$\begin{matrix} \mathbf{P}_{03} & \mathbf{P}_{13} & \mathbf{P}_{23} & \mathbf{P}_{33} \\ \mathbf{P}_{02} & \mathbf{P}_{12} & \mathbf{P}_{22} & \mathbf{P}_{32} \\ \mathbf{P}_{01} & \mathbf{P}_{11} & \mathbf{P}_{21} & \mathbf{P}_{31} \\ \mathbf{P}_{00} & \mathbf{P}_{10} & \mathbf{P}_{20} & \mathbf{P}_{30}. \end{matrix}$$

To determine the 16 unknown coefficients, we write 16 equations, each based on one of the given points:

$$\begin{matrix} \mathbf{P}(0,0)=\mathbf{P}_{00} & \mathbf{P}(0,1/3)=\mathbf{P}_{01} & \mathbf{P}(0,2/3)=\mathbf{P}_{02} & \mathbf{P}(0,1)=\mathbf{P}_{03} \\ \mathbf{P}(1/3,0)=\mathbf{P}_{10} & \mathbf{P}(1/3,1/3)=\mathbf{P}_{11} & \mathbf{P}(1/3,2/3)=\mathbf{P}_{12} & \mathbf{P}(1/3,1)=\mathbf{P}_{13} \\ \mathbf{P}(2/3,0)=\mathbf{P}_{20} & \mathbf{P}(2/3,1/3)=\mathbf{P}_{21} & \mathbf{P}(2/3,2/3)=\mathbf{P}_{22} & \mathbf{P}(2/3,1)=\mathbf{P}_{23} \\ \mathbf{P}(1,0)=\mathbf{P}_{30} & \mathbf{P}(1,1/3)=\mathbf{P}_{31} & \mathbf{P}(1,2/3)=\mathbf{P}_{32} & \mathbf{P}(1,1)=\mathbf{P}_{33}. \end{matrix}$$

Solving, substituting the solutions in Equation (7.46), and simplifying produces the *geometric representation* of the bicubic surface

$$\mathbf{P}(u,w)=(u^3,u^2,u,1)\mathbf{N}\begin{pmatrix} \mathbf{P}_{33} & \mathbf{P}_{32} & \mathbf{P}_{31} & \mathbf{P}_{30} \\ \mathbf{P}_{23} & \mathbf{P}_{22} & \mathbf{P}_{21} & \mathbf{P}_{20} \\ \mathbf{P}_{13} & \mathbf{P}_{12} & \mathbf{P}_{11} & \mathbf{P}_{10} \\ \mathbf{P}_{03} & \mathbf{P}_{02} & \mathbf{P}_{01} & \mathbf{P}_{00} \end{pmatrix}\mathbf{N}^T\begin{pmatrix} w^3 \\ w^2 \\ w \\ 1 \end{pmatrix}, \tag{7.47}$$

where $\mathbf{N}$ is the Hermite matrix of Equation (7.45).

The surface of Equation (7.47) can now be used to predict the value of a pixel as a polynomial interpolation of 16 of its near neighbors. All that is necessary is to substitute $u = 0.5$ and $w = 0.5$. The following *Mathematica* code

```
Clear[Nh,P,U,W];
Nh={{-4.5,13.5,-13.5,4.5},{9,-22.5,18,-4.5},
 {-5.5,9,-4.5,1},{1,0,0,0}};
P={{p33,p32,p31,p30},{p23,p22,p21,p20},
 {p13,p12,p11,p10},{p03,p02,p01,p00}};
U={u^3,u^2,u,1};
W={w^3,w^2,w,1};
u:=0.5;
w:=0.5;
Expand[U.Nh.P.Transpose[Nh].Transpose[W]]
```

does that and produces

$$\begin{aligned} &\mathbf{P}(0.5,0.5)\\ &\quad=0.00390625\mathbf{P}_{00}-0.0351563\mathbf{P}_{01}-0.0351563\mathbf{P}_{02}+0.00390625\mathbf{P}_{03}\\ &\quad-0.0351563\mathbf{P}_{10}+0.316406\mathbf{P}_{11}+0.316406\mathbf{P}_{12}-0.0351563\mathbf{P}_{13}\\ &\quad-0.0351563\mathbf{P}_{20}+0.316406\mathbf{P}_{21}+0.316406\mathbf{P}_{22}-0.0351563\mathbf{P}_{23}\\ &\quad+0.00390625\mathbf{P}_{30}-0.0351563\mathbf{P}_{31}-0.0351563\mathbf{P}_{32}+0.00390625\mathbf{P}_{33}, \end{aligned}$$

where the 16 coefficients are the ones used in Table 7.141.

◇ **Exercise 7.43:** How can this method be used in cases where not all 16 points are known?

◇ **Exercise 7.44:** The center point of the surface is calculated as a weighted sum of the 16 equally-spaced data points. It makes sense to assign small weights to points located away from the center, but our result assigns *negative* weights to eight of the 16 points. Explain the meaning of negative weights and show what role they play in interpolating the center of the surface.

Readers who find it difficult to follow the details above should compare the way two-dimensional polynomial interpolation is presented here to the way it is discussed by [Press et al. 88]. The following quotation is from page 125: "...The formulas that obtain the c's from the function and derivative values are just a complicated linear transformation, with coefficients which, having been determined once, in the mists of numerical history, can be tabulated and forgotten."

7.26 Adaptive Golomb

The principle of run-length encoding is simple. Given a bi-level image, it is scanned row by row and alternating run lengths of black and white pixels are computed. Each run length is in the interval $[1, r]$, where r is the row size. In principle, the run lengths constitute the compressed image, but in practice we need to write them on the compressed stream such that the decoder will be able to read them unambiguously. This is done by replacing each run length by a prefix code and writing the codes on the compressed stream without any separators. This section shows that the Golomb codes (Section 3.24) are ideal for this application. The Golomb codes depend on a parameter m, so we also propose a simple adaptive algorithm for estimating m for each run length in order to get the best compression in a one-pass algorithm.

Figure 7.146 is a simple finite-state machine (see Section 11.8 for more information) that summarizes the process of reading black and white pixels from an image. At any time, the machine can be in one of two states: a 0 (white) or a 1 (black). If the current state is 0 (i.e., the last pixel input was white), then the probability that the next state will be a 1 (the next pixel input will be black) is denoted by α, which implies that the probability that the next state will also be 0 is $1-\alpha$.

Suppose we are at state 0. The probability $p_w(1)$ of a sequence of exactly one white pixel is the probability α of a transition to state 1. Similarly, the probability $p_w(2)$ of a sequence of exactly two white pixels is the probability of returning once to state 0, then switching to state 1. Since these events are independent, the probability is the product $(1-\alpha)\alpha$. In general, the probability $p_w(n)$ of a run length of n white pixels is $(1-\alpha)^{n-1}\alpha$ (this can be proved by induction). Thus, we conclude that $p_w(n)$ and, by symmetry, also $p_b(n) = (1-\beta)^{n-1}\beta$ are distributed geometrically. This analysis is known as the Capon Markov-1 model of the run lengths of a bi-level image [Capon 59] and it justifies (as explained in Section 3.24) the use of Golomb codes to compress the run lengths.

Once this is understood, it is easy to derive a simple, adaptive process that estimates the best Golomb parameter m for each run length, computes the Golomb code for the

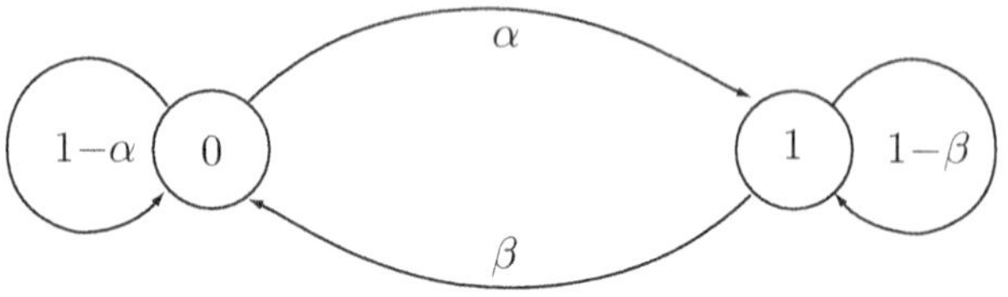

Figure 7.146: A Finite-State Machine for Reading Bi-level Pixels.

run length, and updates a table that's used to obtain better estimates for the remainder of the input.

Each run length is an integer in the interval $[1, r]$, where r is the row size. The first step is to decrement the run length by 1, since the Golomb codes are for the nonnegative (as opposed to the positive) integers. Thus, we denote the modified run length of l pixels of color c by (c, l), where c is either b or w and l is in the interval $[0, r-1]$. A table $S[b{:}w, 0{:}r-1]$ of two rows indexed by b and w and r columns indexed from 0 to $r-1$ is maintained with counts of past run lengths and is used to estimate the probability p of each run length. Assume that the current run length is (c, l), then p is computed as entry $S[c, l]$ divided by the sum of all entries of S. p is then used to compute the best Golomb parameter m as the integer nearest $-1/\log_2(1-p)$. Run length (c, l) is compressed by computing the Golomb code for the integer l with the parameter m, and table S is updated by incrementing element $S[c, l]$ by 1. This process can be reversed by the decoder while maintaining the same table.

There is the question of how to initialize S. In principle, all of its entries should be initialized to zero, but this would produce probabilities of zero, which cannot be used to compute m values. This is the zero-probability problem, introduced in Section 5.14. A reasonable solution is to initialize each table entry to a default value such as 1 and allow the user to change this value according to the nature of the image.

This makes sense because even though bi-level is the simplest image type, there is a wide variety of such images. A page of printed text, for example, normally results, once digitized, in 5–6% of black pixels and with very short runs of those pixels (this figure is used by printer manufacturers to estimate the life of print cartridges). The values of α and β for such a page are about 0.02–0.03 and 0.34–0.35, respectively. In the other extreme, we find traditional woodcuts. When a woodcut is printed and digitized in black and white, it often results in a large majority (and with long runs) of black pixels.

A woodcut is art carved on (and printed directly from) a flat piece of wood. The parts to be printed remain flat with the surface of the wood and the remaining parts should be removed by the artist. A roller is soaked with ink and is rolled on the wood so it covers the surface, but not the carved parts, with ink. A sheet of paper is then placed on the wood and is pressed, either by hand or by a roller, to transfer the ink from the wood. The resulting print on the paper is a mirror image of that on the wood, which makes it especially difficult to include text in a woodcut.

When printed on paper, a woodcut tends to have large black areas because any white (nonprinting) areas have to be removed from the wood, but black areas are those parts of the surface of the wood that are untouched by the artist.

It is possible to have color woodcuts. The artist starts with several flat pieces of wood, each for one color. In the piece for red, for example, the red parts of the image are untouched and the other parts are removed. The same sheet of paper is then pressed to each piece of wood in turn and is left to dry between presses.

Because modern life is fast, many woodcut artists remove the nonprinting areas from the wood by sandblasting instead of slow hand carving (after covering the printing areas with a metal or plastic shield).

The art of carving a woodcut is called xylography.

7.27 PPPM

The reader should review the PPM method, Section 5.14, before reading this section. The PPPM method uses the ideas of MLP (Section 7.25). It is also (remotely) related to the context-based image compression method of Section 7.22.

PPM encodes a symbol by comparing its present context to other similar contexts and selecting the longest match. The context selected is then used to estimate the symbol's probability in the present context. This way of context matching works well for text, where we can expect strings of symbols to repeat exactly, but it does not work well for images, because a digital image is often the result of digitizing an analog image. Assume that the current pixel has intensity 118 and its context is the two neighboring pixels with values 118 and 120. It is possible that 118 was never seen in the past with the context 118, 120 but was seen with contexts 119, 120 and 118, 121. Clearly, these contexts are close enough to the current one to justify using one of them. Once a closely matching context has been found, it is used to estimate the variance (not the probability) of the current prediction error. This idea serves as one principle of the Prediction by Partial Precision Matching (PPPM) method [Howard and Vitter 92a]. The other principle is to use the Laplace distribution to estimate the probability of the prediction error, as done in MLP.

Figure 7.147 shows how prediction is done in PPPM. Pixels are raster-scanned, row by row. The two pixels labeled C are used to predict the one labeled P. The prediction R is simply the rounded average of the two C pixels. Pixels in the top or left edges

o	o	o	o	o
o	S	C	S	o
o	C	P		

Figure 7.147: Prediction and Variance-Estimation Contexts for PPPM.

are predicted by one neighbor only. The top-left pixel of the image is encoded without prediction. After predicting the value of P, the encoder calculates the error E = R − P and uses the Laplace distribution to estimate the probability of the error, as in MLP.

The only remaining point to discuss is how PPPM estimates the variance of the particular Laplace distribution that should be used to obtain the probability of E. PPPM uses the four neighbors labeled C and S in Figure 7.147. These pixels have already been encoded, so their values are known. They are used as the variance-estimation context of P. Assume that the four values are 3, 0, 7, and 5, expressed as 4-bit numbers. They are combined to form the 16-bit key 0011|0000|0111|0101, and the encoder uses a hash table to find all the previous occurrences of this context. If this context occurred enough times in the past (more than the value of a threshold parameter), the statistics of these occurrences are used to obtain a mean m and a variance V. If the context did not occur enough times in the past, the least-significant bit of each of the four values is dropped to obtain the 12-bit key 001|000|011|010, and the encoder hashes this value. Thus, the encoder iterates in a loop until it finds m and V. (It turns out that using the errors of the C and S pixels as a key, instead of their values, produces slightly better compression, so this is what PPPM actually does.)

Once m and V have been obtained, the encoder quantizes V and uses it to select one of 37 Laplace probability tables, as in MLP. The encoder then adds $E+m$ and sends this value to be arithmetically encoded with the probability obtained from the Laplace table. To update the statistics, the PPPM encoder uses a lazy approach. It updates the statistics of the context that is actually used and also updates, if applicable, the context with one additional bit of precision.

One critical point is the number of times a context had to be seen in the past to be considered meaningful and not random. The PPMB method, Section 5.14.4, "trusts" a context if it has been seen twice. For an image, a threshold of 10–15 is more reasonable.

7.28 CALIC

Sections 7.22 through 7.27 describe context-based image compression methods that have one feature in common: They determine the context of a pixel based on some of its neighbor pixels that have already been seen and processed. Normally, these are some of the pixels above and to the left of the current pixel, which leads to asymmetric context. It seems intuitive that a symmetric context, one that predicts the current pixel from pixels all around it, would produce better compression, so attempts have been made to develop image compression methods that employ such contexts.

The MLP method, Section 7.25, provides an interesting twist to the problem of symmetric context. The CALIC method of this section takes a different approach. The name CALIC ([Wu 95] and [Wu 96]) stands for Context-based, Adaptive, Lossless Image Compression. It performs three passes to create a symmetric context around the current pixel, and it uses quantization to reduce the number of possible contexts to something manageable. The method has been developed for compressing grayscale images (where each pixel is a c-bit number representing a shade of gray), but like any other method for grayscale images, it can handle a color image by separating it into three color components and treating each component as a grayscale image.

7.28.1 Three Passes

We start with an image $I[i,j]$ that consists of H rows and W columns of pixels. Both encoder and decoder perform three passes over the image. The first pass calculates averages of pairs of pixels. It looks only at pixels $I[i,j]$ where i and j have the same parity (i.e., both are even or both are odd). The second pass uses these averages to actually encode the same pixels. The third pass uses the same averages plus the pixels of the second pass to encode all pixels $I[i,j]$ where i and j have different parities (one is odd and the other is even).

The first pass calculates the $W/2 \times H/2$ values $\mu[i,j]$ defined by

$$\mu[i,j] = (I[2i,2j] + I[2i+1,2j+1])/2, \text{ for } 0 \le i < H/2,\ 0 \le j < W/2. \tag{7.48}$$

(In the original CALIC papers i and j denote the columns and rows, respectively. We use the standard notation where the first index denotes the rows.) Each $\mu[i,j]$ is therefore the average of two diagonally adjacent pixels. Table 7.148 shows the pixels (in boldface) involved in this computation for an 8×8-pixel image. Each pair that's used to calculate a value $\mu[i,j]$ is connected with an arrow. Notice that the two original pixels cannot be fully reconstructed from the average because 1 bit may be lost by the division by 2 in Equation (7.48).

0,0	0,1	**0,2**	0,3	**0,4**	0,5	**0,6**	0,7
1,0	↘**1,1**	1,2	↘**1,3**	1,4	↘**1,5**	1,6	↘**1,7**
2,0	2,1	**2,2**	2,3	**2,4**	2,5	**2,6**	2,7
3,0	↘**3,1**	3,2	↘**3,3**	3,4	↘**3,5**	3,6	↘**3,7**
4,0	4,1	**4,2**	4,3	**4,4**	4,5	**4,6**	4,7
5,0	↘**5,1**	5,2	↘**5,3**	5,4	↘**5,5**	5,6	↘**5,7**
6,0	6,1	**6,2**	6,3	**6,4**	6,5	**6,6**	6,7
7,0	↘**7,1**	7,2	↘**7,3**	7,4	↘**7,5**	7,6	↘**7,7**

Table 7.148: The 4×4 Values $\mu[i,j]$ for an 8×8-Pixel Image.

The newly calculated values $\mu[i,j]$ are now considered the pixels of a new, small, $W/2\times H/2$-pixel image (a quarter of the size of the original image). This image is raster-scanned, and each of its pixels is predicted by four of its neighbors, three centered above it and one on its left. If $x = \mu[i,j]$ is the current pixel, it is predicted by the quantity

$$\hat{x} = \frac{1}{2}\mu[i,j-1] - \frac{1}{4}\mu[i-1,j-1] + \frac{1}{2}\mu[i-1,j] + \frac{1}{4}\mu[i-1,j+1]. \tag{7.49}$$

(The coefficients 1/2, 1/4, and −1/4, as well as the coefficients used in the other passes, were determined by linear regression, using a set of "training" images. The idea is to find the set of coefficients a_k that gives the best compression for those images, then round them to integer powers of 2, and build them into the algorithm.) The error value $x - \hat{x}$ is then encoded.

The second pass involves the same pixels as the first pass (half the pixels of the original image), but this time each of them is individually predicted. They are raster scanned, and assuming that $x = I[2i, 2j]$ denotes the current pixel, it is predicted using five known neighbor pixels above it and to its left, and three averages μ, known from the first pass, below it and to its right:

$$\begin{aligned}\hat{x} &= 0.9\mu[i,j] + \frac{1}{6}(I[2i+1,2j-1] + I[2i-1,2j-1] + I[2i-1,2j+1]) \\ &\quad - 0.05(I[2i,2j-2] + I[2i-2,2j]) - 0.15(\mu[i,j+1] + \mu[i+1,j]). \qquad (7.50)\end{aligned}$$

Figure 7.149a shows the five pixels (gray dots) and three averages (slanted lines) involved. The task of the encoder is again to encode the error value $x - \hat{x}$ for each pixel x.

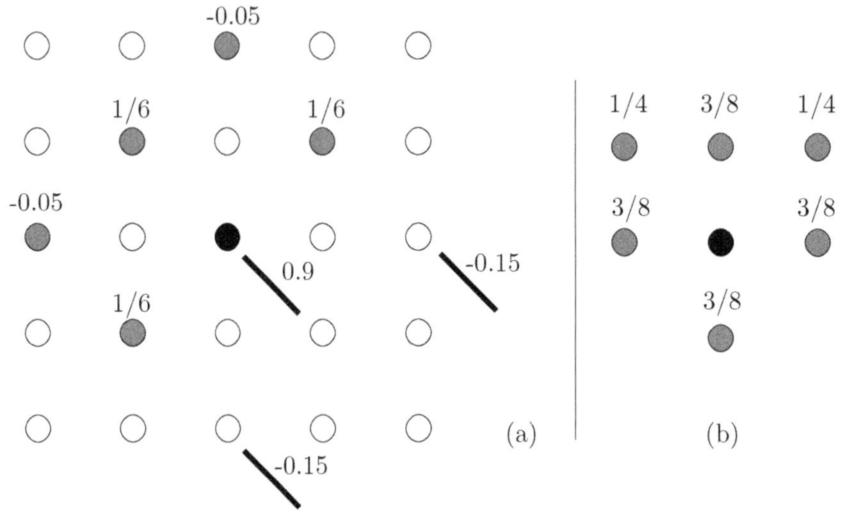

Figure 7.149: Weighted Sums for 360° Contexts.

◇ **Exercise 7.45:** Pixel $I[2i-1, 2j+1]$ is located below $x = I[2i, 2j]$, so how does the decoder know its value when x is decoded?

The third pass involves the remaining pixels:

$$I[2i, 2j+1] \text{ and } I[2i+1, 2j], \quad \text{for } 0 \le i < H/2, \quad 0 \le j < W/2. \qquad (7.51)$$

Each is predicted by an almost symmetric context of six pixels (Figure 7.149b) consisting of all of its four-connected neighbors and two of its eight-connected neighbors. If $x =$

$I[i,j]$ is the current pixel, it is predicted by

$$\begin{aligned}\hat{x} =& \frac{3}{8}\left(I[i,j-1]+I[i-1,j]+I[i,j+1]+I[i+1,j]\right)\\ &-\frac{1}{4}\left(I[i-1,j-1]+I[i-1,j+1]\right).\end{aligned} \tag{7.52}$$

The decoder can mimic this operation, since the pixels below and to the right of x are known to it from the second pass.

Notice that each of the last two passes of the decoder creates half of the image pixels. Thus, CALIC can be considered a progressive method where the two progressive steps increase the resolution of the image.

7.28.2 Context Quantization

In each of the three passes, the error values $x - \hat{x}$ are arithmetically encoded, which means that they have to be assigned probabilities. Assume that pixel x is predicted by the n pixel neighbors $x_1, x_2, \ldots, x_n$. The values of n for the three passes are 4, 8, and 6, respectively. In order to assign a probability to x, the encoder has to count the number of times x was found in the past with every possible n-pixel context. If a pixel is stored as a c-bit number (representing a shade of gray), then the number of possible contexts for a pixel is $2^{n\cdot c}$. Even for $c=4$ (just 16 shades of gray) this number is $2^{8\cdot 4} \approx 4.3$ billion, too large for a practical implementation. CALIC reduces the number of contexts in several steps. It first creates the single n-bit number $t = t_n \ldots t_1$ by

$$t_k = \begin{cases} 0, & \text{if } x_k \ge \hat{x},\\ 1, & \text{if } x_k < \hat{x}.\end{cases}$$

Next it calculates the quantity

$$\Delta = \sum_{k=1}^{n} w_k |x_k - \hat{x}|,$$

where the coefficients w_k were prepared in advance, by using the same set of training images, and are built into the algorithm. The quantity Δ is called the *error strength discriminant* and is quantized to one of L integer values d, where L is typically set to 8. Once $\hat{x}$ and the n neighbors $x_1, x_2, \ldots, x_n$ are known, both t and the quantized value d of Δ can be determined, and they become the indices of the context. This reduces the number of contexts to $L \cdot 2^n$, which is at most $8 \cdot 2^8 = 2{,}048$. The encoder maintains an array S of d rows and t columns where context counts are kept. The following is a summary of the steps performed by the CALIC encoder.

```
For all passes
INITIALIZATION: N(d,t):=1; S(d,t):=0; d=0,1,...,L, t=0,1,...,2^n;
PARAMETERS: a_k and w_k are assigned their values;
for all pixels x in the current pass do
 0: x̂ = Σ_{k=1}^n a_k·x_k;
 1: Δ = Σ_{k=1}^n w_k(x_k − x̂);
```

2: $d =$ `Quantize`(Δ);
3: `Compute` $t = t_n \ldots t_2 t_1$;
4: $\bar{\epsilon} = S(d,t)/N(d,t)$;
5: $\dot{x} = \hat{x} + \bar{\epsilon}$;
6: $\epsilon = x - \dot{x}$;
7: $S(d,t) := S(d,t) + \epsilon$; $N(d,t) := N(d,t) + 1$;
8: <u>if</u> $N(d,t) \geq 128$ <u>then</u>
   $S(d,t) := S(d,t)/2$; $N(d,t) := N(d,t)/2$;
9: <u>if</u> $S(d,t) < 0$ `encode`$(-\epsilon, d)$ <u>else</u> `encode`(ϵ, d);
<u>endfor</u>;
end.

7.29 Differential Lossless Compression

There is always a trade-off between speed and performance, so there is always a demand for fast compression methods as well as for methods that are slow but very efficient. The differential method of this section, due to Sayood and Anderson [Sayood and Anderson 92], belongs to the former class. It is fast and simple to implement, while offering good, albeit not spectacular, performance.

The principle is to compare each pixel p to a *reference pixel*, which is one of its previously-encoded immediate neighbors, and encode p in two parts: a prefix, which is the number of most-significant bits of p that are identical to those of the reference pixel, and a suffix, which is the remaining least-significant bits of p. For example, if the reference pixel is 10110010 and p is 10110100, then the prefix is 5, because the five most-significant bits of p are identical to those of the reference pixel, and the suffix is 00. Notice that the remaining three least-significant bits are 100 but the suffix does not have to include the 1, since the decoder can easily deduce its value.

◇ **Exercise 7.46:** How can the decoder do this?

The prefix in our example is 5, and in general it is an integer in the range $[0, 8]$, and compression can be improved by encoding the prefix further. Huffman coding is a good choice for this purpose, with either a fixed set of nine Huffman codes or with adaptive codes. The suffix can be any number of between zero and eight bits, so there are 256 possible suffixes. Since this number is relatively large, and since we expect most suffixes to be small, it makes sense to write the suffix on the output stream unencoded.

This method encodes each pixel with a different number of bits. The encoder generates bits until it has 8 or more of them, then outputs a byte. The decoder can easily mimic this. All that it has to know is the location of the reference pixel and the Huffman codes. In the example above, if the Huffman code of 6 is, say, 010, the code of p will be the five bits 010|00.

The only point remaining to be discussed is the selection of the reference pixel. It should be close to the current pixel p, and it should be known to the decoder when p is decoded. The rules adopted by the developers of this method for selecting the reference pixel are therefore simple. The very first pixel of an image is written on the output stream unencoded. For every other pixel in the first (top) scan line, the reference pixel is selected as its immediate left neighbor. For the first (leftmost) pixel on subsequent scan lines, the reference pixel is the one above it. For every other pixel, it is possible to select the reference pixel in one of three ways: (1) the pixel immediately to its left; (2) the pixel above it; and (3) the pixel on the left, except that if the resulting prefix is less than a predetermined threshold, the pixel above it.

An example of case 3 is a threshold value of 3. Initially, the reference pixel for p is chosen to be its left neighbor, but if this results in a prefix value of 0, 1, or 2, the reference pixel is changed to the one above p, regardless of the prefix value that is then produced.

This method assumes one byte per pixel (256 colors or grayscale values). If a pixel is expressed by three bytes, the image should be separated into three parts, and the method applied to each part separately.

◇ **Exercise 7.47:** Can this method be used for images with 16 grayscale values (where each pixel is four bits, and a byte contain two pixels)?

"The nerve impulse, including the skeptic waves, will have to jump the tiny gap of the synapse and, in doing so, the dominant thoughts will be less attenuated than the others. In short, if we jump the synapse, too, we will reach a region where we may, for a while at least, detect what we want to hear with less interference from trivial noise."

"Really?" asked Morrison archly. "This notion of differential attenuation is new to me."

—Isaac Asimov, *Fantastic Voyage II: Destination Brain*

7.30 DPCM

The DPCM compression method is a member of the family of differential encoding compression methods, which itself is a generalization of the simple concept of relative encoding (Section 1.3.1). It is based on the well-known fact that neighboring pixels in an image (and also adjacent samples in digitized sound, Section 10.2) are correlated. Correlated values are generally similar, so their differences are small, resulting in compression. Table 7.150 lists 25 consecutive values of $\sin\theta_i$, calculated for θ_i values from 0 to 360° in steps of 15°. The values therefore range from -1 to $+1$, but the 24 differences $\sin\theta_{i+1} - \sin\theta_i$ (also listed in the table) are all in the range $[-0.259, 0.259]$. The average of the 25 values is zero, as is the average of the 24 differences. However, the variance of the differences is small, since they are all closer to their average.

Figure 7.151a shows a histogram of a hypothetical image that consists of 8-bit pixels. For each pixel value between 0 and 255 there is a different number of pixels. Figure 7.151b shows a histogram of the differences of consecutive pixels. It is easy to see that most of the differences (which, in principle, can be in the range $[0, 255]$) are small; only a few are outside the range $[-50, +50]$.

Differential encoding methods calculate the differences $d_i = a_i - a_{i-1}$ between consecutive data items a_i, and encode the d_i's. The first data item, a_0, is either encoded

sin(t) :	0	0.259	0.500	0.707	0.866	0.966	1.000	0.966	
diff :	–	0.259	0.241	0.207	0.159	0.100	0.034	−0.034	
sin(t) :	0.866	0.707	0.500	0.259	0	−0.259	−0.500	−0.707	
diff :	−0.100	−0.159	−0.207	−0.241	−0.259	−0.259	−0.241	−0.207	
sin(t) :	−0.866	−0.966	−1.000	−0.966	−0.866	−0.707	−0.500	−0.259	0
diff :	−0.159	−0.100	−0.034	0.034	0.100	0.159	0.207	0.241	0.259

```
S=Table[N[Sin[t Degree]], {t,0,360,15}]
Table[S[[i+1]]-S[[i]], {i,1,24}]
```

Table 7.150: 25 Sine Values and 24 Differences.

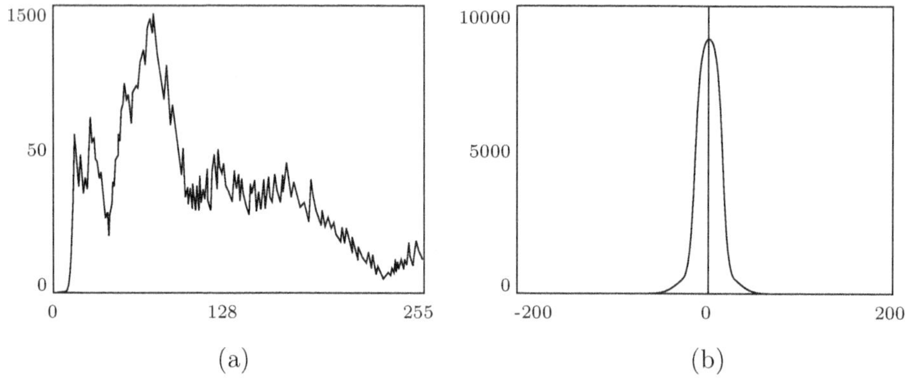

Figure 7.151: A Histogram of an Image and Its Differences.

separately or is written on the compressed stream in raw format. In either case the decoder can decode and generate a_0 in exact form. In principle, any suitable method, lossy or lossless, can be used to encode the differences. In practice, quantization is often used, resulting in lossy compression. The quantity encoded is not the difference d_i but a similar, quantized number that we denote by $\hat{d}_i$. The difference between d_i and $\hat{d}_i$ is the *quantization error* q_i. Thus, $\hat{d}_i = d_i + q_i$.

It turns out that the lossy compression of differences introduces a new problem, namely, the accumulation of errors. This is easy to see when we consider the operation of the decoder. The decoder inputs encoded values of $\hat{d}_i$, decodes them, and uses them to generate "reconstructed" values $\hat{a}_i$ (where $\hat{a}_i = \hat{a}_{i-1} + \hat{d}_i$) instead of the original data values a_i. The decoder starts by reading and decoding a_0. It then inputs $\hat{d}_1 = d_1 + q_1$ and calculates $\hat{a}_1 = a_0 + \hat{d}_1 = a_0 + d_1 + q_1 = a_1 + q_1$. The next step is to input $\hat{d}_2 = d_2 + q_2$ and to calculate $\hat{a}_2 = \hat{a}_1 + \hat{d}_2 = a_1 + q_1 + d_2 + q_2 = a_2 + q_1 + q_2$. The decoded value $\hat{a}_2$ contains the sum of two quantization errors. In general, the decoded value $\hat{a}_n$ equals

$$\hat{a}_n = a_n + \sum_{i=1}^{n} q_i,$$

and includes the sum of n quantization errors. Sometimes, the errors q_i are signed and tend to cancel each other out in the long run. In general, however, this is a problem.

The solution is easy to understand once we realize that the encoder and the decoder operate on different pieces of data. The encoder generates the exact differences d_i from the original data items a_i, while the decoder generates the reconstructed $\hat{a}_i$ using only the quantized differences $\hat{d}_i$. The solution is therefore to modify the encoder to calculate differences of the form $d_i = a_i - \hat{a}_{i-1}$. A general difference d_i is therefore calculated by subtracting the most recent reconstructed value $\hat{a}_{i-1}$ (which both encoder and decoder have) from the current original item a_i.

The decoder now starts by reading and decoding a_0. It then inputs $\hat{d}_1 = d_1 + q_1$ and calculates $\hat{a}_1 = a_0 + \hat{d}_1 = a_0 + d_1 + q_1 = a_1 + q_1$. The next step is to input $\hat{d}_2 = d_2 + q_2$ and calculate $\hat{a}_2 = \hat{a}_1 + \hat{d}_2 = \hat{a}_1 + d_2 + q_2 = a_2 + q_2$. The decoded value $\hat{a}_2$ contains just the single quantization error q_2, and in general, the decoded value $\hat{a}_i$ equals $a_i + q_i$, so it contains just quantization error q_i. We say that the *quantization noise* in decoding $\hat{a}_i$ equals the noise generated when a_i was quantized.

Figure 7.152a summarizes the operations of both encoder and decoder. It shows how the current data item a_i is saved in a storage unit (a delay), to be used for encoding the next item a_{i+1}.

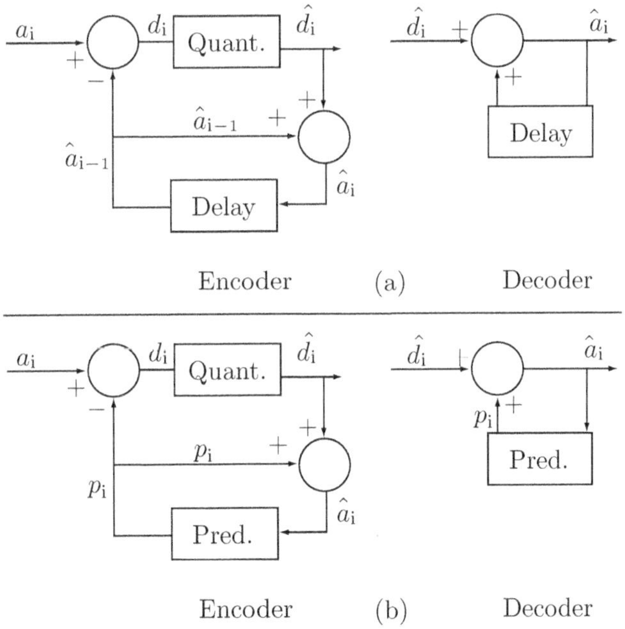

Figure 7.152: (a) A Differential Codec. (b) DPCM.

The next step in developing a general differential encoding method is to take advantage of the fact that the data items being compressed are correlated. This means that in general, an item a_i depends on *several* of its near neighbors, not just on the preceding item a_{i-1}. Better prediction (and, as a result, smaller differences) can therefore be obtained by using N of the previously-seen neighbors to encode the current item a_i (where N is a parameter). We therefore would like to have a function $p_i = f(\hat{a}_{i-1}, \hat{a}_{i-2}, \ldots, \hat{a}_{i-N})$ to predict a_i (Figure 7.152b). Notice that f has to be a function of the $\hat{a}_{i-j}$, not the a_{i-j}, since the decoder has to calculate the same f. Any method using such a predictor is called *differential pulse code modulation*, or DPCM. In practice, DPCM methods are used mostly for audio compression, but are illustrated here in connection with image compression.

The simplest predictor is linear. In such a predictor the value of the current pixel a_i is predicted by a weighted sum of N of its previously-seen neighbors (in the case of an image these are the pixels above it or to its left):

$$p_i = \sum_{j=1}^{N} w_j a_{i-j},$$

where w_j are the weights, which still need to be determined.

Figure 7.153a shows a simple example for the case $N = 3$. Let's assume that a pixel X is predicted by its three neighbors A, B, and C according to the simple weighted sum

$$X = 0.35A + 0.3B + 0.35C. \tag{7.53}$$

Figure 7.153b shows 16 8-bit pixels, part of a bigger image. We use Equation (7.53) to predict the nine pixels at the bottom right. The predictions are shown in Figure 7.153c. Figure 7.153d shows the differences between the pixel values a_i and their predictions p_i.

(a)

B	C
A	X

(b)

101	128	108	110
100	90	95	100
102	80	90	85
105	75	96	91

(c)

	110	108	104
	97	88	95
	95	82	90

(d)

	-20	-13	-4
	-17	2	-10
	-20	14	1

Figure 7.153: Predicting Pixels and Calculating Differences.

The weights used in Equation (7.53) have been selected more or less arbitrarily and are for illustration purposes only. However, they make sense, because they add up to unity.

◇ **Exercise 7.48:** Why should the weights add up to 1? (This is an easy exercise, but it is important, because weights that add up to unity are very common. Such weights are called barycentric.)

In order to determine the best weights, we denote by e_i the prediction error for pixel a_i,

$$e_i = a_i - p_i = a_i - \sum_{j=1}^{N} w_j a_{i-j}, \quad i = 1, 2, \ldots, n,$$

where n is the number of pixels to be compressed (in our example nine of the 16 pixels), and we find the set of weights w_j that minimizes the sum

$$E = \sum_{i=1}^{n} e^2 = \sum_{i=1}^{n} \left[a_i - \sum_{j=1}^{N} w_j a_{i-j} \right]^2 .$$

We denote by $\mathbf{a} = (90, 95, 100, 80, 90, 85, 75, 96, 91)$ the vector of the nine pixels to be compressed. We denote by $\mathbf{b}^{(k)}$ the vector consisting of the kth neighbors of the six pixels. Thus

$$\begin{aligned}
\mathbf{b}^{(1)} &= (100, 90, 95, 102, 80, 90, 105, 75, 96),\\
\mathbf{b}^{(2)} &= (101, 128, 108, 100, 90, 95, 102, 80, 90),\\
\mathbf{b}^{(3)} &= (128, 108, 110, 90, 95, 100, 80, 90, 85).
\end{aligned}$$

The total square prediction error is

$$E = \left| \mathbf{a} - \sum_{j=1}^{3} w_j \mathbf{b}^{(j)} \right|^2 ,$$

where the vertical bars denote the absolute value of the vector between them. To minimize this error, we need to find the linear combination of vectors $\mathbf{b}^{(j)}$ that's closest to $\mathbf{a}$. Readers familiar with the algebraic concept of vector spaces know that this is done by finding the orthogonal projection of $\mathbf{a}$ on the vector space spanned by the $\mathbf{b}^{(j)}$'s, or, equivalently, by finding the difference vector

$$\mathbf{a} - \sum_{j=1}^{3} w_j \mathbf{b}^{(j)}$$

that's orthogonal to all the $\mathbf{b}^{(j)}$'s. Two vectors are orthogonal if their dot product is zero, which produces the M (in our case, 3) equations

$$\mathbf{b}^{(k)} \cdot \left(\mathbf{a} - \sum_{j=1}^{M} w_j \mathbf{b}^{(j)} \right) = 0, \quad \text{for } 1 \le k \le M,$$

or, equivalently,

$$\sum_{j=1}^{M} w_j \left(\mathbf{b}^{(k)} \cdot \mathbf{b}^{(j)} \right) = \left(\mathbf{b}^{(k)} \cdot \mathbf{a} \right), \quad \text{for } 1 \le k < M.$$

The *Mathematica* code of Figure 7.154 produces, for our example, the solutions $w_1 = 0.1691$, $w_2 = 0.1988$, and $w_3 = 0.5382$. Note that they add up to 0.9061, not to 1, and this is discussed in the following exercise.

```
a={90.,95,100,80,90,85,75,96,91};
b1={100,90,95,102,80,90,105,75,96};
b2={101,128,108,100,90,95,102,80,90};
b3={128,108,110,90,95,100,80,90,85};
Solve[{b1.(a-w1 b1-w2 b2-w3 b3)==0,
b2.(a-w1 b1-w2 b2-w3 b3)==0,
b3.(a-w1 b1-w2 b2-w3 b3)==0},{w1,w2,w3}]
```

Figure 7.154: Solving for Three Weights.

◇ **Exercise 7.49:** Repeat this calculation for the six pixels 90, 95, 100, 80, 90, and 85. Discuss your results.

Adaptive DPCM: This variant of DPCM is commonly used for audio compression. In ADPCM the quantization step size adapts to the changing frequency of the sound being compressed. The predictor also has to adapt itself and recalculate the weights according to changes in the input. Several versions of ADPCM exist. A popular version is the IMA ADPCM standard (Section 10.6), which specifies the compression of PCM from 16 down to four bits per sample. ADPCM is fast, but it introduces noticeable quantization noise and achieves unimpressive compression factors of about four.

7.31 Context-Tree Weighting

The context-tree weighting method of Section 5.16 can be applied to images. The method described here [Ekstrand 96] proceeds in five steps as follows:

Step 1, prediction. This step uses differencing to predict pixels and is similar to the lossless mode of JPEG (Section 7.10.5). The four immediate neighbors A, B, C, and D of the current pixel X (Figure 7.155) are used to calculate a linear prediction P of X according to

$$L = aA + bB + cC + dD, \quad P = X - \lfloor L + 1/2 \rfloor, \quad \text{for } a + b + c + d = 1.$$

The four weights should add up to 1 and are selected such that a and c are assigned slightly greater values than b and d. A possible choice is $a = c = 0.3$ and $b = d = 0.2$, but experience seems to suggest that the precise values are not critical.

	B	C	D	
	A	X		

Figure 7.155: Pixel Prediction.

The values of P are mostly close to zero and are often Laplace distributed (Figure 7.143).

Step 2, Gray encoding. Our intention is to apply the CTW method to the compression of grayscale or color images. In the case of a grayscale image we have to separate the bitplanes and compress each individually, as if it were a bi-level image. A color image has to be separated into its three components and each component compressed separately as a grayscale image. Section 7.4.1 discusses *reflected Gray codes* (RGC) and shows how their use preserves pixel correlations after the different bitplanes are separated.

Step 3, serialization. The image is converted to a bit stream that will be arithmetically encoded in step 5. Perhaps the best way of doing this is to scan the image row by row for each bitplane. The bit stream therefore starts with all the bits of the first (least-significant) bitplane, continues with the bits of the second least-significant bitplane, and so on. Each bit in the resulting bit stream has as context (its neighbors on the left) bits that originally were above it or to its left in the two-dimensional bitplane. An alternative is to start with the first rows of all the bitplanes, continue with the second rows, and so on.

Step 4, estimation. The bitstream is read bit by bit, and a context tree is constructed and updated. The weighted probability at the root of the tree is used to predict the current bit, and both the probability and the bit are sent to the arithmetic encoder (Step 5). The tree should be deep, since the prediction done in Step 1 reduces the correlation between pixels. On the other hand, a deep CTW tree slows down the encoder, since more nodes should be updated for each new input bit.

Step 5, encoding. This is done by a standard adaptive arithmetic encoder.

7.32 Block Decomposition

Readers of this chapter may have noticed that most image compression methods have been designed for, and perform best on, continuous-tone images, where adjacent pixels normally have similar intensities or colors. The method described here is intended for the lossless compression of discrete-tone images, be they bi-level, grayscale, or color. Such images are (with few exceptions) artificial, having been obtained by scanning a document, or grabbing a computer screen. The pixel colors of such an image do not vary continuously or smoothly, but have a small set of values, such that adjacent pixels may differ much in intensity or color.

The method works by searching for, and locating, identical blocks of pixels. A copy B of a block A is compressed by preparing the height, width, and location (image coordinates) of A, and compressing those four numbers by means of Huffman codes. The method is called *Flexible Automatic Block Decomposition* (FABD) [Gilbert and Brodersen 98]. Finding identical blocks of pixels is a natural method for image compression, because an image is a two-dimensional structure. The GIF graphics file format (Section 6.21), for example, scans an image row by row to compress it. It is therefore a one-dimensional method, so its efficiency as an image compressor is not very high. The JBIG method of Section 7.14 considers pixels individually, and examines just the local neighborhood of a pixel. It does not attempt to discover correlations in distant parts of the image (at least, not explicitly).

FABD, on the other hand, assumes that identical parts (blocks) of pixels may appear several times in the image. In other words, it assumes that images have a *global two-dimensional redundancy.* It also assumes that large, uniform blocks of pixels will exist in the image. Thus, FABD performs well on images that satisfy these assumptions, such as discrete-tone images. The method scans the image in raster order, row by row, and divides it into (possibly overlapping) sets of blocks. There are three types of blocks namely, copied blocks, solid fill blocks, and punts.

> Basically "punting" is often used as slang (at least in Massachusetts, where I am from) to mean "give up" or do something suboptimal—in my case the punting is a sort of catch-all to make sure that pixels that cannot efficiently take part in a fill or copy block are still coded.
>
> —Jeffrey M. Gilbert

A copied block B is a rectangular part of the image that has been seen before (which is located above, or on the same line but to the left of, the current pixel). It can have any size. A solid fill block is a rectangular uniform region of the image. A punt is any image area that's neither a copied block nor a solid fill one. Each of these three types is compressed by preparing a set of parameters that fully describe the block, and writing their Huffman codes on the compressed stream. Here is a general description of the operations of the encoder.

The encoder proceeds from pixel to pixel. It looks at the vicinity of the current pixel P to see if its future neighbors (those to the right and below P) have the same color. If yes, the method locates the largest uniform block of which P is the top-left corner. Once such a block has been identified, the encoder knows the width and height of the block. It writes the Huffman codes of these two quantities on the compressed stream, followed by the color of the block, and preceded by a code specifying a solid fill block. The four values

fill-block code, width, height, pixel value,

are written, encoded, on the output. Figure 7.156a shows a fill block with dimensions 4×3 at pixel B.

If the near neighbors of P have different values, the encoder starts looking for an identical block among past pixels. It considers P the top-left corner of a block with unspecified dimensions, and it searches pixels seen in the past (those above or to the left of P) to find the largest block A that will match P. If it finds such a block, then P is designated a copy block of A. Since A has already been compressed, its copy block P can be fully identified by preparing its dimensions (width and height) and source location (the coordinates of A). The five quantities

copy-block code, width, height, A_x, A_y,

are written, suitably encoded, on the output. Notice that there is no need to write the coordinates of P on the output, because both encoder and decoder proceed pixel by pixel in raster order. Figure 7.156a shows a copy block with dimensions 3×4 at pixel P. This is a copy of block A whose image coordinates are $(2, 2)$, so the quantities

copy-block code, 3, 4, 2, 2,

should be written, encoded, on the output.

If no identical block A can be found (we propose that blocks should have a certain minimum size, such as 4×4), the encoder marks P and continues with the next pixel. Thus, P becomes a punt pixel. Suppose that P and the four pixels following it are punts, but the next one starts a copy or a fill block. The encoder prepares the quantities

$$\text{punt-block code, 5, } P_1,\ P_2,\ P_3,\ P_4,\ P_5,$$

where the P_i are the punt pixels, and writes them, encoded, on the output. Figure 7.156a shows how the first six pixels of the image are all different and thus form a punt block. The seventh pixel, marked x, is identical to the first pixel, and therefore has a chance of starting a fill or a copy block.

In each of these cases, the encoder marks the pixels of the current block "encoded," and skips them in its raster scan of the image. Figure 7.156a shows how the scan order is affected when a block is identified.

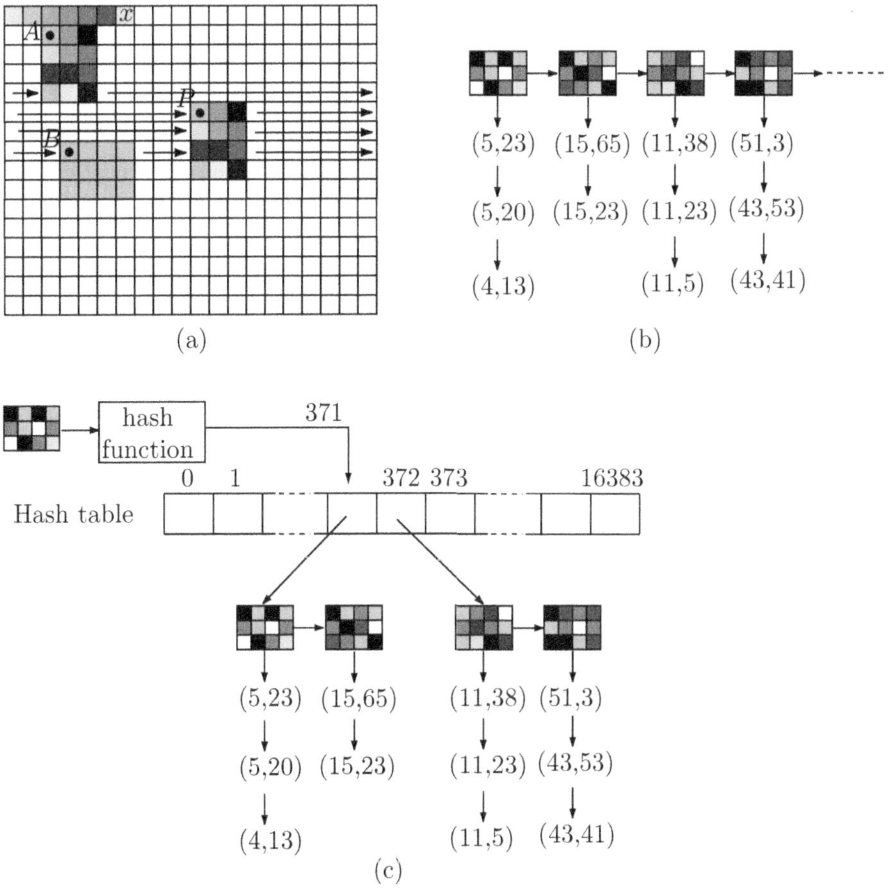

Figure 7.156: FABD Encoding of an Image.

The decoder scans the image in raster order. For each pixel P it inputs the next block code from the compressed stream. This can be the code of a copy block, a fill block, or a punt block. If this is the code of a copied block, the decoder knows that it will be followed by a pair of dimensions and a pair of image coordinates. Once the decoder inputs those, it knows the dimensions of the block and where to copy it from. The block is copied and is anchored with P as its top-left corner. The decoder continues its raster scan, but it skips all pixels of the newly constructed block (and any other blocks constructed previously). If the next block code is that of a fill block or a punt block, the decoder generates the block of pixels, and continues with the raster scan, skipping already-generated pixels. The task of the decoder is therefore very simple, making this method ideal for use by Web browsers, where fast decoding is a must.

It is clear that FABD is a highly asymmetric method. The job of the decoder is mostly to follow pointers and to copy pixels, but the encoder has to identify fill and copy blocks. In principle, the encoder has to scan the entire image for each pixel P, looking for bigger and bigger blocks identical to those that start at P. If the image consists of n pixels, this implies n^2 searches, where each search has to examine several, possibly even many, pixels. For $n = 10^6$, the number of searches is 10^{12} and the number of pixels examined may be 2–3 orders of magnitude bigger. This kind of exhaustive search may take hours on current computers and has to be drastically improved before the method can be considered practical. The discussion here explores ways of speeding up the encoder's search. The following two features are obvious:

1. The search examines only past pixels, not future ones. Searching the entire image has to be done for the last pixels only. There are no searches at all for the first pixel. This reduces the total number of searches from n^2 to $n^2/2$ on average. In fact, the actual implementation of FABD tries to start blocks only at uncoded pixel locations—which reduces the number of searches from n^2 to $n^2/(\text{average block size})^2$.
2. There is no search for the pixels of a solid fill block. The process of identifying such a block is simple and fast.

There still remains a large number of searches, and they are speeded up by the following technique. The minimum block size can be limited to 4×4 pixels without reducing the amount of compression, since blocks smaller than 4×4 aren't much bigger than a single pixel. This suggests constructing a list with all the 4×4 pixel patterns found so far in the image (in reverse order, so that recently-found patterns are placed at the start of the list). If the current pixel is P, the encoder constructs the 4×4 block B starting at P, and searches this list for occurrences of B. The list may be extremely long. For a bi-level image, where each pixel is one bit, the total number of 16-bit patterns is $2^{16} = 65{,}536$. For an image with 8-bit pixels, the total number of such patterns is $2^{8\cdot 16} = 2^{128} \approx 3.4\cdot 10^{38}$. Not every pattern must occur in a given image, but the number of patterns that do occur may still be large, and more improvements are needed. Here are three improvements:

3. The list of patterns should include each pattern found so far *only once*. Our list is now called the main list, and it becomes a list of lists. Each element of the main list starts with a unique 4×4 pattern, followed by a match-list of image locations where this pattern was found so far. Figure 7.156b shows an example. Notice how each pattern points to a match-list of coordinates that starts with recently-found pixels. A new item is always added to the start of a match-list, so those lists are naturally maintained in sorted order. The elements of a match list are sorted by (descending) rows and, within a row, by column.

4. Hashing can be used to locate a 4×4 pattern in the main list. The bits of a pattern are hashed into a 14-bit number that's employed as a pointer (for 1-bit pixels, there are 16 bits in a pattern. The actual implementation of FABD assumes 8-bit pixels, resulting in 128-bit patterns). It points to an element in an array of pointers (the hash table). Following a pointer from the hash table brings the encoder to the start of a match-list. Because of hash collisions, more than one 16-bit pattern may hash into the same 14-bit number, so each pointer in the hash table actually points to a (normally short) list whose elements are match-lists. Figure 7.156c shows an example.
5. If the image has many pixels and large redundancy, some match lists may be long. The total compression time depends heavily on a fast search of the match lists, so it makes sense to limit those searches. A practical implementation may have a parameter k that limits the depth of the search of a match list. Thus, setting $k = 1000$ limits the search of a match list to its 1000 top elements. This reduces the search time while having only a minimal detrimental effect on the final compression ratio. Experience shows that even values as low as $k = 50$ can be useful. Such a value may increase the compression ratio by a few percent, while cutting down the compression time of a typical $1\text{K}\times1\text{K}$ image to just a few seconds. Another beneficial effect of limiting the search depth has to do with hard-to-compress regions in the image. Such regions may be rare, but tend nevertheless to increase the total compression time significantly.

Denoting the current pixel by P, the task of the encoder is to (1) construct the 4×4 block B of which P is the upper-left corner, (2) hash the 16 pixel values of B into a 14-bit pointer, (3) follow the pointer to the hash table and, from there, to a short main list, (4) search this main list linearly, to find a match list that starts with B, and (5) search the first k items in this match list. Each item is the start location of a block that matches P by at least 4×4 pixels. The largest match is selected.

The last interesting feature of FABD has to do with transforming the quantities that should be encoded, before the actual encoding. This is based on the *spatial locality* of discrete-tone images. This property implies that a block will normally be copied from a nearby source block. Also, a given region will tend to have just a few colors, even though the image in general may have many colors.

Thus, spatial locality suggests the use of *relative image coordinates*. If the current pixel is located at, say, $(81, 112)$ and it is a copy of a block located at $(41, 10)$, then the location of the source block is expressed by the pair $(81-41, 112-10)$ of relative coordinates. Relative coordinates are small numbers and are also distributed nonuniformly. Thus, they compress well with Huffman coding.

Spatial locality also suggests the use of *color age*. This is a simple way to assign relative codes to colors. The age of a color C is the number of unique colors located between the current instance of C and its previous instance. As an example, given the sequence of colors

green, yellow, red, red, red, green, red,

their color ages are

?, ?, ?, 0, 0, 2, 1.

It is clear that color ages in an image with spatial locality are small numbers. The first time a color is seen it does not have an age, so it is encoded raw.

Experiments with FABD on discrete-tone images yield compression of between 0.04 bpp (for bi-level images) and 0.65 bpp (for 8-bit images).

7.33 Binary Tree Predictive Coding

Binary tree predictive coding (BTPC) is intended for lossless and lossy compression of all types of images. The method is based on the concept of *image pyramid* and its lossy mode also uses quantization. BTPC was designed to meet the following criteria:

1. It should be able to compress continuous-tone (photographic), discrete-tone (graphical), and mixed images as well as the standard methods for each.
2. The lossy option should not require a fundamental change in the basic algorithm.
3. When the same image is compressed several times, losing more and more data each time, the result should visually appear as a gradual blurring of the decompressed images, not as sudden changes in image quality.
4. A software implementation should be efficient in its time and memory requirements. The decoder's memory requirements should be just a little more than the image size. The decoding time (number of steps) should be a small multiple of the image size.
5. A hardware implementation of both encoder and decoder should allow for fine-grain parallelism. In the ideal case, the hardware has enough processors so that each processor should be able to process one bit of the image.

There are two versions, BTPC1 and BTPC2 [Robinson 97]. The details of both are described here, and the name BTPC is used for features that are common to both.

The main innovation of BTPC is the use of a *binary image pyramid*. The technique repeatedly decomposes the image into two components, a low band L, which is recursively decomposed, and a high band H. The low band is a low-resolution part of the image. The high band contains differences between L and the original image. These differences are later used by the decoder to reconstruct the image from L. If the original image is highly correlated, L will contain correlated (i.e., highly redundant) values, but will still contribute to the overall compression, since it is small. On the other hand, the differences that are the contents of H will be decorrelated (i.e., with little or no redundancy left), will be small numbers, and will have a histogram that peaks around zero. Thus, their entropy will be small, making it possible to compress them efficiently with an entropy coder. Since the difference values in H are small, it is natural to obtain lossy compression by quantizing them, a process that generates many zeros.

The final compression ratio depends on how small L is and how decorrelated the values of H are. The main idea of the binary pyramid used by BTPC is to decompose the original image into two bands L_1 and H_1, decompose L_1 into bands L_2 and H_2, and continue decomposing the low bands until bands L_8 and H_8 are obtained. The eight high bands and the last low band L_8 are written on the compressed stream after being entropy encoded. They constitute the binary image pyramid, and they are used by the BTPC decoder to reconstruct the original image.

It is natural for the encoder to write them in the order L_1, H_1, $H_2, \ldots, H_8$. The decoder, however, needs them in the opposite order, so it makes sense for the encoder to collect these matrices in memory and write them in reverse order. The decoder needs just one memory buffer, the size of the original image, plus some more memory to input a row of H_i from the compressed stream. The elements of the row are used to process pixels in the buffer, and the next row is then input.

If the original image has $2^n \times 2^n = N$ pixels, then band H_1 contains $N/2$ elements, band H_2 has $N/4$ elements, and bands L_8 and H_8 have $N/2^8$ elements each. The total number of values to be encoded is therefore

$$(N/2 + N/2^2 + N/2^3 + \cdots + N/2^8) + N/2^8 = N(1 - 2^{-8}) + N/2^8 = N.$$

If the original image size has $2^{10}\times 2^{10} = 2^{20}$ pixels, then bands L_8 and H_8 have $2^{12} = 4096$ elements each. When dealing with bigger images, it may be better to continue beyond L_8 and H_8 down to matrices of size 2^{10} to 2^{12}.

Figure 7.157 illustrates the details of the BTPC decomposition. Figure 7.157a shows an 8×8 highly correlated grayscale image (note how pixel values grow from 1 to 64). The first low band L_1 is obtained by removing every even pixel on every odd-numbered row and every odd pixel on every even-numbered row. The result (shown in Figure 7.157b) is a rectangular lattice with eight rows and four pixels per column. It contains half the number of pixels in the original image. Since its elements are pixels, we call it a *subsampled band.* The positions of the removed pixels are labeled H_1 and their values are shown in small type in Figure 7.157c. Each H_1 is calculated by subtracting the original pixel value at that location from the average of the L_1 pixels directly above and below it. For example, the H_1 value 3 in the top row, column 2, is obtained by subtracting the original pixel 2 from the average $(10 + 0)/2 = 5$ (we use a simple *edge rule* where any missing pixels along edges of the image are considered zero for the purpose of predicting pixels). The H_1 value -33 at the bottom-left corner is obtained by subtracting 57 from the average $(0 + 49)/2 = 24$. For simplicity, we use just integers in these examples. A practical implementation, however, should deal with real values. Also, the prediction methods actually used by BTPC1 and BTPC2 are more sophisticated than the simple method shown here. They are discussed below. The various H_i bands are called *difference bands.*

Among the difference bands, H_1 is the finest (because it contains the most values), and H_8 is the coarsest. Similarly, L_8 (which is the only subsampled band written on the compressed stream) is the coarsest subsampled band.

Figure 7.157d shows how the second low band L_2 is obtained from L_1 by removing the pixels on even-numbered rows. The result is a square pattern containing half the number of pixels in L_1. The positions of the H_2 values are also shown. Notice that they don't have neighbors above or below, so we use two diagonal neighbors for prediction (again, the actual prediction used by BTPC1 and BTPC2 is different). For example, the H_2 value -5 in Figure 7.158a was obtained by subtracting the original pixel 16 from the prediction $(23+0)/2 = 11$. The next low band, L_3 (Figure 7.157e), is obtained from L_2 in the same way that L_1 is obtained from the original image. Band L_4 (Figure 7.157f) is obtained from L_3 in the same way that L_2 is obtained from L_1. Each band contains half the number of pixels of its predecessor. It is also obvious that the four near neighbors of a value H_i are located in its four corners for even values of i and in its four sides for odd values of i.

◇ **Exercise 7.50:** Calculate the values of L_i and H_i, for $i = 2, 3, 4$. For even values of i use the average of the bottom-left and top-right neighbors for prediction. For odd values of i use the average of the neighbors above and below.

1	2	3	4	5	6	7	8
9	10	11	12	13	14	15	16
17	18	19	20	21	22	23	24
25	26	27	28	29	30	31	32
33	34	35	36	37	38	39	40
41	42	43	44	45	46	47	48
49	50	51	52	53	54	55	56
57	58	59	60	61	62	63	64

(a)

1	H1	3	H1	5	H1	7	H1
H1	10	H1	12	H1	14	H1	16
17	H1	19	H1	21	H1	23	H1
H1	26	H1	28	H1	30	H1	32
33	H1	35	H1	37	H1	39	H1
H1	42	H1	44	H1	46	H1	48
49	H1	51	H1	53	H1	55	H1
H1	58	H1	60	H1	62	H1	64

(b)

1	3	3	2	5	1	7	0
0	10	0	12	0	14	0	16
17	0	19	0	21	0	23	0
0	26	0	28	0	30	0	32
33	0	35	0	37	0	39	0
0	42	0	44	0	46	0	48
49	0	51	0	53	0	55	0
-33	58	-34	60	-35	62	-36	64

(c)

1	.	3	.	5	.	7	.
.	H2	.	H2	.	H2	.	H2
17	.	19	.	21	.	23	.
.	H2	.	H2	.	H2	.	H2
33	.	35	.	37	.	39	.
.	H2	.	H2	.	H2	.	H2
49	.	51	.	53	.	55	.
.	H2	.	H2	.	H2	.	H2

(d)

1	.	H3	.	5	.	H3	.
.	.	.	.	.	.	.	.
H3	.	19	.	H3	.	23	.
.	.	.	.	.	.	.	.
33	.	H3	.	37	.	H3	.
.	.	.	.	.	.	.	.
H3	.	51	.	H3	.	55	.
.	.	.	.	.	.	.	.

(e)

1	.	.	.	5	.	.	.
.	.	.	.	.	.	.	.
.	.	H4	.	.	.	H4	.
.	.	.	.	.	.	.	.
33	.	.	.	37	.	.	.
.	.	.	.	.	.	.	.
.	.	H4	.	.	.	H4	.
.	.	.	.	.	.	.	.

(f)

Figure 7.157: An 8×8 Image and Its First Three L Bands.

1	.	3	.	5	.	7	.
.	0	.	0	.	0	.	-5
17	.	19	.	21	.	23	.
.	0	.	0	.	0	.	-13
33	.	35	.	37	.	39	.
.	0	.	0	.	0	.	-21
49	.	51	.	53	.	55	.
.	-33	.	-34	.	-35	.	-64

(a)

1	.	7	.	5	.	5	.
.	.	.	.	.	.	.	.
15	.	19	.	11	.	23	.
.	.	.	.	.	.	.	.
33	.	0	.	37	.	0	.
.	.	.	.	.	.	.	.
-33	.	51	.	-35	.	55	.
.	.	.	.	.	.	.	.

(b)

1	.	.	.	5	.	.	.
.	.	.	.	.	.	.	.
.	.	0	.	.	.	-5	.
.	.	.	.	.	.	.	.
33	.	.	.	37	.	.	.
.	.	.	.	.	.	.	.
.	.	-33	.	.	.	-55	.
.	.	.	.	.	.	.	.

(c)

Figure 7.158: (a) Bands L_2 and H_2. (b) Bands L_3 and H_3. (c) Bands L_4 and H_4.

◇ **Exercise 7.51:** Use Figure 7.158a to compute the entropy of band H_2.

The next step in BTPC encoding is to turn the binary pyramid into a binary tree, similar to the bintree of Section 7.34.1. This is useful, because many H_i values tend to be zeros, especially when lossy compression is used. If a node v in this tree is zero and all its children are zeros, the children will not be written on the compressed stream and v will contain a special termination code telling the decoder to substitute zeros for the children. The rule used by BTPC to construct the tree tells how to associate a difference value in H_i with its two children in H_{i-1}:

1. If i is even, a value in H_i has one child $i/2$ rows above it and another child located $i/2$ columns to its left in H_{i-1}. Thus, the 0 in H_4 of Figure 7.158c has the two children

7 and 15 in H_3 of Figure 7.158b. Also, the -33 in H_4 has the two children 0 and -33 in H_3
2. For odd i, if the parent (in H_{i-1}) of a value v in H_i is located below v, then the two children of v are located in H_{i-1}, $(i-1)/2$ rows below it and $(i-1)/2$ columns to its left and right. For example, the 5 in H_3 of Figure 7.158b has its parent (the -5 of H_4) below it, so its two children 0 and -5 are located one row below and one column to its left and right in the H_2 band of Figure 7.158a.
3. For odd i, if the parent (in H_{i-1}) of a value v in H_i is to the right of v, then the two children of v are located in H_{i-1}, $(i-1)/2$ rows below it. One is $(i-1)/2$ columns to its right, and the other is $3(i-1)/2$ columns to its right. For example, the -33 of H_3 has its parent (the -33 of H_4) to its right, so its children are the -33 and -34 of H_2. They are located one row below it, and one and three columns to its right.

These rules seem arbitrary, but they have the advantage that the descendants of a difference value form either a square or a rectangle around it. For example, the descendants of the 0 value of H_4 (shown in Figure 7.159 in boldface) are (1) its two children, the 7 and 15 of H_3, (2) their four children, the four top-left zeros of H_2 (shown in italics), and (3) the eight grandchildren in H_1 (shown in small type). These 14 descendants are shown in Figure 7.159.

.	3	7	2	.	.	.	.
0	*0*	0	*0*	.	.	.	.
15	0	**0**	0	.	.	.	.
0	*0*	0	*0*	.	.	.	.
.	.	.	.	.	.	.	.
.	.	.	.	.	.	.	.
.	.	.	.	.	.	.	.
.	.	.	.	.	.	.	.

Figure 7.159: The Fourteen Descendants of the Zero of H_4.

Not all these descendants are zeros, but if they were, the zero value of H_4 would have a special zero-termination code associated with it. This code tells the decoder that this value and all its descendants are zeros and are not written on the compressed stream. BTPC2 adds another twist to the tree. It uses the leaf codeword for left siblings in H_1 to indicate that both siblings are zero. This slightly improves the encoding of H_1.

Experiments show that turning the binary pyramid into a binary tree increases the compression factor significantly.

It should now be clear why BTPC is a natural candidate for implementation on a parallel computer. Each difference value input by the decoder from a difference band H_i is used to compute a pixel in subsampled band L_i, and these calculations can be done in parallel, because they are independent (except for competition for memory accesses).

We now turn to the pixel prediction used by BTPC. The goal is to have simple prediction, using just two or four neighbors of a pixel, while minimizing the prediction error. Figure 7.157 shows how each H_i difference value is located at the center of a group of four pixels that are known to the decoder from the decoding of bands H_{i-1}, H_{i-2}, etc. It is therefore a good idea for the encoder to use this group to predict the

H_i value. Figure 7.160a,b shows that there are three ways to estimate an H_i value at the center X of a group where A and C are two opposite pixels, and B and D are the other two. Two estimations are the linear predictions $(A + C)/2$ and $(B + D)/2$, and the third is the bilinear $(A + B + C + D)/4$. Based on the results of experiments, the developers of BTPC1 decided to use the following rule for prediction (notice that the decoder can mimic this rule):

> **Rule:** If the two extreme (i.e., the largest and smallest) pixels of A, B, C, and D are opposite each other in the prediction square, use the average of the other two opposite pixels (i.e., the middle two of the four values). Otherwise, use the average of the two opposite pixels closest in value to each other.

A		D		A		1		6		1	
	X		D	X	B		X		9	X	5
B		C		C		4		9		3	
	(a)			(b)			(c)			(d)	

Figure 7.160: The Four Neighbors Used to Predict an H_i Value X.

The two extreme values in Figure 7.160c are 1 and 9. They are opposite each other, so the prediction is the average 5 of the other two opposite pixels 4 and 6. In Figure 7.160d, on the other hand, the two extreme values, which are the same 1 and 9, are not opposite each other, so the prediction is the average, 2, of the 1 and 3, since they are the opposite pixels closest in value.

◇ **Exercise 7.52:** It seems that the best prediction is obtained when the encoder tries all three ways to estimate X and selects the best one. Why not use this prediction?

When BTPC2 was developed, extensive experiments were performed, resulting in more sophisticated prediction. BTPC2 selects one of 13 different predictions, depending on the values and positions of the four pixels forming the prediction square. These pixels are assigned names a, b, c, and d in such a way that they satisfy $a < b < c < d$. The 13 cases are summarized in Table 7.162.

As has been mentioned, BTPC has a natural lossy option that quantizes the prediction differences (the H_i values). The main aim of the quantization is to increase the number of zero differences, so it uses a double-size zero zone, illustrated in Figure 7.161. The quantizer step size in this figure is 3, so the values 3, 4, and 5 are quantized to 4 (quantization bin 1). The values -6, -7, and -8 are quantized to -7 (bin -2), but the values 0, 1, and 2 are quantized to 0 (bin 0), as are the values 0, -1, and -2.

The amount of quantization varies from level to level in the pyramid. The first difference band H_1 (the finest one) is the most coarsely quantized, since it is half the image size and since its difference values affect single pixels. Any quantization errors in this level do not propagate to the following difference bands. The main decision in the quantization process is how to vary the quantization step size s from level to level. The principle is to set the step size s_i of the current level H_i such that any value that would

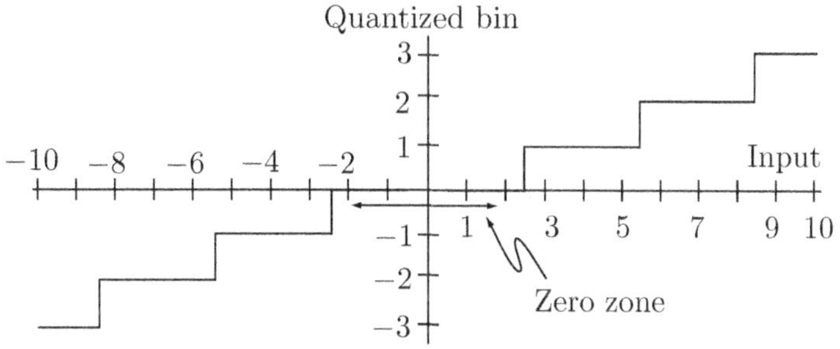

Figure 7.161: Quantization in BTPC.

Num	Name	Pixels	Closest-Opposite-Pair (BTPC1) Prediction	Adaptive (BTPC2) Prediction
0	Flat	a a a a	a	a
1	High point	a a a b	a	a
2	Line	a b b a	a or b	a or b (note1)
3	Aligned edge	a a b b	$(a+b)/2$	a or b (note2)
4	Low	a b b b	b	b
5	Twisted edge	a a b c	$(a+b)/2$	$(a+b)/2$
6	Valley	a b c a	a	a or $(a+b)/2$ (note1)
7	Edge	a b b c	b	b
8	Doubly twisted edge	a b c b	$(a+b)/2$ or $(b+c)/2$	b
9	Twisted edge	a b c c	$(b+c)/2$	$(b+c)/2$
10	Ridge	a c c b	c	c or $(a+b)/2$ (note1)
11	Edge	a b c d	$(b+c)/2$	$(b+c)/2$
12	Doubly twisted edge	a b d c	$(a+c)/2$ or $(b+d)/2$	$(b+c)/2$
13	Line	a c d b	$(a+b)/2$ or $(c+d)/2$	$(a+b)/2$ or $(c+d)/2$ (note1)

Note 1: Flag is needed to indicate choice

Note 2: Depending on other surrounding values

Table 7.162: Thirteen Predictions Used by BTPC2.

be quantized to 0 with the step s_{i-1} of the preceding level using exact prediction (no quantization), is quantized to 0 with the inexact prediction actually obtained by step size s_i. If we can compute the range of prediction error caused by earlier quantization errors, then the appropriate value for s_i is the preceding step size s_{i-1} plus the maximum error in the prediction. Since this type of calculation is time-consuming, BTPC determines the step size s_i for level i by scaling down the preceding step size s_{i-1} by a constant factor a. Thus, $s_i = a\,s_{i-1}$, where the constant a satisfies $a < 1$. Its value was determined by experiment to be in the range 0.75 to 0.8.

The last step in BTPC compression is the entropy coding of L_8 and the eight difference bands H_i. BTPC1 does not include an entropy coder. Its output can be sent to any available adaptive lossless coder, such as Huffman, arithmetic or dictionary-based. BTPC2 includes an integrated adaptive Huffman coder. This coder is reset for each difference band, because each band is quantized differently, so their statistics are different. If the coder is not reset at the start of a band H_i, it may produce bad compression while getting adapted to the statistical model of H_i.

The binary tree structure introduces another feature that should be taken into account. When a leaf is found with a zero-termination code, it means that the current difference value and all its descendants are zero. However, if an interior node with a zero value is found in the binary tree, and if its left child is a leaf, then its right child cannot be a leaf (since otherwise, the parent would have two zero children and would itself be a leaf). As a result, a right child whose parent is zero and whose left sibling is a leaf is special in some sense. Because of this property, BTPC2 uses three adaptive Huffman coders for each difference band, one for left children, another for "normal" right children, and the third for "special" right children.

7.34 Quadtrees

A quadtree compression of a bi-level image is based on the principle of image compression (Section 1.4) which states; If we select a pixel in the image at random, there is a good chance that its immediate neighbors will have the same or similar color. The quadtree method scans the bitmap, area by area, looking for areas composed of identical pixels (uniform areas). This should be compared to RLE image compression (Section 1.4), where only neighbors on the same scan row are checked, even though neighbors on the same column may also be identical or very similar.

The input consists of bitmap pixels, and the output is a tree (a quadtree, where each node is either a leaf or has exactly four children). The size of the quadtree depends on the complexity of the image. For complex images, the tree may be bigger than the original bitmap, resulting in expansion. The method starts by constructing a single node, the root of the final quadtree. It divides the bitmap into four quadrants, each to become a child of the root. A uniform quadrant (one where all the pixels have the same color) is saved as a leaf child of the root. A nonuniform quadrant is saved as an (interior node) child of the root. Any nonuniform quadrants are then recursively divided into four smaller subquadrants that are saved as four sibling nodes of the quadtree. Figure 7.163 shows a simple example.

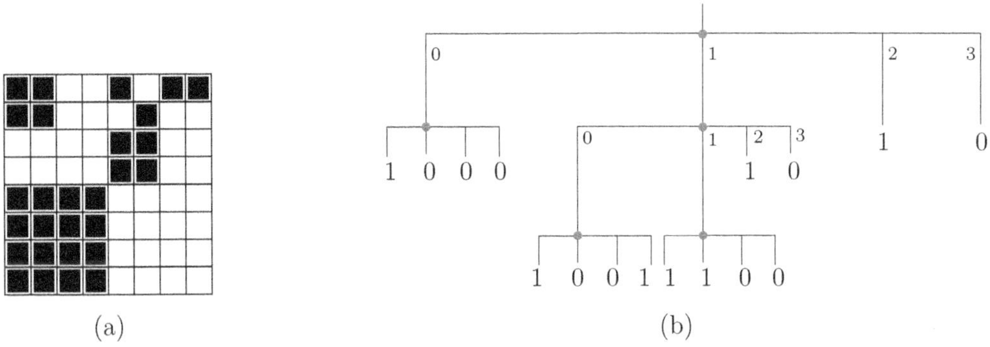

Figure 7.163: A Quadtree.

The 8×8 bitmap in 7.163a produces the 21-node quadtree of 7.163b. Sixteen nodes are leaves (each containing the color of one quadrant, 0 for white, 1 for black), and the other five (the gray circles) are interior nodes containing four pointers each. The quadrant numbering used is $\binom{0\,1}{2\,3}$ (but see Exercise 7.67 for a more natural numbering scheme).

The size of a quadtree depends on the complexity of the image. Assuming a bitmap size of $2^N \times 2^N$, one extreme case is where all the pixels are identical. The quadtree in this case consists of just one node, the root. The other extreme case is where each quadrant, even the smallest one, is nonuniform. The lowest level of the quadtree has, in such a case, $2^N \times 2^N = 4^N$ nodes. The level directly above it has a quarter of that number (4^{N-1}), and the level above that one has 4^{N-2} nodes. The total number of nodes in this case is $4^0+4^1+\cdots+4^{N-1}+4^N = (4^{N+1}-1)/3 \approx 4^N(4/3) \approx 1.33\times 4^N = 1.33(2^N\times 2^N)$. In this worst case the quadtree contains about 33% more nodes than the number of pixels (the bitmap size). Such an image therefore generates considerable expansion when converted to a quadtree.

A nonrecursive approach to generating a quadtree starts by building the complete quadtree assuming that all quadrants are nonuniform and then checking the assumption. Every time a quadrant is tested and found to be uniform, the four nodes corresponding to its four quarters are deleted from the quadtree. This process proceeds from the bottom (the leaves) up towards the root. The main steps are the following:

1. A complete quadtree of height N is constructed. It contains levels $0, 1, \ldots, N$ where level k has 4^k nodes.
2. All $2^N \times 2^N$ pixels are copied from the bitmap into the leaves (the lowest level) of the quadtree.
3. The tree is scanned level by level, from the bottom (level N) to the root (level 0). When level k is scanned, its 4^k nodes are examined in groups of four, the four nodes in each group having a common parent. If the four nodes in a group are leaves and have the same color (i.e., if they represent a uniform quadrant), they are deleted and their parent is changed from an interior node to a leaf having the same color.

It's easy to analyze the time complexity of this approach. A complete quadtree has about 1.33×4^N nodes, and since each is tested once, the number of operations (in step 3) is 1.33×4^N. Step 1 requires 1.33×4^N operations, and Step 2 requires 4^N operations.

The total number of operations is therefore $(1.33+1.33+1)\times 4^N = 3.66\times 4^N$. We are faced with comparing the first method, which requires $(1/3)\times 4^N$ steps, with the second method, which needs 3.66×4^N operations. Since N usually varies in the narrow range 8–12, the difference is not very significant. A similar analysis of storage requirements shows that the first method needs only the memory space required by the final quadtree, whereas the second method uses all the storage needed for a complete quadtree.

The following discussion shows the relation between the positions of pixels in a quadtree and in the image. Imagine a complete quadtree. Its bottom row consists of all the $2^n\times 2^n$ pixels of the image. Suppose we scan these pixels from left to right and number them. We show how the number of a pixel can be used to determine its (x,y) image coordinates.

Each quadrant, subquadrant, or pixel (or, in short, each subsquare) obtained by a quadtree partitioning of an image can be represented by a string of the quaternary (base 4) digits 0, 1, 2, and 3. The longer the string, the smaller the subsquare it represents. We denote such a string by $d_i d_{i-1}\ldots d_0$, where $0\le i\le n$. We assume that the quadrant numbering of Figure 7.186a (Section 7.38) is extended recursively to subsquares of all sizes. Figure 7.186b shows how each of the 16 subquadrants produced from the four original ones is identified by a 2-digit quaternary number. After another subdivision, each of the resulting subsubquadrants is identified by a 3-digit number, and so on. The black area in Figure 7.186c, for example, is identified by the quaternary integer 1032, and the gray area is identified by the integer 2011_4.

Consecutive quaternary numbers are easy to generate. We simply have to increment the digit 3_4 to the 2-digit number 10_4. The first n-digit quaternary numbers are (notice that there are $4\times 2^{2n-2}=2^n\times 2^n$ of them)

$$0,\ 1,\ 2,\ 3,\ 10,\ 11,\ 12,\ 13,\ 20,\ 21,\ 22,\ 23,\ 30,\ 31,\ 32,\ 33,\ 100,\ 101,\ 102,\ 103,\ldots,$$
$$\ldots,130,\ 131,\ 132,\ 133,\ 200,\ \ldots\ ,\underbrace{33\ldots3}_{n}.$$

Figure 7.164a shows a complete quadtree for a $2^2\times 2^2$ image. The 16 pixels constitute the bottom row, and their quaternary numbers are listed. Once we know how to locate a pixel in the image by its quaternary number, we can construct the quadtree with a bottom-up, left-to-right approach. Here are the details. We start with the four pixels with quaternary numbers 00, 01, 02, and 03. They become the bottom-left part of the tree and are numbered 1–4 in Figure 7.164b. We construct their parent node (numbered 5 in Figure 7.164b). If the four pixels are uniform, they are deleted from the quadtree. We do the same with the next group of pixels, whose quaternary numbers are 10, 11, 12, and 13. They become pixels 6–9, with a parent numbered 10 in Figure 7.164b. If they are uniform, they are deleted. This is repeated until four parents, numbered 5, 10, 15, and 20, are constructed. Their parent (numbered 21) is then created, and the four nodes are checked. If they are uniform, they are deleted. The process continues until the root is created, its four children nodes are checked, and, if necessary, deleted. This a recursive approach whose advantage is that no extra memory is needed. Any unnecessary nodes of the quadtree are deleted as soon as they and their parent are created.

Given a square image with $2^n\times 2^n$ pixels, each pixel is identified by an n-digit quaternary number. Given such a number $d_{n-1}d_{n-2}\ldots d_0$, we show how to locate "its"

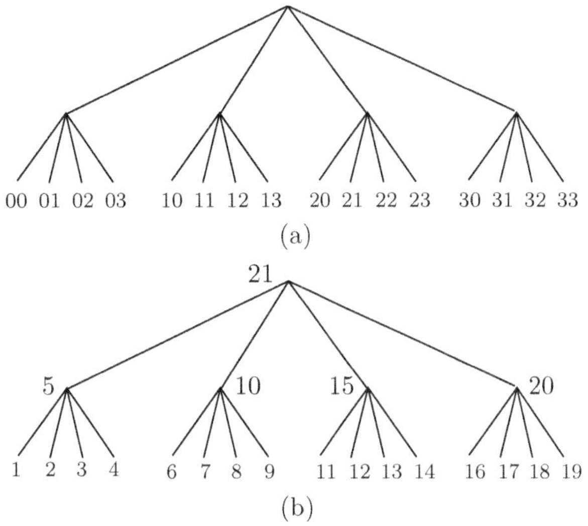

Figure 7.164: A Quadtree for a $2^2 \times 2^2$ Image.

pixel in the image. We assume that a pixel has image coordinates (x, y), with the origin of the coordinate system at the bottom-left corner of the image. We start at the origin and scan the quaternary digits from left to right. A digit of 1 tells us to move up to reach our target pixel. A digit of 2 signals a move to the right, and a digit of 3 directs us to move up and to the right. A digit of 0 corresponds to no movement. The amount of the move halves from digit to digit. The leftmost digit corresponds to a move of 2^{n-1} rows and/or columns, the second digit from the left corresponds to a move of 2^{n-2} rows and/or columns, and so on, until the rightmost digit corresponds to moving just $1\,(=2^0)$ pixels (or none, if this digit is a 0). The pseudo-code of Figure 7.165 summarizes this process

```
x:=0; y:=0;
for k:= n-1 step -1 to 0 do
 if digit(k)=1 or 3 then y := y + 2^k;
 if digit(k)=2 or 3 then x := x + 2^k
endfor;
```

Figure 7.165: Pseudo-Code to Locate a Pixel in an Image.

It should also be noted that quadtrees are a special case of the Hilbert curve, discussed in Section 7.36.

7.34.1 Bintrees

Instead of partitioning the image into quadrants, it can be recursively split in halves. This is the principle of the bintree method. Figure 7.166a–e shows the 8×8 image of Figure 7.163a and the first four steps in its bintree partitioning. Figure 7.166f shows part of the resulting bintree. It is easy to see how the bintree method alternates between vertical and horizontal splits, and how the subimages being generated include all those produced by a quadtree plus other ones. As a compression method, bintree partitioning is less efficient than quadtree, but it may be useful in cases where many subimages are needed. A case in point is the WFA method of Section 7.38. The original method uses a quadtree to partition an image into nonoverlapping subsquares, and compresses the image by matching a subsquare with a linear combination of other (possibly bigger) subsquares. An extension of WFA (page 707) uses bintrees to obtain more subimages and therefore better compression.

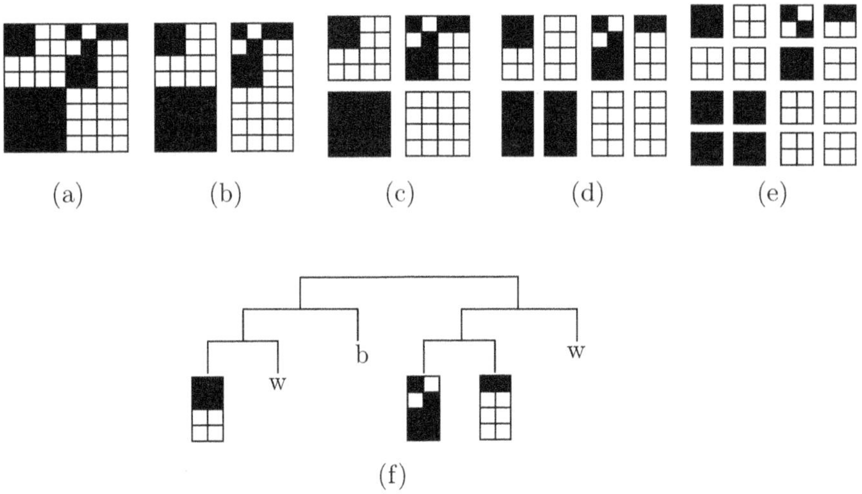

Figure 7.166: A Bintree for an 8×8 Image.

7.34.2 Composite and Difference Values

The average of a set of integers is a good representative of the integers in the set, but is not always an integer. The median of such a set is an integer but is not always a good representative. The progressive image method described in this section, due to K. Knowlton [Knowlton 80], employs the concepts of *composite* and *differentiator* to encode an image progressively. The image is encoded in layers, where early layers consist of a few large, low-resolution blocks, followed by later layers with smaller, higher-resolution blocks. The image is divided into layers using the method of bintrees (Section 7.34.1). The entire image is divided into two vertical halves, each is divided into two horizontal quadrants, and so on.

The first layer consists of a single uniform area, the size of the entire image. We call it the *zeroth* progressive approximation. The next layer is the *first* approximation. It consists of two uniform rectangles (or cells), one above the other, each half the size of the image. We can consider them the two children of the single cell of the zeroth approximation. In the *second* approximation, each of these cells is divided into two children, which are two smaller cells, placed side by side. If the original image has 2^n

pixels, then they (the pixels) are the cells of the nth approximation, which constitutes the last layer.

If the image has 16 grayscales, then each pixel consists of four bits, describing its intensity. The pixels are located at the bottom of the bintree (they are the leaves of the tree), so each cell in the layer immediately above the leaves is the parent of two pixels. We want to represent such a cell by means of two 4-bit numbers. The first number, called the *composite*, should be similar to an average. It should be a representative of both pixels taken as a unit. The second number, called the *differentiator*, should reflect the difference between the pixels. Using the composite and differentiator (which are 4-bit integers) it should be possible to reconstruct the two pixels.

Figure 7.167a,b,c shows how two numbers $v_1 = 30$ and $v_2 = 03$ can be considered the vectors $(3, 0)$ and $(0, 3)$, and how their sum $v_1+v_2 = (3, 3)$ and difference $v_1-v_2 = (3, -3)$ correspond to a 45° rotation of the vectors. The sum can be considered, in a certain sense, the average of the vectors, and the difference can be used to reconstruct them. However, the sum and difference of binary vectors are, in general, not binary. The sum and difference of the vectors $(0, 0, 1, 1)$ and $(1, 0, 1, 0)$, for example, are $(1, 0, 2, 1)$ and $(-1, 0, 0, 1)$.

The method proposed here for determining the composite and differentiator is illustrated by Figure 7.167d. Its 16 rows and 16 columns correspond to the 16 grayscales of our hypothetical image. The diagram is divided into 16 narrow strips of 16 locations each. The strips are numbered 0 through 15, and these numbers (shown in boldface) are the composite values. The 16 locations within each strip are also numbered 0 through 15 (this numbering is shown for strips 3 and 8), and they are the differentiators. Given two adjacent cells, $v_1 = 0011$ and $v_2 = 0100$ in one of the approximations, we use their values as (x, y) table coordinates to determine the composite $3 = 0011$ and differentiator $7 = 0111$ of their parent.

Once a pair of composite and differentiator values are given, Figure 7.167d can be used to reconstruct the two original values. It is obvious that such a diagram can be designed in many different ways, but the important feature of this particular diagram is the way it determines the composite values. They are always close to the true average of v_1 and v_2. The elements along the main diagonal, for example, satisfy $v_1 - v_2$, and the diagram is designed such that the composite values c for these elements are $c = v_1 = v_2$. The diagram is constructed by laying a narrow, 1-unit, strip of length 16 around the bottom-left corner, and adding similar strips that are symmetric about the main diagonal.

Most of the time, the composite value that is determined by this diagram for a pair v_1 and v_2 of 4-bit numbers is their true average, or it differs from the true average by 1. Only in 82 out of the 256 possible pairs (v_1, v_2) does the composite value differ from the true average $(v_1 + v_2)/2$ by more than 1. This is about 32%. The deviations of the composite values from the true averages are shown in Figure 7.167e. The maximum deviation is 4, and it occurs in just two cases.

Since the composite values are so close to the average, they are used to color the cells of the various approximations, which makes for realistic-looking progressive images. This is the main feature of the method.

Figure 7.168 shows the binary tree resulting from the various approximations. The root represents the entire image, and the leaves are the individual pixels. Every pair of

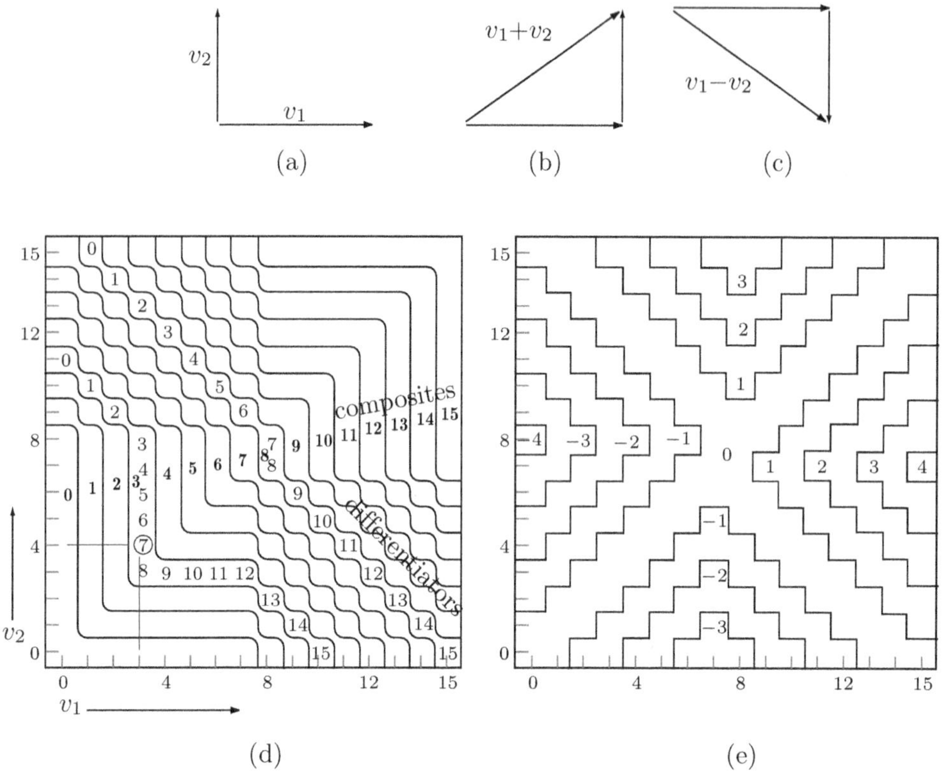

Figure 7.167: Determining Composite and Differentiator Values.

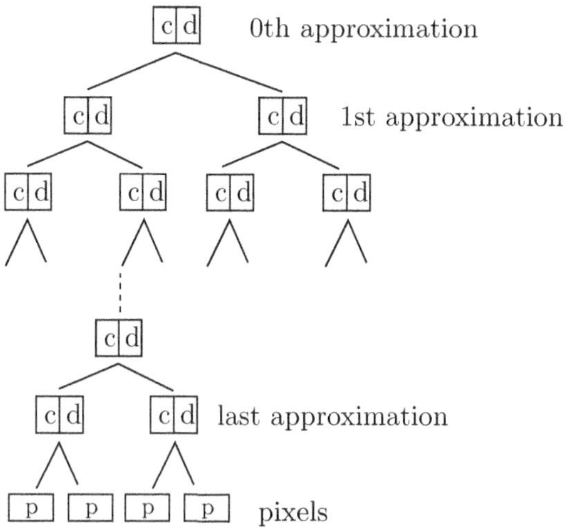

Figure 7.168: Successive Approximation in a Binary Tree.

pixels becomes a pair (c, d) of composite and differentiator values in the level above the leaves. Every pair of composite values in that level becomes a pair (c, d) in the level above it, and so on, until the root becomes one such pair. If the image contains 2^n pixels, then the bottom level (the leaves) contains 2^n nodes. The level above it contains 2^{n-1} nodes, and so on. Displaying the image progressively is done by the progressive decoder in steps as follows:

1. The pair (c_0, d_0) for the root is input and the image is displayed as one large uniform block (0th approximation) of intensity c_0.
2. Values c_0 and d_0 are used to determine, from Figure 7.167d, the two composite values c_{10} and c_{11}. The first approximation, consisting of two large uniform cells with intensities c_{10} and c_{11}, is displayed, replacing the 0th approximation. Two differentiator values d_{10} and d_{11} for the next level (2nd approximation) are input.
3. Values c_{10} and d_{10} are used to determine, from Figure 7.167d, the two composite values c_{20} and c_{21}. Values c_{11} and d_{11} are similarly used to determine the two composite values c_{22} and c_{23}. The second approximation, consisting of four large cells with intensities c_{20}, c_{21}, c_{22}, and c_{23}, is displayed, replacing the 1st approximation. Four differentiator values d_{2i} for the next level (3rd approximation) are input.
4. This process is repeated until, in the last step, the 2^{n-1} composite values for the next to last approximation are determined and displayed. The 2^{n-1} differentiator values for these composites are now input, and each of the 2^{n-1} pairs (c, d) is used to compute a pair of pixels. The 2^n pixels are displayed, completing the progressive generation of the image.

These steps show that the image file should contain the value c_0, the value d_0, the two values d_{10} and d_{11}, the four values d_{20}, d_{21}, d_{22}, and d_{23}, and so on, up to the 2^{n-1} differentiator values for the next to last approximation. The total is one composite value and $2^n - 1$ differentiator values. Thus, the image file contains 2^n 4-bit values, so its size equals that of the original image. The method discussed so far is for progressive image transmission and does not yet provide any compression.

◇ **Exercise 7.53:** Given the eight pixel values 3, 4, 5, 6, 6, 4, 5, and 8, build a tree similar to the one of Figure 7.168 and list the contents of the progressive image file.

The encoder starts with the original image pixels (the leaves of the binary tree) and prepares the tree from bottom to top. The image file, however, should start with the root of the tree, so the encoder must keep all the levels of the tree in memory while the tree is being generated. Fortunately, this does not require any extra memory. The original pixel values are not needed after the first step, and can be discarded, leaving room for the composite and differentiator values. The composite values are used in the second step and can later also be discarded. Only the differentiator values have to be saved for all the steps, but they can be stored in the memory space originally occupied by the image.

It is clear that the first few approximations are too coarse to show any recognizable image. However, they involve very few cells and are computed and displayed quickly. Each approximation has twice the number of cells of its predecessor, so the time it takes to display an approximation increases geometrically. Since the computations involved are simple, we can estimate the time needed to generate and display an approximation by considering the time it takes to input the values needed for the following approximation;

the time for computations can be ignored. Assuming a transmission speed of 28,800 baud, the zeroth approximation (input two 4-bit values) takes $8/28800 \approx 0.00028$ sec, the first approximation (input another two 4-bit values) takes the same time. The second approximation (four 4-bit values) takes 0.000556 sec, and the tenth approximation (input 2^{10} 4-bit numbers) takes 0.14 sec Subsequent approximations take 0.28, 0.56, 1.04, and 2.08 sec.

It is also possible to develop the image progressively in a nonuniform way. The simplest way to encode an image progressively is to compute the differentiator values of each approximation in raster order and write them on the image file in this order. However, if we are interested in the center of the image, we may change this order, as long as we do it in the same way for the encoder and decoder. The encoder may write the differentiator values for the center of the image first, followed by the remaining differentiator values. The decoder should mimic this. It should start each approximation by displaying the center of the image, followed by the rest of the approximation.

The image file generated by this method can be compressed by entropy coding the individual values in it. All the values (except one) in this file are differentiators, and experiments indicate that these values are distributed normally. For 4-bit pixels, differentiator values are in the range 0 through 15, and Table 7.169 lists typical counts and possible Huffman codes for each of the 16 values. This reduces the data from four bits per differentiator to about 2.7 bits/value.

Diff. value	Count	Huffman code
0	27	00000000
1	67	00000001
2	110	0000001
3	204	000001
4	485	00001
5	1564	0001
6	4382	001
7	8704	01
8	10206	11
9	4569	101
10	1348	1001
11	515	10001
12	267	100001
13	165	1000001
14	96	10000001
15	58	10000000

Table 7.169: Possible Huffman Codes for 4-bit Differentiator Values.

Another way to compress the image file is to quantize the original pixels from 16 to 15 intensity levels and use the extra pixel value as a termination code, indicating a uniform cell. When the decoder inputs this value from the image file, it does not split the cell further.

7.34.3 Progressive Bintree Compression

Various techniques for progressive image representation are discussed in Section 7.13. Section 7.34.2 describes an application of bintrees for the progressive transmission of grayscale images. This section (based on [Knowlton 80]) outlines a similar method for bi-level images. We illustrate the method on an image with resolution 384×512. The image is divided into blocks of size 3×2 each. There are $128=2^7$ rows and $256=2^8$ columns of these 6-tuples, so their total number is $2^{15}=32{,}768$.

The encoder constructs a binary tree by dividing the entire image into two horizontal halves, splitting each into two vertical quadrants, and continuing in this way, down to the level of 6-tuples. In practice, this is done from the bottom up, following which the tree is written on the output file top to bottom. Figure 7.170 shows the final tree. Each node is marked as either uniform black (b, with prefix code 10), uniform white ($w=11$), or mixed ($m=0$). A mixed node also contains another prefix code, following the 0, to specify one of five shades of gray. A good set of these codes is

110 for 16.7%, 00 for 33.3%, 01 for 50%, 10 for 66.7%, and 111 for 83.3%.

Five codes are needed, since a group of six pixels can have between zero and six white pixels, corresponding to seven shades of gray. Two shades, namely black and white, do not occur in a mixed 6-tuple, so only the five shades above need be specified. The average size of these codes is 2.4 bits, so a mixed (m) tree node has on average 3.4 bits, the code 0 followed by two or three bits.

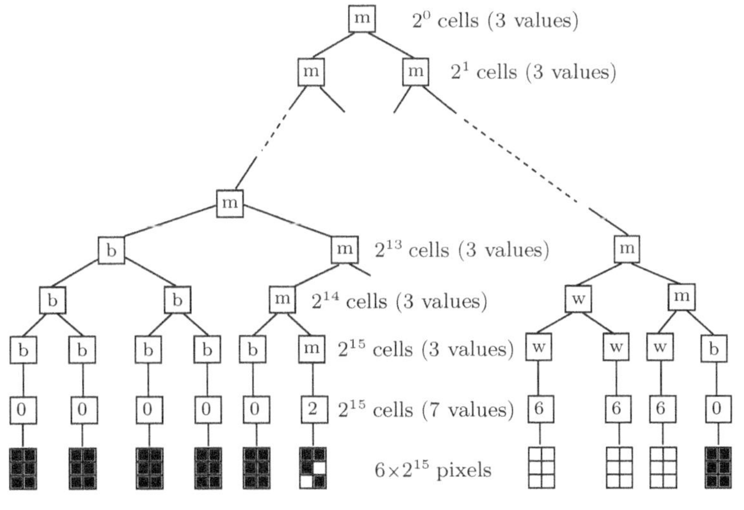

Figure 7.170: A Complete Bintree for a Bi-level Image.

When the decoder inputs one of these levels, it splits each block of the preceding level into two smaller blocks and uses the codes to paint them black, white, or one of five shades of gray.

The bottom two levels of the tree are different. The next to last level contains 2^{15} nodes, one for each 6-tuple. A node in this level contains one of seven codes, specifying the number of white pixels in the 6-tuple. When the decoder gets to this level, each block is already a 6-tuple, and it is painted the right shade of gray, according to the number of the white pixels in the 6-tuple. When the decoder gets to the last level, it already knows how many white pixels each 6-tuple contains. It only needs to be told *where* these white pixels are located in the 6-tuple. This is also done by means of codes. If a 6-tuple contains just one white pixel, that pixel may be located in one of six positions, so six 3-bit codes are needed. A 6-tuple with five white pixels also requires a 3-bit code. If a 6-tuple contains two white pixels, they may be located in the 6-tuple in 15 different ways, so 15 4-bit codes are needed. A 6-tuple with four white pixels also requires a 4-bit code. When a 6-tuple contains three white pixels, they may form 20 configurations, so 20 prefix (variable-length) codes are used, ranging in size from 3 to 5 bits.

◇ **Exercise 7.54:** Show the 15 6-tuples with two white pixels.

The size of the binary tree of Figure 7.170 can now be estimated. The top levels (excluding the bottom three levels) contain between 1 and 2^{14} nodes, for a total of $2^{15} - 1 \approx 2^{15}$. The third level from the bottom contains 2^{15} nodes, so the total number of nodes so far is 2×2^{15}. Each of these nodes contains either a 2-bit code (for uniform b and w nodes) or a 3.4-bit code (for m nodes). Assuming that half the nodes are mixed, the average is 2.7 bits per node. The total number of bits so far is $2{\times}2^{15}{\times}2.7 = 5.4{\times}2^{15}$. This almost equals the original size of the image (which is 6×2^{15}), and we still have two more levels to go!

The next to last level contains 2^{15} nodes with a 3-bit code each (one of seven shades of gray). The bottom level also contains 2^{15} nodes with 3, 4, or 5-bit codes each. Assuming a 4-bit average code size for this level, the total size of the tree is

$$(5.4 + 3 + 4) \times 2^{15} = 12.4 \times 2^{15},$$

more than twice the image size! Compression is achieved by pruning the tree, similar to a quadtree, such that any b or w node located high in the tree becomes a leaf. Experiments indicate a typical compression factor of 6, implying that the tree is reduced in size from twice the image size to 1/6 the image size: a factor of 12. [Knowlton 80] describes a more complex coding scheme that produces typical compression factors of 8.

One advantage of the method proposed here is the fact that it produces gray blocks for tree nodes of type m (mixed). This means that the method can be used even with a bi-level display. A block consisting of black and white pixels looks gray when we watch it from a distance, and the amount of gray is determined by the mixture of black and white pixels in the block. Methods that attempt to get nice-looking gray blocks on a bi-level display are known as *dithering* [Salomon 99].

7.34.4 Compression of N-Tree Structures

Objects encountered in real life are normally three dimensional, although many objects are close to being two- or even one-dimensional. In geometry, objects can have any number of dimensions, although we cannot visualize objects in more than three dimensions. An N-tree data structure is a special tree that stores an N-dimensional object. The most common example is a quadtree (Section 7.34), a popular structure for storing a two-dimensional object such as an image.

An *octree* is the obvious extension of a quadtree to three dimensions [Samet 90a,b]. An octree isn't used for data compression; it is a data structure where a three-dimensional object can be stored. In an octree, a node is either a leaf or has exactly eight children. The object is divided into eight octants, each nonuniform octant is recursively divided into eight suboctants, and the process continues until the subsuboctants reach a certain minimal size. Similar trees (N-trees) can, in principle, be constructed for N-dimensional objects.

The technique described here, due to [Buyanovsky 02], is a simple, fast method to compress an N-tree. It has been developed as part of a software package to handle medical data such as NMR, ultrasound, and CT. Such data consists of a set of three-dimensional medical images taken over a short period of time and is therefore four dimensional, with time as the fourth dimension. It can be stored in a hextree, where each node is either a leaf or has exactly 16 children. This section describes the data compression aspect of the project and uses a quadtree as an example.

Imagine a quadtree with the pixels of an image. For the purpose of compression, we assume that any node A of the quadtree contains (in addition to pointers to the four children) two values: the minimum and maximum pixel values of the nodes of the subtree whose root A is. For example, the quadtree of Figure 7.171 has pixels with values between 0 and 255, so the root of the tree contains the pair $0, 255$. Without compression, pixel values in this quadtree are 8-bit numbers, but the method described here makes it possible to write many of those values on the compressed stream encoded with fewer bits. The leftmost child of the root, node ①, is itself the root of a subtree whose pixel values are in the interval $[15, 255]$. Obviously, this interval cannot be wider than the interval $[0, 255]$ for the entire quadtree, so pixel values in this subtree may, in principle, be encoded in fewer than eight bits. The number of bits required to arithmetically encode those values is $\log_2(255 - 15 + 1) \approx 7.91$, so there is a small gain for the four immediate children of node ①. Similarly, pixel values in the subtree defined by node ② are in the interval $[101, 255]$ and therefore require $\log_2(255 - 101 + 1) \approx 7.276$ bits each when arithmetically encoded. Pixel values in the subtree defined by node ③ are in the narrow interval $[172, 216]$ and therefore require only $\log_2(216 - 172 + 1) \approx 5.49$ bits each. This is achieved by encoding these four numbers on the compressed stream relative to the constant 172. The numbers actually encoded are 44, 0, 9, and 25.

The two nodes marked ⑦ have identical minimum and maximum pixel values, which indicates that these nodes correspond to uniform quadrants of the image and are therefore leaves of the quadtree.

The following point helps to understand how pixel values are encoded. When the decoder starts decoding a subtree, it already knows the values of the maximum and minimum pixels in the subtree because it has already decoded the parent tree of that subtree. For example, when the decoder starts decoding the subtree whose root is ②, it knows that the maximum and minimum pixel values in that subtree are 101 and 255, because it has already decoded the four children of node ①. This knowledge is utilized to further compress a subtree, such as ③, whose four children are pixels.

A group of four pixels is encoded by first encoding the values of the first (i.e., leftmost) two pixels using the required number of bits. For the four children of node ③,

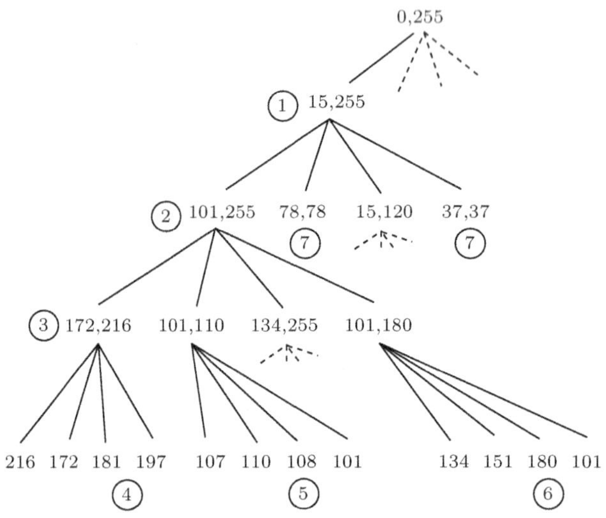

Figure 7.171: A Quadtree with Pixels in the Interval $[0, 255]$.

these are 44 and 0, encoded in 5.49 bits each. The remaining two pixel values may be encoded with fewer bits, and this is done by distinguishing three cases as follows:

Case 1: The first two pixel values are the minimum and maximum of the four. Once the decoder reads and decodes those two values, all it knows about the following two values is that they are between the minimum and the maximum. As a result, the last two values have to be encoded by the encoder with the required number of bits and there is no gain. In our example, the first two values are 216 and 172, so the next two values 181 and 197 (marked by ④) have to be written as the numbers 9 and 25. The encoder encodes the four values 44, 0, 9, and 25 arithmetically in 5.49 bits each. The decoder reads and decodes the first two values and finds out that they are the minimum and maximum, so it knows that this is Case 1 and two more values remain to be read and decoded.

Case 2: One of the first two pixel values is the minimum or the maximum. In this case, one of the remaining two values is the maximum or the minimum, and this is indicated by a 1-bit indicator that's encoded and emitted by the encoder following the second pixel. Consider the four pixel values 107, 110, 108, and 101. They should be encoded in 3.32 bits each [because $\log_2(110 - 101 + 1) \approx 3.32$] relative to 101. The first two values encoded are 6 and 9. After the decoder reads and decodes them, it knows that the maximum (110) is one of them but the minimum (101) is not. Thus, the decoder knows that this is Case 2, and an indicator, followed by one more value, remain to be read. Once the indicator is read, the decoder knows that the minimum is the 4th value and therefore the next value is the 3rd of the four pixels. That value (108, marked by ⑤) is then read and decoded. Compression is increased because the encoder does not need to encode the minimum pixel, 101. Without an indicator, the four values would require $4\times 3.32 = 13.28$ bits. With the indicator, the number of bits required is $2\times 3.32 + 1 + 3.32 = 10.96$, a savings of 2.32 bits or 17.5% of 13.28.

Case 3: None of the first two pixel values is the minimum or the maximum. Thus, one of the remaining two values is the maximum and the other one is the minimum. The encoder writes the first two values (encoded) on the compressed file, followed by a 1-bit indicator that indicates which of the two remaining values is the maximum. Compression is enhanced, since the encoder does not have to write the actual values of the minimum and maximum pixels. After reading and decoding the first two values, the decoder finds out that they are not the minimum and maximum, so this is Case 3, and only an indicator remains to be read and decoded. The four pixel values 134, 151, 180, and 101 serve as an example. The first two values are written (relative to 101) in 6.32 bits each. The 180 (marked by ⑥) is the maximum, so a 1-bit indicator is encoded to indicate that the maximum is the third value. Instead of using $4 \times 6.32 = 25.28$ bits, the encoder uses $2 \times 6.32 + 1 = 13.64$ bits.

The case max = min+1 is especially interesting. In this case $\log_2(\text{max}-\text{min}+1) = 1$, so each pixel value is encoded in one bit on average. The algorithm described here does just that and does not treat this case in any special way. However, in principle, it is possible to handle this case separately, and to encode the four pixel values in fewer than four bits (in 3.8073 bits, to be precise). Here is how.

We denote min and max by n and x, respectively, and explore all the possible combinations of n and x.

If one of the first two pixel values is n and the other one is x, then both encoder and decoder know that this is case 1 above. No indicator is used. Notice that the remaining two pixels can be one of the four pairs (n, n), (n, x), (x, n), and (x, x), so a 1-bit indicator wouldn't be enough to distinguish between them. Thus, the encoder encodes each of the four pixel values with one bit, for a total of four bits.

If the first two pixel values are both n or both x, then this is Case 2 above. There can be two subcases as follows:

Case 2.1: The first two values are n and n. These can be followed by one of the three pairs (n, x), (x, n), or (x, x).

Case 2.2: The first two values are x and x. These can be followed by one of the three pairs (n, x), (x, n), or (n, n).

Thus, once the decoder has read the first two values, it has to identify one of three alternatives. In order to achieve this, the encoder can, in principle, follow the first two values (which are encoded in one bit each) with a special code that indicates one of three alternatives. Such a code can, in principle, be encoded in $-\log_2 3 \approx 1.585$ bits. The total number of bits required, in principle, to encode Case 2 is thus $2 + 1.585 = 3.585$.

In Case 3, none of the first two pixel values is n or x. This, of course, cannot happen when max = min + 1, since in this case each of the four values is either n or x. Case 3 is therefore impossible.

We therefore conclude that in the special case max = min + 1, the four pixel values can be encoded either in 4 bits or in 3.585 bits. On average, the four values can be encoded in fewer than 4 bits, which seems magical! The explanation is that two of the 16 possible combinations never occur. We can think of the four pixel values as a 4-tuple where the elements are either n or x. In general, there are 16 such 4-tuples, but the two cases (n, n, n, n) and (x, x, x, x) cannot occur, which leaves just 14 4-tuples to be encoded. It takes $-\log_2(14) \approx 3.8073$ bits to encode one of 14 binary 4-tuples. Since the leaves of a quadtree tend to occupy much space and since the case max $=$ min $+ 1$ is highly probable in images, we estimate that the special treatment described here may improve compression by 2–3%. The pseudocode listed here, however, does not treat the case max $=$ min $+ 1$ in any special way.

The pseudocode shows how pixel values and indicators are sent to a procedure `encode(value, min, max)` to be arithmetically encoded. This procedure encodes its first parameter in $\log_2(\max - \min + 1)$ bits on average. An indicator is encoded by setting min $= 0$, max $= 1$, and a value of 0 or 1. The indicator bit is therefore encoded in $\log_2(1 - 0 + 1) = 1$ bit on average. (This means that some indicators are encoded in more than one bit, but others are encoded in as few as zero bits! In general, the number of bits spent on encoding an indicator is distributed normally around 1.)

The method is illustrated by the following C-style code:

```
/* Definitions:
encode( value , min , max ); - function of encoding of value
 (output bits stream),
 min<=value<=max, the length of code is Log2(max-min+1) bits;
Log2(N) - depth of pixel level of quadtree.
struct knot_descriptor
{
int min,max; //min,max of the whole sub-plane
int pix ; // value of pixel (in case of pixel level)
int depth ; // depth of knot.
knot_descriptor *square[4];//children's sub-planes
} ;
Compact_quadtree (...) - recursive procedure of quadtree compression.
*/
Compact_quadtree (knot_descriptor *knot, int min, int max)
{
encode( knot->min , min , max ) ;
encode( knot->max , min , max ) ;
if ( knot->min == knot->max ) return ;
min = knot->min ;
max= knot->max ;
if ( knot->depth < Log2(N) )
{
Compact_quadtree(knot->square[ 0 ],min,max) ;
Compact_quadtree(knot->square[ 1 ],min,max) ;
Compact_quadtree(knot->square[ 2 ],min,max) ;
Compact_quadtree(knot->square[ 3 ],min,max) ;
}
else
{ // knot->depth == Log2(N) e pixel level
int slc = 0 ;
encode( (knot->square[ 0 ])->pix , min , max );
if ((knot->square[ 0 ])->pix == min) slc=1;
else if ((knot->square[ 0 ])->pix==max) slc=2;
encode( (knot->square[ 1 ])->pix , min , max );
if ((knot->square[ 1 ])->pix == min) slc |= 1;
else if ((knot->square[ 1 ])->pix==max) slc |= 2;
switch( slc )
{
case 0:
encode(((knot->square[2])->pix==max),0,1);
return ;
case 1:
if ((knot->square[2])->pix==max)
{
```

```
encode(1,0,1);
encode((knot->square[ 3 ])->pix,min,max);
}
else
{
encode(0,0,1);
encode((knot->square[ 2 ])->pix,min,max);
}
return ;
case 2:
if ((knot->square[2])->pix==min)
{
encode(1,0,1);
encode((knot->square[ 3 ])->pix,min,max);
}
else
{
encode(0,0,1);
encode((knot->square[ 2 ])->pix,min,max);
}
return ;
case 3:
encode((knot->square[ 2 ])->pix,min,max);
encode((knot->square[ 3 ])->pix,min,max);
}
}
}
```

An improvement

The following improvement to the method of this section was communicated to me by its originator, Stephan Wolf, who also wrote these paragraphs.

I was surprised by the fact that arithmetic encoding is suggested for the `encode()` function. Arithmetic coding is apparently the best method if individual symbols have different probabilities. But in this case, the paper seems to assume equal probabilities for all the individual pixel values in a particular encoded range.

Thus, I had the idea for a very simple algorithm that uses exactly $\log_2 n$ bits to encode a value for a given zero-relative range of values.

I thought I would wait until I receive your book and see if it describes this algorithm, i.e., if this is an already-known technique.

But I could not find any hints in your book or anywhere else (for example, the Internet) about the algorithm I devised.

So may I present you with this algorithm for revisal. May I also add that the following ideas are all mine:

Given a (zero-relative) range `r` of equally-distributed values `v`, i.e., all values have equal probabilities. In this paper, I denote a range/value pair as a tuple $(\mathtt{r}, \mathtt{v})$. Both encoder and decoder need to already know the range `r` in each step of encoding/decoding an individual value `v`.

As an example, I use the following range/value pairs $(9, 2)$, $(199, 73)$, $(123, 89)$, and $(50, 43)$. The theoretical number of bits required to encode each individual value is $\log_2(r+1)$ which translates (in rounded values using three decimal digits) to $\log_2(9+1) = 3.32$, $\log_2(199+1) = 7.64$, $\log_2(123+1) = 6.95$, and $\log_2(50+1) = 5.67$. Thus, a total maximum of 23.6 bits is required to encode all values in the example. A perfect encoder would therefore use at most 24 bits.

I propose the algorithm `e[n+1]=e[n]*(r[n]+1)+v[n]` to encode the values in the range/value pairs above. The results are

$$\begin{aligned}
\texttt{e[1]} &= 0 \times (9+1) + 2 = 2,\\
\texttt{e[2]} &= 2 \times (199+1) + 73 = 473,\\
\texttt{e[3]} &= 473 \times (123+1) + 89 = 58{,}741,\\
\texttt{e[4]} &= 58{,}741 \times (50+1) + 43 = 2{,}995{,}834.
\end{aligned}$$

This value can be expressed in 22 bits as 1011011011011001111010.

Decoding is quite similar (note that % denotes the modulo operation, i.e., the integral remainder of division)

```
v[n+1] = d[n] % r
d[n+1] = d[n] / r
v[1] = 2,995,834 % (50+1) = 43
d[1] = 2,995,834 / (50+1) = 58,741
v[2] = 58,741 % (123+1) = 89
d[2] = 58,741 / (123+1) = 473
v[3] = 473 % (199+1) = 73
d[3] = 473 / (199+1) = 2
v[4] = 2 % (9+1) = 2
d[4] = 2 / (9+1) = 0
```

Note that any implementation of this encoding/decoding algorithm requires long integer arithmetic, similar to what is normally available in most cryptography libraries.

7.34.5 Prefix Compression

Prefix compression is a variant of quadtrees proposed in [Anedda and Felician 88] (see also Section 11.5.5). We start with a $2^n \times 2^n$ image. Each quadrant in the quadtree of this image is numbered 0, 1, 2, or 3, a two-bit number. Each subquadrant has a two-digit (i.e., a four-bit) number, and each subsubquadrant receives a 3-digit number. As the quadrants get smaller, their numbers get longer. When this numbering scheme is carried down to individual pixels, the number of a pixel turns out to be n digits, or $2n$ bits, long. Prefix compression is designed for bi-level images containing text or diagrams where the number of black pixels is relatively small. It is not suitable for grayscale, color, or any image that contains many black pixels, such as a painting. The method is best explained by an example. Figure 7.172 shows the pixel numbering in an 8×8 image (i.e., $n = 3$) and also a simple 8×8 image consisting of 18 black pixels. Each pixel number is three digits long, and they range from 000 to 333.

The first step is to use quadtree methods to figure the three-digit id numbers of the 18 black pixels. They are 000 101 003 103 030 121 033 122 123 132 210 301 203 303 220 221 222 223.

The next step is to select a prefix value P. We select $P = 2$, a choice that's justified below. The code of a pixel is now divided into P prefix digits followed by $3 - P$ suffix

000	001	010	011	100	101	110	111
002	003	012	013	102	103	112	113
020	021	030	031	120	121	130	131
022	023	032	033	122	123	132	133
200	201	210	211	300	301	310	311
202	203	212	213	302	303	312	313
220	221	230	231	320	321	330	331
222	223	232	233	322	323	332	333

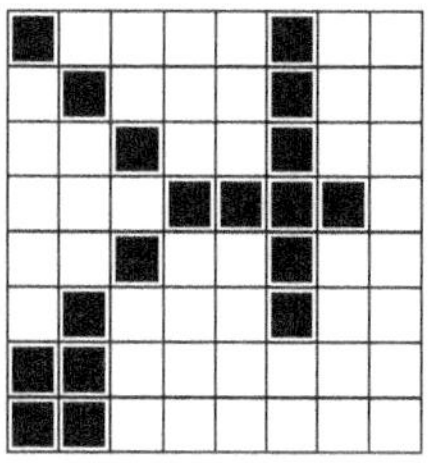

Figure 7.172: Example of Prefix Compression.

digits. The last step goes over the sequence of black pixels and selects all the pixels with the same prefix. The first prefix is 00, so all the pixels that start with 00 are selected. They are 000 and 003. They are removed from the original sequence and are compressed by writing the token 00|1|0|3 on the output stream. The first part of this token is a prefix (00), the second part is a count (1), and the rest are the suffixes of the two pixels having prefix 00. Notice that a count of one implies two pixels. The count is always one less than the number of pixels being counted. Sixteen pixels now remain in the original sequence, and the first of them has prefix 10. The two pixels with this prefix are removed and compressed by writing the token 10|1|1|3 on the output stream. This continues until the original sequence is empty. The final result is the 9-token string

$$00|1|0|3, 10|1|1|3, 03|1|0|3, 12|2|1|2|3, 13|0|2, 21|0|0, 30|1|1|3, 20|0|3, 22|3|0|1|2|3$$

(without the commas). Such a string can be decoded uniquely, since each segment starts with a two-digit prefix, followed by a one-digit count c, followed by $c+1$ one-digit suffixes.

In general, the prefix is P digits long, and the count and each suffix are $n-P$ digits each. The maximum number of suffixes in a segment is therefore 4^{n-P}. The maximum size of a segment is thus $P+(n-P)+4^{n-P}(n-P)$ digits. Each segment corresponds to a different prefix. A prefix has P digits, each between 0 and 3, so the maximum number of segments is 4^P. The entire compressed string therefore occupies at most

$$4^P\left[P+(n-P)+4^{n-P}(n-P)\right]=n\cdot 4^P+4^n(n-P)$$

digits. To find the optimum value of P we differentiate the expression above with respect to P,

$$\frac{d}{dP}\left[n\cdot 4^P+4^n(n-P)\right]=n\cdot 4^P\ln 4-4^n,$$

and set the derivative to zero. The solution is

$$4^P=\frac{4^n}{n\cdot\ln 4},\quad\text{or } P=\log_4\left[\frac{4^n}{n\cdot\ln 4}\right]=\frac{1}{2}\log_2\left[\frac{4^n}{n\cdot\ln 4}\right].$$

For $n=3$ this yields

$$P\approx\frac{1}{2}\log_2\left[\frac{4^3}{3\times 1.386}\right]=\frac{\log_2 15.388}{2}=3.944/2=1.97.$$

This is why $P = 2$ was selected earlier.

A downside of this method is that some pixels may be assigned numbers with different prefixes even though they are near neighbors. This happens when they are located in different quadrants. An example is pixels 123 and 301 of Figure 7.172.

Improvement: The `count` field was arbitrarily set to one digit (two bits). The maximum count is therefore $3\,(= 11_2)$, i.e., four pixels. It is possible to have a variable-size `count` field containing a variable-length code, such as the unary code of Section 3.1. This way a single token could compress any number of pixels.

7.35 Quadrisection

A quadtree exploits the redundancy in an image by examining smaller and smaller quadrants, looking for uniform areas. The method of *quadrisection* [Kieffer et al. 96a,b] is related to quadtrees, because it uses the quadtree principle for dividing an image into subimages. However, quadrisection does not divide an image into four parts, but rather reshapes the image in steps by increasing the number of its rows and decreasing the number of its columns. The method is lossless. It performs well for bi-level images and is illustrated here for such an image, but can also be applied to grayscale (and therefore to color) images.

The method assumes that the original image is a $2^k \times 2^k$ square matrix M_0, and it reshapes M_0 into a sequence of matrices $M_1, M_2, \ldots, M_{k+1}$ with fewer and fewer columns. These matrices naturally have more and more rows, and the quadrisection method achieves compression by searching for and removing duplicate rows. The more rows and the fewer columns a matrix has, the better the chance of having duplicate rows. The end result is a matrix M_{k+1} with one column and not more than 64 rows. This matrix is treated as a short string and is written at the end of the compressed stream, to help in decoding and recovering the original image M_0. The compressed stream must also contain information on how to reconstruct each matrix M_{j-1} from its successor M_j. This information is in the form of an *indicator vector* denoted by I_{j-1}. Thus, the output consists of all the I_j vectors (each arithmetically encoded into a string w_j), followed by the bits of M^T_{k+1} (the transpose of the last, single-column, matrix, in raw format). This string is preceded by a prefix that indicates the value of k, the sizes of all the w_j's, and the size of M^T_{k+1}.

Here is the encoding process in a little more detail. We assume that the original image is a $2^k \times 2^k$ square matrix M_0 of 4^k pixels, each a single bit. The encoder uses an operation called *projection* (discussed below) to construct a sequence of matrices M_1, M_2, up to M_{k+1}, such that M_j has 4^{k-j+1} columns. This implies that M_1 has $4^{k-1+1} = 4^k$ columns and therefore just one row. Matrix M_2 has 4^{k-1} columns and therefore four rows. Matrix M_3 has 4^{k-2} columns, but may have fewer than eight rows, since any duplicate rows are removed from M_2 before M_3 is constructed. The number of rows of M_3 is four times the number of *distinct* rows of M_2. Indicator vector I_2 is constructed at the same time as M_3 to indicate those rows of M_2 that are duplicates. The size of I_2 equals the number of rows of M_2, and the elements of I_2 are nonnegative integers. The decoder uses I_2 to reconstruct M_2 from M_3.

All the I_j indicator vectors are needed by the decoder. Since the size of I_j equals the number of rows of matrix M_j, and since these matrices have more and more rows, the combined sizes of all the indicator vectors I_j may be large. However, most elements of a typical indicator vector I_j are zeros, so these vectors can be highly compressed with arithmetic coding. Also, the first few indicator vectors are normally all zeros, so they may all be replaced by one number indicating how many of them there are.

◇ **Exercise 7.55:** Even though we haven't yet described the details of the method (specifically, the projection operation), it is possible to logically deduce (or guess) the answer to this exercise. The question is We know that images with little or no correlation will not compress. Yet the size of the last matrix, M_{k+1}, is small (64 bits or fewer) regardless of the image being compressed. Can quadrisection somehow compress images that other methods cannot?

The decoder starts by decoding the prefix, to obtain the sizes of all the compressed strings w_j. It then decompresses each w_j to obtain an indicator vector I_j. The decoder knows how many indicator vectors there are, since j goes from 1 to k. After decoding all the indicator vectors, the decoder reads the rest of the compressed stream and considers it the bits of M_{k+1}. The decoder then uses the indicator vectors to reconstruct matrices M_k, M_{k-1}, and so on all the way down to M_0.

The projection operation is the heart of quadrisection. It is described here in three steps.

Step 1: An indicator vector I for a matrix M with r rows contains r components, one for each row of M. Each distinct row of M has a zero component in I, and each duplicate row has a positive component. If, for example, row 8 is a duplicate of the 5th distinct row, then the 8th component of I will be 5. Notice that the 5th distinct row is not necessarily row 5. As an example, the indicator vector for matrix M of Equation (7.54) is $I = (0,0,0,3,0,4,3,1)$:

$$M_3 = \begin{bmatrix} 1001 \\ 1101 \\ 1011 \\ 1011 \\ 0000 \\ 0000 \\ 1011 \\ 1001 \end{bmatrix}. \tag{7.54}$$

Step 2: Given a row vector v of length m, where m is a power of 2, we perform the following: (1) Divide it into $\sqrt{m}$ segments of length $\sqrt{m}$ each. (2) Arrange the segments in a matrix of size $\sqrt{m}\times\sqrt{m}$. (3) Divide it into four quadrants, each a matrix of size $\sqrt{m}/2\times\sqrt{m}/2$. (4) Label the quadrants 1, 2, 3, and 4 according to $\binom{1\,2}{3\,4}$. (5) Unfold each into a vector v_i of length $\sqrt{m}$. As an example, consider the vector

$$v = (0,1,2,3,4,5,6,7,8,9,10,11,12,13,14,15),$$

of length $16 = 2^4$. It first becomes the 4×4 matrix

$$M = \begin{pmatrix} 0 & 1 & 2 & 3 \\ 4 & 5 & 6 & 7 \\ 8 & 9 & 10 & 11 \\ 12 & 13 & 14 & 15 \end{pmatrix}.$$

Partitioning M yields the four 2×2 matrices

$$M_1 = \begin{pmatrix} 0 & 1 \\ 4 & 5 \end{pmatrix}, \quad M_2 = \begin{pmatrix} 2 & 3 \\ 6 & 7 \end{pmatrix}, \quad M_3 = \begin{pmatrix} 8 & 9 \\ 12 & 13 \end{pmatrix}, \quad \text{and } M_4 = \begin{pmatrix} 10 & 11 \\ 14 & 15 \end{pmatrix}.$$

Each is unfolded, to produce the four vectors

$$v_1 = (0,1,4,5), \quad v_2 = (2,3,6,7), \quad v_3 = (8,9,12,13), \text{ and } v_4 = (10,11,14,15).$$

This step illustrates the relation between quadrisection and quadtrees.

Step 3: This finally describes the projection operation. Given an $n\times m$ matrix M where m (the number of columns) is a power of 2, we first construct an indicator vector I for it. Vector I will have n components, for the n rows of M. If M has r distinct rows, vector I will have r zeros corresponding to these rows. We construct the projected matrix M' with $4r$ rows and $m/4$ columns in the following steps: (3.1) Ignore all duplicate rows of M (i.e., all rows corresponding to nonzero elements of I). (3.2) For each of the remaining r distinct rows of M construct four vectors v_i as shown in Step 2 above, and make them into the next four rows of M'. We use the notation

$$M \xrightarrow{I} M',$$

to indicate this projection.

Example: Given the 16×16 matrix M_0 of Figure 7.173 we concatenate its rows to construct matrix M_1 with one row and 256 columns. This is always our starting point, regardless of whether the rows of M_0 are distinct or not. Since M_1 has just one row, its indicator vector is $I_1 = (0)$.

```
0 0 0 0 0 0 0 0 1 0 0 0 0 0 0 0
0 0 0 0 0 0 0 1 1 0 0 0 0 0 0 0
0 0 0 0 0 0 1 1 1 1 0 0 0 0 0 0
0 0 0 0 0 0 1 1 0 1 1 1 0 0 0 0
0 0 0 0 0 0 1 1 1 1 1 0 0 0 0 0
0 0 0 0 1 1 1 1 1 1 0 0 0 0 0 0
0 0 1 1 1 1 1 1 1 1 0 0 0 0 0 0
0 1 1 1 1 1 1 1 1 1 0 0 0 0 0 0
0 1 1 1 1 1 1 1 1 0 0 0 0 0 0 0
1 1 1 1 1 1 1 1 1 0 0 0 0 0 0 0
1 1 1 1 1 1 1 1 0 0 0 0 0 0 0 0
1 1 1 1 1 0 1 1 1 0 0 0 0 0 0 0
1 1 1 1 1 0 0 1 1 0 0 0 0 0 0 0
1 1 1 1 1 1 0 1 1 0 0 0 0 0 0 0
0 1 1 1 1 1 0 1 1 0 0 0 0 0 0 0
0 1 1 1 1 1 0 1 1 1 0 0 1 1 1 1
```

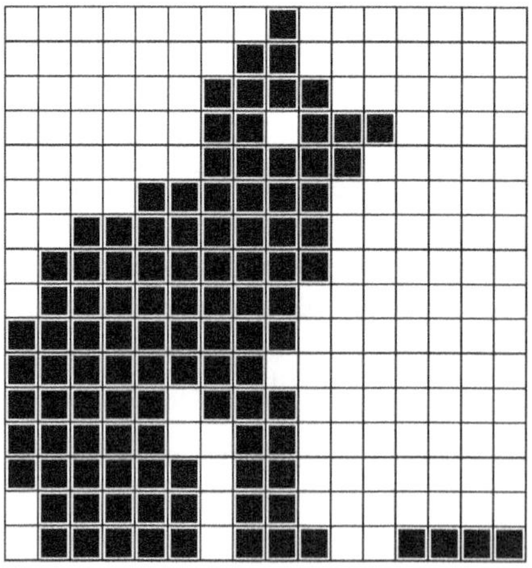

Figure 7.173: A 16×16 Matrix M_0.

Matrix M_2 is easily projected from M_1. It has four rows and 64 columns

$$M_2 = \begin{bmatrix} 0000000000000000100000011000000110000001100001111001111110111111 \\ 100000001000000001100000000111000011100000011000000011000000011000000 \\ 0111111111111111111111111111111011111110011111110101111110101111101 \\ 10000000100000000000000001000000010000000100000001000000011001111 \end{bmatrix}.$$

All four rows of M_2 are distinct, so its indicator vector is $I_2 = (0,0,0,0)$. Notice how each row of M_2 is an 8×8 submatrix of M_0. The top row, for example, is the upper-left 8×8 corner of M_0.

To project M_3 from M_2 we perform Step 2 above on each row of M_2, converting it from a 1×64 to a 4×16 matrix. Thus, matrix M_3 consists of four parts, each 4×16, so its size is 16×16:

$$M_3 = \begin{bmatrix} 0000000000000000 \\ 0000000100110011 \\ 0000000000110111 \\ 0011111111111111 \\ \hline 1000100011000111 \\ 0000000000000000 \\ 1110110011001100 \\ 0000000000000000 \\ \hline 0111111111111111 \\ 1111111111111011 \\ 1111111101110111 \\ 1001110111011101 \\ \hline 1000100000001000 \\ 0000000000000000 \\ 1000100010001100 \\ 0000000000001111 \end{bmatrix}.$$

Notice how each row of M_3 is a 4×4 submatrix of M_0. The top row, for example, is the upper-left 4×4 corner of M_0, and the second row is the upper-second-from-left 4×4 corner. Examining the 16 rows of M_3 results in the indicator vector $I_3 = (0,0,0,0,0,1,0,1,0,0,0,0,0,1,0,0)$ (i.e., rows 6, 8, and 14 are duplicates of row 1). This is how the next projection creates some compression. Projecting M_3 to M_4 is done by ignoring rows 6, 8, and 14. M_4 [Equation (7.55)] therefore has $4\times 13 = 52$ rows and four columns (a quarter of the number of M_3). It therefore has $52\times 4 = 208$ elements instead of 256.

It is clear that with so many short rows, M_4 must have many duplicate rows. An examination indicates only 12 distinct rows, and produces vector I_4:

$$\begin{aligned} I_4 = (&0,1,1,1,1,0,1,0,1,1,2,3,0,3,3,3|0,1,0,4,3,0,3,1| \\ &0,3,3,3,3,3,0,3,3,3,0,3,0,10,3,10|5,1,0,1,5,1,11,1,1,1,4,4). \end{aligned}$$

Projecting M_4 to obtain M_5 is done as before. Matrix M_5 has $4\times 12 = 48$ rows and only one column (a quarter of the number of columns of M_4). This is therefore the final matrix M_{k+1}, whose transpose will be the last part of the compressed stream.

◇ **Exercise 7.56:** Compute matrix M_5.

The compressed stream consists therefore of a prefix, followed by the value of k (= 4), followed by I_1, I_2, I_3, and I_4, arithmetically encoded, followed by the 48 bits of M_5^T. There is no need to encode those last bits, since there may be at most 64 of them (see exercise 7.57). Moreover, since I_1 and I_2 are zeros, there is no need to write them on the output. All that the encoder has to write instead is the number 3 (encoded) to point out that the first indicator vector included in the output is I_3.

$$M_4 = \left[\begin{array}{c}
0000\\ 0000\\ 0000\\ 0000\\ 0000\\ 0001\\ 0000\\ 1111\\ 0000\\ 0000\\ 0001\\ 1111\\ 0011\\ 1111\\ 1111\\ 1111\\ \hline
1010\\ 0000\\ 1101\\ 0011\\ 1111\\ 1000\\ 1111\\ 0000\\ \hline
0111\\ 1111\\ 1111\\ 1111\\ 1111\\ 1111\\ 1110\\ 1111\\ 1111\\ 1111\\ 0101\\ 1111\\ 1011\\ 0101\\ 1111\\ 0101\\ \hline
1010\\ 0000\\ 0010\\ 0000\\ 1010\\ 0000\\ 1011\\ 0000\\ 0000\\ 0000\\ 0011\\ 0011
\end{array}\right]. \tag{7.55}$$

The prefix consists of 4 (k), 3 (index of first nonzero indicator vector), and the encoded lengths of indicator vectors I_3 and I_4. After decoding this, the decoder can decode the two indicator vectors, read the remaining compressed stream as M_5^T, then use all this data to reconstruct M_4, M_3, M_2, M_1, and M_0.

◇ **Exercise 7.57:** Show why M_5 cannot have more than 64 elements.

Extensions: Quadrisection is used to compress two-dimensional data, and can therefore be considered the second method in a succession of three compression methods the first and third of which are *bisection* and *octasection*. Following is a short description of these two methods.

Bisection is an extension (reduction) of quadrisection for the case of one-dimensional data (typical examples are sampled audio, a binary string, or text). We assume that the data consists of a string L_0 of 2^k data items, where an item can be a single bit, an ASCII code, a audio sample, or anything else, but all items have the same size.

The encoder iterates k times, varying j from 1 to k. In iteration j, a list L_j is created by bisecting the elements of the preceding list L_{j-1}. An indicator vector I_j is constructed for L_j. The duplicate elements of L_j are then deleted (this is where we get compression).

Each of the two elements of the first constructed list L_1 is therefore a block of 2^{k-1} data items (half the number of the data items in L_0). Indicator vector I_1 also has two elements. Each element of L_2 is a block of 2^{k-2} data items, but the number of elements of L_2 is not necessarily four. It can be smaller depending on how many distinct elements L_1 has. In the last step, where $j = k$, list L_k is created, where each element is a block of size $2^{k-k} = 1$. Thus, each element of L_k is one of the original data items of L_0. It is not necessary to construct indicator vector I_k (the last indicator vector generated is I_{k-1}).

The compressed stream consists of k, followed by indicator vectors I_1, I_2, through I_{k-1} (compressed), followed by L_k (raw). Since the first few indicator vectors tend to be all zeros, they can be replaced by a single number indicating the index of the first nonzero indicator vector.

Example: The 32-element string L_0 where each data element is a single bit:

$$L_0 = (0,1,1,1,0,0,1,0,1,1,0,0,0,0,1,0,0,1,1,0,1,1,1,1,0,1,0,1,1,0,1,0).$$

The two halves of L_0 are distinct, so L_1 consists of two elements

$$L_1 = (0111001011000010, 0110111101011010),$$

and the first indicator vector is $I_1 = (0,0)$. The two elements of L_1 are distinct, so L_2 has four elements,

$$L_2 = (01110010, 11000010, 01101111, 01011010),$$

and the second indicator vector is $I_2 = (0,0,0,0)$. The four elements of L_2 are distinct, so L_3 has eight elements,

$$L_3 = (0111, 0010, 1100, 0010, 0110, 1111, 0101, 1010),$$

and the third indicator vector is $I_3 = (0,0,0,2,0,0,0,0)$. Seven elements of L_3 are distinct, so L_4 has 14 elements,

$$L_4 = (01, 11, 00, 10, 11, 00, 01, 10, 11, 11, 01, 01, 10, 10),$$

and the fourth indicator vector is $I_4 = (0,0,0,0,2,3,1,4,2,2,1,1,4,4)$. Only four elements of L_4 are distinct, so L_5 has eight elements, $L_5 = (0,1,1,1,0,0,1,0)$.

The output therefore consists of $k = 5$, indicator vectors $I_1 = (0,0)$, $I_2 = (0,0,0,0)$, $I_3 = (0,0,0,2,0,0,0,0)$, and $I_4 = (0,0,0,0,2,3,1,4,2,2,1,1,4,4)$ (encoded), followed by $L_5 = (0,1,1,1,0,0,1,0)$. Since the first nonzero indicator vector is I_3, we can omit both I_1 and I_2 and replace them with the integer 3.

◇ **Exercise 7.58:** Describe the operations of the decoder for this example.

◇ **Exercise 7.59:** Use bisection to encode the 32-bit string

$$L_0 = (0101010101010101\ 1010101010101010).$$

The discussion above shows that if the original data to be compressed is a bit string of length 2^k, then L_{k-1} is a list of pairs of bits. L_{k-1} can therefore have at most four distinct elements, so L_k, whose elements are single bits, can have at most eight elements. This, of course, does not mean that any binary string L_0 can be compressed into eight bits. The reader should bear in mind that the compressed stream must also include the nonzero indicator vectors. If the elements of L_0 are not bits, then L_k could, in principle, be as long as L_0.

Example: A source string $L_0 = (a_1, a_2, \ldots, a_{32})$ where the a_i's are distinct data items, such as ASCII characters. L_1 consists of two elements,

$$L_1 = (a_1a_2\ldots a_{16},\ a_{17}a_{18}\ldots a_{32}),$$

and the first indicator vector is $I_1 = (0,0)$. The two elements of L_1 are distinct, so L_2 has four elements,

$$L_2 = (a_1a_2\ldots a_8,\ a_9a_{10}\ldots a_{16},\ a_{17}a_{18}\ldots a_{24},\ a_{25}a_{26}\ldots a_{32}),$$

and the second indicator vector is $I_2 = (0,0,0,0)$. All four elements of L_2 are distinct, so

$$L_3 = (a_1a_2a_3a_4,\ a_5a_6a_7a_8,\ a_9a_{10}a_{11}a_{12}, \ldots,\ a_{29}a_{30}a_{31}a_{32}).$$

Continuing this way, it is easy to see that all indicator vectors will be zero, and L_5 will have the same 32 elements as L_0. The result will be no compression at all.

If the length L of the original data is not a power of two, we can still use bisection by considering the following: There is some integer k such that $2^{k-1} < L < 2^k$. If L is close to 2^k, we can add $d = 2^k - L$ zeros to the original string L_0, compress it by bisection, and write d on the compressed stream, so the decoder can delete the d zeros. If L is close to 2^{k-1} we divide L_0 into a string L_0^1 with the first 2^{k-1} items, and string L_0^2 with the remaining items. The former string can be compressed with bisection and the latter can be compressed by either appending zeros to it or splitting it recursively.

Octasection is an extension of both bisection and quadrisection for three-dimensional data. Examples of such data are a video (a sequence of images), a grayscale image, and a color image. Such an image can be viewed as a three-dimensional matrix where the third dimension is the bits constituting each pixel (the bitplanes). We assume that the data is a rectangular box P of dimensions $2^{k_1}\times 2^{k_2}\times 2^{k_3}$, where each entry is a data item (a single bit or several bits) and all entries have the same size. The encoder performs k iterations, where $k = \min(k_1, k_2, k_3)$, varying j from 1 to k. In iteration j, a list L_j is created, by subdividing each element of the preceding list L_{j-1} into eight smaller rectangular boxes. An indicator vector I_j is constructed for L_j. The duplicate elements of L_j are then deleted (this is where we get compression).

The compressed stream consists, as before, of all the (arithmetically encoded) indicator vectors, followed by the last list L_k.

7.36 Space-Filling Curves

A space-filling curve [Sagan 94] is a parametric function $\mathbf{P}(t)$ that passes through every point in a given two-dimensional area, normally the unit square, when its parameter t varies in the range $[0, 1]$. For any value t_0, the value of $\mathbf{P}(t_0)$ is a point $[x_0, y_0]$ in the unit square. Mathematically, such a curve is a mapping from the interval $[0, 1]$ to the two-dimensional interval $[0, 1] \times [0, 1]$. To understand how such a curve is constructed it is best to think of it as the limit of an infinite sequence of curves $\mathbf{P}_1(t)$, $\mathbf{P}_2(t), \ldots$, which are drawn inside the unit square, where each curve is derived from its predecessor by a process of *refinement*. The details of the refinement depend on the specific curve. Section 7.37.1 discusses several well-known space-filling curves, among them the Hilbert curve and the Sierpiński curve. Since the sequence of curves is infinite, it is impossible to compute all its components. Fortunately, we are interested in a curve that passes through every pixel in a bitmap, not through every mathematical point in the unit square. Since the number of pixels is finite, it is possible to construct such a curve in practice.

To understand why such curves are useful for image compression, the reader should recall the principle that has been mentioned several times in the past, namely, if we select a pixel in an image at random, there is a good chance that its neighbors will have the same (or similar) colors. Both RLE image compression and the quadtree method are based on this principle, but they are not always efficient, as Figure 7.174 shows. This 8×8 bitmap has two concentrations of pixels, but neither RLE nor the quadtree method compress it very well, since there are no long runs and since the pixel concentrations happen to cross quadrant boundaries.

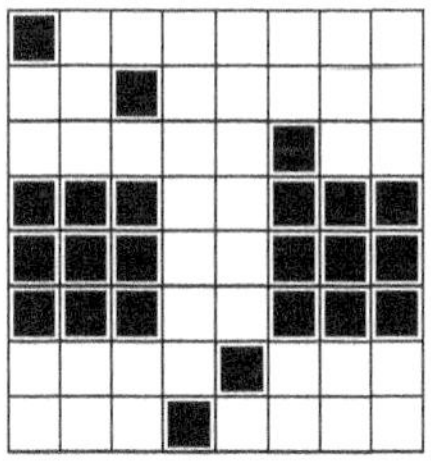

Figure 7.174: An 8×8 Bitmap.

Better compression may be produced by a method that scans the bitmap area by area instead of line by line or quadrant by quadrant. This is why space-filling curves provide a new approach to image compression. Such a curve visits every point in a given area, and does that by visiting all the points in a subarea, then moves to the next subarea and traverses it, and so on. We use the Hilbert curve (Section 7.37.1) as an example. Each curve H_i is constructed by making four copies of the previous curve

H_{i-1}, shrinking them, rotating them, and connecting them. The new curve ends up covering the same area as its predecessor. This is the refinement process for the Hilbert curve.

Scanning an 8×8 bitmap in a Hilbert curve results in the sequence of pixels

(0,0), (0,1), (1,1), (1,0), (2,0), (3,0), (3,1), (2,1),
(2,2), (3,2), (3,3), (2,3), (1,3), (1,2), (0,2), (0,3),
(0,4), (1,4), (1,5), (0,5), (0,6), (0,7), (1,7), (1,6),
(2,6), (2,7), (3,7), (3,6), (3,5), (2,5), (2,4), (3,4),
(4,4), (5,4), (5,5), (4,5), (4,6), (4,7), (5,7), (5,6),
(6,6), (6,7), (7,7), (7,6), (7,5), (6,5), (6,4), (7,4),
(7,3), (7,2), (6,2), (6,3), (5,3), (4,3), (4,2), (5,2),
(5,1), (4,1), (4,0), (5,0), (6,0), (6,1), (7,1), (7,0).

Section 7.37.1 discusses space-filling curves in general and shows methods for fast traversal of some curves. Here we would like to point out that quadtrees (Section 7.34) are a special case of the Hilbert curve, a fact illustrated by six figures in that Section. See also [Prusinkiewicz and Lindenmayer 90] and [Prusinkiewicz et al. 91].

◇ **Exercise 7.60:** Scan the 8×8 bitmap of Figure 7.174 using a Hilbert curve and calculate the runs of identical pixels and compare them to the runs produced by RLE.

There are actually two types of the Hilbert space-filling curve. Relations between these types, Lindemayer's L-systems, and Gray codes can be found in *Hilbert Curve* by Eric Weisstein, available from MathWorld—A Wolfram Web Resource
`http://mathworld.wolfram.com/HilbertCurve.html`

7.37 Hilbert Scan and VQ

The space-filling Hilbert curve (Section 7.37.1) is employed by this method [Sampath and Ansari 93] as a preprocessing step for the lossy compression of images (notice that the authors use the term "Peano scan," but they actually use the Hilbert curve). The Hilbert curve is used to transform the original image into another, highly correlated, image. The compression itself is done by a vector quantization algorithm that exploits the correlation produced by the preprocessing step.

In the preprocessing step, an original image of size $M\times N$ is partitioned into small blocks of $m \times m$ pixels each (with $m = 8$ typically) that are scanned in Hilbert curve order. The result is a linear (one-dimensional) sequence of blocks, which is then rearranged into a new two-dimensional array of size $\frac{M}{m} \times Nm$. This process results in bringing together (or clustering) blocks that are highly correlated. The "distance" between adjacent blocks is now smaller than the distance between blocks that are raster-scan neighbors in the original image.

The distance between two blocks B_{ij} and C_{ij} of pixels is measured by the *mean absolute difference*, a quantity defined by

$$\frac{1}{m^2}\sum_{i=1}^{m}\sum_{j=1}^{m}|B_{ij}-C_{ij}|.$$

It turns out that the mean absolute difference of adjacent blocks in the new, rearranged, image is about half that of adjacent blocks in a raster scan of the original image. The reason for this is that the Hilbert curve is space filling. Points that are nearby on the curve are nearby in the image. Conversely, points that are nearby in the image are normally nearby on the curve. Thus, the Hilbert curve acts as an image transform. This fact is the main innovation of the method described here.

As an example of a Hilbert scan, imagine an image of size $M\times N = 128\times 128$. If we select $m = 8$, we get $16\times 16 = 256$ blocks that are scanned and made into a one-dimensional array. After rearranging, we end up with a new image of size 2×128 blocks or $\frac{128}{8}\times 128\cdot 8 = 16\times 1024$ pixels. Figure 7.175 shows how the blocks are Hilbert scanned to produce the sequence shown in Figure 7.176. This sequence is then rearranged. The top row of the rearranged image contains blocks 1, 17, 18, 2, 3, 4, ..., and the bottom row contains blocks 138, 154, 153, 169, 185, 186, ...

The new, rearranged image is now partitioned into new blocks of $4\times 4 = 16$ pixels each. The 16 pixels of a block constitute a vector. (Notice that the vector has 16 components, each of which can be one or more bits, depending on the size of a pixel.) The LBG algorithm (Section 7.19) is used for the vector quantization of those vectors. This algorithm calls for an initial codebook, so the implementors chose five images, scanned them at a resolution of 256×256, and used them as training images, to generate three codebooks, consisting of 128, 512, and 1024 codevectors, respectively.

The main feature of the particular vector quantization algorithm used here is that the Hilbert scan results in adjacent blocks that are highly correlated. As a result, the LBG algorithm frequently assigns the same codevector to a *run* of blocks, and this fact can be used to highly compress the image. Experiments indicate that as many as 2–10 consecutive blocks (for images with details) and 30–70 consecutive blocks (for images with high spatial redundancy) may participate in such a run. Therefore, the method precedes each codevector with a code indicating the length of the run (one block or several). There are two versions of the method, one with a fixed-length code and the other with a variable-length prefix code.

When a fixed-length code is used preceding each codevector, the main question is the size of the code. A long code, such as six bits, allows for runs of up to 64 consecutive blocks with the same codevector. On the other hand, if the image has small details, the runs would be shorter and some of the 6 bits would be wasted. A short code, such as two bits, allows for runs of up to four blocks only, but fewer bits are wasted in the case of a highly detailed image. A good solution is to write the size of the code (which is typically 2–6 bits) at the start of the compressed stream, so the decoder knows what it is. The encoder can then be very sophisticated, trying various code sizes before settling on one and using it to compress the image.

Using prefix codes can result in slightly better compression, especially if the encoder can perform a two-pass job to determine the frequency of each run before anything is compressed. In such a case, the best Huffman codes can be assigned to the runs, resulting in best compression.

Further improvement can be achieved by a variant of the method that uses *dynamic codebook partitioning*. This is again based on adjacent blocks being very similar. Even if such blocks end up with different codevectors, those codevectors may be very similar. This variant selects the codevectors for the first block in the usual way, using the entire

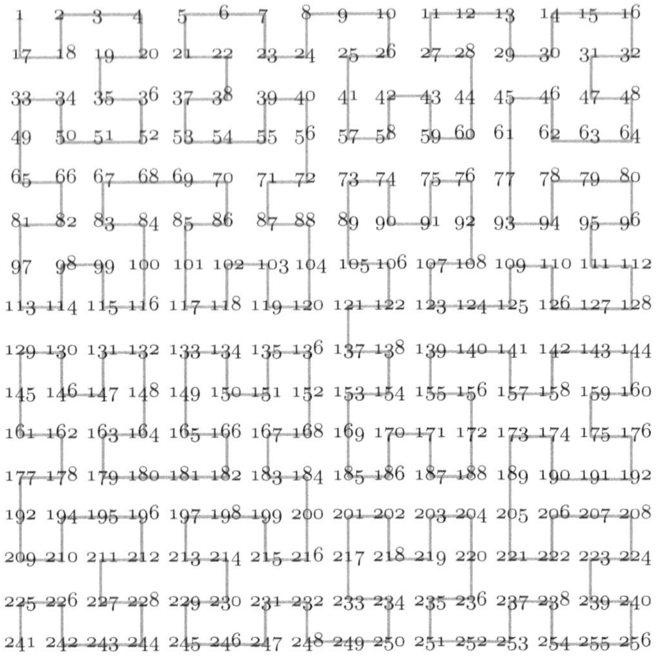

Figure 7.175: Hilbert Scan of 16×16 Blocks.

1, 17, 18, 2, 3, 4, 20, 19, 35, 36, 52, 51, 50, 34, 33, 49,
65, 66, 82, 81, 97, 113, 114, 98, 99, 115, 116, 100, 84, 83, 67, 68,
69, 70, 86, 85, 101, 117, 118, 102, 103, 119, 120, 104, 88, 87, 71, 72,
56, 40, 39, 55, 54, 53, 37, 38, 22, 21, 5, 6, 7, 23, 24, 8,
9, 10, 26, 25, 41, 57, 58, 42, 43, 59, 60, 44, 28, 27, 11, 12,
13, 29, 30, 14, 15, 16, 32, 31, 47, 48, 64, 63, 62, 46, 45, 61,
77, 93, 94, 78, 79, 80, 96, 95, 111, 112, 128, 127, 126, 110, 109, 125,
124, 123, 107, 108, 92, 76, 75, 91, 90, 74, 73, 89, 105, 106, 122, 121,
137, 138, 154, 153, 169, 185, 186, 170, 171, 187, 188, 172, 156, 155, 139, 140,
141, 157, 158, 142, 143, 144, 160, 159, 175, 176, 192, 191, 190, 174, 173, 189,
205, 221, 222, 206, 207, 208, 224, 223, 239, 240, 256, 255, 254, 238, 237, 253,
252, 251, 235, 236, 220, 204, 203, 219, 218, 202, 201, 217, 233, 234, 250, 249,
248, 232, 231, 247, 246, 245, 229, 230, 214, 213, 197, 198, 199, 215, 216, 200,
184, 183, 167, 168, 152, 136, 135, 151, 150, 134, 133, 149, 165, 166, 182, 181,
180, 179, 163, 164, 148, 132, 131, 147, 146, 130, 129, 145, 161, 162, 178, 177,
192, 209, 210, 194, 195, 196, 212, 211, 227, 228, 244, 243, 242, 226, 225, 241

Figure 7.176: The Resulting 256 Blocks.

codebook. It then selects a set of the next best codevectors that could have been used to code this block. This set becomes the *active part* of the codebook. (Notice that the decoder can mimic this selection.) The second block is then compared to the first block and the distance between them measured. If this distance is less than a given threshold, the codevector for the second block is selected from the active set. Since the active set is much smaller than the entire codebook, this leads to much better compression. However, each codevector must be preceded by a bit telling the decoder whether the codevector was selected from the entire codebook or just from the active set.

If the distance is greater than the threshold, a codevector for the second block is selected from the entire codebook, and a new active set is chosen, to be used (hopefully) for the third block.

If the Hilbert scan really ends up with adjacent blocks that are highly correlated, a large fraction of the blocks are coded from the active sets, thereby considerably improving the compression. A typical codebook size is 128–1024 entries, whereas the size of an active set may be just four codevectors. This reduces the size of a codebook pointer from 7–10 bits to 2 bits.

The choice of the threshold for this variant is important. It seems that an adaptive threshold adjustment may work best, but the developers don't indicate how this may be implemented.

7.37.1 Examples

A space-filling curve completely fills up part of space by passing through every point in that part. It does that by changing direction repeatedly. We will only discuss curves that fill up part of the two-dimensional plane, but the concept of a space-filling curve exists for any number of dimensions.

◇ **Exercise 7.61:** Show an example of a space-filling curve in one dimension.

Several such curves are known and all are defined recursively. A typical definition starts with a simple curve C_0, shows how to use it to construct another, more complex curve C_1, and defines the final, space-filling curve as the limit of the sequence of curves $C_0, C_1, \ldots$.

7.37.2 The Hilbert Curve

(This discussion is based on the approach of [Wirth 76].) Perhaps the most familiar of these curves is the Hilbert curve, discovered by the great mathematician David Hilbert in 1891. The Hilbert curve [Hilbert 91] is the limit of a sequence $H_0, H_1, H_2 \ldots$ of curves, some of which are shown in Figure 7.177. They are defined by the following:

0. H_0 is a single point.

1. H_1 consists of four copies of (the point) H_0, connected with three straight segments of length h at right angles to each other. Four orientations of this curve, labeled 1, 2, 3, and 4, are shown in Figure 7.177a.

2. The next curve, H_2, in the sequence is constructed by connecting four copies of different orientations of H_1 with three straight segments of length $h/2$ (shown in bold in Figure 7.177b). Again there are four possible orientations of H_2, and the one shown is #2. It is constructed of orientations 1223 of H_1, connected by segments that go to the

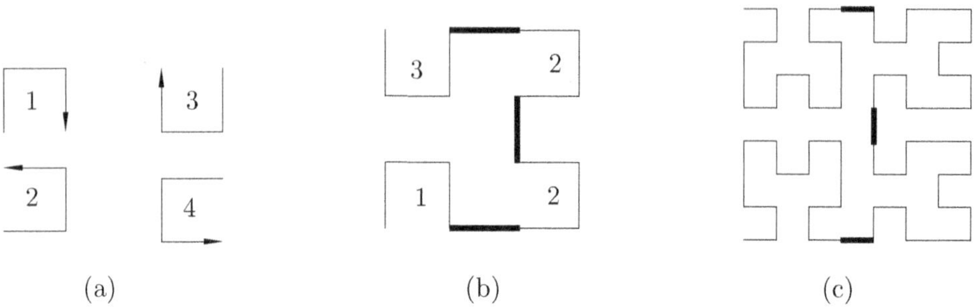

Figure 7.177: Hilbert Curves of Orders 1, 2, and 3.

right, up, and to the left. The construction of the four orientations of H_2 is summarized in Table Ans.41.

Curve H_3 is shown in Figure 7.177c. The particular curve shown is orientation 1223 of H_2.

Figures 7.178, 7.179 and 7.180 show the Hilbert curves of orders 4, 5 and 6. It is easy to see how fast these curves become extremely complex.

7.37.3 The Sierpiński Curve

Another well-known space-filling curve is the Sierpiński curve. Figure 7.181 shows curves S_1 and S_2, and Sierpiński has proved [Sierpiński 12] that the limit of the sequence $S_1, S_2, \ldots$ is a curve that passes through every point of the unit square $[0,1] \times [0,1]$.

To construct this curve, we need to figure out how S_2 is constructed out of four copies of S_1. The first thing that comes to mind is to follow the construction method used for the Hilbert curve, i.e., to take four copies of S_1, eliminate one edge in each, and connect them. This, unfortunately, does not work, because the Sierpiński curve is very different from the Hilbert curve. It is closed, and it has one orientation only. A better approach is to start with four parts that constitute four orientations of one open curve, and connect them with straight segments. The segments are shown dashed in Figure 7.181. Notice how Figure 7.181a is constructed of four orientations of a basic, three-part curve connected by four short, dashed segments. Figure 7.181b is similarly constructed of four orientations of a complex, 15-part curve, connected by the same short, dashed segments. If we denote the four basic curves A, B, C, and D, then the basic construction rule of the Sierpiński curve is S: $\mathrm{A}\searrow\mathrm{B}\swarrow\mathrm{C}\nwarrow\mathrm{D}\nearrow$, and the recursion rules are:

$$
\begin{aligned}
&\mathrm{A}\colon\ \mathrm{A}\searrow\mathrm{B}\rightarrow\rightarrow\mathrm{D}\nearrow\mathrm{A}\\
&\mathrm{B}\colon\ \mathrm{B}\swarrow\mathrm{C}\downarrow\downarrow\mathrm{A}\searrow\mathrm{B}\\
&\mathrm{C}\colon\ \mathrm{C}\nwarrow\mathrm{D}\leftarrow\leftarrow\mathrm{B}\swarrow\mathrm{C}\\
&\mathrm{D}\colon\ \mathrm{D}\nearrow\mathrm{A}\uparrow\uparrow\mathrm{C}\nwarrow\mathrm{D}
\end{aligned}
\tag{7.56}
$$

Figure 7.182 shows the five Sierpiński curves of orders 1 through 5 superimposed on each other.

◇ **Exercise 7.62:** Figure 7.183 shows three iterations of the Peano space-filling curve, developed in 1890. Use the techniques developed earlier for the Hilbert and Sierpiński

Figure 7.178: Hilbert Curve of Order 4.

Figure 7.179: Hilbert Curve of Order 5.

Figure 7.180: Hilbert Curve of Order 6.

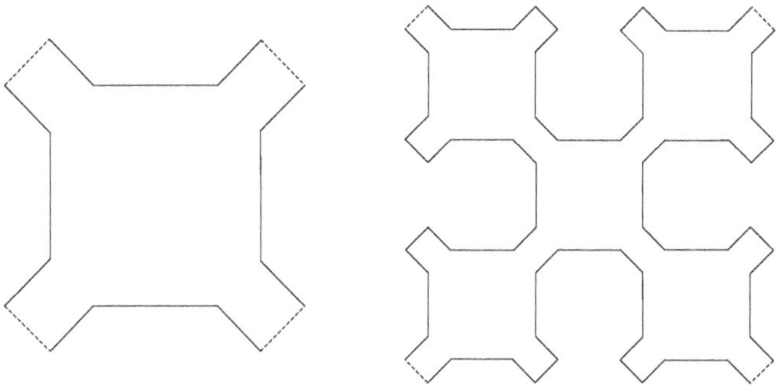

(a) (b)

Figure 7.181: Sierpiński Curves of Orders 1 and 2.

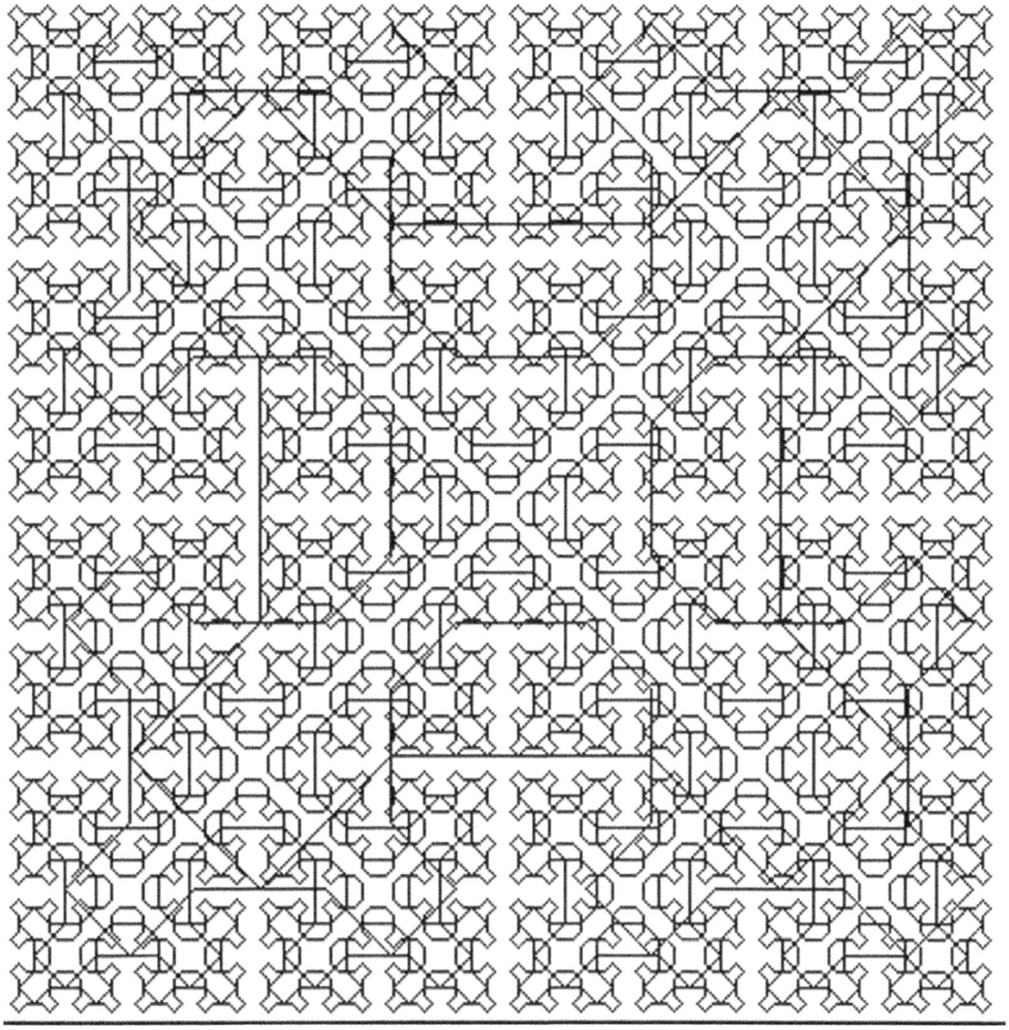

Figure 7.182: Sierpiński Curves of Orders 1–5.

curves, to describe how the Peano curve is constructed. (Hint: The curves shown are P_1, P_2, and P_3. The first curve, P_0, in this sequence is not shown.)

7.37.4 Traversing the Hilbert Curve

Space-filling curves are used in image compression (Section 7.36), which is why it is important to develop methods for a fast traversal of such a curve. Two approaches, both table-driven, are illustrated here for traversing the Hilbert curve.

The first approach [Cole 86] is based on the observation that the Hilbert curve H_i is constructed of four copies of its predecessor H_{i-1} placed at different orientations. A look

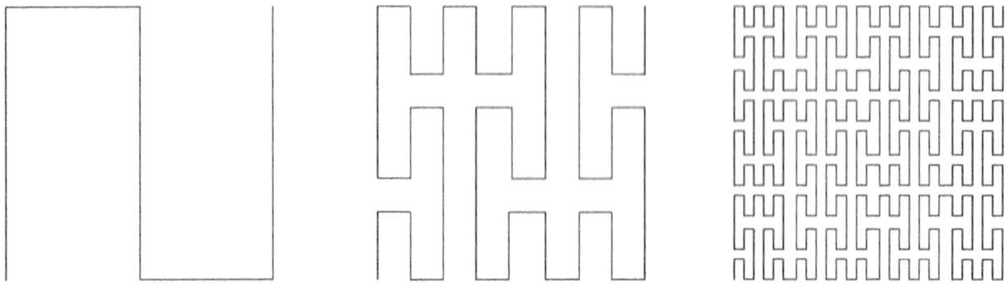

(a) (b) (c)

Figure 7.183: Three Iterations of the Peano Curve.

at Figures 7.177, 7.178, 7.179, and 7.180 should convince the reader that H_i consists of 2^{2i} nodes connected with straight segments. The node numbers therefore vary from 0 to $2^{2i}-1$ and require $2i$ bits each. In order to traverse the curve we need a function that computes the coordinates (x, y) of a node i from the node number i. The (x, y) coordinates of a node in H_i are i-bit numbers.

A look at Figure 7.178 shows how successive nodes are initially located at the bottom-left quadrant, and then move to the bottom-right quadrant, the top-right quadrant, and finally the top-left one. This figure shows orientation #2 of the curve, so we can say that this orientation of H_i traverses quadrants 0, 1, 2, and 3, where quadrants are numbered $\binom{3\,2}{0\,1}$. It is now clear that the two leftmost bits of a node number determine its quadrant. Similarly, the next pair of bits in the node number determine its subquadrant within the quadrant, but here we run into the added complication that each subquadrant is placed at a different orientation in its quadrant. This approach therefore uses Table 7.184 to determine the coordinates of a node from its number.

Bit pair	x	y	Next table	Bit pair	x	y	Next table	Bit pair	x	y	Next table	Bit pair	x	y	Next table
00	0	0	2	00	0	0	1	00	1	1	4	00	1	1	3
01	1	0	1	01	0	1	2	01	0	1	3	01	1	0	4
10	1	1	1	10	1	1	2	10	0	0	3	10	0	0	4
11	0	1	4	11	1	0	3	11	1	0	2	11	0	1	1
(1)				(2)				(3)				(4)			

Table 7.184: Coordinates of Nodes in H_i.

As an example, we compute the xy coordinates of node 109 (the 110th node) of orientation #2 of H_4. The H_4 curve has $2^{2\cdot 4}=256$ nodes, so node numbers are eight bits each, and $109 = 01101101_2$. We start with Table 7.184(1). The two leftmost bits of the node number are 01, and table (1) tells us that the x coordinate start with 1, the y coordinate, with 0, and we should continue with table (1). The next pair of bits is 10, and table (1) tells us that the next bit of x is 1, the next bit of y is 1, and we should stay with table (1). The third pair of bits is 11, so table (1) tells us that the next bit of x is 0, the next bit of y is 1, and we should switch to table (4). The last pair of bits is 01, and table (4) tells us to append 1 and 0 to the coordinates of x and y, respectively. Thus, the coordinates are $x = 1101 = 13$, $y = 0110 = 6$, as can be verified directly from Figure 7.178 (the small circle).

It is also possible to transform a pair of coordinates (x, y), each in the range $[0, 2^i - 1]$, to a node number in H_i by means of Table 7.185.

xy pair	Int. pair	Next table	xy pair	Int. pair	Next table	xy pair	Int. pair	Next table	xy pair	Int. pair	Next table
00	00	2	00	00	1	00	10	3	00	10	4
01	11	4	01	01	2	01	01	3	01	11	1
10	01	1	10	11	3	10	11	2	10	01	4
11	10	1	11	10	2	11	00	4	11	00	3
	(1)			(2)			(3)			(4)	

Table 7.185: Node Numbers in H_i.

◇ **Exercise 7.63:** Use Table 7.185 to compute the node number of the H_4 node whose coordinates are $(13, 6)$.

The second approach to Hilbert curve traversal uses Table Ans.41. Orientation #2 of the H_2 curve shown in Figure 7.177(b) is traversed in order 1223. The same orientation of the H_3 curve of Figure 7.177(c) is traversed in 2114 1223 1223 4332, but Table Ans.41 tells us that 2114 is the traversal order for orientation #1 of H_2, 1223 is the traversal for orientation #2 of H_2, and 4332 is for orientation #3. The traversal of orientation #2 of H_3 is therefore also based on the sequence 1223. Similarly, orientation #2 of H_4 is traversed (Figure 7.178) in the order

1223 2114 2114 3441 2114 1223 1223 4332
2114 1223 1223 4332 3441 4332 4332 1223,

which is reduced to 2114 1223 1223 4332, which in turn is reduced to the same sequence 1223.

The idea is therefore to create the traversal order for orientation #2 of H_i by starting with the sequence 1223 and recursively expanding it $i - 1$ times, using Table Ans.41.

◇ **Exercise 7.64:** (Easy.) Show how to apply this method to traversing orientation #1 of H_i.

A MATLAB function `hilbert.m` to compute the traversal of the curve is available at [Matlab 99]. It was written by Daniel Leo Lau (`dllau@cngr.uky.edu`). The call `hilbert(4)` produces the 4×4 matrix

$$\begin{array}{cccc} 5 & 6 & 9 & 10 \\ 4 & 7 & 8 & 11 \\ 3 & 2 & 13 & 12 \\ 0 & 1 & 14 & 15 \end{array}.$$

7.37.5 Traversing the Peano Curve

The Peano curves P_0, P_1, and P_2 of Figure Ans.40 have 1, 3^2, and 3^4 nodes, respectively. In general, P_n has 3^{2n} nodes, numbered $0, 1, 2, \ldots, 3^{2n}-1$. This suggests that the Peano curve [Peano 90] is somehow based on the number 3, in contrast with the Hilbert curve, which is based on 2. The coordinates of the nodes vary from $(0,0)$ to $(n-1, n-1)$. It turns out that there is a correspondence between the node numbers and their coordinates [Cole 85], which uses base-3 reflected Gray codes (Section 7.4.1).

A reflected Gray code [Gray 53] is a permutation of the i-digit integers such that consecutive integers differ by one digit only. Here is one way to derive these codes for binary numbers. Start with $i = 1$. There are only two 1-bit digits, namely, 0 and 1, and they differ by 1 bit only. To get the RGC for $i = 2$ proceed as follows:

1. Copy the sequence $(0, 1)$.
2. Append (on the left or on the right) a 0 bit to the original sequence and a bit of 1 to the copy. The result is $(00, 01)$, $(10, 11)$.
3. Reflect (reverse) the second sequence. The result is $(11, 10)$.
4. Concatenate the two sequences to get $(00, 01, 11, 10)$.

It is easy to see that consecutive numbers differ by one bit only.

◊ **Exercise 7.65:** Follow the rules above to get the binary RGC for $i = 3$.

Notice that the first and last numbers in an RGC also differ by one bit. RGCs can be created for any number system using the following notation and rules: Let $a = a_1a_2 \cdots a_m$ be a non-negative, base-n integer (i.e., $0 \le a_i < n$). Define the quantity $p_j = (\sum_{i=1}^{j} a_i) \bmod 2$, and denote the base-$n$ RGC of a by $a' = b_1b_2 \cdots b_m$. The digits b_i of a' can be computed by

$$b_1 = a_1; \quad b_i = \begin{cases} a_i & \text{if } p_{i-1} = 0; \\ n-1-a_i & \text{if } p_{i-1} = 1. \end{cases} \text{Odd } n \quad \begin{cases} a_i & \text{if } a_{i-1} \text{ is even}; \\ n-1-a_i & \text{if } a_{i-1} \text{ is odd}. \end{cases} \text{Even } n \qquad i = 2, 3, \ldots, m.$$

[Note that $(a')' = a$ for both even and odd n.] For example, the RGC of the sequence of base-3 numbers (trits) 000, 001, 002, 010, 011, 012, 020, 021, 022, 100, 101... is 000, 001, 002, 012, 011, 010, 020, 021, 022, 122, 121... .

The connection between Peano curves and RGCs is as follows: Let a be a node in the Peano curve P_m. Write a as a ternary (base-3) number with $2m$ trits $a = a_1a_2 \cdots a_{2m}$. Let $a' = b_1b_2 \cdots b_{2m}$ be the RGC equivalent of a. Compute the two numbers $x' = b_2b_4b_6 \cdots b_{2m}$ and $y' = b_1b_3b_5 \cdots b_{2m-1}$. The number x' is the RGC of a number x, and similarly for y'. The two numbers (x, y) are the coordinates of node a in P_m.

7.38 Finite Automata Methods

Finite automata methods are somewhat similar to the IFS method of Section 7.39. They are based on the fact that images used in practice have a certain amount of self-similarity, i.e., it is often possible to find a part of the image that looks the same (or almost the same) as another part, except perhaps for size, brightness, or contrast. It is shown in Section 7.39 that IFS uses affine transformations to match image parts. The methods described here, on the other hand, try to describe a subimage as a weighted sum of other subimages.

Two methods are described in this section, *weighted finite automata* (or WFA) and *generalized finite automata* (or GFA). Both are due to Karel Culik, the former in collaboration with Jarkko Kari, and the latter with Vladimir Valenta. The term "automata" is used because these methods represent the image to be compressed in terms of a graph that is very similar to graphs used to represent finite-state automata (also called finite-state machines, see Section 11.8). The main references are [Culik and Kari 93], [Culik and Kari 94a,b], and [Culik and Kari 95].

7.38.1 Weighted Finite Automata

WFA starts with an image to be compressed. It locates image parts that are identical or very similar to the entire image or to other parts, and constructs a graph that reflects the relationships between these parts and the entire image. The various components of the graph are then compressed and become the compressed image.

The image parts used by WFA are obtained by a quadtree partitioning of the image. Any quadrant, subquadrant, or pixel in the image can be represented by a string of the digits 0, 1, 2, and 3. The longer the string, the smaller the image area (subsquare) it represents. We denote such a string by $a_1a_2\ldots a_k$.

Quadtrees were introduced in Section 7.34. We assume that the quadrant numbering of Figure 7.186a is extended recursively to subquadrants. Figure 7.186b shows how each of the 16 subquadrants produced from the 4 original ones are identified by a 2-digit string of the digits 0, 1, 2, and 3. After another subdivision, each of the resulting subsubquadrants is identified by a 3-digit string, and so on. The black area in Figure 7.186c, for example, is identified by the string 1032.

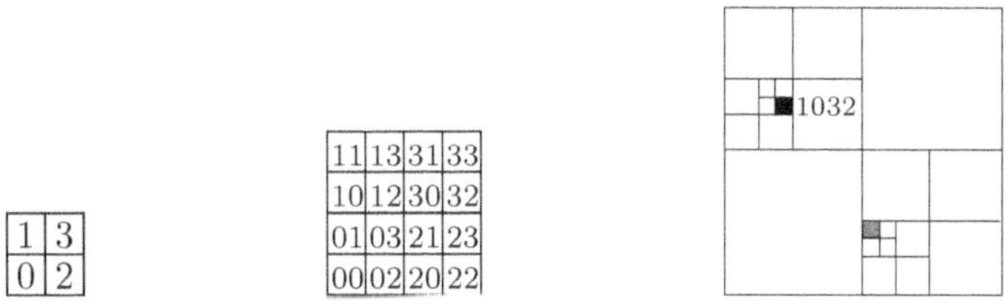

(a) (b) (c)

Figure 7.186: Quadrant Numbering.

◇ **Exercise 7.66:** What string identifies the gray area of Figure 7.186c.

Instead of constantly saying "quadrant, subquadrant, or subsubquadrant," we use the term "subsquare" throughout this section.

◇ **Exercise 7.67:** (Proposed by Karel Culik.) What is special about the particular quadrant numbering of Figure 7.186a?

If the image size is $2^n \times 2^n$, then a single pixel is represented by a string of n digits, and a string $a_1a_2\ldots a_k$ of k digits represents a subsquare of size $2^{n-k} \times 2^{n-k}$ pixels. Once this is grasped, it is not hard to see how any type of image, bi-level, grayscale, or color, can be represented by a graph. Mathematically, a graph is a set of nodes (or vertices) connected by edges. A node may contain data, and an edge may have a label. The graphs used in WFA represent finite automata, so the term *state* is used instead of node or vertex.

The WFA encoder starts by generating a graph that represents the image to be compressed. One state in this graph represents the entire image, and each of the other states represents a subimage. The edges show how certain subimages can be expressed as linear combinations of the entire (scaled) image or of other subimages. Since the subimages are generated by a quadtree, they do not overlap. The basic rule for connecting states with edges is; If quadrant a (where a can be 0, 1, 2, or 3) of subimage i is identical to subimage j, construct an edge from state i to state j, and label it a. In an arbitrary image it is not very common to find two identical subimages, so the rule above is extended to; If subsquare a of subimage i can be obtained by multiplying all the pixels of subimage j by a constant w, construct an edge from state i to state j, label it a, and assign it a weight w. We use the notation $a(w)$ or, if $w = 1$, just a.

A sophisticated encoding algorithm may discover that subsquare a of subimage i can be expressed as the weighted sum of subimages j and k with weights u and w, respectively. In such a case, two edges should be constructed. One from i to j, labeled $a(u)$, and another, from i to k, labeled $a(w)$. In general, such an algorithm may discover that a subsquare of subimage i can be expressed as a weighted sum (or a linear combination) of several subimages, with different weights. In such a case, an edge should be constructed for each term in the sum.

We denote the set of states of the graph by Q and the number of states by m.

Notice that the weights do not have to add up to 1. They can be any real numbers, and can be greater than 1 or negative. Figure 7.187b is the graph of a simple 2×2 chessboard. It is clear that we can ignore subsquares 1 and 2, since they are all white. The final graph includes states for subsquares that are not all white, and it has two states. State 1 is the entire image, and state 2 is an all black subimage. Since subsquares 0 and 3 are identical to state 2, there are two edges (shown as one edge) from state 1 to state 2, where the notation $0, 3(1)$ stands for $0(1)$ and $3(1)$. Since state 2 represents subsquares 0 and 3, it can be named q_0 or q_3.

Figure 7.187c shows the graph of a 2×2 chessboard where quadrant 3 is 50% gray. This is also a two-state graph. State 2 of this graph is an all-black image and is identical to quadrant 0 of the image (we can also name it q_0). This is why there is an edge labeled $0(1)$. State 2 can also generate quadrant 3, if multiplied by 0.5, and this is expressed by the edge $3(0.5)$. Figure 7.187d shows another two-state graph representing the same

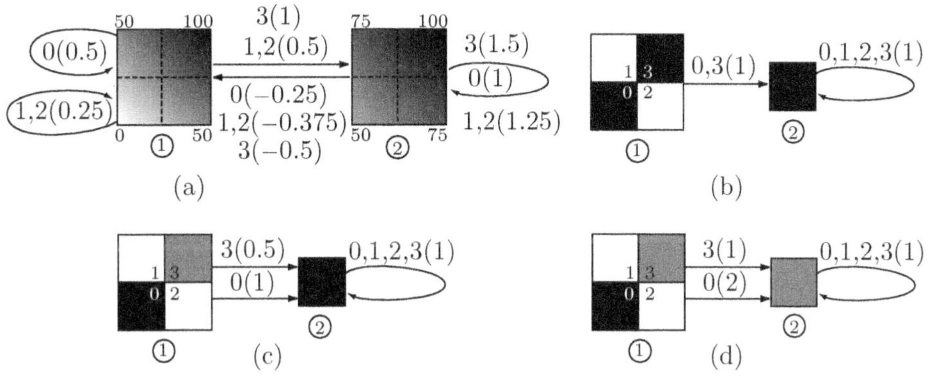

Figure 7.187: Graphs for Simple Images.

image. This time, state 2 (which may be named q_3) is a 50%-gray image, and the weights are different.

Figure 7.187a shows an example of a weighted sum. The image (state 1) varies smoothly from black in the top-right corner to white in the bottom-left corner. The graph contains one more state, state 2, which is identical to quadrant 3 (we can also name it q_3). It varies from black in the top-right corner to 50% gray in the bottom-left corner, with 75% gray in the other two corners. Quadrant 0 of the image is obtained when the entire image is multiplied by 0.5, so there is an edge from state 1 to itself labeled 0(0.5). Quadrants 1 and 2 are identical and are obtained as a weighted sum of the entire image (state 1) multiplied by 0.25 and of state 2 (quadrant 3) multiplied by 0.5. The four quadrants of state 2 are similarly obtained as weighted sums of states 1 and 2.

The average grayness of each state is easy to figure out in this simple case. For example, the average grayness of state 1 of Figure 7.187a is 0.5 and that of quadrant 3 of state 2 is $(1 + 0.75)/2 = 0.875$. The average grayness is used later and is called the *final distribution.*

Notice that the two states of Figure 7.187a have the same size. In fact, nothing has been said so far about the sizes and resolutions of the image and the various subimages. The process of constructing the graph does not depend on the resolution, which is why WFA can be applied to *multiresolution images*, i.e., images that may exist in several different resolutions.

Figure 7.188 is another example. The original image is state 1 of the graph, and the graph has four states. State 2 is quadrant 0 of the image (so it may be called q_0). State 3 represents (the identical) quadrants 1 and 2 of the image (it may be called q_1 or q_2). It should be noted that, in principle, a state of a WFA graph does not have to correspond to any subsquare of the image. However, the recursive inference algorithm used by WFA generates a graph where each state corresponds to a subsquare of the image.

◇ **Exercise 7.68:** In Figure 7.188, state 2 of the graph is represented in terms of itself and of state 4. Show how to represent it in terms of itself and an all-black state.

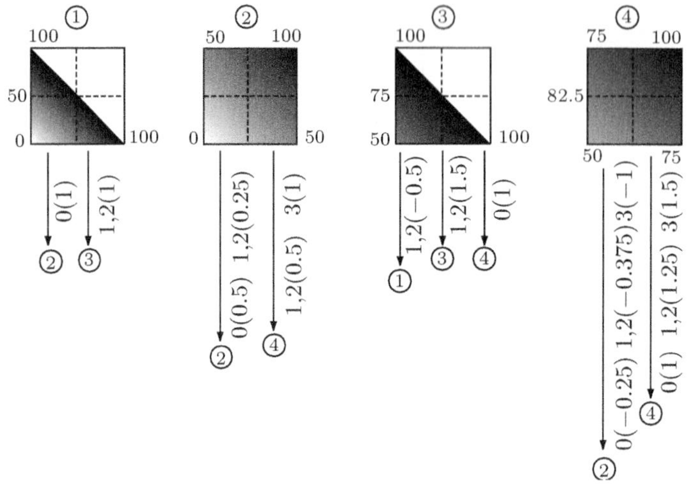

Figure 7.188: A Four-State Graph.

A multiresolution image M may, in principle, be represented by an intensity function $f_M(x, y)$ that gives the intensity of the image at point (x, y). For simplicity, we assume that x and y vary independently in the range $[0, 1]$, that the intensity varies between 0 (white) and 1 (black), and that the bottom-left corner of the image is located at the origin [i.e., it is point $(0, 0)$]. Thus, the subimage of state 2 of Figure 7.188 is defined by the function

$$f(x,y) = \frac{x+y}{2}.$$

◇ **Exercise 7.69:** What function describes the image of state 1 of Figure 7.188?

Since we are dealing with multiresolution images, the intensity function f must be able to generate the image at any resolution. If we want, for example, to lower the resolution to one-fourth the original one, each new pixel should be computed from a set of four original ones. The obvious value for such a "fat" pixel is the average of the four original pixels. This implies that at low resolutions, the intensity function f should compute the average intensity of a region in the original image. In general, the value of f for a subsquare w should satisfy

$$f(w) = \frac{1}{4}\left[f(w0) + f(w1) + f(w2) + f(w3)\right].$$

Such a function is called *average preserving* (ap). It is also possible to increase the resolution, but the details created in such a case would be artificial.

Given an arbitrary image, however, we generally do not know its intensity function. Also, any given image has limited, finite resolution. Thus, WFA proceeds as follows: It starts with a given image with resolution $2^n \times 2^n$. It uses an inference algorithm to construct its graph, then extracts enough information from the graph to be able to reconstruct the image at its original or any lower resolution. The information obtained from the graph is compressed and becomes the compressed image.

The first step in extracting information from the graph is to construct four *transition matrices* W_0, W_1, W_2, and W_3 according to the following rule: If there is an edge labeled q from state i to state j, then element (i,j) of transition matrix W_q is set to the weight w of the edge. Otherwise, $(W_q)_{i,j}$ is set to zero. An example is the four transition matrices resulting from the graph of Figure 7.189. They are

$$W_0=\begin{pmatrix}0.5 & 0\\ 0 & 1\end{pmatrix},\ W_1=\begin{pmatrix}0.5 & 0.25\\ 0 & 1\end{pmatrix},\ W_2=\begin{pmatrix}0.5 & 0.25\\ 0 & 1\end{pmatrix},\ W_3=\begin{pmatrix}0.5 & 0.5\\ 0 & 1\end{pmatrix}.$$

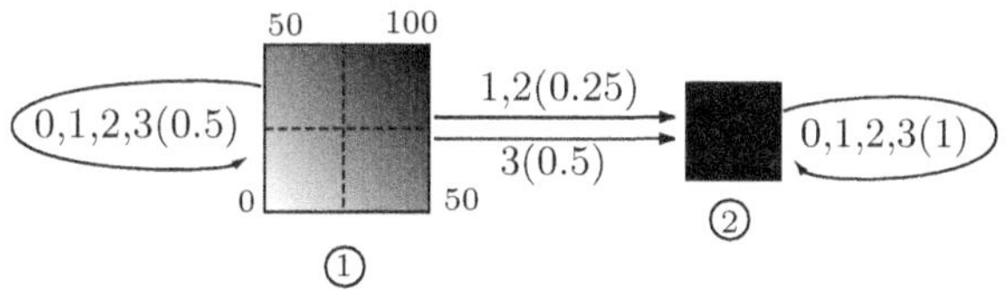

Figure 7.189: A Two-State Graph.

The second step is to construct a column vector F of size m called the *final distribution* (this is not the same as the intensity function f). Each component of F is the average intensity of the subimage associated with one state of the graph. Thus, for the two-state graph of Figure 7.189 we get

$$F=(0.5,1)^T.$$

◇ **Exercise 7.70:** Write the four transition matrices and the final distribution for the graph of Figure 7.188.

In the third step we define quantities $\psi_i(a_1a_2\ldots a_k)$. As a reminder, if the size of the original image is $2^n\times 2^n$, then the string $a_1a_2\ldots a_k$ (where $a_i=0,1,2$, or 3) defines a subsquare of size $2^{n-k}\times 2^{n-k}$ pixels. The definition is

$$\psi_i(a_1a_2\ldots a_k)=(W_{a1}\cdot W_{a2}\cdots W_{ak}\cdot F)_i. \tag{7.57}$$

Thus, $\psi_i(a_1a_2\ldots a_k)$ is the ith element of the column $(W_{a1}\cdot W_{a2}\cdots W_{ak}\cdot F)^T$; it is a number.

Each transition matrix W_a has dimensions $m\times m$ (where m the number of states of the graph) and F is a column of size m. For any given string $a_1a_2\ldots a_k$ there are therefore m numbers $\psi_i(a_1a_2\ldots a_k)$, for $i=1,2,\ldots,m$. We use the two-state graph of Figure 7.189 to calculate some ψ_i's:

$$\psi_i(0)=(W_0\cdot F)_i=\begin{pmatrix}0.5 & 0\\ 0 & 1\end{pmatrix}\begin{pmatrix}0.5\\ 1\end{pmatrix}_i=\begin{pmatrix}0.25\\ 1\end{pmatrix}_i,$$

$$
\begin{aligned}
\psi_i(01) &= (W_0{\cdot}W_1{\cdot}F)_i = \begin{pmatrix} 0.5 & 0 \\ 0 & 1 \end{pmatrix} \begin{pmatrix} 0.5 & 0.25 \\ 0 & 1 \end{pmatrix} \begin{pmatrix} 0.5 \\ 1 \end{pmatrix}_i = \begin{pmatrix} 0.25 \\ 1 \end{pmatrix}_i, \\
\psi_i(1) &= (W_1{\cdot}F)_i = \begin{pmatrix} 0.5 & 0.25 \\ 0 & 1 \end{pmatrix} \begin{pmatrix} 0.5 \\ 1 \end{pmatrix}_i = \begin{pmatrix} 0.5 \\ 1 \end{pmatrix}_i, \qquad (7.58) \\
\psi_i(00) &= (W_0{\cdot}W_0{\cdot}F)_i = \begin{pmatrix} 0.5 & 0 \\ 0 & 1 \end{pmatrix} \begin{pmatrix} 0.5 & 0 \\ 0 & 1 \end{pmatrix} \begin{pmatrix} 0.5 \\ 1 \end{pmatrix}_i = \begin{pmatrix} 0.125 \\ 1 \end{pmatrix}_i, \\
\psi_i(03) &= (W_0{\cdot}W_3{\cdot}F)_i = \begin{pmatrix} 0.5 & 0 \\ 0 & 1 \end{pmatrix} \begin{pmatrix} 0.5 & 0.5 \\ 0 & 1 \end{pmatrix} \begin{pmatrix} 0.5 \\ 1 \end{pmatrix}_i = \begin{pmatrix} 0.375 \\ 1 \end{pmatrix}_i, \\
\psi_i(33\ldots3) &= (W_3{\cdot}W_3\cdots W_3{\cdot}F)_i \\
&= \begin{pmatrix} 0.5 & 0.5 \\ 0 & 1 \end{pmatrix} \begin{pmatrix} 0.5 & 0.5 \\ 0 & 1 \end{pmatrix} \cdots \begin{pmatrix} 0.5 & 0.5 \\ 0 & 1 \end{pmatrix} \begin{pmatrix} 0.5 \\ 1 \end{pmatrix}_i \\
&= \begin{pmatrix} 0 & 1 \\ 0 & 1 \end{pmatrix} \begin{pmatrix} 0.5 \\ 1 \end{pmatrix}_i = \begin{pmatrix} 1 \\ 1 \end{pmatrix}_i.
\end{aligned}
$$

◇ **Exercise 7.71:** Compute ψ_i at the center of the image.

Notice that each $\psi_i(w)$ is identified by two quantities, its subscript i (which corresponds to a state of the graph) and a string $w = a_1a_2\ldots a_k$ that specifies a subsquare of the image. The subsquare can be as big as the entire image or as small as a single pixel. The quantity ψ_i (where no subsquare is specified) is called the *image* of state i of the graph. The name *image* makes sense, since for each subsquare w, $\psi_i(w)$ is a number, so ψ_i consists of several numbers from which the pixels of the image of state i can be computed and displayed. The WFA encoding algorithms described in this section generate a graph where each state is a subsquare of the image. In principle, however, some states of the graph may not correspond to any subsquare of the image. An example is state 2 of the graph of Figure 7.189. It is all black, even though the image does not have an all-black subsquare.

Figure 7.190 shows the image of Figure 7.189 at resolutions 2×2, 4×4, and 256×256.

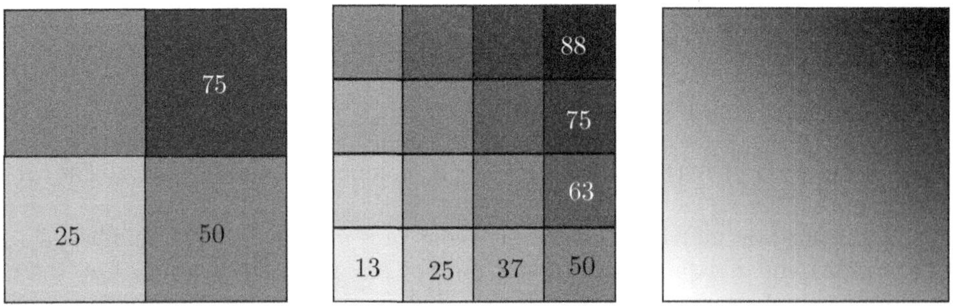

Figure 7.190: Image $f = (i + j)/2$ at Three Resolutions.

◇ **Exercise 7.72:** Use mathematical software to compute a matrix such as those of Figure 7.190.

We now introduce the *initial distribution* $I = (I_1, I_2, \ldots, I_m)$, a row vector with m elements. If we set it to $I = (1, 0, 0, \ldots, 0)$, then the dot product $I\cdot\psi(a_1a_2\ldots a_k)$ [where the notation $\psi(q)$ stands for a vector with the m values $\psi_1(q)$ through $\psi_m(q)$] gives the average intensity $f(a_1a_2\ldots a_k)$ of the subsquare specified by $a_1a_2\ldots a_k$. This intensity function can also be expressed as the matrix product

$$f(a_1a_2\ldots a_k) = I\cdot W_{a1}\cdot W_{a2}\cdots W_{ak}\cdot F. \tag{7.59}$$

In general, the initial distribution specifies the image defined by a WFA as a linear combination $I_1\psi_1 + \cdots + I_n\psi_n$ of "state images." If $I = (1, 0, 0, \ldots, 0)$, then the image defined is ψ_1, the image corresponding to the first state. This is always the case for the image resulting from the inference algorithm described later in this section.

Given a WFA, we can easily find another WFA for the image obtained by zooming to the subsquare with address $a_1a_2\ldots a_k$. We just replace the initial distribution I by $I\cdot W_{a1}\cdot W_{a2}\cdots W_{ak}$.

To prove this, consider the subsquare with address $b_1b_2\ldots b_m$ in the zoomed square. It corresponds to the subsquare in the entire image with address $a_1a_2\ldots a_kb_1b_2\ldots b_m$. Hence, the grayness value computed for it by the original WFA is

$$I\,W_{a1}W_{a2}\ldots W_{ak}W_{b1}W_{b2}\ldots W_{bm}\,F.$$

Using the new WFA for the corresponding subsquare, we get the same value, namely $I'\,W_{b1}W_{b2}\ldots W_{bm}\,F$, where $I' = I\,W_{a1}W_{a2}\ldots W_{ak}$ (proof provided by Karel Culik).

For the ψ_i's computed in Equation (7.58) the dot products of the form $I\psi(a_1a_2\ldots a_k)$ yield

$$\begin{aligned}
f(0) &= I\cdot\psi(0) = (1,0)\begin{pmatrix}0.25\\1\end{pmatrix} = I_1\psi_1(0) + I_2\psi_2(0) = 0.25,\\
f(01) &= I\cdot\psi(01) = 1\times 0.25 + 0\times 1 = 0.25,\\
f(1) &= I\cdot\psi(1) = 0.5,\\
f(00) &= I\cdot\psi(00) = 0.125,\\
f(03) &= I\cdot\psi(03) = 0.375,\\
f(33\ldots3) &= I\cdot\psi(33\ldots3) = 1.
\end{aligned}$$

◇ **Exercise 7.73:** Compute the ψ_i's and the corresponding f values for subsquares 0, 01, 1, 00, 03, and 3 of the five-state graph of Figure 7.188.

Equation (7.57) is the definition of $\psi_i(a_1a_2\ldots a_k)$. It shows that this quantity is the ith element of the column vector $(W_{a1}\cdot W_{a2}\cdots W_{ak}\cdot F)^T$. We now examine the reduced column vector $(W_{a2}\cdots W_{ak}\cdot F)^T$. Its ith element is, according to the definition of ψ_i, the quantity $\psi_i(a_2\ldots a_k)$. Thus, we conclude that

$$\begin{aligned}
&\psi_i(a_1a_2\ldots a_k)\\
&\quad= (W_{a1})_{i,1}\psi_1(a_2\ldots a_k) + (W_{a1})_{i,2}\psi_2(a_2\ldots a_k) + \cdots + (W_{a1})_{i,m}\psi_m(a_2\ldots a_k),
\end{aligned}$$

or, if we denote the string $a_2 \ldots a_k$ by w,

$$\begin{aligned}\psi_i(a_1 w) &= (W_{a1})_{i,1}\psi_1(w) + (W_{a1})_{i,2}\psi_2(w) + \cdots + (W_{a1})_{i,m}\psi_m(w)\\ &= \sum_{j=1}^{m}(W_{a1})_{i,j}\psi_j(w).\end{aligned} \tag{7.60}$$

The quantity $\psi_i(a_1 w)$ (which corresponds to quadrant a_1 of subsquare w) can be expressed as a linear combination of the quantities $\psi_j(w)$, where $j = 1, 2, \ldots, m$. This justifies calling ψ_i the *image* of state i of the graph. We say that subsquare a_1 of image $\psi_i(w)$ can be expressed as a linear combination of images $\psi_j(w)$, for $j = 1, 2, \ldots, m$. This is how the self-similarity of the original image enters the picture.

Equation (7.60) is recursive, since it defines a smaller image in terms of larger ones. The largest image is, of course, the original image, for which w is null (we denote the null string by ϵ). This is where the recursion starts

$$\begin{aligned}\psi_i(a) &= (W_a)_{i,1}\psi_1(\epsilon) + (W_a)_{i,2}\psi_2(\epsilon) + \cdots + (W_a)_{i,m}\psi_m(\epsilon)\\ &= \sum_{j=1}^{m}(W_a)_{i,j}\psi_j(\epsilon), \quad \text{for } a = 0, 1, 2, 3.\end{aligned} \tag{7.61}$$

On the other hand, Equation (7.57) shows that $\psi_i(\epsilon) = F_i$, so Equation (7.61) becomes

$$\psi_i(a) = \sum_{j=1}^{m}(W_a)_{i,j}F_j, \quad \text{for } a = 0, 1, 2, 3. \tag{7.62}$$

It should now be clear that if we know the four transition matrices and the final distribution, we can compute the images $\psi_i(w)$ for strings w of any length (i.e., for subsquares of any size). Once an image $\psi_i(w)$ is known, the average intensity $f(w)$ of subsquare w can be computed by Equation (7.59) (which requires knowledge of the initial distribution).

The problem of representing an image f in WFA is thus reduced to finding vectors I and F and matrices W_0, W_1, W_2, and W_3 that will produce f (or an image close to it). Alternatively, we can find I and images $\psi_i(\epsilon)$ for $i = 1, 2, \ldots, m$.

Here is an informal approach to this problem. We can start with a graph consisting of one state and select this state as the entire image, $\psi_1(\epsilon)$. We now concentrate on quadrant 0 of the image and try to determine $\psi_1(0)$. If quadrant 0 is identical, or similar enough, to (a scaled version of) the entire image, we can write

$$\psi_1(0) = \psi_1(\epsilon) = \sum_{j=1}^{m}(W_0)_{1,j}\psi_j(\epsilon),$$

which is true if the first row of W_0 is $(1, 0, 0, \ldots, 0)$. We have determined the first row of W_0, even though we don't yet know its size (it is m but m hasn't been determined yet). If quadrant 0 is substantially different from the entire image, we add $\psi_1(0)$ to the graph as a new state, state 2 (i.e., we increment m by 1) and call it $\psi_2(0)$. The result is

$$\psi_1(0) = \psi_2(0) = \sum_{j=1}^{m} (W_0)_{1,j}\psi_j(0),$$

which is true if the first row of W_0 is $(0, 1, 0, \ldots, 0)$. We have again determined the first row of W_0, even though we still don't know its size.

Next, we process the remaining three quadrants of $\psi_1(\epsilon)$ and the four quadrants of $\psi_2(0)$. Let's examine, for example, the processing of quadrant three $[\psi_2(03)]$ of $\psi_2(0)$. If we can express it (precisely or close enough) as the linear combination

$$\psi_2(03) = \alpha\psi_1(3) + \beta\psi_2(3) = \sum_j (W_0)_{2,j}\psi_j(3),$$

then we know that the second row of W_0 must be $(\alpha, \beta, 0, \ldots, 0)$. If we cannot express $\psi_2(03)$ as a linear combination of already known ψ's, then we declare it a new state $\psi_3(03)$, and this implies $(W_3)_{2,3} = 1$, and also determines the second row of W_0 to be $(0, 0, 1, 0, \ldots, 0)$.

This process continues until all quadrants of all the ψ_j's have been processed. This normally generates many more ψ_i's, all of which are subsquares of the image.

This intuitive algorithm is now described more precisely, using pseudocode. It constructs a graph from a given multiresolution image f one state at a time. The graph has the minimal number of states (to be denoted by m) but a relatively large number of edges. Since m is minimal, the four transition matrices (whose dimensions are $m \times m$) are small. However, since the number of edges is large, most of the elements of these matrices are nonzero. The image is compressed by writing the transition matrices (in compressed format) on the compressed stream, so the sparser the matrices, the better the compression. Thus, this algorithm does not produce good compression and is described here because of its simplicity. We denote by i the index of the first unprocessed state, and by γ, a mapping from states to subsquares. The steps of this algorithm are as follows:

Step 1: Set $m = 1$, $i = 1$, $F(q_1) = f(\epsilon)$, $\gamma(q_1) = \epsilon$.
Step 2: Process q_i, i.e., for $w = \gamma(q_i)$ and $a = 0, 1, 2, 3$, do:

Step 2a: Start with $\psi_j = f_{\gamma(q_j)}$ for $j = 1, \ldots, m$ and try to find real numbers $c_1, \ldots, c_m$ such that $f_{wa} = c_1\psi_1 + \cdots + c_m\psi_m$. If such numbers are found, they become the m elements of row q_i of transition matrix W_a, i.e., $W_a(q_i, q_j) = c_j$ for $j = 1, \ldots, m$.

Step 2b: If such numbers cannot be found, increment the number of states $m = m + 1$, and set $\gamma(q_m) = wa$, $F(q_m) = f(wa)$ (where F is the final distribution), and $W_a(q_i, q_m) = 1$.

Step 3: Increment the index of the next unprocessed state $i = i + 1$. If $i \le m$, go to Step 2.
Step 4: The final step. Construct the initial distribution I by setting $I(q_1) = 1$ and $I(q_j) = 0$ for $j = 2, 3, \ldots, m$.

[Litow and Olivier 95] presents another inference algorithm that also yields a minimum state WFA and uses only additions and inner products.

The real breakthrough in WFA compression came when a better inference algorithm was developed. This algorithm is the most important part of the WFA method. It generates a graph that may have more than the minimal number of states, but that has a small number of edges. The four transition matrices may be large, but they are sparse, resulting in better compression of the matrices and hence better compression of the image. The algorithm starts with a given finite-resolution image A, and generates its graph by matching subsquares of the image to the entire image or to other subsquares. An important point is that this algorithm can be lossy, with the amount of the loss controlled by a user-defined parameter G. Larger values of G "permit" the algorithm to match two parts of the image even if they poorly resemble each other. This naturally leads to better compression. The metric used by the algorithm to match two images (or image parts) f and g is the square of the L2 metric, a common measure where the distance $d_k(f,g)$ is defined as

$$d_k(f,g) = \sum_{w} \left[f(w) - g(w)\right]^2,$$

where the sum is over all subsquares w.

The algorithm tries to produce a small graph. If we denote the size of the graph by (size A), the algorithm tries to keep as small as possible the value of

$$d_k(f, f_A) + G\cdot(\text{size } A).$$

The quantity m indicates the number of states, and there is a multiresolution image ψ_i for each state $i = 1, 2, \ldots, m$. The pseudocode of Figure 7.191 describes a recursive function $make\_wfa(i,k,max)$ that tries to approximate ψ_i at level k as well as possible by adding new edges and (possibly) new states to the graph. The function minimizes the value of the cost quantity

$$cost = d_k(\psi_i, \psi_i') + G\cdot s,$$

where ψ_i' is the current approximation to ψ_i and s is the increase in the size of the graph caused by adding new edges and states. If $cost > max$, the function returns ∞, otherwise, it returns $cost$.

When the algorithm starts, m is set to 1 and ψ is set to f, where f is the function that needs to be approximated at level k. The algorithm then calls $make\_wfa(1,k,\infty)$, which calls itself. For each of the four quadrants, the recursive call $make\_wfa(i,k,max)$ tries to approximate $(\psi_i)_a$ for $a = 0, 1, 2, 3$ in two different ways, as a linear combination of the functions of existing states (step 1), and by adding a new state and recursively calling itself (steps 2 and 3). The better of the two results is then selected (in steps 4 and 5).

The algorithm constructs an initial distribution $I = (1, 0, \ldots, 0)$ and a final distribution $F_i = \psi_i(\epsilon)$ for all states i.

WFA is a common acronym as, e.g., in "World Fellowship Activities."

function $make\_wfa(i, k, max)$;
If $max < 0$, return ∞;
$cost \leftarrow 0$;
if $k = 0$, $cost \leftarrow d_0(f, 0)$
else do steps 1–5 with $\psi = (\psi_i)_a$ for $a = 0, 1, 2, 3$;

1. Find $r_1, r_2, \ldots r_m$ such that the value of

$$cost1 \leftarrow d_{k-1}(\psi, r_1\psi_1 + \cdots + r_n\psi_n) + G{\cdot}s$$

is small, where s denotes the increase in the size of the graph caused by adding edges from state i to states j with nonzero weights r_j and label a, and d_{k-1} denotes the distance between two multiresolution images at level $k-1$ (quite a mouthful).
2. Set $m_0 \leftarrow m$, $m \leftarrow m+1$, $\psi_m \leftarrow \psi$ and add an edge with label a and weight 1 from state i to the new state m. Let s denote the increase in the size of the graph caused by the new state and new edge.
3. Set $cost2 \leftarrow G{\cdot}s + make\_wfa(m, k-1, min(max - cost, cost1) - G{\cdot}s)$;
4. If $cost2 \le cost1$, set $cost \leftarrow cost + cost2$;
5. If $cost1 < cost2$, set $cost \leftarrow cost + cost1$, remove all outgoing edges from states $m_0+1, \ldots, m$ (added during the recursive call), as well as the edge from state i added in step 2. Set $m \leftarrow m_0$ and add the edges from state i with label a to states $j = 1, 2, \ldots, m$ with weights r_j whenever $r_j \ne 0$.

If $cost \le max$, return($cost$), else return(∞).

Figure 7.191: The WFA Recursive Inference Algorithm.

Next, we discuss how the elements of the graph can be compressed. Step 2 creates an edge whenever a new state is added to the graph. They form a tree, and their weights are always 1, so each edge can be coded in four bits. Each of the four bits indicates which of two alternatives was selected for the label in steps 4–5.

Step 1 creates edges that represent linear combinations (i.e., cases where subimages are expressed as linear combinations of other subimages). Both the weight and the endpoints need be stored. Experiments indicate that the weights are normally distributed, so they can be efficiently encoded with prefix codes. Storing the endpoints of the edges is equivalent to storing four sparse binary matrices, so run length encoding can be used.

The WFA decoder reads vectors I and F and the four transition matrices W_a from the compressed stream and decompresses them. Its main task is to use them to reconstruct the original $2^n \times 2^n$ image A (precisely or approximately), i.e., to compute $f_A(w)$ for all strings $w = a_1a_2\ldots a_n$ of size n. The original WFA decoding algorithm is fast but has storage requirements of order $m4^n$, where m is the number of states of the graph. The algorithm consists of four steps as follows:

Step 1: Set $\psi_p(\epsilon) = F(p)$ for all $p \in Q$.

Step 2: Repeat Step 3 for $i = 1, 2, \ldots, n$.

Step 3: For all $p \in Q$, $w = a_1a_2 \ldots a_{i-1}$, and $a = 0, 1, 2, 3$, compute

$$\psi_p(aw) = \sum_{q \in Q} W_a(p, q) \cdot \psi_q(w).$$

Step 4: For each $w = a_1a_2 \ldots a_n$ compute

$$f_A(w) = \sum_{q \in Q} I(q) \cdot \psi_q(w).$$

This decoding algorithm was improved by Raghavendra Udupa, Vinayaka Pandit, and Ashok Rao [Udupa et al. 99]. They define a new multiresolution function Φ_p for every state p of the WFA graph by

$$\Phi_p(\epsilon) = I(p),$$
$$\Phi_p(wa) = \sum_{q \in Q} (W_a)_{q,p} \Phi_q(w), \text{ for } a = 0, 1, 2, 3,$$

or, equivalently,

$$\Phi_p(a_1a_2 \ldots a_k) = (I\, W_{a1} \ldots W_{ak})_p,$$

where $\Phi_p(wa)$ is the sum of the weights of all the paths with label wa ending in p. The weight of a path starting at state i is the product of the initial distribution of i and the weights of the edges along the path wa.

One result of this definition is that the intensity function f can be expressed as a linear combination of the new multiresolution functions Φ_p,

$$f(w) = \sum_{q \in Q} \Phi_q(w) F(q),$$

but there is a more important observation! Suppose that both $\Phi_p(u)$ and $\psi_p(w)$ are known for all states $p \in Q$ of the graph and for all strings u and w. Then the intensity function of subsquare uw can be expressed as

$$f(uw) = \sum_{p \in Q} \Phi_p(u) \psi_p(w).$$

This observation is used in the improved 6-step algorithm that follows to reduce both the space and time requirements. The input to the algorithm is a WFA A with resolution $2^n \times 2^n$, and the output is an intensity function $f_A(w)$ for every string w up to length n.

Step 1: Select nonnegative integers $n1$ and $n2$ satisfying $n1 + n2 = n$. Set $\Phi_p = I(p)$ and $\psi_p = F(p)$ for all $p \in Q$.

Step 2: For $i = 1, 2, \ldots, n1$ do Step 3.

Step 3: For all $p \in Q$ and $w = a_1a_2 \ldots a_{i-1}$ and $a = 0, 1, 2, 3$, compute

$$\Phi_p(aw) = \sum_{q \in Q} (W_a)_{q,p} \Phi_q(w).$$

Step 4: For $i = 1, 2, \ldots, n2$ do Step 5.
Step 5: For all $p \in Q$ and $w = a_1 a_2 \ldots a_{i-1}$ and $a = 0, 1, 2, 3$, compute

$$\psi_p(aw) = \sum_{q \in Q} (W_a)_{p,q} \psi_q(w).$$

Step 6: For each $u = a_1 a_2 \ldots a_{n1}$ and $w = b_1 b_2 \ldots b_{n2}$, compute

$$f_A(uw) = \sum_{q \in Q} \Phi_q(u) \psi_q(w).$$

The original WFA decoding algorithm is a special case of the above algorithm for $n1 = 0$ and $n2 = n$. For the case where $n = 2l$ is an even number, it can be shown that the space requirement of this algorithm is of order $m4^{n1} + m4^{n2}$. This expression has a minimum when $n1 = n2 = n/2 = l$, implying a space requirement of order $m4^l$.

Another improvement on the original WFA is the use of *bintrees* (Section 7.34.1). This idea is due to Ullrich Hafner [Hafner 95]. Recall that WFA uses quadtree methods to partition the image into a set of nonoverlapping subsquares. These can be called (in common with the notation used by IFS) the *range images*. Each range image is then matched, precisely or approximately, to a linear combination of other images that become the *domain images*. Using bintree methods to partition the image results in finer partitioning and an increase in the number of domain images available for matching. In addition, the compression of the transition matrices and of the initial and final distributions is improved. Each node in a bintree has two children compared to four children in a quadtree. A subimage can therefore be identified by a string of bits instead of a string of the digits 0–3. This also means that there are just two transition matrices W_0 and W_1.

WFA compression works particularly well for color images, since it constructs a single WFA graph for the three color components of each pixel. Each component has its own initial state, but other states are shared. This normally saves many states because the color components of pixels in real images are not independent. Experience indicates that the WFA compression of a typical image may create, say, 300 states for the Y color component, 40 states for I, and only about 5 states for the Q component. Another property of WFA compression is that it is relatively slow for high-quality compression, since it builds a large automaton, but is faster for low-quality compression, since the automaton constructed in such a case is small.

7.38.2 Generalized Finite Automata

The graphs constructed and used by WFA represent finite-state automata with "weighted inputs." Without weights, finite automata specify multiresolution bi-level images. At resolution $2^k \times 2^k$, a finite automaton specifies an image that is black in those subsquares whose addresses are accepted by the automaton. It is well known that every nondeterministic automaton can be converted to an equivalent deterministic one. This is not the case for WFA, where nondeterminism gives much more power. This is why an image with a smoothly varying gray scale, such as the one depicted in Figure 7.189, can be represented by a simple, two-state automaton. This is also why WFA can efficiently compress a large variety of grayscale and color images.

For bi-level images the situation is different. Here nondeterminism might provide more concise description. However, experiments show that a nondeterministic automaton does not improve the compression of such images by much, and is slower to construct than a deterministic one. This is why the *generalized finite automata* (GFA) method was originally developed ([Culik and Valenta 96] and [Culik and Valenta 97a,b]). We also show how it is extended for color images. The algorithm to generate the GFA graph is completely different from the one used by WFA. In particular, it is not recursive (in contrast to the edge-optimizing WFA algorithm of Figure 7.191, which is recursive). GFA also uses three types of transformations, rotations, flips, and complementation, to match image parts.

A GFA graph consists of states and of edges connecting them. Each state represents a subsquare of the image, with the first state representing the entire image. If quadrant q of the image represented by state i is identical (up to a scale factor) to the image represented by state j, then an edge is constructed from i to j and is labeled q. More generally, if quadrant q of state i can be made identical to state j by applying transformation t to the image of j, then an edge is constructed from i to j and is labeled q, t or (q, t). There are 16 transformations, shown in Figure 7.193. Transformation 0 is the identity, so it may be omitted when an edge is labeled. Transformations 1–3 are 90° rotations, and transformations 4–7 are rotations of a reflection of transformation 0. Transformations 8–15 are the reverse videos of 0–7, respectively. Each transformation t is thus specified by a 4-bit number.

Figure 7.192: Image for Exercise 7.74.

Figure 7.194 shows a simple bi-level image and its GFA graph. Quadrant 3 of state 0, for example, has to go through transformation 1 in order to make it identical to state 1, so there is an edge labeled $(3, 1)$ from state 0 to state 1.

◇ **Exercise 7.74:** Construct the GFA of the image of Figure 7.192 using transformations, and list all the resulting edges in a table.

Images used in practice are more complex and less self-similar than the examples shown here, so the particular GFA algorithm discussed here (although not GFA in general) allows for a certain amount of data loss in matching subsquares of the image. The distance $d_k(f, g)$ between two images or subimages f and g of resolution $2^k \times 2^k$ is defined as

$$d_k(f, g) = \frac{\sum_{w=a1a2\ldots ak} |f(w) - g(w)|}{2^k \times 2^k}.$$

The numerator of this expression counts the number of pixels that differ in the two images, while the denominator is the total number of pixels in each image. The distance

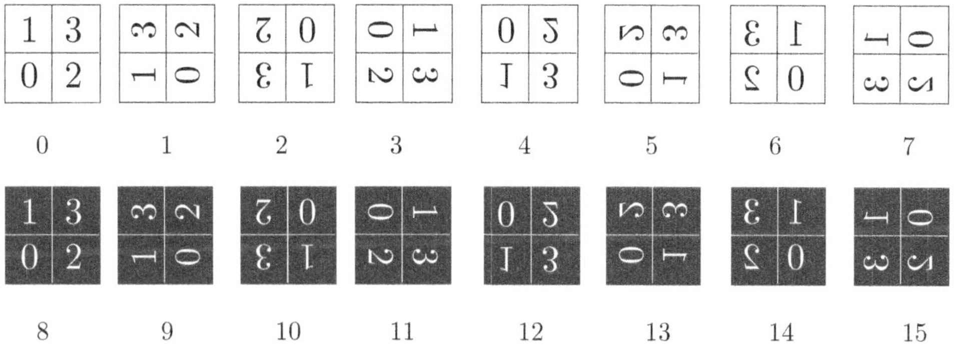

Figure 7.193: Sixteen Image Transformations.

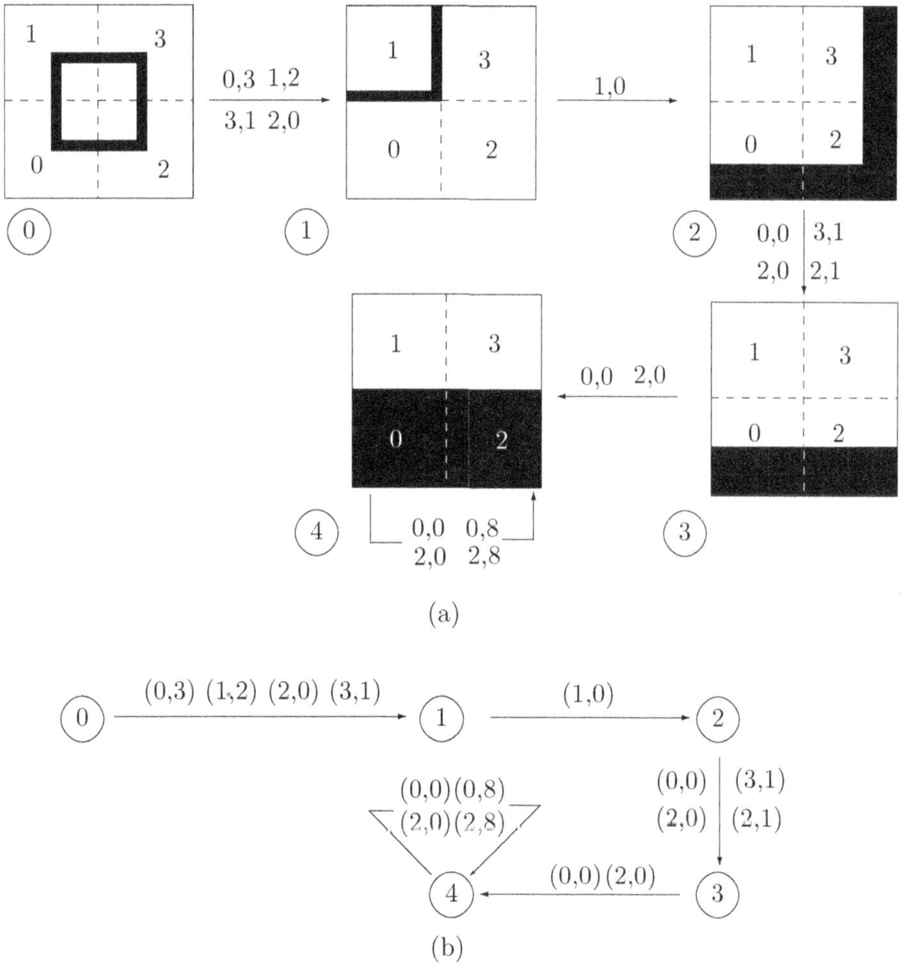

Figure 7.194: A Five-State GFA.

is thus the percentage of pixels that differ in the two images (we assume that 0 and 1 represent white and black pixels, respectively). The four-step algorithm described here for constructing the graph also uses an error parameter input by the user to control the amount of loss.

Step 1: Generate state 0 as the entire image (i.e., subsquare ϵ). We select the initial distribution as $I = (1, 0, 0, \ldots, 0)$, so state 0 will be the only one displayed. The final distribution is a vector of all ones.

Step 2: Process each state as follows: Let the next unprocessed state be q, representing subsquare w. Partition w into its four quadrants $w0$, $w1$, $w2$, and $w3$, and perform Step 3 for each of the four wa's.

Step 3: Denote the image of subsquare wa by ψ'. If $\psi' = 0$, there is no edge from state q with label a. Otherwise, examine all the states generated so far and try to find a state p and a transformation t such that $t(\psi_p)$ (the image of state p transformed by t) will be similar enough to image ψ'. This is expressed by $d_k(\psi', t(\psi_p)) \leq$ error. If such p and t are found, construct an edge $q(a,t)p$. Otherwise, add a new, unprocessed, state r to ψ' and construct a new edge $q(a,0)r$.

Step 4: If there are any unprocessed states left, go to Step 2. Otherwise stop.

Step 3 may involve many searches, since many states may have to be examined up to 16 times until a match is found. Therefore, the algorithm uses two versions for this search, a *first fit* and a *best fit*. Both approaches proceed from state to state, apply each of the 16 transformations to the state, and calculate the distance between each transformed state and ψ'. The first fit approach stops when it finds the first transformed state that fits, while the best fit conducts the full search and selects the state and transformation that yield the best fit. The former is thus faster, while the latter produces better compression.

The GFA graph for a real, complex image may have a huge number of states, so the GFA algorithm discussed here has an option where vector quantization is used. This option reduces the number of states (and thus speeds up the process of constructing the graph) by treating subsquares of size 8×8 differently. When this option is selected, the algorithm does not try to find a state similar to such a small subsquare, but rather encodes the subsquare with vector quantization (Section 7.19). An 8×8 subsquare has 64 pixels, and the algorithm uses a 256-entry codebook to code them. It should be noted that in general, GFA does not use quantization or any specific codebook. The vector quantization option is specific to the implementation discussed here.

After constructing the GFA graph, its components are compressed and written on the compressed stream in three parts. The first part is the edge information created in step 3 of the algorithm. The second part is the state information (the indices of the states), and the third part is the 256 codewords used in the vector quantization of small subsquares. All three parts are arithmetically encoded.

GFA decoding is done as in WFA, except that the decoder has to consider possible transformations when generating new image parts from existing ones.

GFA has been extended to color images. The image is separated into individual bitplanes and a graph is constructed for each. However, common states in these graphs are stored only once, thereby improving compression. A little thinking shows that this works best for color images consisting of several areas with well-defined boundaries (i.e., discrete-tone or cartoon-like images). When such an image is separated into its bitplanes, they tend to be similar. A subsquare q in bitplane 1, for example, may be identical to the same subsquare in, say, bitplane 3. The graphs for these bitplanes can thus share the state for subsquare q.

The GFA algorithm is then applied to the bitplanes, one by one. When working on bitplane b, Step 3 of the algorithm searches through all the states of all the graphs constructed so far for the preceding $b-1$ bitplanes. This process can be viewed in two different ways. The algorithm may construct n graphs, one for each bitplane, and share states between them. Alternatively, it may construct just one graph with n initial states.

Experiments show that GFA works best for images with a small number of colors. Given an image with many colors, quantization is used to reduce the number of colors.

7.39 Iterated Function Systems

Fractals have been popular since the 1970s and have many applications (see [Demko et al. 85], [Feder 88], [Mandelbrot 82], [Peitgen et al. 82], [Peitgen and Saupe 85], and [Reghbati 81] for examples). One such application, relatively underused, is data compression. Applying fractals to data compression is done by means of *iterated function systems*, or IFS. Two references to IFS are [Barnsley 88] and [Fisher 95]. IFS compression can be very efficient, achieving excellent compression factors (32 is not uncommon), but it is lossy and also computationally intensive. The IFS encoder partitions the image into parts called ranges; it then matches each range to some other part called a domain, and produces an *affine transformation* from the domain to the range. The transformations are written on the compressed stream, and they constitute the compressed image. We start with an introduction to two-dimensional affine transformations.

7.39.1 Affine Transformations

In computer graphics, a complete two-dimensional image is built part by part and is normally edited before it is considered satisfactory. Editing is done by selecting a figure (part of the drawing) and applying a transformation to it. Typical transformations (Figure 7.195) are moving or sliding (translation), reflecting or flipping (mirror image), zooming (scaling), rotating, and shearing.

The transformation can be applied to every pixel of the figure. Alternatively, it can be applied to a few key points that completely define the figure (such as the four corners of a rectangle), following which the figure is reconstructed from the transformed key points.

We use the notation $\mathbf{P} = (x, y)$ for a two-dimensional point, and $\mathbf{P}^* = (x^*, y^*)$ for the transformed point. The simplest linear transformation is $x^* = ax + cy$, $y^* = bx + dy$, in which each of the new coordinates is a *linear combination* of the two original coordinates. This transformation can be written $\mathbf{P}^* = \mathbf{PT}$, where $\mathbf{T}$ is the 2×2 matrix $\binom{a\,b}{c\,d}$.

To understand the functions of the four matrix elements, we start by setting $b = c = 0$. The transformation becomes $x^* = ax$, $y^* = dy$. Such a transformation is called *scaling*. If applied to all the points of an object, all the x dimensions are scaled by a factor a, and all the y dimensions are scaled by a factor d. Note that a and d can also be

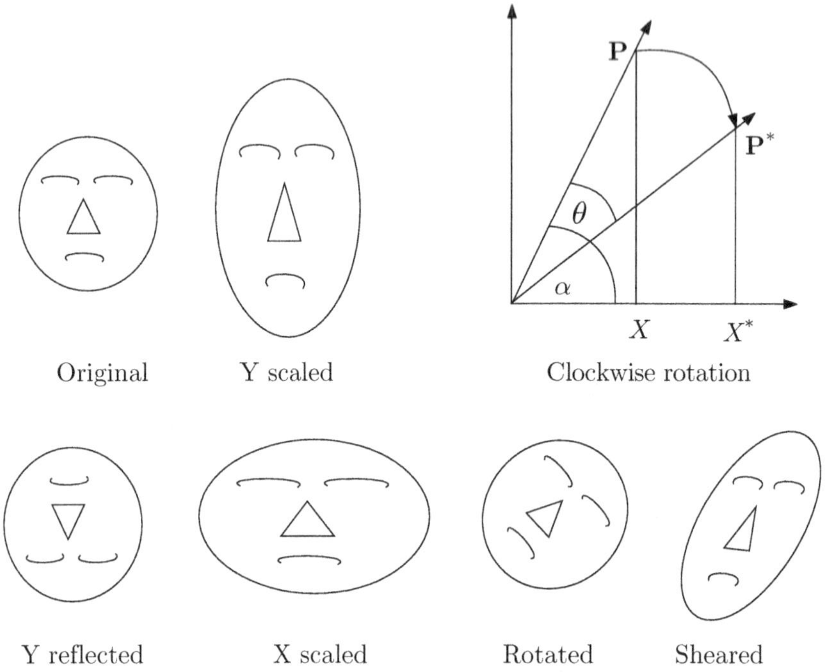

Figure 7.195: Two-Dimensional Transformations.

less than 1, causing shrinking of the object. If any of a or d equals -1, the transformation is a *reflection*. Any other negative values cause both scaling and reflection.

Note that scaling an object by factors of a and d changes its area by a factor of $a \times d$, and that this factor is also the value of the determinant of the scaling matrix $\left(\begin{smallmatrix} a & 0 \\ 0 & d \end{smallmatrix}\right)$.

(Scaling, reflection, and other geometrical transformations can be extended to three dimensions, where they become much more complex.)

Below are examples of matrices for scaling and reflection. In a, the y-coordinates are scaled by a factor of 2. In b, the x-coordinates are reflected. In c, the x dimensions are shrunk to 0.001 their original values. In d, the figure is shrunk to a vertical line:

$$a = \begin{pmatrix} 1 & 0 \\ 0 & 2 \end{pmatrix}, \quad b = \begin{pmatrix} -1 & 0 \\ 0 & 1 \end{pmatrix}, \quad c = \begin{pmatrix} .001 & 0 \\ 0 & 1 \end{pmatrix}, \quad d = \begin{pmatrix} 0 & 0 \\ 0 & 1 \end{pmatrix}.$$

The next step is to set $a = 1$, $d = 1$ (no scaling or reflection), and explore the effect of b and c on the transformations. The transformation becomes $x^* = x + cy$, $y^* = bx + y$. We first select matrix $\left(\begin{smallmatrix} 1 & 1 \\ 0 & 1 \end{smallmatrix}\right)$ and use it to transform the rectangle whose four corners are $(1, 0)$, $(3, 0)$, $(1, 1)$, and $(3, 1)$. The corners are transformed to $(1, 1)$, $(3, 3)$, $(1, 2)$, $(3, 4)$. The original rectangle has been *sheared* vertically and transformed into a parallelogram. A similar effect occurs when we try the matrix $\left(\begin{smallmatrix} 1 & 0 \\ 1 & 1 \end{smallmatrix}\right)$. Thus, the quantities b and c are responsible for *shearing*. Figure 7.196 shows the connection between shearing and the operation of scissors. The word *shearing* comes from the concept of shear in mechanics.

Figure 7.196: Scissors and Shearing.

◇ **Exercise 7.75:** Apply the shearing transformation $\left(\begin{smallmatrix}1 & -1\\ 0 & 1\end{smallmatrix}\right)$ to the four points $(1, 0)$, $(3, 0)$, $(1, 1)$, and $(3, 1)$. What are the transformed points? What geometrical figure do they represent?

The next important transformation is *rotation*. Figure 7.195 illustrates rotation. It shows a point $\mathbf{P}$ rotated clockwise through an angle θ to become $\mathbf{P}^*$. Simple trigonometry yields $x = R\cos\alpha$ and $y = R\sin\alpha$. From this we get the expressions for x^* and y^*:

$$\begin{aligned} x^* &= R\cos(\alpha-\theta) = R\cos\alpha\cos\theta + R\sin\alpha\sin\theta = x\cos\theta + y\sin\theta,\\ y^* &= R\sin(\alpha-\theta) = -R\cos\alpha\sin\theta + R\sin\alpha\cos\theta = -x\sin\theta + y\cos\theta. \end{aligned}$$

Thus the rotation matrix in two dimensions is

$$\begin{pmatrix}\cos\theta & -\sin\theta\\ \sin\theta & \cos\theta\end{pmatrix}, \tag{7.63}$$

which also equals

$$\begin{pmatrix}\cos\theta & 0\\ 0 & \cos\theta\end{pmatrix}\begin{pmatrix}1 & -\tan\theta\\ \tan\theta & 1\end{pmatrix}.$$

This proves that any rotation in two dimensions is a combination of scaling (and, perhaps, reflection) and shearing, an unexpected result that's true for all angles satisfying $\tan\theta \neq \infty$.

Matrix $\mathbf{T}_1$ below rotates anticlockwise. Matrix $\mathbf{T}_2$ reflects about the line $y = x$, and matrix $\mathbf{T}_3$ reflects about the line $y = -x$. Note the determinants of these matrices. In general, a determinant of $+1$ indicates pure rotation, whereas a determinant of -1 indicates pure reflection. (As a reminder, $\det\left(\begin{smallmatrix}a & b\\ c & d\end{smallmatrix}\right) = ad - bc$.)

$$\mathbf{T}_1 = \begin{pmatrix}\cos\theta & \sin\theta\\ -\sin\theta & \cos\theta\end{pmatrix},\quad \mathbf{T}_2 = \begin{pmatrix}0 & 1\\ 1 & 0\end{pmatrix},\quad \mathbf{T}_3 = \begin{pmatrix}0 & -1\\ -1 & 0\end{pmatrix}.$$

A 90° Rotation

In the case of a 90° clockwise rotation, the rotation matrix is

$$\begin{pmatrix}\cos(90) & -\sin(90)\\ \sin(90) & \cos(90)\end{pmatrix} = \begin{pmatrix}0 & -1\\ 1 & 0\end{pmatrix}. \tag{7.64}$$

A point $\mathbf{P} = (x, y)$ is thus transformed to the point $(y, -x)$. For a counterclockwise 90° rotation, (x, y) is transformed to $(-y, x)$. This is called the *negate and exchange* rule.

Translations

Unfortunately, our simple 2×2 matrix cannot generate all the necessary transformations! Specifically, it cannot generate *translation*. This is proved by realizing that any object containing the origin will, after any of the transformations above, still contain the origin (the result of $(0, 0) \times \mathbf{T}$ is $(0, 0)$ for any matrix $\mathbf{T}$).

One way to implement translation (which can be expressed by $x^* = x + m$, $y^* = y + n$), is to generalize our transformations to $\mathbf{P}^* = \mathbf{PT} + (m, n)$, where $\mathbf{T}$ is the familiar 2×2 transformation matrix $\binom{a\,b}{c\,d}$. A more elegant approach, however, is to stay with $\mathbf{P}^* = \mathbf{PT}$ and to generalize $\mathbf{T}$ to the 3×3 matrix

$$\mathbf{T} = \begin{pmatrix} a & b & 0 \\ c & d & 0 \\ m & n & 1 \end{pmatrix}.$$

This approach is called *homogeneous coordinates* and is commonly used in projective geometry. It makes it possible to unify all the two-dimensional transformations within one matrix. Notice that only six of the nine elements of matrix $\mathbf{T}$ are variables. Our points should now be the triplets $\mathbf{P} = (x, y, 1)$.

It is easy to see that the transformations discussed above can change lengths and angles. Scaling changes the lengths of objects. Rotation and shearing change angles. One property that is preserved, though, is parallel lines. A pair of parallel lines will remain parallel after any scaling, reflection, rotation, shearing, and translation. A transformation that preserves parallelism is called *affine*.

The final conclusion of this section is that any affine two-dimensional transformation can be fully specified by only six numbers!

Affine transformations can be defined in different ways. One important definition is that a transformation of points in space is affine if it preserves *barycentric sums* of the points. A barycentric sum of points $\mathbf{P}_i$ has the form $\sum w_i \mathbf{P}_i$, where w_i are numbers and $\sum w_i = 1$, so if $\mathbf{P} = \sum w_i \mathbf{P}_i$ and if $\sum w_i = 1$, then any affine transformation $\mathbf{T}$ satisfies

$$\mathbf{TP} = \sum_1^n w_i \mathbf{TP}_i.$$

7.39.2 IFS Definition

A simple example of IFS is the set of three transformations

$$\mathbf{T}_1 = \begin{pmatrix} .5 & 0 & 0 \\ 0 & .5 & 0 \\ 8 & 8 & 1 \end{pmatrix}, \quad \mathbf{T}_2 = \begin{pmatrix} .5 & 0 & 0 \\ 0 & .5 & 0 \\ 96 & 16 & 1 \end{pmatrix}, \quad \mathbf{T}_3 = \begin{pmatrix} .5 & 0 & 0 \\ 0 & .5 & 0 \\ 120 & 60 & 1 \end{pmatrix}. \tag{7.65}$$

We first discuss the concept of the *fixed point*. Imagine the sequence $\mathbf{P}_1 = \mathbf{P}_0\mathbf{T}_1$, $\mathbf{P}_2 = \mathbf{P}_1\mathbf{T}_1, \ldots$, where transformation $\mathbf{T}_1$ is applied repeatedly to create a sequence of points $\mathbf{P}_1, \mathbf{P}_2, \ldots$. We start with an arbitrary point $\mathbf{P}_0 = (x_0, y_0)$ and examine the limit of the sequence $\mathbf{P}_1 = \mathbf{P}_0\mathbf{T}_1$, $\mathbf{P}_2 = \mathbf{P}_1\mathbf{T}_1, \ldots$. It is easy to show that the limit of $\mathbf{P}_i$ for large i is the fixed point $(2m, 2n) = (16, 16)$ (where m and n are the translation factors of matrix $\mathbf{T}_1$, Equation (7.65)) regardles of the values of x_0 and y_0. Point $(16, 16)$ is called the *fixed point* of $\mathbf{T}_1$, and *it does not depend* on the particular starting point $\mathbf{P}_0$ selected.

Proof: $\mathbf{P}_1 = \mathbf{P}_0\mathbf{T}_1 = (.5x_0{+}8, .5y_0{+}8)$, $\mathbf{P}_2 = \mathbf{P}_1\mathbf{T}_1 = (0.5(0.5x_0{+}8){+}8, .5(0.5y_0{+}8){+}8)$. It is easy to see (and to prove by induction) that $x_i = 0.5^i x_0 + 0.5^{i-1}8 + 0.5^{i-2}8 + \cdots + 0.5^1 8 + 8$. In the limit $x_i = 0.5^i x_0 + 8\sum_{j=0}^{\infty} 0.5^j = 0.5^i x_0 + 8\times 2$, which approaches the limit $8 \times 2 = 16$ for large i regardless of x_0.

Now it is easy to show that for the transformations above, with scale factors of 0.5 and no shearing, each new point in the sequence moves half the remaining distance toward the fixed point. Given a point $\mathbf{P}_i = (x_i, y_i)$, the point midway between $\mathbf{P}_i$ and the fixed point $(16, 16)$ is

$$\left(\frac{x_i + 16}{2}, \frac{y_i + 16}{2}\right) = (0.5x_i + 8, 0.5y_i + 8) = (x_{i+1}, y_{i+1}) = \mathbf{P}_{i+1}.$$

Consequently, for the particular transformations above there is no need to use the transformation matrix. At each step of the iteration, point $\mathbf{P}_{i+1}$ is obtained by $(\mathbf{P}_i + (2m, 2n))/2$. For other transformations, matrix multiplication is necessary to compute point $\mathbf{P}_{i+1}$.

In general, every affine transformation where the scale and shear factors are less than 1 has a fixed point, but it may not be easy to find it.

The principle of IFS is now easy to describe. A set of transformations (an IFS code) is selected. A sequence of points is computed and plotted by starting with an arbitrary point $\mathbf{P}_0$, selecting a transformation from the set at random, and applying it to $\mathbf{P}_0$, transforming it into a point $\mathbf{P}_1$, and then randomly selecting another transformation and applying it to $\mathbf{P}_1$, thereby generating point $\mathbf{P}_2$, and so on.

Every point is plotted on the screen as it is calculated, and gradually, the object begins to take shape before the viewer's eyes. The shape of the object is called the IFS *attractor*, and it depends on the IFS code (the transformations) selected. The shape also depends slightly on the particular selection of $\mathbf{P}_0$. It is best to choose $\mathbf{P}_0$ as one of the fixed points of the IFS code (if they are known in advance). In such a case, all the points in the sequence will lie inside the attractor. For any other choice of $\mathbf{P}_0$, a finite number of points will lie outside the attractor, but eventually they will move into the attractor and stay there.

It is surprising that the attractor does not depend on the precise order of the transformations used. This result has been proved by the mathematician John Elton.

Another surprising property of IFS is that the random numbers used don't have to be uniformly distributed; they can be weighted. Transformation $\mathbf{T}_1$, for example, may be selected at random 50% of the time, transformation $\mathbf{T}_2$, 30%, and $\mathbf{T}_3$, 20%. The shape being generated does not depend on the probabilities, but the computation time does. The weights should add up to 1 (a normal requirement for a set of mathematical weights), and none can be zero.

The three transformations of Equation (7.65) above create an attractor in the form of a Sierpiński triangle (Figure 7.197a). The translation factors determine the coordinates of the three triangle corners. The six transformations of Table 7.198 create an attractor in the form of a fern (Figure 7.197b). The notation used in Table 7.198 is $\binom{a\,b}{c\,d} + \binom{m}{n}$.

The Sierpiński triangle, also known as the Sierpiński gasket (Figure 7.197a), is defined recursively. Start with any triangle, find the midpoint of each edge, connect the three midpoints to obtain a new triangle fully contained in the original one, and cut the new triangle out. The newly created hole now divides the original triangle into three smaller ones. Repeat the process on each of the smaller triangles. At the limit, there is no area left in the triangle. It resembles Swiss cheese without any cheese, just holes.

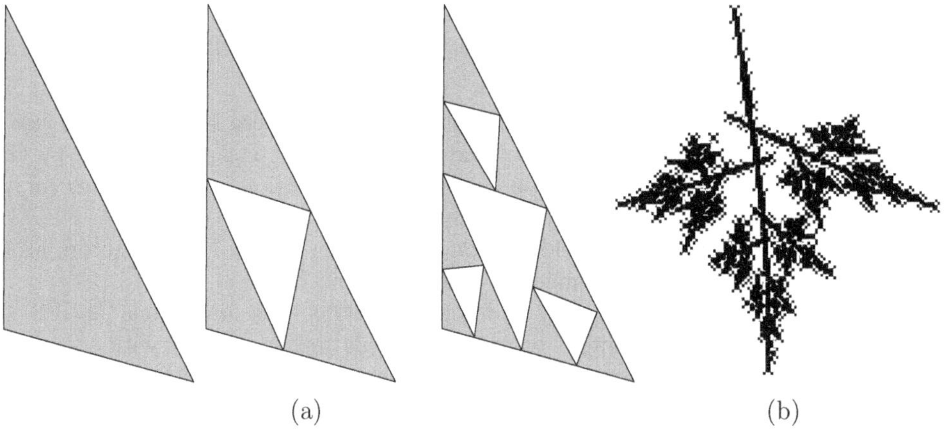

Figure 7.197: (a) Sierpiński Triangle. (b) A Leaf.

The program of Figure 7.199 calculates and displays IFS attractors for any given set of transformations. It runs on the Macintosh computer (because of changes in the Macintosh operating system, this program no longer runs and should be considered pseudocode).

7.39.3 IFS Principles

Before we describe how IFS is used to compress real-life images, let's look at IFS from a different point of view. Figure 7.200 shows three images: a person, the letter "T", and the Sierpiński gasket (or triangle). The first two images are transformed in a special way. Each image is shrunk to half its size, then copied three times, and the three copies are arranged in the shape of a triangle. When this transformation is applied a few times to an image, it is still possible to discern the individual copies of the original image. However, when it is applied many times, the result is the Sierpiński gasket (or something very close to it, depending on the number of iterations and on the resolution of the output device). The point is that each transformation shrinks the image (the transformations

	a	b	c	d	m	n
1:	0	−28	0	29	151	92
2:	−2	37	−31	29	85	103
3:	−1	18	−18	−1	88	147

	a	b	c	d	m	n
4:	64	0	0	64	82	6
5:	0	−80	−22	1	243	151
6:	2	−48	0	50	160	80

Table 7.198: All numbers are shown as integers, but $a, b, c,$ and d should be divided by 100, to make them less than 1. The values m and n are the translation factors.

```
PROGRAM IFS;
USES ScreenIO, Graphics, MathLib;
CONST LB = 5; Width = 490; Height = 285;
(* LB=left bottom corner of window *)
VAR i,k,x0,y0,x1,y1,NumTransf: INTEGER;
Transf: ARRAY[1..6,1..10] OF INTEGER;
Params:TEXT;
filename:STRING;
BEGIN (* main *)
Write('params file='); Readln(filename);
Assign(Params,filename); Reset(Params);
Readln(Params,NumTransf);
FOR i:=1 TO NumTransf DO
Readln(Params,Transf[1,i],Transf[2,i],Transf[3,i],
 Transf[4,i],Transf[5,i],Transf[6,i]);
 OpenGraphicWindow(LB,LB,Width,Height,'IFS shape');
 SetMode(paint);
 x0:=100; y0:=100;
 REPEAT
 k:=RandomInt(1,NumTransf+1);
 x1:=Round((x0*Transf[1,k]+y0*Transf[2,k])/100)+Transf[5,k];
 y1:=Round((x0*Transf[3,k]+y0*Transf[4,k])/100)+Transf[6,k];
 Dot(x1,y1); x0:=x1; y0:=y1;
 UNTIL Button()=TRUE;
  ScBOL; ScWriteStr('Hit a key & close this window to quit');
  ScFreeze;
END.
```

Figure 7.199: Calculate and Display IFS Attractors.

are *contractive*), so the final result does not depend on the shape of the original image. The shape can be that of a person, a letter, or anything else; the final result depends only on the particular transformation applied to the image. A different transformation will create a different result, which again will not depend on the particular image being transformed. Figure 7.200d, for example, shows the results of transforming the letter "T" by reducing it, making three copies, arranging them in a triangle, and flipping the top copy. The final image obtained at the limit, after applying a certain transformation infinitely many times, is called the *attractor* of the transformation. (An attractor is a state to which a dynamical system evolves after a long enough time.)

The following sets of numbers create especially interesting patterns.

1. A frond.

```
5
  0  -28    0   29  151   92
 64    0    0   64   82    6
 -2   37  -31   29   85  103
 17  -51  -22    3  183  148
 -1   18  -18   -1   88  147
```

2. A coastline

```
4
-17  -26   34  -12   84   53
 25  -20   29   17  192   57
 35    0    0   35   49    3
 25   -6    6   25  128   28
```

3. A leaf (Figure 7.197b)

```
4
  2   -7   -2   48  141   83
 40    0   -4   65   88   10
 -2   45  -37   10   82  132
-11  -60  -34   22  237  125
```

4. A Sierpiński Triangle

```
3
 50    0    0   50    0    0
 50    0    0   50  127   79
 50    0    0   50  127    0
```

◇ **Exercise 7.76:** The three affine transformations of example 4 above (the Sierpiński triangle) are different from those of Equation (7.65). What is the explanation?

The result of each transformation is an image containing all the images of all the previous transformations. If we apply the same transformation many times, it is possible to zoom on the result, to magnify it many times, and still see details of the original images. In principle, if we apply the transformation an infinite number of times, the final result will show details at *any* magnification. It will be a fractal.

The case of Figure 7.200c is especially interesting. It seems that the original image is simply shown four times, without any transformations. A little thinking, however, shows that our particular transformation transforms this image to itself. The original image is already the Sierpiński gasket, and it gets transformed to itself because it is self-similar.

◇ **Exercise 7.77:** Explain the geometrical meaning of the combined three affine transformations below and show the attractor they converge to

$$w_1\begin{pmatrix}x\\y\end{pmatrix}=\begin{pmatrix}1/2&0\\0&1/2\end{pmatrix}\begin{pmatrix}x\\y\end{pmatrix},$$
$$w_2\begin{pmatrix}x\\y\end{pmatrix}=\begin{pmatrix}1/2&0\\0&1/2\end{pmatrix}\begin{pmatrix}x\\y\end{pmatrix}+\begin{pmatrix}0\\1/2\end{pmatrix},$$
$$w_3\begin{pmatrix}x\\y\end{pmatrix}=\begin{pmatrix}1/2&0\\0&1/2\end{pmatrix}\begin{pmatrix}x\\y\end{pmatrix}+\begin{pmatrix}1/2\\0\end{pmatrix}.$$

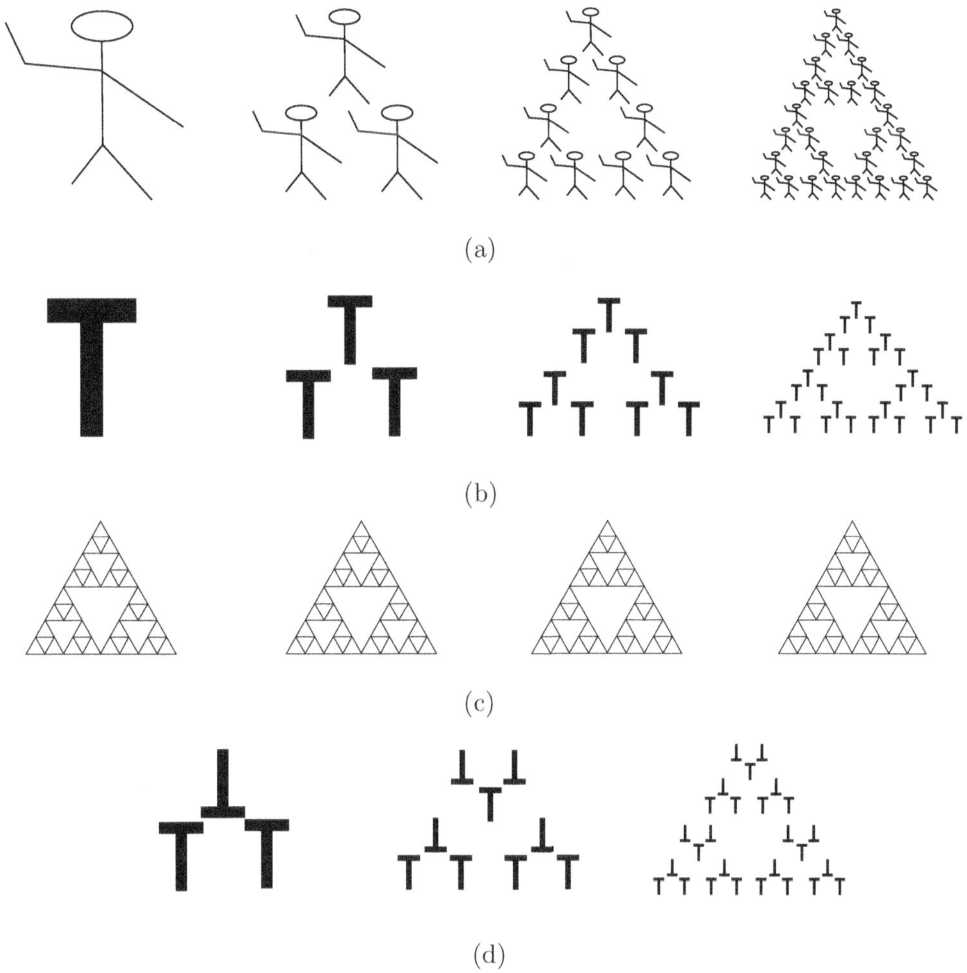

Figure 7.200: Creating the Sierpiński Gasket.

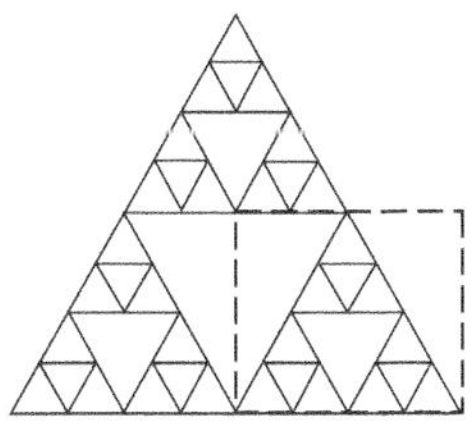

Figure 7.201: A Self-Similar Image.

The Sierpiński gasket is therefore easy to compress because it is self-similar; it is easy to find parts of it that are identical to the entire image. In fact, every part of it is identical to the entire image. Figure 7.201 shows the bottom-right part of the gasket surrounded by dashed lines. It is easy to see the relation between this part and the entire image. Their shapes are identical, up to a scale factor. The size of this part is half the size of the image, and we know where it is positioned relative to the entire image (we can measure the displacement of its bottom-left corner from the bottom-left corner of the entire image).

This points to a possible way to compress real images. If we can divide an image into parts such that each part is identical (or at least very close) to the entire image up to a scale factor, then we can highly compress the image by IFS. All that we need is the scale factor (actually two scale factors, in the x and y directions) and the displacement of each part relative to the entire image [the (x, y) distances between a corner of the part and the same corner of the image]. Sometimes we may find a part of the image that has to be reflected in order to become identical to the entire image. In such a case we also need the reflection coefficients. Thus, we can compress an image by figuring out the transformations that transform each part (called "range") into the entire image. The transformation for each part is expressed by a few numbers, and these numbers become the compressed stream.

It is easy to see that this simple approach will not work for real-life images. Such images are complex, and it is generally impossible to divide such an image into parts that will all be identical (or even very close) to the entire image. A different approach is needed to make IFS practical. The approach used by any practical IFS algorithm is to partition the image into nonoverlapping parts called *ranges*. They can be of any size and shape, but in practice it is easiest to work with squares, rectangles, or triangles. For each range R_i, the encoder has to find a *domain* D_i that's very similar, or even identical in shape, to the range but is bigger. Once such a domain is found, it is easy to figure out the transformation w_i that will transform the domain into the range $R_i = w_i(D_i)$. Two scale factors have to be determined (the scaling is shrinking, since the domain is bigger than the range) as well as the displacement of the domain relative to the range [the (x, y) distances between one corner of the domain and the same corner of the range]. Sometimes, the domain has to be rotated and/or reflected to make it identical to the range, and the transformation should, of course, include these factors as well. This approach to IFS image compression is called PIFS (for partitioned IFS).

7.39.4 IFS Decoding

Before looking into the details of PIFS encoding, let's try to understand how the PIFS decoder works. All that the decoder has is the set of transformations, one per range. It does not know the shapes of any ranges or domains. In spite of this, decoding is very simple. It is based on the fact, mentioned earlier, that a contractive transformation creates a result that does not depend on the shape of the initial image used. We can therefore create any range R_i by applying contractive transformation w_i many times to *any* bigger shape D_i (except an all-white shape).

The decoder therefore starts by setting all the domains to arbitrary shapes (e.g., it can initially set the entire image to black). It then goes into a loop where in each iteration it applies every transformation w_i once. The first iteration applies the transformations to domains D_i that are all black. This creates ranges R_i that may already, after this single iteration, slightly resemble the original ranges. This iteration changes the image from the initial all black to something resembling the original image. In the second iteration the decoder applies again all the w_i transformations, but this time they are applied to domains that are no longer all black. The domains already somewhat resemble the original ones, so the second iteration results in better-looking ranges, and thus in a better image. Experience shows that only 8–10 iterations are normally needed to get a result that closely resembles the original image.

It is important to realize that this decoding process is resolution independent! Normally, the decoder starts with an initial image whose size is identical to that of the original image. It can, however, start with an all-black image of any size. The affine transformations used to encode the original image do not depend on the resolution of the image or on the resolutions of the ranges. Decoding an image at, say, twice its original size will create a large, smooth image, with details not seen in the original and without pixelization (without jagged edges or "fat" pixels). The extra details will, of course, be artificial. They may not be what one would see when looking at the original image through a magnifying glass, but the point is that PIFS decoding is resolution independent; it creates a natural-looking image at any size, and it does not involve pixelization.

The resolution-independent nature of PIFS decoding also means that we have to be careful when measuring the compression performance. After compressing a 64-Kb image into, say, 2 Kb, the compression factor is 32. Decoding the 2-Kb compressed file into a large, 2 Mb, image (with a lot of artificial detail) does not mean that we have changed the compression factor to 2M/2K=1024. The compression factor is still the same 32.

7.39.5 IFS Encoding

PIFS decoding is therefore easy, if somewhat magical, but we still need to see the details of PIFS encoding. The first point to consider is how to select the ranges and find the domains. The following is a straightforward way of doing this.

Suppose that the original image has resolution 512×512. We can select as ranges the nonoverlapping groups of 16×16 pixels. There are $32\times32 = 1024$ such groups. The domains should be bigger than the ranges, so we may select as domains all the 32×32 groups of pixels in the image (they may, of course, overlap). There are $(512-31)\times(512-31) = 231361$ such groups. The encoder should compare each range with all 231,361 domains. Each comparison involves eight steps, because a range may be identical to a rotation or a reflection of a domain (Figure 7.193). The total number of steps in comparing ranges and domains is therefore $1024\times231361\times8 = 1{,}895{,}309{,}312$. If each step takes a microsecond, the total time required is 1,895 seconds, or about 31 minutes.

◇ **Exercise 7.78:** Repeat the computation above for a 256×256 image with ranges of size 8×8 and domains of size 16×16.

If the encoder is looking for a domain to match range R_i and it is lucky to find one that's identical to R_i, it can proceed to the next range. In practice, however, domains identical to a given range are very rare, so the encoder has to compare all $231{,}361\times8$ domains to each range R_i and select the one that's closest to R_i (PIFS is, in general, a lossy compression method). We therefore have to answer two questions: When is a domain identical to a range (remember: they have different sizes) and how do we measure the "distance" between a domain and a range?

To compare a 32×32-pixel domain to a 16×16-pixel range, we can either choose one pixel from each 2×2 square of pixels in the domain (this is called subsampling) or average each 2×2 square of pixels in the domain and compare it to one pixel of the range (averaging).

To decide how close a range R_i is to a domain D_j, we have to use one of several *metrics*. A metric is a function that measures "distance" between, or "closeness" of, two mathematical quantities. Experience recommends the use of the *rms* (root mean square) metric

$$M_{\rm rms}(R_i, D_j) = \sqrt{\sum_{x,y} \left[R_i(x,y) - D_j(x,y)\right]^2}. \tag{7.66}$$

This involves a square root calculation, so a simpler metric may be

$$M_{\rm max}(R_i, D_j) = \max |R_i(x,y) - D_j(x,y)|$$

(the largest difference between a pixel of R_i and a pixel of D_j). Whatever metric is used, a comparison of a range and a domain involves subsampling (or averaging) followed by a metric calculation.

After comparing range R_i to all the (rotated and reflected) domains, the encoder selects the domain with the smallest metric and determines the transformation that will bring the domain to the range. This process is repeated for all ranges.

Even this simple way of matching parts of the image produces excellent results. Compression factors are typically in the 15–32 range and data loss is minimal.

◇ **Exercise 7.79:** What is a reasonable way to estimate the amount of image information lost in PIFS compression?

The main disadvantage of this method of determining ranges and domains is the fixed size of the ranges. A method where ranges can have different sizes may lead to better compression and less loss. Imagine an image of a hand with a ring on one finger. If the ring happens to be inside a large range R_i, it may be impossible to find any domain that will even come close to R_i. Too much data may be lost in such a case. On the other hand, if part of an image is fairly uniform, it may become a large range, since there is a better chance that it will match some domain. Clearly, large ranges are preferable, since the compressed stream contains one transformation per range. Therefore, quadtrees offer a good solution.

Quadtrees: We start with a few large ranges, each a subquadrant. If a range does not match well any domain (the metric between it and any domain is greater than a user-controlled tolerance parameter), it is divided into four subranges, and each is matched separately. As an example, consider a 256×256-pixel image. We can choose for domains all the image squares of size 8, 12, 16, 24, 32, 48, and 64 pixels. We start with ranges that are subquadrants of size 32×32. Each range is compared with domains that are larger than itself (48 or 64) pixels. If a range does not match well, it is divided into four quadrants of size 16×16 each, and each is compared, as a new range, with all domains of sizes 24, 32, 48, and 64 pixels. This process continues until all ranges have been matched to domains. Large ranges result in better compression, but small ranges are easier to match, because they contain few adjacent pixels, and we know from experience that adjacent pixels tend to be highly correlated in real-life images.

7.39.6 IFS for Grayscale Images

Up until now we have assumed that our transformations have to be affine. The truth is that any contractive transformations, even nonlinear ones, can be used for IFS. Affine transformations are used simply because they are linear and therefore computationally simple. Also, up to now we have assumed a monochromatic image, where the only problem is to determine which pixels should be black. IFS can easily be extended to compress grayscale images (and therefore also color images; see below). The problem here is to determine which pixels to paint, and what gray level to paint each.

Matching a domain to a range now involves the intensities of the pixels in both. Whatever metric is employed, it should use only those intensities to determine the "closeness" of the domain and the range. Assume that a certain domain D contains n pixels with gray levels $a_1 \ldots a_n$, and the IFS encoder tries to match D to a range R containing n pixels with gray levels $b_1, \ldots, b_n$. The rms metric, mentioned earlier, works by finding two numbers, r and g (called the contrast and brightness controls), that will minimize the expression

$$Q = \sum_{1}^{n} \left((r \cdot a_i + g) - b_i\right)^2. \tag{7.67}$$

This is done by solving the two equations $\partial Q/\partial r = 0$ and $\partial Q/\partial g = 0$ for the unknowns r and g (see details below). Minimizing Q minimizes the difference in contrast and brightness between the domain and the range. The value of the rms metric is $\sqrt{Q}$ [compare with Equation (7.66)].

When the IFS encoder finally decides which domain to associate with the current range, it has to figure out the transformation w between them. The point is that r and g should be included in the transformation, so that the decoder will know what gray level to paint the pixels when the domain is recreated in successive decoding iterations. It is common to use transformations of the form

$$w\begin{pmatrix} x \\ y \\ z \end{pmatrix} = \begin{pmatrix} a & b & 0 \\ c & d & 0 \\ 0 & 0 & r \end{pmatrix}\begin{pmatrix} x \\ y \\ z \end{pmatrix} + \begin{pmatrix} l \\ m \\ g \end{pmatrix}. \tag{7.68}$$

A pixel (x, y) in the domain D is now given a third coordinate z (its gray level) and is transformed into a pixel (x^*, y^*, z^*) in the range R, where $z^* = z \cdot r + g$. Transformation (7.68) has another property. It is contractive if $r < 1$, regardless of the scale factors.

Any compression method for grayscale images can be extended to color images. It is only necessary to separate the image into three color components (preferably YIQ) and compress each individually as a grayscale image. This is how IFS can be applied to the compression of color images.

The next point that merits consideration is how to write the coefficients of a transformation w on the compressed stream. There are three groups of coefficients, the scale factors a and d, the reflection/rotation factors a, b, c, and d, the displacements l and m, and the contrast/brightness controls r and g. If a domain is twice as large as a range, then the scale factors are always 0.5 and consequently do not have to be written on the compressed stream. If the domains and ranges can have several sizes, then only certain scale factors are possible, and they can be encoded either arithmetically or with some prefix code. The particular rotation or reflection of the domain relative to the range can be coded with three bits, since there are only eight rotations/reflections possible. The displacement can be encoded by encoding the positions and sizes of the domain and range.

The quantities r and g are not distributed in any uniform way, and they are also real (floating-point) numbers that can have many different values. They should thus be quantized, i.e., converted to an integer in a certain range. Experience shows that the contrast r can be quantized into a 4-bit or 5-bit integer (i.e., 16 or 32 contrast values are enough in practice), whereas the brightness g should become a 6-bit or 7-bit integer (resulting in 64 or 128 brightness values).

Here are the details of computing r and g that minimize Q and then calculating the (minimized) Q and the rms metric.

From Equation (7.67), we get

$$\frac{\partial Q}{\partial g}=0\rightarrow\sum 2(r\cdot a_i+g-b_i)=0\rightarrow ng+\sum(r\cdot a_i-b_i)=0,$$
$$g=\frac{1}{n}\left[\sum b_i-r\sum a_i\right], \tag{7.69}$$

and

$$\frac{\partial Q}{\partial r}=0\rightarrow\sum 2(r\cdot a_i+g-b_i)a_i=0\rightarrow\sum(r\cdot a_i^2+g\cdot a_i-a_ib_i)=0,$$
$$r\sum a_i^2+\frac{1}{n}\left[\sum b_i-r\sum a_i\right]\sum a_i-\sum a_ib_i=0,$$
$$r\left[\sum a_i^2-\frac{1}{n}\left(\sum a_i\right)^2\right]=\sum a_ib_i-\frac{1}{n}\sum a_i\sum b_i,$$
$$r=\frac{\sum a_ib_i-\frac{1}{n}\sum a_i\sum b_i}{\sum a_i^2-\frac{1}{n}\left(\sum a_i\right)^2}=\frac{n\sum a_ib_i-\sum a_i\sum b_i}{n\sum a_i^2-\left(\sum a_i\right)^2}. \tag{7.70}$$

From the same Equation (7.67), we also get the minimized Q

$$\begin{aligned}Q&=\sum_1^n(r\cdot a_i+g-b_i)^2=\sum(r^2a_i^2+g^2+b_i^2+2rga_i-2ra_ib_i-2gb_i)\\&=r^2\sum a_i^2+ng^2+\sum b_i^2+2rg\sum a_i-2r\sum a_ib_i-2g\sum b_i.\end{aligned} \tag{7.71}$$

The following steps are needed to calculate the rms metric:

1. Compute the sums $\sum a_i$ and $\sum a_i^2$ for all domains.
2. Compute the sums $\sum b_i$ and $\sum b_i^2$ for all ranges.
3. Every time a range R and a domain D are compared, compute:
   3.1. The sum $\sum a_ib_i$.

3.2. The quantities r and g from Equations (7.69) and (7.70) using the five sums above. Quantize r and g.

3.3. Compute Q from Equation (7.71) using the quantized r, g and the five sums. The value of the rms metric for these particular R and D is $\sqrt{Q}$.

Finally, Figures 7.202 and 7.203 are pseudocode algorithms outlining two approaches to IFS encoding. The former is more intuitive. For each range R it selects the domain that's closest to R. The latter tries to reduce data loss by sacrificing compression ratio. This is done by letting the user specify the minimum number T of transformations to be generated. (Each transformation w is written on the compressed stream using roughly the same number of bits, so the size of that stream is proportional to the number of transformations.) If every range has been matched, and the number of transformations is still less than T, the algorithm continues by taking ranges that have already been matched and partitioning them into smaller ones. This increases the number of transformations but reduces the data loss, since smaller ranges are easier to match with domains.

> He gathered himself together and then banged his fist on the table. "To hell with art, I say."
>
> "You not only say it, but you say it with tiresome iteration," said Clutton severely.
>
> —W. Somerset Maugham, *Of Human Bondage*

7.40 Spatial Prediction

Image compression by spatial prediction is a combination of JPEG and various fractal-based image compression algorithms. The method is described in [Feig et al. 95]. The principle is to select an integer N, partition the input image into blocks of $N \times N$ pixels each, and compress each block with JPEG or with a simple fractal method depending on which is better. Compressing the block with JPEG results in a zigzag sequence of $N \times N$ transform coefficients which are quantized and further compressed as described in Section 7.10.4. In contrast, the fractal compression of a block results in five integers as shown later in this section.

The method described in Section 7.39 is perhaps the simplest fractal-based approach to image compression. It is based on the observation that any part of an image may look very similar to other parts, or to transformed versions of other parts. Thus, a given part A of an image may be very close to a scaled, rotated, and reflected version of part B. The method partitions the image into blocks and tries to match the current block A to a transformed version of a previously-encoded block. If (a transformed version of) a block B provides a sufficiently-close match, the location of B, as well as the transform parameters, are written on the output and become the compressed data of the current block.

```
t:=some default value; [t is the tolerance]
push(entire image); [stack contains ranges to be matched]
repeat
 R:=pop();
 match all domains to R, find the one (D) that's closest to R,
  pop(R);
 if metric(R,D)<t then
  compute transformation w from D to R and output it;
 else partition R into smaller ranges and push them
      into the stack;
 endif;
until stack is empty;
```

Figure 7.202: IFS Encoding: Version I.

```
input T from user;
push(entire image); [stack contains ranges to be matched]
repeat
 for every unmatched R in the stack find the best matching domain D,
  compute the transformation w, and push D and w into the stack;
 if the number of ranges in the stack is <T then
   find range R with largest metric (worst match)
   pop R, D and w from the stack
   partition R into smaller ranges and push them, as unmatched,
    into the stack;
 endif
until all ranges in the stack are matched;
output all transformations w from the stack;
```

Figure 7.203: IFS Encoding: Version II.

The spatial prediction method compresses an image in blocks. It first tries to match the current block to transformed versions of previously-encoded blocks (which serve as predictors). If no good match is found, the current block is compressed with DCT, as in JPEG.

A color image is first separated into its three color components, and each is compressed as a grayscale image. We denote such a grayscale image by I, denote its height by $H = h \cdot N$ and its width by $W = w \cdot N$. The image is partitioned into $h \cdot w$ blocks of $N \times N$ pixels each. The block whose top-left corner is at pixel (i, j) is denoted by B_{ij} and the N^2 pixels of the block are denoted by $b_{ij}(k)$ for $1 \le k \le N^2$. The compressed image and the compressed blocks are denoted by $E(I)$ and $E(B_{ij})$, respectively. The symbols $\tilde{I}$, $\tilde{B}_{ij}$, and $\tilde{b}_{ij}(k)$ denote the decompressed results of the image, its blocks, and their pixels, respectively.

If B_{ij} is the current block to be encoded, it is first compared with (the decompressed versions $\tilde{B}$ of) all the overlapping $N \times N$ blocks in the already-encoded parts of the image, i.e., the parts either to the left of the current block or in rows above it. Figure 7.204 illustrates this process. The figure shows a 32×32-pixel image partitioned into 4×4 blocks of 8×8 pixels each. The current block is $B_{8,16}$ (shown in gray), and the encoder should try to match it with (1) the 25 blocks $\tilde{B}_{ij}$ above it, whose addresses are $(1, 1)$, $(1, 2)$, through $(1, 25)$ and (2) the 18 blocks $\tilde{B}_{ij}$ to its left, whose addresses are $(1, 1)$, $(2, 1)$, through $(9, 1)$ and $(1, 2)$, $(2, 2)$, through $(9, 2)$.

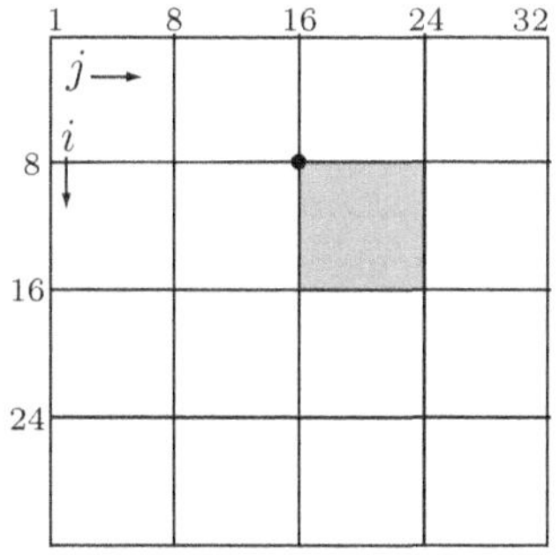

Figure 7.204: Overlapping Blocks For Matching $B_{8,16}$.

Notice that a block B_{ij} has to be matched with decompressed blocks $\tilde{B}_{i'j'}$ for various i' and j' values. This is because the method is lossy and the decoder has access only to the decompressed blocks, not to the original ones. Thus, after compressing a block, the encoder has to decompress it and save the decompressed version for later matchings. Each matching of a block B_{ij} with a block $\tilde{B}_{i'j'}$ is an 8-step process where B_{ij} is compared with rotated and reflected versions of $\tilde{B}_{i'j'}$, as illustrated by Figure 7.205 (compare with Figure 7.193).

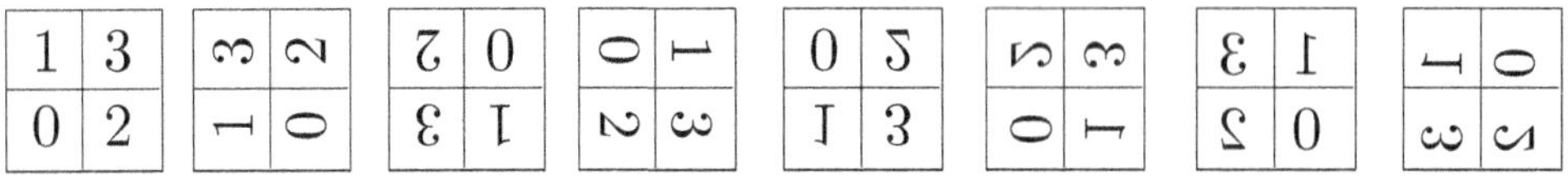

Figure 7.205: Eight Transformations of a Block of Pixels.

Each of the eight comparisons of a pair of blocks is the sum of squares of pixel differences. It has the form

$$\Delta = \sum_{k=1}^{N^2} \left[s \cdot \tilde{b}_{i'j'}(k) + o - b_{ij}(k) \right]^2,$$

where the two parameters s and o are adjusted to obtain the smallest Δ. Once block B_{ij} has been compared eight times with each of the valid predictor blocks $\tilde{B}_{i'j'}$, the s and o values that produce the smallest Δ (to be denoted by Δ^*) are selected.

If Δ^* is less than the threshold T_{ij} for the block, the encoder outputs the quintet (i', j', l, s, o) as the compressed form of B_{ij}. The two indexes i' and j' indicate the location of the predictor block that produced Δ^*, l indicates the number (between 0 and 7) of the best transformation of that block, and s and o are the parameters that produced Δ^* for predictor (i', j') and transformation l. This quintet is output as the compressed version $E(B_{ij})$ of the block.

If Δ^* is not less than the threshold T_{ij} for the block, the encoder applies the JPEG algorithm (DCT, quantization, and encoding) to the block and outputs the encoded, quantized DCT coefficients as the compressed version $E(B_{ij})$ of the block.

In either case, the encoder immediately decodes $E(B_{ij})$ to obtain the decompressed version $\tilde{B}_{ij}$ of the block, to be used later as a predictor block.

In summary, a block B_{ij} of pixels is compressed either by prediction (if the best match yields a Δ^* that is within the threshold T_{ij} of the block) or by DCT. The performance of this algorithm as compared to the performance of pure JPEG therefore depends on the choice of thresholds. A simple choice of thresholds uses average pixel values as follows: Given a block B_{ij} with pixels $b_{ij}(k)$, we denote by $\bar{b}_{ij}$ the average pixel value and define the threshold T_{ij} as the sum of squares of differences

$$T_{i,j} = \sum_{k=1}^{N^2} \left(b_{ij}(k) - \bar{b}_{ij} \right)^2.$$

Tests of this algorithm immediately point out its main advantage; when it is set for coarse quantization (high lossy compression), the resulting compressed/decompressed image has considerably fewer blocky artifacts compared to JPEG.

7.41 Cell Encoding

Imagine an image stored in a bitmap and displayed on a screen. Let's start with the case where the image consists of just text, with each character occupying the same area, say 8×8 pixels (large enough for a 5×7 or a 6×7 character and some spacing). Assuming a set of 256 characters, each cell can be encoded as an 8-bit pointer pointing to a 256-entry table where each entry contains the description of an 8×8 character as a 64-bit string. The compression factor is therefore 64/8, or eight to one. Figure 7.206 shows the letter **H** both as a bitmap and as a 64-bit string.

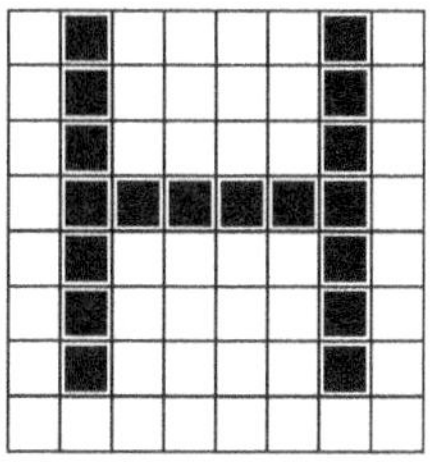

0 1 0 0 0 0 1 0
0 1 0 0 0 0 1 0
0 1 0 0 0 0 1 0
0 1 1 1 1 1 1 0
0 1 0 0 0 0 1 0
0 1 0 0 0 0 1 0
0 1 0 0 0 0 1 0
0 0 0 0 0 0 0 0

Figure 7.206: An 8×8 Letter H.

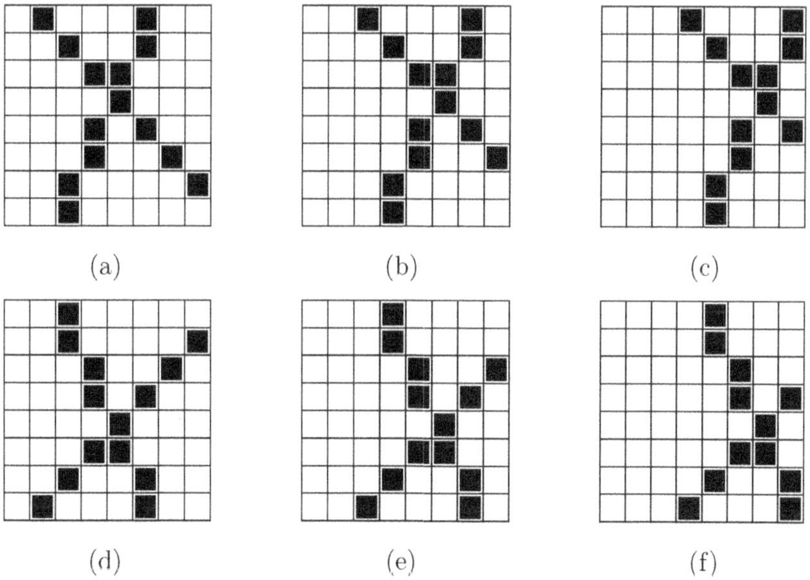

Figure 7.207: Six 8×8 Bitmaps Translated (a–c) and Reflected (d–f).

Cell encoding is not very useful for text (which can always be represented with eight bits per character), but this method can also be extended to an image that consists of straight lines only. The entire bitmap is divided into cells of, say, 8×8 pixels and is scanned cell by cell. The first cell is stored in entry 0 of a table and is encoded (i.e., written on the compressed file) as the pointer 0. Each subsequent cell is searched in the table. If found, its index in the table becomes its code and is written on the compressed file. Otherwise, it is added to the table. With 8×8 cells, each of the 64 pixels can be black or white, so the total number of different cells is $2^{64} \approx 1.8\times 10^{19}$, an immense number. However, some patterns never appear, as they don't represent any possible combination of line segments. Also, many cells are translated or reflected versions of other cells (Figure 7.207). All this brings the total number of distinct cells to just 108 [Jordan and Barrett 74]. These 108 cells can be stored in ROM and used frequently to compress images.

A teacher's purpose is not to create students in his own image, but to develop students who can create their own image.

—Anonymous

8
Wavelet Methods

Back in the early 1800s, the French mathematician Joseph Fourier discovered that any periodic fucntion can be expressed as a (possibly infinite) sum of sines and cosines. This surprising fact is now known as Fourier expansion and it has many applications in engineering, mainly in the analysis of signals. It can isolate the various frequencies that underlie a signal and thereby enable the user to study the signal and also edit it by deleting or adding certain frequencies. The downside of Fourier expansion is that it does not tell us when (at which point or points in time) each frequency is active in a given signal. We therefore say that Fourier expansion offers frequency resolution but no time resolution.

Wavelet analysis (or the wavelet transform) is a successful approach to the problem of analyzing a signal both in time and in frequency. Given a signal that varies with time, we select a time interval, and use the wavelet approach to identify and isolate the frequencies that constitute the signal in that interval. The interval can be wide, in which case we say that the signal is analysed on a large scale. As the time interval gets narrower, the scale of analysis is said to become smaller and smaller. A large scale analysis illustrates the global behavior of the signal, while each small scale analysis illuminates the way the signal behaves at a short interval of time; it is like zooming in the signal in time, instead of in space. Thus, the fundamental idea behind wavelets is to analyze a function or a signal according to scale.

Mathematically, wavelets are functions that satisfy certain requirements. Among these is the requirement that a wavelet integrates to zero. This implies that for each area of the wavelet function above the x axis, there must be an equal area below that axis. Thus, the wavelet function has to wave above and below the x axis, which justifies the name "wave." Other requirements result in functions that are localized in space, thereby suggesting the use of the diminutive "wavelet" instead of "wave."

This chapter starts with a discussion of the Fourier transform and the concepts of the time and frequency domains, which naturally lead to an uncertainty principle. This is followed by a discussion of the continuous wavelet transform. The important part

D. Salomon, G. Motta, *Handbook of Data Compression*, 5th ed.
DOI 10.1007/978-1-84882-903-9_8, © Springer-Verlag London Limited 2010

of the chapter introduces the discrete wavelet transform by using the Haar transform as an example. The simple Haar transform is then extended to the more general filter banks. In order to illustrate the power of the wavelet approach, the chapter describes several important compression methods, such as the Laplacian pyramid, SPIHT, WSQ, and JPEG 2000.

8.1 Fourier Transform

The concept of a transform is familiar to mathematicians. It is a standard mathematical tool used to solve problems in many areas. The idea is to change a mathematical quantity (a number, a vector, a function, or anything else) to another form, where it may look unfamiliar but may have useful properties. The transformed quantity is used to solve a problem or to perform a calculation, and the result is then transformed back to the original form.

A simple, illustrative example is Roman numerals. The ancient Romans presumably knew how to operate on such numbers, but when we have to, say, multiply two Roman numerals, we may find it more convenient to transform them into modern (Arabic) notation, multiply, and then transform the result back into a Roman numeral. Here is a simple example:

$$\text{XCVI} \times \text{XII} \rightarrow 96 \times 12 = 1152 \rightarrow \text{MCLII}.$$

Functions used in science and engineering often use *time* as their parameter. We therefore say that a function $g(t)$ is represented in the *time domain*. A typical function varies over time, which is why we can think of it as being similar to a wave, and we may try to represent it as a wave (or as a combination of waves). When this is done, we denote the resulting function by $G(f)$, where f stands for the frequency of the wave, and we say that the function is represented in the *frequency domain*. The concept of two domains turns out to be useful, since many operations on functions are easy to carry out in the frequency domain. Transforming a function between the time and frequency domains is easy when the function is *periodic*, but it can also be done for certain nonperiodic functions.

Definition: A function $g(t)$ is periodic if there exists a nonzero constant P such that $g(t+P) = g(t)$ for all values of t. The least such P is called the *period* of the function.

Figure 8.1 shows three periodic functions. The function of Figure 8.1a is a square pulse, that of Figure 8.1b is a sine wave, and the function of Figure 8.1c is more complex.

A periodic function has four important attributes: its amplitude, period, frequency, and phase. The amplitude of the function is the maximum value it has in any period. The frequency f is the inverse of the period ($f = 1/P$). It is expressed in cycles per second, or hertz (Hz). The phase is the least understood of the four attributes. It measures the position of the function within a period, and it is easy to visualize when a function is compared to its own copy. Consider the two sinusoids of Figure 8.1b. They are identical but out of phase. One follows the other at a fixed interval called the *phase difference*. We can write them as $g_1(t) = A\sin(2\pi ft)$ and $g_2(t) = A\sin(2\pi ft + \theta)$. The phase difference between them is θ, but we can also say that the first one has no phase,

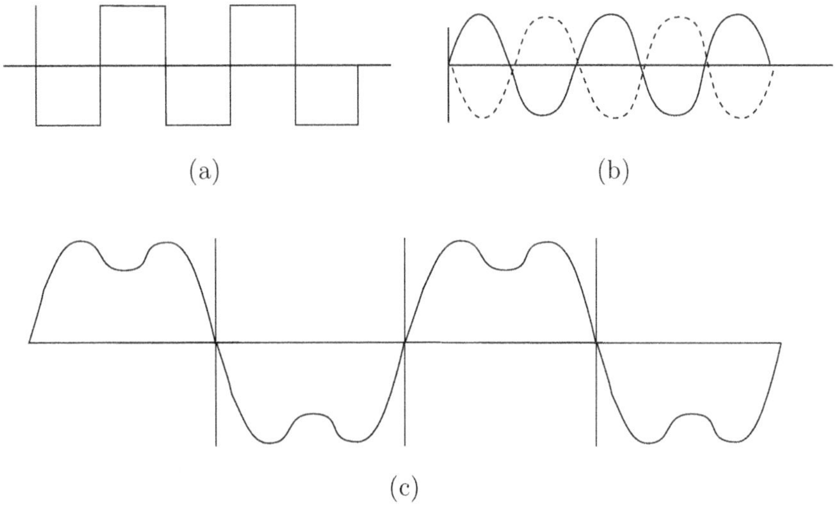

Figure 8.1: Periodic Functions.

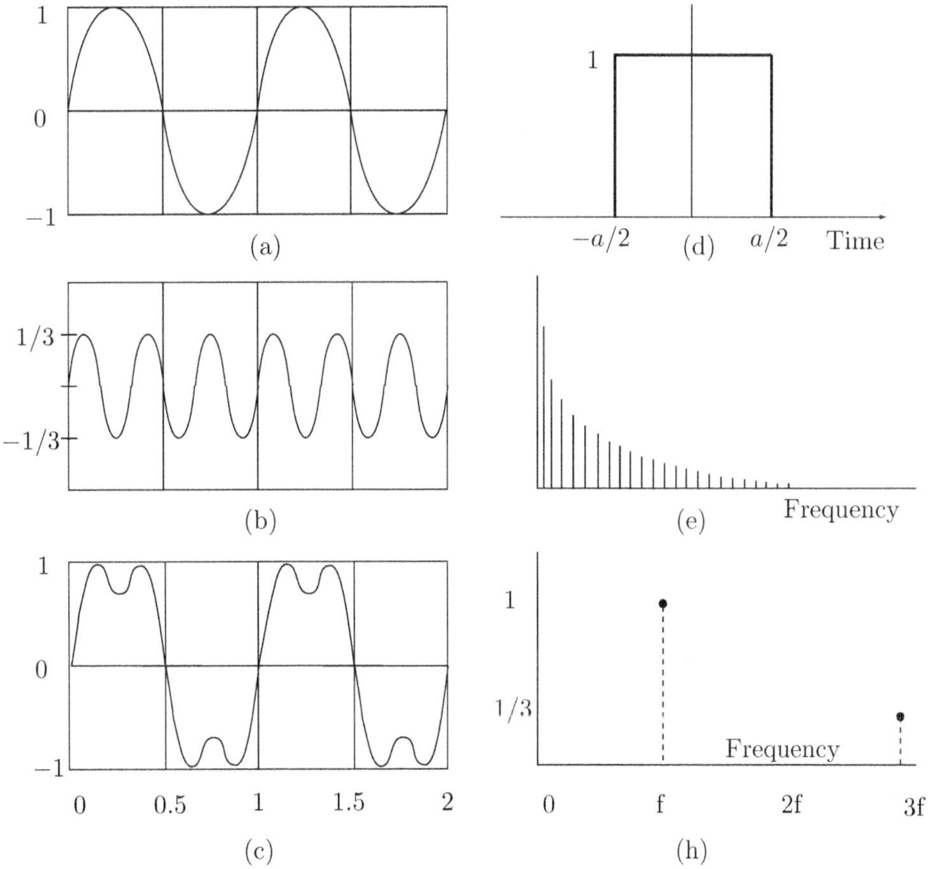

Figure 8.2: Time and Frequency Domains.

while the second one has a phase of θ. (This example also shows that the cosine is a sine function with a phase $\theta = \pi/2$.)

8.2 The Frequency Domain

To understand the concept of frequency domain, let's look at two simple examples. The function $g(t) = \sin(2\pi ft) + (1/3)\sin(2\pi(3f)t)$ is a combination of two sine waves with amplitudes 1 and 1/3, and frequencies f and $3f$, respectively. They are shown in Figure 8.2a,b. Their sum (Figure 8.2c) is also periodic, with frequency f (the smaller of the two frequencies f and $3f$). The frequency domain of $g(t)$ is a function consisting of just the two points $(f,1)$ and $(3f,1/3)$ (Figure 8.2h). It indicates that the original (time domain) function is made up of frequency f with amplitude 1, and frequency $3f$ with amplitude 1/3.

This example is extremely simple, since it involves just two frequencies. When a function involves several frequencies that are integer multiples of some lowest frequency, the latter is called the *fundamental frequency* of the function.

Not every function has a simple frequency domain representation. Consider the single square pulse of Figure 8.2d. Its time domain is

$$g(t) = \begin{cases} 1, & -a/2 \le t \le a/2, \\ 0, & \text{elsewhere}, \end{cases}$$

but its frequency domain is Figure 8.2e. It consists of all the frequencies from 0 to ∞, with amplitudes that drop continuously. This means that the time domain representation, even though simple, consists of all possible frequencies, with lower frequencies contributing more, and higher ones, contributing less and less.

In general, a periodic function can be represented in the frequency domain as the sum of (phase shifted) sine waves with frequencies that are integer multiples (harmonics) of some fundamental frequency. However, the square pulse of Figure 8.2d is not periodic. It turns out that frequency domain concepts can be applied to a nonperiodic function, but only if it is nonzero over a finite range (like our square pulse). Such a function is represented as the sum of (phase shifted) sine waves with all kinds of frequencies, not just harmonics.

The *spectrum* of the frequency domain (sometimes also called the *frequency content* of the function) is the range of frequencies it contains. For the function of Figure 8.2h, the spectrum is the two frequencies f and $3f$. For the one of Figure 8.2e, it is the entire range $[0,\infty]$. The *bandwidth* of the frequency domain is the width of the spectrum. It is $2f$ in our first example, and infinity in the second example.

Another important concept to define is the *dc component* of the function. The time domain of a function may include a component of zero frequency. Engineers call this component the *direct current*, so the rest of us have adopted the term "dc component." Figure 8.3a is identical to Figure 8.2c except that it goes from 0 to 2, instead of from -1 to $+1$. The frequency domain (Figure 8.3b) now has an added point at $(0,1)$, representing the dc component.

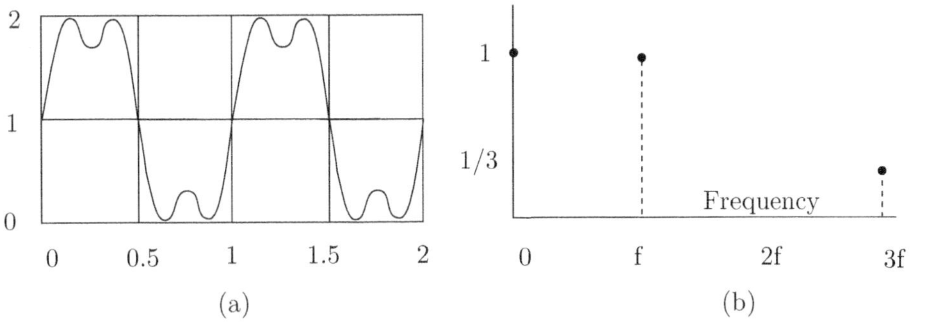

Figure 8.3: Time and Frequency Domains With a dc Component.

Joseph Fourier was born March 21, 1768, in Auxerre, France and died on May 16, 1830. A brilliant mathematician, he developed the partial differential equations governing the steady-state and time-dependent propagations of heat in solid bodies. He solved these equations with an infinite series of trigonometric equations, now known as a Fourier Series. His analytical methods for solving heat transfer problems departed from others of the time by distinguishing between interior and surface conditions, treating each separately. He was awarded the Grand Prize in Mathematics from the Institut de France in 1810 for his work.

The entire concept of the two domains is due to the French mathematician Joseph Fourier. He proved a fundamental theorem which says that every periodic function, real or complex, can be represented as the sum of sine and cosine functions. He also showed how to transform a function between the time and frequency domains. If the shape of the function is far from a uniform wave, its Fourier expansion may include an infinite number of frequencies. For a continuous function $g(t)$, the Fourier transform and its inverse are given by

$$\begin{aligned} G(f) &= \int_{-\infty}^{\infty} g(t)[\cos(2\pi ft) - i\sin(2\pi ft)]\,dt, \\ g(t) &= \int_{-\infty}^{\infty} G(f)[\cos(2\pi ft) + i\sin(2\pi ft)]\,df. \end{aligned}$$

In computer applications, we normally have discrete functions that take just n (equally spaced) values. In such a case the discrete Fourier transform is

$$\begin{aligned} G(f) &= \sum_{t=0}^{n-1} g(t)\left[\cos\left(\frac{2\pi ft}{n}\right) - i\sin\left(\frac{2\pi ft}{n}\right)\right] \\ &= \sum_{t=0}^{n-1} g(t)e^{-ift}, \quad 0 \le f \le n-1, \end{aligned} \tag{8.1}$$

and its inverse

$$\begin{aligned} g(t) &= \frac{1}{n}\sum_{f=0}^{n-1} G(f)\left[\cos\left(\frac{2\pi ft}{n}\right) + i\sin\left(\frac{2\pi ft}{n}\right)\right] \\ &= \sum_{t=0}^{n-1} G(f)e^{ift}, \quad 0 \le t \le n-1. \end{aligned} \tag{8.2}$$

Note that $G(f)$ is complex, so it can be written $G(f) = \mathcal{R}(f) + i\mathcal{I}(f)$. For any value of f, the amplitude (or magnitude) of G is given by $|G(f)| = \sqrt{\mathcal{R}^2(f) + \mathcal{I}^2(f)}$.

A word on terminology. Generally, $G(f)$ is called the Fourier transform of $g(t)$. However, if $g(t)$ is periodic, then $G(f)$ is its Fourier series.

A function $f(t)$ is a *bandpass function* if its Fourier transform $F(\omega)$ is confined to a frequency interval $\omega_1 < |\omega| < \omega_2$, where $\omega_1 > 0$ and ω_2 is finite.

Note how the function in Figure 8.2c, obtained by adding the simple functions in Figure 8.2a,b, starts resembling a square pulse. It turns out that we can bring it closer to a square pulse (like the one in Figure 8.1a) by adding $(1/5)\sin(2\pi(5f)t)$, $(1/7)\sin(2\pi(7f)t)$, and so on. We say that the Fourier series of a square wave with amplitude A and frequency f is the infinite sum

$$A\sum_{k=1,3,5,\ldots}^{\infty}\frac{1}{k}\sin(2\pi kft),$$

where successive terms have smaller and smaller amplitudes.

Here are two examples that show the relation between a function $g(t)$ and its Fourier expansion $G(f)$. The first example is the step function of Figure 8.4a, defined by

$$g(t) = \begin{cases} \pi/2, & \text{for } 2k\pi \le t < (2k+1)\pi, \\ -\pi/2, & \text{for } (2k+1)\pi \le t < (2k+2)\pi, \end{cases}$$

where k is any integer (positive or negative). The Fourier expansion of $g(t)$ is

$$G(f) = 2\sum_{k=0}^{\infty}\frac{\sin[f(2k+1)]}{2k+1} = 2\sin f + \frac{2\sin 3f}{3} + \frac{2\sin 5f}{5} + \cdots.$$

Figure 8.5a shows the first three terms of this series, and Figure 8.5b shows three partial sums, of the first four, eight, and 13 terms. It is obvious that these partial sums quickly approach the original function.

The second example is the sawtooth function of Figure 8.4b, defined by $g(t) = t/2$ for every interval $[2k\pi, (2k+1)\pi)$. Its Fourier expansion is

$$G(f) = \frac{\pi}{2} - \sum_{k=1}^{\infty}\frac{\sin(kf)}{k}.$$

Figure 8.6 shows four partial sums, of the first three, five, nine, and 17 terms, of this expansion.

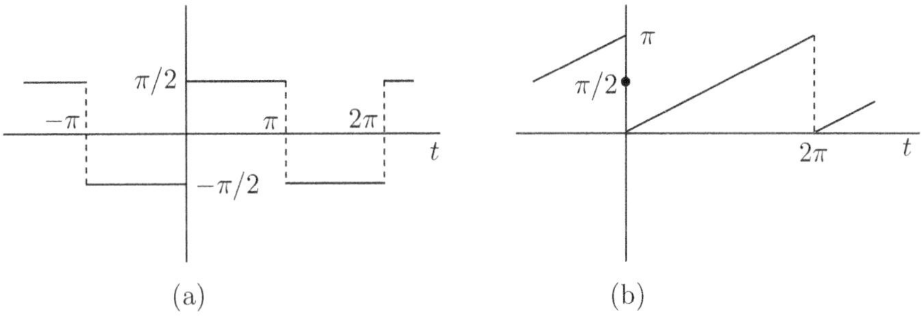

Figure 8.4: Two Functions $g(t)$ to be Transformed.

8.3 The Uncertainty Principle

These examples illustrate an important relation between the time and frequency domains; namely, they are *complementary.* Each of them complements the other in the sense that when one of them is localized the other one must be global. Consider, for example, a pure sine wave. It has one frequency, so it is well localized in the frequency domain. However, it is infinitely long, so in the time domain it is global. On the other hand, a function may be localized in the time domain, such as the single spike of Figure 8.38a, but it will invariably be global in the frequency domain; its Fourier expansion will contain many (possibly even infinitely many) frequencies. This relation between the time and frequency domains makes them complementary, and it is popularly described by the term *uncertainty relation* or *uncertainty principle.*

The Heisenberg uncertainty principle, first recognized in 1927 by Werner Heisenberg, is a very important physical principle. It says that position and momentum are complementary. The better we know the position of a particle of matter, the less certain we are of its momentum. The reason for this relation is the way we measure positions. In order to locate a particle in space, we have to see it (with our eyes or with an instrument). This requires shining light on the particle (either visible light or some other wavelength), and it is this light that disturbs the particle, moves it from its position and thus changes its momentum. We think of light as consisting of small units, photons, that don't have any mass but have momentum. When a photon hits a particle, it gives the particle a "kick," which moves it away. The larger the particle, the smaller the effects of the kick, but in principle, the complementary relation between position and momentum exists even for macroscopic objects.

It is important to realize that the uncertainty principle is part of nature; it is not just a result of our imperfect knowledge or primitive instruments. It is expressed by the relation

$$\Delta x\,\Delta p \geq \frac{\hbar}{2} = \frac{h}{4\pi},$$

where Δx and Δp are the uncertainties in the position and momentum of the particle,

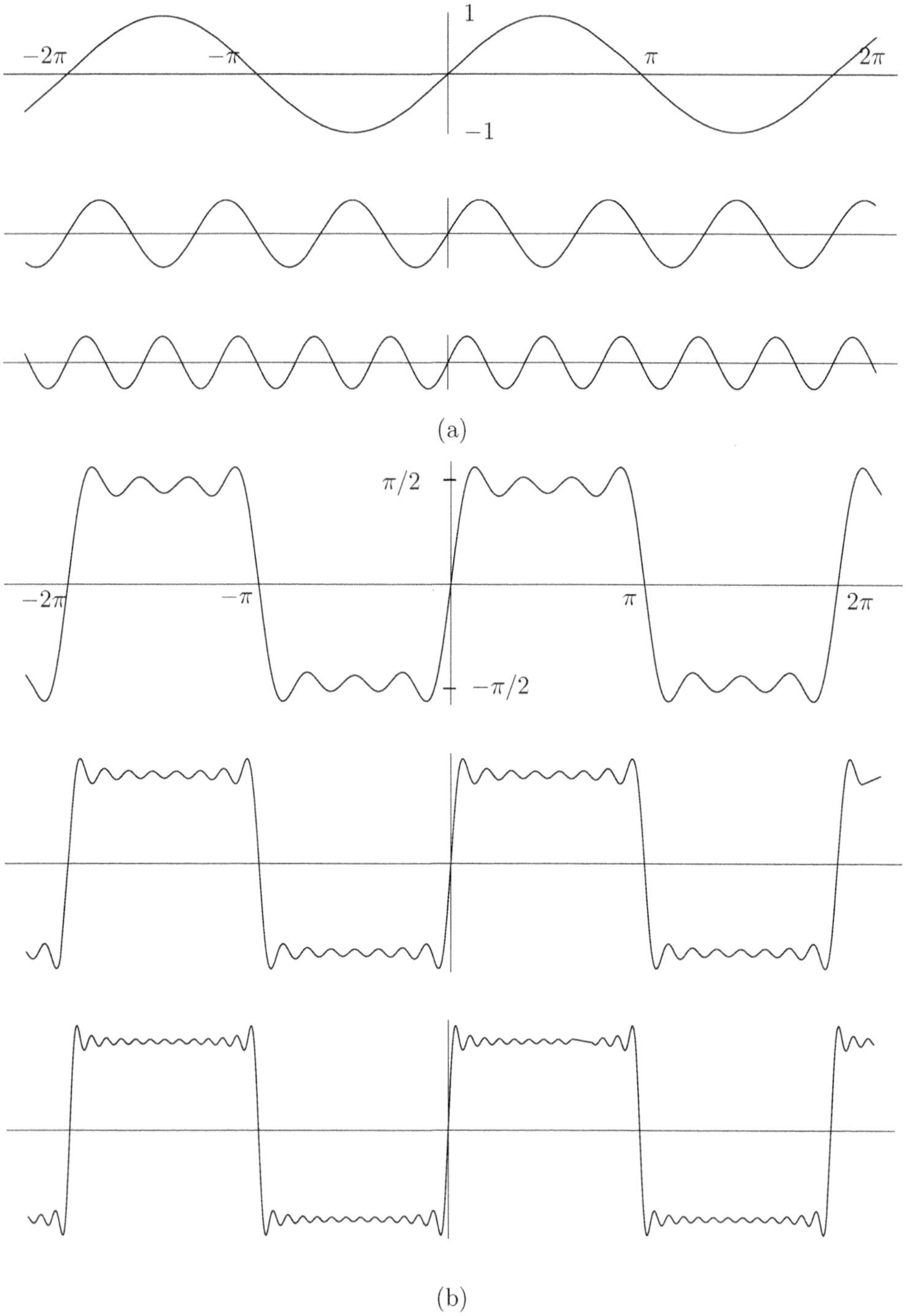

Figure 8.5: The First Three Terms and Three Partial Sums.

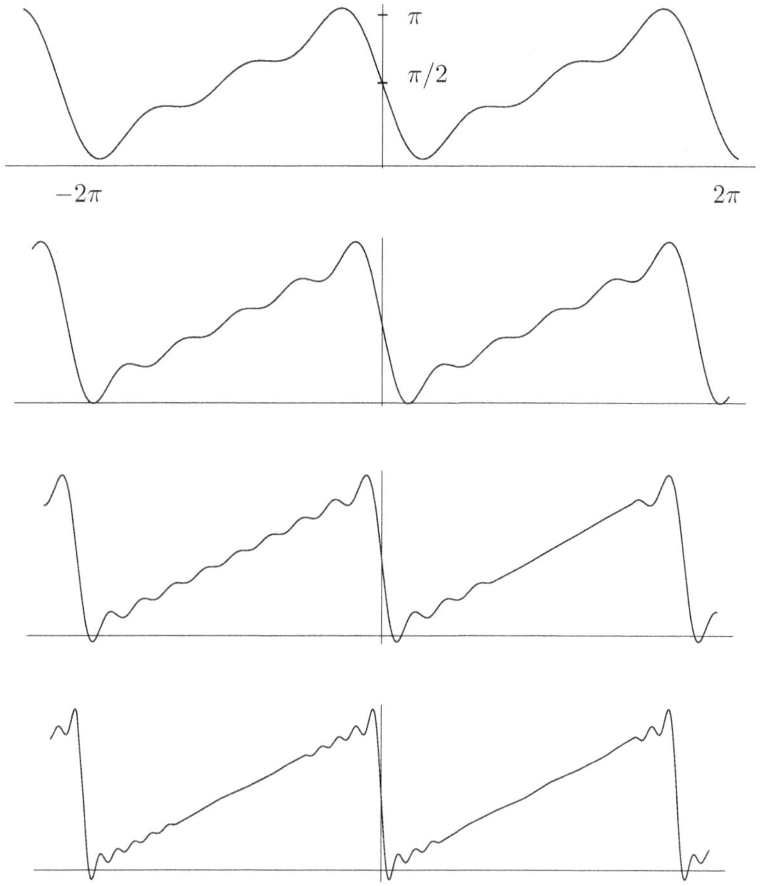

Figure 8.6: Four Partial Sums.

and h is the Planck constant (6.626176×10^{-27}erg·sec). The point is that h is very small, so

$$\hbar \stackrel{\text{def}}{=} \frac{h}{2\pi} = 1.05\times10^{-34} = 0.000000000000000000000000000000000105\,\text{joule}\cdot\text{sec}.$$

(An erg and a joule are units of energy whose precise values are irrelevant for this discussion.) This is why we don't notice the position/momentum uncertainty in everyday life. A simple example should make this clear. Suppose that the mass of a moving ball is 0.15 Kg and we measure its velocity to a precision of 0.05 m/sec (that is about 1 mph). Momentum is the product of mass and velocity, so if the uncertainty in the velocity of the ball is 0.05 m/sec, the uncertainty in its momentum is $\Delta p = 0.15\times0.05 = 0.0075$ Kg·m/sec The uncertainty principle says that $\Delta x\times0.0075 \geq 1.05\times10^{-34}/2$, which implies that the uncertainty in the position of the ball is on the order of 70×10^{-34} m or 7×10^{-30} mm, too small to measure!

The uncertainty principle is a special case of the principle of complementarity. Position and momentum are complementary quantities, but there are other pairs of physical quantities that are complementary, such as time and energy.

While we are discussing physics, here is another similarity between the time and frequency domains and physical reality. Real objects (i.e., objects with mass) have a property called the "wave particle duality." This is true for all objects but is noticeable and can be measured only for microscopic particles. This property means that every object has a wave associated with it. The amplitude of this wave (actually, the square of the absolute value of the amplitude) in a given region describes the probability of finding the object in that region. When we observe the object, it will normally be found in the region where its wave function has the highest amplitude.

The wave associated with an object is generally not spread throughout the entire universe, but is confined to a small region in space. This "localized" wave is called a *wave packet*. Such a packet does not have one particular wavelength, but consists of waves of many wavelengths. The momentum of an object depends on its wavelength, so an object that is localized in space doesn't have a specific, well-defined momentum; it has many momenta. In other words, there's an uncertainty in the momentum of the particle.

Nuggets of Uncertainty
A full discussion of Heisenberg's uncertainty principle may be found in the Appendix. Then again, it may not.
Uncertainty is the very essence of romance. It's what you don't know that intrigues you.

8.4 Fourier Image Compression

We now apply the concepts of time and frequency domains to digital images. Imagine a grayscale photograph scanned line by line. For all practical purposes we can assume that the photograph has infinite resolution (its shades of gray can vary continuously). An ideal scan would therefore result in an infinite sequence of numbers that can be considered the values of a (continuous) intensity function $I(t)$. In practice, we can store only a finite amount of data in memory, so we have to select a finite number of values $I(1)$ through $I(n)$. This process is known as *sampling*.

Intuitively, sampling seems a trade-off between quality and price. The bigger the sample, the better the quality of the final image, but more hardware (more memory and higher screen resolution) is required, resulting in higher costs. This intuitive conclusion, however, is not entirely true. Sampling theory tells us that we can sample an image and reconstruct it later in memory without loss of quality if we can do the following:

1. Transform the intensity function from the time domain $I(t)$ to the frequency domain $G(f)$.
2. Find the maximum frequency f_m.

3. Sample $I(t)$ at a rate slightly higher than $2f_m$ (e.g., if $f_m = 22{,}000$ Hz, generate samples at the rate of 44,100 Hz).
4. Store the sampled values in the bitmap. The resulting image would be equal in quality to the original one on the photograph.

There are two points to consider. The first is that f_m could be infinite. In such a case, a value f_m should be selected such that frequencies that are greater than f_m do not contribute much (have low amplitudes). There is some loss of image quality in such a case. The second point is that the bitmap (and consequently, the resolution) may be too small for the sample generated in step 3. In such a case, a smaller sample has to be taken, again resulting in a loss of image quality.

The result above was proved by Harry Nyquist [Nyquist 28], and the quantity $2f_m$ is called the *Nyquist rate*. It is used in many practical situations. The range of human hearing, for instance, is between 16 Hz and 22,000 Hz. When sound is digitized at high quality (such as music recorded on a CD), it is sampled at the rate of 44,100 Hz. Anything lower than that results in distortions. Section 7.1 discusses the application of this theorem to images (two-dimensional signals).

Why is it that sampling at the Nyquist rate is enough to restore the original signal? It seems that sampling ignores the behavior of the analog signal between samples, and can therefore miss important information. What guarantees that the signal will not go up or down considerably between consecutive samples? In principle, such behavior may happen, but in practice, all analog signals have a frequency response limit because they are created by sources (such as microphone, speaker, or mouth) whose response speed is limited. Thus, the rate at which a real signal can change is limited, thereby making it possible to predict the way it will vary from sample to sample. We say that the *finite bandwidth* of real signals is what makes their digitization possible.

> Fourier is a mathematical poem.
>
> —William Thomson (Lord Kelvin)

The Fourier transform is useful and popular, having applications in many areas. It has, however, one drawback: It shows the frequency content of a function $f(t)$, but it does not specify where (i.e., for what values of t) the function has low or high frequencies. The reason for this is that the basis functions (sine and cosine) used by this transform are infinitely long. They pick up the different frequencies of $f(t)$ regardless of where they are located.

A better transform should specify the frequency content of $f(t)$ as a function of t. Instead of producing a one-dimensional function (in the continuous case) or a one-dimensional array of numbers (in the discrete case), it should produce a two-dimensional function or array of numbers $W(a,b)$ that describes the frequency content of $f(t)$ for different values of t. A column $W(a_i,b)$ of this array (where $i = 1,2,\ldots,n$) lists the frequency spectrum of $f(t)$ for a certain value (or range of values) of t. A row $W(a,b_i)$ contains numbers that describe how much of a certain frequency (or range of frequencies) $f(t)$ has for any given t.

The *wavelet transform* is such a method. It has been developed, researched, and applied to many areas of science and engineering since the early 1980s, although its roots go much earlier. The main idea is to select a mother wavelet, a function that is nonzero in some small interval, and use it to explore the properties of $f(t)$ in that interval. The mother wavelet is then translated to another interval of t and used again in the same way. Parameter b specifies the translation. Different frequency resolutions of $f(t)$ are explored by scaling the mother wavelet with a scale factor a.

Before getting into any details, we illustrate the relation between the "normal" (time domain) representation of a function and its two-dimensional transform by looking at the standard musical notation. This notation, used in the West for hundreds of years, is two-dimensional. Notes are written on a stave in two dimensions, where the horizontal axis denotes the time (from left to right) and the vertical axis denotes the pitch. The higher the note is on the stave, the higher the pitch of the tone played. In addition, the shape of a note indicates its duration. Figure 8.7a shows, from left to right, one whole note (called "C" in the U.S. and "do" in Europe), two half notes, three quarter notes, and two eighth notes. In addition to the stave and the notes, musical notation includes many other symbols and directions from the composer. However, the point is that the same music can also be represented by a one-dimensional function that simply describes the amplitude of the sound as a function of the time (Figure 8.7b). The two representations are mathematically equivalent, but are used differently in practice. The two-dimensional representation is used by musicians to actually perform the music. The one-dimensional representation is used to replay music that has already been performed and recorded.

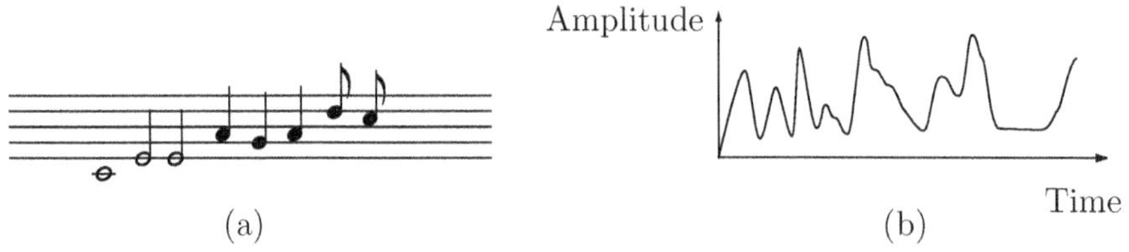

Figure 8.7: Two Representations of Music.

◇ **Exercise 8.1:** Come up with an example of a common notation that has a familiar two-dimensional representation used by humans and an unfamiliar one-dimensional representation used by machines.

The principle of analyzing a function by time and by frequency can be applied to image compression because images contain areas that exhibit "trends" and areas with "anomalies." A trend is an image feature that involves just a few frequencies (it is localized in frequency) but is spread spatially. A typical example is an image area where the brightness varies gradually. An anomaly is an image feature that involves several frequencies but is localized spatially (it is concentrated in a small image area). An example is an edge.

We start by looking at functions that can serve as a wavelet, by defining the continuous wavelet transform (CWT) and its inverse, and illustrating the way the CWT works. We then show in detail how the Haar wavelet can be applied to the compression of images. This naturally leads to the concepts of filter banks and the *discrete wavelet transform* (Section 8.8). The *lifting scheme* for the calculation of the wavelet transform and its inverse are described in Section 8.11. This is followed by descriptions of several compression methods that either employ the wavelet transform or compress the coefficients that result from such a transform.

8.5 The CWT and Its Inverse

The continuous wavelet transform (CWT, [Lewalle 95] and [Rao and Bopardikar 98]) of a function $f(t)$ involves a mother wavelet $\psi(t)$. The mother wavelet can be any real or complex continuous function that satisfies the following properties:

1. The total area under the curve of the function is zero, i.e.,

$$\int_{-\infty}^{\infty} \psi(t)\,dt = 0.$$

2. The total area of $|\psi(t)|^2$ is finite, i.e.

$$\int_{-\infty}^{\infty} |\psi(t)|^2\,dt < \infty.$$

This condition implies that the integral of the square of the wavelet has to exist. We can also say that a wavelet has to be *square integrable*, or that it belongs to the set $\mathcal{L}^2(R)$ of all the square integrable functions.
3. The admissibility condition, discussed below.

Property 1 suggests a function that oscillates above and below the t axis. Such a function tends to have a wavy appearance. Property 2 implies that the *energy* of the function is finite, suggesting that the function is localized in some finite interval and is zero, or almost zero, outside this interval. These properties justify the name "wavelet." An infinite number of functions satisfy these conditions, and some of them have been researched and are commonly used for wavelet transforms. Figure 8.8a shows the Morlet wavelet, defined by

$$\psi(t) = e^{-t^2}\cos\left(\pi t\sqrt{\frac{2}{\ln 2}}\right) \approx e^{-t^2}\cos(2.885\pi t).$$

This is a cosine curve whose oscillations are dampened by the exponential (or Gaussian) factor. More than 99% of its energy is concentrated in the interval $-2.5 \le t \le 2.5$. Figure 8.8b shows the so-called Mexican hat wavelet, defined as

$$\psi(t) = (1 - 2t^2)e^{-t^2}.$$

This is the second derivative of the (negative) Gaussian function $-0.5e^{-t^2}$.

The energy of a function

A function $y = f(x)$ relates each value of the independent variable x with a value of y. Plotting these pairs of values results in a representation of the function as a curve in two dimensions. The *energy* of a function is defined in terms of the area enclosed by this curve and the x axis. It makes sense to say that a curve that stays close to the axis has little energy, while a curve that spends "time" away from the x axis has more energy. Negative values of y push the curve under the x axis, where its area is considered negative, so the energy of $f(t)$ is defined as the area under the curve of the nonnegative function $f(x)^2$. If the function is complex, its absolute value is used in area calculations, so the energy of $f(x)$ is defined as

$$\int_{-\infty}^{\infty} |f(x)|^2 dx.$$

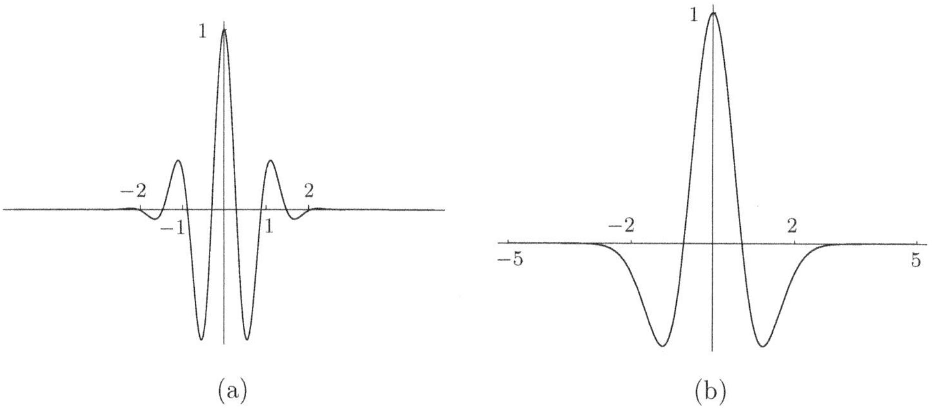

Figure 8.8: The (a) Morlet and (b) Mexican Hat Wavelets.

Once a wavelet $\psi(t)$ has been chosen, the CWT of a square integrable function $f(t)$ is defined as

$$W(a,b) = \int_{-\infty}^{\infty} f(t)\frac{1}{\sqrt{|a|}}\psi^*\left(\frac{t-b}{a}\right) dt. \tag{8.3}$$

The transform is a function of the two real parameters a and b. The $*$ denotes the complex conjugate. If we define a function $\psi_{a,b}(t)$ as

$$\psi_{a,b}(t) = \frac{1}{\sqrt{|a|}}\psi\left(\frac{t-b}{a}\right),$$

we can write Equation (8.3) in the form

$$W(a,b) = \int_{-\infty}^{\infty} f(t)\psi_{a,b}(t)\,dt. \tag{8.4}$$

Mathematically, the transform is the *inner product* of the two functions $f(t)$ and $\psi_{a,b}(t)$. The quantity $1/\sqrt{|a|}$ is a normalizing factor that ensures that the energy of $\psi(t)$ remains independent of a and b, i.e.,

$$\int_{-\infty}^{\infty} |\psi_{a,b}(t)|^2\,dt = \int_{-\infty}^{\infty} |\psi(t)|^2\,dt.$$

For any a, $\psi_{a,b}(t)$ is a copy of $\psi_{a,0}$ shifted b units along the time axis. Thus, b is a *translation parameter.* Setting $b = 0$ shows that

$$\psi_{a,0}(t) = \frac{1}{\sqrt{|a|}}\psi\left(\frac{t}{a}\right),$$

implying that a is a scaling (or a dilation) parameter. Values $a > 1$ stretch the wavelet, while values $0 < a < 1$ shrink it.

Since wavelets are used to transform a function, the inverse transform is needed. We denote by $\Psi(\omega)$ the Fourier transform of $\psi(t)$:

$$\Psi(\omega) = \int_{-\infty}^{\infty} \psi(t)e^{-i\omega t}\,dt.$$

If $W(a,b)$ is the CWT of a function $f(t)$ with a wavelet $\psi(t)$, then the inverse CWT is defined by

$$f(t) = \frac{1}{C}\int_{-\infty}^{\infty}\int_{-\infty}^{\infty} \frac{1}{|a|^2} W(a,b)\psi_{a,b}(t)\,da\,db,$$

where the quantity C is defined as

$$C = \int_{-\infty}^{\infty} \frac{|\Psi(\omega)|^2}{|\omega|}\,d\omega.$$

The inverse CWT exists if C is positive and finite. Since C is defined by means of Ψ, which itself is defined by means of the wavelet $\psi(t)$, the requirement that C be positive and finite imposes another restriction, called the *admissibility condition*, on the choice of wavelet.

The CWT is best thought of as an array of numbers that are inner products of $f(t)$ and $\psi_{a,b}(t)$. The rows of the array correspond to values of a, and the columns are indexed by b. The *inner product* of two functions $f(t)$ and $g(t)$ is defined as

$$\langle f(t), g(t)\rangle = \int_{-\infty}^{\infty} f(t)g^*(t)\,dt,$$

so the CWT is the inner product

$$\langle f(t), \psi_{a,b}(t)\rangle = \int_{-\infty}^{\infty} f(t)\psi_{a,b}(t)\,dt.$$

After this introduction, we are now in a position to explain the intuitive meaning of the CWT. We start with a simple example: the CWT of a sine wave, where the Mexican hat is chosen as the wavelet. Figure 8.9a shows a sine wave with two copies of the wavelet. Copy 1 is positioned at a point where the sine wave has a maximum. At this point there is a good match between the function being analyzed (the sine) and the wavelet. The wavelet *replicates* the features of the sine wave. As a result, the inner product of the sine and the wavelet is a large positive number. In contrast, copy 2 is positioned where the sine wave has a minimum. At that point the wave and the wavelet are almost mirror images of each other. Where the sine wave is positive, the wavelet is negative and vice versa. The product of the wave and the wavelet at this point is negative, and the CWT (the integral or the inner product) becomes a large negative number. In between points 1 and 2, the CWT drops from positive, to zero, to negative.

As the wavelet is translated along the sine wave, from left to right, the CWT alternates between positive and negative and produces the small wave shown in Figure 8.9b. This shape is the result of the close match between the function being analyzed (the sine wave) and the analyzing wavelet. They have similar shapes and also similar frequencies.

We now extend the analysis to cover different frequencies. This is done by scaling the wavelet. Figure 8.9c shows (in part 3) what happens when the wavelet is stretched such that it covers several periods of the sine wave. Translating the wavelet from left to right does not affect the match between it and the sine wave by much, with the result that the CWT varies just a little. The wider the wavelet, the closer the CWT is to a constant. Notice how the amplitude of the wavelet has been reduced, thereby reducing its area and producing a small constant. A similar thing happens in part 4 of the figure, where the wavelet has shrunk and is much narrower than one cycle of the sine wave. Since the wavelet is so "thin," the inner product of it and the sine wave is always a small number (positive or negative) regardless of the precise position of the wavelet relative to the sine wave. We see that translating the wavelet does not much affect its match to the sine wave, resulting in a CWT that is close to a constant.

The results of translating the wavelet, scaling it, and translating again and again are summarized in Figure 8.9d. This is a density plot of the transform function $W(a,b)$ where the horizontal axis corresponds to values of b (translation) and the vertical axis corresponds to values of a (scaling). Figure 8.9e is a contour plot of the same $W(a,b)$. These diagrams show that there is a good match between the function and the wavelet at a certain frequency (the frequency of the sine wave). At other frequencies the match deteriorates, resulting in a transform that gets closer and closer to a constant.

This is how the CWT provides a time-frequency analysis of a function $f(t)$. The result is a function $W(a,b)$ of two variables that shows the match between $f(t)$ and the wavelet at different frequencies of the wavelet and at different times. It is obvious that the quality of the match depends on the choice of wavelet. If the wavelet is very different from $f(t)$ at any frequencies and any times, the values of the resulting $W(a,b)$ will all be small and will not exhibit much variation. As a result, when wavelets are used to

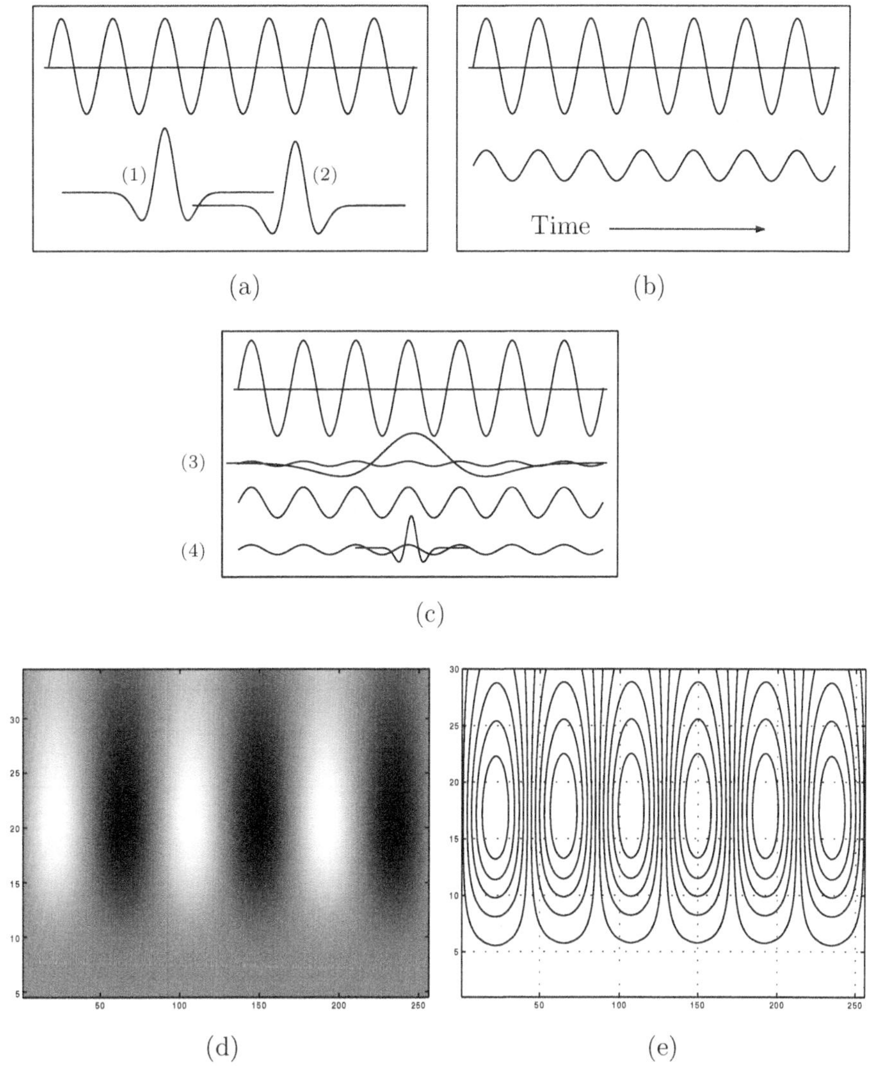

Figure 8.9: The Continuous Wavelet Transform of a Pure Sine Wave.

```
t=linspace(0,6*pi,256);
sinwav=sin(t);
plot(t,sinwav)
cwt=CWT(sinwav,10,'Sombrero');
axis('ij'); colormap(gray);
imagesc(cwt')
x=1:256; y=1:30;
[X,Y]=meshgrid(x,y);
contour(X,Y,cwt',10)
```

```
t=linspace(-10,10,256);
sombr=(1-2*t.^2).*exp(-t.^2);
plot(t,sombr)
```

Code For Figure 8.9.

compress images, different wavelets should be selected for different image types (bi-level, continuous-tone, and discrete-tone), but the precise choice of wavelet is still the subject of much research.

Since both our function and our wavelet are simple functions, it may be possible in this case to calculate the integral of the CWT as an indefinite integral, and come up with a closed formula for $W(a,b)$. In the general case, however, this is impossible, either in practice or in principle, and the calculations have to be done numerically.

The next example is slightly more complex and leads to a better understanding of the CWT. The function being analyzed this time is a sine wave with an accelerated frequency: the so-called *chirp*. It is given by $f(t)=\sin(t^2)$, and it is shown in Figure 8.10a. The wavelet is the same Mexican hat. Readers who have gone over the previous example carefully will have no trouble in interpreting the CWT of this example. It is given by Figure 8.10b,c and shows how the frequency of $f(t)$ increases with time.

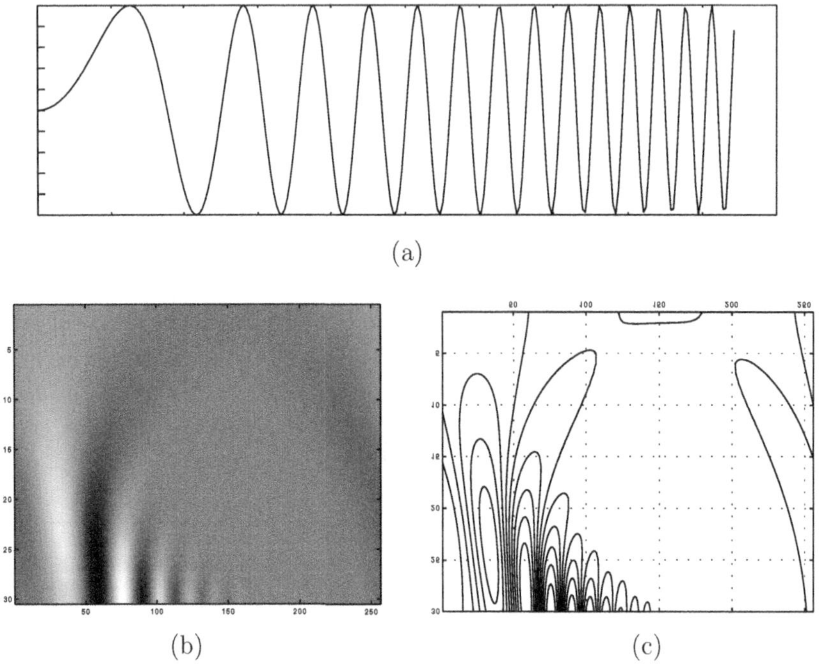

Figure 8.10: The Continuous Wavelet Transform of a Chirp.

These two examples illustrate how the CWT is used to analyze the time frequency of a function $f(t)$. It is clear that the user needs experience both in order to select the right wavelet for the job and in order to interpret the results. However, with the right experience, the CWT can be a powerful analytic tool in the hands of scientists and engineers.

◇ **Exercise 8.2:** Experiment with the CWT by trying to work out the following analysis. Select the function $f(t)=1+\sin(\pi t+t^2)/8$ as your candidate for analysis. This is a sine wave that oscillates between $y = 1 - 1/8$ and $y = 1 + 1/8$ with a frequency $2\pi + t$ that increases with t. As the wavelet, select the Mexican hat. Plot several translated copies of the wavelet against $f(t)$, then use appropriate software, such as *Mathematica* or Matlab, to calculate and display the CWT.

Wavelets are also subject to the uncertainty principle (Section 8.3). The Haar wavelet is very well localized in the time domain but is spread in the frequency domain due to the appearance of sidebands in the Fourier spectrum. In contrast, the Mexican hat wavelet and especially the Morlet wavelet have more concentrated frequencies but are spread over time. An important result of the uncertainty principle is that it is impossible to achieve a complete simultaneous mapping of both time and frequency. Wavelets provide a compromise, or a near optimal solution, to this problem, and this is one feature that makes them superior to Fourier analysis.

We can compare wavelet analysis to looking at a complex object first from a distance, then closer, then through a magnifying glass, and finally through a microscope. When looking from a distance, we see the overall shape of the object, but not any small details. When looking through a microscope, we see small details, but not the overall shape. This is why it is important to analyze in different scales. When we change the scale of the wavelet, we get new information about the function being analyzed.

8.6 The Haar Transform

Alfréd Haar (1885–1933)

Alfréd Haar was born in Budapest and received his higher mathematical training in Göttingen, where he later became a privatdozent. In 1912, he returned to Hungary and became a professor of mathematics first in Kolozsvár and then in Szeged, where he and his colleagues created a major mathematical center.

Haar is best remembered for his work on analysis on groups. In 1932 he introduced an invariant measure on locally compact groups, now called the Haar measure, which allows an analog of Lebesgue integrals to be defined on locally compact topological groups. Mathematical lore has it that John von Neumann tried to discourage Haar in this work because he felt certain that no such measure could exist. The following limerick celebrates Haar's achievement.

Said a mathematician named Haar,
"Von Neumann can't see very far.
He missed a great treasure—
They call it Haar measure—
Poor Johnny's just not up to par."

Information that is being wavelet transformed in practical applications, such as digitized sound and images, is discrete, consisting of individual numbers. This is why the discrete, and not the continuous, wavelet transform is used in practice. The discrete wavelet transform is described in general in Section 8.8, but we precede this discussion by presenting a simple example of this type of transform, namely, the Haar wavelet transform.

The use of the Haar transform for image compression is described here from a practical point of view. We first show how this transform is applied to the compression of grayscale images, then show how this method can be extended to color images. The Haar transform [Stollnitz et al. 96] was introduced in Section 7.7.3.

The Haar transform uses a scale function $\phi(t)$ and a wavelet $\psi(t)$, both shown in Figure 8.11a, to represent a large number of functions $f(t)$. The representation is the infinite sum

$$f(t) = \sum_{k=-\infty}^{\infty} c_k \phi(t-k) + \sum_{k=-\infty}^{\infty} \sum_{j=0}^{\infty} d_{j,k} \psi(2^j t - k),$$

where c_k and $d_{j,k}$ are coefficients to be calculated.

The basic scale function $\phi(t)$ is the unit pulse

$$\phi(t) = \begin{cases} 1, & 0 \le t < 1, \\ 0, & \text{otherwise.} \end{cases}$$

The function $\phi(t-k)$ is a copy of $\phi(t)$, shifted k units to the right. Similarly, $\phi(2t-k)$ is a copy of $\phi(t-k)$ scaled to half the width of $\phi(t-k)$. The shifted copies are used to approximate $f(t)$ at different times t. The scaled copies are used to approximate $f(t)$ at higher resolutions. Figure 8.11b shows the functions $\phi(2^j t - k)$ for $j = 0, 1, 2,$ and 3 and for $k = 0, 1, \ldots, 7$.

The basic Haar wavelet is the step function

$$\psi(t) = \begin{cases} 1, & 0 \le t < 0.5, \\ -1, & 0.5 \le t < 1. \end{cases}$$

From this we can see that the general Haar wavelet $\psi(2^j t - k)$ is a copy of $\psi(t)$ shifted k units to the right and scaled such that its total width is $1/2^j$.

◇ **Exercise 8.3:** Draw the four Haar wavelets $\psi(2^2 t - k)$ for $k = 0, 1, 2,$ and 3.

Both $\phi(2^j t - k)$ and $\psi(2^j t - k)$ are nonzero in an interval of width $1/2^j$. This interval is their *support*. Since this interval tends to be short, we say that these functions have *compact support*.

We illustrate the basic transform on the simple step function

$$f(t) = \begin{cases} 5, & 0 \le t < 0.5, \\ 3, & 0.5 \le t < 1. \end{cases}$$

It is easy to see that $f(t) = 4\phi(t) + \psi(t)$. We say that the original steps (5, 3) have been transformed to the (low resolution) average 4 and the (high resolution) detail 1. Using

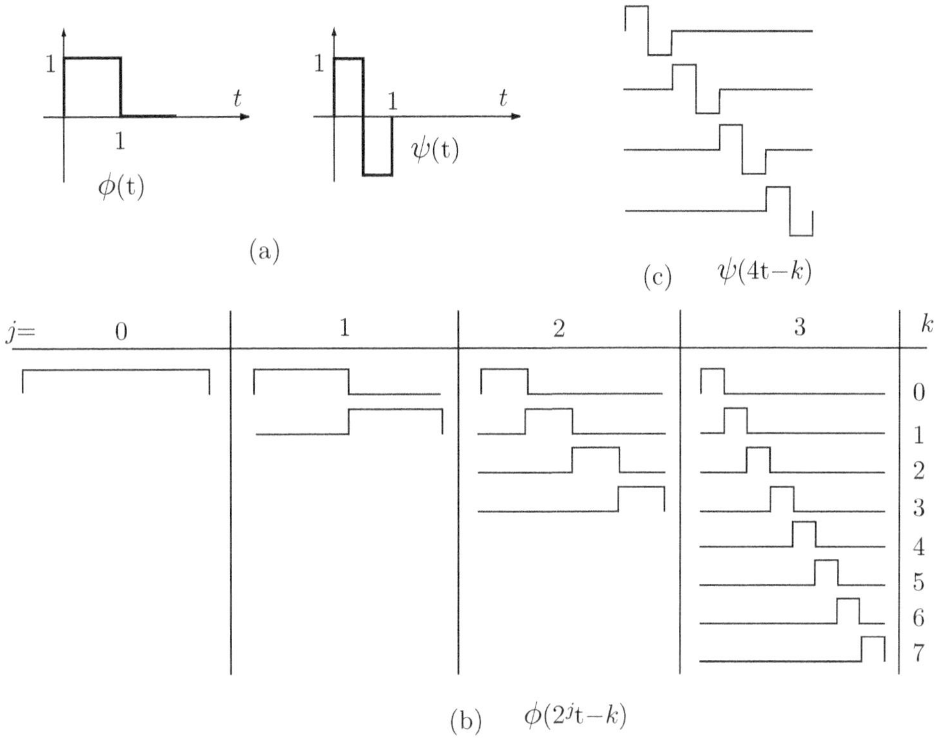

Figure 8.11: The Haar Basis Scale and Wavelet Functions.

matrix notation, this can be expressed (up to a factor of $\sqrt{2}$) as $(5,3)\mathbf{A}_2 = (4,1)$, where $\mathbf{A}_2$ is the order-2 Haar transform matrix of Equation (7.12).

An image is a two-dimensional array of pixel values. To illustrate how the Haar transform is used to compress an image, we start with a single row of pixel values, i.e., a one-dimensional array of n values. For simplicity we assume that n is a power of 2. (We use this assumption throughout this chapter, but there is no loss of generality. If n has a different value, the data can be extended by appending zeros. After decompression, the extra zeros are removed.) Consider the array of eight values $(1,2,3,4,5,6,7,8)$. We first compute the four averages $(1+2)/2 = 3/2$, $(3+4)/2 = 7/2$, $(5+6)/2 = 11/2$, and $(7+8)/2 = 15/2$. It is impossible to reconstruct the original eight values from these four averages, so we also compute the four differences $(1-2)/2 = -1/2$, $(3-4)/2 = -1/2$, $(5-6)/2 = -1/2$, and $(7-8)/2 = -1/2$. These differences are called *detail coefficients*, and in this section the terms "difference" and "detail" are used interchangeably. We can think of the averages as a coarse resolution representation of the original image, and of the details as the data needed to reconstruct the original image from this coarse resolution. If the pixels of the image are correlated, the coarse representation will resemble the original pixels, while the details will be small. This explains why the Haar wavelet compression of images uses averages and details.

It is easy to see that the array $(3/2, 7/2, 11/2, 15/2, -1/2, -1/2, -1/2, -1/2)$ made of the four averages and four differences can be used to reconstruct the original eight

> Prolonged, lugubrious stretches of Sunday afternoon in a university town could be mitigated by attending Sillery's tea parties, to which anyone might drop in after half-past three. Action of some law of averages always regulated numbers at these gatherings to something between four and eight persons, mostly undergraduates, though an occasional don was not unknown.
>
> —Anthony Powell, *A Question of Upbringing*

values. This array has eight values, but its last four components, the differences, tend to be small numbers, which helps in compression. Encouraged by this, we repeat the process on the four averages, the large components of our array. They are transformed into two averages and two differences, yielding $(10/4, 26/4, -4/4, -4/4, -1/2, -1/2, -1/2, -1/2)$. The next, and last, iteration of this process transforms the first two components of the new array into one average (the average of all eight components of the original array) and one difference $(36/8, -16/8, -4/4, -4/4, -1/2, -1/2, -1/2, -1/2)$. The last array is the *Haar wavelet transform* of the original data items.

Because of the differences, the wavelet transform tends to have numbers smaller than the original pixel values, so it is easier to compress using RLE, perhaps combined with move-to-front and Huffman coding. Lossy compression can be obtained if some of the smaller differences are quantized or even completely deleted (changed to zero).

Before we continue, it is interesting (and also useful) to calculate the *complexity* of this transform, i.e., the number of necessary arithmetic operations as a function of the size of the data. In our example we needed $8+4+2=14$ operations (additions and subtractions), a number that can also be expressed as $14 = 2(8-1)$. In the general case, assume that we start with $N = 2^n$ data items. In the first iteration we need 2^n operations, in the second one we need 2^{n-1} operations, and so on, until the last iteration, where $2^{n-(n-1)} = 2^1$ operations are needed. Thus, the total number of operations is

$$\sum_{i=1}^{n} 2^i = \sum_{i=0}^{n} 2^i - 1 = \frac{1-2^{n+1}}{1-2} - 1 = 2^{n+1} - 2 = 2(2^n - 1) = 2(N-1).$$

The Haar wavelet transform of N data items can therefore be performed with $2(N-1)$ operations, so its complexity is $\mathcal{O}(N)$, an excellent result.

It is useful to associate with each iteration a quantity called *resolution*, which is defined as the number of remaining averages at the end of the iteration. The resolutions after each of the three iterations above are $4(=2^2)$, $2(=2^1)$, and $1(=2^0)$. Section 8.6.3 shows that each component of the wavelet transform should be normalized by dividing it by the square root of the resolution. (This is the *orthonormal Haar transform*, also discussed in Section 7.7.3.) Thus, our example wavelet transform becomes

$$\left(\frac{36/8}{\sqrt{2^0}}, \frac{-16/8}{\sqrt{2^0}}, \frac{-4/4}{\sqrt{2^1}}, \frac{-4/4}{\sqrt{2^1}}, \frac{-1/2}{\sqrt{2^2}}, \frac{-1/2}{\sqrt{2^2}}, \frac{-1/2}{\sqrt{2^2}}, \frac{-1/2}{\sqrt{2^2}}\right).$$

If the normalized wavelet transform is used, it can be formally proved that ignoring the smallest differences is the best choice for lossy wavelet compression, since it causes the smallest loss of image information.

The two procedures of Figure 8.12 illustrate how the normalized wavelet transform of an array of n components (where n is a power of 2) can be computed. Reconstructing the original array from the normalized wavelet transform is illustrated by the pair of procedures of Figure 8.13.

These procedures seem at first to be different from the averages and differences discussed earlier. They don't compute averages, because they divide by $\sqrt{2}$ instead of by 2; the first procedure starts by dividing the entire array by $\sqrt{n}$, and the second one ends by doing the reverse. The final result, however, is the same as that shown above. Starting with array $(1,2,3,4,5,6,7,8)$, the three iterations of procedure `NWTcalc` result in the following

$$\left(\frac{3}{\sqrt{2^4}}, \frac{7}{\sqrt{2^4}}, \frac{11}{\sqrt{2^4}}, \frac{15}{\sqrt{2^4}}, \frac{-1}{\sqrt{2^4}}, \frac{-1}{\sqrt{2^4}}, \frac{-1}{\sqrt{2^4}}, \frac{-1}{\sqrt{2^4}}\right),$$
$$\left(\frac{10}{\sqrt{2^5}}, \frac{26}{\sqrt{2^5}}, \frac{-4}{\sqrt{2^5}}, \frac{-4}{\sqrt{2^5}}, \frac{-1}{\sqrt{2^4}}, \frac{-1}{\sqrt{2^4}}, \frac{-1}{\sqrt{2^4}}, \frac{-1}{\sqrt{2^4}}\right),$$
$$\left(\frac{36}{\sqrt{2^6}}, \frac{-16}{\sqrt{2^6}}, \frac{-4}{\sqrt{2^5}}, \frac{-4}{\sqrt{2^5}}, \frac{-1}{\sqrt{2^4}}, \frac{-1}{\sqrt{2^4}}, \frac{-1}{\sqrt{2^4}}, \frac{-1}{\sqrt{2^4}}\right),$$
$$\left(\frac{36/8}{\sqrt{2^0}}, \frac{-16/8}{\sqrt{2^0}}, \frac{-4/4}{\sqrt{2^1}}, \frac{-4/4}{\sqrt{2^1}}, \frac{-1/2}{\sqrt{2^2}}, \frac{-1/2}{\sqrt{2^2}}, \frac{-1/2}{\sqrt{2^2}}, \frac{-1/2}{\sqrt{2^2}}\right).$$

> We spake no word, Tho' each I ween did hear the other's soul.
> Not a wavelet stirred,
> And yet we heard
> The loneliest music of the weariest waves
> That ever roll.
>
> —Abram J. Ryan, *Poems*

8.6.1 Applying the Haar Transform

Once the concept of a wavelet transform is grasped, it's easy to generalize it to a complete two-dimensional image. This can be done in several ways that are discussed in Section 8.10. Here we show two such approaches, called the *standard decomposition* and the *pyramid decomposition*.

The former (Figure 8.15) starts by computing the wavelet transform of every row of the image. This results in a transformed image where the first column contains averages and all the other columns contain differences. The standard algorithm then computes the wavelet transform of every column. This results in one average value at the top-left corner, with the rest of the top row containing averages of differences, and with all other pixel values transformed into differences.

```
procedure NWTcalc(a:array of real, n:int);
 comment n is the array size (a power of 2)
 a:=a/√n comment divide entire array
 j:=n;
 while j≥ 2 do
  NWTstep(a, j);
  j:=j/2;
 endwhile;
end;

procedure NWTstep(a:array of real, j:int);
 for i=1 to j/2 do
  b[i]:=(a[2i-1]+a[2i])/√2;
  b[j/2+i]:=(a[2i-1]-a[2i])/√2;
 endfor;
 a:=b; comment move entire array
end;
```

Figure 8.12: Computing the Normalized Wavelet Transform.

```
procedure NWTreconst(a:array of real, n:int);
 j:=2;
 while j≤n do
  NWTRstep(a, j);
  j:=2j;
 endwhile
 a:=a√n; comment multiply entire array
end;

procedure NWTRstep(a:array of real, j:int);
 for i=1 to j/2 do
  b[2i-1]:=(a[i]+a[j/2+i])/√2;
  b[2i]:=(a[i]-a[j/2+i])/√2;
 endfor;
 a:=b; comment move entire array
end;
```

Figure 8.13: Restoring From a Normalized Wavelet Transform.

The latter method computes the wavelet transform of the image by alternating between rows and columns. The first step is to calculate averages and differences for all the rows (just one iteration, not the entire wavelet transform). This creates averages in the left half of the image and differences in the right half. The second step is to calculate averages and differences (just one iteration) for all the columns, which results in averages in the top-left quadrant of the image and differences elsewhere. Steps 3 and 4 operate on the rows and columns of that quadrant, resulting in averages concentrated in the top-left subquadrant. Pairs of steps are repeatedly executed on smaller and smaller subsquares, until only one average is left, at the top-left corner of the image, and all other pixel values have been reduced to differences. This process is summarized in Figure 8.16.

The transforms described in Section 7.6 are orthogonal. They transform the original pixels into a few large numbers and many small numbers. In contrast, wavelet transforms, such as the Haar transform, are *subband transforms*. They partition the image into regions such that one region contains large numbers (averages in the case of the Haar transform) and the other regions contain small numbers (differences). However, these regions, which are called subbands, are more than just sets of large and small numbers. They reflect different geometrical artifacts of the image. To illustrate this important feature, we examine a small, mostly-uniform image with one vertical line and one horizontal line. Figure 8.14a shows an 8×8 image with pixel values of 12, except for a vertical line with pixel values of 14 and a horizontal line with pixel values of 16.

12	12	12	12	14	12	12	12
12	12	12	12	14	12	12	12
12	12	12	12	14	12	12	12
12	12	12	12	14	12	12	12
12	12	12	12	14	12	12	12
16	16	16	16	14	16	16	16
12	12	12	12	14	12	12	12
12	12	12	12	14	12	12	12

(a)

12	12	13	12	0	0	2	0
12	12	13	12	0	0	2	0
12	12	13	12	0	0	2	0
12	12	13	12	0	0	2	0
12	12	13	12	0	0	2	0
16	16	15	16	0	0	$\underline{2}$	0
12	12	13	12	0	0	2	0
12	12	13	12	0	0	2	0

(b)

12	12	13	12	0	0	2	0
12	12	13	12	0	0	2	0
14	14	14	14	0	0	0	0
12	12	13	12	0	0	2	0
0	0	0	0	0	0	0	0
0	0	0	0	0	0	0	0
$\underline{4}$	$\underline{4}$	$\underline{2}$	$\underline{4}$	0	0	4	0
0	0	0	0	0	0	0	0

(c)

Figure 8.14: An 8×8 Image and Its Subband Decomposition.

Figure 8.14b shows the results of applying the Haar transform once to the rows of the image. The right half of this figure (the differences) is mostly zeros, reflecting the uniform nature of the image. However, traces of the vertical line can easily be seen (the notation $\underline{2}$ indicates a negative difference). Figure 8.14c shows the results of applying the Haar transform once to the columns of Figure 8.14b. The upper-right subband now features traces of the vertical line, whereas the lower-left subband shows traces of the horizontal line. These subbands are denoted by HL and LH, respectively (Figures 8.16 and 8.55, although there is inconsistency in the use of this notation by various authors). The lower-right subband, denoted by HH, reflects diagonal image artifacts (which our example image lacks). Most interesting is the upper-left subband, denoted by LL, that consists entirely of averages. This subband is a one-quarter version of the entire image, containing traces of both the vertical and the horizontal lines.

```
procedure StdCalc(a:array of real, n:int);
 comment array size is nxn (n = power of 2)
 for r=1 to n do NWTcalc(row r of a, n);
 endfor;
 for c=n to 1 do comment loop backwards
  NWTcalc(col c of a, n);
 endfor;
end;
procedure StdReconst(a:array of real, n:int);
 for c=n to 1 do comment loop backwards
  NWTreconst(col c of a, n);
 endfor;
 for r=1 to n do
  NWTreconst(row r of a, n);
 endfor;
end;
```

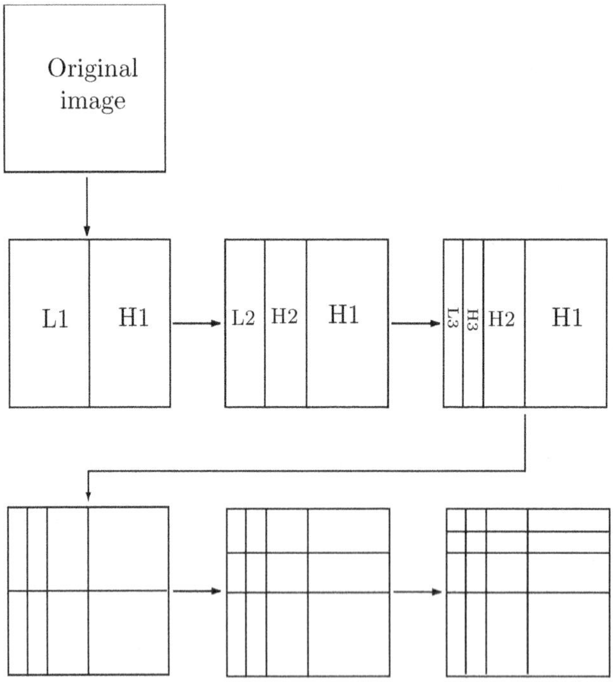

Figure 8.15: The Standard Image Wavelet Transform and Decomposition.

```
procedure NStdCalc(a:array of real, n:int);
 a:=a/√n comment divide entire array
 j:=n;
 while j≥ 2 do
  for r=1 to j do NWTstep(row r of a, j);
  endfor;
  for c=j to 1 do comment loop backwards
   NWTstep(col c of a, j);
  endfor;
  j:=j/2;
 endwhile;
end;
procedure NStdReconst(a:array of real, n:int);
 j:=2;
 while j≤n do
  for c=j to 1 do comment loop backwards
   NWTRstep(col c of a, j);
  endfor;
  for r=1 to j do
   NWTRstep(row r of a, j);
  endfor;
  j:=2j;
 endwhile
 a:=a√n; comment multiply entire array
end;
```

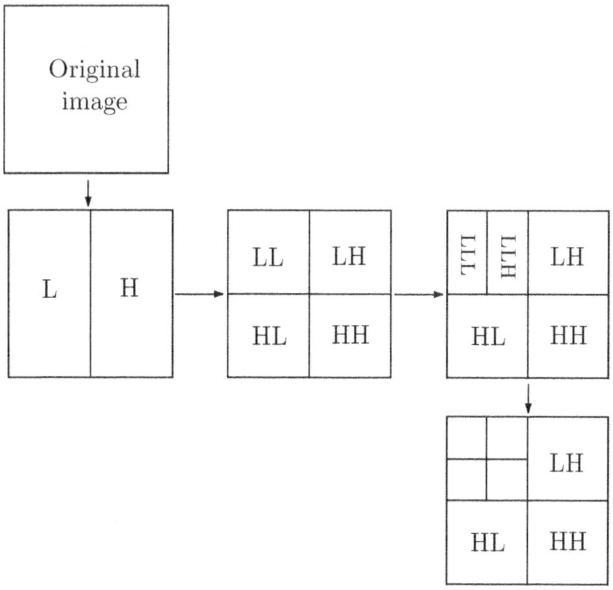

Figure 8.16: The Pyramid Image Wavelet Transform.

◇ **Exercise 8.4:** Construct a diagram similar to Figure 8.14 to show how subband HH reflects diagonal artifacts of the image.

(Artifact: A feature not naturally present, introduced during preparation or investigation.)

Figure 8.55 shows four levels of subbands, where level 1 contains the detailed features of the image (also referred to as the high-frequency or fine-resolution wavelet coefficients) and the top level, level 4, contains the coarse image features (low-frequency or coarse-resolution coefficients). It is clear that the lower levels can be quantized coarsely without much loss of important image information, while the higher levels should be quantized finely. The subband structure is the basis of all the image compression methods that use the wavelet transform.

Figure 8.17 shows typical results of the pyramid wavelet transform. The original image is shown in Figure 8.17a, and Figure 8.17c is a general pyramid decomposition. In order to illustrate how the pyramid transform works, this image consists only of horizontal, vertical, and slanted lines. The four quadrants of Figure 8.17b show smaller versions of the image. The top-left subband, containing the averages, is similar to the entire image, while each of the other three subbands shows image details. Because of the way the pyramid transform is constructed, the top-right subband contains vertical details, the bottom-left subband contains horizontal details, and the bottom-right one contains the details of slanted lines. Figure 8.17c shows the results of repeatedly applying this transform. The image is transformed into subbands of horizontal, vertical, and diagonal details, while the top-left subsquare, containing the averages, is shrunk to a single pixel.

Section 7.6 discusses image transforms. It should be mentioned that there are two main types of transforms, *orthogonal* and *subband*. An orthogonal linear transform is performed by computing the *inner product* of the data (pixel values or sound samples) with a set of *basis functions*. The result is a set of transform coefficients that can later be quantized or compressed with RLE, Huffman coding, or other methods. Several examples of important orthogonal transforms, such as the DCT, the WHT, and the KLT, are described in detail in Section 7.6. The Fourier transform also belongs in this category. It is discussed in Section 8.1.

The other main type of transform is the *subband transform*. It is performed by computing a convolution of the data (Section 8.7) with a set of *bandpass filters*. Each resulting subband encodes a particular portion of the frequency content of the data.

As a reminder, the discrete inner product of the two vectors f_i and g_i is defined by

$$\langle f, g\rangle = \sum_i f_i\, g_i.$$

The discrete convolution h is defined by Equation (8.5):

$$h_i = f \star g = \sum_j f_j\, g_{i-j}. \tag{8.5}$$

(Each element h_i of the discrete convolution h is the sum of products. It depends on i in the special way shown.)

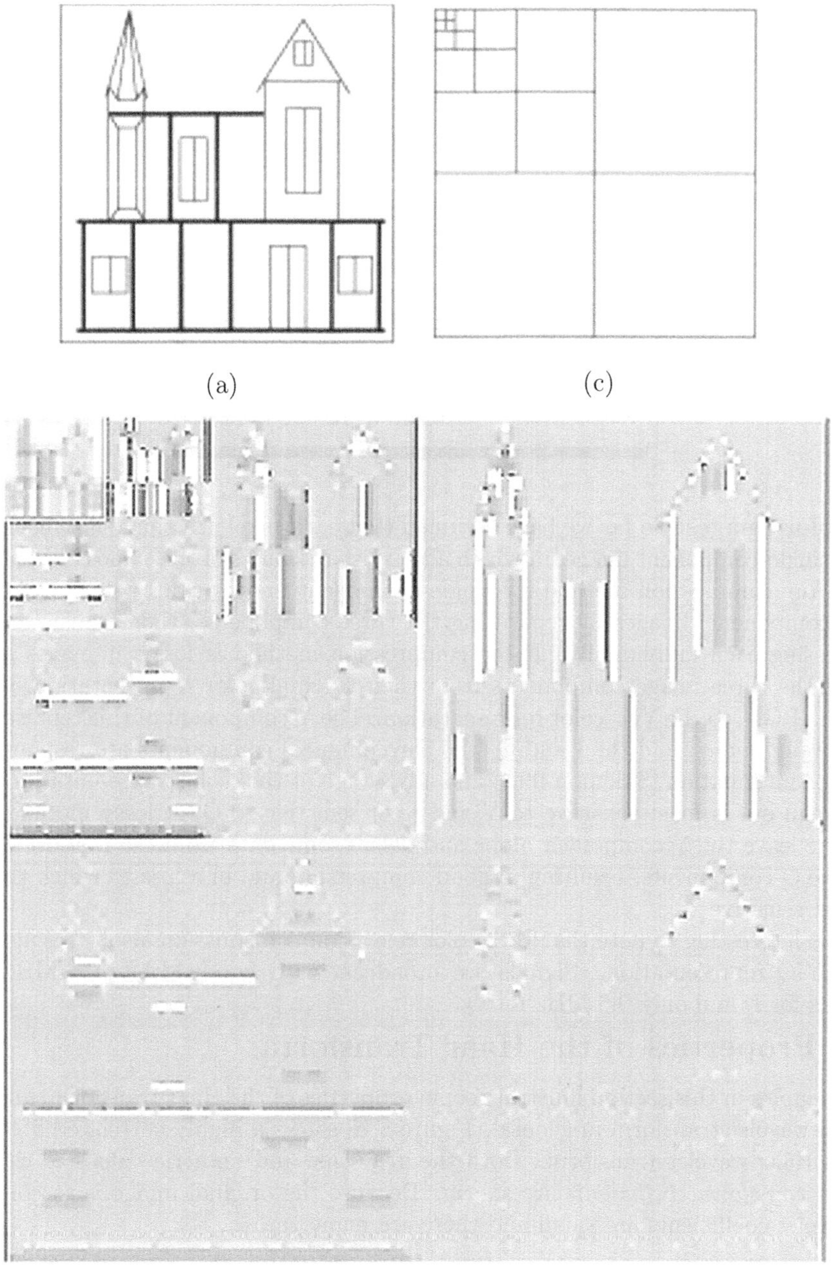

(a) (c)

(b)

Figure 8.17: An Example of the Pyramid Image Wavelet Transform.

Either method, standard or uniform, results in a transformed, although not yet compressed, image that has one average at the top-left corner and smaller numbers, differences, or averages of differences everywhere else. These numbers can be compressed using a combination of methods, such as RLE, move-to-front, and Huffman coding. If lossy compression is acceptable, some of the smallest differences can be quantized or even set to zeros, which creates run lengths of zeros, making the use of RLE even more attractive.

> Whiter foam than thine, O wave,
> Wavelet never wore,
> Stainless wave; and now you lave
> The far and stormless shore —
> Ever — ever — evermore!
> —Abram J. Ryan, *Poems*

Color Images: So far we have assumed that each pixel is a single number (i.e., we have a single-component image, in which all pixels are shades of the same color, normally gray). Any compression method for single-component images can be extended to color (three-component) images by separating the three components, then transforming and compressing each individually. If the compression method is lossy, it makes sense to convert the three image components from their original color representation, which is normally RGB, to the YIQ color representation. The Y component of this representation is called *luminance*, and the I and Q (the chrominance) components are responsible for the color information [Salomon 99]. The advantage of this color representation is that the human eye is most sensitive to Y and least sensitive to Q. A lossy method should therefore leave the Y component alone and delete some data from the I, and more data from the Q components, resulting in good compression and in a loss to which the eye is not that sensitive.

It is interesting to note that U.S. color television transmission also takes advantage of the YIQ representation. Signals are broadcast with bandwidths of 4 MHz for Y, 1.5 MHz for I, and only 0.6 MHz for Q.

8.6.2 Properties of the Haar Transform

The examples in this section illustrate some properties of the Haar transform, and of the discrete wavelet transform in general. Figure 8.18 shows a highly correlated 8×8 image and its Haar wavelet transform. Both the grayscale and numeric values of the pixels and the *transform coefficients* are shown. Because the original image is so correlated, the wavelet coefficients are small and there are many zeros.

◇ **Exercise 8.5:** A glance at Figure 8.18 suggests that the last sentence is wrong. The wavelet transform coefficients listed in the figure are very large compared with the pixel values of the original image. In fact, we know that the top-left Haar transform coefficient should be the average of all the image pixels. Since the pixels of our image have values that are (more or less) uniformly distributed in the interval $[0, 255]$, this average should be around 128, yet the top-left transform coefficient is 1051. Explain this!

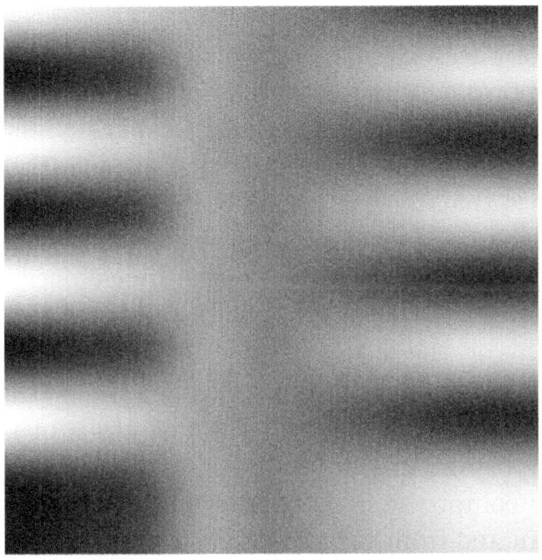

255	224	192	159	127	95	63	32
0	32	64	159	127	159	191	223
255	224	192	159	127	95	63	32
0	32	64	159	127	159	191	223
255	224	192	159	127	95	63	32
0	32	64	159	127	159	191	223
255	224	192	159	127	95	63	32
0	32	64	159	127	159	191	223

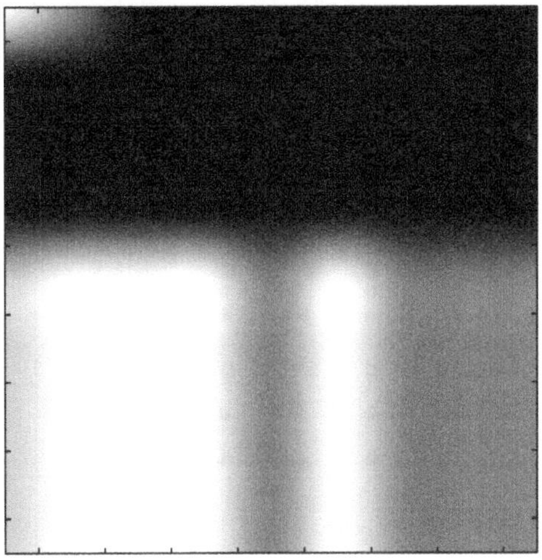

1051	34.0	-44.5	-0.7	-1.0	-62	0	-1.0
0	0.0	0.0	0.0	0.0	0	0	0.0
0	0.0	0.0	0.0	0.0	0	0	0.0
0	0.0	0.0	0.0	0.0	0	0	0.0
48	239.5	112.8	90.2	31.5	64	32	31.5
48	239.5	112.8	90.2	31.5	64	32	31.5
48	239.5	112.8	90.2	31.5	64	32	31.5
48	239.5	112.8	90.2	31.5	64	32	31.5

Figure 8.18: The 8×8 Image Reconstructed in Figure 8.21 and Its Haar Transform.

(a) (b)

Figure 8.19: (a) A 128×128 Image With Activity on the Right. (b) Its Transform.

In a discrete wavelet transform, most of the wavelet coefficients are details (or differences). The details in the lower levels represent the fine details of the image. As we move higher in the subband level, we find details that correspond to coarser image features. Figure 8.19a illustrates this concept. It shows an image that is smooth on the left and has "activity" (i.e., adjacent pixels that tend to be different) on the right. Part (b) shows the wavelet transform of the image. Low levels (corresponding to fine details) have transform coefficients on the right, since this is where the image activity is located. High levels (coarse details) look similar but also have coefficients on the left side, because the image is not completely blank on the left.

The Haar transform is the simplest wavelet transform, but even this simple method illustrates the power of the wavelet transform. It turns out that the low levels of the discrete wavelet transform contain the unimportant image features, so quantizing or discarding these coefficients can lead to lossy compression that is both efficient and of high quality. Often, the image can be reconstructed from very few transform coefficients without any noticeable loss of quality. Figure 8.21a–c shows three reconstructions of the simple 8×8 image of Figure 8.18. They were obtained from only 32, 13, and 5 wavelet coefficients, respectively.

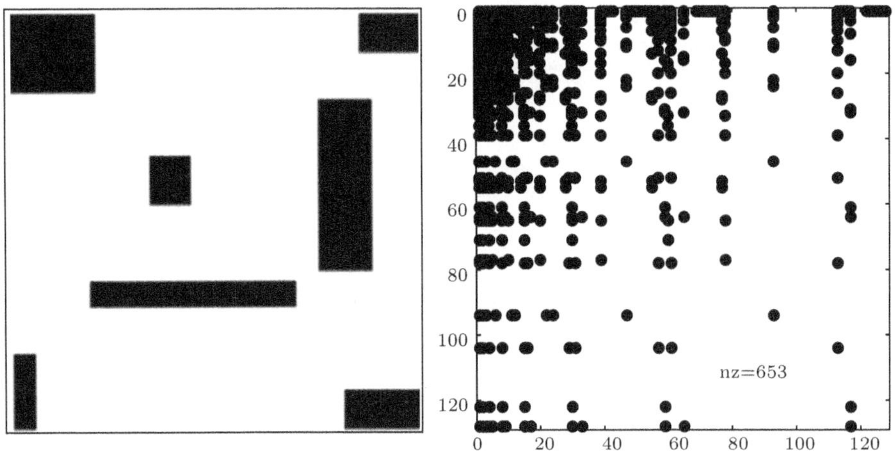

(a) (b)

Figure 8.20: Reconstructing a 128×128 Simple Image From 4% of Its Coefficients.

Figure 8.20 is a similar example. It shows a bi-level image fully reconstructed from just 4% of its transform coefficients (653 coefficients out of 128×128).

Experimenting is the key to understanding these concepts. Proper mathematical software makes it easy to input images and experiment with various features of the discrete wavelet transform. In order to help the interested reader, Figure 8.22 lists a Matlab program that inputs an image, computes its Haar wavelet transform, discards a given percentage of the smallest transform coefficients, then computes the inverse transform to reconstruct the image.

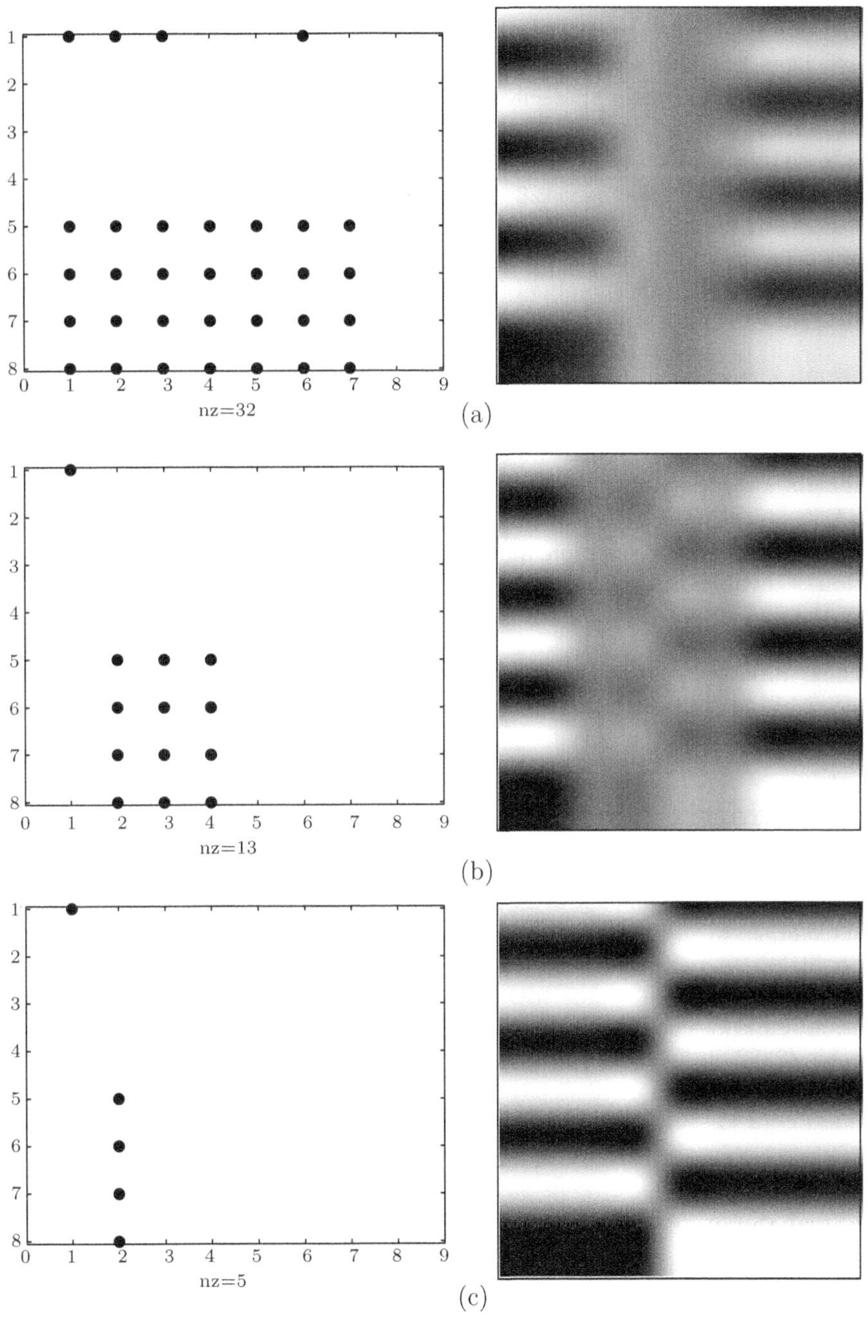

Figure 8.21: Three Lossy Reconstructions of an 8×8 Image.

Lossy wavelet image compression involves the discarding of coefficients, so the concept of *sparseness ratio* is defined to measure the amount of coefficients discarded. Sparseness is defined as the number of nonzero wavelet coefficients divided by the number of coefficients left after some are discarded. The higher the sparseness ratio, the fewer coefficients are left. Higher sparseness ratios lead to better compression but may result in poorly reconstructed images. The sparseness ratio is distantly related to *compression factor*, a compression measure defined in the Introduction.

The line "`filename='lena128'; dim=128;`" contains the image file name and the dimension of the image. The image files used by the authors were in raw form and contained just the grayscale values, each as a single byte. There is no header, and not even the image resolution (number of rows and columns) is included in the file. However, Matlab can read other types of files. The image is assumed to be square, and parameter "`dim`" should be a power of 2. The assignment "`thresh=`" specifies the percentage of transform coefficients to be deleted. This provides an easy way to experiment with lossy wavelet image compression.

File "`harmatt.m`" contains two functions that compute the Haar wavelet coefficients in a matrix form (Section 8.6.3).

(A technical note: A Matlab `m` file may include commands or a function but not both. It may, however, contain more than one function, provided that only the top function is invoked from outside the file. All the other functions must be called from within the file. In our case, function `harmatt(dim)` calls function `individ(n)`.)

◇ **Exercise 8.6:** Use the code of Figure 8.22 (or similar code) to compute the Haar transform of the "Lena" image (Figure 7.55) and reconstruct it three times by discarding more and more detail coefficients.

8.6.3 A Matrix Approach

The principle of the Haar transform is to compute averages and differences. It turns out that this can be done by means of matrix multiplication ([Mulcahy 96] and [Mulcahy 97]). As an example, we look at the top row of the simple 8×8 image of Figure 8.18. Anyone with a little experience with matrices can construct a matrix that when multiplied by this vector creates a vector with four averages and four differences. Matrix A_1 of Equation (8.6) does that and, when multiplied by the top row of pixels of Figure 8.18, generates $(239.5, 175.5, 111.0, 47.5, 15.5, 16.5, 16.0, 15.5)$. Similarly, matrices A_2 and A_3 perform the second and third steps of the transform, respectively. The results are shown in Equation (8.7):

$$A_1=\begin{pmatrix}\frac{1}{2}&\frac{1}{2}&0&0&0&0&0&0\\0&0&\frac{1}{2}&\frac{1}{2}&0&0&0&0\\0&0&0&0&\frac{1}{2}&\frac{1}{2}&0&0\\0&0&0&0&0&0&\frac{1}{2}&\frac{1}{2}\\\frac{1}{2}&-\frac{1}{2}&0&0&0&0&0&0\\0&0&\frac{1}{2}&-\frac{1}{2}&0&0&0&0\\0&0&0&0&\frac{1}{2}&-\frac{1}{2}&0&0\\0&0&0&0&0&0&\frac{1}{2}&-\frac{1}{2}\end{pmatrix},\quad A_1\begin{pmatrix}255\\224\\192\\159\\127\\95\\63\\32\end{pmatrix}=\begin{pmatrix}239.5\\175.5\\111.0\\47.5\\15.5\\16.5\\16.0\\15.5\end{pmatrix},\qquad(8.6)$$

```
clear; % main program
filename='lena128'; dim=128;
fid=fopen(filename,'r');
if fid==-1 disp('file not found')
else img=fread(fid,[dim,dim])'; fclose(fid);
end
thresh=0.0;        % percent of transform coefficients deleted
figure(1), imagesc(img), colormap(gray), axis off, axis square
w=harmatt(dim);  % compute the Haar dim x dim transform matrix
timg=w*img*w';   % forward Haar transform
tsort=sort(abs(timg(:)));
tthresh=tsort(floor(max(thresh*dim*dim,1)));
cim=timg.*(abs(timg) > tthresh);
[i,j,s]=find(cim);
dimg=sparse(i,j,s,dim,dim);
% figure(2) displays the remaining transform coefficients
%figure(2), spy(dimg), colormap(gray), axis square
figure(2), image(dimg), colormap(gray), axis square
cimg=full(w'*sparse(dimg)*w);      % inverse Haar transform
density = nnz(dimg);
disp([num2str(100*thresh) '% of smallest coefficients deleted.'])
disp([num2str(density) ' coefficients remain out of ' ...
 num2str(dim) 'x' num2str(dim) '.'])
figure(3), imagesc(cimg), colormap(gray), axis off, axis square

File harmatt.m with two functions

function x = harmatt(dim)
num=log2(dim);
p = sparse(eye(dim)); q = p;
i=1;
while i<=dim/2;
 q(1:2*i,1:2*i) = sparse(individ(2*i));
 p=p*q; i=2*i;
end
x=sparse(p);

function f=individ(n)
x=[1, 1]/sqrt(2);
y=[1,-1]/sqrt(2);
while min(size(x)) < n/2
 x=[x, zeros(min(size(x)),max(size(x)));...
   zeros(min(size(x)),max(size(x))), x];
end
while min(size(y)) < n/2
 y=[y, zeros(min(size(y)),max(size(y)));...
   zeros(min(size(y)),max(size(y))), y];
end
f=[x;y];
```

Figure 8.22: Matlab Code for the Haar Transform of an Image.

$$A_2=\begin{pmatrix} \frac{1}{2} & \frac{1}{2} & 0 & 0 & 0 & 0 & 0 & 0\\ 0 & 0 & \frac{1}{2} & \frac{1}{2} & 0 & 0 & 0 & 0\\ \frac{1}{2} & -\frac{1}{2} & 0 & 0 & 0 & 0 & 0 & 0\\ 0 & 0 & \frac{1}{2} & -\frac{1}{2} & 0 & 0 & 0 & 0\\ 0 & 0 & 0 & 0 & 1 & 0 & 0 & 0\\ 0 & 0 & 0 & 0 & 0 & 1 & 0 & 0\\ 0 & 0 & 0 & 0 & 0 & 0 & 1 & 0\\ 0 & 0 & 0 & 0 & 0 & 0 & 0 & 1 \end{pmatrix},\quad A_3=\begin{pmatrix} \frac{1}{2} & \frac{1}{2} & 0 & 0 & 0 & 0 & 0 & 0\\ \frac{1}{2} & -\frac{1}{2} & 0 & 0 & 0 & 0 & 0 & 0\\ 0 & 0 & 1 & 0 & 0 & 0 & 0 & 0\\ 0 & 0 & 0 & 1 & 0 & 0 & 0 & 0\\ 0 & 0 & 0 & 0 & 1 & 0 & 0 & 0\\ 0 & 0 & 0 & 0 & 0 & 1 & 0 & 0\\ 0 & 0 & 0 & 0 & 0 & 0 & 1 & 0\\ 0 & 0 & 0 & 0 & 0 & 0 & 0 & 1 \end{pmatrix},$$

$$A_2\begin{pmatrix}239.5\\175.5\\111.0\\47.5\\15.5\\16.5\\16.0\\15.5\end{pmatrix}=\begin{pmatrix}207.5\\79.25\\32.0\\31.75\\15.5\\16.5\\16.0\\15.5\end{pmatrix},\quad A_3\begin{pmatrix}207.5\\79.25\\32.0\\31.75\\15.5\\16.5\\16.0\\15.5\end{pmatrix}=\begin{pmatrix}143.375\\64.125\\32.\\31.75\\15.5\\16.5\\16.\\15.5\end{pmatrix}. \tag{8.7}$$

Instead of calculating averages and differences, all we have to do is construct matrices A_1, A_2, and A_3, multiply them to get $W=A_1A_2A_3$, and apply W to all the columns of the image I by multiplying $W\cdot I$:

$$W\begin{pmatrix}255\\224\\192\\159\\127\\95\\63\\32\end{pmatrix}=\begin{pmatrix} \frac{1}{8} & \frac{1}{8} & \frac{1}{8} & \frac{1}{8} & \frac{1}{8} & \frac{1}{8} & \frac{1}{8} & \frac{1}{8}\\ \frac{1}{8} & \frac{1}{8} & \frac{1}{8} & \frac{1}{8} & \frac{-1}{8} & \frac{-1}{8} & \frac{-1}{8} & \frac{-1}{8}\\ \frac{1}{4} & \frac{1}{4} & \frac{-1}{4} & \frac{-1}{4} & 0 & 0 & 0 & 0\\ 0 & 0 & 0 & 0 & \frac{1}{4} & \frac{1}{4} & \frac{-1}{4} & \frac{-1}{4}\\ \frac{1}{2} & \frac{-1}{2} & 0 & 0 & 0 & 0 & 0 & 0\\ 0 & 0 & \frac{1}{2} & \frac{-1}{2} & 0 & 0 & 0 & 0\\ 0 & 0 & 0 & 0 & \frac{1}{2} & \frac{-1}{2} & 0 & 0\\ 0 & 0 & 0 & 0 & 0 & 0 & \frac{1}{2} & \frac{-1}{2} \end{pmatrix}\begin{pmatrix}255\\224\\192\\159\\127\\95\\63\\32\end{pmatrix}=\begin{pmatrix}143.375\\64.125\\32\\31.75\\15.5\\16.5\\16\\15.5\end{pmatrix}.$$

This, of course, is only half the job. In order to compute the complete transform, we still have to apply W to the rows of the product $W\cdot I$, and we do this by applying it to the columns of the transpose $(W\cdot I)^T$, then transposing the result. Thus, the complete transform is (see line `timg=w*img*w'` in Figure 8.22)

$$I_{\text{tr}}=\left(W(W\cdot I)^T\right)^T=W\cdot I\cdot W^T.$$

The inverse transform is performed by

$$W^{-1}(W^{-1}\cdot I_{\text{tr}}^T)^T=W^{-1}\left(I_{\text{tr}}\cdot(W^{-1})^T\right),$$

and this is where the normalized Haar transform (mentioned on page 752) becomes important. Instead of calculating averages [quantities of the form $(d_i+d_{i+1})/2$] and differences [quantities of the form $(d_i-d_{i+1})/2$], it is better to use the quantities $(d_i+$

$d_{i+1})/\sqrt{2}$ and $(d_i - d_{i+1})/\sqrt{2}$. This results is an *orthonormal* matrix W, and it is well known that the inverse of such a matrix is simply its transpose. Thus, we can write the inverse transform in the simple form $W^T \cdot I_{tr} \cdot W$ [see line `cimg=full(w'*sparse(dimg)*w)` in Figure 8.22].

In between the forward and inverse transforms, some transform coefficients may be quantized or deleted. Alternatively, matrix I_{tr} may be compressed by means of run length encoding and/or Huffman codes.

Function `individ(n)` of Figure 8.22 starts with a 2×2 Haar transform matrix (notice that it uses $\sqrt{2}$ instead of 2) and then uses it to construct as many individual matrices A_i as necessary. Function `harmatt(dim)` combines those individual matrices to form the final Haar matrix for an image of `dim` rows and `dim` columns.

◇ **Exercise 8.7:** Perform the calculation $W \cdot I \cdot W^T$ for the 8×8 image of Figure 8.18.

8.7 Filter Banks

This section uses the matrix approach to the Haar transform to introduce the reader to the idea of *filter banks*. We show how the Haar transform can be interpreted as a bank of two filters, a lowpass and a highpass. We explain the terms "filter," "lowpass," and "highpass" and show how the idea of filter banks leads naturally to the concept of *subband transform*. The Haar transform, of course, is the simplest wavelet transform, so it is used here to illustrate the new concepts. However, using it as a filter bank may not be very efficient. Most practical applications of wavelet filters employ more sophisticated sets of filter coefficients, but they are all based on the concept of filters and filter banks [Strang and Nguyen 96].

> And just like the wavelet that moans on the beach,
> And, sighing, sinks back to the sea,
> So my song—it just touches the rude shores of speech,
> And its music melts back into me.
>
> —Abram J. Ryan, *Poems*

A *filter* is a linear operator defined in terms of its *filter coefficients* $h(0)$, $h(1)$, $h(2), \ldots$. It can be applied to an input vector x to produce an output vector y according to

$$y(n) = \sum_k h(k)x(n-k) = h \star x,$$

where the symbol $\star$ indicates a convolution. Notice that the limits of the sum above have not been stated explicitly. They depend on the sizes of vectors x and h. Since our independent variable is the time t, it is convenient to assume that the inputs (and, consequently, also the outputs) come at all times $t = \ldots, -2, -1, 0, 1, 2, \ldots$. Thus, we use the notation

$$x = (\ldots, a, b, c, d, e, \ldots),$$

where the central value c is the input at time zero $[c = x(0)]$, values d and e are the inputs at times 1 and 2, respectively, and $b = x(-1)$ and $a = x(-2)$. In practice, the inputs are always finite, so the infinite vector x will have only a finite number of nonzero elements.

Deeper insight into the behavior of a linear filter can be gained by considering the simple input $x = (\ldots, 0, 0, 1, 0, 0, \ldots)$. This input is zero at all times except at $t = 0$. It is called a *unit pulse* or a *unit impulse*. Even though the limits of the sum in the convolution have not been specified, it is easy to see that for any n there is only one nonzero term in the sum, so $y(n) = h(n)x(0) = h(n)$. We say that the output $y(n) = h(n)$ at time n is the *response* at time n to the unit impulse $x(0) = 1$. Since the number of filter coefficients $h(i)$ is finite, this filter is a *finite impulse response* or FIR.

Figure 8.23 shows the basic idea of a filter bank. It shows an *analysis bank* consisting of two filters, a lowpass filter H_0 and a highpass filter H_1. The lowpass filter employs convolution to remove the high frequencies from the input signal x and let the low frequencies through. The highpass filter does the opposite. Together, they separate the input into *frequency bands*.

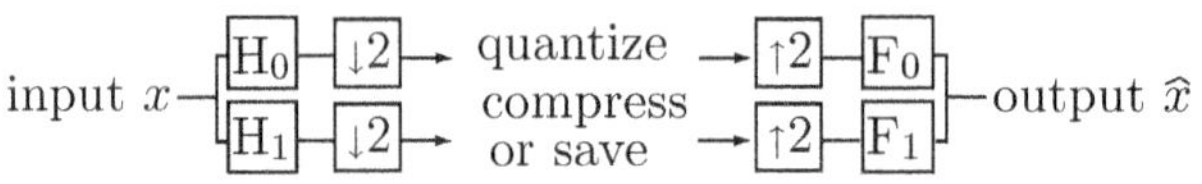

Figure 8.23: A Two-Channel Filter Bank.

The input x can be a one-dimensional signal (a vector of real numbers, which is what we assume in this section) or a two-dimensional signal, an image. The elements $x(n)$ of x are fed into the filters one by one, and each filter computes and outputs one number $y(n)$ in response to $x(n)$. The number of responses is therefore double the number of inputs (because we have two filters); a bad result, since we are interested in data compression. To correct this situation, each filter is followed by a *downsampling* process where the odd-numbered outputs are thrown away. This operation is also called *decimation* and is represented by the boxes marked "↓2". After decimation, the number of outputs from the two filters together equals the number of inputs.

Notice that the filter bank described here, followed by decimation, performs exactly the same calculations as matrix $W = A_1 A_2 A_3$ of Section 8.6.3. Filter banks are just a more general way of looking at the Haar transform (or, in general, at the discrete wavelet transform). We look at this transform as a filtering operation, followed by decimation, and we can then try to find better filters.

The reason for having a bank of filters as opposed to just one filter is that several filters working together, with downsampling, can exhibit behavior that is impossible to obtain with just a single filter. The most important feature of a filter bank is its ability to reconstruct the input from the outputs $H_0 x$ and $H_1 x$, even though each has been decimated.

Downsampling is not time invariant. After downsampling, the output is the even-numbered values $y(0)$, $y(2)$, $y(4)$,..., but if we delay the inputs by one time unit, the new outputs will be $y(-1)$, $y(1)$, $y(3)$,..., and these are different from and independent of the original outputs. These two sequences of signals are two phases of vector y.

The outputs of the analysis bank are called *subband coefficients.* They can be quantized (if lossy compression is acceptable), and they can be compressed by means of RLE, Huffman, arithmetic coding, or any other method. Eventually, they are fed into the *synthesis bank*, where they are first upsampled (by inserting zeros for each odd-numbered coefficient that was thrown away), then passed through the inverse filters F_0 and F_1, and finally combined to form a single output vector $\hat{x}$. The output of each analysis filter (after decimation) is

$$(\downarrow y) = (\ldots, y(-4), y(-2), y(0), y(2), y(4), \ldots).$$

Upsampling inserts zeros for the decimated values, so it converts the output vector above to

$$(\uparrow y) = (\ldots, y(-4), 0, y(-2), 0, y(0), 0, y(2), 0, y(4), 0, \ldots).$$

Downsampling causes loss of data. Upsampling alone cannot compensate for it, because it simply inserts zeros for the missing data. In order to achieve lossless reconstruction of the original signal x, the filters have to be designed such that they compensate for this loss of data. One feature that is commonly used in the design of good filters is *orthogonality.* Figure 8.24 shows a set of orthogonal filters of size 4. The filters of the set are orthogonal because their dot product is zero

$$(a, b, c, d) \cdot (d, -c, b, -a) = 0.$$

Notice how similar H_0 and F_0 are (and also H_1 and F_1). It still remains, of course, to choose actual values for the four *filter coefficients* a, b, c, and d. A full discussion of this is outside the scope of this book, but Section 8.7.1 illustrates some of the methods and rules used in practice to determine the values of various filter coefficients. An example is the Daubechies D4 filter, whose values are listed in Equation (8.11).

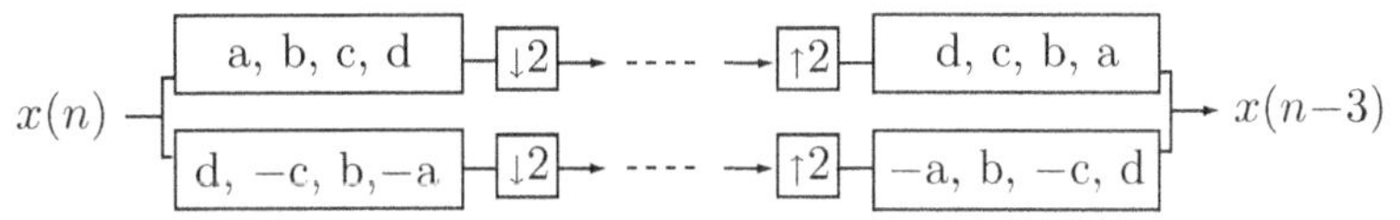

Figure 8.24: An Orthogonal Filter Bank With Four Filter Coefficients.

Simulating the operation of this filter manually shows that the reconstructed input is identical to the original input but lags three time units behind it.

A filter bank can also be *biorthogonal*, a less restricted type of filter. Figure 8.25 shows an example of such a set of filters that can reconstruct a signal exactly. Notice the similarity of H_0 and F_1 and also of H_1 and F_0.

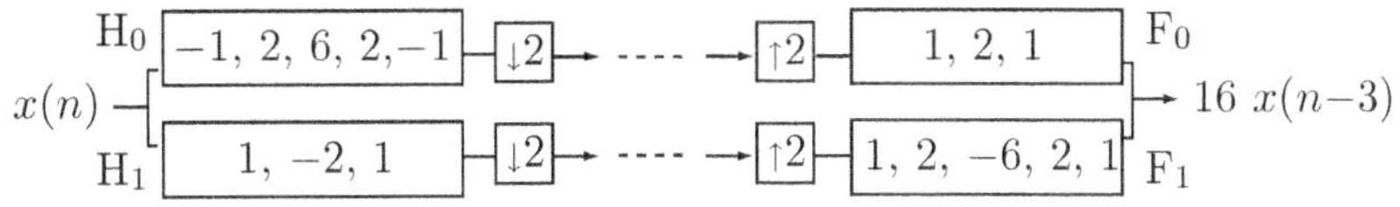

Figure 8.25: A Biorthogonal Filter Bank With Perfect Reconstruction.

We already know, from the discussion in Section 8.6, that the outputs of the lowpass filter H_0 are normally passed through the analysis filter several times, creating shorter and shorter outputs. This recursive process can be illustrated as a tree (Figure 8.26). Since each node of this tree produces half the number of outputs as its predecessor, the tree is called a *logarithmic tree.* Figure 8.26 shows how the scaling function $\phi(t)$ and the wavelet $\psi(t)$ are obtained at the limit of the logarithmic tree. This is the connection between the discrete wavelet transform (using filter banks) and the CWT.

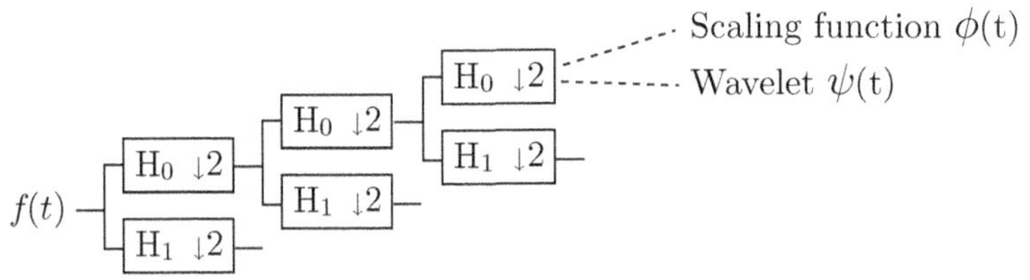

Figure 8.26: Scaling Function and Wavelet as Limits of a Logarithmic Tree.

As we "climb" up the logarithmic tree from level i to the next finer level $i+1$, we compute the new averages from the new, higher-resolution scaling functions $\phi(2^i t-k)$ and the new details from the new wavelets $\psi(2^i t-k)$

$$\begin{array}{rl} \text{signal at level } i \text{ (averages)} \searrow & \\ + & \text{signal at level } i+1. \\ \text{details at level } i \text{ (differences)} \nearrow & \end{array}$$

Each level of the tree corresponds to twice the frequency (or twice the resolution) of the preceding level, which is why the logarithmic tree is also called a *multiresolution tree.* Successive filtering through the tree separates lower and lower frequencies.

People who do quantitative work with sound and music know that two tones at frequencies ω and 2ω sound like the same note and differ only by pitch. The frequency interval between ω and 2ω is divided into 12 subintervals (the so-called *chromatic scale*), but Western music has a tradition of favoring just eight of the twelve tones that result from this division (a *diatonic scale*, made up of seven notes, with the eighth note as the "octave"). This is why the basic frequency interval used in music is traditionally called an *octave.* We therefore say that adjacent levels of the multiresolution tree differ in an octave of frequencies.

In order to understand the meaning of *lowpass* and *highpass* we need to work in the frequency domain, where the convolution of two vectors is replaced by a multiplication of their Fourier transforms. The vector $x(n)$ is in the time domain, so its frequency domain equivalent is its discrete Fourier transform (Equation (8.1))

$$X(\omega) \stackrel{\text{def}}{=} X(e^{i\omega}) = \sum_{-\infty}^{\infty} x(n)e^{-in\omega},$$

which is sometimes written in the z-domain,

$$X(z) = \sum_{-\infty}^{\infty} x(n) z^{-n},$$

where $z \stackrel{\text{def}}{=} e^{i\omega}$. The convolution by h in the time domain now becomes a multiplication by the function $H(\omega) = \sum h(n) e^{-in\omega}$ in the frequency domain, so we can express the output in the frequency domain by

$$Y(e^{i\omega}) = H(e^{i\omega}) X(e^{i\omega}),$$

or, in reduced notation, $Y(\omega) = H(\omega)X(\omega)$, or, in the z-domain, $Y(z) = H(z)X(z)$. When all the inputs are $X(\omega) = 1$, the output at frequency ω is $Y(\omega) = H(\omega)$.

We can now understand the operation of the lowpass Haar filter. It works by averaging two consecutive inputs, so it produces the output

$$y(n) = \frac{1}{2}x(n) + \frac{1}{2}x(n-1). \tag{8.8}$$

This is a convolution with only the two terms $k = 0$ and $k = 1$ in the sum. The filter coefficients are $h(0) = h(1) = 1/2$, and we can call the output a *moving average*, since each $y(n)$ depends on the current input and its predecessor. If the input is the unit impulse $x = (\ldots, 0, 0, 1, 0, 0, \ldots)$, then the output is $y(0) = y(1) = 1/2$, or $y = (\ldots, 0, 0, 1/2, 1/2, 0, \ldots)$. The output values are simply the filter coefficients as we saw earlier.

We can look at this averaging filter as the combination of an identity operator and a delay operator. The output produced by the identity operator equals the current input, while the output produced by the delay is the input one time unit earlier. Thus, we write

$$\text{averaging filter} = \frac{1}{2}(\text{identity}) + \frac{1}{2}(\text{delay}).$$

In matrix notation this can be expressed by

$$\begin{pmatrix} \cdots \\ y(-1) \\ y(0) \\ y(1) \\ \cdots \end{pmatrix} = \begin{pmatrix} \cdots & & & & \\ \frac{1}{2} & \frac{1}{2} & & & \\ & \frac{1}{2} & \frac{1}{2} & & \\ & & \frac{1}{2} & \frac{1}{2} & \\ & & & & \cdots \end{pmatrix} \begin{pmatrix} \cdots \\ x(-1) \\ x(0) \\ x(1) \\ \cdots \end{pmatrix}.$$

The $1/2$ values on the main diagonal are copies of the weight of the identity operator. They all equal to the $h(0)$ Haar filter coefficient. The $1/2$ values on the diagonal below are copies of the weights of the delay operator. They all equal the $h(1)$ Haar filter coefficient. Thus, the matrix is a *constant diagonal* matrix (or a *banded* matrix). A wavelet filter that has a coefficient $h(3)$ would correspond to a matrix where this filter coefficient appears on the second diagonal below the main diagonal. The rule of matrix multiplication produces the familiar convolution

$$y(n) = h(0)x(n) + h(1)x(n-1) + h(2)x(n-2) + \cdots = \sum_k h(k)x(n-k).$$

Notice that the matrix is lower triangular. The upper diagonal, which would naturally correspond to the filter coefficients $h(-1)$, $h(-2)$,..., is zero. All filter coefficients with negative indices must be zero, since such coefficients lead to outputs that precede the inputs in time. In the real world, we are used to a cause preceding its effect, so our finite impulse response filters should also be *causal.*

Summary: A causal FIR filter with $N+1$ filter coefficients $h(0), h(1), \ldots, h(N)$ (a filter with $N+1$ "taps") has $h(i) = 0$ for all negative i and for $i > N$. When expressed in terms of a matrix, the matrix is lower triangular and banded. Such filters are commonly used and are important.

From the Dictionary

Tap (noun).

1. A cylindrical plug or stopper for closing an opening through which liquid is drawn, as in a cask; spigot.
2. A faucet or cock.
3. A connection made at an intermediate point on an electrical circuit or device.
4. An act or instance of wiretapping.

To illustrate the frequency response of a filter we select an input vector of the form

$$x(n) = e^{in\omega} = \cos(n\omega) + i\sin(n\omega), \text{ for } -\infty < n < \infty.$$

This is a complex function whose real and imaginary parts are a cosine and a sine, respectively, both with frequency ω. Recall that the Fourier transform of a pulse contains all the frequencies (Figure 8.2d,e), but the Fourier transform of a sine wave has just one frequency. The smallest frequency is $\omega = 0$, for which the vector becomes $x = (\ldots, 1, 1, 1, 1, 1, \ldots)$. The highest frequency is $\omega = \pi$, where the same vector becomes $x = (\ldots, 1, -1, 1, -1, 1, \ldots)$. The special feature of this input is that the output vector $y(n)$ is a multiple of the input.

For the moving average, the output (filter response) is

$$y(n) = \frac{1}{2}x(n) + \frac{1}{2}x(n-1) = \frac{1}{2}e^{in\omega} + \frac{1}{2}e^{i(n-1)\omega} = \left(\frac{1}{2} + \frac{1}{2}e^{-i\omega}\right)e^{in\omega} = H(\omega)x(n),$$

where $H(\omega) = (\frac{1}{2} + \frac{1}{2}e^{-i\omega})$ is the *frequency response function* of the filter. Since $H(0) = 1/2 + 1/2 = 1$, we see that the input $x = (\ldots, 1, 1, 1, 1, 1, \ldots)$ is transformed to itself. Also, $H(\omega)$ for small values of ω generates output that is very similar to the input. This filter "lets" the low frequencies through, hence the name "lowpass filter." For $\omega = \pi$, the input is $x = (\ldots, 1, -1, 1, -1, 1, \ldots)$ and the output is all zeros (since the average of 1 and -1 is zero). This lowpass filter smooths out the high-frequency regions (the bumps) of the input signal.

Notice that we can write

$$H(\omega) = \left(\cos\frac{\omega}{2}\right) e^{i\omega/2}.$$

When we plot the magnitude $|H(\omega)| = \cos(\omega/2)$ of $H(\omega)$ (Figure 8.27a), it is easy to see that it has a maximum at $\omega = 0$ (the lowest frequency) and two minima at $\omega = \pm\pi$ (the highest frequencies).

The highpass filter uses differences to pick up the high frequencies in the input signal, and reduces or removes the smooth (low frequency) parts. In the case of the Haar transform, the highpass filter computes

$$y(n) = \frac{1}{2}x(n) - \frac{1}{2}x(n-1) = h \star x,$$

where the filter coefficients are $h(0) = 1/2$ and $h(1) = -1/2$, or

$$h = (\ldots, 0, 0, 1/2, -1/2, 0, \ldots).$$

In matrix notation this can be expressed by

$$\begin{pmatrix} \cdots \\ y(-1) \\ y(0) \\ y(1) \\ \cdots \end{pmatrix} = \begin{pmatrix} \cdots & & & & \\ -\frac{1}{2} & \frac{1}{2} & & & \\ & -\frac{1}{2} & \frac{1}{2} & & \\ & & -\frac{1}{2} & \frac{1}{2} & \\ & & & & \cdots \end{pmatrix} \begin{pmatrix} \cdots \\ x(-1) \\ x(0) \\ x(1) \\ \cdots \end{pmatrix}.$$

The main diagonal contains copies of $h(0)$, and the diagonal below contains $h(1)$. Using the identity and delay operator, this can also be written

$$\text{highpass filter} = \frac{1}{2}(\text{identity}) - \frac{1}{2}(\text{delay}).$$

Again selecting input $x(n) = e^{in\omega}$, it is easy to see that the output is

$$y(n) = \frac{1}{2}e^{in\omega} - \frac{1}{2}e^{i(n-1)\omega} = \left(\frac{1}{2} - \frac{1}{2}e^{-i\omega}\right) e^{-i\omega/2} = \sin(\omega/2)\, ie^{-i\omega/2}.$$

This time the highpass response function is

$$H_1(\omega) = \frac{1}{2} - \frac{1}{2}e^{-i\omega} = \frac{1}{2}\left(e^{i\omega/2} - e^{-i\omega/2}\right) e^{-i\omega/2} = \sin(\omega/2)\, e^{-i\omega/2}.$$

The magnitude is $|H_1(\omega)| = |\sin\left(\frac{\omega}{2}\right)|$. It is shown in Figure 8.27b, and it is obvious that it has a minimum for frequency zero and two maxima for large frequencies.

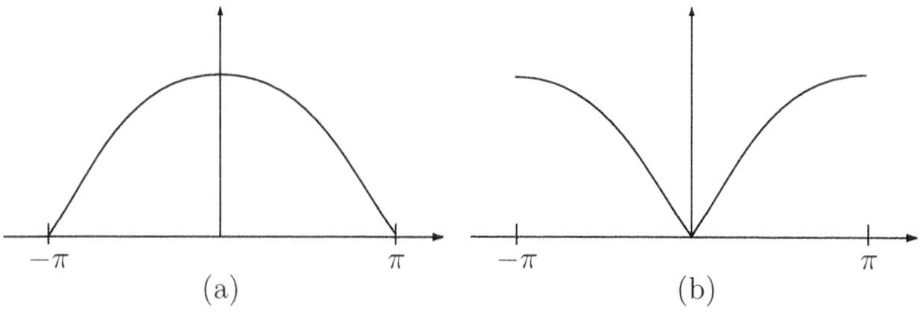

Figure 8.27: Magnitudes of (a) Lowpass and (b) Highpass Filters.

An important property of filter banks is that none of the individual filters are invertible, but the bank as a whole has to be designed such that the input signal could be perfectly reconstructed from the output in spite of the data loss caused by downsampling. It is easy to see, for example, that the constant signal $x = (\ldots, 1, 1, 1, 1, 1, \ldots)$ is transformed by the highpass filter H_1 to an output vector of all zeros. Obviously, there cannot exist an inverse filter H_1^{-1} that will be able to reconstruct the original input from a zero vector. The best that such an inverse transform can do is to use the zero vector to reconstruct another zero vector.

◇ **Exercise 8.8:** Show an example of an input vector x that is transformed by the lowpass filter H_0 to a vector of all zeros.

Summary: The discussion of filter banks in this section should be compared to the discussion of image transforms in Section 7.6. Even though both sections describe transforms, they differ in their approach, since they describe different classes of transforms. Each of the transforms described in Section 7.6 is based on a set of *orthogonal* basis functions (or orthogonal basis images), and is computed as an inner product of the input signal with the basis functions. The result is a set of transform coefficients that are subsequently compressed either losslessly (by RLE or some entropy encoder) or lossily (by quantization followed by entropy coding).

This section deals with *subband transforms* [Simoncelli and Adelson 90], a different type of transform that is computed by taking the *convolution* of the input signal with a set of bandpass filters and decimating the results. Each decimated set of transform coefficients is a subband signal that encodes a specific range of the frequencies of the input. Reconstruction is done by upsampling, followed by computing the inverse transforms, and merging the resulting sets of outputs from the inverse filters.

The main advantage of subband transforms is that they isolate the different frequencies of the input signal, thereby making it possible for the user to precisely control the loss of data in each frequency range. In practice, such a transform decomposes an image into several subbands, corresponding to different image frequencies, and each subband can be quantized differently.

The main disadvantage of this type of transform is the introduction of artifacts, such as aliasing and ringing, into the reconstructed image, because of the downsampling. This is why the Haar transform is unsatisfactory, and most of the research in this field is concerned with finding better sets of filters.

Figure 8.28 illustrates a typical case of a general subband filter bank with N bandpass filters and three stages. Notice how the output of the lowpass filter H_0 of each stage is sent to the next stage for further decomposition, and how the combined output of the synthesis bank of a stage is sent to the top inverse filter of the synthesis bank of the preceding stage.

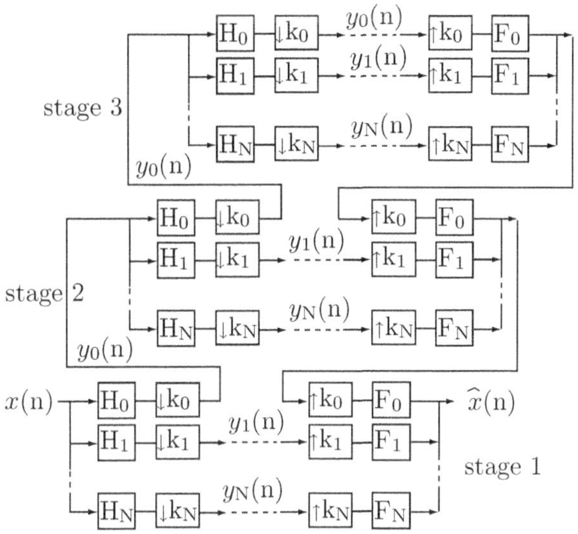

Figure 8.28: A General Filter Bank.

8.7.1 Deriving the Filter Coefficients

After presenting the basic operation of filter banks, the natural question is, How are the filter coefficients derived? A full answer is outside the scope of this book (see, for example, [Akansu and Haddad 92]), but this section provides a glimpse at the rules and methods used to figure out the values of various filter banks.

Given a set of two forward and two inverse N-tap filters H_0 and H_1, and F_0 and F_1 (where N is even), we denote their coefficients by

$$h_0 = \big(h_0(0), h_0(1), \ldots, h_0(N-1)\big), \quad h_1 = \big(h_1(0), h_1(1), \ldots, h_1(N-1)\big),$$
$$f_0 = \big(f_0(0), f_0(1), \ldots, f_0(N-1)\big), \quad f_1 = \big(f_1(0), f_1(1), \ldots, f_1(N-1)\big).$$

The four vectors h_0, h_1, f_0, and f_1 are the *impulse responses* of the four filters. The simplest set of conditions that these quantities have to satisfy is:

1. Normalization: Vector h_0 is normalized (i.e., its length is one unit).
2. Orthogonality: For any integer i that satisfies $1 \leq i < N/2$, the vector formed by the first $2i$ elements of h_0 should be orthogonal to the vector formed by the last $2i$ elements of the same h_0.
3. Vector f_0 is the reverse of h_0.

4. Vector h_1 is a copy of f_0 where the signs of the odd-numbered elements (the first, third, etc.) are reversed. We can express this by saying that h_1 is computed by coordinate multiplication of h_1 and $(-1, 1, -1, 1, \ldots, -1, 1)$.
5. Vector f_1 is a copy of h_0 where the signs of the even-numbered elements (the second, fourth, etc.) are reversed. We can express this by saying that f_1 is computed by coordinate multiplication of h_0 and $(1, -1, 1, -1, \ldots, 1, -1)$.

For two-tap filters, rule 1 implies

$$h_0^2(0) + h_0^2(1) = 1. \tag{8.9}$$

Rule 2 is not applicable because $N = 2$, so $i < N/2$ implies $i < 1$. Rules 3–5 yield

$$f_0 = \big(h_0(1), h_0(0)\big), \quad h_1 = \big(-h_0(1), h_0(0)\big), \quad f_1 = \big(h_0(0), -h_0(1)\big).$$

It all depends on the values of $h_0(0)$ and $h_0(1)$, but the single Equation (8.9) is not enough to determine them. However, it is not difficult to see that the choice $h_0(0) = h_0(1) = 1/\sqrt{2}$ satisfies Equation (8.9).

For four-tap filters, rules 1 and 2 imply

$$h_0^2(0) + h_0^2(1) + h_0^2(2) + h_0^2(3) = 1, \quad h_0(0)h_0(2) + h_0(1)h_0(3) = 0, \tag{8.10}$$

and rules 3–5 yield

$$\begin{aligned} f_0 &= \big(h_0(3), h_0(2), h_0(1), h_0(0)\big), \\ h_1 &= \big(-h_0(3), h_0(2), -h_0(1), h_0(0)\big), \\ f_1 &= \big(h_0(0), -h_0(1), h_0(2), -h_0(3)\big). \end{aligned}$$

Again, Equation (8.10) is not enough to determine four unknowns, and other considerations (plus mathematical intuition) are needed to derive the four values. They are listed in Equation (8.11) (this is the Daubechies D4 filter).

◇ **Exercise 8.9:** Write the five conditions above for an eight-tap filter.

Determining the N filter coefficients for each of the four filters H_0, H_1, F_0, and F_1 depends on $h_0(0)$ through $h_0(N-1)$, so it requires N equations. However, in each of the cases above, rules 1 and 2 supply only $N/2$ equations. Other conditions have to be imposed and satisfied before the N quantities $h_0(0)$ through $h_0(N-1)$ can be determined. Here are some examples:

Lowpass H_0 filter: We want H_0 to be a lowpass filter, so it makes sense to require that the frequency response $H_0(\omega)$ be zero for the highest frequency $\omega = \pi$.

Minimum phase filter: This condition requires the zeros of the complex function $H_0(z)$ to lie on or inside the unit circle in the complex plane.

Controlled collinearity: The linearity of the phase response can be controlled by requiring that the sum

$$\sum_i \big(h_0(i) - h_0(N-1-i)\big)^2$$

be a minimum.

Other conditions are discussed in [Akansu and Haddad 92].

8.8 The DWT

Information that is produced and analyzed in real-life situations is discrete. It comes in the form of numbers, rather than a continuous function. This is why the discrete rather than the continuous wavelet transform is the one used in practice ([Daubechies 88], [DeVore et al. 92], and [Vetterli and Kovacevic 95]). Recall that the CWT [Equation (8.3)] is the integral of the product $f(t)\psi^*(\frac{t-b}{a})$, where a, the scale factor, and b, the time shift, can be any real numbers. The corresponding calculation for the discrete case (the DWT) involves a *convolution*, but experience shows that the quality of this type of transform depends heavily on two factors, the choice of scale factors and time shifts, and the choice of wavelet.

In practice, the DWT is computed with scale factors that are negative powers of 2 and time shifts that are nonnegative powers of 2. Figure 8.29 shows the so-called *dyadic lattice* that illustrates this particular choice. The wavelets used are those that generate orthonormal (or biorthogonal) wavelet bases.

The main thrust in wavelet research has therefore been the search for wavelet families that form orthogonal bases. Of those wavelets, the preferred ones are those that have compact support, because they allow for DWT computations with *finite impulse response* (FIR) filters.

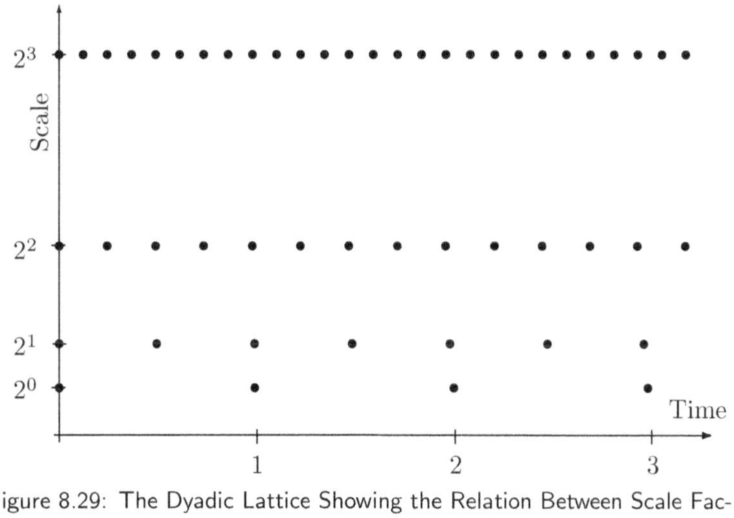

Figure 8.29: The Dyadic Lattice Showing the Relation Between Scale Factors and Time.

The simplest way to describe the discrete wavelet transform is by means of matrix multiplication, along the lines developed in Section 8.6.3. The Haar transform depends on two *filter coefficients* c_0 and c_1, both with a value of $1/\sqrt{2} \approx 0.7071$. The smallest transform matrix that can be constructed in this case is $\binom{1\ \ 1}{1\ -1}/\sqrt{2}$. It is a 2×2 matrix, and it generates two transform coefficients, an average and a difference. (Notice that these are not exactly an average and a difference, because $\sqrt{2}$ is used instead of 2. Better names for them are *coarse detail* and *fine detail*, respectively.) In general, the DWT can use any set of wavelet filters, but it is computed in the same way regardless of the particular filter used.

We start with one of the most popular wavelets, the Daubechies D4. As its name implies, it is based on four filter coefficients c_0, c_1, c_2, and c_3, whose values are listed in Equation (8.11). The transform matrix W is [compare with matrix A_1, Equation (8.6)]

$$W=\begin{pmatrix}
c_0 & c_1 & c_2 & c_3 & 0 & 0 & \ldots & 0\\
c_3 & -c_2 & c_1 & -c_0 & 0 & 0 & \ldots & 0\\
0 & 0 & c_0 & c_1 & c_2 & c_3 & \ldots & 0\\
0 & 0 & c_3 & -c_2 & c_1 & -c_0 & \ldots & 0\\
\vdots & \vdots & & & & \ddots & & \\
0 & 0 & \ldots & 0 & c_0 & c_1 & c_2 & c_3\\
0 & 0 & \ldots & 0 & c_3 & -c_2 & c_1 & -c_0\\
c_2 & c_3 & 0 & \ldots & 0 & 0 & c_0 & c_1\\
c_1 & -c_0 & 0 & \ldots & 0 & 0 & c_3 & -c_2
\end{pmatrix}.$$

When this matrix is applied to a column vector of data items $(x_1, x_2, \ldots, x_n)$, its top row generates the weighted sum $s_1 = c_0x_1 + c_1x_2 + c_2x_3 + c_3x_4$, its third row generates the weighted sum $s_2 = c_0x_3 + c_1x_4 + c_2x_5 + c_3x_6$, and the other odd-numbered rows generate similar weighted sums s_i. Such sums are *convolutions* of the data vector x_i with the four filter coefficients. In the language of wavelets, each of them is called a *smooth coefficient*, and together they are called an H smoothing filter.

In a similar way, the second row of the matrix generates the quantity $d_1 = c_3x_1 - c_2x_2 + c_1x_3 - c_0x_4$, and the other even-numbered rows generate similar convolutions. Each d_i is called a *detail coefficient*, and together they are called a G filter. G is not a smoothing filter. In fact, the filter coefficients are chosen such that the G filter generates small values when the data items x_i are correlated. Together, H and G are called *quadrature mirror filters* (QMF).

The discrete wavelet transform of an image can therefore be viewed as passing the original image through a QMF that consists of a pair of lowpass (H) and highpass (G) filters.

If W is an $n\times n$ matrix, it generates $n/2$ smooth coefficients s_i and $n/2$ detail coefficients d_i. The transposed matrix is

$$W^T=\begin{pmatrix}
c_0 & c_3 & 0 & 0 & \ldots & & & & c_2 & c_1\\
c_1 & -c_2 & 0 & 0 & \ldots & & & & c_3 & -c_0\\
c_2 & c_1 & c_0 & c_3 & \ldots & & & & 0 & 0\\
c_3 & -c_0 & c_1 & -c_2 & \ldots & & & & 0 & 0\\
 & & & & \ddots & & & & & \\
 & & & & & c_2 & c_1 & c_0 & c_3 & 0 & 0\\
 & & & & & c_3 & -c_0 & c_1 & -c_2 & 0 & 0\\
 & & & & & & & c_2 & c_1 & c_0 & c_3\\
 & & & & & & & c_3 & -c_0 & c_1 & -c_2
\end{pmatrix}.$$

It can be shown that in order for W to be orthonormal, the four coefficients have to satisfy the two relations $c_0^2 + c_1^2 + c_2^2 + c_3^2 = 1$ and $c_2c_0 + c_3c_1 = 0$. The other two equations used to calculate the four filter coefficients are $c_3 - c_2 + c_1 - c_0 = 0$ and $0c_3 - 1c_2 + 2c_1 - 3c_0 = 0$. They represent the vanishing of the first two moments of the sequence $(c_3, -c_2, c_1, -c_0)$. The solutions are

$$\begin{aligned} c_0 &= (1+\sqrt{3})/(4\sqrt{2}) \approx 0.48296, \quad & c_1 &= (3+\sqrt{3})/(4\sqrt{2}) \approx 0.8365,\\ c_2 &= (3-\sqrt{3})/(4\sqrt{2}) \approx 0.2241, \quad & c_3 &= (1-\sqrt{3})/(4\sqrt{2}) \approx -0.1294. \end{aligned} \tag{8.11}$$

Using a transform matrix W is conceptually simple, but not very practical, since W should be of the same size as the image, which can be large. However, a look at W shows that it is very regular, so there is really no need to construct the full matrix. It is enough to have just the top row of W. In fact, it is enough to have just an array with the filter coefficients. Figure 8.30 lists Matlab code that performs this calculation. Function `fwt1(dat,coarse,filter)` takes a row vector `dat` of 2^n data items, and another array, `filter`, with filter coefficients. It then calculates the first `coarse` levels of the discrete wavelet transform.

◇ **Exercise 8.10:** Write similar code for the inverse one-dimensional discrete wavelet transform.

Plotting Functions: Wavelets are being used in many fields and have many applications, but the simple test of Figure 8.30 suggests another application, namely, plotting functions. Any graphics program or graphics software package has to include a routine to plot functions. It works by calculating the function at certain points and connecting the points with straight segments. In regions where the function has small curvature (it resembles a straight line) only few points are needed, whereas in areas where the function has large curvature (it changes direction rapidly) more points are required. An ideal plotting routine should therefore be adaptive. It should select the points depending on the curvature of the function.

The curvature, however, may not be easy to compute (it is essentially given by the second derivative of the function) which is why many plotting routines use instead the angle between consecutive segments. Figure 8.31 shows how a typical plotting routine works. It starts with a fixed number (say, 50) of points. This implies 49 straight segments connecting them. Before any of the segments is actually plotted, the routine measures the angles between consecutive segments. If an angle at point $\mathbf{P}_i$ is extreme (close to zero or close to 360°, as it is around points 4 and 10 in the figure), then more points are calculated between points $\mathbf{P}_{i-1}$ and $\mathbf{P}_{i+1}$; otherwise (if the angle is closer to 180°, as, for example, around points 5 and 9 in the figure), $\mathbf{P}_i$ is considered the only point necessary in that region.

Better and faster results may be obtained using a discrete wavelet transform. The function is evaluated at n points (where n, a parameter, is large), and the values are collected in a vector v. A discrete wavelet transform of v is then calculated, to produce n transform coefficients. The next step is to discard m of the smallest coefficients (where m is another parameter). We know, from the previous discussion, that the smallest coefficients represent small details of the function, so discarding them leaves the important details practically untouched. The inverse transform is then performed on the remaining $n - m$ transform coefficients, resulting in $n - m$ new points that are

```
function wc1=fwt1(dat,coarse,filter)
%  The 1D Forward Wavelet Transform
%  dat must be a 1D row vector of size 2^n,
%  coarse is the coarsest level of the transform
%  (note that coarse should be <<n)
%  filter is an orthonormal quadrature mirror filter
%  whose length should be <2^(coarse+1)
n=length(dat); j=log2(n); wc1=zeros(1,n);
beta=dat;
for i=j-1:-1:coarse
  alfa=HiPass(beta,filter);
  wc1((2^(i)+1):(2^(i+1)))=alfa;
  beta=LoPass(beta,filter) ;
end
wc1(1:(2^coarse))=beta;

function d=HiPass(dt,filter) % highpass downsampling
d=iconv(mirror(filter),lshift(dt));
% iconv is matlab convolution tool
n=length(d);
d=d(1:2:(n-1));

function d=LoPass(dt,filter) % lowpass downsampling
d=aconv(filter,dt);
% aconv is matlab convolution tool with time-
% reversal of filter
n=length(d);
d=d(1:2:(n-1));

function sgn=mirror(filt)
% return filter coefficients with alternating signs
sgn=-((-1).^(1:length(filt))).*filt;
```

A simple test of fwt1 is

```
n=16; t=(1:n)./n;
dat=sin(2*pi*t)
filt=[0.4830 0.8365 0.2241 -0.1294];
wc=fwt1(dat,1,filt)
```

which outputs

```
dat=
0.3827  0.7071  0.9239 1.0000 0.9239 0.7071 0.3827 0
-0.3827 -0.7071 -0.9239 -1.0000 -0.9239 -0.7071 -0.3827 0
wc=
1.1365 -1.1365 -1.5685 1.5685 -0.2271 -0.4239 0.2271 0.4239
-0.0281 -0.0818 -0.0876 -0.0421 0.0281 0.0818 0.0876 0.0421
```

Figure 8.30: Code for the One-Dimensional Forward Discrete Wavelet Transform.

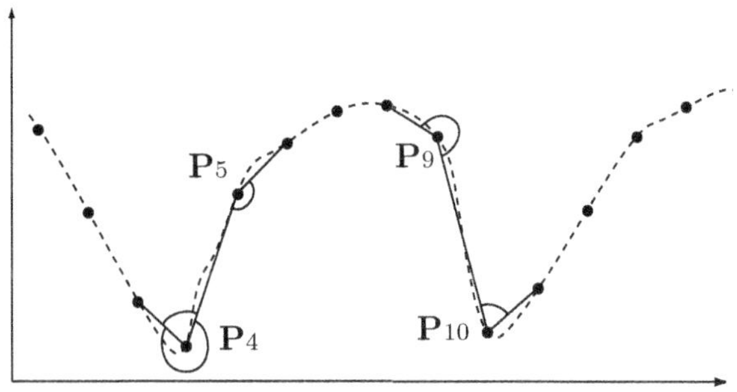

Figure 8.31: Using Angles Between Segments to Add More Points.

then connected with straight segments. The larger m, the fewer segments necessary, but the worse the fit.

Readers who take the trouble to read and understand functions `fwt1` and `iwt1` (Figures 8.30 and 8.32) may be interested in their two-dimensional equivalents, functions `fwt2` and `iwt2`, which are listed in Figures 8.33 and 8.34, respectively, with a simple test routine.

Table 8.35 lists the filter coefficients for some of the most common wavelets currently in use. Notice that each of those sets should still be normalized. Following are the main features of each set:

- The Daubechies family of filters maximize the smoothness of the father wavelet (the scaling function) by maximizing the rate of decay of its Fourier transform.

- The Haar wavelet can be considered the Daubechies filter of order 2. It is the oldest filter. It is simple to work with, but it does not produce best results, since it is not continuous.

- The Beylkin filter places the roots of the frequency response function close to the Nyquist frequency (page 741) on the real axis.

- The Coifman filter (or "Coiflet") of order p (where p is a positive integer) gives both the mother and father wavelets $2p$ zero moments.

- Symmetric filters (symmlets) are the most symmetric compactly supported wavelets with a maximum number of zero moments.

- The Vaidyanathan filter does not satisfy any conditions on the moments but produces exact reconstruction. This filter is especially useful in speech compression.

Figures 8.36 and 8.37 are diagrams of some of those wavelets.

The Daubechies family of wavelets is a set of orthonormal, compactly supported functions where consecutive members are increasingly smoother. Some of them are shown in Figure 8.37. The term *compact support* means that these functions are zero (exactly zero, not just very small) outside a finite interval.

The Daubechies D4 wavelet is based on four coefficients, shown in Equation (8.11). The D6 wavelet is, similarly, based on six coefficients. They are determined by solving six equations, three of which represent orthogonality requirements and the other

```
function dat=iwt1(wc,coarse,filter)
% Inverse Discrete Wavelet Transform
dat=wc(1:2^coarse);
n=length(wc); j=log2(n);
for i=coarse:j-1
 dat=ILoPass(dat,filter)+ ...
  IHiPass(wc((2^(i)+1):(2^(i+1))),filter);
end

function f=ILoPass(dt,filter)
f=iconv(filter,AltrntZro(dt));

function f=IHiPass(dt,filter)
f=aconv(mirror(filter),rshift(AltrntZro(dt)));

function sgn=mirror(filt)
% return filter coefficients with alternating signs
sgn=-((-1).^(1:length(filt))).*filt;

function f=AltrntZro(dt)
% returns a vector of length 2*n with zeros
% placed between consecutive values
n =length(dt)*2; f =zeros(1,n);
f(1:2:(n-1))=dt;
```

A simple test of `iwt1` is

```
n=16; t=(1:n)./n;
dat=sin(2*pi*t)
filt=[0.4830 0.8365 0.2241 -0.1294];
wc=fwt1(dat,1,filt)
rec=iwt1(wc,1,filt)
```

Figure 8.32: Code for the One-Dimensional Inverse Discrete Wavelet Transform.

```
function wc=fwt2(dat,coarse,filter)
% The 2D Forward Wavelet Transform
% dat must be a 2D matrix of size (2^n:2^n),
% "coarse" is the coarsest level of the transform
% (note that coarse should be <<n)
% filter is an orthonormal qmf of length<2^(coarse+1)
q=size(dat); n = q(1); j=log2(n);
if q(1)~=q(2), disp('Nonsquare image!'), end;
wc = dat; nc = n;
for i=j-1:-1:coarse,
 top = (nc/2+1):nc; bot = 1:(nc/2);
 for ic=1:nc,
  row = wc(ic,1:nc);
  wc(ic,bot)=LoPass(row,filter);
  wc(ic,top)=HiPass(row,filter);
 end
 for ir=1:nc,
  row = wc(1:nc,ir)';
  wc(top,ir)=HiPass(row,filter)';
  wc(bot,ir)=LoPass(row,filter)';
 end
nc = nc/2;
end

function d=HiPass(dt,filter) % highpass downsampling
d=iconv(mirror(filter),lshift(dt));
% iconv is matlab convolution tool
n=length(d);
d=d(1:2:(n-1));

function d=LoPass(dt,filter) % lowpass downsampling
d=aconv(filter,dt);
% aconv is matlab convolution tool with time-
% reversal of filter
n=length(d);
d=d(1:2:(n-1));

function sgn=mirror(filt)
% return filter coefficients with alternating signs
sgn=-((-1).^(1:length(filt))).*filt;
```

A simple test of fwt2 and iwt2 is

```
filename='house128'; dim=128;
fid=fopen(filename,'r');
if fid==-1 disp('file not found')
 else img=fread(fid,[dim,dim])'; fclose(fid);
end
filt=[0.4830 0.8365 0.2241 -0.1294];
fwim=fwt2(img,4,filt);
figure(1), imagesc(fwim), axis off, axis square
rec=iwt2(fwim,4,filt);
figure(2), imagesc(rec), axis off, axis square
```

Figure 8.33: Code for the Two-Dimensional Forward Discrete Wavelet Transform.

```
function dat=iwt2(wc,coarse,filter)
% Inverse Discrete 2D Wavelet Transform
n=length(wc); j=log2(n);
dat=wc;
nc=2^(coarse+1);
for i=coarse:j-1,
 top=(nc/2+1):nc; bot=1:(nc/2); all=1:nc;
 for ic=1:nc,
  dat(all,ic)=ILoPass(dat(bot,ic)',filter)'  ...
   +IHiPass(dat(top,ic)',filter)';
 end % ic
 for ir=1:nc,
  dat(ir,all)=ILoPass(dat(ir,bot),filter)  ...
   +IHiPass(dat(ir,top),filter);
 end % ir
nc=2*nc;
end % i

function f=ILoPass(dt,filter)
f=iconv(filter,AltrntZro(dt));

function f=IHiPass(dt,filter)
f=aconv(mirror(filter),rshift(AltrntZro(dt)));

function sgn=mirror(filt)
% return filter coefficients with alternating signs
sgn=-((-1).^(1:length(filt))).*filt;

function f=AltrntZro(dt)
% returns a vector of length 2*n with zeros
% placed between consecutive values
n =length(dt)*2; f =zeros(1,n);
f(1:2:(n-1))=dt;
```

A simple test of fwt2 and iwt2 is

```
filename='house128'; dim=128;
fid=fopen(filename,'r');
if fid==-1 disp('file not found')
 else img=fread(fid,[dim,dim])'; fclose(fid);
end
filt=[0.4830 0.8365 0.2241 -0.1294];
fwim=fwt2(img,4,filt);
figure(1), imagesc(fwim), axis off, axis square
rec=iwt2(fwim,4,filt);
figure(2), imagesc(rec), axis off, axis square
```

Figure 8.34: Code for the Two-Dimensional Inverse Discrete Wavelet Transform.

.099305765374	.424215360813	.699825214057	.449718251149	-.110927598348	-.264497231446
.026900308804	.155538731877	-.017520746267	-.088543630623	.019679866044	.042916387274
-.017460408696	-.014365807969	.010040411845	.001484234782	-.002736031626	.000640485329
		Beylkin			
.038580777748	-.126969125396	-.077161555496	.607491641386	.745687558934	.226584265197
		Coifman 1-tap			
.016387336463	-.041464936782	-.067372554722	.386110066823	.812723635450	.417005184424
-.076488599078	-.059434418646	.023680171947	.005611434819	-.001823208871	-.000720549445
		Coifman 2-tap			
-.003793512864	.007782596426	.023452696142	-.065771911281	-.061123390003	.405176902410
.793777222626	.428483476378	-.071799821619	-.082301927106	.034555027573	.015880544864
-.009007976137	-.002574517688	.001117518771	.000466216960	-.000070983303	-.000034599773
		Coifman 3-tap			
.000892313668	-.001629492013	-.007346166328	.016068943964	.026682300156	-.081266699680
-.056077313316	.415308407030	.782238930920	.434386056491	-.066627474263	-.096220442034
.039334427123	.025082261845	-.015211731527	-.005658286686	.003751436157	.001266561929
-.000589020757	-.000259974552	.000062339034	.000031229876	-.000003259680	-.000001784985
		Coifman 4-tap			
-.000212080863	.000358589677	.002178236305	-.004159358782	-.010131117538	.023408156762
.028168029062	-.091920010549	-.052043163216	.421566206729	.774289603740	.437991626228
-.062035963906	-.105574208706	.041289208741	.032683574283	-.019761779012	-.009164231153
.006764185419	.002433373209	-.001662863769	-.000638131296	.000302259520	.000140541149
-.000041340484	-.000021315014	.000003734597	.000002063806	-.000000167408	-.000000095158
		Coifman 5-tap			
.482962913145	.836516303738	.224143868042	-.129409522551		
		Daubechies 4-tap			
.332670552950	.806891509311	.459877502118	-.135011020010	-.085441273882	.035226291882
		Daubechies 6-tap			
.230377813309	.714846570553	.630880767930	-.027983769417	-.187034811719	.030841381836
.032883011667	-.010597401785				
		Daubechies 8-tap			
.160102397974	.603829269797	.724308528438	.138428145901	-.242294887066	-.032244869585
.077571493840	-.006241490213	-.012580751999	.003335725285		
		Daubechies 10-tap			
.111540743350	.494623890398	.751133908021	.315250351709	-.226264693965	-.129766867567
.097501605587	.027522865530	-.031582039317	.000553842201	.004777257511	-.001077301085
		Daubechies 12-tap			
.077852054085	.396539319482	.729132090846	.469782287405	-.143906003929	-.224036184994
.071309219267	.080612609151	-.038029936935	-.016574541631	.012550998556	.000429577973
-.001801640704	.000353713800				
		Daubechies 14-tap			
.054415842243	.312871590914	.675630736297	.585354683654	-.015829105256	-.284015542962
.000472484574	.128747426620	-.017369301002	-.044088253931	.013981027917	.008746094047
-.004870352993	-.000391740373	.000675449406	-.000117476784		
		Daubechies 16-tap			
.038077947364	.243834674613	.604823123690	.657288078051	.133197385825	-.293273783279
-.096840783223	.148540749338	.030725681479	-.067632829061	.000250947115	.022361662124
-.004723204758	-.004281503682	.001847646883	.000230385764	-.000251963189	.000039347320
		Daubechies 18-tap			
.026670057901	.188176800078	.527201188932	.688459039454	.281172343661	-.249846424327
-.195946274377	.127369340336	.093057364604	-.071394147166	-.029457536822	.033212674059
.003606553567	-.010733175483	.001395351747	.001992405295	-.000685856695	-.000116466855
.000093588670	-.000013264203				
		Daubechies 20-tap			

Table 8.35: Filter Coefficients for Some Common Wavelets (Continues).

-.107148901418	-.041910965125	.703739068656	1.136658243408	.421234534204	-.140317624179
-.017824701442	.045570345896				
Symmlet 4-tap					
.038654795955	.041746864422	-.055344186117	.281990696854	1.023052966894	.896581648380
.023478923136	-.247951362613	-.029842499869	.027632152958		
Symmlet 5-tap					
.021784700327	.004936612372	-.166863215412	-.068323121587	.694457972958	1.113892783926
.477904371333	-.102724969862	-.029783751299	.063250562660	.002499922093	-.011031867509
Symmlet 6-tap					
.003792658534	-.001481225915	-.017870431651	.043155452582	.096014767936	-.070078291222
.024665659489	.758162601964	1.085782709814	.408183939725	-.198056706807	-.152463871896
.005671342686	.014521394762				
Symmlet 7-tap					
.002672793393	-.000428394300	-.021145686528	.005386388754	.069490465911	-.038493521263
-.073462508761	.515398670374	1.099106630537	.680745347190	-.086653615406	-.202648655286
.010758611751	.044823623042	-.000766690896	-.004783458512		
Symmlet 8-tap					
.001512487309	-.000669141509	-.014515578553	.012528896242	.087791251554	-.025786445930
-.270893783503	.049882830959	.873048407349	1.015259790832	.337658923602	-.077172161097
.000825140929	.042744433602	-.016303351226	-.018769396836	.000876502539	.001981193736
Symmlet 9-tap					
.001089170447	.000135245020	-.012220642630	-.002072363923	.064950924579	.016418869426
-.225558972234	-.100240215031	.667071338154	1.088251530500	.542813011213	-.050256540092
-.045240772218	.070703567550	.008152816799	-.028786231926	-.001137535314	.006495728375
.000080661204	-.000649589896				
Symmlet 10-tap					
-.000062906118	.000343631905	-.000453956620	-.000944897136	.002843834547	.000708137504
-.008839103409	.003153847056	.019687215010	-.014853448005	-.035470398607	.038742619293
.055892523691	-.077709750902	-.083928884366	.131971661417	.135084227129	-.194450471766
-.263494802488	.201612161775	.635601059872	.572797793211	.250184129505	.045799334111
Vaidyanathan					

Table 8.35: Continued.

three represent the vanishing of the first three moments. The result is shown in Equation (8.12):

$$
\begin{aligned}
c_0 &= (1+\sqrt{10}+\sqrt{5+2\sqrt{10}})/(16\sqrt{2}) \approx .3326,\\
c_1 &= (5+\sqrt{10}+3\sqrt{5+2\sqrt{10}})/(16\sqrt{2}) \approx .8068,\\
c_2 &= (10-2\sqrt{10}+2\sqrt{5+2\sqrt{10}})/(16\sqrt{2}) \approx .4598,\\
c_3 &= (10-2\sqrt{10}-2\sqrt{5+2\sqrt{10}})/(16\sqrt{2}) \approx -.1350,\\
c_4 &= (5+\sqrt{10}-3\sqrt{5+2\sqrt{10}})/(16\sqrt{2}) \approx -.0854,\\
c_5 &= (1+\sqrt{10}-\sqrt{5+2\sqrt{10}})/(16\sqrt{2}) \approx .0352.
\end{aligned}
\tag{8.12}
$$

Each member of this family has two more coefficients than its predecessor and is smoother. The derivation of these functions is outside the scope of this book and can be found in [Daubechies 88]. They are derived recursively, do not have a closed form, and are non-differentiable at infinitely many points. Truly unusual functions!

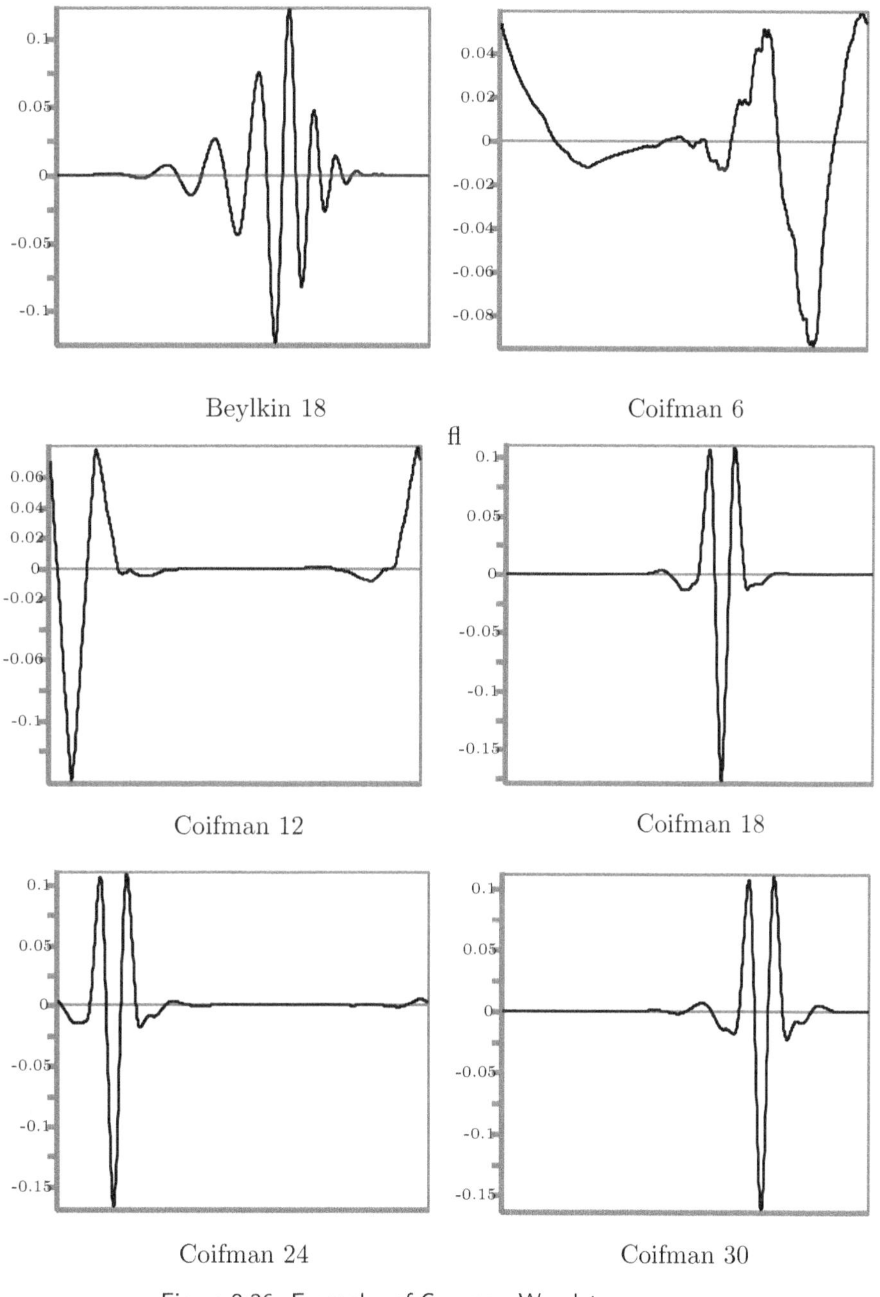

Figure 8.36: Examples of Common Wavelets.

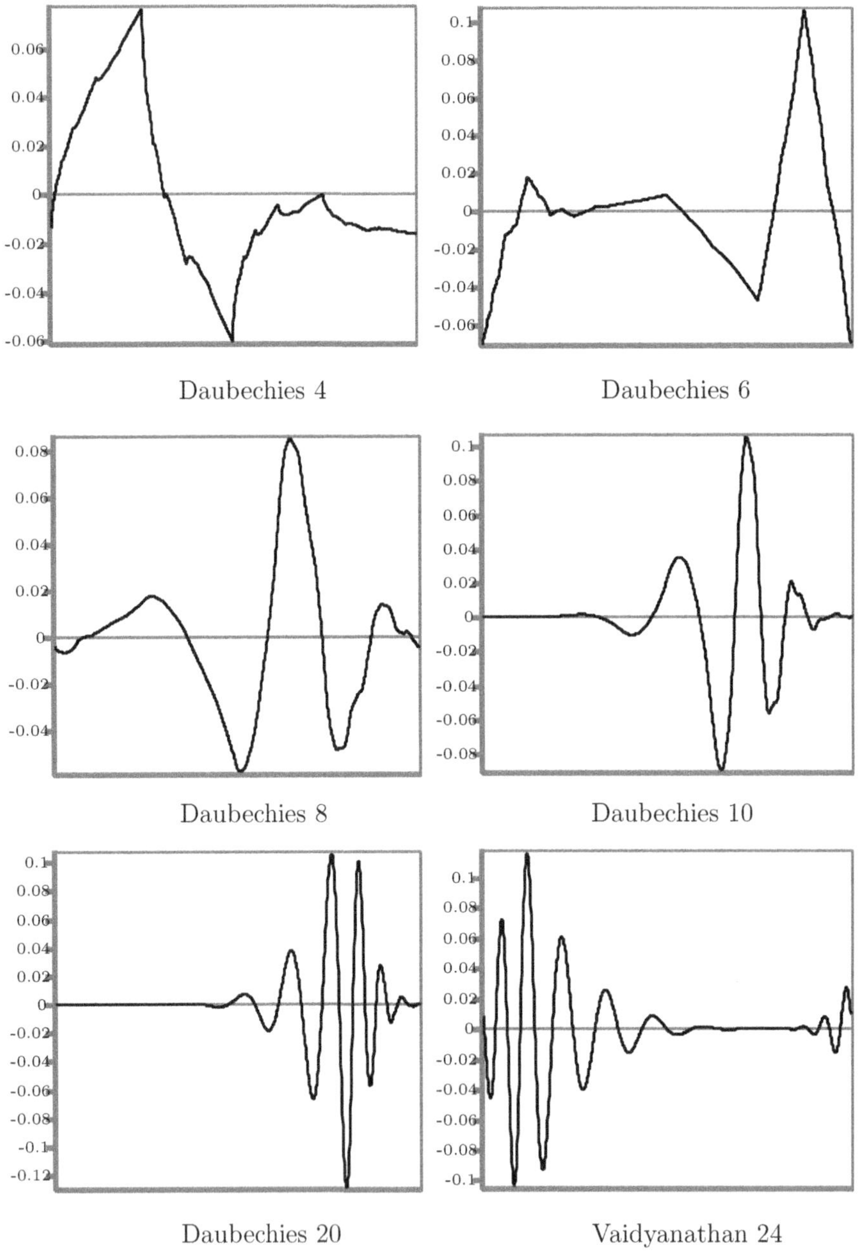

Figure 8.37: Examples of Common Wavelets.

Nondifferentiable Functions

Most functions used in science and engineering are smooth. They have a well-defined direction (or tangent) at every point. Mathematically, we say that they are *differentiable*. Some functions may have a few points where they suddenly change direction, and so do not have a tangent. The Haar wavelet is an example of such a function. A function may have many such "sharp" corners, even infinitely many. A simple example is an infinite square wave. It has infinitely many sharp points, but it does not look strange or unusual, because those points are separated by smooth areas of the function.

What is hard for us to imagine (and even harder to accept the existence of) is a function that is everywhere continuous but is nowhere differentiable! Such a function has no "holes" or "gaps." It continues without interruptions, but it changes its direction sharply at *every* point. It is as if it has a certain direction at point x and a different direction at the point that immediately follows x; except, of course, that a real number x does not have an immediate successor.

Such functions exist. The first was discovered in 1875 by Karl Weierstrass. It is the sum of the infinite series

$$W_{b,w}(t) = \frac{\sum_{n=0}^{\infty} w^n e^{2\pi i b^n t}}{\sqrt{1-w^2}},$$

where $b > 1$ is a real number, w is written either as $w = b^h$, with $0 < h < 1$, or as $w = b^{d-2}$, with $1 < d < 2$, and $i = \sqrt{-1}$. Notice that $W_{b,w}(t)$ is complex; its real and imaginary parts are called the Weierstrass cosine and sine functions, respectively.

Weierstrass proved the unusual behavior of this function, and also showed that for $d < 1$ it is differentiable. In his time, such a function was so contrary to common sense and mathematical intuition that he did not publish his findings. Today, we simply call this function and others like it *fractals*.

◇ **Exercise 8.11:** Use functions `fwt2` and `iwt2` of Figures 8.33 and 8.34 to blur an image. The idea is to compute the 4-step subband transform of an image (thus ending up with 13 subbands), then set most of the transform coefficients to zero and heavily quantize some of the others. This, of course, results in a loss of image information, and in a nonperfectly reconstructed image. The aim of this exercise, however, is to have the inverse transform produce a *blurred image*. This illustrates an important property of the discrete wavelet transform, namely its ability to reconstruct images that degrade gracefully when more and more transform coefficients are zeroed or coarsely quantized. Other transforms, most notably the DCT, may introduce artifacts in the reconstructed image, but this property of the DWT makes it ideal for applications such as fingerprint compression (Section 8.18).

8.9 Multiresolution Decomposition

The main idea of wavelet analysis, illustrated in detail in Section 8.5, is to analyze a function at different scales. A mother wavelet is used to construct wavelets at different scales (dilations) and translate each relative to the function being analyzed. The results of translating a wavelet depend on how much it matches the function being analyzed. Wavelets at different scales (resolutions) produce different results. The principle of multiresolution decomposition, due to Stephane Mallat and Yves Meyer, is to group all the wavelet transform coefficients for a given scale, display their superposition, and repeat for all scales.

Figure 8.39 lists a Matlab function `multres` for this operation, together with a test. Figure 8.38 shows two examples, a single spike and multiple spikes. It is easy to see how the fine-resolution coefficients are concentrated at the values of t that correspond to the spikes (i.e., the high activity areas of the data).

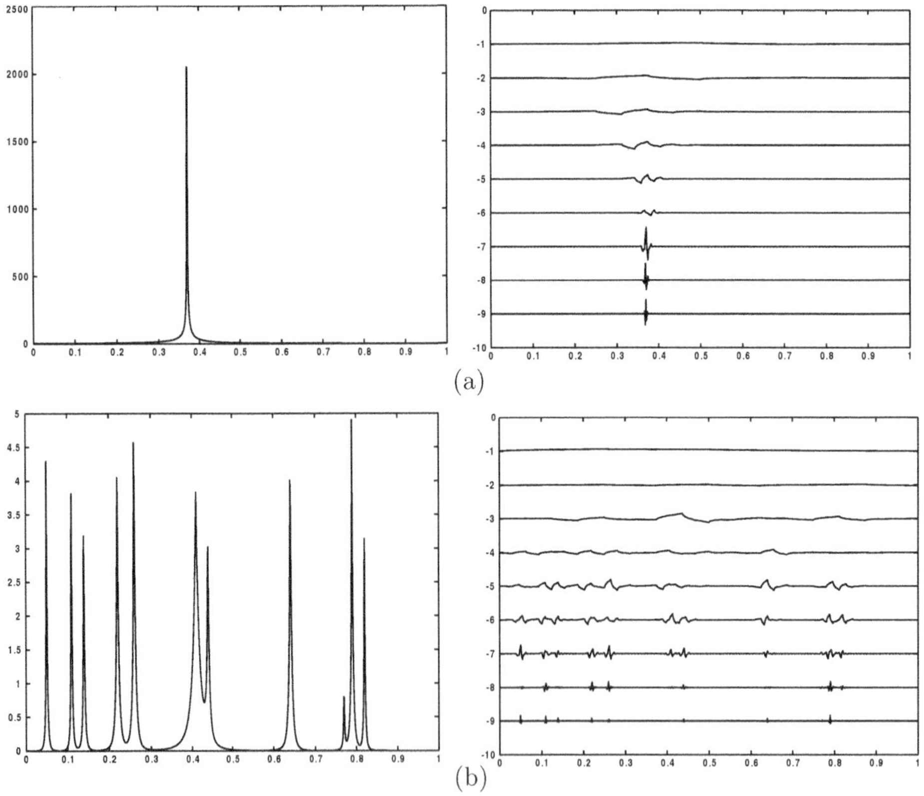

Figure 8.38: Examples of Multiresolution Decomposition. (a) A Spike. (b) Several Spikes.

```
function multres(wc,coarse,filter)
% A multi resolution plot of a 1D wavelet transform
scale=1./max(abs(wc));
n=length(wc); j=log2(n);
LockAxes([0 1 -(j) (-coarse+2)]);
t=(.5:(n-.5))/n;
for i=(j-1):-1:coarse
 z=zeros(1,n);
 z((2^(i)+1):(2^(i+1)))=wc((2^(i)+1):(2^(i+1)));
 dat=iwt1(z,i,filter);
 plot(t,-(i)+scale.*dat);
end
z=zeros(1,n);
z(1:2^(coarse))=wc(1:2^(coarse));
dat=iwt1(z,coarse,filter);
plot(t,(-(coarse-1))+scale.*dat);
UnlockAxes;
```

And a test routine

```
n=1024; t=(1:n)./n;
dat=spikes(t);                    % several spikes
%p=floor(n*.37); dat=1./abs(t-(p+.5)/n); % one spike
figure(1), plot(t,dat)
filt=[0.4830 0.8365 0.2241 -0.1294];
wc=fwt1(dat,2,filt);
figure(2), plot(t,wc)
figure(3)
multres(wc,2,filt);

function dat=spikes(t)
pos=[.05 .11 .14 .22 .26 .41 .44 .64  .77 .79 .82];
hgt=[5  5 4 4.5 5 3.9 3.3 4.6 1.1 5 4];
wth=[.005 .005 .006 .01 .01 .03 .01 .01  .005 .008 .005];
dat=zeros(size(t));
for i=1:length(pos)
dat=dat+hgt(i)./(1+abs((t-pos(i))./wth(i))).^4;
end;
```

Figure 8.39: Matlab Code for the Multiresolution Decomposition of a One-Dimensional Row Vector.

8.10 Various Image Decompositions

Section 8.6.1 shows two ways of applying a discrete wavelet transform to an image in order to partition it into several subbands. This section (based on [Strømme 99], which also contains comparisons of experimental results) discusses seven ways of doing the same thing, each involving a different algorithm and resulting in subbands with different energy compactions. Other decompositions are also possible (Section 8.18 describes a special, symmetric decomposition).

It is important to realize that the wavelet filters and the decomposition method are independent. The discrete wavelet transform of an image can use any set of wavelet filters and can decompose the image in any way. The only limitation is that there must be enough data points in the subband to cover all the filter taps. For example, if a 12-tap Daubechies filter is used, and the image and subband sizes are powers of two, then the smallest subband that can be produced has size 8×8. This is because a subband of size 16×16 is the smallest that can be multiplied by the 12 coefficients of this particular filter. Once such a subband is decomposed, the resulting 8×8 subbands are too small to be multiplied by 12 coefficients and to be decomposed further.

1. Laplacian Pyramid: This technique for image decomposition is described in detail in Section 8.13. Its main feature is progressive image transmission. During decompression and image reconstruction, the user sees small, blurred images that grow and become sharper. The main reference is [Burt and Adelson 83].

The Laplacian pyramid is generated by subtracting an upsampled lowpass version of the image from the original image. The image is partitioned into a Gaussian pyramid (the lowpass subbands) and a Laplacian pyramid that consists of the detail coefficients (the highpass subbands). Only the Laplacian pyramid is needed to reconstruct the image. The transformed image is bigger than the original image, which is the main difference between the Laplacian pyramid decomposition and the pyramid decomposition (method 4 below).

2. Line: This technique is a simpler version of the standard wavelet decomposition (method 5). The wavelet transform is applied to each row of the image, resulting in smooth coefficients on the left (subband L1) and detail coefficients on the right (subband H1). Subband L1 is then partitioned into L2 and H2, and the process is repeated until the entire coefficient matrix is turned into detail coefficients, except the leftmost column, which contains smooth coefficients. The wavelet transform is then applied recursively to the leftmost column, resulting in one smooth coefficient at the top-left corner of the coefficient matrix. This last step may be omitted if the compression method being used requires that image rows be individually compressed (notice the distinction between the wavelet transform and the actual compression algorithm).

This technique exploits correlations only within an image row to calculate the transform coefficients. Also, discarding a coefficient that is located on the leftmost column may affect just a particular group of rows and may this way introduce artifacts into the reconstructed image.

Implementation of this method is simple, and execution is fast, about twice that of the standard decomposition. This type of decomposition is illustrated in Figure 8.40.

It is possible to apply this decomposition to the columns of the image, instead of to the rows. Ideally, the transform should be applied in the direction of highest image redundancy, and experience suggests that for natural images this is the horizontal direction. Thus, in practice, line decomposition is applied to the image rows.

3. Quincunx: Somewhat similar to the Laplacian pyramid, quincunx decomposition proceeds level by level and decomposes subband L_i of level i into subbands H_{i+1} and L_{i+1} of level $i+1$. Figure 8.41 illustrates this type of decomposition. The method is due to Strømme and McGregor [Strømme and McGregor 97], who originally called it nonstandard decomposition. (See also [Starck et al. 98] for a different presentation of this method.) It is efficient and computationally simple. On average, it achieves more than four times the energy compaction of the line method.

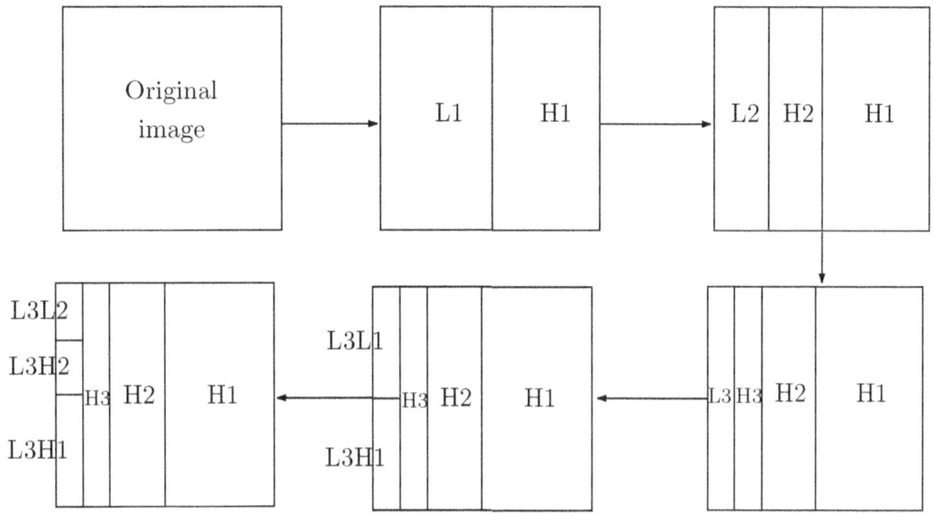

Figure 8.40: Line Wavelet Decomposition.

Quincunx decomposition results in fewer subbands than most other wavelet decompositions, a feature that may lead to reconstructed images with slightly lower visual quality. The method is not used much in practice, but [Strømme 99] presents results that suggest that the quincunx decomposition performs extremely well and may be the best performer in many practical situations.

4. Pyramid: The pyramid decomposition is by far the most common method used to decompose images that are wavelet transformed. It results in subbands with horizontal, vertical, and diagonal image details, as illustrated by Figure 8.17. The three subbands at each level contain horizontal, vertical, and diagonal image features at a particular scale, and each scale is divided by an octave in spatial frequency (division of the frequency by two).

Pyramid decomposition turns out to be a very efficient way of transferring significant visual data to the detail coefficients. Its computational complexity is about 30% higher than that of the quincunx method, but its image reconstruction abilities are higher. The reasons for the popularity of the pyramid method may be that (1) it is symmetrical, (2) its mathematical description is simple, and (3) it was used by the influential paper [Mallat 89].

Figure 8.42 illustrates pyramid decomposition. It is obvious that the first step is identical to that of the quincunx decomposition. However, while the quincunx method leaves the high-frequency subband untouched, the pyramid method resolves it into two bands. On the other hand, pyramid decomposition involves more computations in order to spatially resolve the asymmetric high-frequency band into two symmetric high-frequency and low-frequency bands.

5. Standard: The first step in the standard decomposition is to apply whatever discrete wavelet filter is being used to all the rows of the image, obtaining subbands L_1 and H_1. This is repeated on L_1 to obtain L_2 and H_2, and so on k times. This is followed by a second step where a similar calculation is applied k times to the columns. If $k = 1$,

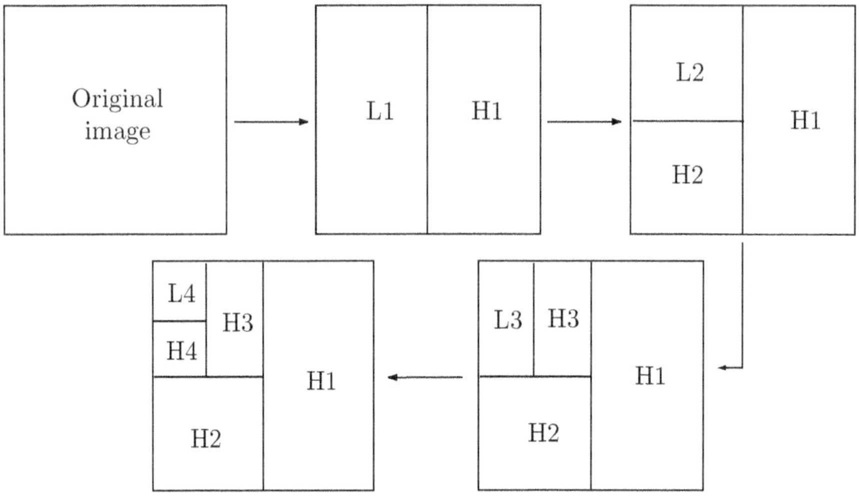

Figure 8.41: Quincunx Wavelet Decomposition.

From the Dictionary

QUINCUNX: An arrangement of five objects in a square or rectangle, one at each corner, and one at the center. The word seems to have originated from Latin around 1640–1650. It stands for "five-twelvths." *Quinc* is a variation of *quinque*, and *unx* or *unc* is a form of *uncia*, meaning twelfth. It was used to indicate a Roman coin worth five-twelfths of an AS and marked with a quincunx of spots. (AS is an ancient Roman unit of weight, equal to about 12 ounces.)

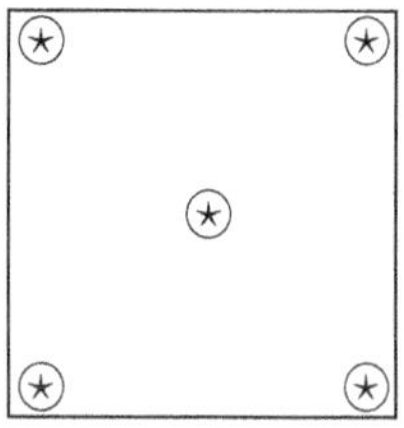

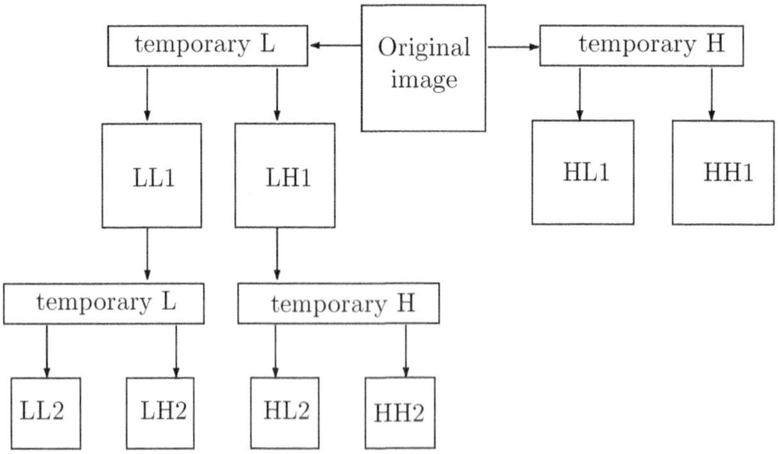

Figure 8.42: Pyramid Wavelet Decomposition.

the decomposition alternates between rows and columns, but k may be greater than 1. The end result is to have one smooth coefficient at the top-left corner of the coefficient matrix. This method is somewhat similar to line decomposition. An important feature of standard decomposition is that when a coefficient is quantized, it may affect a long, thin rectangular area in the reconstructed image. Thus, very coarse quantization may result in artifacts in the reconstructed image in the form of horizontal rectangles.

Standard decomposition has the second-highest reconstruction quality of the seven methods described here. The reason for the improvement compared to the pyramid decomposition may be that the higher directional resolution gives thresholding a better chance to cover larger uniform areas. On the other hand, standard decomposition is computationally more expensive than pyramid decomposition.

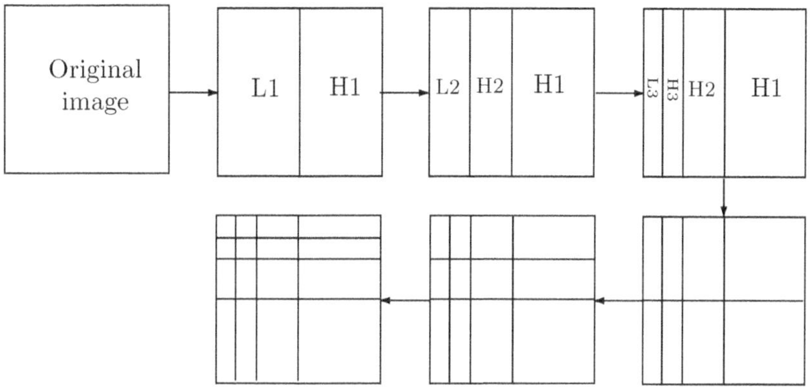

Figure 8.43: Standard Wavelet Decomposition.

6. Uniform Decomposition: This method is also called the *wavelet packet transform.* It is illustrated in Figure 8.44. In the case where the user elects to compute all the levels of the transform, the uniform decomposition becomes similar to the discrete Fourier transform (DFT) in that each coefficient represents a spatial frequency for the entire image. In such a case (where all the levels are computed), the removal of one coefficient in the transformed image affects the entire reconstructed image.

The computational cost of uniform decomposition is very high, since it effectively computes n^2 coefficients for every level of decomposition, where n is the side length of the (square) image. Despite having comparably high average reconstruction qualities, the perceptual quality of the image starts degrading at lower ratios than for the other decomposition methods. The reason for this is the same as for Fourier methods: Because the support for a single coefficient is global, its removal has the effect of blurring the reconstructed image. The conclusion is that the increased computational complexity of uniform decomposition does not result in increased quality of the reconstructed image.

6.1. Full Wavelet Decomposition: (This is a special case of uniform decomposition.) Denote the original image by I_0. We assume that its size is $2^l \times 2^l$. After applying the discrete wavelet transform to it, we end up with a matrix I_1 partitioned into four subbands. The same discrete wavelet transform (i.e., using the same wavelet filters) is then applied recursively to each of the four subbands individually. The result

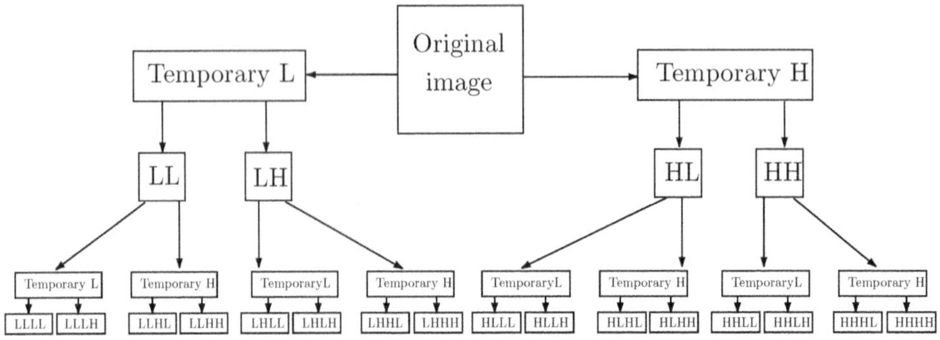

LLLL	LLHL	HLLL	HLHL
LLLH	LLHH	HLLH	HLHH
LHLL	LHHL	HHLL	HHHL
LHLH	LHHH	HHLH	HHHH

Figure 8.44: Uniform Wavelet Decomposition.

is a coefficient matrix I_2 consisting of 16 subbands. When this process is carried out r times, the result is a coefficient matrix consisting of $2^r \times 2^r$ subbands, each of size $2^{l-r} \times 2^{l-r}$. The top-left subband contains the smooth coefficients (depending on the particular wavelet filter used, it may look like a small version of the original image), and the other subbands contain detail coefficients. Each subband corresponds to a frequency band, while each individual transform coefficient corresponds to a local spatial region. By increasing the recursion depth r, we can increase frequency resolution at the expense of spatial resolution.

This type of wavelet image decomposition has been proposed by Wong and Kuo [Wong and Kuo 93], who highly recommend its use. However, it seems that it has been ignored by researchers and implementers in the field of image compression.

7. Adaptive Wavelet Packet Decomposition: The uniform decomposition method is costly in terms of computations, and the adaptive wavelet packet method is potentially even more so. The idea is to skip those subband splits that do not contribute significantly to energy compaction. The result is a coefficient matrix with subbands of different (possibly even many) sizes. The bottom-right part of Figure 8.45 shows such a case (after [Meyer et al. 98], which shows the adaptive wavelet packet transform matrix for the bi-level "mandril" image, Figure 7.56).

The justification for this complex decomposition method is the prevalence of continuous tone (natural) images. These images, which are discussed at the start of Chapter 7, are mostly smooth but normally also have some regions with high frequency data. Such regions should end up as many small subbands (to better enable an accurate spatial frequency representation of the image), with the rest of the image giving rise to a few large subbands. The bottom-left coefficient matrix in Figure 8.45 is an example of a very uniform image, resulting in just 10 subbands. (The test for a split depends on the absolute magnitude of the transform coefficients. Thus, the test can be adjusted so high that very few splits are done.)

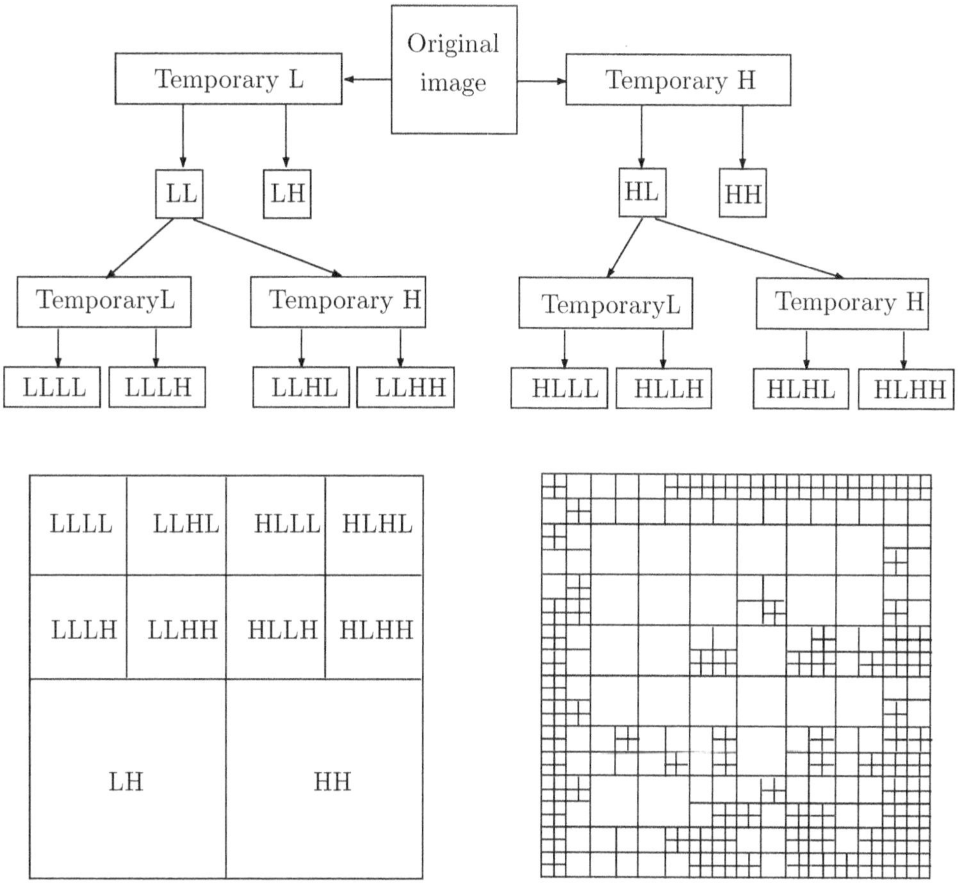

Figure 8.45: Adaptive Wavelet Packet Decomposition.

The downside of this type of decomposition is finding an algorithm that will determine which subband splits can be skipped. Such an algorithm uses entropy calculations and should be efficient. It should identify all the splits that do not have to be performed, and it should identify as many of them as possible. An inefficient algorithm may lead to the split of every subband, thereby performing many unnecessary computations and ending up with a coefficient matrix where every coefficient is a subband, in which case this decomposition reduces to the uniform decomposition.

This type of decomposition has the highest reproduction quality of all the methods discussed here, a feature that may justify the high computational costs in certain special applications. This quality, however, is not much higher than what is achieved with simpler decomposition methods, such as standard, pyramid, or quincunx.

8.11 The Lifting Scheme

The lifting scheme ([Stollnitz et al. 96] and [Sweldens and Schröder 96]) is a novel and useful way of looking at the discrete wavelet transform. It is easy to understand, since it performs all the operations in the time domain, rather than in the frequency domain, and has other advantages as well. This section illustrates the lifting approach using the Haar transform, which is already familiar to the reader, as an example. The following section extends the same approach to other transforms.

The Haar transform, described Section 8.6, is based on the computations of averages and differences (details). Given two adjacent pixels a and b, the principle is to calculate the average $s = (a+b)/2$ and difference $d = b - a$. If a and b are similar, s will be similar to both and d will be small, i.e., require fewer bits to represent. This transform is reversible, because $a = s - d/2$ and $b = s + d/2$, and it can be written using matrix notation as

$$(s,d) = (a,b)\begin{pmatrix} 1/2 & -1 \\ 1/2 & 1 \end{pmatrix} = (a,b)\mathbf{A}, \quad (a,b) = (s,d)\begin{pmatrix} 1 & 1 \\ -1/2 & 1/2 \end{pmatrix} = (s,d)\mathbf{A}^{-1}.$$

Consider a row of 2^n pixel values $s_{n,l}$ for $0 \le l < 2^n$. There are 2^{n-1} pairs of pixels $s_{n,2l}, s_{n,2l+1}$ for $l = 0, 2, 4, \ldots, 2^{n-2}$. Each pair is transformed into an average $s_{n-1,l} = (s_{n,2l} + s_{n,2l+1})/2$ and a difference $d_{n-1,l} = s_{n,2l+1} - s_{n,2l}$. The result is a set s_{n-1} of 2^{n-1} averages and a set d_{n-1} of 2^{n-1} differences.

The same operations can be applied to the 2^{n-1} averages $s_{n-1,l}$ of set s_{n-1}, resulting in 2^{n-2} averages $s_{n-2,l}$ and 2^{n-2} differences $d_{n-2,l}$. After applying these operations n times we end up with a set s_0 consisting of one average $s_{0,0}$, and with n sets of differences $d_{j,l}$ where $j = 0, 1, \ldots, n-1$ and $l = 0, 1, \ldots, 2^j - 1$. Set j consists of j differences, so the total number of differences is

$$\sum_{j=0}^{n-1} 2^j = 2^n - 1.$$

Adding the single average $s_{0,0}$ brings the total number of results to 2^n, the same as the number of original pixel values. Notice that the final average $s_{0,0}$ is the average S of the original 2^n pixel values, so it can be called the DC component of the original values. In fact, if we look at any set s_j of averages $s_{j,l}$, $l = 0, 1, \ldots, 2^j - 1$, we find that its average

$$S = \frac{1}{2^j}\sum_{l=0}^{2^j-1} s_{j,l}$$

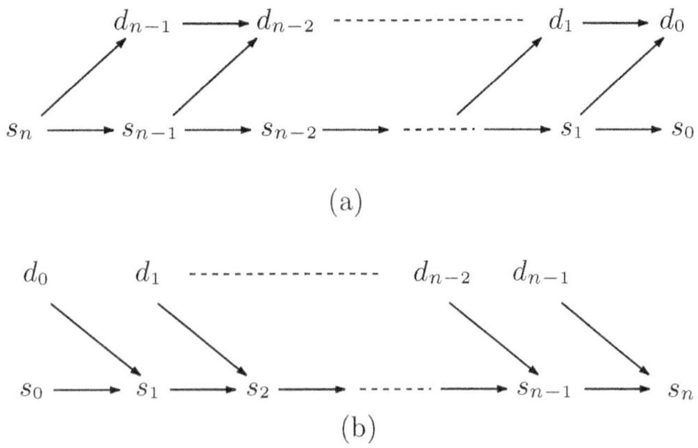

Figure 8.46: (a) The Haar Wavelet Transform and (b) Its Inverse.

is the average of all the original 2^n pixel values. Thus, the average S of set s_j is independent of j. Figure 8.46a,b illustrates the transform and its inverse.

The main idea in the lifting scheme is to perform all the necessary operations without using extra space. The entire transform is performed *in place*, replacing the original image. We start with a pair of consecutive pixels a and b. They are replaced with their average s and difference d by first replacing b with $d = b - a$, then replacing a with $s = a + d/2$ [since $d = b - a$, $a + d/2 = a + (b - a)/2$ equals $(a + b)/2$]. In the C language, this is written

```
b-=a;   a+=b/2;
```

◇ **Exercise 8.12:** Write the reverse operations in C.

This is easy to apply to an entire row of pixels. Suppose that we have a row s_j with 2^j values and we want to transform it to a row s_{j-1} with 2^{j-1} averages and 2^{j-1} differences. The lifting scheme performs this transform in three steps, *split*, *predict*, and *update*.

The split operation splits row s_j into two separate sets denoted by even_{j-1} and odd_{j-1}. The former contains all the even values $s_{j,2l}$, and the latter contains all the odd values $s_{j,2l+1}$. The range of l is from 0 to $2^j - 1$. This kind of splitting into odd and even values is called the *lazy wavelet transform*. We denote it by

$$(\text{even}_{j-1}, \text{odd}_{j-1}) := \text{Split}(s_j).$$

The predict operation uses the even set even_{j-1} to predict the odd set odd_{j-1}. This is based on the fact that each value $s_{j,2l+1}$ in the odd set is adjacent to the corresponding value $s_{j,2l}$ in the even set. Thus, the two values are correlated and either can be used to predict the other. Recall that a general difference $d_{j-1,l}$ is computed as the difference $d_{j-1,l} = s_{j,2l+1} - s_{j,2l}$ between an odd value and an adjacent even value (or between an odd value and its prediction), which is why we can define the prediction operator P as

$$d_{j-1} = \text{odd}_{j-1} - P(\text{even}_{j-1}).$$

The update operation U follows the prediction step. It calculates the 2^{j-1} averages $s_{j-1,l}$ as the sum

$$s_{j-1,l} = s_{j,2l} + d_{j-1,l}/2. \tag{8.13}$$

This operation is formally defined by

$$s_{j-1} = \text{even}_{j-1} + U(d_{j-1}).$$

◇ **Exercise 8.13:** Use Equation (8.13) to show that sets s_j and s_{j-1} have the same average.

The important point to notice is that all three operations can be performed in place. The even locations of set s_j are overwritten with the averages (i.e., with set even_{j-1}), and the odd locations are overwritten with the differences (set odd_{j-1}). The sequence of three operations can be summarized by

$$(\text{odd}_{j-1}, \text{even}_{j-1}) := \text{Split}(s_j); \quad \text{odd}_{j-1} -= P(\text{even}_{j-1}); \quad \text{even}_{j-1} += U(\text{odd}_{j-1});$$

Figure 8.47a is a wiring diagram of this process.

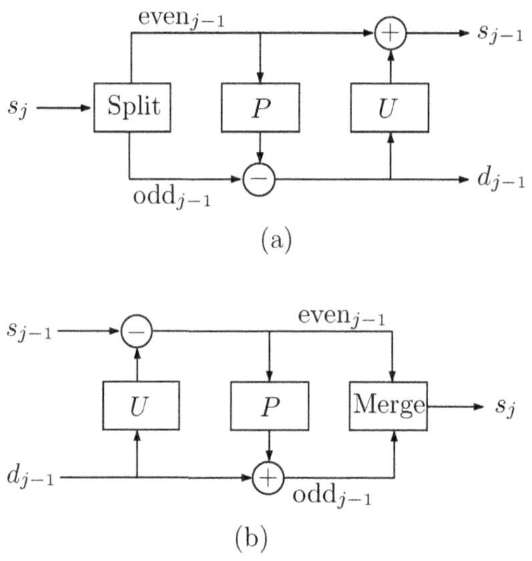

Figure 8.47: The Lifting Scheme. (a) Forward Transform. (b) Reverse Transform.

The reverse transform is similar. It is based on the three operations *undo update*, *undo prediction*, and *merge*.

Given two sets s_{j-1} and d_{j-1}, the "undo update" operation reconstructs the averages $s_{j,2l}$ (the even values of set s_j) by subtracting the update operation. It generates the set even_{j-1} by subtracting the sets $s_{j-1} - U(d_{j-1})$. Written explicitly, this operation becomes

$$s_{j,2l} = s_{j-1,l} - d_{j-1,l}/2, \text{ for } 0 \le l < 2^j.$$

Given the two sets even_{j-1} and d_{j-1}, the "undo predict" operation reconstructs the differences $s_{j,2l+1}$ (the odd values of set s_j) by adding the prediction operator P. It generates the set odd_{j-1} by adding the sets $d_{j-1} + P(\text{even}_{j-1})$. Written explicitly, this operation becomes

$$s_{j,2l+1} = d_{j-1,l} + s_{j,2l}, \text{ for } 0 \le l < 2^j.$$

Now that the two sets even_{j-1} and odd_{j-1} have been reconstructed, they are merged by the "merge" operation into the set s_j. This is the inverse lazy wavelet transform, formally denoted by

$$s_j = \text{Merge}(\text{even}_{j-1}, \text{odd}_{j-1}).$$

It moves the averages and the differences into the even and odd locations of s_j, respectively, without using any extra space. The three operations are summarized by

$$\text{even}_{j-1} -= U(\text{odd}_{j-1}); \quad \text{odd}_{j-1} += P(\text{even}_{j-1}); \quad s_j := \text{Merge}(\text{odd}_{j-1}, \text{even}_{j-1});$$

Figure 8.47b is a wiring diagram of this process.

The wiring diagrams show one of the important features of the lifting scheme, namely its inherent parallelism. Given an SIMD (single instruction, multiple data) computer with 2^n processing units, each unit can be "responsible" for one pixel value. Such a computer executes a single program where each instruction is executed in parallel by all the processing units. Another advantage of lifting is the simplicity of its inverse transform. It is simply the code for the forward transform, run backward. The main advantage of lifting is the fact that it is easy to extend. It has been presented here for the Haar transform, where averages and differences are calculated in a simple way. It can be extended to more complex cases, where the prediction and update operations are more complex.

8.11.1 The Linear Wavelet Transform

The reason for extending the lifting scheme beyond the Haar transform is that this transform does not produce high-quality results, since it uses such simple prediction. Recall that the "predict" operation uses the even set even_{j-1} to predict the odd set odd_{j-1}. This gives accurate prediction only in the (very rare) cases where the two sets are identical. We can say that the Haar transform eliminates order-zero correlation between pixels by using an order-one predictor. The Haar transform also preserves the average of all the pixels of the image, and this average can be called the *order-zero moment* of the image.

Better compression can be achieved by transforms that use better predictors, predictors that exploit correlations between several neighbor pixels and also preserve higher-order moments of the image. The predictor and update operations described here are of order two. This implies that the predictor will provide exact prediction if the image pixels vary linearly, and the update operation will preserve the first two (order-zero and order-one) moments. The principle is easy to describe. We again concentrate on a row s_j of pixel values. An odd-numbered value $s_{j,2l+1}$ is predicted as the average of its two immediate neighbors $s_{j,2l}$ and $s_{j,2l+2}$. To be able to reconstruct $s_{j,2l+1}$, we have to calculate a detail value $d_{j,l}$ that is no longer a simple difference but is given by

$$d_{j,l} = s_{j,2l+1} - \frac{1}{2}(s_{j,2l} + s_{j,2l+2}).$$

(This notation assumes that every pixel has two immediate neighbors. This is not true for pixels located on the boundaries of the image, so edge rules will have to be developed and justified.)

Figure 8.48 illustrates the meaning of the detail values in this case. Figure 8.48a shows the values of nine hypothetical pixels numbered 0 through 8. Figure 8.48b shows straight segments connecting the even-numbered pixels. This is the reason for the name "linear transform." In Figure 8.48c the odd-numbered pixels are predicted by these straight segments, and Figure 8.48d shows (in dashed bold) the difference between each odd-numbered pixel and its prediction. The equation of a straight line is $y = ax + b$, a polynomial of degree 1, which is why we can think of the detail values as the amount by which the pixel values deviate locally from a degree-1 polynomial. If pixel x had value $ax + b$, all the detail values would be zero. This is why we can say that the detail values capture the high frequencies of the image.

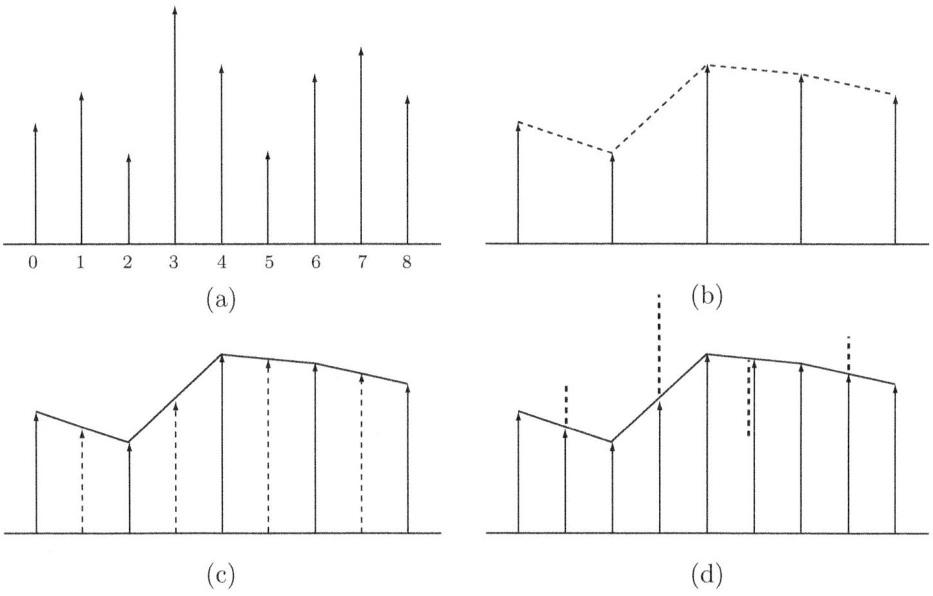

Figure 8.48: Linear Prediction.

The "update" operation reconstructs the averages $s_{j-1,l}$ from the averages $s_{j,l}$ and the differences $d_{j-1,l}$. In the case of the Haar transform, this operation is defined by Equation (8.13): $s_{j-1,l} = s_{j,2l} + d_{j-1,l}/2$. In the case of the linear transform the operation is somewhat more complicated. We derive the update operation for the linear transform using the requirement that it preserves the zeroth-order moment of the image, i.e., the average of the averages $s_{j,l}$ should not depend on j. We try an update of the form

$$s_{j-1,l} = s_{j,2l} + A(d_{j-1,l-1} + d_{j-1,l}), \tag{8.14}$$

where A is an unknown coefficient. The sum of $s_{j-1,l}$ now becomes

$$\sum_l s_{j-1,l} = \sum_l s_{j,2l} + 2A\sum_l d_{j-1,l} = (1-2A)\sum_l s_{j,2l} + 2A\sum_l s_{j,2l+1},$$

so the choice $A = 1/4$ results in

$$\sum_{l=0}^{2^{j-1}-1} s_{j-1,l} = \left(1-\frac{1}{2}\right)\sum_{l=0}^{2^j-1} s_{j,2l} + \frac{2}{4}\sum_{l=0}^{2^j-1} s_{j,2l+1} = \frac{1}{2}\sum_{l=0}^{2^j-1} s_{j,l}.$$

Comparing this with Equation (8.15) shows that the zeroth-order moment of the image is preserved by the update operation of Equation (8.14). A direct check also verifies that

$$\sum_l l\, s_{j-1,l} = \frac{1}{2}\sum_l l\, s_{j,l},$$

which shows that this update operation also preserves the first-order moment of the image and is therefore of order 2.

Equation (8.15), duplicated below, is derived in the answer to Exercise 8.13:

$$\sum_{l=0}^{2^{j-1}-1} s_{j-1,l} = \sum_{l=0}^{2^{j-1}-1} (s_{j,2l} + d_{j-1,l}/2) = \frac{1}{2}\sum_{l=0}^{2^{j-1}-1} (s_{j,2l} + s_{j,2l+1}) = \frac{1}{2}\sum_{l=0}^{2^j-1} s_{j,l}. \quad (8.15)$$

The inverse linear transform reconstructs the even and odd average values by

$$s_{j,2l} = s_{j-1,l} - \frac{1}{4}(d_{j-1,l-1} + d_{j-1,l}), \text{ and } s_{j,2l+1} = d_{j,l} + \frac{1}{2}(s_{j,2l} + s_{j,2l+2}),$$

respectively.

Figure 8.49 summarizes the linear transform. The top row shows the even and odd averages $s_{j,l}$. The middle row shows how a detail value $d_{j-1,l}$ is calculated as the difference between an odd average $s_{j,2l+1}$ and half the sum of its two even neighbors $s_{j,2l}$ and $s_{j,2l+2}$. The bottom row shows how the next set s_{j-1} of averages is calculated. Each average $s_{j-1,l}$ in this set is the sum of an even average $s_{j,2l}$ and one-quarter of the two detail values $d_{j-1,l-1}$ and $d_{j-1,l}$. The figure also illustrates the main feature of the lifting scheme, namely how the even averages $s_{j,2l}$ are replaced by the next set of averages $s_{j-1,l}$ and how the odd averages $s_{j,2l+1}$ are replaced by the detail values $d_{j-1,l}$ (the dashed arrows indicate items that are not moved). All these operations are performed in place.

8.11.2 Interpolating Subdivision

The method of interpolating subdivision starts with a set s_0 of pixels where pixel $s_{0,k}$ (for $k = 0, 1, \ldots$) is stored in location k of an array a. It creates a set s_1 (twice the size of s_0) of pixels $s_{1,k}$ such that the even-numbered pixels $s_{1,2k}$ are simply the even-numbered

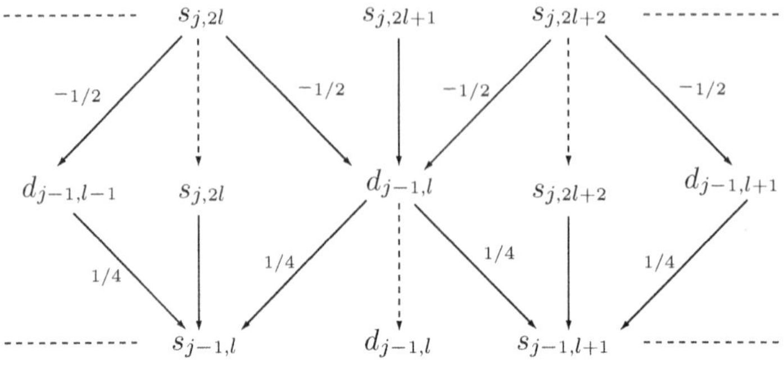

Figure 8.49: Summary of the Linear Wavelet Transform.

pixels $s_{0,k}$ and each of the odd-numbered pixels $s_{1,2k+1}$ is obtained by interpolating some of the pixels (odd and/or even) of set s_0. The new content of array a is

$$s_{1,0} = s_{0,0},\ s_{1,1},\ s_{1,2} = s_{0,1},\ s_{1,3},\ s_{1,4} = s_{0,2},\ s_{1,5},\ldots,s_{1,2k} = s_{0,k},\ s_{1,2k+1},\ldots.$$

The original elements $s_{0,k}$ of s_0 are now stored in a in locations $2k = k{\cdot}2^1$. We say that set s_1 was created from s_0 by a process of subdivision (or refinement).

Next, set s_2 (twice the size of s_1) is created in the same way from s_1. The even-numbered pixels $s_{2,2k}$ are simply the even-numbered pixels $s_{1,k}$, and each of the odd-numbered pixels $s_{2,2k+1}$ is obtained by interpolating some of the pixels (odd and/or even) of set s_1. The new content of array a becomes

$$s_{2,0} = s_{1,0} = s_{0,0},\ s_{2,1},\ s_{2,2} = s_{1,1},\ s_{2,3},\ s_{2,4} = s_{1,2} = s_{0,1},\ s_{2,5},\ldots$$
$$s_{2,2k} = s_{1,k},\ s_{2,2k+1},\ldots,s_{2,4k} = s_{1,2k} = s_{0,k}, s_{2,4k+1},\ldots.$$

The original elements $s_{0,k}$ of s_0 are now stored in a in locations $4k = k{\cdot}2^2$.

In a general subdivision step, a set s_j of pixel values $s_{j,k}$ is used to construct a new set s_{j+1}, twice as large. The even-numbered pixels $s_{j+1,2k}$ are simply the even-numbered pixels $s_{j,k}$, and each of the odd-numbered pixels $s_{j+1,2k+1}$ is obtained by interpolating some of the pixels (odd and/or even) of set s_j. The original elements $s_{0,k}$ of s_0 are now stored in a in locations $k{\cdot}2^j$.

We employ linear interpolation to illustrate this refinement process. Each of the odd-numbered pixels $s_{1,2k+1}$ is calculated as the average of the two pixels $s_{0,k}$ and $s_{0,k+1}$ of set s_0. In general, we get

$$s_{j+1,2k} = s_{j,k}, \quad s_{j+1,2k+1} = \frac{1}{2}\left(s_{j,k} + s_{j,k+1}\right). \tag{8.16}$$

Figure 8.50a–d shows several steps in this process, starting with a set s_0 of four pixels (it is useful to visualize each pixel as a two-dimensional point). It is obvious that the pixel values converge to the *polyline* that passes through the original four points.

Given an image where the pixel values go up and down linearly (in a polyline), we can compress it by selecting the pixels at the corners of the polyline (i.e., those pixels

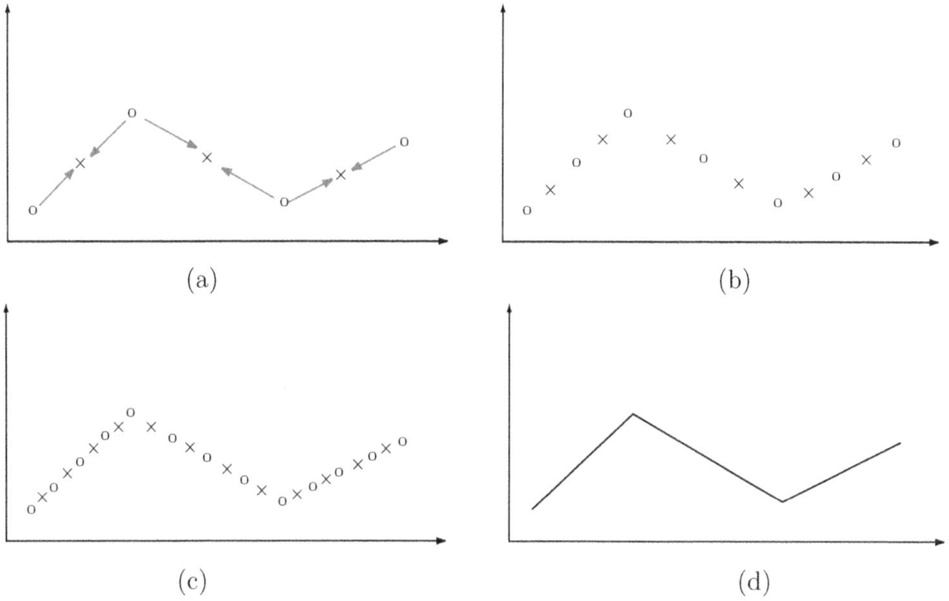

Figure 8.50: An Example of a Linear Subdivision.

where the value changes direction) and writing them on the compressed file. The image could then be perfectly reconstructed to any resolution by using linear subdivision to compute as many pixels as needed between corner pixels. Most images, however, feature more complex behavior of pixel values, so more complex interpolation is needed.

We now show how to extend linear subdivision to *polynomial subdivision*. Instead of computing an odd-numbered pixel $s_{j+1,2k+1}$ as the average of its two immediate level-j neighbors $s_{j,k}$ and $s_{j,k+1}$, we calculate it as a weighted sum of its **four** immediate level-j neighbors $s_{j,k-1}$, $s_{j,k}$, $s_{j,k+1}$, and $s_{j,k+2}$. It is obvious that the two closer neighbors $s_{j,k}$ and $s_{j,k+1}$ should be assigned more weight than the two extreme neighbors $s_{j,k-1}$ and $s_{j,k+2}$, but what should the weights be? The answer is given by Equation (7.45) (Section 7.25.4), which shows that the degree-3 (cubic) polynomial $\mathbf{P}(t)$ that interpolates four arbitrary points $\mathbf{P}_1$, $\mathbf{P}_2$, $\mathbf{P}_3$, and $\mathbf{P}_4$ is given by

$$\mathbf{P}(t) = (t^3, t^2, t, 1)\begin{pmatrix} -4.5 & 13.5 & -13.5 & 4.5 \\ 9.0 & -22.5 & 18 & -4.5 \\ -5.5 & 9.0 & -4.5 & 1.0 \\ 1.0 & 0 & 0 & 0 \end{pmatrix}\begin{pmatrix} \mathbf{P}_1 \\ \mathbf{P}_2 \\ \mathbf{P}_3 \\ \mathbf{P}_4 \end{pmatrix}. \tag{8.17}$$

We are going to place the new odd-numbered pixel $s_{j+1,2k+1}$ in the middle of the group of its four level-j neighbors, so it makes sense to assign it the value of the interpolating polynomial in the middle of its interval, i.e., $\mathbf{P}(0.5)$. Calculated from Equation (8.17), this value is

$$\mathbf{P}(0.5) = -0.0625\mathbf{P}_1 + 0.5625\mathbf{P}_2 + 0.5625\mathbf{P}_3 - 0.0625\mathbf{P}_4.$$

(Notice that the four weights add up to one. This is an example of barycentric functions. Also, see Exercise 7.44 for an explanation of the negative weights.)

The subdivision rule in this case is [by analogy with Equation (8.16)]

$$s_{j+1,2k} = s_{j,k}, \quad s_{j+1,2k+1} = \mathbf{P}_{3,j,k-1}(0.5), \tag{8.18}$$

where the notation $\mathbf{P}_{3,j,k-1}(t)$ indicates the degree-3 interpolating polynomial for the group of four level-j pixels that starts at $s_{j,k-1}$. We define the *order* of this subdivision to be 4 (the number of interpolated pixels). Since linear subdivision interpolates two pixels, its order is 2. Figure 8.51 shows three levels of pixels generated by linear (8.51a) and by cubic (8.51b) subdivisions.

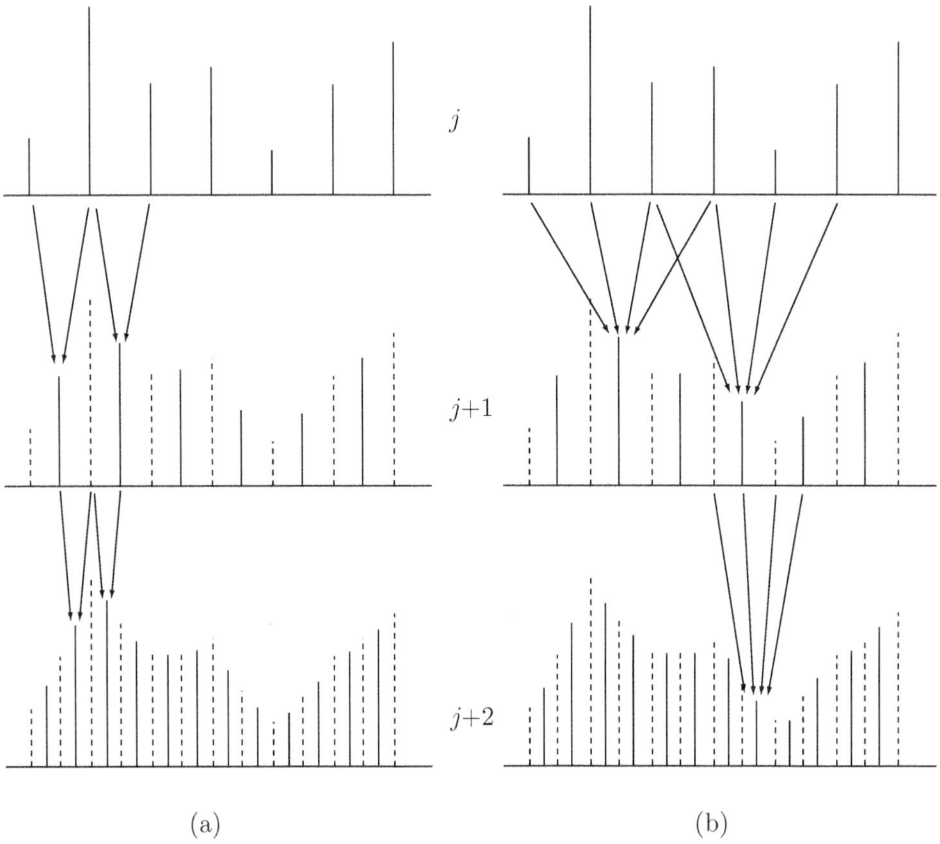

Figure 8.51: Linear and Cubic Subdivisions.

It is now obvious how interpolating subdivision can be extended to higher orders. Select an even integer n, derive the degree-$(n-1)$ polynomial $\mathbf{P}_{n-1,j,k-(n/2-1)}(t)$ that will interpolate the n level-j pixels

$$s_{j,k-(n/2-1)},\ s_{j,k-n/2},\ s_{j,k-n/2+1},\ldots,s_{j,k},\ s_{j,k+1},\ldots,s_{j,k+n/2},$$

calculate the midpoint $\mathbf{P}_{n-1,j,k-(n/2-1)}(0.5)$, and generate the level-$(j+1)$ pixels according to

$$s_{j+1,2k} = s_{j,k}, \quad s_{j+1,2k+1} = \mathbf{P}_{n-1,j,k-(n/2-1)}(0.5), \tag{8.19}$$

◇ **Exercise 8.14:** Compute the midpoint $\mathbf{P}(0.5)$ of the degree-5 interpolating polynomial for six points $\mathbf{P}_1$ through $\mathbf{P}_6$ as a function of the points.

8.11.3 Scaling Functions

The scaling functions $\phi_{j,k}(x)$ have been mentioned in Section 8.6, in connection with the Haar transform. Here we show how they are constructed for an interpolating subdivision. Each coefficient $s_{j,k}$ computed in level j has a scaling function $\phi_{j,k}(x)$ associated with it, which is defined as follows: Select values n, j, and k. Set pixel $s_{j,k}$ to 1 and all other $s_{j,i}$ to 0 (this can be expressed as $s_{j,k} = \delta_{0,k}$). Use a subdivision scheme (based on n points) to compute levels $(j+1)$, $(j+2)$, etc. Each level has twice the number of pixels as its predecessor. In the limit, we end up with an infinite number of pixels. We can view these pixels as real numbers and define the range of the scaling function $\phi_{j,k}(x)$ as these numbers. Each pair of values j and k defines a different scaling function $\phi_{j,k}(x)$, but we can intuitively see that the scaling functions depend on j and k in simple ways. The shape of $\phi_{j,k}(x)$ does not depend on k, since we calculate ϕ by setting $s_{j,k} = 1$ and all other $s_{j,i} = 0$. Thus, the function $\phi_{j,8}(x)$ is a shifted copy of $\phi_{j,7}(x)$, a twice shifted copy of $\phi_{j,6}(x)$, etc. In general, we may write $\phi_{j,k}(x) = \phi_{j,0}(x-k)$. To understand the dependence of $\phi_{j,k}(x)$ on j, the reader should recall the following sentence (from page 804):

"The original elements $s_{0,k}$ of s_0 are now stored in a in locations $k{\cdot}2^j$."

This implies that if we select a small value for j, we end up with a wide scaling function $\phi_{j,k}(x)$. In general, we have

$$\phi_{j,k}(x) = \phi_{0,0}(2^j x - k) \stackrel{\text{def}}{=} \phi(2^j x - k),$$

implying that all the scaling functions for different values of j and k are translations and dilations (scaling) of $\phi(x)$, the fundamental solution of the subdivision process. $\phi(x)$ is shown in Figure 8.52 for $n = 2$, 4, 6, and 8.

The main properties of $\phi(x)$ are compact support and smoothness (it can be nonzero only inside the interval $[-(n-1), n-1]$), but it also satisfies a *refinement relation* of the form

$$\phi(x) = \sum_{l=-n}^{n} h_l \phi(2x - l),$$

where h_l are called the *filter coefficients*. A change of variables allows us to write the same refinement relation in the form

$$\phi_{j,k}(x) = \sum_l h_{l-2k} \phi_{j+1,l}(x).$$

The odd-numbered filter coefficients are the coefficients of the interpolating polynomial at its midpoint $\mathbf{P}_{n-1,j,k-(n/2-1)}(0.5)$. The even-numbered coefficients are zero except

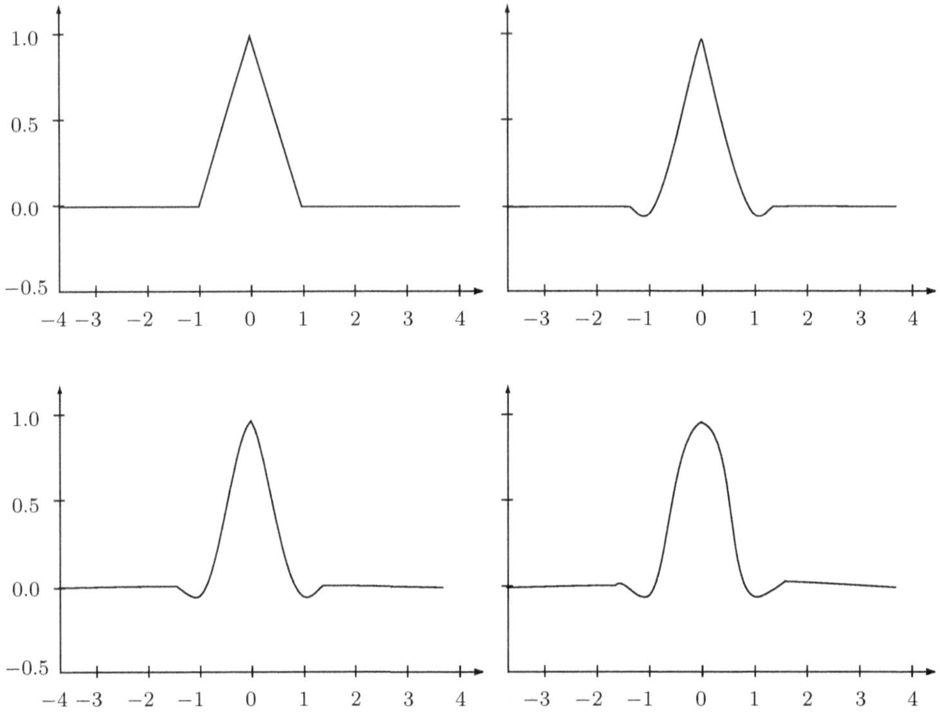

Figure 8.52: Scaling Functions $\phi_{j,k}(x)$ for n=2, 4, 6, and 8.

for $l = 0$ (this property can be expressed as $h_{2l} = \delta_{0,l}$). The general expression for h_k is

$$h_k = \begin{cases} k \text{ even}, & \delta_{k,0}, \\ k \text{ odd}, & (-1)^{d+k} \frac{\Pi_{i=0}^{2d-1}(i-d+1/2)}{(k+1/2)(d+k)!(d-k-1)!}, \end{cases}$$

where $d = n/2$. For linear subdivision, the filter coefficients are $(1/2, 1, 1/2)$. For cubic interpolation the eight filter coefficients are $(-1/16, 0, 9/16, 1, 9/16, 0, -1/16)$. The filter coefficients are useful, since they allow us to write the interpolating subdivision in the form

$$s_{j+1,l} = \sum_k h_{l-2k} s_{j,k}.$$

8.12 The IWT

The DWT is simple but has an important drawback, namely, it uses noninteger filter coefficients, which is why it produces noninteger transform coefficients. There are several ways to modify the basic DWT such that it produces integer transform coefficients. This section describes a simple *integer wavelet transform* (IWT) that can be used to decompose an image in any of the ways described in Section 8.10. The transform is reversible, i.e., the image can be fully reconstructed from the (integer) transform coefficients. This IWT can be used to compress the image either lossily (by quantizing the transform coefficients) or losslessly (by entropy coding the transform coefficients).

The simple principle of this transform is illustrated here for the one-dimensional case. Given a data vector of N integers x_i, where $i = 0, 1, \ldots, N-1$, we define $k = N/2$ and we compute the transform vector y_i by calculating the odd and even components of y separately. We first discuss the case where N is even. The $N/2$ odd components y_{2i+1} (where $i = 0, 1, \ldots, k-1$) are calculated as differences of the x_i's. They become the detail (high frequency) transform coefficients. Each of the even components y_{2i} (where i varies in the same range $[0, k-1]$) is calculated as a weighted average of five data items x_i. These $N/2$ numbers become the low-frequency transform coefficients, and are normally transformed again, into $N/4$ low-frequency and $N/4$ high-frequency coefficients.

The basic rule for the odd transform coefficients is

$$y_{2i+1} = -\frac{1}{2}x_{2i} + x_{2i+1} - \frac{1}{2}x_{2i+2},$$

except the last coefficient, where $i = k-1$, which is computed as the simple difference $y_{2k-1} = x_{2k-1} - x_{2k-2}$. This can be summarized as

$$y_{2i+1} = \begin{cases} x_{2i+1} - (x_{2i} + x_{2i+2})/2, & \text{for } i = 0, 1, \ldots, k-2, \\ x_{2i+1} - x_{2i}, & \text{for } i = k-1. \end{cases} \tag{8.20}$$

The even transform coefficients are calculated as the weighted average

$$y_{2i} = -\frac{1}{8}x_{2i-2} + \frac{1}{4}x_{2i-1} + \frac{3}{4}x_{2i} + \frac{1}{4}x_{2i+1} - \frac{1}{8}x_{2i+2},$$

except the first coefficient, where $i = 0$, which is calculated as

$$y_0 = \frac{3}{4}x_0 + \frac{1}{2}x_1 - \frac{1}{4}x_2.$$

In practice, this calculation is done by computing each even coefficient y_{2i} in terms of x_{2i} and the two odd coefficients y_{2i-1} and y_{2i+1}. This can be summarized by

$$y_{2i} = \begin{cases} x_{2i} + y_{2i+1}/2, & \text{for } i = 0, \\ x_{2i} + (y_{2i-1} + y_{2i+1})/4, & \text{for } i = 1, 2, \ldots, k-1. \end{cases} \tag{8.21}$$

The inverse transform is easy to figure out. It uses the transform coefficients y_i to calculate data items z_i that are identical to the original x_i. It first computes the even elements

$$z_{2i} = \begin{cases} y_{2i} - y_{2i+1}/2, & \text{for } i = 0, \\ y_{2i} - (y_{2i-1} + y_{2i+1})/4, & \text{for } i = 1, 2, \ldots, k-1, \end{cases} \tag{8.22}$$

and then the odd elements

$$z_{2i+1} = \begin{cases} y_{2i+1} + (z_{2i} + z_{2i+2})/2, & \text{for } i = 0, 1, \ldots, k-2, \\ y_{2i+1} + z_{2i}, & \text{for } i = k-1. \end{cases} \tag{8.23}$$

Now comes the interesting part. The transform coefficients calculated by Equations (8.20) and (8.21) are generally nonintegers, because of the divisions by 2 and 4. The same is true for the reconstructed data items of Equations (8.22) and (8.23). The main feature of the particular IWT described here is the use of *truncation*. Truncation, denoted by the "floor" symbols $\lfloor$ and $\rfloor$, is used to produce integer transform coefficients y_i **and** also integer reconstructed data items z_i. Equations (8.20) through (8.23) are modified to

$$\begin{aligned}
y_{2i+1} &= \begin{cases} x_{2i+1} - \lfloor (x_{2i} + x_{2i+2})/2 \rfloor, & \text{for } i = 0, 1, \ldots, k-2, \\ x_{2i+1} - x_{2i}, & \text{for } i = k-1. \end{cases} \\
y_{2i} &= \begin{cases} x_{2i} + \lfloor y_{2i+1}/2 \rfloor, & \text{for } i = 0, \\ x_{2i} + \lfloor (y_{2i-1} + y_{2i+1})/4 \rfloor, & \text{for } i = 1, 2, \ldots, k-1. \end{cases} \\
z_{2i} &= \begin{cases} y_{2i} - \lfloor y_{2i+1}/2 \rfloor, & \text{for } i = 0, \\ y_{2i} - \lfloor (y_{2i-1} + y_{2i+1})/4 \rfloor, & \text{for } i = 1, 2, \ldots, k-1, \end{cases} \\
z_{2i+1} &= \begin{cases} y_{2i+1} + \lfloor (z_{2i} + z_{2i+2})/2 \rfloor, & \text{for } i = 0, 1, \ldots, k-2, \\ y_{2i+1} + z_{2i}, & \text{for } i = k-1. \end{cases}
\end{aligned} \tag{8.24}$$

Because of truncation, some information is lost when the y_i are calculated. However, truncation is also used in the calculation of the z_i, which restores the lost information. Thus, Equation (8.24) is a true forward and inverse IWT that reconstructs the original data items exactly.

◇ **Exercise 8.15:** Given the data vector $x = (112, 97, 85, 99, 114, 120, 77, 80)$, use Equation (8.24) to calculate its forward and inverse integer wavelet transforms.

The same concepts can be applied to the case where the number N of data items is odd. We first define k by $N = 2k + 1$, then define the forward and inverse integer transforms by

$$\begin{aligned}
y_{2i+1} &= x_{2i+1} - \lfloor (x_{2i} + x_{2i+2})/2 \rfloor, \text{ for } i = 0, 1, \ldots, k-1, \\
y_{2i} &= \begin{cases} x_{2i} + \lfloor y_{2i+1}/2 \rfloor, & \text{for } i = 0, \\ x_{2i} + \lfloor (y_{2i-1} + y_{2i+1})4 \rfloor, & \text{for } i = 1, 2, \ldots, k-1, \\ x_{2i} + \lfloor y_{2i-1}/2 \rfloor, & \text{for } i = k. \end{cases} \\
z_{2i} &= \begin{cases} y_{2i} - \lfloor y_{2i+1}/2 \rfloor, & \text{for } i = 0, \\ y_{2i} - \lfloor (y_{2i-1} + y_{2i+1})/4 \rfloor, & \text{for } i = 1, 2, \ldots, k-1, \\ y_{2i} - \lfloor y_{2i-1}/2 \rfloor, & \text{for } i = k, \end{cases} \\
z_{2i+1} &= y_{2i+1} + \lfloor (z_{2i} + z_{2i+2})/2 \rfloor, \text{ for } i = 0, 1, \ldots, k-1.
\end{aligned}$$

Notice that the IWT produces a vector y_i where the detail coefficients and the weighted averages are interleaved. The algorithm should be modified to place the averages in the first half of y and the details in the second half.

The extension of this transform to the two-dimensional case is obvious. The IWT is applied to the rows and the columns of the image using any of the image decomposition methods discussed in Section 8.10.

8.13 The Laplacian Pyramid

The main feature of the Laplacian pyramid method [Burt and Adelson 83] is progressive compression. The decoder inputs the compressed stream section by section, and each section improves the appearance on the screen of the image-so-far. The method uses both prediction and transform techniques, but its computations are simple and local (i.e., there is no need to examine or use values that are far away from the current pixel). The name "Laplacian" comes from the field of image enhancement, where it is used to indicate operations similar to the ones used here. We start with a general description of the method.

We denote by $g_0(i,j)$ the original image. A new, reduced image g_1 is computed from g_0 such that each pixel of g_1 is a weighted sum of a group of 5×5 pixels of g_0. Image g_1 is computed [see Equation (8.25)] such that it has half the number of rows and half the number of columns of g_0, so it is one-quarter the size of g_0. It is a blurred (or lowpass filtered) version of g_0. The next step is to expand g_1 to an image $g_{1,1}$ the size of g_0 by interpolating pixel values [Equation (8.26)]. A difference image (also called an error image) L_0 is calculated as the difference $g_0 - g_{1,1}$, and it becomes the bottom level of the Laplacian pyramid. The original image g_0 can be reconstructed from L_0 and $g_{1,1}$, and also from L_0 and g_1. Since g_1 is smaller than $g_{1,1}$, it makes sense to write L_0 and g_1 on the compressed stream. The size of L_0 equals that of g_0, and the size of g_1 is 1/4 of that, so it seems that we have generated expansion, but in fact, compression is achieved, because the error values in L_0 are decorrelated to a high degree, and so are small (and therefore have small variance and low entropy) and can be represented with fewer bits than the original pixels in g_0.

In order to achieve progressive representation of the image, only L_0 is written on the output, and the process is repeated on g_1. A new, reduced image g_2 is computed from g_1 (the size of g_2 is 1/16 that of g_0). It is expanded to an image $g_{2,1}$, and a new difference image L_1 is computed as the difference $g_1 - g_{2,1}$ and becomes the next level, above L_0, of the Laplacian pyramid. The final result is a sequence L_0, $L_1,\ldots$, L_{k-1}, L_k where the first k items are difference images and the last one, L_k, is simply the (very small) reduced image g_k. These items constitute the Laplacian pyramid. They are written on the compressed stream in reverse order, so the decoder inputs L_k first, uses it to display a small, blurred image, inputs L_{k-1}, reconstructs and displays g_{k-1} (which is four times bigger), and repeats until g_0 is reconstructed and displayed. This process can be summarized by

$$\begin{aligned} g_k &= L_k, \\ g_i &= L_i + \text{Expand}(g_{i+1}) = L_i + g_{i+1,1}, \quad \text{for } i = k-1, k-2, \ldots, 2, 1, 0. \end{aligned}$$

The user sees small, blurred images that grow and become sharper. The decoder can be modified to expand each intermediate image g_i (before it is displayed) several times, by interpolating pixel values, until it gets to the size of the original image g_0. This way, the user sees an image that progresses from very blurred to sharp, while remaining the same size. For example, image g_3, which is 1/64th the size of g_0, can be brought to that size by expanding it three times, yielding the chain

$$g_{3,3} = \text{Expand}(g_{3,2}) = \text{Expand}(g_{3,1}) = \text{Expand}(g_3).$$

In order for all the intermediate g_i and L_i images to have well-defined dimensions, the original image g_0 should have $R = M_R 2^M + 1$ rows and $C = M_C 2^M + 1$ columns, where M_R, M_C, and M are integers. Selecting, for example, $M_R = M_C$ results in a square image. Image g_1 has dimensions $(M_R 2^{M-1} + 1) \times (M_C 2^{M-1} + 1)$, and image g_p has dimensions $(M_R 2^{M-p} + 1) \times (M_C 2^{M-p} + 1)$. An example is $M_R = M_C = 1$ and $M = 8$. The dimensions of the original image are $(2^8+1) \times (2^8+1) = 257 \times 257$ and the reduced images g_1 through g_5 have dimensions $(2^7+1) \times (2^7+1) = 129 \times 129$, 65×65, 33×33, 17×17, and 9×9, respectively.

◇ **Exercise 8.16:** Calculate the dimensions of the first six images g_0 through g_5 for the case $M_C = 3$, $M_R = 4$, and $M = 5$.

We now turn to the details of reducing and expanding images. Reducing an image g_{p-1} to an image g_p of dimensions $R_p \times C_p$ (where $R_p = M_R 2^{M-p} + 1$, and $C_p = M_C 2^{M-p} + 1$) is done by

$$g_p(i,j) = \sum_{m=-2}^{2} \sum_{n=-2}^{2} w(m,n) g_{p-1}(2i+m, 2j+n), \tag{8.25}$$

where $i = 0, 1, \ldots, C_p - 1$, $j = 0, 1, \ldots, R_p - 1$, and p (the level) varies from 1 to $k-1$. Each pixel of g_p is a weighted sum of 5×5 pixels of g_{p-1} with weights $w(m,n)$ that are the same for all levels p. Figure 8.53a illustrates this process in one dimension. It shows how each pixel in a higher level of the pyramid is generated as a weighted sum of five pixels from the level below it, and how each level has (about) half the number of pixels of its predecessor. Figure 8.53b is in two dimensions. It shows how a pixel in a high level is obtained from 25 pixels located one level below. Some of the weights are also shown. Notice that in this case each level has about 1/4 the number of pixels of its predecessor.

The weights $w(m,n)$ (also called the generating kernel) are determined by first separating each in the form $w(m,n) = \hat{w}(m)\hat{w}(n)$, where the functions $\hat{w}(m)$ should be normalized, i.e.,

$$\sum_{m=-2}^{2} \hat{w}(m) = 1,$$

and symmetric, $\hat{w}(m) = \hat{w}(-m)$. These two constraints are not enough to determine $\hat{w}(m)$, so we add a third one, called *equal contribution*, that demands that all the pixels at a given level contribute the same total weight (= 1/4) to pixels at the next higher

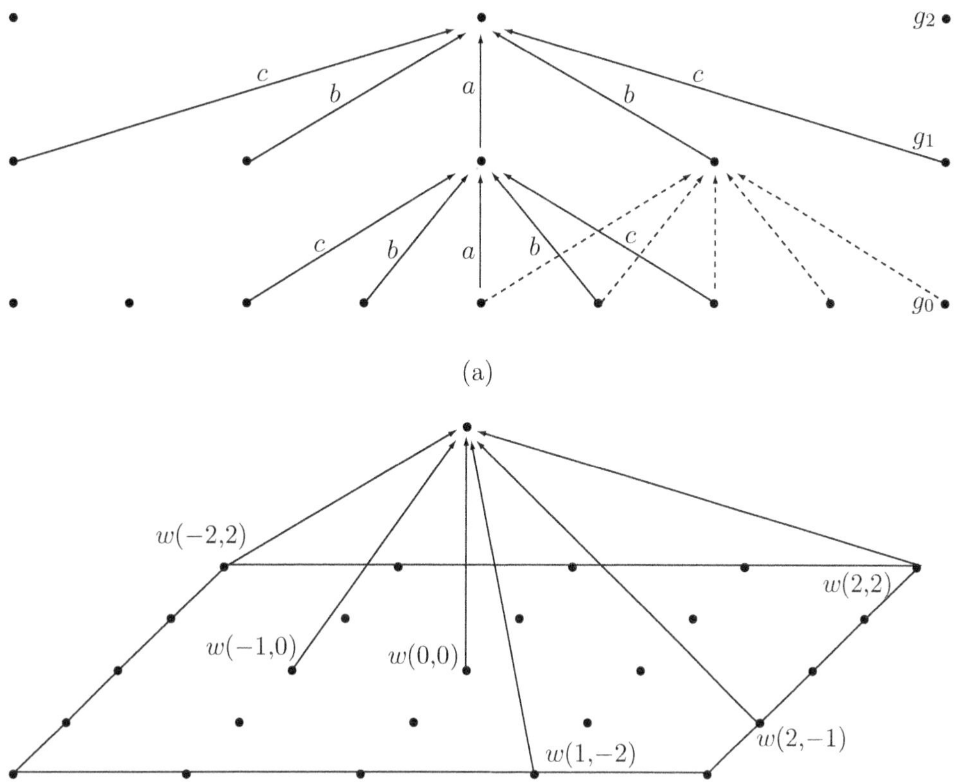

Figure 8.53: Illustrating Reduction.

level. If we set $\hat{w}(0) = a$, $\hat{w}(-1) = \hat{w}(1) = b$, and $\hat{w}(-2) = \hat{w}(2) = c$, then the three constraints are satisfied if

$$w(0) = a, \quad \hat{w}(-1) = \hat{w}(1) = 0.25, \quad \hat{w}(-2) = \hat{w}(2) = 0.25 - a/2.$$

(The reader should compare this to the discussion of interpolating polynomials in Section 7.25.4.)

Experience recommends setting $a = 0.6$, which yields (see values of $w(m,n)$ in Table 8.54)

$$\hat{w}(0) = 0.6, \quad \hat{w}(-2) = \hat{w}(2) = 0.25, \quad \hat{w}(-1) = \hat{w}(1) = 0.25 - 0.3 = -0.05.$$

The same weights used in reducing images are also used in expanding them. Expanding an image g_p of dimensions $R_p \times C_p$ (where $R_p = M_R 2^{M-p} + 1$, and $C_p = M_C 2^{M-p} + 1$) to an image $g_{p,1}$ that has the same dimensions as g_{p-1} (i.e., is four times bigger than g_p) is done by

$$g_{p,1}(i,j) = 4 \sum_{m=-2}^{2} \sum_{n=-2}^{2} w(m,n) g_p\big([i-m]/2, [j-n]/2\big), \tag{8.26}$$

where $i = 0, 1, \ldots, C_p - 1$, $j = 0, 1, \ldots, R_p - 1$, and the sums include only terms for which both $(i-m)/2$ and $(j-n)/2$ are integers. As an example we compute the single pixel

$$g_{1,1}(4,5) = 4 \sum_{m=-2}^{2} \sum_{n=-2}^{2} w(m,n) g_1\big([4-m]/2, [5-n]/2\big).$$

Of the 25 terms of this sum, only six, namely those with $m = -2, 0, 2$ and $n = -1, 1$, satisfy the condition above and are included. The six terms correspond to

$$(m,n) = (-2,-1),\ (-2,1),\ (0,-1),\ (0,1),\ (2,-1),\ (2,1),$$

and the sum is

$$\begin{aligned} 4\big[&w(-2,-1)g_1(3,3) + w(-2,1)g_1(3,2) + w(0,-1)g_1(2,3) \\ &+ w(0,1)g_1(2,2) + w(2,-1)g_1(1,3) + w(2,1)g_1(1,2)\big]. \end{aligned}$$

	−0.05	0.25	0.6	0.25	−0.05
−0.05	0.0025	−0.0125	−0.03	−0.01	0.0025
0.25		0.0625	0.15	0.0625	−0.0125
0.6			0.36	0.15	−0.03
0.25				0.0625	−0.0125
−0.05					0.0025

Table 8.54: Values of $w(m,n)$ for $a = 0.6$.

A lossy version of the Laplacian pyramid can be obtained by quantizing the values of each L_i image before it is encoded and written on the compressed stream.

8.14 SPIHT

Section 8.6 shows how the Haar transform can be applied several times to an image, creating regions (or subbands) of averages and details. The Haar transform is simple, and better compression can be achieved by other wavelet filters. It seems that different wavelet filters produce different results depending on the image type, but it is currently not clear what filter is the best for any given image type. Regardless of the particular filter used, the image is decomposed into subbands, such that lower subbands correspond to higher image frequencies (they are the highpass levels) and higher subbands correspond to lower image frequencies (lowpass levels), where most of the image energy is concentrated (Figure 8.55). This is why we can expect the detail coefficients to get smaller as we move from high to low levels. Also, there are spatial similarities among the subbands (Figure 8.17b). An image part, such as an edge, occupies the same spatial position in each subband. These features of the wavelet decomposition are exploited by the SPIHT (set partitioning in hierarchical trees) method [Said and Pearlman 96].

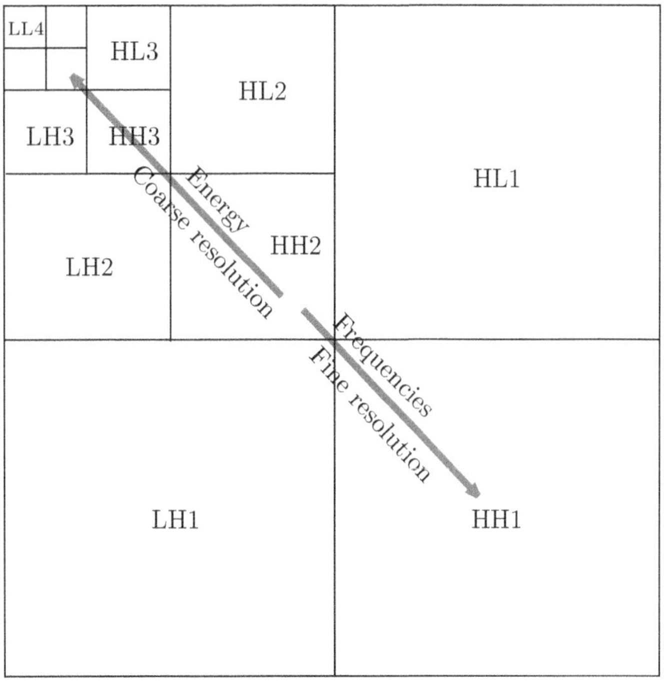

Figure 8.55: Subbands and Levels in Wavelet Decomposition.

SPIHT was designed for optimal progressive transmission, as well as for compression. One of the important features of SPIHT (perhaps a unique feature) is that at any point during the decoding of an image, the quality of the displayed image is the best that can be achieved for the number of bits input by the decoder up to that moment.

Another important SPIHT feature is its use of embedded coding. This feature is defined as follows: If an (embedded coding) encoder produces two files, a large one of size M and a small one of size m, then the smaller file is identical to the first m bits of the larger file.

The following example aptly illustrates the meaning of this definition. Suppose that three users wait for you to send them a certain compressed image, but they need different image qualities. The first one needs the quality contained in a 10 Kb file. The image qualities required by the second and third users are contained in files of sizes 20 Kb and 50 Kb, respectively. Most lossy image compression methods would have to compress the same image three times, at different qualities, to generate three files with the right sizes. SPIHT, on the other hand, produces one file, and then three chunks—of lengths 10 Kb, 20 Kb, and 50 Kb, all starting at the beginning of that file—can be sent to the three users, thereby satisfying their needs.

We start with a general description of SPIHT. We denote the pixels of the original image $\mathbf{p}$ by $p_{i,j}$. Any set $\mathbf{T}$ of wavelet filters can be used to transform the pixels to wavelet coefficients (or transform coefficients) $c_{i,j}$. These coefficients constitute the transformed image $\mathbf{c}$. The transformation is denoted by $\mathbf{c} = \mathbf{T}(\mathbf{p})$. In a progressive transmission method, the decoder starts by setting the reconstruction image $\hat{\mathbf{c}}$ to zero. It then inputs (encoded) transform coefficients, decodes them, and uses them to generate an improved reconstruction image $\hat{\mathbf{c}}$, which, in turn, is used to produce a better image $\hat{\mathbf{p}}$. We can summarize this operation by $\hat{\mathbf{p}} = \mathbf{T}^{-1}(\hat{\mathbf{c}})$.

The main aim in progressive transmission is to transmit the most important image information first. This is the information that results in the largest reduction of the distortion (the difference between the original and the reconstructed images). SPIHT uses the mean squared error (MSE) distortion measure [Equation (7.2)]

$$D_{\text{mse}}(\mathbf{p}-\hat{\mathbf{p}}) = \frac{|\mathbf{p}-\hat{\mathbf{p}}|^2}{N} = \frac{1}{N}\sum_i\sum_j (p_{i,j}-\hat{p}_{i,j})^2,$$

where N is the total number of pixels. An important consideration in the design of SPIHT is the fact that this measure is invariant to the wavelet transform, a feature that allows us to write

$$D_{\text{mse}}(\mathbf{p}-\hat{\mathbf{p}}) = D_{\text{mse}}(\mathbf{c}-\hat{\mathbf{c}}) = \frac{|\mathbf{p}-\hat{\mathbf{p}}|^2}{N} = \frac{1}{N}\sum_i\sum_j (c_{i,j}-\hat{c}_{i,j})^2. \tag{8.27}$$

Equation (8.27) shows that the MSE decreases by $|c_{i,j}|^2/N$ when the decoder receives the transform coefficient $c_{i,j}$ (we assume that the decoder receives the exact value of the coefficient, i.e., there is no loss of precision due to limitations imposed by computer arithmetic). It is now clear that the largest coefficients $c_{i,j}$ (largest in absolute value, regardless of their signs) contain the information that reduces the MSE distortion most, so a progressive encoder should send those coefficients first. This is an important principle of SPIHT.

Another principle is based on the observation that the most significant bits of a binary integer whose value is close to maximum tend to be ones. This suggests that the most significant bits contain the most important image information, and that they should be sent to the decoder first (or written first on the compressed stream).

The progressive transmission method used by SPIHT incorporates these two principles. SPIHT sorts the coefficients and transmits their most significant bits first. To simplify the description, we first assume that the sorting information is explicitly transmitted to the decoder; the next section shows an efficient way to code this information.

We now show how the SPIHT encoder uses these principles to progressively transmit the wavelet coefficients to the decoder (or write them on the compressed stream), starting with the most important information. We assume that a wavelet transform has already been applied to the image (SPIHT is a coding method, so it can work with any wavelet transform) and that the transformed coefficients $c_{i,j}$ are already stored in memory. The coefficients are sorted (ignoring their signs), and the sorting information is contained in an array m such that array element $m(k)$ contains the (i, j) coordinates of a coefficient $c_{i,j}$, and such that $|c_{m(k)}| \geq |c_{m(k+1)}|$ for all values of k. Table 8.56 lists hypothetical values of 16 coefficients. Each is shown as a 16-bit number where the most significant bit (bit 15) is the sign and the remaining 15 bits (numbered 14 through 0, top to bottom) constitute the magnitude. The first coefficient $c_{m(1)} = c_{2,3}$ is $s1aci\ldots r$ (where s, a, etc., are bits). The second one $c_{m(2)} = c_{3,4}$ is $s1bdj\ldots s$, and so on.

k		1	2	3	4	5	6	7	8	9	10	11	12	13	14	15	16
	sign	s	s	s	s	s	s	s	s	s	s	s	s	s	s	s	s
msb	14	1	1	0	0	0	0	0	0	0	0	0	0	0	0	0	0
	13	a	b	1	1	1	1	0	0	0	0	0	0	0	0	0	0
	12	c	d	e	f	g	h	1	1	1	0	0	0	0	0	0	0
	11	i	j	k	l	m	n	o	p	q	1	0	0	0	0	0	0
	$\vdots$	$\vdots$	$\vdots$														$\vdots$
lsb	0	r	s	t	u	v	w	x	y		...		...		...		z
$m(k) = i,j$		2,3	3,4	3,2	4,4	1,2	3,1	3,3	4,2	4,1	...		...		...		4,3

Table 8.56: Transform Coefficients Ordered by Absolute Magnitudes.

The sorting information that the encoder has to transmit is the sequence $m(k)$, or

$$(2,3),\ (3,4),\ (3,2),\ (4,4),\ (1,2),\ (3,1),\ (3,3),\ (4,2),\ldots,(4,3).$$

In addition, it has to transmit the 16 signs, and the 16 coefficients in order of significant bits. A direct transmission would send the 16 numbers

$$ssssssssssssssss,\quad 1100000000000000,\quad ab11110000000000,$$
$$cdefgh1110000000,\quad ijklmnopq1000000,\ \ldots,\ rstuvwxy\ldots z,$$

but this is clearly wasteful. Instead, the encoder goes into a loop, where in each iteration it performs a *sorting step* and a *refinement step*. In the first iteration it transmits the number $l = 2$ (the number of coefficients $c_{i,j}$ in our example that satisfy $2^{14} \leq |c_{i,j}| < 2^{15}$) followed by the two pairs of coordinates $(2,3)$ and $(3,4)$ and by the signs of the first two coefficients. This is done in the first sorting pass. This information enables the decoder to construct approximate versions of the 16 coefficients as follows: Coefficients $c_{2,3}$ and $c_{3,4}$ are constructed as the 16-bit numbers $s100\ldots0$. The remaining 14 coefficients are constructed as all zeros. This is how the most significant bits of the largest coefficients are transmitted to the decoder first.

The next step of the encoder is the refinement pass, but this is not performed in the first iteration.

In the second iteration the encoder performs both passes. In the sorting pass it transmits the number $l = 4$ (the number of coefficients $c_{i,j}$ in our example that satisfy $2^{13} \le |c_{i,j}| < 2^{14}$), followed by the four pairs of coordinates $(3,2)$, $(4,4)$, $(1,2)$, and $(3,1)$ and by the signs of the four coefficients. In the refinement step it transmits the two bits a and b. These are the 14th most significant bits of the two coefficients transmitted in the previous iteration.

The information received so far enables the decoder to improve the 16 approximate coefficients constructed in the previous iteration. The first six become

$$c_{2,3} = s1a0\ldots0,\; c_{3,4} = s1b0\ldots0,\; c_{3,2} = s0100\ldots0,$$
$$c_{4,4} = s0100\ldots0,\; c_{1,2} = s0100\ldots0,\; c_{3,1} = s0100\ldots0,$$

and the remaining 10 coefficients are not changed.

◇ **Exercise 8.17:** Perform the sorting and refinement passes of the next (third) iteration.

The main steps of the SPIHT encoder should now be easy to understand. They are as follows:

Step 1: Given an image to be compressed, perform its wavelet transform using any suitable wavelet filter, decompose it into transform coefficients $c_{i,j}$, and represent the resulting coefficients with a fixed number of bits. (In the discussion that follows we use the terms *pixel* and *coefficient* interchangeably.) We assume that the coefficients are represented as 16-bit signed-magnitude numbers. The leftmost bit is the sign, and the remaining 15 bits are the magnitude. (Notice that the sign-magnitude representation is different from the 2's complement method, which is used by computer hardware to represent signed numbers.) Such numbers can have values from $-(2^{15}-1)$ to $2^{15}-1$. Set n to $\lfloor \log_2 \max_{i,j}(c_{i,j}) \rfloor$. In our case n will be set to $\lfloor \log_2(2^{15}-1) \rfloor = 14$.

Step 2: Sorting pass: Transmit the number l of coefficients $c_{i,j}$ that satisfy $2^n \le |c_{i,j}| < 2^{n+1}$. Follow with the l pairs of coordinates and the l sign bits of those coefficients.

Step 3: Refinement pass: Transmit the nth most significant bit of all the coefficients satisfying $|c_{i,j}| \ge 2^{n+1}$. These are the coefficients that were selected in previous sorting passes (not including the immediately preceding sorting pass).

Step 4: Iterate: Decrement n by 1. If more iterations are needed (or desired), go back to Step 2.

The last iteration is normally performed for $n = 0$, but the encoder can stop earlier, in which case the least important image information (some of the least significant bits of all the wavelet coefficients) will not be transmitted. This is the natural lossy option of SPIHT. It is equivalent to scalar quantization, but it produces better results than what is usually achieved with scalar quantization, since the coefficients are transmitted in sorted order. An alternative is for the encoder to transmit the entire image (i.e., all the bits of all the wavelet coefficients) and the decoder can stop decoding when the reconstructed image reaches a certain quality. This quality can either be predetermined by the user or automatically determined by the decoder at run time.

8.14.1 Set Partitioning Sorting Algorithm

The method as described so far is simple, since we have assumed that the coefficients had been sorted before the loop started. In practice, the image may have 1K$\times$1K pixels or more; there may be more than a million coefficients, so sorting all of them is too slow. Instead of sorting the coefficients, SPIHT uses the fact that sorting is done by comparing two elements at a time, and each comparison results in a simple yes/no result. Therefore, if both encoder and decoder use the same sorting algorithm, the encoder can simply send the decoder the sequence of yes/no results, and the decoder can use those to duplicate the operations of the encoder. This is true not just for sorting but for any algorithm based on comparisons or on any type of branching.

The actual algorithm used by SPIHT is based on the realization that there is really no need to sort *all* the coefficients. The main task of the sorting pass in each iteration is to select those coefficients that satisfy $2^n \le |c_{i,j}| < 2^{n+1}$. This task is divided into two parts. For a given value of n, if a coefficient $c_{i,j}$ satisfies $|c_{i,j}| \ge 2^n$, then we say that it is *significant*; otherwise, it is called *insignificant*. In the first iteration, relatively few coefficients will be significant, but their number increases from iteration to iteration, because n keeps getting decremented. The sorting pass has to determine which of the significant coefficients satisfies $|c_{i,j}| < 2^{n+1}$ and transmit their coordinates to the decoder. This is an important part of the algorithm used by SPIHT.

The encoder partitions all the coefficients into a number of sets T_k and performs the significance test

$$\max_{(i,j)\in T_k} |c_{i,j}| \ge 2^n\ ?$$

on each set T_k. The result may be either "no" (all the coefficients in T_k are insignificant, so T_k itself is considered insignificant) or "yes" (some coefficients in T_k are significant, so T_k itself is significant). This result is transmitted to the decoder. If the result is "yes," then T_k is partitioned by both encoder and decoder, using the same rule, into subsets and the same significance test is performed on all the subsets. This partitioning is repeated until all the significant sets are reduced to size 1 (i.e., they contain one coefficient each, and that coefficient is significant). This is how the significant coefficients are identified by the sorting pass in each iteration.

The significance test performed on a set T can be summarized by

$$S_n(T) = \begin{cases} 1, & \max_{(i,j)\in T} |c_{i,j}| \ge 2^n, \\ 0, & \text{otherwise.} \end{cases} \tag{8.28}$$

The result, $S_n(T)$, is a single bit that is transmitted to the decoder. The result of each significance test becomes a single bit written on the compressed stream, which is why the number of tests should be minimized. To achieve this goal, the sets should be created and partitioned such that sets expected to be significant will be large and sets that are expected to be insignificant will contain just one element.

8.14.2 Spatial Orientation Trees

The sets T_k are created and partitioned using a special data structure called a *spatial orientation tree*. This structure is defined in a way that exploits the spatial relationships between the wavelet coefficients in the different levels of the subband pyramid. Experience has shown that the subbands in each level of the pyramid exhibit spatial similarity (Figure 8.17b). Any special features, such as a straight edge or a uniform region, are visible in all the levels at the same location.

The spatial orientation trees are illustrated in Figure 8.57a,b for a 16×16 image. The figure shows two levels, level 1 (the highpass) and level 2 (the lowpass). Each level is divided into four subbands. Subband LL2 (the lowpass subband) is divided into four groups of 2×2 coefficients each. Figure 8.57a shows the top-left group, and Figure 8.57b shows the bottom-right group. In each group, each of the four coefficients (except the top-left one, marked in gray) becomes the root of a spatial orientation tree. The arrows show examples of how the various levels of these trees are related. The thick arrows indicate how each group of 4×4 coefficients in level 2 is the parent of four such groups in level 1. In general, a coefficient at location (i,j) in the image is the parent of the four coefficients at locations $(2i,2j)$, $(2i+1,2j)$, $(2i,2j+1)$, and $(2i+1,2j+1)$.

The roots of the spatial orientation trees of our example are located in subband LL2 (in general, they are located in the top-left LL subband, which can be of any size), but any wavelet coefficient, except the gray ones on level 1 (also except the leaves), can be considered the root of some spatial orientation subtree. The leaves of all those trees are located on level 1 of the subband pyramid.

In our example, subband LL2 is of size 4×4, so it is divided into four 2×2 groups, and three of the four coefficients of a group become roots of trees. Thus, the number of trees in our example is 12. In general, the number of trees is 3/4 the size of the highest LL subband.

Each of the 12 roots in subband LL2 in our example is the parent of four children located on the same level. However, the children of these children are located on level 1. This is true in general. The roots of the trees are located on the highest level, and their children are on the same level, but from then on, the four children of a coefficient on level k are themselves located on level $k-1$.

We use the terms *offspring* for the four children of a node, and *descendants* for the children, grandchildren, and all their descendants. The set partitioning sorting algorithm uses the following four sets of coordinates:

1. $\mathcal{O}(i,j)$: the set of coordinates of the four offspring of node (i,j). If node (i,j) is a leaf of a spatial orientation tree, then $\mathcal{O}(i,j)$ is empty.
2. $\mathcal{D}(i,j)$: the set of coordinates of the descendants of node (i,j).
3. $\mathcal{H}(i,j)$: the set of coordinates of the roots of all the spatial orientation trees (3/4 of the wavelet coefficients in the highest LL subband).
4. $\mathcal{L}(i,j)$: The difference set $\mathcal{D}(i,j)-\mathcal{O}(i,j)$. This set contains all the descendants of tree node (i,j), except its four offspring.

The spatial orientation trees are used to create and partition the sets T_k. The set partitioning rules are as follows:

1. The initial sets are $\{(i,j)\}$ and $\mathcal{D}(i,j)$, for all $(i,j)\in\mathcal{H}$ (i.e., for all roots of the spatial orientation trees). In our example there are 12 roots, so there will initially be 24

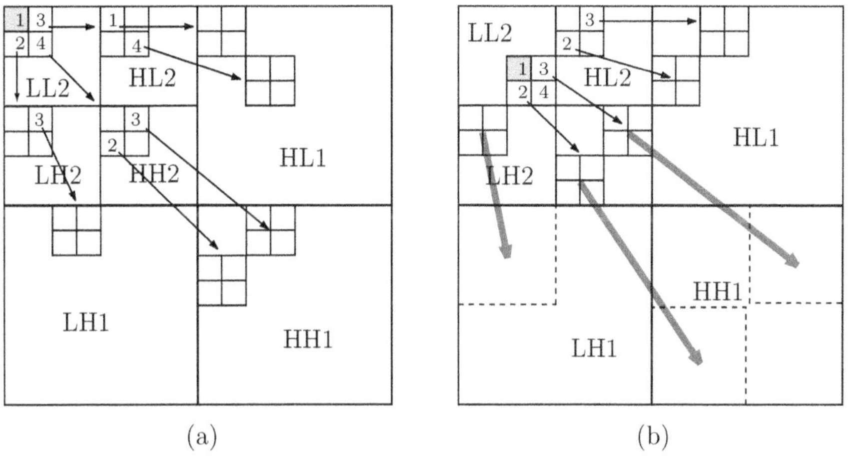

Figure 8.57: Spatial Orientation Trees in SPIHT.

sets: 12 sets, each containing the coordinates of one root, and 12 more sets, each with the coordinates of all the descendants of one root.
2. If set $\mathcal{D}(i,j)$ is significant, then it is partitioned into $\mathcal{L}(i,j)$ plus the four single-element sets with the four offspring of (i,j). In other words, if any of the descendants of node (i,j) is significant, then its four offspring become four new sets and all its other descendants become another set (to be significance tested in rule 3).
3. If $\mathcal{L}(i,j)$ is significant, then it is partitioned into the four sets $\mathcal{D}(k,l)$, where (k,l) are the offspring of (i,j).

Once the spatial orientation trees and the set partitioning rules are understood, the coding algorithm can be described.

8.14.3 SPIHT Coding

It is important to have the encoder and decoder test sets for significance in the same way. The coding algorithm therefore uses three lists called *list of significant pixels* (LSP), *list of insignificant pixels* (LIP), and *list of insignificant sets* (LIS). These are lists of coordinates (i,j) that in the LIP and LSP represent individual coefficients, and in the LIS represent either the set $\mathcal{D}(i,j)$ (a type A entry) or the set $\mathcal{L}(i,j)$ (a type B entry).

The LIP contains coordinates of coefficients that were insignificant in the previous sorting pass. In the current pass they are tested, and those that test significant are moved to the LSP. In a similar way, sets in the LIS are tested in sequential order, and when a set is found to be significant, it is removed from the LIS and is partitioned. The new subsets with more than one coefficient are placed back in the LIS, to be tested later, and the subsets with one element are tested and appended to the LIP or the LSP, depending on the results of the test. The refinement pass transmits the nth most significant bit of the entries in the LSP.

Figure 8.58 shows this algorithm in detail. Figure 8.59 is a simplified version, for readers who are intimidated by too many details.

The decoder executes the detailed algorithm of Figure 8.58. It always works in *lockstep* with the encoder, but the following notes shed more light on its operation:

1. Initialization: Set n to $\lfloor \log_2 \max_{i,j}(c_{i,j}) \rfloor$ and transmit n. Set the LSP to empty. Set the LIP to the coordinates of all the roots $(i,j) \in \mathcal{H}$. Set the LIS to the coordinates of all the roots $(i,j) \in \mathcal{H}$ that have descendants.
2. Sorting pass:
   - 2.1 for each entry (i,j) in the LIP do:
     - 2.1.1 output $S_n(i,j)$;
     - 2.1.2 if $S_n(i,j) = 1$, move (i,j) to the LSP and output the sign of $c_{i,j}$;
   - 2.2 for each entry (i,j) in the LIS do:
     - 2.2.1 if the entry is of type A, then
       - output $S_n(\mathcal{D}(i,j))$;
       - if $S_n(\mathcal{D}(i,j)) = 1$, then
         - * for each $(k,l) \in \mathcal{O}(i,j)$ do:
           - · output $S_n(k,l)$;
           - · if $S_n(k,l) = 1$, add (k,l) to the LSP, output the sign of $c_{k,l}$;
           - · if $S_n(k,l) = 0$, append (k,l) to the LIP;
         - * if $\mathcal{L}(i,j) \neq 0$, move (i,j) to the end of the LIS, as a type-B entry, and go to step 2.2.2; else, remove entry (i,j) from the LIS;
     - 2.2.2 if the entry is of type B, then
       - output $S_n(\mathcal{L}(i,j))$;
       - if $S_n(\mathcal{L}(i,j)) = 1$, then
         - * append each $(k,l) \in \mathcal{O}(i,j)$ to the LIS as a type-A entry:
         - * remove (i,j) from the LIS:
3. Refinement pass: for each entry (i,j) in the LSP, except those included in the last sorting pass (the one with the same n), output the nth most significant bit of $|c_{i,j}|$;
4. Loop: decrement n by 1 and go to step 2 if needed.

Figure 8.58: The SPIHT Coding Algorithm.

1. Set the threshold. Set LIP to all root nodes coefficients. Set LIS to all trees (assign type D to them). Set LSP to an empty set.
2. Sorting pass: Check the significance of all coefficients in LIP:
   - 2.1 If significant, output 1, output a sign bit, and move the coefficient to the LSP.
   - 2.2 If not significant, output 0.
3. Check the significance of all trees in the LIS according to the type of tree:
   - 3.1 For a tree of type D:
     - 3.1.1 If it is significant, output 1, and code its children:
       - 3.1.1.1 If a child is significant, output 1, then a sign bit, add it to the LSP
       - 3.1.1.2 If a child is insignificant, output 0 and add the child to the end of LIP.
       - 3.1.1.3 If the children have descendants, move the tree to the end of LIS as type L, otherwise remove it from LIS.
     - 3.1.2 If it is insignificant, output 0.
   - 3.2 For a tree of type L:
     - 3.2.1 If it is significant, output 1, add each of the children to the end of LIS as an entry of type D and remove the parent tree from the LIS.
     - 3.2.2 If it is insignificant, output 0.
4. Loop: Decrement the threshold and go to step 2 if needed.

Figure 8.59: A Simplified SPIHT Coding Algorithm.

1. Step 2.2 of the algorithm evaluates all the entries in the LIS. However, step 2.2.1 appends certain entries to the LIS (as type-B) and step 2.2.2 appends other entries to the LIS (as type-A). It is important to realize that all these entries are also evaluated by step 2.2 in the same iteration.
2. The value of n is decremented in each iteration, but there is no need to bring it all the way to zero. The loop can stop after any iteration, resulting in lossy compression. Normally, the user specifies the number of iterations, but it is also possible to have the user specify the acceptable amount of distortion (in units of MSE), and the encoder can use Equation (8.27) to decide when to stop the loop.
3. The encoder knows the values of the wavelet coefficients $c_{i,j}$ and uses them to calculate the bits S_n (Equation (8.28)), which it transmits (i.e., writes on the compressed stream). These bits are input by the decoder, which uses them to calculate the values of $c_{i,j}$. The algorithm executed by the decoder is that of Figure 8.58 but with the word "output" changed to "input."
4. The sorting information, previously denoted by $m(k)$, is recovered when the coordinates of the significant coefficients are appended to the LSP in steps 2.1.2 and 2.2.1. This implies that the coefficients indicated by the coordinates in the LSP are sorted according to

$$\lfloor \log_2 |c_{m(k)}| \rfloor \geq \lfloor \log_2 |c_{m(k+1)}| \rfloor,$$

for all values of k. The decoder recovers the ordering because its three lists (LIS, LIP, and LSP) are updated in the same way as those of the encoder (remember that the decoder works in lockstep with the encoder). When the decoder inputs data, its three lists are identical to those of the encoder at the moment it (the encoder) output that data.
5. The encoder starts with the wavelet coefficients $c_{i,j}$; it never gets to "see" the actual image. The decoder, however, has to display the image and update the display in each iteration. In each iteration, when the coordinates (i,j) of a coefficient $c_{i,j}$ are moved to the LSP as an entry, it is known (to both encoder and decoder) that $2^n \leq |c_{i,j}| < 2^{n+1}$. As a result, the best value that the decoder can give the coefficient $\hat{c}_{i,j}$ that is being reconstructed is midway between 2^n and $2^{n+1} = 2\times 2^n$. Thus, the decoder sets $\hat{c}_{i,j} = \pm 1.5\times 2^n$ (the sign of $\hat{c}_{i,j}$ is input by the decoder just after the insertion). During the refinement pass, when the decoder inputs the actual value of the nth bit of $c_{i,j}$, it improves the value 1.5×2^n by either adding 2^{n-1} to it (if the input bit was a 1) or subtracting 2^{n-1} from it (if the input bit was a 0). This way, the decoder can improve the appearance of the image (or, equivalently, reduce the distortion) during *both* the sorting and refinement passes.

It is possible to improve the performance of SPIHT by entropy coding the encoder's output, but experience shows that the added compression gained in this way is minimal and does not justify the additional expense of both encoding and decoding time. It turns out that the signs and the individual bits of coefficients output in each iteration are uniformly distributed, so entropy coding them does not produce any compression. The bits $S_n(i,j)$ and $S_n(\mathcal{D}(i,j))$, on the other hand, are distributed nonuniformly and may gain from such coding.

8.14.4 Example

We assume that a 4×4 image has already been transformed, and the 16 coefficients are stored in memory as 6-bit signed-magnitude numbers (one sign bit followed by five magnitude bits). They are shown in Figure 8.60, together with the single spatial orientation tree. The coding algorithm initializes LIP to the one-element set $\{(1,1)\}$, the LIS to the set $\{\mathcal{D}(1,1)\}$, and the LSP to the empty set. The largest coefficient is 18, so n is set to $\lfloor \log_2 18 \rfloor = 4$. The first two iterations are shown.

18	6	8	−7
3	−5	13	1
2	1	−6	3
2	−2	4	−2

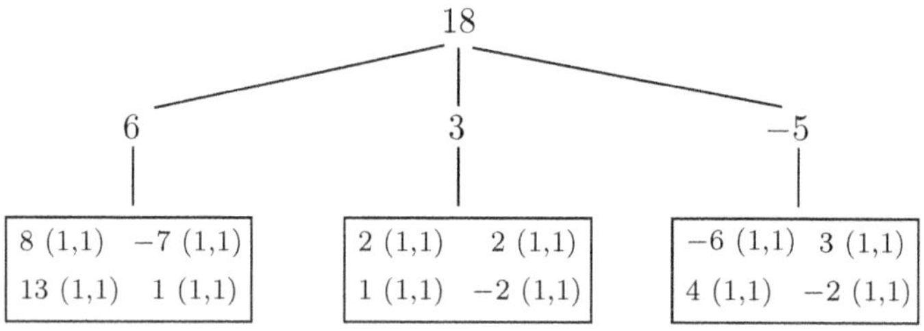

Figure 8.60: Sixteen Coefficients and One Spatial Orientation Tree.

Sorting Pass 1:
$2^n = 16$.
Is (1,1) significant? yes: output a 1.
LSP $= \{(1,1)\}$, output the sign bit: 0.
Is $\mathcal{D}(1,1)$ significant? no: output a 0.
LSP $= \{(1,1)\}$, LIP $= \{\}$, LIS $= \{\mathcal{D}(1,1)\}$.
Three bits output.
Refinement pass 1: no bits are output (this pass deals with coefficients from sorting pass $n-1$).
Decrement n to 3.
Sorting Pass 2:
$2^n = 8$.
Is $\mathcal{D}(1,1)$ significant? yes: output a 1.
Is $(1,2)$ significant? no: output a 0.
Is $(2,1)$ significant? no: output a 0.
Is $(2,2)$ significant? no: output a 0.
LIP $= \{(1,2),(2,1),(2,2)\}$, LIS $= \{\mathcal{L}(1,1)\}$.
Is $\mathcal{L}(1,1)$ significant? yes: output a 1.
LIS $= \{\mathcal{D}(1,2),\mathcal{D}(2,1),\mathcal{D}(2,2)\}$.
Is $\mathcal{D}(1,2)$ significant? yes: output a 1.
Is $(1,3)$ significant? yes: output a 1.
LSP $= \{(1,1),(1,3)\}$, output sign bit: 1.
Is $(2,3)$ significant? yes: output a 1.
LSP $= \{(1,1),(1,3),(2,3)\}$, output sign bit: 1.
Is $(1,4)$ significant? no: output a 0.
Is $(2,4)$ significant? no: output a 0.
LIP $= \{(1,2),(2,1),(2,2),(1,4),(2,4)\}$,

LIS = $\{\mathcal{D}(2,1), \mathcal{D}(2,2)\}$.
Is $\mathcal{D}(2,1)$ significant? no: output a 0.
Is $\mathcal{D}(2,2)$ significant? no: output a 0.
LIP = $\{(1,2),(2,1),(2,2),(1,4),(2,4)\}$,
LIS = $\{\mathcal{D}(2,1), \mathcal{D}(2,2)\}$,
LSP = $\{(1,1),(1,3),(2,3)\}$.
Fourteen bits output.
Refinement pass 2: After iteration 1, the LSP included entry $(1,1)$, whose value is $18 = 10010_2$.
One bit is output.
Sorting Pass 3:
$2^n = 4$.
Is $(1,2)$ significant? yes: output a 1.
LSP = $\{(1,1),(1,3),(2,3),(1,2)\}$, output a sign bit: 1.
Is $(2,1)$ significant? no: output a 0.
Is $(2,2)$ significant? yes: output a 1.
LSP = $\{(1,1),(1,3),(2,3),(1,2),(2,2)\}$, output a sign bit: 0.
Is $(1,4)$ significant? yes: output a 1.
LSP = $\{(1,1),(1,3),(2,3),(1,2),(2,2),(1,4)\}$, output a sign bit: 1.
Is $(2,4)$ significant? no: output a 0.
LIP = $\{(2,1),(2,4)\}$.
Is $D(2,1)$ significant? no: output a 0.
Is $D(2,2)$ significant? yes: output a 1.
Is $(3,3)$ significant? yes: output a 1.
LSP = $\{(1,1),(1,3),(2,3),(1,2),(2,2),(1,4),(3,3)\}$, output a sign bit: 0.
Is $(4,3)$ significant? yes: output a 1.
LSP = $\{(1,1),(1,3),(2,3),(1,2),(2,2),(1,4),(3,3),(4,3)\}$, output a sign bit: 1.
Is $(3,4)$ significant? no: output a 0.
LIP = $\{(2,1),(2,4),(3,4)\}$.
Is $(4,4)$ significant? no: output a 0.
LIP = $\{(2,1),(2,4),(3,4),(4,4)\}$.
LIP = $\{(2,1),(3,4),(3,4),(4,4)\}$, LIS = $\{\mathcal{D}(2,1)\}$,
LSP = $\{(1,1),(1,3),(2,3),(1,2),(2,2),(1,4),(3,3),(4,3)\}$.
Sixteen bits output.
Refinement Pass 3:
After iteration 2, the LSP included entries $(1,1)$, $(1,3)$, and $(2,3)$, whose values are $18 = 10010_2$, $8 = 1000_2$, and $13 = 1101_2$. Three bits are output
After two iterations, a total of 37 bits has been output.

8.14.5 QTCQ

Closely related to SPIHT, the QTCQ (quadtree classification and trellis coding quantization) method [Banister and Fischer 99] uses fewer lists than SPIHT and explicitly forms classes of the wavelet coefficients for later quantization by means of the ACTCQ and TCQ (arithmetic and trellis coded quantization) algorithms of [Joshi, Crump, and Fischer 93].

The method uses the spatial orientation trees originally developed by SPIHT. This type of tree is a special case of a quadtree (Section 7.34). The encoding algorithm is iterative. In the nth iteration, if any element of this quadtree is found to be significant, then the four highest elements in the tree are defined to be in class n. They also become roots of four new quadtrees. Each of the four new trees is tested for significance, moving down each tree until all the significant elements are found. All the wavelet coefficients declared to be in class n are stored in a *list of pixels* (LP). The LP is initialized with all the wavelet coefficients in the lowest frequency subband (LFS). The test for significance is performed by the function $S_T(k)$, which is defined by

$$S_T(k) = \begin{cases} 1, & \max_{(i,j)\in k} |C_{i,j}| \geq T, \\ 0, & \text{otherwise,} \end{cases}$$

where T is the current threshold for significance and k is a tree of wavelet coefficients. The QTCQ encoding algorithm uses this test, and is listed in Figure 8.61.

The QTCQ decoder is similar. All the outputs in Figure 8.61 should be replaced by inputs, and ACTCQ encoding should be replaced by ACTCQ decoding.

```
1. Initialization:
    Initialize LP with all C_{i,j} in LFS,
    Initialize LIS with all parent nodes,
    Output n = ⌊log_2(max |C_{i,j}/q)⌋.
    Set the threshold T = q2^n, where q is a quality factor.
2. Sorting:
    for each node k in LIS do
      output S_T(k)
      if S_T(k) = 1 then
       for each child of k do
        move coefficients to LP
        add to LIS as a new node
       endfor
       remove k from LIS
      endif
    endfor
3. Quantization: For each element in LP,
     quantize and encode using ACTCQ.
     (use TCQ step size Δ = α · q).
4. Update: Remove all elements in LP. Set T = T/2. Go to step 2.
```

Figure 8.61: QTCQ Encoding.

The QTCQ implementation, as described in [Banister and Fischer 99], does not transmit the image progressively, but the authors claim that this property can be added to it.

8.15 CREW

The CREW method (compression with reversible embedded wavelets) was developed in 1994 by A. Zandi at Ricoh Silicon Valley for the high-quality lossy and lossless compression of medical images. It was later realized that he independently developed a method very similar to SPIHT (Section 8.14), which is why the details of CREW are not described here. The interested reader is referred to [Zandi et al. 95], but more recent and detailed descriptions can be found at [CREW 00].

8.16 EZW

The SPIHT method is in some ways an extension of EZW, so this method, whose full name is "embedded coding using zerotrees of wavelet coefficients," is described here by outlining its principles and showing an example. Some of the details, such as the relation between parents and descendants in a spatial orientation tree, and the meaning of the term "embedded," are described in Section 8.14.

The EZW method, as implemented in practice, starts by performing the 9-tap symmetric quadrature mirror filter (QMF) wavelet transform [Adelson et al. 87]. The main loop is then repeated for values of the threshold that are halved at the end of each iteration. The threshold is used to calculate a *significance map* of significant and insignificant wavelet coefficients. Zerotrees are used to represent the significance map in an efficient way. The main steps are as follows:

1. Initialization: Set the threshold T to the smallest power of 2 that is greater than $\max_{(i,j)}|c_{i,j}|/2$, where $c_{i,j}$ are the wavelet coefficients.
2. Significance map coding: Scan all the coefficients in a predefined way and output a symbol when $|c_{i,j}| > T$. When the decoder inputs this symbol, it sets $c_{i,j} = \pm 1.5T$.
3. Refinement: Refine each significant coefficient by sending one more bit of its binary representation. When the decoder receives this, it increments the current coefficient value by $\pm 0.25T$.
4. Set $T = T/2$, and go to step 2 if more iterations are needed.

A wavelet coefficient $c_{i,j}$ is considered insignificant with respect to the current threshold T if $|c_{i,j}| \le T$. The zerotree data structure is based on the following well-known experimental result: If a wavelet coefficient at a coarse scale (i.e., high in the image pyramid) is insignificant with respect to a given threshold T, then all of the coefficients of the same orientation in the same spatial location at finer scales (i.e., located lower in the pyramid) are very likely to be insignificant with respect to T.

In each iteration, all the coefficients are scanned in the order shown in Figure 8.62a. This guarantees that when a node is visited, all its parents will already have been scanned. The scan starts at the lowest frequency subband LL_n, continues with subbands HL_n, LH_n, and HH_n, and drops to level $n-1$, where it scans HL_{n-1}, LH_{n-1}, and HH_{n-1}. Each subband is fully scanned before the algorithm proceeds to the next subband.

Each coefficient visited in the scan is classified as a zerotree root (ZTR), an isolated zero (IZ), positive significant (POS), or negative significant (NEG). A zerotree root is a coefficient that is insignificant and all its descendants (in the same spatial orientation tree) are also insignificant. Such a coefficient becomes the root of a zerotree. It is encoded with a special symbol (denoted by ZTR), and the important point is that its descendants don't have to be encoded in the current iteration. When the decoder inputs a ZTR symbol, it assigns a zero value to the coefficients and to all its descendants in the spatial orientation tree. Their values get improved (refined) in subsequent iterations. An isolated zero is a coefficient that is insignificant but has some significant descendants. Such a coefficient is encoded with the special IZ symbol. The other two classes are coefficients that are significant and are positive or negative. The flowchart of Figure 8.62b illustrates this classification. Notice that a coefficient is classified into one of five classes, but the fifth class (a zerotree node) is not encoded.

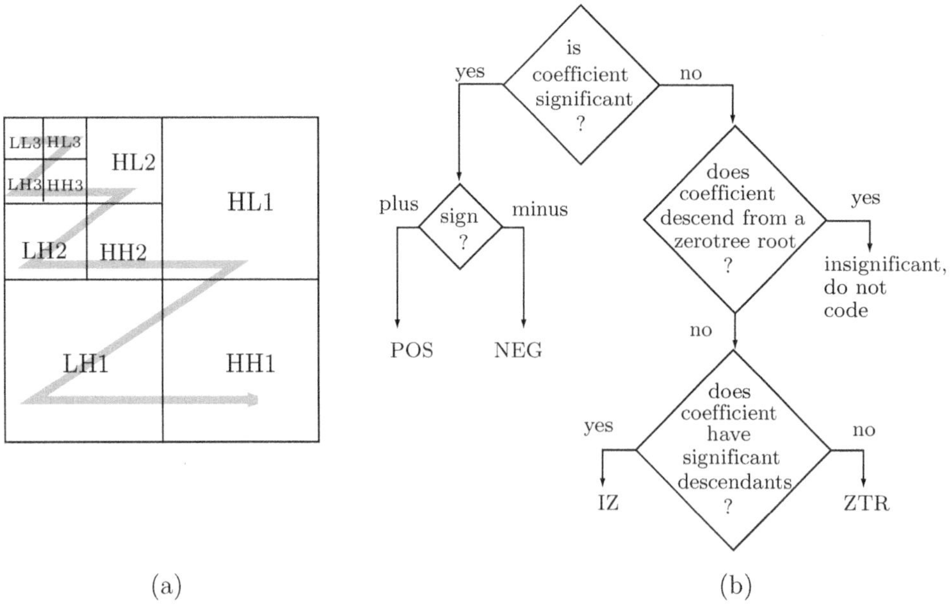

Figure 8.62: (a) Scanning a Zerotree. (b) Classifying a Coefficient.

Coefficients in the lowest pyramid level don't have any children, so they cannot be the roots of zerotrees. Thus, they are classified into isolated zero, positive significant, or negative significant.

The zerotree can be viewed as a structure that helps to find insignificance. Most methods that try to find structure in an image try to find significance. The IFS method of Section 7.39, for example, tries to locate image parts that are similar up to size and/or transformation, and this is much harder than to locate parts that are insignificant relative to other parts.

Two lists are used by the encoder (and also by the decoder, which works in lockstep) in the scanning process. The *dominant list* contains the coordinates of the coefficients that have not been found to be significant. They are stored in the order scan, by pyramid levels, and within each level by subbands. The *subordinate list* contains the *magnitudes* (not coordinates) of the coefficients that have been found to be significant. Each list is scanned once per iteration.

An iteration consists of a *dominant pass* followed by a *subordinate pass*. In the dominant pass, coefficients from the dominant list are tested for significance. If a coefficient is found significant, then (1) its sign is determined, (2) it is classified as either POS or NEG, (3) its magnitude is appended to the subordinate list, and (4) it is set to zero in memory (in the array containing all the wavelet coefficients). The last step is done so that the coefficient does not prevent the occurrence of a zerotree in subsequent dominant passes at smaller thresholds.

Imagine that the initial threshold is $T = 32$. When a coefficient $c_{i,j} = 63$ is encountered in the first iteration, it is found to be significant. Since it is positive, it is encoded as POS. When the decoder inputs this symbol, it does not know its value, but it knows that the coefficient is positive significant, i.e., it satisfies $c_{i,j} > 32$. The decoder also knows that $c_{i,j} \leq 64 = 2\times 32$, so the best value that the decoder can assign the coefficient is $(32+64)/2 = 48$. The coefficient is then set to 0, so subsequent iterations will not identify it as significant.

We can think of the threshold T as an indicator that specifies a bit position. In each iteration, the threshold indicates the next less significant bit position. We can also view T as the current quantization width. In each iteration that width is divided by 2, so another less significant bit of the coefficients becomes known.

During a subordinate pass, the subordinate list is scanned and the encoder outputs a 0 or a 1 for each coefficient to indicate to the decoder how the magnitude of the coefficient should be improved. In the example of $c_{i,j} = 63$, the encoder transmits a 1, indicating to the decoder that the actual value of the coefficient is greater than 48. The decoder uses that information to improve the magnitude of the coefficient from 48 to $(48+64)/2 = 56$. If a 0 had been transmitted, the decoder would have refined the value of the coefficient to $(32+48)/2 = 40$.

The string of bits generated by the encoder during the subordinate pass is entropy encoded using a custom version of adaptive arithmetic coding (Section 5.10).

At the end of the subordinate pass, the encoder (and, in lockstep, also the decoder) sorts the magnitudes in the subordinate list in decreasing order.

The encoder stops the loop when a certain condition is met. The user may, for example, specify the desired bitrate (number of bits per pixel). The encoder knows the image size (number of pixels), so it knows when the desired bitrate has been reached or exceeded. The advantage is that the compression ratio is known (in fact, it is determined by the user) in advance. The downside is that too much information may be lost if the compression ratio is too high, thereby leading to a poorly reconstructed image. It is also possible for the encoder to stop in the midst of an iteration, when the exact bitrate specified by the user has been reached. However, in such a case the last codeword may be incomplete, leading to wrong decoding of one coefficient.

The user may also specify the bit budget (the size of the compressed stream) as a stopping condition. This is similar to specifying the bitrate. An alternative is for the user to specify the maximum acceptable distortion (the difference between the original and the compressed image). In such a case, the encoder iterates until the threshold becomes 1, and the decoder stops decoding when the maximum acceptable distortion has been reached.

8.16.1 Example

This example follows the one in [Shapiro 93]. Figure 8.63a shows three levels of the wavelet transform of an 8×8 image. The largest value is 63, so the initial threshold can be anywhere in the range $(31, 64]$. We set it to 32. Table 8.63b lists the results of the first dominant pass.

63	−34	49	10	7	13	−12	7
−31	23	14	−13	3	4	6	−1
15	14	3	−12	5	−7	3	9
−9	−7	−14	8	4	−2	3	2
−5	9	−1	47	4	6	−2	2
3	0	−3	2	3	−2	0	4
2	−3	6	−4	3	6	3	6
5	11	5	6	0	3	−4	4

(a)

Subband	Coeff. value	Symbol	Reconstr. value	Note
LL3	63	POS	48	1
HL3	−34	NEG	−48	
LH3	−31	IZ	0	2
HH3	23	ZTR	0	3
HL2	49	POS	48	
HL2	10	ZTR	0	4
HL2	14	ZTR	0	
HL2	−13	ZTR	0	
LH2	15	ZTR	0	
LH2	14	IZ	0	5
LH2	−9	ZTR	0	
LH2	−7	ZTR	0	
HL1	7	Z	0	
HL1	13	Z	0	
HL1	3	Z	0	
HL1	4	Z	0	
LH1	−1	Z	0	
LH1	47	POS	48	6
LH1	−3	Z	0	
LH1	−2	Z	0	

(b)

Figure 8.63: An EZW Example: Three Levels of an 8×8 Image.

Notes:

1. The top-left coefficient is 63. It is greater than the threshold, and it is positive, so a POS symbol is generated and is transmitted by the encoder (and the 63 is changed to 0). The decoder assigns this POS symbol the value 48, the midpoint of the interval $[32, 64)$.

2. The coefficient 31 is insignificant with respect to 32, but it is not a zerotree root, since one of its descendants (the 47 in LH1) is significant. The 31 is therefore an isolated zero (IZ).

3. The 23 is less than 32. Also, all its descendants (the 3, −12, −14, and 8 in HH2, and all of HH1) are insignificant. The 23 is therefore a zerotree root (ZTR). As a result, no symbols will be generated by the encoder in the dominant pass for its descendants (this is why none of the HH2 and HH1 coefficients appear in the table).

4. The 10 is less than 32, and all its descendants (the −12, 7, 6, and −1 in HL1) are also less than 32. Thus, the 10 becomes a zerotree root (ZTR). Notice that the −12 is greater, in absolute value, than the 10, but is still less than the threshold.
5. The 14 is insignificant with respect to the threshold, but one of its children (they are −1, 47, −3, and 2) is significant. Thus, the 14 becomes an IZ.
6. The 47 in subband LH1 is significant with respect to the threshold, so it is coded as POS. It is then changed to zero, so that a future pass (with a threshold of 16) will code its parent, 14, as a zerotree root.

Four significant coefficients were transmitted during the first dominant pass. All that the decoder knows about them is that they are in the interval $[32, 64)$. They will be refined during the first subordinate pass, so the decoder will be able to place them either in $[32, 48)$ (if it receives a 0) or in $[48, 64)$ (if it receives a 1). The encoder generates and transmits the bits "1010" for the four significant coefficients 63, 34, 49, and 47. Thus, the decoder refines them to 56, 40, 56, and 40, respectively.

In the second dominant pass, only those coefficients not yet found to be significant are scanned and tested. The ones found significant are treated as zero when the encoder checks for zerotree roots. This second pass ends up identifying the −31 in LH3 as NEG, the 23 in HH3 as POS, the 10, 14, and −3 in LH2 as zerotree roots, and also all four coefficients in LH2 and all four in HH2 as zerotree roots. The second dominant pass stops at this point, since all other coefficients are known to be insignificant from the first dominant pass.

The subordinate list contains, at this point, the six magnitudes 63, 49, 34, 47, 31, and 23. They represent the 16-bit-wide intervals $[48, 64)$, $[32, 48)$, and $[16, 31)$. The encoder outputs bits that define a new subinterval for each of the three. At the end of the second subordinate pass, the decoder could have identified the 34 and 47 as being in different intervals, so the six magnitudes are ordered as 63, 49, 47, 34, 31, and 23. The decoder assigns them the refined values 60, 52, 44, 36, 28, and 20. (End of example.)

8.17 DjVu

Image compression methods are normally designed for one type of image. JBIG (Section 7.14), for example, was designed for bi-level images, the FABD block decomposition method (Section 7.32) is intended for the compression of discrete-tone images, and JPEG (Section 7.10) works best on continuous-tone images. Certain images, however, combine the properties of all three image types. An important example of such an image is a scanned document containing text, line drawings, and regions with continuous-tone pictures, such as paintings or photographs. Libraries all over the world are currently digitizing their holdings by scanning and storing them in compressed format on disks and CD-ROMs. Organizations interested in making their documents available to the public (such as a mail-order firm with a colorful catalog, or a research institute with scientific papers) also have collections of documents. They can all benefit from an efficient lossy compression method that can highly compress such documents. Viewing such a document is normally done in a web browser, so such a method should feature fast decoding. Such a method is DjVu (pronounced "déjà vu"), from AT&T laboratories ([ATT 96] and [Haffner et al. 98]). We start with a short summary of its performance.

DjVu routinely achieves compression factors as high as 1000—which is 5 to 10 times better than competing image compression methods. Scanned pages at 300 dpi in full color are typically compressed from 25 Mb down to 30–60 Kb with excellent quality. Black-and-white pages become even smaller, typically compressed to 10–30 Kb. This creates high-quality scanned pages whose size is comparable to that of an average HTML page.

For color documents with both text and pictures, DjVu files are typically 5–10 times smaller than JPEG at similar quality. For black-and-white pages, DjVu files are typically 10–20 times smaller than JPEG and five times smaller than GIF. DjVu files are also about five times smaller than PDF files (Section 11.13) produced from scanned documents.

To help users read DjVu-compressed documents in a web browser, the developers have implemented a decoder in the form of a plug-in for standard web browsers. With this decoder (freely available for all popular platforms) it is easy to pan the image and zoom on it. The decoder also uses little memory, typically 2 Mbyte of RAM for images that normally require 25 Mbyte of RAM to fully display. The decoder keeps the image in RAM in a compact form, and decompresses, in real time, only the part that is actually displayed on the screen.

The DjVu method is progressive. The viewer sees an initial version of the document very quickly, and the visual quality of the display improves progressively as more bits arrive. For example, the text of a typical magazine page appears in just three seconds over a 56 Kbps modem connection. In another second or two, the first versions of the pictures and backgrounds appear. The final, full-quality, version of the page is completed after a few more seconds.

We next outline the main features of DjVu. The main idea is that the different elements of a scanned document, namely text, drawings, and pictures, have different perceptual characteristics. Digitized text and line drawings require high spatial resolution but little color resolution. Pictures and backgrounds, on the other hand, can be coded at lower spatial resolution, but with more bits per pixel (for high color resolution). We know from experience that text and line diagrams should be scanned and displayed at 300 dpi or higher resolution, since at any lower resolution text is barely legible and lines lose their sharp edges. Also, text normally uses one color, and drawings use few colors. Pictures, on the other hand, can be scanned and viewed at 100 dpi without much loss of picture detail if adjacent pixels can have similar colors (i.e., if the number of available colors is large).

DjVu therefore starts by decomposing the document into three components: mask, foreground, and background. The background component contains the pixels that constitute the pictures and the paper background. The mask contains the text and the lines in bi-level form (i.e., one bit per pixel). The foreground contains the color of the mask pixels. The background is a continuous-tone image and can be compressed at the low resolution of 100 dpi. The foreground normally contains large uniform areas and is also compressed as a continuous-tone image at the same low resolution. The mask is left at 300 dpi but can be efficiently compressed, because it is bi-level. The background and foreground are compressed with a wavelet-based method called IW44 ("IW" stands for "integer wavelet"), while the mask is compressed with JB2, a version of JBIG2 (Section 7.15) developed at AT&T.

The decoder decodes the three components, increases the resolution of the background and foreground components back to 300 dpi, and generates each pixel in the final decompressed image according to the mask. If a mask pixel is 0, the corresponding image pixel is taken from the background. If the mask pixel is 1, the corresponding image pixel is generated in the color of the foreground pixel.

The rest of this section describes the image separation method used by DjVu. This is a multiscale bicolor clustering algorithm based on the concept of *dominant colors*. Imagine a document containing just black text on a white background. In order to obtain best results, such a document should be scanned as a grayscale, antialiased image. This results in an image that has mostly black and white pixels, but also gray pixels of various shades located at and near the edges of the text characters. It is obvious that the dominant colors of such an image are black and white. In general, given an image with several colors or shades of gray, its two dominant colors can be identified by the following algorithm:

Step 1: Initialize the background color b to white and the foreground color f to black.
Step 2: Loop over all the pixels of the image. For each pixel, calculate the distances f and b between the pixel's color and the current foreground and background colors. Select the shorter of the two distances and flag the pixel as either f or b accordingly.
Step 3: Calculate the average color of all the pixels that are flagged f. This becomes the new foreground color. Do the same for the background color.
Step 4: Repeat steps 2 and 3 until the two dominant colors converge (i.e., until they vary by less than the value of a preset threshold).

This algorithm is simple, and it converges rapidly. However, a real document seldom has two dominant colors, so DjVu uses two extensions of this algorithm. The first extension is called *block bicolor clustering*. It divides the document into small rectangular blocks, and executes the clustering algorithm above in each block to get two dominant colors. Each block is then separated into a foreground and a background using these two dominant colors. This extension is not completely satisfactory because of the following problems involving the block size:

1. A block should be small enough to have a dominant foreground color. Imagine, for instance, a red word in a region of black text. Ideally, we want such a word to be in a block of its own, with red as one of the dominant colors of the block. If the blocks are large, this word may end up being a small part of a block whose dominant colors will be black and white; the red will effectively disappear.
2. On the other hand, a small block may be located completely outside any text area. Such a block contains just background pixels and should not be separated into background and foreground. A block may also be located completely inside a large character. Such a block is all foreground and should not be separated. In either case, this extension of the clustering algorithm will not find meaningfully dominant colors.
3. Experience shows that in a small block this algorithm does not always select the right colors for foreground and background.

The second extension is called *multiscale bicolor clustering*. This is an iterative algorithm that starts with a grid of large blocks and applies the block bicolor clustering algorithm to each. The result of this first iteration is a pair of dominant colors for each large block. In the second and subsequent iterations, the grid is divided into smaller and smaller blocks, and each is processed with the original clustering algorithm. However, this algorithm is slightly modified, since it now attracts the new foreground and background colors of an iteration not toward black and white but toward the two dominant colors found in the previous iteration. Here is how it works:

1. For each small block b, identify the larger block B (from the preceding iteration) inside which b is located. Initialize the background and foreground colors of b to those of B.
2. Loop over the entire image. Compare the color of a pixel to the background and foreground colors of its block and tag the pixel according to the smaller of the two distances.
3. For each small block b, calculate a new background color by averaging (1) the colors of all pixels in b that are tagged as background, and (2) the background color that was used for b in this iteration. This is a weighted average where the pixels of (1) are given a weight of 80%, and the background color of (2) is assigned a weight of 20%. The new foreground color is calculated similarly.
4. Steps 2 and 3 are repeated until both background and foreground converge.

The multiscale bicolor clustering algorithm then divides each small block b into smaller blocks and repeats.

This process is usually successful in separating the image into the right foreground and background components. To improve it even more, it is followed by a variety of filters that are applied to the various foreground areas. They are designed to find and eliminate the most obvious mistakes in the identification of foreground parts.

déjà vu—French for "already seen."
Vuja De—The feeling you've never been here.

8.18 WSQ, Fingerprint Compression

Most of us don't realize it, but fingerprints are "big business." The FBI started collecting fingerprints in the form of inked impressions on paper cards back in 1924, and today they have about 200 million cards, occupying an acre of filing cabinets in the J. Edgar Hoover building in Washington, D.C. (The FBI, like many of us, never throws anything away. They also have many "repeat customers," which is why "only" about 29 million out of the 200 million cards are distinct; these are the ones used for running background checks.) What's more, these cards keep accumulating at a rate of 30,000–50,000 new cards per day (this is per day, not per year)! There's clearly a need to digitize this collection, so it will occupy less space and will lend itself to automatic search and classification. The main problem is size (in bits). When a typical fingerprint card is scanned at 500 dpi, with eight bits/pixel, it results in about 10 Mb of data. Thus, the total size of the digitized collection would be more than 2000 terabytes (a terabyte is 2^{40} bytes); huge even by current (2006) standards.

◇ **Exercise 8.18:** Apply these numbers to estimate the size of a fingerprint card.

Compression is therefore a must. At first, it seems that fingerprint compression must be lossless because of the small but important details involved. However, lossless image compression methods produce typical compression ratios of 0.5, whereas in order to make a serious dent in the huge amount of data in this collection, compressions of about 1 bpp or better are needed. What is needed is a lossy compression method that results in graceful degradation of image details, and does not introduce any artifacts into the reconstructed image. Most lossy image compression methods involve the loss of small details and are therefore unacceptable, since small fingerprint details, such as sweat pores, are admissible points of identification in court. This is where wavelets come into the picture. Lossy wavelet compression, if carefully designed, can satisfy the criteria above and result in efficient compression where important small details are preserved or are at least identifiable. Figure 8.64a,b (obtained, with permission, from Christopher M. Brislawn), shows two examples of fingerprints and one detail, where ridges and sweat pores can clearly be seen.

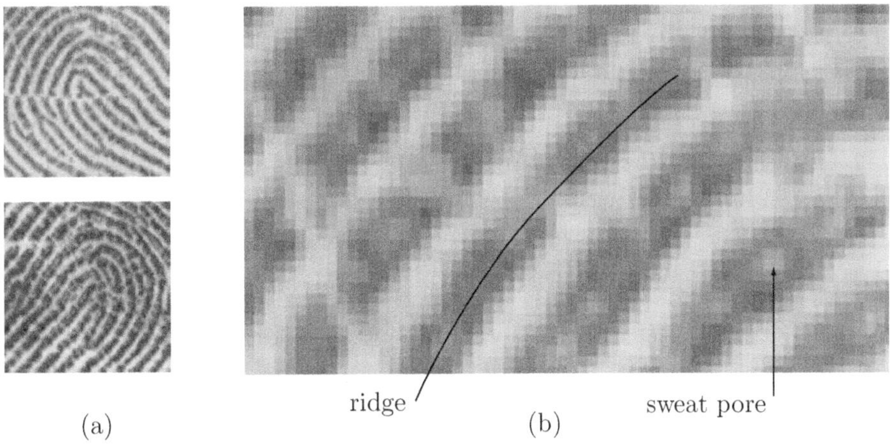

Figure 8.64: Examples of Scanned Fingerprints (courtesy Christopher Brislawn).

Compression is also necessary, because fingerprint images are routinely sent between law enforcement agencies. Overnight delivery of the actual card is too slow and risky (there are no backup cards), and sending 10 Mb of data through a 9600 baud modem takes about three hours.

The method described here [Bradley, Brislawn, and Hopper 93] has been adopted by the FBI as its standard for fingerprint compression [Federal Bureau of Investigations 93]. It involves three steps: (1) a discrete wavelet transform, (2) adaptive scalar quantization of the wavelet transform coefficients, and (3) a two-pass Huffman coding of the quantization indices. This is the reason for the name *wavelet/scalar quantization*, or WSQ. The method typically produces compression factors of about 20. Decoding is the opposite of encoding, so WSQ is a symmetric compression method.

The first step is a symmetric discrete wavelet transform (SWT) using the symmetric filter coefficients listed in Table 8.65 (where $\mathcal{R}$ indicates the real part of a complex number). They are symmetric filters with seven and nine impulse response taps, and they depend on the two numbers x_1 (real) and x_2 (complex). The final standard adopted by the FBI uses the values

$$x_1 = A + B - \frac{1}{6}, \quad x_2 = \frac{-(A+B)}{2} - \frac{1}{6} + \frac{i\sqrt{3}(A-B)}{2},$$

where

$$A = \left(\frac{-14\sqrt{15}+63}{1080\sqrt{15}}\right)^{1/3}, \quad \text{and } B = \left(\frac{-14\sqrt{15}-63}{1080\sqrt{15}}\right)^{1/3}.$$

Tap	Exact value	Approximate value
$h_0(0)$	$-5\sqrt{2}x_1(48\|x_2\|^2 - 16\mathcal{R}x_2 + 3)/32$	0.852698790094000
$h_0(\pm1)$	$-5\sqrt{2}x_1(8\|x_2\|^2 - \mathcal{R}x_2)/8$	0.377402855612650
$h_0(\pm2)$	$-5\sqrt{2}x_1(4\|x_2\|^2 + 4\mathcal{R}x_2 - 1)/16$	−0.110624404418420
$h_0(\pm3)$	$-5\sqrt{2}x_1(\mathcal{R}x_2)/8$	−0.023849465019380
$h_0(\pm4)$	$-5\sqrt{2}x_1/64$	0.037828455506995
$h_1(-1)$	$\sqrt{2}(6x_1 - 1)/16x_1$	0.788485616405660
$h_1(-2,0)$	$-\sqrt{2}(16x_1 - 1)/64x_1$	−0.418092273222210
$h_1(-3,1)$	$\sqrt{2}(2x_1 + 1)/32x_1$	−0.040689417609558
$h_1(-4,2)$	$-\sqrt{2}/64x_1$	0.064538882628938

Table 8.65: Symmetric Wavelet Filter Coefficients for WSQ.

The wavelet image decomposition is different from those discussed in Section 8.10. It can be called symmetric and is shown in Figure 8.66. The SWT is first applied to the image rows and columns, resulting in $4\times4 = 16$ subbands. The SWT is then applied in the same manner to three of the 16 subbands, decomposing each into 16 smaller subbands. The last step is to decompose the top-left subband into four smaller ones.

The larger subbands (51–63) contain the fine-detail, high-frequency information of the image. They can later be heavily quantized without loss of any important information (i.e., information needed to classify and identify fingerprints). In fact, subbands 60–63 are completely discarded. Subbands 7–18 are important. They contain that portion of the image frequencies that corresponds to the ridges in a fingerprint. This information is important and should be quantized lightly.

The transform coefficients in the 64 subbands are floating-point numbers to be denoted by a. They are quantized to a finite number of floating-point numbers that are denoted by $\hat{a}$. The WSQ encoder maps a transform coefficient a to a quantization index p (an integer that is later mapped to a code that is itself Huffman encoded). The index p can be considered a pointer to the quantization bin where a lies. The WSQ decoder receives an index p and maps it to a value $\hat{a}$ that is close, but not identical, to a. This

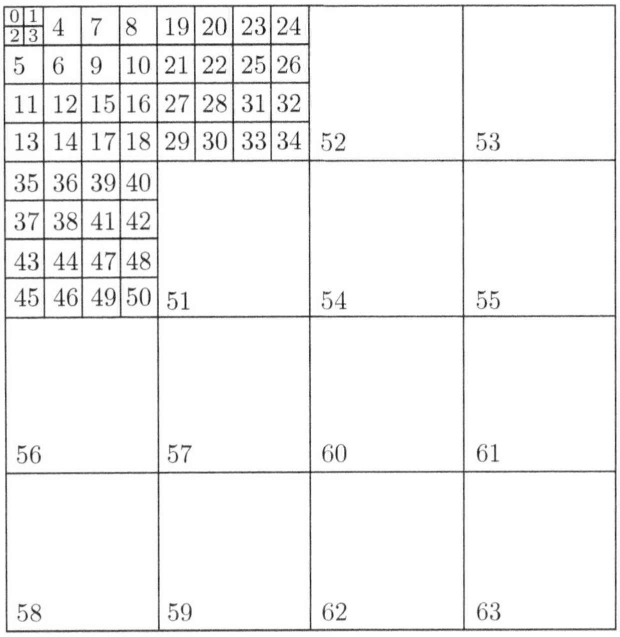

Figure 8.66: Symmetric Image Wavelet Decomposition.

is how WSQ loses image information. The set of $\hat{a}$ values is a discrete set of floating-point numbers called the *quantized wavelet coefficients*. The quantization depends on parameters that may vary from subband to subband, since different subbands have different quantization requirements.

Figure 8.67 shows the setup of quantization bins for subband k. Parameter Z_k is the width of the zero bin, and parameter Q_k is the width of the other bins. Parameter C is in the range $[0,1]$. It determines the reconstructed value $\hat{a}$. For $C=0.5$, for example, the reconstructed value for each quantization bin is the center of the bin. Equation (8.29) shows how parameters Z_k and Q_k are used by the WSQ encoder to quantize a transform coefficient $a_k(m,n)$ (i.e., a coefficient in position (m,n) in subband k) to an index $p_k(m,n)$ (an integer), and how the WSQ decoder computes a quantized coefficient $\hat{a}_k(m,n)$ from that index:

$$p_k(m,n)=\begin{cases}\left\lfloor\frac{a_k(m,n)-Z_k/2}{Q_k}\right\rfloor+1, & a_k(m,n)>Z_k/2,\\ 0, & -Z_k/2\le a_k(m,n)\le Z_k/2,\\ \left\lceil\frac{a_k(m,n)+Z_k/2}{Q_k}\right\rceil+1, & a_k(m,n)<-Z_k/2,\end{cases}$$

$$\hat{a}_k(m,n)=\begin{cases}(p_k(m,n)-C)Q_k+Z_k/2, & p_k(m,n)>0,\\ 0, & p_k(m,n)=0,\\ (p_k(m,n)+C)Q_k-Z_k/2, & p_k(m,n)<0.\end{cases} \tag{8.29}$$

The final standard adopted by the FBI uses the value $C = 0.44$ and determines the bin widths Q_k and Z_k from the variances of the coefficients in the different subbands in the following steps:

Step 1: Let the width and height of subband k be denoted by X_k and Y_k, respectively. We compute the six quantities

$$\begin{aligned} W_k &= \left\lfloor \frac{3X_k}{4} \right\rfloor, \quad H_k = \left\lfloor \frac{7Y_k}{16} \right\rfloor, \\ x_{0k} &= \left\lfloor \frac{X_k}{8} \right\rfloor, \quad x_{1k} = x_{0k} + W_k - 1, \\ y_{0k} &= \left\lfloor \frac{9Y_k}{32} \right\rfloor, \quad y_{1k} = y_{0k} + H_k - 1. \end{aligned}$$

Step 2: Assuming that position $(0,0)$ is the top-left corner of the subband, we use the subband region from position (x_{0k}, y_{0k}) to position (x_{1k}, y_{1k}) to estimate the variance σ_k^2 of the subband by

$$\sigma_k^2 = \frac{1}{W \cdot H - 1} \sum_{n=x_{0k}}^{x_{1k}} \sum_{m=y_{0k}}^{y_{1k}} \left(a_k(m,n) - \mu_k\right)^2,$$

where μ_k denotes the mean of $a_k(m,n)$ in the region.

Step 3: Parameter Q_k is computed by

$$q\,Q_k = \begin{cases} 1, & 0 \le k \le 3, \\ \dfrac{10}{A_k \log_e(\sigma_k^2)}, & 4 \le k \le 59, \text{ and } \sigma_k^2 \ge 1.01, \\ 0, & 60 \le k \le 63, \text{ or } \sigma_k^2 < 1.01, \end{cases}$$

where q is a proportionality constant that controls the bin widths Q_k and thereby the overall level of compression. The procedure for computing q is complex and will not be described here. The values of the constants A_k are

$$A_k = \begin{cases} 1.32, & k = 52, 56, \\ 1.08, & k = 53, 58, \\ 1.42, & k = 54, 57, \\ 1.08, & k = 55, 59, \\ 1, & \text{otherwise.} \end{cases}$$

Notice that the bin widths for subbands 60–63 are zero. As a result, these subbands, containing the finest detail coefficients, are simply discarded.

Step 4: The width of the zero bin is set to $Z_k = 1.2Q_k$.

The WSQ encoder computes the quantization indices $p_k(m,n)$ as shown, then maps them to the 254 codes shown in Table 8.68. These values are encoded with Huffman codes (using a two-pass process), and the Huffman codes are then written on the compressed stream. A quantization index $p_k(m,n)$ can be any integer, but most are small and there are many zeros. Thus, the codes of Table 8.68 are divided into three groups. The first

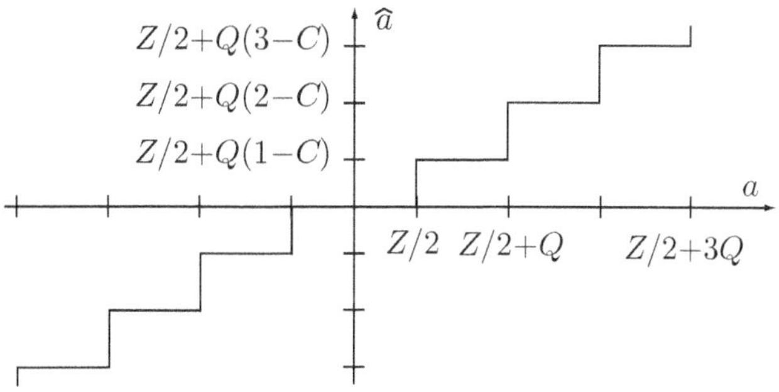

Figure 8.67: WSQ Scalar Quantization.

group consists of 100 codes (codes 1 through 100) for run lengths of 1 to 100 zero indices. The second group is codes 107 through 254. They specify small indices, in the range $[-73, +74]$. The third group consists of the six *escape* codes 101 through 106. They indicate large indices or run lengths of more than 100 zero indices. Code 180 (which corresponds to an index $p_k(m,n) = 0$) is not used, because this case is really a run length of a single zero. An escape code is followed by the (8-bit or 16-bit) raw value of the index (or size of the run length). Here are some examples:

An index $p_k(m,n) = -71$ is coded as 109. An index $p_k(m,n) = -1$ is coded as 179. An index $p_k(m,n) = 75$ is coded as 101 (escape for positive 8-bit indices) followed by 75 (in eight bits). An index $p_k(m,n) = -259$ is coded as 104 (escape for negative large indices) followed by 259 (the absolute value of the index, in 16 bits). An isolated index of zero is coded as 1, and a run length of 260 zeros is coded as 106 (escape for large run lengths) followed by 260 (in 16 bits). Indices or run lengths that require more than 16 bits cannot be encoded, but the particular choice of the quantization parameters and the wavelet transform virtually guarantee that large indices will never be generated.

> O'Day figured that that was more than he'd had the right to expect under the circumstances. A fingerprint identification ordinarily required ten individual points—the irregularities that constituted the art of fingerprint identification—but that number had always been arbitrary. The inspector was certain that Cutter had handled this computer disk, even if a jury might not be completely sure, if that time ever came.
>
> —Tom Clancy, *Clear and Present Danger*

The last step is to prepare the Huffman code tables. They depend on the image, so they have to be written on the compressed stream. The standard adopted by the FBI specifies that subbands be grouped into three blocks and all the subbands in a group use the same Huffman code table. This facilitates progressive transmission of the image. The first block consists of the low- and mid-frequency subbands 0–18. The second and third blocks contain the highpass detail subbands 19–51 and 52–59, respectively (recall that subbands 60–63 are completely discarded). Two Huffman code tables are prepared, one for the first block and the other for the second and third blocks.

A Huffman code table for a block of subbands is prepared by counting the number of times each of the 254 codes of Table 8.68 appears in the block. The counts are used to determine the length of each code and to construct the Huffman code tree. This is a

Code	Index or run length
1	run length of 1 zeros
2	run length of 2 zeros
3	run length of 3 zeros
⋮	
100	run length of 100 zeros
101	escape code for positive 8-bit index
102	escape code for negative 8-bit index
103	escape code for positive 16-bit index
104	escape code for negative 16-bit index
105	escape code for zero run, 8-bit
106	escape code for zero run, 16-bit
107	index value −73
108	index value −72
109	index value −71
⋮	
179	index value −1
180	unused
181	index value 1
⋮	
253	index value 73
254	index value 74

Table 8.68: WSQ Codes for Quantization Indices and Run Lengths.

two-pass job (one pass to determine the code tables and another pass to encode), and it is done in a way similar to the use of the Huffman code by JPEG (Section 7.10.4).

8.19 JPEG 2000

This section was originally written in mid-2000 and was slightly improved in early 2003.

The data compression field is very active, with new approaches, ideas, and techniques being developed and implemented all the time. JPEG (Section 7.10) is widely used for image compression but is not perfect. The use of the DCT on 8×8 blocks of pixels results sometimes in a reconstructed image that has a blocky appearance (especially when the JPEG parameters are set for much loss of information). This is why the JPEG committee has decided, as early as 1995, to develop a new, wavelet-based standard for the compression of still images, to be known as JPEG 2000 (or JPEG Y2K). Perhaps the most important milestone in the development of JPEG 2000 occurred in December 1999, when the JPEG committee met in Maui, Hawaii and approved the first committee draft of Part 1 of the JPEG 2000 standard. At its Rochester meeting in August 2000, the JPEG committee approved the final draft of this International Standard. In December 2000 this draft was finally accepted as a full International Standard by the ISO and

ITU-T. Following is a list of areas where this new standard is expected to improve on existing methods:

- High compression efficiency. Bitrates of less than 0.25 bpp are expected for highly detailed grayscale images.
- The ability to handle large images, up to $2^{32} \times 2^{32}$ pixels (the original JPEG can handle images of up to $2^{16} \times 2^{16}$).
- Progressive image transmission (Section 7.13). The proposed standard can decompress an image progressively by SNR, resolution, color component, or region of interest.
- Easy, fast access to various points in the compressed stream.
- The decoder can pan/zoom the image while decompressing only parts of it.
- The decoder can rotate and crop the image while decompressing it.
- Error resilience. Error-correcting codes can be included in the compressed stream, to improve transmission reliability in noisy environments.

The main sources of information on JPEG 2000 are [JPEG 00] and [Taubman and Marcellin 02]. This section, however, is based on [ISO/IEC 00], the final committee draft (FCD), released in March 2000. This document defines the compressed stream (referred to as the *bitstream*) and the operations of the decoder. It contains informative sections about the encoder, but any encoder that produces a valid bitstream is considered a valid JPEG 2000 encoder.

One of the new, important approaches to compression introduced by JPEG 2000 is the "compress once, decompress many ways" paradigm. The JPEG 2000 encoder selects a maximum image quality Q and maximum resolution R, and it compresses an image using these parameters. The decoder can decompress the image at any image quality up to and including Q and at any resolution less than or equal to R. Suppose that an image I was compressed into B bits. The decoder can extract A bits from the compressed stream (where $A < B$) and produce a lossy decompressed image that will be identical to the image obtained if I was originally compressed lossily to A bits.

In general, the decoder can decompress the entire image in lower quality and/or lower resolution. It can also decompress parts of the image (regions of interest) at either maximum or lower quality and resolution. Even more, the decoder can *extract* parts of the compressed stream and assemble them to create a new compressed stream without having to do any decompression. Thus, a lower-resolution and/or lower-quality image can be created without the decoder having to decompress anything. The advantages of this approach are (1) it saves time and space and (2) it prevents the buildup of image noise, common in cases where an image is lossily compressed and decompressed several times.

JPEG 2000 also makes it possible to crop and transform the image. When an image is originally compressed, several regions of interest may be specified. The decoder can access the compressed data of any region and write it as a new compressed stream. This necessitates some special processing around the region's borders, but there is no need to decompress the entire image and recompress the desired region. In addition, mirroring (flipping) the image or rotating it by 90°, 180°, and 270° can be carried out almost entirely in the compressed stream without decompression.

Image progression is also supported by JPEG 2000 and has four aspects: quality, resolution, spatial location, and component.

As the decoder inputs more data from the compressed stream it improves the quality of the displayed image. An image normally becomes recognizable after only 0.05 bits/pixel have been input and processed by the decoder. After 0.25 bits/pixel have been input and processed, the partially decompressed image looks smooth, with only minor compression artifacts. When the decoder starts, it uses the first data bytes input to create a small version (a thumbnail) of the image. As more data is read and processed, the decoder adds more pixels to the displayed image, thereby increasing its size in steps. Each step increases the image by a factor of 2 on each side. This is how resolution is increased progressively. This mode of JPEG 2000 corresponds roughly to the hierarchical mode of JPEG.

Spatial location means that the image can be displayed progressively in raster order, row by row. This is useful when the image has to be decompressed and immediately sent to a printer. Component progression has to do with the color components of the image. Most images have either one or three color components, but four components, such as CMYK, are also common. JPEG 2000 allows for up to 16,384 components. Extra components may be overlays containing text and graphics. If the compressed stream is prepared for progression by component, then the image is first decoded as grayscale, then as a color image, then the various overlays are decompressed and displayed to add realism and detail.

The four aspects of progression may be mixed within a single compressed stream. A particular stream may be prepared by the encoder to create, for example, the following result. The start of the stream may contain data enough to display a small, grayscale image. The data that follows may add color to this small image, followed by data that increases the resolution several times, followed, in turn, by an increase in quality until the user decides that the quality is high enough to permit printing the image. At that point, the resolution can be increased to match that of the available printer. If the printer is black and white, the color information in the remainder of the compressed image can be skipped. All this can be done while only the data required by the user needs be input and decompressed. Regardless of the original size of the compressed stream, the compression ratio is effectively increased if only part of the compressed stream needs be input and processed.

How does JPEG 2000 work? The following paragraph is a short summary of the algorithm. Certain steps are described in more detail in the remainder of this section.

If the image being compressed is in color, it is divided into three components. Each component is partitioned into rectangular, nonoverlapping regions called *tiles*, that are compressed individually. A tile is compressed in four main steps. The first step is to compute a wavelet transform that results in subbands of wavelet coefficients. Two such transforms, an integer and a floating point, are specified by the standard. There are $L+1$ resolution levels of subbands, where L is a parameter determined by the encoder. In step two, the wavelet coefficients are quantized. This is done if the user specifies a target bitrate. The lower the bitrate, the coarser the wavelet coefficients have to be quantized. Step three uses the MQ coder (an encoder similar to the QM coder, Section 5.11) to arithmetically encode the wavelet coefficients. The EBCOT algorithm [Taubman 99] has been adopted for the encoding step. The principle of EBCOT is to divide each subband into blocks (termed *code-blocks*) that are coded individually. The bits resulting from coding several code-blocks become a *packet* and the packets are the components of the bitstream. The last step is to construct the bitstream. This step places the packets, as well as many *markers*, in the bitstream. The markers can be used by the decoder to skip certain areas of the bitstream and to reach certain points quickly. Using markers, the decoder can, e.g., decode certain code-blocks before others, thereby displaying certain regions of the image before other regions. Another use of the markers is for the decoder to progressively decode the image in one of several ways. The bitstream is organized in *layers*, where each layer contains higher-resolution image information. Thus, decoding the image layer by layer is a natural way to achieve progressive image transmission and decompression.

Before getting to the details, the following is a short history of the development effort of JPEG 2000. The history of JPEG 2000 starts in 1995, when Ricoh Inc. submitted the CREW algorithm (compression with reversible embedded wavelets; Section 8.15) to the ISO/IEC as a candidate for JPEG-LS (Section 7.11). CREW was not selected as the algorithm for JPEG-LS, but was sufficiently advanced to be considered a candidate for the new method then being considered by the ISO/IEC. This method, later to become known as JPEG 2000, was approved as a new, official work item, and a working group (WG1) was set for it in 1996. In March 1997, WG1 called for proposals and started evaluating them. Of the many algorithms submitted, the WTCQ method (wavelet trellis coded quantization) performed the best and was selected in November 1997 as the reference JPEG 2000 algorithm. The WTCQ algorithm includes a wavelet transform and a quantization method.

In November 1998, the EBCOT algorithm was presented to the working group by its developer David Taubman and was adopted as the method for encoding the wavelet coefficients. In March 1999, the MQ coder was presented to the working group and was adopted by it as the arithmetic coder to be used in JPEG 2000. During 1999, the format of the bitstream was being developed and tested, with the result that by the end of 1999 all the main components of JPEG 2000 were in place. In December 1999, the working group issued its committee draft (CD), and in March 2000, it has issued its final committee draft (FCD), the document on which this section is based. In August 2000, the JPEG group met and decided to approve the FCD as a full International Standard in December 2000. This standard is now known as ISO/IEC-15444, and its formal description is available (in 13 parts, of which part 7 has been abandoned) from the ISO, ITU-T, and certain national standards organizations.

This section continues with details of certain conventions and operations of JPEG 2000. The goal is to illuminate the key concepts in order to give the reader a general understanding of this new international standard.

Color Components: A color image consists of three color components. The first step of the JPEG 2000 encoder is to transform the components by means of either a reversible component transform (RCT) or an irreversible component transform (ICT). Each transformed component is then compressed separately.

If the image pixels have unsigned values (which is the normal case), then the component transform (either RCT or ICT) is preceded by a DC level shifting. This process translates all pixel values from their original, unsigned interval $[0, 2^s - 1]$ (where s is the pixels' depth) to the signed interval $[-2^{s-1}, 2^{s-1} - 1]$ by subtracting 2^{s-1} from each value. For $s=4$, e.g., the $2^4=16$ possible pixel values are transformed from the interval $[0,15]$ to the interval $[-8,+7]$ by subtracting $2^{4-1}=8$ from each value.

The RCT is a decorrelating transform. It can only be used with the integer wavelet transform (which is reversible). Denoting the pixel values of image component i (after a possible DC level shifting) by $I_i(x,y)$ for $i=0,1$, and 2, the RCT produces new values $Y_i(x,y)$ according to

$$\begin{aligned} Y_0(x,y) &= \left\lfloor \frac{I_0(x,y)+2I_1(x,y)+I_2(x,y)}{4} \right\rfloor, \\ Y_1(x,y) &= I_2(x,y)-I_1(x,y), \\ Y_2(x,y) &= I_0(x,y)-I_1(x,y). \end{aligned}$$

Notice that the values of components Y_1 and Y_2 (but not Y_0) require one more bit than the original I_i values.

The ICT is also a decorrelating transform. It can only be used with the floating-point wavelet transform (which is irreversible). The ICT is defined by

$$\begin{aligned} Y_0(x,y) &= 0.299I_0(x,y)+0.587I_1(x,y)+0.144I_2(x,y), \\ Y_1(x,y) &= -0.16875I_0(x,y)-0.33126I_1(x,y)+0.5I_2(x,y), \\ Y_2(x,y) &= 0.5I_0(x,y)-0.41869I_1(x,y)-0.08131I_2(x,y). \end{aligned}$$

If the original image components are red, green, and blue, then the ICT is very similar to the YCbCr color representation (Section 9.2).

Tiles: Each (RCT or ICT transformed) color component of the image is partitioned into rectangular, nonoverlapping tiles. Since the color components may have different resolutions, they may use different tile sizes. Tiles may have any size, up to the size of the entire image (i.e., one tile). All the tiles of a given color component have the same size, except those at the edges. Each tile is compressed individually.

Figure 8.69a shows an example of image tiling. JPEG 2000 allows the image to have a vertical offset at the top and a horizontal offset on the left (the offsets can be zero). The origin of the tile grid (the top-left corner) can be located anywhere inside the intersection area of the two offsets. All the tiles in the grid are the same size, but those located on the edges, such as T0, T4, and T19 in the figure, have to be truncated. Tiles are numbered in raster order.

The main reason for having tiles is to enable the user to decode parts of the image (regions of interest). The decoder can identify each tile in the bitstream and decompress just those pixels included in the tile. Figure 8.69b shows an image with an aspect ratio (height/width) of 16 : 9 (the aspect ratio of high-definition television, HDTV; Section 9.3.1). The image is tiled with four tiles whose aspect ratio is 4:3 (the aspect ratio of current, analog television), such that tile T0 covers the central area of the image. This makes it easy to crop the image from the original aspect ratio to the 4:3 ratio.

Wavelet Transform: Two wavelet transforms are specified by the standard. They are the (9,7) floating-point wavelet (irreversible) and the (5,3) integer wavelet (reversible). Either transform allows for progressive transmission, but only the integer transform can produce lossless compression.

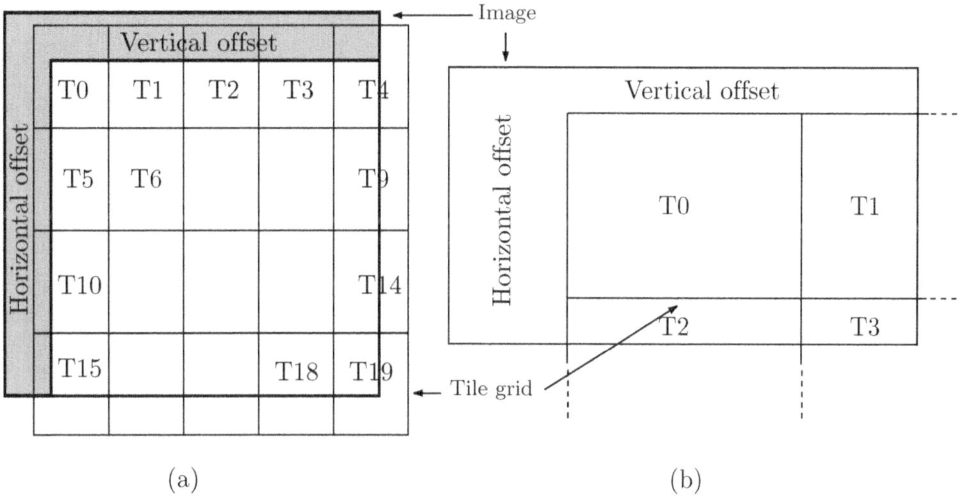

Figure 8.69: JPEG 2000 Image Tiling.

We denote a row of pixels in a tile by $P_k, P_{k+1}, \ldots, P_m$. Because of the nature of the wavelet transforms used by JPEG 2000, a few pixels with indices less than k or greater than m may have to be used. Therefore, before any wavelet transform is computed for a tile, the pixels of the tile may have to be extended. The JPEG 2000 standard specifies a simple extension method termed *periodic symmetric extension*, which is illustrated by Figure 8.70. The figure shows a row of seven symbols "ABCDEFG" and how they are reflected to the left and to the right to extend the row by l and r symbols, respectively. Table 8.71 lists the minimum values of l and r as functions of the parity of k and m and of the particular transform used.

$$\overleftarrow{\quad l \quad} \qquad\qquad \overrightarrow{\quad r \quad}$$
$$\ldots\text{DEFGFEDCB}|\underset{\substack{\uparrow\\ k}}{\text{A}}\text{BCDEF}\underset{\substack{\uparrow\\ m}}{\text{G}}|\text{FEDCBABCD}\ldots$$

Figure 8.70: Extending a Row of Pixels.

k	l (5,3)	l (9,7)	m	r (5,3)	r (9,7)
even	2	4	odd	2	4
odd	1	3	even	1	3

Table 8.71: Minimum Left and Right Extensions.

We now denote a row of extended pixels in a tile by $P_k, P_{k+1}, \ldots, P_m$. Since the pixels have been extended, index values below k and above m can be used. The (5,3) integer wavelet transform computes wavelet coefficients $C(i)$ by first computing the odd values $C(2i+1)$ and then using them to compute the even values $C(2i)$. The computations are

$$C(2i+1) = P(2i+1) - \left\lfloor \frac{P(2i)+P(2i+2)}{2} \right\rfloor, \quad \text{for } k-1 \le 2i+1 < m+1,$$
$$C(2i) = P(2i) + \left\lfloor \frac{C(2i-1)+C(2i+1)+2}{4} \right\rfloor, \quad \text{for } k \le 2i < m+1.$$

The (9,7) floating-point wavelet transform is computed by executing four "lifting" steps followed by two "scaling" steps on the extended pixel values P_k through P_m. Each step is performed on all the pixels in the tile before the next step starts. Step 1 is performed for all i values satisfying $k-3 \le 2i+1 < m+3$. Step 2 is performed for all i such that $k-2 \le 2i < m+2$. Step 3 is performed for $k-1 \le 2i+1 < m+1$. Step 4 is performed for $k \le 2i < m$. Step 5 is done for $k \le 2i+1 < m$. Finally, step 6 is executed for $k \le 2i < m$. The calculations are

$$\begin{aligned} C(2i+1) &= P(2i+1) + \alpha[P(2i)+P(2i+2)], && \text{step 1}\\ C(2i) &= P(2i) + \beta[C(2i-1)+C(2i+1)], && \text{step 2}\\ C(2i+1) &= C(2i+1) + \gamma[C(2i)+C(2i+2)], && \text{step 3}\\ C(2i) &= C(2i) + \delta[C(2i-1)+C(2i+1)], && \text{step 4}\\ C(2i+1) &= -K \times C(2i+1), && \text{step 5}\\ C(2i) &= (1/K) \times C(2i), && \text{step 6} \end{aligned}$$

where the five constants (wavelet filter coefficients) used by JPEG 2000 are given by $\alpha = -1.586134342$, $\beta = -0.052980118$, $\gamma = 0.882911075$, $\delta = 0.443506852$, and $K = 1.230174105$.

These one-dimensional wavelet transforms are applied L times, where L is a parameter (either user-controlled or set by the encoder), and are interleaved on rows and columns to form L levels (or resolutions) of subbands. Resolution $L-1$ is the original image (resolution 3 in Figure 8.72a), and resolution 0 is the lowest-frequency subband. The subbands can be organized in one of three ways, as illustrated in Figure 8.72a–c.

Quantization: Each subband can have a different quantization step size. Each wavelet coefficient in the subband is divided by the quantization step size and the result is truncated. The quantization step size may be determined iteratively in order to achieve a target bitrate (i.e., the compression factor may be specified in advance by the user) or in order to achieve a predetermined level of image quality. If lossless compression is desired, the quantization step is set to 1.

Precincts and Code-Blocks: Consider a tile in a color component. The original pixels are wavelet transformed, resulting in subbands of L resolution levels. Figure 8.73a shows one tile and four resolution levels. There are three subbands in each resolution level (except the lowest level). The total size of all the subbands equals the size of the tile. A grid of rectangles known as *precincts* is now imposed on the entire image as shown in Figure 8.73b. The origin of the precinct grid is anchored at the top-left corner of the image and the dimensions of a precinct (its width and height) are powers of 2. Notice that subband boundaries are generally not identical to precinct boundaries. We now examine the three subbands of a certain resolution and pick three precincts located in the same

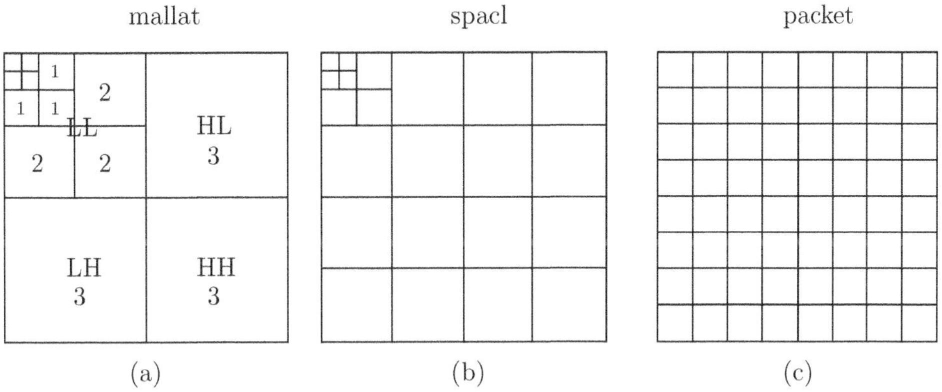

Figure 8.72: JPEG 2000 Subband Organization.

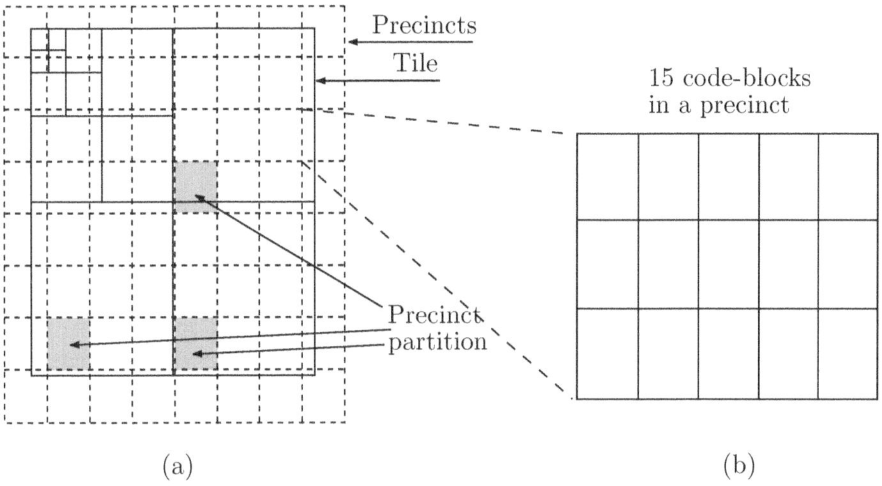

Figure 8.73: Subbands, Precincts, and Code-Blocks.

regions in the three subbands (the three gray rectangles in Figure 8.73a). These three precincts constitute a *precinct partition.* The grid of precincts is now divided into a finer grid of *code-blocks,* which are the basic units to be arithmetically coded. Figure 8.73b shows how a precinct is divided into 15 code-blocks. Thus, a precinct partition in this example consists of 45 code-blocks. A code-block is a rectangle of size 2^{xcb} (width) by 2^{ycb} (height), where $2 \leq xcb,\ ycb \leq 10$, and $xcb + ycb \leq 12$.

We can think of the tiles, precincts, and code-blocks as coarse, medium, and fine partitions of the image, respectively. Partitioning the image into smaller and smaller units helps in (1) creating memory-efficient implementations, (2) streaming, and (3) allowing easy access to many points in the bitstream. It is expected that simple JPEG 2000 encoders would ignore this partitioning and have just one tile, one precinct, and one code-block. Sophisticated encoders, on the other hand, may end up with a large number of code-blocks, thereby allowing the decoder to perform progressive decompression, fast streaming, zooming, panning, and other special operations while decoding only parts of the image.

Entropy Coding: The wavelet coefficients of a code-block are arithmetically coded by bitplane. The coding is done from the most-significant bitplane (containing the most important bits of the coefficients) to the least-significant bitplane. Each bitplane is scanned as shown in Figure 8.74. A context is determined for each bit, a probability is estimated from the context, and the bit and its probability are sent to the arithmetic coder.

Many image compression methods work differently. A typical image compression algorithm may use several neighbors of a pixel as its context and encode the pixel (not just an individual bit) based on the context. Such a context can include only pixels that will be known to the decoder (normally pixels located above the current pixel or to its left). JPEG 2000 is different in this respect. It encodes individual bits (this is why it uses the MQ coder, which encodes bits, not numbers) and it uses symmetric contexts. The context of a bit is computed from its eight near neighbors. However, since at decoding time the decoder will not know all the neighbors, the context cannot use the values of the neighbors. Instead, it uses the *significance* of the neighbors. Each wavelet coefficient has a 1-bit variable (a flag) associated with it, which indicates its significance. This is the *significance state* of the coefficient. When the encoding of a code-block starts, all its wavelet coefficients are considered insignificant and all the significance states are cleared.

Some of the most-significant bitplanes may be all zeros. The number of such bitplanes is stored in the bitstream, for the decoder's use. Encoding starts from the first bitplane that is not identically zero. That bitplane is encoded in one pass (a *cleanup pass*). Each of the less-significant bitplanes following it is encoded in three passes, referred to as the *significance propagation pass*, the *magnitude refinement pass*, and the *cleanup pass*. Each pass divides the bitplane into stripes that are each four rows high (Figure 8.74). Each stripe is scanned column by column from left to right. Each bit in the bitplane is encoded in one of the three passes. As mentioned earlier, encoding a bit involves (1) determining its context, (2) estimating a probability for it, and (3) sending the bit and its probability to the arithmetic coder.

The first encoding pass (significance propagation) of a bitplane encodes all the bits that belong to wavelet coefficients satisfying (1) the coefficient is insignificant and (2) at least one of its eight nearest neighbors is significant. If a bit is encoded in this pass and if the bit is 1, its wavelet coefficient is marked as significant by setting its significance state to 1. Subsequent bits encoded in this pass (and the following two passes) will consider this coefficient significant.

It is clear that for this pass to encode *any* bits, some wavelet coefficients must be declared significant before the pass even starts. This is why the first bitplane that is being encoded is encoded in just one pass known as the cleanup pass. In that pass, all the bits of the bitplane are encoded. If a bit happens to be a 1, its coefficient is declared significant.

The second encoding pass (magnitude refinement) of a bitplane encodes all bits of wavelet coefficients that became significant in a *previous* bitplane. Thus, once a coefficient becomes significant, all its less-significant bits will be encoded one by one, each one in the second pass of a different bitplane.

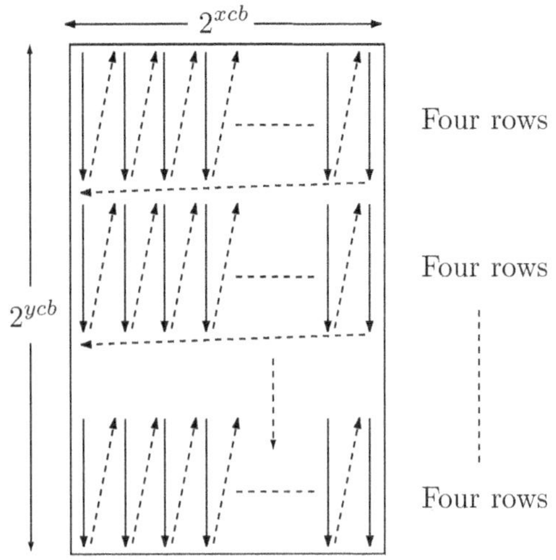

Figure 8.74: Stripes in a Code-Block.

The third and final encoding pass (cleanup) of a bitplane encodes all the bits not encoded in the first two passes. Let's consider a coefficient C in a bitplane B. If C is insignificant, its bit in B will not be encoded in the second pass. If all eight near neighbors of C are insignificant, the bit of C in bitplane B will not be encoded in the first pass either. That bit will therefore be encoded in the third pass. If the bit happens to be 1, C will become significant (the encoder will set its significance state to 1).

Wavelet coefficients are signed integers and have a sign bit. In JPEG 2000, they are represented in the *sign-magnitude* method. The sign bit is 0 for a positive (or a zero) coefficient and 1 for a negative coefficient. The magnitude bits are the same, regardless of the sign. If a coefficient has one sign bit and eight bits of magnitude, then the value $+7$ is represented as $0|00000111$ and -7 is represented as $1|00000111$. The sign bit of a coefficient is encoded following the first 1 bit of the coefficient.

The behavior of the three passes is illustrated by a simple example. We assume four coefficients with values $10 = 0|00001010$, $1 = 0|00000001$, $3 = 0|00000011$, and $-7 = 1|00000111$. There are eight bitplanes numbered 7 through 0 from left (most significant) to right (least significant). The bitplane with the sign bits is initially ignored. The first four bitplanes 7–4 are all zeros, so encoding starts with bitplane 3 (Figure 8.75). There is just one pass for this bitplane—the cleanup pass. One bit from each of the four coefficients is encoded in this pass. The bit for coefficient 10 is 1, so this coefficient is declared significant (its remaining bits, in bitplanes 2, 1, and 0, will be encoded in pass 2). Also the sign bit of 10 is encoded following this 1. Next, bitplane 2 is encoded. Coefficient 10 is significant, so its bit in this bitplane (a zero) is encoded in pass 2. Coefficient 1 is insignificant, but one of its near neighbors (the 10) is significant, so the bit of 1 in bitplane 2 (a zero) is encoded in pass 1. The bits of coefficients 3 and -7 in this bitplane (a zero and a one, respectively) are encoded in pass 3. The sign bit of coefficient -7 is encoded following the 1 bit of that coefficient. Also, coefficient -7 is declared significant.

Bitplane 1 is encoded next. The bit of coefficient 10 is encoded in pass 2. The bit of coefficient 1 is encoded in pass 1, same as in bitplane 2. The bit of coefficient 3, however, is encoded in pass 1 since its near neighbor, the −7, is now significant. This bit is 1, so coefficient 3 becomes significant. This bit is the first 1 of coefficient 3, so the sign of 3 is encoded following this bit.

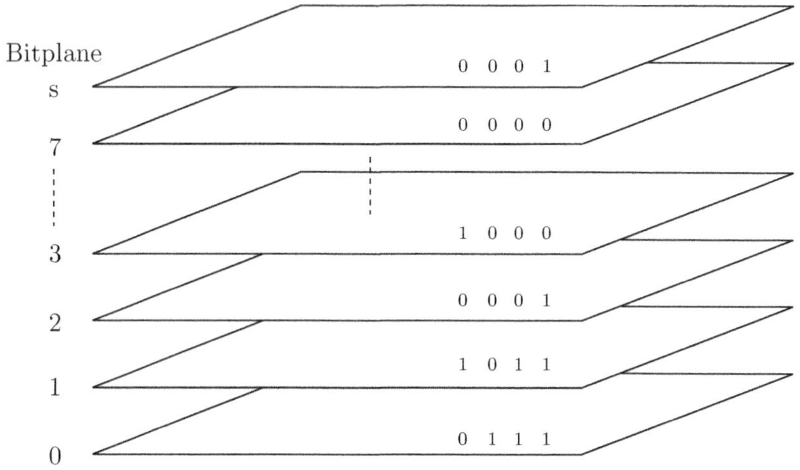

Figure 8.75: Bitplanes of Four Coefficients in a Code-Block.

◇ **Exercise 8.19:** Describe the encoding order of the last bitplane.

	Coefficients			
Bitplane	10	1	3	−7
sign	0	0	0	1
7	0	0	0	0
6	0	0	0	0
5	0	0	0	0
4	0	0	0	0
3	1	0	0	0
2	0	0	0	1
1	1	0	1	1
0	0	1	1	1

(a)

		Coefficients			
Bitplane	Pass	10	1	3	−7
3	cleanup	1+	0	0	0
2	significance		0		
2	refinement	0			
2	cleanup			0	1−
1	significance		0	1+	
1	refinement	1			1
1	cleanup				
0	significance		1+		
0	refinement	0		1	1
0	cleanup				

(b)

Table 8.76: Encoding Four Bitplanes of Four Coefficients.

The context of a bit is determined in a different way for each pass. Here we show how it is determined for the significance propagation pass. The eight near neighbors of the current wavelet coefficient X are used to determine the context used in encoding each bit of X. One of nine contexts is selected and is used to estimate the probability of the bit being encoded. However, as mentioned earlier, it is the current *significance states* of the eight neighbors, not their values, that are used in the determination. Figure 8.77 shows the names assigned to the significance states of the eight neighbors. Table 8.78 lists the criteria used to select one of the nine contexts. Notice that those criteria depend on which of the four subbands (HH, HL, LH, or LL) are being encoded. Context 0 is selected when coefficient X has no significant near neighbors. Context 8 is selected when all eight neighbors of X are significant.

D_0	V_0	D_1
H_0	X	H_1
D_2	V_1	D_3

Figure 8.77: Eight Neighbors.

The JPEG 2000 standard specifies similar rules for determining the context of a bit in the refinement and cleanup passes, as well as for the context of a sign bit. Context determination in the cleanup pass is identical to that of the significance pass, with the difference that run-length coding is used if all four bits in a column of a stripe are insignificant and each has only insignificant neighbors. In such a case, a single bit is encoded, to indicate whether the column is all zero or not. If not, then the column has a run of (between zero and three) zeros. In such a case, the length of this run is encoded. This run must, of course, be followed by a 1. This 1 does not have to encoded since its presence is easily deduced by the decoder. Normal (bit by bit) encoding resumes for the bit following this 1.

LL and LH Subbands (vertical highpass)			HL Subband (horizontal highpass)			HH Subband (diagonal highpass)		Context
$\sum H_i$	$\sum V_i$	$\sum D_i$	$\sum H_i$	$\sum V_i$	$\sum D_i$	$\sum (H_i+V_i)$	$\sum D_i$	
2				2			≥ 3	8
1	≥ 1		≥ 1	1		≥ 1	2	7
1	0	≥ 1	0	1	≥ 1	0	2	6
1	0	0	0	1	0	≥ 2	1	5
0	2		2	0		1	1	4
0	1		1	0		0	1	3
0	0	≥ 2	0	0	≥ 2	≥ 2	0	2
0	0	1	0	0	1	1	0	1
0	0	0	0	0	0	0	0	0

Table 8.78: Nine Contexts for the Significance Propagation Pass.

Once a context has been determined for a bit, it is used to estimate a probability for encoding the bit. This is done using a probability estimation table, similar to Table 5.65. Notice that JBIG and JBIG2 (Sections 7.14 and 7.15, respectively) use similar tables, but they use thousands of contexts, in contrast with the few contexts (nine or fewer) used by JPEG 2000.

Packets: After all the bits of all the coefficients of all the code-blocks of a precinct partition have been encoded into a short bitstream, a header is added to that bitstream, thereby turning it into a packet. Figure 8.73a,b shows a precinct partition consisting of three precincts, each divided into 15 code-blocks. Encoding this partition therefore results in a packet with 45 encoded code-blocks. The header contains all the information needed to decode the packet. If all the code-blocks in a precinct partition are identically zero, the body of the packet is empty. Recall that a precinct partition corresponds to the same spatial location in three subbands. As a result, a packet can be considered a quality increment for one level of resolution at a certain spatial location.

Layers: A layer is a set of packets. It contains one packet from each precinct partition of each resolution level. Thus, a layer is a quality increment for the entire image at full resolution.

Progressive Transmission: This is an important feature of JPEG 2000. The standard provides four ways of progressively transmitting and decoding an image: by resolution, quality, spatial location, and component. Progression is achieved simply by storing the packets in a specific order in the bitstream. For example, quality (also known as SNR) progression can be achieved by arranging the packets in layers, within each layer by component, within each component by resolution level, and within each resolution level by precinct partition. Resolution progression is achieved when the packets are arranged by precinct partition (innermost nesting), layer, image component, and resolution level (outermost nesting).

When an image is encoded, the packets are placed in the bitstream in a certain order, corresponding to a certain progression. If a user or an application require a different progression (and thus a different order of the packets), it should be easy to read the bitstream, identify the packets, and rearrange them. This process is known as *parsing*, and is an easy task because of the many *markers* embedded in the bitstream. There are different types of marker and they are used for different purposes. Certain markers identify the type of progression used in the bitstream, and others contain the lengths of all the packets. Thus, the bitstream can be parsed without having to decode any of it.

A typical example of parsing is printing a color image on a grayscale printer. In such a case, there is no point in sending the color information to the printer. A parser can use the markers to identify all the packets containing color information and discard them. Another example is decreasing the size of the bitstream (by increasing the amount of image loss). The parser has to identify and discard the layers that contribute the least to the image quality. This is done repeatedly, until the desired bitstream size is achieved.

The parser can be part of an *image server*. A client sends a request to such a server with the name of an image and a desired attribute. The server executes the parser to obtain the image with that attribute and transmits the bitstream to the client.

Regions of Interest: A client may want to decode just part of an image—a region of interest (ROI). The parser should be able to identify the parts of the bitstream that correspond to the ROI and transmit just those parts. An ROI may be specified at compression time. In such a case, the encoder identifies the code-blocks located in the ROI and writes them early in the bitstream. In many cases, however, an ROI is specified to a parser after the image has been compressed. In such a case, the parser may use the tiles to identify the ROI. Any ROI will be contained in one or several tiles, and the parser can identify those tiles and transmit to the client a bitstream consisting of just those tiles. This is an important application of tiles. Small tiles make it possible to specify small ROIs, but result in poor compression. Experience suggests that tiles of size 256×256 have a negligible negative impact on the compression ratio and are usually small enough to specify ROIs.

Since each tile is compressed individually, it is easy to find the tile's information in the bitstream. Each tile has a header and markers that simplify this task. Any parsing that can be done on the entire image can also be performed on individual tiles. Other tiles can either be ignored or can be transmitted at a lower quality.

Alternatively, the parser can handle small ROIs by extracting from the bitstream the information for individual code-blocks. The parser has to (1) determine what code-blocks contain a given pixel, (2) find the packets containing those code-blocks, (3) decode the packet headers, (4) use the header data to find the code-block information within each packet, and (5) transmit just this information to the decoder.

Summary: Current experimentation indicates that JPEG 2000 performs better than the original JPEG, especially for images where very low bitrates (large compression factors) or very high image quality are required. For lossless or near-lossless compression, JPEG 2000 offers only modest improvements over JPEG.

Discover the power of wavelets! Wavelet analysis, in contrast to Fourier analysis, uses approximating functions that are localized in both time and frequency space. It is this unique characteristic that makes wavelets particularly useful, for example, in approximating data with sharp discontinuities.

—Wolfram Research

9 Video Compression

Sound recording and the movie camera were among the greatest inventions of Thomas Edison. They were later united when "talkies" were developed, and they are still used together in video recordings. This unification is one reason for the popularity of movies and video. With the rapid advances in computers in the 1980s and 1990s came multimedia applications, where images and sound are combined in the same file. Such files tend to be large, which is why compressing them became an important field of research.

This chapter starts with basic discussions of analog and digital video, continues with the principles of video compression, and concludes with a description of several compression methods designed specifically for video, namely MPEG-1, MPEG-4, H.261, H.264, and VC-1.

9.1 Analog Video

An analog video camera converts the image it "sees" through its lens to an electric voltage (a signal) that varies with time according to the intensity and color of the light emitted from the different image parts. Such a signal is called *analog*, because it is analogous (proportional) to the light intensity. The best way to understand this signal is to see how a television receiver responds to it.

From the Dictionary

Analog (adjective).

Being a mechanism that represents data by measurement of a continuously variable quantity (as electrical voltage).

D. Salomon, G. Motta, *Handbook of Data Compression*, 5th ed.
DOI 10.1007/978-1-84882-903-9_9,

9.1.1 The CRT

Modern televisions are mostly digital and use an array of LEDs to generate the picture. Older televisions were analog and were based on a CRT (cathode ray tube, Figure 9.1a). A CRT is a glass tube with a familiar shape. In the back it has an electron gun (the cathode) that emits a stream of electrons. Its front surface is positively charged, so it attracts the electrons (which have a negative electric charge). The front is coated with a phosphor compound that converts the kinetic energy of the electrons hitting it to light. The flash of light lasts only a fraction of a second, so in order to get a constant display, the picture has to be refreshed several times a second. The actual refresh rate depends on the *persistence* of the compound (Figure 9.1b). For certain types of work, such as architectural drawing, long persistence is acceptable. For animation, short persistence is a must.

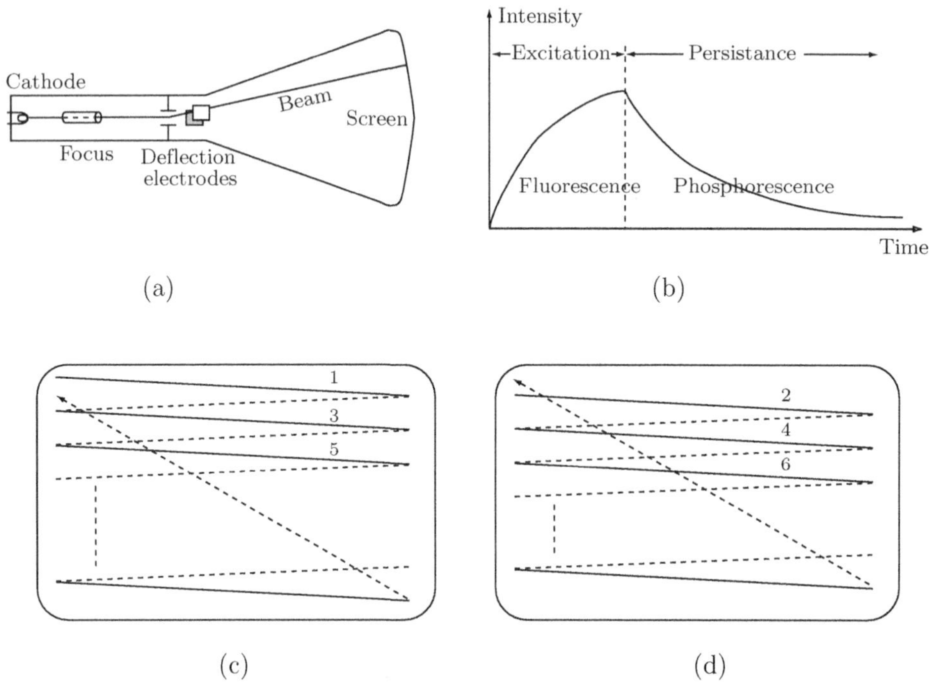

Figure 9.1: (a) CRT Operation. (b) Persistence. (c) Odd Scan Lines. (d) Even Scan Lines.

The early pioneers of motion pictures found, after much experimentation, that the minimum refresh rate required for smooth animation is 15 pictures (or frames) per second (fps), so they adopted 16 fps as the refresh rate for their cameras and projectors. However, when movies began to film fast action (such as in westerns), the motion pictures industry decided to increased the refresh rate to 24 fps, a rate that is used to this day. At a certain point it was discovered that this rate can artificially be doubled, to 48 fps (which produces smoother animation), by projecting each frame twice. This is done by employing a double-blade rotating shutter in the movie projector. The shutter exposes a picture, covers it, and exposes it again, all in 1/24 of a second, thereby achieving an effective refresh rate of 48 fps. Modern movie projectors have very bright lamps and can even use a triple-blade shutter, for an effective refresh rate of 72 fps.

The frequency of electric current in Europe is 50 Hz, so television standards used there, such as PAL and SECAM, employ a refresh rate of 25 fps. This is convenient for transmitting a movie on television. The movie, which was filmed at 24 fps, is shown at 25 fps, an undetectable difference.

The frequency of electric current in the United States is 60 Hz, so when television arrived, in the 1930s, it used a refresh rate of 30 fps. When color was added, in 1953, that rate was decreased by 1%, to 29.97 fps, because of the need for precise separation of the video and audio signal carriers. Because of interlacing, a complete television picture consists of two frames, so a refresh rate of 29.97 pictures per second requires a rate of 59.94 frames per second.

It turns out that the refresh rate for television should be higher than the rate for movies. A movie is normally watched in darkness, whereas television is watched in a lighted room, and human vision is more sensitive to flicker under conditions of bright illumination. This is why 30 (or 29.97) fps is better than 25.

The electron beam can be turned off and on very rapidly. It can also be deflected horizontally and vertically by two pairs (X and Y) of electrodes. Displaying a single point on the screen is done by turning the beam off, moving it to the part of the screen where the point should appear, and turning it on. This is done by special hardware in response to the analog signal received by the television set.

The signal instructs the hardware to turn the beam off, move it to the top-left corner of the screen, turn it on, and sweep a horizontal line on the screen. While the beam is swept horizontally along the top scan line, the analog signal is used to adjust the beam's intensity according to the image parts being displayed. At the end of the first scan line, the signal instructs the television hardware to turn the beam off, move it back and slightly down, to the start of the third (not the second) scan line, turn it on, and sweep that line. Moving the beam to the start of the next scan line is known as a *retrace*. The time it takes to retrace is the *horizontal blanking time*.

This way, one field of the picture is created on the screen line by line, using just the odd-numbered scan lines (Figure 9.1c). At the end of the last line, the signal contains instructions for a frame retrace. This turns the beam off and moves it to the start of the next field (the second scan line) to scan the field of even-numbered scan lines (Figure 9.1d). The time it takes to do the vertical retrace is the *vertical blanking time*. The picture is therefore created in two fields that together constitute a *frame*. The picture is said to be *interlaced*.

This process is repeated several times each second, to refresh the picture. This order of scanning (left to right, top to bottom, with or without interlacing) is called *raster scan*. The word raster is derived from the Latin *rastrum*, meaning rake, since this scan is done in a pattern similar to that left by a rake on a field.

A consumer television set uses one of three international standards. The standard used in the United States is called NTSC (National Television Standards Committee), although the new digital standard (Section 9.3.1) is fast becoming popular. NTSC specifies a television transmission of 525 lines (today, this would be $2^9 = 512$ lines, but since television was developed before the advent of computers with their preference for binary numbers, the NTSC standard has nothing to do with powers of two). Because of vertical blanking, however, only 483 lines are visible on the screen. Since the aspect ratio (width/height) of a television screen is 4:3, each line has a size of $\frac{4}{3}483 = 644$ pixels. The resolution of a standard television set is therefore 483×644. This may be considered at best medium resolution. (This is the reason why text is so hard to read on a standard television.)

◇ **Exercise 9.1:** (Easy.) What would be the resolution if all 525 lines were visible on the screen?

The aspect ratio of 4:3 was selected by Thomas Edison when he built the first movie cameras and projectors, and was adopted by early television in the 1930s. In the 1950s, after many tests on viewers, the movie industry decided that people prefer larger aspect ratios and started making wide-screen movies, with aspect ratios of 1.85 or higher. Influenced by that, the developers of digital video opted (Section 9.3.1) for the large aspect ratio of 16:9. Exercise 9.4 compares the two aspect ratios, and Table 9.2 lists some common aspect ratios of television and film.

Image formats	Aspect ratio
NTSC, PAL, and SECAM TV	1.33
16 mm and 35 mm film	1.33
HDTV	1.78
Widescreen film	1.85
70 mm film	2.10
Cinemascope film	2.35

Table 9.2: Aspect Ratios of Television and Film.

The concept of *pel aspect ratio* is also useful and should be mentioned. We usually think of a pel (or a pixel) as a mathematical dot, with no dimensions and no shape. In practice, however, pels are printed or displayed, so they have shape and dimensions. The use of a shadow mask (see below) creates circular pels, but computer monitors normally display square or rectangular pixels, thereby creating a crisp, sharp image (because square or rectangular pixels completely fill up space). MPEG-1 (Section 9.6) even has a parameter `pel_aspect_ratio`, whose 16 values are listed in Table 9.26.

It should be emphasized that analog television does not display pixels. When a line is scanned, the beam's intensity is varied continuously. The picture is displayed line by line, but each line is continuous. Consequently, the image displayed by analog television is sampled only in the vertical dimension.

NTSC also specifies a refresh rate of 59.94 (or 60/1.001) frames per second and can be summarized by the notation 525/59.94/2:1, where the 2:1 indicates interlacing. The notation 1:1 indicates *progressive scanning* (not the same as progressive image compression). The PAL television standard (phase alternate line), used in Europe and Asia, is summarized by 625/50/2:1. The quantity $262.5 \times 59.94 = 15734.25$ kHz is called the *line rate* of the 525/59.94/2:1 standard. This is the product of the frame size (number of lines per frame) and the refresh rate.

It should be mentioned that NTSC and PAL are standards for color encoding. They specify how to encode the color into the analog black-and-white video signal. However, for historical reasons, television systems using 525/59.94 scanning normally employ NTSC color coding, whereas television systems using 625/50 scanning normally employ PAL color coding. This is why 525/59.94 and 625/50 are loosely called NTSC and PAL, respectively.

A word about color: Many color CRTs use the *shadow mask* technique (Figure 9.3). They have three guns emitting three separate electron beams. Each beam is associated with one color, but the beams themselves, of course, consist of electrons and do not have any color. The beams are adjusted such that they always converge a short distance behind the screen. By the time they reach the screen they have diverged a bit, and they strike a group of three different (but very close) points called a *triad*.

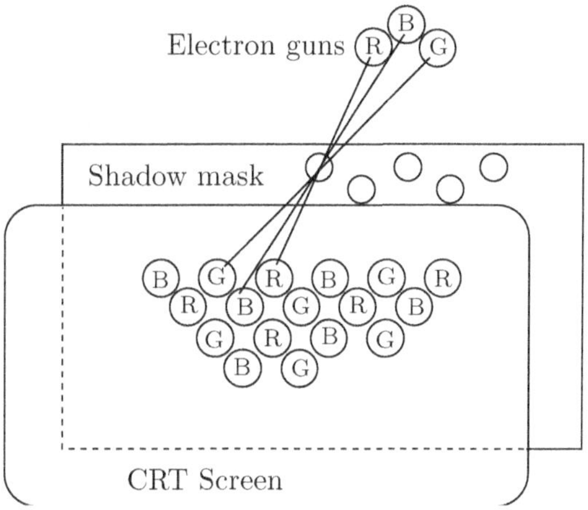

Figure 9.3: A Shadow Mask.

The screen is coated with dots made of three types of phosphor compounds that emit red, green, and blue light, respectively, when excited. At the plane of convergence there is a thin, perforated metal screen: the shadow mask. When the three beams converge at a hole in the mask, they pass through, diverge, and hit a triad of points coated with different phosphor compounds. The points glow at the three colors, and the observer sees a mixture of red, green, and blue whose precise color depends on the intensities of the three beams. When the beams are deflected a little, they hit the mask and are absorbed. After some more deflection, they converge at another hole and hit the screen at another triad.

At a screen resolution of 72 dpi (dots per inch) we expect 72 ideal, square pixels per inch of screen. Each pixel should be a square of side $25.4/72 \approx 0.353$ mm. However, as Figure 9.4a shows, each triad produces a wide circular spot, with a diameter of 0.63 mm, on the screen. These spots highly overlap, and each affects the perceived colors of its neighbors.

When watching television, we tend to position ourselves at a distance from which it is comfortable to watch. When watching from a greater distance, we miss some details,

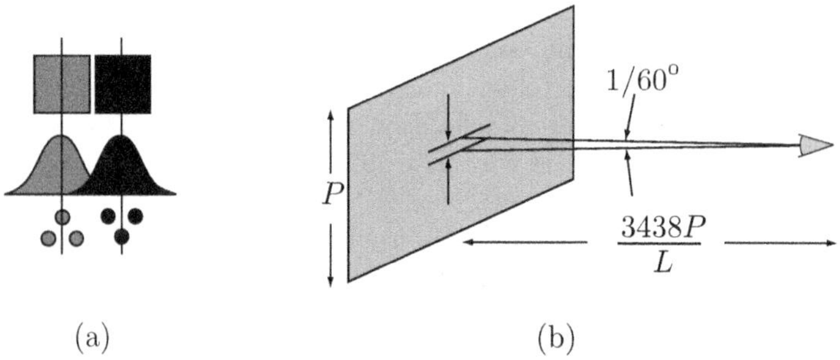

Figure 9.4: (a) Square and Circular Pixels. (b) Comfortable Viewing Distance.

and when watching closer, the individual scan lines are visible. Experiments show that the comfortable viewing distance is determined by the rule: The smallest detail that we want to see should subtend an angle of about one minute of arc $(1/60)^\circ$. We denote by P the height of the image and by L the number of scan lines. The relation between degrees and radians is $360^\circ = 2\pi$ radians. Combining this with Figure 9.4b produces the expression

$$\frac{P/L}{\text{Distance}} = \left(\frac{1}{60}\right)^\circ = \frac{2\pi}{360\cdot 60} = \frac{1}{3438},$$

or

$$\text{Distance} = \frac{3438P}{L}. \tag{9.1}$$

If $L = 483$, the comfortable distance is $7.12P$. For $L = 1080$, Equation (9.1) suggests a distance of $3.18P$.

◇ **Exercise 9.2:** Measure the height of the image on your television set and calculate the comfortable viewing distance from Equation (9.1). Compare it to the distance you actually use.

> All of the books in the world contain no more information than is broadcast as video in a single large American city in a single year. Not all bits have equal value.
>
> —Carl Sagan

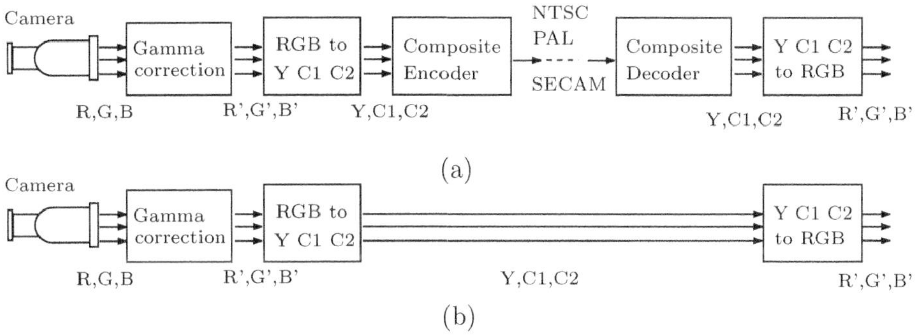

Figure 9.5: (a) Composite and (b) Component Television Transmission.

9.2 Composite and Components Video

The common television receiver found in many homes receives from the transmitter a composite signal, where the luminance and chrominance components [Salomon 99] are multiplexed. This type of signal was designed in the early 1950s, when color was added to television transmissions. The basic black-and-white signal becomes the luminance (Y) component, and two chrominance components $C1$ and $C2$ are added. Those can be U and V, Cb and Cr, I and Q, or any other chrominance components. Figure 9.5a shows the main components of a transmitter and a receiver using a composite signal. The main point is that only one signal is needed. If the signal is sent on the air, only one frequency is needed. If it is sent on a cable, only one cable is used.

> Television, a medium. So called because it is neither rare nor well done.
> —Proverb

NTSC uses the YIQ components, which are defined by

$$\begin{aligned}
Y &= 0.299R' + 0.587B' + 0.114B',\\
I &= 0.596R' - 0.274G' - 0.322B'\\
&= -(\sin 33°)U + (\cos 33°)V,\\
Q &= 0.211R' - 0.523G' + 0.311B'\\
&= (\cos 33°)U + (\sin 33°)V.
\end{aligned}$$

At the receiver, the gamma-corrected $R'G'B'$ components are extracted using the inverse transformation

$$\begin{aligned}
R' &= Y + 0.956I + 0.621Q,\\
G' &= Y - 0.272I - 0.649Q,\\
B' &= Y - 1.106I + 1.703Q.
\end{aligned}$$

PAL uses the basic YUV color space, defined by

$$\begin{aligned}
Y &= 0.299R' + 0.587G' + 0.114B',\\
U &= -0.147R' - 0.289G' + 0.436B' = 0.492(B' - Y),\\
V &= 0.615R' - 0.515G' - 0.100B' = 0.877(R' - Y),
\end{aligned}$$

whose inverse transform is

$$\begin{aligned}
R' &= Y + 1.140V,\\
G' &= Y - 0.394U - 0.580V,\\
B' &= Y - 2.030U.
\end{aligned}$$

(The YUV color space was originally introduced to preserve compatibility between black-and-white and color television standards.)

SECAM uses the composite color space $YDrDb$, defined by

$$\begin{aligned}
Y &= 0.299R' + 0.587G' + 0.114B',\\
Db &= -0.450R' - 0.833G' + 1.333B' = 3.059U,\\
Dr &= -1.333R' + 1.116G' - 0.217B' = -2.169V.
\end{aligned}$$

The inverse transformation is

$$\begin{aligned}
R' &= Y - 0.526Dr,\\
G' &= Y - 0.129Db + 0.268Dr,\\
B' &= Y + 0.665Db.
\end{aligned}$$

Composite video is cheap but has problems such as cross-luminance and cross-chrominance artifacts in the displayed image. High-quality video systems often use *component video*, where three cables or three frequencies carry the individual color components (Figure 9.5b). A common component video standard is the ITU-R recommendation 601, which adopts the YCbCr color space (page 844). In this standard, the luminance Y has values in the range $[16, 235]$, whereas each of the two chrominance components has values in the range $[16, 240]$ centered at 128, which indicates zero chrominance.

9.3 Digital Video

Digital video is the case where a (digital) camera generates a digital image, i.e., an image that consists of pixels. Many people may intuitively feel that an image produced in this way is inferior to an analog image. An analog image seems to have infinite resolution, whereas a digital image has a fixed, finite resolution that cannot be increased without loss of image quality. In practice, however, the high resolution of analog images is not an advantage, because we view them on a television screen or a computer monitor in a certain, fixed resolution. Digital video, on the other hand, has the following important advantages:

1. It can be easily edited. This makes it possible to produce special effects. Computer-generated images, such as spaceships or cartoon characters, can be combined with real-life action to produce complex, realistic-looking effects. The images of an actor in a movie can be edited to make him look young at the beginning and old later. Software for editing digital video is available for most computer platforms. Users can edit a video file and attach it to an email message, thereby creating *vmail.* Multimedia applications, where text, audio, still images, and video are integrated, are common today and involve the editing of video.
2. It can be stored on any digital medium, such as disks, flash memories, CD-ROMs, or DVDs. An error-correcting code can be added, if needed, for increased reliability. This makes it possible to duplicate a long movie or transmit it between computers without loss of quality (in fact, without a single bit getting corrupted). In contrast, analog video is typically stored on tape, each copy of which is slightly inferior to the original, and the medium is subject to wear.
3. It can be compressed. This allows for more storage (when video is stored on a digital medium) and also for fast transmission. Sending compressed video between computers makes video telephony, video chats, and video conferencing possible. Transmitting compressed video also makes it possible to increase the capacity of television cables and thus carry more channels in the same cable.

Digital video is, in principle, a sequence of images, called frames, displayed at a certain *frame rate* (so many frames per second, or fps) to create the illusion of animation. This rate, as well as the image size and pixel depth, depend heavily on the application. Surveillance cameras, for example, use the very low frame rate of five fps, while HDTV displays 25 fps. Table 9.6 shows some typical video applications and their video parameters.

The table illustrates the need for compression. Even the most economic application, a surveillance camera, generates $5\times 640\times 480\times 12 = 18{,}432{,}000$ bits per second! This is equivalent to more than 2.3 million bytes per second, and this information has to be saved for at least a few days before it can be deleted. Most video applications also involve audio. It is part of the overall video data and has to be compressed with the video image.

◇ **Exercise 9.3:** What video applications do not include sound?

A complete piece of video is sometimes called a *presentation.* It consists of a number of *acts*, where each act is broken down into several *scenes.* A scene is made of several *shots* or *sequences* of action, each a succession of *frames* (still images), where there is

Application	Frame rate	Resolution	Pixel depth
Surveillance	5	640×480	12
Video telephony	10	320×240	12
Multimedia	15	320×240	16
Analog TV	25	640×480	16
HDTV (720p)	60	1280×720	24
HDTV (1080i)	60	1920×1080	24
HDTV (1080p)	30	1920×1080	24

Table 9.6: Video Parameters for Typical Applications.

a small change in scene and camera position between consecutive frames. Thus, the five-step hierarchy is

$$\text{piece}\to\text{act}\to\text{scene}\to\text{sequence}\to\text{frame}.$$

9.3.1 High-Definition Television

The NTSC standard was created in the 1930s, for black-and-white television transmissions. NTSC stands for National Television Standards Committee. This is a standard that specifies the shape of the signal generated and sent by a television transmitter. The signal is analog, with amplitude that goes up and down during each scan line in response to the dark and bright parts of the line. When color was incorporated in this standard, in 1953, it had to be added such that black-and-white television sets would be able to display the color signal in black and white. The result was phase modulation of the black-and-white carrier, a kludge (television engineers jokingly refer to it as NSCT "never the same color twice").

With the explosion of computers and digital equipment in the 1980s and 1990s came the realization that a digital signal is a better, more reliable way of transmitting images over the air. In such a signal, the image is sent pixel by pixel, where each pixel is represented by a number specifying its color. The digital signal is still a wave, but the amplitude of the wave no longer represents the image. Rather, the wave is *modulated* to carry binary information. The term modulation means that something in the wave is modified to distinguish between the zeros and ones being sent. An FM digital signal, for example, modifies (modulates) the frequency of the wave. This type of wave uses one frequency to represent a binary 0 and another to represent a 1. The DTV (Digital TV) standard uses a modulation technique called 8-VSB (for *vestigial sideband*), which provides robust and reliable terrestrial transmission. The 8-VSB modulation technique allows for a broad coverage area, reduces interference with existing analog broadcasts, and is itself immune from interference.

History of DTV: The Advanced Television Systems Committee (ATSC), established in 1982, is an international organization that develops technical standards for advanced video systems. Even though these standards are voluntary, they are generally adopted by the ATSC members and other manufacturers. There are currently about eighty ATSC member companies and organizations, which represent the many facets of the television, computer, telephone, and motion picture industries.

The ATSC Digital Television Standard adopted by the United States Federal Communications Commission (FCC) is based on a design by the Grand Alliance (a coalition of electronics manufacturers and research institutes) that was a finalist in the first round of DTV proposals under the FCC's Advisory Committee on Advanced Television Systems (ACATS). The ACATS is composed of representatives of the computer, broadcasting, telecommunications, manufacturing, cable television, and motion picture industries. Its mission is to assist in the adoption of an HDTV transmission standard and to promote the rapid implementation of HDTV in the U.S.

The ACATS announced an open competition: Anyone could submit a proposed HDTV standard, and the best system would be selected as the new television standard for the United States. To ensure fast transition to HDTV, the FCC promised that every television station in the nation would be temporarily lent an additional channel of broadcast spectrum.

The ACATS worked with the ATSC to review the proposed DTV standard, and gave its approval to final specifications for the various parts—audio, transport, format, compression, and transmission. The ATSC documented the system as a standard, and ACATS adopted the Grand Alliance system in its recommendation to the FCC in late 1995.

In late 1996, corporate members of the ATSC had reached an agreement on the DTV standard (Document A/53) and asked the FCC to approve it. On December 31, 1996, the FCC formally adopted every aspect of the ATSC standard except for the video formats. These video formats nevertheless remain a part of the ATSC standard, and are expected to be used by broadcasters and by television manufacturers in the foreseeable future.

HDTV Specifications: The NTSC standard in use since the 1930s specifies an interlaced image composed of 525 lines where the odd numbered lines $(1, 3, 5, \ldots)$ are drawn on the screen first, followed by the even numbered lines $(2, 4, 6, \ldots)$. The two fields are woven together and are drawn in 1/30 of a second, allowing for 30 screen refreshes each second. In contrast, a noninterlaced picture displays the entire image line by line. This *progressive scan* type of image is what's used by today's computer monitors.

The digital television sets that have been available since mid 1998 use an aspect ratio of 16/9 and can display both the interlaced and progressive-scan images in several different resolutions—one of the best features of digital video. These formats include 525-line progressive-scan (525P), 720-line progressive-scan (720P), 1050-line progressive-scan (1050P), and 1080-interlaced (1080I), all with square pixels.

Our present, analog, television sets cannot deal with the new, digital signal broadcast by television stations, but inexpensive converters will be available (in the form of a small box that can comfortably sit on top of a television set) to translate the digital signals to analog ones (and lose image information in the process).

The NTSC standard calls for 525 scan lines and an aspect ratio of 4/3. This implies $\frac{4}{3} \times 525 = 700$ pixels per line, yielding a total of $525 \times 700 = 367{,}500$ pixels on the screen. (This is the theoretical total, in practice only 483 lines are actually visible.) In comparison, a DTV format calling for 1080 scan lines and an aspect ratio of 16/9 is equivalent to 1920 pixels per line, bringing the total number of pixels to $1080 \times 1920 = 2{,}073{,}600$, about 5.64 times more than the NTSC interlaced standard.

◇ **Exercise 9.4:** The NTSC aspect ratio is $4/3 = 1.33$ and that of DTV is $16/9 = 1.77$. Which one looks better?

In addition to the 1080×1920 DTV format, the ATSC DTV standard calls for a lower-resolution format with just 720 scan lines, implying $\frac{16}{9} \times 720 = 1280$ pixels per line. Each of these resolutions can be refreshed at one of three different rates: 60 frames/second (for live video) and 24 or 30 frames/second (for material originally produced on film). The refresh rates can be considered *temporal resolution*. The result is a total of six different formats. Table 9.7 summarizes the screen capacities and the necessary transmission rates of the six formats. With high-resolution and 60 frames per second the transmitter must be able to send 2,985,984,000 bits/sec (about 356 Mbyte/sec), which is why this format uses compression. (It uses MPEG-2. Other video formats can also use this compression method.) The fact that DTV can have different spatial and temporal resolutions allows for trade-offs. Certain types of video material (such as a fast-moving horse or a car race) may look better at high refresh rates even with low spatial resolution, while other material (such as museum-quality paintings) should ideally be watched in high resolution even with low refresh rates.

	total #	refresh rate		
lines × pixels	of pixels	24	30	60
1080 × 1920	2,073,600	1,194,393,600	1,492,992,000	2,985,984,000
720 × 1280	921,600	530,841,600	663,552,000	1,327,104,000

Table 9.7: Resolutions and Capacities of Six DTV Formats.

Digital Television (DTV) is a broad term encompassing all types of digital transmission. HDTV is a subset of DTV indicating 1080 scan lines. Another type of DTV is standard definition television (SDTV), which has picture quality slightly better than a good analog picture. (SDTV has resolution of 640×480 at 30 frames/sec and an aspect ratio of 4:3.) Since generating an SDTV picture requires fewer pixels, a broadcasting station will be able to transmit multiple channels of SDTV within its 6 MHz allowed frequency range. HDTV also incorporates Dolby Digital sound technology to bring together a complete presentation.

Figure 9.8 shows the most important resolutions used in various video systems. Their capacities range from 19,000 pixels to more than two million pixels.

> Sometimes cameras and television are good to people and sometimes they aren't. I don't know if it's the way you say it, or how you look.
>
> —Dan Quayle

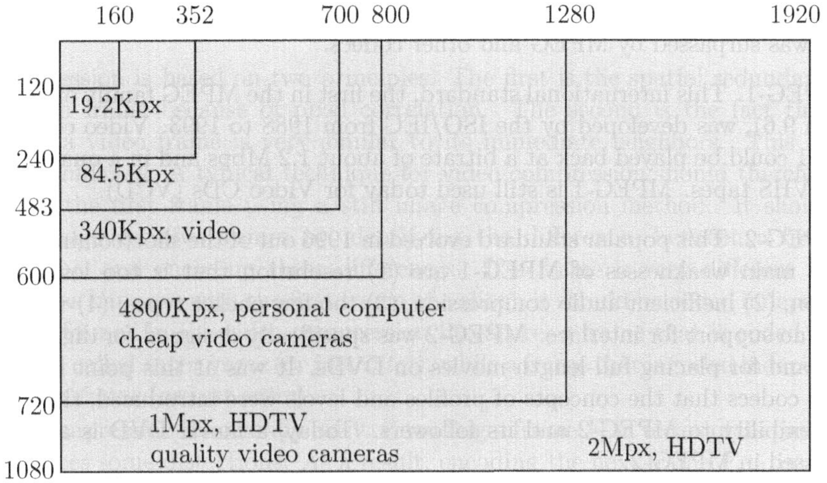

Figure 9.8: Various Video Resolutions.

9.4 History of Video Compression

The search for better and faster compression algorithms is driven by two basic facts of life, the limited bandwidth of our communication channels and the restricted amount of storage space available. Even though both commodities are being expanded all the time, they always remain in short supply because (1) the bigger the storage available, the bigger the data files we want to store and save and (2) the faster our communication channels, the more data we try to deliver through them. Nowhere is this tendency more evident than in the area of video. Video data has the two attributes of being voluminous and being in demand. People generally prefer images over text, and video over still images.

In uncompressed form, digital video, both standard definition (SD) and high definition (HD), requires huge quantities of data and high delivery rates (bitrates). A typical two-hour movie requires, in uncompressed format, a total of 110–670 GB, which is way over the capacity of today's DVDs. Streaming such a movie over a channel to be viewed at real time requires a data transfer rate of 125–750 Mbps. At the time of writing (late 2008), typical communication channels used in broadcasting can deliver only 1–3 Mbps for SD and 10–15 Mbps for HD, a small fraction of the required bandwidth. This is why video compression is crucial to the development of video and has been so for more than two decades. The following is a short historical overview of the main milestones of this field.

- DVI. There seems to be a consensus among experts that the field of video compression had its origin in 1987, when *Digital Video Interactive* (DVI) made its appearance. This video compression method [Golin 92] was developed with the video consumer market in mind, and its developers hoped to see it institutionalized as a standard. DVI was never adopted as a standard by any international body. In 1992 it metamorphosed and became known as Indeo, a software product by Intel, where it remained proprietary. It in the order in which they are generated by the encoder (and input by the decoder). Figure 9.9b shows the same sequence in the order in which the frames are output by the decoder and displayed. The frame labeled 2 should be displayed after frame 5, so each frame should have two time stamps, its coding time and its display time.

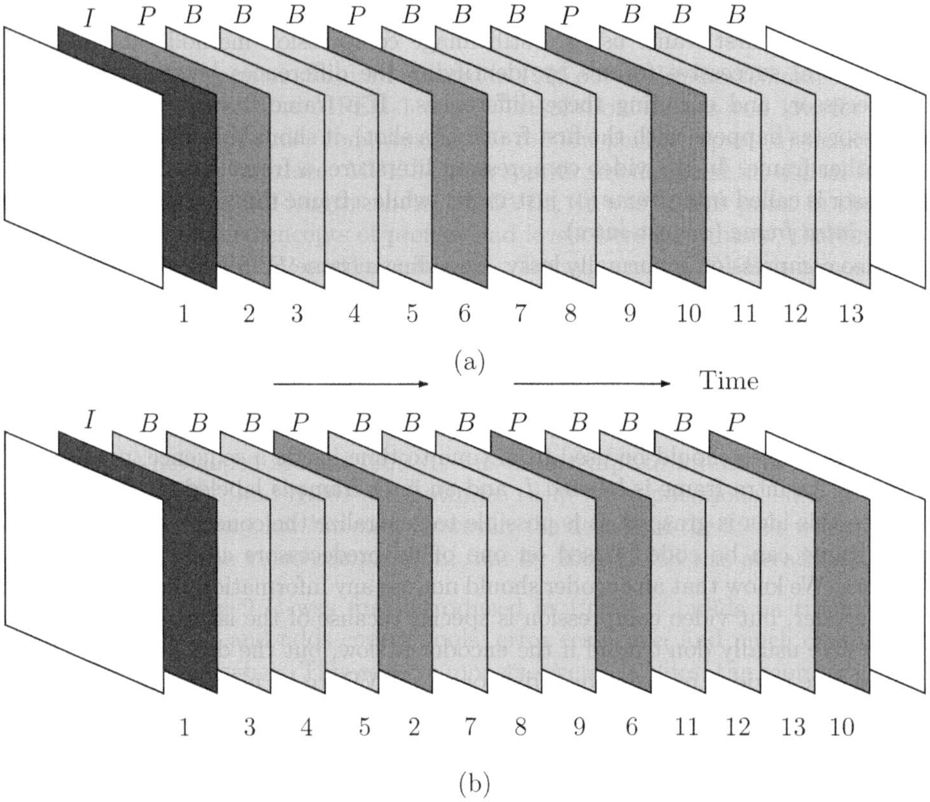

Figure 9.9: (a) Coding Order. (b) Display Order.

We start with a few intuitive video compression methods.

Subsampling: The encoder selects every other frame and writes it on the compressed stream. This yields a compression factor of 2. The decoder inputs a frame and duplicates it to create two frames.

Differencing: A frame is compared to its predecessor. If the difference between them is small (just a few pixels), the encoder encodes the pixels that are different by writing three numbers on the compressed stream for each pixel: its image coordinates, and the difference between the values of the pixel in the two frames. If the difference between the frames is large, the current frame is written on the output in raw format. Compare this method with relative encoding, Section 1.3.1.

A lossy version of differencing looks at the amount of change in a pixel. If the difference between the intensities of a pixel in the preceding frame and in the current frame is smaller than a certain (user controlled) threshold, the pixel is not considered different.

Block Differencing: This is a further improvement of differencing. The image is divided into blocks of pixels, and each block B in the current frame is compared with the corresponding block P in the preceding frame. If the blocks differ by more than a certain amount, then B is compressed by writing its image coordinates, followed by the values of all its pixels (expressed as differences) on the compressed stream. The advantage is that the block coordinates are small numbers (smaller than a pixel's coordinates), and these coordinates have to be written just once for the entire block. On the downside, the values of all the pixels in the block, even those that haven't changed, have to be written on the output. However, since these values are expressed as differences, they are small numbers. Consequently, this method is sensitive to the block size.

Motion Compensation: Anyone who has watched movies knows that the difference between consecutive frames is small because it is the result of moving the scene, the camera, or both between frames. This feature can therefore be exploited to achieve better compression. If the encoder discovers that a part P of the preceding frame has been rigidly moved to a different location in the current frame, then P can be compressed by writing the following three items on the compressed stream: its previous location, its current location, and information identifying the boundaries of P. The following discussion of motion compensation is based on [Manning 98].

In principle, such a part can have any shape. In practice, we are limited to equal-size blocks (normally square but can also be rectangular). The encoder scans the current frame block by block. For each block B it searches the preceding frame for an identical block C (if compression is to be lossless) or for a similar one (if it can be lossy). Finding such a block, the encoder writes the difference between its past and present locations on the output. This difference is of the form

$$(C_x - B_x, C_y - B_y) = (\Delta x, \Delta y),$$

so it is called a *motion vector*. Figure 9.10a,b shows an example where the seagull sweeps to the right while the rest of the image slides slightly to the left because of camera movement.

Motion compensation is effective if objects are just translated, not scaled or rotated. Drastic changes in illumination from frame to frame also reduce the effectiveness of this method. In general, motion compensation is lossy. The following paragraphs discuss the main aspects of motion compensation in detail.

Frame Segmentation: The current frame is divided into equal-size nonoverlapping blocks. The blocks may be squares or rectangles. The latter choice assumes that motion in video is mostly horizontal, so horizontal blocks reduce the number of motion vectors without degrading the compression ratio. The block size is important, because large blocks reduce the chance of finding a match, and small blocks result in many motion vectors. In practice, block sizes that are integer powers of 2, such as 8 or 16, are used, since this simplifies the software.

Search Threshold: Each block B in the current frame is first compared to its counterpart C in the preceding frame. If they are identical, or if the difference between them is less than a preset threshold, the encoder assumes that the block hasn't been

Figure 9.10: Motion Compensation.

moved.

Block Search: This is a time-consuming process, and so has to be carefully designed. If B is the current block in the current frame, then the previous frame has to be searched for a block identical to or very close to B. The search is normally restricted to a small area (called the *search area*) around B, defined by the *maximum displacement* parameters dx and dy. These parameters specify the maximum horizontal and vertical distances, in pixels, between B and any matching block in the previous frame. If B is a square with side b, the search area will contain $(b+2dx)(b+2dy)$ pixels (Figure 9.11) and will consist of $(2dx+1)(2dy+1)$ distinct, overlapping $b\times b$ squares. The number of candidate blocks in this area is therefore proportional to $dx\cdot dy$.

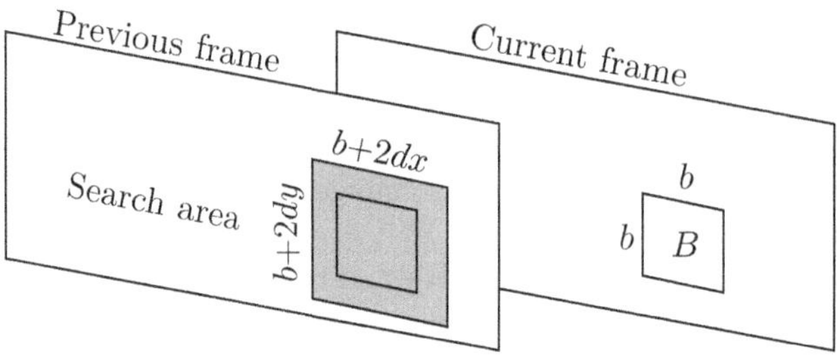

Figure 9.11: Search Area.

Distortion Measure: This is the most sensitive part of the encoder. The distortion measure selects the best match for block B. It has to be simple and fast, but also reliable. A few choices—similar to the ones of Section 7.19—are discussed here.

The *mean absolute difference* (or *mean absolute error*) calculates the average of the absolute differences between a pixel B_{ij} in B and its counterpart C_{ij} in a candidate block C:

$$\frac{1}{b^2}\sum_{i=1}^{b}\sum_{j=1}^{b}|B_{ij}-C_{ij}|.$$

This involves b^2 subtractions and absolute value operations, b^2 additions, and one division. This measure is calculated for each of the $(2dx+1)(2dy+1)$ distinct, overlapping $b\times b$ candidate blocks, and the smallest distortion (say, for block C_k) is examined. If it is smaller than the search threshold, then C_k is selected as the match for B. Otherwise, there is no match for B, and B has to be encoded without motion compensation.

◇ **Exercise 9.5:** How can such a thing happen? How can a block in the current frame match nothing in the preceding frame?

The *mean square difference* is a similar measure, where the square, rather than the absolute value, of a pixel difference is calculated:

$$\frac{1}{b^2}\sum_{i=1}^{b}\sum_{j=1}^{b}(B_{ij}-C_{ij})^2.$$

The *pel difference classification* (PDC) measure counts how many differences $|B_{ij}-C_{ij}|$ are smaller than the PDC parameter p.

The *integral projection* measure computes the sum of a row of B and subtracts it from the sum of the corresponding row of C. The absolute value of the difference is added to the absolute value of the difference of the columns sum:

$$\sum_{i=1}^{b}\left|\sum_{j=1}^{b}B_{ij}-\sum_{j=1}^{b}C_{ij}\right|+\sum_{j=1}^{b}\left|\sum_{i=1}^{b}B_{ij}-\sum_{i=1}^{b}C_{ij}\right|.$$

Suboptimal Search Methods: These methods search some, instead of all, the candidate blocks in the $(b+2dx)(b+2dy)$ area. They speed up the search for a matching block, at the expense of compression efficiency. Several such methods are discussed in detail in Section 9.5.1.

Motion Vector Correction: Once a block C has been selected as the best match for B, a motion vector is computed as the difference between the upper-left corner of C and the upper-left corner of B. Regardless of how the matching was determined, the motion vector may be wrong because of noise, local minima in the frame, or because the matching algorithm is not perfect. It is possible to apply smoothing techniques to the motion vectors after they have been calculated, in an attempt to improve the matching. Spatial correlations in the image suggest that the motion vectors should also be correlated. If certain vectors are found to violate this, they can be corrected.

This step is costly and may even backfire. A video presentation may involve slow, smooth motion of most objects, but also swift, jerky motion of some small objects. Correcting motion vectors may interfere with the motion vectors of such objects and cause distortions in the compressed frames.

Coding Motion Vectors: A large part of the current frame (perhaps close to half of it) may be converted to motion vectors, which is why the way these vectors are encoded is crucial; it must also be lossless. Two properties of motion vectors help in encoding them: (1) They are correlated and (2) their distribution is nonuniform. As we scan the frame block by block, adjacent blocks normally have motion vectors that don't differ by much; they are correlated. The vectors also don't point in all directions. There are often one or two preferred directions in which all or most motion vectors point; the vectors are nonuniformly distributed.

No single method has proved ideal for encoding the motion vectors. Arithmetic coding, adaptive Huffman coding, and various prefix codes have been tried, and all seem to perform well. Here are two different methods that may perform better:

1. Predict a motion vector based on its predecessors in the same row and its predecessors in the same column of the current frame. Calculate the difference between the prediction and the actual vector, and Huffman encode it. This algorithm is important. It is used in MPEG and other compression methods.
2. Group the motion vectors in blocks. If all the vectors in a block are identical, the block is encoded by encoding this vector. Other blocks are encoded as in 1 above. Each encoded block starts with a code identifying its type.

Coding the Prediction Error: Motion compensation is lossy, since a block B is normally matched to a somewhat different block C. Compression can be improved by coding the difference between the current uncompressed and compressed frames on a block by block basis and only for blocks that differ much. This is usually done by transform coding. The difference is written on the output, following each frame, and is used by the decoder to improve the frame after it has been decoded.

9.5.1 Suboptimal Search Methods

Video compression requires many steps and computations, so researchers have been looking for optimizations and faster algorithms, especially for steps that involve many calculations. One such step is the search for a block C in the previous frame to match a given block B in the current frame. An exhaustive search is time-consuming, so it pays to look for suboptimal search methods that search just some of the many overlapping candidate blocks. These methods do not always find the best match, but can generally speed up the entire compression process while incurring only a small loss of compression efficiency.

Signature-Based Methods: Such a method performs a number of steps, restricting the number of candidate blocks in each step. In the first step, all the candidate blocks are searched using a simple, fast distortion measure such as pel difference classification. Only the best matched blocks are included in the next step, where they are evaluated by a more restrictive distortion measure, or by the same measure but with a smaller parameter. A signature method may involve several steps, using different distortion measures in each.

Distance-Diluted Search: We know from experience that fast-moving objects look blurred in an animation, even if they are sharp in all the frames. This suggests a way to lose data. We may require a good block match for slow-moving objects, but allow for a worse match for fast-moving ones. The result is a block matching algorithm that searches all the blocks close to B, but fewer and fewer blocks as the search gets farther away from B. Figure 9.12a shows how such a method may work for maximum displacement parameters $dx = dy = 6$. The total number of blocks C being searched goes from $(2dx+1)\cdot(2dy+1) = 13 \times 13 = 169$ to just 65, less than 39%!

Locality-Based Search: This method is based on the assumption that once a good match has been found, even better matches are likely to be located near it (remember

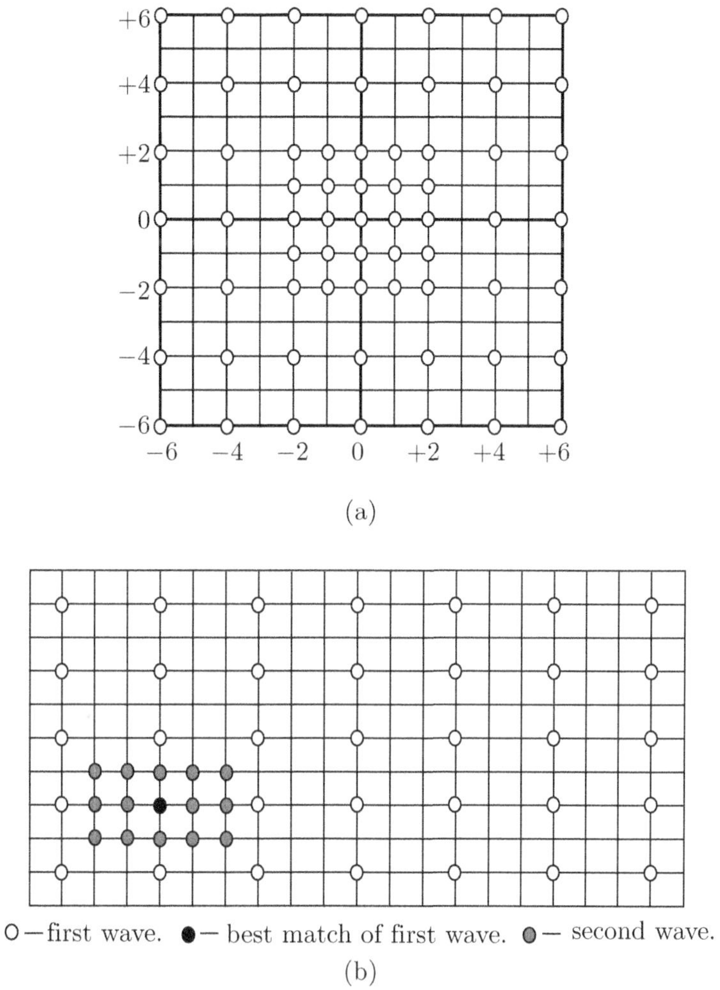

Figure 9.12: (a) Distance-Diluted Search for $dx = dy = 6$. (b) A Locality Search.

that the blocks C searched for matches highly overlap). An obvious algorithm is to start searching for a match in a sparse set of blocks, then use the best-matched block C as the center of a second wave of searches, this time in a denser set of blocks. Figure 9.12b shows two waves of search, the first considers widely-spaced blocks, selecting one as the best match. The second wave searches every block in the vicinity of the best match.

Quadrant Monotonic Search: This is a variant of a locality-based search. It starts with a sparse set of blocks C that are searched for a match. The distortion measure is computed for each of those blocks, and the result is a set of distortion values. The idea is that the distortion values increase as we move away from the best match. By examining the set of distortion values obtained in the first step, the second step may predict where the best match is likely to be found. Figure 9.13 shows how a search of a

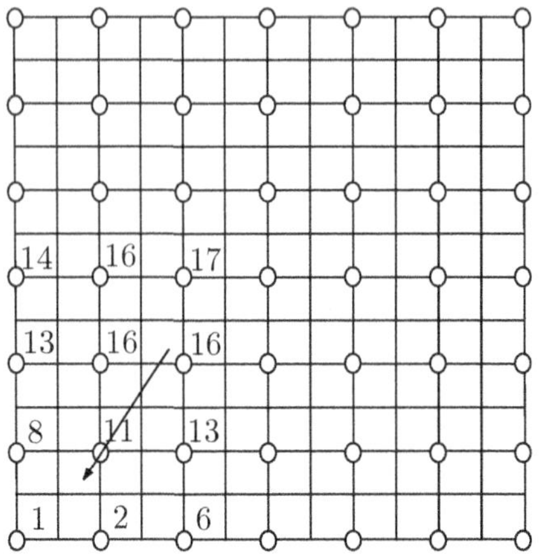

Figure 9.13: Quadrant Monotonic Search.

region of 4×3 blocks suggests a well-defined direction in which to continue searching.

This method is less reliable than the previous ones because the direction proposed by the set of distortion values may lead to a local best block, whereas *the* best block may be located elsewhere.

Dependent Algorithms: As mentioned earlier, motion in a frame is the result of either camera movement or object movement. If we assume that objects in the frame are bigger than a block, we conclude that it is reasonable to expect the motion vectors of adjacent blocks to be correlated. The search algorithm can therefore start by estimating the motion vector of a block B from the motion vectors that have already been found for its neighbors, then improve this estimate by comparing B to some candidate blocks C. This is the basis of several *dependent algorithms*, which can be spatial or temporal.

Spatial dependency: In a spatial dependent algorithm, the neighbors of a block B in the current frame are used to estimate the motion vector of B. These, of course, must be neighbors whose motion vectors have already been computed. Most blocks have eight neighbors each, but considering all eight may not be the best strategy (also, when a block B is considered, only some of its neighbors may have their motion vectors already computed). If blocks are matched in raster order, then it makes sense to use one, two, or three previously-matched neighbors, as shown in Figure 9.14a,b,c. Because of symmetry, however, it is better to use four symmetric neighbors, as in Figure 9.14d,e. This can be done by a three-pass method that scans blocks as in Figure 9.14f. The first pass scans all the blocks shown in black (one-quarter of the blocks in the frame). Motion vectors for those blocks are calculated by some other method. Pass two scans the blocks shown in gray (25% of the blocks) and estimates a motion vector for each using the motion vectors of its four corner neighbors. The white blocks (the remaining 50%) are scanned in the third pass, and the motion vector of each is estimated using the motion vectors of its neighbors on all four sides. If the motion vectors of the neighbors are very different, they should not be used, and a motion vector for block B is computed with a different method.

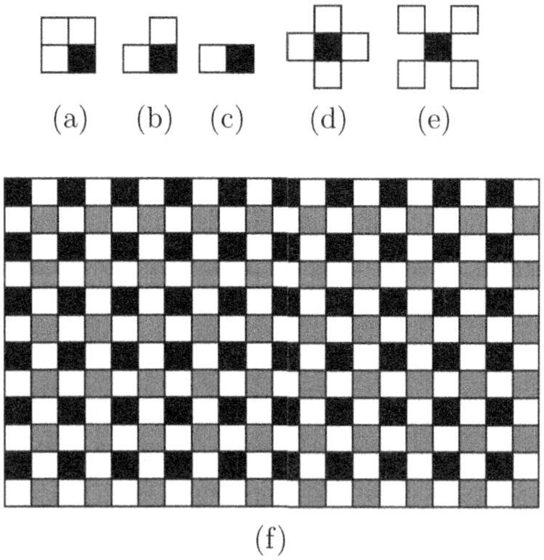

Figure 9.14: Strategies for Spatial Dependent Search Algorithms.

Temporal dependency: The motion vector of block B in the current frame can be estimated as the motion vector of the same block in the previous frame. This makes sense if we can assume uniform motion. After the motion vector of B is estimated this way, it should be improved and corrected using other methods.

More Quadrant Monotonic Search Methods: The following suboptimal block matching methods use the main assumption of the quadrant monotonic search method.

Two-Dimensional Logarithmic Search: This multistep method reduces the search area in each step until it shrinks to one block. We assume that the current block B is located at position (a, b) in the current frame. This position becomes the initial center of the search. The algorithm uses a distance parameter d that defines the search area. This parameter is user-controlled with a default value. The search area consists of the $(2d+1)\times(2d+1)$ blocks centered on the current block B.

Step 1: A step size s is computed by

$$s = 2^{\lfloor \log_2 d \rfloor - 1},$$

and the algorithm compares B with the five blocks at positions (a, b), $(a, b+s)$, $(a, b-s)$, $(a+s, b)$, and $(a-s, b)$ in the previous frame. These five blocks form the pattern of a plus sign "+".

Step 2: The best match among the five blocks is selected. We denote the position of this block by (x, y). If $(x, y) = (a, b)$, then s is halved (this is the reason for the name *logarithmic*). Otherwise, s stays the same, and the center (a, b) of the search is moved to (x, y).

Step 3: If $s = 1$, then the nine blocks around the center (a, b) of the search are searched, and the best match among them becomes the result of the algorithm. Otherwise the algorithm goes to Step 2.

Any blocks that need be searched but are outside the search area are ignored and are not used in the search. Figure 9.15 illustrates the case where $d = 8$. For simplicity we assume that the current block B has frame coordinates $(0,0)$. The search is limited to the (17×17)-block area centered on block B. Step 1 computes

$$s = 2^{\lfloor \log_2 8 \rfloor -1} = 2^{3-1} = 4,$$

and searches the five blocks (labeled 1) at locations $(0,0)$, $(4,0)$, $(-4,0)$, $(0,4)$, and $(0,-4)$. We assume that the best match of these five is at $(0,4)$, so this becomes the new center of the search, and the three blocks (labeled 2) at locations $(4,-4)$, $(4,4)$, and $(8,0)$ are searched in the second step.

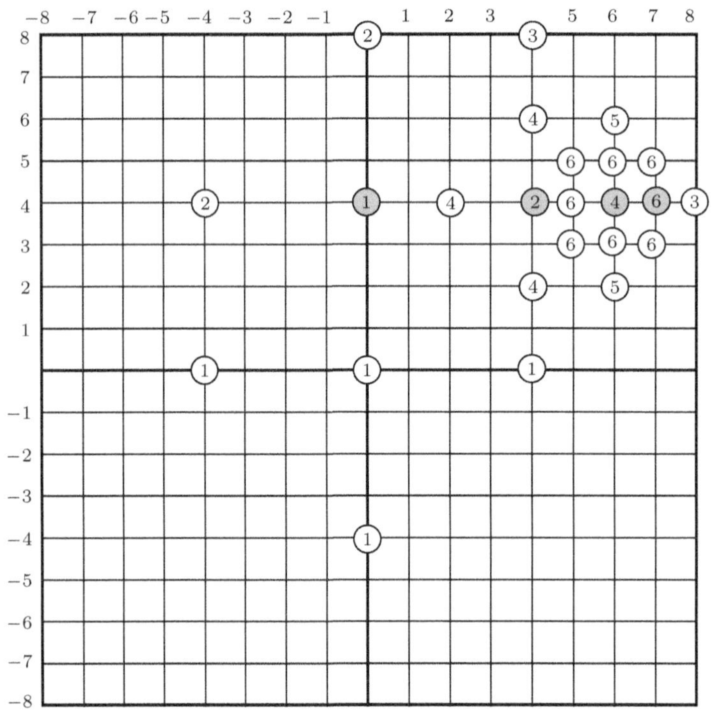

Figure 9.15: The Two-Dimensional Logarithmic Search Method.

Assuming that the best match among these three is at location $(4,4)$, the next step searches the two blocks labeled 3 at locations $(8,4)$ and $(4,8)$, the block (labeled 2) at $(4,4)$, and the "1" blocks at $(0,4)$ and $(4,0)$.

◇ **Exercise 9.6:** Assuming that $(4,4)$ is again the best match, use the figure to describe the rest of the search.

Three-Step Search: This is somewhat similar to the two-dimensional logarithmic search. In each step it tests eight blocks, instead of four, around the center of search, then halves the step size. If $s=3$ initially, the algorithm terminates after three steps, hence its name.

Orthogonal Search: This is a variation of both the two-dimensional logarithmic search and the three-step search. Each step of the orthogonal search involves a horizontal and a vertical search. The step size s is initialized to $\lfloor (d+1)/2 \rfloor$, and the block at the center of the search and two candidate blocks located on either side of it at a distance of s are searched. The location of smallest distortion becomes the center of the vertical search, where two candidate blocks above and below the center, at distances of s, are searched. The best of these locations becomes the center of the next search. If the step size s is 1, the algorithm terminates and returns the best block found in the current step. Otherwise, s is halved, and a new, similar set of horizontal and vertical searches is performed.

One-at-a-Time Search: In this type of search there are again two steps, a horizontal and a vertical. The horizontal step searches all the blocks in the search area whose y coordinates equal that of block B (i.e., that are located on the same horizontal axis as B). Assuming that block H has the minimum distortion among those, the vertical step searches all the blocks on the same vertical axis as H and returns the best of them. A variation repeats this on smaller and smaller search areas.

Cross Search: All the steps of this algorithm, except the last one, search the five blocks at the edges of a multiplication sign "×". The step size is halved in each step until it gets down to 1. At the last step, the plus sign "+" is used to search the areas located around the top-left and bottom-right corners of the preceding step.

This has been a survey of quadrant monotonic search methods. We follow with an outline of two advanced search methods.

Hierarchical Search Methods: Hierarchical methods take advantage of the fact that block matching is sensitive to the block size. A hierarchical search method starts with large blocks and uses their motion vectors as starting points for more searches with smaller blocks. Large blocks are less likely to stumble on a local maximum, while a small block generally produces a better motion vector. A hierarchical search method is therefore computationally intensive, and the main point is to speed it up by reducing the number of operations. This can be done in several ways as follows:

1. In the initial steps, when the blocks are still large, search just a sample of blocks. The resulting motion vectors are not the best, but they are only going to be used as starting points for better ones.
2. When searching large blocks, skip some of the pixels of a block. The algorithm may, for example, examine just one-quarter of the pixels of the large blocks, one-half of the pixels of smaller blocks, and so on.
3. Select the block sizes such that the block used in step i is divided into several (typically four or nine) blocks used in the following step. This way, a single motion vector computed in step i can be used as an estimate for several better motion vectors in step $i+1$.

Multidimensional Search Space Methods: These methods are more complex. When searching for a match for block B, such a method looks for matches that are rotations or zooms of B, not just translations.

A multidimensional search space method may also find a block C that matches B but has different lighting conditions. This is useful when an object moves among areas that are illuminated differently. All the methods discussed so far compare two blocks by comparing the luminance values of corresponding pixels. Two blocks B and C that contain the same objects but differ in luminance would be declared different by such methods.

When a multidimensional search space method finds a block C that matches B but has different luminance, it may declare C the match of B and append a luminance value to the compressed frame B. This value (which may be negative) is added by the decoder to the pixels of the decompressed frame, to bring them back to their original values.

A multidimensional search space method may also compare a block B to rotated versions of the candidate blocks C. This is useful if objects in the video presentation may be rotated in addition to being moved. The algorithm may also try to match a block B to a block C containing a scaled version of the objects in B. If, for example, B is of size 8×8 pixels, the algorithm may consider blocks C of size 12×12, shrink each to 8×8, and compare it to B.

This kind of block search requires many extra operations and comparisons. We say that it increases the size of the *search space* significantly, hence the name *multidimensional search space*. It seems that at present there is no multidimensional search space method that can account for scaling, rotation, and changes in illumination and also be fast enough for practical use.

> Television? No good will come of this device. The word is half Greek and half Latin.
>
> —C. P. Scott

9.6 MPEG

Starting in 1988, the MPEG project was developed by a group of hundreds of experts under the auspices of the ISO (International Standardization Organization) and the IEC (International Electrotechnical Committee). The name MPEG is an acronym for Moving Pictures Experts Group. MPEG is a method for video compression, which involves the compression of digital images and sound, as well as synchronization of the two. There currently are several MPEG standards. MPEG-1 is intended for intermediate data rates, on the order of 1.5 Mbit/sec MPEG-2 is intended for high data rates of at least 10 Mbit/sec. MPEG-3 was intended for HDTV compression but was found to be redundant and was merged with MPEG-2. MPEG-4 is intended for very low data rates of less than 64 Kbit/sec. A third international body, the ITU-T, has been involved in the design of both MPEG-2 and MPEG-4. This section concentrates on MPEG-1 and discusses only its image compression features.

The formal name of MPEG-1 is the international standard for moving picture video compression, IS11172-2. Like other standards developed by the ITU and ISO, the document describing MPEG-1 has *normative* and *informative* sections. A normative section is part of the specification of the standard. It is intended for implementers, is written in a precise language, and should be strictly adhered to when implementing the standard on actual computer platforms. An informative section, on the other hand, illustrates concepts discussed elsewhere, explains the reasons that led to certain choices and decisions, and contains background material. An example of a normative section is the various tables of variable codes used in MPEG. An example of an informative section is the algorithm used by MPEG to estimate motion and match blocks. MPEG does not require any particular algorithm, and an MPEG encoder can use any method to match blocks. The section itself simply describes various alternatives.

The discussion of MPEG in this section is informal. The first subsection (main components) describes all the important terms, principles, and codes used in MPEG-1. The subsections that follow go into more details, especially in the description and listing of the various parameters and variable-length codes.

The importance of a widely accepted standard for video compression is apparent from the fact that many manufacturers (of computer games, DVD movies, digital television, and digital recorders, among others) implemented MPEG-1 and started using it even before it was finally approved by the MPEG committee. This also was one reason why MPEG-1 had to be frozen at an early stage and MPEG-2 had to be developed to accommodate video applications with high data rates.

There are many sources of information on MPEG. [Mitchell et al. 97] is one detailed source for MPEG-1, and the MPEG consortium [MPEG 98] contains lists of other resources. In addition, there are many Web pages with descriptions, explanations, and answers to frequently asked questions about MPEG.

To understand the meaning of the words "intermediate data rate" we consider a typical example of video with a resolution of 360×288, a depth of 24 bits per pixel, and a refresh rate of 24 frames per second. The image part of this video requires $360\times 288\times 24\times 24 = 59{,}719{,}680$ bits/s. For the audio part, we assume two sound tracks (stereo sound), each sampled at 44 kHz with 16-bit samples. The data rate is $2\times 44{,}000\times 16 = 1{,}408{,}000$ bits/s. The total is about 61.1 Mbit/s and this is supposed to be compressed by MPEG-1 to an intermediate data rate of about 1.5 Mbit/s (the size of the sound track alone), a compression factor of more than 40! Another aspect is the decoding speed. An MPEG-compressed movie may end up being stored on a CD-ROM or DVD and has to be decoded and played in real time.

MPEG uses its own vocabulary. An entire movie is considered a *video sequence*. It consists of *pictures*, each having three *components*, one luminance (Y) and two chrominance (Cb and Cr). The luminance component (Section 7.3) contains the black-and-white picture, and the chrominance components provide the color hue and saturation (see [Salomon 99] for a detailed discussion). Each component is a rectangular array of *samples*, and each row of the array is called a *raster line*. A *pel* is the set of three samples. The eye is sensitive to small spatial variations of luminance, but is less sensitive to similar changes in chrominance. As a result, MPEG-1 samples the chrominance components at half the resolution of the luminance component. The term *intra* is used, but *inter* and *nonintra* are used interchangeably.

The input to an MPEG encoder is called the *source data*, and the output of an MPEG decoder is the *reconstructed data*. The source data is organized in packs (Figure 9.16b), where each pack starts with a start code (32 bits) followed by a header, ends with a 32-bit end code, and contains a number of packets in between. A packet contains compressed data, either audio or video. The size of a packet is determined by the MPEG encoder according to the requirements of the storage or transmission medium, which is why a packet is not necessarily a complete video picture. It can be any part of a video picture or any part of the audio.

The MPEG decoder has three main parts, called *layers*, to decode the audio, the video, and the system data. The system layer reads and interprets the various codes and headers in the source data, and routes the packets to either the audio or the video layers (Figure 9.16a) to be buffered and later decoded. Each of these two layers consists of several decoders that work simultaneously.

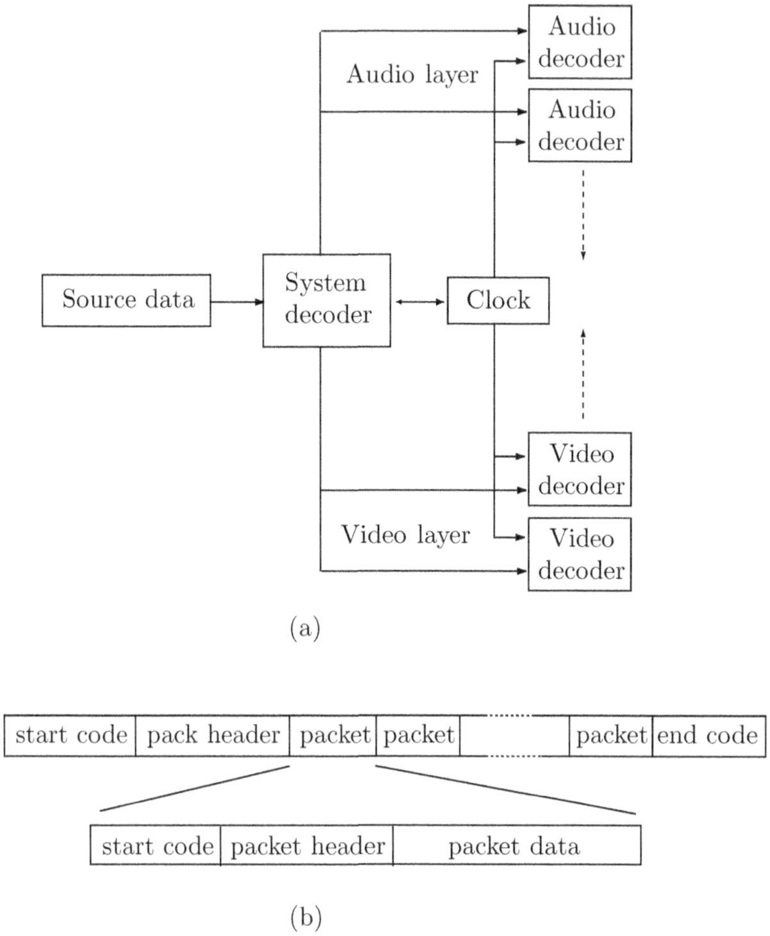

Figure 9.16: (a) MPEG Decoder Organization. (b) Source Format.

The MPEG home page is at `http://www.chiariglione.org/mpeg/index.htm`.

9.6.1 MPEG-1 Main Components

MPEG uses I, P, and B pictures, as discussed in Section 9.5. They are arranged in groups, where a group can be open or closed. The pictures are arranged in a certain order, called the *coding order*, but (after being decoded) they are output and displayed in a different order, called the *display order*. In a closed group, P and B pictures are decoded only from other pictures in the group. In an open group, they can be decoded from pictures outside the group. Different regions of a B picture may use different pictures for their decoding. A region may be decoded from some preceding pictures, from some following pictures, from both types, or from none. Similarly, a region in a P picture may use several preceding pictures for its decoding, or use none at all, in which case it is decoded using MPEG's intra methods.

The basic building block of an MPEG picture is the *macroblock* (Figure 9.17a). It consists of a 16×16 block of luminance (grayscale) samples (divided into four 8×8 blocks) and two 8×8 blocks of the matching chrominance samples. The MPEG compression of a macroblock consists mainly in passing each of the six blocks through a discrete cosine transform, which creates decorrelated values, then quantizing and encoding the results. It is very similar to JPEG compression (Section 7.10), the main differences being that different quantization tables and different code tables are used in MPEG for intra and nonintra, and the rounding is done differently.

A picture in MPEG is organized in slices, where each slice is a contiguous set of macroblocks (in raster order) that have the same grayscale (i.e., luminance component). The concept of slices makes sense because a picture may often contain large uniform areas, causing many contiguous macroblocks to have the same grayscale. Figure 9.17b shows a hypothetical MPEG picture and how it is divided into slices. Each square in the picture is a macroblock. Notice that a slice can continue from scan line to scan line.

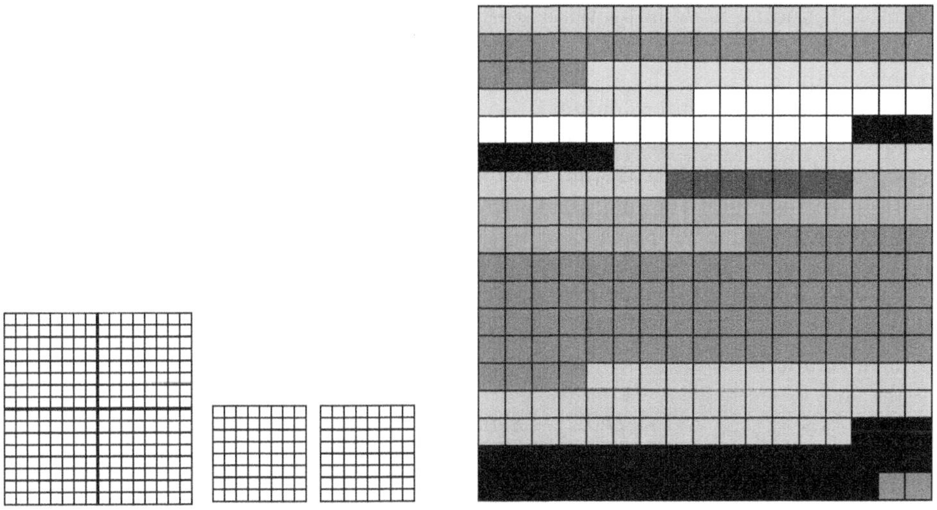

Luminance Cb Cr

(a) (b)

Figure 9.17: (a) A Macroblock. (b) A Possible Slice Structure.

◇ **Exercise 9.7:** How many samples are there in the hypothetical MPEG picture of Figure 9.17b?

When a picture is encoded in nonintra mode (i.e., it is encoded by means of another picture, normally its predecessor), the MPEG encoder generates the differences between the pictures, then applies the DCT to the differences. In such a case, the DCT does not contribute much to the compression, because the differences are already decorrelated. Nevertheless, the DCT is useful even in this case, since it is followed by quantization, and the quantization in nonintra coding can be quite deep.

The precision of the numbers processed by the DCT in MPEG also depends on whether intra or nonintra coding is used. MPEG samples in intra coding are 8-bit unsigned integers, whereas in nonintra they are 9-bit signed integers. This is because a sample in nonintra is the difference of two unsigned integers, and may therefore be negative. The two summations of the two-dimensional DCT, Equation (7.16), can at most multiply a sample by $64 = 2^6$ and may therefore result in an $8+6 = 14$-bit integer (see Exercise 7.19 for a similar case). In those summations, a sample is multiplied by cosine functions, which may result in a negative number. The result of the double sum is therefore a 15-bit signed integer. This integer is then multiplied by the factor $C_iC_j/4$ which is at least $1/8$, thereby reducing the result to a 12-bit signed integer.

This 12-bit integer is then quantized by dividing it by a quantization coefficient (QC) taken from a quantization table. The result is, in general, a noninteger and has to be rounded. It is in quantization and rounding that information is irretrievably lost. MPEG specifies default quantization tables, but custom tables can also be used. In intra coding, rounding is done in the normal way, to the nearest integer, whereas in nonintra, rounding is done by truncating a noninteger to the nearest smaller integer. Figure 9.18a,b shows the results graphically. Notice the wide interval around zero in nonintra coding. This is the so-called *dead zone*.

The quantization and rounding steps are complex and involve more operations than just dividing a DCT coefficient by a quantization coefficient. They depend on a scale factor called `quantizer_scale`, an MPEG parameter that is an integer in the interval $[1, 31]$. The results of the quantization, and hence the compression performance, are sensitive to the value of `quantizer_scale`. The encoder can change this value from time to time and has to insert a special code in the compressed stream to indicate this.

We denote by DCT the DCT coefficient being quantized, by Q the QC from the quantization table, and by $QDCT$ the quantized value of DCT. The quantization tule for intra coding is

$$QDCT = \frac{(16\times DCT) + \text{Sign}(DCT)\times\texttt{quantizer\_scale}\times Q}{2\times\texttt{quantizer\_scale}\times Q}, \tag{9.2}$$

where the function Sign(DCT) is the sign of DCT, defined by

$$\text{Sign}(DCT) = \begin{cases} +1, & \text{when } DCT > 0,\\ 0, & \text{when } DCT = 0,\\ -1, & \text{when } DCT < 0. \end{cases}$$

The second term of Equation (9.2) is called the *rounding term* and is responsible for the special form of rounding illustrated by Figure 9.18a. This is easy to see when we consider the case of a positive DCT. In this case, Equation (9.2) is reduced to the simpler expression

$$QDCT = \frac{(16\times DCT)}{2\times\texttt{quantizer\_scale}\times Q} + \frac{1}{2}.$$

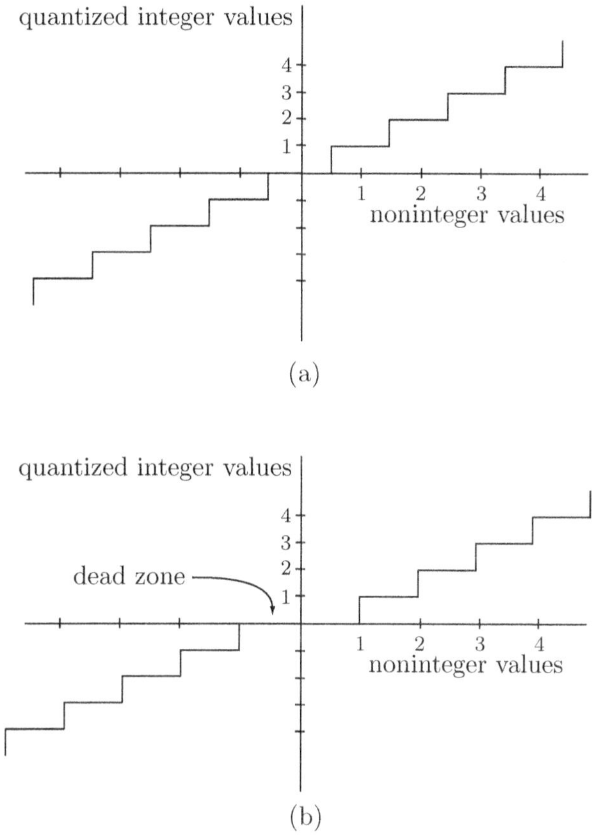

Figure 9.18: Rounding of Quantized DCT Coefficients.
(a) For Intra Coding. (b) For Nonintra Coding.

The rounding term is eliminated for nonintra coding, where quantization is done by

$$QDCT = \frac{(16\times DCT)}{2\times\texttt{quantizer\_scale}\times Q}. \qquad (9.3)$$

Dequantization, which is done by the decoder in preparation for the IDCT, is the inverse of quantization. For intra coding it is done by

$$DCT = \frac{(2\times QDCT)\times\texttt{quantizer\_scale}\times Q}{16}$$

(notice that there is no rounding term), and for nonintra it is the inverse of Equation (9.2)

$$DCT = \frac{\big((2\times QDCT) + Sign(QDCT)\big)\times\texttt{quantizer\_scale}\times Q}{16}.$$

The precise way to compute the IDCT is not specified by MPEG. This can lead to distortions in cases where a picture is encoded by one implementation and decoded by another, where the IDCT is done differently. In a chain of inter pictures, where each picture is decoded by means of its neighbors, this can lead to accumulation of errors, a phenomenon known as *IDCT mismatch.* This is why MPEG requires periodic intra coding of every part of the picture. This *forced updating* has to be done at least once for every 132 P pictures in the sequence. In practice, forced updating is rare, since I pictures are fairly common, occurring every 10 to 15 pictures.

The quantized numbers $QDCT$ are Huffman coded, using the nonadaptive Huffman method and Huffman code tables that were computed by gathering statistics from many training image sequences. The particular code table being used depends on the type of picture being encoded. To avoid the zero probability problem (Section 5.14), all the entries in the code tables were initialized to 1 before any statistics were collected.

Decorrelating the original pels by computing the DCT (or, in the case of inter coding, by calculating pel differences) is part of the statistical model of MPEG. The other part is the creation of a symbol set that takes advantage of the properties of Huffman coding. Section 5.2 explains that the Huffman method becomes inefficient when the data contains symbols with large probabilities. If the probability of a symbol is 0.5, it should ideally be assigned a 1-bit code. If the probability is higher, the symbol should be assigned a shorter code, but the Huffman codes are integers and hence cannot be shorter than one bit. To avoid symbols with high probability, MPEG uses an alphabet where several old symbols (i.e., several pel differences or quantized DCT coefficients) are combined to form one new symbol. An example is run lengths of zeros. After quantizing the 64 DCT coefficients of a block, many of the resulting numbers are zeros. The probability of a zero is therefore high and can easily exceed 0.5. The solution is to deal with runs of consecutive zeros. Each run becomes a new symbol and is assigned a Huffman code. This method creates a large number of new symbols, and many Huffman codes are needed as a result. Compression efficiency, however, is improved.

Table 9.19 is the default quantization coefficients table for the luminance samples in intra coding. The MPEG documentation "explains" this table by saying, "This table has a distribution of quantizing values that is roughly in accord with the frequency response of the human eye, given a viewing distance of approximately six times the screen width and a 360×240 pel picture." Quantizing in nonintra coding is completely different, since the quantities being quantized are pel differences, and they do not have any spatial frequencies. This type of quantization is done by dividing the DCT coefficients of the differences by 16 (the default quantization table is thus flat), although custom quantization tables can also be specified.

In an I picture, the DC coefficients of the macroblocks are coded separately from the AC coefficients, similar to what is done in JPEG (Section 7.10.4). Figure 9.20 shows how three types of DC coefficients, for the Y, Cb, and Cr components of the I picture, are encoded separately in one stream. Each macroblock consists of four Y blocks, one Cb, and one Cr block, so it contributes four DC coefficients of the first type, and one coefficient of each of the other two types. A coefficient DC_i is first used to calculate a difference $\Delta DC = DC_i - P$ (where P is the previous DC coefficient *of the same type*), and then the difference is encoded by coding a size category followed by bits for the magnitude and sign of the difference. The size category is the number of bits required to encode the sign and magnitude of the difference ΔDC. Each size category is assigned a code. Three steps are needed to encode a DC difference ΔDC as follows (1) The

8	16	19	22	26	27	29	34
16	16	22	24	27	29	34	37
19	22	26	27	29	34	34	38
22	22	26	27	29	34	37	40
22	26	27	29	32	35	40	48
26	27	29	32	35	40	48	58
26	27	29	34	38	46	56	69
27	29	35	38	46	56	69	83

Table 9.19: Default Luminance Quantization Table for Intra Coding.

size category is first determined and its code is emitted, (2) if ΔDC is negative, a 1 is subtracted from its 2's complement representation, and (3) the *size* least-significant bits of the difference are emitted. Table 9.21 summarizes the size categories, their codes, and the range of differences ΔDC for each size. Notice that the size category of zero is defined as 0. This table should be compared with Table 7.65, which lists the corresponding codes used by JPEG.

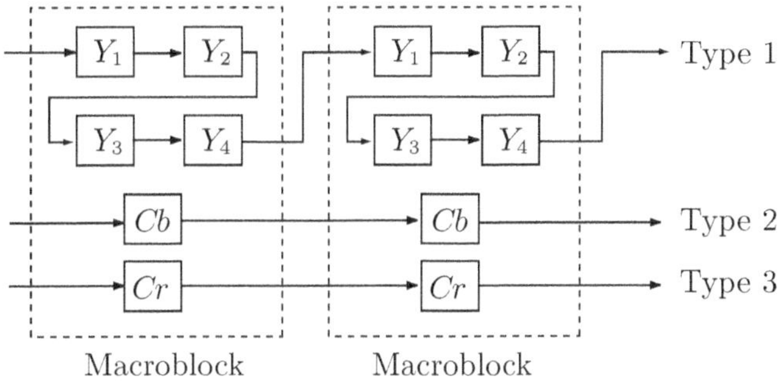

Figure 9.20: The Three Types of DC Coefficients.

Y codes	C codes	size	magnitude range
100	00	0	0
00	01	1	$-1,1$
01	10	2	$-3\cdots-2,2\cdots3$
101	110	3	$-7\cdots-4,4\cdots7$
110	1110	4	$-15\cdots-8,8\cdots15$
1110	11110	5	$-31\cdots-16,16\cdots31$
11110	111110	6	$-63\cdots-32,32\cdots63$
111110	1111110	7	$-127\cdots-64,64\cdots127$
1111110	11111110	8	$-255\cdots-128,128\cdots255$

Table 9.21: Codes for DC Coefficients (Luminance and Chrominance).

Examples: (1) A luminance ΔDC of 5. The number 5 can be expressed in three bits, so the size category is 3, and code 101 is emitted first. It is followed by the three least-significant bits of 5, which are 101. (2) A chrominance ΔDC of -3. The number 3 can be expressed in 2 bits, so the size category is 2, and code 10 is first emitted. The difference -3 is represented in twos complement as ...11101. When 1 is subtracted from this number, the 2 least-significant bits are 00, and this code is emitted next.

◇ **Exercise 9.8:** Compute the code of luminance $\Delta DC = 0$ and the code of chrominance $\Delta DC = 4$.

The AC coefficients of an I picture (intra coding) are encoded by scanning them in the zigzag order shown in Figure 1.8b. The resulting sequence of AC coefficients consists of nonzero coefficients and run lengths of zero coefficients. A *run-level* code is output for each nonzero coefficient C, where *run* refers to the number of zero coefficients preceding C, and *level* refers to the absolute size of C. Each run-level code for a nonzero coefficient C is followed by the 1-bit sign of C (1 for negative and 0 for positive). The run-level code for the last nonzero coefficient is followed by a special 2-bit end-of-block (EOB) code. Table 9.23 lists the EOB code and the run-level codes for common values of runs and levels. Combinations of runs and levels that are not in the table are encoded by the escape code, followed by a 6-bit code for the run length and an 8- or 16-bit code for the level.

Figure 9.22a shows an example of an 8×8 block of quantized coefficients. The zigzag sequence of these coefficients is

$$127,\ 0,\ 0,\ -1,\ 0,\ 2,\ 0,\ 0,\ 0,\ 1,$$

where 127 is the DC coefficient. Thus, the AC coefficients are encoded by the three run level codes $(2, -1)$, $(1, 2)$, $(3, 1)$, followed by the EOB code. Table 9.23 shows that the codes are (notice the sign bits following the run-level codes)

$$0101\,1|000110\,0|00111\,0|10$$

(without the vertical bars).

(a)

127	0	2	0	0	0	0	0
0	0	0	0	0	0	0	0
−1	0	0	0	0	0	0	0
1	0	0	0	0	0	0	0
0	0	0	0	0	0	0	0
0	0	0	0	0	0	0	0
0	0	0	0	0	0	0	0
0	0	0	0	0	0	0	0

(b)

118	2	0	0	0	0	0	0
0	0	0	0	0	0	0	0
−2	0	0	−1	0	0	0	0
0	0	0	0	0	0	0	0
0	0	0	0	0	0	0	0
0	0	0	0	0	0	0	0
0	0	0	0	0	0	0	0
0	0	0	0	0	0	0	0

Figure 9.22: Two 8×8 Blocks of DCT Quantized Coefficients.

0/1	1s (first)	2	1/1	011s	4
0/1	11s (next)	3	1/2	0001 10s	7
0/2	0100 s	5	1/3	0010 0101 s	9
0/3	0010 1s	6	1/4	0000 0011 00s	11
0/4	0000 110s	8	1/5	0000 0001 1011 s	13
0/5	0010 0110 s	9	1/6	0000 0000 1011 0s	14
0/6	0010 0001 s	9	1/7	0000 0000 1010 1s	14
0/7	0000 0010 10s	11	1/8	0000 0000 0011 111s	16
0/8	0000 0001 1101 s	13	1/9	0000 0000 0011 110s	16
0/9	0000 0001 1000 s	13	1/10	0000 0000 0011 101s	16
0/10	0000 0001 0011 s	13	1/11	0000 0000 0011 100s	16
0/11	0000 0001 0000 s	13	1/12	0000 0000 0011 011s	16
0/12	0000 0000 1101 0s	14	1/13	0000 0000 0011 010s	16
0/13	0000 0000 1100 1s	14	1/14	0000 0000 0011 001s	16
0/14	0000 0000 1100 0s	14	1/15	0000 0000 0001 0011 s	17
0/15	0000 0000 1011 1s	14	1/16	0000 0000 0001 0010 s	17
0/16	0000 0000 0111 11s	15	1/17	0000 0000 0001 0001 s	17
0/17	0000 0000 0111 10s	15	1/18	0000 0000 0001 0000 s	17
0/18	0000 0000 0111 01s	15	2/1	0101 s	5
0/19	0000 0000 0111 00s	15	2/2	0000 100s	8
0/20	0000 0000 0110 11s	15	2/3	0000 0010 11s	11
0/21	0000 0000 0110 10s	15	2/4	0000 0001 0100 s	13
0/22	0000 0000 0110 01s	15	2/5	0000 0000 1010 0s	14
0/23	0000 0000 0110 00s	15	3/1	0011 1s	6
0/24	0000 0000 0101 11s	15	3/2	0010 0100 s	9
0/25	0000 0000 0101 10s	15	3/3	0000 0001 1100 s	13
0/26	0000 0000 0101 01s	15	3/4	0000 0000 1001 1s	14
0/27	0000 0000 0101 00s	15	4/1	0011 0s	6
0/28	0000 0000 0100 11s	15	4/2	0000 0011 11s	11
0/29	0000 0000 0100 10s	15	4/3	0000 0001 0010 s	13
0/30	0000 0000 0100 01s	15	5/1	0001 11s	7
0/31	0000 0000 0100 00s	15	5/2	0000 0010 01s	11
0/32	0000 0000 0011 000s	16	5/3	0000 0000 1001 0s	14
0/33	0000 0000 0010 111s	16	6/1	0001 01s	7
0/34	0000 0000 0010 110s	16	6/2	0000 0001 1110 s	13
0/35	0000 0000 0010 101s	16	6/3	0000 0000 0001 0100 s	17
0/36	0000 0000 0010 100s	16	7/1	0001 00s	7
0/37	0000 0000 0010 011s	16	7/2	0000 0001 0101 s	13
0/38	0000 0000 0010 010s	16	8/1	0000 111s	8
0/39	0000 0000 0010 001s	16	8/2	0000 0001 0001 s	13
0/40	0000 0000 0010 000s	16			

Table 9.23: Variable-Length Run-Level Codes (Part 1).

9/1	0000 101s	8	18/1	0000 0001 1010 s	13
9/2	0000 0000 1000 1s	14	19/1	0000 0001 1001 s	13
10/1	0010 0111 s	9	20/1	0000 0001 0111 s	13
10/2	0000 0000 1000 0s	14	21/1	0000 0001 0110 s	13
11/1	0010 0011 s	9	22/1	0000 0000 1111 1s	14
11/2	0000 0000 0001 1010 s	17	23/1	0000 0000 1111 0s	14
12/1	0010 0010 s	9	24/1	0000 0000 1110 1s	14
12/2	0000 0000 0001 1001 s	17	25/1	0000 0000 1110 0s	14
13/1	0010 0000 s	9	26/1	0000 0000 1101 1s	14
13/2	0000 0000 0001 1000 s	17	27/1	0000 0000 0001 1111 s	17
14/1	0000 0011 10s	11	28/1	0000 0000 0001 1110 s	17
14/2	0000 0000 0001 0111 s	17	29/1	0000 0000 0001 1101 s	17
15/1	0000 0011 01s	11	30/1	0000 0000 0001 1100 s	17
15/2	0000 0000 0001 0110 s	17	31/1	0000 0000 0001 1011 s	17
16/1	0000 0010 00s	11	EOB	10	2
16/2	0000 0000 0001 0101 s	17	ESC	0000 01	6
17/1	0000 0001 1111 s	13			

Table 9.23: Variable-Length Run-Level Codes (Part 2).

◇ **Exercise 9.9:** compute the zigzag sequence and run-level codes for the AC coefficients of Figure 9.22b.

◇ **Exercise 9.10:** Given a block with 63 zero AC coefficients how is it coded?

A peculiar feature of Table 9.23 is that it lists two codes for run-level $(0,1)$. Also, the first of those codes (labeled "first") is $1s$, which may conflict with the EOB code. The explanation is that the second of those codes (labeled "next"), $11s$, is normally used, and this causes no conflict. The first code, $1s$, is used only in nonintra coding, where an all-zero DCT coefficients block is coded in a special way.

The discussion so far has concentrated on encoding the quantized DCT coefficients for intra coding (I pictures). For nonintra coding (i.e., P and B pictures) the situation is different. The process of predicting a picture from another picture already decorrelates the samples, and the main advantage of the DCT in nonintra coding is quantization. Deep quantization of the DCT coefficients increases compression, and even a flat default quantization table (that does not take advantage of the properties of human vision) is effective in this case. Another feature of the DCT in nonintra coding is that the DC and AC coefficients are not substantially different, since they are the DCT transforms of differences. There is, therefore, no need to separate the encoding of the DC and AC coefficients.

The encoding process starts by looking for runs of macroblocks that are completely zero. Such runs are encoded by a macroblock address increment. If a macroblock is not all zeros, some of its six component blocks may still be completely zero. For such macroblocks the encoder prepares a *coded block pattern* (cbp). This is a 6-bit variable where each bit specifies whether one of the six blocks is completely zero or not. A zero block is identified as such by the corresponding cbp bit. A nonzero block is encoded using the codes of Table 9.23. When such a nonzero block is encoded, the encoder knows that it cannot be all zeros. There must be at least one nonzero coefficient among the 64 quantized coefficients in the block. If the first nonzero coefficient has a run-level code of $(0, 1)$, it is coded as "1s" and there is no conflict with the EOB code since the EOB code cannot be the first code in such a block. Any other nonzero coefficients with a run-level code of $(0, 1)$ are encoded using the "next" code, which is "11s".

9.6.2 MPEG-1 Video Syntax

Some of the many parameters used by MPEG to specify and control the compression of a video sequence are described in this section in detail. Readers who are interested only in the general description of MPEG may skip this section. The concepts of video sequence, picture, slice, macroblock, and block have already been discussed. Figure 9.24 shows the format of the compressed MPEG stream and how it is organized in six layers. Optional parts are enclosed in dashed boxes. Notice that only the video sequence of the compressed stream is shown; the system parts are omitted.

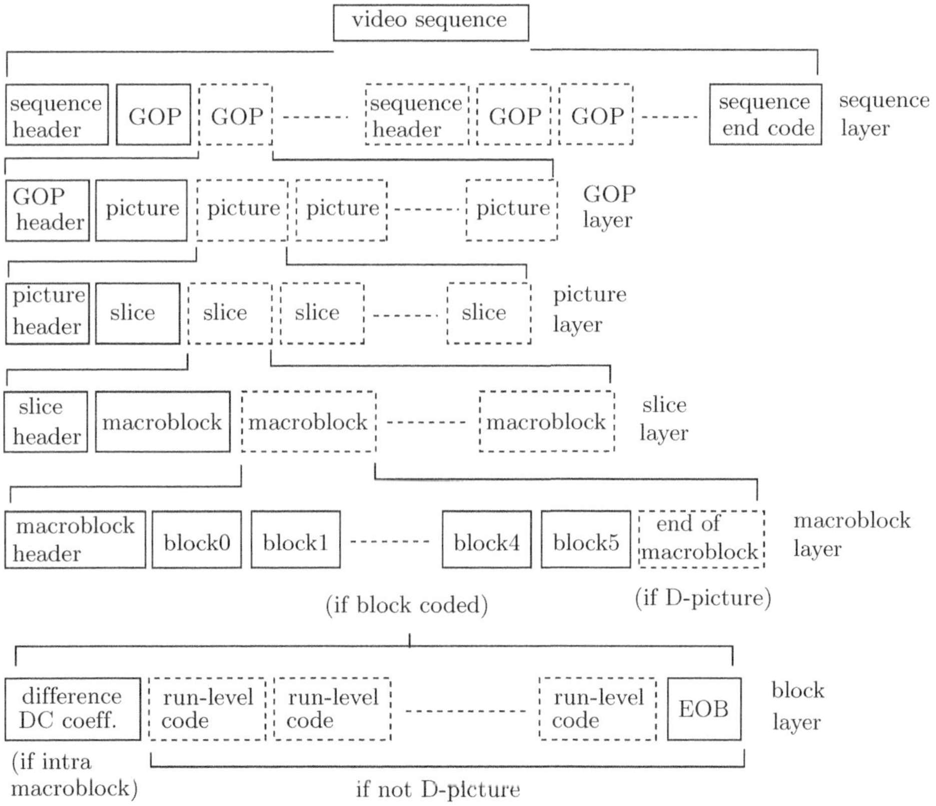

Figure 9.24: The Layers of a Video Stream.

The video sequence starts with a sequence header, followed by a group of pictures (GOP) and optionally by more GOPs. There may be other sequence headers followed by more GOPs, and the sequence ends with a sequence-end-code. The extra sequence headers may be included to help in random access playback or video editing, but most of the parameters in the extra sequence headers must remain unchanged from the first header.

A group of pictures (GOP) starts with a GOP header, followed by one or more pictures. Each picture in a GOP starts with a picture header, followed by one or more slices. Each slice, in turn, consists of a slice header followed by one or more macroblocks of encoded, quantized DCT coefficients. A macroblock is a set of six 8×8 blocks, four blocks of luminance samples and two blocks of chrominance samples. Some blocks may be completely zero and may not be encoded. Each block is coded in intra or nonintra. An intra block starts with a difference between its DC coefficient and the previous DC coefficient (of the same type), followed by run-level codes for the nonzero AC coefficients and zero runs. The EOB code terminates the block. In a nonintra block, both DC and AC coefficients are run-level coded.

It should be mentioned that in addition to the I, P, and B picture types, there exists in MPEG a fourth type, a D picture (for *DC coded*). Such pictures contain only DC coefficient information; no run-level codes or EOB are included. However, D pictures are not allowed to be mixed with the other types of pictures, so they are rare and will not be discussed here.

The headers of a sequence, GOP, picture, and slice all start with a byte-aligned 32-bit start code. In addition to these video start codes there are other start codes for the system layer, user data, and error tagging. A start code starts with 23 zero bits, followed by a single bit of 1, followed by a unique byte. Table 9.25 lists all the video start codes. The "sequence.error" code is for cases where the encoder discovers unrecoverable errors in a video sequence and cannot encode it as a result. The run-level codes have variable lengths, so some 0 bits normally have to be appended to the video stream before a start code, to make sure the code starts on a byte boundary.

Start code	Hex	Binary
extension.start	000001B5	00000000 00000000 00000001 10110101
GOP.start	000001B8	00000000 00000000 00000001 10111000
picture.start	00000100	00000000 00000000 00000001 00000000
reserved	000001B0	00000000 00000000 00000001 10110000
reserved	000001B1	00000000 00000000 00000001 10110001
reserved	000001B6	00000000 00000000 00000001 10110110
sequence.end	000001B7	00000000 00000000 00000001 10110111
sequence.error	000001B4	00000000 00000000 00000001 10110100
sequence.header	000001B3	00000000 00000000 00000001 10110011
slice.start.1	00000101	00000000 00000000 00000001 00000001
...	...	
slice.start.175	000001AF	00000000 00000000 00000001 10101111
user.data.start	000001B2	00000000 00000000 00000001 10110010

Table 9.25: MPEG Video Start Codes.

Code	height/width	Video source
0000	forbidden	
0001	1.0000	computers (VGA)
0010	0.6735	
0011	0.7031	16:9, 625 lines
0100	0.7615	
0101	0.8055	
0110	0.8437	16:9 525 lines
0111	0.8935	
1000	0.9157	CCIR Rec, 601, 625 lines
1001	0.9815	
1010	1.0255	
1011	1.0695	
1100	1.0950	CCIR Rec. 601, 525 lines
1101	1.1575	
1110	1.2015	
1111	reserved	

Table 9.26: MPEG Pel Aspect Ratio Codes.

Code	nominal picture rate	Typical applications
0000		forbidden
0001	23.976	Movies on NTSC broadcast monitors
0010	24	Movies, commercial clips, animation
0011	25	PAL, SECAM, generic 625/50Hz component video
0100	29.97	Broadcast rate NTSC
0101	30	NTSC profession studio, 525/60Hz component video
0110	50	noninterlaced PAL/SECAM/625 video
0111	59.94	Noninterlaced broadcast NTSC
1000	60	Noninterlaced studio 525 NTSC rate
1001		
...		
1111	reserved	

Table 9.27: MPEG Picture Rates and Typical Applications.

`horizontal_size` ≤ 768 pels.
`vertical_size` ≤ 576 lines.
number of macroblocks ≤ 396.
(number of macroblocks)$\times$`picture_rate` $\leq 396\times 25$.
`picture_rate` ≤ 30 pictures per second.
`vbv_buffer_size` ≤ 160.
`bit_rate`≤ 4640.
`forward_f_code` ≤ 4.
`backward_f_code` ≤ 4.

Table 9.28: Constrained Parameters Bounds.

Video Sequence Layer: This layer starts with start code 000001B3, followed by nine fixed-length data elements. The parameters `horizontal_size` and `vertical_size` are 12-bit parameters that define the width and height of the picture. Neither is allowed to be zero, and `vertical_size` must be even. Parameter `pel_aspect_ratio` is a 4-bit parameter that specifies the aspect ratio of a pel. Its 16 values are listed in Table 9.26. Parameter `picture_rate` is a 4-bit parameter that specifies one of 16 picture refresh rates. Its eight nonreserved values are listed in Table 9.27. The 18-bit `bit_rate` data element specifies the compressed data rate to the decoder (in units of 400 bits/s). This parameter has to be positive and is related to the true bit rate R by

$$\texttt{bit\_rate} = \lceil R/400 \rceil.$$

Next comes a `marker_bit`. This single bit of 1 prevents the accidental generation of a start code in cases where some bits get corrupted. Marker bits are common in MPEG. The 10-bit `vbv_buffer_size` follows the marker bit. It specifies to the decoder the lower bound for the size of the compressed data buffer. The buffer size, in bits, is given by

$$B = 8\times 2048\times \texttt{vbv\_buffer\_size},$$

and is a multiple of 2K bytes. The `constrained_parameter_flag` is a 1-bit parameter that is normally 0. When set to 1, it signifies that some of the other parameters have the values listed in Table 9.28.

The last two data elements are 1-bit each and control the loading of the intra and nonintra quantization tables. When `load_intra_quantizer_matrix` is set to 1, it means that it is followed by the 64 8-bit QCs of the `intra_quantizer_matrix`. Similarly, `load_non_intra_quantizer_matrix` signals whether the quantization table `non_intra_quantizer_matrix` follows it or whether the default should be used.

GOP Layer: This layer starts with nine mandatory elements, optionally followed by extensions and user data, and by the (compressed) pictures themselves.

The 32-bit group start code 000001B8 is followed by the 25-bit `time_code`, which consists of the following six data elements: `drop_frame_flag` (one bit) is zero unless the picture rate is 29.97 Hz; `time_code_hours` (five bits, in the range $[0, 23]$), data elements `time_code_minutes` (six bits, in the range $[0, 59]$), and `time_code_seconds` (six bits, in the same range) indicate the hours, minutes, and seconds in the time interval from the start of the sequence to the display of the first picture in the GOP. The 6-bit `time_code_pictures` parameter indicates the number of pictures in a second. There is a `marker_bit` between `time_code_minutes` and `time_code_seconds`.

Following the `time_code` there are two 1-bit parameters. The flag `closed_gop` is set if the GOP is closed (i.e., its pictures can be decoded without reference to pictures from outside the group). The `broken_link` flag is set to 1 if editing has disrupted the original sequence of groups of pictures.

Picture Layer: Parameters in this layer specify the type of the picture (I, P, B, or D) and the motion vectors for the picture. The layer starts with the 32-bit `picture_start_code`, whose hexadecimal value is 00000100. It is followed by a 10-bit `temporal_reference` parameter, which is the picture number (modulo 1024) in the sequence. The next parameter is the 3-bit `picture_coding_type` (Table 9.29), and this is followed by the 16-bit vbv_delay that tells the decoder how many bits must be in the compressed data buffer before the picture can be decoded. This parameter helps prevent buffer overflow and underflow.

Code	picture type
000	forbidden
001	I
010	P
011	B
100	D
101	reserved
...	...
111	reserved

Table 9.29: Picture Type Codes.

If the picture type is P or B, then this is followed by the forward motion vectors scale information, a 3-bit parameter called forward_f_code (see Table 9.34). For B pictures, there follows the backward motion vectors scale information, a 3-bit parameter called backward_f_code.

Slice Layer: There can be many slices in a picture, so the start code of a slice ends with a value in the range $[1, 175]$. This value defines the macroblock row where the slice starts (a picture can therefore have up to 175 rows of macroblocks). The horizontal position where the slice starts in that macroblock row is determined by other parameters.

The quantizer_scale (five bits) initializes the quantizer scale factor, discussed earlier in connection with the rounding of the quantized DCT coefficients. The extra_bit_slice flag following it is always 0 (the value 1 is reserved for future ISO standards). Following this, the encoded macroblocks are written.

Macroblock Layer: This layer identifies the position of the macroblock relative to the position of the current macroblock. It codes the motion vectors for the macroblock, and identifies the zero and nonzero blocks in the macroblock.

Each macroblock has an address, or index, in the picture. Index values start at 0 in the upper-left corner of the picture and continue in raster order. When the encoder starts encoding a new picture, it sets the macroblock address to -1. The macroblock_address_increment parameter contains the amount needed to increment the macroblock address in order to reach the macroblock being coded. This parameter is normally 1. If macroblock_address_increment is greater than 33, it is encoded as a sequence of macroblock_escape codes, each incrementing the macroblock address by 33, followed by the appropriate code from Table 9.30.

The macroblock_type is a variable-size parameter, between 1 and 6 bits long, whose values are listed in Table 9.31. Each value of this variable-size code is decoded into five bits that become the values of the following five flags:

1. macroblock_quant. If set, a new 5-bit quantization scale is sent.
2. macroblock_motion_forward. If set, a forward motion vector is sent.
3. macroblock_motion_backward. If set, a backward motion vector is sent.

Increment value	macroblock address increment	Increment value	macroblock address increment
1	1	19	0000 00
2	011	20	0000 0100 11
3	010	21	0000 0100 10
4	0011	22	0000 0100 011
5	0010	23	0000 0100 010
6	0001 1	24	0000 0100 001
7	0001 0	25	0000 0100 000
8	0000 111	26	0000 0011 111
9	0000 011	27	0000 0011 110
10	0000 1011	28	0000 0011 101
11	0000 1010	29	0000 0011 100
12	0000 1001	30	0000 0011 011
13	0000 1000	31	0000 0011 010
14	0000 0111	32	0000 0011 001
15	0000 0110	33	0000 0011 000
16	0000 0101 11	stuffing	0000 0001 111
17	0000 0101 10	escape	0000 0001 000
18	0000 0101 01		

Table 9.30: Codes for Macroblock Address Increment.

	Code	quant	forward	backward	pattern	intra
I	1	0	0	0	0	1
	01	1	0	0	0	1
	001	0	1	0	0	0
	01	0	0	0	1	0
	00001	1	0	0	1	0
P	1	0	1	0	1	0
	00010	1	1	0	1	0
	00011	0	0	0	0	1
	000001	1	0	0	0	1
	0010	0	1	0	0	0
	010	0	0	1	0	0
	10	0	1	1	0	0
	0011	0	1	0	1	0
	000011	1	1	0	1	0
B	011	0	0	1	1	0
	000010	1	0	1	1	0
	11	0	1	1	1	0
	00010	1	1	1	1	0
	00011	0	0	0	0	1
	000001	1	0	0	0	1
D	1	0	0	0	0	1

Table 9.31: Variable Length Codes for Macroblock Types.

cbp dec.	cbp binary	block # 012345	cbp code	cbp dec.	cbp binary	block # 012345	cbp code
0	000000	......	forbidden	32	100000	c.....	1010
1	000001	c	01011	33	100001	c....c	0010100
2	000010	....c.	01001	34	100010	c...c.	0010000
3	000011	cc	001101	35	100011	c...cc	00011100
4	000100	...c..	1101	36	100100	c..c..	001110
5	000101	...c.c	0010111	37	100101	c..c.c	00001110
6	000110	...cc.	0010011	38	100110	c..cc.	00001100
7	000111	...ccc	00011111	39	100111	c..ccc	000000010
8	001000	..c...	1100	40	101000	c.c...	10000
9	001001	..c..c	0010110	41	101001	c.c..c	00011000
10	001010	..c.c.	0010010	42	101010	c.c.c.	00010100
11	001011	..c.cc	00011110	43	101011	c.c.cc	00010000
12	001100	..cc..	10011	44	101100	c.cc..	01110
13	001101	..cc.c	00011011	45	101101	c.cc.c	00001010
14	001110	..ccc.	00010111	46	101110	c.ccc.	00000110
15	001111	..cccc	00010011	47	101111	c.cccc	000000110
16	010000	.c....	1011	48	110000	cc....	10010
17	010001	.c...c	0010101	49	110001	cc...c	00011010
18	010010	.c..c.	0010001	50	110010	cc..c.	00010110
19	010011	.c..cc	00011101	51	110011	cc..cc	00010010
20	010100	.c.c..	10001	52	110100	cc.c..	01101
21	010101	.c.c.c	00011001	53	110101	cc.c.c	00001001
22	010110	.c.cc.	00010101	54	110110	cc.cc.	00000101
23	010111	.c.ccc	00010001	55	110111	cc.ccc	000000101
24	011000	.cc...	001111	56	111000	ccc...	01100
25	011001	.cc..c	00001111	57	111001	ccc..c	00001000
26	011010	.cc.c.	00001101	58	111010	ccc.c.	00000100
27	011011	.cc.cc	000000011	59	111011	ccc.cc	000000100
28	011100	.ccc..	01111	60	111100	cccc..	111
29	011101	.ccc.c	00001011	61	111101	cccc.c	01010
30	011110	.cccc.	00000111	62	111110	ccccc.	01000
31	011111	.ccccc	000000111	63	111111	cccccc	001100

Table 9.32: Codes for Macroblock Address Increment.

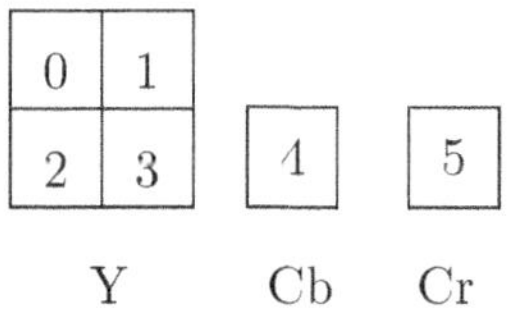

Figure 9.33: Indexes of Six Blocks in a Macroblock.

4. macroblock_pattern. If set, the coded_block_pattern code (variable length) listed in Table 9.32 follows to code the six pattern_code bits of variable cbp discussed earlier (blocks labeled "." in the table are skipped, and blocks labeled c are coded). These six bits identify the six blocks of the macroblock as completely zero or not. The correspondence between the six bits and blocks is shown in Figure 9.33, where block 0 corresponds to the most-significant bit of the pattern_code.
5. macroblock_intra. If set, the six blocks of this macroblock are coded as intra.

These five flags determine the rest of the processing steps for the macroblock.

Once the pattern_code bits are known, blocks that correspond to 1 bits are encoded.

Block Layer: This layer is the lowest in the video sequence. It contains the encoded 8×8 blocks of quantized DCT coefficients. The coding depends on whether the block contains luminance or chrominance samples and on whether the macroblock is intra or nonintra. In nonintra coding, blocks that are completely zero are skipped; they don't have to be encoded.

The macroblock_intra flag gets its value from macroblock_type. If it is set, the DC coefficient of the block is coded separately from the AC coefficients.

9.6.3 Motion Compensation

An important element of MPEG is *motion compensation*, which is used in inter coding only. In this mode, the pels of the current picture are predicted by those of a previous reference picture (and, possibly, by those of a future reference picture). Pels are subtracted, and the differences (which should be small numbers) are DCT transformed, quantized, and encoded. The differences between the current picture and the reference picture are normally caused by motion (either camera motion or scene motion), so best prediction is obtained by matching a region in the current picture with a different region in the reference picture. MPEG does not require the use of any particular matching algorithm, and any implementation can use its own method for matching macroblocks (see Section 9.5 for examples of matching algorithms). The discussion here concentrates on the operations of the decoder.

Differences between consecutive pictures may also be caused by random noise in the video camera, or by variations of illumination, which may change brightness in a nonuniform way. In such cases, motion compensation is not used, and each region ends up being matched with the same spatial region in the reference picture.

If the difference between consecutive pictures is caused by camera motion, one motion vector is enough for the entire picture. Normally, however, there is also scene motion and movement of shadows, so a number of motion vectors are needed, to describe the motion of different regions in the picture. The size of those regions is critical. A large number of small regions improves prediction accuracy, whereas the opposite situation simplifies the algorithms used to find matching regions and also leads to fewer motion vectors and sometimes to better compression. Since a macroblock is such an important unit in MPEG, it was also selected as the elementary region for motion compensation.

Another important consideration is the precision of the motion vectors. A motion vector such as $(15, -4)$ for a macroblock M typically means that M has been moved from the reference picture to the current picture by displacing it 15 pels to the right and four pels up (a positive vertical displacement is down). The components of the vector are in units of pels. They may, however, be in units of half a pel, or even smaller. In MPEG-1, the precision of motion vectors may be either full-pel or half-pel, and the encoder signals this decision to the decoder by a parameter in the picture header (this parameter may be different from picture to picture).

It often happens that large areas of a picture move at identical or at similar speeds, and this implies that the motion vectors of adjacent macroblocks are correlated. This is the reason why the MPEG encoder encodes a motion vector by subtracting it from the motion vector of the preceding macroblock and encoding the difference.

A P picture uses an earlier I picture or P picture as a reference picture. We say that P pictures use forward motion-compensated prediction. When a motion vector `MD` for a macroblock is determined (`MD` stands for *motion displacement*, since the vector consists of two components, the horizontal and the vertical displacements), MPEG denotes the motion vector of the preceding macroblock in the slice by `PMD` and computes the difference `dMD`=`MD`–`PMD`. `PMD` is reset to zero at the start of a slice, after a macroblock is intra coded, when the macroblock is skipped, and when parameter `block_motion_forward` is zero.

The 1-bit parameter `full_pel_forward_vector` in the picture header defines the precision of the motion vectors (1=full-pel, 0=half-pel). The 3-bit parameter `forward_f_code` defines the range.

A B picture may use forward or backward motion compensation. If both are used, then two motion vectors are calculated by the encoder for the current macroblock, and each vector is encoded by first computing a difference, as in P pictures. Also, if both motion vectors are used, then the prediction is the average of both. Here is how the prediction is done in the case of both forward and backward prediction. Suppose that the encoder has determined that macroblock M in the current picture matches macroblock MB in the following picture and macroblock MF in the preceding picture. Each pel $M[i,j]$ in macroblock M in the current picture is predicted by computing the difference

$$M[i,j]-\frac{MF[i,j]+MB[i,j]}{2},$$

where the quotient of the division by 2 is rounded to the nearest integer.

Parameters `full_pel_forward_vector` and `forward_f_code` have the same meaning as for a P picture. In addition, there are the two corresponding parameters (for backward) `full_pel_backward_vector` and `backward_f_code`.

The following rules apply when coding motion vectors in B pictures:

1. The quantity `PMD` is reset to zero at the start of a slice and after a macroblock is skipped.
2. In a skipped macroblock, `MD` is predicted from the preceding macroblock in the slice.
3. When `motion_vector_forward` is zero, the forward `MD`s are predicted from the preceding macroblock in the slice.
4. When `motion_vector_backward` is zero, the backward `MD`s are predicted from the preceding macroblock in the slice.
5. A B picture is predicted by means of I and P pictures but not by another B picture.
6. The two components of a motion vector (the motion displacements) are computed to a precision of either full-pel or half-pel, as specified in the picture header.

Displacement Principal and Residual: The two components of a motion vector are motion displacements. Each motion displacement has two parts, a principal and a residual. The principal part is denoted by dMD_p. It is a signed integer that is given by the product

$$dMD_p = \texttt{motion\_code} \times f,$$

where (1) `motion_code` is an integer parameter in the range $[-16, +16]$, included in the compressed stream by means of a variable-length code, and (2) f, the scaling factor, is a power of 2 given by

$$f = 2^{rsize}.$$

The integer *rsize* is simply `f_code` $-$ 1, where `f_code` is a 3-bit parameter with values in the range $[1, 7]$. This implies that *rsize* has values $[0, 6]$ and f is a power of 2 in the range $[1, 2, 4, 8, 16, 32, 64]$.

The residual part is denoted by `r` and is a positive integer, defined by

$$\texttt{r} = |dMD_p| - |dMD|.$$

After computing `r`, the encoder encodes it by concatenating the ones-complement of `r` and `motion_r` to the variable-length code for parameter `motion_code`. The parameter `motion_r` is related to `r` by

$$\texttt{motion\_r} = (f - 1) - \texttt{r},$$

and f should be chosen as the smallest value that satisfies the following inequalities for the largest (positive or negative) differential displacement in the entire picture

$$-(16 \times f) \leq \texttt{dMD} < (16 \times f).$$

Once f has been selected, the value of parameter `motion_code` for the differential displacement of a macroblock is given by

$$\texttt{motion\_code} = \frac{\texttt{dMD} + \text{Sign}(\texttt{dMD}) \times (f - 1)}{f},$$

where the quotient is rounded such that `motion_code`$\times f \geq$ `dMD`.

Table 9.34 lists the header parameters for motion vector computation ("p" stands for picture header and "mb" stands for macroblock header). Table 9.35 lists the generic and full names of the parameters mentioned here, and Table 9.36 lists the range of values of `motion_r` as a function of `f_code`.

9.6.4 Pel Reconstruction

The main task of the MPEG decoder is to reconstruct the pels of the entire video sequence. This is done by reading the codes of a block from the compressed stream, decoding them, dequantizing them, and calculating the IDCT. For nonintra blocks in P and B pictures, the decoder has to add the motion-compensated prediction to the results of the IDCT. This is repeated six times (or fewer, if some blocks are completely zero)

Header parameter	set in header	number of bits	displacement parameter
full_pel_forward_vector	p	1	precision
forward_f_code	p	3	range
full_pel_backward_vector	p	1	precision
backward_f_code	p	3	range
motion_horizontal_forward_code	mb	vlc	principal
motion_horizontal_forward_r	mb	forward_r_size	residual
motion_vertical_forward_code	mb	vlc	principal
motion_vertical_forward_r	mb	forward_r_size	residual

Table 9.34: Header Parameters for Motion Vector Computation.

Generic name	Full name	Range
full_pel_vector	full_pel_forward_vector full_pel_backward_vector	0,1
f_code	forward_f_code backward_f_code	1–7
r_size	forward_r_size backward_r_size	0–6
f	forward_f backward_f	1,2,4,8,16,32,64
motion_code	motion_horizontal_forward_code motion_vertical_forward_code motion_horizontal_backward_code motion_vertical_backward_code	$-16 \rightarrow +16$
motion_r	motion_horizontal_forward_r motion_vertical_forward_r motion_horizontal_backward_r motion_vertical_backward_r	$0 \rightarrow (f-1)$
r	compliment_horizontal_forward_r compliment_vertical_forward_r compliment_horizontal_backward_r compliment_vertical_backward_r	$0 \rightarrow (f-1)$

Table 9.35: Generic Names for Motion Displacement Parameters.

f_code	r_size	f	$f-1$	r	motion_r
1	0	1	0		0
2	1	2	1	0,1	1,0
3	2	4	3	0–3	3,. . . ,0
4	3	8	7	0–7	7,. . . ,0
5	4	16	15	0–15	15,. . . ,0
6	5	32	31	0–31	31,. . . ,0
7	6	64	63	0–63	63,. . . ,0

Table 9.36: Range of Values of motion_r as a Function of f_code.

for the six blocks of a macroblock. The entire sequence is decoded picture by picture, and within each picture, macroblock by macroblock.

It has already been mentioned that the IDCT is not rigidly defined in MPEG, which may lead to accumulation of errors, called *IDCT mismatch*, during decoding.

For intra-coded blocks, the decoder reads the differential code of the DC coefficient and uses the decoded value of the previous DC coefficient (of the same type) to decode the DC coefficient of the current block. It then reads the run-level codes until an EOB code is encountered, and decodes them, generating a sequence of 63 AC coefficients, normally with few nonzero coefficients and runs of zeros between them. The DC and 63 AC coefficients are then collected in zigzag order to create an 8×8 block. After dequantization and inverse DCT calculation, the resulting block becomes one of the six blocks that make up a macroblock (in intra coding all six blocks are always coded, even those that are completely zero).

For nonintra blocks, there is no distinction between DC and AC coefficients and between luminance and chrominance blocks. They are all decoded in the same way.

9.7 MPEG-4

MPEG-4 is a new standard for audiovisual data. Although video and audio compression is still a central feature of MPEG-4, this standard includes much more than just compression of the data. As a result, MPEG-4 is huge and this section can only describe its main features. No details are provided. We start with a bit of history (see also Section 9.9).

The MPEG-4 project started in May 1991 and initially aimed at finding ways to compress multimedia data to very low bitrates with minimal distortions. In July 1994, this goal was significantly altered in response to developments in audiovisual technologies. The MPEG-4 committee started thinking of future developments and tried to guess what features should be included in MPEG-4 to meet them. A call for proposals was issued in July 1995 and responses were received by October of that year. (The proposals were supposed to address the eight major functionalities of MPEG-4, listed below.) Tests of the proposals were conducted starting in late 1995. In January 1996, the first verification model was defined, and the cycle of calls for proposals—proposal implementation and verification was repeated several times in 1997 and 1998. Many proposals were accepted for the many facets of MPEG-4, and the first version of MPEG-4 was accepted and approved in late 1998. The formal description was published in 1999 with many amendments that keep coming out.

At present (mid-2006), the MPEG-4 standard is designated the ISO/IEC 14496 standard, and its formal description, which is available from [ISO 03], consists of 17 parts plus new amendments. More readable descriptions can be found in [Pereira and Ebrahimi 02] and [Symes 03].

MPEG-1 was originally developed as a compression standard for interactive video on CDs and for digital audio broadcasting. It turned out to be a technological triumph but a visionary failure. On the one hand, not a single design mistake was found during the implementation of this complex algorithm and it worked as expected. On the other hand, interactive CDs and digital audio broadcasting have had little commercial success, so MPEG-1 is used today for general video compression. One aspect of MPEG-1 that was supposed to be minor, namely MP3, has grown out of proportion and is commonly used today for audio (Section 10.14). MPEG-2, on the other hand, was specifically designed for digital television and this standard has had tremendous commercial success.

The lessons learned from MPEG-1 and MPEG-2 were not lost on the MPEG committee members and helped shape their thinking for MPEG-4. The MPEG-4 project started as a standard for video compression at very low bitrates. It was supposed to deliver reasonable video data in only a few thousand bits per second. Such compression is important for video telephones, video conferences or for receiving video in a small, handheld device, especially in a mobile environment, such as a moving car. After working on this project for two years, the committee members, realizing that the rapid development of multimedia applications and services will require more and more compression standards, have revised their approach. Instead of a compression standard, they decided to develop a set of tools (a toolbox) to deal with audiovisual products in general, at present and in the future. They hoped that such a set will encourage industry to invest in new ideas, technologies, and products in confidence, while making it possible for consumers to generate, distribute, and receive different types of multimedia data with ease and at a reasonable cost.

Traditionally, methods for compressing video have been based on pixels. Each video frame is a rectangular set of pixels, and the algorithm looks for correlations between pixels in a frame and between frames. The compression paradigm adopted for MPEG-4, on the other hand, is based on objects. (The name of the MPEG-4 project was also changed at this point to "coding of audiovisual objects.") In addition to producing a movie in the traditional way with a camera or with the help of computer animation, an individual generating a piece of audiovisual data may start by defining objects, such as a flower, a face, or a vehicle, and then describing how each object should be moved and manipulated in successive frames. A flower may open slowly, a face may turn, smile, and fade, a vehicle may move toward the viewer and appear bigger. MPEG-4 includes an object description language that provides for a compact description of both objects and their movements and interactions.

Another important feature of MPEG-4 is *interoperability*. This term refers to the ability to exchange any type of data, be it text, graphics, video, or audio. Obviously, interoperability is possible only in the presence of standards. All devices that produce data, deliver it, and consume (play, display, or print) it must obey the same rules and read and write the same file structures.

During its important July 1994 meeting, the MPEG-4 committee decided to revise its original goal and also started thinking of future developments in the audiovisual field and of features that should be included in MPEG-4 to meet them. They came up with eight points that they considered important functionalities for MPEG-4.

1. **Content-based multimedia access tools**. The MPEG-4 standard should provide tools for accessing and organizing audiovisual data. Such tools may include indexing, linking, querying, browsing, delivering files, and deleting them. The main tools currently in existence are listed later in this section.

2. **Content-based manipulation and bitstream editing**. A syntax and a coding scheme should be part of MPEG-4. The idea is to enable users to manipulate and edit compressed files (bitstreams) without fully decompressing them. A user should

be able to select an object and modify it in the compressed file without decompressing the entire file.

3. **Hybrid natural and synthetic data coding**. A natural scene is normally produced by a video camera. A synthetic scene consists of text and graphics. MPEG-4 recognizes the need for tools to compress natural and synthetic scenes and mix them interactively.

4. **Improved temporal random access**. Users may often want to access part of the compressed file, so the MPEG-4 standard should include tags to make it easy to quickly reach any point in the file. This may be important when the file is stored in a central location and the user is trying to manipulate it remotely, over a slow communications channel.

5. **Improved coding efficiency**. This feature simply means improved compression. Imagine a case where audiovisual data has to be transmitted over a low-bandwidth channel (such as a telephone line) and stored in a low-capacity device such as a smartcard. This is possible only if the data is well compressed, and high compression rates (or equivalently, low bitrates) normally involve a trade-off in the form of smaller image size, reduced resolution (pixels per inch), and lower quality.

6. **Coding of multiple concurrent data streams**. It seems that future audiovisual applications will allow the user not just to watch and listen but also to interact with the image. As a result, the MPEG-4 compressed stream can include several views of the same scene, enabling the user to select any of them to watch and to change views at will. The point is that the different views may be similar, so any redundancy should be eliminated by means of efficient compression that takes into account identical patterns in the various views. The same is true for the audio part (the soundtracks).

7. **Robustness in error-prone environments**. MPEG-4 must provide error-correcting codes for cases where audiovisual data is transmitted through a noisy channel. This is especially important for low-bitrate streams, where even the smallest error may be noticeable and may propagate and affect large parts of the audiovisual presentation.

8. **Content-based scalability**. The compressed stream may include audiovisual data in fine resolution and high quality, but any MPEG-4 decoder should be able to decode it at low resolution and low quality. This feature is useful in cases where the data is decoded and displayed on a small, low-resolution screen, or in cases where the user is in a hurry and prefers to see a rough image rather than wait for a full decoding.

Once the eight fundamental functionalities above have been identified and listed, the MPEG-4 committee started the process of developing separate tools to satisfy these functionalities. This is an ongoing process that continues to this day and will continue in the future. An MPEG-4 author faced with an application has to identify the requirements of the application and select the right tools. It is now clear that compression is a central requirement in MPEG-4, but not the only requirement, as it was for MPEG-1 and MPEG-2.

An example may serve to illustrate the concept of natural and synthetic objects. In a news session on television, a few seconds may be devoted to the weather. The viewers see a weather map of their local geographic region (a computer-generated image) that may zoom in and out and pan. Graphic images of sun, clouds, rain drops, or a rainbow (synthetic scenes) appear, move, and disappear. A person is moving, pointing, and talking (a natural scene), and text (another synthetic scene) may also appear from time to time. All those scenes are mixed by the producers into one audiovisual presentation that's compressed, transmitted (on television cable, on the air, or into the Internet), received by computers or television sets, decompressed, and displayed (consumed).

In general, audiovisual content goes through three stages: production, delivery, and consumption. Each of these stages is summarized below for the traditional approach and for the MPEG-4 approach.

Production. Traditionally, audiovisual data consists of two-dimensional scenes; it is produced with a camera and microphones and consists of natural objects. All the mixing of objects (composition of the image) is done during production. The MPEG-4 approach is to allow for both two-dimensional and three-dimensional objects and for natural and synthetic scenes. The composition of objects is explicitly specified by the producers during production by means of a special language. This allows later editing.

Delivery. The traditional approach is to transmit audiovisual data on a few networks, such as local-area networks and satellite transmissions. The MPEG-4 approach is to let practically any data network carry audiovisual data. Protocols exist to transmit audiovisual data over any type of network.

Consumption. Traditionally, a viewer can only watch video and listen to the accompanying audio. Everything is precomposed. The MPEG-4 approach is to allow the user as much freedom of composition as possible. The user should be able to interact with the audiovisual data, watch only parts of it, interactively modify the size, quality, and resolution of the parts being watched, and be as active in the consumption stage as possible.

Because of the wide goals and rich variety of tools available as part of MPEG-4, this standard is expected to have many applications. The ones listed here are just a few important examples.

1. Streaming multimedia data over the Internet or over local-area networks. This is important for entertainment and education.

2. Communications, both visual and audio, between vehicles and/or individuals. This has military and law enforcement applications.

3. Broadcasting digital multimedia. This, again, has many entertainment and educational applications.

4. Context-based storage and retrieval. Audiovisual data can be stored in compressed form and retrieved for delivery or consumption.

5. Studio and television postproduction. A movie originally produced in English may be translated to another language by dubbing or subtitling.

6. Surveillance. Low-quality video and audio data can be compressed and transmitted from a surveillance camera to a central monitoring location over an inexpensive, slow communications channel. Control signals may be sent back to the camera through the same channel to rotate or zoom it in order to follow the movements of a suspect.

7. Virtual conferencing. This time-saving application is the favorite of busy executives.

Our short description of MPEG-4 concludes with a list of the main tools specified by the MPEG-4 standard.

Object descriptor framework. Imagine an individual participating in a video conference. There is an MPEG-4 object representing this individual and there are video and audio streams associated with this object. The object descriptor (OD) provides information on elementary streams available to represent a given MPEG-4 object. The OD also has information on the source location of the streams (perhaps a URL) and on various MPEG-4 decoders available to consume (i.e., display and play sound) the streams. Certain objects place restrictions on their consumption, and these are also included in the OD of the object. A common example of a restriction is the need to pay before an object can be consumed. A video, for example, may be watched only if it has been paid for, and the consumption may be restricted to streaming only, so that the consumer cannot copy the original movie.

Systems decoder model. All the basic synchronization and streaming features of the MPEG-4 standard are included in this tool. It specifies how the buffers of the receiver should be initialized and managed during transmission and consumption. It also includes specifications for timing identification and mechanisms for recovery from errors.

Binary format for scenes. An MPEG-4 scene consists of objects, but for the scene to make sense, the objects must be placed at the right locations and moved and manipulated at the right times. This important tool (BIFS for short) is responsible for describing a scene, both spatially and temporally. It contains functions that are used to describe two-dimensional and three-dimensional objects and their movements. It also provides ways to describe and manipulate synthetic scenes, such as text and graphics.

MPEG-J. A user may want to use the Java programming language to implement certain parts of an MPEG-4 content. MPEG-J allows the user to write such MPEGlets and it also includes useful Java APIs that help the user interface with the output device and with the networks used to deliver the content. In addition, MPEG-J also defines a delivery mechanism that allows MPEGlets and other Java classes to be streamed to the output separately.

Extensible MPEG-4 textual format. This tool is a format, abbreviated XMT, that allows authors to exchange MPEG-4 content with other authors. XMT can be described as a framework that uses a textual syntax to represent MPEG-4 scene descriptions.

Transport tools. Two such tools, MP4 and FlexMux, are defined to help users transport multimedia content. The former writes MPEG-4 content on a file, whereas the latter is used to interleave multiple streams into a single stream, including timing information.

Video compression. It has already been mentioned that compression is only one of the many goals of MPEG-4. The video compression tools consist of various algorithms that can compress video data to bitrates between 5 kbits/sec (very low bitrate, implying low-resolution and low-quality video) and 1 Gbit/sec. Compression methods vary from very lossy to nearly lossless, and some also support progressive and interlaced video. Many MPEG-4 objects consist of polygon meshes, so most of the video compression tools are designed to compress such meshes. Section 11.11 is an example of such a method.

Robustness tools. Data compression is based on removing redundancies from the original data, but this also renders the data more vulnerable to errors. All methods for error detection and correction are based on increasing the redundancy of the data. MPEG-4 includes tools to add robustness, in the form of error-correcting codes, to the compressed content. Such tools are important in applications where data has to be transmitted through unreliable lines. Robustness also has to be added to very low bitrate MPEG-4 streams because these suffer most from errors.

Fine-grain scalability. When MPEG-4 content is streamed, it is sometimes desirable to first send a rough image, and then improve its visual quality by adding layers of extra information. This is the function of the fine-grain scalability (FGS) tools.

Face and body animation. Often, an MPEG-4 file contains human faces and bodies, and they have to be animated. The MPEG-4 standard therefore provides tools for constructing and animating such surfaces.

Speech coding. Speech may often be part of MPEG-4 content, and special tools are provided to compress it efficiently at bitrates from 2 Kbit/sec up to 24 Kbit/sec. The main algorithm for speech compression is CELP, but there is also a parametric coder.

Audio coding. Several algorithms are available as MPEG-4 tools for audio compression. Examples are (1) advanced audio coding (AAC, Section 10.15), (2) transform-domain weighted interleave vector quantization (Twin VQ, can produce low bitrates such as 6 kbit/sec/channel), and (3) harmonic and individual lines plus noise (HILN, a parametric audio coder).

Synthetic audio coding. Algorithms are provided to generate the sound of familiar musical instruments. They can be used to generate synthetic music in compressed format. The MIDI format, popular with computer music users, is also included among these tools. Text-to-speech tools allow authors to write text that will be pronounced when the MPEG-4 content is consumed. This text may include parameters such as pitch contour and phoneme duration that improve the speech quality.

9.8 H.261

In late 1984, the CCITT (currently the ITU-T) organized an expert group to develop a standard for visual telephony for ISDN services. The idea was to send images and sound between special terminals, so that users could talk and see each other. This type of application requires sending large amounts of data, so compression became an important consideration. The group eventually came up with a number of standards, known as the H series (for video) and the G series (for audio) recommendations, all operating at speeds of $p\times 64$ Kbit/sec for $1 \le p \le 30$. These standards are known today under the umbrella name of $p \times 64$ and are summarized in Table 9.37. This section provides a short summary of the H.261 standard [Liou 91].

Members of the $p\times 64$ also participated in the development of MPEG, so the two methods have many common elements. There is, however, an important difference between them. In MPEG, the decoder must be fast, since it may have to operate in real time, but the encoder can be slow. This leads to very asymmetric compression, and the encoder can be hundreds of times more complex than the decoder. In H.261, both encoder and decoder operate in real time, so both have to be fast. Still, the H.261 standard defines only the data stream and the decoder. The encoder can use any method as long as it creates a valid compressed stream. The compressed stream is organized in layers, and macroblocks are used as in MPEG. Also, the same 8×8 DCT and the same zigzag order as in MPEG are used. The intra DC coefficient is quantized by always

Standard	Purpose
H.261	Video
H.221	Communications
H.230	Initial handshake
H.320	Terminal systems
H.242	Control protocol
G.711	Companded audio (64 Kbits/s)
G.722	High quality audio (64 Kbits/s)
G.728	Speech (LD-CELP @16kbits/s)

Table 9.37: The $p\times 64$ Standards.

dividing it by 8, and it has no dead zone. The inter DC and all AC coefficients are quantized with a dead zone.

Motion compensation is used when pictures are predicted from other pictures, and motion vectors are coded as differences. Blocks that are completely zero can be skipped within a macroblock, and variable-length codes that are very similar to those of MPEG (such as run-level codes), or are even identical (such as motion vector codes) are used. In all these aspects, H.261 and MPEG are very similar.

There are, however, important differences between them. H.261 uses a single quantization coefficient instead of an 8×8 table of QCs, and this coefficient can be changed only after 11 macroblocks. AC coefficients that are intra coded have a dead zone. The compressed stream has just four layers, instead of MPEG's six. The motion vectors are always full-pel and are limited to a range of just ± 15 pels. There are no B pictures, and only the immediately preceding picture can be used to predict a P picture.

9.8.1 H.261 Compressed Stream

H.261 limits the image to just two sizes, the common intermediate format (CIF), which is optional, and the quarter CIF (QCIF). These are shown in Figure 9.38a,b. The CIF format has dimensions of 288×360 for luminance, but only 352 of the 360 columns of pels are actually coded, creating an active area of 288×352 pels. For chrominance, the dimensions are 144×180, but only 176 columns are coded, for an active area of 144×176 pels. The QCIF format is one-fourth of CIF, so the luminance component is 144×180 with an active area of 144×176, and the chrominance components are 72×90, with an active area of 72×88 pels.

The macroblocks are organized in groups (known as groups of blocks or GOB) of 33 macroblocks each. Each GOB is a 3×11 array (48×176 pels), as shown in Figure 9.38c. A CIF picture consists of 12 GPBs, and a QCIF picture is three GOBs, numbered as in Figure 9.38d.

Figure 9.39 shows the four H.261 video sequence layers. The main layer is the picture layer. It start with a picture header, followed by the GOBs (12 or 3, depending on the image size). The compressed stream may contain as many pictures as necessary. The next layer is the GOB, which consists of a header, followed by the macroblocks. If the motion compensation is good, some macroblocks may be completely zero and may be skipped. The macroblock layer is next, with a header that may contain a motion vector, followed by the blocks. There are six blocks per macroblock, but blocks that are

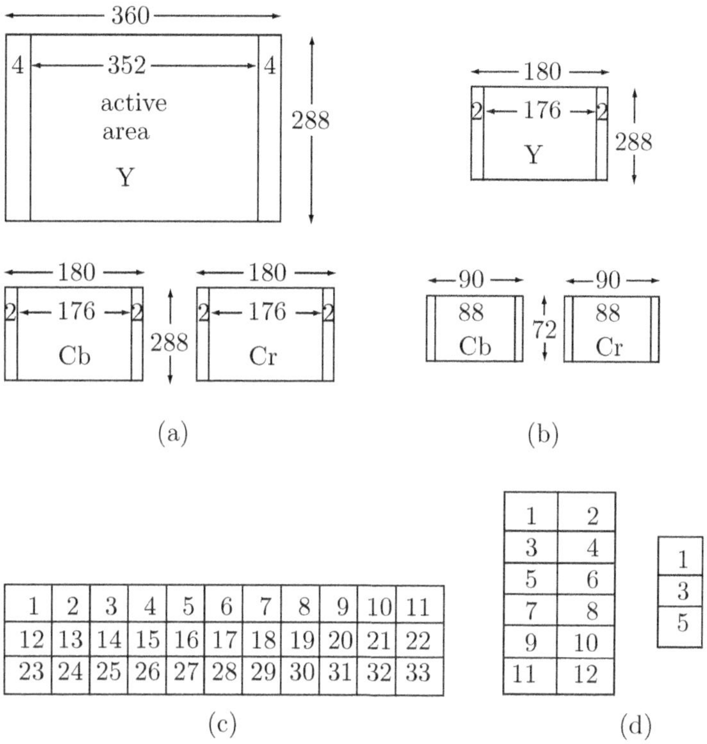

Figure 9.38: (a) CIF. (b) QCIF. (c) GOB.

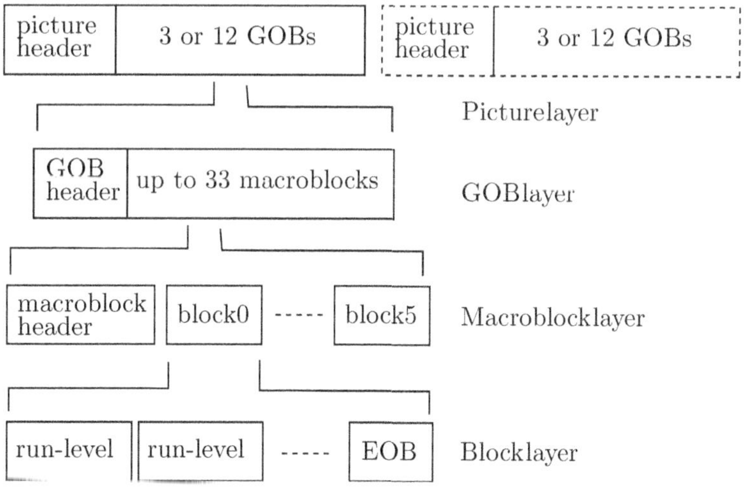

Figure 9.39: H.261 Video Sequence Layers.

zero are skipped. The block layer contains the blocks that are coded. If a block is coded, its nonzero coefficients are encoded with a run-level code. A 2-bit EOB code terminates each block.

9.9 H.264

The last years of the 20th century witnessed an unprecedented and unexpected progress in computing power, video applications, and network support for video data. Both producers and consumers of video felt the need for an advanced video codec to replace the existing video compression standards H.261, H.262, and H.263. The last of these standards, H.263, was developed around 1995 and in 2001 was already outdated. The two groups responsible for developing video compression standards, ISO-MPEG and ITU-VCEG (motion pictures experts group and video coding experts group), felt that the new standard should offer (1) increased compression efficiency, (2) support for special video applications such as videoconferencing, DVD storage, video broadcasting, and streaming over the Internet, and (3) greater reliability. In 2001, in response to this demand, the ITU started two projects. The first, short term, project was an attempt to add extra features to H.263. This project resulted in version 2 of this standard. The second, long term, project was an effort to develop a new standard for efficient (i.e., low bitrate) video compression. This project received the codename H.26L. In the same year, the ITU established a new study group, SG 16, whose members came from MPEG and VCEG, and whose mission was to study new methods, select promising ones, and implement, test, and approve the new standard. The new standard, H.264, was approved in May 2003 [ITU-T264 02], and a corrigendum was added and approved in May 2004. This standard, which is now part of the huge MPEG-4 project, is known by several names. The ITU calls it ITU-T Recommendation H.264, advanced video coding for generic audiovisual services. Its official title is advanced video coding (AVC). The ISO considers it part 10 of the MPEG-4 standard. Regardless of its title, this standard (referred to in this section as H.264) is the best video codec available at the time of writing (mid 2006).

The following is a list of some of the many references currently available for H.264. The official standard is [H.264Draft 06]. As usual with such standards, this document describes only the format of the compressed stream and the operation of the decoder. As such, it is difficult to follow. The main text is [Richardson 03], a detailed book on H.264 and MPEG-4. The integer transform of H.264 is surveyed in [H.264Transform 06]. Many papers and articles are available online at [H.264Standards 06]. Two survey papers by Iain Richardson [H.264PaperIR 06] and Rico Malvar [H.264PaperRM 06] are also widely available and serve as introductions to this topic. The discussion in this section follows both [H.264Draft 06] and [Richardson 03].

Principles of operation

The H.264 standard does not specify the operations of the encoder, but it makes sense to assume that practical implementations of the encoder and decoder will consist of the building blocks shown in Figures 9.40 and 9.41, respectively. Those familiar with older video codecs will notice that the main components of H.264 (prediction, transform, quantization, and entropy encoding) are not much different from those of its predecessors MPEG-1, MPEG-2, MPEG-4, H.261, and H.263. What makes H.264 new and original are the details of those components and the presence of the only new component, the filter. Because of this similarity, the discussion here concentrates on those features of H.264 that distinguish it from its predecessors. Readers who are interested in more details should study the principles of MPEG-1 (Section 9.6) before reading any further.

The input to a video encoder is a set of video frames (where a frame may consist of progressive or interlaced video). Each frame is encoded separately, and the encoded frame is referred to as a coded picture. There are several types of frames, mostly I, P, and B as discussed at the start of Section 9.5, and the order in which the frames are encoded may be different from the order in which they have to be displayed (Figure 9.9). A frame is broken up into slices, and each slice is further partitioned into macroblocks as illustrated in Figure 9.17.

In H.264, slices and macroblocks also have types. An I slice may have only I-type macroblocks, a P slice may have P and I macroblocks, and a B slice may have B and I macroblocks (there are also slices of types SI and SP). Figure 9.40 shows the main steps of the H.264 encoder. A frame F_n is predicted (each macroblock in the frame is predicted by other macroblocks in the same frame or by macroblocks from other frames). The predicted frame P is subtracted from the original F_n to produce a difference D_n, which is transformed (in the box labeled T), filtered (in Q), reordered, and entropy encoded. What is new in the H.264 encoder is the reconstruction path.

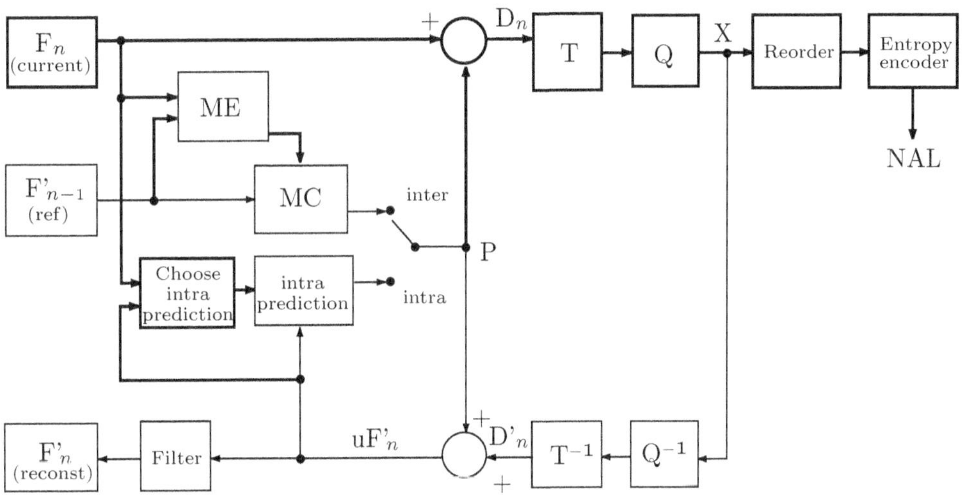

Figure 9.40: H.264 Encoder.

Even a cursory glance at Figure 9.40 shows that the encoder consists of two data paths, a "forward" path (mostly from left to right, shown in thick lines) and a "reconstruction" path (mostly from right to left, shown in thin lines). The decoder (Figure 9.41) is very similar to the encoder's reconstruction path and has most of the latter's components.

The main part of the encoder is its forward path. The next video frame to be compressed is denoted by F_n. The frame is partitioned into macroblocks of 16×16 pixels each, and each macroblock is encoded in intra or inter mode. In either mode, a prediction macroblock P is constructed based on a reconstructed video frame. In the intra mode, P is constructed from previously-encoded samples in the current frame n. These samples are decoded and reconstructed, becoming uF'_n in the figure. In the inter mode, P is constructed by motion-compensated prediction from one or several reference frames (F'_{n-1} in the Figure). Notice that the prediction for each macroblock may be based on one or two frames that have already been encoded and reconstructed (these may be past or future frames). The prediction macroblock P is then subtracted from the current macroblock to produce a residual or difference macroblock $\mathrm{D_n}$. This macroblock is transformed and quantized to produce a set X of quantized transform coefficients. The coefficients are reordered in zigzag and entropy encoded into a short bit string. This string, together with side information for the decoder (also entropy encoded, it includes the macroblock prediction mode, quantizer step size, motion vector information specifying the motion compensation, as well as other items), become the compressed stream which is passed to a network abstraction layer (NAL) for transmission outside the computer or proper storage. NAL consists of units, each with a header and a raw byte sequence payload (RBSP) that can be sent as packets over a network or stored as records that constitute the compressed file.

Each slice starts with a header that contains the slice type, the number of the coded picture that includes the slice, and other information. The slice data consists mostly of coded macroblocks, but may also have skip indicators, indicating macroblocks that have been skipped (not encoded).

Slices of type *I* (intra) contain only *I* macroblocks (macroblocks that are predicted from previous macroblocks in the same slice). This type of slice is used in all three profiles (profiles are discussed below).

Slices of type *P* (predicted) contain *P* and/or *I* macroblocks. The former type is predicted from one reference picture. This type of slice is used in all three profiles.

Slices of type *B* (bipredictive) contain *B* and/or *I* macroblocks. The former type is predicted from a list of reference pictures. This type of slice is used in the main and extended profiles.

Slices of type *SP* (switching P) are special. They facilitate switching between coding streams and they contain extended *P* and/or *I* macroblocks. This type of slice is used only in the extended profile.

Slices of type *SI* (switching I) are also special and contain extended *SI* macroblocks. This type of slice is used only in the extended profile.

The encoder has a special reconstruction path whose purpose is to reconstruct a frame for encoding of further macroblocks. The main step in this path is to decode the quantized macroblock coefficients X. They are rescaled in box Q^{-1} and inverse transformed in T^{-1}, resulting in a difference macroblock $\mathrm{D'_n}$. Notice that this macroblock is different from the original difference macroblock $\mathrm{D_n}$ because of losses introduced by the quantization. We can consider $\mathrm{D'_n}$ a distorted version of $\mathrm{D_n}$. The next step creates a reconstructed macroblock $\mathrm{uF'_n}$ (a distorted version of the original macroblock) by adding the prediction macroblock P to $\mathrm{D'_n}$. Finally, a filter is applied to a series of macroblocks $\mathrm{uF'_n}$ in order to soften the effects of blocking. This creates a reconstructed reference frame $\mathrm{F'_n}$.

The decoder (Figure 9.41) inputs a compressed bitstream from the NAL. The first two steps (entropy decoding and reordering) produce a set of quantized coefficients X. Once these are rescaled and inverse transformed they result in a $\mathrm{D'_n}$ identical to the $\mathrm{D'_n}$ of the encoder). Using the header side information from the bitstream, the decoder constructs a prediction macroblock P, identical to the original prediction P created by

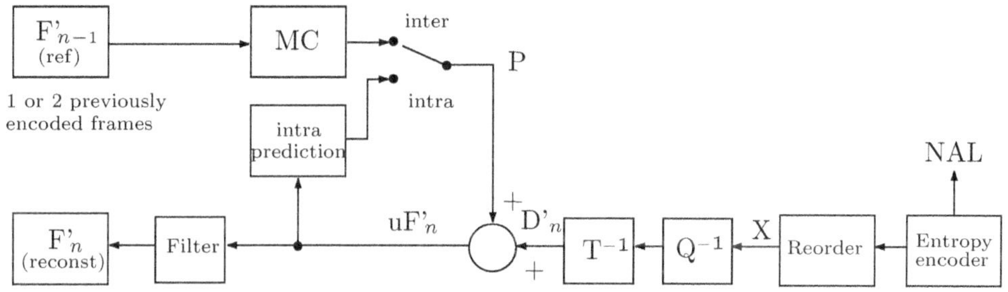

Figure 9.41: H.264 Decoder.

the encoder. In the next step, P is added to D'_n to produce uF'_n. In the final step, uF'_n is filtered to create the decoded macroblock F'_n.

Figures 9.40 and 9.41 and the paragraphs above make it clear that the reconstruction path in the encoder has an important task. It ensures that both encoder and decoder use identical reference frames to create the prediction P. It is important for the predictions P in encoder and decoder to be identical, because any changes between them tend to accumulate and lead to an increasing error or "drift" between the encoder and decoder.

The remainder of this section discusses several important features of H.264. We start with the profiles and levels. H.264 defines three profiles, baseline, main, and extended. Each profile consists of a certain set of encoding functions, and each is intended for different video applications. The baseline profile is designed for videotelephony, video conferencing, and wireless communication. It has inter and intra coding with I and P slices and performs the entropy encoding with context-adaptive variable-length codes (CAVLC). The main profile is designed to handle television broadcasting and video storage. It can deal with interlaced video, perform inter coding with B slices, and encode macroblocks with context-based arithmetic coding (CABAC). The extended profile is intended for streaming video and audio. It supports only progressive video and has modes to encode SP and SI slices. It also employs data partitioning for greater error control. Figure 9.42 shows the main components of each profile and makes it clear that the baseline profile is a subset of the extended profile.

This section continues with the main components of the baseline profile. Those who are interested in the details of the main and extended profiles are referred to [Richardson 03].

The baseline profile

This profile handles I and P slices. The macroblocks in an I slice are intra coded, meaning that each macroblock is predicted from previous macroblocks in the same slice. The macroblocks in a P slice can either be skipped, intra coded, or inter coded. The latter term means that a macroblock is predicted from the same macroblock in previously-coded pictures. H.264 uses a tree-structured motion compensation algorithm based on motion vectors as described in Section 9.5.

Once a macroblock F_n has been predicted, the prediction is subtracted from F_n to obtain a macroblock D_n of residuals. The residuals are transformed with a 4×4 integer transform, similar to the DCT. The transform coefficients are reordered (collected

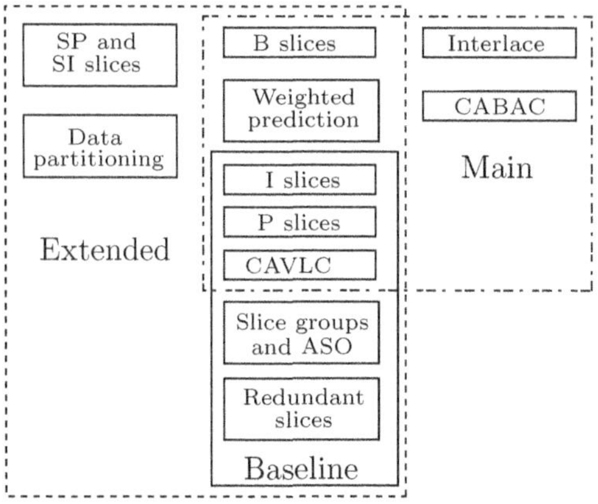

Figure 9.42: H.264 Profiles.

in zigzag order), quantized, and encoded by a context-adaptive variable-length code (CAVLC). Other information that has to go on the compressed stream is encoded with a Golomb code.

Inter prediction. The model for inter prediction is based on several frames that have already been encoded. The model is similar to that employed by older MPEG video standards, with the main differences being (1) it can deal with block sizes from 16×16 video samples down to 4×4 samples and (2) it uses fine-resolution motion vectors.

The tree-structured motion compensation algorithm starts by splitting the luma component of a 16×16 macroblock into partitions and computing a separate motion vector for each partition. Thus, each partition is motion compensated separately. A macroblock can be split in four different ways as illustrated in Figure 9.43a. The macroblock is either left as a single 16×16 block, or is split in one of three ways into (1) two 8×16 partitions, (2) two 16×8 partitions, or (3) four 8×8 partitions. If the latter split is chosen, then each partition may be further split in one of four ways as shown in Figure 9.43b.

The partition sizes listed here are for the luma color component of the macroblock. Each of the two chroma components (Cb and Cr) is partitioned in the same way as the luma component, but with half the vertical and horizontal sizes. As a result, when a motion vector is applied to the chroma partitions, its vertical and horizontal components have to be halved.

It is obvious that the partitioning scheme and the individual motion vectors have to be sent to the decoder as side information. There is therefore an important trade-off between the better compression that results from small partitions and the larger side information that small partitions require. The choice of partition size is therefore important, and the general rule is that large partitions should be selected by the encoder for uniform (or close to uniform) regions of a video frame while small partitions perform better for regions with much video noise.

Intra prediction. In this mode, the luma component of a macroblock either remains a single 16×16 block or is split into small 4×4 partitions. In the former case, the standard specifies four methods (summarized later) for predicting the macroblock. In

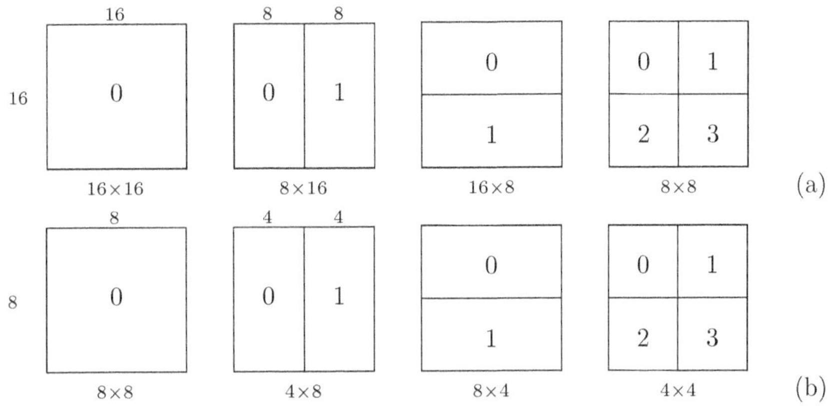

Figure 9.43: H.264 Macroblock Partitions.

the latter case, the 16 elements of a 4×4 partition are predicted in one of nine modes from 12 elements of three previously-encoded partitions to its left and above it. These 12 elements are shown in the top-left part of Figure 9.44 where they are labeled A through H, I through L, and M. Either the x or the y coordinate of these elements is -1. Notice that these elements have already been encoded and reconstructed, which implies that they are also available to the decoder, not just to the encoder. The 16 elements of the partition to be predicted are labeled `a`–`p`, and their x and y coordinates are in the interval $[0, 3]$. The remaining nine parts of Figure 9.44 show the nine prediction modes available to the encoder. A sophisticated encoder should try all nine modes, predict the 4×4 partition nine times, and select the best prediction. The mode number for each partition should be written on the compressed stream for the use of the decoder. Following are the details of each of the nine modes.

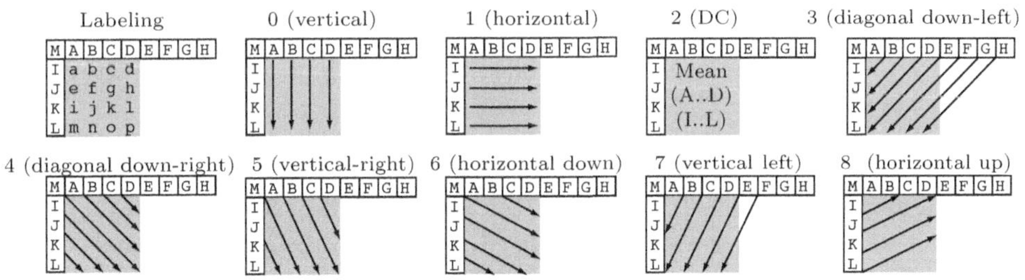

Figure 9.44: Nine Luma Prediction Modes.

Mode 0 (vertical) is simple. Elements `a`, `e`, `i`, and `m` are set to `A`. In general $p(x, y) = p(x, -1)$ for $x, y = 0, 1, 2,$ and 3.

Mode 1 (horizontal) is similar. Elements `a`, `b`, `c`, and `d` are set to `I`. In general $p(x, y) = p(-1, y)$ for $x, y = 0, 1, 2,$ and 3.

Mode 2 (DC). Each of the 16 elements is set to the average $(\mathtt{I}+\mathtt{J}+\mathtt{K}+\mathtt{L}+\mathtt{A}+\mathtt{B}+\mathtt{C}+\mathtt{D}+4)/8$. If any of `A`, `B`, `C`, and `D` haven't been previously encoded, then each of the

16 elements is set to the average $(\mathtt{I}+\mathtt{J}+\mathtt{K}+\mathtt{L}+2)/4$. Similarly, if any of I, J, K, and L haven't been previously encoded, then each of the 16 elements is set to the average $(\mathtt{A}+\mathtt{B}+\mathtt{C}+\mathtt{D}+2)/4$. In the case where some of A, B, C, and D and some of I, J, K, and L haven't been previously encoded, each of the 16 elements is set to 128. This mode can always be selected and used by the encoder.

Mode 3 (diagonal down-left). This mode can be used only if all eight elements A through H have been encoded. Element p is set to $(\mathtt{G}+3\mathtt{H}+2)/4$, and elements $p(x,y)$ where $x\neq 3$ or $y\neq 3$ are set to $[p(x+y,-1)+2p(x+y+1,-1)+p(x+y+2,-1)+2]/4$. As an example, element g, whose coordinates are $(2,1)$, is set to the average $[p(3,-1)+2p(4,-1)+p(5,-1)+2]/4=[\mathtt{D}+2\mathtt{E}+\mathtt{F}+2]/4$.

Mode 4 (diagonal down-right). This mode can be used only if elements A through D, M, and I through L have been encoded. The four diagonal elements a, f, k, and p are set to $[\mathtt{A}+2\mathtt{M}+\mathtt{I}+2]/4$. Elements $p(x,y)$ where $x>y$ are set to $[p(x-y-2,-1)+2p(x-y-1,-1)+p(x-y,-1)+2]/4$. Elements where $x<y$ are set to a similar average but with $y-x$ instead of $x-y$. Example. Element h, whose coordinates are $(3,1)$, is set to $[p(0,-1)+2p(1,-1)+p(2,-1)+2]/4=[\mathtt{A}+2\mathtt{B}+\mathtt{C}+2]/4$.

Mode 5 (vertical right). This mode can be used only if elements A through D, M, and I through L have been encoded. We set variable z to $2x-y$. If z is even (equals 0, 2, 4, or 6), element $p(x,y)$ is set to $[p(x-y/2-1,-1)+p(x-y/2,-1)+1]/2$. If z equals 1, 3, or 5, element $p(x,y)$ is set to $p[(x-y/2-2,-1)+2p(x-y/2-1,-1)+p(x-y/2,-1)+2]/4$. If $z=-1$, element $p(x,y)$ is set to $[\mathtt{A}+2\mathtt{M}+\mathtt{I}+2]/4$. Finally, if $z=-2$ or -3, element $p(x,y)$ is set to $[p(-1,y-1)+2p(-1,y-2)+p(-1,y-3)+2]/4$.

Mode 6 (horizontal down). This mode can be used only if elements A through D, M, and I through L have been encoded. We set variable z to $2y-x$. If z is even (equals 0, 2, 4, or 6), element $p(x,y)$ is set to $[p(-1,y-x/2-1)+p(-1,y-x/2)+1]/2$. If z equals 1, 3, or 5, element $p(x,y)$ is set to $[p(-1,y-x/2-2)+2p(-1,y-x/2-1)+p(-1,y-x/2)+2]/4$. If $z=-1$, element $p(x,y)$ is set to $[\mathtt{A}+2\mathtt{M}+\mathtt{I}+2]/4$. Finally, if $z=-2$ or -3, element $p(x,y)$ is set to $[(p(x-1,-1)+2p(x-2,-1)+p(x-3,-1)+2]/4$.

Mode 7 (vertical left). This mode can be used only if elements A through H have been encoded. If y equals 0 or 2, element $p(x,y)$ is set to $[p(x+y/2,-1)+p(x+y/2+1,-1)+1]/2$. If y equals 1 or 3, element $p(x,y)$ is set to $[p(x+y/2,-1)+2p(x+y/2+1,-1)+p(x+y/2+2,-1)+2]/4$.

Mode 8 (horizontal up). This mode can be used only if elements I through L have been encoded. We set z to $x+2y$. If z is even (equals 0, 2, or 4), element $p(x,y)$ is set to $[p(-1,y+x/2-1)+p(-1,y+x/2)+1]/2$. If z equals 1 or 3, element $p(x,y)$ is set to $[p(-1,y+x/2)+2p(-1,y+x/2+1)+p(-1,y+x/2+2)+2]/4$. If z equals 5, $p(x,y)$ is set to $[\mathtt{K}+3\mathtt{L}+2]/4$. If $z>5$, $p(x,y)$ is set to L.

If the encoder decides to predict an entire 16×16 macroblock of luma components without splitting it, the standard specifies four prediction modes where modes 0, 1, and 2 are identical to the corresponding modes for 4×4 partitions described above and mode 3 specifies a special linear function of the elements above and to the left of the 16×16 macroblock.

There are also four prediction modes for the 4×4 blocks of chroma components.

Deblocking filter. Individual macroblocks are part of a video frame. The fact that each macroblock is predicted separately introduces blocking distortions, which is why H.264 has a new component, the deblocking filter, designed to reduce these distortions. The filter is applied after the inverse transform. Naturally, the inverse transform is computed by the decoder, but the reader should recall, with the help of Figure 9.40, that in H.264 the inverse transform is also performed by the reconstruction path of the encoder (box T^{-1}, before the macroblock is reconstructed and stored for later predictions). The filter improves the appearance of decoded video frames by smoothing the edges of blocks. Without the filter, decompressed blocks often feature artifacts due to small differences between the edges of adjacent blocks.

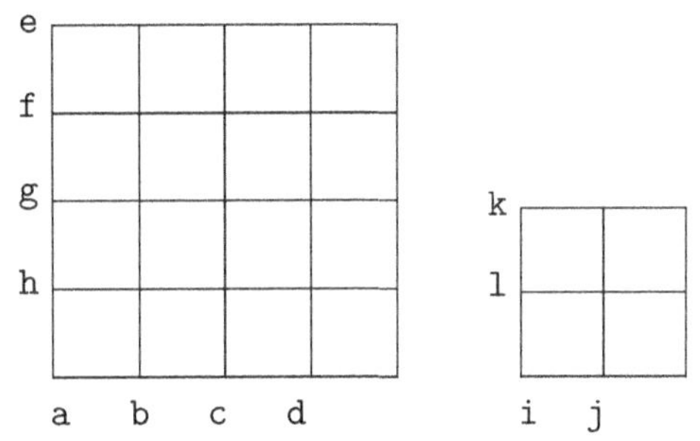

Figure 9.45: Edge Designation in a Macroblock.

The filter is applied to the horizontal or vertical edges of 4×4 blocks of a macroblock, except for edges located on the boundary of a slice, in the following order (Figure 9.45):

1. Filter the four vertical boundaries `a`, `b`, `c`, and `d` of the 16×16 luma component of the macroblock.

2. Filter the four horizontal boundaries `e`, `f`, `g`, and `h` of the 16×16 luma component of the macroblock.

3. Filter the two vertical boundaries `i` and `j` of each 8×8 luma component of the macroblock.

4. Filter the two horizontal boundaries `k` and `l` of each 8×8 luma component of the macroblock.

The filtering operation is an n-tap discrete wavelet transform (Section 8.7) where n can be 3, 4, or 5. Figure 9.46 shows eight samples `p0` through `p3` and `q0` through `q3` being filtered, four on each side of the boundary (vertical and horizontal) between macroblocks. Samples `p0` and `q0` are on the boundary. The filtering operation can have different strengths, and it affects up to three samples on each side of the boundary. The strength of the filtering is an integer between 0 (no filtering) and 4 (maximum filtering), and it depends on the current quantizer, on the decoding modes (inter or intra) of adjacent macroblocks, and on the gradient of video samples across the boundary.

The H.264 standard defines two threshold parameters α and β whose values depend on the average quantizer parameters of the two blocks `p` and `q` in question. These parameters help to determine whether filtering should take place. A set of eight video samples `pi qi` is filtered only if $|\texttt{p0} - \texttt{q0}| < \alpha$, $|\texttt{p1} - \texttt{p0}| < \beta$, and $|\texttt{q1} - \texttt{q0}| \leq \beta$.

When the filtering strength is between 1 and 3, the filtering process proceeds as follows: Samples `p1`, `p0`, `q0`, and `q1` are processed by a 4-tap filter to produce `p'0` and `q'0`. If $|\texttt{p2} - \texttt{p0}| < \beta$, another 4-tap filter is applied to `p2`, `p1`, `p0`, and `q0` to produce `p'1`. If $|\texttt{q2} - \texttt{q0}| < \beta$, a third 4-tap filter is applied to `q2`, `q1`, `q0`, and `p0` to produce

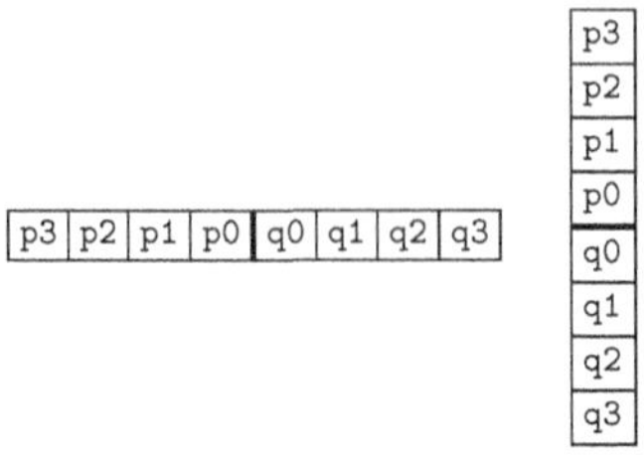

Figure 9.46: Horizontal and Vertical Boundaries.

q'1 (the last two steps are performed for the luma component only). When the filtering strength is 4, the filtering process is similar, but longer. It has more steps and employs 3-, 4-, and 5-tap filters.

Transform. Figures 9.40 and 9.41 show that blocks D_n of residuals are transformed (in box T) by the encoder and reverse transformed (box T^{-1}) by both decoder and the reconstruction path of the encoder. Most of the residuals are transformed in H.264 by a special 4×4 version of the discrete cosine transform (DCT, Section 7.8). Recall that an $n \times n$ block results, after being transformed by the DCT, in an $n \times n$ block of transform coefficients where the top-left coefficient is DC and the remaining $n^2 - 1$ coefficients are AC. In H.264, certain DC coefficients are tranformed by the Walsh-Hadamard transform (WHT, Section 7.7.2). The DCT and WHT transforms employed by H.264 are special and have the following features:

1. They use only integers, so there is no loss of accuracy.
2. The core parts of the transforms use only additions and subtractions.
3. Part of the DCT is scaling multiplication and this is integrated into the quantizer, thereby reducing the number of multiplications.

Given a 4×4 block **X** of residuals, the special DCT employed by H.264 can be written in the form

$$\mathbf{Y} = \mathbf{A}\,\mathbf{X}\,\mathbf{A}^T = \begin{bmatrix} a & a & a & a \\ b & c & -c & -b \\ a & -a & -a & a \\ c & -b & b & -c \end{bmatrix} \mathbf{X} \begin{bmatrix} a & b & a & c \\ a & c & -a & -b \\ a & -c & -a & b \\ a & -b & a & -c \end{bmatrix},$$

where

$$a = \frac{1}{2}, \quad b = \sqrt{\frac{1}{2}}\cos\frac{\pi}{8}, \quad \text{and} \quad c = \sqrt{\frac{1}{2}}\cos\frac{3\pi}{8}.$$

This can also be written in the form

$$\begin{aligned}\mathbf{Y} =& (\mathbf{CXC}^T) \otimes \mathbf{E} \\ =& \left\{ \begin{bmatrix} 1 & 1 & 1 & 1 \\ 1 & d & -d & -1 \\ 1 & -1 & -1 & 1 \\ d & -1 & 1 & -d \end{bmatrix} \mathbf{X} \begin{bmatrix} 1 & 1 & 1 & d \\ 1 & d & -1 & -1 \\ 1 & -d & -1 & 1 \\ 1 & -1 & 1 & -d \end{bmatrix} \right\} \otimes \begin{bmatrix} a^2 & ab & a^2 & ab \\ ab & b^2 & ab & b^2 \\ a^2 & ab & a^2 & ab \\ ab & b^2 & ab & b^2 \end{bmatrix},\end{aligned}$$

where $(\mathbf{CXC}^T)$ is a core transform (where $d = c/b \approx 0.414$), matrix **E** consists of scaling factors, and the $\otimes$ operation indicates scalar matrix multiplication (pairs of corresponding matrix elements are multiplied).

In order to simplify the computations and ensure that the transform remains orthogonal, the following approximate values are actually used

$$a = \frac{1}{2}, \quad b = \sqrt{\frac{2}{5}}, \quad d = \frac{1}{2},$$

bringing this special forward transform to the form

$$\mathbf{Y} = \left\{ \begin{bmatrix} 1 & 1 & 1 & 1 \\ 2 & 1 & -1 & -2 \\ 1 & -1 & -1 & 1 \\ 1 & -2 & 2 & -1 \end{bmatrix} \mathbf{X} \begin{bmatrix} 1 & 2 & 1 & 1 \\ 1 & 1 & -1 & -2 \\ 1 & -1 & -1 & 2 \\ 1 & -2 & 1 & -1 \end{bmatrix} \right\} \otimes \begin{bmatrix} a^2 & ab/2 & a^2 & ab/2 \\ ab/2 & b^2/4 & ab/2 & b^2/4 \\ a^2 & ab/2 & a^2 & ab/2 \\ ab/2 & b^2/4 & ab/2 & b^2/4 \end{bmatrix}.$$

Following the transform, each transform coefficient Y_{ij} is quantized by the simple operation $Z_{ij} = \text{round}(Y_{ij}/Q_{\text{step}})$ where Q_{step} is the quantization step. The rounding operation is also special. The encoder may decide to use "floor" or "ceiling" instead of rounding. The H.264 standard specifies 52 quantization steps Q_{step} that range from 0.625 up to 224 depending on the quantization parameter QP whose values range from 0 to 51. Table 9.47 lists these values and shows that every increment of 6 in QP doubles Q_{step}. The wide range of values of the latter quantity allows a sophisticated encoder to precisely control the trade-off between low bitrate and high quality. The values of QP and Q_{step} may be different for the luma and chroma transformed blocks.

QP	0	1	2	3	4	5	6	7	8	9	10	11	12	...
Q_{step}	0.625	0.6875	0.8125	0.875	1	1.125	1.25	1.375	1.625	1.75	2	2.25	2.5	...

QP	...	18	...	24	...	30	...	36	...	42	...	48	...	51
Q_{step}		5		10		20		40		80		160		224

Table 9.47: Quantization Step Sizes.

If a macroblock is encoded in 16×16 intra prediction mode, then each 4×4 block of residuals in the macroblock is first transformed by the core transform described above, and then the 4×4 Walsh-Hadamard transform is applied to the resulting DC coefficients as follows

$$\mathbf{Y}_D = \frac{1}{2} \begin{bmatrix} 1 & 1 & 1 & 1 \\ 1 & 1 & -1 & -1 \\ 1 & -1 & -1 & 1 \\ 1 & -1 & 1 & 1 \end{bmatrix} \mathbf{W}_D \begin{bmatrix} 1 & 1 & 1 & 1 \\ 1 & 1 & -1 & -1 \\ 1 & -1 & -1 & 1 \\ 1 & -1 & 1 & -1 \end{bmatrix},$$

where $\mathbf{W}_D$ is a 4×4 block of DC coefficients.

After a macroblock is predicted and the residuals are transformed and quantized, they are reordered. Each 4×4 block of quantized transform coefficients is scanned in zigzag order (Figure 1.8b). If the macroblock is encoded in intra mode, there will be 16 DC coefficients (the top-left coefficient of each 4×4 luma block). These 16 coefficients are arranged in their own 4×4 block which is scanned in zigzag. The remaining 15 AC coefficients in each 4×4 luma block are also scanned in the same zigzag order. The same reordering is applied to the smaller chroma blocks.

Entropy coding

H.264 involves many items—counts, parameters, and flags—that have to be encoded. The baseline profile employs variable-length codes and a special context adaptive arithmetic coding (CABAC) scheme to encode parameters such as macroblock type, reference frame index, motion vectors differences, and differences of quantization parameters. The discussion in this section concentrates on the encoding of the zigzag sequence of quantized transform coefficients. This sequence is encoded in a special context-adaptive variabe-length coding scheme (CAVLC) based on Golomb codes (Section 3.24).

CAVLC is used to encode reordered zigzag sequences of either 4×4 or 2×2 blocks of quantized transform coefficients. The method is designed to take advantage of the following features of this special type of data.

1. Such a sequence is sparse, has runs of zeros, and normally ends with such a run.

2. Among the nonzero coefficients, such a sequence has many +1s and −1s and these are referred to as trailing ±1s.

3. The number of nonzero coefficients in such a sequence is often similar to the corresponding numbers in neighboring blocks. Thus, these numbers are correlated across adjacent blocks.

4. The first (DC) coefficient tends to be the largest one in the sequence, and subsequent nonzero coefficients tend to get smaller.

CAVLC starts by scanning and analyzing the sequence of 16 coefficients. This step assigns values to the six encoding parameters `coeff_token`, `trailing_ones_sign_flag`, `level_prefix`, `level_suffix`, `total_zeros`, and `run_before`. These parameters are then used in five steps to select tables of variable-length codes and codes within those tables. The codes are concatenated to form a binary string that is later formatted by the NAL stage and is finally written on the compressed stream. The five encoding steps are as follows:

Step 1. Parameter `coeff_token` is the number of nonzero coefficients and trailing ±1s among the 16 coefficients in the sequence. The number of nonzero coefficients can be from 0 to 16. The number of trailing ±1s can be up to the number of nonzero coefficients, but `coeff_token` takes into account (i.e., it signals) up to three trailing ±1s. Any trailing ±1s beyond the first three are encoded separately as other nonzero coefficients. If the number of nonzero coefficients is zero, there can be no trailing ±1s. If the number of nonzero coefficients is one, there can be zero or one trailing ±1s. If the number of nonzero coefficients is two, the number of trailing ±1s can be 0, 1, or 2. If the number of nonzero coefficients is greater than two, there can be between zero and three trailing ±1s that are signaled by `coeff_token`. There can be between zero and 16 nonzero coefficients, so the total number of values of parameter `coeff_token` is $1+2+3+14\times4=62$. Thus, the H.264 standard should specify a table of 62 variable-length codes to encode the value of `coeff_token`. However, because the number of nonzero coefficients is correlated across neighboring blocks, the standard specifies four such tables (table 9–5 on page 159 of [H.264Draft 06]).

To select one of the four tables, the encoder counts the numbers nA and nB of nonzero coefficients in the blocks to the left of and above the current block (if such blocks exist, they have already been encoded). A value nC is computed from nA and nB as follows. If both blocks exist, then nC = round[(nA + nB)/2]. If only the left block exists, then nC = nA. If only the block above exists, then nC = nB. Otherwise nC = 0. The resulting value of nC is therefore in the interval [0, 16] and it is used to select one of four tables as listed in Table 9.48.

nC:	0	1	2	3	4	5	6	7	8	9	...
Table:	1	1	2	2	3	3	3	3	4	4	...

Table 9.48: nC Values to Select a Table.

Step 2. Parameter `trailing_ones_sign_flag` is encoded as a single bit for each trailing ±1 signaled by `coeff_token`. The trailing ±1s are scanned in reverse zigzag order, and the (up to three) sign bits that are generated are appended to the binary string generated so far.

Step 3. The next set of bits appended to the binary string encodes the values of the remaining nonzero coefficients (i.e, the nonzero coefficients except those trailing ±1s that were signaled by `coeff_token`). These nonzero coefficients are scanned in reverse zigzag order, and each is encoded in two parts, as a prefix (`level_prefix`) and suffix (`level_suffix`). The prefix is an integer between 0 and 15 and is encoded with a variable-length code specified by the standard (table 9–6 on page 162 of [H.264Draft 06]). The length of the suffix part is determined by parameter suffixLength. This parameter can have values between 0 and 6 and is adapted while the nonzero coefficients are being located and encoded. The principle of adapting suffixLength is to match the suffix of a coefficient to the magnitudes of the recently-encoded nonzero coefficients. The following rules govern the way suffixLength is adapted:

1. Parameter suffixlength is initialized to 0 (but if there are more than 10 nonzero coefficients and fewer than three trailing ±1s, it is initialized to 1).

2. The next coefficient in reverse zigzag order is encoded.

3. If the magnitude of that coefficient is greater than the current threshold, increment suffixLength by 1. This also selects another threshold. The values of the thresholds are listed in Table 9.49.

suffixLength:	0	1	2	3	4	5	6
Threshold:	0	3	6	12	24	48	na

Table 9.49: Thresholds to Increment suffixLength.

Step 4. Parameter `total_zeros` is the number of zeros preceding the last nonzero coefficient. (This number does not include the last run of zeros, the run that follows the last nonzero coefficient.) `total_zeros` is an integer in the interval [0, 15] and it is encoded with a variable-length code specified by the standard (tables 9–7 and 9–8 on page 163 of [H.264Draft 06]). This parameter and the next one (`run_before`) together are sufficient to encode all the runs of zeros up to the last nonzero coefficient.

Step 5. Parameter `run_before` is the length of a run of zeros preceding a nonzero coefficient. This parameter is determined and encoded for each nonzero coefficient (except, of course, the first one) in reverse zigzag order. The length of such a run can be zero, but its largest value is 14 (this happens in the rare case where the DC coefficient is followed by a run of 14 zeros, which is followed in turn by a nonzero coefficient). While the encoder encodes these run lengths, it keeps track of their total length and terminates this step when the total reaches `total_zeros` (i.e., when no more zeros remain to be encoded). The standard specifies a table of variable-length codes (table 9–10 on page 164 of [H.264Draft 06]) to encode these run lengths.

Once the five steps have encoded all the nonzero coefficients and all the runs of zeros between them, there is no need to encode the last run of zeros, the one (if any) that follows the last nonzero coefficient, because the length of this run will be known to the decoder once it has decoded the entire bit string.

The conscientious reader should consult [Richardson 03] for examples of encoded sequences and also compare the CAVLC method of encoding a sequence of DCT transform coefficients with the (much simpler) way similar sequences are encoded by JPEG (Section 7.10.4).

9.10 H.264/AVC Scalable Video Coding

The Joint Video Team, an organization that combines members of the ITU-T Video coding Group (VCEG) and members of the ISO/IEC Moving Picture expert Group (MPEG), has extended H.264/AVC with an Amendment defining a scalable video encoder called *Scalable Video Coding* or *SVC*. The purpose of this extension is to specify a compressed bitstream that supports temporal, spatial, and quality scalable video coding, while retaining a base layer that is still backward compatible with H.264/AVC.

A bitstream is called *multi-layer* or *scalable* if parts of the bitstream can be discarded to generate a substream that is still valid and decodable by a target decoder, but with a reconstruction quality that is lower than the quality obtained by decoding the entire bitstream. A bitstream that does not have this property is called *single-layer*.

A scalable video bitstream can be easily adapted to the capacity of a transmission channel, to the computational power of a decoder, or to the resolution of a display device. Furthermore, since the layers in a scalable bitstream can be easily prioritized, assigning unequal error protection to the layers can improve error resilience and gracefully degrade the quality of the reconstructed signal in the presence of channel error and packet loss.

Scalable video compression has not been introduced by SVC. In fact, almost all the recent international coding standards such as MPEG-2 Video, H.263, and MPEG-4 Visual already support the most useful forms of scalability. Unfortunately, the lack of coding efficiency and the increased complexity of the scalable tools offered by these standards has limited the deployment of this technology.

In order to be useful, a scalable encoder should generate a bitstream substantially more compact than the independent compression of all its component layers, so to provide a convenient alternative to *simulcast*, the simultaneous broadcasting of all the layers independently compressed. Furthermore, the increase in decoder complexity should be lower than the complexity of *transcoding* the bitstream, the conversion of a layer into another directly in the digital domain. It is obvious that the scalable compression of a layer other than the base layer will, in general, be slightly worse than its non-scalable compression. However, it is important to design the system so that performance loss is kept to a minimum.

Designed according to these principles, the Scalable Video Coding extension to H.264/AVC provides temporal, spatial, and quality scalability with the mechanisms outlined in the following paragraphs.

Temporal Scalability: Video coding standards are traditionally "decoder" or "bitstream" standards. The format of the bitstream and the behavior of a reference decoder are specified by the standard, while leaving the implementors complete freedom in the design of the encoder. For this reason, a standard will specify a tool only when strictly necessary to achieve a given functionality. Following this basic guideline, SVC supports temporal scalability without adding any new tool to H.264/AVC.

Syntactically, the only extension is the signalling of a *temporal layer identification* T. T is an integer that starts from 0 (to signal the pictures belonging to the base layer) and grows as the temporal resolution of a layer increases. The temporal layer identification is necessary because a decoder that wants to decode a temporal layer $T = t$ should be able to isolate and decode all pictures with $T \leq t$ and discard any remaining picture. The temporal layer identification is contained in a header that is peculiar to SVC, the *SVC NAL unit header.* This additional header is stored in the so-called *prefix NAL units.* Prefix NAL units can be discarded in order to generate a base layer with NALs that are fully H.264/AVC compliant.

H.264/AVC already allows a pretty flexible temporal scalability with the use of hierarchical prediction. Figures 9.50, 9.51, and 9.52 show three possible alternatives that implement dyadic and non-dyadic temporal scalability with different delays. The arrows indicate the reference pictures used in the encoding of each picture.

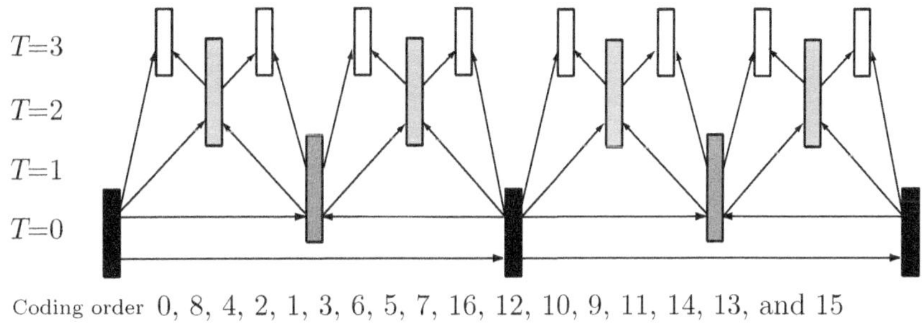

Figure 9.50: Dyadic Hierarchical Prediction with Four Temporal Layers.

The hierarchical structure of Figure 9.50 uses a GOP (group of pictures) of eight frames to specify a four-layer dyadic scheme. In a dyadic scheme, each layer doubles the temporal resolution of the layer preceding it. The gray and white frames in the figure are B-pictures, respectively, used and unused as reference in the motion compensated prediction. Figure 9.51 depicts a three-layer non-dyadic scheme.

The hierarchical structures depicted in Figures 9.50 and 9.51 are capable of generating a temporally scalable bitstream with a compression of the upper layers that is

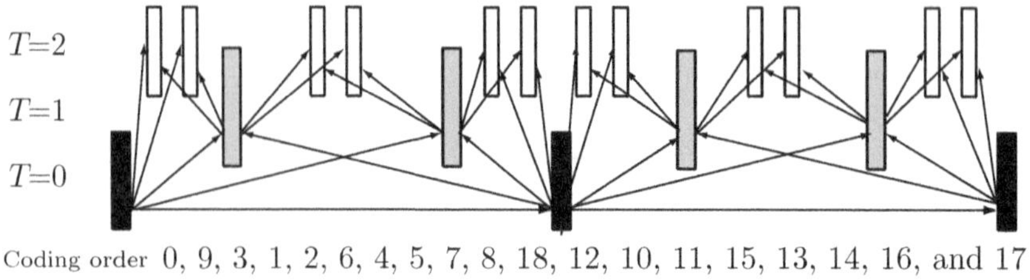

Figure 9.51: Non-dyadic Hierarchical Prediction with Three Temporal Layers.

substantially better than what could be achieved with a single layer "IPPP" coding scheme. This is apparently counterintuitive, since we have said before that scalability slightly penalizes compression efficiency. The reason for this apparent contradiction is found in the delay introduced by these schemes. The hierarchical structures depicted in Figure 9.50 and Figure 9.51 introduce delays of seven and eight pictures, respectively. These delays are the "price" paid for the better compression.

It is important to notice though that temporal scalability can be implemented with arbitrary delays, all we have to do is to appropriately constrain the prediction structure. Figure 9.52 shows a zero-delay dyadic scheme that does not use B-pictures. A temporally scalable prediction that is implemented via a low-delay scheme will eventually exhibit a reduced coding efficiency.

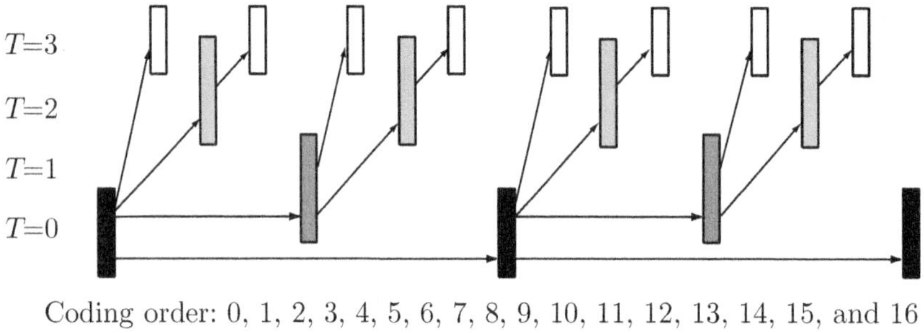

Figure 9.52: Low-delay Dyadic Hierarchical Prediction with Four Temporal Layers.

Spatial Scalability: With a mechanism similar to the temporal layer identification described above, SVC introduces the concept of *dependency identifier*, D. The dependency identifier describes the relationship between spatial layers and is also stored in a prefix NAL unit. The value of D is zero for the base layer ($T = 0$) and increases by 1 with each spatial layer.

In SVC, each layer can use an independent motion compensation. Additionally, SVC exploits the correlation of layers with the following three inter-layer prediction mechanisms:

- *Inter-Layer Motion Prediction:* The partitioning used in a block in the reference layer is upsampled and used to infer the partitioning of the corresponding macroblock in the enhancement layer. Motion vectors from the block in the reference layer are eventually scaled before being used in the enhancement layer.

- *Inter-Layer Residual Prediction:* It can be used on all inter-coded macroblocks and uses the upsampled residual of the corresponding reference block. Upsampling is performed with a bilinear filter.

- *Inter-Layer Intra Prediction:* The prediction signal of intra coded blocks in the reference layer is upsampled and used to predict the corresponding macroblock in the enhancement layer.

In general, the coding loss introduced by spatial scalability is observable only in the enhancement layers, since the base layer is often encoded independently by a regular H.264/AVC encoder. However, since enhancement layers are predicted starting from the base layer, joint optimization of all layers is known to reduce coding loss at the cost of more complex encoding. With the use of joint optimization, the bit rate increase due to spatial scalability can be as low as 10%.

The main restriction imposed by spatial scalability is that horizontal and vertical resolutions of the video sequence cannot decrease from one layer to the next. The pictures in an enhancement layer are allowed to have the same resolution as the pictures in the reference layer. This is a feature that enables scalable encoding of video formats with dimensions that are not necessarily multiples of each other. For example, SVC can encode a base layer representing a video sequence in QVGA (Quarter VGA) format, with pictures having 320×240 pixels. The first enhancement layer can encode the same sequence, but at a higher resolution, for example, a VGA version of the video with pictures of 640×480 pixels. A second enhancement layer can then be used to represent the original video in WVGA (Wide VGA) format, at 800×480 pixels (Figure 9.53). In this case, the second enhancement layer will only need to encode columns that have been cropped out to obtain the VGA sequence from the WVGA original.

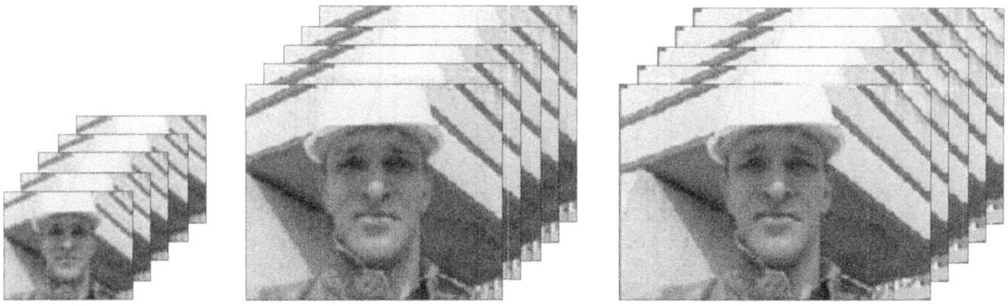

Base layer T=0 QVGA (320×240) | Enhancement layer T=1 VGA (640×480) | Enhancement layer T=2 WVGA (800×480)

Figure 9.53: Spatial Layers Encoding QVGA, VGA, and WVGA Video Sequences.

In SVC, the coding order of a picture cannot change from one layer to another. However, lower layers are not required to contain information on all pictures. This option allows combination of spatial and temporal scalability.

Quality Scalability: A coarse form of quality scalability can be easily implemented as a special case of spatial scalability, where the reference and the enhancement layers have identical size and resolution but different quantization parameters. While simple, this solution has the major drawback that only a prefixed number of quality levels can be reconstructed, one for each layer present in the bitstream. SVC adds to this possibility a more refined form of quality scalability in which a part of the packets that constitute the enhancement layer can be discarded in order to match a target bit rate that is between the rate of the reference layer and the rate of the enhancement layer. Enhancement layer's packets can be randomly discarded until the target bit rate is reached. A better alternative, which is also supported by the SVC syntax, is assigning priority level to packets, so that packets can be selectively discarded in order of increasing priority.

A critical issue in the design of a quality scalable video encoder is the decision of which layer should be used in the motion estimation. The highest coding efficiency is obtained when motion estimation is performed on the pictures with the best quality, and so, on the highest quality enhancement layer. However, discarding packets from the enhancement layer compromises the quality of the reference frames used by the decoder in the motion compensation. The difference between the reference frames used by the encoder in the motion estimation and the reference frames used by the decoder in the motion compensation generates an error that is known as *drift*. Using only the base layer in the motion estimation prevents drift while reducing coding effectiveness.

SVC compromises between these two solutions by introducing the concept of *key pictures*. A key picture is one in which only the base layer is used in the motion compensated prediction. All non-key pictures will instead use the enhancement layer. Key pictures control drift by periodically resynchronizing encoder and decoder. Figure 9.54 shows an example where key pictures (marked in gray) are periodically inserted in the base and in the enhancement layer. The arrows represent the pictures used in the motion compensated prediction. The bit rate increase due to the introduction of quality scalability is typically less than 10%.

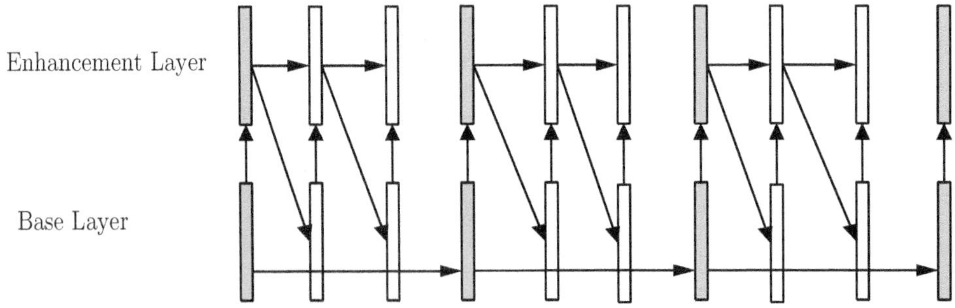

Figure 9.54: Key Pictures (Gray) Used to Control Drift in Quality Scalable Video Coding.

A detailed description of the H.264/AVC scalable video coding extension can be found in [Schwarz et. al. 07] and [Segall and Sullivan 07], both published in a special issue of the *IEEE Transaction on Circuits and Systems for Video Technology* dedicated to SVC.

9.11 VC-1

Political and social revolutions are common and some consider them the engine that drives human history, but scientific and technical progress is mostly evolutional. Among the very few scientific revolutions that immediately spring to mind are relativity, quantum mechanics, and Newton's theory of gravity (in physics), Natural selection and the genetic code (in biology), the atomic hypothesis (in chemistry), and information theory. As in other scientific fields, most progress in data compression is evolutionary and VC-1, the video codec described here, is no exception. It represents an evolution of the successful DCT-based video codec design that is found in popular video codecs such as H.261, H.263, MPEG-1, MPEG-2, and MPEG-4 Part 2.

Like most video codecs that preceded it, VC-1 is a hybrid codec that is based on the two pillars of video compression, namely a transform (often the DCT, Section 7.8) combined with motion compensated prediction (Section 9.5). The former is used to reduce spatial redundancy, while the latter is used to reduce temporal redundancy. Like other video codecs, it extends these techniques and adds other "bells and whistles." One of the chief goals of the VC-1 project was to develop an algorithm that can encode interlaced video without first converting it to progressive, and this feature, among others, has made VC-1 attractive to video professionals.

VC-1 (the name stands for video codec) is a video codec specification that was adopted in 2006 as a standard by the Society of Motion Picture and Television Engineers (SMPTE). SMPTE (located at [SMPTE 08]) is the pre-eminent society of film and video experts, with members in 85 countries worldwide. Similar to the ITU, the standards that SMPTE publishes are voluntary, but are nevertheless widely adopted, implemented, and employed by companies doing business in video, motion pictures, and digital cinema. Within SMPTE, the VC-1 standard was developed by C42, the Video Compression Technology Committee, which is responsible for all video technologies in SMPTE.

VC-1 has been implemented by Microsoft as Microsoft Windows Media Video (WMV) 9 [WMV 08]. The formal title of VC-1/WMV standard is SMPTE 421M [SMPTE-421M 08] and there are two documents (SMPTE RP227 and SMPTE RP228) that accompany it. They describe VC-1 transport (details about carrying VC-1 elementary streams in MPEG-2 Program and Transport Streams) and conformance (the test procedures and criteria for determining conformance to the 421M specifications). As usual, the formal standard specifies only the syntax of the compressed stream and the operations of the decoder; the precise steps of the encoder are left unspecified, in the hope that innovative programmers will come up with clever algorithms and shortcuts to improve the encoder's performance and speed it up.

References. [SMPTE-421M 08] and [Lee and Kalva 08] are the two main references. The former is the formal specification of this standard, which many people may find cryptic, while the latter may include too much detail for most readers. Those interested in short, easy-to-understand overviews may want to consult [WMV review 08] and [WikiVC-1 08].

VC-1 adoption. Notwithstanding its tender age, VC-1 has been favorably received and adopted by the digital video industry. The main reasons for its popularity are its performance combined with the fact that it is well documented, extremely stable, and easily licensable. Of special notice are the following adopters:

- High-definition DVD. The DVD forum (`dvdforum.org`) has mandated VC-1, H.264, and MPEG-2 for the HD DVD format.

- The Blu-ray Disc Association (`blu-raydisc.com`) has mandated the same three codecs for their Blu-ray Disc format (so named because of its use of blue laser).

- The FVD (forward versatile disc) standard from Taiwan has adopted VC-1 as its only mandated video codec. (FVD is the Taiwanese alternative for the high-definition storing optical media based on red laser technology.)

- Several DSP and chip manufacturers have plans (or even devices) to implement VC-1 in hardware.

- VC-1 is already used for professional video broadcast. Several companies are already offering products, ranging from encoders and decoders to professional video test equipment, that use VC-1.

- The Digital Living Network Alliance (DLNA, `dlna.org`) is a group of 245 companies that agreed to adopt compatible standards and make compatible equipment in the areas of consumer electronics, computer, and mobile devices. VC-1 is one of several compression formats adopted by the DLNA.

- The acronym `DVB-H` stands for Digital Video Broadcasting - Handheld (`dvb-h.org`). This is a set of specifications for bringing broadcast services to battery-powered handheld receivers. Again, VC-1 is one of the compression formats adopted by this standard.

- VC-1 has been designated by Microsoft as the Xbox 360 video game console's official video codec. Thus, game developers may utilize VC-1 for any video included with games.

- The FFmpeg project (`http://ffmpeg.org/`) includes a free VC-1 decoder (FFmpeg is a computer program that can record, convert, and stream digital audio and video in numerous formats.)

9.11.1 Overview

VC-1 is complex. It includes many parameters, features, and processes. The description here attempts to cover the general design and the main parts of this algorithm. The only items that are described in detail are the pixel interpolation and the transform. The formal definition of VC-1 simply lists the steps and parameters used by these items, but the discussion here tries to explain why these specific steps and constants were chosen by the codec's developers.

Most video codecs are designed for a specific application and a particular bitrate, but VC-1 is intended for use in a wide variety of video applications, and must therefore perform well at bitrates that may vary from very low (low-resolution, 160×120 pixels video, requiring about 10 Kbps) to very high ($2{,}048 \times 1{,}536$ pixels, or about 135 Mbps, for high-definition video). It is expected that a careful implementation of the VC-1 encoder would be up to three times as efficient as MPEG-2 and up to twice as good as MPEG-4 Part 2.

As has already been mentioned, the main encoding techniques used by VC-1 are a (16-bit integer) transform and block-based motion compensation. These are optimized in several ways and, with the help of several innovations, result in a fast, efficient video

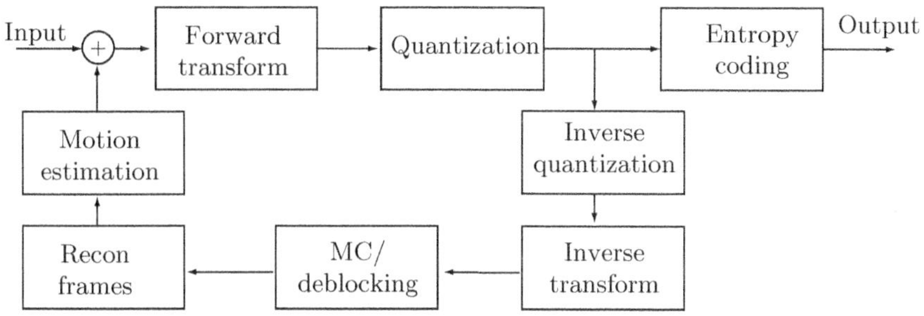

Figure 9.55: VC-1 Main Block Diagram.

codec. Figure 9.55 is a block diagram of the main components of the VC-1 encoder (where MC stands for motion compensation).

The figure illustrates the main steps in encoding a frame. Blocks of prediction residuals are transformed, the transformation coefficients are quantized, entropy encoded, and output. The motion compensation part involves inverse quantization, deblocking, reconstructing each frame, predicting the motion (the differences between this frame and its predecessor/seccessor), and using the prediction to improve the encoding of blocks in the next frame.

The experienced reader will notice that the various MPEG methods feature similar block diagrams. Thus, the superior performance of VC-1 stems from its innovations, the main ones of which are listed here:

- Adaptive Block-Size Transform. The JPEG standard (Section 7.10) applies the DCT to blocks of 8×8 pixels, and these dimensions have become a tradition in image and video compression. Users of JPEG know from long experience that setting the JPEG parameters to maximum compression often results in compression artifacts (known as blocking or ringing, see Figure 7.57) along block boundaries. This phenomenon is caused by differences between pixel blocks due to the short range of pixel correlation. Thus, image regions with low pixel correlation compress better when the DCT (or any other transform) is applied to smaller blocks.

VC-1 can transform a block of 8×8 pixels in one of the following ways (Figure 9.56): As a single 8×8 block, as two 8×4 blocks, as two 4×8 blocks, or as four 4×4 blocks. The encoder decides which transform size to use for any given block (although the definition of the standard does not specify how this decision should be made).

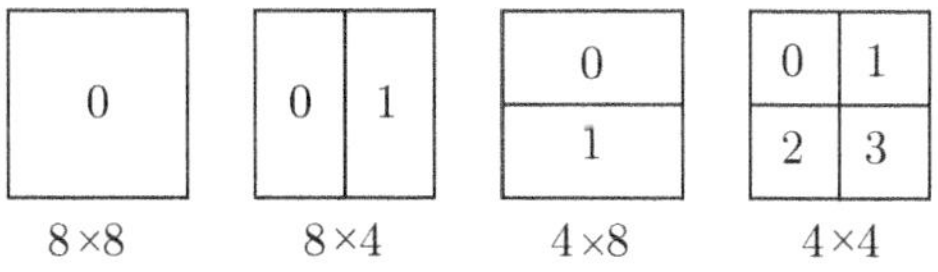

Figure 9.56: Variable-Size Transformed Blocks.

■ A specially-developed integer transform is employed by VC-1. The integer nature of the transform makes it easy to implement the VC-1 decoder in 16-bit architectures. This features makes VC-1 attractive to hardware implementors, especially for use in a variety of DSP (digital signal processing) devices. The transform matrices were derived in a special way to allow for fast, easy decoding.

■ Section 9.5 discusses motion compensation. The encoder computes a prediction of a video frame and uses it to predict the next frame. Naturally, the decoder should be able to mimic this operation, which is why Figure 9.55 shows that the encoder computes the prediction by performing the same operations (inverse quantization, inverse transform, deblocking, and reconstructing the frame from its constituent blocks) as the decoder. Typically, each 8×8 block or each 16×16 macroblock of pixels is motion compensated individually to generate a motion vector. The vector specifies how much the encoder predicts the image of the block (or macroblock) to be shifted in the x and y directions in the corresponding (macro)block in the reference frame.

The motion compensation component employs two sets of filters, an approximate bicubic filter with four taps and a bilinear filter with two taps. Filtering the pixels of a block sometimes allows the decoder to isolate the color of a part of a pixel (a sub-pixel) as small as 1/4 of a pixel. The encoder uses the block size, sub-pixel data, and filter type to determine one of four motion compensation modes for each block. These modes improve compression as well as simplify the decoder's implementation.

■ Blocking artifacts have already been mentioned. They are the result of discontinuities along block boundaries, caused by the differences between quantization and dequantization. These artifacts are especially noticeable (and annoying) when the codec's parameters are set for maximum compression (i.e., coarse quantization). The VC-1 encoder has an in-loop deblocking filter that tries to remove these discontinuities. This filter needs to know the block size of the transform. Also, the VC-1 encoder can overlap the transform between blocks in order to reduce annoying blocking artifacts. This technique is known as overlapped transform (OLT).

■ Many video transmissions are interlaced (see Section 9.1.1 for interlacing). VC-1 includes an innovative algorithm that uses data from both sets of scan lines to improve the prediction of the encoder's motion compensation component. This algorithm is referred to as interlace coding.

■ VC-1 supports five types of video frames. Three of these are the standard I frame (a key frame or intra; a frame that's encoded independently of its neighbors), P frame (predictive, a frame encoded based on its predecessor), and B frame (bidirectional, a frame encoded based on its predecessors and successors). These types are covered in more detail in Section 9.5. One of the innovations included in VC-1 is advanced B-frame encoding. This feature consists of several optimizations that make B-frame encoding more efficient.

The other two frame types are S (skipped) and BI. The former indicates a frame that is so similar to its reference frame that there is no need to encode it. When the decoder identifies an S frame, it simply repeats its reference frame. The latter type is a B frame with only I macroblocks. When there is a complete change of scene in a video stream, a B frame tends to result in a large number of I macroblocks. In such a case, the special BI frame type occupies fewer bits than a standard B frame.

It should be noted that video codecs normally collect several consecutive frames into a group of frames (GOP). In VC-1, the number of frames in a GOP is variable.

- It has been known for a while that motion compensation has its drawbacks, one of which is that it cannot efficiently model fading. Often, a succession of video frames contains regions that fade to or from black, and these are encoded very inefficiently because of the nature of motion compensation. The fading compensation feature of VC-1 detects fades and employs methods other than motion compensation to improve encoding in such cases. This feature also handles other global illumination changes.

- The coefficients produced by the transform are quantized by dividing them by a quantization parameter (QP) which itself is transmitted to the decoder in the compressed stream. Quantization is the step where information is irretrievably lost but much compression is gained.

  Many compression algorithms based on quantization employ a single QP for the entire image (in case of image compression) or for an entire video frame (for video compression). VC-1 supports differential quantization (or dquant), where several quantization steps are applied to a frame. This innovation is based on the fact that different macroblocks require different grades of quantization, depending on how many nonzero AC coefficients remain after the macroblock is transformed. The simplest form of differential quantization uses two quantization levels and is termed bi-level dquant.

- The audio component of VC-1 can produce bitrates of 128–768 Kbps and is hybrid; it employs different algorithms for speech and music.

- From its inception, VC-1 was designed to perform well on a wide variety of video data. To this end, the designers included three profiles, each with several levels. Each level is geared toward a range of video applications and each determines the features a VC-1 encoder would need (its mathematical complexity) to process a video stream generated by those applications. Table 9.57 lists the profiles and their levels and Table 9.58 lists the features and innovations included in each profile.

9.11.2 The Encoder

This section is a high-level summary of the operations of the encoder. It should again be stressed that the formal definition of the VC-1 standard does not specify how each step is to be performed and leaves the details to the ingenuity of implementors. The following is a list of the main steps performed by the encoder:

- A video frame is read from the input and is decomposed into macroblocks and blocks (this is common in video encoders).

- For an intra-coded block:
  1. Transform the block with an 8×8 transform.
  2. The transform can overlap neighboring intra-coded blocks.
  3. Quantize the 64 transform coefficients. DC/AC prediction can be applied to the quantized coefficients to reduce redundancy even more.
  4. Encode the DC coefficient with a variable-length code. Scan the AC coefficients in zigzag and encode the run-lengths and nonzero coefficients with another variable-length code.

Profile	Level	Max bitrate	Representative resolutions by frame rate
Simple	Low	96 Kbps	176 × 144 @ 15 Hz (QCIF)
	Medium	384 Kbps	240 × 176 @ 30 Hz
			352 × 288 @ 15 Hz (CIF)
Main	Low	2 Mbps	320 × 240 @ 24 Hz (QVGA)
	Medium	10 Mbps	720 × 480 @ 30 Hz (480p)
			720 × 576 @ 25 Hz (576p)
	High	20 Mbps	1920 × 1080 @ 30 Hz (1080p)
Advanced	L0	2 Mbps	352 × 288 @ 30 Hz (CIF)
	L1	10 Mbps	720 × 480 @ 30 Hz (NTSC-SD)
			720 × 576 @ 25 Hz (PAL-SD)
	L2	20 Mbps	720 × 480 @ 60 Hz (480p)
			1280 × 720 @ 30 Hz (720p)
	L3	45 Mbps	1920 × 1080 @ 24 Hz (1080p)
			1920 × 1080 @ 30 Hz (1080i)
			1280 × 720 @ 60 Hz (720p)
	L4	135 Mbps	1920 × 1080 @ 60 Hz (1080p)
			2048 × 1536 @ 24 Hz

Table 9.57: VC-1 Profiles and Levels.

Feature	Simple	Main	Adv.
Baseline intra frame compression	Yes	Yes	Yes
Variable-sized transform	Yes	Yes	Yes
16-bit transform	Yes	Yes	Yes
Overlapped transform	Yes	Yes	Yes
4 motion vector per macroblock	Yes	Yes	Yes
1/4 pixel luminance motion compensation	Yes	Yes	Yes
1/4 pixel chrominance motion compensation	No	Yes	Yes
Start codes	No	Yes	Yes
Extended motion vectors	No	Yes	Yes
Loop filter	No	Yes	Yes
Dynamic resolution change	No	Yes	Yes
Adaptive macroblock quantisation	No	Yes	Yes
B frames	No	Yes	Yes
Intensity compensation	No	Yes	Yes
Range adjustment	No	Yes	Yes
Field and frame coding modes	No	No	Yes
GOP Layer	No	No	Yes
Display metadata	No	No	Yes

Table 9.58: VC-1 Profiles and Included Features.

- For an inter-coded block:

1. Predict the block from the corresponding motion-compensated block in the previous frame. If the frame is bidirectional, the block can also be predicted from the motion-compensated block in the next frame.
2. Predict the motion vectors themselves from the motion vectors of neighboring blocks and encode the motion vector differences with a variable-length code.
3. Apply one of the four sizes of the transform to the prediction error of the block.
4. Quantize the resulting transform coefficients, scan the results in a zigzag pattern, and encode the run-lengths and nonzero coefficients with a variable-length code.

Figure 9.59 shows how an 8×8 intra block is encoded after an overlap operation. Figure 9.60 shows the main steps in the encoding of an inter block.

9.11.3 Comparison of Features

Today (early 2009) there are quite a few competing algorithms, implementations, and standards for video compression. Many experts may agree that the most important participants in this field are MPEG-2 (Section 9.6), H.264/AVC (Section 9.9), and the newcomer VC-1. This short section discusses ways to compare the main features of these standards.

What important features of a codec should be included in such a comparison? One way to compare codecs is to list the features each has or is missing. A more useful comparison should try to compare (1) the quality of the decompressed data, (2) the compression ratios (or equivalently, bitrates) achieved, and (3) the time requirements for both encoding and decoding of each implementation.

Quality comparison. The algorithms being compared include quantization as an important encoding step and are therefore lossy. The decompressed data is not identical to the original and may even contain visible distortions and compression artifacts. It is therefore important to have a well-defined process that can measure the quality of a given compression. Notice that the quality depends on the specific implementation, and not just on the algorithm. A poor implementation may result in poor compression quality regardless of the sophistication of the underlying algorithm.

The chief aim of a lossy video encoder is to lose only data that is perceptually irrelevant; data whose absence would not be noticed by a viewer. However, viewers (and people in general) rarely agree on the quality of video (or on the quality of anything else for that matter). Thus, any test for compression quality that relies on the subjective judgment of people has to be carefully planned, has to involve several judges, and should be repeated several times, with several different video streams screened under different conditions.

Because of these difficulties, some testers rely mostly on objective measures of quality, measures that yield a single number that depends on the differences between the original and decompressed video streams. One such measure is the peak signal to noise ratio (PSNR). The PSNR for two $N \times M$ images (original and reconstructed) with n-bit pixels is defined by

$$\text{PSNR}_{\text{dB}} = 10 \log_{10} \frac{(2^n - 1)N \times M}{\sum_{i=0}^{N-1} \sum_{j=0}^{M-1} (P_{i,j} - Q_{i,j})^2}, \tag{9.4}$$

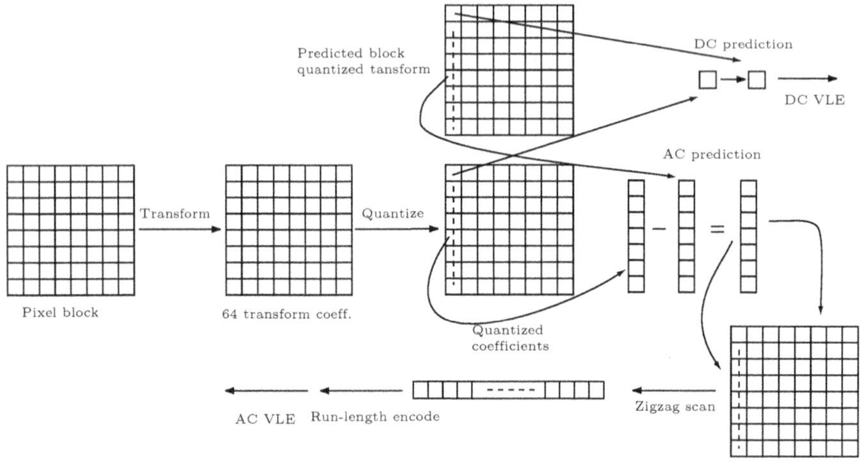

Figure 9.59: Encoding an Intra Block.

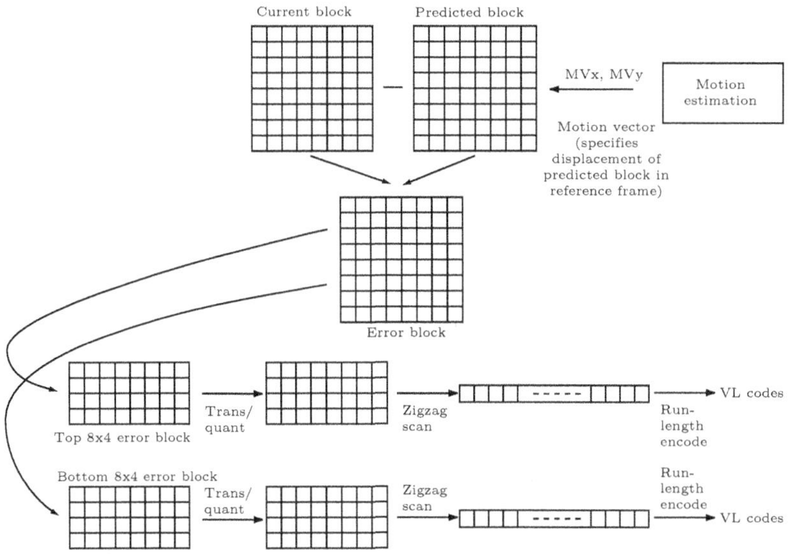

Figure 9.60: Encoding an Inter Block.

where $P_{i,j}$ and $Q_{i,j}$ are the original and reconstructed values of pixel (i,j), respectively. This expression is slightly different from Equation (7.3), but it is also dimensionless.

The PSNR is the ratio of the maximum value of a signal (255 in the case of 8-bit pixels) to the quantization noise. Thus, the lower the PSNR, the noisier the signal. Lower compression ratios (i.e, higher bitrates) generally correspond to less noise and therefore higher PSNR. For the purposes of measuring compression quality, it therefore makes sense to plot the PSNRs for many bitrates. Such a graph (which is a rate-distortion curve) would illustrate the behavior of the codec over a wide range of bitrates (or equivalently, a range of encoder settings).

When objective measures, such as the PSNR, alone are deemed insufficient, the testers may resort to subjective measures of quality that involve human judgment. One such test, comparing codecs for the next-generation optical disc standard, was conducted in 2002 by the DVD forum. Viewers from several interested organizations were asked to rate each video clip for (1) resolution, (2) noise, and (3) overall impression, on a scale of 1 to 5 for each measure. The aim was to select an algorithm and an implementation, so several video clips—each compressed and decompressed by MPEG-2, MPEG-4, H.264, and WMV9—were anonymously screened to the participants. The results confirmed VC-1 (in its WMV implementation) as the leader in all three categories.

Other independent subjective quality tests, performed by *DV Magazine* (located at `dv.com/magazine`), Tandberg television (`tandbergtv.com`), *C'T Magazine* (German, `heise.de/ct`), and European Broadcasting Union (EBU, `www.ebu.ch`) also confirmed VC-1 as a leader.

Complexity comparison. Currently, there are many simple devices that can decompress video and send it to a screen in real time. The fastest way to do this is to implement the decoder in hardware, which is why a video codec whose decoder is simple to implement would be a winner in any complexity comparison. Here, VC-1 (and also H.264) are often losers, because they are more complex than MPEG-2.

The remainder of this section provides more details on the most important building blocks and features of VC-1.

9.11.4 The Input and Output Streams

- VC-1 expects its input stream to be in the YUV 4:2:0 format. YUV is a color space (Section 9.1.1) that is related to the familiar RGB by the simple expressions

$$\begin{aligned} Y &= 0.299R' + 0.587G' + 0.114B', \\ U &= -0.147R' - 0.289G' + 0.436B' = 0.492(B' - Y), \\ V &= 0.615R' - 0.515G' - 0.100B' = 0.877(R' - Y), \end{aligned}$$

whose inverse transform is

$$\begin{aligned} R' &= Y + 1.140V, \\ G' &= Y - 0.394U - 0.580V, \\ B' &= Y - 2.030U. \end{aligned}$$

The Y component of YUV is called luma (brightness) and the U and V components are color differences or chroma (they specify the color). The PAL television standard also uses YUV.

The three components of the RGB color space designate the intensities of the primary colors. However, lossy image and video codecs should lose information to which the eye is not sensitive, which is why such codecs prefer a color space that has luma (or luminance, Section 7.10.1) as a component. The human visual system is much more sensitive to variations in brightness than in color, which is why a video codec can lose more information (by quantization) in the chroma components, while keeping the luma component intact or almost so. Losing information is done either by quantization or by subsampling (deleting one or two color components in a certain percentage of the pixels).

Notations such as 4:2:0 refer to subsampling pixels (or decimating the chroma components, see for example [Poynton 08]). This is illustrated in Figure 9.61. The 4:4:4 notation means no subsampling. Every pixel contains all three color components. 4:2:2 is the case where the chroma components are subsampled horizontally by a factor of 2. Similarly, a horizontal subsampling by a factor of 4 is indicated by 4:1:1. Finally, 4:2:0 is the case where pixels are subsampled in both directions by a factor of 2. This is equivalent to saying that the pixels retain only their luma component and a virtual chroma pixel is positioned between the rows and columns of the real pixels. There are actually three versions of 4:2:0 that differ in the precise positioning of the virtual chroma components.

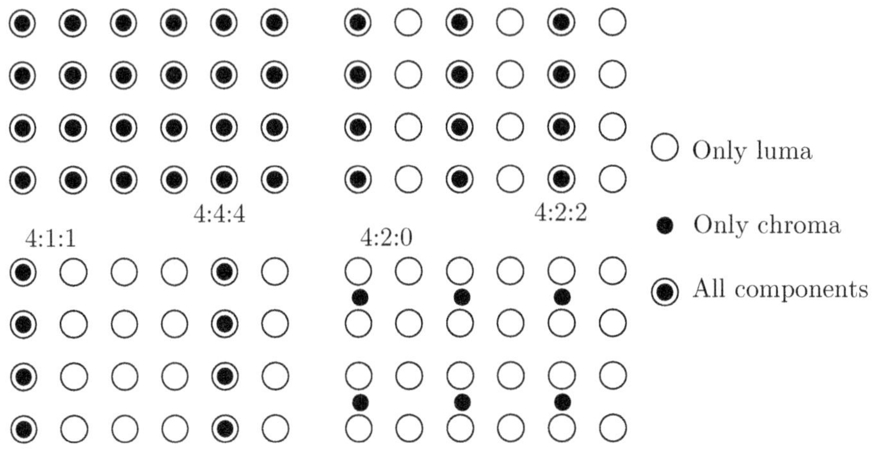

Figure 9.61: Subsampling Pixels.

The compressed stream output by VC-1 is also in YUV 4:2:0 format. It features a hierarchy of the following layers: sequence, picture, entry point, slice, macroblock, and block (the slice and entry point layers exist only in the advanced profile). A sequence is a set of pictures, a picture is a set of slices, and a slice consists of rows of macroblocks. The entry point layer is very useful. It makes it possible to enter the compressed stream at any point and also to skip parts of the video.

9.11.5 Profiles and Levels

As is the norm with important compression standards, the operations of the VC-1 encoder are not described in the official publications. This allows implementors to come up with original, sophisticated, and fast algorithms for the encoder's basic building blocks—the prediction (including both intra and inter prediction, where inter prediction means the motion estimation and compensation), the transform, and the entropy coding. A general VC-1 encoder can therefore be quite complex, which is one reason why the designers incorporated the concepts of profiles and levels in VC-1.

A profile is a precisely defined subset of a compression standard. It includes some of the tools, algorithms, parameters, and syntax of the overall standard, and it restricts the values that parameters may take. A profile may further be divided into levels which are subsubsets of the standard.

The VC-1 definition specifies three profiles, simple, main, and advanced. The simple profile has two levels and is intended for low bitrate applications, such as mobile video communication devices, security cameras, and software for video playback on personal computers. The main profile is intended for higher bitrate internet applications, such as video delivery over IP. This profile has three levels. The advanced profile is aimed for broadcast applications, such as digital television, DVDs, of HDTV. This profile supports interlaced video and contains five levels. Thus, someone trying to implement VC-1 for a mobile telephone or a portable video player may decide to include only the features of the simple profile, while engineers designing a new DVD format might consider to go all the way with a VC-1 encoder operating in the advanced profile.

Each profile has levels that place constraints on the values of the parameters used in the profile. In this way, a level limits the video resolution, frame rate, bitrate, buffer sizes, and the range of motion vectors.

9.11.6 Motion Compensation

As with other video encoders, the motion compensation step is the most elaborate and time consuming part of the encoder. The principle of motion compensation is to locate, for each block B in the current frame F, the most similar block S among the reference frames of F. Once found, S is subtracted from B and the difference, which is small, is transformed, quantized, and entropy encoded.

Motion compensation is beneficial because a video frame tends to be similar to its immediate predecessors and successors. Often, the only difference between a frame and the preceding frame is the location of certain objects that have moved sightly between the frames. Thus, the chance of finding blocks similar to the current block increases when the blocks are small. Similarly, a large number of reference frames also increases the chance of finding good matches. As usual, however, there is a tradeoff. Smaller blocks (and therefore a larger number of blocks) and many reference frames imply more work and increased complexity for the encoder. VC-1 allows for only two reference frames and a maximum of two block sizes.

When a frame of type inter is encoded, its macroblocks are motion compensated with either 16×16 blocks and one motion vector (MV) for each block, or with four 8×8 blocks, each with its own MV. Notice that the MVDs (MV differences) have to transmitted to the decoder. An MV specifies the xy coordinates of the matched block S relative to the current block B. Normally, the coordinates are measured in pixels, but VC-1 allows for coordinates whose units are $1/2$ pixels or even $1/4$ pixels (and this choice has also to be transmitted to the decoder).

Intensity Compensation is one of the many innovations and complexities introduced in VC-1. This is a special mode that the encoder enters when it decides that the effectiveness of motion compensation can be improved by scaling the original color. In this mode, the encoder scales the luma and chroma color components of the reference frame before using them in motion compensation.

9.11.7 Pixel Interpolation

The principles of motion compensation are discussed in Section 9.5. The idea is to predict a block from the corresponding blocks in previous (and sometimes also future) frames. Modern video codecs extend this idea by artificially increasing the resolution of each block before it is predicted. This is done by interpolating the values of pixels. In its simplest form, interpolation starts with a group of 2×2 pixels, computes their average, and considers this average a new, fractional, pixel at the center of the group. In VC-1, pixels can be interpolated to half-pixel or even quarter-pixel, as illustrated by Figure 9.62a,b. The black squares in the figure are pixels, the white squares are half pixels, and the circles are quarter pixels.

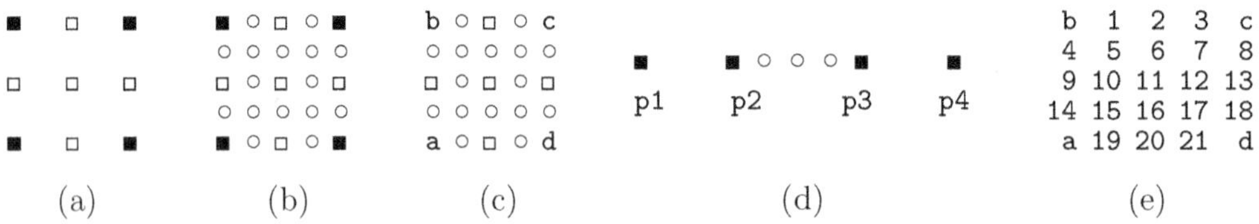

Figure 9.62: Fractional Pixels.

VC-1 uses bilinear and bicubic pixel interpolation methods. Instead of listing the complex rules and interpolation formulas employed by VC-1, this short section tries to explain how the formulas and the many constants they include were derived.

We start with bilinear interpolation (see [Salomon 06] for more information and figures). Figure 9.62c shows a group of 2×2 pixels labeled a through d, with five half pixels and 16 quarter pixels to be linearly interpolated. We imagine that the four original pixels are equally-spaced three-dimensional points (where the values of the pixels are the z coordinates of the points) and we use linear operations to construct a rectangular surface patch whose four corners are these points. We connect points a and d with the straight line $L_1(u) = \mathtt{a}(1-u) + \mathtt{d}u$ and then connect points b and c with the straight line $L_2(u) = \mathtt{b}(1-u) + \mathtt{c}u$. Notice that $L_1(0) = \mathtt{a}$, $L_1(1) = \mathtt{d}$, and similarly for L_2. The entire bilinear surface $P(u,w)$ is obtained by selecting a value for parameter w, selecting the two points $P_1 = L_1(w)$ and $P_2 = L_2(w)$, and connecting P_1 and P_2 with a straight line $P(u,w) = P_1(1-w) + P_2 w$. The entire surface (Figure 9.63 is an example) is obtained when u and w vary independently in the interval $[0,1]$.

We now consider the pixel grid of Figure 9.62c a two-dimensional coordinate system with point a as its origin, parameter u as the x axis, and parameter w as the y axis. With this notation, it is easy to see that the half pixels (from top to bottom and left to right) are the five points $P(.5,1)$, $P(0,.5)$, $P(.5,.5)$, $P(1,.5)$, and $P(.5,0)$. The two quarter pixels on the top row are the surface points $P(.25,1)$ and $P(.75,1)$, and the remaining 14 quarter pixels have similar coordinates.

Figure 9.64 lists Mathematica code for bilinear interpolation, as well as a few results for half- and quarter pixels. It is easy to see how some fractional pixels depend on only two of the four original pixels, while others depend on all four. Notice that the formal definition of the VC-1 standard employs integers (followed by a right shift) instead of the real numbers listed in our figure. For point $P(0.5, 0.25)$, for example, a VC-1

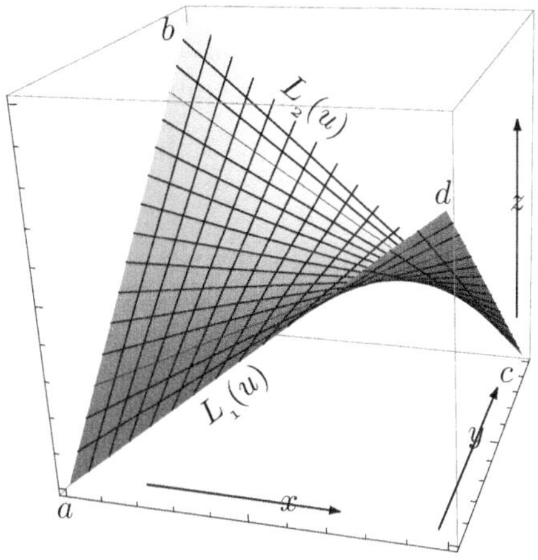

Figure 9.63: A Bilinear Surface.

implementation computes the integer expression $(6a + 2b + 2c + 6d + 2) >> 4$ instead of the equivalent real expression $0.375a + 0.125b + 0.125c + 0.375d$.

```
bil[u_,w_]:=a(1-u)(1-w)+b(1-u)w+d(1-w)u+c u w;
bil[.25, .25]
bil[.75, .5]
bil[.25, 1]
bil[.5, .25]

0.5625a+0.1875b+0.0625c+0.1875d
0.125a+0.125b+0.375c+0.375d
0.75b+0.25c
0.375a+0.125b+0.125c+0.375d
```

Figure 9.64: Code For Bilinear Interpolation.

The principles of bicubic interpolation (we are interested in the one-dimensional case) are discussed in Section 7.25.5. The idea is to start with the values of four consecutive pixels, consider them the y coordinates of four equally-spaced two-dimensional points $p1$ through $p4$, and construct a polynomial $P(t)$ that passes through the points when t varies from 0 to 1. This polynomial is given by Equation (7.45), duplicated here

$$\mathbf{G}(t) = (t^3, t^2, t, 1) \begin{pmatrix} -4.5 & 13.5 & -13.5 & 4.5 \\ 9.0 & -22.5 & 18 & -4.5 \\ -5.5 & 9.0 & -4.5 & 1.0 \\ 1.0 & 0 & 0 & 0 \end{pmatrix} \begin{pmatrix} p1 \\ p2 \\ p3 \\ p4 \end{pmatrix}. \tag{7.45}$$

The four original points are obtained for t values 0, 1/3, 2/3, and 1 and Figure 9.62d makes it clear that we are interested in the three pixels that are equally spaced between

$p2$ and $p3$. These correspond to t values 0.44444, 0.5, and 0.55555, respectively. The Mathematica code of Figure 9.65 shows how this computation is carried out. In order to speed up the decoder, the definition of VC-1 employs integers and a right shift instead of the real numbers listed in the figure.

```
g[t_]:={t^3, t^2, t, 1}.
{{-4.5,13.5,-13.5,4.5},{9.,-22.5,18.,-4.5},
{-5.5,9.,-4.5,1.},{1.,0,0,0}}.{p1,p2,p3,p4};
g[0.44444]
g[0.5]
g[0.55555]

-0.0617277p1+0.740754p2+0.370355p3-0.0493812 p4
-0.0625p1+0.5625p2+0.5625p3-0.0625p4
-0.0493846p1+0.37039p2+0.740724p3-0.0617293p4
```

Figure 9.65: Code For Bicubic Interpolation.

Section 7.25.7 shows how cubic interpolation can be extended to two dimensions, but VC-1 specifies this process differently. The VC-1 encoder computes the bicubic interpolation of half- and quarter pixels in several steps. First, this type of interpolation is performed horizontally to compute the three pixels 1, 2, 3 between `b` and `c` (Figure 9.62e), using, in addition to `b` and `c`, the values of the pixel to the left of `b` (as our $p1$) and the pixel to the right of `c` (as our $p4$). Next, the three pixels 19, 20, and 21 between `a` and `d` are computed in a similar way. Using the values of 1 (and the corresponding pixel in the 5×5 grid above) and 19 (and the corresponding pixel in the grid below), it is possible to interpolate for pixels 5, 10, and 15. The values of 6, 11, and 16, and 7, 12, and 17 are similarly computed. Finally, the values of 4, 9, and 14 are computed from pixels `b` (and the one above it) and `a` (and the one below it), and similarly for 8, 13, and 18. Another tough job for the VC-1 implementor!

9.11.8 Deblocking Filters

The VC-1 encoder transforms blocks of pixels and it has long been known that this introduces boundary artifacts into the decompressed frame. Imagine two adjacent identical pixels with a value of, say, 100. Assume further that a block boundary passes between the pixels. After being transformed and quantized by the encoder and later reverse-transformed and dequantized by the decoder, one pixel may end up as 102, while the other may become 98. The difference between an original and a reconstructed pixel is 2, but the difference between the two reconstructed pixels is double that. Blocking artifact can also be caused by a poor quantization of the DC coefficient, causing the intensity of a decoded block to become lower or higher than the intensities of its neighbors.

VC-1 introduces two innovative techniques to reduce the effect of blocking artifacts. They are referred to as overlapped transform (OLT) and in-loop deblocking filter (ILF).

OLT acts as a smoothing filter. It is designed to perform best when two adjacent 8×8 blocks differ much in their quality. Imagine a block P on the left and a block Q on the right with a vertical boundary between them (alternatively, P may be directly above

Q, with a horizontal boundary between them). If P and Q differ in their qualities (one is high quality and the other is low quality), then the encoder applies the OLT filter to smooth their boundary. This operation is carried out only on intra-coded blocks, which are always 8×8. The formal definition places many restrictions on the OLT filter, so the encoder has to signal to the decoder the presence of this operation.

Once the encoder decides to apply OLT to two adjacent blocks (the codec definition, as usual, does not specify how the encoder should make this decision), it starts with the four edge pixels P1, P0, Q0, and Q1 shown in Figure 9.66 (for two blocks with a horizontal boundary, the figure should be rotated 90°). For each of the eight rows of the blocks, the encoder computes new values for these pixels, based on their original values and on certain rounding parameters, and substitutes the new values for the old ones. The decoder performs the reverse operations. Specifically, the computation carried out by the decoder is

$$\begin{bmatrix} P1' \\ P0' \\ Q0' \\ Q1' \end{bmatrix} = \left\{ \begin{bmatrix} 7 & 0 & 0 & 1 \\ -1 & 7 & 1 & 1 \\ 1 & 1 & 7 & -1 \\ 1 & 0 & 0 & 7 \end{bmatrix} \begin{bmatrix} P1 \\ P0 \\ Q0 \\ Q1 \end{bmatrix} + \begin{bmatrix} r0 \\ r1 \\ r0 \\ r1 \end{bmatrix} \right\} \gg 3,$$

where the notation $\gg$ means a right shift and the two rounding parameters $r0$ and $r1$ take the values $(3, 4)$ for even-numbered rows of pixels and $(4, 3)$ for odd-numbered rows.

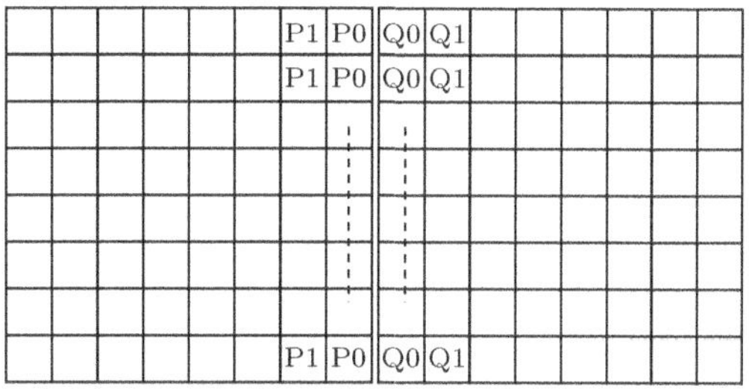

Figure 9.66: Edge Pixels For OLT.

When two adjacent blocks have similar qualities, the VC-1 standard recommends the use of ILF. Notice that an in-loop filter improves prediction since it filters frames that will be used for reference (this is different than a post-filter). For blocks in I and B frames, the ILF filter should be applied to all four boundaries of the 8×8 blocks in the following way. First, ILF should be applied to every 8th and 9th horizontal rows of the frame, and then to every 8th and 9th vertical columns. If the rows are numbered 0 through $N-1$, then ILF should be applied to rows $(7, 8)$, $(15, 16)$,..., $\big((8(N-1)-1, 8(N-1)\big)$, and similarly for the columns (Figure 9.67). Those rows and columns form the block boundaries in the frame.

For blocks in P frames, the situation is more complex, because blocks can be of four different sizes, as illustrated by Figure 9.69a. The encoder therefore has to work harder to identify the pixels that constitute block boundaries, as shown in part (b) of

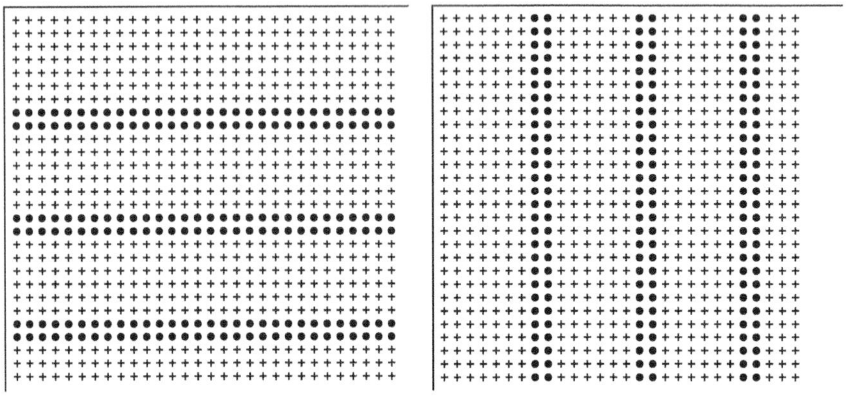

Figure 9.67: ILF Boundary Rows and Columns For I and B Frames.

the figure. The figure shows examples where filtering between neighboring blocks (or subblocks) does or does not occur. If the motion vectors of adjacent blocks are different, then the boundary between them is always filtered. Thus, our example assumes that the motion vectors of the blocks are identical. The shaded blocks or subblocks in the figure are those that have at least one nonzero coefficient, while the clear blocks or subblocks have no transform coefficients. The thick lines designate boundaries that should be filtered. The order of filtering is important and is specified in detail in the VC-1 definition. The figure illustrates only horizontal block neighbors. The handling of vertical block neighbors is similar.

Once the encoder identifies the block boundaries that should be filtered, the filtering operation can proceed. Figure 9.68 illustrates this process. Filtering is always done on a segment of 2×4 (horizontal) or 4×2 (vertical) pixels. Thus, if a boundary is eight pixels long, its filtering becomes a two-step process. In each four-pixel segment, the third pair of pixels is filtered first and the result of this operation determines whether the other three pairs should also be filtered.

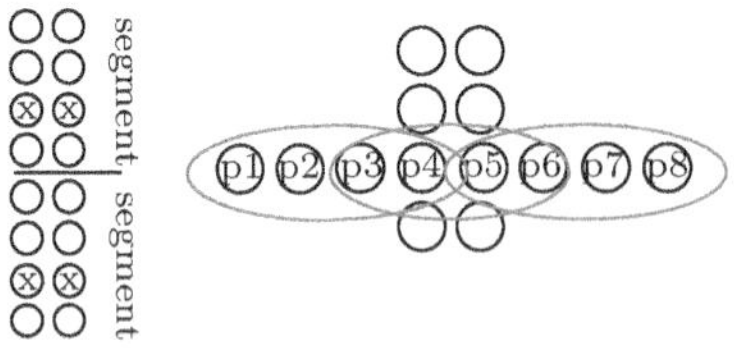

Figure 9.68: ILF Filtering.

The rest of the filtering operation differs slightly for P frames and for the other frame types. It is described here in essence but not in detail. The third pair and its six near neighbors are split into three groups as shown in the figure and several expressions involving differences of pairs of pixels in each group are computed. The idea is to discover any blocking effects caused by differences between pixels on both sides of the boundary. These differences are examined to see whether the remaining three pairs should also

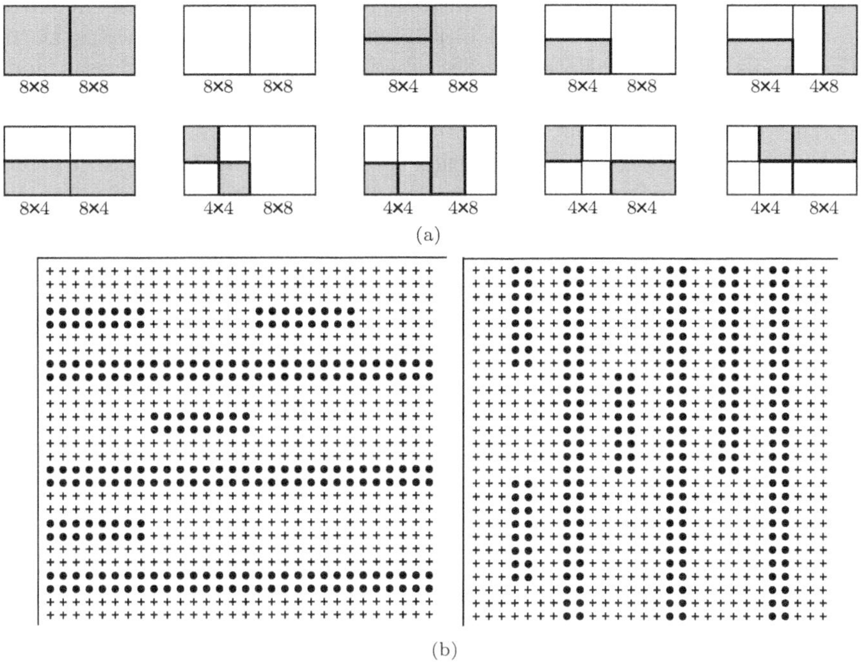

Figure 9.69: ILF Boundary Rows and Columns For P Frames.

be filtered. The two pixels p4 and p5 that are closest to the boundary are updated according to

$$d = (p4 - p5)/2, \quad p4 = p4 - d, \quad p5 = p5 + d.$$

If the decision is made to filter the other three pairs, they are updated in a similar way. In each pair, pixel differences are computed and the two pixels closest to the boundary are updated.

9.11.9 The Transform

The transform is a central component of VC-1. It is an integer transform that is based on the discrete cosine transform (DCT). Clearly, an integer transform is easier to implement and is faster to execute. Its precision, however, depends on the size of the integers in question, but each color component of a pixel is typically only eight bits long, so a 16-bit transform provides enough precision to generate intermediate results that are much bigger than an original pixel.

It has already been mentioned that the transform operates on four different block sizes, as shown in Figure 9.56. This is one of the many innovations of VC-1 and it is an option of the encoder (that is signaled at the sequence layer). Blocks are first motion compensated and then transformed. The intra blocks (blocks in macroblocks in I frames) always use the 8×8 transform, while the inter blocks (in the other types of frames) may use any of the four transform sizes (and this information is stored in the picture, macroblock, or block layers). When the transform size appears in the picture layer, all the blocks in the picture use the same transform size. When the size appear at the macroblock layer, all the blocks of the macroblock use the same transform size.

The details of this transform are described here, based on [Srinivasan and Regunathan 05], but the interested reader should first become familiar with the concept of orthogonal transforms (Section 7.7) and especially with the matrix of Equation (7.6) and with the DCT matrix in one dimension for $n = 8$, Table 7.38, shown here as matrix (9.5) (unnormalized) and (9.6) (normalized). Notice that the odd-numbered rows of this matrix have only three distinct elements and the even-numbered rows consist of another set of four distinct elements.

$$\begin{bmatrix} 1. & 1. & 1. & 1. & 1. & 1. & 1. & 1. \\ 0.981 & 0.831 & 0.556 & 0.195 & -0.195 & -0.556 & -0.831 & -0.981 \\ 0.924 & 0.383 & -0.383 & -0.924 & -0.924 & -0.383 & 0.383 & 0.924 \\ 0.831 & -0.195 & -0.981 & -0.556 & 0.556 & 0.981 & 0.195 & -0.831 \\ 0.707 & -0.707 & -0.707 & 0.707 & 0.707 & -0.707 & -0.707 & 0.707 \\ 0.556 & -0.981 & 0.195 & 0.831 & -0.831 & -0.195 & 0.981 & -0.556 \\ 0.383 & -0.924 & 0.924 & -0.383 & -0.383 & 0.924 & -0.924 & 0.383 \\ 0.195 & -0.556 & 0.831 & -0.981 & 0.981 & -0.831 & 0.556 & -0.195 \end{bmatrix}. \tag{9.5}$$

$$\begin{bmatrix} 0.3536 & 0.3536 & 0.3536 & 0.3536 & 0.3536 & 0.3536 & 0.3536 & 0.3536 \\ 0.4904 & 0.4157 & 0.2778 & 0.0975 & -0.0975 & -0.2778 & -0.4157 & -0.4904 \\ 0.4619 & 0.1913 & -0.1913 & -0.4619 & -0.4619 & -0.1913 & 0.1913 & 0.4619 \\ 0.4157 & -0.0975 & -0.4904 & -0.2778 & 0.2778 & 0.4904 & 0.0975 & -0.4157 \\ 0.3536 & -0.3536 & -0.3536 & 0.3536 & 0.3536 & -0.3536 & -0.3536 & 0.3536 \\ 0.2778 & -0.4904 & 0.0975 & 0.4157 & -0.4157 & -0.0975 & 0.4904 & -0.2778 \\ 0.1913 & -0.4619 & 0.4619 & -0.1913 & -0.1913 & 0.4619 & -0.4619 & 0.1913 \\ 0.0975 & -0.2778 & 0.4157 & -0.4904 & 0.4904 & -0.4157 & 0.2778 & -0.0975 \end{bmatrix}. \tag{9.6}$$

The first thing that comes to mind, when trying to derive an integer version of the DCT, is to round off the elements of Equation (9.5) to the nearest integer, but it is easy to see that this would simply result in a matrix of zeros and 1's, similar to the basic orthogonal transform matrix of Equation (7.6). It would still be an orthogonal transform, but without the useful energy compaction properties of the DCT (however, note that energy compaction is not a critical issue here, because the DCT operates on the residuals of the prediction). Thus, a different approach is needed. The designers of the VC-1 transform started with the following considerations:

- Video encoding can sometimes be slow, but decoding must be done at real time. Video decoding is often done by small, specialized processors that are embedded in DVD players, digital televisions, home theaters, handheld mp3 and video devices, and other devices and instruments designed for video playback. Currently, most such devices are based on 16-bit architectures.

- It is known from long experience with video compression that the inverse transform is generally responsible for 10–30% of the CPU time used by a video decoder. A single video frame of 1920 × 1080 pixel (the so-called HDTV 1080p) consists of 32,400 8 × 8 blocks that have to be inverse transformed. At 30 frames per second, the decoder has to inverse-transform 972,000 blocks each second; a tough job for even fast processors.

These two considerations imply that a 16-bit integer transform can be very useful if it also offers energy compaction equivalent to that of the DCT. The integer transform developed for VC-1 allows for fast decoding. The decoder performs the inverse transform using just 16-bit integer arithmetic. The encoder, on the other hand, can compute the forward transform with fixed-point (scaled integer) or even floating-point arithmetic. (The encoder can also benefit from a simple transform, because it has to try all the valid block sizes, such as 8×8 and 4×8, in order to choose the best.)

A full understanding of this transform starts with the concept of *norm*. In mathematics, the term "norm" is defined for certain mathematical objects that are more complex than a single number. Intuitively, the norm of an object is a quantity that describes the length, magnitude, or extent of the object. Norms are defined for complex numbers, quaternions, vectors, matrices, and other objects that are not just numbers. Often, several different norms are defined for the same type of object.

The norm of a matrix can be defined in several ways, but perhaps the most common definition is the Frobenius norm (older texts refer to it as the Hilbert-Schmidt norm or the Schur norm). Given an $n \times m$ matrix A with elements a_{ij}, its Frobenius norm is

$$||A|| = \left[\sum_{i=1}^{n}\sum_{j=1}^{m} |a_{ij}|^2\right]^{1/2}.$$

Similarly, the norm of a vector is the square root of the sum of the squares of its elements. These norms are very similar to the concept of energy, defined in Section 7.6.

Given a vector $V = (2, 2, 2)$ and a 3×3 matrix A all whose elements are 10, the result of the product $V \cdot A$ is the vector $W = (60, 60, 60)$, and it is easy to verify that $||W|| = ||V|| \cdot ||A|| = 60\sqrt{3}$. We therefore conclude that multiplying a vector by a matrix A scales its norm by (approximately) a factor of $||A||$.

The two-dimensional DCT (Equation (7.16), duplicated here) can be written in matrix notation as the product of a row of m functions $\cos[(2y+1)j\pi/2m]$, an $n \times m$ matrix of pixels p_{xy}, and a column of n functions $\cos[(2x+1)i\pi/2n]$. Thus

$$\begin{aligned} G_{ij} &= \sqrt{\frac{2}{m}}\sqrt{\frac{2}{n}}C_iC_j\sum_{x=0}^{n-1}\sum_{y=0}^{m-1} p_{xy}\cos\left[\frac{(2y+1)j\pi}{2m}\right]\cos\left[\frac{(2x+1)i\pi}{2n}\right] \\ &= \left[\text{row of } \cos\left[\frac{(2y+1)j\pi}{2m}\right]\right] \begin{bmatrix} p_{00} & & \\ & \ddots & \\ & & p_{nm} \end{bmatrix} \begin{bmatrix} \text{column} \\ \text{of} \\ \cos\left[\frac{(2x+1)i\pi}{2n}\right] \end{bmatrix}, \end{aligned} \quad (9.7)$$

Each coefficient G_{ij} is therefore the product of a row of cosines, a block of pixels, and a column of cosines. An entire block of transform coefficients G_{ij} is therefore the product of the one-dimensional DCT matrix (Equation (7.14)), the corresponding block of pixels, and the transpose of the same DCT matrix. Our task is to replace the DCT matrix with a matrix T of integers, and the earlier discussion of norms makes it clear that the resulting transform coefficients G_{ij} will be bigger than the original pixels by a factor of approximately $||T||^2$.

We denote $L \stackrel{\text{def}}{=} \log_2 ||T||$. We assume that the prediction residuals are 9-bit signed integers, i.e., in the interval $[-256, 255]$. We require that the transform coefficients be 12-bit signed integers. Transforming a block of pixels therefore increases the magnitude of a pixel by three bits. Because transforming is done by two matrix multiplications (by the transform matrix T and its transpose), we conclude that each matrix multiplication increases the magnitude of pixels by 1.5 bits on average.

To a large extent, decoding is the reverse of encoding, but the integer nature of the transform creates a problem. Decoding starts with 12-bit signed transform coefficients and has to end up with 9-bit signed integer pixels, but decoding is done by two matrix multiplications which *increase* the size of the numbers involve. Thus, each matrix multiplication should be followed by a right shift (denoted by $>>$) which shrinks the results from 12 bits to 10.5 bits and then to nine bits. Assuming that decoding starts with a matrix A of 12-bit transform coefficients, the first step is to multiply $B = T \cdot A$ and shift $C = B >> n$, where the value of n should be such that the elements of C will be 10.5 bits long on average. The second step is another multiplication $D = C \cdot T'$ followed by another right shift $E = D >> m$, where m should be chosen such that the elements of E are 9-bit integers.

This sequence of transformations and magnitude changes is achieved if we require that $||T||^2 = 2^{n+m}$, which implies $2L = n + m$. Also, the dynamic range (the range of values) of matrix B should be $10.5 + L$ and the range of D should be $9 + 2L - n$. These conditions should be achieved within the constraints of 16-bit arithmetic, which results in the following relations

$$2L = n + m, \quad 10.5 + L \le 16, \quad 9 + 2L - n \le 16.$$

One solution of these relations is $L \le 5.5$ (which implies $||T||^2 = 2^{11} = 2{,}048$) and $m = 6$.

These values are enough to place a limit on the DC coefficient of the transform. Equation (9.7) implies that the DC coefficient G_{00} is obtained by multiplying the pixels by the top row of the transform matrix, a row whose elements are identical. If we denote the (yet unknown) element on the top row of our 8×8 integer transform matrix T by d_8, then the norm of that row is $\sqrt{8d_8^2}$. This norm must be less than or equal the norm of T, so it must satisfy $8d_8^2 \le 2{,}048$ or $d_8 \le 16$. For the 4×4 transform matrix, where each row is half the size of the 8×8 matrix, the element of the top row is denoted by d_4 and the norm requirement is $d_4 = \sqrt{2}d_8$. It is now easy to check every pair of integers (d_4, d_8) for the constraints $d_8 \le 16$ and $d_4 = \sqrt{2}d_8$, and a quick search verifies that the only pairs that come within 5% of the constraints are $(7, 5)$, $(10, 7)$, $(17, 12)$ and $(20, 14)$.

The next task is to determine the remaining seven rows of the 8×8 integer transform matrix. We first examine the even-numbered rows and it is easy to see that they contain only four distinct elements that we denote by $C1$ through $C4$. These rows form the 4×8 matrix

$$\begin{bmatrix} C1 & C2 & C3 & C4 & -C4 & -C3 & -C2 & -C1 \\ C2 & -C4 & -C1 & -C3 & C3 & C1 & C4 & -C2 \\ C3 & -C1 & C4 & C2 & -C2 & -C4 & C1 & -C3 \\ C4 & -C3 & C2 & -C1 & C1 & -C2 & C3 & -C4 \end{bmatrix}. \tag{9.8}$$

The following conditions were set by experts in order to determine the four elements (the curious reader might want to glance at Section 8.7.1, for a list of similar conditions used to determine the values of wavelet filter coefficients):

1. The rows of T must be orthogonal. We know that each even-numbered row is orthogonal to the odd-numbered rows, because of the properties of the cosine function. Thus, we have to choose the four elements such that the four even-numbered rows will be orthogonal.

2. The vector $(C1, C2, C3, C4)$ should correlate well with the special DCT vector $(0.4904, 0.4157, 0.2778, 0.0975)$. The term "correlate" refers to the cosine of the angle between the vectors, as measured by their dot product. The DCT vector above was obtained in the following way. The DCT matrix of Equation (9.6) is normalized. Normalization is done by dividing each row by its length, so it becomes a vector of length 1. When this is done to the second row, it becomes the unit vector
$(0.4904, 0.4157, 0.2778, 0.0975, -0.0975, -0.2778, -0.4157, -0.4904)$ and the DCT vector above is the first half of this (its length is therefore 0.5).

3. The norm $C1^2 + C2^2 + C3^3 + C4^2$ of the four elements should match that of d_8, i.e., it should be as close as possible to $d_8^2 + d_8^2 + d_8^2 + d_8^2 = 4d_8^2$. The four choices of d_8 listed earlier yield the four values 100, 196, 576, and 784 for $4d_8^2$.

How can we derive the set of four integers that come closest to satisfying the three conditions above? The simplest approach is to use brute force (i.e., the high speed of our computers). We loop over all the possible sets of four positive integers and select the sets that satisfy $C1^2 + C2^2 + C3^3 + C4^2 \leq 800$. (There is a very large number of such sets, but it is finite.) For each set, we check to see if the four rows of matrix (9.8) are orthogonal. If so, we compute the dot product of condition 2 above and save it. After all possible sets of four integers have been examined in this way, we select the set whose dot product is the maximal. This turns out to be the set $(16, 15, 9, 4)$, whose norm is 578 (implying that $d_4 = 17$ and $d_8 = 12$), very close to the ideal 576. The correlation of condition 2 is 0.9984, also very encouraging.

We next examine the odd-numbered rows of Equation (9.6). We already know that $d_8 = 12$, so these four rows become the 4×8 matrix

$$\begin{bmatrix} 12 & 12 & 12 & 12 & 12 & 12 & 12 & 12 \\ C6 & C6 & -C6 & -C5 & -C5 & -C6 & C6 & C5 \\ 12 & -12 & -12 & 12 & 12 & -12 & -12 & 12 \\ C6 & -C5 & C5 & -C6 & -C6 & C5 & -C5 & C6 \end{bmatrix}$$

and require the derivation of only two elements, denoted by $C5$ and $C6$. The norm $C5^2 + C6^2$ of these elements should match (similar to condition 3 above) that of d_4, i.e., it should be as close as possible to $d_8^2 + d_8^2 = 288$, and this is satisfied, to within 1%, by the pair $(16, 6)$.

The 8×8 transform matrix T_8 has now been fully determined and is given by

$$T_8 = \begin{bmatrix} 12 & 12 & 12 & 12 & 12 & 12 & 12 & 12 \\ 16 & 15 & 9 & 4 & -4 & -9 & -15 & -16 \\ 16 & 6 & -6 & -16 & -16 & -6 & 6 & 16 \\ 15 & -4 & -16 & -9 & 9 & 16 & 4 & -15 \\ 12 & -12 & -12 & 12 & 12 & -12 & -12 & 12 \\ 9 & -16 & 4 & 15 & -15 & -4 & 16 & -9 \\ 6 & -16 & 16 & -6 & -6 & 16 & -16 & 6 \\ 4 & -9 & 15 & -16 & 16 & -15 & 9 & -4 \end{bmatrix}.$$

The 4×4 integer transform is derived in a similar way, but is much simpler. We already know that the transform matrix should look like the 4×4 matrix of Equation (7.6) and that d_4 equals 17. Thus, we are looking for a matrix of the form

$$\begin{bmatrix} 17 & 17 & 17 & 17 \\ D1 & D2 & -D2 & -D1 \\ 17 & -17 & -17 & 17 \\ D2 & -D1 & D1 & -D2 \end{bmatrix}.$$

This matrix is inherently orthogonal for any choice of elements $D1$ and $D2$, but the best choices are those that satisfy the norm condition $D1^2+D2^2 = d_4^2+d_4^2 = 578$, and a direct search yields the pair $D1 = 22$ and $D2 = 10$, that satisfy $D1^2 + D2^2 = 22^2 + 10^2 = 584$. Thus, the complete 4×4 integer transform used by VC-1 is

$$T_4 = \begin{bmatrix} 17 & 17 & 17 & 17 \\ 22 & 10 & -10 & -22 \\ 17 & -17 & -17 & 17 \\ 10 & -22 & 22 & -10 \end{bmatrix}.$$

Once the two integer transform matrices have been constructed, we are ready for the transform itself. As has already been mentioned, the forward transform is computed by the encoder, which doesn't always have to work at real time. Thus, the forward transform can be performed with fixed-point (scaled integer) or even floating-point arithmetic. We start with two normalization vectors

$$c_4 = \left(\frac{8}{289}, \frac{8}{292}, \frac{8}{289}, \frac{8}{292}\right), \quad c_8 = \left(\frac{8}{288}, \frac{8}{289}, \frac{8}{292}, \frac{8}{289}, \frac{8}{288}, \frac{8}{289}, \frac{8}{292}, \frac{8}{289}\right).$$

These are used to construct the four normalization matrices $N_{4\times4}$, $N_{8\times4}$, $N_{4\times8}$, and $N_{8\times8}$ by means of outer products (also known as tensor products) of the vectors c_4 and c_8. Thus, for example, $N_{8\times4} = c_8 \otimes c_4$.

The outer product is a vector operation whereby a matrix is constructed from the elements of two vectors. Given the two vectors $a = (a_1, a_2, \ldots, a_m)$ and $b = (b_1, b_2, \ldots, b_n)$, their outer product $a \otimes b$ is the $m \times n$ matrix

$$\begin{bmatrix} a_1b_1 & a_1b_2 & \dots & a_1b_n \\ a_2b_1 & a_2b_2 & \dots & a_2b_n \\ & & \ddots & \\ a_mb_1 & a_mb_2 & \dots & a_mb_n \end{bmatrix}.$$

In contrast, the more familiar inner product (or dot product) operates on the elements of two equal-length vectors to produce a single number.

The forward transform starts with a block D of pixels and produces a block $\hat{D}$ of transform coefficients. The transform is defined by $\hat{D} = (T_i D T_j') \circ N_{i\times j}$ where i and j take the values 4 or 8, and the notation "$\circ$" means an entry-wise multiplication of two matrices. Thus, depending on the transform block size, there are the following four cases:

$$\hat{D} = (T_4 D T_4') \circ N_{4\times4},\ \hat{D} = (T_8 D T_4') \circ N_{8\times4},\ \hat{D} = (T_4 D T_8') \circ N_{4\times8},\ \hat{D} = (T_8 D T_8') \circ N_{8\times8}.$$

The inverse transform starts with a block of 12-bit signed (i.e., in the interval $[-2{,}048, +2{,}047]$), quantized transform coefficients $D_{i\times j}$ (where i and j are 4 or 8) and results in a block $R_{i\times j}$ of 10-bit signed integers (i.e., in the interval $[-512, 511]$) that become the reconstructed pixels. The transform consists of the two steps

$$E_{i\times j} = (D_{i\times j} \cdot T_j + \mathbf{4}) >> 3, \quad R_{i\times j} = (T_i' \cdot E_{i\times j} + C_i \otimes \mathbf{1}_j + \mathbf{64}) >> 7,$$

where $E_{i\times j}$ are intermediate matrices whose entries are 13-bit signed integers (i.e., in the interval $[-4{,}096, +4{,}095]$), $\mathbf{1}_i$ is a row of i 1's, C_j are the columns $C_4 = (0,0,0,0)'$ and $C_8 = (0,0,0,0,1,1,1,1)'$, and the notation "$>>$" means an arithmetic (sign-preserving) right shift, performed on all the entries of a matrix. The boldface **4** and **64** refer to adding these constants to all the elements of a matrix.

9.11.10 Quantization and Scan

VC-1 uses both uniform and non-uniform quantization schemes. The uniform quantization method treats the DC coefficients of intra blocks differently from the other transform coefficients. These DC coefficients are quantized by dividing them by a parameter DCStepSize, that is computed as follows

$$\text{DCStepSize} = \begin{cases} \text{MQUANT} = 1, 2, & 2 \cdot \text{MQUANT}, \\ \text{MQUANT} = 3, 4, & 8, \\ \text{MQUANT} \geq 5, & 6 + \text{MQUANT}/2, \end{cases}$$

where MQUANT is a parameter derived in a complex way from the syntax elements DQBILEVEL, MQDIFF, PQUANT (the latter parameter is the picture quantizer scale parameter, which itself is derived in a complex way from the Picture Quantizer Index, PQINDEX), ALTPQUANT and ABSMQ. All other transform coefficients are quantized by dividing them by $2 \cdot \text{MQUANT}$ (except in some cases, not listed here, where the quantization parameter is $2 \cdot \text{MQUANT} + \text{HALFQP}$).

The non-uniform quantization method also treats the DC coefficients of intra blocks differently from the other transform coefficients. This coefficient is quantized as in the uniform method, while all other transform coefficients are quantized by dividing them by first subtracting MQUANT and then dividing by $2 \cdot \text{MQUANT}$ (except in some cases, not listed here, where the transform parameter is divided by $2 \cdot \text{MQUANT} + \text{HALFQP}$).

The decoder performs the reverse operations, then rounds the results to the nearest integer and, if necessary, also truncates them to 12 bits, signed, so they lie in the interval [−2,048, +2,047].

Once the transform coefficients have been quantized, they are collected in a zigzag scan of the block. As most other features of VC-1, the zigzag scan depends on syntax parameters and can be done in different ways.

Intra-coded blocks are scanned in normal, horizontal, or vertical mode depending on parameter ACPRED and on the DC prediction direction (which can be top or left) as follows. If ACPRED = 0, perform normal scan. If ACPRED = 1 and prediction direction is top, perform horizontal scan. If ACPRED = 1 and prediction direction is left, perform vertical scan. Figure 9.70 shows the normal and horizontal zigzag scans and cell numbering within an 8×8 block.

0	1	2	3	4	5	6	7
8	9	10	11	12	13	14	15
16	17	18	19	20	21	22	23
24	25	26	27	28	29	30	31
32	33	34	35	36	37	38	39
40	41	42	43	44	45	46	47
48	49	50	51	52	53	54	55
56	57	58	59	60	61	62	63

Normal scan: 0, 8, 1, 2, 9, 16, 24, 17, 10, 3, 4, 11, 18, 25, 32, 40, 33, 48, 26, 19, 12, 5, 6, 13, 20, 27, 34, 41, 56, 49, 57, 42, 35, 28, 21, 14, 7, 15, 22, 29, 36, 43, 50, 58, 51, 59, 44, 37, 30, 23, 31, 38, 45, 52, 60, 53, 61, 46, 39, 47, 54, 62, 55, 63.

Horizontal scan: 0, 1, 8, 2, 3, 9, 16, 24, 17, 10, 4, 5, 11, 18, 25, 32, 40, 48, 33, 26, 19, 12, 6, 7, 31, 20, 27, 34, 41, 56, 49, 57, 42, 35, 28, 21, 14, 15, 22, 29, 36, 43, 50, 58, 51, 44, 37, 30, 23, 31, 38, 45, 52, 59, 60, 53, 46, 39, 47, 54, 61, 62, 55, 63.

Figure 9.70: VC-1 Normal and Horizontal Zigzag Scans.

For inter-coded blocks, there are four block sizes and VC-1 defines two scan patterns for each, one for the simple and main profiles, and the other for the advanced profile; an implementor's nightmare!

9.11.11 Entropy Coding

The step following quantization is the entropy coding of the quantized transform coefficients. This is also very complex because many different tables of indexes and of variable-length codes are used, depending on the profile/level and the type of block being coded (intra or inter). Instead of listing the many rules, syntax parameters, and tables, we show a typical example that illustrates the basic process. Figure 9.71 shows an 8×8 block of quantized transform coefficients, where only eight of the 64 coefficients are nonzero. The block is scanned in basic zigzag pattern that brings the nonzero coefficients to the front, creates several short runs of zeros between them, and ends with a long run of zeros. Each nonzero coefficient C is first converted to a triplet (`run`, `level`, `flag`), where `run` is the length of the run of zeros preceding C, `level` is the (signed) value of C, and `flag` is a single bit, that is zero except for the last triplet.

Thus, the eight triplets for our example are $(1, 1, 0)$, $(0, -2, 0)$, $(2, 5, 0)$, $(1, -1, 0)$, $(1, 1, 0)$, $(0, -3, 0)$, $(0, 4, 0)$, and $(1, 1, 1)$.

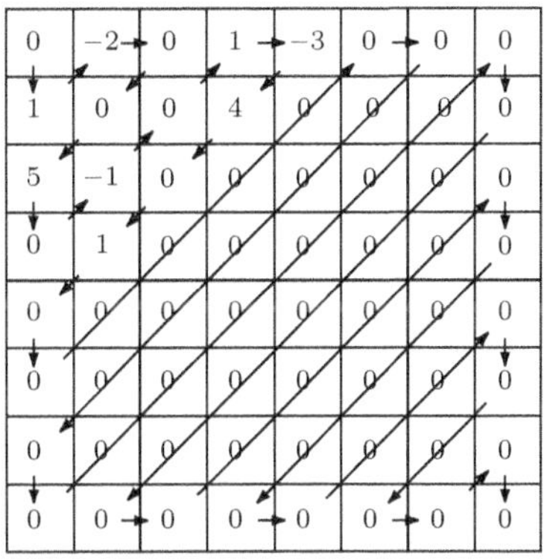

Figure 9.71: Example of Variable-Length Coding of a Block.

VC-1 employs 65 tables of variable-length codes. In the formal specification of this codec [SMPTE-421M 08], these tables are numbered from 168 to 232 and most are used to encode transform coefficients. Tables 175–190 and 219–232 contain variable-length codes for high-level inter and intra-coded blocks. Tables 191 through 204 are similarly used to encode low-level transform coefficients, and tables 205 through 218 specify codes for mid-level. Each set consists of tables with indexes and tables with the actual codes. For our example, we use the tables for low-motion, specifically table 199 for indexes of triplets of the form $(\texttt{run}, \texttt{level}, 0)$, table 200 for an index for the last triplet, whose format is $(\texttt{run}, \texttt{level}, 1)$, and table 198 for the actual variable-length codes. Table 9.72 lists parts of these three code tables.

The first triplet is $(1, 1, 0)$. We search table 199 for run 1 and level 1, and find index 14. Entry 14 of table 198 specifies the 4-bit code $11_{10} = 1011$. Thus, the encoded stream starts with 1011|0, where the single zero implies a positive nonzero coefficient. The second triplet is $(0, -2, 0)$. We search table 199 for run 0 and level 2, and find index 1. Entry 1 of table 198 specifies the 5-bit code $20_{10} = 10100$. The encoded stream becomes 1011|0|10100|1, where the last 1 implies a negative coefficient. The third triplet is $(2, 5, 0)$. We search table 199 for run 2 and level 5, and find index 27. Entry 27 of table 198 specifies the 14-bit code $1522_{10} = 00010111110010$. The encoded stream becomes 1011|0|10100|1|00010111110010|0. This process continues until the last triplet $(1, 1, 1)$. Since the flag is 1 (indicating the last triplet), we search table 200 for an index. The table yields index 86 for run 1 and level 1, so we look at entry 86 of table 198 and find the 4-bit code $5 = 0101$. The complete encoded bitstream is therefore

1011|0|10100|1|00010111110010|0|1011|1|1011|0|0010111|1|01111111|0|0101|0,

and this encodes eight 12-bit signed nonzero quantized transform coefficients.

Index	VLC Code	Size
0	4	3
1	20	5
2	23	7
3	127	8
4	340	9
5	498	10
...	...	
14	11	4
...	...	
27	1522	14
...	...	
86	5	4
...	...	
147	5524	13

Table 198

Index	Run	Level
0	0	1
1	0	2
2	0	3
3	0	4
4	0	5
5	0	6
...	...	
14	1	1
...	...	
27	2	5
...	...	
80	29	1

Table 199

Index	Run	Level
81	0	1
82	0	2
83	0	3
84	0	4
85	0	5
86	1	1
...	...	
103	7	2

Table 200

Table 9.72: Tables of Indexes and Variable-Length Codes.

In seven to ten years video traffic on the Internet
will exceed data and voice traffic combined.

—Bob Metcalfe

10
Audio Compression

Text does not occupy much space in the computer. An average book, consisting of a million characters, can be stored uncompressed in about 1 Mbyte, because each character of text occupies one byte (the Colophon at the end of the book illustrates this with data collected from the book itself).

◇ **Exercise 10.1:** It is a handy rule of thumb that an average book occupies about a million bytes. Explain why this makes sense.

In contrast, images occupy much more space, lending another meaning to the phrase "a picture is worth a thousand words." Depending on the number of colors used in an image, a single pixel occupies between one bit and three bytes. Thus, a 4 Mpixel picture taken by a typical current (2006) digital camera occupies between 512 Kbytes and 12 Mbytes before compression. With the advent of powerful, inexpensive personal computers in the 1980s and 1990s came multimedia applications, where text, images, movies, and *audio* are stored in the computer, and can be uploaded, downloaded, displayed, edited, and played back. The storage requirements of sound are smaller than those of images or video, but greater than those of text. This is why audio compression has become important and has been the subject of much research and experimentation throughout the 1990s.

Two important features of audio compression are (1) it can be lossy and (2) it requires fast decoding. Text compression must be lossless, but images and audio can lose much data without a noticeable degradation of quality. Thus, there are both lossless and lossy audio compression algorithms. It only rarely happens that a user will want to read text while it is decoded and decompressed, but this is common with audio. Often, audio is stored in compressed form and has to be decompressed in real-time when the user wants to listen to it. This is why most audio compression methods are asymmetric. The encoder can be slow, but the decoder has to be fast. This is also why audio compression methods are not dictionary based. A dictionary-based compression method may have many advantages, but fast decoding is not one of them.

D. Salomon, G. Motta, *Handbook of Data Compression*, 5th ed.
DOI 10.1007/978-1-84882-903-9_10,

The field of audio compression is rapidly developing and more and more references become available. Out of this huge crowd, reference [wiki.audio 06] should especially be noted. It should be kept up-to-date and provide the reader with the latest on techniques, resources, and links to works in this important field. An important survey paper on lossless audio compression is [Hans and Schafer 01].

This chapter starts with a short introduction to sound and digitized audio. It then discusses those properties of the human auditory system (ear and brain) that make it possible to delete much audio information without adversely affecting the quality of the reconstructed audio. The chapter continues with a survey of several lossy and lossless audio compression methods. Most methods are general and can compress any audio data. Other methods are aimed at compressing speech. Of special note is the detailed description of the three lossy methods (layers) of audio compression used by the MPEG-1 and MPEG-2 standards (Section 10.14).

10.1 Sound

To most of us, sound is a very familiar phenomenon, because we hear it all the time. Nevertheless, when we try to define sound, we find that we can approach this concept from two different points of view, and we end up with two definitions, as follows:

An intuitive definition: Sound is the sensation detected by our ears and interpreted by our brain in a certain way.

A scientific definition: Sound is a physical disturbance in a medium. It propagates in the medium as a pressure wave by the movement of atoms or molecules.

We normally hear sound as it propagates through the air and hits the diaphragm in our ears. However, sound can propagate in many different media. Marine animals produce sounds underwater and respond to similar sounds. Hitting the end of a metal bar with a hammer produces sound waves that propagate through the bar and can be detected at the other end. Good sound insulators are rare, and the best insulator is vacuum, where there are no particles to vibrate and propagate the disturbance.

Sound can also be considered a wave, even though its frequency may vary all the time. It is a longitudinal wave, one where the disturbance is in the direction of the wave itself. In contrast, electromagnetic waves and ocean waves are transverse waves. Their undulations are perpendicular to the direction of the wave.

Like any other wave, sound has three important attributes, its speed, amplitude, and period. The frequency of a wave is not an independent attribute; it is the number of periods that occur in one time unit (one second). The unit of frequency is the hertz (Hz). The speed of sound depends mostly on the medium it passes through, and on the temperature. In air, at sea level (one atmospheric pressure), and at 20° Celsius (68° Fahrenheit), the speed of sound is 343.8 meters per second (about 1128 feet per second).

The human ear is sensitive to a wide range of sound frequencies, normally from about 20 Hz to about 22,000 Hz, depending on a person's age and health. This is the range of *audible frequencies*. Some animals, most notably dogs and bats, can hear higher frequencies (ultrasound). A quick calculation reveals the wavelengths associated with audible frequencies. At 22,000 Hz, each wavelength is about 1.56 cm long, whereas at 20 Hz, a wavelength is about 17.19 meters long.

◇ **Exercise 10.2:** Verify these calculations.

The amplitude of sound is also an important property. We perceive it as loudness. We sense sound when air molecules strike the diaphragm of the ear and apply pressure to it. The molecules move back and forth tiny distances that are related to the *amplitude*, not to the period of the sound. The period of a sound wave may be several meters, yet an individual molecule in the air may move just a millionth of a centimeter in its oscillations. With very loud noise, an individual molecule may move about one thousandth of a centimeter. A device to measure noise levels should therefore be based on a sensitive diaphragm where the pressure of the sound wave is sensed and is converted to electrical voltage, which in turn is displayed as a numeric value.

The problem with measuring noise intensity is that the ear is sensitive to a very wide range of sound levels (amplitudes). The ratio between the sound level of a cannon at the muzzle and the lowest level we can hear (the threshold of hearing) is about 11–12 orders of magnitude. If we denote the lowest audible sound level by 1, then the cannon noise would have a magnitude of 10^{11}! It is inconvenient to deal with measurements in such a wide range, which is why the units of sound loudness use a *logarithmic scale*. The (base-10) logarithm of 1 is 0, and the logarithm of 10^{11} is 11. Using logarithms, we only have to deal with numbers in the range 0 through 11. In fact, this range is too small, and we typically multiply it by 10 or by 20, to get numbers between 0 and 110 or 220. This is the well-known (and sometimes confusing) decibel system of measurement.

◇ **Exercise 10.3:** (For the mathematically weak.) What exactly is a logarithm?

The decibel (dB) unit is defined as the base-10 logarithm of the ratio of two physical quantities whose units are powers (energy per time). The logarithm is then multiplied by the convenient scale factor 10. (If the scale factor is not used, the result is measured in units called "Bel." The Bel, however, was dropped long ago in favor of the decibel.) Thus, we have

$$\text{Level} = 10\log_{10}\frac{P_1}{P_2}\ \text{dB},$$

where P_1 and P_2 are measured in units of power such as watt, joule/sec, gram·cm/sec, or horsepower. This can be mechanical power, electrical power, or anything else. In measuring the loudness of sound, we have to use units of acoustical power. Since even loud sound can be produced with very little energy, we use the microwatt (10^{-6} watt) as a convenient unit.

From the Dictionary

Acoustics: (1) The science of sound, including the generation, transmission, and effects of sound waves, both audible and inaudible. (2) The physical qualities of a room or other enclosure (such as size, shape, amount of noise) that determine the audibility and perception of speech and music within the room.

The decibel is the logarithm of a ratio. The numerator, P_1, is the power (in microwatts) of the sound whose intensity level is being measured. It is convenient to select as the denominator the number of microwatts that produce the faintest audible sound (the threshold of hearing). This number is shown by experiment to be 10^{-6} microwatt $= 10^{-12}$ watt. Thus, a stereo unit that produces 1 watt of acoustical power has an intensity level of

$$10\log\frac{10^6}{10^{-6}} = 10\log\left(10^{12}\right) = 10\times 12 = 120\text{ dB}$$

(this happens to be about the threshold of feeling; see Figure 10.1), whereas an earphone producing 3×10^{-4} microwatt has a level of

$$10\log\frac{3\times 10^{-4}}{10^{-6}} = 10\log\left(3\times 10^{2}\right) = 10\times(\log 3 + \log 100) = 10\times(0.477+2) \approx 24.77\text{ dB}.$$

In the field of electricity, there is a simple relation between (electrical) power P and pressure (voltage) V. Electrical power is the product of the current and voltage $P = I\cdot V$. The current, however, is proportional to the voltage by means of Ohm's law $I = V/R$ (where R is the resistance). We can therefore write $P = V^2/R$, and use pressure (voltage) in our electric decibel measurements.

In practical acoustics work, we don't always have access to the source of the sound, so we cannot measure its electrical power output. In practice, we may find ourselves in a busy location, holding a sound decibel meter in our hands, trying to measure the noise level around us. The decibel meter measures the pressure Pr applied by the sound waves on its diaphragm. Fortunately, the acoustical power per area (denoted by P) is proportional to the square of the sound pressure Pr. This is because sound power P is the product of the pressure Pr and the speed v of the sound, and because the speed can be expressed as the pressure divided by the specific impedance of the medium through which the sound propagates. This is why sound loudness is commonly measured in units of dB SPL (sound pressure level) instead of sound power. The definition is

$$\text{Level} = 10\log_{10}\frac{P_1}{P2} = 10\log_{10}\frac{Pr_1^2}{Pr_2^2} = 20\log_{10}\frac{Pr_1}{Pr_2}\text{ dB SPL}.$$

The zero reference level for dB SPL becomes 0.0002 dyne/cm^2, where the dyne, a small unit of force, is about 0.0010197 grams. Since a dyne equals 10^{-5} newtons, and since one centimeter is 0.01 meter, that zero reference level (the threshold of hearing) equals 0.00002 newton/meter^2. Table 10.2 shows typical dB values in both units of power and SPL.

The sensitivity of the human ear to sound level depends on the frequency. Experiments indicate that people are more sensitive to (and therefore more annoyed by) high-frequency sounds (which is why sirens have a high pitch). It is possible to modify the dB SPL system to make it more sensitive to high frequencies and less sensitive to low frequencies. This is called the dBA standard (ANSI standard S1.4-1983). There are also dBB and dBC standards of noise measurement. (Electrical engineers use also decibel standards called dBm, dBm0, and dBrn; see, for example, [Shenoi 95].)

Because of the use of logarithms, dB measures don't simply add up. If the first trumpeter starts playing his trumpet just before the concert, generating, say, a 70 dB noise level, and the second trombonist follows on his trombone, generating the same sound level, then the (poor) listeners hear twice the noise intensity, but this corresponds

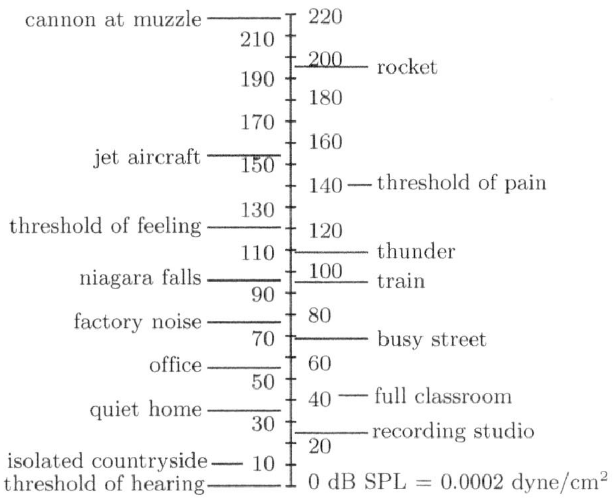

Figure 10.1: Common Sound Levels in dB SPL Units.

watts	dB	pressure n/m^2	dB SPL	source
30000.0	165	2000.0	160	jet
300.0	145	200.0	140	threshold of pain
3.0	125	20.0	120	factory noise
0.03	105	2.0	100	highway traffic
0.0003	85	0.2	80	appliance
0.000003	65	0.02	60	conversation
0.00000003	45	0.002	40	quiet room
0.0000000003	25	0.0002	20	whisper
0.000000000001	0	0.00002	0	threshold of hearing

Table 10.2: Sound Levels in Power and Pressure Units.

to just 73 dB, not 140 dB. To show why this is true we notice that if

$$10\log\left(\frac{P_1}{P_2}\right) = 70,$$

then

$$10\log\left(\frac{2P_1}{P_2}\right) = 10\left[\log_{10}2 + \log\left(\frac{P_1}{P_2}\right)\right] = 10(0.3 + 70/10) = 73.$$

Doubling the noise level increases the dB level by 3 (if SPL units are used, the 3 should be doubled to 6).

◇ **Exercise 10.4:** Two sound sources A and B produce dB levels of 70 and 79 dB, respectively. How much louder is source B compared to A?

10.2 Digital Audio

Much as an image can be digitized and broken up into pixels, where each pixel is a number, sound can also be digitized and broken up into numbers. When sound is played into a microphone, it is converted into a voltage that varies continuously with time. Figure 10.3 shows a typical example of sound that starts at zero and oscillates several times. Such voltage is the *analog* representation of the sound. Digitizing sound is done by measuring the voltage at many points in time, translating each measurement into a number, and writing the numbers on a file. This process is called *sampling*. The sound wave is sampled, and the samples become the digitized sound. The device used for sampling is called an analog-to-digital converter (ADC).

The difference between a sound wave and its samples can be compared to the difference between an analog clock, where the hands seem to move continuously, and a digital clock, where the display changes abruptly every second.

Since the audio samples are numbers, they are easy to edit. However, the main use of an audio file is to play it back. This is done by converting the numeric samples back into voltages that are continuously fed into a speaker. The device that does that is called a digital-to-analog converter (DAC). Intuitively, it is clear that a high sampling rate would result in better sound reproduction, but also in many more samples and therefore bigger files. Thus, the main problem in audio sampling is how often to sample a given sound.

Figure 10.3a shows what may happen if the sampling rate is too low. The sound wave in the figure is sampled four times, and all four samples happen to be identical. When these samples are used to play back the sound, the result is silence. Figure 10.3b shows seven samples, and they seem to "follow" the original wave fairly closely. Unfortunately, when they are used to reproduce the sound, they produce the curve shown in dashed. There simply are not enough samples to reconstruct the original sound wave.

The solution to the sampling problem is to sample sound at a little over the *Nyquist frequency* (page 741), which is twice the maximum frequency contained in the sound. Thus, if a sound contains frequencies of up to 2 kHz, it should be sampled at a little more than 4 kHz. (A detail that's often ignored in the literature is that signal reconstruction is perfect only if something called the Nyquist-Shannon sampling theorem [Wikipedia 03] is used.) Such a sampling rate guarantees true reproduction of the sound. This is illustrated in Figure 10.3c, which shows 10 equally-spaced samples taken over four periods. Notice that the samples do not have to be taken from the maxima or minima of the wave; they can come from any point. Section 7.1 discusses the application of sampling to pixels in images (two-dimensional signals).

The range of human hearing is typically from 16–20 Hz to 20,000–22,000 Hz, depending on the person and on age. When sound is digitized at high fidelity, it should therefore be sampled at a little over the Nyquist rate of $2\times 22000 = 44000$ Hz. This is why high-quality digital sound is based on a 44,100-Hz sampling rate. Anything lower than this rate results in distortions, while higher sampling rates do not produce any improvement in the reconstruction (playback) of the sound. We can consider the sampling rate of 44,100 Hz a lowpass filter, since it effectively removes all the frequencies above 22,000 Hz.

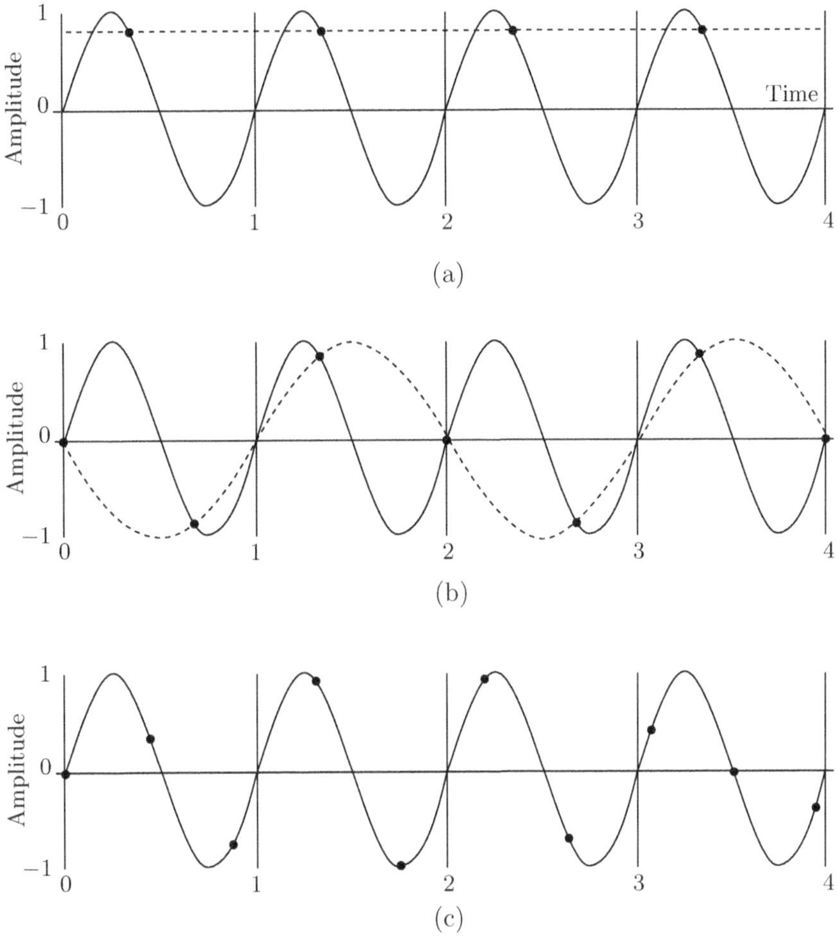

Figure 10.3: Sampling a Sound Wave.

Many low-fidelity applications sample sound at 11,000 Hz, and the telephone system, originally designed for conversations, not for digital communications, samples sound at only 8 kHz. Thus, any frequency higher than 4000 Hz gets distorted when sent over the phone, which is why it is hard to distinguish, on the phone, between the sounds of `f` and `s`. This is also why, when someone gives you an address over the phone you should ask, "Is it H street, as in EFGH?" Often, the answer is, "No, this is Eighth street, as in sixth, seventh, eighth."

> The meeting was in Mr. Rogers' law office, at 1415 *H* Street. My slip of paper said 1415 *8th* Street. (The address had been given over the telephone.)
> —Richard P. Feynman, *What Do YOU Care What Other People Think?*

The second problem in sound sampling is the sample size. Each sample becomes a number, but how large should this number be? In practice, samples are normally either 8 or 16 bits, although some high-quality sound cards that are available for many computer platforms may optionally use 32-bit samples. Assuming that the highest voltage in a sound wave is 1 volt, an 8-bit sample can distinguish voltages as low as $1/256 \approx 0.004$ volt, or 4 millivolts (mv). A quiet sound, generating a wave lower than 4 mv, would be sampled as zero and played back as silence. In contrast, with a 16-bit sample it is possible to distinguish sounds as low as $1/65,536 \approx 15$ microvolt (μv). We can think of the sample size as a quantization of the original audio data. Eight-bit samples are more coarsely quantized than 16-bit samples. As a result, they produce better compression but poorer audio reconstruction (the reconstructed sound features only 256 different amplitudes).

◇ **Exercise 10.5:** Suppose that the sample size is one bit. Each sample has a value of either 0 or 1. What would we hear when these samples are played back?

Audio sampling is also called *pulse code modulation* (PCM). We have all heard of AM and FM radio. These terms stand for *amplitude modulation* and *frequency modulation*, respectively. They indicate methods to modulate (i.e., to include binary information in) continuous waves. The term *pulse modulation* refers to techniques for converting a continuous wave to a stream of binary numbers (audio samples). Possible pulse modulation methods include pulse amplitude modulation (PAM), pulse position modulation (PPM), pulse width modulation (PWM), and pulse number modulation (PNM). [Pohlmann 85] is a good source of information on these methods. In practice, however, PCM has proved the most effective form of converting sound waves to numbers. When stereo sound is digitized, the PCM encoder multiplexes the left and right sound samples. Thus, stereo sound sampled at 22,000 Hz with 16-bit samples generates 44,000 16-bit samples per second, for a total of 704,000 bits/sec, or 88,000 bytes/sec.

10.2.1 Digital Audio and Laplace Distribution

The Laplace distribution has already been mentioned in Section 7.25. This short section explains the relevance of this distribution to audio compression. A large audio file with a long, complex piece of music tends to have all the possible values of audio samples. Consider the simple case of 8-bit audio samples, which have values in the interval $[0, 255]$. A large audio file, with millions of audio samples, will tend to have many audio samples concentrated around the center of this interval (around 128), fewer large samples (close to the maximum 255), and few small samples (although there may be many audio samples of 0, because many types of sound tend to have periods of silence). The distribution of the samples may have a maximum at its center and another spike at 0. Thus, the audio samples themselves do not normally have a simple distribution.

However, when we examine the differences of adjacent samples, we observe a completely different behavior. Consecutive audio samples tend to be correlated, which is why the differences of consecutive samples tend to be small numbers. Experiments with many types of sound indicate that the distribution of audio differences resembles the Laplace distribution. The following experiment reinforces this conclusion. Figure 10.4a shows the distribution of about 2900 audio samples taken from a real audio file (file `a8.wav`, where the single English letter `a` is spoken). It is easy to see that the distribution has a peak at 128, but is very rough and does not resemble either the Gaussian or the Laplacian distributions. Also, there are very few samples smaller than 75 and greater than 225 (because of the small size of the file and its particular, single-letter, content).

However, part b of the figure is different. It displays the distribution of the differences and it is clearly smooth, narrow, and has a sharp peak at 0 (in the figure, the peak is at 256 because indexes in Matlab must be nonnegative, which is why an artificial 256 is added to the indexes of differences).

Parts c,d of the figure show the distribution of about 2900 random numbers. The numbers themselves have a flat (although nonsmooth) distribution, while their differences have a very rough distribution, centered at 0 but very different from Laplace (only about 24 differences are zero).

The distribution of the differences is not flat; it resembles a very rough Gaussian distribution. Given random integers R_i between 0 and n, the differences of consecutive numbers can be from $-n$ to $+n$, but these differences occur with different probabilities. Extreme differences are rare. The only way to obtain a difference of n is to have an $R_{i+1} = 0$ follow an $R_i = n$ and subtract $n - 0$. Small differences, on the other hand, have higher probabilities because they can be obtained in more cases. A difference of 2 is obtained by subtracting $56 - 54$, but also $57 - 55$, $58 - 56$ and so on.

The conclusion is that differences of consecutive correlated values tend to have a narrow, peaked distribution, resembling the Laplace distribution. This is true for the differences of audio samples as well as for the differences of consecutive pixels of an image. A compression algorithm may take advantage of this fact and encode the differences with variable-length codes that have a Laplace distribution. A more sophisticated version may compute differences between actual values (audio samples or pixels) and their predicted values, and then encode the (Laplace distributed) differences. Two such methods are image MLP (Section 7.25) and FLAC (Section 10.10).

10.3 The Human Auditory System

The frequency range of the human ear is from about 20 Hz to about 20,000 Hz, but the ear's sensitivity to sound is not uniform. It depends on the frequency, and experiments indicate that in a quiet environment the ear's sensitivity is maximal for frequencies in the range 2 kHz to 4 kHz. Figure 10.5a shows the *hearing threshold* for a quiet environment (see also Figure 10.67).

◇ **Exercise 10.6:** Come up with an appropriate way to conduct such experiments.

It should also be noted that the range of the human voice is much more limited. It is only from about 500 Hz to about 2 kHz (Section 10.8).

The existence of the hearing threshold suggests an approach to lossy audio compression. Just delete any audio samples that are below the threshold. Since the threshold depends on the frequency, the encoder needs to know the frequency spectrum of the sound being compressed at any time. The encoder therefore has to save several of the previously input audio samples at any time ($n-1$ samples, where n is either a constant or a user-controlled parameter). When the current sample is input, the first step is to transform the most-recent n samples to the frequency domain (Section 8.2). The result is a number m of values (called *signals*) that indicate the strength of the sound at m

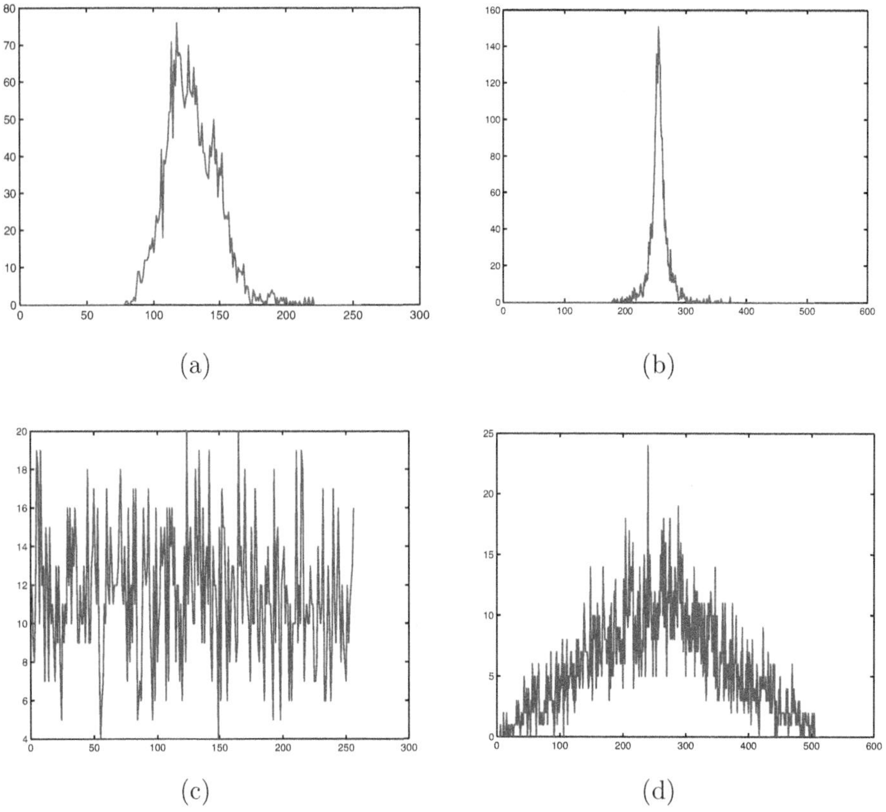

Figure 10.4: Distribution of Audio Samples, Random Numbers, and Differences.

```
% File 'LaplaceWav.m'
filename='a8.wav';
dim=2950; % size of file a8.wav
dist=zeros(256,1); ddist=zeros(512,1);
fid=fopen(filename,'r');
buf=fread(fid,dim,'uint8'); %input unsigned integers
 for i=46:dim % skip .wav file header
  x=buf(i)+1; dif=buf(i)-buf(i-1)+256;
  dist(x)=dist(x)+1; ddist(dif)=ddist(dif)+1;
 end
figure(1), plot(dist), colormap(gray) %dist of audio samples
figure(2), plot(ddist), colormap(gray) %dist of differences
dist=zeros(256,1); ddist=zeros(512,1); % clear buffers
buf=randint(dim,1,[0 255]); % many random numbers
 for i=2:dim
  x=buf(i)+1; dif=buf(i)-buf(i-1)+256;
  dist(x)=dist(x)+1; ddist(dif)=ddist(dif)+1;
 end
figure(3), plot(dist), colormap(gray) %dist of random numbers
figure(4), plot(ddist), colormap(gray) %dist of differences
```

Code for Figure 10.4.

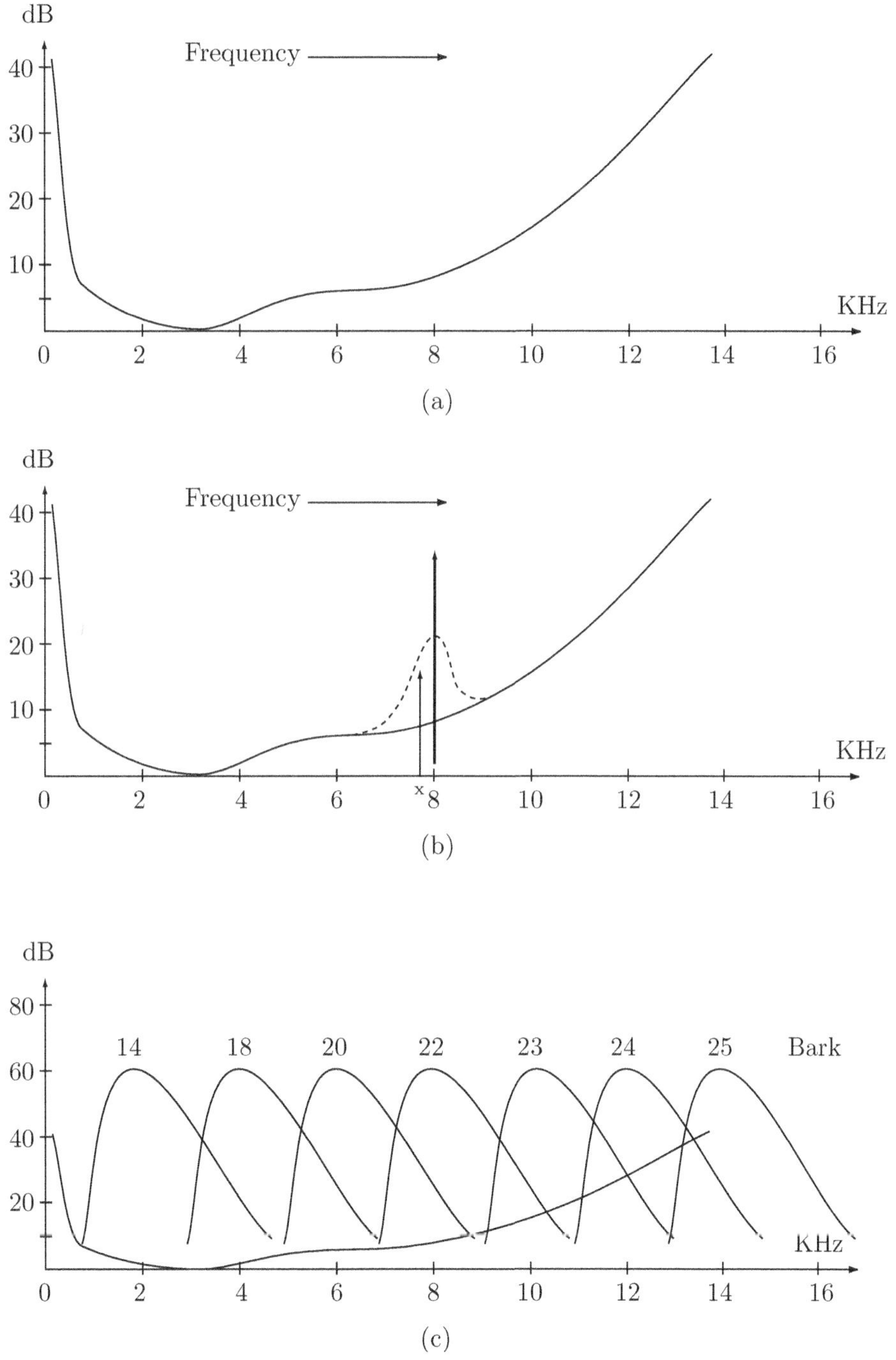

Figure 10.5: Threshold and Masking of Sound.

different frequencies. If a signal for frequency f is smaller than the hearing threshold at f, it (the signal) should be deleted.

In addition to this, two more properties of the human hearing system are used in audio compression. They are *frequency masking* and *temporal masking*.

Frequency masking (also known as *auditory masking*) occurs when a sound that we can normally hear (because it is loud enough) is masked by another sound with a nearby frequency. The thick arrow in Figure 10.5b represents a strong sound source at 800 kHz. This source raises the normal threshold in its vicinity (the dashed curve), with the result that the nearby sound represented by the arrow at "x", a sound that would normally be audible because it is above the threshold, is now masked, and is inaudible (see also Figure 10.67). A good lossy audio compression method should identify this case and delete the signals corresponding to sound "x", because it cannot be heard anyway. This is one way to lossily compress sound.

The frequency masking (the width of the dashed curve of Figure 10.5b) depends on the frequency. It varies from about 100 Hz for the lowest audible frequencies to more than 4 kHz for the highest. The range of audible frequencies can therefore be partitioned into a number of *critical bands* that indicate the declining sensitivity of the ear (rather, its declining resolving power) for higher frequencies. We can think of the critical bands as a measure similar to frequency. However, in contrast to frequency, which is absolute and has nothing to do with human hearing, the critical bands are determined according to the sound perception of the ear. Thus, they constitute a perceptually uniform measure of frequency. Table 10.6 lists 27 approximate critical bands.

Another way to describe critical bands is to say that because of the ear's limited perception of frequencies, the threshold at a frequency f is raised by a nearby sound only if the sound is within the critical band of f. This also points the way to designing a practical lossy compression algorithm. The audio signal should first be transformed into its frequency domain, and the resulting values (the frequency spectrum) should be divided into subbands that resemble the critical bands as much as possible. Once this is done, the signals in each subband should be quantized such that the quantization noise (the difference between the original sound sample and its quantized value) should be inaudible.

band	range	band	range	band	range
0	0–50	9	800–940	18	3280–3840
1	50–95	10	940–1125	19	3840–4690
2	95–140	11	1125–1265	20	4690–5440
3	140–235	12	1265–1500	21	5440–6375
4	235–330	13	1500–1735	22	6375–7690
5	330–420	14	1735–1970	23	7690–9375
6	420–560	15	1970–2340	24	9375–11625
7	560–660	16	2340–2720	25	11625–15375
8	660–800	17	2720–3280	26	15375–20250

Table 10.6: Twenty-Seven Approximate Critical Bands.

Yet another way to look at the concept of critical bands is to consider the human auditory system a filter that lets through only frequencies in the range (bandpass) of 20 Hz to 20000 Hz. We visualize the ear–brain system as a collection of filters, each with a different bandpass. The bandpasses are called critical bands. They overlap and they have different widths. They are narrow (about 100 Hz) at low frequencies and become wider (to about 4–5 kHz) at high frequencies.

The width of a critical band is called its size. The widths of the critical bands introduce a new unit, the *Bark* (after H. G. Barkhausen) such that one Bark is the width (in Hz) of one critical band. The Bark is defined as

$$1\text{ Bark} = \begin{cases} \frac{f}{100}, & \text{for frequencies } f < 500\text{ Hz},\\ 9 + 4\log_2\left(\frac{f}{1000}\right), & \text{for frequencies } f \geq 500\text{ Hz}. \end{cases}$$

Figure 10.5c shows some critical bands, with Barks between 14 and 25, positioned above the threshold.

Heinrich Georg Barkhausen

Heinrich Barkhausen was born on December 2, 1881, in Bremen, Germany. He spent his entire career as a professor of electrical engineering at the Technische Hochschule in Dresden, where he concentrated on developing electron tubes. He also discovered the so-called "Barkhausen effect," where acoustical waves are generated in a solid by the movement of domain walls when the material is magnetized. He also coined the term "phon" as a unit of sound loudness. The institute in Dresden was destroyed, as was most of the city, in the famous fire bombing in February 1945. After the war, Barkhausen helped rebuild the institute. He died on February 20, 1956.

Temporal masking may occur when a strong sound A of frequency f is preceded or followed in time by a weaker sound B at a nearby (or the same) frequency. If the time interval between the sounds is short, sound B may not be audible. Figure 10.7 illustrates an example of temporal masking (see also Figure 10.68). The threshold of temporal masking due to a loud sound at time 0 goes down, first sharply, then slowly. A weaker sound of 30 dB will not be audible if it occurs 10 ms before or after the loud sound, but will be audible if the time interval between the sounds is 20 ms.

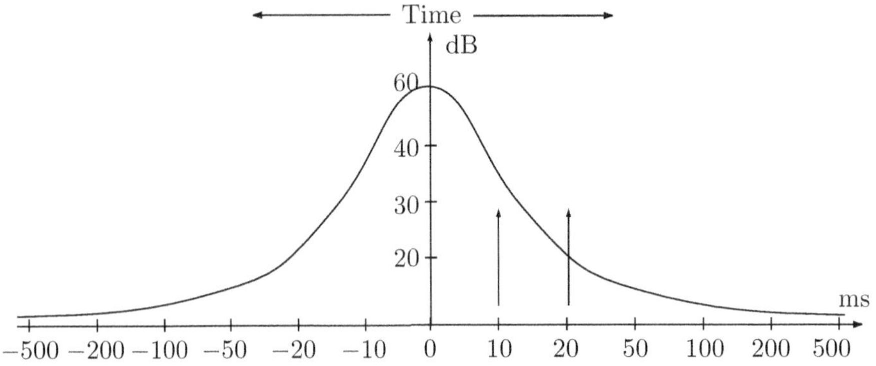

Figure 10.7: Threshold and Masking of Sound.

10.3.1 Conventional Methods

Conventional compression methods, such as RLE, statistical, and dictionary-based, can be used to losslessly compress audio data, but the results depend heavily on the specific sound. Some sounds may compress well under RLE but not under a statistical method. Other sounds may lend themselves to statistical compression but may expand when processed by a dictionary method. Here is how sounds respond to each of the three classes of compression methods.

RLE may work well when the sound contains long runs of identical samples. With 8-bit samples this may be common. Recall that the difference between the two 8-bit samples n and $n+1$ is about 4 mv. A few seconds of uniform music, where the wave does not oscillate more than 4 mv, may produce a run of thousands of identical samples. With 16-bit samples, long runs may be rare and RLE, consequently, ineffective.

Statistical methods assign variable-length codes to the samples according to their frequency of occurrence. With 8-bit samples, there are only 256 different samples, so in a large audio file, the samples may sometimes have a flat distribution. Such a file will therefore not respond well to Huffman coding (See Exercise 5.7). With 16-bit samples there are more than 65,000 sample values, so they may sometimes feature skewed probabilities (i.e., some samples may occur very often, while others may be rare). Such a file may therefore compress better with arithmetic coding, which works well even for skewed probabilities.

Dictionary-based methods expect to find the same phrases again and again in the data. This happens with text, where certain strings may repeat often. Sound, however, is an analog signal and the particular samples generated depend on the precise way the

ADC works. With 8-bit samples, for example, a wave of 8 mv becomes a sample of size 2, but waves very close to that, say, 7.6 mv or 8.5 mv, may become samples of different sizes. This is why parts of speech that sound the same to us, and should therefore have become identical phrases, end up being digitized slightly differently, and go into the dictionary as different phrases, thereby reducing compression. Dictionary-based methods are not well suited for sound compression.

> I don't like the sound of that sound.
>
> —Heather Graham as Judy Robinson in *Lost in Space* (1998)

10.3.2 Lossy Audio Compression

It is possible to get better audio compression by developing lossy methods that take advantage of our perception of sound, and discard data to which the human ear is not sensitive. This is similar to lossy image compression, where visual data to which the human eye is not sensitive is discarded. In both cases we use the fact that the original information (image or sound) is analog and has already lost some quality when digitized. Losing some more data, if done carefully, may not significantly affect the played-back sound, and may therefore be indistinguishable from the original. We briefly describe two approaches, *silence compression* and *companding*.

The principle of silence compression is to treat small samples as if they were silence (i.e., as samples of 0). This generates run lengths of zero, so silence compression is actually a variant of RLE, suitable for sound compression. This method uses the fact that some people have less sensitive hearing than others, and will tolerate the loss of sound that is so quiet they may not hear it anyway. Audio files containing long periods of low-volume sound will respond to silence compression better than other files with high-volume sound. This method requires a user-controlled parameter that specifies the largest sample that should be suppressed. Two other parameters are also necessary, although they may not have to be user-controlled. One specifies the shortest run length of small samples, typically 2 or 3. The other specifies the minimum number of consecutive large samples that should terminate a run of silence. For example, a run of 15 small samples, followed by two large samples, followed by 13 small samples may be considered one silence run of 30 samples, whereas the runs 15, 3, 13 may become two distinct silence runs of 15 and 13 samples, with nonsilence in between.

Companding (short for "compressing/expanding") uses the fact that the ear requires more precise samples at low amplitudes (soft sounds), but is more forgiving at higher amplitudes. A typical ADC used in sound cards for personal computers converts voltages to numbers linearly. If an amplitude a is converted to the number n, then amplitude $2a$ will be converted to the number $2n$. A compression method using companding examines every sample in the sound file, and employs a nonlinear formula to reduce the number of bits devoted to it. For 16-bit samples, for example, a companding encoder may use a formula as simple as

$$\text{mapped} = 32{,}767\left(2^{\frac{\text{sample}}{65536}} - 1\right) \tag{10.1}$$

to reduce each sample. This formula maps the 16-bit samples nonlinearly to 15-bit numbers (i.e., numbers in the range [0, 32767]) such that small samples are less affected than large ones. Table 10.8 illustrates the nonlinearity of this mapping. It shows eight pairs of samples, where the two samples in each pair differ by 100. The two samples of the first pair get mapped to numbers that differ by 34, whereas the two samples of the last pair are mapped to numbers that differ by 65. The mapped 15-bit numbers can be decoded back into the original 16-bit samples by the inverse formula

$$\text{Sample} = 65{,}536 \log_2 \left(1 + \frac{\text{mapped}}{32{,}767}\right). \tag{10.2}$$

Sample	Mapped	Diff	Sample	Mapped	Diff
100 →	35	34	30,000 →	12,236	47
200 →	69		30,100 →	12,283	
1,000 →	348	35	40,000 →	17,256	53
1,100 →	383		40,100 →	17,309	
10,000 →	3,656	38	50,000 →	22,837	59
10,100 →	3,694		50,100 →	22,896	
20,000 →	7,719	43	60,000 →	29,040	65
20,100 →	7,762		60,100 →	29,105	

Table 10.8: 16-Bit Samples Mapped to 15-Bit Numbers.

Reducing 16-bit numbers to 15 bits doesn't produce much compression. Better compression can be achieved by substituting a smaller number for 32,767 in equations (10.1) and (10.2). A value of 127, for example, would map each 16-bit sample into an 8-bit one, yielding a compression ratio of 0.5. However, decoding would be less accurate. A 16-bit sample of 60,100, for example, would be mapped into the 8-bit number 113, but this number would produce 60,172 when decoded by Equation (10.2). Even worse, the small 16-bit sample 1000 would be mapped into 1.35, which has to be rounded to 1. When Equation (10.2) is used to decode a 1, it produces 742, significantly different from the original sample. The amount of compression should therefore be a user-controlled parameter, and this is an interesting example of a compression method where the compression ratio is known in advance!

In practice, there is no need to go through Equations (10.1) and (10.2), since the mapping of all the samples can be prepared in advance in a table. Both encoding and decoding are therefore fast.

Companding is not limited to Equations (10.1) and (10.2). More sophisticated methods, such as μ-law and A-law, are commonly used and have been designated international standards.

10.4 WAVE Audio Format

WAVE (or simply Wave) is the native file format employed by the Windows operating system for storing digital audio data. Audio has become as important as images and video, which is why modern operating systems support a native format for audio files. The popularity of Windows and the vast amount of software available for the PC platform have guaranteed the popularity of the Wave format, hence this short description.

First, three technical notes. (1) The Wave file format is native to Windows and therefore runs on Intel processors which use the little endian byte order. (2) Wave files normally contain strings of text for specifying cue points, labels, notes, and other information. Strings whose size is not known in advance are stored in the so-called Pascal format where the first byte specifies the number of the ASCII text bytes in the string. (3) The WAV (or .WAV) file format is a special case of Wave where no compression is used.

A recommended reference for Wave and other important file formats is [Born 95]. However, it is always a good idea to search the Internet for new references.

The organization of a Wave file is based on the standard RIFF structure. (RIFF stands for Resource Interchange File Format. This is the basis of several important file formats.) The file consists of chunks, where each chunk starts with a header and may contain subchunks as well as data. For example, the fmt␣ and data chunks are actually subchunks of the RIFF chunk. New types of chunks may be added in the future, leading to a situation where existing software does not recognize new chunks. Thus, the general rule is to skip any unrecognizable chunks. The total size of a chunk must be an even number of bytes (a multiple of two bytes), which is why a chunk may end with one byte of zero padding.

The first chunk of a Wave file is RIFF. It start with the 4-byte string RIFF, followed by the size of the remaining part of the file (in four bytes), followed by string WAVE. This is immediately followed by the format and data subchunks. The former is listed in Table 10.9.

The main specifications included in the format chunk are as follows:

- Compression Code: Wave files may be compressed. A special case is .WAV files, which are not compressed. The compression codes are listed in Table 10.10.

- Number of Channels: The number of audio channels encoded in the data chunk. A value of 1 indicates a mono signal, 2 indicates stereo, and so on.

- Sample Rate: The number of audio samples per second. This value is unaffected by the number of channels.

- Average Bytes Per Second: How many bytes of wave data must be sent to a D/A converter per second in order to play the wave file? This value is the product of the sample rate and block align.

- Block Align: The number of bytes per sample slice. This value is not affected by the number of channels.

- Significant Bits Per Sample: The number of bits per audio sample, most often 8, 16, 24, or 32. If the number of bits is not a multiple of 8, then the number of bytes used

Byte #	Description
0–3	`fmt␣`
4–7	Length of subchunk (24+extra format bytes)
8–9	16-bit compression code
10–11	Channel numbers (1=Mono, 2=Stereo)
12-15	Sample Rate (Binary, in Hz)
16–19	Bytes Per Second
20–21	Bytes Per Sample: 1=8 bit Mono, 2=8 bit Stereo or 16 bit Mono, 4=16 bit Stereo
22-23	Bits Per Sample
24–	Extra format bytes

Table 10.9: A `Format` Chunk.

Code	(Hex)	Compression
0	0000	Unknown
1	0001	PCM/uncompressed
2	0002	Microsoft ADPCM
6	0006	ITU G.711 A-law
7	0007	ITU G.711 μ-law
17	0011	IMA ADPCM
20	0016	ITU G.723 ADPCM (Yamaha)
49	0031	GSM 6.10
64	0040	ITU G.721 ADPCM
80	0050	MPEG
65,536	FFFF	Experimental

Table 10.10: Wave Compression Codes.

per sample is rounded up to the nearest byte size and the unused bytes are set to 0 and ignored.

▪ Extra Format Bytes: The number of additional format bytes that follow. These bytes are used for nonzero compression codes. Their meaning depends on the specific compression used. If this value is not even (a multiple of 2), a single byte of zero-padding should be added to the end of this data to align it, but the value itself should remain odd.

The data chunk starts with the string `data`, which is followed by the size of the data that follows (as a 32-bit integer). This is followed by the data bytes (audio samples).

Table 10.11 lists the first 80 bytes of a simple `.WAV` file.

The file starts with the identifying string `RIFF`. The next four bytes are the length, which is $16e6_{16} = 5{,}862_{10}$ bytes. This number is the length of the entire file minus the eight bytes for the `RIFF` and length. The strings `WAVE` and `fmt␣` follow.

The second row starts with the length $10_{16} = 24_{10}$ of the format chunk. This indicates no extra bytes. The next two bytes are 1 (compression code of uncompressed), followed by another two bytes of 1 (mono audio, only one channel). The audio sampling

	Hexadecimal	ASCII
00	52 49 46 46 E6 16 00 00 57 41 56 45 66 6D 74 20	RIFF....WAVEfmt␣
10	10 00 00 00 01 00 01 00 10 27 00 00 20 4E 00 00	.........'...N..
20	02 00 10 00 64 61 74 61 C2 16 00 00 8B FC 12 F8	....data........
30	B9 F8 EB F8 25 F8 31 F3 67 EE 66 ED 80 EB F1 E7	....%.1.g.f.....
40	3D EB 5C 09 C3 1B 44 E9 EF FF 18 39 35 0E 5B 0A	=.\...D....95.[.

Table 10.11: Example of a `.WAV` File.

rate is $2{,}710_{16} = 10{,}000_{10}$, and the bytes per second are $4e20_{16} = 20{,}000_{10}$.

The third row starts with bytes per sample (two bytes of 2, indicating 8-bit stereo or 16-bit mono). The next two bytes are the bits per sample (16 in our case). We now know that this audio is mono with 16-bit audio samples. The data subchunk follows immediately. It starts with the string `data`, followed by the four bytes $16c2_{16}$ indicating that 5826_{10} bytes of audio data follow. The first four such bytes (two audio samples) are `8B FC 12 F8`.

10.5 μ-Law and A-Law Companding

These two international standards, formally known as recommendation G.711, are documented in [ITU-T 89]. They employ logarithm-based functions to encode audio samples for ISDN (integrated services digital network) digital telephony services, by means of nonlinear quantization. The ISDN hardware samples the voice signal from the telephone 8,000 times per second, and generates 14-bit samples (13 for A-law). The method of μ-law companding is used in North America and Japan, and A-law is used elsewhere. The two methods are similar; they differ mostly in their quantizations (midtread vs. midriser).

Experiments (documented in [Shenoi 95]) indicate that the low amplitudes of speech signals contain more information than the high amplitudes. This is why nonlinear quantization makes sense. Imagine an audio signal sent on a telephone line and digitized to 14-bit samples. The louder the conversation, the higher the amplitude, and the bigger the value of the sample. Since high amplitudes are less important, they can be coarsely quantized. If the largest sample, which is $2^{14} - 1 = 16{,}383$, is quantized to 255 (the largest 8-bit number), then the compression factor is $14/8 = 1.75$. When decoded, a code of 255 will become very different from the original 16,383. We say that because of the coarse quantization, large samples end up with high quantization noise. Smaller samples should be finely quantized, so they end up with low quantization noise.

The μ-law encoder inputs 14-bit samples and outputs 8-bit codewords. The A-law inputs 13-bit samples and also outputs 8-bit codewords. The telephone signals are sampled at 8 kHz (8,000 times per second), so the μ-law encoder receives $8{,}000 \times 14 = 112{,}000$ bits/sec. At a compression factor of 1.75, the encoder outputs 64,000 bits/sec. The G.711 standard [G.711 72] also specifies output rates of 48 Kbps and 56 Kbps.

The μ-law encoder receives a 14-bit *signed* input sample x. Thus, the input is in the range $[-8,192, +8,191]$. The sample is normalized to the interval $[-1, +1]$, and the encoder uses the logarithmic expression

$$\mathrm{sgn}(x)\frac{\ln(1+\mu|x|)}{\ln(1+\mu)},\quad \text{where } \mathrm{sgn}(x)=\begin{cases}+1, & x>0,\\ 0, & x=0,\\ -1, & x<0\end{cases}$$

(and μ is a positive integer), to compute and output an 8-bit code in the same interval $[-1,+1]$. The output is then scaled to the range $[-256,+255]$. Figure 10.12 shows this output as a function of the input for the three μ values 25, 255, and 2555. It is clear that large values of μ cause coarser quantization for larger amplitudes. Such values allocate more bits to the smaller, more important, amplitudes. The G.711 standard recommends the use of $\mu=255$. The diagram shows only the nonnegative values of the input (i.e., from 0 to 8191). The negative side of the diagram has the same shape but with negative inputs and outputs.

The A-law encoder uses the similar expression

$$\begin{cases}\mathrm{sgn}(x)\dfrac{A|x|}{1+\ln(A)}, & \text{for } 0\le|x|<\frac{1}{A},\\[2mm] \mathrm{sgn}(x)\dfrac{1+\ln(A|x|)}{1+\ln(A)}, & \text{for } \frac{1}{A}\le|x|<1.\end{cases}$$

The G.711 standard recommends the use of $A=87.6$.

The following simple examples illustrate the nonlinear nature of the μ-law. The two (normalized) input samples 0.15 and 0.16 are transformed by μ-law to outputs 0.6618 and 0.6732. The difference between the outputs is 0.0114. On the other hand, the two input samples 0.95 and 0.96 (bigger inputs but with the same difference) are transformed to 0.9908 and 0.9927. The difference between these two outputs is 0.0019; much smaller.

Bigger samples are decoded with more noise, and smaller samples are decoded with less noise. However, the signal-to-noise ratio (SNR, Section 7.4.2) is constant because both the μ-law and the SNR use logarithmic expressions.

Logarithms are slow to compute, so the μ-law encoder performs much simpler calculations that produce an approximation. The output specified by the G.711 standard is an 8-bit codeword whose format is shown in Figure 10.13.

Bit P in Figure 10.13 is the sign bit of the output (same as the sign bit of the 14-bit signed input sample). Bits S2, S1, and S0 are the segment code, and bits Q3 through Q0 are the quantization code. The encoder determines the segment code by (1) adding a bias of 33 to the absolute value of the input sample, (2) determining the bit position of the most significant 1-bit among bits 5 through 12 of the input, and (3) subtracting 5 from that position. The 4-bit quantization code is set to the four bits following the bit position determined in step 2. The encoder ignores the remaining bits of the input sample, and it inverts (1's complements) the codeword before it is output.

We use the input sample -656 as an example. The sample is negative, so bit P becomes 1. Adding 33 to the absolute value of the input yields $689=0001010110001_2$ (Figure 10.14). The most significant 1-bit in positions 5 through 12 is found at position 9. The segment code is thus $9-5=4$. The quantization code is the four bits 0101 at positions 8–5, and the remaining five bits 10001 are ignored. The 8-bit codeword (which is later inverted) becomes

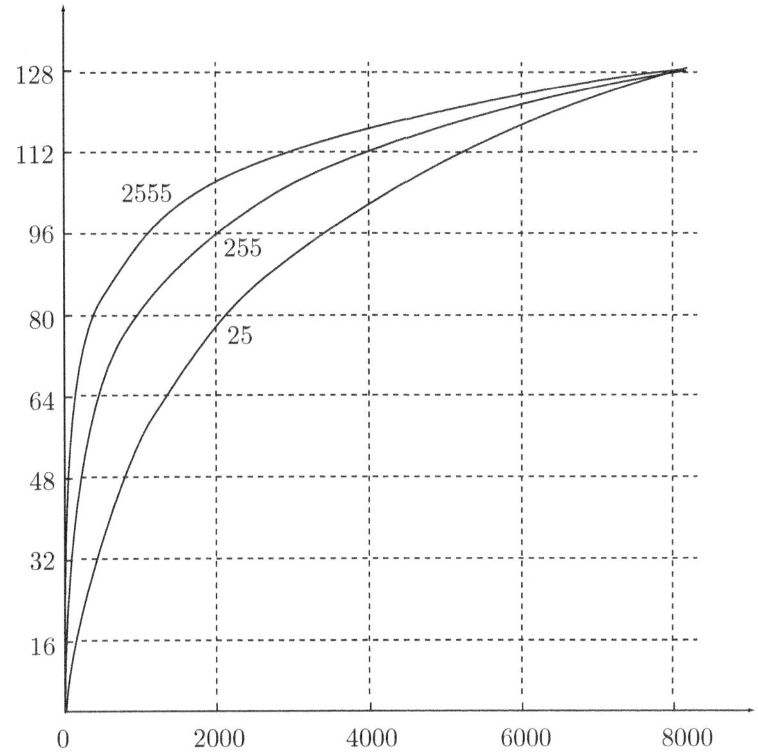

Figure 10.12: The μ-Law for μ Values of 25, 255, and 2555.

```
dat=linspace(0,1,1000);
mu=255;
plot(dat*8159,128*log(1+mu*dat)/log(1+mu));
```

Matlab code for Figure 10.12. Notice how the input is normalized to the range $[0,1]$ before the calculations, and how the output is scaled from the interval $[0,1]$ to $[0,128]$.

P	S2	S1	S0	Q3	Q2	Q1	Q0

Figure 10.13: G.711 μ-Law Codeword.

				Q3	Q2	Q1	Q0					
0	0	0	1	0	1	0	1	1	0	0	0	1
12	11	10	9	8	7	6	5	4	3	2	1	0

Figure 10.14: Encoding Input Sample -656.

P	S2	S1	S0	Q3	Q2	Q1	Q0
1	1	0	0	0	1	0	1

The μ-law decoder inputs an 8-bit codeword and inverts it. It then decodes it as follows:

1. Multiply the quantization code by 2 and add 33 (the bias) to the result.
2. Multiply the result by 2 raised to the power of the segment code.
3. Decrement the result by the bias.
4. Use bit P to determine the sign of the result.

Applying these steps to our example produces

1. The quantization code is $101_2 = 5$, so $5 \times 2 + 33 = 43$.
2. The segment code is $100_2 = 4$, so $43 \times 2^4 = 688$.
3. Decrement by the bias $688 - 33 = 655$.
4. Bit P is 1, so the final result is -655. Thus, the quantization error (the noise) is 1; very small.

Figure 10.15a illustrates the nature of the μ-law midtread quantization. Zero is one of the valid output values, and the quantization steps are centered at the input value of 0. The steps are organized in eight segments of 16 steps each. The steps within each segment have the same width, but they double in width from one segment to the next. If we denote the segment number by i (where $i = 0, 1, \ldots, 7$) and the width of a segment by k (where $k = 1, 2, \ldots, 16$), then the middle of the tread of each step in Figure 10.15a (i.e., the points labeled x_j) is given by

$$x(16i + k) = T(i) + k \times D(i), \tag{10.3}$$

where the constants $T(i)$ and $D(i)$ are the initial value and the step size for segment i, respectively. They are given by

i	0	1	2	3	4	5	6	7
$T(i)$	1	35	103	239	511	1055	2143	4319
$D(i)$	2	4	8	16	32	64	128	256

Table 10.16 lists some values of the breakpoints (points x_j) and outputs (points y_j) shown in Figure 10.15a.

The operation of the A-law encoder is similar, except that the quantization (Figure 10.15b) is of the midriser variety. The breakpoints x_j are given by Equation (10.3), but the initial value $T(i)$ and the step size $D(i)$ for segment i are different from those used by the μ-law encoder and are given by

i	0	1	2	3	4	5	6	7
$T(i)$	0	32	64	128	256	512	1024	2048
$D(i)$	2	2	4	8	16	32	64	128

Table 10.17 lists some values of the breakpoints (points x_j) and outputs (points y_j) shown in Figure 10.15b.

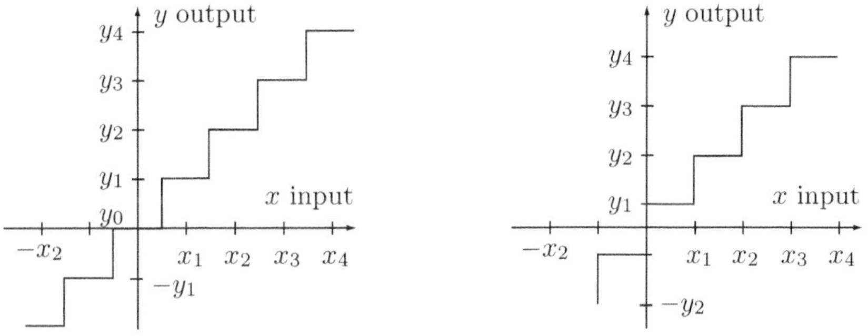

(a) (b)

Figure 10.15: (a) μ-Law Midtread Quantization. (b) A-Law Midriser Quantization.

segment 0		segment 1		...	segment 7	
break points	output values	break points	output values		break points	output values
	$y_0 = 0$		$y_{16} = 33$	...		$y_{112} = 4191$
$x_1 = 1$		$x_{17} = 35$			$x_{113} = 4319$	
	$y_1 = 2$		$y_{17} = 37$	...		$y_{113} = 4447$
$x_2 = 3$		$x_{18} = 39$			$x_{114} = 4575$	
	$y_2 = 4$		$y_{18} = 41$	...		$y_{114} = 4703$
$x_3 = 5$		$x_{19} = 43$			$x_{115} = 4831$	
	$y_3 = 6$		$y_{19} = 45$	...		$y_{115} = 4959$
$x_4 = 7$		$x_{20} = 47$			$x_{116} = 5087$	
...				...		
...				...		
$x_{15} = 29$		$x_{31} = 91$			$x_{127} = 7903$	
	$y_{15} = 28$		$y_{31} = 93$	...		$y_{127} = 8031$
$x_{16} = 31$		$x_{32} = 95$			$x_{128} = 8159$	

Table 10.16: Specification of the μ-Law Quantizer.

The A-law encoder generates an 8-bit codeword with the same format as the μ-law encoder. It sets the P bit to the sign of the input sample. It then determines the segment code in the following steps:

1. Determine the bit position of the most significant 1-bit among the seven most significant bits of the input.
2. If such a 1-bit is found, the segment code becomes that position minus 4. Otherwise, the segment code becomes zero.

The 4-bit quantization code is set to the four bits following the bit position determined in step 1, or to half the input value if the segment code is zero. The encoder ignores the remaining bits of the input sample, and it inverts bit P and the even-numbered bits of the codeword before it is output.

segment 0		segment 1		...	segment 7	
break points	output values	break points	output values		break points	output values
$x_0 = 0$		$x_{16} = 32$			$x_{112} = 2048$	
	$y_1 = 1$		$y_{17} = 33$	...		$y_{113} = 2112$
$x_1 = 2$		$x_{17} = 34$			$x_{113} = 2176$	
	$y_2 = 3$		$y_{18} = 35$	...		$y_{114} = 2240$
$x_2 = 4$		$x_{18} = 36$			$x_{114} = 2304$	
	$y_3 = 5$		$y_{19} = 37$	...		$y_{115} = 2368$
$x_3 = 6$		$x_{19} = 38$			$x_{115} = 2432$	
	$y_4 = 7$		$y_{20} = 39$	...		$y_{116} = 2496$
...				...		
...				...		
$x_{15} = 30$		$x_{31} = 62$			$x_{128} = 4096$	
	$y_{16} = 31$		$y_{32} = 63$	...		$y_{127} = 4032$

Table 10.17: Specification of the A-Law Quantizer.

The A-law decoder decodes an 8-bit codeword into a 13-bit audio sample as follows:

1. It inverts bit P and the even-numbered bits of the codeword.

2. If the segment code is nonzero, the decoder multiplies the quantization code by 2 and increments this by the bias (33). The result is then multiplied by 2 and raised to the power of the (segment code minus 1). If the segment code is 0, the decoder outputs twice the quantization code, plus 1.

3. Bit P is then used to determine the sign of the output.

Normally, the output codewords are generated by the encoder at the rate of 64 Kbps. The G.711 standard [G.711 72] also provides for two other rates, as follows:

1. To achieve an output rate of 48 Kbps, the encoder masks out the two least-significant bits of each codeword. This works, because $6/8 = 48/64$.

2. To achieve an output rate of 56 Kpbs, the encoder masks out the least-significant bit of each codeword. This works, because $7/8 = 56/64 = 0.875$.

This applies to both the μ-law and the A-law. The decoder typically fills up the masked bit positions with zeros before decoding a codeword.

10.6 ADPCM Audio Compression

Compression is possible only because audio, and therefore audio samples, tend to have redundancies. Adjacent audio samples tend to be similar in much the same way that neighboring pixels in an image tend to have similar colors. The simplest way to exploit this redundancy is to subtract adjacent samples and code the differences, which tend to be small integers. Any audio compression method based on this principle is called DPCM (differential pulse code modulation). Such methods, however, are inefficient, because they do not adapt themselves to the varying magnitudes of the audio stream. Better results are achieved by an adaptive version, and any such version is called ADPCM [ITU-T 90].

ADPCM employs linear prediction (this is also commonly used in predictive image compression). It uses the previous sample (or several previous samples) to predict the current sample. It then computes the difference between the current sample and its prediction, and quantizes the difference. For each input sample X[n], the output C[n] of the encoder is simply a certain number of quantization levels. The decoder multiplies this number by the quantization step (and may add half the quantization step, for better precision) to obtain the reconstructed audio sample. The method is efficient because the quantization step is updated all the time, by both encoder and decoder, in response to the varying magnitudes of the input samples. It is also possible to modify adaptively the prediction algorithm.

Various ADPCM methods differ in the way they predict the current audio sample and in the way they adapt to the input (by changing the quantization step size and/or the prediction method).

In addition to the quantized values, an ADPCM encoder can provide the decoder with *side information*. This information increases the size of the compressed stream, but this degradation is acceptable to the users, because it makes the compressed audio data more useful. Typical applications of side information are (1) help the decoder recover from errors and (2) signal an entry point into the compressed stream. An original audio stream may be recorded in compressed form on a medium such as a CD-ROM. If the user (listener) wants to listen to song 5, the decoder can use the side information to quickly find the start of that song.

Figure 10.18a,b shows the general organization of the ADPCM encoder and decoder. Notice that they share two functional units, a feature that helps in both software and hardware implementations. The adaptive quantizer receives the difference D[n] between the current input sample X[n] and the prediction Xp[n − 1]. The quantizer computes and outputs the quantized code C[n] of X[n]. The same code is sent to the adaptive dequantizer (the same dequantizer used by the decoder), which produces the next dequantized difference value Dq[n]. This value is added to the previous predictor output Xp[n − 1], and the sum Xp[n] is sent to the predictor to be used in the next step.

Better prediction would be obtained by feeding the actual input X[n] to the predictor. However, the decoder wouldn't be able to mimic that, since it does not have X[n]. We see that the basic ADPCM encoder is simple, and the decoder is even simpler. It inputs a code C[n], dequantizes it to a difference Dq[n], which is added to the preceding predictor output Xp[n − 1] to form the next output Xp[n]. The next output is also fed into the predictor, to be used in the next step.

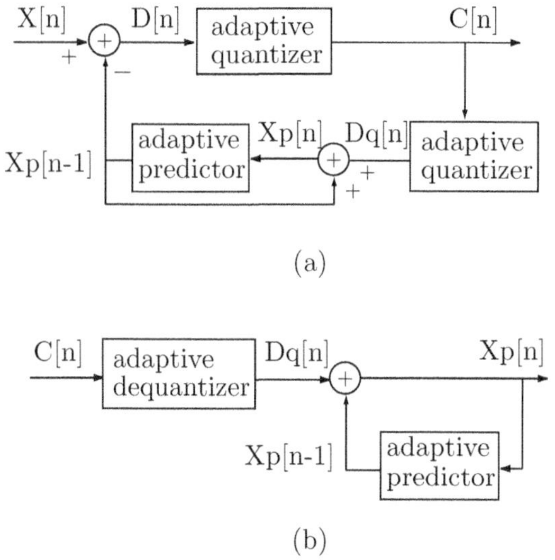

Figure 10.18: (a) ADPCM Encoder and (b) Decoder.

The remainder of this section describes the particular ADPCM algorithm adopted by the Interactive Multimedia Association [IMA 06]. The IMA is a consortium of computer hardware and software manufacturers, established to develop standards for multimedia applications. The goal of the IMA in developing its audio compression standard was to have a public domain method that is simple and fast enough such that a 20-MHz 386-class personal computer would be able to decode, in real time, sound recorded in stereo at 44,100 16-bit samples per second (this is 88,200 16-bit samples per second).

The encoder quantizes each 16-bit audio sample into a 4-bit code. The compression factor is therefore a constant 4.

The "secret" of the IMA algorithm is the simplicity of its predictor. The predicted value $\mathrm{Xp}[\mathrm{n}-1]$ that is output by the predictor is simply the decoded value Xp[n] of the preceding input X[n]. The predictor just stores Xp[n] for one cycle (one audio sample interval), then outputs it as $\mathrm{Xp}[\mathrm{n}-1]$. It does not use any of the preceding values Xp[i] to obtain better prediction. Thus, the predictor is not adaptive (but the quantizer is). Also, no side information is generated by the encoder.

Figure 10.19a is a block diagram of the IMA quantizer. It is both simple and adaptive, varying the quantization step size based on both the current step size and the previous quantizer output. The adaptation is done by means of two table lookups, so it is fast. The quantizer outputs 4-bit codes where the leftmost bit is a sign and the remaining three bits are the number of quantization levels computed for the current audio sample. These three bits are used as an index to the first table. The item found in this table serves as an index adjustment to the second table. The index adjustment is added to a previously stored index, and the sum, after being checked for proper range, is used as the index for the second table lookup. The sum is then stored, and it becomes the stored index used in the next adaptation step. The item found in the second table becomes the new quantization step size. Figure 10.19b illustrates this process, and

Tables 10.21 and 10.22 list the two tables. Table 10.20 shows the 4-bit output produced by the quantizer as a function of the sample size. For example, if the sample is in the range $[1.5ss, 1.75ss)$, where ss is the step size, then the output is 0|110.

Table 10.21 adjusts the index by bigger steps when the quantized magnitude is bigger. Table 10.22 is constructed such that the ratio between successive entries is about 1.1.

> ADPCM: Short for Adaptive Differential Pulse Code Modulation, a form of pulse code modulation (PCM) that produces a digital signal with a lower bit rate than standard PCM. ADPCM produces a lower bit rate by recording only the difference between samples and adjusting the coding scale dynamically to accommodate large and small differences.
>
> —From `Webopedia.com`

10.7 MLP Audio

Note. The MLP audio compression method described in this section is different from and unrelated to the MLP (multilevel progressive) image compression method of Section 7.25. The identical acronyms are an unfortunate coincidence.

> Ambiguity refers to the property of words, terms, and concepts, (within a particular context) as being in undefined, undefinable, or otherwise vague, and thus having an unclear meaning. A word, phrase, sentence, or other communication is called "ambiguous" if it can be interpreted in more than one way.
>
> —From `http://en.wikipedia.org/wiki/Disambiguation`

Meridian [Meridian 03] is a British company specializing in high-quality audio products, such as CD and DVD players, loudspeakers, radio tuners, and surround sound stereo amplifiers. Good-quality digitized sound, such as found in an audio CD, employs two channels (stereo sound), each sampled at 44.1 kHz with 16-bit samples (Section 10.2). A typical high-quality digitized sound, on the other hand, may use six channels (i.e., the sound is originally recorded by six microphones, for surround sound), sampled at the high rate of 96 kHz (to ensure that all the nuances of the performance can be delivered), with 24-bit samples (to get the highest possible dynamic range). This kind of audio data is represented by $6\times 96{,}000\times 24 = 13.824$ Mbps (that's megabits, not megabytes). In contrast, a DVD (digital versatile disc) holds 4.7 Gbytes, which at 13.824 Mbps in uncompressed form translates to only 45 minutes of playing. (Recall that even CDs, which have a much smaller capacity, hold 74 minutes of play time. This is an industry standard.) Also, the maximum data transfer rate for DVD-A (audio) is 9.6 Mbps, much lower than 13.824 Mbps. It is obvious that compression is the key to achieving a practical DVD-A format, but the high quality (as opposed to just good quality) requires lossless compression.

The algorithm that has been selected as the compression standard for DVD-A (audio) is MLP (Meridian Lossless Packing). This algorithm is patented, and some of its details are still kept proprietary, which is reflected in the information provided in this

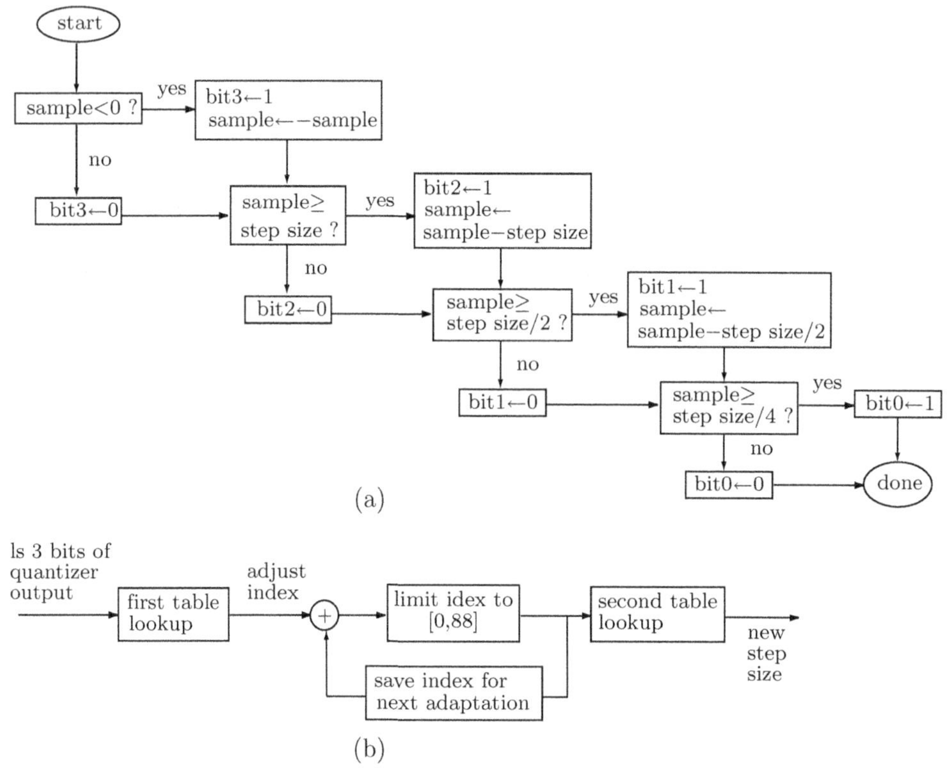

Figure 10.19: (a) IMA ADPCM Quantization. (b) Step Size Adaptation.

If sample is in range	4-Bit quant	If sample is in range	4-Bit quant
$[1.75ss, \infty)$	0|111	$[-\infty, -1.75ss)$	1|111
$[1.5ss, 1.75ss)$	0|110	$[-1.75ss, -1.5ss)$	1|110
$[1.25ss, 1.5ss)$	0|101	$[-1.5ss, -1.25ss)$	1|101
$[1ss, 1.25ss)$	0|100	$[-1.25ss, -1ss)$	1|100
$[.75ss, 1ss)$	0|011	$[-1ss, -.75ss)$	1|011
$[.5ss, .75ss)$	0|010	$[-.75ss, -.5ss)$	1|010
$[.25ss, .5ss)$	0|001	$[-.5ss, -.25ss)$	1|001
$[0, .25ss)$	0|000	$[-.25ss, 0)$	1|000

Table 10.20: Step Size and 4-Bit Quantizer Outputs.

three bits quantized magnitude	index adjust
000	−1
001	−1
010	−1
011	−1
100	2
101	4
110	6
111	8

Table 10.21: First Table for IMA ADPCM.

Index	Step Size	Index	Step Size	Index	Step Size	Index	Step Size
0	7	22	60	44	494	66	4,026
1	8	23	66	45	544	67	4,428
2	9	24	73	46	598	68	4,871
3	10	25	80	47	658	69	5,358
4	11	26	88	48	724	70	5,894
5	12	27	97	49	796	71	6,484
6	13	28	107	50	876	72	7,132
7	14	29	118	51	963	73	7,845
8	16	30	130	52	1,060	74	8,630
9	17	31	143	53	1,166	75	9,493
10	19	32	157	54	1,282	76	10,442
11	21	33	173	55	1,411	77	11,487
12	23	34	190	56	1,552	78	12,635
13	25	35	209	57	1,707	79	13,899
14	28	36	230	58	1,878	80	15,289
15	31	37	253	59	2,066	81	16,818
16	34	38	279	60	2,272	82	18,500
17	37	39	307	61	2,499	83	20,350
18	41	40	337	62	2,749	84	22,358
19	45	41	371	63	3,024	85	24,633
20	50	42	408	64	3,327	86	27,086
21	55	43	449	65	3,660	87	29,794
						88	32,767

Table 10.22: Second Table for IMA ADPCM.

section. The term "packing" has a dual meaning. It refers to (1) removing redundancy from the original data in order to "pack" it densely, and (2) the audio samples are encoded in packets.

MLP operates by reducing or completely removing redundancies in the digitized audio, without any quantization or other loss of data. Notice that high-quality audio formats such as 96 kHz with 24-bit samples carry more information than is strictly necessary for the human listener (or more than is available from modern microphone and converter techniques). Thus, such audio formats contain much redundancy and can be compressed efficiently. MLP can handle up to 63 audio channels and sampling rates of up to 192 kHz.

The main features of MLP are as follows:

1. At least 4 bits/sample of compression for both average and peak data rates.
2. Easy transformation between fixed-rate and variable-rate data streams.
3. Careful and economical handling of mixed input sample rates.
4. Simple, fast decoding.
5. It is cascadable. An audio stream can be encoded and decoded multiple times in succession, and the output will always be an exact copy of the original. With MLP, there is no generation loss.

The term "variable data rate" is important. An uncompressed data stream consists of audio samples, and each second of sound requires the same number of samples. Such a stream has a fixed data rate. In contrast, the compressed stream generated by a lossless audio encoder has a variable data rate; each second of sound occupies a different number of bits in this stream, depending on the nature of the sound. A second of silence occupies very few bits, whereas a second of random sound will not compress and will require the same number of bits in the compressed stream as in the original file. Most lossless audio compression methods are designed to reduce the average data rate, but MLP has the important feature that it reduces the instantaneous peak data rate by a known amount. This feature makes it possible to record 74 min of any kind of nonrandom sound on a 4.7-Gbyte DVD-A.

In addition to being lossless (which means that the original data is delivered bit-for-bit at the playback), MLP is also robust. It does not include any error-correcting code but has error-protection features. It uses check bits to make sure that each packet decompressed by the decoder is identical to that compressed by the encoder. The compressed stream contains restart points, placed at intervals of 10–30 ms. When the decoder notices an error, it simply skips to the next restart point, with a minimal loss of sound. This is another meaning of the term "high-quality sound." For the first time, a listener hears exactly what the composer/performer intended—bit-for-bit and note-for-note.

With lossy audio compression, the amount of compression is measured by the number of bits per second of sound in the compressed stream, regardless of the audio-sample size. With lossless compression, a large sample size (which really means more least-significant bits), must be losslessly compressed, so it increases the size of the compressed stream, but the extra LSBs typically have little redundancy and are therefore harder to compress. This is why lossless audio compression should be measured by the number of bits saved in each audio sample—a relative measure of compression.

MLP reduces the audio samples from their original size (typically 24 bits) depending on the sampling rate. For average data rates, the reduction is as follows: For sampling rates of 44.1 kHz and 48 kHz, a sample is reduced by 5 to 11 bits. At 88.2 kHz and 96 kHz, the reduction increases to 9 to 13 bits. At 192 kHz, MLP can sometimes reduce a sample by 14 bits. Even more important are the savings for peak data rates. They are 4 bits for 44.1 kHz, 8 bits for 96 kHz, and 9 bits for 192 kHz samples. These peak data rate savings amount to a virtual guarantee, and they are one of the main reasons for the adoption of MLP as the DVD-A compression standard.

The remainder of this section covers some of the details of MLP. It is based on [Stuart et al. 99]. The techniques used by MLP to compress audio samples include:

1. It looks for blocks of consecutive audio samples that are small, i.e., have several most-significant 0 bits. This is known as "dead air." The LSBs of such samples are removed, which is equivalent to shifting the samples to the right.
2. It identifies channels that do not use their full bandwidth.
3. It identifies and removes correlation between channels.
4. It removes correlation between consecutive audio samples in each channel.
5. It uses buffering to smooth the production rate of the compressed output.

The main compression steps are (1) lossless processing, (2) lossless matrixing, (3) lossless IIR filtering, (4) entropy encoding, and (5) FIFO buffering of the output. Lossless processing refers to identifying blocks of audio samples with unused capacity. Such samples are shifted. The term "matrixing" refers to reducing interchannel correlations by means of an affine transformation matrix. The IIR filter decorrelates the audio samples in each channel (intrachannel decorrelation) by predicting the next sample from its predecessors and producing a prediction difference. The differences are then compressed by an entropy encoder. The result is a compressed stream for each channel, and those streams are FIFO buffered to smooth the rate at which the encoded data is output. Finally, the output from the buffer is divided into packets, and check bits and restart points are added to each packet for error protection.

An audio file with several channels is normally obtained by recording sound with several microphones placed at different locations. Each microphone outputs an analog signal that's digitized and becomes an audio channel. Thus, the audio samples in the various channels are correlated, and reducing or removing this correlation is an important step, termed "matrixing," in MLP compression. Denoting audio sample i of channel j by a_{ij}, matrixing is done by subtracting from a_{ij} a linear combination of audio samples a_{ik} from all other channels k. Thus,

$$a_{ij} \leftarrow a_{ij} - \sum_{k \neq j} a_{ik} m_{kj},$$

where column j of matrix m corresponds to audio channel j. This linear combination can also be viewed as a weighted sum where each audio sample a_{ik} is multiplied by a weight m_{kj}. It can also be interpreted as an affine transformation of the vector of audio samples $(a_{i,1}, a_{i,2}, \ldots, a_{i,j-1}, a_{i,j+1}, \ldots, a_{i,n})$.

The next MLP encoding step is prediction. The matrixing step may remove all or part of the correlation between audio channels, but the individual samples in each channel are still correlated; a sample tends to be similar to its near neighbors, and the samples in an audio channel make up what is called *quasi-periodic components*. MLP uses linear prediction where a special IIR filter is applied to the data to remove the correlated or timbral portion of the signal. The filter leaves a low-amplitude residual signal that is aperiodic and resembles noise. This residual signal represents sound components such as audio transients, phase relationships between partial harmonics, and even extra rosin on strings (no kidding).

The entropy encoding step follows the matrixing (removing interchannel correlations) and prediction (removing intra-channel correlations). This step gets rid of any remaining correlations between the original audio samples. Several entropy encoders are built into the MLP encoder, and it may select any of them to encode any audio channel. Generally, MLP assumes that audio signals have a Laplacian distribution about zero, and its entropy encoders assign variable-length codes to the audio samples based on this assumption (recall that the MLP image compression method of Section 7.25 also uses this distribution, which is displayed graphically in Figure 7.143b).

Buffering is the next step in MLP encoding. It smooths out the variations in bitrate caused by variations in the sound that's being compressed. Passages of uniform sound or even silence may alternate with complex, random-like sound. When compressing such sound, the MLP encoder may output a few bits for each second of the uniform passages (which compress well) followed by many bits for each second for those parts of the sound that don't compress well. Buffering ensures that the encoder outputs data to the compressed stream at a constant rate. This may not be very important when the compressed stream is a file, but it can make a big difference when the compressed audio is transmitted and is supposed to be received by an MLP decoder and played at real time without pauses.

In addition to the compressed data, the compressed stream includes instructions to the decoder and CRC check bits. The input audio file is divided into blocks that are typically 40–160 samples long each. Blocks are assembled into packets whose length varies between 640 and 2,560 samples and is controlled by the user. Each packet is self-checked and contains restart information. If the decoder senses bad data and cannot continue decompressing a packet, it simply skips to the start of the next packet, with a typical loss of just 7 ms of play time—hardly noticeable by a listener.

MLP also has lossy options. One such option is to reduce all audio samples from 24 to 22 bits. Another is to pass some of the audio channels through a lowpass filter. This tends to reduce the entropy of the audio signal, making it possible for the MLP encoder to produce better compression. Such options may be important in cases where more than 74 min of audio should be recorded on one DVD-A.

10.8 Speech Compression

Certain audio codecs are designed specifically to compress speech signals. Such signals are audio and are sampled like any other audio data, but because of the nature of human speech, they have properties that can be exploited for efficient compression. [Jayant 97] covers several types of speech codecs, but more information on this topic is available on the World Wide Web in researchers' sites. This section starts by discussing the properties of human speech and continues with a description of various speech codecs.

10.8.1 Properties of Speech

We produce sound by forcing air from the lungs through the vocal cords into the vocal tract (Figure 10.23). The air ends up escaping through the mouth, but the sound is generated in the vocal tract (which extends from the vocal cords to the mouth, and in an average adult it is 17 cm long) by vibrations in much the same way as air passing through a flute generates sound. The pitch of the sound is controlled by varying the shape of the vocal tract (mostly by moving the tongue) and by moving the lips. The intensity (loudness) is controlled by varying the amount of air sent from the lungs. Humans are much slower than computers or other electronic devices, and this is also true with regard to speech. The lungs operate slowly, and the vocal tract changes shape slowly, so the pitch and loudness of speech vary slowly. When speech is captured by a microphone and is sampled, we find that adjacent samples are similar, and even samples separated by 20 ms are strongly correlated. This correlation is the basis of speech compression.

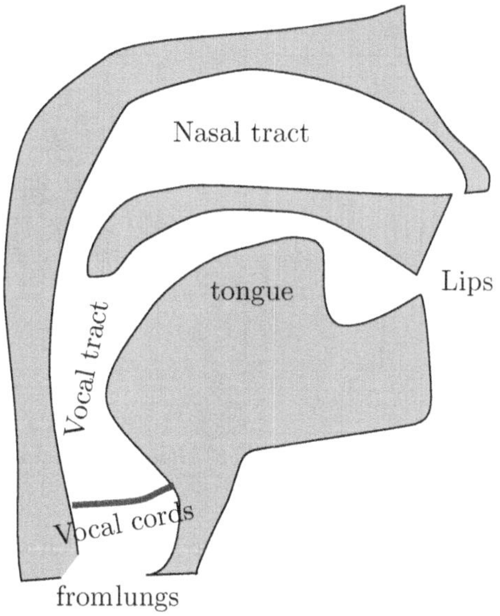

Figure 10.23: A Cross Section of the Human Head.

The vocal cords can open and close, and the opening between them is called the glottis. The movements of the glottis and vocal tract give rise to different types of sound. The three main types are as follows:

1. Voiced sounds. These are the sounds we make when we talk. The vocal cords vibrate, which opens and closes the glottis, thereby sending pulses of air at varying pressures to the tract, where it is shaped into sound waves. Varying the shape of the vocal cords and their tension changes the rate of vibration of the glottis and therefore controls the pitch of the sound. Recall that the ear is sensitive to sound frequencies of from 16 Hz to about 20,000–22,000 Hz. The frequencies of the human voice, on the other hand, are much more restricted and are generally in the range of 500 Hz to about 2 kHz. This is equivalent to time periods of 2 ms to 20 ms, and to a computer, such

periods are very long. Thus, voiced sounds have long-term periodicity, and this is the key to good speech compression. Figure 10.24a is a typical example of waveforms of voiced sound.

2. Unvoiced sounds. These are sounds that are emitted and can be heard, but are not parts of speech. Such a sound is the result of holding the glottis open and forcing air through a constriction in the vocal tract. When an unvoiced sound is sampled, the samples show little correlation and are random or close to random. Figure 10.24b is a typical example of the waveforms of unvoiced sound.

3. Plosive sounds. These result when the glottis closes, the lungs apply air pressure on it, and it suddenly opens, letting the air escape suddenly. The result is a popping sound.

There are also combinations of the above three types. An example is the case where the vocal cords vibrate and there is also a constriction in the vocal tract. Such a sound is called fricative.

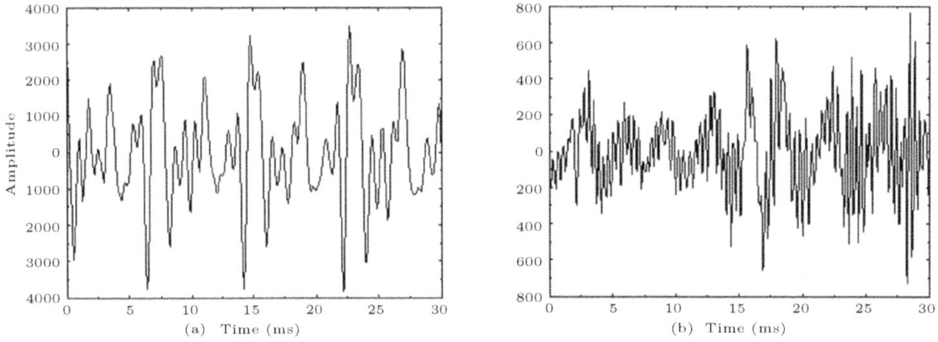

Figure 10.24: (a) Voiced and (b) Unvoiced Sound Waves.

Speech codecs. There are three main types of speech codecs. Waveform speech codecs produce good to excellent speech after compressing and decompressing it, but generate bitrates of 10–64 kbps. Source codecs (also called vocoders) generally produce poor to fair speech but can compress it to very low bitrates (down to 2 kbps). Hybrid codecs are combinations of the former two types and produce speech that varies from fair to good, with bitrates between 2 and 16 kbps. Figure 10.25 illustrates the speech quality versus bitrate of these three types.

10.8.2 Waveform Codecs

This type of codec does not attempt to predict how the original sound was generated. It only tries to produce, after decompression, audio samples that are as close to the original ones as possible. Thus, such codecs are not designed specifically for speech coding and can perform equally well on all kinds of audio data. As Figure 10.25 illustrates, when such a codec is forced to compress sound to less than 16 kbps, the quality of the reconstructed sound drops significantly.

The simplest waveform encoder is pulse code modulation (PCM). This encoder simply quantizes each audio sample. Speech is typically sampled at only 8 kHz. If

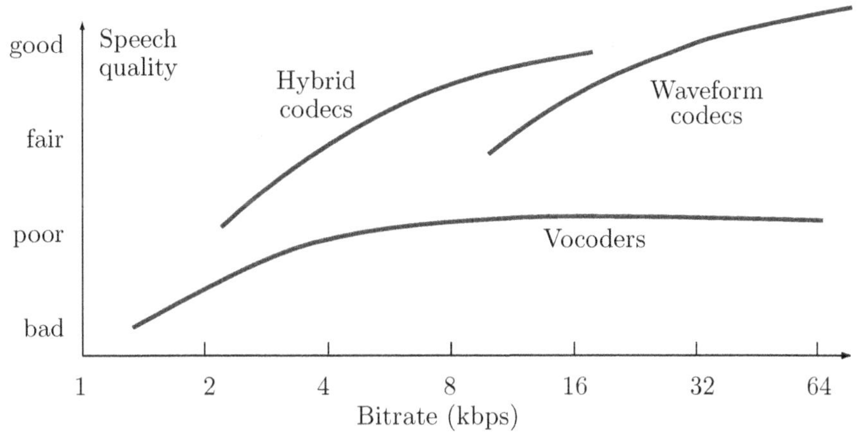

Figure 10.25: Speech Quality Versus Bitrate for Speech Codecs.

each sample is quantized to 12 bits, the resulting bitrate is $8k \times 12 = 96$ kbps and the reproduced speech sounds almost natural. Better results are obtained with a logarithmic quantizer, such as the μ-law and A-law companding methods (Section 10.5). They quantize audio samples to varying numbers of bits and may compress speech to 8 bits per sample on average, thereby resulting in a bitrate of 64 kbps, with very good quality of the reconstructed speech.

A differential PCM (DPCM; Section 10.6) speech encoder uses the fact that the audio samples of voiced speech are correlated. This type of encoder computes the difference between the current sample and its predecessor and quantizes the difference. An adaptive version (ADPCM; Section 10.6) may compress speech at good quality down to a bitrate of 32 kbps.

Waveform coders may also operate in the frequency domain. The subband coding algorithm (SBC) transforms the audio samples to the frequency domain, partitions the resulting coefficients into several critical bands (or frequency subbands), and codes each subband separately with ADPCM or a similar quantization method. The SBC decoder decodes the frequency coefficients, recombines them, and performs the inverse transformation to (lossily) reconstruct audio samples. The advantage of SBC is that the ear is sensitive to certain frequencies and less sensitive to others (Section 10.3, especially Table 10.6). Subbands of frequencies to which the ear is less sensitive can therefore be coarsely quantized without loss of sound quality. This type of coder typically produces good reconstructed speech quality at bitrates of 16–32 kbps. They are, however, more complex to implement than PCM codecs and may also be slower.

The adaptive transform coding (ATC) speech compression algorithm transforms audio samples to the frequency domain with the discrete cosine transform (DCT). The audio file is divided into blocks of audio samples and the DCT is applied to each block, resulting in a number of frequency coefficients. Each coefficient is quantized according to the frequency to which it corresponds. Good quality reconstructed speech can be achieved at bitrates as low as 16 kbps.

10.8.3 Source Codecs

In general, a source encoder uses a mathematical model of the source of data. The model depends on certain parameters, and the encoder uses the input data to compute those parameters. Once the parameters are obtained, they are written (after being suitably encoded) on the compressed stream. The decoder inputs the parameters and employs the mathematical model to reconstruct the original data. If the original data is audio, the source coder is called vocoder (from "vocal coder"). The vocoder described in this section is the linear predictive coder (LPC, see [Rabiner and Schafer 78]). Figure 10.26 shows a simplified model of speech production. Part (a) illustrates the process in a person, whereas part (b) shows the corresponding LPC mathematical model.

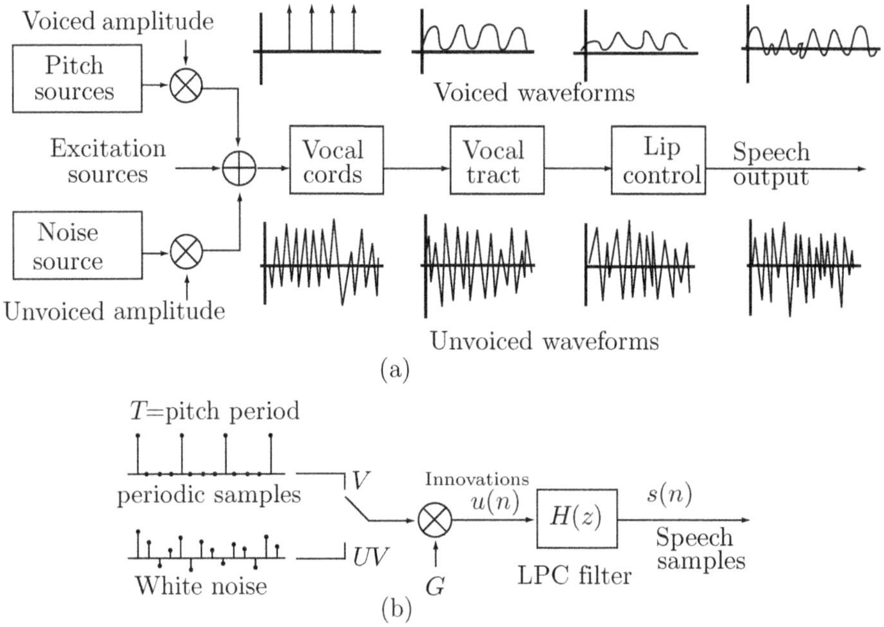

Figure 10.26: Speech Production: (a) Real. (b) LPC Model.

In this model, the output is the sequence of speech samples $s(n)$ coming out of the LPC filter (which corresponds to the vocal tract and lips). The input $u(n)$ to the model (and to the filter) is either a train of pulses (when the sound is voiced speech) or white noise (when the sound is unvoiced speech). The quantities $u(n)$ are also termed *innovation*. The model illustrates how samples $s(n)$ of speech can be generated by mixing innovations (a train of pulses and white noise). Thus, it represents mathematically the relation between speech samples and innovations. The task of the speech encoder is to input samples $s(n)$ of actual speech, use the filter as a mathematical function to determine an equivalent sequence of innovations $u(n)$, and output the innovations in compressed form. The correspondence between the model's parameters and the parts of real speech is as follows:

1. Parameter V (voiced) corresponds to the vibrations of the vocal cords. UV expresses the unvoiced sounds.

2. T is the period of the vocal cords vibrations.

3. G (gain) corresponds to the loudness or the air volume sent from the lungs each second.

4. The innovations $u(n)$ correspond to the air passing through the vocal tract.

5. The symbols $\otimes$ and $\oplus$ denote amplification and combination, respectively.

The main equation of the LPC model describes the output of the LPC filter as

$$H(z) = \frac{1}{1 + a_1 z^{-1} + a_2 z^{-2} + \cdots + a_{10} z^{-10}},$$

where z is the input to the filter [the value of one of the $u(n)$]. An equivalent equation describes the relation between the innovations $u(n)$ on the one hand and the 10 coefficients a_i and the speech audio samples $s(n)$ on the other hand. The relation is

$$u(n) = s(n) + \sum_{i=1}^{10} a_i s(n-i). \tag{10.4}$$

This relation implies that each number $u(n)$ input to the LPC filter is the sum of the current audio sample $s(n)$ and a weighted sum of the 10 preceding samples.

The LPC model can be written as the 13-tuple

$$\mathbf{A} = (a_1, a_2, \ldots, a_{10}, G, V/UV, T),$$

where V/UV is a single bit specifying the source (voiced or unvoiced) of the input samples. The model assumes that $\mathbf{A}$ stays stable for about 20 ms, then gets updated by the audio samples of the next 20 ms. At a sampling rate of 8 kHz, there are 160 audio samples $s(n)$ every 20 ms. The model computes the 13 quantities in $\mathbf{A}$ from these 160 samples, writes $\mathbf{A}$ (as 13 numbers) on the compressed stream, then repeats for the next 20 ms. The resulting compression factor is therefore 13 numbers for each set of 160 audio samples.

It's important to distinguish the operation of the encoder from the diagram of the LPC's mathematical model depicted in Figure 10.26b. The figure shows how a sequence of innovations $u(n)$ generates speech samples $s(n)$. The encoder, however, starts with the speech samples. It inputs a 20-ms sequence of speech samples $s(n)$, computes an equivalent sequence of innovations, compresses them to 13 numbers, and outputs the numbers after further encoding them. This repeats every 20 ms.

LPC encoding (or analysis) starts with 160 sound samples and computes the 10 LPC parameters a_i by minimizing the energy of the innovation $u(n)$. The energy is the function

$$f(a_1, a_2, \ldots, a_{10}) = \sum_{i=0}^{159} u^2(n)$$

and its minimum is computed by differentiating it 10 times, with respect to each of its 10 parameters and setting the derivatives to zero. The 10 equations

$$\frac{\partial f}{\partial a_i}=0, \quad \text{for} \quad i=1,2,\ldots,10$$

can be written in compact notation

$$\begin{bmatrix} R(0) & R(1) & R(2) & R(3) & R(4) & R(5) & R(6) & R(7) & R(8) & R(9)\\ R(1) & R(0) & R(1) & R(2) & R(3) & R(4) & R(5) & R(6) & R(7) & R(8)\\ R(2) & R(1) & R(0) & R(1) & R(2) & R(3) & R(4) & R(5) & R(6) & R(7)\\ R(3) & R(2) & R(1) & R(0) & R(1) & R(2) & R(3) & R(4) & R(5) & R(6)\\ R(4) & R(3) & R(2) & R(1) & R(0) & R(1) & R(2) & R(3) & R(4) & R(5)\\ R(5) & R(4) & R(3) & R(2) & R(1) & R(0) & R(1) & R(2) & R(3) & R(4)\\ R(6) & R(5) & R(4) & R(3) & R(2) & R(1) & R(0) & R(1) & R(2) & R(3)\\ R(7) & R(6) & R(5) & R(4) & R(3) & R(2) & R(1) & R(0) & R(1) & R(2)\\ R(8) & R(7) & R(6) & R(5) & R(4) & R(3) & R(2) & R(1) & R(0) & R(1)\\ R(9) & R(8) & R(7) & R(6) & R(5) & R(4) & R(3) & R(2) & R(1) & R(0) \end{bmatrix} \begin{bmatrix} a_1\\ a_2\\ a_3\\ a_4\\ a_5\\ a_6\\ a_7\\ a_8\\ a_9\\ a_{10} \end{bmatrix} = \begin{bmatrix} -R(1)\\ -R(2)\\ -R(3)\\ -R(4)\\ -R(5)\\ -R(6)\\ -R(7)\\ -R(8)\\ -R(9)\\ -R(10) \end{bmatrix},$$

where $R(k)$, which is defined as

$$R(k)=\sum_{n=0}^{159-k} s(n)s(n+k),$$

is the autocorrelation function of the samples $s(n)$. This system of 10 algebraic equations is easy to solve numerically, and the solutions are the 10 LPC parameters a_i. The remaining three parameters, V/UV, G, and T, are determined from the 160 audio samples. If those samples exhibit periodicity, then T becomes that period and the 1-bit parameter V/UV is set to V. If the 160 samples do not feature any well-defined period, then T remains undefined and V/UV is set to UV. The value of G is determined by the largest sample.

LPC decoding (or synthesis) starts with a set of 13 LPC parameters and computes 160 audio samples as the output of the LPC filter by

$$H(z)=\frac{1}{1+a_1z^{-1}+a_2z^{-2}+\cdots+a_{10}z^{-10}}.$$

These samples are played at 8,000 samples per second and result in 20 ms of (voiced or unvoiced) reconstructed speech.

The 2.4-kbps version of the LPC vocoder goes through one more step to encode the 10 parameters a_i. It converts them to 10-line spectrum pair (LSP) parameters, denoted by ω_i, by means of the relations

$$\begin{aligned} P(z)&=1+(a_1-a_{10})z^{-1}+(a_2-a_9)z^{-2}+\cdots+(a_{10}-a_1)z^{-10}-z^{-11}\\ &=(1-z^{-1})\Pi_{k=2,4,\ldots,10}(1-2\cos\omega_k z^{-1}+z^{-2}),\\ Q(z)&=1+(a_1+a_{10})z^{-1}+(a_2+a_9)z^{-2}+\cdots+(a_{10}+a_1)z^{-10}+z^{-11}\\ &=(1+z^{-1})\Pi_{k=1,3,\ldots,9}(1-2\cos\omega_k z^{-1}+z^{-2}). \end{aligned}$$

The ten LSP parameters satisfy $0 < \omega_1 < \omega_2 < \cdots < \omega_{10} < \pi$. Each is quantized to either three or four bits, as shown here, for a total of 34 bits.

ω_1	ω_2	ω_3	ω_4	ω_5	ω_6	ω_7	ω_8	ω_9	ω_{10}
3	4	4	4	4	3	3	3	3	3

The gain parameter G is encoded in seven bits, the period T is quantized to six bits, and the V/UV parameter requires just one bit. Thus, the 13 LPC parameters are quantized to 48 bits and provide enough data for a 20-ms frame of decompressed speech. Each second of speech has 50 frames, so the bitrate is $50 \times 48 = 2.4$ kbps. Assuming that the original speech samples were eight bits each, the compression factor is $(8{,}000\times8)/2{,}400 = 26.67$—very impressive.

10.8.4 Hybrid Codecs

This type of speech codec combines features from both waveform and source codecs. The most popular hybrid codecs are Analysis-by-Synthesis (AbS) time-domain algorithms. Like the LPC vocoder, these codecs model the vocal tract by a linear prediction filter, but use an excitation signal instead of the simple, two-state voice-unvoice model to supply the $u(n)$ (innovation) input to the filter. Thus, an AbS encoder starts with a set of speech samples (a frame), encodes them similar to LPC, decodes them, and subtracts the decoded samples from the original ones. The differences are sent through an error minimization process that outputs improved encoded samples. These samples are again decoded, subtracted from the original samples, and new differences computed. This is repeated until the differences satisfy a termination condition. The encoder then proceeds to the next set of speech samples (next frame).

One of the best-known AbS codecs is CELP, an acronym that stands for code-excited linear predictive. The 4.8-kbps CELP codec is similar to the LPC vocoder, with the following differences:

1. The frame size is 30 ms (i.e., 240 speech samples per frame and 33.3 frames per second of speech).
2. The innovation $u(n)$ is coded directly.
3. A pitch prediction is included in the algorithm.
4. Vector quantization is used.

Each frame is encoded to 144 bits as follows: The LSP parameters are quantized to a total of 34 bits. The pitch prediction filter becomes a 48-bit number. Codebook indices consume 36 bits. Four gain parameters require 5 bits each. One bit is used for synchronization, four bits encode the forward error control (FEC, similar to CRC), and one bit is reserved for future use. At 33.3 frames per second, the bitrate is therefore $33.3 \times 144 = 4{,}795 \approx 4.8$ kbps

There is also a conjugate-structured algebraic CELP (CS-ACELP) that uses 10 ms frames (i.e., 80 samples per frame) and encodes the LSP parameters with a two-stage vector quantizer. The gains (two per frame) are also encoded with vector quantization.

10.9 Shorten

Shorten is a simple, special-purpose, lossless compressor for waveform files. Any file whose data items (which are referred to as *samples*) go up and down as in a wave can be efficiently compressed by this method. Its developer [Robinson 94] had in mind applications to speech compression (where audio files with speech are distributed on a CD-ROM), but any other waveform files can be similarly compressed. The compression performance of Shorten isn't as good as that of mp3, but Shorten is lossless. Shorten performs best on files with low-amplitude and low-frequency samples, where it yields compression factors of 2 or better. It has been implemented on UNIX and on MS-DOS and is freely available at [Softsound 03].

Shorten encodes the individual samples of the input file by partitioning the file into blocks, predicting each sample from some of its predecessors, subtracting the prediction from the sample, and encoding the difference with a special variable-length code. It also has a lossy mode, where samples are quantized before being compressed. The algorithm has been implemented by its developer and can input files in the audio formats ulaw (Section 10.5), s8, u8, s16 (this is the default input format), u16, s16x, u16x, s16hl, u16hl, s16lh, and u16lh, where "s" and "u" stand for "signed" and "unsigned," respectively, a trailing "x" specifies byte-mapped data, "hl" implies that the high-order byte of a sample is followed in the file by the low-order byte, and "lh" signifies the reverse.

An entire file is encoded in blocks, where the block size (typically 128 or 256 samples) is specified by the user and has a default value of 256. (At a sampling rate of 16 kHz, this is equivalent to 16 ms of sound.) The samples in the block are first converted to integers with an expected mean of 0. The idea is that the samples within each block have the same spectral characteristic and can therefore be predicted accurately. Some audio files consist of several interleaved channels, so Shorten starts by separating the channels in each block. Thus, if the file has two channels and the samples are interleaved as L_1, R_1, L_2, R_2, and so on up to L_b, R_b, the first step creates the two blocks $(L_1, L_2, \ldots, L_b)$ and $(R_1, R_2, \ldots, R_b)$ and each block is then compressed separately. In practice, blocks that correspond to audio channels are often highly correlated, so sophisticated methods, such as MLP (Section 10.7), try to remove interblock correlations before tackling the samples within a block.

Once a block has been constructed, its samples are predicted and difference values are computed. A predicted value $\hat{s}(t)$ for the current sample $s(t)$ is computed from the p immediately-preceding samples by a linear combination (see also Section 7.30)

$$\hat{s}(t) = \sum_{i=1}^{p} a_i s(t-i).$$

If the prediction is done properly, then the difference (also termed error or residual) $e(t) = s(t) - \hat{s}(t)$ will almost always be a small (positive or negative) number. The simplest type of wave is stationary. In such a wave, a single set of coefficients a_i always produces the best prediction. Naturally, most waves are not stationary and should select a different set of a_i coefficients to predict each sample. Such selection can be done in different ways, involving more and more neighbor samples, and this results in predictors of different orders.

A zeroth-order predictor simply predicts each sample $s(t)$ as zero. A first-order predictor (Figure 10.27a) predicts each $s(t)$ as its predecessor $s(t-1)$. Similarly, a second-order predictor (Figure 10.27b) computes a straight segment (a linear function or a degree-1 polynomial) from $s(t-2)$ to $s(t-1)$ and continues it to predict $s(t)$. Extending this idea to one more point, a third-order predictor (Figure 10.27c) computes a degree-2 polynomial (a conic section) that passes through the three points $s(t-3)$, $s(t-2)$, and $s(t-1)$ and extrapolates it to predict $s(t)$. In general, an nth-order predictor computes a degree-$(n-1)$ polynomial that passes through the n points $s(t-n)$ through $s(t-1)$ and extrapolates it to predict $s(t)$. We show how to compute second- and third-order predictors.

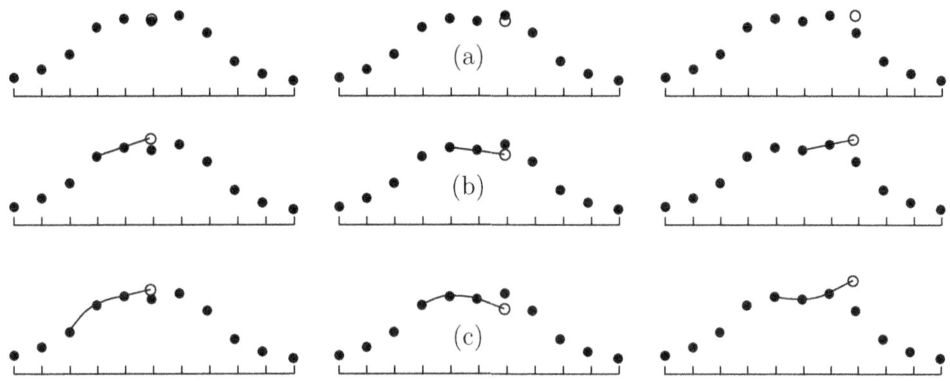

Figure 10.27: Predictors of Orders 1, 2, and 3.

Given the two points $P_2=(t-2,s_2)$ and $P_1=(t-1,s_1)$, we can write the parametric equation of the straight segment connecting them as

$$L(u)=(1-u)P_2+u\,P_1=(1-u)(t-2,s_2)+u(t-1,s_1)=(u+t-2,(1-u)s_2+u\,s_1).$$

It's easy to see that $L(0)=P_2$ and $L(1)=P_1$. Extrapolating to the next point, at $u=2$, yields $L(2)=(t,2s_1-s_2)$. Using our notation, we conclude that the second-order predictor predicts sample $s(t)$ as the linear combination $2s(t-1)-s(t-2)$.

For the third-order predictor, we start with the three points $P_3=(t-3,s_3)$, $P_2=(t-2,s_2)$, and $P_1=(t-1,s_1)$. The degree-2 polynomial that passes through those points is given by the uniform quadratic Lagrange interpolation polynomial (see, for example, [Salomon 06] p. 78, Equation 3.12.)

$$\begin{aligned}L(u)&=[u^2,u,1]\begin{bmatrix}\frac{1}{2}&-1&\frac{1}{2}\\-\frac{3}{2}&2&-\frac{1}{2}\\1&0&0\end{bmatrix}\begin{bmatrix}P_3\\P_2\\P_1\end{bmatrix}\\&=\left[\frac{u^2}{2}-\frac{3u}{2}+1\right]P_3+(-u^2+2u)P_2+\left[\frac{u^2}{2}-\frac{u}{2}\right]P_1.\end{aligned}$$

It is easy to verify that $L(0)=P_3$, $L(1)=P_2$, and $L(2)=P_1$. Extrapolating to $u=3$ yields $L(3)=3P_1-3P_2+P_3$. When this is translated to our samples, the result is

$\hat{s}(t) = 3s(t-1) - 3s(t-2) + s(t-3)$. The first four predictors are summarized as

$$\begin{aligned}
\hat{s}_0(t) &= 0,\\
\hat{s}_1(t) &= s(t-1),\\
\hat{s}_2(t) &= 2s(t-1) - s(t-2),\\
\hat{s}_3(t) &= 3s(t-1) - 3s(t-2) + s(t-3).
\end{aligned} \tag{10.5}$$

These predictors can now be used to compute the error (or difference) values for the first four orders:

$$\begin{aligned}
e_0(t) &= s(t) - \hat{s}_0(t) = s(t),\\
e_1(t) &= s(t) - \hat{s}_1(t) = s(t) - s(t-1) = e_0(t) - e_0(t-1),\\
e_2(t) &= s(t) - \hat{s}_2(t) = s(t) - 2s(t-1) + s(t-2)\\
&= [s(t) - s(t-1)] - [s(t-1) - s(t-2)] = e_1(t) - e_1(t-1),\\
e_3(t) &= s(t) - \hat{s}_3(t) = s(t) - 3s(t-1) + 3s(t-2) - s(t-3)\\
&= [s(t) - 2s(t-1) + s(t-2)] - [s(t-1) - 2s(t-2) + s(t-3)]\\
&= e_2(t) - e_2(t-1).
\end{aligned} \tag{10.6}$$

This computation is recursive but it involves only three steps, it is arithmetically simple, and does not require any multiplications (see also Equation (10.8)).

For maximum compression, it is possible to compute all four predictors and their errors and select the smallest error. However, experience gained by the developer of the method indicates that even a zeroth-order predictor results in typical compression of 48%, and going all the way to third-order prediction improves this only to 58%. For most cases, there is therefore no need to use higher-order predictors, and the precise predictor used should be determined by compression quality versus run time considerations. The default mode of Shorten uses linear (second-order) prediction.

The error (or difference) values are assumed decorrelated and are replaced by variable-length codes. Strong correlation between the original samples implies that most error values are small or even zero, and only few are large. They are also signed. Experiments suggest that the distribution of the errors is very close to the well-known Laplace distribution (Figure 7.143b), suggesting that the variable-length codes used to encode the differences should follow this distribution. (The developer of this method claims that the error introduced by coding the difference values according to the Laplace distribution, instead of using the actual probability distribution of the differences, is only 0.004 bits.)

The codes selected for Shorten have been developed by Robert F. Rice and are known as the Rice codes ([Rice 79], [Rice 91], and [Fenwick 96a]). They are a special case of the Golomb code (Section 3.24) and are also related to the subexponential code of Section 7.24.1. A Rice code depends on the choice of a base n and is computed in the following steps: (1) Separate the sign bit from the rest of the number. This becomes the most-significant bit of the Rice code. (2) Separate the n LSBs. They become the LSBs of the Rice code. (3) Code the remaining bits in unary and make this the middle part of the Rice code. (If the remaining bits are, say, 11, then the unary code is either three zeros followed by a 1 or three 1's followed by a 0.) Thus, this code is computed with a few logical operations, which is faster than computing a Huffman code, which requires sliding down the Huffman tree while collecting the individual bits of the code. This feature is especially important for the decoder, which has to be simple and fast. Table 10.28 shows examples of this code for $n = 2$ (the column labeled "No. of zeros" lists the number of zeros in the unary part of the code).

i	Binary	Sign	LSB	No. of Zeros	Code	i	Binary	Sign	LSB	No. of Zeros	Code
0	0	0	00	0	0|1|00						
1	1	0	01	0	0|1|01	−1	1	1	01	0	1|1|01
2	10	0	10	0	0|1|10	−2	10	1	10	0	1|1|10
3	11	0	11	0	0|1|11	−3	11	1	11	0	1|1|11
4	100	0	00	1	0|01|00	−4	100	1	00	1	1|01|00
5	101	0	01	1	0|01|01	−5	101	1	01	1	1|01|01
6	110	0	10	1	0|01|10	−6	110	1	10	1	1|01|10
7	111	0	11	1	0|01|11	−7	111	1	11	1	1|01|11
8	1000	0	00	2	0|001|00	−8	1000	1	00	2	1|001|00
11	1011	0	11	2	0|001|11	−11	1011	1	11	2	1|001|11
12	1100	0	00	3	0|0001|00	−12	1100	1	00	3	1|0001|00
15	1111	0	11	3	0|0001|11	−15	1111	1	11	3	1|0001|11

Table 10.28: Various Positive and Negative Rice Codes.

The Rice code, similar to the Huffman code, assigns an integer number of bits to each difference value, so in general, it is not an entropy code. However, the developer of this method claims that the error introduced by using these codes is only 0.12 bits per difference value.

Rice codes are ideal for data items with a Laplace distribution, but other prefix codes exist that are easier to construct and to decode and that may, under certain circumstances, outperform the Rice codes. Table 10.29 lists three such codes. The "pod" code, due to Robin Whittle [firstpr 06], codes the number zero with the single bit 1, and codes the binary number $1\underbrace{b...b}_{k}$ as $\underbrace{0...0}_{k+1}1\underbrace{b...b}_{k}$. In two cases, the pod code is one bit longer than the Rice code, in four cases it has the same length, and in all other cases it is shorter than the Rice codes. The Elias Gamma code [Fenwick 96a] is identical to the pod code minus its leftmost zero. It is therefore shorter, but does not include a code for zero. The biased Elias Gamma code corrects this fault in an obvious way but at the cost of making some codes one bit longer.

There remains the question of what base value n to select for the Rice codes. The base determines how many low-order bits of a difference value are included directly in the Rice code, and this is linearly related to the variance of the difference values. The developer provides the formula $n = \log_2[\log(2)E(|x|)]$, where $E(|x|)$ is the expected value of the differences. This value is the sum $\sum |x|p(x)$ taken over all possible differences x.

Number		Pod	Elias	Biased Elias
Dec	Binary		Gamma	Gamma
0	00000	1		1
1	00001	01	1	010
2	00010	0010	010	011
3	00011	0011	011	00100
4	00100	000100	00100	00101
5	00101	000101	00101	00110
6	00110	000110	00110	00111
7	00111	000111	00111	0001000
8	01000	00001000	0001000	0001001
9	01001	00001001	0001001	0001010
10	01010	00001010	0001010	0001011
11	01011	00001011	0001011	0001100
12	01100	00001100	0001100	0001101
13	01101	00001101	0001101	0001110
14	01110	00001110	0001110	0001111
15	01111	00001111	0001111	000010000
16	10000	0000010000	000010000	000010001
17	10001	0000010001	000010001	000010010
18	10010	0000010010	000010010	000010011

Table 10.29: Pod, Elias Gamma, and Biased Elias Gamma Codes.

The steps described so far result in lossless coding of the audio samples. Sometimes, a certain loss in the samples is acceptable if it results in significantly better compression. Shorten offers two options of lossy compression by quantizing the original audio samples before they are compressed. The quantization is done separately for each segment. One lossy option encodes every segment at the same bitrate (the user specifies the maximum bitrate), and the other option maintains a user-specified signal-to-noise ratio in each segment (the user specifies the minimum acceptable signal-to-noise ratio in dB).

Tests of Shorten indicate that it generally performs better and faster than UNIX compress and gzip, but that the lossy options are slow.

10.10 FLAC

The name FLAC is an acronym that stands for free lossless audio compression. FLAC was especially designed for audio compression and it also supports streaming and archival of audio data. FLAC is the brainchild of Josh Coalson who developed it in 1999 based on ideas from Shorten. He then started the FLAC project on the well-known sourceforge Web site [sourceforge.flac 06] by releasing his reference implementation. Since then many developers have contributed to improving the reference implementation and to writing alternative implementations. The FLAC project, administered and coordinated by Josh Coalson, maintains the software and provides a reference codec and input plugins for several popular audio players.

FLAC is free both in the sense that it is available at no cost and that its specifications and format are fully open to the public and can be used for any purpose. This includes the FLAC source code, which is available under open-source licenses. Neither the FLAC format nor any of the implemented encoding/decoding algorithms are covered by any known patents. Thus, FLAC is one of the first open and free lossless audio compression methods. It is possible that Squish (or Ogg Squish, by Chris Montgomery, [Ogg squish 06]), holds the title of "first," though it is not widely used.]

The FLAC project consists of (1) the format of the compressed stream, (2) reference encoders and decoders in library form, (3) a command-line software FLAC codec, (4) metaflac, a command-line metadata editor for FLAC files, and (5) input plugins for various music players.

The reference implementation of FLAC compiles on many platforms and supports many operating systems. Examples are Windows, BeOS, OS/2, and most Unix and Unix-like systems (including Linux, *BSD, Solaris, and Mac OS X).

SourceForge.net is the world's largest Open Source software development web site, hosting more than 100,000 projects and over 1,000,000 registered users with a centralized resource for managing projects, issues, communications, and code. SourceForge.net has the largest repository of Open Source code and applications available on the Internet, and hosts more Open Source development products than any other site or network worldwide. SourceForge.net provides a wide variety of services to projects we host, and to the Open Source community.

SourceForge.net provides free hosting to Open Source software development projects. The essence of the Open Source development model is the rapid creation of solutions within an open, collaborative environment. Collaboration within the Open Source community (developers and end users) promotes a higher standard of quality, and helps to ensure the long-term viability of both data and applications.

SourceForge.net is owned by the Open Source Technology Group (OSTG).

From `http://sourceforge.net/docs/about`

FLAC compresses the audio input block by block and is based on prediction and Rice codes. Thus, it somewhat resembles Shorten (Section 10.9). A block of audio samples is compressed by predicting the samples, subtracting each audio sample from its prediction, and encoding the difference with a Rice code (Sections 3.24 and 10.9). The compressed stream consists of the Rice codes, a few parameters specifying the prediction, and metadata information. The latter includes information such as the audio sampling rate, the number of audio channels, the minimum and maximum data rates, the minimum and maximum block sizes, the MD5 signature of the unencoded audio data, padding, seek tables, tags, cuesheets, and application-specific data. Users who need custom metadata can define their own format and request a metadata ID from the FLAC developers at [flacID 06].

FLAC does not forbid copy-prevention schemes such as DRM from appearing in FLAC streams, it just does not make any provisions for copy prevention in the format. It is possible, say, for Apple to encrypt a FLAC stream with FairPlay and store it in an `m4a` container.

Many audiophiles who are concerned about high audio quality are suspicious of lossy compression methods such as mp3 or AAC, and it is true that lossy methods are sensitive to certain types of sound and may sometimes produce reconstructed sound of inferior quality. Such users/listeners prefer FLAC, and they have made this format popular. As a result, audio equipment manufacturers have started supporting this format in hardware. Currently (mid 2006), many home stereo, car stereo, portable, and handheld audio devices can play music in this format. Software is also available on many computer platforms and on most operating systems—including Windows, Unix (Linux, *BSD, Solaris, OS X, IRIX), BeOS, OS/2, and Amiga—to encode and decode sound in FLAC. A list of available devices and software is maintained at [flac.devices 06].

Because of its lossless nature, FLAC is limited to only integer audio samples. There are audio encoders that can input audio samples in floating-point, but floating-point computations often return slightly different results on different computers, because of differences in word size and in the way ALU circuits handle rounding errors. Restricting FLAC to integer audio samples ensures that a correct FLAC implementation will be able to fully and truly reconstruct any audio file.

FLAC can handle audio sampling rates over a wide range. Any sampling rates of 1 Hz to 1,048,570 Hz (in 1-Hz increments) can be input and compressed correctly. FLAC is designed to handle up to eight audio channels (that can be grouped in stereo or as in 5.1 channel surround, page 1083) and take advantage of interchannel correlations to improve compression.

FLAC Goals. Being open source means that anyone can volunteer to help in the development of FLAC. Thus, it is important to have a well-defined set of goals for this project. Developers should keep these goals in mind when trying to improve, modify, or maintain any part of FLAC, but the project administrator may modify the goals from time to time, to reflect changes in thinking and users' needs. The current goals are:

- FLAC is and should stay an open format with an open-source reference implementation.
- FLAC is and should stay lossless. Future developers should not add lossy compression even as an option.
- The compression efficiency of FLAC should be on par or better than other lossless codecs.
- Decoding should be fast. A FLAC decoder should work at least in real time (i.e., decode the audio while it is played) even on average-speed hardware.
- FLAC should support fast sample-accurate seeking. A user listening to decoded audio should be able to skip to any point in the audio with ease.
- Being lossless, FLAC should allow for the playback of consecutive audio streams without gaps.
- No copy protection of any kind is allowed in a FLAC compressed stream.

These goals imply the main features of FLAC, which are the following:

- Lossless. There are many excellent lossy audio compression methods, which is why FLAC's developers decided to design it as lossless and keep it lossless. The decoded audio should be identical to the original input, and each implementation has to be tested to verify this property. FLAC has a "verify" option (`-V`) to verify the output while encoding.

The decoder runs in parallel with the encoder, and the decoder's output is compared with the original input. The software signals an error if a mismatch is found.

Each frame in the compressed stream includes a 16-bit CRC of the frame data for detecting (but not correcting) transmission errors. It has already been mentioned that the header of the compressed file contains the MD5 signature of the original (uncompressed) data. This can be used by the decoder to verify its output and can also be employed by users and implementors to test an implementation.

- Fast. FLAC was designed for fast decoding. This is important in audio compression because the decoder often has to output the audio in real time, to be decoded and played simultaneously. FLAC is therefore an asymmetric compression method. Decoding is fast because it is simple (much less computationally intensive than perceptual decoders) and uses only integer operations. Hardware support: Because of FLAC's free reference implementation and low decoding complexity, FLAC is currently the only lossless audio codec that has any kind of hardware support.

- Streamable. The FLAC compressed stream consists of frames, where each frame contains all the data necessary to decode the frame. Decoding a frame does not require any previous frames. The FLAC compressed stream includes sync codes and CRCs (similar to MPEG and other formats) that make it easy for the decoder to quickly skip to any point in the compressed stream.

- Seekable. Being able to skip to any point in the audio is useful for playback as well as for audio editing applications.

- Flexible metadata. The FLAC compressed stream starts with metadata blocks that include useful information. Users can design their own metadata blocks, for special and private information, and can apply to the FLAC project for block IDs.

- Suitable for archiving. An archive can benefit from compressed data, but it requires lossless compression. Lossless compression is also handy for converting between formats. The fact that FLAC is open source also encourages users to select FLAC for archiving audio.

- Convenient CD archiving. One of the metadata blocks on the FLAC compressed stream has a "cue sheet" for storing a CD table-of-contents and all track and index points. In a typical application, a CD is converted to a single file and its cue sheet is extracted. The file is then compressed into a single FLAC file with a metadata block containing the extracted cue sheet. If the original CD is damaged, the FLAC file with the cue sheet can be converted back and burned to become an exact copy of the CD.

- Error resistant. Many lossless codecs do not include any features for data reliability or integrity. A single error in a compressed stream may prevent the decoder from continuing. The FLAC compressed stream, in contrast, consists of frames, each representing a small fraction of a second of audio. If an error damages a frame, the decoder can still locate the next frame and continue. This limits any damages caused to the audio by errors during transmission or storage; an important, practical feature.

Before we get to the details of the FLAC algorithm, here are the definitions of the two key concepts, block and frame. A block in FLAC is a number of consecutive audio samples. If the input has more than one audio channel, a block consists of one subblock for each channel. Thus, a subblock is a set of contiguous audio samples from the same channel. The block size is actually the size of a subblock. Thus, if the block size is 4,608 and there are two audio channels, then the actual size of the block is $4{,}608 \times 2 = 9{,}216$ samples. The block size is important, and it affects the efficiency, speed, and reliability of a FLAC codec. A small block represents a very small fraction of a second of audio. In case of an error during decoding, only that small period of time is lost. On the downside, small blocks imply many blocks, which presents more work for both the encoder and decoder. Also, each block becomes a frame in the compressed stream and a frame has a header. A large number of headers results in more overhead which leads to less compression. Another consideration is that large blocks feature reduced correlation between their audio samples and thus lead to inefficient prediction and compression.

In a typical case where the sampling rate is 44.1 kHz and linear prediction is used, the optimal block size is 2,000–6,000 samples and the default value in such a case is 4,608 samples (except when the user-controlled option `-l` is set to zero, in which case the block size becomes 1,152).

A frame is written on the FLAC output file for each block of the input file. The frame starts with a header which is followed by a number of subframes. Each subframe starts with its own header, followed by Rice codes for encoded audio samples from the same channel. A frame consists of one subframe for each audio channel, and each subframe consists of the same number of (Rice encoded) audio samples.

A FLAC output stream starts with the four-byte identifying string `fLaC`. This is followed by the STREAMINFO metadata block. Other metadata blocks may optionally follow. They include all the side information that the decoder and end-user need. Metadata blocks can be of any length, and new ones can be defined by users. Thus, a FLAC decoder may not recognize certain metadata blocks and is allowed to skip them silently. The remainder of the stream consists of frames (there must be at least one such frame).

The encoder. The FLAC encoder encodes each block of audio samples separately. The first step is to optionally exploit the inter-channel correlation of audio samples. If the input is stereo (left and right channels), it is decorrelated by means of the usual lossless transformation mid = (left + right)/2 and side = left − right. FLAC has two user-selected options for this decorrelation. Option `-m` directs the encoder to encode both the left and right channels and the mid and side channels, and then select the better (i.e., smaller) result. Option `-M` tells the encoder to switch between left-right and mid-side adaptively and always select the smaller frame.

The next encoder step is modeling. Given a block of audio samples a_i (from the left, right, mid, side, or any other channel), the encoder tries to find a function $f(t)$ such that $f(t_i)$ is close to a_i for all values of i and for certain values t_i. The function (termed the approximation function or the fit) is fully specified by a few parameters which are sent to the decoder as side information. The encoder then subtracts $a_i - f(t_i)$ to obtain a difference (or residue) e_i. If the function closely fits the audio samples, the residues will be small integers (and may, of course, be negative).

FLAC employs four methods for constructing an approximating function (although only the last two are general and provide compression).

- Verbatim. This method always predicts a zero audio sample. The difference (or residue) is therefore the original audio sample. It is pointless to replace the audio samples with a Rice code, which is why this predictor does not encode the samples and provides no compression. Verbatim is essentially a zero-order predictor of the audio samples. The advantage of the verbatim predictor is fast decoding. If the decoder is told that a certain block or subblock employs this predictor, the decoder simply inputs the audio samples from the compressed stream without any decoding. Verbatim is the baseline against which the other predictors are measured. A good FLAC encoder should compress each subblock twice, first with the prediction method selected by the user, and then with the verbatim predictor. The smaller result should be written on the compressed stream. The verbatim predictor should do at least as well as any other predictor when the original audio is random or close to random.

- Constant. Sound is a wave. We hear sound when air vibrations strike our ear. Such vibrations are converted by a microphone to an electric wave (voltage that varies with time), and this wave is sampled to produce the audio samples. In the absence of sound, a microphone outputs a constant voltage (normally zero but possibly nonzero), which becomes a long sequence of identical audio samples; a digital silence. A sophisticated FLAC encoder should check each subblock for digital silence before it is encoded. When such a subblock is found, it is compressed by run-length encoding.

- Fixed linear predictor. This is a simple predictor that fits a polynomial to the audio samples (this method is selected if the FLAC user-controlled option `-l` is set to zero). Prediction by polynomial fit is faster than LPC (the next method) but usually results in files being 5-10% larger. The polynomial fit is similar to that employed by Shorten (zero-order to third-order prediction, Section 10.9), but also includes a fourth-order polynomial. The only side information included in the compressed stream is the order of prediction, a 3-bit integer with values of 0, 1, 2, 3, or 4. Fourth-order polynomial prediction is discussed in Section 10.10.1.

- FIR Linear prediction. This is a more sophisticated method that's also referred to as general linear predictive coding (LPC) (see Section 7.30) and [Rabiner and Schafer 78]). This predictor is used if option `-l` is set to a value between 1 and 32. This value becomes the maximum order of the LPC. Generally, the larger the maximum order, the more accurate the prediction (fit) provided by LPC, but the slower the encoder. However, increasing the order (normally above 9) brings diminishing returns and may even degrade the LPC fit and thereby decrease compression. LPC is described in Section 10.10.2, and the following is quoted from the official FLAC site

> The reference encoder uses the Levinson-Durbin algorithm for calculating the LPC coefficients from the autocorrelation coefficients, and the coefficients are quantized before computing the residual. Whereas encoders such as Shorten use a fixed quantization for the entire input, FLAC allows the quantized coefficient precision to vary from subframe to subframe. The FLAC reference encoder estimates the optimal precision to use based on the block size and dynamic range of the original signal.

Another difference between the last two methods is that the parameter for polynomial fit can be written on the compressed stream as a 3-bit number, whereas the size of the LPC parameters depends on the sample size (number of bits per audio sample) and the maximum LPC order.

Residue encoding. Once the audio samples for a subblock have been computed, they are subtracted from the original samples to obtain the residues. It has been observed experimentally (see Section 10.2.1 for such an experiment) that the residues often have a Laplacian distribution (Equation (7.40) and Figure 7.143b), and it is known that the Rice codes (Section 10.9) are ideal for such a distribution. Thus, FLAC encodes the residues with Rice codes. Recall that each original block of audio samples becomes a frame on the compressed stream and each subblock becomes a subframe. Most of the content of a subframe are the Rice codes.

The Rice codes depend on the choice of a single parameter m, and the best value for the parameter depends on the precise shape of the Laplace distribution. The FLAC encoder estimates m based on the statistical distribution of the residues, by $m = \log_2\big((\log_e 2)E[|r_i|]\big)$ where E is the expected value of the residues r_i (this estimate was originally proposed for Shorten in [Robinson 94]).

Even more, the encoder can vary the Rice parameter inside a subframe if the distribution of the Rice codes varies significantly. The residues in a subframe can be broken up into partitions where each partition uses a different Rice parameter. The user-controlled option `-r` can be assigned values `m` and `n` (for min and max). The FLAC encoder then tries to divide each subframe into 2^m partitions, then into 2^{m+1} partitions, and so on, up to 2^n partitions. In each try, partitions are assigned different values of the Rice parameter, and the encoder measures the final size of the subframe. When all the sizes are known, the encoder selects the partitioning scheme that produces the smallest subframe. This process is slow and becomes even slower when the user selects values `m` and `n` that are very different.

The FLAC literature recommends to set both `m` and `n` to 2, unless the block size is large, in which case the recommended values are `m` = `n` where n satisfies blocksize/$(2^n) = 128$. Values `m` = `0` and `n` = `16` result in maximum optimization, but encoding is slow.

The last encoding step is framing. The encoder prepares the frames and writes each on the compressed stream. Each frame contains a frame header, the Rice codes, and the frame footer.

Format of FLAC Output. FLAC may be modified in the future, so the format of its output stream contains some empty, reserved spaces. In the future, these spaces may contain a FLAC version number and other information relating to as-yet nonexistent features. Certain fields are limited to only certain bit patterns, while other patterns are invalid. This contributes to the integrity of the output. All the numbers included in the output are integers and are in big-endian format. The output of FLAC starts with the marker `fLaC`, followed by one or more metadata blocks. The STREAMINFO metadata is mandatory and there may be up to 128 different metadata blocks. Currently, the following types of metadata are defined:

> The term Big Endian means that the high-order byte (the big end) of a number or a string is stored in memory at the lowest address (it comes first). For example, given the 4-byte number $b_3b_2b_1b_0$, if the most-significant byte b_3 is stored at address A, then the least-significant byte b_0 will be stored at address $A + 3$.

- STREAMINFO. This metadata block contains information about the entire output stream, such as the sampling rate, number of audio channels, and the total number of samples. STREAMINFO must be the first metadata block.

- APPLICATION. A metadata block for use by third-party applications. The only mandatory field is a 32-bit identifier. Anyone can apply to the FLAC development team for such an ID. The remainder of an APPLICATION block is specified by the registered application.

- PADDING. This is a special type of metadata block that allows for an arbitrary amount of padding. The content of such a block is meaningless. This type of metadata is used in cases where the user plans to edit metadata after encoding. The user can instruct the encoder to reserve a PADDING block of sufficient size for later inclusion of metadata.

- SEEKTABLE. A metadata block with a table for storing seek points. There can be only one SEEKTABLE in an output stream, but the table can have any number of seek points. Seeking a particular point in a compressed audio file is especially useful. It is possible to skip to any given sample in a FLAC stream without a seek table, but the time it takes to skip is unpredictable because the bitrate may vary widely within a stream. Including seek points in the output stream significantly reduces the skip time. Each seek point occupies 18 bytes, so 1% resolution within a stream adds $100 \times 18 = 1800$ bytes to the output.

- VORBIS-COMMENT. This block implements the Vorbis comment specification. It contains a list of human-readable name/value pairs encoded in UTF-8. The VORBIS-COMMENT metadata is the only officially-supported tagging mechanism in FLAC. There may be only one VORBIS-COMMENT block in an output stream.

- CUESHEET. This useful type of metadata block serves for data that can be used in a cue sheet. It supports track and index points that are compatible with the CD digital audio disc standard, as well as other CD-DA metadata. The CUESHEET block is especially useful for backing up CD-DA discs, but it seems that its normal use is as a general purpose cueing mechanism for audio playback.

The metadata blocks are followed by the frames where each frame starts with a header. The frame header starts with a sync code, and contains the side information needed by the decoder, such as the sampling rate, number of bits per sample, number of audio channels, a frame/sample number, and an 8-bit CRC of the frame header itself.

The 14-bit sync code 11111111111110 starts each frame. This code is important because the decoder may have to start decoding inside the compressed stream and must therefore have a way to locate the start of a frame. Unfortunately, the Rice codes inside a frame feature arbitrary bit patterns that may look like a sync code. Thus, when the decoder finds the bit pattern of a sync, it has to read the rest of the header and verify its CRC to make sure that this is a frame header.

The sampling rate, number of bits per sample, and number of audio channels are the same for the entire compression job, but they appear in the header of every frame because the decoder may have to start decoding in midstream and may not have a chance to read the STREAMINFO metadata block located at the start of the output stream.

The frame/sample number in the header is either the number of the first sample in the frame (in cases where frames have different sizes) or the number of the frame itself (in cases where frames have fixed sizes). The sync code, CRC, and frame/sample number allow resynchronization of the decoder in case of errors. They also allow seeking even if no seek points have been specified.

Because the same parameters have to appear in every frame header, the header is merely overhead and it adversely affects the compression performance. To keep the headers small, at least in those cases where common parameters are used, FLAC employs lookup tables for the most-commonly-used values of certain parameters. For example, the sampling rate is stored in the frame header as a 4-bit code. Eight of the 16 possible codes (codes 4 through 11) specify the commonly-used sampling rates of 8, 16, 22.05, 24, 32, 44.1, 48, and 96 kHz. Three more codes direct the decoder to find the sampling rate at the end of the frame header. They are the following:

- Code 1100, find 8-bit sampling rate (in kHz) in end of header.
- Code 1101, find 16-bit sampling rate (in Hz) in end of header.
- Code 1110, find 16-bit sampling rate (in tens of Hz) in end of header.

Code 1111 is invalid (to reduce the chance of sync fooling), and codes 0–3 are reserved. The same method is used to specify the block size and bits per sample.

The (Rice encoded) subframes follow the header. Each subframe has its own header (with the attributes of the subframe, such as prediction method and order, and residual coding parameters), and there is one subframe for each audio channel. Notice that the subframes are not interleaved. The decoder therefore needs to reserve buffers large enough to read and store the subframes of a frame, and then loop over the buffers and decode a Rice code from each subframe in turn.

The subframes are followed by a pad of zero bits, if necessary, to complete the last byte of the last subframe (recall that the Rice codes are variable-length, so the total length of a subframe may not be an integer multiple of eight bits).

The frame footer contains a 16-bit CRC of the entire encoded frame, for robust error detection. If the reference decoder detects a CRC error, it generates a silent block. Other decoders may display an error message.

10.10.1 Fourth-Order Polynomial Prediction

Section 10.9 develops the expressions for linear predictors of orders 0 through 3 (see also Figure 10.27). Extending these concepts to a 4th-order linear predictor is straightforward. We start with the four points $P_4=(t-4,s_4)$, $P_3=(t-3,s_3)$, $P_2=(t-2,s_2)$, and $P_1=(t-1,s_1)$ and construct a degree-3 polynomial that passes through those points (the points are selected such that their x coordinates correspond to time and their y coordinates are audio samples). A natural choice is the nonuniform cubic Lagrange interpolation polynomial $Q_{3nu}(t)=\sum_{i=0}^{3}P_{i+1}L_i^3(t)$ whose coefficients are given by (see, for example, [Salomon 99] p. 204, Equations 4.17 and 4.18)

$$L_i^3(t)=\frac{\prod_{j\neq i}^{3}(t-t_j)}{\prod_{j\neq i}^{3}(t_i-t_j)},\quad \text{for}\quad 0\le i\le 3.$$

The *Mathematica* code of Figure 10.30 performs the computations and produces

$$Q(t)=-\frac{1}{6}(t-1)(t-2)(t-3)P_4+\frac{1}{2}t(t-2)(t-3)P_3-\frac{1}{2}t(t-1)(t-3)P_2+\frac{1}{6}t(t-1)(t-2)P_1.$$

It is easy to verify that $Q(0) = P_4$, $Q(1) = P_3$, $Q(2) = P_2$, and $Q(3) = P_1$. Extrapolating to $t = 4$ yields $Q(4) = 4P_1 - 6P_2 + 4P_3 - P_4$, and when this is translated to audio samples the result is

$$\hat{s}_4(t) = 4s(t-1) - 6s(t-2) + 4s(t-3) - s(t-4) \tag{10.7}$$

[compare with Equation (10.5)]. When this prediction is subtracted from the current audio sample $s(t)$, the residue is

$$e_4(t) = s(t) - \hat{s}_4(t) = s(t) - 4s(t-1) + 6s(t-2) - 4s(t-3) + s(t-4). \tag{10.8}$$

This is a simple arithmetic expression that involves only four additions and subtractions.

```
(* Uniform Cubic Lagrange polynomial for 4th-order prediction in FLAC *)
Clear[Q,t]; t0=0; t1=1; t2=2; t3=3;
Q[t_] := Plus @@ {
((t-t1)(t-t2)(t-t3))/((t0-t1)(t0-t2)(t0-t3))P4,
((t-t0)(t-t2)(t-t3))/((t1-t0)(t1-t2)(t1-t3))P3,
((t-t0)(t-t1)(t-t3))/((t2-t0)(t2-t1)(t2-t3))P2,
((t-t0)(t-t1)(t-t2))/((t3-t0)(t3-t1)(t3-t2))P1}
```

Figure 10.30: Code for a Lagrange Polynomial.

◇ **Exercise 10.7:** Check the performance of the 4th-order prediction developed here. Select four correlated items and compare the prediction of Equation (10.7) to the actual value of the next correlated item.

10.10.2 Linear Predictive Coding (LPC)

Given a set of correlated values $\{a_i\}$, linear predictive coding (LPC) is a sophisticated method to predict any set element a_i from its n immediate predecessors a_{i-1} through a_{i-n}, where n is a parameter. The idea is (see also Section 7.30) to try various sets of coefficients c_j and select the set that minimizes the difference

$$d_i = \left[a_i - \sum_{j=1}^{n} c_j a_{i-j}\right]^2. \tag{10.9}$$

One reference for LPC is [Rabiner and Schafer 78].

First, some mathematical background. In statistics and probability we deal with random variables and define useful quantities such as average and variance. Given a random variable V that takes values v_i, we denote the probability that V will have a specific value v by $P(V = v)$. The average of V is $(\sum_i^n v_i)/n$, but statisticians also define a related quantity E called the expectation (or expected value) of V. The expectation $E[V]$ of random variable V is the sum of all the values v_i of V, each multiplied by its probability. Thus,

$$E[V] = \sum_i v_i P(V = v_i).$$

The average and expectation of a random variable are often, but not always, the same. In many cases, the probability of a value equals its frequency of occurrence. When this is true, then the average and expectation are the same. Otherwise, they are different.

The autocorrelation $R_V(d)$ of a random variable V is the correlation of V with a copy of itself, shifted d positions. For example, given an array $a = (a_1, a_2, \ldots, a_n)$ of n values, we construct the two arrays $x = (a_1, a_2, \ldots, a_{n-1})$ and $y = (a_2, a_3, \ldots, a_n)$ of $n-1$ values each, and compute the Pearson correlation coefficient R of x and y. This becomes the autocorrelation $R_a(1)$ of array a with a shift of 1 position. Those unfamiliar with correlation may consult any text on probability and/or statistics. The short document [corr.pdf 02] may also be of help. What is important for the purpose of this discussion is that (under certain assumptions) the autocorrelation and expectation of a random variable are related by $R_V(k) = E[V_i V_{i+k}]$.

Now, back to linear predictive coding. In order to find the set of coefficients $\{c_j\}$ that minimizes Equation (10.9), we need to differentiate the expected value of d_i with respect to a coefficient c_p, set the derivative to zero, and do this for all possible values of p, from 1 to n. The result is

$$-2\left[\left(a_i - \sum_{j=1}^{n} c_j a_{i-j}\right) a_{i-p}\right] = 0, \quad \text{for} \quad 1 \le p \le n,$$

or

$$\sum_{j=1}^{n} c_j E[a_{i-j}a_{i-p}] = E[a_i a_{i-p}], \quad \text{for} \quad 1 \le p \le n.$$

Replacing the expectations with autocorrelation coefficients yields the system of n linear equations

$$\begin{bmatrix} R(0) & R(1) & R(2) & \ldots & R(n-1) \\ R(1) & R(0) & R(1) & \ldots & R(n-2) \\ R(2) & R(1) & R(0) & \ldots & R(n-3) \\ \vdots & \vdots & \vdots & & \vdots \\ R(n-1) & R(n-2) & R(n-3) & \ldots & R(0) \end{bmatrix} \begin{bmatrix} c_1 \\ c_2 \\ c_3 \\ \vdots \\ c_n \end{bmatrix} = \begin{bmatrix} R(1) \\ R(2) \\ R(3) \\ \vdots \\ R(n) \end{bmatrix},$$

with the n coefficients c_j as the unknowns. This can be written compactly as $\mathbf{R}C = P$ and can easily be solved by Gaussian elimination or by inverting matrix $\mathbf{R}$. The point is that matrix inversion requires in general $O(n^3)$ operations, but ours is not a general case, because our matrix $\mathbf{R}$ is special. It is easy to see that each diagonal of $\mathbf{R}$ consists of identical elements. Such a matrix is called a Toeplitz matrix, after its originator, Otto Toeplitz, and it can be inverted by a number of specialized, efficient algorithms. In addition, our $\mathbf{R}$ is also symmetric.

FLAC employs a recursive, efficient method, known as the Levinson-Durbin algorithm, to solve this system of n equations. This algorithm was first proposed by Norman Levinson in 1947 [Levinson 47], improved by J. Durbin in 1960 [Durbin 60], and further improved by several researchers. In its present form it requires only $3n^2$ multiplications. A description of this algorithm is outside the scope of this book, but can be found, together with a derivation, in [Parsons 87].

Otto Toeplitz (1881–1940) came from a Jewish family that produced several teachers of mathematics. Both his father, Emil Toeplitz, and his grandfather, Julius Toeplitz, taught mathematics in a Gymnasium and they also both published mathematics papers. Otto was brought up in Breslau and attended a Gymnasium in that city. His family background made it natural that he also should study mathematics.

During his long, productive carrer, Toeplitz studied algebraic geometry, integral equations, and spectral theory. He worked also on summation processes, infinite-dimensional spaces and functions on them. In the 1930s he developed a general theory of infinite dimensional spaces and criticized Banach's work as being too abstract.

Toeplitz operators and Toeplitz matrices bear his name.

We would like to thank Josh Coalson for his help with this section.

10.11 WavPack

(This section written by David Bryant, WavPack's developer.)

WavPack [WavPack 06] is a completely open, multiplatform audio compression algorithm and software that supports three compression modes, lossless, high-quality lossy, and a unique hybrid compression mode. It handles integer audio samples up to 32-bits wide and also 32-bit IEEE floating-point data [IEEE754 85]. The input stream is partitioned by WavPack into blocks that can be either mono or stereo and are generally 0.5 seconds long (but the length is actually flexible). Blocks may be combined in sequence by the encoder to handle multichannel audio streams. All audio sampling rates are supported by WavPack in all its modes.

WavPack defaults to the lossless mode. In this mode, the audio samples are simply compressed at their full resolution and no information is discarded. WavPack generally provides a reasonable compromise between compression ratio and encoding/decoding speed. However, for specific applications, an alternate compromise may be more desirable. For this reason, WavPack incorporates an optional "fast" mode that is fast and entails only a small penalty in compression performance and a "high" mode that aims for maximum compression (at a somewhat slower pace).

The lossy mode employs no subband coding or psychoacoustic noise masking. Instead, it is based on variable quantization in the time domain combined with mild noise shaping. This mode can operate from bitrates as low as 2.22 bits per sample up to fully lossless and offers considerably more flexibility and performance than the similar, but much simpler, ADPCM (Section 10.6). This makes the lossy mode ideal for high-quality audio encoding where the data storage or bandwidth requirements of the lossless mode might be prohibitive.

Finally, the hybrid mode combines the lossless and lossy modes into a single operation. Instead of a single output file being generated from a source, two files are generated. The first, with the extension `wv`, is a smaller lossy file that can be played on its own. The second is a "correction" file (`wvc`) that, when combined with the first, provides lossless restoration of the original audio. The two files together have essentially the same size as the output produced by the pure lossless mode, so the hybrid mode generates very little additional overhead.

Efficient hardware decoding was among the design goals of WavPack, and this weighed heavily (along with patent considerations) in the design decisions. As a result, even the most demanding WavPack mode will easily decode CD quality audio in real time on an ARM7 processor running at 75 MHz (and the default modes use considerably less than that).

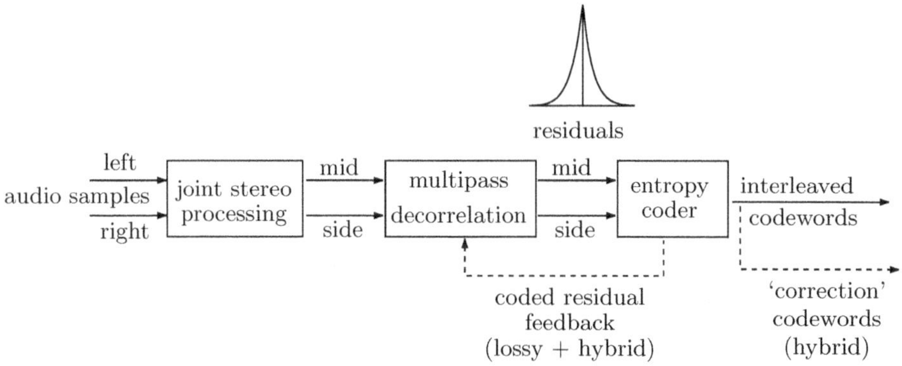

Figure 10.31: The WavPack Encoder.

Figure 10.31 is a block diagram of the WavPack encoder. Its main parts are the joint-stereo processing (removing inter-channel correlations), multipass decorrelation (removing intra-channel correlations between neighboring audio samples), and the entropy coder.

Decorrelation

The decorrelation passes are virtually identical in all three modes of WavPack. The first step in the decorrelation process is to convert the left and right stereo channels to the standard difference and average (also referred to as side and mid) channels by $s(k) = l(k) - r(k)$ and $s'(k) = \text{int}\big((l(k)+r(k))/2\big)$. Notice that the integer division by 2 discards the least-significant bit of the sum $l(k)+r(k)$, but this bit can be reconstructed from the difference, because a sum $A + B$ and a difference $A - B$ have the same least-significant bit.

The second decorrelation step is prediction. WavPack performs multiple passes of very simple, single-tap linear predictive filters that employ the sign-sign LMS method for adaptation. (The sign-sign LMS method [adaptiv9 06] is a variation on the standard least-mean-squares method.) Having only one adjustable weighing factor per filter eliminates problems of instability and non-convergence that can occur when multiple taps are updated from a single result. The overall computational complexity of the process is simply controlled by the total number of filtering passes made. The default case performs five passes, as this provides a good compromise of compression vs. speed. In the "fast" mode, only two passes are made, while the "high" mode incorporates the maximum allowed 16 passes.

There are actually 13 variations of the filter depending on what sample value (or linear function of two sample values) is used as the input value $u(k)$ to the prediction filter. These variations are identified by a parameter named "term." In the case of a single channel, filtering depends on the value of this parameter as follows

$$u(k) = \begin{cases} 1 \le \text{term} \le 8, & s(k - \text{term}), \\ \text{term} = 17, & 2s(k-1) - s(k-2), \\ \text{term} = 18, & \bigl(3s(k-1) - s(k-2)\bigr)/2.0, \end{cases}$$

where $s(k)$ is the sample to be predicted and the recent sample history is $s(k-1)$, $s(k-2),\ldots,\ s(k-8)$.

For the dual channel case (which may be left and right or mid and side) there are three more filter variations available that use cross-channel information

$$\begin{aligned} &\texttt{if}(\text{term} = -1),\ u(k) = s'(k),\ u'(k) = s(k-1), \\ &\texttt{if}(\text{term} = -2),\ u(k) = s'(k-1),\ u'(k) = s(k), \\ &\texttt{if}(\text{term} = -3),\ u(k) = s'(k-1),\ u'(k) = s(k-1). \end{aligned}$$

Once $u(k)$ has been determined, an error (or residual) value $e(k)$ is computed as the difference $e(k) = s(k) - w(k)u(k)$, where $w(k)$ is the current filter weight value. The weight generally varies between +1 for highly correlated samples and 0 for completely uncorrelated samples. It can also become negative if there is a negative correlation between the sample values. Negative correlation would exist if $(-1 \times u(k))$ was better than $u(k)$ as a predictor of $s(k)$. This often happens with samples with large amounts of high frequencies.

Finally, the weight is updated for the next sample based on the signs of the filter input and the residual $w(k+1) = w(k) + d \cdot \text{sgn}(u(k)) \cdot \text{sgn}(e(k))$, where parameter d is the step-size of the adaptation, a small positive integer. (The `sgn` function used by WavPack returns +1, 0, or −1.) The rule for adapting $w(k)$ implies that if either $u(k)$ or $e(k)$ is zero, the weight is not updated. In cases where the input to the filter comes from the same channel as the sample being predicted (i.e., where "term" is positive), the magnitude of the weight is self limiting and is therefore allowed to exceed ±1. However, in the cross-channel cases (where "term" is negative), the weight could grow indefinitely, and so is clipped to ±1 after each adaptation.

Here's an example of an adaptation step. Assuming that the input $u(k)$ to the predictor is 105, sample $s(k)$ to be predicted is 100, the current filter weight $w(k)$ is 0.8, and the current stepsize d is 0.01, we first compute the residual $e(k) = s(k) - w(k) \cdot u(k) = 100 - 105 \cdot 0.8 = 16$ and then update the weight

$$\begin{aligned} w(k+1) &= w(k) + s \cdot \text{sgn}(e(k)) \cdot \text{sgn}(u(k)) \\ &= 0.8 + 0.01 \cdot \text{sgn}(100) \cdot \text{sgn}(16) \\ &= 0.8 + 0.01 \cdot 1 \cdot 1 = 0.81. \end{aligned}$$

We can see that if the original weight had been more positive (greater than 0.8), the prediction would have been closer to $s(k)$ and the residual would have been smaller. Therefore, we increase the weight in this iteration.

The magnitude of the stepsize is a compromise between fast adaptation and noisy operation around the optimum value. In other words, if the stepsize is large, the filter will adjust quickly to sudden changes in the signal, but the weight will jump around a lot when it is near the correct value. Too small a stepsize will cause the filter to adjust too slowly to rapidly changing audio.

In the lossless mode, the results of the decorrelation (the residuals $e(k)$) are simply passed to the entropy coder for exact translation. In the lossy and hybrid modes, in contrast, the decorrelated values may be coded inexactly based on a defined maximum allowable error. In this situation, the actual value coded is returned from the entropy coder and this value is used to update the filter weight and generate the reconstructed sample value. This is required because during subsequent decoding the exact sample values will not be available, and we must operate the encoder with only the information that we will have at that time.

Implementation

All sample data is presented to the decorrelator as 32-bit signed integers, even if the source is smaller. This is done to allow common routines, but is also required from an arithmetic standpoint because, for example, 16-bit data might clip at some steps if restricted to 16-bit storage. The routines are designed in such a way that they will not overflow on sample data with up to 25 bits of significance. This allows for standard 24-bit audio and the 25-bit significand of 32-bit IEEE data. Any significance beyond this must be handled by side channel data.

The filter weights are represented in 16-bit signed integers with 10 bits of fraction (for example $1024 = 000001\underbrace{0\ldots00}_{10}{}_2$ represents 1.0), and the adjusting step-size may range from 1 to 7 units of that resolution. For input values that fit in 16-bit signed integers, it is possible to perform the rounded weight multiply with a single 16-bit $\times$ 16-bit = 32-bit operation by `prediction = (sample * weight + 512) >> 10;`

Since "sample" and "weight" both fit in 16-bit integers, the signed product will fit in 32 bits. We add the 512 for rounding and then shift the result to the right 10 places to compensate for the 10 bits of fraction in the weight representation. This operation is very efficiently implemented in modern processors.

For those cases where the sample does not fit in 16 bits, we need more than 32 bits in the intermediate result, and several alternative methods are provided in the source code. If a multiply operation is provided with 40 bits of product resolution (as in the ColdFire EMAC or several Texas Instruments DSP chips), then the above operation will work (although some scaling of the inputs may be required). If only 32-bit results are provided, it is possible to perform separate operations for the lower and upper 16 bits of the input sample and then combine the results with shifts and an addition. This has been efficiently implemented with the MMX extensions to the `x86` instruction set. Finally, some processors provide hardware-assisted double-precision floating-point arithmetic which can easily perform the operation with the required precision.

As mentioned previously, multiple passes of the decorrelation filter are used. For the standard operating modes (fast, default, and high) stepsize is always 2 (which represents about 0.002) and the following terms are employed, in the order shown:

$$\begin{aligned}\text{fast mode terms} &= \{17, 17\},\\ \text{default mode terms} &= \{18, 18, 2, 3, -2\},\\ \text{high mode terms} &= \{18, 18, 2, 3, -2, 18, 2, 4, 7, 5, 3, 6, 8, -1, 18, 2\}.\end{aligned}$$

For single channel operation, the negative term passes (i.e. passes with cross-channel correlation) are skipped. Note, however, that the decoder is not fixed to any of these configurations because the number of decorrelation passes, as well as other information such as what terms and stepsize to use in each pass, is sent to the decoder as side information at the beginning of each block.

For lossless processing, it is possible to perform each pass of the decorrelation in a separate loop acting on an array of samples. This allows the process to be implemented in small, highly optimized routines. Unfortunately, this technique is not possible in the lossy mode, because the results of the inexact entropy coding must be accounted for at each sample. However, on the decoding side, this technique may be used for both lossless and lossy modes (because they are essentially identical from the decorrelator's viewpoint).

Generally, the operations are not parallelizable because the output from one pass is the input to the next pass. However, in the two-channel mode the two channels may be performed in parallel for the positive term values (and again this has been implemented in MMX instructions).

Entropy Coder

The decorrelation step generates residuals $e(k)$ that are signed numbers (generally small). Recall that decorrelation is a multistep process, where the differences $e(k)$ computed by a step become the input of the next step. Because of this feature of WavPack, we refer to the final outputs $e(k)$ of the decorrelation process as residuals. The residuals are compressed by the entropy encoder. The sequence of residuals is best characterized as a two-sided geometric distribution (TSGD) centered at zero (see Figure 3.49 for the standard, one-sided geometric distribution). The Golomb codes (Section 3.24) provide a simple method for optimally encoding similar one-sided geometric distributions. This code depends on a single positive integer parameter m. To Golomb encode any non-negative integer n, we first divide n into a magnitude `mag` $= \text{int}(n/m)$ and a remainder `rem` $= n \bmod m$, then we code the magnitude as a unary prefix (i.e. `mag` 1's followed by a single 0) and follow that with an adjusted-binary code representation of the remainder. If m is an integer power of 2 ($m = 2^k$), then the remainder is simply encoded as the binary value of `rem` using k bits. If m is not an integer power of 2, then the remainder is coded in either k or $k+1$ bits where $2^k < m < 2^{k+1}$. Because the values at the low side of the range are more probable than values at the high side, the shorter codewords are reserved for the smaller values.

Table 10.32 lists the adjusted binary codes for m values 6 through 11.

The Rice codes are a common simplification of this method, where m is a power of 2. This eliminates the need for the adjusted binary codes and eliminates the division

remainder to code	when $m = 6$	when $m = 7$	when $m = 8$	when $m = 9$	when $m = 10$	when $m = 11$
0	00	00	000	000	000	000
1	01	010	001	001	001	001
2	100	011	010	010	010	010
3	101	100	011	011	011	011
4	110	101	100	100	100	100
5	111	110	101	101	101	1010
6		111	110	110	1100	1011
7			111	1110	1101	1100
8				1111	1110	1101
9					1111	1110
10						1111

Table 10.32: Adjusted Binary Codes for Six m Values.

required by the general Golomb codes (because division by a power of 2 is implemented as a shift). Rice codes are commonly used by lossless audio compression algorithms to encode prediction errors. However, Rice codes are less efficient when the optimum m is midway between two consecutive powers of 2. They are also less suitable for lossy encoding because of discontinuities resulting from large jumps in the value of m. For these reasons Rice codes were not chosen.

Before a residual $e(k)$ is encoded, it is converted to a nonnegative value by means of `if` $e(k) < 0,\ e(k) = -\big(e(k)+1\big)$. The original sign bit of $e(k)$ is appended to the end of the coded value. Note that the sign bit is always written on the compressed stream, even when the value to be encoded is zero, because -1 is also coded as zero. This is less efficient for very small encoded values where the probability of the value zero is significantly higher than neighboring values. However, we know from long experience that an audio file may have long runs of consecutive zero audio samples (silence), but relatively few isolated zero audio samples. Thus, our chosen method is slightly more efficient when zero is not significantly more common than its neighbors. We show later that WavPack encodes runs of zero residuals with an Elias code.

The simplest method for determining m is to divide the residuals into blocks and determine the value of m that encodes the block in the smallest number of bits. Compression algorithms that employ this method have to send the value of m to the decoder as side information. It is included in the compressed stream before the encoded samples. In keeping with the overall philosophy of WavPack, we instead implement a method that dynamically adjusts m at each audio sample.

The discussion on page 164 (and especially Equation (3.12)) shows that the best value for m is the median of the recent sample values. We therefore attempt to adapt m directly from the current residual $e(k)$ using the simple expression

$$\texttt{if}\ \big(e(k) \geq m(k)\big)\ \texttt{then}\ m(k+1) = m(k) + \mathrm{int}\left[\frac{m(k)+127}{128}\right],$$

$$\texttt{else}\; m(k+1) = m(k) - \text{int}\left[\frac{m(k)+126}{128}\right]. \qquad (10.10)$$

The idea is that the current median $m(k)$ will be incremented by a small amount when a residual $e(k)$ occurs above it and will be decremented by an almost identical amount when the residual occurs below it. The different offsets (126 and 127) for the two cases were selected so that the median value can move up from 1, but not go below 1. This works very well for large values, but because the median is simply an integer, it tends to be "jumpy" at small values because it must move at least one unit in each step. For this reason the actual implementation adds four bits of fraction to the median that are simply ignored when the value is used in a comparison. This way, it can move smoothly at smaller values.

It is interesting to note that the initial value of the median is not crucial. Regardless of its initial value, m will move towards the median of the values it is presented with. In WavPack, m is initialized at the start of each block to the value that it left off on the previous block, but even if it started at 1 it would still work (but would take a while to catch up with the median).

There is a problem with using an adaptive median value in this manner. Sometimes, a residual will come along that is so big compared to the median that a Golomb code with a huge number of 1's will be generated. For example, if m was equal to 1 and a residual of 24,000 occurred, then the Golomb code for the sample would result in $\texttt{mag} = \texttt{int}(24{,}000/1) = 24{,}000$ and $\texttt{rem} = 24{,}000 \bmod 1 = 0$ and would therefore start with 24,000 1's, the equivalent of 3,000 bytes of all 1's. To prevent such a situation and avoid ridiculously long Golomb codes, WavPack generates only Golomb codes with 15 or fewer consecutive 1's. If a Golomb code for a residual $e(k)$ requires k consecutive 1's, where k is greater than or equal 16, then WavPack encodes $e(k)$ by generating 16 1's, followed by the Elias gamma code of k (Code C_1 of Table 3.20). When the residuals are distributed geometrically, this situation is exceedingly rare, so it does not affect efficiency, but it does prevent extraordinary and rarely-occurring cases from drastically reducing compression.

Recall that each encoded value is followed by the sign bit of the original residual $e(k)$. Thus, an audio sample of zero is inefficiently encoded as the two bits 00. Because of this, WavPack employs a special method to encode runs of consecutive zero residuals. A run of k zeros is encoded in the same Elias gamma code preceded by a single 1 to distinguish it from a regular code (this is done only when the medians are very small to avoid wasting a bit for every sample.)

One source of less than optimum efficiency with the method described so far is that sometimes the data is not distributed in an exactly geometrical manner. The decay for higher values may be faster or slower than exponential, either because the decorrelator is not perfect or the audio data has uncommon distribution characteristics. In these cases, it may be that the dynamically estimated median is not a good value for m, or it may be that no value of m can produce optimal codes. For these cases, a further refinement has been added that I have named Recursive Golomb Coding.

In this method, we calculate the first median as in Equation (10.10) and denote it by m. However, for values above the median we subtract m from the value and calculate a new median, called m', that represents the 3/4 population point of samples. Finally, we calculate a third median (m'') for the sample values that lie above the second median (m'), and we use this for all remaining partitions (Figure 10.33). The first three coding zones still represent 1/2, 1/4, and 1/8 of the residuals, but they don't necessarily span the same number of possible values as they would with a single m. However, this is not an issue for the decoder because it knows all the medians and therefore all the span sizes.

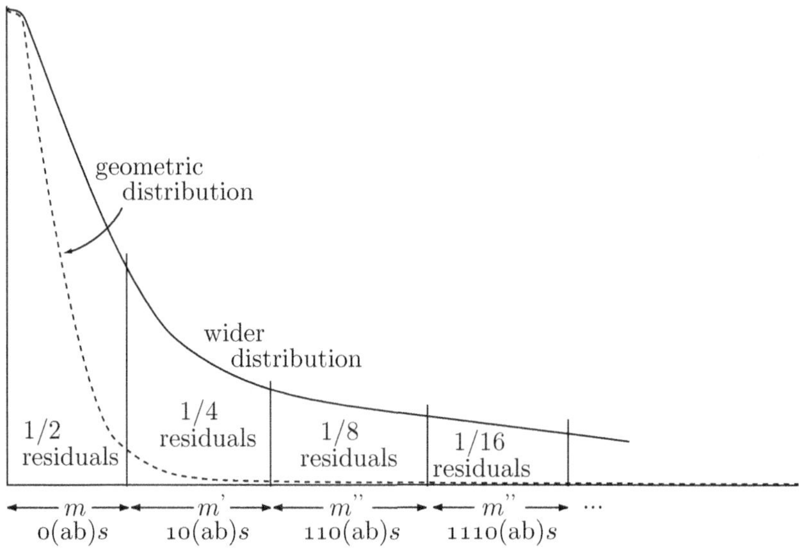

Figure 10.33: Median Zones for Recursive Golomb Coding.

In practice, recursive Golomb coding works like this. If the residual being coded is in the first zone, fine. If not, then subtract m from the residual and pretend we're starting all over. In other words, first we divide the residuals into two equal groups (under m and over m). Then, we subtract m from the "over m" group (so they start at zero again) and divide those into two groups. Then, we do this once more. This way, the first several zones are guaranteed to have the proper distribution because we calculate the second and third medians directly instead of assuming that they're equal in width to the first one. Table 10.34 lists the first few zones where (ab) represents the adjusted-binary encoding of the remainder (residual modulo m) and S represents the sign bit.

range	prob.	coding
$0 \le \text{residual} < m$	1/2	$0(ab)S$
$m \le \text{residual} < m + m'$	1/4	$10(ab)S$
$m + m' \le \text{residual} < m + m' + m''$	1/8	$110(ab)S$
$m + m' + m'' \le \text{residual} < m + m' + 2m''$	1/16	$1110(ab)S$
...		

Table 10.34: Probabilities for Recursive Golomb Coding.

Lossy and hybrid coding

Recursive Golomb coding allows arbitrary TSGD integers to be simply and efficiently coded, even when the distribution is not perfectly exponential and even when it contains runs of zeros. For the implementation of lossy encoding, it is possible to simply transmit the unary magnitude and the sign bit, leaving out the remainder data altogether. On the decode side, the value at the center of the specified magnitude range is chosen as the output value. This results in an average of three bits per residual as illustrated by Table 10.35. The table is based on the infinite sum

$$1+\sum_{n=1}^{\infty}\frac{n}{2^n},$$

where the 1 represents the sign bit and each term in the sum is a probability multiplied by the number of bits for the probability. The infinite sum adds up to two bits, and the total after adding the sign bit is three bits.

	prob.	bits sent	# of bits	(# of bits) × prob.
	1/2	$0S$	2	2/2
	1/4	$10S$	3	3/4
	1/8	$110S$	4	4/8
	1/16	$1110S$	5	5/16
	1/32	$11110S$	6	6/32
	1/64	$111110S$	7	7/64
	...	...	...	...
total	1			3

Table 10.35: Three Bits Per Residual.

This works fine. However, a lower minimum bitrate is desired and so a new scheme is employed that splits the samples 5:2 instead of down the middle at 1:1. In other words, a special "median" is found that has five residuals lower for every two that are higher. This is accomplished easily by modifying the procedure above to

$$\begin{aligned}\texttt{if}\ \big(s(k)>m(k)\big)\ \texttt{then}\ m(k+1)&=m(k)+5\cdot\mathrm{int}\left[\frac{m(k)+127}{128}\right],\\ \texttt{else}\ m(k+1)&=m(k)-2\cdot\mathrm{int}\left[\frac{m(k)+125}{128}\right].\end{aligned}$$

The idea is that the median is bumped up five units every time the value is higher, and bumped down two units when it is lower. If there are two ups for every five downs, the median stays stationary. Because the median moves down a minimum of two units, we consider 1 or 2 to be the minimum (it will not go below this).

Now the first region contains 5/7 of the residuals, the second has 10/49, the third 20/343, and so on. Using these enlarged median zones, the minimum bitrate drops from

3 to 2.4 bits per residual, as shown by the following infinite sum

$$1+\sum_{n=1}^{\infty}\frac{5n\times 2^{n-1}}{7^n}.$$

There is an efficiency loss with these unary codes because the number of 1's and 0's in the datastream is no longer equal. There is a single 0 for each residual, but since there are only 1.4 total bits per residual (not counting the sign bit), there must be only an average of 0.4 1's per residual. In other words, the probability for 0's is 5/7 and the probability for 1's is 2/7. Using the $\log_2$ method to determine the ideal number of bits required to represent those symbols, we obtain the numbers listed in Table 10.36a and derive an optimum multibit encoding scheme (Table 10.36b).

symbol	freq	$\log_2(1/\text{freq})$
0	5/7	0.485427
1	2/7	1.807355

(a)

input sequence	freq	ideal # bits	output sequence
00	25/49	0.970854	0
01	10/49	2.292782	10
1	14/49	1.807355	11

(b)

Table 10.36: Optimum Translation of Unary Magnitudes.

	input sequence	freq	input bits	net* input	output sequence	output bits	net* output
	00	25/49	2	50/49	0	1	25/49
	01	10/49	2	20/49	10	2	20/49
	1	14/49	1	14/49	11	2	28/49
totals		1		84/49			73/49

*The "net" values are the frequency multiplied by the number of bits.

Table 10.37: Net Savings from Unary Translation.

Table 10.37 shows that the average number of unary magnitude bits transmitted is reduced by the translation to $(73/84)\times 1.4 = 1.21666 \approx 1.22$ bits per sample. We use the 1.4 figure because only the unary magnitude data is subject to this translation; adding back the untranslated sign bit gets us to about 2.22 bits per sample. This represents the encoder running "flat out" lossy (i.e., no remainder data sent) however, the method is so efficient that it is used for all modes. In the lossless case, the complete adjusted-binary representation of the remainder is inserted between the magnitude data and the terminating sign bit.

In practice, we don't normally want the encoder to be running at its absolute minimum lossy bitrate. Instead, we would generally like to be somewhere in-between full lossy mode and lossless mode so that we send some (but not all) of the remainder value. Specifically, we want to have a maximum allowed error that can be adjusted during encoding to give us a specific signal-to-noise ratio or constant bitrate. Sending just enough data to reach a specified error limit for each sample (and no more) is optimal for this kind of quantization because the perceived noise is equivalent to the RMS level of the error values.

To accomplish this, we have the encoder calculate the range of possible sample values that satisfy the magnitude information that was just sent. If this range is smaller than the desired error limit, then we are finished and no additional data is sent for that sample. Otherwise, successive bits are sent to the datastream, each one narrowing the possible range by half (i.e., 0 = lower half and 1 = upper half) until the size of the range becomes less than the error limit. This binary process is done in lockstep by the decoder, to remain synchronized with the bitstream.

At this point it should be clear how the hybrid lossless mode splits the bitstream into a lossy part and a correction part. Once the specified error limit has been met, the numeric position of the sample value in the remaining range is encoded using the adjusted binary code described above, but this data goes into the "correction" stream rather than the main bitstream. Again, on decoding, the process is exactly reversed so the decoder can stay synchronized with the two bitstreams.

With respect to the entropy coder, the overhead associated with having two files instead of one is negligible. The only cost is the difference resulting from coding the lossy portion of the remainder with the binary splitting method instead of the adjusted binary method. The rest of the cost comes on the decorrelator side where it must deal with a more noisy signal.

More information on the theory behind and the implementation of WavPack can be found in [WavPack 06].

10.12 Monkey's Audio

Monkey's audio is a fast, efficient, free, lossless audio compression algorithm and implementation (for the Windows operating system) by Matt Ashland [monkeyaudio 06]. This method also has the following important features:

- Error detection. The implementation incorporates CRCs in the compressed stream to detect virtually all errors.

- Tagging support. The algorithm inserts tags in the compressed stream. This makes it easy for a user to manage and catalog an entire collection of monkey-compressed audio files.

- External support. Monkey's audio can also be used as a front-end to a command-line encoder. The current GUI tool allows different command line encoders to be used by creating a simple XML script that describes how to pass parameters and what return values to look for.

- The source code is freely available. Software developers can use it (modified or not) in other programs without restrictions.

The first step in monkey's audio is to transform the left and right stereo channels to X (mid or average) and Y (side or difference) channels according to $X = (L+R)/2$ and $Y = L - R$. The inter correlation between L and R implies that the mid channel X is similar to both L and R, while the side channel Y consists of small numbers. Notice that the elements (transform coefficients) of X and Y are integers but are no longer audio samples. Also, the transform is reversible even though the division by 2 causes the loss of the least-significant bit (LSB) of X. This bit can be reconstructed by the decoder because it is identical to the LSB of Y. (Notice that the LSBs of any sum $A + B$ and difference $A - B$ are the same.)

The next step is prediction. This step exploits the intra correlation between consecutive elements in the mid and side channels. Monkey's audio employs a novel method for linear prediction that consists of a fixed first-order predictor followed by multiple adaptive offset filters. There are varying degrees of neural net filtering, depending on the compression level—up to three consecutive neural-net filters with 1024 sample windows at the highest compression.

In its last step, monkey's audio encodes the residuals with a range-style coder (Section 5.10.1).

The compressed stream is organized in frames, where each frame has a 31-bit CRC (a standard 32-bit CRC with the most-significant bit truncated). The encoder also computes an MD5 checksum for the entire file. The decoder starts by verifying this checksum. A failed verification indicates an error, and the decoder locates the error by verifying the CRCs of the individual frames.

> The best lossless audio compressor around... written by the best monkey around. Great job, Matt!
>
> —From `http://www.ashlands.net/Links.htm`.

10.13 MPEG-4 Audio Lossless Coding (ALS)

MPEG-4 Audio Lossless Coding (ALS) is the latest addition to the family of MPEG-4 audio codecs. ALS can input floating-point audio samples and is based on a combination of linear prediction (both short-term and long-term), multichannel coding, and efficient encoding of audio residues by means of Rice codes and block codes (the latter are also known as block Gilbert-Moore codes, or BGMC [Gilbert and Moore 59] and [Reznik 04]). Because of this organization, ALS is not restricted to the encoding of audio signals, and can efficiently and losslessly compress other types of one-dimensional, fixed-size, correlated signals, such as medical (ECG-EKG and EEG) and seismic data. Because of its lossless nature and high compression gain, its developers envision the following audio compression applications for ALS:

- Archival (audio archives in broadcasting, sound studios, record labels, and libraries)
- Studio operations (storage, collaborative working, and digital transfer)

- High-resolution disc formats (CD, DVD, and future formats)
- Internet distribution of audio files
- Online music stores (downloading purchased music)
- Portable music players (an especially popular application)

Currently, not all these applications employ ALS, because many believe that audio compression should always be lossy and there is no point in preserving every bit of audio, because the ear cannot distinguish between reconstructed lossy and reconstructed lossless sounds. There is an ongoing debate about the advantage of lossless audio compression over lossy methods and whether lossless audio compression is really necessary. An Internet search returns many strong opinions, ranging from "Not another losslessphile. When will you people realize that there is no point in lossless audio compression?" to "I spent all of last summer converting my 200 Gb audio collection to a lossy format, and now I have to redo it in lossless because of the low quality of mp3." Here are two more opinions:

"While trying to encode lossless some of it I came up with enormous files. For example with only 50 minutes of music, the Bach Inventions and Sinfonias with Kenneth Gilbert in Archiv ended up around 350 Mb with Ape. Apparently the sound of cembalo [harpsichord] is so rich in harmonics that it creates very complex sound waves (with a very high entropy indicator—my apologies, I'm translating a Greek term). That's why many cembalo-solo mp3 files sound not-so-pleasant in mp3 (while other instruments with a more 'rounded' sound like the flute, sound just fine)."

"For me actually harpsichord mp3 sounds better than other instruments simply because it is rich in harmonics, decoding to mp3 cuts off a lot of high frequencies, harpsichord has a lot so it still sounds good, other instruments, for example piano, sound much worse, human voice is very vulnerable to cutting too many high frequencies as well. Besides, you have to know how to make a good mp3. Some 256K/s bitrate mp3s are horrible while sometimes a 160K/s bitrate mp3 sounds really good. Of course, such low bitrates are almost always too low for orchestral music, in this case only lossless compression is enough (if you have a decent CD player and good ripping software of course)."

The main reference for MPEG-4 ALS is [mpeg-4.als 06], a Web site maintained by Tilman Liebchen, a developer of this algorithm and the editor of the MPEG-4 ALS standard. MPEG-4 ALS (in this section we will refer to it simply as ALS) supports any audio sampling rates, audio samples of up to 32 bits (including samples in 32-bit floating-point format), and up to 2^{16} audio channels. Tests performed by the ALS development team are described in [Liebchen et al. 05] and indicate that ALS, even in its simplest modes, outperforms FLAC.

First, a bit of history. In July 2002, the MPEG committee issued a call for proposals for a lossless audio coding algorithm. By December 2002, seven organizations have submitted codecs that met the basic requirements. These were evaluated in early 2003 in terms of compression efficiency, complexity, and flexibility. In March 2003, the codec proposed by researchers from the Technical University of Berlin was selected as the first working draft. The following two years saw further development, followed by implementation and testing. This work culminated in the technical specifications of ALS, which were finalized in July 2005, and adopted by the MPEG committee and later by the ISO. The formal designation of ALS is the ISO/IEC 14496-3:2005/Amd 2:2006 international standard.

The organization of the ALS encoder and decoder is shown in Figure 10.38. The original audio data is divided into frames, each frame is divided into audio channels, and each channel may be divided into blocks. The encoder may modify block sizes if this improves compression. Audio channels may be joined by either subtracting pairs of adjacent channels or by multi-channel coding, in order to reduce inter-channel redundancy. The audio samples in each block are predicted by a combination of short-term and long-term predictors to generate residues. The residues are encoded by Rice codes and block Gilbert-Moore codes that are written on the output stream. This stream also includes tags that allow random access of the audio at virtually any point. Another useful optional feature is a CRC checksum for better integrity of the output.

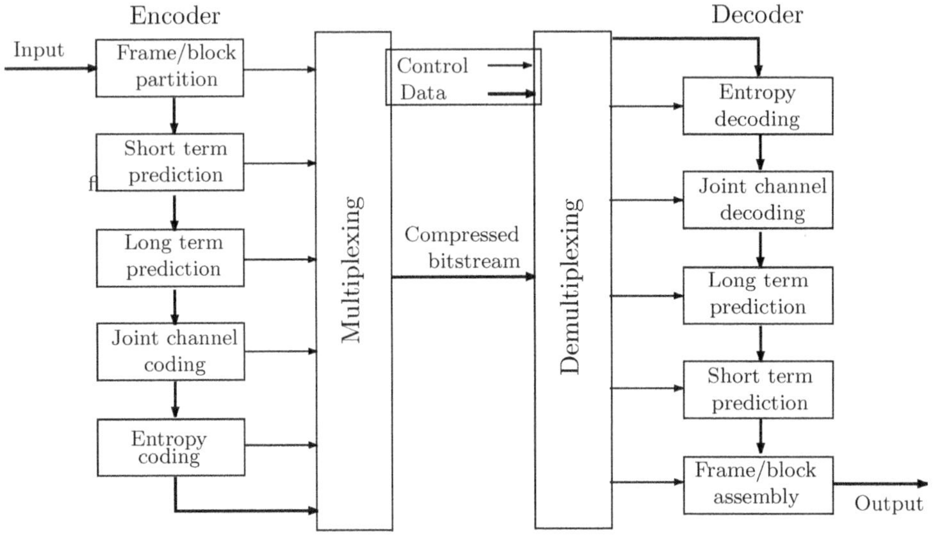

Figure 10.38: Organization of the ALS Encoder and Decoder.

Prediction. Linear prediction is commonly used in lossless audio encoders and its principles have been mentioned elsewhere in this book (see Sections 7.30, 10.9, and 10.10.1). A finite-impulse-response (FIR) linear predictor of order K predicts the current audio sample $x(n)$ from its K immediate predecessors by computing the linear sum

$$\hat{x}(n)=\sum_{k=1}^{K}h_k x(n-k),$$

where h_k are coefficients to be determined. The residue $e(n)$ is computed as the difference $x(n)-\hat{x}(n)$ and is later encoded. If the prediction is done properly, the residues are decorrelated and are also small numbers, ideal for later encoding with a variable-length code.

The decoder obtains $e(n)$ by reading its code and decoding it. It then computes $\hat{x}(n)$ from the coefficients h_k. Thus, the decoder needs to know these coefficients and this can be achieved in three ways as follows:

1. The coefficients are always the same. This is termed nth-order prediction and is discussed in Section 10.9 for $n = 0$, 1, 2, and 3 and in Section 10.10.1 for $n = 4$.

2. The coefficients are computed based on the K audio samples preceding $x(n)$ (this is referred to as forward prediction) and may also depend on some residues preceding $e(n)$ (backward prediction). These samples and residues are known to the decoder and it can compute the coefficients in lockstep with the encoder.

3. The coefficients are computed by the encoder based on all the audio samples in the current block. When the decoder gets to decoding audio sample $x(n)$, it has only the samples preceding it and not those following $x(n)$. Therefore, the coefficients have to be transmitted in the bitstream, as side information, for the use of the decoder. Naturally, they have to be efficiently encoded.

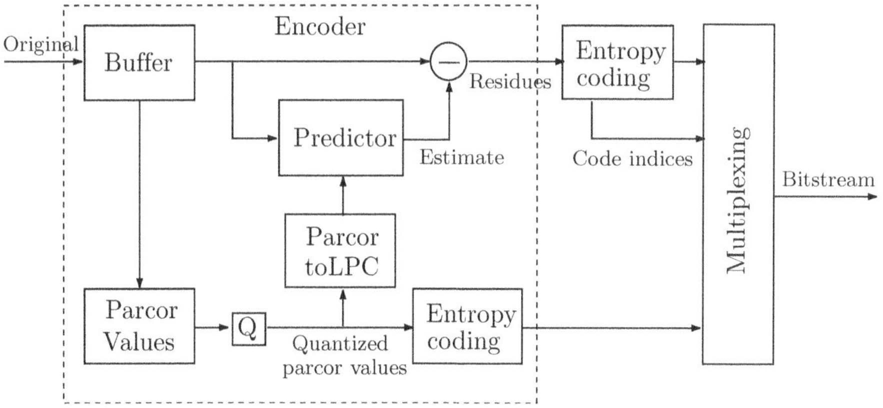

Figure 10.39: Forward Adaptive Prediction in ALS.

ALS adopts forward-adaptive prediction as in 3 above with a prediction-order K of up to 1023. Figure 10.39 is a block diagram of the prediction steps of the ALS encoder. An entire block of audio samples is read into a buffer, and the encoder executes the Levinson-Durbin algorithm (as also does FLAC, Section 10.10). This algorithm (see, for example, [Parsons 87] for a derivation) is recursive, where each iteration increments the order of the filter. The final order K becomes known in the last iteration. The algorithm produces a set of filter coefficients h_i that minimizes the variance of the residues for the block.

The Levinson-Durbin algorithm is adaptive because it computes a different set of filter coefficients and a different order K for each block, depending on the varying statistics of the audio samples and on the block size. The result is a set of filter coefficients that not only produce the smallest residues, but are themselves small, easy to compress, and therefore minimize the amount of side information in the bitstream.

[The value of the predictor order K is important. A large K reduces the variance of the prediction error and results in a smaller bitrate R_e for the encoded residues. On the other hand, a large prediction order implies more filter coefficients and consequently a higher bitrate R_c of the encoded coefficients (the side information). Thus, an important aspect of the Levinson-Durbin algorithm is to determine the value K that will minimize the total bitrate $R(K) = R_e(K) + R_c(K)$. The algorithm estimates both $R_e(K)$ and $R_c(K)$ in each iteration and stops when their sum no longer decreases.]

While computing the filter coefficients h_i in an iteration, the Levinson-Durbin algorithm also computes a set of constants k_i that are referred to as partial correlation (parcor) coefficients (also known as reflection coefficients, see [Rabiner and Schafer 78]). These coefficients have an interesting property. As long as their magnitudes are less than 1, they produce a stable filter. Therefore, it is preferable to quantize the parcor coefficients k_i and use the quantized parcors to compute the desired coefficients h_i. The quantized parcors are compressed and multiplexed into the bitstream to be used later by the decoder.

The parcors are now used to compute a set of filter coefficients h_i that are in turn used to predict all the audio samples in the block. The predicted estimates are subtracted from the actual samples, and the resulting residues are encoded and are also multiplexed into the bitstream.

The ALS decoder (Figure 10.40) is much simpler because it doesn't have to compute any coefficients. For each block, it demultiplexes the encoded parcors from the bitstream, decodes them, and converts them to filter coefficients h_i that are then used to compute the estimates $\hat{x}(n)$ for the entire block. Each encoded residue is then read from the bitstream, decoded, and converted to an audio sample. The complexity of computations in the decoder depends mostly on the order K of the prediction and this order varies from block to block.

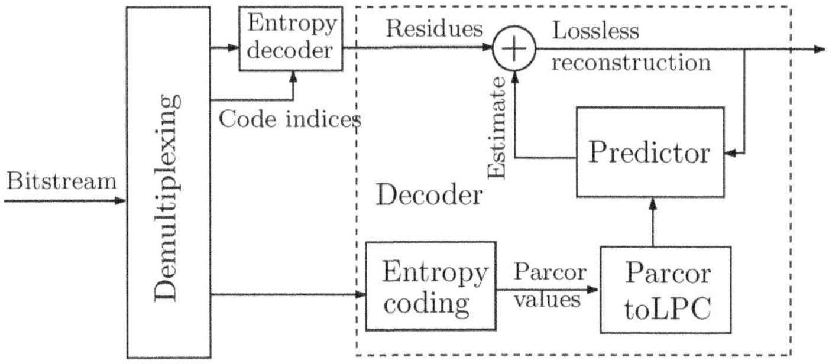

Figure 10.40: Decoding of Predictions in ALS.

Quantization. ALS quantizes the parcors r_i, not the filter coefficients h_i, because it is known (see, for example, [Kleijn and Paliwal 95]) that the latter are very sensitive to even small quantization errors. In order to end up with a stable filter, the parcors are quantized to the interval $[-1, 1]$ and the quantization method takes into account two important facts: (1) The parcors are less sensitive to quantization errors than the filter coefficients, but parcors close to $+1$ or -1 are sensitive. (2) The first two parcors r_1 and r_2, are normally close to -1 and $+1$, respectively, while the remaining parcors are normally smaller. Thus, the main quantization problem is with regard to the first two parcors and is solved by the simple companding expression

$$C(r) = -1 + \sqrt{2(r+1)}.$$

Figure 10.41 shows the behavior of this compander. Function $C(r)$ is close to vertical at $r = -1$, implying that it is very sensitive to small values of r around -1. Thus, $C(r)$ is an ideal compander for r_1. Its complement, $-C(-r)$, is close to vertical at $r = +1$ and is therefore an ideal candidate to compand r_2. In order to simplify the computations, $+C(-r_2)$ is actually used to compand r_2, leading to a companded value that is around -1. After companding, both parcors are quantized to seven bit values (a sign bit and six bits of magnitude) a_1 and a_2 using a simple, uniform quantizer. The final results are

$$a_1 = \left\lfloor 64\left(-1 + \sqrt{2(r_1+1)}\right)\right\rfloor, \quad a_2 = \left\lfloor 64\left(-1 + \sqrt{2(-r_2+1)}\right)\right\rfloor.$$

The remaining parcors r_i $(i > 2)$ are not companded, but are quantized to seven bits by the uniform quantizer $a_i = \lfloor 64r_i \rfloor$. The 7-bit quantized parcors a_i are signed, which places them in the interval $[-64, +63]$. This interval is centered at zero, but the average of the a_is may be nonzero. To shrink them even further, the a_is are re-centered around their most probable values and are then encoded with Rice codes. The result is that each quantized recentered parcor can be encoded with about four bits, without any noticeable degradation of the reconstructed audio.

The parcor to LPC conversion shown in Figures 10.39 and 10.40 uses integer computations and can therefore be reconstructed to full precision by both encoder and decoder.

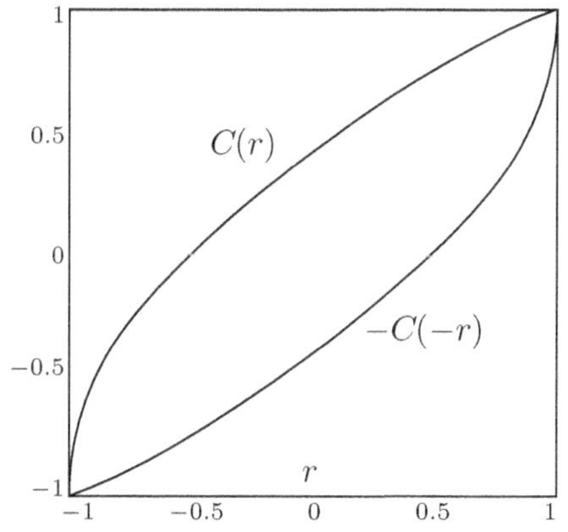

Figure 10.41: Compander Functions $C(r)$ and $-C(-r)$.

Block Size Switching. As mentioned earlier, the input is divided into fixed-size frames. The frame size can be selected by the user depending on the audio sampling rate. For example, if the user prefers 43 ms of audio in a frame, then the frame size should be set to 2,048 samples if the sampling rate is 48 kHz, or 4,096 samples if the sampling rate is 96 kHz. Initially, each frame is one block, but a sophisticated encoder may subdivide a frame into blocks in order to adapt the linear prediction to variations in the input audio. This is referred to as block switching. In general, long blocks with high predictor orders are preferable for stationary segments of audio, while short blocks with lower orders are suitable for transient audio segments.

A frame of size N can be subdivided into blocks of sizes $N/2$, $N/4$, $N/8$, $N/16$, and $N/32$, and the only restriction is that in each subdivision step a block of size n should be divided into two blocks of size $n/2$ each. Figure 10.42 shows some examples of valid (a–d) and invalid (e, f) block subdivisions.

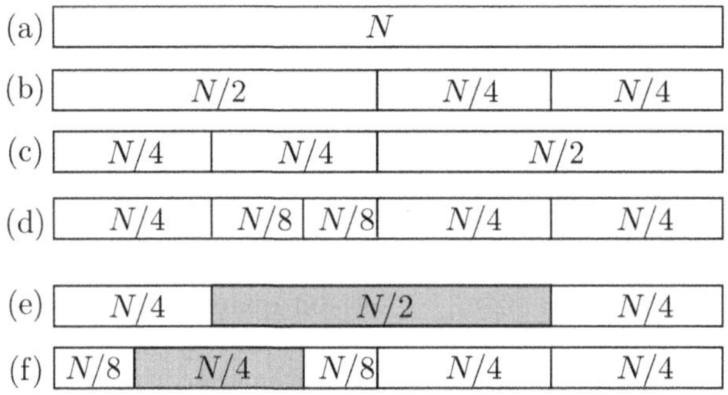

Figure 10.42: Block Subdivision Examples.

The ALS standard does not specify how an encoder is to determine block switching. Methods may range from no block switching, to switching by estimating audio signal characteristics, to complex algorithms that try many block decompositions and select the best one. The block subdivision (BS) scheme selected by the endocder has to be transmitted to the decoder as side information, but it does not increase the work of the decoder since the decoder still has to process the same number of audio samples per frame regardless of how the frame has been subdivided. Thus, block switching can be employed by a sophisticated encoder to significantly increase compression, without degrading decoder performance.

Because of the rule governing block switching, the BS side information sent to the decoder is very small and does not exceed 31 bits. The idea is to start with one bit that specifies whether the size-N frame has been partitioned into two blocks of size $N/2$ each. If this bit is 1, it is followed by two bits that indicate whether any of these blocks has been partitioned into two blocks of size $N/4$ each. If either of the two bits is 1, they are followed by four bits that indicate which of the four size-$N/4$ blocks, if any, has been partitioned into two blocks of size $N/8$. The four bits may be followed by eight bits (for any blocks of size $N/16$) and then by another 16 bits (for any blocks of size $N/32$), for a total of at most 31 bits.

As an example, the block structure of Figure 10.42c is described by the 31-bit BS 1|10|0100|00000000|0000000000000000. The simplest representation of BS is to always send the 31 bits to the decoder, but a small gain may be achieved by preceding BS with two "count" bits that specify its length. Here is how this can be done. Our BS consists of five fields of which the first is a single bit and must always exist. The remaining four fields are needed only if the the first field is 1. Thus, two count bits preceding the first bit of BS are sufficient to specify how many of the remaining four fields follow the first field. For example, $xx0$ means any two count bits followed by a first field of 0, indicating no block switching, whereas 101|10|0100 indicates two fields (two bits and four bits) following the first bit of 1 (the count field is $10_2 = 2$).

Random Access. A practical, user-friendly audio compression algorithm should allow the user to skip forward and back to any desired point in the bitstream and decode the audio from that point. This feature, referred to as random access or seeking, is implemented in ALS by means of random-access frames. Normally, the first K audio samples of a frame are predicted by samples from the preceding frame, which prevents the decoder from jumping to the start of a frame and decoding it without first decoding the previous frame. A random-access frame is a special frame that can be decoded without decoding the preceding frame. Assuming that a frame corresponds to 50 ms of audio, if every 10th frame is generated by the encoder as a random-access frame, the decoder can skip to points in the audio with a precision of 500 ms, more than acceptable for normal audio listening, editing, and streaming applications. The size of a frame and the distance between random-access frames are user-controlled parameters. The latter can be from 1 to 255 frames.

The only problem with random access in ALS is the prediction of the first K audio samples of a random-access frame. A Kth-order linear predictor normally predicts the first K samples of a frame using as many samples as needed from the previous frame, but in a random-access frame, the ALS predictor uses only samples from the current frame and has to predict, for example, the 4th sample from the first three samples of the frame. This is termed progressive prediction. The first sample cannot be predicted. The second sample is predicted from the first (first-order prediction), the third sample is predicted from the first two (second-order), and so on until sample $K+1$ is predicted from its K predecessors in the same frame. Figure 10.43 summarizes this process. Part (a) shows a typical sequence of audio samples. Part (b) shows the result of continuous (non-progressive) prediction, where the first K samples of a random-access frame are not predicted (so the residues equal the original samples). Part (c) illustrates the results of progressive prediction. It shows how the first residue of the random-access frame is identical to the first sample, but successive residues get smaller and smaller.

Long-Term Prediction. A pixel in a digital image is correlated with its near neighbors and an audio sample is similarly correlated with its neighbor samples, but there is a difference between correlation in images and in sound. Many musical instruments generate sound that consists of a fundamental frequency and higher harmonics that are multiples of that frequency. The concept of harmonics is easy to grasp when we consider a string instrument (but is also valid for wind and precussion instruments). A string is attached to the instrument at both ends. When plucked, the string tends to vibrate as shown in Figure 10.44a where it takes the form of half a wave. The fundamental frequency (or first harmonics) is therefore associated with a wavelength that is twice the length of the string. However, part of the time the string also vibrates as in part (b) of the figure, where its center is stationary. This creates the second harmonics frequency, whose associated wavelength equals the length of the string. If the string vibrates as in part (c) of the figure, it generates the third harmonics, with a wavelength that equals two-thirds of the string. In general, the wavelength λ of the nth harmonics is related to the length L of the string by $\lambda = (2/n)L$, which means that the frequency f of

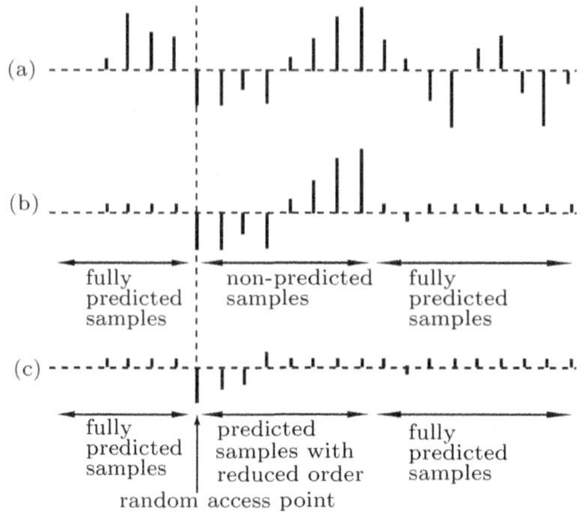

Figure 10.43: Progressive Prediction in ALS.

the nth harmonics is given by $f = s/\lambda = (s/2L)n$, where s is the speed of the wave (notice that the speed depends on the medium of the string, but not on the wavelength or frequency). Thus, the frequencies of higher harmonics are integer multiples of the fundamental frequency.

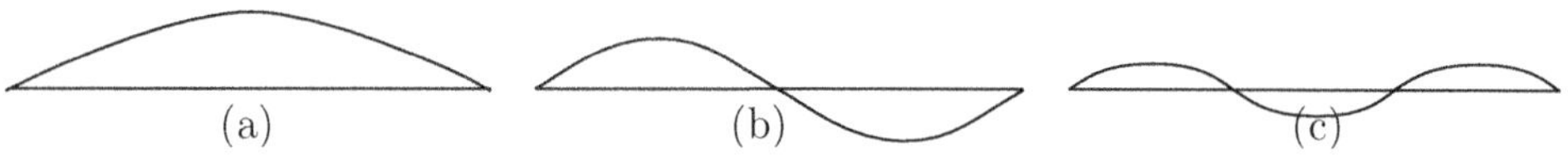

Figure 10.44: Fundamental and Harmonic Frequencies.

We normally don't hear the harmonics as separate tones, mostly because they have increasingly lower amplitudes than the fundamental frequency, but also because our ear-brain system gets used to them from an early age and it combines them with the fundamental frequency. However, higher harmonics are normally present in sound generated by speech or by musical instruments, and they contribute to the richness of the audio we hear. A "pure" tone, with only a fundamental frequency and without any higher frequencies, can be generated by electronic devices, but it sounds artificial, thin, and uninteresting.

Thus, harmonics exist in many audio streams and they contribute to the correlation between audio samples. However, it is easy to see that this correlation is long range. The frequency of the middle C musical note is 261 Hz. At a sampling rate of 48 kHz, a complete sine wave of middle C corresponds to $48{,}000/261 = 184$ audio samples, and at a sampling rate of 192 kHz, a complete sine wave corresponds to $192{,}000/261 = 736$ audio samples. Thus, there may be some correlation between two audio samples separated by hundreds of samples, and ALS tries to take advantage of this correlation.

ALS employs a long-term predictor (LTP) that starts with a residue $e(n)$ and computes a new residue $\tilde{e}(n)$ by subtracting from $e(n)$ the contributions of five distant but related residues $e(n-r+j)$. Parameter r is the lag. It has to be determined by the encoder depending on the current frequency of the audio input and the sampling rate. Index j varies over five values. The rule of computation is

$$\tilde{e}(n) = e(n) - \left(\sum_{j=-2}^{2} \gamma_{r+j} \cdot e(n - r + j) \right),$$

where the five coefficients γ are the quantized gain values. Both r and the five gain values have to be determined by the encoder (using an unspecified algorithm) and sent to the decoder as side information. The resulting long-term residue $\tilde{e}(n)$ is then encoded instead of the short-term residual $e(n)$. [Notice that $\tilde{e}(n)$ is used instead of $e(n)$ also for multi-channel prediction.]

The decoder computes a short-term residue $e(n)$ from a long-term residue $\tilde{e}(n)$ by means of

$$e(n) = \tilde{e}(n) + \left(\sum_{j=-2}^{2} \gamma_{r+j} \cdot e(n - r + j) \right),$$

and then employs $e(n)$ and LPC to determine the next audio sample.

Joint Channel Coding. ALS can support up to 2^{16} audio channels, which is why it is important to exploit any possible correlations and dependencies between channels. Two schemes for multichannel coding are implemented as part of ALS and either can be used.

1. Difference Coding. Often, two audio channels $x_1(n)$ and $x_2(n)$ are known or suspected to be correlated. To take advantage of this, many compression methods compute the difference $d(n) = x_1(n) - x_2(n)$ and encode either the pair $[d(n), x_1(n)]$ or the pair $[d(n), x_2(n)]$. An even better approach is to use LPC to compute and encode residues for the three pairs $00 = [x_1(n), x_2(n)]$, $01 = [d(n), x_1(n)]$, and $10 = [d(n), x_2(n)]$ and select the smallest of the three residues. If this is done on a sample by sample basis, then the decoder has to be told which of the three pairs were selected for audio sample i. This information requires two bits, but only three of the four 2-bit combinations are used. Therefore, this side information should be compressed. We can consider it a sequence of trits (ternary digits), one trit for each audio sample. At the end of a block, the encoder tries to compress the sequence for the block with either run length or by counting the number of occurrences of 00, 01, and 10, and assigning them three appropriate Huffman codes.

2. Multichannel Coding. A volcano in a remote south sea island starts showing signs of an imminent eruption. Geologists rush to the site and place many seismometers on and around the mountain. The signals generated by the seismometers are correlated and serve as an example of many channels of correlated samples. Another example of many correlated channels is biomedical signals obtained from a patient by several measuring devices.

The principle used by ALS to code many correlated channels is to select one channel as a reference r and encode it independently by predicting its samples, computing residues $e^r(n)$, and encoding them. Any other channel c is coded by computing residues $e^c(n)$ and then making each residue smaller by subtracting from it a linear combination of three corresponding residues (three taps) $e^r(n-1)$, $e^r(n)$, and $e^r(n+1)$ from the reference channel. The computation is

$$\hat{e}^c(n) = e^c(n) - \left(\sum_{j=-1}^{1} \gamma_j \cdot e^r(n+j)\right),$$

where the three gain coefficients γ_j are computed by solving the system of equations

$$\gamma = \mathbf{X^{-1}} \cdot \mathbf{y}, \qquad (10.11)$$

where

$$\begin{aligned}
\gamma &= (\gamma_{-1}, \gamma_0, \gamma_{+1})^T,\\
\mathbf{X} &= \begin{bmatrix} \mathbf{e}^{r\,T}_{-1} \cdot \mathbf{e}^r_{-1} & \mathbf{e}^{r\,T}_{-1} \cdot \mathbf{e}^r_0 & \mathbf{e}^{r\,T}_{-1} \cdot \mathbf{e}^r_{+1}\\ \mathbf{e}^{r\,T}_{-1} \cdot \mathbf{e}^r_0 & \mathbf{e}^{rT}_0 \cdot \mathbf{e}^r_0 & \mathbf{e}^{rT}_0 \cdot \mathbf{e}^r_{+1}\\ \mathbf{e}^{r\,T}_{-1} \cdot \mathbf{e}^r_{+1} & \mathbf{e}^{r\,T}_{-1} \cdot \mathbf{e}^r_{+1} & \mathbf{e}^{r\,T}_{+1} \cdot \mathbf{e}^r_{+1} \end{bmatrix},\\
\mathbf{y} &= \left(\mathbf{e}^{cT} \cdot \mathbf{e}^r_{-1}, \mathbf{e}^{cT} \cdot \mathbf{e}^r_0, \mathbf{e}^{cT} \cdot \mathbf{e}^r_{+1}\right),\\
\mathbf{e}^c &= \left(e^c(0), e^c(1), e^c(2), \ldots, e^c(N-1)\right)^T,\\
\mathbf{e}^r_{-1} &= \left(e^r(0), e^r(1), e^r(2), \ldots, e^r(N-1)\right)^T,\\
\mathbf{e}^r_{+1} &= \left(e^r(1), e^r(2), e^r(3), \ldots, e^r(N)\right)^T.
\end{aligned}$$

(Where N is the frame size and T denotes a transpose.) The resulting residues $\hat{e}^c(n)$ are then encoded and written on the bitstream.

The decoder reconstructs the original residual signal by applying the reverse operation

$$e^c(n) = \hat{e}^c(n) + \left(\sum_{j=-1}^{1} \gamma_j \cdot e^r(n+j)\right).$$

There is also a mode of multichannel coding that employs six taps for long-range prediction. Residues $e^c(n)$ of a coding channel are computed and are then made smaller by subtracting a linear combination of six residues from the reference channel. These include the three residues $e^r(n-1)$, $e^r(n)$, and $e^r(n+1)$ that correspond to $e^c(n)$ and another set of three consecutive residues at a distance of r from n, where the lag parameter r is to be estimated by the encoder by cross correlation between the coding channel and the reference channel. The computation is

$$\hat{e}^c(n) = e^c(n) - \left(\sum_{j=-1}^{1} \gamma_j \cdot e^r(n+j) + \sum_{j=-1}^{1} \gamma_{r+j} \cdot e^r(n+r+j)\right).$$

The six gain parameters γ can be obtained by minimizing the energy of the six subtracted residues, similar to Equation (10.11). The decoder, as usual, reconstructs the original residues by the similar computation

$$e^c(n) = \hat{e}^c(n) + \left(\sum_{j=-1}^{1} \gamma_j \cdot e^r(n+j) + \sum_{j=-1}^{1} \gamma_{r+j} \cdot e^r(n+r+j) \right).$$

Once the reconstructed residual signal $e^c(n)$ is obtained by the decoder, it is used by short-term and/or long-term linear prediction to obtain the reconstructed audio sample.

Encoding the Residues. Two modes, simple and advanced, are available to the ALS encoder for encoding the residues $e(n)$. In the simple mode, only Rice codes are used. The enocder has to determine the Rice parameter, and a sophisticated encoder selects the best parameter by computing the distribution of the residues. One option in the simple mode is to use the same Rice parameter for all the residues in a block. An alternative is to partition a block into four parts and use a different Rice parameter to encode the residues in each part. The Rice parameters have to be sent to the decoder as side information as shown in Figure 10.39.

The advanced mode determines a value $e_{\max}$ and encodes each residue $e(n)$ according to its size relative to $e_{\max}$ as shown in Figure 10.45. Residues that satisfy $|e(n)| < e_{\max}$ are located in the central region of the residue distribution and are encoded by a special version of block Gilbert-Moore codes (BGMC, [Gilbert and Moore 59] and [Reznik 04]). Each residue is split into a most-significant part that is coded with BGMC and a least-significant part that is written in raw format (fixed-length) on the bitstream. Reference [Reznik 04] has more information on how the encoder can select reasonable values for $e_{\max}$ and on the number of least-significant bits. Residues that satisfy $|e(n)| \geq e_{\max}$ are located in the tails of the residue distribution. They are first recentered by $e_t(n) = e(n) - e_{\max}$ and then encoded in Rice codes as in the simple mode.

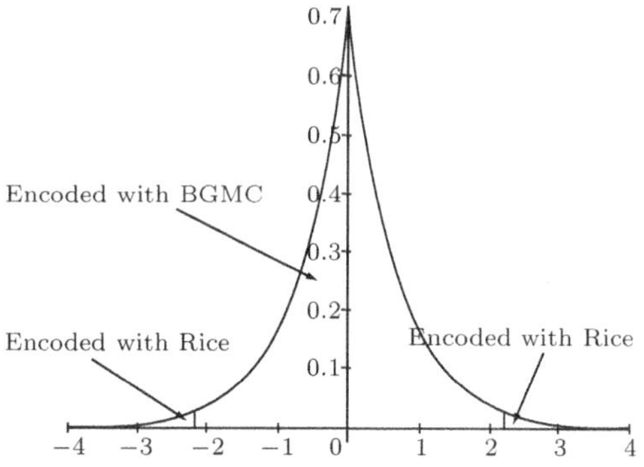

Figure 10.45: Advanced Encoding of Residues.

Floating-Point Audio Samples. It has been mentioned earlier that ALS supports audio signals in floating-point format, specifically, in the IEEE 32-bit format [IEEE754 85]. The encoder separates the sequence **X** of floating-point samples of a block into the sum of an integer sequence **Y** and a sequence **Z** of small floating-point numbers. The sequence $\mathbf{Y}$ of integer values is compressed in the normal manner, while the sequence $\mathbf{Z}$ is compressed by an algorithm termed masked Lempel-Ziv tool, a variant of LZW.

The original sequence $\mathbf{X}$ of floating-point audio samples is separated in the form $\mathbf{X} = A \otimes \mathbf{Y} + \mathbf{Z}$, where $\mathbf{X}$, $\mathbf{Y}$, and $\mathbf{Z}$ are vectors, $\otimes$ denotes multiplication with rounding, and A is a common multiplier, a floating-point number in the interval $[1,2)$. The main problem is to select a common multiplier that will minimize the floating-point values in vector $\mathbf{Z}$. The ALS standard proposes an approach to this problem, but does not guarantee that this approach is the best one. For a detailed discussion of this point as well as masked Lempel-Ziv tool, see [Liebchen et al. 05].

10.14 MPEG-1/2 Audio Layers

The video compression aspects of the MPEG-1 and MPEG-2 standards are described in Section 9.6. Its audio compression principles are discussed here. Readers are advised to read Sections 10.2 and 10.3 before trying to tackle this material. Some references for this section are [Brandenburg and Stoll 94], [Pan 95], [Rao and Hwang 96], and [Shlien 94].

The formal name of MPEG-1 is *the international standard for moving picture video compression, IS 11172*. It consists of five parts, of which part 3 [ISO/IEC 93] is the definition of the audio compression algorithm. Like other standards developed by the ITU and ISO, the document describing MPEG-1 has *normative* and *informative* sections. A normative section is part of the standard specification. It is intended for implementers, it is written in a precise language, and it should be strictly followed in implementing the standard on actual computer platforms. An informative section, on the other hand, illustrates concepts that are discussed elsewhere, explains the reasons that led to certain choices and decisions, and contains background material. An example of a normative section is the tables of various parameters and of the Huffman codes used in MPEG audio. An example of an informative section is the algorithm used by MPEG audio to implement a psychoacoustic model. MPEG does not require any particular algorithm, and an MPEG encoder can use any method to implement the model. This informative section simply describes various alternatives.

The MPEG-1 and MPEG-2 (or in short, MPEG-1/2) audio standard specifies three compression methods called *layers* and designated I, II, and III. All three layers are part of the MPEG-1 standard. A movie compressed by MPEG-1 uses only one layer, and the layer number is specified in the compressed stream. Any of the layers can be used to compress an audio file without any video. The functional modules of the lower layers are also used by the higher layers, but the higher layers have additional features that result in better compression. An interesting aspect of the design of the standard is that the layers form a hierarchy in the sense that a layer-III decoder can also decode audio files compressed by layers I or II.

The result of having three layers was an increasing popularity of layer III. The encoder is extremely complex, but it produces excellent compression, and this, combined with the fact that the decoder is much simpler, has produced in the late 1990s an explosion of what is popularly known as `mp3` audio files. It is easy to legally and freely obtain a layer-III decoder and much music that is already encoded in layer III. So far, this has been a big success of the audio part of the MPEG project.

The MPEG audio standard [ISO/IEC 93] starts with the normative description of the format of the compressed stream for each of the three layers. It follows with a normative description of the decoder. The description of the encoder (it is different for the three layers) and of two psychoacoustic models follows and is informative; any encoder that generates a correct compressed stream is a valid MPEG encoder. There are appendices (annexes) discussing related topics such as error protection.

In contrast with MPEG video, where many information sources are available, there is relatively little in the technical literature about MPEG audio. In addition to the references elsewhere in this section, the reader is referred to the MPEG consortium [MPEG 00]. This site contains lists of other resources, and is updated from time to time. Another resource is the audio engineering society (AES). Most of the ideas and techniques used in the MPEG audio standard (and also in other audio compression methods) were originally published in the many conference proceedings of this organization. Unfortunately, these are not freely available and can be purchased from the AES.

The history of MPEG-1 audio starts in December 1988, when a group of experts met for the first time in Hanover, Germany, to discuss the topic. During 1989, no fewer than 14 algorithms were proposed by these experts. They were eventually merged into four groups, known today as ASPEC, ATAC, MUSICAM, and SB-ADPCM. These were tested and compared in 1990, with results that led to the idea of adopting the three layers. Each of the three layers is a different compression method, and they increase in complexity (and in performance) from layer I to layer III. The first draft of the ISO MPEG-1 audio standard (standard 11172 part 3) was ready in December 1990. The three layers were implemented and tested during the first half of 1991, and the specifications of the first two layers were frozen in June 1991. Layer III went through further modifications and tests during 1991, and the complete MPEG audio standard was ready and was sent by the expert group to the ISO in December 1991, for its approval. It took the ISO/IEC almost a year to examine the standard, and it was finally approved in November 1992.

When a movie is digitized, the audio part may often consist of two sound tracks (stereo sound), each sampled at 44.1 kHz with 16-bit samples. The audio data rate is therefore $2\times 44{,}100\times 16 = 1{,}411{,}200$ bits/sec; close to 1.5 Mbits/sec. In addition to the 44.1 kHz sampling rate, the MPEG standard allows sampling rates of 32 kHz and 48 kHz. An important feature of MPEG audio is its compression ratio, which the standard specifies in advance! The standard calls for a compressed stream with one of several bitrates (Table 10.46) ranging from 32 to 224 kbps per audio channel (there normally are two channels, for stereo sound). Depending on the original sampling rate, these bitrates translate to compression factors of from 2.7 (low) to 24 (impressive)! The reason for specifying the bitrates of the compressed stream is that the stream also includes compressed data for the video and system parts. (These parts are mentioned in Section 9.6 and are not the same as the three audio layers.)

The principle of MPEG audio compression is quantization. The values being quantized, however, are not the audio samples but numbers (called signals) taken from the frequency domain of the sound (this is discussed in the next paragraph). The fact that the compression ratio (or equivalently, the bitrate) is known to the encoder means that

	Bitrate (Kbps)		
Index	Layer I	Layer II	Layer III
0000	free format	free format	free format
0001	32	32	32
0010	64	48	40
0011	96	56	48
0100	128	64	56
0101	160	80	64
0110	192	96	80
0111	224	112	96
1000	256	128	112
1001	288	160	128
1010	320	192	160
1011	352	224	192
1100	384	256	224
1101	416	320	256
1110	448	384	320
1111	forbidden	forbidden	forbidden

Table 10.46: Bitrates in the Three Layers.

the encoder knows at any time how many bits it can allocate to the quantized signals. Thus, the (adaptive) *bit allocation algorithm* is an important part of the encoder. This algorithm uses the known bitrate and the frequency spectrum of the most recent audio samples to determine the size of the quantized signals such that the quantization noise (the difference between an original signal and a quantized one) will be inaudible (i.e., will be below the *masked threshold*, a concept discussed in Section 10.3).

The psychoacoustic models use the frequency of the sound that is being compressed, but the input stream consists of audio samples, not sound frequencies. The frequencies have to be computed from the samples. This is why the first step in MPEG audio encoding is a discrete Fourier transform, where a set of 512 consecutive audio samples is transformed to the frequency domain. Since the number of frequencies can be huge, they are grouped into 32 equal-width frequency subbands (layer III uses different numbers but the same principle). For each subband, a number is obtained that indicates the intensity of the sound at that subband's frequency range. These numbers (called *signals*) are then quantized. The coarseness of the quantization in each subband is determined by the masking threshold in the subband and by the number of bits still available to the encoder. The masking threshold is computed for each subband using a psychoacoustic model.

MPEG uses two psychoacoustic models to implement frequency masking and temporal masking. Each model describes how loud sound masks other sounds that happen to be close to it in frequency or in time. The model partitions the frequency range into 24 critical bands and specifies how masking effects apply within each band. The masking effects depend, of course, on the frequencies and amplitudes of the tones. When the sound is decompressed and played, the user (listener) may select any playback amplitude, which is why the psychoacoustic model has to be designed for the worst case. The masking effects also depend on the nature of the source of the sound being compressed. The source may be tone-like or noise-like. The two psychoacoustic models employed by MPEG are based on experimental work done by researchers over many years.

The decoder must be fast, since it may have to decode the entire movie (video and audio) at real time, so it must be simple. As a result it does not use any psychoacoustic model or bit allocation algorithm. The compressed stream must therefore contain all the information that the decoder needs for dequantizing the signals. This information (the size of the quantized signals) must be written by the encoder on the compressed stream, and it constitutes overhead that should be subtracted from the number of remaining available bits.

Figure 10.47 is a block diagram of the main components of the MPEG audio encoder and decoder. The ancillary data is user-definable and would normally consist of information related to specific applications. This data is optional.

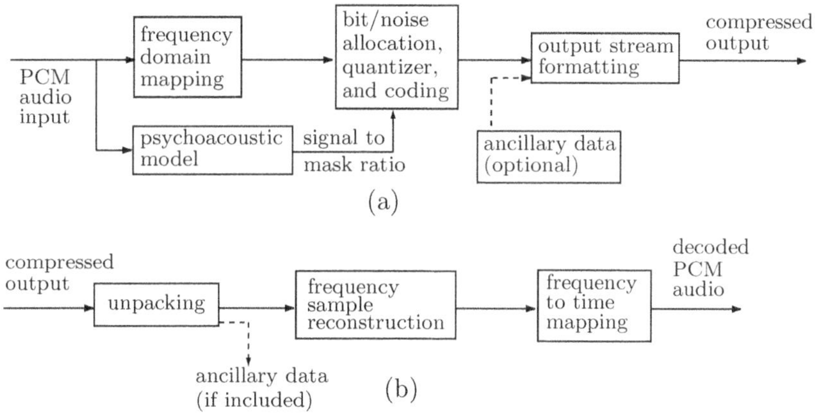

Figure 10.47: MPEG Audio: (a) Encoder and (b) Decoder.

10.14.1 Frequency Domain Coding

The first step in encoding the audio samples is to transform them from the time domain to the frequency domain. This is done by a bank of *polyphase filters* that transform the samples into 32 equal-width frequency subbands. The filters were designed to provide fast operation combined with good time and frequency resolutions. As a result, their design involved three compromises.

The first compromise is the equal widths of the 32 frequency bands. This simplifies the filters but is in contrast to the behavior of the human auditory system, whose sensitivity is frequency-dependent. Ideally, the filters should divide the input into the critical bands discussed in Section 10.3. These bands are formed such that the perceived loudness of a given sound and its audibility in the presence of another, masking, sound are consistent within a critical band, but different across these bands. Unfortunately, each of the low-frequency subbands overlaps several critical bands, with the result that the bit allocation algorithm cannot optimize the number of bits allocated to the quantized signal in those subbands. When several critical bands are covered by a subband X, the bit allocation algorithm selects the critical band with the least noise masking and uses that critical band to compute the number of bits allocated to the quantized signals in subband X.

The second compromise involves the inverse filter bank, the one used by the decoder. The original time-to-frequency transformation involves loss of information (even before any quantization). The inverse filter bank therefore receives data that is slightly bad, and uses it to perform the inverse frequency-to-time transformation, resulting in more distortions. Therefore, the design of the two filter banks (for direct and inverse transformations) had to use compromises to minimize this loss of information.

The third compromise has to do with the individual filters. Adjacent filters should ideally pass different frequency ranges. In practice, they have considerable frequency overlap. Sound of a single, pure, frequency can therefore penetrate through two filters and produce signals (that are later quantized) in two of the 32 subbands instead of in just one subband.

The polyphase filter bank uses (in addition to other intermediate data structures) a buffer X with room for 512 input samples. The buffer is a FIFO queue and always contains the most-recent 512 samples input. Figure 10.48 shows the five main steps of the polyphase filtering algorithm.

1. Shift in 32 new input samples into FIFO buffer X

2. Window samples: $Z_i = C_i \times X_i$, for $i = 0, 1, \ldots, 511$

3. Partial computation: $Y_i = \sum_{j=0}^{7} Z_{i+64j}$, for $i = 0, 1, \ldots, 63$

4. Compute 32 signals: $S_i = \sum_{k=0}^{63} M_{i,k} \times Y_k$, for $i = 0, 1, \ldots, 31$

5. Output the 32 subband signals S_i

Figure 10.48: Polyphase Filter Bank.

The next 32 audio samples read from the input are shifted into the buffer. Thus, the buffer always holds the 512 most recent audio samples. The signals $S_t[i]$ for the 32 subbands are computed by

$$S_t[i] = \sum_{k=0}^{63} \sum_{j=0}^{7} M_{i,k} \big(C[k+64j] \times X[k+64j]\big), \; i = 0, \ldots, 31. \tag{10.12}$$

The notation $S_t[i]$ stands for the signal of subband i at time t. Vector C contains 512 coefficients of the analysis window and is fully specified by the standard. Some of the subband filter coefficients C_i are listed in Table 10.49a. M is the analysis matrix defined by

$$M_{i,k} = \cos\left(\frac{(2i+1)(k-16)\pi}{64}\right),\ i=0,\ldots,31,\ k=0,\ldots,63. \tag{10.13}$$

Notice that the expression in parentheses in Equation (10.12) does not depend on i, while $M_{i,k}$ in Equation (10.13) does not depend on j. (This matrix is a variant of MDCT, which is a modified version of the well-known DCT matrix.) This feature is a compromise that results in fewer arithmetic operations. In fact, the 32 signals $S_t[i]$ are computed by only $512 + 32 \times 64 = 2560$ multiplications and $64 \times 7 + 32 \times 63 = 2464$ additions, which come to about 80 multiplications and 80 additions per signal. Another point worth mentioning is the decimation of samples (Section 8.7). The entire filter bank produces 32 output signals for 32 input samples. Since each of the 32 filters produces 32 signals, its output has to be decimated, retaining only one signal per filter.

Figure 10.51a,b illustrates graphically the operations performed by the encoder and decoder during the polyphase filtering step. Part (a) of the figure shows how the X buffer holds 64 segments of 32 audio samples each. The buffer is shifted one segment to the right before the next segment of 32 new samples is read from the input stream and is entered on the left. After multiplying the X buffer by the coefficients of the C window, the products are moved to the Z vector. The contents of this vector are partitioned into segments of 64 numbers each, and the segments are added. The result is vector Y, which is multiplied by the MDCT matrix, to produce the final vector of 32 subband signals.

Part (b) of the figure illustrates the operations performed by the decoder. A group of 32 subband signals is multiplied by the IMDCT matrix $N_{i,k}$ to produce the V vector, consisting of two segments of 32 values each. The two segments are shifted into the V FIFO buffer from the left. The V buffer has room for the last 16 V vectors (i.e., 16×64, or 1024, values). A new 512-entry U vector is created from 32 alternate segments in the V buffer, as shown. The U vector is multiplied by the 512 coefficients D_i of the synthesis window (similar to the C_i coefficients of the analysis window used by the encoder), to create the W vector. Some of the D_i coefficients are listed in Table 10.49b. This vector is divided into 16 segments of 32 values each and the segments added. The result is 32 reconstructed audio samples. Figure 10.50 is a flowchart illustrating this process. The IMDCT synthesis matrix $N_{i,k}$ is given by

$$N_{i,k} = \cos\left(\frac{(2k+1)(i+16)\pi}{64}\right),\ i=0,\ldots,63,\ k=0,\ldots,31.$$

The subband signals computed by the filtering stage of the encoder are collected and packaged into *frames* containing 384 signals (in layer I) or 1152 signals (in layers II and III) each. The signals in a frame are then scaled and quantized according to the psychoacoustic model used by the encoder and the bit allocation algorithm. The quantized values, together with the scale factors and quantization information (the number of quantization levels in each subband) are written on the compressed stream (layer III also uses Huffman codes to encode the quantized values even further).

10.14.2 Format of Compressed Data

In layer I, each frame consists of 12 signals per subband, for a total of $12 \times 32 = 384$ signals. In layers II and III, a frame contains 36 signals per subband, for a total of

i	C_i	i	C_i	i	C_i
0	0.000000000	252	−0.035435200	506	−0.000000477
1	−0.000000477	253	−0.035586357	507	−0.000000477
2	−0.000000477	254	−0.035694122	508	−0.000000477
3	−0.000000477	255	−0.035758972	509	−0.000000477
4	−0.000000477	256	−0.035780907	510	−0.000000477
5	−0.000000477	257	−0.035758972	511	−0.000000477
⋮		⋮			

(a)

i	D_i	i	D_i	i	D_i
0	0.000000000	252	−1.133926392	506	0.000015259
1	−0.000015259	253	−1.138763428	507	0.000015259
2	−0.000015259	254	−1.142211914	508	0.000015259
3	−0.000015259	255	−1.144287109	509	0.000015259
4	−0.000015259	256	−1.144989014	510	0.000015259
5	−0.000015259	257	−1.144287109	511	0.000015259
⋮		⋮			

(b)

Table 10.49: Coefficients of (a) the Analysis and (b) the Synthesis Window.

1. Input 32 new signals S_i, for $i = 0, \ldots, 31$

2. Shift* the FIFO buffer $V_i = V_{i-64}$, $i = 1023, \ldots, 64$

3. Multiply $V_i = \sum_{k=0}^{31} N_{ik} S_k$, $i = 0, \ldots, 63$

4. Construct 512 values in vector U
for $i = 0$ to 7 do
 for $j = 0$ to 31 do
  $U_{j+64i} = V_{j+128i}$
  $U_{32+j+64i} = V_{96+j+128i}$

5. Construct W. For $i = 0$ to 511 do, $W_i = U_i \times D_i$

6. Compute 32 samples. For $j = 0$ to 31 do, $S_j = \sum_{i=0}^{15} W_{j+32i}$

7. Output the 32 audio samples S_j

* V is initialized to all zeros at startup.

Figure 10.50: Reconstructing Audio Samples.

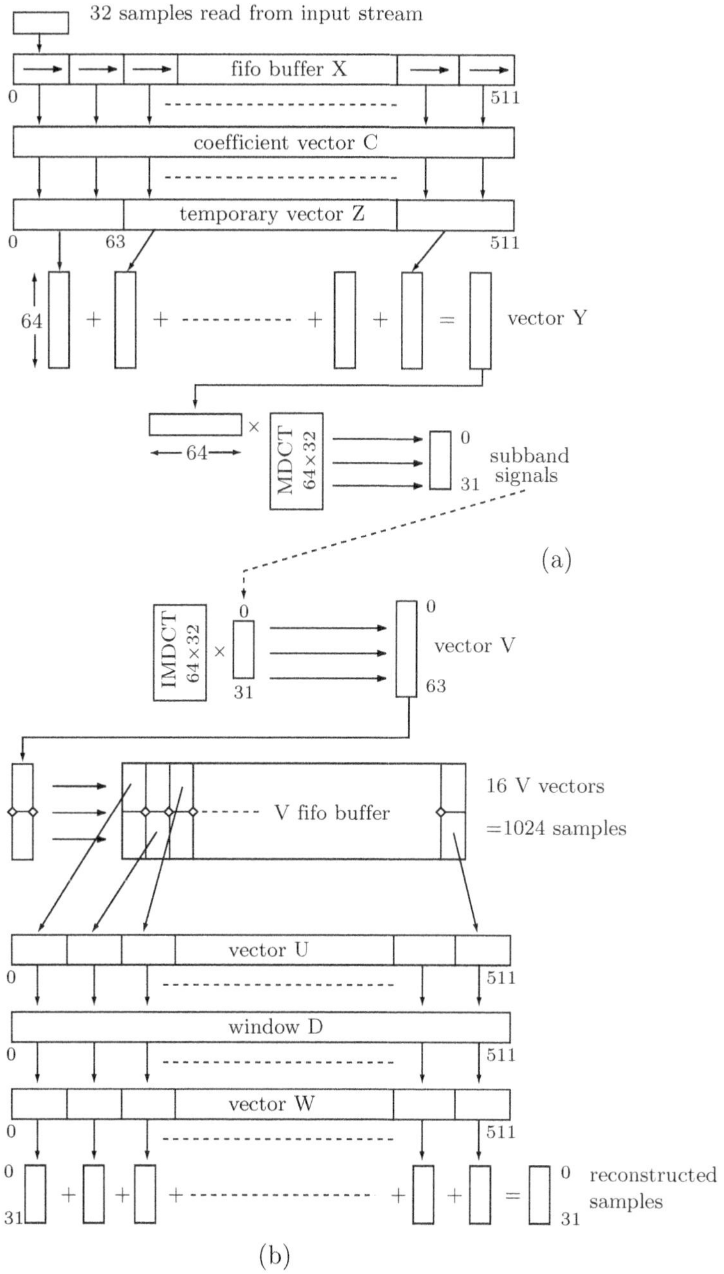

Figure 10.51: MPEG Audio: (a) Encoder and (b) Decoder.

1152 signals. The signals in the frame are quantized (this is the important step where compression is achieved) and written on the compressed stream together with other information.

Each frame written on the output starts with a 32-bit header whose format is identical for the three layers. The header contains a synchronization code (12 1-bits) and 20 bits of coding parameters listed below. If error protection is used, the header is immediately followed by a 16-bit CRC check word (Section 6.32) that uses the generating polynomial $CRC_{16}(x) = x^{16}+x^{15}+x^2+1$. Next comes a frame of the quantized signals, followed by an (optional) ancillary data block. The formats of the last two items depend on the layer.

In layer I, the CRC 16-bit code is computed from the last 16 bits (bits 16–31) of the header and from the bit allocation information. In layer II, the CRC code is computed from these bits plus the scalefactor selection information. In layer III, the CRC code is computed from the last 16 bits of the header and also from either bits 0–135 of the audio data (in single-channel mode) or bits 0–255 of the audio data (in the other three modes, see field 8 below).

The synchronization code is used by the decoder to verify that what it is reading is, in fact, the header. This code is a string of 12 bits of 1, so the entire format of the compressed stream had to be designed to avoid an accidental occurrence of a string of twelve 1's elsewhere.

The remaining 20 bits of the header are divided into 12 fields as follows:

Field 1. An ID bit whose value is 1 (this indicates that MPEG is used). A value of 0 is reserved and is currently unused.

Field 2. Two bits to indicate the layer number. Valid values are 11—layer I, 10—layer II, and 01—layer III. The value 00 is reserved.

Field 3. An error protection indicator bit. A value of 0 indicates that redundancy has been added to the compressed stream to help in error detection.

Field 4. Four bits to indicate the bitrate (Table 10.46). A zero index indicates a "fixed" bitrate, where a frame may contain an extra slot, depending on the padding bit (field 6).

Field 5. Two bits to indicate one of three sampling frequencies. The three values are 00—44.1 kHz, 01—48 kHz, and 10—32 kHz. The value 11 is reserved.

Field 6. One bit to indicate whether padding is used. Padding may add a slot (slots are discussed in Section 10.14.3) to the compressed stream after a certain number of frames, to make sure that the total size of the frames either equals or is slightly less than the sum

$$\sum_{\text{first frame}}^{\text{current frame}} \frac{\text{frame-size} \times \text{bitrate}}{\text{sampling frequency}},$$

where frame-size is 384 signals for layer I and 1152 signals for layers II and III. The following algorithm may be used by the encoder to determine whether or not padding is necessary.

```
for first audio frame:
  rest:=0;
  padding:=no;
```

```
for each subsequent audio frame:
  if layer=I
    then dif:=(12 × bitrate) modulo (sampling-frequency)
    else dif:=(144 × bitrate) modulo (sampling-frequency);
  rest:=rest−dif;
  if rest<0 then
     padding:=yes;
     rest:=rest+(sampling-frequency)
  else padding:=no;
```

This algorithm has a simple interpretation. A frame is divided into N or $N+1$ slots, where N depends on the layer. For layer I, N is given by

$$N = 12\times\frac{\text{bitrate}}{\text{sampling frequency}}.$$

For layers II and III, it is given by

$$N = 144\times\frac{\text{bitrate}}{\text{sampling frequency}}.$$

If this does not produce an integer, the result is truncated and padding is used.

Padding is also mentioned in Section 10.14.3

Field 7. One bit for private use of the encoder. This bit will not be used by ISO/IEC in the future.

Field 8. A two-bit stereo mode field. Values are 00—stereo, 01—joint-stereo (intensity-stereo and/or ms-stereo), 10—dual-channel, and 11—single-channel.

Stereo information is encoded in one of four modes: stereo, dual channel, joint stereo, and ms-stereo. In the first two modes, samples from the two stereo channels are compressed and written on the output. The encoder does not check for any correlations between the two. The stereo mode is used to compress the left and right stereo channels, while the dual channel mode is used to compress different sets of audio samples, such as a bilingual broadcast. The joint stereo mode exploits redundancies between the left and right channels, since many times they are identical, similar, or differ by a small time lag. The ms-stereo mode ("ms" stands for "middle-side") is a special case of joint stereo, where two signals, a middle value M_i and a side value S_i, are encoded instead of the left and right audio channels L_i and R_i. The middle-side values are computed by the following sum and difference

$$L_i = \frac{M_i + S_i}{\sqrt{2}}, \text{ and } R_i = \frac{M_i - S_i}{\sqrt{2}}.$$

Field 9. A two-bit mode extension field. This is used in the joint-stereo mode. In layers I and II the bits indicate which subbands are in intensity-stereo. All other subbands are coded in "stereo" mode. The four values are:

00—subbands 4–31 in intensity-stereo, bound=4.

01—subbands 8–31 in intensity-stereo, bound=8.

10—subbands 12–31 in intensity-stereo, bound=12.

11—subbands 16–31 in intensity-stereo, bound=16.

In layer III these bits indicate which type of joint stereo coding method is applied. The values are:

00—intensity-stereo off, ms-stereo off.

01—intensity-stereo on, ms-stereo off.

10—intensity-stereo off, ms-stereo on.

11—intensity-stereo on, ms-stereo on.

Field 10. Copyright bit. If the compressed stream is copyright protected, this bit should be 1.

Field 11. One bit indicating original/copy. A value of 1 indicates an original compressed stream.

Field 12. A 2-bit emphasis field. This indicates the type of de-emphasis that is used. The values are 00—none, 01—50/15 microseconds, 10—reserved, and 11 indicates CCITT J.17 de-emphasis.

Layer I: The 12 signals of each subband are scaled such that the largest one becomes 1 (or very close to 1, but not greater than 1). The psychoacoustic model and the bit allocation algorithm are invoked to determine the number of bits allocated to the quantization of each scaled signal in each subband (or, equivalently, the number of quantization levels). The scaled signals are then quantized. The number of quantization levels, the scale factors, and the 384 quantized signals are then placed in their areas in the frame, which is written on the output. Each bit allocation item is four bits, each scale factor is six bits, and each quantized sample occupies between two and 15 bits in the frame.

The number l of quantization levels and the number q of bits per quantized value are related by $2^q = l$. The bit allocation algorithm uses tables to determine q for each of the 32 subbands. The 32 $(q-1)$ values are then written, as 4-bit numbers, on the frame, for the use of the decoder. Thus, the 4-bit value 0000 read by the decoder from the frame for subband s indicates to the decoder that the 12 signals of s have been quantized coarsely to one bit each, while the value 1110 implies that the 16-bit signals have been finely quantized to 15 bits each. The value 1111 is not used, to avoid conflict with the synchronization code. The encoder decides many times to quantize all the 12 signals of a subband s to zero, and this is indicated in the frame by a different code. In such a case, the 4-bit value input from the frame for s is ignored by the decoder.

◇ **Exercise 10.8:** Explain why the encoder may decide to quantize 12 consecutive signals of subband s to zero.

The decoder multiplies the dequantized signal values by the scale factors found in the frame. There is one scale factor for the 12 signals of a subband. This scale factor is selected by the encoder from a table of 63 scale factors specified by the standard (Table 10.52). The scale factors in the table increase by a factor of $\sqrt[3]{2}$.

◇ **Exercise 10.9:** How does this increase translate to an increase in the decibel level?

Quantization is performed by the following simple rule: If the bit allocation algorithm allocates b bits to each quantized value, then the number n of quantization levels is determined by $b = \log_2(n+1)$, or $2^b = n+1$. The value $b = 3$, for example, results

Index	scalefactor	Index	scalefactor	Index	scalefactor
0	2.00000000000000	21	0.01562500000000	42	0.00012207031250
1	1.58740105196820	22	0.01240157071850	43	0.00009688727124
2	1.25992104989487	23	0.00984313320230	44	0.00007689947814
3	1.00000000000000	24	0.00781250000000	45	0.00006103515625
4	0.79370052598410	25	0.00620078535925	46	0.00004844363562
5	0.62996052494744	26	0.00492156660115	47	0.00003844973907
6	0.50000000000000	27	0.00390625000000	48	0.00003051757813
7	0.39685026299205	28	0.00310039267963	49	0.00002422181781
8	0.31498026247372	29	0.00246078330058	50	0.00001922486954
9	0.25000000000000	30	0.00195312500000	51	0.00001525878906
10	0.19842513149602	31	0.00155019633981	52	0.00001211090890
11	0.15749013123686	32	0.00123039165029	53	0.00000961243477
12	0.12500000000000	33	0.00097656250000	54	0.00000762939453
13	0.09921256574801	34	0.00077509816991	55	0.00000605545445
14	0.07874506561843	35	0.00061519582514	56	0.00000480621738
15	0.06250000000000	36	0.00048828125000	57	0.00000381469727
16	0.04960628287401	37	0.00038754908495	58	0.00000302772723
17	0.03937253280921	38	0.00030759791257	59	0.00000240310869
18	0.03125000000000	39	0.00024414062500	60	0.00000190734863
19	0.02480314143700	40	0.00019377454248	61	0.00000151386361
20	0.01968626640461	41	0.00015379895629	62	0.00000120155435

Table 10.52: Layers I and II Scale Factors.

in $n = 7$ quantization values. Figure 10.53a,b shows examples of such quantization. The input signals being quantized have already been scaled to the interval $[-1, +1]$, and the quantization is midtread. For example, all input values in the range $[-1/7, -3/7]$ are quantized to 010 (dashed lines in the figure). Dequantization is also simple. The quantized value 010 is always dequantized to $-2/7$. Notice that quantized values range from 0 to $n-1$. The value $n = 11\ldots1$ is never used, in order to avoid a conflict with the synchronization code.

Table 10.54 shows that the bit allocation algorithm for layer I can select the size of the quantized signals to be between 0 and 15 bits. For each of these sizes, the table lists the 4-bit allocation code that the encoder writes on the frame for the decoder's use, the number q of quantization levels (between 0 and $2^{15}-1 = 32{,}767$, and the signal-to-noise ratio in decibels.

In practice, the encoder scales a signal S_i by a scale factor scf that is determined from Table 10.52, and quantizes the result by computing

$$S_{qi} = \left(A\left[\frac{S_i}{scf}\right] + B\right)\Bigg|_N,$$

where A and B are the constants listed in Table 10.55 (the three entries flagged by asterisks are used by layer II but not by layer I) and N is the number of bits needed to encode the number of quantization levels. (The vertical bar with subscript N means: Take the N most-significant bits.) In order to avoid conflicts with the synchronization

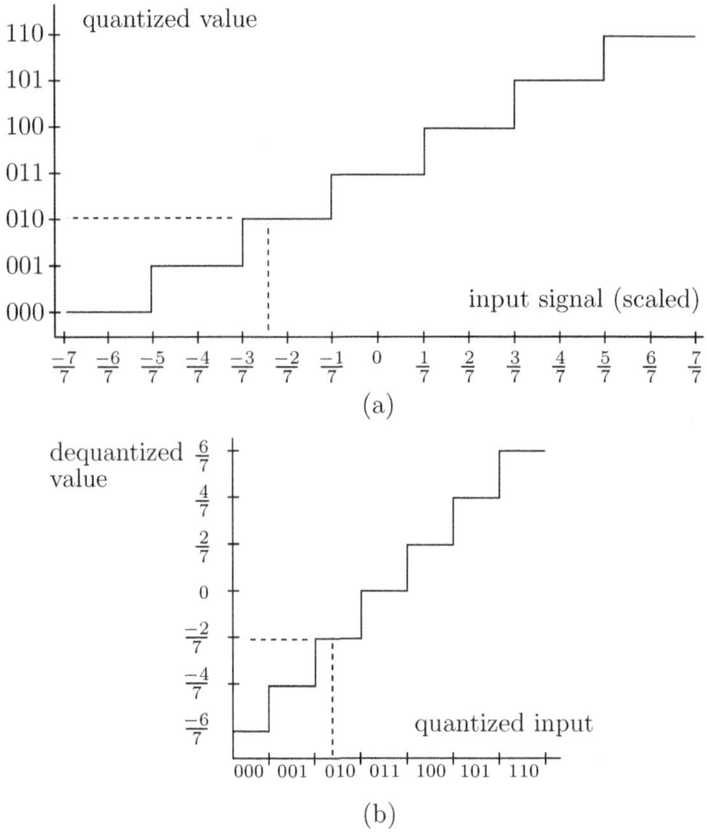

Figure 10.53: Examples of (a) Quantizer and (b) Dequantizer.

code, the most-significant bit of the quantized value S_{qi} is inverted before that quantity is written on the output.

Layer II Format: This is an extension of the basic method outlined for layer I. Each frame now consists of 36 signals per subband, and both the bit allocation data and the scale factor information are coded more efficiently. Also, quantization can be much finer and can have up to $2^{16} - 1 = 65{,}535$ levels. A frame is divided into three parts, numbered 0, 1, and 2. Each part resembles a layer-I frame and contains 12 signals per subband. The bit allocation data is common to the three parts, but the information for the scale factors is organized differently. It can be common to all three parts, it can apply to just two of the three parts, or it can be specified for each part separately. This information consists of a 2-bit scale factor selection information (sfsi). This number indicates whether one, two, or three scale factors per subband are written in the frame, and how they are applied.

The bit allocation section of the frame is encoded more efficiently by limiting the choice of quantization levels for the higher subbands and the lower bitrates. Instead of the four bits per subband used by layer I to specify the bit allocation choice, the number of bits used by layer II varies from 0 to 4 depending on the subband number. The MPEG

bit alloc	4-bit code	number of levels	SNR (dB)
0	0000	0	0.00
2	0001	3	7.00
3	0010	7	16.00
4	0011	15	25.28
5	0100	31	31.59
6	0101	63	37.75
7	0110	127	43.84
8	0111	255	49.89
9	1000	511	55.93
10	1001	1023	61.96
11	1010	2047	67.98
12	1011	4095	74.01
13	1100	8191	80.03
14	1101	16383	86.05
15	1110	32767	92.01
invalid	1111		

Table 10.54: Bit Allocation and Quantization in Layer I.

number of levels	A	B
3	0.750000000	−0.250000000
5*	0.625000000	−0.375000000
7	0.875000000	−0.125000000
9*	0.562500000	−0.437500000
15	0.937500000	−0.062500000
31	0.968750000	−0.031250000
63	0.984375000	−0.015625000
127	0.992187500	−0.007812500
255	0.996093750	−0.003906250
511	0.998046875	−0.001953125
1023	0.999023438	−0.000976563
2047	0.999511719	−0.000488281
4095	0.999755859	−0.000244141
8191	0.999877930	−0.000122070
16383	0.999938965	−0.000061035
32767	0.999969482	−0.000030518
65535*	0.999984741	−0.000015259

Table 10.55: Quantization Coefficients in Layers I and II.

standard contains a table that the encoder searches by subband number and bitrate to find the number of bits.

Quantization is similar to layer I, except that layer II can sometimes pack three consecutive quantized values in a single codeword. This can be done when the number of quantization levels is a power of 2, and it reduces the number of wasted bits.

10.14.3 Encoding: Layers I and II

The MPEG standard specifies a table of scale factors. For each subband, the encoder compares the largest of the 12 signals to the values in the table, finds the next largest value, and uses the table index of that value to determine the scale factor for the 12 signals. In layer II, the encoder determines three scale factors for each subband, one for each of the three parts. It calculates the difference between the first two and the last two and uses the two differences to decide whether to encode one, two, or all three scale factors. This process is described in Section 10.14.4.

The bit allocation information in layer II uses 2–4 bits. The *scale factor select information* (sfsi) is two bits. The scale factor itself uses six bits. Each quantized signal uses 2–16 bits, and there may be ancillary data.

The standard describes two psychoacoustic models. Each produces a quantity called the *signal to mask ratio* (SMR) for each subband. The bit allocation algorithm uses this SMR and the SNR from Table 10.54 to compute the *mask to noise* ratio (MNR) as the difference

$$\text{MNR} = \text{SMR} - \text{SNR dB}.$$

The MNR indicates the discrepancy between waveform error and perceptual measurement, and the idea is that the subband signals can be compressed as much as the MNR. As a result, each iteration of the bit allocation loop determines the minimum MNR of all the subbands. The basic principle of bit allocation is to minimize the MNR over a frame while using not more than the number of bits B_f available for the frame.

In each iteration, the algorithm computes the number of bits B_f available to encode a frame. This is computed from the sample rate (number of samples input per second, normally $2\times 44{,}100$) and the bitrate (number of bits written on the compressed output per second). The calculation is

$$\text{frames/sec} = \frac{\text{samples/second}}{\text{samples/frame}}, \quad B_f = \frac{\text{bits/second}}{\text{frames/second}} = \frac{\text{bitrate} \times \text{samples per frame}}{\text{sampling rate}}.$$

Thus, B_f is measured in bits/frame. The frame header occupies 32 bits, and the CRC, if used, requires 16 bits. The bit allocation data is four bits per subband. If ancillary data are used, its size is also determined. These amounts are subtracted from the value of B_f computed earlier.

The main step of the bit allocation algorithm is to maximize the minimum MNR for all subbands by assigning the remaining bits to the scale factors and the quantized signals. In layer I, the scale factors take six bits per subband, but in layer II there are several ways to encode them (Section 10.14.4).

The main bit allocation step is a loop. It starts with allocating zero bits to each of the subbands. If the algorithm assigns zero bits to a particular subband, then no bits will be needed for the scale factors and the quantized signals. Otherwise, the number of bits assigned depends on the layer number and on the number of scale factors (1, 2, or 3) encoded for the subband. The algorithm computes the SNR and MNR for each subband and searches for the subband with the lowest MNR whose bit allocation has not yet reached its maximum limit. The bit allocation for that subband is incremented by one level, and the number of extra bits needed is subtracted from the balance of available bits. This is repeated until the balance of available bits reaches zero or until all the subbands have reached their maximum limit.

The format of the compressed output is shown in Figure 10.56. Each frame consists of between two and four parts. The frame is organized in slots, where a slot size is 32 bits in layer I, and 8 bits in layers II and III. Thus, the number of slots is $B_f/32$ or $B_f/8$. If the last slot is not full, it is padded with zeros.

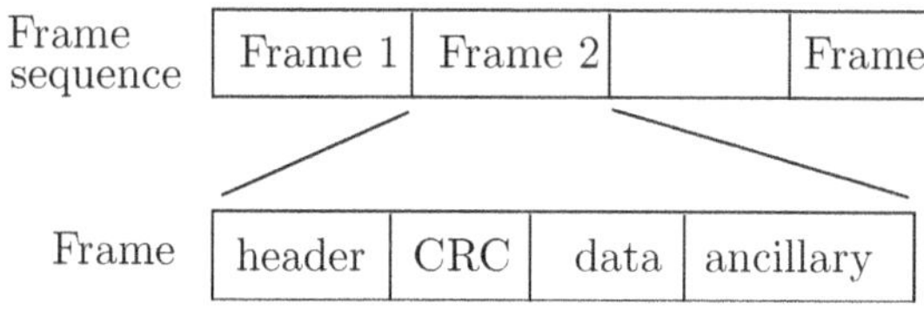

Figure 10.56: Format of Compressed Output.

The relation between sampling rate, frame size, and number of slots is easy to visualize. Typical layer I parameters are (1) a sampling rate of 48,000 samples/sec, (2) a bitrate of 64,000 bits/sec, and (3) 384 quantized signals per frame. The decoder has to decode 48,000/384 = 125 frames per second. Thus, each frame has to be decoded in 8 ms. In order to output 125 frames in 64,000 bits, each frame must have $B_f = 512$ bits available to encode it. The number of slots per frame is thus $512/32 = 16$.

A similar example assumes (1) a sampling rate of 32,000 samples/sec, (2) a bitrate of 64,000 bits/sec, and (3) 384 quantized signals per frame. The decoder has to decode 32,000/384 = 83.33 frames per second. Thus, each frame has to be decoded in 12 ms. In order to output 83.33 frames in 64000 bits, each frame must have $B_f = 768$ bits available to encode it. Thus, the number of slots per frame is $768/32 = 24$.

◇ **Exercise 10.10:** Do the same calculation for a sampling rate of 44,100 samples/s.

10.14.4 Encoding: Layer II

A frame in layer II consists of 36 subband signals, organized in 12 granules as shown in Figure 10.57. Three scale factors $scf1$, $scf2$, and $scf3$ are computed for each subband, one scale factor for each group of 12 signals. This is done in six steps as follows:

Step 1: The maximum of the absolute values of these 12 signals is determined.
Step 2: This maximum is compared with the values in column "scalefactors" of Table 10.52, and the smallest entry that is greater than the maximum is noted.
Step 3: The value in column "Index" of that entry is denoted by scf_i.
Step 4: After repeating steps 1–3 for $i = 1, 2, 3$, two differences, $D_1 = scf1 - scf2$ and $D_2 = scf2 - scf3$, are computed.

Step 5: Two "class" values, for D_1 and D_2, are determined by

$$\text{Class}_i = \begin{cases} 1, & \text{if } Di \le -3, \\ 2, & \text{if } -3 < Di < 0, \\ 3, & \text{if } Di = 0, \\ 4, & \text{if } 0 < Di < 3, \\ 5, & \text{if } Di \ge 3. \end{cases}$$

Step 6: Depending on the two classes, three scale factors are determined from Table 10.58, column "s. fact. used." The values 1, 2, and 3 in this column stand for the first, second, and third scale factors, respectively, within a frame. The value 4 means the maximum of the three scale factors. The column labeled "trans. patt." indicates those scale factors that are actually written on the compressed stream.

Table 10.59 lists the four values of the 2-bit *scale factor select information* (scfsi).

As an example, suppose that the two differences D_1 and D_2 between the three scale factors A, B, and C of the three parts of a certain subband are classified as $(1, 1)$. Table 10.58 shows that the transmission factor is 123 and the select information is 0. The top rule in Table 10.59 shows that scfsi of 0 means that each of the three scale factors is encoded separately in the output stream as a 6-bit number. (This rule is used by both encoder and decoder.) The three scale factors occupy, in this case, 18 bits, so there is no savings compared to layer I encoding.

We next assume that the two differences are classified as $(1, 3)$. Table 10.58 shows that the scale factor is 122 and the select information is 3. The bottom rule in Table 10.59 shows that a scfsi of 3 means that only scale factors A and B need be encoded (since $B = C$), occupying just 12 bits. The decoder assigns the first six bits as the value of A, and the following six bits as the values of both B and C. Thus, the redundancy in the scale factors has been translated to a small savings.

◇ **Exercise 10.11:** How are the three scale factors encoded when the two differences are classified as $(3, 2)$?

Quantization in layer II is similar to that of layer I with the difference that the two constants C and D of Table 10.60 are used instead of the constants A and B. Another difference is the use of grouping. Table 10.60 shows that grouping is required when the number of quantization levels is 3, 5, or 9. In any of these cases, three signals are combined into one codeword, which is then quantized. The decoder reconstructs the three signals $s(1)$, $s(2)$, and $s(3)$ from the codeword w by

$$\begin{aligned} & w = 0, \\ & \text{for } i = 0 \text{ to } 2, \\ & \quad s(i) = w \bmod \text{nos}, \\ & \quad w = w \div \text{nos}, \end{aligned}$$

where "mod" is the modulo function, "÷" denotes integer division, and "nos" denotes the number of quantization steps.

Figure 10.61a,b shows the format of a frame in both layers I and II. In layer I (part (a) of the figure), there are 384 signals per frame. Assuming a sampling rate of 48,000

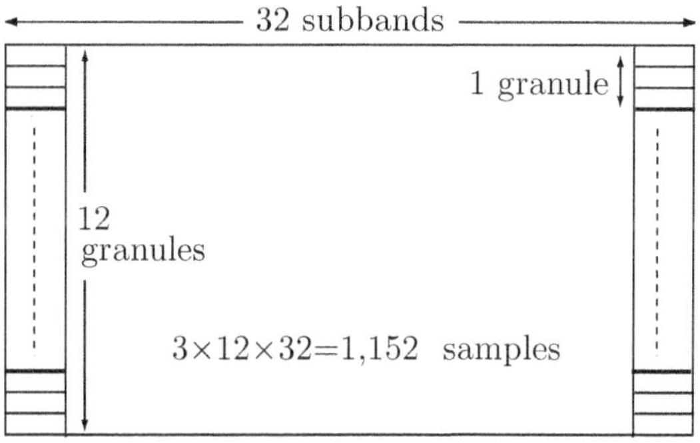

Figure 10.57: Organization of Layer II Subband Signals.

class 1	class 2	s. fact used	trans. patt.	s. fact. select	class 1	class 2	s. fact used	trans. patt.	s. fact. select
1	1	123	123	0	3	4	333	3	2
1	2	122	12	3	3	5	113	13	1
1	3	122	12	3	4	1	222	2	2
1	4	133	13	3	4	2	222	2	2
1	5	123	123	0	4	3	222	2	2
2	1	113	13	1	4	4	333	3	2
2	2	111	1	2	4	5	123	123	0
2	3	111	1	2	5	1	123	123	0
2	4	444	4	2	5	2	122	12	3
2	5	113	13	1	5	3	122	12	3
3	1	111	1	2	5	4	133	13	3
3	2	111	1	2	5	5	123	123	0
3	3	111	1	2					

Table 10.58: Layer II Scale Factor Transmission Patterns.

scsfi	# of coded scale factors	decoding scale factor
0 (00)	3	*scf1, scf2, scf3*
1 (01)	2	1st is *scf1* and *scf2* 2nd is *scf3*
2 (10)	1	one scale factor
3 (11)	2	1st is *scf1* 2nd is *scf2* and *scf3*

Table 10.59: Layer II Scale Factor Select Information.

number of steps	C	D	grouping	samples/code	bits/code
3	1.3333333333	0.5000000000	yes	3	5
5	1.6000000000	0.5000000000	yes	3	7
7	1.1428571428	0.2500000000	no	1	3
9	1.7777777777	0.5000000000	yes	3	10
15	1.0666666666	0.1250000000	no	1	4
31	1.0322580645	0.0625000000	no	1	5
63	1.0158730158	0.0312500000	no	1	6
127	1.0078740157	0.0156250000	no	1	7
255	1.0039215686	0.0078125000	no	1	8
511	1.0019569471	0.0039062500	no	1	9
1023	1.0009775171	0.0019531250	no	1	10
2047	1.0004885197	0.0009765625	no	1	11
4095	1.0002442002	0.0004882812	no	1	12
8191	1.0001220852	0.0002441406	no	1	13
16383	1.0000610388	0.0001220703	no	1	14
32767	1.0000305185	0.0000610351	no	1	15
65535	1.0000152590	0.0000305175	no	1	16

Table 10.60: Quantization Classes for Layer II.

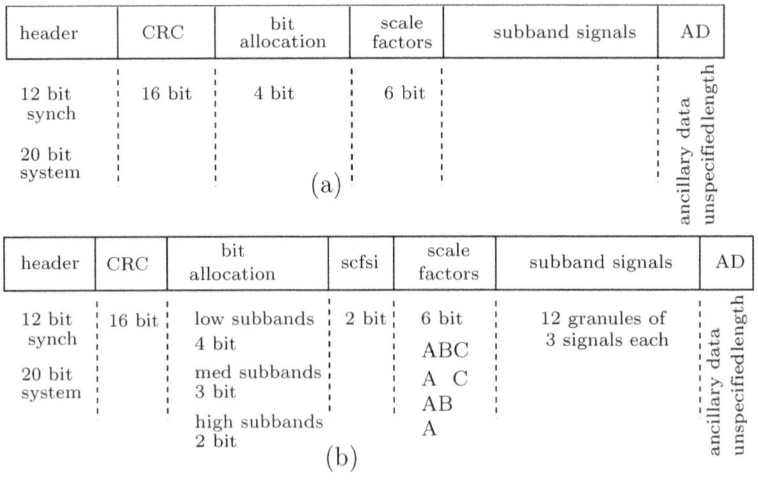

Figure 10.61: Organization of Layers I and II Output Frames.

samples/sec and a bitrate of 64,000 bits/sec, such a frame must be fully decoded in 80 ms, for a decoding rate of 125 frames/sec.

◇ **Exercise 10.12:** What is a typical frame rate for layer II?

10.14.5 Psychoacoustic Models

The task of a psychoacoustic model is to make it possible for the encoder to easily decide how much quantization noise to allow in each subband. This information is then used by the bit allocation algorithm, together with the number of available bits, to determine the number of quantization levels for each subband. The MPEG audio standard specifies two psychoacoustic models. Either model can be used with any layer, but only model II generates the specific information needed by layer III. In practice, model I is the only one used in layers I and II. Layer III can use either model, but it achieves better results when using model II.

The MPEG standard allows considerable freedom in the way the models are implemented. The sophistication of the model that is actually implemented in a given MPEG audio encoder depends on the desired compression factor. For consumer applications, where large compression factors are not critical, the psychoacoustic model can be completely eliminated. In such a case, the bit allocation algorithm does not use an SMR (signal to mask ratio). It simply assigns bits to the subband with the minimum SNR (signal to noise ratio).

A complete description of the models is outside the scope of this book and can be found in the text of the MPEG audio standard [ISO/IEC 93], pp. 109–139. The main steps of the two models are as follows:

1. A Fourier transform is used to convert the original audio samples to their frequency domain. This transform is separate and different from the polyphase filters because the models need finer frequency resolution in order to accurately determine the masking threshold.
2. The resulting frequencies are grouped into critical bands, not into the same 32 subbands used by the main part of the encoder.
3. The spectral values of the critical bands are separated into tonal (sinusoid-like) and nontonal (noise-like) components.
4. Before the noise masking thresholds for the different critical bands can be determined, the model applies a masking function to the signals in the different critical bands. This function has been determined empirically, by experimentation.
5. The model computes a masking threshold for each subband.
6. The SMR (signal to mask ratio) is calculated for each subband. It is the signal energy in the subband divided by the minimum masking threshold for the subband. The set of 32 SMRs, one per subband, constitutes the output of the model.

10.14.6 Encoding: Layer III

Layer III employs a much more refined and complex algorithm than the first two layers. This is reflected in the compression factors, which are much higher. The difference between layer III and layers I and II starts at the very first step, filtering. The same polyphase filter bank (Table 10.49a) is used, but it is followed by a modified version of the discrete cosine transform. The MDCT corrects some of the errors introduced by the polyphase filters and also subdivides the subbands to bring them closer to the critical bands. The layer III decoder has to use the inverse MDCT, so it has to work harder. The MDCT can be performed on either a short block of 12 samples (resulting in six transform coefficients) or a long block of 36 samples (resulting in 18 transform coefficients). Regardless of the block size chosen, consecutive blocks transformed by the MDCT have considerable overlap, as shown in Figure 10.62. In this figure, the blocks are shown above the thick line, and the resulting groups of 18 or 6 coefficients are below the line. The long blocks produce better frequency spectrum for stationary sound (sound where adjacent samples don't differ much), while the short blocks are preferable when the sound varies often.

The MDCT uses n input samples x_k (where n is either 36 or 12) to obtain $n/2$ (i.e., 18 or 6) transform coefficients S_i. The transform and its inverse are given by

$$S_i = \sum_{k=0}^{n-1} x_k \cos\left(\frac{\pi}{2n}\left[2k+1+\frac{n}{2}\right](2i+1)\right), \; i=0,1,\ldots,\frac{n}{2}-1, \tag{10.14}$$

$$x_k = \sum_{i=0}^{n/2-1} S_i \cos\left(\frac{\pi}{2n}\left[2k+1+\frac{n}{2}\right](2i+1)\right), \; k=0,1,\ldots,n-1. \tag{10.15}$$

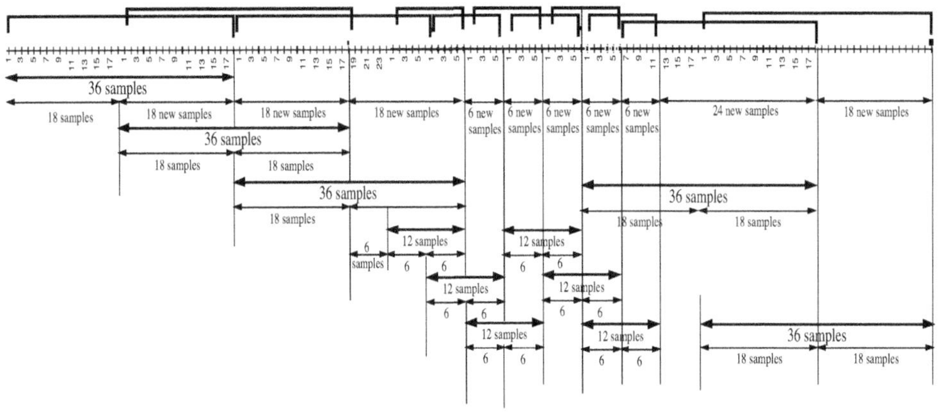

Figure 10.62: Overlapping MDCT Windows.

The size of a short block is one-third that of a long block, so they can be mixed. When a frame is constructed, the MDCT can use all long blocks, all short blocks (three times as many), or long blocks for the two lowest-frequency subbands and short blocks for the remaining 30 subbands. This is a compromise where the long blocks provide finer frequency resolution for the lower frequencies, where it is most useful, and the short blocks maintain better time resolution for the high frequencies.

Since the MDCT provides better frequency resolution, it has to result in poorer time resolution because of the uncertainty principle (Section 8.3). What happens in practice is that the quantization of the MDCT coefficients causes errors that are spread over time and cause audible distortions that manifest themselves as preechoes (read: "pre-echoes").

The psychoacoustic model used by layer III has extra features that detect conditions for preechoes. In such cases, layer III uses a complex bit allocation algorithm that borrows bits from the pool of available bits in order to temporarily increase the number of quantization levels and thus reduce preechoes. Layer III can also switch to short MDCT blocks, thereby reducing the size of the time window, if it "suspects" that conditions are favorable for preechoes.

(The layer-III psychoacoustic model calculates a quantity called "psychoacoustic entropy" (PE), and the layer-III encoder "suspects" that conditions are favorable for preechoes if PE > 1800.)

The MDCT coefficients go through some processing to remove artifacts caused by the frequency overlap of the 32 subbands. This is called *aliasing reduction.* Only the long blocks are sent to the aliasing reduction procedure. The MDCT uses 36 input samples to compute 18 coefficients, and the aliasing reduction procedure uses a butterfly operation between two sets of 18 coefficients. This operation is illustrated graphically in Figure 10.63a with a C-language code listed in Figure 10.63b. Index i in the figure is the distance from the last line of the previous block to the first line of the current block. Eight butterflies are computed, with different values of the weights cs_i and ca_i that are given by

$$cs_i = \frac{1}{\sqrt{1+c_i^2}}, \quad ca_i = \frac{c_i}{\sqrt{1+c_i^2}}, \quad i = 0, 1, \ldots, 7.$$

The eight c_i values specified by the standard are −0.6, −0.535, −0.33, −0.185, −0.095, −0.041, −0.0142, and −0.0037. Figures 10.63c,d show the details of a single butterfly for the encoder and decoder, respectively.

Quantization in layer III is nonuniform. The quantizer raises the values to be quantized to 3/4 power before quantization. This provides a more consistent SNR. The decoder reverses this operation by dequantizing a value and raising it to 4/3 power. The quantization is performed by

$$is(i) = \text{nint}\left[\left(\frac{xr(i)}{\text{quant}}\right)^{3/4} - 0.0946\right], \tag{10.16}$$

where $xr(i)$ is the absolute value of the signal of subband i, "quant" is the quantization step size, "nint" is the nearest-integer function, and $is(i)$ is the quantized value. As in layers I and II, the quantization is midtread, i.e., values around 0 are quantized to 0 and the quantizer is symmetric about 0.

In layers I and II, each subband can have its own scale factor. Layer III uses *bands* of scale factors. These bands cover several MDCT coefficients each, and their widths are close to the widths of the critical bands. There is a noise allocation algorithm that selects values for the scale factors.

Layer III uses Huffman codes to compress the quantized values even further. The encoder produces 18 MDCT coefficients per subband. It sorts the resulting 576 coefficients (= 18×32) by increasing order of frequency (for short blocks, there are three sets of coefficients within each frequency). Notice that the 576 MDCT coefficients correspond to 1152 transformed audio samples. The set of sorted coefficients is divided into

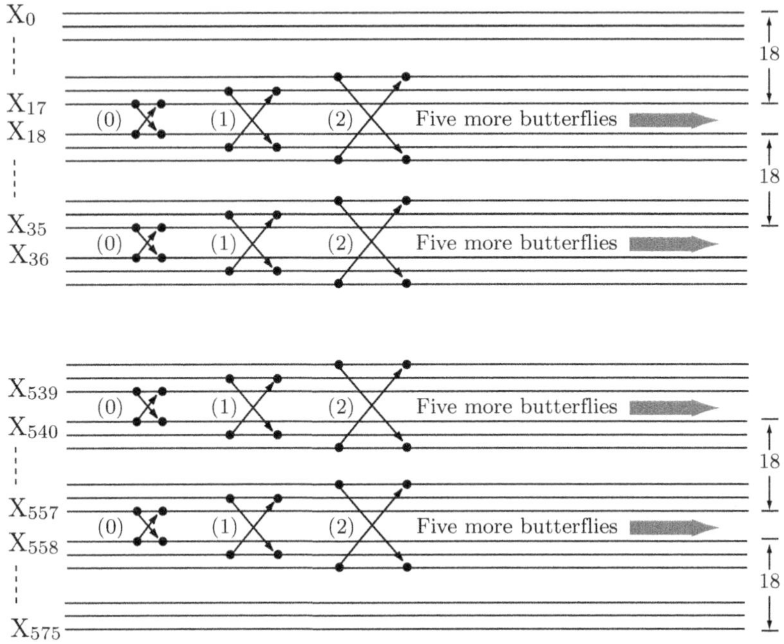

(a)

```
for(sb=1; sb<32; sb++)
 for(i=0; i<8; i++){
   xar[18*sb-1-i]=xr[18*sb-1-i]cs[i]-xr[18*sb+i]ca[i]
   xar[18*sb+i]=xr[18*sb+i]cs[i]+xr[18*sb-1-i]ca[i]
                   }
```

(b)

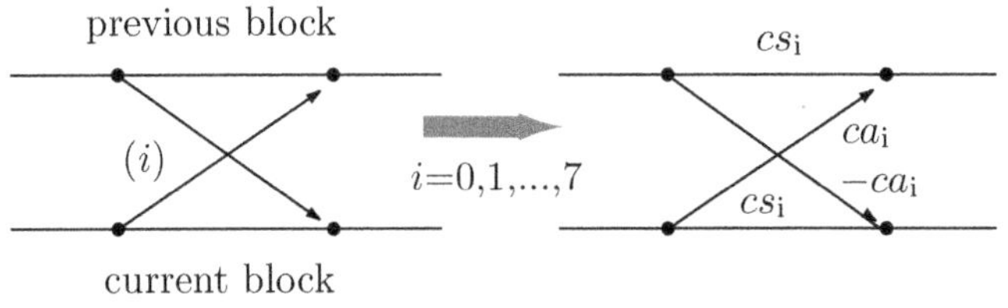

(c)

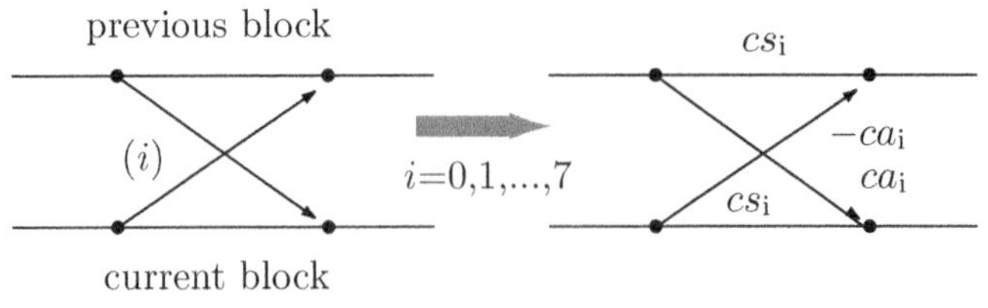

(d)

Figure 10.63: Layer III Aliasing Reduction Butterfly.

three regions, and each region is encoded with a different set of Huffman codes. This is because the values in each region have a different statistical distribution. The values for the higher frequencies tend to be small and have runs of zeros, whereas the values for the lower frequencies tend to be large. The code tables are provided by the standard (32 tables on pages 54–61 of [ISO/IEC 93]). Dividing the quantized values into three regions also helps to control error propagation.

Starting at the values for the highest frequencies, where there are many zeros, the encoder selects the first region as the continuous run of zeros from the highest frequency. It is rare, but possible, not to have any run of zeros. The run is limited to an even number of zeros. This run does not have to be encoded, since its value can be deduced from the sizes of the other two regions. Its size, however, should be even, since the other two regions code their values in even-numbered groupings.

The second region consists of a continuous run of the three values -1, 0, and 1. This is called the "count1" region. Each Huffman code for this region encodes four consecutive values, so the number of codes must be $3^4 = 81$. The length of this region must, of course, be a multiple of 4.

The third region, known as the "big values" region, consists of the remaining values. It is (optionally) further divided into three subregions, each with its own Huffman code table. Each Huffman code encodes two values.

The largest Huffman code table specified by the standard has 16×16 codes. Larger values are encoded using an escape mechanism.

A frame F in layer III is organized as follows: It starts with the usual 32-bit header, which is followed by the optional 16-bit CRC. This is followed by 136 bits (for single channel) and 256 bits (for dual channel) of side information. The size of side information for MPEG/Audio Layer III, for single and dual channel, are listed in Table 10.64. The last part of the frame is the main data. The side information is followed by a segment of main data (the side information contains, among other data, the length of the segment) but the data in this segment does not have to be that of frame F! The segment may contain main data from several frames because of the encoder's use of a *bit reservoir*.

The concept of the bit reservoir has already been mentioned. The encoder can borrow bits from this reservoir when it decides to increase the number of quantization levels because it suspects preechoes. The encoder can also donate bits to the reservoir when it needs fewer than the average number of bits to encode a frame. Borrowing, however, can be done only from past donations; the reservoir cannot have a negative number of bits.

The side information of a frame includes a 9-bit pointer that points to the start of the main data for the frame, and the entire concept of main data, fixed-size segments, pointers, and bit reservoirs is illustrated in Figure 10.65. In this diagram, frame 1 needed only about half of its bit allocation, so it left the other half in the reservoir, where it was eventually used by frame 2. Frame 2 needed a little additional space in its "own" segment, leaving the rest of the segment in the reservoir. This was eventually used by frames 3 and 4. Frame 3 did not need any of its own segment, so the entire segment was left in the reservoir, and was eventually used by frame 4. That frame also needed some of its own segment, and the rest was used by frame 5.

Bit allocation in layer III is similar to that in the other layers, but includes the added complexity of noise allocation. The encoder (Figure 10.66) computes the bit allocation,

Field name	Single channel	Dual channel
main Data	9	9
private bits=	5	3
scfsi:	4	8
part2 3length =	12×2	12×4
big values =	9×2	9×4
global gain =	8×2	8×4
scalefac compress =	4×2	4×4
preflag:	1×2	1×4
scalefac scale:	1×2	1×4
count1table Select =	1×2	1×4
window Switching flag =	1×2	1×4
Additional info =	22×2	22×4
(dependent on window switching flag)		
Total (bits)	136	256

Table 10.64: Size of Side Info for MPEG Audio Layer III.

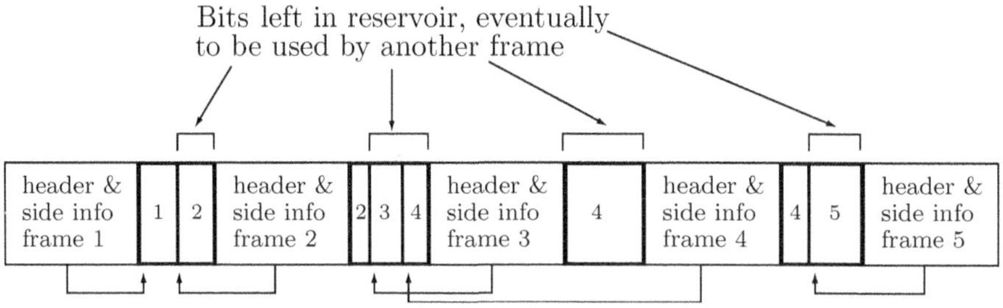

Figure 10.65: Layer III Compressed Stream.

performs the actual quantization of the subband signals, encodes them with the Huffman codes, and counts the total number of bits generated in this process. This is the bit allocation inner loop. The noise allocation algorithm (also called analysis-by-synthesis procedure) becomes an outer loop where the encoder calculates the quantization noise (i.e., it dequantizes and reconstructs the subband signals and computes the differences between each signal and its reconstructed counterpart). If it finds that certain scalefactor bands have more noise than what the psychoacoustic model allows, the encoder increases the number of quantization levels for these bands, and repeats the process. This process terminates when any of the following three conditions becomes true:

1. All the scalefactor bands have the allowed noise or less.
2. The next iteration would require a requantization of ALL the scalefactor bands.
3. The next iteration would need more bits than are available in the bit reservoir.

◇ **Exercise 10.13:** Layer III is extremely complex and hard to implement. In view of this, how is it that there are so many free and low-cost programs available to play `.mp3` audio files on all computer platforms?

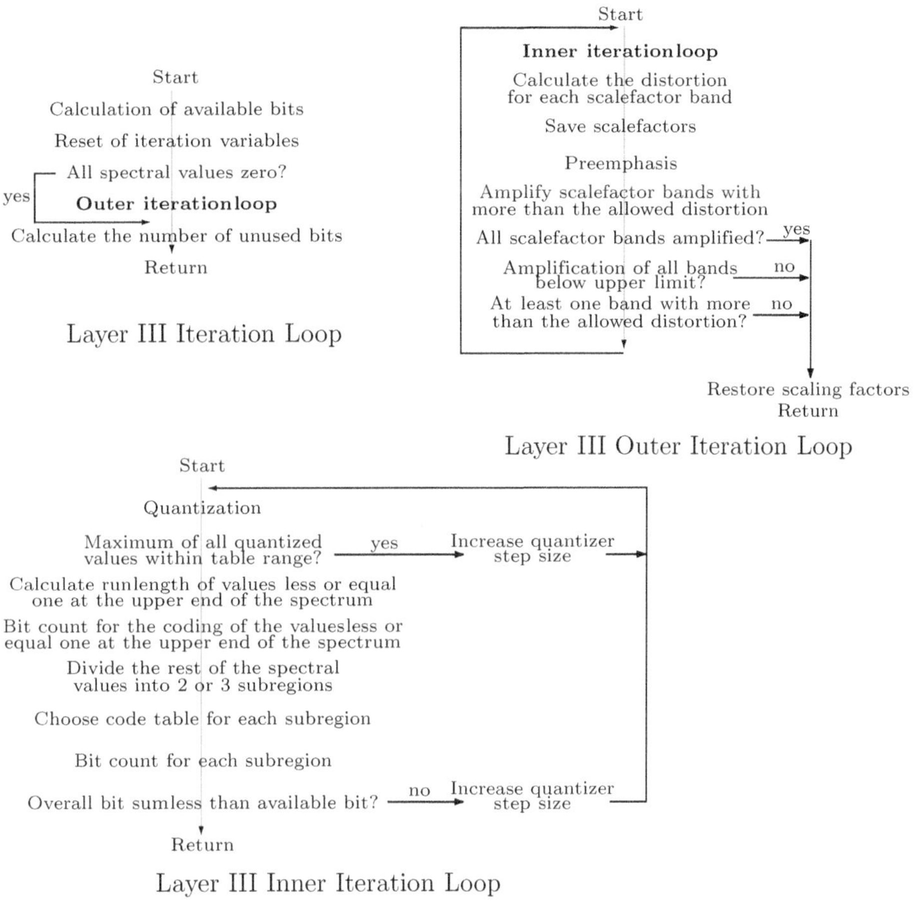

Figure 10.66: Layer III Iteration Loop.

10.15 Advanced Audio Coding (AAC)

The development, by Philips and Sony, of the familiar compact disc (CD), started around 1974. In June 1980, the two companies agreed on a common CD standard and in 1981 this standard was approved by the Digital Audio Disc committee. (Notice the spelling `disc`, as opposed to a magnetic `disk`.) The standard, which has been used ever since, includes hardware and software specifications of the signal format, disc material and dimensions, error-correcting code, and many other features. What strikes us as unexpected and unusual, though, is that our music CDs record the audio in uncompressed format. The raw audio samples, typically 44,100 16-bit samples per second of audio, are written on the disc with a sophisticated error-correcting code, but without any attempt to compress them.

Current CDs have a capacity of about 80 minutes of sound. This can accommodate many songs, and is enough for a concerto or two and for most symphonies (although

Mahler's third symphony, at about 100 minutes, is a notable exception that comes to mind). An opera, however, requires two or more CDs (Wagner's Ring Cycle, which clocks in at 18 hours, requires 14–16 CDs), as also do the collected works of many composers. Thus, compression can be quite useful. Currently, the ever popular mp3 format routinely achieves compression factors of 18–20 and can reduce a music library of 100 CDs to just five or six CDs and eventually to just a single DVD.

Considering all this, it is natural to wonder about the lack of audio compression in the original CD standard. The best explanation is that engineers simply did not believe that audio can be compressed lossily and efficiently, to perhaps 20% of its original size, without a noticeable loss of quality. A common term used by audio experts in the 1970s and 80s was "golden ears," which refers to audiophiles who claim they can distinguish between live music and recorded music, especially music played from a compressed format. Thus, when the Moving Pictures Experts Group (MPEG) came into being, it concentrated on video compression. It was only in 1988 that several audio pioneers formed the audio group within MPEG (see also page 1031) and started the process that brought us the three MPEG-1 and MPEG-2 layers, followed by the advanced audio compression (AAC) standard.

The AAC compression standard was originally part of the MPEG-2 project, and was later augmented and extended as part of MPEG-4. Some important references for this efficient and complex method are [Bosi and Goldberg 03], [Brandenburg 99], [Pereira and Ebrahimi 02], and [ISO/IEC 03] (the latter is the formal specification of the standard and is not very readable). AAC is the result of an intensive international effort involving companies and individuals. The project started in November 1994, when a number of proposals were submitted to the MPEG-2 committee. The committee came up with a preliminary structure of AAC as a set of modules that constitute independent parts of the overall system. Each module was developed, implemented, and tested separately before becoming part of the final product. The main modules are the following:

- Gain control
- Filterbank
- Prediction
- Quantization and coding
- Noiseless Huffman coding
- Bitstream multiplexing
- Temporal noise shaping (TNS)
- Mid/side (M/S) stereo coding
- Intensity stereo coding

These modules are listed in Figures 10.72 and 10.73 and most of them are described in this section.

Today, virtually all producers and consumers of audio agree that audio compression (both lossy and lossless) is effective and is very useful (notice the popularity of mp3 players). Even more, those familiar with the MPEG committees and the way they operate agree that the MPEG process of advertising for algorithms, selecting promising candidates, and implementing and testing them, has resulted in the best audio compression methods, while also encouraging compatibility among many different types of audio equipment.

Another aspect of the success of MPEG is that the main methods currently used for audio compression came from this organization and not from commercial developers who try to lock users into proprietary algorithms and software. The chief reason for the success of MPEG audio is its development process. The entire process of selecting, testing, and adopting a compression standard is open and competitive. The MPEG audio committee includes academic researchers and industry representatives. They meet periodically to set goals for the next method, to advertise those goals, to receive, study, and discuss ideas from many contributors, and to implement and test the most promising ideas.

The first success of the MPEG audio group came in 1991 with MPEG-1. This early video compression standard already incorporated the three layers discussed in Section 10.14. MPEG-2 was finalized in 1997. It added a few minor enhancements to the three layers, but its main contribution to audio compression was the early version of the advanced audio compression (AAC) standard [ISO/IEC 03].

[Layer 1 of MPEG-1/2 is rarely used. Layer 2 is used for DAB (digital audio broadcasting) in Europe, in audio for video, and in broadcast delivery systems. Layer 3 is, of course, widely used—under the name mp3, derived from its file extension—in consumer products as well as in broadcast codecs.]

Thus, the full name of mp3 is MPEG-1 and MPEG-2 layer 3. A somewhat shorter name is MPEG-1/2 layer 3. It has long been observed by users and readers that the term "layers" is confusing and unfortunate. Perhaps "levels" or "versions" would have been a better choice. There is also often a confusion of MPEG-2 with level 2.

MPEG-3 was intended for HDTV compression but was found to be redundant (it was very similar to MPEG-2) and was merged with MPEG-2 (Section 9.6).

The next MPEG design, MPEG-4, was completed in 1999 (Section 9.7). It incorporated several enhancements to AAC and added the AAC-LD (low delay, Section 10.15.3) module.

Work on MPEG-7 is currently underway, but many curious souls ask the natural question, What about MPEGs 5 and 6? The official MPEG literature says nothing about this jump in numbering, but it seems (at least this is what one participant in the MPEG meetings recalls) that someone proposed to continue the integer sequence 1, 2, and 4 with 8 (because these are the first powers of 2 and are written in binary as $10...0_2$). In response, someone else proposed (perhaps as a joke) to select 7 instead of 8 because MPEG-7 is supposed to be so different from its predecessors and because 7 is a lucky number.

Why is 7 considered a lucky number? No one can say for sure, but here are seven examples that justify this title:

- The Pythagoreans called 7 the perfect number. They called 3 and 4 (the triangle and the square) the perfect figures.

- The ancients knew seven planets, the sun, the moon, Mercury, Venus, Mars, Jupiter and Saturn. Similarly, there were seven wonders in the ancient world.

- In ancient writings and lore, the number 7 plays an important role. The Bible mentions this number many times (the 7 days of the week, the 7 sabbatical years, the 7 years of famine, the 7 years of plenty, the 7 years occupied in the building of King Solomon's Temple). The Arabians had seven holy temples. In Persian mysteries there were seven spacious caverns through which the aspirants had to pass. The Goths had seven deities, as did the Romans, from whose names are derived our days of the week. This preoccupation with 7 has found its way to modern times in the form of Snow White and the seven dwarfs, and other works of literature.

- The seven deadly sins (introduced by St. Gregory The Great around 600) are pride, envy, anger, avarice, sadness, gluttony, and lust.

- The Trumpton fire brigade (Pugh, Pugh, Barney, Magrew, Cuthburt, Dibble and Grub). Trumpton was a stop-motion children's television show in 1967.

- Up until the 15th century, the world was believed to have seven seas. The Red sea, the Mediterranean, the Persian gulf, the Black sea, the Adriatic sea, the Caspian sea, and the Indian ocean.

- In Chinese culture, the 7th day of the first moon of the lunar year is known as Human's Day and is celebrated as the universal birthday of all humans.

The scientists, engineers, and experts who contributed to the three layers of MPEG-1 and MPEG-2 have hit on a practical and efficient approach to audio compression, the so-called perceptual coding. The idea is to identify those parts of an audio stream that are not fully perceived by the ear/brain system and quantize them or even delete them completely. What is actually quantized is frequency coefficients (often referred to as spectral coefficients) and not the audio samples themselves. It is this principle of acoustic masking that made first mp3 and later, AAC, possible.

The human ear (Section 10.3) is very sensitive, but it is not a precision instrument. Its sensitivity depends heavily on the frequency of the sound (as well as on other factors such as age and environment). Figure 10.67 shows the normal threshold of the ear (a curve determined by many sensitive experiments) and how it depends on the frequency (see also Figure 10.5a). Notice that the frequency scale is logarithmic because of its wide interval (from 20 Hz to 20 kHz). Any sound below this threshold is inaudible, and the figure shows that the threshold is high at both low and high frequencies. In order for a sound at these frequencies to be audible, it has to be loud; its amplitude has to exceed the threshold. The threshold is lowest at the frequency range of 3–5 kHz, indicating that the ear is very sensitive at this range and can detect very faint sounds.

The point is that the normal threshold can be momentarily perturbed by loud sounds. The figure shows how a loud sound at a frequency of 300 Hz raises the normal threshold for frequencies from 80 Hz to about 1 kHz. The result is that the sound labeled "x," at 150 Hz, which is above the normal threshold, is now located below the new, perturbed threshold and is therefore masked; it cannot be heard and its audio samples can be deleted (see also Figure 10.5b).

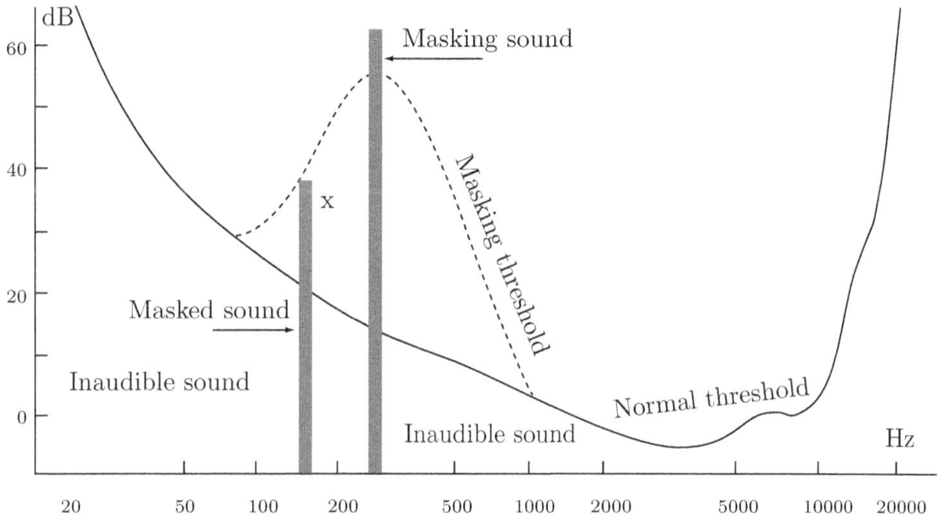

Figure 10.67: Audio Frequency Masking.

Time is also a factor in acoustic masking. Any perturbation of the normal threshold is temporary and normally decays in less than a second. Figure 10.68 shows a loud, 60-dB sound (at frequency f) that lasts 5 ms. A new threshold starts at 60 dB and decays almost completely in about 500 ms. Any sounds at or around frequency f that happen to be below this threshold are inaudible. Sound "x" at 30 dB is inaudible because it occurs only 5 ms after the masking sound, but the same 30-dB sound "y" occurring 10 ms later is fully audible. The ear becomes overwhelmed by the loud sound to such an extent that it cannot perceive other sounds for a while.

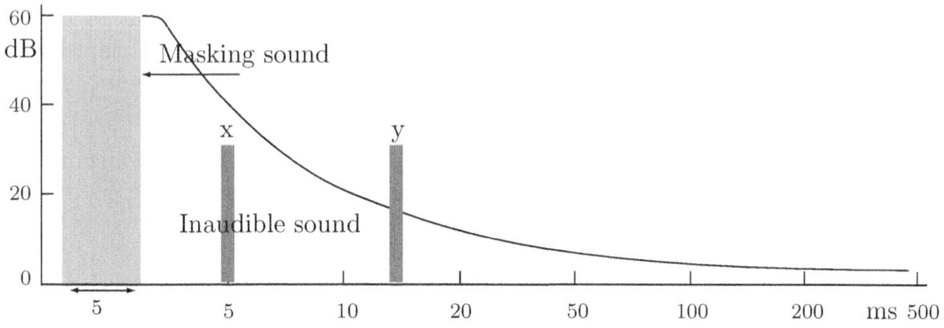

Figure 10.68: Audio Masking in Time.

Figure 10.69 illustrates the concept of acoustic masking in both frequency and time. A masking sound (in the form of a rectangle) creates a surface that is spread over neighboring frequencies and over time. Any sounds with amplitudes and frequencies under this surface are masked.

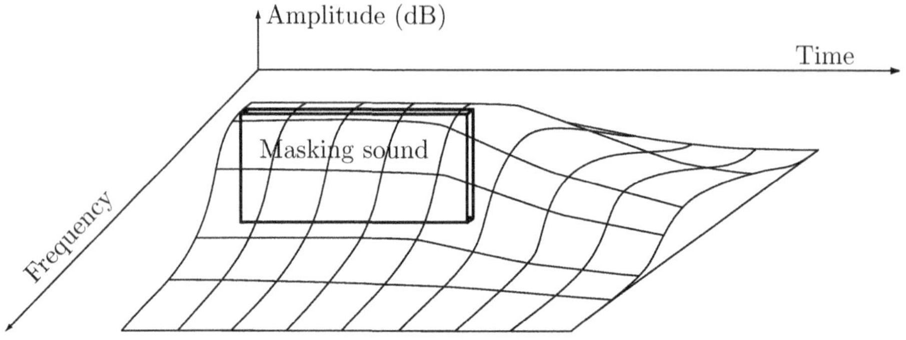

Figure 10.69: Audio Masking In Frequency and Time.

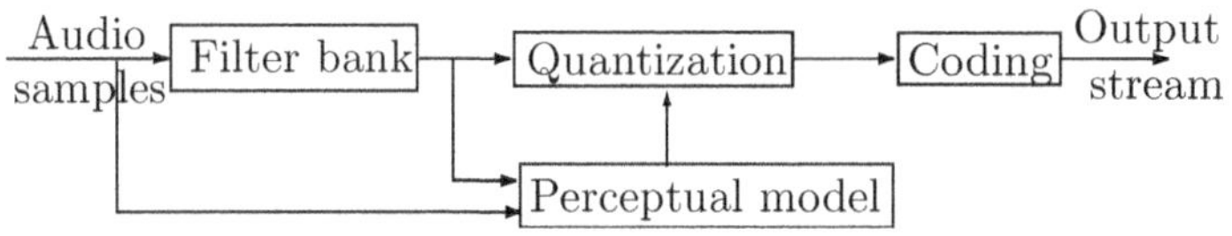

Figure 10.70: Block Diagram of a Perceptual Encoder.

Based on this behavior of the ear, a time/frequency (T/F) lossy audio compression algorithm works in the following steps (Figure 10.70):

- The encoder goes over the audio samples in the input stream and groups them in overlapping ranges (windows) of samples. Each window is then filtered to convert its audio samples to frequency coefficients. Each coefficient corresponds to a small interval of frequencies (a frequency band) and indicates the intensity of the sound in the band.
- Use the frequency coefficients to identify the masking sounds.
- A perceptual (or psychoacoustic) model then estimates the height of the mask curve at each frequency band. This height is determined by the normal threshold and by perturbations to it, caused by masking sounds at various frequencies and in various times in the past.
- When the updated threshold curve is ready, it is used to determine which frequency coefficients are unimportant. These coefficients are now quantized.
- The quantized coefficients are encoded by a variable-length code.
- The codes are written on the output stream together with other side information needed by the decoder.

An important point is that audio masking depends on the frequency of sound, but the data to be compressed is in the form of audio samples. Thus, the first task of the encoder is to take a set of audio samples and compute the frequencies of the sound wave expressed by them. (In practice, intervals of frequencies, known as frequency bands, are identified and one frequency (or spectral) coefficient is computed per band. The bands must not exceed the width of the ear's critical bands discussed in Section 10.3.) This is done by means of a bank of polyphase filters. The way this is done in layer 3 (mp3) is discussed in Section 10.14.1 and particularly in Figure 10.48. In mp3, the most-recent 512 audio samples are stored in a buffer and are used to compute spectral coefficients for 32 frequency bands. Each spectral coefficient indicates the amplitude of the audio (specifically, that part of the audio described by the 512 samples) in one frequency band. The next 32 audio samples are then shifted into the buffer and the process is repeated. AAC performs a similar computation, but uses a buffer with 2048 audio samples, computes 1024 spectral coefficients, each indicating the amplitude of the audio in a 23.4-Hz-wide frequency band, and shifts the next 1024 samples into the buffer.

[The spectral coefficients are later used by the decoder to reconstruct the audio samples, but these samples are different from the original ones because the spectral coefficients are quantized to achieve compression.]

The next component of mp3 is a psychoacoustic model (Section 10.14.5). It determines the height of the masking threshold for every frequency band. However, the model is not specified in detail and the idea is that each implementation selects its own model and the success of a particular implementation depends on the complexity of the model and its resemblance to the actual behavior of the ear/brain system. The AAC standard specification similarly employs a psychoacoustic model without specifying its precise operation. (See flowchart on page 144 of [ISO/IEC 03].)

For each frequency band, the encoder compares the spectral coefficient to the threshold of the band. If the spectral coefficient is smaller than the threshold, the coefficient is quantized. This is the main source of compression in the three layers of MPEG-1/2 as well as in AAC, and it is lossy. The amount of quantization depends on how much smaller the coefficient is than the threshold. The amount of quantization is crucial. If a coefficient is quantized too much, the decoder would generate audio samples much different from the original, resulting in perceptible noise and poor audio quality. On the other hand, the result of insufficient (too fine) quantization is low compression that doesn't achieve the target bitrate specified by the user.

AAC employs several tools that improve quantization as follows:

- Temporal noise shaping (TNS) is a sophisticated algorithm that minimizes the effect of temporal spread. This improves mostly the quantization (and hence the compression) of voice signals.
- A prediction module improves the performance of the quantizer in cases where the original audio features patterns, such as high tonality (sinusoid-like sound).
- Perceptual noise shaping (PNS) results in a finer control of quantization, which saves bits and improves compression.

The next step is coding. The quantized frequency coefficients are coded. This step improves compression because the coefficients are replaced by variable-length codes, but this added compression is lossless (the AAC literature refers to it as "noiseless"). Both mp3 and AAC use Huffman codes. The former encoding process is described in Section 10.14.6. The latter uses 12 Huffman code tables. To select a code table, the encoder selects a group of two or four consecutive quantized frequency coefficients and selects a code table depending on the group size and on the largest absolute value of the coefficients in the group. Huffman codes for coefficients up to ± 16 are provided, but there is also an escape mechanism that allows coding of coefficients of up to $\pm 8{,}191$.

The last step is to collect the Huffman codes, add flags, control codes, and other information needed by the decoder, and multiplex all this to create the final output stream.

Two unusual features of mp3 are the bit reservoir and the joint stereo mode. The user specifies the quality of the compression by specifying a bitrate. A bitrate of 64 Kb/sec, for example, directs the encoder to compress each second of audio to only 64 K bits. For stereo sound, a sampling rate of 44,100 samples per second results in 88,200 samples (equivalent to 176,400 bytes or 1,411,200 bits) per second. Compressing 1,411,200 bits into 65,536 bits implies a compression factor of about 21.5. This is impressive, but may lead to degredation of the reconstructed sound. Thus, mp3 maintains a bit reservoir. Each time a set of coefficients is quantized, encoded, and output, the encoder figures out how many bits are allowed for the set by the user-specified bitrate and how many bits were actually used. Any "unused" bits are added to the bit reservoir to be used for future sets if needed. If too many bits were used for the compression of the current set, the extra bits are subtracted from the reservoir. If the reservoir doesn't have enough bits, the encoder repeats its operation with coarser quantization, which saves bits but results in bad reconstructed sound. The joint stereo mode takes advantage of the correlation between the left and right stereo audio channels.

Thus, AAC shares the basic features of mp3, with the following improvements:

- A larger, improved filter bank that computes 1,024 spectral coefficients from each window of 2,048 audio samples, resulting in narrow frequency bands and a frequency resolution much finer than that of mp3.
- Temporal noise shaping (TNS), a new algorithm that minimizes the effect of temporal masking. This is especially useful for the compression of voice signals.
- A prediction module improves quantization for periodic audio or audio with patterns.
- Perceptual noise shaping (PNS) provides fine control of quantization which leads to better compression.

In April 2003, Apple Computer brought public attention to AAC by announcing that its iTunes and iPod products would support audio in MPEG-4 AAC format (older iPods require a firmware update). Customers can download music from the iTunes Music Store in a protected version of the format (it uses the file extensions `m4a` and `m4p`). This is why AAC has become so associated with Apple hardware and software that many mistakenly believe that AAC stands for "Apple Audio Codec."

MP4 also refers to Møller-Plesset perturbation theory of the fourth order.

10.15.1 Details of AAC

We start with the details of AAC as defined in MPEG-2. Section 10.15.2 discusses the features added to AAC by MPEG-4. Figures 10.72 and 10.73 are block diagrams of the AAC encoder and decoder, respectively.

AAC supports three *profiles*. These are different configurations of the basic algorithm, and they offer different trade-offs between compression efficiency and encoder

index	profile
0	Main
1	Low Complexity (LC)
2	Scalable Sampling Rate (SSR)
3	Reserved

Table 10.71: The Three AAC Profiles.

complexity. A 2-bit profile index is written on the compressed stream (Table 10.71) to indicate the profile to the decoder.

Index 0. Main profile. In this profile, AAC offers the best compression for any sampling rate and bitrate. All the modules, routines, and tools of the AAC specification (except the gain control tool) are employed. This is the normal profile and it makes sense when enough memory and processing power are available (as is common in current computers).

Index 1. Low complexity profile (LC). The gain control tool and prediction are not used. Also, TNS order is limited to 12. Otherwise, this profile is a subset of the main profile, which means that a compressed stream generated by LC can be decoded by the main profile.

Index 2. Scalable sampling rate (SSR). This profile is designed to become simplified (to scale down in complexity) when the original audio data consists of limited frequencies (reduced audio bandwidth). The gain-control tool is required in SSR and is used only by SSR (but is not used in the lowest of the four PQF subbands). The prediction and coupling channels are not used. Also, the TNS order and bandwidth are limited. A compressed stream generated by LC can be decoded by SSR, but the resulting audio will be limited to (approximately) 5 kHz because the lowest band of the first filterbank is not used. SSR is simpler than the other profiles.

Index 3 is reserved for a future profile.

Gain Control. This tool is a preprocessor and is used only in the SSR profile. It consists of modules for filtering, gain detection, and gain modification. Gain control starts by filtering the input (audio samples) into four 6-kHz-wide frequency bands. This filtering is done by a bank of polyphase quadrature filters (PQF). The frequency coefficients in each frequency band are examined by the gain detectors, searching for rapid variations in energy (i.e., consecutive coefficients with very different sizes). The gain modifiers use these results to modify the coefficients so as to compress the dynamics of the input audio.

In the AAC decoder, gain control performs the same operations in reverse order; it becomes a postprocessor. The modification are reversed, to restore the original dynamics of the audio signal, and the inverse PQF filterbank is used to generate audio samples.

The coefficients of the four PQF frequency bands are computed by

$$h_i = \frac{1}{4}\cos\left[\frac{(2i+1)(2n+5)\pi}{16}\right] Q(n), \quad \text{for} \quad 0 \le n \le 95, \quad \text{and} \quad 0 \le i \le 3,$$

where the first 48 $Q(n)$ values are listed in Table 10.74 and the remaining 48 values are given by $Q(n) = Q(95 - n)$ for $48 \le n \le 95$.

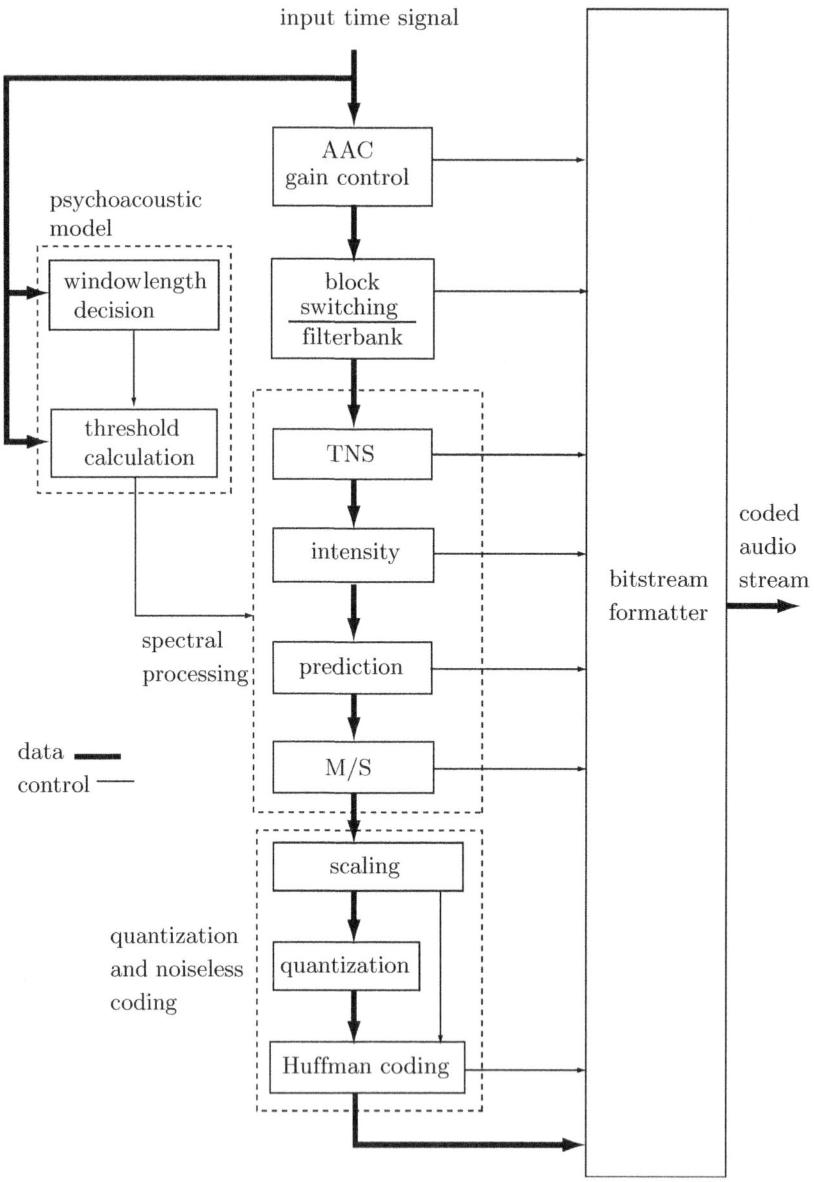

Figure 10.72: AAC Encoder.

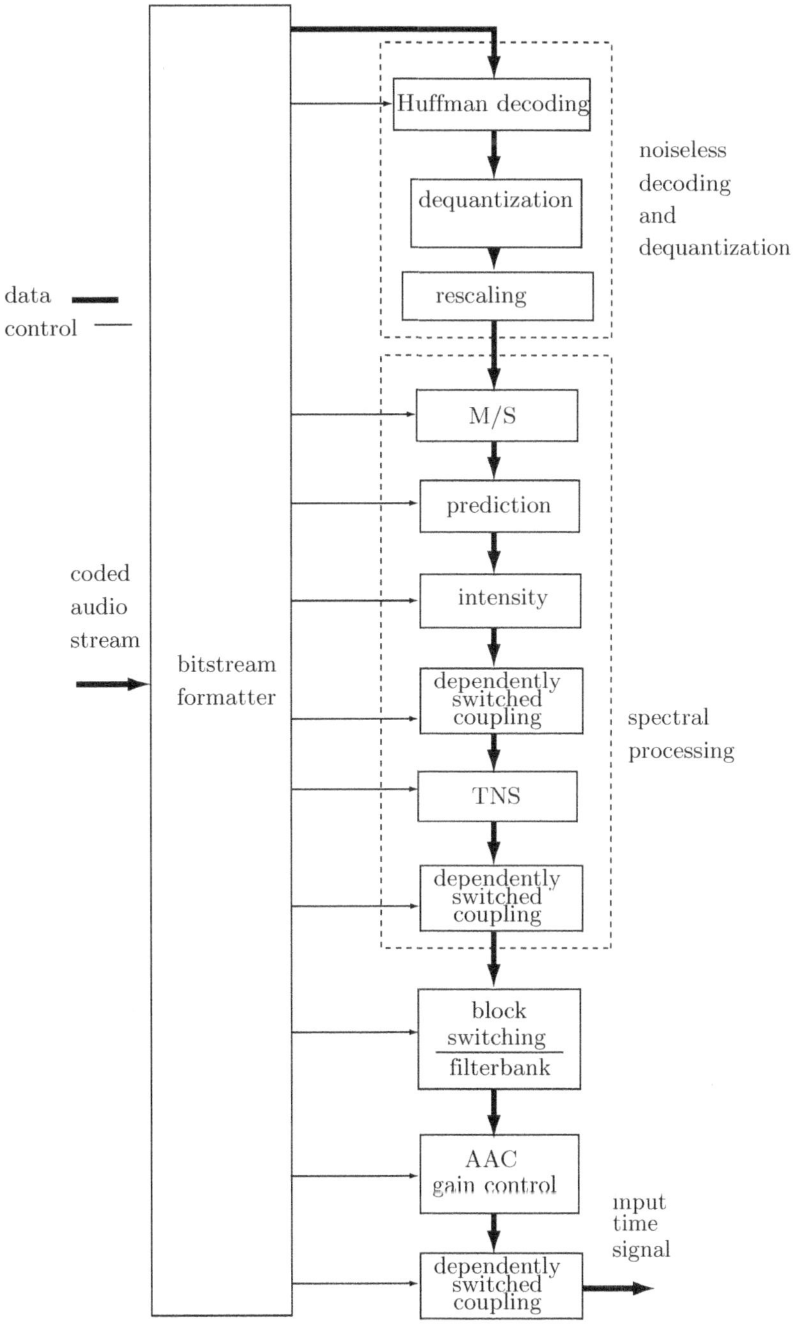

Figure 10.73: AAC Decoder.

j	Q(j)	j	Q(j)
0	9.7655291007575512E-05	24	-2.2656858741499447E-02
1	1.3809589379038567E-04	25	-6.8031113858963354E-03
2	9.8400749256623534E-05	26	1.5085400948280744E-02
3	-8.6671544782335723E-05	27	3.9750993388272739E-02
4	-4.6217998911921346E-04	28	6.2445363629436743E-02
5	-1.0211814095158174E-03	29	7.7622327748721326E-02
6	-1.6772149340010668E-03	30	7.9968338496132926E-02
7	-2.2533338951411081E-03	31	6.5615493068475583E-02
8	-2.4987888343213967E-03	32	3.3313658300882690E-02
9	-2.1390815966761882E-03	33	-1.4691563058190206E-02
10	-9.5595397454597772E-04	34	-7.2307890475334147E-02
11	1.1172111530118943E-03	35	-1.2993222541703875E-01
12	3.9091309127348584E-03	36	-1.7551641029040532E-01
13	6.9635703420118673E-03	37	-1.9626543957670528E-01
14	9.5595442159478339E-03	38	-1.8073330670215029E-01
15	1.0815766540021360E-02	39	-1.2097653136035738E-01
16	9.8770514991715300E-03	40	-1.4377370758549035E-02
17	6.1562567291327357E-03	41	1.3522730742860303E-01
18	-4.1793946063629710E-04	42	3.1737852699301633E-01
19	-9.2128743097707640E-03	43	5.1590021798482233E-01
20	-1.8830775873369020E-02	44	7.1080020379761377E-01
21	-2.7226498457701823E-02	45	8.8090632488444798E-01
22	-3.2022840857588906E-02	46	1.0068321641150089E+00
23	-3.0996332527754609E-02	47	1.0737914947736096E+00

Table 10.74: The First 48 $Q(n)$ Coefficients.

Figure 10.75 shows the main components of the gain control encoder (part a) and decoder (part b).

Filtering. The AAC encoder transforms the audio samples into frequency coefficients by means of a modified DCT (MDCT) transform. The process is similar to that employed by the three layers of MPEG-1 and MPEG-2, but with higher resolution and with improvements. At any given time, the encoder transforms a set of consecurive audio samples referred to as a window. There are two types of windows, long and short. A long window consists of 2,048 consecutive samples and is transformed to produce 1,024 frequency coefficients (a long transform). A short window is 256 samples long and results in 128 coefficients (a short transform). Once the coefficients of a window have been computed, the encoder shifts the audio samples in the buffer by half the window size. Thus, the windows overlap by 50% of their size.

Long windows produce many coefficients, so each coefficient corresponds to a narrow frequency band. This leads to precise quantization and thus to a more meaningful loss of data. The data lost in quantization corresponds to those parts of the audio that are not perceived by the ear/brain system. Thus, long windows make sense in those parts of the audio that are either tonal or low frequency. On the other hand, atonal or high-frequency audio varies rapidly, which is why such parts of the audio input lend themselves to better compression with short windows.

It is up to the AAC encoder to analyze the input, identify its stationary and transient parts, and use this knowledge to decide when and how often to switch filtering windows. In order to smooth the transition between long and short windows, AAC defines three types of long windows as follows:

- `LONG_WINDOW`. This is the normal type of long windows, Many consecutive long windows may be used, they overlap by 1,024 audio samples and each produces 1,024 frequency coefficients.

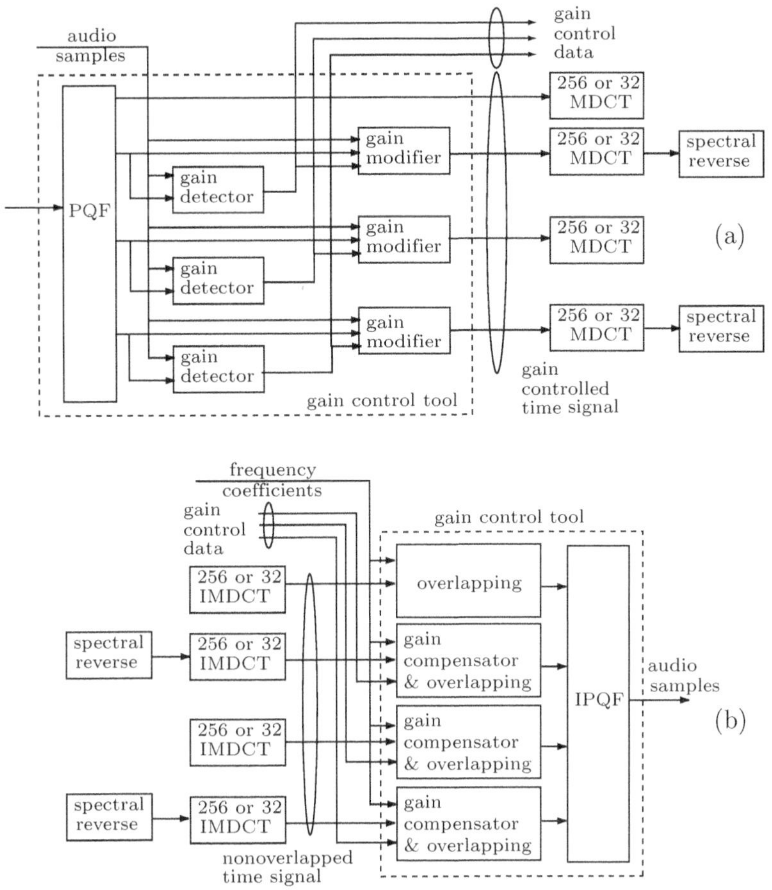

Figure 10.75: AAC Gain Control (a) Encoder and (b) Decoder.

- LONG_START_WINDOW. When the encoder decides to switch from a long window to a short one, it uses one window of this type. It has an overlap of 1,024 audio samples with the long window that precedes it and of 128 samples with the short window that follows it. It produces 1,024 frequency coefficients.

- LONG_STOP_WINDOW. This type of long window is used when the encoder decides to switch from a set of short windows (they always come in sets of eight) to a long window. This type of window has an overlap of 128 audio samples with the short window that precedes it and an overlap of 1,024 samples with the long window that follows it. It also produces 1,024 frequency coefficients.

The width of the frequency bands depends on the sampling rate (number of audio samples per second per audio channel), the number of channels (two for stereo), and the number of frequency bands. The AAC literature provides specifications and tables for the following 12 sampling rates (in Hz): 96,000, 88,200, 64,000, 48,000, 44,100, 32,000, 24,000, 22,050, 16,000, 12,000, 11,025, and 8,000. Any other sampling rates must use one of the 12 sets of specifications and tables listed in Table 10.76. For a sampling rate of 48,000 Hz and two stereo channels, each channel is sampled at 24,000 Hz. A long window generates 1,024 frequency coefficients, so each coefficient corresponds to a 24,000/1,024 ≈ 23.4 kHz frequency band.

Frequency	Specs & tables
$f \geq 92,017$	96,000
$92,017 > f \geq 75,132$	88,200
$75,132 > f \geq 55,426$	64,000
$55,426 > f \geq 46,009$	48,000
$46,009 > f \geq 37,566$	44,100
$37,566 > f \geq 27,713$	32,000
$27,713 > f \geq 23,004$	24,000
$23,004 > f \geq 18,783$	22,050
$18,783 > f \geq 13,856$	16,000
$13,856 > f \geq 11,502$	12,000
$11,502 > f \geq 9,391$	11,025
$9,391 > f$	8,000

Table 10.76: 12 Sets of Frequency Specifications.

Figure 10.77a shows a typical sequence of three long windows, spanning a total of 4,096 audio samples. Part (b) of the figure shows a `LONG_START_WINDOW`, followed by eight short windows, followed by a `LONG_STOP_WINDOW`.

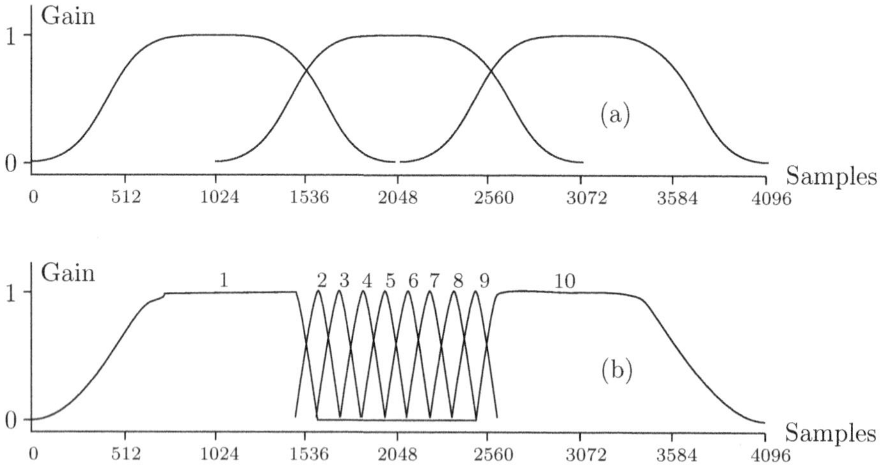

Figure 10.77: Overlapping Long and Short Windows for AAC Filtering.

The MDCT calculated by the encoder is specified by [compare with Equation (10.14)]

$$X_{i,k} = 2\sum_{n=0}^{N-1} z_{i,n}\cos\left[\frac{2\pi}{N}(n+n_0)(k+1/2)\right], \quad \text{for} \quad 0 \leq k < N/2,$$

where $z_{i,n}$ is an audio sample, i is the block index (see "grouping and interleaving" below), n is the audio sample in the current window, k is the index of the frequency coefficients X, N is the window size (2,048 or 256), and $n_0 = (N/2{+}1)/2$. Each frequency coefficient $X_{i,k}$ is computed by a loop that covers the entire current window (window i) and iterates over all N audio samples in that window. However, the index k of the frequency coefficients varies in the interval $[0, N/2)$, so only 1,024 or 128 coefficients are computed.

The inverse MDCT (IMDCT) computed by the decoder is given by [compare with Equation (10.15)]

$$x_{i,n} = \frac{2}{N}\sum_{k=0}^{\frac{N}{2}-1} spec[i][k]\cos\left[\frac{2\pi}{N}(n+n_0)(k+1/2)\right], \quad \text{for} \quad 0 \le n < N,$$

where i is the block (i.e., window) index, n is the index of the audio sample in the current window, k is the index of the frequency coefficients *spec*, N is the window size (2,048 or 256), and $n_0 = (N/2+1)/2$. A close look at this computation shows that N values of $x_{i,n}$ are computed, and each is obtained as a sum of $N/2$ frequency coefficients.

Quantization in AAC is, like many other features of AAC, an improvement over the three layers. The quantization rule is [compare to Equation (10.16)]

$$is(i) = \text{sign}(x(i))\,\text{nint}\left[\left(\frac{|x(i)|}{2^{\text{sf}/4}}\right)^{3/4} + 0.4054\right], \qquad (10.17)$$

where "sign" is the sign bit of the coefficient $x(i)$, "nint" stands for "nearest integer," and sf relates to the scale factor (it is the quantizer step size). In addition, the quantized values are limited to a maximum of 8,191. The idea is to apply coarser quantization to large coefficients, because large coefficients can lose more bits with less effect on the reconstructed audio. Thus, the *Mathematica* code

```
lst = {1., 10., 100., 1000., 10000.};
Table[lst[[i]] - lst[[i]]^0.75, {i, 1, 5}]
```

results in 0, 4.37659, 68.3772, 822.172, and 9,000, thereby illustrating how large coefficients are quantized more than small ones. The five elements of list `lst` are quantized to 1, 5.62341, 31.6228, 177.828, and 1,000. The specific nonlinearity power-constant of 3/4 is not magical and was obtained as a result of many tests and attempts to fine-tune the encoder. A power-constant of 0.5, for example, would have resulted in the more pronounced quantization 0, 6.83772, 90, 968.377, and 9900. Similarly, the "magic" constant 0.4054 is the result of experiments (in early versions of AAC and in mp3 this constant is −0.0946).

The quantization rule of Equation (10.17) involves scale factors. Every frequency coefficient is scaled, during quantization, by a quantity related to its scalefactor. The scalefactors are unsigned 8-bit integers. When the 1,024 coefficients of a long window are quantized, they are grouped into scalefactor bands, where each band is a set of spectral coefficients that are scaled by one scalefactor. The number of coefficients in a band is a multiple of 4 with a maximum of 32. This restriction makes it possible to Huffman-code sets of four consecutive coefficients. Table 10.78 lists the 49 scalefactor bands for the three types of long windows and for sampling rates of 44.1 and 48 kHz. For example, scalefactor band 20 starts at frequency coefficient 132 and ends at coefficient 143, for a total of 12 coefficients. The last scalefactor band includes the 96 coefficients 928 to 1,023. The AAC standard specifies other scalefactor bands for short windows and for other sampling rates.

#	from	#	from	#	from	#	from
0	0	25	216	1	4	26	240
2	8	27	264	3	12	28	292
4	16	29	320	5	20	30	352
6	24	31	384	7	28	32	416
8	32	33	448	9	36	34	480
10	40	35	512	11	48	36	544
12	56	37	576	13	64	38	608
14	72	39	640	15	80	40	672
16	88	41	704	17	96	42	736
18	108	43	768	19	120	44	800
20	132	45	832	21	144	46	864
22	160	47	896	23	176	48	928
24	196					to	1023

Table 10.78: Scalefactor Bands for Long Windows at 44.1 and 48 kHz.

AAC parameter `global_gain` is an unsigned 8-bit integer with the value of the scalefactor of the first band. This value is computed by the psychoacoustic model depending on the masking sounds found in the current window. The scalefactors of the other bands are determined by computing values in increments of 1.5 dB. The scalefactors are compressed differentially by computing the difference scalefactor(i) − scalefactor($i-1$) and replacing it with a Huffman code. The fact that the scalefactors of consecutive bands increase by increments of 1.5 dB implies that the difference between consecutive scalefactors cannot exceed 120. Table 10.79 lists some of the Huffman codes used to compress the differences of the scalefactors. The code lengths vary from a single bit to 19 bits.

Noiseless Coding. Once the frequency coefficients have been quantized, they are replaced by Huffman codes. This increases compression but is lossless, hence the name "noiseless." There are 12 Huffman codebooks (Table 10.80), although one is a pseudo-table, for coding runs of zero coefficients. A block of 1,024 quantized coefficients is divided into sections where each section contains one or several scalefactor bands. The same Huffman codebook is used to code all the coefficients of a section, but different sections can use different code tables. The length of each section (in units of scalefactor bands) and the index of the Huffman codebook used to code the section must be included in the output stream as side information for the decoder.

The idea in sectioning is to employ an adaptive algorithm in order to minimize the total number of bits of the Huffman codes used to encode a block. Thus, in order to determine the sections for a block, the encoder has to execute a complex, greedy-type, merge algorithm that starts by trying the largest number of sections (where each section is one scalefactor band) and using the Huffman codebook with the smallest possible index. The algorithm proceeds by tentatively merging sections. Two merged sections stay merged if this reduces the total bit count. If the two sections that are being tentatively merged use different Huffman codebooks, the resulting section must use the codebook with the higher index.

index	length	code	index	length	code
0	18	3ffe8	61	4	a
1	18	3ffe6	62	4	c
2	18	3ffe7	63	5	1b
3	18	3ffe5	64	6	39
4	19	7fff5	65	6	3b
5	19	7fff1	66	7	78
6	19	7ffed	67	7	7a
7	19	7fff6	68	8	f7
8	19	7ffee	69	8	f9
9	19	7ffef	70	9	1f6
10	19	7fff0	71	9	1f9
⋮			⋮		
57	5	1a	118	19	7ffec
58	4	b	119	19	7fff4
59	3	4	120	19	7fff3
60	1	0			

Table 10.79: Huffman Codes for Differences of Scalefactors.

Huffman codebook 0 is special. It is an escape mechanism that's used for sections where all the quantized coefficients are zero

Grouping and interleaving. A long window produces 1,024 coefficients and a short window produces only 128 coefficients. However, short windows always come in sets of eight, so they also produce 1,024 coefficients, organized as an 8×128 matrix. In order to further increase coding efficiency, the eight short windows can be grouped in such a way that coefficients within a group share scalefactors (they have only one set of scalefactors). Each group is a set of adjacent windows. For example, the first three windows may become group 0, the next window may become group 1 (a single-window group), the next two windows may form group 2, and the last two windows may constitute group 3. Each window now has two indexes, the group index and the window index within the group. Indexes start at zero. Thus, window `[0][2]` is the third window (index 2) of the first group (index 0). Each window has several scalefactor bands, and each band is a set of coefficients. A particular coefficient `c` is therefore identified by four indexes, group (`g`), window within the group (`w`), scalefactor band within the window (`b`), and coefficient within a band (`k`). Thus, `c[g][w][b][k]` is a four-dimensional array where `k` is the fastest index.

Interleaving is the process of interchanging the order of scalefactor bands and windows; a coefficient `c[g][w][b][k]` becomes `c[g][b][w][k]`. The result of interleaving is that frequency coefficients that belong to the same scalefactor band but to different block types (i.e., coefficients that should be quantized with the same scalefactors), are put together in one (bigger) scalefactor band. This has the advantage of combining all zero sections due to band-limiting within each group.

Once all the sections and groups have been determined and interleaving is complete, the coefficients are coded by replacing each coefficient with a Huffman code taken from one of the codebooks. Table 10.80 lists all 12 Huffman codebooks. For each book, the table lists its index (0–11), the maximum tuple size (up to two or four frequency coefficients can be coded by one Huffman code), the maximum absolute value of the coefficients that can be encoded by the codebook, and signed/unsigned information. A signed codebook encodes only positive coefficients. An unsigned codebook provides codes for unsigned values of frequency coefficients. Thus, if coefficient 5 is to be encoded by such a codebook to the 10-bit Huffman code `3f0`, then its sign (0) is written on the compressed stream following this code. Similarly, coefficient −5 is coded to the same 10-bit Huffman code and is later followed by a sign bit of 1. If two or four coefficients are encoded by a single Huffman code, then their sign bits follow this code in the compressed stream. There are two codebooks for each maximum absolute value, each for a different probability distribution. The better fit is always chosen.

	Tuple size	Max. abs. value	Signed
0		0	
1	4	1	yes
2	4	1	yes
3	4	2	no
4	4	2	no
5	2	4	yes
6	2	4	yes
7	2	7	no
8	2	7	no
9	2	12	no
10	2	12	no
11	2	16 (ESC)	no

Table 10.80: 12 Huffman Codebooks.

The first step in encoding the next tuple of frequency coefficients is to select a Huffman codebook (its index is then written on the compressed stream for the decoder's use). Knowing the index of the codebook (1–11), Table 10.80 specifies the tuple size (two or four consecutive coefficients to be encoded). If the coefficients do not exceed the maximum size allowed by the codebook (1, 2, 4, 7, 12, or 16, depending on the codebook), the encoder uses the values of the coefficients in the tuple to compute an index to the codebook. The codebook is then accessed to provide the Huffman code for the n-tuple of coefficients. The Huffman code is written on the output stream, and if the codebook is unsigned (codebook index 3, 4, or 7–11), the Huffman code is followed by two or four sign bits of the coefficients. Decoding a Huffman code is done in the reverse steps, except that the index of the Huffman codebook is read by the decoder from the compressd stream. The decoding steps are listed, as C code, in Figure 10.81. The figure uses the following conventions:

- `unsigned` is the boolean value in column 4 of Table 10.80.
- `dim` is the tuple size of codebook, listed in the second column of Table 10.80.
- `lav` is the maximum absolute value in column 3 of Table 10.80.
- `idx` is the codeword index.

The figure shows how `unsigned`, `dim`, `lav`, and `idx` are used to compute either the four coefficients `w`, `x`, `y`, and `z` or the two coefficients `y` and `z`. If an unsigned codebook was used, the two or four sign bits are read from the compressed stream and are attached to the newly-computed coefficients.

```
if (unsigned) {
  mod = lav + 1;
  off = 0;
}
else {
  mod = 2*lav + 1;
  off = lav;
}
if (dim == 4) {
  w = INT(idx/(mod*mod*mod)) - off;
  idx -= (w+off)*(mod*mod*mod)
  x = INT(idx/(mod*mod)) - off;
  idx -= (x+off)*(mod*mod)
  y = INT(idx/mod) - off;
  idx -= (y+off)*mod
  z = idx - off;
}
else {
  y = INT(idx/mod) - off;
  idx -= (y+off)*mod
  z = idx - off;
}
```

Figure 10.81: Decoding Frequency Coefficients.

If the coefficients exceed the maximum size allowed by the codebook, each is encoded with an escape sequence. A Huffman code representing an escape sequence is constructed by the following rule: Start with N 1's followed by a single 0. Follow with a binary value (called `escapeword`) in $N+4$ bits, and interpret the entire sequence of $N+1+(N+4)$ bits as the number $2^{N+4}+$ `escapeword`. Thus, the escape sequence 01111 corresponds to $N=0$ and a 4-bit `escapeword` of 15. Its value is therefore $2^{0+4}+15=31$. The escape sequence 1011111 corresponds to $N=1$ and a 5-bit `escapeword` of 31. Its value is therefore $2^{1+4}+31=63$. Escape sequences are limited to 21 bits, which is why the largest escape sequence starts with eight 1's, followed by a single 0, followed by $8+4=12$ 1's. It corresponds to $N=8$ and an $(8+4)$-bit `escapeword` of $2^{12}-1$. Its value is therefore $2^{8+4}+(2^{12}-1)=8{,}191$.

Table 10.82 lists parts of the first Huffman Codebook. The "length" field is the length of the correpsonding code, in bits.

Temporal Noise Shaping (TNS). This module of AAC addresses the problem of preechos (read: "pre-echoes"), a problem created when the audio signal to be compressed is transient and varies rapidly. The problem is a mismatch between the masking threshold of the ear (as predicted by the psychoacoustic model) and the quantization noise (the difference between a frequency coefficient and its quantized value). The AAC

index	length	codeword	index	length	codeword
0	11	7f8	41	5	14
1	9	1f1	42	7	65
2	11	7fd	43	5	16
3	10	3f5	44	7	6d
4	7	68	45	9	1e9
5	10	3f0	46	7	63
6	11	7f7	47	9	1e4
7	9	1ec	48	7	6b
8	11	7f5	49	5	13
9	10	3f1	50	7	71
10	7	72	51	9	1e3
⋮		⋮			
38	7	62	79	9	1f7
39	5	12	80	11	7f4
40	1	0			

Table 10.82: Huffman Codebook 1 (Partial).

encoder (as well as the three layers of MPEG-1 and MPEG-2) generates a quantization noise that's evenly distributed in each filterbank window, whereas the masking threshold varies significantly even during the short time period that corresponds to one window.

A long filterbank window consists of 2,048 audio samples, so it represents a very short time. A typical sampling rate is 44,100 samples/second, so 2,048 samples correspond to 0.46 seconds, a very short time. If the original sound (audio signal) doesn't vary much during this time period, the AAC encoder will determine the correct masking threshold and will perform the correct quantization. However, if the audio signal varies considerably during this period, the quantization is wrong, because the quantization noise is evenly distributed in each filterbank window. The TNS module alleviates the preechoes problem by reshaping and controlling the structure of the quantization noise over time (within each transform window). This is done by applying a filtering process to parts of the frequency data of each channel.

Perhaps the best way to understand the principle of operation of TNS is to consider the following duality. Most audio signals feature audio samples that are correlated, similar to adjacent pixels in an image. Such audio can therefore be compressed either by predicting audio samples or by transforming the audio samples to frequency coefficients and quantizing the latter. In transient sounds, the audio samples are not correlated, which suggests the following opposite (or dual) approach to compressing such sounds. Either transform the audio samples to frequency coefficients and predict the next coefficient from its n immediate predecessors, or encode the audio samples directly (i.e., by variable-length codes or some entropy encoder).

Prediction. This tool improves compression by predicting a frequency coefficient from the corresponding coefficients in the two preceding windows and quantizing and encoding the prediction difference. Prediction makes sense for items that are correlated. In AAC, short windows are used when the encoder decides that the sound being compressed is non-stationary. Therefore, prediction is used only on the 1,024 frequency coefficients of long windows.

AAC prediction is second-order, similar to what is shown in Figure 10.27b, but is adaptive. The prediction algorithm is therefore complex and requires many computations. The AAC encoder is often complex and its speed and memory requirements are normally not crucial. The decoder, however, should be fast. This is why AAC has features that limit the complexity of prediction. The main features are: (1) Prediction is used only in the main profile (Table 10.71). (2) In each long window, prediction is performed on those frequency coefficients that correspond to low frequencies, but is stopped at a certain maximum frequency that depends on the sampling rate (Table 10.83). (3) Also, the prediction algorithm depends on variables that are stored internally as 16-bit floating-point numbers (called truncated f.p. in the IEEE floating-point standard) instead of full 32-bit floating-point numbers. This leads to substantial savings in memory space.

Sampling	sf bands	# of predictors	Max. Frequency
96,000	33	512	24,000.00
88,200	33	512	22,050.00
64,000	38	664	20,750.00
48,000	40	672	15,750.00
44,100	40	672	14,470.31
32,000	40	672	10,500.00
24,000	41	652	7,640.63
22,050	41	652	7,019.82
16,000	37	664	5,187.50
12,000	37	664	3,890.63
11,025	37	664	3,574.51
8,000	34	664	2,593.75

Table 10.83: Upper Frequency Limit For Prediction.

The table indicates, for example, that at the 48-kHz sampling rate, prediction can be used in the first 40 scalefactor bands (bands 0 through 39) and stops at the first frequency coefficient that corresponds to a frequency greater than or equal to 15.75 kHz. [Column 3 of Table 10.83, (the number of predictors), is not discussed here.]

Sometimes, prediction does not work; the prediction error is greater than the coefficient being predicted. Therefore, prediction is done twice for each scalefactor band in the current window. First, the coefficients in the band are predicted and the prediction errors are quantized and encoded. Then the coefficients themselves are quantized and encoded without prediction. The encoder selects the method that produces fewer bits and writes the bits on the output stream, preceded by a flag that indicates whether or not prediction was used for that scalefactor band.

The quantization rule of Equation (10.17) generates intermediate results that are real numbers, then converts the final result to the nearest integer (nint). Different word lengths and ALU circuits in different computers may therefore cause slightly different numerical results, and the chance of this happening increases when prediction is used.

This is why the AAC prediction mechanism uses rules for prediction reset. The predictors of both the encoder and decoder (which may run on different computers) are periodically reset to well-defined initial states, which helps avoid differences between them.

10.15.2 MPEG-4 Extensions to AAC

The main features and operation of AAC have been determined as part of MPEG-2. The MPEG-4 project added to AAC several algorithms that contribute to compression efficiency and offer new, extended features. The main enhancements are (1) perceptual noise substitution (PNS), (2) long-term prediction (LTP), (3) transform-domain weighted interleave vector quantization (TwinVQ) coding kernel, (4) long-delay AAC (AAC-LD), (5) error-resilience (ER) module, (6) scalable audio coding, and (7) fine-grain scalability (FGS) mode. Some of these features are discussed here.

Perceptual Noise Substitution. A microphone converts sound waves in the air to an electric signal that varies with time; a waveform. AAC is normally a lossy compression method. It quantizes the frequency coefficients to achieve most of its compression efficiency (the remaining compression is achieved by the Huffman coding of the coefficients and is lossless). If the quantization step is skipped, AAC becomes lossless (although it loses most of its compression ability) and the AAC decoder reconstructs the original sound waveform (except for small changes due to the limited nature of machine arithmetic). If the quantization step is not skipped, the decoder generates a waveform that resembles the original.

Perceptual noise substitution (PNS) is a novel approach to audio compression where part of the original sound is converted by the PNS encoder to one number, thereby achieving compression. This number, however, is not enough for the decoder to reconstruct the original audio waveform. Instead, the decoder generates audio that is perceived by the ear as the original sound, even though it corresponds to a different waveform. PNS is especially suited to noiselike audio, where it achieves high compression factors and reconstructed sound of intermediate to good quality.

PNS is based on the fact that what the ear perceives in the presence of noiselike audio does not depend on the precise waveform of the audio as much as on how its frequency varies with time (its spectral fine structure). PNS is optionally included in the AAC encoder to analyze the frequency coefficients before they are quantized, in order to determine the noiselike parts of the input. Once a scalefactor band of frequency coefficients is discovered to be noiselike, it is not quantized and Huffman coded but instead is sent to the PNS encoder to be processed. The PNS encoder simply writes a flag on the output stream to indicate that this scalefactor band is processed by PNS. The flag is followed by the total noise power of the set of coefficients (the sum of the squares of the coefficients, suitably Huffman coded). When the AAC decoder reads the compressed stream and identifies the flag, it invokes the PNS decoder. This decoder inputs the total power P and tries various sets of pseudorandom numbers until it finds a set whose total power equals P. The numbers in this set are then sent to the AAC decoder as if they were dequantized frequency coefficients.

It is easy to see why PNS achieves excellent compression for the noiselike parts of the input audio. What is harder to grasp is how a set of random numbers is decoded to generate noiselike sound that the ear cannot distinguish from the original sound.

Long-Term Prediction is a technique that discovers and exploits a special redundancy in the original audio samples, redundancy that's related to (a normally invisible) periodicity in parts of the audio. The AAC encoder works as usual, it prepares windows of frequency coefficients C, and quantizes and encodes each window. Before the Huffman codes are set to the output stream, however, the AAC encoder invokes the LTP module which performs the following steps:

- It acts temporarily as an AAC decoder. It decodes the window and reconstructs the audio samples.
- It compares these samples to the original samples.
- The comparison results in delay and gain parameters that are then used to predict the current window from its predecessors.
- The predicted samples are filtered to become frequency coefficients.
- These coefficients are subtracted from the coefficients of set C to become a set of residues.
- Either the residues or the coefficients of set C (whichever results in the smaller number of bits) are Huffman coded and are sent to the output stream.

Twin-Vector Quantization (VQ). This quantization algorithm was developed as part of MPEG-4 for use in the scalable audio coder. (MPEG-4 offers a choice of several audio codecs such as AAC, scalable, HILN, ALS, CELP, and HVXC. The last two are speech coders and ALS is lossless.) Because of the success of TwinVQ, it is sometimes used in AAC as an alternative to the original AAC quantization.

TwinVQ is a two-step process. The first step (spectral normalization) rescales the frequency coefficients to a desired range of amplitudes and computes several signal parameters that are later used by the decoder. The second stage interleaves the normalized coefficients, arranges them in so-called subvectors, and applies weighted vector quantization to the subvectors. (Vector quantization is discussed in Section 7.19.)

10.15.3 AAC-LD (Low Delay)

People like to communicate. This fact becomes obvious when we consider the success of the many inventions that allow and improve communications. Technologies such as telegraph, telephone, radio, and Internet email and telephony have all been successful. An important and obvious aspect of communication is a conversation. No one likes to talk to the walls. When people talk they expect quick response. Even a short delay in a response is annoying, and a long delay may kill a conversation. The high speed of electromagnetic waves and of electrons in wires prevents delays in local telephone calls and in two-way radio conversations. However, long-distance telephone calls that are routed over one or more communications satellites may cause a noticeable delay (close to a second) between the speaker and the listener.

When future manned spaceships reach beyond the orbit of the moon (about 1.2–1.5 light seconds away), the delay caused by the finite speed of light would be annoying. At the distance of Mars (about four light minutes) normal conversation between the crew and Earth would be impossible.

> There was a chorus of "good-byes," and the vision screen went blank. How strange to think, Poole told himself, that all this had happened more than an hour ago; by now his family would have dispersed again and its members would be miles from home. But in a way that time lag, though it could be frustrating, was also a blessing in disguise. Like every man of his age, Poole took it for granted that he could talk instantly, to anyone on Earth, whenever he pleased. Now that this was no longer true, the psychological impact was profound. He had moved into a new dimension of remoteness, and almost all emotional links had been stretched beyond the yield point.
>
> —Arthur C. Clarke, *2001: A Space Odyssey*, (1968)

Various studies of the effects of conversational delays on people have concluded that delays of up to 0.1 sec are unnoticeable, delays of up to 0.25 sec are acceptable, and longer delays are normally annoying. It is generally agreed that a delay of 0.15 sec is the maximum for "good" interactivity.

There is, however, the problem of echoes and it is especially pronounced in Internet telephony and computer chats. In a computer chat, person A speaks into a microphone hooked to his computer, the chat program digitizes the sound and sends it to the receiving computer where it is converted to an analog voltage and is fed into a speaker. This causes a delay (if the chat includes video as well, the data has to be compressed before it is transmitted and decompressed at the destination, causing a longer delay). At the receiver, person B hears the message, but his microphone hears it too and immediately transmits it back to A, who hears it as a faint echo. Using earphones instead of a speaker eliminates this problem, but in conference calls it is natural to use speakers.

The presence of echoes reduces the quality of remote conversations and is, of course, undesirable. Audio engineers and telecommunications researchers have experimented with the combined effect of delays and echoes on the responses of test volunteers and have come up with conclusions and recommendations (see [G131 06]).

This is why MPEG-4 has added a low delay module to AAC. This module can handle both speech and music with high compression, high-quality reconstruction, and short delays. One advantage of AAC-LD is scalability. When the user permits high bitrates (i.e., low compression), the quality of the audio reconstructed by this module gets higher and can easily become indistinguishable from lossless.

The total delay caused by AAC is a sum of several delays caused by high sampling rates (many audio samples per second), filtering (the filterbanks are designed for high resolution), window switching, and handling of the bit reservoir. Window switching causes delays because it requires a look-ahead process to determine the properties of future data (audio samples that haven't been examined yet). This is one reason why it is so difficult to implement a good quality AAC encoder and why many "cheap" AAC encoders produce low compression.

Each of these sources of delay has been examined and modified in AAC-LD. The main features of this variant are as follows:

- It limits the sampling rate to only 48 kHz and uses frame sizes of either 480 or 512 samples. This reduces the window size to half its normal length.

- Windows are not switched. Preecho artifacts are reduced by TNS.

- The "window shape" of the frequency filterbank is now adaptive. In AAC, the shape is a wide sine curve, but AAC-LD dynamically switches a window to a shape that has a lower overlap between the bands (Figure 10.84).
- The bit reservoir is either eliminated altogether or its use is limited.

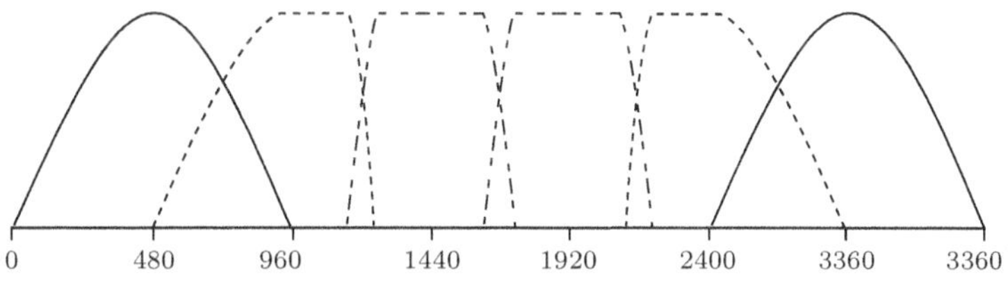

Figure 10.84: Window Shapes in AAC-LD.

These features significantly reduce any delays to well below 0.1 sec, while the decrease in compression relative to AAC is only moderate. The bitrate is increased by about 8 kbit per stereo channel, not very significant.

A battery of tests was perormed to compare the performance of mp3 (single channel) to AAC-LD. The results indicate that AAC-LD outperforms mp3 for half the tests while being as good as mp3 for the other half.

10.15.4 AAC Tests

Testing audio compression methods is especially important and complex. An algorithm for compressing text is easy to test, but methods (especially lossily) for compressing images or audio are difficult to test because these types of data (especially audio) are subject to opinion. A human tester has to be exposed to an original image or an audio sequence (before it is compressed) and to the same data after its compression and decompression. This kind of double test is blind (the tester is not told which data is original and which is reconstructed) and has to be repeated several times (because even a person who decides at random will achieve a 50% success on average in identifying data) and with several volunteer testers, both experienced and inexperienced (because we all perceive images, speech, and music in different ways). Extensive and successful tests for AAC were conducted at the BBC in England, the CBC and CRC (Communications Research Centre) in Canada, and NHK in Japan.

Once the MPEG-4 group decided on the component algorithms of AAC, they were implemented and tested. The aim of the entire project was to create an audio compression method that will be able to compress audio at a bitrate of 64 kbps and generate reconstructed sound that's indistinguishable from the original. Not all sounds are the same. Certain sounds lend themselves to efficient compression, while others are close to random and defy any attempts to compress them. The same sound sequence may have easy and difficult segments, so the essence of the tests was to identify sounds that pose a challenge to the AAC encoder and use them in the tests.

A rigorous definition of the term "indistinguishable" is provided by the ITU-R publication TG-10 [ITU/TG10 91].

The tests proved that AAC behaves at least as good as other audio compressors for low bitrates (in the range of 16 to 64 kbps) and is definitely superior to other compressors at bitrates higher than 64 kbps.

Two series of tests were conducted in 1996–97. The first in 1996 at the BBC in England and in the Japan broadcasting corporation [NHK 06]. The second took place in 1997 at the communications research center [CRC 98] in Canada.

The first set of tests employed a group of 23 reliable expert listeners at the BBC and 16 reliable expert listeners at the NHK. The tests compared 10 audio samples, each compressed in AAC and in MPEG-2 layer 2. The tests were of the type known as triple-stimulus/hidden-reference/double-blind, which is specified in [ITU-R/BS1116 97]. A set of 94 audio samples was originally chosen and was later reduced to 10 excerpts judged critical. They are listed in Table 10.85. The results (which are discussed in depth on page 360 of [Bosi and Goldberg 03]) indicated the superiority of AAC compared to layer 2.

#	Name	Description
1	Cast	Castanets panned across the front, noise in surround
2	Clarinet	Clarinet in front, theater foyer ambience, rain in surround
3	Eliot	Female and male speech in a restaurant, chamber music
4	Glock	Glockenspiel and timpani
5	Harp	Harpsichord
6	Manc	Orchestra, strings, cymbals, drums, horns
7	Pipe	Pitch pipe
8	Station	Male voice with steam-locomotive effects
9	Thal	Piano front left, sax in front right, female voice in center
10	Tria	Triangle

Table 10.85: AAC Tests in 1996 (After [Bosi and Goldberg 03]).

The second set of tests has been well documented and seems to have been very extensive. The tests started with a selection of appropriate audio material. A panel of three experts was charged with this task. Over several months they collected and listened to 80 sounds obtained from (1) past tests, (2) the CRC archives, and (3) the private collections of the experts themselves. In addition, all those who contributed AAC implementations to be tested were invited to supply audio excerpts that they felt were difficult to compress.

Each of the 80 items was then compressed at 96 kbps and decompressed by each of 17 AAC codecs used in the tests, resulting in 1,360 audio sequences. The panel of experts listened to all 1,360 sequences and reduced their numbers to 340 ($= 20 \times 17$) that they felt would resist compression and present the AAC encoders with considerable challenges.

The final selection step reduced the number of audio items tested from 20 to just eight, and these eight items were the ones used in the tests. They included complex sounds such as arpeggios, music of a bowed double bass, a muted trumpet, a song by Suzanne Vega, and a mixture of music and rain.

The first mp3 ever recorded was the song *Tom's Diner* by Suzanne Vega [suzannevega 06] written by her (about Tom's restaurant in New York city) in 1983. As a result, she is sometimes known as the mother of the mp3.

In 1993, Karlheinz Brandenburg headed a team of programmers/engineers who implemented mp3 for Fraunhofer-Gesellschaft and patented their software. As a result, Brandenburg is sometimes called the father of the mp3. The history of mp3 is well documented. Two of the many available sources are [h2g2 06] and [MPThree 06].

Brandenburg used the *Tom's Diner* song as his first audio test item for mp3, because the unusual vocal clarity of this piece allowed him to accurately perceive any audio degradation after this audio file was compressed and decompressed.

"I was ready to fine-tune my compression algorithm," Brandenburg recalls. "Somewhere down the corridor a radio was playing *Tom's Diner*. I was electrified. I knew it would be nearly impossible to compress this warm a capella voice."

Because the song depends on very subtle nuances of Vega's inflection, the algorithm would have to be very, very good to identify the most important parts of the audio and discard the rest. So Brandenburg tested each refinement of his implementation with this song. He ended up listening to the song thousands of times, and the result was a fast, efficient, and robust mp3 implementation.

Once the test objects (the audio items) were ready, the test subjects (the persons making the judgments) were selected. a total of 24 listeners were selected, including seven musicians (performers, composers, students), six audio engineers, three audio professionals, three piano tuners, two programmers, and three persons chosen at random.

The judges concluded that the quality of the audio test pieces in AAC (at 96 kbps) were comparable to that produced by layer 2 at 192 kbps and by layer 3 at 128 kbps. Based on these tests, the CRC announced that AAC had passed the goal set by the ITU of "indistinguishable quality" of compressed sound.

As a result of those tests, audio compressed by AAC at 128 kbps is currently considered the best choice for lossy audio compression and is the choice of many users, including discriminating classical music lovers and critics.

10.16 Dolby AC-3

Dolby Laboratories develops and delivers products and technologies that make the entertainment experience more realistic and immersive. For four decades Dolby has been at the forefront of defining high-quality audio and surround sound in cinema, broadcast, home audio systems, cars, DVDs, headphones, games, televisions, and personal computers. Based in San Francisco with European headquarters in England, the company has entertainment industry liaison offices in New York and Los Angeles, and licensing liaison offices in London, Shanghai, Beijing, Hong Kong, and Tokyo. (Quoted from [Dolby 06].)

Dolby Laboratories was founded by Ray Dolby, who started his career in high school, when he went to work part-time for Ampex Corporation in Redwood City, California. While still in college, he joined the small team of Ampex engineers dedicated to inventing the world's first practical video tape recorder, which was introduced in 1956; his focus was the electronics.

Upon graduation from Stanford University in 1957, Dolby was awarded a Marshall Fellowship to Cambridge University in England. After six years at Cambridge leading to a Ph.D. in physics, Dolby worked in India for two years as a United Nations Adviser to the Central Scientific Instruments Organization. He returned to England in 1965 to found his own company, Dolby Laboratories, Inc. in London. Always a United States corporation, the company moved its headquarters to San Francisco in 1976. (Quoted from [Dolby 06].)

AC-3, also known as Dolby Digital, stands for Dolby's third-generation audio coder. AC-3 is a perceptual audio codec based on the same principles as the three MPEG-1/2 layers and AAC. This short section concentrates on the special features of AC-3 and what distinguishes it from other perceptual codecs. Detailed information can be found in [Bosi and Goldberg 03]. AC-3 was approved by the United States Advanced Television Systems Committee (ATSC) in 1994. The formal specification can be found at [ATSC 06] and may also be available from the authors.

AC-3 was developed to combine low bitrates with an excellent quality of the reconstructed sound. It is used in several important video applications such as the following:

- HDTV (in North America). HDTV (high-definition television) is a standard for high-resolution television. Currently, HDTV is becoming popular and its products are slowly replacing the existing older television formats, such as NTSC and PAL.

- DVD-video. This is a standard developed by the DVD Forum for storing full-length digital movies on DVDs. The standard allows the use of either MPEG-1 or MPEG-2 for compressing the video, but the latter is more common. Commercial DVD-Video players are currently very popular. Such a device connects to a television set just like a videocassette player. The Digital-Video format includes a Content Scrambling System (CSS) to prevent unauthorized copying of DVDs.

- DVB. The Digital Video Broadcasting (DVB) project is an industry-led consortium of over 270 broadcasters, manufacturers, network operators, software developers, regulatory bodies, and others in over 35 countries committed to designing global standards for the global delivery of digital television and data services. Services using DVB standards are available on every continent with more than 110 million DVB receivers deployed. (Quoted from [DVB 06].)

- Various digital cable and satellite transmissions.

What distinguishes AC-3 from other perceptual coders is that it has originally been designed to support several audio channels. Old audio equipment utilized just one audio channel (so-called mono or monophonic). The year 1957 marked the debut of the stereo long-play (LP) phonograph record, which popularized stereo (more accurately, stereophonic) sound. Stereo requires two audio channels, left and right.

The Quadraphonic format (quad) appeared in the early 1970s. It consists of matrix encoding of four channels of audio data within a two-channel recording. It allows a two-channel recording to contain four channels of audio that are played on four speakers. Quad never became very popular because of the cost of new amplifiers, receivers, and additional speakers.

In the mid 1970s, Dolby Labs unveiled a new surround sound process that was easy to adapt for home use. The development of the HiFi stereo VCR and stereo television broadcasting in the 1980s provided an additional chance for surround sound to become popular. An important reason for the popularity of surround sound is the ability to add Dolby surround processors to existing stereo receivers; there is no need to buy all new equipment.

The Dolby surround method involves encoding four channels of audio—front left, center, front right, and rear surround—into a two-channel signal. A decoding circuit then separates the four channels and sends them to the appropriate speakers, the left, right, rear, and phantom center (center channel is derived from the L/R front channels). The result is a balanced listening environment where the chief sounds come from the left and right channels, the vocal or dialog audio emanates from the center phantom channel, and the ambience or effects information comes from behind the listener.

Thus, the number of audio channels used in sound equipment has risen steadily over the years, and it was only natural for Dolby to come up with an audio compression method that supports up to five channels. Bitrates range from 32 to 640 kbps, depending on the number of channels (Table 10.86, where "code" is the bitrate code sent to the decoder). More accurately, AC-3 supports from one to 5.1 audio channels. The unusual designation 5.1 means five channels plus a special low-frequency-effects (LFE) channel (referred to as a 0.1 channel) which provides an economical and practical way of achieving extra bass impact. The eight channel configurations supported by AC-3 are listed in Table 10.87 where "code" is a code sent to the decoder and "config" is the notation used by Dolby to indicate the channel configuration.

In addition to the compressed audio, the output stream of AC-3 may include auxiliary data such as language identifiers, copyright notices, time stamps, and other control information. AC-3 also supports so-called listener features that include downmixing (to reduce the number of channels, for those listeners who lack advanced listening equipment), dialog normalization, compatibility with Dolby surround, bitstreams for the

code	bitrate	code	bitrate
2	32	20	160
4	40	22	192
6	48	24	224
8	56	26	256
10	64	28	320
12	80	30	384
14	96	32	448
16	112	34	512
18	128	36	640

Table 10.86: Bitrates (kbps) in AC-3.

code	config	#	channel	speakers	code	config	#	channel	speakers
000	1 + 1	2	C1, C2		100	2/1	3	L, R, S	
001	1/0	1	C		101	3/1	4	L, C, R, S	
010	2/0	2	L, R		110	2/2	4	L, R, SL, SR	
110	3/0	3	L, C, R	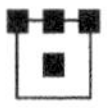	111	3/2	5	L, C, R, SL, SR	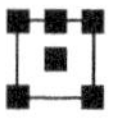

Table 10.87: AC-3 Channel Configurations.

hearing impaired, and dynamic range control. The last feature is especially useful. The person encoding audio with AC-3 can specify range control parameters for individual segments of the audio. The AC-3 decoder uses these parameters to adjust the level of the decoded audio on a block by block basis (where a block consists of a few ms of audio, depending on the sampling rate), thereby guaranteeing that the audio being played back will always have the full dynamic range.

The first step of the AC-3 encoder is to group the audio samples (which can be up to 24 bits each) in blocks of 512 samples each. Each block is assessed to determine whether its audio is tonal or transient. If the audio is transient, the block size is cut to 256 audio samples. Each block is mapped onto the frequency domain by a modified DCT. The number of frequency coefficients computed for the block is half the block size. Six blocks make up a window, and block sizes in the window can be switched between long and short, similar to the way it is done in AAC.

Once the frequency coefficients have been computed, the encoder uses them in a process termed rematrixing to perform multichannel coding. The frequency coefficients resulting from the MDCT are real numbers and are stored in floating-point format, with a mantissa and an exponent. The former includes the significant digits of the coefficient, while the latter indicates its size. Thus, the two floating-point numbers 0.12345678×2^{-10} and $0.12345678\times2^{+10}$ have identical mantissas that constitute eight significant digits, but have wildly different sizes indicated by their different exponents.

Compression is achieved by coding the exponents and quantizing the mantissas (the amount of quantization depends on the bit allocation routine). The former compression is lossless, whereas the latter is lossy. The bit allocation routine applies a psychoacoustic model to determine the appropriate precision needed for the mantissas.

The last step of the encoder is to combine (multiplex) the encoded exponents and quantized mantissas with control information such as block switching flags, coupling parameters, rematrixing flags, exponent stategy bits, dither flags, and bit allocation parameters to create the final output stream.

Thomas Dolby (born Thomas Morgan Robertson, on 14 October 1958, in London) is a British musician. His father was a professor of Greek history and in his youth he lived in Greece, Italy, France, and other countries of Mediterranean Europe. The "Dolby" nickname comes from the name Dolby Laboratories, and was given to him by friends impressed with his studio tinkering. Dolby Laboratories was reportedly very displeased with Robertson using the company name as his own stage name and sued him, trying to stop him from using the name Dolby entirely. Eventually, they succeeded in restricting him from using the word Dolby in any context other than with the name Thomas. (See [Thomas Dolby 06]).

I don't have an English accent because this is what English sounds like when spoken properly.

—James Carr

11
Other Methods

Previous chapters discuss the main classes of compression methods: RLE, statistical methods, and dictionary-based methods. There are data compression methods that are not easy to classify and do not clearly belong in any of the classes discussed so far. A few such methods are described here.

- The Burrows-Wheeler method (Section 11.1) starts with a string S of n symbols and scrambles (i.e., permutes) them into another string L that satisfies two conditions: (1) Any area of L will tend to have a concentration of just a few symbols. (2) It is possible to reconstruct the original string S from L.

- The technique of *symbol ranking* (Section 11.2) uses context to rank symbols rather than assign them probabilities.

- ACB is a new method, based on an associative dictionary (Section 11.3). It has features that relate it to the traditional dictionary-based methods as well as to the symbol ranking method.

- Section 11.4 is a description of the sort-based context similarity method. This method uses the context of a symbol in a way reminiscent of ACB. It also assigns ranks to symbols, and this feature relates it to the Burrows-Wheeler method and also to symbol ranking.

- The special case of *sparse binary strings* is discussed in Section 11.5. Such strings can be compressed very efficiently due to the large number of consecutive zeros they contain.

- Compression methods that are based on words rather than individual symbols are the subject of Section 11.6.

- Textual image compression is the topic of Section 11.7. When a printed document has to be saved in the computer, it has to be scanned first, a process that converts it

D. Salomon, G. Motta, *Handbook of Data Compression*, 5th ed.
DOI 10.1007/978-1-84882-903-9_11,

into an image typically containing millions of pixels. The complex method described here has been developed for this kind of data, which forms a special type of image, a *textual image*. Such an image is made of pixels, but most of the pixels are grouped to form characters, and the number of different groups is not large.

- The FHM method (for Fibonacci, Huffman, and Markov) is an unusual, special-purpose method for the compression of curves.

- Dynamic Markov coding uses finite-state machines to estimate the probability of symbols, and arithmetic coding to actually encode them. This is a compression method for two-symbol (binary) alphabets.

- Sequitur, Section 11.10, is a method designed for the compression of semistructured text. It is based on context-free grammars.

- Section 11.11 is a detailed description of edgebreaker, a highly original method for compressing the connectivity information of a triangle mesh. This method and its various extensions may become the standard for compressing polygonal surfaces, one of the most common surface types used in computer graphics. Edgebreaker is an example of a *geometric compression* method.

- Sections 11.12 and 11.12.1 describe two algorithms, SCSU and BOCU-1, for the compression of Unicode-based documents.

- Section 11.13 is a short summary of the compression methods used by the popular Portable Document Format (PDF) by Adobe.

- Section 11.14 covers a little-known variant of data compression, namely how to compress the differences between two files. We all have cellular telephones, and we take these handy little devices for granted. However, a cell phone is a complex device that is driven by a computer and has an operating system. Naturally, the software inside the telephone has to be updated from time to time, but sending a large executable file to many thousands of telephones is a major communication headache. It is better to isolate the differences between the old and the new software, compress these differences, and send the compressed file to all the telephones.

- Section 11.15 on hyperspectral data compression treats the topic of hyperspectral data. Such data is similar to a digital image, but each "pixel" consists of many (hundred to thousands) of components.

- The commercial compression software known as stuffit has been around since 1987. The methods and algorithms it employs are proprietary, but some information exists in various patents. The new Section 11.16 is an attempt to describe what is publicly known about this software and how it works.

11.1 The Burrows-Wheeler Method

Most compression methods operate in the *streaming mode*, where the codec inputs a byte or several bytes, processes them, and continues until an end-of-file is sensed. The Burrows-Wheeler (BW) method, described in this section [Burrows and Wheeler 94], works in a *block mode*, where the input stream is read block by block and each block is encoded separately as one string. The method is therefore referred to as *block sorting*. The BW method is general purpose, it works well on images, sound, and text, and can achieve very high compression ratios (1 bit per byte or even better).

The method was developed by Michael Burrows and David Wheeler in 1994, while working at DEC Systems Research Center in Palo Alto, California. This interesting and original transform is based on a previously unpublished transform originated by Wheeler in 1983. After its publication, the BW method was further improved by several researchers, among them Ziya Arnavut, Michael Schindler, and Peter Fenwick. The main idea of the BW method is to start with a string S of n symbols and to scramble them into another string L that satisfies two conditions:

1. Any region of L will tend to have a concentration of just a few symbols. Another way of saying this is, if a symbol s is found at a certain position in L, then other occurrences of s are likely to be found nearby. This property means that L can easily and efficiently be compressed with the move-to-front method (Section 1.5), perhaps in combination with RLE. This also means that the BW method will work well only if n is large (at least several thousand symbols per string).
2. It is possible to reconstruct the original string S from L (a little more data may be needed for the reconstruction, in addition to L, but not much).

The mathematical term for scrambling symbols is *permutation*, and it is easy to show that a string of n symbols has $n!$ (pronounced "n factorial") permutations. This is a large number even for relatively small values of n, so the particular permutation used by BW has to be carefully selected. The BW codec proceeds in the following steps:

1. String L is created, by the encoder, as a permutation of S. Some more information, denoted by I, is also created, to be used later by the decoder in step 3.
2. The encoder compresses L and I and writes the results on the output stream. This step typically starts with RLE, continues with move-to-front coding, and finally applies Huffman coding.
3. The decoder reads the output stream and decodes it by applying the same methods as in 2 above but in reverse order. The result is string L and variable I.
4. Both L and I are used by the decoder to reconstruct the original string S.

> I do hate sums. There is no greater mistake than to call arithmetic an exact science. There are permutations and aberrations discernible to minds entirely noble like mine; subtle variations which ordinary accountants fail to discover; hidden laws of number which it requires a mind like mine to perceive. For instance, if you add a sum from the bottom up, and then from the top down, the result is always different.
>
> —Mrs. La Touche, *Mathematical Gazette*, v. 12 (1924)

The first step is to understand how string L is created from S, and what information needs to be stored in I for later reconstruction. We use the familiar string `swiss miss` to illustrate this process.

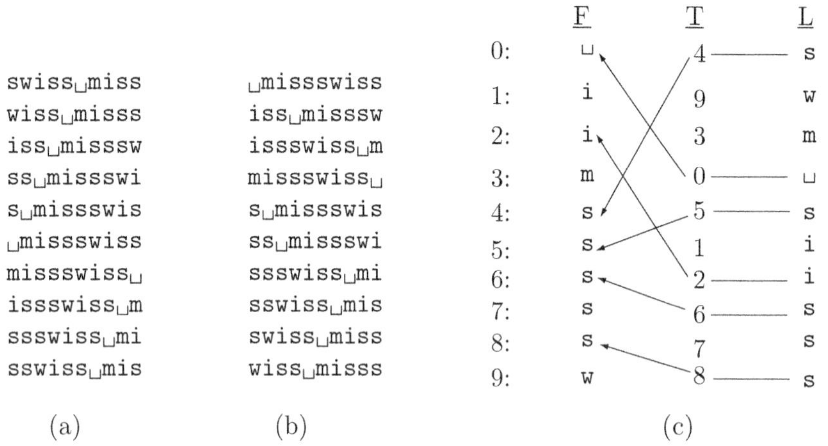

Figure 11.1: Principles of BW Compression.

Given an input string of n symbols, the encoder constructs an $n \times n$ matrix where it stores string S in the top row, followed by $n-1$ copies of S, each cyclically shifted (rotated) one symbol to the left (Figure 11.1a). The matrix is then sorted lexicographically by rows (see Section 6.4 for lexicographic order), producing the sorted matrix of Figure 11.1b. Notice that every row and every column of each of the two matrices is a permutation of S and thus contains all n symbols of S. The permutation L selected by the encoder is the **last column** of the sorted matrix. In our example this is the string `swm siisss`. The only other information needed to eventually reconstruct S from L is the row number of the original string in the sorted matrix, which in our example is 8 (row and column numbering starts from 0). This number is stored in I.

It is easy to see why L contains concentrations of identical symbols. Assume that the words `bail`, `fail`, `hail`, `jail`, `mail`, `nail`, `pail`, `rail`, `sail`, `tail`, and `wail` appear somewhere in S. After sorting, all the permutations that start with `il` will appear together. All of them contribute an `a` to L, so L will have a concentration of `a`'s. Also, all the permutations starting with `ail` will end up together, contributing to a concentration of the letters `bfhjmnprstw` in one region of L.

We can now characterize the BW method by saying that it uses sorting to group together symbols based on their contexts. However, the method considers context on only one side of each symbol.

◇ **Exercise 11.1:** The last column, L, of the sorted matrix contains concentrations of identical characters, which is why L is easy to compress. However, the first column, F, of the same matrix is even easier to compress, since it contains runs, not just concentrations, of identical characters. Why select column L and not column F?

Notice also that the encoder does not actually have to construct the two $n \times n$ matrices (or even one of them) in memory. The practical details of the encoder are discussed in Section 11.1.2, as well as the compression of L and I, but let's first see how the decoder works.

The decoder reads a compressed stream, decompresses it using Huffman and move-to-front (and perhaps also RLE), and then reconstructs string S from the decompressed L in three steps:

1. The first column of the sorted matrix (column F in Figure 11.1c) is constructed from L. This is a straightforward process, since F and L contain the same symbols (both are permutations of S) and F is sorted. The decoder simply sorts string L to obtain F.
2. While sorting L, the decoder prepares an auxiliary array T that shows the relations between elements of L and F (Figure 11.1c). The first element of T is 4, implying that the first symbol of L (the letter "s") is located in position 4 of F. The second element of T is 9, implying that the second symbol of L (the letter "w") is located in position 9 of F, and so on. The contents of T in our example are $(4, 9, 3, 0, 5, 1, 2, 6, 7, 8)$.
3. String F is no longer needed. The decoder uses L, I, and T to reconstruct S according to

$$\begin{aligned} &\mathrm{S}[n-1-i] \leftarrow \mathrm{L}[\mathrm{T}^i[\mathrm{I}]], \quad \text{for } i = 0, 1, \ldots, n-1, \\ &\text{where } \mathrm{T}^0[j] = j, \text{ and } \mathrm{T}^{i+1}[j] = \mathrm{T}[\mathrm{T}^i[j]]. \end{aligned} \tag{11.1}$$

Here are the first two steps in this reconstruction:

$$\texttt{S[10-1-0]=L[T}^0\texttt{[I]]=L[T}^0\texttt{[8]]=L[8]=s,}$$
$$\texttt{S[10-1-1]=L[T}^1\texttt{[I]]=L[T[T}^0\texttt{[I]]]=L[T[8]]=L[7]=s.}$$

◇ **Exercise 11.2:** Complete this reconstruction.

Before getting to the details of the compression, it may be interesting to understand why Equation (11.1) reconstructs S from L. The following arguments explain why this process works:

1. T is constructed such that $\mathrm{F}[\mathrm{T}[i]] = \mathrm{L}[i]$ for $i = 0, \ldots, n$.
2. A look at the sorted matrix of Figure 11.1b shows that in each row i, symbol L[i] precedes symbol $\mathrm{F}[i]$ in the original string S (the word *precedes* has to be understood as *precedes cyclically*). Specifically, in row I (8 in our example), L[I] cyclically precedes F[I], but F[I] is the first symbol of S, so L[I] is the *last* symbol of S. The reconstruction starts with L[I] and reconstructs S from right to left.
3. $\mathrm{L}[i]$ precedes $\mathrm{F}[i]$ in S for $i = 0, \ldots, n-1$. Therefore $\mathrm{L}[\mathrm{T}[i]]$ precedes $\mathrm{F}[\mathrm{T}[i]]$, but $\mathrm{F}[\mathrm{T}[i]] = \mathrm{L}[i]$. The conclusion is that $\mathrm{L}[\mathrm{T}[i]]$ precedes $\mathrm{L}[i]$ in S.
4. The reconstruction therefore starts with $\mathrm{L}[\mathrm{I}] = \mathrm{L}[8] = \texttt{s}$ (the last symbol of S) and proceeds with $\mathrm{L}[\mathrm{T}[\mathrm{I}]] = \mathrm{L}[\mathrm{T}[8]] = \mathrm{L}[7] = \texttt{s}$ (the next-to-last symbol of S). This is why Equation (11.1) correctly describes the reconstruction.

11.1.1 Compressing L

Compressing L is based on its main attribute, namely, it contains concentrations (although not necessarily runs) of identical symbols. Using RLE makes sense, but only as a first step in a multistep compression process. The main step in compressing L should use the move-to-front method (Section 1.5). This method is applied to our example L=`swm␣siisss` as follows:

1. Initialize A to a list containing our alphabet A=(␣, `i`, `m`, `s`, `w`).
2. For $i := 0, \ldots, n-1$, encode symbol L_i as the number of symbols preceding it in A, and then move symbol L_i to the beginning of A.
3. Combine the codes of step 2 in a list C, which will be further compressed with Huffman or arithmetic coding.

The results are summarized in Figure 11.2a. The final list of codes is the 10-element array $C = (3, 4, 4, 3, 3, 4, 0, 1, 0, 0)$, illustrating how any concentration of identical symbols produces small codes. The first occurrence of `i` is assigned code 4 but the second occurrence is assigned code 0. The first two occurrences of `s` get code 3, but the next one gets code 1.

L	A	Code	C	A	L
s	␣imsw	3	3	␣imsw	s
w	s␣imw	4	4	s␣imw	w
m	ws␣im	4	4	ws␣im	m
␣	mws␣i	3	3	mws␣i	␣
s	␣mwsi	3	3	␣mwsi	s
i	s␣mwi	4	4	s␣mwi	i
i	is␣mw	0	0	is␣mw	i
s	is␣mw	1	1	is␣mw	s
s	si␣mw	0	0	si␣mw	s
s	si␣mw	0	0	si␣mw	s
	(a)			(b)	

Figure 11.2: Encoding/Decoding L by Move-to-Front.

It is interesting to compare the codes in C, which are integers in the range $[0, n-1]$, with the codes obtained without the extra step of "moving to front." It is easy to encode L using the three steps above but without moving symbol L_i to the beginning of A. The result is $C' = (3, 4, 2, 0, 3, 1, 1, 3, 3, 3)$, a list of integers *in the same range* $[0, n-1]$. This is why applying move-to-front is not enough. Lists C and C' contain elements in the same range, but the elements of C are smaller on average. They should therefore be further encoded using Huffman coding or some other statistical method. Huffman codes for C can be assigned assuming that code 0 has the highest probability and code $n-1$, the smallest probability.

In our example, a possible set of Huffman codes is 0—0, 1—10, 2—110, 3—1110, 4—1111. Applying this set to C yields "1110|1111|1111|1110|1110|1111|0|10|0|0"; 29 bits. (Applying it to C' yields "1110|1111|110|0|1110|10|10|1110|1110|1110"; 32 bits.) Our original 10-character string `swiss miss` has thus been coded using 2.9 bits/character, a very good result. It should be noted that the Burrows-Wheeler method can easily achieve better compression than that when applied to longer strings (thousands of symbols).

◇ **Exercise 11.3:** Given the string S=`sssssssssh` calculate string L and its move-to-front compression.

Decoding C is done with the inverse of move-to-front. We assume that the alphabet list A is available to the decoder (it is either the list of all possible bytes or it is written by the encoder on the output stream). Figure 11.2b shows the details of decoding C = $(3,4,4,3,3,4,0,1,0,0)$. The first code is 3, so the first symbol in the newly constructed L is the *fourth* one in A, or "`s`". This symbol is then moved to the front of A, and the process continues.

11.1.2 Implementation Hints

Since the Burrows-Wheeler method is efficient only for long strings (at least thousands of symbols), any practical implementation should allow for large values of n. The maximum value of n should be so large that two $n \times n$ matrices would not fit in the available memory (at least not comfortably), and all the encoder operations (preparing the permutations and sorting them) should be done with one-dimensional arrays of size n. In principle, it is enough to have just the original string S and the auxiliary array T in memory. [Manber and Myers 93] and [McCreight 76] discuss the data structures used in this implementation.

String S contains the original data, but surprisingly, it also contains all the necessary permutations. Since the only permutations we need to generate are rotations, we can generate permutation i of matrix 11.1a by scanning S from position i to the end, then continuing cyclically from the start of S to position $i-1$. Permutation 5, for example, can be generated by scanning substring $(5,9)$ of S (`␣miss`), followed by substring $(0,4)$ of S (`swiss`). The result is `␣missswiss`. The first step in a practical implementation would thus be to write a procedure that takes a parameter i and scans the corresponding permutation.

Any method used to sort the permutations has to compare them. Comparing two permutations can be done by scanning them in S, without having to move symbols or create new arrays.

Once the sorting algorithm determines that permutation i should be in position j in the sorted matrix (Figure 11.1b), it sets T[i] to j. In our example, the sort ends up with T $= (5,2,7,6,4,3,8,9,0,1)$.

◇ **Exercise 11.4:** Show how how T is used to create the encoder's main output, L and I.

Implementing the decoder is straightforward, because there is no need to create $n \times n$ matrices. The decoder inputs bits that are Huffman codes. It uses them to create the codes of C, decompressing each as it is created, with inverse move-to-front, into the next symbol of L. When L is ready, the decoder sorts it into F, generating array T in the process. Following that, it reconstructs S from L, T, and I. Thus, the decoder needs at most three structures at any time, the two strings L and F (having typically one byte per symbol), and the array T (with at least two bytes per pointer, to allow for large values of n).

We describe a block-sorting,lossless data compression algorithm, and our implementation of that algorithm. We compare the performance of our implementation with widely available data compressors running on the same hardware.

M. Burrows and D. J. Wheeler, May 10, 1994

11.2 Symbol Ranking

Like so many other ideas in the realm of information and data, the idea of text compression by symbol ranking is due to Claude Shannon, the creator of information theory. In his classic paper on the information content of English text [Shannon 51] he describes a method for experimentally determining the entropy of such texts. In a typical experiment, a passage of text has to be predicted, character by character, by a person (the examinee). In one version of the method the examinee predicts the next character and is then told by the examiner whether the prediction was correct or, if it was not, what the next character is. In another version, the examinee has to continue predicting until he obtains the right answer. The examiner then uses the number of wrong answers to estimate the entropy of the text.

As it turned out, in the latter version of the test, the human examinees were able to predict the next character in one guess about 79% of the time and rarely needed more than 3–4 guesses. Table 11.3 shows the distribution of guesses as published by Shannon.

# of guesses	1	2	3	4	5	>5
Probability	79%	8%	3%	2%	2%	5%

Table 11.3: Probabilities of Guesses of English Text.

The fact that this probability is so skewed implies low entropy (Shannon's conclusion was that the entropy of English text is in the range of 0.6–1.3 bits per letter), which in turn implies the possibility of very good compression.

The symbol ranking method of this section [Fenwick 96] is based on the latter version of the Shannon test. The method uses the context C of the current symbol S (the N symbols preceding S) to prepare a list of symbols that are likely to follow C. The list is arranged from most likely to least likely. The position of S in this list (position numbering starts from 0) is then written by the encoder, after being suitably encoded, on the output stream. If the program performs as well as a human examinee, we can expect 79% of the symbols being encoded to result in 0 (first position in the ranking list), creating runs of zeros, which can easily be compressed by RLE.

The various context-based methods described elsewhere in this book, most notably PPM, use context to estimate symbol probabilities. They have to generate and output escape symbols when switching contexts. In contrast, symbol ranking does not estimate probabilities and does not use escape symbols. The absence of escapes seems to be the main feature contributing to the excellent performance of the method. Following is an outline of the main steps of the encoding algorithm.

Step 0: The *ranking index* (an integer counting the position of S in the ranked list) is set to 0.
Step 1: An LZ77-type dictionary is used, with a search buffer containing text that has already been input and encoded, and with a look-ahead buffer containing new, unprocessed text. The most-recent text in the search buffer becomes the *current context* C. The leftmost symbol, R, in the look-ahead buffer (immediately to the right of C) is the *current symbol.* The search buffer is scanned from right to left (from recent to older text) for strings matching C. This process is very similar to the one described in Section 6.19 (LZP compression). The longest match is selected (if there are several longest matches, the most recent one is selected). The match length, N, becomes the *current order.* The symbol P following the matched string (i.e., immediately to the right of it) is examined. This is the symbol ranked first by the algorithm. If P is identical to R, the search is over and the algorithm outputs the ranking index (which is currently 0).
Step 2: If P is different from R, the ranking index is incremented by 1, P is declared *excluded,* and the other order-N matches, if any, are examined in the same way. Assume that Q is the symbol following such a match. If Q is in the list of excluded symbols, then it is pointless to examine it, and the search continues with the next match. If Q has not been excluded, it is compared with R. If they are identical, the search is over, and the encoding algorithm outputs the ranking index. Otherwise the ranking index is incremented by 1, and Q is excluded.
Step 3: If none of the order-N matches is followed by a symbol identical to R, the order of the match is decremented by 1, and the search buffer is again scanned from right to left (from more recent text to older text) for strings of size $N-1$ that match C. For each failure in this scan, the ranking index is incremented by 1, and Q is excluded.
Step 4: When the match order gets all the way down to 0, symbol R is compared with symbols in a list containing the entire alphabet, again using exclusions and incrementing the ranking index. If the algorithm gets to this step, it will find R in this list, and will output the current value of the ranking index (which will then normally be a large number).

Some implementation details are discussed here.

1. Implementing exclusion. When a string S that matches C is found, the symbol P immediately to the right of S is compared with R. If P and R are different, P should be declared excluded. This means that any future occurrences of P should be ignored. The first implementation of exclusion that comes to mind is a list to which excluded symbols are appended. Searching such a list, however, is time consuming, and it is possible to do much better.

The method described here uses an array `excl` indexed by the alphabet symbols. If the alphabet consists, for example, of just the 26 letters, the array will have 26 locations indexed `a` through `z`. Figure 11.4 shows a simple implementation that requires just one step to determine whether a given symbol is excluded. Assume that the current context

C is the string "...abc". We know that the c will remain in the context even if the algorithm has to go down all the way to order-1. The algorithm therefore prepares a pointer to c (to be called the *context index*). Assume that the scan finds another string abc, followed by a y, and compares it to the current context. They match, but they are followed by different symbols. The decision is to exclude y, and this is done by setting array element excl[y] to the context index (i.e., to point to c). As long as the algorithm scans for matches to the same context C, the context index will stay the same. If another matching string abc is later found, also followed by y, the algorithm compares excl[y] to the context index, finds that they are equal, so it knows that y has already been excluded. When switching to the next current context there is no need to initialize or modify the pointers in array excl.

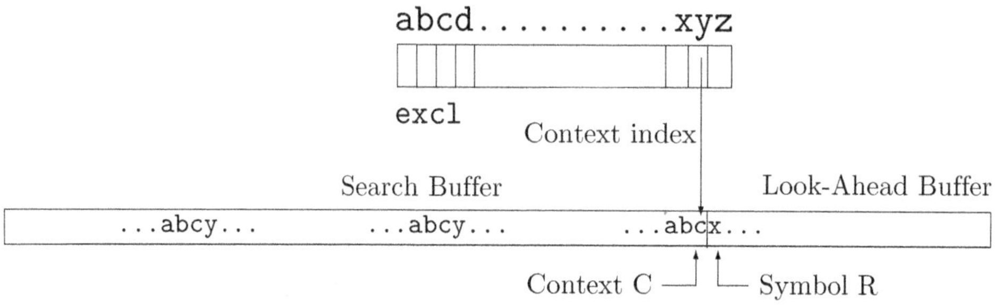

Figure 11.4: Exclusion Mechanism.

2. It has been mentioned earlier that scanning and finding matches to the current context C is done by a method similar to the one used by LZP. The reader should review Section 6.19 before reading ahead. Recall that N (the order) is initially unknown. The algorithm has to scan the search buffer and find the longest match to the current context. Once this is done, the length N of the match becomes the current order. The process therefore starts by hashing the two rightmost symbols of the current context C and using them to locate a possible match.

Figure 11.5 shows the current context "...amcde". We assume that it has already been matched to some string of length 3 (i.e., a string "...cde"), and we try to match it to a longer string. The two symbols "de" are hashed and produce a pointer to string "lmcde". The problem is to compare the current context to "lmcde" and find whether and by how much they match. This is done by the following three rules.

Rule 1: Compare the symbols preceding (i.e., to the left of) cde in the two strings. In our example they are both m, so the match is now of size 4. Repeat this rule until it fails. It determines the order N of the match. Once the order is known, the algorithm may have to decrement it later and compare shorter strings. In such a case, this rule has to be modified. Instead of comparing the symbols *preceding* the strings, it should compare the leftmost symbols of the two strings.

Rule 2: (We are still not sure whether the two strings are identical.) Compare the middle symbols of the two strings. In our case, since the strings have a length of 4, this would be either the c or the d. If the comparison fails, the strings are different. Otherwise, Rule 3 is used.

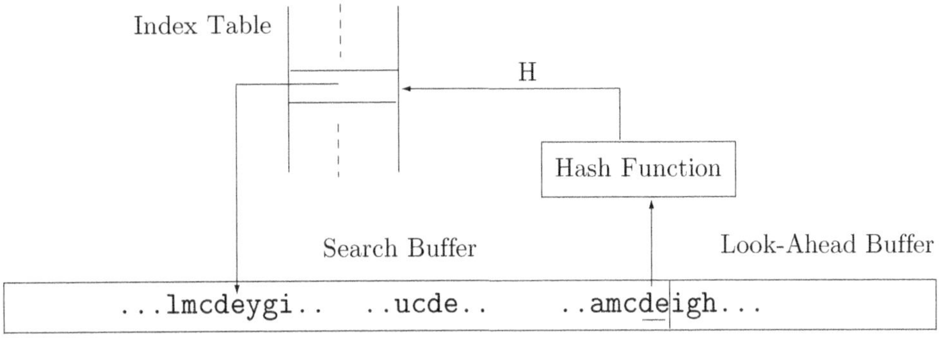

Figure 11.5: String Search and Comparison Method.

Rule 3: Compare the strings symbol by symbol to finally determine whether they are identical.

It seems unnecessarily cumbersome to go through three rules when only the third one is really necessary. However, the first two rules are simple, and they identify 90% of the cases where the two strings are different. Rule 3, which is slow, has to be applied only if the first two rules have not identified the strings as different.

3. If the encoding algorithm has to decrement the order all the way down to 1, it faces a special problem. It can no longer hash two symbols. Searching for order-1 matching strings (i.e., single symbols) therefore requires a different method which is illustrated by Figure 11.6. Two linked lists are shown, one linking occurrences of `s` and the other linking occurrences of `i`. Notice how only certain occurrences of `s` are linked, while others are skipped. The rule is to skip an occurrence of `s` which is followed by a symbol that has already been seen. Thus, the first occurrences of `si`, `ss`, `s␣`, and `sw` are linked, whereas other occurrences of `s` are skipped.

The list linking these occurrences of `s` starts empty and is built gradually, as more text is input and is moved into the search buffer. When a new context is created with `s` as its rightmost symbol, the list is updated. This is done by finding the symbol to the right of the new `s`, say `a`, scanning the list for a link `sa`, deleting it if found (not more than one may exist), and linking the current `s` to the list.

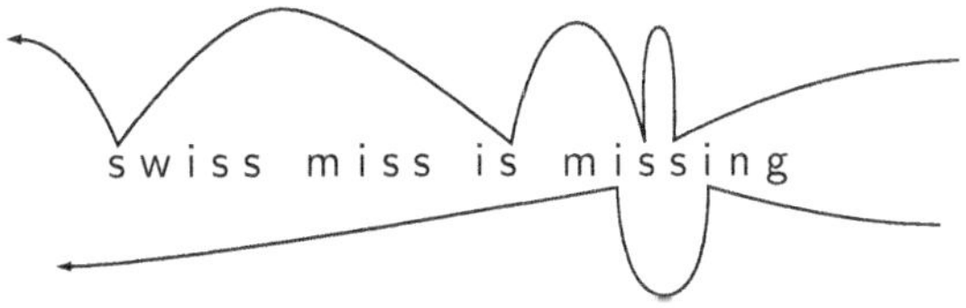

Figure 11.6: Context Searching for Order-1.

This list makes it easy to search and find all occurrences of the order-1 context `s` that are followed by different symbols (i.e., with exclusions).

Such a list should be constructed and updated for each symbol in the alphabet. If the algorithm is implemented to handle 8-bit symbols, then 256 such lists are needed and have to be updated.

The implementation details above show how complex this method is. It is slow, but it produces excellent compression.

11.3 ACB

Not many details are available of the actual implementation of ACB, an original, highly efficient text compression method by George Buyanovsky. The only documentation currently available is in Russian [Buyanovsky 94] and is outdated. (An informal interpretation in English, by Leonid Broukhis, is available at `http://www.cbloom.com/news/leoacb.html`.) The name ACB stands for Associative Coder (of) Buyanovsky. We start with an example and follow with some features and a variant. The precise details of the ACB algorithm, however, are still unknown. The reader should also consult [Buyanovsky 94], [Fenwick 96], and [Lambert 99].

Assume that the text "`...swiss␣miss␣is␣missing...`" is part of the input stream. The method uses an LZ77-type sliding buffer where we assume that the first seven symbols have already been input and are now in the search buffer. The look-ahead buffer starts with the string "`iss␣is...`".

`...swiss␣m`	`iss␣is␣missing...`	...← text to be read.

While text is input and encoded, all contexts are placed in a dictionary, each with the text following it. This text is called the *content* string of the context. The six entries (context|content) that correspond to the seven rightmost symbols in the search buffer are shown in Table 11.7a. The dictionary is then sorted by contexts, *from right to left*, as shown in Table 11.7b. Both the contexts and contents are unbounded. They are assumed to be as long as possible but may include only symbols from the search buffer since the look-ahead buffer is unknown to the decoder. This way both encoder and decoder can create and update their dictionaries in lockstep.

(From the Internet.) ACB - Associative coder of Buyanovsky. The usage of the ACB algorithm ensures a record compression coefficient.

11.3.1 The Encoder

The current context `...swiss␣m` is matched by the encoder to the dictionary entries. The best match is between entries 2 and 3 (matching is from right to left). We arbitrarily assume that the match algorithm selects entry 2 (obviously, the algorithm does not make arbitrary decisions and is the same for encoder and decoder). The current content `iss...` is also matched to the dictionary. The best content match is to entry 6. The four symbols `iss␣` match, so the output is (6 − 2,4,`i`), a triplet that compresses the *five* symbols `iss␣i`. The first element of the triplet is the distance `d` between the best content and best context matches (it can be negative). The second element is the number `l` of

(a)		(b)
...s|wiss␣m	1	...swiss␣|m
...sw|iss␣m	2	...swi|ss␣m
...swi|ss␣m	3	...s|wiss␣m
...swis|s␣m	4	...swis|s␣m
...swiss|␣m	5	...swiss|␣m
...swiss␣|m	6	...sw|iss␣m

Table 11.7: Six Contexts and Contents.

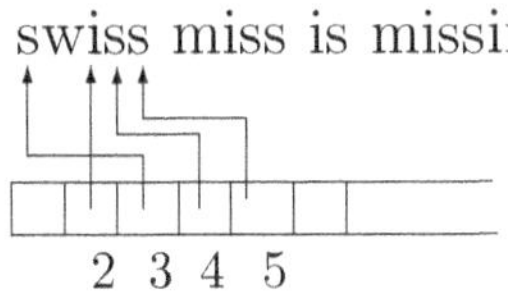

Figure 11.8: Dictionary Organization.

(a)		(b)
...s|wiss␣miss␣i	1	...swiss␣miss␣|i
...sw|iss␣miss␣i	2	...swiss␣|miss␣i
...swi|ss␣miss␣i	3	...swiss␣mi|ss␣i
...swis|s␣miss␣i	4	...swi|ss␣miss␣i
...swiss|␣miss␣i	5	...swiss␣m|iss␣i
...swiss␣|miss␣i	6	...s|wiss␣miss␣i
...swiss␣m|iss␣i	7	...swiss␣mis|s␣i
...swiss␣mi|ss␣i	8	...swis|s␣miss␣i
...swiss␣mis|s␣i	9	...swiss␣miss|␣i
...swiss␣miss|␣i	10	...swiss|␣miss␣i
...swiss␣miss␣|i	11	...sw|iss␣miss␣i

Table 11.9: Eleven Contexts and Their Contents.

symbols matched (hopefully large, but could also be zero). The third element is the first unmatched symbol in the look-ahead buffer (in the spirit of LZ77). The five compressed symbols are appended to the "content" fields of *all* the dictionary entries (Table 11.9a) and are also shifted into the search buffer. These symbols also cause five entries to be added to the dictionary, which is shown, re-sorted, in Table 11.9b.

The new sliding buffer is

`...swiss␣miss␣i|s␣missing...|...` ← text to be read.

The best context match is between entries 2 and 3 (we arbitrarily assume that the match algorithm selects entry 3). The best content match is entry 8. The six symbols `s␣miss` match, so the output is $(8-3,6,\texttt{i})$, a triplet that compresses seven symbols. The seven symbols are appended to the "content" field of every dictionary entry and are also shifted into the search buffer. Seven new entries are added to the dictionary, which is shown in Table 11.10a (unsorted) and 11.10b (sorted).

(a) context	(a) content		(b) context	(b) content
...s	wiss␣miss␣is␣missi	1	...swiss␣miss␣is␣	missi
...sw	iss␣miss␣is␣missi	2	...swiss␣miss␣	is␣missi
...swi	ss␣miss␣is␣missi	3	...swiss␣	miss␣is␣missi
...swis	s␣miss␣is␣missi	4	...swiss␣miss␣i	s␣missi
...swiss	␣miss␣is␣missi	5	...swiss␣miss␣is␣mi	ssi
...swiss␣	miss␣is␣missi	6	...swiss␣mi	ss␣is␣missi
...swiss␣m	iss␣is␣missi	7	...swi	ss␣miss␣is␣missi
...swiss␣mi	ss␣is␣missi	8	...swiss␣miss␣is␣m	issi
...swiss␣mis	s␣is␣missi	9	...swiss␣m	iss␣is␣missi
...swiss␣miss	␣is␣missi	10	...s	wiss␣miss␣is␣missi
...swiss␣miss␣	is␣missi	11	...swiss␣miss␣is	␣missi
...swiss␣miss␣i	s␣missi	12	...swiss␣miss␣is␣mis	si
...swiss␣miss␣is	␣missi	13	...swiss␣mis	s␣is␣missi
...swiss␣miss␣is␣	missi	14	...swis	s␣miss␣is␣missi
...swiss␣miss␣is␣m	issi	15	..swiss␣miss␣is␣miss	i
...swiss␣miss␣is␣mi	ssi	16	...swiss␣miss	␣is␣missi
...swiss␣miss␣is␣mis	si	17	...swiss	␣miss␣is␣missi
..swiss␣miss␣is␣miss	i	18	...sw	iss␣miss␣is␣missi

(a) (b)

Table 11.10: Eighteen Contexts and Their Contents.

The new sliding buffer is

`...swiss␣miss␣is␣missi|ng...|` ...← text to be read.

(Notice that each sorted dictionary is a permutation of the text symbols in the search buffer. This feature of ACB resembles the Burrows-Wheeler method, Section 11.1.)

The best context match is now entries 6 or 7 (we assume that 6 is selected), but there is no content match, since no content starts with an `n`. No symbols match, so the output is (0,0,`n`), a triplet that compresses the single symbol `n` (it actually generates expansion). This symbol should now be added to the dictionary and also shifted into the search buffer. (End of example.)

◇ **Exercise 11.5:** Why does this triplet have a first element of zero?

11.3.2 The Decoder

The ACB decoder builds and updates the dictionary in lockstep with the encoder. At each step, the encoder and decoder have the same dictionary (same contexts and contents). The difference between them is that the decoder does not have the data in the look-ahead buffer. The decoder does have the data in the search buffer, though, and uses it to find the best context match at, say, dictionary entry `t`. This is done before the decoder inputs anything. It then inputs a triplet (`d`,`l`,`x`) and adds the distance `d` to `t` to find the best content match `c`. The decoder then simply copies `l` symbols from the content part of entry `c`, appends symbol `x`, and outputs the resulting string to the decompressed stream. This string is also used to update the dictionary.

Notice that the content part of entry `c` may have fewer than `l` symbols. In this case, the decoding becomes somewhat more complicated and resembles the LZ77 example (from Section 6.3)

`···|alf␣eastman␣easily␣yells␣A|AAAAAAAAAA|AAAAAH...`

(The authors are indebted to Donna Klaasen for pointing this out.)

A modified version of ACB writes pairs (distance, match length) on the compressed stream instead of triplets. When the match length `l` is zero, the raw symbol code (typically ASCII or 8 bits) is written, instead of a pair. Each output, a pair or raw code, must now be preceded by a flag indicating its type.

The dictionary may be organized as a list of pointers to the search buffer. Figure 11.8 shows how dictionary entry 4 points to the second `s` of `swiss`. Following this pointer, it is easy to locate both the context of entry 4 (the search buffer to the left of the pointer, the past text) and its content (that part of the search buffer to the right of the pointer, the future text).

Part of the excellent performance of ACB is attributed to the way it encodes the distances `d` and match lengths `l`, which are its main output. Unfortunately, the details of this are unknown.

It is clear that ACB is somewhat related to both LZ77 and LZ78. What is not immediately obvious is that ACB is also related to the symbol-ranking method (Section 11.2). The distance `d` between the best-content and best-context entries can be regarded a measure of ranking. In this sense ACB is a *phrase-ranking* compression method.

11.3.3 A Variation

Here is a variation of the basic ACB method that is slower, requiring an extra sort for each match, but is more efficient. We assume the string

`|...your␣swiss␣mis|s␣is␣mistress...|...` ← text to be read.

in the search and look-ahead buffers. We denote this string by S. Part of the current dictionary (sorted by context, as usual) is shown in Table 11.11a, where the first eight and the last five entries are from the current search buffer `your swiss mis`, and the middle ten entries are assumed to be from older data.

All dictionary entries whose context fields agree with the search buffer by at least `k` symbols—where `k` is a parameter, set to 9 in our example—are selected and become the *associative list*, shown in Table 11.11b. Notice that these entries agree with the search buffer by ten symbols, but we assume that `k` has been set to 9. All the entries in the associative list have identical, `k`-symbol contexts and represent dictionary entries with contexts similar to the search buffer (hence the name "associative").

The associative list is now sorted in ascending order by the **contents**, producing Table 11.12a. It is now obvious that S can be placed between entries 4 and 5 of the

```
        ...your␣|swiss␣mis
  ...your␣swiss␣|mis
...your␣swiss␣mi|s
     ...your␣swi|ss␣mis
 ...your␣swiss␣m|is
           ...yo|ur␣swiss␣mis
         ...your|␣swiss␣mis
       ...your␣s|wiss␣mis
    ...young␣mis|creant...
  ...unusual␣mis|fortune...
    ...plain␣mis|ery...
 ...no␣swiss␣mis|spelled␣it␣so..
 ...no␣swiss␣mis|s␣is␣mistaken..
 ...or␣swiss␣mis|read␣it␣to...
..your␣swiss␣mis|s␣is␣missing...
...his␣swiss␣mis|s␣is␣here...
 ...my␣swiss␣mis|s␣is␣trouble...
   ...always␣mis|placed␣it...
    ...your␣swis|s␣mis
   ...your␣swiss|␣mis
          ...you|r␣swiss␣mis
      ...your␣sw|iss␣mis
            ...y|our␣swiss␣mis
```

(a)

```
swiss␣mis|spelled␣it␣so.
swiss␣mis|s␣is␣mistaken.
swiss␣mis|read␣it␣to...
swiss␣mis|s␣is␣missing..
swiss␣mis|s␣is␣here...
swiss␣mis|s␣is␣trouble..
```

(b)

Table 11.11: (a) Sorted Dictionary. (b) Associative List.

```
1 swiss␣mis|read␣it␣to...
2 swiss␣mis|s␣is␣here...
3 swiss␣mis|s␣is␣missing...
4 swiss␣mis|s␣is␣mistaken..
5 swiss␣mis|s␣is␣trouble...
6 swiss␣mis|spelled␣it␣so..
```

(a)

```
4. swiss mis|s is mistaken..
S. swiss mis|s is mistress..
5. swiss mis|s is trouble...
```

(b)

Table 11.12: (a) Sorted Associative List. (b) Three Lines.

```
4. xx...x0zz...z0A
S. xx...x0zz...z1B
5. xx...x1CC...
```

(a)

```
4. xx...x0CC...
S. xx...x1zz...z0B
5. xx...x1zz...z1A
```

(b)

```
4. xx...x0CC...
S. xx...x1zz...z1B
5. xx...x1zz...z0A
```

(c)

Table 11.13: (a, b) Two Possibilities, and (c) One Impossibility, of Three Lines.

sorted list (Table 11.12b).

Since each of these three lines is sorted, we can temporarily forget that they consist of characters, and simply consider them three sorted bit-strings that can be written as in Table 11.13a. The `xx...x` bits are the part were all three lines agree (the string `swiss␣mis|s␣is␣`), and the `zz...z` bits are a further match between entry 4 and the look-ahead buffer (the string `mist`). All that the encoder has to output is the index 4, the underlined bit (which we denote by `b` and which may, of course, be a zero), and the length `l` of the `zz...z` string. The encoder's output is thus the triplet (4,`b`,`l`).

In our example S agrees best with the entry preceding it. In some cases it may best agree with the entry following it, as in Table 11.13b (where bit `b` is shown as zero).

◇ **Exercise 11.6:** Show why the configuration of Table 11.13c is impossible.

The decoder maintains the dictionary in lockstep with the encoder, so it can create the same associative list, sort it, and use the identical parts (the intersection) of entries 4 and 5 to identify the `xx...x` string. It then uses `l` to identify the `zz...z` part in entry 4 and generates the bit-string `xx...x`$\tilde{\mathtt{b}}$`zz...zb` (where $\tilde{\mathtt{b}}$ is the complement of `b`) as the decompressed output of the triplet (4,`b`,`l`).

This variant can be further improved (producing better but slower compression) if instead of `l`, the encoder generates the number `q` of $\tilde{\mathtt{b}}$ bits in the `zz...z` part. This improves compression since $\mathtt{q} \le \mathtt{l}$. The decoder then starts copying bits from the `zz...z` part of entry 4 until it finds the $(\mathtt{q}+1)$st occurrence of $\tilde{\mathtt{b}}$, which it ignores. Example: if $\mathtt{b} = 1$ and the `zz...z` part is `01011110001011` (preceded by $\tilde{\mathtt{b}} = 0$ and followed by $\mathtt{b} = 1$) then $\mathtt{q} = 6$. The three lines are shown in Table 11.14. It is easy to see how the decoder can create the 14-bit `zz...z` part by copying bits from entry 4 until it finds the seventh 0, which it ignores. The encoder's output is thus the (encoded) triplet $(4, 1, 6)$ instead of $(4, 1, 14)$. Writing the value 6 (encoded) instead of 14 on the compressed stream improves the overall compression performance somewhat.

```
         zz..........z
4. xx...x0|01011110001011|0A
S. xx...x0|01011110001011|1B
5. xx...x1 CC...
```

Table 11.14: An Example.

Another possible improvement is to delete any identical entries in the sorted associative list. This technique may be called *phrase exclusion*, in analogy with the exclusion techniques of PPM and the symbol-ranking method. In our example, Table 11.12a, there are no identical entries, but had there been any, exclusion would have reduced the number of entries to fewer than 6.

◇ **Exercise 11.7:** How would this improve compression?

The main strength of ACB stems from the way it operates. It selects dictionary entries with contexts that are similar to the current context (the search buffer), then sorts the selected entries by content and selects the best content match. This is slow and also requires a huge dictionary (a small dictionary would not provide good matches) but results in excellent context-based compression without the need for escape symbols or any other "artificial" device.

11.3.4 Context Files

An interesting feature of ACB is its ability to create and use *context files*. When a file `abc.ext` is compressed, the user may specify the creation of a context file called, for example, `abc.ctx`. This file contains the final dictionary generated during the compression of `abc.ext`. The user may later compress another file `lmn.xyz` asking ACB to use `abc.ctx` as a context file. File `lmn.xyz` will be compressed using the dictionary of `abc.ext`. Following this, ACB will replace the contents of `abc.ctx`. Instead of the original dictionary, it will now contain the dictionary of `lmn.xyz` (which was not used for the actual compression of `lmn.xyz`). If the user wants to keep the original contents of `abc.ctx`, its attributes can be set to "read only." Context files can be very useful, as the following examples illustrate.

1. A writer emails a large manuscript to an editor. Because of its size, the manuscript file should be sent compressed. The first time this is done, the writer asks ACB to create a context file, then emails both the compressed manuscript and the context file to the editor. Two files need be emailed, so compression doesn't do much good this first time.

The editor decompresses the manuscript using the context file, reads it, and responds with proposed modifications to the manuscript. The writer modifies the manuscript, compresses it again with the same context file, and emails it, this time without the context file. The writer's context file has now been updated, so the writer cannot use it to decompress what he has just emailed (but then he doesn't need to). The editor still has the original context file, so he can decompress the second manuscript version, during which process ACB creates a new context file for the editor's use next time.

2. The complete collection of detective stories by a famous author should be compressed and saved as an archive. Since all the files are detective stories and are all by the same author, it makes sense to assume that they feature similar writing styles and therefore similar contexts. One story is selected to serve as a "training" file. It is compressed and a context file created. This context file is permanently saved and is used to compress and decompress all the other files in the archive.

3. A shareware author writes an application `abc.exe` that is used (and paid for) by many people. The author decides to make version 2 available. He starts by compressing the old version while creating a context file `abc.ctx`. The resulting compressed file is not needed and is immediately deleted. The author then uses `abc.ctx` as a context file to compress his version 2, and then deletes `abc.ctx`. The result is a compressed (i.e., small) file, containing version 2, which is placed on the internet, to be downloaded by users of version 1. Anyone who has version 1 can download the result and decompress it. All they need is to compress their version 1 in order to obtain a context file, then use that context file to decompress what has been downloaded.

> This algorithm ... is simple for software and hardware implementations.
>
> —George Buyanovsky

11.4 Sort-Based Context Similarity

The idea of context similarity is a "relative" of the symbol ranking method of Section 11.2 and of the Burrows-Wheeler method (Section 11.1). In contrast to the Burrows-Wheeler method, the context similarity method of this section is adaptive.

The method uses context similarity to sort previously seen contexts by reverse lexicographic order. Based on the sorted sequence of contexts, a *rank* is assigned to the next symbol. The ranks are written on the compressed stream and are later used by the decoder to reconstruct the original data. The compressed stream also includes each of the distinct input symbols in raw format, and this data is also used by the decoder.

The Encoder: The encoder reads the input symbol by symbol and maintains a sorted list of (context, symbol) pairs. When the next symbol is input, the encoder inserts a new pair into the proper place in the list, and uses the list to assign a rank to the symbol. The rank is written by the encoder on the compressed stream and is sometimes followed by the symbol itself in raw format. The operation of the encoder is best illustrated by an example. Suppose that the string bacacaba has been input so far, and the next symbol (still unread by the encoder) is denoted by x. The current list is shown in Table 11.15 where λ stands for the empty string.

#	context	symbol
0	λ	b
1	ba	c
2	bacacaba	x
3	baca	c
4	bacaca	b
5	b	a
6	bacacab	a
7	bac	a
8	bacac	a

Table 11.15: The Sorted List for bacacaba.

Each of the nine entries in the list consists of a context and the symbol that followed the context in the input stream (except entry 2, where the input is still unknown). The list is sorted by contexts, but in reverse order. The empty string is assumed, by definition, to be less than any other string. It is followed by all the contexts that end with an "a" (there are four of them), and they are sorted according to the second symbol from the right, then the third symbol, and so on. These are followed by all the contexts that end with "b", then the ones that end with "c", and so on. The current context bacacaba happens to be number 2 in this list. Once the decoder has decoded the first eight symbols, it will have the same nine-entry list available.

The encoder now ranks the contexts in the list according to how similar they are to context 2. It is clear that context 1 is the most similar to 2, since they share two symbols. Thus, the ranking starts with context 1. Context 1 is then compared to the remaining seven contexts 0 and 3–8. This context ends with an "a", so is similar to contexts 3 and 4. We select context 3 as the most similar to 1, since it is shorter than

4. This rule of selecting the shortest context is arbitrary. It simply guarantees that the decoder will be able to construct the same ranking. The ranking so far is $1 \to 3$. Context 3 is now compared to the remaining six contexts. It is clear that it is most similar to context 4, since they share the last symbol. The ranking so far is therefore $1 \to 3 \to 4$. Context 4 is now compared to the remaining five contexts. These contexts do not share any suffixes with 4, so the shortest one, context 0, is selected as the most similar. The ranking so far is $1 \to 3 \to 4 \to 0$. Context 0 is now compared to the remaining four contexts. It does not share any suffix with them, so the shortest of the four, context 5, is selected. The ranking so far is $1 \to 3 \to 4 \to 0 \to 5$.

◇ **Exercise 11.8:** Continue this process.

As the answer to this exercise shows, the final ranking of the contexts is

$$\begin{array}{cccccccc} 1 \to & 3 \to & 4 \to & 0 \to & 5 \to & 6 \to & 7 \to & 8\ . \\ \text{c} & \text{c} & \text{b} & \text{b} & \text{a} & \text{a} & \text{a} & \text{a} \end{array} \qquad (11.2)$$

This ranking of contexts is now used by the encoder to assign a rank to the next symbol x. The encoder inputs x and compares it, from left to right, to the symbols shown in Equation (11.2). The rank of x is one more than the number of distinct symbols that are encountered in the comparison. Thus, if x is the symbol "c", it is found immediately in Equation (11.2), there are no distinct symbols, and "c" is assigned a rank of 1. If x is "b", the encoder encounters only one distinct symbol, namely "c", before it gets to "b", so the rank of "b" becomes 2. If x is "a", its rank becomes 3, and if x is a different symbol, its rank is 4 [one more than the number of distinct symbols in Equation (11.2)]. The encoder writes the rank on the compressed stream, and if the rank is 4, the encoder also writes the actual symbol in raw format, following the rank.

We now show the first few steps in encoding the input string bacacaba. The first symbol "b" is written on the output in raw format. It is not assigned any rank. The encoder (and also the decoder, in lockstep) constructs the one-entry table (λ b). The second symbol, "a", is input. Its context is "b", and the encoder inserts entry (b a) into the list. The new list is shown in Table 11.16a with x denoting the new input symbol, since this symbol is not yet known to the decoder. The next five steps are summarized in Table 11.16b–f.

(a)

0	λ	b
1	b	x

(b)

0	λ	b
1	ba	x
2	b	a

(c)

0	λ	b
1	ba	c
2	b	a
3	bac	x

(d)

0	λ	b
1	ba	c
2	baca	x
3	b	a
4	bac	a

(e)

0	λ	b
1	ba	c
2	baca	c
3	b	a
4	bac	a
5	bacac	x

(f)

0	λ	b
1	ba	c
2	baca	c
3	bacaca	x
4	b	a
5	bac	a
6	bacac	a

Table 11.16: Constructing the Sorted Lists for bacacaba.

Equation (11.3) lists the different context rankings.

$$
\begin{array}{lll}
0\,, & 0\rightarrow 2\,, & 0\rightarrow 2\rightarrow 1\,,\\
\text{b} & \text{b}\quad\text{a} & \text{b}\quad\text{a}\quad\text{c}\\
1\rightarrow 0\rightarrow 3\rightarrow 4\,, & \multicolumn{2}{l}{4\rightarrow 0\rightarrow 3\rightarrow 1\rightarrow 2\,,}\\
\text{c}\quad\text{a}\quad\text{a}\quad\text{a} & \multicolumn{2}{l}{\text{a}\quad\text{b}\quad\text{a}\quad\text{c}\quad\text{c}}\\
\multicolumn{3}{l}{2\rightarrow 1\rightarrow 0\rightarrow 4\rightarrow 5\rightarrow 6\,.}\\
\multicolumn{3}{l}{\text{c}\quad\text{c}\quad\text{b}\quad\text{a}\quad\text{a}\quad\text{a}}
\end{array}
\qquad (11.3)
$$

With this information, it is easy to manually construct the compressed stream. The first symbol, "b", is output in raw format. The second symbol, "a", is assigned rank 1 and is also output following its rank. The first "c" is assigned rank 2 and is also output (each distinct input symbol is output raw, following its rank, the first time it is read and processed). The second "a" is assigned rank 2, because there is one distinct symbol ("b") preceding it in the list of context ranking. The second "c" is assigned rank 1.

◇ **Exercise 11.9:** Complete the output stream.

◇ **Exercise 11.10:** Practice your knowledge of the encoder on the short input string `ubladiu`. Show the sorted contexts and the context ranking after each symbol is input. Also show the output produced by the encoder.

The Decoder: Once the operation of the encoder is understood, it is clear that the decoder can mimic the encoder. It can construct and maintain the table of sorted contexts as it reads the compressed stream, and use the table to regenerate the original data. The decoding algorithm is shown in Figure 11.17.

```
Input the first item. This is a raw symbol. Output it.
while not end-of-file
 Input the next item. This is the rank of a symbol.
 If this rank is > the total number of distinct symbols seen so far
   then Input the next item. This is a raw symbol. Output it.
   else Translate the rank into a symbol using the current
        context ranking. Output this symbol.
 endif
 The string that has been output so far is the current context.
  Insert it into the table of sorted contexts.
endwhile
```

Figure 11.17: The Decoding Algorithm.

The Data Structure: Early versions of the context sorting method used a binary decision tree to store the various contexts. This was slow, so the length of the contexts had to be limited to eight symbols. The new version, described in [Yokoo 99a], uses a *prefix list* as the data structure, and is fast enough to allow contexts of unlimited length. We denote the input symbols by s_i. Let $S[1\ldots n]$ be the string $s_1s_2\ldots s_n$ of n symbols. We use the notation $S[i\ldots j]$ to denote the substring $s_i\ldots s_j$. If $i>j$, then $S[i\ldots j]$ is the empty string λ.

As an example, consider the 9-symbol string $S[1\ldots 9]=$ `yabrecabr`. Table 11.18a lists the ten prefixes of this string (including the empty prefix) sorted in reverse lexicographic order. Table 11.18b considers the prefixes, contexts and lists the ten (context, symbol) pairs. This table illustrates how to insert the next prefix, which consists of the next input symbol s_{10} appended to the current context `yabrecabr`. If s_{10} is not any of the rightmost symbols of the prefixes (i.e., if it is not any of `abcery`), then s_{10} determines the position of the next prefix. For example, if s_{10} is "x", then prefix `yabrecabrx` should be inserted between `yabr` and `y`. If, on the other hand, s_{10} is one of `abcery`, we compare `yabrecabr`s_{10} to the prefixes that precede it and follow it in the table, until we find the first match.

(a)		(b)	
$S[1\ldots 0]=$	λ	λ	`y`
$S[1\ldots 7]=$	`yabreca`	`yabreca`	`b`
$S[1\ldots 2]=$	`ya`	`ya`	`b`
$S[1\ldots 8]=$	`yabrecab`	`yabrecab`	`r`
$S[1\ldots 3]=$	`yab`	`yab`	`r`
$S[1\ldots 6]=$	`yabrec`	`yabrec`	`a`
$S[1\ldots 5]=$	`yabre`	`yabre`	`c`
$S[1\ldots 9]=$	`yabrecabr`	`yabrecabr`	$s_{10}\uparrow\downarrow$
$S[1\ldots 4]=$	`yabr`	`yabr`	`e`
$S[1\ldots 1]=$	`y`	`y`	`a`

Table 11.18: (a) Sorted List for `yabrecabr`. (b) Inserting the Next Prefix.

For example, if s_{10} is "e", then comparing `yabrecabre` with the prefixes that follow it (`yabr` and `y`) will not find a match, but comparing it with the preceding prefixes will match it with `yabre` in one step. In such a case (a match found while searching up), the rule is that `yabrecabre` should become the predecessor of `yabre`. Similarly, if s_{10} is "a", then comparing `yabrecabra` with the preceding prefixes will find a match at `ya`, so `yabrecabra` should become the predecessor of `ya`. If s_{10} is "r", then comparing `yabrecabrr` with the prefixes following it will match it with `yabr`, and the rule in this case (a match found while searching down) is that `yabrecabrr` should become the successor of `yabr`.

Once this is grasped, the prefix list data structure is easy to understand. It is a doubly-linked list where each node is associated with an input prefix $S[1\ldots i]$ and contains the integer i and three pointers. Two pointers (*pred* and *succ*) point to the predecessor and successor nodes, and the third one (*next*) points to the node associated with prefix $S[1\ldots i+1]$. Figure 11.19 shows the pointers of a general node representing substring $S[1\ldots \mathrm{P}.index]$. If a node corresponds to the entire input string $S[1\ldots n]$, then its *next* field is set to the null pointer *nil*. The prefix list is initialized to a special node H that represents the empty string. Some list operations are simplified if the list ends with another special node T. Figure 11.20 shows the prefix list for `yabrecabr`.

Here is how the next prefix is inserted into the prefix list. We assume that the list already contains all the prefixes of string $S[1\ldots i]$ and that the next prefix $S[1\ldots i+1]$ should now be inserted. If the newly input symbol s_{i+1} precedes or succeeds all the symbols seen so far, then the node representing $S[1\ldots i+1]$ should be inserted at the start of the list (i.e., to the right of special node H) or at the end of the list (i.e., to the left of special node T), respectively. Otherwise, if s_{i+1} is not included in $S[1\ldots i]$, then there is a unique position Q that satisfies

$$S[Q.index] < s_{i+1} < S[Q.succ.index]. \tag{11.4}$$

The new node for $S[1\ldots i+1]$ should be inserted between the two nodes pointed to by Q and $Q.succ$.

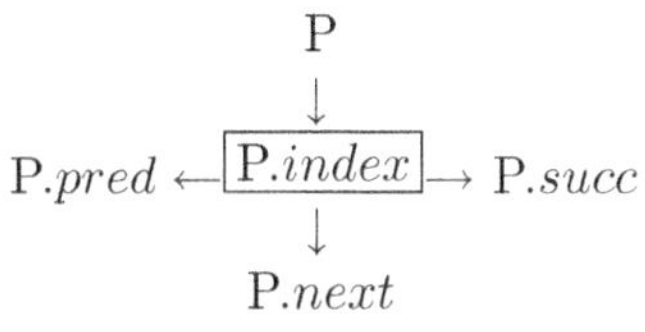

Figure 11.19: Node Representing $S[1\ldots \mathrm{P}.index]$.

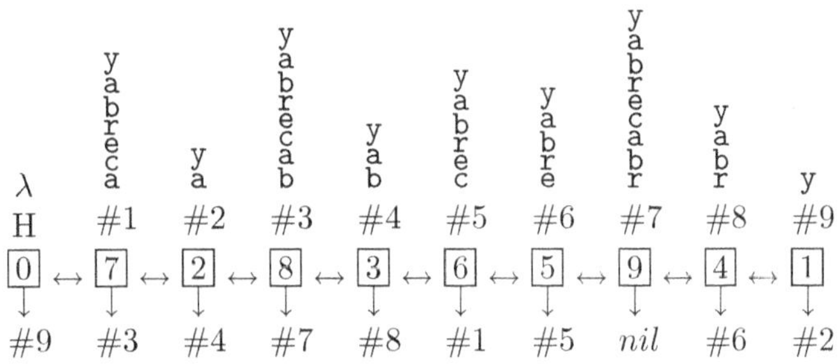

Figure 11.20: Prefix List for `yabrecabr`.

If symbol s_{i+1} has already appeared in $S[1\ldots i]$, then the inequalities in Equation (11.4) may become equalities. In this case, the immediate predecessor or successor of $S[1\ldots i+1]$ has the same last symbol as s_{i+1}. If this is true for the immediate successor (i.e., if the immediate successor $S[1\ldots j+1]$ of $S[1\ldots i+1]$ satisfies $s_{j+1} = s_{i+1}$), then $S[1\ldots j]$ precedes $S[1\ldots i]$. The node corresponding to $S[1\ldots j]$ should be the first node that satisfies $s_{j+1} = s_{i+1}$ in traversing the list from the current node to the start. We can test whether the following symbol matches s_{i+1} by following the pointer *next*.

Conversely, if the last symbol s_{j+1} of the immediate successor $S[1\ldots j+1]$ of $S[1\ldots i+1]$ equals s_{i+1}, then the node for $S[1\ldots j]$ should be the first one satisfying $s_{j+1} = s_{i+1}$ when the list is traversed from the current node to the last node. In either case we start from the current node and search forward and backward, looking for the node for $S[1\ldots j]$ by comparing the last symbols s_{j+1} with s_{i+1}. Once the node for $S[1\ldots j]$ has been located, the node for $S[1\ldots j+1]$ can be reached in one step by following the *next* pointer. The new node for $S[1\ldots i+1]$ should be inserted adjacent to the node for $S[1\ldots j+1]$.

[Yokoo 99a] has time complexity analysis and details about this structure.

The context sorting method was first published in [Yokoo 96] with analysis and evaluation added in [Yokoo 97]. It was further developed and improved by its developer, Hidetoshi Yokoo. It has been communicated to the authors in [Yokoo 99b].

11.5 Sparse Strings

Regardless of what the input data represents—text, binary, images, or anything else—we can think of the input stream as a string of bits. If most of the bits are zeros, the string is *sparse*. Sparse strings can be compressed very efficiently, and this section describes methods developed specifically for this task. Before getting to the individual methods it may be useful to convince the reader that sparse strings are not a theoretical concept but do occur commonly in practice. Here are some examples.

1. A drawing. Imagine a drawing, technical or artistic, done with a black pen on white paper. If the drawing is not very complex, most of it remains white. When such a drawing is scanned and digitized, most of the resulting pixels are white, and the percentage of black ones is small. The resulting bitmap is an example of a sparse string.
2. A bitmap index for a large data base. Imagine a large data base of text documents. A bitmap for such a data base is a set of bitstrings (or bitvectors) that makes it easy to identify all the documents where a given word w appears. To implement a bitmap, we first have to prepare a list of all the distinct words w_j in all the documents. Suppose that there are W such words. The next step is to go over each document d_i and prepare a bit-string D_i that is W bits long, containing a 1 in position j if word w_j appears in document d_i. The bitmap is the set of all those bit-strings. Depending on the database, such bit-strings may be sparse.

(Indexing large databases is an important operation, since a computerized database should be easy to search. The traditional method of indexing is to prepare a *concordance*. Originally, the word concordance referred to any comprehensive index of the Bible, but today there are concordances for the collected works of Shakespeare, Wordsworth, and many others. A computerized concordance is called an *inverted file*. Such a file includes one entry for each term in the documents constituting the database. An entry is a list of pointers to all the occurrences of the term, similar to an index of a book. A pointer may be the number of a chapter, of a section, a page, a page-line pair, of a sentence, or even of a word. An inverted file where pointers point to individual words is considered *fine grained*. An inverted file where they point to, say, a chapter is considered *coarse grained*. A fine-grained inverted file may seem preferable, but it must use large pointers, and as a result, it may turn out to be so large that it may have to be stored in compressed form.)

3. Sparse strings have also been mentioned in Section 7.10.3, in connection with JPEG.

The methods described here (except prefix compression, Section 11.5.5) are due to [Fraenkel and Klein 85]. Section 3.24 discusses the Golomb codes and illustrates their application to sparse strings.

11.5.1 OR-ing Bits

This method starts with a sparse string L_1 of size n_1 bits. In the first step, L_1 is divided into k substrings of equal size. In each substring all bits are logically OR-ed, and the results (one bit per substring) become string L_2, which will be compressed in step 2. All zero substrings of L_1 are now deleted. Here is an example of a sparse, 64-bit string L_1, which we divide into 16 substrings of size 4 each:

$$L_1 = 0000|0000|0000|0100|0000|0000|0000|1000|0000$$
$$|0000|0000|0000|0010|0000|0000|0000.$$

After ORing each 4-bit substring we get the 16-bit string $L_2 = 0001|0001|0000|1000$.

In step 2, the same process is applied to L_2, and the result is the 4-bit string $L_3 = 1101$, which is short enough so no more compression steps are needed. After deleting all zero substrings in L_1 and L_2, we end up with the three short strings

$$L_1 = 0100|1000|0010, \qquad L_2 = 0001|0001|1000, \qquad L_3 = 1101.$$

The output stream consists of seven 4-bit substrings instead of the original 16! (A few more numbers are needed, to indicate how long each substring is.)

The decoder works differently (this is an asymmetric compression method). It starts with L_3 and considers each of its 1-bits a pointer to a substring of L_2 and each of its 0-bits a pointer to a substring of all zeros that is not stored in L_2. This way, string L_2 can be reconstructed from L_3, and string L_1, in turn, can be reconstructed from L_2. Figure 11.21 illustrates this process. The substrings shown in square brackets are the ones not contained in the compressed stream.

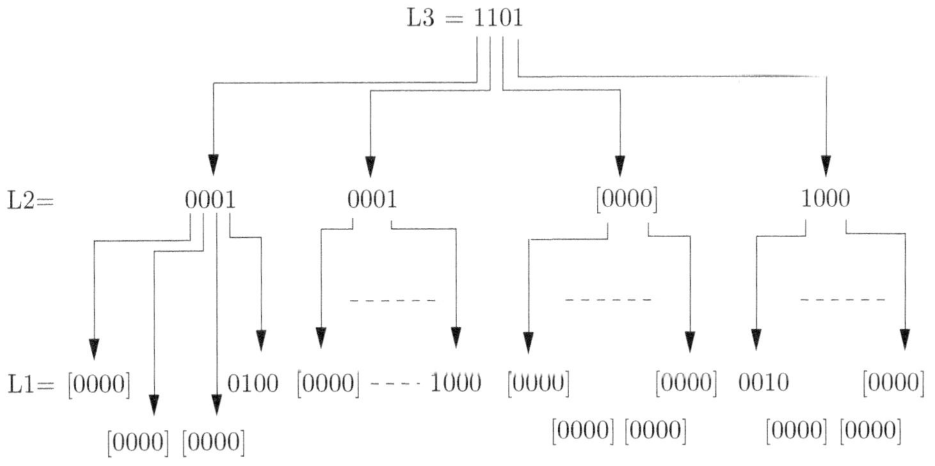

Figure 11.21: Reconstructing L_1 from L_3.

◇ **Exercise 11.11:** This method becomes highly inefficient for strings that are not sparse, and may easily result in expansion. Analyze the worst case, where every group of L_1 is nonzero.

11.5.2 Variable-Length Codes

We start with an input stream that is a sparse string L of n bits. We divide it into groups of l bits each, and assign each group a variable-length code. Since a group of l bits can have one of 2^l values, we need 2^l codes. Since L is sparse, most groups will consist of l zeros, implying that the variable-length code assigned to the group of l zeros (the zero group) should be the shortest (perhaps just one bit). The other $2^l - 1$ variable-length codes can be assigned arbitrarily, or according to the frequencies of occurrence of the groups. The latter choice requires an extra pass over the input stream to compute the frequencies. In the ideal case, where all the groups are zeros, and each is coded as one bit, the output stream will consist of n/l bits, yielding a compression ratio of $1/l$. This shows that in principle, the compression ratio can be improved by increasing l, but in practice, large l means many codes, which, in turn, increases the code size and decreases the compression ratio for an "average" string L.

A better approach can be developed once we realize that a sparse input stream must contain **runs** of zero groups. A run of zero groups may consist of one, two, or up to n/l such groups. It is possible to assign variable-length codes to the runs of zero groups, as well as to the nonzero groups, and Table 11.22 illustrates this approach for the case of 16 groups. The trouble is that there are $2^l - 1$ nonzero groups and n/l possible run lengths of zero groups. Normally, n is large and l is small, so n/l is large. If we increase l, then n/l gets smaller but $2^l - 1$ gets bigger. Thus, we always end up with many codes, which implies long codes.

Size of run length	Run of zeros	Nonzero group
1	0000	1 0001
2	0000 0000	2 0010
3	0000 0000 0000	3 0011
⋮	⋮ ⋮	⋮
16	0000 0000 ... 0000	15 1111
	(a)	(b)

Table 11.22: (a) n/l Run lengths. (b) $2^l - 1$ Nonzero Groups.

A more promising approach is to divide the run lengths (which are integers between 1 and n/l) into classes, assign one variable-length code C_i to each class i, and assign a two-part code to each run length. Imagine a run of r zero groups, where r happens to be in class i, and happens to be the third one in this class. When a run of r zero groups is found in the input stream, the code of r is written to the output stream. Such a code consists of two parts: The first is the class code C_i, and the second is 2, the position of r in its class (positions are numbered from zero). Experience with algorithm design and binary numbers suggests the following definition of classes: A run length of r zero groups is in class i if $2^{i-1} \le r < 2^i$, where $i = 1, 2, \ldots, \lfloor \log_2(n/l) \rfloor$. This definition implies that the position of r in its class is $m = r - 2^{i-1}$, a number that can be written in $(i-1)$ bits. Table 11.23 shows the four classes for the case $n/l = 16$ (16 groups). Notice how the numbers m are written as $(i-1)$-bit numbers, so for $i = 1$ (the first class), no m is necessary. The variable-length Huffman codes C_i shown in the table are for illustration purposes only and are based on the (arbitrary) assumption that the most common run lengths are 5, 6, and 7.

Run length	Code	$r - 2^{i-1}$	$i-1$	Huffman code|m
1	C_1	$1 - 2^{1-1} = 0$	0	00010
2	C_2	$2 - 2^{2-1} = 0$	1	00011|0
3	C_2	$3 - 2^{2-1} = 1$	1	0010|1
4	C_3	$4 - 2^{3-1} = 0$	2	0011|00
5	C_3	$5 - 2^{3-1} = 1$	2	010|01
6	C_3	$6 - 2^{3-1} = 2$	2	011|10
7	C_3	$7 - 2^{3-1} = 3$	2	1|11
8	C_4	$8 - 2^{4-1} = 0$	3	00001|000
9	C_4	$9 - 2^{4-1} = 1$	3	000001|001
⋮				⋮
15	C_4	$15 - 2^{4-1} = 7$	3	000000000001|111

Table 11.23: $\log_2(n/l)$ Classes of Run Lengths.

It is easy to see from the table that a run of 16 zero groups (which corresponds to an input stream of all zeros) does not belong in any of the classes. It should therefore be assigned a special variable-length code. The total number of variable-length codes required in this approach is therefore $2^l - 1$ (for the nonzero groups) plus $\lfloor \log_2(n/l) \rfloor$ (for the run lengths of zero groups) plus 1 for the special case where all the groups are zero. A typical example is a 1-megabit input stream (i.e., $n = 2^{20}$). Assuming $l = 8$, the number of codes is $2^8 - 1 + \log_2(2^{20}/8) + 1 = 256 - 1 + 17 + 1 = 273$. With $l = 4$ the number of codes is $2^4 - 1 + \log_2(2^{20}/4) + 1 = 16 - 1 + 18 + 1 = 34$; much smaller, but more codes are required to encode the same input stream.

The operation of the decoder is straightforward. It reads the next variable-length code, which represents either a nonzero group of l bits, or a run of r zero groups, or an input stream of all zeros. In the first case, the decoder creates the nonzero group. In the second case, the code tells the decoder the class number i. The decoder then reads the next $i-1$ bits to get the value of m, and computes $r = m + 2^{i-1}$ as the size of the run length of zero groups. The decoder then creates a run of r zero groups. In the third case the decoder creates a stream of n zero bits.

Example: An input stream of size $n = 64$ divided into 16 groups of $l = 4$ bits each. The number of codes needed is $2^4 - 1 + \log_2(64/4) + 1 = 16 - 1 + 4 + 1 = 20$. We arbitrarily assume that each of the 15 nonzero groups occurs with probability 1/40, and that the probability of occurrence of runs in the four classes are 6/40, 8/40, 6/40, and 4/40, respectively. The probability of occurrence of a run of 16 groups is assumed to be 1/40. Table 11.24 shows possible codes for each nonzero group and for each class of runs of zero groups. The code for 16 zero groups is 00000 (corresponds to the 16 in italics).

Nonzero groups								Classes	
1	111111	5	01111	9	00111	13	00011	C_1	110
2	111110	6	01110	10	00110	14	00010	C_2	10
3	111101	7	01101	11	00101	15	00001	C_3	010
4	111100	8	01100	12	00100	*16*	00000	C_4	1110

Table 11.24: Twenty Codes.

◇ **Exercise 11.12:** Encode the input stream

0000|0000|0000|0100|0000|0000|0000|1000|0000|0000|0000|0000|0010|0000|0000|0000

using the codes of Table 11.24.

11.5.3 Variable-Length Codes for Base 2

Classes defined as in the preceding section require a set of $(2^l-1)+\lfloor\log_2(n/l)\rfloor+1$ codes. The method is efficient but slow, since the code for a run of zero groups involves both a class code C_i and the quantity m. In this section, we look at a way to handle runs of zero groups by defining codes R_1, R_2, R_4, $R_8,\ldots$ for run lengths of $1,2,4,8,\ldots,2^i$ zero groups. The binary representation of the number 17, e.g., is 10001, so a run of 17 zero groups would be coded as R_{16} followed by R_1. Since run lengths can be from 1 to n/l, the number of codes R_i required is $1+(n/l)$, more than before. Experience shows, however, that long runs are rare, so the Huffman codes assigned to R_i should be short for small values of i, and long for large i's. In addition to the R_i codes, we still need 2^l-1 codes for the nonzero groups. In the case $n=64$, $l=4$ we need 15 codes for the nonzero groups and 7 codes for R_1 through R_{64}. An example is illustrated in Table 11.25 where all the codes for nonzero groups are 5 bits long and start with 0, while the seven R_i codes start with 1 and are variable-length.

Nonzero groups				Run lengths	
1	0 0001	9	0 1001	R_1	1 1
2	0 0010	10	0 1010	R_2	1 01
3	0 0011	11	0 1011	R_4	1 001
4	0 0100	12	0 1100	R_8	1 00011
5	0 0101	13	0 1101	R_{16}	1 00010
6	0 0110	14	0 1110	R_{32}	1 00001
7	0 0111	15	0 1111	R_{64}	1 00000
8	0 1000				

Table 11.25: Codes for Base-2 R_i.

The R_i codes don't have to correspond to powers of 2. They may be based on 3, 4, or even larger integers. Let's take a quick look at octal R_i codes (8-based). They are denoted by R_1, R_8, $R_{64}, \ldots$. To encode a run length of 17 zero groups ($= 21_8$) we need **two** copies of the code for R_8, followed by the code for R_1. The number of R_i codes is smaller, but some may have to appear several (up to 7) times.

The general rule is; Suppose that the R_i codes are based on the number B. If we identify a run of R zero groups, we first have to express R in base B, then create copies of the R_i codes according to the digits of that number. If $R = d_3d_2d_1$ in base B, then the coded output for run length R should consist of d_1 copies of R_1, followed by d_2 copies of R_2 and by d_3 copies of R_3.

◇ **Exercise 11.13:** Encode the 64-bit input stream of Exercise 11.12 using the codes of Table 11.25.

11.5.4 Fibonacci-Based Variable-Length Codes

The codes R_i used in the previous section are based on powers of 2, since any positive integer can be expressed in this base using just the digits 0 and 1. It turns out that the well-known Fibonacci numbers also have this property. Any positive integer R can be expressed as $R = b_1F_1 + b_2F_2 + b_3F_3 + b_4F_5 + \cdots$ (that is b_4F_5, not b_4F_4), where the F_i are the Fibonacci numbers $1, 2, 3, 5, 8, 13, \ldots$ and the b_i are binary digits. The Fibonacci numbers grow more slowly than the powers of 2, implying that more of them are needed to express a given run length R of zero groups. However, this representation has the interesting property that the string $b_1b_2\ldots$ does not contain any adjacent 1's ([Knuth 73], ex. 34, p. 85). If the representation of R in this base consists of d digits, at most $\lceil d/2 \rceil$ codes F_i would actually be needed to code a run length of R zero groups. As an example, the integer 33 equals the sum $1 + 3 + 8 + 21$, so it is expressed in the Fibonacci base as the 7-bit number 1010101. A run length of 33 zero groups is therefore coded, in this method, as the four codes F_1, F_3, F_8, and F_{21}. (See also Section 3.20.)

Table 11.26 is an example of Fibonacci codes for the run length of zero groups. Notice that with seven Fibonacci codes we can express only runs of up to $1 + 2 + 3 + 5 + 8 + 13 + 21 = 53$ groups. Since we want up to 64 groups, we need one more code. Table 11.26 thus has eight codes, compared to seven in Table 11.25.

Nonzero groups				Run lengths	
1	0 0001	9	0 1001	F_1	1 1
2	0 0010	10	0 1010	F_2	1 01
3	0 0011	11	0 1011	F_3	1 001
4	0 0100	12	0 1100	F_5	1 00011
5	0 0101	13	0 1101	F_8	1 00010
6	0 0110	14	0 1110	F_{13}	1 00001
7	0 0111	15	0 1111	F_{21}	1 00000
8	0 1000			F_{34}	1 000000

Table 11.26: Codes for Fibonacci-Based F_i.

◇ **Exercise 11.14:** Encode the 64-bit input stream of Exercise 11.12 using the codes of Table 11.26.

This section and the previous one suggest that any number system can be used to construct codes for the run lengths of zero groups. However, number systems based on binary digits (see Section 3.19) are preferable, since certain codes can be omitted in such a case, and no code has to be duplicated. Another possibility is to use number systems where certain combinations of digits are impossible. Here is an example, also based on Fibonacci numbers.

The well-known recurrence relation these numbers satisfy is $F_i = F_{i-1} + F_{i-2}$. It can be written

$$F_{i+2} = F_{i+1} + F_i = (F_i + F_{i-1}) + F_i = 2F_i + F_{i-1}.$$

The numbers produced by this relation can also serve as the basis for a number system that has two interesting properties: (1) Any positive integer can be expressed using just the digits 0, 1, and 2. (2) Any digit of 2 is followed by a 0.

The first few numbers produced by this relation are 1, 3, 7, 17, 41, 99, 239, 577, 1393 and 3363. It is easy to verify the following examples:

1. $7000_{10} = 2001002001$ (since $7000_{10} = 2 \times 3363 + 239 + 2 \times 17 + 1$.
2. $168_{10} = 111111$.
3. $230_{10} = 201201$.
4. $271_{10} = 1001201$.

Thus, a run of 230 zero groups can be compressed by generating two copies of the Huffman code of 99, followed by the Huffman code of 17, by two copies of the code of 7, and by the code of 1. Another possibility is to assign each of the base numbers 1, 3, 7, etc., two Huffman codes, one for two copies and the other one, for a single copy.

11.5.5 Prefix Compression

The principle of the prefix compression method [Salomon 00] is to assign an address to each bit of 1 in the sparse string, divide each address into a prefix and a suffix, then select all the 1-bits whose addresses have the same prefix and write them on the compressed file by writing the common prefix, followed by all the different suffixes. Compression will be achieved if we end up with many 1-bits having the same prefix. In order for several 1-bits to have the same prefix, their addresses should be close. However, in a long, sparse string, the 1-bits may be located far apart. Prefix compression tries to bring together 1-bits that are separated in the string, by breaking up the string into equal-size segments that are then placed one below the other, effectively creating a sparse *matrix* of bits. It is easy to see that 1-bits that are widely separated in the original string may get closer in this matrix. As an example consider a binary string of 2^{20} bits (a megabit). The maximum distance between bits in the original string is about a million, but when the string is rearranged as a matrix of dimensions $2^{10} \times 2^{10} = 1024 \times 1024$, the maximum distance between bits is only about a thousand.

Our problem is to assign addresses to matrix elements such that (1) the address of a matrix element is a single number and (2) elements that are close will be assigned addresses that do not differ by much. The usual way of referring to matrix elements is by row and column. We can create a one-number address by concatenating the row and column numbers of an element, but this is unsatisfactory. Consider, for example, the two matrix elements at positions $(1, 400)$ and $(2, 400)$. They are certainly close neighbors, but their numbers are 1,400 and 2,400, not very close!

We therefore use a different method. We think of the matrix as a digital image where each bit becomes a pixel (white for a 0 and black for a 1) and we require that this image be of size $2^n \times 2^n$ for some integer n. This normally necessitates extending the original string with 0 bits until its size becomes an even power of 2 (2^{2n}). The original size of the string should therefore be written, in raw form, on the compressed file for the use of the decompressor. If the string is sparse, the corresponding image has few black pixels. Those pixels are assigned addresses using the concept of a *quadtree.*

To understand how this is done, let's assume that an image of size $2^n \times 2^n$ is given. We divide it into four quadrants and label them $\binom{0\,1}{2\,3}$. Notice that these are 2-bit numbers. Each quadrant is subsequently divided into four subquadrants labeled in the same way. Thus, each subquadrant gets a four-bit (two-digit) number. This process continues recursively, and as the subsubquadrants get smaller, their numbers get longer. When this numbering scheme is carried down to individual pixels, the number of a pixel turns out to be $2n$ bits long. Figure 7.172, duplicated here, shows the pixel numbering in a $2^3 \times 2^3 = 8 \times 8$ image and also a simple image consisting of 18 black pixels. Each pixel number is six bits (three digits) long, and they range from 000 to 333. The original string being used to create this image is

1000010001000100001001000001111000100100010001001100000011000000

(where the six trailing zeros, or some of them, have been added to make the string size an even power of two).

000	001	010	011	100	101	110	111
002	003	012	013	102	103	112	113
020	021	030	031	120	121	130	131
022	023	032	033	122	123	132	133
200	201	210	211	300	301	310	311
202	203	212	213	302	303	312	313
220	221	230	231	320	321	330	331
222	223	232	233	322	323	332	333

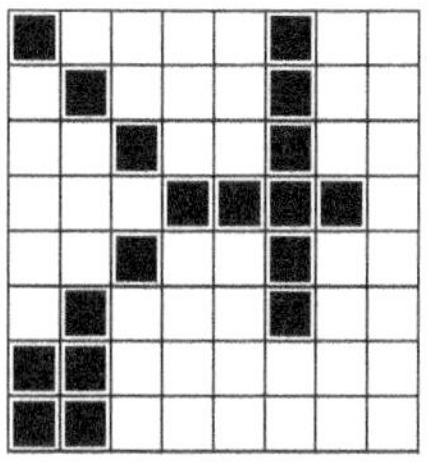

Figure 7.172 (duplicate): Example of Prefix Compression.

The first step is to use quadtree methods to figure out the three-digit id numbers of the 18 black pixels. They are 000, 101, 003, 103, 030, 121, 033, 122, 123, 132, 210, 301, 203, 303, 220, 221, 222, and 223.

The next step is to select a *prefix* value. For our example we select $P = 2$, a choice that is justified below. The code of a pixel is now divided into P prefix digits followed by $3 - P$ suffix digits. The last step goes over the sequence of black pixels and selects all the pixels with the same prefix. The first prefix is 00, so all the pixels that start with 00 are selected. They are 000 and 003. They are removed from the original sequence and are compressed by writing the token 00|1|0|3 on the output stream. The first part of this token is a prefix (00), the second part is a count (1), and the rest are the suffixes of the two pixels having prefix 00. Notice that a count of 1 implies two pixels. The count is always one less than the number of pixels being counted. Sixteen pixels now remain in the original sequence, and the first of them has prefix 10. The two pixels with this prefix, namely 101 and 103, are removed and compressed by writing the token 10|1|1|3 on the output stream. This continues until the original sequence becomes empty. The final result is the nine-token string

00|1|0|3 10|1|1|3 03|1|0|3 12|2|1|2|3 13|0|2 21|0|0 30|1|1|3 20|0|3 22|3|0|1|2|3,

or in binary,

0000010011 0100010111 0011010011 011010011011 01110010
10010000 1100010111 10000011 10101100011011

(without the spaces). Such a string can be decoded uniquely, since each token starts with a two-digit prefix, followed by a one-digit count c, followed by $(c+1)$ 1-digit suffixes. Preceding the tokens, the compressed file should contain, in raw form, the value of n, the original size of the sparse string, and the value of P that was used in the compression. Once this is understood, it becomes obvious that decompression is trivial. The decompressor reads the values of n and P, and these two numbers are all it needs to read and decode all the tokens unambiguously.

◇ **Exercise 11.15:** Can adjacent pixels have different prefixes?

Notice that our example results in expansion because our binary string is short and therefore not sparse. A sparse string has to be at least tens of thousands of bits long.

In general, the prefix is P digits (or $2P$ bits) long, and the count and each suffix are $n-P$ digits each. The maximum number of suffixes in a token is therefore 4^{n-P} and the maximum size of a token is $P+(n-P)+4^{n-P}(n-P)$ digits. Each token corresponds to a different prefix. A prefix has P digits, each between 0 and 3, so the maximum number of tokens is 4^P. Thus, the entire compressed string occupies at most

$$4^P\left[P+(n-P)+4^{n-P}(n-P)\right]=n\cdot 4^P+4^n(n-P)$$

digits. To find the optimum value of P we differentiate this expression with respect to P,

$$\frac{d}{dP}\left[n\cdot 4^P+4^n(n-P)\right]=n\cdot 4^P\ln 4-4^n,$$

and set the derivative to zero. The solution is

$$4^P=\frac{4^n}{n\cdot\ln 4},\quad \text{or } P=\log_4\left[\frac{4^n}{n\cdot\ln 4}\right]=\frac{1}{2}\log_2\left[\frac{4^n}{n\cdot\ln 4}\right].$$

For $n=3$ this yields

$$P\approx\frac{1}{2}\log_2\left[\frac{4^3}{3\times 1.386}\right]=\frac{\log_2 15.388}{2}=3.944/2=1.97.$$

This is why $P = 2$ was selected in our example. A practical compression program should contain a table with P values precalculated for all expected values of n. Table 11.27 shows such values for $n = 1, 2, \ldots, 12$.

n:	1	2	3	4	5	6	7	8	9	10	11	12
P:	0.76	1.26	1.97	2.76	3.60	4.47	5.36	6.26	7.18	8.10	9.03	9.97

Table 11.27: Dependence of P on n.

```
Clear[t]; t=Log[4]; (* natural log *)
Table[{n,N[0.5 Log[2,4^n/(n t)],3]}, {n,1,12}]//TableForm
```

Mathematica Code for Table 11.27.

This method for calculating the optimum value of P is based on the worst case. It uses the maximum number of suffixes in a token, but many tokens have just a few suffixes. It also uses the maximum number of prefixes, but in a sparse string many prefixes may not occur. Experiments indicate that for large sparse strings (corresponding to n values of 9–12), better compression is obtained if the P value selected is one less than the value listed in Table 11.27. However, for short sparse strings, the values of Table 11.27 are optimum. Selecting, for example, $P = 1$ instead of $P = 2$ for our short example (with $n = 3$) results in the four tokens

0|3|00|03|30|33 1|5|01|03|21|22|23|32 2|5|10|03|20|21|22|23 3|1|01|03,

which require a total of 96 bits, more than the 90 bits required for the choice $P = 2$.

In our example the `count` field can go up to 3, which means that an output token, whose format is `prefix|count|suffixes`, can compress at most four pixels. A better choice may be to encode the `count` field so its length can vary. Even the simple unary code might produce good results, but a better choice may be a Huffman code where small counts are assigned short codes. Such a code may be constructed based on distribution of values for the `count` field determined by several "training" sparse strings. If the `count` field is encoded, then a token may have any size. The compressor therefore has to generate the current token in a short array of bytes and write the "full" bytes on the compressed file, moving the last byte, which is normally only partly filled, to the start of the array before generating the next token. The very last byte is written to the compressed file with trailing zeros.

◊ **Exercise 11.16:** Does it make sense to also encode the `prefix` and `suffix` fields?

One of the principles of this method is to bring the individual bits of the sparse string closer together by rearranging the one-dimensional string into a two-dimensional array. Thus, it seems reasonable to try to improve the method by rearranging the string into a three-dimensional array, a cube, or a rectangular box, bringing the bits even closer. If the size of the string is 2^{3n}, each dimension of the array will be of size 2^n. For $n = 7$, the maximum distance between bits in the string is $2^{37} = 2^{21} \approx 2$ million, while the distance between them in the three-dimensional cube is just $2^7 = 128$, a considerable reduction!

The cube can be partitioned by means of an *octree*, where each of the eight octants is identified by a number in the range 0–7 (a 3-bit number). When this partitioning is carried out recursively all the way to individual pixels, each pixel is identified by an n-digit number, where each digit is in the range 0–7.

◇ **Exercise 11.17:** Is it possible to improve the method even more by rearranging the original string into a four-dimensional hypercube?

David Salomon has an article on a somewhat more esoteric problem: the compression of sparse strings. While this isn't a general-purpose one-size-fits-all algorithm, it does show you how to approach a more specialized but not uncommon problem. Prefix Compression of Sparse Binary Strings `http://www.acm.org/crossroads/xrds6-3/prefix.html`. (From *Dr. Dobbs Journal*, March 2000.)

Decoding: The decoder starts by reading the value of n and constructing a matrix M of $2^n \times 2^n$ zeros. We assume that the rows and columns of M are numbered from 0 to $2^n - 1$. It then reads the next token from the compressed file and reconstructs the addresses (we'll call them *id numbers*) of the pixels (the 1-bits) included in the token. This task is straightforward and does not require any special algorithms. Next, the decoder has to convert each id into a row and column numbers. A recursive algorithm for that is described here. Once the row and column of a pixel are known, the decoder sets that location in matrix M to 1. The last task is to "unfold" M into a single bit string, and delete artificially appended zeros, if any, from the end.

Our recursive algorithm assumes that the individual digits of the id number are stored in an array `id`, with the most significant digit stored at location 0. The algorithm starts by setting the two variables `R1` and `R2` to 0 and $2^n - 1$, respectively. They denote the range of row numbers in M. Each step of the algorithm examines the next less-significant digit of the id number in array `id` and updates `R1` or `R2` such that the distance between them is halved. After n recursive steps, they meet at a common value which is the row number of the current pixel. Two more variables, `C1` and `C2`, are initialized and updated similarly, to become the column number of the current pixel in M. Figure 11.28 is a pseudocode algorithm of a recursive procedure `RowCol` that executes this algorithm. A main program that performs the initialization and invokes `RowCol` is also listed. Notice that the symbol ÷ stands for integer division, and that the algorithm can easily be modified to the case where the row and column numbers start from 1.

```
procedure RowCol(ind,R1,R2,C1,C2: integer);
case ind of
0: R2:=(R1+R2)÷2; C2:=(C1+C2)÷2;
1: R2:=(R1+R2)÷2; C1:=((C1+C2)÷2) + 1;
2: R1:=((R1+R2)÷2) + 1; C2:=(C1+C2)÷2;
3: R1:=((R1+R2)÷2) + 1; C1:=((C1+C2)÷2) + 1;
endcase;
if ind≤n then RowCol(ind+1,R1,R2,C1,C2);
end RowCol;

main program
integer ind, R1, R2, C1, C2;
integer array id[10];
bit array M[2^n,2^n];
ind:=0; R1:=0; R2:=2^n − 1; C1:=0; C2:=2^n − 1;
RowCol(ind, R1, R2, C1, C2);
M[R1,C1]:=1;
end;
```

Figure 11.28: Recursive Procedure RowCol.

11.6 Word-Based Text Compression

All the data compression methods mentioned in this book operate on small alphabets. A typical alphabet may consist of the two binary digits, the sixteen 4-bit pixels, the 7-bit ASCII codes, or the 8-bit bytes. In this section, we consider the application of known methods to large alphabets that consist of *words*.

It is not clear how to define a word in cases where the input stream consists of the pixels of an image, so we limit our discussion to text streams. In such a stream a word is defined as a maximal string of either alphanumeric characters (letters and digits) or other characters (punctuations and spaces). We denote by `A` the alphabet of all the alphanumeric words and by `P`, that of all the other words. One consequence of this definition is that in any text stream—whether the source code of a computer program, a work of fiction, or a restaurant menu—words from `A` and `P` strictly *alternate*. A simple example is the C-language source line

```
␣␣for␣(␣short␣i=0;␣i␣<␣npoints;␣i++␣)•
```

where • indicates the end-of-line character (CR, LF, or both). This line can easily be broken up into the 15-word alternating sequence

“␣␣” “for” “␣(␣” “short” “␣” “i” “=” “0” “;␣” “i” “␣<␣” “npoints” “;␣” “i” “++)•”.

Clearly, the size of a word alphabet can be very large and may for all practical purposes be considered infinite. This implies that a method that requires storing the entire alphabet in memory cannot be modified to deal with words as the basic units (symbols) of compression.

◇ **Exercise 11.18:** What is an example of such a method?

A minor point to keep in mind is that short input streams tend to have a small number of distinct words, so when an existing compression method is modified to operate on words, care should be taken to make sure it still operates efficiently on small quantities of data.

Any compression method based on symbol frequencies can be modified to compress words if an extra pass is added, where the frequency of occurrence of all the words in the input is counted. However, such a modification is impractical for the following reasons:

1. A two-pass method is inherently slow.
2. The information gathered by the first pass has to be included in the compressed stream, because the decoder needs it. This degrades compression even if that information is included in compressed form.

It therefore makes more sense to come up with *adaptive* versions of existing methods. Such a version should start with an empty database (dictionary or frequency counts) and should add words to it as they are found in the input stream. When a new word is input, the raw, uncompressed ASCII codes of the individual characters in the word should be output, preceded by an escape code. In fact, it is even better to use some simple compression scheme to compress the ASCII codes. Such a version should also take advantage of the alternating nature of the words in the input stream.

11.6.1 Word-Based Adaptive Huffman Coding

This is a modification of the character-based adaptive Huffman coding (Section 5.3). Two Huffman trees are maintained, for the two alphabets `A` and `P`, and the algorithm alternates between them. Figure 11.29 lists the main steps of the algorithm.

The main problems with this method are the following:

1. In what format to output new words. A new word can be written on the output stream, following the escape code, using the ASCII codes of its characters. However, since a word normally consists of several characters, a better idea is to code it by using the original, character-based, adaptive Huffman method. Thus, the word-based adaptive Huffman algorithm "contains" a character-based adaptive Huffman algorithm that is invoked from time to time. This point is critical, since a short input stream normally contains a high percentage of new words. Writing their raw codes on the output stream may degrade the overall compression performance considerably.
2. What to do when the encoder runs out of memory because of large Huffman trees. A good solution is to delete nodes from the tree (and rearrange the tree after such deletions, so it remains a Huffman tree) instead of deleting the entire tree. The best nodes to delete are those whose counts are so low that their Huffman codes are longer than the codes they would be assigned if they were seen for the first time. If there are just a few (or no) such nodes, then nodes with low frequency counts should be deleted.

Experience shows that word-based adaptive Huffman coding produces better compression than the character-based version but is slower, since the Huffman trees tend to get big, thereby slowing down the search and update operations.
3. The first word in the input stream may be either alphanumeric or other. Thus, the compressed stream should start with a flag indicating the type of this word.

```
repeat
 input an alphanumeric word W;
 if W is in the A-tree then
  output code of W;
  increment count of W;
 else
  output an A-escape;
  output W (perhaps coded);
  add W to the A-tree with a count of 1;
  Increment the escape count
 endif;
 rearrange the A-tree if necessary;
 input an ``other'' word P;
 if P is in the P-tree then
 ...
 ...  code similar to the above
 ...
until end-of-file.
```

Figure 11.29: Word-Based Adaptive Huffman Algorithm.

11.6.2 Word-Based LZW

Word-based LZW is a modification of the character-based LZW method (Section 6.13). The number of words in the input stream is not known beforehand and may also be very large. As a result, the LZW dictionary cannot be initialized to all the possible words, as is done in the character-based original LZW method. The main idea is to start with an empty dictionary (actually two dictionaries, an A-dictionary and a P-dictionary) and use escape codes.

> You watch your phraseology!
>
> —Paul Ford as Mayor Shinn in *The Music Man* (1962)

Each phrase added to a dictionary consists of two strings, one from `A` and the other from `P`. All phrases where the first string is from `A` are added to the `A`-dictionary. All those where the first string is from `P` are added to the `P`-dictionary. The advantage of having two dictionaries is that phrases can be numbered starting from 1 in each dictionary, which keeps the phrase numbers small. Notice that two different phrases in the two dictionaries can have the same number, since the decoder knows whether the next phrase to be decoded comes from the `A`- or the `P`-dictionary. Figure 11.30 is a general algorithm, where the notation "`S,W`" stands for string `W` appended to string `S`.

Notice the line "`output an escape followed by the text of W;`". Instead of writing the raw code of `W` on the output stream, it is again possible to use (character-based) LZW to code it.

```
 S:=empty string;
repeat
 if currentIsAlph then input alphanumeric word W
                  else input non-alphanumeric word W;
 endif;
 if W is a new word then
  if S is not the empty string then output string # of S; endif;
  output an escape followed by the text of W;
  S:=empty string;
 else
  if startIsAlph then search A-dictionary for string S,W
                 else search P-dictionary for string S,W;
  endif;
  if S,W was found then S:=S,W
  else
    output string numer of S;
    add S to either the A- or the P-dictionary;
    startIsAlph:=currentIsAlph;
    S:=W;
  endif;
 endif;
 currentIsAlph:=not currentIsAlph;
until end-of-file.
```

Figure 11.30: Word-Based LZW.

11.6.3 Word-Based Order-1 Prediction

English grammar imposes obvious correlations between consecutive words. It is common, for example, to find the pairs of words `the boy` or `the beauty` in English text, but rarely a pair such as `the went`. This reflects the basic syntax rules governing the structure of a sentence, and similar rules should exist in other languages as well. A compression algorithm using order-1 prediction can therefore be very successful when applied to an input stream that obeys strict syntax rules. Such an algorithm [Horspool and Cormack 92] should maintain an appropriate data structure for the frequencies of all the pairs of alphanumeric words seen so far. Assume that the text "$\ldots P_i\, A_i\, P_j$" has recently been input, and the next word is A_j. The algorithm should get the frequency of the pair (A_i, A_j) from the data structure, compute its probability, send it to an arithmetic encoder together with A_j, and update the count of (A_i, A_j). Notice that there are no obvious correlations between consecutive punctuation words, but there may be some correlations between a pair (P_i, A_i) or (A_i, P_j). An example is a punctuation word that contains a period, which usually indicates the end of a sentence, suggesting that the next alphanumeric word is likely to start with an uppercase letter, and to be an article. Figure 11.31 is a basic algorithm in pseudocode, implementing these ideas. It tries to discover correlations only between alphanumeric words.

Since this method uses an arithmetic encoder to encode words, it is natural to extend it to use the same arithmetic encoder, applied to individual characters, to encode the raw text of new words.

```
prevW:=escape;
repeat
 input next punctuation word WP and output its text;
 input next alphanumeric word WA;
 if WA is new then
  output an escape;
  output WA arithmetically encoded by characters;
  add AW to list of words;
  set frequency of pair (prevW,WA) to 1;
  increment frequency of the pair (prevW,escape) by 1;
 else
  output WA arithmetically encoded;
  increment frequency of the pair (prevW,WA);
 endif;
prevW:=WA;
until end-of-file.
```

Figure 11.31: Word-Based Order-1 Predictor.

When I use a word it means just what I choose it to mean—neither more nor less.

—Humpty Dumpty

11.6.4 Syllable-Based Compression

Words in most languages are constructed from letters, but there is an intermediate linguistic form between letters and words, namely the syllable. The following is an intuitive definition: A syllable is a unit of pronunciation that consists of a single vowel sound, with or without surrounding consonants, and forming all or a part of a word. Thus, syllables can be considered the phonological building blocks of words.

A word may be a monosyllable (it is monosyllabic), a disyllable, a trisyllable, or a polysyllable. A language may have long monosyllables. Examples in English are *squirreled* (from the squirrel's habit of storing up gathered nuts and seeds for winter use), *broughamed* (or *broughammed*, meaning "travelled by brougham"), and *scroonched* (a variant of *scrunched*, meaning *squeezed*). The opposite is also true. Short words may consist of several syllables. The following English words—*exciting*, *inviting*, and *delightful*—have three syllables each.

Syllabification is the process of separating a word into its syllables.

Haiku is a poetic form originated in Japan and now popular in many languages. Haiku is perhaps the simplest, shortest literary form. It consists of three lines, with five, seven, and five syllables, respectively. The original Japanese haiku are based on sound units known as "on" that are subtly different from syllables. With the worldwide popularity of haiku came a relaxation of rules and today we find poems that are referred to as haiku and are based on many different rules (for example, three lines with up to 17 syllables, three lines with three, six, and three syllables, and three lines with between 10 and 14 syllables).

The Haiku Society of America [haiku-usa 09] is a nonprofit organization founded in 1968 to promote the writing and appreciation of haiku in English.

The following haiku are: on the left, a traditional one and on the right, a free one.

Syllable compression for common applications I like this notion	Modern technology Owes ecology An apology
—D. Salomon	—Alan M. Eddison

This short section is based on several publications available in [Lansky 09]. The reason for considering syllables in data compression is that in morphologically-rich languages, such as German, Czech, and Russian, many words are formed from a single root syllable (a stem). The following are just a few of the many English words that start with *work*:

workability, works, worked, working, workable, workableness, workably, workaday(s), workaholic(s), workaholism, workalike, workaway, workbag(s), workbench, workboard, workboat(s), workbook(s), workbox(es), workday(s),

and there are many more words that contain "work." In many languages, negations and antonyms of a word are created by appending a prefix or prepending a suffix to the original word. Such prefixes and suffixes may themselves be common syllables in the language.

The fact that a syllable may appear in several words suggests that syllables repeat more than words. The King James version of the Bible (part of the Canterbury Corpus) is a good example. It contains 767,857 words, of which 13,455 (or 1.8%) are distinct, but it also consists of 1,073,882 syllables, of which only 5,604 (or about 0.5%) are distinct. Data compression algorithms look for repeating patterns in the data, which is why syllables may be better than words as a basic unit for compression.

The developers of this approach to compression have implemented and tested a syllable-based modification of LZW (referred to as LZWL) and a syllable-based modification of adaptive Huffman coding (termed HuffSyllable or HS).

LZWL. Recall that the original LZW (Section 6.13) is based on the following principle. The encoder inputs symbols one by one and accumulates them in a string `I`. After each symbol is input and is concatenated to `I`, the dictionary is searched for string `I`. As long as `I` is found in the dictionary, the process continues. At a certain point, adding the next symbol `x` causes the search to fail; string `I` is in the dictionary but string `Ix` (symbol `x` concatenated to `I`) is not. At this point the encoder (1) outputs the dictionary pointer that points to string `I`, (2) saves string `Ix` (which is now called a *phrase*) in the next available dictionary entry, and (3) initializes string `I` to symbol `x`.

LZWL is based on the same principle, except that the basic symbols are syllables. The number of syllables in a language is normally very large, so LZWL starts with a dictionary that contains many of the most-common syllables. These syllables depend on the language and the type of text (such as literature, legal documents, scientific papers, and government forms) being compressed, which is why the LZWL encoder and decoder should have access to several dictionaries (or should have them built-in).

A dictionary of syllables can be prepared from a set of training documents. Once a language and a data type are selected, a large number of documents of that type is collected and analyzed. Each document is read, its words are broken up into individual syllables, and the number of occurrences of each syllable is counted. The syllables are sorted by their counts, and a number of the most-common syllables are stored in the dictionary. Care should be taken to avoid common syllables that appear in one document only, because such syllables tend to be rare in the language as a whole.

It has been verified experimentally that the choice of training documents is critical to the performance of LZWL, especially when the files being compressed are small. The training documents should be typical in the sense that they should include most of the syllables that exist in the selected language and data type, while not including rare syllables. In addition, the dictionary size is important. In one extreme case the initial dictionary contains many syllables. This causes each new syllable added to the dictionary to have a large pointer. If the initial dictionary was too large, it may happen that many syllables in it would never be used, while the useful syllables added to the dictionary would have large pointers. In the opposite extreme case, a small initial dictionary implies many new syllables added to it, and each has to be encoded character by character.

> ...the professor had some pretty peculiar talents. For instance, he could instantly reverse the syllables in a phrase and repeat them backward.
>
> Yoko Ogawa, *The Housekeeper and the Professor*, (2003)

HuffSyllable (HS). This syllable-based compression method is a modification of the adaptive Huffman method (Section 5.3). The words in the input data are broken up into syllables of different types (such as lowercase, uppercase, mixed, numeric, and other) and an adaptive Huffman tree is constructed and maintained for each type.

In a general encoding step, the encoder uses the previous syllable (and sometimes also the latest letter syllable) to predict the type of the current syllable `K` (Table 11.32). If the type of `K` is different from that predicted, an escape symbol is emitted. In any case, syllable `K` is encoded by the Huffman tree that corresponds to its type. The decoder uses the same table to predict the type of the next syllable. When the decoder does not input an escape symbol, it uses its prediction to select a Huffman tree, decode the syllable, and update the tree. If an escape symbol is present, it tells the decoder which Huffman tree to select.

Previous syllable	Predicted
Lowercase	Lowercase
Uppercase	Uppercase
Mixed	Lowercase
Other (w/o a period) if the latest letter syllable was not uppercase	Lowercase
Other (with a period) if the latest letter syllable was not uppercase	Mixed
Other, if the latest letter syllable was uppercase	Uppercase

Table 11.32: Predicting the Next Syllable.

11.7 Textual Image Compression

All the methods described so far assume that the input stream is either a computer file or resides in memory. Life, however, isn't always so simple, and sometimes the data to be compressed consists of a printed document that includes text, perhaps in several columns, and rules (horizontal and vertical). The method described here cannot deal with images very well, so we assume that the input documents do not include any images. The document may be in several languages and fonts, and may contain musical notes or other notation instead of plain text. It may also be handwritten, but the method described here works best with printed material, since handwriting normally has too much variation. Examples of such data are (1) rare books and important original historical documents that are deteriorating because of old age or mishandling, (2) old library catalog cards about to be discarded because of automation, and (3) typed manuscripts that are of interest to scholars. In many of these cases, it is important to preserve *all* the information on the document, not just the text. This includes the original fonts, margin notes, and various smudges, fingerprints, and other stains.

Before any processing by computer, the document has, of course, to be scanned and converted into black and white pixels. Such a scanned document is called a *textual image*, because it is text described by pixels. In the discussion below, this collection of pixels is called *the input* or the *original image*. The scanning resolution should be as high as possible, and this raises the question of compression. Even at the low resolution of 300 dpi, an 8.5×11" page with 1-inch margins on all sides has a printed area of $6.5 \times 9 = 58.5$ square inches, which translates to $58.5 \times 300^2 = 5.265$ million pixels. At 600 dpi (medium resolution) such a page is converted to about 21 billion pixels. Compression makes even more sense if the document contains lots of blank space, since in such a case most of the pixels will be white, resulting in excellent compression.

One approach to this problem is OCR (optical character recognition). Existing OCR software uses sophisticated algorithms to recognize the shape of printed characters and output their ASCII codes. If OCR is used, the compressed file should include the ASCII codes, each with a pair of (x, y) coordinates specifying its position on the page.

◇ **Exercise 11.19:** The (x, y) coordinates may specify the position of a character with respect to an origin, possibly at the top-left or the bottom-left corner of the page. What may be a better choice for the coordinates?

OCR may be a good solution in cases where the entire text is in one font, and there is no need to preserve stains, smudges, and the precise shape of badly printed characters. This makes sense for documents such as old technical manuals that are not quite obsolete and might be needed in the future. However, if the document contains several fonts, OCR software often does a poor job. It also cannot handle accents, images, stains, musical notes, hieroglyphs, or anything other than text.

(It should be mentioned that recent releases of some OCR packages, such as Xerox Techbridge, *do* handle accents. The Adobe Acrobat Capture application goes even further. It inputs a scanned page and converts it to PDF format. Internally it represents the page as a collection of recognized glyphs and unrecognized bitmaps. As a result, it is capable of producing a PDF file that, when printed or viewed may be almost indistinguishable from the original.)

Facsimile compression (Section 5.7) can be used, but does not produce the best results, since it is based on RLE and does not pay any attention to the text itself. A document where letters repeat all the time and another one where no character appears twice may end up being compressed by the same amount.

The method described here [Witten et al. 92] is complex and requires several steps, but is general, it preserves the entire document, and results in excellent compression. Compression factors of 25 are not uncommon. The method can also be easily modified to include lossy compression as an option, in which case it may produce compression factors of 100 [Witten et al. 94]. An additional reference is [Constantinescu and Arps 97].

The principle of the method is to separate the pixels representing text from the rest of the document. The text is then compressed with a method that counts symbol frequencies and assigns them probabilities, while the rest of the document—which typically consists of random pixels and may be considered "noise"—is compressed by another, more appropriate, method. Here is a summary of the method (the reader is referred to [Witten et al. 94] for the full details).

The encoder starts by identifying the lines of text. It then scans each line, identifying the boundaries of individual characters. The encoder does not attempt to actually recognize the characters. It treats each connected set of pixels as a character, called a *mark*. Often, a mark is a character of text, but it may also be part of a character. The letter `i`, for example, is made up of two unconnected parts, the stem and the dot, so each becomes a mark. Something like ö becomes three marks. This way, the algorithm does not need to know anything about the languages, fonts, or accents used in the text. The method works even if the text consists of "exotic" characters, musical notes, or hieroglyphs. Figure 11.33 shows examples of three marks and some specks. A human can easily recognize the marks as the letters PQR, but software would have a hard time at this, especially since some pixels are missing.

Very small marks (less than the size of a period) are left in the input and are not further processed. Each mark above a certain size is compared to a library of previously found marks (called symbols). If the mark is identical to one of the symbols, its pixels are removed from the original textual image (the input). If the mark is "close enough" to one of the library symbols, then the difference between the mark and the symbol is left in the input (it becomes part of what is called the *residue*) and the rest is removed. If the mark is sufficiently different from all the library symbols, it is added to the library as a new symbol and all its pixels are removed from the input. In each of these cases

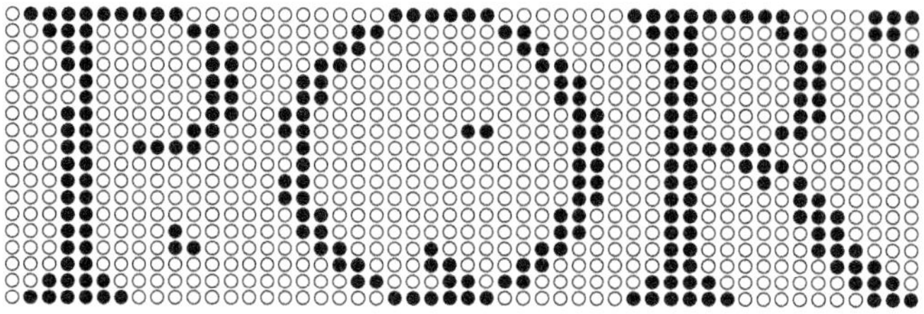

Figure 11.33: Marks and Specks.

the encoder generates the triplet

$$(\#\text{ of symbol in the library, } x,\ y),$$

which is later compressed. The quantities x and y are the horizontal and vertical distances (measured in pixels) between the bottom-left corner of the mark and the bottom-right corner of its predecessor; they are *offsets*. The first mark on a print line normally has a large negative x offset, since it is located way to the left of its predecessor.

The case where a mark is "sufficiently close" to a library symbol is important. In practice, this usually means that the mark and the symbol describe the same character but there are small differences between them due to poor printing or bad alignment of the document during scanning. The pixels that constitute the difference are therefore normally in the form of a *halo* around the mark. The residue is thus made up of halos (which are recognizable or almost recognizable as "ghost" characters) and specks and stains that are too small to be included in the library. Considered as an image, the residue is therefore fairly random (and therefore poorly compressible), because it does not satisfy the condition "the near neighbors of a pixel should have the same value as the pixel itself."

Mark: A visible indication made on a surface.

When the entire input (the scanned document) has been scanned in this way, the encoder selects all the library symbols that matched just one mark and returns their pixels to the input. They become part of the residue. (This step is omitted when the lossy compression option is used.) The encoder is then left with the symbol library, the string of symbol triplets, and the residue. Each of these is compressed separately.

The decoder first decompresses the library and the list of triplets. This is fast and normally results in text that can immediately be displayed and interpreted by the user. The residue is then decompressed, adding pixels to the display and bringing it to its original form. This process suggests a way to implement lossy compression. Just ignore the residue (and omit the step above of returning once-used symbols to the residue). This speeds up both compression and decompression, and significantly improves the compression performance. Experiments show that the residue, even though made up of relatively few pixels, may occupy up to 75% of the compressed stream, since it is random and therefore compresses poorly.

The actual algorithm is very complex, because it has to identify the marks and decide whether a mark is close enough to any library symbol. Here, however, we will discuss just the way the library, triplets, and residue are compressed.

The number of symbols in the library is encoded first, by using one of the prefix codes of Section 3.1. Each symbol is then encoded in two parts. The first encodes the height and depth of the symbol (using the same code as for the number of symbols); the second encodes the pixels of the symbol using the two-level context-based image compression method of Section 7.22.

The triplets are encoded in three parts. The first part is the list of symbol numbers, which is encoded in a modified version of PPM (Section 5.14). The original PPM method was designed for an alphabet whose size is known in advance, but the number of marks in a document is unknown in advance and can be very large (especially if the document consists of more than one page). The second part is the list of x offsets, and the third part, the list of y offsets. They are encoded with adaptive arithmetic coding.

The x and y offsets are the horizontal and vertical distances (measured in pixels) between the bottom-right corner of one mark and the bottom-left corner of the next. In a neatly printed page, such as this one, all characters on a line, except those with descenders, are vertically aligned at their bottoms, which means that most y offsets will be either zero or small numbers. If a proportional font is used, then the horizontal gaps between characters in a word are also identical, resulting in x offsets that are also small numbers. The first character in a word has an x offset whose size is the interword space. In a neatly printed page all interword spaces on a line should be the same, although those on other lines may be different.

All this means that many values of x will appear several times in the list of x values, and the same for y values. What is more, if an x value of, say, 3 is found to be associated with symbol s, there is a good chance that other occurrences of the same s will have an x offset of 3. This argument suggests the use of an adaptive compression method for compressing the lists of x and y offsets.

The method that's actually used, inputs the next triplet (s, x, y) and checks to see whether symbol s was seen in the past followed by the same offset x. If yes, then offset x is encoded with a probability

$$\frac{\text{the number of times this } x \text{ was seen associated with this } s}{\text{the number of times this } s \text{ was seen}},$$

and the count (s, x) incremented by 1. If x hasn't been seen associated with this s, then the algorithm outputs an escape code, and assigns x the probability

$$\frac{\text{the number of times this } x \text{ was seen}}{\text{the total number of } x \text{ offsets seen so far}},$$

(disregarding any associated symbols). If this particular value of x has never been seen in the past, the algorithm outputs a second escape code and encodes x with the same prefix code used for the number of library symbols (and for the height and width of a symbol). The y value of the triplet is then encoded in the same way.

Compressing the residue presents a special problem, since viewed as an image, the residue is random and therefore incompressible. However, viewed as text, the residue consists mostly of halos around characters (in fact, most of it may be legible, or close to legible), which suggests the following approach:

The encoder compresses the library and triplets, and writes them on the compressed stream, followed by the compressed residue. The decoder reads the library and triplets, and uses them to decode the *reconstructed text.* Only then does it read and decompress the residue. Thus, both encoder and decoder have access to the reconstructed text when they encode and decode the residue, and this fact is exploited to compress the residue! The two-level context-based image compression method of Section 7.22 is employed, but with a twist.

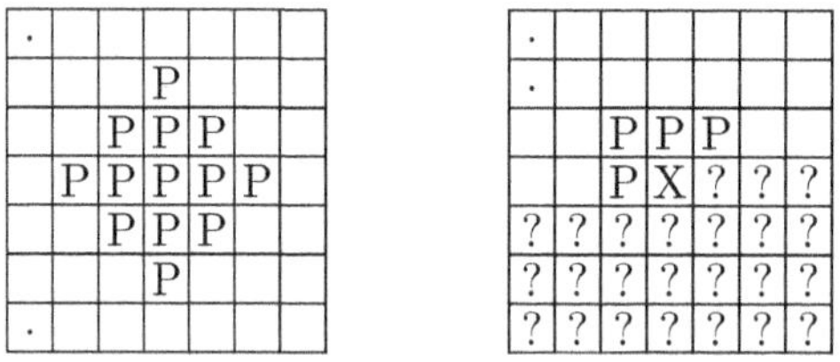

(a) (b)

Figure 11.34: (a) Clairvoyant Context. (b) Secondary Context.

A table of size 2^{17} is used to accumulate frequency counts of 17-bit contexts (in practice, it is organized as a binary search tree or a hash table). The residue pixels are scanned row by row. The first step in encoding a pixel at residue position (r, c) is to go to the same position (r, c) *in the reconstructed text* (which is, of course, an image made of pixels) and use the 13-pixel context shown in Figure 11.34a to generate a 13-bit index. The second step is to use the four-pixel context of Figure 11.34b *on the pixels of the residue* to add four more bits to this index. The final 17-bit index is then used to compute a probability for the current pixel based on the pixel's value (0 or 1) and on the counts found in the table for this 17-bit index. The point is that the 13-bit part of the index is based on pixels that *follow* the current pixel in the reconstructed text. Such a context is normally impossible (it is called *clairvoyant*) but can be used in this case, because the reconstructed text is known to the decoder. This is an interesting variation on the theme of compressing random data.

Even with this clever method, the residue still takes up a large part (up to 75%) of the compressed stream. After some experimentation, the developers realized that it is not necessary to compress the residue at all! Instead, the original image (the input) can be compressed and decompressed using the method above, and this gives better results, even though the original image is less sparse than the residue, because it (the original image) is not as random as the residue.

This approach, of compressing the original image instead of the residue also means that the residue isn't necessary at all. The encoder does not need to reserve memory space for it and to actually create the pixels, which speeds up encoding.

Thus, the compressed stream consists of two parts, the symbol library and triplets (which are decoded to form the reconstructed text), followed by the entire input in compressed form. The decoder decompresses the first part, displays it so the user can read it immediately, then uses it to decompress the second part. Another advantage of this method is that as pixels of the original image are decompressed and become known, they can be displayed, replacing the pixels of the reconstructed text and thereby improving the decompressed image viewed by the user in *real time*. The decoder is therefore able to display an approximate image very quickly, and then clean it up row by row, while the user is watching, until the final image is displayed.

As has been mentioned, the details of this method are complex, because they involve a pattern recognition process in addition to the encoding and decoding algorithms. Here are some of the complexities involved.

Identifying and extracting a mark from the document is done by scanning the input from left to right and top to bottom. The first nonwhite pixel found is the top-left pixel of a mark. This pixel is used to trace the entire boundary of the mark, a complex process that involves an algorithm similar to those used in computer graphics to fill an area. The main point is that the mark may have an inside boundary as well as an outside one (think of the letters `O`, `P`, `Q`, and Φ) and there may be other marks nested inside it.

◇ **Exercise 11.20:** It seems that no letter of the Latin alphabet consists of two nested parts. What are examples of marks that contain other, smaller marks nested within them?

Tracing the boundary of a mark also involves the question of connectivity. When are pixels considered connected? Figure 11.35 illustrates the concepts of 4- and 8-connectivity and makes it clear that the latter method should be used, because the former may miss letter segments that we normally consider connected.

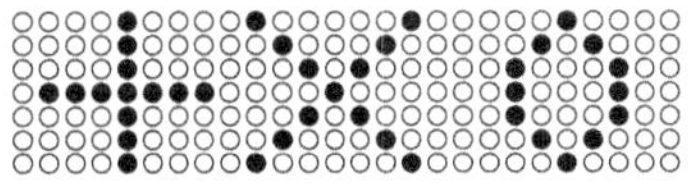

Figure 11.35: 4- and 8-Connectivity.

Comparing a mark to library symbols is the next complex problem. When is a mark considered "sufficiently close" to a library symbol? It is not enough to simply ignore small areas of pixels where the two differ, as this may lead to identifying, for example, an `e` with a `c`. A more sophisticated method is needed, based on pattern recognition techniques. It is also important to speed up the process of comparing a mark to a symbol, since a mark has to be compared to all the library symbols before it is considered a new one. An algorithm is needed that will find out quickly whether the mark and the symbol are too different. This algorithm may use clues such as large differences in the height and width of the two, or in their total areas or perimeters, or in the number of black pixels of each.

When a mark is determined to be sufficiently different from all the existing library symbols, it is added to the library and becomes a symbol. Other marks may be found in the future that are close enough to this symbol and end up being associated with it. They should be "remembered" by the encoder. The encoder therefore maintains a list attached to each library symbol, containing the marks associated with the symbol. When the entire input has been scanned, the library contains the first version of each symbol, along with a list of marks that are similar to it. To achieve better compression, each symbol is now replaced with an average of all the marks in its list. A pixel in this average is set to black if it is black in more than half the marks in the list. The averaged symbols not only result in better compression but also look better in the reconstructed text, making it possible to use lossy compression more often. In principle, any change in a symbol should result in modifications to the residue, but we already know that in practice the residue does not have to be maintained at all.

All these complexities make textual images a complex, interesting example of a special-purpose data compression method and show how much can be gained from a systematic approach in which every idea is implemented and experimented with before it is rejected, accepted, or improved upon.

11.8 Dynamic Markov Coding

Dynamic Markov coding is an adaptive, two-stage statistical compression method due to G. V. Cormack and R. N. Horspool [Cormack and Horspool 87] (see also [Yu 96] for an implementation). Stage 1 employs a finite-state machine to estimate the probability of the next symbol. Stage 2 is an arithmetic encoder that performs the actual compression. Recall that the PPM method (Section 5.14) works similarly. Finite automata (also called finite-state machines) are discussed in many texts.

> Three centuries after Hobbes, automata are multiplying with an agility that no vision formed in the seventeenth century could have foretold.
>
> —George Dyson, *Darwin Among the Machines* (1997)

A finite-state machine can be used in data compression as a model, to compute probabilities of input symbols. Figure 11.36a shows the simplest model, a one-state machine. The alphabet consists of the three symbols `a`, `b`, and `c`. Assume that the input stream is the 600-symbol string `aaabbcaaabbc`.... Each time a symbol is input, the machine outputs its estimated probability, updates its count, and stays in its (only) state. Each of the three probabilities is initially set to 1/3 and gets very quickly updated to its correct value (300/600 for `a`, 200/600 for `b`, and 100/600 for `c`) because of the regularity of this particular input. Assuming that the machine always uses the correct probabilities, the entropy (number of bits per input symbol) of this first example is

$$-\frac{300}{600}\log_2\left(\frac{300}{600}\right)-\frac{200}{600}\log_2\left(\frac{200}{600}\right)-\frac{100}{600}\log_2\left(\frac{100}{600}\right)\approx 1.46.$$

State: mode or condition of being (a state of readiness)

Figure 11.36b illustrates a two-state model. When an `a` is input by state 1, it gets counted, and the machine switches to state 2. When state 2 inputs an `a` or a `b` it counts them and switches back to state 1 (this model will stop prematurely if state 2 inputs a `c`). Each sextet of symbols read from the input stream switches between the two states four times as follows:

$$a\xrightarrow{2}a\xrightarrow{1}a\xrightarrow{2}b\xrightarrow{1}b\xrightarrow{1}c\xrightarrow{1}.$$

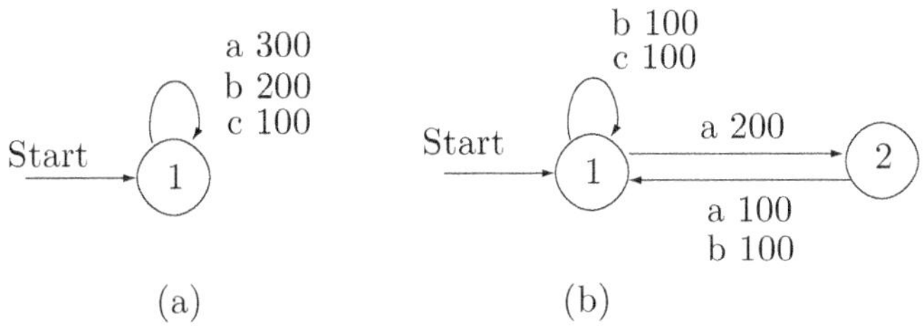

Figure 11.36: Finite-State Models for Data Compression.

State 1 accumulates 200 counts for a, 100 counts for b, and 100 counts for c. State 2 accumulates 100 counts for a and 100 counts for b. State 1 thus handles 400 of the 600 symbols, and state 2, the remaining 200. The probability of the machine being in state 1 is therefore $400/600 = 4/6$, and that of its being in state 2 is $2/6$.

The entropy of this model is calculated separately for each state, and the total entropy is the sum of the individual entropies of the two states, weighted by the probabilities of the states. State 1 has "a", "b", and "c" coming out of it with probabilities $2/6$, $1/6$, and $1/6$, respectively. State 2 has "a" and "b" coming out of it, each with probability $1/6$. Thus, the entropies of the two states are

$$
\begin{aligned}
-\frac{100}{600}\log_2\left(\frac{100}{600}\right)-\frac{200}{600}\log_2\left(\frac{200}{600}\right)-\frac{100}{600}\log_2\left(\frac{100}{600}\right) &\approx 1.3899 \quad \text{(state 1)},\\
-\frac{100}{600}\log_2\left(\frac{100}{600}\right)-\frac{100}{600}\log_2\left(\frac{100}{600}\right) &\approx 0.8616 \quad \text{(state 1)}.
\end{aligned}
$$

The total entropy is therefore $1.3899 \times 4/6 + 0.8616 \times 2/6 = 1.21$.

Assuming that the arithmetic encoder works at or close to the entropy, this two-state model encodes a symbol in 1.21 bits, compared to 1.46 bits/symbol for the previous, one-state model. This is how a finite-state machine with the right states can be used to produce good probability estimates (good predictions) for compressing data.

The natural question at this point is, given a particular input stream, how do we find the particular finite-state machine that will feature the smallest entropy for that stream. A simple, brute force approach is to try all the possible finite-state machines, pass the input stream through each of them, and measure the results. This approach is impractical, since there are n^{na} n-state machines for an alphabet of size a. Even for the smallest alphabet, with two symbols, this number grows exponentially: one 1-state machine, 16 2-state machines, 729 3-state machines, and so on.

Clearly, a clever approach is needed, where the algorithm can start with a simple one-state machine, and adapt it to the particular input data by adding states as it goes along, based on the counts accumulated at any step. This is the approach adopted by the DMC algorithm.

11.8.1 The DMC Algorithm

This algorithm was originally developed for binary data (i.e., a two-symbol alphabet). Common examples of binary data are machine code (executable) files, images (both monochromatic and color), and sound. Each state of the finite-state DMC machine (or DMC model; in this section the words "machine" and "model" are used interchangeably) reads a bit from the input stream, assigns it a probability based on what it has counted in the past, and switches to one of two other states depending on whether the input bit was 1 or 0. The algorithm starts with a small machine (perhaps as simple as just one state) and adds states to it based on the input. Thus, it is adaptive. As soon as a new state is added to the machine, it starts counting bits and using them to compute probabilities for 0 and 1. Even in this simple form, the machine can grow very large and quickly fill up the entire available memory. One advantage of dealing with binary data is that the arithmetic encoder can be made very efficient if it has to deal with just two symbols.

In its original form, the DMC algorithm does not compress text very well. Recall that compression is done by reducing redundancy, and that the redundancy of a text file is featured in the text characters, not in the individual bits. It is possible to extend DMC to handle the 128 ASCII characters by implementing a finite-state machine with more complex states. A state in such a machine should input an ASCII character, assign it a probability based on what characters were counted in the past, and switch to one of 128 other states depending on what the input character was. Such a machine would grow to consume even more memory space than the binary version.

The DMC algorithm has two parts; the first is concerned with computing probabilities, and the second is concerned with adding new states to the existing machine. The first part calculates probabilities by counting, for each state S, how many zeros and ones were input in that state. Assume that in the past the machine was in state S several times, and it input a 0 s_0 times and a 1 s_1 times while in this state (i.e., it switched out of state S s_0 times on the 0 output, and s_1 times on the 1 output; Figure 11.37a). The simplest way to assign probabilities to the two bits is by defining the following:

$$\text{The probability that a 0 will be input while in state } S \text{ is } \frac{s_0}{s_0+s_1},$$
$$\text{The probability that a 1 will be input while in state } S \text{ is } \frac{s_1}{s_0+s_1}.$$

But this, of course, raises the *zero-probability problem*, since either s_0 or s_1 may be zero. The solution adopted by DMC is to assign probabilities that are always nonzero and that depend on a positive integer parameter c. The definitions are the following:

$$\text{The probability that a 0 will be input while in state } S \text{ is } \frac{s_0+c}{s_0+s_1+2c},$$
$$\text{The probability that a 1 will be input while in state } S \text{ is } \frac{s_1+c}{s_0+s_1+2c}.$$

Assigning small values to c implies that small values of s_0 and s_1 will affect the probabilities significantly. This is done when the user feels that the distributions of the two bits in the data can be "learned" fast by the model. If the data is such that it takes longer to adapt to the correct bit distributions, larger values of c can lead to better compression. Experience shows that for very large input streams, the precise value of c does not make much difference.

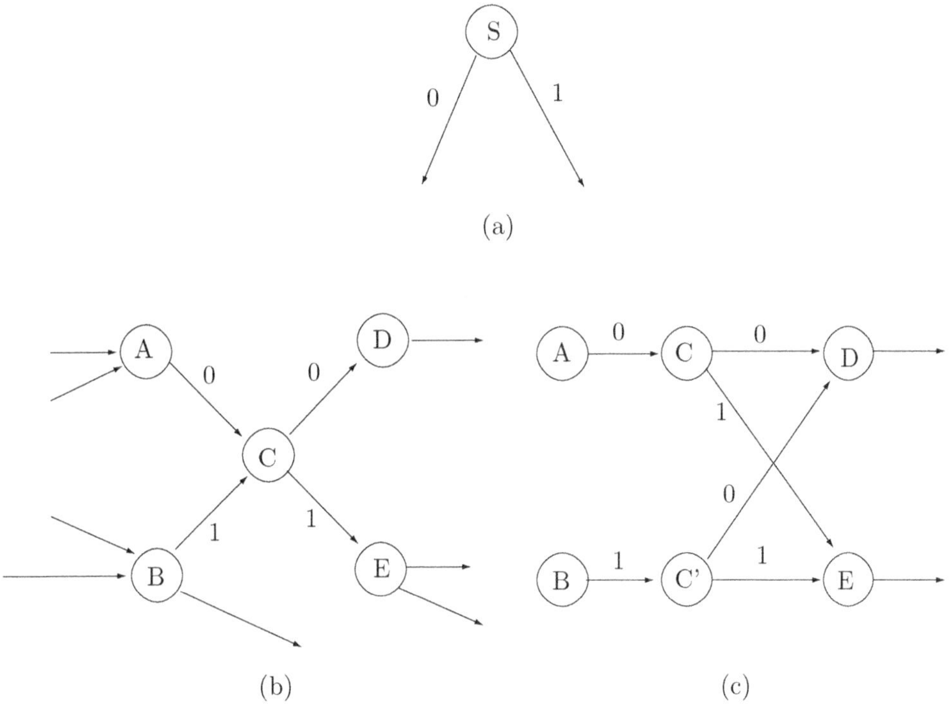

Figure 11.37: The Principles of DMC.

◇ **Exercise 11.21:** Why is there c in the numerator but $2c$ in the denominator of the two probabilities?

The second part of the DMC algorithm is concerned with how to add a new state to the machine. Consider the five states shown in Figure 11.37b, which may be part of a large finite-state DMC model. When a 0 is input while in state A, or when a 1 is input while in state B, the machine switches to state C. The next input bit switches it to either D or E. When switching to D, e.g., some information is lost, since the machine does not "remember" whether it got there from A or from B. This information may be important if the input bits are correlated (i.e., if the probabilities of certain bit patterns are much different from those of other patterns). If the machine is currently in state A, it will get to D if it inputs 00. If it is in state B, it will get to D if it inputs 10. If the probabilities of the input patterns 00 and 10 are very different, the model may compute better probabilities (may produce better predictions) if it knew whether it came to D from A or from B.

The central idea of DMC is to compare the counts of the transitions $A \to C$ and $B \to C$, and if they are significantly different, to create a copy of state C, call it C', and place the copy such that $A \to C \to (D, E)$ but $B \to C' \to (D, E)$ (Figure 11.37c). This copying process is called *cloning*. The machine becomes more complex but can now keep better counts (counts that depend on the specific correlations between A and D, A and E, B and D, and B and E) and, as a result, compute better probabilities. Even adding one state may improve the probability estimates significantly since it may "bring to light" correlations between a state preceding A and a state following D. In general, the more states that are added by cloning, the easier it is for the model to "learn" about correlations (even long range ones) between the input bits.

Once the new state C' is created, the original counts of state C should be divided between C and C'. Ideally, they should be divided in proportion to the counts of the transitions $A \to C \to (D, E)$ and $B \to C \to (D, E)$, but these counts are not available (in fact, the cloning is done precisely in order to have these counts in the future). The next-best thing is to divide the new counts in proportion to the counts of the transitions $A \to C$ and $B \to C$.

An interesting point is that unnecessary cloning does not do much harm. It increases the size of the finite-state machine by one state, but the computed probabilities will not get worse. (Since the machine now has one more state, each state will be visited less often, which will lead to smaller counts and will therefore amplify small fluctuations in the distribution of the input bits, but this is a minor disadvantage.)

All this suggests that cloning be performed as early as possible, so we need to decide on the exact rule(s) for cloning. A look at Figure 11.37b shows that cloning should be done only when both transitions $A \to C$ and $B \to C$ have high counts. If both have low counts, there is "not enough statistics" to justify cloning. If A has a high count, and B has a low count, not much will be gained by cloning C, since B is not very active. This suggests that cloning should be done when both A and B have high counts and one of them has a much higher count than the other. Therefore, the DMC algorithm uses two parameters $C1$ and $C2$, and the following rule:

> If the current state is A and the next state is C, then C is a candidate for cloning, and should be cloned if the count for the transition $A \to C$ is greater than $C1$ **and** the total counts for all the other transitions $X \to C$ are greater than $C2$ (X stands for all the states feeding C, except the current state A).

The choice of values for $C1$ and $C2$ is critical. Small values result in fast cloning of states. This implies better compression, because the model "learns" the correlations in the data faster, but also more memory usage, thereby increasing the chance of running out of memory while there is still much data to be compressed. Large values have the opposite effect. Any practical implementation should therefore let the user specify the values of the two parameters. It also makes sense to start with small values and increase them gradually as compression goes along. This enables the model to "learn" fast initially, and also delays the moment when the algorithm runs out of memory.

◇ **Exercise 11.22:** Figure 11.38 shows part of a finite-state DMC model. State A switches to D when it inputs a 1, so D is a candidate for cloning when A is the current state. Assuming that the algorithm has decided to clone D, show the states after the cloning.

Figure 11.39 is a simple example that shows six steps of adding states to a hypothetical DMC model. The model starts with the single state 0 whose two outputs loop back and become its inputs (they are *reflexive*). In 11.39b a new state, state 1, is added to the 1-output of state 0. We use the notation $0, 1 \to 1$ (read: state 0 output 1 goes to new state 1) to indicate this operation. In 11.39c the operation $0, 0 \to 2$ adds a new state 2. In 11.39d,e,f states 3, 4, and 5 are added by the operations $1, 1 \to 3$; $2, 1 \to 4$; and $0, 0 \to 5$. Figure 11.39f, for example, was constructed by adding state 5 to output

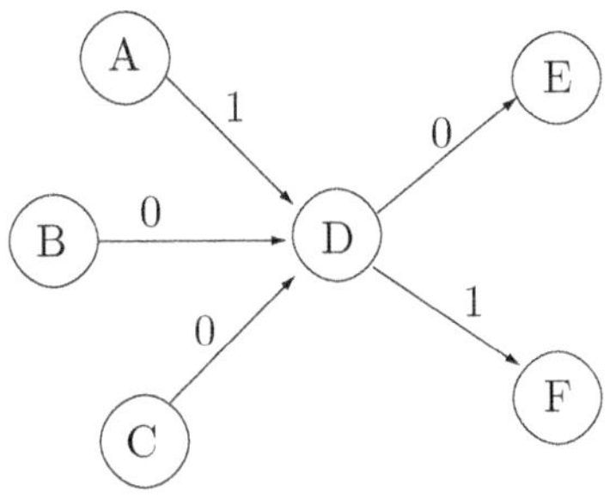

Figure 11.38: State D Is a Candidate.

0 of state 0. The two outputs of state 5 are determined by examining the 0-output of state 0. Since this output used to go to state 2, the new 0-output of state 5 goes to state 2. Also, since this output used to go to state 2, the new 1-output of state 5 becomes a copy of the 1-output of state 2, which is why it goes to state 4.

◇ **Exercise 11.23:** Draw the DMC model after the operation $1, 1 \to 6$.

11.8.2 DMC Start and Stop

When the DMC algorithm starts, it needs to have only one state that switches to itself when either 0 or 1 are input, as shown in Figure 11.40a. This state is cloned many times and may grow very fast to become a complex finite-state machine with many thousands of states. This way to start the DMC algorithm works well for binary input. However, if the input consists of nonbinary symbols, an appropriate initial machine, one that takes advantage of possible correlations between the individual bits of a symbol, may lead to much better compression. The tree of Figure 11.40b is a good choice for 4-bit symbols, because each level corresponds to one of the four bits. If there is, for example, a large probability that a symbol $01xx$ will have the form $011x$ (i.e., a 01 at the start of a symbol will be followed by another 1), the model will discover it very quickly and will clone the state marked in the figure. A similar complete binary tree, but with 255 states instead of 15, may be appropriate as an initial model in cases where the data consists of 8-bit symbols. More complex initial machines may even take advantage of correlations between the last bit of an input symbol and the first bit of the next symbol. One such model, a *braid* designed for 3-bit input symbols, is shown in Figure 11.40c.

Any practical implementation of DMC should address the question of memory usage. The number of states can grow very rapidly and fill up any amount of available memory. The simplest solution is to continue, once memory is full, without cloning. A better solution is to discard the existing model and start afresh. This has the advantage that new states being cloned will be based on new correlations discovered in the data, and old correlations will be forgotten. An even better solution is to always keep the k most recent input symbols in a circular queue and use them to build a small initial model when the old one is discarded. When the algorithm resumes, this small initial model will let it take advantage of the recently discovered correlations, so the algorithm will not have to "relearn" the data from scratch. This method minimizes the loss of compression that results from discarding the old model.

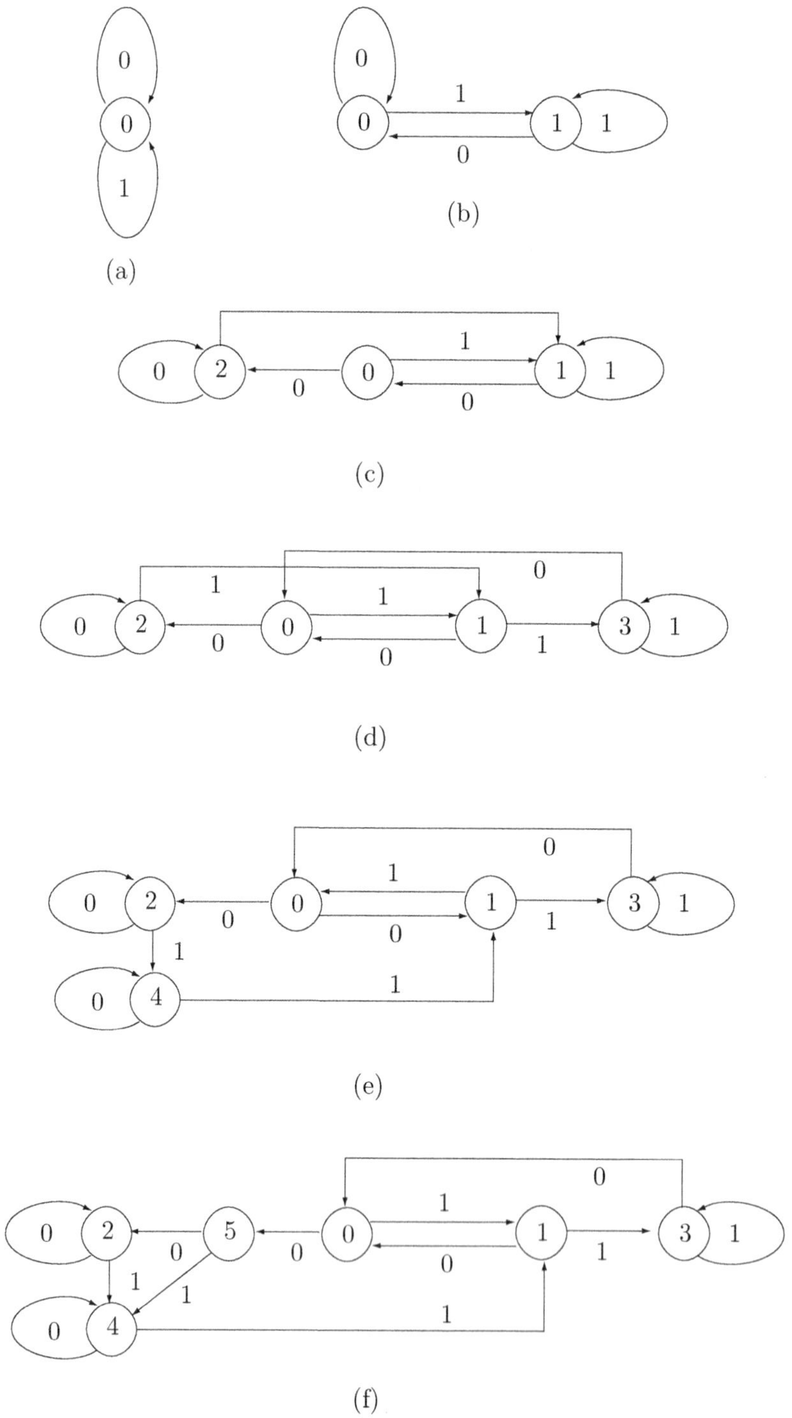

Figure 11.39: First Six States.

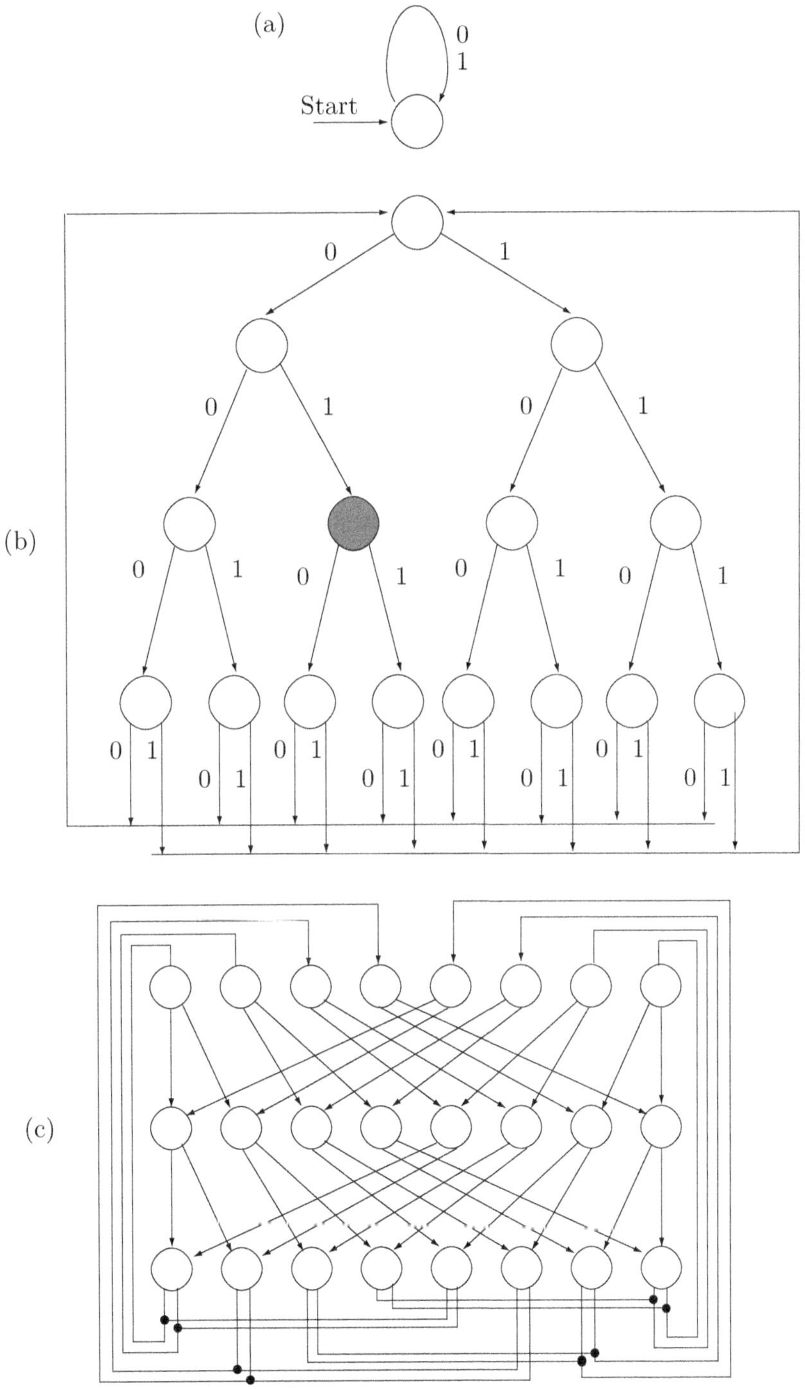

Figure 11.40: Initial DMC Models.

◊ **Exercise 11.24:** How does the loss of compression depend on the value of k?

The main principle of DMC, the rule of cloning, is based on intuition, not on theory. Consequently, the main justification of DMC is that it works! It produces excellent compression, comparable to that achieved by PPM, while being faster.

11.9 FHM Curve Compression

The name FHM is an acronym that stands for Fibonacci, Huffman, and Markov. This method is a modification of the *multiring chain coding* method (see [Freeman 61] and [Wong and Koplowitz 92]), and it uses Fibonacci numbers [Vorobev 83] to construct squares of specific sizes around the current point, such that the total number of choices at any point in the compression is exactly 256.

The method is designed to compress curves, and it is especially suited for the compression of digital signatures. Such a signature is executed, at a point-of-sale or during parcel delivery, with a stylus on a special graphics tablet that breaks the signature curve into a large sequence of points. For each point $\mathbf{P}_i$ such a tablet records its coordinates, the time it took the user to move the stylus to $\mathbf{P}_i$ from $\mathbf{P}_{i-1}$, and the angle and pressure of the stylus at $\mathbf{P}_i$. This information is then kept in an archive and can be used later by a sophisticated algorithm to compare with future signatures. Since the sequence can be large (five items per point and hundreds of points), it should be archived in compressed form.

In the present discussion, we consider only the compression of the coordinates. The resulting compressed curve is very close to the original curve but any time, angle, and pressure information is lost. The method is based on the following two ideas:

1. A straight line is fully defined by its two endpoints, so any interior point can be ignored. In regions where the curve that is being compressed has small curvature (i.e., it is close to a straight line) certain points that have been digitized by the tablet can be ignored without affecting the shape of the curve.
2. At any point in the compression process the curve is placed inside a grid centered on the current *anchor point* A (the grid has the same coordinate size used by the tablet). The next anchor point B is selected such that (1) the straight segment AB is as long as possible, (2) the curve in the region AB is close to a straight line, and (3) B can be chosen from among 256 grid points. Any points that have been originally digitized by the tablet between A and B are deleted, and B becomes the current anchor point. Notice that in general B is not any of the points originally digitized. Thus, the method replaces the original set of points with the set of anchor points, and replaces the curve with the set of straight segments connecting the anchor points. In regions of large curvature the anchor points are close together. The first anchor point is the first point digitized by the tablet.

Figure 11.41a shows a 5×5 grid S_1 centered on the current anchor point X. There are 16 points on the circumference of S_1, and eight of them are marked with circles. We select the first point Y that's on the curve but is outside the grid, and connect points X and Y with a straight segment (the arrow in the figure). We select the two marked points nearest the arrow and construct the triangle shown in dashed. There are no points inside this triangle, which means that the part of the curve inside S_1 is close to a straight line. Before selecting the next anchor point, we try the next larger grid, S_2 (Figure 11.41b). This grid has 32 points on its circumference, and 16 of them are marked. As before, we select the first point Z located on the curve but outside S_2, and connect points X and Z with a straight segment (the arrow in the figure). We select the two marked points nearest the arrow and construct the triangle (in dashed). This time there is one point (Y) on the curve which is outside the triangle. This means that the part of the curve inside S_2 is not sufficiently close to a straight line. We therefore go back to S_1 and select point P (the marked point nearest the arrow) as the next anchor point. The distance between the next anchor and the true curve is therefore less than one grid unit. Point P is encoded, the grids are centered on P, and the process continues.

Figure 11.41c shows all six grids used by this method. They are denoted by S_1, S_2, S_3, S_5, S_8, and S_{13}, with 8, 16, 24, 40, 64, and 104 marked points, respectively. The total number of marked points is thus 256.

◇ **Exercise 11.25:** Where do the Fibonacci numbers come into play?

Since the next anchor point P can be one of 256 points, it can be written on the compressed stream in eight bits. Both encoder and decoder should have a table or a rule telling them where each of the 256 points are located relative to the current anchor point. Experience shows that some of the 256 points are selected more often than others, which suggests a Huffman code to encode them. The 104 points on the border of S_{13}, for example, are selected only when the curve has a long region (26 coordinate units or more) where the curve is close to a straight line. These points should therefore be assigned long Huffman codes. A practical way to determine the frequency of occurrence of each of the 256 points is to *train* the algorithm on many actual signatures. Once the table of 256 Huffman codes has been determined, it is built into both encoder and decoder.

Now for Markov. A Markov chain (or a Markov model) is a sequence of values where each value depends on one of its predecessors, not necessarily the one immediately preceding it, but not on any other value (in a k-order Markov model, a value depends on k of its past neighbors). Working with real signatures shows that the number of direction reversals (or near reversals) is small compared to the size of the signature. This implies that if a segment L_i between two anchor points goes in a certain direction, there is a good chance that the following segment L_{i+1} will point in a slightly different, but not very different, direction. Thus, segment L_{i+1} depends on segment L_i but not on L_{i-1} or preceding segments. The sequence of segments L_i is therefore a Markov chain, a fact that suggests using a different Huffman code table for each segment. A segment goes from one anchor point to the next. Since there are 256 different anchor points, a segment can point in one of 256 directions. We denote the directions by D_0, D_1,...,D_{255}. We now need to construct 256 tables with 256 Huffman codes each; a total of $2^{16} = 65{,}536$ codes!

To calculate the Huffman code table for a particular direction D_i we need to analyze many signatures and count how many times a segment pointing in direction D_i is preceded by a segment pointing in direction D_0, how many times it is preceded by a segment pointing in direction D_1, and so on. In general, code j in the Huffman code table for direction D_i depends on the *conditional probability* $P(D_j|D_i)$ that a segment

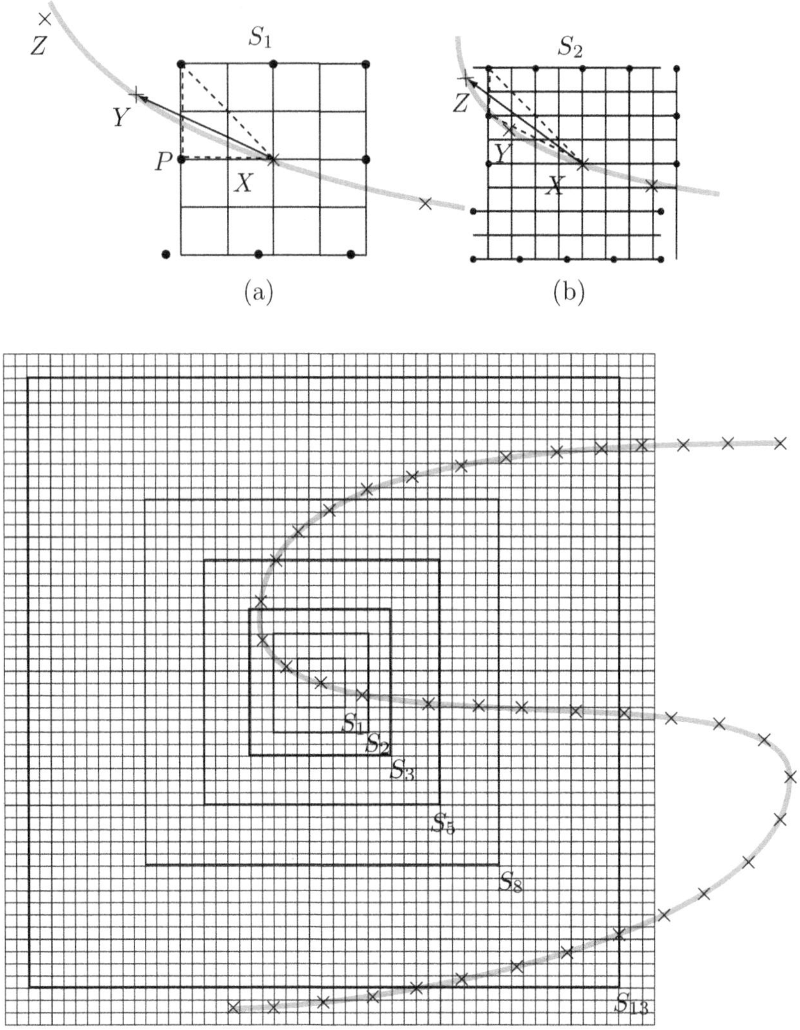

(c)

Figure 11.41: FHM Compression of a Curve.

pointing in direction D_i is preceded by a segment pointing in direction D_j. Allocating 2^{16} locations for tables is not unusual in current (2006) applications, and the main point is that the decoder can mimic all the encoder's operations.

◇ **Exercise 11.26:** Estimate the conditional probability $P(D_i|D_i)$.

◇ **Exercise 11.27:** Estimate the compression ratio of this method.

11.10 Sequitur

Sequitur is based on the concept of context-free grammars, so we start with a short review of this field. A (natural) language starts with a small number of building blocks (letters and punctuation marks) and uses them to construct words and sentences. A sentence is a finite sequence (a string) of symbols that obeys certain grammar rules, and the number of valid sentences is, for all practical purposes, unlimited. Similarly, a formal language uses a small number of symbols (called *terminal symbols*) from which valid sequences can be constructed. Any valid sequence is finite, the number of valid sequences is normally unlimited, and the sequences are constructed according to certain rules (sometimes called *production rules*).

The rules can be used to construct valid sequences and also to determine whether a given sequence is valid. A production rule consists of a nonterminal symbol on the left and a string of terminal and nonterminal symbols on the right. The nonterminal symbol on the left becomes the name of the string on the right. In general, the right-hand side may contain several alternative strings, but the rules generated by sequitur have just a single string.

The entire field of formal languages and grammars is based on the pioneering work of Noam Chomsky in the 1950s [Chomsky 56]. The BNF notation, used to describe the syntax of programming languages, is based on the concept of production rules, as are also L Systems [Salomon 99].

We use lowercase letters to denote terminal symbols and uppercase letters for the nonterminals. Suppose that the following production rules are given: `A` $\rightarrow$ `ab`, `B` $\rightarrow$ `Ac`, `C` $\rightarrow$ `BdA`. With these rules we can generate the valid strings `ab` (an application of the nonterminal `A`), `abc` (an application of `B`), `abcdab` (an application of `C`), as well as many others. Alternatively, we can verify that the string `abcdab` is valid since we can write it as `AcdA`, rewrite this as `BdA`, and replace this with `C`. It is clear that the production rules reduce the redundancy of the original sequence, so they can serve as the basis of a compression method.

In a context free grammar, the production rules do not depend on the context of a symbol. There are also context-sensitive grammars.

Sequitur (from the Latin for "it follows") is based on the concept of context-free grammars. It considers the input stream a valid sequence in some formal language. It reads the input symbol by symbol and uses repeated phrases in the input data to build a set of context-free production rules. Each repetition results in a rule, and is replaced by the name of the rule (a nonterminal symbol), thereby resulting in a shorter representation. Generally, a set of production rules can be used to generate many valid sequences, but the production rules produced by sequitur are not general. They can be used only to reconstruct the original data. The production rules themselves are not much smaller than the original data, so sequitur has to go through one more step, where it compresses the production rules. The compressed production rules become the compressed stream, and the sequitur decoder applies the rules (after decompressing them) to reconstruct the original data.

If the input is a typical text in a natural language, the top-level rule becomes very long, typically 10–20% of the size of the input, and the other rules are short, with typically 2–3 symbols each.

Figure 11.42 shows three examples of short input sequences and grammars. The input sequence S on the left of Figure 11.42a is already a (one-rule) grammar. However, it contains the repeated phrase bc, so this phrase becomes a production rule whose name is the nonterminal symbol A. The result is a two-rule grammar, where the first rule is the input sequence with its redundancy removed, and the second rule is short, replacing the *digram* bc with the single nonterminal symbol A. (The reader should review the discussion of digram encoding in Section 1.3.)

Input	Grammar
S → abcdbc	S → aAdA A → bc
(a)	
S → abcdbcabcdbc	S → AA A → aBdB B → bc
(b)	
S → abcdbcabcdbc	S → AA A → abcdbc
	S → CC A → bc B → aA C → BdA
(c)	

Figure 11.42: Three Input Sequences and Grammars.

Figure 11.42b is an example of a grammar where rule A includes rule B. The input S is considered a one-rule grammar. It has redundancy, so each occurrence of abcdbc is replaced with A. Rule A still has redundancy because of a repetition of the phrase bc, which justifies the introduction of another rule B.

Sequitur constructs its grammars by observing two principles (or enforcing two *constraints*) that we denote by $p1$ and $p2$. Constraint $p1$ states: No pair of adjacent symbols will appear more than once in the grammar (this can be rephrased as; Every digram in the grammar is unique). Constraint $p2$ says: Every rule should be used more than once. This ensures that rules are useful. A rule that occurs just once is useless and should be deleted.

The sequence of Figure 11.42a contains the digram bc twice, so $p1$ requires the creation of a new rule (rule A). Once this is done, digram bc occurs just once, inside rule A. Figure 11.42c shows how the two constraints can be violated. The first grammar of Figure 11.42c contains two occurrences of bc, thereby violating $p1$. The second grammar contains rule B, which is used just once. It is easy to see how removing B reduces the size of the grammar.

◇ **Exercise 11.28:** Show this!

The sequitur encoder constructs the grammar rules while enforcing the two constraints at all times. If constraint $p1$ is violated, the encoder generates a new production rule. When $p2$ is violated, the useless rule is deleted. The encoder starts by setting rule S to the first input symbol. It then goes into a loop where new symbols are input and appended to S. Each time a new symbol is appended to S, the symbol and its predecessor become the current digram. If the current digram already occurs in the grammar, then $p1$ has been violated, and the encoder generates a new rule with the current digram on the right-hand side and with a new nonterminal symbol on the left. The two occurrences of the digram are replaced by this nonterminal.

Figure 11.43 illustrates the operation of the encoder on the input sequence abcdbcabcd. The leftmost column shows the new symbol being input, or an action being taken to enforce one of the two constraints. The next two columns from the left list the input so far and the grammar created so far. The last two columns list duplicate digrams and any underused rules. The last line of Figure 11.43a shows how the input symbol c creates a repeat digram bc, thereby triggering the generation of a new rule A. Notice that the appearance of a new, duplicate digram does not always generate a new rule. If, for example, the new, duplicate digram xy is created, but is found to be the right-hand side of an existing rule A $\rightarrow$ xy, then xy is replaced by A and there is no need to generate a new rule. This is illustrated in Figure 11.43b, where a third occurrence of bc is found. No new rule is generated, and the existing rule A is appended to S. This creates a new, duplicate digram aA in S, so a new rule B $\rightarrow$ aA is generated in Figure 11.43c. Finally, Figure 11.43d illustrates how enforcing $p2$ results in a new rule (rule C) whose right-hand side consists of three symbols. This is how rules that are longer than a digram can be generated.

◇ **Exercise 11.29:** Why is rule B removed in Figure 11.43d?

One more detail, namely rule utilization, still needs to be discussed. When a new rule X is generated, the encoder also generates a counter associated with X, and initializes the counter to the number of times X is used (a new rule is normally used twice when it is first generated). Each time X is used in another rule Y, the encoder increments X's counter by 1. When Y is deleted, the counter for X is decremented by 1. If X's counter reaches 1, rule X is deleted.

As mentioned earlier, the grammar (the set of production rules) is not much smaller than the original data, so it has to be compressed. An example is file book1 of the Calgary Corpus (Table Intro.4), which is 768,771 bytes long. The sequitur encoder generates for this file a grammar where the first rule (rule S) has 131,416 symbols, and each of the other 27,364 rules has 1.967 symbols on average. Thus, the size of the grammar is 185,253 symbols, or about 24% the size of the original data; not very impressive. There is also the question of the names of the nonterminal symbols. In the case of book1 there are 27,365 rules (including rule S), so 27,365 names are needed for the nonterminal symbols.

The method described here for compressing the grammar has two parts. The first part uses arithmetic coding to compress the individual grammar symbols. The second part provides an elegant way of handling the names of the many nonterminal symbols.

Part 1 employs adaptive arithmetic coding (Section 5.10) with an order-0 model. Constraint $p1$ ensures that no digram appears twice in the grammar, so there is no ad-

New symbol or action	the string so far	resulting grammar	duplicate digrams	unused rules
a	a	S → a		
b	ab	S → ab		
c	abc	S → abc		
d	abcd	S → abcd		
b	abcdb	S → abcdb		
c	abcdbc	S → abcdbc	bc	
enforce *p*1		S → aAdA A → bc		
(a)				
a	abcdbca	S → aAdAa A → bc		
b	abcdbcab	S → aAdAab A → bc		
c	abcdbcabc	S → aAdAabc A → bc	bc	
enforce *p*1		S → aAdAaA A → bc	aA	
(b)				
enforce *p*1	abcdbcabc	S → BdAB A → bc B → aA		
(c)				
d	abcdbcabcd	S → BdABd A → bc B → aA	Bd	
enforce *p*1		S → CAC A → bc B → aA C → Bd		B
enforce *p*2		S → CAC A → bc C → aAd		
(d)				

Figure 11.43: A Detailed Example of the Sequitur Encoder (After [Nevill-Manning and Witten 97]).

vantage to using higher-order models to estimate symbol probabilities while compressing a grammar. This is also the reason why compression methods that use high-order models, such as PPM (Section 5.14), would not do a better job in this case. Applying this method to the grammar of `book1` yields compression of 3.49 bpc; not very good.

Part 2 eliminates the need to compress the names of the nonterminal symbols. The number of terminal symbols (normally letters and punctuation marks) is relatively small. This is typically the set of 128 ASCII characters. The number of nonterminals, on the other hand, can be huge (27,365 for `book1`). The solution is to not assign explicit names to the nonterminals. The encoder sends (i.e., writes on the compressed stream) the sequence of input symbols, and whenever it generates a rule, it sends enough information so the decoder can reconstruct the rule. Rule `S` represents the entire input, so this method is equivalent to sending rule `S` and sending other rules as they are generated.

When a nonterminal is found by the encoder while sending rule `S`, it is handled in one of three ways, depending on how many times it has been encountered in the past. The first time a nonterminal is found in `S`, its contents (i.e., the right-hand side of the rule) are sent. The decoder does not even know that it is receiving symbols that constitute a rule. The second time the nonterminal is found, a pair (pointer, count) is sent. The pointer is the offset of the nonterminal from the start of rule `S`, and the counter is the length of the rule. (This is similar to the tokens generated by LZ77, Section 6.3.) The decoder uses the pair to form a new rule that it can use later. The name of the rule is simply its serial number, and rules are numbered by both encoder and decoder in the same way. On the third and subsequent occurrences of the nonterminal, its serial number is sent by the encoder and is identified by the decoder.

This way, the first two times a rule is encountered, its name (i.e., its serial number) is not sent, a feature that greatly improves compression. Also, instead of sending the grammar to the decoder rule by rule, a rule is sent only when it is needed for the first time. Using arithmetic coding together with part 2 to compress `book1` yields compression of 2.82 bpc.

◇ **Exercise 11.30:** Show the information sent to the decoder for the input string `abcdbcabcdbc` (Figure 11.42b).

Detailed information about the actual implementation of sequitur can be found in [Nevill-Manning 96].

One advantage of sequitur is that every rule is used more than once. This is in contrast to some dictionary-based methods that add (to the dictionary) strings that may never occur in the future and thus may never be used.

Once the principles of sequitur are understood, it is easy to see that it performs best when the data to be compressed consists of identical strings that are adjacent. In general, identical strings are not adjacent in the input stream, but there is one type of data, namely *semistructured text*, where identical strings are many times also adjacent. Semistructured text is defined as data that is human readable and also suitable for machine processing. A common example is HTML. An HTML file consists of text with markup tags embedded. There is a small number of different tags, and they have to conform to certain rules. Thus, the tags can be considered highly structured, in contrast with the text, which is unstructured and free. The entire HTML file is therefore semistructured. Other examples of semistructured text are forms, email messages, and data bases. When a form is stored in a computer, some fields are fixed (these are the highly structured parts) and other parts have to be filled out by the user (with unstructured, free text). An email message includes several fixed parts in addition to the text of the message. The same is true for a database. Sequitur was used by its developers to compress two large genealogical data bases [Nevill-Manning and Witten 97], resulting in compression ratios of 11–13%.

11.11 Triangle Mesh Compression: Edgebreaker

Polygonal surfaces are commonly used in computer graphics. Such a surface is made up of flat polygons, and is therefore very simple to construct, save in memory, and render. A surface is normally rendered by simulating the light reflected from it (although some surfaces are rendered by simulating light that they emit or transmit). Since a polygonal surface is made of flat polygons, it looks angular and unnatural when rendered. There are, however, simple methods (such as Gouraud shading and Phong shading, see, e.g., [Salomon 99]) that smooth the reflection over the polygons, resulting in a realistic-looking smooth, curved surface. This fact, combined with the simplicity of the polygonal surface, has made this type of surface very common.

Any flat polygon can be used in a polygonal surface, but triangles are common, since a triangle is always flat (other polygons have to be tested for flatness before they can be included in such a surface). This is why a polygonal surface is normally a triangle mesh. Such a mesh is fully represented by the coordinates of its vertices (the triangle corners) and by its connectivity information (the edges connecting the vertices). Since polygonal surfaces are so common, compressing a triangle mesh is a practical problem. The edgebreaker algorithm described here [Rossignac 98] is an efficient method that can compress the connectivity information of a triangle mesh to about two bits per triangle; an impressive result (the list of vertex coordinates is compressed separately).

Mathematically, the connectivity information is a graph called the *incidence graph*. Edgebreaker is therefore an example of a *geometric compressor*. It can compress certain types of geometries. For most triangle meshes, the number of triangles is roughly twice the number of vertices. As a result, the incidence graph is typically twice as big as the list of vertex coordinates. The list of vertex coordinates is also easy to compress, since it consists of triplets of numbers (integer or real), but it is not immediately clear how to efficiently compress the geometric information included in the incidence graph. This is why edgebreaker concentrates on compressing the connectivity information.

The edgebreaker encoder encodes the connectivity information of a triangle mesh by traversing it triangle by triangle, and assigning one of five codes to each triangle. The code expresses the topological relation between the current triangle and the boundary of the remaining mesh. The code is appended to a history list of triangle codes, and the current triangle is then removed. Removing a triangle may change the topology of the remaining mesh. In particular, if the remaining mesh is separated into two regions with just one common vertex, then each region is compressed separately. Thus, the method is recursive, and it uses a stack to save one edge of each of the regions waiting to be encoded. When the last triangle of a region is removed, the encoder pops the stack and starts encoding another region. If the stack is empty (there are no more regions to compress), the algorithm terminates. The decoder uses the list of triangle codes to reconstruct the connectivity, following which, it uses the list of vertex coordinates to assign the original coordinates to each vertex.

The Encoding Algorithm: We assume that the original triangle mesh is homeomorphic to half a sphere, that is, it is a single region, and it has a boundary B that is a closed polygonal curve without self-intersections. The boundary is called a *loop*. One edge on the boundary is selected and is denoted by g (this is the current *gate*). Since the gate is on the boundary, it is an edge of just one triangle. Depending on the local topology of g and its triangle, the triangle is assigned one of the five codes C, L, E, R, or S. The triangle is then removed (which changes the boundary B) and the next gate is selected among one of the edges on the boundary. The next gate is an edge of the new current triangle, adjacent to the previous one.

The term *homeomorphism* is a topological concept that refers to intrinsic topological equivalence. Two objects are homeomorphic if they can be deformed into each other by a continuous, invertible mapping. Homeomorphism ignores the geometric details of the objects and also the space in which they are embedded, so the deformation can be performed in a higher-dimensional space. Examples of homeomorphic objects are (1) mirror images, (2) a Möbius strip with an even number of half-twists and another Möbius strip with an odd number of half-twists, (3) a donut and a ring.

Figure 11.44e demonstrates how removing a triangle (the one marked X) may convert a simple mesh into two regions sharing a common vertex but not a common edge. The boundary of the new mesh now intersects itself, so the encoder divides it into two regions, each with a simple, non-self-intersecting boundary. An edge on the boundary of one region is pushed into the stack, to be used later, and the encoder continues with the triangles of the remaining region. If the original mesh is large, many regions may be formed during encoding. After the last triangle in a mesh has been assigned a code and has been removed (the code of the last triangle in a mesh is always E), the encoder pops the stack, and starts on another region. Once a code is determined for a triangle, it is appended to a compression history H. This history is a string whose elements are the five symbols C, L, E, R, and S, where each symbol is encoded by a prefix code. Surprisingly, this history is all that the decoder needs to reconstruct the connectivity of the original mesh.

Next, we show how the five codes are assigned to the triangles. Let **v** be the third vertex of the current triangle (the triangle whose outer edge is the current gate g, marked with an arrow). If **v** hasn't been visited yet, the triangle is assigned code C. Vertex **v** of Figure 11.44a hasn't been visited yet, since none of the triangles around it has been removed. The current triangle (marked X) is therefore assigned code C. A triangle is assigned code L (left) if its third vertex (the one that does not bound the gate) is exterior (located on the boundary) and immediately precedes g (Figure 11.44b). Similarly, a triangle is assigned code R (right) if its third vertex is exterior and immediately follows g (Figure 11.44d). If the third vertex **v** is exterior but does not immediately precede or follow the current gate, the triangle is assigned code S (Figure 11.44e). Finally,

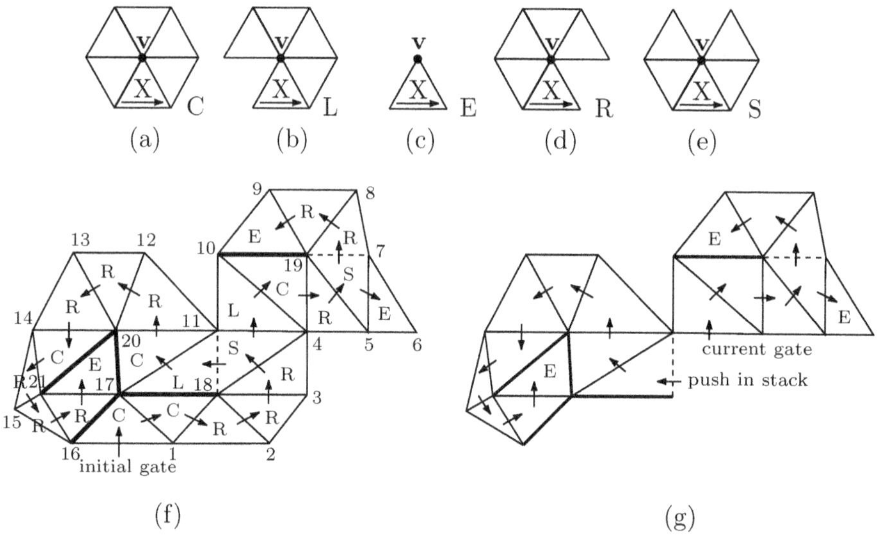

Figure 11.44: The Five Codes Assigned to Triangles.

if the triangle is the last one in its region (i.e., if vertex **v** immediately precedes and immediately follows g), the triangle is assigned code E (Figure 11.44c).

Figure 11.44f shows an example of a mesh with 24 triangles. The initial gate (selected arbitrarily) is indicated. The arrows show the order in which the triangles are visited. The vertices are numbered in the order in which they are added to the list **P** of vertex coordinates. The decoder reconstructs the mesh in the same order as the encoder, so it knows how to assign the vertices the same serial numbers. The decoder then uses the list **P** of vertex coordinates to assign the actual coordinates to all the vertices. The third set of coordinates in **P**, for example, is assigned to the vertex labeled 3.

Initially, the entire mesh is one region. Its boundary is a 16-segment polygonal curve that does not intersect itself. During encoding, as the encoder removes more and more triangles, the mesh degenerates into three regions, so two edges (the ones shown in dashed lines) are pushed into the stack and are popped out later. Figure 11.44g shows the first two regions and the last triangles of the three regions. As the encoder moves from triangle to triangle, all the interior edges (except the five edges shown in thick lines) become gates. The encoder produces a list **P** of the coordinates of all 21 vertices, as well as the compression history

$$H = \text{CCRRRSLCRSERRELCRRRCRRRE}.$$

Notice that there are three triangles with a code of E. They are the last triangles visited in the three regions. Also, each C triangle corresponds to an interior vertex (of which there are five in our example).

The list **P** of vertex coordinates is initialized to the coordinates of the vertices on the boundary of the mesh (if the mesh has a boundary; a mesh that is homeomorphic to a sphere does not have a boundary curve). In our example, these are vertices 1–16. During the main loop, while the encoder examines and removes triangles, it appends (an interior) vertex to **P** for each triangle with a code of C. In our example, these are the five interior vertices 17–21.

A few more terms, as well as some notation, have to be specified before the detailed steps of the encoding algorithm can be listed. The term *half-edge* is a useful topological concept. This is a directed edge together with one of the two triangles incident on it. Figure 11.45a shows two triangles X and Y with a common edge. The common edge together with triangle X is one half-edge, and the same common edge together with triangle Y is another half-edge pointing in the opposite direction. An exterior edge is associated with a single half-edge. The following terms are associated with a half-edge h (Figure 11.45b,c):

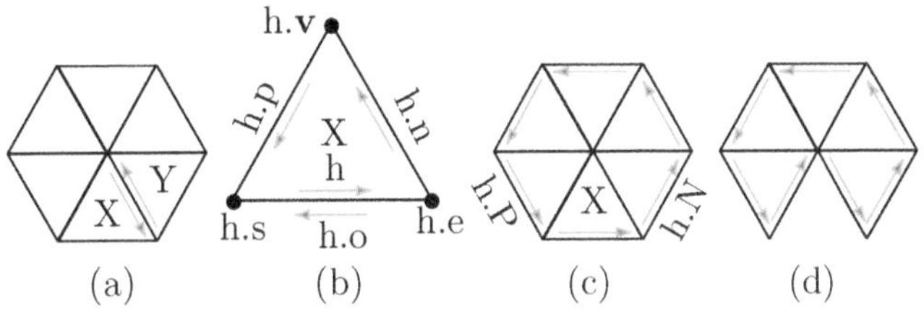

Figure 11.45: Manipulating Half-Edges.

1. The start vertex of h is denoted by h.s.
2. The end vertex of h is denoted by h.e.
3. The third vertex of X (the one that does not bound h) is denoted by h.v.
4. The half-edge that follows h in the triangle X is denoted by h.n.
5. The half-edge that precedes h in the triangle X is denoted by h.p.
6. The half-edge that is the opposite of h is denoted by h.o.
7. The half-edge that follows h in the boundary of the mesh is denoted by h.N.
8. The half-edge that precedes h in the boundary is denoted by h.P.

Figure 11.45c shows a simple mesh composed of six triangles. Six half-edges are shown along the boundary of the mesh. While the encoder visits triangles and removes them, the boundary of the mesh changes. When a triangle is removed, a half-edge that used to be on the boundary disappears, and two half-edges that used to be interior are now positioned on the boundary. This is illustrated in Figure 11.45d. To help the encoder manipulate the half-edges, they are linked in a list. This is a doubly linked, cyclic list where each half-edge points both to its successor and to its predecessor.

The algorithm presented here uses simple notation to indicate operations on the list of half-edges. The notation h.x = y indicates that field h.x of half-edge h should be set to point to y. The algorithm also uses two types of binary flags. Flags **v**.m mark each previously visited vertex **v**. They are used to distinguish between C and S triangles without having to traverse the boundary. Flags h.m mark each half-edge h located on the boundary of the remaining portion of the mesh. These flags are used during S operations to simplify the process of finding the half-edge b such that g.**v** is b.**e**. The last notational element is the vertical bar, used to denote concatenation. Thus, H=H|C indicates the concatenation of code C to the history so far, and **P**=**P**|**v** indicates the operation of appending vertex **v** to the list of vertex coordinates **P**.

The edgebreaker encoding algorithm can now be described in detail. It starts with an initialization step where (1) the first gate is selected and a pointer to it is pushed into the stack, (2) all the half-edges along the boundary are identified, marked, and linked in a doubly-linked, cyclic list, and (3) the list of coordinate vertices **P** is initialized. **P** is initialized to the coordinates of all the vertices on the boundary of the mesh (if the mesh has a boundary) starting from the end-vertex of the gate.

The main loop iterates on the triangles, starting from the triangle associated with the first gate. Each triangle is assigned a code and is removed. Depending on the code, the encoder executes one of five cases (recall that B stands for the boundary of the current region). The loop can stop only when a triangle is assigned code E (i.e., it is the last in its region). The routine that handles case E tries to pop the stack for a pointer to the gate of the next region. If the stack is empty, all the regions that constitute the mesh have been encoded, so the routine stops the encoding loop. Here is the main loop:

```
if not g.v.m then case C  % v not marked
 else if g.p==g.P         % left edge of X is in B
   then if g.n==g.N then case E else case L endif;
   else if g.n==g.N then case R else case S endif;
     endif;
endif;
```

The details of the five cases are shown here, together with diagrams illustrating each case.

Case C: Figure 11.47 shows a simple region with six half-edges on its boundary, linked in a list. The bottom triangle gets code C, it is removed, and the list loses one half-edge and gains two new ones.

Case L: Figure 11.48 is a simple example of this case. The left edge of the bottom triangle is part of the boundary of the region. The triangle is removed, and the boundary is updated.

Case R: Figure 11.49 shows an example of this case. The right edge of the current triangle is part of the boundary of the region. The triangle is removed, and the boundary is updated.

Case S: Figure 11.50 shows a simple example. The upper triangle is missing, so removing the bottom triangle converts the original mesh to two regions. The one on the left is pushed into the stack, and the encoder continues with the region on the right.

Case E: Figure 11.46 shows the last triangle of a region. It is only necessary to unmark its three edges. The stack is then popped for the gate to the next region. If the stack is empty (no more regions), the encoder stops.

Compressing the History: The history string H consists of just five types of symbols, so it can easily and efficiently be compressed with prefix codes. An efficient method is a two-pass compression job where the first pass counts the frequency of each symbol and the second pass does the actual compression. In between the passes, a set of five Huffman codes is computed and is written at the start of the compressed stream. Such a method, however, is slow. Faster (and only slightly less efficient) results can be obtained by selecting the following fixed set of five prefix codes: Assign a 1-bit code to

C and 3-bit codes to each of the other four symbols. A possible choice is the set 0, 100, 101, 110, and 111 to code C, S, R, L, and E, respectively.

```
H=H|E;                        % append E to history
g.m=0; g.n.m=0; g.p.m=0;      % unmark edges
if StackEmpty then stop
 else PopStack; g=StackTop;% start on next region
endif
```

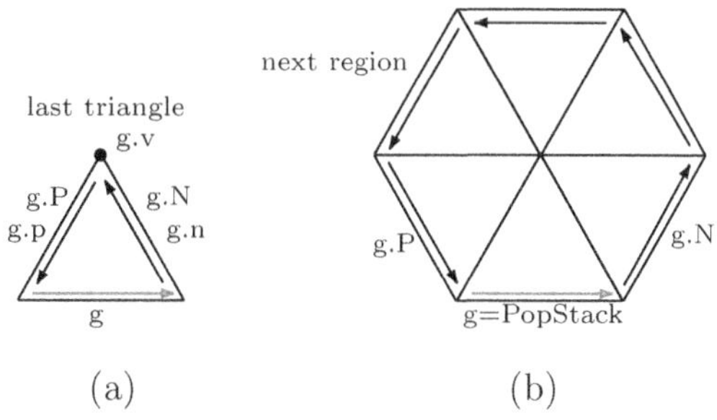

Figure 11.46: Handling Case E.

The average code size is not hard to estimate. The total number of bits used to encode the history H is $c = |\mathrm{T}| - |\mathrm{C}| = |\mathrm{C}| + 3(|\mathrm{S}| + |\mathrm{L}| + |\mathrm{R}| + |\mathrm{E}|)$. We denote by T the set of triangles in the mesh, by $\mathrm{V_i}$ the set of interior vertices of the mesh, and by $\mathrm{V_e}$ the set of exterior vertices. The size of $\mathrm{V_i}$ is the size of C $|\mathrm{C}| = |\mathrm{V_i}|$. We also use Euler's equation for simple meshes, $t - e + v = 1$, where t is the number of triangles, e is the number of edges, and v is the number of vertices. All this is combined to yield

$$|\mathrm{S}| + |\mathrm{L}| + |\mathrm{R}| + |\mathrm{E}| = |\mathrm{T}| - |\mathrm{C}| = |\mathrm{T}| - |\mathrm{V_i}|,$$

and $c = |\mathrm{V_i}| + 3(|\mathrm{T}| - |\mathrm{V_i}|)$ or $c = 2|\mathrm{T}| + (|\mathrm{T}| - 2|\mathrm{V_i}|)$. Since $|\mathrm{T}| - 2|\mathrm{V_i}| = |\mathrm{V_e}| - 2$ we get $c = 2|\mathrm{T}| + |\mathrm{V_e}| - 2$.

The result is that for simple meshes, with a simple, short initial boundary, we have $|\mathrm{V_e}| \ll |\mathrm{V_i}|$ (the mesh is interior heavy), so $c \approx 2|\mathrm{T}|$. The length of the history is two bits per triangle.

A small mesh may have a relatively large number of exterior edges and may thus be exterior heavy. Most triangles in such a mesh get a code of R, so the five codes above should be changed. The code of R should be 1 bit long, and the other four codes should be three bits each. This set of codes yields $c = 3|\mathrm{T}| - 2|\mathrm{R}|$. If most triangles have a code of R, then $|\mathrm{T}| \approx |\mathrm{R}|$ and $c \approx 1$; even more impressive.

We conclude that the set of five prefix codes should be selected according to the ratio $|\mathrm{V_e}|/|\mathrm{V_i}|$.

The Decoder: Edgebreaker is an asymmetric compression method. The operation of the decoder is very different from that of the encoder and requires two passes, preprocessing and generation. The preprocessing pass determines the numbers of triangles, edges, and vertices, as well as the offsets for the S codes. The generation pass creates the

```
H=H|C;                          % append C to history
P=P|g.v;                        % append v to P
g.m=0; g.p.o.m=1;               % update flags
g.n.o.m=1; g.v.m=1;
g.p.o.P=g.P; g.P.N=g.p.o;       % fix link 1
g.p.o.N=g.n.o; g.n.o.P=g.p.o;   % fix link 2
g.n.o.N=g.N; g.N.P=g.n.o;       % fix link 3
g=g.n.o; StackTop=g;            % advance gate
```

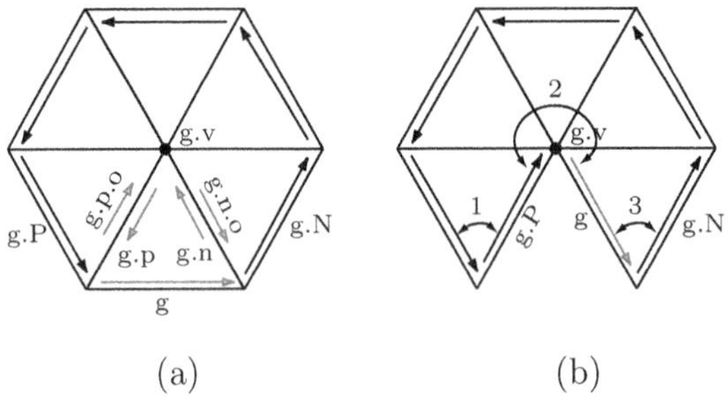

Figure 11.47: Handling Case C.

```
H=H|L;                              % append L to history
g.m=0; g.P.m=0; g.n.o.m=1;          % update flags
g.P.P.n=g.n.o; g.n.o.P=g.P.P;       % fix link 1
g.n.o.N=g.N; g.N.P=g.n.o;           % fix link 2
g=g.n.o; StackTop=g;                % advance gate
```

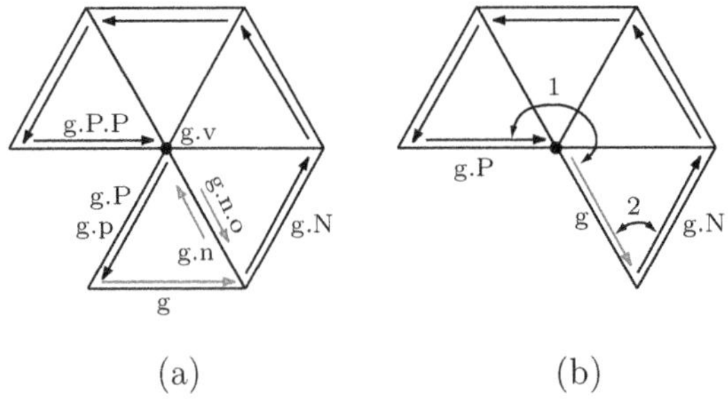

Figure 11.48: Handling Case L.

```
H=H|R;                      % append R to history
g.m=0; g.N.m=0; g.p.o.m=1;  % update flags
g.N.N.P=g.p.o; g.p.o.N=g.N.N.  % fix link 1
g.p.o.P=g.P; g.P.N=g.p.o;      % fix link 2
g=g.p.o; StackTop=g;         % advance gate
```

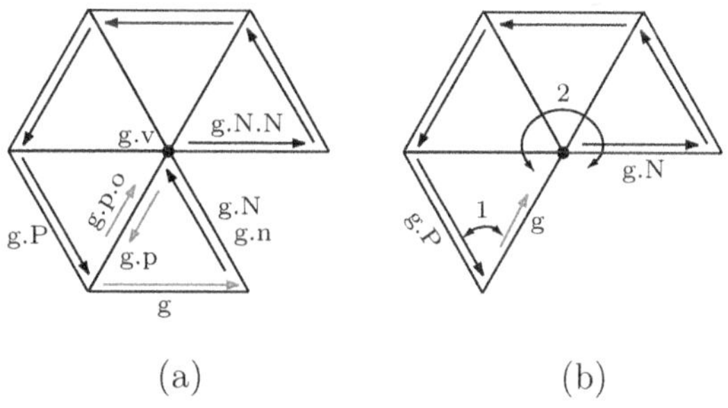

Figure 11.49: Handling Case R.

```
H=H|S;                       % append S to history
g.m=0; g.n.o.m=1; g.p.o.m=1; % update flags
b=g.n;                       % initial candidate for b
while not b.m do b=b.o.p;
                         % turn around v to marked b
g.P.N=g.p.o; g.p.o.P=g.P.        % fix link 1
g.p.o.N=b.N; b.N.P=g.p.o;        % fix link 2
b.N=g.n.o; g.n.o.P=b;            % fix link 3
g.n.o.N=g.N; g.N.P=g.n.o;        % fix link 4
StackTop=g.p.o; PushStack; % save new region
g=g.n.o; StackTop=g;            % advance gate
```

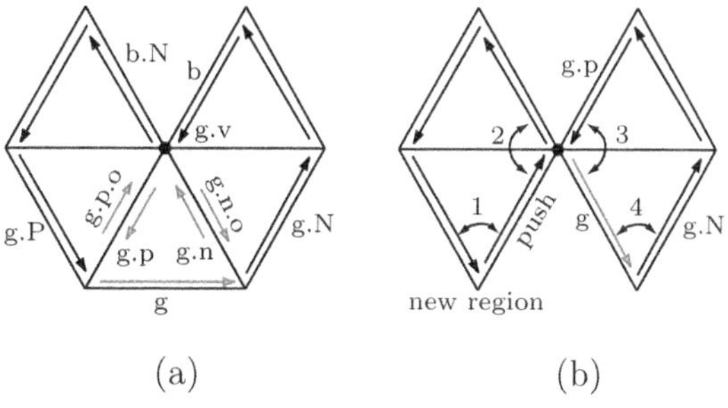

Figure 11.50: Handling Case S.

triangles in the order in which they were removed by the encoder. This pass determines the labels of the three vertices of each triangle and stores them in a table.

The preprocessing pass reads the codes from the history H and performs certain actions for each of the five types of codes. It uses the following variables and data structures:

1. The triangle count t. This is incremented for each code, so it tracks the total number of triangles.
2. The value of $|\mathrm{S}| - |\mathrm{E}|$ is tracked in d. Once the operation of the encoder is fully understood, it should be clear that codes S and E act as pairs of balanced brackets and that the last code in H is always E. After reading this last E, the value of d becomes negative.
3. The vertex counter c is incremented each time a C code is found in the history H. The current value of c becomes the label of the next vertex g.**v**.
4. Variable e tracks the value of $3|\mathrm{E}| + |\mathrm{L}| + |\mathrm{R}| - |\mathrm{C}| - |\mathrm{S}|$. Its final value is $|\mathrm{V_e}|$. When an S code is read, the current value of e is pushed by the decoder into a stack.
5. The number of S codes is tracked by variable s. When e is pushed into the stack, s is used to relate e to the corresponding S.
6. A stack into which pairs (e, s) are pushed when an S code is found. The last pair is popped when an E code is read from H. It is used to compute the offset.
7. A table O of offsets.

All variables are initialized to zero. The stack and table O start empty.

The operations performed by the preprocessing pass for each type of code are as follows:

S code: $e- = 1$; $s+ = 1$; push(e, s); $d+ = 1$;
E code: $e+ = 3$; $(e', s') =$ popstack; $O[s'] = e = e' - 2$; $d- = 1$; if $d < 0$, stop.
C code: $e- = 1$; $c+ = 1$;
L code: $e+ = 1$;
R code: $e+ = 1$;

In addition, t is incremented for each code read. At the end of this pass, t contains the total number $|\mathrm{T}|$ of triangles, c is the number $|\mathrm{V_i}|$ of interior vertices, e is set to the number $|\mathrm{V_e}|$ of exterior vertices, and table O contains the offsets, sorted in the order in which codes S occur in H.

Two points should be explained in connection with these operations. The first is why the final value of e is $|\mathrm{V_e}|$. The calculation of e uses H to find out how many edges were added to the boundary B or deleted from it by the encoder. This number is then used to determine the initial size of B. Recall that the encoder deletes two edges from B and adds one edge to it while processing an R or an L code. Processing an E code by the encoder involves removing three edges.

◇ **Exercise 11.31:** How does the number of edges change when the encoder processes a C code or an S code?

These edge-count changes imply that the initial number of edges (and thus also the initial number of vertices) in the boundary is $3|\mathrm{E}| + |\mathrm{L}| + |\mathrm{R}| - |\mathrm{C}| - |\mathrm{S}|$. This is why e tracks this number.

The second point is the offsets. We know that S and E codes act as paired brackets in H. Any substring in H that starts with an S and ends with the corresponding E contains the connectivity information for a region of the original mesh. We also know that e tracks the value of $3|\text{E}| + |\text{L}| + |\text{R}| - |\text{C}| - |\text{S}|$ for the substring of H that has already been read and processed. Therefore, the difference between the values of e at the E operation and at the corresponding S operation is the number of vertices in the boundary of that region. We subtract 2 from this value in order not to count the two vertices of the gate g as part of the offset.

For each S code, the value of e is pushed into the stack. When the corresponding E code is read from H, the stack is popped and is subtracted from the current value of e.

The preprocessing pass is illustrated by applying it to the history

$$\text{H} = \text{CCRRRSLCRSERRELCRRRCRRRE}$$

that was generated by the encoder from the mesh of Figure 11.44f. Table 11.51 lists the values of all the variables involved after each code is read from H. The pair (e', s') is the contents of the top of the decoder's stack. The two offsets 1 and 6 are stored in table O. The final value of e is 16 (the number of exterior vertices or, equivalently, the number of vertices in B). The final value of c is 5. This is the number of interior vertices. The final value of t is, of course, 24, the number of triangles.

	C	C	R	R	R	S	L	C	R	S	E	R	R	E	L	C	R	R	R	C	R	R	R	E
t	1	2	3	4	5	6	7	8	9	10	11	12	13	14	15	16	17	18	19	20	21	22	23	24
d	0	0	0	0	0	1	1	1	1	2	1	1	1	0	0	0	0	0	0	0	0	0	0	-1
c	1	2	2	2	2	2	2	3	3	3	3	3	3	3	3	4	4	4	4	5	5	5	5	5
e	-1	-2	-1	0	1	0	1	0	1	0	3	4	5	8	9	8	9	10	11	10	11	12	13	16
s	0	0	0	0	0	1	1	1	1	2	2	2	2	2	2	2	2	2	2	2	2	2	2	2
e',s'						0,1	0,1	0,1	0,1	0,2	0,1	0,1	0,1											
$O[s']$											1		6											

Table 11.51: An Example of the Preprocessing Pass.

The generation pass starts with an initialization phase where it does the following:

1. Starts an empty table TV of triangle vertices, where each entry will contain the labels of the three vertices of a triangle.
2. Initializes a vertex counter c to $|\text{V}_\text{e}|$, so that references to exterior vertices precede references to interior ones.
3. Constructs a circular, doubly linked list of $|\text{V}_\text{e}|$ edges, where each node corresponds to an edge G and contains a pointer G.P to the preceding node, a pointer G.N to the successor node, and an integer label G.e that identifies the end vertex of edge G. The labels are integers increasing from 1 to $|\text{V}_\text{e}|$.
4. Creates a stack of references to edges and initializes it to a single entry that refers to the first edge G (the gate) in the boundary B. Notice that uppercase letters are used for the edges to distinguish them from the half-edges used by the encoder.

5. Sets a triangle counter $t = 0$ and a counter $s = 0$ of S operations.

The pass then reads the history H symbol by symbol. For each symbol it determines the labels of the three vertices of the current triangle X and stores them in TV[t]. It also updates B, G, and the stack, if necessary. Once the current gate, G, is known, it determines two of the three vertices of the current triangle. They are G.P.**e** and G.**e**. Determining the third vertex depends on the current code read from H. The actions for each of the five codes are shown here (the notation x++ means return the current value of x, then increment it, while ++x means increment x , then return its new value):

C code: TV[$++t$]=(G.P.**e**, G.**e**, $++c$);
New Edge A; A. **e**=c;
G.P.N=A; A.P=G.P;
A.N=G; G.P=A;

R Code: TV[$++t$]=(G.P.**e**, G.**e**, G.N.**e**);
G.P.N=G.N; G.N.P=G.P;
G=G.N;

L Code: TV[$++t$]=(G.P.**e**, G.**e**, G.P.P.**e**);
G.P=G.P.P; G.P.P.N=G;

E Code: TV[$++t$]=(G.P.**e**, G.**e**, G.N.**e**);
G=PopStack;

S Code: D=G.N; repeat D=D.N; O[$++s$] times;
TV[$++t$]=(G.P.**e**, G.**e**, D.**e**);
New Edge A; A.**e**=D.**e**;
G.P.N=A; A.P=G.P;
PopStack; Push A;
A.N=D.N; D.N.P=A;
G.P=D; D.N=G;
Push G;

These actions are illustrated here for the history

$$\text{H} = \text{CCRRRSLCRSERRELCRRRCRRRE},$$

originally generated by the encoder from the mesh of Figure 11.44f. The preprocessing pass computes $|\text{V}_\text{e}| = 16$, so we start with an initial boundary of 16 edges and of 16 vertices that we label 1 through 16. The first edge (the one that is associated with vertex 1) is our initial gate G. Variable c is set to 16. The first C code encountered in H sets entry TV[1] to the three labels 16 (G.P.**e**), 1 (G.**e**), and 17 (the result of $++c$). The reader should use Figure 11.44f to verify that these are, in fact, the vertices of the first triangle processed and removed by the encoder. A new edge, A, is also created, and 17 is stored as its label. The new edge A is inserted before G by updating the pointers as follows: G.P.N=A, A.P=G.P, A.N=G, and G.P=A. The second C code creates triangle $(17, 1, 18)$ and inserts another new edge, with label 18, before G. (The reader should again verify that $(17, 1, 18)$ are the vertices of the second triangle processed and removed by the encoder.) The first R code creates triangle $(18, 1, 2)$, deletes gate G from the boundary, and declares the edge labeled 2 the current gate.

◇ **Exercise 11.32:** Show the result of processing the second and third R codes.

The first S code skips the six vertices (six, because O[1] = 6) 5, 6, 7, 8, 9, and 10. It then determines that triangle 6 has vertices $(18, 4, 11)$, and splits the boundary into the two regions $(11, 12, 13, 14, 15, 16, 17, 18)$ and $(4, 5, 6, 7, 8, 9, 10)$. The bottom of the stack points to edge $(8, 11)$, the first edge in the first region. The top of the stack points to edge $(11, 4)$, the first edge in the second region. The L code creates triangle $(11, 4, 10)$ and deletes the last edge of the second region. At this point the current edge G is edge $(10, 4)$.

[Rossignac 98] has more details of this interesting and original method, including extensions of edgebreaker for meshes with holes in them, and for meshes without a boundary (meshes that are homeomorphic to a sphere).

11.12 SCSU: Unicode Compression

The ASCII code is old, having been designed in the early 1960s. With the advent of inexpensive laser and inkjet printers and high-resolution displays, it has become possible to display and print characters of any size and shape. As a result, the ASCII code, with its 128 characters, no longer satisfies the needs of modern computing. Starting in 1991, the Unicode consortium (whose members include major computer corporations, software producers, database vendors, research institutions, international agencies, various user groups, and interested individuals) has proposed and designed a new character coding scheme that satisfies the demands of current hardware and software. Information about Unicode is available at [Unicode 03].

The Unicode Standard assigns a number, called a code point, to each character (code element). A code point is listed in hexadecimal with a "U+" preceding it. Thus, the code point U+0041 is the number $0041_{16} = 65_{10}$. It represents the character "A" in the Unicode Standard.

The Unicode Standard also assigns a unique name to each character. Code element U+0041, for example, is assigned the name "LATIN CAPITAL LETTER A" and U+0A1B is assigned the name "GURMUKHI LETTER CHA."

An important feature of the Unicode Standard is the way it groups related codes. A group of related characters is referred to as a script, and such characters are assigned consecutive codes; they become a contiguous area or a region of Unicode. If the characters are ordered in the original script (such as A–Z in the Latin alphabet and α through ω in Greek), then their Unicodes reflect that order. Region sizes vary greatly, depending on the script.

Most Unicode code points are 16-bit (2-byte) numbers. There are 64K (or 65,536) such codes, but Unicode reserves 2,048 of the 16-bit codes to extend this set to 32-bit codes (thereby adding about 1.4 million surrogate code pairs). Most of the characters in common use fit into the first 64K code points, a region of the codespace that's called the basic multilingual plane (BMP). There are about 6,700 unassigned code points for future expansion in the BMP, plus over 870,000 unused supplementary code points in the other regions of the codespace. More characters are under consideration for inclusion in future versions of the standard.

Unicode starts with the set of 128 ASCII codes U+0000 through U+007F and continues with Greek, Cyrillic, Hebrew, Arabic, Indic, and other scripts. These are followed by symbols and punctuation marks, diacritics, mathematical symbols, technical symbols, arrows, dingbats, and so forth. The codespace continues with Hiragana, Katakana, and Bopomofo. The unified Han ideographs are followed by the complete set of modern Hangul. Toward the end of the BMP is a range of code points reserved for private use, followed by a range of compatibility characters. The compatibility characters are character variants that are encoded only to enable transcoding to earlier standards and old implementations that happen to use them.

The Unicode Standard also reserves code points for private use. Anyone can assign these codes privately for their own characters and symbols or use them with specialized fonts. There are 6,400 private-use code points on the BMP and another 131,068 supplementary private-use code points elsewhere in the codespace.

Version 3.2 of Unicode specifies codes for 95,221 characters from the world's alphabets, ideograph sets, and symbol collections. The current version is 5 and its detailed specifications should be published in late 2006.

The method described in this section is due to [Wolf et al. 00]. It is a standard compression scheme for Unicode, abbreviated SCSU. It compresses strings of code points. Like any compression method, it works by removing redundancy from the original data. The redundancy in Unicode stems from the fact that typical text in Unicode tends to have characters located in the same region in the Unicode codespace. Thus, a text using the basic Latin character set consists mostly of code points of the form U+00xx. These can be compressed to one byte each. A text in Arabic tends to use just Arabic characters, which start at U+0600. Such text can be compressed by specifying a start address and then converting each code point to its distance (or offset) from that address. This introduces the concept of a window. The distance should be just one byte because a 2-byte distance replacing a 2-byte code results in no compression. This kind of compression is called the single-byte mode. A 1-byte offset suggests a window size of 256, but we'll see why the method uses windows of half that size. In practice, there may be complications, as the following three examples demonstrate:

1. A string of text in a certain script may have punctuation marks embedded in it, and these have code points in a different region. A single punctuation mark inside a string can be written in raw form on the compressed stream and also requires a special tag to indicate a raw code. The result is a slight expansion.

2. The script may include hundreds or even thousands of characters. At a certain point, the next character to be compressed may be too far from the start address, so a new start address (a new window) has to be specified just for the next character. This is done by a nonlocking-shift tag.

3. Similarly, at a certain point, the characters being compressed may all be in a different window, so a locking-shift tag has to be inserted, to indicate the start address of the new window.

As a result, the method employs tags, implying that a tag has to be at least a few bits, which raises the question of how the decoder distinguishes tags from compressed characters. The solution is to limit the window size to 128. There are 8 static and 8 dynamic windows (Tables 11.52 and 11.53, respectively, where CJK stands for Chinese, Japanese, and Korean). The start positions of the latter can be changed by tags.

n	Start	Major area
0	0000	Quoting tags in single-byte mode
1	0080	Latin1 supplement
2	0100	Latin Extended-A
3	0300	Combining diacritical marks
4	2000	General punctuation marks
5	2080	Currency symbols
6	2100	Letterlike symbols and number forms
7	3000	CJK symbols and punctuation

Table 11.52: Static Windows.

n	Start	Major area
0	0080	Latin1 supplement
1	00C0	Latin1 supp. + Latin Extended-A
2	0400	Cyrillic
3	0600	Arabic
4	0900	Devanagari
5	3040	Hiragana
6	30A0	Katakana
7	FF00	Fullwidth ASCII

Table 11.53: Dynamic Windows (Default Positions).

SCSU employs the following conventions:

1. Each tag is a byte in the interval [0x00,0x1F], except that the ASCII codes for CR (0x0D), LF (0x0A), and TAB (or HT 0x09) are not used for tags. There can therefore be $32 - 3 = 29$ tags (but we'll see that more values are reserved for tags in the Unicode mode). The tags are used to indicate a switch to another window, a repositioning of a window, or an escape to an uncompressed (raw) mode called the Unicode mode.

2. A code in the range U+0020 through U+007F is compressed by eliminating its most-significant eight zeros. It becomes a single byte.

3. Any other codes are compressed to a byte in the range 0x80 through 0xFF. These indicate offsets in the range 0x00 through 0xEF (0 through 127) in the current window.

Example: The string 041C, 043E, 0441, 002D, 043A, 0562, and 000D is compressed to the bytes 12, 9C, BE, C1, 2D, BA, 1A, 02, E2, and 0D. The tag byte 12 indicates the window from 0x0400 to 0x047F. Code 041C is at offset 1C from the start of that window, so it is compressed to the byte 1C + 80 = 9C. Code 043E is at offset 3E, so it is compressed to 3E + 80 = BE. Code 0441 is similarly compressed to C1. Code 002D (the ASCII code of a hyphen) is compressed (without any tags) to its least-significant 8 bits 2D. (This is less than 0x80, so the decoder does not get confused.) Code 043A is compressed in the current window to BA, but compressing code 0562 must be done in window [0x0500,0x057F] and must therefore be preceded by tag 1A (followed by index 02) which selects this window. The offset of code 0562 in the new window is 62, so it is compressed to byte E2. Finally, code 000D (CR) is compressed to its eight least-significant bits 0D without an additional tag.

Tag 12 is called SC2. It indicates a locking shift to dynamic window 2, which starts at 0x0400 by default. Tag 1A is called SD2 and indicates a repositioning of window 2. The byte 02 that follows 1A is an index to Table 11.54 and changes the window's start position by $2\times 80_{16} = 100_{16}$, so the window moves from the original 0x0400 to 0x0500.

X	Offset[X]	Comments
00	reserved	for internal use
01–67	$X\times 80$	half-blocks from U+0080 to U+3380
68–A7	$X\times 80 + AC00$	half-blocks from U+E000 to U+FF80
A8–F8		reserved for future use
F9	00C0	Latin1 characters + half of Extended-A
FA	0250	IPA extensions
FB	0370	Greek
FC	0530	Armenian
FD	3040	Hiragana
FE	30A0	Katakana
FF	FF60	Halfwidth Katakana

Table 11.54: Window Offsets.

We start with the details of the single-byte mode. This mode is the one in effect when the SCSU encoder starts. Each 16-bit code is compressed in this mode to a single byte. Tags are needed from time to time and may be followed by up to two arguments, each a byte. This mode continues until one of the following is encountered: (1) end-of-input, (2) an SCU tag, or (3) an SQU tag. Six types of tags (for a total of 27 different tags) are used in this mode as follows.

1. SQU (0E). This tag (termed quote Unicode) is a temporary (nonlocking) shift to Unicode mode. This tag is followed by the two bytes of a raw code.

2. SCU (0F). This tag (change to Unicode) is a permanent (locking) shift to Unicode mode. This is used for a string of consecutive characters that belong to different scripts and are therefore in different windows.

3. SQn (01–08). This tag (quote from window n) is a nonlocking shift to window n. It quotes (i.e., writes in raw format) the next code, so there is no compression. The value of n (between 0 and 7) is determined by the tag (between 1 and 8). This tag must be followed by a byte used as an offset to the selected window. If the byte is in the interval 00 to 7F, static window n should be selected. If it is in the range 80 to FF, dynamic window n should be selected. This tag quotes one code, then switches back to the single-byte mode. For example, SQ3 followed by 14 selects offset 14 in static window 3, so it quotes code 0300 + 14 = 0314. Another example is SQ4 followed by 8A. This selects offset $8A - 80 = 0A$ in dynamic window 4 (which normally starts at 0900, but could be repositioned), so it quotes code $0900 + 0A = 090A$.

4. SCn (10–17). This tag (change to window n) is a locking shift to window n.

5. SDn (18–1F). This tag (define window n) repositions window n and makes it the current window. The tag is followed by a one-byte index to Table 11.54 that indicates the new start address of window n.

6. SDX (0B). This is the "define extended" tag. It is followed by two bytes denoted by H and L. The three most-significant bits of H determine the static window to be selected, and the remaining 13 bits of H and L become one integer N that determines the start address of the window as $10000 + 80\times N$ (hexadecimal).

Tag SQ0 is important. It is used to flag code points whose most-significant byte may be confused with a tag (i.e., it is in the range 00 through 1F). When encountering such a byte, the decoder should be able to tell whether it is a tag or the start of a raw code. As a result, when the encoder inputs a code that starts with such a byte, it writes it on the output in raw format (quoted), preceded by an SQ0 tag.

Next comes the Unicode mode. Each character is written in this mode in raw form, so there is no compression (there is even slight expansion due to the tags required). Once this mode is selected by an SCU tag, it stays in effect until the end of the input or until a tag that selects an active window is encountered. Four types of tags are used in this mode as follows:

1. UQU (F0). This tag quotes a Unicode character. The two bytes following the tag are written on the output in raw format.

2. UCn (E0–E7). This tag is a locking shift to single-byte mode and it also selects window n.

3. UDn (E8–EF). Define window n. This tag is followed by a single-byte index. It selects window n and repositions it according to the start positions of Table 11.54.

4. UDX (F1). Define extended window. This tag (similar to SDX) is followed by two bytes denoted by H and L. The three most-significant bits of H determine the dynamic window to be selected, and the remaining 13 bits of H and L become an integer N that determines the start address of the window by $10000 + 80\times N$ (hexadecimal).

The four types of tags require 18 tag values, but almost all the possible 29 tag values are used by the single-byte mode. As a result, the Unicode mode uses tag values that are valid code points. Byte E0, for example, is tag UC0, but is also the most-significant half of a valid code point (in fact, it is the most-significant half of 256 valid code points). The encoder therefore reserves these 18 values (plus a few more for future use) for tags. When the encoder encounters any character whose code starts with one of those values, the character is written in raw format (preceded by a UQU tag). Such cases are not common because the reserved values are taken from the private-use area of Unicode, and this area is rarely used.

SCSU also specifies ways to compress Unicode surrogates. With 16-bit codes, there can be 65,536 codes. However, $800_{16} = 2048_{10}$ 16-bit codes have been reserved for an extension of Unicode to 32-bit codes. The 400_{16} codes U+D800 through U+DBFF are reserved as high surrogates, and the 400_{16} codes U+DC00 through U+DFFF are reserved as low surrogates. This allows for an additional $400\times 400 = 100,000_{16}$ 32-bit codes. A 32-bit code is known as a surrogate pair and can be encoded in SCSU in one of several ways, three of which are shown here:

1. In Unicode mode, in raw format (four bytes).

2. In single-byte mode, with each half quoted. Thus, SQU, H1, L1, SQU, H2, L2.

3. Also in single-byte mode, as a single byte, by setting a dynamic window to the appropriate position with an SDX or UDX tag.

The 2-code sequence U+FEFF (or its reversed counterpart U+FFFE) occurs very rarely in text files, so it serves as a signature, to identify text files in Unicode. This sequence is known as a *byte order mark* or BOM. SCSU recommends several ways of compressing this signature, and an encoder can select any of those.

11.12.1 BOCU-1: Unicode Compression

The acronym BOCU stands for binary-ordered compression for Unicode. BOCU is a simple compression method for Unicode-based files [BOCU 01]. Its main feature is preserving the binary sort order of the code points being compressed. Thus, if two code points x and y are compressed to a and b and if $x < y$, then $a < b$.

The basic BOCU method is based on differencing (Section 1.3.1). The previous code point is subtracted from the current code point to yield a difference value. Consecutive code points in a document are normally similar, so most differences are small and fit in a single byte. However, because a code point in Unicode 2.0 (published in 1996) is in the range U+000000 to U+10FFFF (21-bit codes), the differences can, in principle, be any numbers in the interval $[-10\text{FFFF}, 10\text{FFFF}]$ and may require up to three bytes each. This basic method is enhanced in two ways.

The first enhancement improves compression in small alphabets. In Unicode, most small alphabets start on a 128-byte boundary, although the alphabet size may be more than 128 symbols. This suggests that a difference be computed not between the current and previous code values but between the current code value and the value in the middle of the 128-byte segment where the previous code value is located. Specifically, the difference is computed by subtracting a *base value* from the current code point. The base value is obtained from the previous code point as follows. If the previous code value is in the interval xxxx00 to xxxx7F (i.e., its seven least-significant bits are 0 to 127), the base value is set to xxxx40 (the seven LSBs are 64), and if the previous code point is in the range xxxx80 to xxxxFF (i.e., its seven least-significant bits are 128 to 255), the base value is set to xxxxC0 (the seven LSBs are 192). This way, if the current code point is within 128 positions of the base value, the difference is in the range $[-128, +127]$ which makes it fit in one byte.

The second enhancement has to do with remote symbols. A document in a non-Latin alphabet (where the code points are very different from the ASCII codes) may use spaces between words. The code point for a space is the ASCII code 20_{16}, so any pair of code points that includes a space results in a large difference. BOCU therefore computes a difference by first computing the base values of the three previous code points, and then subtracting the smallest base value from the current code point.

BOCU-1 is the version of BOCU that's commonly used in practice [BOCU 02]. It differs from the original BOCU method by using a different set of byte value ranges and by encoding the ASCII control characters U+0000 through U+0020 with byte values 0 through 20_{16}, respectively. These features make BOCU-1 suitable for compressing input files that are MIME (text) media types.

> Il faut avoir beaucoup étudié pour savoir peu (it is necessary to study much in order to know little).
>
> —Montesquieu (Charles de Secondat), *Pensées diverses*

11.13 Portable Document Format (PDF)

The first digital computers were developed in the 1940s during and after the Second World War, as a fast and flexible tool used mostly for code breaking. Already in the early 1950s, there was a great variety of computer models, made by diverse manufacturers such as Philco, Fairchild, NCR, and RCA. It is no wonder that these early computers did not follow any standards of data organization and were incompatible. The introduction of the ASCII code, in 1967 (it was last updated in 1986), did much to standardize digital data. For the first time, it became relatively easy to transfer data between computers—first on magnetic tapes and disks, and later through networks. With the advent of multimedia applications in the 1980s it became possible to represent, store, and edit any type of data, text, images, video, and audio in digital form in a computer. Users immediately felt the need for a standard that will make it easy to create, edit, print, and transfer multimedia documents between computers.

In 1991, John Warnock, a cofounder of Adobe Inc., came up with an outline of a standard that would enable computer users everywhere to do just that. His ideas became the basis of the portable document format (PDF) standard that is currently very popular. PDF was first introduced to the public in 1992 at the COMDEX conference, and became commercially available, under the tradename *Acrobat*, in 1993. The advantages of this technique were immediately noticed by commercial and private computer users, and already in 1994 the United States government adopted it to distribute tax forms to the public. The appearance of the free *Acrobat reader* software in 1995 did much to increase the popularity of PDF, and more was done by extensions to the basic standard. Features such as color, plugins for Internet browsers, and double-byte character codes, have gradually been added to PDF and in 1999 PDF became an ANSI standard. Today, there are several PDF versions including PDF/A (for archiving, published by the ISO in 2005), PDF/E (for engineering), and PDF/X (for printing). A general reference for PDF is [adobepdf 06].

A striking proof of the immense popularity of PDF came in 2000, when *Riding the Bullet*, a book by Stephen King, was downloaded 400,000 times within the first 24 hours of its becoming available as an e-book.

The main features of the PDF format are as follows:

- Any application can send its output to Acrobat software that converts it to PDF. Thus, it is easy to create documents in this format.

- Once a PDF document has been created, it can be sent to another computer platform where it can be read, edited, and printed.

- Acrobat software makes it possible to edit PDF documents by adding bookmarks and comments, drawing markups, measuring dimensions, touching up text, and cropping pages.

- Integrity. A PDF document looks exactly like the original. All the details of text, drawings, color graphics, and photographs are fully preserved.

- Security. It is possible to encrypt a PDF document with a password so that only authorized persons can read it. It is also possible to make the document read-only, requiring a password to copy its data or modify it. Password-protected digital signatures can also be included.

- Easy search. A PDF document can be searched for words or phrases. It is also possible to add bookmarks and skip to any bookmark quickly and easily.

- Open format. The details of the PDF format are available at [adobepdf 06].

- Small file size. PDF uses compression algorithms to compress the text and images in a document. This important feature is a special favorite of PDF users who love the small storage space occupied by large documents and the short time it takes to transfer them between computers.

The output file generated by a compressor is binary, but PDF can optionally encode such a file in ASCII base-85, a method developed by Adobe Inc., so it looks like a text file and consists of 7-bit ASCII codes. This process produces five ASCII codes for every four bytes of binary data and thereby causes expansion (by a factor of $5/4 = 1.25$). Given a group of four bytes b_1, b_2, b_3, and b_4, the encoder generates five output bytes c_1 through c_5 from the relation

$$b_1\times 256^3 + b_2\times 256^2 + b_3\times 256^1 + b_4 = c_1\times 85^4 + c_2\times 85^3 + c_3\times 85^2 + c_4\times 85^1 + c_5.$$

This relation can be interpreted as follows. A byte consists of eight bits, so it has values in the range 0 through 255. The four input bytes are considered the digits of a base-256 integer that is converted to a base-85 integer. Each of the resulting five bytes c_i therefore has values in the range 0 through 84 and is later converted to an ASCII character by adding 33. The resulting characters have codes from 33 (ASCII character `!`) up to 117 (ASCII character `u`). If all five output bytes are zeros, they are represented by ASCII code 122 (character `z`). The last group of input bytes may have only one, two, or three bytes, and is complemented to four bytes by adding 0 bytes as needed.

In comparison with the properties above, the compression features of PDF seem simple. The algorithms used by PDF to compress the various types of data are either well-known, widely-used methods or versions of such methods. PDF employs LZW (Section 6.13) to compress text, graphics, and images. Starting with PDF version 1.2, Flate (a variant of Deflate, Section 6.25) is also used for the same types of data. JPEG (Section 7.10) is used to compress color and grayscale images. Starting with PDF 1.5, JPEG 2000 (Section 8.19) is also used for the same purpose. Monochrome (bi-level) images can also be compressed in PDF by fax compression (group 3 or group 4, Section 5.7) or run-length encoding (Section 1.2). Starting with PDF 1.4, JBIG2 (Section 7.15) is also used for the same purpose. Acrobat software normally selects the appropriate compression method automatically, but users can, in principle, select or disable any compression algorithm. Disabling JPEG and JBIG2 may be important in certain applications because these methods are normally used for lossy compression and should be avoided in applications where data loss is prohibitive.

The run-length encoding used by PDF produces sequences of bytes where the first byte of a sequence specifies a length and may be followed by up to 128 bytes of data. If the length is in the range $[0, 127]$, it specifies no runs. In this case, the length byte is followed by length + 1 (i.e., between one and 128) data bytes. If the length is in the range $[129, 255]$, it implies that the single byte following it constitutes a run of

(257 – length) identical bytes (i.e., a run of length two to 128 bytes). A length of 128 specifies end-of-data (EOD).

In the best case (data that consists of one long run), this method produces a sequence of two bytes for each set of 128 identical bytes, which is equivalent to a compression factor of 64. In the worst case (no runs), each sequence of 128 data bytes is encoded into 129 bytes, thereby causing slight expansion.

11.14 File Differencing

The term *file differencing* refers to any method that locates and compresses the differences between two files. Imagine a file `A` with two copies that are kept by two users. When a copy is updated by one user, it should be sent to the other user, to keep the two copies identical. Instead of sending a copy of `A` which may be big, a much smaller file containing just the differences, in compressed format, can be sent and used at the receiving end to update the copy of `A`.

In the professional literature, this kind of compression is often called "differencing." We use "file differencing" because the term "differencing" is used in this book (Section 1.3.1) to describe a completely different compression method.

> Differencing: To cause to differ, to make different.

More formally, file differencing is concerned with the compression of a *target* data set given a *reference* data set. Applications of this compression technique include software distribution and updates (or *patching*), revision control systems, compression of backup files, and archival of multiple versions of data.

When used for the compression of backup files or in a revision control system, multiple versions of the same file can be stored compactly by storing the most recent version and the differences with the version immediately precedent. If necessary, the differences can be used to retrieve older versions. In these applications the difference (sometimes called *reverse delta*) is applied to retrieve an older version. The opposite happens in the case of a software distribution system, where the difference (or *patch*) is applied in order to update the software from an original (older) version.

If the reference and the target files are sufficiently similar, file differencing is able to generate compressed files that are orders of magnitude smaller than what is achievable with ordinary compression techniques.

File differencing is particularly useful for the distribution of software updates over the Internet. Even more relevant is its application to the patching and update of wireless mobile devices like cellular phones and PDAs, where the capacity of the wireless link is limited.

> I don't paint things. I only paint the difference between things.
> —Henri Matisse

11.14.1 UNIX diff

The earliest approach to file differencing uses a combination of operations that `APPEND`, `DELETE` and `CHANGE` lines of text to transform a file into another. A popular UNIX tool, `diff`, is based on this paradigm [Hunt 76]. Given two text files, `diff` generates a minimal set of line changes in the form of commands. The commands, applied in sequence, transform the first file into the second. A lossless compressor (`gzip`, for example) can be used to further reduce the size of the commands that `diff` outputs. `diff` generates commands in a human-readable format. Optionally, it can generate batch commands that can be fed directly to a text editor like `ed`.

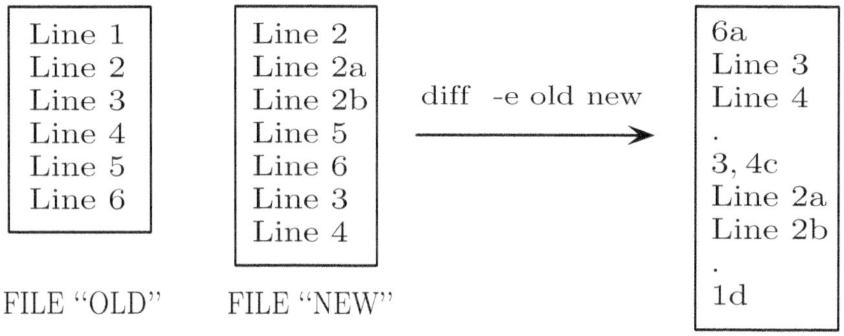

Figure 11.55: UNIX Command `diff`.

Figure 11.55 shows the output of `diff` applied to the text files `OLD` and `NEW`. The optional switch `-e` is used to obtain output that is compatible with the text editor `ed`. The output consists of line numbers, commands, and parameters. The dot "." ends a sequence of parameters. The command `a` is used to append new lines of text, `d` deletes one or more lines, and `c` replaces the content of one or more lines. With the file `OLD` opened in the editor `ed`, the command `6a` positions the cursor on the sixth line and appends the text lines "`Line 3`" and "`Line 4`". The next command, `3,4c` changes the third and the fourth line into "`Line 2a`" and "`Line 2b`", respectively. Finally, `1d` deletes the first line and completes the transformation. The individual effect of each command can be seen in Figure 11.56.

The commands issued by `diff` append, delete, and change entire text lines, so `diff` fails to capture character-based differences. An even bigger limitation preventing `diff` from achieving higher compression is its inability to copy patterns that are out of order or are repeated multiple times in the target file. For example, in Figure 11.56 we can see how "`Line 3`" and "`Line 4`" are appended at the end of the file with their text and explicitly sent as a parameter of the command `6a`. `diff` does not use the fact that two identical lines are already present in the original file and could be copied at the end of it with a command having a more compact representation.

The core algorithm used by `diff` is based on the solution of the Longest Common Subsequence problem [Cormen et al. 01]. Specific optimizations are introduced to contain memory usage and achieve good performance on typical text files. Variants and improvements to the original algorithm have been described in [Myers 86].

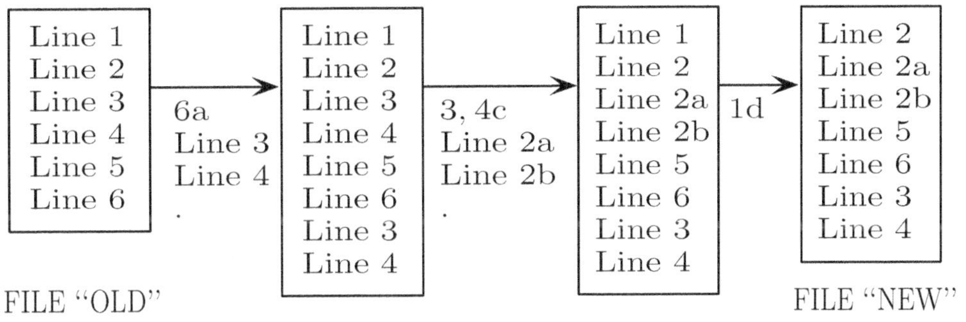

Figure 11.56: Patching With the Editor ed.

11.14.2 File Differencing: VCDIFF

The VCDIFF method described here is due to [Korn et al. 02]. We start with a source file. The file is copied, and the copy is modified to become a target file. The task is to use both files to encode the differences between them in compressed form, such that anyone who has the source file and the encoded differences will be able to reconstruct the target file. The principle is to append the target file to the source file, to pass the combined file through LZ77 or a similar method, and to start writing the compressed file when reaching the target file (i.e., when the start of the target file reaches the boundary between the look-ahead buffer and the search buffer). Even without any details, it is immediately clear that compression is a special case of file differencing. If there is no source file, then file differencing reduces to an LZ77 compression of the target file.

When the LZ77 process reaches the target file, it is compressed very efficiently because the source file has already been fully read into the search buffer. Recall that the target file is a modified version of the source file. Thus, many parts of the target file come from the source file and will therefore be found in the search buffer, which results in excellent compression. If the source file is too big to fully fit in the search buffer, both source and target files have to be segmented, and the difference between each pair of segments should be compressed separately.

The developers of VCDIFF propose a variant of the old LZ77 algorithm that compresses the differences between the source and target files and creates a compressed "delta" file with three types of instructions: `ADD`, `RUN`, and `COPY`. We imagine the source file, with bytes denoted by $S_0, S_1, \ldots, S_{s-1}$, immediately followed by the target file, with bytes $T_0, T_1, \ldots, T_{t-1}$. Both files are stored in a buffer U, so the buffer index of S_i is i and the buffer index of T_j is $s+j$. The VCDIFF encoder has a nontrivial job and is not described in [Korn et al. 02]. The encoder's task is to scan U, find matches between source and target strings, and create a delta file with delta instructions. Any encoder that creates a valid delta file is considered a compliant VCDIFF encoder. The VCDIFF decoder, on the other hand, is straightforward and employs the delta instructions to generate the target file as follows:

`ADD` has two arguments, a length x and a sequence of x bytes. The bytes are appended by the decoder to the target file that's being generated.

`RUN` is a special case of `ADD` where the x bytes are identical. It has two arguments, a

length x and a single byte b. The decoder executes RUN by appending x occurrences of b to the target file.
COPY also has two arguments, a length x and an index p in buffer U. The decoder locates the substring of length x that starts at index p in U and appends a copy to the target file.

The following example is taken from [Korn et al. 02]. Assume that the source and target files are the strings

$$abcdefghijklmnop$$
$$abcdwxyzefghefghefghefghzzzz$$

Then the differences between them can be expressed by the five delta instructions

```
COPY    4,0
ADD     4,wxyz
COPY    4,4
COPY   12,24
RUN     4,z
```

which are easily decoded. The decoder places the source file in its buffer, then reads and executes the delta instructions to create the target file in the same buffer, immediately following the source file. The first instruction tells the decoder to copy the four bytes starting at buffer index 0 (i.e., the first four source symbols) to the target. The second instruction tells it to append the four bytes $wxyz$ to the target. The third instruction refers to the four bytes that start at buffer index 4 (i.e., the string $efgh$). These are appended to the target file. At this point, the decoder's buffer consists of the 28 bytes

$$abcdefghijklmnop|abcdwxyzefgh$$

so the fourth instruction, which refers to the 12-byte starting at index 24, is special. The decoder first copies the four bytes $efgh$ that start at index 24 (this increases the size of the buffer to 32 bytes), then copies the four bytes from index 28 to index 31, then repeats this once more to copy a total of 12 bytes. (Notice how the partially-created target file is used by this instruction to append data to itself.) The last instruction appends four bytes of z to the buffer, thereby completing the target file.

A delta file starts with a header that contains various details about the rest of the file. The header is followed by windows, where each window consists of the delta instructions for one segment of the source and target files. If the files are small enough to fit in one buffer, then the delta file has just one window. Each window starts with several items that specify the format of the delta instructions, followed by the delta instructions themselves. For better compression, the instructions are encoded in three arrays, as shown in the remainder of this section.

A COPY instruction has two arguments, of which the second one is an index. These indexes are the locations of matches, which is why they are often correlated. Recall that a match is the same string found in the source and target files. Therefore, if a match occurs in buffer index p, chances are that the next match will be found at buffer index

$p+e$, where e is a small positive number. This is why VCDIFF writes the indexes of consecutive matches in a special array (`addr`), separate from their instructions, and in relative format (each index is written relative to its predecessor).

The number of possible delta instructions is vast, but experience indicates that a small number of those instructions are used most of the time, while the rest are rarely used. As a result, a special "instruction code table" with 256 entries has been specified to efficiently encode the delta instructions. Each entry in the table corresponds to a commonly-used delta instruction or a pair of consecutive instructions, and the delta file itself has an array (`inst`) with indexes to the table instead of to the actual instructions. The table is fixed and is built into both encoder and decoder, so it does not have to be written on the delta file (special applications may require different tables, and those have to be written on the file). The instruction code table enhances compression of the delta file, but requires a sophisticated encoder. If the encoder needs to use a certain delta instruction that's not in the table, it has to change its strategy and use instead two (or more) instructions from the table.

Each entry in the instruction code table consists of two triplets (or, equivalently, six fields): inst1, size1, mode1, and inst2, size2, and mode2. Each triplet indicates a delta instruction, a size, and a mode. If the "inst" field is 0, then the triplet indicates no delta instruction. The "size" field is the length of the data associated with the instruction (`ADD` and `RUN` instructions have data, but this data is stored in a separate array). The "mode" field is used only for `COPY` instructions, which have several modes.

To summarize, the delta instructions are written on the delta file in three arrays. The first array (`data`) has the data values for the `ADD` and `RUN` instructions. This information is not compressed. The second array (`inst`) is the instructions themselves, encoded as pointers to the instruction code table. This array is a sequence of triplets (index, size1, size2) where "index" is an index to the instruction code table (indicating one or two delta instructions), and the two (optional) sizes preempt the sizes included in the table. The third array (`addr`) contains the indexes for the `COPY` instructions, encoded in relative format.

11.14.3 Zdelta

Zdelta is a file differencing algorithm developed by Dimitre Trendafilov, Nasir Memon and Torsten Suel [Trendafilov et al. 02]. Zdelta adapts the compression library `zlib` to the problem of differential file compression. Similarly to VCDIFF, zdelta represents the target file by combining copies from both the reference and the already compressed target file. A Huffman encoder is used to further compress this representation.

Copies from the reference and target files are found with the use of two hash tables. If the reference file is sufficiently small, the corresponding hash table is fully built in advance by hashing each sequence of three consecutive characters (3-grams) in the reference file. In the hash table, each 3-gram is associated with the position of its first character in the file. Large reference files require the use of a window to contain memory usage. The hash table for to the target file is constructed during the compression. As in `zlib`, hash entries are never deleted individually. The hash table is flushed periodically to make space for new entries.

A `COPY` command has four parameters: the length of the copy, the offset from one of several pointers, the pointer itself, and the direction of the offset. One of the most important differences from VCDIFF is the use of multiple pointers to specify the location of the copy. The use of multiple pointers allows a more compact encoding. Zdelta maintains and updates independently one or more pointers in the reference file and an implicit pointer in the target file. The implicit pointer in the target file always points to the beginning of the section that is about to be compressed. Offsets can be specified from any pointer, but a more compact encoding is achieved by always preferring the pointer that generates the smaller offset.

The position of each pointer represents a guess (prediction) of the location of the next match. The offset (i.e., the most accurate of these guesses) can be interpreted as the prediction error.

After each match, the pointers in the reference buffer are moved according to a predetermined strategy that aims at predicting the position of the next copy. In [Trendafilov et al. 02] experiments show that the best performances can been achieved by maintaining only two pointers in the reference file and one in the target. Adding more pointers complicates encoding and brings only a very modest compression gain. Only one pointer in the reference file is updated after a match. If the offset of the current match is smaller than a given amount, the pointer used for that offset (the one closest to the match) is moved to the end of the copy. This is done to anticipate the location of the next copy in the case of similar files. Otherwise, the other pointer is moved to the end of the copy. This strategy takes into account the possibility of an isolated match. If zdelta cannot find a match of at least three characters, one character is emitted as a literal and the matching process restarts from the next character in the file.

The implementation described in [Trendafilov et al. 02] allows matches of up to 1026 characters and offsets in the range $[0, 32766]$. Zdelta tries to reuse as much as possible the library `zlib`. Unfortunately, the Huffman encoder used in `zlib` allows only codes for lengths of $[0, 255]$ characters. Zdelta overcomes this limitation by representing the length l of a match as $L = (l + 3) + 256 \times c$ and encoding L and c separately. When three pointers are used, it is possible to encode in a single code word, called *zdelta flag*, the parameter c, the pointer, and the sign of the offset.

The encoding used in [Trendafilov et al. 02] is listed in Table 11.57 and uses 20 different codes since there are four possible values for the parameter c and five combinations of pointer and offset direction. The direction of the offset can be positive or negative for pointers in the reference buffer but only negative for the implicit pointer in the target buffer. Zdelta uses three different Huffman trees: one for the offsets, one for the literals, and the lengths, and one for the flags.

c (lengths)	$-ptr_{\text{target}}$	$+ptr_{\text{ref}(1)}$	$-ptr_{\text{ref}(1)}$	$+ptr_{\text{ref}(2)}$	$-ptr_{\text{ref}(2)}$
0 (3–258)	code 1	code 2	code 3	code 4	code 5
1 (259–514)	code 6	code 7	code 8	code 9	code 10
2 (515–770)	code 11	code 12	code 13	code 14	code 15
3 (771–1026)	code 16	code 17	code 18	code 19	code 20

Table 11.57: Zdelta Encoding for the Flags.

A greedy strategy is used to search for matches. All matches found are compared according to their length, and longer matches are preferred to shorter ones. If a match offset is larger than a given constant, its length is penalized so that a slightly shorter but closer match can be selected instead. Closer matches are always preferred because their offset can be encoded more compactly.

The best match is not emitted right away. Its length is saved in a variable l_{prev} and compared to the best match starting from the next character. The longer of the two matches is emitted. If the second of the two is selected, the extra character is emitted as a literal. This strategy is borrowed from `zlib`, and it is known to achieve good compression because it mitigates the greediness of the match selection by deferring the choice of the longer match.

A hash table with overpopulated buckets slows down execution time without any substantial improvement in the compression. The occurrence of extremely large buckets can be due to a faulty hash function (and so to a design error) or to many repetitions of the same pattern in the files. Since zdelta relies on the hash function used by `zlib`, it is likely that a large bucket is caused by many repetitions of the same pattern. This happens for example when a file contains many consecutive repetitions of the same character. Zdelta prevents this problem by limiting the number of elements searched in a given hash bucket to 1,024.

Zdelta reuses many functions from `zlib`. The main differences are:

- An additional hash table maintains pointers to the reference data set.
- An extra Huffman tree is used to encode the zdelta flags.
- Compression is performed in a single step, with the reference file entirely available from the beginning.

11.14.4 Exediff

A differential file compression algorithm based on the operations of COPY and ADD works well in the case of text files because insertion and deletion of new material are the operations typically performed on text. Furthermore, in text files, changes tend to be localized. Material may be added, deleted, or moved, but in general most of the text remains the same. The situation is quite different in the case of executable code (binary files). Executable code uses relative or absolute references to objects (functions, variables, etc.), so a change in the source code, such as the addition of a new object, or its relocation, will affect references in sections of the file that have not been changed explicitly. For example, if new code is added to a binary file, as in Figure 11.58, all relative offsets crossing the inserted section are likely to change. The same will happen to absolute references following the insertion point. As a consequence, a small change in the source program is likely to result in changes spread throughout the entire executable file.

When discussing differential compression of executable code, it is helpful to make a distinction between the two kinds of file differences typically encountered:

- *Primary Changes.* Changes that are caused by modifications of the source code such as the addition of a new function, the deletion of an object, or a change to the structure of a control loop.
- *Secondary Changes.* That refer to changes in the values assumed by pointers, offsets, and references introduced as a consequence of a primary change.

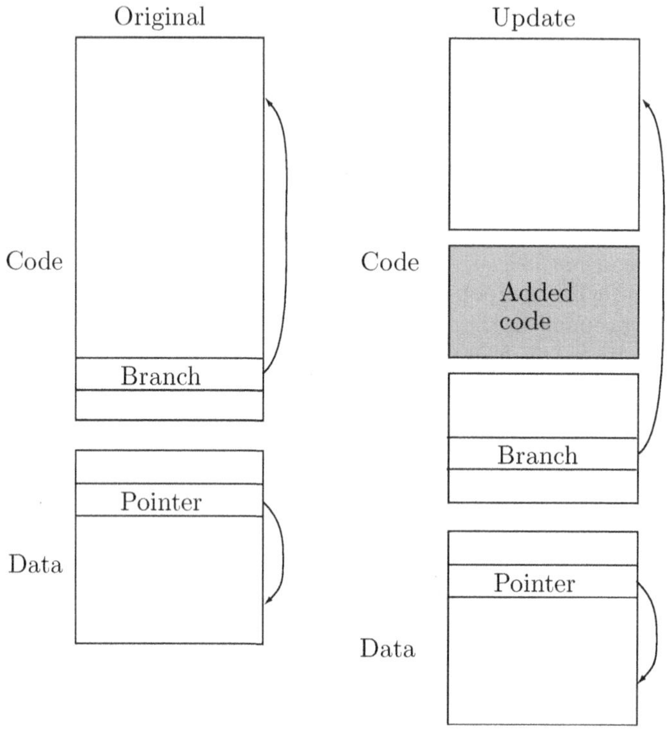

Figure 11.58: Changes in Executable Files After the Addition of New Code.

Because of the secondary changes, sections of the file that have not been explicitly modified may not match. If we assume that the compressor has no access to the original and to the target source programs, a brute-force approach relies on a full disassembling of the reference and of the target files. Corresponding objects can be matched, the references that have changed from one version to another adjusted, and finally a file differencing algorithm like VCDIFF or zdelta can be used.

This solution is very effective and allows compression two to five times better than ordinary file differencing algorithms. However, it has the disadvantage of being complex and platform (or, depending on the implementation, even compiler and linker) dependent. Furthermore, disassembling a binary file can be an extremely hard task, complicated by variable-length instructions, compiler optimizations, human intervention, and data structures mixed with the instructions.

A more interesting solution is proposed by Brenda Baker, Udi Manber, and Robert Muth in [Baker et al. 99]. It uses two algorithms called *exediff* and *exepatch*. The former generates a patch, given an original and an updated executable code. The latter applies the patch to the original code in order to retrieve the update.

Exediff uses a lossy transform to reduce the effect of the secondary changes in the executable code. It iterates two operations called *pre-matching* and *value recovery* until the size of the patch converges to a minimum and cannot be further reduced. Pre-matching is based on a solution of the Longest Common Subsequence algorithm described in [Cormen et al. 01]. The lossy transform relies on the detailed knowledge of the architecture since it involves locating the instructions and the pointers in the binary file. Exediff has been developed and tested on a 64-bit Alpha microprocessor, however the algorithm can be adapted to other architectures. The implementation on other architectures can be greatly complicated by the presence of variable-length instructions. However, exediff does not assume access to the source code.

The lossy transform described in [Baker et al. 99] assumes that the code being compressed has been developed for a DEC UNIX Alpha. DEC UNIX Alpha executables are stored using an object file format called ECOFF and are composed of three sections:

1. *Text Segment.* Containing instructions and read-only constants.
2. *Data Segment.* Containing uninitialized data structures.
3. *Bss Segment.* Containing zero initialized data structures.

Pre-matching and value recovery are applied only to the Data Segment and to the executable code in the Text Segment. In [Baker et al. 99] it is assumed that all instructions in the Text Segment are 64 bits long and consist of an *operand code*, three *registers*, and an *immediate* value: *opcode*, reg_1, reg_2, reg_3, and *immediate*. The values for reg_i indicate the index of a general-purpose or a special-purpose register.

The transform locates machine instructions, inspects the three registers reg_1 through reg_3 and replaces the index of the general-purpose registers with a default value called ϵ. The indexes of the special-purpose registers are left unchanged. The *immediate* operand is replaced by a flag indicating whether its value is negative or not: If *immediate* is non-negative, it is set to POS; otherwise, it is set to NEG. Similar operations are performed on the pointers in the Data Segment. A 64-bit pointer to an instruction in the Text Segment is replaced by the keyword "$TEXT$" followed by the hashing of the first three opcodes at the destination of the pointer. A pointer to a byte b in the Data Segment is replaced by the keyword "$DATA$" followed by the value of b. Pointers to other objects are replaced by the keyword "$OTHER$".

By replacing the items subject to secondary changes with hash values, flags, and keywords, the transform attempts to remove the effect of the secondary changes while preserving valuable information that can be used in the matching performed by the Longest Common Subsequence algorithm. LCS aligns the transformed files in an attempt to find the primary changes. The common subsequences correspond to matches between the transformed original and the transformed upgrade file. Matches will retain the same order in the two files and they will never cross.

The LCS partitions the two files into sections as follows:

1. *Matched and equal.* These sections represent executable code that has not changed between the original and the update.

2. *Matched but not equal.* After the lossy transform, the sections in the original and the update match. These matching sections are likely to represent secondary changes. Some of the matches will be spurious and will be eliminated by the iterative process.

3. *Unmatched sections.* They represent code that has changed between the two versions. Unmatched sections correspond to primary changes.

Matched sections are reconstructed by exepatch with the use of `COPY` commands. Unmatched sections will consist of inserted material and will translate to `ADD` instructions. Sections that are matched but unequal (because of the secondary changes) will be copies but they need an extra step in order to faithfully reflect the content of the update file.

Exediff uses value recovery to predict the content of the lossy transformed pointers, registers, and immediates. Exepatch performs a similar prediction, so value recovery must rely on information common to both encoder and decoder. Values that cannot be predicted correctly are called *Unrecoverable Matching Items* and are explicitly stored in the patch. Values are recovered in sequence, starting from the small to the high offsets in the upgrade file. Recovered values are used in successive recoveries to determine a cascade effect.

Value recovery is based on a set of chained heuristics that depend on the domain knowledge and on the specific features of the type of item being predicted (register, pointer, immediate value, etc.. . .). The heuristics are applied in a predetermined sequence known to both exediff and exepatch.

The following recovery schemes are described in [Baker et al. 99]:

1. *Match Value.* The value is already matching and needs no further prediction.

2. *Translate Address.* The value represents the address of an object within a file and that address can be computed by using the relative offset of the two matches. By address we mean any kind of reference like an offset, the index of a table, a file position, etc.

3. *Equal Value.* An identical value has been successfully recovered in a previous match. The previously recovered value can be reused in the current match.

4. *Close Value.* The value can be computed from a numerically close value recovered in a previous match.

The lossy nature of the transform used in the pre-matching may generate spurious matches. Spurious matches increase the size of the patch because they cause value recovery to fail. Values that cannot be recovered have to be sent explicitly in the patch as unrecoverable matched items. This problem is solved in [Baker et al. 99] with an iterative method that performs the pre-matching, computes the "benefit" of each match, eliminates the matches that compromise compression, and repeats the process until no further reduction of the patch size is achievable.

11.14.5 BSDiff

BSDiff is an algorithm that addresses the problem of differential file compression of executable code while maintaining a platform-independent approach.

BSDiff has been created by Colin Percival while working on FreeBSD Update as a diversion from his Doctoral research. During this period, Colin wrote several versions of BSDiff (up to version 4.0) before realizing, four months later, that it was possible to improve BSDiff's matching algorithm and write a Doctoral Thesis about it and the related matching problems [Percival 06]. The algorithm version described in the Thesis (BSDiff 6) is based on the same principles as BSDiff 4 but uses a more sophisticated matching algorithm. Unless otherwise specified, in the following we focus on the description of BSDiff 4, since BSDiff 6 is not yet available to the public.

While originally designed to solve the problem of binary security update for FreeBSD UNIX [Percival 03a], the most recent release of BSDiff 4 (version 4.3) has been ported on several platforms and is available on FreeBSD, NetBSD, OpenBSD, Darwin, Debian,

Gentoo, OS X, and Windows. BSDiff is currently used by FreeBSD and OS X to distribute binary security updates and, with some modifications, by the Mozilla project to speed-up the download of FireFox updates.

The algorithm that BSDiff 4 [Percival 03b] uses is based on the following observations:

1. Regions of executable code affected by secondary changes (i.e., not involved in a modification of the source program, See 11.14.4) will present sparse differences. Secondary changes will constitute only a small portion of the compiled code, and references are likely to change only in one or two bytes.

2. Since data and code are usually moved around in blocks, there will be many references affected by secondary changes that need to be adjusted by the same amount.

When sections of the original and the update files are matched against each other with an approximate matching algorithm, the positions in which the matching sections do not match will be sparse and the bytewise differences between these positions will be highly compressible.

BSDiff 4 scans both the original and the update file and builds an index with a hash table or with a similar method. Then, by using the index, it finds a set of exactly matching regions. These regions are further extended forward and backward by allowing mismatches. The set of approximately matching regions will roughly correspond to secondary changes and to unmodified sections of code. Regions in the update files for which no approximate match can be found correspond to primary changes. With this set of approximate matches, BSDiff 4 creates a patch file consisting of:

1. A *control* section, containing `ADD` and `INSERT` instructions. Each `ADD` instruction specifies an offset in the original file and the number of bytes being copied from the original file. An `INSERT` instruction specifies the number of bytes that have to be inserted in the update file. Inserted bytes are stored together in another section.

2. A *difference* section, containing the bytewise differences between the approximate matches.

3. A section containing the bytes that were not part of the approximate matches. These bytes will be inserted in the update file by the `INSERT` instructions in the control section.

The concatenation of these three sections is slightly bigger than the update file. However, the control section and the bytewise differences are highly compressible, so BSDiff 4 uses `bzip2` to effectively reduce the size of the patch. `bzip2` is called independently for each section. The three sections exhibit different statistics, and independent compression often provides a substantial improvement over compressing everything at once. For the same reason, the bytes that are added by the `INSERT` instruction are not interspersed with the instruction codes in the control section but are instead grouped into a separate section.

> Birds of a feather compress better together.
>
> —Colin Percival, *Matching with Mismatches and Assorted Applications*, (2006)

A decoder program called *bspatch* decompresses the patch file and applies in sequence the instructions contained in the control section. For each `ADD` instruction, bspatch copies bytes from the original file to the update. Copies represent sections that have been approximately matched, so bspatch must retrieve an equal amount of bytes from the difference section of the patch and bytewise add these bytes to the copied section. For each `INSERT`, bspatch will merely retrieve the specified number of bytes from the third section of the patch and insert them in the update file.

BSDiff 4 and exediff achieve comparable compression on a set of reference files [Percival 03b]. This result is remarkable since BSDiff 4 is platform-independent. Even better compression is achieved by BSDiff 6, the version described in [Percival 06]. While relying upon the same basic principles of encoding the locations of approximately matching regions and their sparse differences, BSDiff 6 uses a different and more sophisticated matching algorithm. On the same set of reference files, BSDiff 6 improves the compression of BSDiff 4 by about 10% to 20%.

We would like to thank Colin Percival for his comments on this section.

11.15 Hyperspectral Data Compression

A digital image consists of pixels. In a monochromatic (bi-level) image, a pixel can have one of two colors, so it is represented by one bit. In a color image, a pixel represented by k bits can have one of 2^k colors. A typical value is $k = 24$, where a pixel can have approximately 16.78 million colors. Current inexpensive display monitors for personal computers can display this number of colors. It seems that such a large number of colors in a single image would be sufficient for any purpose, but life isn't that simple. There is a large (and growing) field of applications that require images where each pixel is represented by hundreds or even thousands of bits. Such a large set of data is no longer referred to as an image, but is termed *hyperspectral data.* Following are a few examples of applications that require hyperspectral data.

1. Spy satellites (officially referred to as reconnaissance satellites or recon sats). It is not enough to have a high-resolution camera mounted on a satellite, taking pictures, and transmitting them to Earth. The enemy can easily hide a tank by covering it with branches and leaves. The enemy can mislead us by making tanks, airplanes, and armored vehicles from inflatable rubber and plastic. When a camera takes a picture, it registers the visible light reflected from the objects it sees. In order to distinguish between light reflected from steel and light reflected from rubber (or between light reflected from normal tree branches and light reflected from tree branches placed over a tank) the camera in a spy satellite has to look at the ground in more than just visible light. It has to measure the radiation reflected from each point on the ground in many wavelengths.

A wavelength is a real number, not an integer, so in principle there is an infinite number of wavelengths in even a very narrow part of the electromagnetic spectrum. Naturally, a practical camera cannot measure the reflection intensity of radiation at the precise wavelength of 423.708235 nm. It may measure the radiation in, say, the narrow frequency range of 420–430 nm, and such a range is called a *band.* A typical spy camera consists of a set of sensitive sensors that can measure and record radiation in perhaps 250 frequency bands normally located between 400 nm and 2,400 nm. This includes the visible range (400–700 nm) and part of the infrared range of the electromagnetic spectrum and may contain information that's much more useful than visible light alone. (The infrared range is very wide and includes wavelengths of up to 1 mm.)

As an example, the AVIRIS sensor (airborne visible/infrared imaging spectrometer) consists of three sensors of 64 frequency bands each plus a fourth sensor with 32 bands, for a total of 224 bands. Other sensors have different numbers of bands ranging from hundreds to thousands.

The eye cannot perceive infrared radiation (see discussion of human vision on page 526), which is why such an image must be painted artificial colors. A trained person looking at a high-resolution image in artificial colors can immediately tell the difference between a rubber tank and a steel tank, between real tree branches and branches hiding a rocket launcher. If more detailed analysis is required, only a few bands may be displayed at a time and painted artificially.

Thus, each "pixel" in the image taken by such a camera consists of 250 numbers, each an integer of at least 16 bits. The resolution of the camera (the size of a pixel on the ground) must be high. The resolution of modern spy satellites is kept secret, but is conjectured to be around 10 cm. As a result, the amount of data collected by a spy satellite for one square kilometer is $10{,}000^2\times 250\times 2 = 5\times 10^{10}$ bytes; huge, but crucial for successful battlefield surveillance!

2. Remote sensing. This term refers to the use of a sensor to remotely gather data about a certain environment. The data may be optical (reflection intensities at many wavelengths), acoustic (intensity of sound waves at various frequencies), or the concentrations of chemicals. The following are typical examples of remote sensors:

- Radar is used to measure range and velocity of well-defined targets such as an aircraft or blurred targets such as a cloud of water vapor.

- Radar altimeters (both microwaves and laser) mounted on satellites are used to map the entire Earth (to a high precision) as well as features on the sea floor (to a much lower precision).

- Lidar, radiometers, and photometers measure the concentrations of various chemicals. This application provides important information on chemical concentrations in the atmosphere.

- Oil and mineral companies are constantly on the lookout for new deposits. They mount sensors in earth observation satellites and measure the sunlight reflected from large regions of the Earth in attempts to locate natural resources.

- Oceanographers use sonar to "look" deep underwater for interesting objects, marine species, and changes in the seabed. This is a hyperspectral application because different sound frequencies penetrate the water and are reflected from objects in different ways.

- Geologists use seismometers to "see" inside the ground, either to locate resources or to study geologic formations. The principle is to create acoustic waves underground with an explosion, and it makes sense to create several explosions with varying powers and measure the acoustic waves created by each.

- Medical imaging. In addition to X rays and magnetic resonance imaging, modern medicine can see inside our bodies with ultrasonography. High-frequency sound waves (typically 2 to 10 MHz) are directed into an area of interest in the body and are reflected by tissues in different ways to produce an image on a display monitor. This type of imaging is often used to visualize the fetus during pregnancy.

■ Many scientists are interested in information such as the health of crops in many areas or the spread of plankton in the oceans. Such information can be part of a long range study of climate changes and global warming.

In all these cases, the result is a large amount of data organized in three dimensions. Two dimensions are spatial (each pixel has a location and is identified by a pair (x, y) of coordinates) and the third dimension is spectral (the bands of a pixel, normally indexed by λ or by t). An image has a resolution, but hyperspectral data has both spatial resolution (the size of a pixel) and spectral resolution (the number of spectral bands of a pixel). Data with two spectral bands per pixel is called dual band. Data with three to several (perhaps 8–10) bands is termed multispectral, and data with more bands is hyperspectral. Figure 11.59 shows a typical organization of hyperspectral data. Each plane is a band and it consists of rows and columns of pixels. The smallest units are called pixels even though they are not always visual units.

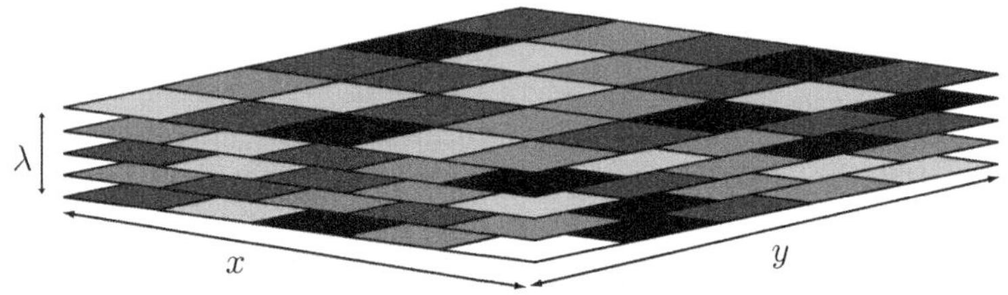

Figure 11.59: Organization of Hyperspectral Data.

We can interpret each plane of Figure 11.59 as an image (pixels displayed in spatial relationship to one another) and each column as a spectrum (variations within pixels as a function of wavelength). A column can also be considered a point in n-dimensional space.

Before we talk about compression of hyperspectral data, here are a few words about its processing. Most algorithms for processing hyperspectral data are concerned with anomaly detection or material identification, but the first step in processing is to remove the effects of varying illumination from a block of data. A satellite may collect data over a certain region for several hours, and the light reflected from the ground during this time varies as the sun moves across the sky. This variation has to be removed before any other processing can be done.

Because of the vast size of a block of hyperspectral data, compression is important. An overview of various methods for hyperspectral data compression can be found in [Motta et al. 06]. The data can be compressed either at the source (the satellite, seismometer, or other sensor) or at the sink (the computer to which it is sent and where it is stored and processed). A satellite may be out of touch with ground stations several times during each orbit, so it may spend these times compressing data on board. When communication resumes, the satellite may stop compressing and start sending the compressed data to Earth. A seismometer, on the other hand, should be portable, lightweight, and inexpensive, so it makes sense for it to simply transmit its data in raw format as soon as it is collected, to a remote computer where it will be stored, compressed, and processed.

Figure 11.59 shows that compressing hyperspectral data is similar to image compression with the added feature of inter-band correlation. An image can be considered hyperspectral data with one band, and compressing an image exploits the correlation of the pixels in this band (intra-band correlation). When several bands exist, a compression method should also take advantage of the correlation of pixels across neighboring bands (inter-band correlation). It is well known that in a color image there is correlation between the color planes. The difference between an image and hyperspectral data is that the bands in the latter type are narrow, so the correlation is higher in the spectral domain than in the spatial domain.

More often than not, compression should be lossless, because certain applications (especially spying and medical) may depend crucially on the values of a few individual pixels. In fact, important data items may sometimes be smaller than a pixel. Lossy compression of hyperspectral data may make sense in cases where such data is sent for an initial evaluation. Sometimes, a potential buyer may want to examine a block of data before buying it, and may not mind looking at data that is compressed lossily. Alternatively, certain clients may be attracted by the low cost of lossy-compressed hyperspectral data. When lossy compression is chosen for hyperspectral data, it is important to develop a suitable distortion measure that will guarantee the quality of the compressed data. The guiding principle is that applying any processing algorithm to the lossy data and the lossless data will produce results that differ only in the details, not in their principal features.

Given a block of hyperspectral data to be compressed, we denote a general pixel before the compression by $B(x, y, t)$ and the same pixel after lossy compression and decompression by $\hat{B}(x, y, t)$. The simplest distortion measure is maximum absolute distortion (MAD). It guarantees that the absolute value of the difference $B(x, y, t) - \hat{B}(x, y, t)$ does not exceed a certain distance parameter d. The downside of this distortion measure is that small pixels may become relatively more distorted than large ones; MAD doesn't take into account the absolute values of individual pixels.

A better distortion measure is percentage maximum absolute distortion (PMAD), where the absolute difference $|B(x, y, t) - \hat{B}(x, y, t)|$ is always maintained at or below $p \times B(x, y, t)$ where the percentage parameter p is in the range $[0, 1]$. Other measures are possible.

Given a suitable distortion measure, Figure 11.60a is a block diagram of the main components of a lossy encoder for hyperspectral data. The compressor itself can be any standard algorithm—such as a transform, vector quantization, or prediction—that has been extended to take advantage of the three-dimensional correlations in the data. The most important feature of the encoder is the feedback of information from the compressor to the distortion measure. The latter component examines the information and decides whether more compression is needed. The preprocessing stage is optional. It involves a simple process, such as band reordering, that improves the main compression that follows it. Preprocessing must be reversible, because the decoder (Figure 11.60b) has to perform postprocessing. Notice how the side information is sent to the decoder by interleaving it with the compressed data in the output stream.

The remainder of this section discusses the extensions of several standard compression techniques to hyperspectral data.

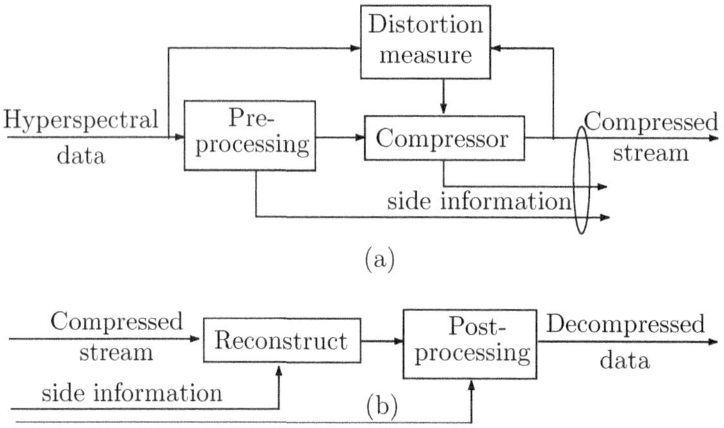

Figure 11.60: Organization of a Codec for Hyperspectral Data.

11.15.1 Predictive Methods

In a predictive method, the encoder predicts the current data item x from several of its predecessors. The prediction $\hat{x}$ is subtracted from x to form a residual e, which is then entropy encoded. If the prediction is done properly and if the individual data items are correlated, the residuals $e = x - \hat{x}$ will normally be small integers which are easy to encode by variable-length codes. Notice that only predecessors of x can be used in the prediction because the decoder has to perform the same prediction and it has access only to those data items that have already been decoded (the predecessors).

Many lossless audio compression methods use this approach. Shorten (Section 10.9) employs order-n linear predictors on its input (a one-dimensional sequence of audio samples) to predict the current sample $s(t)$. The first four linear predictors, corresponding to n values 0 through 3, are given by Equation (10.5), duplicated here.

$$\begin{aligned}\hat{s}_0(t) &= 0,\\ \hat{s}_1(t) &= s(t-1),\\ \hat{s}_2(t) &= 2s(t-1) - s(t-2),\\ \hat{s}_3(t) &= 3s(t-1) - 3s(t-2) + s(t-3).\end{aligned} \tag{10.5}$$

When a predictive method is applied to the compression of images, it predicts a pixel by computing a weighted sum of its near neighbors, assigning more weight to nearby neighbors. Figure 7.72 shows how JPEG-LS (Section 7.11) predicts a pixel x from the three neighbors c, b, and d above it and the neighbor a to its left. These neighbors will be known to the decoder when it gets to decoding x. The context-tree weighting method (Section 7.31) similarly predicts a pixel X from the same four neighbors (Figure 7.155) by means of a weighted sum. If we consider an image a two-dimensional array of pixels arranged in rows and columns, we can express such prediction with the notation

$$\hat{x}[i,j] = a\,x[i-1,j-1] + b\,x[i-1,j] + c\,x[i-1,j+1] + d\,x[i,j-1], \tag{11.5}$$

where the four weights a through d should add up to 1.

It is now clear how linear prediction can be extended to hyperspectral data. A block of data is considered a three-dimensional rectangular box, where each pixel $B[x,y,t]$ has three coordinates and is predicted by pixels above it, to its left, and close to it in the preceding band (band $t-1$). If the encoder scans the hyperspectral data band by band, then Equation (11.5) can be extended to

$$\begin{aligned}\hat{B}[x,y,t] =&\Big(B[x-1,y-1,t]+B[x,y-1,t]+B[x+1,y-1,t]+B[x-1,y,t]\\&+B[x,y-1,t-1]+B[x-1,y,t-1]\\&+B[x,y,t-1]+B[x+1,y,t-1]+B[x,y+1,t-1]\Big)/9\end{aligned}$$

(Figure 11.61a, where each pixel receives a weight of 1/9). Notice that any pixel in band $t-1$ is known to the decoder and can therefore be used for the prediction. An alternative is for the encoder to scan the data row by row, and for each row band by band. In such a case, not all the pixels in band $t-1$ are available for prediction, and Equation (11.5) can be extended to

$$\begin{aligned}\hat{B}[x,y,t] =&\Big(B[x-1,y-1,t]+B[x,y-1,t]+B[x+1,y-1,t]+B[x-1,y,t]\\&+B[x-1,y-1,t-1]+B[x,y-1,t-1]+B[x+1,y-1,t-1]\\&+B[x-1,y,t-1]+B[x,y,t-1]+B[x+1,y,t-1]\Big)/10\end{aligned}$$

(Figure 11.61b, where each pixel receives a weight of 1/10). It is also possible to assign different weights to the pixels used in the prediction. The weights should add up to unity and should reflect the distance of a predicting pixel from the predicted pixel $B[x,y,t]$.

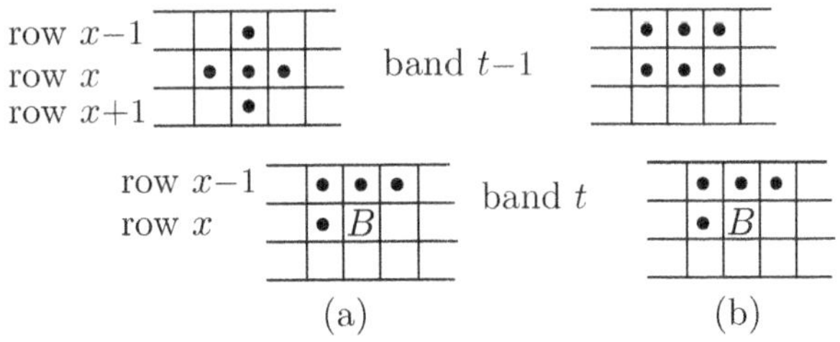

Figure 11.61: Two Alternative Predictions for Hyperspectral Data.

A practical example of this approach to hyperspectral data compression is the three-dimensional adaptive differential pulse code modulation (ADPCM) algorithm described in [Roger and Cavenor 96]. The two authors have tried 25 configurations of pixels for prediction and have experimentally selected the five listed in Table 11.62.

The prefixes SA, SE, and SS refer to spatial, spectral, and mixed predictors, respectively. The last three predictors use parameters that were obtained by minimizing

Name	Pixels used
SA-2RC	$(B[x-1,y,t]+B[x,y-1,t])/2$
SS-1	$B[x,y-1,t]+B[x,y,t-1]-B[x,y-1,t-1]$
SE-01B	$a+b\,B[x,y,t-1]$
SE-02B	$a+b\,B[x,y,t-1]+c\,B[x,y,t-2]$
SS-01	$a+b\,B[x,y-1,t]+c\,B[x,y,t-1]+d\,B[x,y-1,t-1]$

Table 11.62: Five Prediction Configurations for Hyperspectral Data.

the variance of the prediction errors in each row of pixels. Assuming that there are n pixels in each row, the variance of row x is the sum

$$\sum_{y=1}^{n}\left(\hat{B}[x,y,t]-B[x,y,t]\right)^2.$$

This sum depends on the values of a, b, c, and d, and the set of four parameters that results in the smallest sum (as computed by least-squares minimization) is chosen for pixel prediction for row x. This set is also sent to the decoder as side information, and another set is computed for the next row.

The residuals resulting from this prediction are encoded by Rice codes. This type of code was selected experimentally by the developers of ADPCM as the one that produced the best results.

11.15.2 Three-Dimensional DCT

Section 7.8 is a detailed discussion of the discrete cosine transform (DCT). It explains the origins of this important transform and shows how it can be applied to the compression of digital images. Even a cursory glance at Equation (7.16) shows that the DCT is two-dimensional, which immediately suggests the possibility of extending it to three dimensions and applying it to the compression of hyperspectral data. The extension is straightforward. The forward DCT in three dimensions (3DCT) becomes

$$\begin{aligned}G_{ijk}=&\sqrt{\frac{2^3}{n^3}}C_iC_jC_k\sum_{x=0}^{n-1}\sum_{y=0}^{n-1}\sum_{z=0}^{n-1}p_{xyz}\\&\cos\left[\frac{(2x+1)i\pi}{2n}\right]\cos\left[\frac{(2y+1)j\pi}{2n}\right]\cos\left[\frac{(2z+1)k\pi}{2n}\right],\end{aligned}\qquad(11.6)$$

for $0\le i,j,k\le n-1$ and for

$$C_f=\begin{cases}\frac{1}{\sqrt{2}}, & f=0\\ 1\ , & f>0\end{cases}.$$

And the inverse three-dimensional DCT is

$$p_{xyz}=\sqrt{\frac{2^3}{n^3}}\sum_{i=0}^{n-1}\sum_{j=0}^{n-1}\sum_{k=0}^{n-1}C_iC_jC_kG_{ijk}$$

$$\cos\left[\frac{(2x+1)i\pi}{2n}\right]\cos\left[\frac{(2y+1)j\pi}{2n}\right]\cos\left[\frac{(2z+1)k\pi}{2n}\right], \tag{11.7}$$

for $0 \le x, y, z \le n-1$.

Those familiar with the principles of JPEG (Section 7.10) will find it easy to visualize its extension to three-dimensional data. Simply partition a large set of hyperspectral data into cubes of $8\times8\times8$ pixels each, apply the 3DCT to each cube, collect the resulting transform coefficients in a zigzag sequence, quantize them, and encode the results with an entropy coder such as Huffman code. This works, but two points should be taken into consideration.

The first point has to do with the correlation between bands. Depending on the nature of the hyperspectral data, the user may know or suspect that the inter-band correlation is weaker than the intra-band correlation. In such a case, the cubes that are transformed and quantized may contain fewer bands than spatial dimensions and become rectangular boxes. Such a box may, for example, have eight spatial dimensions but only four spectral dimensions. Equations (11.6) and (11.7) should be modified accordingly. The sum on z should go from 0 to 3, the cosines should have either 2×8 or 2×4 in their denominators, and the constant should be $\sqrt{2^3/8\times8\times4}\approx0.177$.

The second point has to do with the zigzag sequence. Figure 11.63 shows simple *Mathematica* code that computes the 3DCT of a 4×4×4 block of random integers. The constant becomes $\sqrt{2^3/4\times4\times4}\approx0.3536$.

```
(* 3D DCT for hyperspectral data *)
Clear[Pixl, DCT];
Cr[i_]:=If[i==0,1/Sqrt[2],1];
DCT[i_,j_,k_]:=(Sqrt[2]/32) Cr[i]Cr[j]Cr[k]Sum[Pixl[[x+1,y+1,z+1]]
Cos[(2x+1)i Pi/8]Cos[(2y+1)j Pi/8]Cos[(2z+1)k Pi/8],
{x, 0, 3}, {y, 0, 3}, {z, 0, 3}];
Pixl = Table[Random[Integer, {30, 60}],{4},{4},{4}];
Table[Round[DCT[m,n,p]],{m,0,3},{n,0,3},{p,0,3}];
MatrixForm[%]
```

Figure 11.63: Three-Dimensional DCT Applied to Correlated Data.

Three tests were performed, with the random data in the intervals $[30,60]$, $[30,160]$, and $[30,260]$. The constant has been set such that the DC coefficient would be the average of the random data items. These sets of random data exhibit less and less correlation as their variances increase. The resulting 4x4x4 cubes of transform coefficients (rounded to the nearest integer and with overbars for minus signs) are listed in Figure 11.64.

In all three tests, it is obvious that the resulting transform coefficients are getting bigger as the random numbers deviate from their average and thus become less correlated. Also, the zigzag sequence for the 3DCT is more complex than in the two-dimensional DCT. The coefficients around the top-left corner of each of the four planes are large and should be collected first. In addition, the coefficients tend to get smaller as we move from plane to plane away from the DC coefficient. As a result, a sequence

$$\begin{bmatrix} 46 & 0 & 0 & \bar{2} \\ \bar{3} & 1 & \bar{2} & 0 \\ 2 & 0 & 0 & 0 \\ 0 & \bar{1} & 0 & 3 \end{bmatrix} \begin{bmatrix} \bar{1} & 0 & 1 & \bar{1} \\ 2 & \bar{2} & 0 & 1 \\ 1 & 0 & 1 & 0 \\ \bar{1} & 1 & 0 & 1 \end{bmatrix} \begin{bmatrix} 0 & 1 & 0 & \bar{1} \\ 0 & 0 & 0 & 0 \\ \bar{1} & \bar{2} & 0 & 1 \\ 1 & 2 & 0 & \bar{1} \end{bmatrix} \begin{bmatrix} \bar{1} & 1 & 1 & 0 \\ 0 & 1 & 1 & 1 \\ 0 & 1 & 1 & \bar{1} \\ 0 & 2 & 2 & \bar{1} \end{bmatrix}.$$

$$\begin{bmatrix} 95 & 8 & 0 & \bar{9} \\ 4 & 0 & 5 & 0 \\ 3 & \bar{1} & 1 & 3 \\ \bar{2} & 4 & 1 & 1 \end{bmatrix} \begin{bmatrix} \bar{6} & 3 & 5 & \bar{4} \\ 3 & \bar{2} & 1 & 10 \\ 1 & \bar{9} & \bar{6} & \bar{4} \\ \bar{6} & \bar{5} & 1 & 7 \end{bmatrix} \begin{bmatrix} \bar{2} & 1 & 6 & 1 \\ \bar{6} & 4 & \bar{2} & 0 \\ 5 & \bar{1} & \bar{1} & \bar{3} \\ 6 & \bar{3} & 4 & \bar{2} \end{bmatrix} \begin{bmatrix} \bar{4} & 2 & \bar{2} & 1 \\ \bar{3} & 13 & \bar{5} & \bar{1} \\ \bar{2} & 4 & 4 & \overline{12} \\ 1 & 10 & 4 & 0 \end{bmatrix}.$$

$$\begin{bmatrix} 137 & \bar{5} & \overline{14} & 8 \\ \overline{21} & 10 & 1 & 17 \\ 3 & \bar{8} & \bar{6} & 13 \\ 4 & \bar{4} & 0 & \overline{10} \end{bmatrix} \begin{bmatrix} \overline{11} & 10 & \overline{10} & 3 \\ 4 & 14 & 5 & 3 \\ \bar{6} & 3 & 8 & \overline{18} \\ \bar{6} & 8 & 2 & 12 \end{bmatrix} \begin{bmatrix} 14 & 16 & 2 & 4 \\ 7 & \overline{11} & 12 & 8 \\ \overline{11} & \bar{3} & 9 & \bar{5} \\ 1 & 2 & 12 & \bar{4} \end{bmatrix} \begin{bmatrix} 3 & 8 & \bar{9} & \bar{9} \\ \bar{1} & 7 & 11 & 18 \\ \bar{4} & \bar{7} & 1 & 8 \\ \bar{7} & 3 & 3 & \bar{1} \end{bmatrix}.$$

Figure 11.64: Three-Dimensional DCT Coefficients.

such as the one shown here may make sense.

$$\begin{bmatrix} 1 & 2 & 6 & 7 \\ 3 & 5 & 8 & 23 \\ 4 & 9 & 22 & 53 \\ 10 & 21 & 54 & 61 \end{bmatrix} \begin{bmatrix} 11 & 12 & 16 & 24 \\ 13 & 15 & 25 & 30 \\ 14 & 26 & 29 & 56 \\ 27 & 28 & 55 & 62 \end{bmatrix} \begin{bmatrix} 17 & 18 & 33 & 34 \\ 19 & 32 & 35 & 40 \\ 31 & 36 & 39 & 57 \\ 37 & 38 & 58 & 63 \end{bmatrix} \begin{bmatrix} 20 & 41 & 45 & 46 \\ 42 & 44 & 47 & 52 \\ 43 & 48 & 51 & 59 \\ 49 & 50 & 60 & 64 \end{bmatrix}.$$

Another possibility is the traditional zigzag pattern where each move is repeated in all the bands

$$\begin{bmatrix} 1 & 5 & 21 & 25 \\ 9 & 17 & 29 & 49 \\ 13 & 33 & 45 & 53 \\ 37 & 41 & 57 & 61 \end{bmatrix} \begin{bmatrix} 2 & 6 & 22 & 26 \\ 10 & 18 & 30 & 50 \\ 14 & 34 & 46 & 54 \\ 38 & 42 & 58 & 62 \end{bmatrix} \begin{bmatrix} 3 & 7 & 23 & 27 \\ 11 & 19 & 31 & 51 \\ 15 & 35 & 47 & 55 \\ 39 & 43 & 59 & 63 \end{bmatrix} \begin{bmatrix} 4 & 8 & 24 & 28 \\ 12 & 20 & 32 & 52 \\ 16 & 36 & 48 & 56 \\ 40 & 44 & 60 & 64 \end{bmatrix}.$$

This approach to the compression of hyperspectral data is a direct extension of the concepts behind JPEG. A similar application of this transform has been proposed by [Li and Furht 03] for the compression of two-dimensional images. A low-power VLSI implementation of the 3DCT has been proposed by [Saponara et al. 03], and a fast algorithm for this transform has been developed by [Boussakta and Alshibami 04].

Variations on this approach are possible and have been tried. Glen Abousleman has developed a combination of 3DCT followed by trellis-coded-quantization (TCQ) specifically to compress hyperspectral images taken by satellites. The hyperspectral data is partitioned into cubes of $8 \times 8 \times 8$ pixels each, the 3DCT is applied to each cube, resulting in a cube of transform coefficients. Coefficients located in the same position in each cube (like-coefficients) are collected into a sequence and each sequence is then encoded using TCQ. The number of sequences is $8^3 = 512$, and each sequence consists of $n/512$ pixels, where n is the total size of the hyperspectral data. Details of the TCQ method and its codebook design can be found in [Abousleman 06].

11.15.3 Vector Quantization

Like transform and predictive methods, vector quantization can also be applied to the compression of hyperspectral data. Since hyperspectral data exhibits strong correlation in the spectral domain, a natural choice is to consider each pixel a vector to which quantization is applied. Once the vector has been quantized and replaced by an index, encoding that exploits spatial correlation can be used to further reduce the size of this representation. The main issues prevent a direct application of this idea are:

- Vector quantization is lossy. Hyperspectral data is collected at great cost which is why it is necessary to preserve its full information content. Applications that use this data rely on the highest possible quality. The loss introduced by a vector quantizer has statistical nature. Information that is statistically rare is discarded to favor vectors that occur often. However, applications such as target detection aim at locating objects with a rare spectral signature and it often happens that the object the algorithm wants to locate has sub-pixel physical dimensions. For example, if a tank in a battlefield is imaged by a sensor in which a pixel covers a 20×20 meter area, the tank will occupy only a fraction of a pixel. The spectral signature of the tank will be (linearly) combined with the signature of the ground. Since the ground is statistically dominant, a vector quantizer will encode it well, but will miss the signature of the tank.

- A hyperspectral pixel can have from hundreds to thousands of components. The dimension of each pixel is expected to grow with the next generation of sensors. Data size grows linearly with the spectral resolution and quadratically with the spatial resolution, so increasing the spectral resolution is the first option when designing a new imager. The design of a vector quantizer for highly-dimensional vectors is extremely demanding and can quickly become computationally infeasible. Furthermore, vector quantizers become more and more inefficient when the vector dimension increases [Kendall 61].

A vector quantizer can be used to perform lossless compression if the quantization error is encoded in the bitstream together with the quantization indexes. However, since the quantization error is a vector that has the same size as the input, the first issue can be solved with this method only if the vector that is decomposed into quantization index and error can be encoded more compactly than the original input.

A vector quantizer can be simplified by structuring the codebook. Traditional methods use *trees*, *partitioning*, *residual coding*, and *trellises*. A structured vector quantizer trades some compression (or quality, for a given compression ratio) for speed. A structured codebook may be easier to design and search.

Since spectral signatures are expected to increase in the future, a partitioned vector quantizer offers the advantage of scalability. In a partitioned vector quantizer, the input vectors are subdivided into a fixed number of sub-vectors, each quantized independently. Longer vectors are easily accommodated by increasing the number of partitions, since the complexity of design and search grows linearly with the number of partitions.

The sub-vectors are quantized independently, so it is clear that partitioning cannot be better than quantizing the entire vector. A partitioned vector quantizer does not take advantage of the correlation that exists between partitions of the same vector. There is also the issue of deciding how wide each partition should be. Partitioning a spectral signature vector into equally-sized partitions would be highly sub-optimal due to the fact that vector components have very different statistics, with mean, variance and range varying wildly.

The *Locally Optimal Partitioned Vector Quantizer* (LPVQ) introduced by Giovanni Motta, Francesco Rizzo, and James Storer [Motta et al. 06] addresses these two issues by

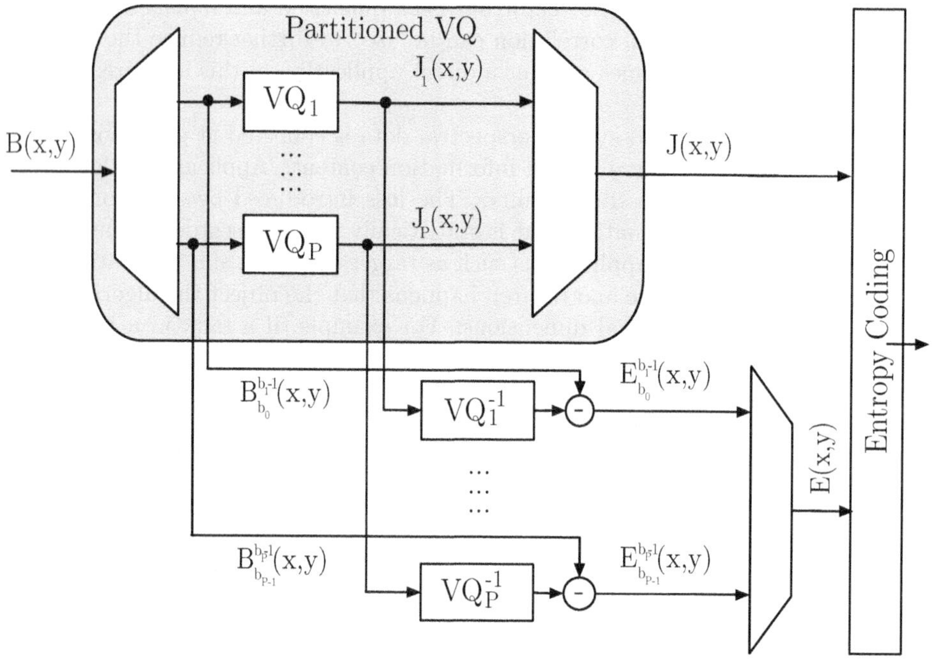

Figure 11.65: Locally Optimal Partitioned Vector Quantizer (Encoder).

exploiting correlation between quantization indices and by determining at design-time the (locally) optimal partition of the spectral signatures for a given data set.

Figure 11.65 shows the LPVQ encoder. An input spectral signature $B(x, y)$ is partitioned into P disjoint sub-vectors having boundaries $b_0, b_1, \cdots, b_{P-1}, b_P$. Sub-vectors are independently quantized by $VQ_1, VQ_2, \cdots, VQ_P$. Each quantizer outputs an index $J_i(x, y)$. Indices are reconstructed by the inverse quantizers VQ_i^{-1}, and the lossy compressed sub-vectors are subtracted to determine the quantization error $E(x, y) = B(x, y) - \hat{B}(x, y)$. Finally, an entropy encoder removes redundancies from the quantization indices and from the quantization error.

If the application allows the use of a near-lossless compressor, a small, controlled quantization error can be introduced before the residual is entropy coded. An interesting feature of LPVQ is the possibility of tightly bounding the error on a pixel-by-pixel basis. A number of experiments with several error metrics are described in [Motta et al. 06].

If the vectors in the P codebooks are sorted according to their energy, the LPVQ index planes will resemble P grayscale images. The index planes retain many features of the original data and the image-like nature suggests an encoding inspired by JPEG-LS (Section 7.11), the ISO/JPEG standard for lossless coding of natural images. The three-dimensional nature of the LPVQ index planes, however, permits the use of a three-dimensional causal context, which JPEG-LS does not provide. The statistics of the index planes show that, even after quantization, the correlation between the planes is stronger than the spatial correlation.

The entropy coding of the residual error is performed by an arithmetic encoder. Each spectral component and each quantization index has a different statistical model. This method is based on the assumption that errors for different spectral components and different quantization classes have slightly different probability distributions.

The P codebooks are designed by using an adaptation of the Linde-Buzo-Gray (or LBG, Section 7.19) algorithm [Linde, Buzo, and Gray 80]. The partitioning algorithm generalizes LBG by observing that, once the partition boundaries are kept fixed, distortion measures that are additive with respect to the vector components (like the *Mean Squared Error*, for example) can be minimized independently for each partition by applying the optimality conditions on the centroids and on the cells. Similarly, when the centroids and the cells are held fixed, the (locally optimal) partitions' boundaries can be determined in a greedy fashion. So the design algorithm starts from equally-sized partitions and iterates the optimality conditions for the centroids, the cells, and the boundaries. This design converges to locally optimal centroids and vector boundaries.

Besides competitive compression, LPVQ has the advantage that the quantization indices retain important information about the original scene. Quantization indices constitute only a small fraction of the bitstream, but they can be used to browse the image and select regions of interest.

Another feature of the LPVQ compressor is to optionally produce a tightly-bound quantization error that's controlled on a pixel-by-pixel basis. Due to the hierarchical structure of the data, it is possible to perform pure-pixel classification and target detection directly in the compressed domain, with considerable speed-up and memory savings. A suitable algorithm is described in [Motta et al. 06].

11.16 Stuffit

These words are written in January 2009, a month that marks the 25th anniversary of the Macintosh computer. Both the architecture and the operating system of this computer were revolutionary and it immediately attracted a large number of software developers and many devoted users. Among them was the New-York high-school student Raymond Lau [Lau 09] who, in the summer of 1987, came up with a data compression utility that he termed stuffit. It very quickly proved its value and became a popular, de-facto standard for compressing files on the Macintosh. I was one of its many users in those early days, and I remember well how it would start compressing a file with both LZW and RLE. At a certain point, stuffit could see which of these methods was better for the file and would abandon the other method.

We start with a historical overview of the development of stuffit [stuffit-wiki 09]. Initially, Lau sold stuffit as shareware, but after version 1.5.1 he realized he couldn't continue to develop and market a successful product while also going to college (he had already been admitted to MIT). Thus, in 1998, Aladdin systems was formed to develop and market stuffit. Ray participated in the development of stuffIt for a few years while

> This blew my mind as I had never seen a computer could be accessed and used instantly with no prior knowledge! I knew from that moment that this was the way all personal computers would be controlled from then on.
>
> —Unknown

attending MIT's undergraduate program. Aladdin sold StuffIt Classic as shareware and StuffIt Deluxe as a commercial package. The latter version had extra functions, compression methods, and finder integration in the form of a "magic menu."

In 1996, after releasing a few updates with Aladdin, Ray Lau decided to follow other pursuits, with the result that stuffit started losing market share to its main rival Compact Pro. In order to compete, Aladdin decided to release the free stuffit expander, thereby allowing anyone to freely decompress stuffit files. This single decision returned stuffit to its former place as the main data compression utility for the Macintosh platform. (It is interesting to note that Wolfram research and Adobe adopted similar tactics and gave us the free Mathematica player and Acrobat reader.) Throughout the 1990s, stuffit had very little competition. New, fast and efficient compression utilities such as ZIP (Section 6.25), RAR (Section 6.22), and 7z (Section 6.26) were implemented for Windows and had little effect on Macintosh users.

The Macintosh new operating system, OS X, was introduced in 2001, but Aladdin was late to respond to it, thereby losing popularity and market share to Unix utilities such as gzip and tar. The new stuffit deluxe 7.0 for OS X was finally announced in September 2002. In addition to LZW, this version supported PPM (Section 5.14) and the Burrows-Wheeler transform (Section 11.1). It also supported Unix and Windows file attributes and offered optional encryption. JPEG compression was added in 2005. Files compressed by stuffit X versions have the extension `.sitx` to distinguish them from the original `.sit` extension.

(A `.sitx` file has a built-in error-correcting code and offers optional encryption. The new blockmode makes it possible to compress a set of files as a single `.sitx` file. An interesting feature is duplicate folding. If several files with identical data forks are added to an archive, only one copy of the data is compressed and saved. For each of the other files, stuffit saves only the header information and a pointer to the unique copy of the compressed data.)

In July 2004 Aladdin was forced to change its name because of a trademark lawsuit with Aladdin Knowledge Systems. Thus, the short-lived Allume systems was born. In 2005, Allume Systems was acquired by Smith Micro Software, but the development of stuffit remained in the same location in Watsonville, California.

In addition to the original Macintosh implementation, there are currently stuffit implementations for Windows, Linux, and Sun Solaris.

Recently, new versions of stuffit have been issued annually, a behavior pattern which irritates some devoted users of this software. The details of its operations remain a trade secret, leading several users to complain that stuffit takes their money without offering anything new. (However, the remainder of this section documents new features that have been added to stuffit 9, hopefully satisfying any frustrated users.)

> I got irritated at this product years ago where every "new release" was just fixing one set of bugs and as they introduced new ones; some of which were actually modifying files and permissions in ways that harmed your system. No real new functionality has been seen for going on 10 years, yet they charge for each new "update."
>
> StuffIt compression is far more effective than that produced by zip. I generate large datasets in my work. I tried an example just now. StuffIt compressed the folder down to 1.7 MB. Tar and gzip compressed it to 4.4 MB. Zip compressed it to 9.2 MB....
>
> I do not bother to upgrade every year, but you can upgrade cheaply from any prior version. The new version is much better than version 11.
>
> I just bought this thing 6 months ago—and Smith Micro wants another $30—for what? Required bug fixes? There are no "new" features that warrant a full version number, much less an upgrade fee higher than most full prices. (Quick Look = quick money).
>
> Smith Micro in keeping with its tradition, has submitted its new and little-improved addition for its annual fee assessment. Maybe Smith Micro can simply start submitting it with my property taxes to make it easier for them.
>
> I find that version 12 of the Stuffit Suite is excellent. Stuffit Expander always works for me, unless the file is damaged.
>
> —Users' comments found in
> `http://www.versiontracker.com/dyn/moreinfo/macosx/180&page=1`

Technical Overview

Because of its proprietary nature, many details of stuffit compression are a trade secret, which is why the description here is incomplete. However, I feel that the description is still useful because it reveals the design philosophy of stuffit, and illustrates its main components. Some of the material was sent to me by Smith Micro and the rest comes from three patents applied for by this company.

Popular compression software packages normally employ a single algorithm. Thus, Zip (Section 6.25) is based on Deflate and Unix compress uses LZC, a variation of LZW. The PAQ algorithm (Section 5.15) goes a step further. It employs several predictors and computes a weighted prediction average for each input bit. Stuffit, on the other hand, supports several compression algorithms and analyzes the input data to decide which algorithm(s) to use for each part of the input. Stuffit even goes to the trouble of decompressing or partly decompressing the input (or part of it) if it is already in compressed form, and then recompressing it with more efficient algorithms.

A good example of the success of this approach is PDF. A PDF file (Section 11.13) may consist of text, fonts, images (raster graphics for photographs and scanned documents), vector graphics (for illustrations and designs that consist of shapes and lines), and formatting information. Later PDF versions also allow objects such as support links, forms, JavaScript, and other types of embedded contents. Most parts of a PDF file are already in compressed form, so passing it through a compressor does not reduce its size. However, when stuffit identifies an input file as PDF, it parses the file, separates the text, images, and formatting information, and processes each separately. Text is decompressed, preprocessed, and then recompressed. Each raster image is identified (as raw, JPEG, GIF, TIFF, or other types), decompressed, and then compressed with a different compression method according to its type (monochrome, grayscale, or color). Any unknown parts of the input are then compressed with one of stuffit's general compressors, which are a custom LZ algorithm, the Burrows-Wheeler transform (Section 11.1), and several PPM variants. Following the compression, stuffit can optionally encrypt the output with DES and add a layer of error-correcting code. This process is illustrated in Figure 11.66.

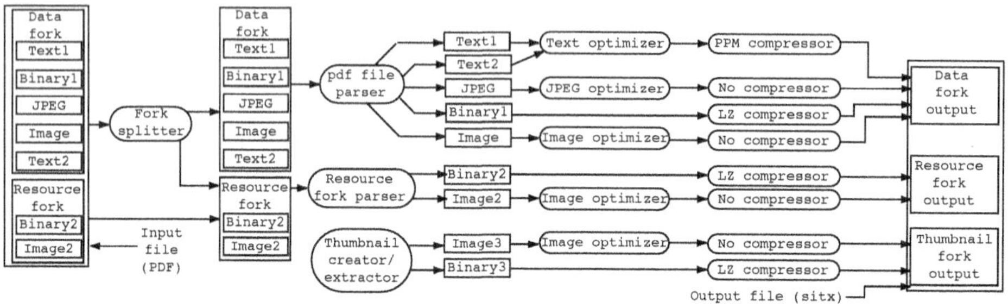

Figure 11.66: Compression of a PDF File by Stuffit.

The following list indicates the processing (or preprocessing) done by stuffit for common types of data.

- JPG. The last step of JPEG compression is to encode the 64 quantized transform coefficients of a block with special Huffman codes (Section 7.10.4). Stuffit undoes this step and compresses the coefficients with a better algorithm, thereby reducing the size of the original JPEG file by 20–30%. Notice that the stuffit decompressor must end up with a standard JPEG file, so it starts by decoding the transform coefficients, and then encoding them with the Huffman codes used by JPEG. (This is true for most of the data types listed here.)

- JLS. This method (JPEG-LS, Section 7.11) is based on prediction by context. On identifying such an image, stuffit completely decompresses it and then compresses it with its lossless image compressor.

- J2K (JPEG 2000, Section 8.19). Stuffit again completely decompresses such an image and then encodes it with its proprietary image compressor.

- BMP (Section 1.4.4). BMP is the native format for image files in the Microsoft Windows operating system. It is based on RLE and is not particularly efficient. Stuffit completely decompresses such an image and then encodes it with its proprietary lossless image compressor.

- GIF (Section 6.21). The Graphics Interchange Format, is a graphics file format that employs LZW for compression. Stuffit undoes any compression of such an image and encodes it with a better compressor.

- TIFF. The Tagged Image File Format, originally developed by Aldus and now owned by Adobe, is another graphics file format. It offers LZW compression as an option and can act as a container for JPEG- and RLE-compressed images. Stuffit improves TIFF compression with its own custom image compressor.

- PPM, PBM, and PGM. These are intermediate graphics formats used in the conversion of many little-known graphics file formats via pbmplus, the Portable Bitmap

Utilities. These formats are mainly available under UNIX and on Intel-based PCs. Stuffit again recompresses such files with its proprietary lossless image encoder.

- PSD (Photoshop image files). Stuffit parses these files and compresses each layer separately.

- PNG (portable network graphics, Section 6.27). This algorithm was developed specifically to allow the use of patent-free compressed images on the Internet. Stuffit undoes the original PNG compression and encodes the result with its lossless image compressor.

- ZIP. Stuffit decompresses ZIP files and then checks the resulting data, which may consist of several different types of files. Stuffit then compresses each of these files separately according to its type.

- PDF. The compressed parts of a PDF file are decompressed and each is compressed with an appropriate encoder.

- MP3. Stuffit undoes the last step of mp3 compression (encoding, Section 10.14.6) and reencodes it with its own encoder.

We conclude this discussion with some details of the inner workings of stuffit, using JPEG as an example. If stuffit identifies the input file as JPEG, it starts by reading, parsing, and encoding the JPEG header (Figure 11.67). The header is encoded with a generic compressor that is a mixed-model statistical compressor employing order-0 context with weight 1 and order-1 context with weight 2. While parsing the header, stuffit isolates the JPEG parameters, verifies them, and stores them for later use. If the file contains any non-standard tables, they are also stored by stuffit for later use. The stuffit encoder now reserves enough memory to store two rows of JPEG 8×8 blocks and it starts reading the blocks.

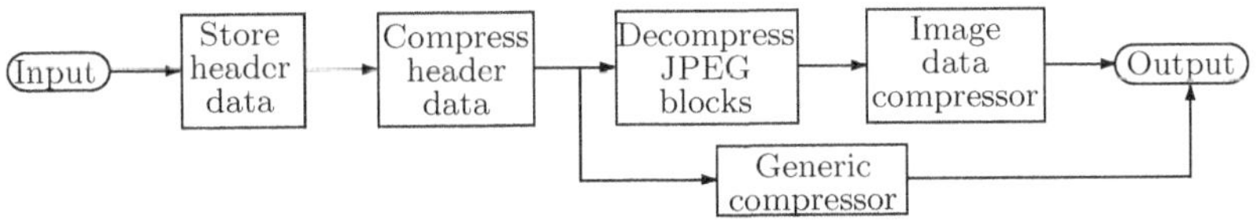

Figure 11.67: JPEG Compressor Structure.

Each block has been originally DCT-transformed by JPEG, quantized, zigzagged, and encoded with Huffman codes. Stuffit undoes the encoding and zigzagging, and ends up with an 8×8 block of quantized transform coefficients. The block is compared to the block above it and to the block on its left. If our block is identical to either of those, it is declared a reference block, and appropriate information is written on the output, to enable the stuffit decoder to recreate the block from its identical twin (Figure 11.68).

If the block is not a reference block, it is encoded efficiently by stuffit. First, its DC coefficient is encoded (Figure 11.69) by a predictive model that uses four DC coefficients as predictors. Two of the four are the DC coefficients of the block above and the block to the left of the current block. The other two are the DC coefficients of the block

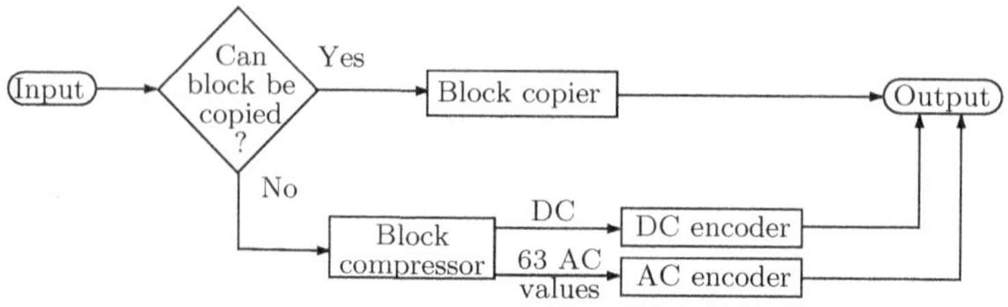

Figure 11.68: JPEG Block Compressor.

below and the block to the right of the current block. These two blocks are not yet known to the stuffit encoder, so instead of using their DC coefficients, it predicts those coefficients from the DC coefficients of the block to the left and below (or, if this block is not available, from the block on the left) and the block to the right and above (or, if this block is not available, from the block above). Once the four predictors have been determined, they are assigned different weights, their weighted average is subtracted from the DC coefficient of the current block, and the residual is encoded by a context-based compressor that uses as context the difference between the four predictors and the sparsity of the blocks to the left and above the current block.

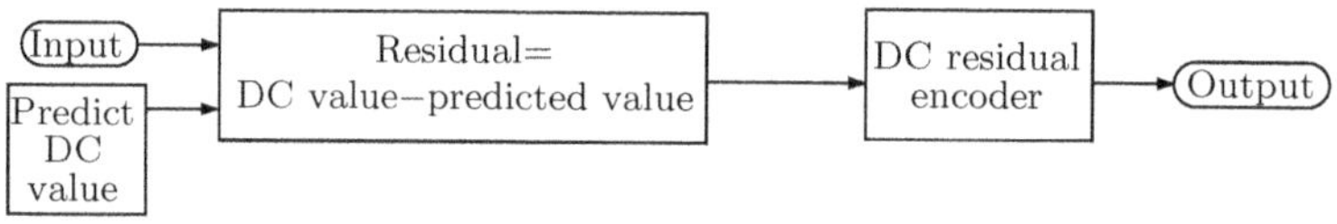

Figure 11.69: DC Encoder.

Next, the 63 AC coefficients are encoded using two predictors. An up-predictor is computed based on the block above and a left-predictor is computed based on the block to the left. These predictors are continually updated based on the already-encoded AC coefficients. Notice that this can be followed in lockstep by the decoder. The AC coefficients are scanned in raster order, not in a zigzag, and are encoded by a context-based zero/nonzero encoder that employs a mixed model based on the following contexts: the block type, the position within the block, values from the current block, the quantization table, the two predictors, and for Cr blocks also the corresponding values from the Cb block.

Encoding an AC coefficient starts with a zero/nonzero flag (Figure 11.70). If the coefficient is zero, its flag is all that is needed. If it is nonzero, its absolute value and sign are encoded separately. Three different encoders are used for this purpose, one for AC coefficients on the top row, one for coefficients on the leftmost column, and a third encoder for the remaining coefficients. Each encoder is context based using the contexts listed above, which are assigned different weights. The encoder counts the values already encountered for each context and uses this information to adjust the weights.

This is how stuffit can compress a JPEG file. In principle, any file compressed by a lossy method can be recompressed in this way. If the original method was inefficient, a

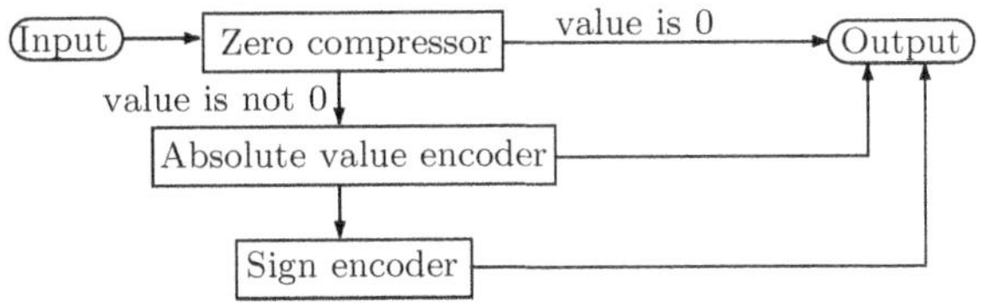

Figure 11.70: AC Encoder.

partial decompression of the last, lossless stages followed by efficient compression, may result in a smaller file.

If the input was compressed losslessly, it can be completely decompressed, and stuffit can then employ its own, efficient lossless encoder. Here is how such an encoder works (the following makes sense especially when the input is an image or video).

After decompressing the input, the encoder parses it into blocks of the following types: solid, reference, natural, or artificial. In a solid block, all the pixels have the same color. This is obviously easy to compress. A reference block is identical to one or more of the preceding blocks, so very little information has to be sent to the decoder for such a block. A natural block depicts a natural image, and an artificial block contains a discrete-tone image. The tests for solid and reference blocks are obvious. The test for a natural/artificial block starts with pixel comparisons. Each pixel in the block is compared with its immediate neighbors above and to the left. If the pixel is identical to either of the two neighbors, a count variable C is incremented by 1. If the ratio C/B (where B is the number of the pixels in the block) is greater than $100/256 \approx 0.39$, the block is considered artificial and is compressed by a proprietary context-based encoder optimized for discrete-tone images.

The specific method used depends on whether the block is in grayscale or color, but either encoder employs an interesting feature termed pixel marking. If a pixel is identical to another pixel, it is marked and a special code is then output to indicate this fact. If a pixel is different from any other pixel, it is compressed depending on its context. Natural blocks are similarly compressed, except that a predictive encoder is used instead of a context-based encoder. Several predictors, all based on neighbors above and to the left of the current pixel, are used and the final prediction is a weighted average of all the predictors. The weights are adjusted depending on how well they predicted neighboring pixels in the past, and this adjustment is done such that the decoder can do it in lockstep. For each block, a code is inserted into the compressed stream to indicate its type and thus the method used to compress it.

Another test for a natural/artificial block is based on the number of unique colors in the block. This number is counted and is divided by the total number of pixels in the block. If the ratio is below a certain threshold, the block is deemed artificial.

The prediction errors (residuals) for the individual pixels are encoded, but if the image is in color, the residuals of the three colors of a pixel are correlated, which suggests that the differences of the residuals, rather than the residuals themselves, should be encoded. Thus, if P_0, P_1, and P_2 are the residuals of the three colors of a pixel, then the encoder first computes $T_0 = P_0$, $T_1 = P_1 - P_0$, and $T_2 = P_2/2 - P_1/2$, and then encodes the three T_i's.

We would like to thank Darryl Lovato of Smith Micro (the current owner of stuffit) for his help with this section.

Age cannot wither her, nor custom stale Her infinite variety.
—William Shakespeare, *Antony and Cleopatra*

Appendix A
Information Theory

This appendix serves as a brief introduction to information theory, the foundation of many techniques used in data compression. The two most important terms covered here are entropy and redundancy (see also Section 2.3 for an alternative discussion of these concepts).

A.1 Information Theory Concepts

We intuitively know what information is. We constantly receive and send information in the form of text, sound, and images. We also feel that information is an elusive nonmathematical quantity that cannot be precisely defined, captured, or measured. The standard dictionary definitions of information are (1) knowledge derived from study, experience, or instruction; (2) knowledge of a specific event or situation; intelligence; (3) a collection of facts or data; (4) the act of informing or the condition of being informed; communication of knowledge.

Imagine a person who does not know what information is. Would those definitions make it clear to them? Unlikely.

The importance of information theory is that it quantifies information. It shows how to measure information, so that we can answer the question "how much information is included in this piece of data?" with a precise number! Quantifying information is based on the observation that the information content of a message is equivalent to the amount of *surprise* in the message. If I tell you something that you already know (for example, "you and I work here"), I haven't given you any information. If I tell you something new (for example, "we both received a raise"), I have given you some information. If I tell you something that really surprises you (for example, "only I received a raise"), I have given you more information, regardless of the number of words I have used, and of how you feel about my information.

D. Salomon, G. Motta, *Handbook of Data Compression*, 5th ed.
DOI 10.1007/978-1-84882-903-9_BM2, © Springer-Verlag London Limited 2010

We start with a simple, familiar event that's easy to analyze, namely the toss of a coin. There are two results, so the result of any toss is initially uncertain. We have to actually throw the coin in order to resolve the uncertainty. The result is heads or tails, which can also be expressed as a yes or no, or as a 0 or 1; a bit.

A single bit resolves the uncertainty in the toss of a coin. What makes this example important is the fact that it can easily be generalized. Many real-life problems can be resolved, and their solutions expressed, by means of several bits. The principle of doing so is to find the minimum number of yes/no questions that must be answered in order to arrive at the result. Since the answer to a yes/no question can be expressed with one bit, the number of questions will equal the number of bits it takes to express the information contained in the result.

A slightly more complex example is a deck of 64 playing cards. For simplicity let's ignore their traditional names and numbers and simply number them 1 to 64. Consider the event of person `A` drawing one card and person `B` having to guess what it was. The guess is a number between 1 and 64. What is the minimum number of yes/no questions that are needed to guess the card? Those who are familiar with the technique of *binary search* know the answer. Using this technique, `B` should divide the interval 1–64 in two, and should start by asking "is the result between 1 and 32?" If the answer is no, then the result is in the interval 33 to 64. This interval is then divided by two and `B`'s next question should be "is the result between 33 and 48?" This process continues until the interval selected by `B` reduces to a single number.

It does not take much to see that exactly six questions are necessary to get at the result. This is because 6 is the number of times 64 can be divided in half. Mathematically, this is equivalent to writing $6 = \log_2 64$. This is why the *logarithm* is the mathematical function that quantifies information.

Another approach to the same problem is to ask the question; Given a nonnegative integer N, how many digits does it take to express it? The answer, of course, depends on N. The greater N, the more digits are needed. The first 100 nonnegative integers (0 to 99) can be expressed by two decimal digits. The first 1000 such integers can be expressed by three digits. Again it does not take long to see the connection. The number of digits required to represent N equals approximately $\log N$. The base of the logarithm is the same as the base of the digits. For decimal digits, use base 10; for binary digits (bits), use base 2. If we agree that the number of digits it takes to express N is proportional to the information content of N, then again the logarithm is the function that gives us a measure of the information.

◇ **Exercise A.1:** What is the precise size, in bits, of the binary integer i?

Here is another approach to quantifying information. We are familiar with the ten decimal digits. We know that the value of a digit in a number depends on its position. Thus, the value of the digit 4 in the number 14708 is 4×10^3, or 4000, since it is in position 3 (positions are numbered from right to left, starting from 0). We are also familiar with the two binary digits (bits) 0 and 1. The value of a bit in a binary number similarly depends on its position, except that powers of 2 are used. Mathematically, there is nothing special about 2 or 10. We can use the number 3 as the basis of our arithmetic. This would require the three digits, 0, 1, and 2 (we might call them *trits*). A trit t at position i would have a value of $t \times 3^i$.

◇ **Exercise A.2:** Actually, there is something special about 10. We use base-10 numbers because we have ten fingers. There is also something special about the use of 2 as the basis for a number system. What is it?

Given a decimal (base 10) or a ternary (base 3) number with k digits, a natural question is; how much information is included in this k-digit number? We answer this by determining the number of bits it takes to express the given number. Assuming that the answer is x, then $10^k - 1 = 2^x - 1$. This is because $10^k - 1$ is the largest k-digit decimal number and $2^x - 1$ is the largest x-bit binary number. Solving the equation above for x as the unknown is easily done using logarithms and yields

$$x = k\frac{\log 10}{\log 2}.$$

We can use any base for the logarithm, as long as we use the same base for $\log 10$ and $\log 2$. Selecting base 2 simplifies the result, which becomes $x = k\log_2 10 \approx 3.32k$. This shows that the information included in one decimal digit equals that contained in about 3.32 bits. In general, given numbers in base n, we can write $x = k\log_2 n$, which expresses the fact that the information included in one base-n digit equals that included in $\log_2 n$ bits.

◇ **Exercise A.3:** How many bits does it take to express the information included in one trit?

We now turn to a transmitter, a piece of hardware that can transmit data over a communications line (a channel). In practice, such a transmitter sends binary data (a modem is a good example). However, in order to obtain general results, we assume that the data is a string made up of occurrences of the n symbols a_1 through a_n. Such a set is an n-symbol alphabet. Since there are n symbols, we can think of each as a base-n digit, which means that it is equivalent to $\log_2 n$ bits. As far as the hardware is concerned, this means that it must be able to transmit at n discrete levels.

If the transmitter takes $1/s$ time units to transmit a single symbol, then the speed of the transmission is s symbols per time unit. A common example is $s = 28800$ baud (*baud* is the term for "bits per second"), which translates to $1/s \approx 34.7$ μsec (where the Greek letter μ stands for "micro" and 1 μsec $= 10^{-6}$ sec). In one time unit, the transmitter can send s symbols, which as far as information content is concerned, is equivalent to $s\log_2 n$ bits. We denote by $H = s\log_2 n$ the amount of information, measured in bits, transmitted in each time unit.

The next step is to express H in terms of the probabilities of occurrence of the n symbols. We assume that symbol a_i occurs in the data with probability P_i. The sum of the probabilities equals, of course, unity: $P_1+P_2+\cdots+P_n = 1$. In the special case where all n probabilities are equal, $P_i = P$, we get $1 = \sum P_i = nP$, implying that $P = 1/n$, and resulting in $H = s\log_2 n = s\log_2(1/P) = -s\log_2 P$. In general, the probabilities are different, and we want to express H in terms of all of them. Since symbol a_i occurs a fraction P_i of the time in the data, it occurs on the average sP_i times each time unit, so its contribution to H is $-sP_i\log_2 P_i$. The sum of the contributions of all n symbols to H is therefore $H = -s\sum_1^n P_i\log_2 P_i$.

As a reminder, H is the amount of information, in bits, sent by the transmitter in one time unit. The amount of information contained in one base-n symbol is therefore H/s (because it takes time $1/s$ to transmit one symbol), or $-\sum_1^n P_i \log_2 P_i$. This quantity is called the *entropy* of the data being transmitted. In analogy we can define the entropy of a single symbol a_i as $-P_i \log_2 P_i$. This is the smallest number of bits needed, on average, to represent the symbol.

(Information theory was developed, in the late 1940s, by Claude Shannon, of Bell Labs, and he chose the term *entropy* because this term is used in thermodynamics to indicate the amount of disorder in a physical system.)

Since I think it is better to take the names of such quantities as these, which are important for science, from the ancient languages, so that they can be introduced without change into all the modern languages, I propose to name the magnitude S the *entropy* of the body, from the Greek word "trope" for "transformation." I have intentionally formed the word "entropy" so as to be as similar as possible to the word "energy" since both these quantities which are to be known by these names are so nearly related to each other in their physical significance that a certain similarity in their names seemed to me advantageous.

—Rudolph Clausius, 1865 (translated by Hans C. von Baeyer)

The entropy of data depends on the individual probabilities P_i and it is at its maximum (see Exercise A.4) when all n probabilities are equal. This fact is used to define the redundancy R in the data. It is defined as the difference between a symbol set's largest possible entropy and its actual entropy. Thus

$$R = \left[-\sum_1^n P \log_2 P\right] - \left[-\sum_1^n P_i \log_2 P_i\right] = \log_2 n + \sum_1^n P_i \log_2 P_i.$$

Thus, the test for fully compressed data (no redundancy) is $\log_2 n + \sum_1^n P_i \log_2 P_i = 0$.

◇ **Exercise A.4:** Analyze the entropy of a two-symbol set.

Given a string of characters, the probability of a character can be determined by counting the frequency of the character and dividing by the length of the string. Once the probabilities of all the characters are known, the entropy of the entire string can be calculated. With current availability of powerful mathematical software, it is easy to calculate the entropy of a given string. The *Mathematica* code

```
Frequencies[list_]:=Map[{Count[list,#],#}&, Union[list]];
Entropy[list_]:=-Plus @@ N[# Log[2,#]]& @
 (First[Transpose[Frequencies[list]]]/Length[list]);
```

```
Characters["swiss miss"]
Entropy[%]
```

does that and shows that, for example, the entropy of the string `swiss␣miss` is 1.96096.

The main theorem proved by Shannon says essentially that a message of n symbols can, on average, be compressed down to nH bits, but not further. It also says that almost optimal compressors (called *entropy encoders*) exist, but does not show how to construct them. Arithmetic coding (Section 5.9) is an example of an entropy encoder, as are also the dictionary-based algorithms of Chapter 6 (but the latter require huge quantities of data to perform at the entropy level).

You have two chances—
One of getting the germ
And one of not.
And if you get the germ
You have two chances—
One of getting the disease
And one of not.
And if you get the disease
You have two chances—
One of dying
And one of not.
And if you die—
Well, you still have two chances.

—Unknown

Appendix A.1.1 Algorithmic Information Content

Consider the following three sequences:

$$S_1 = 100100100100100100100100100100100100\ldots,$$
$$S_2 = 010110110110101101101011011011011011\ldots,$$
$$S_3 = 011100100111100100001011000000111011111\ldots.$$

The first sequence, S_1, is just a repetition of the simple pattern 100. S_2 is less regular. It can be described as a 01, followed by r_1 repetitions of 011, followed by another 01, followed by r_2 repetitions of 011, etc., where $r_1 = 3$, $r_2 = 2$, $r_3 = 4$, and the other r_i's are not shown. S_3 is more difficult to describe, since it does not seem to have any apparent regularity; it looks random. Notice that the meaning of the ellipsis is clear in the case of S_1 (just repeat the pattern 100), less clear in S_2 (what are the other r_i's?), and completely unknown in S_3 (is it random?).

We now assume that these sequences are very long (say, 999,999 bits each), and each continues "in the same way." How can we define the complexity of such a binary sequence, at least qualitatively? One way to do so, called the Kolmogorov-Chaitin complexity (KCC), is to define the complexity of a binary string S as the length, in bits, of the shortest computer program that, when executed, generates S (display it, print it, or write it on file). This definition is also called the *algorithmic information content* of string S.

A computer program P_1 to generate string S_1 could just loop 333,333 times and print 100 in each iteration. Alternatively, the program could loop 111,111 times and print 100100100 in each iteration. Such a program is very short (especially when compared with the length of the sequence it generates), concurring with our intuitive feeling that S_1 has low complexity.

A program P_2 to generate string S_2 should know the values of all the r_i's. They could either be built in or input by the user at run time. The program initializes a variable i to 1. It then prints "01", loops r_i times printing "011" in each iteration, increments i by 1, and repeats this behavior until 999,999 bits have been printed. Such a program is longer than P_1, thereby reflecting our intuitive feel that S_2 is more complex than S_1.

A program P_3 to generate S_3 should (assuming that we cannot express this string in any regular way) simply print all 999,999 bits of the sequence. Such a program is as long as the sequence itself, implying that the KCC of S_3 is as large as S_3.

Using this definition of complexity, Gregory Chaitin showed (see [Chaitin 77] or [Chaitin 97]) that most binary strings of length n are random; their complexities are close to n. However, the "interesting" (or "practical") binary strings, those that are used in practice to represent text, images, and sound, and are compressed all the time, are similar to S_2. They are not random. They exhibit some regularity, which makes it possible to compress them. Very regular strings, such as S_1, are rare in practice.

Algorithmic information content is a measure of the amount of information included in a message. It is related to the KCC and is different from the way information is measured in information theory. Shannon's information theory defines the amount of information in a string by considering the amount of *surprise* this information contains when revealed. Algorithmic information content, on the other hand, measures information that has already been revealed. An example may serve to illustrate this difference. Imagine two persons A (well-read, sophisticated, and knowledgeable) and B (inexperienced and naive), reading the same story. There are few surprises in the story for A. He has already read many similar stories and can predict the development of the plot, the behavior of the characters, and even the end. The opposite is true for B. As he reads, he is surprised by the (to him) unexpected twists and turns that the story takes and by the (to him) unpredictable behavior of the characters. The question is; How much information does the story really contain?

Shannon's information theory tells us that the story contains less information for A than for B, since it contains fewer surprises for A than for B. Recall that A's mind already has memories of similar plots and characters. As they read more and more, however, both A and B become more and more familiar with the plot and characters, and therefore less and less surprised (although at different rates). Thus, they get less and less (Shannon's type of) information. At the same time, as more of the story is revealed to them, their minds' complexities increase (again at different rates). Thus, they get more algorithmic information content. The sum of Shannon's information and KCC is therefore constant (or close to constant).

This example suggests a way to measure the information content of the story in an absolute way, regardless of the particular reader. It is the *sum* of Shannon's information and the KCC. This measure has been proposed by the physicist Wojciech Zurek [Zurek 89], who termed it "physical entropy."

Information's pretty thin stuff unless mixed with experience.

—Clarence Day *The Crow's Nest*

Answers to Exercises

> A bird does not sing because he has an answer, he sings because he has a song.
>
> —Chinese Proverb

Intro.1: abstemious, abstentious, adventitious, annelidous, arsenious, arterious, facetious, sacrilegious.

Intro.2: When a software house has a popular product they tend to come up with new versions. A user can update an old version to a new one, and the update usually comes as a compressed file on a floppy disk. Over time the updates get bigger and, at a certain point, an update may not fit on a single floppy. This is why good compression is important in the case of software updates. The time it takes to compress and decompress the update is unimportant since these operations are typically done just once. Recently, software makers have taken to providing updates over the Internet, but even in such cases it is important to have small files because of the download times involved.

1.1: (1) ask a question, (2) absolutely necessary, (3) advance warning, (4) boiling hot, (5) climb up, (6) close scrutiny, (7) exactly the same, (8) free gift, (9) hot water heater, (10) my personal opinion, (11) newborn baby, (12) postponed until later, (13) unexpected surprise, (14) unsolved mysteries.

1.2: A reasonable way to use them is to code the five most-common strings in the text. Because irreversible text compression is a special-purpose method, the user may know what strings are common in any particular text to be compressed. The user may specify five such strings to the encoder, and they should also be written at the start of the output stream, for the decoder's use.

1.3: 6,8,0,1,3,1,4,1,3,1,4,1,3,1,4,1,3,1,2,2,2,2,6,1,1. The first two are the bitmap resolution (6×8). If each number occupies a byte on the output stream, then its size is 25 bytes, compared to a bitmap size of only 6×8 bits $= 6$ bytes. The method does not work for small images.

1.4: RLE of images is based on the idea that adjacent pixels tend to be identical. The last pixel of a row, however, has no reason to be identical to the first pixel of the next row.

1.5: Each of the first four rows yields the eight runs 1,1,1,2,1,1,1,eol. Rows 6 and 8 yield the four runs 0,7,1,eol each. Rows 5 and 7 yield the two runs 8,eol each. The total number of runs (including the eol's) is thus 44.

When compressing by columns, columns 1, 3, and 6 yield the five runs 5,1,1,1,eol each. Columns 2, 4, 5, and 7 yield the six runs 0,5,1,1,1,eol each. Column 8 gives 4,4,eol, so the total number of runs is 42. This image is thus "balanced" with respect to rows and columns.

1.6: The result is five groups as follows:

$$\begin{aligned}
W_1 \text{ to } W_2 &:00000, 11111,\\
W_3 \text{ to } W_{10} &:00001, 00011, 00111, 01111, 11110, 11100, 11000, 10000,\\
W_{11} \text{ to } W_{22} &:00010, 00100, 01000, 00110, 01100, 01110,\\
&11101, 11011, 10111, 11001, 10011, 10001,\\
W_{23} \text{ to } W_{30} &:01011, 10110, 01101, 11010, 10100, 01001, 10010, 00101,\\
W_{31} \text{ to } W_{32} &:01010, 10101.
\end{aligned}$$

1.7: The seven codes are

$$0000, 1111, 0001, 1110, 0000, 0011, 1111,$$

forming a string with six runs. Applying the rule of complementing yields the sequence

$$0000, 1111, 1110, 1110, 0000, 0011, 0000,$$

with *seven* runs. The rule of complementing does not always reduce the number of runs.

1.8: As "11 22 90 00 00 33 44". The 00 following the 90 indicates no run, and the following 00 is interpreted as a regular character.

1.9: The six characters "`123ABC`" have ASCII codes 31, 32, 33, 41, 42, and 43. Translating these hexadecimal numbers to binary produces "`00110001 00110010 00110011 01000001 01000010 01000011`".
The next step is to divide this string of 48 bits into 6-bit blocks. They are 001100=12, 010011=19, 001000=8, 110011=51, 010000=16, 010100=20, 001001=9, and 000011=3. The character at position 12 in the BinHex table is "`-`" (position numbering starts at zero). The one at position 19 is "`6`". The final result is the string "`-6)c38*$`".

1.10: Exercise A.1 shows that the binary code of the integer i is $1 + \lfloor \log_2 i \rfloor$ bits long. We add $\lfloor \log_2 i \rfloor$ zeros, bringing the total size to $1 + 2\lfloor \log_2 i \rfloor$ bits.

1.11: Table Ans.1 summarizes the results. In (a), the first string is encoded with $k = 1$. In (b) it is encoded with $k = 2$. Columns (c) and (d) are the encodings of the second string with $k = 1$ and $k = 2$, respectively. The averages of the four columns are 3.4375, 3.25, 3.56, and 3.6875; very similar! The move-ahead-k method used with small values of k does not favor strings satisfying the concentration property.

a	abcdmnop	0	a	abcdmnop	0	a	abcdmnop	0	a	abcdmnop	0
b	abcdmnop	1	b	abcdmnop	1	b	abcdmnop	1	b	abcdmnop	1
c	bacdmnop	2	c	bacdmnop	2	c	bacdmnop	2	c	bacdmnop	2
d	bcadmnop	3	d	cbadmnop	3	d	bcadmnop	3	d	cbadmnop	3
d	bcdamnop	2	d	cdbamnop	1	m	bcdamnop	4	m	cdbamnop	4
c	bdcamnop	2	c	dcbamnop	1	n	bcdmanop	5	n	cdmbanop	5
b	bcdamnop	0	b	cdbamnop	2	o	bcdmnaop	6	o	cdmnbaop	6
a	bcdamnop	3	a	bcdamnop	3	p	bcdmnoap	7	p	cdmnobap	7
m	bcadmnop	4	m	bacdmnop	4	a	bcdmnopa	7	a	cdmnopba	7
n	bcamdnop	5	n	bamcdnop	5	b	bcdmnoap	0	b	cdmnoapb	7
o	bcamndop	6	o	bamncdop	6	c	bcdmnoap	1	c	cdmnobap	0
p	bcamnodp	7	p	bamnocdp	7	d	cbdmnoap	2	d	cdmnobap	1
p	bcamnopd	6	p	bamnopcd	5	m	cdbmnoap	3	m	dcmnobap	2
o	bcamnpod	6	o	bampnocd	5	n	cdmbnoap	4	n	mdcnobap	3
n	bcamnopd	4	n	bamopncd	5	o	cdmnboap	5	o	mndcobap	4
m	bcanmopd	4	m	bamnopcd	2	p	cdmnobap	7	p	mnodcbap	7
	bcamnopd			mbanopcd			cdmnobpa			mnodcpba	
	(a)			(b)			(c)			(d)	

Table Ans.1: Encoding With Move-Ahead-k.

1.12: Table Ans.2 summarizes the decoding steps. Notice how similar it is to Table 1.16, indicating that move-to-front is a symmetric data compression method.

Code input	A (before adding)	A (after adding)	Word
0the	()	(the)	the
1boy	(the)	(the, boy)	boy
2on	(boy, the)	(boy, the, on)	on
3my	(on, boy, the)	(on, boy, the, my)	my
4right	(my, on, boy, the)	(my, on, boy, the, right)	right
5is	(right, my, on, boy, the)	(right, my, on, boy, the, is)	is
5	(is, right, my, on, boy, the)	(is, right, my, on, boy, the)	the
2	(the, is, right, my, on, boy)	(the, is, right, my, on, boy)	right
5	(right, the, is, my, on, boy)	(right, the, is, my, on, boy)	boy
	(boy, right, the, is, my, on)		

Table Ans.2: Decoding Multiple-Letter Words.

	Prob.	Steps				Final
1.	0.25	1	1			:11
2.	0.20	1	0			:101
3.	0.15	1	0			:100
4.	0.15	0	1			:01
5.	0.10	0	0	1		:001
6.	0.10	0	0	0	0	:0001
7.	0.05	0	0	0	0	:0000

Table Ans.3: Shannon-Fano Example.

5.1: Subsequent splits can be done in different ways, but Table Ans.3 shows one way of assigning Shannon-Fano codes to the 7 symbols.
The average size in this case is $0.25 \times 2 + 0.20 \times 3 + 0.15 \times 3 + 0.15 \times 2 + 0.10 \times 3 + 0.10 \times 4 + 0.05 \times 4 = 2.75$ bits/symbols.

5.2: The entropy is $-2(0.25 \times \log_2 0.25) - 4(0.125 \times \log_2 0.125) = 2.5$.

5.3: Figure Ans.4a,b,c shows the three trees. The codes sizes for the trees are

$$(5+5+5+5{\cdot}2+3{\cdot}3+3{\cdot}5+3{\cdot}5+12)/30 = 76/30,$$
$$(5+5+4+4{\cdot}2+4{\cdot}3+3{\cdot}5+3{\cdot}5+12)/30 = 76/30,$$
$$(6+6+5+4{\cdot}2+3{\cdot}3+3{\cdot}5+3{\cdot}5+12)/30 = 76/30.$$

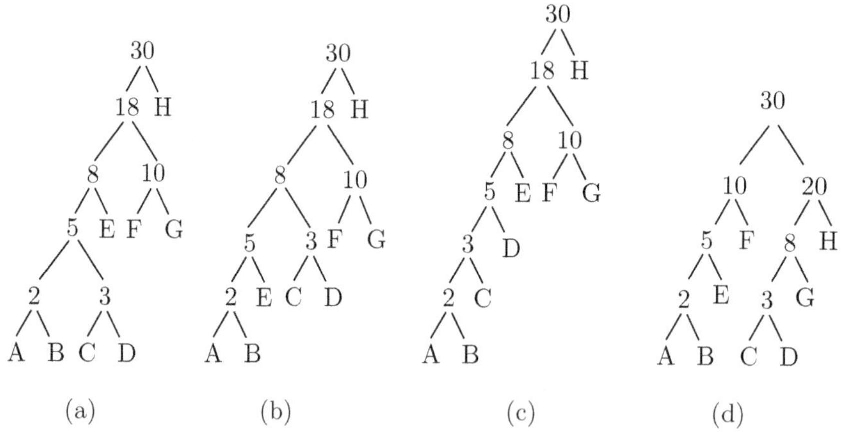

Figure Ans.4: Three Huffman Trees for Eight Symbols.

5.4: After adding symbols A, B, C, D, E, F, and G to the tree, we were left with the three symbols ABEF (with probability 10/30), CDG (with probability 8/30), and H (with probability 12/30). The two symbols with lowest probabilities were ABEF and

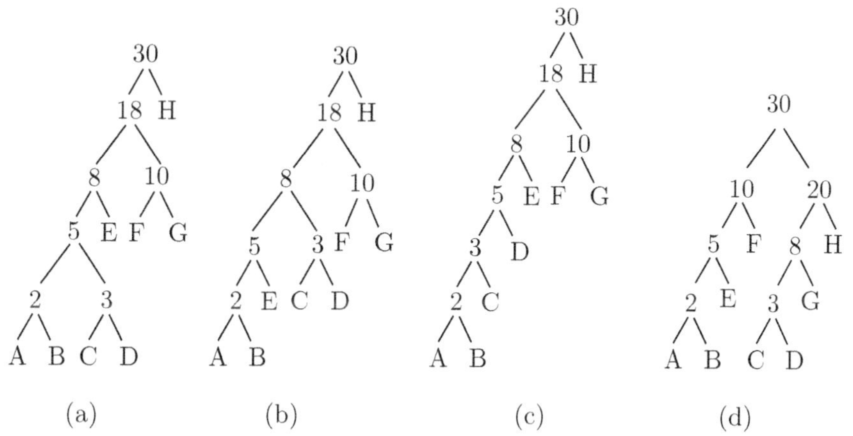

Figure Ans.5: Three Huffman Trees for Eight Symbols.

CDG, so they had to be merged. Instead, symbols CDG and H were merged, creating a non-Huffman tree.

5.5: The second row of Table Ans.6 (due to Guy Blelloch) shows a symbol whose Huffman code is three bits long, but for which $\lceil -\log_2 0.3 \rceil = \lceil 1.737 \rceil = 2$.

P_i	Code	$-\log_2 P_i$	$\lceil -\log_2 P_i \rceil$
.01	000	6.644	7
*.30	001	1.737	2
.34	01	1.556	2
.35	1	1.515	2

Table Ans.6: A Huffman Code Example.

5.6: The explanation is simple. Imagine a large alphabet where all the symbols have (about) the same probability. Since the alphabet is large, that probability will be small, resulting in long codes. Imagine the other extreme case, where certain symbols have high probabilities (and, therefore, short codes). Since the probabilities have to add up to 1, the rest of the symbols will have low probabilities (and, therefore, long codes). We therefore see that the size of a code depends on the probability, but is indirectly affected by the size of the alphabet.

5.7: Figure Ans.7 shows Huffman codes for 5, 6, 7, and 8 symbols with equal probabilities. In the case where n is a power of 2, the codes are simply the fixed-sized ones. In other cases the codes are very close to fixed-length. This shows that symbols with equal probabilities do not benefit from variable-length codes. (This is another way of saying that random text cannot be compressed.) Table Ans.8 shows the codes, their average sizes and variances.

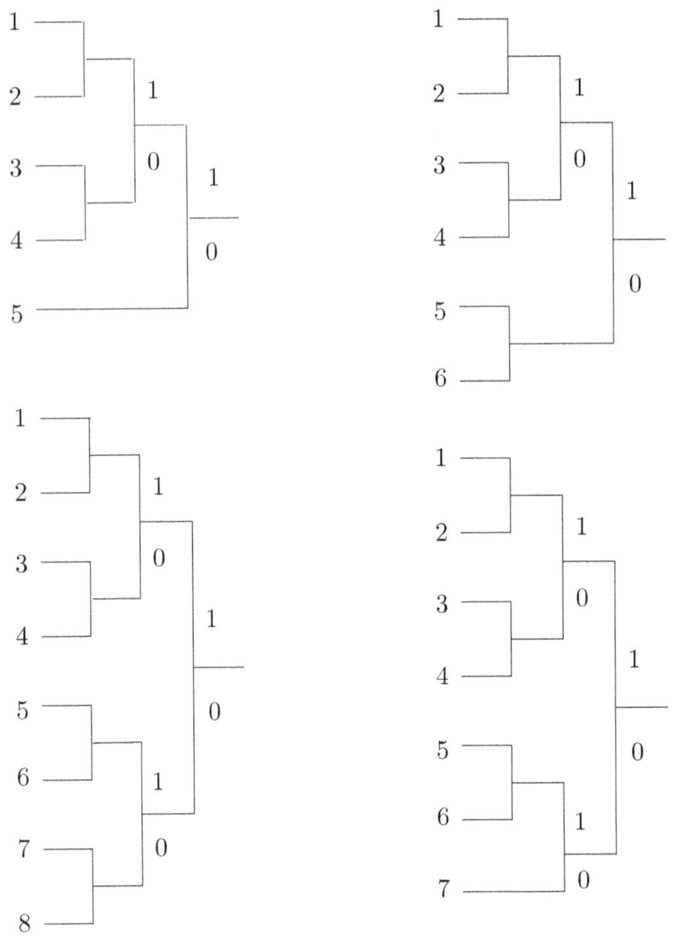

Figure Ans.7: Huffman Codes for Equal Probabilities.

n	p	a_1	a_2	a_3	a_4	a_5	a_6	a_7	a_8	Avg. size	Var.
5	0.200	111	110	101	100	0				2.6	0.64
6	0.167	111	110	101	100	01	00			2.672	0.2227
7	0.143	111	110	101	100	011	010	00		2.86	0.1226
8	0.125	111	110	101	100	011	010	001	000	3	0

Table Ans.8: Huffman Codes for 5–8 Symbols.

5.8: It increases exponentially from 2^s to $2^{s+n} = 2^s \times 2^n$.

5.9: The binary value of 127 is 01111111 and that of 128 is 10000000. Half the pixels in each bitplane will therefore be 0 and the other half, 1. In the worst case, each bitplane will be a checkerboard, i.e., will have many runs of size one. In such a case, each run requires a 1-bit code, leading to one codebit per pixel per bitplane, or eight codebits per pixel for the entire image, resulting in no compression at all. In comparison, a Huffman code for such an image requires just two codes (since there are just two pixel values) and they can be one bit each. This leads to one codebit per pixel, or a compression factor of eight.

5.10: The two trees are shown in Figure 5.16c,d. The average code size for the binary Huffman tree is

$$1\times0.49+2\times0.25+5\times0.02+5\times0.03+5\times.04+5\times0.04+3\times0.12=2\,\text{bits/symbol},$$

and that of the ternary tree is

$$1\times0.26+3\times0.02+3\times0.03+3\times0.04+2\times0.04+2\times0.12+1\times0.49=1.34\,\text{trits/symbol}.$$

5.11: Figure Ans.9 shows how the loop continues until the heap shrinks to just one node that is the single pointer 2. This indicates that the total frequency (which happens to be 100 in our example) is stored in `A[2]`. All other frequencies have been replaced by pointers. Figure Ans.10a shows the heaps generated during the loop.

5.12: The final result of the loop is

1	2	3	4	5	6	7	8	9	10	11	12	13	14
[*2*]	100	*2*	*2*	*3*	*4*	*5*	*3*	*4*	*6*	*5*	*7*	*6*	*7*

from which it is easy to figure out the code lengths of all seven symbols. To find the length of the code of symbol 14, e.g., we follow the pointers 7, 5, 3, 2 from `A[14]` to the root. Four steps are necessary, so the code length is 4.

5.13: The code lengths for the seven symbols are 2, 2, 3, 3, 4, 3, and 4 bits. This can also be verified from the Huffman code-tree of Figure Ans.10b. A set of codes derived from this tree is shown in the following table:

Count:	25	20	13	17	9	11	5
Code:	01	11	101	000	0011	100	0010
Length:	2	2	3	3	4	3	4

5.14: A symbol with high frequency of occurrence should be assigned a shorter code. Therefore, it has to appear high in the tree. The requirement that at each level the frequencies be sorted from left to right is artificial. In principle, it is not necessary, but it simplifies the process of updating the tree.

1	2	3	4	5	6	7	8	9	10	11	12	13	14
[*7*	*11*	*6*	*8*	*9*]	24	14	25	20	*6*	17	*7*	*6*	*7*

1	2	3	4	5	6	7	8	9	10	11	12	13	14
[*11*	*9*	*8*	*6*]		24	14	25	20	*6*	17	*7*	*6*	*7*

1	2	3	4	5	6	7	8	9	10	11	12	13	14
[*11*	*9*	*8*	*6*]	17+14	24	14	25	20	*6*	17	*7*	*6*	*7*

1	2	3	4	5	6	7	8	9	10	11	12	13	14
[*5*	*9*	*8*	*6*]	31	24	*5*	25	20	*6*	*5*	*7*	*6*	*7*

1	2	3	4	5	6	7	8	9	10	11	12	13	14
[*9*	*6*	*8*	*5*]	31	24	*5*	25	20	*6*	*5*	*7*	*6*	*7*

1	2	3	4	5	6	7	8	9	10	11	12	13	14
[*6*	*8*	*5*]		31	24	*5*	25	20	*6*	*5*	*7*	*6*	*7*

1	2	3	4	5	6	7	8	9	10	11	12	13	14
[*6*	*8*	*5*]	20+24	31	24	*5*	25	20	*6*	*5*	*7*	*6*	*7*

1	2	3	4	5	6	7	8	9	10	11	12	13	14
[*4*	*8*	*5*]	44	31	*4*	*5*	25	*4*	*6*	*5*	*7*	*6*	*7*

1	2	3	4	5	6	7	8	9	10	11	12	13	14
[*8*	*5*	*4*]	44	31	*4*	*5*	25	*4*	*6*	*5*	*7*	*6*	*7*

1	2	3	4	5	6	7	8	9	10	11	12	13	14
[*5*	*4*]		44	31	*4*	*5*	25	*4*	*6*	*5*	*7*	*6*	*7*

1	2	3	4	5	6	7	8	9	10	11	12	13	14
[*5*	*4*]	25+31	44	31	*4*	*5*	25	*4*	*6*	*5*	*7*	*6*	*7*

1	2	3	4	5	6	7	8	9	10	11	12	13	14
[*3*	*4*]	56	44	*3*	*4*	*5*	*3*	*4*	*6*	*5*	*7*	*6*	*7*

1	2	3	4	5	6	7	8	9	10	11	12	13	14
[*4*	*3*]	56	44	*3*	*4*	*5*	*3*	*4*	*6*	*5*	*7*	*6*	*7*

1	2	3	4	5	6	7	8	9	10	11	12	13	14
[*3*]		56	44	*3*	*4*	*5*	*3*	*4*	*6*	*5*	*7*	*6*	*7*

1	2	3	4	5	6	7	8	9	10	11	12	13	14
[*3*]	56+44	56	44	*3*	*4*	*5*	*3*	*4*	*6*	*5*	*7*	*6*	*7*

1	2	3	4	5	6	7	8	9	10	11	12	13	14
[*2*]	100	*2*	*2*	*3*	*4*	*5*	*3*	*4*	*6*	*5*	*7*	*6*	*7*

Figure Ans.9: Sifting the Heap.

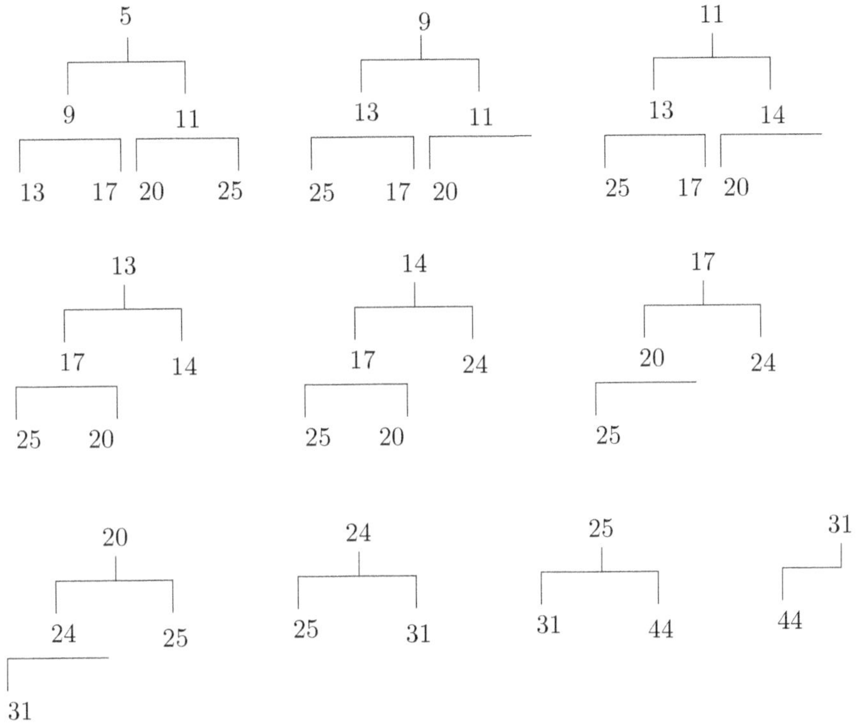

(a)

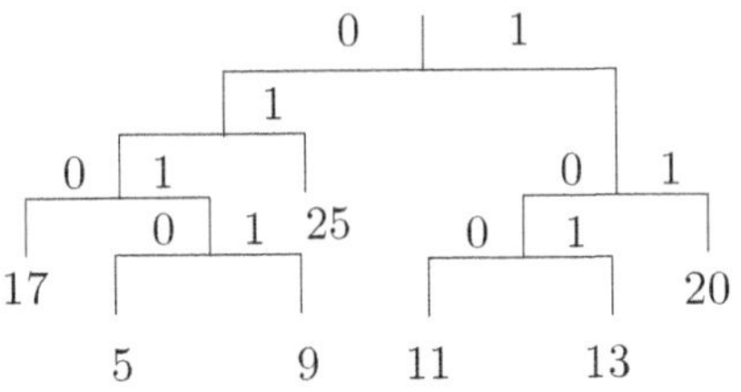

(b)

Figure Ans.10: (a) Heaps. (b) Huffman Code-Tree.

5.15: Figure Ans.11 shows the initial tree and how it is updated in the 11 steps (a) through (k). Notice how the *esc* symbol gets assigned different codes all the time, and how the different symbols move about in the tree and change their codes. Code 10, e.g., is the code of symbol "i" in steps (f) and (i), but is the code of "s" in steps (e) and (j). The code of a blank space is 011 in step (h), but 00 in step (k).

The final output is: "`s0i00r100␣1010000d011101000`". A total of $5\times 8+22=62$ bits. The compression ratio is thus $62/88\approx 0.7$.

5.16: A simple calculation shows that the average size of a token in Table 5.25 is about nine bits. In stage 2, each 8-bit byte will be replaced, on average, by a 9-bit token, resulting in an expansion factor of $9/8=1.125$ or 12.5%.

5.17: The decompressor will interpret the input data as 111110 0110 11000 0..., which is the string `XRP`....

5.18: Because a typical fax machine scans lines that are about 8.2 inches wide ($\approx$ 208 mm), so a blank scan line produces 1,664 consecutive white pels.

5.19: These codes are needed for cases such as example 4, where the run length is 64, 128, or any length for which a make-up code has been assigned.

5.20: There may be fax machines (now or in the future) built for wider paper, so the Group 3 code was designed to accommodate them.

5.21: Each scan line starts with a white pel, so when the decoder inputs the next code it knows whether it is for a run of white or black pels. This is why the codes of Table 5.31 have to satisfy the prefix property in each column but not between the columns.

5.22: Imagine a scan line where all runs have length one (strictly alternating pels). It's easy to see that this case results in expansion. The code of a run length of one white pel is 000111, and that of one black pel is 010. Two consecutive pels of different colors are thus coded into 9 bits. Since the uncoded data requires just two bits (01 or 10), the compression ratio is $9/2=4.5$ (the compressed stream is 4.5 times longer than the uncompressed one; a large expansion).

5.23: Figure Ans.12 shows the modes and the actual code generated from the two lines.

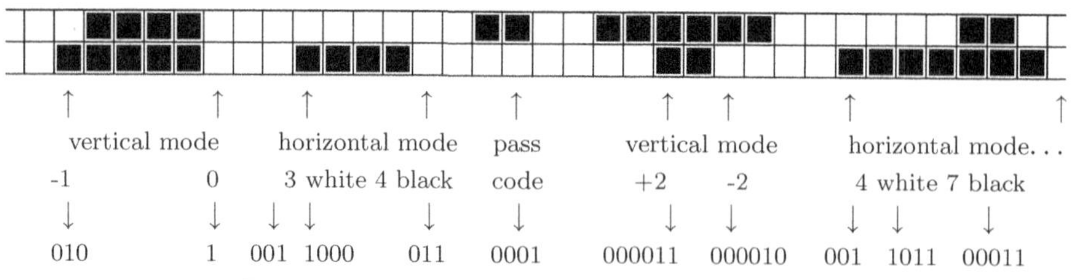

Figure Ans.12: Two-Dimensional Coding Example.

Initial tree

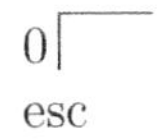

(a). Input: s. Output: 's'.
$esc\, s_1$

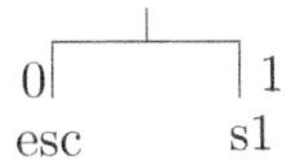

(b). Input: i. Output: 0'i'.
$esc\, i_1\, 1\ s_1$

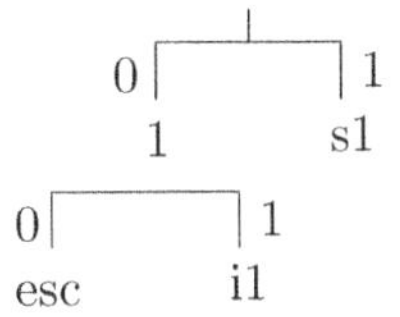

(c). Input: r. Output: 00'r'.
$esc\, r_1\, 1\, i_1\, 2\, s_1 \rightarrow$
$esc\, r_1\, 1\, i_1\, s_1\, 2$

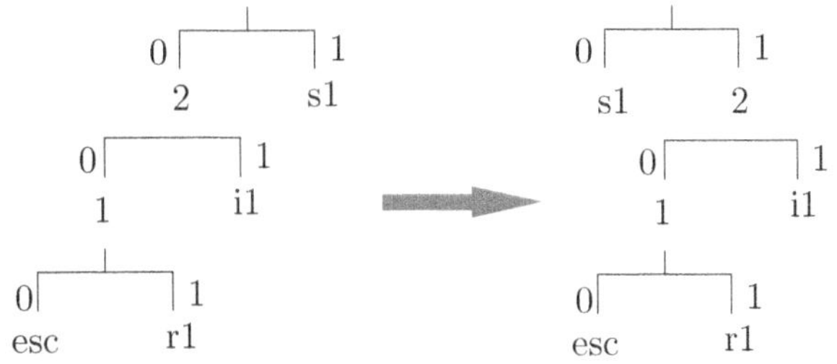

(d). Input: ␣. Output: 100'␣'.
$esc_{\sqcup 1}\, 1\, r_1\, 2\, i_1\, s_1\, 3 \rightarrow$
$esc_{\sqcup 1}\, 1\, r_1\, s_1\, i_1\, 2\, 2$

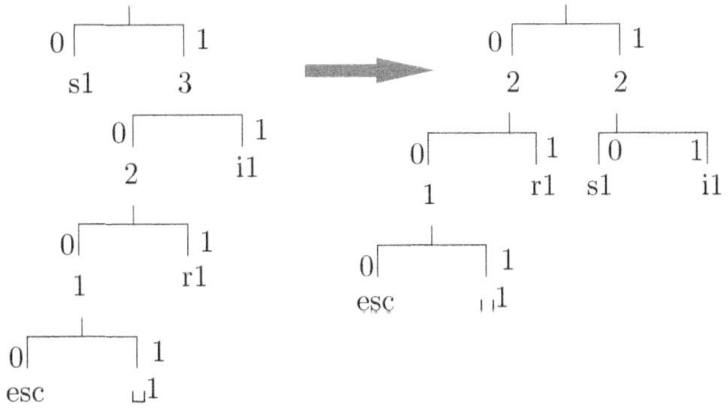

Figure Ans.11: Exercise 5.15. Part I.

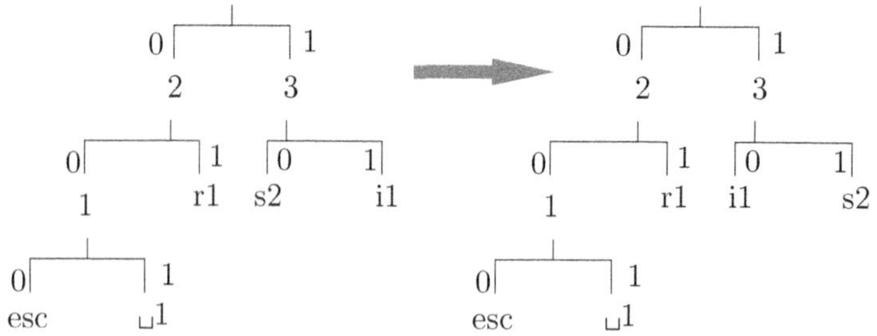

(e). Input: s. Output: 10.
$esc\,\sqcup_1\,1\,r_1\,s_2\,i_1\,2\,3 \rightarrow$
$esc\,\sqcup_1\,1\,r_1\,i_1\,s_2\,2\,3$

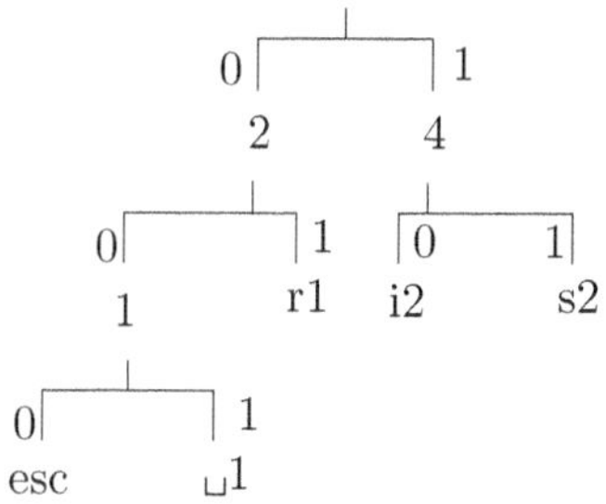

(f). Input: i. Output: 10.
$esc\,\sqcup_1\,1\,r_1\,i_2\,s_2\,2\,4$

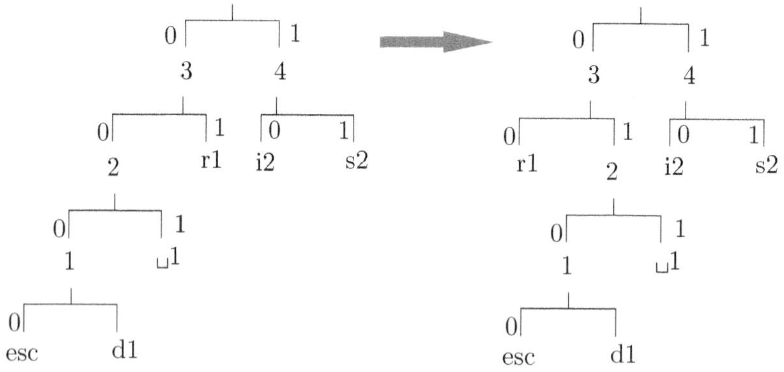

(g). Input: d. Output: 000‘d’.
$esc\,d_1\,1\,\sqcup_1\,2\,r_1\,i_2\,s_2\,3\,4 \rightarrow$
$esc\,d_1\,1\,\sqcup_1\,r_1\,2\,i_2\,s_2\,3\,4$

Figure Ans.11: Exercise 5.15. Part II.

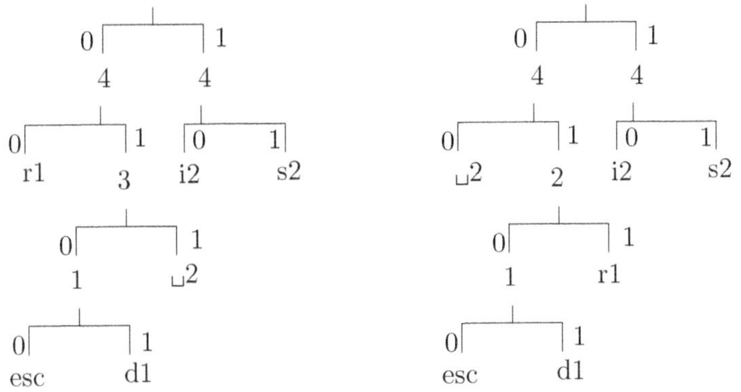

(h). Input: ␣. Output: 011.
$esc\, d_1\, 1\, {\sqcup}_2\, r_1\, 3\, i_2\, s_2\, 4\, 4 \rightarrow$
$esc\, d_1\, 1\, r_1\, {\sqcup}_2\, 2\, i_2\, s_2\, 4\, 4$

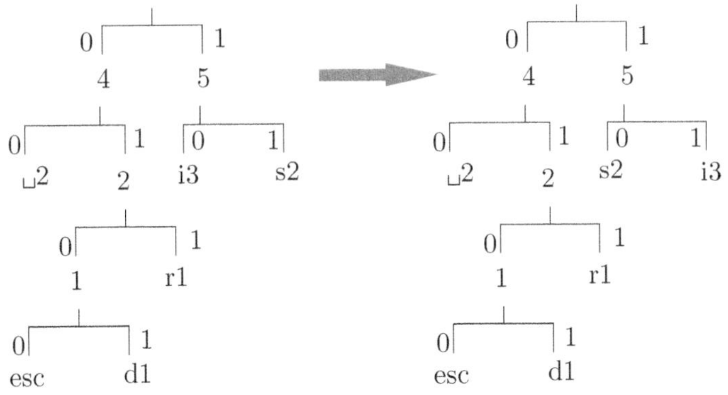

(i). Input: i. Output: 10.
$esc\, d_1\, 1\, r_1\, {\sqcup}_2\, 2\, i_3\, s_2\, 4\, 5 \rightarrow$
$esc\, d_1\, 1\, r_1\, {\sqcup}_2\, 2\, s_2\, i_3\, 4\, 5$

Figure Ans.11: Exercise 5.15. Part III.

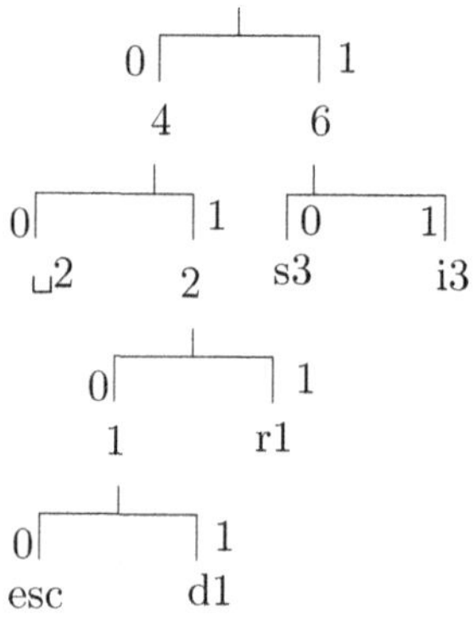

(j). Input: s. Output: 10.
$esc\, d_1\, 1\, r_1\, \sqcup_2\, 2\, s_3\, i_3\, 4\, 6$

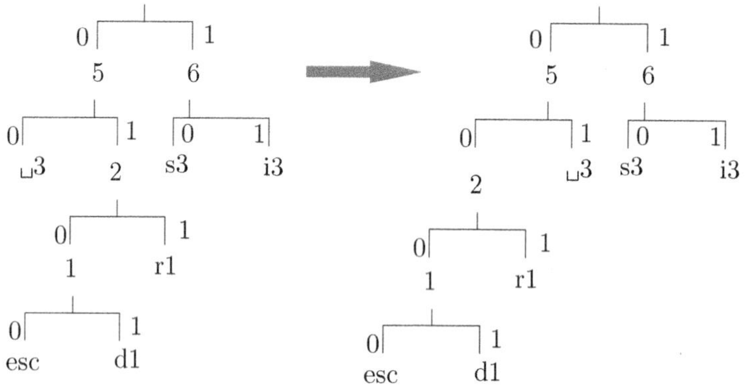

(k). Input: ␣. Output: 00.
$esc\, d_1\, 1\, r_1\, \sqcup_3\, 2\, s_3\, i_3\, 5\, 6 \rightarrow$
$esc\, d_1\, 1\, r_1\, 2\, \sqcup_3\, s_3\, i_3\, 5\, 6$

Figure Ans.11: Exercise 5.15. Part IV.

5.24: Table Ans.13 shows the steps of encoding the string $a_2a_2a_2a_2$. Because of the high probability of a_2 the low and high variables start at very different values and approach each other slowly.

a_2	$0.0 + (1.0 - 0.0) \times 0.023162{=}0.023162$
	$0.0 + (1.0 - 0.0) \times 0.998162{=}0.998162$
a_2	$0.023162 + .975 \times 0.023162{=}0.04574495$
	$0.023162 + .975 \times 0.998162{=}0.99636995$
a_2	$0.04574495 + 0.950625 \times 0.023162{=}0.06776322625$
	$0.04574495 + 0.950625 \times 0.998162{=}0.99462270125$
a_2	$0.06776322625 + 0.926859375 \times 0.023162{=}0.08923124309375$
	$0.06776322625 + 0.926859375 \times 0.998162{=}0.99291913371875$

Table Ans.13: Encoding the String $a_2a_2a_2a_2$.

5.25: It can be written either as 0.1000...or 0.0111... .

5.26: In practice, the eof symbol has to be included in the original table of frequencies and probabilities. This symbol is the last to be encoded, and the decoder stops when it detects an eof.

5.27: The encoding steps are simple (see first example on page 265). We start with the interval $[0, 1)$. The first symbol a_2 reduces the interval to $[0.4, 0.9)$. The second one, to $[0.6, 0.85)$, the third one to $[0.7, 0.825)$ and the eof symbol, to $[0.8125, 0.8250)$. The approximate binary values of the last interval are 0.1101000000 and 0.1101001100, so we select the 7-bit number 1101000 as our code.

The probability of $a_2a_2a_2$eof is $(0.5)^3 \times 0.1 = 0.0125$, but since $-\log_2 0.0125 \approx 6.322$ it follows that the practical minimum code size is 7 bits.

5.28: Perhaps the simplest way to do this is to compute a set of Huffman codes for the symbols, using their probabilities. This converts each symbol to a binary string, so the input stream can be encoded by the QM-coder. After the compressed stream is decoded by the QM-decoder, an extra step is needed, to convert the resulting binary strings back to the original symbols.

5.29: The results are shown in Tables Ans.14 and Ans.15. When all symbols are LPS, the output C always points at the bottom $A(1 - Qe)$ of the upper (LPS) subinterval. When the symbols are MPS, the output always points at the bottom of the lower (MPS) subinterval, i.e., 0.

5.30: If the current input bit is an LPS, A is shrunk to Qe, which is always 0.5 or less, so A always has to be renormalized in such a case.

5.31: The results are shown in Tables Ans.16 and Ans.17 (compare with the answer to exercise 5.29).

Symbol	C	A
Initially	0	1
s1 (LPS)	$0+1(1-0.5)=0.5$	$1\times 0.5=0.5$
s2 (LPS)	$0.5+0.5(1-0.5)=0.75$	$0.5\times 0.5=0.25$
s3 (LPS)	$0.75+0.25(1-0.5)=0.875$	$0.25\times 0.5=0.125$
s4 (LPS)	$0.875+0.125(1-0.5)=0.9375$	$0.125\times 0.5=0.0625$

Table Ans.14: Encoding Four Symbols With $Qe=0.5$.

Symbol	C	A
Initially	0	1
s1 (MPS)	0	$1\times(1-0.1)=0.9$
s2 (MPS)	0	$0.9\times(1-0.1)=0.81$
s3 (MPS)	0	$0.81\times(1-0.1)=0.729$
s4 (MPS)	0	$0.729\times(1-0.1)=0.6561$

Table Ans.15: Encoding Four Symbols With $Qe=0.1$.

Symbol	C	A	Renor. A	Renor. C
Initially	0	1		
s1 (LPS)	$0+1-0.5=0.5$	0.5	1	1
s2 (LPS)	$1+1-0.5=1.5$	0.5	1	3
s3 (LPS)	$3+1-0.5=3.5$	0.5	1	7
s4 (LPS)	$7+1-0.5=6.5$	0.5	1	13

Table Ans.16: Renormalization Added to Table Ans.14.

Symbol	C	A	Renor. A	Renor. C
Initially	0	1		
s1 (MPS)	0	$1-0.1=0.9$		
s2 (MPS)	0	$0.9-0.1=0.8$		
s3 (MPS)	0	$0.8-0.1=0.7$	1.4	0
s4 (MPS)	0	$1.4-0.1=1.3$		

Table Ans.17: Renormalization Added to Table Ans.15.

5.32: The four decoding steps are as follows:

Step 1: $C = 0.981$, $A = 1$, the dividing line is $A(1 - Qe) = 1(1 - 0.1) = 0.9$, so the LPS and MPS subintervals are $[0, 0.9)$ and $[0.9, 1)$. Since C points to the upper subinterval, an LPS is decoded. The new C is $0.981 - 1(1 - 0.1) = 0.081$ and the new A is $1\times 0.1 = 0.1$.
Step 2: $C = 0.081$, $A = 0.1$, the dividing line is $A(1 - Qe) = 0.1(1 - 0.1) = 0.09$, so the LPS and MPS subintervals are $[0, 0.09)$ and $[0.09, 0.1)$, and an MPS is decoded. C is unchanged and the new A is $0.1(1 - 0.1) = 0.09$.
Step 3: $C = 0.081$, $A = 0.09$, the dividing line is $A(1 - Qe) = 0.09(1 - 0.1) = 0.0081$, so the LPS and MPS subintervals are $[0, 0.0081)$ and $[0.0081, 0.09)$, and an LPS is decoded. The new C is $0.081 - 0.09(1 - 0.1) = 0$ and the new A is $0.09\times 0.1 = 0.009$.
Step 4: $C = 0$, $A = 0.009$, the dividing line is $A(1 - Qe) = 0.009(1 - 0.1) = 0.00081$, so the LPS and MPS subintervals are $[0, 0.00081)$ and $[0.00081, 0.009)$, and an MPS is decoded. C is unchanged and the new A is $0.009(1 - 0.1) = 0.00081$.

5.33: In practice, an encoder may encode texts other than English, such as a foreign language or the source code of a computer program. Acronyms, such as QED and abbreviations, such as qwerty, are also good examples. Even in English texts there are many examples of a `q` not followed by a `u`, such as in words transliterated from other languages, most commonly Arabic and Chinese. Examples are `suq` (market) and `qadi` (a Moslem judge). See [PQ-wiki 09] for more examples.

5.34: The number of order-2 and order-3 contexts for an alphabet of size $2^8 = 256$ is $256^2 = 65,536$ and $256^3 = 16,777,216$, respectively. The former is manageable, whereas the latter is perhaps too big for a practical implementation, unless a sophisticated data structure is used or unless the encoder gets rid of older data from time to time.

For a small alphabet, larger values of N can be used. For a 16-symbol alphabet there are $16^4 = 65,536$ order-4 contexts and $16^6 = 16,777,216$ order-6 contexts.

5.35: A practical example of a 16-symbol alphabet is a color or grayscale image with 4-bit pixels. Each symbol is a pixel, and there are 16 different symbols.

5.36: An object file generated by a compiler or an assembler normally has several distinct parts including the machine instructions, symbol table, relocation bits, and constants. Such parts may have different bit distributions.

5.37: The alphabet has to be extended, in such a case, to include one more symbol. If the original alphabet consisted of all the possible 256 8-bit bytes, it should be extended to 9-bit symbols, and should include 257 values.

5.38: Table Ans.18 shows the groups generated in both cases and makes it clear why these particular probabilities were assigned.

5.39: The `d` is added to the order-0 contexts with frequency 1. The escape frequency should be incremented from 5 to 6, bringing the total frequencies from 19 up to 21. The probability assigned to the new `d` is therefore $1/21$, and that assigned to the escape is $6/21$. All other probabilities are reduced from $x/19$ to $x/21$.

Context	f	p
abc→x	10	10/11
Esc	1	1/11

Context		f	p
abc→	a_1	1	1/20
→	a_2	1	1/20
→	a_3	1	1/20
→	a_4	1	1/20
→	a_5	1	1/20
→	a_6	1	1/20
→	a_7	1	1/20
→	a_8	1	1/20
→	a_9	1	1/20
→	a_{10}	1	1/20
Esc		10	10/20
Total		20	

Table Ans.18: Stable vs. Variable Data.

5.40: The new `d` would require switching from order 2 to order 0, sending two escapes that take 1 and 1.32 bits. The `d` is now found in order-0 with probability 1/21, so it is encoded in 4.39 bits. The total number of bits required to encode the second `d` is therefore $1 + 1.32 + 4.39 = 6.71$, still greater than 5.

5.41: The first three cases don't change. They still code a symbol with 1, 1.32, and 6.57 bits, which is less than the 8 bits required for a 256-symbol alphabet without compression. Case 4 is different since the `d` is now encoded with a probability of 1/256, producing 8 instead of 4.8 bits. The total number of bits required to encode the `d` in case 4 is now $1 + 1.32 + 1.93 + 8 = 12.25$.

5.42: The final trie is shown in Figure Ans.19.

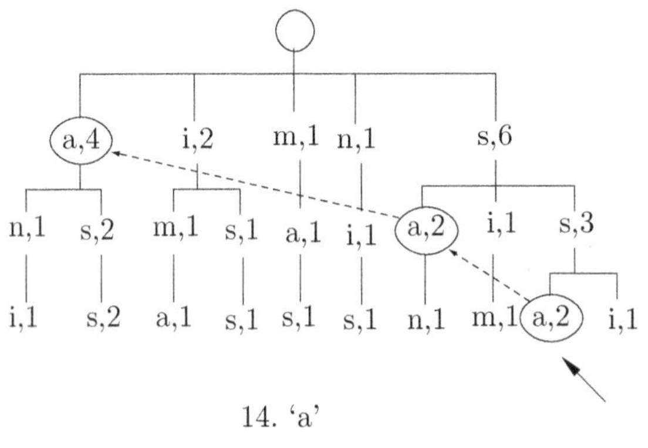

Figure Ans.19: Final Trie of `assanissimassa`.

5.43: This probability is, of course

$$1 - P_e(b_{t+1} = 1|b_1^t) = 1 - \frac{b + 1/2}{a + b + 1} = \frac{a + 1/2}{a + b + 1}.$$

5.44: For the first string the single bit has a suffix of 00, so the probability of leaf 00 is $P_e(1,0) = 1/2$. This is equal to the probability of string 0 without any suffix. For the second string each of the two zero bits has suffix 00, so the probability of leaf 00 is $P_e(2,0) = 3/8 = 0.375$. This is greater than the probability 0.25 of string 00 without any suffix. Similarly, the probabilities of the remaining three strings are $P_e(3,0) = 5/8 \approx 0.625$, $P_e(4,0) = 35/128 \approx 0.273$, and $P_e(5,0) = 63/256 \approx 0.246$. As the strings get longer, their probabilities get smaller but they are greater than the probabilities without the suffix. Having a suffix of 00 thus increases the probability of having strings of zeros following it.

5.45: The four trees are shown in Figure Ans.20a–d. The weighted probability that the next bit will be a zero given that three zeros have just been generated is 0.5. The weighted probability to have two consecutive zeros given the suffix 000 is 0.375, higher than the 0.25 without the suffix.

6.1: The size of the output stream is $N[48 - 28P] = N[48 - 25.2] = 22.8N$. The size of the input stream is, as before, $40N$. The compression factor is therefore $40/22.8 \approx 1.75$.

6.2: The list has up to 256 entries, each consisting of a byte and a count. A byte occupies eight bits, but the counts depend on the size and content of the file being compressed. If the file has high redundancy, a few bytes may have large counts, close to the length of the file, while other bytes may have very low counts. On the other hand, if the file is close to random, then each distinct byte has approximately the same count.

Thus, the first step in organizing the list is to reserve enough space for each "count" field to accommodate the maximum possible count. We denote the length of the file by L and find the positive integer k that satisfies $2^{k-1} < L \leq 2^k$. Thus, L is a k-bit number. If k is not already a multiple of 8, we increase it to the next multiple of 8. We now denote $k = 8m$, and allocate m bytes to each "count" field.

Once the file has been input and processed and the list has been sorted, we examine the largest count. It may be large and may occupy all m bytes, or it may be smaller. Assuming that the largest count occupies n bytes (where $n \leq m$), we can store each of the other counts in n bytes.

When the list is written on the compressed file as the dictionary, its length s is first written in one byte. s is the number of distinct bytes in the original file. This is followed by n, followed by s groups, each with one of the distinct data bytes followed by an n-byte count. Notice that the value n should be fairly small and should fit in a single byte. If n does not fit in a single byte, then it is greater than 255, implying that the largest count does not fit in 255 bytes, implying in turn a file whose length L is greater than $2^{255} \approx 10^{76}$ bytes.

An alternative is to start with s, followed by n, followed by the s distinct data bytes, followed by the $n \times s$ bytes of counts. The last part could also be in compressed

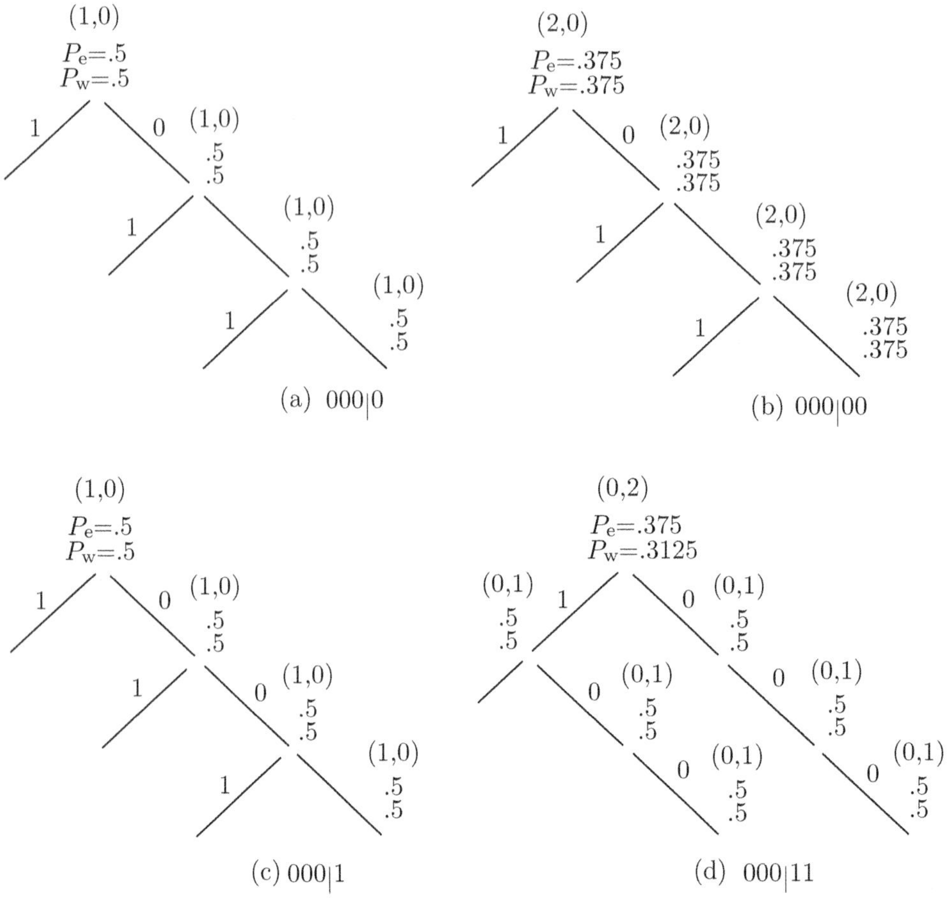

Figure Ans.20: Context Trees For 000|0, 000|00, 000|1, and 000|11.

form, because only a few largest counts will occupy all n bytes. Most counts may be small and occupy just one or two bytes, which implies that many of the $n \times s$ count bytes will be zero, resulting in high redundancy and therefore good compression.

6.3: The straightforward answer is The decoder doesn't know but it does not need to know. The decoder simply reads tokens and uses each offset to locate a string of text without having to know whether the string was a first or a last match.

6.4: The next step matches the space and encodes the string ␣e.

`sir␣sid`	`␣eastman␣easily␣`	⇒	(4,1,e)
`sir␣sid␣e`	`astman␣easily␣te`	⇒	(0,0,a)

and the next one matches nothing and encodes the `a`.

6.5: The first two characters `CA` at positions 17–18 are a repeat of the `CA` at positions 9–10, so they will be encoded as a string of length 2 at offset $18 - 10 = 8$. The next two characters AC at positions 19–20 are a repeat of the string at positions 8–9, so they will be encoded as a string of length 2 at offset $20 - 9 = 11$.

6.6: The decoder interprets the first 1 of the end marker as the start of a token. The second 1 is interpreted as the prefix of a 7-bit offset. The next 7 bits are 0, and they identify the end marker as such, since a "normal" offset cannot be zero.

6.7: This is straightforward. The remaining steps are shown in Table Ans.21

Dictionary		Token	Dictionary		Token
15	␣t	(4, t)	21	␣si	(19,i)
16	e	(0, e)	22	c	(0, c)
17	as	(8, s)	23	k	(0, k)
18	es	(16,s)	24	␣se	(19,e)
19	␣s	(4, s)	25	al	(8, l)
20	ea	(4, a)	26	s(eof)	(1, (eof))

Table Ans.21: Next 12 Encoding Steps in the LZ78 Example.

6.8: Table Ans.22 shows the last three steps.

p_src	3 chars	Hash index	P	Output	Binary output
11	h t	7	any→11	h	01101000
12	␣th	5	5→12	4,7	0000|0011|00000111
16	ws			ws	01110111|01110011

Table Ans.22: Last Steps of Encoding that thatch thaws.

The final compressed stream consists of 1 control word followed by 11 items (9 literals and 2 copy items)
0000010010000000|01110100|01101000|01100001|01110100|00100000|0000|0011
|00000101|01100011|01101000|0000|0011|00000111|01110111|01110011.

6.9: An example is a compression utility for a personal computer that maintains all the files (or groups of files) on the hard disk in compressed form, to save space. Such a utility should be transparent to the user; it should automatically decompress a file every time it is opened and automatically compress it when it is being closed. In order to be transparent, such a utility should be fast, with compression ratio being only a secondary feature.

6.10: Table Ans.23 summarizes the steps. The output emitted by the encoder is
97 (a), 108 (l), 102 (f), 32 (␣), 101 (e), 97 (a), 116 (t), 115 (s), 32 (␣), 256 (al), 102 (f), 265 (alf), 97 (a),
and the following new entries are added to the dictionary
(256: al), (257: lf), (258: f␣), (259: ␣e), (260: ea), (261: at), (262: ts),
(263: s␣), (264: ␣a), (265: alf), (266: fa), (267: alfa).

I	in dict?	new entry	output	I	in dict?	new entry	output
a	Y			s␣	N	263-s␣	115 (s)
al	N	256-al	97 (a)	␣	Y		
l	Y			␣a	N	264-␣a	32 (␣)
lf	N	257-lf	108 (l)	a	Y		
f	Y			al	Y		
f␣	N	258-f␣	102 (f)	alf	N	265-alf	256 (al)
␣	Y			f	Y		
␣e	N	259-␣e	32 (w)	fa	N	266-fa	102 (f)
e	Y			a	Y		
ea	N	260-ea	101 (e)	al	Y		
a	Y			alf	Y		
at	N	261-at	97 (a)	alfa	N	267-alfa	265 (alf)
t	Y			a	Y		
ts	N	262-ts	116 (t)	a,eof	N		97 (a)
s	Y						

Table Ans.23: LZW Encoding of "alf eats alfalfa".

6.11: The encoder inputs the first a into I, searches and finds a in the dictionary. It inputs the next a but finds that Ix, which is now aa, is not in the dictionary. The encoder thus adds string aa to the dictionary as entry 256 and outputs the token 97 (a). Variable I is initialized to the second a. The third a is input, so Ix is the string aa, which is now in the dictionary. I becomes this string, and the fourth a is input. Ix is now aaa which is not in the dictionary. The encoder thus adds string aaa to the dictionary as entry 257 and outputs 256 (aa). I is initialized to the fourth a. Continuing this process is straightforward.

The result is that strings aa, aaa, aaaa,... are added to the dictionary as entries 256, 257, 258,..., and the output is

97 (a), 256 (aa), 257 (aaa), 258 (aaaa),...

The output consists of pointers pointing to longer and longer strings of as. The first k pointers thus point at strings whose total length is $1+2+\cdots+k=(k+k^2)/2$.

Assuming an input stream that consists of one million as, we can find the size of the compressed output stream by solving the quadratic equation $(k+k^2)/2=1000000$ for the unknown k. The solution is $k\approx 1414$. The original, 8-million bit input is thus compressed into 1414 pointers, each at least 9-bit (and in practice, probably 16-bit) long. The compression factor is thus either $8\text{M}/(1414\times 9)\approx 628.6$ or $8\text{M}/(1414\times 16)\approx 353.6$.

This is an impressive result but such input streams are rare (notice that this particular input can best be compressed by generating an output stream containing just "1000000 a", and without using LZW).

6.12: We simply follow the decoding steps described in the text. The results are:
1. Input 97. This is in the dictionary so set I=a and output a. String ax needs to be saved in the dictionary but x is still unknown.

2. Input 108. This is in the dictionary so set J=`l` and output `l`. Save `al` in entry 256. Set I=`l`.
3. Input 102. This is in the dictionary so set J=`f` and output `f`. Save `lf` in entry 257. Set I=`f`.
4. Input 32. This is in the dictionary so set J=␣ and output ␣. Save `f␣` in entry 258. Set I=␣.
5. Input 101. This is in the dictionary so set J=`e` and output `e`. Save `␣e` in entry 259. Set I=`e`.
6. Input 97. This is in the dictionary so set J=`a` and output `a`. Save `ea` in entry 260. Set I=`a`.
7. Input 116. This is in the dictionary so set J=`t` and output `t`. Save `at` in entry 261. Set I=`t`.
8. Input 115. This is in the dictionary so set J=`s` and output `s`. Save `ts` in entry 262. Set I=`t`.
9. Input 32. This is in the dictionary so set J=␣ and output ␣. Save `s␣` in entry 263. Set I=␣.
10. Input 256. This is in the dictionary so set J=`al` and output `al`. Save `␣a` in entry 264. Set I=`al`.
11. Input 102. This is in the dictionary so set J=`f` and output `f`. Save `alf` in entry 265. Set I=`f`.
12. Input 265. This has just been saved in the dictionary so set J=`alf` and output `alf`. Save `fa` in dictionary entry 266. Set I=`alf`.
13. Input 97. This is in the dictionary so set J=`a` and output `a`. Save `alfa` in entry 267 (even though it will never be used). Set I=`a`.
14. Read eof. Stop.

6.13: We assume that the dictionary is initialized to just the two entries (1: `a`) and (2:`␣b`). The encoder outputs

1 (`a`), 2 (`b`), 3 (`ab`), 5(`aba`), 4(`ba`), 7 (`bab`), 6 (`abab`), 9 (`ababa`), 8 (`baba`),...

and adds the new entries (3: `ab`), (4: `ba`), (5: `aba`), (6: `abab`), (7: `bab`), (8: `baba`), (9: `ababa`), (10: `ababab`), (11: `babab`),... to the dictionary. This regular behavior can be analyzed and the kth output pointer and dictionary entry predicted, but the effort is probably not worth it.

6.14: The answer to exercise 6.11 shows the relation between the size of the compressed file and the size of the largest dictionary string for the "worst case" situation (input that creates the longest strings). For a 1 Mbyte input stream, there will be 1,414 strings in the dictionary, the largest of which is 1,414 symbols long.

6.15: This is straightforward (Table Ans.24) but not very efficient since only one two-symbol dictionary phrase is used.

6.16: Table Ans.25 shows all the steps. In spite of the short input, the result is quite good (13 codes to compress 18-symbols) because the input contains concentrations of `a`s and `b`s.

Step	Input	Output	S	Add to dict.	S'
	swiss miss				
1	s	115	s	—	s
2	w	119	w	256-sw	w
3	i	105	i	257-wi	i
4	s	115	s	258-is	s
5	s	115	s	259-ss	s
6	-	32	␣	260-s␣	␣
7	m	109	m	261-␣m	m
8	is	258	is	262-mis	is
9	s	115	s	263-iss	s

Table Ans.24: LZMW Compression of "swiss miss".

Step	Input	Output	S	Add to dict.	S'
	yabbadabbadabbadoo				
1	y	121	y	—	y
2	a	97	a	256-ya	a
3	b	98	b	257-ab	b
4	b	98	b	258-bb	b
5	a	97	a	259-ba	a
6	d	100	a	260-ad	a
7	ab	257	ab	261-dab	ab
8	ba	259	ba	262-abba	ba
9	dab	261	dab	263-badab	dab
10	ba	259	ba	264-dabba	ba
11	d	100	d	265-bad	d
12	o	111	o	266-do	o
13	o	111	o	267-o	o

Table Ans.25: LZMW Compression of "yabbadabbadabbadoo".

6.17: 1. The encoder starts by shifting the first two symbols xy to the search buffer, outputting them as literals and initializing all locations of the index table to the null pointer.
2. The current symbol is a (the first a) and the context is xy. It is hashed to, say, 5, but location 5 of the index table contains a null pointer, so P is null. Location 5 is set to point to the first a, which is then output as a literal. The data in the encoder's buffer is shifted to the left.
3. The current symbol is the second a and the context is ya. It is hashed to, say, 1, but location 1 of the index table contains a null pointer, so P is null. Location 1 is set to point to the second a, which is then output as a literal. The data in the encoder's buffer is shifted to the left.
4. The current symbol is the third a and the context is aa. It is hashed to, say, 2, but

location 2 of the index table contains a null pointer, so P is null. Location 2 is set to point to the third a, which is then output as a literal. The data in the encoder's buffer is shifted to the left.
5. The current symbol is the fourth a and the context is aa. We know from step 4 that it is hashed to 2, and location 2 of the index table points to the third a. Location 2 is set to point to the fourth a, and the encoder tries to match the string starting with the third a to the string starting with the fourth a. Assuming that the look-ahead buffer is full of as, the match length L will be the size of that buffer. The encoded value of L will be written to the compressed stream, and the data in the buffer shifted L positions to the left.
6. If the original input stream is long, more a's will be shifted into the look-ahead buffer, and this step will also result in a match of length L. If only n as remain in the input stream, they will be matched, and the encoded value of n output.

The compressed stream will consist of the three literals x, y, and a, followed by (perhaps several values of) L, and possibly ending with a smaller value.

6.18: T percent of the compressed stream is made up of literals, some appearing consecutively (and thus getting the flag "1" for two literals, half a bit per literal) and others with a match length following them (and thus getting the flag "01", one bit for the literal). We assume that two thirds of the literals appear consecutively and one third are followed by match lengths. The total number of flag bits created for literals is thus

$$\frac{2}{3}T \times 0.5 + \frac{1}{3}T \times 1.$$

A similar argument for the match lengths yields

$$\frac{2}{3}(1-T) \times 2 + \frac{1}{3}(1-T) \times 1$$

for the total number of the flag bits. We now write the equation

$$\frac{2}{3}T \times 0.5 + \frac{1}{3}T \times 1 + \frac{2}{3}(1-T) \times 2 + \frac{1}{3}(1-T) \times 1 = 1,$$

which is solved to yield $T = 2/3$. This means that if two thirds of the items in the compressed stream are literals, there would be 1 flag bit per item on the average. More literals would result in fewer flag bits.

6.19: The first three 1's indicate six literals. The following 01 indicates a literal (b) followed by a match length (of 3). The 10 is the code of match length 3, and the last 1 indicates two more literals (x and y).

7.1: An image with no redundancy is not always random. The definition of redundancy (Section A.1) tells us that an image where each color appears with the same frequency has no redundancy (statistically) yet it is not necessarily random and may even be interesting and/or useful.

7.2: Figure Ans.26 shows two 32×32 matrices. The first one, a, with random (and therefore decorrelated) values and the second one, b, is its inverse (and therefore with correlated values). Their covariance matrices are also shown and it is obvious that matrix $\text{cov}(a)$ is close to diagonal, whereas matrix $\text{cov}(b)$ is far from diagonal. The Matlab code for this figure is also listed.

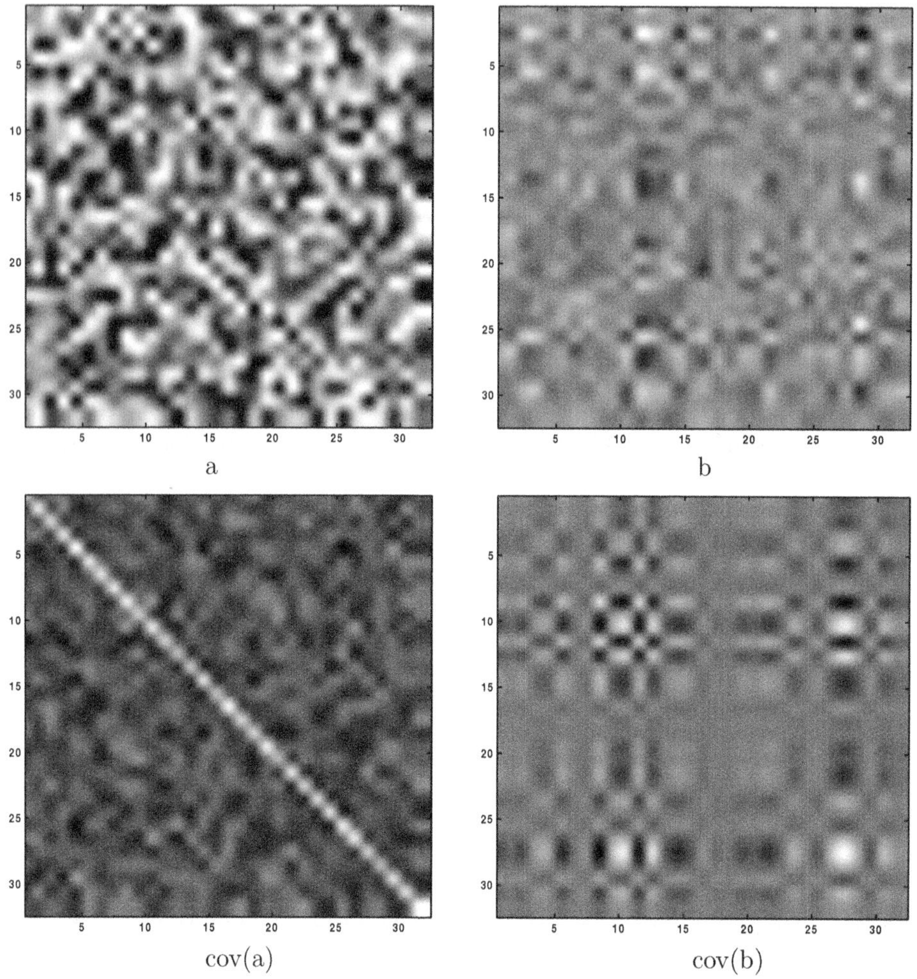


```
a=rand(32); b=inv(a);
figure(1), imagesc(a), colormap(gray); axis square
figure(2), imagesc(b), colormap(gray); axis square
figure(3), imagesc(cov(a)), colormap(gray); axis square
figure(4), imagesc(cov(b)), colormap(gray); axis square
```

Figure Ans.26: Covariance Matrices of Correlated and Decorrelated Values.

43210	Gray	43210	Gray	43210	Gray	43210	Gray
00000	00000	0**1000**	01100	**10000**	11000	1**1000**	10100
00001	00001	01001	01101	10001	11001	11001	10101
000**10**	00011	010**10**	01111	100**10**	11011	110**10**	10111
00011	00010	01011	01110	10011	11010	11011	10110
00**100**	00110	01**100**	01010	10**100**	11110	11**100**	10010
00101	00111	01101	01011	10101	11111	11101	10011
001**10**	00101	011**10**	01001	101**10**	11101	111**10**	10001
00111	00100	01111	01000	10111	11100	11111	10000

```
a=linspace(0,31,32); b=bitshift(a,-1);
b=bitxor(a,b); dec2bin(b)
```

Table Ans.27: First 32 Binary and Gray Codes.

7.3: The results are shown in Table Ans.27 together with the Matlab code used to calculate it.

7.4: One feature is the regular way in which each of the five code bits alternates periodically between 0 and 1. It is easy to write a program that will set all five bits to 0, will flip the rightmost bit after two codes have been calculated, and will flip any of the other four code bits in the middle of the period of its immediate neighbor on the right. Another feature is the fact that the second half of the table is a mirror image of the first half, but with the most significant bit set to one. After the first half of the table has been computed, using any method, this symmetry can be used to quickly calculate the second half.

7.5: Figure Ans.28 is an *angular code wheel* representation of the 4-bit and 6-bit RGC codes (part a) and the 4-bit and 6-bit binary codes (part b). The individual bitplanes are shown as rings, with the most significant bits as the innermost ring. It is easy to see that the maximum angular frequency of the RGC is half that of the binary code and that the first and last codes also differ by just one bit.

7.6: If pixel values are in the range $[0, 255]$, a difference $(P_i - Q_i)$ can be at most 255. The worst case is where all the differences are 255. It is easy to see that such a case yields an RMSE of 255.

7.7: The code of Figure 7.16 yields the coordinates of the rotated points

$$(7.071, 0),\ (9.19, 0.7071),\ (17.9, 0.78),\ (33.9, 1.41),\ (43.13, -2.12)$$

(notice how all the y coordinates are small numbers) and shows that the cross-correlation drops from 1729.72 before the rotation to -23.0846 after it. A significant reduction!

7.8: Figure Ans.29 shows the 64 basis images and the Matlab code to calculate and display them. Each basis image is an 8×8 matrix.

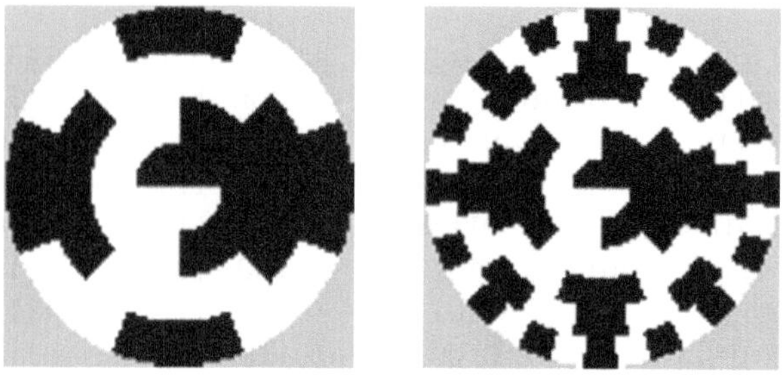

(a)

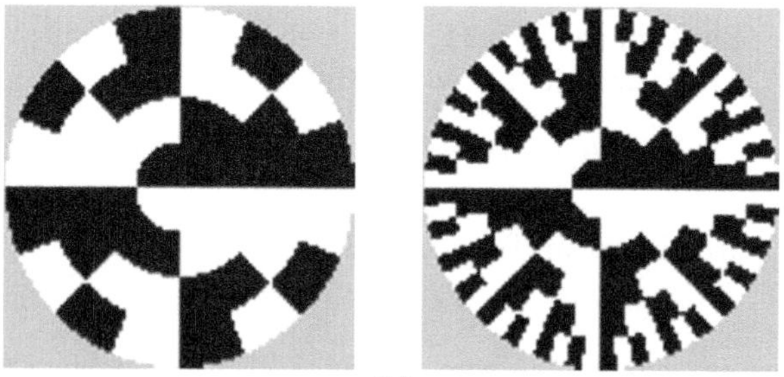

(b)

Figure Ans.28: Angular Code Wheels of RGC and Binary Codes.

7.9: $\mathbf{A}_4$ is the 4×4 matrix

$$\mathbf{A}_4=\begin{pmatrix} h_0(0/4) & h_0(1/4) & h_0(2/4) & h_0(3/4)\\ h_1(0/4) & h_1(1/4) & h_1(2/4) & h_1(3/4)\\ h_2(0/4) & h_2(1/4) & h_2(2/4) & h_2(3/4)\\ h_3(0/4) & h_3(1/4) & h_3(2/4) & h_3(3/4)\end{pmatrix}=\frac{1}{\sqrt{4}}\begin{pmatrix} 1 & 1 & 1 & 1\\ 1 & 1 & -1 & -1\\ \sqrt{2} & -\sqrt{2} & 0 & 0\\ 0 & 0 & \sqrt{2} & -\sqrt{2}\end{pmatrix}.$$

Similarly, $\mathbf{A}_8$ is the matrix

$$\mathbf{A}_8=\frac{1}{\sqrt{8}}\begin{pmatrix} 1 & 1 & 1 & 1 & 1 & 1 & 1 & 1\\ 1 & 1 & 1 & 1 & -1 & -1 & -1 & -1\\ \sqrt{2} & \sqrt{2} & -\sqrt{2} & -\sqrt{2} & 0 & 0 & 0 & 0\\ 0 & 0 & 0 & 0 & \sqrt{2} & \sqrt{2} & -\sqrt{2} & -\sqrt{2}\\ 2 & -2 & 0 & 0 & 0 & 0 & 0 & 0\\ 0 & 0 & 2 & -2 & 0 & 0 & 0 & 0\\ 0 & 0 & 0 & 0 & 2 & -2 & 0 & 0\\ 0 & 0 & 0 & 0 & 0 & 0 & 2 & -2\end{pmatrix}.$$

7.10: The average of vector $\mathbf{w}^{(i)}$ is zero, so Equation (7.13) yields

$$(\mathbf{W}\cdot\mathbf{W}^T)_{jj}=\sum_{i=1}^{k}w_j^{(i)}w_j^{(i)}=\sum_{i=1}^{k}\left(w_j^{(i)}-0\right)^2=\sum_{i=1}^{k}\left(c_i^{(j)}-0\right)^2=k\,\text{Variance}(\mathbf{c}^{(j)}).$$

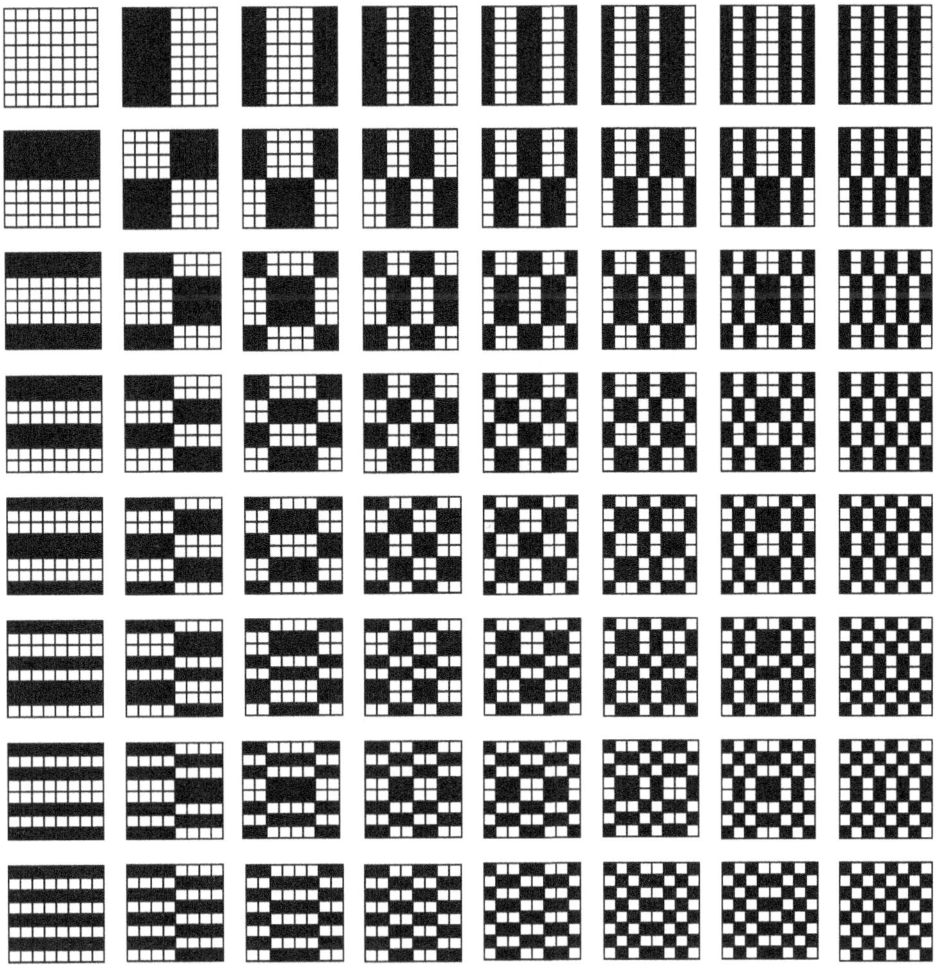

```
M=3; N=2^M; H=[1 1; 1 -1]/sqrt(2);
for m=1:(M-1) % recursion
  H=[H H; H -H]/sqrt(2);
end
A=H';
map=[1 5 7 3 4 8 6 2]; % 1:N
for n=1:N, B(:,n)=A(:,map(n)); end;
A=B;
sc=1/(max(abs(A(:))).^2); % scale factor
for row=1:N
  for col=1:N
    BI=A(:,row)*A(:,col).'; % tensor product
    subplot(N,N,(row-1)*N+col)
    oo=round(BI*sc); % results in -1, +1
    imagesc(oo), colormap([1 1 1; .5 .5 .5; 0 0 0])
    drawnow
  end
end
```

Figure Ans.29: The 8×8 WHT Basis Images and Matlab Code.

7.11: The *Mathematica* code of Figure 7.22 produces the eight coefficients 140, −71, 0, −7, 0, −2, 0, and 0. We now quantize this set coarsely by clearing the last two nonzero weights −7 and −2, When the IDCT is applied to the sequence 140, −71, 0, 0, 0, 0, 0, 0, it produces 15, 20, 30, 43, 56, 69, 79, and 84. These are not identical to the original values, but the maximum difference is only 4; an excellent result considering that only two of the eight DCT coefficients are nonzero.

7.12: The eight values in the top row are very similar (the differences between them are either 2 or 3). Each of the other rows is obtained as a right-circular shift of the preceding row.

7.13: It is obvious that such a block can be represented as a linear combination of the patterns in the leftmost column of Figure 7.40. The actual calculation yields the eight weights 4, 0.72, 0, 0.85, 0, 1.27, 0, and 3.62 for the patterns of this column. The other 56 weights are zero or very close to zero.

7.14: The arguments of the cosine functions used by the DCT are of the form $(2x+1)i\pi/16$, where i and x are integers in the range $[0,7]$. Such an argument can be written in the form $n\pi/16$, where n is an integer in the range $[0,15\times 7]$. Since the cosine function is periodic, it satisfies $\cos(32\pi/16)=\cos(0\pi/16)$, $\cos(33\pi/16)=\cos(\pi/16)$, and so on. As a result, only the 32 values $\cos(n\pi/16)$ for $n=0,1,2,\ldots,31$ are needed. We are indebted to V. Saravanan for pointing out this feature of the DCT.

7.15: Figure 7.54 shows the results (that resemble Figure 7.40) and the Matlab code. Notice that the same code can also be used to calculate and display the DCT basis images.

7.16: First figure out the zigzag path manually, then record it in an array `zz` of structures, where each structure contains a pair of coordinates for the path as shown, e.g., in Figure Ans.30.

(0,0)	(0,1)	(1,0)	(2,0)	(1,1)	(0,2)	(0,3)	(1,2)
(2,1)	(3,0)	(4,0)	(3,1)	(2,2)	(1,3)	(0,4)	(0,5)
(1,4)	(2,3)	(3,2)	(4,1)	(5,0)	(6,0)	(5,1)	(4,2)
(3,3)	(2,4)	(1,5)	(0,6)	(0,7)	(1,6)	(2,5)	(3,4)
(4,3)	(5,2)	(6,1)	(7,0)	(7,1)	(6,2)	(5,3)	(4,4)
(3,5)	(2,6)	(1,7)	(2,7)	(3,6)	(4,5)	(5,4)	(6,3)
(7,2)	(7,3)	(6,4)	(5,5)	(4,6)	(3,7)	(4,7)	(5,6)
(6,5)	(7,4)	(7,5)	(6,6)	(5,7)	(6,7)	(7,6)	(7,7)

Figure Ans.30: Coordinates for the Zigzag Path.

If the two components of a structure are `zz.r` and `zz.c`, then the zigzag traversal can be done by a loop of the form :

```
for (i=0; i<64; i++){
row:=zz[i].r; col:=zz[i].c
...data_unit[row][col]...}
```

7.17: The third DC difference, 5, is located in row 3 column 5, so it is encoded as 1110|101.

7.18: Thirteen consecutive zeros precede this coefficient, so Z = 13. The coefficient itself is found in Table 7.65 in row 1, column 0, so R = 1 and C = 0. Assuming that the Huffman code in position (R, Z) = (1, 13) of Table 7.68 is 1110101, the final code emitted for 1 is 1110101|0.

7.19: This is shown by multiplying the largest four n-bit number, $\underbrace{11\ldots1}_{n}$ by 4, which is easily done by shifting it 2 positions to the left. The result is the $n+2$-bit number $\underbrace{11\ldots1}_{n}00$.

7.20: They make for a more natural progressive growth of the image. They make it easier for a person watching the image develop on the screen to decide if and when to stop the decoding process and accept or discard the image.

7.21: The only specification that depends on the particular bits assigned to the two colors is Equation (7.26). All the other parts of JBIG are independent of the bit assignment.

7.22: For the 16-bit template of Figure 7.101a the relative coordinates are

$$A_1 = (3, -1), \quad A_2 = (-3, -1), \quad A_3 = (2, -2), \quad A_4 = (-2, -2).$$

For the 13-bit template of Figure 7.101b the relative coordinates of A_1 are $(3, -1)$. For the 10-bit templates of Figure 7.101c,d the relative coordinates of A_1 are $(2, -1)$.

7.23: Transposing S and T produces better compression in cases where the text runs vertically.

7.24: It may simply be too long. When compressing text, each symbol is normally 1-byte long (two bytes in Unicode). However, images with 24-bit pixels are very common, and a 16-pixel block in such an image is 48-bytes long.

7.25: If the encoder uses a $(2, 1, k)$ general unary code, then the value of k should also be included in the header.

7.26: Going back to step 1 we have the same points participate in the partition for each codebook entry (this happens because our points are concentrated in four distinct regions, but in general a partition $P_i^{(k)}$ may consist of different image blocks in each iteration k). The distortions calculated in step 2 are summarized in Table Ans.32. The average distortion $D_i^{(1)}$ is

$$D^{(1)} = (277+277+277+277+50+50+200+117+37+117+162+117)/12 = 163.17,$$

much smaller than the original 603.33. If step 3 indicates no convergence, more iterations should follow (Exercise 7.27), reducing the average distortion and improving the values of the four codebook entries.

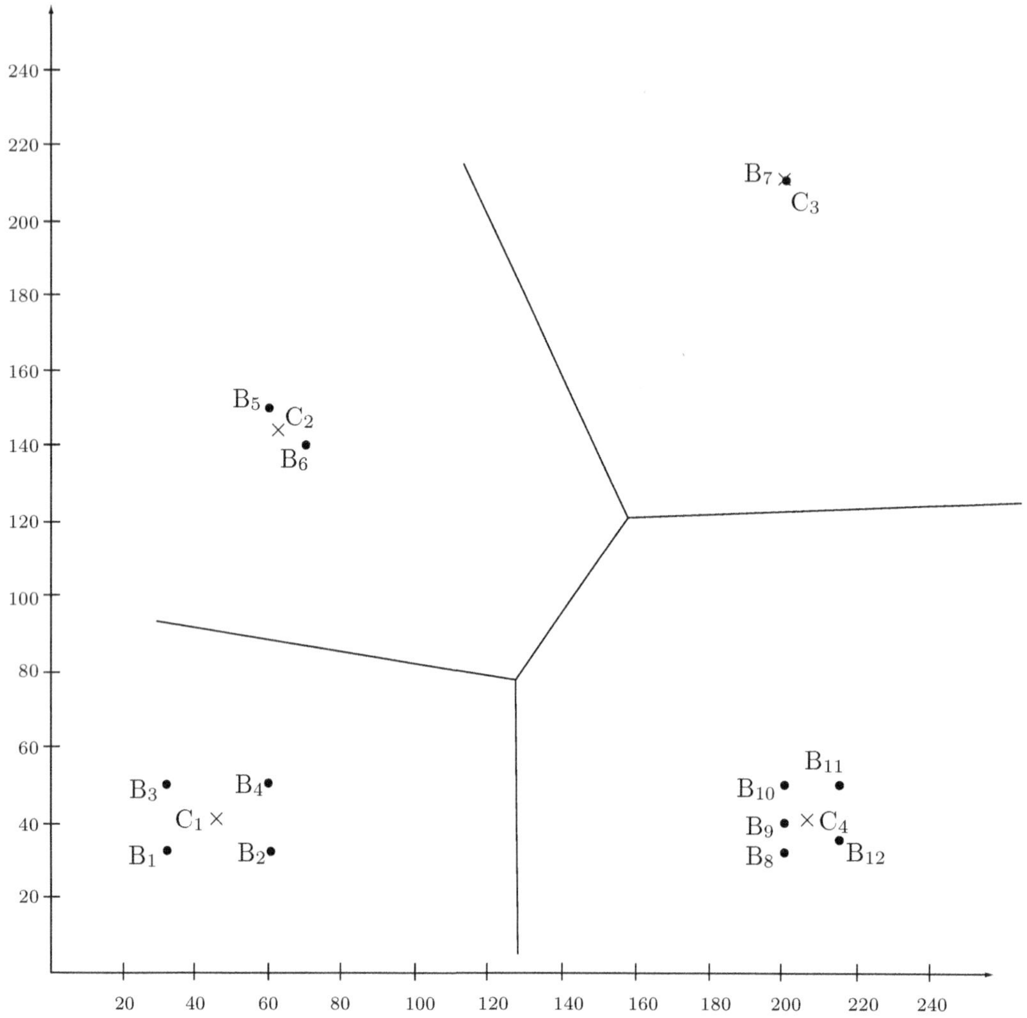

Figure Ans.31: Twelve Points and Four Codebook Entries $C_i^{(1)}$.

$$
\begin{array}{lrr}
\text{I:} & (46-32)^2+(41-32)^2=277, & (46-60)^2+(41-32)^2=277,\\
 & (46-32)^2+(41-50)^2=277, & (46-60)^2+(41-50)^2=277,\\
\text{II:} & (65-60)^2+(145-150)^2=\ \ 50, & (65-70)^2+(145-140)^2=\ \ 50,\\
\text{III:} & (210-200)^2+(200-210)^2=200, & \\
\text{IV:} & (206-200)^2+(41-32)^2=117, & (206-200)^2+(41-40)^2=\ \ 37,\\
 & (206-200)^2+(41-50)^2=117, & (206-215)^2+(41-50)^2=162,\\
 & (206-215)^2+(41-35)^2=117. &
\end{array}
$$

Table Ans.32: Twelve Distortions for $k=1$.

7.27: Each new codebook entry $C_i^{(k)}$ is calculated, in step 4 of iteration k, as the average of the block images comprising partition $P_i^{(k-1)}$. In our example the image blocks (points) are concentrated in four separate regions, so the partitions calculated for iteration $k = 1$ are the same as those for $k = 0$. Another iteration, for $k = 2$, will therefore compute the same partitions in its step 1 yielding, in step 3, an average distortion $D^{(2)}$ that equals $D^{(1)}$. Step 3 will therefore indicate convergence.

7.28: Monitor the compression ratio and delete the dictionary and start afresh each time compression performance drops below a certain threshold.

7.29: Step 4: Point $(2, 0)$ is popped out of the GPP. The pixel value at this position is 7. The best match for this point is with the dictionary entry containing 7. The encoder outputs the pointer 7. The match does not have any concave corners, so we push the point on the right of the matched block, $(2, 1)$, and the point below it, $(3, 0)$, into the GPP. The GPP now contains points $(2, 1)$, $(3, 0)$, $(0, 2)$, and $(1, 1)$. The dictionary is updated by appending to it (at location 18) the block $\begin{array}{|c|}\hline 4\\\hline 7\\\hline\end{array}$.

Step 5: Point $(1, 1)$ is popped out of the GPP. The pixel value at this position is 5. The best match for this point is with the dictionary entry containing 5. The encoder outputs the pointer 5. The match does not have any concave corners, so we push the point to the right of the matched block, $(1, 2)$, and the point below it, $(2, 1)$, into the GPP. The GPP contains points $(1, 2)$, $(2, 1)$, $(3, 0)$, and $(0, 2)$. The dictionary is updated by appending to it (at locations 19, 20) the two blocks $\begin{array}{|c|}\hline 2\\\hline 5\\\hline\end{array}$ and $\begin{array}{|c|c|}\hline 4 & 5\\\hline\end{array}$.

7.30: The mean and standard deviation are $\bar{p} = 115$ and $\sigma = 77.93$, respectively. The counts become $n^+ = n^- = 8$, and Equations (7.34) are solved to yield $p^+ = 193$ and $p^- = 37$. The original block is compressed to the 16 bits

$$\begin{pmatrix} 1 & 0 & 1 & 1 \\ 1 & 0 & 0 & 1 \\ 1 & 1 & 0 & 1 \\ 0 & 0 & 0 & 0 \end{pmatrix},$$

and the two 8-bit values 37 and 193.

7.31: Table Ans.33 summarizes the results. Notice how a 1-pixel with a context of 00 is assigned high probability after being seen 3 times.

#	Pixel	Context	Counts	Probability	New counts
5	0	10=2	1,1	1/2	2,1
6	1	00=0	1,3	3/4	1,4
7	0	11=3	1,1	1/2	2,1
8	1	10=2	2,1	1/3	2,2

Table Ans.33: Counts and Probabilities for Next Four Pixels.

7.32: Such a thing is possible for the encoder but not for the decoder. A compression method using "future" pixels in the context is useless because its output would be impossible to decompress.

7.33: The model used by FELICS to predict the current pixel is a second-order Markov model. In such a model the value of the current data item depends on just two of its past neighbors, not necessarily the two immediate ones.

7.34: The two previously seen neighbors of P=8 are A=1 and B=11. P is thus in the central region, where all codes start with a zero, and L=1, H=11. The computations are straightforward:

$$k = \lfloor \log_2(11 - 1 + 1) \rfloor = 3, \qquad a = 2^{3+1} - 11 = 5, \qquad b = 2(11 - 2^3) = 6.$$

Table Ans.34 lists the five 3-bit codes and six 4-bit codes for the central region. The code for 8 is thus 0|111.

The two previously seen neighbors of P=7 are A=2 and B=5. P is thus in the right outer region, where all codes start with 11, and L=2, H=7. We are looking for the code of $7 - 5 = 2$. Choosing $m = 1$ yields, from Table 7.135, the code 11|01.

The two previously seen neighbors of P=0 are A=3 and B=5. P is thus in the left outer region, where all codes start with 10, and L=3, H=5. We are looking for the code of $3 - 0 = 3$. Choosing $m = 1$ yields, from Table 7.135, the code 10|100.

Pixel P	Region code	Pixel code
1	0	0000
2	0	0010
3	0	0100
4	0	011
5	0	100
6	0	101
7	0	110
8	0	111
9	0	0001
10	0	0011
11	0	0101

Table Ans.34: The Codes for a Central Region.

7.35: Because the decoder has to resolve ties in the same way as the encoder.

7.36: The weights have to add up to 1 because this results in a weighted sum whose value is in the same range as the values of the pixels. If pixel values are, for example, in the range $[0, 15]$ and the weights add up to 2, a prediction may result in values of up to 30.

7.37: Each of the three weights 0.0039, −0.0351, and 0.3164 is used twice. The sum of the weights is therefore 0.5704, and the result of dividing each weight by this sum is 0.0068, −0.0615, and 0.5547. It is easy to verify that the sum of the renormalized weights $2(0.0068-0.0615+0.5547)$ equals 1.

7.38: An archive of static images is an example where this approach is practical. NASA has a large archive of images taken by various satellites. They should be kept highly compressed, but they never change, so each image has to be compressed only once. A slow encoder is therefore acceptable but a fast decoder is certainly handy. Another example is an art collection. Many museums have already scanned and digitized their collections of paintings, and those are also static.

7.39: Such a polynomial depends on three coefficients **b**, **c**, and **d** that can be considered three-dimensional points, and any three points are on the same plane.

7.40: This is straightforward

$$\begin{aligned}\mathbf{P}(2/3) =&(0,-9)(2/3)^3+(-4.5,13.5)(2/3)^2+(4.5,-3.5)(2/3)\\ =&(0,-8/3)+(-2,6)+(3,-7/3)\\ =&(1,1)=\mathbf{P}_3.\end{aligned}$$

7.41: We use the relations $\sin 30^\circ=\cos 60^\circ=.5$ and the approximation $\cos 30^\circ=\sin 60^\circ\approx .866$. The four points are $\mathbf{P}_1=(1,0)$, $\mathbf{P}_2=(\cos 30^\circ,\sin 30^\circ)=(.866,.5)$, $\mathbf{P}_3=(.5,.866)$, and $\mathbf{P}_4=(0,1)$. The relation $\mathbf{A}=\mathbf{N}\cdot\mathbf{P}$ becomes

$$\begin{pmatrix}\mathbf{a}\\ \mathbf{b}\\ \mathbf{c}\\ \mathbf{d}\end{pmatrix}=\mathbf{A}=\mathbf{N}\cdot\mathbf{P}=\begin{pmatrix}-4.5 & 13.5 & -13.5 & 4.5\\ 9.0 & -22.5 & 18 & -4.5\\ -5.5 & 9.0 & -4.5 & 1.0\\ 1.0 & 0 & 0 & 0\end{pmatrix}\begin{pmatrix}(1,0)\\ (.866,.5)\\ (.5,.866)\\ (0,1)\end{pmatrix}$$

and the solutions are

$$\begin{aligned}\mathbf{a}&=-4.5(1,0)+13.5(.866,.5)-13.5(.5,.866)+4.5(0,1)=(.441,-.441),\\ \mathbf{b}&=19(1,0)-22.5(.866,.5)+18(.5,.866)-4.5(0,1)=(-1.485,-0.162),\\ \mathbf{c}&=-5.5(1,0)+9(.866,.5)-4.5(.5,.866)+1(0,1)=(0.044,1.603),\\ \mathbf{d}&=1(1,0)-0(.866,.5)+0(.5,.866)-0(0,1)=(1,0).\end{aligned}$$

Thus, the PC is $\mathbf{P}(t)=(.441,-.441)t^3+(-1.485,-0.162)t^2+(0.044,1.603)t+(1,0)$. The midpoint is $\mathbf{P}(.5)=(.7058,.7058)$, only 0.2% away from the midpoint of the arc, which is at $(\cos 45^\circ,\sin 45^\circ)\approx(.7071,.7071)$.

7.42: The new equations are easy enough to set up. Using *Mathematica*, they are also easy to solve. The following code

```
Solve[{d==p1,
```

```
a al^3+b al^2+c al+d==p2,
a be^3+b be^2+c be+d==p3,
a+b+c+d==p4},{a,b,c,d}];
ExpandAll[Simplify[%]]
```

(where `al` and `be` stand for α and β, respectively) produces the (messy) solutions

$$
\begin{aligned}
\mathbf{a} &= -\frac{\mathbf{P}_1}{\alpha\beta} + \frac{\mathbf{P}_2}{-\alpha^2+\alpha^3+\alpha\beta-\alpha^2\beta} + \frac{\mathbf{P}_3}{\alpha\beta-\beta^2-\alpha\beta^2+\beta^3} + \frac{\mathbf{P}_4}{1-\alpha-\beta+\alpha\beta},\\
\mathbf{b} &= \mathbf{P}_1\left(-\alpha+\alpha^3+\beta-\alpha^3\beta-\beta^3+\alpha\beta^3\right)/\gamma + \mathbf{P}_2\left(-\beta+\beta^3\right)/\gamma\\
&\quad + \mathbf{P}_3\left(\alpha-\alpha^3\right)/\gamma + \mathbf{P}_4\left(\alpha^3\beta-\alpha\beta^3\right)/\gamma,\\
\mathbf{c} &= -\mathbf{P}_1\left(1+\frac{1}{\alpha}+\frac{1}{\beta}\right) + \frac{\beta\mathbf{P}_2}{-\alpha^2+\alpha^3+\alpha\beta-\alpha^2\beta}\\
&\quad + \frac{\alpha\mathbf{P}_3}{\alpha\beta-\beta^2-\alpha\beta^2+\beta^3} + \frac{\alpha\beta\mathbf{P}_4}{1-\alpha-\beta+\alpha\beta},\\
\mathbf{d} &= \mathbf{P}_1,\\
&\text{where}\quad \gamma = (-1+\alpha)\alpha(-1+\beta)\beta(-\alpha+\beta).
\end{aligned}
$$

From here, the basis matrix immediately follows

$$
\begin{pmatrix}
-\frac{1}{\alpha\beta} & \frac{1}{-\alpha^2+\alpha^3\alpha\beta-\alpha^2\beta} & \frac{1}{\alpha\beta-\beta^2-\alpha\beta^2+\beta^3} & \frac{1}{1-\alpha-\beta+\alpha\beta}\\
\frac{-\alpha+\alpha^3+\beta-\alpha^3\beta-\beta^3+\alpha\beta^3}{\gamma} & \frac{-\beta+\beta^3}{\gamma} & \frac{\alpha-\alpha^3}{\gamma} & \frac{\alpha^3\beta-\alpha\beta^3}{\gamma}\\
-\left(1+\frac{1}{\alpha}+\frac{1}{\beta}\right) & \frac{\beta}{-\alpha^2+\alpha^3+\alpha\beta-\alpha^2\beta} & \frac{\alpha}{\alpha\beta-\beta^2-\alpha\beta^2+\beta^3} & \frac{\alpha\beta}{1-\alpha-\beta+\alpha\beta}\\
1 & 0 & 0 & 0
\end{pmatrix}.
$$

A direct check, again using *Mathematica*, for $\alpha = 1/3$ and $\beta = 2/3$, reduces this matrix to matrix $\mathbf{N}$ of Equation (7.45).

7.43: The missing points will have to be estimated by interpolation or extrapolation from the known points before our method can be applied. Obviously, the fewer points are known, the worse the final interpolation. Note that 16 points are necessary, because a bicubic polynomial has 16 coefficients.

7.44: Figure Ans.35a shows a diamond-shaped grid of 16 equally-spaced points. The eight points with negative weights are shown in black. Figure Ans.35b shows a cut (labeled xx) through four points in this surface. The cut is a curve that passes through pour data points. It is easy to see that when the two exterior (black) points are raised, the center of the curve (and, as a result, the center of the surface) gets lowered. It is now clear that points with negative weights push the center of the surface in a direction opposite that of the points.

Figure Ans.35c is a more detailed example that also shows why the four corner points should have positive weights. It shows a simple symmetric surface patch that

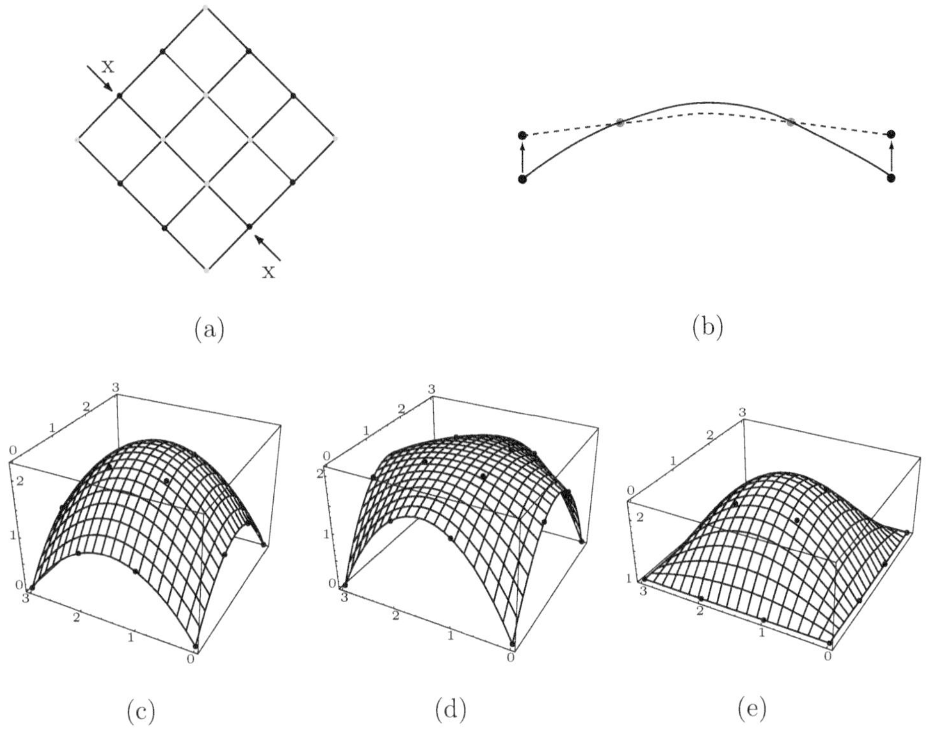

Figure Ans.35: An Interpolating Bicubic Surface Patch.

```
Clear[Nh,p,pnts,U,W];
p00={0,0,0}; p10={1,0,1}; p20={2,0,1}; p30={3,0,0};
p01={0,1,1}; p11={1,1,2}; p21={2,1,2}; p31={3,1,1};
p02={0,2,1}; p12={1,2,2}; p22={2,2,2}; p32={3,2,1};
p03={0,3,0}; p13={1,3,1}; p23={2,3,1}; p33={3,3,0};
Nh={{-4.5,13.5,-13.5,4.5},{9,-22.5,18,-4.5},
 {-5.5,9,-4.5,1},{1,0,0,0}};
pnts={{p33,p32,p31,p30},{p23,p22,p21,p20},
 {p13,p12,p11,p10},{p03,p02,p01,p00}};
U[u_]:={u^3,u^2,u,1};   W[w_]:={w^3,w^2,w,1};
(* prt [i] extracts component i from the 3rd dimen of P *)
prt[i_]:=pnts[[Range[1,4],Range[1,4],i]];
p[u_,w_]:={U[u].Nh.prt[1].Transpose[Nh].W[w],
 U[u].Nh.prt[2].Transpose[Nh].W[w], \
 U[u].Nh.prt[3].Transpose[Nh].W[w]};
 g1=ParametricPlot3D[p[u,w], {u,0,1},{w,0,1},
 Compiled->False, DisplayFunction->Identity];
g2=Graphics3D[{AbsolutePointSize[2],
 Table[Point[pnts[[i,j]]],{i,1,4},{j,1,4}]}];
Show[g1,g2, ViewPoint->{-2.576, -1.365, 1.718}]
```

Code For Figure Ans.35

interpolates the 16 points

$$\begin{aligned}
&\mathbf{P}_{00}=(0,0,0), &&\mathbf{P}_{10}=(1,0,1), &&\mathbf{P}_{20}=(2,0,1), &&\mathbf{P}_{30}=(3,0,0),\\
&\mathbf{P}_{01}=(0,1,1), &&\mathbf{P}_{11}=(1,1,2), &&\mathbf{P}_{21}=(2,1,2), &&\mathbf{P}_{31}=(3,1,1),\\
&\mathbf{P}_{02}=(0,2,1), &&\mathbf{P}_{12}=(1,2,2), &&\mathbf{P}_{22}=(2,2,2), &&\mathbf{P}_{32}=(3,2,1),\\
&\mathbf{P}_{03}=(0,3,0), &&\mathbf{P}_{13}=(1,3,1), &&\mathbf{P}_{23}=(2,3,1), &&\mathbf{P}_{33}=(3,3,0).
\end{aligned}$$

We first raise the eight boundary points from $z = 1$ to $z = 1.5$. Figure Ans.35d shows how the center point $\mathbf{P}(.5, .5)$ gets lowered from $(1.5, 1.5, 2.25)$ to $(1.5, 1.5, 2.10938)$. We next return those points to their original positions and instead raise the four corner points from $z = 0$ to $z = 1$. Figure Ans.35e shows how this raises the center point from $(1.5, 1.5, 2.25)$ to $(1.5, 1.5, 2.26563)$.

7.45: The decoder knows this pixel since it knows the value of average $\mu[i-1, j] = 0.5(I[2i-2, 2j] + I[2i-1, 2j+1])$ and since it has already decoded pixel $I[2i-2, 2j]$

7.46: The decoder knows how to do this because when the decoder inputs the 5, it knows that the difference between p (the pixel being decoded) and the reference pixel starts at position 6 (counting from the left). Since bit 6 of the reference pixel is 0, that of p must be 1.

7.47: Yes, but compression would suffer. One way to apply this method is to separate each byte into two 4-bit pixels and encode each pixel separately. This approach is bad since the prefix and suffix of a 4-bit pixel may often consist of more than four bits. Another approach is to ignore the fact that a byte contains two pixels, and use the method as originally described. This may still compress the image, but is not very efficient, as the following example illustrates.

Example: The two bytes 1100|1101 and 1110|1111 represent four pixels, each differing from its immediate neighbor by its least-significant bit. The four pixels therefore have similar colors (or grayscales). Comparing consecutive pixels results in prefixes of 3 or 2, but comparing the two bytes produces the prefix 2.

7.48: The weights add up to 1 because this produces a value X in the same range as A, B, and C. If the weights were, for instance, 1, 100, and 1, X would have much bigger values than any of the three pixels.

7.49: The four vectors are

$$\begin{aligned}
\mathbf{a} &= (90, 95, 100, 80, 90, 85),\\
\mathbf{b}^{(1)} &= (100, 90, 95, 102, 80, 90),\\
\mathbf{b}^{(2)} &= (101, 128, 108, 100, 90, 95),\\
\mathbf{b}^{(3)} &= (128, 108, 110, 90, 95, 100),
\end{aligned}$$

and the code of Figure Ans.36 produces the solutions $w_1 = 0.1051$, $w_2 = 0.3974$, and $w_3 = 0.3690$. Their total is 0.8715, compared with the original solutions, which added up to 0.9061. The point is that the numbers involved in the equations (the elements of the four vectors) are not independent (for example, pixel 80 appears in **a** and in $\mathbf{b}^{(1)}$) except for the last element (85 or 91) of **a** and the first element 101 of $\mathbf{b}^{(2)}$, which are independent. Changing these two elements affects the solutions, which is why the solutions do not always add up to unity. However, compressing nine pixels produces solutions whose total is closer to one than in the case of six pixels. Compressing an entire image, with many thousands of pixels, produces solutions whose sum is very close to 1.

```
a={90.,95,100,80,90,85};
b1={100,90,95,100,80,90};
b2={100,128,108,100,90,95};
b3={128,108,110,90,95,100};
Solve[{b1.(a-w1 b1-w2 b2-w3 b3)==0,
b2.(a-w1 b1-w2 b2-w3 b3)==0,
b3.(a-w1 b1-w2 b2-w3 b3)==0},{w1,w2,w3}]
```

Figure Ans.36: Solving for Three Weights.

7.50: Figure Ans.37a,b,c shows the results, with all H_i values shown in small type. Most H_i values are zero because the pixels of the original image are so highly correlated. The H_i values along the edges are very different because of the simple edge rule used. The result is that the H_i values are highly decorrelated and have low entropy. Thus, they are candidates for entropy coding.

1	.	3	.	5	.	7	.
.	0	.	0	.	0	.	-5
17	.	19	.	21	.	23	.
.	0	.	0	.	0	.	-13
33	.	35	.	37	.	39	.
.	0	.	0	.	0	.	-21
49	.	51	.	53	.	55	.
.	-33	.	-34	.	-35	.	-64

(a)

1	.	7	.	5	.	5	.
.	.	.	.	.	.	.	.
15	.	19	.	11	.	23	.
.	.	.	.	.	.	.	.
33	.	0	.	37	.	0	.
.	.	.	.	.	.	.	.
-33	.	51	.	-35	.	55	.
.	.	.	.	.	.	.	.

(b)

1	.	.	.	5	.	.	.
.	.	.	.	.	.	.	.
.	.	0	.	.	.	-5	.
.	.	.	.	.	.	.	.
33	.	.	.	37	.	.	.
.	.	.	.	.	.	.	.
.	.	-33	.	.	.	-55	.
.	.	.	.	.	.	.	.

(c)

Figure Ans.37: (a) Bands L_2 and H_2. (b) Bands L_3 and H_3. (c) Bands L_4 and H_4.

7.51: There are 16 values. The value 0 appears nine times, and each of the other seven values appears once. The entropy is therefore

$$-\sum p_i \log_2 p_i = -\frac{9}{16}\log_2\left(\frac{9}{16}\right) - 7\frac{1}{16}\log_2\left(\frac{1}{16}\right) \approx 2.2169.$$

Not very small, because seven of the 16 values have the same probability. In practice, values of an H_i difference band tend to be small, are both positive and negative, and are concentrated around zero, so their entropy is small.

7.52: Because the decoder needs to know how the encoder estimated X for each H_i difference value. If the encoder uses one of three methods for prediction, it has to precede each difference value in the compressed stream with a code that tells the decoder which method was used. Such a code can have variable size (for example, 0, 10, 11) but even adding just one or two bits to each prediction reduces compression performance significantly, because each H_i value needs to be predicted, and the number of these values is close to the size of the image.

7.53: The binary tree is shown in Figure Ans.38. From this tree, it is easy to see that the progressive image file is 3 6|5 7|7 7 10 5.

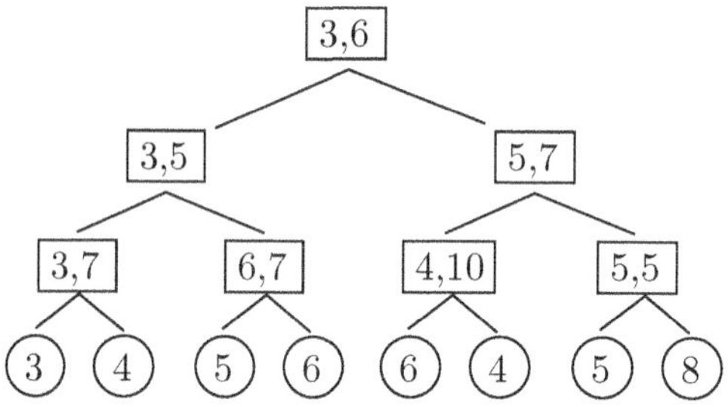

Figure Ans.38: A Binary Tree for an 8-Pixel Image.

7.54: They are shown in Figure Ans.39.

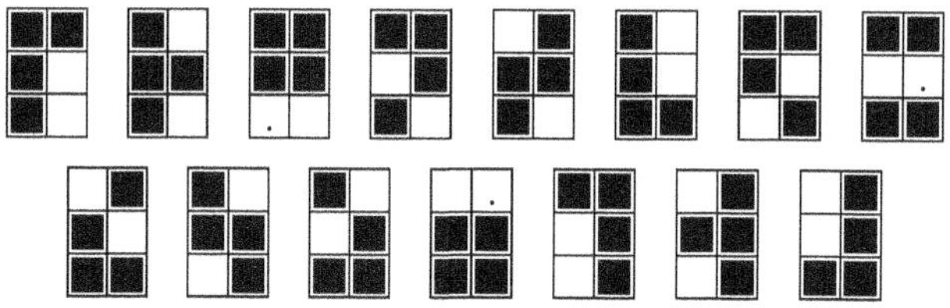
Figure Ans.39: The 15 6-Tuples With Two White Pixels.

7.55: No. An image with little or no correlation between the pixels will not compress with quadrisection, even though the size of the last matrix is always small. Even without knowing the details of quadrisection we can confidently state that such an image will produce a sequence of matrices M_j with few or no identical rows. In the extreme case, where the rows of any M_j are all distinct, each M_j will have four times the number of rows of its predecessor. This will create indicator vectors I_j that get longer and longer, thereby increasing the size of the compressed stream and reducing the overall compression performance.

7.56: Matrix M_5 is just the concatenation of the 12 distinct rows of M_4

$$M_5^T = (0000|0001|1111|0011|1010|1101|1000|0111|1110|0101|1011|0010).$$

7.57: M_4 has four columns, so it can have at most 16 distinct rows, implying that M_5 can have at most $4\times 16 = 64$ elements.

7.58: The decoder has to read the entire compressed stream, save it in memory, and start the decoding with L_5. Grouping the eight elements of L_5 yields the four distinct elements 01, 11, 00, and 10 of L_4, so I_4 can now be used to reconstruct L_4. The four zeros of I_4 correspond to the four distinct elements of L_4, and the remaining 10 elements of L_4 can be constructed from them. Once L_4 has been constructed, its 14 elements are grouped to form the seven distinct elements of L_3. These elements are 0111, 0010, 1100, 0110, 1111, 0101, and 1010, and they correspond to the seven zeros of I_3. Once L_3 has been constructed, its eight elements are grouped to form the four distinct elements of L_2. Those four elements are the entire L_2 since I_2 is all zero. Reconstructing L_1 and L_0 is now trivial.

7.59: The two halves of L_0 are distinct, so L_1 consists of the two elements

$$L_1 = (0101010101010101, 1010101010101010),$$

and the first indicator vector is $I_1 = (0, 0)$. The two elements of L_1 are distinct, so L_2 has the four elements

$$L_2 = (01010101, 01010101, 10101010, 10101010),$$

and the second indicator vector is $I_2 = (0, 1, 0, 2)$. Two elements of L_2 are distinct, so L_3 has the four elements $L_3 = (0101, 0101, 1010, 1010)$, and the third indicator vector is $I_3 = (0, 1, 0, 2)$. Again two elements of L_3 are distinct, so L_4 has the four elements $L_4 = (01, 01, 10, 10)$, and the fourth indicator vector is $I_4 = (0, 1, 0, 2)$. Only two elements of L_4 are distinct, so L_5 has the four elements $L_5 = (0, 1, 1, 0)$.

The output thus consists of $k = 5$, the value 2 (indicating that I_2 is the first nonzero vector) I_2, I_3, and I_4 (encoded), followed by $L_5 = (0, 1, 1, 0)$.

7.60: Using a Hilbert curve produces the 21 runs 5, 1, 2, 1, 2, 7, 3, 1, 2, 1, 5, 1, 2, 2, 11, 7, 2, 1, 1, 1, 6. RLE produces the 27 runs 0, 1, 7, eol, 2, 1, 5, eol, 5, 1, 2, eol, 0, 3, 2, 3, eol, 0, 3, 2, 3, eol, 0, 3, 2, 3, eol, 4, 1, 3, eol, 3, 1, 4, eol.

7.61: A straight line segment from a to b is an example of a one-dimensional curve that passes through every point in the interval a, b.

7.62: The key is to realize that P_0 is a single point, and P_1 is constructed by connecting nine copies of P_0 with straight segments. Similarly, P_2 consists of nine copies of P_1, in different orientations, connected by segments (the dashed segments in Figure Ans.40).

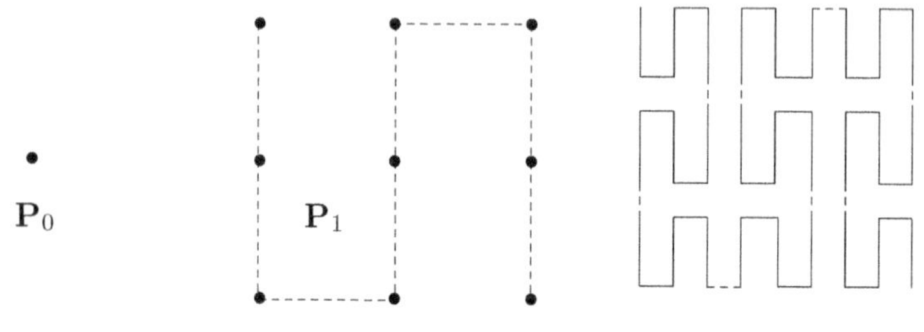

(a) (b) (c)

Figure Ans.40: The First Three Iterations of the Peano Curve.

7.63: Written in binary, the coordinates are (1101, 0110). We iterate four times, each time taking 1 bit from the x coordinate and 1 bit from the y coordinate to form an (x, y) pair. The pairs are 10, 11, 01, 10. The first one yields [from Table 7.185(1)] 01. The second pair yields [also from Table 7.185(1)] 10. The third pair [from Table 7.185(1)] 11, and the last pair [from Table 7.185(4)] 01. Thus, the result is $01|10|11|01 = 109$.

7.64: Table Ans.41 shows that this traversal is based on the sequence 2114.

$$\begin{array}{l} 1\colon 2\uparrow\ 1\rightarrow 1\downarrow\ 4 \\ 2\colon 1\rightarrow 2\uparrow\ 2\leftarrow 3 \\ 3\colon 4\downarrow\ 3\leftarrow 3\uparrow\ 2 \\ 4\colon 3\leftarrow 4\downarrow\ 4\rightarrow 1 \end{array}$$

Table Ans.41: The Four Orientations of H_2.

7.65: This is straightforward

$$\begin{aligned}(00,01,11,10) &\rightarrow (000,001,011,010)(100,101,111,110)\\ &\rightarrow (000,001,011,010)(110,111,101,100)\\ &\rightarrow (000,001,011,010,110,111,101,100).\end{aligned}$$

7.66: The gray area of Figure 7.186c is identified by the string 2011.

7.67: This particular numbering makes it easy to convert between the number of a subsquare and its image coordinates. (We assume that the origin is located at the bottom-left corner of the image and that image coordinates vary from 0 to 1.) As an example, translating the digits of the number 1032 to binary results in (01)(00)(11)(10). The first bits of these groups constitute the x coordinate of the subsquare, and the second bits constitute the y coordinate. Thus, the image coordinates of subsquare 1032 are $x = .0011_2 = 3/16$ and $y = .1010_2 = 5/8$, as can be directly verified from Figure 7.186c.

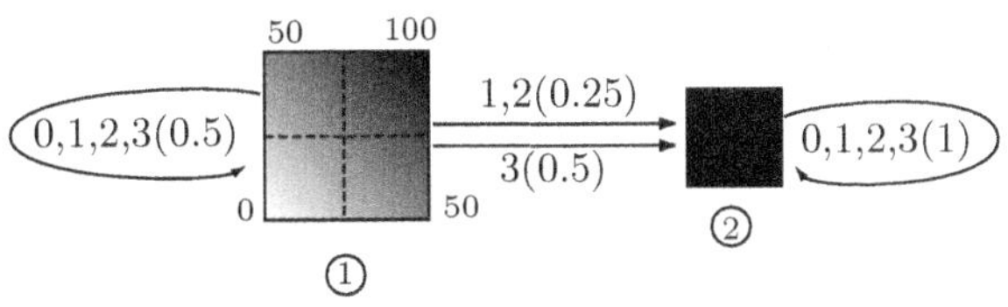

Figure Ans.42: A Two-State Graph.

7.68: This is shown in Figure Ans.42.

7.69: This image is described by the function

$$f(x,y)=\begin{cases} x+y, & \text{if } x+y\le 1,\\ 0, & \text{if } x+y>1.\end{cases}$$

7.70: The graph has five states, so each transition matrix is of size 5×5. Direct computation from the graph yields

$$W_0=\begin{pmatrix} 0 & 1 & 0 & 0 & 0\\ 0 & 0.5 & 0 & 0 & 0\\ 0 & 0 & 0 & 0 & 1\\ 0 & -0.5 & 0 & 0 & 1.5\\ 0 & -0.25 & 0 & 0 & 1\end{pmatrix},\quad W_3=\begin{pmatrix} 0 & 0 & 0 & 0 & 0\\ 0 & 0 & 0 & 0 & 1\\ 0 & 0 & 0 & 0 & 0\\ 0 & 0 & 0 & 0 & 0\\ 0 & -1 & 0 & 0 & 1.5\end{pmatrix},$$

$$W_1=W_2=\begin{pmatrix} 0 & 0 & 1 & 0 & 0\\ 0 & 0.25 & 0 & 0 & 0.5\\ 0 & 0 & 0 & 1.5 & 0\\ 0 & 0 & -0.5 & 1.5 & 0\\ 0 & -0.375 & 0 & 0 & 1.25\end{pmatrix}.$$

The final distribution is the five-component vector

$$F=(0.25,0.5,0.375,0.4125,0.75)^T.$$

7.71: One way to specify the center is to construct string $033\ldots3$. This yields

$$\begin{aligned}\psi_i(03\ldots3)&=(W_0\cdot W_3\cdots W_3\cdot F)_i\\ &=\begin{pmatrix}0.5 & 0\\ 0 & 1\end{pmatrix}\begin{pmatrix}0.5 & 0.5\\ 0 & 1\end{pmatrix}\cdots\begin{pmatrix}0.5 & 0.5\\ 0 & 1\end{pmatrix}\begin{pmatrix}0.5\\ 1\end{pmatrix}_i\\ &=\begin{pmatrix}0.5 & 0\\ 0 & 1\end{pmatrix}\begin{pmatrix}0 & 1\\ 0 & 1\end{pmatrix}\begin{pmatrix}0.5\\ 1\end{pmatrix}_i=\begin{pmatrix}0.5\\ 1\end{pmatrix}_i.\end{aligned}$$

```
dim=256;
for i=1:dim
 for j=1:dim
   m(i,j)=(i+j-2)/(2*dim-2);
 end
end
m
```

Figure Ans.43: Matlab Code for a Matrix $m_{i,j} = (i + j)/2$.

7.72: Figure Ans.43 shows Matlab code to compute a matrix such as those of Figure 7.190.

7.73: A direct examination of the graph yields the ψ_i values

$$\begin{aligned}
\psi_i(0) &= (W_0 \cdot F)_i = (0.5, 0.25, 0.75, 0.875, 0.625)_i^T,\\
\psi_i(01) &= (W_0 \cdot W_1 \cdot F)_i = (0.5, 0.25, 0.75, 0.875, 0.625)_i^T,\\
\psi_i(1) &= (W_1 \cdot F)_i = (0.375, 0.5, 0.61875, 0.43125, 0.75)_i^T,\\
\psi_i(00) &= (W_0 \cdot W_0 \cdot F)_i = (0.25, 0.125, 0.625, 0.8125, 0.5625)_i^T,\\
\psi_i(03) &= (W_0 \cdot W_3 \cdot F)_i = (0.75, 0.375, 0.625, 0.5625, 0.4375)_i^T,\\
\psi_i(3) &= (W_3 \cdot F)_i = (0, 0.75, 0, 0, 0.625)_i^T,
\end{aligned}$$

and the f values

$$\begin{aligned}
&f(0) = I \cdot \psi(0) = 0.5, \quad f(01) = I \cdot \psi(01) = 0.5, \quad f(1) = I \cdot \psi(1) = 0.375,\\
&f(00) = I \cdot \psi(00) = 0.25, \quad f(03) = I \cdot \psi(03) = 0.75, \quad f(3) = I \cdot \psi(3) = 0.
\end{aligned}$$

7.74: Figure Ans.44a,b shows the six states and all 21 edges. We use the notation $i(q,t)j$ for the edge with quadrant number q and transformation t from state i to state j. This GFA is more complex than pervious ones since the original image is less self-similar.

7.75: The transformation can be written $(x, y) \to (x, -x + y)$, so $(1, 0) \to (1, -1)$, $(3, 0) \to (3, -3)$, $(1, 1) \to (1, 0)$ and $(3, 1) \to (3, -2)$. Thus, the original rectangle is transformed into a parallelogram.

7.76: The explanation is that the two sets of transformations produce the same Sierpiński triangle but at different sizes and orientations.

7.77: All three transformations shrink an image to half its original size. In addition, w_2 and w_3 place two copies of the shrunken image at relative displacements of $(0, 1/2)$ and $(1/2, 0)$, as shown in Figure Ans.45. The result is the familiar Sierpiński gasket but in a different orientation.

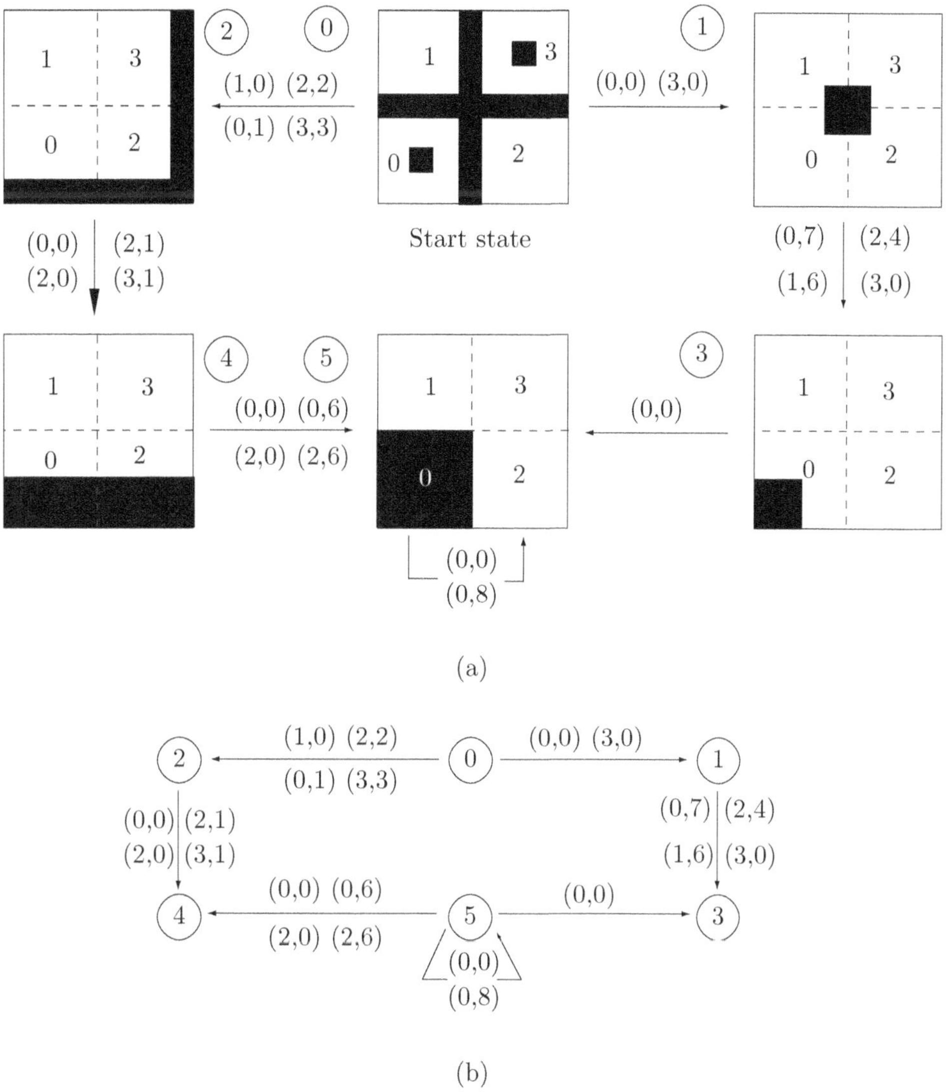

0(0,0)1	0(3,0)1	0(0,1)2	0(1,0)2	0(2,2)2	0(3,3)2	1(0,7)3
1(1,6)3	1(2,4)3	1(3,0)3	2(0,0)4	2(2,0)4	2(2,1)4	2(3,1)4
3(0,0)5	4(0,0)5	4(0,6)5	4(2,0)5	4(2,6)5	5(0,0)5	5(0,8)5

Figure Ans.44: A GFA for Exercise 7.74.

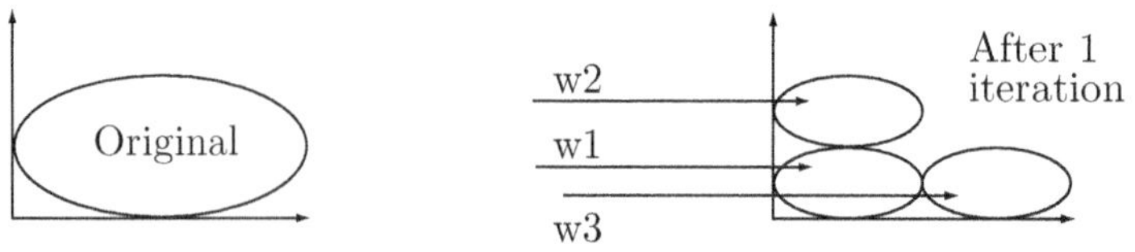

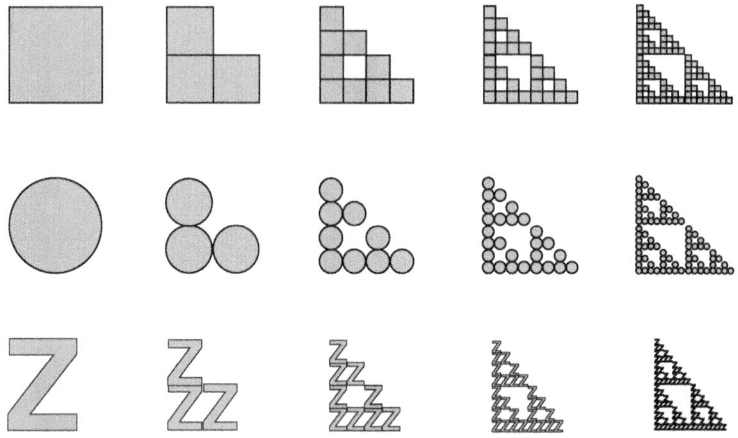

Figure Ans.45: Another Sierpiński Gasket.

7.78: There are $32\times32 = 1,024$ ranges and $(256-15)\times(256-15) = 58,081$ domains. Thus, the total number of steps is $1,024\times58,081\times8 = 475,799,552$, still a large number. PIFS is therefore computationally intensive.

7.79: Suppose that the image has G levels of gray. A good measure of data loss is the difference between the value of an average decompressed pixel and its correct value, expressed in number of gray levels. For large values of G (hundreds of gray levels) an average difference of $\log_2 G$ gray levels (or fewer) is considered satisfactory.

8.1: A written page is such an example. A person can place marks on a page and read them later as text, mathematical expressions, and drawings. This is a two-dimensional representation of the information on the page. The page can later be scanned by, e.g., a fax machine, and its contents transmitted as a one-dimensional stream of bits that constitute a different representation of the same information.

8.2: Figure Ans.46 shows $f(t)$ and three shifted copies of the wavelet, for $a=1$ and $b=2$, 4, and 6. The inner product $W(a,b)$ is plotted below each copy of the wavelet. It is easy to see how the inner products are affected by the increasing frequency.

The table of Figure Ans.47 lists 15 values of $W(a,b)$, for $a=1$, 2, and 3 and for $b=2$ through 6. The density plot of the figure, where the bright parts correspond to large values, shows those values graphically. For each value of a, the CWT yields values that drop with b, reflecting the fact that the frequency of $f(t)$ increases with t. The five values of $W(1,b)$ are small and very similar, while the five values of $W(3,b)$ are larger

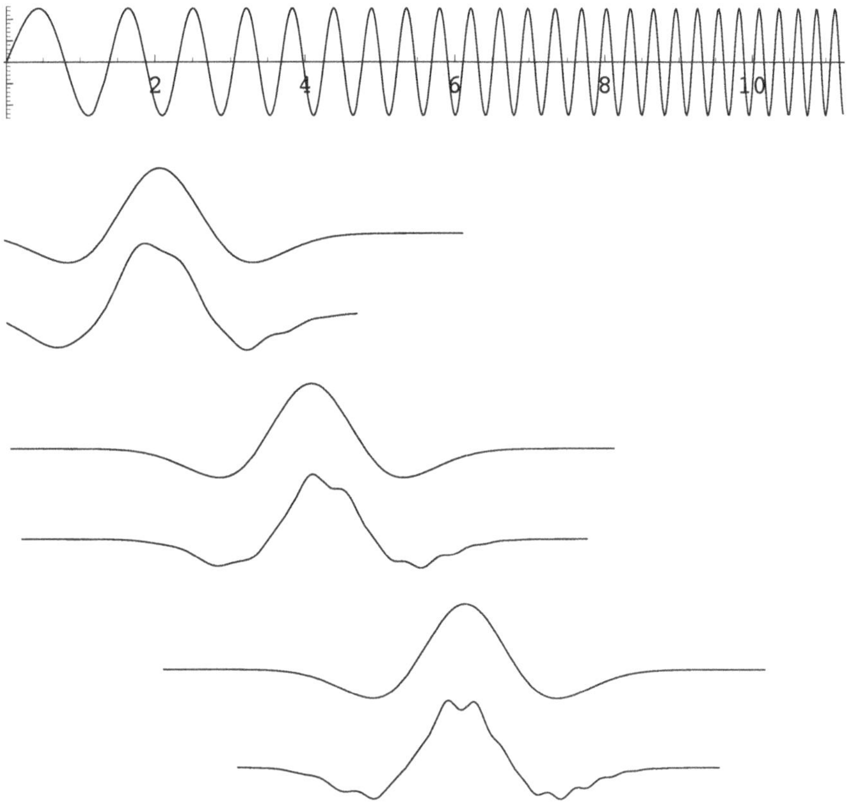

Figure Ans.46: An Inner Product for $a = 1$ and $b = 2, 4, 6$.

a	$b = 2$	3	4	5	6
1	0.032512	0.000299	1.10923×10^{-6}	2.73032×10^{-9}	8.33866×10^{-11}
2	0.510418	0.212575	0.0481292	0.00626348	0.00048097
3	0.743313	0.629473	0.380634	0.173591	0.064264

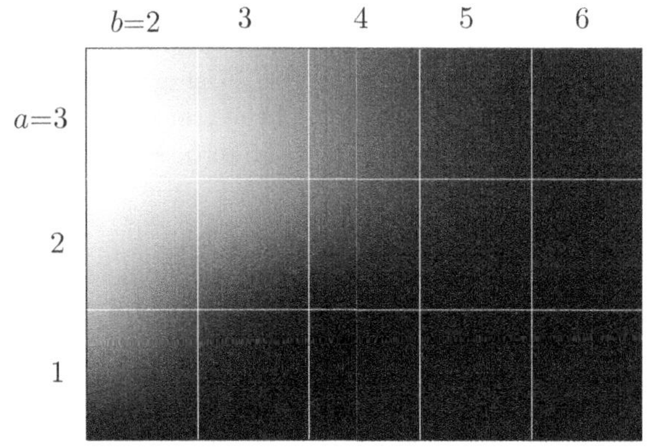

Figure Ans.47: Fifteen Values and a Density Plot of $W(a, b)$.

and differ more. This shows how scaling the wavelet up makes the CWT more sensitive to frequency changes in $f(t)$.

8.3: Figure 8.11c shows these wavelets.

8.4: Figure Ans.48a shows a simple, 8×8 image with one diagonal line above the main diagonal. Figure Ans.48b,c shows the first two steps in its pyramid decomposition. It is obvious that the transform coefficients in the bottom-right subband (HH) indicate a diagonal artifact located above the main diagonal. It is also easy to see that subband LL is a low-resolution version of the original image.

$$\begin{array}{cccccccc}
12&16&12&12&12&12&12&12\\
12&12&16&12&12&12&12&12\\
12&12&12&16&12&12&12&12\\
12&12&12&12&16&12&12&12\\
12&12&12&12&12&16&12&12\\
12&12&12&12&12&12&16&12\\
12&12&12&12&12&12&12&16\\
12&12&12&12&12&12&12&12
\end{array}
\qquad
\begin{array}{cccc|cccc}
14&12&12&12&\underline{4}&0&0&0\\
12&14&12&12&0&4&0&0\\
12&14&12&12&0&\underline{4}&0&0\\
12&12&14&12&0&0&4&0\\
12&12&14&12&0&0&\underline{4}&0\\
12&12&12&14&0&0&0&4\\
12&12&12&14&0&0&0&\underline{4}\\
12&12&12&12&0&0&0&0
\end{array}
\qquad
\begin{array}{cccc|cccc}
13&13&12&12&\underline{2}&2&0&0\\
12&13&13&12&0&\underline{2}&2&0\\
12&12&13&13&0&0&\underline{2}&2\\
12&12&12&13&0&0&0&\underline{2}\\
\hline
2&\underline{2}&0&0&\underline{4}&\underline{4}&0&0\\
0&2&\underline{2}&0&0&\underline{4}&\underline{4}&0\\
0&0&2&\underline{2}&0&0&\underline{4}&\underline{4}\\
0&0&0&2&0&0&0&\underline{4}
\end{array}$$

(a) (b) (c)

Figure Ans.48: The Subband Decomposition of a Diagonal Line.

8.5: The average can easily be calculated. It turns out to be 131.375, which is exactly 1/8 of 1051. The reason the top-left transform coefficient is eight times the average is that the Matlab code that did the calculations uses $\sqrt{2}$ instead of 2 (see function `individ(n)` in Figure 8.22).

8.6: Figure Ans.49a–c shows the results of reconstruction from 3277, 1639, and 820 coefficients, respectively. Despite the heavy loss of wavelet coefficients, only a very small loss of image quality is noticeable. The number of wavelet coefficients is, of course, the same as the image resolution $128 \times 128 = 16,384$. Using 820 out of 16,384 coefficients corresponds to discarding 95% of the smallest of the transform coefficients (notice, however, that some of the coefficients were originally zero, so the actual loss may amount to less than 95%).

8.7: The Matlab code of Figure Ans.50 calculates W as the product of the three matrices A_1, A_2, and A_3 and computes the 8×8 matrix of transform coefficients. Notice that the top-left value 131.375 is the average of all the 64 image pixels.

8.8: The vector $x = (\ldots, 1, -1, 1, -1, 1, \ldots)$ of alternating values is transformed by the lowpass filter H_0 to a vector of all zeros.

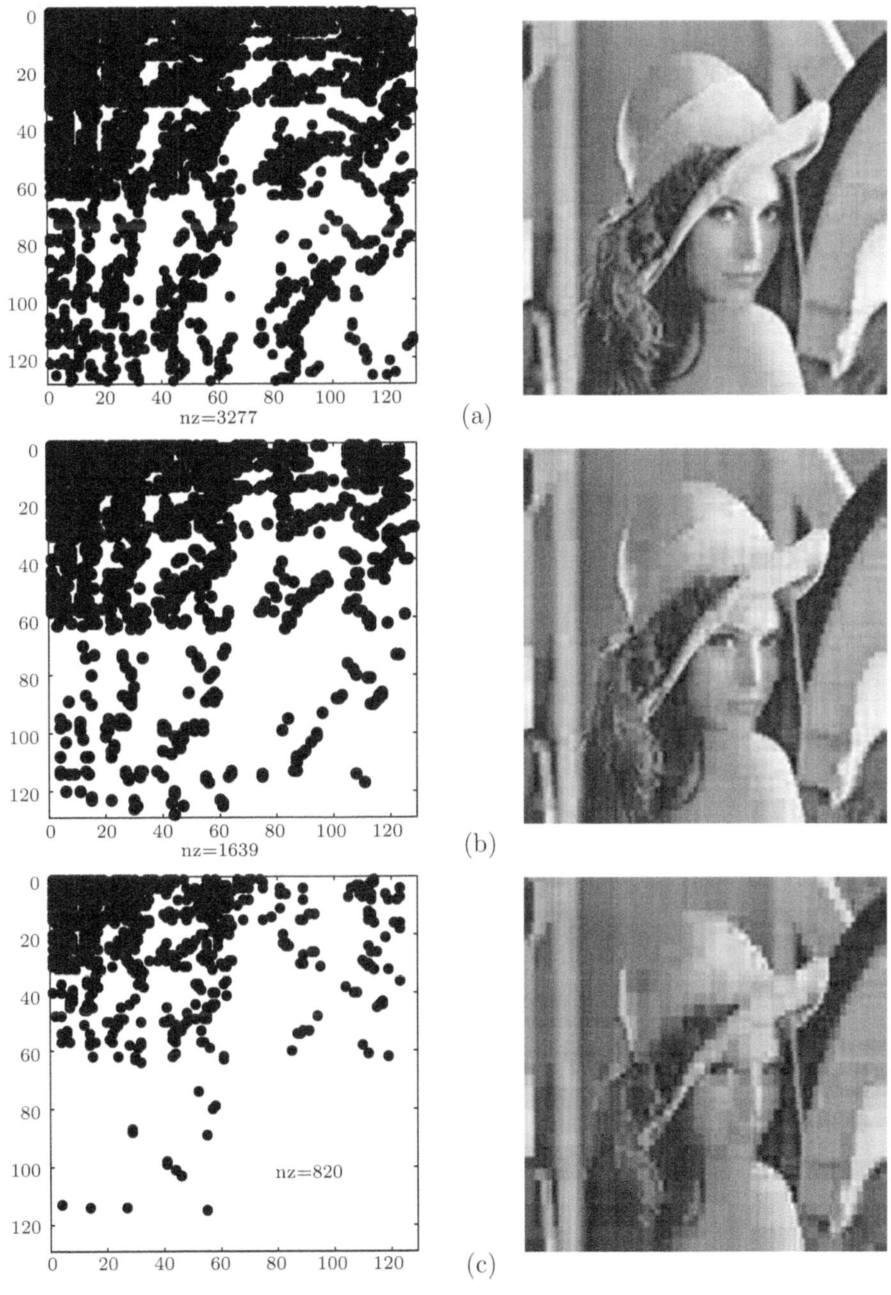

Figure Ans.49: Three Lossy Reconstructions of the 128×128 Lena Image.

```
clear
a1=[1/2 1/2 0 0 0 0 0 0; 0 0 1/2 1/2 0 0 0 0;
 0 0 0 0 1/2 1/2 0 0; 0 0 0 0 0 0 1/2 1/2;
 1/2 -1/2 0 0 0 0 0 0; 0 0 1/2 -1/2 0 0 0 0;
 0 0 0 0 1/2 -1/2 0 0; 0 0 0 0 0 0 1/2 -1/2];
% a1*[255; 224; 192; 159; 127; 95; 63; 32];
a2=[1/2 1/2 0 0 0 0 0 0; 0 0 1/2 1/2 0 0 0 0;
 1/2 -1/2 0 0 0 0 0 0; 0 0 1/2 -1/2 0 0 0 0;
 0 0 0 0 1 0 0 0; 0 0 0 0 0 1 0 0;
 0 0 0 0 0 0 1 0; 0 0 0 0 0 0 0 1];
a3=[1/2 1/2 0 0 0 0 0 0; 1/2 -1/2 0 0 0 0 0 0;
 0 0 1 0 0 0 0 0; 0 0 0 1 0 0 0 0;
 0 0 0 0 1 0 0 0; 0 0 0 0 0 1 0 0;
 0 0 0 0 0 0 1 0; 0 0 0 0 0 0 0 1];
w=a3*a2*a1;
dim=8; fid=fopen('8x8','r');
img=fread(fid,[dim,dim])'; fclose(fid);
w*img*w' % Result of the transform
```

131.375	4.250	−7.875	−0.125	−0.25	−15.5	0	−0.25
0	0	0	0	0	0	0	0
0	0	0	0	0	0	0	0
0	0	0	0	0	0	0	0
12.000	59.875	39.875	31.875	15.75	32.0	16	15.75
12.000	59.875	39.875	31.875	15.75	32.0	16	15.75
12.000	59.875	39.875	31.875	15.75	32.0	16	15.75
12.000	59.875	39.875	31.875	15.75	32.0	16	15.75

Figure Ans.50: Code and Results for the Calculation of Matrix W and Transform $W \cdot I \cdot W^T$.

8.9: For these filters, rules 1 and 2 imply

$$\begin{gathered}
h_0^2(0)+h_0^2(1)+h_0^2(2)+h_0^2(3)+h_0^2(4)+h_0^2(5)+h_0^2(6)+h_0^2(7)=1,\\
h_0(0)h_0(2)+h_0(1)h_0(3)+h_0(2)h_0(4)+h_0(3)h_0(5)+h_0(4)h_0(6)+h_0(5)h_0(7)=0,\\
h_0(0)h_0(4)+h_0(1)h_0(5)+h_0(2)h_0(6)+h_0(3)h_0(7)=0,\\
h_0(0)h_0(6)+h_0(1)h_0(7)=0,
\end{gathered}$$

and rules 3–5 yield

$$\begin{aligned}
f_0&=\big(h_0(7),h_0(6),h_0(5),h_0(4),h_0(3),h_0(2),h_0(1),h_0(0)\big),\\
h_1&=\big(-h_0(7),h_0(6),-h_0(5),h_0(4),-h_0(3),h_0(2),-h_0(1),h_0(0)\big),\\
f_1&=\big(h_0(0),-h_0(1),h_0(2),-h_0(3),h_0(4),-h_0(5),h_0(6),-h_0(7)\big).
\end{aligned}$$

The eight coefficients are listed in Table 8.35 (this is the Daubechies D8 filter).

8.10: Figure Ans.51 lists the Matlab code of the inverse wavelet transform function `iwt1(wc,coarse,filter)` and a test.

```
function dat=iwt1(wc,coarse,filter)
% Inverse Discrete Wavelet Transform
dat=wc(1:2^coarse);
n=length(wc); j=log2(n);
for i=coarse:j-1
 dat=ILoPass(dat,filter)+ ...
  IHiPass(wc((2^(i)+1):(2^(i+1))),filter);
end

function f=ILoPass(dt,filter)
f=iconv(filter,AltrntZro(dt));

function f=IHiPass(dt,filter)
f=aconv(mirror(filter),rshift(AltrntZro(dt)));

function sgn=mirror(filt)
% return filter coefficients with alternating signs
sgn=-((-1).^(1:length(filt))).*filt;

function f=AltrntZro(dt)
% returns a vector of length 2*n with zeros
% placed between consecutive values
n =length(dt)*2; f =zeros(1,n);
f(1:2:(n-1))=dt;
```

Figure Ans.51: Code for the 1D Inverse Discrete Wavelet Transform.

A simple test of `iwt1` is

```
n=16; t=(1:n)./n;
dat=sin(2*pi*t)
filt=[0.4830 0.8365 0.2241 -0.1294];
wc=fwt1(dat,1,filt)
rec=iwt1(wc,1,filt)
```

8.11: Figure Ans.52 shows the result of blurring the "lena" image. Parts (a) and (b) show the logarithmic multiresolution tree and the subband structure, respectively. Part (c) shows the results of the quantization. The transform coefficients of subbands 5–7 have been divided by two, and all the coefficients of subbands 8–13 have been cleared. We can say that the blurred image of part (d) has been reconstructed from the coefficients of subbands 1–4 (1/64th of the total number of transform coefficients) and half of the coefficients of subbands 5–7 (half of 3/64, or 3/128). On average, the image has been reconstructed from $5/128 \approx 0.039$ or 3.9% of the transform coefficients. Notice that the Daubechies D8 filter was used in the calculations. Readers are encouraged to use this code and experiment with the performance of other filters.

8.12: They are written in the form `a-=b/2; b+=a;`.

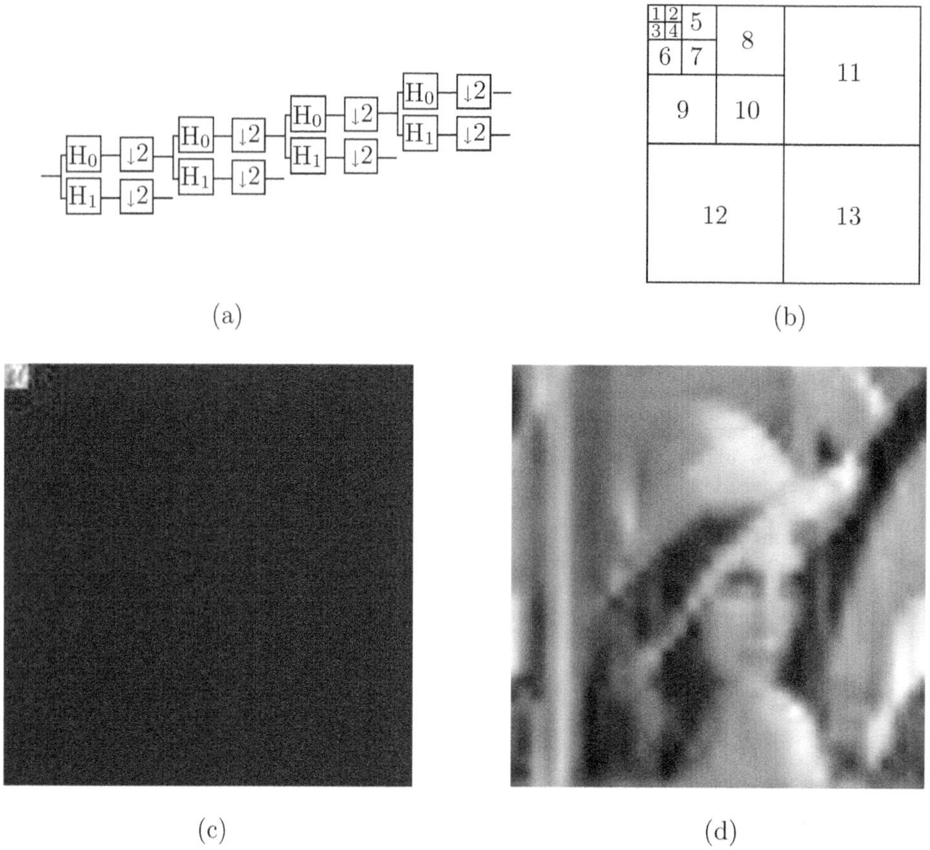

Figure Ans.52: Blurring as a Result of Coarse Quantization.

```
clear, colormap(gray);
filename='lena128'; dim=128;
fid=fopen(filename,'r');
img=fread(fid,[dim,dim])';
filt=[0.23037,0.71484,0.63088,-0.02798, ...
 -0.18703,0.03084,0.03288,-0.01059];
fwim=fwt2(img,3,filt);
figure(1), imagesc(fwim), axis square
 fwim(1:16,17:32)=fwim(1:16,17:32)/2;
 fwim(1:16,33:128)=0;
 fwim(17:32,1:32)=fwim(17:32,1:32)/2;
 fwim(17:32,33:128)=0;
 fwim(33:128,:)=0;
figure(2), colormap(gray), imagesc(fwim)
rec=iwt2(fwim,3,filt);
figure(3), colormap(gray), imagesc(rec)
```

Code for Figure Ans.52.

8.13: We sum Equation (8.13) over all the values of l to get

$$\sum_{l=0}^{2^{j-1}-1} s_{j-1,l} = \sum_{l=0}^{2^{j-1}-1} (s_{j,2l} + d_{j-1,l}/2) = \frac{1}{2}\sum_{l=0}^{2^{j-1}-1} (s_{j,2l} + s_{j,2l+1}) = \frac{1}{2}\sum_{l=0}^{2^{j}-1} s_{j,l}. \quad \text{(Ans.1)}$$

Therefore, the average of set s_{j-1} equals

$$\frac{1}{2^{j-1}}\sum_{l=0}^{2^{j-1}-1} s_{j-1,l} = \frac{1}{2^{j-1}}\frac{1}{2}\sum_{l=0}^{2^{j}-1} s_{j,l} = \frac{1}{2^{j}}\sum_{l=0}^{2^{j}-1} s_{j,l}$$

the average of set s_j.

8.14: The code of Figure Ans.53 produces the expression

$$0.0117\mathbf{P}_1 - 0.0977\mathbf{P}_2 + 0.5859\mathbf{P}_3 + 0.5859\mathbf{P}_4 - 0.0977\mathbf{P}_5 + 0.0117\mathbf{P}_6.$$

```
Clear[p,a,b,c,d,e,f];
p[t_]:=a t^5+b t^4+c t^3+d t^2+e t+f;
Solve[{p[0]==p1, p[1/5.]==p2, p[2/5.]==p3,
p[3/5.]==p4, p[4/5.]==p5, p[1]==p6}, {a,b,c,d,e,f}];
sol=ExpandAll[Simplify[%]];
Simplify[p[0.5] /.sol]
```

Figure Ans.53: Code for a Degree-5 Interpolating Polynomial.

8.15: The Matlab code of Figure Ans.54 does that and produces the transformed integer vector $y = (111, -1, 84, 0, 120, 25, 84, 3)$. The inverse transform generates vector z that is identical to the original data x. Notice how the detail coefficients are much smaller than the weighted averages. Notice also that Matlab arrays are indexed from 1, whereas the discussion in the text assumes arrays indexed from 0. This causes the difference in index values in Figure Ans.54.

8.16: For the case $M_C = 3$, the first six images g_0 through g_5 will have dimensions

$$(3 \cdot 2^5 + 1 \times 4 \cdot 2^5 + 1) = 97 \times 129, \quad 49 \times 65, \quad 25 \times 33, \quad 13 \times 17, \text{ and } 7 \times 9.$$

8.17: In the sorting pass of the third iteration the encoder transmits the number $l = 3$ (the number of coefficients $c_{i,j}$ in our example that satisfy $2^{12} \le |c_{i,j}| < 2^{13}$), followed by the three pairs of coordinates $(3,3)$, $(4,2)$, and $(4,1)$ and by the signs of the three coefficients. In the refinement step it transmits the six bits *cdefgh*. These are the 13th most significant bits of the coefficients transmitted in all the previous iterations.

```
clear;
N=8; k=N/2;
x=[112,97,85,99,114,120,77,80];
% Forward IWT into y
for i=0:k-2,
 y(2*i+2)=x(2*i+2)-floor((x(2*i+1)+x(2*i+3))/2);
end;
y(N)=x(N)-x(N-1);
y(1)=x(1)+floor(y(2)/2);
for i=1:k-1,
 y(2*i+1)=x(2*i+1)+floor((y(2*i)+y(2*i+2))/4);
end;
% Inverse IWT into z
z(1)=y(1)-floor(y(2)/2);
for i=1:k-1,
 z(2*i+1)=y(2*i+1)-floor((y(2*i)+y(2*i+2))/4);
end;
for i=0:k-2,
 z(2*i+2)=y(2*i+2)+floor((z(2*i+1)+x(2*i+3))/2);
end;
z(N)=y(N)+z(N-1);
```

Figure Ans.54: Matlab Code for Forward and Inverse IWT.

The information received so far enables the decoder to further improve the 16 approximate coefficients. The first nine become

$$c_{2,3} = s1ac0\ldots0,\ c_{3,4} = s1bd0\ldots0,\ c_{3,2} = s01e00\ldots0,$$
$$c_{4,4} = s01f00\ldots0,\ c_{1,2} = s01g00\ldots0,\ c_{3,1} = s01h00\ldots0,$$
$$c_{3,3} = s0010\ldots0,\ c_{4,2} = s0010\ldots0,\ c_{4,1} = s0010\ldots0,$$

and the remaining seven are not changed.

8.18: The simple equation $10\times2^{20}\times8 = (500x)\times(500x)\times8$ is solved to yield $x^2 = 40$ square inches. If the card is square, it is approximately 6.32 inches on a side. Such a card has 10 rolled impressions (about 1.5×1.5 each), two plain impressions of the thumbs (about 0.875×1.875 each), and simultaneous impressions of both hands (about 3.125×1.875 each). All the dimensions are in inches.

8.19: The bit of 10 is encoded, as usual, in pass 2. The bit of 1 is encoded in pass 1 since this coefficient is still insignificant but has significant neighbors. This bit is 1, so coefficient 1 becomes significant (a fact that is not used later). Also, this bit is the first 1 of this coefficient, so the sign bit of the coefficient is encoded following this bit. The bits of coefficients 3 and -7 are encoded in pass 2 since these coefficients are significant.

9.1: It is easy to calculate that $525 \times 4/3 = 700$ pixels.

9.2: The vertical height of the picture on the author's 27 in. television set is 16 in., which translates to a viewing distance of $7.12 \times 16 = 114$ in. or about 9.5 feet. It is easy to see that individual scan lines are visible at any distance shorter than about 6 feet.

9.3: Three common examples are: (1) Surveillance camera, (2) an old, silent movie being restored and converted from film to video, and (3) a video presentation taken underwater.

9.4: The golden ratio $\phi \approx 1.618$ has traditionally been considered the aspect ratio that is most pleasing to the eye. This suggests that 1.77 is the better aspect ratio.

9.5: Imagine a camera panning from left to right. New objects will enter the field of view from the right all the time. A block on the right side of the frame may therefore contain objects that did not exist in the previous frame.

9.6: Since $(4,4)$ is at the center of the "+", the value of s is halved, to 2. The next step searches the four blocks labeled 4, centered on $(4,4)$. Assuming that the best match is at $(6,4)$, the two blocks labeled 5 are searched. Assuming that $(6,4)$ is the best match, s is halved to 1, and the eight blocks labeled 6 are searched. The diagram shows that the best match is finally found at location $(7,4)$.

9.7: This figure consists of 18×18 macroblocks, and each macroblock constitutes six 8×8 blocks of samples. The total number of samples is therefore $18\times18\times6\times64 = 124,416$.

9.8: The size category of zero is 0, so code 100 is emitted, followed by zero bits. The size category of 4 is 3, so code 110 is first emitted, followed by the three least-significant bits of 4, which are 100.

9.9: The zigzag sequence is

$$118, 2, 0, -2, \underbrace{0, \ldots, 0}_{13}, -1, 0, \ldots.$$

The run-level pairs are $(0,2)$, $(1,-2)$, and $(13,-1)$, so the final codes are (notice the sign bits following the run-level codes)

$$0100\,0|000110\,1|00100000\,1|10,$$

(without the vertical bars).

9.10: There are no nonzero coefficients, no run-level codes, just the 2-bit EOB code. However, in nonintra coding, such a block is encoded in a special way.

10.1: An average book may have 60 characters per line, 45 lines per page, and 400 pages. This comes to $60 \times 45 \times 400 = 1,080,000$ characters, requiring one byte of storage each.

10.2: The period of a wave is its speed divided by its frequency. For sound we get

$$\frac{34380\,\text{cm/s}}{22000\,\text{Hz}} = 1.562 \text{ cm}, \qquad \frac{34380}{20} = 1719 \text{ cm}.$$

10.3: The (base-10) logarithm of x is a number y such that $10^y = x$. The number 2 is the logarithm of 100 since $10^2 = 100$. Similarly, 0.3 is the logarithm of 2 since $10^{0.3} = 2$. Also, The base-b logarithm of x is a number y such that $b^y = x$ (for any real $b > 1$).

10.4: Each doubling of the sound intensity increases the dB level by 3. Therefore, the difference of 9 dB $(3 + 3 + 3)$ between A and B corresponds to three doublings of the sound intensity. Thus, source B is $2\cdot2\cdot2 = 8$ times louder than source A.

10.5: Each 0 would result in silence and each sample of 1, in the same tone. The result would be a nonuniform buzz. Such sounds were common on early personal computers.

10.6: Such an experiment should be repeated with several persons, preferably of different ages. The person should be placed in a sound insulated chamber, and a pure tone of frequency f should be played. The amplitude of the tone should be gradually increased from zero until the person can just barely hear it. If this happens at a decibel value d, point (d, f) should be plotted. This should be repeated for many frequencies until a graph similar to Figure 10.5a is obtained.

10.7: We first select identical items. If all $s(t - i)$ equal s, Equation (10.7) yields the same s. Next, we select values on a straight line. Given the four values a, $a + 2$, $a + 4$, and $a + 6$, Equation (10.7) yields $a + 8$, the next linear value. Finally, we select values roughly equally-spaced on a circle. The y coordinates of points on the first quadrant of a circle can be computed by $y = \sqrt{r^2 - x^2}$. We select the four points with x coordinates 0, $0.08r$, $0.16r$, and $0.24r$, compute their y coordinates for $r = 10$, and substitute them in Equation (10.7). The result is 9.96926, compared to the actual y coordinate for $x = 0.32r$ which is $\sqrt{r^2 - (0.32r)^2} = 9.47418$, a difference of about 5%. The code that did the computations is shown in Figure Ans.55.

```
(* Points on a circle. Used in exercise to check
 4th-order prediction in FLAC *)
r = 10;
ci[x_] := Sqrt[100 - x^2];
ci[0.32r]
4ci[0] - 6ci[0.08r] + 4ci[0.16r] - ci[0.24r]
```

Figure Ans.55: Code for Checking 4th-Order Prediction.

10.8: Imagine that the sound being compressed contains one second of a pure tone (just one frequency). This second will be digitized to 44,100 consecutive samples per channel. The samples indicate amplitudes, so they don't have to be the same. However, after filtering, only one subband (although in practice perhaps two subbands) will have nonzero signals. All the other subbands correspond to different frequencies, so they will have signals that are either zero or very close to zero.

10.9: Assuming that a noise level P_1 translates to x decibels

$$20\log\left(\frac{P_1}{P_2}\right) = x \text{ dB SPL},$$

results in the relation

$$20\log\left(\frac{\sqrt[3]{2}P_1}{P_2}\right) = 20\left[\log_{10}\sqrt[3]{2} + \log\left(\frac{P_1}{P_2}\right)\right] = 20(0.1 + x/20) = x + 2.$$

Thus, increasing the sound level by a factor of $\sqrt[3]{2}$ increases the decibel level by 2 dB SPL.

10.10: For a sampling rate of 44,100 samples/sec, the calculations are similar. The decoder has to decode 44,100/384 $\approx$ 114.84 frames per second. Thus, each frame has to be decoded in approximately 8.7 ms. In order to output 114.84 frames in 64,000 bits, each frame must have $B_f = 557$ bits available to encode it. Thus, the number of slots per frame is 557/32 $\approx$ 17.41. Thus, the last (18th) slot is not full and has to padded.

10.11: Table 10.58 shows that the scale factor is 111 and the select information is 2. The third rule in Table 10.59 shows that a scfsi of 2 means that only one scale factor was coded, occupying just six bits in the compressed output. The decoder assigns these six bits as the values of all three scale factors.

10.12: Typical layer II parameters are (1) a sampling rate of 48,000 samples/sec, (2) a bitrate of 64,000 bits/sec, and (3) 1,152 quantized signals per frame. The decoder has to decode 48,000/1152 = 41.66 frames per second. Thus, each frame has to be decoded in 24 ms. In order to output 41.66 frames in 64,000 bits, each frame must have B_f = 1,536 bits available to encode it.

10.13: A program to play `.mp3` files is an MPEG layer III *decoder*, not an encoder. Decoding is much simpler since it does not use a psychoacoustic model, nor does it have to anticipate preechoes and maintain the bit reservoir.

11.1: Because the original string S can be reconstructed from L but not from F.

11.2: A direct application of Equation (11.1) eight more times produces:

```
S[10-1-2]=L[T²[I]]=L[T[T¹[I]]]=L[T[7]]=L[6]=i;
S[10-1-3]=L[T³[I]]=L[T[T²[I]]]=L[T[6]]=L[2]=m;
S[10-1-4]=L[T⁴[I]]=L[T[T³[I]]]=L[T[2]]=L[3]=␣;
```

```
S[10-1-5]=L[T⁵[I]]=L[T[T⁴[I]]]=L[T[3]]=L[0]=s;
S[10-1-6]=L[T⁶[I]]=L[T[T⁵[I]]]=L[T[0]]=L[4]=s;
S[10-1-7]=L[T⁷[I]]=L[T[T⁶[I]]]=L[T[4]]=L[5]=i;
S[10-1-8]=L[T⁸[I]]=L[T[T⁷[I]]]=L[T[5]]=L[1]=w;
S[10-1-9]=L[T⁹[I]]=L[T[T⁸[I]]]=L[T[1]]=L[9]=s;
```

The original string `swiss miss` is indeed reproduced in S from right to left.

11.3: Figure Ans.56 shows the rotations of S and the sorted matrix. The last column, L of Ans.56b happens to be identical to S, so S=L=`sssssssssh`. Since A=(`s`,`h`), a move-to-front compression of L yields $C = (1, 0, 0, 0, 0, 0, 0, 0, 0, 1)$. Since C contains just the two values 0 and 1, they can serve as their own Huffman codes, so the final result is 1000000001, 1 bit per character!

(a)	(b)
sssssssssh	hsssssssss
ssssssssh s	shssssssss
sssssssshss	sshsssssss
ssssssshsss	ssshssssss
sssssshssss	sssshsssss
ssssshsssss	ssssshssss
sssshssssss	sssssshsss
ssshsssssss	ssssssshss
sshssssssss	sssssssshs
hsssssssss	sssssssssh

Figure Ans.56: Permutations of "sssssssssh".

11.4: The encoder starts at T[0], which contains 5. The first element of L is thus the last symbol of permutation 5. This permutation starts at position 5 of S, so its last element is in position 4. The encoder thus has to go through symbols S[T[i-1]] for $i = 0, \ldots, n-1$, where the notation $i-1$ should be interpreted cyclically (i.e., $0-1$ should be $n-1$). As each symbol S[T[i-1]] is found, it is compressed using move-to-front. The value of I is the position where T contains 0. In our example, T[8]=0, so I=8.

11.5: The first element of a triplet is the distance between two dictionary entries, the one best matching the content and the one best matching the context. In this case there is no content match, no distance, so any number could serve as the first element, 0 being the best (smallest) choice.

11.6: because the three lines are sorted in ascending order. The bottom two lines of Table 11.13c are not in sorted order. This is why the `zz...z` part of string S must be preceded and followed by complementary bits.

11.7: The encoder places S between two entries of the sorted associative list and writes the (encoded) index of the entry above or below S on the compressed stream. The fewer the number of entries, the smaller this index, and the better the compression.

11.8: Context 5 is compared to the three remaining contexts 6, 7, and 8, and it is most similar to context 6 (they share a suffix of "b"). Context 6 is compared to 7 and 8 and, since they don't share any suffix, context 7, the shorter of the two, is selected. The remaining context 8 is, of course, the last one in the ranking. The final context ranking is

$$1 \rightarrow 3 \rightarrow 4 \rightarrow 0 \rightarrow 5 \rightarrow 6 \rightarrow 7 \rightarrow 8.$$

11.9: Equation (11.3) shows that the third "a" is assigned rank 1 and the "b" and "a" following it are assigned ranks 2 and 3, respectively.

11.10: Table Ans.57 shows the sorted contexts. Equation (Ans.2) shows the context ranking at each step.

$$\begin{array}{l} \underset{u}{0}\,, \qquad \underset{u}{0} \rightarrow \underset{b}{2}\,, \qquad \underset{l}{1} \rightarrow \underset{b}{3} \rightarrow \underset{u}{0}\,, \\ \underset{u}{0} \rightarrow \underset{l}{2} \rightarrow \underset{a}{3} \rightarrow \underset{b}{4}\,, \qquad \underset{l}{2} \rightarrow \underset{a}{4} \rightarrow \underset{d}{1} \rightarrow \underset{b}{5} \rightarrow \underset{u}{0}\,, \\ \underset{i}{3} \rightarrow \underset{a}{5} \rightarrow \underset{l}{2} \rightarrow \underset{b}{6} \rightarrow \underset{d}{5} \rightarrow \underset{u}{0}\,. \end{array} \tag{Ans.2}$$

The final output is "u 2 b 3 l 4 a 5 d 6 i 6." Notice that each of the distinct input symbols appears once in this output in raw format.

(a)

0	λ	u
1	u	x

(b)

0	λ	u
1	ub	x
2	u	b

(c)

0	λ	u
1	ub	l
2	ubl	x
3	u	b

(d)

0	λ	u
1	ubla	x
2	ub	l
3	ubl	a
4	u	b

(e)

0	λ	u
1	ubla	d
2	ub	l
3	ublad	x
4	ubl	a
5	u	b

(f)

0	λ	u
1	ubla	d
2	ub	l
3	ublad	i
4	ubladi	x
5	ubl	a
6	u	b

Table Ans.57: Constructing the Sorted Lists for `ubladiu`.

11.11: All n_1 bits of string L_1 need be written on the output stream. This already shows that there is going to be no compression. String L_2 consists of n_1/k 1's, so all of it has to be written on the output stream. String L_3 similarly consists of n_1/k^2 1's, and so on. Thus, the size of the output stream is

$$n_1 + \frac{n_1}{k} + \frac{n_1}{k^2} + \frac{n_1}{k^3} + \cdots + \frac{n_1}{k^m} = n_1 \frac{k^{m+1} - 1}{k^m(k-1)},$$

for some value of m. The limit of this expression, when $m \to \infty$, is $n_1 k/(k-1)$. For $k=2$ this equals $2n_1$. For larger values of k this limit is always between n_1 and $2n_1$.

For the curious reader, here is how the sum above is computed. Given the series

$$S = \sum_{i=0}^{m} \frac{1}{k^i} = 1 + \frac{1}{k} + \frac{1}{k^2} + \frac{1}{k^3} + \cdots + \frac{1}{k^{m-1}} + \frac{1}{k^m},$$

we multiply both sides by $1/k$

$$\frac{S}{k} = \frac{1}{k} + \frac{1}{k^2} + \frac{1}{k^3} + \cdots + \frac{1}{k^m} + \frac{1}{k^{m+1}} = S + \frac{1}{k^{m+1}} - 1,$$

and subtract

$$\frac{S}{k}(k-1) = \frac{k^{m+1}-1}{k^{m+1}} \to S = \frac{k^{m+1}-1}{k^m(k-1)}.$$

11.12: The input stream consists of:
1. A run of three zero groups, coded as 10|1 since 3 is in second position in class 2.
2. The nonzero group 0100, coded as 111100.
3. Another run of three zero groups, again coded as 10|1.
4. The nonzero group 1000, coded as 01100.
5. A run of four zero groups, coded as 010|00 since 4 is in first position in class 3.
6. 0010, coded as 111110.
7. A run of two zero groups, coded as 10|0.

The output is thus the 31-bit string 1011111001010110001000111110100.

11.13: The input stream consists of:
1. A run of three zero groups, coded as $R_2 R_1$ or 101|11.
2. The nonzero group 0100, coded as 00100.
3. Another run of three zero groups, again coded as 101|11.
4. The nonzero group 1000, coded as 01000.
5. A run of four zero groups, coded as $R_4 = 1001$.
6. 0010, coded as 00010.
7. A run of two zero groups, coded as $R_2 = 101$.

The output is thus the 32-bit string 10111001001011101000100100010101.

11.14: The input stream consists of:
1. A run of three zero groups, coded as F_3 or 1001.
2. The nonzero group 0100, coded as 00100.
3. Another run of three zero groups, again coded as 1001.
4. The nonzero group 1000, coded as 01000.
5. A run of four zero groups, coded as $F_3 F_1 = 1001|11$.
6. 0010, coded as 00010.
7. A run of two zero groups, coded as $F_2 = 101$.

The output is thus the 32-bit string 10010010010010100010011100010101.

11.15: Yes, if they are located in different quadrants or subquadrants. Pixels 123 and 301, for example, are adjacent in Figure 7.172 but have different prefixes.

11.16: No, because all prefixes have the same probability of occurrence. In our example the prefixes are four bits long and all 16 possible prefixes have the same probability because a pixel may be located anywhere in the image. A Huffman code constructed for 16 equally-probable symbols has an average size of four bits per symbol, so nothing would be gained. The same is true for suffixes.

11.17: This is possible, but it places severe limitations on the size of the string. In order to rearrange a one-dimensional string into a four-dimensional cube, the string size should be 2^{4n}. If the string size happens to be $2^{4n}+1$, it has to be extended to $2^{4(n+1)}$, which increases its size by a factor of 16. It is possible to rearrange the string into a rectangular box, not just a cube, but then its size will have to be of the form $2^{n1}2^{n2}2^{n3}2^{n4}$ where the four ni's are integers.

11.18: The LZW algorithm, which starts with the entire alphabet stored at the beginning of its dictionary, is an example of such a method. However, an adaptive version of LZW can be designed to compress words instead of individual characters.

11.19: A better choice for the coordinates may be relative values (or *offsets*). Each (x, y) pair may specify the position of a character relative to its predecessor. This results in smaller numbers for the coordinates, and smaller numbers are easier to compress.

11.20: There may be such letters in other, "exotic" alphabets, but a more common example is a rectangular box enclosing text. The four rules that constitute such a box should be considered a mark, but the text characters inside the box should be identified as separate marks.

11.21: This guarantees that the two probabilities will add up to 1.

11.22: Figure Ans.58 shows how state A feeds into the new state D' which, in turn, feeds into states E and F. Notice how states B and C haven't changed. Since the new state D' is identical to D, it is possible to feed A into either D or D' (cloning can be done in two different but identical ways). The original counts of state D should now be divided between D and D' in proportion to the counts of the transitions $A \to D$ and $B, C \to D$.

11.23: Figure Ans.59 shows the new state 6 after the operation $1, 1 \to 6$. Its 1-output is identical to that of state 1, and its 0-output is a copy of the 0-output of state 3.

11.24: A precise answer requires many experiments with various data files. A little thinking, though, shows that the larger k, the better the initial model that is created when the old one is discarded. Larger values of k thus minimize the loss of compression. However, very large values may produce an initial model that is already large and cannot grow much. The best value for k is therefore one that produces an initial model large enough to provide information about recent correlations in the data, but small enough so it has room to grow before it too has to be discarded.

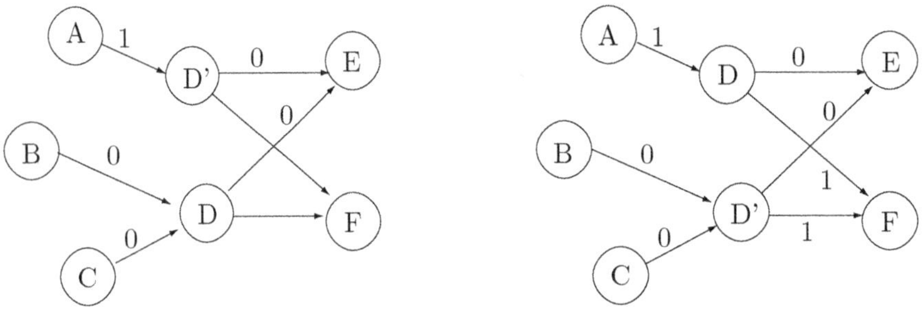

Figure Ans.58: New State D' Cloned.

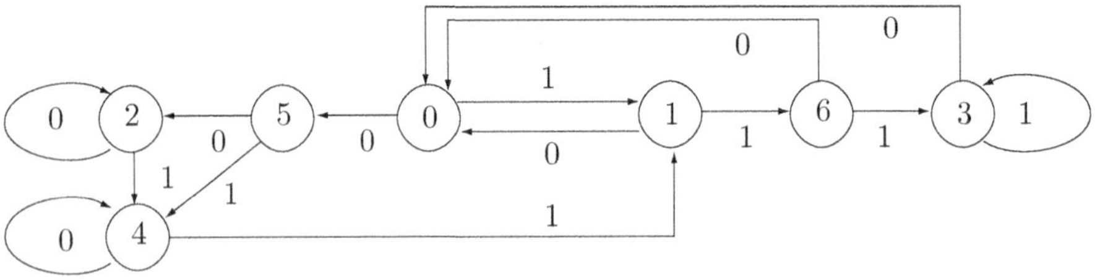

Figure Ans.59: State 6 Added.

11.25: The number of marked points can be written $8(1 + 2 + 3 + 5 + 8 + 13) = 256$ and the numbers in parentheses are the Fibonacci numbers.

11.26: The conditional probability $P(D_i|D_i)$ is very small. A segment pointing in direction D_i can be preceded by another segment pointing in the same direction only if the original curve is straight or very close to straight for more than 26 coordinate units (half the width of grid S_{13}).

11.27: We observe that a point has two coordinates. If each coordinate occupies eight bits, then the use of Fibonacci numbers reduces the 16-bit coordinates to an 8-bit number, a compression ratio of 0.5. The use of Huffman codes can typically reduce this 8-bit number to (on average) a 4-bit code, and the use of the Markov model can perhaps cut this by another bit. The result is an estimated compression ratio of $3/16 = 0.1875$. If each coordinate is a 16-bit number, then this ratio improves to $3/32 = 0.09375$.

11.28: The resulting, shorter grammar is shown in Figure Ans.60. It is one rule and one symbol shorter.

Input	Grammar
S → abcdbcabcdbc	S → CC
	A → bc
	C → aAdA

Figure Ans.60: Improving the Grammar of Figure 11.42.

11.29: Because generating rule `C` has made rule `B` underused (i.e., used just once).

11.30: Rule `S` consists of two copies of rule `A`. The first time rule `A` is encountered, its contents `aBdB` are sent. This involves sending rule `B` twice. The first time rule `B` is sent, its contents `bc` are sent (and the decoder does not know that the string `bc` it is receiving is the contents of a rule). The second time rule `B` is sent, the pair $(1,2)$ is sent (offset 1, count 2). The decoder identifies the pair and uses it to set up the rule $1 \rightarrow$ `bc`. Sending the first copy of rule `A` therefore amounts to sending `abcd`$(1,2)$. The second copy of rule `A` is sent as the pair $(0,4)$ since `A` starts at offset 0 in `S` and its length is 4. The decoder identifies this pair and uses it to set up the rule $2 \rightarrow$ `a`$\boxed{1}$`d`$\boxed{1}$. The final result is therefore `abcd`$(1,2)(0,4)$.

11.31: In each of these cases, the encoder removes one edge from the boundary and inserts two new edges. There is a net gain of one edge.

11.32: They create triangles $(18,2,3)$ and $(18,3,4)$, and reduce the boundary to the sequence of vertices

$$(4,5,6,7,8,9,10,11,12,13,14,15,16,17,18).$$

A.1: It is $1+\lfloor \log_2 i \rfloor$ as can be seen by simple experimenting.

A.2: The integer 2 is the smallest integer that can serve as the basis for a number system.

A.3: Replacing 10 by 3 we get $x = k\log_2 3 \approx 1.58k$. A trit is therefore worth about 1.58 bits.

A.4: We assume an alphabet with two symbols a_1 and a_2, with probabilities P_1 and P_2, respectively. Since $P_1 + P_2 = 1$, the entropy of the alphabet is $-P_1 \log_2 P_1 - (1 - P_1)\log_2(1 - P_1)$. Table 2.2 shows the entropies for certain values of the probabilities. When $P_1 = P_2$, at least 1 bit is required to encode each symbol, reflecting the fact that the entropy is at its maximum, the redundancy is zero, and the data cannot be compressed. However, when the probabilities are very different, the minimum number of bits required per symbol drops significantly. We may not be able to develop a compression method using 0.08 bits per symbol but we know that when $P_1 = 99\%$, this is the theoretical minimum.

> A writer is someone who can make a riddle out of an answer.
>
> Karl Kraus

Bibliography

All URLs have been checked and updated as of early July 2009. Any broken links reported to the authors will be added to the errata list in the book's Web site.

The main event in the life of the data compression community is the annual data compression conference (DCC, see Joining the Data Compression Community) whose proceedings are published by the IEEE. The editors have traditionally been James Andrew Storer and Martin Cohn. Instead of listing many references that differ only by year, we start this bibliography with a generic reference to the DCC, where "XX" is the last two digits of the conference year.

Storer, James A., and Martin Cohn (eds.) (annual) *DCC 'XX: Data Compression Conference*, Los Alamitos, CA, IEEE Computer Society Press.

3R (2006) is `http://f-cpu.seul.org/whygee/ddj-3r/ddj-3R.html`.

7z (2006) is `http://www.7-zip.org/sdk.html`.

Abramson, N. (1963) *Information Theory and Coding*, New York, McGraw-Hill.

Abousleman, Glen P. (2006) "Coding of Hyperspectral Imagery with Trellis-Coded Quantization," in G. Motta, F. Rizzo, and J. A. Storer, editors, *Hyperspectral Data Compression*, New York, Springer Verlag.

Acharya, Tinku and Joseph F. JáJá (1995) "Enhancing LZW Coding Using a Variable-Length Binary Encoding," Tech Report TR 1995-70, Institute for Systems Research, University of Maryland.

Acharya, Tinku and Joseph F. JáJá (1996) "An On-Line Variable Length Binary Encoding of Text," *Information and Computer Science*, **94**:1–22.

acronyms (2006) is
`http://acronyms.thefreedictionary.com/Refund-Anticipated+Return`.

adaptiv9 (2006) is `http://www.weizmann.ac.il/matlab/toolbox/filterdesign/` file `adaptiv9.html`.

Adelson, E. H., E. Simoncelli, and R. Hingorani (1987) "Orthogonal Pyramid Transforms for Image Coding," *Proceedings SPIE*, vol. 845, Cambridge, MA, pp. 50–58, October.

adobepdf (2006) is `http://www.adobe.com/products/acrobat/adobepdf.html`.

afb (2006) is `http://www.afb.org/prodProfile.asp?ProdID=42`.

Ahlswede, Rudolf, Te Sun Han, and Kingo Kobayashi (1997) "Universal Coding of Integers and Unbounded Search Trees," *IEEE Transactions on Information Theory*, **43**(2):669–682.

Ahmed, N., T. Natarajan, and R. K. Rao (1974) "Discrete Cosine Transform," *IEEE Transactions on Computers*, **C-23**:90–93.

Akansu, Ali, and R. Haddad (1992) *Multiresolution Signal Decomposition*, San Diego, CA, Academic Press.

amu (2006) is `http://ifa.amu.edu.pl/~kprzemek`.

Anderson, K. L., et al., (1987) "Binary-Image-Manipulation Algorithm in the Image View Facility," *IBM Journal of Research and Development*, **31**(1):16–31, January.

Anedda, C. and L. Felician (1988) "P-Compressed Quadtrees for Image Storing," *The Computer Journal*, **31**(4):353–357.

Anh, Vo Ngoc and Alistair Moffat (2005) "Inverted Index Compression Using Word-Aligned Binary Codes," *Information Retrieval*, **8**:151–166.

Apostolico, Alberto and A. S. Fraenkel (1987) "Robust Transmission of Unbounded Strings Using Fibonacci Representations," *IEEE Transactions on Information Theory*, **33**(2):238–245, March.

ascii-wiki (2009) is `http://en.wikipedia.org/wiki/ASCII`.

ATSC (2006) is `http://www.atsc.org/standards/a_52b.pdf`.

ATT (1996) is `http://www.djvu.att.com/`.

Baker, Brenda, Udi Manber, and Robert Muth (1999) "Compressing Differences of Executable Code," in *ACM SIGPLAN Workshop on Compiler Support for System Software (WCSSS '99)*.

Banister, Brian, and Thomas R. Fischer (1999) "Quadtree Classification and TCQ Image Coding," in Storer, James A., and Martin Cohn (eds.) (1999) *DCC '99: Data Compression Conference*, Los Alamitos, CA, IEEE Computer Society Press, pp. 149–157.

Barf (2008) is `http://www.cs.fit.edu/~mmahoney/compression/barf.html`.

Barnsley, M. F., and Sloan, A. D. (1988) "A Better Way to Compress Images," *Byte Magazine*, pp. 215–222, January.

Barnsley, M. F. (1988) *Fractals Everywhere*, New York, Academic Press.

Bass, Thomas A. (1992) *Eudaemonic Pie*, New York, Penguin Books.

Bauer, Rainer and Joachim Hagenauer (2001) "On Variable-Length Codes for Iterative Source/Channel-Decoding," *Proceedings of the IEEE Data Compression Conference, 2001*, Snowbird, UT, pp. 273–282.

Bell, Timothy C. (1986) "Better OPM/L Text Compression," *IEEE Transactions on Communications*, **34**(12):1176–1182, December.

Bell, Timothy C. (1987) *A Unifying Theory and Improvements for Existing Approaches to Text Compression*, PhD thesis, Department of Computer Science, University of Canterbury, Christchurch, New Zealand.

Bell, Timothy C., John J. Cleary, and Ian Hugh Witten (1990) *Text Compression*, Prentice-Hall, Englewood Cliffs, N.J.

Bentley, J. L. and A. C. Yao (1976) "An Almost Optimal Algorithm for Unbounded Searching," *Information Processing Letters*, **5**(3):82–87.

Bentley, J. L. et al. (1986) "A Locally Adaptive Data Compression Algorithm," *Communications of the ACM*, **29**(4):320–330, April.

Berger, T. and R. Yeung (1990) "Optimum '1'-Ended Binary Prefix Codes," *IEEE Transactions on Information Theory*, **36**(6):1435–1441, November.

Bergmans (2008) is `http://www.maximumcompression.com`.

Berstel, Jean and Dominique Perrin (1985) *Theory of Codes*, Orlando, FL, Academic Press.

Blackstock, Steve (1987) "LZW and GIF Explained," available from
`http://www.cis.udel.edu/~amer/CISC651/lzw.and.gif.explained.html`.

Bloom, Charles R. (1996) "LZP: A New Data Compression Algorithm," in *Proceedings of Data Compression Conference*, J. Storer, editor, Los Alamitos, CA, IEEE Computer Society Press, p. 425.

Bloom, Charles R. (1998) "Solving the Problems of Context Modeling," available for ftp from `http://www.cbloom.com/papers/ppmz.zip`.

BOCU (2001) is `http://icu-project.org/docs/papers/binary_ordered_compression_for_unicode.html`.

BOCU-1 (2002) is `http://www.unicode.org/notes/tn6/`.

Boldi, Paolo and Sebastiano Vigna (2004a), "The WebGraph Framework I: Compression Techniques," in *Proceedings of the 13th International World Wide Web Conference (WWW 2004)*, pages 595–601, New York, ACM Press.

Boldi, Paolo and Sebastiano Vigna (2004b), "The WebGraph Framework II: Codes for the World-Wide Web," in *Data Compression Conference, DCC 2004*.

Boldi, Paolo and Sebastiano Vigna (2005), "Codes for the World-Wide Web," *Internet Mathematics*, **2**(4):405–427.

Bookstein, Abraham and S. T. Klein (1993) "Is Huffman Coding Dead?" *Proceedings of the 16th Annual International ACM SIGIR Conference on Research and Development in Information Retrieval*, pp. 80–87. Also published in *Computing*, **50**(4):279–296, 1993, and in *Proceedings of the Data Compression Conference, 1993*, Snowbird, UT. p. 464.

Born, Günter (1995) *The File Formats Handbook*, London, New York, International Thomson Computer Press.

Bosi, Marina, and Richard E. Goldberg (2003) *Introduction To Digital Audio Coding and Standards*, Boston, MA, Kluwer Academic.

Boussakta, Said, and Hamoud O. Alshibami (2004) "Fast Algorithm for the 3-D DCT-II," *IEEE Transactions on Signal Processing*, **52**(4).

Bradley, Jonathan N., Christopher M. Brislawn, and Tom Hopper (1993) "The FBI Wavelet/Scalar Quantization Standard for Grayscale Fingerprint Image Compression," *Proceedings of Visual Information Processing II*, Orlando, FL, SPIE vol. 1961, pp. 293–304, April.

Brandenburg, Karlheinz, and Gerhard Stoll (1994) "ISO-MPEG-1 Audio: A Generic Standard for Coding of High-Quality Digital Audio," *Journal of the Audio Engineering Society*, **42**(10):780–792, October.

Brandenburg, Karlheinz (1999) "MP3 and AAC Explained," *The AES 17th International Conference*, Florence, Italy, Sept. 2–5. Available at `http://www.cselt.it/mpeg/tutorials.htm`.

Brent, R. P. (1987) "A linear Algorithm for Data Compression," *Australian Computer Journal*, **19**(2):64–68, May.

Brislawn, Christopher, Jonathan Bradley, R. Onyshczak, and Tom Hopper (1996) "The FBI Compression Standard for Digitized Fingerprint Images," in *Proceedings SPIE*, Vol. 2847, Denver, CO, pp. 344–355, August.

Britanak, Vladimir, Patrick C. Yip, and Kamisetty Ramamohan Rao (2006) *Discrete Cosine and Sine Transforms: General Properties, Fast Algorithms and Integer Approximations*, St. Louis, MO, Academic Press, 2006

Brittain, Nathanael J. and Mahmoud R. El-Sakka (2005) "Grayscale Two-Dimensional Lempel-Ziv Encoding,"*International Conference on Image Analysis and Recognition, ICIAR'2005*, LNCS 3656, pp. 328–334.

Brittain, Nathanael and Mahmoud R. El-Sakka (2007) "Grayscale True Two-Dimensional Dictionary-Based Image Compression," *Journal of Visual Communication and Image Representation*, **18**(1):35–44, February.

BSDiff (2005) is `http://www.daemonology.net/bsdiff/bsdiff-4.3.tar.gz`

Burrows, Michael, et al. (1992) *On-line Data Compression in a Log-Structured File System*, Digital, Systems Research Center, Palo Alto, CA.

Burrows, Michael, and D. J. Wheeler (1994) *A Block-Sorting Lossless Data Compression Algorithm*, Digital Systems Research Center Report 124, Palo Alto, CA, May 10.

Burt, Peter J., and Edward H. Adelson (1983) "The Laplacian Pyramid as a Compact Image Code," *IEEE Transactions on Communications*, **COM-31**(4):532–540, April.

Buttigieg, Victor (1995a) *Variable-Length Error-Correcting Codes*, PhD thesis, University of Manchester.

Buttigieg, Victor and P. G. Farrell (1995b) "A Maximum A-Posteriori (MAP) Decoding Algorithm For Variable-Length Error-Correcting Codes," *Codes and Cyphers: Cryptography and Coding IV*, Essex, England, Institute of Mathematics and Its Applications, pp. 103–119.

Buyanovsky, George (1994) "Associative Coding" (in Russian), *Monitor*, Moscow, #8, 10–19, August. (Hard copies of the Russian source and English translation are available from the authors of this book. Send requests to the authors' email address found in the Preface.)

Buyanovsky, George (2002) Private communications (`buyanovsky@home.com`).

Cachin, Christian (1998) "An Information-Theoretic Model for Steganography," in *Proceedings of the Second International Workshop on Information Hiding*, D. Aucsmith, ed. vol. 1525 of *Lecture Notes in Computer Science*, Berlin, Springer-Verlag, pp. 306–318.

Calgary (2006) is `ftp://ftp.cpsc.ucalgary.ca/pub/projects/` directory `text.compression.corpus`.

Calgary challenge (2008) is `http://mailcom.com/challenge`.

Campos, Arturo San Emeterio (2006) *Range coder*, in `http://www.arturocampos.com/ac_range.html`.

Canterbury (2006) is `http://corpus.canterbury.ac.nz`.

Capocelli, Renato (1989) "Comments and Additions to 'Robust Transmission of Unbounded Strings Using Fibonacci Representations'," *IEEE Transactions on Information Theory*, **35**(1):191–193, January.

Capocelli, R. and A. De Santis (1992) "On The Construction of Statistically Synchronizable Codes," *IEEE Transactions on Information Theory*, **38**(2):407–414, March.

Capocelli, R. and A. De Santis (1994) "Binary Prefix Codes Ending in a '1'," *IEEE Transactions on Information Theory*, **40**(4):1296–1302, July.

Capon, J. (1959) "A Probabilistic Model for Run-length Coding of Pictures," *IEEE Transactions on Information Theory*, **5**(4):157–163, December.

Carpentieri, B., M.J. Weinberger, and G. Seroussi (2000) "Lossless Compression of Continuous-Tone Images," *Proceedings of the IEEE*, **88**(11):1797–1809, November.

Chaitin, Gregory J. (1966) "On the Lengths of Programs for Computing Finite Binary Sequences," *Journal of the ACM*, **13**(4):547–569, October.

Chaitin, Gregory J. (1977) "Algorithmic Information Theory," *IBM Journal of Research and Development*, **21**:350–359, July.

Chaitin, Gregory J. (1997) *The Limits of Mathematics*, Singapore, Springer-Verlag.

Chen, Wen-Hsiung, C. Harrison Smith, and S. C. Fralick (1977) "A Fast Computational Algorithm For the Discrete Cosine Transform," *IEEE Transactions on Communications*, **25**(9):1004–1009, September.

Chomsky, N. (1956) "Three Models for the Description of Language," *IRE Transactions on Information Theory*, **2**(3):113–124.

Choueka Y., Shmuel T. Klein, and Y. Perl (1985) "Efficient Variants of Huffman Codes in High Level Languages," *Proceedings of the 8th ACM-SIGIR Conference*, Montreal, pp. 122–130.

Cleary, John G., and I. H. Witten (1984) "Data Compression Using Adaptive Coding and Partial String Matching," *IEEE Transactions on Communications*, **COM-32**(4):396–402, April.

Cleary, John G., W. J. Teahan, and Ian H. Witten (1995) "Unbounded Length Contexts for PPM," *Data Compression Conference, 1995*, 52–61.

Cleary, John G. and W. J. Teahan (1997) "Unbounded Length Contexts for PPM," *The Computer Journal*, **40**(2/3):67–75.

codethatword (2007) is `http://www.codethatword.com/`.

Cole, A. J. (1985) "A Note on Peano Polygons and Gray Codes," *International Journal of Computer Mathematics*, **18**:3–13.

Cole, A. J. (1986) "Direct Transformations Between Sets of Integers and Hilbert Polygons," *International Journal of Computer Mathematics*, **20**:115–122.

Constantinescu, C., and J. A. Storer (1994a) "Online Adaptive Vector Quantization with Variable Size Codebook Entries," *Information Processing and Management*, **30**(6)745–758.

Constantinescu, C., and J. A. Storer (1994b) "Improved Techniques for Single-Pass Adaptive Vector Quantization," *Proceedings of the IEEE*, **82**(6):933–939, June.

Constantinescu, C., and R. Arps (1997) "Fast Residue Coding for Lossless Textual Image Compression," in *Proceedings of the 1997 Data Compression Conference*, J. Storer, ed., Los Alamitos, CA, IEEE Computer Society Press, pp. 397–406.

Cormack G. V., and R. N. S. Horspool (1987) "Data Compression Using Dynamic Markov Modelling," *The Computer Journal*, **30**(6):541–550.

Cormen, Thomas H., Charles E. Leiserson, Ronald L. Rivest and Clifford Stein (2001) *Introduction to Algorithms*, 2nd Edition, MIT Press and McGraw-Hill.

corr.pdf (2002) is `http://www.davidsalomon.name/DC2advertis/Corr.pdf`.

CRC (1998) Soulodre, G. A., T. Grusec, M. Lavoie, and L. Thibault, "Subjective Evaluation of State-of-the-Art 2-Channel Audio Codecs," *Journal of the Audio Engineering Society*, **46**(3):164–176, March.

CREW 2000 was `www.crc.ricoh.com/CREW/` but it no longer exists.

Crochemore, Maxime, Filippo Mignosi, Antonio Restivo, and Sergio Salemi (2000) "Data Compression Using Antidictionaries," *Proceedings of the IEEE*, **88**(11):1756–1768 (a special issue on lossless data compression, edited by J. Storer).

Crochemore, Maxime and G. Navarro (2002) "Improved Antidictionary Based Compression," *XII International Conference of the Chilean Computer Science Soc. (SCCC'02)*, pp. 7–13, November.

Crocker, Lee Daniel (1995) "PNG: The Portable Network Graphic Format," *Dr. Dobb's Journal of Software Tools*, **20**(7):36–44.

Culik, Karel II, and J. Kari (1993) "Image Compression Using Weighted Finite Automata," *Computer and Graphics*, **17**(3):305–313.

Culik, Karel II, and J. Kari (1994a) "Image-Data Compression Using Edge-Optimizing Algorithm for WFA Inference," *Journal of Information Processing and Management*, **30**(6):829–838.

Culik, Karel II, and Jarkko Kari (1994b) "Inference Algorithm for Weighted Finite Automata and Image Compression," in *Fractal Image Encoding and Compression*, Y. Fisher, editor, New York, NY, Springer-Verlag.

Culik, Karel II, and Jarkko Kari (1995) "Finite State Methods for Compression and Manipulation of Images," in *DCC '96, Data Compression Conference*, J. Storer, editor, Los Alamitos, CA, IEEE Computer Society Press, pp. 142–151.

Culik, Karel II, and V. Valenta (1996) "Finite Automata Based Compression of Bi-Level Images," in Storer, James A. (ed.), *DCC '96, Data Compression Conference*, Los Alamitos, CA, IEEE Computer Society Press, pp. 280–289.

Culik, Karel II, and V. Valenta (1997a) "Finite Automata Based Compression of Bi-Level and Simple Color Images," *Computer and Graphics*, **21**:61–68.

Culik, Karel II, and V. Valenta (1997b) "Compression of Silhouette-like Images Based on WFA," *Journal of Universal Computer Science*, **3**:1100–1113.

curiosities (2008) is `http://www.maximumcompression.com/compression_fun.php`.

curio.strange (2008) is `http://www.maximumcompression.com/strange.rar`.

curio.test (2008) is `http://www.maximumcompression.com/test.rar`.

dartmouth (2006) is
`http://www.math.dartmouth.edu/~euler/correspondence/letters/OO0765.pdf`.

Dasarathy, Belur V. (ed.) (1995) *Image Data Compression: Block Truncation Coding (BTC) Techniques*, Los Alamitos, CA, IEEE Computer Society Press.

Daubechies, Ingrid (1988) "Orthonormal Bases of Compactly Supported Wavelets," *Communications on Pure and Applied Mathematics*, **41**:909–996.

Davidson, Michael and Lucian Ilie (2004) "Fast Data Compression with Antidictionaries," *Fundamenta Informaticae*, **64**(1–4):119–134, July.

Davisson, I. D. (1966) "Comments on 'Sequence Time Coding for Data Compression'," *Proceedings of the IEEE*, **54**:2010, December.

Deflate (2003) is `http://www.zlib.net/`.

della Porta, Giambattista (1558) *Magia Naturalis*, Naples, first edition, four volumes 1558, second edition, 20 volumes 1589. Translated by Thomas Young and Samuel Speed, *Natural Magick by John Baptista Porta, a Neopolitane*, London 1658.

Demko, S., L. Hodges, and B. Naylor (1985) "Construction of Fractal Objects with Iterated Function Systems," *Computer Graphics*, **19**(3):271–278, July.

DeVore, R., et al. (1992) "Image Compression Through Wavelet Transform Coding," *IEEE Transactions on Information Theory*, **38**(2):719–746, March.

Dewitte, J., and J. Ronson (1983) "Original Block Coding Scheme for Low Bit Rate Image Transmission," in *Signal Processing II: Theories and Applications—Proceedings of EUSIPCO 83*, H. W. Schussler, ed., Amsterdam, Elsevier Science Publishers B. V. (North-Holland), pp. 143–146.

Dolby (2006) is `http://www.dolby.com/`.

donationcoder (2006) is
`http://www.donationcoder.com/Reviews/Archive/ArchiveTools/index.html`.

Durbin J. (1960) "The Fitting of Time-Series Models," *JSTOR: Revue de l'Institut International de Statistique*, **28**:233–344.

DVB (2006) is `http://www.dvb.org/`.

Ekstrand, Nicklas (1996) "Lossless Compression of Gray Images via Context Tree Weighting," in Storer, James A. (ed.), *DCC '96: Data Compression Conference*, Los Alamitos, CA, IEEE Computer Society Press, pp. 132–139, April.

Elias, P. (1975) "Universal Codeword Sets and Representations of the Integers," *IEEE Transactions on Information Theory*, **IT-21**(2):194–203, March.

Emilcont (2008) is `http://www.freewebs.com/emilcont/`.

Even, S. and M. Rodeh (1978) "Economical Encoding of Commas Between Strings," *Communications of the ACM*, **21**(4):315–317, April.

Faller N. (1973) "An Adaptive System for Data Compression," in *Record of the 7th Asilomar Conference on Circuits, Systems, and Computers*, pp. 593–597.

Fang I. (1966) "It Isn't ETAOIN SHRDLU; It's ETAONI RSHDLC," *Journalism Quarterly*, **43**:761–762.

Fano, R. M. (1949) "The Transmission of Information," Research Laboratory for Electronics, MIT, Tech Rep. No. 65.

Feder, Jens (1988) *Fractals*, New York, Plenum Press.

Federal Bureau of Investigation (1993) *WSQ Grayscale Fingerprint Image Compression Specification, ver. 2.0*, Document #IAFIS-IC-0110v2, Criminal Justice Information Services, February.

Feig, E., and E. Linzer (1990) "Discrete Cosine Transform Algorithms for Image Data Compression," in *Proceedings Electronic Imaging '90 East*, pp. 84–87, Boston, MA.

Feig, Ephraim, Heidi Peterson, and Viresh Ratnakar (1995) "Image Compression Using Spatial Prediction," *International Conference on Acoustics, Speech, and Signal Processing, ICASSP-95*, **4**:2339–2342, May.

Feldspar (2003) is http://zlib.net/feldspar.html.

Fenwick, P. (1996) *Symbol Ranking Text Compression*, Tech. Rep. 132, Dept. of Computer Science, University of Auckland, New Zealand, June.

Fenwick, Peter (1996a) "Punctured Elias Codes for variable-length coding of the integers," Technical Report 137, Department of Computer Science, The University of Auckland, December. This is also available online at www.firstpr.com.au/audiocomp/lossless/TechRep137.pdf.

Fenwick, P. (2002), "Variable-Length Integer Codes Based on the Goldbach Conjecture, and Other Additive Codes," *IEEE Transactions on Information Theory*, **48**(8):2412–2417, August.

Ferguson, T. J. and J. H. Rabinowitz (1984) "Self-Synchronizing Huffman codes," *IEEE Transactions on Information Theory*, **30**(4):687–693, July.

Fiala, E. R., and D. H. Greene (1989), "Data Compression with Finite Windows," *Communications of the ACM*, **32**(4):490–505.

Fibonacci (1999) is file Fibonacci.html in http://www-groups.dcs.st-and.ac.uk/~history/References/.

FIPS197 (2001) *Advanced Encryption Standard*, FIPS Publication 197, November 26, 2001. Available from http://csrc.nist.gov/publications/fips/fips197/fips-197.pdf.

firstpr (2006) is http://www.firstpr.com.au/audiocomp/lossless/#rice.

Fisher, Yuval (ed.) (1995) *Fractal Image Compression: Theory and Application*, New York, Springer-Verlag.

flac.devices (2006) is http://flac.sourceforge.net/links.html#hardware.

flacID (2006) is http://flac.sourceforge.net/id.html.

Fox, E. A., et al. (1991) "Order Preserving Minimal Perfect Hash Functions and Information Retrieval," *ACM Transactions on Information Systems*, **9**(2):281–308.

Fraenkel, A. S., and S. T. Klein (1985) "Novel Compression of Sparse Bit-Strings—Preliminary Report," in A. Apostolico and Z. Galil, eds. *Combinatorial Algorithms on Words*, Vol. 12, NATO ASI Series F:169–183, New York, Springer-Verlag.

Fraenkel, A. S. and Shmuel T. Klein (1985a) "Robust Universal Complete Codes as Alternatives to Huffman Codes," Tech. Report CS85-16, Dept. of Applied Mathematics, Weizmann Institute of Science, October.

Fraenkel, Aviezri S. and Shmuel T. Klein (1990) "Bidirectional Huffman Coding," *The Computer Journal*, **33**:296–307.

Fraenkel, Aviezri S. and Shmuel T. Klein (1996) "Robust Universal Complete Codes for Transmission and Compression," *Discrete Applied Mathematics*, **64**(1):31–55, January.

Frank, Amalie J., J. D. Daniels, and Diane R. Unangst (1980) "Progressive Image Transmission Using a Growth-Geometry Coding," *Proceedings of the IEEE*, **68**(7):897–909, July.

Freeman, G. H. (1991) "Asymptotic Convergence of Dual-Tree Entropy Codes," *Proceedings of the Data Compression Conference (DCC '91)*, pp. 208–217.

Freeman, G. H. (1993) "Divergence and the Construction of Variable-to-Variable-Length Lossless Codes by Source-Word Extensions," *Proceedings of the Data Compression Conference (DCC '93)*, pp. 79–88.

Freeman, H. (1961) "On The Encoding of Arbitrary Geometric Configurations," *IRE Transactions on Electronic Computers*, **EC-10**(2):260–268, June.

G131 (2006) ITU-T Recommendation G.131, *Talker echo and its control.*

G.711 (1972) is `http://en.wikipedia.org/wiki/G.711`.

Gage, Philip (1994) "A New Algorithm for Data Compression," *C/C++ Users Journal*, **12**(2)23–28, Feb 01. This is available online at `http://www.ddj.com/cpp/184402829;jsessionid=LGSEIODZNDHKIQSNDLRSKHSCJUNN2JVN?_requestid=927467`.

Gallager, Robert G., and David C. van Voorhis (1975) "Optimal Source Codes for Geometrically Distributed Integer Alphabets," *IEEE Transactions on Information Theory*, **IT-21**(3):228–230, March.

Gallager, Robert G. (1978) "Variations On a Theme By Huffman," *IEEE Transactions on Information Theory*, **IT-24**(6):668–674, November.

Gardner, Martin (1972) "Mathematical Games," *Scientific American*, **227**(2):106, August.

Gemstar (2006) is
`http://www.macrovision.com/products/ce_manufacturers/ipg_ce/vcr_plus.htm`

Gersho, Allen, and Robert M. Gray (1992) *Vector Quantization and Signal Compression*, Boston, MA, Kluwer Academic Publishers.

Gharavi, H. (1987) "Conditional Run-Length and Variable-Length Coding of Digital Pictures," *IEEE Transactions on Communications*, **COM-35**(6):671–677, June.

Gilbert, E. N., and E. F. Moore (1959) "Variable Length Binary Encodings," *Bell System Technical Journal, Monograph 3515*, **38**:933–967, July.

Gilbert, Jeffrey M., and Robert W. Brodersen (1998) "A Lossless 2-D Image Compression Technique for Synthetic Discrete-Tone Images," in *Proceedings of the 1998 Data Compression Conference*, J. Storer, ed., Los Alamitos, CA, IEEE Computer Society Press, pp. 359–368, March. This is also available from URL
`http://bwrc.eecs.berkeley.edu/Publications/1999/A_lossless_2-D_image_compression_technique/JMG_DCC98.pdf`.

Gilchrist (2008) is http://compression.ca/act/.

Girod, Bernd (1999) "Bidirectionally Decodable Streams of Prefix Code-Words," *IEEE Communications Letters*, **3**(8):245–247, August.

Givens, Wallace (1958) "Computation of Plane Unitary Rotations Transforming a General Matrix to Triangular Form," *Journal of the Society for Industrial and Applied Mathematics*, **6**(1):26–50, March.

Glusker, Mark, David M. Hogan, and Pamela Vass (2005) "The Ternary Calculating Machine of Thomas Fowler," *IEEE Annals of the History of Computing*, **27**(3):4–22, July.

Golin, S. (1992) "DVI Image Compression, Second Generation," *Proceedings SPIE Conference, Image Processing Algorithms*, **1657**:106–113, February.

Golomb, Solomon W. (1966) "Run-Length Encodings," *IEEE Transactions on Information Theory*, **IT-12**(3):399–401.

Gonzalez, Rafael C., and Richard E. Woods (1992) *Digital Image Processing*, Reading, MA, Addison-Wesley.

Gottlieb, D., et al. (1975) *A Classification of Compression Methods and their Usefulness for a Large Data Processing Center*, Proceedings of National Computer Conference, **44**:453–458.

Gray, Frank (1953) "Pulse Code Communication," United States Patent 2,632,058, March 17.

Greek-Unicode (2007) is http://www.unicode.org/charts/PDF/U0370.pdf.

Grimm, R. E. (1973) "The Autobiography of Leonardo Pisano," *Fibonacci Quarterly*, **11**(1):99–104, February.

H.264Draft (2006) is ftp://standards.polycom.com/JVT_Site/draft_standard/ file JVT-G050r1.zip. This is the H.264 draft standard.

H.264PaperIR (2006) is http://www.vcodex.com/h264.html (papers by Iain Richardson).

H.264PaperRM (2006) is http://research.microsoft.com/apps/pubs/, file default.aspx?id=77882 (a paper by Rico Malvar).

H.264Standards (2006) is ftp://standards.polycom.com/JVT_Site/ (the H.264 standards repository).

H.264Transform (2006) ftp://standards.polycom.com/JVT_Site/2002_01_Geneva/ files JVT-B038r2.doc and JVT-B039r2.doc (the H.264 integer transform).

h2g2 (2006) is http://www.bbc.co.uk/dna/h2g2/A406973.

Haffner, Patrick, et al. (1998) "High-Quality Document Image Compression with DjVu," *Journal of Electronic Imaging*, **7**(3):410–425, SPIE. This is also available from http://citeseer.ist.psu.edu/old/523172.html.

Hafner, Ullrich (1995) "Asymmetric Coding in (m)-WFA Image Compression," Report 132, Department of Computer Science, University of Würzburg, December.

haiku-usa (2009) is `http://www.hsa-haiku.org/index.htm`.

Hans, Mat and R. W. Schafer (2001) "Lossless Compression of Digital Audio," *IEEE Signal Processing Magazine*, **18**(4):21–32, July.

Haralambous, Yannis (2007) *Fonts and Encodings*, (Translated by P. Scott Horne), Sebastopol, CA, O'Reilly Associates.

Havas, G., et al. (1993) *Graphs, Hypergraphs and Hashing*, in *Proceedings of the International Workshop on Graph-Theoretic Concepts in Computer Science (WG'93)*, Berlin, Springer-Verlag.

Heath, F. G. (1972) "Origins of the Binary Code," *Scientific American*, **227**(2):76, August.

Hilbert, D. (1891) "Ueber stetige Abbildung einer Linie auf ein Flächenstück," *Math. Annalen*, **38**:459–460.

Hirschberg, D., and D. Lelewer (1990) "Efficient Decoding of Prefix Codes," *Communications of the ACM*, **33**(4):449–459.

Horspool, N. R. (1991) "Improving LZW," in *Proceedings of the 1991 Data Compression Conference*, J. Storer, ed., Los Alamitos, CA, IEEE Computer Society Press, pp .332–341.

Horspool, N. R., and G. V. Cormack (1992) "Constructing Word-Based Text Compression Algorithms," in *Proceedings of the 1992 Data Compression Conference*, J. Storer, ed., Los Alamitos, CA, IEEE Computer Society Press, PP. 62–71, April.

Horstmann (2006) `http://www.horstmann.com/bigj/help/windows/tutorial.html`.

Howard, Paul G. (1998) "Interleaving Entropy Codes," *Proceedings Compression and Complexity of Sequences 1997*, Salerno, Italy, pp. 45–55, June.

Howard, Paul G., and J. S. Vitter (1992a) "New Methods for Lossless Image Compression Using Arithmetic Coding," *Information Processing and Management*, **28**(6):765–779.

Howard, Paul G., and J. S. Vitter (1992b) "Error Modeling for Hierarchical Lossless Image Compression," in *Proceedings of the 1992 Data Compression Conference*, J. Storer, ed., Los Alamitos, CA, IEEE Computer Society Press, pp. 269–278.

Howard, Paul G., and J. S. Vitter (1992c) "Practical Implementations of Arithmetic Coding," in *Image and Text Compression*, J. A. Storer, ed., Norwell, MA, Kluwer Academic Publishers, PP. 85–112. Also available from URL
`http://www.cs.duke.edu/~jsv/Papers/catalog/node85.html`.

Howard, Paul G., and J. S. Vitter, (1993) "Fast and Efficient Lossless Image Compression," in *Proceedings of the 1993 Data Compression Conference*, J. Storer, ed., Los Alamitos, CA, IEEE Computer Society Press, pp. 351–360.

Howard, Paul G., and J. S. Vitter (1994a) "Fast Progressive Lossless Image Compression," Proceedings of the Image and Video Compression Conference, *IS&T/SPIE 1994*

Symposium on Electronic Imaging: Science & Technology, 2186, San Jose, CA, pp. 98–109, February.

Howard, Paul G., and J. S. Vitter (1994b) "Design and Analysis of Fast text Compression Based on Quasi-Arithmetic Coding," *Journal of Information Processing and Management*, **30**(6):777–790. Also available from URL
`http://www.cs.duke.edu/~jsv/Papers/catalog/node91.html`.

Huffman, David (1952) "A Method for the Construction of Minimum Redundancy Codes," *Proceedings of the IRE*, **40**(9):1098–1101.

Hunt, James W. and M. Douglas McIlroy (1976) "An Algorithm for Differential File Comparison," Computing Science Technical Report No. 41, Murray Hill, NJ, Bell Labs, June.

Hunter, R., and A. H. Robinson (1980) "International Digital Facsimile Coding Standards," *Proceedings of the IEEE*, **68**(7):854–867, July.

Hutter (2008) is `http://prize.hutter1.net`.

hydrogenaudio (2006) is `www.hydrogenaudio.org/forums/`.

IA-32 (2006) is `http://en.wikipedia.org/wiki/IA-32`.

IBM (1988) *IBM Journal of Research and Development*, #6 (the entire issue).

IEEE754 (1985) ANSI/IEEE Standard 754-1985, "IEEE Standard for Binary Floating-Point Arithmetic."

IMA (2006) is `www.ima.org/`.

ISO (1984) "Information Processing Systems-Data Communication High-Level Data Link Control Procedure-Frame Structure," IS 3309, 3rd ed., October.

ISO (2003) is `http://www.iso.ch/`.

ISO/IEC (1993) International Standard IS 11172-3 "Information Technology, Coding of Moving Pictures and Associated Audio for Digital Storage Media at up to about 1.5 Mbits/s—Part 3: Audio."

ISO/IEC (2000), International Standard IS 15444-1 "Information Technology—JPEG 2000 Image Coding System." This is the FDC (final committee draft) version 1.0, 16 March 2000.

ISO/IEC (2003) International Standard ISO/IEC 13818-7, "Information technology, Generic coding of moving pictures and associated audio information, Part 7: Advanced Audio Coding (AAC)," 2nd ed., 2003-08-01.

ITU-R/BS1116 (1997) ITU-R, document BS 1116 "Methods for the Subjective Assessment of Small Impairments in Audio Systems Including Multichannel Sound Systems," Rev. 1, Geneva.

ITU-T (1989) CCITT Recommendation G.711: "Pulse Code Modulation (PCM) of Voice Frequencies."

ITU-T (1990), Recommendation G.726 (12/90), *40, 32, 24, 16 kbit/s Adaptive Differential Pulse Code Modulation (ADPCM).*

ITU-T (1994) ITU-T Recommendation V.42, Revision 1 "Error-correcting Procedures for DCEs Using Asynchronous-to-Synchronous Conversion."

ITU-T264 (2002) ITU-T Recommendation H.264, ISO/IEC 11496-10, "Advanced Video Coding," Final Committee Draft, Document JVT-E022, September.

ITU/TG10 (1991) ITU-R document TG-10-2/3-E "Basic Audio Quality Requirements for Digital Audio Bit-Rate Reduction Systems for Broadcast Emission and Primary Distribution," 28 October.

Jakobsson, Matti (1985) "Compression of Character Strings by an Adaptive Dictionary," *BIT*, **25**(4):593–603, december.

Jayant N. (ed.) (1997) *Signal Compression: Coding of Speech, Audio, Text, Image and Video*, Singapore, World Scientific Publications.

JBIG (2003) is `http://www.jpeg.org/jbig/index.html`.

JBIG2 (2003) is `http://www.jpeg.org/jbig/jbigpt2.html`.

JBIG2 (2006) is `http://jbig2.com/`.

Jordan, B. W., and R. C. Barrett (1974) "A Cell Organized Raster Display for Line Drawings," *Communications of the ACM*, **17**(2):70–77.

Joshi, R. L., V. J. Crump, and T. R. Fischer (1993) "Image Subband Coding Using Arithmetic and Trellis Coded Quantization," *IEEE Transactions on Circuits and Systems Video Technology*, **5**(6):515–523, December.

JPEG 2000 Organization (2000) is `http://www.jpeg.org/JPEG2000.htm`.

Karp, R. S. (1961) "Minimum-Redundancy Coding for the Discrete Noiseless Channel," *Transactions of the IRE*, **7**:27–38.

Kendall, Maurice G. (1961) *A Course in the Geometry of n-Dimensions*, New York, Hafner.

Kiely, A. (2004) "Selecting the Golomb Parameter in Rice Coding," IPN (Interplanetary Network) Progress Report, **42–159**:1–8, November 15.

KGB (2008) is `http://kgbarchiver.net/`.

Kieffer, J., G. Nelson, and E-H. Yang (1996a) "Tutorial on the quadrisection method and related methods for lossless data compression." Available at URL `http://www.ece.umn.edu/users/kieffer/index.html`.

Kieffer, J., E-H. Yang, G. Nelson, and P. Cosman (1996b) "Lossless compression via bisection trees," at `http://www.ece.umn.edu/users/kieffer/index.html`.

Kleijn, W. B., and K. K. Paliwal (1995) *Speech Coding and Synthesis*, Elsevier, Amsterdam.

Knowlton, Kenneth (1980) "Progressive Transmission of Grey-Scale and Binary Pictures by Simple, Efficient, and Lossless Encoding Schemes," *Proceedings of the IEEE*, **68**(7):885–896, July.

Knuth, Donald E. (1973) *The Art of Computer Programming*, Vol. 1, 2nd Ed., Reading, MA, Addison-Wesley.

Knuth, Donald E. (1985) "Dynamic Huffman Coding," *Journal of Algorithms*, **6**:163–180.

Knuth, D. E. (1986) *Computers and Typesetting*, Reading, Mass., Addison Wesley.

Korn D., et al. (2002) "The VCDIFF Generic Differencing and Compression Data Format," RFC 3284, June 2002, available on the Internet as text file "rfc3284.txt".

Kraft, L. G. (1949) *A Device for Quantizing, Grouping, and Coding Amplitude Modulated Pulses*, Master's Thesis, Department of Electrical Engineering, MIT, Cambridge, MA.

Krichevsky, R. E., and V. K. Trofimov (1981) "The Performance of Universal Coding," *IEEE Transactions on Information Theory*, **IT-27**:199–207, March.

Laković, Ksenija and John Villasenor (2002) "On Design of Error-Correcting Reversible Variable Length Codes," *IEEE Communications Letters*, **6**(8):337–339, August.

Laković, Ksenija and John Villasenor (2003) "An Algorithm for Construction of Efficient Fix-Free Codes," *IEEE Communications Letters*, **7**(8):391–393, August.

Lam, Wai-Man and Sanjeev R. Kulkarni (1996) "Extended Synchronizing Codewords for Binary Prefix Codes," *IEEE Transactions on Information Theory*, **42**(3):984–987, May.

Lambert, Sean M. (1999) "Implementing Associative Coder of Buyanovsky (ACB) Data Compression," M.S. thesis, Bozeman, MT, Montana State University (available from Sean Lambert at `sum1els@mindless.com`).

Langdon, Glen G., and J. Rissanen (1981) "Compression of Black White Images with Arithmetic Coding," *IEEE Transactions on Communications*, **COM-29**(6):858–867, June.

Langdon, Glen G. (1983) "A Note on the Ziv-Lempel Model for Compressing Individual Sequences," *IEEE Transactions on Information Theory*, **IT-29**(2):284–287, March.

Langdon, Glenn G. (1983a) "An Adaptive Run Length Coding Algorithm," *IBM Technical Disclosure Bulletin*, **26**(7B):3783–3785, December.

Langdon, Glen G. (1984) *On Parsing vs. Mixed-Order Model Structures for Data Compression*, IBM research report RJ-4163 (46091), January 18, 1984, San Jose, CA.

Lansky (2009) is `http://www.ksi.mff.cuni.cz/~lansky/SC/`.

Larsson and Moffat (2000) "Off-Line Dictionary-Based Compression," *Proceedings of the IEEE*, **88**(11)1722–1732. An earlier, shorter version was published in *Proceedings of the Conference on Data Compression* 1999, pages 296–305. An implementation is available at `http://www.bic.kyoto-u.ac.jp/pathway/rwan/software/restore.html`

Lau (2009) is http://www.raylau.com/biography.html.

Lawrence, John C. (1977) "A New Universal Coding Scheme for the Binary Memoryless Source," *IEEE Transactions on Information Theory*, **23**(4):466–472, July.

LBE (2007) is http://in.geocities.com/iamthebiggestone/how_lbe_works.htm

Lee, J. B. and H. Kalva (2008) *The VC-1 and H.264 Video Compression Standards for Broadband Video Services*, New York, Springer Verlag.

Lempel A. and J. Ziv (1976) "On the Complexity of Finite Sequences," *IEEE Transactions on Information Theory*, **22**(1):75–81.

Levenstein (2006) is http://en.wikipedia.org/wiki/Levenstein_coding.

Levinson, N. (1947) "The Weiner RMS Error Criterion in Filter Design and Prediction," *Journal of Mathematical Physics*, **25**:261–278.

Lewalle, Jacques (1995) "Tutorial on Continuous Wavelet Analysis of Experimental Data" available at http://www.ecs.syr.edu/faculty/lewalle/tutor/tutor.html.

Li, Xiuqi and Borko Furht (2003) "An Approach to Image Compression Using Three-Dimensional DCT," *Proceeding of The Sixth International Conference on Visual Information System 2003 (VIS2003)*, September 24–26.

Liebchen, Tilman et al. (2005) "The MPEG-4 Audio Lossless Coding (ALS) Standard - Technology and Applications," AES 119th Convention, New York, October 7–10, 2005. Available at URL
http://www.nue.tu-berlin.de/forschung/projekte/lossless/mp4als.html

Liefke, Hartmut and Dan Suciu (1999) "XMill: an Efficient Compressor for XML Data," *Proceedings of the ACM SIGMOD Symposium on the Management of Data, 2000*, pp. 153–164. Available at http://citeseer.ist.psu.edu/327443.html.

Linde, Y., A. Buzo, and R. M. Gray (1980) "An Algorithm for Vector Quantization Design," *IEEE Transactions on Communications*, **COM-28**:84–95, January.

Liou, Ming (1991) "Overview of the p×64 kbits/s Video Coding Standard," *Communications of the ACM*, **34**(4):59–63, April.

Litow, Bruce, and Olivier de Val (1995) "The Weighted Finite Automaton Inference Problem," Technical Report 95-1, James Cook University, Queensland.

Loeffler, C., A. Ligtenberg, and G. Moschytz (1989) "Practical Fast 1-D DCT Algorithms with 11 Multiplications," in *Proceedings of the International Conference on Acoustics, Speech, and Signal Processing (ICASSP '89)*, pp. 988–991.

Lynch, T. J. (1966) "Sequence Time Coding for Data Compression," *Proceedings of the IEEE*, **54**:1490–1491, October.

Lyon, Richard F. (2009) "A Brief History of Pixel," available online from
http://www.foveon.com/files/ABriefHistoryofPixel2.pdf.

LZX (2009) is http://xavprods.free.fr/lzx/.

Mahoney (2008) is http://www.cs.fit.edu/~mmahoney/compression.

Mahoney, Michael S. (1990) "Goldbach's Biography" in Charles Coulston Gillispie *Dictionary of Scientific Biography*, New York, NY, Scribner's, 14 vols. 1970–1990.

Mallat, Stephane (1989) "A Theory for Multiresolution Signal Decomposition: The Wavelet Representation," *IEEE Transactions on Pattern Analysis and Machine Intelligence*, **11**(7):674–693, July.

Manber, U., and E. W. Myers (1993) "Suffix Arrays: A New Method for On-Line String Searches," *SIAM Journal on Computing*, **22**(5):935–948, October.

Mandelbrot, B. (1982) *The Fractal Geometry of Nature*, San Francisco, CA, W. H. Freeman.

Manning (1998), is file compression/adv08.html at http://www.newmediarepublic.com/dvideo/.

Marking, Michael P. (1990) "Decoding Group 3 Images," *The C Users Journal* pp. 45–54, June.

Matlab (1999) is http://www.mathworks.com/.

McConnell, Kenneth R. (1992) *FAX: Digital Facsimile Technology and Applications*, Norwood, MA, Artech House.

McCreight, E. M (1976) "A Space Economical Suffix Tree Construction Algorithm," *Journal of the ACM*, **32**(2):262–272, April.

McMillan, Brockway (1956) "Two Inequalities Implied by Unique Decipherability," *IEEE Transactions on Information Theory*, **2**(4):115–116, December.

Meridian (2003) is http://www.meridian-audio.com/.

Meyer, F. G., A. Averbuch, and J.O. Strömberg (1998) "Fast Adaptive Wavelet Packet Image Compression," *IEEE Transactions on Image Processing*, **9**(5) May 2000.

Miano, John (1999) *Compressed Image File Formats*, New York, ACM Press and Addison-Wesley.

Miller, V. S., and M. N. Wegman (1985) "Variations On a Theme by Ziv and Lempel," in A. Apostolico and Z. Galil, eds., NATO ASI series Vol. F12, *Combinatorial Algorithms on Words*, Berlin, Springer, pp. 131–140.

Mitchell, Joan L., W. B. Pennebaker, C. E. Fogg, and D. J. LeGall, eds. (1997) *MPEG Video Compression Standard*, New York, Chapman and Hall and International Thomson Publishing.

MNG (2003) is http://www.libpng.org/pub/mng/spec/.

Moffat, Alistair (1990) "Implementing the PPM Data Compression Scheme," *IEEE Transactions on Communications*, **COM-38**(11):1917–1921, November.

Moffat, Alistair (1991) "Two-Level Context Based Compression of Binary Images," in *Proceedings of the 1991 Data Compression Conference*, J. Storer, ed., Los Alamitos, CA, IEEE Computer Society Press, pp. 382–391.

Moffat, Alistair, Radford Neal, and Ian H. Witten (1998) "Arithmetic Coding Revisited," *ACM Transactions on Information Systems*, **16**(3):256–294, July.

Moffat, Alistair and Lang Stuiver (2000) "Binary Interpolative Coding for Effective Index Compression," *Information Retrieval*, **3**(1):25–47, July.

Moffat, Alistair (2000) "Compressing Integer Sequences and Sets," in *Encyclopedia of Algorithms*, Ming-Yang Kao (Ed.), Springer Verlag.

Moffat, Alistair and Andrew Turpin (2002) *Compression and Coding Algorithms*, Norwell, MA, Kluwer Academic.

Moffat, Alistair and Vo Ngoc Anh (2006) "Binary Codes for Locally Homogeneous Sequences," *Information Processing Letters*, **99**(5):175–180, September.

monkeyaudio (2006) is `http://www.monkeysaudio.com/index.html`.

morse-tape (2006) is `http://memory.loc.gov/mss/mmorse/071/071009/0001d.jpg`.

Motta, Giovanni, James A. Storer, and Bruno Carpentieri (2000) "Lossless Image Coding via Adaptive Linear Prediction and Classification," *Proceedings of the IEEE*, **88**(11):1790–1796, November.

Motta, G., F. Rizzo, and J. A. Storer, eds. (2006) *Hyperspectral Data Compression*, New York, Springer Verlag.

Motte, Warren F. (1998) *Oulipo, A Primer of Potential Literature*, Normal, Ill, Daleky Archive Press.

MPEG (1998), is `http://www.mpeg.org/`.

mpeg-4.als (2006) is
`http://www.nue.tu-berlin.de/forschung/projekte/lossless/mp4als.html`.

MPThree (2006) is `http://inventors.about.com/od/mstartinventions/a/`, file `MPThree.htm`.

Mulcahy, Colm (1996) "Plotting and Scheming with Wavelets," *Mathematics Magazine*, **69**(5):323–343, December. See also `http://www.spelman.edu/~colm/csam.ps`.

Mulcahy, Colm (1997) "Image Compression Using the Haar Wavelet Transform," *Spelman College Science and Mathematics Journal*, **1**(1):22–31, April. Also available at URL `http://www.spelman.edu/~colm/wav.ps`. (It has been claimed that any smart 15-year-old could follow this introduction to wavelets.)

Murray, James D., and William vanRyper (1994) *Encyclopedia of Graphics File Formats*, Sebastopol, CA, O'Reilly and Assoc.

Myers, Eugene W. (1986) "An $O(ND)$ Difference Algorithm and its Variations," *Algorithmica*, **1**(2):251–266.

Nakamura, Hirofumi and Sadayuki Murashima (1991) "The Data Compression Based on Concatenation of Frequentative Code Neighbor," *Proceedings of the 14th Symposium on Information Theory and its Applications (SITA '91)*, (Ibusuki, Japan), pp. 701–704, December 11–14 (in Japanese).

Nakamura, Hirofumi and Sadayuki Murashima (1996) "Data Compression by Concatenations of Symbol Pairs," *Proceedings of the IEEE International Symposium on Information Theory and its Applications*, (Victoria, BC, Canada), pp. 496–499, September.

Netravali, A. and J. O. Limb (1980) "Picture Coding: A Preview," *Proceedings of the IEEE*, **68**:366–406.

Nevill-Manning, C. G. (1996) "Inferring Sequential Structure," Ph.D. thesis, Department of Computer Science, University of Waikato, New Zealand.

Nevill-Manning, C. G., and Ian H. Witten (1997) "Compression and Explanation Using Hierarchical Grammars," *The Computer Journal*, **40**(2/3):104–116.

NHK (2006) is `http://www.nhk.or.jp/english/`.

Nix, R. (1981) "Experience With a Space Efficient Way to Store a Dictionary," *Communications of the ACM*, **24**(5):297–298.

ntfs (2006) is `http://www.ntfs.com/`.

Nyquist, Harry (1928) "Certain Topics in Telegraph Transmission Theory," *AIEE Transactions*, **47**:617–644.

Ogg squish (2006) is `http://www.xiph.org/ogg/flac.html`.

Okumura, Haruhiko (1998) is `http://oku.edu.mie-u.ac.jp/~okumura/` directories `compression/history.html` and `compression.html`.

Osterberg, G. (1935) "Topography of the Layer of Rods and Cones in the Human Retina," *Acta Ophthalmologica*, (suppl. 6):1–103.

Ota, Takahiro and Hiroyoshi Morita (2007) "On the Construction of an Antidictionary of a Binary String with Linear Complexity," *IEICE Transactions on Fundamentals of Electronics, Communications and Computer Sciences*, **E90-A**(11):2533–2539, November.

Paeth, Alan W. (1991) "Image File Compression Made Easy," in *Graphics Gems II*, James Arvo, editor, San Diego, CA, Academic Press.

Parsons, Thomas W. (1987) *Voice and Speech Processing*, New York, McGraw-Hill.

Pan, Davis Yen (1995) "A Tutorial on MPEG/Audio Compression," *IEEE Multimedia*, **2**:60–74, Summer.

Pasco, R. (1976) "Source Coding Algorithms for Fast Data Compression," Ph.D. dissertation, Dept. of Electrical Engineering, Stanford University, Stanford, CA.

pass (2006) is `http://pass.maths.org/issue2/xfile/`.

patents (2006) is `www.ross.net/compression/patents.html`.

PDF (2001) *Adobe Portable Document Format Version 1.4*, 3rd ed., Reading, MA, Addison-Wesley, December.

Peano, G. (1890) "Sur Une Courbe Qui Remplit Toute Une Aire Plaine," *Math. Annalen*, **36**:157–160.

PeaZip (2008) is `http://sourceforge.net/projects/peazip/`.

Peitgen, H. -O., et al. (eds.) (1982) *The Beauty of Fractals*, Berlin, Springer-Verlag.

Peitgen, H. -O., and Dietmar Saupe (1985) *The Science of Fractal Images*, Berlin, Springer-Verlag.

Pennebaker, William B., and Joan L. Mitchell (1988a) "Probability Estimation for the Q-coder," *IBM Journal of Research and Development*, **32**(6):717–726.

Pennebaker, William B., Joan L. Mitchell, et al. (1988b) "An Overview of the Basic Principles of the Q-coder Adaptive Binary Arithmetic Coder," *IBM Journal of Research and Development*, **32**(6):737–752.

Pennebaker, William B., and Joan L. Mitchell (1992) *JPEG Still Image Data Compression Standard*, New York, Van Nostrand Reinhold.

Percival, Colin (2003a) "An Automated Binary Security Update System For FreeBSD," *Proceedings of BSDCon '03*, PP. 29–34.

Percival, Colin (2003b) "Naive Differences of Executable Code," Computing Lab, Oxford University. Available from
`http://www.daemonology.net/papers/bsdiff.pdf`

Percival, Colin (2006) "Matching with Mismatches and Assorted Applications," Ph.D. Thesis (pending paperwork). Available at URL
`http://www.daemonology.net/papers/thesis.pdf`.

Pereira, Fernando and Touradj Ebrahimi (2002) *The MPEG-4 Book*, Upper Saddle River, NJ, Prentice-Hall.

Phillips, Dwayne (1992) "LZW Data Compression," *The Computer Application Journal* Circuit Cellar Inc., **27**:36–48, June/July.

Pigeon, Steven (2001a) "Start/Stop Codes," *Proceedings of the Data Compression Conference (DCC '01)*, p. 511. Also available at
`http://www.stevenpigeon.org/Publications/publications/ssc_full.pdf`.

Pigeon, Steven (2001b) *Contributions to Data Compression*, PhD Thesis, University of Montreal (in French). Available from
`http://www.stevenpigeon.org/Publications/publications/phd.pdf`. The part on taboo codes is "Taboo Codes, New Classes of Universal Codes," and is also available at `www.iro.umontreal.ca/~brassard/SEMINAIRES/taboo.ps`. A new version has been submitted to *SIAM Journal of Computing*.

PKWare (2003) is `http://www.pkware.com`.

PNG (2003) is `http://www.libpng.org/pub/png/`.

Pohlmann, Ken (1985) *Principles of Digital Audio*, Indianapolis, IN, Howard Sams & Co.

Poynton (2008) `http://www.poynton.com/PDFs/Chroma_subsampling_notation.pdf`.

PQ-wiki (2009) is http://en.wikipedia.org/wiki/List_of_English_words_containing_Q_not_followed_by_U.

Press, W. H., B. P. Flannery, et al. (1988) *Numerical Recipes in C: The Art of Scientific Computing*, Cambridge, UK, Cambridge University Press. (Also available at URL http://www.nr.com/.)

Prusinkiewicz, P., and A. Lindenmayer (1990) *The Algorithmic Beauty of Plants*, New York, Springer-Verlag.

Prusinkiewicz, P., A. Lindenmayer, and F. D. Fracchia (1991) "Synthesis of Space-Filling Curves on the Square Grid," in *Fractals in the Fundamental and Applied Sciences*, Peitgen, H.-O., et al. (eds.), Amsterdam, Elsevier Science Publishers, pp. 341–366.

Pylak, Paweł (2003) "Efficient Modification of LZSS Compression Algorithm," *Annales UMCS Informatica*, section AI, pp. 61–72, Lublin, Poland. Also available online at www.annales.umcs.lublin.pl/AI/2003/07.pdf.

quicktimeAAC (2006) is http://www.apple.com/quicktime/technologies/aac/.

Rabbani, Majid, and Paul W. Jones (1991) *Digital Image Compression Techniques*, Bellingham, WA, Spie Optical Engineering Press.

Rabiner, Lawrence R. and Ronald W. Schafer (1978) *Digital Processing of Speech Signals*, Englewood Cliffs, NJ, Prentice-Hall Series in Signal Processing.

Ramabadran, Tenkasi V., and Sunil S. Gaitonde (1988) "A Tutorial on CRC Computations," *IEEE Micro*, pp. 62–75, August.

Ramstad, T. A., et al (1995) *Subband Compression of Images: Principles and Examples*, Amsterdam, Elsevier Science Publishers.

Randall, Keith, Raymie Stata, Rajiv Wickremesinghe, and Janet L. Wiener (2001) "The LINK Database: Fast Access to Graphs of the Web." *Research Report 175*, Compaq Systems Research Center, Palo Alto, CA.

Rao, K. R., and J. J. Hwang (1996) *Techniques and Standards for Image, Video, and Audio Coding*, Upper Saddle River, NJ, Prentice Hall.

Rao, K. R., and P. Yip (1990) *Discrete Cosine Transform—Algorithms, Advantages, Applications*, London, Academic Press.

Rao, Kamisetty Ramamohan and Patrick C. Yip, editors (2000) *The Transform and Data Compression Handbook*, Boca Raton, FL, CRC Press.

Rao, Raghuveer M., and Ajit S. Bopardikar (1998) *Wavelet Transforms: Introduction to Theory and Applications*, Reading, MA, Addison-Wesley.

rarlab (2006) is http://www.rarlab.com/.

Ratushnyak (2008) is http://www.geocities.com/SiliconValley/Bay/1995/texts22.html.

Reghbati, H. K. (1981) "An Overview of Data Compression Techniques," *IEEE Computer*, **14**(4):71–76.

Reznik, Yuriy (2004) "Coding Of Prediction Residual In MPEG-4 Standard For Lossless Audio Coding (MPEG-4 ALS)," *Proceedings of the IEEE International Conference on Acoustics, Speech, and Signal Processing, 2004*, (ICASSP apos;04), **3**(17–21):1024–1027, May.

RFC1945 (1996) *Hypertext Transfer Protocol—HTTP/1.0*, available at URL `http://www.faqs.org/rfcs/rfc1945.html`.

RFC1950 (1996) *ZLIB Compressed Data Format Specification version 3.3*, is `http://www.ietf.org/rfc/rfc1950`.

RFC1951 (1996) *DEFLATE Compressed Data Format Specification version 1.3*, is `http://www.ietf.org/rfc/rfc1951`.

RFC1952 (1996) *GZIP File Format Specification Version 4.3*. Available in PDF format at URL `http://www.gzip.org/zlib/rfc-gzip.html`.

RFC1962 (1996) *The PPP Compression Control Protocol (CCP)*, available from many sources.

RFC1979 (1996) is `http://www.faqs.org/rfcs/rfc1979.html` (*PPP Deflate Protocol*).

RFC2616 (1999) *Hypertext Transfer Protocol – HTTP/1.1*. Available in PDF format at URL `http://www.faqs.org/rfcs/rfc2616.html`.

In computer network engineering, a Request for Comments (RFC) is a memorandum published by the Internet Engineering Task Force (IETF) describing methods, behaviors, research, or innovations applicable to the working of the Internet and Internet-connected systems.

Through the Internet Society, engineers and computer scientists may publish discourse in the form of an RFC, either for peer review or simply to convey new concepts, information, or (occasionally) engineering humor. The IETF adopts some of the proposals published as RFCs as Internet standards.

The acronym RFC also stands for Radio frequency choke, Remote function call, Request for change, and Regenerative fuel cell.

—From `http://en.wikipedia.org/`

Rice, Robert F. (1979) "Some Practical Universal Noiseless Coding Techniques," Jet Propulsion Laboratory, JPL Publication 79-22, Pasadena, CA, March.

Rice, R. F. (1991) "Some Practical Universal Noiseless Coding Techniques—Part III. Module PSI14.K," Jet Propulsion Laboratory, JPL Publication 91-3, Pasadena, CA, November.

Richardson, Iain G. (2003) *H.264 and MPEG-4 Video Compression Video Coding for Next-generation Multimedia*, Chichester, West Sussex, UK, John Wiley and Sons,

Rissanen, J. J. (1976) "Generalized Kraft Inequality and Arithmetic Coding," *IBM Journal of Research and Development*, **20**:198–203, May.

Robinson, John A. (1997) "Efficient General-Purpose Image Compression with Binary Tree Predictive Coding," *IEEE Transactions on Image Processing*, **6**(4):601–607 April.

Robinson, P. and D. Singer (1981) "Another Spelling Correction Program," *Communications of the ACM*, **24**(5):296–297.

Robinson, Tony (1994) "Simple Lossless and Near-Lossless Waveform Compression," Technical Report CUED/F-INFENG/TR.156, Cambridge University, December. Also at `http://citeseerx.ist.psu.edu/viewdoc/download?doi=10.1.1.53.7337&rep=rep1&type=ps`.

Rodeh, Michael, Vaughan R. Pratt and Shimon Even (1981) "Linear Algorithm for Data Compression via String Matching," Journal of the ACM, **28**(1):16–24, January.

Rodriguez, Karen (1995) "Graphics File Format Patent Unisys Seeks Royalties from GIF Developers," *InfoWorld*, January 9, **17**(2):3.

Roetling, P. G. (1976) "Halftone Method with Edge Enhancement and Moiré Suppression," *Journal of the Optical Society of America*, **66**:985–989.

Roetling, P. G. (1977) "Binary Approximation of Continuous Tone Images," *Photography Science and Engineering*, **21**:60–65.

Roger, R. E., and M. C. Cavenor (1996) "Lossless Compression of AVIRIS Images," *IEEE Transactions on Image Processing*, **5**(5):713–719, May.

Rokicki, Tomas (1985) "Packed (PK) Font File Format," *TUGboat*, **6**(3):115–120. Available at `http://www.tug.org/TUGboat/Articles/tb06-3/tb13pk.pdf`.

Rokicki, Tomas (1990) "GFtoPK, v. 2.3," available at `http://tug.org/texlive/devsrc/Build/source/texk/web2c/gftopk.web`.

Ronson, J. and J. Dewitte (1982) "Adaptive Block Truncation Coding Scheme Using an Edge Following Algorithm," in *Proceedings of the IEEE International Conference on Acoustics, Speech, and Signal Processing*, Piscataway, NJ, IEEE Press, pp. 1235–1238.

Rossignac, J. (1998) "Edgebreaker: Connectivity Compression for Triangle Meshes," GVU Technical Report GIT-GVU-98-35, Atlanta, GA, Georgia Institute of Technology.

Rubin, F. (1979) "Arithmetic Stream Coding Using Fixed Precision Registers," *IEEE Transactions on Information Theory*, **25**(6):672–675, November.

Rudner, B. (1971) "Construction of Minimum-Redundancy Codes with an Optimum Synchronization Property," *IEEE Transactions on Information Theory*, **17**(4):478–487, July.

Sacco, William, et al. (1988) *Information Theory, Saving Bits*, Providence, RI, Janson Publications.

Sagan, Hans (1994) *Space-Filling Curves*, New York, Springer-Verlag.

Said, A. and W. A. Pearlman (1996), "A New Fast and Efficient Image Codec Based on Set Partitioning in Hierarchical Trees," *IEEE Transactions on Circuits and Systems for Video Technology*, **6**(6):243–250, June.

Salomon, David (1999) *Computer Graphics and Geometric Modeling*, New York, Springer.

Salomon, David (2000) "Prefix Compression of Sparse Binary Strings," *ACM Crossroads Magazine*, **6**(3), February.

Salomon, David (2003) *Data Privacy and Security*, New York, Springer.

Salomon, David (2006) *Curves and Surfaces for Computer Graphics*, New York, Springer.

Samet, Hanan (1990a) *Applications of Spatial Data Structures: Computer Graphics, Image Processing, and GIS*, Reading, MA, Addison-Wesley.

Samet, Hanan (1990b) *The Design and Analysis of Spatial Data Structures*, Reading, MA, Addison-Wesley.

Sampath, Ashwin, and Ahmad C. Ansari (1993) "Combined Peano Scan and VQ Approach to Image Compression," *Image and Video Processing*, Bellingham, WA, SPIE vol. 1903, pp. 175–186.

Saponara, Sergio, Luca Fanucci, Pierangelo Terren (2003) "Low-Power VLSI Architectures for 3D Discrete Cosine Transform (DCT)," in *Midwest Symposium on Circuits and Systems (MWSCAS)*.

Sayood, Khalid and K. Anderson (1992) "A Differential Lossless Image Compression Scheme," *IEEE Transactions on Signal Processing*, **40**(1):236–241, January.

Sayood, Khalid (2005) *Introduction to Data Compression*, 3rd Ed., San Francisco, CA, Morgan Kaufmann.

Schalkwijk, J. Pieter M. "An Algorithm for Source Coding," *IEEE Transactions on Information Theory*, **18**(3):395–399, May.

Schindler, Michael (1998) "A Fast Renormalisation for Arithmetic Coding," a poster in the Data Compression Conference, 1998, available at URL
`http://www.compressconsult.com/rangecoder/`.

Schwarz, Heiko, Detlev Marpe, and Thomas Wiegand (2007) "Overview of the Scalable Video Coding Extension of the H.264/AVC Standard," *IEEE Transaction on Circuits and Systems for Video Technology*, **17**(9):1103–1120, September.

Segall, C. Andrew and Gary J. Sullivan (2007) "Spatial Scalability Within the H.264/AVC Scalable Video Coding Extension," *IEEE Transaction on Circuits and Systems for Video Technology*, **17**(9):1121–1135, September.

seul.org (2006) is
`http://f-cpu.seul.org/whygee/phasing-in_codes/PhasingInCodes.nb`.

SHA256 (2002) *Secure Hash Standard*, FIPS Publication 180-2, August 2002. Available at `csrc.nist.gov/publications/fips/fips180-2/fips180-2.pdf`.

Shannon, Claude E. (1948), "A Mathematical Theory of Communication," *Bell System Technical Journal*, **27**:379–423 and 623–656, July and October,

Shannon, Claude (1951) "Prediction and Entropy of Printed English," *Bell System Technical Journal*, **30**(1):50–64, January.

Shapiro, J. (1993) "Embedded Image Coding Using Zerotrees of Wavelet Coefficients," *IEEE Transactions on Signal Processing*, **41**(12):3445–3462, October.

Shenoi, Kishan (1995) *Digital Signal Processing in Telecommunications*, Upper Saddle River, NJ, Prentice Hall.

Shlien, Seymour (1994) "Guide to MPEG-1 Audio Standard," *IEEE Transactions on Broadcasting*, **40**(4):206–218, December.

Sieminski, A. (1988) "Fast Decoding of the Huffman Codes," *Information Processing Letters*, **26**(5):237–241.

Sierpiński, W. (1912) "Sur Une Nouvelle Courbe Qui Remplit Toute Une Aire Plaine," *Bull. Acad. Sci. Cracovie*, **Serie A**:462–478.

Simoncelli, Eero P., and Edward. H. Adelson (1990) "Subband Transforms," in *Subband Coding*, John Woods, ed., Boston, MA, Kluwer Academic Press, pp. 143–192.

Sloane, Neil J. A. (2006) "The On-Line Encyclopedia of Integer Sequences," at `www.research.att.com/~njas/sequences/`.

Smith, Alvy Ray (1984) "Plants, Fractals and Formal Languages," *Computer Graphics*, **18**(3):1–10.

Smith, Alvy Ray (2009) "A Pixel Is Not A Little Square," available from `ftp://ftp.alvyray.com/Acrobat/6_Pixel.pdf`.

SMPTE (2008) is `http://www.smpte.org/`.

SMPTE-421M (2008) is `http://www.smpte.org/standards`.

softexperience (2006) is `http://peccatte.karefil.com/software/Rarissimo/RarissimoEN.htm`.

Softsound (2003) was `http://www.softsound.com/Shorten.html` but try also URL `http://mi.eng.cam.ac.uk/reports/ajr/TR156/tr156.html`.

sourceforge.flac (2006) is `http://sourceforge.net/projects/flac`.

squeezemark (2009) is `http://www.squeezechart.com/`.

Srinivasan, Sridhar and Shankar Regunathan (2005) "Computationally Efficient Transforms for Video Coding," *IEEE International Conference on Image Processing, ICIP 2005*, **2**:325–328, 11–14 Sept.

Starck, J. L., F. Murtagh, and A. Bijaoui (1998) *Image Processing and Data Analysis: The Multiscale Approach*, Cambridge, UK, Cambridge University Press.

Stollnitz, E. J., T. D. DeRose, and D. H. Salesin (1996) *Wavelets for Computer Graphics*, San Francisco, CA, Morgan Kaufmann.

Storer, James A. and T. G. Szymanski (1982) "Data Compression via Textual Substitution," *Journal of the ACM*, **29**:928–951.

Storer, James A. (1988) *Data Compression: Methods and Theory*, Rockville, MD, Computer Science Press.

Storer, James A., and Martin Cohn (eds.) (annual) *DCC 'XX: Data Compression Conference*, Los Alamitos, CA, IEEE Computer Society Press.

Storer, James A., and Harald Helfgott (1997) "Lossless Image Compression by Block Matching," *The Computer Journal*, **40**(2/3):137–145.

Stout, Quentin F. (1980) "Improved Prefix Encodings of the Natural Numbers," *IEEE Transactions on Information Theory*, **26**(5):607–609, September.

Strang, Gilbert, and Truong Nguyen (1996) *Wavelets and Filter Banks*, Wellesley, MA, Wellesley-Cambridge Press.

Strang, Gilbert (1999) "The Discrete Cosine Transform," *SIAM Review*, **41**(1):135–147.

Strømme, Øyvind, and Douglas R. McGregor (1997) "Comparison of Fidelity of Reproduction of Images After Lossy Compression Using Standard and Nonstandard Wavelet Decompositions," in *Proceedings of The First European Conference on Signal Analysis and Prediction (ECSAP 97)*, Prague, June.

Strømme, Øyvind (1999) *On The Applicability of Wavelet Transforms to Image and Video Compression*, Ph.D. thesis, University of Strathclyde, February.

Stuart, J. R. et al. (1999) "MLP Lossless Compression," *AES 9th Regional Convention, Tokyo*. Available at `http://www.meridian-audio.com/w_paper/mlp_jap_new.PDF`.

stuffit-wiki (2009) is `http://en.wikipedia.org/wiki/Stuffit`.

suzannevega (2006) is `http://www.suzannevega.com/suzanne/funfacts/music.aspx`.

Swan, Tom (1993) *Inside Windows File Formats*, Indianapolis, IN, Sams Publications.

Sweldens, Wim and Peter Schröder (1996), *Building Your Own Wavelets At Home*, SIGGRAPH 96 Course Notes. Available on the WWW.

Symes, Peter D. (2003) *MPEG-4 Demystified*, New York, NY, McGraw-Hill Professional.

T-REC-h (2006) is `http://www.itu.int/rec/T-REC-h`.

Takishima, Y., M. Wada, and H. Murakami, (1995) "Reversible Variable-Length Codes," *IEEE Transactions on Communications*, **43**(2,3,4):158–162, Feb./Mar./Apr.

Tanenbaum, Andrew S. (2002) *Computer Networks*, Upper Saddle River, NJ, Prentice Hall.

Taubman, David (1999) "High Performance Scalable Image Compression with EBCOT," *IEEE Transactions on Image Processing*, **9**(7):1158–1170.

Taubman, David S., and Michael W. Marcellin (2002) *JPEG 2000, Image Compression Fundamentals, Standards and Practice*, Norwell, MA, Kluwer Academic.

Teuhola, J. (1978) "A Compression Method for Clustered Bit-Vectors," *Information Processing Letters*, **7**:308–311, October.

Thomborson, Clark, (1992) "The V.42bis Standard for Data-Compressing Modems," *IEEE Micro*, pp. 41–53, October.

Thomas Dolby (2006) is http://www.thomasdolby.com/.

Tischer, Peter (1987) "A Modified LZW Data Compression Scheme," *Australian computer science communications*, **9**(1):262–272.

Tjalkens, T. and Frans M. Willems (1992) "A Universal Variable-to-Fixed Length Source Code Based on Lawrence's Algorithm," *IEEE Transactions on Information Theory*, **38**(2):247–253, March.

Trendafilov, Dimitre, Nasir Memon, and Torsten Suel (2002) "Zdelta: An Efficient Delta Compression Tool," Technical Report TR-CIS-2002-02, New York, NY, Polytechnic University.

Tsai, C. W. and J. L. Wu (2001a) "On Constructing the Huffman-Code Based Reversible Variable Length Codes," *IEEE Transactions on Communications*, **49**(9):1506–1509, September.

Tsai, Chien-Wu and Ja-Ling Wu (2001b) "Modified Symmetrical Reversible Variable-Length Code and its Theoretical Bounds," *IEEE Transactions on Information Theory*, **47**(6):2543–2548, September.

Tunstall, B. P., (1967) "Synthesis of Noiseless Compression Codes," Ph.D. dissertation, Georgia Institute of Technology, Atlanta, GA.

UCLC (2008) is http://uclc.info.

UDA (2008) is http://wex.cn/dwing/mycomp.htm.

Udupa, Raghavendra U., Vinayaka D. Pandit, and Ashok Rao (1999), Private Communication.

uklinux (2007) is http://www.bckelk.uklinux.net/menu.html.

Ulam (2006) is Weisstein, Eric W. "Ulam Sequence." From MathWorld–A Wolfram Web Resource. http://mathworld.wolfram.com/UlamSequence.html.

Unicode (2003) is http://unicode.org/.

Unisys (2003) is http://www.unisys.com.

unrarsrc (2006) is http://www.rarlab.com/rar/unrarsrc-3.5.4.tar.gz.

UPX (2003) is http://upx.sourceforge.net/.

UTF16 (2006) is http://en.wikipedia.org/wiki/UTF-16.

utm (2006) is http://primes.utm.edu/notes/conjectures/.

Vetterli, M., and J. Kovacevic (1995) *Wavelets and Subband Coding*, Englewood Cliffs, NJ, Prentice-Hall.

Vitter, Jeffrey S. (1987) "Design and Analysis of Dynamic Huffman Codes," *Journal of the ACM*, **34**(4):825–845, October.

Volf, Paul A. J. (1997) "A Context-Tree Weighting Algorithm for Text Generating Sources," in Storer, James A. (ed.), *DCC '97: Data Compression Conference*, Los Alamitos, CA, IEEE Computer Society Press, pp. 132–139, (Poster).

Vorobev, Nikolai N. (1983) in Ian N. Sneddon (ed.), and Halina Moss (translator), *Fibonacci Numbers*, New Classics Library.

Wallace, Gregory K. (1991) "The JPEG Still Image Compression Standard," *Communications of the ACM*, **34**(4):30–44, April.

Wang, Muzhong (1988) "Almost Asymptotically Optimal Flag Encoding of the Integers," *IEEE Transactions on Information Theory*, **34**(2):324–326, March.

Watson, Andrew (1994) "Image Compression Using the Discrete Cosine Transform," *Mathematica Journal*, **4**(1):81–88.

WavPack (2006) is `http://www.wavpack.com/`.

Weinberger, M. J., G. Seroussi, and G. Sapiro (1996) "LOCO-I: A Low Complexity, Context-Based, Lossless Image Compression Algorithm," in *Proceedings of Data Compression Conference*, J. Storer, editor, Los Alamitos, CA, IEEE Computer Society Press, pp. 140–149.

Weinberger, M. J., G. Seroussi, and G. Sapiro (2000) "The LOCO-I Lossless Image Compression Algorithm: Principles and Standardization Into JPEG-LS," *IEEE Transactions on Image Processing*, **9**(8):1309–1324, August.

Welch, T. A. (1984) "A Technique for High-Performance Data Compression," *IEEE Computer*, **17**(6):8–19, June.

Wen, Jiangtao and John D. Villasenor (1998) "Reversible Variable Length Codes for Efficient and Robust Image and Video Coding," *Data Compression Conference*, pp. 471–480, Snowbird, UT, March–April.

wiki.audio (2006) is `http://en.wikipedia.org/wiki/Audio_data_compression`.

wikiHutter (2008) is `http://en.wikipedia.org/wiki/Hutter_Prize`.

wikiPAQ (2008) is `http://en.wikipedia.org/wiki/PAQ`.

Wikipedia (2003) is file `Nyquist-Shannon_sampling_theorem` in `http://www.wikipedia.org/wiki/`.

WikiVC-1 (2008) is `http://en.wikipedia.org/wiki/VC-1`.

Willems, F. M. J. (1989) "Universal Data Compression and Repetition Times," *IEEE Transactions on Information Theory*, **IT-35**(1):54–58, January.

Willems, F. M. J., Y. M. Shtarkov, and Tj. J. Tjalkens (1995) "The Context-Tree Weighting Method: Basic Properties," *IEEE Transactions on Information Theory*, **IT-41**:653–664, May.

Williams, Ross N. (1991a) *Adaptive Data Compression*, Boston, MA, Kluwer Academic Publishers.

Williams, Ross N. (1991b) "An Extremely Fast Ziv-Lempel Data Compression Algorithm," in *Proceedings of the 1991 Data Compression Conference*, J. Storer, ed., Los Alamitos, CA, IEEE Computer Society Press, pp. 362–371.

Williams, Ross N. (1993), "A Painless Guide to CRC Error Detection Algorithms," available from http://ross.net/crc/download/crc_v3.txt.

WinAce (2003) is http://www.winace.com/.

windots (2006) is http://www.uiciechi.it/vecchio/cnt/schede/windots-eng.html.

Wirth, N. (1976) *Algorithms + Data Structures = Programs*, 2nd ed., Englewood Cliffs, NJ, Prentice-Hall.

Witten, Ian H., Radford M. Neal, and John G. Cleary (1987) "Arithmetic Coding for Data Compression," *Communications of the ACM*, **30**(6):520–540.

Witten, Ian H. and Timothy C. Bell (1991) "The Zero-Frequency Problem: Estimating the Probabilities of Novel Events in Adaptive Text Compression," *IEEE Transactions on Information Theory*, **IT-37**(4):1085–1094.

Witten, Ian H., T. C. Bell, M. E. Harrison, M. L. James, and A. Moffat (1992) "Textual Image Compression," in *Proceedings of the 1992 Data Compression Conference*, J. Storer, ed., Los Alamitos, CA, IEEE Computer Society Press, pp. 42–51.

Witten, Ian H., T. C. Bell, H. Emberson, S. Inglis, and A. Moffat, (1994) "Textual image compression: two-stage lossy/lossless encoding of textual images," *Proceedings of the IEEE*, **82**(6):878–888, June.

Witten, Ian H, A. Moffat, and T. Bell (1999) *Managing Gigabytes: Compressing and Indexing Documents and Images*, 2nd ed., San Francisco, Morgan Kaufmann.

WMV (2008) is
http://www.microsoft.com/windows/windowsmedia/forpros/codecs/video.aspx.

WMV review (2008) is http://www.microsoft.com/windows/windowsmedia/howto/articles/vc1techoverview.aspx.

Wolf, Misha et al. (2000) "A Standard Compression Scheme for Unicode," Unicode Technical Report #6, available at http://unicode.org/unicode/reports/tr6/index.html.

Wolff, Gerry (1999) is http://www.cognitionresearch.org.uk/sp.htm.

Wong, Kwo-Jyr, and C. C. Jay Kuo (1993) "A Full Wavelet Transform (FWT) Approach to Image Compression," *Image and Video Processing*, Bellingham, WA, SPIE vol. 1903:153–164.

Wong, P. W., and J. Koplowitz (1992) "Chain Codes and Their Linear Reconstruction Filters," *IEEE Transactions on Information Theory*, **IT-38**(2):268–280, May.

Wright, E. V. (1939) *Gadsby*, Los Angeles, Wetzel. Reprinted by University Microfilms, Ann Arbor, MI, 1991.

Wu, Xiaolin (1995), "Context Selection and Quantization for Lossless Image Coding," in James A. Storer and Martin Cohn (eds.), *DCC '95, Data Compression Conference*, Los Alamitos, CA, IEEE Computer Society Press, p. 453.

Wu, Xiaolin (1996), "An Algorithmic Study on Lossless Image Compression," in James A. Storer, ed., *DCC '96, Data Compression Conference*, Los Alamitos, CA, IEEE Computer Society Press.

XMill (2003) is http://sourceforge.net/projects/xmill/.

XML (2003) is http://www.xml.com/.

Yamamoto, Hirosuke (2000) "A New Recursive Universal Code of the Positive Integers," *IEEE Transactions on Information Theory*, **46**(2):717–723, March.

Yamamoto, Hirosuke and Hiroshi Ochi (1991) "A New Asymptotically Optimal Code for the Positive Integers," *IEEE Transactions on Information Theory*, **37**(5):1420–1429, September.

Yokoo, Hidetoshi (1991) "An Improvement of Dynamic Huffman Coding with a Simple Repetition Finder," *IEEE Transactions on Communications*, **39**(1):8–10, January.

Yokoo, Hidetoshi (1996) "An Adaptive Data Compression Method Based on Context Sorting," in *Proceedings of the 1996 Data Compression Conference*, J. Storer, ed., Los Alamitos, CA, IEEE Computer Society Press, pp. 160–169.

Yokoo, Hidetoshi (1997) "Data Compression Using Sort-Based Context Similarity Measure," *The Computer Journal*, **40**(2/3):94–102.

Yokoo, Hidetoshi (1999a) "A Dynamic Data Structure for Reverse Lexicographically Sorted Prefixes," in *Combinatorial Pattern Matching, Lecture Notes in Computer Science 1645*, M. Crochemore and M. Paterson, eds., Berlin, Springer Verlag, pp. 150–162.

Yokoo, Hidetoshi (1999b) Private Communication.

Yoo, Youngjun, Younggap Kwon, and Antonio Ortega (1998) "Embedded Image-Domain Adaptive Compression of Simple Images," in *Proceedings of the 32nd Asilomar Conference on Signals, Systems, and Computers*, Pacific Grove, CA, Nov. 1998.

Young, D. M. (1985) "MacWrite File Format," *Wheels for the Mind*, **1**:34, Fall.

Yu, Tong Lai (1996) "Dynamic Markov Compression," *Dr Dobb's Journal*, pp. 30–31, January.

Zalta, Edward N. (1988) "Are Algorithms Patentable?" *Notices of the American Mathematical Society*, **35**(6):796–799.

Zandi A., J. Allen, E. Schwartz, and M. Boliek, (1995), "CREW: Compression with Reversible Embedded Wavelets," in James A. Storer and Martin Cohn (eds.) *DCC '95: Data Compression Conference*, Los Alamitos, CA, IEEE Computer Society Press, pp. 212–221, March.

Zhang, Manyun (1990) *The JPEG and Image Data Compression Algorithms* (dissertation).

Zeckendorf, E. (1972) "Représentation des Nombres Naturels par Une Somme de Nombres de Fibonacci ou de Nombres de Lucas," *Bull. Soc. Roy. Sci. Liège*, **41**:179–182.

Zeilberger, D. (1993) "Theorems for a Price: Tomorrow's Semi-Rigorous Mathematical Culture," *Notices of the American Mathematical Society*, **40**(8):978–981, October. Reprinted in *Mathematical Intelligencer*, **16**(4):11–14 (Fall 1994).

Zipf's Law (2007) is http://en.wikipedia.org/wiki/Zipf's_law

Ziv, Jacob, and A. Lempel (1977) "A Universal Algorithm for Sequential Data Compression," *IEEE Transactions on Information Theory*, **IT-23**(3):337–343.

Ziv, Jacob and A. Lempel (1978) "Compression of Individual Sequences via Variable-Rate Coding," *IEEE Transactions on Information Theory*, **IT-24**(5):530–536.

zlib (2003) is `http://www.zlib.org/zlib_tech.html`.

Zurek, Wojciech (1989) "Thermodynamic Cost of Computation, Algorithmic Complexity, and the Information Metric," *Nature*, **341**(6238):119–124, September 14.

Yet the guide is fragmentary, incomplete, and in no sense a bibliography. Its emphases vary according to my own indifferences and ignorance as well as according to my own sympathies and knowledge.

J. Frank Dobie, *Guide to Life and Literature of the Southwest* (1943)

Glossary

7-Zip. A file archiver with high compression ratio. The brainchild of Igor Pavlov, this free software for Windows is based on the LZMA algorithm. Both LZMA and `7z` were designed to provide high compression, fast decompression, and low memory requirements for decompression. (See also LZMA.)

AAC. A complex and efficient audio compression method. AAC is an extension of and the successor to mp3. Like mp3, AAC is a time/frequency (T/F) codec that employs a psychoacoustic model to determine how the normal threshold of the ear varies in the presence of masking sounds. Once the perturbed threshold is known, the original audio samples are converted to frequency coefficients which are quantized (thereby providing lossy compression) and then Huffman encoded (providing additional, lossless, compression).

AC-3. A perceptual audio coded designed by Dolby Laboratories to support several audio channels.

ACB. A very efficient text compression method by G. Buyanovsky (Section 11.3). It uses a dictionary with unbounded contexts and contents to select the context that best matches the search buffer and the content that best matches the look-ahead buffer.

Adaptive Compression. A compression method that modifies its operations and/or its parameters according to new data read from the input stream. Examples are the adaptive Huffman method of Section 5.3 and the dictionary-based methods of Chapter 6. (See also Semiadaptive Compression, Locally Adaptive Compression.)

Affine Transformations. Two-dimensional or three-dimensional geometric transformations, such as scaling, reflection, rotation, and translation, that preserve parallel lines (Section 7.39.1).

Alphabet. The set of all possible symbols in the input stream. In text compression, the alphabet is normally the set of 128 ASCII codes. In image compression it is the set of values a pixel can take (2, 16, 256, or anything else). (See also Symbol.)

ALPC. ALPC (*adaptive linear prediction and classification*), is a lossless image compression algorithm based on a linear predictor whose coefficients are computed for each pixel individually in a way that can be mimicd by the decoder.

ALS. MPEG-4 Audio Lossless Coding (ALS) is the latest addition to the family of MPEG-4 audio codecs. ALS can handle integer and floating-point audio samples and is based on a combination of linear prediction (both short-term and long-term), multichannel coding, and efficient encoding of audio residues by means of Rice codes and block codes.

Antidictionary. (See DCA.)

ARC. A compression/archival/cataloging program written by Robert A. Freed in the mid 1980s (Section 6.24). It offers good compression and the ability to combine several files into an archive. (See also Archive, ARJ.)

Archive. A set of one or more files combined into one file (Section 6.24). The individual members of an archive may be compressed. An archive provides a convenient way of transferring or storing groups of related files. (See also ARC, ARJ.)

Arithmetic Coding. A statistical compression method (Section 5.9) that assigns one (normally long) code to the entire input stream, instead of assigning codes to the individual symbols. The method reads the input stream symbol by symbol and appends more bits to the code each time a symbol is input and processed. Arithmetic coding is slow, but it compresses at or close to the entropy, even when the symbol probabilities are skewed. (See also Model of Compression, Statistical Methods, QM Coder.)

ARJ. A free compression/archiving utility for MS/DOS (Section 6.24), written by Robert K. Jung to compete with ARC and the various PK utilities. (See also Archive, ARC.)

ASCII Code. The standard character code on all modern computers (although Unicode is becoming a competitor). ASCII stands for American Standard Code for Information Interchange. It is a $(1+7)$-bit code, with one parity bit and seven data bits per symbol. As a result, 128 symbols can be coded. They include the uppercase and lowercase letters, the ten digits, some punctuation marks, and control characters. (See also Unicode.)

Bark. Unit of critical band rate. Named after Heinrich Georg Barkhausen and used in audio applications. The Bark scale is a nonlinear mapping of the frequency scale over the audio range, a mapping that matches the frequency selectivity of the human ear.

BASC. A compromise between the standard binary (β) code and the Elias gamma codes. (See also RBUC.)

Bayesian Statistics. (See Conditional Probability.)

Bi-level Image. An image whose pixels have two different colors. The colors are normally referred to as black and white, "foreground" and "background," or 1 and 0. (See also Bitplane.)

Benchmarks. (See Compression Benchmarks.)

BinHex. A file format for reliable file transfers, designed by Yves Lempereur for use on the Macintosh computer (Section 1.4.3).

Bintrees. A method, somewhat similar to quadtrees, for partitioning an image into nonoverlapping parts. The image is (horizontally) divided into two halves, each half is divided (vertically) into smaller halves, and the process continues recursively, alternating between horizontal and vertical splits. The result is a binary tree where any uniform part of the image becomes a leaf. (See also Prefix Compression, Quadtrees.)

Bitplane. Each pixel in a digital image is represented by several bits. The set of all the kth bits of all the pixels in the image is the kth bitplane of the image. A bi-level image, for example, consists of one bitplane. (See also Bi-level Image.)

Bitrate. In general, the term "bitrate" refers to both bpb and bpc. However, in audio compression, this term is used to indicate the rate at which the compressed stream is read by the decoder. This rate depends on where the stream comes from (such as disk, communications channel, memory). If the bitrate of an MPEG audio file is, e.g., 128 Kbps. then the encoder will convert each second of audio into 128 K bits of compressed data, and the decoder will convert each group of 128 K bits of compressed data into one second of sound. Lower bitrates mean smaller file sizes. However, as the bitrate decreases, the encoder must compress more audio data into fewer bits, eventually resulting in a noticeable loss of audio quality. For CD-quality audio, experience indicates that the best bitrates are in the range of 112 Kbps to 160 Kbps. (See also Bits/Char.)

Bits/Char. Bits per character (bpc). A measure of the performance in text compression. Also a measure of entropy. (See also Bitrate, Entropy.)

Bits/Symbol. Bits per symbol. A general measure of compression performance.

Block Coding. A general term for image compression methods that work by breaking the image into small blocks of pixels, and encoding each block separately. JPEG (Section 7.10) is a good example, because it processes blocks of 8×8 pixels.

Block Decomposition. A method for lossless compression of discrete-tone images. The method works by searching for, and locating, identical blocks of pixels. A copy B of a block A is compressed by preparing the height, width, and location (image coordinates) of A, and compressing those four numbers by means of Huffman codes. (See also Discrete-Tone Image.)

Block Matching. A lossless image compression method based on the LZ77 sliding window method originally developed for text compression. (See also LZ Methods.)

Block Truncation Coding. BTC is a lossy image compression method that quantizes pixels in an image while preserving the first two or three *statistical moments*. (See also Vector Quantization.)

BMP. BMP (Section 1.4.4) is a palette-based graphics file format for images with 1, 2, 4, 8, 16, 24, or 32 bitplanes. It uses a simple form of RLE to compress images with 4 or 8 bitplanes.

BOCU-1. A simple algorithm for Unicode compression (Section 11.12.1).

BSDiff. A file differencing algorithm created by Colin Percival. The algorithm addresses the problem of differential file compression of executable code while maintaining a platform-independent approach. BSDiff combines matching with mismatches and entropy coding of the differences with `bzip2`. BSDiff's decoder is called bspatch. (See also Exediff, File differencing, UNIX diff, VCDIFF, and Zdelta.)

Burrows-Wheeler Method. This method (Section 11.1) prepares a string of data for later compression. The compression itself is done with the move-to-front method (Section 1.5), perhaps in combination with RLE. The BW method converts a string S to another string L that satisfies two conditions:

1. Any region of L will tend to have a concentration of just a few symbols.

2. It is possible to reconstruct the original string S from L (a little more data may be needed for the reconstruction, in addition to L, but not much).

CALIC. A context-based, lossless image compression method (Section 7.28) whose two main features are (1) the use of three passes in order to achieve symmetric contexts and (2) context quantization, to significantly reduce the number of possible contexts without degrading compression.

CCITT. The International Telegraph and Telephone Consultative Committee (Comité Consultatif International Télégraphique et Téléphonique), the old name of the ITU, the International Telecommunications Union. The ITU is a United Nations organization responsible for developing and recommending standards for data communications (not just compression). (See also ITU.)

Cell Encoding. An image compression method where the entire bitmap is divided into cells of, say, 8×8 pixels each and is scanned cell by cell. The first cell is stored in entry 0 of a table and is encoded (i.e., written on the compressed file) as the pointer 0. Each subsequent cell is searched in the table. If found, its index in the table becomes its code and it is written on the compressed file. Otherwise, it is added to the table. In the case of an image made of just straight segments, it can be shown that the table size is just 108 entries.

CIE. CIE is an abbreviation for Commission Internationale de l'Éclairage (International Committee on Illumination). This is the main international organization devoted to light and color. It is responsible for developing standards and definitions in this area. (See Luminance.)

Circular Queue. A basic data structure (Section 6.3.1) that moves data along an array in circular fashion, updating two pointers to point to the start and end of the data in the array.

Codec. A term used to refer to both encoder and decoder.

Codes. A code is a symbol that stands for another symbol. In computer and telecommunications applications, codes are virtually always binary numbers. The ASCII code is the defacto standard, although the new Unicode is used on several new computers and the older EBCDIC is still used on some old IBM computers. (See also ASCII, Unicode.)

Compösite and Difference Values. A progressive image method that separates the image into layers using the method of *bintrees*. Early layers consist of a few large, low-resolution blocks, followed by later layers with smaller, higher-resolution blocks. The main principle is to transform a pair of pixels into two values, a composite and a differentiator. (See also Bintrees, Progressive Image Compression.)

Compress. In the large UNIX world, `compress` is commonly used to compress data. This utility uses LZW with a growing dictionary. It starts with a small dictionary of just 512 entries and doubles its size each time it fills up, until it reaches 64K bytes (Section 6.14). (See also LZT.)

Compression Benchmarks. In order to prove its value, an algorithm has to be implemented and tested. Thus, every researcher, programmer, and developer compares a new algorithm to older, well-established and known methods, and draws conclusions about its performance. Such comparisons are known as benchmarks.

Compression Factor. The inverse of compression ratio. It is defined as

$$\text{compression factor} = \frac{\text{size of the input stream}}{\text{size of the output stream}}.$$

Values greater than 1 indicate compression, and values less than 1 imply expansion. (See also Compression Ratio.)

Compression Gain. This measure is defined as

$$100 \log_e \frac{\text{reference size}}{\text{compressed size}},$$

where the reference size is either the size of the input stream or the size of the compressed stream produced by some standard lossless compression method.

Compression Ratio. One of several measures that are commonly used to express the efficiency of a compression method. It is the ratio

$$\text{compression ratio} = \frac{\text{size of the output stream}}{\text{size of the input stream}}.$$

A value of 0.6 indicates that the data occupies 60% of its original size after compression. Values greater than 1 mean an output stream bigger than the input stream (negative compression).

Sometimes the quantity $100 \times (1 - \text{compression ratio})$ is used to express the quality of compression. A value of 60 means that the output stream occupies 40% of its original size (or that the compression has resulted in a savings of 60%). (See also Compression Factor.)

Conditional Image RLE. A compression method for grayscale images with n shades of gray. The method starts by assigning an n-bit code to each pixel depending on its near neighbors. It then concatenates the n-bit codes into a long string, and calculates run lengths. The run lengths are encoded by prefix codes. (See also RLE, Relative Encoding.)

Conditional Probability. We tend to think of probability as something that is built into an experiment. A true die, for example, has probability of 1/6 of falling on any side, and we tend to consider this an intrinsic feature of the die. Conditional probability is a different way of looking at probability. It says that knowledge affects probability. The main task of this field is to calculate the probability of an event A given that another event, B, is known to have occurred. This is the conditional probability of A (more precisely, the probability of A conditioned on B), and it is denoted by $P(A|B)$. The field of conditional probability is sometimes called *Bayesian statistics*, since it was first developed by the Reverend Thomas Bayes, who came up with the basic formula of conditional probability.

Context. The N symbols preceding the next symbol. A context-based model uses context to assign probabilities to symbols.

Context-Free Grammars. A formal language uses a small number of symbols (called *terminal symbols*) from which valid sequences can be constructed. Any valid sequence is finite, the number of valid sequences is normally unlimited, and the sequences are constructed according to certain rules (sometimes called *production rules*). The rules can be used to construct valid sequences and also to determine whether a given sequence is valid. A production rule consists of a nonterminal symbol on the left and a string of terminal and nonterminal symbols on the right. The nonterminal symbol on the left becomes the name of the string on the right. The set of production rules constitutes the grammar of the formal language. If the production rules do not depend on the context of a symbol, the grammar is context-free. There are also context-sensitive grammars. The sequitur method of Section 11.10 is based on context-free grammars.

Context-Tree Weighting. A method for the compression of bitstrings. It can be applied to text and images, but they have to be carefully converted to bitstrings. The method constructs a context tree where bits input in the immediate past (context) are used to estimate the probability of the current bit. The current bit and its estimated probability are then sent to an arithmetic encoder, and the tree is updated to include the current bit in the context. (See also KT Probability Estimator.)

Continuous-Tone Image. A digital image with a large number of colors, such that adjacent image areas with colors that differ by just one unit appear to the eye as having continuously varying colors. An example is an image with 256 grayscale values. When adjacent pixels in such an image have consecutive gray levels, they appear to the eye as a continuous variation of the gray level. (See also Bi-level image, Discrete-Tone Image, Grayscale Image.)

Continuous Wavelet Transform. An important modern method for analyzing the time and frequency contents of a function $f(t)$ by means of a wavelet. The wavelet is itself a function (which has to satisfy certain conditions), and the transform is done by multiplying the wavelet and $f(t)$ and computing the integral of the product. The wavelet is then translated, and the process is repeated. When done, the wavelet is scaled, and the entire process is carried out again in order to analyze $f(t)$ at a different scale. (See also Discrete Wavelet Transform, Lifting Scheme, Multiresolution Decomposition, Taps.)

Convolution. A way to describe the output of a linear, shift-invariant system by means of its input.

Correlation. A statistical measure of the linear relation between two paired variables. The values of R range from -1 (perfect negative relation), to 0 (no relation), to $+1$ (perfect positive relation).

CRC. CRC stands for *Cyclical Redundancy Check* (or *Cyclical Redundancy Code*). It is a rule that shows how to obtain vertical check bits from all the bits of a data stream (Section 6.32). The idea is to generate a code that depends on all the bits of the data stream, and use it to detect errors (bad bits) when the data is transmitted (or when it is stored and retrieved).

CRT. A CRT (cathode ray tube) is a glass tube with a familiar shape. In the back it has an electron gun (the cathode) that emits a stream of electrons. Its front surface is positively charged, so it attracts the electrons (which have a negative electric charge). The front is coated with a phosphor compound that converts the kinetic energy of the electrons hitting it to light. The flash of light lasts only a fraction of a second, so in order to get a constant display, the picture has to be refreshed several times a second.

Data Compression Conference. A scientific meeting of researchers and developers in the area of data compression. The DCC takes place every year in Snowbird, Utah, USA. The time is normally the second half of March and the conference lasts three days.

Data Structure. A set of data items used by a program and stored in memory such that certain operations (for example, finding, adding, modifying, and deleting items) can be performed on the data items fast and easily. The most common data structures are the array, stack, queue, linked list, tree, graph, and hash table. (See also Circular Queue.)

DCA. DCA stands for data compression with antidictionaries. An antidictionary contains bitstrings that do not appear in the input. With the help of an antidictionary, both encoder and decoder can predict with certainty the values of certain bits, which can then be eliminated from the output, thereby causing compression. (See also Antidictionary.)

Decibel. A logarithmic measure that can be used to measure any quantity that takes values over a very wide range. A common example is sound intensity. The intensity (amplitude) of sound can vary over a range of 11–12 orders of magnitude. Instead of using a linear measure, where numbers as small as 1 and as large as 10^{11} would be needed, a logarithmic scale is used, where the range of values is $[0, 11]$.

Decoder. A decompression program (or algorithm).

Deflate. A popular lossless compression algorithm (Section 6.25) used by Zip and gzip. Deflate employs a variant of LZ77 combined with static Huffman coding. It uses a 32-Kb-long sliding dictionary and a look-ahead buffer of 258 bytes. When a string is not found in the dictionary, its first symbol is emitted as a literal byte. (See also Gzip, Zip.)

Dictionary-Based Compression. Compression methods (Chapter 6) that save pieces of the data in a "dictionary" data structure. If a string of new data is identical to a piece that is already saved in the dictionary, a pointer to that piece is output to the compressed stream. (See also LZ Methods.)

Differential Image Compression. A lossless image compression method where each pixel p is compared to a *reference pixel*, which is one of its immediate neighbors, and is then encoded in two parts: a prefix, which is the number of most significant bits of p that are identical to those of the reference pixel, and a suffix, which is (almost all) the remaining least significant bits of p. (See also DPCM.)

Digital Video. A form of video in which the original image is generated, in the camera, in the form of pixels. (See also High-Definition Television.)

Digram. A pair of consecutive symbols.

Discrete Cosine Transform. A variant of the discrete Fourier transform (DFT) that produces just real numbers. The DCT (Sections 7.8, 7.10.2, and 11.15.2) transforms a set of numbers by combining n numbers to become an n-dimensional point and rotating it in n-dimensions such that the first coordinate becomes dominant. The DCT and its inverse, the IDCT, are used in JPEG (Section 7.10) to compress an image with acceptable loss, by isolating the high-frequency components of an image, so that they can later be quantized. (See also Fourier Transform, Transform.)

Discrete-Tone Image. A discrete-tone image may be bi-level, grayscale, or color. Such images are (with some exceptions) artificial, having been obtained by scanning a document, or capturing a computer screen. The pixel colors of such an image do not vary continuously or smoothly, but have a small set of values, such that adjacent pixels may differ much in intensity or color. Figure 7.59 is an example of such an image. (See also Block Decomposition, Continuous-Tone Image.)

Discrete Wavelet Transform. The discrete version of the continuous wavelet transform. A wavelet is represented by means of several filter coefficients, and the transform is carried out by matrix multiplication (or a simpler version thereof) instead of by calculating an integral. (See also Continuous Wavelet Transform, Multiresolution Decomposition.)

DjVu. Certain images combine the properties of all three image types (bi-level, discrete-tone, and continuous-tone). An important example of such an image is a scanned document containing text, line drawings, and regions with continuous-tone pictures, such as paintings or photographs. DjVu (pronounced "déjà vu") is designed for high compression and fast decompression of such documents.

It starts by decomposing the document into three components: mask, foreground, and background. The background component contains the pixels that constitute the pictures and the paper background. The mask contains the text and the lines in bi-level form (i.e., one bit per pixel). The foreground contains the color of the mask pixels. The background is a continuous-tone image and can be compressed at the low resolution of 100 dpi. The foreground normally contains large uniform areas and is also compressed as a continuous-tone image at the same low resolution. The mask is left at 300 dpi but can be efficiently compressed, since it is bi-level. The background and foreground are compressed with a wavelet-based method called IW44, while the mask is compressed with JB2, a version of JBIG2 (Section 7.15) developed at AT&T.

DPCM. DPCM compression is a member of the family of differential encoding compression methods, which itself is a generalization of the simple concept of relative encoding (Section 1.3.1). It is based on the fact that neighboring pixels in an image (and also adjacent samples in digitized sound) are correlated. (See also Differential Image Compression, Relative Encoding.)

Embedded Coding. This feature is defined as follows: Imagine that an image encoder is applied twice to the same image, with different amounts of loss. It produces two files, a large one of size M and a small one of size m. If the encoder uses embedded coding, the smaller file is identical to the first m bits of the larger file.

The following example aptly illustrates the meaning of this definition. Suppose that three users wait for you to send them a certain compressed image, but they need different image qualities. The first one needs the quality contained in a 10 Kb file. The image qualities required by the second and third users are contained in files of sizes 20 Kb and 50 Kb, respectively. Most lossy image compression methods would have to compress the same image three times, at different qualities, to generate three files with the right sizes. An embedded encoder, on the other hand, produces one file, and then three chunks—of lengths 10 Kb, 20 Kb, and 50 Kb, all starting at the beginning of the file—can be sent to the three users, satisfying their needs. (See also SPIHT, EZW.)

Encoder. A compression program (or algorithm).

Entropy. The entropy of a single symbol a_i is defined (in Section A.1) as $-P_i \log_2 P_i$, where P_i is the probability of occurrence of a_i in the data. The entropy of a_i is the smallest number of bits needed, on average, to represent symbol a_i. Claude Shannon, the creator of information theory, coined the term *entropy* in 1948, because this term is used in thermodynamics to indicate the amount of disorder in a physical system. (See also Entropy Encoding, Information Theory.)

Entropy Encoding. A lossless compression method where data can be compressed such that the average number of bits/symbol approaches the entropy of the input symbols. (See also Entropy.)

Error-Correcting Codes. The opposite of data compression, these codes detect and correct errors in digital data by increasing the redundancy of the data. They use check bits or parity bits, and are sometimes designed with the help of generating polynomials.

EXE Compressor. A compression program for compressing EXE files on the PC. Such a compressed file can be decompressed and executed with one command. The original EXE compressor is LZEXE, by Fabrice Bellard (Section 6.29).

Exediff. A differential file compression algorithm created by Brenda Baker, Udi Manber, and Robert Muth for the differential compression of executable code. Exediff is an iterative algorithm that uses a lossy transform to reduce the effect of the secondary changes in executable code. Two operations called *pre-matching* and *value recovery* are iterated until the size of the patch converges to a minimum. Exediff's decoder is called exepatch. (See also BSdiff, File differencing, UNIX diff, VCDIFF, and Zdelta.)

EZW. A progressive, embedded image coding method based on the zerotree data structure. It has largely been superseded by the more efficient SPIHT method. (See also SPIHT, Progressive Image Compression, Embedded Coding.)

Facsimile Compression. Transferring a typical page between two fax machines can take up to 10–11 minutes without compression, This is why the ITU has developed several standards for compression of facsimile data. The current standards (Section 5.7) are T4 and T6, also called Group 3 and Group 4, respectively. (See also ITU.)

FELICS. A Fast, Efficient, Lossless Image Compression method designed for grayscale images that competes with the lossless mode of JPEG. The principle is to code each pixel with a variable-length code based on the values of two of its previously seen neighbor pixels. Both the unary code and the Golomb code are used. There is also a progressive version of FELICS (Section 7.24). (See also Progressive FELICS.)

FHM Curve Compression. A method for compressing curves. The acronym FHM stands for Fibonacci, Huffman, and Markov. (See also Fibonacci Numbers.)

Fibonacci Numbers. A sequence of numbers defined by

$$F_1 = 1, \quad F_2 = 1, \quad F_i = F_{i-1} + F_{i-2}, \quad i = 3, 4, \ldots.$$

The first few numbers in the sequence are 1, 1, 2, 3, 5, 8, 13, and 21. These numbers have many applications in mathematics and in various sciences. They are also found in nature, and are related to the golden ratio. (See also FHM Curve Compression.)

File Differencing. A compression method that locates and compresses the differences between two slightly different data sets. The decoder, that has access to one of the two data sets, can use the differences and reconstruct the other. Applications of this compression technique include software distribution and updates (or *patching*), revision control systems, compression of backup files, archival of multiple versions of data. (See also VCDIFF.) (See also BSdiff, Exediff, UNIX diff, VCDIFF, and Zdelta.)

FLAC. An acronym for free lossless audio compression, FLAC is an audio compression method, somewhat resembling Shorten, that is based on prediction of audio samples and encoding of the prediction residues with Rice codes. (See also Rice codes.)

Fourier Transform. A mathematical transformation that produces the frequency components of a function (Section 8.1). The Fourier transform shows how a periodic function can be written as the sum of sines and cosines, thereby showing explicitly the frequencies "hidden" in the original representation of the function. (See also Discrete Cosine Transform, Transform.)

Gaussian Distribution. (See Normal Distribution.)

GFA. A compression method originally developed for bi-level images that can also be used for color images. GFA uses the fact that most images of interest have a certain amount of self-similarity (i.e., parts of the image are similar, up to size, orientation, or brightness, to the entire image or to other parts). GFA partitions the image into subsquares using a quadtree, and expresses relations between parts of the image in a graph.

The graph is similar to graphs used to describe finite-state automata. The method is lossy, because parts of a real image may be very (although not completely) similar to other parts. (See also Quadtrees, Resolution Independent Compression, WFA.)

GIF. An acronym that stands for Graphics Interchange Format. This format (Section 6.21) was developed by Compuserve Information Services in 1987 as an efficient, compressed graphics file format that allows for images to be sent between computers. The original version of GIF is known as GIF 87a. The current standard is GIF 89a. (See also Patents.)

Golomb Code. The Golomb codes consist of an infinite set of *parametrized prefix codes*. They are the best ones for the compression of data items that are distributed geometrically. (See also Unary Code.)

Gray Codes. These are binary codes for the integers, where the codes of consecutive integers differ by one bit only. Such codes are used when a grayscale image is separated into bitplanes, each a bi-level image. (See also Grayscale Image,)

Grayscale Image. A continuous-tone image with shades of a single color. (See also Continuous-Tone Image.)

Growth Geometry Coding. A method for progressive lossless compression of bi-level images. The method selects some *seed* pixels and applies geometric rules to grow each seed pixel into a pattern of pixels. (See also Progressive Image Compression.)

GS-2D-LZ. GS-2D-LZ stands for Grayscale Two-Dimensional Lempel-Ziv Encoding (Section 7.18). This is an innovative dictionary-based method for the lossless compression of grayscale images.

Gzip. Popular software that implements the Deflate algorithm (Section 6.25) that uses a variation of LZ77 combined with static Huffman coding. It uses a 32 Kb-long sliding dictionary, and a look-ahead buffer of 258 bytes. When a string is not found in the dictionary, it is emitted as a sequence of literal bytes. (See also Zip.)

H.261. In late 1984, the CCITT (currently the ITU-T) organized an expert group to develop a standard for visual telephony for ISDN services. The idea was to send images and sound between special terminals, so that users could talk and see each other. This type of application requires sending large amounts of data, so compression became an important consideration. The group eventually came up with a number of standards, known as the H series (for video) and the G series (for audio) recommendations, all operating at speeds of $p\times 64$ Kbit/sec for $p = 1, 2, \ldots, 30$. These standards are known today under the umbrella name of $p \times 64$.

H.264. A sophisticated method for the compression of video. This method is a successor of H.261, H.262, and H.263. It has been approved in 2003 and employs the main building blocks of its predecessors, but with many additions and improvements.

HD Photo. A compression standard (algorithm and file format) for continuous-tone images. HD Photo was developed from Windows Media Photo, a Microsoft compression algorithm. HD Photo follows the basic steps of JPEG, but employs an integer transform instead of the DCT. (See also JPEG, JPEG XR.)

Halftoning. A method for the display of gray scales in a bi-level image. By placing groups of black and white pixels in carefully designed patterns, it is possible to create the effect of a gray area. The trade-off of halftoning is loss of resolution. (See also Bi-level Image, Dithering.)

Hamming Codes. A type of error-correcting code for 1-bit errors, where it is easy to generate the required parity bits.

Hierarchical Progressive Image Compression. An image compression method (or an optional part of such a method) where the encoder writes the compressed image in layers of increasing resolution. The decoder decompresses the lowest-resolution layer first, displays this crude image, and continues with higher-resolution layers. Each layer in the compressed stream uses data from the preceding layer. (See also Progressive Image Compression.)

High-Definition Television. A general name for several standards that are currently replacing traditional television. HDTV uses digital video, high-resolution images, and aspect ratios different from the traditional 3:4. (See also Digital Video.)

Huffman Coding. A popular method for data compression (Section 5.2). It assigns a set of "best" variable-length codes to a set of symbols based on their probabilities. It serves as the basis for several popular programs used on personal computers. Some of them use just the Huffman method, while others use it as one step in a multistep compression process. The Huffman method is somewhat similar to the Shannon-Fano method. It generally produces better codes, and like the Shannon-Fano method, it produces best code when the probabilities of the symbols are negative powers of 2. The main difference between the two methods is that Shannon-Fano constructs its codes top to bottom (from the leftmost to the rightmost bits), while Huffman constructs a code tree from the bottom up (builds the codes from right to left). (See also Shannon-Fano Coding, Statistical Methods.)

Hutter prize. The Hutter prize (Section 5.13 tries to encourage the development of new methods for text compression.

Hyperspectral data. A set of data items (called pixels) arranged in rows and columns where each pixel is a vector. An example is an image where each pixel consists of the radiation reflected from the ground in many frequencies. We can think of such data as several image planes (called bands) stacked vertically. Hyperspectral data is normally large and is an ideal candidate for compression. Any compression method for this type of data should take advantage of the correlation between bands as well as correlations between pixels in the same band.

Information Theory. A mathematical theory that quantifies information. It shows how to measure information, so that one can answer the question; How much information is included in a given piece of data? with a precise number! Information theory is the creation, in 1948, of Claude Shannon of Bell labs. (See also Entropy.)

Interpolating Polynomials. Given two numbers a and b we know that $m = 0.5a + 0.5b$ is their average, since it is located midway between a and b. We say that the average is an *interpolation* of the two numbers. Similarly, the weighted sum $0.1a +$

0.9b represents an interpolated value located 10% away from b and 90% away from a. Extending this concept to points (in two or three dimensions) is done by means of *interpolating polynomials*. Given a set of points, we start by fitting a parametric polynomial $\mathbf{P}(t)$ or $\mathbf{P}(u, w)$ through them. Once the polynomial is known, it can be used to calculate interpolated points by computing $\mathbf{P}(0.5)$, $\mathbf{P}(0.1)$, or other values.

Interpolative Coding. An algorithm that assigns dynamic variable-length codes to a strictly monotonically increasing sequence of integers (Section 3.28).

ISO. The International Standards Organization. This is one of the organizations responsible for developing standards. Among other things it is responsible (together with the ITU) for the JPEG and MPEG compression standards. (See also ITU, CCITT, MPEG.)

Iterated Function Systems (IFS). An image compressed by IFS is uniquely defined by a few affine transformations (Section 7.39.1). The only rule is that the scale factors of these transformations must be less than 1 (shrinking). The image is saved in the output stream by writing the sets of six numbers that define each transformation. (See also Affine Transformations, Resolution Independent Compression.)

ITU. The International Telecommunications Union, the new name of the CCITT, is a United Nations organization responsible for developing and recommending standards for data communications (not just compression). (See also CCITT.)

JBIG. A special-purpose compression method (Section 7.14) developed specifically for progressive compression of bi-level images. The name JBIG stands for Joint Bi-Level Image Processing Group. This is a group of experts from several international organizations, formed in 1988 to recommend such a standard. JBIG uses multiple arithmetic coding and a resolution-reduction technique to achieve its goals. (See also Bi-level Image, JBIG2.)

JBIG2. A recent international standard for the compression of bi-level images. It is intended to replace the original JBIG. Its main features are

1. Large increases in compression performance (typically 3–5 times better than Group 4/MMR, and 2–4 times better than JBIG).

2. Special compression methods for text, halftones, and other bi-level image parts.

3. Lossy and lossless compression modes.

4. Two modes of progressive compression. Mode 1 is quality-progressive compression, where the decoded image progresses from low to high quality. Mode 2 is content-progressive coding, where important image parts (such as text) are decoded first, followed by less important parts (such as halftone patterns).

5. Multipage document compression.

6. Flexible format, designed for easy embedding in other image file formats, such as TIFF.

7. Fast decompression. In some coding modes, images can be decompressed at over 250 million pixels/second in software.

(See also Bi-level Image, JBIG.)

JFIF. The full name of this method (Section 7.10.7) is JPEG File Interchange Format. It is a graphics file format that makes it possible to exchange JPEG-compressed images between different computers. The main features of JFIF are the use of the YCbCr triple-component color space for color images (only one component for grayscale images) and the use of a *marker* to specify features missing from JPEG, such as image resolution, aspect ratio, and features that are application-specific.

JPEG. A sophisticated lossy compression method (Section 7.10) for color or grayscale still images (not movies). It works best on continuous-tone images, where adjacent pixels have similar colors. One advantage of JPEG is the use of many parameters, allowing the user to adjust the amount of data loss (and thereby also the compression ratio) over a very wide range. There are two main modes: lossy (also called baseline) and lossless (which typically yields a 2:1 compression ratio). Most implementations support just the lossy mode. This mode includes progressive and hierarchical coding.

The main idea behind JPEG is that an image exists for people to look at, so when the image is compressed, it is acceptable to lose image features to which the human eye is not sensitive.

The name JPEG is an acronym that stands for Joint Photographic Experts Group. This was a joint effort by the CCITT and the ISO that started in June 1987. The JPEG standard has proved successful and has become widely used for image presentation, especially in Web pages. (See also JPEG-LS, MPEG.)

JPEG-LS. The lossless mode of JPEG is inefficient and often is not even implemented. As a result, the ISO decided to develop a new standard for the lossless (or near-lossless) compression of continuous-tone images. The result became popularly known as JPEG-LS. This method is not simply an extension or a modification of JPEG. It is a new method, designed to be simple and fast. It does not employ the DCT, does not use arithmetic coding, and applies quantization in a limited way, and only in its near-lossless option. JPEG-LS examines several of the previously-seen neighbors of the current pixel, uses them as the *context* of the pixel, employs the context to predict the pixel and to select a probability distribution out of several such distributions, and uses that distribution to encode the prediction error with a special Golomb code. There is also a run mode, where the length of a run of identical pixels is encoded. (See also Golomb Code, JPEG.)

JPEG XR. See HD Photo.

> As for my mother, perhaps the Ambassador had not the type of mind towards which she felt herself most attracted. I should add that his conversation furnished so exhaustive a glossary of the superannuated forms of speech peculiar to a certain profession, class and period.
>
> —Marcel Proust, *Within a Budding Grove* (1913–1927)

Kraft-MacMillan Inequality. A relation (Section 2.5) that says something about unambiguous variable- codes. Its first part states: Given an unambiguous variable-length code, with n codes of lengths L_i, then

$$\sum_{i=1}^{n} 2^{-L_i} \leq 1.$$

[This is Equation (2.3).] The second part states the opposite, namely, given a set of n positive integers $(L_1, L_2, \ldots, L_n)$ that satisfy Equation (2.3), there exists an unambiguous variable-length code such that L_i are the sizes of its individual codes. Together, both parts state that a code is unambiguous if and only if it satisfies relation (2.3).

KT Probability Estimator. A method to estimate the probability of a bitstring containing a zeros and b ones. It is due to Krichevsky and Trofimov. (See also Context-Tree Weighting.)

Laplace Distribution. A probability distribution similar to the normal (Gaussian) distribution, but narrower and sharply peaked. The general Laplace distribution with variance V and mean m is given by

$$L(V, x) = \frac{1}{\sqrt{2V}} \exp\left(-\sqrt{\frac{2}{V}}|x - m|\right).$$

Experience seems to suggest that the values of the residues computed by many image compression algorithms are Laplace distributed, which is why this distribution is employed by those compression methods, most notably MLP. (See also Normal Distribution.)

Laplacian Pyramid. A progressive image compression technique where the original image is transformed to a set of difference images that can later be decompressed and displayed as a small, blurred image that becomes increasingly sharper. (See also Progressive Image Compression.)

LHArc. This method (Section 6.24) is by Haruyasu Yoshizaki. Its predecessor is LHA, designed jointly by Haruyasu Yoshizaki and Haruhiko Okumura. These methods are based on adaptive Huffman coding with features drawn from LZSS.

Lifting Scheme. A method for computing the discrete wavelet transform in place, so no extra memory is required. (See also Discrete Wavelet Transform.)

Locally Adaptive Compression. A compression method that adapts itself to local conditions in the input stream, and varies this adaptation as it moves from area to area in the input. An example is the move-to-front method of Section 1.5. (See also Adaptive Compression, Semiadaptive Compression.)

Lossless Compression. A compression method where the output of the decoder is identical to the original data compressed by the encoder. (See also Lossy Compression.)

Lossy Compression. A compression method where the output of the decoder is different from the original data compressed by the encoder, but is nevertheless acceptable to a user. Such methods are common in image and audio compression, but not in text compression, where the loss of even one character may result in wrong, ambiguous, or incomprehensible text. (See also Lossless Compression, Subsampling.)

LPVQ. An acronym for Locally Optimal Partitioned Vector Quantization. LPVQ is a quantization algorithm proposed by Giovanni Motta, Francesco Rizzo, and James Storer [Motta et al. 06] for the lossless and near-lossless compression of hyperspectral data. Spectral signatures are first partitioned in sub-vectors on unequal length and independently quantized. Then, indices are entropy coded by exploiting both spectral and spatial correlation. The residual error is also entropy coded, with the probabilities conditioned by the quantization indices. The locally optimal partitioning of the spectral signatures is decided at design time, during the training of the quantizer.

Luminance. This quantity is defined by the CIE (Section 7.10.1) as radiant power weighted by a spectral sensitivity function that is characteristic of vision. (See also CIE.)

LZ Methods. All dictionary-based compression methods are based on the work of J. Ziv and A. Lempel, published in 1977 and 1978. Today, these are called LZ77 and LZ78 methods, respectively. Their ideas have been a source of inspiration to many researchers, who generalized, improved, and combined them with RLE and statistical methods to form many commonly used adaptive compression methods, for text, images, and audio. (See also Block Matching, Dictionary-Based Compression, Sliding-Window Compression.)

LZAP. The LZAP method (Section 6.16) is an LZW variant based on the following idea: Instead of just concatenating the last two phrases and placing the result in the dictionary, place all prefixes of the concatenation in the dictionary. The suffix AP stands for All Prefixes.

LZARI. An improvement on LZSS, developed in 1988 by Haruhiko Okumura. (See also LZSS.)

LZB. LZB is the result of evaluating and comparing several data structures and variable-length codes with an eye to improving the performance of LZSS.

LZC. See Compress, LZT.

LZFG. This is the name of several related methods (Section 6.10) that are hybrids of LZ77 and LZ78. They were developed by Edward Fiala and Daniel Greene. All these methods are based on the following scheme. The encoder produces a compressed file with tokens and literals (raw ASCII codes) intermixed. There are two types of tokens, a *literal* and a *copy*. A literal token indicates that a string of literals follow, a copy token points to a string previously seen in the data. (See also LZ Methods, Patents.)

LZJ. LZJ (Section 6.17) is an interesting LZ variant. It stores in its dictionary, which can be viewed either as a multiway tree or as a forest, *every* phrase found in the input. If a phrase if found n times in the input, only one copy is stored in the dictionary.

LZMA. LZMA (Lempel-Ziv-Markov chain-Algorithm) is one of the many LZ77 variants. Developed by Igor Pavlov, this algorithm, which is used in his popular `7z` software, is based on a large search buffer, a hash function that generates indexes, somewhat similar to LZRW4, and two search methods. The fast method uses a hash-array of lists of indexes and the normal method uses a hash-array of binary decision trees. (See also 7-Zip.)

LZMW. A variant of LZW, the LZMW method (Section 6.15) works as follows: Instead of adding I plus one character of the next phrase to the dictionary, add I plus the entire next phrase to the dictionary. (See also LZW.)

LZP. An LZ77 variant developed by C. Bloom (Section 6.19). It is based on the principle of context prediction that says "if a certain string `abcde` has appeared in the input stream in the past and was followed by `fg...`, then when `abcde` appears again in the input stream, there is a good chance that it will be followed by the same `fg...`." (See also Context.)

LZPP. LZPP is a modern, sophisticated algorithm that extends LZSS in several directions. LZPP identifies several sources of redundancy in the various quantities generated and manipulated by LZSS and exploits these sources to obtain better overall compression. (See also LZSS.)

LZR. LZR is a variant of the basic LZ77 method, where the lengths of both the search and look-ahead buffers are unbounded.

LZSS. This version of LZ77 (Section 6.4) was developed by Storer and Szymanski in 1982 [Storer 82]. It improves on the basic LZ77 in three ways: (1) it holds the look-ahead buffer in a circular queue, (2) it implements the search buffer (the dictionary) in a binary search tree, and (3) it creates tokens with two fields instead of three. (See also LZ Methods, LZARI, LZPP, SLH.)

LZT. LZT is an extension of UNIX compress/LZC. The major innovation of LZT is the way it handles a full dictionary. (See also Compress.)

LZW. This is a popular variant (Section 6.13) of LZ78, developed by Terry Welch in 1984. Its main feature is eliminating the second field of a token. An LZW token consists of just a pointer to the dictionary. As a result, such a token always encodes a string of more than one symbol. (See also Patents.)

LZWL. A syllable-based variant of LZW. (See also Syllable-Based Compression.)

LZX. LZX is an LZ77 variant for the compression of cabinet files (Section 6.8).

LZY. LZY (Section 6.18) is an LZW variant that adds one dictionary string per input character and increments strings by one character at a time.

MLP. A progressive compression method for grayscale images. An image is compressed in levels. A pixel is predicted by a symmetric pattern of its neighbors from preceding levels, and the prediction error is arithmetically encoded. The Laplace distribution is used to estimate the probability of the error. (See also Laplace Distribution, Progressive FELICS.)

MLP Audio. The new lossless compression standard approved for DVD-A (audio) is called MLP. It is the topic of Section 10.7.

MNP5, MNP7. These have been developed by Microcom, Inc., a maker of modems, for use in its modems. MNP5 (Section 5.4) is a two-stage process that starts with run-length encoding, followed by adaptive frequency encoding. MNP7 (Section 5.5) combines run-length encoding with a two-dimensional variant of adaptive Huffman.

Model of Compression. A model is a method to "predict" (to assign probabilities to) the data to be compressed. This concept is important in statistical data compression. When a statistical method is used, a model for the data has to be constructed before compression can begin. A simple model can be built by reading the entire input stream, counting the number of times each symbol appears (its frequency of occurrence), and computing the probability of occurrence of each symbol. The data stream is then input again, symbol by symbol, and is compressed using the information in the probability model. (See also Statistical Methods, Statistical Model.)

One feature of arithmetic coding is that it is easy to separate the statistical model (the table with frequencies and probabilities) from the encoding and decoding operations. It is easy to encode, for example, the first half of a data stream using one model, and the second half using another model.

Monkey's audio. Monkey's audio is a fast, efficient, free, lossless audio compression algorithm and implementation that offers error detection, tagging, and external support.

Move-to-Front Coding. The basic idea behind this method (Section 1.5) is to maintain the alphabet A of symbols as a list where frequently occurring symbols are located near the front. A symbol `s` is encoded as the number of symbols that precede it in this list. After symbol `s` is encoded, it is moved to the front of list A.

MPEG. This acronym stands for Moving Pictures Experts Group. The MPEG standard consists of several methods for the compression of video, including the compression of digital images and digital sound, as well as synchronization of the two. There currently are several MPEG standards. MPEG-1 is intended for intermediate data rates, on the order of 1.5 Mbit/sec. MPEG-2 is intended for high data rates of at least 10 Mbit/sec. MPEG-3 was intended for HDTV compression but was found to be redundant and was merged with MPEG-2. MPEG-4 is intended for very low data rates of less than 64 Kbit/sec. The ITU-T, has been involved in the design of both MPEG-2 and MPEG-4. A working group of the ISO is still at work on MPEG. (See also ISO, JPEG.)

Multiresolution Decomposition. This method groups all the discrete wavelet transform coefficients for a given scale, displays their superposition, and repeats for all scales. (See also Continuous Wavelet Transform, Discrete Wavelet Transform.)

Multiresolution Image. A compressed image that may be decompressed at any resolution. (See also Resolution Independent Compression, Iterated Function Systems, WFA.)

Normal Distribution. A probability distribution with the well-known bell shape. It is found in many places in both theoretical models and real-life situations. The normal distribution with mean m and standard deviation s is defined by

$$f(x) = \frac{1}{s\sqrt{2\pi}} \exp\left\{ -\frac{1}{2}\left(\frac{x-m}{s}\right)^2 \right\}.$$

PAQ. An open-source, high-performance compression algorithm and free software (Section 5.15) that features sophisticated prediction combined with adaptive arithmetic encoding.

Patents. A mathematical algorithm can be patented if it is intimately associated with software or firmware implementing it. Several compression methods, most notably LZW, have been patented (Section 6.34), creating difficulties for software developers who work with GIF, UNIX `compress`, or any other system that uses LZW. (See also GIF, LZW, Compress.)

Pel. The smallest unit of a facsimile image; a dot. (See also Pixel.)

phased-in codes. Phased-in codes (Section 2.9) are a minor extension of fixed-length codes and may contribute a little to the compression of a set of consecutive integers by changing the representation of the integers from fixed n bits to either n or $n-1$ bits.

Phrase. A piece of data placed in a dictionary to be used in compressing future data. The concept of phrase is central in dictionary-based data compression methods since the success of such a method depends a lot on how it selects phrases to be saved in its dictionary. (See also Dictionary-Based Compression, LZ Methods.)

Pixel. The smallest unit of a digital image; a dot. (See also Pel.)

PKZip. A compression program for MS/DOS (Section 6.24) written by Phil Katz who has founded the PKWare company which also markets the PKunzip, PKlite, and PKArc software (`http://www.pkware.com`).

PNG. An image file format (Section 6.27) that includes lossless compression with Deflate and pixel prediction. PNG is free and it supports several image types and number of bitplanes, as well as sophisticated transparency.

Portable Document Format (PDF). A standard developed by Adobe in 1991–1992 that allows arbitrary documents to be created, edited, transferred between different computer platforms, and printed. PDF compresses the data in the document (text and images) by means of LZW, Flate (a variant of Deflate), run-length encoding, JPEG, JBIG2, and JPEG 2000.

PPM. A compression method that assigns probabilities to symbols based on the context (long or short) in which they appear. (See also Prediction, PPPM.)

PPPM. A lossless compression method for grayscale (and color) images that assigns probabilities to symbols based on the Laplace distribution, like MLP. Different contexts of a pixel are examined, and their statistics used to select the mean and variance for a particular Laplace distribution. (See also Laplace Distribution, Prediction, PPM, MLP.)

Prediction. Assigning probabilities to symbols. (See also PPM.)

Prefix Compression. A variant of quadtrees, designed for bi-level images with text or diagrams, where the number of black pixels is relatively small. Each pixel in a $2^n \times 2^n$ image is assigned an n digit, or $2n$ bit, number based on the concept of quadtrees. Numbers of adjacent pixels tend to have the same prefix (most-significant bits), so the common prefix and different suffixes of a group of pixels are compressed separately. (See also Quadtrees.)

Prefix Property. One of the principles of variable-length codes. It states; Once a certain bit pattern has been assigned as the code of a symbol, no other codes should start with that pattern (the pattern cannot be the *prefix* of any other code). Once the string 1, for example, is assigned as the code of a_1, no other codes should start with 1 (i.e., they all have to start with 0). Once 01, for example, is assigned as the code of a_2, no other codes can start with 01 (they all should start with 00). (See also Variable-Length Codes, Statistical Methods.)

Progressive FELICS. A progressive version of FELICS where pixels are encoded in levels. Each level doubles the number of pixels encoded. To decide what pixels are included in a certain level, the preceding level can conceptually be rotated 45° and scaled by $\sqrt{2}$ in both dimensions. (See also FELICS, MLP, Progressive Image Compression.)

Progressive Image Compression. An image compression method where the compressed stream consists of "layers," where each layer contains more detail of the image. The decoder can very quickly display the entire image in a low-quality format, and then improve the display quality as more and more layers are being read and decompressed. A user watching the decompressed image develop on the screen can normally recognize most of the image features after only 5–10% of it has been decompressed. Improving image quality over time can be done by (1) sharpening it, (2) adding colors, or (3) increasing its resolution. (See also Progressive FELICS, Hierarchical Progressive Image Compression, MLP, JBIG.)

Psychoacoustic Model. A mathematical model of the sound masking properties of the human auditory (ear brain) system.

QIC-122 Compression. An LZ77 variant that has been developed by the QIC organization for text compression on 1/4-inch data cartridge tape drives.

QM Coder. This is the arithmetic coder of JPEG and JBIG. It is designed for simplicity and speed, so it is limited to input symbols that are single bits and it employs an approximation instead of exact multiplication. It also uses fixed-precision integer arithmetic, so it has to resort to *renormalization* of the probability interval from time to time, in order for the approximation to remain close to the true multiplication. (See also Arithmetic Coding.)

Quadrisection. This is a relative of the quadtree method. It assumes that the original image is a $2^k \times 2^k$ square matrix M_0, and it constructs matrices M_1, $M_2, \ldots, M_{k+1}$ with fewer and fewer columns. These matrices naturally have more and more rows, and quadrisection achieves compression by searching for and removing duplicate rows. Two closely related variants of quadrisection are bisection and octasection (See also Quadtrees.)

Quadtrees. This is a data compression method for bitmap images. A quadtree (Section 7.34) is a tree where each leaf corresponds to a uniform part of the image (a quadrant, subquadrant, or a single pixel) and each interior node has exactly four children. (See also Bintrees, Prefix Compression, Quadrisection.)

Quaternary. A base-4 digit. It can be 0, 1, 2, or 3.

RAR. RAR An LZ77 variant designed and developed by Eugene Roshal. RAR is extremely popular with Windows users and is available for a variety of platforms. In addition to excellent compression and good encoding speed, RAR offers options such as error-correcting codes and encryption. (See also Rarissimo.)

Rarissimo. A file utility that's always used in conjunction with RAR. It is designed to periodically check certain source folders, automatically compress and decompress files found there, and then move those files to designated target folders. (See also RAR.)

RBUC. A compromise between the standard binary (β) code and the Elias gamma codes. (See also BASC.)

Recursive range reduction (3R). Recursive range reduction (3R) is a simple coding algorithm that offers decent compression, is easy to program, and its performance is independent of the amount of data to be compressed.

Relative Encoding. A variant of RLE, sometimes called *differencing* (Section 1.3.1). It is used in cases where the data to be compressed consists of a string of numbers that don't differ by much, or in cases where it consists of strings that are similar to each other. The principle of relative encoding is to send the first data item a_1 followed by the differences $a_{i+1} - a_i$. (See also DPCM, RLE.)

Reliability. Variable-length codes and other codes are vulnerable to errors. In cases where reliable storage and transmission of codes are important, the codes can be made more reliable by adding check bits, parity bits, or CRC (Section 5.6). Notice that reliability is, in a sense, the opposite of data compression, because it is achieved by increasing redundancy. (See also CRC.)

Resolution Independent Compression. An image compression method that does not depend on the resolution of the specific image being compressed. The image can be decompressed at any resolution. (See also Multiresolution Images, Iterated Function Systems, WFA.)

Rice Codes. A special case of the Golomb code. (See also Golomb Codes.)

RLE. RLE stands for run-length encoding. This is a general name for methods that compress data by replacing a run of identical symbols with a single code, or token, that encodes the symbol and the length of the run. RLE sometimes serves as one step in a multistep statistical or dictionary-based method. (See also Relative Encoding, Conditional Image RLE.)

Scalar Quantization. The dictionary definition of the term "quantization" is "to restrict a variable quantity to discrete values rather than to a continuous set of values." If the data to be compressed is in the form of large numbers, quantization is used to convert them to small numbers. This results in (lossy) compression. If the data to be compressed is analog (e.g., a voltage that changes with time), quantization is used to digitize it into small numbers. This aspect of quantization is used by several audio compression methods. (See also Vector Quantization.)

SCSU. A compression algorithm designed specifically for compressing text files in Unicode (Section 11.12).

SemiAdaptive Compression. A compression method that uses a two-pass algorithm, where the first pass reads the input stream to collect statistics on the data to be compressed, and the second pass performs the actual compression. The statistics (model) are included in the compressed stream. (See also Adaptive Compression, Locally Adaptive Compression.)

Semistructured Text. Such text is defined as data that is human readable and also suitable for machine processing. A common example is HTML. The sequitur method of Section 11.10 performs especially well on such text.

Shannon-Fano Coding. An early algorithm for finding a minimum-length variable-length code given the probabilities of all the symbols in the data (Section 5.1). This method was later superseded by the Huffman method. (See also Statistical Methods, Huffman Coding.)

Shorten. A simple compression algorithm for waveform data in general and for speech in particular (Section 10.9). Shorten employs linear prediction to compute residues (of audio samples) which it encodes by means of Rice codes. (See also Rice codes.)

Simple Image. A simple image is one that uses a small fraction of the possible grayscale values or colors available to it. A common example is a bi-level image where each pixel is represented by eight bits. Such an image uses just two colors out of a palette of 256 possible colors. Another example is a grayscale image scanned from a bi-level image. Most pixels will be black or white, but some pixels may have other shades of gray. A cartoon is also an example of a simple image (especially a cheap cartoon, where just a few colors are used). A typical cartoon consists of uniform areas, so it may use a small number of colors out of a potentially large palette. The EIDAC method of Section 7.16 is especially designed for simple images.

SLH. SLH is a variant of LZSS. It is a two-pass algorithm where the first pass employs a hash table to locate the best match and to count frequencies and the second pass encodes the offsets and the raw symbols with Huffman codes prepared from the frequencies counted by the first pass.

Sliding Window Compression. The LZ77 method (Section 6.3) uses part of the previously seen input stream as the dictionary. The encoder maintains a window to the input stream, and shifts the input in that window from right to left as strings of symbols are being encoded. The method is therefore based on a *sliding window*. (See also LZ Methods.)

Space-Filling Curves. A space-filling curve (Section 7.36) is a function $\mathbf{P}(t)$ that goes through every point in a given two-dimensional region, normally the unit square, as t varies from 0 to 1. Such curves are defined recursively and are used in image compression.

Sparse Strings. Regardless of what the input data represents—text, binary, images, or anything else—we can think of the input stream as a string of bits. If most of the bits are zeros, the string is *sparse*. Sparse strings can be compressed very efficiently by specially designed methods (Section 11.5).

Spatial Prediction. An image compression method that is a combination of JPEG and fractal-based image compression.

SPIHT. A progressive image encoding method that efficiently encodes the image after it has been transformed by any wavelet filter. SPIHT is embedded, progressive, and has a natural lossy option. It is also simple to implement, fast, and produces excellent results for all types of images. (See also EZW, Progressive Image Compression, Embedded Coding, Discrete Wavelet Transform.)

Statistical Methods. These methods (Chapter 5) work by assigning variable-length codes to symbols in the data, with the shorter codes assigned to symbols or groups of symbols that appear more often in the data (have a higher probability of occurrence). (See also Variable-Length Codes, Prefix Property, Shannon-Fano Coding, Huffman Coding, and Arithmetic Coding.)

Statistical Model. (See Model of Compression.)

String Compression. In general, compression methods based on strings of symbols can be more efficient than methods that compress individual symbols (Section 6.1).

Stuffit. Stuffit (Section 11.16) is compression software for the Macintosh platform, developed in 1987. The methods and algorithms it employs are proprietary, but some information exists in various patents.

Subsampling. Subsampling is, possibly, the simplest way to compress an image. One approach to subsampling is simply to ignore some of the pixels. The encoder may, for example, ignore every other row and every other column of the image, and write the remaining pixels (which constitute 25% of the image) on the compressed stream. The decoder inputs the compressed data and uses each pixel to generate four identical pixels of the reconstructed image. This, of course, involves the loss of much image detail and is rarely acceptable. (See also Lossy Compression.)

SVC. SVC (scalable video coding) is an extension to H.264 that supports temporal, spatial, and quality scalable video coding, while retaining a base layer that is still backward compatible with the original H.264/AVC standard.

Syllable-Based Compression. An approach to compression where the basic data symbols are syllables, a syntactic unit between letters and words. (See also LZWL.)

Symbol. The smallest unit of the data to be compressed. A symbol is often a byte but may also be a bit, a trit $\{0, 1, 2\}$, or anything else. (See also Alphabet.)

Symbol Ranking. A context-based method (Section 11.2) where the context C of the current symbol S (the N symbols preceding S) is used to prepare a list of symbols that are likely to follow C. The list is arranged from most likely to least likely. The position of S in this list (position numbering starts from 0) is then written by the encoder, after being suitably encoded, on the output stream.

Taps. Wavelet filter coefficients. (See also Continuous Wavelet Transform, Discrete Wavelet Transform.)

TAR. The standard UNIX archiver. The name TAR stands for Tape ARchive. It groups a number of files into one file without compression. After being compressed by the UNIX `compress` program, a TAR file gets an extension name of `tar.z`.

Textual Image Compression. A compression method for hard copy documents containing printed or typed (but not handwritten) text. The text can be in many fonts and may consist of musical notes, hieroglyphs, or any symbols. Pattern recognition techniques are used to recognize text characters that are identical or at least similar. One copy of each group of identical characters is kept in a library. Any leftover material is considered residue. The method uses different compression techniques for the symbols and the residue. It includes a lossy option where the residue is ignored.

Time/frequency (T/F) codec. An audio codec that employs a psychoacoustic model to determine how the normal threshold of the ear varies (in both time and frequency) in the presence of masking sounds.

Token. A unit of data written on the compressed stream by some compression algorithms. A token consists of several fields that may have either fixed or variable sizes.

Transform. An image can be compressed by transforming its pixels (which are correlated) to a representation where they are *decorrelated*. Compression is achieved if the new values are smaller, on average, than the original ones. Lossy compression can be achieved by quantizing the transformed values. The decoder inputs the transformed values from the compressed stream and reconstructs the (precise or approximate) original data by applying the opposite transform. (See also Discrete Cosine Transform, Fourier Transform, Continuous Wavelet Transform, Discrete Wavelet Transform.)

Triangle Mesh. Polygonal surfaces are very popular in computer graphics. Such a surface consists of flat polygons, mostly triangles, so there is a need for special methods to compress a triangle mesh. One such a method is edgebreaker (Section 11.11).

Trit. A ternary (base 3) digit. It can be 0, 1, or 2.

Tunstall codes. Tunstall codes are a variation on variable-length codes. They are fixed-size codes, each encoding a variable-length string of data symbols.

Unary Code. A way to generate variable-length codes of the integers in one step. The unary code of the nonnegative integer n is defined (Section 3.1) as $n-1$ 1's followed by a single 0 (Table 3.1). There is also a general unary code. (See also Golomb Code.)

Unicode. A new international standard code, the Unicode, has been proposed, and is being developed by the international Unicode organization (`www.unicode.org`). Unicode uses 16-bit codes for its characters, so it provides for $2^{16} = 64\text{K} = 65{,}536$ codes. (Notice that doubling the size of a code much more than doubles the number of possible codes. In fact, it *squares* the number of codes.) Unicode includes all the ASCII codes in addition to codes for characters in foreign languages (including complete sets of Korean, Japanese, and Chinese characters) and many mathematical and other symbols. Currently, about 39,000 out of the 65,536 possible codes have been assigned, so there is room for adding more symbols in the future.

The Microsoft Windows NT operating system has adopted Unicode, as have also AT&T Plan 9 and Lucent Inferno. (See also ASCII, Codes.)

UNIX diff. A file differencing algorithm that uses `APPEND`, `DELETE` and `CHANGE` to encode the differences between two text files. `diff` generates an output that is human-readable or, optionally, it can generate batch commands for a text editor like `ed`. (See also BSdiff, Exediff, File differencing, VCDIFF, and Zdelta.)

V.42bis Protocol. This is a standard, published by the ITU-T (page 248) for use in fast modems. It is based on the older V.32bis protocol and is supposed to be used for fast transmission rates, up to 57.6K baud. The standard contains specifications for data compression and error correction, but only the former is discussed, in Section 6.23.

V.42bis specifies two modes: a *transparent* mode, where no compression is used, and a *compressed* mode using an LZW variant. The former is used for data streams that don't compress well, and may even cause expansion. A good example is an already compressed file. Such a file looks like random data, it does not have any repetitive patterns, and trying to compress it with LZW will fill up the dictionary with short, two-symbol, phrases.

Variable-Length Codes. These are used by statistical methods. Many codes of this type satisfy the prefix property (Section 2.1) and should be assigned to symbols based on the probability distribution of the symbols. (See also Prefix Property, Statistical Methods.)

VC-1. VC-1 (Section 9.11) is a hybrid video codec, developed by SMPTE in 2006, that is based on the two chief principles of video compression, a transform and motion compensation. This codec is intended for use in a wide variety of video applications and promises to perform well at a wide range of bitrates from 10 Kbps to about 135 Mbps.

VCDIFF. A method for compressing the differences between two files. (See also BSdiff, Exediff, File differencing, UNIX diff, and Zdelta.)

Vector Quantization. This is a generalization of the scalar quantization method. It is used for both image and audio compression. In practice, vector quantization is commonly used to compress data that has been digitized from an analog source, such as sampled sound and scanned images (drawings or photographs). Such data is called *digitally sampled analog data* (DSAD). (See also Scalar Quantization.)

Video Compression. Video compression is based on two principles. The first is the spatial redundancy that exists in each video frame. The second is the fact that very often, a video frame is very similar to its immediate neighbors. This is called *temporal redundancy*. A typical technique for video compression should therefore start by encoding the first frame using an image compression method. It should then encode each successive frame by identifying the differences between the frame and its predecessor, and encoding these differences.

Voronoi Diagrams. Imagine a petri dish ready for growing bacteria. Four bacteria of different types are simultaneously placed in it at different points and immediately start multiplying. We assume that their colonies grow at the same rate. Initially, each colony consists of a growing circle around one of the starting points. After a while some of them meet and stop growing in the meeting area due to lack of food. The final result is that the entire dish gets divided into four areas, one around each of the four starting points, such that all the points within area i are closer to starting point i than to any other start point. Such areas are called *Voronoi regions* or *Dirichlet Tessellations*.

WavPack. WavPack is an open, multiplatform audio compression algorithm and software that supports three compression modes, lossless, high-quality lossy, and a unique hybrid mode. WavPack handles integer audio samples up to 32-bits wide and also 32-bit IEEE floating-point audio samples. It employs an original entropy encoder that assigns variable-length Golomb codes to the residuals and also has a recursive Golomb coding mode for cases where the distribution of the residuals is not geometric.

WFA. This method uses the fact that most images of interest have a certain amount of self-similarity (i.e., parts of the image are similar, up to size or brightness, to the entire image or to other parts). It partitions the image into subsquares using a quadtree, and uses a recursive inference algorithm to express relations between parts of the image in a graph. The graph is similar to graphs used to describe finite-state automata. The method is lossy, since parts of a real image may be very similar to other parts. WFA is a very efficient method for compression of grayscale and color images. (See also GFA, Quadtrees, Resolution-Independent Compression.)

WSQ. An efficient lossy compression method specifically developed for compressing fingerprint images. The method involves a wavelet transform of the image, followed by scalar quantization of the wavelet coefficients, and by RLE and Huffman coding of the results. (See also Discrete Wavelet Transform.)

XMill. Section 6.28 is a short description of XMill, a special-purpose compressor for XML files.

Zdelta. A file differencing algorithm developed by Dimitre Trendafilov, Nasir Memon and Torsten Suel. Zdelta adapts the compression library `zlib` to the problem of differential file compression. zdelta represents the target file by combining copies from both the reference and the already compressed target file. A Huffman encoder is used to further compress this representation. (See also BSdiff, Exediff, File differencing, UNIX diff, and VCDIFF.)

Zero-Probability Problem. When samples of data are read and analyzed in order to generate a statistical model of the data, certain contexts may not appear, leaving entries with zero counts and thus zero probability in the frequency table. Any compression method requires that such entries be somehow assigned nonzero probabilities.

Zip. Popular software that implements the Deflate algorithm (Section 6.25) that uses a variant of LZ77 combined with static Huffman coding. It uses a 32-Kb-long sliding dictionary and a look-ahead buffer of 258 bytes. When a string is not found in the dictionary, its first symbol is emitted as a literal byte. (See also Deflate, Gzip.)

The expression of a man's face is commonly a
help to his thoughts, or glossary on his speech.
—Charles Dickens, *Life and Adventures of Nicholas Nickleby* (1839)

Joining the Data Compression Community

Those interested in a personal touch can join the "DC community" and communicate with researchers and developers in this growing area in person by attending the Data Compression Conference (DCC). It has taken place, mostly in late March, every year since 1991, in Snowbird, Utah, USA, and it lasts three days. Detailed information about the conference, including the organizers and the geographical location, can be found at `http://www.cs.brandeis.edu/~dcc/`.

In addition to invited presentations and technical sessions, there is a poster session and "Midday Talks" on issues of current interest.

The poster session is the central event of the DCC. Each presenter places a description of recent work (including text, diagrams, photographs, and charts) on a 4-foot-wide by 3-foot-high poster. They then discuss the work with anyone interested, in a relaxed atmosphere, with refreshments served. The Capocelli prize is awarded annually for the best student-authored DCC paper. This is in memory of Renato M. Capocelli.

The program committee reads like a who's who of data compression, but the two central figures are James Andrew Storer and Michael W. Marcellin.

The conference proceedings are published by the IEEE Computer Society and are distributed prior to the conference; an attractive feature. A complete bibliography (in bibTEX format) of papers published in past DCCs can be found at `http://liinwww.ira.uka.de/bibliography/Misc/dcc.html`.

What greater thing is there for two human souls than to feel
that they are joined... to strengthen each other... to
be one with each other in silent unspeakable memories.

—George Eliot

Index

The index caters to those who have already read the book and want to locate a familiar item, as well as to those new to the book who are looking for a particular topic. We have included any terms that may occur to a reader interested in any of the topics discussed in the book (even topics that are just mentioned in passing). As a result, even a quick glancing over the index gives the reader an idea of the terms and topics included in the book. Notice that the index items "data compression" and "image compression" have only general subitems such as "logical," "lossless," and "bi-level." No specific compression methods are listed as subitems.
we have attempted to make the index items as complete as possible, including middle names and dates. Any errors and omissions brought to my attention are welcome. They will be added to the errata list and will be included in any future editions.

C

G

I

J

M

P

R

Indexing requires decision making of a far higher order than computers are yet capable of.
—The Chicago Manual of Style, 13th ed. (1982)

Colophon

This volume is an extension of *Data Compression: The Complete Reference*, whose first edition appeared in 1996. The book was designed by the authors and was typeset with the TeX typesetting system developed by D. Knuth. The text and tables were done with Textures and TeXshop on a Macintosh computer. The figures were drawn in Adobe Illustrator. Figures that required calculations were computed either in Mathematica or Matlab, but even those were "polished" in Adobe Illustrator. The following facts illustrate the amount of work that went into the book:

- The book (including the auxiliary material located in the book's Web site) contains about 523,000 words, consisting of about 3,081,000 characters (big, even by the standards of Marcel Proust). However, the size of the auxiliary material collected in the author's computer and on his shelves while working on the book is about 10 times bigger than the entire book. This material includes articles and source codes available on the Internet, as well as many pages of information collected from various sources.
- The text is typeset mainly in font cmr10, but about 30 other fonts were used.
- The raw index file has about 5150 items.
- There are about 1300 cross references in the book.

You can't just start a new project in Visual Studio/Delphi/whatever, then add in an ADPCM encoder, the best psychoacoustic model, some DSP stuff, Levinson Durbin, subband decomposition, MDCT, a Blum-Blum-Shub random number generator, a wavelet-based brownian movement simulator and a Feistel network cipher using a cryptographic hash of the Matrix series, and expect it to blow everything out of the water, now can you?

Anonymous, found in [hydrogenaudio 06]

GPSR Compliance

The European Union's (EU) General Product Safety Regulation (GPSR) is a set of rules that requires consumer products to be safe and our obligations to ensure this.

If you have any concerns about our products, you can contact us on ProductSafety@springernature.com

In case Publisher is established outside the EU, the EU authorized representative is:

Springer Nature Customer Service Center GmbH
Europaplatz 3
69115 Heidelberg, Germany

Batch number: 10755974

Printed by Printforce, the Netherlands